Handbook of Neurochemistry and Molecular Neurobiology

Behavioral Neurochemistry, Neuroendocrinology and Molecular Neurobiology

Abel Lajtha(Ed.)

Handbook of Neurochemistry and Molecular Neurobiology

Behavioral Neurochemistry, Neuroendocrinology and Molecular Neurobiology

Volume Editor: Jeff Blaustein

With 130 Figures and 7 Tables

Springer

Editor-in-Chief
Abel Lajtha
Director
Center for Neurochemistry
Nathan S. Kline Institute for Psychiatric Research
140 Old Orangeburg Road
Orangeburg
New York, 10962
USA

Volume Editor
Jeffrey D. Blaustein
Center for Neuroendocrine Studies
135 Hicks Way
University of Massachusetts
Amherst, MA 01003-9271
USA

Library of Congress Control Number: 2006922553

ISBN 13: 978-0-387-30362-8

Additionally, the whole set will be available upon completion under ISBN-13: 978-0-387-35443-9
The electronic version of the whole set will be available under ISBN-13: 978-0-387-30426-7
The print and electronic bundle of the whole set will be available under ISBN-13: 978-0-387-35478-1

springer.com

Printed on acid-free paper

SPIN: 11417767 2109 – 5 4 3 2 1 0

Preface

Behavioral neuroscience was not covered extensively in the second edition of the Handbook of Neurochemistry, published in 1983. That was nearly a decade before the formation of the international society, which named itself after this discipline, the International Society for Behavioral Neuroscience, and it was even longer before the inception of the Society for Behavioral Neuroendocrinology, which focuses on a subfield of behavioral neuroscience. The progress that has been made in the study of the cellular and molecular underpinnings of behavior was almost unimaginable in 1983. The field has prospered thanks to development in novel drugs, genetic models, and related molecular techniques, neuroanatomical techniques, including *in situ* hybridization histochemistry, new immunocytochemical techniques, real time PCR, microarrays, and more sophisticated behavioral analysis.

This volume is filled with a tremendous amount of history that documents the coming of age of behavioral neuroscience. Learning the history and following the development of a field are often an important part of understanding an area of science, and many of the authors have elaborated extensively on the history of their field. Though behavioral neuroscience has advanced tremendously in recent years, impediments to progress still remain in this field. For example, behavior still occasionally takes a back seat to the study of simpler physiological endpoints, such as the control of ovulation. Yet, it is the more complex regulation of behavior and the interactions of the environment on it that allow for fertilization, without which, ovulation would be irrelevant. In his memorable biography of the field of hormones and behavior, Frank Beach (1981) explained some of the background for bias against studies of behavior. He also provided what is perhaps one of the most noteworthy examples of this bias against studying behavioral endpoints. In 1935, Edward Dempsey, working in William C. Young's group, made the heretic proposal that sequential exposure to estradiol and progesterone was necessary to induce the expression of estrous behaviors in female guinea pigs. This made no sense at the time since it was widely believed that the source of progesterone was the corpus luteum, most definitely formed after ovulation, long after the animal's estrous behavior had commenced. Beach recounts that Edgar Allen, a co-discoverer of estradiol, suggested that Young would be "well-advised to give up behavior and return to his more promising early studies on physiology of the epididymis." It was to be another 30 years before novel biochemical procedures would be developed, which would prove Dempsey, Young and colleagues correct; progesterone was being secreted from another source before the formation of the corpus luteum. Behavioral studies had indeed informed physiology.

Throughout this volume, you will find examples of dogmas that ultimately did not hold water (the timing of progesterone secretion during the estrous cycle just discussed is but one example). Time and time again, throughout the history of science, scientists who have questioned dogma have been subjected to ridicule or derision for their actions. Win or lose, the dogma fight is always worth waging, and it is good science. If the dogma represents truth, it will stand; if not, it will eventually topple, but usually not without significant battle. I dedicate this volume to all scientists who have at one point or another challenged dogma in their work.

In this volume, I have collected in one place the expertise of numerous authorities in the diverse field of behavioral neuroscience. A volume of this size does not allow for an exhaustive treatment of the neurochemistry, neuroendocrinology, and molecular neurobiology of behavior; rather it is a sampling of some fascinating areas within the realm of behavioral neuroscience. Moreover, because of my personal bias, these areas of behavioral neuroscience often have an important, well-developed, endocrine slant. To appreciate how much space a comprehensive treatment of the entire field would require, consider that a comprehensive coverage of the subfield, the relatively narrow field of behavioral neuroendocrinology, was recently accomplished admirably in a discipline-defining, five-volume, and nearly 4,000 page work, which was edited by Donald W. Pfaff et al.

It is impossible to acknowledge here each of the authors and all of the important findings of the fields that they represent. Suffice it to say that tremendous progress has been made in the study of reproductive behaviors, affiliative and aggressive behaviors, bird song, sex differences, the hypothalamo-pituitary-adrenal axis, stress, ingestive behaviors, fear, cognitive function, reward, rhythms, and sleep. The book starts at the beginning with reproduction and ends appropriately with sleep. Between those two basic, life-generating and restorative activities, tremendous progress in all of these fields is described.

Recent changes in the funding climate have affected many areas within behavioral sciences. For example, the National Institutes of Health have shifted emphasis to more translational research; in some cases, at the expense of more basic research. This volume is filled with examples of basic research that have led to a better understanding of the human brain and behavior. I hope that it serves as testament to the indispensable value of basic research, as well as translational research, in behavioral neuroscience.

Frank Beach ended his biography of the field of behavioral endocrinology with the following passage: "Scientists with doctorates in psychology study development of progesterone receptors in neurons of the rat hypothalamus while other investigators initially trained in pharmacology invent elegant behavioral measures of sexual motivation in the estrous female. These developments appear to represent more than a mere borrowing of techniques by one discipline from another. Instead they seem to reflect progress toward recognition of common goals and shared theoretical interests. If such indeed is the case, behavioral endocrinology may well be a discipline *in statu nascendi*," that is, a discipline in a state of being born. He could have said the same about behavioral neuroscience. I submit that the discipline of behavioral neuroscience has become a fully developed discipline with investigators answering questions that truly run the gamut from molecular to behavioral and all levels in between.

I am immensely grateful to all who contributed to this volume. Writing a comprehensive review of a field, even one's specialty, is time-consuming and laborious, and it invariably takes time away from other worthy tasks. My job of convincing colleagues to contribute to this volume was made relatively easy because previous versions of the Handbook of Neurochemistry have been well received, and have often served as landmark volumes in their respective fields. It is my hope that the authors will be well compensated for their work with the satisfaction of knowing that their reviews will be read and that, as their areas evolve, progress can be updated and followed in the electronic version of the Handbook.

Jeffrey D. Blaustein
Amherst, USA
April 2006

Table of Contents

Contributors

L. Alanko
Institute of Biomedicine/Physiology, University of Helsinki, Helsinki, Finland

H. E. Albers
Departments of Biology and Psychology and Center for Behavioral Neuroscience, Georgia State University, Atlanta, GA 30303, USA

G. F. Ball
Department of Psychological and Brain Sciences, Johns Hopkins University, 3400 N. Charles Street, Baltimore, MD 21218-2686, USA

J. Balthazart
Center for Cellular and Molecular Neurobiology, Research Group in Behavioral Neuroendocrinology, University of Liège, 1 Boulevard de l'Hopital (Bat. L36), B-4000 Liège 1, Belgium

M. J. Baum
Dept. of Biology Boston University Boston, MA 02215, USA

J. B. Becker
Department of Psychology. Neuroscience Program and Reproductive Sciences Program, University of Michigan, 525 East University Ave., Ann Arbor, MI 48109, USA

S. C. Benoit
Department of Psychiatry, University of Cincinnati, USA

J. D. Blaustein
Center for Neuroendocrine Studies and Psychology Department, University of Massachusetts, Amherst, MA, USA

S. Chattarji
National Centre for Biological Sciences, Tata Institute of Fundamental Research, Bangalore 560065, India

D. Choi
Department of Psychiatry, University of Cincinnati, Cincinnati, OH 45237-0506, USA

D. F. Clayton
School of Molecular and Cellular Biology, Beckman Institute, and Institute for Genomic Biology, University of Illinois, Urbana-Champaign, IL 61801, USA

D. J. Clegg
Department of Psychiatry, University of Cincinnati, USA

M. A. Cooper
Department of Psychology and Center for Behavioral Neuroscience, Georgia State University, Atlanta, GA 30303, USA

P. Dash
The Vivian L. Smith Center for Neurologic Research Department of Neurobiology and Anatomy The University of Texas Medical School at Houston, Houston, Texas 77030, USA

G. J. de Vries
Center for Neuroendocrine Studies and Department of Psychology University of Massachusetts, Amherst, MA 01003, USA

J. F. DeBold
Department of Psychology, Tufts University, Medford and Boston, Massachusetts, USA

G. E. Demas
Department of Biology, Program in Neural Science and Center for the Integrative Study of Animal Behavior, Indiana University, Bloomington, IN 47405, USA

J. Diorio
Douglas Hospital Research Centre, Montreal, Canada

J. M. Dominguez
Department of Psychology Florida State University
Tallahassee, FL 32306-1270, USA

N. K. Evanson
Department of Psychiatry, University of Cincinnati,
Cincinnati, OH 45237-0506, USA

S. P. Faccidomo
Department of Psychology, Tufts University, Medford
and Boston, Massachusetts, USA

H. F. Figueiredo
Department of Psychiatry, University of Cincinnati,
Cincinnati, OH 45237-0506, USA

E. W. Fish
Department of Psychology, Tufts University, Medford
and Boston, Massachusetts, USA

A. Furay
Department of Psychiatry, University of Cincinnati,
Cincinnati, OH 45237-0506, USA

E. A. D. Hammock
Department of Psychiatry and Behavioral Sciences,
Emory University, Atlanta, GA 30329. Center for
Behavioral Neuroscience, Atlanta, GA 30329.
Yerkes National Primate Research Center, Atlanta,
GA 30329, USA

I. Hellstrom
McGill Program for the Study of Behavior, Genes and
Environment, McGill University, Montreal, Canada

J. P. Herman
Departments of Psychiatry and Cell Biology,
Neurobiology and Anatomy, University of Cincinnati,
Cincinnati, OH 45237-0506, USA

E. M. Hull
Department of Psychology Florida State University
Tallahassee, FL 32306-1270, USA

A. K. Johnson
Departments of Psychology, Pharmacology and
Exercise Science, and the Cardiovascular Center,
University of Iowa, Iowa City, IA 52242, USA

J. E. LeDoux
Center for Neural Science, New York University, USA

T. M. Lee
Department of Psychology and Neuroscience
Program, University of Michigan, Ann Arbor,
48109-1043, USA

J. S. Lonstein
Program in Neuroscience and Department of
Psychology, Giltner Hall, Michigan State University,
East Lansing, MI 48824, USA

V. N. Luine
Department of Psychology, Hunter College of C.U.N.Y.,
695 Park Ave., New York, NY 10021, USA

S. K. Mani
Department of Molecular and Cellular Biology, Baylor
College of Medicine, Houston, TX, USA

M. M. McCarthy, PhD
Departments of Physiology and Psychiatry, University
of Maryland, Baltimore – School of Medicine. 655
W. Baltimore St., Baltimore, MD 21201, USA

E. L. McDearmon
Howard Hughes Medical Institute

B. S. McEwen
Harold and Margaret Milliken Hatch Laboratory
of Neuroendocrinology, The Rockefeller University,
New York, NY 10021, USA

M. J. Meaney
McGill Program for the Study of Behavior, Genes and
Environment, McGill University, Montreal Canada.
Douglas Hospital Research Centre, 6875 boul. LaSalle,
Montreal, Québec, Canada H4H 1R3

R. L. Meisel
Department of Psychological Sciences and
Neuroscience Program, Purdue University, 703 Third St.,
West Lafayette, IN 47907, USA

P. Micevych
Dept of Neurobiology, Laboratory of
Neuroendocrinology of the Brain Research Institute,
David Geffen School of Medicine at UCLA,
Los Angeles, CA 90095-1763, USA

K. A. Miczek
Department of Psychology, Psychiatry, Pharmacology,
and Neuroscience Tufts University Medford and Boston,
Massachusetts, USA

B. H. Miller
The Scripps Research Institute, Jupiter, FL 33458, USA

A. N. Moore
The Vivian L. Smith Center for Neurologic Research Department of Neurobiology and Anatomy The University of Texas Medical School at Houston, Houston, Texas 77030, USA

J. I. Morrell
Center for Molecular and Behavioral Neuroscience, 197 University Ave, Rutgers University, Newark, NJ 07102, USA

N. K. Mueller
Department of Psychiatry, University of Cincinnati, Cincinnati, OH 45237-0506, USA

J. W. Muschamp
Department of Psychology, Florida State University Tallahassee, FL 32306-1270, USA

E. B. Nelson
Department of Psychiatry, University of Cincinnati, Cincinnati, OH 45237-0506, USA

M. M. Ostrander
Department of Psychiatry, University of Cincinnati, Cincinnati, OH 45237-0506, USA

T. Porkka-Heiskanen
Institute of Biomedicine/Physiology, University of Helsinki, Helsinki, Finland

G. E. Schafe
Department of Psychology & Interdisciplinary Neuroscience Program, Yale University. Department of Psychology, Yale University, 2 Hillhouse Avenue, New Haven, CT 06520, USA

K. Sinchak
Dept of Psychology, Center for Neuroendocrine Studies and Neuroscience and Behavior Program, University of Massachusetts, Amherst, MA, USA

L. Smale
Department of Psychology, Michigan State University, E. Lansing, MI 48824, USA

K. K. Soma
Departments of Psychology and Zoology, Graduate Program in Neuroscience, University of British Columbia, Vancouver, BC Canada V6T 1Z4

A. Steiger
Max Planck Institute of Psychiatry, Munich, Germany

D. Stenberg
Institute of Biomedicine/Physiology, University of Helsinki, Helsinki, Finland

M. Szyf
McGill Program for the Study of Behavior, Genes and Environment, McGill University, Montreal, Canada. Department of Pharmacology, McGill University, Montreal Canada

J. S. Takahashi
Howard Hughes Medical Institute, Department of Neurobiology & Physiology, Northwestern University, Evanston, IL 60208, USA

M. Tauchi
Department of Psychiatry, University of Cincinnati, Cincinnati, OH 45237-0506, USA

R. L. Thunhorst
Departments of Psychology, Pharmacology and Exercise Science, and the Cardiovascular Center, University of Iowa, Iowa City, IA 52242, USA

Y. M. Ulrich-Lai
Department of Psychiatry, University of Cincinnati, Cincinnati, OH 45237-0506, USA

I. C. G. Weaver
McGill Program for the Study of Behavior, Genes and Environment, McGill University, Montreal, Canada. Douglas Hospital Research Centre, Montreal, Canada. Department of Pharmacology, McGill University, Montreal, Canada

S. C. Woods
Department of Psychiatry, University of Cincinnati, USA

T. Wu
McGill Program for the Study of Behavior, Genes and Environment, McGill University, Montreal Canada. Douglas Hospital Research Centre, Montreal, Canada. Department of Pharmacology, McGill University, Montreal, Canada

L. J. Young
Department of Psychiatry and Behavioral Sciences, Emory University, Atlanta, GA 30329. Center for Behavioral Neuroscience, Atlanta, GA 30329. Yerkes National Primate Research Center, Atlanta, GA 30329, USA

R. Zhang
Department of Psychiatry, University of Cincinnati, Cincinnati, OH 45237-0506, USA

1 Neuroendocrinology of Male Reproductive Behavior

M. J. Baum

Abstract: Berthold, working in the mid-nineteenth century, first published data linking the endocrine secretions of the rooster testes to the display of masculine courtship behavior. Since then hundreds of experiments have been published showing that testosterone is the endocrine signal produced by the Leydig cells of the testes, which in male mammals contributes both to the activation of mating behavior in adulthood and to the organization during perinatal life of neural mechanisms that control this behavior. The broader topic of the neural basis of male sex behavior has recently been reviewed (Hull et al., 2002). Therefore, the present review will concentrate selectively on neuroendocrine variables that control the adult activation as well as the perinatal development of brain mechanisms controlling male-typical sexual motivation, courtship, penile erection, and coital behaviors, with an emphasis on common mammalian models including the rat, mouse, hamster, ferret, and monkey. A common theme to all of these studies is that testosterone exerts its actions in the neural systems controlling male-typical sexual behavior both by acting directly via neural androgen receptors and after neural aromatization to estradiol or 5α–reduction to dihydrotestosterone. Estradiol, in turn, affects neural morphology and function via estradiol receptors of the alpha- or beta-subtypes, whereas dihydrotestosterone, like testosterone, acts via androgen receptors. The evidence reviewed indicates that there are many similarities and a few differences among mammalian species, including higher primates, in the principles of neuroendocrine regulation that control the development and expression of male sexual behavior.

List of Abbreviations: AOB, accessory olfactory bulb; AR, androgen receptor; ArKO, aromatase knockout (mouse); ATD, 1,4,6-androstatriene-3,17-dione (aromatase blocker); BNST, bed nucleus of the stria terminalis; CA, cyproterone acetate; CBP, cAMP response element binding protein; COX-2, cyclooxygenase 2; CPP, conditioned place preference; DA, dopamine; DHT, dihydrotestosterone; DHTP, dihydrotestosterone propionate; DNA, deoxyribonucleic acid; DOPAC, 3,4-dihydroxy-phenylacetic acid; E, estradiol; E25, embryonic day 25; EB, estradiol benzoate; ER, estrogen receptor; F1, first generation; GABA, gamma amino butyric acid; GnRH, gonadotrophin releasing hormone; HIV/AIDS, human immunodeficiency virus/ acquired immunodeficiency syndrome; INA-3, third interstitial nucleus of the anterior hypothalamus; IR, immunoreactivity; LS, lumbar spinal cord; MeA, anterior amygdaloid nucleus; MN-POA/AH, (sexually dimorphic) male nucleus of the preoptic area/anterior hypothalamus (ferret); MPOA/AH, medial preoptic area/anterior hypothalamus; mRNA, messenger ribonucleic acid; MSH, melanocyte stimulating hormone; NO, nitric oxide; NOS, nitric oxide synthase; OHF, hydroxyflutamide; P, progesterone; PGE2, prostaglandin E2; POE, parent of origin effect; POM, sexually dimorphic medial preoptic nucleus (quail); PR, progesterone receptor; sc, subcutaneous; SDN, sexually dimorphic nucleus; SPFp, subparafascicular nucleus ; SRC-1, steroid receptor co-activator 1; Sry, sex determining region of the Y chromosome; T, testosterone; Tfm, testicular feminization; TH, tyrosine hydroxylase; TP, testosterone propionate; VMH, ventromedial hypothalamic nucleus; VNO, vomeronasal organ; WT, wild type

1 Introduction

Contemporary textbooks of behavioral endocrinology (Nelson, 2000; Becker et al., 2002) list the first formal experiment in this field as having been carried out at the University of Gottingen, Germany by Arnold A. Berthold (Berthold, 1849), who demonstrated that castrating young male chickens prevented the development of crowing, sexual behavior, and secondary sex characteristics, including a red comb. By contrast, castrated chickens that were implanted with a testis (either one of their own or from another male) showed normal male-typical development of a comb along with courtship and sexual behaviors. The implanted testes were vascularized and they produced sperm along with an endocrine product that stimulated the observed behavioral changes, including mating. Many years later it was found that testosterone (T), secreted by Leydig cells of the testes, was the relevant endocrine signal that caused all of these behavioral and somatic effects. These early observations laid the groundwork for a myriad of experiments, conducted mainly in avian and mammalian species, on the actions of T in both the nervous system and in other androgen-sensitive target tissues including the prostate gland and penis. Androgens can act directly in the brain to facilitate the expression of numerous social behaviors, including mating. Through its action on another androgen-sensitive tissue, the penis, androgen also indirectly affects the patterning of mating behavior. This chapter concentrates on the actions of T, and its metabolites estradiol (E) and 5α–dihydrotestosterone (DHT), in the adult and/or fetal brain that contribute to male-typical sex partner preference, sexual arousal including penile erection, and the control of courtship and mating behaviors in male mammals. The reversible, adult (so-called activational), the permanent perinatal (so-called organizational), and the pubertal actions of testosterone and its neural metabolites on neural mechanisms controlling male-typical sexual behavior are considered. Experimental findings from several vertebrate species including rat, ferret, mouse, hamster, quail, monkey, and human are used to illustrate the relevant principles of neuroendocrine regulation. Extensive reviews of the literature on the neuroendocrine and neurochemical regulation of male sexual behavior (Hull et al., 2002), as well as of brain and behavioral sexual differentiation (Wallen and Baum, 2002; De Vries and Simerly, 2002), appeared in 2002. I therefore concentrate here on studies concerning the neuroendocrine regulation of masculine sexual behavior that were either not highlighted in those reviews or have been published since 2001.

2 Methods of Studying Appetitive versus Consummatory Components of Male Sexual Behavior as well as Erectile Function

A distinction between the neuroendocrine mechanisms controlling appetitive and consummatory components of male-typical sexual behaviors is emphasized throughout this review. Much research using animal models has concentrated on the different neuroendocrine mechanisms controlling sexual motivation, penile erection, and mating behavior per se. The most common model system is the male rat in which mating performance has been observed during interactions with an estrous female. In this situation, a receptive female is placed in a chamber with the male being tested, and the observer records the occurrence of mounts with pelvic thrusting, penile intromissions, and ejaculations over time. In rats, as in many rodent species, the male displays a series of discrete mounts of the female partner that are accompanied by very brief penile erection, pelvic thrusting, and intromission into the vagina. The male dismounts after each intromission. After 5 to 15 such mounts with intromission, the male ejaculates a copulatory plug composed of secretions of the prostate, seminal vesicle, and coagulating gland and sperm from the testes. Deposition of this copulatory plug against the female's cervix ensures that sperm will pass into the uterus, thereby increasing the likelihood that fertilization of ova will occur in the female's fallopian tubes. Receipt of a minimal number of intromissions from the male over a particular period insures the activation of a neuroendocrine reflex in female rodents, which is needed to stimulate prolactin secretion from the pituitary gland. Prolactin then stimulates the corpora lutea of the ovaries to produce the progesterone needed to establish pregnancy (Erskine, 1995).

Significant species variations exist in the behavioral patterns displayed by male vertebrates during mating. For example, many species of fish perform stereotyped courtship movements that entice the female to deposit eggs in a nest, whereupon the male positions himself over these eggs and deposits his sperm without any physical contact with the female. Male birds exhibit a wide variety of courtship displays, including vocalization (crowing or singing), strutting, mounting, and deposition of sperm through direct cloacal contact with the female. The pattern of mating displayed by the male rhesus monkey resembles that shown by the male rat, as described earlier. By contrast, the male ferret mates by grasping the female's neck, mounting, and exhibiting episodes of pelvic thrusting (accompanied by penile erection). Once the erect penis is inserted into the female's vagina, thrusting ceases, and the intromission is maintained for up to 1.5 h, even after ejaculation occurs. This intromissive stimulation activates a neuroendocrine reflex in females leading to the pituitary secretion of luteinizing hormone (LH) and subsequent ovulation (Carroll et al., 1985). Specific patterns of masculine courtship and coital behavior have evolved in different species (Dewsbury, 1972) to maximize the chances of reproductive success.

Recording the frequency or timing the duration of neck grip, mounting, intromission, and ejaculation provides a useful index of masculine coital performance. However, these variables provide only partial insight into an animal's level of sexual motivation. Sexual motivation is a construct that refers to the inclination of an individual to seek out and approach a partner for the purpose of mating. Masculine sexual motivation has been studied in several different ways. These include (Stone et al., 1935) monitoring the willingness of male rats to cross an electrified grid to gain access to an estrous female, as well as latencies of males to approach an estrous female tethered in the goal box of a straight runway (Beach and Jordan, 1956; Bolles et al., 1968; Lopez et al., 1999). Another approach has been to require rats to press a lever in a Skinner box to gain access to an estrous female (Beck, 1971). A shortcoming of this method is that subjects' operant responses occurred at a low rate under a continuous reinforcement schedule, and when the female became available, the resulting sexual interaction disrupted subjects' operant lever-pressing behavior. Everitt et al. (1987) improved on this procedure by providing male rats with a conditioned secondary reinforcer (a red light) that was initially associated with access to an estrous female that dropped into the male's compartment from an overhead location. This procedure led to high levels of lever pressing by male subjects in order to illuminate the conditioned stimulus. Although some interesting data were obtained using this method, it also had certain disadvantages in that considerable pretraining was needed for subjects to acquire the task, and various experimental manipulations could influence task performance by affecting subjects' motivation to lever press for the conditioned stimulus as opposed to the unconditioned sexual stimuli.

The method of measuring runway-approach latencies is insensitive and the method of training animals to press a lever for access to a goal stimulus is tedious. As a compromise, numerous investigators have chosen to assess the subjects' preference to approach and interact with any one of two different social stimuli that are tethered to the opposite ends of a 3-compartment box or in the goal boxes of a T- or Y-shaped maze. This method has also been used to provide a choice between volatile odors from anesthetized conspecifics and physical access to these animals. Such an approach has been used to assess the preference of rats (Vega Matuszczyk et al., 1988), hamsters (Johnson and Tiefer, 1972), ferrets (Stockman et al., 1985), and mice (Bakker et al., 2002) for same-sex versus opposite-sex conspecifics. This method has also been extensively used (Winslow et al., 1993) to establish the roles of vasopressin and oxytocin in monogamous pair bonding in male and female prairie voles, respectively. The motivation of male ferrets to approach same-sex conspecifics in a T-maze was studied after the placement of lesions in the sexually dimorphic preoptic/anterior hypothalamic region (Paredes and Baum, 1995). More recently, this method was adapted so that volatile odors from same-sex versus opposite-sex conspecifics could be presented in an air-tight Y-maze, with the aim of establishing the role of body odorants in heterosexual mate recognition in ferrets of both sexes (Kelliher and Baum, 2001). Avian species use visual signals to identify preferred mating partners, and the time that male quail spend approaching a window to view a receptive female has been taken as an index of their sexual motivation (Balthazart et al., 1998).

Sexual motivation in male rats and mice has been assessed using bilevel test chambers. Males' level-changing behavior increased significantly when subjects were tested in the presence of an estrous—as opposed to an anestrous—stimulus female (Mendelson and Pfaus, 1989). Anosmic male rats showed less level changing in the presence of an estrous female (Van Furthand Van Ree, 1996b), implying that males' motivation to approach the female resulted from their attraction to volatile estrous odors. Male rats showed significantly less level-changing behavior within a few minutes following ejaculation with an estrous female (i.e., during the postejaculatory interval) (Van Furth and Van Ree, 1996a), providing further evidence that this behavior is a useful index of the males' sexual motivation. Male rats that were treated with dopamine (DA) receptor blockers also showed significantly less level-changing behavior in the presence of an estrous female (Pfaus and Phillips, 1991), implying that the activation of DA neurons normally augments masculine sexual motivation and reward.

A simple method for assessing the rewarding characteristics of drugs of abuse (e.g., heroin, cocaine, amphetamine) has been to demonstrate that the administration of a particular drug in a compartment (distinctive because of its color, odor, and/or floor texture), which was initially not preferred by a subject, causes the subject to prefer that compartment as a result of repeated pairing of its physical features with receipt of the drug (Mucha et al., 1982). Male rats learn a conditioned place preference (CPP) for the opportunity to mate with an estrous female (Miller and Baum, 1987). This type of CPP was less evident in male subjects that had no control over the rate at which an estrous female allowed mounting and intromission behaviors to occur (Martinez and Paredes, 2001).

In addition to the above-mentioned methods for measuring sexual motivation and mating performance, per se, several model systems have been established in which to study the neuroendocrine regulation of penile erection in mammals. In many commonly studied laboratory species, penile erections can be easily observed and counted "in copula" with the use of a mirror for ventral viewing during mating sessions with a female conspecific. More detailed measurements of erectile function, including monitoring blood pressure increments in the penile corpora cavernosa that are associated with erection induced by electrical stimulation of the cavernosal nerve, have been made in studies using anesthetized male rats (Lugg et al., 1995; Marin et al., 1999) and rabbits (Traish et al., 1999). Numerous experiments have also been conducted in which erectile function was studied "ex copula" in awake male rats during whole body restraint and retraction of the penile foreskin coupled with continuous pressure at the base of the penis (Hart, 1967). Finally, erection has also been studied ex copula by placing an awake male rat down wind from an estrous female whereupon noncontact, psychogenic erections induced by volatile chemosensory signals from the female are observed using a ventrally placed mirror (Sachs, 1997).

3 Activation of Male-typical Sexual Behavior and Penile Erection by Testosterone and its Neural Metabolites, Estradiol and Dihydrotestosterone

3.1 Effects of Castration and Systemic Administration of Steroids or Antagonist Drugs

With one notable exception (discussed later in detail), studies conducted over the past 75 years have established the indispensable role of the testicular steroid hormone, T, in promoting the activation of both appetitive and consummatory components of sexual behavior among infra primate mammalian species (Hull et al., 2002). Thus, castration of male rodents (e.g., rats, mice, hamsters, and guinea pigs) inevitably leads to a steady decline over a period of several weeks in ejaculation, followed by a decline in mounting and approach of an estrous female. Administration of T immediately after castration will maintain high levels of male-typical mating behavior, and this treatment will readily restore mating in castrated males even when treatment is begun many months after the postcastration disappearance of sexual behavior. Evidence will be considered suggesting that T, either acting itself or after metabolism in the brain into E or DHT, plays a permissive role in the sense that it facilitates the display of appetitive and consummatory sexual behaviors in response to olfactory, visual, and/or genital–somatosensory stimuli derived from a sexually receptive estrous female. In the absence of circulating T, these same stimuli lack the ability to elicit sexual behaviors. Adult male rats normally have circulating levels of T that range from 1 to 3 ng/ml; however, sexual behavior can be readily maintained in castrated rats by s.c. administration of very low doses of T that result in plasma levels of <1.0 ng/ml (Damassa et al., 1977). Higher circulating levels of T are required to restore mating in long-term castrate males. Male mammals normally begin displaying mating behavior around the time of puberty, when testicular production of T and the production of mature sperm is established (55–65 days postnatal in male rats). The first display of sexual behavior can be significantly advanced by daily administration of testosterone priopionate (TP) to prepubertal male rats (Stone, 1940; Baum, 1972), and the doses of TP required for these effects are considerably higher than those needed to maintain or restore mating in adult castrated males. In male hamsters that were castrated either prepubertally or after the age of puberty, s.c. administration of different doses of T (pellets given s.c.) activated all aspects of mating in the adult, but not the prepubertal males (Meek et al., 1997), suggesting that males' ability to respond behaviorally to T increases as they pass through the age of puberty.

Early in the history of contemporary behavioral neuroendocrinology, Beach argued that the activational effect of T on male rats' sexual behavior reflected the increased size and sensitivity of somatosensory receptors on cornified papillae of the glans penis (Beach and Levinson, 1950). Evidence that this peripheral action of T—presumably acting via the androgen receptor (AR) agonist actions of its metabolite, DHT—cannot account for the behavioral effects of T, came from the observation (McDonald et al., 1970; Feder, 1971; Whalen and Luttge, 1971) that administration of DHT to castrated male rats failed to activate mating behavior, whereas it stimulated the growth of the prostate gland and other accessory sex organs, including the penis, up to levels characteristic of testes-intact controls. By the early 1970s, it also became apparent that

DHT and T are specifically bound, with high affinity, in the male rat hypothalamus (Kato and Onouchi, 1973). It was subsequently shown (Vreeburg et al., 1975) that specific, high-affinity hypothalamic binding of the androgen metabolite, E, also occurs in male rats. In subsequent years the AR and the ERα were cloned, and mRNAs for each type of receptor were found to be expressed in the hypothalamus and amygdala (among other regions) of male rats (Simerly et al., 1990). Early studies also established that T can be aromatized to E (Naftolin et al., 1972) and reduced to DHT (Denef et al., 1973) in the rat hypothalamus and amygdala. These facts strongly pointed to a neural action of T and/or its androgenic or estrogenic metabolites in the activation of masculine sexual behavior.

In contrast to the inability of systemic treatment with DHT, by itself, to activate mating in male rats, two studies (Davidson, 1969; Sodersten, 1973a) demonstrated that s.c. injections of estradiol benzoate (EB) into castrated male rats restored appreciable levels of male sexual behavior, especially mounting and intromission, along with an occasional ejaculation response. Administration of EB prepubertally to male rats activated precocious expression of mounting and intromission behaviors even more readily than a prepubertal treatment with TP (Baum, 1972). As shown in *Figure 1-1*, combined s.c. administration of 2 μg EB and 200 μg DHTP activated all components of sexual behavior in castrated male rats as readily as

Figure 1-1

Combined s.c. administration to castrated male rats of estradiol benzoate (EB)+dihydrotestosterone propionate (DHTP) duplicated the activational effects of testosterone propionate (TP) on sexual behavior. The percentage of males showing intromission behavior with an estrous female is shown before (pre) and after the onset of daily injections of different hormones. Similar data (not shown) were obtained for the occurrence of ejaculation. Group Ns are given in parentheses. Adapted from Baum and vreeburg (1973)

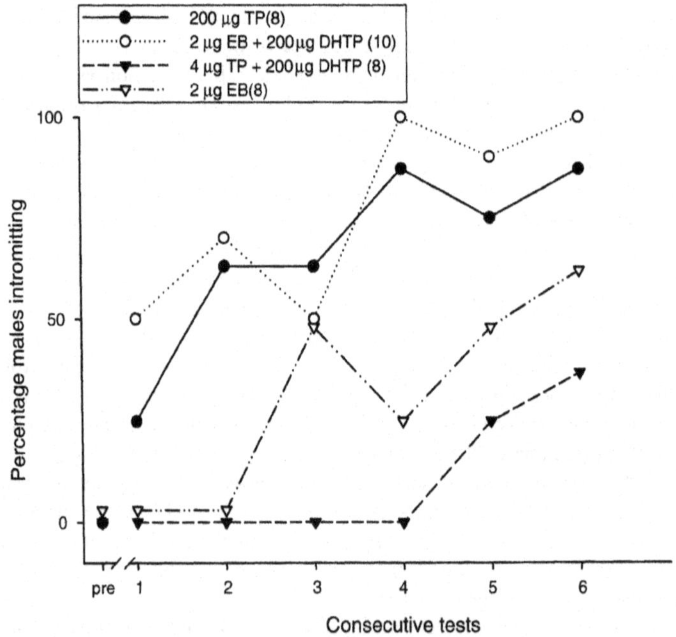

200 μg TP itself. By contrast, combining a low dose of TP (4 μg/rat/day) with the same high dose of DHTP stimulated only low levels of mating behavior (Baum and Vreeburg, 1973). Similar results were also obtained by other investigators (Larsson et al., 1973; Feder et al., 1974). An obvious explanation for the

combined action of E and DHT was that E acted on neural mechanisms that control appetitive and pre-ejaculatory components of the behavior, whereas DHT acted in the penis to promote the genital sensory inputs needed in order for intromissions to trigger ejaculation. Interestingly, the ability of s.c. injections of DHT to augment mounting behavior in castrates that were given EB concurrently was not reduced by pudendal nerve transaction (Lodder and Baum, 1977). Pudendal nerve transaction interrupted sensory inputs from the penis to the spinal cord; thus, the result obtained implies that there are central activating effects of DHT on the expression of sexual behavior in addition to any facilitation of mating due to a stimulation of penile sensory responsiveness.

A wealth of information (Hull et al., 2002) has appeared assessing the respective roles of E and DHT in the activation of sexual behavior in a variety of vertebrate species. For example, in male hamsters the conversion of T to E may not be required for the activation of sexual behavior in that s.c. administration of the aromatase inhibitor, Fadrozole, failed to inhibit the expression of any aspect of male sexual behavior (Cooper et al., 2000). This outcome contrasts with a similar study in male rats (Bonsall et al., 1992) in which systemic Fadrozole treatment dramatically inhibited sexual behavior. Similar effects had been obtained in early studies (Christensen and Clemens, 1975; Beyer et al., 1976) in response to the administration of other aromatase inhibitors to male rats. Likewise, s.c. treatment with Fadrozole reduced the number of anticipatory level changes shown by male rats in anticipation of the presentation of an estrous female (Roselli et al., 2003), in addition to reducing mating behavior. In the latter study, s.c. administration of E reversed these inhibitory effects of Fadrozole treatment, further affirming the essential role of E as a mediator of T's activational effect on appetitive as well as some consummatory components (mounting, intromission) of mating in the male rat.

It must be noted that despite the wealth of data pointing to a role of E metabolites of T in the activation of sexual motivation and aspects of mating behavior in the male rat, there are several studies that continue to implicate T in the control of these behaviors. Thus, systemic injections with the potent AR blocker, Flutamide, to castrated, T-treated rats blocked the restoration of mounting, intromission, and ejaculation behaviors (Gladue and Clemens, 1980b), although only intromission and ejaculatory responses were blocked by Flutamide treatment of male rats in other studies (Sodersten et al., 1975; Gray, 1977). As DHT, given by itself, fails to augment even mounting behavior in rats, these workers concluded that T may act at ARs to facilitate appetitive courtship responses and consummatory components of this behavior. This view is supported by another study (McGinnis and Dreifuss, 1989), which compared the nuclear occupation of AR and ER in several hypothalamic and amygdaloid regions of castrated male rats that were treated with E, DHT, T, or E + DHT. Although a low dose of E + DHT treatment resulted in nuclear AR and ER occupation levels that were comparable to those of T-treated castrates, these particular doses of E and DHT failed to activate even mounting behavior in castrated rats. Treatment with higher doses of E, either alone or combined with DHT, activated higher levels of mounting, with some intromission and ejaculation behavior. Based on these results, the authors argued that T itself, and not E, is primarily responsible for the activation of male sex behavior in rats.

There is a small literature suggesting that progesterone (P), acting via neural progestin receptors (PR), may contribute in male vertebrates to the activation of male-typical courtship and mounting behaviors. When given in moderate doses, P activated all facets of sexual behavior in castrated male lizards (Young et al., 1991; Lindzey and Crews, 1992), although treatment with T had greater and longer-lasting stimulatory effects on males' courtship behaviors than P (Sakata et al., 2003). Likewise, administration of P to castrated male rats partially duplicated the activational effects of T on mating behavior, and this effect of P was blocked by concurrent administration of a PR antagonist, suggesting that the effects of P were mediated by its action at PR as opposed to AR (Witt et al., 1995). Gonadally intact male mice with a homozygous null mutation of the PR showed significantly lower mount frequencies than WT controls, and heterozygous mutant males were less responsive to the activational effect of T following castration (Phelps et al., 1998). This latter finding points to the interesting possibility that PR activation may somehow contribute to the activational effects of T on the expression of male-typical sexual behaviors. More research will be required to characterize the nature of any such relationship.

3.2 Species and Genotype Variations in the Effects of Castration and Hormone Replacement on Male Sexual Behavior

With the exception of one particular hybrid mouse strain (details given later), male rodents reliably display a progressive loss in all aspects of sexual motivation, mating, and erectile function within a few weeks after castration and the resultant deprivation of circulating T, regardless of whether or not subjects had received coital experience prior to castration (Hull et al., 2002). Considerably more variability in the display of sexual behavior has been reported in males of two carnivore species, the dog (Hart, 1968; Beach, 1970) and cat (Rosenblatt and Aronson, 1958), especially when castration was carried out after subjects had received mating experience. In these instances, mating persisted in some individuals for many months after castration. By contrast, in another carnivore, the ferret, castration reliably led to a disappearance of mating within one month in all males tested (Lambert and Baum, 1991). Perhaps not surprisingly, castration of male rhesus monkeys failed to disrupt mating performance in some individuals. In contrast to the rat, administration of DHTP duplicated the facilitatory effect of TP on ejaculation capacity in castrated rhesus monkeys (Phoenix, 1974) whereas administration of EB failed to influence mating in castrated monkeys (Phoenix, 1978; Michael et al., 1990). This observation raised doubts about a possible contribution of T aromatization to the activation of sexual behavior in male primates. More recently, however, it was reported (Zumpe et al., 1993) that systemic administration of the aromatase inhibitor, Fadrozole, to gonadally intact male cynomolgus monkeys significantly reduced all aspects of mating, including ejaculation after 2 weeks of treatment. This effect of Fadrozole treatment was partially reversed by concurrent treatment with E, raising the possibility that aromatization of T to E plays some role in the activation of sexual behavior in male primates just as it does in several species of rodent.

Adult castration of human males had variable effects on sexual behavior in different individuals (Heim and Hursch, 1979), although average levels of erectile function, sexual fantasies, and intercourse were reliably reduced in groups of men within 6 weeks after the inhibition of testicular T production by administration of a GnRH antagonist drug (Bagatell et al., 1994). These behavioral effects were reversed by concurrent treatment with T plus the GnRH antagonist. However, administering an aromatase inhibitor failed to reduce this latter activational effect of T treatment, providing no evidence that the conversion of T to E contributes to sexual arousal or performance in adult men. This conclusion must be tempered, however, by the absence of any independent confirmation that the particular aromatase inhibitor administered by these investigators had its intended effects on the neural conversion of T into E. Indeed, a more recent clinical study (Carani et al., 1999) of a single man with a spontaneous mutation in the *Cyp* 19 aromatase gene suggests that the E may contribute to sexual arousal and performance in adult men. This individual had a male-typical gender identity and a heterosexual orientation (he was married to and reported having sex with a woman). Administration of T over several months failed to influence the incidence of sexual intercourse, masturbation, or erotic fantasies; however, each of these variables was increased after an equivalent period of E treatment. More clinical studies are needed to determine conclusively whether the aromatization of T to E plays an essential role in the regulation of sexual motivation and mating performance in higher primates including man. Regardless of the outcome of such studies, the available comparative evidence supports the early suggestion of Frank Beach (1942b) that the degree to which males of different species depend upon gonadal hormones for its ability to display courtship and mating behaviors is inversely correlated with the degree of cerebral cortex development. An exception to this rule is now discussed.

Male mice, like other rodents, typically stop pursuing and mating with an estrous female within a few weeks after castration. However, McGill and Tucker (1964) initially made the serendipitous observation that the male F1 hybrid offspring produced by mating a female C57BL/6J with a male DBA/2J mouse (B6D2F1 males) were surprisingly resistant to the effects of castration on the display of all aspects of male sexual behavior, including ejaculation. Subsequent studies of this particular F1 hybrid strain of male mice (McGill and Haynes, 1973; Clemens et al., 1988; Coquelin, 1991) showed that all aspects of sexual behavior continued to be expressed after castration over the subsequent lifespan (upto 2 years) in some individuals. Several studies have explored a possible endocrine basis for the persistence of mating capacity in B6D2F1 mice, following castration. Castration in these animals, as in other genotypes, caused a profound reduction in circulating levels of T (Clemens et al., 1988; Sinchak et al., 1996) and in removal of the adrenal glands, in addition, castration of B6D2F1 male mice failed to disrupt the postcastration maintenance of mating behavior (Thompson et al., 1976). Some evidence raises the possibility that the persistence of sexual behavior in castrated B6D2F1 males depends on the activational effects of E, which continues to circulate in very low concentrations in this as well as other mouse strains even after castration (Sinchak et al., 1996). Administration of the aromatase inhibitor, ATD, to castrated B6D2F1 significantly reduced the expression of sexual behavior (Sinchak et al., 1989). No explanation was provided for why B6D2F1 males, unlike other mouse strains, are so highly responsive to the activating effects of the low circulating levels of E that persist after castration. One hint of a possible mechanism stems from the early observation (McGill and Manning, 1976) that there is a parent of origin effect (POE) in the ability of mating to persist in male mice following castration. Thus, this persistence of mating capacity was seen only in the F1 male offspring of a female C57BL/6J with a male DBA/2J; male F1 offspring of the reverse parentage (female DBA/2J × male C57BL/6J) stopped mating within a few weeks after castration. Many POEs have been attributed to the differential expression (genomic imprinting) of autosomal genes, depending on whether particular alleles are inherited from the mother or the father of the individual being studied (Isles et al., 2002). Future studies of the possible contributions of a small set of imprinted genes to the activational effects of E on neural mechanisms controlling sexual arousal and mating performance could provide insights into the mechanism that allows B6D2F1 male mice to continue to display sexual behavior after castration.

3.3 Effects of Hypothalamic Implantation of T, E, DHT or Antagonist Drugs on the Expression of Sexual Behavior

Several early experiments (Brookhard and Dey, 1941; Hillarp et al., 1954; Soulairac and Soulairac, 1956) established that destructive lesions of the medial preoptic area/anterior hypothalamus (MPOA/AH) disrupted the display of sexual behavior in male rats and guinea pigs. These studies, which appeared well before the advent of knowledge about the hypothalamic localization of T-metabolic enzymes or AR and ER, led shortly thereafter to a preliminary study (Fisher, 1956) that showed that intra-hypothalamic implantation of T-activated sexual behavior in castrated male rats. More than a decade later several studies were published that corroborated and extended these findings. Davidson (1966) reported that small amounts of TP, implanted into the anterior and posterior hypothalamus, were effective in restoring sexual behavior in castrated male rats. Subsequently, more detailed studies showed that implanting TP in the MPOA/AH more effectively restored sexual behavior than implanting TP in the posterior hypothalamus (Lisk, 1967; Johnston and Davidson, 1972; Kierniesky and Gerall, 1973). Such activational effects of TP implanted directly into the MPOA/AH of castrated rats could not be explained by leakage of T into the general circulation (Smith et al., 1977). Activational effects of MPOA/AH implants of TP on masculine courtship behavior were also reported in several other species including the domestic fowl (Barfield, 1969), the ring dove (Barfield, 1971), the Barbary dove (Hutchison, 1974), the quail (Balthazart and Surlemont, 1990), the mouse (Matochik et al., 1994), and the ferret (Tang and Sisk, 1991), thereby illustrating an activational effect of this steroid on male sexual behavior in a range of avian and mammalian species.

As already explained, combined systemic injections of E and DHT activated mating behavior in castrated male rats as effectively as injections of T (Baum and Vreeburg, 1973; Larsson et al., 1973). Several studies implicate E, formed locally in the MPOA/AH from circulating T, as a sex steroid that acts on neurons in this particular brain region to activate masculine sexual behavior. Thus, implants of E into the MPOA/AH restored sexual behavior in castrated male rats (Christensen and Clemens, 1974; Davis and Barfield, 1979), thereby duplicating the effect of T when implanted into this region. The results of these experiments are complemented by studies in which infusion of aromatase-inhibiting drugs such as ATD (Christensen and Clemens, 1975) or Fadrozole (Clancy et al., 1995) into the MPOA/AH of male rats inhibited the ability of exogenous or endogenous T to activate male-typical sexual behavior. Likewise, when the aromatase inhibitor, ATD, was bilaterally implanted into the sexually dimorphic nucleus (POM) of the MPOA/AH, the activation of copulatory behavior induced by a systemic T treatment in castrated male quail was completely blocked (Balthazart et al., 1990). This inhibition of coital behavior was accompanied by an inhibition of the aromatase activity in the MPOA/AH. Moreover, implants of a synthetic estrogen (diethylstilbestrol) into the MPOA/AH restored copulatory behavior in castrated quail, whereas implantation of a nonaromatizable androgen (methyltrienolone) lacked consistent activational effects on courtship behavior (Balthazart et al., 1990). These workers also found that the activational effects of systemic T treatment were blocked by POM implants of either an antiandrogen (Flutamide) or an antiestrogen (Tamoxifen). The authors concluded that estrogenic metabolites of T are produced in the sexually dimorphic POM of male quail, where they selectively activate the expression of sexual behavior. Two additional studies point to a similar conclusion in male rats. In one study (Vagell and McGinnis, 1997), infusion of the aromatase inhibitor, Fadrozole, into the lateral ventricle, blocked the ability of systemic T to restore ejaculation. This inhibition of T-induced mating by central administration of Fadrozole was reversed in other castrated males that also received an s.c. capsule containing a low dose of E. In another study (Clancy et al., 2000), systemic administration of Fadrozole to gonadally intact male rats effectively reduced mounting and ejaculation; however, bilateral implantation of E directly into the MPOA/AH effectively reversed the inhibitory effect of systemic Fadrozole on mating performance. These results confirm the specific role of T aromatization in the activation of mating performance, including mounting, intromission, and ejaculation. Interestingly, however, administration of a low systemic dosage of E to castrated males, by itself, failed to activate sexual behavior (Vagell and McGinnis, 1997). This suggests that the activational effects of E on mating may depend on the concurrent, synergistic action of an androgen, acting via neural androgen receptors. In contrast to its effect on mating performance per se, icv administration of Fadrozole failed to attenuate the activational effect of T on males' preference to approach an estrous as opposed to an anestrous stimulus female. This latter result suggests that males' motivation to seek out an estrous female may depend on the effects of T itself, as opposed to its estrogenic metabolite.

An initial attempt to attenuate the activational effect of T on ejaculation in castrated male rats by implanting the antiandrogen, cyproterone, into the MPOA/AH failed; if anything, implanted males showed higher levels of mating than cholesterol-implanted controls (Block and Davidson, 1971). Several subsequent studies using more potent AR receptor antagonists, including flutamide or its active metabolite, hydroxyflutamide (OHF), have provided more definitive evidence of a role of T, and/or DHT, acting via AR, in promoting aspects of sexual partner preference and mating performance in male rats. In an earlier study (McGinnis et al., 1996), implantation of OHF into the MPOA/AH prevented the restoration of sexual behavior (mounting and ejaculation) in castrated rats that received s.c. T capsules concurrently. In a subsequent study (McGinnis et al., 2002), this inhibitory effect of OHF was most reliably obtained when implants were placed in the rostral portion of the POA. Implanting OHF into the posterior POA did not disrupt T-induced mating behavior, whereas it did reduce males' preference to seek out an estrous as opposed to an anestrous female. This latter effect may have resulted from a leakage of OHF into caudal regions of the hypothalamus. Indeed, in a more recent study (Harding and McGinnis, 2004) implantation of OHF into the ventromedial hypothalamic nucleus (VMH), but not the MPOA/AH, significantly reduced the preference of castrated, T-treated male rats to approach an estrous versus an anestrous female and blocked the establishment of a CPP response for the opportunity to mate with an estrous female. As in previous studies by this group, implantation of OHF into the MPOA/AH attenuated the activational effect of T on mating behavior, whereas it failed to affect either partner preference or males' ability to learn a CPP response for access to an estrous female. Further evidence of the role of a male's VMH in male-typical sexual motivation was provided by the observation that bilateral implants of TP into this brain region stimulated approach to an estrous as opposed to an anestrous female without facilitating mating behavior per se (Harding and McGinnis, 2003).

As already explained, the enzyme 5α–reductase is expressed in the rat hypothalamus and in other subcortical brain regions (Denef et al., 1973), raising the question of whether any of the reported effects of AR blockade (after hypothalamic administration of OHF) reflect an antagonism of DHT action or that of T itself. In one study (Butera and Czaja, 1989), s.c. administration of DHT when combined with implantation of either T or of DHT into the MPOA activated equivalent levels of mating behavior in castrated male rats. In the majority of studies, however, neither systemic administration of DHT nor intra-hypothalamic implants of this androgen succeeded in activating sexual behavior. Thus, systemic administration of

DHT (McDonald et al., 1970; Feder, 1971; Whalen and Luttge, 1971), a nonaromatizable synthetic AR agonist, R1881 (Baum et al., 1987), or direct administration of DHT into the MPOA/AH (Johnston and Davidson, 1972; Baum et al., 1982b) failed to activate mating behavior in castrated male rats, suggesting that any contribution of hypothalamic AR activation to the expression of masculine sexual behavior is caused by T itself. Further support for this conclusion is also drawn from the observation (Bradshaw et al., 1981) that concurrent administration of a 5α-reductase inhibitor, 17β-testosterone carboxylic acid, failed to attenuate mating induced in castrated male rats by T, even though this treatment did significantly reduce T-stimulated increments in prostate and seminal vesicle weights.

Studies conducted in males of several different infraprimate mammalian species, including the rat (Carr et al., 1965; Kelliher et al., 1999), mouse (Wysocki et al., 1982), hamster (Powers et al., 1979), and ferret (Kelliher and Baum, 2001) have implicated odor cues emitted from the urine and/or scent glands of estrous females in male-typical patterns of sex partner preference and masculine sexual arousal. Extensive neuroanatomical evidence (Wood, 1997) has established the existence of projections from odor receptors in the vomeronasal organ (VNO) to the accessory olfactory bulb (AOB). Mitral cells from the AOB project to the medial amygdala, which in turn sends projections to the principal nucleus of the bed nucleus of the stria terminalis (BNST) and the MPOA. Afferents from the main olfactory bulb also gain access to this VNO projection to the hypothalamus via axonal projections to the anterior cortical amygdaloid nucleus (Scalia and Winans, 1975). By monitoring the expression of the immediate-early gene, c-fos, it has been possible to show that body odors (e.g., urine and vaginal secretions in rats and hamsters, respectively) activate neurons in each of the above-mentioned brain regions projecting to the MPOA/AH (Fiber et al., 1993; Bressler and Baum, 1996). Experiments in male hamsters (Fiber and Swann, 1996) and male rats (Paredes et al., 1998a) showed that administration of T enhanced the number of Fos-immunoreactive (IR) cells in the VNO-projection circuit that were induced by exposure to odors from opposite-sex conspecifics. Wood and Newman (1995b) provided evidence that such steroidal enhancement of olfactory processing contributes to the activation of sexual behavior in castrated male hamsters. Unilateral implantation of T into the MPOA/BNST successfully activated mating in subjects in which the ipsilateral olfactory bulb was intact; removal of the olfactory bulb ipsilateral to the T implant prevented the activation of sexual behavior. It is noteworthy that previous work by these same investigators (Wood and Newman, 1993) had demonstrated that many of the Fos-IR neurons activated in the MPOA and BNST of male hamsters after they mated with a female also coexpressed AR. More recent studies carried out in rat (Greco et al., 1998a, 1998b) and mouse (Shah et al., 2004) showed that mating-induced Fos expressing neurons located in several extra-hypothalamic sites including the cortical–medial amygdala express AR and/or ERα.

3.4 Extra-Hypothalamic Sites of Sex-Steroid Activation of Male Sexual Behavior

The suggestion (Wood, 1997; Newman, 2002) that T or its metabolites E and/or DHT act in the medial amygdala (MeA) to facilitate the transmission of olfactory information to the hypothalamus, with a resultant stimulation of approach, sexual arousal, and mounting of an estrous female is supported by numerous experiments in which sex hormones have been implanted directly into this temporal region of male rodents. Unilateral implants of T into the MeA stimulated approach and mounting of an estrous female in male hamsters (Wood and Newman, 1995a), and the facilitation of sexual behavior seen after such implants was not further enhanced by combined MeA and MPOA/AH implants of T (Coolen and Wood, 1999). These results point to a distributed network of steroid-sensitive neurons (Cottingham and Pfaff, 1986) that independently amplify and transmit sexually relevant chemosensory inputs. As already described, olfactory bulbectomy carried out ipsilaterally, but not contralateral to, a unilateral T implant into the MPOA and/or BNST, disrupted mating behavior in castrated male hamsters (Wood and Newman, 1995b). When T was implanted unilaterally into the MeA instead of the hypothalamus of castrated male hamsters, surgical removal of the olfactory bulb that was either ipsilateral or contralateral to the implant blocked the activation of sexual behavior (Wood and Coolen, 1997), suggesting that bilateral olfactory inputs are required at the level of the MeA where they are amplified and transmitted into the hypothalamus by steroidal actions. Implantation of T into the MeA of castrated male rats augmented the display of penile erections shown in response to odors emitted from estrous females (Bialy and Sachs, 2002), suggesting that the steroidal enhancement of olfactory processing may also facilitate sexual arousal. Implantation of E, but not DHT, into the MeA of castrated male hamsters duplicated the activational effects of T on the display of approach and mounting of an estrous female, but not intromission and ejaculation (Wood, 1996). Likewise, implantation of E into the MeA of castrated male rats stimulated mounting behavior (Rasia-Filho et al., 1991) and implantation of E into the MeA of gonadally intact male rats that received the aromatase inhibitor, fadrozole, s.c., stimulated mounting, but not intromission and ejaculation (Huddleston et al., 2003). One interpretation of these results is that estrogen-sensitive neurons in the MeA respond to E, formed via the local aromatization of T, to facilitate the sexual arousal induced by olfactory signals from estrous females. Whereas neither systemic (McDonald et al., 1970) nor intracranial (MPOA or MeA) (Johnston and Davidson, 1972; Wood, 1996) administration of DHT by itself stimulated sexual behavior in castrated male rats, implanting DHT into the MeA of castrated male rats given concurrent s.c. implants of E, stimulated all aspects of sexual behavior including ejaculation (Baum et al., 1982b). No such stimulation was seen in groups of castrated, E-primed males that received DHT implants in the MPOA/AH. These results are consistent with the suggestion (Baum and Vreeburg, 1973) that both estrogenic and 5α reduced androgenic metabolites of T contribute to the activation of mating behavior in the male rat by acting in the brain. Further evidence that DHT affects sexual behavior by acting neurally, as opposed to solely at the level of penile sensory receptors stems from the observation (Lodder and Baum, 1977) that denervating the penis by cutting the pudendal nerves failed to diminish the ability of DHT, given concurrently with E to castrated male rats, to stimulate mounting behavior. In the event that DHT (as opposed to T) is the behaviorally active AR ligand that normally facilitates mating behavior in gonadally intact animals, it would appear that it occurs in brain regions other than the MPOA/AH. It is possible that the synergistic action of E and DHT reflects an up-regulation of AR synthesis (Handa et al., 1987), and a resultant enhancement of the behavioral effects of DHT (or of T) at these ARs in sites such as the MeA or BNST. In this regard it is interesting to note that the distribution of AR was significantly reduced in the BNST of male mice bearing a null mutation for ERα (Wersinger et al., 1997).

The MeA and BNST (also referred to as the "extended amygdala" (Alheid and Heimer, 1988)) contain steroid-sensitive neurons that amplify olfactory inputs to the MPOA/AH, which are thought to enhance masculine sexual motivation and sexual arousal including penile erection. On the other hand, somatosensory inputs associated with achieving intromission with an estrous female are apparently conveyed to the MPOA/AH via a polysynaptic pathway that originates in laminae 7 and 10 of lumbar spinal cord segments L3 and L4 (Ju et al., 1987) and includes the parvocellular subparafascicular nucleus (SPFp) of the posterior thalamus (also known as the central tegmental field) (Simerly and Swanson, 1986). Significant increases in the number of Fos-IR neurons were noted in the SPFp of male rats after they achieved intromissions, with a maximal level of Fos seen after a series of intromissions leading to ejaculation (Baum and Everitt, 1992; Veening and Coolen, 1998). Many of the same neurons in the SPFp that express Fos after intromissions and ejaculation coexpress AR (Greco et al., 1998b), raising the question of whether androgenic facilitation of mating may depend, in part, on a steroidal action in these neurons. Indeed, earlier studies (Brackett and Edwards, 1984; Maillard and Edwards, 1991) showed that thalamic lesions that included the SPFp disrupted the expression of sexual behavior in male rats as reliably as lesions of the MPOA/AH.

Coolen and coworkers (Truitt and Coolen, 2002; Truitt et al., 2003) provided evidence that a cluster of lumbar spinal (LS) neurons, which project to the thalamic SPFp nucleus may comprise an "ejaculation generator" in male rats. Mating with ejaculation stimulated Fos-IR in neurons of the LS (Truitt and Coolen, 2001), and in another study (Greco et al., 1999) many of the mating-induced Fos-IR neurons in the LS, which were retrogradely labeled by fluorogold injected into the thalamic SPFp, also coexpressed AR. Targeted destruction of spinal LS neurons in male rats by intrathecal administration of the neurotoxin, saporin (which was conjugated to substance P, thereby causing it to bind with high affinity to neurokinin receptors expressed on LS neurons), eliminated ejaculatory capacity without disrupting males' capacity to mount and intromit with estrous females (Truitt and Coolen, 2002). These results point to a spinal mechanism whereby AR activation by either T and/or DHT may facilitate the processing of penile sensory inputs associated with erection and intromissions that lead to ejaculation.

3.5 MPOA/AH Sex Steroid–Dopamine Relationships in the Control of Male Sexual Behavior

Several lines of pharmacological and neurochemical evidence have established a link between the activational effects of T (perhaps acting via its aromatized metabolite, E) on mating and a facilitation of activity in dopaminergic neurons that are located in the dorso-lateral hypothalamus and in zona incerta, and that terminate in hypothalamic regions including the MPOA/AH and the paraventricular nucleus (Dahlstrom and Fuxe, 1964). The earliest evidence of such a link derived from a study (Malmnas, 1977) showed that systemic injections of the nonspecific dopamine (DA) agonist, apomorphine, reversed the postcastration decline in mating performance otherwise seen in male rats. Similar results were later obtained by administering apomorphine either systemically or into the MPOA/AH (Scaletta and Hull, 1990). Conversely, either systemic (Baum and Starr, 1980) or intra MPOA/AH (Pehek et al., 1988) administration of mixed D1 and D2 type DA receptor antagonist drugs attenuated steroid-activated sexual behavior in male rats.

Establishing a direct link between the action of T, sexually arousing signals from an estrous female, and neurochemical indices of DA function in the MPOA/AH of male subjects has been a difficult task. Early studies (Simpkins et al., 1980, 1983; Gunnet et al., 1986) suggested that DA turnover in the MPOA/AH, which was measured by the rate of decline of DA concentrations after systemic treatment with the tyrosine hydroxylase inhibitor, α–methylparatyrosine, was reduced in castrated male rats by T treatment. Another study (Baum et al., 1986) revealed no effect of castration, with or without T replacement, on the concentrations of DA, its neural metabolite DOPAC, or the ratio of DOPA/DA (an index of DA release) in male rats. Similar negative results were obtained in groups of castrated males that were killed 10 min after exposure to the sight, sound, and odors of an estrous female. Subsequently, different groups of investigators used in vivo electrochemical recordings (Mas et al., 1990) and in vivo microdialysis (Pfaus et al., 1990; Pleim et al., 1990) to document significant increases in the extracellular levels of DA in the nucleus accumbens of gonadally intact male rats while they engaged in mating behavior with an estrous female. Increased DA release in the nucleus accumbens was also seen in males that were simply exposed to odors from estrous females (Mitchell and Gratton, 1991, 1992). Likewise, in vivo microdialysis studies subsequently established that extracellular levels of DOPAC (Fumero et al., 1994) as well as DA itself (Hull et al., 1995; Sato et al., 1995) were significantly increased in the MPOA of gonadally intact and castrated, T-treated male rats while they were exposed to an estrous female. Likewise, exposure to odors emitted from an estrous female significantly augmented both DA and DOPAC concentrations in the paraventricular nucleus of gonadally intact male rats (Melis et al., 2003). There was a close correlation between increases in MPOA levels of DA that were shown in response to distal cues from estrous females, and the subsequent display of mating behavior when given access to an estrous female (Hull et al., 1995). Castrated males given no T replacement rarely mated with estrous females, and these males did not show increases in MPOA levels of DA in response to the presence of a female (Hull et al., 1995). In a subsequent study (Putnam et al., 2001) a close correlation was seen between the ability of T replacement to restore mating behavior in castrated male rats and to allow exposure to distal cues from an estrous female to elevate extracellular DA levels in the MPOA. Castrated male rats only mustered a significant MPOA DA response to the presence of an estrous female if they were treated with T or with E + DHT. Castrated males given only E showed no such DA response, although they did display mounts and intromissions (no ejaculation) when given access to a female (Putnam et al., 2003). This result implies that AR activation is required in order for signals from an estrous female to augment DA release from terminals in the male's MPOA. As already explained, implanting either T or DHT into the MeA augmented mating behavior in male hamsters and rats, respectively, and it is possible that these behavioral responses reflect an activation of DA release in the MPOA that is driven by inputs to this hypothalamic region from the MeA. Evidence supporting this conclusion derives from the observations (Dominguez et al., 2001) that excitotoxic lesions of the medial amygdala disrupted mating in male rats, whereas infusion of a DA agonist into the MPOA restored this behavior. Also, lesions of the MeA blocked the increases in extracellular DA levels in the MPOA otherwise observed in male rats after they were exposed to distal signals from an estrous female.

The studies just summarized support the notion that a steroid-induced facilitation of the responsiveness of hypothalamic and/or ventral tegmental DA neurons to olfactory cues from estrous females contributes to sexual arousal and the display of male-typical sexual behavior. Further support for this view stems from a study (Szczypka et al., 1998) using male mice in which a null mutation of the tyrosine hydroxylase (TH) gene eliminated the capacity to produce DA and in which TH expression was selectively rescued in noradrenergic neurons by daily injections of the DA precursor, L-DOPA. These authors noted that mutant DA-/-males, unlike wild type controls, showed high levels of mounting behavior as well as aggression toward either female or male cagemates immediately after the daily injection of L-DOPA. Castration of mutant males led to a reduction in the display of both mounting and aggressive behaviors. Subsequent administration of T stimulated higher levels of mounting in DA-/-mutants than in wild type control males (all of which continued to receive daily injections of L-DOPA). These results suggest that nondopaminergic neurochemical mechanisms must exist, which mediate at least some of the facilitatory effects of T and/or E on masculine sexual motivation and mating performance.

3.6 Androgenic Facilitation of Penile Erectile Function

A complete description of the neural innervation of the penis and the neurochemical control of erectile function can be found in the recent review by Hull et al. (2002). The present review is confined to a consideration of the role of T and its androgenic metabolite, DHT, acting via ARs, in the regulation of erectile functions in rat and rabbit.

Electrical stimulation of the MPOA/AH reliably elicits penile erection in rat (Giuliano et al., 1996) and monkey (Maclean et al., 1959), due to a complex modulation of parasympathetic (tumescence promoting) and sympathetic (tumescence inhibiting) inputs to the smooth muscle lining of the corpora caveronsa of the penile shaft as well as the corpus spongiosum that surrounds the urethra and extends distally into the glans penis. Erection can also be induced by electrical stimulation of the cavernosal nerve in rat (Lugg et al., 1995; Marin et al., 1999) and rabbit (Traish et al., 1999), and in both models castration reduced the stimulation-induced erection (indexed by increases in intracavernosal blood pressure), whereas T replacement reversed this effect. A similar modulation of stimulation-induced penile erection was observed in male mice after the administration of the melanocortin 4 receptor agonist, THIQ, which is a synthetic analog of the natural ligand for melanocortin receptors, $\alpha-$MSH (Van der Ploeg et al., 2002). Systemic injections of the melanocortin receptor agonist into wild type male mice enhanced erections induced by cavernous nerve stimulation; however, no such enhancement was seen in male mice with a null mutation of the melanocortin 4 receptor. More research is needed to determine what, if any, interaction exists between the facilitatory actions of androgen and melanocortin receptor activation on erectile responsiveness.

A hormonal modulation of erectile function was also observed in an early study (Hart, 1967), which showed that severing the spinal cord of male rats at the midthoracic level greatly increased the display of penile erections as well as dorsal flips when subjects were tested in the ex copula restraint paradigm with constant pressure being applied to the base of the penis. As in the electrical stimulation paradigm, castration reduced the occurrence of erections in these spinal rats, and administration of T reversed this effect, suggesting that the steroid acted at sites distal to the level of the spinal section to enhance erectile function. Subsequent studies established that ex copula erections displayed by male rats tested either in the restraint paradigm (Hart, 1973; Gray et al., 1980; Meisel et al., 1984) or in the noncontact paradigm in the presence of odors from an estrous female (Manzo et al., 1999) were reliably activated in castrated rats by administration of either T or DHT, but not by E. As shown in ➋ *Figure 1-2*, administration of the 5α-reductase inhibiting drug, 17β-testosterone carboxylic acid, significantly attenuated the activational effect of T on erection frequency (restraint paradigm) in castrated rats (Bradshaw et al., 1981). This effect was reversed by administering DHT, suggesting that this metabolite of T plays an essential role in facilitating erectile function, probably acting via androgen receptors (AR) that are expressed in the corpora cavernosa (Takane et al., 1991).

Several lines of evidence suggest that DHT facilitates erectile function by augmenting the activity of nitric oxide synthase (NOS; both the neural and endothelial isoenzymes), which catalyzes the formation of nitric oxide (NO) in nerve terminals in the endothelial lining of the penile corpora caversosa. In one study

Figure 1-2

Administration of the 5α-reductase inhibitor, 17β-testosterone carboxylic acid (17β–C) significantly attenuated the activational effect of a low dose (8 μg/100 g body weight) of testosterone propionate (TP) on the display of *ex copula* penile erections in castrated male rats. The frequency (Freq) of erections is shown 4 and 2 days prior to as well as 4–13 days after castration and the onset of daily s.c. injections of the different hormone treatments or sesame oil vehicle. Group Ns are given in parentheses. Adapted from Bradshaw et al. (1981)

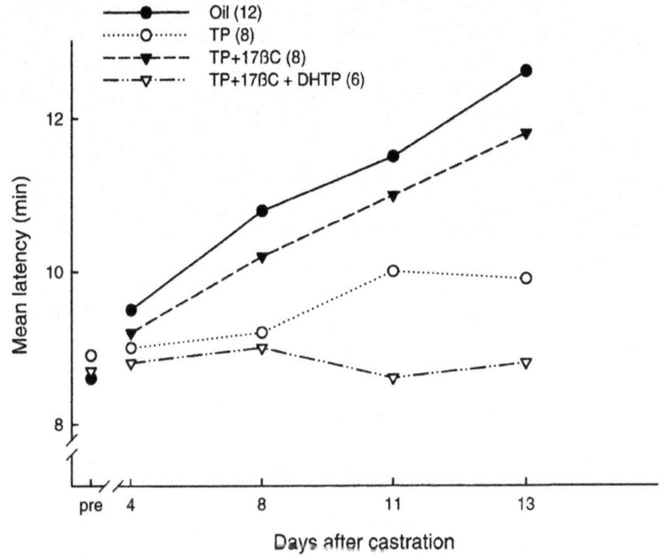

(Lugg et al., 1995) castration of male rats significantly reduced NOS activity in penile cytosols, whereas administration of T or DHT reversed this effect. Combined treatment with T and the 5α-reductase inhibiting drug, Finasteride, duplicated the effect of castration on NOS activity, suggesting that the stimulatory effects of T on enzyme activity are mediated by DHT. This profile of steroidal regulation of NOS activity was paralleled by the effects of these treatments on the maximal degree of intracavernosal pressure shown in response to electrical stimulation of the cavernosal nerve. Again, blocking 5α-reductase activity completely blocked the stimulatory effect of T on stimulation-induced erection (Lugg et al., 1995). In another study (Marin et al., 1999) electrical stimulation of the cavernosal nerve stimulated erection in anesthetized male rats, and augmented the extracellular levels of NO, measured by in vivo electrochemistry, in the corpora cavernosa. Again, castration reduced stimulation-induced erectile responses and associated levels of NO, and administration of T to castrated rats reversed these reductions. Western blot analysis showed that castration reduced the content of both the neural and endothelial isoforms of NOS in the corpora cavernosa, and administration of T reversed this effect of castration (Marin et al., 1999).

4 Sex Steroids, Acting Perinatally in the Male Brain, "Organize" Neural Mechanisms Controlling Male-Typical Mating Behavior

Much evidence (Arnold, 2002; Wallen and Baum, 2002) indicates that the differentiation of male-typical courtship behaviors in mammalian species results from the perinatal neural actions of T, and/or its neural metabolite, E, after the fetal differentiation of testes in response to the expression of the Y chromosome *Sry* gene in the embryonic genital ridge. In this respect, the hormonal control of the development of neural mechanisms controlling masculine sexual behavior parallels the androgenic control of male-typical differentiation of the internal and external genital structures (Jost, 1970). Several studies have recently explored the possibility that genes expressed on the Y chromosome may also contribute directly to male-typical aspects of brain and behavioral sexual differentiation (Arnold et al., 2004). These studies used mice in which the *Sry* gene was deleted from the Y chromosome and in some cases was expressed via an autosome. By expressing *Sry* via an autosome it was possible to generate XY⁻*Sry* (Sry deleted from the Y chromosome and instead expressed off an autosome) and XX*Sry* (Sry expressed off an autosome) mice in which testes developed as in normal XY males. Also, XY⁻ mice (*Sry* deleted from the Y chromosome without autosomal expression) with ovaries (females) were generated along with normal XX females. Sexually dimorphic aspects of male-typical courtship behavior, including time spent sniffing an anesthetized female and the display of mounting and thrusting behaviors, were associated with the presence of testes (males) and ovaries (females) as opposed to the presence or absence of Y chromosome genes other than *Sry* (De Vries et al., 2002). There were no differences in any aspect of male-typical courtship behaviors between males (mice with testes) in which Y chromosome genes were present (i.e., XY and XY⁻*Sry*) and those in which they were absent (i.e., XX*Sry*). By contrast, XY males that possessed Y chromosome genes in addition to *Sry* had a more masculine vasopressin innervation of the lateral septal region than XX males. These results suggest that purely endocrine, as opposed to direct genetic factors, controls the differentiation of the male-typical neural mechanisms controlling courtship and mating behaviors. The available evidence suggests that the critical endocrine signal controlling behavioral masculinzation is T, secreted perinatally from the testes. As reviewed later, this androgen organizes neural mechanisms controlling male-typical behavior via either a direct action at neural AR or after its neural aromatization to E and subsequent actions at neural receptors ERα or ERβ.

4.1 Contribution of Testosterone (T) Acting via Androgen Receptor (AR)

Naturally occurring mutations in the androgen receptor gene (AR) have been studied in humans, rats, and mice for many years. The resultant androgen insensitivity or testicular feminization (tfm) syndrome involves an absence of normal male-typical external and internal genital development in genetic (XY) male individuals, combined with an inhibition of internal female-typical duct development due to the secretion of Mullerian inhibiting hormone by the fetal testes. These tfm individuals, who are typically assigned the female gender at birth, are reared as girls, and live as women in adulthood, present a complex profile of psychosexual characteristics. In woman with complete androgen insensitivity syndrome who had undergone castration and surgical vaginal enlargement followed by estrogen therapy as teenagers there was subsequently a high level of satisfaction with their female assignment and sex of rearing coupled with a heterosexual orientation (Wisniewski et al., 2000). The majority of the 20 androgen insensitive women studied also reported that they were satisfied with their sexual and romantic lives. In an early study (Beach and Buehler, 1977) tfm male rats showed deficient pelvic thrusting, intromissive, and ejaculatory behaviors when tested while gonadally intact or after treatment with TP. By contrast, Olsen (1979) reported that administration of TP, EB, or EB + DHT activated mounting, pelvic thrusting, and intromission-like behaviors but not ejaculation in tfm male rats after castration in adulthood. One inference drawn from these observations is that the brain mechanisms controlling the motivation to seek out, mount, and intromit with a female do not rely on AR activation, either perinatally or in adulthood. Instead, E formed perinatally and/or in adulthood via the neural aromatization of T may promote a degree of male-typical brain sexual differentiation that enables male rats that lack AR expression to show these components of sexual behavior.

In an early study (Olsen, 1992) tfm male mice showed no male-typical mating behavior in tests with estrous females, although mounting behavior was displayed in 25% of these same mice when they were treated with EB or with EB + DHT. More recently it was found (Bodo and Rissman, 2004) that tfm male mice showed mounting and pelvic thrusting behavior as readily as WT control males and females following adult gonadectomy and E treatment, whereas the preference to approach soiled bedding from an estrous female as opposed to a gonadally intact male was significantly reduced in tfm compared with WT males. Transgenic mice have been produced with a null mutation of the AR gene (Sato et al., 2004). As in tfm males, male-typical external genital development was completely inhibited in AR null mutant (ARKO) males. When tested with estrous females, gonadally intact ARKO males, like tfm males (Olsen, 1992), showed no sexual behavior. Following gonadectomy and treatment with E, however, approximately 40% of ARKO males displayed mounting, pelvic thrusting, and intromission-like behaviors toward estrous females whereas 80% of WT control males showed these behaviors following castration. Neither WT controls nor ARKO males displayed lordosis behavior in response to mounts from a stud male after castration and E treatment. This latter result implies that the perinatal defeminization of the male brain occurred despite the absence of AR signaling, presumably as a result of the actions of estrogen formed perinatally via the neural aromatization of T. Interestingly, perinatal administration of DHT to WT mice significantly enhanced their subsequent capacity to display male-typical sexual behaviors (including ejaculation-like behavioral responses) toward estrous females after gonadectomy and treatment with either DHT or E in adulthood (Sato et al., 2004). However, this behavioral facilitation of perinatal DHT treatment was absent in ARKO females, leading the authors to conclude that AR activation normally contributes to the perinatal differentiation of neural circuits that control male-typical mating behaviors. As will be seen, a body of other recent evidence suggests that E receptor α (ERα) signaling in the perinatal brain also makes an essential contribution to the differentiation of male-typical sexual behavior.

4.2 Contribution of E, Aromatized in Brain from T, Acting via E Receptor (ER) α and β

In early studies (Booth, 1977a; Sodersten, 1978), neonatal administration of the anti-estrogen MER-25 to male rats led to later deficits in their capacity to achieve ejaculation even though their mounting and intromission behaviors were normal. A similar behavioral profile was observed in male mice with a null mutation of ERα. Thus, homozygous null mutant ERα–knockout (ERαKO) male mice that were gonadectomized and treated with TP showed significant reductions in their capacity to achieve ejaculation, even though they often pursued and mounted, and occasionally intromitted with estrous females (Wersinger et al., 1997; Ogawa et al., 1998; Rissman et al., 1999). Mount latencies were lengthened in ERαKO males, suggesting a deficit in male-typical sexual motivation. Confirmation of this impression was obtained in additional studies (Wersinger and Rissman, 2000a; Imwalle et al., 2002), in which ERαKO male mice, unlike WT controls, spent very little time investigating chemosensory cues emitted from estrous females, even though they clearly were able to detect these signals. Still further evidence of a deficit in motivation, as opposed to an inability to show the motor patterns associated with mating, was provided by the report (Wersinger and Rissman, 2000b) that systemic administration of the dopamine receptor agonist, apomorphine, reliably stimulated all aspects of mating, including ejaculation, in ERαKO males, which had been castrated and given T in adulthood.

A second type of E receptor, ER-β [Kuiper, 1966 #221], was discovered shortly after ERαKO mice were first produced. A homozygous mutation of the ER-β led to the creation of ERβKO male mice which, when tested while gonadally intact with estrous females, showed no deficits in mating behavior (Ogawa et al., 1999), and indeed, were fertile (Krege et al., 1998). However, a recent developmental study (Temple et al., 2004) showed that gonadally intact ERβKO males first mated with estrous females at an older age than WT control males. Thus, estrogenic signaling via ERβ may control the timing of behavioral puberty. Perhaps not surprisingly, a double null mutation of both types of estrogen receptor gene eliminated mating behavior in male mice tested while gonadally intact (Ogawa et al., 2000). However, after castration and s.c. treatment with T plus icv administration of the dopamine agonist, apomorphine, 75% of double mutant ERαβKO males displayed mounting behavior, and 50% showed intromission behavior, although none ejaculated (Burns-Cusato et al., 2004). These results are consistent with the conclusion that estrogen signaling, primarily acting via the ERα, contributes to either the organization during perinatal development or to the activation of neural circuits that control sex discrimination, heterosexual partner preference, or sexual arousal. Insofar as the mutations in the AR as well as both types of ER are present in transgenic mice from the earliest stages of development right on into adulthood (the time when sex behavior tests are administered), it is impossible, based on these model systems, to distinguish between possible organizational versus activational deficiencies in steroid hormone signaling to explain observed deficits in the mating behavior of different mutant mice. A future solution to this problem may be the development of "inducible knockout" transgenic mice in which AR or ER genes are only rendered functionless under the influence of a drug (e.g., tetracycline) or some other pharmacological or environmental stimulus. In lieu of such models, the availability of transgenic mice with a null mutation of the *Cyp*19 gene (Fisher et al., 1998), which encodes the enzyme for estrogen biosynthesis, aromatase, has provided an attractive model in which to study the contribution of perinatal estrogen actions on the differentiation of brain mechanisms controlling male-typical mating behavior. Males that are homozygous for a null mutation of the aromatase gene, *Cyp*19, (ArKO) are thought to be fetally deprived of estrogenic stimulation, because any maternal estrogen that might otherwise reach the brain would be bound with high affinity by α-feto protein (Savu et al., 1981), and the endogenous production of brain E via neural aromatization of T (Wozniak et al., 1992) would be absent due to the ArKO mutation (Bakker et al., 2004a). In an initial study (Honda et al., 1998), ArKO males that were gonadally intact showed significantly longer mount latencies and frequencies than WT control males in 30-min. tests. In a subsequent study (Matsumoto et al., 2003), fertility as well as the display of intromissive and ejaculatory behaviors was reduced, though not eliminated, in ArKO male mice. A similar deficiency in fertility among ArKO male mice was also reported by another group of investigators (Toda et al., 2001). Interestingly, this deficit in fertility was reversed in ArKO males that received neonatal injections of E, suggesting that the observed deficits in male sexual behavior and resultant fertility resulted from the lack of the perinatal, organizational actions of estrogens formed via neural aromatization as opposed to a deficiency in the activational actions of estrogens on the adult expression of sexual behavior. ArKO males, when tested while gonadally intact or after castration and treatment with T, did not prefer to approach (Y-maze tests) volatile body odors emitted from an estrous female as opposed to a stimulus male (Bakker et al., 2002). Again, as others had reported for gonadally intact males, deficits in mounting, intromission, and ejaculation were seen in these castrated, T-treated ArKO males. In a subsequent study (Bakker et al., 2004b), administration of E + DHT to gonadally intact ArKO male mice in adulthood stimulated the expression of all aspects of male sexual behavior above the level shown by intact ArKO males given no supplemental hormone treatments. This implies that some proportion of the deficit in male sexual behavior among ArKO males resulted from the absence of an activational (adult) synthesis and action of E. However, even after E + DHT treatment, ArKO males intromitted and ejaculated less frequently than WT controls. In addition, even after adult treatment with EB, gonadally intact ArKO males approached volatile body odors emitted from estrous females in Y-maze tests. These deficits in olfactory investigation as well as actual coital performance among ArKO males treated in adulthood with estrogen imply that estrogens, acting at some earlier (possibly perinatal) period in the subjects' lives, played an essential role in the process of male-typical psychosexual differentiation. Definitive proof of this hypothesis awaits a study to determine whether the perinatal administration of E to ArKO male mice will reverse their deficient olfactory and mating behavior, as assessed in adulthood after treatment with E.

4.3 Possible Interactions of AR and ERα in the Organization of Male Sexual Behavior

Although there are gaps in the database on the effects of AR, ER, and aromatase null mutations on the display of male-typical mating behavior in mice, the available results reviewed earlier are consistent with the previous suggestion (Baum et al., 1990a, 1990b; Tobet and Baum, 1987) that E, formed in the fetal male brain from circulating T, sensitizes the nervous system to the subsequent fetal and/or immediate postnatal and neonatal actions of T itself in completing the masculinization of neural circuits that control male-typical partner preference and mating behavior. Other than several reports (Vale et al., 1973; Manning and McGill, 1974; Gandelman and Kozak, 1988) that neonatal administration of TP to female mice enhanced their later capacity to show male-typical intromission- and ejaculation-like behaviors, there are no pharmacological experiments on mice that have systematically explored the effects of perinatal manipulation of AR- or ER-activation on the later expression of these sexual behaviors. Thus, the only evidence for this proposed sequential action of both E (acting via ERα) and T (acting via AR) comes from previous studies using ferrets. In ferrets, circulating T levels are consistently greater in males than in females over the last quarter of the 41-day gestation (Krohmer and Baum, 1989), again within 2 h after birth (Erskine et al., 1988) and then intermittently over the first 3 postnatal weeks (Erskine and Baum, 1982). High levels of hypothalamic aromatase activity (Tobet et al., 1985; Krohmer and Baum, 1989; Weaver and Baum, 1991) as well as ERα (Tobet et al., 1993) and AR (Vito et al., 1985) are present in both sexes over this extended perinatal period. Although males have relatively high levels of circulating T, beginning as early as embryonic day 25 (E25) and extending to a few hours after birth, neither transplacental administration of TP to female ferrets over days E16–34 followed by a single injection of TP on postnatal day 3 (Baum, 1976) nor transplacental treatment of females with T over days E30–41 (Tobet and Baum, 1987) significantly augmented their capacity to display male-typical sexual behavior in adulthood. Instead, the available evidence (Baum et al., 1990b) suggests that sex steroid exposure over this period sensitizes the developing nervous system to a second phase of sex hormone action, which begins shortly after birth and extends to postnatal day 20. Recent studies on relative contributions of perinatal AR- and ER-activation in organizing male-typical coital capacity in mice have only pointed to an important role of each type of steroid in the control of this process. By contrast, the administration during specific perinatal periods of T, E, or of drugs that either block androgen receptors (Flutamide) or inhibit aromatase activity (1,4,6-androstatriene-3,17-dione; ATD) has allowed the specification of the sequential roles of ER and AR activation in the organization of neural mechanisms controlling ferrets' male-typical sexual behavior. Thus, depriving male ferrets of estrogens between days E30 and E41 via maternal ovariectomy and transplacental ATD treatment reduced their later ability to display male-typical mating behavior (Tobet and Baum, 1987), whereas treatment with the AR antagonist, Flutamide, had no such effect, even though it attenuated penile development. These findings imply that estrogenic metabolites of T play an important role in the initial perinatal masculinization of neural mechanisms controlling male sexual behavior. By contrast, during the second, neonatal period of steroidal masculinization of behavior it appears that T itself, acting via AR, plays a critical role. Thus T itself, when administered to female ferrets over postnatal days 0–15 was significantly more effective than treatment with either E or DHT in masculinizing male sexual behavior (Baum et al., 1982a). Further support for a neonatal role of T itself in this regard stems from the observation (Baum et al., 1983) that treating female ferrets over postnatal days 0–15 with either the aromatase inhibitor, ATD or the 5α reductase blocker, testosterone 17β–carboxylic acid, failed to disrupt the normal male-typical differentiation of sexual behavior capacity.

The duration of gestation in commonly studied rodents such as rats and mice is considerably shorter (22 and 19 days, respectively) than that of the ferret, a carnivore (41 days). The longer duration (~36 days in ferrets versus ~16 days in rats) of a critical period of development for neural tissues controlling male-typical brain and behavioral sexual differentiation has perhaps made it experimentally more feasible to identify separate, sequential periods of estrogen and androgen actions in the ferret than in rodent models. Estrogenic metabolites of T, formed perinatally, contribute to the capacity of male rats to ejaculate. Thus, neonatal administration of the aromatase blocker, ATD (Bakker et al., 1993b), or the anti-estrogen, MER-25 (Booth, 1977b) to male rats attenuated their later ability to achieve ejaculation after a series of mounts with intromission. Combined pre- and neonatal treatment with ATD also attenuated ejaculatory capacity, whereas prenatal ATD alone had no such treatment (Brand and Slob, 1991; Houtsmuller et al., 1994). These drug treatments had no detrimental effect on penile development in males. Thus, their disruptive effect on ejaculatory capacity cannot be attributed to a reduction in penile sensory functions during a mounting series. Castration of male rats within a day after birth, as opposed to 7–15 days postnatally, disrupted intromission and ejaculation without affecting mounting behavior (Beach et al., 1969). These behavioral effects were correlated with reduced penile development in neonatally castrated males. Likewise, disruptions in intromission and ejaculatory capacity were obtained in male rats after prenatal treatment with the antiandrogen, Flutamide (Clemens et al., 1978; Brand and Slob, 1991; Dominguez et al., 2002; Castro et al., 2003). These behavioral effects of prenatal Flutamide, like those of neonatal castration, were correlated with disruptions in normal AR-dependent penile development. A similar peripheral as opposed to central neural explanation, likely applies to the deficits in intromissive and ejaculatory capacity reported in androgen-insensitive tfm rats (Beach and Buehler, 1977; Olsen, 1979).

In contrast to the data implicating neonatal estrogen biosynthesis in the differentiation of neural mechanisms controlling intromissive and/or ejaculatory behavioral capacity, the evidence suggesting that either ER or AR, activated either sequentially or over the same perinatal periods, is required for the development of male-typical mounting capacity is controversial. In female rats (Beach, 1942a; Emery and Sachs, 1975) as in female mice (Wersinger et al., 1997), surprisingly high levels of mounting behavior toward a stimulus female were observed after adult treatment with ovarian hormones, sometimes including intromission- and ejaculation-like behaviors. The presence of this capacity in female rats may reflect the fact that male and female rat pups have similar circulating levels of T over days E16, E17, E20, E21, and E22, with males having significantly higher levels only on days E18 and E19 (Weisz and Ward, 1980; Baum et al., 1991).

Although essentially all female rats will display mounting after ovariectomy and TP treatment in adulthood, the level of this behavior is typically significantly lower than the level displayed by male rats following adult castration and TP treatment (Whalen and Edwards, 1967; Dominguez et al., 2002; Amateau and McCarthy, 2004). There are conflicting reports on whether prenatal treatment of male rats with the aromatase inhibitor, ATD, reduces later mounting behavior. In one study (Gladue and Clemens, 1980a), prenatal ATD reduced males' later mounting capacity whereas in two other reports (Houtsmuller et al., 1994; Dominguez et al., 2002) no such effect of ATD was obtained in either male or female rats. Likewise, prenatal treatment with an AR blocker, Flutamide or cyproterone acetate (CA), had different effects on rats' mounting capacity in different studies. In some instances, prenatal Flutamide (Stewart et al., 1971; Clemens et al., 1978; Castro et al., 2003) or CA (Ward and Renz, 1972) treatment reduced later mounting capacity in both male and female rats, whereas in other studies (Brand et al., 1991; Dominguez et al., 2002) no such effect was obtained. Taken together, these results provide no conclusive evidence that fetal availability of T, or its estrogenic metabolites, contributes to the development of the mounting capacity that is seen in rats of both sexes. They also provide little evidence that the brief (days E18–19) sex difference in circulating T can account for the fact that male rats display higher levels of mounting behavior than females following treatment with T or E in adulthood.

In the rat, as in the ferret, males are exposed to significantly higher levels of T than females, beginning 2h after birth (Corbier et al., 1978; Baum et al., 1988) and again over the first 10 postnatal days (Resko et al., 1968; Pang et al., 1979). Attempts to demonstrate that neonatal administration of TP to female rats enhances later mounting capacity have met with mixed outcomes. Thus in two studies (Hendricks, 1969; Sodersten, 1973b) neonatal TP treatment significantly augmented mounting capacity following ovariectomy and TP treatment later in life, whereas in other experiments (Whalen and Edwards, 1967; Pfaff and Sigmond, 1971) no such effect was found. Castration carried out on the day of birth (more than 4h postpartum) failed to reduce the capacity of male rats to display mounting behavior after TP treatment later in life (Grady et al., 1965; Whalen and Edwards, 1967). Important, however, castration within minutes after birth caused a significant reduction, though not an elimination, of T-induced mounting behavior in male rats (Roffi et al., 1987). Immediate postnatal injection of T after castration reversed the disruptive effect of this treatment on later mounting potential, suggesting that the immediate postparturient rise in circulating T normally augments the sensitivity of neural mechanisms that control mounting to the activational effects of later T exposure.

Whereas there is general agreement that neonatal inhibition of E's availability or neural action disrupts later ejaculatory capacity in male rats (Booth, 1977b; Bakker et al., 1993a), there is less conclusive evidence that the action of E in the postnatal male rat brain contributes to the masculinization of mounting behavior capacity. Thus, there is a discrepancy in the outcomes of experiments assessing the ability of neonatal EB treatment to augment later mounting capacity in female or neonatally castrated male rats. No effects were found in two studies (Whalen and Edwards, 1967; Hendricks and Weltin, 1976), whereas an enhancement of later mounting capacity was induced by neonatal EB treatment in two other studies (Mullins and Levine, 1968; Sodersten and Hansen, 1978). Administration of the aromatase blocker, ATD, beginning 3–9h after birth, in addition to dramatically reducing ejaculatory capacity in male rats, significantly reduced their mounting and intromission behavior later in life (Bakker et al., 1993a). In another study (Bakker et al., 1993b), however, the same ATD treatment slightly augmented the later display of mounting and intromissive behaviors in male rats that were tested late as opposed to early in the dark phase of the daily day/night cycle. Studies are needed to determine whether inhibiting the aromatization of the immediate postparturient T peak would disrupt later mounting potential as reliably as castration within minutes after birth. The observation (Rhoda et al., 1984; Amateau et al., 2004) that levels of E in the hypothalamus and preoptic area were significantly higher in male than in female rats killed 1–2h after birth points to a potential contribution in males of this steroid, formed via neural aromatization of the postparturient surge in T secretion, to the masculinization of brain mechanisms controlling mounting behavior.

4.4 Steroid Receptor Coactivators, GABA, and Prostaglandin-E2 as Possible Mediators of Sex Hormone-Dependent Behavioral Masculinization

Steroid hormones, including T and E, affect neural development and function by binding to specific intracellular ARs and ERs, respectively. In order for steroid–receptor complexes to interact effectively with DNA to promote gene transcription, an additional class of proteins referred to as nuclear receptor coactivators is thought to be required (Shibata et al., 1997). On the presumption that either ER or AR, or both receptors, normally contribute in male rats to behavioral masculinization during neonatal life, McCarthy and coworkers have carried out studies to explore the possible contribution of two such steroid coactivators to this process. In an initial study (Auger et al., 2000), antisense oligonucleotides to the steroid receptor coactivator-1 (SRC-1) mRNA were infused on the day of birth into the hypothalamus of either male or neonatally TP-treated female rats. This treatment dramatically augmented the ability of these groups of rats to display lordosis behavior after adult treatment with ovarian hormones, suggesting that the SRC-1 protein is essential for the normal defeminizing actions of E in male rats during neonatal life. By contrast, antisense to SRC-1 failed to reduce the incidence of later mounting behavior in these same subjects, suggesting that the neuroendocrine mechanism controlling behavioral defeminization and masculinization in male rats is fundamentally different. The cAMP response element-binding protein (CBP) may also function as a nuclear receptor coactivator by interacting with SRC-1, thereby enhancing steroid receptor action at hormone response elements on target DNA. The levels of CBP in the basomedial hypothalamus were significantly higher in male than female rats over the first 10 postnatal days (Auger et al., 2002). As with antisense to SRC-1, neonatal hypothalamic infusions of antisense to CBP into neonatally TP-treated female rats blocked the normal defeminizing action of this neonatal steroid treatment without reducing the capacity of these same females to display mounting behavior in adulthood after treatment with testosterone (Auger et al., 2002). Again, this divergence in the behavioral effects of neonatal antisense treatments to two steroid receptor coactivators reinforces the view that the neural mechanisms controlling these two types of behavior are different in many important ways (Goy, 1970), possibly including timing of their perinatal development.

Significantly higher levels of the neurotransmitter, γ-aminobutyric acid (GABA), were reported in the hypothalamus and limbic structures of male versus female rats killed over the first 5 postnatal days (Davis et al., 1999), raising the possibility that excitatory inputs from GABA neurons may mediate some of the effects of elevated ER- or AR-mediated neural signaling in the neonatal male rat brain contributing to the development of male-typical features of behavioral sexual differentiation (McCarthy et al., 1997; Perrot-Sinal et al., 2001). Indeed, neonatal intrahypothalamic infusion of antisense oligonucleotides to glutamic acid decarboxylase mRNA (the rate limiting enzyme controlling GABA synthesis) into female rats that had received neonatal injections of TP reduced their later display of intromission-like behavior in tests with estrous females (Davis et al., 2000). In another study (Amateau and McCarthy, 2004), prostaglandin-E2 (PGE2) was implicated in the estrogen-dependent signaling that has been hypothesized to contribute to the differentiation of masculine sexual behavior in male rats. Bilateral intracerebroventricular (icv) infusions of PGE2 made within 6h after birth and again 24h later significantly enhanced the capacity of female rats to display mounting and intromission-like behaviors in adult tests (after ovariectomy and T treatment) with estrous females. Males in which PGE2 synthesis was inhibited neonatally by s.c. injections of indomethacin, a potent cyclooxygenase-2 (COX-2) blocker, later showed significant reductions in mounting and intromission behaviors. This behavioral effect of indomethacin was correlated with a reduction in a marker for dendritic spine density in the preoptic area. Administering E on the day of birth and again on postnatal day 1 caused a significant rise in COX-2 mRNA levels in the preoptic area when female subjects were killed on postnatal day 3, suggesting that the higher levels of E present in male versus female hypothalamus shortly after birth (Rhoda et al., 1984; Amateau et al., 2004) may stimulate PGE2 synthesis and male-typical aspects of preoptic area development. Finally, chronic perinatal exposure of developing male rats' pups to aspirin, a COX-2 inhibitor that was placed in their mothers' drinking water, led to significant reductions in mounting and intromission behaviors displayed later in two adult tests followed by a recovery of function in a third test. We need more research to determine whether in vivo inhibition of E biosynthesis by administering an aromatase inhibitor immediately after birth causes reductions in PGE2 synthesis that are correlated with deficits in masculine sexual behavior in adulthood. It is also important to show that these effects of neonatal aromatase inhibition are reversed by concomitant administration of either E or of PGE2.

5 Relationship Between Sexually Dimorphic Hypothalamic Morphology and Male-Typical Sexual Behavior and Partner Preference

Gorski et al. (1978) first reported the existence of a sexually dimorphic nucleus (SDN) of the rat POA/AH. The SDN is 3–4 times larger in volume in male than in female rats. In light of the large literature (Hull et al., 2002) showing that destructive lesions of the medial POA/AH disrupt masculine sexual behavior, it was widely presumed that the behavioral deficits observed could be attributed to the destruction of the SDN. Indeed, Houtsmuller et al. (1994) reported that perinatal exposure to an aromatase inhibiting drug, ATD, reduced the volume of the SDN in male rats, and this reduction was correlated with subjects expressing less masculine mating behavior. Likewise, a strain (Noble) of male rats with significantly larger volumes of the SDN displayed significantly higher levels of mating, including ejaculation, than another strain (Wistar–Furth) with a smaller SDN (Lephart et al., 2001). These correlations not withstanding, bilateral destructive lesions restricted to the male rat SDN caused either no (Arendash and Gorski, 1983) or minimal (De Jonge et al., 1989) disruption of sexual behavior. In the latter study, disruptive effects of SDN lesions were most evident in male rats that lacked mating experience at the time the lesions were placed. There is also a sexually dimorphic male nucleus of the POA/AH in male ferrets, which is not present in females (Tobet et al., 1986). In contrast to the rat example, neonatal administration of TP to female ferrets augmented their later capacity to display male-typical mating behaviors without exerting any organizational effect on the morphology of the POA/AH (Cherry et al., 1991). Again, as in the male rat, bilateral lesions that were restricted to the male ferret's male nucleus of the POA/AH failed to disrupt male sexual behavior (Cherry and Baum, 1990). Lesions of the male gerbil's sexually dimorphic nucleus disrupted mating performance in gonadally intact subjects, although the deficits were abolished after castration and treatment with T (Commins and Yahr, 1984). Taken together, these various studies provide no consistent evidence that sexually dimorphic subdivisions of the medial POA/AH play an essential role in controlling the expression of male-typical mating behaviors in mammalian species.

Whereas the capacity to display *consummatory* components of masculine sexual behavior has not been reliably linked to the presence of a male-typical SDN, several lines of evidence have linked *appetitive* features of masculine sexual behavior, including males' preference to seek out a female as opposed to another male conspecific, to the function of sexually dimorphic portions of the POA/AH. An early linkage of this sort was obtained in ferrets (Cherry and Baum, 1990) in which discrete electrolytic lesions of the males' sexually dimorphic POA/AH caused them to show female-like patterns of approach behavior to stud males in L-maze tests. Following gonadectomy and adult treatment with increasing dosages of EB, female ferrets showed a dose-dependent reduction in approach latencies to the stimulus male, whereas male subjects approached this stimulus at a constant speed regardless of estrogen dosage (Baum et al., 1985). As shown in ➊ *Figure 1-3*, discrete bilateral lesions of the sexually dimorphic POA/AH caused castrated male ferrets, like control females, to show progressively shorter approach latencies to a stud male in response to increasing doses of EB (Cherry and Baum, 1990). Two subsequent studies using ferrets showed that either excitotoxic (Paredes and Baum, 1995) or electrolytic (Kindon et al., 1996) lesions centered in the males' sexually dimorphic POA/AH caused them to prefer to approach and interact with a stud male as opposed to a sexually receptive female when tested with these two types of stimuli in a T-maze. Again, castrated ferrets used in both of these studies received E at the time of testing. Likewise, as shown in ➊ *Figure 1-4*, castrated adult male rats that were treated with either T showed a robust preference to approach a stud male as opposed to

Figure 1-3
Descrete bilateral lesions of the sexually dimorphic male nucleus of the medial preoptic area/anterior hypothalamus (Lesion male) caused castrated adult male ferrets to show a female-typical profile of approach latencies in an L-shaped maze to a sexually active male ferret in response to s.c. injections of increasing doses of estradiol benzoate (EB). Approach latency results from both pre- and postoperative tests are shown for sham operated female (Fem), sham-operated male, and MN lesioned male ferrets. Data are expressed as mean +/−SEM; ns are given in parentheses. * Significant effect of EB dose by 1-way ANOVA. Adapted from Cherry and Baum (1990)

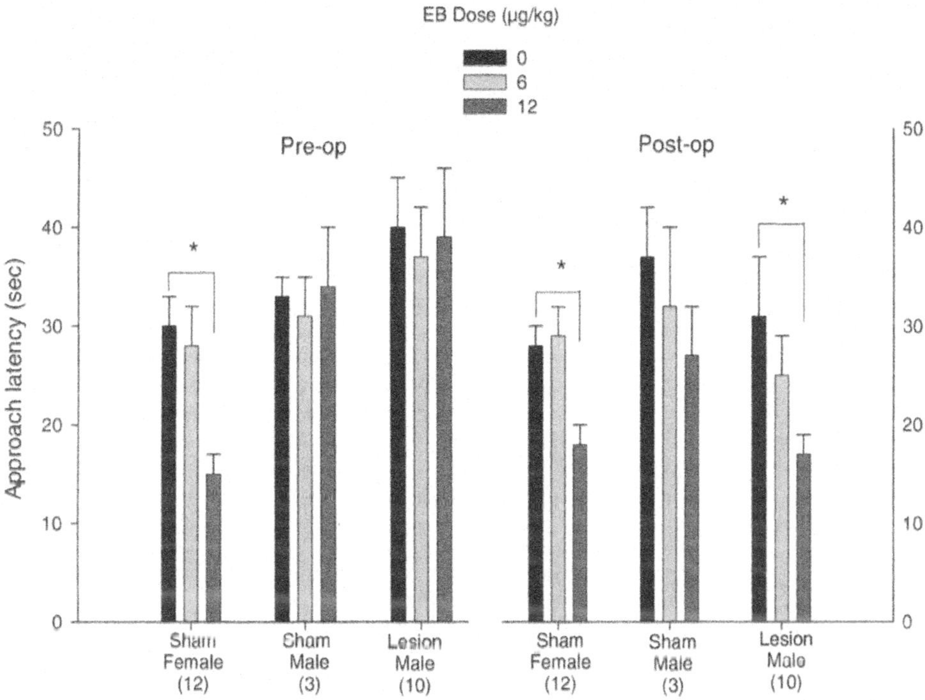

an estrous female following the placement of electrolytic lesions in the POA/AH (Paredes et al., 1998b). Sham-operated control males preferred to approach the estrous female. The placement of POA/AH lesions in female ferrets (Paredes and Baum, 1995) or in female rats (Paredes et al., 1998b) had no effect on their male-oriented partner preference. Taken together, these results suggest that a male-typical preference to approach female, as opposed to male conspecifics, depends on yet-to-be understood characteristics of the sexually dimorphic structures of the male's POA/AH.

Two additional lines of correlative evidence support the view that male-typical features (e.g., larger volume of the nucleus, larger cell bodies, larger dendritic arbor, high number of aromatase-expressing neurons (Tobet and Hanna, 1997) of the sexually dimorphic subdivision of the POA/AH) somehow contribute to the male-typical preference of male mammals to seek out female conspecifics, as opposed to other males, for the purpose of mating. The first of these examples stems from the early observation (Perkins et al., 1992) that ~8% of rams prefer to approach and mount other rams as opposed to ewes when given a choice between these alternatives. In sheep, as in rat and ferret, a sexually dimorphic group of cells exists in the POA/AH (ovine SDN), which is significantly larger in volume in males than in females (Roselli et al., 2004b). The volume of this ovine SDN was twice as great in female-oriented than in male-oriented rams (Roselli et al., 2004b), raising the question of whether there is a causal link between the size of the ram's sexually dimorphic POA/AH nucleus and the motivation to seek out an opposite-sex mating partner.

Figure 1-4
Bilateral lesions of the medial preoptic area/anterior hypothalamus (Bilateral) caused castrated, testosterone-treated male rats to switch their preference from a tethered estrous female (to female) to a tethered sexually active male (to male). Preference scores were calculated by subtracting the number of seconds that male subjects spent in a 3-compartment box with the tethered male from the time spent with the receptive female (negative numbers reflect time spent with the stimulus male). * Significant switch in male rats' partner preference following placement of a bilateral MPOA/AH lesion. Group Ns are given in parentheses. Adapted from Paredes et al. (1998b)

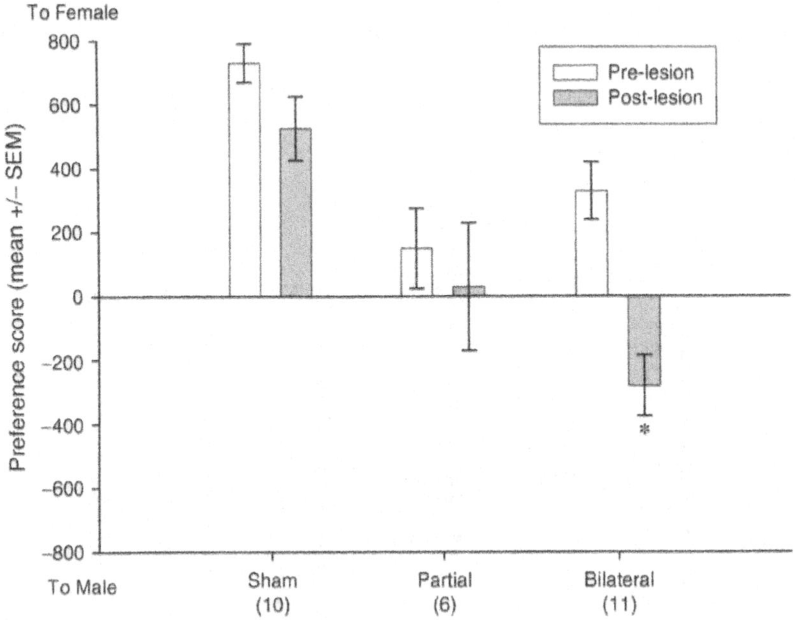

Three independent groups of investigators (Allen et al., 1989; LeVay, 1991; Byne et al., 2000) have published results showing that the volume of the third interstitial nucleus of the anterior hypothalamus (INAH3) is significantly greater in (presumptive) heterosexual men than women. In a widely publicized study (LeVay, 1991), it was reported that the volume of INAH3 was significantly greater in heterosexual than in homosexual men. The homosexual men whose brains were analyzed in this study had nearly all died of HIV/AIDS whereas the heterosexual subjects typically died of non-HIV related causes. This difference in cause of death led some critics to attribute the differences in INAH3 volume observed between homosexual versus heterosexual men to the presence of HIV infection in the homosexual sample as opposed to a difference in sexual orientation. A more recent study (Byne et al., 2001) suggested, however, that there was no effect of HIV infection on the dimensions of INAH3 in either heterosexual or homosexual male subjects. These investigators also reported only a nonsignificant trend for INAH3 volume to be greater in heterosexual as compared with homosexual men. Thus, the original report of LeVay (LeVay, 1991), although intriguing, still awaits full replication. Certainly, the recent report (Roselli et al., 2004b) of an analogous difference in the volume of the male sheep's sexually dimorphic POA/AH, depending on whether rams preferred to approach and mount other males over females, increases the importance of obtaining additional data on the dimensions of the human male's INAH3 as a function of sexual orientation.

There are no empirical data that establish whether fetal differences in T exposure cause the above-mentioned differences between men and women or between heterosexual and homosexual men in INAH3 volume. Nor is any information available about the respective roles of AR- or ER-mediation of any such organizational effects of fetal T exposure in the human nervous system. By contrast, in both rat (Jacobson et al., 1981) and ferret (Park et al., 1997), experimental data have established that perinatal administration of T to females organizes male-typical size as well as phenotypic characteristics of sexually dimorphic structures in the MPOA/AH. Additional studies in both species have established that male-typical characteristics of POA/AH morphology depends on the perinatal neural aromatization of T, secreted by the testes, into E. Thus, neonatal treatment of male rats with the ER antagonist, tamoxifen, significantly reduced SDN volume (Dohler et al., 1984), and transplacental administration of the aromatase blocker, ATD, blocked the differentiation of the MN in the POA/AH of male ferrets (Tobet et al., 1986; Cherry et al., 1990). Subgroups of rams that preferred to mount females had significantly higher levels of aromatase activity (Resko et al., 1996) and aromatase mRNA (Roselli et al., 2004b) as well as estradiol receptor content (Perkins et al., 1995) in the sexually dimorphic portion of the POA/AH than did rams that preferred to mount other males. These correlative results raise the questions of whether variations in fetal estrogen signaling account for variations in the dimensions of the male-typical ovine SDN and whether these variations, in turn, control heterosexual versus homosexual partner orientation in rams (Roselli et al., 2004a). In the sheep model, as in other model systems in which the development and adult control of appetitive as well as consummatory components of masculine sexual behavior are studied, the future challenge is to understand how neuroendocrine signals organize the perinatal development and modulate the adult functions of male-typical circuits that mediate the sensory (olfactory, visual, auditory) inputs, which control these two kinds of behavior.

Acknowledgment

Preparation of this review was supported in part by funding from NIH grants HD21094 and HD44897.

References

Alheid GF, Heimer L. 1988. New perspectives in basal forebrain organization of special relevance for neuropsychiatric disorders: the striatopallidal, amygdaloid, and corticopetal components of substantia innominata. Neuroscience 27: 1-39.

Allen LS, Hines M, Shryne JE, Gorski RA. 1989. Two sexually dimorphic cell groups in the human brain. J Neurosci 9: 497-506.

Amateau SK, McCarthy MM. 2004. Induction of PGE (2) by estradiol mediates developmental masculinization of sex behavior. Nat Neurosci 7: 643-650.

Amateau SK, Alt JJ, Stamps CL, McCarthy MM. 2004. Brain estradiol content in newborn rats: sex differences, regional heterogeneity, and possible de novo synthesis by the female telencephalon. Endocrinology 145: 2906-2917.

Arendash GW, Gorski RA. 1983. Effects of discrete lesions of the sexually dimorphic nucleus of the preoptic area or other medial preoptic regions on the sexual behavior of male rats. Brain Res Bull 10: 147-154.

Arnold AP. 2002. Concepts of genetic and hormonal induction of vertebrate sexual differentiation in the twentieth century, with special reference to the brain. Hormones, Brain and Behavior. Pfaff DW, Arnold AP, Etgen AM, Fahrbach SE, Rubin RT, editors. San Diego: Academic; pp. 105-135.

Arnold AP, Xu J, Grisham W, Chen X, Kim YH, et al. 2004. Minireview: Sex chromosomes and brain sexual differentiation. Endocrinology 145: 1057-1062.

Auger AP, Tetel MJ, McCarthy MM. 2000. Steroid receptor coactivator-1 (SRC-1) mediates the development of sex-specific brain morphology and behavior. Proc Natl Acad Sci USA 97: 7551-7555.

Auger AP, Perrot-Sinal TS, Auger CJ, Ekas LA, Tetel MJ, et al. 2002. Expression of the nuclear receptor coactivator, cAMP response element-binding protein, is sexually dimorphic and modulates sexual differentiation of neonatal rat brain. Endocrinology 143: 3009-3016.

Bagatell CJ, Heiman JR, Rivier JE, Bremner WJ. 1994. Effects of endogenous testosterone and estradiol on sexual behavior in normal young men. J Clin Endocrinol Metab 78: 711-716.

Bakker J, Brand T, van Ophemert J, Slob AK. 1993a. Hormonal regulation of adult partner preference behavior in neonatally ATD-treated male rats. Behav Neurosci 107: 480-487.

Bakker J, van Ophemert J, Slob AK. 1993b. Organization of partner preference and sexual behavior and its nocturnal rhythmicity in male rats. Behav Neurosci 107: 1049-1058.

Bakker J, Honda S, Harada N, Balthazart J. 2002. Sexual partner preference requires a functional aromatase (cyp19) gene in male mice. Horm Behav 42: 158-171.

Bakker J, Baillien M, Honda S, Harada N, Balthazart J. 2004a. Relationships between aromatase activity in the brain and gonads and behavioural deficits in homozygous and heterozygous aromatase knockout mice. J Neuroendocrinol 16: 483-490.

Bakker J, Honda S, Harada N, Balthazart J. 2004b. Restoration of male sexual behavior by adult exogenous estrogens in male aromatase knockout mice. Horm Behav 46: 1-10.

Balthazart J, Surlemont C. 1990. Androgen and estrogen action in the preoptic area and activation of copulatory behavior in quail. Physiol Behav 48: 599-609.

Balthazart J, Foidart A, Hendrick JC. 1990. The induction by testosterone of aromatase activity in the preoptic area and activation of copulatory behavior. Physiol Behav 47: 83-94.

Balthazart J, Absil P, Gerard M, Appeltants D, Ball GF. 1998. Appetitive and consummatory male sexual behavior in Japanese quail are differentially regulated by subregions of the preoptic medial nucleus. J Neurosci 18: 6512-6527.

Barfield RJ. 1969. Activation of copulatory behavior by androgen implanted into the preoptic area of the male fowl. Horm Behav 1: 37-52.

Barfield RJ. 1971. Activation of sexual and aggressive behavior by androgen implanted into the male ring dove brain. Endocrinology 89: 1470-1476.

Baum MJ. 1972. Precocious mating in male rats following treatment with androgen or estrogen. J Comp Physiol Psychol 78: 356-367.

Baum MJ. 1973. Hormonal stimulation of precocious mating in male rats without antecedent effects on sexual clasping or ambulation. Physiol Behav 10: 137-140.

Baum MJ. 1976. Effects of testosterone propionate administered perinatally on sexual behavior of female ferrets. J Comp Physiol Psychol 90: 399-410.

Baum MJ, Vreeburg JT. 1973. Copulation in castrated male rats following combined treatment with estradiol and dihydrotestosterone. Science 182: 283-285.

Baum MJ, Starr MS. 1980. Inhibition of sexual behavior by dopamine antagonist or serotonin agonist drugs in castrated male rats given estradiol or dihydrotestosterone. Pharmacol Biochem Behav 13: 57-67.

Baum MJ, Everitt BJ. 1992. Increased expression of c-fos in the medial preoptic area after mating in male rats: role of afferent inputs from the medial amygdala and midbrain central tegmental field. Neuroscience 50: 627-646.

Baum MJ, Stockman ER, Lundell LA. 1985. Evidence of proceptive without receptive defeminization in male ferrets. Behav Neurosci 99: 742-750.

Baum MJ, Melamed E, Globus M. 1986. Dissociation of the effects of castration and testosterone replacement on sexual behavior and neural metabolism of dopamine in the male rat. Brain Res Bull 16: 145-148.

Baum MJ, Kingsbury PA, Erskine MS. 1987. Failure of the synthetic androgen 17 beta-hydroxy-17 alpha-methyl-estra-4,9,11-triene-3-one (methyltrienolone, R1881) to duplicate the activational effect of testosterone on mating in castrated male rats. J Endocrinol 113: 15-20.

Baum MJ, Woutersen PJ, Slob AK. 1991. Sex difference in whole-body androgen content in rats on fetal days 18 and 19 without evidence that androgen passes from males to females. Biol Reprod 44: 747-751.

Baum MJ, Gallagher CA, Martin JT, Damassa DA. 1982a. Effects of testosterone, dihydrotestosterone, or estradiol administered neonatally on sexual behavior of female ferrets. Endocrinology 111: 773-780.

Baum MJ, Tobet SA, Starr MS, Bradshaw WG. 1982b. Implantation of dihydrotestosterone propionate into the lateral septum or medial amygdala facilitates copulation in castrated male rats given estradiol systemically. Horm Behav 16: 208-223.

Baum MJ, Canick JA, Erskine MS, Gallagher CA, Shim JH. 1983. Normal differentiation of masculine sexual behavior in male ferrets despite neonatal inhibition of brain aromatase or 5-alpha-reductase activity. Neuroendocrinology 36: 277-284.

Baum MJ, Carroll RS, Cherry JA, Tobet SA. 1990a. Steroidal control of behavioural, neuroendocrine and brain sexual differentiation: studies in a carnivore, the ferret. J Neuroendocrinology 2: 401-418.

Baum MJ, Erskine MS, Kornberg E, Weaver CE. 1990b. Prenatal and neonatal testosterone exposure interact to affect differentiation of sexual behavior and partner preference in female ferrets. Behav Neurosci 104: 183-198.

Baum MJ, Brand T, Ooms M, Vreeburg JT, Slob AK. 1988. Immediate postnatal rise in whole body androgen content in male rats: correlation with increased testicular content and reduced body clearance of testosterone. Biol Reprod 38: 980-986.

Beach FA. 1942a. Male and female mating behavior in prepuberally castrated female rats treated with androgens. Endocrinology 31: 373-378.

Beach FA. 1942b. Central nervous mechanisms involved in the reproductive behavior of vertebrates. Psychol Bull 39: 200-226.

Beach FA. 1970. Coital behavior in dogs. VI. Long-term effects of castration upon mating in the male. J Comp Physiol Psychol 70: 1-32.

Beach FA, Levinson G. 1950. Effects of androgen on the glans penis and mating behavior of castrated male rats. J Exp Zool 114: 159-168.

Beach FA, Jordan L. 1956. Effects of sexual reinforcement upon the performance of male rats in a straight runway. J Comp Physiol Psychol 49: 105-110.

Beach FA, Buehler MG. 1977. Male rats with inherited insensitivity to androgen show reduced sexual behavior. Endocrinology 100: 197-200.

Beach FA, Noble RG, Orndoff RK. 1969. Effects of perinatal androgen treatment on responses of male rats to gonadal hormones in adulthood. J Comp Physiol Psychol 68: 490-497.

Beck J. 1971. Instrumental conditioned reflexes with sexual reinforcement in rats. Acta Neurobiol Exp (Wars) 31: 251-252.

Becker JB, Breedloue SM, Clews D, McCarthy MM, editors. 2002. Behavioral Endocrinology 2nd edition. Cambridge: MET press.

Berthold AA. 1849. Transplantation der Hoden. Arch Anat Physiol Wissensch Med 16: 42-49.

Beyer C, Morali G, Naftolin F, Larsson K, Perez P. 1976. Effect of some antiestrogens and aromatase inhibitors on androgen induced sexual behavior in castrated male rats. Horm Behav 7: 353-363.

Bialy M, Sachs BD. 2002. Androgen implants in medial amygdala briefly maintain noncontact erection in castrated male rats. Horm Behav 42: 345-355.

Block GJ, Davidson JM. 1971. Behavioral and somatic responses to the antiandrogen cyproterone. Horm Behav 2: 11-25.

Bodo CCA, Rissman EF. 2004. Normal masculine behavior does not require the activation of the androgen receptor in mice. Horm Behav 46: 85, abstract.

Bolles RC, Rapp HM, White GC. 1968. Failure of sexual activity to reinforce female rats. J Comp Physiol Psychol 65: 311-313.

Bonsall RW, Clancy AN, Michael RP. 1992. Effects of the nonsteroidal aromatase inhibitor, Fadrozole, on sexual behavior in male rats. Horm Behav 26: 240-254.

Booth JE. 1977a. Sexual behaviour of neonatally castrated rats injected during infancy with oestrogen and dihydrotestosterone. J Endocrinol 72: 135-141.

Booth JE. 1977b. Sexual behavior of male rats injected with anti-estrogen MER-25 during infancy. Physiol Behav 19: 35-39.

Brackett NL, Edwards DA. 1984. Medial preoptic connections with the midbrain tegmentum are essential for male sexual behavior. Physiol Behav 32: 79-84.

Bradshaw WG, Baum MJ, Awh CC. 1981. Attenuation by a 5 alpha-reductase inhibitor of the activational effect of testosterone propionate on penile erections in castrated male rats. Endocrinology 109: 1047-1051.

Brand T, Slob AK. 1991. Neonatal organization of adult partner preference behavior in male rats. Physiol Behav 49: 107-111.

Brand T, Kroonen J, Mos J, Slob AK. 1991. Adult partner preference and sexual behavior of male rats affected by perinatal endocrine manipulations. Horm Behav 25: 323-341.

Bressler SC, Baum MJ. 1996. Sex comparison of neuronal Fos immunoreactivity in the rat vomeronasal projection circuit after chemosensory stimulation. Neuroscience 71: 1063-1072.

Brookhard JM, Dey FL. 1941. Reduction of sexual behavior in male guinea pigs by hypothalamic lesions. Am J Physiol 133: 551-554.

Burns-Cusato M, Scordalakes EM, Rissman EF. 2004. Of mice and missing data: what we know and need to learn about sexual behavior in male mice. Physiol Behav in press.

Butera PC, Czaja JA. 1989. Activation of sexual behavior in male rats by combined subcutaneous and intracranial treatments of 5 alpha-dihydrotestosterone. Horm Behav 23: 92-105.

Byne W, Lasco MS, Kemether E, Shinwari A, Edgar MA, et al. 2000. The interstitial nuclei of the human anterior hypothalamus: an investigation of sexual variation in volume and cell size, number and density. Brain Res 856: 254-258.

Byne W, Tobet S, Mattiace LA, Lasco MS, Kemether E, et al. 2001. The interstitial nuclei of the human anterior hypothalamus: an investigation of variation with sex, sexual orientation, and HIV status. Horm Behav 40: 86-92.

Carani C, Rochira V, Faustini- Fustini M, Balestrieri A, Granata AR. 1999. Role of oestrogen in male sexual behaviour: insights from the natural model of aromatase deficiency. Clin Endocrinol (Oxf) 51: 517-524.

Carr WJ, Loeb LS, Dissinger ML. 1965. Responses of Rats to Sex Odors. J Comp Physiol Psychol 59: 370-377.

Carroll RS, Erskine MS, Doherty PC, Lundell LA, Baum MJ. 1985. Coital stimuli controlling luteinizing hormone secretion and ovulation in the female ferret. Biol Reprod 32: 925-933.

Castro JM, Ward OB, Bartke A. 2003. Play, copulation, anatomy, and testosterone in gonadally intact male rats prenatally exposed to flutamide. Physiol Behav 79: 633-641.

Cherry JA, Baum MJ. 1990. Effects of lesions of a sexually dimorphic nucleus in the preoptic/anterior hypothalamic area on the expression of androgen- and estrogen-dependent sexual behaviors in male ferrets. Brain Res 522: 191-203.

Cherry JA, Basham ME, Baum MJ. 1991. Neonatal testosterone masculinizes sexual behavior without affecting the morphology of the dorsal preoptic/anterior hypothalamic area of female ferrets. Brain Res 546: 321-328.

Cherry JA, Basham ME, Weaver CE, Krohmer RW, Baum MJ. 1990. Ontogeny of the sexually dimorphic male nucleus in the preoptic/anterior hypothalamus of ferrets and its manipulation by gonadal steroids. J Neurobiol 21: 844-857.

Christensen LW, Clemens LG. 1974. Intrahypothalamic implants of testosterone or estradiol and resumption of masculine sexual behavior in long-term castrated male rats. Endocrinology 95: 984-990.

Christensen LW, Clemens LG. 1975. Blockade of testosterone-induced mounting behavior in the male rat with intracranial application of the aromatization inhibitor, androst-1,4,6,-triene-3,17-dione. Endocrinology 97: 1545-1551.

Clancy AN, Zumpe D, Michael RP. 1995. Intracerebral infusion of an aromatase inhibitor, sexual behavior and brain estrogen receptor-like immunoreactivity in intact male rats. Neuroendocrinology 61: 98-111.

Clancy AN, Zumpe D, Michael RP. 2000. Estrogen in the medial preoptic area of male rats facilitates copulatory behavior. Horm Behav 38: 86-93.

Clemens LG, Gladue BA, Coniglio LP. 1978. Prenatal endogenous androgenic influences on masculine sexual behavior and genital morphology in male and female rats. Horm Behav 10: 40-53.

Clemens LG, Wee BE, Weaver DR, Roy EJ, Goldman BD, et al. 1988. Retention of masculine sexual behavior following castration in male B6D2F1 mice. Physiol Behav 42: 69-76.

Commins D, Yahr P. 1984. Lesions of the sexually dimorphic area disrupt mating and marking in male gerbils. Brain Res Bull 13: 185-193.

Coolen LM, Wood RI. 1999. Testosterone stimulation of the medial preoptic area and medial amygdala in the control of male hamster sexual behavior: redundancy without amplification. Behav Brain Res 98: 143-153.

Cooper TT, Clancy AN, Karom M, Moore TO, Albers HE. 2000. Conversion of testosterone to estradiol may not be necessary for the expression of mating behavior in male Syrian hamsters (Mesocricetus auratus). Horm Behav 37: 237-245.

Coquelin A. 1991. Persistent sexual behavior in castrated, recombinant inbred mice. Biol Reprod 45: 680-684.

Corbier P, Kerdelhue B, Picon R, Roffi J. 1978. Changes in testicular weight and serum gonadotropin and testosterone levels before, during, and after birth in the perinatal rat. Endocrinology 103: 1985-1991.

Cottingham SL, Pfaff DW. 1986. Interconnectedness of steroid hormone-binding neurons: Existence and implications. Current Topics in Neuroendocrinology 7: 223-249.

Dahlstrom A, Fuxe K. 1964. Evidence for the existence of monoamine-containing neurons in the central nervous system. I. Demonstration of monoamines in the cell bodies of brainstem neurons. Acta Physiol Scand 62: 1-55.

Damassa DA, Smith ER, Tennent B, Davidson JM. 1977. The relationship between circulating testosterone levels and male sexual behavior in rats. Horm Behav 8: 275-286.

Davidson JM. 1966. Activation of the male rat's sexual behavior by intracerebral implantation of androgen. Endocrinology 79: 783-794.

Davidson JM. 1969. Effects of estrogen on the sexual behavior of male rats. Endocrinology 84: 1365-1372.

Davis PG, Barfield RJ. 1979. Activation of masculine sexual behavior by intracranial estradiol benzoate implants in male rats. Neuroendocrinology 28: 217-227.

Davis AM, Grattan, DR, McCarthy MM. 2000. Decreasing GAD neonatally attenuates steroid-induced sexual differentiation of the rat brain. Behav Neurosci 114: 923-933.

Davis AM, Ward SC, Selmanoff M, Herbison AE, McCarthy MM. 1999. Developmental sex differences in amino acid neurotransmitter levels in hypothalamic and limbic areas of rat brain. Neuroscience 90: 1471-1482.

De Jonge FH, Louwerse AL, Ooms MP, Evers P, Endert E, et al. 1989. Lesions of the SDN-POA inhibit sexual behavior of male Wistar rats. Brain Res Bull 23: 483-492.

De Vries GJ, Simerly RB. 2002. Anatomy, development, and function of sexually dimorphic neural circuits in the mammalian brain. Hormones, Brain and Behavior. Pfaff DW, Arnold AP, Etgen AM, Fahrbach SE, Rubin RT, editors. San Diego: Academic Press; pp. 137-192.

De Vries GJ, Rissman EF, Simerly RB, Yang LY, Scordalakes EM, et al. 2002. A model system for study of sex chromosome effects on sexually dimorphic neural and behavioral traits. J Neurosci 22: 9005-9014.

Denef C, Magnus C, McEwen BS. 1973. Sex differences and hormonal control of testosterone metabolism in rat pituitary and brain. J Endocrinol 59: 605-621.

Dewsbury DA. 1972. Patterns of copulatory behavior in male mammals. Q Rev Biol 47: 1-33.

Dohler KD, Srivastava SS, Shryne JE, Jarzab B, Sipos A, et al. 1984. Differentiation of the sexually dimorphic nucleus in the preoptic area of the rat brain is inhibited by postnatal treatment with an estrogen antagonist. Neuroendocrinology 38: 297-301.

Dominguez J, Riolo JV, Xu Z, Hull EM. 2001. Regulation by the medial amygdala of copulation and medial preoptic dopamine release. J Neurosci 21: 349-355.

Dominguez E, Portillo W, Baum MJ, Bakker J, Paredes RG. 2002. Effect of prenatal androgen receptor antagonist or aromatase inhibitor on the differentiation of neuronal Fos responses to estrous female odors in the rat accessory olfactory system. Physiol Behav 75: 337-346.

Emery DE, Sachs BD. 1975. Ejaculatory pattern in female rats without androgen treatment. Science 190: 484-486.

Erskine MS. 1995. Prolactin release after mating and genitosensory stimulation in females. Endocr Rev 16: 508-528.

Erskine MS, Baum MJ. 1982. Plasma concentrations of testosterone and dihydrotestosterone during perinatal development in male and female ferrets. Endocrinology 111: 767-772.

Erskine MS, Tobet SA, Baum MJ. 1988. Effect of birth on plasma testosterone, brain aromatase activity, and hypothalamic estradiol in male and female ferrets. Endocrinology 122: 524-530.

Everitt BJ, Fray P, Kostarczyk E, Taylor S, Stacey P. 1987. Studies of instrumental behavior with sexual reinforcement in male rats (Rattus norvegicus): I. Control by brief visual stimuli paired with a receptive female. J Comp Psychol 101: 395-406.

Feder HH. 1971. The comparative actions of testosterone propionate and 5-androstan-17-ol-3-one propionate on the reproductive behaviour, physiology and morphology of male rats. J Endocrinol 51: 241-252.

Feder HH, Naftolin F, Ryan KJ. 1974. Male and female sexual responses in male rats given estradiol benzoate and 5 alpha-androstan-17 beta-ol-3-one propionate. Endocrinology 94: 136-141.

Fiber JM, Swann JM. 1996. Testosterone differentially influences sex-specific pheromone-stimulated Fos expression in limbic regions of Syrian hamsters. Horm Behav 30: 455-473.

Fiber JM, Adames P, Swann JM. 1993. Pheromones induce c-fos in limbic areas regulating male hamster mating behavior. Neuroreport 4: 871-874.

Fisher AE. 1956. Maternal and sexual behavior induced by intracranial chemical stimulation. Science 124: 228-229.

Fisher CR, Graves KH, Parlow AF, Simpson ER. 1998. Characterization of mice deficient in aromatase (ArKO) because of targeted disruption of the cyp19 gene. Proc Natl Acad Sci USA 95: 6965-6970.

Fumero B, Fernandez-Vera JR, Gonzalez-Mora JL, Mas M. 1994. Changes in monoamine turnover in forebrain areas associated with masculine sexual behavior: a microdialysis study. Brain Res 662: 233-239.

Gandelman R, Kozak M. 1988. Neonatal testosterone administration, but not in utero contiguity to males, augments the display of male sexual behavior by testosterone- treated adult female mice. Physiol Behav 42: 453-456.

Giuliano F, Rampin O, Brown K, Courtois F, Benoit G, et al. 1996. Stimulation of the medial preoptic area of the hypothalamus in the rat elicits increases in intracavernous pressure. Neurosci Lett 209: 1-4.

Gladue BA, Clemens LG. 1980a. Masculinization diminished by disruption of prenatal estrogen biosynthesis in male rats. Physiol Behav 25: 589-593.

Gladue BA, Clemens LG. 1980b. Flutamide inhibits testosterone-induced masculine sexual behavior in male and female rats. Endocrinology 106: 1917-1922.

Gorski RA, Gordon JH, Shryne JE, Southam AM. 1978. Evidence for a morphological sex difference within the medial preoptic area of the rat brain. Brain Res 148: 333-346.

Goy RW. 1970. Experimental control of psychosexuality. Philos Trans R Soc Lond B Biol Sci 259: 149-162.

Grady KL, Phoenix CH, Young WC. 1965. Role of the developing rat testis in differentiation of the neural tissues mediating mating behavior. J Comp Physiol Psychol 59: 176-182.

Gray GD. 1977. Differential effects of the antiandrogen flutamide on aspects of sexual behavior in castrated, androgen-treated male rats. Psychoneuroendocrinology 2: 315-320.

Gray GD, Smith ER, Davidson JM. 1980. Hormonal regulation of penile erection in castrated male rats. Physiol Behav 24: 463-468.

Greco B, Edwards DA, Michael RP, Clancy AN. 1998a. Androgen receptors and estrogen receptors are colocalized in male rat hypothalamic and limbic neurons that express Fos immunoreactivity induced by mating. Neuroendocrinology 67: 18-28.

Greco B, Edwards DA, Zumpe D, Michael RP, Clancy AN. 1998b. Fos induced by mating or noncontact sociosexual interaction is colocalized with androgen receptors in neurons within the forebrain, midbrain, and lumbosacral spinal cord of male rats. Horm Behav 33: 125-138.

Greco B, Edwards DA, Michael RP, Zumpe D, Clancy AN. 1999. Colocalization of androgen receptors and mating-induced FOS immunoreactivity in neurons that project to the central tegmental field in male rats. J Comp Neurol 408: 220-236.

Gunnet JW, Lookingland KJ, Moore KE. 1986. Comparison of the effects of castration and steroid replacement on incertohypothalamic dopaminergic neurons in male and female rats. Neuroendocrinology 44: 269-275.

Handa RJ, Roselli CE, Horton L, Resko JA. 1987. The quantitative distribution of cytosolic androgen receptors in microdissected areas of the male rat brain: effects of estrogen treatment. Endocrinology 121: 233-240.

Harding SM, McGinnis MY. 2003. Effects of testosterone in the VMN on copulation, partner preference, and vocalizations in male rats. Horm Behav 43: 327-335.

Harding SM, McGinnis MY. 2004. Androgen receptor blockade in the MPOA or VMN: effects on male sociosexual behaviors. Physiol Behav 81: 671-680.

Hart BL. 1967. Testosterone regulation of sexual reflexes in spinal male rats. Science 155: 1283-1284.

Hart BL. 1968. Role of prior experience in the effects of castration on sexual behavior of male dogs. J Comp Physiol Psychol 66: 719-725.

Hart BL. 1973. Effects of testosterone propionate and dihydrotestosterone on penile morphology and sexual reflexes of spinal male rats. Horm Behav 4: 239-246.

Heim N, Hursch CJ. 1979. Castration for sex offenders: treatment or punishment? A review and critique of recent European literature. Arch Sex Behav 8: 281-304.

Hendricks SE. 1969. Influence of neonatally administered hormones and early gonadectomy on rats' sexual behavior. J Comp Physiol Psychol 69: 408-413.

Hendricks SE, Weltin M. 1976. Effect of estrogen given during various periods of prepubertal life on the sexual behavior of rats. Physiol Psychol 4: 105-110.

Hillarp NA, Olivecrona H, Silfverskiold W. 1954. Evidence for the participation of the preoptic area in male mating behaviour. Experientia 10: 224-225.

Honda S, Harada N, Ito S, Takagi Y, Maeda S. 1998. Disruption of sexual behavior in male aromatase-deficient mice lacking exons 1 and 2 of the cyp19 gene. Biochem Biophys Res Commun 252: 445-449.

Houtsmuller EJ, Brand T, de Jonge FH, Joosten RN, Poll van de NE, et al. 1994. SDN-POA volume, sexual behavior, and partner preference of male rats affected by perinatal treatment with ATD. Physiol Behav 56: 535-541.

Huddleston GG, Michael RP, Zumpe D, Clancy AN. 2003. Estradiol in the male rat amygdala facilitates mounting but not ejaculation. Physiol Behav 79: 239-246.

Hull EM, Du J, Lorrain DS, Matuszewich L. 1995. Extracellular dopamine in the medial preoptic area: implications for sexual motivation and hormonal control of copulation. J Neurosci 15: 7465-7471.

Hull EM, Meisel RL, Sachs BD. 2002. Male sexual behavior. Hormones, Brain and Behavior. Pfaff DW, Arnold AP, Etgen AM, Fahrhant SE, Rubin RT, editors. San Diego: Academic Press; pp. 3-137.

Hutchison JB. 1974. Post-castration decline in behavioural responsiveness to intrahypothalamic androgen in doves. Brain Res 81: 169-181.

Imwalle DB, Scordalakes EM, Rissman EF. 2002. Estrogen receptor alpha influences socially motivated behaviors. Horm Behav 42: 484-491.

Isles AR, Baum MJ, Ma D, Szeto A, Keverne EB, et al. 2002. A possible role for imprinted genes in inbreeding avoidance and dispersal from the natal area in mice. Proc R Soc Lond B Biol Sci 269: 665-670.

Jacobson CD, Csernus VJ, Shryne JE, Gorski RA. 1981. The influence of gonadectomy, androgen exposure, or a gonadal graft in the neonatal rat on the volume of the sexually dimorphic nucleus of the preoptic area. J Neurosci 1: 1142-1147.

Johnson WA, Tiefer L. 1972. Sexual preferences in neonatally castrated male golden hamsters. Physiol Behav 9: 213-217.

Johnston P, Davidson JM. 1972. Intracerebral androgen and sexual behavior in the male rat. Horm Behav 3: 345-357.

Jost A. 1970. Hormonal factors in the sex differentiation of the mammalian foetus. Philos Trans R Soc Lond B Biol Sci 259: 119-130.

Ju G, Melander T, Ceccatelli S, Hokfelt T, Frey P. 1987. Immunohistochemical evidence for a spinothalamic pathway co-containing cholecystokinin- and galanin-like immunoreactivities in the rat. Neuroscience 20: 439-456.

Kato J, Onouchi T. 1973. 5Alpha-dihydrotestosterone "receptor" in the rat hypothalamus. Endocrinol Jpn 20: 429-432.

Kelliher KR, Baum MJ. 2001. Naris occlusion eliminates heterosexual partner selection without disrupting coitus in feuets of both sexes. J Neurosci 21: 5832-5840.

Kelliher KR, Liu YC, Baum MJ, Sachs BD. 1999. Neuronal Fos activation in olfactory bulb and forebrain of male rats having erections in the presence of inaccessible estrous females. Neuroscience 92: 1025-1033.

Kierniesky NC, Gerall AA. 1973. Effects of testosterone propionate implants in the brain on the sexual behavior and peripheral tissue of the male rat. Physiol Behav 11: 633-640.

Kindon HA, Baum MJ, Paredes RJ. 1996. Medial preoptic/anterior hypothalamic lesions induce a female-typical profile of sexual partner preference in male ferrets. Horm Behav 30: 514-527.

Krege JH, Hodgin JB, Couse JF, Enmark E, Warner M, et al. 1998. Generation and reproductive phenotypes of mice lacking estrogen receptor beta. Proc Natl Acad Sci USA 95: 15677-15682.

Krohmer RW, Baum MJ. 1989. Effect of sex, intrauterine position and androgen manipulation on the development of brain aromatase activity in fetal ferrets. J Neuroendocrinology 1: 265-271.

Lambert GM, Baum MJ. 1991. Reciprocal relationships between pulsatile androgen secretion and the expression of mating behavior in adult male ferrets. Horm Behav 25: 382-393.

Larsson K, Sodersten P, Beyer C. 1973. Induction of male sexual behaviour by oestradiol benzoate in combination with dihydrotestosterone. J Endocrinol 57: 563-564.

Lephart ED, Call SB, Rhees RW, Jacobson NA, Weber KS, et al. 2001. Neuroendocrine regulation of sexually dimorphic brain structure and associated sexual behavior in male rats is genetically controlled. Biol Reprod 64: 571-578.

Le Vay S. 1991. A difference in hypothalamic structure between heterosexual and homosexual men. Science 253: 1034-1037.

Lindzey J, Crews D. 1992. Interactions between progesterone and androgens in the stimulation of sex behaviors in male little striped whiptail lizards, Cnemidophorus inornatus. Gen Comp Endocrinol 86: 52-58.

Lisk RD. 1967. Neural localization for androgen activation of copulatory behavior in the male rat. Endocrinology 80: 754-761.

Lodder J, Baum MJ. 1977. Facilitation of mounting behavior by dihydrotestosterone propionate in castrated estradiol benzoate-treated male rats following pudendectomy. Behav Biol 20: 141-148.

Lopez HH, Olster DH, Ettenberg A. 1999. Sexual motivation in the male rat: the role of primary incentives and copulatory experience. Horm Behav 36: 176-185.

Lugg JA, Rajfer J, Gonzalez- Cadavid NF. 1995. Dihydrotestosterone is the active androgen in the maintenance of nitric oxide-mediated penile erection in the rat. Endocrinology 136: 1495-1501.

Maclean PD, Robinson BW, Ploog DW. 1959. Experiments on localization of genital function in the brain. Trans Am Neurol Assoc 84: 105-109.

Maillard CA, Edwards DA. 1991. Excitotoxin lesions of the zona incerta/lateral tegmentum continuum: effects on male sexual behavior in rats. Behav Brain Res 46: 143-149.

Malmnas CO. 1977. Dopaminergic reversal of the decline after castration of rat copulatory behaviour. J Endocrinol 73: 187-188.

Manning A, McGill TE. 1974. Neonatal androgen and sexual behavior in female house mice. Horm Behav 5: 19-31.

Manzo J, Cruz MR, Hernandez ME, Pacheco P, Sachs BD. 1999. Regulation of noncontact erection in rats by gonadal steroids. Horm Behav 35: 264-270.

Marin R, Escrig A, Abreu P, Mas M. 1999. Androgen-dependent nitric oxide release in rat penis correlates with levels of constitutive nitric oxide synthase isoenzymes. Biol Reprod 61: 1012-1016.

Martinez I, Paredes RG. 2001. Only self-paced mating is rewarding in rats of both sexes. Horm Behav 40: 510-517.

Mas M, Gonzalez-Mora JL, Louilot A, Sole C, Guadalupe T. 1990. Increased dopamine release in the nucleus accumbens of copulating male rats as evidenced by in vivo voltammetry. Neurosci Lett 110: 303-308.

Matochik JA, Sipos ML, Nyby JG, Barfield RJ. 1994. Intracranial androgenic activation of male-typical behaviors in house mice: motivation versus performance. Behav Brain Res 60: 141-149.

Matsumoto T, Honda S, Harada N. 2003. Alteration in sex-specific behaviors in male mice lacking the aromatase gene. Neuroendocrinology 77: 416-424.

McCarthy MM, Davis AM, Mong JA. 1997. Excitatory neurotransmission and sexual differentiation of the brain. Brain Res Bull 44: 487-495.

McDonald P, Beyer C, Newton F, Brien B, Baker R, et al. 1970. Failure of 5alpha-dihydrotestosterone to initiate sexual behaviour in the castrated male rat. Nature 227: 964-965.

McGill TE, Tucker GR. 1964. Genotype and Sex Drive in Intact and in Castrated Male Mice. Science 145: 514-515.

McGill TE, Haynes CM. 1973. Heterozygosity and retention of ejaculatory reflex after castration in male mice. J Comp Physiol Psychol 84: 423-429.

McGill TE, Manning A. 1976. Genotype and retention of the ejaculatory reflex in castrated male mice. Anim Behav 24: 507-518.

McGinnis MY, Dreifuss RM. 1989. Evidence for a role of testosterone-androgen receptor interactions in mediating masculine sexual behavior in male rats. Endocrinology 124: 618-626.

McGinnis MY, Williams GW, Lumia AR. 1996. Inhibition of male sex behavior by androgen receptor blockade in preoptic area or hypothalamus, but not amygdala or septum. Physiol Behav 60: 783-789.

McGinnis MY, Montana RC, Lumia AR. 2002. Effects of hydroxyflutamide in the medial preoptic area or lateral septum on reproductive behaviors in male rats. Brain Res Bull 59: 227-234.

Meek LR, Romeo RD, Novak CM, Sisk CL. 1997. Actions of testosterone in prepubertal and postpubertal male hamsters: dissociation of effects on reproductive behavior and brain androgen receptor immunoreactivity. Horm Behav 31: 75-88.

Meisel RL, O'Hanlon JK, Sachs BD. 1984. Differential maintenance of penile responses and copulatory behavior by gonadal hormones in castrated male rats. Horm Behav 18: 56-64.

Melis MR, Succu S, Mascia MS, Cortis L, Argiolas A. 2003. Extra-cellular dopamine increases in the paraventricular nucleus of male rats during sexual activity. Eur J Neurosci 17: 1266-1272.

Mendelson SD, Pfaus JG. 1989. Level searching: a new assay of sexual motivation in the male rat. Physiol Behav 45: 337-341.

Michael RP, Zumpe D, Bonsall RW. 1990. Estradiol administration and the sexual activity of castrated male rhesus monkeys (Macaca mulatta). Horm Behav 24: 71-88.

Miller RL, Baum MJ. 1987. Naloxone inhibits mating and conditioned place preference for an estrous female in male rats soon after castration. Pharmacol Biochem Behav 26: 781-789.

Mitchell JB, Gratton A. 1991. Opioid modulation and sensitization of dopamine release elicited by sexually relevant stimuli: a high speed chronoamperometric study in freely behaving rats. Brain Res 551: 20-27.

Mitchell JB, Gratton A. 1992. Mesolimbic dopamine release elicited by activation of the accessory olfactory system: a high speed chronoamperometric study. Neurosci Lett 140: 81-84.

Mucha RF, van der Kooy D, O'Shaughnessy M, Bucenieks P. 1982. Drug reinforcement studied by the use of place conditioning in rat. Brain Res 243: 91-105.

Mullins RF, Levine S. 1968. Hormonal determinants during infancy of adult sexual behavior in the female rat. Physiol Behav 3: 333-338.

Naftolin F, Ryan KJ, Petro Z. 1972. Aromatization of androstenedione by the anterior hypothalamus of adult male and female rats. Endocrinology 90: 295-298.

Nelson RJ. 2005. Introduction to the behavioral endocrinology, 3rd edition. Sunderland: Sinaner Association.

Newman SW. 2002. Pheromonal signals access the medial extended amygdala: One node in a proposed social behavior network. Hormones, Brain and Behavior. Pfaff DW,

Arnold AP, Etgen AM, Fahrbach SE, Rubin RT, editors. San Diego: Elsevier Science; pp. 17-32.

Ogawa S, Washburn TF, Taylor J, Lubahn DB, Korach KS, et al. 1998. Modifications of testosterone-dependent behaviors by estrogen receptor-alpha gene disruption in male mice. Endocrinology 139: 5058-5069.

Ogawa S, Chan J, Chester AE, Gustafsson JA, Korach KS, et al. 1999. Survival of reproductive behaviors in estrogen receptor beta gene- deficient (betaERKO) male and female mice. Proc Natl Acad Sci USA 96: 12887-12892.

Ogawa S, Chester AE, Hewitt SC, Walker VR, Gustafsson JA, et al. 2000. Abolition of male sexual behaviors in mice lacking estrogen receptors alpha and beta (alpha beta ERKO). Proc Natl Acad Sci USA 97: 14737-14741.

Olsen KL. 1979. Induction of male mating behavior in androgen-insensitive (tfm) and Normal (King-Holtzman) male rats: effect of testosterone propionate, estradiol benzoate, and dihydrotestosterone. Horm Behav 13: 66-84.

Olsen K. 1992. Genetic influences on sexual behavior differentiation. Handbook of behavioral neurobiology, sexual differentiation. Gerall AA, Moltz H, Ward IL, editors. New York: Plenum; pp. 1-40.

Pang SF, Caggiula AR, Gay VL, Goodman RL, Pang CS. 1979. Serum concentrations of testosterone, oestrogens, luteinizing hormone and follicle-stimulating hormone in male and female rats during the critical period of neural sexual differentiation. J Endocrinol 80: 103-110.

Paredes RG, Baum MJ. 1995. Altered sexual partner preference in male ferrets given excitotoxic lesions of the preoptic area/anterior hypothalamus. J Neurosci 15: 6619-6630.

Paredes RG, Lopez ME, Baum MJ. 1998a. Testosterone augments neuronal Fos responses to estrous odors throughout the vomeronasal projection pathway of gonadectomized male and female rats. Horm Behav 33: 48-57.

Paredes RG, Tzschentke T, Nakach N. 1998b. Lesions of the medial preoptic area/anterior hypothalamus (MPOA/AH) modify partner preference in male rats. Brain Res 813: 1-8.

Park JJ, Baum MJ, Tobet SA. 1997. Sex difference and steroidal stimulation of galanin immunoreactivity in the ferret's dorsal preoptic area/anterior hypothalamus. J Comp Neurol 389: 277-288.

Pehek EA, Warner RK, Bazzett TJ, Bitran D, Band LC, et al. 1988. Microinjection of cis-flupenthixol, a dopamine antagonist, into the medial preoptic area impairs sexual behavior of male rats. Brain Res 443: 70-76.

Perkins A, Fitzgerald JA, Price EO. 1992. Luteinizing hormone and testosterone response of sexually active and inactive rams. J Anim Sci 70: 2086-2093.

Perkins A, Fitzgerald JA, Moss GE. 1995. A comparison of LH secretion and brain estradiol receptors in heterosexual and homosexual rams and female sheep. Horm Behav 29: 31-41.

Perrot-Sinal TS, Davis AM, Gregerson KA, Kao JP, McCarthy MM. 2001. Estradiol enhances excitatory gamma-aminobutyric [corrected] acid-mediated calcium signaling in neonatal hypothalamic neurons. Endocrinology 142: 2238-2243.

Pfaff DW, Zigmond RE. 1971. Neonatal androgen effects on sexual and non-sexual behavior of adult rats tested under various hormone regimes. Neuroendocrinology 7: 129-145.

Pfaus JG, Phillips AG. 1991. Role of dopamine in anticipatory and consummatory aspects of sexual behavior in the male rat. Behav Neurosci 105: 727-743.

Pfaus JG, Damsma G, Nomikos GG, Wenkstern DG, Blaha CD, et al. 1990. Sexual behavior enhances central dopamine transmission in the male rat. Brain Res 530: 345-348.

Phelps SM, Lydon JP, O'Malley BW, Crews D. 1998. Regulation of male sexual behavior by progesterone receptor, sexual experience, and androgen. Horm Behav 34: 294-302.

Phoenix CH. 1974. Effects of dihydrotestosterone on sexual behavior of castrated male rhesus monkeys. Physiol Behav 12: 1045-1055.

Phoenix CH. 1978. Steroids and sexual behavior in castrated male rhesus monkeys. Horm Behav 10: 1-9.

Pleim ET, Matochik JA, Barfield RJ, Auerbach SB. 1990. Correlation of dopamine release in the nucleus accumbens with masculine sexual behavior in rats. Brain Res 524: 160-163.

Powers JB, Fields RB, Winans SS. 1979. Olfactory and vomeronasal system participation in male hamsters' attraction to female vaginal secretions. Physiol Behav 22: 77-84.

Putnam SK, Sato S, Hull EM. 2003. Effects of testosterone metabolites on copulation and medial preoptic dopamine release in castrated male rats. Horm Behav 44: 419-426.

Putnam SK, Du J, Sato S, Hull EM. 2001. Testosterone restoration of copulatory behavior correlates with medial preoptic dopamine release in castrated male rats. Horm Behav 39: 216-224.

Rasia-Filho AA, Peres TM, Cubilla-Gutierrez FH, Lucion AB. 1991. Effect of estradiol implanted in the corticomedial amygdala on the sexual behavior of castrated male rats. Braz J Med Biol Res 24: 1041-1049.

Resko JA, Feder HH, Goy RW. 1968. Androgen concentrations in plasma and testis of developing rats. J Endocrinol 40: 485-491.

Resko JA, Perkins A, Roselli CE, Fitzgerald JA, Choate JV, et al. 1996. Endocrine correlates of partner preference behavior in rams. Biol Reprod 55: 120-126.

Rhoda J, Corbier P, Roffi J. 1984. Gonadal steroid concentrations in serum and hypothalamus of the rat at birth: aromatization of testosterone to 17 beta-estradiol. Endocrinology 114: 1754-1760.

Rissman EF, Wersinger SR, Fugger HN, Foster TC. 1999. Sex with knockout models: behavioral studies of estrogen receptor alpha. Brain Res 835: 80-90.

Roffi J, Chami F, Corbier P, Edwards DA. 1987. Testicular hormones during the first few hours after birth augment the tendency of adult male rats to mount receptive females. Physiol Behav 39: 625-628.

Roselli CE, Cross E, Poonyagariyagorn HK, Stadelman HL. 2003. Role of aromatization in anticipatory and consummatory aspects of sexual behavior in male rats. Horm Behav 44: 146-151.

Roselli CE, Larkin K, Schrunk JM, Stormshak F. 2004a. Sexual partner preference, hypothalamic morphology and aromatase in rams. Physiol Behav 83: 233-245.

Roselli CE, Larkin K, Resko JA, Stellflug JN, Stormshak F. 2004b. The volume of a sexually dimorphic nucleus in the ovine medial preoptic area/anterior hypothalamus varies with sexual partner preference. Endocrinology 145: 478-483.

Rosenblatt JS, Aronson LR. 1958. The decline of sexual behavior in male cats after castration with special reference to the role prior sexual experience. Behaviour 12: 285-338.

Sachs BD. 1997. Erection invoked in male rats by airborne scent from estrous females. Physiol Behav 62: 921-924.

Sakata JT, Woolley SC, Gupta A, Crews D. 2003. Differential effects of testosterone and progesterone on the activation and retention of courtship behavior in sexual and parthenogenetic whiptail lizards. Horm Behav 43: 523-530.

Sato Y, Wada H, Horita H, Suzuki N, Shibuya A, et al. 1995. Dopamine release in the medial preoptic area during male copulatory behavior in rats. Brain Res 692: 66-70.

Sato T, Matsumoto T, Kawano H, Watanabe T, Uematsu Y, et al. 2004. Brain masculinization requires androgen receptor function. Proc Natl Acad Sci USA 101: 1673-1678.

Savu L, Benassayag C, Vallette G, Christeff N, Nunez E. 1981. Mouse alpha 1-fetoprotein and albumin. A comparison of their binding properties with estrogen and fatty acid ligands. J Biol Chem 256: 9414-9418.

Scaletta LL, Hull EM. 1990. Systemic or intracranial apomorphine increases copulation in long-term castrated male rats. Pharmacol Biochem Behav 37: 471-475.

Scalia F, Winans SS. 1975. The differential projections of the olfactory bulb and accessory olfactory bulb in mammals. J Comp Neurol 161: 31-55.

Shah NM, Pisapia DJ, Maniatis S, Mendelsohn MM, Nemes A, et al. 2004. Visualizing sexual dimorphism in the brain. Neuron 43: 313-319.

Shibata H, Spencer TE, Onate SA, Jenster G, Tsai SY, et al. 1997. Role of co-activators and co-repressors in the mechanism of steroid/thyroid receptor action. Recent Prog Horm Res 52: 141-164; discussion 164–165.

Simerly RB, Swanson LW. 1986. The organization of neural inputs to the medial preoptic nucleus of the rat. J Comp Neurol 246: 312-342.

Simerly RB, Chang C, Muramatsu M, Swanson LW. 1990. Distribution of androgen and estrogen receptor mRNA-containing cells in the rat brain: an in situ hybridization study. J Comp Neurol 294: 76-95.

Simpkins JW, Kalra PS, Kalra SP. 1980. Inhibitory effects of androgens on preoptic area dopaminergic neurons in castrate rats. Neuroendocrinology 31: 177-181.

Simpkins JW, Kalra SP, Kalra PS. 1983. Variable effects of testosterone on dopamine activity in several microdissected regions in the preoptic area and medial basal hypothalamus. Endocrinology 112: 665-669.

Sinchak K, Clemens LG. 1989. Expression of the ejaculatory reflex in castrated male B6D2F1 hybrid house mouse is reduced by the aromatase inhibitor ATD. Society for Neuroscience Abstracts 15: 155.16.

Sinchak K, Roselli CE, Clemens LG. 1996. Levels of serum steroids, aromatase activity, and estrogen receptors in preoptic area, hypothalamus, and amygdala of B6D2F1 male house mice that differ in the display of copulatory behavior after castration. Behav Neurosci 110: 593-602.

Smith ER, Damassa DA, Davidson JM. 1977. Plasma testosterone and sexual behavior following intracerebral implantation of testosterone propionate in the castrated male rat. Horm Behav 8: 77-87.

Sodersten P. 1973a. Estrogen-activated sexual behavior in male rats. Horm Behav 4: 247-256.

Sodersten P. 1973b. Increased mounting behavior in the female rat following a single neonatal injection of testosterone propionate. Horm Behav 4: 1-17.

Sodersten P. 1978. Effects of anti-oestrogen treatment of neonatal male rats on lordosis behaviour and mounting behaviour in the adult. J Endocrinol 76: 241-249.

Sodersten P, Hansen S. 1978. Effects of castration and testosterone, dihydrotestosterone or oestradiol replacement treatment in neonatal rats on mounting behaviour in the adult. J Endocrinol 76: 251-260.

Sodersten P, Gray G, Damassa DA, Smith ER, Davidson JM. 1975. Effects of a non-steroidal antiandrogen on sexual behavior and pituitary-gonadal function in the male rat. Endocrinology 97: 1468-1475.

Soulairac A, Soulairac ML. 1956. [Effects of hypothalamic lesions on the sex behavior & genital tract in the male rat]. Ann Endocrinol (Paris) 17: 731-745.

Stewart J, Pottier J, Kaczender-Henrik. 1971. Male copulatory behavior in the female rat after perinatal treatment with an antiandrogenic steroid. Horm Behav 2: 247-254.

Stockman ER, Callaghan RS, Baum MJ. 1985. Effects of neonatal castration and testosterone treatment on sexual partner preference in the ferret. Physiol Behav 34: 409-414.

Stone CP. 1940. Precocious copulatory activity produced in male rats by subcutaneous injections of testosterone propionate. Endocrinology 26: 511-515.

Stone CP, Barker RG, Tomlin MI. 1935. Sexual drive in potent and impotent male rats as measured by the Columbia obstruction apparatus. J Genet Psychol 47: 33-48.

Szczypka MS, Zhou QY, Palmiter RD. 1998. Dopamine-stimulated sexual behavior is testosterone dependent in mice. Behav Neurosci 112: 1229-1235.

Takane KK, Husmann DA, McPhaul MJ, Wilson JD. 1991. Androgen receptor levels in the rat penis are controlled differently in distinctive cell types. Endocrinology 128: 2234-2238.

Tang YP, Sisk CL. 1991. Testosterone in MPOA elicits behavioral but not neuroendocrine responses in ferrets. Brain Res Bull 26: 373-378.

Temple JL, Scordalakes EM, Bodo C, Gustafsson JA, Rissman EF. 2004. Lack of functional estrogen receptor beta gene disrupts pubertal male sexual behavior. Horm Behav in press.

Thompson ML, McGill TE, McIntosh SM, Manning A. 1976. Effects of adrenalectomy on the sexual behaviour of castrated and intact BDF1 mice. Anim Behav 24: 519-522.

Tobet SA, Baum MJ. 1987. Role for prenatal estrogen in the development of masculine sexual behavior in the male ferret. Horm Behav 21: 419-429.

Tobet SA, Hanna IK. 1997. Ontogeny of sex differences in the mammalian hypothalamus and preoptic area. Cell Mol Neurobiol 17: 565-601.

Tobet SA, Zahniser DJ, Baum MJ. 1986. Differentiation in male ferrets of a sexually dimorphic nucleus of the preoptic/anterior hypothalamic area requires prenatal estrogen. Neuroendocrinology 44: 299-308.

Tobet SA, Chickering TW, Fox TO, Baum MJ. 1993. Sex and regional differences in intracellular localization of estrogen receptor immunoreactivity in adult ferret forebrain. Neuroendocrinology 58: 316-324.

Tobet SA, Shim JH, Osiecki ST, Baum MJ, Canick JA. 1985. Androgen aromatization and 5 alpha-reduction in ferret brain during perinatal development: effects of sex and testosterone manipulation. Endocrinology 116: 1869-1877.

Toda K, Okada T, Takeda K, Akira S, Saibara T, et al. 2001. Oestrogen at the neonatal stage is critical for the reproductive ability of male mice as revealed by supplementation with 17beta-oestradiol to aromatase gene (Cyp19) knock-out mice. J Endocrinol 168: 455-463.

Traish AM, Park K, Dhir V, Kim NN, Moreland RB, et al. 1999. Effects of castration and androgen replacement on erectile function in a rabbit model. Endocrinology 140: 1861-1868.

Truitt WA, Coolen LM. 2001. Ejaculation-induced neural activation in lumbar spinal cord of the rat. Horm Behav 39: 352 (abstract).

Truitt WA, Coolen LM. 2002. Identification of a potential ejaculation generator in the spinal cord. Science 297: 1566-1569.

Truitt WA, Shipley MT, Veening JG, Coolen LM. 2003. Activation of a subset of lumbar spinothalamic neurons after copulatory behavior in male but not female rats. J Neurosci 23: 325-331.

Vagell ME, McGinnis MY. 1997. The role of aromatization in the restoration of male rat reproductive behavior. J Neuroendocrinol 9: 415-421.

Vale JR, Ray D, Vale CA. 1973. The interaction of genotype and exogenous neonatal androgen and estrogen: sex behavior in female mice. Dev Psychobiol 6: 319-327.

Ploeg Van der LH, Martin WJ, Howard AD, Nargund RP, Austin CP, et al. 2002. A role for the melanocortin 4 receptor in sexual function. Proc Natl Acad Sci USA 99: 11381-11386.

Van Furth WR, Van Ree JM. 1996a. Appetitive sexual behavior in male rats: 2. sexual reward and level-changing behavior. Physiol Behav 60: 1007-1012.

Van Furth WR, Van Ree JM. 1996b. Appetitive sexual behavior in male rats: 1. The role of olfaction in level-changing behavior. Physiol Behav 60: 999-1005.

Veening JG, Coolen LM. 1998. Neural activation following sexual behavior in the male and female rat brain. Behav Brain Res 92: 181-193.

Vega Matuszczyk J, Fernandez-Guasti A, Larsson K. 1988. Sexual orientation, proceptivity, and receptivity in the male rat as a function of neonatal hormonal manipulation. Horm Behav 22: 362-378.

Vito CC, Baum MJ, Bloom C, Fox TO. 1985. Androgen and estrogen receptors in perinatal ferret brain. J Neurosci 5: 268-274.

Vreeburg JT, Schretlen PJ, Baum MJ. 1975. Specific, high-affinity binding of 17beta-estradiol in cytosols from several brain regions and pituitary of intact and castrated adult male rats. Endocrinology 97: 969-977.

Wallen K, Baum MJ. 2002. Masculinization and defeminization in altricial and precocial mammals: Comparative aspects of steroid hormone action. Hormones, Brain and Behavior. Pfaff DW, Arnold AP, Etgen AM, Fahrbach SE, Rubin RT, editors. San Diego: Academic Press; pp. 385-423.

Ward IL, Renz FJ. 1972. Consequences of perinatal hormone manipulation on the adult sexual behavior of female rats. J Comp Physiol Psychol 78: 349-355.

Weaver CE, Baum MJ. 1991. Differential regulation of brain aromatase by androgen in adult and fetal ferrets. Endocrinology 128: 1247-1254.

Weisz J, Ward IL. 1980. Plasma testosterone and progesterone titers of pregnant rats, their male and female fetuses, and neonatal offspring. Endocrinology 106: 306-316.

Wersinger SR, Rissman EF. 2000a. Oestrogen receptor alpha is essential for female-directed chemo-investigatory behaviour but is not required for the pheromone-induced luteinizing hormone surge in male mice. J Neuroendocrinol 12: 103-110.

Wersinger SR, Rissman EF. 2000b. Dopamine activates masculine sexual behavior independent of the estrogen receptor alpha. J Neurosci 20: 4248-4254.

Wersinger SR, Sannen K, Villalba C, Lubahn DB, Rissman EF, et al. 1997. Masculine sexual behavior is disrupted in male and female mice lacking a functional estrogen receptor alpha gene. Horm Behav 32: 176-183.

Whalen RE, Edwards DA. 1967. Hormonal determinants of the development of masculine and feminine behavior in male and female rats. Anat Rec 157: 173-180.

Whalen RE, Luttge WG. 1971. Testosterone, androstenedione, and dihydrotestosterone: Effects on mating behavior of male rats. Horm Behav 2: 117-125.

Winslow JT, Hastings N, Carter CS, Harbaugh CR, Insel TR. 1993. A role for central vasopressin in pair bonding in monogamous prairie voles. Nature 365: 545-548.

Wisniewski AB, Migeon CJ, Meyer-Bahlburg HF, Gearhart JP, Berkovitz GD, et al. 2000. Complete androgen insensitivity syndrome: long-term medical, surgical, and psychosexual outcome. J Clin Endocrinol Metab 85: 2664-2669.

Witt DM, Young LJ, Crews D. 1995. Progesterone modulation of androgen-dependent sexual behavior in male rats. Physiol Behav 57: 307-313.

Wood RI. 1996. Estradiol, but not dihydrotestosterone, in the medial amygdala facilitates male hamster sex behavior. Physiol Behav 59: 833-841.

Wood RI. 1997. Thinking about networks in the control of male hamster sexual behavior. Horm Behav 32: 40-45.

Wood RI, Coolen LM. 1997. Integration of chemosensory and hormonal cues is essential for sexual behaviour in the male Syrian hamster: role of the medial amygdaloid nucleus. Neuroscience 78: 1027-1035.

Wood RI, Newman SW. 1993. Mating activates androgen receptor-containing neurons in chemosensory pathways of the male Syrian hamster brain. Brain Res 614: 65-77.

Wood RI, Newman SW. 1995a. The medial amygdaloid nucleus and medial preoptic area mediate steroidal control of sexual behavior in the male Syrian hamster. Horm Behav 29: 338-353.

Wood RI, Newman SW. 1995b. Integration of chemosensory and hormonal cues is essential for mating in the male Syrian hamster. J Neurosci 15: 7261-7269.

Wozniak A, Hutchison RE, Hutchison JB. 1992. Localisation of aromatase activity in androgen target areas of the mouse brain. Neurosci Lett 146: 191-194.

Wysocki CJ, Nyby J, Whitney G, Beauchamp GK, Katz Y. 1982. The vomeronasal organ: primary role in mouse chemosensory gender recognition. Physiol Behav 29: 315-327.

Young LJ, Greenberg N, Crews D. 1991. The effects of progesterone on sexual behavior in male green anole lizards (Anolis carolinensis). Horm Behav 25: 477-488.

Zumpe D, Bonsall RW, Michael RP. 1993. Effects of the nonsteroidal aromatase inhibitor, fadrozole, on the sexual behavior of male cynomolgus monkeys (Macaca fascicularis). Horm Behav 27: 200-215.

2 Neurochemistry of Male Sexual Behavior

E. M. Hull · J. M. Dominguez · J. W. Muschamp

Abstract: Steroid hormones regulate male sexual behavior, in part, by orchestrating the production and release of neurotransmitters in brain areas that control sexual motivation and copulatory performance. Major facilitative neurotransmitters are dopamine, glutamate, and, in some cases, norepinephrine. Serotonin, GABA, and opioid peptides provide primarily inhibitory influence, although low levels of some of these neurotransmitters, acting via specific receptor subtypes, may facilitate certain aspects of behavior. Other neuropeptides, including oxytocin, orexin/hypocretin, α-melanocyte stimulating hormone, gonadotropin releasing hormone, and galanin-like peptide may facilitate mating, whereas neuropeptide Y, prolactin, corticotropin releasing hormone, and angiotensin II may have inhibitory effects. Brain areas that are especially important for the control of male sexual behavior include the medial preoptic area, the medial amygdala, the paraventricular nucleus of the hypothalamus, the mesocorticolimbic dopamine tract, several midbrain and brainstem nuclei, and nuclei in the spinal cord that control the erectile and ejaculatory reflexes. Some dopaminergic and serotonergic receptor subtypes promote parasympathetically mediated erection, and other subtypes facilitate sympathetically mediated seminal emission and ejaculation.

List of Abbreviations: 5-HT, 5-hydroxytryptamine, serotonin; 5-HT_{1A}, 5-HT_{1B}, 5-HT_{2C}, Serotonin receptor subtypes; 5,7-DHT, 5,7-Dihydroxytryptamine, serotonin neurotoxin; α_1, α_2, Alpha noradrenergic receptors; α-MSH, Alpha-melanocyte stimulating hormone; AChE, Acetylcholinesterase; ACTH, Adrenocorticotropic hormone; ADP, Adenosine 5′-diphosphate; AMPA, alpha-amino-3-hydroxy-5-methyl-4-isoxazolepropionic acid, specific agonist for a type of glutamate receptor; AR, Androgen receptor; BL, Baseline; BS, Bulbospongiosus muscle of the penis; CCK, Cholecystokinin; cGMP, Cyclic guanosine monophosphate; CNQX, 6-cyano-7-nitroquinoxaline-2,3-dione, an AMPA/kainate receptor antagonist; CRH, Corticotropin releasing hormone; D_1, D_2, D_3, D_4, D_5, Dopamine receptor subtypes; D5KO, Mouse with deletion of the D_5 dopamine receptor; DA, Dopamine; EEG, Electroencephalogram; EMG, Electromyogram; eNOS, Endothelial nitric oxide synthase; ER, Estrogen receptor; ERα KO, Mouse with deletion of the ERα estrogen receptor; FSH, Follicle stimulating hormone; GABA, Gamma-aminobutyric acid; GAD, L-Glutamic acid decarboxylase; GALP, Galanin-like peptide; $G_{I/O}$, GTP binding protein that couples receptors to inhibition of adenylyl cyclase or other effects; GTP, Guanosine triphosphate; icv, Intracerebroventricular; IP_3, Inositol triphosphate; L5-S1, Fifth lumbar to first sacral vertebrae; L-DOPA, 3,4-dihydroxy-L-phenylalanine, precursor of dopamine; LHA, Lateral hypothalamic area; L-NAME, N-Nitro-L-arginine methyl ester; L-NMMA, N(G)-monomethyl L-arginine ; MC, melanocortin; MPOA, Medial preoptic area; NAc, Nucleus accumbens; NE, norepinephrine; NMDA, N-methyl-D-aspartic acid; nNOS, Neuronal nitric oxide synthase; NO, Nitric oxide; NOS, Nitric oxide synthase; NOS-ir, Nitric oxide synthase immunoreactivity; nPGi, Nucleus paragigantocellularis; NPY, Neuropeptide Y; ODQ, 1H-[1,2,4]Oxadiazole[4,3-a]quinoxalin-1-one, inhibitor of guanylyl cyclase; *p*CPA, *p*-chlorophenylalanine; PDE5, Phosphodiesterase 5; PGE_2, Prostaglandin E_2; PKA, Protein kinase A; PKG, Protein kinase G; POMC, Proopiomelanocortin; PVN, Paraventricular nucleus; SPFp, Subparafascicular nucleus, parvocellular division; SSRI, Selective serotonin reuptake inhibitor; tfm, Testicular feminization mutation; THP, Tetrahydrothienopyridine; TRH, Thyrotropin releasing hormone; VIP, Vasoactive intestinal peptide; VNX, Vomeronasal organ removal; VTA, Ventral tegmental area

1 Introduction

Sexual behavior and the courtship that precedes it are regulated by steroid hormones, which orchestrate the production and release of numerous neurotransmitters and neuropeptides that interact in specific brain areas to elicit the appropriately timed and coordinated behavioral response. In this chapter, we summarize recent contributions to our understanding of how neurotransmitters and neuropeptides in the central nervous system interact to regulate male sexual behavior. We begin with a brief description of copulatory behaviors of the most commonly studied species. Next, we consider the major neurotransmitters and neuropeptides that have been shown to influence male sexual behavior. For each, we describe the effects of systemically or intraventricularly administered drugs, followed by consideration of the role of each neurotransmitter in specific brain areas. We conclude with a summary and a set of questions that may guide future research.

2 Patterns of Sexual Behavior of Male Mammals

2.1 Precopulatory Behaviors

Males advertise their fitness by using species-specific displays. They also assess the female's desirability and willingness to mate by investigating her face and anogenital region. Chemosensory, visual, and auditory stimuli may arouse him and signal the female's receptivity. A female may exhibit proceptive behaviors, such as hopping and darting, ending with an abrupt stop in front of the male. However, a female may reject the male's advances by running away, keeping her hindquarters and tail low, or kicking him.

2.2 Copulatory Behavior

2.2.1 Mounts

A receptive female typically remains immobile while the male clasps her flanks with his forepaws and initiates several shallow thrusts with his pelvis, while his penis is at least partially erect. The female may stand in a rigid lordosis posture, with her back flat or concave and her tail deflected to expose her vagina.

2.2.2 Intromissions

Intromission is the definitive event of copulation. If the male locates the female's vagina, he will perform a deeper, intravaginal thrust, followed by a rapid, springing dismount. The deep thrust and springing dismount are reliably associated with penile insertion (Pollak and Sachs, 1975) and are typically used as the measure of intromission in most rodents. Each intromission in rats lasts about half a second. In mice and some other species, the male thrusts repeatedly throughout the intromission (McGill, 1962), and in ungulates and canids, the male may ejaculate immediately after vaginal intromission (Bermant et al., 1969; Lott, 1981). Following an intromission, the male typically grooms his genitals.

In some species, increased numbers of intromissions preceding ejaculation may promote fertility by increasing the number of sperm in the ejaculate (Adler and Toner, 1986; Toner and Adler, 1986; Toner et al., 1987) or facilitating sperm transport in the female (Adler and Toner, 1986). Females of some species require a sufficient amount of vaginocervical stimulation to ovulate or to trigger a progestational state (reviewed in Dewsbury, 1990).

2.2.3 Ejaculation

Males of most mammalian species ejaculate only after receiving stimuli from multiple intromissions. Ejaculation in rats can be recognized behaviorally by a longer, deeper thrust followed by a slow dismount. Skeletal and striated perineal muscles usually contract rhythmically during ejaculation. Such muscle contractions are associated with orgasm in human males and females.

2.3 Postejaculatory Behavior and Satiety

Following ejaculation, males typically groom themselves and enter a period of sexual quiescence, which may last for as little as 30 s in Syrian hamsters or as long as hours or days in other species (Dewsbury, 1972); in rats, it is usually 5–10 min. During the first 50–75% of this interval, the absolute refractory interval, the male will not copulate again, regardless of the amount of stimulation. Ultrasonic vocalizations may occur during this period, as well as a pattern of synchronized EEG waves, similar to those of sleep (Barfield and Geyer, 1975). During the remainder of the postejaculatory interval (the relative refractory period), the male may resume copulation with a novel female. Most male rats reach sexual satiety after 7–8 ejaculations and may not copulate again for 1–3 days. Recent data suggest that a reduction in androgen receptors in the medial preoptic area and related areas may give rise to satiety (Fernandez-Guasti and Rodriguez-Manzo, 2003).

2.4 Sexual Motivation versus Performance

Sometimes it is useful to distinguish between the motivation for sexual activity and the actual performance of the behavior. Appetitive behaviors are more variable than the relatively stereotypic consummatory behaviors and are thought to reflect an underlying motivational state. Sexual motivation may include sensory and motor processes that are sensitized by gonadal hormones, as well as more general activational processes that can energize other behaviors as well. It is best to use tests that are designed specifically to assess motivation and that do not confound possible changes in sensory, motor, or learning ability with motivation per se. Some of the tests that have been used for this purpose include the following: (1) the X (or cross) maze, in which a female is in a goal box at the end of one of four arms and a male or other goal objects occupy other goal boxes, or the other goal boxes remain empty; (2) the bilevel apparatus, in which a male changes levels apparently in search of a female, following a previous sexual encounter in the apparatus; (3) place-preference test, in which the male spends more time in an area in which he has previously copulated; (4) time spent near an estrous versus nonestrous female; and (5) lever pressing for access to a receptive female. However, several of these tests are subject to confounding of motor and motivational effects. It is difficult to design tests with no possible confounds, so it is helpful to test separately for drug effects on sensory or motor function.

2.5 Sexual Experience

Previous sexual experience confers increased sexual "efficiency." Experienced males require less time to initiate copulation and to ejaculate (Larsson, 1959; Dewsbury, 1969; Pfaus and Wilkins, 1995; Bialy et al., 2000). They are also less impaired by castration, various brain lesions, and exposure to a novel environment (Rosenblatt and Aronson, 1958; Lisk and Heimann, 1980; Arendash and Gorski, 1983; Meredith, 1986; Saito and Moltz, 1986; Kondo, 1992; Bialy and Beck, 1993; Claro et al., 1995; Pfaus and Wilkins, 1995; Bialy et al., 2000). Ejaculation activated more neurons (i.e., immunoreactive for Fos—the protein product of the immediate-early gene *c-fos*) in the medial preoptic area (MPOA) (Lumley and Hull, 1999) and in the shell of the nucleus accumbens (Lopez and Ettenberg, 2002a) in sexually experienced than in sexually naïve males. Therefore, increased sensitivity to sexual stimuli may be one mechanism that confers greater sexual efficiency and resistance to disruptions in experienced males.

2.6 Sexual Behavior Throughout the Lifespan: Puberty and Aging

The onset of copulatory ability is variable, with some male rats beginning to mount as early as 40 days of age, and the first ejaculations occurring between 48 and 75 days (reviewed in Meisel and Sachs, 1994). Prepubertal castration prevents the development of sexual behavior (Larsson, 1967), and exogenous testosterone or estrogens can hasten its onset (Baum, 1972; Södersten et al., 1985). However, the natural onset of copulatory behavior in male rats occurs several days before the pubertal rise in testosterone

(Södersten et al., 1977; Sachs and Meisel, 1979). Furthermore, the maturation or innervation of genital structures was not the decisive factor, as spinally transected male rats showed ex copula erections and penile anteroflexions 10 days before spinally intact animals (Sachs and Meisel, 1979).

Unlike rats, male hamsters begin to copulate only after the pubertal testosterone increase begins, although before it peaks (Romeo et al., 2002). Lack of testosterone during puberty resulted in impaired copulation after testosterone was replaced in adulthood, compared with castrates with testosterone replacement during puberty (Schultz et al., 2004). The odor of a receptive female activated Fos expression in the MPOA of male hamsters even before puberty (Romeo et al., 1998); however, increased dopamine (DA) activity in the MPOA (measured as increased DOPAC, a DA metabolite) in response to female odor occurred only after puberty (Schultz et al., 2003). Therefore, a hormonal factor is apparently required for this critical area to become fully responsive to the female's stimuli. Sisk and her colleagues have suggested that puberty is a second organizational period for gonadal hormones to alter neural processing in areas that regulate sexual behavior (Romeo et al., 2002; Sisk et al., 2003; Schultz et al., 2004).

The ability of aged males to copulate is also highly variable. Sexual behavior of aged male monkeys was not correlated with testosterone levels (Chambers and Phoenix, 1981; Chambers et al., 1981), nor did exogenous testosterone improve behavior in castrated or intact old monkeys, compared with young males (Phoenix and Chambers, 1986, 1988). In old male mice, testosterone levels, gonad weights, and spermatogenesis were lower than in young animals; however, some of the aged males could copulate and others could not, despite similar testosterone levels (Bronson and Desjardins, 1982). Neither decreased testosterone levels nor decreased nuclear androgen receptors (ARs) in the MPOA could account for the inability of old male rats to ejaculate (Chambers et al., 1991). However, a later study found that nuclear estrogen receptors (ERs) in the amygdala were lower in old than in young males, and that testosterone replacement did not restore ERs to the levels of young males (Roselli et al., 1993). The authors suggested that old males' ability to bind estradiol, derived from testosterone, may limit their androgen responsiveness.

2.7 Measures of Penile Function

2.7.1 In Copula Measures

The most direct way to study erectile function is to observe the penis during sexual behavior. Although this may be easy in some species, it is difficult in rodents, because erections are very brief and usually obscured from view. Erection, intromission, and ejaculation are usually inferred from the male's behavior. However, brain lesions or drug administration could affect ejaculation without obvious changes in behavior. Experimenters may inspect the female's vagina for sperm immediately after copulation or examine the cage floor for coagulated semen. However, males often eat their ejaculate while grooming their genitalia immediately after ejaculation (Orbach, 1961). Paradigms have been developed for monitoring genital reflexes ex copula. However, in copula and ex copula erections may differ in their physiological, CNS, neurochemical, or hormonal control (Sachs, 2000). Thus, one should be cautious when predicting the effects of a drug on copulation, based on ex copula tests. It is possible to measure physiological responses during copulation, using chronic implantation of electrodes in striated perineal muscles (Holmes et al., 1991) or telemetric measures of penile pressures (Giuliano et al., 1994). These methods are technically difficult, but do provide physiological measurements during normal sexual behavior.

2.7.2 Ex Copula Measures

2.7.2.1 Spontaneous Erections Various drugs, administered either systemically or into specific CNS sites, can increase the number of erections elicited outside of a sexual context. These drug-induced erections usually involve extension of the engorged glans beyond the sheath and are often accompanied by genital grooming. The obvious advantages of this model are its ease and simplicity. However, it may not reliably predict the effects of a drug on copulatory performance.

2.7.2.2 Noncontact Erections Another model for studying erection is noncontact erection. Male rats show penile erections in response to volatile odors from an estrous female, even in the absence of physical contact (Sachs et al., 1994; Sachs, 1997; Kondo et al., 1999). This may be used as a model for psychogenic erections in humans. This response is produced by supraspinal mechanisms rather than spinal reflex pathways. Discrete brain lesions have resulted in differential effects on copulatory performance and erectile capability, as judged by noncontact erections (Liu et al., 1997).

2.7.2.3 Touch-Based Erections Although reflexive touch-based erections can be elicited in dogs (Hart, 1967; Beach, 1984), penile erection is inhibited by tactile stimulation of the penis in rats (Hart, 1968b). However, erections can be elicited in a male rat, restrained on his back, with the penile sheath retracted, exposing the glans penis. A series of erections typically ensue, usually accompanied by penile anteroflexions (previously called "flips"). Three gradations of glans erection can be observed—in the first, the penile body elongates and rises; in the second, the glans is engorged and slightly flared; and in the third, the glans flares into a cup. The cup is used during copulation to hold the ejaculated semen around the female's cervix, where it coagulates to form a copulatory plug. Without this plug, the semen can seep out of the vagina, greatly impairing fertility (Sachs, 1982). The sensory stimulus responsible for these erections may be the continuing pressure of the sheath around the base of the penis. Ex copula seminal emission occasionally occurs in this context.

2.7.2.4 Urethrogenital Reflex Sexual reflexes can be studied in anesthetized, spinally transected male and female rats, using urethral distension or stimulation of the dorsal nerve of the penis (McKenna et al., 1991). In male rats, the urethrogenital reflex comprises clonic contractions of the perineal muscles, rhythmic firing of the cavernous nerve, penile erections, and ejaculation. Therefore, the urethrogenital reflex is used as a model for both erection and ejaculatory reflexes.

3 Neurotransmitters and Neuromodulators

3.1 General Considerations

Steroid hormones regulate sexual behavior primarily through slow, genomically mediated processes. However, sexual behavior requires rapid processing of sensory input and motor output to produce intricate moment-to-moment interactions with a partner. Therefore, the slow hormonal effects must be translated into rapid neuronal responses. A major effect of hormones is to bias sensory inputs to favor sexually relevant stimuli and to prime central integrative areas and appropriate motor outputs to orchestrate integrated sexual responses. Major targets of steroid hormones are enzymes that synthesize (or catabolize) neurotransmitters or neuromodulators, mechanisms that regulate neurotransmitter release, neurotransmitter receptors, or second messenger systems. Drugs that influence one or more of these targets may be used to study the influence of a neurotransmitter on behavior. Because no drug is entirely specific for its major effect, it is important to block the effects of an agonist with a relevant antagonist or to administer a drug of a different class that targets the same receptor or enzyme.

There are advantages to systemic or intracerebroventricular (icv) drug administration. First, the brain areas in which a given neurotransmitter influences a behavior may not be known a priori. Second, a given neurotransmitter may act synergistically in several central and/or peripheral sites. Finally, drugs being developed for possible clinical use must be effective with systemic administration. However, drugs may also have competing effects at multiple sites. Therefore, studies employing both systemic and site-directed routes of administration should be used to fully explore a drug's effects on behavior.

3.2 Specific Neurotransmitters

3.2.1 Dopamine

3.2.1.1 Effects of Systemically or Intraventricularly Administered Drugs The catecholamines DA, norepinephrine (NE), and epinephrine are formed from the amino acid tyrosine; tyrosine hydroxylase, the rate-limiting enzyme, produces L-DOPA, which in turn is decarboxylated by aromatic amino acid decarboxylase to form DA. In noradrenergic neurons, dopamine β-hydroxylase converts DA to NE. In a few neurons, NE can be converted to epinephrine by phenylethanolamine *N*-methyl transferase. Two major groups of dopamine-containing cells are the A9 group in the substantia nigra of the midbrain and the A10 group in the adjacent ventral tegmental area (VTA). A9 axons ascend to the caudate-putamen (dorsal striatum) via the nigrostriatal tract, and A10 axons ascend via the mesocorticolimbic tract to the nucleus accumbens (ventral striatum), prefrontal cortex, and other limbic structures (reviewed in Fallon, 1986). There are several dopaminergic cell groups in the hypothalamus, including the A14 periventricular group, which lies along the third ventricle and sends axons laterally into the adjacent MPOA and anterior hypothalamus (reviewed in Moore and Lookingland, 1995).

As long ago as the late 1960s, DA was reported to facilitate sexual function. L-DOPA administered to Parkinsonian patients often increased libido and sexual potency (Barbeau, 1969; Bowers et al., 1971), and the classic DA agonist apomorphine has more recently been used to treat erectile dysfunction (Lal et al., 1984, 1987; O'Sullivan and Hughes, 1998). A sublingual form of apomorphine appears to be clinically effective, with fewer side effects (Dula et al., 2000; reviewed in Padma-Nathan and Giuliano, 2001; Giuliano and Allard, 2002).

In rodents, systemic administration of DA agonists facilitates male sexual behavior (reviewed in (Bitran and Hull, 1987; Melis and Argiolas, 1995; Giuliano and Allard, 2001). L-DOPA and/or apomorphine elicited copulation in sexually sluggish males and increased the numbers, while decreasing the latencies, of ejaculations. Apomorphine also elicited copulation in sexually satiated male rats (Mas et al., 1995b; Rodriguez-Manzo, 1999) and in socially stressed mice (Sugiura et al., 1997) and rats (Niikura et al., 2002). DA releasers decreased mount and intromission latencies in male rats that were sexually sluggish because of damage to the medial frontal cortex (Ågmo and Villalpando, 1995). Systemically administered apomorphine partially restored copulation in short-term (Malmnas, 1976) and long-term (Scaletta and Hull, 1990) castrated rats. Apomorphine also elicited copulation in mice lacking the gene for estrogen receptor alpha (ERα); such mice usually show little sexual behavior (Wersinger and Rissman, 2000). Therefore, stimulation of DA receptors appears to bypass the usual requirement for the organizational and activational effects of ERα, at least in some animals. Apomorphine also elicited ex copula penile erections in mice (Rampin et al., 2003). This effect was blocked by the DA antagonist haloperidol, which crosses the blood–brain barrier, but not by domperidone, which does not. In hamsters, apomorphine produced only slight decreases in ejaculation latency and postejaculatory interval; however, it also abolished long intromissions (a sign of approaching sexual satiety) in 60% of the animals (Arteaga et al., 2002). The effects of systemically administered dopamine agonists are typically dose-dependent; low doses facilitate, but high doses inhibit copulation, at least in part by eliciting stereotypic behaviors (Foreman and Hall, 1987).

Daily injections of amphetamine in sexually naïve male rats facilitated copulation on their first sexual behavior test (Fiorino and Phillips, 1999). Similar injections of amphetamine resulted in sensitization of locomotor responsiveness to amphetamine. Therefore, the (presumably dopaminergic) mechanisms that produce drug sensitization apparently "cross-sensitize" to a natural motivated behavior. Similar mechanisms may underlie the enhanced copulatory ability of sexually experienced animals and their resistance to various surgical and environmental insults.

The consistent facilitative effects of DA agonists suggest that DA antagonists, which block the effects of endogenous DA, would impair sexual behavior. As expected, copulation by both experienced (Ahlenius and Larsson, 1984a; Ågmo and Fernandez, 1989; Pfaus and Phillips, 1989; Ahlenius and Larsson, 1990) and inexperienced (Ågmo and Picker, 1990) male rats was impaired by systemically administered DA antagonists. These inhibitory effects ranged from failure to copulate to increased intromission and ejaculation latencies. Touch-based erections were also decreased by haloperidol, which crosses the blood–brain barrier, but not by domperidone, which does not (Pehek et al., 1988a). Mild tail pinch can sometimes induce sexually sluggish males to copulate. The facilitative effects of tail pinch were blocked by both a DA and an opioid antagonist in sexually sluggish castrates maintained on a suboptimal testosterone regimen (Leyton and Stewart, 1996). Therefore, tail pinch may have activated copulation by increasing the activity of midbrain DA systems via an opioid mechanism. On subsequent days when tail pinch was no longer applied, the facilitation was still observed, suggesting that any temporary increases in motor activity did not explain the increased responsiveness.

A DA antagonist has also inhibited sexual motivation. Sexually naïve male rats received ten daily trials on which they ran down a runway to be near an estrous or a nonestrous female on different days (Lopez and Ettenberg, 2000). They then received either vehicle or one of three doses of haloperidol before being allowed to copulate to one ejaculation; the drug did not significantly affect any copulatory measure. Two days later, animals previously treated with vehicle ran faster after sexual experience than before and ran faster for the estrous than for the nonestrous female. Males that had received the low and moderate doses of haloperidol showed no change in running speed after the sexual experience and showed no preference for the estrous over the nonestrous female. Males that had received the highest dose ran more slowly after sexual experience than before and did not prefer the estrous female. They also showed more retreats as they approached the goalbox, suggesting that it was aversive. The drug effects on motivation do not appear to be confounded with motoric side effects. First, the runway test occurred 2 days after the drug was given, which should have allowed sufficient time for the drug to be metabolized before that test. Second, the drug did not significantly affect copulation on the experience day, and all males ejaculated. Thus, the experience-induced enhancement of sexual motivation was blocked by a systemically administered DA antagonist.

A similar experiment from the same lab provided additional support for dopaminergic influence on sexual motivation. Sexually naïve males were first exposed for 4 min to an estrous female, a nonestrous female, or an empty goal box, through a perforated barrier (Lopez and Ettenberg, 2001). They then received one of two doses of haloperidol or vehicle before traversing the runway. The drug did not affect run times to the nonestrous female or the empty goal box, suggesting that motoric ability was not impaired. However, the drug slowed run times to the estrous female, compared to vehicle-treated animals, and resulted in similar run times to all three targets.

Two additional experiments from that lab tested sexual motivation using conditioned olfactory stimuli (Lopez and Ettenberg, 2002b). Males had five copulatory experiences in the presence of either orange or almond odor and experienced the other odor when alone for a similar amount of time. After the 10 days of conditioning, all males had faster run times to the goal box scented with the odor that had been associated with copulation. An additional group of males was subjected to the same conditioning protocol, but received one of three doses of haloperidol or vehicle before the final runway test. As expected, animals receiving vehicle injections showed an increase in running speed to the odor associated with copulation and a decrease in running speed to the control odor. There were no differences in run times to the two odors in animals treated with the low and middle doses of haloperidol and no changes relative to their baselines. The highest dose of haloperidol slowed running to both odors, suggesting an impairment of motor ability. However, the selective inhibition of the conditioned preference by the two lower doses, without slowing running, compared with baseline, suggests that a DA antagonist can inhibit motivational processes independently of motor impairment. One study did report a lack of effect of dopaminergic drugs on a measure of sexual motivation in sexually inexperienced male rats (Ågmo, 2003). In that experiment, the DA agonists apomorphine and amphetamine did not affect the amount of time spent in front of a receptive female behind a barrier. In addition, only a high dose of the DA antagonist *cis*-flupenthixol, which also slowed motor behavior, decreased this measure. However, it is not clear that motoric effects explained the apparent decrease in motivation by the high dose, because, as the author noted, no more motor activity was required to stand in front of the estrous female than in front of the male.

Mice with a deletion of the gene for tyrosine hydroxylase only in dopamine-, and not in norepinephrine-containing neurons, required daily injections of L-DOPA in order to survive (Szczypka et al., 1998). After their daily L-DOPA injections, the DA-deficient males mounted estrous females, as well as other males and nonestrous females. Following castration, they continued to mount receptive females for 21 days, whereas wild-type castrates mounted for only 7 days. In addition, DA-deficient castrates were more sensitive to both L-DOPA and testosterone than were wild-type castrates, in that lower doses of each were sufficient to restore copulation. The authors suggested that DA receptors might have been coupled to second messenger systems more efficiently in the DA-deficient animals.

DA receptors are divided into two families. The D_1 family consists of D_1 and D_5 receptors, whereas the D_2 family comprises D_2, D_3, and D_4 receptors. The D_1 family members stimulate adenylyl cyclase, whereas the D_2-like receptors inhibit adenylyl cyclase and influence certain ion channels and the phosphoinositide pathway. DA and apomorphine stimulate both D_1 and D_2 families; however, drugs that are relatively selective for members of one family have been used to analyze the roles of the two subtypes in various aspects of copulation.

There is evidence for a facilitative role of D_2-like receptors in the control of ejaculation. Ejaculation latencies and the numbers of intromissions preceding ejaculation were decreased by a D_3-selective agonist (7-OH-DPAT) and two D_2/D_3 agonists (SND 919 and B-HT 920); however, this effect was observed only in sexually active male rats, not in sexually inactive animals (Ferrari and Giuliani, 1995, 1996a; Giuliani and Ferrari, 1996). The pro-ejaculatory effects of 7-OH-DPAT (Ahlenius and Larsson, 1995) and of SND 919 (Ferrari and Giuliani, 1994) were blocked by a D_2/D_3 antagonist (eticlopride), confirming their receptor specificity. However, the muscarinic antagonist atropine also inhibited the effects of both a dopaminergic agonist and a muscarinic agonist (Zarrindast et al., 1994), suggesting that cholinergic muscarinic receptors may contribute to the pro-ejaculatory effects of D_2-like agonists. The D_1 agonist SKF 81292 facilitated copulatory ability in the DA-deficient mice mentioned earlier, but the D_2 agonist quinpirole inhibited their copulation by increasing stereotypic behavior (Szczypka et al., 1998). A different D_1 agonist, SKF 38393, also increased sexual motivation, measured as the time spent in a goal compartment with a receptive female (Beck et al., 2002). This effect reflected an increase in the number of copulatory behaviors while in the female's compartment. Neither the latency to run to the female nor the run duration was affected, suggesting that the effect on sexual motivation was independent of any motor confounds. The D_5 receptor is a member of the D_1-like receptor family. Mice with deletion of the D_5 receptor (D5KO) showed a place preference when ejaculation was paired with the experimental context, but not when intromissions alone were paired with the context (Kudwa et al., 2005). Wild-type (WT) males showed a preference for the context paired with either intromissions alone or ejaculation. Therefore, D_5 receptors may contribute to the rewarding aspects of intromission.

Some of the reported effects of D_1- and D_2-like agonists on ex copula genital reflexes appear to be contradictory. A presumed-selective D_2 agonist in one series of experiments elicited erections in a neutral arena, whereas a D_1 agonist inhibited those erections (Zarrindast et al., 1992). However, the D_2/D_3 antagonist eticlopride, administered with either cocaine or presumed-selective D_2 agonists, actually increased the number of drug-induced erections, suggesting that stimulation of D_2-like receptors inhibits, rather than stimulates, erections (Ferrari and Giuliani, 1996b). Cocaine increases extracellular levels of DA by inhibiting reuptake, which results in stimulation of both D_1- and D_2-like receptors. It seems likely that the presumed-selective D_2-like agonists also stimulated substantial activity at D_1 receptors. Stimulation of D_2-like receptors also inhibits touch-based erections in restrained male rats. Either systemic or intra-medial preoptic area administration of the D_2/D_3 agonist quinelorane decreased touch-based erections (Bitran et al., 1989). As is discussed later, intense stimulation of D_2-like receptors in the MPOA apparently shifts autonomic influence to favor sympathetically mediated seminal emission and ejaculation, and to inhibit parasympathetically mediated erection. However, the selective D_2/D_3 agonists 7-OH-DPAT and B-HT 920 evoked erections in sexually naïve male rats, tested in a neutral arena, and also elicited stretching, yawning, and sedation (Ferrari et al., 2002). Similarly, the D_4 agonist ABT-724 elicited erections in rats; its effects were blocked by the primarily D_2 agonist haloperidol and the primarily D_4 agonist clozapine, but not by domperidone, which does not cross the blood–brain barrier (Brioni et al., 2004). It is not clear whether these divergent findings resulted from different dose ranges or different test regimens (touch-based erections vs. drug-induced erections in a neutral arena).

The facilitative effects of D_1 stimulation appear to be evolutionarily conserved. A D_1 agonist (SKF 81297) systemically administered to two species of lizards (little striped whiptail lizards and parthenogenetic desert grassland lizards) increased the proportion of animals that mated after gonadectomy and decreased mount latencies (Woolley et al., 2001). Conversely, a DA receptor antagonist inhibited courtship behavior in castrated, testosterone-implanted leopard geckos (reviewed in Woolley et al., 2004). Similarly, a D_1 agonist (SKF 38393) in quail increased both appetitive (the amount of time near a window through which a female could be seen) and copulatory behaviors (the numbers of mount attempts and of cloacal contact movements) (Balthazart et al., 1997). Conversely, the D_1 antagonist SCH 23390 decreased the numbers of approaches to the female's window and mount attempts. D_2-selective drugs yielded generally opposite results—the D_2 agonist quinpirole decreased both appetitive and copulatory behaviors, and the D_2 antagonist spiperone increased mount attempts but decreased approaches to the window. Somewhat earlier research from the same lab found that the classic DA agonist apomorphine inhibited both appetitive and copulatory behaviors, apparently by stimulating primarily D_2 receptors (Castagna et al., 1997). Intense stereotyped pecking elicited by the two higher doses of apomorphine could explain their inhibition of sexual behavior; however, the lowest dose also inhibited sexual behavior but did not significantly increase pecking. The indirect DA agonists nomifensine (which inhibits reuptake) and amfonelic acid (which primarily increases release) yielded mixed facilitative and inhibitory effects. Two recent articles from the same lab reported inhibition of sexual behavior by intraventricular injections of DA itself and suggested that the inhibitory effects were mediated by DA acting at α-2 noradrenergic receptors, rather than DA receptors (Cornil et al., 2004, 2005). However, the problem of cross-reactivity with α-2 receptors was probably avoided in the earlier studies, which used selective DA agonists and antagonists.

In summary, convergent data support the suggestion that systemically administered DA agonists facilitate multiple aspects of male sexual behavior in most species, including sexual motivation, genital reflexes, and copulatory behavior, and that DA antagonists produce the opposite pattern of results. DA agonists have restored sexual behavior in stressed or sexually satiated males, in castrates, in males with lesions of the prefrontal cortex, in males lacking ERα, and in DA-deficient males. DA agonists have also facilitated ex copula erections, although the relative importance of D_1 and D_2 families of receptors for ex copula erections is unclear. Conversely, DA antagonists have impaired both conditioned and unconditioned sexual motivation, often without confounding motor effects, have inhibited genital reflexes, and have decreased the incidence, rate, and efficiency of copulation. There is also conservation of D_1-like receptors' facilitative effects in reptiles, birds, and mammals.

A critical review recently suggested that DA has no role in sexual motivation because some of the studies reporting inhibitory effects of DA antagonists either measured approach to a female, which can confound motor and motivational factors, or studied sexually experienced males (Paredes and Ågmo, 2004). However, several of the studies that measured approach to a female specifically ruled out motor effects. Furthermore, the motivation of experienced males is nevertheless motivation and should be studied. DA antagonists have inhibited both the initial conditioning and the later expression of the enhanced motivation, without confounding motoric effects. Although no procedure has been devised to measure motivation selectively, without any possible sensory, learning, or motoric confounds, the preponderance of evidence supports the suggestion that DA contributes to sexual motivation, as well as to genital reflexes and copulatory performance.

3.2.1.2 Role of DA in Specific Brain Areas *Medial Preoptic Area.* The MPOA is a major integrative site for male sexual behavior. It has reciprocal connections with virtually all sensory modalities, allowing it to modify the processing of its sensory inputs (Simerly and Swanson, 1986). Both the MPOA and its afferent connections contain steroid receptors, which may bias inputs to favor sexually relevant stimuli. It has efferent connections to hypothalamic, midbrain, and brain stem areas that control autonomic or somatomotor patterns and motivational processes. Dopaminergic drugs microinjected into the MPOA influence copulation, genital reflexes, and sexual motivation. Dopamine innervation is from the periventricular system, whose cell bodies lie along the third ventricle and send widely branching axons laterally into the MPOA and anterior hypothalamus (reviewed in Moore and Lookingland, 1995).

Effects of drug microinjections into the MPOA. Microinjections of the classic D_1/D_2 agonist apomorphine into the MPOA facilitated copulation in intact male rats (Hull et al., 1986) and in long-term castrates (Scaletta and Hull, 1990), and increased the numbers of touch-based erections and seminal emissions

(Pehek et al., 1989a). Similarly, administration of the DA reuptake inhibitor bupropion into the MPOA by reverse dialysis increased both touch-based and noncontact erections (Adachi et al., 2003). Conversely, copulation was impaired by the D_1/D_2 antagonist *cis*-flupenthixol microinjected into the MPOA (Warner et al., 1991); *cis*-flupenthixol also decreased touch-based genital reflexes (Pehek et al., 1988b) and decreased the number of choices of a female in an X-maze, without affecting running speed or the number of no-choice trials (Warner et al., 1991). There was site specificity for the effects of both apomorphine and *cis*-flupenthixol, with microinjections anterior, dorsal, or lateral to the MPOA being ineffective. There was also behavioral specificity, with no effect of apomorphine or *cis*-flupenthixol on eating, drinking, locomotion, or rearing in the home cage, although flupenthixol slightly increased duration of inactivity (Warner et al., 1991). A different DA antagonist, haloperidol, increased intromission latency and decreased the number of ejaculations in a bilevel apparatus, although motoric slowing may have contributed to these deficits (Pfaus and Phillips, 1991). More recently, MPOA microinjections of apomorphine fully restored the ability to copulate in males whose sexual behavior was severely impaired by large excitotoxic lesions of the amygdala, which provides a major input to the MPOA (Dominguez et al., 2001). There were also trends toward apomorphine-induced increases in numbers of intromissions and ejaculations in sham animals. Therefore, a major consequence of amygdala lesions may be impaired DA activity in the MPOA.

6-OHDA neurotoxic lesions of MPOA DA neurons impaired copulation only during the first 24 h after the lesion (Bazzett et al., 1992) or if a subthreshold dose of a DA synthesis inhibitor was administered before testing (Bitran et al., 1988a). The lesions produced only a 23% depletion of DA, probably because of the sparse DA transporters in the MPOA, which are needed to transport the toxin into the axon terminals. Therefore, increased DA synthesis occurred in the remaining neurons during the first 24 h postlesion and was able to restore copulation to normal. This effect was specific for DA, because the 6-OHDA was coadministered with an NE transporter inhibitor, desipramine, and NE levels were not affected.

As with systemic injections, MPOA microinjections of a D_1 agonist facilitated copulation and touch-based erections, whereas stimulation of D_2 receptors yielded biphasic effects (See ➋ *Figure 2-1*). A low dose of quinelorane, a D_2/D_3 agonist, decreased the latency to the first touch-based genital reflex, without affecting the numbers of erections, anteroflexions, or seminal emissions (Bazzett et al., 1991). Therefore, a small increase in dopaminergic stimulation may disinhibit reflexes via D_2-like receptors. However, a high dose of quinelorane, or of the D_1 antagonist SCH-23390, decreased the number of erections but increased the number of seminal emissions. Thus, intense stimulation of D_2-like receptors or antagonism of D_1-like receptors appears to increase sympathetic facilitation of seminal emission, while decreasing parasympathetic stimulation of erection. Conversely, the D_1 full agonist tetrahydrothienopyridine (THP) increased touch-based erections but inhibited seminal emission (Markowski et al., 1994). The D_1 partial agonist SKF 38393 was ineffective (unpublished observations). The suggestion that D_1 and D_2 receptors have opposing effects on autonomic function is supported by experiments using low and high doses of apomorphine, together with D_1 and D_2 antagonists. The low dose of apomorphine (1 μg) increased touch-based erections, but not seminal emissions; this effect was blocked by both D_1 (SCH-23390) and D_2 (raclopride) antagonists (Hull et al., 1992). These data suggest that both disinhibition by D_2 receptors and facilitation by D_1 receptors contributed to apomorphine's increase. Unexpectedly, a high dose (10 μg) of apomorphine did not increase erections significantly; however, coadministration of the high dose with a D_2 antagonist unmasked the facilitative effects mediated by D_1 receptors. The high dose of apomorphine did increase seminal emissions, and that effect was blocked by a high dose of the D_2 antagonist, but not by the D_1 antagonist.

The association of erections with D_1 receptor stimulation in the MPOA and of ejaculation with intense D_2 stimulation was supported in copulation tests. Microinjections of the D_1 agonist THP into the MPOA increased copulatory efficiency, resulting in more ejaculations in the 30-min test (Markowski et al., 1994). Conversely, microinjection of the D_1 antagonist SCH-23390 increased intromission latencies but decreased the threshold for ejaculation (Hull et al., 1989). It also decreased sexual motivation, measured as the choice of the female in an X-maze, without impairing motoric activity (Moses et al., 1995). A high dose of the D_2/D_3 agonist quinelorane in the MPOA also increased intromission latency and decreased ejaculation threshold (Hull et al., 1989; Moses et al., 1995). It did not affect the percentage of X-maze trials on which the male chose the female's goal box, but it did increase the latency to reach the female's goal box (Moses et al., 1995). However, this delay did not result from slowed motor function, but rather from apparent reluctance

Figure 2-1
Model showing possible effects of D_1 versus D_2 stimulation in the MPOA, as a result of a sexually exciting stimulus and/or sexual activity (reviewed in Dominguez and Hull, 2005). In this model, a low-threshold mechanism mediated by D_2 receptors disinhibits genital reflexes. A moderate threshold mechanism facilitates penile erections, via activation of D_1 receptors. A high threshold mechanism, activated by intense stimulation of D_2 receptors, facilitates seminal emissions and inhibits erections. These mechanisms may be activated successively by increasing levels or duration of DA activity (figure is from Dominguez and Hull, 2005, with permission)

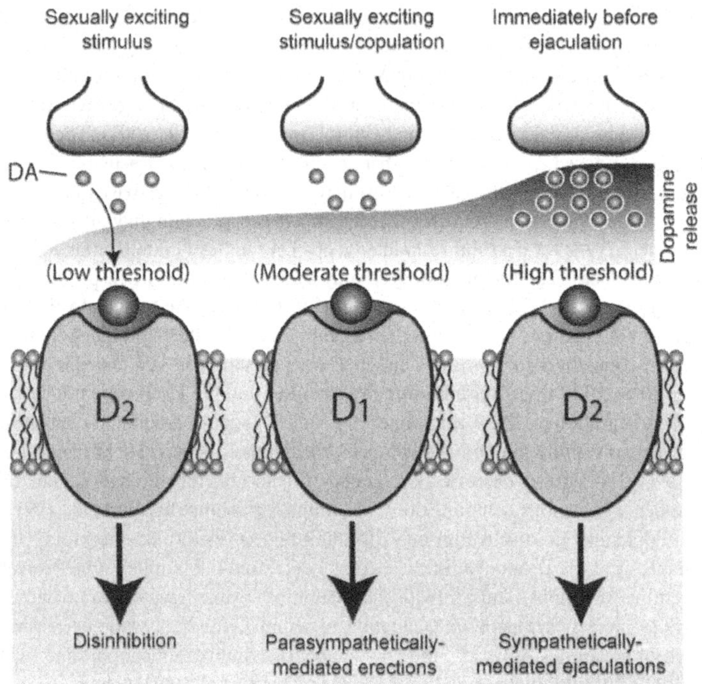

to cross the line in front of the female's compartment, beyond which the male would be placed into that compartment. In summary, D_1 and D_2 receptors in the MPOA appear to exert opposing effects on copulation and genital reflexes. Stimulation of D_1 receptors facilitates parasympathetically mediated erections and copulatory performance, whereas intense activation of D_2 receptors promotes sympathetically mediated ejaculation (See *Figure 2-1*).

Intra- and extracellular recordings in slice preparations from the MPOA of Japanese quail showed that bath applications of DA inhibited most cells (52–80%), but excited a few (10–24%). Neither D_1 nor D_2 antagonists blocked these effects; however, they were blocked by α_1 or α_2 antagonists, respectively (Cornil et al., 2002). However, DA was not being converted to NE to produce the effects, as dopamine-β-hydroxylase inhibitors did not block DA's effects. Therefore, in the quail MPOA DA affects neuronal activity via cross talk with NE receptors.

Chemical changes detected by microdialysis or from tissue punches. There is a close correlation between extracellular MPOA DA levels and sexual behavior in male rats. DA levels increased as soon as an estrous female was introduced across a perforated barrier; DA levels remained high or increased further, when the female was placed into the male's compartment and the animals were allowed to copulate (Hull et al., 1995; Sato et al., 1995) (See *Figure 2-2*). The recent presence of T is required for both the DA response to the

Figure 2-2
Levels of extracellular DA in the MPOA of male rats during baseline (BL), a 6-min precopulatory exposure to an estrous female (PRECOP), and three 6-min copulation samples (COP). Gonadally intact male rats showed an increase in extracellular DA during precopulatory exposure to an inaccessible estrous female, and all intact males copulated; males castrated 2 weeks before showed no DA release in response to the female, and none copulated. Values are expressed as mean ± SEM, *P < 0.05, compared to final baseline for intact males or for one-week vehicle-treated castrates that copulated; **P < 0.01, compared to final baseline for intact males or for one-week vehicle-treated castrates that copulated; +P < 0.05, compared to baseline for testosterone-treated castrates; #P < 0.05, compared to final baseline for vehicle-treated castrates that failed to copulate (figure is from Hull et al., 1995, with permission)

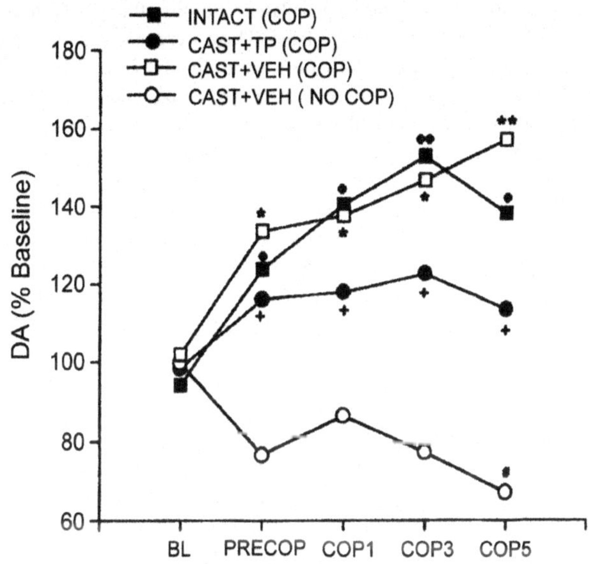

female and the ability to copulate. All gonadally intact males, as well as two-thirds of males castrated one week before the test, showed a precopulatory increase in DA and copulated when the female was placed with the male (Hull et al., 1995) (See *Figure 2-2*). The remaining third of the 1-week castrates, and all 2-week castrates failed to show the DA response to the female and failed to copulate. There was behavioral specificity, in that neither presentation of a male behind the barrier nor voluntary exercise in a running wheel increased DA levels (Hull et al., 1995); in addition, eating a palatable food did not increase levels of the DA metabolite DOPAC in an earlier study (Hull et al., 1993). There was also site specificity, in that no DA response to the female was observed in probes anterior or lateral to the MPOA. Because the increase in DA occurred before copulation began, it is evident that the increase was not caused by copulation, but was more likely associated with sexual motivation.

In the earlier studies described above, basal levels of extracellular DA were not measured; therefore, it was unclear whether castration also affected basal levels or only the response to a female. Subsequently, the no-net-flux technique was used to measure basal DA levels in castrates and in gonadally intact males (Du et al., 1998). In this technique, known amounts of DA are added to the dialysate; if there is more DA in the dialysate than in the brain, some of it will diffuse out of the dialysate into the brain, and the loss can be detected (Olson and Justice, 1993). Conversely, if there is less DA in the dialysate, or none, as in standard dialysis, DA will diffuse into the dialysate from the brain, and the gain can be detected. A regression line is drawn, and the point at which it crosses from loss to gain in the dialysate represents the basal extracellular DA level. In castrates, the basal extracellular DA levels were only about one-fourth those of gonadally intact males (0.3 vs. 1.3 nM) (Du et al., 1998). Therefore, the deficit in extracellular DA in castrates was a general one, not only a lack of response to a female. However, systemic injections of amphetamine evoked greater DA release in castrates than in intact males, and tissue punches from castrates contained more DA than did those from intact males (Du et al., 1998). Therefore, DA synthesis and storage were at least as great in castrates as in intact males, but castrates were unable to release their abundant stores.

Further evidence of a correlation between DA release and the copulatory ability of castrates was provided by a study of restoration with 2-, 5-, or 10-day regimens of testosterone treatment (Putnam et al., 2001). None of the 2-day testosterone-treated castrates showed a DA response to the female, and none was able to copulate. Eight of nine 5-day testosterone-treated castrates showed a DA response and those eight were able to copulate, with five also able to ejaculate. All of the 10-day testosterone-treated castrates showed a precopulatory DA response, and all copulated. There were numerous correlations between DA levels and copulatory measures. Therefore, both postcastration loss of copulation and its restoration following testosterone replacement were closely correlated with female-elicited DA release in the MPOA.

Testosterone is primarily a prohormone, which is converted to either estradiol or dihydrotestosterone in target tissues. Its metabolites were differentially effective in restoring basal and female-stimulated DA release, as well as copulation, in long-term castrates (Putnam et al., 2003). Daily injections of estradiol restored high basal levels of DA, but not the increase in response to a female. All estradiol-treated males intromitted, but none showed a behavioral ejaculation pattern. Castrates treated with dihydrotestosterone or oil vehicle had very low basal DA levels and did not show a female-stimulated increase; no dihydrotestosterone- or oil-treated castrate copulated. Although dihydrotestosterone was ineffective by itself, when it was administered with estradiol, the combination restored basal and female-stimulated DA responses, as well as copulation.

Although copulatory measures were positively correlated with extracellular DA levels, copulation was negatively associated with tissue (stored) DA levels (Putnam et al., 2005). Dihydrotestosterone- and oil-treated animals could synthesize and store DA, but had impaired release in basal conditions and in response to a female. An important factor regulating DA release in the MPOA is nitric oxide (NO). Both basal (Lorrain and Hull, 1993) and copulation-induced (Lorrain et al., 1996) DA release in the MPOA depended, in part, on NO. In addition, immunoreactivity for nitric oxide synthase (NOS-ir) in the medial preoptic nucleus is positively regulated by testosterone and estradiol (Hadeishi and Wood, 1996; Ceccatelli et al., 1996; Du and Hull, 1999; Scordalakes et al., 2002; Putnam et al., 2005) (but see Singh et al., 2000). A deficiency in NO may have led to the impaired DA release in the dihydrotestosterone- and oil-treated castrates described earlier. Those groups had less NOS-ir than did the three groups that were able to copulate (estradiol, estradiol+dihydrotestosterone, and testosterone) (Putnam et al., 2005). Therefore, testosterone may maintain both basal and female-stimulated DA release by upregulating NOS in the MPOA.

The major input that directly stimulates MPOA DA release in response to a female is from the medial amygdala, which in turn receives input from both the main olfactory and vomeronasal systems, as well as genitosensory information. Chemical stimulation of the medial amygdala increased DA release in the MPOA (Dominguez and Hull, 2001), and microinjection of the DA agonist apomorphine into the MPOA restored copulatory ability that had been severely impaired by amygdala lesions (Dominguez et al., 2001). Basal DA levels in the MPOA were normal in males with small radio-frequency lesions of the medial amygdala, but there was no DA increase in response to an estrous female or during copulation (Dominguez et al., 2001) (See ➲ *Figure 2-3*). In addition, males with lesions achieved only half the number of ejaculations in a 30-min test as did sham lesion animals. Therefore, as with estradiol treatment of castrates (Putnam et al., 2003), normal basal levels of MPOA DA were sufficient for suboptimal copulation. However, an increase in DA before and during copulation appears to contribute to copulatory ability. There are no DA-containing neurons in the medial amygdala of rats; therefore, amygdala efferents apparently synapse, directly or indirectly, on dopaminergic cell bodies or terminals in the MPOA.

The signal relayed by the amygdala to the MPOA apparently arises from the olfactory bulbs. Male hamsters with either sham bulbectomy or unilateral bulbectomy contralateral to the MPOA microdialysis site showed increases in extracellular DA in the MPOA when presented with an estrous female (Triemstra

Figure 2-3

Lesions of the MeA inhibit the release of DA in the MPOA resulting from exposure to an estrous female and copulation. Levels represent percent changes from baseline (BL) in response to precopulatory exposure to an estrous female (PRE), during copulation (C1–C3), and after the female was removed (POST). Extracellular levels of DA significantly increased during the precopulatory and copulatory stages of testing for animals with sham lesions but not for animals with MeA lesions. The baseline value used for computation was obtained by dividing the value of the last baseline by the mean of all three baselines. Values are expressed as mean ± SEM, *P < 0.05; **P < 0.01 (figure is from Dominguez et al., 2001, with permission)

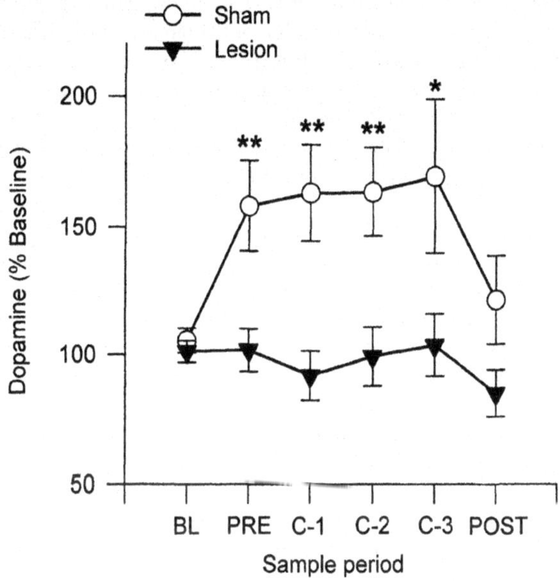

et al., 2005). However, either bilateral or ipsilateral bulbectomy blocked the DA response and impaired copulatory behavior, similar to the results obtained by Dominguez et al. (2001). In another study of sexually mature male hamsters, pheromones from estrous females elicited an apparent increase in DA release, inferred from an increase in the DA metabolite DOPAC, in tissue punches from the MPOA, suggesting release of DA and subsequent uptake and metabolism (Schultz et al., 2003). Juveniles showed no such increase and were unable to copulate. Therefore, a maturational process rendered the processing of sensory stimuli effective in postpubertal males.

Immunocytochemistry. Copulation to one ejaculation elicited more Fos immunoreactivity (Fos-ir) in the medial preoptic nucleus of sexually experienced males than in naïve males, even though the experienced males had fewer intromissions before ejaculating (Lumley and Hull, 1999). The increased Fos-ir was inhibited by systemic injection of a D_1 antagonist in the naïve males, but not in experienced animals. This suggests that D_1 receptors may contribute to in the enhanced processing of sexual stimuli in the MPOA that results from sexual experience.

In summary, DA is released in the MPOA as soon as a male rat detects the presence of an estrous female, and it may increase further during copulation. Both basal and female-stimulated DA release depends on testosterone or estradiol, which in turn increases NOS. The signal that a female is present is conveyed by the medial amygdala. Normal basal levels of MPOA DA are sufficient for suboptimal copulation, but the increase in response to a female is associated with enhanced copulatory ability. Microinjection data show that stimulation of DA receptors in the MPOA facilitates several measures of sexual behavior, including sexual motivation, genital reflexes, and copulation.

Paraventricular Nucleus of the Hypothalamus. The paraventricular nucleus (PVN) is an important integrative site for autonomic and neuroendocrine functions. It consists of a parvocellular division, which projects to several brain areas and the spinal cord, and a magnocellular division, which releases oxytocin and vasopressin from the posterior pituitary (reviewed in Swanson and Sawchenko, 1980). The parvocellular division receives input from the MPOA and from periventricular DA neurons.

Effects of drug microinjections into the PVN. Drug-induced erections were elicited by microinjection of the DA agonist apomorphine (Melis et al., 1987), the D_2 agonist quinpirole, or oxytocin (Argiolas et al., 1987) into the PVN; these effects were blocked by icv microinjections of an oxytocin antagonist (Melis et al., 1989). Microinjections of apomorphine (Pehek et al., 1989a) or the D_2/D_3 agonist quinelorane (Eaton et al., 1991) into the PVN also increased the numbers of touch-based erections and ex copula seminal emissions. Furthermore, PVN apomorphine increased intracavernous pressure in anesthetized rats; that increase was enhanced by systemic administration of the monoamine oxidase B inhibitor, selegiline (Allard et al., 2002), which would have decreased degradation of DA. In summary, stimulation of PVN D_2 receptors apparently increases oxytocin release, which then promotes drug-induced and noncontact erections (reviewed in Argiolas and Melis, 1995, 2004).

Chemical changes detected by microdialysis or voltammetry. Extracellular DA levels increased in the PVN during display of noncontact erections by male rats; PVN DA levels increased even more during copulation (Melis et al., 2003).

Mesocorticolimbic DA Tract. DA-containing cell bodies in the VTA give rise to axons of the mesocorticolimbic DA tract that innervate the nucleus accumbens (NAc), prefrontal cortex, and other forebrain regions (Domesick, 1988). This system is activated before and/or during a number of motivated behaviors, including eating, drinking, sexual activity, drug self-administration, and intracranial self-stimulation (Ikemoto and Panksepp, 1999; Kiyatkin, 2002). There is disagreement about whether this system is more important for reward (Wise and Bozarth, 1984), behavioral activation (Koob, 1992; Wilson et al., 1995; Salamone and Correa, 2002), "wanting" [as opposed to "liking" (Kelley and Berridge, 2002)], or incentive learning (Di Chiara, 1995; Spanagel and Weiss, 1999). However, it clearly is necessary for appetitive behaviors.

Effects of drug microinjections into the VTA or NAc. Stimulation of autoreceptors in the VTA with apomorphine delayed the onset and slowed the rate of copulation (Hull et al., 1990). Similar microinjections slowed running speed in all four arms of an X-maze and increased the number of trials on which the male did not leave the start area; however, the percentage of trials on which the male chose the female's goal box was not affected, nor were ex copula genital reflexes (Hull et al., 1991). Therefore, inhibiting DA activity in the mesocorticolimbic tract decreased behavioral activation but did not specifically affect sexual motivation or genital reflexes.

Microinjection of drugs into the NAc also primarily affected activity measures. The latency to begin copulating was speeded by amphetamine and delayed by *cis*-flupenthixol or by the DA neurotoxin 6-OHDA (reviewed in Everitt, 1990). Amphetamine in the NAc also increased responding for a secondary reinforcer that had previously been paired with copulation (Everitt, 1990). The D_2/D_3 agonist quinelorane increased the number of trials on which the male did not leave the start area but did not affect the percentage of trials on which he chose the female's goal box or his copulatory behavior when he reached the female (Moses et al., 1995). It is not clear whether quinelorane's effects were caused by stimulation of terminal autoreceptors or postsynaptic receptors. However, drug injections into either the VTA or the NAc have primarily affected motoric activation, with little influence specifically on sexual motivation or on copulatory measures.

Chemical changes detected by microdialysis or voltammetry. Exposure of male rats to the odor of an estrous, but not a nonestrous, female increased extracellular DA in the NAc (Pfaus et al., 1990; Louilot et al., 1991; Damsma et al., 1992; Wenkstern et al., 1993; Fumero et al., 1994). DA levels were also increased during copulation (Mas et al., 1990; Pleim et al., 1990; Wenkstern et al., 1993; Fumero et al., 1994) (See ➋ *Figure 2-4*). The DA response to the female's odor occurred on the male's first exposure to the odor (Wenkstern et al., 1993) and was lessened by systemic injections of the opioid antagonist naloxone, which may have blocked the disinhibition of DA neurons by endogenous opioids (Mitchell and Gratton, 1991). More recently, a neutral odor that had been paired with copulation enhanced the release of DA in the NAc of male rats (Gelez et al., 2005).

Finer temporal analysis, obtained with either in vivo voltammetry (Mas et al., 1990; Blackburn et al., 1992) or microdialysis with capillary chromatography (Lorrain et al., 1999), revealed that NAc DA rose with

Figure 2-4
Levels of extracellular DA, DOPAC, and HVA in the nucleus accumbens and striatum of sexually experienced male rats; microdialysis samples were collected during baseline when the male was in the sex chamber alone, during exposure to a female behind a barrier, during copulation, and after the female was removed from the chamber. Arrows indicate the first sample collected after the male was exposed to a new condition. Levels of DA increased in the presence of sexually exciting stimuli and rose sharply during copulation (figure is from Pfaus et al., 1990, with permission)

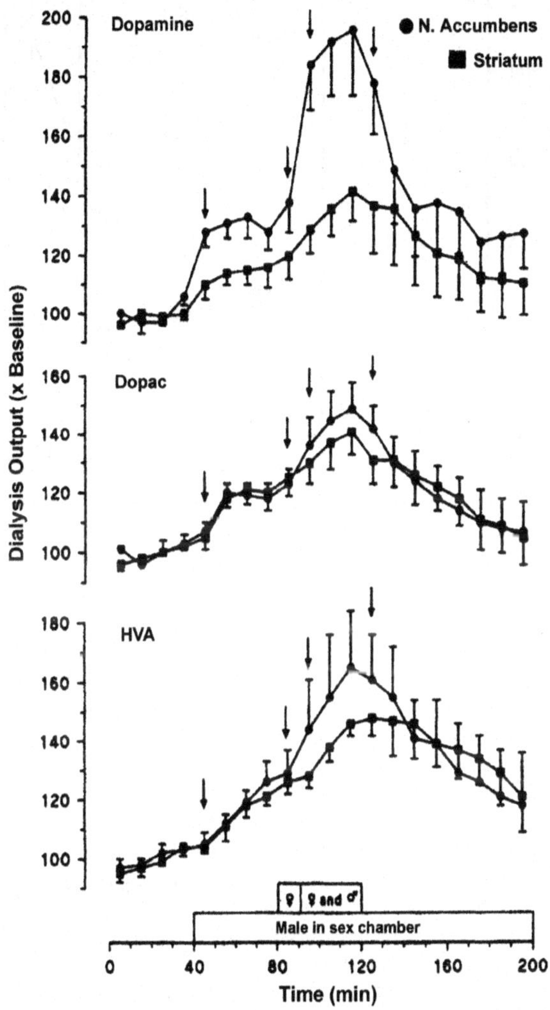

the initial presentation of the female, increased further during copulation, fell following each ejaculation, and increased again when copulation resumed. Reverse dialysis of serotonin (5-HT) into the lateral hypothalamic area (LHA) decreased basal levels of DA in the NAc and prevented the DA response to the female (Lorrain et al., 1999) (See *Figure 2-5*). Because 5-HT is increased in the LHA with ejaculation (Lorrain et al., 1997), one factor contributing to the postejaculatory interval of quiescence may be decreased NAc DA, because of 5-HT release in the LHA.

Sexually satiated male rats showed a small increase in extracellular DA in the NAc when they were presented with a novel estrous female behind a barrier; DA levels increased significantly when the barrier

Figure 2-5

Levels of extracellular DA in the nucleus accumbens of six male rats before and during 40 min of 5-HT perfusion into the anterior lateral hypothalamic area (LHA). Perfusion of 5-HT into the (LHA) decreased basal DA levels and inhibited the mating-induced DA release (*P < 0.05 relative to baseline) (figure is from Lorrain et al., 1999, with permission)

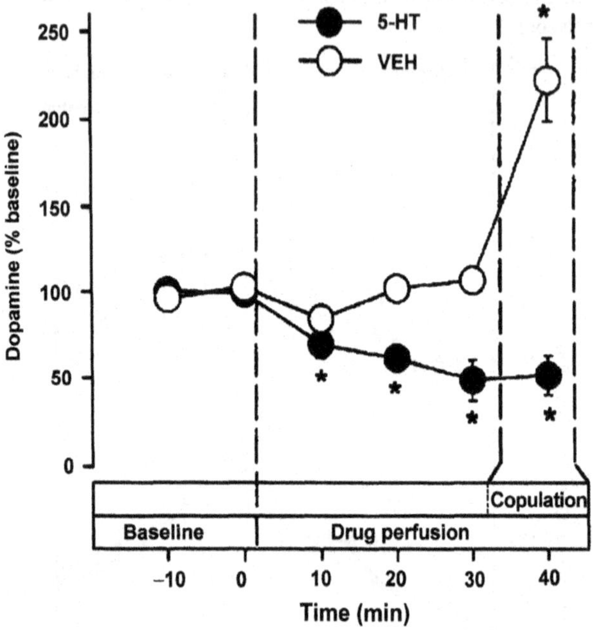

was removed and the animals copulated (Fiorino et al., 1997). In addition, sensitization to the motor-activating effects of amphetamine, after repeated systemic injections of amphetamine, also increased the NAc DA response to a female (Fiorino and Phillips, 1999). Therefore, there is evidence for cross-sensitization between a psychostimulant drug and a natural motivated behavior. However, prenatal stress decreased the number of male rats that were able to copulate, but did not affect the NAc DA response to an inaccessible female or DA levels during copulation (Wang et al., 1995).

Immunocytochemistry. Exposure of male quail to a receptive female for 60 min increased Fos-ir in tyrosine hydroxylase-containing neurons in the VTA, but not in the substantia nigra (Charlier et al., 2005). Therefore, dopaminergic neurons were activated in the VTA of male quail during copulation.

Nigrostriatal DA Tract. The nigrostriatal tract arises from cell bodies in the substantia nigra of the midbrain (A9 cell group) and innervates the dorsal striatum (caudate-putamen). Nigrostriatal activity enhances responsiveness to external stimuli (Robbins and Everitt, 1992).

Chemical changes detected by microdialysis or voltammetry. DA was released in the dorsal striatum only during copulation, in contrast to the DA release in the NAc and MPOA before, as well as during, copulation (Damsma et al., 1992) (See *Figure 2-4*). Therefore, nigrostriatal DA appears to be more important for motoric activation than for sexual motivation.

Spinal Cord. The spinal cord contains somatic and autonomic nuclei that control the striated and hemodynamic mediators of erection, ejaculation, and detumescence. They also provide the initial processing of sensory information from the genitals. Both descending axons from the brain and local reflex circuits regulate these processes (reviewed in Giuliano and Rampin, 2000).

Effects of intrathecal (around the spinal cord) drug administration. Intrathecal injections of the DA agonist apomorphine decreased the numbers of intromissions preceding ejaculation, suggesting that ejaculation was facilitated by stimulation of spinal cord DA receptors (Pehek et al., 1989b). However, intrathecal apomorphine also increased interintromission interval, ejaculation latency, and postejaculatory interval. The highest dose of intrathecal apomorphine also decreased touch-based erections and anteroflexions but did not alter seminal emissions (Pehek et al., 1989b). These data contrast markedly with the effects of systemically administered apomorphine, which increased erections at low doses, inhibited erections at high doses, and facilitated seminal emissions at both low and high doses (Pehek et al., 1988a). In anesthetized rats, intrathecal apomorphine increased intracavernous pressure (Giuliano et al., 2001a). The explanation for these seemingly incongruent results is unclear.

Immunocytochemistry. DA-containing fibers and terminals are found in essentially all laminae throughout the spinal cord (Ridet et al., 1992; Holstege et al., 1996). D_2 receptors identified by immunohistochemistry and in situ hybridization are located in the parasympathetic area of the lumbosacral spinal cord and are abundant in the dorsomedial and dorsolateral motoneurons that innervate the bulbospongiosus and ischiocavernosus muscles of the penis (van Dijken et al., 1996).

3.2.2 Norepinephrine

Cell bodies of noradrenergic neurons in the CNS are primarily located in the locus coeruleus (A6) and in nuclei (A1–5,7) throughout the medulla and pons (reviewed in Kuhar et al., 1999). The locus coeruleus sends axons via the dorsal noradrenergic bundle to the cerebral cortex and hippocampus; the remaining cell groups project via the ventral noradrenergic bundle to the hypothalamus, amygdala, and septum. NE also serves as the major postganglionic neurotransmitter for the sympathetic nervous system. Noradrenergic receptors are divided into two classes. α_1 receptors activate phospholipase C to increase intracellular calcium; α_2 receptors inhibit adenylyl cyclase and are located on presynaptic terminals. β receptors are linked to stimulation of adenylyl cyclase.

3.2.2.1 Effects of Systemically or Intraventricularly Administered Drugs or Lesions of Cell Groups There are inconsistent results from experiments using noradrenergic drugs and lesions of NE pathways. Electrolytic lesions of the ascending dorsal noradrenergic bundle in the midbrain decreased postejaculatory intervals and increased the number of ejaculations (Clark, 1975, 1980). However, cytotoxic 6-OHDA lesions of the same area did not significantly affect copulation (Clark, 1980). Electrolytic lesions of the locus coeruleus or systemic injection of an NE synthesis inhibitor impaired copulation in one experiment (McIntosh and Barfield, 1984), but not in a more recent one (Fernandez-Guasti and Rodriguez-Manzo, 1997). However, locus coeruleus lesions blocked the facilitative effect of introduction of a new female to a sexually satiated male and decreased the facilitative effects of the 5-HT_{1A} agonist 8-OH-DPAT in both satiated and nonsatiated males (Fernandez-Guasti and Rodriguez-Manzo, 1997).

Presynaptic α_2 autoreceptors produce negative feedback on NE release. Systemic administration of the α_2 antagonist yohimbine increased mounts and intromissions by sexually sluggish males (Clark et al., 1984; Smith et al., 1987a), castrated males (Clark et al., 1985b), and males with an anesthetized penis (Clark et al., 1984), and also reversed sexual satiety (Rodriguez-Manzo and Fernandez-Guasti, 1994, 1995; Rodriguez-Manzo, 1999). However, yohimbine's reversal of satiety was inhibited by the DA antagonist haloperidol, suggesting that activation of DA receptors mediated that effect (Rodriguez-Manzo, 1999). The more selective α_2 antagonist delaquamine increased mounting by castrates and facilitated copulation in sexually naïve males (Tallentire et al., 1996). Copulation was also facilitated by two other α_2 antagonists, rauwalscine and phentolamine (Clark, 1995). In hamsters, too, yohimbine increased the numbers of ejaculations and of long intromissions (usually a sign of approaching satiety) (Arteaga et al., 2002). Therefore, blocking α_2 autoreceptors, and thereby increasing NE release, has enhanced sexual performance in both rats and hamsters. Conversely, the α_2 agonist clonidine inhibited both copulation and touch-based ex copula erections in rats (Clark et al., 1985b; Clark and Smith, 1990). However, copulation was inhibited by high doses of yohimbine (Sala et al., 1990), and the NE precursor dl-threodihydroxyphenylserine increased latencies to mount and intromit (Ågmo and Villalpando, 1995). Thus, increasing NE activity can either increase or decrease sexual behavior. The facilitative effects of moderate doses of yohimbine may be mediated by an increase in NE release, which would stimulate α_1 receptors; consistent with this suggestion, the α_1 antagonists prazosin or methoxamine impaired copulation (Clark et al., 1985b, 1987). Yohimbine also facilitated erections in men with moderate erectile dysfunction; it was more effective in men with less severe dysfunction and higher levels of testosterone (Guay et al., 2002).

As the major postganglionic neurotransmitter of the sympathetic nervous system, NE inhibits erection. In contrast to its facilitative effects on copulation, yohimbine inhibited ex copula touch-based erections and seminal emissions (Smith et al., 1987b). However, touch-based erections were also inhibited by the α_2 agonist clonidine (Clark and Smith, 1990). Because penile reflexes were inhibited by both an agonist and an antagonist of α_2 receptors, it seems likely that low to moderate levels of NE facilitate erections, whereas high levels are inhibitory.

Most systemically administered drugs affect both central and peripheral receptors that regulate genital reflexes. However, guanethidine depletes peripheral, but not central, noradrenergic nerve terminals. A single injection of guanethidine increased the number of anteroflexions in restrained supine males; however, repeated treatments over 4 to 8 weeks decreased both touch-based erections and seminal emissions (Stefanick et al., 1985). Therefore, the sympathetic nervous system may provide some facilitation of erection, as demonstrated by Giuliano and colleagues (1997, 2000), in addition to its major roles in seminal emission and detumescence. The mixed excitatory and inhibitory influences of NE on genital reflexes may be relevant to human sexual function, as reviewed by Bancroft and Janssen (2000).

3.2.2.2 Role of NE in Specific Brain Areas *Medial Preoptic Area. Effects of drug microinjections into the MPOA.* NE microinjected into the MPOA facilitated both sexual arousal and copulatory performance; conversely, both the β-noradrenergic antagonist propranolol and the α antagonist phenoxybenzamine inhibited copulatory behavior (Mallick et al., 1996). The α_2 agonist clonidine, which would have stimulated autoreceptors and thereby decreased NE release, decreased the number of animals copulating and increased interintromission intervals and ejaculation latencies; however, intromissions preceding ejaculation were decreased (Clark, 1991). Furthermore, MPOA microinjections of the α_2 antagonist yohimbine (which would have disinhibited NE release) blocked the inhibitory effects of systemically injected clonidine (Clark, 1991). Therefore, copulation appears to be facilitated by stimulation of α_2 receptors in the MPOA.

Spinal Cord. Effects of intrathecal drug administration. Intrathecal injections of NE increased the frequency of pelvic thrusting during intromission, whereas similar administration of either the α_2 agonist clonidine or the β agonist isoproterenol decreased the frequency of thrusting (Hernandez et al., 1994). Clonidine's inhibitory effect may have resulted from stimulation of α_2 autoreceptors, which would have decreased NE release. These results are consistent with the findings noted earlier that α_1 receptors probably mediate NE's facilitation of copulation.

Penile detumescence is maintained by sympathetic noradrenergic nerves, which stimulate α_1 receptors on smooth muscle cells in the corpora cavernosa. This activates the inositol triphosphate (IP_3) second messenger cascade and increases intracellular calcium both from intracellular stores and through plasma membrane calcium channels. The increased calcium then activates a series of processes that result in smooth muscle contraction, decreased inflow of blood, and detumescence (Christ, 1995; Andersson and Stief, 1997). α_1 receptors may contribute to detumescence indirectly by increasing sensitivity of smooth muscle to calcium, via the Rho-kinase pathway (Somlyo and Somlyo, 2000). A Rho-kinase antagonist relaxed smooth muscle in strips of human or rabbit corpus cavernosum that had been contracted by the α receptor agonist phenylephrine (Rees et al., 2001). This antagonist also stimulated penile erection in rats (Chitaley et al., 2001). Therefore, Rho-kinase antagonists may provide a new type of treatment for erectile dysfunction.

Immunocytochemistry. Neurons and fibers containing immunoreactivity for α_{2a} and α_{2c} adrenoceptor subtypes were found in the intermediolateral cell column, the dorsal gray commissure, and the ventral horn of the T12-L2 and L5-S1 spinal cord. There were only immunoreactive fibers, not cell bodies, in the dorsal horn. Neurons in the autonomic nuclei that were retrogradely labeled from the penis were immunoreactive for α_{2a} and α_{2c} adrenoceptor subtypes and were closely apposed by α_{2a} and α_{2c} immunoreactive fibers (Yaîci et al., 2002). Therefore, pre- and postsynaptic α_2 adrenoceptors may provide an intraspinal modulation of the noradrenergic control of the autonomic outflow to the penis.

3.2.3 Serotonin

Serotonin (5-hydroxytryptamine, 5-HT) is synthesized from tryptophan, beginning with tryptophan hydroxylase, the rate-limiting enzyme, which produces 5-hydroxytryptophan, which is in turn converted to 5-HT by aromatic amino acid decarboxylase. There are nine groups of serotonergic neurons (B1–9) scattered throughout the midbrain, pons, and medulla. Serotonergic axons ascend and descend from these nuclei to nearly every part of the brain and spinal cord. However, some serotonergic cell bodies lie outside these nuclei, and the majority of neurons within the nuclei do not contain serotonin (Hillegaart, 1991; reviewed in Frazer and Hensler, 1999).

3.2.3.1 Effects of Systemically or Intraventricularly Administered Drugs or Lesions of Cell Groups Most serotonergic influences on sexual behavior are inhibitory. The use of selective serotonin reuptake inhibitor (SSRI) antidepressants often results in difficulty achieving orgasm or ejaculation and sometimes in decreased sexual interest as well (Seagraves, 1990; reviewed in Rosen et al., 1999). The time course of the inhibitory sexual effects is usually similar to that of the antidepressant effects. The physiological changes that mediate the sexual impairments are unclear, but they have been thought to include increased prolactin secretion, decreased NO production, anticholinergic effects (Rosen et al., 1999), or decreased sensitivity of the penis (Yilmaz et al., 1999). The ability of SSRIs to inhibit ejaculation has resulted in their use to treat premature ejaculation (Haensel et al., 1995; Kim and Seo, 1998; Waldinger et al., 1998; McMahon and Touma, 1999; Yilmaz et al., 1999); however, undesirable side effects have included anorgasmia and decreased sexual interest (McMahon and Touma, 1999).

In rats, too, chronic administration of the SSRI fluoxetine decreased both sexual motivation and the ability to ejaculate (Taylor et al., 1996; Vega Matuszcyk et al., 1998; Cantor et al., 1999; Frank et al., 2000). The inhibition of ejaculation was reversed by oxytocin injections, but sexual motivation, measured as level changes in a bilevel apparatus, was not improved by oxytocin (Cantor et al., 1999). Hamsters also showed ejaculatory impairment following chronic administration of a 5-HT reuptake inhibitor, clomipramine (Boscarino and Parfitt, 2002). A possible mechanism of the ejaculatory impairment may be a decrease in the pressure response of the seminal vesicle to splanchnic nerve stimulation (Hsieh et al., 1998). Most studies of acute, as opposed to chronic, administration of fluoxetine did not report inhibition of ejaculation (Vega Matuszcyk et al., 1998; Cantor et al., 1999; Mos et al., 1999). However, one study found that acute injections slowed copulation and that this slowing could be reversed by amantadine, which releases NE (Yells et al., 1995). Therefore, 5-HT and NE influences may interact in the regulation of ejaculation. Acute injections of fluoxetine have also inhibited one measure of sexual motivation (level changing in search of a female) but not the amount of time spent near a receptive female (Vega Matuszcyk et al., 1998). Injections of 5-hydroxytryptophan, the 5-HT precursor, also increased intromission and ejaculation latencies and increased the number of intromissions preceding ejaculation (Ahlenius and Larsson, 1991). Similarly, intrathecal (around the spinal cord) administration of 5-HT blocked the urethrogentital reflex, which was restored by systemic administration of the 5-HT antagonist methysergide (Marson and McKenna, 1992).

Conversely, copulation has been facilitated by drugs that deplete 5-HT (Salis and Dewsbury, 1971; Ahlenius et al., 1971; Mitler et al., 1972) and by lesions of the medial raphe nucleus, a major source of 5-HT in the brain; lesions of the dorsal raphe nucleus were ineffective (Albinsson et al., 1996; Kondo and Yamanouchi, 1997). Facilitation was found most often in animals that were initially noncopulators (Salis and Dewsbury, 1971; Ginton, 1976; Dallo, 1977), in castrates maintained on suboptimal testosterone replacement (Malmnas and Meyerson, 1971; Södersten et al., 1976, 1977; Larsson et al., 1978; Rodriguez et al., 1984), or in sexually naïve males (Rodriguez et al., 1984). In addition, 5-HT depletion caused by lesions of the median and pontine raphe nuclei increased ex copula genital reflexes (Monaghan et al., 1993), and neurotoxic destruction of 5-HT terminals in the brain and spinal cord disinhibited the urethrogenital reflex, a model of orgasm (Marson and McKenna, 1994a). Inhibition of 5-HT synthesis by *p*-chlorophenylalanine (*p*CPA) not only decreased the number of touch-based erections, but also decreased the latency to the first erection (Matsumoto et al., 1997). *p*CPA-treated males also had more noncontact erections in the presence of a female, but not when they were tested alone (Matsumoto et al., 1997). Therefore, 5-HT appears to inhibit erections in response to a female, and to have mixed, but mostly inhibitory, effects on touch-based erections.

Although 5-HT's effects on sexual behavior are generally inhibitory, stimulation of the 5-HT_{1A} receptor subtype facilitates ejaculation. The classic 5-HT_{1A} agonist 8-OH-DPAT [8-hydroxy-2-(di-*n*-propylamino) tetralin] decreased both ejaculation latency and the number of intromissions preceding ejaculation (Ahlenius et al., 1981; Ahlenius and Larsson, 1984b; Ahlenius et al., 1989; Schnur et al., 1989; Rehman et al., 1999); indeed, some rats ejaculated on their first intromission (Ahlenius et al., 1989; Ahlenius and Larsson, 1991; Haensel and Slob, 1997). Another 5-HT_{1A} agonist, flesinoxan, also facilitated ejaculation, although no animals ejaculated on the first intromission (Haensel and Slob, 1997). The opposite influences of 8-OH-DPAT, compared to 5-HT itself, and the fact that some 5-HT_{1A} receptors are autoreceptors on serotonergic cell bodies and dendrites suggested that 8-OH-DPAT might facilitate sexual behavior by inhibiting 5-HT release. However, the effects of 8-OH-DPAT were not altered by lesions of serotonergic nuclei, suggesting that postsynaptic receptors mediated those effects (Fernandez-Guasti and Escalante, 1991). Furthermore, microinjections of 8-OH-DPAT into terminal fields (MPOA and NAc), where all 5-HT_{1A} receptors are postsynaptic, also facilitated sexual behavior (Fernandez-Guasti et al., 1992; Matuszewich et al., 1999). The facilitative effects of 8-OH-DPAT are seen only in sexually experienced animals with normal levels of testosterone. 8-OH-DPAT did not facilitate copulation in castrates with subnormal testosterone replacement, and in those with threshold testosterone levels, only sexually experienced males were able to copulate after 8-OH-DPAT administration (Rowland and Houtsmuller, 1998). The drug had a strong effect in males with high levels of testosterone.

One possible mechanism of 8-OH-DPAT's effects is increased noradrenergic activity in some brain area(s). The percentage of sexually satiated males that copulated on the following day was increased, not only by 8-OH-DPAT, but also by the α_2-adrenergic receptor antagonist yohimbine and the opioid antagonist naloxone (Rodriguez-Manzo and Fernandez-Guasti, 1994, 1995). In addition, neurotoxic lesions of noradrenergic neurons blocked the facilitation by 8-OH-DPAT and by naloxone, but not that of yohimbine, in satiated males (Rodriguez-Manzo and Fernandez-Guasti, 1995). Similar lesions in nonsatiated males decreased 8-OH-DPAT's effects (Fernandez-Guasti and Rodriguez-Manzo, 1997).

8-OH-DPAT's most common effect is facilitation of ejaculation, which is promoted by the sympathetic nervous system. Studies of ex copula reflexes are consistent with a prosympathetic, anti-parasympathetic influence. In male rats touch-based erections, cups, and anteroflexions were inhibited by 8-OH-DPAT, and the time and number of intromissions preceding ejaculation were decreased in copula (Rehman et al., 1999). In monkeys, a moderate dose of 8-OH-DPAT decreased erections in the presence of an inaccessible female (Pomerantz et al., 1993a) and facilitated ejaculation in copula, although a high dose delayed ejaculation (Pomerantz et al., 1993b). Therefore, stimulation of 5-HT_{1A} receptors may shift autonomic balance to favor ejaculation, in addition to facilitating copulation in previously noncopulating males. However, there is some species specificity in 8-OH-DPAT's effects. In male ferrets, low doses had no effect, and the doses that had facilitated ejaculation in male rats inhibited copulation in ferrets (Paredes et al., 1994). Furthermore, some of the effects of 8-OH-DPAT are mediated by D_2 receptors, rather than by 5-HT_{1A} receptors. Intraventricular delivery of 8-OH-DPAT induced rhythmic contractions of the bulbospongiosus (BS) muscles, which mediate the expulsion of semen (Clement et al., 2006). This effect was blocked by two D_2 receptor antagonists (raclopride and spiperone), but not by a 5-HT_{1A} receptor antagonist (WAY100635). Additional support for a major role of D_2 receptors came from the effectiveness of the D_2 agonist quinelorane in stimulating these contractions. A similar finding is described later, in which raclopride, coadministered with 8-OH-DPAT into the MPOA, partially blocked the effects of 8-OH-DPAT on copulation.

Stimulation of other 5-HT receptor subtypes generally inhibits copulation. In rats, ejaculation was inhibited by the 5-HT_{1B} agonist anpirtoline, whose effects were blocked by selective 5-HT_{1B} antagonists (Hillegaart and Ahlenius, 1998). Ejaculation latency was lengthened by coadministration of the SSRI citalopram and a 5-HT_{1A} antagonist, as expected, and that effect was blocked by a 5-HT_{1B} antagonist (Ahlenius and Larsson, 1999). A 5-HT_{1B} antagonist also blocked the increase in ejaculation latency that resulted from systemic administration of the 5-HT precursor 5-hydroxytryptophan (plus a peripheral decarboxylase inhibitor to restrict the increase in 5-HT to the CNS) (Ahlenius and Larsson, 1998). Several serotonergic drugs produced less sexual impairment in 5-HT_{1B} knockout mice than in wild-type mice, consistent with an inhibitory effect of the 5-HT_{1B} receptor (Rodriguez-Manzo et al., 2002b). Therefore, 5-HT_{1A} and 5-HT_{1B} receptors appear to have opposite effects on ejaculation, with 5-HT_{1A} receptors facilitating, and 5-HT_{1B} receptors inhibiting, ejaculation.

However, in two studies of mice both the 5-HT_{1A} agonist 8-OH-DPAT and two 5-HT_{1B} agonists inhibited one or more components of sexual behavior. 8-OH-DPAT decreased the time spent near an estrous female (Popova and Amstislavskaya, 2002), as well as copulatory performance (Rodriguez-Manzo et al., 2002b). The 5-HT_{1B} agonist CGS-12066A also decreased the time spent near the estrous female (Popova and Amstislavskaya, 2002), and the 5-HT_{1B} agonist TFMPP inhibited copulation. These data suggest that both subtypes of 5-HT receptor impair sexual behavior. On the other hand, 5-HT_{1B} knockout mice required more stimulation to achieve ejaculation, suggesting that 5-HT_{1B} receptors may facilitate ejaculation; however, the knockout mice were less sensitive to the inhibitory effects of 5-hydroxytryptophan, the 5-HT precursor (Rodriguez-Manzo et al., 2002b). The authors suggested that compensatory mechanisms may have rendered the knockout mice less sensitive to the inhibitory effects of high levels of 5-HT.

Whereas stimulation of 5-HT_{1A} receptors facilitates ejaculation in male rats and may inhibit erection, 5-HT_{2C} agonists appear to produce the opposite pattern of effects. The 5-$HT_{2C/1D}$ agonist mCPP increased drug-induced and noncontact erections in male monkeys, but decreased the number of males that initiated copulation and that ejaculated (Pomerantz et al., 1993a). mCPP also increased erections in rats (Berendsen et al., 1990; Millan et al., 1997). When these studies were conducted mCPP was classified as a 5-$HT_{1C/1D}$ agonist. More recently the 5-HT_{1C} receptor was renamed the 5-HT_{2C} receptor (Hoyer et al., 1994). Therefore, we have changed the nomenclature from the original publication to be consistent with the current classification. mCPP also increased firing of the cavernous nerves to the penis and increased intracavernous pressure in anesthetized rats (Steers and de Groat, 1989) and produced drug-induced erections in awake rats (Berendsen et al., 1990). Erections in awake rats were also increased by a more selective 5-HT_{2C} agonist (RO60-0175) (Millan et al., 1997). Therefore, stimulation of 5-HT_{1A} receptors appears to promote sympathetically mediated ejaculation, whereas activation of 5-HT_{2C} receptors facilitates parasympathetic effects on erection, but may inhibit ejaculation.

A relatively nonselective 5-HT_2 agonist, DOI, has been reported to inhibit copulation (Foreman et al., 1989; Gonzalez et al., 1994; Klint and Larsson, 1995; Padoin and Lucion, 1995), and a 5-HT_2 antagonist facilitated copulation (Gonzalez et al., 1994). To summarize, 5-HT_{1B}, 5-HT_{2C}, and other 5-HT_2 agonists inhibit ejaculation and may impair copulation, although 5-HT_{2C} agonists may also facilitate erection. On the other hand, stimulation of 5-HT_{1A} receptors facilitates ejaculation and may inhibit erection.

3.2.3.2 Role of 5-HT in Specific Brain Areas *Medial Preoptic Area. Effects of drug microinjections into the MPOA.* In keeping with 5-HT's general inhibitory role in male sexual behavior, large doses of 5-HT microinjected into the MPOA impaired copulation (Fernandez-Guasti et al., 1992; Verma et al., 1989); similarly, a 5-HT_{1B} agonist delayed ejaculation (Fernandez-Guasti et al., 1992). Consistent with the facilitative effects of systemically administered 5-HT_{1A} agonists, 8-OH-DPAT microinjections into the MPOA facilitated ejaculation (Fernandez-Guasti and Escalante, 1991). Administration of 8-OH-DPAT by reverse dialysis into the MPOA also facilitated copulation (Matuszewich et al., 1999) and increased extracellular levels of both DA and 5-HT (Lorrain et al., 1998). 8-OH-DPAT's facilitative effects were partially blocked by the D_2 antagonist raclopride, but not by the 5-HT_{1A} antagonist pMPPI, suggesting that stimulation of D_2 receptors, as a result of the elevated DA levels, mediated much of the facilitation (Matuszewich et al., 1999).

Microdialysis. The 5-HT metabolite 5-hydroxyindole acetic acid (5-HIAA) was increased in dialysates from the POA of male rats after ejaculation (Fumero et al., 1994; Mas et al., 1995a); 5-HT itself was increased in tissue punches from the POA of males that had ejaculated (Mas et al., 1987). Mas and his colleagues suggested that 5-HT increases in the POA after ejaculation might contribute to postejaculatory quiescence. However, a more recent analysis of 5-HT itself concluded that 5-HT levels in the MPOA and, more laterally, in the POA were constant throughout copulation and the postejaculatory interval (Lorrain et al., 1997). However, there were increases of 5-HT in the anterior LHA at the time of ejaculation. The increase in 5-HIAA that was observed by Fumero et al. (1994) and Mas et al. (1995a) may have reflected diffusion from the adjacent LHA. Therefore, it seems improbable that 5-HT in the MPOA contributes to postejaculatory quiescence.

Lateral Hypothalamic Area. Effects of drug microinjections into the lateral hypothalamic area (LHA). Microinjection into the anterior LHA of a SSRI delayed the onset of copulation and delayed ejaculation after the animals did begin to copulate (Lorrain et al., 1997). Thus, the impairment by SSRIs of sexual motivation and ejaculation (reviewed in Rosen et al., 1999) may be mediated, in part, by increased 5-HT in the LHA.

Changes detected by microdialysis. Extracellular 5-HT increased in the LHA when male rats ejaculated (Lorrain et al., 1997). Reverse dialysis of 5-HT into the LHA decreased basal DA levels in the NAc and prevented the DA increase usually seen during copulation (Lorrain et al., 1999) (See ➋ *Figure 2-5*). The inhibition of NAc DA release produced by 5-HT in the LHA may contribute to the decreased sexual motivation associated with clinical use of SSRI antidepressants (reviewed in Rosen et al., 1999).

Mesocorticolimbic Dopamine Tract. Microinjection of 5-HT into the NAc inhibited ejaculation (i.e., it increased the numbers of mounts, intromissions, and time preceding ejaculation); those same measures were decreased by the 5-HT_{1A} agonist 8-OH-DPAT (Hillegaart et al., 1991). Microinjection of these compounds into the dorsal striatum did not affect copulation. The authors suggested that 8-OH-DPAT's effects might have resulted from antagonism of 5-HT_2 receptors.

Paraventricular Nucleus. Axons from the PVN project to the nucleus paragigantocellularis (nPGi) (Bancila et al., 2002), which is a major source of inhibition of genital reflexes and copulation (see section on nPGi later). PVN axon terminals formed close appositions to serotonergic neurons in the nPGi that projected to the lumbosacral spinal cord and indirectly to the corpus cavernosum. Therefore, the PVN can provide control over inhibitory serotonergic projections to the spinal areas that control erection.

Nucleus Paragigantocellularis of the Medulla. There is tonic inhibitory control of the spinal mechanisms that control erection. Spinal transection in rats (Hart, 1968a; Sachs and Garinello, 1979; Sachs and Garinello, 1980), mice (Sachs, 1980), and dogs (Hart and Kitchell, 1966; Hart, 1967) increases the number or intensity of erections or urethrogenital reflexes or decreases the amount of stimulation required to elicit those responses. A major source of that inhibition is the nucleus paragigantocellularis (nPGi) of the medulla in the ventrolateral medulla. Neurons in the nPGi project directly to pudendal motor neurons, sympathetic and parasympathetic preganglionic neurons, and the interneuronal regions of the medial gray lumbosacral cord (Marson et al., 1992; Hermann et al., 2003).

Effects of drug microinjections into the nPGi. Administration of the serotonergic neurotoxin 5,7 dihydroxytryptamine into the PVN decreased the descending inhibition in UG reflex tests (Marson and McKenna, 1994a). Therefore, serotonergic input from the nPGi to the spinal cord tonically inhibits genital reflexes.

Immunocytochemistry. Approximately 78% of nPGi neurons that project to the ipsilateral lumbosacral spinal cord contain 5-HT (Marson and McKenna, 1992). These serotonergic neurons receive direct input from the PVN (Bancila et al., 2002).

Spinal Cord. Effects of intrathecally administered drugs. Typically, the urethrogenital reflex can be elicited only after transection of the spinal cord (McKenna et al., 1991), ablation of the nPGi (Marson and McKenna, 1990; Marson et al., 1992), or stimulation of the MPOA (Marson and McKenna, 1994b). However, the urethrogenital reflex could be elicited without any of these measures following either intraventricular or intrathecal administration of the 5-HT neurotoxin 5,7-DHT (Marson and McKenna, 1994a). Similarly, lesions of the median raphe nuclei, which, together with the nPGi, provide a major source of serotonergic axons to the spinal cord, significantly increased the numbers of touch-based anteroflexions (penile body erections) and cups (intense glans erections) (Monaghan et al., 1993). Conversely, application of 5-HT to the spinal cord suppressed the urethrogenital reflex in rats with transected spinal cords, and the 5-HT antagonist methysergide prevented 5-HT's effect (Marson and McKenna, 1992). Therefore, descending serotonergic input to the spinal cord appears to mediate much of the tonic inhibition of genital reflexes.

In copulation tests, intrathecal application of either 5-HT or thyrotropin releasing hormone (TRH) produced only slight increases in mount and intromission latencies. However, coadministration of 5-HT and TRH markedly increased mount and intromission latencies (Hansen et al., 1983). Intrathecal administration of TRH alone inhibited touch-based penile reflexes, decreasing both the number of responders and the numbers of reflexes, and increasing reflex latencies (Holmes et al., 1997). TRH and 5-HT are colocalized in some neurons that descend from the medulla to the spinal cord (Hokfelt et al., 1986). Furthermore, methiothepin, a 5-HT antagonist, partially blocked some of TRH's inhibitory effects (Holmes et al., 2001). Therefore, either 5-HT or TRH alone may inhibit copulation and genital reflexes, but their effects may be enhanced by coadministration or endogenous corelease.

Intrathecal administration of the 5-HT_{1A} agonist 8-OH-DPAT facilitated copulation, producing decreased ejaculation latencies, intercopulatory intervals, and numbers of intromissions preceding ejaculation (Lee et al., 1990). In dramatic contrast, intrathecal injections of 8-OH-DPAT decreased the numbers of male rats that displayed touch-based erections and seminal emissions (Lee et al., 1990). The basis for the apparent discrepancy between in copula and ex copula measures is not clear.

Immunocytochemistry. Spinal 5-HT_{2C} receptors appear to facilitate erection. Consistent with a proerectile effect, 5-HT_{2C} receptor immunoreactivity was observed on motor neurons of the sacral parasympathetic nucleus, the dorsal gray commissure, and the motor neurons of the ventral horn (Bancila et al., 1999). In addition, 5-HT_{2C} receptor immunoreactivity was present in all neurons that were retrogradely labeled from the corpus cavernosum to the sacral parasympathetic nucleus and the dorsal gray commissure of L5-L6, as well as in ventral horn motoneurons retrogradely labeled from the ischiocavernosus and BS muscles of the penis. Therefore, the neurons that promote erection possess 5-HT_{2C} receptors.

3.2.4 Acetylcholine

Acetylcholine (ACh) is synthesized from choline and acetyl coenzyme A by choline acetyltransferase. There are eight major groups of cholinergic cell bodies (Ch 1–8) in the basal forebrain and brainstem (reviewed in Mesulam, 1995). ACh is also the neurotransmitter of peripheral motor nerves, preganglionic autonomic neurons, and postganglionic parasympathetic neurons. There are five types of muscarinic receptors, coupled to either $G_{q/11}$ or $G_{I/O}$ G-proteins (Alexander and Peters, 2000). They have either excitatory or inhibitory effects, depending on the tissue (Taylor and Brown, 1999). Nine subtypes of nicotinic receptors are excitatory, acting on sodium channels (Taylor and Brown, 1999; Alexander and Peters, 2000).

3.2.4.1 Effects of Systemically or Intraventricularly Administered Drugs or Lesions of Cell Groups Slowing the degradation of ACh (using the acetylcholinesterase antagonist physostigmine) or administration of the muscarinic agonist pilocarpine increased ejaculations in male rats; this effect was blocked by atropine, a muscarinic antagonist (Zarrindast et al., 1992). Pilocarpine also evoked ex copula erections in a neutral test cage (Maeda et al., 1990). The muscarinic agonist oxotremorine facilitated copulation in both sexually experienced and naïve castrates with testosterone replacement, but not in castrates without testosterone (Retana-Marquez and Velazquez-Moctezuma, 1997). A more recent study found that intraperitoneal administration of muscarine to spinally transected rats increased the numbers of penile reflexes, decreased their latencies, and increased their duration (Vargas et al., 2004). In addition, six of the nine animals ejaculated. The authors suggested that muscarinic receptors facilitate both erectile and ejaculatory functions mediated by the spinal cord. In summary, systemically administered muscarinic drugs appear to facilitate copulation and ex copula erections. However, because systemically administered cholinergic drugs affect striated muscles and autonomic targets, as well as central regulatory areas, it is difficult to determine whether their effects are centrally or peripherally mediated.

3.2.4.2 Role of ACh in Specific Brain Areas *Medial Preoptic Area. Effects of drug microinjections into the MPOA.* Microinjections of either the nonspecific ACh agonist carbachol or the muscarinic agonist oxotremorine facilitated ejaculation by decreasing the numbers of intromissions preceding ejaculation (Hull et al., 1988a). This facilitation was blocked by the muscarinic antagonist scopolamine, which also decreased the numbers of animals that copulated, when administered alone (Hull et al., 1988b).

Immunocytochemistry. Immunoreactivity for acetylcholinesterase (AChE), the degradative enzyme, was more dense in the sexually dimorphic area of male gerbils, compared with females (Commins and Yahr, 1984). AChE immunoreactivity decreased after castration and was restored by testosterone replacement, suggesting that ACh in the MPOA may contribute to hormonal regulation of male sexual behavior.

3.2.5 Glutamate

Glutamate is the major excitatory neurotransmitter in the CNS. However, relatively few studies have examined the role of glutamate in the regulation of male sexual behavior. Glutamate receptors are grouped into two distinct classes, the ionotropic and metabotropic receptors. The ionotropic receptors contain cation-specific ion channels and are further subdivided into AMPA (α-amino-3-hydroxy-5-methyl-4-isooxazole-propionic acid), kainate, and NMDA (N-methyl-D-aspartate) receptors. The metabotropic glutamate receptors, on the other hand, are coupled to GTP binding proteins (G-proteins) and modulate the production of intracellular messengers.

3.2.5.1 Effects of Systemically or Intraventricularly Administered Drugs or Lesions of Cell Groups Because of the ubiquity of glutamate in the nervous system and its ability to produce seizures and neurotoxicity, there have been no studies on the effects of systemically administered glutamatergic agonists on male sexual behavior. However, administration of the NMDA antagonist MK-801 impaired copulation in both naive and sexually experienced males and prevented the facilitation due to repeated noncopulatory exposures to a female (Powell et al., 2003).

3.2.5.2 Role of Glutamate in Specific Brain Areas *Medial Preoptic Area. Effects of drug microinjections into the MPOA.* One brain region where increased glutamate appears to facilitate sexual behavior is the MPOA. Microinjections of glutamate into the MPOA of anesthetized rats increased erectile responses (Giuliano et al., 1996) and the urethrogenital reflex, a model of orgasm (Marson and McKenna, 1994b). Conversely, microinjections of the NMDA receptor antagonist MK-801 into the MPOA of both sexually naïve and experienced male rats decreased the numbers of mounts, intromissions, and ejaculations (Dominguez et al., 2003; Vigdorchik et al., 2003). MPOA microinjections of MK-801 in sexually naïve males before each of seven noncopulatory exposures to an estrous female prevented the facilitation of such exposures on a drug-free test on day 8 (Vigdorchik et al., 2003). Microinjections of MK-801 also inhibited 50-kHz ultrasonic vocalizations evoked in anticipation of a receptive female (Brudzynski and Pniak, 2002). Therefore, endogenous stimulation of NMDA receptors in the MPOA facilitates copulation in both naïve and experienced male rats, and is permissive for the sensitization to female stimuli induced by repeated exposures to a female.

Chemical changes detected by microdialysis. A recent microdialysis study found mating-induced increases in extracellular glutamate in the MPOA of male rats (Dominguez et al., 2006) (See ➋ *Figure 2-6*). Levels of glutamate increased to nearly 170% of baseline during mounts and intromissions, and nearly 300% during ejaculations. Glutamate levels decreased back to baseline in the first postejaculatory sample; the magnitude of this fall was highly correlated with the length of the postejaculatory interval of quiescence, suggesting that glutamate in the MPOA might also play a role in regulating the postejaculatory interval. In the same study, reverse dialysis of a cocktail of glutamate uptake inhibitors into the MPOA before and during mating increased extracellular glutamate (~280% of BL), increased the number of ejaculations, decreased ejaculation latency, and decreased the postejaculatory latency to resume copulation (Dominguez et al., 2006).

Immunocytochemistry. It is not clear which mechanisms in the MPOA are altered by glutamate. However, glutamatergic stimulation of NMDA receptors appears to be important. Nearly 100% of cells in the MPOA containing mating-induced Fos-ir also contained NMDA receptors; additionally, mating increased phosphorylation of NMDA receptors, suggesting that NMDA receptors were indeed activated by mating (Dominguez et al., 2003). Finally, microinjections of the NMDA receptor antagonist MK-801 into the MPOA of male rats reduced the number of cells containing mating-induced Fos-ir (Dominguez and Coolen, 2004). Together, these studies indicate that glutamatergic stimulation in the MPOA facilitates male sexual behavior.

■ Figure 2-6

Mating increased extracellular glutamate in the MPOA of male rats. This increase was highest (~300% of baseline, BL) in samples collected when the animals ejaculated (EJAC); levels then dropped precipitously in samples collected during the postejaculatory interval (PEI). Levels returned to BL after mating, when the female was removed (POST). Values are expressed as mean ± SEM, *P < 0.05; **P < 0.001 (figure is from Dominguez et al., 2006, with permission)

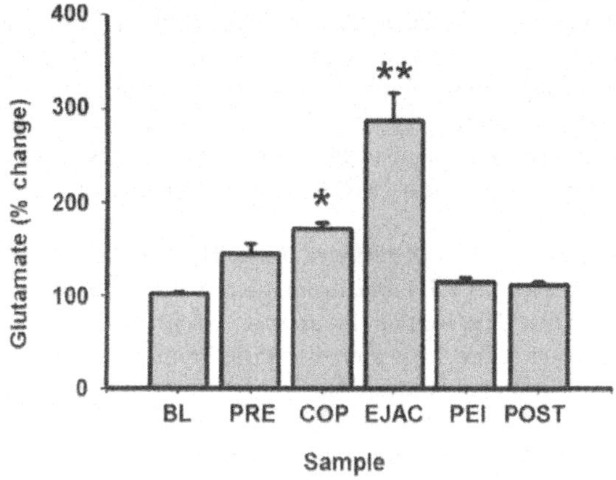

Paraventricular Nucleus. Effects of drug microinjections into the PVN. Another region in the hypothalamus where glutamate might facilitate male sexual behavior is the PVN. Like in the MPOA, microinjections of glutamate agonists into the PVN facilitated erections, whereas antagonists inhibited this response. Specifically, microinjections of the glutamate agonist NMDA into the PVN increased erectile response; this effect was blocked by the NMDA antagonist MK-801 (dizocilpine) and by the NOS antagonist L-NAME (Melis et al., 1997). Therefore, glutamate, by opening calcium channels associated with NMDA receptors, may lead to calcium activation of calmodulin, which then activates NOS. Microinjections of MK-801 into the PVN also inhibited noncontact erections and copulation; an AMPA receptor antagonist (CNQX) had a similar effect, but to a lesser extent (Melis et al., 2004).

Chemical changes detected by microdialysis. Levels of extracellular glutamate and aspartic acid in the PVN increased during exposure to a receptive female and increased even further during copulation (Melis et al., 2004), suggesting that increased glutamate in the PVN may enhance male sexual behavior.

3.2.6 Gamma Amino Butyric Acid

Gamma amino butyric acid (GABA) is the major inhibitory neurotransmitter in the brain (reviewed in Paul, 1995). It is formed from glutamate by glutamic acid dehydroxylase (GAD). Stimulation of $GABA_A$ receptors opens chloride channels, thereby hyperpolarizing neurons. $GABA_B$ receptors inhibit several intracellular processes by acting through the $G_{i/o}$ G-protein.

3.2.6.1 Effects of Systemically Administered Drugs Systemically administered $GABA_A$ (Ågmo et al., 1997) and $GABA_B$ (Paredes and Ågmo, 1995) agonists impaired copulation without significant effects on motor activity. The $GABA_A$ agonist produced similar effects in rabbits (Paredes et al., 1998). However, a different $GABA_A$ agonist, as well as a GABA synthesis inhibitor and an inhibitor of GABA catabolism all impaired social and drinking behaviors, as well as sexual behavior, suggesting that these drugs had nonspecific effects on motivated behaviors (Paredes et al., 1997). Furthermore, some of the inhibitory effects on copulation may have resulted from decreased EMG activity in ischiocavernosus muscles during thrusting, (Paredes et al., 1993). The ischiocavernosus muscle is necessary for erection; therefore, the impairment of copulation may have been due to a peripheral, rather than central effect. Furthermore, ex copula genital reflexes were inhibited by a $GABA_B$, but not a $GABA_A$, receptor agonist (Leipheimer and Sachs, 1988).

3.2.6.2 Role of GABA in Specific Brain Areas *Medial Preoptic Area. Effects of drug microinjections into the MPOA.* MPOA microinjections of the GABAergic drugs muscimol and ethanolamine-O-sulfate decreased the numbers of male rats that mounted, intromitted, or ejaculated; conversely, the $GABA_A$ antagonists bicuculline or 3-mercaptopropionic acid (3-MPA) dramatically shortened both ejaculation latencies and postejaculatory intervals and decreased the numbers of intromissions preceding ejaculation (Fernandez-Guasti et al., 1985, 1986). However, MPOA microinjections of the same two antagonists did not restore copulation in satiated males, suggesting that sexual satiety is governed by mechanisms separate from those that regulate the postejaculatory interval.

Paraventricular Nucleus. Microinjections of the $GABA_A$ agonist muscimol into the PVN inhibited erections elicited by apomorphine, oxytocin, or NMDA (Melis et al., 2000a). The $GABA_B$ agonist baclofen was ineffective.

Spinal Cord. Effects of intrathecally administered drugs. The $GABA_B$ agonist baclofen, intrathecally administered into the lumbosacral (L5-S1) spinal cord, decreased the numbers of touch-based erections and increased latency to the first glans erection (Bitran et al., 1988b). The highest dose completely blocked penile reflexes. However, none of the doses prevented males from copulating to ejaculation. In copula ejaculation shortly before the reflex test decreased latency to the first touch-based erection in saline controls and blocked the inhibitory effects of the two lower doses of baclofen, but not the highest dose. Therefore, different neural mechanisms appear to control copulation and ex copula reflexes. In contrast to the inhibitory effects of baclofen on ex copula reflexes, intrathecal administration of the $GABA_A$ agonist THIP produced only slight inhibitory effects at the highest dose. Therefore, ex copula reflexes are inhibited by spinal $GABA_B$, but not $GABA_A$, receptors.

Chemical changes detected by analysis of cerebrospinal fluid. Following ejaculation, the concentration of GABA in cerebrospinal fluid of male rats increased by more than 1,000%; aspartate and glutamate increased by about 200% (Quershi and Södersten, 1986). Although it is possible that such an increase in a major inhibitory neurotransmitter contributes to the absolute refractory interval, as noted earlier, ex copula reflexes can be elicited soon after an ejaculation. Therefore, any contribution of GABA to the PEI is probably mediated by brain, not spinal cord mechanisms.

3.2.7 Nitric Oxide

NO is a short-lived soluble gas that is produced by the enzyme NOS in the conversion of L-arginine to citrulline. NO acts both peripherally and centrally to promote vasodilation, erection, and other parasympathetic functions. Both neuronal (nNOS) and endothelial (eNOS) isoforms of NOS are present in central structures that regulate male sexual behavior, as well as in the corpus cavernosum of the penis and dorsal penile nerve. The major mediator of NO's effects is soluble guanylyl cyclase, which produces cGMP. The activity of cGMP is terminated by phosphodiesterase 5 (PDE 5).

3.2.7.1 Role of NO in Penile Erection nNOS in parasympathetic nerves produces NO, which directly promotes smooth muscle relaxation; it may also stimulate production of NO by eNOS in endothelial cells. The cascade of nNOS and eNOS activation can amplify the effects of neuronal stimulation. NO diffuses into smooth muscle in the corpus cavernosum and activates guanylyl cyclase, producing cGMP, which in turn activates protein kinase G (PKG) and, to a lesser extent, protein kinase A (PKA). These enzymes phosphorylate various proteins, including phospholamban, which normally inhibits the calcium pump that sequesters calcium into the sarcoplasmic reticulum (Horowitz et al., 1996). However, phosphorylation inhibits this inhibitory effect, resulting in greater sequestration and less intracellular calcium, resulting in muscle relaxation. PKG may also activate ion channels that extrude calcium from the cell, open potassium channels to hyperpolarize smooth muscle, and inhibit the IP3 contractile pathway (Andersson and Wagner, 1995; Miller et al., 1995; Bivalacqua et al., 1998, 2000). Because PDE 5 terminates the activity of cGMP, inhibitors of this enzyme prolong the effects of cGMP, thereby promoting erection.

Testosterone and dihydrotestosterone upregulate both nNOS and eNOS (Lugg et al., 1995; Zvara et al., 1995; Reilly et al., 1997; Marin et al., 1999; Baba et al., 2000). Castration decreased NOS in the penis and epididymis, but actually increased NOS in the seminal vesicles and prostate, an effect that was reversed by exogenous testosterone (Chamness et al., 1995). Furthermore, fertility in male rats was impaired by systemic administration of the NO precursor L-arginine, although sexual behavior was not affected (Ratnasooriya and Dharmasiri, 2001). Therefore, excessive NO in some portions of the reproductive tract may impair certain aspects of ejaculation or sperm motility.

3.2.7.2 Effects of Systemically Administered Drugs Sildenafil (Viagra), vardenafil (Levitra), and tadalafil (Cialis) inhibit PDE 5 and are used to treat erectile dysfunction (Goldstein et al., 1998; Christiansen et al., 2000; Giuliano et al., 2000; Meuleman et al., 2001; Porst et al., 2001). Sildenafil had no effect on erectile function in healthy men but did decrease the time between ejaculation and a subsequent erection (Aversa et al., 2000). In rats, too, sildenafil increased the number of ex copula erections, as well as genital grooming and homosexual mounting (Ferrari et al., 2002). Vardenafil administered to anesthetized rats was more effective than sildenafil in increasing intracavernous pressure (Giuliano et al., 2003).

Systemic administration of NOS inhibitors decreased the numbers of sexually experienced (Hull et al., 1994; Bialy et al., 1996) and inexperienced (Benelli et al., 1995) males that could ejaculate and increased the number of nonintromissive mounts. The numbers of noncontact erections (Melis and Argiolas, 1997) and touch-based erections (Hull et al., 1994) were decreased by NOS inhibitors, but the number of seminal emissions was increased (Hull et al., 1994). This pattern suggests a facilitation of parasympathetic activity and inhibition of sympathetic activity by NO. NOS inhibitors did not affect sexual motivation, measured as either percent choice of a female in an X-maze (Hull et al., 1994) or increased mount latency (Bialy et al., 1996). However, another study reported decreases in precoital activity and numbers of males that mounted (Ratnasooriya et al., 2000). In the latter study, the NOS inhibitor did not affect the numbers of spermatozoa, but did decrease the number of pregnancies and the number of implanted embryos per pregnancy. A possible explanation for the impaired fertility may be that NO must be injected by the sperm into the egg in order to release internal stores of calcium, which then trigger cell division (Kuo et al., 2000). nNOS "knockout" mice had normal erectile function, resulting from increased production of eNOS (Burnett et al., 1996). In fact, the knockout mice were able to ejaculate after fewer mounts and intromissions (Kriegsfeld et al., 1999), again suggesting that NO normally activates parasympathetic, and inhibits sympathetic, function. Therefore, NO may work to prevent "premature ejaculation."

Systemic administration of the precursor L-arginine or of NO donors has produced inconsistent results. L-arginine increased the number of sexually naïve male rats that were able to copulate and enhanced the performance of experienced males (Benelli et al., 1995). L-arginine also increased production of NO in the corpus cavernosum (Escrig et al., 1999). However, sodium nitroprusside, an NO donor, increased intracavernosal pressure in untreated castrates, but not in testosterone-replaced males (Reilly et al., 1997). In addition, L-arginine had only marginal facilitative effects on touch-based erections (Hull et al., 1994) and failed to increase intracavernosal pressure elicited by electrical stimulation in castrates, with or without testosterone replacement (Reilly et al., 1997). Therefore, most gonadally intact males may have sufficient NOS activity for normal copulation and may not benefit from exogenous precursor or NO donors.

3.2.7.3 Role of NO in Specific Brain Areas *Medial Preoptic Area. Effects of drug microinjections or reverse dialysis into the MPOA.* Reverse dialysis of the NO precursor, L-arginine, into the MPOA increased the rate of mounting, and the NOS inhibitor N(G)-monomethyl L-arginine (L-NMMA) decreased mount rate (Sato et al., 1998). L-NMMA microinjections decreased the number of sexually naïve males that copulated, although it did not affect experienced males; it also increased the number of ex copula seminal emissions (Moses and Hull, 1994, 1999). More recently, microinjections of the NOS inhibitor N(G)-nitro-L-arginine methyl ester (L-NAME) into the MPOA prevented copulation in sexually naïve males and decreased intromissions and ejaculations in experienced males (Lagoda et al., 2004). In addition, L-NAME microinjections in naïve males, before each of seven exposures to an inaccessible female, prevented the facilitation produced by such exposures, which was observed in saline-treated males, compared to nonexposed males (Lagoda et al., 2004).

The most common second messenger for NO is cGMP, although ADP-ribosylation and S-nitrosylation of intracellular proteins may mediate some of NO's effects. Reverse dialysis of a membrane-permeable cGMP analog, 8-bromo-cGMP, increased DA release in the MPOA, whereas an inhibitor of soluble guanylyl cyclase, ODQ, decreased DA levels (Sato and Hull, 2006). In addition, ODQ was able to block the increase elicited by an NO donor, sodium nitroprusside, whereas NOS inhibitor, L-NAME, was not able to block the increase produced by 8-bromo-cGMP, indicating that cGMP was "downstream" from NO. Finally, ODQ blocked the copulation-induced increase in DA and inhibited copulation, whereas 8-bromo-cGMP increased DA release and facilitated sexual behavior (Sato and Hull, 2006). In addition, production of cGMP was decreased in castrated male rats, as were NMDA-induced increases in plasma gonadotropin releasing hormone (GnRH) and in cGMP in MPOA slices (Pu et al., 1996). Therefore, the NO->cGMP pathway in the MPOA is important for male sexual behavior and neuroendocrine function.

Activins are peptides that are primarily known for their stimulation of follicle stimulating hormone (FSH) in the anterior pituitary and in ovarian and testicular function (Gaddy-Kurten et al., 1995). However, they are also expressed in the brain. Adult activin receptor type II knockout mice have numerous copulatory deficits, including delayed and slowed copulation and fewer copulatory behaviors (Ma et al., 2005). The behavioral impairment was traced to deficient NOS activity in the MPOA, but not in the rest of the hypothalamus or the cortex. These animals had no defects in olfactory or motor functions and no penile malformations. This study provides additional support for the importance of MPOA NOS for male sexual behavior.

Immunocytochemistry. MPOA NOS is upregulated as a result of sexual experience. There was more nNOS-ir, as well as increased nNOS protein in Western blots from the MPOA of males that had copulated previously, whether or not they also copulated on the day they were euthanized (Dominguez et al., in press). The increased DA release occasioned by increased NO in experienced animals may account for at least some of their greater copulatory efficiency and resistance to impairments by stress, lesions, or castration.

MPOA NOS is also hormonally regulated. nNOS-ir in the MPN of male hamsters (Hadeishi and Wood, 1996) and rats (Du and Hull, 1999; Putnam et al., 2005) was decreased by castration. Testosterone replacement for 2, 5, or 10 days in castrated rats produced progressive increases in NOS-ir and improvement in copulation; greater NOS-ir density was correlated with shorter mount latencies (Sato et al., 2005). NOS-ir was restored by estradiol or testosterone replacement, but not by dihydrotestosterone, which cannot be aromatized to estradiol (Putnam et al., 2005). Estrogen receptor-α knockout mice (ERαKO) had less nNOS-ir in the MPN than did wild-type mice or mice that lacked AR because of the testicular feminization mutation (tfm) (Scordalakes et al., 2002). Therefore, aromatization of testosterone to estradiol, which would then bind to ERα, probably mediates the hormonal upregulation of NOS-ir. nNOS is colocalized with gonadal steroid receptors in mice (Scordalakes et al., 2002), rats (Sato et al., 2005), and hamsters (Hadeishi and Wood, 1996). In contrast to the decreases in NOS-ir in castrates in the studies mentioned here, there is one report of castration-induced increases in NOS-ir and activity and decreases produced by either testosterone or dihydrotesterone (Singh et al., 2000). The reason for this discrepancy is not clear, but may be a function of the more posterior location examined by Singh and colleagues or their use of an antibody that labeled very few cells. In summary, copulation is enhanced and GnRH release is increased by NO in the MPOA. Testosterone, probably via aromatization to estradiol and activation of ERα, increases NOS production in the MPN, although apparently not in more posterior areas.

3.2.7.3 Paraventricular Nucleus *Effects of Drugs Administered into the PVN.* Administration of the NOS inhibitor L-NAME into the PVN decreased the number of noncontact erections and impaired copulation in male rats (Melis et al., 1998). Reverse dialysis of a different NOS inhibitor, L-NMMA into the PVN decreased touch-based erections, and the NO precursor L-arginine increased erections; however, neither drug affected copulation (Sato et al., 1999). These authors pointed out that similar administration of L-NMMA into the MPOA did decrease the rate of mounting (Sato et al., 1998). Therefore, NO antagonists in the PVN have consistently inhibited noncontact and touch-based erections, but manipulations of NO have had inconsistent effects on copulation.

Chemical Changes Detected by Microdialysis. Increases in production of NO, inferred from increases in NO_2 and NO_3, were detected in PVN microdialysate during both copulation and noncontact erections

(Melis et al., 1998; Melis et al., 1999a). These increases were blocked by injections of hemoglobin (an NO scavenger) into the ventricle, but not by injections into the PVN; hemoglobin in the PVN also failed to inhibit noncontact erections (Melis et al., 1998). Therefore, NO may work intracellularly in the PVN, but it appears to have additional effects in other areas that regulate erections. Microinjections of morphine into the PVN inhibited both copulation and the copulation-induced increase in NO (Melis et al., 1999a). Reverse dialysis of L-arginine into the PVN increased both NO production and touch-based erections, and the NOS inhibitor L-NMMA had the opposite effects (Sato et al., 1999). Both noncontact erections and NO production were increased by microinjections of apomorphine or the D_2 agonist LY-171555, but not by the D_1 agonist SKF 38393 (Melis et al., 1996). The D_2 antagonist L-sulpiride, but not the D_1 antagonist SCH-23390, blocked the increase in NO production during exposure to the estrous female. However, the increase in erections was blocked by both antagonists, suggesting that both D_1 and D_2 receptors contribute to erectile function. On the other hand, a later study from the same lab reported no effect of PVN administration of either a D_1 or D_2 antagonist, or of an oxytocin antagonist, on noncontact erections or on the production of NO in the PVN, although both were reduced by an NMDA antagonist (Melis et al., 2000c). In addition, the NMDA microinjections that elicited erections (see earlier) also increased NO in the PVN (Melis et al., 1997).

NO production was prevented by microinjection of omega-conotoxin, a potent inhibitor of N-type voltage-activated calcium channels; omega-conotoxin also prevented apomorphine- and oxytocin-induced erections and (Succu et al., 1998). (Intracellular calcium is needed to activate NOS.) Administration of the NO donors sodium nitroprusside or hydroxylamine overcame the effects of the omega-conotoxin; apparently, the NO donors bypassed the need for activation of N-type channels that were indirectly activated by apomorphine and oxytocin. NMDA's facilitation of NO production and erections also was not blocked by omega-conotoxin, because NMDA stimulates ligand-activated calcium channels, not voltage-gated channels.

3.2.7.4 Spinal Cord Intrathecal administration of an NO donor, a cGMP analog, or the PDE 5 inhibitor sildenafil enhanced the increase in intracavernous pressure produced by electrical stimulation of the MPOA; this enhancement was reduced by the NOS inhibitor L-NAME (Sato et al., 2001). These treatments did not affect blood pressure or responses to cavernous nerve stimulation. Therefore, NO and cGMP in the spinal cord can increase erectile funtion.

3.2.8 Opioids

The three major classes of endogenous opioids are endorphins, enkephalins, and dynorphins (reviewed in Mains and Eipper, 1999). β-endorphin and several other neuroactive peptides are synthesized from proopiomelanocortin (POMC) in the hypothalamus and in the anterior and intermediate lobes of the pituitary. Met- and leu-enkephalin are produced from proenkephalin and are distributed throughout the central and peripheral nervous systems. Dynorphins, made from prodynorphin, include four major classes, dynorphins A and B and neoendorphins α and β, which also have a wide distribution. A more recently discovered peptide, orphanin FQ (also called nociceptin), binds to an "orphan" receptor that is structurally similar to other opioid receptors. Besides the orphanin FQ receptor, there are three major types of opioid receptors, μ (activated primarily by β-endorphin), δ (activated by enkephalins), and κ (activated by dynorphins) (Mains and Eipper, 1999). Neuropeptides, including opioids, are usually colocalized with classical transmitters and are released during high neural activity.

3.2.8.1 Effects of Systemically or Intraventricularly Administered Drugs Endogenous opioids and exogenous opiates affect male sexual behavior in a dose-dependent manner (reviewed in Hull et al., 2002). Opioids and opiates are generally thought to inhibit sexual behavior because of the facilitative effects of antagonists in sexually sluggish males (Gessa et al., 1979), sexually satiated males (Rodriguez-Manzo and Fernandez-Guasti, 1995), or sexually naïve males tested in a novel environment (Pfaus and Wilkins, 1995). The μ antagonist naloxone also facilitated copulatory behaviors in male quail, but did not affect anticipatory behaviors (Riters et al., 1999). However, the opiate antagonists naloxone and naltrexone have also produced inhibitory effects, including lengthened postejaculatory intervals (Sachs et al., 1981; van Furth and van Ree,

1994; Ågmo et al., 1994b) and decreased ability of mild tail pinch to activate copulation in castrated rats with subnormal testosterone replacement (Leyton and Stewart, 1996).

In addition to the reports of facilitation of copulation by opioid antagonists, opioid agonists have been found to inhibit sexual behavior. Either morphine or the κ agonists U-50,488 or bremazocine decreased the number of animals that copulated and inhibited the behavior of those that did copulate (Ågmo and Paredes, 1988; Leyton and Stewart, 1992; Ågmo et al., 1994b). However, the inhibitory effects of opioid antagonists on active copulators noted above suggest that at least a minimal amount of stimulation of opioid receptors can facilitate copulation, but an excess may impair it. Inhibition of genital reflexes may account for some of the inhibitory effects of systemically administered morphine. All doses of morphine inhibited touch-based erections and seminal emissions (Gomez-Marrero et al., 1988). However, touch-based erections were also inhibited by naloxone in those experiments, again suggesting that some opioid activity can facilitate sexual function. Naloxone also inhibited anticipatory level changing in a bilevel apparatus (van Furth and van Ree, 1994; van Furth et al., 1994; van Furth and van Ree, 1996), suggesting that sexual motivation is also facilitated by at least minimal endogenous opioid activity. Facilitation of sexual behavior by opioids may result from disinhibition of the mesolimbic DA system, as there are μ receptors on inhibitory GABAergic neurons in the VTA.

Stimulation of κ receptors also impairs sexual behavior. Dynorphins are the endogenous ligands of κ receptors, and they are upregulated after castration (Molineaux et al., 1986; Almeida et al., 1987). Systemic treatment with the κ agonist U-50,448 decreased the number of male rats that copulated and increased ejaculation latencies in the 12% of animals that remained competent (Leyton and Stewart, 1992). Selective κ receptor blockade by pretreatment with nor-binaltorphimine placed into the VTA, NAc, or MPOA, all partially restored decrements in copulation and female-directed investigative behavior induced by systemic U-50,448. Although the mechanism behind the MPOA effect is unclear, it seems likely that the behavioral impairments that follow κ receptor activation in VTA may be due to inhibition of DA neurons there (Margolis et al., 2003, 2005), particularly a subpopulation having projections to the prefrontal cortex (Margolis et al., 2006). Similarly, inasmuch as κ receptor signaling in the NAc is aversive (Carlezon et al., 2006), this would seem to offset natural reinforcers like sex.

Expression of the final class of endogenous opioid, the δ receptor ligand enkephalin, is downregulated in the ventromedial nucleus of the hypothalamus following castration and is restored by estradiol treatment (Hammer et al., 1993). An agonist at δ receptors had modest inhibitory effects on copulation (Ågmo and Paredes, 1988). Increasing endogenous enkephalins by intraventricular injections of an enkephalinase inhibitor decreased the time and number of intromissions preceding ejaculation, and increased mount and intromission latencies (Ågmo et al., 1994a).

In summary, low levels of opioid activity may increase genital reflexes and sexual motivation; however, excess opioid activity can impair sexual behavior, with some measures being more sensitive than others are to such interference. Because testosterone and estradiol downregulate opioids, it is tempting to speculate that opioid peptides may be one mediator of the impairments seen following castration.

3.2.8.2 Role of Opioids in Specific Brain Areas *Medial Preoptic Area. Effects of drug microinjections into the MPOA.* As with systemic manipulations, studies administering opiates or opioids into the MPOA support the proposal that low doses of agonists facilitate, and high doses inhibit copulation. The lowest doses of the μ agonist morphine and of the κ agonist dynorphin (1–13) decreased ejaculation latency and the numbers of intromissions preceding ejaculation, whereas animals treated with the highest dose of morphine failed to resume copulation after the second ejaculation (Band and Hull, 1990). The selective μ agonist morphiceptin increased mount and intromission latencies but had no effect on copulatory performance, touch-based erections, sexual motivation, or locomotion (Matuszewich et al., 1995). However, microinjection of β-endorphin into the MPOA delayed copulation and inhibited its performance without impairing grooming or drinking of sweetened solutions (Hughes et al., 1987; van Furth et al., 1995). In support of the importance of low levels of endogenous opioids, microinjections of naloxone into the MPOA prevented induction of sexual reinforcement (Ågmo and Gomez, 1993).

Immunocytochemistry. Endocytosis of μ opioid receptors in the MPOA was observed within 30 min after copulation; endocytosis was still apparent 6 h later (Coolen et al., 2004). Receptor internalization was prevented by the opioid antagonist naloxone. Similar endocytosis was elicited by microinjections of a μ agonist, and copulation increased the number of Fos-ir in neurons that contained μ receptors. However, pretreatment with naloxone did not prevent the Fos response in μ receptor-containing neurons, suggesting that the Fos response was not a result of μ activation during mating.

Paraventricular Nucleus. Unilateral administration of morphine into the PVN before presentation of an inaccessible estrous female prevented noncontact erections as well as the increases in production of NO observed with control injections (Melis et al., 1999a).

Ventral Tegmental Area. GABAergic neurons in the VTA contain μ receptors; stimulation of these μ receptors disinhibits DA neurons of the mesocorticolimbic tract (Chiodo, 1988; Johnson and North, 1992). The opioid antagonist naloxone, microinjected into the VTA, decreased anticipatory level changing in a bilevel apparatus but did not affect sexual performance; β-endorphin microinjections had no effect (van Furth and van Ree, 1996). However, microinjections of low doses of morphine into the VTA, but not the substantia nigra, increased female-directed behavior and the proportions of animals displaying mounts, intromissions, and ejaculations (Mitchell and Stewart, 1990). Therefore, stimulation of opioid receptors in the VTA may disinhibit DA neurons there and increase behavioral activation.

Spinal Cord. Effects of Intrathecally Administered Drugs. Intrathecal infusions of naloxone decreased the number of intromissions preceding ejaculation, and intrathecal morphine had the opposite effect (Wiesenfeld-Hallin and Sodersten, 1984). These treatments did not affect other measures of copulation. The authors suggested that opioids influence the strength of the sensory signal from each intromission.

3.2.8.3 Effects of Sexual Experience or Hormones on Opioid Peptides or Precursor Leu- and met-enkephalins and opioid octapeptide were increased in tissue from the cortex, hypothalamus, and midbrain 24 h after copulation either to one ejaculation or to exhaustion, with no significant difference between the two groups (Rodriguez-Manzo et al., 2002a). Met-enkephalin and opioid octapeptide remained elevated in hypothalamic tissue even 48 h after ejaculation. In addition, sexually inactive males had higher enkephalin content in the hypothalamus than did sexually active males, consistent with an inhibitory effect of relatively high opioid levels.

Production of POMC, the precursor of β-endorphin and other neuroactive peptides, is under hormonal control in the mediobasal hypothalamus, which contains the arcuate nucleus, the major population of POMC-producing neurons in the forebrain. There is first a decrease in β-endorphin, measured with radioimmunoassy immediately following castration (Wardlaw and Blum, 1990), followed by a prolonged increase (Wardlaw, 1986). Either 5-α-dihydrotestosterone or estradiol was able to abolish postcastration increases in β-endorphin. These effects were replicated at the level of POMC message, with 4-week castrates showing elevated POMC mRNAs compared to testosterone replaced animals (Blum et al., 1989). Therefore, there is negative regulation of POMC by both estradiol and dihydrotestosterone. The mechanism underlying this effect must be indirect, as only a small portion of POMC neurons expresses ER or AR (Fodor and Delemarre-van de Waal, 2001).

3.2.9 α-Melanocyte Stimulating Hormone and Related Peptides

α-Melanocyte stimulating hormone (α-MSH) is derived from the precursor POMC and is secreted into the systemic circulation from the intermediate lobe of the pituitary. It and related melanocortin peptides are also expressed in numerous CNS and peripheral structures. In addition to α-MSH, these peptides include adrenocorticotropic hormone (ACTH) and β-lipotropin, which are released from the anterior pituitary; the opioid β-endorphin, discussed earlier; and other variants formed by posttranslational processing. They bind to G-protein-coupled receptors (MC-1 to MC-5), all of which activate adenylyl cyclase (reviewed in Giuliano, 2004).

3.2.9.1 Effects of Systemically or Intraventricularly Administered Drugs Intraventricularly administered α-MSH and ACTH were reported as early as the 1960s to increase erections in dogs, cats, rats, mice, and rabbits (reviewed in Argiolas, 1999; Giuliano, 2004). They have also decreased the numbers of mounts and intromissions preceding ejaculation and ejaculation latency. Injection of either α-MSH or ACTH into the third ventricle elicited penile erections and stretching and yawning in rats; the stretching and yawning were blocked by an MC-4 receptor antagonist, but penile erection was not (Argiolas et al., 2000), suggesting that a different MC receptor mediated facilitation of erection (but see later). However, erections evoked by ACTH were inhibited by either intraventricular or systemic administration of the NOS anatagonist L-NAME, suggesting that the facilitative effects of intraventricular ACTH were mediated by increased central production of NO (Poggioli et al., 1995). Microinjections of ACTH (1–24) into the periventricular hypothalamus, including the PVN, elicited drug-induced erections (Argiolas et al., 2000). Similar injections into the POA, caudate nucleus, or hippocampus were ineffective.

There is currently a resurgence of interest in these peptides, and several synthetic peptides that are selective for subtypes of melanocortin (MC) receptors have been produced. Two of these synthetic peptides have increased erections and sexual interest in men—Melanotan-II, also called MT-II (Dorr et al., 1996; Wessels et al., 1998, 2000; Van der Ploeg et al., 2002) and PT-141 (Molinoff et al., 2003)—with nausea, yawning, stretching, and decreased appetite reported as side effects of MT-II. MT-II is a selective melanocortin-4 (MC-4) agonist that has also increased erections in rats; intrathecal administration was most effective, but intraventricular and intravenous injections also yielded positive effects (Wessels et al., 2003). The effects of MT-II were blocked by the MC-3/MC-4 receptor antagonist SHU 9119, confirming its receptor specificity. Direct intracavernosal injections were ineffective, suggesting that MC receptors in the spinal cord and brain mediated the proerectile activity. Intravenous injections of MT-II also increased cavernosal pressure in rabbits, and, as with rats, isolated cavernosal tissue was not responsive (Vemulapalli et al., 2001). The effects of MT-II were abolished by both SHU 9119 and the NOS inhibitor L-NAME, indicating that the proerectile effects of central MC-3/MC-4 receptors are mediated by release of NO. L-NAME also blocked the proerectile effects of intraventricular ACTH (Poggioli et al., 1995) and α-MSH (Mizusawa et al., 2002) in rats. In addition, MT-II facilitated copulation and augmented erections elicited by stimulation of the cavernous nerve in wild-type, but not MC-4 receptor knockout mice (Van der Ploeg et al., 2002). MC-4 receptors were expressed on nerve fibers and mechanoreceptors in both rat and human penis, but not in cavernosal smooth muscle; they were also expressd in rat spinal cord, brain stem, hypothalamus, and pelvic gangion (Van der Ploeg et al., 2002). Intranasal administration of the MC-4 agonist PT-141 also elicited erections in rats and in men with erectile dysfunction (Molinoff et al., 2003). Finally, a different, highly selective, MC-4 receptor agonist, THIQ, administered either systemically or intraventricularly, increased intracavernous pressure and elicited reflexive erections in rats; these effects were blocked by a synthetic MC-4-preferring antagonist, MBP10, as well as by the endogenous nonselective antagonist, agouti-related protein (Martin et al., 2002).

3.2.9.2 Effects of Hormones on α-MSH Measurements of hypothalamic α-MSH by radioimmunoassay show a testosterone-reversible decrease in the peptide 3 weeks following castration (Scimonelli and Celis, 1987). This is in contrast to the castration-induced increase in the precursor, POMC, as well as β-endorphin, as noted in the preceding section. In this way, castration reduced levels of a prosexual neuropeptide, even while increasing expression of its encoding gene. The simplest explanation that can reconcile these findings is that posttranslational processing of POMC may be differentially regulated with regard to its neuroactive cleavage products: In the absence of gonadal steroids, β-endorphin synthesis proceeds unimpeded and in a manner that is positively correlated with the abundance of POMC transcripts. Synthesis of α-MSH, on the other hand, appears to be inhibited and results in a much lower peptide-to-transcript ratio. This creates a circumstance in which castrates experience both an excess of the sexually inhibitory β-endorphin, and an insufficiency in the prosexual α-MSH.

In summary, α-MSH and ACTH, as well as the synthetic MC-3 and MC-4 receptor agonists, can promote erection by acting synergistically at nerve endings in the penis, the lower spinal cord, and several brain areas. Some of these effects are mediated by NO.

3.2.10 Oxytocin

The nonapeptide oxytocin is synthesized in the PVN and in the supraoptic nucleus. Magnocellular PVN neurons descend to the posterior pituitary, from which they release oxytocin into the circulation.

Parvocellular PVN neurons release oxytocin into numerous CNS structures. Circulating oxytocin stimulates smooth muscle and thereby facilitates seminal emission and, in humans, contributes to orgasm. In female mammals, it stimulates parturition and milk letdown. Plasma oxytocin was increased by copulation in naïve, but not sexually experienced male rats (Hillegaart et al., 1998). In the naïve males, plasma oxytocin was correlated with intensity of copulation.

3.2.10.1 Effects of Systemically or Intraventricularly Administered Drugs Systemically administered oxytocin reduced ejaculation latency and postejaculatory interval in male rats (Arletti et al., 1985). This facilitative effect extends to older (~20 months) rats that were "sexually sluggish" (Arletti et al., 1990). In these animals, oxytocin treatment reduced mount, intromission, and ejaculation latencies, as well as postejaculatory intervals. In addition, oxytocin increased the number of animals able to initiate additional bouts of copulation after the first ejaculation. Systemically administered oxytocin also restored copulation in male rats whose copulatory behavior had been impaired by chronic fluoxetine (Cantor et al., 1999). Facilitative effects of low doses of oxytocin have also been observed in dominant squirrel monkeys (Winslow and Insel, 1991). Conversely, intraventricular infusions of a peptide antagonist at oxytocin receptors dose-dependently impaired or abolished copulation in sexually experienced rats (Argiolas et al., 1988a). However, high doses of oxytocin have inhibited copulation in prairie voles (Mahalati et al., 1991) and rats (Stoneham et al., 1985; reviewed in Witt, 1995).

Intraventricular injections of oxytocin also increased spontaneous erections, and this increase was blocked by an oxytocin antagonist or an anticholinergic drug (atropine), but not by a DA antagonist (Argiolas et al., 1986, 1988b). The oxytocin antagonist also blocked erections elicited by the DA agonist apomorphine (Argiolas et al., 1988b), suggesting that DA is "upstream" of oxytocin in the control of drug-induced erections; ACh may, in turn, be "downstream" from oxytocin. In addition, an oxytocin antagonist, administered intraventricularly, inhibited noncontact erections, suggesting a role of oxytocin in the control of erections elicited by normal physiological stimuli, as well as drug-induced erections (Melis et al., 1999a). Finally, oxytocin injected systemically in 7-day-old neonatal rats elicited the rhythmic ejaculatory pattern characteristic of adult male rats (Carro-Juarez and Rodriguez-Manzo, 2005). The ejaculatory pattern matured over a period from postnatal day 2 until it was fully mature on day 28.

Presumptive evidence for the role of oxytocin in male sex behavior also comes from experiments in which levels of oxytocin in the CSF were seen to increase markedly following copulation (Hughes et al., 1987). Although the change in oxytocin level was significant by 5 min after ejaculation, it did not peak for another 15 min. This is likely due to the poor spatial and temporal resolution of the sampling method in which CSF was tapped rather than dialyzed.

3.2.10.2 Role of Oxytocin in Specific Brain Areas *Paraventricular Nucleus. Effects of drug microinjections into the PVN.* Microinjection of oxytocin into the PVN elicited drug-induced erections (Melis et al., 1989). However, an oxytocin antagonist administered into the PVN failed to inhibit those erections, although icv administration of the antagonist was effective (Melis et al., 1989). This suggests that axons from the PVN, terminating perhaps in the hippocampus (Melis et al., 1992), facilitate noncontact erections.

Immunocytochemistry. Fos-ir in the parvocellular PVN was increased by both intromission and ejaculation, with ejaculation inducing 2.5 times as many Fos-ir neurons (Witt and Insel, 1994). Ejaculation also elicited more Fos-ir in the magnocellular PVN than did intromission, although there were fewer labeled cells than in the parvocellular region. In half of the neurons, Fos was colocalized with oxytocin, even in noncopulating males. In fact, copulation did not alter the percentage of double-labeled cells, except in the most caudal portion of the PVN, where one-third of the lateral parvocellular neurons were double labeled after ejaculation, but no cells were double labeled in noncopulating animals. On the other hand, use of an antibody that is more selective for Fos, as opposed to Fos plus Fos-related antigens (Fras), revealed no double labeling of Fos and oxytocin following ejaculation in male rats (L. Coolen, personal communication). In gerbils, PVN Fos expression was not increased by exposure to an arena previously associated with copulation or by copulation itself (Heeb and Yahr, 1996). However, a more recent study found that copulation did increase double labeling for oxytocin and Fos-ir (Nishitani et al., 2004). Fos-ir was also increased in oxytocin-containing cells in sexually experienced males, but not in naïve males, following exposure to an anesthetized female. Variability in Fos-ir in nonoxytocin cells may have contributed to a lack of statistical significance in the earlier studies that found no Fos increases.

Spinal Cord. Effects of intrathecally administered drugs. Lumbosacral, but not thoracolumbar, administration of oxytocin dose-dependently increased intracavernous pressure (Giuliano et al., 2001b). An oxytocin antagonist or pelvic nerve section blocked oxytocin's proerectile effects. Intrathecal administration of an NO inhibitor blocked the proerectile effects of intrathecally administered oxytocin, suggesting that the effects of oxytocin in the spinal cord are mediated by NO interneurons. Vasopressin, a nonapeptide that differs from oxytocin by only one amino acid, was ineffective, as was systemically administered oxytocin. Oxytocin's facilitative effects were not inhibited by blockade of striated muscle activation, suggesting that oxytocin's effects were mediated by the parasympathetic nervous system, rather than by striated penile muscles. Oxytocin release in the spinal cord also promotes seminal emission in rats. Excitotoxic lesions of the parvocellular PVN significantly reduced oxytocin content in the lower lumbar spinal cord and decreased the weight of seminal plugs following ejaculation, but did not affect measures of copulatory behavior (Ackerman et al., 1997). Together, these data provide compelling evidence for a role of oxytocin-containing axons from the PVN to the lumbosacral cord in both erection and seminal emission.

Expression of mRNA or immunocytochemistry for oxytocin. The PVN sends oxytocinergic axons to the thoracolumbar spinal cord, where they form dense clusters around sympathetic preganglionic neurons in the intermediolateral cell column (Swanson and McKellar, 1979; Millan et al., 1984). Oxytocinergic axons also make synaptic contacts on preganglionic neurons in the sacral parasympathetic nucleus (Swanson and McKellar, 1979; Schoenen et al., 1985; Tang et al., 1998) as well as the dorsal horn and the dorsal gray commissure (Véronneau-Longueville et al., 1999). These contacts provide a mechanism for oxytocin to regulate both parasympathetic facilitation of erections and sympathetic regulation of seminal emission and ejaculation.

Sexually impotent male rats had less expression of oxytocin mRNA in the PVN, and greater expression of mRNA for proenkephalin and prodynorphin, than did sexually competent males (Arletti et al., 1997). This finding fits well with the observation that morphine in the PVN inhibited NO production and copulation (Melis et al., 1999a). In addition, NOS is colocalized with oxytocin in the PVN (Yamada et al., 1996). Activation of these neurons facilitates ex copula erections and may contribute to copulation. The oxytocin is released in the hippocampus, where it promotes erections (Melis et al., 1992), and in the spinal cord, where it promotes both erection and seminal emission (Ackerman et al., 1997; Giuliano et al., 2001b).

3.2.11 Gonadotropin-Releasing Hormone

GnRH is a decapeptide that is produced in the mediobasal hypothalamus, lamina terminalis, and MPOA (Barry et al., 1985). It is released in the median eminence and carried via the hypophyseal portal system to the anterior pituitary, where it stimulates gonadotropin secretion. GnRH also acts as a neuromodulator in the MPOA and amygdala, as well as several other brain areas (Barry et al., 1985; Kostarczyk, 1986). Female pheromones elicit GnRH release in mice and hamsters (Meredith and Fernandez-Fewell, 1994; reviewed in Meredith, 1998; Westberry and Meredith, 2003).

3.2.11.1 Effects of Systemically or Intraventricularly Administered Drugs Experiments manipulating GnRH have provided inconsistent results. One study, by Moss and his colleagues, found that GnRH decreased the intromission and ejaculation latencies in castrated male rats that were maintained on a low dose of testosterone; GnRH was ineffective in gonadally intact males (Moss et al., 1975). However, Myers and Baum found that only intact males, but not testosterone-replaced castrates, were facilitated (Myers and Baum, 1980). In fact, the postejaculatory interval in testosterone-treated castrates was actually lengthened by GnRH. Phoenix and Chambers also found lengthened postejaculatory intervals in old rhesus monkeys after GnRH administration (Phoenix and Chambers, 1990). Ryan and Frankel found no effects of GnRH in male rats (Ryan and Frankel, 1978), but Dorsa and Smith reported increased mount rates in male rats with anesthetized genitalia (Dorsa and Smith, 1980), and Boyd and Moore observed decreased intromission and ejaculation latencies in gray-tailed voles (Boyd and Moore, 1985). Intraventricular administration of GnRH promoted copulation in sexually naïve male hamsters whose vomeronasal organs had been removed (VNX) (Fernandez-Fewell and Meredith, 1995). VNX male hamsters are usually able to copulate only if they have had previous sexual experience; therefore, GnRH administration compensated for the lack of sexual experience. Intraventricular GnRH also increased female pheromone-induced Fos-ir in the MPOA of intact male hamsters and in sexually experienced VNX hamsters (Westberry and Meredith, 2003). GnRH increased pheromone-induced Fos-ir in the medial amygdala, but not the MPOA, of sexually naïve VNX males. GnRH improved sexual function in hypogonadal men with decreased LH release (Mortimer et al., 1974). However, it led to little (Benkert, 1975; Ehrensing et al., 1981) or no (Perras et al., 2001) improvement in men with erectile dysfunction.

Chronic administration of a synthetic GnRH agonist ([6-D-(2-napthyl)-alanine]GnRH) led to decreased testosterone levels and a disruption of copulation in gonadally intact male rats (Dorsa et al., 1981). This inhibitory effect probably occurred because continuous high doses of GnRH inhibit LH release and gonadal function; normally endogenous GnRH is released in a pulsatile fashion (Knobil, 1980; Crowley et al., 1985). Therefore, GnRH agonists have been suggested as potential antifertility drugs (Brodie and Crowley, 1984). However, they can also inhibit sexual behavior, probably because of the decrease in testosterone.

3.2.12 Galanin and Galanin-Like Peptide

Galanin is known primarily for its role in stimulating feeding. However, it has also produced apparently contradictory effects on sexual behavior.

3.2.12.1 Effects of Intraventricularly Administered Drugs Intraventricular microinjections of galanin decreased the numbers of mounts and intromissions in male rats (Poggioli et al., 1992). Those behaviors were increased by a galanin antagonist (galantide) (Benelli et al., 1994). However, intraventricular injections of a galanin antagonist in male ferrets decreased the time they spent near a receptive female, although it did not influence copulation (Park and Baum, 1999). It is not clear whether differences in site, species, dosage, or other factors account for these contradictory findings. Galanin-like peptide (GALP) is derived from a different gene but shares partial sequence homology with galanin. Intraventricular administration of GALP facilitated copulation in male rats, whereas similar injections of galanin inhibited the same measures that GALP had facilitated (Fraley et al., 2004). GALP stimulates GnRH secretion; however, even castrated males showed facilitative effects of GALP, suggesting that the behavioral effects were not mediated by increases in testosterone secretion.

1.3.2.12.2 Role of Galanin in Specific Brain Areas *Medial Preoptic Area.* In contrast to the apparently inhibitory effects of intraventricular galanin, microinjections of galanin into the MPOA of castrated, testosterone-replaced rats facilitated their mating behavior (Bloch et al., 1993). In addition, galanin neurons in the MPOA of male rats were activated selectively by ejaculation (Bakker et al., 2002).

Subparafascicular Nucleus of the Thalamus. The parvocellular portion of the subparafascicular nucleus of the thalamus (SPFp) comprises medial and lateral subdivisions. The medial subdivision contains numerous galanin-ir fibers (Veening and Coolen, 1998). Mating induces Fos in the medial subdivision, and galanin-ir fibers surround the Fos-positive nuclei. A major input to the medial SPFp is from laminae 7 and 10 of the lumbosacral spinal cord (Gréco et al., 1999; Truitt et al., 2003; Coolen et al., 2003), and the SPFp has bidirectional connections with forebrain areas that control sexual behavior (Coolen et al., 2003). Many of these connections also contained AR and mating-induced Fos (Gréco et al., 1998). Thus, galanin may be used to relay ejaculation-related somatosensory information to the SPFp, which relays it to forebrain areas related to ejaculation.

Spinal Cord. Neurons that project to the SPFp originate in segments L3 and L4 of the lumbar spinal cord (Wells and Coolen, 2001; Truitt and Coolen, 2002) and contain both galanin and CCK (Ju et al., 1987).

They are believed to relay somatosensory information from the genitals to the brain. Galanin- and CCK-containing neurons showed increased Fos-ir only in males that ejaculated, and not in those that only mounted or intromitted (except for a few Fos-ir neurons in one male that only intromitted) (Truitt and Coolen, 2002). In addition, ejaculatory ability of male rats was severely impaired by lesions of galanin-containing neurons in L3 and L4 (Truitt and Coolen, 2002). Therefore, these galanin-containing neurons are critical for eliciting ejaculation, not just relaying ejaculation-specific sensory information to the brain.

3.2.13 Orexin/Hypocretin

Orexin/hypocretin neurons are located in the perifornical lateral hypothalamus. This peptide is better known for its facilitation of feeding and prolonging wakefulness. However, recent studies suggest that it is important for several other functions, including male sexual behavior.

3.2.13.1 Role of Orexin/Hypocretin in Specific Brain Areas *Medial Preoptic Area.* Microinjection of orexin/hypocretin into the MPOA of male rats facilitated copulation, decreasing mount, intromission, and ejaculation latencies, while also increasing frequencies for these copulatory measures (Gulia et al., 2003).

Lateral Hypothalamus. Either exposure to a receptive female or copulation increased the number of orexin/hypocretin neurons that were Fos-ir, compared to exposure to another male or placement into an empty cage, respectively (Muschamp et al., 2004). Orexin/hypocretin is hormonally regulated. The number of orexin/hypocretin immunopositive cells near the perifornical lateral hypothalamus declined markedly at 28, but not 7 or 14 days following castration (Muschamp et al., 2004). Findings for the 28-day group were replicated using Western immunoblot, in which a $\sim$20% decrease in hypothalamic orexin/hypocretin was seen, compared to sham-treated controls. It is unknown, however, which gonadal steroid mediates this effect. Whereas a small population ($<$5%) of orexin/hypocretin neurons in the dorsomedial hypothalamus coexpress ERα, the majority of cells express neither this receptor, nor nuclear AR, suggesting indirect hormonal regulation of orexin/hypocretin transcription (Muschamp, unpublished data).

Ventral Tegmental Area. Although orexin/hypocretin is likely to enhance sexual performance by its actions at the MPOA, its effects on sexual motivation may be exerted at the VTA, as orexin/hypocretin seems to influence natural reward (Harris et al., 2005; Thorpe et al., 2005). Intra-VTA orexin/hypocretin increased NAc DA efflux (Narita et al., 2006), and orexin/hypocretin peptides increased VTA DA neuron activity in vitro (Korotkova et al., 2003) and in vivo (Muschamp, unpublished data).

3.2.14 Other Peptides

3.2.14.1 Vasoactive Intestinal Peptide Vasoactive Intestinal Peptide (VIP) stimulates adenylyl cyclase in the corpus cavernosum, which contributes to smooth muscle relaxation and penile erection (reviewed in Aoki et al., 1994; Bivalacqua et al., 2000). VIP and ACh antagonists reduced erection elicited by pelvic nerve stimulation; however, administration of VIP and ACh together did not elicit a full erectile response (Suh et al., 1995). Therefore, although both VIP and ACh contribute to erection, they are not the principal mediators of that effect. VIP may also facilitate copulation. Systemically administered VIP reduced intromission latencies and interintromission intervals in castrated male rats that were maintained on suboptimal testosterone replacement; these effects were blocked by coadministration of a VIP antagonist (Gozes et al., 1989).

3.2.14.2 Prostaglandin E_2 Injection of prostaglandin E_2 (PGE_2) into the fourth ventricle of male rats decreased the postejaculatory interval (reviewed in Moltz, 1990; Blumberg, 1991). This treatment also increased basal forebrain temperature, measured in the MPOA. Similarly, microinjection of PGE_2 into the MPOA of castrates with subthreshold testosterone replacement activated copulation, with 40% of the treated males ejaculating within 30 min (Clemens and Gladue, 1977). MPOA temperature normally increases during copulation and decreases following ejaculation (Blumberg and Moltz, 1987). It is not clear whether PGE_2 facilitates copulation directly, or by increasing brain temperature. Manipulations of PGE_2 during early development also influence masculinization of rats (Amateau and McCarthy, 2004; Todd et al., 2005).

3.2.14.3 Neuropeptide Y Neuropeptide Y (NPY) is distributed widely throughout the brain and is often colocalized with NE (Chronwall et al., 1985). Intraventricular administration of NPY stimulated feeding in male rats but inhibited copulation (Clark et al., 1985a; Poggioli et al., 1990). Conversely, intraventricular administration of an NPY antagonist facilitated copulation (Poggioli et al., 1994). However, systemic administration of NPY did not affect reflexive erections (Clark et al., 1985a).

3.2.14.4 Prolactin Numerous studies in the 1980s reported inhibitory effects of chronic prolactinemia, resulting from implantation of excess pituitary glands or prolactin-secreting tumors, from drugs that stimulate prolactin release, or by injection of prolactin itself (reviewed in Hull et al., 2002). Increased mount and ejaculation latencies and general slowing of copulation were the most commonly reported effects. However, not all studies found inhibitory effects of prolactin, and the mechanism by which chronic elevations of prolactin may impair sexual behavior is not clear.

3.2.14.5 Corticotropin-Releasing Hormone Corticotropin-Releasing Hormone (CRH) injected into the third ventricle increased latencies to mount, intromit, and ejaculate and increased the numbers of mounts and intromissions preceding ejaculation (Sirinathsinghji, 1987). These effects were attenuated by coadministration of the opioid antagonist naloxone. CRH can also inhibit GnRH release, perhaps by acting through endogenous opioids (Almeida et al., 1988). CRH promotes freezing and flight behaviors, suggesting that it facilitates a constellation of behaviors that are appropriate to stressful situations.

3.2.14.6 Cholecystokinin Cholecystokinin (CCK) has produced mixed effects on male sexual behavior. In one study, systemic administration of sulfated CCK decreased both the number of intromissions preceding ejaculation and ejaculation latency; the CCK antagonist proglumide reversed these effects (Pfaus and Phillips, 1987). However, other studies found that systemic (Linden et al., 1987; Bloch et al., 1988) or intraventricular (Bloch et al., 1988; Dornan and Malsbury, 1989) administration of CCK was without effect. However, the two main subtypes of CCK receptor (Altar, 1989) may have opposite effects (Dauge et al., 1989; Minabe et al., 1991). Therefore, lack of significant effects of CCK may result from opposing excitatory and inhibitory effects.

Microinjection of a cholecystokinin A (CCK_A) antagonist into the posteromedian portion of the NAc attenuated the facilitation that resulted from electrical stimulation of the VTA (Markowski and Hull, 1995). A similar result was observed with either a CCK_A or CCK_B antagonist microinjected into the anterolateral NAc (Markowski and Hull, 1995). In the posteromedian NAc, CCK is located in dopaminergic terminals, but it is in nondopaminergic terminals in the anterolateral portion (Hokfelt et al., 1980). Therefore, CCK in the NAc may potentiate the facilitative effects of mesocorticolimbic DA activity.

3.2.14.7 Angiotensin II Microinjections of angiotensin II into the third ventricle increased intromission latencies, numbers of intromissions preceding ejaculation, and postejaculatory intervals (Clark, 1989). The dose that produced the greatest inhibition of sexual behavior also increased drinking during the test. A more recent study also reported inhibitory effects of angiotensin II and those effects were blocked by an angiotensin I receptor antagonist administered in the drinking water (Keaton and Clark, 1998). However, it is not clear if the inhibition of sexual behavior was a direct effect or a result of increased thirst.

3.2.14.8 Hexarelin Analog Peptides Microinjections of hexarelin analog peptides into the PVN elicited erections, probably by stimulating receptors on oxytocin-containing cell bodies, which would increase Ca^{2+} influx and activate NOS (Melis et al., 2000b, 2001; reviewed in Argiolas and Melis, 2004).

4 Summary

Steroid hormones orchestrate the production and release of neurotransmitters in key CNS sites that control sexual motivation and copulatory performance. Dopamine, glutamate, and, in some cases, norepinephrine are the major facilitative neurotransmitters, and serotonin, GABA, and opioid peptides provide primarily inhibitory influence. However, activation of specific receptor subtypes may oppose the effects of other receptor subtypes that are activated by the same neurotransmitter. For example, some dopaminergic and serotonergic receptor subtypes promote parasympathetically mediated erection, and other subtypes facilitate sympathetically mediated seminal emission and ejaculation. The neuropeptides oxytocin, orexin/hypocretin, α-melanocyte-stimulating hormone, gonadotropin-releasing hormone, and galanin-like peptide may facilitate mating, whereas NPY, prolactin, corticotropin-releasing hormone and angiotensin II usually have inhibitory effects. The brain areas that are especially important for the control of male sexual behavior are the MPOA, the medial amygdala, the PVN of the hypothalamus, the mesocorticolimbic dopamine tract, several midbrain and brainstem nuclei, and nuclei in the spinal cord that control the erectile and ejaculatory reflexes.

5 Unanswered Questions and Future Directions

Answers to previous questions inevitably lead to new questions that can guide future research. Some of these questions include: Which neurotransmitters, and which receptors, mediate the behavioral effects in the various sensory, integrative, and motor control sites? Which neuropeptides are coreleased with classic neurotransmitters, in which brain areas, and under which conditions? Do alterations in neurotransmitter release primarily reflect changes in cell firing or influences on presynaptic terminals? Which (combinations of) neurotransmitters mediate long-term changes in sensitivity, as opposed to immediate responses to stimuli? Which ion channels or second messenger systems elicit the immediate and long-term effects of neurotransmitters in brain areas that respond to sexually relevant stimuli? What changes in neurotransmitter function or gene expression mediate the facilitative effects of sexual experience?

References

Ackerman AE, Lange GM, Clemens LG. 1997. Effects of paraventricular lesions on sex behavior and seminal emissions in male rats. Physiol Behav 63: 49-53.

Adachi H, Sato Y, Kato R, Hisasue S, Suzuki K, et al. 2003. Direct evidence of facilitative actions of dopamine in the medial preoptic area on reflexive and noncontact erections in male rats. J Urol 169: 386-389.

Adler NT, Toner JP. 1986. The effects of copulatory behavior on sperm transport and fertility in rats. Ann NY Acad Sci 474: 21-32.

Ågmo A. 2003. Lack of opioid or dopaminergic effects on unconditioned sexual incentive motivation in male rats. Behav Neurosci 117: 55-68.

Ågmo A, Fernandez H. 1989. Dopamine and sexual behavior in the male rat: a reevaluation. J Neural Transm 77: 21-37.

Ågmo A, Gomez M. 1993. Sexual reinforcement is blocked by infusion of naloxone into the medial preoptic area. Behav Neurosci 107: 812-818.

Ågmo A, Paredes R. 1988. Opioids and sexual behavior in the male rat. Pharmacol Biochem Behav 30: 1021-1034.

Ågmo A, Picker Z. 1990. Catecholamines and the initiation of sexual behavior in male rats without sexual experience. Pharmacol Biochem Behav 35: 327-334.

Ågmo A, Villalpando A. 1995. Central nervous stimulants facilitate sexual behavior in male rats with medial prefrontal cortex lesions. Brain Res 696: 187-193.

Ågmo A, Gomez M, Irazabal Y. 1994a. Enkephalinase inhibition facilitates sexual behavior in the male rat but does not produce conditioned place preference. Pharmacol Biochem Behav 47: 771-778.

Ågmo A, Paredes RG, Contreras JL. 1994b. Opioids and sexual behaviors in the male rabbit: The role of central and peripheral opioid receptors. J Neural Transm Gen Sec 97: 211-223.

Ågmo A, Paredes RG, Sierra L, Garces I. 1997. The inhibitory effects on sexual behavior and ambulatory activity of the mixed $GABA_A/GABA_B$ agonist progabide are differentially blocked by GABA receptor antagonists. Psychopharmacology 129: 27-34.

Ahlenius S, Larsson K. 1984a. Apomorphine and haloperidol-induced effects on male rat sexual behavior: no evidence for actions due to stimulation of central dopamine autoreceptors. Pharmacol Biochem Behav 21: 463-466.

Ahlenius S, Larsson K. 1984b. Lisuride, LY-141865, and 8-OH-DPAT facilitate male rat sexual behavior via a non-dopaminergic system. Psychopharmacology 83: 330-334.

Ahlenius S, Larsson K. 1990. Effects of selective D_1 and D_2 antagonists on male rat sexual behavior. Experentia 46: 1026-1028.

Ahlenius S, Larsson K. 1991. Opposite effects of 5-methoxy-N, N-di-methyl-tryptamine and 5-hydroxytryptophan on male rat sexual behavior. Pharmacol Biochem Behav 38: 201-205.

Ahlenius S, Larsson K. 1995. Effects of the dopamine D_3 receptor ligand 7-OH-DPAT on male rat ejaculatory behavior. Pharmacol Biochem Behav 51: 545-547.

Ahlenius S, Larsson K. 1998. Evidence for an involvement of 5-HT_{1B} receptors in the inhibition of male rat ejaculatory behavior produced by 5-HTP. Psychopharmacology 137: 374-382.

Ahlenius S, Larsson K. 1999. Synergistic actions of the 5-HT_{1A} receptor antagonist WAY-100635 and citalopram on male rat ejaculatory behavior. Eur J Pharm 379: 1-6.

Ahlenius S, Eriksson H, Larsson K, Modigh K, Södersten P. 1971. Mating behavior in the male rat treated with p-chlorophenylalanine methyl ester alone and in combination with pargyline. Psychopharmacologia 20: 383-388.

Ahlenius S, Larsson K, Arvidsson LE. 1989. Effects of stereoselective 5-HT_{1A} agonists on male rat sexual behavior. Pharmacol Biochem Behav 33: 691-695.

Ahlenius S, Larsson K, Svensson L, Hjorth S, Carlsson A, et al. 1981. Effects of a new type of 5-HT receptor agonist on male rat sexual behavior. Pharmacol Biochem Behav 15: 785-792.

Albinsson A, Andersson G, Andersson K, Vega-Matuszczyk J, Larsson K. 1996. The effects of lesions in the mesencephalic raphe systems on male rat sexual behavior and locomotor activity. Behav Brain Res 80: 57-63.

Alexander SPH, Peters JA, Editors 2000. Trends in pharmacological sciences receptor and ion channel nomenclature receptor and ion channel nomenclature Supplement 11, Elsevier.

Allard J, Bernabe J, Derdinger F, Alexantre L, McKenna K, et al. 2002. Selegiline enhances erectile activity induced by dopamine injection in the paraventricular nucleus of the hypothalamus in anesthetized rats. Int J Impot Res 14: 518-522.

Almeida OF, Nikolarakis KE, Herz A. 1987. Significance of testosterone in regulating hypothalamic content and in vitro release of beta-endorphin and dynorphin. J Neurochem 49: 742-747.

Almeida OF, Nikolarakis KE, Herz A. 1988. Evidence for the involvement of endogenous opioids in the inhibition of luteinizing hormone by corticotropin-releasing factor. Endocrinology 122: 1034-1041.

Altar CA. 1989. Cholecystokinin receptor subtypes and neuromodulation. Prog Neuropsychopharmacol Biol Psychiatry 13: 381-393.

Amateau SK, McCarthy MM. 2004. Induction of PGE2 by estradiol mediates developmental masculinization of sex behavior. Nature Neurosci 7: 643-650.

Andersson KE, Stief CG. 1997. Neurotransmission and the contraction and relaxation of penile erectile tissues. World J Urol 15: 14-20.

Andersson KE, Wagner G. 1995. Physiology of penile erection. Physiol Rev 75: 191-236.

Aoki H, Matsuzaka J, Yeh KH, Sato F, Fujioka T, et al. 1994. Involvement of vasoactive intestinal peptide (VIP) as a humoral mediator of penile erectile function in the dog. J Androl 15: 174-182.

Arendash GW, Gorski RA. 1983. Effects of discrete lesions of the sexually dimorphic nucleus of the preoptic area or other medial preoptic regions on the sexual behavior of male rats. Brain Res Bull 10: 147-154.

Argiolas A. 1999. Neuropeptides and sexual behaviour. Neurosci Biobehav Rev 23: 1127-1142.

Argiolas A, Melis MR. 1995. Neuromodulation of penile erection: an overview of the role of neurotransmitters and neuropeptides. Prog Neurobiol 47: 235-255.

Argiolas A, Melis MR. 2004. The role of oxytocin and the paraventricular nucleus in the sexual behavior of male mammals. Physiol Behav 83: 309-317.

Argiolas A, Collu M, Gessa GL, Melis MR, Serra G. 1988a. The oxytocin antagonist d(CH2)5Tyr(Me)-Orn8-vasotocin inhibits male copulatory behaviour in rats. Eur J Pharmacol 149: 389-392.

Argiolas A, Melis MR, Gessa GL. 1986. Oxytocin: an extremely potent inducer of penile erection and yawning in male rats. Eur J Pharmacol 130: 265-272.

Argiolas A, Melis MR, Gessa GL. 1988b. Yawning and penile erection: central dopamine-oxytocin-adrenocorticotropin connection. Ann NY Acad Sci 525: 330-337.

Argiolas A, Melis MR, Mauri A, Gessa GL. 1987. Paraventricular nucleus lesion prevents yawning and penile erection induced by apomorphine and oxytocin, but not by ACTH in rats. Brain Res 421: 349-352.

Argiolas A, Melis MR, Murgia S, Schioth HB. 2000. ACTH- and alpha-MSH-induced grooming, stretching, yawning and penile erection in male rats: site of action in the brain and role of melanocortin receptors. Brain Res Bull 51: 425-431.

Arletti R, Bazzani C, Castelli M, Bertolini A. 1985. Oxytocin improves male copulatory performance in rats. Horm Behav 19: 14-20.

Arletti R, Benelli A, Bertolini A. 1990. Sexual behavior of aging male rats is stimulated by oxytocin. Eur J Pharmacol 179: 377-381.

Arletti R, Calza L, Giardino L, Benelli A, Cavazzuti E, et al. 1997. Sexual impotence is associated with a reduced production of oxytocin and with an increased production of opioid peptides in the paraventricular nucleus of male rats. Neurosci Lett 233: 65-68.

Arteaga M, Motte-Lara J, Velazquez-Moctezuma J. 2002. Effects of yohimbine and apomorphine on the male sexual behaviour pattern of the golden hamster (Mesocricetus auratus). Eur Neuropsychopharmacol 12: 39-45.

Aversa A, Mazzilli F, Rossi T, Delfino M, Isidori AM, et al. 2000. Effects of sildenafil (Viagra) administration on seminal parameters and post-ejaculatory refractory time in normal males. Hum Reprod 15: 131-134.

Baba K, Yajima M, Carrier S, Akkus E, Reman J, et al. 2000. Effect testosterone on the number of NADPH diaphorase-stained nerve fibers in the rat corpus cavernosum-dorsal nerve. Urology 56: 533-538.

Bakker J, Woodley SK, Kelliher KR, Baum MJ. 2002. Sexually dimorphic activation of galanin neurones in the ferret's dorsomedial preoptic area/anterior hypothalamus after mating. J Neuroendocrinol 14: 116-125.

Balthazart J, Castagna C, Ball GF. 1997. Differential effects of D_1 and D_2 dopamine-receptor agonists and antagonists on appetitive and consummatory aspects of male sexual behavior in Japanese quail. Physiol Behav 62: 571-580.

Bancila M, Giuliano F, Rampin O, Mailly P, Brisorgueil MJ, et al. 2002. Evidence for a direct projection from the paraventricular nucleus of the hypothalamus to putative serotoninergic neurons of the nucleus paragigantocellularis involved in the control of erection in rats. Eur J Neurosci 16: 1240-1248.

Bancila M, Verge D, Rampin O, Backstrom JR, Sanders-Bush E, et al. 1999. 5-Hydroxytryptamine2C receptors on spinal neurons controlling penile erection in the rat. Neuroscience 92: 1523-1537.

Bancroft J, Janssen E. 2000. The dual control model of male sexual response: a theoretical approach to centrally mediated erectile dysfunction. Neurosci Biobehav Rev 24: 571-579.

Band LC, Hull EM. 1990. Morphine and dynorphin (1–13) microinjected into the medial preoptic area and nucleus accumbens: effects on sexual behavior in male rats. Brain Res 524: 77-84.

Barbeau A. 1969. L-DOPA therapy in Parkinson's disease: a critical review of nine years' experience. J Can Med Assoc 101: 791-800.

Barfield RJ, Geyer LA. 1975. The ultrasonic postejaculatory vocalization and the postejaculatory refractory period of the male rat. J Comp Physiol Psychol 88: 723-734.

Barry J, Hoffman GE, Wray S. 1985. LHRH containing systems. Bjorklund A, Hokfelt T, Handbook of Chemical Neuroanatomy. Elsevier; Amsterdam: pp. 116-215.editors.

Baum MJ. 1972. Precocious mating in male rats following treatment with androgen or estrogen. J Comp Physiol Psychol 78: 356-367.

Bazzett TJ, Eaton RC, Thompson JT, Markowski VP, Lumley LA, et al. 1991. Dose dependent D_2 effects on genital reflexes after MPOA injections of quinelorane and apomorphine. Life Sci 48: 2309-2315.

Bazzett T, Lumley L, Bitran D, Markowski V, Warner R, et al. 1992. Male rat copulation following 6-OHDA lesions of the medial preoptic area: resistance to repeated administration and rapid behavioral recovery. Brain Res 580: 164-170.

Beach FA. 1984. Hormonal modulation of genital reflexes in male and masculinized female dogs. Behav Neurosci 98: 325-332.

Beck J, Bialy M, Kostowski W. 2002. Effects of D(1) receptor agonist SKF 38393 on male rat sexual behavior and post-copulatory departure in the goal compartment–runway paradigm. Physiol Behav 76: 91-97.

Benelli A, Arletti R, Bertolini A, Menozzi B, Basaglia R, et al. 1994. Galantide stimulates sexual behavior in male rats. Eur J Pharmacol 260: 279-282.

Benelli A, Bertolini A, Poggioli R, Cavazzuti E, Calza L, et al. 1995. Nitric oxide is involved in male sexual behavior of rats. Eur J Pharmacol 294: 505-510.

Benkert O. 1975. Effects of hypothalamic releasing hormones in depression and sexual impotence. Neuropsychopharmacology. Boissier JR, Hippius H, Pichot P, editors. Amsterdam: Excerpta Medica; pp. 663-671.

Berendsen HH, Jenck F, Broekkamp CL. 1990. Involvement of $5\text{-}HT_{1C}$-receptors in drug-induced penile erections in rats. Psychopharmacology 101: 57-61.

Bermant G, Clegg MT, Beamer W. 1969. Copulatory behaviour of the ram, Ovis aries: I. A normative study. Anim Behav 17: 700-705.

Bialy M, Beck J. 1993. The influence of vibrissae removal on copulatory behavior in male rats. Acta Neurobiol Exp 53: 415-419.

Bialy M, Beck J, Abramczyk P, Trzebski A, Przybylski J. 1996. Sexual behavior in male rats after nitric oxide synthesis inhibition. Physiol Behav 60: 139-143.

Bialy M, Rydz M, Kacamarek L. 2000. Precontact 50-kHz vocalizations in male rats during acquisition of sexual experience. Behav Neurosci 114: 983-990.

Bitran D, Hull EM. 1987. Pharmacological analysis of male rat sexual behavior. Neurosci Biobehav Rev 11: 365-389.

Bitran D, Hull EM, Holmes GM, Lookingland KJ. 1988a. Regulation of male rat copulatory behavior by preoptic incertohypothalamic dopamine neurons. Brain Res Bull 20: 323-331.

Bitran D, Miller SA, McQuade DB, Leipheimer RE, Sachs BD. 1988b. Inhibition of sexual reflexes by lumbosacral injection of a $GABA_B$ agonist in the male rat. Pharmacol Biochem Behav 31: 657-666.

Bitran D, Thompson JT, Hull EM, Sachs BD. 1989. Quinelorane (LY163502), a D_2 dopamine receptor agonist, facilitates seminal emission but inhibits penile erection in the rat. Pharmacol Biochem Behav 34: 453-458.

Bivalacqua TJ, Champion HC, Hellstrom WJG, Kadowitz PJ. 2000. Pharmacotherapy for erectile dysfunction. Trends Pharmacol Sci 21: 484-489.

Bivalacqua TJ, Rajasekaran M, Champion HC, Wang R, Sikka SC, et al. 1998. The influence of castration on pharmacologically induced penile erection in the cat. J Androl 19: 551-557.

Blackburn JR, Pfaus JG, Phillips AG. 1992. Dopamine functions in appetitive and defensive behaviours. Prog Neurobiol 39: 247-279.

Bloch GJ, Babcock AM, Gorski RA, Micevych PE. 1988. Effects of cholecystokinin on male copulatory behavior and lordosis behavior in male rats. Physiol Behav 43: 351-357.

Bloch GJ, Butler PC, Kohlert JG, Bloch DA. 1993. Microinjection of galanin into the medial preoptic nucleus facilitates copulatory behavior in the male rat. Physiol Behav 54: 615-624.

Blum M, Roberts JL, Wardlaw SL. 1989. Androgen regulation of proopiomelanocortin gene expression and peptide content in the basal hypothalamus. Endocrinology 124: 2283-2288.

Blumberg MS. 1991. Prostaglandin E accelerates sexual behavior in male rats. Physiol Behav 50: 95-99.

Blumberg MS, Moltz H. 1987. Hypothalamic temperature and the 22 kHz vocalization of the male rat. Physiol Behav 40: 637-640.

Boscarino BT, Parfitt DB. 2002. Chronic oral administration of clomipramine decreases sexual behavior in the male Syrian hamster (Mesocricetus auratus). Physiol Behav 75: 361-366.

Bowers MB Jr, Van Woert M, Davis L. 1971. Sexual behavior during L-dopa treatment for Parkinsonism. Am J Psychiatry 127: 1691-1693.

Boyd SK, Moore FL. 1985. Luteinizing hormone-releasing hormone facilitates the display of sexual behavior in male voles (Microtus canicaudus). Horm Behav 19: 252-264.

Brioni JD, Moreland RB, Cowart M, Hsieh GC, Stewart AO, et al. 2004. Activation of dopamine D4 receptors by ABT-724 induces penile erection in rats. Proc Natl Acad Sci USA 101: 6758-6763.

Brodie TD, Crowley WF. 1984. Neuroendocrine control of reproduction and its manipulation with LHRH and its analogs. Trends Neurosci 7: 340-342.

Bronson FH, Desjardins C. 1982. Endocrine responses to sexual arousal in male mice. Endocrinology 111: 1286-1291.

Brudzynski SM, Pniak A. 2002. Social contacts and production of 50-kHz short ultrasonic calls in adult rats. J Comp Psychol 116: 73-82.

Burnett AL, Nelson RJ, Calvin DC, Liu JX, Demas GE, et al. 1996. Nitric oxide-dependent penile erection in mice lacking neuronal nitric oxide synthase. Mol Med 2: 288-296.

Cantor JM, Binik YM, Pfaus JG. 1999. Chronic fluoxetine inhibits sexual behavior in the male rat: reversal with oxytocin. Psychopharmacology 144: 355-362.

Carlezon WA Jr, Beguin C, Di Nieri JA, Baumann MH, Richards MR, et al. 2006. Depressive-like effects of the kappa-opioid receptor agonist salvinorin A on behavior and neurochemistry in rats. J Pharmacol Exp Ther 316: 440-447.

Carro-Juarez M, Rodriguez-Manzo G. 2005. Evidence for the presence and functioning of the spinal generator for ejaculation in the neonatal male rat. Int J Impot Res 17: 270-276.

Castagna C, Ball GF, Balthazart J. 1997. Effects of dopamine agonists on appetitive and consummatory male sexual behavior in Japanese quail. Pharmacol Biochem Behav 58: 403-414.

Ceccatelli S, Grandison L, Scott RE, Pfaff DW, Kow LM. 1996. Estradiol regulation of nitric oxide synthase mRNAs in rat hypothalamus. Neuroendocrinology 64: 357-363.

Chambers KC, Phoenix CH. 1981. Diurnal patterns of testosterone, dihydrotestosterone, estradiol, and cortisol in serum of rhesus males: relationship to sexual behavior in aging males. Horm Behav 15: 416-426.

Chambers KC, Hess DL, Phoenix CH. 1981. Relationship of free and bound testosterone to sexual behavior in old rhesus males. Physiol Behav 27: 615-620.

Chambers KC, Thornton JE, Roselli CE. 1991. Age-related deficits in brain androgen binding and metabolism, testosterone, and sexual behavior of male rats. Neurobiol Aging 12: 123-130.

Chamness SL, Ricker DD, Crone JK, Dembeck CL, Maguire MP, et al. 1995. The effect of androgen on nitric oxide synthase in the male reproductive tract of the rat. Fertil Steril 63: 1101-1107.

Charlier TD, Ball GF, Balthazart J. 2005. Sexual behavior activates the expression of the immediate early genes c-fos and Zenk (egr-1) in catecholaminergic neurons of male Japanese quail. Neuroscience 131: 13-30.

Chiodo LA. 1988. Dopamine-containing neurons in the mammalian central nervous system: electrophysiology and pharmacology. Neurosci Biobehav Rev 12: 49-91.

Chitaley K, Wingard CJ, Webb RC, Branam H, Stopper VS, et al. 2001. Antagonism of Rho-kinase stimulates rat penile erection via a nitric oxide-independent pathway. Nature Med 7: 119-122.

Christ GJ. 1995. The penis as a vascular organ: the importance of corporal smooth muscle tone in the control of erection. Urol Clin North Am 22: 727-745.

Christiansen E, Guirguis WR, Cox D, Osterloh IH. 2000. Long-term efficacy and safety of oral Viagra (sildenafil citrate) in men with erectile dysfunction and the effect of randomized treatment withdrawal. Intl J Impot Res 12: 177-182.

Chronwall BM, Di Maggio DA, Massari VJ, Pickel VM, O'Donohue TL. 1985. The anatomy of neuropeptide Y-containing neurons in rat brain. Neuroscience 15: 1159-1181.

Clark JT. 1989. A possible role for angiotensin II in the regulation of male sexual behavior in rats. Physiol Behav 45: 221-246.

Clark JT. 1991. Suppression of copulatory behavior in male rats following central administration of clonidine. Neuropharmacology 30: 373-382.

Clark JT. 1995. Sexual arousal and performance are modulated by adrenergic-neuropeptide–steroid interactions. The Pharmacology of Sexual Function and Dysfunction: Proceedings of the Esteve Foundation Symposium VI, Son Vida, Mallorca, 9–12 October 1994. Bancroft J. editor. Amsterdam: Excerpta Medica; pp. 55-68.

Clark JT, Smith ER. 1990. Clonidine suppresses copulatory behavior and erectile reflexes in male rats: lack of effect of naloxone pretreatment. Neuroendocrinology 51: 357-364.

Clark JT, Kalra PS, Kalra SP. 1985a. Neuropeptide Y stimulates feeding but inhibits sexual behavior in rats. Endocrinology 117: 2435-2442.

Clark JT, Kalra SP, Kalra PS. 1987. Effects of a selective alpha-1-adrenoceptor agonist, methoxamine, on sexual behavior and penile reflexes. Physiol Behav 40: 747-753.

Clark JT, Smith ER, Davidson JM. 1984. Enhancement of sexual motivation in male rats by yohimbine. Science 225: 847-849.

Clark JT, Smith ER, Davidson JM. 1985b. Evidence for the modulation of sexual behavior by α-adrenoceptors in male rats. Neuroendocrinology 41: 36-43.

Clark TK. 1975. Sexual inhibition is reduced by rostral midbrain lesions in the male rat. Science 190: 169-171.

Clark TK. 1980. Male rat sexual behavior compared after 6-OHDA and electrolytic lesions in the dorsal NA bundle region of the midbrain. Brain Res 202: 429-443.

Claro F, Segovia S, Guilamon A, Del Abril A. 1995. Lesions in the medial posterior region of the BST impair sexual behavior in sexually experienced and inexperienced male rats. Brain Res Bull 36: 1-10.

Clemens LG, Gladue BA. 1977. Effect of prostaglandin E2 on masculine sexual behaviour in the rat. J Endocrinol 75: 383-389.

Clement P, Bernabe J, Kia HK, Alexandre L, Giuliano F. 2006. D_2-like receptors mediate the expulsion phase of ejaculation elicited by 8-hydroxy-2-(di-N-propylamino)tetralin in rats. J Pharmacol Exp Ther 316: 830-834.

Commins D, Yahr P. 1984. Acetylcholinesterase activity in the sexually dimorphic area of the gerbil brain: sex differences and influences of adult gonadal steroids. J Comp Neurol 224: 123-131.

Coolen LM, Fitzgerald ME, Yu L, Lehman MN. 2004. Activation of mu opioid receptors in the medial preoptic area following copulation in male rats. Neuroscience 124: 11-21.

Coolen LM, Veening JG, Wells AB, Shipley MT. 2003. Afferent connections of the parvocellular subparafascicular thalamic nucleus in the rat: evidence for functional subdivisions. J Comp Neurol 463: 132-156.

Cornil CA, Balthazart J, Motte P, Massotte L, Seutin V. 2002. Dopamine activates noradrenergic receptors in the preoptic area. J Neurosci 22: 9320-9330.

Cornil CA, Dejace C, Ball GF, Balthazart J. 2005. Dopamine modulates male sexual behavior in Japanese quail in part via actions on noradrenergic receptors. Behav Brain Res 163: 42-57.

Cornil CA, Seutin V, Balthazart J. 2004. Interaction of dopamine with noradrenergic receptors in the preoptic area and sexual behavior in male quail. Horm Behav 46: 128-129.

Crowley WF Jr, Filicori M, Spratt DI, Santoro NF. 1985. The physiology of gonadotropin-releasing hormone (GnRH) secretion in men and women. Recent Prog Horm Res 41: 473-531.

Dallo J. 1977. Effect of two brain serotonin depletors on the sexual behavior of male rats. Pol J Pharmacol Pharm 29: 247-251.

Damsma G, Pfaus JG, Wenkstern D, Philips AG, Fibiger HC. 1992. Sexual behavior increases dopamine transmission in the nucleus accumbens and striatum of male rats: a comparison with novelty and locomotion. Behav Neurosci 106: 181-191.

Dauge V, Steimes P, Derrien M, Beau N, Roques BP, et al. 1989. CCK8 effects on motivational and emotional states of rats involve CCKA receptors of the postero-median part of the nucleus accumbens. Pharmacol Biochem Behav 34: 157-163.

Dewsbury DA. 1969. Copulatory behavior of rats (Rattus norvegicus) as a function of prior copulatory experience. Animal Behav 17: 217-223.

Dewsbury DA. 1972. Patterns of copulatory behavior in male mammals. Q Rev Biol 47: 1-33.

Dewsbury DA. 1990. Modes of estrus induction as a factor in studies of the reproductive behavior of rodents. Neurosci Biobehav Rev 14: 147-155.

Di Chiara G. 1995. The role of dopamine in drug abuse viewed from the perspective of its role in motivation. Drug Alcohol Depend 38: 95

Domesick VB. 1988. Neuroanatomical organization of dopamine neurons in the ventral tegmental area. Ann NY Acad Sci 537: 10-26.

Dominguez JM, Hull EM. 2001. Stimulation of the medial amygdala enhances medial preoptic dopamine release: implications for male rat sexual behavior. Brain Res 917: 225-229.

Dominguez JM, Balfour ME, Coolen LM. 2003. Copulation-induced activation of NMDA receptor containing neurons in the medial preoptic nucleus. Horm Behav 44: 46

Dominguez JM, Brann JH, Gil M, Hull EM. In press. Sexual experience increases nitric oxide synthase in the medial preoptic area of male rats, Behav Neurosci.

Dominguez JM, Coolen LM. 2004. Activation of NMDA receptors in the medial preoptic nucleus is essential for the expression of Fos and male sexual behavior. Horm Behav 46: 129

Dominguez JM, Gil M, Hull EM. 2006. Preoptic glutamate facilitates male sexual behavior. J Neurosci 26: 1699-1703.

Dominguez J, Riolo JV, Xu Z, Hull EM. 2001. Regulation by the medial amygdala of copulation and medial preoptic dopamine release. J Neurosci 21: 349-355.

Dornan WA, Malsbury CW. 1989. Peptidergic control of male rat sexual behavior: the effects of intracerebral injections of substance P and cholecystokinin. Physiol Behav 46: 547-556.

Dorr RT, Lines R, Livine N, Brooks C, Xiang L, et al. 1996. Evaluation of melanotan-II, a superpotent cyclic melanotropic peptide in a pilot phase-I clinical study. Life Sci 58: 1777-1784.

Dorsa DM, Smith ER. 1980. Facilitation of mounting behavior in male rats by intracranial injections of luteinizing hormone-releasing hormone. Peptides 1: 147-155.

Dorsa DM, Smith ER, Davidson JM. 1981. Endocrine and behavioral effects of continuous exposure of male rats to a potent luteinizing hormone-releasing hormone (LHRH) agonist: evidence for central nervous system actions of LHRH. Endocrinology 109: 729-735.

Du J, Hull EM. 1999. Effects of testosterone on neuronal nitric oxide synthase and tyrosine hydroxylase. Brain Res 836: 90-98.

Du J, Lorrain DS, Hull EM. 1998. Castration decreases extracellular, but increases intracellular, dopamine in medial preoptic area of male rats. Brain Res 782: 11-17.

Dula E, Keating W, Siami PF, Edmonds A, O'Neil J, et al. 2000. Efficacy and safety of fixed-dose and dose-optimization regimens of sublingual apomorphine versus placebo in men with erectile dysfunction. The Apomorphine Study Group. Urology 56: 130-135.

Eaton RC, Markowski VP, Lumley LA, Thompson JT, Moses J, et al. 1991. D_2 receptors in the paraventicular nucleus regulate genital responses and copulation in male rats. Pharm Biochem Behav 39: 177-181.

Ehrensing RH, Kastin AJ, Schally AV. 1981. Behavioral and hormonal effects of prolonged high doses of LHRH in male impotency. Peptides 1: 147-155.

Escrig A, Gonzalez-Mora JL, Mas M. 1999. Nitric oxide release in penile corpora cavernosa in a rat model of erection. J Physiol 516: 261-269.

Everitt BJ. 1990. Sexual motivation: a neural and behavioral analysis of the mechanisms underlying appetitive and copulatory responses of male rats. Neurosci Biobehav Rev 14: 217-232.

Fallon JH. 1986. Topographic organization of ascending dopaminergic projections. Ann NY Acad Sci 537: 1-9.

Fernandez-Fewell GD, Meredith M. 1995. Facilitation of mating behavior in male hamsters by LHRH and AcLHRH5–10: interaction with the vomeronasal system. Physiol Behav 57: 213-221.

Fernandez-Guasti A, Escalante AL. 1991. Role of presynaptic serotonergic receptors on the mechanism of action of 5-HT_{1A} and 5-HT_{1B} agonists on masculine sexual behaviour: physiological and pharmacological implications. J Neural Transm 85: 95-107.

Fernandez-Guasti A, Escalante AL, Ahlenius S, Hillegaart V, Larsson K. 1992. Stimulation of 5-HT_{1A} and 5-HT_{1B} receptors in brain regions and its effects on male rat sexual behavior. Eur J Pharmacol 210: 121-129.

Fernandez-Guasti A, Larsson K, Beyer C. 1985. Comparison of the effects of different isomers of bicuculline infused in the preoptic area on male rat sexual behavior. Experientia 41: 1414-1416.

Fernandez-Guasti A, Larsson K, Beyer C. 1986. GABAergic control of masculine sexual behavior. Pharmacol Biochem Behav 24: 1065-1070.

Fernandez-Guasti A, Rodriguez-Manzo G. 1997. 8-OH-DPAT and male rat sexual behavior: partial blockade by noradrenergic lesion and sexual exhaustion. Pharmacol Biochem Behav 56: 111-116.

Fernandez-Guasti A, Rodriguez-Manzo G. 2003. Pharmacological and physiological aspects of sexual exhaustion in male rats. Scand J Psychol 44: 257-263.

Ferrari F, Giuliani D. 1994. The selective D_2 dopamine receptor antagonist eticlopride counteracts the ejaculatio praecox induced by the selective D_2 dopamine agonist SND 919 in the rat. Life Sci 55: 1155-1162.

Ferrari F, Giuliani D. 1995. Sexual attraction and copulation in male rats: effects of the dopamine agonist SND 919. Pharmacol Biochem Behav 50: 29-34.

Ferrari F, Giuliani D. 1996a. Behavioral effects induced by the dopamine D_3 agonist 7-OH-DPAT in sexually-active and -inactive male rats. Neuropharmacology 35: 279-284.

Ferrari F, Giuliani D. 1996b. Influence of eticlopride on cocaine and DA D_2 agonist-induced behavioral effects in rats. Pharmacol Biochem Behav 53: 525-530.

Ferrari F, Ottani A, Giuliani D. 2002. Influence of sildenafil on central dopamine-mediated behaviour in male rats. Life Sci 70: 1501-1508.

Fiorino DF, Phillips AG. 1999. Facilitation of sexual behavior and enhanced dopamine efflux in the nucleus accumbens of male rats after D-amphetamine-induced behavioral sensitization. J Neurosci 19: 456-463.

Fiorino DF, Coury A, Phillips AG. 1997. Dynamic changes in nucleus accumbens dopamine efflux during the Coolidge effect in male rats. J Neurosci 17: 4849-4855.

Fodor M, Delemarre-van de Waal HA. 2001. Are POMC neurons targets for sex steroids in the arcuate nucleus of the rat? Neuroreport 12: 3989-3991.

Foreman MM, Hall JL. 1987. Effects of D_2-dopaminergic receptor stimulation on male rat sexual behavior. J Neural Transm 68: 153-170.

Foreman MM, Hall JL, Love RL. 1989. The role of the 5-HT_2 receptor in the regulation of sexual performance of male rats. Life Sci 45: 1263-1270.

Fraley GS, Thomas-Smith SE, Acohido BV, Steiner RA, Clifton DK. 2004. Stimulation of sexual behavior in the male rat by galanin-like peptide. Horm Behav 46: 551-557.

Frank JL, Hendricks SE, Olson CH. 2000. Multiple ejaculations and chronic fluoxetine: effects on male rat copulatory behavior. Pharmacol Biochem Behav 66: 337-342.

Frazer A, Hensler JG. 1999. Serotonin. Basic Neurochemistry, 6th ed. Siegel GJ, Agranoff BW, Albers RW, Fisher SK, Uhler MD, editors. Philadelphia: Lippencott-Raven Press; pp. 263-292.

Fumero B, Fernandez-Vera JR, Gonzalez-Mora JL, Mas M. 1994. Changes in monoamine turnover in forebrain areas associated with masculine sexual behavior: a microdialysis study. Brain Res 662: 233-239.

Gaddy-Kurten D, Tsuchida K, Vale W. 1995. Activins and the receptor serine kinase superfamily. Recent Prog Horm Res 50: 109-129.

Gelez H, Avila GAC, Woehrling D, Rajabi H, Pfaus JG. 2005. A neutral odor previously paired with copulation enhances dopamine release in the nucleus accumbens of male rats. Horm Behav 48: 102

Gessa GL, Paglietti E, Quarantotti BP. 1979. Induction of copulatory behavior in sexually inactive rats by naloxone. Science 204: 203-205.

Ginton A. 1976. Copulation in noncopulators: effect of PCPA in male rats. Pharmacol Biochem Behav 4: 357-359.

Giuliani D, Ferrari F. 1996. Differential behavioral response to dopamine D_2 agonists by sexually naïve, sexually active, and sexually inactive male rats. Behav Neurosci 110: 802-808.

Giuliano F. 2004. Control of penile erection by the melanocortinergic system: experimental evidences and therapeutic perspectives. J Androl 25: 683-694.

Giuliano F, Allard J. 2001. Dopamine and sexual function. Int J Impot Res 13 Suppl 3: S18-S28.

Giuliano F, Allard J. 2002. Apomorphine SL (Uprima): preclinical and clinical experiences learned the first central nervous system-acting ED drug. Int J Impot Res 14 Suppl 1: S53-S56.

Giuliano F, Rampin O. 2000. Central neural regulation of penile erection. Neurosci Biobehav Rev 24: 517-533.

Giuliano F, Allard J, Rampin O, Droupy S, Benoit G, 2001a. Spinal proerectile effect of apomorphine in the anesthetized rat. Int J Impot Res 13: 110-115.

Giuliano F, Bernabe J, Alexandre L, Niewoehner U, Haning H, et al. 2003. Pro-erectile effect of vardenafil: in vitro experiments in rabbits and in vivo comparison with sildenafil in rats. Eur Urol 44: 731-736.

Giuliano F, Bernabé J, Brown K, Droupy S, Benoit G, et al. 1997. Erectile response to hypothalamic stimulation in rats: role of peripheral nerves. Am J Physiol 273: R1990-R1997.

Giuliano F, Bernabe J, McKenna K, Longueville F, Rampin O. 2001b. Spinal proerectile effect of oxytocin in anesthetized rats. Am J Physiol Regul Integr Comp Physiol 280: R1870-R1877.

Giuliano F, Bernabe J, Rampin O, Courtois F, Benoit G, et al. 1994. Telemetric monitoring of intracavernous pressure in freely moving rats during copulation. J Urol 152: 1271-1274.

Giuliano F, Montorsi F, Mirone V, Rossi D, Sweeney M. 2000. Switching from intracavernous prostaglandin E1 injections to oral sildenafil citrate in patients with erectile dysfunction: results of a multicenter European study. J Urol 164: 708-711.

Giuliano F, Rampin O, Brown K, Courtois F, Benoit G, et al. 1996. Stimulation of the medial preoptic area of the hypothalamus in the rat elicits increases in intracavernous pressure. Neurosci Lett 209: 1-4.

Goldstein I, Lue TF, Padma-Nathan H, Rosen RC, Steers WD, et al. 1998. Oral sildenafil in the treatment of erectile dysfunction. N Engl J Med 338: 1397-1404.

Gomez-Marrero J, Feria M, Mas M. 1988. Stimulation of opioid receptors suppresses penile erectile reflexes and seminal emission in rats. Pharmacol Biochem Behav 31: 393-396.

Gonzalez MI, Farabollini F, Albonetti E, Wilson CA. 1994. Interactions between 5-hydroxytryptamine (5-HT) and testosterone in the control of sexual and nonsexual behaviour in male and female rats. Pharm Biochem Behav 47: 591-601.

Gozes I, Meltzer E, Rubinrout S, Brenneman DE, Fridkin M. 1989. Vasoactive intestinal peptide potentiates sexual behavior: inhibition by a novel antagonist. Endocrinology 125: 2945-2949.

Gréco B, Edwards DA, Michael RP, Zumpe D, Clancy AN. 1999. Colocalization of androgen receptors and mating-induced FOS immunoreactivity in neurons that project to the central tegmental field in male rats. J Comp Neurol 408: 220-236.

Gréco B, Edwards DA, Zumpe D, Clancy AN. 1998. Androgen receptor and mating-induced Fos immunoreactivity are co-localized in limbic and midbrain neurons that project to the male rat medial preoptic area. Brain Res 781: 15-24.

Guay AT, Spark RF, Jacoobson J, Murray FT, Geisser ME. 2002. Yohimbine treatment of organic erectile dysfunction in a dose-escalation trial. Int J Impot Res 14: 25-31.

Gulia KK, Mallick HN, Kumar VM. 2003. Orexin A (hypocretin-1) application at the medial preoptic area potentiates male sexual behavior in rats. Neuroscience 116: 921-923.

Hadeishi Y, Wood RI. 1996. Nitric oxide synthase in mating behavior circuitry of male Syrian hamster brain. J Neurobiol 30: 480-492.

Haensel SM, Slob AK. 1997. Flesinoxan: a prosexual drug for male rats. Eur J Pharmacol 330: 1-9.

Haensel SM, Rowland DL, Slob AK. 1995. Serotonergic drugs and masculine sexual behavior in laboratory rats and men. The Pharmacology of Sexual Function and Dysfunction: Proceedings of the Esteve Foundation Symposium VI, Son Vida, Mallorca, 9–12 October 1994. Bancroft J. editor. Amsterdam: Excerpta Medica; pp. 235-247.

Hammer RP Jr, Bogic L, Handa RJ. 1993. Estrogenic regulation of proenkephalin mRNA expression in the ventromedial hypothalamus of the adult male rat. Brain Res Mol Brain Res 19: 129-134.

Hansen S, Svensson L, Hokfelt T, Everitt BJ. 1983. 5-Hydroxytryptamine-thyrotropin releasing hormone interactions in the spinal cord: effects on parameters of sexual behaviour in the male rat. Neurosci Lett 42: 299-304.

Harris GC, Wimmer M, Aston-Jones G. 2005. A role for lateral hypothalamic orexin neurons in reward seeking. Nature 437: 556-559.

Hart BL. 1967. Sexual reflexes and mating behavior in the male dog. J Comp Physiol Psychol 64: 388-399.

Hart BL. 1968a. Alteration of quantitative aspects of sexual reflexes in spinal male dogs by testosterone. J Comp Physiol Psychol 66: 726-730.

Hart BL. 1968b. Sexual reflexes and mating behavior in the male rat. J Comp Physiol Psychol 65: 453-460.

Hart BL, Kitchell RL. 1966. Penile erection and contraction of penile muscles in the spinal and intact dog. Am J Physiol 210: 257-262.

Heeb MM, Yahr P. 1996. c-fos immunoreactivity in the sexually dimorphic area of the hypothalamus and related brain regions of male gerbils after exposure to sex-related stimuli or performance of specific sexual behaviors. Neuroscience 72: 1049-1071.

Hermann GE, Holmes GM, Rogers RC, Beattie MS, Bresnahan JC. 2003. Descending spinal projections from the rostral gigantocellular reticular nuclei complex. J Comp Neurol 455: 210-221.

Hernandez L, Gonzalez M, Oropeza MV, Guevara MA, Cervantes M, et al. 1994. Effects of intrathecal administration of adrenergic agonists on the frequency of copulatory pelvic thrusting of the male rat. Arch Med Res 25: 419-425.

Hillegaart V. 1991. Functional topography of brain serotonergic pathways in the rat. Acta Physiol Scand Suppl 598: 1-54.

Hillegaart V, Ahlenius S. 1998. Facilitation and inhibition of male rat ejaculatory behaviour by the respective 5-HT_{1A} and 5-HT_{1B} receptor agonists 8-OH-DPAT and anpirtoline, as evidenced by use of the corresponding new and selective receptor antagonists NAD-299 and NAS-181. Br J Pharm 125: 1733-1743.

Hillegaart V, Ahlenius S, Larsson K. 1991. Region-selective inhibition of male rat sexual behavior and motor performance by localized forebrain 5-HT injections: a comparison with effects produced by 8-OH-DPAT. Behav Brain Res 42: 169-180.

Hillegaart V, Alster P, Uvnas-Moberg K, Ahlenius S. 1998. Sexual motivation promotes oxytocin secretion in male rats. Peptides 19: 39-45.

Hokfelt T, Fried G, Hansen S, Holets V, Lundberg JM, et al. 1986. Neurons with multiple messengers: distribution and possible functional significance. Prog Brain Res 65: 115-137.

Hokfelt T, Skirboll L, Rehfeld JF, Goldstein M, Markey K, et al. 1980. A subpopulation of mesencephalic dopamine neurons projecting to limbic areas contains a cholecystokinin-like peptide: evidence from immunohistochemistry combined with retrograde tracing. Neuroscience 5: 2093-2124.

Holmes GM, Chapple WD, Leipheimer RE, Sachs BD. 1991. Electromyographic analysis of male rat perineal muscles during copulation and reflexive erections. Physiol Behav 49: 1235-1246.

Holmes GM, Rogers RC, Bresnahan JC, Beattie MS. 1997. Differential effects of intrathecal thyrotropin-releasing hormone (TRH) on perineal reflexes in male rats. Physiol Behav 61: 57-63.

Holmes GM, Stephens RL, Breshnahan JC, Beattie MS. 2001. Intrathecal methiothepin and thyrotropin-releasing hormone (TRH): effects on penile erection. Physiol Behav 73: 59-64.

Holstege JC, Van Dijken H, Buijs RM, Goedknegt H, Gosens T, et al. 1996. Distribution of dopamine immunoreactivity in the rat, cat and monkey spinal cord. J Comp Neurol 37: 631-652.

Horowitz A, Menice CB, Laporte R, Morgan KG. 1996. Mechanisms of smooth muscle contraction. Physiol Rev 76: 967-1003.

Hoyer D, Clarke DE, Fozard JR, Hartig PR, Martin GR, et al. 1994. International Union of Pharmacology classification of receptors for 5-hydroxytryptamine (serotonin). Pharmacol Rev 46: 157-203.

Hsieh JT, Chang HC, Law HS, Hsieh CH, Cheng JT. 1998. In vivo evaluation of serotonergic agents and alpha-adrenergic blockers on premature ejaculation by inhibiting the seminal vesicle pressure response to electrical nerve stimulation. Brit J Urol 82: 237-240.

Hughes AM, Everitt BJ, Herbert J. 1987. Selective effects of beta-endorphin infused into the hypothalamus, preoptic area and bed nucleus of the stria terminalis on the sexual and ingestive behaviour of male rats. Neuroscience 23: 1063-1073.

Hull EM, Bazzett TJ, Warner RK, Eaton RC, Thompson JT. 1990. Dopamine receptors in the ventral tegmental area modulate male sexual behavior in rats. Brain Res 512: 1-6.

Hull EM, Bitran D, Pehek EA, Holmes GM, Warner RK, et al. 1988a. Brain localization of cholinergic influence on male sex behavior in rats: antagonists. Pharmacol Biochem Behav 31: 169-174.

Hull EM, Bitran D, Pehek EA, Warner RK, Band LC. 1986. Dopaminergic control of male sex behavior in rats: effects of an intracerebrally infused agonist. Brain Res 370: 73-81.

Hull EM, Du J, Lorrain DS, Matuszewich L. 1995. Extracellular dopamine in the medial preoptic area: implications for sexual motivation and hormonal control of copulation. J Neurosci 15: 7465-7471.

Hull EM, Eaton RC, Markowski VP, Moses J, Lumley LA, et al. 1992. Opposite influence of medial preoptic D_1 and D_2 receptors on genital reflexes: implications for copulation. Life Sci 51: 1705-1713.

Hull EM, Eaton RC, Moses J, Lorrain D. 1993. Copulation increases dopamine activity in the medial preoptic area of male rats. Life Sci 52: 935-940.

Hull EM, Lumley LA, Matuszewich L, Dominguez J, Moses J, et al. 1994. The roles of nitric oxide in sexual function of male rats. Neuropharmacology 33: 1499-1504.

Hull EM, Meisel RL, Sachs BD. 2002. Male sexual behavior. Hormones, Brain and Behavior. Pfaff DW, Arnold AP, Etgen AM, Fahrbach SE, Rubin RT, editors. San Diego: Academic; pp. 3-137.

Hull EM, Pehek EA, Bitran D, Holmes GM, Warner RK, et al. 1988b. Brain localizations of cholinergic influence on male sex behavior in rats: agonists. Pharm Biochem Behav 31: 175-178.

Hull EM, Warner RK, Bazzett TJ, Eaton RC, Thompson JT, et al. 1989. D_2/D_1 ratio in the medial preoptic area affects copulation of male rats. J Pharmacol Exp Ther 251: 422-427.

Hull EM, Weber MS, Eaton RC, Dua R, Markowski VP, et al. 1991. Dopamine receptors in the ventral tegmental area affect motor, but not motivational or reflexive, components of copulation in male rats. Brain Res 554: 72-76.

Ikemoto S, Panksepp J. 1999. The role of nucleus accumbens dopamine in motivated behavior: a unifying interpretation with special reference to reward-seeking. Brain Res Rev 31: 6-41.

Johnson SW, North RA. 1992. Two types of neurone in the rat ventral tegmental area and their synaptic inputs. J Physiol 450: 455-468.

Ju G, Ceccatelli S, Hokfelt T, Frey P. 1987. Immunohystochemical evidence for a spinothalamic pathway co-containing cholecystokinin- and galanin-like immunoreactivities in the rat. Neuroscience 20: 439-456.

Keaton AK, Clark JT. 1998. Effects of angiotensin II on sexual function, blood pressure, and fluid intake are differentially affected by AT-1 receptor blockade. Physiol Behav 64: 339-346.

Kelley AE, Berridge KC. 2002. The neuroscience of natural rewards: relevance to addictive drugs. J Neurosci 22: 3306-3311.

Kim SC, Seo KK. 1998. Efficacy and safety of fluoxetine, sertraline and clomipramine in patients with premature ejaculation: a double-blind, placebo controlled study. J Urol 159: 425-427.

Kiyatkin EA. 2002. Dopamine in the nucleus accumbens: cellular actions, drug- and behavior-associated fluctuations, and a possible role in an organism's adaptive activity. Behav Brain Res 137: 27-46.

Klint T, Larsson K. 1995. Clozapine acts as a 5-HT_2 antagonist by attenuating DOI-induced inhibition of male rat sexual behaviour. Psychopharmacology 119: 291-294.

Knobil E. 1980. The neuroendocrine control of the menstrual cycle. Recent Prog Horm Res 36: 53-88.

Kondo Y. 1992. Lesions of the medial amygdala produce severe impairment of copulatory behavior in sexually inexperienced male rats. Physiol Behav 51: 939-943.

Kondo Y, Yamanouchi K. 1997. Potentiation of ejaculatory activity by median raphe nucleus lesions in male rats: effect of p-chlorophenylalanine. Endocr J 44: 873-879.

Kondo Y, Tomihara K, Sakuma Y. 1999. Sensory requirements for noncontact penile erection in the rat. Behav Neurosci 113: 1062-1070.

Koob GF. 1992. Drugs of abuse: anatomy, pharmacology and function of reward pathways. Trends Pharmacol Sci 13: 177-184.

Korotkova TM, Sergeeva OA, Eriksson KS, Haas HL, Brown RE. 2003. Excitation of ventral tegmental area dopaminergic and nondopaminergic neurons by orexins/hypocretins. J Neurosci 23: 7-11.

Kostarczyk EM. 1986. The amygdala and male reproductive functions: I. Anatomical and endocrine bases. Neurosci Biobehav Rev 10: 67-77.

Kriegsfeld LJ, Demas GE, Huang PL, Burnett AL, Nelson RJ. 1999. Ejaculatory abnormalities in mice lacking the gene for endothelial nitric oxide synthase (eNOS-/-). Physiol Behav 67: 561-566.

Kudwa AE, Dominguez-Salazar E, Cabrera DM, Sibley DR, Rissman EF. 2005. Dopamine D5 receptor modulates male and female sexual behavior in mice. Psychopharmacology 180: 206-214.

Kuhar MJ, Couceyro PR, Lambert PD. 1999. Catecholamines. Basic Neurochemistry: Molecular, Cellular and Medical Aspects, 6th edition. Siegel GJ, Agranoff BW, Albers RW, Fisher SK, Uhler MD, editors. Philadelphia: Lippencott-Raven Publishers; pp. 243-262.

Kuo RC, Baxter GT, Thompson SH, Stricker SA, Patton C, et al. 2000. NO is necessary and sufficient for egg activation at fertilization. Nature 406: 633-636.

Lagoda G, Muschamp JM, Vigdorchik A, Hull EM. 2004. A nitric oxide synthase inhibitor in the medial preoptic area inhibits copulation and stimulus sensitization in male rats. Behav Neurosci 118: 1317-1323.

Lal S, Ackman D, Thavundayil JX, Kiely ME, Etienne P. 1984. Effect of apomorphine, a dopamine receptor agonist, on penile tumescence in normal subjects. Prog Neuropsychopharmacol Biol Psychiatry 8: 695-699.

Lal S, Laryea E, Thavundayil JX, Nair NP, Negrete J, et al. 1987. Apomorphine-induced penile tumescence in impotent patients: preliminary findings. Prog Neuropsychopharmacol Biol Psychiatry 11: 235-242.

Larsson K. 1959. Experience and maturation in the development of sexual behaviour in the male puberty rat. Behaviour 14: 101-107.

Larsson K. 1967. Testicular hormone and developmental changes in mating behavior of the male rat. J Comp Physiol Psychol 63: 223-230.

Larsson K, Füxe K, Everitt BJ, Holmgren M, Södersten P. 1978. Sexual behavior in male rats after intracerebral injection of 5,7-dihydroxytryptamine. Brain Res 141: 293-303.

Lee RL, Smith ER, Mas M, Davidson JM. 1990. Effects of intrathecal administration of S-OH-DPAT on genital reflexes and mating behavior in male rats. Physiol Behav 47: 665-669.

Leipheimer RE, Sachs BD. 1988. GABAergic regulation of penile reflexes and copulation in rats. Physiol Behav 42: 351-357.

Leyton M, Stewart J. 1992. The stimulation of central kappa opioid receptors decreases male sexual behavior and locomotor activity. Brain Res 594: 56-74.

Leyton M, Stewart J. 1996. Acute and repeated activation of male sexual behavior by tail pinch: opioid and dopaminergic mechanisms. Physiol Behav 60: 77-85.

Linden A, Hansen S, Bednar I, Forsberg G, Södersten P, et al. 1987. Sexual activity increases plasma concentrations of cholecystokinin octapeptide and offsets hunger in male rats. J Endocrinol 115: 91-95.

Lisk RD, Heimann J. 1980. The effects of sexual experience and frequency of testing on retention of copulatory behavior following castration in the male hamster. Behav Neural Biol 28: 156-171.

Liu YC, Salamone JD, Sachs BD. 1997. Lesions in medial preoptic area and bed nucleus of stria terminalis: differential effects on copulatory behavior and noncontact erection in male rats. J Neurosci 17: 5245-5253.

Lopez HH, Ettenberg A. 2000. Haloperidol challenge during copulation prevents subsequent increase in male sexual motivation. Pharmacol Biochem Behav 67: 387-393.

Lopez HH, Ettenberg A. 2001. Dopamine antagonism attenuates the unconditioned incentive value of estrous female cues. Pharmacol Biochem Behav 68: 411-416.

Lopez HH, Ettenberg A. 2002a. Exposure to female rats produces differences in c-fos induction between sexually-naïve and experienced male rats. Brain Res 947: 57-66.

Lopez HH, Ettenberg A. 2002b. Sexually conditioned incentives: attenuation of motivational impact during dopamine receptor antagonism. Pharmacol Biochem Behav 72: 65-72.

Lorrain DS, Hull EM. 1993. Nitric oxide increases dopamine and serotonin release in the medial preoptic area. Neuroreport 5: 87-89.

Lorrain DS, Matuszewich L, Friedman RD, Hull EM. 1997. Extracellular serotonin in the lateral hypothalamic area is increased during postejaculatory interval and impairs copulation in male rats. J Neurosci 17: 9361-9366.

Lorrain DS, Matuszewich L, Howard RV, Du J, 1996. Nitric oxide promotes medial preoptic dopamine release during male rat copulation. Neuroreport 8: 31-34.

Lorrain DS, Matuszewich L, Hull EM. 1998. 8-OH-DPAT influences extracellular levels of serotonin and dopamine in the medial preoptic area of male rats. Brain Res 790: 217-223.

Lorrain DS, Riolo JV, Matuszewich L, Hull EM. 1999. Lateral hypothalamic serotonin inhibits nucleus accumbens dopamine: implications for sexual refractoriness. J Neurosci 19: 7648-7652.

Lott DF. 1981. Sexual behavior and intersexual strategies in American bison. Z Tierpsychol 56: 97-114.

Louilot A, Gonzalez-Mora JL, Guadalupe T, Mas M. 1991. Sex-related olfactory stimuli induce a selective increase in dopamine release in the nucleus accumbens of male rats: a voltammetric study. Brain Res 553: 313-317.

Lugg JA, Rajfer J, Gonzalez-Cadavid NF. 1995. Dihydrotestosterone is the active androgen in the maintenance of nitric oxide-mediated penile erection in the rat. Endocrinology 136: 1495-1501.

Lumley LA, Hull EM. 1999. Effects of a D_1 antagonist and of sexual experience on copulation-induced Fos-like immunoreactivity in the medial preoptic nucleus. Brain Res 829: 55-68.

Ma X, Reyna A, Mani SK, Matzuk MM, Kumar TR. 2005. Impaired male sexual behavior in activin receptor type II knockout mice. Biol Reprod 73: 1182-1190.

Maeda N, Matsuoka N, Yamaguchi I. 1990. Septohippocampal cholinergic pathway and penile erections induced by dopaminergic and cholinergic stimulants. Brain Res 537: 163-168.

Mahalati K, Okanoya K, Witt DM, Carter CS. 1991. Oxytocin inhibits male sexual behavior in prairie voles. Pharmacol Biochem Behav 39: 219-222.

Mains RE, Eipper BA. 1999. Peptides. Basic Neurochemistry: Molecular, Cellular and Medical Aspects, 6th ed. Siegel GJ, Agranoff BW, Albers RW, Fisher SK, Uhler ME, editors. Lippencott-Raven Publishers; Philadelphia: pp. 363-382.

Mallick H, Manchanda SK, Kumar VM. 1996. Beta-adrenergic modulation of male sexual behavior elicited from the medial preoptic area in rats. Behav Brain Res 74: 181-187.

Malmnas CO. 1976. The significance of dopamine, versus other catecholamines, for L-dopa induced facilitation of sexual behavior in the castrated male rat. Pharmacol Biochem Behav 4: 521-526.

Malmnas CO, Meyerson BJ. 1971. p-Chlorophenylalanine and copulatory behaviour in the male rat. Nature 232: 398-400.

Margolis EB, Hjelmstad GO, Bonci A, Fields HL. 2003. Kappa-opioid agonists directly inhibit midbrain dopaminergic neurons. J Neurosci 23: 9981-9986.

Margolis EB, Hjelmstad GO, Bonci A, Fields HL. 2005. Both kappa and mu opioid agonists inhibit glutamatergic input to ventral tegmental area neurons. J Neurophysiol 93: 3086-3093.

Margolis EB, Lock H, Chefer VI, Shippenberg TS, Hjelmstad GO, et al. 2006. κ opioids selectively control dopaminergic neurons projecting to the prefrontal cortex. Proc Natl Acad Sci USA 103: 2938-2942.

Marin R, Escrig A, Abreu P, Mas M. 1999. Androgen-dependent nitric oxide release in rat penis correlates with levels of constitutive nitric oxide synthase isoenzymes. Biol Reprod 61: 1012-1016.

Markowski VP, Hull EM. 1995. Cholecystokinin modulates mesolimbic dopaminergic influences on male rat copulatory behavior. Brain Res 699: 266-274.

Markowski VP, Eaton RC, Lumley LA, Moses J, Hull EM. 1994. A D_1 agonist in the MPOA area facilitates copulation of male rats. Pharmacol Biochem Behav 47: 483-486.

Marson L, McKenna KE. 1990. The identification of a brainstem site controlling spinal sexual reflexes in male rats. Brain Res 515: 303-308.

Marson L, McKenna KE. 1992. A role for 5-hydroxytryptamine in descending inhibition of spinal sexual reflexes. Exp Brain Res 88: 313-320.

Marson L, McKenna KE. 1994a. Serotonergic neurotoxic lesions facilitate male sexual reflexes. Pharmacol Biochem Behav 47: 883-888.

Marson L, McKenna KE. 1994b. Stimulation of the hypothalamus initiates the urethrogenital reflex in male rats. Brain Res 638: 103-108.

Marson L, List MS, McKenna KE. 1992. Lesions of the nucleus paragigantocellularis alter ex copula penile reflexes. Brain Res 592: 187-192.

Martin WJ, McGowan E, Cashen DE, Gantert LT, Drisko JE, et al. 2002. Activation of melanocortin MC(4) receptors increases erectile activity in rats ex copula. Eur J Pharmacol 454: 71-79.

Mas M, Fernandez-Vera JR, Gonzalez-Mora JL. 1995a. Neurochemical correlates of sexual exhaustion and recovery as assessed by in vivo microdialysis. Brain Res 675: 13-19.

Mas M, Fumero B, Perez-Rodriguez I. 1995b. Induction of mating behavior by apomorphine in sexually sated rats. Eur J Pharmacol 280: 331-334.

Mas M, Gonzalez-Mora JL, Louilot A, Sole C, Guadalupe T. 1990. Increased dopamine release in the nucleus accumbens of copulating male rats as evidenced by in vivo voltammetry. Neurosci Lett 110: 303-308.

Mas M, Rodriguez del Castillo A, Guerra M, Davidson JM, Battaner E. 1987. Neurochemicals correlates of male sexual behavior. Physiol Behav 41: 341-345.

Matsumoto T, Kondo Y, Sachs BD, Yamanouchi I. 1997. Effects of p-chlorophenylalanine on reflexive and noncontact penile erections in male rats. Physiol Behav 61: 165-168.

Matuszewich L, Lorrain DS, Trujillo R, Dominguez J, Putnam SK, et al. 1999. Partial antagonism of 8-OH-DPAT's effects on male rat sexual behavior with a D_2, but not a 5-HT_{1A}, antagonist. Brain Res 820: 55-62.

Matuszewich L, Ormsby JL, Moses J, Lorrain DS, Hull EM. 1995. Effects of morphiceptin in the medial preoptic area on male sexual behavior. Psychopharmacology 122: 330-335.

McGill TE. 1962. Sexual behavior in three inbred strains of mice. Behaviour 19: 341-350.

McIntosh TK, Barfield RJ. 1984. Brain monoaminergic control of male reproductive behavior. III. Norepinephrine and the postejaculatory refractory period. Behav Brain Res 12: 275-281.

McKenna KE, Chung SK, McVary KT. 1991. A model for the study of sexual function in anesthetized male and female rats. Am J Physiol 30: R1276-R1285.

McMahon CG, Touma K. 1999. Treatment of premature ejaculation with paroxetine hydrochloride. Int J Impot Res 11: 241-245.

Meisel RL, Sachs BD. 1994. The physiology of male sexual behavior. Physiology of Reproduction, 2nd ed. Knobil E, Neill JD, editors. New York: Raven; pp. 3-106.

Melis MR, Argiolas A. 1995. Dopamine and sexual behavior. Neurosci Biobehav Rev 19: 19-38.

Melis MR, Argiolas A. 1997. Role of central nitric oxide in the control of penile erection and yawning. Prog Neuropsychopharmacol Biol Psychiatry 21: 899-922.

Melis MR, Argiolas A, Gessa GL. 1987. Apomorphine-induced penile erection and yawning: site of action in brain. Brain Res 415: 98-104.

Melis MR, Argiolas A, Gessa GL. 1989. Evidence that apomorphine induces penile erection and yawning by releasing oxytocin in the central nervous system. Eur J Pharmacol 164: 565-572.

Melis MR, Spano MS, Succu S, Argiolas A. 1999a. The oxytocin antagonist d(CH2)5Tyr(Me)2-Orn8-vasotocin reduces non-contact penile erections in male rats. Neurosci Lett 265: 171-174.

Melis MR, Spano MS, Succu S, Argiolas A. 2000a. Activation of gamma-aminobutyric acid(A) receptors in the paraventricular nucleus of the hypothalamus reduces apomorphine-, N-methyl-D-aspartic acid- and oxytocin-induced penile erection and yawning in male rats. Neurosci Lett 281: 127-130.

Melis MR, Spano MS, Succu S, Locatelli V, Torsello A, et al. 2000b. EP 60761- and EP 50885-induced penile erection: structure–activity studies and comparison with apomorphine, oxytocin and N-methyl-D-aspartic acid. Int J Impot Res 12: 255-262.

Melis MR, Stancampiano R, Argiolas A. 1992. Hippocampal oxytocin mediates apomorphine-induced penile erection and yawning. Pharmacol Biochem Behav 42: 61-66.

Melis MR, Succu S, Argiolas A. 1996. Dopamine agonists increase nitric oxide production in the paraventricular nucleus of the hypothalamus: correlation with penile erection and yawning. Eur J Neurosci 8: 2056-2063.

Melis MR, Succu S, Iannucci U, Argiolas A. 1997. N-methyl-D-aspartic acid-induced penile erection and yawning: role of hypothalamic paraventricular nitric oxide. Eur J Pharmacol 328: 115-123.

Melis MR, Succu S, Mascia MS, Cortis L, Argiolas A. 2003. Extra-cellular dopamine increases in the paraventricular nucleus of the hypothalamus: correlation with penile erection and yawning. Eur J Neurosci 17: 1266-1272.

Melis MR, Succu S, Mascia MS, Cortis L, Argiolas A. 2004. Extracellular excitatory amino acids increase in the paraventricular nucleus of male rats during sexual activity: main role of N-methyl-d-aspartic acid receptors in erectile function. Eur J Neurosci 19: 2569-2575.

Melis MR, Succu S, Mauri A, Argiolas A. 1998. Nitric oxide production is increased in the paraventricular nucleus of the hypothalamus of male rats during noncontact penile erections and copulation. Eur J Neurosci 10: 1968-1974.

Melis MR, Succu S, Spano MS, Argiolas A. 2000c. Effect of excitatory amino acid, dopamine, and oxytocin receptor antagonists on noncontact penile erections and paraventricular nitric oxide production in male rats. Behav Neurosci 114: 849-857.

Melis MR, Succu S, Spano MS, Deghenghi R, Argiolas A. 2001. EP 91073 prevents EP 80661-induced penile erection: new evidence for the existence of specific EP peptide receptors mediating penile erection. Neuropharmacology 41: 254-262.

Meredith M. 1986. Vomeronasal organ removal before sexual experience impairs male hamster mating behavior. Physiol Behav 36: 737-743.

Meredith M. 1998. Vomeronasal, olfactory, hormonal convergence in the brain. Cooperation or coincidence? Ann NY Acad Sci 855: 349-361.

Meredith M, Fernandez-Fewell G. 1994. Vomeronasal system, LHRH, and sex behaviour. Psychoneuroendocrinology 19: 657-672.

Mesulam MM. 1995. Structure and function of cholinergic pathways in the cerebral cortex, limbic system, basal ganglia, and thalamus of the human brain. Psychopharmacology: The Fourth Generation of Progress. Bloom FE, Kupfer DJ, editors. New York: Raven; pp. 135-146.

Meuleman E, Cuzin B, Opsomer RJ, Hartmann U, Bailey MJ, et al. 2001. A dose-escalation study to assess the efficacy and safety of sildenafil citrate in men with erectile dysfunction. BJU Int 87: 75-81.

Millan MJ, Millan MH, Czlonkowski A, Herz A. 1984. Vasopressin and oxytocin in the rat spinal cord: distribution and origins in comparison to [Met]enkephalin, dynorphin and related opioids and their irresponsiveness to stimuli modulating neurohypophyseal secretion. Neuroscience 13: 179-187.

Millan MJ, Peglion JL, Lavielle G, Perrin-Monneyron S. 1997. 5-HT_{2C} receptors mediate penile erections in rats, actions of novel and selective agonists and antagonists. Eur J Pharmacol 325: 9-12.

Miller MA, Morgan RJ, Thompson CS, Mikhailidis DP, Jeremy JY. 1995. Effects of papaverine and vasointestinal polypeptide on penile and vascular cAMP and cGMP in control and diabetic animals: an in vitro study. Int J Impot Res 7: 91-100.

Minabe Y, Ashby CR Jr, Wang RY. 1991. The CCK-A receptor antagonist devazepide but not the CCK-B antagonist L-365,260 reverses the effects of chronic clozapine and haloperidol on midbrain dopamine neurons. Brain Res 549: 151-154.

Mitchell JB, Gratton A. 1991. Opioid modulation and sensitization of dopamine release elicited by sexually relevant stimuli: a high speed chronoamperometric study in freely behaving rats. Brain Res 551: 20-27.

Mitchell JB, Stewart J. 1990. Facilitation of sexual behaviors in the male rat associated with intra-VTA injections of opiates. Pharmacol Biochem Behav 35: 643-650.

Mitler MM, Morden B, Levine S, Dement W. 1972. The effects of parachlorophenylalanine on the mating behavior of male rats. Physiol Behav 8: 1147-1150.

Mizusawa H, Hedlund P, Andersson KE. 2002. Alpha-melanocyte stimulating hormone and oxytocin induced penile erections, and intracavernous pressure increases in the rat. J Urol 167: 757-760.

Molineaux CJ, Hassen AH, Rosenberger JG, Cox BM. 1986. Response of rat pituitary anterior lobe prodynorphin products to changes in gonadal steroid environment. Endocrinology 119: 2297-2305.

Molinoff PB, Shadiack AM, Earle D, Diamond LE, Quon CY. 2003. PT-141: a melanocortin agonist for the treatment of sexual dysfunction. Ann NY Acad Sci 994: 96-102.

Moltz H. 1990. E-series prostaglandins and arginine vasopressin in the modulation of male sexual behavior. Neurosci Biobehav Rev 14: 109-115.

Monaghan EP, Arjomand J, Breedlove SM. 1993. Brain lesions affect penile reflexes. Horm Behav 27: 122-131.

Moore KE, Lookingland KJ. 1995. Dopaminergic neuronal systems in the hypothalamus. Psychopharmacology: The Fourth Generation of Progress. Bloom FE, Kupfer DJ, editors. New York: Raven; pp. 245-246.

Mortimer CH, McNeilly AS, Fisher RA, Murray MAF, Besser GM. 1974. Gonadotrophin-releasing hormone therapy in hypogonadal males with hypothalamic or pituitary dysfunction. Br Med J 4: 617-621.

Mos J, Mollet I, Tolboom JT, Waldinger MD, Olivier B. 1999. A comparison of the effects of different serotonin reuptake blockers on sexual behaviour of the male rat. Euro Neuropsychopharmacol 9: 123-135.

Moses J, Hull EM. 1994. Inhibition of nitric oxide synthase in the MPOA increases seminal emissions in restrained supine rats and reduces the incidence of copulation in the copulation test. Conf Reprod Behav Abst 26: 37.

Moses J, Hull EM. 1999. A nitric oxide synthesis inhibitor administered into the medial preoptic area increases seminal emissions in an ex copula reflex test. Pharmacol Biochem Behav 63: 345-348.

Moses J, Loucks JA, Watson HL, Matuszewich L, Hull EM. 1995. Dopaminergic drugs in the medial preoptic area and nucleus accumbens: effects on motor activity, sexual motivation, and sexual performance. Pharmacol Biochem Behav 51: 681-686.

Moss RL, McCann SM, Dudley CA. 1975. Releasing hormones and sexual behavior. Prog Brain Res 42: 37-46.

Muschamp JW, Lagoda G, Hull EM. 2004. Increased Fos immunoreactivity in hypocretin/orexin neurons following copulation and sexual stimuli in male rats. Horm Behav 46: 96-97.

Myers BM, Baum MJ. 1980. Facilitation of copulatory performance in male rats by naloxone: effects of hypophysectomy, 17 alpha-estradiol, and luteinizing hormone releasing hormone. Pharmacol Biochem Behav 12: 365-370.

Narita M, Nagumo Y, Hashimoto S, Khotib J, Miyatake M, et al. 2006. Direct involvement of orexinergic systems in the activation of the mesolimbic dopamine pathway and related behaviors induced by morphine. J Neurosci 26: 398-405.

Niikura S, Yokoyama O, Komatsu K, Yotsuyanagi S, Mizuno T, et al. 2002. A causative factor of copulatory disorder in rats following social stress. J Urol 168: 843-849.

Nishitani S, Moriya T, Kondo Y, Sakuma Y, Shinohara K. 2004. Induction of Fos immunoreactivity in oxytocin neurons in the paraventricular nucleus after female odor exposure in male rats: effects of sexual experience. Cell Mol Neurobiol 24: 283-291.

Olson RJ, Justice JB Jr. 1993. Quantitative microdialysis under transient conditions. Anal Chem 65: 1017-1022.

Orbach J. 1961. Spontaneous ejaculation in rat. Science 134: 1072-1073.

O'Sullivan JD, Hughes AJ. 1998. Apomorphine-induced penile erections in Parkinson's disease. Mov Disord 13: 536-539.

Padma-Nathan H, Giuliano F. 2001. Oral drug therapy for erectile dysfunction. Urol Clin North Am 28: 321-334.

Padoin MJ, Lucion AB. 1995. The effect of testosterone and DOI (1-(2,5-dimethoxy-4-iodophenyl)-2-aminopropane) on male sexual behavior of rats. Eur J Pharmacol 277: 1-6.

Paredes RG, Ågmo A. 1995. The $GABA_B$ antagonist CGP 35348 inhibits the effects of baclofen on sexual behavior and motor coordination. Brain Res Bull 36: 495-497.

Paredes RG, Ågmo A. 2004. Has dopamine a physiological role in the control of sexual behavior? A critical review of the evidence. Prog Neurobiol 73: 179-226.

Paredes RG, Contreras JL, Ågmo A. 1998a. GABAergic drugs and sexual behaviour in the rabbit: evidence for species-specific effects. J Psychopharmacol 12: 186-191.

Paredes RG, Holmes GM, Sachs BD, Ågmo A. 1993b. Electromyographic activity of rat ischiocavenosus muscles during copulation after treatment with a GABA- transaminase inhibitor. Behav Neurol Biol 60: 118-122.

Paredes RG, Karam P, Highland L, Ågmo A. 1997. GABAergic drugs and socio-sexual behavior. Pharmacol Biochem Behav 58: 291-298.

Paredes RG, Kica E, Baum MJ. 1994. Differential effects of the serotonin1A agonist, 8-OH-DPAT, on masculine and feminine sexual behavior of the ferret. Psychopharmacology 114: 591-596.

Park JJ, Baum MJ. 1999. Intracerebroventricular infusion of the galanin antagonist M40 attenuates heterosexual partner preferences in ferrets. Behav Neurosci 113: 391-400.

Paul SM. 1995. GABA and glycine. Psychopharmacology: The Fourth Generation of Progress. Bloom FE, Kupfer DJ, editors. New York: Raven; pp. 87-94.

Pehek EA, Thompson JT, Hull EM. 1989a. The effects of intracranial administration of the dopamine agonist apomorphine on penile reflexes and seminal emission in the rat. Brain Res 500: 325-332.

Pehek EA, Thompson JT, Hull EM. 1989b. The effects of intrathecal administration of the dopamine agonist apomorphine on penile reflexes and copulation in the male rat. Psychopharmacology 99: 304-308.

Pehek EA, Thompson JT, Eaton RC, Bazzett TJ, Hull EM. 1988a. Apomorphine and haloperidol, but not domperidone, affect penile reflexes in rats. Pharmacol Biochem Behav 31: 201-208.

Pehek EA, Warner RK, Bazzett T, Bitran D, Band LC, et al. 1988b. Microinjection of *cis*-flupenthixol, a dopamine antagonist, into the medial preoptic area impairs sexual behavior of male rats. Brain Res 443: 70-76.

Perras B, Smolnik R, Fehm HL, Born J. 2001. Signs of sexual behaviour are not increased after subchronic treatment with LHRH in young men. Psychoneuroendocrinology 26: 1-15.

Pfaus JG, Phillips AG. 1987. Cholecystokinin facilitates ejaculation in male rats: blockade with proglumide and apomorphine. Eur J Pharmacol 141: 331-338.

Pfaus JG, Phillips AG. 1989. Differential effects of dopamine receptor antagonists on the sexual behavior of male rats. Psychopharmacology 98: 363-368.

Pfaus JG, Phillips AG. 1991. Role of dopamine in anticipatory and consummatory aspects of sexual behavior in the male rat. Behav Neurosci 105: 727-743.

Pfaus JG, Wilkins MF. 1995. A novel environment disrupts copulation in sexually naïve but not experienced male rats: reversal with naloxone. Physiol Behav 57: 1045-1049.

Pfaus JG, Damsma G, Nomikos GG, Wenkstern DG, Blaha CD, et al. 1990. Sexual behavior enhances central dopamine transmission in the rat. Brain Res 530: 345-348.

Phoenix CH, Chambers KC. 1986. Threshold for behavioral response to testosterone in old castrated male rhesus monkeys. Biol Reprod 35: 918-926.

Phoenix CH, Chambers KC. 1988. Testosterone therapy in young and old rhesus males that display low levels of sexual activity. Physiol Behav 43: 479-484.

Phoenix CH, Chambers KC. 1990. Sexual performance of old and young male rhesus macaques following treatment with GnRH. Physiol Behav 47: 513-517.

Pleim ET, Matochik JA, Barfield RJ, Auerbach SB. 1990. Correlation of dopamine release in the nucleus accumbens with masculine sexual behavior in rats. Brain Res 524: 160-163.

Poggioli R, Benelli A, Arletti R, Menozzi B, Bertolini A. 1994. Benextramine, an NPY antagonist, improves sexual behavior in male rats. Neuropeptides 26: 355-358.

Poggioli R, Benelli A, Arletti R, Cavazzuti E, Bertolini A. 1995. Nitric oxide is involved in the ACTH-induced behavioral syndrome. Peptides 16: 1263-1268.

Poggioli R, Rasori E, Bertolini A. 1992. Galanin inhibits sexual behavior in male rats. Eur J Pharmacol 213: 87-90.

Poggioli R, Vergoni AV, Marrama D, Giuliani D, Bertolini A. 1990. NPY-induced inhibition of male copulatory activity is a direct behavioural effect. Neuropeptides 16: 169-172.

Pollak EI, Sachs BD. 1975. Masculine sexual behavior and morphology: paradoxical effects of perinatal androgen treatment in male and female rats. Behav Biol 13: 401-411.

Pomerantz SM, Hepner BC, Wertz JM. 1993a. 5-HT_{1A} and 5-$HT_{1C/1D}$ receptor agonists produce reciprocal effects on male sexual behavior of rhesus monkeys. Eur J Pharmacol 243: 227-234.

Pomerantz SM, Hepner BC, Wertz JM. 1993b. Serotonergic influences on male sexual behavior of rhesus monkeys: effects of serotonin agonists. Psychopharmacology 111: 47-54.

Popova NK, Amstislavskaya TG. 2002. Involvement of the 5-HT(1A) and 5-HT(1B) serotonergic receptor subtypes in sexual arousal in male mice. Psychoneuroendocrinology 27: 609-618.

Porst H, Rosen R, Padma-Nathan H, Goldstein I, Giuliano F, et al. 2001. The efficacy and tolerability of vardenafil, a new, oral, selective phosphodiesterase type 5 inhibitor, in patients with erectile dysfunction: the first at-home clinical trial. Int J Impot Res 13: 192-199.

Pu S, Xu B, Kalra SP, Kalra PS. 1996. Evidence that gonadal steroids modulate nitric oxide efflux in the medial preoptic area: effects of N-methyl-D-aspartate and correlation with luteinizing hormone secretion. Endocrinology 137: 1949-1955.

Putnam SK, Du J, Hull EM. 2001. Testosterone restoration of copulation and medial preoptic dopamine release in castrated male rats: 2-, 5-, and 10-day treatments. Horm Behav 39: 216-224.

Putnam SK, Sato S, Hull EM. 2003. Hormonal maintenance of copulation in castrates: association with extracellular dopamine in MPOA. Horm Behav 44: 419-426.

Putnam SK, Sato S, Hull EM. 2005. Effects of testosterone metabolites on copulation, medial preoptic dopamine content, and nitric oxide synthase. Horm Behav 47: 513-522.

Quershi GA, Södersten P. 1986. Sexual activity alters the concentration of amino acids in cerebrospinal fluid of male rat. Neurosci Lett 70: 374-378.

Rampin O, Jerome N, Suaudeau C. 2003. Proerectile effects of apomorphine in mice. Life Sci 72: 2329-2336.

Ratnasooriya WD, Dharmasiri MG. 2001. L-arginine, the substrate of nitric oxide synthase, inhibits fertility of male rats. Asian J Androl 3: 97-103.

Ratnasooriya WD, Dharmasiri MG, Wadsworth RM. 2000. Reduction in libido and fertility of male rats by administration of the nitric oxide (NO) synthase inhibitor N-nitro-L-arginine methylester. Int J Androl 23: 187-191.

Rees RW, Ralph DJ, Royle M, Moncada S, Cellek S. 2001. Y-27632, an inhibitor of Rho-kinase, antagonizes noradrenergic contractions in the rabbit and human corpus cavernosum. Brit J Pharmacol 133: 455-458.

Rehman J, Kaynan A, Christ G, Valcic M, Maayani S, et al. 1999. Modification of sexual behavior of Long–Evans male rats by drugs acting on the 5-HT_{1A} receptor. Brain Res 821: 414-425.

Reilly CM, Zamorano P, Stopper VS, Mills TM. 1997. Androgenic regulation of NO availability in rat penile erection. J Androl 18: 110-115.

Retana-Marquez S, Velazquez-Moctezuma J. 1997. Cholinergic-androgenic interaction in the regulation of male sexual behavior in rats. Pharmacol Biochem Behav 56: 373-378.

Ridet JL, Sandillon F, Rajaofetra N, Geffard M, Privat A. 1992. Spinal dopaminergic system of the rat: light and electron microscopic study using an antiserum against dopamine, with particular emphasis on synaptic incidence. Brain Res 598: 233-241.

Riters LV, Absil P, Balthazart J. 1999. Effects of naloxone on the acquisition and expression of appetitive and consummatory sexual behavior in male Japanese quail. Physiol Behav 66: 763-773.

Robbins TW, Everitt BJ. 1992. Functions of dopamine in the dorsal and ventral striatum. Semin Neurosci 4: 119-128.

Rodriguez M, Castro R, Hernandez G, Mas M. 1984. Different roles of catecholaminergic and serotoninergic neurons of the medial forebrain bundle on male rat sexual behavior. Physiol Behav 33: 5-11.

Rodriguez-Manzo G. 1999. Yohimbine interacts with the dopaminergic system to reverse sexual satiation: further evidence for a role of sexual motivation in sexual exhaustion. Eur J Pharmacol 372: 1-8.

Rodriguez-Manzo G, Fernandez-Guasti A. 1994. Reversal of sexual exhaustion by serotonergic and noradrenergic agents. Behav Brain Res 62: 127-134.

Rodriguez-Manzo G, Fernandez-Guasti A. 1995. Participation of the central noradrenergic system in the reestablishment of copulatory behavior of sexually exhausted rats by yohimbine, naloxone, and 8-OH-DPAT. Brain Res Bull 38: 399-404.

Rodriguez-Manzo G, Asai M, Fernandez-Guasti A. 2002a. Evidence for changes in brain enkephalin contents associated to male rat sexual activity. Behav Brain Res 131: 47-55.

Rodriguez-Manzo G, Lopez-Rubalcava C, Hen R, Fernandez-Guasti A. 2002b. Participation of 5-HT(1B) receptors in the inhibitory actions of serotonin on masculine sexual behaviour of mice: pharmacological analysis in 5-HT(1B) receptor knock-out mice. Br J Pharmacol 136: 1127-1134.

Romeo RD, Parfitt DB, Richardson HN, Sisk CL. 1998. Pheromones elicit equivalent levels of Fos-immunoreactivity in prepubertal and adult male Syrian hamsters. Horm Behav 34: 48-55.

Romeo RD, Richardson HN, Sisk CL. 2002. Puberty and the maturation of the male brain and sexual behavior: recasting a behavioral potential. Neurosci Biobehav Rev 26: 381-391.

Roselli CE, Thornton JE, Chambers KC. 1993. Age-related deficits in brain estrogen receptors and sexual behavior of male rats. Behav Neurosci 107: 202-209.

Rosen RC, Lane RM, Menza M. 1999. Effects of SSRIs on sexual function: a critical review. J Clin Psychopharmacol 19: 67-85.

Rosenblatt JS, Aronson LR. 1958. The decline of sexual behavior in male cats after castration with special reference to the role of prior sexual experience. Behaviour 12: 285-338.

Rowland DL, Houtsmuller EJ. 1998. 8-OH-DPAT interacts with sexual experience and testosterone to affect ejaculatory response in rats. Pharmacol Biochem Behav 60: 143-149.

Ryan EL, Frankel AI. 1978. Studies on the role of the medial preoptic area in sexual behavior and hormonal response to sexual behavior in the mature male laboratory rat. Biol Reprod 19: 971-983.

Sachs BD. 1980. Sexual reflexes of spinal male house mice. Physiol Behav 24: 489-492.

Sachs BD. 1982. Role of striated penile muscles in penile reflexes, copulation, and induction of pregnancy in the rat. J Reprod Fertil 66: 433-443.

Sachs BD. 1997. Erection evoked in male rats by airborne scent from estrous females. Physiol Behav 62: 921-924.

Sachs BD. 2000. Contextual approaches to the physiology and classification of erectile function, erectile dysfunction, and sexual arousal. Neurosci Biobehav Rev 24: 541-560.

Sachs BD, Garinello LD. 1979. Spinal pacemaker controlling sexual reflexes in male rats. Brain Res 171: 152-156.

Sachs BD, Garinello LD. 1980. Hypothetical spinal pacemaker regulating penile reflexes in rats: evidence from transection of spinal cord and dorsal penile nerves. J Comp Physiol Psychol 94: 530-535.

Sachs BD, Meisel RL. 1979. Pubertal development of penile reflexes and copulation in male rats. Psychoneuroendocrinology 4: 287-296.

Sachs BD, Akasofu K, Citron JH, Daniels SB, Natoli JH. 1994. Noncontact stimulation from estrous females evokes penile erection in rats. Physiol Behav 55: 1073-1079.

Sachs BD, Valcourt RJ, Flagg HC. 1981. Copulatory behavior and sexual reflexes of male rats treated with naloxone. Pharmacol Biochem Behav 14: 251-253.

Saito TR, Moltz H. 1986. Copulatory behavior of sexually naïve and sexually experienced male rats following removal of the vomeronasal organ. Physiol Behav 37: 507-510.

Sala M, Braida D, Leone MP, Calcaterra P, Monti S, et al. 1990. Central effect of yohimbine on sexual behavior in the rat. Physiol Behav 47: 165-173.

Salamone JD, Correa M. 2002. Motivational views of reinforcement: implications for understanding the behavioral functions of nucleus accumbens dopamine. Behav Brain Res 137: 3-25.

Salis PS, Dewsbury DA. 1971. p-Chlorophenylalanine facilitates copulatory behaviour in male rats. Nature 232: 400-401.

Sar M, Stumpf WE. 1975. Distribution of androgen-concentrating neurons in rat brain. Anatomical Neuroendocrinology. Stumpf WE, Grant LD, editors. Basel: Karger; pp. 120-133.

Sato S, Hull EM. 2006. The nitric oxide-guanosine 3′, 5′-cyclic monophosphate pathway regulates dopamine efflux in the medial preoptic area and copulation in male rats. Neuroscience 139: 417-428.

Sato S, Braham CS, Putnam SK, Hull EM. 2005. Neuronal nitric oxide synthase and gonadal steroid interaction in the MPOA of male rats: co-localization and testosterone-induced restoration of copulation and nNOS-immunoreactivity. Brain Res 1043: 205-213.

Sato Y, Christ GJ, Horita H, Adachi H, Suzuki N, et al. 1999. The effects of alterations in nitric oxide levels in the paraventricular nucleus on copulatory behavior and reflexive erections in male rats. J Urol 162: 2182-2185.

Sato Y, Horita H, Kurohata T, Adachi H, Tsukamoto T. 1998. Effect of the nitric oxide level in the medial preoptic area on male copulatory behavior in rats. Am J Physiol 274: R243-R247.

Sato Y, Wada H, Horita H, Suzuki N, Shibuya A, et al. 1995. Dopamine release in the medial preoptic area during male copulatory behavior in rats. Brain Res 692: 66-70.

Sato Y, Zhao W, Christ GJ. 2001. Central modulation of the NO–cGMP pathway affects the MPOA-induced intracavernous pressure response. Am J Physiol Regul Integr Comp Physiol 281: R269-R278.

Scaletta LL, Hull EM. 1990. Systemic or intracranial apomorphine increases copulation in long-term castrated male rats. Pharmacol Biochem Behav 37: 471-475.

Schnur SL, Smith ER, Lee RL, Mas M, Davidson JM. 1989. A component analysis of the effects of DPAT on male rat sexual behavior. Physiol Behav 45: 897-901.

Schoenen J, Lotstra F, Vierendeels G, Reznik M, Vanderhaeghen JJ. 1985. Substance P, enkephalins, somatostatin, cholecystokinin, oxytocin, and vasopressin in human spinal cord. Neurology 35: 881-890.

Schultz KM, Richardson HN, Romeo RD, Morris JA, Lookingland KJ, et al. 2003. Medial preoptic area dopaminergic responses to female pheromones develop during puberty in the male Syrian hamster. Brain Res 988: 139-145.

Schultz KM, Richardson HN, Zehr JL, Osetek AJ, Menard TA, et al. 2004. Gonadal hormones masculinize and defiminize reproductive behaviors during puberty in the male Syrian hamster. Horm Behav 45: 242-249.

Scimonelli TN, Celis ME. 1987. Effect of gonadectomy and reposition of gonadal steroids on alpha-melanocyte-stimulating hormone concentration in discrete hypothalamic areas during a 24-hour period. Neuroendocrinology 45: 441-445.

Scordalakes EM, Shetty SJ, Rissman EF. 2002. Roles of estrogen receptor alpha and androgen receptor in the regulation of neuronal nitric oxide synthase. J Comp Neurol 453: 336-344.

Seagraves RT. 1990. Effects of psychotropic drugs on human erection and ejaculation. Arch Gen Psychiatry 46: 275-284.

Simerly RB, Swanson LW. 1986. The organization of neural inputs to the medial preoptic nucleus of the rat. J Comp Neurol 246: 312-342.

Singh R, Pervin S, Shryne J, Gorski R, Chaudhuri G. 2000. Castration increases and androgens decrease nitric oxide synthase activity in the brain: physiologic implications. Proc Natl Acad Sci USA 97: 3672-3677.

Sirinathsinghji DJ. 1987. Inhibitory influence of corticotropin releasing factor on components of sexual behaviour in the male rat. Brain Res 407: 185-190.

Sisk CL, Schultz KM, Zehr JL. 2003. Puberty: a finishing school for male social behavior. Ann NY Acad Sci 1007: 189-198.

Smith ER, Lee RL, Schnur SL, Davidson JM. 1987a. Alpha 2-adrenoceptor antagonists and male sexual behavior: I. Mating behavior. Physiol Behav 41: 7-14.

Smith ER, Lee RL, Schnur SL, Davidson JM. 1987b. Alpha 2-adrenoceptor antagonists and male sexual behavior: II. Erectile and ejaculatory reflexes. Physiol Behav 41: 15-19.

Södersten P, Damassa DA, Smith ER. 1977. Sexual behavior in developing male rats. Horm Behav 8: 320-341.

Södersten P, Eneroth P, Mode A, Gustafsson JA. 1985. Mechanisms of androgen-activated sexual behaviour in rats. Neurobiology. Gilles R, Balthazart J, editors. Berlin: Springer-Verlag; pp. 48-59.

Södersten P, Larsson K, Ahlenius S, Engel J. 1976. Stimulation of mounting behavior but not lordosis behavior in ovariectomized female rats by *p*-chlorophenylalanine. Pharmacol Biochem Behav 5: 329-333.

Somlyo AP, Somlyo AV. 2000. Signal transduction by G-proteins, Rho-kinase and protein phosphatase to smooth muscle and non-muscle myosin II. J Physiol 522: 177-185.

Spanagel R, Weiss F. 1999. The dopamine hypothesis of reward: past and current status. Trends Neurosci 22: 521-527.

Steers WD, de Groat WC. 1989. Effects of m-chlorophenylpiperazine on penile and bladder function in rats. Am J Physiol 257: R1441-R1449.

Stefanick ML, Smith ER, Szumowski DA, Davidson JM. 1985. Reproductive physiological behavior in the male rat following acute and chronic peripheral adrenergic depletion by guanethidine. Pharmacol Biochem Behav 23: 55-63.

Stoneham MD, Everitt BJ, Hansen S, Lightman SL, Todd K. 1985. Oxytocin and sexual behaviour in the male rat and rabbit. J Endocrinol 107: 97-106.

Succu S, Spano MS, Melis MR, Argiolas A. 1998. Different effects of omega-conotoxin on penile erection, yawning and paraventricular nitric oxide in male rats. Eur J Pharmacol 359: 19-26.

Sugiura K, Yoshimura H, Yokoyama M. 1997. An animal model of copulatory disorder induced by social stress in male mice: effects of apomorphine and L-dopa. Psychopharmacology 133: 249-255.

Suh JK, Mun KH, Cho CK, Shin HC, Kim YS, et al. 1995. Effect of vasoactive intestinal peptide and acetylcholine on penile erection in the rat in vivo. Int J Impot Res 7: 111-118.

Swann JM. 1997. Gonadal steroids regulate behavioral responses to pheromones by actions on a subdivision of the medial preoptic nucleus. Brain Res 750: 189-194.

Swanson LW, McKellar S. 1979. The distribution of oxytocin- and neurophysin-stained fibers in the spinal cord of the rat and monkey. J Comp Neurol 188: 87-106.

Swanson LW, Sawchenko PE. 1980. Paraventricular nucleus: a site for the integration of neuroendocrine and autonomic mechanisms. Neuroendocrinology 31: 410-417.

Szczypka MS, Zhou QY, Palmiter RD. 1998. Dopamine-stimulated sexual behavior is testosterone dependent in mice. Behav Neurosci 112: 1229-1235.

Tallentire D, McRae G, Spedding R, Clark R, Vickery B. 1996. Modulation of sexual behaviour in the rat by a potent and selective 2-adrenoceptor antagonist, delequamine (RS-15385–197). Br J Pharm 118: 63-72.

Tang Y, Rampin O, Calas A, Facchinetti P, Giuliano F. 1998. Oxytocinergic and serotoninergic innervation of identified lumbosacral nuclei controlling penile erection in the male rat. Neuroscience 82: 241-254.

Taylor G, Bardgett M, Csernansky J, Early T, Haller J, et al. 1996. Male reproductive systems under chronic fluoxetine or trimipramine treatment. Physiol Behav 59: 479-485.

Taylor P, Brown JH. 1999. Acetycholine. Basic Neurochemistry: Molecular, Cellular and Medical Aspects, 6th ed. Siegel GJ, Agranoff BW, Albers RW, Fisher SK, Uhler MD, editors. Philadelphia: Lippencott-Raven Publishers; pp. 213-242.

Thorpe AJ, Cleary JP, Levine AS, Kotz CM. 2005. Centrally administered orexin A increases motivation for sweet pellets in rats. Psychopharmacology 182: 75-83.

Todd BJ, Schwarz JM, McCarthy MM. 2005. Prostaglandin-E2: a point of divergence in estradiol-mediated sexual differentiation. Horm Behav 48: 512-521.

Toner JP, Adler NT. 1986. The pre-ejaculatory behavior of male and female rats affects the number of sperm in the vagina and uterus. Physiol Behav 36: 363-367.

Toner JP, Attas AI, Adler NT. 1987. Transcervical sperm transport in the rat: the roles of pre-ejaculatory behavior and copulatory plug fit. Physiol Behav 39: 371-375.

Triemstra JL, Nagatani S, Wood RI. 2005. Chemosensory cues are essential for mating-induced dopamine release in MPOA of male Syrian hamsters. Neuropsychopharmacology 30: 1436-1442.

Truitt WA, Coolen LM. 2002. Identification of a potential ejaculation generator in the spinal cord. Science 297: 1566-1569.

Truitt WA, Shipley MT, Veening JG, Coolen LM. 2003. Activation of a subset of lumbar spinothalamic neurons after copulatory behavior in male but not female rats. J Neurosci 23: 325-331.

Van der Ploeg LHT, Martin WJ, Howard AD, Nargund RP, Austin CP, et al. 2002. A role for the melanocortin 4 receptor in sexual function. Proc Natl Acad Sci USA 99: 11381-11386.

van Dijken H, Dijk J, Voom P, Holstege JC. 1996. Localization of dopamine D2 receptor in rat spinal cord identified with immunocytochemistry and in situ hybridization. Eur J Neurosci 8: 621-628.

van Furth WR, van Ree JM. 1994. Endogenous opioids and sexual motivation and performance during the light phase of the diurnal cycle. Brain Res 636: 175-179.

van Furth WR, van Ree JM. 1996. Sexual motivation: involvement of endogenous opioids in the ventral tegmental area. Brain Res 729: 20-28.

van Furth WR, van Ernst MG, van Ree JM. 1995. Opioids and sexual behavior of male rats: involvement of the medial preoptic area. Behav Neurosci 109: 123-134.

van Furth WR, Wolterink-Donselaar IG, van Ree JM. 1994. Endogenous opioids are differentially involved in appetitive and consummatory aspects of sexual behavior of male rats. Am J Physiol 266: R606-R613.

Vargas VM, Torres D, Corona F, Vergara M, Gomez LE, et al. 2004. Cholinergic facilitation of erection and ejaculation in spinal cord-transected rats. Int J Impot Res 16: 86-90.

Veening JG, Coolen LM. 1998. Neural activation following sexual behavior in the male and female rat brain. Behav Brain Res 92: 181-193.

Vega Matuszcyk J, Larsson K, Eriksson E. 1998. The selective serotonin reuptake inhibitor fluoxetine reduces sexual motivation in male rats. Pharmacol Biochem Behav 60: 527-532.

Vemulapalli R, Kurowski S, Salisbury B, Parker E, Davis H. 2001. Activation of central melanocortin receptors by MT-II increases cavernosal pressure in rabbits by the neuronal release of NO. Br J Pharmacol 134: 1705-1710.

Verma S, Chhina GS, Kumar MV, Singh B. 1989. Inhibition of male sexual behavior by serotonin application in the medial preoptic area. Physiol Behav 46: 327-330.

Véronneau-Longueville F, Rampin O, Freund-Mercier MJ, Tang Y, Calas A, et al. 1999. Oxytocinergic innervation of automatic nuclei controlling penile erection in the rat. Neuroscience 93: 1437-1447.

Vigdorchik A, Lagoda G, Hull EM. 2003. A glutamate receptor antagonist microinjected into the MPOA impairs male sexual behavior. Horm Behav 44: 82

Waldinger MD, Hengeveld MW, Zwinderman AH, Olivier B. 1998. Effects of SSRI antidepressants on ejaculation: a double-blind, randomized, placebo-controlled study with fluoxetine, fluvoxamine, paroxetine, and sertraline. J Clin Psychopharmacol 18: 274-281.

Wardlaw SL. 1986. Regulation of beta-endorphin, corticotropin-like intermediate lobe peptide, and alpha-melanotropin-stimulating hormone in the hypothalamus by testosterone. Endocrinology 119: 19-24.

Wardlaw SL, Blum M. 1990. Biphasic effect of orchiectomy on pro-opiomelanocortin gene expression in the hypothalamus. Neuroendocrinology 52: 521-526.

Wang CT, Huang RL, Tai MY, Tsai YF, Peng MT. 1995. Dopamine release in the nucleus accumbens during sexual behavior in prenatally stressed adult male rats. Neurosci Lett 200: 29-32.

Warner RK, Thompson JT, Markowski VP, Loucks JA, Bazzett TJ, et al. 1991. Microinjection of the dopamine antagonist *cis*-flupenthixol into the MPOA impairs copulation, penile reflexes and sexual motivation in male rats. Brain Res 540: 177-182.

Wells AB, Coolen LM. 2001. Lumbar spinal cord efferents directly contact thalamic neurons that project to MPOA or BNST. Soc Behav Neuroendocrinol Abst 5: 354.

Wenkstern D, Pfaus JG, Fibiger HC. 1993. Dopamine transmission increases in the nucleus accumbens of male rats during their first exposure to sexually receptive female rats. Brain Res 618: 41-46.

Wersinger SR, Rissman EF. 2000. Dopamine activates masculine sexual behavior independent of the estrogen receptor alpha. J Neurosci 20: 4248-4254.

Wessels H, Fuciarelli K, Hansen J, Hadley ME, Hruby VJ, et al. 1998. Synthetic melanotropic peptide initiates erections in men with psychogenic erectile dysfunction: double-blind, placebo controlled crossover study. J Urol 160: 389-393.

Wessels H, Gralnek D, Dorr R, Hruby VJ, Hadley ME, et al. 2000. Effect of an alpha-melanocyte stimulating hormone analog on penile erection and sexual desire in men with organic erectile dysfunction. Urology 56: 641-646.

Wessels H, Hruby VJ, Hackett J, Han G, Balse-Sriinivasan P, et al. 2003. Ac-Nle-c[Asp-His-DPhe-Arg-Trp-Lys]-NH2 induces penile erection via brain and spinal melanocortin receptors. Neuroscience 118: 755-762.

Westberry J, Meredith M. 2003. The influence of chemosensory input and gonadotropin releasing hormone on mating behavior circuits in male hamsters. Brain Res 974: 1-16.

Wiesenfeld-Hallin Z, Sodersten P. 1984. Spinal opiates affect sexual behaviour in rats. Nature 309: 257-258.

Wilson C, Nomikos GG, Collu M, Fibiger HC. 1995. Dopaminergic correlates of motivated behavior: importance of drive. J Neurosci 15: 5169-5178.

Winslow JT, Insel TR. 1991. Social status in pairs of male squirrel monkeys determines the behavioral response to central oxytocin administration. J Neurosci 24: 761-765.

Wise RA, Bozarth MA. 1984. Brain reward circuitry: four circuits "wired" in apparent series. Brain Res Bull 12: 203-208.

Witt DM. 1995. Oxytocin and rodent sociosexual responses: from behavior to gene expression. Neurosci Biobehav Rev 19: 315-324.

Witt DM, Insel TR. 1994. Increased Fos expression in oxytocin neurons following masculine sexual behavior. J Neuroendocrinol 6: 13-18.

Woolley SC, Sakata JT, Crews D. 2004. Evolutionary insights into the regulation of courtship behavior in male amphibians and reptiles. Physiol Behav 83: 347-360.

Woolley SC, Sakata JT, Gupta A, Crews D. 2001. Evolutionary changes in dopaminergic modulation of courtship behaviors in Cnemidophorus whiptail lizards. Horm Behav 40: 483-489.

Yaîci ED, Rampin O, Calas A, Jestin A, McKenna KE, et al. 2002. Alpha(2a) and alpha(2c) adrenoceptors on spinal neurons controlling penile erection. Neuroscience 114: 945-960.

Yamada K, Emson P, Hokfelt T. 1996. Immunohistochemical mapping of nitric oxide synthase in the rat hypothalamus and colocalization with neuropeptides. J Chem Neuroanat 10: 295-316.

Yells DP, Prendergast MA, Hendricks SE, Miller ME. 1995. Monoaminergic influences on temporal patterning of sexual behavior in male rats. Physiol Behav 58: 847-852.

Yilmaz U, Tatlisen A, Turan H, Arman F, Ekmekcioglu O. 1999. The effects of fluoxetine on several neurophysiological variables in patients with premature ejaculation. J Urol 161: 107-111.

Zarrindast MR, Mamanpush SM, Rashidy-Pour A. 1994. Morphine inhibits dopaminergic and cholinergic induced ejaculation in rats. Gen Pharmacol 25: 803-808.

Zarrindast MR, Shokravi S, Samini M. 1992. Opposite influences of dopaminergic receptor subtypes on penile erection. Gen Pharmacol 23: 671-675.

Zvara P, Sioufi R, Schipper HM, Begin LR, Brock GB. 1995. Nitric oxide mediated erectile activity is a testosterone dependent event: A rat erection model. Int J Impot Res 7: 209-219.

3 Feminine Sexual Behavior from Neuroendocrine and Molecular Neurobiological Perspectives

J. D. Blaustein · S. K. Mani

1 Introduction: Why Study Feminine Sexual Behavior in Rodents from Neuroendocrine and Molecular Neuroendocrine Perspectives?

Feminine sexual behavior in animals is studied by researchers in a variety of fields for an assortment of reasons. While some are interested in feminine sexual behavior from a strictly behavioral, comparative, or evolutionary perspective, others use it as a biologically significant behavior worthy of study to understand neuroendocrine, cellular, and molecular mechanisms. In this review, we will focus on the background and the research that have contributed to an understanding of the neuroendocrine, cellular, and molecular neuroendocrine underpinnings of this complex set of behaviors.

The neuroendocrine regulation of feminine sexual behaviors has long been used as a model to probe the cellular processes by which hormones act in the brain to result in changes in behavior. There are several reasons why this particular set of behavior lends itself particularly well to this analysis. Feminine sexual behavior has reflexive components that are readily studied and quantified. Motivational components are under hormonal regulation, as are sensory and motor aspects. Finally, the behaviors are influenced by an integration of hormonal and environmental factors. Because of the richness of the literature on the neuroendocrine and molecular neurobiology of feminine sexual behavior in rats, guinea pigs, and mice, most of the work to be discussed will focus on those species.

2 Description of Behaviors

A variety of terms have been used to describe the various components of feminine sexual behavior. Beach (1976) proposed that feminine sexual behavior comprises three basic types of behaviors: receptivity, proceptivity, and attractivity. Receptivity in rodents is defined as the reflexive postural responses by the female to sexual contact. In female rats, sexual receptivity, exemplified by lordosis, is accompanied by immobility, rear leg extension, dorsiflexion of the spine, elevation of the head, and tail deviation (Pfaff et al., 1973). Proceptive behaviors include a number of behaviors that indicate the female's willingness to mate and are usually followed by mounts by the male. These behaviors include, for example, ear-wiggling and darting and hopping. Attractivity, or the elements that make the female attractive to the male, comprises both behavioral and nonbehavioral aspects. These may include olfactory cues and visual displays.

More recently, Blaustein and Erskine (2002) suggested another set of categories of sexual behavior to highlight the active contribution of the female. These include copulatory, paracopulatory, and progestative behaviors. Copulatory behaviors are the behaviors that result in successful transfer of sperm from the male to the female. Although, like the term receptivity, this refers mainly to lordosis, this term emphasizes active participation by the female. Additional copulatory behaviors include those postural adjustments that occur during lordosis, and which are required to facilitate intromission by the male (Adler et al., 1977).

Paracopulatory behaviors are species-typical behaviors displayed by females, which presumably arouse the male and stimulate him to mount. In addition to Beach's term, proceptivity (Beach, 1976), these behaviors have been referred to as precopulatory (Madlafousek and Hlinak, 1977) or soliciting (Erskine, 1989) behaviors. Paracopulatory behaviors shown by estrous female rats include ear-wiggling (Beach, 1976; Madlafousek and Hlinak, 1977), presenting postures (Emery and Moss, 1984a, b), a rapid sequence of approach toward, orientation to, and withdrawal from a sexually active male (darting and hopping) (McClintock and Adler, 1978), and production of ultrasonic vocalizations (White and Barfield, 1989). The frequency and rate at which these paracopulatory behaviors are expressed are influenced by many factors, including hormone levels, the rate of copulation, and a variety of experimental conditions (Erskine, 1989).

The third component of feminine sexual behavior, progestative behaviors, includes those behaviors that increase the probability of occurrence of pregnancy (Blaustein and Erskine, 2002). These behaviors include primarily the female's pacing of sexual interactions, through approaches toward and withdrawals from the male (Bermant, 1961; Peirce and Nuttall, 1961; Krieger et al., 1976; Gilman and Hitt, 1978; Erskine, 1985). Also included are behaviors that regulate the frequency and timing of intromissions and ejaculations and include the female's selection of males that are ready to ejaculate (McClintock and Anisko, 1982) and the female's postejaculatory interval, which enhances sperm transport by preventing removal of the copulatory plug (McClintock et al., 1982). These short-term behavioral adjustments in the timing of sexual interactions, in addition to adoption of a lordosis posture that allows for a sufficient level of vaginocervical stimulation (VCS) (Erskine et al., 2004), increase the probability of reproductive success.

3 Basic Neuroendocrinology of Feminine Sexual Behavior

Many behavioral tests have been used to assess the different aspects of sexual behaviors in female rodents. Each technique for assessing a particular aspect of sexual behavior fills in a different piece of the puzzle. Each technique has its advantages, and each has its disadvantages, which will be discussed, along with their neuroendocrine underpinnings.

3.1 Copulatory Behaviors

In many species, including rats and mice, lordosis in response to mounts by a sexually vigorous male is assessed as an index of copulatory (or receptive) behavior. Often, the lordosis quotient, the percentage of mounts that result in lordosis, is the only metric used. In rats, the lordosis rating, a rating on a 0–3 scale reflecting the intensity of the response, is often used as well (Hardy and Debold, 1971; Hardy and DeBold, 1972). In guinea pigs and hamsters, species with a sustained and obvious lordosis posture, lordosis duration is typically used, often in response to manual palpation (Young, 1969; Noble, 1973) (often with a soft brush in hamsters).

Although it took many decades and a multitude of experiments to fully understand the basic hormonal underpinnings of feminine sexual behavior in rats and guinea pigs, it is clear that during the estrous cycle, the sequential secretion of estradiol and progesterone from the ovaries results in a period of sexual behavior that is linked to the time of ovulation (Collins et al., 1938; Boling and Blandau, 1939; Powers, 1970; Barfield and Lisk, 1974) (➋ *Figure 3-1*). After the period of sexual behavior (heat; behavioral estrus) terminates, sexual receptivity is not seen until the proestrous stage of the reproductive cycle returns, with the next episode of secretion of estradiol followed by progesterone. Many of the experiments on the neuroendocrinology of feminine sexual behavior were limited to assessing influences only on receptivity, so a bit of caution is warranted in generalizations to other aspects of sexual behavior.

Removal of the ovaries causes an immediate decline in circulating ovarian hormones, and consequently cessation of the expression of feminine sexual behaviors (Dempsey et al., 1936; Boling and Blandau, 1939). Results of experiments in which rats or guinea pigs were ovariectomized at precise times during the estrous cycle demonstrate that the hormonal requirements for the induction of optimal sexual behavior include a sufficient period of estradiol priming followed by exposure to progesterone (Powers, 1970; Joslyn et al., 1971). Although far less work has been done in mice, it is clear that many of the same neuroendocrine principles apply to that species as well (Ring, 1944; Edwards, 1970; Thompson and Edwards, 1971; Mani et al., 1997; Rissman et al., 1997a, b), albeit with some notable differences, which will be discussed later (see ➋ *section 3.5*).

Estradiol and progesterone are both necessary and sufficient for optimal levels of sexual behaviors. Rats (Davidson et al., 1968), guinea pigs (Crowley et al., 1978), hamsters (Carter et al., 1976), and mice (Mani et al., 1997) respond to treatment with estradiol alone under some conditions. However, it is sequential treatment with estradiol and progesterone that typically results in expression of feminine sexual behavior that most resembles that seen in estrous-cycling rodents (Dempsey et al., 1936; Collins et al., 1938; Boling and Blandau, 1939; Beach, 1942; Edwards, 1970; Tennent et al., 1980). Increasing doses of estradiol used for priming allow lower levels of progesterone to be used to facilitate sexual behavior, and increasing doses of progesterone allow for lower priming doses of estradiol (Whalen, 1974).

Often after exposure to progesterone, rats (Powers and Moreines, 1976; Blaustein and Wade, 1977; Morin, 1977), hamsters (Carter et al., 1976), guinea pigs (Dempsey et al., 1936; Goy et al., 1966; Zucker, 1966, 1968), and mice (Edwards et al., 1968) become refractory to further stimulation of sexual behavior by either progesterone alone or, in some cases, estradiol and progesterone. Although progesterone desensitizes

Figure 3-1
Schematic diagram depicting the relationship between estradiol and progesterone during the estrous cycle of female rats. Although there are differences in the regulation of the estrous cycle among the species discussed in this review—guinea pigs, rats, hamsters, and mice—most of the elements of hormonal changes around the time of expression of sexual behaviors are similar. Note that the increase in progesterone around the time of metestrus/estrus is due to secretion from the corpus luteum, which is short-lived, unless the female rat becomes pregnant or pseudopregnant

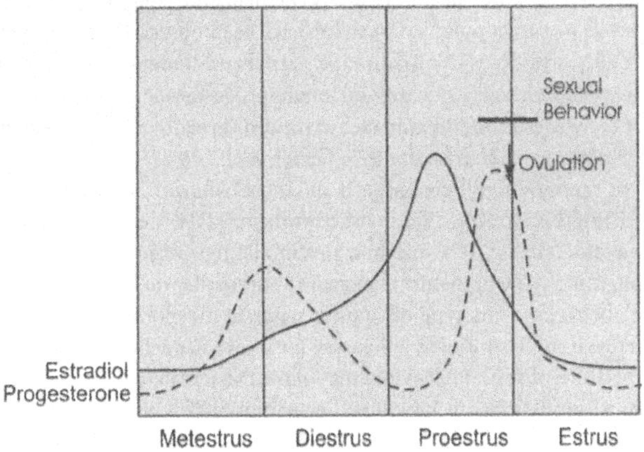

response to itself in ovariectomized rodents receiving exogenous hormones, and it has generally been accepted that progesterone inhibits sexual behavior in guinea pigs under some circumstances (Goy et al., 1966; Feder et al., 1968), its role in termination of sexual behavior during the estrous cycle (Sodersten and Hansen, 1977; Hansen and Sodersten, 1978; Sodersten and Eneroth, 1981) and pregnancy (Baum et al., 1979; Blaustein and Feder, 1979a) of rats is unclear. As will be discussed later (see *section 4.2.6.2*), it has been suggested that it is not progesterone per se that is the proximate cause of heat termination. Rather, it is the secondary influence of progesterone and other factors on the regulation of neural progestin receptors (PRs) that seems to be critical to heat termination and the ensuing refractoriness to further stimulation by progesterone (also known as sequential inhibition).

Estradiol and progesterone are not the only sex hormones involved in regulation of sexual behaviors. In fact, dihydrotestosterone is inhibitory to the induction of sexual receptivity (Baum et al., 1974; Tobet and Baum, 1982; Blasberg et al., 1998), and it has been suggested as one of the factors that determines the duration of the period of sexual receptivity during the estrous cycle (Erskine, 1983).

The time course of estradiol and progesterone action on sexual behavior is of great interest, particularly inasmuch as it provides additional clues to the underlying cellular processes by which hormones act in the brain. Although estradiol priming of response to progesterone generally takes about a day (Green et al., 1970; Feder and Marrone, 1977), an intravenous injection of progesterone facilitates the expression of lordosis within an hour of injection (Meyerson, 1972; Kubli-Garfias and Whalen, 1977; McGinnis et al., 1981; Glaser et al., 1983). Therefore, it is clear that progesterone induces its neuronal changes associated with sexual behavior considerably more rapidly than estradiol, which suggests possible differences in the cellular mechanisms of each of the hormones.

3.2 Paracopulatory Behaviors

In some studies in rats and hamsters, paracopulatory behaviors, such as darting and hopping, ear-wiggling (Gerall et al., 1972; Pfaus et al., 2004), and ultrasound production (Barfield and Schoelch Krieger, 1977; McIntosh et al., 1977; Geyer and Barfield, 1978; Geyer et al., 1978; Floody et al., 1979), have also been assessed. Sometimes, rejection behaviors, presumably indicative of aversive components of sexual interaction, are also quantified.

Paracopulatory behaviors are generally thought to be dependent on progesterone. Although estradiol alone induces paracopulatory (Fadem et al., 1979; Gorzalka and Moe, 1994), as well as copulatory (Boling and Blandau, 1939), behaviors in ovariectomized rats, in most cases, progesterone of adrenal origin has been shown to be necessary for the paracopulatory, but not the copulatory behaviors (Tennent et al., 1980; Gorzalka and Moe, 1994). However, with extremely high doses of estradiol, paracopulatory behaviors can be seen even in ovariectomized–adrenalectomized rats (Zemlan and Adler, 1977). It should be noted that on occasion, very high levels of paracopulatory behavior have been observed in ovariectomized–adrenalectomized rats, even with quite low doses of estradiol (SF Farrell and JD Blaustein, unpublished observations), suggesting that paracopulatory behaviors are not totally progesterone dependent. While lordosis can be observed within 1 h of progesterone treatment, maximum levels of paracopulatory behaviors are seen within about 2 h of treatment (Fadem et al., 1979; Glaser et al., 1983).

The importance of examining different aspects of sexual behavior is illustrated in a recent pharmacological experiment. Using this strategy, Pfaus and coworkers (2004) demonstrated a selective stimulatory effect of PT-141, a peptide analog of α melanocyte-stimulating hormone, which binds to melanocortin 2 receptors on the soliciting (paracopulatory) behaviors, but neither on lordosis (copulatory behavior) nor pacing (progestative) behavior. This type of finding suggests that lordosis, although critical for vaginal intromission and useful in some studies as a bioassay for particular cellular processes, cannot be used as a proxy for the full repertoire of sexual behaviors that allow the female to also initiate and pace the timing of sexual contact. Thus, limitations of individual behavioral tests highlight a need for a combination of behavioral tests required to elucidate the complexities of feminine sexual behavior.

3.3 Progestative Behaviors: Pacing

While indices of copulatory and paracopulatory behaviors are useful parameters, feminine sexual behavior is far more complex than these. Many studies of feminine sexual behavior have been performed in small arenas in which the male, not the female, is the principal controller of the rate of sexual stimulation. In contrast, in a natural or semi-natural condition, the solicitation behavior of the female activates the male pursuit and consequently the type and timing of mating stimulation (McClintock and Adler, 1978; McClintock and Anisko, 1982). Several methods are used to allow and assess the pacing of sexual interactions by the female. These include mating in bilevel chambers in which the female can move between the two levels to pace the rate of sexual interaction (Mendelson and Pfaus, 1989; Pfaus et al., 1999), an arena with a partition with a small hole through which the female, but not the male, can enter and exit (Erskine et al., 1989; Paredes and Alonso, 1997), or a bar-pressing situation in which the female must press a lever to gain access to the male (Bermant, 1961; Bermant and Westbrook, 1966; Matthews et al., 1997).

The role of estradiol and progesterone in pacing behavior has been studied. In general, nonreceptive rats do not pace (Brandling-Bennett et al., 1999). Although paracopulatory behaviors are dose dependent on progesterone, recent data suggest that pacing is unaffected by the dose of progesterone (Brandling-Bennett et al., 1999). However, it should be noted that in some situations, dose dependence of contact–return latencies on progesterone has been observed (Gilman and Hitt, 1978; Fadem et al., 1979). The reasons for the discrepancy in the role of progesterone are unclear, but it may relate to the specific procedures used.

In a direct comparison of paced mating during the estrous cycle and after injection of 10 μg estradiol benzoate followed by 500 μg progesterone in ovariectomized rats (Zipse et al., 2000), return latencies after mounts or intromissions, but not ejaculations, were considerably longer in the ovariectomized, hormone-replaced rats than in ovary-intact rats. Although this issue of timing of rate of copulation is of tremendous importance, at this time, there is virtually no understanding of the cellular mechanisms that underlie these acute, transient changes in the female's behavior.

It must be kept in mind that not all effects of steroid hormones on feminine sexual behavior are due to effects of the hormones in the brain. For example, the systemic administration of the estrogen antagonist, ICI-182,780, which does not seem to cross the blood-brain-barrier (Wade et al., 1993), was injected in rats treated with estradiol + progesterone (Clark et al., 2003). While the antagonist was without effect on the expression of lordosis in response to mating stimulation by male rats, it lengthened the return latencies after intromission and ejaculations in a paced mating situation. Although the mechanism by which this occurs is not known, it is unlikely to be by an action on the brain. The possibility has been suggested that peripheral, nociceptive changes are involved (Clark et al., 2003).

3.4 Sexual Motivation

While assessment of naturally occurring paracopulatory behaviors, such as darting and hopping, and ear-wiggling, goes beyond mere recording of the lordosis response (Erskine, 1989), there is no evidence that these are indices of the female's motivation to mate. Because hormones influence the female's motivation to mate, techniques have been developed to assess this more complex aspect of feminine sexual behavior. These include, for example, tests of partner preference, the increasing barrier method, and tests of conditioned place preference.

Assessing sexual motivation often involves determining whether the female will seek proximity to a sexually active male (Meyerson and Lindstrom, 1973; Clark et al., 1981a; Edwards and Pfeifle, 1983). In addition, it can be determined if the female displays a partner preference for a sexually active male over another stimulus animal (Meyerson and Lindstrom, 1973; Edwards and Pfeifle, 1983). In this situation, females are placed in a three-compartment chamber in which two stimulus animals are positioned in different compartments, and the length of time the female spends with each animal is recorded. Preference for a sexually active male is interpreted as an increase in her sexual motivation. In general, rats in the proestrus/estrus stage of the estrous cycle, or rats injected with estradiol, testosterone, or estradiol + progesterone, are more likely to approach sexually active males (Meyerson and Lindstrom, 1973; Edwards and Pfeifle, 1983; De Jonge et al., 1986; Clark et al., 2004), demonstrating an influence of these hormones on the female's sexual motivation.

The female's motivation is influenced by the ability of the male to copulate with her. Females spend less time with males that are allowed to copulate with them than with males that are not allowed to mate. In tests with two sexually active and interactive stimulus males, ovariectomized rats treated with estradiol and progesterone avoid contact with the males, and spend more time in a neutral compartment. In addition, when given a choice between a castrated and an intact male, the females spend more time with the castrated male than with the intact male. However, females exposed to males in a situation that restricts their sexual interactions (for instance, by keeping the males in a wire cage or by using a vaginal mask to prevent intromissions) spend more time with sexually active males than with castrated, inactive males (Meyerson and Lindstrom, 1973; Broekman et al., 1988). These results suggest that females initially seek out contact with sexually active males, but that the mating stimulation they receive results in subsequent avoidance of this stimulation. This avoidance has also been observed in female rats, where intensive mounting accompanied by intromission and ejaculation by stimulus males during mating, not only reduces the probability and the intensity of lordosis response, but also enhances the rejection behaviors by the female (Hardy and Debold, 1971).

Stimulation by a male is reinforcing only when the females are allowed to pace the sexual interaction (Gilman and Westbrook, 1978), consistent with the idea that paced mating is a rewarding aspect of sexual behavior (Paredes and Vazquez, 1999; Martinez and Paredes, 2001). Finally, when a contact situation allows for pacing, lack of preference for a sexually active male is demonstrated in her pacing behavior, spending a lot of time away from the male (Clark et al., 2004). Therefore, the specific type of mating stimulation provided can have dramatic influences on the female's apparent motivational state, and the particular testing situation used must be kept in mind in interpretation of each experiment.

There are many inconsistencies in published reports of experiments, which have assessed motivation through the use of partner-preference tests (Oldenburger et al., 1992). Clark et al. (2004) recently published an informative paper that elaborates the inconsistencies in results of experiments that have used this technique. They point out the importance of allowing or not allowing contact with the male have on the results of partner-preference tests. They argue for the importance of having a contact and no-contact test when assessing sexual motivation using partner-preference tests.

Sexual motivation can also be assessed by the increasing barrier method in which an electric grid is positioned between the starting cage and the goal cage containing a stimulus male (McDonald and Meyerson, 1973). The grid current is increased across trials, and the number of grid crossings and the maximum current exposure to gain contact with the male are measured as indices of motivation. In general, proestrous–estrous female rats and ovariectomized rats injected with estradiol are more willing to overcome the aversive stimulus and perform a higher number of grid crossings when approaching a sexually active male (Meyerson and Lindstrom, 1973), consistent with the idea that hormonal changes influence the female's motivation for sexual behavior.

Recently, conditioned place preference has been used extensively to assess the female's motivation for various aspects of the mating situation (Oldenburger et al., 1992). For example, it has been used to establish that estradiol– or estradiol + progesterone-treated female rats prefer a compartment, which has previously been associated with sexual interaction. They prefer, however, a compartment associated with a caged male to one in which copulation has been allowed. Similar to what was seen in partner-preference experiments, females develop a conditioned place preference for compartments in which they have been allowed to pace the mating, but not places where rate of copulation has been determined by the male (Martinez and Paredes, 2001). Furthermore, estradiol alone is not sufficient to induce the reward state that leads to a conditioned place preference; progesterone is necessary (Paredes and Alonso, 1997; Paredes and Vazquez, 1999; Gonzalez-Flores et al., 2004a).

3.5 Mice: A Different Pattern

The feminine sexual behavior of mice differs from rats, guinea pigs, and hamsters in a fundamental way. Although other species tend to respond to appropriate hormonal treatment at first exposure after ovariectomy, mice generally do not (Thompson and Edwards, 1971; Mani et al., 1996; Mani et al., 1997; Rissman et al., 1997a). In fact, mice often require numerous exposures to estradiol and progesterone before they express high levels of sexual receptivity, and as would be expected, this varies with strain of mouse (Thompson and Edwards, 1971) and housing conditions (Laroche et al., 2005). Interestingly, weekly hormone treatments are not effective without being accompanied by the experience of behavioral testing (Thompson and Edwards, 1971).

Most research on mice assesses sexual receptivity by way of the expression of lordosis in response to mounts by a sexually vigorous male. However, in one attempt to determine the presence of other behaviors using a semi-natural environment (Garey et al., 2002), a darting behavior was reported after mounting, that the authors suggested resembled a paracopulatory behavior. Likewise, after being mounted, the female was reported to dart to another part of the arena and then return, similar to pacing in rats.

4 Cellular Mechanisms of Hormone Action

As demonstrated by the preceding discussion, a good deal is known about the hormonal regulation of paracopulatory and progestational behaviors and on motivational aspects of sexual behaviors. However, the majority of research on cellular and molecular underpinnings of sexual behavior has been on the simplest, most reductionist measure of sexual receptivity, the lordosis response. Much of the discussion regarding cellular mechanisms of feminine sexual behavior will be focused on this behavior, but information about other aspects of sexual behavior will be added when known.

It was originally proposed (Gorski et al., 1968; Jensen et al., 1968) that steroid hormone receptors act by a two-step mechanism that involved steroid hormone binding to the receptor in the cytoplasm of cells, the receptor undergoing a conformational change, allowing it to be translocated to the cell nucleus. This model has morphed and become far more complex in the past four decades. The two-step model was modified in 1984 (King and Greene, 1984; Welshons et al., 1984), when it was reported that both unoccupied and occupied estrogen receptors (ERs) are present almost exclusively in cell nuclei. It was suggested that the high concentration of unoccupied receptors typically found in the cytoplasm (usually in the cytosol—the high-speed supernatant after cell fractionation) was a procedural artifact. It should be noted that the presence of unoccupied receptor in cytoplasm remains somewhat controversial, seen in some situations (e.g., Blaustein, 1994), not others.

What started out as a very simple model of hormone binding to receptor protein and inducing changes in gene expression has become more elaborate over the years (➔ *Figure 3-2*). A variety of protein–protein interactions is involved in steroid hormone regulation of the transcriptional machinery. This includes association with various heat shock proteins (Toft et al., 1987; Pratt and Toft, 1997; Gehring, 1998) and coregulators (McKenna et al., 1999; Tetel, 2000). Furthermore, much is now known about the nature of

Figure 3-2

Simplified, schematic representation of mechanism of action of steroid hormone receptors. Unoccupied receptors are present inactive complexes associated with heat shock proteins and chaperone proteins (not shown). In the classical, ligand-dependent pathway of activation of receptors by steroid hormones, binding of cognate hormone to receptors results in conformational change, dissociation of heat shock proteins, dimerization of the receptor, and binding to target DNA sequences (hormone response elements). The ligand-induced conformational change facilitates recruitment of cofactors and other basal TFs to the promoters of target genes, producing a transcriptionally active complex that can direct target gene transcription. Protein kinases and PPs present in the cell contribute to receptor activation by altering the phosphorylation of both receptor and coregulators (e.g., SRC-1 and cAMP-binding protein). Compounds such as neurotransmitters and growth factors can activate second messengers and protein kinase pathways in a rapid manner to activate receptors and/or coactivators in a ligand-independent pathway. In addition, although not depicted here, rapid effects of hormone and metabolites may involve direct actions on membrane receptor sites, GPCRs, ion channels, and second messenger systems

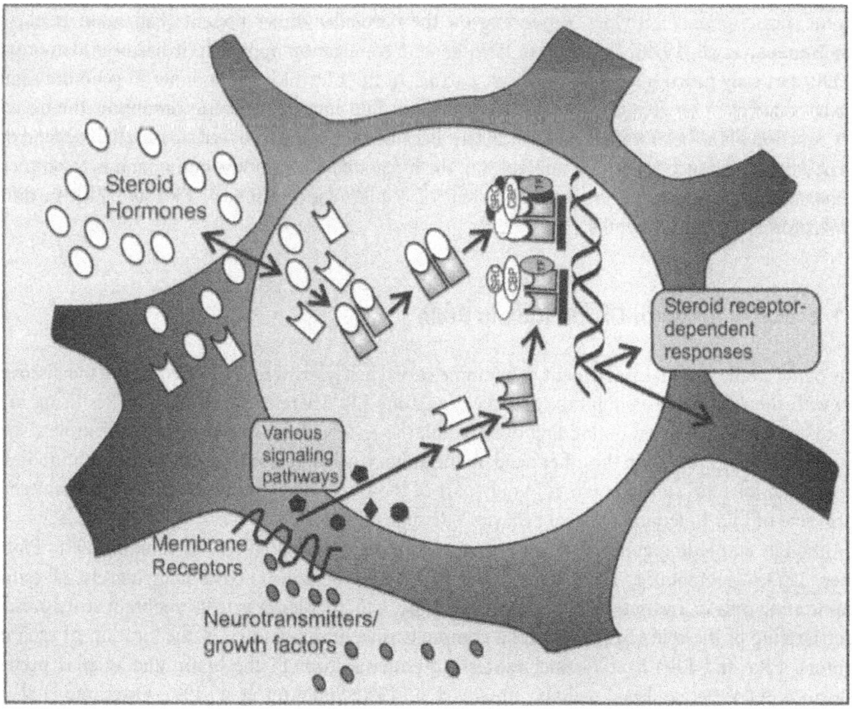

hormone response elements, often found in the promoter region of target genes. Coregulators (coactivators and corepressors) have been demonstrated to form a functional link between the activated steroid receptors and the transcriptional machinery for efficient transcriptional regulation. The modular arrangement of these complexes provides potential for different activators to assemble diverse configurations of regulatory complexes at the promoters of genes (McKenna et al., 1999; Glass and Rosenfeld, 2000; McKenna and OMalley, 2002). In vitro studies suggest that the relative expression levels of the coactivators and corepressors determine cell-specific, appropriate, and graded responses to steroid hormones (Lonard and O'Malley, 2005).

It also was originally believed that steroid receptors required hormone ligand for activation, which is of course how steroid receptors originally received their names. However, it is now well accepted that receptors can be activated by a variety of intracellular signaling pathways in the absence of hormone by a process called ligand-independent activation (Power et al., 1991a, b; Blaustein, 2004b). This will be discussed later in this chapter (➋ *Section 7.1*).

4.1 Estrogen Receptors

Nearly 20 years after an estrogen-specific binding protein that met the criteria for a receptor was first documented (Gorski et al., 1968; Jensen et al., 1968), and its presence confirmed in the brain (Eisenfeld, 1969), the first ER gene, *ERα*, was cloned and sequenced (Greene et al., 1986). A second ER gene (*ERβ*) was discovered (Kuiper et al., 1996) 20 years later, and its protein was also found in the brain. ERβ shares a high level of sequence homology with ERα in some portions of the protein. Although both receptors bind to estradiol with similar binding affinity, they may differ in their affinity for other estrogens (Kuiper et al., 1997).

Depending on the ligand and the estrogen response elements present in cells, ERα and ERβ can have different transcriptional activities (Paech et al., 1997; Hyder et al., 1999; Zou et al., 1999). Furthermore, ERα and ERβ may heterodimerize or homodimerize, then bind to estrogen response elements, and have different transcriptional activities, depending on the particular dimer present (Pettersson et al., 1997, 1998; Tremblay et al., 1999). Using an in vitro genetic conjugation approach, it has been demonstrated that ERα not only functions as the dominant partner in the ERα/ERβ heterodimer in genomic signaling pathways dependent on estrogen response elements, but ERα homodimers also can mimic the transregulatory function of the heterodimer (Li et al., 2004). Because the responses of cells to estradiol depend on the presence and abundance of each ER subtype and the integrated effects of the homo- and heterodimers, an understanding of the distribution and the extent of cellular coexpression of ERα and ERβ in different neuroanatomical areas is essential.

4.1.1 Estrogen Receptor Distribution in Brain

Many of the studies on ER involvement in feminine sexual behavior were performed before the discovery of ERβ, with the reasonable assumption that there was one ER. Therefore, in earlier studies using in vitro receptor assays, which relied on binding of estradiol, the results obtained represented a complex, variable mixture of both receptors. On the other hand, in immunocytochemical studies, the original antibodies used were specific for ERα, so the results told only part of the story, albeit an important part, considering the critical role of ERα in feminine sexual behavior (Rissman et al., 1997a).

Although autoradiographic work and in vitro binding assays (Pfaff, 1968; Stumpf, 1968; Pfaff and Keiner, 1973; Warembourg, 1977; Rainbow et al., 1982c) demonstrated a high density of estradiol-concentrating cells in a variety of hypothalamic, limbic, and, to a lesser extent, midbrain structures, a full understanding of the role of ERs requires a comprehensive understanding of the location of each of the receptors. ERα and ERβ have distinct expression patterns through the brain, and each is present in abundance in a variety of brain regions (Simerly et al., 1990; Shughrue et al., 1997; Osterlund et al., 1998;

Shughrue et al., 1998a; Greco et al., 2001). In some cases, both are present in the same locations or individual neurons (Greco et al., 2001; Devidze et al., 2005), but not in others.

Although not localized exclusively to these regions, ERα mRNA or protein is expressed most densely throughout the bed nucleus of stria terminalis, in the posterodorsal part and cortical nucleus of the amygdala, medial and periventricular preoptic area, arcuate nucleus, ventromedial and periventricular hypothalamus, the ventral part of the lateral septum, and the midbrain central gray (Simerly et al., 1990; DonCarlos et al., 1991; Shughrue et al., 1997; Merchenthaler et al., 2004).

Abundant ERβ mRNA or protein has been reported in hypothalamic and limbic regions, which are known to contain high concentrations of ERα and PRs, as well as in other regions known to express little or no ERα or PRs. Although there is overlap of the areas expressing ERβ and ERα mRNA (Shughrue et al., 1996, 1997) and/or protein (Li et al., 1997; Shughrue and Merchenthaler, 2001; Merchenthaler et al., 2004), in the bed nucleus of stria terminalis, most amygdaloid nuclei, and medial and periventricular preoptic area, distribution of ERβ mRNA or protein extends to some regions containing no detectable or low levels of ERα (Shughrue and Merchenthaler, 2001; Merchenthaler et al., 2004). Some of the most striking differences in distribution between ERα and ERβ are found in the paraventricular and supraoptic nuclei, which are rich in ERβ mRNA with almost no detectable ERα mRNA (Shughrue et al., 1996, 1998a; Simonian and Herbison, 1997; Alves et al., 1998; Laflamme et al., 1998), as well as the dentate gyrus, isocortex, accessory olfactory nucleus, anterodorsal portion of the preoptic nucleus, and cerebellum. Most importantly, for a discussion of hormonal regulation of feminine sexual behavior, ERβ is abundant in the medial preoptic area, bed nucleus of stria terminalis, medial amygdala, ventromedial nucleus of the hypothalamus (VMN), midbrain central gray, and ventral tegmentum in rats (Shughrue and Merchenthaler, 2001; Merchenthaler et al., 2004), and it is present in all of these regions in mice as well (Mitra et al., 2003).

Since ERs can exist as homodimers or heterodimers, it is essential to know the coexpression patterns of the two receptors at the cellular level, not just at the level of whole brain areas. The highest proportion of cells coexpressing ERα and ERβ are within the principal nucleus of the stria terminalis, the posterodorsal medial amygdala, and the periventricular preoptic nucleus (Shughrue et al., 1998b; Greco et al., 2001). In the anteroventral periventricular area, 83% of the cells expressing ERα protein also coexpress ERβ mRNA. However, many cells containing only ERα or ERβ are also found in these regions. Because genes may be differentially regulated by the particular combinations of the ER dimers, changes in the intracellular coexpression may influence the moment-to-moment response of particular ER-dependent genes.

4.1.2 Role of Transcription Factor (Direct Genomic) Estrogen Receptors in Sexual Behavior

4.1.2.1 Interference with Estrogen Receptors The necessity of ERs for action of estradiol on sexual receptivity has been tested directly in two ways; using estrogen antagonists, which block the binding of estradiol to ERs, and in ER gene-disrupted mice (ERKOs, ER knockouts).

4.1.2.2 Estrogen Antagonists Receptor-blocking experiments demonstrated the necessity of the presence of unoccupied ERs and of accumulation of bound receptors in mediating the effects of estradiol on sexual behavior. Although the first estrogen antagonists used did not differentiate the two forms of ERs, treatment with a variety of antiestrogens that block the binding of estradiol to ERs, also blocks the effects of estradiol on the induction of sexual behavior (Landau, 1977; Roy and Wade, 1977; Walker and Feder, 1977a; Wade and Blaustein, 1978; Etgen, 1979; Vagell and McGinnis, 1996).

Antiestrogens can act as both estrogen antagonists and estrogen agonists, depending on the cell type and the context. For example, the antiestrogen tamoxifen and some other antiestrogens are antiestrogenic on breast tissue (Jordan, 1995), but estrogenic on bone metabolism (Jordan, 1999). Likewise, antiestrogens may block effects of estradiol on feminine sexual behavior (Roy and Wade, 1977), but have estrogenic effects on other behaviors [e.g., feeding; Roy and Wade (1976)] or neuronal endpoints. Furthermore, antiestrogens can act both as antagonists or agonists on the same end-points, including feminine sexual behavior, depending on timing of administration and dose (Walker and Feder, 1977b; Wade and Blaustein, 1978; Wilcox and Feder, 1983; Etgen and Shamamian, 1986). While the data currently available deal with the effects of antiestrogens on ERs in general, it will be interesting to learn the effects of ERα- versus ERβ-specific antagonists and agonists on feminine sexual behavior.

4.1.2.3 Estrogen Receptor Gene-Disrupted Mice Because the two forms of ERs are made from independent genes, knockout strains of mice have been developed in which the gene for each (Lubahn et al., 1993; Krege et al., 1998) or both (Couse and Korach, 1999) of the receptors is disrupted. Targeted disruption of the *ERα* gene completely eliminates hormonal induction of feminine sexual behavior (Rissman et al., 1997a; Ogawa et al., 1998). However, disruption of the ERβ is without apparent effect in ovariectomized, hormone-injected mice (Kudwa and Rissman, 2003), but extends the period of behavioral estrus and enhances receptivity in estrous cycling mice (Ogawa et al., 1999). Double knockout mice (αβERKOs) with disruption of both ERs are not only infertile, but also show decreased levels of sexual receptivity confirming the critical role of ERα in copulatory behavior in mice (Kudwa and Rissman, 2003). Interestingly, an important role for ERβ in defeminization of the brain and behavior has recently been suggested (Kudwa et al, 2005).

While the knockout strains of mice provide powerful models for the study of the role of particular proteins in behavior, they suffer from all of the well-known shortcomings of this technique. The observed effects are a result of loss of receptors during early development, as well as absence during adulthood. In addition, some knockouts are not true null mutations. In fact, the first ERαKO mouse developed (Lubahn et al., 1993) had a residual ER fragment containing the AF2 region of the gene (Couse et al., 1997), and it has been shown that the transactivational activity can be as high as 75% of the full ERα in one cell type (Kos et al., 2002) and varies considerably by cell type. While this fragment does not allow for estradiol facilitation of feminine sexual behavior under typical hormonal priming conditions (Rissman et al., 1997a), under some conditions there is, in fact, a moderate amount of induction of neural PRs that could be mediated by this fragment (Moffatt et al., 1998; Kudwa and Rissman, 2003).

4.1.3 Necessity of Long-Term Retention of Estrogen Receptors

Is it necessary that estradiol remain bound to ERs in neurons for the full period of priming of sexual behavior, or is a short exposure sufficient? Although it was originally proposed that estradiol is needed only to trigger the early events necessary for subsequent response to progesterone (McEwen et al., 1975), later work showed that a small amount of estradiol remained bound to ERs in vivo a day after estradiol injection (Blaustein et al., 1979), the time at which animals express sexual behavior in response to progesterone treatment. Treatment with estrogen antagonists, long after estradiol injection, eliminates this remaining bound estradiol, inhibiting the expression of sexual behavior (Blaustein and Wade, 1977; Walker and Feder, 1979; Blaustein and Feder, 1979a). Therefore, it appears that estradiol must remain bound to activated ERs and interact with ER elements on particular genes around the time that sexual behavior is expressed. The reason for this undoubtedly relates to half-lives of transcripts and relevant proteins directed by the occupied ERs, but this area of research has not progressed to a point at which this type of question about molecular neuroendocrinology can be answered.

4.1.4 Involvement of Coregulators of Estrogen Receptors

Although a great deal of work has been done on the role of coregulators in modulation of transcriptional activity of ERs in vitro, very little has been done within the context of ER action in behavior. However, work in rats using antisense oligonucleotides to steroid receptor coactivator (SRC) mRNA reveals that SRC-1 and CBP (cAMP response element binding protein [CREB] binding protein) act together to modulate the induction of sexual receptivity by estradiol (Molenda et al., 2002), as well as progesterone (Molenda-

Fugueira et al., 2006). Likewise, others (Apostolakis et al., 2002) have shown the importance of SRC-1 and SRC-2 in the cellular action of estradiol in the induction of feminine sexual behavior in rats and mice.

4.1.5 Transcriptional Networks Involved in the Ventromedial Nucleus of the Hypothalamus

A novel approach has recently been used to characterize the network of genes involved in hormonal regulation of lordosis. Making the assumption that the ventrolateral aspect of the VMN is an essential component in the display of hormonally induced lordosis, individual neurons were probed for the expression of various mRNAs (Devidze et al., 2005). Real-time quantitative PCR was used to determine expression patterns for mRNAs for ERα, ERβ, oxytocin receptor, and four protein kinase C (PKC) isozymes. This type of experiment, of course, generates a tremendous amount of data. However, the most interesting results are obtained from analysis of patterns of expression of multiple transcripts. A marked sex difference was observed in the expression of four different PKC isozymes in cells expressing ERα, ERβ, and oxytocin receptors. However, there are some important limitations to this single-cell approach. Lesions of this particular neuroanatomical area do not always lead to severe deficits in estradiol plus progesterone-induced sexual behavior (➋ *Section 5*), and of course, there is no way of determining if the neurons selected are actually involved in the particular behavior or physiological end-point of interest. Nevertheless, this novel approach promises to be a strong one in developing a catalog of the expression patterns associated with neurons in brain areas involved in feminine sexual behavior.

4.1.6 Regulation of Neural Estrogen Receptors by Ovarian Hormones

In general, estradiol downregulates ERα mRNA (Lauber et al., 1990; Simerly and Young, 1991), protein (Yuri and Kawata, 1991; Meredith et al., 1994; Osterlund et al., 1998), and [^{3}H]estradiol binding (Brown et al., 1996), but there are a number of discrepancies in the literature. Estradiol downregulates ERα mRNA levels in the arcuate nucleus and ventrolateral–VMN (Simerly and Young, 1991; Lauber et al., 1991). Greco et al. (2001) reported a decrease in ERα protein in the posterodorsal medial amygdala and paraventricular nucleus, but not the principal nucleus of bed nucleus of stria terminalis. However, estradiol treatment caused a decrease in ERα-ir in the guinea pig ventrolateral hypothalamus and the bed nucleus of stria terminalis, but not the periventricular preoptic area (DonCarlos et al., 1995). In related work in guinea pigs, estradiol downregulated ERα-ir in all brain areas studied, with the exception of the medial preoptic nucleus (Meredith et al., 1994). In recent work, it has been reported that estradiol increases ERα mRNA expression in some individual neurons of the ventrolateral–VMN (Devidze et al., 2005). Unfortunately, it is not known what accounts for these differences, but they could relate to specific physiological context or phenotypic differences in different cells.

Estradiol seems to either downregulate ERβ or have no effect, mainly based on the particular neuroanatomical area in which ERβ is assessed. One study reported a decrease in ERβ mRNA in the paraventricular nucleus, but not other areas (Patisaul et al., 1999), while another (Greco et al., 2001) reported a decrease in ERβ-ir in the periventricular preoptic area and the bed nucleus of stria terminalis, but not the paraventricular nucleus and posterodorsal medial amygdala. Shima et al. (2003) reported decreases in ERβ mRNA in a variety of areas including the amygdala, but not the medial preoptic area, ventromedial hypothalamus, or bed nucleus of stria terminalis. In contrast, a decrease in ERβ mRNA in the posterodorsal medial amygdala, but an increase in the arcuate nucleus, has been reported (Osterlund et al., 1998). In the recent study of mRNA within single cells in the ventrolateral aspect of the VMN discussed in the previous section, no change in ERβ mRNA expression was observed after estradiol treatment (Devidze et al., 2005) As with ERα, it is not known what accounts for these inconsistencies among studies. Nevertheless, it is reasonable to conclude that estradiol downregulates ERβ in some areas of the brain, and not others, and that the specifics depend on physiological and specific cellular context.

One situation in which the regulation of ERs by steroid hormones would be expected to be robust is after ovariectomy. Ovariectomy typically decreases the response of rats and guinea pigs to hormonal induction of sexual behavior (Beach and Orndoff, 1974; Clark et al., 1981b; Czaja et al., 1985; Delville and Blaustein, 1989). Long-term ovariectomy (Parsons et al., 1979; Clark et al., 1981b; Delville and Blaustein, 1989) results in a decrease in the concentration of estradiol-induced, hypothalamic PRs in rats, presumably secondary to diminished responsiveness to estradiol. Although typically no change in brain ERs has been reported after long-term ovariectomy (Parsons et al., 1979; Clark et al., 1981b), one study suggests that long-term ovariectomy results in an increase in ERα immunoreactivity (Liposits et al., 1990). Although this apparent increase in ERs after ovariectomy is consistent with the finding of downregulation of ER mRNA and proteins by acute estradiol (Simerly and Young, 1991; Meredith et al., 1994; DonCarlos et al., 1995), it does not explain the change in response to estradiol and progesterone, which typically results from ovariectomy. The fact that long-term ovariectomy, which decreases behavioral response to estradiol and progesterone, suppresses estradiol induction of PRs in the hypothalamus suggests a relationship of behavioral response to PR induction. The fact that ERs do not seem to be decreased by long-term ovariectomy suggests that regulation of response may occur at one of the many other steps involved in estradiol action or in subpopulations of ERα-containing neurons.

Although there have not been any studies to date that have investigated the effects of progesterone specifically on each form of ER, ligand-binding studies, which assess both ERα and ERβ, have typically found that progesterone downregulates ERs in hypothalamus (Smanik et al., 1983; Attardi, 1984; Blaustein and Brown, 1984).

Modulation of the concentration of steroid hormone receptors should be thought of as a means by which sensitivity to a particular class of hormone is modulated. Changes in the concentration of a particular receptor should result in changes in sensitivity to the hormone. Unfortunately, much of the immunocytochemical work that has been published has reported the number of cells, which immunostain positively for a particular receptor. However, "number of cells" can be influenced by sensitivity of the procedure used, and often tells us little about the concentration of protein in particular, relevant cells. While we are still a long way off from being able to assay concentrations of receptor protein in individual relevant cells, optical density measurements can provide indications of relative levels after particular treatments, as long as the quantitative aspects of the analysis are interpreted cautiously.

4.1.7 Another Neurally Active Form of Estradiol?

Finally, yet another twist in the ever-more complex story of ER action in the brain comes from reports that 17α-estradiol is an active estrogen in the brain (MacLusky et al., 2005; Toran-Allerand et al., 2005), while all along has been assumed that only the 17β-isomer was active. The mechanism by which the 17α-isomer exerts its effects is unclear, and it may involve a newly characterized receptor, ERX (Toran-Allerand et al., 2002, 2005) or a membrane receptor derived from the transcription-factor ER (Wade et al., 2001). Nevertheless, it is clear that 17α-estradiol induces functional changes in spatial memory and in synaptogenesis (MacLusky et al., 2005), and has neuroprotective effects in the brain (Green et al., 1997). It has also been suggested that 17α-estradiol is synthesized in the brain, as it is present even in ovariectomized–adrenalectomized rats (Toran-Allerand et al., 2005). Although potentially of great interest, earlier work failed to find an effect of this isomer on sexual receptivity (MacLusky et al., 1985).

4.2 Progestin Receptors

Although difficult to imagine considering the progress that has been made in the past three decades, early attempts to extend the intracellular receptor model of steroid hormone mechanisms to progesterone action in the brain were unsuccessful. There were numerous controversial and conflicting reports regarding the presence of neural PRs, based on in vitro binding techniques (Seiki and Hattori, 1973; Atger et al., 1974; Seiki et al., 1977) and in vivo unsaturable [^{3}H]progesterone uptake (Whalen and Luttge, 1971; Wade and

Feder, 1972; Luttge et al., 1974). While [^{3}H] progesterone or its metabolites were shown to concentrate in guinea pig brain in autoradiographic studies (Sar and Stumpf, 1973; Warembourg, 1978b), attempts to characterize the receptors biochemically were mostly unsuccessful (McEwen, 1976; Marrone and Feder, 1977). The development of R 5020 (promegestone), a synthetic progestin with a higher affinity for PRs than progesterone, was an important step in providing assays to identify and characterize neural PRs (Kato and Onouchi, 1977; Blaustein and Wade, 1978; Kato et al., 1978; MacLusky and McEwen, 1978; Moguilewsky and Raynaud, 1979b; Blaustein and Feder, 1979c).

4.2.1 Progestin Receptor Distribution in Brain

While estradiol priming dramatically increases the concentration of PRs (assayed by in vitro binding) in some neuroanatomical areas, including the hypothalamus and medial preoptic area of rats and guinea pigs (MacLusky and McEwen, 1978; Moguilewsky and Raynaud, 1979b; Blaustein and Feder, 1979c; Balthazart et al., 1980; Roselli and Snipes, 1983; Fraile et al., 1987) and the midbrain of guinea pigs (Blaustein and Feder, 1979c), PRs in other areas are mostly unaffected (MacLusky and McEwen, 1978, 1980). The punch microdissection technique provided slightly better spatial resolution than the grosser dissection done earlier (Parsons et al., 1982b; Thornton et al., 1986). With this technique, brain regions with the highest abundance of estradiol-induced PRs included the arcuate nucleus, the VMN, the periventricular preoptic area, the periventricular hypothalamus, the suprachiasmatic preoptic area, and the medial preoptic area, all areas that also have a high concentration of ERs (Rainbow et al., 1982c). Evidence obtained with the autoradiographic technique (Sar and Stumpf, 1973; Warembourg, 1978b) was consistent with the presence of PRs in the areas in which estradiol induces the highest level of PRs.

Despite the progress, a longstanding question that has remained unanswered is whether the PRs observed in many sites throughout the brain using in vitro binding assays represent a low level in all cells, or a few, scattered cells with an abundance of PRs (Blaustein and Olster, 1989). The development of a sensitive technique, immunocytochemistry (Warembourg et al., 1986; Blaustein et al., 1988; DonCarlos et al., 1989), provided much better anatomical resolution. Using this technique, the sites with the highest number of estradiol-induced PR-immunoreactivity (PRir) neurons in guinea pigs were found to be the bed nucleus of arcuate nucleus, periventricular preoptic region, medial preoptic nucleus, medial preoptic area, and the ventrolateral hypothalamic area. This technique also revealed PRir and ERir cells extending well beyond the Nissl-defined nuclei within the ventromedial/ventrolateral hypothalamus, a distribution that varies considerably from species to species (➋ *Figure 3-3*). Areas with lower numbers of PRir cells include the bed nucleus of stria terminalis, paraventricular nucleus, and lateral hypothalamus. Even within most of the PR-rich areas, relatively few neurons are darkly stained. Although immunocytochemistry is not quantitative in a linear sense, this suggests that relatively few neurons have a high concentration of PRs. Consistent with the idea that most effects of progesterone require priming with estradiol, estradiol-induced PRir is found only in cells containing ERir (Blaustein and Turcotte, 1989; Warembourg et al., 1989).

The neuroanatomical sites containing high concentrations of PR-containing cells are in reasonable agreement in guinea pigs and rats with a few exceptions, including the following. In rats, many of the PRir cells lateral to the ventrolateral–VMN are located in a somewhat less dorsal position. PRs, assessed by either immunocytochemistry (Numan et al., 1999; Greco et al., 2001), autoradiography (Sar, 1988), or in situ hybridization histochemistry (Hagihara et al., 1992) are also seen in the parts of the amygdala in rats. In guinea pigs, PRir is only seen in the amygdala after colchicine treatment (Blaustein and Olster, 1993), which disrupts axonal transport.

Although there are PRs in the cerebral cortex and cerebellum, these PRs do not seem to be induced by estradiol. In fact, early immunocytochemical (Warembourg et al., 1986; Blaustein et al., 1988) and autoradiographic experiments (Warembourg, 1978a) failed to demonstrate any PR-containing neurons in these areas, suggesting that, while the PR concentration is sufficient to assay by in vitro binding techniques, the cellular concentration of PRs may be too low to be detected immunochemically with the procedures that were used. However, it should be noted that in many of the studies, antibody concentrations were

Figure 3-3
Reconstruction of the location of PRir cells at two levels of the ventrolateral hypothalamus in guinea pigs in which sexual behavior was induced by implantation of an estradiol cannula in the rostral region ventrolateral hypothalamus, followed by progesterone injection. The VLN of guinea pig is considered to be analogous to the VMNvl of rats. Note that the effective cannula placements induced PRir in regions clearly outside the Nissl-defined VMN and VLN (arrows). VMN = ventromedial nucleus of the hypothalamus, VLN = ventrolateral nucleus of the hypothalamus; f = fornix (from Delville and Blaustein, 1991)

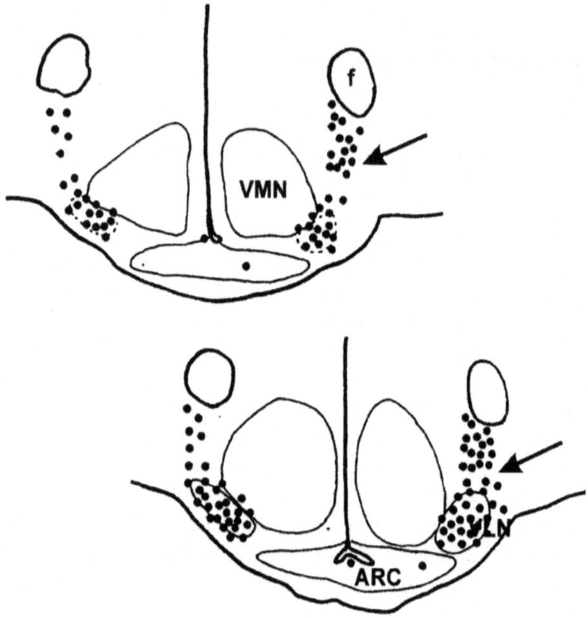

titrated, so that only estradiol-induced PRs were visible (Blaustein et al., 1988). If this titration is not done, many more PRir cells are identified (Greco et al., 2001).

4.2.2 Progestin Receptors and Facilitation of Feminine Sexual Behavior

PRs are generally believed to be essential in mediating the effect of progesterone on facilitation of feminine sexual behavior ,although some exceptions have been reported (Frye and Vongher, 2001), presumably by an interaction of activated hormone–receptor complexes with the genome (Blaustein and Olster, 1989; Blaustein and Erskine, 2002). This hypothesis predicts that sensitivity to progesterone is determined by the concentration of unoccupied PRs available in specific neurons involved in progesterone-facilitated sexual behavior, and response is dependent on the presence of an adequate concentration of activated PRs after progesterone treatment in those cells. An increase in the concentration of PRs after estradiol priming increases the sensitivity of target cells for progesterone, presumably by increasing the concentration of receptors available to become activated in response to progesterone. Similarly, a decreased concentration of unoccupied PRs results in decreased sensitivity to progesterone. Although a number of studies also suggest nongenomic effects of progestins on feminine sexual behaviors (e.g., Debold and Frye, 1994), these are not inconsistent with a PR-mediated, genomic action as well.

Although estradiol has numerous effects in the brain, one of the ways in which estradiol prepares the brain for response to progesterone is by increasing the concentration of PRs in relevant neurons. Estradiol injection results in an increase in the concentration of PRs in the hypothalamus and preoptic area within about a day in both guinea pigs (Blaustein and Feder, 1979c) and rats (Moguilewsky and Raynaud, 1979b; Parsons et al., 1980). The concentration of PRs first increases noticeably at about the same time as the first behavioral response to progesterone (Green et al., 1970; Feder and Marrone, 1977; Parsons et al., 1980). Behavioral response to progesterone is transient; response declines as the concentration of PRs in those areas declines. The timing of this decline is dependent upon the species, dose of estradiol, and mode of administration of the progesterone (Blaustein and Feder, 1979c; Parsons et al., 1980; Clark et al., 1982). Although most work has been done in ovariectomized animals in which hormone levels can be controlled, the concentration of hypothalamic unoccupied PRs in rats increases during proestrus, after peak estradiol levels and prior to, and presumably in preparation for, the preovulatory release of progesterone (McGinnis et al., 1981).

4.2.3 Interference with Progestin Receptors

4.2.3.1 Progestin Antagonists Treatment with progestin antagonists by system injection or by intracerebroventricular or intrahypothalamic administration inhibits the facilitation of sexual behavior by progesterone in rats, guinea pigs, and mice (Brown and Blaustein, 1984c; Etgen and Barfield, 1986; Richmond and Clemens, 1986; Mani et al., 1994a). In guinea pigs, RU486-induced inhibition of progesterone-facilitated, feminine sexual behavior was overcome by a high dose of progesterone, but not by cortisol (Brown and Blaustein, 1986), supporting the idea that inhibition was specific to PRs, and not due to antiglucorticoid (Moguilewsky and Philibert, 1984) or ERβ antagonist (Sathya et al., 2002) activity. Like estrogen antagonists that can have both antagonistic and agonistic effects, progestin antagonists have tissue-specific, agonist-like effects under particular conditions (Vegeto et al., 1992). One study suggested that the progesterone antagonist, RU486 (Pleim et al., 1990) can have progesterone-like effects on feminine sexual behavior under some conditions.

Recent in vitro studies on the structural and functional properties of PRs using isolated domains of the PRs have shed light on the mechanisms by which the RU486 acts as an agonist. By binding to the ligand-binding domain of the PR, RU486 induces conformational changes that not only prevent PR interactions with its coactivators, but also facilitate the interactions with and recruitment of corepressors, resulting in active repression of target gene expression (Allan et al., 1992; Jackson et al., 1997; Wagner et al., 1998a). The partial agonist property of RU486 is mediated exclusively through the PR-B isoform (discussed in *Section 4.2.4*) via its N-terminal domain (Meyer et al., 1990) and is dependent on the promoter context and the ratio of the coactivators and corepressors present in the cell. However, the regions of the PR-B domain and the actual mechanism by which RU486 functions as a partial agonist remain unknown.

4.2.3.2 Antisense Oligonucleotides Further evidence for the importance of PRs in the regulation of progesterone-facilitated sexual behavior comes from intraventricular (Mani et al., 1994c) or intrahypothalamic (Pollio et al., 1993; Ogawa et al., 1994) infusions of antisense oligonucleotides to PR mRNA, which inhibit PR synthesis. The results of these studies support the idea that facilitation of copulatory behavior, as well as paracopulatory behavior (Ogawa et al., 1994) by progesterone in rats, is mediated predominantly by genomic activation of neural PRs in the brain, including the ventromedial hypothalamic area (Pollio et al., 1993; Ogawa et al., 1994). Interestingly, the antisense oligonucleotides to PR mRNA decreased the levels of estadiol-induced cytosol PRs by half, a reduction sufficient to abolish lordosis response, suggesting the requirement of a threshold concentration of PRs for the facilitation of sexual behavior. A reduction in PR concentration would induce hyposensitivity and/or nonresponsiveness to progesterone, as noted later.

4.2.3.3 Progestin Receptor Gene Disruption A mutant strain of mouse with a targeted disruption of the PR gene, progesterone receptor knockouts (PRKOs) (Lydon et al., 1995), is completely unresponsive to the effect of progesterone in the facilitation of feminine sexual behavior (Mani et al., 1996 but *cf.* Frye and Vongher, 2001). In a parallel study of hypothalamic, unoccupied PRs in PRKO mice, a 70% reduction in estradiol-induced PRs was seen in the PRKO females, while heterozygous females had a 40% decrease in estradiol-induced PRs, compared with wild-type controls. It cannot be determined if the binding in the PRKO represents a residual low level of high-affinity binding (i.e., PRs) or the low-affinity binder that binds [^{3}H]progestins in vitro. However, it is clear that the PRKOs at least have a greatly reduced level of estradiol-induced brain PRs. Additional experiments using mice with targeted disruption of each PR subtype will be discussed later (➲ *Section 4.2.4.*).

4.2.3.4 Estrogen Antagonists and Agonists and Progestin Receptors As would be expected, work with estrogen antagonists is consistent with an obligatory role for estradiol-induced PRs in the facilitation of sexual behavior. In most cases, antiestrogen treatments that inhibit priming actions of estradiol on sexual behavior block induction of estradiol of PRs in the rat hypothalamus (Roy et al., 1979; Etgen and Shamamian, 1986). However, as was discussed earlier with respect to progesterone antagonists, estrogen antagonists (also referred to as selective ER modulators) also have estrogenic actions in some cases. Enclomiphene, an estrogen antagonist, substitutes for estradiol in its early priming effects on sexual behavior in guinea pigs, and it also results in an increase in the concentration of PRs in the hypothalamus (Wilcox and Feder, 1983). Likewise, the estrogen antagonist nafoxidine has estrogen antagonist effects in some situations, but agonist effects in others (Wade and Blaustein, 1978).

Interestingly, some xenoestrogens that have estrogen agonist properties can also act on other receptors. For example, the polychlorinated insecticide 1-(*o*-chlorophenyl)-1(*p*-chlorophenyl)-2,2,2-trichloroethane (*o,p*-DDT), has the estrogenic property of inducing behavioral response to progesterone in rats (Etgen, 1982). However, *o,p*-DDT also binds to PRs, suggesting that *o,p*-DDT can bind to ERs and induce PRs and then also bind to these PRs (Brown and Blaustein, 1984a).

4.2.4 Multiple Progestin Receptor Isoforms

The complexity of PR-mediated effects is augmented by the existence of two naturally occurring ligand-binding forms of the PR, PR-A and PR-B. Unlike the case with the two forms of ERs, the two PR isoforms are transcribed from the same gene and arise as the result of alternative initiation of translation from a single mRNA transcript (Conneely et al., 1989) or by alternative transcription from two distinct estrogen-inducible promoters on that gene (Kastner et al., 1990; Kraus et al., 1993). The production of these isoforms is conserved in a variety of vertebrate species, including rodents, primates, and humans (Lessey et al., 1983; Conneely et al., 1989; Duffy et al., 1997; Giangrande and McDonnell, 1999), rabbits being the only exception with only one form of PR (Loosfelt et al., 1984; Savouret et al., 1991). The isoforms differ only at the amino terminus, with PR-B containing an additional stretch of amino acids (128–165), known as the B-upstream sequence (BUS). The size variation in the isoforms is also reflected in their migratory profiles with the heavier form of the receptor (PR-B) typically migrating as a 110–120 kDa protein on Western immunoblots, and the smaller form (PR-A) typically migrating as an 80–95 kDa protein (e.g., Ilenchuk and Walters, 1987).

The BUS region of the PR also contains the transactivation function (AF-3) and is thought to be responsible for the divergent transactivational properties that have been reported for the isoforms, both in vitro and in vivo (Giangrande and McDonnell, 1999; Mulac-Jericevic and Conneely, 2004; Fernandez-Valdivia et al., 2005). The ability of PR-A and PR-B to form active homo- (A:A, B:B) or hetero-dimers (A:B) that bind DNA response elements, each with a distinct transactivational property (depending upon the absence or presence of the BUS domain) together with the differential coregulator recruitment, could expand the functional repertoire of progesterone responses and potentially dictate how a regulated expression of the two isoforms is critical for the spatial and the temporal responses to progesterone in vivo. Such a contribution by the distinct dimer ratios regulating progesterone effects is evident in vivo, where the ratios of the individual isoforms in reproductive tissues are known to change as a consequence of developmental (Kato et al., 1994; Beyer et al., 2002; Sakamoto et al., 2003; Aupperlee et al., 2005) and hormonal status (Mangal et al., 1997; Cheng et al., 2005). An imbalance of the native ratio of A to B form by the introduction of a PR-A transgene in mice results in developmental abnormalities of mammary glands, providing evidence that an aberration in the mechanisms regulating the differential expression of PR-A and PR-B can have major implications in physiology (Shyamala et al., 1998).

Although often ignored in work on the effects of progesterone in the central nervous system, these isoforms are of great importance. Experiments in vitro have suggested that some genes are preferentially activated by PR-B (Tora et al., 1988; Kastner et al., 1990; Vegeto et al., 1993), while some genes are equally responsive to PR-A and PR-B. In yet other cases, PR-A is an inhibitory regulator of PR-B (Vegeto et al., 1993). The isoforms also respond differently to progestin antagonists (Giangrande and McDonnell, 1999). While antagonist-bound PR-A remains inactive, antagonist-bound PR-B can function as a strongly active transcription factor (TF) by modulating intracellular phosphorylation pathways (Beck et al., 1993; Musgrove et al., 1993; Sartorius et al., 1994). Studies suggest differential subcellular localization of PR-A and PR-B in a mammalian cell line (Lim et al., 1999), with unoccupied PR-A being more nuclear in its localization than PR-B.

As with the finding of two forms of ERs, it is essential to keep in mind the presence of the two isoforms in interpretation of experiments on PRs. As progestins bind equally to both isoforms, and most antibodies are not selective for a particular isoform, nearly all experiments on PRs would have assayed a sum of both isoforms. However, a few studies have looked at the presence of PR-A and PR-B in the brain. Kato et al. (1993, 1994) first provided evidence for two distinct mRNA transcripts in the brain for PR-A and PR-B that seem to be regulated differentially. While PR mRNA transcripts are present throughout the brain, the PR-B form is differentially distributed, and is present in very high levels in cerebral cortex and in much lower levels in hypothalamus/preoptic area. In a more recent study (Camacho-Arroyo et al., 1998), both PR mRNA isoforms were induced and downregulated by estradiol and progesterone, in the hypothalamus. However, in the preoptic area, such a regulation was seen only for PR-B, not PR-A. In contrast, in the hippocampus, PR-A, but not PR-B, was induced by estradiol, and progesterone was without effect. Guerra-Araiza and coworkers (2000) also reported that during the estrous cycle, the PR-B isoform predominated in the hypothalamus, the preoptic area, and the frontal cerebral cortex, and the pattern of regulation differed among these areas. In contrast, no change was observed in the hippocampus. In monkey hypothalamus (Bethea and Widmann, 1998), the relative concentrations of the two isoforms are similar and do not appear to be differentially affected by hormonal treatments.

Interestingly, Guerra-Araiza et al. (2003) also examined the protein levels of the PR isoforms using western immunoblotting and found no downregulation of PR-B protein by progesterone in the preoptic area, contrary to their published findings on PR-B mRNA. This discrepancy between the mRNA and the protein levels could be due to the nature of antibodies used for protein detection by western immunoblotting. The available antibodies for PR bind to either PR-A and PR-B, or PR-B alone, and the extent of antigen–antibody reaction depends upon the epitope on the PR protein against which the antibodies are produced. Thus, reports of changes in isoform levels in neuronal tissue using western immunoblotting studies are antibody specific and need to be cautiously interpreted. Moreover, while the studies describe the spatiotemporal expression of one isoform over the other in the brain, the contribution of the individual isoforms in feminine sex behavior was not answered.

As mentioned earlier, recent in vivo studies using transgenic and knockout strains of mice demonstrate that normal tissue responses to progesterone are dependent upon appropriate levels of PR-A and PR-B (Shyamala et al., 1998; Mulac-Jericevic et al., 2000; Chou et al., 2003). The generation of PR isoform specific knockouts, PR-AKO and PR-BKO, has facilitated the analysis of spatiotemporal expression of the PR isoforms and their contributions to the reproductive actions of progesterone. The studies reveal that PR-A and PR-B isoforms modulate distinct, but partially overlapping, subset of reproductive functions of progesterone by regulation of diverse subsets of progesterone-responsive target genes in a cell- and tissue-specific manner (Mulac-Jericevic et al., 2000; Conneely et al., 2003; Mulac-Jericevic and Conneely, 2004; Fernandez-Valdivia et al., 2005). These knockouts have also elucidated the selective contribution of individual isoforms in progesterone-dependent and -independent facilitation of feminine sexual behavior. PR-AKO mice, containing the PR-B isoform alone, displayed significantly reduced levels of progesterone-facilitated lordosis response compared with their wild-type littermates. Interestingly, PR-BKO mice, containing the PR-A isoform alone, displayed higher levels of lordosis response than the PR-AKO mice, though not to the same level as their wild-type littermates. Western immunoblotting and normalization of the isoforms from the hypothalamus showed no evidence of upregulation of the detectable isoform upon ablation of the other. The studies demonstrate that the PR-A isoform plays a predominant role in mediating

Progesteron effects on feminine sexual behavior. The expression of the dominant PR-B isoform alone is insufficient to mediate progesterone-facilitated lordosis in mice. Furthermore, the observations also suggest that, while PR-A can mediate some of the effects of progesterone on lordosis, it is inadequate to mediate the full magnitude of the response, suggesting that the functional participation of both the isoforms, probably via heterodimerization of the two isoforms, contributes to the expression of the lordosis response.

4.2.5 Lack of Role of Progestin Receptors in Progesterone-Independent Sexual Behavior (Estrogen Heats)

Because estradiol binds to PRs in rats in vivo with an affinity 1% that of progesterone (MacLusky and McEwen, 1980; Parsons et al., 1984), it was suggested that induction of feminine sexual behavior in the absence of progesterone ("estrogen heats" or progesterone-independent sexual receptivity) might be referable to in vivo interaction of estradiol with intracellular PRs in addition to ERs. However, this idea has been ruled out by two major lines of evidence. First, a progesterone antagonist that blocks progesterone-facilitated sexual behavior does not block estradiol-induced or estradiol-facilitated sexual behavior (Blaustein et al., 1987). Second, estradiol induces lordosis, albeit at a lower level, in PRKO mice, mice without functional PRs, at the same level as in wild-type mice, demonstrating that the ability of PR-deficient mice to respond and exhibit lordosis is not compromised (Mani et al., 1996). As will be discussed in the section on the VMN (➋ *Section 5*), the two modes of induction of sexual behavior are dissociable in other ways as well, so it should not be expected that they induce sexual behavior through a common mechanism.

4.2.6 Cellular Mechanisms Regulating Estrous Duration and the Refractory Period

4.2.6.1 Downregulation of PRs Leads to Estrous Termination The period of sexual receptivity typical for each species is rather tightly regulated. For example, in estrous-cycling guinea pigs, as after estradiol and progesterone treatments, sexual receptivity lasts about 8 h (Young, 1969). Similarly, rats remain sexually receptive for approximately 14 h (Blandau et al., 1941), but in ovariectomized rats, the duration is very dependent on hormonal treatment and the manner in which heat duration is defined (Pfaus et al., 2000). We suggested that this timing of sexual receptivity is referable to the regulation of occupied PRs in the neurons of the hypothalamus and the preoptic area (Blaustein and Feder, 1980). Injection of a behaviorally effective dose of progesterone in estradiol-primed guinea pigs and rats causes the rapid occupation (and presumably activation) of brain PRs, including in the hypothalamus and the preoptic area (Blaustein and Feder, 1980; McGinnis et al., 1981; Rainbow et al., 1982b). In rats, the preovulatory secretion of progesterone during the estrous cycle binds to hypothalamic PRs (Rainbow et al., 1982b). In guinea pigs, the transient presence of occupied PRs after progesterone injection returns to baseline levels by about 12 h after injection and correlates well with the expression of sexual behavior (Blaustein and Feder, 1980). This temporal concordance first suggested that the termination of sexual behavior is due to loss of intracellular occupied PRs in relevant neurons.

Physiological manipulations that extend the duration of the period of sexual behavior also increase the duration of occupied hypothalamic PRs, consistent with the hypothesis that the maintenance of sexual behavior is dependent upon occupied PRs. For example, a supplemental injection of estradiol in conjunction with a facilitating injection of progesterone in ovariectomized guinea pigs extends heat duration (Joslyn and Feder, 1971; Blaustein, 1982b), and prolongs the duration of occupied, hypothalamic PRs, presumably by increasing the concentration of PRs available to bind progesterone (Blaustein, 1982b). Similarly, supplemental progesterone treatment that prolongs the duration of occupied PRs (Brown and Blaustein, 1985) increases the duration of behavioral estrus.

Although a decrease in blood levels of progesterone and termination of behavioral estrus are often coincident, they are not inextricably linked (Feder et al., 1968; Blaustein and Feder, 1980). While supplemental progesterone administration elevates blood levels of progesterone and prolongs heat duration for 2–3 h in guinea pigs (Morin and Feder, 1973; Brown and Blaustein, 1985), heat nevertheless terminates even in the presence of elevated progesterone levels. These treatments prolong the retention of occupied PRs for a short time (Brown and Blaustein, 1985). Failure to prolong retention of occupied PRs for longer durations, leading to the termination of behavioral estrus, may be due to the decline in PR concentration. Therefore, the loss of behavioral response can be caused by either a declining concentration of hypothalamic unoccupied PRs or the absence of a sufficient level of progesterone to interact with the threshold level of unoccupied receptors. The decline in the concentration of unoccupied PRs, in turn, can be caused by loss of estradiol priming with its resulting decline in the concentration of unoccupied PRs, previous exposure to progesterone with its resulting downregulation of PRs, or both. Considering the complexity of these interactions, it is not surprising that conflicting results concerning progesterone's role in estrous termination have been obtained, some results suggesting a role for progesterone (Barfield and Lisk, 1974; Powers and Moreines, 1976), and some suggesting no role (Hansen and Sodersten, 1978).

In an attempt to tease apart a possible active effect of progesterone on termination of behavioral estrus from the effects of declining estradiol levels, Wallen and Thornton (1979) induced sexual behavior in guinea pigs by daily injections of estradiol alone. When estradiol injections were stopped and the animals treated with progesterone or oil, sexual behavior did not terminate more rapidly in the progesterone-treated animals. The conclusion that termination of heat is not due to progesterone action may have relevance to the process of termination of sexual receptivity induced by estradiol alone. However, because estradiol-induced sexual behavior does not seem to be attributable to interaction with PRs (➋ *Section 4.2.5*), this experiment does not shed light on progesterone-facilitated sexual receptivity.

Besides the concordance between the occupation of PRs and the maintenance of progesterone-facilitated sexual behavior, a causal relationship was shown by injection of the progestin antagonist, RU486, during the period of sexual behavior. The progestin antagonist causes the rapid displacement of hypothalamic, progesterone-occupied PRs, and it shortens the duration of behavioral estrus in guinea pigs (Brown and Blaustein, 1986) and rats (Brown et al., 1987). This loss of neural PRs appears to involve their degradation by the 26S proteosome, as treatment with a proteosome inhibitor prevents the degradation, while stabilizing the concentration of PRs in the hypothalamic and the preoptic areas (Gonzalez-Flores et al., 2004b)(➋ *Section 4.2.6.2*).

Thus, by a variety of experimental paradigms in which estrous duration was either extended or abbreviated, the relationship between the retention of occupied neural PRs and the duration of behavioral estrus has been confirmed. Although the evidence suggests that the accumulation of occupied PRs is essential, the sequelae of PR action, whatever they may be, are essential as well. Thus, levels of other proteins, peptides, or neurotransmitters, endpoints of PR action, should also correlate well with facilitation of sexual behavior, heat duration, termination, and lack of further responsiveness.

4.2.6.2 Downregulation of Progestin Receptors and the Progesterone-Induced Refractory Period Following termination of behavioral estrus in guinea pigs (Goy et al., 1966) and in some circumstances in rats (Nadler, 1970; Blaustein and Wade, 1977; Gonzalez-Flores et al., 2004b), a refractory period occurs during which animals are not responsive to a second exposure to progesterone. Like heat termination, this seems to be due to downregulation of PRs by progesterone. First, during the refractory period, the concentration of hypothalamic PRs is depressed (Moguilewsky and Raynaud, 1979a; Blaustein and Feder, 1979b; Parsons et al., 1981; Gonzalez-Flores et al., 2004b), especially in the ventrolateral hypothalamus of guinea pigs (Blaustein and Turcotte, 1990), and progesterone treatment results in low levels of occupied PRs (Blaustein and Feder, 1980; Blaustein, 1982a). Second, hormonal treatments that influence responsiveness to progesterone also induce correlated changes in the concentration of PRs. Supplemental estradiol priming (Nadler, 1970; Joslyn and Feder, 1971; Blaustein and Wade, 1977; Shivers et al., 1980) offsets the progesterone-induced refractory period, so that animals regain responsiveness to the second progesterone injection. In guinea pigs, the supplemental estradiol injection also offsets the decrease in the concentration of unoccupied PRs, resulting in high levels of occupied PRs in response to progesterone (Blaustein, 1982b). Third, despite the presence of low PR levels, the refractory period can be overcome by a large dose of progesterone (Hansen and Sodersten, 1979; Blaustein, 1982b). This high dose increases progesterone-occupied, hypothalamic PRs, while a typical, behaviorally ineffective, lower dose does not (Blaustein, 1982a). Therefore, under a variety of conditions, there is a strong relationship between the level of hypothalamic PRs, which become occupied after progesterone treatment, and the expression of lordosis in response to progesterone.

Heat termination and the progesterone-induced refractory period may both be due to a progesterone-induced decrease in the concentration of unoccupied PRs with consequent loss of occupied PRs (heat termination) or failure to accumulate an adequate concentration of occupied PRs in response to progesterone (refractory period). Indeed, pharmacological agents that prevent degradation of the PRs by inhibiting 26S proteosome activity within the hypothalamus and preoptic areas, not only stabilize the concentration of PRs, but they also prevent the progesterone-induced refractory period in female rats, confirming that the behavioral refractoriness is causally related to the downregulation of PRs (Gonzalez-Flores et al., 2004b).

The results of some pharmacological experiments support the hypothesis that the progesterone-induced refractory period is caused by desensitization specifically to progesterone, rather than by inhibition of action of estradiol. In rats that are refractory to progesterone, the serotonin antagonist, methysergide (Rodriguez-Sierra and Davis, 1977; Gilchrist and Blaustein, 1984), facilitates sexual behavior, even in progesterone-unresponsive rats. These earlier researches suggested that estradiol priming itself is unaffected by the intervening progesterone injection, and that animals seem to be specifically desensitized to progesterone, a finding consistent with a role for downregulation of PRs during the refractory period. However, in a more recent research (Gonzalez-Mariscal et al., 1993), in which the progestin, norgestrol, was used to induce "sequential inhibition," an opposite conclusion was arrived at. In this experiment, it was reported that animals became refractory to gonadotropin-releasing hormones, prostaglandin E_2 and cAMP, as well as progesterone. Gonzalez-Mariscal et al. first suggested that this cannot be explained by depletion of PRs, but by progestins having a depressive effect on the nervous system. On the other hand, in a subsequent experiment (Beyer et al., 1997), it was demonstrated that a progestin antagonist blocks the facilitation by each of these compounds, providing evidence that each may facilitate sexual receptivity by activation of PRs, a concept that will be discussed later in this chapter (➋ *Section 9*) (Mani et al., 1994a, 1996).

The hypothesis that progesterone induces the refractory period by downregulation of PRs was challenged by the finding that when the protein synthesis inhibitor, anisomycin, was injected into rats around the time of the first progesterone injection, the response to the second progesterone injection appeared to be restored (Parsons and McEwen, 1981). However, although inhibition of protein synthesis rapidly inhibits ongoing sexual behaviors (Brown et al., 1987), it delays the termination of behavioral estrus, resulting in rats still being sexually receptive 24 h after the progesterone injection (Blaustein et al., 1982). Therefore, inhibition of protein synthesis can delay heat termination. Although the cellular basis for the delay is not known, it may be due to interference with the process of downregulation of PRs, perhaps by interference with the 26S proteosome. The suggestion that the process of downregulation of PRs results in heat termination and then persists in causing a refractory period may explain the difficulties in demonstrating a refractory period in rats. The ease of overcoming the refractory period with progesterone should be related to the ease of extending heat duration with progesterone. In fact, guinea pigs have a strong refractory period (Goy et al., 1966), they respond to supplemental progesterone with only a minor increase in heat duration (Morin and Feder, 1973; Brown and Blaustein, 1985), and they show dramatic downregulation of PRs with moderate doses of progesterone (Blaustein and Feder, 1979b). Conversely, in those rats in which the refractory period is only seen with very specific hormonal conditions like low estradiol and high progesterone (Blaustein and Wade, 1977), heat duration is readily extended by a variety of progesterone treatments (Hansen and Sodersten, 1978; Dudley et al., 1980; Brown and Blaustein, 1984b), and downregulation of PRs is far less extreme (Schwartz et al., 1979; Moguilewsky and Raynaud, 1979a; Parsons et al., 1981).

Questions have been raised about whether or not rats have a progesterone-induced refractory period during their estrous cycle. In fact, rats become hyposensitive to progesterone after termination of behavioral estrus during the estrous cycle as well as in the ovariectomized, hormonally treated model. A typical dose (0.5 mg) of progesterone is only effective in inducing sexual behavior during the proestrous stage of the estrous cycle (Sodersten and Hansen, 1979). The fact that in ovariectomized animals, a large dose of progesterone is effective immediately after heat termination (Hansen and Sodersten, 1979), but a low dose is not (Zucker, 1967; Powers and Zucker, 1969; Sodersten and Hansen, 1979) suggests that the animals are in fact hyposensitive to progesterone at the time of heat termination.

Another cellular event that may be related to estrous termination is the downregulation of ERs by progesterone under some conditions (Attardi, 1981; Smanik et al., 1983; Blaustein and Brown, 1984; Brown and MacLusky, 1994; Bethea et al., 1996). While this presumably decreases the effectiveness of estradiol, downregulation of ERs should then decrease the synthesis of PRs, which would then further contribute to the downregulation of PRs.

To summarize, the evidence suggests that termination of the period of sexual receptivity after progesterone treatment results from downregulation of PRs in relevant cells, and the ensuing period of hyposensitivity to progesterone is a consequence of this downregulation. Fewer PRs are available to bind progesterone, so animals tend to be hyposensitive to progesterone. Thus, the refractory period can be seen as a result of the cellular processes that lead to heat termination. Although this has not been studied yet, both may be referable to actions of the 26S proteosome on PRs.

4.3 Membrane Estrogen and Progestin Receptors (Nongenomic and Indirect Genomic)

Although the emphasis of this chapter is mostly on the genomic mechanisms involved in hormonal regulation of sexual behavior, the idea that membrane ERs and/or PRs are involved in mediating the effects of estradiol and progesterone on neuronal function (Watson and Gametchu, 1999; Kelly and Levin, 2001; Vasudevan et al., 2001; Ronnekleiv and Kelly, 2005) and sexual behavior (Debold and Frye, 1994) has received a great deal of support. There are at least two classes of membrane ERs—those made from the same genes that direct synthesis of the TF ERs (Levin, 1999; Razandi et al., 1999; Wade et al., 2001; Levin, 2002; Mhyre et al., 2006) and those potentially made from distinct genes (Ronnekleiv and Kelly, 2005).

In vitro, both the *ERα* and *ERβ* genes are capable of directing the synthesis of receptors that become associated with membranes, and are capable of signaling through the mitogen-activated protein kinase (MAPK), extracellular signal-regulated kinase (ERK) pathway by coupling, directly or indirectly, to G proteins (Driggers and Segars, 2002). Phosphatidyl inositol-3 kinase (Malyala et al., 2005) is also activated by estradiol binding to membrane ERs or other transmembrane G-protein coupled receptors (GPCRs). The activation of signaling pathways by estradiol appears to be tissue specific. Such regulation has been observed in activation of PKC by estradiol in the preoptic area, but not in the hypothalamus and the cortex of female rat brain slices (Ansonoff and Etgen, 1998). Activated signaling pathways influence not only the phosphorylation state of multiple nonER TFs [e.g., Fos, cAMP response element binding (CREB), activator protein-1 (AP-1)], but also the TF-ERs and their coactivators (Zhang and Trudeau, 2006).

Interestingly, membrane signaling also has an impact on the activation of nuclear ERs, resulting in recruitment of coactivator proteins and enhanced transcription of the receptors (Levin, 2005). In addition, a distinct class of GPCR, GPR30, has been shown to have the binding and the signaling characteristics of a membrane ER (Filardo and Thomas, 2005; Thomas et al., 2005). A novel GPCR for progesterone, totally unrelated to the nuclear PR, has been recently identified and demonstrated to play a role in progesterone-induced oocyte maturation in teleosts (Zhu et al., 2003a, b). Human PRs have been shown to mediate rapid progesterone effects via the activation of Src/Ras/Raf/MAPK signaling pathway by a direct interaction with the Src homology 3 domain of Src tyrosine kinases through a proline-rich (PXXPXR) motif located in the N-terminal domain of the PR (Boonyaratanakornkit et al., 2001).

The fact that some membrane receptors are derived from the same gene as the TF receptors has bearing on many of the earlier studies in which antagonists, antisense oligonucleotides, or targeted gene disruption were used to assess the involvement of steroid receptors acting as TFs. If the same ER and PR genes that direct synthesis of the receptors that act as TFs also direct synthesis of membrane receptors, then these manipulations could have influenced this type of membrane receptor as well as the targeted, TF receptors. Antiestrogens have been shown to block some of the rapid effects of estradiol on electrophysiological response (Lagrange et al., 1997). The possibility must be considered that the membrane receptor that is being blocked is a derivative product of the same *ERα* or *ERβ* genes that code for the TF receptors (Wade et al., 2001). ERα (Blaustein, 1992; Blaustein et al., 1992; Wagner et al., 1998b; Watson and Gametchu, 1999; Clarke et al., 2000; Milner et al., 2001) and PRs (Blaustein et al., 1988) have been found in extranuclear locations, including distal dendrites, axon terminals, and synaptic densities in immunocytochemical studies (Blaustein et al., 1992; Blaustein, 1994). These findings are consistent with the idea that some extranuclear receptors are synthesized from the same *ERα* and/or *ERβ* genes.

Rats do not have to be exposed to estradiol continuously during the priming period in order to express sexual behavior. Two brief pulses of a very low dose of estradiol are more effective in inducing feminine sexual behavior in response to progesterone than a single large injection or continuous exposure (Sodersten et al., 1981; Parsons et al., 1982a; Clark and Roy, 1983; Wilcox et al., 1984). The behavioral effects of each pulse can be blocked by a protein synthesis inhibitor (Parsons and McEwen, 1981) or pentobarbital anesthesia (Roy et al., 1985). The particular proteins modulated in response to each injection may differ (Jones et al., 1986). Very little is known about the cellular basis for this understudied, enhanced response to pulsed exposure to estradiol. However, estradiol conjugated to bovine serum albumin (BSA), which makes the compound impermeable to cell membranes, can substitute for estradiol-17β in this two-pulse procedure (Kow et al., 2005). Interestingly, and at odds with previous work done in neuroblastoma cells (Vasudevan et al., 2001), the estradiol–BSA conjugate could also substitute for the second pulse of estradiol. The actions of the estradiol–BSA conjugate can be mimicked by compounds that activate either protein kinase A (PKA) or PKC, suggesting involvement of these two signaling pathways in the action of estradiol on feminine sexual behavior, as has been shown for the action of estradiol on neuronal electrophysiology (Kelly and Wagner, 1999). Although of great interest, experiments using BSA conjugates of steroid hormones must be interpreted very cautiously because of stability problems (Stevis et al., 1999) and because the position of the BSA on the steroid can have tremendous effects on function (Temple and Wray, 2005).

5 Neuroanatomy: Role of the Ventromedial Hypothalamus

In the vast majority of work on cellular and molecular underpinnings of feminine sexual behavior, one brain area, the VMN, and often specifically the ventrolateral aspect of this nucleus has been studied exclusively (VMNvl). This review is not focused on the neuroanatomical substrates of feminine sexual behavior, and the involvement of other important areas, including the preoptic area, midbrain central gray, and medial amygdala will not be discussed here [see reviews by Pfaff et al. (1994) and Blaustein and Erskine (2002) for more extensive discussion of other areas]. However, it is important to understand the reasons why a particular neuroanatomical area, the VMN, has been singled out as the most important area in studies of the cellular and molecular underpinnings and the limitations of the data in support of this idea.

Recent emphasis in this field has been on understanding the cellular and the molecular processes involved in hormonal regulation of sexual behaviors, with little emphasis on refining our understanding of the neuroanatomical substrates for these behaviors. It has become axiomatic that the VMN is the most critical site for hormonal regulation of feminine sexual behavior. In a recent review, the question was raised of whether there is strong evidence to support this idea (Blaustein and Erskine, 2002). Unfortunately, while this is not currently an area of intense investigation, we still do not have the full answers to some of the original questions. The issues to be considered concerning the possible critical role of the VMN will be summarized below.

1. In many cases, rather gross procedures were used that included, but were not limited to lesion of the VMNvl. As we have discussed earlier (Blaustein and Erskine, 2002), some of the earliest lesion studies implicated an area within the mediobasal hypothalamus in the regulation of feminine sexual behavior. In these studies, large hypothalamic lesions were often used, and in fact deficits in sexual behavior were seen. However, the fine points of many of these studies have been lost through the years. As is often cited, Goy and Phoenix (1963) did in fact report that lesions of the ventromedial hypothalamic area decreased lordosis in estradiol plus progesterone-treated, ovariectomized female guinea pigs. However, the lesions in that study included a wide variety of structures within the hypothalamic region, including the periventricular anterior nucleus, the anterior hypothalamic nucleus, and other areas. In referring to the four animals with the most severe deficit in behavior, the authors noted that "the common area of destruction for these brains may be described as lying with the lateral extensions of the arcuate nucleus

and ventrolateral portions of the ventromedial." The lesions from the four poorest responders appear to be in the area adjacent to the VMN, the precise region of highest density of ERir and PRir neurons in this area. The results support a role for the steroid–receptor-rich area in this region, so it is not accurate for this study to be cited in support of the involvement of the VMN specifically.

2. In many cases, in summaries of earlier work, the effects of estradiol have not been distinguished from those of estradiol and progesterone. The results of studies investigating the effects of lesions of the VMN on estradiol only versus estradiol + progesterone-induced sexual behaviors differ. In the work with most disparate results of the two treatments, Mathews and Edwards (1977a, b) reported that after some VMN lesions in ovariectomized rats, deficits were seen in estradiol-alone induced sexual behavior, but not in estradiol + progesterone-induced sexual behavior. This pattern of interference with estradiol-induced lordosis, but not estradiol + progesterone-induced, was also seen after sagittal knife cuts were made laterally to the VMN (Mathews and Edwards, 1977a) or anterior hypothalamus (Pfeifle et al., 1980). Although these data actually suggest a dissociation in the neuroanatomical substrates of estradiol-only versus estradiol + progesterone-facilitated sexual receptivity, other studies by Clark and coworkers (1981a) suggested that VMN lesions inhibit copulatory and progestative behaviors (pacing), induced by either estradiol alone or estradiol and progesterone. The possibility of slight differences in the extent of the lesions cannot be eliminated, especially because the lesions of the earlier researches were quite small, and the lesions made in much of the more recent researches have been larger. However, a later study found that VMN lesions inhibited time that estradiol + progesterone-injected females spent with a male, but did not inhibit their lordosis (Emery and Moss, 1984a). Nevertheless, these studies suggest that specific sites within the VMN itself, and perhaps its surrounding might be differentially involved in estradiol-induced versus estradiol + progesterone-induced sexual behavior. In related work, quite large VMN lesions resulted in near-total loss of estradiol and progesterone-induced lordosis, while lesions that spare 30–50% of the nucleus resulted in no decrement at all (Richmond and Clemens, 1988).

   To summarize, although not conclusive, the data suggest that the neuroanatomical sites mediating estradiol-induced sexual behavior may not be identical to estradiol + progesterone-induced sexual behavior. Therefore, caution must be exercised in extrapolating from the pharmacological induction of sexual behavior (estradiol only) to the more physiological means (estradiol + progesterone).

3. The VMN is not part of the actual reflex arc in estrogen only–induced lordosis (progesterone independent). Pfaff and Sakuma (1979) followed the time course of loss of sexual receptivity after VMN lesions in ovariectomized rats in which lordosis was induced by daily estradiol injections. Lordosis declined over the course of 36–60 h, suggesting that, while this area is critical to estradiol modulation of sexual receptivity, it is not likely to be a part of the actual reflex arc. In contrast, disruption of lordosis occurs within 15 min of lesion in hamsters treated with estradiol and progesterone (Gibson and Floody, 1998), suggesting that the VMN is part of the reflex arc in this particular situation. Besides the obvious species difference, an important difference between the two studies is that the latter was a study of progesterone-facilitated sexual behavior, and the former was a study of estradiol-induced sexual behavior. This illustrates the possibility that the neuroanatomical substrates by which sexual behavior is induced by either estradiol versus estradiol + progesterone are not necessarily identical.

   In a study in which pseudorabies virus (PRV) was used to label transneuronally the neurons innervating the lumbar epaxial muscles involved in the lordosis response (Daniels and Flanagan-Cato, 2000), many of the PRV-labeled neurons in the ventromedial hypothalamic area were outside of the core of the VMN. This study supports the suggestion that many of the relevant neurons involved in the lordosis reflex lie outside of the Nissl-defined VMN.

4. Hormone implant studies can only provide clues to the relevant sites. Many experiments zeroed in on the VMN as the most sensitive site of action for estradiol priming of sexual behavior in ovariectomized rats (Barfield and Chen, 1977; Davis and Barfield, 1979; Rubin and Barfield, 1980; Davis et al., 1982) and hamsters (Debold et al., 1982; Floody et al., 1987). While the results of many experiments point to the VMN as the most sensitive forebrain site for estradiol action, it must be noted that most other areas have not been as extensively studied as this nucleus. Results of experiments using localized implants of

antiestrogens (Howard et al., 1984; Meisel et al., 1987), protein synthesis inhibitors (Rainbow et al., 1982a; Glaser and Barfield, 1984; Meisel and Pfaff, 1985), and transcription inhibitors (Yahr and Ulibarri, 1986) confirm that a site in or near the VMN is a critical site involved in regulation of sexual behavior by estradiol. However, it is critical to note that compounds implanted or infused into the brain do not remain precisely where they are placed, so the results should not be interpreted as having pin-point accuracy. Rather than concluding that the VMN itself is the active site, a more conservative interpretation is that the VMN and/or adjacent areas are involved in the regulation.

In related work in guinea pigs, small, dilute estradiol-containing cannulae were implanted directed at the ventrolateral hypothalamus, the region analogous to the VMNvl in rats (Delville and Blaustein, 1991). Guinea pigs were injected with progesterone, tested for lordosis responses, and brains immunostained for the expression of PRs. Due to the particular technique used, the presence of PRir is dependent upon exposure to estradiol, its presence in a cell served as a bioassay for spread of estradiol to other cells capable of expressing estradiol-induced PRir. Only cannulae located at the rostral and the ventral aspect of the ventrolateral hypothalamus induced behavioral response to progesterone, and the expression of estradiol-induced PRir in the rostral–ventral ventrolateral hypothalamus correlated with response to progesterone. Thus, stimulation of a population of ovarian steroid hormone-sensitive neurons within the ventrolateral hypothalamus by very low levels of estradiol may be sufficient to induce sexual behavior in response to progesterone in female guinea pigs. Many of the cells in which PRir is induced are located outside of the actual ventrolateral (or ventromedial) nucleus of the hypothalamus (*Figure 3-4*). This is in agreement with the suggestion that the defined nucleus itself may not be the critical site of action of estradiol and progesterone on feminine sexual behavior. Rather, receptor-containing cells in the adjacent area could be equally or more important.

In earlier work on the site of action of progesterone, the general area of the mediobasal hypothalamus was singled out as an important site for progesterone facilitation of sexual receptivity in rats (Powers, 1972) and guinea pigs (Morin and Feder, 1974), although many other areas have also been

Figure 3-4

Photomicrograph of ERα-ir in the rostral ventromedial hypothalamic area of an ovariectomized female rat showing that many of the ERα-ir cells are not in the actual ventromedial hypothalamic nucleus; rather they are in the surrounding (arrows point to general area). VMN = ventromedial nucleus of hypothalamus; VMNvl = ventrolateral aspect of ventromedial nucleus of the hypothalamus; arc = arcuate nucleus; IIIV = third ventricle

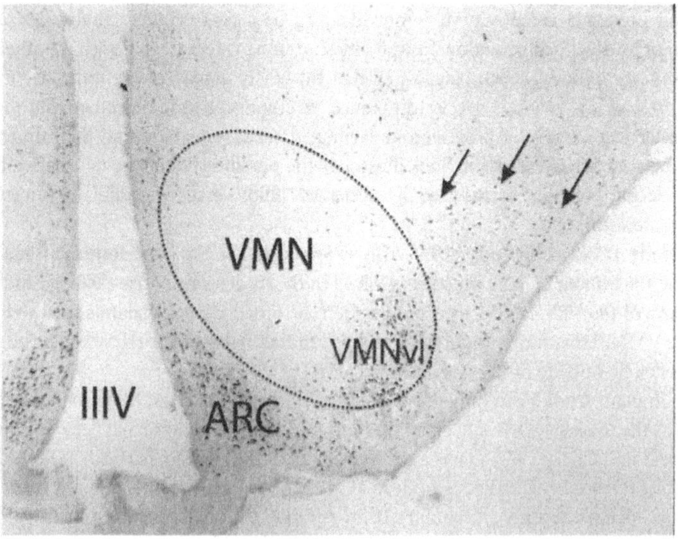

suggested (Ross et al., 1971; Ward et al., 1975; Luttge and Hughes, 1976; Yanase and Gorski, 1976; Franck and Ward, 1981; Rodriguez-Sierra and Komisaruk, 1982; Tennent et al., 1982). Later work using more refined intracranial implants of progesterone in rats (Rubin and Barfield, 1983) and hamsters (Debold and Malsbury, 1989) or a progesterone antagonist in rats (Etgen and Barfield, 1986) suggested that stimulation of the VMN is sufficient for facilitation of sexual behavior. These implant experiments are not inconsistent with the lesion experiments described earlier, which showed that estradiol + progesterone-induced sexual receptivity often remains even after VMN lesions, because the progesterone may diffuse to the adjacent area, where many steroid-responsive cells are found, and which may be a critical site of action.

As with the search for the site of facilitation of sexual receptivity, the work by Rubin and Barfield (1984) suggested that progesterone induces a refractory period in rats at the same site as it facilitates the expression of feminine sexual behavior—the VMN. This is consistent with our earlier reanalysis (Blaustein and Brown, 1985) of a report of Morin and Feder (1974), which suggested that progesterone implants in the mediobasal hypothalamus of guinea pigs at sites that cause facilitation of sexual behavior, also cause refractoriness to progesterone. Therefore, both the facilitation and the refractory period induced by progesterone can occur after implantation of progesterone in the same neuroanatomical site, perhaps in the same neurons. This interpretation is consistent with the PR model for the cellular mechanism of action of progesterone in feminine sexual behavior discussed earlier.

5. The constellation of sexual behaviors has seldom been studied after VMN lesions. Some apparent effects on lordosis may be secondary to motivational changes. VMN lesions inhibit pacing behaviors (Clark et al., 1981a). Similarly, although VMN lesions decrease contact with males when estradiol + progesterone-treated females are allowed to pace the contacts, they are often capable of displaying lordosis when mounted (Emery and Moss, 1984a). It was argued that the decrease in lordosis seen in other studies may be due to avoidance of mounting by the lesioned rats (Emery and Moss, 1984a). The possibility must be considered that this issue influenced the results of some lesion experiments, and is consistent with the earlier interpretation of La Vaque and Rodgers (1975), who studied the effects of lesions in gonadally intact female rats. While this discussion does not take away the importance of the VMN in feminine sexual behavior, it emphasizes the importance of distinguishing the effects of lordosis from other aspects of sexual behavior, and once again, of distinguishing estradiol-only induced sexual behavior and estradiol + progesterone-induced sexual behavior.

In spite of the limitations of these studies, there appears to be a number of conclusions that can be drawn about the role of the VMN:

1. The VMN and/or a nearby population of cells is involved in the action of estradiol on sexual receptivity.
2. Estradiol-induced sexual behavior is more adversely affected by small, VMN lesions than is sexual receptivity induced by estradiol and progesterone, suggesting that the VMN core may be less critical in the presence of progesterone.
3. Likewise, although the VMN may not be part of the actual reflex arc for estradiol-only induced sexual behavior, it may be part of the reflex arc for progesterone-facilitated sexual behavior.
4. Lesions of the VMN always reduce sexual behaviors, but they rarely eliminate them, perhaps because either the entire steroid receptor-containing region associated with the ventromedial hypothalamic area, or an as-yet-undefined subpopulation of cells within this region, is critical for hormonal regulation of sexual behavior.
5. Hormone implants in the area of the VMN are effective in inducing the expression of some aspects of sexual behavior.
6. The VMN is likely to be involved in motivational aspects of sexual behavior.

Therefore, studies of the cellular and molecular neuroendocrinology of this general area are warranted, but it could be argued that it is premature to investigate neurons within only the VMN to the exclusion of steroid-sensitive neurons in the adjacent areas that could be as—or more—important with the assumption that these are the neurons specifically involved in lordosis.

6 Intracellular Signaling Pathways Involved in Feminine Sexual Behavior

Regulation of feminine sexual behavior appears to involve more than steroid hormones acting via rapid nongenomic and slower genomic pathways (*Figure 3-5*). Diffusible molecules like nitric oxide (NO) have been shown to be involved in lordosis response in female rats. While NO has been demonstrated to be a mediator of gonadotropin releasing hormone (GnRH) release and of subsequent estrous behavior (Mani et al., 1994b), a role for NO in the facilitating effects mediated by α1-adrenoceptors (Chu and Etgen, 1999) and progestins on lordosis has also been reported (Gonzalez-Flores and Etgen, 2004). There are also reports

Figure 3-5

Crosstalk between signal transduction pathways operating in feminine sexual behavior. The schematic diagram presents several of the potential signal transduction pathways operating in the hypothalamus and preoptic areas. (1) Progesterone effects mediated by the "classical" genomic mechanism involving the intracellular receptors play a predominant role in signal transduction. Progesterone, binding to the classical intracellular PR, allosterically activates it to promote interactions with nuclear coactivators. (2) Progesterone effects mediated by the second messenger cAMP/cGMP, activates the PKA-mediated signaling, leading to the activation of MAP kinases and the phosphorylation of nuclear TFs, nuclear PRs, CREB, and/or its associated coactivator protein, CBP. (3) Progesterone, acting through the membrane via the cAMP/PKA/MAPK/DARPP-32 pathway, results in decreased dephosphorylation of PR and/or its associated coactivators. (4) Dopamine and mating stimuli (VCS) stimulate cAMP release and PKA activation. PKA-mediated pathway phosphorylates DARPP-32, which inhibits PP-1, leading to the activation of CREB/PR/coactivators. PRs are critical for this pathway denoting a crosstalk between dopamine and progesterone-initiated pathways. (5) Progesterone can also activate PKC pathway leading to the signal transmission to the nucleus and interaction with nuclear TFs. (6) Similarly, neuropeptides, GnRH and PGE$_2$, through various receptor-mediated mechanisms and/or second messengers (e.g., NO, cGMP, cAMP) transmit signals to the nuclear TFs to facilitate sexual behavior. The allosteric activation and the phosphorylation of various TFs work synergistically to promote progesterone receptor-mediated gene function. In addition, interactions between the signal pathways may serve as amplification mechanism and multiple upstream regulators converge on common effectors such as PRs/coactivators and/or CREB to regulate transcription of immediate early genes (*c-fos, c-jun, Egr*) and/or later genes to facilitate feminine sexual behavior

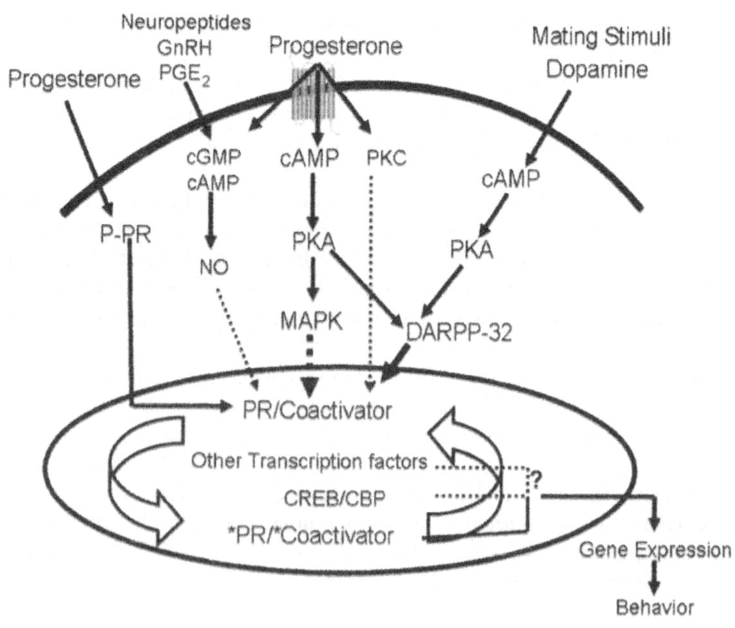

of NO involvement in melanocortin facilitation of lordosis in female rats (Nocetto et al., 2004). The involvement of NO in feminine sex behavior appears to be modulated by gonadal hormones, as correlated by the expression of nNOS in the estradiol-primed hypothalamic neurons (Rachman et al., 1998). NO regulation of sex behavior also appears to be species-specific, because estradiol-regulated nNOS increase has not been observed in avian species (Panzica et al., 2006).

A variety of compounds that activate several second and third messenger systems and stimulate protein kinases within the neurons also facilitate feminine sex behavior in estradiol-primed rats, adding another level of complexity to the mechanisms involved. These compounds include adenosine and guanosine nucleotides and their second messengers, cyclic adenosine monophosphate (cAMP) and cyclic guanosine monophosphate (cGMP) (Beyer et al., 1981; Limbird, 1981; Whalen and Lauber, 1986; Chu and Etgen, 1997) in addition to GnRH, prostaglandin E2, neurotransmitters, and neuropeptides [see Micevych and Sinchak (2006) in this volume]. Phosphodiesterase inhibitors that increase cAMP by interfering with the hydrolysis of cAMP, potentiate the effects of subthreshold doses of GnRH and progesterone in the facilitation of lordosis (Beyer and Gonzalez-Mariscal, 1986). Studies on the role of cGMP in lordosis response have identified a role for protein kinase G in the signaling cascade. Interestingly, PRs are also critical in this cGMP-mediated response suggesting the existence of crosstalk between the two (Chu and Etgen, 1997). Recent studies on progesterone-mediated signal transduction pathways also indicate rapid elevations in hypothalamic cAMP levels and PKA (cAMP-dependent PKA) activity by progesterone in estrogen-primed rats and mice. Inhibitors of PKA reduced progesterone-facilitated sexual behavior in estradiol benzoate (EB)-primed rats. This signaling cascade also includes a third messenger, dopamine and cAMP-regulated phosphoprotein-32 (DARPP-32), the absence of which renders both rats and mice incapable of expressing the progesterone-facilitated lordosis response (Mani et al., 2000). However, progesterone was ineffective in facilitating lordosis in the absence of PRs (Mani et al., 1994c, 1996). Thus, rapid membrane-initiated signaling cascades crosstalk with the genomic pathways involving intracellular PRs to enhance or bring about facilitation of sex behavior.

Two important factors to be borne in mind, when discussing the involvement of the multitude of signaling pathways and steroid hormones in feminine sex behavior, are that the pathways share signaling components and are not unidirectional. For example, involvement of MAPK pathway in progestin-facilitation of lordosis has been demonstrated in rats (Etgen and Acosta-Martinez, 2003; Gonzalez-Flores et al., 2004c). Not only does the MAPK-activated pathway facilitate the expression of lordosis, but it also subsequently promotes termination of behavioral estrous by targeting the PR for degradation by the 26S proteasome (Camacho-Arroyo et al., 2002; Gonzalez-Flores et al., 2004b). Furthermore, the MAPK pathway is also involved in δ opioid actions on lordosis (Acosta-Martinez et al., 2006). While the functional role of multiple signaling pathways can be explained by their ability to relay, amplify, and integrate signals from a variety of extracellular stimuli, it still remains unclear how the same MAPK pathway can both activate and inhibit the same behavioral endpoint. How does the neuron differentiate between the stimuli to activate as opposed to inhibit feminine sexual behavior?

Cellular signaling mechanisms that alter the levels of cAMP, PKA, MAPK, or calcium can increase phosphorylation of CREB (Ca^{2+}/CBP) or other TFs, for example, activating transcription factor-1 (ATF-1), enabling them to become transcriptional regulators (Armstrong and Montminy, 1993). Estrogen and progesterone are known to influence gene expression by activating signaling pathways involving PKA and PKC in rats (Petitti and Etgen, 1989, 1990; Kow et al., 1994). In several regions of the rat brain lacking the classical steroid receptors, estradiol causes a rapid increase in p–CREB immunostaining, with no concomitant increases in CREB protein or mRNA levels (Gu et al., 1996; Zhou et al., 1996). Progesterone, on the other hand, appears to have a bimodal effect on the phosphorylation of CREB, bringing about a rapid decrease followed by an increase (Gu et al., 1996). These rapid effects on CREB phosphorylation also appear to be nuclear receptor mediated; estrogen antagonists and progestin antagonists block the hormonal effects on CREB phosphorylation, again suggesting crosstalk between the distinct signaling pathways.

In addition, protein kinases also have effects on ion channels (Mermelstein et al., 1996; Rupprecht and Holsboer, 1999). Steroid hormones, growth factors, and neurotransmitters are known to activate intracellular cascades leading to increased p-CREB expression, binding of p-CREB to its coactivator, CBP, and inducing transcription of immediate early genes containing cAMP response element-sequences, as *c-fos* and *c-jun* (Schule and Evans, 1991). These genes encode the TFs Fos and Jun that can form hetero- or homodimers and regulate downstream gene expression by acting on target AP-1 DNA recognition sequences near promoter elements. AP-1 elements can substitute for hormone response elements in the steroid hormone regulation of gene transcription (Kushner et al., 1994). Evidence from in vitro experiments indicates that ERs and PRs can also mediate transcription of genes controlled by an AP-1 enhancer element (Umayahara et al., 1994; Webb et al., 1995). Furthermore, it has also been demonstrated that the two distinct classes of TFs, steroid hormone receptors and AP-1 complexes, interact to modulate each other's activity (Shemshedini et al., 1991; Alkhalaf and Murphy, 1992). In addition, recent studies have also suggested that nuclear receptor coregulators may also integrate steroid hormone signaling through CBP (Torchia et al., 1997; Mahajan and Samuels, 2000; Xu et al., 2000a). Functional cooperation between MAPK cascade-mediated phosphorylation of coactivator SRC-1 and CBP has been demonstrated in the activation nuclear PRs in vitro (Rowan et al., 2000a). Thus, second messenger systems, including CaMK (calcium/calmodulin-dependent protein kinase) and PKC modulate gene expression via multiple TFs or transcription coactivators, providing an alternative pathway to the genome (Watters et al., 1997; Watters and Dorsa, 1998; Sweatt, 2004).

There is no doubt that multiple intra- and intercellular mechanisms coexist to ensure that the female is in behavioral estrus at the most appropriate time. However, several conundrums still exist. With all these kinases and coactivator proteins in the same cells, how does the cell/neuron keep the pathways distinct from one another? How do the neurons/cells distinguish different signals that use similar signaling molecules (e.g., MAPK)? How do these various mechanisms coexist, yet ensure that the neurons respond to the correct extracellular signals to control genomic and behavioral response? What mechanisms exist to prevent crosstalk between the pathways that could potentially have opposite behavioral endpoints?

7 Crosstalk Between Neurotransmitters and Steroid Hormone Receptors: Activation of Steroid Receptors by Afferent Input

7.1 Ligand-Independent Activation of PRs

During the first 25 years of studies of steroid receptors, it had been assumed that the steroid hormones were necessary for activation of the receptor for binding of receptors to their cognate ligands. However, we now know that steroid hormone receptors can be activated by a means other than steroid hormones, and a comprehensive understanding of feminine sexual behavior has to consider the potential roles of these other routes to activation. In 1991, Power et al. discovered that PRs and other steroid hormone receptors could be activated by dopamine and dopaminergic agonists in vitro (Power et al., 1991b). The process by which dopamine activates the receptors is via a second messenger signaling pathway rather than a direct effect of the agonist on the PR. This is referred to as ligand-independent activation or hormone-independent activation. The idea that steroid receptors can be activated indirectly via membrane-related events has now been supported by a wide variety of work (Blaustein, 2004b).

In the first test of this model on behavior, it was shown that the intracerebroventricular administration of a D_1-specific, dopamine agonist substitutes for progesterone in the facilitation of sexual behavior in estradiol-primed rats (Mani et al., 1994a). This finding confirmed and extended earlier work in which dopaminergic agonists were infused into the hypothalamus and preoptic area (Foreman and Moss, 1979). However, like facilitation of sexual behavior by progesterone, the facilitation of sexual behavior by dopaminergic agonists was blocked by progesterone antagonists or antisense oligonucleotides directed at the PR mRNA administered directly to the brain, as well as by a D_1-dopamine antagonist (Mani et al., 1994c). The inability of PRKO to exhibit D_1 agonist-facilitated sexual behavior, while their wild-type littermates responded to the agonist with lordosis (Mani et al., 1996), provides definitive evidence for the obligatory role of PRs as transcriptional mediators for dopamine-initiated pathways in sexual behavior. A similar relationship holds for sexual behavior facilitated by GnRH (Mani et al., 1995; Beyer et al., 1997), δ-opioid receptors (Acosta-Martinez et al., 2006), as well as prostaglandin E2 (Beyer et al., 1997); each can be blocked by treatment with a progesterone antagonist. Likewise, NO (Mani et al., 1994b), cAMP (Beyer et al., 1997), cGMP (Chu et al., 1999), and MAPK (Acosta-Martinez et al., 2006) may each facilitate sexual behavior via activation of PRs. Thus, dopamine and other neurotransmitters may activate feminine sexual behavior by indirect activation of neural PRs.

7.2 A Mechanism for Ligand-Independent Activation

Protein phosphorylation that is common to the pathways and molecular mechanisms through which neurotransmitters and steroid hormones produce their biological effects is under investigation. The regulatory mechanisms governing a variety of cellular processes in target cells are not only dependent upon the state of intracellular phosphorylation of the receptor, but are also dependent on the dynamic balance between cellular protein kinases and protein phosphatases (PPs). This has been found to be the case with PRs, where the equilibrium between transcriptionally active and inactive forms of the receptor is under the regulation of kinases and phosphatases (Denner et al., 1990; Shenolikar and Nairn, 1991).

Modulation of protein kinases and PPs in phosphorylation and signal transduction mechanisms occurs in the mammalian brain, a tissue having an abundance of kinases and phosphatases (Nairn et al., 1985). Neuronal phosphoproteins, like neurotransmitters and cyclic nucleotides, are components of the signal transduction pathway in the nervous system (Shenolikar and Nairn, 1991; Watanabe et al., 2001; Svenningsson et al., 2004), and can be phosphorylated/dephosphorylated in response to extracellular stimuli; such dynamic covalent modification is evident in modulation of the activity of PP1 and PP2. Several studies have documented that dopamine signaling through the D_1 subclass of dopamine receptors increases the level of second messenger cAMP and activates cAMP-dependent PKA in the neostriatum. Increased PKA activity leads to the phosphorylation of the neuronal phosphoproteins, DARPP-32, and/or inhibitor-1 (I-1). The phosphoprotein I-1 is closely related structurally, enzymologically, and functionally to DARPP-32 (Shenolikar and Nairn, 1991). In its phosphorylated state, DARPP-32 and/or I-1, by inhibiting the activity of PP-1, increases the state of phosphorylation of many substrate proteins, leading to induction of physiological responses. Thus, the PR is one of the potential substrate proteins phosphorylated by DARPP-32. The involvement of DARPP-32 and/or I-1 as the downstream mediators in the dopamine–PR interactions was investigated in the facilitation of sexual behavior in female rats and mice.

Antisense oligonucleotides to DARPP-32 administered into the third cerebral ventricle inhibited D_1 agonist- and progesterone-facilitated sexual receptivity in estradiol-primed female rats (Mani et al., 2000). Sexual receptivity facilitated by either a D_1 agonist or by progesterone after estradiol priming was also inhibited in mice carrying a null mutation for DARPP-32 gene. However, I-1 null mutant mice exhibited no deleterious defects in D_1 agonist- and progesterone-facilitated lordosis response compared with their wild-type littermates, revealing that involvement of the DARPP-32 pathway was specific. Also, increased immunoreactive phospho-DARPP-32 cells in the PR-containing areas of the rat hypothalamus have been observed after genitosensory stimulation (Meredith et al., 1998).

Similar to dopamine effects in the neostriatum, the D_1 agonist, as well as progesterone, significantly increase hypothalamic cAMP levels and PKA activities and enhance the phosphorylation of DARPP-32 on threonine-34. Increases induced by a D_1 dopamine agonist are inhibited by the D_1 subclass dopamine antagonist, SCH 23390. This suggests that the increases are due to the effects of dopamine initiated at its membrane receptor. The progesterone-induced increase, however, was not inhibited by SCH 23390, suggesting that the increases observed are due to the direct effects of progesterone and not secondary to modulation of dopamine or dopamine receptors by progesterone (Mani et al., 2000). Rp-cAMPS, a compound that blocks the cAMP signal transduction cascade by inhibiting PKA, inhibited D_1 agonist, and progesterone facilitated sexual receptivity in estradiol-primed female rats. While these observations suggest that DARPP-32 activation is an obligatory step in PR regulation of sexual receptivity, the sequence of events leading to the activation of PR (from the DARPP-32 phosphorylation step) is complicated and remains to be completely defined. It is likely that the mechanisms include not only a direct decrease in dephosphorylation (activation) of PRs, but also enhanced phosphorylation of PR-associated coactivators leading to rapid efficient transcriptional activation (Font de and Brown, 2000; Rowan et al., 2000a, b).

Now that it is known that steroid hormone receptors can be activated by afferent input via neurotransmitters, we suggest that results of earlier experiments on facilitation of sexual behaviors by various drugs should be interpreted cautiously. The assumption is often made that the drugs are stimulating or antagonizing output of steroid hormone-sensitive systems. The existence of ligand-independent activation of steroid hormone receptors suggests that some of the drugs may act by altering the activation state of PRs as well as by more traditional mechanisms.

Studies on the regulation of GnRH during the rat and the mouse estrous cycles are consistent with the idea that activation of PRs may be an intermediate step in the mechanism by which some neurotransmitters and other factors regulate GnRH. Blockade of PRs or inhibition of PR synthesis blocks some of the effects of estradiol on GnRH regulation. It has been hypothesized that a neural consequence of estradiol action facilitates GnRH secretion via ligand-independent activation of PRs (Chappell et al., 1999; Chappell and Levine, 2000; Xu et al., 2000b).

In 1986, Whalen and Lauber proposed the novel hypothesis that many drugs that substitute for progesterone do so by elevating neuronal levels of cGMP. Furthermore, Beyer and his colleagues (1981) showed that drugs that increase cAMP may substitute for progesterone. Experiments discussed earlier are consistent with these ideas and suggest that some of the drugs that influence feminine sexual behavior do so by modulating second messenger systems and then by secondarily activating PRs (and perhaps other TFs).

The influence of afferent input on steroid receptors could be involved in the regulation of reproductive physiology. For example, VCS causes a variety of neuroendocrine changes in female rats, including the induction of pseudopregnancy (Gunnet and Freeman, 1983; Erskine, 1995; Erskine et al., 2004), increases in luteinizing hormone (LH) release (Moss et al., 1977), increases in lordosis intensity (Diakow, 1975), and subsequently termination of the period of sexual behavior (Blandau et al., 1941). Similarly, olfactory stimuli associated with mating have profound effects on reproductive physiology, and some of these could be mediated via afferent influences onto steroid-sensitive neurons.

8 Influence of Mating Stimulation on Long-Term and Short-Term Changes in Sexual Behavior and Influences on the Brain

It is generally accepted that fluctuations in hormone levels are the principal factors responsible for the changes in sexual responsiveness over the estrous cycle and in ovariectomized female rats. This is based on a long history of experiments demonstrating the effects of presence or absence of particular hormones on sexual behavior. Although it is well established that the hormonal changes underlying the onset of sexual behavior are an increase in estradiol followed by an increase in progesterone levels, this is not always the case. Many other factors associated with the mating situation result in both short- and long-term modulations of sexual behavior.

8.1 Short-Term Slowing of Sexual Response by Vaginocervical Stimulation: Pacing

During paced mating, the likelihood that a female will leave the compartment containing a male and the latency for her to return to the male's compartment after receipt of a copulatory stimulus is influenced by the specific type of stimulus that precedes the response. Withdrawals from the male occur more often, and the latency of the female's return to the male's cage is longer following an intromission or an ejaculation, than after a mount without intromissive stimulation (Bermant, 1961; Bermant and Westbrook, 1966; Erskine, 1985). Therefore, mounts, which include VCS, result in transient inhibition of sexual responsiveness or a short-term delay in the rate of copulation. Over the course of a test, the latency to return to the male's compartment (and consequently, the inter–intromission interval) increases as a function of the number of prior intromissions (Coopersmith et al., 1996). Therefore, allowing the female to pace sexual contacts with males causes a progressive decline in the rate at which females receive intromissions.

8.2 Short-Term and Long-Term Enhancement of Sexual Responding by Mating Stimuli

In contrast to the rapid and transient inhibitory influences of intromissive stimuli on pacing, in some cases, intromissive stimulation enhances the reflexive component of sexual behavior both in the short and the longer terms. As an example of the short-term change, while female rats that are sexually receptive respond to appropriate stimulation of the flanks and the perineum with the expression of the lordosis response, the presence of VCS by penile intromission results in a greater intensity of lordosis (Diakow, 1975). Similarly, VCS administered by a glass probe show an increased lordosis reflex in response to manual palpation of the flanks and the perineum (Komisaruk and Diakow, 1973). Ovariectomized rats, whether treated with subthreshold doses of estradiol or not, express enhanced lordosis to manual palpation immediately following experimenter-induced VCS (Komisaruk and Diakow, 1973). This acute facilitation of the lordosis response by VCS persists for at least 60 min (Rodriguez-Sierra et al., 1975). It should be noted that knife cuts parasagittal to the VMN that inhibit estradiol + progesterone-induced lordosis in response to a male, are without effect on cervical probing–induced lordosis (Pfeifle et al., 1980), potentially suggesting a different neural substrate for lordosis that is stimulated in this manner.

As an example of longer-term enhancement of sexual behavior, when ovariectomized rats are exposed to sexually active males for 15 min periods every 30 min (Foreman and Moss, 1977; Rajendren et al., 1990; Rajendren and Moss, 1993; Auger et al., 1997), the level of their sexual receptivity increases over time. In most earlier reports of mating enhancement of sexual receptivity, rats that had been ovariectomized, but not adrenalectomized, were used (Clemens et al., 1969; Hardy and Debold, 1973; Rajendren et al., 1990). Because adrenalectomy was reported to prevent this increase (Larsson et al., 1974), and because mere handling can substitute for mating (Hardy and Debold, 1973), the increase seen in those experiments was thought to be due to nonspecific causes, perhaps progesterone release from the adrenal (Feder and Ruf, 1969) in response to a minor stressor, rather than a specific response to afferent influences from the mating situation to the brain. It should be noted, however, that a small amount of enhancement had been seen in ovariectomized–adrenalectomized rats injected with large doses of estradiol (Zemlan and Adler, 1977).

In subsequent experiments in which the adrenal gland was removed as well, it was determined that repeated mating stimulation enhances sexual behavior in the absence of ovarian and adrenal secretions (Auger et al., 1997). In studies of the relative contribution of VCS as compared with flank and perineal stimulation provided by the male (Bennett et al., 2001), the first set of results published suggested that the mating enhancement is dependent in part upon VCS. Elimination of VCS (by placing tape over the vagina) blocked most of the enhancement. However, experimenter-induced VCS had only a very small influence on subsequent responding (Bennett et al., 2001). Nevertheless, on the basis of that work and work demonstrating the importance of intromissive stimuli/VCS on initiation of pregnancy and on abbreviation of estrus, a great deal of attention was paid to the cellular changes that accompanied intromissive/VCS specifically.

Surprisingly, a more recent study (Ghavami et al., 2005) suggests that VCS may not in fact be the proximate cause of mating enhancement in the repetitive mating situation. Mounts without intromissions were as effective in enhancing sexual behavior as mounts with intromission. Furthermore, this enhancement is not limited to lordosis; dramatic increases in paracopulatory behaviors are observed as well. While rats both receiving and not receiving intromissions show high levels of paracopulatory behaviors (ear-wiggling, darting, and hopping), those receiving intromissive stimuli predictably express higher levels of negative behaviors, such as rejection.

The conflicting results on the necessity of intromissive stimuli for enhancement of sexual behaviors in a repetitive mating situation are difficult to reconcile. There were, however, differences in procedures between the earlier and the later studies. For example, in the first study, ovariectomized–adrenalectomized rats were given salt replacement, but not corticosterone, while in the later study, the rats were implanted with a corticosterone pellet. It is not known at this time if there is an interaction with the presence of corticosterone replacement that could explain the differences. It should, however, be emphasized that there are complex interactions between adrenal hormones and ovarian hormones in the regulation of sexual responsiveness (DeCatanzaro and Gorzalka, 1979; DeCatanzaro et al., 1981; Gorzalka and Moe, 1994) and neuronal response to mating stimulation (Cameron and Erskine, 2003). In some neuroanatomical areas, for example, ovariectomized–adrenalectomized rats show Fos expression in fewer cells in response to mounts without intromission than rats that are only ovariectomized, but expression in more cells in response to intromissions. Cameron and Erskine (2003) suggest that adrenal secretions may decrease sensitivity to low levels of mating stimulation. Regardless of the basis for the conflicting reports, it is now clear that mounting without intromission is a sufficient stimulus for enhancement of mating in estradiol-treated rats in ovariectomized–adrenalectomized rats in the absence of progesterone. This makes logical sense, because intromissive stimulation is rare in the early stages of this mating enhancement procedure, because at the start, rats are not expressing lordosis in response to mounts by the male.

Based on the finding that VCS is not necessary for mating enhancement of sexual receptivity, it can be predicted that the pelvic nerve does not mediate enhancement of sexual behaviors by mating stimulation. Rather it is likely to be mediated by the pudendal nerve, which innervates the perineal region, and whose sensory field is increased dramatically by ovarian hormone treatment (Komisaruk et al., 1972; Adler et al., 1977). In related work in rats at the proestrous stage of the estrous cycle, it has been shown that in some neuroanatomical areas, such as the preoptic area, medial amygdala, and bed nucleus of stria terminalis, mounts with and without intromissions are similar in their ability to induce Fos expression (Rowe and Erskine, 1993). In others, such as the VMNvl, paraventricular nucleus of the hypothalamus, and midbrain, mounts with or without intromissions had similar effects (Rowe and Erskine, 1993). A similar increase was seen after mounts without intromission in hormonally treated, ovariectomized rats in the central tegmental field and paraventricular nucleus of the hypothalamus (Polston and Erskine, 1995). While pelvic neurectomy blocked Fos expression in the preoptic area, medial amygdala, and bed nucleus of stria terminalis, it was without effect in the VMNvl, paraventricular nucleus of the hypothalamus, and midbrain (Rowe and Erskine, 1993).

Much of the earlier work in this field showing enhancement of sexual responding as a consequence of nonintromissive mating and even handling was dismissed as being due to nonspecific influences on the adrenal gland. The fact that enhancement of sexual behavior occurs even in rats without ovaries and adrenals suggests that adrenal involvement may not have been an adequate explanation for some of these results. In fact, in some cases, enhancement by repeated testing could be blocked by removal of the adrenals (Larsson et al., 1974). However, that is not the case for the results presented earlier. Most importantly, it is clear that there are numerous factors other than hormones that influence the moment-to-moment levels of sexual responding. As Hardy and Debold wrote (1973), "Thus, in spite of the clear dose–response relationship between level of hormone administration and degree of display of lordosis, variables other than hormonal dosage exert an important influence on the sexual behavior of the female rat." While these have been less well studied than the mechanisms mediating overall response to the hormones, because they can have a dramatic influence on fertilization, they are just as important as the more easily studied overall changes in lordosis quotients as a function of hormone dose.

8.3 Longer Latency Inhibition of Sexual Behavior by Mating Stimuli: Heat Abbreviation

The receipt of intromissions and ejaculations during mating has another longer latency influence on sexual behavior. Females that receive intromissive stimulation during mating express copulatory behaviors for a shorter length of time than do females who have not received VCS [rats (Boling and Blandau, 1939; Lodder and Zeilmaker, 1976; Reading and Blaustein, 1984; Erskine, 1985; Pfaus et al., 2000), guinea pigs (Goldfoot and Goy, 1970; Roy et al., 1993), and hamsters (Carter and Schein, 1971; Carter, 1972, 1973; Carter et al., 1976; Ramos and Debold, 1999)]. Estrous duration in unmated cycling female rats assessed the presence of lordosis in response to manual palpation as approximately 19 h, while the females' receipt of intromissive, but not mounts-without-intromissive, stimulation significantly reduced the length of estrus to 16 h (Lodder and Zeilmaker, 1976). In rats, pacing of sexual contacts increases the effectiveness of intromissive stimulation on reducing estrous duration (Erskine and Baum, 1982; Erskine, 1985; Coopersmith et al., 1996). Excessive intromissions reduce subsequent sexual receptivity (Hardy and DeBold, 1972), and experimenter-induced VCS resulted in some active rejection behaviors as well (Rodriguez-Sierra et al., 1975; Pfaus et al., 2000). In some cases, some of the inhibitory effects of intromissions on subsequent sexual receptivity may be dependent upon trauma to the vagina, vulva, and cervix (VanderSchoot et al., 1992).

Although earlier studies defined the period of sexual receptivity as the time during which a female responds to mating stimulation with lordosis, another definition is that the period of sexual receptivity actually ends when the female either passively avoids the male in a pacing situation and/or demonstrates active rejection behaviors. For example, in a study of the effects of administration of 50 applications of VCS with a glass probe on subsequent sexual behavior, (Pfaus et al., 2000), rats displayed both passive (avoiding the male) and active rejection behavior toward the male, long before the levels of lordosis had declined.

8.4 Vaginocervical Stimulation Versus Perineal Stimulation

VCS, usually the result of penile intromission by the male, is an important component of the stimulation received by female rats during mating. VCS influences hormone-sensitive processes, including LH release (Moss et al., 1977) and the twice daily surges of prolactin that then result in pseudopregnancy (Gunnet and Freeman, 1983). In the short term, it prolongs lordosis responses (Diakow, 1975), and in the long term, it decreases lordosis responding, increases rejection (Hardy and DeBold, 1972; Pfaus et al., 2000), and causes abbreviation of the period of sexual receptivity (Blandau et al., 1941; Reading and Blaustein, 1984; Pfaus et al., 2000), especially when mating stimulation is paced by the female (Erskine et al., 1989).

The results of many experiments suggest that experimentally administered VCS in rats can sometimes mimic the effects of intromissions by a male rat on reproductive physiology and behavior. This finding is interesting, because the experimental probes that are used are typically smooth glass or plastic, and they provide pressure directly to the cervix with only mild distension pressure on the vaginal wall [and in some cases, electrical stimulation of the cervix has been used; e.g.,Gorospe and Freeman (1981)]. In contrast, the penis of rats is covered with keratinous spines, which likely cause much more intense stimulation of the vaginal wall than a smooth probe (Taylor et al., 1983; Sachs et al., 1984). Furthermore, it is not actually known if the penis contacts the cervix directly during an intromission. It is likely, however, that stimulation would result either from ejaculation or deposition of the seminal plug. Although it is unclear whether experimentally induced VCS provides the same crucial element of genitosensory stimulation provided by an intromission, experimenter-administered VCS causes longitudinal stretching of the vaginal wall, which causes sufficient vaginal stimulation to induce immobilization similar to that induced by an intromission (Komisaruk and Larsson, 1971). Therefore, experimentally administered VCS may induce some of the same physiological changes as intromission. It should be considered that experimenter-induced VCS may also be a better proxy for ejaculations than for intromissions, although it may have some common features of both.

The pelvic nerve mediates mating-induced heat abbreviation in rats, because pelvic neurectomy blocks the heat abbreviating effects of intromissions (Lodder and Zeilmaker, 1976). In contrast, in guinea pigs, transection of the pelvic, pudendal, and genitofemoral nerves (Slimp, 1977) is without effect. It has been argued that there is a nonneural route of travel for the information from the genitals to the brain in guinea pigs. However, because the vagus nerve also conveys sensory information from the vagina/cervix to the brain (Cueva-Rolon et al., 1996; Komisaruk et al., 1996), this neural route has to be considered as well before a neural connection is dismissed.

9 Mechanisms of Changes in Sexual Responsiveness: Mating Enhancement Requires Ligand-Independent Activation of Progestin Receptors

The mating enhancement of sexual receptivity seen in ovariectomized–adrenalectomized rats is dependent on PRs, because it is blocked by progesterone antagonists (Auger et al., 1997). Because mating-related stimuli induce release of dopamine, as well as other neurotransmitters (Vathy and Etgen, 1989; Mermelstein and Becker, 1995; Kohlert et al., 1997; Matuszewich et al., 2000; Paredes and Agmo, 2004) in the forebrain, we have suggested that the enhancement is referable to ligand-independent activation of PRs, perhaps secondary to the release of dopamine or another neurotransmitter. This represents a physiological application of the pharmacological, ligand-independent activation of PRs by dopamine agonists discussed earlier.

Moss and collaborators (Moss et al., 1977; Foreman and Moss, 1979; Rajendren et al., 1990, 1991; Dudley et al., 1992; Rajendren and Moss, 1993) suggested an alternate pathway for at least some of the effects of mating on enhancement of sexual behavior. They suggested that an accessory olfactory system pathway, involving the vomeronasal organ, medial amygdala, and the VMN and release of GnRH, is involved in this potentiation. While the olfactory pathway may also be involved in the mechanism by which mating enhances sexual receptivity, the mechanism seems to be independent of the ligand-independent activation of PRs described earlier (Bennett et al., 2002).

We have proposed that heat abbreviation by intromissive stimuli may be a consequence of down-regulation of PRs induced by ligand-independent activation of PRs. This is, however, just a hypothesis at this time.

The question has arisen recently as to whether the facilitation of sexual behavior that has been attributed to ligand-independent activation of PRs is actually due to increases in neural progesterone synthesis. We now know that progesterone is synthesized in the brain (Jung-Testas et al., 1989; Baulieu et al., 1996; Guennoun et al., 1997), but little is known about the regulation of this synthesis. The fact that estradiol induces progesterone synthesis in the hypothalamus (Micevych et al., 2003; Sinchak et al., 2003) and that it increases mRNA for 3β-hydroxysteroid dehydrogenase (Soma et al., 2005), however, raises the possibility that some afferent influences on PRs could be attributable to increases in neuroprogesterone synthesis, and therefore ligand-dependent activation rather than ligand-independent activation. One report provided indirect evidence that this is not the case (Auger et al., 2000). Furthermore, we recently clamped progesterone synthesis with a steroid hormone synthesis inhibitor and found that this does not mitigate the effects of mating on activation of PRs. This suggests that an increase in neural progesterone synthesis is not the mechanism by which the social environment/mating stimulation activates PRs (Lawrence et al., 2005). Nevertheless, further study is essential to support or refute this alternate hypothesis. In addition, much more has to be learned about the factors involved in the regulation of neural progesterone synthesis.

10 Sites of Integration Between Information About Social Environment and Hormonal Milieu

In general, mating stimulation and/or experimenter-applied VCS induces Fos-ir expression in the medial preoptic area, ventromedial hypothalamic area, bed nucleus of stria terminalis, paraventricular nucleus, medial amygdala, accessory olfactory bulb, midbrain central grey, and central tegmental field (Dudley et al., 1992; Erskine, 1993; Pfaus et al., 1993; Rowe and Erskine, 1993; Tetel et al., 1993; Wersinger et al., 1993; Dudley and Moss, 1994). However, many of the cells that respond to mating-related stimulation also contain ERs (Tetel et al., 1994a; Calizo and Flanagan-Cato, 2003) and/or PRs (Auger et al., 1996), suggesting that they are part of the neuronal substrate for integration of hormonal signals with afferent input from the social environment. Extensive coexpression of Fos-ir with ERαir occurs in the medial preoptic area, bed nucleus of stria terminalis, posterodorsal medial amygdala, midbrain central gray (Tetel et al., 1994a), and ventromedial hypothalamic area (Tetel et al., 1994a; Calizo and Flanagan-Cato, 2003). Although other areas were not investigated, extensive coexpression of Fos-ir with PRir is seen in the medial preoptic area, ventromedial hypothalamic area, and the arcuate nucleus (Auger et al., 1996; Blaustein and Greco, 2002). Surprisingly, in the ventromedial hypothalamic area and bed nucleus of stria terminalis, hormonal priming with estradiol and progesterone decreases the number of Fos-ir cells after a moderate amount of VCS (Tetel et al., 1994b; Pfaus et al., 1996). In contrast, hormonal treatment increased Fos expression in other areas, including the medial preoptic area and posterodorsal medial amygdala (Pfaus et al., 1996).

In some, but not all, cells, VCS-induced Fos expression in estradiol-primed, ovariectomized rats, is dependent on PRs, suggesting that ligand-independent activation of PRs is an obligatory step in the induction of this immediate early gene (Auger et al., 1997; Blaustein and Greco, 2002). Although progestin antagonists block VCS-induced Fos expression in the medial preoptic area, medial bed nucleus of stria terminalis, and caudal ventromedial hypothalamic area, these were without effect in other areas, including the medial amygdala, dorsomedial hypothalamus, and paraventricular nucleus. Interestingly, a progestin antagonist blocks VCS-induced Fos in the rostral medial preoptic area, but not the caudal (Blaustein and Greco, 2002). Collectively, the data support the interpretation that in cells expressing PRs, VCS induction of Fos expression requires functional PRs. The mechanism is likely to be ligand-independent activation of those PRs by the afferent input.

Olfactory stimuli are of great importance to reproduction and sexual behavior. However, unlike VCS-induced Fos expression (Auger et al., 1996), the neurons in which Fos is expressed after exposure to bedding soiled by male rats do not contain PRs (Bennett et al., 2002). Furthermore, unlike the case with VCS-induced Fos expression (Auger et al., 1997; Blaustein and Greco, 2002), injection of a progesterone antagonist does not block male odor-induced Fos expression. While olfactory stimulation, like genital stimulation, enhances sexual responding (Rajendren et al., 1990), it seems to do so by a mechanism that does not require PRs.

11 Integration of Approaches

The goal of much of the work in this field is to understand the regulation of feminine sexual behavior from genes to behavior and everything in between. The integration of studies directed at cellular, molecular, and organismic levels is an iterative process. It is important to attempt to complete the synthesis of behavior to genes, and, for example, to determine which genes are influenced by behaviorally relevant hormonal treatments in neurons that are likely to be involved in the behavior. However, as we study the mechanisms of, for example, copulatory behavior, we must keep in mind that the findings may or may not generalize to paracopulatory and/or progestative behaviors. We have to keep in mind that a particular mating situation may establish a reward state, and another may not. How does that difference in procedure alter the results and the conclusions of the experiments? We have to keep in mind that many experiments have looked at estradiol-only induced sexual behavior, and there are fundamental differences between this and estradiol- + progesterone-induced sexual behavior. If a particular study looked at the estradiol-only induced sexual behavior, does it generalize to progesterone facilitated? Is the neural substrate for both hormones one and the same?

A great deal of progress has been made at the molecular, cellular, and organismic levels. However, as we continue to fully integrate our understanding of the multiple levels, it is essential to remember that some of our assumptions may be false. We must constantly challenge the assumptions of the field (Blaustein, 2004a). Some of them are surely wrong. Just as we have learned over the past 30 years that the receptor does not always reside in the subcellular locations where we originally thought it did, and just as we have learned that steroid hormone receptors can be activated by multiple signaling routes besides steroid hormone ligand, we have to remain open minded to the possibility that the behavior is not modified in precisely the ways and the anatomical locations that we have always believed. We have to remain objective in assessing the possibility that the cells adjacent to the VMN are as important, or even more important than, the cells within the VMN. We also have to remain neutral in assessing the possibility that a reductionist approach that can help us understand some of the molecular aspects of complex behaviors may fall short without integrating what we learn about the complexities of regulation of the behaviors at the organismic level. Ultimately, an open mind to all levels of investigation will provide us the opportunity to develop a comprehensive understanding of how hormones regulate a set of behaviors, how the nervous system works to modify the expression of a behavior, and how the environment can influence both brain and behavior. Finally, there are numerous topics that could not be covered or covered in depth in this chapter.

Finally, it cannot be overemphasized that numerous relevant areas that contribute to the overall understanding of feminine sexual behavior were not covered in this chapter. Some of these include the influences of stress on sexual behaviors, the interactions of other hormones, such as glucocorticoids and thyroid hormones with sex steroid hormones and their receptors, the influence of metabolic changes on sexual behavior, and the relationships between the cellular mechanisms regulating sexual behavior with those regulating ovulation. Although this chapter has been quite selective in discussing work from just a few species, and by necessity interesting work on numerous other species has been ignored, the species covered are those for which most information is available on cellular and molecular neuroendocrinology of feminine sexual behavior. The ultimate goal, which cannot be achieved in a review like this, is to integrate all of these issues at all levels, and in many species paying attention to all of the complexities of sexual behavior in a grand synthesis.

Acknowledgments

Work from the authors' laboratories was supported by the following grants from the National Institutes of Health: NS 19327 (JDB) and MH57442 and MH63954 (SKM). We are grateful to Dr Mary Erskine for her contributions to an earlier comprehensive review of feminine sexual behavior (Blaustein and Erskine, 2002), which provided some of the ideas discussed in this chapter.

References

Acosta-Martinez M, Gonzalez-Flores O, Etgen AM. 2006. The role of progestin receptors and the mitogen-activated protein kinase pathway in delta opioid receptor facilitation of female reproductive behaviors. Horm Behav 49: 458-462.

Adler NT, Davis PG, Komisaruk BR. 1977. Variation in the size and sensitivity of a genital sensory field in relation to the estrous cycle in rats. Horm Behav 9: 334-344.

Alkhalaf M, Murphy LC. 1992. Regulation of c-jun and jun-B by progestins in T-47D human breast cancer cells. Mol Endocrinol 6: 1625-1633.

Allan GF, Leng XH, Tsai SY, Weigel NL, Edwards DP, et al. 1992. Hormone and antihormone induce distinct conformational changes which are central to steroid receptor activation. J Biol Chem 267: 19513-19520.

Alves SE, Lopez V, McEwen BS, Weiland NG. 1998. Differential colocalization of estrogen receptor beta (ER beta) with oxytocin and vasopressin in the paraventricular and supraoptic nuclei of the female rat brain: An immunocytochemical study. Proc Natl Acad Sci USA 95: 3281-3286.

Ansonoff MA, Etgen AM. 1998. Estradiol elevates protein kinase C catalytic activity in the preoptic area of female rats. Endocrinology 139: 3050-3056.

Apostolakis EM, Ramamurphy M, Zhou D, Onate S, O'Malley BW. 2002. Acute disruption of select steroid receptor coactivators prevents reproductive behavior in rats and unmasks genetic adaptation in knockout mice. Mol Endocrinol 16: 1511-1523.

Armstrong RC, Montminy MR. 1993. Transsynaptic control of gene expression. Annu Rev Neurosci 16: 17-29.

Atger M, Baulieu EE, Milgrom E. 1974. An investigation of progesterone receptors in guinea pig vagina, uterine cervix, mammary glands, pituitary and hypothalamus. Endocrinology 94: 161-167.

Attardi B. 1981. Facilitation and inhibition of the estrogen-induced luteinizing hormone surge in the rat by progesterone: Effects on cytoplasmic and nuclear estrogen receptors in the hypothalamus-preoptic area, pituitary, and uterus. Endocrinology 108: 1487-1496.

Attardi B. 1984. Progesterone modulation of the luteinizing hormone surge: Regulation of hypothalamic and pituitary progestin receptors. Endocrinology 115: 2113-2122.

Auger AP, La Riccia LM, Moffatt CA, Blaustein JD. 2000. Progesterone, but not progesterone-independent activation of progestin receptors by a mating stimulus, rapidly decreases progestin receptor immunoreactivity in female rat brain. Horm Behav 37: 135-144.

Auger AP, Moffatt CA, Blaustein JD. 1996. Reproductively-relevant stimuli induce Fos-immunoreactivity within progestin receptor-containing neurons in localized regions of female rat forebrain. J Neuroendocrinol 8: 831-838.

Auger AP, Moffatt CA, Blaustein JD. 1997. Progesterone-independent activation of rat brain progestin receptors by reproductive stimuli. Endocrinology 138: 511-514.

Aupperlee MD, Smith KT, Kariagina A, Haslam SZ. 2005. Progesterone receptor isoforms A and B: Temporal and spatial differences in expression during murine mammary gland development. Endocrinology 146: 3577-3588.

Balthazart J, Blaustein JD, Cheng MF, Feder HH. 1980. Hormones modulate the concentration of cytoplasmic progestin receptors in the brain of male ring doves (*Streptopelia risoria*). J Endocrinol 86: 251-261.

Barfield MA, Lisk RD. 1974. Relative contributions of ovarian and adrenal progesterone to the timing of heat in the 4-day cyclic rat. Endocrinology 94: 571-575.

Barfield RJ, Chen JJ. 1977. Activation of estrous behavior in ovariectomized rats by intracerebral implants of estradiol benzoate. Endocrinology 101: 1716-1725.

Barfield RJ, Schoelch Krieger M. 1977. Ejaculatory and post-ejaculatory behavior of male and female rats: Effects of sex hormones and electric shock. Physiol Behav 19: 203-208.

Baulieu EE, Schumacher M, Koenig H, Jung-Testas I, Akwa Y. 1996. Progesterone as a neurosteroid: Actions within the nervous system. Cell Mol Neurobiol 16: 143-154.

Baum MJ, De Greef WJ, Kloet GA, Schretlen PJ. 1979. Evidence that a factor besides progesterone, prolactin, or plasma-estradiol-binding protein inhibits estrogen-induced sexual receptivity in pregnant rats. J Comp Physiol Psychol 93: 278-294.

Baum MJ, Sodersten P, Vreeburg JTM. 1974. Mounting and receptive behavior in the ovariectomized female rat: Influence of estradiol, dihydrotestosterone, and genital anesthetization. Horm Behav 5: 175-190.

Beach FA. 1942. Importance of progesterone to induction of sexual receptivity in spayed female rats. Proc Soc Exp Biol Med 51: 369-371.

Beach FA. 1976. Sexual attractivity, proceptivity, and receptivity in female mammals. Horm Behav 7: 105-138.

Beach FA, Orndoff RK. 1974. Variation in the responsiveness of female rats to ovarian hormones as a function of preceding hormonal deprivation. Horm Behav 5: 201-205.

Beck CA, Weigel NL, Moyer ML, Nordeen SK, Edwards DP. 1993. The progesterone antagonist RU486 acquires agonist activity upon stimulation of cAMP signaling pathways. Proc Natl Acad Sci USA 90: 4441-4445.

Bennett AL, Blasberg ME, Blaustein JD. 2001. Sensory cues mediating mating-induced potentiation of sexual receptivity in female rats. Horm Behav 40: 77-83.

Bennett AL, Greco B, Blasberg ME, Blaustein JD. 2002. Response to male odours in progestin receptor- and oestrogen receptor-containing cells in female rat brain. J Neuroendocrinol 14: 442-449.

Bermant G. 1961. Response latencies of female rats during sexual intercourse. Science 133: 1771-1773.

Bermant G, Westbrook WH. 1966. Peripheral factors in the regulation of sexual contact by female rats. J Comp Physiol Psychol 61: 244-250.

Bethea CL, Widmann AA. 1998. Differential expression of progestin receptor isoforms in the hypothalamus, pituitary, and endometrium of *Rhesus macaques*. Endocrinology 139: 677-687.

Bethea CL, Brown NA, Kohama SG. 1996. Steroid regulation of estrogen and progestin receptor messenger ribonucleic acid in monkey hypothalamus and pituitary. Endocrinology 137: 4372-4383.

Beyer C, Gonzalez-Mariscal G. 1986. Elevation in hypothalamic cyclic AMP as a common factor in the facilitation of lordosis in rodents: A working hypothesis. Ann N Y Acad Sci 474: 270-281.

Beyer C, Canchola E, Larsson K. 1981. Facilitation of lordosis behavior in the ovariectomized estrogen primed rat by dibutyryl cAMP. Physiol Behav 26: 249-251.

Beyer C, Gonzalez-Flores O, Gonzalez-Mariscal G. 1997. Progesterone receptor participates in the stimulatory effect of LHRH, prostaglandin E_2, and cyclic AMP on lordosis and proceptive behaviours in rats. J Neuroendocrinol 9: 609-614.

Beyer C, Damm N, Brito V, Kuppers E. 2002. Developmental expression of progesterone receptor isoforms in the mouse midbrain. Neuroreport 13: 877-880.

Blandau RJ, Boling JL, Young WC. 1941. The length of heat in the albino rat as determined by the copulatory response. Anat Rec 79: 453-463.

Blasberg ME. Robinson S, Henderson LP, Clark AS. 1998. Inhibition of estrogen-induced sexual receptivity by androgens: Role of the androgen receptor. Horm Behav 34: 283-293.

Blaustein JD. 1982a. Progesterone in high doses may overcome progesterone's desensitization effect on lordosis by translocation of hypothalamic progestin receptors. Horm Behav 16: 175-190.

Blaustein JD. 1982b. Alteration of sensitivity to progesterone facilitation of lordosis in guinea pigs by modulation of hypothalamic progestin receptors. Brain Res 243: 287-300.

Blaustein JD. 1992. Cytoplasmic estrogen receptors in rat brain: Immunocytochemical evidence using three antibodies with distinct epitopes. Endocrinology 131: 1336-1342.

Blaustein JD. 1994. Estrogen receptors in neurons: New subcellular locations and functional implications. Endocrine J 2: 249-258.

Blaustein JD. 2004a. Can you teach an old dogma new tricks? Endocrinology 145: 1055-1056.

Blaustein JD. 2004b. Minireview: Neuronal steroid hormone receptors: They're not just for hormones anymore. Endocrinology 145: 1075-1081.

Blaustein JD, Brown TJ. 1984. Progesterone decreases the concentration of hypothalamic and anterior pituitary estrogen receptors in ovariectomized rats. Brain Res 304: 225-236.

Blaustein JD, Brown TJ. 1985. Neural progestin receptors: Regulation of progesterone-facilitated sexual behaviour in female guinea pigs. Comparative Physiology and Biochemistry Current Topics and Trends. Volume C: Comparative Neurobiology. Gilles R, Balthazart J, editors. Berlin Heidelberg: Springer-Verlag; pp. 60-76.

Blaustein JD, Erskine MS. 2002. Feminine sexual behavior: Cellular integration of hormonal and afferent information in the rodent forebrain. Hormones, Brain and Behavior. Pfaff DW, Arnold AP, Etgen AM, Fahrbach SE, Rubin RT, editors. New York: Academic Press; pp. 139-214.

Blaustein JD, Feder HH. 1979a. Progesterone, at plasma levels lower than those of mid-pregnancy, decreases sexual behavior in ovariectomized rats. Physiol Behav 23: 1099-1104.

Blaustein JD, Feder HH. 1979b. Cytoplasmic progestin receptors in female guinea pig brain and their relationship to refractoriness in expression of female sexual behavior. Brain Res 177: 489-498.

Blaustein JD, Feder HH. 1979c. Cytoplasmic progestin receptors in guinea pig brain: Characteristics and relationship to the induction of sexual behavior. Brain Res 169: 481-497.

Blaustein JD, Feder HH. 1980. Nuclear progestin receptors in guinea pig brain measured by and in vitro exchange assay after hormonal treatments that affect lordosis. Endocrinology 106: 1061-1069.

Blaustein JD, Greco B. 2002. A progestin antagonist blocks vaginocervical stimulation-induced Fos expression in neurones containing progestin receptors in the rostral medial preoptic area. J Neuroendocrinol 14: 109-115.

Blaustein JD, Olster DH. 1989. Gonadal steroid hormone receptors and social behaviors. Molecular and Cellular Basis of Social Behavior in Vertebrates. Balthazart J. editor. Berlin: Springer-Verlag; pp. 31-104.

Blaustein JD, Olster DH. 1993. Colchicine-induced accumulation of estrogen receptor and progestin receptor immunoreactivity in atypical areas in guinea-pig brain. J Neuroendocrinol 5: 63-70.

Blaustein JD, Turcotte JC. 1989. Estradiol-induced progestin receptor immunoreactivity is found only in estrogen receptor-immunoreactive cells in guinea pig brain. Neuroendocrinology 49: 454-461.

Blaustein JD, Turcotte JC. 1990. Down-regulation of progestin receptors in guinea pig brain: New findings using an immunocytochemical technique. J Neurobiol 21: 675-685.

Blaustein JD, Wade GN. 1977. Sequential inhibition of sexual behavior by progesterone in female rats: Comparison with a synthetic antiestrogen. J Comp Physiol Psychol 91: 752-760.

Blaustein JD, Wade GN. 1978. Progestin binding by brain and pituitary cell nuclei and female rat sexual behavior. Brain Res 140: 360-367.

Blaustein JD, Brown TJ, Reading DS. 1982. Failure of a protein synthesis inhibitor to block progesterone's desensitization action on lordosis in female rats. Physiol Behav 29: 475-481.

Blaustein JD, Finkbohner R, Delville Y. 1987. Estrogen-induced and estrogen-facilitated female sexual behavior is not mediated by progestin receptors. Neuroendocrinology 45: 152-159.

Blaustein JD, King JC, Toft DO, Turcotte J. 1988. Immunocytochemical localization of estrogen-induced progestin receptors in guinea pig brain. Brain Res 474: 1-15.

Blaustein JD, Lehman MN, Turcotte JC, Greene G. 1992. Estrogen receptors in dendrites and axon terminals in the guinea pig hypothalamus. Endocrinology 131: 281-290.

Blaustein JD, Dudley SD, Gray JM, Roy EJ, Wade GN. 1979. Long-term retention of estradiol by brain cell nuclei and female rat sexual behavior. Brain Res 103: 355-359.

Boling JL, Blandau RJ. 1939. The estrogen-progesterone induction of mating responses in the spayed female rat. Endocrinology 25: 359-364.

Boonyaratanakornkit V, Scott MP, Ribon V, Sherman L, Anderson SM, et al. 2001. Progesterone receptor contains a proline-rich motif that directly interacts with SH3 domains and activates c-Src family tyrosine kinases. Mol Cell 8: 269-280.

Brandling-Bennett EM, Blasberg ME, Clark AS. 1999. Paced mating behavior in female rats in response to different hormone priming regimens. Horm Behav 35: 144-154.

Broekman M, de Bruin M, Smeenk J, Slob AK, van der Schoot P. 1988. Partner preference behavior of estrous female rats affected by castration of tethered male incentives. Horm Behav 22: 324-337.

Brown TJ, Blaustein JD. 1984a. 1-(o-Chlorophenyl)-1(p-Chlorophenyl)2,2,2,-trichlorothane induces functional progestin receptors in the rat hypothalamus and pituitary gland. Endocrinology 115: 1052-2058.

Brown TJ, Blaustein JD. 1984b. Supplemental progesterone delays heat termination and loss of progestin receptors from hypothalamic cell nuclei in female guinea pigs. Neuroendocrinology 39: 384-391.

Brown TJ, Blaustein JD. 1984c. Inhibition of sexual behavior in female guinea pigs by a progestin receptor antagonist. Brain Res 301: 343-349.

Brown TJ, Blaustein JD. 1985. Loss of hypothalamic nuclear-bound progestin receptors: Factors involved and the relationship to heat termination in female guinea pigs. Brain Res 358: 180-190.

Brown TJ, Blaustein JD. 1986. Abbreviation of the period of sexual behavior in female guinea pigs by the progesterone antagonist, RU 486. Brain Res 373: 103-113.

Brown TJ, Mac Lusky NJ. 1994. Progesterone modulation of estrogen receptors in microdissected regions of the rat hypothalamus. Mol Cell Neurosci 5: 283-290.

Brown TJ, Moore MJ, Blaustein JD. 1987. Maintenance of progesterone-facilitated sexual behavior in female rats requires continued hypothalamic protein synthesis and nuclear progestin receptor occupation. Endocrinology 121: 298-304.

Brown TJ, Scherz B, Hochberg RB, Mac Lusky NJ. 1996. Regulation of estrogen receptor concentrations in the rat brain: Effects of sustained androgen and estrogen exposure. Neuroendocrinology 63: 53-60.

Calizo LH, Flanagan-Cato LM. 2003. Hormonal-neural integration in the female rat ventromedial hypothalamus: Triple labeling for estrogen receptor-alpha, retrograde tract tracing from the periaqueductal gray, and mating-induced Fos expression. Endocrinology 144: 5430-5440.

Camacho-Arroyo I, Guerra Araiza C, Cerbon MA. 1998. Progesterone receptor isoforms are differentially regulated by sex steroids in the rat forebrain. Neuroreport 9: 3993-3996.

Camacho-Arroyo I, Villamar-Cruz O, Gonzalez-Arenas A, Guerra-Araiza C. 2002. Participation of the 26S proteasome in the regulation of progesterone receptor concentrations in the rat brain. Neuroendocrinology 76: 267-271.

Cameron N, Erskine MS. 2003. c-fos expression in the forebrain after mating in the female rat is altered by adrenalectomy. Neuroendocrinology 77: 305-313.

Carter CS. 1972. Postcopulatory sexual receptivity in the female hamster: The role of the ovary and adrenal. Horm Behav 3: 261-265.

Carter CS. 1973. Stimuli contributing to the decrement in sexual receptivity of female golden hamsters (*Mesocricetus auratus*). Anim Behav 21: 827-834.

Carter CS, Schein MW. 1971. Sexual receptivity and exhaustion in the female golden hamster. Horm Behav 2: 191-200.

Carter CS, Landauer MR, Tierney BM, Jones T. 1976. Regulation of female sexual behavior in the golden hamster: Behavioral effects of mating and ovarian hormones. J Comp Physiol Psychol 90: 839-850.

Chappell PE, Levine JE. 2000. Stimulation of gonadotropin-releasing hormone surges by estrogen. I. Role of hypothalamic progesterone receptors. Endocrinology 141: 1477-1485.

Chappell PE, Schneider JS, Kim P, Xu M, Lydon JP, et al. 1999. Absence of gonadotropin surges and gonadotropin-releasing hormone self-priming in ovariectomized (OVX), estrogen (E-2)-treated, progesterone receptor knockout (PRKO) mice. Endocrinology 140: 3653-3658.

Cheng G, Li Y, Omoto Y, Wang Y, Berg T, et al. 2005. Differential regulation of estrogen receptor (ER)alpha and ERbeta in primate mammary gland. J Clin Endocrinol Metab 90: 435-444.

Chou YC, Uehara N, Lowry JR, Shyamala G. 2003. Mammary epithelial cells of PR-A transgenic mice exhibit distinct alterations in gene expression and growth potential associated with transformation. Carcinogenesis 24: 403-409.

Chu HP, Etgen AM. 1997. A potential role of cyclic GMP in the regulation of lordosis behavior of female rats. Horm Behav 32: 125-132.

Chu HP, Etgen AM. 1999. Ovarian hormone dependence of alpha(1)-adrenoceptor activation of the nitric oxide-cGMP pathway: Relevance for hormonal facilitation of lordosis behavior. J Neurosci 19: 7191-7197.

Chu HP, Morales JC, Etgen AM. 1999. Cyclic GMP may potentiate lordosis behaviour by progesterone receptor activation. J Neuroendocrinol 11: 107-113.

Clark AS, Roy EJ. 1983. Behavioral and cellular responses to pulses of low doses of estradiol-17. Physiol Behav 30: 561-565.

Clark CR, Mac Lusky NJ, Naftolin F. 1982. Oestrogen induction of progestin receptors in the rat brain and pituitary gland: Quantitative and kinetic aspects. J Endocrinol 93: 339-353.

Clark AS, Pfeifle JK, Edwards DA. 1981a. Ventromedial hypothalamic damage and sexual proceptivity in female rats. Physiol Behav 27: 597-602.

Clark AS, Kelton MC, Guarraci FA, Clyons EQ. 2004. Hormonal status and test condition, but not sexual experience, modulate partner preference in female rats. Horm Behav 45: 314-323.

Clark CR, MacLusky NJ, Parsons B, Naftolin F. 1981b. Effects of estrogen deprivation on brain estrogen and progestin receptor levels and the activation of female sexual behavior. Horm Behav 15: 289-298.

Clark AS, Guarraci FA, Megroz AB, Porter DM, Henderson LP. 2003. The display of sexual behaviors by female rats administered ICI 182,780. Horm Behav 43: 454-464.

Clarke CH, Norfleet AM, Clarke MSF, Watson CS, Cunningham KA, et al. 2000. Perimembrane localization of the estrogen receptor alpha protein in neuronal processes of cultured hippocampal neurons. Neuroendocrinology 71: 34-42.

Clemens LG, Hiroi M, Gorski RA. 1969. Induction and facilitation of female mating behavior in rats treated neonatally with low doses of testosterone propionate. Endocrinology 84: 1430-1438.

Collins VJ, Boling JI, Dempsey EW, Young WC. 1938. Quantitative studies of experimentally induced sexual receptivity in the spayed guinea pig. Endocrinology 23: 188-196.

Conneely OM, Mulac-Jericevic B, Lydon JP. 2003. Progesterone-dependent regulation of female reproductive activity by two distinct progesterone receptor isoforms. Steroids 68: 771-778.

Conneely OM, Kettleberger DM, Tsai MJ, Schrader WT, O'Malley BW. 1989. The chicken progesterone receptor A and B isoforms are products of an alternate translation initiation event. J Biol Chem 264: 14062-14064.

Coopersmith CB, Candurra C, Erskine MS. 1996. Effects of paced mating and intromissive stimulation on feminine sexual behavior and estrus termination in the cycling rat. J Comp Psychol 110: 176-186.

Couse JF, Korach KS. 1999. Estrogen receptor null mice: What have we learned and where will they lead us? Endocr Rev 20: 358-417.

Couse JF, Lindzey J, Grandien K, Gustafsson JA, Korach KS. 1997. Tissue distribution and quantitative analysis of estrogen receptor-alpha (ER alpha) and estrogen receptor-beta (ER beta) messenger ribonucleic acid in the wild-type and ER alpha-knockout mouse. Endocrinology 138: 4613-4621.

Crowley WR, Nock BL, Feder HH. 1978. Facilitation of lordosis behavior by clonidine in female guinea pigs. Pharmacol Biochem Behav 8: 207-209.

Cueva-Rolon R, Sansone G, Bianca R, Gomez LE, Beyer C, et al. 1996. Vagotomy blocks responses to vaginocervical stimulation after genitospinal neurectomy in rats. Physiol Behav 60: 19-24.

Czaja JA, Butera PC, McCaffrey TA. 1985. Duration of hormonal deprivation: Influences on physiological and behavioral responsiveness to estradiol. Horm Behav 19: 52-63.

Daniels D, Flanagan-Cato LM. 2000. Functionally-defined compartments of the lordosis neural circuit in the ventromedial hypothalamus in female rats. J Neurobiol 45: 1-13.

Davidson JM, Rodgers CH, Smith ER, Bloch GJ. 1968. Stimulation of female sex behavior in adrenalectomized rats with estrogen alone. Endocrinology 82: 193-195.

Davis PG, Barfield RJ. 1979. Activation of masculine sexual behavior by intracranial estradiol benzoate implants in male rats. Neuroendocrinology 28: 217-227.

Davis PG, Krieger MS, Barfield RJ, McEwen BS, Pfaff DW. 1982. The site of action in intrahypothalamic estrogen implants in feminine sexual behavior: An autoradiographic analysis. Endocrinology 111: 1581-1586.

De Jonge FH, Eerland EMJ, Poll Van De NE. 1986. The influence of estrogen, testosterone and progesterone on partner preference, receptivity and proceptivity. Physiol Behav 37: 885-891.

Debold JF, Frye CA. 1994. Genomic and non-genomic actions of progesterone in the control of female hamster sexual behavior. Horm Behav 28: 445-453.

Debold JF, Malsbury CW. 1989. Facilitation of sexual receptivity by hypothalamic and midbrain implants of progesterone in female hamsters. Physiol Behav 46: 655-660.

Debold JF, Malsbury CW, Harris VS, Malenka R. 1982. Sexual receptivity: Brain sites of estrogen action in female hamsters. Physiol Behav 29: 589-593.

DeCatanzaro D, Gorzalka BB. 1979. Effects of dexamethasone, corticosterone, and ACTH on lordosis in ovariectomized and adrenalectomized-ovariectomized rats. Pharmacol Biochem Behav 12: 201-206.

DeCatanzaro D, Knipping RP, Gorzalka BB. 1981. Antagonism of estrogen-induced lordosis by corticosterone in adrenalectomized-ovariectomized female rats and mice. Pharmacol Biochem Behav 15: 761-766.

Delville Y, Blaustein JD. 1989. Long-term ovariectomy and hormone-induced sexual behavior, progestin receptors, and hypothalamic morphology in female rats. Horm Behav 23: 269-278.

Delville Y, Blaustein JD. 1991. A site for estradiol priming of progesterone-facilitated sexual receptivity in the ventrolateral hypothalamus of female guinea pigs. Brain Res 559: 191-199.

Dempsey EW, Hertz R, Young WC. 1936. The experimental induction of oestrus (sexual receptivity) in the normal and ovariectomized guinea pig. Am J Physiol 116: 201-209.

Denner LA, Weigel NL, Maxwell BL, Schrader WT, O'Malley BW. 1990. Regulation of progesterone receptor-mediated transcription by phosphorylation. Science 250: 1740-1743.

Devidze N, Mong JA, Jasnow AM, Kow LM, Pfaff DW. 2005. Sex and estrogenic effects on coexpression of mRNAs in single ventromedial hypothalamic neurons. Proc Natl Acad Sci USA 102: 14446-14451.

Diakow C. 1975. Motion picture analysis of rat mating behavior. J Comp Physiol Psychol 88: 704-712.

DonCarlos LL, Greene GL, Morrell JI. 1989. Estrogen plus progesterone increases progestin receptor immunoreactivity in the brain of ovariectomized guinea pigs. Neuroendocrinology 50: 613-623.

DonCarlos LL, Monroy E, Morrell JI. 1991. Distribution of estrogen receptor-immunoreactive cells in the forebrain of the female guinea pig. J Comp Neurol 305: 591-612.

DonCarlos LL, Mcabee M, Ramerquinn DS, Stancik DM. 1995. Estrogen receptor mRNA levels in the preoptic area of neonatal rats are responsive to hormone manipulation. Dev Brain Res 84: 253-260.

Driggers PH, Segars JH. 2002. Estrogen action and cytoplasmic signaling pathways. Part II: The role of growth factors and phosphorylation in estrogen signaling. Trends Endocrinol Metab 13: 422-428.

Dudley CA, Moss RL. 1994. Lesions of the accessory olfactory bulb decrease lordotic responsiveness and reduce mating-induced c-fos expression in the accessory olfactory system. Brain Res 642: 29-37.

Dudley CA, Cooper KJ, Moss RL. 1980. Progesterone-induced mating behavior in rats on the day following spontaneous heat. Physiol Behav 25: 759-762.

Dudley CA, Rajendren G, Moss RL. 1992. Induction of FOS immunoreactivity in central accessory olfactory structures of the female rat following exposure to conspecific males. Mol Cell Neurosci 3: 360-369.

Duffy DM, Wells TR, Haluska GJ, Stouffer RL. 1997. The ratio of progesterone receptor isoforms changes in the monkey corpus luteum during the luteal phase of the menstrual cycle. Biol Reprod 57: 693-699.

Edwards DA. 1970. Induction of estrus in female mice: Estrogen–progesterone interactions. Horm Behav 1: 299-304.

Edwards DA, Pfeifle JK. 1983. Hormonal control of receptivity, proceptivity and sexual motivation. Physiol Behav 30: 437-443.

Edwards DA, Whalen RE, Nadler RD. 1968. Induction of estrus: Estrogen–progesterone interactions. Physiol Behav 3: 29-33.

Eisenfeld AJ. 1969. Hypothalamic oestradiol-binding macromolecules. Nature 224: 1202-1203.

Emery DE, Moss RL. 1984a. Lesions confined to the ventromedial hypothalamus decrease the frequency of coital contacts in female rats. Horm Behav 18: 313-329.

Emery DE, Moss RL. 1984b. p-Chlorophenylalanin alters pacing of copulation in female rats. Pharmacol Biochem Behav 20: 337-341.

Erskine MS. 1983. Effects of an anti-androgen and 5α-reductase inhibitors on estrus duration in the cycling female rat. Physiol Behav 30: 519-524.

Erskine MS. 1985. Effects of paced coital stimulation on estrous duration in intact cycling rats and ovariectomized and ovariectomized–adrenalectomized hormone-primed rats. Behav Neurosci 99: 151-161.

Erskine MS. 1989. Solicitation behavior in the estrous female rat: A review. Horm Behav 23: 473-502.

Erskine MS. 1993. Mating-induced increases in FOS protein in preoptic area and medial amygdala of cycling female rats. Brain Res Bull 32: 447-451.

Erskine MS. 1995. Prolactin release after mating and genitosensory stimulation in females. Endoc Rev 16: 508-528.

Erskine MS, Baum MJ. 1982. Effects of paced coital stimulation on termination of estrus and brain indoleamine levels in female rats. Pharmacol Biochem Behav 17: 857-861.

Erskine MS, Kornberg E, Cherry JA. 1989. Paced copulation in rats: Effects of intromission frequency and duration on luteal activation and estrus length. Physiol Behav 45: 33-39.

Erskine MS, Lehmann ML, Cameron NM, Polston EK. 2004. Co-regulation of female sexual behavior and pregnancy induction: An exploratory synthesis. Behav Brain Res 153: 295-315.

Etgen AM. 1979. Antiestrogens: Effects of tamoxifen, nafoxidine, and CI-628 on sexual behavior, cytoplasmic receptors, and nuclear binding of estrogen. Horm Behav 13: 97-112.

Etgen AM. 1982. 1-(o-Chlorophenyl)-1-(p-chlorophenyl) 2,2,2-trichloroethane: A probe for studying estrogen and progestin receptor mediation of female sexual behavior and neuroendocrine responses. Endocrinology 111: 1498-1504.

Etgen AM, Acosta-Martinez M. 2003. Participation of growth factor signal transduction pathways in estradiol facilitation of female reproductive behavior. Endocrinology 144: 3828-3835.

Etgen AM, Barfield RJ. 1986. Antagonism of female sexual behavior with intracerebral implants of antiprogestin RU 38486: Correlation with binding to neural progestin receptors. Endocrinology 119: 1610-1617.

Etgen AM, Shamamian P. 1986. Regulation of estrogen-stimulated lordosis behavior and hypothalamic progestin receptor induction by antiestrogens in female rats. Horm Behav 20: 166-180.

Fadem BH, Barfield RJ, Whalen RE. 1979. Dose–response and time–response relationships between progesterone and the display of patterns of receptive and proceptive behavior in the female rat. Horm Behav 13: 40-48.

Feder HH, Marrone BL. 1977. Progesterone: Its role in the central nervous system as a facilitator and inhibitor of sexual behavior and gonadotropin release. Ann N Y Acad Sci 286: 331-354.

Feder HH, Ruf KB. 1969. Stimulation of progesterone release and estrous behavior by ACTH in ovariectomized rodents. Endocrinology 84: 171-172.

Feder HH, Resko JA, Goy RW. 1968. Progesterone concentrations in the arterial plasma of guinea pigs during the oestrous cycle. J Endocrinol 40: 505-513.

Fernandez-Valdivia R, Mukherjee A, Mulac-Jericevic B, Conneely OM, De Mayo FJ, et al. 2005. Revealing progesterone's role in uterine and mammary gland biology: Insights from the mouse. Semin Reprod Med 23: 22-37.

Filardo EJ, Thomas P. 2005. GPR30: A seven-transmembrane-spanning estrogen receptor that triggers EGF release. Trends Endocrinol Metab 16: 362-367.

Floody OR, Blinn NE, Lisk RD, Vomachka AJ. 1987. Localization of hypothalamic sites for the estrogen-priming of sexual receptivity in female hamsters. Behav Neurosci 101: 309-314.

Floody OR, Merkle DA, Cahill TJ, Shopp GM. 1979. Gonadal hormones stimulate ultrasound production by female hamsters. Horm Behav 12: 172-184.

Font de MJ, Brown M. 2000. AIB1 is a conduit for kinase-mediated growth factor signaling to the estrogen receptor. Mol Cell Biol 20: 5041-5047.

Foreman MM, Moss RL. 1977. Effects of subcutaneous injection and intrahypothalamic infusion of releasing hormones upon lordotic response to repetitive coital stimulation. Horm Behav 8: 219-234.

Foreman MM, Moss RL. 1979. Role of hypothalamic dopaminergic receptors in the control of lordosis behavior in the female rat. Physiol Behav 22: 282-289.

Fraile IG, Pfaff DW, McEwen BS. 1987. Progestin receptors with and without estrogen induction in male and female hamster brain. Neuroendocrinology 45: 487-491.

Franck JAE, Ward IL. 1981. Intralimbic progesterone and methysergide facilitate lordotic behavior in estrogen-primed female rats. Neuroendocrinology 32: 50-56.

Frye CA, Vongher JM. 2001. Progesterone and 3 alpha,5 alpha-THP enhance sexual receptivity in mice. Behav Neurosci 115: 1118-1128.

Garey J, Kow LM, Huynh W, Ogawa S, Pfaff DW. 2002. Temporal and spatial quantitation of nesting and mating behaviors among mice housed in a semi-natural environment. Horm Behav 42: 294-306.

Gehring U. 1998. Steroid hormone receptors and heat shock proteins. Vitamins and Hormones—Advances in Research and Applications, Vol 54. Litwack G, editor. San Diego: Academic Press Inc; pp. 167-205.

Gerall AA, Dunlap JL, Hendricks SE. 1972. Effect of ovarian secretions on female potentiality in the rat. J Comp Physiol Psychol 82: 449-465.

Geyer LA, Barfield RJ. 1978. Influence of gonadal hormones and sexual behavior on ultrasonic vocalization in rats: I. Treatment of females. J Comp Physiol Psychol 92: 438-446.

Geyer LA, McIntosh TK, Barfield RJ. 1978. Effects of ultrasonic vocalizations and male's urine on female rat readiness to mate. J Comp Physiol Psychol 92: 457-462.

Ghavami G, Farrell SF, Laroche J, Blaustein JD. 2005. Enhancement of feminine rat sexual behavior by non-intromissive, mating stimulation via ligand-independent activation of progestin receptors. Abstract Viewer/Itinerary Planner. Washington, DC: Society for Neurosience, online. Program no. 320.14.

Giangrande PH, McDonnell DP. 1999. The A and B isoforms of the human progesterone receptor: Two functionally different transcription factors encoded by a single gene. Recent Prog Horm Res 54: 291-313.

Gibson BM, Floody OR. 1998. Time course of VMN lesion effects on lordosis and ultrasound production in hamsters. Behav Neurosci 112: 1236-1246.

Gilchrist S, Blaustein JD. 1984. The desensitization effect of progesterone on female rat sexual behavior is not due to interference with estrogen priming. Physiol Behav 32: 879-882.

Gilman DP, Hitt JC. 1978. Effects of gonadal hormones on pacing of sexual contacts by female rats. Behav Biol 24: 77-87.

Gilman DP, Westbrook WH. 1978. Mating preference and sexual reinforcement in female rats. Physiol Behav 20: 11-14.

Glaser JH, Barfield RJ. 1984. Blockade of progesterone-activated estrous behavior in rats by intracerebral anisomycin is site specific. Neuroendocrinology 38: 337-343.

Glaser JH, Rubin BS, Barfield RJ. 1983. Onset of the receptive and proceptive components of feminine sexual behavior in rats following the intravenous administration of progesterone. Horm Behav 17: 18-27.

Glass CK, Rosenfeld MG. 2000. The coregulator exchange in transcriptional functions of nuclear receptors. Gene Develop 14: 121-141.

Goldfoot DA, Goy RW. 1970. Abbreviation of behavioral estrus in guinea pigs by coital and vagino-cervical stimulation. J Comp Physiol Psychol 72: 426-434.

Gonzalez-Flores O, Etgen AM. 2004. The nitric oxide pathway participates in estrous behavior induced by progesterone and some of its ring A-reduced metabolites. Horm Behav 45: 50-57.

Gonzalez-Flores O, Shu J, Camacho-Arroyo I, Etgen AM. 2004c. Regulation of lordosis by cyclic 3′,5′-guanosine monophosphate, progesterone, and its 5 alpha-reduced metabolites involves mitogen-activated protein kinase. Endocrinology 145: 5560-5567.

Gonzalez-Flores O, Guerra-Araiza C, Cerbon M, Camacho-Arroyo I, Etgen AM. 2004b. The 26S proteasome participates in the sequential inhibition of estrous behavior induced by progesterone in rats. Endocrinology 145: 2328-2336.

Gonzalez-Flores O, Camacho FJ, Dominguez-Salazar E, Ramirez-Orduna JM, Beyer C, et al. 2004a. Progestins and place preference conditioning after paced mating. Horm Behav 46: 151-157.

Gonzalez-Mariscal G, Melo AI, Beyer C. 1993. Progesterone, but not LHRH or prostaglandin E_2, induces sequential inhibition of lordosis to various lordogenic agents. Neuroendocrinology 57: 940-945.

Gorospe WC, Freeman ME. 1981. The effects of various methods of cervical stimulation on continuation of prolactin surges in rats. Proc Soc Exp Biol Med 167: 78-82.

Gorski J, Toft D, Shyamala G, Smith D, Notides A. 1968. Hormone receptors: Studies on the interaction of estrogen with the uterus. Recent Prog Horm Res 24: 45-80.

Gorzalka BB, Moe IV. 1994. Adrenal role in proceptivity and receptivity induced by two modes of estradiol treatment. Physiol Behav 55: 29-34.

Goy RW, Phoenix CH. 1963. Hypothalamic regulation of female sexual behaviour, establishment of behavioural oestrus in spayed guinea-pigs following hypothalamic lesions. J Reprod Fertil 5: 23-40.

Goy RW, Phoenix CH, Young WC. 1966. Inhibitory action of the corpus luteum on the hormonal induction of estrous behavior in the guinea pig. Gen Comp Endocrinol 6: 267-275.

Greco B, Allegretto EA, Tetel MJ, Blaustein JD. 2001. Coexpression of ER beta with ER alpha and progestin receptor proteins in the female rat forebrain: Effects of estradiol treatment. Endocrinology 142: 5172-5181.

Green PS, Bishop J, Simpkins JW. 1997. 17 Alpha-estradiol exerts neuroprotective effects on SK-N-SH cells. J Neurosci 17: 511-515.

Green R, Luttge WG, Whalen RE. 1970. Induction of receptivity in ovariectomized female rats by a single intravenous injection of estradiol-17β. Physiol Behav 5: 137-141.

Greene GL, Gilna P, Waterfield M, Baker A, Hort Y, et al. 1986. Sequence and expression of human estrogen receptor complementary DNA. Science 231: 1150-1154.

Gu GB, Rojo AA, Zee MC, Yu JH, Simerly RB. 1996. Hormonal regulation of CREB phosphorylation in the anteroventral periventricular nucleus. J Neurosci 16: 3035-3044.

Guennoun R, Schumacher M, Robert F, Delespierre B, Gouezou M, et al. 1997. Neurosteroids: Expression of functional 3 beta-hydroxysteroid dehydrogenase by rat sensory neurons and Schwann cells. Eur J Neurosci 9: 2236-2247.

Guerra-Araiza C, Cerbon MA, Morimoto S, Camacho-Arroyo I. 2000. Progesterone receptor isoforms expression pattern in the rat brain during the estrous cycle. Life Sci 66: 1743-1752.

Guerra-Araiza C, Villamar-Cruz O, Gonzalez-Arenas A, Chavira R, Camacho-Arroyo I. 2003. Changes in progesterone receptor isoforms content in the rat brain during the oestrous cycle and after oestradiol and progesterone treatments. J Neuroendocrinol 15: 984-990.

Gunnet JW, Freeman ME. 1983. The mating-induced release of prolactin: A unique neuroendocrine response. Endocr Rev 4: 44-61.

Hagihara K, Hirata S, Osada T, Hirai M, Kato J. 1992. Distribution of cells containing progesterone receptor messenger RNA in the female rat diencephalon and telencephalon—an in situ hybridization study. Mol Brain Res 14: 239-249.

Hansen S, Sodersten P. 1978. Effects of subcutaneous implants of progesterone on the induction and duration of sexual receptivity in ovariectomized rats. J Endocrinol 77: 373-379.

Hansen S, Sodersten P. 1979. Reversal of progesterone inhibition of sexual behaviour in ovariectomized rats by high doses of progesterone. J Endocrinol 80: 381-388.

Hardy DF, De Bold J. 1972. Effects of coital stimulation upon behavior of the female rat. J Comp Physiol Psychol 78: 400-408.

Hardy DF, Debold JF. 1971. Effects of mounts without intromission upon the behavior of female rats during the onset of estrogen-induced heat. Physiol Behav 7: 643-645.

Hardy DF, Debold JF. 1973. Effects of repeated testing on sexual behavior of the female rat. J Comp Physiol Psychol 85: 195-202.

Howard SB, Etgen AM, Barfield RJ. 1984. Antagonism of central estrogen action by intracerebral implants of tamoxifen. Horm Behav 18: 256-266.

Hyder SM, Chiappetta C, Stancel GM. 1999. Interaction of human estrogen receptors alpha and beta with the same naturally occurring estrogen response elements. Biochem Pharmacol 57: 597-601.

Ilenchuk TT, Walters MR. 1987. Rat uterine progesterone receptor analyzed by [^{3}H]R5020 photoaffinity labeling: Evidence that the A and B subunits are not equimolar. Endocrinology 120: 1449-1456.

Jackson TA, Richer JK, Bain DL, Takimoto GS, Tung L, et al. 1997. The partial agonist activity of antagonist-occupied steroid receptors is controlled by a novel hinge domain-binding coactivator L7/SPA and the corepressors N-CoR or SMRT. Mol Endocrinol 11: 693-705.

Jensen EV, Suzuki T, Kawasima T, Stumpf WE, Jungblut PW, et al. 1968. A two-step mechanism for the interaction of estradiol with rat uterus. Proc Natl Acad Sci USA 59: 632-638.

Jones KJ, Chikaraishi DM, Harrington CA, McEwen BS, Pfaff DW. 1986. In situ hybridization of estradiol-induced changes in ribosomal RNA levels in rat brain. Mol Brain Res 1: 145-152.

Jordan VC. 1995. Alternate antiestrogens and approaches to the prevention of breast cancer. J Cell Biochem 22: 51-57.

Jordan VC. 1999. Targeted antiestrogens to prevent breast cancer. Trends Endocrinol Metab 10: 312-317.

Joslyn WD, Feder HH. 1971. Facilitatory and inhibitory effects of supplementary estradiol benzoate given to ovariectomized, estrogen-primed guinea pigs. Horm Behav 2: 307-314.

Joslyn WD, Wallen K, Goy RW. 1971. Cyclic changes in sexual response to exogenous progesterone in female guinea pigs. Physiol Behav 7: 915-917.

Jung-Testas I, Hu ZY, Baulieu EE, Robel P. 1989. Neurosteroids—Biosynthesis of pregnenolone and progesterone in primary cultures of rat glial cells. Endocrinology 125: 2083-2091.

Kastner P, Krust A, Turcotte B, Stropp U, Tora L, et al. 1990. Two distinct estrogen-regulated promoters generate transcripts encoding the two functionally different human progesterone receptor forms A and B. EMBO J 9: 1603-1614.

Kato J, Onouchi T. 1977. Specific progesterone receptors in the hypothalamus and anterior hypophysis of the rat. Endocrinology 101: 920-928.

Kato J, Onouchi T, Okinaga S. 1978. Hypothalamic and hypophysial progesterone receptors: Estrogen-priming effect, differential localization, 5α-dihydroprogesterone binding, and nuclear receptors. J Steroid Biochem 9: 419-427.

Kato J, Hirata S, Nozawa A, Mouri N. 1993. The ontogeny of gene expression of progestin receptors in the female rat brain. J Steroid Biochem Mol Biol 47: 173-182.

Kato J, Hirata S, Nozawa A, Yamadamouri N. 1994. Gene expression of progesterone receptor isoforms in the rat brain. Horm Behav 28: 454-463.

Kelly MJ, Levin ER. 2001. Rapid actions of plasma membrane estrogen receptors. Trends Endocrinol Metab 12: 152-156.

Kelly MJ, Wagner EJ. 1999. Estrogen modulation of G-protein-coupled receptors. Trends Endocrinol Metab 10: 369-374.

King WJ, Greene GL. 1984. Monoclonal antibodies localize oestrogen receptor in the nuclei of target cells. Nature 307: 745-747.

Kohlert JG, Rowe RK, Meisel RL. 1997. Intromissive stimulation from the male increases extracellular dopamine release from fluoro-gold-identified neurons within the midbrain of female hamsters. Horm Behav 32: 143-154.

Komisaruk BR, Diakow C. 1973. Lordosis, reflex intensity in rats in relation to the estrous cycle, ovariectomy, estrogen administration and mating behavior. Endocrinology 93: 548-557.

Komisaruk BR, Larsson K. 1971. Suppression of a spinal and a cranial reflex by vaginal or rectal probing in rats. Brain Res 35: 231-235.

Komisaruk BR, Adler NT, Hutchinson J. 1972. Genital sensory field: Enlargement by estrogen treatment in female rats. Science 178: 1295-1298.

Komisaruk BR, Bianca R, Sansone G, Gomez LE, Cueva Rolon R, et al. 1996. Brain-mediated responses to vaginocervical stimulation in spinal cord-transected rats: Role of the vagus nerves. Brain Res 708: 128-134.

Kos M, Denger S, Reid G, Korach KS, Gannon F. 2002. Down but not out? A novel protein isoform of the estrogen receptor alpha is expressed in the estrogen receptor alpha knockout mouse. J Mol Endocrinol 29: 281-286.

Kow LM, Brown HE, Pfaff DW. 1994. Activation of protein kinase C in the hypothalamic ventromedial nucleus or the midbrain central gray facilitates lordosis. Brain Res 660: 241-248.

Kow LM, Easton A, Pfaff DW. 2005. Acute estrogen potentiates excitatory responses of neurons in rat hypothalamic ventromedial nucleus. Brain Res 1043: 124-131.

Kraus WL, Montano MM, Katzenellenbogen BS. 1993. Cloning of the rat progesterone receptor gene 5′-region and identification of two functionally distinct promoters. Mol Endocrinol 7: 1603-1616.

Krege JH, Hodgin JB, Couse JF, Enmark E, Warner M, et al. 1998. Generation and reproductive phenotypes of mice lacking estrogen receptor beta. Proc Natl Acad Sci USA 95: 15677-15682.

Krieger MS, Orr D, Perper T. 1976. Temporal patterning of sexual behavior in the female rat. Behav Biol 18: 379-386.

Kubli-Garfias C, Whalen RE. 1977. Induction of lordosis behavior in female rats by intravenous administration of progestins. Horm Behav 9: 380-386.

Kudwa AE, Bodo C, Gustafsson JA, Rissman EF. 2005. A previously uncharacterized role for estrogen receptor beta: defeminization of male brain and behavior. Proc Natl Acad Sci USA 102: 4608-4612.

Kudwa AE, Rissman EF. 2003. Double oestrogen receptor alpha and beta knockout mice reveal differences in neural oestrogen-mediated progestin receptor induction and female sexual behaviour. J Neuroendocrinol 15: 978-983.

Kuiper GG, Carlsson B, Grandien K, Enmark E, Haggblad J, et al. 1997. Comparison of the ligand binding specificity and transcript tissue distribution of estrogen receptors alpha and beta. Endocrinology 138: 863-870.

Kuiper GGJM, Enmark E, Peltohuikko M, Nilsson S, Gustafsson JA. 1996. Cloning of a novel estrogen receptor expressed in rat prostate and ovary. Proc Natl Acad Sci USA 93: 5925-5930.

Kushner PJ, Baxter JD, Duncan KG, Lopez GN, Schaufele F, et al. 1994. Eukaryotic regulatory elements lurking in plasmid DNA: The activator protein-1 site in pUC. Mol Endocrinol 8: 405-407.

La Vaque TJ, Rodgers CH. 1975. Recovery of mating behavior in the female rat following VMH lesions. Physiol Behav 14: 59-63.

Laflamme N, Nappi RE, Drolet G, Labrie C, Rivest S. 1998. Expression and neuropeptidergic characterization of estrogen receptors (ERalpha and ERbeta) throughout the rat brain: Anatomical evidence of distinct roles of each subtype. J Neurobiol 36: 357-378.

Lagrange AH, Ronnekleiv OK, Kelly MJ. 1997. Modulation of G protein-coupled receptors by an estrogen receptor that activates protein kinase A. Mol Pharmacol 51: 605-612.

Landau IT. 1977. Relationships between the effects of the antiestrogen, CI-628, on sexual behavior, uterine growth, and cell nuclear estrogen retention after estradiol-17b-benzoate administration in the ovariectomized rat. Brain Res 133: 119-138.

Laroche J, Gasbarro L, Blaustein JD. 2005. Environmental influences on female mouse sexual behavior. Horm Behav 48:110; Abstract no. 142.

Larsson K, Feder HH, Komisaruk BR. 1974. Role of the adrenal glands, repeated matings and monoamines in lordosis behavior of rats. Pharmacol Biochem Behav 2: 685-692.

Lauber AH, Mobbs CV, Muramatsu M, Pfaff DW. 1991. Estrogen receptor messenger RNA expression in rat hypothalamus as a function of genetic sex and estrogen dose. Endocrinology 129: 3180-3186.

Lauber AH, Romano GJ, Mobbs CV, Pfaff DW. 1990. Estradiol regulation of estrogen receptor messenger ribonucleic acid in rat mediobasal hypothalamus—an in situ hybridization study. J Neuroendocrinol 2: 605-611.

Lawrence RC, Laroche J, Blaustein JD. 2005. Ligand-independent activation of progestin receptors, not neural progesterone synthesis, mediates mating-enhanced receptivity in female rats. Society for Neurosience, online. Washington, DC: 2005 Abstract Viewer/Itinerary Planner. Program no. 320.13.

Lessey BA, Alexander PS, Horwitz KB. 1983. The subunit structure of human breast cancer progesterone receptors: Characterization by chromatography and photoaffinity labeling. Endocrinology 112: 1267-1274.

Levin ER. 1999. Cellular functions of the plasma membrane estrogen receptor. Trends Endocrinol Metab 10: 374-377.

Levin ER. 2002. Cellular functions of plasma membrane estrogen receptors. Steroids 67: 471-475.

Levin ER. 2005. Integration of the extranuclear and nuclear actions of estrogen. Mol Endocrinol 19: 1951-1959.

Li X, Schwartz PE, Rissman EF. 1997. Distribution of estrogen receptor-beta-like immunoreactivity in rat forebrain. Neuroendocrinology 66: 63-67.

Li X, Huang J, Yi P, Bambara RA, Hilf R, et al. 2004. Single-chain estrogen receptors (ERs) reveal that the ERalpha/beta heterodimer emulates functions of the ERalpha dimer in genomic estrogen signaling pathways. Mol Cell Biol 24: 7681-7694.

Lim CS, Baumann CT, Htun H, Xian WJ, Irie M, et al. 1999. Differential localization and activity of the A- and B-forms of the human progesterone receptor using green fluorescent protein chimeras. Mol Endocrinol 13: 366-375.

Limbird LE. 1981. Activation and attenuation of adenylate cyclase. The role of GTP-binding proteins as macromolecular messengers in receptor–cyclase coupling. Biochem J 195: 1-13.

Liposits Z, Kallo I, Coen CW, Paull WK, Flerko B. 1990. Ultrastructural analysis of estrogen receptor immunoreactive neurons in the medial preoptic area of the female rat brain. Histochemistry 93: 233-239.

Lodder J, Zeilmaker GH. 1976. Role of pelvic nerves in the postcopulatory abbreviation of behavioral estrus in female rats. J Comp Physiol Psychol 90: 925-929.

Lonard DM, O'Malley BW. 2005. Expanding functional diversity of the coactivators. Trends Biochem Sci 30: 126-132.

Loosfelt H, Logeat F, Vu Hai MT, Milgrom E. 1984. The rabbit progesterone receptor. Evidence for a single steroid-binding subunit and characterization of receptor mRNA. J Biol Chem 259: 14196-14202.

Lubahn DB, Moyer JS, Golding TS, Couse JF, Korach KS, et al. 1993. Alteration of reproductive function but not prenatal sexual development after insertional disruption of the mouse estrogen receptor gene. Proc Natl Acad Sci USA 90 (23, DEC 1): 11162-11166.

Luttge WG, Hughes JR. 1976. Intracerebral implantation of progesterone: Reexamination of the brain sites responsible for facilitation of sexual receptivity in estrogen-primed ovariectomized rats. Physiol Behav 17: 771-775.

Luttge WM, Wallis CJ, Hall NR. 1974. Effects of pre- and post-treatment with unlabeled steroids on the in vivo uptake of [^{3}H]progestins in selected brain regions, uterus and plasma of the female mouse. Brain Res 71: 105-115.

Lydon JP, De Mayo FJ, Funk CR, Mani SK, Hughes AR, et al. 1995. Mice lacking progesterone receptor exhibit pleiotropic reproductive abnormalities. Genes Dev 9: 2266-2278.

MacLusky NJ, McEwen BS. 1978. Oestrogen modulates progestin receptor concentrations in some rat brain regions but not in others. Nature 274: 276-278.

MacLusky NJ, McEwen BS. 1980. Progestin receptors in rat brain: Distribution and properties of cytoplasmic progestin-binding sites. Endocrinology 106: 192-202.

MacLusky NJ, Luine VN, Hajszan T, Leranth C. 2005. The 17alpha and 17beta isomers of estradiol both induce rapid spine synapse formation in the CA1 hippocampal subfield of ovariectomized female rats. Endocrinology 146: 287-293.

MacLusky NJ, Krey LC, Parsons B, Merriam GR, Loriaux DL, et al. 1985. Are catechol oestrogens obligatory mediators of oestrogen action in the central nervous system? II. Potencies of natural and synthetic oestrogens for induction of gonadotrophin release and female sexual behaviour in the rat. J Endocrinol 110: 499-505.

Madlafousek J, Hlinak Z. 1977. Sexual behavior in the female laboratory rat: Inventory, patterning and measurement. Anim Behav 63: 129-174.

Mahajan MA, Samuels HH. 2000. A new family of nuclear receptor coregulators that integrate nuclear receptor signaling through CREB-binding protein. Mol Cell Biol 20: 5048-5063.

Malyala A, Kelly MJ, Ronnekleiv OK. 2005. Estrogen modulation of hypothalamic neurons: Activation of multiple signaling pathways and gene expression changes. Steroids 70: 397-406.

Mangal RK, Wiehle RD, Poindexter AN, Weigel NL. 1997. Differential expression of uterine progesterone receptor forms A and B during the menstrual cycle. J Steroid Biochem Mol Biol 63: 195-202.

Mani SK, Blaustein JD, O'Malley BW. 1997. Progesterone receptor function from a behavioral perspective. Horm Behav 31: 244-255.

Mani SK, Allen JMC, Clark JH, O'Malley BW. 1995. Progesterone receptor involvement in the LHRH-facilitated sexual behavior of female rats. Abstr 77th Annu Meeting Endocr Soc 77: P1-440.

Mani SK, Allen JMC, Clark JH, Blaustein JD, O'Malley BW. 1994a. Convergent pathways for steroid hormone- and neurotransmitter-induced rat sexual behavior. Science 265: 1246-1249.

Mani SK, Allen JMC, Lydon JP, Mulac-Jericevic B, Blaustein JD, et al. 1996. Dopamine requires the unoccupied progesterone receptor to induce sexual behavior in mice. Mol Endocrinol 10: 1728-1737.

Mani SK, Allen JMC, Rettori V, McCann SM, O'Malley BW, et al. 1994b. Nitric oxide mediates sexual behavior in female rats. Proc Natl Acad Sci USA 91: 6468-6472.

Mani SK, Blaustein JD, Allen JM, Law SW, O'Malley BW, et al. 1994c. Inhibition of rat sexual behavior by antisense oligonucleotides to the progesterone receptor. Endocrinology 135: 1409-1414.

Mani SK, Fienberg AA, Ocallaghan JP, Snyder GL, Allen PB, et al. 2000. Requirement for DARPP-32 in progesterone-facilitated sexual receptivity in female rats and mice. Science 287: 1053-1056.

Marrone BL, Feder HH. 1977. Characteristics of (^{3}H) estrogen and (^{3}H) progestin uptake and effects of progesterone on (^{3}H) estrogen uptake in brain, anterior pituitary and peripheral tissues of male and female guinea pigs. Biol Reprod 17: 42-57.

Martinez I, Paredes RG. 2001. Only self-paced mating is rewarding in rats of both sexes. Horm Behav 40: 510-517.

Mathews D, Edwards DA. 1977a. Involvement of the ventromedial and anterior hypothalamic nuclei in the hormonal induction of receptivity in the female rat. Physiol Behav 19: 319-326.

Mathews D, Edwards DA. 1977b. The ventromedial nucleus of the hypothalamus and the hormonal arousal of sexual behaviors in the female rat. Horm Behav 8: 40-51.

Matthews TJ, Grigore M, Tang L, Doat M, Kow LM, et al. 1997. Sexual reinforcement in the female rat. J Exp Anal Behav 68: 399-410.

Matuszewich L, Lorrain DS, Hull EM. 2000. Dopamine release in the medial preoptic area of female rats in response to hormonal manipulation and sexual activity. Behav Neurosci 114: 772-782.

McClintock MK, Adler NT. 1978. The role of the female during copulation in wild and domestic Norway rats (*Rattus norvegicus*). Behaviour 67: 67-96.

McClintock MK, Anisko JJ. 1982. Group mating among Norway rats I. Sex differences in the pattern and neuroendocrine consequences of copulation. Anim Behav 30: 398-409.

McClintock MK, Toner JP, Adler NT, Anisko JJ. 1982. Postejaculatory quiescence in female and male rats: Consequences for sperm transport during group mating. J Comp Physiol Psychol 11: 268-287.

McDonald PG, Meyerson BJ. 1973. The effect of oestradiol, testosterone and dihydrotestosterone on sexual motivation in the ovariectomized female rat. Physiol Behav 11: 515-520.

McEwen BS. 1976. Interactions between hormones and nerve tissue. Sci Am 235: 48-58.

McEwen BS, Pfaff DW, Chaptal C, Luine V. 1975. Brain cell nuclear retention of [^{3}H]estradiol doses able to promote lordosis: Temporal and regional aspects. Brain Res 86: 155-161.

McGinnis MY, Parsons B, Rainbow TC, Krey LC, McEwen BS. 1981. Temporal relationship between cell nuclear progestin receptor levels and sexual receptivity following intravenous progesterone administration. Brain Res 218: 365-371.

McIntosh TK, Barfield RJ, Geyer LA. 1977. Ultrasonic vocalizations facilitate sexual behaviour of female rats. Nature 272: 163-164.

McKenna NJ, OMalley BW. 2002. Minireview: Nuclear receptor coactivators—An update. Endocrinology 143: 2461-2465.

McKenna NJ, Lanz RB, O'Malley BW. 1999. Nuclear receptor coregulators: Cellular and molecular biology. Endocr Rev 20: 321-344.

Meisel RL, Pfaff DW. 1985. Specificity and neural sites of action of anisomycin in the reduction or facilitation of female sexual behavior in rats. Horm Behav 19: 237-251.

Meisel RL, Dohanich GP, McEwen BS, Pfaff DW. 1987. Antagonism of sexual behavior in female rats by ventromedial hypothalamic implants of antiestrogen. Neuroendocrinology 45: 201-207.

Mendelson SD, Pfaus JG. 1989. Level searching: A new assay of sexual motivation in the male rat. Physiol Behav 45: 337-341.

Merchenthaler I, Lane MV, Numan S, Dellovade TL. 2004. Distribution of estrogen receptor alpha and beta in the mouse central nervous system: In vivo autoradiographic and immunocytochemical analyses. J Comp Neurol 473: 270-291.

Meredith JM, Auger CJ, Blaustein JD. 1994. Down-regulation of estrogen receptor immunoreactivity by 17 beta-estradiol in the guinea pig forebrain. J Neuroendocrinol 6: 639-648.

Meredith JM, Moffatt CA, Auger AP, Snyder GL, Greengard P, et al. 1998. Mating-related stimulation induces phosphorylation of dopamine- and cyclic AMP-regulated phosphoprotein-32 in progestin receptor-containing areas in the female rat brain. J Neurosci 18: 10189-10195.

Mermelstein PG, Becker JB. 1995. Increased extracellular dopamine in the nucleus accumbens and striatum of the female rat during paced copulatory behavior. Behav Neurosci 109: 354-365.

Mermelstein PG, Becker JB, Surmeier DJ. 1996. Estradiol reduces calcium currents in rat neostriatal neurons via a membrane receptor. J Neurosci 16: 595-604.

Meyer ME, Pornon A, Ji JW, Bocquel MT, Chambon P, et al. 1990. Agonistic and Antagonistic Activities of RU486 on the Functions of the Human Progesterone Receptor. EMBO J 9: 3923-3932.

Meyerson B. 1972. Latency between intravenous injection of progestins and the appearance of estrous behavior in estrogen-treated ovariectomized rats. Horm Behav 3: 1-9.

Meyerson BJ, Lindstrom LH. 1973. Sexual motivation in the female rat. Acta Physiol Scand Suppl 389: 1-80.

Mhyre AJ, Shapiro RA, Dorsa DM. 2006. Estradiol reduces nonclassical transcription at cyclic adenosine 3′,5′-monophosphate response elements in glioma cells expressing estrogen receptor alpha. Endocrinology 147: 1796-1804.

Micevych PE, Sinchak K. 2006. The neurochemistry of limbinc-hypothalamic circuits regulating sexual receptivity. Behavioral Neurochemistry, Neuroendocrinology and Molecular Neurobiology. Blaustein JD, editor. Springer, p. in press.

Micevych P, Sinchak K, Mills RH, Tao L, La Polt P, Lu JKH. 2003. The luteinizing hormone surge is preceded by an estrogen-induced increase of hypothalamic progesterone in ovariectomized and adrenalectomized rats. Neuroendocrinology 78: 29-35.

Milner TA, McEwen BS, Hayashi S, Li CJ, Reagan LP, et al. 2001. Ultrastructural evidence that hippocampal alpha estrogen receptors are located at extranuclear sites. J Comp Neurol 429: 355-371.

Mitra SW, Hoskin E, Yudkovitz J, Pear L, Wilkinson HA, et al. 2003. Immunolocalization of estrogen receptor beta in the mouse brain: Comparison with estrogen receptor alpha. Endocrinology 144: 2055-2067.

Moffatt CA, Rissman EF, Shupnik MA, Blaustein JD. 1998. Induction of progestin receptors by estradiol in the forebrain of estrogen receptor-alpha gene-disrupted mice. J Neurosci 18: 9556-9563.

Moguilewsky M, Philibert D. 1984. RU 38486: Potent antiglucocorticoid activity correlated with strong binding to the cytosolic glucocorticoid receptor followed by an impaired activation. J Steroid Biochem 20: 271-276.

Moguilewsky M, Raynaud JP. 1979a. The relevance of hypothalamic and hypophyseal progestin receptor regulation in the induction and inhibition of sexual behavior in the female rat. Endocrinology 105: 516-522.

Moguilewsky M, Raynaud JP. 1979b. Estrogen-sensitive progestin-binding sites in the female rat brain and pituitary. Brain Res 164: 165-175.

Molenda HA, Griffin AL, Auger AP, McCarthy MM, Tetel MJ. 2002. Nuclear receptor coactivators modulate hormone-dependent gene expression in brain and female reproductive behavior in rats. Endocrinology 143: 436-444.

Molenda-Fugueira HA, Williams CA, Griffin AL, Rutledge EM, Blaustein JD, Tetel MJ. 2006. Nuclear receptor coactivators function in estrogen receptor- and progestin receptor-dependent aspects of sexual behavior in female rats. Horm Behav 50: 2006-2012.

Morin LP. 1977. Progesterone: Inhibition of rodent sexual behavior. Physiol Behav 18: 701-715.

Morin LP, Feder HH. 1973. Multiple progesterone injections and the duration of estrus in ovariectomized guinea pigs. Physiol Behav 11: 861-865.

Morin LP, Feder HH. 1974. Hypothalamic progesterone implants and facilitation of lordosis behavior in estrogen-primed ovariectomized guinea pigs. Brain Res 70: 81-93.

Moss RL, Dudley CA, Schwartz NB. 1977. Coitus-induced release of luteinizing hormone in the proestrous rat: Fantasy or fact? Endocrinology 100: 394-397.

Mulac-Jericevic B, Conneely OM. 2004. Reproductive tissue selective actions of progesterone receptors. Reproduction 128: 139-146.

Mulac-Jericevic B, Mullinax RA, De Mayo FJ, Lydon JP, Conneely OM. 2000. Subgroup of reproductive functions of progesterone mediated by progesterone receptor-B isoform. Science 289: 1751-1754.

Musgrove EA, Hamilton JA, Lee CS, Sweeney KJ, Watts CK, et al. 1993. Growth factor, steroid, and steroid antagonist regulation of cyclin gene expression associated with changes in T-47D human breast cancer cell cycle progression. Mol Cell Biol 13: 3577-3587.

Nadler RD. 1970. A biphasic influence of progesterone on sexual receptivity of spayed female rats. Physiol Behav 5: 95-97.

Nairn AC, Hemming HC Jr, Greengard P. 1985. Protein kinases in the brain. Annu Rev Biochem 54: 931-976.

Noble RG. 1973. Sexual arousal of the female hamster. Physiol Behav 10: 973-975.

Nocetto C, Cragnolini AB, Schioth HB, Scimonelli TN. 2004. Evidence that the effect of melanocortins on female sexual behavior in preoptic area is mediated by the MC3 receptor; Participation of nitric oxide. Behav Brain Res 153: 537-541.

Numan M, Roach JK, Delcerro MCR, Guillamon A, Segovia S, et al. 1999. Expression of intracellular progesterone receptors in rat brain during different reproductive states, and involvement in maternal behavior. Brain Res 830: 358-371.

Ogawa S, Olazabal UE, Parhar IS, Pfaff DW. 1994. Effects of intrahypothalamic administration of antisense DNA for progesterone receptor mRNA on reproductive behavior and progesterone receptor immunoreactivity in female rat. J Neurosci 14: 1766-1774.

Ogawa S, Chan J, Chester AE, Gustafsson JA, Korach KS, et al. 1999. Survival of reproductive behaviors in estrogen receptor beta gene-deficient (Beta ERKO) male and female mice. Proc Natl Acad Sci USA 96: 12887-12892.

Ogawa S, Eng V, Taylor J, Lubahn DB, Korach KS, et al. 1998. Roles of estrogen receptor alpha gene expression in reproduction-related behaviors in female mice. Endocrinology 139: 5070-5081.

Oldenburger WP, Everitt BJ, De Jonge FH. 1992. Conditioned place preference induced by sexual interaction in female rats. Horm Behav 26: 214-228.

Osterlund M, Kuiper GG, Gustafsson JA, Hurd YL. 1998. Differential distribution and regulation of estrogen receptor-alpha and -beta mRNA within the female rat brain. Mol Brain Res 54: 175-180.

Paech K, Webb P, Kuiper GG, Nilsson S, Gustafsson JA, et al. 1997. Differential ligand activation of estrogen receptors ER alpha and ER beta at AP1 sites. Science 277: 1508-1510.

Panzica GC, Viglietti-Panzica C, Sica M, Gotti S, Martini M, et al. 2006. Effects of gonadal hormones on central nitric oxide producing systems. Neuroscience 138: 987-995.

Paredes RG, Agmo A. 2004. Has dopamine a physiological role in the control of sexual behavior? A critical review of the evidence. Prog Neurobiol 73: 179-226.

Paredes RG, Alonso A. 1997. Sexual behavior regulated (paced) by the female induces conditioned place preference. Behav Neurosci 111: 123-128.

Paredes RG, Vazquez B. 1999. What do female rats like about sex? Paced mating. Behav Brain Res 105: 117-127.

Parsons B, McEwen BS. 1981. Sequential inhibition of sexual receptivity by progesterone is prevented by a protein synthesis inhibitor and is not causally related to decreased levels of hypothalamic progestin receptors in the female rat. J Neurosci 1: 527-531.

Parsons B, McEwen BS, Pfaff DW. 1982a. A discontinuous schedule of estradiol treatment is sufficient to activate progesterone-facilitated feminine sexual behavior and to increase cytosol receptors for progestins in the hypothalamus of the rat. Endocrinology 110: 613-624.

Parsons B, McGinnis MY, McEwen BS. 1981. Sequential inhibition by progesterone: Effects on sexual receptivity and associated changes in brain cytosol progestin binding in the female rat. Brain Res 221: 149-160.

Parsons B, Rainbow Tc, MacLusky NJ, McEwen BS. 1982b. Progestin receptor levels in rat hypothalamic and limbic nuclei. J Neurosci 2: 1446-1452.

Parsons B, Rainbow TC, Snyder L, McEwen BS. 1984. Progesterone-like effects of estradiol on reproductive behavior and hypothalamic progestin receptors in the female rat. Neuroendocrinology 39: 25-30.

Parsons B, Mac Lusky NJ, Krey L, Pfaff DW, McEwen BS. 1980. The temporal relationship between estrogen-inducible progestin receptors in the female rat brain and the time course of estrogen activation of mating behavior. Endocrinology 107: 774-779.

Parsons B, Mac Lusky NJ, Krieger MS, McEwen B, Pfaff DW. 1979. The effects of long-term estrogen exposure on the induction of sexual behavior and measurements of brain estrogen and progestin receptors in female the rat. Horm Behav 13: 301-313.

Patisaul HB, Whitten PL, Young LJ. 1999. Regulation of estrogen receptor beta mRNA in the brain: Opposite effects of 17 beta-estradiol and the phytoestrogen, coumestrol. Mol Brain Res 67: 165-171.

Peirce JT, Nuttall RL. 1961. Self-paced sexual behavior in the femal rat. J Comp Physiol Psychol 54: 310-313.

Petersen DN, Tkalcevic GT, Koza Taylor PH, Turi TG, Brown TA. 1998. Identification of estrogen receptor beta(2), a functional variant of estrogen receptor beta expressed in normal rat tissues. Endocrinology 139: 1082-1092.

Petitti N, Etgen AM. 1989. Progesterone depression of norepinephrine-stimulated cAMP accumulation in hypothalamic slices. Mol Brain Res 5: 109-119.

Petitti N, Etgen AM. 1990. Alpha-1-adrenoceptor augmentation of beta-stimulated cAMP formation is enhanced by estrogen and reduced by progesterone in rat hypothalamic slices. J Neurosci 10: 2842-2849.

Pettersson K, Grandien K, Kuiper GG, Gustafsson JA. 1997. Mouse estrogen receptor beta forms estrogen response element-binding heterodimers with estrogen receptor alpha. Mol Endocrinol 11: 1486-1496.

Pfaff DW. 1968. Uptake of ^{3}H-estradiol by the female rat brain. An autoradiographic study. Endocrinology 83: 1149-1155.

Pfaff D, Keiner M. 1973. Atlas of estradiol-concentrating cells in the central nervous system of the female rat. J Comp Neurol 151: 121-158.

Pfaff DW, Sakuma Y. 1979. Deficit in the lordosis reflex of female rats caused by lesions in the ventromedial nucleus of the hypothalamus. J Physiol 288: 203-210.

Pfaff D, Lewis C, Diakow C, Keiner M. 1973. Neurophysiological analysis of mating behavior responses as hormone-sensitive reflexes. Progress in Physiological Psychology. Vol. 5. Stellar E, Sprague JM, editors. New York: Academic Press; pp. 253-297.

Pfaff DW, Schwartz-Giblin S, McCarthy MM, Kow L-M. 1994. Cellular and molecular mechanisms of female reproductive behaviors. The Physiology of Reproduction. Second Edition. Knobil E, Neill JD, editors. New York: Raven Press Ltd; pp. 107-220.

Pfaus JG, Smith WJ, Coopersmith CB. 1999. Appetitive and consummatory sexual behaviors of female rats in bilevel chambers—I. A correlational and factor analysis and the effects of ovarian hormones. Horm Behav 35: 224-240.

Pfaus JG, Smith WJ, Byrne N, Stephens G. 2000. Appetitive and consummatory sexual behaviors of female rats in bilevel chambers II. Patterns of estrus termination following vaginocervical stimulation. Horm Behav 37: 96-107.

Pfaus JG, Kleopoulos SP, Mobbs CV, Gibbs RB, Pfaff DW. 1993. Sexual stimulation activates c-fos within estrogen-concentrating regions of the female rat forebrain. Brain Res 624: 253-267.

Pfaus JG, Marcangione C, Smith WJ, Manitt C, Abillamaa H. 1996. Differential induction of Fos in the female rat brain following different amounts of vaginocervical stimulation: Modulation by steroid hormones. Brain Res 741: 314-330.

Pfaus JG, Shadiack A, Van Soest T, Tse M, Molinoff P. 2004. Selective facilitation of sexual solicitation in the female rat by a melanocortin receptor agonist. PNAS 101: 10201-10204.

Pfeifle JK, Shivers M, Edwards DA. 1980. Parasagittal hypothalamic knife cuts and sexual receptivity in the female rat. Physiol Behav 24: 145-150.

Pleim ET, Cailliau PJ, Weinstein MA, Etgen AM, Barfield RJ. 1990. Facilitation of receptive behavior in estrogen-primed female rats by the anti-progestin, RU486. Horm Behav 24: 301-310.

Pollio G, Xue P, Zanisi M, Nicolin A, Maggi A. 1993. Antisense oligonucleotide blocks progesterone-induced lordosis behavior in ovariectomized rats. Mol Brain Res 19: 135-139.

Polston EK, Erskine MS. 1995. Patterns of induction of the immediate-early genes c-fos and egr-1 in the female rat brain following differential amounts of mating stimulation. Neuroendocrinology 62: 370-384.

Power RF, Lydon JP, Conneely OM, O'Malley BW. 1991a. Dopamine activation of an orphan of the steroid receptor superfamily. Science 252: 1546-1548.

Power RF, Mani SK, Codina J, Conneely OM, O'Malley BW. 1991b. Dopaminergic and ligand-independent activation of steroid hormone receptors. Science 254: 1636-1639.

Powers JB. 1970. Hormonal control of sexual receptivity during the estrous cycle of the rat. Physiol Behav 5: 831-835.

Powers JB. 1972. Facilitation of lordosis in ovariectomized rats by intracerebral implants. Brain Res 48: 311-325.

Powers JB, Moreines J. 1976. Progesterone: Examination of its postulated inhibitory actions on lordosis during the rat estrous cycle. Physiol Behav 17: 493-498.

Powers J, Zucker I. 1969. Sexual receptivity in pregnant and pseudopregnant rats. Endocrinology 84: 820-827.

Pratt WB, Toft DO. 1997. Steroid receptor interactions with heat shock protein and immunophilin chaperones. Endocr Rev 18: 306-360.

Rachman IM, Unnerstall JR, Pfaff DW, Cohen RS. 1998. Estrogen alters behavior and forebrain c-fos expression in ovariectomized rats subjected to the forced swim test. Proc Natl Acad Sci USA 95: 13941-13946.

Rainbow TC, McGinnis MY, Davis PG, McEwen BS. 1982a. Application of anisomycin to the lateral ventromedial nucleus of the hypothalamus inhibits the activation of sexual behavior by estradiol and progesterone. Brain Res 233: 417-423.

Rainbow TC, McGinnis MY, Krey LC, McEwen BS. 1982b. Nuclear progestin receptors in rat brain and pituitary. Neuroendocrinology 34: 426-432.

Rainbow TC, Parsons B, MacLusky NJ, McEwen BS. 1982c. Estradiol receptor levels in rat hypothalamic and limbic nuclei. J Neurosci 2: 1439-1445.

Rajendren G, Moss RL. 1993. The role of the medial nucleus of amygdala in the mating-induced enhancement of lordosis in female rats—the interaction with luteinizing hormone-releasing hormone neuronal system. Brain Res 617: 81-86.

Rajendren G, Dudley CA, Moss RL. 1990. Role of the vomeronasal organ in the male-induced enhancement of sexual receptivity in female rats. Neuroendocrinology 52s: 368-372.

Rajendren G, Dudley CA, Moss RL. 1991. Role of the ventromedial nucleus of hypothalamus in the male-induced enhancement of lordosis in female rats. Physiol Behav 50: 705-710.

Ramos SM, Debold JF. 1999. Protein synthesis in the medial preoptic area is important for the mating-induced decrease in estrus duration in hamsters. Horm Behav 35: 177-185.

Razandi M, Pedram A, Greene GL, Levin ER. 1999. Cell membrane and nuclear estrogen receptors (ERs) originate from a single transcript: Studies of ER alpha and ER beta expressed in Chinese hamster ovary cells. Mol Endocrinol 13: 307-319.

Reading DS, Blaustein JD. 1984. The relationship between heat abbreviation and neural progestin receptors in female rats. Physiol Behav 32: 973-981.

Richmond G, Clemens LG. 1986. Cholinergic mediation of feminine sexual receptivity: Demonstration of progesterone independence using a progestin receptor antagonist. Brain Res 373: 159-163.

Richmond G, Clemens L. 1988. Ventromedial hypothalamic lesions and cholinergic control of female sexual behavior. Physiol Behav 42: 179-182.

Ring JR. 1944. The estrogen-progesterone induction of sexual receptivity in the spayed female mouse. Endocrinology 34: 269-275.

Rissman EF, Wersinger SR, Taylor JA, Lubahn DB. 1997b. Estrogen receptor function as revealed by knockout studies: Neuroendocrine and behavioral aspects. Horm Behav 31: 232-243.

Rissman EF, Early AH, Taylor JA, Korach KS, Lubahn DB. 1997a. Estrogen receptors are essential for female sexual receptivity. Endocrinology 138: 507-510.

Rodriguez-Sierra JF, Davis GA. 1977. Progesterone does not inhibit lordosis through interference with estrogen priming. Life Sci 22: 373-378.

Rodriguez-Sierra JF, Komisaruk BR. 1982. Common hypothalamic sites for activation of sexual receptivity in female rats by LHRH, PGE2 and progesterone. Neuroendocrinology 35: 363-369.

Rodriguez-Sierra JF, Crowley WR, Komisaruk BR. 1975. Vaginal stimulation in rats induces prolonged lordosis responsiveness and sexual receptivity. J Comp Physiol Psychol 89: 79-85.

Ronnekleiv OK, Kelly MJ. 2005. Diversity of ovarian steroid signaling in the hypothalamus. Front Neuroendocrinol 26: 65-84.

Roselli CE, Snipes CA. 1983. Cytoplasmic progesterone receptors in the hypothalamus-preoptic area of the mouse: Effect of estrogen priming. J Steroid Biochem 19: 1571-1575.

Ross J, Claybaugh C, Clemens LG, Gorski RA. 1971. Short latency of estrous behavior with intracerebral gonadal hormones in ovariectomized rats. Endocrinology 89: 32-38.

Rowan BG, Weigel NL, O'Malley BW. 2000b. Phosphorylation of steroid receptor coactivator-1—Identification of the phosphorylation sites and phosphorylation through the mitogen-activated protein kinase pathway. J Biol Chem 275: 4475-4483.

Rowan BG, Garrison N, Weigel NL, O'Malley BW. 2000a. 8-Bromo-cyclic AMP induces phosphorylation of two sites in SRC-1 that facilitate ligand-independent activation of the chicken progesterone receptor and are critical for functional cooperation between SRC-1 and CREB binding protein. Mol Cell Biol 20: 8720-8730.

Rowe DW, Erskine MS. 1993. c-Fos proto-oncogene activity induced by mating in the preoptic area, hypothalamus and amygdala in the female rat—role of afferent input via the pelvic nerve. Brain Res 621: 25-34.

Roy EJ, Wade GN. 1976. Estrogenic effects of an antiestrogen, MER-25, on eating and body weight in rats. J Comp Physiol Psychol 90: 156-166.

Roy EJ, Wade GN. 1977. Binding of [^{3}H]estradiol by brain cell nuclei and female rat sexual behavior: Inhibition by antiestrogens. Brain Res 126: 73-87.

Roy EJ, Lynn DM, Clark AS. 1985. Inhibition of sexual receptivity by anesthesia during estrogen priming. Brain Res 337: 163-166.

Roy EJ, Mac Lusky NJ, McEwen BS. 1979. Antiestrogen inhibits the induction of progestin receptors by estradiol in the hypothalamus–preoptic area and pituitary. Endocrinology 104: 1333-1336.

Roy MM, Goldstein KL, Williams C. 1993. Estrus termination following copulation in female guinea pigs. Horm Behav 27: 397-402.

Rubin BS, Barfield RJ. 1980. Priming of estrous responsiveness by implants of 17b-estradiol in the ventromedial hypothalamic nucleus of female rats. Endocrinology 106: 504-509.

Rubin BS, Barfield RJ. 1983. Progesterone in the ventromedial hypothalamus facilitates estrous behavior in ovariectomized, estrogen-primed rats. Endocrinology 113: 797-804.

Rubin BS, Barfield RJ. 1984. Progesterone in the ventromedial hypothalamus of ovariectomized, estrogen-primed rats inhibits subsequent facilitation of estrous behavior by systemic progesterone. Brain Res 294: 1-8.

Rupprecht R, Holsboer F. 1999. Neuroactive steroids: Mechanisms of action and neuropsychopharmacological perspectives. Trends Neurosci 22: 410-416.

Sachs BD, Glater GB, O'Hanlon JK. 1984. Morphology of the erect glans penis in rats under various gonadal hormone conditions. Anat Rec 210: 45-52.

Sakamoto H, Shikimi H, Ukena K, Tsutsui K. 2003. Neonatal expression of progesterone receptor isoforms in the cerebellar Purkinje cell in rats. Neurosci Lett 343: 163-166.

Sar M. 1988. Distribution of progestin-concentrating cells in rat brain: Colocalization of [^{3}H]ORG 2058, a synthetic progestin, and antibodies to tyrosine hydroxylase in hypothalamus by combined autoradiography and immunocytochemistry. Endocrinology 123: 1110-1118.

Sar M, Stumpf WE. 1973. Neurons of the hypothalamus concentrate [^{3}H]progesterone or its metabolites. Science 183: 1266-1268.

Sartorius CA, Groshong SD, Miller LA, Powell RL, Tung L, et al. 1994. New T47D breast cancer cell lines for the independent study of progesterone b- and a-receptors: Only antiprogestin-occupied b-receptors are switched to transcriptional agonists by cAMP. Cancer Res 54: 3868-3877.

Sathya G, Jansen MS, Nagel SC, Cook CE, McDonnell DP. 2002. Identification and characterization of novel estrogen receptor-beta-sparing antiprogestins. Endocrinology 143: 3071-3082.

Savouret JF, Bailly A, Misrahi M, Rauch C, Redeuilh G, et al. 1991. Characterization of the hormone responsive element involved in the regulation of the progesterone receptor gene. EMBO J 10: 1875-1883.

Schule R, Evans RM. 1991. Cross-coupling of signal transduction pathways: Zinc finger meets leucine zipper. Trends Genet 7: 377-381.

Schwartz SM, Blaustein JD, Wade GN. 1979. Inhibition of estrous behavior by progesterone in rats: Role of neural estrogen and progestin receptors. Endocrinology 105: 1078-1082.

Seiki K, Hattori M. 1973. In vivo uptake of progesterone by the hypothalamus and pituitary of the female ovariectomized rat and its relationship to cytoplasmic progesterone-binding protein. Endocrinology 20: 111-119.

Seiki K, Haruki Y, Imanishi Y, Enomoto T. 1977. Further evidence of the presence of progesterone-binding proteins in female rat hypothalamus. Endocrinol Jpn 24: 233-238.

Shemshedini L, Knauthe R, Sassonecorsi P, Pornon A, Gronemeyer H. 1991. Cell-specific inhibitory and stimulatory effects of Fos and Jun on transcription activation by nuclear receptors. EMBO J 10: 3839-3849.

Shenolikar S, Nairn AC. 1991. Protein phosphatases: Recent progress. Adv Second Messenger Phosphoprotein Res 23: 1-121.

Shima N, Yamaguchi Y, Yuri K. 2003. Distribution of estrogen receptor beta mRNA-containing cells in ovariectomized and estrogen-treated female rat brain. Anat Sci Int 78: 85-97.

Shivers BD, Harlan RE, Parker CR, Moss RL. 1980. Sequential inhibitory effect of progesterone on lordotic responsiveness in rats: Time course, estrogenic nullification, and actinomycin-D insensitivity. Biol Reprod 23: 963-973.

Shughrue PJ, Merchenthaler I. 2001. Distribution of estrogen receptor beta immunoreactivity in the rat central nervous system. J Comp Neurol 436: 64-81.

Shughrue PJ, Komm B, Merchenthaler I. 1996. The distribution of estrogen receptor-beta mRNA in the rat hypothalamus. Steroids 61: 678-681.

Shughrue PJ, Lane MV, Merchenthaler I. 1997. Comparative distribution of estrogen receptor-alpha and -beta mRNA in the rat central nervous system. J Comp Neurol 388: 507-525.

Shughrue PJ, Scrimo PJ, Merchenthaler I. 1998b. Evidence for the colocalization of estrogen receptor-beta mRNA and estrogen receptor-alpha immunoreactivity in neurons of the rat forebrain. Endocrinology 139: 5267-5270.

Shughrue PJ, Lane MV, Scrimo PJ, Merchenthaler I. 1998a. Comparative distribution of estrogen receptor-alpha (ER-alpha) and beta (ER-beta) mRNA in the rat pituitary, gonad, and reproductive tract. Steroids 63: 498-504.

Shyamala G, Yang X, Silberstein G, Barcellos Hoff MH, Dale E. 1998. Transgenic mice carrying an imbalance in the native ratio of A to B forms of progesterone receptor exhibit developmental abnormalities in mammary glands. Proc Natl Acad Sci USA 95: 696-701.

Simerly RB, Young BJ. 1991. Regulation of estrogen receptor messenger ribonucleic acid in rat hypothalamus by sex steroid hormones. Mol Endocrinol 5: 424-432.

Simerly RB, Chang C, Muramatsu M, Swanson LW. 1990. Distribution of androgen and estrogen receptor messenger RNA-containing cells in the rat brain—an in situ hybridization study. J Comp Neurol 294: 76-95.

Simonian SX, Herbison AE. 1997. Differential expression of estrogen receptor alpha and beta immunoreactivity by oxytocin neurons of rat paraventricular nucleus. J Neuroendocrinol 9: 803-806.

Sinchak K, Mills RH, Tao L, La Polt P, Lu JK, et al. 2003. Estrogen induces de novo progesterone synthesis in astrocytes. Dev Neurosci 25: 343-348.

Slimp JC. 1977. Reduction of genital sensory input and sexual behavior of female guinea pigs. Physiol Behav 18: 1027-1031.

Smanik EJ, Young HK, Muldoon TG, Mahesh VB. 1983. Analysis of the effect of progesterone in vivo on estrogen receptor distribution in the rat anterior pituitary and hypothalamus. Endocrinology 113: 15-22.

Sodersten P, Eneroth P. 1981. Evidence that progesterone does not inhibit the induction of sexual receptivity by oestradiol-17b in the rat. J Endocrinol 89: 63-69.

Sodersten P, Hansen S. 1977. Effects of oestradiol and progesterone on the induction and duration of sexual receptivity in cyclic female rats. J Endocrinol 74: 477-485.

Sodersten P, Hansen S. 1979. Induction of sexual receptivity by oestradiol benzoate in cyclic female rats: Influence of ovarian secretions before injection of oestradiol benzoate. J Endocrinol 80: 389-395.

Sodersten P, Eneroth P, Hansen S. 1981. Induction of sexual receptivity in ovariectomized rats by pulse administration of oestradiol-17β. J Endocrinol 89: 55-62.

Soma KK, Sinchak K, Lakhter A, Schlinger BA, Micevych PE. 2005. Neurosteroids and female reproduction: Estrogen increases 3beta-HSD mRNA and activity in rat hypothalamus. Endocrinology 146: 4386-4390.

Stevis PE, Deecher DC, Suhadolnik L, Mallis LM, Frail DE. 1999. Differential effects of estradiol and estradiol–BSA conjugates. Endocrinology 140: 5455-5458.

Stumpf WE. 1968. Estradiol-concentrating neurons: Topography in the hypothalamus by dry-mount autoradiography. Science 162: 1001-1003.

Svenningsson P, Nishi A, Fisone G, Girault JA, Nairn AC, et al. 2004. DARPP-32: An integrator of neurotransmission. Annu Rev Pharmacol Toxicol 44: 269-296.

Sweatt JD. 2004. Mitogen-activated protein kinases in synaptic plasticity and memory. Curr Opin Neurobiol 14: 311-317.

Taylor GT, Komitowski D, Weiss J. 1983. Light and scanning electron microscopic study of testosterone-restored penile papillae in castrated rats. Anat Rec 205: 277-286.

Temple JL, Wray S. 2005. Bovine serum albumin-estrogen compounds differentially alter gonadotropin-releasing hormone-1 neuronal activity. Endocrinology 146: 558-563.

Tennent BJ, Smith ER, Davidson JM. 1980. The effects of estrogen and progesterone on female rat proceptive behavior. Horm Behav 14: 65-75.

Tennent BJ, Smith ER, Davidson JM. 1982. Effects of progesterone implants in the habenula and midbrain on proceptive and receptive behavior in the female rat. Horm Behav 16: 352-373.

Tetel MJ. 2000. Nuclear receptor coactivators in neuroendocrine function. J Neuroendocrinol 12: 927-932.

Tetel MJ, Celentano DC, Blaustein JD. 1994a. Intraneuronal convergence of tactile and hormonal stimuli associated with female reproduction in rats. J Neuroendocrinol 6: 211-216.

Tetel MJ, Getzinger MJ, Blaustein JD. 1994b. Estradiol and progesterone influence the response of ventromedial hypothalamic neurons to tactile stimuli associated with female reproduction. Brain Res 646: 267-272.

Tetel MJ, Getzinger MJ, Blaustein JD. 1993. Fos expression in the rat brain following vaginal-cervical stimulation by mating and manual probing. J Neuroendocrinol 5: 397-404.

Thomas P, Pang Y, Filardo EJ, Dong J. 2005. Identity of an estrogen membrane receptor coupled to a G protein in human breast cancer cells. Endocrinology 146: 624-632.

Thompson ML, Edwards DA. 1971. Experiential and strain determinants of the estrogen-progesterone induction of sexual receptivity in spayed female mice. Horm Behav 2: 299-305.

Thornton JE, Nock B, McEwen BS, Feder HH. 1986. Estrogen induction of progesterone receptors in microdissected hypothalamic and limbic nuclei of female guinea pigs. Neuroendocrinology 43: 182-188.

Tobet SA, Baum MJ. 1982. Implantation of dihydrotestosterone propionate into the lateral septum inhibits sexual receptivity in estrogen-primed, ovariectomized rats. Neuroendocrinology 34: 333-338.

Toft DO, Sullivan WP, McCormick D, Riehl M. 1987. Heat shock proteins and steroid hormone receptors. Biochemical Actions of Hormones. Academic Press; New York: pp. 293-316.

Tora L, Gronemeyer H, Turcotte B, Gaub M-P, Chambon P. 1988. The N-terminal region of the chicken progesterone receptor specifies target gene activation. Nature 333: 185-188.

Toran-Allerand CD, Tinnikov AA, Singh RJ, Nethrapalli IS. 2005. 17Alpha-estradiol: A brain-active estrogen? Endocrinology 146: 3843-3850.

Toran-Allerand CD, Guan XP, Mac Lusky NJ, Horvath TL, Diano S, et al. 2002. ER-X: A novel, plasma membrane-associated, putative estrogen receptor that is regulated during development and after ischemic brain injury. J Neurosci 22: 8391-8401.

Torchia J, Rose DW, Inostroza J, Kamei Y, Westin S, et al. 1997. The transcriptional co-activator p/CIP binds CBP and mediates nuclear-receptor function. Nature 387: 677-684.

Tremblay GB, Tremblay A, Labrie F, Giguere V. 1999. Dominant activity of activation function 1 (AF-1) and differential stoichiometric requirements for AF-1 and -2 in the estrogen receptor alpha–beta heterodimeric complex. Mol Cell Biol 19: 1919-1927.

Umayahara Y, Kawamori R, Watada H, Imano E, Iwama N, et al. 1994. Estrogen regulation of the insulin-like growth factor I gene transcription involves an AP-1 enhancer. J Biol Chem 269: 16433-16442.

Vagell ME, McGinnis MY. 1996. Effects of the antiestrogen, RU 58668, on female sexual behavior in rats. Brain Res Bull 41: 121-124.

Vander Schoot P, Vanophemert J, Baumgarten R. 1992. Copulatory stimuli in rats induce heat abbreviation through effects on genitalia but not through effects on central nervous mechanisms supporting the steroid hormone-induced sexual responsiveness. Behav Brain Res 49: 213-223.

Vasudevan N, Kow L-M, Pfaff DW. 2001. Early membrane estrogenic effects required for full expression of slower genomic actions in a nerve cell line. Proc Natl Acad Sci USA 98: 12267-12271.

Vathy I, Etgen AM. 1989. Hormonal activation of female sexual behavior is accompanied by hypothalamic norepinephrine release. J Neuroendocrinol 1: 383-388.

Vegeto E, Allan GF, Schrader WT, Tsai MJ, McDonnell DP, et al. 1992. The mechanism of RU 486 antagonism is dependent on the conformation of the carboxy-terminal tail of the human progesterone receptor. Cell 69: 703-713.

Vegeto E, Shahbaz MM, Wen DX, Goldman ME, O'Malley BW, et al. 1993. Human progesterone receptor A form is a cell- and promoter-specific repressor of human progesterone receptor B function. Mol Endocrinol 7: 1244-1255.

Wade GN, Blaustein JD. 1978. Effects of an antiestrogen on neural estradiol binding and on behaviors in female rats. Endocrinology 102: 245-251.

Wade GN, Feder HH. 1972. [1,2-3]Progesterone uptake by guinea pig brain and uterus: Differential localization, time-course of uptake and metabolism, and effects of age, sex, estrogen-priming and competing steroids. Brain Res 45: 525-543.

Wade GN, Blaustein JD, Gray JM, Meredith JM. 1993. ICI-182,780—a pure antiestrogen that affects behaviors and energy balance in rats without acting in the brain. Am J Physiol 265: R1392-R1398.

Wade CB, Robinson S, Shapiro RA, Dorsa DM. 2001. Estrogen receptor (ER)alpha and ERbeta exhibit unique pharmacologic properties when coupled to activation of the mitogen-activated protein kinase pathway. Endocrinology 142: 2336-2342.

Wagner CK, Silverman AJ, Morrell JI. 1998b. Evidence for estrogen receptor in cell nuclei and axon terminals within the lateral habenula of the rat: Regulation during pregnancy. J Comp Neurol 392: 330-342.

Wagner BL, Norris JD, Knotts TA, Weigel NL, McDonnell DP. 1998a. The nuclear corepressors NCoR and SMRT are key regulators of both ligand- and 8-bromo-cyclic AMP-dependent transcriptional activity of the human progesterone receptor. Mol Cell Biol 18: 1369-1378.

Walker WA, Feder HH. 1977a. Anti-estrogen effects on estrogen accumulation in brain cell nuclei: Neurochemical correlates of estrogen action of female sexual behavior in guinea pigs. Brain Res 134: 467-478.

Walker WA, Feder HH. 1977b. Inhibitory and facilitatory effects of various anti-estrogens on the induction of female sexual behavior by estradiol benzoate in guinea pigs. Brain Res 134: 455-465.

Walker WA, Feder HH. 1979. Long term effects of estrogen action are crucial for the display of lordosis in female guinea pigs: Antagonism by antiestrogens and correlations with in vitro cytoplasmic binding activity. Endocrinology 104: 89-96.

Wallen K, Thornton JE. 1979. Progesterone and duration of heat in estrogen-treated ovariectomized guinea pigs. Physiol Behav 22: 95-97.

Ward IL, Crowley WR, Zemlan FP, Margules DL. 1975. Monoaminergic mediation of female sexual behavior. J Comp Physiol Psychol 88: 53-61.

Warembourg M. 1977. Radiographic localization of estrogen-concentrating cells in the brain and pituitary of the guinea pig. Brain Res 123: 357-362.

Warembourg M. 1978b. Radioautographic study of the brain and pituitary after [^{3}H]progesterone injection into estrogen-primed ovariectomized guinea pigs. Neurosci Lett 7: 1-5.

Warembourg M. 1978a. Uptake of ^{3}H labeled synthetic progestin by rat brain and pituitary. A radioautography study. Neurosci Lett 9: 329-332.

Warembourg M, Jolivet A, Milgrom E. 1989. Immunohistochemical evidence of the presence of estrogen and progesterone receptors in the same neurons of the guinea pig hypothalamus and preoptic area. Brain Res 480: 1-15.

Warembourg M, Logeat F, Milgrom E. 1986. Immunocytochemical localization of progesterone receptor in the guinea pig central nervous system. Brain Res 384: 121-131.

Watanabe T, Huang HB, Horiuchi A, Silva EFD, Hsieh Wilson L, et al. 2001. Protein phosphatase 1 regulation by inhibitors and targeting subunits. Proc Natl Acad Sci USA 98: 3080-3085.

Watson CS, Gametchu B. 1999. Membrane-initiated steroid actions and the proteins that mediate them. Proc Soc Exp Biol Med 220: 9-19.

Watters JJ, Dorsa DM. 1998. Transcriptional effects of estrogen on neuronal neurotensin gene expression involve cAMP/protein kinase A-dependent signaling mechanisms. J Neurosci 18: 6672-6680.

Watters JJ, Campbell JS, Cunningham MJ, Krebs EG, Dorsa DM. 1997. Rapid membrane effects of steroids in neuroblastoma cells: Effects of estrogen on mitogen activated protein kinase signalling cascade and c-fos immediate early gene transcription. Endocrinology 138: 4030-4033.

Webb P, Lopez GN, Uht RM, Kushner PJ. 1995. Tamoxifen activation of the estrogen receptor/AP-1 pathway: Potential origin for the cell-specific estrogen-like effects of antiestrogens. Mol Endocrinol 9: 443-456.

Welshons WV, Lieberman ME, Gorski J. 1984. Nuclear localization of unoccupied oestrogen receptors. Nature 307: 747-749.

Wersinger SR, Baum MJ, Erskine MS. 1993. Mating-induced FOS-like immunoreactivity in the rat forebrain—a sex comparison and a dimorphic effect of pelvic nerve transection. J Neuroendocrinol 5: 557-568.

Whalen RE. 1974. Estrogen-progesterone induction of mating in female rats. Horm Behav 5: 157-162.

Whalen RE, Lauber AH. 1986. Progesterone substitutes: cGMP mediation. Neurosci Biobehav Rev 10: 47-53.

Whalen RE, Luttge WG. 1971. Differential localization of progesterone uptake in brain, role of sex, estrogen pretreatment and adrenalectomy. Brain Res 33: 147-155.

White NR, Barfield RJ. 1989. Playback of female rat ultrasonic vocalizations during sexual behavior. Physiol Behav 45: 229-233.

Wilcox JN, Feder HH. 1983. Long-term priming with a low dosage of estradiol benzoate or an antiestrogen (Enclomiphene) increases nuclear progestin receptor levels in brain. Brain Res 266: 243-251.

Wilcox JN, Barclay SR, Feder HH. 1984. Administration of estradiol-17B in pulses to guinea pigs: Self-priming effects of estrogen on brain tissues mediating lordosis. Physiol Behav 32: 483-488.

Xu M, Urban JH, Hill JW, Levine JE. 2000b. Regulation of hypothalamic neuropeptide YY1 receptor gene expression during the estrous cycle: Role of progesterone receptors. Endocrinology 141: 3319-3327.

Xu JM, Liao L, Ning C, Yoshida-Komiya H, Deng CX, et al. 2000a. The steroid receptor coactivator SRC-3 (P/CIP/RAC3/AIB1/ACTR/TRAM-1) is required for normal growth, puberty, female reproductive function, and mammary gland development. Proc Natl Acad Sci USA 97: 6379-6384.

Yahr P, Ulibarri C. 1986. Estrogen induction of sexual behavior in female rats and synthesis of polyadenylated messenger RNA in the ventromedial nucleus of the hypothalamus. Mol Brain Res 1: 153-165.

Yanase M, Gorski RA. 1976. Sites of estrogen and progesterone facilitation of lordosis behavior in the spayed rat. Biol Reprod 15: 536-543.

Young WC. 1969. Psychobiology of sexual behavior in the guinea pig. Advances in the Study of Behavior. Lehrman DS, Hinde RA, Shaw E, editors. Academic Press; pp. 1-110.

Yuri K, Kawata M. 1991. The effect of estrogen on the estrogen receptor-immunoreactive cells in the Rat medial preoptic nucleus. Brain Res 548: 50-54.

Zemlan FP, Adler NT. 1977. Hormonal control of female sexual behavior in the rat. Horm Behav 9: 345-357.

Zhang D, Trudeau VL. 2006. Integration of membrane and nuclear estrogen receptor signaling. Comp Biochem. Physiol A Mol Integr Physiol 144: 306-315.

Zhou Y, Watters JJ, Dorsa DM. 1996. Estrogen rapidly induces the phosphorylation of the cAMP response element binding protein in rat brain. Endocrinology 137: 2163-2166.

Zhu Y, Bond J, Thomas P. 2003a. Identification, classification, and partial characterization of genes in humans and other vertebrates homologous to a fish membrane progestin receptor. Proc Natl Acad Sci USA 100: 2237-2242.

Zhu Y, Rice CD, Pang YT, Pace M, Thomas P. 2003b. Cloning, expression, and characterization of a membrane progestin receptor and evidence it is an intermediary in meiotic maturation of fish oocytes. Proc Natl Acad Sci USA 100: 2231-2236.

Zipse LR, Brandling-Bennett EM, Clark AS. 2000. Paced mating behavior in the naturally cycling and the hormone-treated female rat. Physiol Behav 70: 205-209.

Zou AH, Marschke KB, Arnold KE, Berger EM, Fitzgerald P, et al. 1999. Estrogen receptor beta activates the human retinoic acid receptor alpha-1 promoter in response to tamoxifen and other estrogen receptor antagonists, but not in response to estrogen. Mol Endocrinol 13: 418-430.

Zucker I. 1966. Facilitatory and inhibitory effects of progesterone on sexual responses of spayed guinea pigs. J Comp Physiol Psychol 3: 376-381.

Zucker I. 1967. Actions of progesterone in the control of sexual receptivity of the spayed female rat. J Comp Physiol Psychol 63: 313-316.

Zucker I. 1968. Biphasic effects of progesterone on sexual receptivity in the female guinea pig. J Comp Physiol Psychol 63: 472-478.

4 The Neurochemistry of Limbic-Hypothalamic Circuits Regulating Sexual Receptivity

Paul Micevych · Kevin Sinchak

Abstract: Female sexual receptivity is a vital component of reproduction. In many species, females allow sexual contact only during a period surrounding ovulation to maximize procreative success. In rats, the species most commonly studied, sexual receptivity can be determined by measuring the lordosis reflex. The lordosis reflex is elicited by appropriate hormonal priming: estradiol and progesterone in the intact female and stimulation of mechanoreceptors in the flank, perineum and area around the tail. Thus, lordosis is a global reflex that involves integration of information from a large number of brain areas to elicit this behavior. The critical sites for integration of hormonal with introceptive and extroceptive inputs are located in a circuit that spans the limbic system and hypothalamus. Using this highly reproducible and quantifiable behavior, lordosis, investigators have characterized the neurochemistry of neural circuits that control sexual receptivity. Since the hypothalamus and limbic systems are phylogenetically ancient parts of the brain, it is likely that information learned from rat model will be directly translatable to understand the neurochemistry of human sexual receptivity. A majority of studies have focused on hypothalamic regions to elucidate roles for a divergent population of neurotransmitters (i.e., acetylcholine, glutamate, dopamine, serotonin, norepinephrine, nitric oxide, prostaglandins) and neuropeptides (i.e., β-endorphin, enkephalin. α-MSH, ACTH, neuropeptide Y, oxytocin, galanin, vasopressin, TRH, leptin, PACAP and insulin growth factor-1) in the regulation of the lordosis as a measure of sexual receptivity. A striking discovery of this research is that central control of this lordosis involves a disproportionately large number of neuropeptides. This is reflected in this chapter, especially in terms of opioids. Opioids and opioid receptors are the best studied of the neuropeptides in terms of steroid regulation and behavioral actions. Although we have learned a great deal about the neurochemistry of lordosis, much work remains to be done before a synthetic overview emerges.

List of Abbreviations: 3β-HSD, 3β-hydroxysteroid dehydrogenase; 5-HT1-5, serotonin receptor 1–5; αDHP, 5α-dihydroprogesterone; ACTH, Adrenocorticotropin hormone; ADX, Adrenalectomized; AGT, aminoglutethimide; AP-5, 5-amino-phosphonoheptanoic acid; AP-7, 7-amino-phosphonoheptanoic acid; ARC, arcuate nucleus of the hypothalamus; AVP, arginine vasopressin; AVPV, anteroventral paraventricular nucleus; BSA, bovine serum albumin; BST, bed nucleus of the stria terminalis; cAMP, cyclic 3′,5′-adenosine monophosphate; CCK, Cholecystokinin; CCK-1, Cholecystokinin receptor 1; CCK-2, cholecystokinin receptor 2; CCK-A, cholecystokinin receptor A; CCK-B, cholecystokinin receptor B; cGMP, cyclic 3′,5′-guanosine monophosphate; ChAT, choline acetyltransferase; CRF, corticotrophin releasing factor; CTOP, H-D-Phe-*Cys*-Tyr-D-Trp-Orn-Thr-Pen-Thr-NH_2; DA, Dopamine; DAMGO, H-Tyr-D-Ala-Gly-N-Met-Phe-glycinol-enkephalin; DOP, δ-opioid receptor; DPDPE, ([D-Pen2, D-Pen2]-enkephalin), a selective DOP agonist; EB, 17 β-estradiol benzoate; EOP, endogenous opioid peptides; EP 24-15, zinc metalloendopeptidase EC 34.24.15; ERα, estrogen receptor-α; ERβ, estrogen receptor-β; GABA, γ-amino butyric acid; GAD, glutamic acid decarboxylase; GALP, galanin-like peptide; GnRH, gonadotropin releasing hormone; GPCR, G protein coupled receptor; Icv, Intracerebroventricular; IGF-1, insulin-like growth factor-1; GluR, ionotropic glutamate receptor; KOP, κ-opioid receptor; LH, luteinizing hormone; LHRH/GnRH, luteinizing hormone releasing hormone/gonadotropin releasing hormone; L-NAME, *N*-Nitro-L-Arginine-Methyl-Ester, inhibitor of NOS; mAchR, muscarinic acetylcholine receptor; MC, Melanocortin; MC1-5, melanocortin receptor 1–5; MeApd, medial amygdaloid nucleus, posterodorsal part; MOP, μ-opioid receptor; MPN, medial preoptic nucleus; MSG, monosodium glutamate; MSH, melanocyte stimulating hormone; nAchR, nicotinic acetylcholine receptor; NK-1, neurokinin-1 receptor (substance P receptor); NMDA, *N*-methyl-D-aspartic-acid; NO, nitric oxide; NOP, nociceptin opioid receptor; NOS, nitric oxide synthase; NPY, neuropeptide Y; OFQ/N, orphanin FQ/nociception; OVX, ovariectomized/ovariectomy; OVX/ADX, ovariectomized and adrenalectomized; PAC1, pituitary adenylate cyclase 1 receptor; PACAP, pituitary adenylate cyclase-activating polypeptide; PAG, periaquaductal gray (also midbrain central gray-MCG); PGE2, prostaglandin E2; PKA, protein kinase A; POMC, proopiomelanocortin; PPT, β-preprotachykinin; PRKO, progesterone receptor knock out; V1a, vasopressin-1a receptor; VMH, ventromedial nucleus of the hypothalamus

1 Introduction

Once it was realized that the brain was involved in the regulation of sexual receptivity, it became clear that steroid hormones altered the activity of specific CNS circuits that regulated this behavior. Early studies showed that treatment with estradiol or estradiol + progesterone altered neuronal activity (Innes and Michal, 1970; Kubo et al., 1975; Bueno and Pfaff, 1976; Yamada and Nishida, 1978). The steroid environment imposes "states" on the brain. These states alter the activity of lordosis regulating circuits allowing them to respond to inputs with appropriate behavioral responses. These behavioral states are produced by steroid hormones that bathe the brain to coordinate the activity of inhibitory and facilitative lordosis circuits. Estradiol is obligatory for the expression of lordosis. However, its initial behavioral actions are to inhibit the expression of lordosis by activation of inhibitory circuits to allow for estradiol and progesterone to bring it "on line" with proper timing of the facilitative circuits. The discovery that the brain had a plentitude of neurotransmitters, and especially peptide neuromodulators, created an explosion of neurochemical examination of hypothalamic function in the 1980s. These led to studies aimed at determining the role of various extracellular messengers in the CNS control of sexual receptivity including nitric oxide (NO) and neurosteroids. Many of these studies infused a peptide or neurotransmitter into the III ventricle, and tested lordosis. Such studies were important because they relatively rapidly assessed which messengers needed further study. A number of groups chose to focus their attention on specific messengers or classes of messengers. These studies usually involved testing the actions of steroid hormones on messenger concentrations and receptors. For catecholamines, high pressure liquid chromatography was a major advance. For neuropeptides, the application of radioimmunoassay (RIA) to measure levels and various techniques to assess mRNA levels were a great boon. Soon there was an understanding about steroid-induced changes of neuropeptide levels, neuropeptide mRNA expression, and receptor mRNA expression with appropriate temporal and spatial resolution.

The plethora of transmitters/modulators and receptors associated with the regulation of sexual receptivity is highly redundant, yet every perturbation of the circuit had dramatic effects on the behavior. These outcomes were not always consistent with a highly redundant system. The question remains, why are there so many transmitters/modulators regulating sexual receptivity? Is there a hierarchical steroid activation of neurochemical circuits that regulate sexual receptivity or is there an accretion of actions with each transmitter/modulator contributing a proportionate part of the regulation leading to the full display of sexual receptivity? Another possibility is that the circuits have neurochemical nodes—constrictions through which information must flow to influence behavior. There is evidence that the medial amygdala, medial preoptic nucleus (MPN) and ventromedial nucleus of the hypothalamus (VMH) are discrete, morphologically identified nodes. Considerable data have been collected to support all these hypotheses, but a definitive answer has not emerged. Another important issue is the disparity between the pharmacological or acute molecular manipulations and the genetic manipulations. Often acute treatments dramatically amplify or block sexual receptivity. These and other questions should be kept in mind as this review is read. Sometimes the answers will be evident. In other cases the answers are currently not clear because the data are incomplete.

In this chapter, as appropriate, the methodology used to assess sexual receptivity in the rodent models will be discussed, since this often determines the outcome. The actions of estradiol and estradiol + progesterone in driving the behavior will be mentioned in addition to a review of extracellular CNS messengers implicated in the control of sexual receptivity.

2 Lordosis

Female sexual behavior has been divided into three components: attractivity, proceptivity, and receptivity [Beach, 1976; note the use of terms with slightly different emphasis by Blaustein and Mani (2006) in a chapter in this volume: copulatory, paracopulatory, and progestative, originally defined by Blaustein and Erskine, 2002]. In most laboratories, the female sexual behaviors that are studied are proceptivity and receptivity. Proceptive behaviors—designed to entice the male—are solicitation, hopping, darting, and ear-wiggling. Beach defined sexual receptivity as a lordosis reflex. The stereotypic lordosis reflex is an arching of the back, elevation of the hindquarters, dorsoflexion of the tail, and extension of the neck (*Figure 4-1*). Beach proposed the lordosis quotient (number of lordosis divided by the number of mounts × 100) as a quantifiable measure of sexual receptivity (Beach, 1948; Beach and Leboeuf, 1967), and most studies have used this measure.

Figure 4-1
Photograph of two rats of the Long-Evans strain displaying sexual behavior. The male (*top*) has mounted the female (*bottom*) and achieved intromission (insertion of the penis into the vagina). In response to the mounting of the male, the female has assumed the stereotypical mating posture, the lordosis reflex, characterized by extension of the neck, arching of the back, elevation of the hindquarters, and dorsoflexion of the tail, which allows the male to intromit. The lordosis reflex in the female rat is dependent on the exposure of the neural circuits that regulate this behavior to ovarian hormones. Photograph by John T. Clark, Meharry Medical College, with permission

The lordosis reflex is elicited by appropriate hormonal priming: estradiol and progesterone in the intact female and stimulation of mechanoreceptors in the flank, perineum, and area around the tail. Under natural conditions this tactile stimulation is provided by a male. Although this is a global reflex requiring a large number of brain areas to elicit this behavior, the critical sites for integration of hormonal with introceptive and extroceptive inputs are located in a circuit that spans the limbic system and hypothalamus (see Micevych and Ulibarri, 1992 for review; *Figure 4-2*). Over the years, the underlying circuit that controls the lordosis reflex has been expanded to include the posterodorsal medial amygdala (MeApd), bed nucleus of the stria terminalis (BST), MPN, arcuate nucleus, and VMH. The VMH is considered to be the origin of the final common pathway from the integrative diencephalon to the periaquaductal grey (PAG), reticular formation, and vestibular nuclei. These regions provide descending projections to the spinal cord to activate medial lying dorsal horn motoneurons that innervate trunk and neck musculature producing the lordosis posture (reviewed in Pfaff et al., 1994).

Another method of testing sexual receptivity has been proposed as a more natural situation, and is termed "pacing" (Mendelson and Gorzalka, 1987; Pfaus et al., 1999); and references therein; also (Erskine,

Figure 4-2

Diagrammatic representation of the basic neural circuit regulating the lordosis reflex. Estradiol acts at several levels within the circuit (*) to allow sensory input from the accessory olfactory system and tactile stimulation from the perineum and flanks to elicit the sexual receptive behavior. The primary integration of sensory, hormonal, and metabolic states occurs in the hypothalamus (*shaded*). Secondary integration of reproductive relevant olfactory and hormonal information occurs in the MeApd and BST, and in the PAG, which receives ascending peripheral tactile stimulation from the spinal cord, hormonal information, and the positive output from the VMH. Descending information is relayed through the vestibular nuclei and the reticular formation before stimulating medial lamina IX motoneurons activating epaxial muscles resulting in lordosis

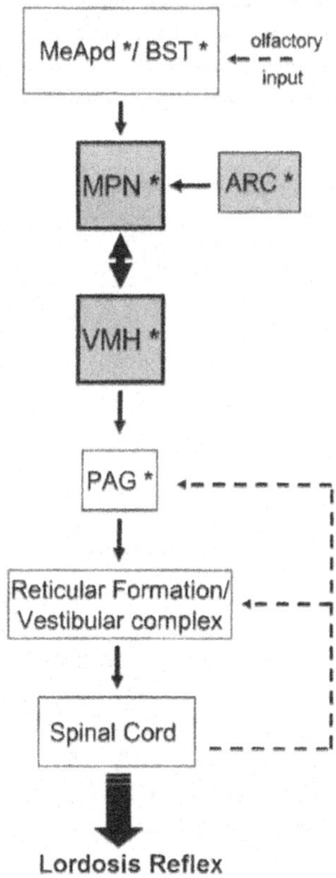

1987; Yang and Clemens, 1997). In the pacing model, females are allowed to approach the male or escape from him—thereby controlling the level of activity. Since this method has not been used to test most of the transmitters and modulators (but see Pfaus and Pfaff, 1992b) we will refer to it only when results with this paradigm conflict with results for the more traditional method of determining a lordosis quotient, the lordosis reflex.

3 Hormonal Modulation of Lordosis Reflex

Estradiol is essential for the display of the lordosis reflex. Estradiol actions in the brain increase the probability of reproductive behavior. Traditionally, estradiol actions that induce sexual receptivity have been thought to be mediated by nuclear receptors that alter the rate of transcription and translation.

Indeed, the induction of sexual receptivity was demonstrated to be dependent on protein synthesis in the brain (Ho et al., 1973; Rainbow et al., 1980a). More recently, both intracellular, nuclear receptors and membrane associated receptors have been implicated in the regulation of lordosis (Vasudevan et al., 2001). Regardless of whether genomic, nongenomic, or nongenomic to genomic signaling occurs, estrogen receptor-α (ERα) is generally considered mediate sexual receptivity relevant signaling (Rissman et al., 1997, 1999; Ogawa et al., 1998; reviewed in Rissman et al., 1999; Micevych et al., 2003b).

In gonadally intact rats, sexual receptivity is tightly regulated by the sequential release of estradiol and progesterone from the ovary. In ovariectomized (OVX) rats, sexual receptivity can be induced by either estradiol only or the sequential treatment with estradiol and progesterone. Although both steroid treatments induce sexual receptivity, several characteristics of the resulting behavior suggest that the mechanisms are different: (1) the dose of estradiol-only needed to induce sexual receptivity is higher than that needed when estradiol is supplemented with progesterone (Pfaff, 1970; reviewed in Clemens and Weaver, 1985); (2) repetitive estradiol treatment needed for maximal sexual receptivity results in a ramping increase of sexual receptivity as measured by lordosis quotient, whereas repeated estradiol + progesterone treatments produces a constant level of lordosis (Sodersten and Eneroth, 1981; Bloch et al., 1987); (3) progesterone receptors are not required for estradiol-only facilitation of lordosis (Mani et al., 1997) as evidenced by the fact that progesterone receptor antagonists do not disrupt estradiol-only induced sexual receptivity (Blaustein et al., 1987); (4) estradiol-only induced lordosis has a later onset and longer window of sexual receptivity compared with estradiol + progesterone (Quadagno et al., 1972; reviewed in Clemens and Weaver, 1985). In summary, progesterone subsequent to estradiol treatment transiently augments the estrogenic induction of the lordosis and eventually inhibits many of the estradiol-induced effects terminating the behavior and "resetting" the behavioral state of the female rat (Sodersten and Eneroth, 1981). In estradiol-only treated rats, the reversal of the behavioral state can be induced by the estradiol priming "wearing off" or induced by levels of vaginocervical stimulation that produces a pregnant or pseudopregnant state (Erskine et al., 2004).

Following estradiol treatment, sexual receptivity cannot be facilitated for almost a day. Progesterone begins facilitating lordosis approximately 20 h after estradiol treatment, and this estradiol + progesterone induced sexual receptivity becomes maximal approximately 24 h after estradiol (Sinchak and Micevych, 2001b). In estradiol-only animals, the dose of estradiol required is higher when the receptivity is facilitated with subsequent progesterone. Either a single large dose of estradiol in a single bolus injection or smaller repeated injections are used to facilitate sexual receptivity. However, the onset of sexual receptivity is delayed approximately 48 h after initial treatment (Boling and Blandau, 1939).

The control of female reproduction requires coordination of sexual receptivity with production of a viable oocyte. The primary stimulus regulating reproductive behavior and ovulation is the increasing levels of estradiol that peak on proestrus. Interestingly, in the intact animal, the rise of ovarian progesterone is coincident with the luteinizing hormone (LH) surge and after the LH surge (Moss, 1974; Sodersten and Eneroth, 1981). In the intact female, both the behavior and LH surge appear dependent on the actions of progesterone (Brom and Schwartz, 1968; Ferin et al., 1969; Labhsetwar, 1970; Rao and Mahesh, 1986; Mahesh and Brann, 1998; Chappell and Levine, 2000). Moreover, progesterone that originates from the ovary and adrenal glands is not involved, since estradiol facilitates lordosis in OVX and adrenalectomized (OVX/ADX) rats (Bloch et al., 1987). One possible explanation is that estradiol induces progesterone synthesis in the hypothalamus that activates reproductive behavior as it has shown to trigger the LH surge (Micevych and Ulibarri, 1992; Micevych et al., 2003a). This locally produced neuroprogesterone would activate estradiol-induced progesterone receptors in the hypothalamus stimulating behavior before the levels of peripheral progesterone are significantly elevated. To directly test this hypothesis, OVX/ADX rats were treated with estradiol (10 μg 17 β-estradiol benzoate, EB) and then with free estradiol (50 μg 17 β-estradiol) 48 h after the first estradiol injection. This treatment induces both proceptive and receptive behaviors (Parsons et al., 1984; ➋ *Figure 4-3*). Treatment with a progesterone (and glucocorticoid) receptor antagonist, RU486, prevented proceptive behaviors, but not lordosis (➋ *Figure 4-3a*; Mills et al., 2003). Similarly, blockade of steroidogenesis of neuroprogesterone with either aminoglutethamide (AGT; icv; ➋ *Figure 4-3b*) or trilostane blocked proceptivity but not sexual receptivity, i.e., lordosis (➋ *Figure 4-3c*; Mills et al., 2003). These data indicate that neuroprogesterone may have a role in initiating proceptive

Figure 4-3

Effects of blocking the synthesis of progesterone and progesterone receptors in the CNS on sexual behavior. Blocking either progesterone synthesis (a and b) or activation of progesterone receptors (c) reduces expression of proceptive behaviors, but has no effect on expression of lordosis OVX/ADX rats treated weekly with 10 μg 17β-estradiol benzoate (EB) and then 4 h before the test with 50 μg 17β-estradiol. Animals were tested 53–56 h after the initial EB injection for sexual receptivity, as measured by lordosis quotient (LQ) and expression of proceptive behaviors (proceptivity scale). Treatments to block progesterone synthesis or progesterone receptors were delivered subcutaneously and started just before the EB treatment and on the following mornings before testing. Aminoglutethamide (AGT, 10 mg per treatment) blocks P450 side chain cleavage that converts cholesterol to pregnenolone (a); trilostane (TRI) blocks the enzyme 3β-hydrysteroid dehyrdrogenase (16.5 mg per treatment; 3β-HSD) which converts pregnenolone to progesterone (b). RU486 (5 mg per treatment) is a progesterone receptor antagonist (c). * = significantly less than control treatment within behavior group (Mills et al., 2003)

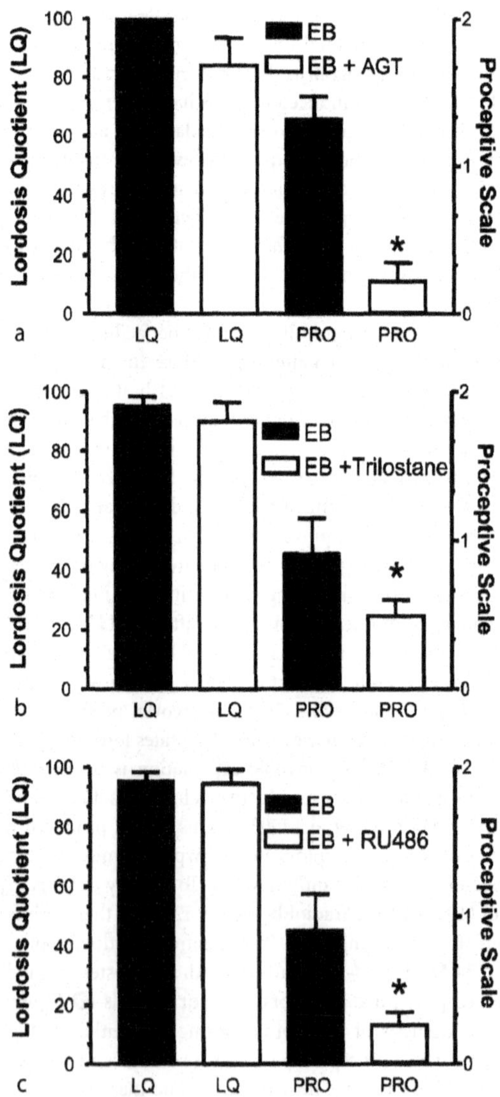

behaviors, but the lordosis reflex is not dependent on de novo synthesis of neuroprogesterone. The conclusion that can be drawn from these results is that neither progesterone nor progesterone receptors are needed to induce lordosis with estradiol alone, suggesting that estradiol activates a circuit that is distinct from the one activated by estradiol + progesterone as previously suggested [Mani et al., 1994a, c, 1997; see also Blaustein and Mani (2006) in this volume].

Progesterone has another important function vis-a-vis receptive behavior; it resets the lordosis regulating circuits in the brain. Sequential treatment of OVX animals with estradiol and progesterone facilitates lordosis and then terminates the behavior (Goy et al., 1966; Nadler, 1970; Feder and Marrone, 1977; Marrone et al., 1977; Morin, 1977; Schwartz et al., 1979). This relatively sharp cessation of lordosis is not seen in OVX animals made receptive by estradiol alone. Perhaps more importantly, females treated with 3–5 μg EB once in every 7–10 days have an increased lordosis quotient with each subsequent treatment until maximally receptive (Bloch et al., 1987; Butler et al., 2001). However, repeated treatments with 2 μg EB produce constant minimum levels of lordosis behavior (Sinchak and Micevych, 2001a), and subsequent progesterone treatment induces maximal receptivity. One intriguing idea is that progesterone induces "sequential inhibition" by promoting the degradation of estradiol-induced progesterone receptors (Blaustein and Mani, 2006). Interestingly, Gonzalez-Flores et al. (2004) recently demonstrated that blocking 26S-proteosome activity was associated with a lack of progesterone-induced sequential inhibition of lordosis reflex, and it prevented the loss of progesterone receptor immunoreactivity.

Studies over the past 40 years have suggested that sexual receptivity is not only complex in terms of the regions that are activated by ovarian steroids but is exceedingly more complicated in terms of the neurochemistry of this behavior. A cornucopia of extracellular signaling molecules has been identified and examined for their roles in modulating the lordosis reflex, or in some cases, pacing of sexually receptive behavior of the female.

Neuropeptides are disproportionately represented on the list of lordosis-promoting or lordosis-inhibiting molecules. The reasons for this are not entirely clear, but may reflect the large number of neuropeptides and their preferential distribution in the hypothalamus. Levels of most of these molecules are regulated by ovarian steroids. In many cases, the receptors for these molecules have also been characterized in terms of their response to estradiol and estradiol + progesterone.

4 Opioid Regulation of Lordosis

Arguably the best studied neurochemical mediators of lordosis are members of the family of endogenous opioid peptides (EOP). The EOP and their receptor systems are involved in all aspects of reproduction. EOP are distributed throughout limbic system and hypothalamic sites that regulate the lordosis reflex (➋ *Figure 4-2*; Lee and Smith, 1980; Patterson et al., 1983; Smith et al., 1983; Law and Loh, 1999; Sinchak et al., 2006). These steroid receptive circuits are important not only for reproductive behavior in females, but also for integrative control of neuroendocrine regulation of reproduction by the hypothalamo-pituitary-gonadal axis. Opioid receptors are part of the superfamily of seven membrane pass, G protein-coupled receptors (Chen et al., 1993a, b; Reisine and Bell, 1993; Thompson et al., 1993; Mollereau et al., 1994; Wang et al., 1994a, b; Lachowicz et al., 1995; Dhawan et al., 1996), which include the classical opioid receptors μ-, δ-, and κ-opioid receptors (MOP, DOP, and KOP, respectively)—all blockable by the opioid receptor antagonist naloxone—and the opioid receptor-like receptor-1 (NOP, also known as ORL-1; Mollereau et al., 1994; Lachowicz et al., 1995). NOP shares a significant sequence homology with the other opioid receptors but is not antagonized by naloxone (Darland et al., 1998; reviewed in Meunier, 1997).

The early studies of opioid regulation of reproduction produced results that appeared incompatible with one another. The chronic use of heroin or morphine inhibited sexual behavior in female dogs, mice, monkeys, and chimpanzees, indicating that the global effect of opioid tone is inhibiting to reproductive behavior (Plant and Pierce, 1928; Tatum et al., 1929; Ko, 1934,1935; Seevers, 1936; Eddy and Reid, 1939; Spragg, 1940). However, further studies showed that chronic opioid administration did not preclude the dogs and rats from becoming pregnant and producing healthy young (Plant and Pierce, 1928; Myers and Flynn, 1931), indicating that estrous cyclicity and reproductive behavior were not eliminated, as previously assumed. There were several factors in these early studies that made interpretation of the findings difficult. First, opioid treatment that produces tolerence state for opioids that may permit "normal" reproduction to occur, whereas acute treatments may have rapid transient effects on behavior. Second, agonists and antagonists used to investigate the role of opioids acted on multiple opioid receptor subtypes. Third, in the majority of these studies, drugs were given either peripherally or icv thereby activating multiple sites within the CNS. Fourth, the opioid receptors are promiscuous and will bind multiple opiates and opioids (Lee and Smith, 1980; Patterson et al., 1983; Smith et al., 1983; Law and Loh, 1999).

Despite these issues, the inhibition of sexual receptivity remained a consistent finding, especially following acute, global activation of opioid systems (reviewed in Pfaus and Gorzalka, 1987a). Yet with time, data suggested an additional, facilitatory role in sexual receptivity (Imura et al., 1985, 1986; Suda et al., 1986; Pfaus and Gorzalka, 1987b; Torii and Kubo, 1994; Torii et al., 1995, 1996, 1997, 1999). In this review, we will categorize opioid regulation of sexual receptivity by opioid receptor subtype.

4.1 MOP

The most well understood mechanism for MOP regulation of lordosis has been worked out in the MPN. MOP is expressed throughout the rostrocaudal extent of the medial preoptic area, especially in the dorsal aspect of the medial part of the MPN (Eckersell et al., 1998; Sinchak and Micevych, 2001a). In sexually receptive females, acute activation of MOP in the MPN inhibits lordosis (Wiesner and Moss, 1984; Sirinathsinghji, 1986; Pfaus and Pfaff, 1992b; Sinchak and Micevych, 2001a; Acosta-Martinez and Etgen, 2002a). Further, infusion of the general opioid receptor antagonist, naloxone, into the MPN of steroid primed nonreceptive females facilitated lordosis (Acosta-Martinez and Etgen, 2002a), indicating that steroid induction of sexual receptivity was blocked by MOP in the MPN.

Estradiol rapidly activates MOP in the MPN. This activation is coincident with the nonreceptive phase following initial estradiol exposure. The specific activation of MOP following estradiol treatment has been followed using the receptor internalization assay (Micevych et al., 1997; Eckersell et al., 1998; Mills et al., 2004; reviewed in Sinchak and Micevych, 2003). Estradiol, acting through ERα (within 30 min), rapidly induces activation of MOP in the MPN that lasts for at least 30 h before returning to basal levels by 48 h (Eckersell et al., 1998; Micevych et al., 2000, 2003b; Sinchak and Micevych, 2001a). This pattern of MOP activation in the MPN parallels the facilitation of lordosis in estradiol-only treated rats. Females treated with estradiol are not receptive at 30 h after treatment (Sinchak et al., 1997; Sinchak and Micevych, 2001a), but do exhibit lordosis approximately 48 h after estradiol treatment (reviewed in Clemens and Weaver, 1985), which coincides with the relief of MOP-mediated inhibition. Progesterone treatment 26 h after priming with physiological levels of estradiol facilitates lordosis 30 h after estradiol treatment and reverses estradiol-induced MOP internalization/activation (Sinchak and Micevych, 2001a). The steroid regulation of MOP activation in the MPN appears to be mediated by a multisynaptic circuit involving NPY and β-endorphin neurons in the arcuate nucleus, as discussed in the NPY section.

Further evidence that MOP in the MPN are relevant to the regulation of sexual receptivity in the intact rats is provided by the pattern of MOP activation during the estrus cycle (Sinchak and Micevych, 2003). MOR internalization/activation in the MPN is increased whereas sexual receptivity is low, on diestrus 1 and 2 when, estradiol is rising, and decreased when receptivity is elevated as on the afternoon and evening of proestrus after progesterone levels rise. After sexual receptivity is terminated, MOR internalization is again increased on the morning of estrus, suggesting that MOR in the MPN may also regulate the termination of sexual receptivity.

The activity of MOP in the MPN is important for integrating gonadal steroid information that regulates the timing of sexual receptivity—both the onset and termination. Antagonizing MOP in the MPN at the time of estradiol treatment prevents full progesterone facilitation of lordosis (Torii and Kubo, 1994; Torii et al., 1995, 1996, 1999). In mice that are MOP deficient, estradiol + progesterone do not fully induce the lordosis reflex or lordosis score (Sinchak et al., 2005). These behavioral studies indicate that the estradiol activates this lordosis inhibiting circuit early and that this MOP activity is important for estradiol actions on other lordosis facilitative circuits and the expression of maximal sexual receptivity. This MOP activation may be one of the neural states that is induced by estradiol. This inhibitory tone over the lordosis circuit is maintained until either progesterone deactivates the MOP, which switches the state by allowing the coordination of circuits controlling the LH surge (ovulation) and sexual behaviors. If progesterone is not present, high levels of estradiol alter these neural circuits such that the female is eventually sexually receptive. Then copulatory behavior and/or olfactory cues from a male induce the LH surge and ovulation (Matt et al., 1987; Day et al., 1988).

The expression of MOP in the VMH remains controversial. A number of immunocytochemical and in situ studies have not demonstrated MOP in the VMH of rats or mice (Mansour et al., 1994a, b; Minami et al., 1994; Micevych et al., 1997; Eckersell et al., 1998), whereas others indicate that MOP may be expressed at very low levels (Bunzow et al., 1995; Loughlin et al., 1995; Acosta-Martinez and Etgen, 2002a). This controversy is born out in behavioral studies. Infusion of either morphine or the selective MOP agonist, H-Tyr-D-Ala-Gly-N-Met-Phe-glycinol-enkephalin (DAMGO), into the region of the VMH inhibited lordosis (Vathy et al., 1991; Acosta-Martinez and Etgen, 2002a). However, other studies indicate that the role of MOP in the VMH is either nonexistent or has little influence on lordosis (Sinchak et al., 1997; Acosta-Martinez and Etgen, 2002b; Sinchak and Micevych, 2003). One explanation is that site-specific injections aimed at the VMH may activate MOP in the arcuate nucleus, which affects the lordosis reflex through a multisynaptic mechanism that involves β-endorphin neurons or MOP that are located outside of the VMH (Bouret et al., 1999; reviewed below in NPY section).

The PAG receives projections from the VMH and is an important part of the motoric pathway for the facilitation of lordosis (Pfaff and Sakuma, 1979; Sakuma and Pfaff, 1979, 1980a, b , 1982; Sakuma and Akaishi, 1987; Dornan et al., 1990; Hennessey et al., 1990; Canteras et al., 1994; Daniels et al., 1999; Flanagan et al., 2006). The PAG expresses MOP, and site-specific injections of β-endorphin, an endogenous MOP ligand, inhibited lordosis. The inhibition was naloxone reversible (Sirinathsinghji, 1984, 1985). More recent studies using selective MOP ligands, endomorphin-1 and endomorphin-2, did not confirm the earlier studies (Acosta-Martinez and Etgen, 2002a). Although it is possible that β-endorphin may be acting through DOP (Law and Loh, 1999) in the PAG, the most parsimonious explanation is that the endomorphins were rapidly degraded before behavioral testing (Sinchak and Micevych, 2001a).

4.2 DOP

DOPs are also expressed throughout the limbic-hypothalamic lordosis regulating circuit. In the MPN, DOP binding has been reported (McLean et al., 1986; Eckersell and Micevych, 1997). The pattern of DOP immunoreactivity is diffuse and located on small fibers suggesting that DOP may be located on afferent fibers (Sinchak et al., 2004b). This is corroborated by a lack of DOP mRNA in the MPN (Desjardins et al., 1990; Mansour et al., 1994a, b). Interestingly, neither DOP binding levels nor patterns of immunoreactivity are sensitive to the steroid milieu (Acosta-Martinez and Etgen, 2002b; Sinchak et al., 2004b). Although, icv infusion of DOP agonists facilitate lordosis in estradiol primed female rats (Pfaus and Gorzalka, 1987b; Pfaus and Pfaff, 1992a; Acosta-Martinez and Etgen, 2002b), activation of DOP facilitates or inhibits lordosis depending on the site of activation. In the VMH, [D-Pen2, D-Pen2]-enkephalin (DPDPE), a selective DOP agonist moderately facilitates female sexual receptivity, whereas in the MPN DOP activation inhibits lordosis (Micevych et al., 2002; Acosta-Martinez and Etgen, 2002b; Sinchak and Micevych, 2003; Sinchak et al., 2004b).

Based on the presence of DOP binding and immunoreactivity in the MPN this is another region of where DOP may regulate lordosis. Infusion of DPDPE into the MPN-inhibited lordosis in sexually receptive animals and was blocked by pretreatment with naltrindole, a DOP antagonist (Sinchak et al., 2004b). This rapid inhibition of lordosis was similar to the activation of MOP in the MPN (Sinchak and Micevych, 2001a). Thus, both MOP and DOP in the MPN inhibit lordosis. In estradiol-primed female rats, infusion of naloxone, which antagonizes classical opioid receptors, into the MPN facilitates lordosis (Acosta-Martinez and Etgen, 2002a). Selective MOP antagonism did not facilitate lordosis in nonreceptive female rats primed with estradiol (Acosta-Martinez and Etgen, 2002a). Thus, the inactivation of both MOP and DOP in the medial preoptic area is required to facilitate lordosis (Sinchak et al.,2004b). The MPN contains the endogenous ligands for both the DOP and the MOP, enkephalin and β-endorphin, which are capable of binding and activating both the receptors (Lee and Smith, 1980; Patterson et al., 1983; Smith et al., 1983; Law and Loh, 1999).

In the VMH, DPDPE moderately facilitates female sexual receptivity (Micevych et al., 2002; Acosta-Martinez and Etgen, 2002b; Sinchak and Micevych, 2003). Levels of sexual receptivity achieved by activation of DOP in the VMH are well below the sexual receptivity achieved by icv injection of DPDPE, suggesting that the VMH is not the only site at which DOP facilitates lordosis.

4.3 KOP

Endogenous KOP ligands (i.e., dynorphin 1–17 (dynorphin A), dynorphin B (rimorphin), leumorphin, a/β-neoendorphin) are produced from the endoproteolysis of prodynorphin (reviewed in Waldhoer et al., 2004). Although these KOP-activating peptides are marginally distributed in the limbic-hypothalamic lordosis regulating circuit, their role or that of KOP in the regulation of lordosis has received little attention. Intracerebroventricular infusion of dynorphin A has been reported to have either no effect or facilitate lordosis (Pfaus and Gorzalka, 1987b; Pfaus and Pfaff,1992a). Similarly, activation of KOP in the PAG was also ineffective in modulating the lordosis reflex (Sirinathsinghji, 1985). Based on these results there is consensus that KOP has a minor role, if any, in regulating sexual receptivity.

4.4 NOP

NOP and its endogenous ligand, orphanin FQ-nociceptin (OFQ/N, also known as nociceptin), comprise a more recently characterized endogenous opioid system (Mollereau et al., 1994; Wang et al., 1994a; Lachowicz et al., 1995; Brit J. Pharm, 2003; Neubig et al., 2003). The amino acid sequences for NOP and its endogenous ligand are highly homologous to KOP and dynorphin A, respectively (Fukuda et al., 1994; Mollereau et al., 1994; Wang et al., 1994a; Lachowicz et al., 1995; Meunier et al., 1995; Lee et al., 1997; Meunier, 1997; Reinscheid et al., 1998). Though it may be tempting to think that KOP actions on lordosis may be due to dynorphin A acting at NOP, NOP exhibits little affinity for endogenous ligands of MOP, DOP, or KOP (Pfaus and Pfaff, 1992a; Fukuda et al., 1994; Mollereau et al., 1994; Wang et al., 1994a, Lachowicz et al., 1995; Lee et al., 1997). Naloxone does not antagonize NOP, but does antagonize dynorphin A actions. The only known endogenous agonist for NOP is OFQ/N (Meunier et al., 1995; Dooley and Houghten, 1996; Reinscheid et al., 1996; Shimohigashi et al., 1996; Ardati et al., 1997; Butour et al., 1997; Guerrini et al., 1997; Meunier, 1997; Reinscheid et al., 1998), suggesting that dynorphin A is not the ligand for NOP.

In the female rat, NOP and OFQ/N are expressed in the limbic–hypothalamic lordosis regulating circuit: MeApd, BST, arcuate nucleus, and MPN (Neal et al., 1999a, b; Sinchak et al., 2006). In the VMH, only the NOP are expressed, but OFQ/N is transported to the nucleus in afferent fibers (Sinchak et al., 1997, 2006). Ovarian steroids regulate the mRNA expression of both OFQ/N and NOP mRNA expression and the functional coupling of NOP to its G protein that is temporally congruent with the facilitation of lordosis (Micevych and Quesada, unpublished observations; Sinchak et al., 2006).

Infusion of OFQ/N into the VMH facilitates lordosis in female rats primed with a physiological dose of estradiol (Micevych et al., 1996; Sinchak et al., 1997; Dewing et al., 2005; Sinchak et al., in press). The sequential exposure to estradiol and progesterone is important for increasing both the ligand and receptor expression and then activation of the OFQ/N in the VMH to facilitate lordosis. Estradiol priming increases the expression of NOP in the VMH, but does not induce NOP activation (Sinchak et al., 1997; Sinchak and Micevych, 2003; Dewing et al., 2005). Subsequent progesterone treatment that facilitates lordosis activates NOP, presumably by the release of endogenous OFQ/N (Sinchak and Micevych, 2003). Potential brain regions known to regulate lordosis that express OFQ/N mRNA and project to the VMH include the MeApd and the MPN. OFQ/N mRNA expression in these regions is upregulated by ovarian steroids (Sinchak et al., 2006).

As in the VMH, infusion of OFQ/N into the MPN facilitates the lordosis reflex of rats primed with physiological levels of estradiol (Sinchak et al., 2003; Dewing et al., 2005; Sinchak et al., in press). NOP activation is enhanced by ovarian steroid regulation of NOP expression and an increase in GPCR-G protein coupling in the MPN. In the medial part of the MPN, estradiol slightly increases NOP expression; however, estradiol and progesterone are needed to significantly increase NOP expression (Sinchak et al., 2006). This finding is consistent with results that the number of NOP binding sites in medial preoptic area, which contains the MPN, is increased by estradiol and progesterone treatment, but not by estradiol alone (Micevych and Quesada, Submitted). Further, estradiol and progesterone treatment increases the coupling of NOP to its G protein in the medial preoptic area. Thus, the upregulation and activation of the NOP system in the MPN by estradiol and progesterone is temporally associated with the onset of sexual receptivity. At this time, the relationship between the activation of NOP and the inactivation of MOP in the MPN needed for sexual receptivity is not clear. Further experiments are needed to understand these interactions.

The arcuate nucleus has recently emerged as a node for steroid regulation of lordosis (Mills et al., 2004). Recent studies suggest two mechanisms through which OFQ/N in the arcuate nucleus may facilitate sexual receptivity. First, OFQ/N that is released in the MPN may originate in the arcuate nucleus (Sinchak et al., 2003, 2004a). Second, OFQ/N hyperpolarizes β-endorphinergic neurons (Wagner et al., 1998), which are activated by estradiol to inhibit lordosis (Nguyen et al., 2006). Thus, progesterone facilitation of lordosis may be through inhibition of β-endorphin neurons that project to the MPN.

The estradiol–progesterone–EOP interaction appears to be designed to provide a timing mechanism that ensures reproductive success. Estradiol initiates a series of transcriptional events, especially the expression of progesterone receptors. Simultaneously, estradiol initiates a series of rapid actions that lead to an inhibition of the lordosis regulating circuits. Normally, the increasing levels of circulating progesterone block the MOR inhibition and activates facilitatory circuits among which is NOP release in the VMH and MPN to facilitate lordosis (Sinchak et al., 1997, 2003; Dewing et al., 2005; Sinchak et al., in press).

5 Neuropeptide Y

NPY infused icv robustly induces increased food intake and acutely inhibits the lordosis reflex in rodents (Clark et al., 1984, 1985; Stanley and Leibowitz, 1985; Kalra et al., 1988; Kulkosky et al., 1988; Corp et al., 2001; Mills et al., 2004). NPY, a member of the pancreatic polypeptide family, is highly conserved among the mammalian species (Tatemoto, 1982; Tatemoto et al., 1982; Adrian et al., 1983; Lundberg et al., 1984; Sabatino et al., 1987; Mercer et al., 1996). There are at least six known NPY receptor subtypes that have been characterized (Michel et al., 1998; Brit J. Pharmacol, 2003). Although there have been relatively few studies linking NPY and central regulation of lordosis, the similarity of NPY and β-endorphin inhibition of the lordosis reflex suggested that a study of these two neuropeptides may yield a more clear understanding of estradiol regulation of the limbic-hypothalamic lordosis regulating circuit. Moreover, it suggested that the arcuate nucleus may be an important, if overlooked node of estradiol and progesterone action. A series of studies have characterized NPY as an integral part of a neural circuit that involves the arcuate nucleus through which estradiol, by a multisynaptic circuit, rapidly induces and maintains the activation of MOP in the MPN that is ultimately important for ovarian steroid for the facilitation of lordosis (Eckersell et al., 1998; Sinchak and Micevych, 2001a; Mills et al., 2004; Sinchak et al., 2005).

Within the arcuate nucleus, NPY neurons express estrogen receptors, and estradiol regulates the expression of NPY (Sar et al., 1990; Baskin et al., 1995), indicating a direct mechanism for estradiol regulation of these arcuate nucleus neurons. Accordingly, estradiol induces the release of NPY that stimulates β-endorphin neurons by the NPY-Y1 receptor which project to the MPN to release β-endorphin and activate MOP (Mills et al., 2004). This is supported by observations that NPY-Y1 receptors in the arcuate nucleus and MOP in the MPN have a similar time course of estradiol-induced activation (Mills et al., 2004). Further, the estradiol-induced internalization of activated NPY-Y1 receptor and MOR is blocked by treatment with an NPY-Y1 receptor antagonist. Subsequent progesterone treatment that facilitates lordosis attenuates activation/internalization of NPY-Y1 in the arcuate nucleus and MOP in the MPN (Sinchak and Micevych, 2001a; Mills et al., 2004). The progesterone inactivation of the MOP and NPY-Y1 receptors is reversed by NPY treatment, indicating that both the estradiol and the progesterone actions on MOR are mediated by the release of NPY in the arcuate nucleus (Mills et al., 2004). The acute NPY-induced (icv) inhibition of lordosis in estradiol + progesterone treated animals is reversed by pretreating with the MOP specific antagonist CTOP (Mills et al., 2004), indicating that the inhibitory action of NPY on lordosis is through the activation of MOP in the MPN (Mills et al., 2004). Thus, ovarian regulation of the MOP lordosis inhibitory circuit in the MPN is by the regulation of the accurate nucleus β-endorphin neuron activity by NPY.

6 Adrenocorticotropic Hormone/Melanocyte Stimulating Hormone

Proopiomelanocortin (POMC) is a precursor from which α-MSH (melanocyte stimulating hormone), β-MSH, γ-MSH, ACTH (adrenocorticotropic hormone), and β-endorphin are produced by posttranslational splicing (Uhler et al., 1983). The opioid product, β-endorphin, has been discussed separately. Both α-MSH and ACTH have implicated the regulation of sexual receptivity (Pfaus and Gorzalka, 1987a; Dornan and Malsbury, 1989; Gonzalez et al., 1999; Scimonelli et al., 2000; Pfaus et al., 2004). Melanocortin effects are mediated through a family of G protein-coupled seven membrane span receptors (reviewed in Catania et al., 2004 and Getting, 2006), which have a small second extracellular loop and short amino and carboxy terminal ends. To date 5 melanocortin (MC) receptors have been cloned and termed MC1 through MC5. These GPCRs are coupled to several intracellular signaling pathways including activation of adenylyl cyclase, and inositol triphosphate leading to increased cAMP levels and intracellular calcium levels, respectively (Wikberg et al., 2000).

The MC1 receptor is distributed in the skin where it mediates pigmentation and cutaneous neuroimmunomodulation. The MC2 receptor is distinguished from the other MC receptors because it binds ACTH, but not the other melanocortins (α-MSH, β-MSH, and γ-MSH). The MC3 receptor is mainly found in the brain (Roselli-Rehfuss et al., 1993), and the MC4 receptor is exclusively found in the brain (Mountjoy et al., 1994). The MC4 receptor subtype appears to mediate the effects of the melanocortins on eating behavior and body weight (Huszar et al., 1997). The MC5 receptor is also found in the brain and a wide range of peripheral tissues. This receptor has been implicated in regulating exocrine gland secretion (Chen et al., 1997).

ACTH and α-MSH act on similar MC receptors, α-MSH on most MC receptors and ACTH on MC2 (Chhajlani and Wikberg, 1992; Mountjoy et al., 1992; Chhajlani et al., 1993). α-MSH behaves like serotonin: facilitating lordosis reflex when animals are lowly receptive and inhibiting lordosis in highly receptive animals (Thody and Wilson, 1983; Raible and Gorzalka, 1986; Cragnolini et al., 2000). Both α-MSH, a pan-MC receptor agonist, and γ-MSH, a selective MC3 agonist, have shown to regulate lordosis in ovariectomized, estradiol primed rats after infusion into the VMH and the medial proptic area (Gonzalez et al., 1993, 1996; Cragnolini et al., 2000; Nocetto et al., 2004). Evidence has accumulated that the mechanism of this action is through the MC3 receptor that mediates stimulation of NO generation. Pretreatment of estradiol-primed rats with L-NAME, an inhibitor of nitric oxide synthase (NOS) completely abolished the stimulatory actions of both α-MSH and γ-MSH (Nocetto et al., 2004). On the other hand, PT-141, an analog of α-MSH that binds to the MC1 and MC2 receptors increased solicitation behaviors without altering lordosis reflex or pacing behavior (Pfaus et al., 2004), suggesting that the melanocortin action on sexual behavior may be differentiated according to the receptors activated. MC1 and MC2 receptors appear to be involved with regulation of proceptivity (anticipatory phase) and MC-3 regulates receptivity (consummatory phase).

7 Oxytocin

The classic neurohormone oxytocin is usually associated with the posterior pituitary and with a peripheral site of action. In the female, these actions are mainly on uterine smooth muscle (parturition) and milk ejection (lactation). Cells in the supraoptic nucleus and paraventricular nucleus of the hypothalamus synthesize oxytocin and paraventricular oxytocinergic neurons have extensive CNS projections. Estradiol increases oxytocin levels (Jirikowski et al., 1988) and release (Caldwell et al., 1996a, b) in the medial preoptic area and medial basal hypothalamus. Estradiol also acts postsynaptically in the oxytocin system to increase oxytocin receptor expression (McCarthy, 1994a; Bale and Dorsa, 1995a, b; Bale et al., 1995a, b). The estradiol-induced increase in oxytocin receptor was shown to be due to a nongenomic to genomic signaling mechanism involving rapid activation of protein kinase C and protein kinase A (Bale and Dorsa, 1997, 1998; Bale et al., 2001).

A role of oxytocin in facilitating lordosis was demonstrated in the mid 1980s by Cort Pedersen and his colleagues (Caldwell et al., 1984, 1986). This action of oxytocin has been studied by other groups as well (Arletti and Bertolini, 1985; Gorzalka and Lester, 1987; Benelli et al., 1994; Bale et al., 2001), who established that oxytocin facilitates both estradiol and estradiol + progesterone induced lordosis reflex (Caldwell et al., 1984, 1986, 1992, 1994; Gorzalka and Lester, 1987; Schumacher et al., 1989; Schulze and Gorzalka, 1991). Oxytocin acts in both the medial preoptic area and the VMH to exert its effects by activating oxytocin receptors (Caldwell et al., 1990, 1994; Schumacher et al., 1990; Coirini et al., 1991; Johnson et al., 1991; Witt and Insel, 1991; McCarthy, 1994a; Pedersen and Boccia, 2002). Estradiol conjugated to bovine serum albumin (BSA), to prevent activation of intracellular ERs (Ke and Ramirez, 1990) along with oxytocin stimulated lordosis reflex after infusion into the medial preoptic area or the mediobasal hypothalamus (Caldwell and Moe, 1999). Additionally, estradiol induction of oxytocin mRNA levels is not as dramatic as estradiol effects on peptide content (Caldwell et al., 1989; Amico et al., 1995), and estradiol increases the affinity of oxytocin receptors (Caldwell et al., 1994), suggesting that estradiol's primary action on the lordosis regulating oxytocin system may be to rapidly influence oxytocin receptors. Further work is needed to demonstrate whether this is due to a change in the affinity of the oxytocin receptor for its G proteins in addition to altered affinity for oxytocin.

8 Vasopressin

Central administration of arginine vasopressin (AVP) has been shown to both facilitate and inhibit lordosis (Sodersten et al., 1983, 1985; Caldwell et al., 1986; Pedersen and Boccia, 2006). The inhibitory actions of AVP appear to be mediated through the V1a receptor subtype. Antagonism of the V1a receptor facilitates lordosis (Caldwell et al., 1990; Pedersen and Boccia, 2006) and appears to be acting in the medial preoptic area (Caldwell et al., 1990, 1994; Pedersen and Boccia, 2006). The inhibitory effect of AVP may be acting through reducing the activity of the oxytocin system (see oxytocin section), since pretreatment with a selective oxytocin receptor antagonist inhibits the facilitation induced by the selective V1a antagonist (Pedersen and Boccia, 2006). The inhibition of behavior by AVP was produced within 15 min, but the facilitation produced by the V1a antagonist had an onset that was much delayed, ranging from 90 min to 4 and 6 h (Caldwell et al., 1990; Pedersen and Boccia, 2006).

9 Cholecystokinin

Like most neuropeptides, CCK was implicated in the control of sexual receptivity in the 1980s (Pfaus et al., 1986; Bloch et al., 1988). These initial studies examined the effects of peripherally administered CCK on the lordosis reflex. CCK was shown to be both lorodosis-promoting and lordosis-inhibiting: facilitating lordosis in rats, which were slightly receptive and inhibiting lordosis in highly receptive rats (Bloch et al., 1987). Subsequent studies defined the distribution of CCK in the limbic-hypothalamic, lordosis-regulating circuit (Micevych et al., 1986, 1987, 1988a, b; see Micevych and Ulibarri, 1992 for review). The expression of CCK is sexually dimorphic (Micevych et al., 1987, 1988a) and regulated by estradiol in both males and females (Micevych et al., 1986, 1988b; Oro et al., 1988). Although there is a slight sexually dimorphic distribution in gonadectomized rats, the primary stimulus for CCK expression is the elevated level of estradiol and testosterone (which is metabolized to estradiol) at puberty (Micevych et al., 1994; Holland et al., 1996). It was determined that a physiologic dose of estradiol (2 μg estradiol benzoate) given every 4 days was more effective at increasing CCK mRNA expression than bolus doses (50 μg EB) or continuously released estradiol from a 10-mm Silastic capsule (Micevych et al., 1996). This 2 μg dose of EB yields peak levels of circulating estradiol that mimic peak proestrous levels (Micevych et al., 1996; Asarian and Geary, 1999). In terms of lordosis, this estradiol treatment produces nonreceptive animals that can be reliably made maximally receptive with progesterone treatment (Sinchak and Micevych, 2001a). These estradiol + progesterone treated animals do not display a ramping of receptivity that was observed by Bloch et al. (1987) with estradiol only treatment (Bloch et al., 1987).

The facilitatory and inhibitory actions of peripherally administered CCK were explained on the basis of the distribution of CCK receptor subtypes. In the MPN, CCK-1 (also known as CCK-A) receptors mediate the facilitation of lordosis (Dornan et al., 1989; Holland et al., 1997). In the VMH, CCK-2 (also known as CCK-B) receptors mediate the inhibitory actions of CCK on lordosis (Bloch et al., 1989; Dornan et al., 1989). In addition, CCK binding in the VMH is modulated by estradiol and varies across the estrous cycle (Akesson et al., 1987). In the VMH, estradiol decreases the number of CCK binding, and this down-regulation is dependent on CCK innervation of the VMH (Schumacher et al., 1991). There are no CCK cell bodies in the VMH (Micevych et al., 1987) but there is a dense CCK terminal field (Fulwiler and Saper, 1985), suggesting that the decrease in receptors is secondary to the estradiol induced release of CCK. Such a rapid action of estradiol on CCK release from hypothalamic fragments was demonstrated in vitro (Micevych et al., 1988b). An hour of estradiol treatment dramatically augmented the potassium-evoked release of CCK. The time course of estradiol regulation of CCK binding levels in the VMH suggest that CCK is involved in the post-estradiol delay in response, 0–20 h after exogenous estradiol treatment of an OVX rat and diestrus in the estrous cycling rat, and in the cessation of lordosis following the period of receptivity. CCK receptor levels are low following estradiol, and return to basal levels by 48 h after estradiol treatment, and it is at this time that CCK is able to inhibit lordosis (Ulibarri and Micevych, 1993). Using microdialysis to measure extracellular levels of CCK in the medial preoptic area, in situ hybridization histochemistry for CCK mRNA and behavioral assessment of lordosis, Micevych and Sinchak (2001) demonstrated that the estradiol-induced synthesis of CCK precedes release, and that release remains basal for approximately 24 h after 2 μg EB treatment. During the time when CCK release is basal in the medial preoptic area, the rats do not respond to EB + progesterone treatment. When CCK release is elevated, the animals are receptive, suggesting a correlation between release of CCK in the medial preoptic area and receptivity. Support for this idea was evidenced by an antisense oligodeoxynuceotide study that demonstrated that blocking the expression of CCK-1 receptors in the medial preoptic area prevented the estradiol-induced lordosis reflex (Holland et al., 1997). When the treatments were stopped, and the animals were allowed to express CCK receptors, the rats previously treated with CCK-1 antisense deoxyoligonucleotide displayed full sexual receptivity.

The distributions of MOR and CCK largely overlap, suggesting that these two neuropeptides interact, except in the MPN (Micevych and Sinchak, 2001). MOR agonists inhibit the potassium-evoked release of CCK from hypothalamus (Micevych et al., 1982, 1985) and selective DOR agonists inhibit the estradiol-induced expression of CCK mRNA (Eckersell and Micevych, 1997), suggesting that activation of different EOP circuits regulate CCK expression or release. It is tempting to speculate that EOPs may regulate sexual receptivity by modulating the expression and release of CCK. Preliminary studies suggest that such an interaction may not be direct, since MOR protein and mRNA are not expressed in CCK-expressing neurons. A possible interaction may involve glutaminergic neurons based on the dependence of estradiol induced CCK expression on glutamate neurotransmission (unpublished observations).

10 Gonadotropin Releasing Hormone/Luteinizing Hormone Releasing Hormone

The distribution of GnRH cell bodies is scattered throughout the hypothalamus, depending on the species in question. In the rat and mouse, the majority of cell bodies is localized in the diagonal band of Broca and extends in a scattered band back into the medial preoptic area in an inverted "Y" when viewed in a horizontal plane (reviewed in Barry et al., 1985). GnRH was shown to facilitate lordosis in nonreceptive rats treated with estrone (Pfaff, 1973; Pfaff and Keiner, 1973; Moss and McCann, 1973, 1975). GnRH was effective in the medial preoptic area and the arcuate nucleus but not the lateral hypothalamic area (Moss and Foreman, 1976). GnRH was also effective in other species (Alderete et al., 1980; Kendrick and Dixson, 1985; Mackay-Sim and Rose, 1986; Maney et al., 1997; Schiml and Rissman, 2000). Whereas GnRH increased the lordosis reflex, it did not increase sexual preference of estradiol-treated rats (Dudley and Moss, 1985) suggesting that only sexual receptivity is regulated by GnRH. In addition to hypothalamic sites, GnRH stimulates lordosis after injection into the PAG (Sakuma and Pfaff, 1980b; Riskind and Moss, 1983).

Reports questioning GnRH's efficacy may be explained by the fact that GnRH may need to be metabolized to an active fragment. As early as 1983, Dudley et al. (1983) reported that the lordosis-facilitating activity of GnRH was contained in the carboxy-terminal half of the decapeptide. However, the Ac-$GnRH_{6-10}$ has not been demonstrated in the brain (see Wu et al., 2006 and references therein). More recent work has demonstrated that the GnRH peptide is metabolized by the zinc metalloendopeptidase EC 3.4.24.15 (EP 24-15), which cleaves the decapeptide in half to produce $LHRH_{1-5}$ (Camargo et al., 1982; Krause et al., 1982). This enzyme has been shown to have a hypothalamic distribution and activity (Healy and Orlowski, 1992; Wu et al., 1997). In a study comparing GnRH and the $LHRH_{1-5}$ fragment, both peptides facilitated lordosis (Wu et al., 2006). Immunoneutralization of EP24-15 resulted in the inhibition of the LHRH-facilitated lordosis but did not inhibit $LHRH_{1-5}$ facilitated lordosis. The GnRH antagonist, antide, was capable of inhibiting GnRH facilitated lordosis, without affecting $LHRH_{1-5}$ facilitated lordosis, Suggesting that both GnRH and $LHRH_{1-5}$ affect sexual receptivity.

11 Substance P and other Tachykinins

Substance P is often associated with nociception, but its distribution in the limbic-hypothalamic lordosis regulating circuit suggests that this neuropeptide is also involved in the regulation of sexual receptivity. Substance P is a member of the tachykinin family of peptides. Mammalian tachykinins comprise a family of peptides now known to contain the products of three genes, *Tac1*, *Tac2*, and *Tac4* in mouse and rat (recently reviewed in Page, 2004; Patacchini et al., 2004; Pennefather et al., 2004). In addition to substance P, the family includes neurokinin A and its N-terminally extended analogs, neuropeptide K, and neuropeptide γ, all derived from the *Tac1* gene, neurokinin B from the *Tac2* gene, and the hemokinins and endokinins derived from the *Tac4* gene. In terms of specificity, substance P, and the mouse and rat hemokinins, endokinin A and B, are tachykinin neurokinin1 (NK1) receptor-preferring (Zhang et al., 2000; Kurtz et al., 2002; Page, 2004; Patacchini et al., 2004). Neurokinin A and its N-terminally extended analogs are tachykinin NK2 receptor-preferring, and neurokinin B is tachykinin NK3 receptor-preferring (Maggi, 1995; reviewed in Pennefather et al., 2004). Estradiol regulates NK1 receptor mRNA in vitro (Villablanca and Hanley, 1997).

Most studies have focused on substance P in the VMH and its descending projections. The expression of substance P gene, β-preprotachykinin (PPT, *Tac1*) is increased by estradiol treatment (Brown et al., 1990; Priest et al., 1995). The estradiol action was likely to be directly on PPT mRNA expressing cells, since 43% of the substance P immunoreactive cells were estradiol receptive in the ventrolateral part of the VMH (Akesson and Micevych, 1988). Interestingly, a constant level of estradiol replacement with an estradiol-filled Silastic capsule did not increase β-PPT mRNA levels (Romano et al., 1989), suggesting that a more natural pulsatile estradiol replacement may be needed to drive gene expression rather than a static replacement paradigm. Further support for this idea was provided by findings that the percentage of progesterone receptor/substance P immunoreactive cells in the ventrolateral nucleus of the hypothalamus of guinea pigs, an area analogous to the VMH in rats, was significantly higher in the estradiol pulse-treated compared with estradiol capsule-implanted animals (Olster and Blaustein, 1992). The estradiol pulse treatment was also more effective at inducing the display of lordosis reflex in guinea pigs.

Substance P injection into the PAG facilitates estradiol-induced lordosis behavior in OVX rats (Dornan et al., 1987). These authors blocked the action of endogenous substance P with antisera directed against the peptide attenuating sexual receptivity. Retrograde tract tracing from the dorsal PAG revealed cell bodies throughout the extent of the VMH with the majority of the substance P cells that project to the PAG localized in the caudal two-thirds of the VMH. Approximately 17% of the substance P-immunoreactive cells of the ventrolateral part of the VMH projected to the dorsal PAG (Dornan et al., 1990). These results provide morphological evidence for a substance P projection from the VMH, where substance P expression is regulated by estradiol, to an area where substance P has been demonstrated to facilitate lordosis behavior.

More recently this idea of a lordosis-relevant substance P projection from the VMH to the PAG has been questioned. An analysis of pseudorabies virus-labeled neurons after injection into the lordosis-relevant, lumbar, epaxial muscles demonstrated that the vast majority of labeled neurons, both within and subjacent to the VMH, did not contain ERα (Daniels and Flanagan-Cato, 2000) the predominant form of ER in the VMH (Shughrue et al., 1997). Furthermore, these lordosis-relevant projection neurons and the population of ERα-containing neurons were largely segregated topographically. Consistent with these findings, previous studies have found estradiol binding within only a minority of the VMH projection neurons (Morrell and Pfaff, 1982; Akesson et al., 1994), whereas a significant population of substance P neurons are estradiol receptive. A significant population (28%), however, of neurons labeled with pseudorabies virus were shown to express NK1 receptors indicating that estradiol information from estradiol-receptive substance P neurons can influence VMH output neurons (Daniels et al., 2003) and suggest that the VMH is an important site for the substance P regulation of lordosis.

12 Corticotropin Releasing Factor

CRF has a CNS distribution that is more extensive than its role, regulating ACTH release, would suggest (Olschowka and Jacobowitz, 1983; Swanson et al., 1983). In the hypothalamus, the primary locus of CRF mRNA is the paraventricular nucleus (PVN); (Swanson et al., 1987; Matthews et al., 1991). Exogenous CRF powerfully inhibits the lordosis reflex in sexually receptive rats in the medial preoptic area and the PAG (Sirinathsinghji et al., 1983; Sirinathsinghji, 1985, 1986). This is consistent with the idea that severe stress negatively regulates the lordosis reflex, since the dose of CRF needed to inhibit lordosis also induced behaviors associated with stress. Note that not all stress is detrimental for reproduction. Psychosocial stress facilitates female sexual receptivity (Williams et al., 1992). Indeed, mild stress facilitates the expression of preproenkephalin (PPE) mRNA in the VMH (Sinchak et al., 2000), an event correlated with an enhanced lordosis reflex (Lauber et al., 1990). Consistent with this idea, estradiol and progesterone have been reported to reduce the negative actions of mild (restraint) stress on the lordosis reflex (White and Uphouse, 2004).

13 Galanin

Galanin is widely distributed throughout the limbic-hypothalamic, lordosis-regulating circuit (see Bedecs et al., 1995 for review) and the distribution of galanin-immunoreactive neurons is sexually dimorphic. Castrated males have more galanin cells in the MPN compared with OVX females (Bloch et al., 1993). Galanin immunoreactive neurons are estradiol receptive and galanin expression is increased by either estradiol or testosterone treatment (Bloch et al., 1992, 1993), suggesting that galanin may also mediate estradiol's actions on lordosis. Indeed, microinjections of galanin into the medial preoptic area facilitate female sexual receptivity in estradiol primed rats (Bloch et al., 1996, 1998).

More recently, a galanin-related peptide, galanin-like peptide (GALP) has been described. GALP is coded by a separate gene but has amino acid sequence homology with galanin (Ohtaki et al., 1999). Although, GALP stimulates male sexual behavior (Fraley et al., 2004), to date, this peptide has not been shown to modulate lordosis.

14 Serotonin

The initial observation that serotonin inhibits sexual receptivity (Meyerson, 1964) has been supported by over 40 years of intense research. In general, treatments that decrease monoamine concentrations increase the lordosis reflex and treatments that increase serotonin decrease lordosis. However, in the final analysis there is little evidence that endogenous serotonin increases the lordosis reflex calling into question the physiological relevance of these observations (Luine and Paden, 1982; Luine et al., 1984; Frankfurt et al., 1985).

As with most transmitters/modulators, serotonin acts at several receptor subtypes that can have diametrically opposite actions. Vis-a-vis lordosis, the 5-HT1, 5-HT2, and 5-HT3 families have been well studied (see Uphouse, 2000 for review). The 5HT2 and 5-HT3 receptors appear to increase the lordosis reflex, and 5-HT1 and specifically the 5-HT1_A receptors inhibit lordosis (Mendelson and Gorzalka, 1985; Hunter et al., 1985; Fernandez-Guasti et al., 1987; James et al., 1989; Uphouse et al., 1991, 1992; Aiello-Zaldivar et al., 1992; Hebert et al., 1995; Gonzalez et al., 1997; Wolf et al., 1998b, 1999; Jackson and Etgen, 2001; Uphouse and Wolf, 2004). Unlike other transmitters/modulators that have different actions on the lordosis reflex depending on the areas exposed, the 5-HT1 receptor is associated with the inhibition of lordosis in sites including the VMH, medial preoptic area, ventrolateral PAG, and medial raphe (Uphouse et al., 1992, 1993, 1996; Caldwell and Albers, 2002). The 5-HT1 is an autoreceptor located on the soma and dendrites of serotonin neurons that reduces neuronal activity and serotonin release (Sprouse and Aghajanian, 1987; Hjorth and Sharp, 1991; Adell et al., 1993; Hjorth et al., 1995). An interesting hypothesis is that the serotonergic system mediates the lordosis reflex in response to acute environmental cues (see Uphouse, 2000 for review). Under this scenario, the increase of serotonin following stress (Kirby et al., 1997; Maswood et al., 1998; Uphouse and Wolf, 2004) reduces lordosis through the 5-HT1_A receptors; activation of 5-HT-2_A/2_C may prevent disruption of the behavior in response to mild stress. Thus, dual activation of 5-HT1_A and 5-HT2_A/2_C enables a female to modify her behavior in response to environmental circumstances. Pharmacological activation of 5-HT1 or 5-HT2 receptors fits this schema (Hunter et al., 1985; Mendelson and Gorzalka, 1990; Gorzalka et al., 1997, 1998, 1999; Caldwell and Albers, 2002 see Mendelson, 1992 for review; but see Wolf et al., 1998a). However, manipulations of the endogenous serotonin system are not congruent with this hypothesis. The level of serotonin in the mediobasal hypothalamus declines near the time of sexual receptivity (Farmer et al., 1996; Maswood et al., 1999), suggesting that endogenous serotonin is tonically inhibitory but, estradiol, which stimulates lordosis, does not reduce ^{3}H-OH-DPAT binding (Clarke and Maayani, 1990; Mendelson and McEwen, 1991; Frankfurt et al., 1994; Gonzalez et al., 1997). Unlike the modest effects of estradiol on 5-HT1_A receptors, 5-HT2_A/2_C receptors are increased with estradiol treatment (Sumner and Fink, 1997; Sumner et al., 1999). Thus, a clear picture of the role of endogenous serotonin in the regulation of female sexual receptivity has not emerged.

15 Dopamine

The hypothalamus has a large number of DA neurons, rivaling the better known population in the substantia nigra. The best known DA populations are the tuberoinfudibular neurons located in the arcuate nucleus and the adjacent periventricular nucleus, the incertohypothalamic DA system clustered in the medial zona incerta, and a rostral AVPV group (Hokfelt et al., 1976; Zoli et al., 1993). Although this molecule and its receptors have been extensively studied in nigrostriatal neurons, its role in the regulation of female sexual receptivity remains complex and controversial. DA was implicated in regulating lordosis reflex in 1975 with the publication of the observation that apomorphine stimulated sexual receptivity (Hamburger-Bar and Rigter, 1975; Foreman and Moss, 1979). Studies from Gorski and Yanase (1981) to Fabre-Nys et al. (2003) conclude that there is a powerful interaction of DA with estradiol stimulation, such that DA facilitates lordosis reflex when administered at the beginning of estradiol priming and inhibits the behavior when given later. One attractive hypothesis for these results is that DA facilitates proceptive behaviors inhibiting receptivity at the same time (Everitt et al., 1974; Caggiula et al., 1979; Grierson et al., 1988; Wilson et al., 1991; Meisel et al., 1993; Mermelstein and Becker, 1995; Pfaus et al., 1995; Kohlert et al., 1997).

DA actions are mediated through a family of GPCR receptors, which is the subject of an excellent recent review (see Smythies, 2005). It is clear that the stimulatory actions of DA are mediated by the D1-like receptors, D1a and D1b (which are also known as D1 and D5) (Mani et al., 1996, 2001; Kudwa et al., 2005). The inhibition of lordosis behavior is mediated by D2-like receptors: D2, D3, D4 (Melis and Argiolas, 1995).

Several reports strongly suggest that DA mediates the hedonic aspects of sexual behavior (Fiorino and Phillips, 1999; Becker et al., 2001; Bradley and Meisel, 2001; Bradley et al., 2005). This action of DA is beyond the scope of the present review and involves the mesolimbic DA system and the nucleus accumbens (Fiorino and Phillips, 1999). The role of DA in activating progesterone receptors and sexual receptivity will be extensively discussed in the chapter by Blaustein and Mani in this volume (2006).

16 Norepinepherine

Norepinephrine is a biogenic amine that is synthesized from the amino acid tyrosine by three enzymes. The first, tyrosine hydroxylase, converts tyrosine to dihydroxyphenylalanine, which is then converted to DA by dihydroxyphenylalanine decarboxylase. Dopamine-β hydroxylase converts DA to norepinephrine, which is metabolized to epinephrine by phenylethanolamine *N*-methyl-transferase. Norepinephrine acts through the adrenergic family of receptors that are composed of two types of receptors, α and β-adrenoceptors. These two families are divided into a number of subtypes: α_{1a}-, α_{1b}-, α_2-, β_1-, β_2-, and β_3-adrenoceptors (Bylund et al., 1994; Alexander et al., 2004). The adrenergic receptors are members of the seven transmembrane GPCRs that are activated by either epinephrine or norepinephrine. The hypothalamus receives noradrenergic innervation from the ventral noradrenergic bundle (lateral tegmental tract) from the locus coeruleus.

Norepinephrine both inhibits and facilitates lordosis depending on the hormonal status of the female and the receptor subtype activated. Early studies demonstrated that drugs that decreased norepinephrine activity or transmission, through α-adrenergic receptors in general, facilitated lordosis in the estradiol-primed rats and those that increased norepinephrine transmission inhibited lordosis (Ahlenius et al., 1972a, b; Everitt et al., 1975; Herndon et al., 1978). Facilitation of lordosis is mediated by α_1-noradrenergic receptors, and inihibitory effects of norepinephrine appear to be through α_2-noradrenergice receptors. Systemic infusions of an α_2-adrenergic receptor agonist, clonidine, inhibited lordosis (Davis and Kohl, 1977). This effect appears to be localized to the medial preoptic area, since site-specific infusion of either epinephrine or clonidine (α_2-adrenergic receptor agonists) mimicked the effects of norepinephrine, whereas α_1- and β-adrenergic receptor agonists had no effect on lordosis (Caldwell and Clemens, 1986).

In contrast, a number of other studies indicate that endogenous norepinephrine is important for the facilitation of lordosis, but these effects are mediated through the α_1-adrenergic receptor. Bilateral lesions of the ventral noradrenergic bundle inhibited lordosis in cycling or estradiol + progesterone primed female rats, which was reversed by ventricular infusions of an α_1-adrenergic receptor agonist (Hansen et al., 1980, 1981; Davis et al., 1991). Although these lesions decreased norepinephrine concentrations in both the MPN and VMH, norepinephrine does not appear to act in either the MPN or VMH, since infusion of an α_1-adrenergic receptor agonist in either of these nuclei after ventral noradrenergic bundle did not restore lordosis (Davis et al., 1991). Furthermore, site-specific noradrenergic lesions of either the MPN or VMH did not affect lordosis in estradiol + progesterone treated rats (Davis et al., 1991). However, others indicate that the VMH is the site of action of norepinephrine for α_1-adrenergic receptor facilitation of lordosis (Fernandez-Guasti et al., 1985a, b; Etgen, 1990; Kow et al., 1992). Using more selective ligands, it was shown that the α_1-adrenergic receptors in the VMH facilitated the lordosis reflex (Etgen, 1990). Furthermore, α_{1b}-adrenergic receptor, which is upregulated by estradiol in the hypothalamus, is responsible for the noradrenergic facilitation (Petitti et al., 1992). The role of norepinephrine in the facilitation of lordosis is complex and involves the behavioral state of the animal. Although estradiol+ progesterone treatment that facilitates lordosis increases stimulus-evoked norepinephrine release in the VMH of anesthetized rats (Vathy and Etgen, 1988), norepinephrine release in the anterior region of the VMH was not increased in steroid-primed rats, unless she received an intromission from the stimulus male rat (Etgen and Morales, 2002). Moreover, norepinephrine levels in the VMH do not rise in steroid-primed females that are highly receptive with vaginal masks that block penile penetration (Etgen and Morales, 2002). These data indicate that basal norepinephrine release in the VMH is important for the facilitation of lordosis, and the increased release in response to vaginocervical stimulation of steroid primed female rats is a part of the steroid responsive pathway that conveys somatosensory, vaginocervical stimulatory information to be integrated in the hypothalamus that may be involved in the induction of pregnancy or psuedopregnant state (reviewed in

Flanagan et al., 1993; Pfaus et al., 1993; Tetel et al., 1993, 1994; Auger et al., 1996; Erskine and Hanrahan, 1997; Etgen and Morales, 2002; Erskine et al., 2004).

17 Acetylcholine

Acetylcholine is the endogenous neurotransmitter that is synthesized from choline and acetyl coenzyme A by a reaction that is catalyzed by choline acetyltransferase (ChAT). Acetylcholine acts through two types of receptors that are named for binding properties to muscarin (muscarinic: mAChR) and nicotine (nicotinic; nAChR; Alexander et al., 2004). The mAChR is a member of the G protein-coupled receptor family and has five distinct subtypes [M_1–M_5; (Caulfield and Birdsall, 1998)]. The nAChR is a member of the superfamily of "cys-loop" ligand-gated, ion channel receptors composed of five subunits that determine receptor properties including the channel open-time and desensitization (Karlin and Akabas, 1995; Itier and Bertrand, 2001; Quick and Lester, 2002; reviewed in Lindstrom, 2003).

Lindstrom and Meyerson (1967) initially demonstrated that activation of central cholinergic systems inhibited sexual receptivity in estradiol+ progesterone primed female rats. Subsequent studies demonstrated that the acute activation of cholinergic systems facilitates sexual receptivity in female rats and hamsters (Clemens et al., 1980, 1981; Dohanich and Clemens, 1981; Dohanich et al., 1990). Global antagonism or blocking choline reuptake of the muscarinic receptors inhibits lordosis in gonadally intact cycling rats or OVX, steroid-primed female rats (Clemens et al., 1980, 1981, 1989; Kaufman et al., 1988; Menard and Dohanich, 1989). Although early studies suggested that the medial preoptic area was important for muscarinic facilitation of lordosis (Clemens et al., 1980, 1981; Dohanich and Clemens, 1981), these weak effects were demonstrated to be due to the leakage of drugs into the ventricular system and interacting with cholinergic receptors elsewhere in the brain (Dohanich et al., 1984). Further evidence contradicting a role of mAChR in the medial preoptic area was a decrease in the receptor binding by estradiol (Dohanich et al., 1982). Based on mAChR distribution and estradiol regulation of mAChR binding (Rainbow et al., 1980b; Dohanich et al., 1982), ACh actions on lordosis were localized to the VMH and PAG (Dohanich and Clemens, 1981; Dohanich et al., 1984; Richmond and Clemens, 1986; Kaufman et al., 1988). The PAG has estradiol-regulated mAChR (Meyers and Clemens, 1985) and an intensive population of cholinergic neurons providing a local network that may act to augment VMH cholinergic effects on lordosis (Satoh et al., 1983).

The nature of the mAChR subtype facilitating lordosis has not been elucidated. Pharmacological manipulations, have ruled out the M1 mAChR subtype (Kow and Pfaff, 1985; Dohanich et al., 1991). The M3 receptor subtype, located in the hypothalamus in higher concentrations than other mAChR subtypes (Wall et al., 1991; Zubieta and Frey, 1993), appears to be at least partially responsible for the steroid facilitation of lordosis, but it does not account for the entire range of ACh actions in the VMH (Kow et al., 1995). The roles of the other mAChR subtypes present in the hypothalamus require further investigation (Ehlert and Tran, 1990; Vilaro et al., 1990).

Global activation of the nicotinic receptor system by nicotine produces moderate facilitation of lordosis in estradiol-primed females, which could be reversed by pretreatment with a nicotinic receptor antagonist (Fuxe et al., 1977; Weaver and Clemens, 1987). However, icv injection of mecamylamine, a nicotinic antagonist, did not affect lordosis in estradiol+ progesterone primed animals, indicating that nicotinic systems are not critically important for the acute regulation of sexual receptivity (Weaver and Clemens, 1987). Thus, the main effects that acetylcholine has on sexual receptivity are through the activation of muscarinic receptors in the VMH and PAG but the mAChR subtype has not been conclusively characterized.

18 Glutamate

In the CNS, the major excitatory transmitter is glutamate. Thus, it is not surprising that glutamate and its receptors are involved in the regulation of the lordosis reflex. Glutamate receptors are broadly divided into ionotropic and metabotropic groups (Seeburg, 1993; Brann, 1995). Ionotropic glutamate receptors (GluR): are segregated based on their pharmacology into AMPA, kainate, and *N*-methyl-*D*-aspartic-acid (NMDA) type receptors, whose subunits are coded by six gene families (Dingledine et al., 1999). There are eight metabotropic glutamate receptors (mGluR; Baskys et al., 2005). The metabotropic glutamate receptors are a part of the GPCR superfamily. A further segregation is possible based on pharmacology, signal transduction, and sequence homology into group I: mGluR1 and mGluR5 and their splice variants; and group II: mGluR2 and mGluR3. The group III receptors are mGluR4, mGluR6, mGluR8 and their splice variants (Baskys et al., 2005). A role for glutamenergic transmission was proposed in 1980 by the results of Rodriguez-Sierra et al. (1980). These authors reported that neonatal administration of *L*-monosodium glutamate (MSG) prevented the estradiol+ progesterone induced lordosis reflex in the adult. Other workers have suggested that the glutamate effect is mediated through the release of GnRH (Hsu et al., 1993; Bailey et al., 2006). More recently, icv administration of a selective NMDA receptor agonist facilitated lordosis, and the antagonist 7-amino-phosphonoheptanoic acid (AP-7), blocked the effect (Gargiulo et al., 1992; Gargiulo and Donoso, 1995). As with a number of neuropeptides, site of action is important to understand glutamate's role in regulating lordosis. Microinjections of NMDA into the mediobasal hypothalamus significantly reduced lordosis in estradiol+ progesterone primed rats (McCarthy et al., 1991a). The NMDA antagonist 5-amino-phosphonoheptanoic acid (AP-5) attenuated lordosis following microinjection into the medial preoptic area. These results suggest that glutamate facilitates lordosis in the medial preoptic area and inhibits lordosis in the mediobasal hypothalamus, suggesting a complex microcircuitry regulating sexual receptivity in these regions. How glutamate interacts with other neurotransmitters or neuropeptides in these regions that control sexual receptivity is largely unknown. An understanding of chemical signaling regulating sexual receptivity will certainly need to include this information.

19 γ-Amino Butyric Acid

Whereas glutamate is the primary excitatory transmitter, GABA is the primary inhibitory transmitter in the CNS. Initially, GABA was thought to inhibit the lordosis reflex based on studies using systemic injections of GABA, or $GABA_A$, or $GABA_B$ agonists (Qureshi et al., 1988; Agmo et al., 1989). Like opioids, CCK and glutamate, eventually GABA was shown to have different actions in the medial preoptic area compared with the mediobasal hypothalamus and these effects are, as expected, opposite to those described for glutamate. Exogenous GABA inhibits lordosis in the medial preoptic area and facilitates lordosis in the mediobasal hypothalamus in estradiol or estradiol+ progesterone primed rats (McCarthy et al., 1990). Moreover, a good correlation exists between endogenous levels of GABA and sexual receptivity: elevated GABA levels in the hypothalamus and lower levels in the preoptic area are found in receptive females (McCarthy et al., 1991). Microinfusion of the $GABA_A$ antagonist, bicuculline, into the PAG of steroid-primed female rats resulted in a significant decrease in lordosis and proceptive behaviors and muscimol, a $GABA_A$ agonist, increased lordosis (McCarthy et al., 1991b; Frye and DeBold, 1992; Frye et al., 1993). That endogenous GABA is involved in regulating sexual receptivity was further evidenced by blocking the expression of the GABA synthetic enzymes glutamic acid decarboxylase, GAD 67 and GAD 65 (McCarthy, 1994b). These studies recapitulated the idea that increased GABA activity in the mediobasal hypothalamus and PAG facilitate lordosis behavior, since infusion of antisense oligodeoxynucleotide to GAD 67 significantly reduced lordosis for 24 h. The antisense oligodeoxynucleotide to GAD 65 reduced the behavior for 48 h. By 5 days following the antisense oligodeoxynucleotide treatments, the lordosis quotients had returned to normal.

20 Other Mediators

20.1 Prostaglandins

The role of prostaglandins in CNS regulation of female sexual receptivity is not clear. Implants of prostaglandin E2 (PGE2) into the medial preoptic area-anterior hypothalamic area facilitate sexual receptivity in OVX, estradiol-primed rats (Hall and Luttge, 1977, 1978). Interestingly, PGE2 was also reported to induce lordosis in OVX rats without steroid priming (Rodriguez-Sierra and Komisaruk, 1978). In more recent studies, the actions of prostaglandins were shown to be dependent on activation of the progesterone receptor since the antiprogesterone receptor RU486 blocked the lordosis stimulating action (Beyer et al., 1997). In guinea pigs, however, prostaglandins E1 and F2α inhibited estradiol+progesterone induced receptivity (Irving et al., 1981), which appears to be a species difference (Marrone et al., 1979). More recently, the induction of lordosis with estradiol and progesterone metabolites (i.e., 5α-dihydroprogesterone (αDHP), allopregnanolone, epipregnanolone) was shown to be dependent on prostaglandin synthesis (Beyer et al., 1999).

20.2 Insulin-Like Growth Factor-1

IGF-1 is a protein related to insulin (Rinderknecht and Humbel, 1978) and an important mediator of growth hormone action (Daughaday and Rotwein, 1989; Jones and Clemmons, 1995). Although IGF-1 is predominantly produced in the liver, and can cross the blood–brain barrier, IGF-1 is also produced by cells in the brain (Bondy and Cheng, 2004; Kaur et al., 2006). IGF-1 has powerful effects on brain development and function (Lee et al., 1996; Fushimi and Shirabe, 2004; Xing et al., 2006; reviewed in Aberg et al., 2006). Estradiol regulates circulating (Zapf et al., 1995; Rempel and Clapper, 2002; Hilleson-Gayne and Clapper, 2005) and brain levels (Shingo and Kito, 2003; Mendez et al., 2005; Darnaudery et al., 2006) of IGF-1. Estradiol increases the expression of IGF-1, its cognate receptor and IGF-1 binding proteins (Dickson et al., 1986; Pons and Torres-Aleman, 1993; Wimalasena et al., 1993; Sahlin et al., 1994). It has been proposed that estradiol actions in the CNS may involve interactions with IGF-1 (Duenas et al., 1996; Garcia-Segura et al., 1996; Fernandez-Galaz et al., 1999; Quesada and Etgen, 2001, 2002; Quesada and Micevych, 2004).

Apropos of sexual receptivity, icv administration of IGF-1 facilitated the lordosis reflex without estradiol priming (Apostolakis et al., 2000). Acute blockade of IGF-1 receptors with a selective competitive, antagonist, JB-1 (Pietrzkowski et al., 1992), did not reduce lordosis, but treatment with JB-1 during estradiol priming attenuated lordosis quotient and lordosis score (a measure of the "quality of lordosis") (Quesada and Etgen, 2002). It was hypothesized that the estradiol-IGF-1 interaction modulates α_1-adrenergic signaling based on the results that IGF-1 potentiates α_1-adrenergic receptor signaling in hypothalamic slices from estradiol-primed females (Quesada and Etgen, 2001). Blockade of IGF-1 receptor during estradiol priming prevents this potentiation of receptor signaling and expression of α_{1B}-adrenergic receptors in the hypothalamus. Further evidence for this hypothesis is that the attenuation of estradiol-induced lordosis reflex with JB-1 was reversed by 8-bromo-cGMP, which stimulates an intracellular signaling cascade that facilitates lordosis (Chu and Etgen, 1997; Chu et al., 1999).

20.3 Leptin

A hormone produced by adipose tissue that has been implicated in the regulation of food intake and energy regulation and puberty facilitates the lordosis reflex in ad libitum fed hamsters and increased food-deprivation induced attenuation of lordosis (Campfield et al., 1996; Wade et al., 1997). On the other hand, leptin did not affect female sexual receptivity in either lean or obese Zucker (*fa/fa*) rats (Fox and Olster, 2000). However, these authors found a reduction in proceptive behaviors in both the lean and obese Zucker rats. In Syrian hamsters, that are made anestrus due to fasting, leptin restored estrous cyclicity (Schneider et al., 1998). This restoration was due to leptin stimulation of metabolic fuel oxidation. These data provide an important neurochemical link between sexual behavior and metabolic state.

20.4 Nitric Oxide

NO is a gaseous extracellular messenger that increases cGMP levels. In addition to an extensive distribution in peripheral tissues, NO is widely distributed in the brain (McCall and Vallance, 1992). NO is formed by the conversion of arginine to citrulline by nitric oxide synthase (NOS). Estradiol which increases the expression of NOS in the hypothalamus and areas involved with lordosis, suggesting that estradiol promotes NO generation and cGMP accumulation (Okamura et al., 1994; Ceccatelli et al., 1996; Pu et al., 1998; Rachman et al., 1998). Additionally, progesterone alone was shown to increase guanylyl cyclase activity in peripheral tissues, suggesting an NO-mediated action (Vesely, 1979; Vesely and Hill, 1980). An interesting correlation: cGMP levels are highest in the afternoon of proestrus (Kimura et al., 1980). Thus, it was not surprising that manipulations that increase brain NO and cGMP levels facilitate lordosis behavior (Fernandez-Guasti et al., 1983; Chu and Etgen, 1997; Chu et al., 1999; Gonzalez-Flores et al., 2004).

It has been hypothesized that estradiol induced NO acts by the release of GnRH (LHRH) to facilitate the lordosis reflex (Mani et al., 1994b). Blocking NOS prevented estradiol+progesterone induced lordosis and this blockade was reversed by the treatment of the rats with GnRH. NO has also been reported to mediate the actions of a number of extracellular messengers on GnRH (LHRH) secretion including: glutamate (Mahachoklertwattana et al., 1994; Rettori et al., 1994), leptin (Yu et al., 1997), oxytocin (Rettori et al., 1997), and α1-adrenoceptors (Canteros et al., 1995, 1996; Kamat et al., 1995), suggesting a point of convergence of these lordosis-relevant circuits.

20.5 Thyroid Releasing Hormone

Thyroid releasing hormone has not been well studied; however reports indicate that this neuropeptide stimulates lordosis following injections into the PAG, and stimulates lumbosacral spinal motoneurons that may be involved in innervating epaxial musculature needed for the lordosis reflex (Ogawa et al., 1992; Kow and Pfaff, 1996).

20.6 Pituitary Adenylate Cyclase-Activating Polypeptide

PACAP is a member of the secretin/glucagon/vasoactive intestinal peptide (VIP) family that, as its name implies, increases the levels of cAMP (cyclic adenosine monophosphate; reviewed in Arimura and Shioda, 1995; and Sherwood et al., 2000). Localization of PACAP in the CNS and in the VMH (Ishihara et al., 1992; Jaworski and Proctor, 2000) suggested that this peptide may mediate steroid actions on the lordosis reflex. Indeed, the administration of PACAP into the VMH facilitated estradiol-induced lordosis (Apostolakis et al., 2004, 2005). These studies also demonstrated that these actions are mediated through PACAP activation of the selective PAC1 receptor (Arimura and Shioda, 1995). The facilitation of lordosis was dependent on estradiol-induced progesterone receptors and the cAMP/protein kinase A (PKA)/DARPP32 pathway (Apostolakis et al., 2005). Interestingly, PACAP did not rescue sexual receptivity in progesterone receptor knock out mice (PRKO; Apostolakis et al., 2004) suggesting that PACAP may be involved in modulating lordosis, but is not the final signaling molecule in this circuit.

21 Conclusions

As is evidenced by this review, much has been accomplished describing the action of a constellation of neurotransmitters and neuropeptides in regulating sexual receptivity. This has been paralleled by a tremendous progress in defining steroid action on the expression and release of endogenous extracellular messengers that regulate the lordosis reflex. One common theme that has emerged from these studies is that neurotransmitters and neuropeptides "substitute" for progesterone; however, what is usually meant by these statements is that neurochemicals facilitate estradiol-induced lordosis behavior. All too often, the other critical effects of progesterone, the termination of lordosis and the resetting of the lordosis regulating circuitry are ignored. In order to truly appreciate the action of progesterone, we need to develop an understanding of both limbs of progesterone action.

Although there is universal agreement that the lordosis reflex is dependent on estradiol action in the brain, less progress has been made in understanding which of the myriad of estradiol actions are critical for the expression of lordosis. In our opinion, this is due to the global nature of the estradiol action. Investigators have developed a catalog of estrogen actions on gene expression, transmitter and peptide release and intracellular signaling. Indeed, many of these studies have carefully demonstrated the temporal and spatial characteristics of estradiol action on the neurochemistry of lordosis regulating circuits.

Despite these gains, in some sense, we are still very far from an integrated view of the neurochemistry controlling the lordosis reflex. This review has concentrated on the neurochemical and pharmacological aspects of circuits that regulate sexual receptivity, and it is clear that the progress is substantial. These studies have provided an excellent underpinning for the next series of investigations. To date, studies have established a framework of neurochemical puzzle pieces and have begun to arrange the pieces into a neuroanatomical substrate providing a general picture of the circuits involved in regulating lordosis behavior. Within the broad picture, however, many parts of the puzzle are still not firmly in place.

Several guiding principles have emerged from the past work: (1) the lordosis regulating circuit is organized into nodes (the most prominent are the MPN, VMH, and PAG) that integrate hormonal, sensory, behavioral, and metabolic information; and (2) although all the nodes are exposed to estrogens at the same time, they are activated in-series to produce sexual receptive behavior that is coordinated with ovulation to maximize the potential for reproduction. For example, it is clear that the MPN is activated early by estradiol and is inhibitory to lordosis. The VMH appears to "come on-line" after the MPN and facilitates lordosis. The problems are within the nodes. What is the microcircuitry that dictates the output from the nodes? Are neurochemical circuits arranged in series, in parallel, or a mixture of the two relationships? Can inputs carrying information from different modalities substitute for others? For example, if the peripheral steroid signal is weak can metabolic and sensory information that food is plentiful and the days are long activate the circuit and elicit sexual receptivity? Or vice versa, despite strong hormonal input, in the face of imminent threats can unfavorable sensory input prevent the display of sexual receptivity. Indeed, there is evidence that this may happen. What is not known is how these circuits are organized to produce this effect. Working out microcircuitry underlying a global behavior such as the lordosis reflex is extremely difficult and is not easily approachable with standard tract tracing/immunohistochemical techniques. One method that may aid elucidation of microcircuitry relies on membrane receptor internalization. Agonist-induced receptor internalization is a universal receptor phenomenon, which can be used to identify activated neurons with neurochemical specificity. Indeed, receptor internalization has been used to begin to determine the neurochemical signature of estradiol and estradiol+progesterone stimulation in lordosis regulating circuits. As more reagents become available, this technique in combination with other techniques holds great promise for revealing microcircuitry within lordosis regulating nodes, and may possibly shed light on the basic problem of the difference between estradiol- only and estradiol+progesterone facilitated lordosis, which is another unresolved issue.

Another issue that must be resolved to elucidate CNS regulation of sexual receptivity is the relationship between neurochemical actions and the timing of sexual receptivity. Many of the studies reviewed here examined short-term effects of transmitters and peptides. Often this short-term regulation is only a part of the story, and may not even be predictive of the long-term consequences for regulation of lordosis. For example, opioid actions at the MOP acutely inhibit lordosis but removing MOP action prevents the full display of sexually receptive behavior, indicating that inhibition may be a necessary component of the circuitry regulating the lordosis reflex. Thus, to develop a comprehensive understanding of both the circuitry and the interaction of transmitters and peptides (microcircuitry) that regulate sexual receptivity requires further concerted effort to match the tremendous progress made to date.

Acknowledgments

The authors thank Ms. Amy Christensen for her help in the preparation of this chapter. We wish to acknowledge the support of NIH grants NS039495, DA13185, and HD042635 that funded some of the research present in this chapter.

References

Aberg ND, Brywe KG, Isgaard J. 2006. Aspects of growth hormone and insulin-like growth factor-I related to neuroprotection, regeneration, and functional plasticity in the adult brain. Scientific WorldJournal 6: 53-80.

Acosta-Martinez M, Etgen AM. 2002a. Activation of mu-opioid receptors inhibits lordosis behavior in estrogen and progesterone-primed female rats. Horm Behav 41: 88-100.

Acosta-Martinez M, Etgen AM. 2002b. The role of delta-opioid receptors in estrogen facilitation of lordosis behavior. Behav Brain Res 136: 93-102.

Adell A, Carceller A, Artigas F. 1993. In vivo brain dialysis study of the somatodendritic release of serotonin in the Raphe nuclei of the rat: effects of 8-hydroxy-2-(di-n-propylamino) tetralin. J Neurochem 60: 1673-1681.

Adrian TE, Allen JM, Bloom SR, Ghatei MA, Rossor MN, et al. 1983. Neuropeptide Y distribution in human brain. Nature 306: 584-586.

Agmo A, Soria P, Paredes R. 1989. GABAergic drugs and lordosis behavior in the female rat. Horm Behav 23: 368-380.

Ahlenius S, Engel J, Eriksson H, Modigh K, Sodersten P. 1972a. Importance of central catecholamines in the mediation of lordosis behaviour in ovariectomized rats treated with estrogen and inhibitors of monoamine synthesis. J Neural Transm 33: 247-255.

Ahlenius S, Engel J, Eriksson H, Sodersten P. 1972b. Effects of tetrabenazine on lordosis behaviour and on brain monoamines in the female rat. J Neural Transm 33: 155-162.

Aiello-Zaldivar M, Luine V, Frankfurt M. 1992. 5, 7-DHT facilitated lordosis: effects of 5-HT agonists. Neuroreport 3: 542-544.

Akesson TR, Micevych PE. 1988. Estrogen concentration by substance P-immunoreactive neurons in the medial basal hypothalamus of the female rat. J Neurosci Res 19: 412-419.

Akesson TR, Ulibarri C, Truitt S. 1994. Divergent axon collaterals originate in the estrogen receptive ventromedial nucleus of hypothalamus in the rat. J Neurobiol 25: 406-414.

Akesson TR, Mantyh PW, Manthyh CR, Matt DW, Micevych PE. 1987. Estrous cyclicity of 125I-cholecystokinin octapeptide binding in the ventromedial hypothalamic nucleus. Neuroendocrinology 45: 257-262.

Alderete MR, Tokarz RR, Crews D. 1980. Luteinizing hormone-releasing hormone and thyrotropin-releasing hormone induction of female sexual receptivity in the lizard, Anolis carolinensis. Neuroendocrinology 30: 200-205.

Alexander SP, Mathie A, Peters JA. 2004. Guide to receptors and channels, 1st edn. Br J Pharmacol 141(Suppl 1): S1-126.

Amico JA, Crowley RS, Insel TR, Thomas A, O'Keefe JA. 1995. Effect of gonadal steroids upon hypothalamic oxytocin expression. Adv Exp Med Biol 395: 23-35.

Apostolakis EM, Garai J, Lohmann JE, Clark JH, O'Malley BW. 2000. Epidermal growth factor activates reproductive behavior independent of ovarian steroids in female rodents. Mol Endocrinol 14: 1086-1098.

Apostolakis EM, Lanz R, O'Malley BW. 2004. Pituitary adenylate cyclase-activating peptide: a pivotal modulator of steroid-induced reproductive behavior in female rodents. Mol Endocrinol 18: 173-183.

Apostolakis EM, Riherd DN, O'Malley BW. 2005. PAC1 receptors mediate pituitary adenylate cyclase-activating polypeptide- and progesterone-facilitated receptivity in female rats. Mol Endocrinol 19: 2798-2811.

Ardati A, Henningsen RA, Higelin J, Reinscheid RK, Civelli O, et al. 1997. Interaction of [H-3]orphanin FQ and I-125-Tyr14-orphanin FQ with the orphanin FQ receptor: Kinetics and modulation by cations and guanine nucleotides. Mol Pharmacol 51: 816-824.

Arimura A, Shioda S. 1995. Pituitary adenylate cyclase activating polypeptide (PACAP) and its receptors: neuroendocrine and endocrine interaction. Front Neuroendocrinol 16: 53-88.

Arletti R, Bertolini A. 1985 Oxytocin stimulates lordosis behavior in female rats. Neuropeptides 6: 247-253.

Asarian L, Geary N. 1999. Cyclic estradiol treatment phasically potentiates endogenous cholecystokinin's satiating action in ovariectomized rats. Peptides 20: 445-450.

Auger AP, Moffatt CA, Blaustein JD. 1996. Reproductively-relevant stimuli induce Fos-immunoreactivity within progestin receptor-containing neurons in localized regions of female rat forebrain. J Neuroendocrinol 8: 831-838.

Bailey JD, Centers A, Jennes L. 2006. Expression of AMPA receptor subunits (GluR1-GluR4) in gonadotrophin-releasing hormone neurones of young and middle-aged persistently oestrous rats during the steroid-induced luteinising hormone surge. J Neuroendocrinol 18: 1-12.

Bale TL, Dorsa DM. 1995a. Sex differences in and effects of estrogen on oxytocin receptor messenger ribonucleic acid expression in the ventromedial hypothalamus. Endocrinology 136: 27-32.

Bale TL, Dorsa DM. 1995b. Regulation of oxytocin receptor messenger ribonucleic acid in the ventromedial hypothalamus by testosterone and its metabolites. Endocrinology 136: 5135-5138.

Bale TL, Dorsa DM. 1997. Cloning, novel promoter sequence, and estrogen regulation of a rat oxytocin receptor gene. Endocrinology 138: 1151-1158.

Bale TL, Dorsa DM. 1998. NGF, cyclic AMP, and phorbol esters regulate oxytocin receptor gene transcription in SK-N-SH and MCF7 cells. Brain Res Mol Brain Res 53: 130-137.

Bale TL, Dorsa DM, Johnston CA. 1995a. Oxytocin receptor mRNA expression in the ventromedial hypothalamus during the estrous cycle. J Neurosci 15: 5058-5064.

Bale TL, Pedersen CA, Dorsa DM. 1995b. CNS oxytocin receptor mRNA expression and regulation by gonadal steroids. Adv Exp Med Biol 395: 269-280.

Bale TL, Davis AM, Auger AP, Dorsa DM, McCarthy MM. 2001. CNS region-specific oxytocin receptor expression: importance in regulation of anxiety and sex behavior. J Neurosci 21: 2546-2552.

Barry J, Hoffman G, Wray S. 1985. LHRH-containing systems. Handbook of Chemical Neuroanatomy, Vol. 4. Bjorklund A. Hokfelt T. Eds. New York: Elsevier; pp. 162-215.

Baskin DG, Norwood BJ, Schwartz MW, Koerker DJ. 1995. Estradiol inhibits the increase of hypothalamic neuropeptide Y messenger ribonucleic acid expression induced by weight loss in ovariectomized rats. Endocrinology 136: 5547-5554.

Baskys A, Fang L, Bayazitov I. 2005. Activation of neuroprotective pathways by metabotropic group I glutamate receptors: a potential target for drug discovery? Ann N Y Acad Sci 1053: 55-73.

Beach F. 1948. Hormones Behavior. New York: Paul & Hoeber.

Beach FA. 1976. Sexual attractivity, proceptivity, and receptivity in female mammals. Horm Behav 7: 105-138.

Beach FA, Leboeuf BJ. 1967. Coital behavior in dogs. 1. Preferential mating in the bitch. Anim Behav 15: 546-558.

Becker JB, Rudick CN, Jenkins WJ. 2001. The role of dopamine in the nucleus accumbens and striatum during sexual behavior in the female rat. J Neurosci 21: 3236-3241.

Bedecs K, Berthold M, Bartfai T. 1995. Galanin-10 years with a neuroendocrine peptide. Int J Biochem Cell Biol 27: 337-349.

Benelli A, Poggioli R, Luppi P, Ruini L, Bertolini A, et al. 1994. Oxytocin enhances, and oxytocin antagonism decreases, sexual receptivity in intact female rats. Neuropeptides 27: 245-250.

Beyer C, Gonzalez-Flores O, Gonzalez-Mariscal G. 1997. Progesterone receptor participates in the stimulatory effect of LHRH, prostaglandin E2, and cyclic AMP on lordosis and proceptive behaviours in rats. J Neuroendocrinol 9: 609-614.

Beyer C, Gonzalez-Flores O, Ramirez-Orduna JM, Gonzalez-Mariscal G. 1999. Indomethacin inhibits lordosis induced by ring A-reduced progestins: possible role of 3alpha-oxoreduction in progestin-facilitated lordosis. Horm Behav 35: 1-8.

Blaustein JD, Erskine MS. 2002. Feminine sexual behavior: cellular integration of hormonal and afferent information in the rodent forebrain. Hormones, Brain and Behavior, Vol 1. Pfaff DW, Ed. New York: Academic Press; pp. 139-214.

Blaustein JD, Mani SK. 2006. Feminine sexual behavior from neuroendocrine and molecular neuroendocrine perspectives. Springer; in press. Handbook of Neurochemistry and Molecular Biology. Vol 21. Blaustein JD, Ed. Behavioral Neurochemistry, Neuroendocrinology and Molecular Neurobiology (Latha, A, Series Editor), New York: Springer; in Press.

Blaustein JD, Finkbohner R, Delville Y. 1987. Estrogen-induced and estrogen-facilitated female rat sexual behavior is not mediated by progestin receptors. Neuroendocrinology 45: 152-159.

Bloch GJ, Butler PC, Kohlert JG. 1996. Galanin microinjected into the medial preoptic nucleus facilitates female- and male-typical sexual behaviors in the female rat. Physiol Behav 59: 1147-1154.

Bloch GJ, Eckersell C, Mills R. 1993. Distribution of galanin-immunoreactive cells within sexually dimorphic components of the medial preoptic area of the male and female rat. Brain Res 620: 259-268.

Bloch GJ, Babcock AM, Gorski RA, Micevych PE. 1987. Cholecystokinin stimulates and inhibits lordosis behavior in female rats. Physiol Behav 39: 217-224.

Bloch GJ, Babcock AM, Gorski RA, Micevych PE. 1988. Effects of cholecystokinin on male copulatory behavior and lordosis behavior in male rats. Physiol Behav 43: 351-357.

Bloch GJ, Butler PC, Eckersell CB, Mills RH. 1998. Gonadal steroid-dependent GAL-IR cells within the medial preoptic nucleus (MPN) and the stimulatory effects of GAL within the MPN on sexual behaviors. Ann N Y Acad Sci 863: 188-205.

Bloch GJ, Kurth SM, Akesson TR, Micevych PE. 1992. Estrogen-concentrating cells within cell groups of the medial preoptic area: sex differences and co-localization with galanin-immunoreactive cells. Brain Res 595: 301-308.

Bloch GJ, Dornan WA, Babcock AM, Gorski RA, Micevych PE. 1989. Effects of site-specific CNS microinjection of cholecystokinin on lordosis behavior in the male rat. Physiol Behav 46: 725-730.

Boling FL, Blandau RJ. 1939. The estrogen-progesterone induction of mating responses in the spayed female rat. Endocrinology 25: 359-364.

Bondy CA, Cheng CM. 2004. Signaling by insulin-like growth factor 1 in brain. Eur J Pharmacol 490: 25-31.

Bouret S, Prevot V, Croix D, Jegou S, Vaudry H, et al. 1999. Mu-opioid receptor mRNA expression in proopiomelanocortin neurons of the rat arcuate nucleus. Brain Res Mol Brain Res 70: 155-158.

Bradley KC, Meisel RL. 2001. Sexual behavior induction of c-Fos in the nucleus accumbens and amphetamine-stimulated locomotor activity are sensitized by previous sexual experience in female Syrian hamsters. J Neurosci 21: 2123-2130.

Bradley KC, Haas AR, Meisel RL. 2005. 6-Hydroxydopamine lesions in female hamsters (*Mesocricetus auratus*) abolish the sensitized effects of sexual experience on copulatory interactions with males. Behav Neurosci 119: 224-232.

Brann DW. 1995. Glutamate: a major excitatory transmitter in neuroendocrine regulation. Neuroendocrinology 61: 213-225.

Brit J Pharmacol 2003. Nomenclature guidelines for authors. Br J Pharmacol 138: 725–729.

Brom GM, Schwartz NB. 1968. Acute changes in the estrous cycle following ovariectomy in the golden hamster. Neuroendocrinology 3: 366-377.

Brown ER, Harlan RE, Krause JE. 1990. Gonadal steroid regulation of substance P (SP) and SP-encoding messenger ribonucleic acids in the rat anterior pituitary and hypothalamus. Endocrinology 126: 330-340.

Bueno J, Pfaff DW. 1976. Single unit recording in hypothalamus and preoptic area of estrogen-treated and untreated ovariectomized female rats. Brain Res 101: 67-78.

Bunzow JR, Zhang G, Bouvier C, Saez C, Ronnekleiv OK, et al. 1995. Characterization and distribution of a cloned rat mu-opioid receptor. J Neurochem 64: 14-24.

Butler PC, Mills RH, Bloch GJ. 2001. Inhibition of lordosis behavior in male and female rats by androgens and progesterone. Horm Behav 40: 384-395.

Butour JL, Moisand C, Mazarguil H, Mollereau C, Meunier JC. 1997. Recognition and activation of the opioid receptor-like ORL 1 receptor by nociceptin, nociceptin analogs and opioids. Eur J Pharmacol 321: 97-103.

Bylund DB, Eikenberg DC, Hieble JP, Langer SZ, Lefkowitz RJ, et al. 1994. International Union of Pharmacology nomenclature of adrenoceptors. Pharmacol Rev 46: 121-136.

Caggiula AR, Herndon JG Scanlon R, Jr, Greenstone D, Bradshaw W, et al. 1979. Dissociation of active from immobility components of sexual behavior in female rats by central 6-hydroxydopamine: implications for CA involvement in sexual behavior and sensorimotor responsiveness. Brain Res 172: 505-520.

Caldwell HK, Albers HE. 2002. The effects of serotonin agonists on the hypothalamic regulation of sexual receptivity in Syrian hamsters. Horm Behav 42: 78-84.

Caldwell JD, Clemens LG. 1986. Norepinephrine infusions into the medial preoptic area inhibit lordosis behavior. Pharmacol Biochem Behav 24: 1015-1023.

Caldwell JD, Moe BD. 1999. Conjugated estradiol increases female sexual receptivity in response to oxytocin infused into the medial preoptic area and medial basal hypothalamus. Horm Behav 35: 38-46.

CaldwellJD, Pedersen CA, Prange A Jr. 1984. Oxytocin facilitates sexual behavior in estrogen-treated ovariectomized rats. J Steroid Biochem 20(6B): 1510.

Caldwell JD, Prange AJ Jr, Pedersen CA. 1986. Oxytocin facilitates the sexual receptivity of estrogen-treated female rats. Neuropeptides 7: 175-189.

Caldwell JD, Walker CH, Pedersen CA, Mason GA. 1992. Effects of steroids and mating on central oxytocin receptors. Ann N Y Acad Sci 652: 433-436.

Caldwell JD, Barakat AS, Smith DD, Hruby VJ, Pedersen CA. 1990. A uterotonic antagonist blocks the oxytocin-induced facilitation of female sexual receptivity. Brain Res 512: 291-296.

Caldwell JD, Johns JM, Faggin BM, Senger MA, Pedersen CA. 1994. Infusion of an oxytocin antagonist into the medial preoptic area prior to progesterone inhibits sexual receptivity and increases rejection in female rats. Horm Behav 28: 288-302.

Caldwell JD, Brooks PJ, Jirikowski GF, Barakat S, Lund PK, et al. 1989. Estrogen alters oxytocin mRNA levels in the preoptic area. J Neuroendocrinol 1: 273-278.

Caldwell JD, Morris MA, Walker CH, Carr RB, Faggin BM, et al. 1996a. Estradiol conjugated to BSA releases oxytocin from synaptosome-containing homogenates from the medial preoptic area-hypothalamus. Horm Metab Res 28: 119-121.

Caldwell JD, Walker CH, O'Rourke ST, Faggin BM, Morris M, et al. 1996b. Analogies between oxytocin systems of the uterus and brain. Horm Metab Res 28: 65-74.

Camargo AC, Spadaro AC, Martins AR, Greene LJ. 1982. Brain endo-oligopeptidase B: inactivation of LH-RH by hydrolysis of the Pro9-Gly-NH2(10) peptide bond. Braz J Med Biol Res 15: 239-245.

Campfield LA, Smith FJ, Burn P. 1996. The OB protein (leptin) pathway–a link between adipose tissue mass and central neural networks. Horm Metab Res 28: 619-632.

Canteras NS, Simerly RB, Swanson LW. 1994. Organization of projections from the ventromedial nucleus of the hypothalamus: a Phaseolus vulgaris-leucoagglutinin study in the rat. J Comp Neurol 348: 41-79.

Canteros G, Rettori V, Franchi A, Genaro A, Cebral E, et al. 1995. Ethanol inhibits luteinizing hormone-releasing hormone (LHRH) secretion by blocking the response of LHRH neuronal terminals to nitric oxide. Proc Natl Acad Sci USA 92: 3416-3420.

Canteros G, Rettori V, Genaro A, Suburo A, Gimeno M, et al. 1996. Nitric oxide synthase content of hypothalamic explants: increase by norepinephrine and inactivated by NO and cGMP. Proc Natl Acad Sci USA 93: 4246-4250.

Catania A, Gatti S, Colombo G, Lipton JM. 2004. Targeting melanocortin receptors as a novel strategy to control inflammation. Pharmacol Rev 56: 1-29.

Caulfield MP, Birdsall NJ. 1998. International Union of Pharmacology. XVII. Classification of muscarinic acetylcholine receptors. Pharmacol Rev 50: 279-290.

Ceccatelli S, Grandison L, Scott RE, Pfaff DW, Kow LM. 1996. Estradiol regulation of nitric oxide synthase mRNAs in rat hypothalamus. Neuroendocrinology 64: 357-363.

Chappell PE, Levine JE. 2000. Stimulation of gonadotropin-releasing hormone surges by estrogen. I. Role of hypothalamic progesterone receptors. Endocrinology 141: 1477-1485.

Chen W, Kelly MA, Opitz-Araya X, Thomas RE, Low MJ, et al. 1997. Exocrine gland dysfunction in MC5-R-deficient mice: evidence for coordinated regulation of exocrine gland function by melanocortin peptides. Cell 91: 789-798.

Chen Y, Mestek A, Liu J, Yu L. 1993b. Molecular cloning of a rat kappa opioid receptor reveals sequence similarities to the mu and delta opioid receptors. Biochem J 295: 625-628.

Chen Y, Mestek A, Liu J, Hurley JA, Yu L. 1993a. Molecular cloning and functional expression of a mu-opioid receptor from rat brain. Mol Pharmacol 44: 8-12.

Chhajlani V, Wikberg JE. 1992. Molecular cloning and expression of the human melanocyte stimulating hormone receptor cDNA. FEBS Lett 309: 417-420.

Chhajlani V, Muceniece R, Wikberg JE. 1993. Molecular cloning of a novel human melanocortin receptor. Biochem Biophys Res Commun 195: 866-873.

Chu HP, Etgen AM. 1997. A potential role of cyclic GMP in the regulation of lordosis behavior of female rats. Horm Behav 32: 125-132.

Chu HP, Morales JC, Etgen AM. 1999. Cyclic GMP may potentiate lordosis behaviour by progesterone receptor activation. J Neuroendocrinol 11: 107-113.

Clark JT, Kalra PS, Kalra SP. 1985. Neuropeptide Y stimulates feeding but inhibits sexual behavior in rats. Endocrinology 117: 2435-2442.

Clark JT, Kalra PS, Crowley WR, Kalra SP. 1984. Neuropeptide Y and human pancreatic polypeptide stimulate feeding behavior in rats. Endocrinology 115: 427-429.

Clarke WP, Maayani S. 1990. Estrogen effects on 5-HT1A receptors in hippocampal membranes from ovariectomized rats: functional and binding studies. Brain Res 518: 287-291.

Clemens LG, Weaver DR. 1985. The role of gonadal hormone in the activation of feminine sexual behavior. Handbook of Behavioral Neurobiology, Vol. 7. Adler N, Pfaff D, Goy RW, Eds. New York: Plenum Press; pp. 183-227.

Clemens LG, Barr P, Dohanich GP. 1989. Cholinergic regulation of female sexual behavior in rats demonstrated by manipulation of endogenous acetylcholine. Physiol Behav 45: 437-442.

Clemens LG, Dohanich GP, Witcher JA. 1981. Cholinergic influences on estrogen-dependent sexual behavior in female rats. J Comp Physiol Psychol 95: 763-770.

Clemens LG, Humphrys RR, Dohanich GP. 1980. Cholinergic brain mechanisms and the hormonal regulation of female sexual behavior in the rat. Pharmacol Biochem Behav 13: 81-88.

Coirini H, Schumacher M, Flanagan LM, McEwen BS. 1991. Transport of estrogen-induced oxytocin receptors in the ventromedial hypothalamus. J Neurosci 11: 3317-3324.

Corp ES, Greco B, Powers JB, Marin Bivens CL, Wade GN. 2001. Neuropeptide Y inhibits estrous behavior and stimulates feeding via separate receptors in Syrian hamsters. Am J Physiol Regul Integr Comp Physiol 280: R1061-1068.

Cragnolini A, Scimonelli T, Celis ME, Schioth HB. 2000. The role of melanocortin receptors in sexual behavior in female rats. Neuropeptides 34: 211-215.

Daniels D, Flanagan-Cato LM. 2000. Functionally-defined compartments of the lordosis neural circuit in the ventromedial hypothalamus in female rats. J Neurobiol 45: 1-13.

Daniels D, Miselis RR, Flanagan-Cato LM. 1999. Central neuronal circuit innervating the lordosis-producing muscles defined by transneuronal transport of pseudorabies virus. J Neurosci 19: 2823-2833.

Daniels D, Miselis RR, Flanagan-Cato LM. 2003. Hypothalamic co-localization of substance P receptor and transneuronal tracer from the lordosis-relevant lumbar epaxial muscles in the female rat. Neurosci Lett 338: 111-114.

Darland T, Heinricher MM, Grandy DK. 1998. Orphanin FQ/nociceptin: a role in pain and analgesia, but so much more. Trends Neurosci 21: 215-221.

Darnaudery M, Perez-Martin M, Belizaire G, Maccari S, Garcia-Segura LM. 2006. Insulin-like growth factor 1 reduces age-related disorders induced by prenatal stress in female rats. Neurobiol Aging 27: 119-127.

Daughaday WH, Rotwein P. 1989. Insulin-like growth factors I and II. Peptide, messenger ribonucleic acid and gene structures, serum, and tissue concentrations. Endocr Rev 10: 68-91.

Davis BL, Manzanares J, Lookingland KJ, Moore KE, Clemens LG. 1991. Noradrenergic innervation to the VMN or MPN is not necessary for lordosis. Pharmacol Biochem Behav 39: 737-742.

Davis GA, Kohl R. 1977. The influence of alpha-receptors on lordosis in the female rat. Pharmacol Biochem Behav 6: 47-53.

Day JR, Morales TH, Lu JK. 1988. Male stimulation of luteinizing hormone surge, progesterone secretion and ovulation in spontaneously persistent-estrous, aging rats. Biol Reprod 38: 1019-1026.

Desjardins GC, Brawer JR, Beaudet A. 1990. Distribution of mu, delta, and kappa opioid receptors in the hypothalamus of the rat. Brain Res 536: 114-123.

Dewing P, Sinchak K, Cook M, Micevych P.2005. Orphanin FQ/ Nociceptin facilitation of lordosis in the medial preoptic area and ventromedial nucleus of the hypothalamus. In Society for Neuroscience Meetings, Washington D.C. Abst 320.17.

Dhawan BN, Cesselin F, Raghubir R, Reisine T, Bradley PB, et al. 1996. International union of pharmacology. 12. Classification of opioid receptors. Pharmacol Rev 48: 567-592.

Dickson RB, Huff KK, Spencer EM, Lippman ME. 1986. Induction of epidermal growth factor-related polypeptides by 17 beta-estradiol in MCF-7 human breast cancer cells. Endocrinology 118: 138-142.

Dingledine R, Borges K, Bowie D, Traynelis SF. 1999. The glutamate receptor ion channels. Pharmacol Rev 51: 7-61.

Dohanich GP, Clemens LG. 1981. Brain areas implicated in cholinergic regulation of sexual behavior. Horm Behav 15: 157-167.

Dohanich GP, McMullan DM, Brazier MM. 1990. Cholinergic regulation of sexual behavior in female hamsters. Physiol Behav 47: 127-131.

Dohanich GP, Barr PJ, Witcher JA, Clemens LG. 1984. Pharmacological and anatomical aspects of cholinergic activation of female sexual behavior. Physiol Behav 32: 1021-1026.

Dohanich GP, McMullan DM, Cada DA, Mangum KA. 1991. Muscarinic receptor subtypes and sexual behavior in female rats. Pharmacol Biochem Behav 38: 115-124.

Dohanich GP, Witcher JA, Weaver DR, Clemens LG. 1982. Alteration of muscarinic binding in specific brain areas following estrogen treatment. Brain Res 241: 347-350.

Dooley CT, Houghten RA. 1996. Orphanin FQ: receptor binding and analog structure activity relationships in rat brain. Life Sci 59: L23-29.

Dornan WA, Malsbury CW. 1989. Peptidergic control of male rat sexual behavior: The effects of intracerebral injections of substance P and cholecystokinin. Physiol Behav 46: 547-556.

Dornan WA, Akesson TR, Micevych PE. 1990. A substance P projection from the VMH to the dorsal midbrain central gray: Implication for lordosis. Brain Res Bull 25: 791-796.

Dornan WA, Malsbury CW, Penney RB. 1987. Facilitation of lordosis by injection of substance P into the midbrain central gray. Neuroendocrinology 45: 498-506.

Dornan WA, Bloch GJ, Priest CA, Micevych PE. 1989. Microinjection of cholecystokinin into the medial preoptic nucleus facilitates lordosis behavior in the female rat. Physiol Behav 45: 969-974.

Dudley CA, Moss RL. 1985. LHRH and mating behavior: sexual receptivity versus sexual preference. Pharmacol Biochem Behav 22: 967-972.

Dudley CA, Vale W, Rivier J, Moss RL. 1983. Facilitation of sexual receptivity in the female rat by a fragment of the LHRH decapeptide, Ac-LHRH. Neuroendocrinology 36(6): 486-488.

Duenas M, Torres-Aleman I, Naftolin F, Garcia-Segura LM. 1996. Interaction of insulin-like growth factor-I and estradiol signaling pathways on hypothalamic neuronal differentiation. Neuroscience 74: 531-539.

Dumont Y, Jacques D, St Pierre JA, Tong Y, Parker R, Herzog H & Quirion R. 2000. Neuropeptide Y, peptide YY and pancreatic polypeptide receptor proteins and mRNAs in mammalian brains. Handbook of Chemical Neuroanatomy, Vol 16. Peptide Receptor, Part 1 Quirion, R., Bjorklund, A & Hokfelt, T. ed. London, U.K.: Elsevier. pp 375-475.

Eckersell CB, Micevych PE. 1997. Opiate receptors modulate estrogen-induced cholecystokinin and tachykinin but not enkephalin mRNA levels in the limbic system and hypothalamus. Neuroscience 80: 473-485.

Eckersell CB, Popper P, Micevych PE. 1998. Estrogen-induced alteration of mu-opioid receptor immunoreactivity in the medial preoptic nucleus and medial amygdala. J Neurosci 18: 3967-3976.

Eddy N, Reid J. 1939. Studies of morphine, codeine, and their derivatives VIII. Dihydromorphine (paramorphan), dihydromorphinone (dilaudid), and dihydrocodeine (dicodid). J Pharmacol Exp Ther 52: 468-493.

Ehlert FJ, Tran LP. 1990. Regional distribution of M1, M2 and non-M1, non-M2 subtypes of muscarinic binding sites in rat brain. J Pharmacol Exp Ther 255: 1148-1157.

Erskine MS. 1987. Serum 5 alpha-androstane-3 alpha, 17 beta-diol increases in response to paced coital stimulation in cycling female rats. Biol Reprod 37: 1139-1148.

Erskine MS, Hanrahan SB. 1997. Effects of paced mating on c-fos gene expression in the female rat brain. J Neuroendocrinol 9: 903-912.

Erskine MS, Lehmann ML, Cameron NM, Polston EK. 2004. Co-regulation of female sexual behavior and pregnancy induction: an exploratory synthesis. Behav Brain Res 153: 295-315.

Etgen AM. 1990. Intrahypothalamic implants of noradrenergic antagonists disrupt lordosis behavior in female rats. Physiol Behav 48: 31-36.

Etgen AM, Morales JC. 2002. Somatosensory stimuli evoke norepinephrine release in the anterior ventromedial hypothalamus of sexually receptive female rats. J Neuroendocrinology 14: 213-218.

Everitt BJ, Fuxe K, Hokfelt T. 1974. Inhibitory role of dopamine and 5-hydroxytryptamine in the sexual behaviour of female rats. Eur J Pharmacol 29: 187-191.

Everitt BJ, Fuxe K, Hokfelt FT, Jonsson G. 1975. Role of monoamines in the control by hormones of sexual receptivity in the female rat. J Comp Physiol Psychol 89: 556-572.

Fabre-Nys C, Chesneau D, de la Riva C, Hinton MR, Locatelli A, et al. 2003. Biphasic role of dopamine on female sexual behaviour via D2 receptors in the mediobasal hypothalamus. Neuropharmacology 44: 354-366.

Farmer CJ, Isakson TR, Coy DJ, Renner KJ. 1996. In vivo evidence for progesterone dependent decreases in serotonin release in the hypothalamus and midbrain central grey: relation to the induction of lordosis. Brain Res 711: 84-92.

Feder HH, Marrone BL. 1977. Progesterone: Its role in the central nervous system as a facilitator and inhibitor of sexual behavior and gonadotropin release. Ann NY Acad Sci 286: 331-354.

Ferin M, Tempone A, Zimmering PE, Van de Wiele RL. 1969. Effect of antibodies to 17beta-estradiol and progesterone on the estrous cycle of the rat. Endocrinology 85: 1070-1078.

Fernandez-Galaz MC, Naftolin F, Garcia-Segura LM. 1999. Phasic synaptic remodeling of the rat arcuate nucleus during the estrous cycle depends on insulin-like growth factor-I receptor activation. J Neurosci Res 55: 286-292.

Fernandez-Guasti A, Larsson K, Beyer C. 1985a. Potentiative action of alpha- and beta-adrenergic receptor stimulation in inducing lordosis behavior. Pharmacol Biochem Behav 22: 613-617.

Fernandez-Guasti A, Larsson K, Beyer C. 1985b. Prevention of progesterone-induced lordosis behavior by alpha or beta adrenergic antagonists in ovariectomized estrogen-primed rats. Pharmacol Biochem Behav 22: 279-282.

Fernandez-Guasti A, Rodriguez-Manzo G, Beyer C. 1983. Effect of guanine derivatives on lordosis behavior in estrogen primed rats. Physiol Behav 31: 589-592.

Fernandez-Guasti A, Ahlenius S, Hjorth S, Larsson K. 1987. Separation of dopaminergic and serotonergic inhibitory mechanisms in the mediation of estrogen-induced lordosis behaviour in the rat. Pharmacol Biochem Behav 27: 93-98.

Fiorino DF, Phillips AG. 1999. Facilitation of sexual behavior and enhanced dopamine efflux in the nucleus accumbens of male rats after D-amphetamine-induced behavioral sensitization. J Neurosci 19: 456-463.

Flanagan LM, Lee B, Calizo L. 2006. Co-localization of midbrain projections, progestin receptors, and mating-induced fos in the hypothalamic ventromedial nucleus of the female rat. Horm Behav: 50: 52-60.

Flanagan LM, Pfaus JG, Pfaff DW, McEwen BS. 1993. Induction of FOS immunoreactivity in oxytocin neurons after sexual activity in female rats. Neuroendocrinology 58: 352-358.

Foreman MM, Moss RL. 1979. Role of hypothalamic dopaminergic receptors in the control of lordosis behavior in the female rat. Physiol Behav 22: 283-289.

Fox AS, Olster DH. 2000. Effects of intracerebroventricular leptin administration on feeding and sexual behaviors in lean and obese female Zucker rats. Horm Behav 37: 377-387.

Fraley GS, Thomas-Smith SE, Acohido BV, Steiner RA, Clifton DK. 2004. Stimulation of sexual behavior in the male rat by galanin-like peptide. Horm Behav 46: 551-557.

Frankfurt M, McKittrick CR, Mendelson SD, McEwen BS. 1994. Effect of 5,7-dihydroxytryptamine, ovariectomy and gonadal steroids on serotonin receptor binding in rat brain. Neuroendocrinology 59: 245-250.

Frankfurt M, Renner K, Azmitia E, Luine V. 1985. Intrahypothalamic 5,7-dihydroxytryptamine: temporal analysis of effects on 5-hydroxytryptamine content in brain nuclei and on facilitated lordosis behavior. Brain Res 340: 127-133.

Frye CA, De Bold JF. 1992. Muscimol facilitates sexual receptivity in hamsters when infused into the ventral tegmentum. Pharmacol Biochem Behav 42: 879-887.

Frye CA, Mermelstein PG, De Bold JF. 1993. Bicuculline infused into the hamster ventral tegmentum inhibits, while sodium valproate facilitates, sexual receptivity. Pharmacol Biochem Behav 46: 1-8.

Fukuda K, Kato S, Mori K, Nishi M, Takeshima H, et al. 1994. cDNA Cloning and Regional Distribution of a Novel Member of the Opioid Receptor Family. FEBS Lett 343: 42-46.

Fulwiler C, Saper CB. 1985. Cholectocystokinin-immunoreactive innervation of the ventrome-dial hypothalamusin the rat: Possible substrate for the autonomic regulation of the feeding. Neurosci Lett 53: 289-296.

Fushimi S, Shirabe T. 2004. Expression of insulin-like growth factors in remyelination following ethidium bromide-induced demyelination in the mouse spinal cord. Neuropathology 24: 208-218.

Fuxe K, Everett B, Holkfelt T. 1977. Enhancement of sexual behavior in the female rat by nicotine. Pharmacol Biochem Behav 7: 147-151.

Garcia-Segura LM, Duenas M, Fernandez-Galaz MC, Chowen JA, Argente J, et al. 1996. Interaction of the signalling pathways of insulin-like growth factor-I and sex steroids in the neuroendocrine hypothalamus. Horm Res 46: 160-164.

Gargiulo PA, Donoso AO. 1995. Interaction between glutamate and luteinizing hormone-releasing hormone (LHRH) in lordosis behavior and luteinizing hormone release (LH): further studies on NMDA receptor mediation. Physiol Behav 58: 169-173.

Gargiulo PA, Munoz V, Donoso AO. 1992. Inhibition by N-methyl-D-aspartic acid (NMDA) receptor antagonist of lordosis behavior induced by estrogen followed by progesterone or luteinizing hormone-releasing hormone (LHRH) in the rat. Physiol Behav 52: 737-739.

Getting SJ. 2006. Targeting melanocortin receptors as potential novel therapeutics. Pharmacol Ther 111: 1-15.

Gonzalez-Flores O, Shu J, Camacho-Arroyo I, Etgen AM. 2004. Regulation of lordosis by cyclic 3′,5′-guanosine monophosphate, progesterone, and its 5alpha-reduced metabolites involves mitogen-activated protein kinase. Endocrinology 145: 5560-5567.

Gonzalez F, Chang L, Horab T, Stanczyk FZ, Crickard K, et al. 1999. Adrenal dynamic responses to physiologic and pharmacologic adrenocorticotropic hormone stimulation before and after ovarian steroid modulation in women with polycystic ovary syndrome. Fertil Steril 71: 439-444.

Gonzalez MI, Vaziri S, Wilson CA. 1996. Behavioral effects of alpha-MSH and MCH after central administration in the female rat. Peptides 17: 171-177.

Gonzalez MI, Celis ME, Hole DR, Wilson CA. 1993. Interaction of oestradiol, alpha-melanotrophin and noradrenaline within the ventromedial nucleus in the control of female sexual behaviour. Neuroendocrinology 58: 218-226.

Gonzalez MI, Greengrass P, Russell M, Wilson CA. 1997. Comparison of serotonin receptor numbers and activity in specific hypothalamic areas of sexually active and inactive female rats. Neuroendocrinology 66: 384-392.

Gorski RA, Yanase M. 1981. Estrogen facilitation of lordosis behavior in the female rat. Exp Brain Res (Suppl)3: 222-237.

Gorzalka BB, Lester GL. 1987. Oxytocin-induced facilitation of lordosis behaviour in rats is progesterone-dependent. Neuropeptides 10: 55-65.

Gorzalka BB, Brotto LA, Hong JJ. 1999. Corticosterone regulation of 5-HT2A receptor-mediated behaviors: attenuation by melatonin. Physiol Behav 67: 439-442.

Gorzalka BB, Hanson LA, Brotto LA. 1998. Chronic stress effects on sexual behavior in male and female rats: mediation by 5-HT2A receptors. Pharmacol Biochem Behav 61: 405-412.

Gorzalka BB, Heddema GM, Lester GL, Hanson LA. 1997. beta-endorphin inhibits and facilitates lordosis behaviour in rats depending on ventricular site of administration. Neuropeptides 31: 517-521.

Goy RW, Phoenix CH, Young WC. 1966. Inhibitory action in the corpus luteum on the hormonal induction of estrous behavior in the guinea pig. Gen Comp Endocrinol 6: 267-275.

Grierson JP, James MD, Pearson JR, Wilson CA. 1988. The effect of selective D1 and D2 dopaminergic agents on sexual receptivity in the female rat. Neuropharmacology 27: 181-189.

Guerrini R, Calo G, Rizzi A, Bianchi C, Lazarus LH, et al. 1997. Address and message sequences for the nociceptin receptor: A structure-activity study of nociceptin-(1–13)-peptide amide. J Med Chem 40: 1789-1793.

Hall NR, Luttge WG. 1977. Diencephalic sites responsive to prostaglandin E2 facilitation of sexual receptivity in estrogen-primed ovariectomized rats. Brain Res Bull 2: 203-207.

Hall NR, Luttge WG. 1978. Effects of prostaglandin synthesis inhibitor, indomethacin on estrogen- and estrogen plus progesterone-induced sexual receptivity in ovariectomized rats. Pharmacol Biochem Behav 8: 597-602.

Hamburger-Bar R, Rigter H. 1975. Apomorphine: facilitation of sexual behaviour in female rats. Eur J Pharmacol 32: 357-360.

Hansen S, Stanfield E, Everitt B. 1980. The role of ventral bundle noradrenergic neurons in sensory components of sexual behaviour and coitus-induced pseudopregnancy. Nature 286: 152-154.

Hansen S, Stanfield EJ, Everitt BJ. 1981. The effects of lesions of lateral tegmental noradrenergic neurons on components of sexual behavior and pseudopregnancy in female rats. Neuroscience 6: 1105-1117.

Healy DP, Orlowski M. 1992. Immunocytochemical localization of endopeptidase 24.15 in rat brain. Brain Res 571: 121-128.

Hebert TJ, Menard CS, Dohanich GP. 1995. Inhibition of lordosis in female hamsters and rats by 8-OH-DPAT treatment. Physiol Behav 57: 523-527.

Hennessey AC, Camak L, Gordon F, Edwards DA. 1990. Connections between the pontine central gray and the ventromedial hypothalamus are essential for lordosis in female rats. Behav Neurosci 104: 477-488.

Herndon JG Jr, Caggiula AR, Sharp D, Ellis D, Redgate E. 1978. Selective enhancement of the lordotic component of female sexual behavior in rats following destruction of central catecholamine-containing systems. Brain Res 141: 137-151.

Hilleson-Gayne CK, Clapper JA. 2005. Effects of decreased estradiol-17beta on the serum and anterior pituitary IGF-I system in pigs. J Endocrinol 187: 369-378.

Hjorth S, Sharp T. 1991. Effect of the 5-HT1A receptor agonist 8-OH-DPAT on the release of 5-HT in dorsal and median raphe-innervated rat brain regions as measured by in vivo microdialysis. Life Sci 48: 1779-1786.

Hjorth S, Suchowski CS, Galloway MP. 1995. Evidence for 5-HT autoreceptor-mediated, nerve impulse-independent, control of 5-HT synthesis in the rat brain. Synapse 19: 170-176.

Ho GK, Quadagno DM, Cooke PH, Gorski RA. 1973. Intracranial implants of actinomycin-D: effects on sexual behavior and nucleolar ultrastructure in the rat. Neuroendocrinology 13: 47-55.

Hokfelt T, Johansson O, Fuxe K, Goldstein M, Park D. 1976. Immunohistochemical studies on the localization and distribution of monoamine neuron systems in the rat brain. I. Tyrosine hydroxylase in the mes- and diencephalon. Med Biol 54: 427-453.

Holland KL, Abelson L, Micevych PE. 1996. Estrogen stimulates cholecystokinin and preproenkephalin mRNA in the female limbic-hypothalamic circuit during the peripubertal period. Abstr Soc Neurosci Vol. 22: 561

Holland KL, Popper P, Micevych PE. 1997. Infusion of CCK-A receptor mRNA antisense oligodeoxynucleotides inhibits lordosis behavior. Physiol Behav 62: 537-543.

Hsu C, Lee JN, Ho ML, Cheng BH, Li PH, et al. 1993. The facilitatory effect of N-methyl-D-aspartate on sexual receptivity in female rats through GnRH release. Acta Endocrinol (Copenh) 128: 385-388.

Hunter AJ, Hole DR, Wilson CA. 1985. Studies into the dual effects of serotonergic pharmacological agents on female sexual behaviour in the rat: preliminary evidence that endogenous 5HT is stimulatory. Pharmacol Biochem Behav 22: 5-13.

Huszar D, Lynch CA, Fairchild-Huntress V, Dunmore JH, Fang Q, et al. 1997. Targeted disruption of the melanocortin-4 receptor results in obesity in mice. Cell 88: 131-141.

Imura H, Nakao K, Yanaihara N, Katsuura G, Nakamura M. 1986. Potent action of leumorphin on consummatory behaviors in rats: comparison with other opioid peptides. Neuroendocrinology 44: 142-148.

Imura H, Kato Y, Nakai Y, Nakao K, Tanaka I, et al. 1985. Endogenous opioids and related peptides: from molecular biology to clinical medicine. The Sir Henry Dale lecture for 1985. J Endocrinol 107: 147-157.

Innes DL, Michal EK. 1970. Effects of progesterone and estrogen on the electrical activity of the limbic system. J Exp Zool 175: 487-492.

Irving SM, Goy RW, Haning RV, Davis GA. 1981. Prostaglandins, clonidine and sexual receptivity in the guinea pig. Brain Res 204: 65-77.

Ishihara T, Shigemoto R, Mori K, Takahashi K, Nagata S. 1992. Functional expression and tissue distribution of a novel receptor for vasoactive intestinal polypeptide. Neuron 8: 811-819.

Itier V, Bertrand D. 2001. Neuronal nicotinic receptors: from protein structure to function. FEBS Lett 504: 118-125.

Jackson A, Etgen AM. 2001. Estrogen modulates 5-HT1A agonist inhibition of lordosis behavior but not binding of [H-3]-8-OH-DPAT. Pharmacol Biochem Behav 68: 221-227.

James MD, Hole DR, Wilson CA. 1989. Differential involvement of 5-hydroxytryptamine (5HT) in specific hypothalamic areas in the mediation of steroid-induced changes in gonadotrophin release and sexual behaviour in female rats. Neuroendocrinology 49: 561-569.

Jaworski DM, Proctor MD. 2000. Developmental regulation of pituitary adenylate cyclase-activating polypeptide and PAC(1) receptor mRNA expression in the rat central nervous system. Brain Res Dev Brain Res 120: 27-39.

Jirikowski GF, Caldwell JD, Pedersen CA, Stumpf WE. 1988. Estradiol influences oxytocin-immunoreactive brain systems. Neuroscience 25: 237-248.

Johnson AE, Coirini H, Insel TR, McEwen BS. 1991. The regulation of oxytocin receptor binding in the ventromedial hypothalamic nucleus by testosterone and its metabolites. Endocrinology 128: 891-896.

Jones JI, Clemmons DR. 1995. Insulin-like growth factors and their binding proteins: biological actions. Endocr Rev 16: 3-34.

Kalra SP, Clark JT, Sahu A, Dube MG, Kalra PS. 1988. Control of feeding and sexual behaviors by neuropeptide Y: physiological implications. Synapse 2: 254-257.

Kamat A, Yu WH, Rettori V, McCann SM. 1995. Glutamic acid induces luteinizing hormone releasing hormone release via alpha receptors. Brain Res Bull 37: 233-235.

Karlin A, Akabas MH. 1995. Toward a structural basis for the function of nicotinic acetylcholine receptors and their cousins. Neuron 15: 1231-1244.

Kaufman LS, McEwen BS, Pfaff DW. 1988. Cholinergic mechanisms of lordotic behavior in rats. Physiol Behav 43: 507-514.

Kaur C, Sivakumar V, Dheen ST, Ling EA. 2006. Insulin-like growth factor I and II expression and modulation in amoeboid microglial cells by lipopolysaccharide and retinoic acid. Neuroscience 138: 1233-1244.

Ke FC, Ramirez VD. 1990. Binding of progesterone to nerve cell membranes of rat brain using progesterone conjugated to ^{125}I-bovine serum albumin as a ligand. J Neurochem 54: 467-472.

Kendrick KM, Dixson AF. 1985. Luteinizing hormone releasing hormone enhances proceptivity in a primate. Neuroendocrinology 41: 449-453.

Kimura F, Kawakami M, Nakano H, McCann SM. 1980. Changes in adenosine 3′,5′-monophosphate and guanosine 3′,5′-monophosphate concentrations in the anterior pituitary and hypothalamus during the rat estrous cycle and effects of administration of sodium pentobarbital in proestrus. Endocrinology 106: 631-635.

Kirby LG, Chou-Green JM, Davis K, Lucki I. 1997. The effects of different stressors on extracellular 5-hydroxytryptamine and 5-hydroxyindoleacetic acid. Brain Res 760: 218-230.

Ko K. 1934. The influence of opium alkaloids on the sexual cycle of female mice. Taiwan Igk Zeitschr 33: 50-62.

Ko K. 1935. The influence of opium alkaloids on the sexual cycle of female mice. Jpn J Med Sci IV Pharmacol 8: 97-108.

Kohlert JG, Rowe RK, Meisel RL. 1997. Intromissive stimulation from the male increases extracellular dopamine release from fluoro-gold-identified neurons within the midbrain of female hamsters. Horm Behav 32: 143-154.

Kow LM, Pfaff DW. 1985. Estrogen effects on neuronal responsiveness to electrical and neurotransmitter stimulation: an in vitro study on the ventromedial nucleus of the hypothalamus. Brain Res 347: 1-10.

Kow LM, Pfaff DW. 1996. Thyrotropin-releasing hormone (TRH) has independent excitatory and modulatory actions on lamina IX neurons of lumbosacral spinal cord slices from adult rats. Peptides 17: 131-138.

Kow LM, Weesner GD, Pfaff DW. 1992. Alpha 1-adrenergic agonists act on the ventromedial hypothalamus to cause neuronal excitation and lordosis facilitation: electrophysiological and behavioral evidence. Brain Res 588: 237-245.

Kow LM, Tsai YF, Weiland NG, McEwen BS, Pfaff DW. 1995. In vitro electro-pharmacological and autoradiographic analyses of muscarinic receptor subtypes in rat hypothalamic ventromedial nucleus: implications for cholinergic regulation of lordosis. Brain Res 694: 29-39.

Krause JE, Advis JP, McKelvy JF. 1982. Characterization of the site of cleavage of luteinizing hormone-releasing hormone under conditions of measurement in which LHRH degradation undergoes physiologically related change. Biochem Biophys Res Commun 108: 1475-1481.

Kubo K, Gorski RA, Kawakami M. 1975. Effects of estrogen on neuronal excitability in the hippocampal-septal-hypothalamic system. Neuroendocrinology 18: 176-191.

Kudwa AE, Dominguez-Salazar E, Cabrera DM, Sibley DR, Rissman EF. 2005. Dopamine D5 receptor modulates male and female sexual behavior in mice. Psychopharmacology (Berl) 180: 206-214.

Kulkosky PJ, Glazner GW, Moore HD, Low CA, Woods SC. 1988. Neuropeptide Y: behavioral effects in the golden hamster. Peptides 9: 1389-1393.

Kurtz MM, Wang R, Clements MK, Cascieri MA, Austin CP, et al. 2002. Identification, localization and receptor characterization of novel mammalian substance P-like peptides. Gene 296: 205-212.

Labhsetwar AP. 1970. Role of estrogens in ovulation: A study using the estrogen-antagonist, I.C.I. 46,474. Endocrinology 87: 542-551.

Lachowicz JE, Shen Y, Monsma FJ Jr, Sibley DR. 1995. Molecular cloning of a novel G protein-coupled receptor related to the opiate receptor family. J Neurochem 64: 34-40.

Lauber AH, Romano GJ, Mobbs CV, Howells RD, Pfaff DW. 1990. Estradiol induction of proenkephalin messenger RNA in hypothalamus: dose-response and relation to reproductive behavior in the female rat. Brain Res Mol Brain Res 8: 47-54.

Law PY, Loh HH. 1999. Regulation of opioid receptor activities. J Pharmacol Exp Ther 289: 607-624.

Lee K, Nicholson JR, McKnight AT. 1997. Nociceptin hyperpolarises neurones in the rat ventromedial hypothalamus. Neurosci Lett 239: 37-40.

Lee NM, Smith AP. 1980. A protein-lipid model of the opiate receptor. Life Sci 26: 1459-1464.

Lee WH, Wang GM, Seaman LB, Vannucci SJ. 1996. Coordinate IGF-I and IGFBP5 gene expression in perinatal rat brain after hypoxia-ischemia. J Cereb Blood Flow Metab 16: 227-236.

Lindstrom JM. 2003. Nicotinic acetylcholine receptors of muscles and nerves: comparison of their structures, functional roles, and vulnerability to pathology. Ann N Y Acad Sci 998: 41-52.

Lindstrom LH, Meyerson BJ. 1967. The effect of pilocarpine, oxotremorine and arecoline in combination with methylatropine or atropine on hormone activated oestrous behaviour in ovariectomized rats. Psychopharmacologia 11: 405-413.

Loughlin SE, Leslie FM, Fallon JJ. 1995. Endogenous opioid systems. The Rat Nervous System. Paxinos G. Ed. San Diego: Academic Press; pp. 975-1001.

Luine VN, Paden CM. 1982. Effects of monoamine oxidase inhibition on female sexual behavior, serotonin levels and type A and B monoamine oxidase activity. Neuroendocrinology 34: 245-251.

Luine VN, Renner KJ, Frankfurt M, Azmitia EC. 1984. Facilitated sexual behavior reversed and serotonin restored by raphe nuclei transplanted into denervated hypothalamus. Science 226: 1436-1439.

Lundberg JM, Terenius L, Hokfelt T, Tatemoto K. 1984. Comparative immunohistochemical and biochemical analysis of pancreatic polypeptide-like peptides with special reference to presence of neuropeptide Y in central and peripheral neurons. J Neurosci 4: 2376-2386.

Mackay-Sim A, Rose JD. 1986. Removal of the vomeronasal organ impairs lordosis in female hamsters: effect is reversed by luteinising hormone-releasing hormone. Neuroendocrinology 42: 489-493.

Maggi CA. 1995. The mammalian tachykinin receptors. Gen Pharmacol 26: 911-944.

Mahesh VB, Brann DW. 1998. Regulation of the preovulatory gonadotropin surge by endogenous steroids. Steroids 63: 616-629.

Mahachoklertwattana P, Black SM, Kaplan SL, Bristow JD, Grumbach MM. 1994. Nitric oxide synthesized by gonadotropin-releasing hormone neurons is a mediator of N-methyl-D-aspartate (NMDA)-induced GnRH secretion. Endocrinology 135: 1709-1712.

Maney DL, Richardson RD, Wingfield JC. 1997. Central administration of chicken gonadotropin-releasing hormone-II enhances courtship behavior in a female sparrow. Horm Behav 32: 11-18.

Mani SK, Blaustein JD, O'Malley BW. 1997. Progesterone receptor function from a behavioral perspective. Horm Behav 31: 244-255.

Mani SK, Mitchell A, O'Malley BW. 2001. Progesterone receptor and dopamine receptors are required in Delta(9)-tetrahydrocannabinol modulation of sexual receptivity in female rats. Proc Natl Acad Sci USA 98: 1249-1254, 1241.

Mani SK, Allen JM, Clark JH, Blaustein JD, O'Malley BW. 1994a. Convergent pathways for steroid hormone- and neurotransmitter-induced rat sexual behavior. Science 265: 1246-1249.

Mani SK, Allen JM, Valeria R, McCann SM, O'Malley BW, et al. 1994b. Nitric Oxide mediates sexual behavior in female rats. Proc Natl Acad Sci USA 91: 6468-6472.

Mani SK, Blaustein JD, Allen JM, Law SW, O'Malley BW, et al. 1994c. Inhibition of rat sexual behavior by antisense oligonucleotides to the progesterone receptor. Endocrinology 135: 1409-1414.

Mani SK, Allen JM, Lydon JP, Mulac-Jericevic B, Blaustein JD, et al. 1996. Dopamine requires the unoccupied progesterone receptor to induce sexual behavior in mice [published erratum appears in Mol. Endocrinol 1997 Apr;11(4):423]. Mol Endocrinol 10: 1728-1737.

Mansour A, Fox CA, Thompson RC, Akil H, Watson SJ. 1994b. Mu-opioid receptor mRNA expression in the rat CNS: comparison to mu-receptor binding. Brain Res 643: 245-265.

Mansour A, Fox CA, Burke S, Meng F, Thompson RC, et al. 1994a. Mu, delta, and kappa opioid receptor mRNA expression in the rat CNS: an in situ hybridization study. J Comp Neurol 350: 412-438.

Marrone BL, Rodriguez-Sierra JF, Feder HH. 1977. Lordosis: inhibiting effects progesterone in the female rat. Horm Behav 8: 391-402.

Marrone BL, Rodriguez-Sierra JF, Feder HH. 1979. Differential effects of prostaglandins on lordosis behavior in female guinea pigs and rats. Biol Reprod 20: 853 861.

Maswood N, Caldarola-Pastuszka M, Uphouse L. 1998. Functional integration among 5-hydroxytryptamine receptor families in the control of female rat sexual behavior. Brain Res 802: 98-103.

Maswood S, Truitt W, Hotema M, Caldarola-Pastuszka M, Uphouse L. 1999. Estrous cycle modulation of extracellular serotonin in mediobasal hypothalamus: role of the serotonin transporter and terminal autoreceptors. Brain Res 831: 146-154.

Matt DW, Coquelin A, Lu JK. 1987. Neuroendocrine control of luteinizing hormone secretion and reproductive function in spontaneously persistent-estrous aging rats. Biol Reprod 37: 1198-1206.

Matthews SG, Heavens RP, Sirinathsinghji DJ. 1991. Cellular localization of corticotropin releasing factor mRNA in the ovine brain. Brain Res Mol Brain Res 11: 171-176.

McCall T, Vallance P. 1992. Nitric oxide takes centre-stage with newly defined roles. Trends Pharmacol Sci 13: 1-6.

McCarthy MM, Curran GH, Feder HH. 1991a. Excitatory amino acid modulation of lordosis in the rat. Neurosci Lett 126: 94-97.

McCarthy MM, Malik KF, Feder HH. 1990. Increased GABAergic transmission in medial hypothalamus facilitates lordosis but has the opposite effect in preoptic area. Brain Res 507: 40-44.

McCarthy MM, Pfaff DW, Schwartz-Giblin S. 1991b. Midbrain central gray GABAA receptor activation enhances, and blockade reduces, sexual behavior in the female rat. Exp Brain Res 86: 108-116.

McCarthy M, Kleopoulos St, Mobbs C, Pfaff D. 1994a. Infusion of antisense oligodeoxynucleotides to the oxytocin receptor in the ventromedial hypothalamus reduces estrogen-induced sexual receptivity and oxytocin receptor binding in the female rat. Neuronal Reg 59: 432-440.

McCarthy MM, Masters DB, Rimvall K, Schwartz-Giblin S, Pfaff DW. 1994b. Intracerebral administration of antisense oligodeoxynucleotides to GAD_{65} and GAD_{67} mRNAs modulate reproductive behavior in the female rat. Brain Res 636: 209-220.

McCarthy MM, Masters DB, Fiber JM, López-Colomé A, Beyer C, et al. 1991. GABAergic Control of Receptivity in the Female Rat. Neuroendocrinology 53: 473-479.

McLean S, Rothman RB, Herkenham M. 1986. Autoradiographic localization of mu- and delta-opiate receptors in the forebrain of the rat. Brain Res 378: 49-60.

Meisel RL, Camp DM, Robinson TE. 1993. A microdialysis study of ventral striatal dopamine during sexual behavior in female Syrian hamsters. Behav Brain Res 55: 151-157.

Melis MR, Argiolas A. 1995. Dopamine and sexual behavior. Neurosci Biobehav Rev 19: 19-38.

Menard CS, Dohanich GP. 1989. Scopolamine inhibition of lordosis in naturally cycling female rats. Physiol Behav 45: 819-823.

Mendelson SD. 1992. A review and reevaluation of the role of serotonin in the modulation of lordosis behavior in the female rat. Neurosci Biobehav Rev 16: 309-350.

Mendelson SD, Gorzalka BB. 1985. A facilitatory role for serotonin in the sexual behavior of the female rat. Pharmacol Biochem Behav 22: 1025-1033.

Mendelson SD, Gorzalka BB. 1987. An improved chamber for the observation and analysis of the sexual behavior of the female rat. Physiol Behav 39: 67-71.

Mendelson SD, Gorzalka BB. 1990. Sex differences in the effects of 1-(m-trifluoromethylphenyl) piperazine and 1-(m-chlorophenyl) piperazine on copulatory behavior in the rat. Neuropharmacology 29: 783-786.

Mendelson SD, McEwen BS. 1991. Autoradiographic analyses of the effects of restraint-induced stress on 5-HT1A, 5-HT1C and 5-HT2 receptors in the dorsal hippocampus of male and female rats. Neuroendocrinology 54: 454-461.

Mendez P, Azcoitia I, Garcia-Segura LM. 2005. Interdependence of oestrogen and insulin-like growth factor-I in the brain: potential for analysing neuroprotective mechanisms. J Endocrinol 185: 11-17.

Mercer JG, Lawrence CB, Atkinson T. 1996. Hypothalamic NPY and CRF gene expression in the food-deprived Syrian hamster. Physiol Behav 60: 121-127.

Mermelstein PG, Becker JB. 1995. Increased extracellular dopamine in the nucleus accumbens and striatum the female rat during paced copulatory behavior. Behav Neurosci 109: 354-365.

Meunier JC. 1997. Nociceptin/orphanin FQ and the opioid receptor-like ORL1 receptor. Eur J Pharmacol 340: 1-15.

Meunier JC, Mollereau C, Toll L, Suaudeau C, Moisand C, et al. 1995. Isolation and structure of the endogenous agonist of opioid receptor-like ORL1 receptor. Nature 377: 532-535.

Meyers T, Clemens LG.1985. Dissociation of cholinergic facilitation of feminine sexual behavior and estrogen induction of muscarinic cholinergic receptors. In Conference on Reproductive Behavior, Pacific Grove, CA.

Meyerson BJ. 1964. The Effect Of Neuropharmacological Agents On Hormone-Activated Estrus Behaviour In Ovariectomised Rats. Arch Int Pharmacodyn Ther 150: 4-33.

Micevych P, Sinchak K. 2001. Estrogen and endogenous opioids regulate CCK in reproductive circuits. Peptides 22: 1235-1244.

Micevych P, Chaban V, Quesada A, Sinchak K. 2002. Oestrogen modulates cholecystokinin: opioid interactions in the nervous system. Pharmacol Toxicol 91: 387-397.

Micevych P, Sinchak K, Mills RH, Tao L, LaPolt P, et al. 2003a. The luteinizing hormone surge is preceded by an estrogen-induced increase of hypothalamic progesterone in ovariectomized and adrenalectomized rats. Neuroendocrinology 78: 29-35.

Micevych PE, Ulibarri C. 1992. Development of the limbic-hypothalamic cholecystokinin circuit: A model of sexual differentiation. Dev Neurosci 14: 11-34.

Micevych PE, Akesson T, Elde R. 1988a. Distribution of cholecystokinin-immunoreactive cell bodies in the male and female rat: II. Bed nucleus of the stria terminalis and amygdala. J Comp Neurol 269: 381-391.

Micevych PE, Matt DW, Go VL. 1988b. Concentrations of cholecystokinin, substance P, and bombesin in discrete regions of male and female rat brain: Sex differences and estrogen effects. Exp Neurol 100: 416-425.

Micevych PE, Yaksh TL, Go VLW. 1982. Opiate-mediated inhibition of the release of cholecystokinin and substance P, but not neurotensin from cat hypothalamic slices. Brain Res 250: 283-289.

Micevych PE, Eckersell CB, Brecha N, Holland K. 1997. Estrogenic modulation of opiate and cholecystokinin systems in the limbic-hypothalamic circuit. Brain Res Bull 44: 335-343.

Micevych PE, Eckersell CB, Holland KL, Smith A. 1996. Induction of CCK mRNA levels in the limbic-hypothalamic circuit: Time course and site-specific effects of estrogen. J Neurobiol 30: 465-479.

Micevych PE, Park SS, Akesson TR, Elde R. 1987. Distribution of cholecystokinin-immunoreactive cell bodies in the male and female rat: I. Hypothalamus. J Comp Neurol 255: 124-136.

Micevych PE, Rissman EF, Gustafsson JA, Sinchak K. 2003b. Estrogen receptor-alpha is required for estrogen-induced mu-opioid receptor internalization. J Neurosci Res 71: 802-810.

Micevych PE, Rissman EF, Macias L, Sinchak K. 2000. Estrogen receptor α deficient mice do not internalized mu-opioid receptors in response to estrogen. In Society of Behavioral Neuroendocrinology Madrid Spain 129-130.

Micevych PE, Stroink A, Yaksh T, Go VL. 1986. Immunochemical studies of substance P and cholecystokinin octapeptide recovery in dorsal horn following unilateral lumbosacral ganglionectomy. Somatosens Res 3: 239-260.

Micevych PE, Yaksh TL, Go VW, Finkelstein JA. 1985. Effect of opiates on the release of cholecystokinin from in vitro hypothalamus and frontal cortex of Zucker lean (Fa/-) and obese (fa/fa) rats. Brain Res 337: 382-385.

Micevych PE, Abelson L, Fok H, Ulibarri C, Priest CA. 1994. Gonadal steroid control of preprocholecystokinin mRNA expression in the limbic-hypothalamic circuit: Comparison of adult with neonatal steroid treatments. J Neurosci Res 38: 386-398.

Michel MC, Beck-Sickinger A, Cox H, Doods HN, Herzog H, et al. 1998. XVI. International Union of Pharmacology recommendations for the nomenclature of neuropeptide Y, peptide YY, and pancreatic polypeptide receptors. Pharmacol Rev 50: 143-150.

Mills RH, Sinchak K, Mann DM, R, Wright R, Micevych P. 2003. Neuroprogesterone regulation of female reproductive behavior. In Society for Neuroscience, New Orleans, LA, USA. Abst 726.10

Mills RH, Sohn RK, Micevych PE. 2004. Estrogen-induced mu-opioid receptor internalization in the medial preoptic nucleus is mediated via neuropeptide Y-Y1 receptor activation in the arcuate nucleus of female rats. J Neurosci 24: 947-955.

Minami M, Onogi T, Toya T, Katao Y, Hosoi Y, et al. 1994. Molecular cloning and in situ hybridization histochemistry for rat mu-opioid receptor. Neurosci Res 18: 315-322.

Mollereau C, Parmentier M, Mailleux P, Butour JL, Moisand C, et al. 1994. ORL1, a novel member of the opioid receptor family. Cloning, functional expression and localization. FEBS Lett 341: 33-38.

Morin LP. 1977. Progesterone: inhibition of rodent behavior. Physiol Behav 18: 701-715.

Morrell JI, Pfaff DW. 1982. Characterization of estrogen-concentrating hypothalamic neurons by their axonal projections. Science 217: 1273-1276.

Moss RL. 1974. Relationship between the central regulation of gonadotropins and mating behavior in female rats. Adv Behav Biol 11: 55-76.

Moss RL, Foreman MM. 1976. Potentiation of lordosis behavior by intrahypothalamic infusion of synthetic luteinizing hormone-releasing hormone. Neuroendocrinology 20: 176-181.

Moss RL, McCann SM. 1973. Induction of mating behavior in rats by luteinizing hormone-releasing factor. Science 181: 177-179.

Moss RL, McCann SM. 1975. Action of luteinizing hormone-releasing factor (lrf) in the initiation of lordosis behavior in the estrone-primed ovariectomized female rat. Neuroendocrinology 17: 309-318.

Mountjoy KG, Robbins LS, Mortrud MT, Cone RD. 1992. The cloning of a family of genes that encode the melanocortin receptors. Science 257: 1248-1251.

Mountjoy KG, Mortrud MT, Low MJ, Simerly RB, Cone RD. 1994. Localization of the melanocortin-4 receptor (MC4-R) in neuroendocrine and autonomic control circuits in the brain. Mol Endocrinol 8: 1298-1308.

Myers H, Flynn J. 1931. The effect of chronic morphine poisoning upon growth, the oestrous cycle, and fertility of the white rat. J Pharmacol Exp Ther 41: 317-331.

Nadler RD. 1970. A biphasic influence of progesterone on sexual receptivity of spayed female rats. Physiol Behav 5: 95-97.

Neal CR, Mansour A, Reinscheid R, Nothacker HP, Civelli O, et al. 1999a. Opioid receptor-like (ORL1) receptor distribution in the rat central nervous system: Comparison of ORL1 receptor mRNA expression with I-125-[(14)Tyr]-orphanin FQ binding. J Comp Neurol 412: 563-605.

Neal CR, Mansour A, Reinscheid R, Nothacker HP, Civelli O, et al. 1999b. Localization of orphanin FQ (nociceptin) peptide and messenger RNA in the central nervous system of the rat. J Comp Neurol 406: 503-547.

Neubig RR, Spedding M, Kenakin T, Christopoulos A. 2003. International Union of Pharmacology Committee on Receptor Nomenclature and Drug Classification. XXXVIII. Update on terms and symbols in quantitative pharmacology. Pharmacol Rev 55: 597-606.

Nguyen Y, Sinchak K, Dewing P, Micevych P. 2006. Orphanin FQ/nociceptin regulation of steroid activation of μ-opioid receptors and sexual behavior in rat models. J Investig Med 54(Suppl 1): #171.

Nocetto C, Cragnolini AB, Schioth HB, Scimonelli TN. 2004. Evidence that the effect of melanocortins on female sexual behavior in preoptic area is mediated by the MC3 receptor; Participation of nitric oxide. Behav Brain Res 153: 537-541.

Ogawa S, Kow LM, Pfaff DW. 1992. Effects of Lordosis-Relevant Neuropeptides on Midbrain Periaqueductal Gray Neuronal Activity Invitro. Peptides 13: 965-975.

Ogawa S, Eng V, Taylor J, Lubahn DB, Korach KS, et al. 1998. Roles of estrogen receptor alpha gene expression in reproduction-related behaviors in female mice. Endocrinology 139: 5070-5081.

Ohtaki T, Kumano S, Ishibashi Y, Ogi K, Matsui H, et al. 1999. Isolation and cDNA cloning of a novel galanin-like peptide (GALP) from porcine hypothalamus. J Biol Chem 274: 37041-37045.

Okamura H, Yokosuka M, McEwen BS, Hayashi S. 1994. Colocalization of NADPH-diaphorase and estrogen receptor immunoreactivity in the rat ventromedial hypothalamic nucleus: stimulatory effect of estrogen on NADPH-diaphorase activity. Endocrinology 135: 1705-1708.

Olschowka JA, Jacobowitz DM. 1983. The coexistence and release of bovine pancreatic polypeptide-like immunoreactivity from noradrenergic superior cervical ganglia neurons. Peptides 4: 231-238.

Olster DH, Blaustein JD. 1992. Estradiol pulses induce progestin receptors selectively in substance P-immunoreactive neurons in the ventrolateral hypothalamus of female guinea pigs. J Neurobiol 23: 293-301.

Oro AE, Simerly RB, Swanson LW. 1988. Estrous cycle variations in levels of cholecystokinin immunoreactivity within cells of three interconnected sexually dimorphic forebrain nuclei. Neuroendocrinology 47: 225-235.

Page NM. 2004. Hemokinins and endokinins. Cell Mol Life Sci 61: 1652-1663.

Parsons B, Rainbow T, McEewn B. 1984. Organizational effects of testosterone via aromatization on female reproductive behavior and neural proogestine receptors in the rat brain. Endocrinology 115: 1412-1417.

Patacchini R, Lecci A, Holzer P, Maggi CA. 2004. Newly discovered tachykinins raise new questions about their peripheral roles and the tachykinin nomenclature. Trends Pharmacol Sci 25: 1-3.

Patterson SJ, Robson LE, Kosterlitz HW. 1983. Classification of opiate receptors. Br Med Bull 39: 31-36.

Pedersen CA, Boccia ML. 2002. Oxytocin maintains as well as initiates female sexual behavior: effects of a highly selective oxytocin antagonist. Horm Behav 41: 170-177.

Pedersen CA, Boccia ML. 2006. Vasopressin interactions with oxytocin in the control of female sexual behavior. Neuroscience 139: 843-851.

Pennefather JN, Lecci A, Candenas ML, Patak E, Pinto FM, et al. 2004. Tachykinins and tachykinin receptors: a growing family. Life Sci 74: 1445-1463.

Petitti N, Karkanias GB, Etgen AM. 1992. Estradiol selectively regulates alpha 1B-noradrenergic receptors in the hypothalamus and preoptic area. J Neurosci 12: 3869-3876.

Pfaff DW. 1970. Nature of sex hormone effects on rat sex behavior: Specificity of effects and individual patterns of response. J Comp Physiol Psychol 73: 349-358.

Pfaff DW. 1973. Luteinizing hormone-releasing factor potentiates lordosis behavior in hypophysectomized ovariectomized female rats. Science 182: 1148-1149.

Pfaff DW, Keiner M. 1973. Atlas of estradiol-concentrating neurons in the central nervous system of the female rat. J Comp Neurol 151: 121-158.

Pfaff DW, Sakuma Y. 1979. Facilitation of the lordosis reflex of female rats from the ventromedial nucleus of the hypothalamus. J Physiol 288: 189-202.

Pfaff DW, Schwartz-Giblin S, McCarthy M, Kow LM. 1994. Cellular and molecular mechanisms of female reproductive behaviors. The Physiology of Reproduction, Vol. 2. Knobil E, Neill, JD, Eds. New York: Raven Press, Ltd; pp. 107-220.

Pfaus JG, Gorzalka BB. 1987a. Opioids and sexual behavior. Neurosci Biobehav Rev 11: 1-34.

Pfaus JG, Gorzalka BB. 1987b. Selective activation of opioid receptors differentially affects lordosis behavior in female rats. Peptides 8: 309-317.

Pfaus JG, Pfaff DW. 1992a. μ-, δ-, and κ-opioid receptor agonists selectively modulate sexual behaviors in the female rat: Differential dependence on progesterone. Horm Behav 26: 457-473.

Pfaus JG, Pfaff DW. 1992b. Mu-, delta-, and kappa-opioid receptor agonists selectively modulate sexual behaviors in the female rat: differential dependence on progesterone. Horm Behav 26: 457-473.

Pfaus JG, Smith WJ, Coopersmith CB. 1999. Appetitive and consummatory sexual behaviors of female rats in bilevel chambers. I. A correlational and factor analysis and the effects of ovarian hormones. Horm Behav 35: 224-240.

Pfaus JG, Pendleton N, Gorzalka BB. 1986. Dual effect of morphiceptin on lordosis behavior: possible mediation by different opioid receptor subtypes. Pharmacol Biochem Behav 24: 1461-1464.

Pfaus JG, Damsma G, Wenkstern D, Fibiger HC. 1995. Sexual activity increases dopamine transmission in the nucleus accumbens and striatum of female rats. Brain Res 693: 21-30.

Pfaus JG, Kleopoulos SP, Mobbs CV, Gibbs RB, Pfaff DW. 1993. Sexual stimulation activates c-fos within estrogen-concentrating regions of the female rat forebrain. Brain Res 624: 253-267.

Pfaus JG, Shadiack A, Van Soest T, Tse M, Molinoff P. 2004. Selective facilitation of sexual solicitation in the female rat by a melanocortin receptor agonist. Proc Natl Acad Sci USA 101: 10201-10204.

Pietrzkowski Z, Wernicke D, Porcu P, Jameson BA, Baserga R. 1992. Inhibition of cellular proliferation by peptide analogues of insulin-like growth factor 1. Cancer Res 52: 6447-6451.

Plant O, Pierce I. 1928. Studies of chronic morphine poisoning in dogs. I General symptoms and behavior during addiction and withdrawal. J Pharmacol Exp Ther 33: 329-357.

Pons S, Torres-Aleman I. 1993. Estradiol modulates insulin-like growth factor I receptors and binding proteins in neurons from the hypothalamus. J Neuroendocrinol 5: 267-271.

Priest CA, Vink KL, Micevych PE. 1995. Temporal regulation by estrogen of beta-preprotachykinin mRNA expression in the rat ventromedial nucleus of the hypothalamus. Brain Res Mol Brain Res 28: 61-71.

Pu S, Kalra PS, Kalra SP. 1998. Ovarian steroid-independent diurnal rhythm in cyclic GMP/nitric oxide efflux in the medial preoptic area: possible role in preovulatory and ovarian steroid-induced LH surge. J Neuroendocrinol 10: 617-625.

Quadagno DM, McCullough J, Langan R. 1972. The effect of varying amounts of exogenous estradiol benzoate on estrous behavior in the rat. Horm Behav 3: 175-179.

Quesada A, Etgen AM. 2001. Insulin-like growth factor-1 regulation of alpha(1)-adrenergic receptor signaling is estradiol dependent in the preoptic area and hypothalamus of female rats. Endocrinology 142: 599-607.

Quesada A, Etgen AM. 2002. Functional interactions between estrogen and insulin-like growth factor-i in the regulation of alpha(1B)-adrenoceptors and female reproductive function. J Neurosci 22: 2401-2408.

Quesada A, Micevych PE. 2004. Estrogen interacts with the IGF-1 system to protect nigrostriatal dopamine and maintain motoric behavior after 6-hydroxdopamine lesions. J Neurosci Res 75: 107-116.

Quick MW, Lester RA. 2002. Desensitization of neuronal nicotinic receptors. J Neurobiol 53: 457-478.

Qureshi GA, Bednar I, Forsberg BG, Sodersten P. 1988. GABA inhibits sexual behaviour in female rats. Neuroscience 27: 169-174.

Rachman IM, Unnerstall JR, Pfaff DW, Cohen RS. 1998. Regulation of neuronal nitric oxide synthase mRNA in lordosis-relevant neurons of the ventromedial hypothalamus following short-term estrogen treatment. Brain Res Mol Brain Res 59: 105-108.

Raible LH, Gorzalka BB. 1986. Short and long term inhibitory actions of alpha-melanocyte stimulating hormone on lordosis in rats. Peptides 7: 581-586.

Rainbow T, Davis P, McEwen B. 1980a. Anisomycin inhibits the activation of sexual behavior by estradiol and progesterone. Brain Res 194: 548-555.

Rainbow TC, Degroff V, Luine VN, McEwen BS. 1980b. Estradiol 17 beta increases the number of muscarinic receptors in hypothalamic nuclei. Brain Res 198: 239-243.

Rao IM, Mahesh VB. 1986. Role of progesterone in the modulation of the preovulatory surge of gonadotropins and ovulation in the pregnant mare's serum gonadotropin-primed immature rat and the adult rat. Biol Reprod 35: 1154-1161.

Reinscheid RK, Ardati A, Monsma FJ, Civelli O. 1996. Structure-Activity Relationship Studies on the Novel Neuropeptide Orphanin FQ. J Biol Chem 271: 14163-14168.

Reinscheid RK, Higelin J, Henningsen RA, Monsma FJ, Civelli O. 1998. Structures that delineate orphanin FQ and dynorphin A pharmacological selectivities. J Biol Chem 273: 1490-1495.

Reisine T, Bell GI. 1993. Molecular Biology of Opioid Receptors. Trends Neurosci 16: 506-510.

Rempel LA, Clapper JA. 2002. Administration of estradiol-17beta increases anterior pituitary IGF-I and relative amounts of serum and anterior pituitary IGF-binding proteins in barrows. J Anim Sci 80: 214-224.

Rettori V, Canteros G, McCann SM. 1997. Interaction between NO and oxytocin: influence on LHRH release. Braz J Med Biol Res 30: 453-457.

Rettori V, Kamat A, McCann SM. 1994. Nitric oxide mediates the stimulation of luteinizing-hormone releasing hormone release induced by glutamic acid in vitro. Brain Res Bull 33: 501-503.

Richmond G, Clemens LG. 1986. Evidence for involvement of midbrain central gray in cholinergic mediation of female sexual receptivity in rats. Behav Neurosci 100: 376-380.

Rinderknecht E, Humbel RE. 1978. The amino acid sequence of human insulin-like growth factor I and its structural homology with proinsulin. J Biol Chem 253: 2769-2776.

Riskind P, Moss RL. 1983. Midbrain LHRH infusions enhance lordotic behavior in ovariectomized estrogen-primed rats independently of a hypothalamic responsiveness to LHRH. Brain Res Bull 11: 481-485.

Rissman EF, Wersinger SR, Fugger HN, Foster TC. 1999. Sex with knockout models: behavioral studies of estrogen receptor alpha. Brain Res 835: 80-90.

Rissman EF, Early AH, Taylor JA, Korach KS, Lubahn DB. 1997. Estrogen receptors are essential for female sexual receptivity. Endocrinology 138: 507-510.

Rodriguez-Sierra JF, Komisaruk BR. 1978. Lordosis induction in the rat by prostaglandin E2 systemically or intracranially in the absence of ovarian hormones. Prostaglandins 15: 513-524.

Rodriguez-Sierra JF, Sridaran R, Blake CA. 1980. Monosodium glutamate disruption of behavioral and endocrine function in the female rat. Neuroendocrinology 31: 228-235.

Romano GJ, Bonner TI, Pfaff DW. 1989. Preprotachykinin gene expression in the mediobasal hypothalamus of estrogen-treated and ovariectomized control rats. Exp Brain Res 76: 21-26.

Roselli-Rehfuss L, Mountjoy KG, Robbins LS, Mortrud MT, Low MJ, et al. 1993. Identification of a receptor for gamma melanotropin and other proopiomelanocortin peptides in the hypothalamus and limbic system. Proc Natl Acad Sci USA 90: 8856-8860.

Sabatino FD, Murnane JM, Hoffman RA, McDonald JK. 1987. Distribution of neuropeptide Y-like immunoreactivity in the hypothalamus of the adult golden hamster. J Comp Neurol 257: 93-104.

Sahlin L, Norstedt G, Eriksson H. 1994. Estrogen regulation of the estrogen receptor and insulinlike growth factor-I in the rat uterus: a potential coupling between effects of estrogen and IGF-I. Steroids 59: 421-430.

Sakuma Y, Akaishi T. 1987. Cell. size, projection path, and localization of estrogen-sensitive neurons in the rat ventromedial hypothalamus. J Neurophysiol 57: 1148-1159.

Sakuma Y, Pfaff D. 1979. Mesencephalic mechanisms of integration of female reproductive behavior in the rat. Am J Physiol 237: R285-290.

Sakuma Y, Pfaff DW. 1980a. Cells of origin of medullary projections in central gray of rat mesencephalon. J Neurophysiol 44: 1002-1011.

Sakuma Y, Pfaff DW. 1980b. LH-RH in the mesencephalic central grey can potentiate lordosis reflex of female rats. Nature 283: 566-567.

Sakuma Y, Pfaff DW. 1982. Properties of ventromedial hypothalamic neurons with axons to midbrain central gray. Exp Brain Res 46: 292-300.

Sar M, Sahu A, Crowley WR, Kalra SP. 1990. Localization of neuropeptide-Y immunoreactivity in estradiol-concentrating cells in the hypothalamus. Endocrinology 127: 2752-2756.

Satoh K, Armstrong DM, Fibiger HC. 1983. A comparison of the distribution of central cholinergic neurons as demonstrated by acetylcholinesterase pharmacohistochemistry and choline acetyltransferase immunohistochemistry. Brain Res Bull 11: 693-720.

Schiml PA, Rissman EF. 2000. Effects of gonadotropin-releasing hormones, corticotropin-releasing hormone, and vasopressin on female sexual behavior. Horm Behav 37: 212-220.

Schneider JE, Goldman MD, Tang S, Bean B, Ji H, et al. 1998. Leptin indirectly affects estrous cycles by increasing metabolic fuel oxidation. Horm Behav 33: 217-228.

Schulze HG, Gorzalka BB. 1991. Oxytocin effects on lordosis frequency and lordosis duration following infusion into the medial pre-optic area and ventromedial hypothalamus of female rats. Neuropeptides 18: 99-106.

Schumacher M, Coirini H, Frankfurt M, McEwen BS. 1989. Localized actions of progesterone in hypothalamus involve oxytocin. Proc Natl Acad Sci USA 86: 6798-6801.

Schumacher M, Coirini H, McEwen BS, Zaborszky L. 1991. Binding of [3 H] cholecystokinin in the ventromedial hypothalamus is modulated by an afferent brainstem projection but not by ovarian steroids. Brain Res 564: 102-108.

Schumacher M, Coirini H, Pfaff DW, McEwen BS. 1990. Behavioral effects of progesterone associated with rapid modulation of oxytocin receptors. Science 250: 691-694.

Schwartz SM, Blaustein JD, Wade GN. 1979. Inhibition of estrogen behavior by progesterone in rats: Role of neural estrogen and progestin receptors. Endocrinology 105: 1078-1082.

Scimonelli T, Medina F, Wilson C, Celis ME. 2000. Interaction of alpha-melanotropin (alpha-MSH) and noradrenaline in the median eminence in the control of female sexual behavior. Peptides 21: 219-223.

Seeburg P. 1993. The TiPS/TINS lecture: The molecular biology of mammalian glutamate receptor channels. Trends Pharmacol Sci 14: 297-303.

Seevers M. 1936. Opiate addiction in the monkey. I. Methods of study. J Pharmacol Exp Ther 56: 147-156.

Sherwood NM, Krueckl SL, McRory JE. 2000. The origin and function of the pituitary adenylate cyclase-activating polypeptide (PACAP)/glucagon superfamily. Endocr Rev 21: 619-670.

Shimohigashi Y, Hatano R, Fujita T, Nakashima R, Nose T, et al. 1996. Sensitivity of opioid receptor-like receptor ORL1 for chemical modification on nociceptin, a naturally occurring nociceptive peptide. J Biol Chem 271: 23642-23645.

Shingo AS, Kito S. 2003. Estrogen induces insulin-like growth factor-1 mRNA expression in the immortalized hippocampal cell: determination by quantitative real-time polymerase chain reaction. Neurochem Res 28: 1379-1383.

Shughrue PJ, Lane MV, Merchenthaler I. 1997. Comparative distribution of estrogen receptor-alpha and -beta mRNA in the rat central nervous system. J Comp Neurol 388: 507-525.

Sinchak K, Micevych PE. 2001a. Progesterone blockade of estrogen activation of μ-opioid receptors regulates reproductive behavior. J Neurosci 21: 5723-5729.

Sinchak K, Micevych PE. 2001b. Progesterone blockade of estrogen activation of mu-opioid receptors regulates reproductive behavior. J Neurosci 21: 5723-5729.

Sinchak K, Micevych P. 2003. Visualizing activation of opioid circuits by internalization of G protein-coupled receptors. Mol Neurobiol 27: 197-222.

Sinchak K, Romeo HE, Micevych PE. 2006. Site-specific estrogen and progestin regulation of orphanin FQ/nociceptin and nociceptin opioid receptor mRNA expression in the female rat limbic hypothalamic system. J Comp Neurol 496: 252-268.

Sinchak K, Dewing P, Cook M, Micevych P. Submitted. Release of orphanin FQ/Nociceptin in the medial preoptic nucleus and ventromedial nucleus of the hypothalamus facilitates lordosis. in press.

Sinchak K, Hendricks DG, Baroudi R, Micevych PE. 1997. Orphanin FQ/nociceptin in the ventromedial nucleus facilitates lordosis in female rats. Neuroreport 8: 3857-3860.

Sinchak K, Hipschman S, Cook M, Micevych P. 2004a. Nociceptin/orphanin FQ afferents of the medial preoptic nucleus. In Society for Neuroscience, San Diego. Program No. 423.417.

Sinchak K, Mills RH, Eckersell CB, Micevych PE. 2004b. Medial preoptic area delta-opioid receptors inhibit lordosis. Behav Brain Res 155: 301-306.

Sinchak K, Shahedi K, Dewing P, Micevych P. 2005. Sexual receptivity is reduced in the female mu-opioid receptor knockout mouse. Neuroreport 16: 1697-1700.

Sinchak K, Yang W, Lino S, Micevych P. 2003. Orphanin FQ infusion into the medial preoptic nucleus facilitates lordosis in estrogen primed female rats, New Orleans, LA. 727.712.

Sinchak K, Eckersell C, Quezada V, Norell A, Micevych P. 2000. Preproenkephalin mRNA levels are regulated by acute stress and estrogen stimulation. Physiol Behav 69: 425-432.

Sirinathsinghji DJ. 1984. Modulation of lordosis behavior of female rats by naloxone, beta-endorphin and its antiserum in the mesencephalic central gray: possible mediation via GnRH. Neuroendocrinology 39: 222-230.

Sirinathsinghji DJ. 1985. Modulation of lordosis behaviour in the female rat by corticotropin releasing factor, beta-endorphin and gonadotropin releasing hormone in the mesencephalic central gray. Brain Res 336: 45-55.

Sirinathsinghji DJS. 1986. Regulation of lordosis behavior in the female rat by corticotropin-releasing factor, beta-endorphin/corticotropin and luteinizing hormone-releasing hormone neuronal systems in the medial preoptic area. Brain Res 375: 149-156.

Sirinathsinghji DJ, Whittington PE, Audsley A, Fraser HM. 1983. β-Endorphin regulates lordosis in female rats by modulating Lh-Rh release. Nature 301: 62-64.

Smith A, Lee N, Loh H. 1983. The multiple-site beta-endorphin receptor. Trends Pharmacol Sci 4: 163-164.

Smythies J. 2005. Section II. The dopamine system. Int Rev Neurobiol 64: 123-172.

Sodersten P, Eneroth P. 1981. Serum levels of oestradiol-17 beta and progesterone in relation to receptivity in intact and ovariectomized rats. J Endocrinol 89: 45-54.

Sodersten P, De Vries GJ, Buijs RM, Melin P. 1985. A daily rhythm in behavioral vasopressin sensitivity and brain vasopressin concentrations. Neurosci Lett 58: 37-41.

Sodersten P, Henning M, Melin P, Ludlin S. 1983. Vasopressin alters female sexual behaviour by acting on the brain independently of alterations in blood pressure. Nature 301: 608-610.

Spragg S. 1940. Morphine addiction in chimpanzees. Comp Psychol Monogr 15: 132-190.

Sprouse JS, Aghajanian GK. 1987. Electrophysiological responses of serotoninergic dorsal raphe neurons to 5-HT1A and 5-HT1B agonists. Synapse 1: 3-9.

Stanley BG, Leibowitz SF. 1985. Neuropeptide Y injected in the paraventricular hypothalamus: a powerful stimulant of feeding behavior. Proc Natl Acad Sci USA 82: 3940-3943.

Suda M, Nakao K, Sakamoto M, Morii N, Sugawara A, et al. 1986. Changes in the immunoreactivities of an opioid peptide leumorphin in the hypothalamus and anterior pituitary during the estrous cycle of the rat and their relation to sexual behavior. Brain Res 374: 236-243.

Sumner BE, Fink G. 1997. The density of 5-hydoxytryptamine2A receptors in forebrain is increased at pro-oestrus in intact female rats. Neurosci Lett 234: 7-10.

Sumner BE, Grant KE, Rosie R, Hegele-Hartung C, Fritzemeier KH, et al. 1999. Effects of tamoxifen on serotonin transporter and 5-hydroxytryptamine(2A) receptor binding sites and mRNA levels in the brain of ovariectomized rats with or without acute estradiol replacement. Brain Res Mol Brain Res 73: 119-128.

Swanson LW, Mogenson GJ, Simerly RB, Wu M. 1987. Anatomical and electrophysiological evidence for a projection from the medial preoptic area to the 'mesencephalic and subthalamic locomotor regions' in the rat. Brain Res 405: 108-122.

Swanson LW, Sawchenko PE, Rivier J, Vale WW. 1983. Organization of ovine corticotropin-releasing factor immunoreactive cells and fibers in the rat brain: an immunohistochemical study. Neuroendocrinology 36: 165-186.

Tatemoto K. 1982. Isolation and characterization of peptide YY (PYY), a candidate gut hormone that inhibits pancreatic exocrine secretion. Proc Natl Acad Sci USA 79: 2514-2518.

Tatemoto K, Carlquist M, Mutt V. 1982. Neuropeptide Y-a novel brain peptide with structural similarities to peptide YY and pancreatic polypeptide. Nature 296: 659-660.

Tatum A, Seevers M, Collins K. 1929. Morphine addiction and its physiological interpretation based on experimental evidence. J Pharmacol Exp Ther 36: 477-510.

Tetel MJ, Celentano DC, Blaustein JD. 1994. Intraneuronal convergence of tactile and hormonal stimuli associated with female reproduction in rats. J Neuroendocrinol 6: 211-216.

Tetel MJ, Getzinger MJ, Blaustein JD. 1993. Fos expression in the rat brain following vaginal-cervical stimulation by mating and manual probing. J Neuroendocrinol 5: 397-404.

Thody AJ, Wilson CA. 1983. Melanocyte stimulating hormone and the inhibition of sexual behaviour in the female rat. Physiol Behav 31: 67-72.

Thompson RC, Mansour A, Akil H, Watson SJ. 1993. Cloning and Pharmacological Characterization of a Rat Mu-Opioid Receptor. Neuron 11: 903-913.

Torii M, Kubo K. 1994. The effects of intraventricular injection of beta-endorphin on initial estrogen action to induce lordosis behavior. Physiol Behav 55: 157-162.

Torii M, Kubo K, Sasaki T. 1995. Naloxone and initial estrogen action to induce lordosis in ovariectomized rats: the effect of a cut between the septum and preoptic area. Neurosci Lett 195: 167-170.

Torii M, Kubo K, Sasaki T. 1996. Influence of opioid peptides on the priming action of estrogen on lordosis in ovariectomized rats. Neurosci Lett 212: 68-70.

Torii M, Kubo K, Sasaki T. 1997. Differential effects of beta-endorphin and Met- and Leu-enkephalin on steroid hormone-induced lordosis in ovariectomized female rats. Pharmacol Biochem Behav 58: 837-842.

Torii M, Kubo K, Sasaki T. 1999. Facilitatory and inhibitory effects of beta-endorphin on lordosis in female rats: relation to time of administration. Horm Behav 35. 271-278.

Uhler M, Herbert E, D'Eustachio P, Ruddle FD. 1983. The mouse genome contains two nonallelic pro-opiomelanocortin genes. J Biol Chem 258: 9444-9453.

Ulibarri C, Micevych PE. 1993. Role of perinatal estrogens in sexual differentiation of the inhibition of lordosis by exogenous cholecystokinin. Physiol Behav 54: 95-100.

Uphouse L. 2000. Female gonadal hormones, serotonin, and sexual receptivity. Brain Res Brain Res Rev 33: 242-257.

Uphouse L, Wolf A. 2004. WAY100635 and female rat lordosis behavior. Brain Res 1013: 260-263.

Uphouse L, Caldarola-Pastuszka M, Montanez S. 1992. Intracerebral actions of the 5-HT1A agonists, 8-OH-DPAT and buspirone and of the 5-HT1A partial agonist/antagonist, NAN-190, on female sexual behavior. Neuropharmacology 31: 969-981.

Uphouse L, Caldarola-Pastuszka M, Moore N. 1993. Inhibitory effects of the 5-HT1A agonists, 5-hydroxy- and 5-methoxy-(3-di-n-propylamino)chroman, on female lordosis behavior. Neuropharmacology 32: 641-651.

Uphouse L, Salamanca S, Caldarola-Pastuszka M. 1991. Gender and estrous cycle differences in the response to the 5-HT1A agonist 8-OH-DPAT. Pharmacol Biochem Behav 40: 901-906.

Uphouse L, Andrade M, Caldarola-Pastuszka M, Jackson A. 1996. 5-HT1A receptor antagonists and lordosis behavior. Neuropharmacology 35: 489-495.

Vasudevan N, Kow LM, Pfaff DW. 2001. Early membrane estrogenic effects required for full expression of slower genomic actions in a nerve cell line. Proc Natl Acad Sci USA 98: 12267-12271.

Vathy I, Etgen AM. 1988. Ovarian steroids and hypothalamic norepinephrine release: studies using in vivo brain microdialysis. Life Sci 43: 1493-1499.

Vathy I, Plas van der J, Vincent PA, Etgen AM. 1991. Intracranial dialysis and microinfusion studies suggest that morphine acts in the ventromedial hypothalamus to inhibit female rat sexual behavior. Horm Behav 25: 354-366.

Vesely DL. 1979. Testosterone and its precursors and metabolites enhance guanylate cyclase activity. Proc Natl Acad Sci USA 76: 3491-3494.

Vesely DL, Hill DE. 1980. Estrogens and progesterone increase fetal and maternal guanylate cyclase activity. Endocrinology 107: 2104-2109.

Vilaro MT, Palacios JM, Mengod G. 1990. Localization of m5 muscarinic receptor mRNA in rat brain examined by in situ hybridization histochemistry. Neurosci Lett 114: 154-159.

Villablanca AC, Hanley MR. 1997. 17beta-estradiol stimulates substance P receptor gene expression. Mol Cell Endocrinol 135: 109-117.

Wade GN, Lempicki RL, Panicker AK, Frisbee RM, Blaustein JD. 1997. Leptin facilitates and inhibits sexual behavior in female hamsters. Am J Physiol 272: R1354-1358.

Wagner EJ, Ronnekleiv OK, Grandy DK, Kelly MJ. 1998. The peptide orphanin FQ inhibits beta-endorphin neurons and neurosecretory cells in the hypothalamic arcuate nucleus by activating an inwardly-rectifying K+ conductance. Neuroendocrinology 67: 73-82.

Waldhoer M, Bartlett SE, Whistler JL. 2004. Opioid receptors. Annu Rev Biochem 73: 953-990.

Wall SJ, Yasuda RP, Li M, Wolfe BB. 1991. Development of an antiserum against m3 muscarinic receptors: distribution of m3 receptors in rat tissues and clonal cell lines. Mol Pharmacol 40: 783-789.

Wang JB, Johnson PS, Imai Y, Persico AM, Ozenberger BA, et al. 1994a. cDNA cloning of an orphan opiate receptor gene family member and its splice variant. FEBS Lett 348: 75-79.

Wang JB, Johnson PS, Persico AM, Hawkins AL, Griffin CA, et al. 1994b. Human mu opiate receptor. cDNA and genomic clones, pharmacologic characterization and chromosomal assignment. FEBS Lett 338: 217-222.

Weaver DR, Clemens LG. 1987. Nicotinic cholinergic influences on sexual receptivity in female rats. Pharmacol Biochem Behav 26: 393-400.

White S, Uphouse L. 2004. Estrogen and progesterone dosedependently reduce disruptive effects of restraint on lordosis behavior. Horm Behav 45: 201-208.

Wiesner JB, Moss. 1984. Beta-endorphin suppression of lordosis behavior lack of effect of peripherally-administered naloxone. Life Sci 34: 1455-1462.

Wikberg JE, Muceniece R, Mandrika I, Prusis P, Lindblom J, et al. 2000. New aspects on the melanocortins and their receptors. Pharmacol Res 42: 393-420.

Williams GW, McGinnis MY, Lumia AR. 1992. The effects of olfactory bulbectomy and chronic psychosocial stress on serum glucocorticoids and sexual behavior in female rats. Physiol Behav 52: 755-760.

Wilson CA, Thody AJ, Hole DR, Grierson JP, Celis ME. 1991. Interaction of estradiol, alpha-melanocyte-stimulating hormone, and dopamine in the regulation of sexual receptivity in the female rat. Neuroendocrinology 54: 14-22.

Wimalasena J, Meehan D, Dostal R, Foster JS, Cameron M, et al. 1993. Growth factors interact with estradiol and gonadotropins in the regulation of ovarian cancer cell growth and growth factor receptors. Oncol Res 5: 325-337.

Witt DM, Insel TR. 1991. A selective oxytocin antagonist attenuates progesterone facilitation of female sexual behavior. Endocrinology 128: 3269-3276.

Wolf A, Caldarola-Pastuszka M, Uphouse L. 1998a. Facilitation of female rat lordosis behavior by hypothalamic infusion of 5-HT(2A/2C) receptor agonists. Brain Res 779: 84-95.

Wolf A, Caldarola-Pastuszka M, De Lashaw M, Uphouse L. 1999. 5-HT2C receptor involvement in female rat lordosis behavior. Brain Res 825: 146-151.

Wolf A, Jackson A, Price T, Trevino A, Caldarola-Pastuszka M, et al. 1998b. Attenuation of the lordosis-inhibiting effects of 8-OH-DPAT by TFMPP and quipazine. Brain Res 804: 206-211.

Wu TJ, Glucksman MJ, Roberts JL, Mani SK. 2006. Facilitation of Lordosis in Rats by a Metabolite of Luteinizing Hormone Releasing Hormone (LHRH). Endocrinology 147: 2544-2549.

Wu TJ, Pierotti AR, Jakubowski M, Sheward WJ, Glucksman MJ, et al. 1997. Endopeptidase EC 3.4.24.15 presence in the rat median eminence and hypophysial portal blood and its modulation of the luteinizing hormone surge. J Neuroendocrinol 9: 813-822.

Xing C, Yin Y, He X, Xie Z. 2006. Effects of insulin-like growth factor 1 on voltage-gated ion channels in cultured rat hippocampal neurons. Brain Res 1072: 30-35.

Yamada Y, Nishida E. 1978. Effects of estrogen and adrenal androgen on unit activity of the rat brain. Brain Res 142: 187-190.

Yang LY, Clemens LG. 1997. Function of intromissions on intromission-return latency of female rats during paced sexual behavior. Physiol Behav 61: 889-894.

Yu WH, Walczewska A, Karanth S, McCann SM. 1997. Nitric oxide mediates leptin-induced luteinizing hormone-releasing hormone (LHRH) and LHRH and leptin-induced

LH release from the pituitary gland. Endocrinology 138: 5055-5058.

Zapf J, Hauri C, Futo E, Hussain M, Rutishauser J, et al. 1995. Intravenously injected insulin-like growth factor (IGF) I/ IGF binding protein-3 complex exerts insulin-like effects in hypophysectomized, but not in normal rats. J Clin Invest 95: 179-186.

Zhang Y, Lu L, Furlonger C, Wu GE, Paige CJ. 2000. Hemokinin is a hematopoietic-specific tachykinin that regulates B lymphopoiesis. Nat Immunol 1: 392-397.

Zoli M, Agnati LF, Tinner B, Steinbusch HW, Fuxe K. 1993. Distribution of dopamine-immunoreactive neurons and their relationships to transmitter and hypothalamic hormone-immunoreactive neuronal systems in the rat mediobasal hypothalamus. A morphometric and microdensitometric analysis. J Chem Neuroanat 6: 293-310.

Zubieta JK, Frey KA. 1993. Autoradiographic mapping of M3 muscarinic receptors in the rat brain. J Pharmacol Exp Ther 264: 415-422.

5 Neuroendocrinology and Neurochemistry of Maternal Motivation and Behavior

J. S. Lonstein · J. I. Morrell

Abstract: Successful reproduction requires transmission of traits from one generation to the next by ensuring that offspring carrying the genetic basis of these traits survive to reproductive age. Maternal caregiving is the most crucial behavioral pattern ensuring the survival of young in most mammals. Expression of maternal behavior requires complex interactions between internal and external sensory cues, numerous endocrine systems, and the brain. We herein summarize the components of maternal care, and the hormonal, sensory, and neural underpinnings of its expression, with the laboratory rat model as an exemplar. The impetus to act maternally is similar to other goal-directed behaviors, in that it requires an appropriate motivational state. We emphasize how endocrine systems during pregnancy, parturition, and lactation may interact with traditional neural motivation systems, including the mesolimbic and nigrosriatal dopamine systems, to facilitate the transition from indifference to infants to a state of irresistible nurturance toward them.

List of Abbreviations: 6-OHDA, 6 hydroxydopamine; AMYbl, basolateral nucleus of the amygdala; AVP, arginine vasopression; BNST, bed nucleus of the stria terminalis; CCK, cholecystokinin; CNS, central nervous system; $cPAG_l$, caudal lateral subregion of the periaqueductal gray; $cPAG_{vl}$, caudal ventrolateral subregion of the periaquaductal gray; D1, dopamine receptor subtype 1; D2, dopamine receptor subtype 2; D3, Dopamine receptor subtype 3; DA, dopamine; DAergic, dopaminergic ; DOPAC, dihydroxyphenylacetic acid; DHT, dihydrotestosterone; EB, estradiol benzoate; ER, estrogen receptor; ERα, estrogen receptor alpha; ERβ, estrogen receptor beta; Fos, protein product of the immediate early gene *c-fos*; FSH, follicle stimulating hormone; GABA, γ-aminobutyric acid; $GABA_A$, γ-aminobutyric acid receptor subtype A; h, hours; Hbc, habenular complex; H-O, hysterectomy-ovariectomy ; ICV, intracerebroventricular ; IR, immunoreactive; Lhb, lateral subdivision of the habenular complex; LH, luteinizing hormone; lPOA, lateral preoptic area; LS, lateral septum; MeA, medial nucleus of the amygdala; mPOA, medial preoptic area ; MRI, magnetic resonance imaging; mRNA, messenger ribonucleic acid; N, number of animals; NA, nucleus accumbens; OT, oxytocin; OTR, oxytocin receptor; PAG, periaqueductal gray; PFC, prefrontal cortex; POA, preoptic area; PR, progestin receptor; PRL, prolactin; PRL-R, prolactin receptor; PVN, paraventricular nucleus of the hypothalamus; s.c., subcutaneous; SEM, standard error of the mean; SN, substantia nigra; SNpc, substantia nigra pars compacta; T, testosterone; TH-ir, tyrosine hydroxylase immunoreactivity; V1a, vasopressin receptor 1[a]; vBNST, ventral area of bed nucleus of the stria terminalis; VMN, ventromedial nucleus of the hypothalamus; VMH, ventromedial hypothalamus; VTA, ventral tegmental area

1 Introduction

One of the most dramatic changes in behavior by females of many mammalian species is the transition from a nonmaternal state existing before and during pregnancy, to a state of high maternal responsiveness that emerges after parturition. This transition from infant rejection to infant acceptance requires a complex milieu of hormones and other neurochemicals. The largest and most detailed literature on the biology underlying parental behavior has been generated using the laboratory rat model and has been reviewed many times with great completeness and care by Rosenblatt et al. (1979), Fahrbach and Pfaff (1982), Stern (1989), Numan (1988, 1994), Rosenblatt (1995), and most recently in an entire book devoted to the subject by Numan and Insel (2003). Our goals in this chapter are to highlight recent findings on the neural and hormonal regulation of maternal behavior in the rat. While we provide some context for these findings, we refer our readers to prior reviews of the extensive literature on this topic for additional details. We have attempted throughout to suggest future directions for the field and in particular, we examine the generally less investigated topic of maternal motivation.

Scientists examining maternal behavior often seek a reductionist, mechanistic model of each of the broad components of this behavior and their behavioral subroutines. Any behavioral subroutine is worth studying, but considering these subroutines within an ethologically relevant context is particularly powerful, because it allows a more complete understanding of how the brain supports a seamless behavioral sequence. This sequence emerges from a complex pattern of activity across neural structures with a particular neurochemistry. Most components of the circuit can be determined to be necessary or not, sufficient or not. A component that is "not necessary" can nonetheless be normally used in the circuit, but redundant pathways may exist so that the behavior can continue to be displayed in the absence of a particular structure. Additionally, structures that are "not necessary" in one ethological context might, in fact, provide valuable flexibility so that normal behavior is maintained regardless of the demands of the environment.

We believe that the study of maternal behavior in the rat has two broad goals. First, maternal behavior can be considered a prototype complex behavior. We know it is important, because it is critical for the continuance of a very large number of vertebrate species. The assumption is that understanding the components of maternal behavior allows insight into how the integrated patterns of brain function ultimately yield other complex behaviors. Secondly, we propose that the information gathered in the rat will validly generalize to the fundamental neurobiological underpinnings of parental behavior in humans. Although obvious to most neuroscientists, we wish to emphasize that even human parenting behavior has its roots in central nervous system function, with cultural and "sentient" influences on parenting unique to humans layered on top of these fundamental biological roots. While most humans accept that our behavior is based in genetic and central nervous system (CNS) function, human maternal/parental behavior is often thought to function independently of these genetic and neural underpinning. That is, "true maternal behavior" is something of pure emotion and soul, and not definable by the cold light of science. We submit that, on the contrary, understanding the fundamental neuroanatomy and neurochemistry of this behavior is possible, and will ultimately be of great societal benefit.

2 Components of Maternal Care

Animal models of parental behavior define it operationally and in a species-specific framework. In rats, only the female cares for the pups and our operational definitions of parental behavior in the rat are referred to as maternal behavior. Operationally, defining complex behaviors in animal models is a typical approach to examining these behaviors. What is our definition of maternal behavior? The two aspects are (1) the outwardly visible behavioral components, or consummatory aspects, and (2) the motivation to perform maternal behavior, or the appetitive aspects. We define expression of the behavior by the retrieval, nest building, and pup care components that are detailed immediately below. We define the motivational components of maternal behavior in ➋ *Section 5* below through examining maternal motivation using measures typical of other behavioral and pharmacological examinations of motivation.

2.1 Active Maternal Behaviors

The consummatory aspects of maternal behavior in rodents encompass two general categories of behavior, active maternal behaviors and inactive maternal behaviors (see Stern, 1996; Lonstein and Fleming, 2001). Parturition behaviors are the first active maternal behaviors dams display toward pups and they begin with dams licking and pulling at the emerging fetuses to extract them from the vagina. Vigorous licking and gentle nibbling ensues, which removes pups from their amniotic sac, cleans them of amniotic fluid, and separates them from their umbilical cord and placenta (Hollaway et al., 1980). These oral behaviors by the dam not only help extract the neonates from their prenatal encumbrances, but also stimulate their respiration and general activity that leads to the first bout of nursing that may begin within minutes after completion of parturition (Hollaway et al., 1980). After parturition, one of the most recognizable and easily studied active maternal behaviors is the retrieval of displaced pups. In most cases, dams orient toward the pups and often sniff them (the appetitive response) prior to gently picking each one up with her mouth and transporting it (the consummatory response). Although retrieval of pups is very often studied in the laboratory, it may be rare in a natural setting unless the pups are older and more likely to wander from the nest, or if there is a disturbance to the nest site (Brewster and Leon, 1980).

Once pups are collected in a central location, such as the nest, dams spend a great deal of time actively hovering over the litter while performing other activities including self-grooming and licking the pups. Self-grooming during this time cannot be considered a maternal behavior because there is no known advantage for the offspring, although the mother's prepartum licking of her ventrum is important for her mammary gland development (Roth and Rosenblatt, 1968). In contrast, the frequent maternal licking of the offspring serves numerous purposes for the pups, not only cleaning them, but also stimulating excretion if licking is directed at their anogenital region. Maternal licking also activates them behaviorally, which facilitates their search for a nipple and subsequent suckling (Stern and Johnson, 1989). Licking the pups is also beneficial to the dam because the ingestion of pup's excreta provides her with additional water and electrolytes during this metabolically challenging period (Friedman et al., 1981).

Other active maternal behaviors differ from retrieval and licking because they are not directed toward the pups, but may still be necessary for their survival. Nest construction involves the dam using her mouth, snout, and paws to carry or push material to a central location and then shaping the material to form a structure with compact walls and a depression in the center. Nest construction is not exclusive to the lactating rat, but maternal nests are typically larger and more tightly constructed than those of nonmaternal animals. Construction of a nest is certainly important because it not only provides the locale where most mother–infant interactions occur, but is also important for thermoregulation of the dam and litter (e.g., Leon and Woodside, 1983).

Heightened aggression also emerges during lactation and is, of course, not directed toward offspring, but rather toward potentially threatening conspecifics. Maternal aggression serves to protect the dam's most valuable resources, the litter and nest site (Lonstein and Gammie, 2002). Increased foraging and food intake are also associated with lactation, and are necessary for the high metabolic demands of lactation (e.g., Fleming, 1976).

2.2 Inactive Maternal Behaviors

Interspersed with the display of these active maternal behaviors are prolonged periods of inactive nursing, when milk is transferred to the infants. When a sufficient number of pups suckle, dams undergo a transition from a highly active state to one of relative quiescence. During this quiescence, she will frequently assume a distinctive nursing posture, termed kyphosis (Stern, 1996), characterized by limb rigidity and an upward flexion of the spinal column. This posture provides sufficient room for pups to breathe, move, and suckle while underneath the dam. Kyphosis is not the only nursing posture that dams can assume. They can also lie prone on top of the litter with little limb support or nurse them while lying on their side in a supine position. Only during these periods of quiescent nursing does milk letdown occur (Voloschin and Tremezani, 1984), and during early lactation it is more likely to occur when dams are displaying kyphosis than other nursing postures (Lonstein et al., 1998).

2.3 Sensory Influences on Maternal Care

The sensory determinants of active and inactive maternal behaviors in rats are well understood and have been reviewed in detail previously (Stern, 1989, 1996; Numan, 1994). Dams normally use all sensory modalities to fine-tune their maternal responding, but do not require every type of sensory input to display maternal behavior. For example, eliminating the ability of dams to see and hear their pups does not abolish retrieval or any other maternal behavior (see Stern, 1989). How the sense of taste influences maternal behavior has not been examined in detail, but one could imagine that it might be involved in maternal licking of pups. Anesthetization of the tongue decreases maternal licking (Stern and Johnson, 1989), but because this procedure eliminates both taste and somatosensory inputs to the tongue, it is unclear if both lingual senses are important.

Olfactory input is critical for maternal behavior in mice (Gandelman et al., 1972), but its effects on maternal behavior in rats depend on their reproductive state. When virgin female rats have their olfactory bulbs or other areas of their olfactory system destroyed, maternal responding is facilitated, presumably because they no longer detect what is thought to be the repellant odors emanating from the neonates (Fleming and Rosenblatt, 1974a, b). The effects of postpartum olfactory bulbectomy on maternal behavior vary. If bulbectomy is performed during pregnancy, severe disruptions in maternal behavior may be observed following parturition in rats (Schwartz and Rowe, 1976; Kolunie and Stern, 1995), although some studies report virtually no effects (Fleming and Rosenblatt, 1974a; Numan and Numan, 1995; Walsh et al., 1996). If bulbectomy does impair maternal behavior, it might be particularly true in primiparous females without prior maternal experience or in previously cannibalistic females (Fleming and Rosenblatt, 1974a,b; Schwartz and Rowe, 1976; Kolunie and Stern, 1995). The relative importance of nonvolatile and volatile odors in maternal care has been carefully examined. Destruction of just the bed nucleus of the accessory olfactory tract, which primarily receives and transmits nonvolatile (pheromonal) information, facilitates maternal responding in virgin females and suggests that nonvolatile cues from pups inhibit the behavior (Del Cerro et al., 1991). Once dams are maternal, though, disruption of nonvolatile olfactory inputs by severing the vomeronasal organ has little effect on maternal behavior in rats (Jirik-Babb et al., 1984; Kolunie and Stern, 1995). Presumed reduction of volatile inputs with postpartum intranasal infusion of zinc sulfate also has little effect (Benuck and Rowe, 1975; Jirik-Babb et al., 1984; Kolunie and Stern, 1995). Therefore, it must be the case that either: (1) olfaction has little role in the postpartum maintenance of maternal behavior, (2) if it does, then volatile or nonvolatile olfactory input can singularly sustain normal maternal behavior, or (3) any olfactory bulbectomy effects on postpartum maternal behavior may be due to surgical or neurological factors other than disruption of olfactory inputs.

The sense of touch has been extensively studied for its role in maternal behaviors in rats (Stern, 1996), and in contrast to any other sensory modality, it is absolutely critical. Reduction or elimination of tactile inputs to the perioral region through anesthetization of the mystacial pads, or lesions of the infraorbital nerve subserving the mystacial pads and surrounding area, impairs or abolishes retrieval and licking of pups (Kenyon et al., 1983; Stern and Johnson, 1989). Analogously, quiescent nursing behavior is impaired or absent if somatosensory inputs to the dam's ventrum are reduced. This can be accomplished by anesthetizing her ventrum or removing her nipples. This can also be accomplished by manipulating the pups (by heating or cooling them, or anesthetizing them periorally or completely) so they cannot root for a nipple and/or suckle (Stern, 1996). In sum, maternal behavior is surely influenced by all sensory modalities, but any individual sensory modality other than somatosensation seems to be dispensable.

3 Peripheral Hormones Associated with the Onset and Maintenance of Maternal Behavior

3.1 Pregnancy and Parturition – the Optimal Endocrine Basis for Maternal Behavior and Maternal Motivation

The parturient rat immediately expresses complete maternal behavior at the moment her pups are born. While this immediate onset of the expression of maternal behavior is understood to require particular hormones, as discussed in detail below, once the hormonal induction of the expression of maternal behavior has occurred, the expression of maternal behavior does not require hormones. This rapid onset and complete behavioral expression at parturition are the basis to compare behavioral outcomes of other endocrine/behavioral models. Reductionist experiments using models with deliberately limited endocrine millieu typically demonstrate a slower onset of maternal behavior (1–4 days after presentation of foster pups), and may not produce all components of the behavior. Whether and to what extent hormones have a role in the onset or maintenance of maternal motivation has been examined much less than the hormonal basis of the expression of maternal behavior.

3.1.1 Endocrine Sources During Pregnancy and Lactation

Our greatest insights into hormonal regulation of maternal behavior have come from considering the hormones of pregnancy and the postpartum period. During this period the pituitary, ovary, and uterus are in unique endocrine states. Notable during pregnancy are the corpora lutea of pregnancy, the decidua (uterine mucosa) of the uterus, the placenta, and the fetus (Gibori et al., 1988; Talamantes and Ogren, 1988). Unique to the postpartum period are the postpartum estrus, the corpora lutea of lactation, the process of lactation, and the presence of the pups as endocrine/neuroendocrine stimuli (Tucker, 1988). If fertilization occurs during the postpartum estrus, the second pregnancy contemporaneous with lactation is another unique feature of this period.

3.1.2 Endocrine Models for the Examination of Maternal Behavior

Correlating hormone levels in the pregnant and parturient female with the expression of maternal behavior was an important initial model, often followed by studies eliminating the hormone sources. Such reductionist experiments to explore the causal links between hormones and behavior are limited by the endocrine and physiological demands of pregnancy, parturition, and lactation. That is, removal of glands may alter maternal behavior because of a disruption in these other processes. For example, the maintenance of pregnancy requires pituitary hormones during the first half of gestation and ovarian steroid hormone production throughout (Heap and Flint, 1986), and these glands cannot be removed without precipitating miscarriage. Similarly, the ovaries and pituitary are required for sustained lactation. Normal adrenal function greatly facilitates pregnancy, parturition, and lactation, but these can continue without adrenal hormones (Thoman and Levine, 1970; Thoman et al., 1970).

A very fruitful model for understanding the hormone dependency of the natural onset of maternal behavior after pregnancy is the hysterectomy–ovariectomy (H–O) model in which the uterus (including fetuses) and/or the ovaries are removed (Rosenblatt, 1995). Together with the use of foster pups provided by other parturient females, this model avoids the limitations of pregnancy and parturition. Nearly three decades of work with this model has uncovered the role of individual hormones or groups of hormone regimes, in the hormonally regulated onset of maternal behavior (Siegel and Rosenblatt, 1975a, b; Rosenblatt et al., 1979, 1998). For similar reasons, the H–O model has also been very useful in examining the neural basis of maternal behavior (Corodimas et al., 1992, 1993).

Virgin females and males, and even juvenile rats, can be stimulated to perform some maternal behaviors simply by continuously exposing them to pups. Pup-induced maternal behavior has historically been referred to as concaveation, or more commonly, pup-sensitized maternal behavior. This model has also been used to consider the hormonal basis of maternal behavior, as the onset of sensitized maternal behavior can be shortened by hormones (Rosenblatt, 1967; Rosenblatt and Ceus, 1998; Olazabal et al., 2002).

3.1.3 Minimum Endocrine and Physiological Elements Necessary for the Onset and Maintenance of Maternal Behavior

Very early studies established that the onset of maternal behavior in parturient females requires ovarian steroids (Terkel and Rosenblatt, 1972; Rosenblatt et al., 1979; Fahrbach and Pfaff, 1982). Less clear was the role of the pituitary, with some studies indicating that the pituitary was not necessary (Rosenblatt et al., 1979) and others indicating that pituitary prolactin was active in the rapid onset of maternal behavior (Moltz et al., 1970; Fahrbach and Pfaff, 1982; Numan and Insel, 2003). Prepartum adrenal hormones are also probably not absolutely necessary (Rosenblatt et al., 1979). Neither the pituitary, ovary, nor adrenal glands are required for the postpartum maintenance of the expression of maternal behavior once it is established, although subtle changes in maternal behavior may exist after their removal (Rosenblatt et al., 1979; Fahrbach and Pfaff, 1982; Numan 1988, 1994; Rees et al., 2004).

The H–O model revealed that as pregnancy progresses, the female will display maternal behavior more and more rapidly after the presentation of pups (Rosenblatt et al., 1979; Numan, 1988, 1994; Rosenblatt, 1995). While nulliparious or early pregnant females require as much as 6 days of exposure to pups before the expression of maternal behavior, females after 17–19th day of pregnancy are fully maternal within 1 day of pup exposure, depending on the additional hormone regime (Rosenblatt et al., 1998). Thus, the hormonal milieu of the progression of pregnancy is a critical basis for the onset of maternal behavior.

Early work has also ruled out parturition and lactation as requirements for maternal behavior. Removal of the fetuses and placenta at the later stages of pregnancy (by Caesarean procedures), leaving the uterus and ovary intact, demonstrated that the sensory processes of parturition are not required for the onset of maternal behavior (Moltz et al., 1966). Nor is the capacity to lactate necessary for the hormonal onset or the maintenance of maternal behavior, as neither galactophone transaction nor the more radical mammectomy (Moltz et al., 1967) decrease maternal behavior, although the patterns of mother–litter interactions are altered (Woodside and Popeski, 1999; Stern and Keer, 2002).

3.1.4 Postpartum Estrus and Consequences

The parturient female rat enters postpartum estrus approximately 6–12 h after the delivery of the last pup, and reaches a peak in lordosis responding at about 9 h after the last pup (Ying et al., 1973; Connor and Davis, 1980a). This behavioral peak is preceded by a peak in estradiol and is coincident with rising progesterone levels (Connor and Davis, 1980b). If mating does not occur or if the female does not become pregnant, the suckling stimulus inhibits further ovarian cycles. The larger the litter, the longer this period of inhibition, with a very large litter increasing the period up to 35 days (Van Der Shoot et al., 1978; Lindblom et al., 1985). Lactation is maintained by steroid hormones from the corpus luteum of the postpartum estrus and hormones from the pituitary.

Resumption of estrous cycles and sexual behavior occurs late in the postpartum period, with rising estradiol and falling progesterone and prolactin levels. This is around the time of weaning, which in the laboratory is postpartum day 21–28, but in the wild this can be later, because food supply and litter size also influence the timing of postpartum resumption of estrous cyclicity (Reisbick et al., 1975; Woodside et al., 2000).

If the female becomes pregnant at the postpartum estrus, there is a delay in the implantation of the embryos proportional to the number of suckling neonates in her newly born litter (Mantalenakis and Ketchel, 1966). This suckling-induced delay in implantation is caused by lack of FSH stimulation and the resultant low estradiol release, and is not related to prolactin release (Rosenblatt et al., 1979; Cummings and Laws, 1981). This reproductive state is probably common in feral rats and yet it is rarely used as a laboratory model. Thus in the wild, the corpora lutea of the estrous cycle or even pseudopregnancy is never formed and the corpora lutea of pregnancy is constantly present (Gibori et al., 1988).

3.2 Estrogen

3.2.1 Estrogens During Pregnancy and Postpartum

During pregnancy, circulating levels of estrogens are generally low and similar to diestrus until about day 16 when a gradual increase commences and continues until parturition (Shaikh, 1971; Garland et al., 1987). With the exception of a brief increase with the postpartum estrus, circulating estrogens are low during most of lactation, slowly rises across days 10–15, and rise more as weaning approaches (Smith and Neill, 1977; Taya and Greenwald, 1982).

The luteal cells of the corpora lutea of the ovary are the source of estradiol during pregnancy and lactation in the rat (Niswender et al., 1994). During the estrous cycle and pregnancy, this is the corpora lutea resultant from the initial ovulatory events that lead to pregnancy. During lactation this corpora lutea "of pregnancy" regresses and the corpora lutea of the postpartum ovulation becomes functionally dominant (Taya and Greeenwald, 1982). During early pregnancy, the corpora lutea produces both the steroidogenic substrate for estrogen synthesis, and estradiol itself, while during the second half of pregnancy, the placenta becomes the source of the androgens that are the steroidogenic substrate for estradiol produced by the corpous luteum (Gibori et al., 1979; Gibori and Sridaran, 1981; Jones et al., 1987).

3.2.2 Estrogenic Influences on Maternal Behavior

One of the best understood hormonal requirements for the onset of maternal behavior is estradiol (Siegel and Rosentblatt, 1975a, b; Rosenblatt, 1995). During pregnancy, estradiol primes the female for the immediate onset of maternal behavior at parturition. The rapid onset of maternal behavior can be stimulated by either the first 10–13 days of pregnancy, or a treatment regime of estradiol for a similar period, followed by a single (triggering) dose of estradiol if progesterone is absent (Rosenblatt et al., 1979). While estradiol certainly affects the periphery, particularly the uterus and mammary glands, a key influence for the induction of behavior is on the components of the neural circuits that mediate maternal motivation and expression of maternal behavior (Rosenblatt et al., 1994). Regardless of the preparation used to demonstrate maternal behavior, estradiol always induces the onset of maternal behavior more rapidly (Rosenblatt et al., 1979).

An interesting corollary of the late pregnancy production of androgens by the placenta is that the circulating levels of androgens (T and DHT) rise from very low levels of 0.4–0.8 ng/ml early in pregnancy to a sustained and remarkable level of 2.8 ng/ml during the second half of pregnancy (Gibori et al., 1979). Circulating androgens in the cycling female range from 0.2 ng/ml to 0.4–0.5 ng/ml preceding the LH surge (Rush and Blake, 1982; Jones et al., 1987), and testosterone in the male rat is between 1.1 and 2.3 ng/ml (Keating and Tcholakian, 1979). In this context, the estrous cycle has much lower levels of androgens, and the peak values of the late pregnant female are remarkable. No studies have examined whether androgens during the second half of pregnancy in rats play any role in the hormonal onset of maternal behavior at parturition. Androgen levels similar to those in the rat are also observed in late pregnancy in the rabbit, where the hormone increases nest building (Gonzalez-Mariscal et al., 2003).

3.3 Progesterone

3.3.1 Progesterone During Pregnancy and Postpartum

During the estrous cycle, pregnancy, and lactation in rats, the corpora lutea of the ovary are also the source of progesterone (Niswender et al., 1994). This capacity is maintained by the luteotrophic action of prolactin and related hormones, the locally generated estradiol of the corpora lutea, and also by the luteinizing hormone (LH) (Gibori and Sridaran, 1981; Gibori et al., 1988). During most of pregnancy, beginning as early as day 5, progesterone levels are as high or up to tenfold higher than peak values during the estrous cycle. This is one of the most remarkable features of the endocrine milieu of the pregnant female. A sharp reduction in progesterone begins on day 19 of pregnancy, followed by a descent to very low levels at parturition (Grota and Eik-Nes, 1967; Sanyal, 1978; Garland et al., 1987; Rosenblatt et al., 1994). During lactation, progesterone levels rise rapidly during the first 5 days, peak at around day 10 to levels similar to proestrus, and then fall by day 21 to lower levels similar to diestrus (Smith et al., 1975; Smith and Neill, 1977; Taya and Greenwald, 1982).

3.3.2 Progesterone Influences on Maternal Behavior

While progesterone is critical for the maintenance of pregnancy, its withdrawal at the end of pregnancy is necessary for the hormonal onset of maternal behavior (Rosenblatt et al., 1979). Experiments with the H–O model have shown that progesterone prevents the inappropriate display of maternal behavior during pregnancy, and that prepartum withdrawal of progesterone allows and facilitates the immediate onset of maternal behavior (Bridges et al., 1978; Siegel and Rosenblatt, 1978). In the absence of estradiol, neither progesterone alone nor withdrawal of progesterone stimulates maternal behavior (Doerr et al., 1981).

Although lactation can continue after ovariectomy, its normal maintenance requires progesterone. There is no evidence, however, that postpartum levels of progesterone influence the robust maternal behavior displayed during this period (Moltz et al., 1969b). The mechanism by which progesterone is alternatively inhibitory to maternal behavior during the prepartum period, and then permissive during the postpartum period, is not understood.

3.4 Prolactin

3.4.1 Prolactin and Related Hormones During Pregnancy and Postpartum

In the rat, prolactin and three functionally related hormones, decidual luteotrophin and placental lactogens I and II, are important during pregnancy (Robertson et al., 1982; Gibori et al., 1988; Tucker, 1988; Bridges, 1990, 1996). These hormones, along with LH, maintain steroidogenesis in the corpus luteum of pregnancy, and prepare the mammary glands for lactation (Smith et al., 1975; Freeman, 1988; Gibori et al. 1988; Niswender et al., 1994). In spite of structural differences, all these protein hormones bind to and exert their effects via the short or long form of the prolactin receptor (PRL-R). These receptors are expressed and differentially regulated in the corpus luteum, mammary gland, and brain (Gibori et al., 1988; Clarke and Linzer, 1993; Grattan, 2001; Bakowska and Morrell, 2003). Furthermore, all these hormones access the brain via an active receptor-transport mechanism in the choroid plexus (Walsh et al., 1987; Bridges et al., 1996).

The anterior pituitary is the source of prolactin, which is the functionally important hormone of this group during the estrous cycle and in the initial days of pregnancy (Freeman, 1988). Shortly after pregnancy begins, pituitary prolactin is secreted in a pattern of a two daily surges, with nocturnal and diurnal peaks (Taya and Sasamoto, 1981). Early in pregnancy, decidual luteotrophin is secreted by the decidua of the uterus, and by 6 days after insemination, pituitary prolactin is no longer required to maintain pregnancy (Gibori et al., 1988). However, the anterior pituitary continues to secrete prolactin for the first 12 days of pregnancy (Freeman, 1988). After midpregnancy, placental lactogens, first I and then II, become the predominant forms of this hormone group in circulation. During this period, decidual luteotrophin disappears and serum prolactin of pituitary origin is substantially lower than that in cycling females (Garland et al., 1987). Throughout the second half of pregnancy, placental lactogens inhibit prolactin secretion from the anterior pituitary through feedback on the brain. This inhibition ceases in the last few days of pregnancy and there is a prepartum surge of prolactin (Gibori et al., 1988; Tucker, 1988; Fliestra and Voogt, 1997).

After parturition, prolactin levels rise by day 4 of lactation and remain high until day 10, after which they decline abruptly and remain low until estrous cycling resumes. During the initial 10 days of lactation, prolactin levels are 2–3 times higher than the maximum of the estrous cycle (Smith et al., 1975; Taya and Sasamoto, 1981; Taya and Greenwald, 1982). During lactation, both prolactin and LH regulate production of progesterone from the corpus luteum (Taya and Greenwald, 1982), and prolactin has an important role in milk production of the mammary gland.

The importance of suckling for the release of prolactin can be most distinctly seen if a female rat is separated from and then reunited with her pups, upon which there is a rapid and sharp increase in circulating prolactin. The female housed constantly with her pups does not show the same stimulus-related secretion, but nonetheless, these lactating females do have generally high levels of prolactin in the first 10 days of lactation followed by a substantial decline (Taya and Sasamoto, 1981; Mattheij et al., 1985; Freeman, 1988). While lactation and suckling-induced prolactin release can be prolonged if younger pups are substituted for the natural pups approaching weaning, if litters are larger or if food is restricted, even under these conditions the period of lactation is finite (Mattheij et al., 1984; Tucker, 1988; Woodside et al., 2000).

The remarkable power of the pup as a sensory stimulus for prolactin release is seen even in the absence of the ability to nurse pups. Galactophore (milk duct) transection prevents suckling-induced milk delivery, but because the sensory capacity of the nipple is intact, both prolactin levels and anovulation are maintained in the postpartum period (Woodside and Popeski, 1999). There are also elevated circulating levels of prolactin in virgin females that have been induced into maternal responsiveness by pup exposure. If foster pups of naturally advancing age are offered to the maternal virgins, they have a prolactin pattern similar to the natural postpartum dam (Marinari and Moltz, 1978).

3.4.2 Prolactin Influences on Maternal Behavior

Prolactin was proposed in the early literature to be a facilitator in the hormonal onset of maternal behavior if estradiol was provided and progesterone withdrawn (Moltz et al., 1970; Numan and Insel, 2003). A surprisingly long period elapsed before Bridges and colleagues demonstrated that peripheral administration (via ectopic pituitary graphs) or a lengthy series of peripheral injections of prolactin will stimulate a fairly rapid onset of maternal behavior in inexperienced ovariectomized and hypophysectomized virgin females (Bridges et al., 1985). This was also contingent upon treatment with estradiol and progesterone. Using a similar steroid treatment, Bridges et al. (1990) demonstrated that an intracerebroventricular (ICV) infusion of prolactin will also stimulate the onset of maternal behavior.

Bakowska and Morrell have similar conclusions from a different model, a pregnant female H–O on day 16 of pregnancy (Bakowska, 1998; unpublished). This model was selected to consider the impact of the elevated levels of prolactin receptors (PRL-R) in the brain of late pregnancy females, compared with cycling virgin females. These females received only a single, very low dose of estradiol to maintain high levels of PRL-R, once the hormones of pregnancy were removed by the H–O procedure. This dose of estradiol does not support the initiation of maternal behavior in this H–O preparation (Rosenblatt et al., 1998) (➋ *Figure 5-1*).

Prolactin substantially facilitated the onset of maternal behavior. All females infused with prolactin became maternal, while no more than 40% of females without prolactin were maternal after 5 days of pup exposure. This H–O model required less prolactin and less estradiol priming than in the virgin, suggesting that the hormones of pregnancy were facilitatory to the role of prolactin in the onset of maternal behavior, possibly by affecting the prolactin or PRL-R.

3.5 Other Participants in the Processes of Pregnancy, Parturition, and Lactation

As we develop a better understanding of the endocrine factors regulating physiological processes concurrent with maternal behavior, it is worthwhile to consider that some of these factors might also influence maternal behavior, or promote the co-ordination of maternal behavior with these physiological processes. Other unproven but possible candidates for such behavioral regulation are prostaglandins and relaxin (Sherwood et al., 1980; Sherwood, 1988). Relaxin is particularly intriguing. The rising level of relaxin in the latter part of pregnancy correlates with the onset of maternal behavior, and it has a close relationship with the secretion of oxytocin, which participates in the expression of maternal behavior. While the major source of relaxin is the corpus lutea of pregnancy, relaxin receptor expression in brain, relaxin synthesis by the brain, and activation of neurons by relaxin have also been demonstrated (Sherwood, 1988; Osheroff and Phillips, 1991; Gunnersen et al., 1995; Heine et al., 1997; McKinley et al., 1997). Relaxin may act in the brain to mediate increased water intake during the second half of pregnancy (Hornsby et al., 2001) and the timing of birth in rats (Summerlee et al., 1998).

Other yet unexplored endocrine links to maternal behavior may be found as our knowledge of pregnancy and lactation grows. For example, our understanding of the regulatory and synthetic capacity of the placenta is still emerging, and we now know that it has both serotonin receptors and transporters (Shearman et al., 1998), D2 dopamine receptors (Lee et al., 1999), glutamate transporters (Matthews et al.,

Figure 5-1
Females were hysterectomized and ovariectomized on day 16 of pregnancy, and administered one s.c. injection of estradiol (1.25 μg/kg). After five ICV infusions of 50 μg of prolactin (n = 16) or vechicle (n = 10) the animals were tested with foster pups for 5 days without any further infusions or injections. Foster pups were fed and cared for by their biological dams. Estradiol treatment as in Rosenblatt et al. (1998); prolactin regime as in Bridges et al. (1990). Cannula placement was verified postmortem. *P < 0.05. (Bakowska and Morrell, unpublished; Bakowska Doctoral Thesis, Rutgers University, 1997)

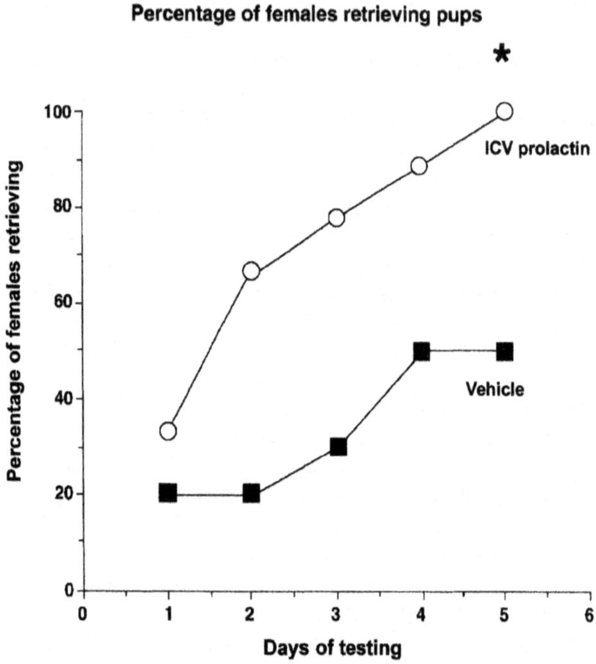

1998), leptin receptors (Smith and Waddell, 2002), and it produces grehlin (Gualillo et al., 2001) and nitric oxide synthase (Casado et al., 1997). In another example, the extinction of suckling-induced prolactin release in the later part of the lactational period can also be delayed by hysterectomy (Kanyicska et al., 1990), suggesting a potential role for the uterus during the postpartum period.

3.6 The Brain as a Possible Source of Hormones that Regulate Maternal Behavior and its Neuroendocrine Processes

The majority of work on the hormones that mediate or support maternal behavior has focused on peripheral sources of these hormones. In certain cases, however, the primary source of these hormones is the brain. For example, oxytocin facilitates maternal behavior when infused into the cerebral ventricles, or into some specific brain areas (see *Section 4*), but there is little evidence that oxytocin secreted into the periphery by the pituitary gland has any effect on maternal behavior. In this case, it is probable that oxytocin facilitates maternal behavior after release from neuronal endings within the brain, in a rich and well-described network of locations (Buijs et al., 1985; Russell et al., 2003). There is evidence that the brain also makes estrogens, prolactin and related compounds, opioids, and other substances that might have a role in mediating behavior (Schlinger and Arnold, 1983; Devito et al., 1992; Emanuele, 1992; Clapp et al., 1994; Gunnersen et al., 1995; Wagner and Morrell, 1996; Balthazart and Ball, 1998; Amateau et al., 2004).

It is interesting to note that these substances may have regulatory roles in maternal behavior that are more dependent on their central sources than we currently understand. It is possible that peripheral and central sources of these hormones work in concert to influence behavior and coordinate endocrine and physiological processes during pregnancy, parturition, and lactation. Grattan and his colleagues have suggested that prolactin may coordinate behavioral, neuroendocrine, and physiological processes during pregnancy and lactation (Grattan, 2001), and we propose that this may occur by the action of prolactin and related compounds of both brain and peripheral origin.

4 Neural Expression of Maternal Behavior

4.1 Fundamentals of the Traditional Maternal Behavior Circuit – the Preoptic Area

The most thoroughly investigated neural site necessary for the display of maternal behaviors in rats is the POA. Fisher (1956) provided the first evidence that the POA might be involved in maternal care and reported that maternal behaviors could be elicited in male rats when testosterone was infused in the POA. Later studies expanded upon these results to show that pre- or postpartum destruction of the POA severely disrupts maternal behaviors, particularly retrieval of pups and nest building (see Numan and Insel, 2003). The importance of the POA in maternal behaviors is not restricted to pregnant or parturient females, as POA lesions also impair the onset or maintenance of maternal responding in sensitized nulliparous female rats (Numan and Insel, 2003). Importantly these females do not have a general oral impairment and continue to carry food, eat, and drink. The POA projections necessary for maternal behaviors exit this structure dorsolaterally, and severing these projections also impairs retrieval and nest building (Terkel et al., 1979; Numan and Callahan, 1980; Numan et al., 1990). Interestingly, dams with severed dorsolateral POA connections will still approach and sniff pups as rapidly as nonlesioned controls (e.g., Terkel et al., 1979; Numan and Callahan, 1980), indicating that their motivation to seek out and investigate pups remains intact. These dorsolateral fibers likely come from many places in the brain to provide input to the mPOA necessary for maternal behaviors, but some important sources of afferents that likely reach the POA include the raphe nuclei, the ventral tegmental area (VTA), the locus coeruleus, the interpeduncular nucleus, and the nucleus of the solitary tract (Numan et al., 1990).

The effect of POA lesions on other maternal behaviors is somewhat unclear. In many cases, nursing the pups is disrupted or absent after destruction of the POA or its projections. However, this is not always the case (e.g., Terkel et al., 1979; Jacobson et al., 1980; Miceli et al., 1983; Numan, 1990; Numan et al., 1990). As noted by Stern and Johnson (1989), nursing is unlikely to occur if dams cannot gather enough pups in one place and receive sufficient suckling stimulation necessary to nurse. Therefore, deficits in nursing after POA lesions may in some cases be secondary to the deficit in retrieval. Licking of the pups is not often examined in these studies, but when quantified, it has ranged from little or no impairment after postpartum severing the lateral connections of the POA (Terkel et al., 1979) to a pronounced deficit after prepartum POA lesioning (Lee et al., 2000).

The POA is part of what is known as the POA/hypothalmic "locomotor" region, a diffuse area where electrical stimulation elicits locomotion in rats (Sinnamon, 1993). These cells are estrogen-sensitive (Takeo and Sakuma, 1995) and project to subthalamic areas and the pedunculopontine nucleus of the midbrain known to be involved in locomotor activity (Swanson et al., 1987). These are presumably the components of the POA projections necessary for the motor control of maternal behavior.

Lesioning the medial POA (mPOA) and lateral POA (lPOA) produce similar effects on maternal behaviors. It is thought that the mPOA is critical for the integration of hormonal and sensory information necessary for the onset and later display of maternal behaviors, but that its afferents synapse onto the lPOA before projecting to areas of the brain necessary for the motor performance of maternal behaviors (Numan et al., 1988). Because the POA is sensitive to a variety of gonadal and pituitary hormones (see ➋ *Section 5.4* below), receives olfactory inputs from the amygdala, and somatosensory information from the trigeminal system (Li et al., 1997), it is perfectly situated to act as a hormone–sensory–motor integration site for the performance of maternal behaviors.

4.2 Other Components of the Maternal Behavior Circuit

In addition to the POA, numerous other neural sites have been examined for their role in maternal behaviors. Only those with a clear and indispensable role in a component of maternal care are presented below. Components of the nigrostriatal and mesolimbic systems involved in maternal care are discussed separately in ➋ *Section 6*.

4.2.1 Ventral Areas of the Bed Nucleus of the Stria Terminalis

Immediately adjacent to the POA is the bed nucleus of the stria terminalis (BNST), which among other functions is integral for the display of many social behaviors in animals because of its ability to relay information between the amygdala and the hypothalamus. As noted by Numan and Numan (1996), previous experimental manipulations of the POA often included ventral areas of the BNST (vBNST – generally corresponding to portions of the anterior dorsomedial, fusiform, and subcommissural nuclei; Swanson, 1998), and it was possible that the vBNST itself had a role in maternal care. Neurotoxic lesions of the vBNST affect maternal behavior as do POA lesions, including a disruption of retrieval and nest building, and a concomitant reduction in nursing, likely the result of the inability to collect the litter in one place (Numan and Numan, 1996). It is not known if the vBNST acts as a separate unit to influence active maternal behaviors or rather as part of a continuous population of neurons in this general area of the basomedial forebrain necessary for the display of maternal care (Numan and Numan, 1996).

4.2.2 Paraventricular Nucleus (PVN)

The PVN of the hypothalamus contains a large number of oxytocin- and vasopressin-synthesizing cells. In accordance with the facilitatory role of these neuropeptides on the onset of maternal behavior (Pedersen et al., 1982), prepartum electrolytic lesions of the PVN and surrounding hypothalamus induce cannibalism at parturition in some females, and also disrupt maternal retrieving, nest building, and nursing (Insel and Harbaugh, 1989). It is not yet known if cells in this structure, rather than fibers passing though it, are responsible for the disruption in the onset of maternal behavior. It is also difficult to determine whether lesioning of peptidergic or nonpeptidergic cells produces these effects. Postpartum electrolytic, radiofrequency, or neurotoxic lesions of the PVN, or knifecuts lateral to it, also impair retrieval and nest building (Numan and Corodimas, 1985; Olazabal and Ferreira, 1997), although not in all cases (Insel and Harbaugh, 1989; Consiglio and Lucion, 1996). Interestingly, PVN lesions apparently have no effect on the duration of nursing behavior (Numan and Corodimas, 1985; Insel and Harbaugh, 1989; Olazabal and Ferreira, 1997), but recent data have shown that central oxytocin (OT) antagonism in lactating rats reduced the dam's licking of pups and their display of kyphosis (Champagne et al., 2001; Pedersen and Boccia, 2003), suggesting that OT originating from the PVN may be more important for these behaviors than previously suggested.

4.2.3 Habenula

The habenular complex (Hbc) has a unique role in the regulation of maternal behavior in rats. Radiofrequency or excitotoxic lesions of the Hbc during midpregnancy produce effects on maternal behavior similar to POA lesions – disruption of retrieval, nest construction, and nursing (Corodimas et al., 1992, 1993; Matthews-Felton et al., 1995). In contrast, postpartum excitotoxic Hbc lesions have either little or only transient effects on maternal behavior, indicating a particular importance of the Hbc for the onset of maternal behavior, but not its maintenance (Corodimas et al., 1993). It has been found that the lateral subdivision of the habenula (Lhb) is responsible for these effects (Corodimas et al., 1993; Matthews-Felton et al., 1995, Felton et al., 1998). While implants of estradiol into the Lhb are not sufficient to facilitate the onset of maternal behavior (Matthews-Felton et al., 1999), a population of neurons in the Lhb that express estradiol receptors suggests at least that some hormonal influence on either maternal behavior or postpartum estrus is possible (Wagner et al., 1998). However, Lhb lesions prior to exposure to pups also disrupt the acquisition of retrieval and nestbuilding in sensitized nulliparous females (Matthews-Felton et al., 1998). Therefore, the Lhb is in all models important for the onset of maternal behavior, but not necessarily related to the mechanisms by which hormones directly influence this onset. Because Lhb lesions produce effects similar to that of mPOA lesions, it is tempting to suggest that the Lhb effects are associated with impairments in mPOA inputs or outputs. Even though the Lhb does not project to or receive a projection from the mPOA directly, it does project to and from the lPOA, through which communication with the mPOA is possible (Wayner et al., 1983; Felton et al., 1999).

4.2.4 Periaqueductal Gray (PAG)

Studies of the neural control of retrieval and nest building in rats demonstrate a critical role of the mPOA, but as noted above, nursing behavior in these studies is not consistently affected by mPOA lesions. This suggests that neural sites other than the mPOA are more important for quiescent nursing. Nothing is known about what neural sites mediate the suckling-induced quiescence necessary for nursing, but the midbrain PAG is necessary for the postural component of kyphosis. Interactions with pups that include prolonged periods of suckling-induced kyphosis elicit expression of Fos, the protein product of the immediate-early gene *c-fos*, in a specific area of the PAG, the caudal lateral and ventrolateral regions ($cPAG_{l,vl}$) (Lonstein and Stern, 1997a, b). Electrolytic lesions of the $cPAG_{l,vl}$ impair the ability of dams to display kyphosis, but have little effect on any other maternal behavior. Lesioned dams do become quiescent and nurse, however, but do so in postures less effective in providing milk to young pups such as laying prone on top of the pups or nursing them in a supine position (Lonstein and Stern, 1997a; Lonstein et al., 1998). All of these effects are specific to the $cPAG_{l,vl}$ because lesions of the dorsolateral PAG do not impair kyphosis (Lonstein and Stern, 1998). Suckling appears to increase GABAergic input to the $cPAG_{l,vl}$ to elicit kyphosis, and infusion of the $GABA_A$ receptor antagonist bicuculline into the $cPAG_{l,vl}$ prevents the ability of suckling to produce this posture (Salzberg et al., 2002). There are dense reciprocal projections between the $cPAG_{l,vl}$ and the POA, some of which express Fos during the display of maternal behavior (Numan and Numan, 1997). These sites may reciprocally inhibit each other, such that kyphosis is inhibited at inappropriate times when dams need to perform active maternal behaviors, and active maternal behaviors are inhibited when dams are showing suckling-induced quiescence and kyphosis (Lonstein and Stern, 1997a).

4.2.5 Septum

Large lesions of the septum performed pre- or postpartum impair maternal care in parturient females, as well as in sensitized virgins. To what degree remains unclear. One study reports that LS lesions disrupt only the efficiency of retrieval, with dams picking up pups, but carrying them for prolonged periods of time and dropping them in locations other than the nest (Fleischer and Slotnick, 1978). In other cases, such lesions completely abolished maternal behavior (Flannelly et al., 1986; Novakova et al., 1993). It is difficult to interpret these results, partly because in some cases the lesions included parts of the ventral striatum, BNST, and/or the dorsal POA, which in itself could impair maternal care. As suggested by Numan and Insel (2003), cases of disorganized retrieval by septal-lesioned females may indicate that this structure is involved in behavioral inhibition and termination, and its integrity is necessary for the smooth transition between sub-components of each retrieval (e.g., picking up pups, transporting pups, dropping pups, searching for the next pup). Consistent with a role in the organization of maternal behaviors, cells of the lateral septum express the immediate-early gene, c-fos, after the display of maternal behavior (Lonstein et al., 1998; Stack and Numan, 2000), and this neural activity requires a projection from the POA (Numan and Numan, 1997; Stack et al., 2002).

4.3 Brain Regions that Inhibit Maternal Behavior in the Nonmaternal State

4.3.1 Amygdala

As noted above (➒ *Section 2*), olfaction is generally unnecessary for the postpartum display of maternal behavior in rats and inhibits the behavior in virgin females. The main and accessory olfactory bulbs densely project to the cortical and medial nuclei of the amygdala, which project heavily to the POA (Simerly and Swanson, 1986). If olfactory information accesses the POA via the amygdala to inhibit maternal responding in nulliparous females, it is possible that destroying the amygdala facilitates maternal behavior. This is exactly what occurs after electrolytic or excitotoxic lesions of the corticomedial amygdala, or of just the medial amygdala (MeA) (Fleming et al., 1980; Numan et al., 1993). Conversely, stimulation of the MeA delays the onset of maternal behavior (Morgan et al., 1999). Although MeA-lesioned females no longer avoid pups, they may not immediately act maternally toward them and still require 2–3 days of exposure to become fully maternal (Fleming et al., 1980), suggesting that inhibitory inputs also come from other neural sites. An interesting study by Sheehan et al. (2000) examined how the nonmaternal virgin female rat brain responds to pup stimuli, with the goal of finding areas that cause such stimuli to be aversive, and found that the MeA showed one of the greatest increases in Fos expression. However, postpartum and sensitized maternal rats also show high levels of Fos expression in the MeA (Fleming et al., 1994; Numan and Numan, 1994, 1995; Lonstein et al., 1998; Kalinichev et al., 2000), albeit at a somewhat lesser magnitude than nonmaternal virgin females (Sheehan et al., 2000).

4.3.2 Anterior and Mediobasal Hypothalamus

The amygdala may also inhibit maternal behavior by stimulating other areas of the brain that can inhibit maternal responding, and the mediobasal hypothalamus is one such site. Lesions of the mediobasal hypothalamus (the anterior hypothalamus, dorsomedial hypothalamus, or ventromedial hypothalamus [VMH]) facilitate maternal response in steroid-treated nulliparous female rats (Bridges et al., 1999; Sheehan et al., 2001). These lesions do not facilitate the behavior in females not given steroid hormones (Bridges et al., 1999) and the VMH must only then inhibit maternal response under certain hormonal conditions, such as pregnancy. In support, nonmaternal female rats show elevated Fos expression in the anterior hypothalamus and ventromedial nucleus in response to pups whereas maternal females do not (Sheehan et al., 2000). Other than the requirement for ovarian hormones to be present, little is known about the neurochemistry underlying the inhibition of maternal behavior by the mediobasal hypothalamus, with the exception that neuropeptide K infused into the VMH can inhibit the behavior (Sheehan and Numan, 1997). Evidence that these mediobasal hypothalamic areas receive an input from the MeA to inhibit maternal behavior is provided by the finding that MeA lesions reduce Fos expression in these sites of nonmaternal females after exposure to pups (Sheehan et al., 2001).

Areas of the brain associated with fear and anxiety, such as the PAG and septum, have also been suggested to be involved in the inhibition of maternal behavior (Numan and Insel, 2003). The novel stimuli eminating from pups are thought to be anxiety-generating in maternally inexperienced female rats and reduced neophobia of pups may be a prerequisite for females to act maternally (Fleming and Luebke, 1981; Hard and Hansen, 1985; Lonstein, 2005). Nonmaternal female rats do show high levels of Fos expression in the PAG and septum when exposed to neonates (Sheehan et al., 2000), but more work is needed to determine if these sites inhibit maternal care.

4.4 Changes in Neural Hormone Receptor Expression Across Pregnancy and Lactation

4.4.1 Estrogen Receptors

Estradiol is thought to be an important hormone for the onset of maternal responding and numerous studies have examined changes in ER expression in the rat brain across pregnancy and lactation. The pattern of these changes is site-specific, but probably not surprisingly, does not have a simple relationship to the onset of maternal behavior in the parturient rat. Estrogen receptor (ER) binding increases in the medial preoptic nucleus and MeA during pregnancy, and at the same time, decreases in other areas of the POA (Giordano et al., 1991). This pattern is most readily seen as preparing this region for the onset of maternal behavior. In contrast, a different pattern emerges in the VMN, where ER binding decreases during midpregnancy and rises again at the time of parturition (Giordano et al., 1991). This region is important for sexual behavior in the female, which is absent during pregnancy (Rosenblatt et al., 1998) but reemerges during the postpartum estrus. Changes in ER binding in the VMN support the idea that this area participates in the absence of sexual behavior during pregnancy and its reemergence after parturition.

Examining the number of neurons that express the mRNA (Wagner and Morrell, 1995) and immunoreactivity (Wagner and Morrell, 1996) of the estrogen receptor alpha isoform (ERα) has provided some insight into the transcriptional and translational regulation underlying the binding data. In general, changes in ERα mRNA expression in the mPOA are antecedent to the increases in binding, with peaks on day 8 of pregnancy, and then again on day 22 of pregnancy. The number of neurons in the mPOA expressing ERα immunoreactivity also increases on day 8 and day 16, accompanied by a shift in the expression per neuron such that more neurons express more protein on days 16–22 than before or after this period. Thus, increased estrogen binding in the mPOA is supported by increased mRNA and protein for the receptor. In the VMN, mRNA and immunoreactive protein are low during pregnancy, and increase just before parturition, consistent with the binding pattern. The patterns of neurons that express mRNA and immunoreactivity of ERα, and the amount per neuron, are the summation of both alterations in synthesis and degradation of the ER-α mRNA and protein. While they seem to generally support the binding patterns, additional levels of regulation are probably also involved, such that the stability of the mRNA and the immunoreactive protein may also change during these periods.

Expression of the more recently discovered beta isoform of ER also differs between virgin, pregnant, and lactating rats in sites relevant to maternal behavior. The number of cells in the POA expressing ERβ mRNA is lower in late-pregnant and lactating rats compared with proestrous virgins, although females at the end of pregnancy have very high ERβ expression per cell. Similarly, the number of ERβ-expressing cells is low in the MeA in lactating females, although they express more per cell (Greco et al., 2003). The functional relationships between ERα and ERβ are only now emerging, as are the functions of the ERβ (Matthews and Gustafsson, 2003).

4.4.2 Progestin Receptors

The only detailed study of nuclear progestin receptor (PR) expression during pregnancy and lactation in rats providing cellular resolution is by Numan and colleagues (1999). In both the POA and VMN, expression of PR-immunoreactivity is high in virgin females, but falls precipitously soon after insemination. PR levels then rise before parturition and fall during early lactation. A similar, but not statistically significant, pattern of change in PR expression was found in the vBNST and MeA. It is interesting that the low levels of PR after parturition are due to contact with pups, because their removal increases PR levels in these sites.

As noted above, progesterone has a dual effect on maternal behavior, first synergizing with estradiol during pregnancy to promote parental responding, but then having to be withdrawn for the behavior to emerge at parturition. How changes in the PR levels in sites involved in maternal behavior mediate the onset of the behavior is unknown and counterintuitive because the high sensitivity to progesterone afforded by high levels of PR in these sites just before parturition might be expected to inhibit maternal responding, not promote it. One might also expect that recently inseminated females, which have very low PR expression, would be highly maternal, but this has not been demonstrated to be the case. Additional levels of regulation within a larger circuit must be interfacing with the PR system and is clearly an area for future attention.

4.4.3 Prolactin Receptors

The localization and concentration of prolactin receptors (PRL-R) in the brain have been examined numerous times in pregnant and lactating rats. PRL-R can be found in two forms Ouhtit et al., 1993, a long form that is regulated during pregnancy and lactation, and a short form that is not. In homogenates of the entire forebrain, mRNA for the long form increases during midpregnancy and is high through most of lactation (Sugiyama et al., 1994). In just the hypothalamus, the greatest long-form PRL-R mRNA expression is found in lactating dams (3–7 days postpartum), compared with late-pregnant or cycling females (Torner et al., 2002). In situ hybridization, which allows for cell-specific resolution, shows that mRNA for the long form of the PRL-R can be found in almost every site involved in maternal behavior. In the POA, PRL-R mRNA generally increases across pregnancy, peaks at parturition, and then falls as lactation progresses (Bakowska and Morrell, 1997; Pi and Grattan, 1999a; Mann and Bridges, 2001, 2002). PRL-R mRNA also rises in the dorsal and ventral BNST as pregnancy progresses (Bakowska and Morrell, 1997; Mann and Bridges, 2001, 2002), and lactating rats have greater long-form PRL-R mRNA in the VMH than diestrous virgins (Pi and Grattan, 1999b). The absence of changes in PRL-R expression in these brain regions has also been reported (Augustine et al., 2003). Other sites important for maternal behavior with numerous cells expressing mRNA for the long-form of the PRL-R include the MeA, lateral septum, anterior hypothalamus, and the ventrolateral PAG (Bakowska and Morrell, 1997), but it is not known if its expression changes across pregnancy and lactation in these sites. Presumably, when changes in PRL-R are found, they are due to changes in circulating hormones, as expression of PRL-R in the hypothalamus is reduced by ovariectomy, and restored by chronic estradiol or progesterone (Mustafa et al., 1995; Pi et al., 1999c, 2001, 2003). In some subdivisions of the POA and VMH, this PRL-R expression requires that dams have relatively recent contact with suckling pups (Sugiyama et al., 1996; Pi and Voogt, 2000, 2001). Exposure to neonatal pups may also alter PRL-R expression in the choroid plexus of virgin female rats (Sugiyama et al., 1994, 1996) and in the homogenized cerebrum of male rats (Sakaguchi et al., 1996).

Expression of mRNA for the short form can be found in many of the same sites (Bakowska and Morrell, 2003), but it is found in low abundance compared with the long form. In regions where both forms of the receptor are found, however, there is an additional level of complexity, as neurons with both forms exist. For example, in the paraventricular and supraoptic nuclei of the hypothalamus, nearly all the neurons have a low level of expression of the short form, while the distribution of neurons that express the long form of the receptor only substantially overlaps with oxytocin-producing neurons (Bakowska and Morrell, 1997, 2003). The differential regulatory roles of these two forms of the receptor in the brain are unknown. Two points are of interest. First, the activities of the short and long form are likely to interact in complex ways (Bole-Feysot et al., 1998), including the possibility that the short form inhibits the activity of the long form. Secondly, both forms may have a role in the regulation of neurogenesis during pregnancy (Shingo et al., 2003), which could be involved in changes in behaviors specific to the postpartum period.

Notably, in every study that includes analysis of the choroid plexus, this was the location of the largest amount of either mRNA for the long or the short form of the prolactin receptor or its immunoreactivity (Bakowska and Morrell, 1997, 2003). Every cell in the choroid plexus expresses high amounts of mRNA for both forms of the receptor. This appears to be the transcriptional evidence for the receptor-mediated mechanism for the transport of prolactin and prolactin-like substances from the peripheral blood into the brain (Walsh et al., 1987).

4.4.4 Oxytocin Receptors

Oxytocin is released most likely from within the brain, rather than from the pituitary, to facilitate maternal behavior in rats (Pedersen, 1997). Oxytocin receptor (OTR) expression can be found in many places of the female rat brain involved in maternal behavior, and it changes across pregnancy and lactation in a manner that might modulate the onset and maintenance of maternal care. Discrepancies between studies using different methods to visualize OTR, however, make interpretation of these data somewhat difficult (see Young et al., 1997 for discussion). Insel (1986) first reported that OTR binding was greater in the dorsolateral BNST of lactating dams (1–6 days postpartum) than that in midpregnant females and in ovariectomized females treated with estradiol than in untreated females. Furthermore, binding was highest earlier in lactation than later (Insel, 1990). Differences between groups were not found in the VMH, and the figures reveal no noticeable differences in the BNST, septum, or POA (Insel, 1986). Additionally, significantly more OTR binding was found in the VMH on the day of parturition than in pregnancy or other times during lactation (Insel, 1990; Bale et al., 1995; Young et al., 1997). OTR mRNA shows a different pattern, with recently parturient females having greater expression in the VMH, lateral septum, and POA, but probably unexpectedly, not the BNST (Bale et al., 1995; Young et al., 1997). Consistent with this finding, analysis of OTR binding in the POA by Pedersen et al. (1994) showed increases during parturition compared with days 15–17 of pregnancy or postpartum days 5–7. OTR binding was also higher during parturition in the VTA (Pedersen et al., 1994).

The significance of changes in OTR expression in the dorsolateral BNST or VMH on the day of parturition is unknown. Day 4 postpartum lesions of the dorsolateral BNST have little effect on maternal behavior (Numan and Numan, 1997), although the very early postpartum increase in OTR expression here suggests that prepartum lesions of this area might effectively disrupt the onset of maternal behavior. Alternatively, changes in OTR expression on the day of parturition might be more closely related to the postpartum estrus or the onset of lactation rather than maternal behavior, as maternally sensitized females do not have increased OTR binding in either the BNST or VMH (Insel, 1990), and some BNST cells are responsive to suckling and are involved in milk letdown (Lambert et al., 1993).

OTR expression is sensitive to ovarian hormones, which may be responsible for the observed changes in OTR expression in parturient females. Estradiol is a potent stimulator of OTR expression in many areas, including the BNST, POA, LS, and VMH (Insel, 1986; Schumacher et al., 1989; Tribollet et al., 1990; Bale and Dorsa, 1995; Bale et al., 1995; Kremarik et al., 1995; Quinones-Jenab, 1997), and increases OTR affinity for OT in the POA (Caldwell et al., 1994). The addition of progesterone in some cases further enhances it (Schumacher et al., 1989; Coirini et al., 1991), but chronic progesterone can suppress basal and estradiol-induced OTR expression in some areas of the brain, including the VMH (Patchev et al., 1993; Bale et al., 1995).

As noted above, previous research had suggested little or no role for OT in maternal behavior after the behavior was established around the time of parturition, but this issue has recently been reexamined in detail. Francis et al. (2000) found that some lactating rats have greater OTR binding in the POA and dorsolateral BNST than other females, and that these differences were related to the higher levels of maternal licking these females received when they themselves were neonates. Furthermore, when such females are virgins, they more rapidly become maternally sensitized (Champagne et al., 2001). These differences in OTR expression appear to be related to differences between females in their postpartum maternal behavior because ventricular infusion of an OT antagonist reduces pup licking in dams with higher OTR expression (Champagne et al., 2001).

4.4.5 Conclusions

Region-specific changes in the expression of hormone receptors have complex relationships to the events of pregnancy and the onset of maternal behavior. One hypothesis is that these changes prepare the brain to rapidly induce and maintain maternal behavior because the correlations of these events and the behavioral outcomes are strong. However, this hypothesis must account mechanistically for the disjointed time course because some of the greatest changes in receptor measures occur well before the behavioral events. Causal linkages between changes in the expression of these receptors and the onset of maternal behavior will probably require examining both region- and temporal-specific regulation of expression using methods that are only now emerging in molecular biology. An understanding of how these region-specific changes in receptors occur and are sustained is also needed. Estrogen and progestin receptors are regulated by circulating steroid hormone levels in several models (Parsons et al., 1982; DonCarlos et al., 1989; Lisciotto and Morrell, 1993; Morrell et al., 1995). The region-specific regulation of these receptors in the brain may involve additional regulation such as afferent signals by neurotransmitters as well as differences in transcriptional and translational regulation within each region.

The case of ER regulation across pregnancy exemplifies a theme that must be constantly revisited when examining the data describing changes in hormone levels, neuroendocrine factor levels, and receptor levels proposed to be functionally related to changes in maternal behavior or its physiological correlates. There are many well-known changes between the second half of pregnancy and the immediate postpartum period that may be regulating components of the neural circuit mediating maternal responsiveness. These correlations can only serve as the basis for a search for causality. It is also the case that there are less well-known changes in hormone regulation that may be associated with maternal care; for example, the work of Gibori and coworkers demonstrating that very high levels of androgens are circulating in the later part of pregnancy (Gibori et al., 1979). Also, consider that other processes, such as food intake, lactogenesis, and responsiveness to adrenal steroids, are altered during this time. All these changes have a temporal coincidence with the changes in the brain appearing to be necessary for the onset of maternal responsiveness. Proving some causal linkages and revealing their mechanisms are important future directions.

4.5 Effects of Site-Specific Application of Hormones on Maternal Behavior

4.5.1 Estradiol

Many sites in the female rat brain express ERs and may be targets for estradiol's facilitatory effects on maternal behavior. Numan et al. (1977) addressed this question by implanting crystalline estradiol benzoate (EB) into the POA of 16-day pregnant rats immediately after hysterectomy and ovariectomy. Two days later, the implants were removed and the females were exposed to neonates. Females with EB implants into the POA showed a significantly reduced onset of maternal behavior, from a median latency of 2 days after a subcutaneous injection of EB to an immediate onset of the behavior. Because undiluted steroid could spread considerably from the cannulation site, Fahrbach and Pfaff (1986) reexamined this question using POA implants of unconjugated estradiol diluted with cholesterol. They found that diluted estradiol implants shortened the latency of nulliparous ovariectomized females to retrieve pups, but had no facilitatory effects on the efficiency of retrieval or nest building. Similar effects of estradiol implants have been found in female rats hysterectomized and ovariectomized during late pregnancy (Felton et al., 1999) and in males (Rosenblatt and Ceus, 1998). Implantation of estradiol in other sites in the brain does not produce these facilitatory effects (Numan et al., 1977; Matthews-Felton et al., 1999).

Even though estradiol hastens the onset of maternal behavior when infused into the POA, antagonism of ERs in the POA via implants of the selective ER modulator tamoxifen at the end of pregnancy does not impair the onset of postpartum maternal behavior (Ahdieh et al., 1987). Nonetheless, tamoxifen implants into the POA somewhat reduced the percentage of females showing maternal care (by ~33%) if litters were delivered by caesarian section (Ahdieh et al., 1987), which suggests an interaction between estradiol and the experience of parturition to facilitate the onset of maternal behavior.

4.5.2 Progesterone

There has only been one study examining where progesterone might act in the brain to inhibit maternal behavior. Numan (1978) showed that H–O pregnant females given exogenous estradiol were almost immediately parental 48 h later, but not if also injected subcutaneously with progesterone. Unilateral or bilateral implants of progesterone into the POA, VMH, raphe nuclei, or the area of the ventromedial substantia nigra of similarly treated females did not inhibit maternal behavior. Numan (1978) suggested that either the site where progesterone is acting was not examined in the study, or that it acts on multiple sites simultaneously to produce its inhibitory effects. This remains a major area of inquiry within the field.

4.5.3 Prolactin

Infusion of ovine prolactin into the POA also facilitates maternal behavior, shortening the latency of steroid-primed females to act maternally from almost 6 days to approximately 1 day (Bridges et al., 1990). Infusion of neither growth hormone nor LH produces this effect (Bridges et al., 1997). Notably, the dose of ovine PRL infused into the POA is lower than the doses necessary to stimulate parental behavior when infused into the cerebral ventricles (Bridges et al., 1990), suggesting that the POA is particularly sensitive to the influence of PRL on maternal care. Bridges also demonstrated that rat placental lactogen I facilitates maternal responding when infused into the POA (Bridges et al., 1997), suggesting that the pituitary is not the only source of prolactogenic hormones that could be important for the onset of maternal behavior. Whether PRL or plactental lactogens act similarly on other areas of the brain is unknown.

4.5.4 Oxytocin

The POA is also a target for oxytocin's effects on maternal behavior. Infusion of oxytocin into the POA facilitates maternal responding (Fahrbach et al., 1985), whereas infusion of an oxytocin antagonist into the POA during parturition greatly delays or completely eliminates retrieval of pups, and delays the latency to crouch over pups (Pedersen et al., 1994). Similar to POA lesions, impairments in nursing behavior were found and may have been consequent to the impairments in retrieval. These behavioral impairments were also found after OT antagonist infusion into the VTA (Fahrbach et al., 1985; Pedersen et al., 1994). As with most other hormones and neuropeptides, other neural areas where oxytocin might produce its effects on maternal care have not been investigated.

4.5.5 Other Neuropeptides

Other neuropeptides have been infused into specific sites and their effects on maternal behavior examined. Infusion of a vasopressin V1a receptor antagonist into the POA during parturition in rats impairs maternal behavior, but not as effectively as an OT antagonist (Pedersen et al., 1994). Unlike the effects of an OT antagonist, V1a antagonism in the VTA has no effects (Pedersen et al., 1994). Administration of morphine into the POA also impairs the onset of maternal behavior in estradiol-primed nulliparious female rats, as well as ongoing maternal behavior in postpartum females (Rubin and Bridges, 1984). Because morphine in the POA also impairs the display of parent-like behavior in juvenile rats (Wellman et al., 1997), its role is not likely related to hormonal influences on the POA that promote parental care. The PAG is also sensitive to opioid effects on maternal responding, and unilateral infusion of the opioid receptor antagonist naloxone into the dorsolateral PAG reverses the impairing effects of acute systemic morphine on postpartum maternal behavior (Miranda-Paiva et al., 2003). The neuropeptide cholecystokinin (CCK) hastens the onset of maternal behavior in estrogen-primed nulliparous female rats when injected peripherally (Linden et al., 1989), but has also been reported to be ineffective after peripheral or ICV infusion (Mann et al., 1995). CCK can act as an opioid antagonist and when infused into the POA, it blocks the impairments in maternal behavior after POA injection of morphine (Mann et al., 1995). The tachykinin, neuropeptide K, delays the display of maternal behavior in pregnancy-terminated, estradiol-treated female rats when infused into VMH (Sheehan and Numan, 1997), which so far, is the only neurochemical known to act in the VMH to impede this behavior.

5 Studies of Maternal Motivation

Hormones acting on the brain surely affect both the appetitive and consummatory aspects of maternal care, but there has been relatively little study of maternal motivation in rodents. Of course, dams will only express the components of maternal care if motivation is sufficiently high, and therefore, motivation is an absolute prerequisite for the behavior. Early studies employed methods commonly used at the time to evaluate other motivational states (e.g., hunger, thirst, libido) to evaluate the effectiveness of pups as an incentive for parturient female rats to learn mazes (Simmons, 1924) or cross electrified grids (Moss, 1924; Nissen, 1930) to gain access to neonates. As noted previously (Fahrbach and Pfaff, 1982), these studies were fraught with methodological problems that render it difficult to determine how maternal motivation compared with other motivating factors.

More recent studies also based on the premise that lactating rats will readily overcome adversity to retrieve offspring have examined retrieval of pups from unfamiliar and potentially anxiety-generating environments as a measure of maternal motivation. Lactating female rats and mice are more likely than virgins to retrieve pups from an unfamiliar straight alley or T-maze extension of their home cage (Gandelman et al., 1970; Bridges et al., 1972; Stern and Mackinnon, 1976). The fact that most maternally sensitized rats of both sexes will not retrieve pups from a T-maze (Bridges et al., 1972) is likely due to their hormonal milieu. Estradiol and progesterone are each anxiolytic in rats (e.g., Mora et al., 1996), and maternally sensitized female rats are as likely as parturient females to retrieve pups from a T-maze extension of the home cage only if administered estradiol benzoate before sensitization (Stern and Mackinnon, 1976).

Operant chambers requiring rodents to press a bar or make other behavioral responses to obtain a reward have been used extensively to evaluate the reinforcing properties of many types of stimuli, including infants. In the first such study (Wilsoncroft, 1969), recently parturient female rats learned to bar-press for access to a neonate, which was invariably retrieved to the subjects' nest box. Pups were apparently highly reinforcing and some dams would bar-press for hours, gaining access to and retrieving hundreds of pups within the 3-h test session. A study in a strain of nulliparous female mice that are spontaneously maternal provided somewhat similar results (Van Hemel, 1973). When compared with lactating mice, however, nulliparous maternal females were less motivated to bar-press for a pup reward (Hauser and Gandelman, 1985).

Circulating hormones are an important influence on females' motivation to bar-press for pups (Hauser and Gandelman, 1985). High levels of bar-pressing in female mice are observed only soon after parturition, suggesting that hormonal fluctuations occurring immediately before or at this time influence maternal motivation. However, hysterectomy and caesarian removal of pups just before parturition did not prevent high levels of operant response in the dams (Hauser and Gandelman, 1985). As noted above, hysterectomy and removal of fetuses during pregnancy produces hormonal changes mimicking those occurring at parturition (Rosenblatt and Siegel, 1975). If the ovaries and the uterus and fetuses are removed, mouse dams did not bar-press at high rates for infant reward (Hauser and Gandelman, 1985). Furthermore, the withdrawal of progesterone is not only necessary for the onset of maternal behavior in mice and rats (Bridges, 1996), but also the females' operant responding for pups, as daily injections of progesterone prevent bar-pressing for pups in pregnancy terminated females (Hauser and Gandelman, 1985). Hauser and Gandelman, (1985) also showed that the reinforcing quality of mouse pups decreases as they age, consistent with the decline in maternal behavior that occurs as lactation progresses and that older pups are less attractive to lactating females than younger ones (Reisbick et al., 1975; Stern and Mackinnon, 1978; Mayer and Rosenblatt, 1980).

Hormonal influences on maternal motivation in an operant paradigm have also been examined in common marmosets (*Callithrix jacchus*). Unlike mice, frequency of bar-pressing for the opportunity to view an artificial marmoset infant and hear infant cries increases not immediately before parturition, but rather during late pregnancy, and continues throughout early lactation. This is also presumably due to changes in ovarian hormones, because virgin marmosets treated with a regimen of estradiol and progesterone mimicking that of late pregnancy also bar-press at high rates for infant-related stimuli (Pryce et al., 1993).

More recently, an operant paradigm has been used to examine the effects of destroying various neural sites on maternal motivation (Lee et al., 2000). Such an examination could discern whether neural lesions impair maternal behavior by altering maternal motivation to be in contact with pups (i.e., the appetitive component) or by altering other factors necessary for the execution of maternal behaviors (i.e., the consummatory component). In multiparous female rats, lesions of the POA and basolateral amygdala (AMYbl) significantly reduced bar-pressing for pups, whereas lesions of the nucleus accumbens (NA) did not. However, all lesions disrupted the performance of maternal behavior when tested in the home cage. POA lesions also impaired bar-pressing for pups and home-cage maternal behavior in sensitized virgins. The reinforcing aspect of pups is apparently the contact that dams have with them after they are delivered through the reward bin, because even unlesioned dams will not bar-press for pups if they cannot thereafter make physical contact with them. In light of this, Lee et al. (2000) acknowledge that POA lesions may impair operant response for pups because they also impair the unconditioned maternal responses (retrieval, licking) that allow pups to be rewarding. Nonetheless, these results suggest that the neural sites mediating the reinforcing properties of pups differ in some ways than those necessary for expression of maternal behavior, with the POA and AMYbl necessary for both factors, but the NA unnecessary for maternal motivation.

Females' preference for locations or objects associated by residual sensory cues with pups also provides insight into the cues relevant for maternal motivation and responsiveness. Females with greater maternal motivation would be expected to show a higher preference for these pup-associated locations or objects. Not surprisingly, late-pregnant and parturient rats prefer a chamber containing bedding soiled by a lactating dam and her litter compared with a clean chamber (Bauer, 1983; Fleming et al., 1989). Virgin females do not show this preference, but a regimen of ovarian hormones that elicits maternal behavior in virgins also causes them to prefer the soiled bedding (Fleming et al., 1989). This paradigm shows that the chemosensory stimuli provided by pups are a sufficient representation of the motivating stimulus (pup) once the female enters late pregnancy, and that even before parturition she is motivationally prepared to respond to pups.

There are two common methods for determining the motivation or drive an animal has for an unconditioned stimulus, operant responding as described above, and place-preference conditioning. Place-preference conditioning is widely used in research on behavioral and neural processes involved in reward and reinforcement (Tzschentke, 1998). After a conditioning period when the rat learns to associate an unconditioned stimulus with a conditioned context (usually multiple neutral environmental cues in a chamber), preference for the conditioned cues is tested in the absence of the unconditioned stimulus. The environmental cues are neutral, that is not derived from the unconditioned stimulus. The time the rat spends in the chambers with the stimulus-associated cues is thought to indicate the motivation or desire for the stimulus. ➲ *Figure 5-2* allows us to consider some key differences in these two paradigms.

Figure 5-2
Schematic representation of the components of a behavioral sequence that can be identified in motivated, complex, naturally-occurring behaviors (B. Mattson, Doctoral Thesis, Rutgers University 2002). The two components discussed in the adjacent section are the motivational aspects and the expression of the behavior, with emphasis on their step-wise role in a complete behavioral sequence, particularly including consideration of the effect of achieving the stimulus on subsequence seeking of the stimulus

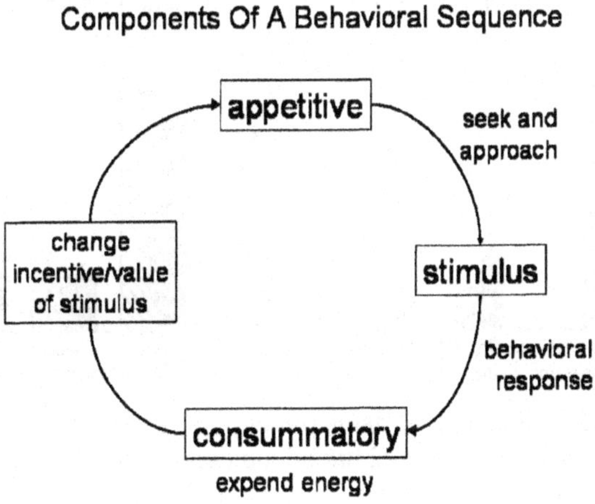

In the case of operant conditioning, the subject obtains and interacts with the stimulus, and so every operant response after the initial contact with the stimulus is one in which the incentive value of the stimulus is altered after interaction with the stimulus. While operant conditioning allows an examination of satiety, it is a paradigm in which motivation is likely altered by delivery of the unconditioned stimulus. Place-preference conditioning never provides the stimulus at any point during testing and so the entire test examines the initial appetitive state. The two methods have been fruitfully applied to explore motivation for pharmacological and natural rewards and data from both approaches together provides insight into issues of motivation.

The first use of a conditioned place-preference to study maternal motivation demonstrated that up to day 8 postpartum, lactating female rats preferred a pup-associated environmental context (Fleming et al., 1994). This response required that dams be deprived of pups for 23 h before testing, after which they were much more likely to show a preference for the pup-associated chamber than nondeprived dams, similar to the effects of deprivation on motivation for other types of rewards, such as food (Tzschentke, 1998). Morrell and colleagues further demonstrated that an equally strong preference for pup-associated environmental context occurred at both day 8 and 16 postpartum, using a similarly long pup deprivation period before conditioning with pups and before testing (➋ *Figures 5-3* and ➋ *5-4*) (Wansaw et al., 2002, 2003a,b). One interpretation of these and the Fleming et al. (1994) data is that the need to nurse pups after such prolonged deprivation is reinforcing simply due to the relief from milk-engorged mammary glands provided by the pups. This would not be interpretable as a demonstration of maternal motivation per se.

Wansaw and Morrell are currently examining how these responses may be related to the different amounts of pup deprivation (Wansaw et al., 2002, 2003a,b). On day 8, postpartum females prefer pups over familiar neutral objects even if deprived of their pups for as little as 15 min or as long as 6 h before

Figure 5-3

Place preference of day 8 or day 16 postpartum dams conditioned and tested after 22 h of pup deprivation. Dams were conditioned to associate neutral context cues with either pups or pup-sized objects. The graph on the left represents the percentage of individuals in each group that met a very conservative quantitative criterion to be categorized as preferring one cue association or the other (Mattson et al., 2001). The graph on the right represents the proportion of the total test time of 60min that the females spent in one of the three chambers – chamber associated with pups, a chamber associated with neutral pup sized objects, or a neutral empty chamber without conditioning. Both groups of dams showed a statistically significant effect of conditioning, and a preference for cues-associated with pups over those associated with objects ($P < 0.05$)

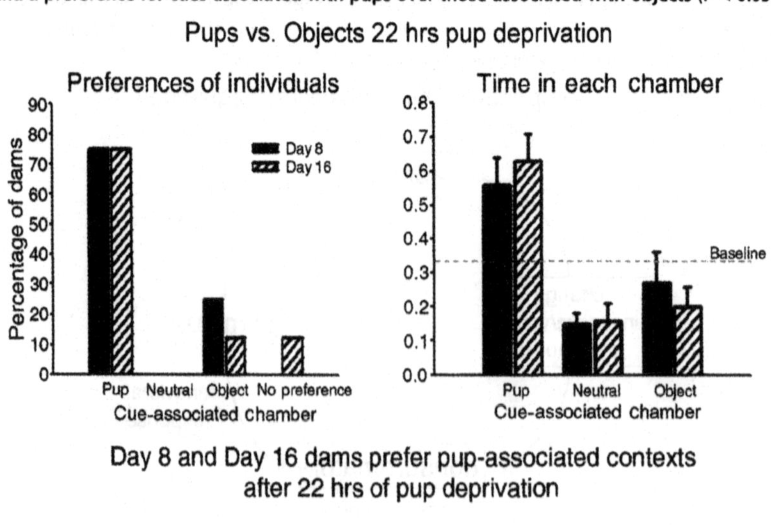

Figure 5-4
Place preference of day 8 or day 16 postpartum dams conditioned and tested after less than 6 h of pup deprivation. Separate groups of dams were deprived of their pups for either 15 min, 2 h or 6 h, and then were conditioned to associate neutral context cues with either pups or pup-sized objects. Both groups of dams showed a statistically significant effect of conditioning. On day 8, dams showed a statistically significant preference for a pup associated context, while on day 16, a preference for cues-associated with pups over those associated with objects emerged (P<0.05). Further details of graphs as in *Figure 5-3*

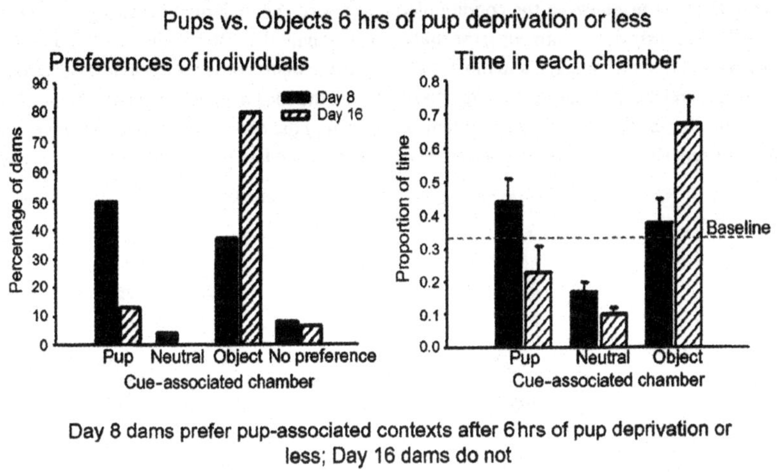

conditioning and testing. In contrast, by the later part of the postpartum period, similarly-treated females no longer preferred pups and instead exhibit a preference for neutral familiar objects. These data demonstrate that maternal motivation toward pups is waning as the postpartum period progresses. The experiments using very short pup deprivation times to reveal pup-associated place preference in the early but not the late postpartum period are interpretable as a demonstration of maternal motivation since the complication (confound) with any physiological need to nurse is avoided.

The reward salience of pups across the postpartum period has been further explored with a place-preference paradigm contrasting the response of cues associated with pups versus the highly rewarding neuroactive drug, cocaine (Mattson et al., 2001, 2003). Early in the postpartum period, the majority of the females preferred cues associated with pups over cues associated with cocaine (*Figure 5-5*). Later in the postpartum period, the preference for cues associated with cocaine becomes the dominant response. This paradigm highlights the strength of the pup stimulus as a reward and provides another illustration of the waning of maternal motivation in the late postpartum period.

Formation of an association between pups and pup associated context requires that dams received the full complement of sensory cues from pups while interacting with them. If dams receive only distal cues from pups are made anosmic, or have their ventrum or mystacial pads anesthetized just before the interaction with the litter during the conditioning phase, they do not form a conditioned place preference (Magnusson and Fleming, 1995). The hormones of pregnancy are likely important for conditioned place preference for pup-associated cues in postpartum dams, as virgin females given a regimen of estradiol and progesterone that induced maternal responding do show a place-preference for the pup-associated chamber (Fleming et al., 1994). The neurochemical underpinnings necessary to form the association between the pups and the environment, or at least necessary to display this preference, involves dopaminergic neurotransmission and can be disrupted by dopamine antagonism during the exposure phase even at doses that did not disrupt mother–litter interactions (Fleming et al., 1994).

Mattson et al. (2003) examined the expression of homecage maternal behavior in females that had been tested for their preference for cues associated with pups or cocaine. Regardless of their

Figure 5-5

Postpartum females were conditioned to associate pups with neutral cues in one chamber and on alternate days conditioned to associate the effects of a subcutaneous injection of cocaine with another set of neutral cues in a different chamber (Mattson et al., 2001, 2003). Summary graph comparing the percentage of females that preferred cues associated with pups, cocaine, or a neutral chamber. Three groups were conditioned: early postpartum conditioned days 4–7, tested day 8; mid postpartum conditioned days 6–9, tested day 10; late postpartum conditioned days 12–15 tested day 16. All groups demonstrated a statistically significant conditioning effect over baseline response to the conditioning apparatus. The majority of females in the early group preferred the chamber associated with pups, the mid group was bimodal with half the population choosing pup and half cocaine associated chambers, and most late postpartum dams chose the cocaine-associated chamber. The number of late and mid postpartum females that preferred the cocaine-associated chamber was significantly greater than the number of early postpartum females preferring the cocaine-associated chamber. Significantly more postpartum females preferred pup-associated cues compared to late-postpartum females ($P < 0.05$)

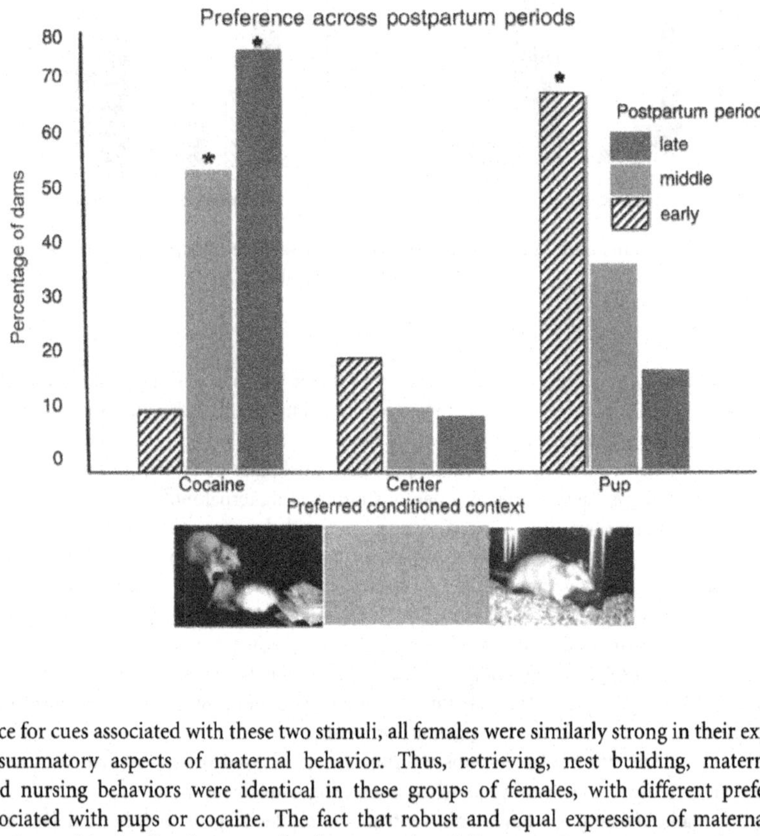

preference for cues associated with these two stimuli, all females were similarly strong in their expression of the consummatory aspects of maternal behavior. Thus, retrieving, nest building, maternal aggression, and nursing behaviors were identical in these groups of females, with different preferences for cues associated with pups or cocaine. The fact that robust and equal expression of maternal behavior could be detected in females that were also demonstrating different preferences suggests that the motivation to seek the pup stimulus (the appetitive component) can be regulated by processes that are, in part, dissociable from the processes underlying the expression of maternal behavior (consummatory components).

To explore the neural substrate that supports maternal motivation, Morrell and colleagues have examined the brains of the females after determining their preference for cues associated with pups, compared with controls using cFos immunoreactive protein as a marker for neural activity *Figures 5-6* and 5-7 demonstrate that the mPOA has a statistically significant increase in the number of cFos immunoreactive neurons when the dam is demonstrating a preference for cues associated with the pup,

Figure 5-6
Place-preference conditioning was used to determine the preference of day 10 postpartum maternal, lactating dams for a distinctive environment associated with pups. Control subjects received the same chamber exposure but no unconditioned stimuli. Animals were perfused after 2 h test exposure to the conditioning apparatus and brain sections prepared with cFos immunocytochemistry (Smith et al., 2003; Mattson and Morrell, 2005). Top – anatomically exact Neurolucida-generated illustration of a representative cross-section in the mPOA for each group. A single dot represents a c-Fos-IR neuron; larger dots represent the c-Fos-IR neurons in the medial preoptic area. Below – photomicrograph of a field in the same region. There was a much greater density of Fos-IR neurons in the mPOA of dams that preferred cues associated with pups (right photo and Neurolucida drawing) than that in control subjects (left photo and Neurolucida drawing)

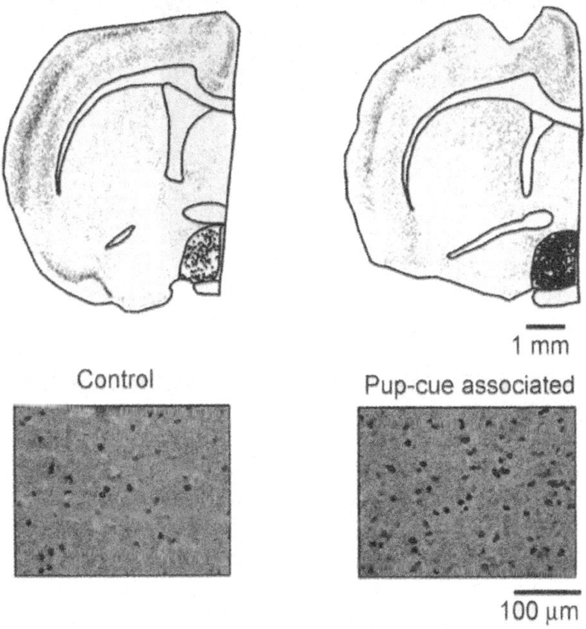

that is when the salience of the pup stimulus is high and that the dam is maternally motivated. Control areas as such as the pyriform cortex or the dorsal striatum showed no such differences, supporting the hypothesis that these neurons are engaged by the preference state of the female, and the salience of the pup-associated cues. Other brain regions including the prefrontal cortex (PFC), NA, and AMYbl also had the basolateral nucleus of the amygdala also had populations that were engaged with the preference for pup-associated cues. These data are the first to demonstrate an engagement of neurons in the mPOA for the purely appetitive components of maternal behavior, in the absence of the pup stimulus (Smith et al., 2003; Mattson and Morrell, 2005).

We know from a wealth of data that the mPOA is key to the expression of maternal behavior (Numan and Insel, 2003). The data presented above extend our understanding of the role of the mPOA in the motivational aspects of maternal behavior by demonstrating a neuronal activation in this region with a purely appetitive measure of motivational status. These data accord well with demonstration that the mPOA is involved in operant responding for delivery of pups (Lee et al., 1999). Thus, the mPOA may have a combined role in recognizing the pup as a highly salient reinforcer and in focusing the consummatory aspects of the behavior. How neuronal subsets within the mPOA carry out this dual function remains to be determined.

Figure 5-7
Histograms from each group represented in *Figure 5-6* **showing the mean (±SEM) number of c-Fos-IR neurons, with one group preferring pup-associated cues and control. Asterisk indicates a significant difference from the control group, P < 0.05. The average number of Fos-immunoreactive neurons in the mPOA was 40% greater in pup-associated, cue-preferring dams (black bar; 92 SEM 2.25) than in controls (grey bar; 54 SEM1.04), and differences were statistically significant, P < 0.001**

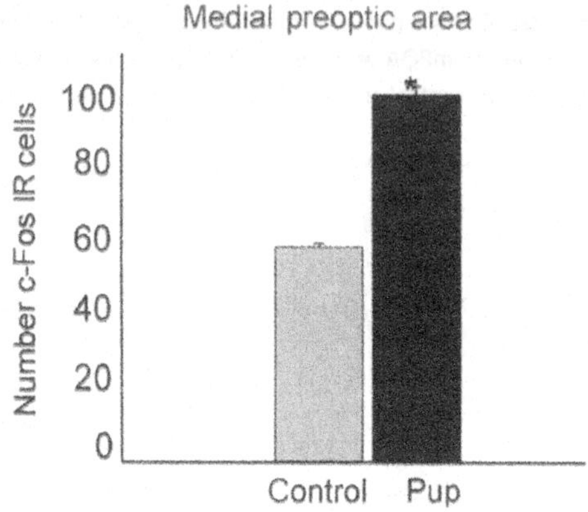

6 Neural Motivation Systems: Interfaces with Maternal Behavior Circuit and Effects of Hormones

The motivation to act maternally is presumed to be regulated by the same neural substrates underlying other types of motivation in rats. Here, we present a summary of this neural substrate and how components of this network are influenced by hormones in virgin rats, with the expectation that this information can be extrapolated to the conditions of pregnancy and lactation.

6.1 Brief Description of Mesolimbic and Nigrostriatal Systems and their Anatomy/Neurochemistry

The basal ganglia is comprised of striatal, pallidal, and associated mesencephalic structures (Heimer et al., 1995). The striatum has a dorsal and ventral subdivision, and the latter includes the NA which has two distinct subdivisions, with the core being more similar to the dorsal striatum, while the shell is more limbic-related (Zaborszky et al., 1985; Zahm and Brog, 1992; Pennartz et al., 1994; Groenewegen et al., 1996). Both striatal subdivisions receive glutaminergic afferents from cortex and thalamus, dopamineric afferents from the mescencephalon, and are constituted of GABAergic projection neurons. The pars compacta of the substantia nigra and the VTA contain DAergic neurons that give rise to the well-known nigrostriatal and mesocorticolimbic projections to the striatum, PFC, and other areas of the forebrain (Heimer et al., 1995). Major efferents targets of the dorsal striatum include the globus pallidus and substantia nigra pars reticulata, while major efferents targets of the ventral striatum include the ventral pallidum and the VTA (Heimer et al., 1995). Generally, the dorsal striatum and related structures are considered important for the programming and execution of voluntary movement, while the ventral striatum and related structures are considered more involved in reward-based responding or motivational processes (Groenewegen et al., 2003; Kelley, 2004). The dorsal structures of the system, however, also probably participate in certain aspects of reward-based responding (Voorn et al., 2004).

The mesolimbic DAergic projections to the NA, PFC, and amygdala are the best understood structures that mediate appetitive/motivational processes (Koob et al., 1987; Kalivas and Barnes, 1994; Wise, 1998;

Robbins and Everitt, 1999; Spanagel and Weiss, 1999). The NA is critical for the mediation of reward processes, containing separable neuronal populations responsive to pharmacological and natural rewards (Carelli et al., 2000; Neisewander et al., 2000; Carelli and Wondolowski, 2003), including pups (Fleming and Walsh, 1994; Lonstein et al., 1998; Stack et al., 2002). The functional repertoire of the medial PFC includes cognitive complexity and emotional tone, making it an excellent candidate to initiate the efferent activity related to motivated behavior (Isaac et al., 1989; Birrell and Brown, 2000; Franklin and Druhan, 2000a, b; Neisewander et al., 2000; Brown and Bowman, 2002). The literature suggesting that the PFC participates in the expression of maternal behavior is limited (Numan and Insel, 2003), although recent functional MRI work in humans shows increased responses in the anterior cingulate of mothers after hearing infant cries (Lorberbaum et al., 1999). Because the infralimbic, prelimbic, and cingulate cortex are engaged by natural as well as pharmacological stimuli (Schroeder et al., 1999; Mattson and Morrell, 2005), these regions may have a general reward response/recognition capacity.

Also important are afferent projections from the PFC, hippocampus, and amygdala *to* the NA. Both the GABAergic and glutametergic projections of these structures are of functional significance in reward processes (Pifl et al., 1993; Cornish et al., 1999; Kalivas and Nakamura, 1999; Li et al., 1999; Rosenkranz and Grace, 2001). These neural substrates are activated by many stimuli, including pharmacologic and natural rewards (Everitt and Wolf, 2002; Kelley and Berridge, 2002), and these diverse reward processes are thought to be mediated by specialized subcircuits or channels within and between these neural components (Pennartz et al., 1994; Ikemoto and Panksepp, 1999; Carelli et al., 2000; Carelli and Wondolowski, 2003).

The basolateral amygdala is well established to be involved in conditioned reward and reinforcement processes (Hiroi and White, 1991; Brown and Fibiger, 1993; Parkinson et al., 2001; Fuchs et al., 2004). Furthermore, it is engaged by both pharmacological and natural rewards, including pup stimuli (Brown et al., 1992; Schroeder et al., 1999; Neisewander et al., 2000; Mattson and Morrell, 2005).

6.1.1 Additional Areas of Note for Maternal Motivation

The mPOA is involved in the response to natural rewards in rats, including chocolate, sexual partners, and offspring (Balthazart et al., 1998; Hull et al., 1999; Schroeder et al., 1999; Lee et al., 2000; Smith et al., 2002; Numan and Insel, 2003; Mattson and Morrell, 2005). There is also a strong case that the mPOA has a remarkably analogous dual functional role for the appetitive and consummatory aspects of male sexual behavior and for maternal behavior (Balthazart et al., 1998; Hull et al., 1999; Numan and Insel, 2003). Known neuroanatomical connections of the mPOA connect it with components of the reward system discussed above and to the lesser understood motor regions of the mesencephalon, suggesting that it has sufficient network connections to support an integrated appetitive/consummatory function (Fahrbach et al., 1986; Simerly and Swanson, 1986, 1988; Swanson, 1988–1989; Corodimas and Morrell, 1990; Saper, 2000).

6.1.2 Reward System Hierarchy

Based on classical models of organization, the neural structures involved in motivation appear to be hierarchically organized (Groenewegen et al., 1996; Groenewegen and Uylings, 2000; Saper, 2000; Eichenbaum and Cohen, 2001). The associated figure presents one example of the possible efferent flow of activity through the neural components of such a neural circuit (➋ *Figure 5-8*). This allows us to focus on the components of the system that are additional to the well studied mesolimbic DAergic projections, which are afferent to cortex and accumbens.

The sensory stimulation provided by natural rewards enters through particular and limited channels (sensory afferents), which vary depending upon the physical nature of the natural stimulus. In the case of the pup as natural stimulus, somatosensory and olfactory components are notable. This sensory information must reach the cortex, after thalamic or other (olfactory) processing, but it is also possible that such sensory information could enter this circuit at several additional points, for example, including the amygdala, POA, and the VTA, independent of thalamic processing. Responding to motivating stimuli

Figure 5-8
Schematic diagram of selected components of the neural circuit that mediates the expression of maternal behavior. Focus is on the efferent outflow of the system's components, presumably mediating the active expression of the behavioral sequences

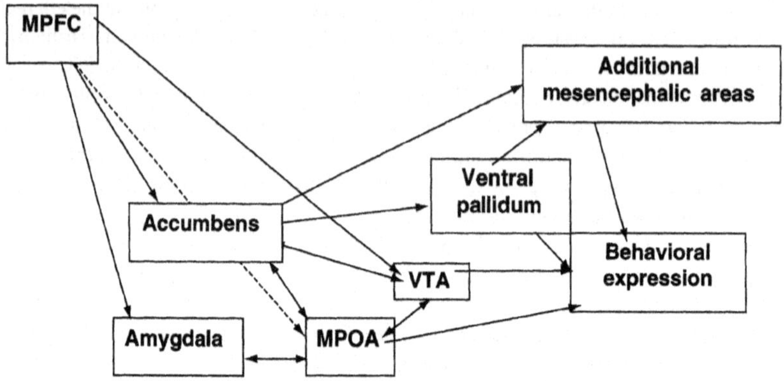

must initially involve the PFC, which soon after receiving stimulus-specific sensory information and association memories via projections from other cortical areas including the sensory and parietal association areas, initiates a descending cascade. These corticofugal projections from the medial PFC to the NA, VTA, basolateral amygdala, and mPOA, are glutamatergic (Park et al., 2002; McFarland et al., 2003). The behavioral expression of motivation or reward response is then due to subsequent outflow of limbic and POA structures via the ventral pallidum and mesencephalic regions, and these efferents are either glutametergic or GABAergic. The consideration of reward based responses, in addition to the classically considered dopamine mediated responses, must include processes regulated by glutamate and GABA. These are generally less frequently examined in the context of maternal motivation or expression of maternal behavior and provide an important area for future focus.

6.2 Role of Dopamine and the Nigrostriatal and Mesolimbic Systems on Maternal Behavior and Motivation

Dopamine is the most thoroughly examined neurotransmitter with regard to maternal behavior in rodents. The general conclusion from this body of work is that increased DAergic neurotransmission in the nigrostriatal, and particularly mesolimbic, systems is necessary for active components of maternal behavior such as retrieval and licking the pups, whereas its inhibition is required for quiescent nursing. This has been gleaned after systemic injections of DAergic drugs. Injection of the mixed DA receptor antagonist haloperidol eliminates retrieval and licking of pups in most dams (Giordano et al., 1990; Stern and Taylor, 1991), and its effects can be reversed with concomitant treatment with the DA agonist apomorphine (Giordano et al., 1990). Notably, even though haloperidol-treated dams do not retrieve pups, they may readily approach and sniff them, and leave the nest for food (Giordano et al., 1990), indicating that these impairments are not due to generalized motor impairments. The fact that dams may still remain in contact with scattered pups and even nurse them (Giordano et al., 1990) provides evidence that the dams remain motivated to act maternally. Haloperidol-treated dams also become quiescent and begin to nurse sooner, and nurse for longer durations, which results in greater milk letdown and litter weight gains (Stern and Taylor, 1991). In fact, rats do not even need to be maternal or lactating for DA inhibition and pup stimulation to elicit kyphosis (Stern, 1991). The D1 and D2 receptors appear to be equally involved in mediating these effects and antagonists to either receptor produce these changes in maternal behaviors (Hansen et al., 1991b; Silva et al., 2001; Byrnes et al., 2002). These DA antagonist effects on maternal behaviors are not exclusive to rats, as haloperidol produces similar effects in both sexes of the biparental prairie vole (Lonstein, 2002). Given all of these results, Numan and Insel (2003) noted that it is surprising that a lack of maternal care has not been even anecdotally reported in female mice with a selective deletion of dopamine receptors.

Although dams given DA antagonists still readily make contact with pups, these antagonists also produce deficits in maternal motivation that can be observed under some testing conditions. Muzzling of dams prevents their perioral contact with pups, but does not prevent them from attempting to make contact with pups with their snouts and paws (Stern and Johnson, 1989). Low doses of haloperidol that do not disrupt maternal retrieval or licking in unmuzzled dams reduce pushing and handling of pups in muzzled dams, suggesting that a basal level of DAergic neurotransmission is necessary for continued maternal motivation to contact inaccessible pups, and possibly accessible pups as well (Stern and Keer, 1999).

DA antagonists act centrally to produce their effects on maternal behavior and DA antagonists that act only peripherally have no effect (Stern and Taylor, 1991). This is in support of the early literature that showed various degrees of disrupted maternal care after chemical depletion of neural catecholmines with ventricular injection of 6-hydroxydopamine (6-OHDA) (Sorenson and Gordon, 1975; Rosenberg et al., 1977; Piccirillo et al., 1980) or lesions of the ascending catecholaminergic projections (Steele et al., 1979; Bridges et al., 1982). The first work examining a particular role for the nigrostriatal systems in maternal behavior was by Numan and Nagle (1983). Large, but not small, bilateral lesions of the substantia nigra and surrounding area disrupted all components of maternal behavior, but also were generally debilitating. A unilateral knifecut of the lateral projections of the POA combined with a lesion of the SN ipsilateral or contralateral to the knifecut produced similar effects, implying that projections between the POA and SN are necessary for maternal care. Maternal behavior in dams with knifecuts and lesions slowly recovered within a few days after surgery, however, concomitant with an improvement in lactational performance and overall health. Therefore, the nigrostriatal system is probably not absolutely necessary for the display of maternal behavior, and postsurgical compensatory mechanisms can supplant any functions that this system normally contributes.

Compared with the nigrostriatal system, the role of the mesolimbic system in maternal behavior has received much more attention. Large premating (Gaffori and LeMoal, 1979) or postpartum (Numan and Smith, 1984) radiofrequency or electrolytic lesions of the VTA and surrounding area severely disrupted all components of maternal care, but similar to dams that receive POA lesions, VTA-lesioned dams still approach and sniff the pups. Because dams with bilateral VTA lesions were often physically debilitated, Numan and Smith (1984) conducted further studies in which postpartum dams received a knifecut of the lateral connections of the mPOA or the lPOA, along with a unilateral VTA lesion on the ipsilateral or contralateral side. Maternal behavior was significantly impaired after all knifecuts, although dams with the mPOA knifecuts and contralateral VTA lesion were the most impaired. Nursing behavior was not more impaired than that produced from other knifecuts/lesions and POA–VTA projections appear to be more likely involved in active maternal behaviors than nursing. A role specifically for the DAergic projection of the VTA was provided by Hansen et al. (1991b) who demonstrated that postpartum 6-OHDA lesions of the VTA eliminate retrieval in most dams, but leave nest building and nursing intact (Hansen et al., 1991b). Preparturitional lesions completely abolished all maternal care, but dams were also physically debilitated (Hansen et al., 1991b). When lesions were made before mating, avoiding postpartum physical debilitation, there were no significant effects on retrieval, although it was severely impaired a few dams (Hansen et al., 1991b). The normal display in most dams may be due to compensation and recovery that might have occurred during the 8 weeks between surgery and parturition.

The role of the termination points of the mesolimbic DA system in maternal behavior has also been examined. Interactions with pups elicit Fos expression in the shell region of the NA (Lonstein et al., 1998; Stack et al., 2002) and this maximal Fos response requires integrity of the mPOA (Stack et al., 2002). Electrolytic lesions of the ventral striatum performed during midpregnancy have little effect on mother–litter interactions immediately after parturition (Lee et al., 1999), but NA-lesioned dams show impairments in all aspects of maternal behavior if tested repeatedly thereafter (Smith and Holland, 1975; Lee et al., 2000; Stack et al., 2001). These effects are probably due to DAergic neurotransmission in the NA because selective postpartum 6-OHDA lesions of the ventral striatum produce similar results

(Hansen et al., 1991a; Hansen, 1994). However, when 6-OHDA-lesioned dams were separated from the pups for 4–6 h prior to testing, retrieval was completely normal (Hansen, 1994), and increasing maternal motivation through separation from pups can apparently override the effects of impaired mesolimbic neurotransmission. It also indicates that sensory cues from pups increase DAergic activity in the ventral striatum in a manner that elicits maternal responses. Indeed, interactions with pups increase extracellular DA release in the NA (Hansen et al., 1993), particularly during bouts of pup licking (Champagne et al., 2004). Furthermore, DA release begins to rise *before* dams begin licking (Champagne et al., 2004) and DA release in the NA might be involved in the dam's transition from one maternal behavior to another. Importantly, 6-OHDA or electrolytic lesions of the dorsomedial striatum, which receives its primary projections from the SN, do not affect maternal behavior (Kirkby, 1967; Numan and Nagle, 1983; Hansen et al., 1991a).

Reducing DAergic neurotransmission in the ventral striatum by directly infusing the DA antagonist *cis*-flupenthixol impairs retrieval and enhances quiescent nursing (Keer and Stern, 1999), with infusions the NA shell producing the greatest effects (Keer and Stern, 1999). However, similar effects have also been found after infusions of the D2 receptor antagonist pimozide into the NA core (Silva et al., 2003). Pharmacological increases in DA receptor activity also impair maternal care. NA core infusion of the DA reuptake inhibitor cocaine reduces retrieval of pups from a runway extension of the home cage and impairs nest construction, with lesser effects found with NA shell infusions (Vernotica et al., 1999). These results are consistent with the inefficient retrieving of pups and reduced time spent in the nest by female mice with a deletion of the DA transporter, which chronically elevates extracellular DA (Spielewoy et al., 2000). Competing behaviors (burying, rearing, and self-grooming) were increased in these mice and thus the impairments may not be specific to maternal behavior. Systemic injection of the DA agonist apomorphine also impairs retrieval, concomitant with increased competing behaviors (Stern and Protomastro, 2000). Therefore, substantially decreased or increased DAergic neurotransmission in the NA and elsewhere is detrimental to active maternal behaviors such as retrieval.

Two recent studies examined intracellular DAergic activity in the striatum of maternal rats, with the aim of correlating changes in DAergic activity with maternal state, but were somewhat inconclusive. Olazabal et al. (2004) examined virgin females without exposure to pups, maternally-sensitized virgins, virgins that did not show maternal behavior after exposure to pups, and day 4 postpartum dams, to find that DAergic activity in the NA was similar between groups. They also found that DAergic activity in the NA was higher in adult rats than in juveniles, which may be associated with age-related differences in the propensity to spontaneously display maternal behavior. In another study, Lonstein et al. (2003) examined DAergic activity in tissue samples from the NA of virgin females or females that were sacrificed during early- and late-pregnancy, soon after parturition, or early and late-lactation, and found no differences between groups. However, in the dorsolateral striatum, DAergic activity was highest during late pregnancy and soon after parturition. Because this area of the striatum is electrophysiologically responsive to perioral stimulation (Mittler et al., 1994), changes in DAergic activity here might be involved in the onset of oral maternal behaviors emerging at parturition (Lonstein et al., 2003).

6.3 Changes in Dopamine Receptors Across Pregnancy and Lactation

There has been limited examination of neurotransmitter receptor or transporter expression across pregnancy and lactation, although naturally-occurring changes in the distribution or numbers of these proteins are likely involved in changes in maternal responsiveness. Dopamine receptors have been examined using in vitro autoradiography during early and late pregnancy, with reference to diestrous females and males (Bakowska and Morrell, 1995). Compared with males, late-pregnant females showed fewer D_1 and D_2 receptors in the striatum and NA. Late-pregnant females also showed fewer D_2 receptors in the olfactory tuberacle than females in either diestrus or early pregnancy. There were no differences in many other regions including the SN and VTA, or regions best understood to participate in the mediation of maternal behavior such as the mPOA and MeA. In general, the decreases in late pregnancy were 11–27% compared with other groups. The magnitude of these changes is less than typical after pharmacological intervention or with pathology, but similar to the down-regulation of the receptors seen with other physiological events such as aging or estrogen treatment. During late pregnancy, voluntary activity decreases markedly, and grooming of the ventral body surface increases. Changes in the motivational state of the female and preparation for new motor patterns are also underway. Perhaps these region-specific cases of down-regulation are related to increased dopamine secretion that then participates in these preparatory events (Bakowska and Morrell, 1995).

6.4 Influences of Ovarian and Pituitary Hormones on Nigrostriatal Neurotransmission

DA neurotransmission within the nigrostriatal, and particularly mesolimbic, systems is strongly implicated in the control of maternal behavior and motivation. How ovarian and pituitary hormones alter DA neurotransmission across pregnancy and lactation in a manner that alters maternal responsiveness is generally unknown, but insight may be gained from the literature investigating hormone effects on these DA systems in nonpregnant and nonlactating rats.

Elevated levels of circulating estrogens established naturally or exogenously generally increase DA neurotransmission in the striatum (Becker and Ramirez, 1981; Crowley et al.,1978a; Tansey et al., 1983; Becker and Beer, 1986; Becker, 1990; Dluzen and Ramirez, 1990; Morissette et al., 1990b; Fernandez-Ruiz et al., 1991; McDermott, 1993; Pasqualini et al., 1995). Similarly, behavioral responses controlled by the nigrostriatal system, including amphetamine-, cocaine-, or electrical stimulation-induced locomotor behavior, are also potentiated by estrogens (e.g., Becker, 1990; Sell et al., 2000). Both the SN and striatum are sensitive to estrogens, and infusion into either site affects DA release in the striatum (Pasqualini et al., 1995), as well as postural and motor regulation in rats (Joyce and Van Hartesveldt, 1984; Roy et al., 1990). However, there are also many negative findings on estrogen's ability to increase nigrostriatal DA function (Gordon et al., 1977; Lofstrom, 1977; Tyler et al., 1979; Euvard et al., 1980; Dupont et al., 1981; Di Paolo et al., 1982a; Fields and Gordon, 1982; Kazandjian et al., 1988; Pasqualini et al., 1995; McDermott et al., 1997; Zsarnovszky et al., 2000) and one can conjecture that methodological differences in dose or length of estrogen treatment could explain the discrepancies.

Estrogenic effects on striatal DA neurotransmission likely result from changes that occur both presynaptically and postsynaptically. For example, physiological increases in circulating estradiol increase presynaptic DA release (Becker, 1990; Xiao and Becker, 1998; Xiao et al., 2003). Estradiol also increases the density of DA receptors in the striatum (Di Paolo et al., 1982a, 1988; Hruska and Pitman, 1982; Hurska and Nowak, 1988), although a decrease in expression or no changes have also been reported (Levesque et al., 1992; Lammers et al., 1999). The DA transporter, which allows DA reuptake from the synapse and termination of DA neurotransmission, also rapidly increases in number in the striatum and substantia nigra pars compacta (SNpc) after treatment with estradiol (Morissette et al., 1990; Morissette and Di Paolo, 1993; Bosse et al., 1997), or during proestrus (Morissette and Di Paolo, 1993; Bosse et al., 1997). It has also been reported that estrogens decrease DA transporter activity in the SNpc (Bosse et al., 1997) and striatum (Disshon et al., 1998; Rehavi et al., 1998).

Less is known about the effects of progesterone on the nigrostriatal system, but it appears to have dose- and time-specific effects. Low doses of progesterone reduce DA release from striatal tissue, whereas higher doses increase it (Becker et al., 1984; Dluzen and Ramirez, 1989a, b; Cabrera et al., 1993), and progesterone increases striatal DA release 2 and 12 h after injection, but reduces it 24 h later (Dluzen and Ramirez, 1984). Interestingly, pulsatile progesterone is more effective than continuous exposure in stimulating DA release (Dluzen and Ramirez, 1987), possibly relevant to different times of the reproductive cycle when the hormone can be either transiently released (as during estrous) or chronically released (as during pregnancy). It is unclear what the effects of progesterone are on striatal DA transporter activity (Morissette and Di Paolo, 1993; Rehavi et al., 1998).

The mechanism by which ovarian hormones influence the nigrostriatal DA system is unclear because this system does not appear to express nuclear ERs or estradiol-induced progestin receptors (e.g., Sar, 1988; Kritzer, 1997; Mufson et al., 1999; Creutz and Kritzer, 2002; Lonstein and Blaustein, 2004). Instead, ovarian hormones probably act rapidly, at least in part, via membrane-bound receptors on either DAergic cell bodies in the mesencephalon and/or locally within the striatum to affect DA release (Chiodo and Caggiula, 1980; Dluzen and Ramirez, 1989a, b; Becker, 1990; Castner et al., 1993; Mermelstein et al., 1996; Ramirez and Zheng, 1996; Xiao and Becker, 1998; Xiao et al., 2003). Nonetheless, some long-term and presumably genomically-based effects of estradiol may also occur (e.g., Chiodo and Caggiula, 1980). It may do this by acting on non-DAergic neurons in the SN that do express ER. In fact, the nigrostriatal system contains a rich GABAergic network originating in the pars reticulata that inhibits DA release by acting directly on SNpc neurons, and by sending projections to the striatum where with local interneurons, it can influence DA release (e.g., Kilpatrick et al., 1980; Kubota et al., 1987; Reid et al., 1990; Tepper et al., 1995; Paladini et al., 1999; Rodriguez and Gonzalez-Hernandez, 1999). Estradiol reduces GABAergic activity in these systems (Gordon et al., 1977; Tyler et al., 1979; McGinnis et al., 1980; Nicoletti et al., 1982, 1985).

Pituitary hormones may also increase nigrostriatal DAergic activity and intracerebroventricular infusion of oxytocin or arginine-vasopressin (AVP) increase striatal DA release (Schulz et al., 1979), although it is unknown where they act to produce this effect. The effects of prolactin in this system have also been investigated, but are highly discrepant (Drago et al., 1981; Chen and Ramirez, 1982; Laping and Ramirez, 1988; Cebeira et al., 1991; Hernandez et al., 1994).

In sum, ovarian and pituitary hormones alter DAergic activity in the nigrostriatal system. Although some of the literature may be contradictory, estradiol seems particularly potent in increasing mesolimbic DA function. Along with contributory effects of progesterone, OT, and PRL, nigrostriatal DAergic activity is likely to be increased by the end of pregnancy when maternal behavior emerges (Lonstein et al., 2004).

6.5 Influences of Ovarian and Pituitary Hormones on Mesolimbic Neurotransmission

Unfortunately, the literature on the effects of hormones on mesolimbic DA neurotransmission in nonpregnant and nonlactating rat is frequently inconsistent and provides only limited insight for the field of maternal care. Cells of the NA do express ER, but not nearly as many as compared with ER-rich areas such as the medial preoptic nucleus (Pfaff and Keiner, 1973; Mufson et al., 1999). Furthermore, the VTA has not been reported to express ER at all, although the excitability of these cells is altered by estradiol (Sakamoto et al., 1993). Nevertheless, ovariectomy reduces DA content in the VTA (Russo et al., 2003) and potassium-stimulated DA release in NA is highest during diestrus when circulating hormones are low, and reuptake is enhanced during proestrus when they are high (Thompson and Moss, 1997). However, DA turnover in the NA has been reported to be highest (Shimizu and Bray, 1993) and extracellular DA the lowest, during proestrus (Shimizu and Bray, 1993). Virtually opposite results have also been reported (Fernandez-Ruiz et al., 1991; Kazandjian et al., 1988). Functionally, there is no influence of the stage of the estrous cycle on the ability of haloperidol to induce catalepsy (Kazandjian et al., 1988), but estrogen-treated female rats do sensitize to cocaine faster than those not given the hormone (Hu and Becker, 2003). While a single injection of estradiol has no effect on DAergic activity in any site measured (Lofstrom et al., 1977; Crowley et al., 1978b), a few days of treatment can either decrease DA turnover in the NA (Shimizu and Bray, 1993), increase DA synthesis in the NA (Hernandez et al., 1991), or have no effects (Gordon et al., 1977). In cases where estrogens reduce evoked DA release from the NA, it may slow clearance of the neurotransmitter from the synapse (Thompson and Moss, 1994; Thompson, 1999). A few days of estradiol also increases the number of cells expressing TH-ir in interfascicular region, but not paranigral region, of the VTA (Zsarnovszky et al., 2000). An even longer (2-week) regimen of estradiol reduces DA content in NA and VTA, but has no effects on turnover (Dupont et al., 1981), although this chronic regimen reduces DA content in the NA of hypophysectomized females (Di Paolo et al., 1982b), pointing to an influence of PRL. Note that very short-term effects of estradiol can also be found, with direct infusion into NA rapidly increasing DA release (Thompson and Moss, 1994). Similar to the nigrostriatal system, the GABAergic innervation of the VTA likely contributes to estradiol's effects on DAergic activity in the mesolimbic system (Gordon et al., 1977; McGinnis et al., 1980).

Chronic estradiol also influences DA receptor expression, having little effect on D1 and D2 receptor expression in either the VTA or NA, but possibly influencing the D3 receptor in the VTA (Lammers et al.,

1999; Zhou et al., 2002). D2/D3 receptor-mediated G-protein activation is reduced in VTA, but not the NA, after 1 week of EB treatment (Febo et al., 2003). The density or affinity of DA transporter sites in the NA is not modified by 2 weeks of EB treatment (Morissette and Di Paolo, 1993; Thompson et al., 2000), but is reduced in the NA shell after 3 weeks of estrogen treatment (Rehavi et al., 1998). Three months after ovariectomy, DA transporter binding is decreased in NA and its mRNA is increased in the VTA and these effects cannot be reversed with 2 weeks of estradiol treatment (Bosse et al., 1997).

There is little known about progesterone's effects on mesolimbic DA neurotransmission. DAergic cells of the VTA do not express nuclear progestin receptors (Sar, 1988; Lonstein and Blaustein, 2004), but may express PR mRNA (Curran-Rauhut and Pedersen, 2002), and there have been no reports of nuclear PR in the NA. Instead, membrane-bound progestin receptors may influence mesolimbic function. It is also likely that progesterone influences mesolimbic DA activity indirectly through non-DA cells that project to this system or by acting through its metabolites on the benzodiazepine binding site of the $GABA_A$ receptor (Frye, 2001). In any case, progesterone prevents estradiol's effects on DA turnover in NA (Shimizu and Bray, 1993), but it has also been reported to have no effects (Crowley et al., 1978b). Two weeks of progesterone is without apparent effect on the density of DA transporter sites in the NA (Morissette and Di Paolo, 1993), but 3 weeks of progesterone treatment reduced DA transporter binding in the NA (Rehavi et al., 1998).

Oxytocin binding sites are found in the NA (De Kloet et al., 1986), but apparently not the VTA. OT receptors are likely present in the VTA, though, because infusion of oxytocin here induces grooming (Kaltwasser and Crawley, 1987), though increased mesolimbic DA release (Drago et al., 1986; Stivers et al., 1988). Antagonism of OT in the VTA of parturient dams also inhibits maternal response (Pedersen et al., 1994), but there has apparently been no examination of mesolimbic DA release after OT infusion into the VTA or NA. Intracerebroventricular prolactin decreases DAergic activity and D1 receptor density in the limibic forebrain of male rats (Hernandez et al., 1994), and increases DOPAC output when infused into the NA, but not VTA (Chen and Ramirez, 1988). While cells expressing mRNA for either the long or short form were not found in the VTA (Bakowska and Morrell, 1997, 2003), fibers immunoreactive for the prolactin receptor are found in this region (Roky et al., 1996).

7 The Future – Functional Integration

It is clear that a substantial amount of information exists regarding the endocrinology, neuroendocrinology, neuroanatomy, and neurochemistry underlying the onset and maintenance of maternal behavior in rats. It is also clear that the volume of information dramatically decreases as one progresses from the endocrinology, to the neuroendocrinology, to the neurochemistry of this behavior. As noted above, future areas of focus that might be particularly fruitful to our understanding of maternal care might include a better comprehension of how progesterone both inhibits and facilitates the onset of maternal behavior, how the numerous areas of the brain involved in this behavior interact as a larger neural circuit, particularly to behavioral outflow, and how the many less-explored hormones secreted during pregnancy and parturition might influence maternal care and the coordination of its behavioral and physiological aspects. Significant focus on what hormones and neural sites are involved in maternal motivation, which is critical for the initiation and maintenance of the expression of the behavior, could be very fruitful. The literature detailing the effects of hormones on the nigrostriatal, and particularly the mesolimbic, DA systems points to these pathways as malleable substrates for hormone-induced changes in maternal motivation and behavior. The details of how pregnancy, parturition, and lactation modify these systems require examination.

Advances in our understanding of the hormonal and neurochemical basis of maternal behavior have been made with the use of relatively new genetic knockout technology. One of the earliest such models was the *fosB* gene knock out mouse, which has severe deficits in postpartum maternal behavior (Brown et al., 1996). Currently, many models with targeted disruption of specific genes can be found in the literature. These include knockout models of either form of the ER, the progestin receptor, the nitric oxide synthase gene, prolactin receptor genes, a variety of genes related to neurotransmitter systems, and genes that are part of other transcription systems or imprinting processes (Brown et al., 1996; Gammie and Nelson, 1999; Ogawa et al., 1998; Kelly et al., 2001; Leckman and Herman, 2002; Schneider et al., 2003). Some of these gene disruptions interfere with the behavior or closely related functions (e.g., ER or prolactin receptor knockouts), while others facilitate it (e.g., progestin receptor knockout facilitates male parental behavior). At this stage in the development of the technology, all of these models completely inactivate the genes throughout the entire organism and during its entire development and adult life. Data gathered by these methods is rich and intriguing, but its interpretation is complex. As elegantly discussed by Rene Hen and his colleagues (Gingrich and Hen, 2000), in addition to simple loss of function, these models are also subject to a wide variety of plastic or compensatory changes leading to the altered functional state of the knockout animal. Advances in technology allowing for temporal- and region-specific knockout (referred to as conditional knockouts), or in the future, knockin models (Tonegawa et al., 2003; Champtiaux and Changeux, 2004), will be improved and valuable tools to explore the basis of maternal behavior.

Many studies have fruitfully used visualization of immediate-early genes or their products (e.g., *c-fos*) as a tool to survey the populations of neurons activated during maternal behavior (Numan and Insel, 2003). This approach, however, has little temporal resolution. Notable limits on progress in understanding maternal behavior can be seen in the field from the lack of understanding of the real time events and the coordinated and sequenced activity across neural components necessary for maternal behavior. Particularly useful would be a functional circuit analysis with electrophysiological approaches, with the goal of moving toward the more complex analysis currently underway in such structures as the hippocampus, amygdala, and isocortex (Pare and Collins, 2000; Buzsaki, 2004).

Functional imaging may also offer a means of simultaneously assessing the entire neural circuit mediating maternal behavior. The first paper dealing with stimulus responses of postpartum females using the tool of functional magnetic resonance imaging has examined the response of postpartum females to sucking by pups, or to the psychostimulant cocaine (Ferris et al., 2005). Relationships between outcomes using these different approaches in rats should provide fruitful links to data gathered from humans expressing parental behaviors and motivation. Expanding the use of these more recently-devised approaches may also help the field in our quest to adapt what we learn from maternal behavior and consider it as a broader model. That is, a prototypical complex behavior from which we may learn generalizable, fundamental concepts that will inform models of other complex behaviors.

Parental behavior directed toward offspring can uniquely help or materially harm the young and with life-long consequences. As researchers, struggling with the biological fundamentals of our models, we must not loose sight of the ideas that our understanding from the rodent model of maternal behavior may inform the human condition to benefit both parents and offspring during this critical period.

Acknowledgments

The authors would like to thank Dr. Jill B. Becker for helpful comments on some sections of this chapter.

References

Ahdieh HB, Mayer AD, Rosenblatt JS. 1987. Effects of brain antiestrogen implants on maternal behavior and on postpartum estrus in pregnant rats. Neuroendocrinology 46(6): 522-531.

Amateau SK, Alt JJ, Stamps CL, McCarthy MM. 2004. Brain estradiol content in newborn rats: sex differences, regional heterogeneity, and possible de novo synthesis by the female telencephalon. Endocrinology 145: 2906-2917.

Augustine RA, Kokay IC, Andrews ZB, Ladyman SR, Grattan DR, 2003. Quantitation of prolactin receptor mRNA in the maternal rat brain during pregnancy and lactation. J Mol Endocrinol 31(1): 221-232.

Bakowska JC. 1998. Prolactinergic and dopaminergic components of the neural chemistry that underlies maternal behavior in the rat. PhD Thesis Rutgers, Newark NJ: The State University of New Jersey. pp. 1-199.

Bakowska JC, Morrell JI. 1995. Quantitative autoradiographic analysis of D1 and D2 dopamine receptors in rat brain in early and late pregnancy. Brain Res 703 (1–2): 191-200.

Bakowska JC, Morrell JI. 1997. Atlas of the neurons that express mRNA for the long form of the prolactin receptor in the forebrain of the female rat. J Comp Neurol 386(2): 161-177.

Bale TL, Dorsa DM. 1995. Sex differences in and effects of estrogen on oxytocin receptor messenger ribonucleic acid expression in the ventromedial hypothalamus. Endocrinology 136: 27-32.

Bakowska JC, Morrell JI. 2003. Neurons that express the mRNA for the short form of the prolactin receptor. Molecular Brain Research 116: 50-58

Bale TL, Pedersen CA, Dorsa DM. 1995. CNS oxytocin receptor mRNA expression and regulation by gonadal steroids. Adv Exp Med Biol 395: 269-280.

Balthazart J, Ball GF. 1998. New insights into the regulation and function of brain estrogen synthase (aromatase). Trends Neurosci 21: 243-249.

Balthazart J, Absil P, Gerard M, Appeltants D, Ball GF. 1998. Appetitive and consummatory male sexual behavior in Japanese quail are differentially regulated by subregions of the preoptic medial nucleus. J Neurosci 18: 6512-6527.

Bauer JH. 1983. Effects of maternal state on the responsiveness to nest odors of hooded rats. Physiol Behav 30(2): 229-232.

Becker JB. 1990. Estrogen rapidly potentiates amphetamine-induced striatal dopamine release and rotational behavior during microdialysis. Neurosci Lett 118(2): 169-171.

Becker JB, Beer ME. 1986. The influence of estrogen on nigrostriatal dopamine activity: behavioral and neurochemical evidence for both pre- and postsynaptic components. Behav Brain Res 19(1): 27-33.

Becker JB, Beer ME, Robinson TE. 1984. Striatal dopamine release stimulated by amphetamine or potassium: influence of ovarian hormones and the light–dark cycle. Brain Res 311: 157-160.

Becker JB, Ramirez VD. 1981. Sex differences in the amphetamine stimulated release of catecholamines from rat striatal tissue in vitro. Brain Res 204: 361-372.

Benuck I, Rowe FA. 1975. Centrally and peripherally induced anosmia: influences on maternal behavior in lactating female rats. Physiol Behav 14: 439-447.

Birrell JM, Brown VJ. 2000. Medial frontal cortex mediates perceptual attentional set shifting in the rat. J Neurosci 20: 4320-4324.

Bole-Feysot C, Goffin V, Edery M, Binart N, Kelly PA. 1998. Prolactin (PRL) and its receptor: actions, signal transduction pathways, and phenotypes observed in PRL receptor knockout mice. Endocr Rev 19: 225-268.

Bosse R, Rivest R, Di Paolo T. 1997. Ovariectomy and estradiol treatment affect the dopamine transporter and its gene expression in the rat brain. Brain Res Mol Brain Res 46(1–2): 343-346.

Brewster J, Leon M. 1980. Relocation of the site of mother–young contact: maternal transport behavior in Norway rats. J Comp Physiol Psychol 94: 69-79.

Bridges RS. 1990. Endocrine regulation of parental behavior in rodents. Mammalian parenting biochemical, neurobiological, and behavioral determinants. Krasnegor NA, Bridges RS, editors. New York: Oxford Press; pp. 93-117.

Bridges RS. 1996. Biochemical basis of parental behavior in the rat. Parental care – evolution, mechanisms, and adaptive significance. Advances in the study of behavior, vol. 25. Rosenblatt JS, Snowdon CT, editors. Academic Press; New York: pp. 215-242.

Bridges RS, Clifton DK, Sawyer CH. 1982. Postpartum luteinizing hormone release and maternal behavior in the rat after late-gestational depletion of hypothalamic norepinephrine. Neuroendocrinology 34(4): 286-291.

Bridges RS, Di Biase R, Loundes DD, Doherty PC. 1985. Prolactin stimulation of maternal behavior in female rats. Science 782–784.

Bridges RS, Mann PE, Coppeta JS. 1999. Hypothalamic involvement in the regulation of maternal behaviour in the rat: inhibitory roles for the ventromedial hypothalamus and the dorsal/anterior hypothalamic areas. J Neuroendocrinol 11(4): 259-266.

Bridges RS, Numan M, Ronsheim PM, Mann PE, Lupini CE. 1990. Central prolactin infusions stimulate maternal behavior in steroid-treated, nulliparous female rats. Proc Natl Acad Sci USA 87(20): 8003-8007.

Bridges RS, Rosenblatt JS, Feder HH. 1978. Serum progesterone concentrations and maternal behavior in rats after pregnancy termination: behavioral stimulation after progesterone withdrawal and inhibition of progesterone maintenance. Endocrinology 102: 258-267.

Bridges RS, Robertson MC, Shiu RPC, Friesen HG, Stuer AM, et al. 1996. Endocrine communication between conceptus and mother: placental lactogen stimulation of maternal behavior. Neuroendocrinology 64: 57-64.

Bridges RS, Robertson MC, Shiu RP, Sturgis JD, Henriquez BM, et al. 1997. Central lactogenic regulation of maternal behavior in rats: steroid dependence, hormone specificity, and behavioral potencies of rat prolactin and rat placental lactogen I. Endocrinology 138(2): 756-763.

Bridges R, Zarrow MX, Gandelman R, Denenberg VH. 1972. Differences in maternal responsiveness between lactating and sensitized rats. Dev Psychobiol 5: 123-127.

Brown VJ, Bowman EM. 2002. Rodent models of prefrontal cortical function. Trends Neurosci 25: 340-343.

Brown EE, Fibiger HC. 1993. Differential effects of excitotoxic lesions of the amygdala on cocaine-induced conditioned locomotion and conditioned place preference. Psychopharmacology 113: 123-130.

Brown EE, Robertson GS, Fibiger HC. 1992. Evidence for conditional neuronal activation following exposure to a cocaine-paired environment: role of forebrain limbic structures. J Neurosci 12: 4112-4121.

Brown JR, Ye H., Bronson RT, Dikkes P, Greenberg ME. 1996 A defect in nurturing in mice lacking the immediate early gene *fosB*. Cell 86: 297-309.

Buijs RM, DeVries GJ, van Leeuwne FW. 1985. The distribution and synaptic release of oxytocin in the central nervous system. Oxytocin: clinical and laboratory studies. Amico JA, Robinson AG, editors. Amsterdam, New York, Oxford: Elsevier Science Publishers BV.

Buzsaki G. 2004. Large-scale recording of neuronal ensembles. Nat Neurosci 7: 446-451.

Byrnes EM, Rigero BA, Bridges RS. 2002. Dopamine antagonists during parturition disrupt maternal care and the retention of maternal behavior in rats. Pharmacol Biochem Behav 73: 869-875.

Cabrera R, Diaz A, Pinter A, Belmar J. 1993. In vitro progesterone effects on 3H-dopamine release from rat corpus striatum slices obtained under difference endocrine conditions. Life Sci 53: 1767-1777.

Caldwell JD, Walker CH, Pedersen CA, Barakat AS, Mason GA. 1994. Estrogen increases affinity of oxytocin receptors in the medial preoptic area-anterior hypothalamus. Peptides 15: 1079-1084.

Carelli RM, Ijames SG, Crumling AJ. 2000. Evidence that separate neural circuits in the nucleus accumbens encode cocaine versus "natural" (water and food) reward. J Neurosci 20: 4255-4266.

Carelli RM, Wondolowski J. 2003. Selective encoding of cocaine versus natural rewards by nucleus accumbens neurons is not related to chronic drug exposure. J Neurosci 23: 11214-11223.

Casado M, Diaz-Guerra MJ, Rodrigo J, Fernandez AP, Bosca L, et al. 1997. Expression of the calcium-independent cytokine-inducible (iNOS) isoform of nitric oxide synthase in rat placenta. Biochem J 324: 201-207.

Castner SA, Xiao L, Becker JB, 1993. Sex differences in striatal dopamine: in vivo microdialysis and behavioral studies. Brain Res 610(1): 127-134.

Cebeira M, Hernandez ML, Rodriguez de Fonseca F, de Miguel R, Fernandez-Ruiz JJ, et al. 1991. Lack of effect of prolactin on the dopaminergic receptor sensitivity in striatal and limbic areas after exprimentally-induced alterations in its peripheral levels. Life Sci 48: 531-541.

Champagne F, Diorio J, Sharma S, Meaney MJ. 2001. Naturally occurring variations in maternal behavior in the rat are associated with differences in estrogen-inducible central oxytocin receptors. Proc Natl Acad Sci USA 98: 12736-12741.

Champagne FA, Chretien P, Stevenson CW, Zhang TY, Gratton A, Meaney MJ. 2004. Variations in nucleus accumbens dopamine associated with individual differences in maternal behavior in the rat. J Neurosci 24: 4113-4123.

Champtiaux N, Changeux JP. 2004 Knockout and knockin mice to investigate the role of nicotinic receptors in the central nervous system. Prog Brain Res 145: 235-251.

Chen YF, Ramirez VD. 1982. Prolactin stimulates dopamine release from males but not female rats striatal tissue superfuse in vitro. Endocrinology 111: 1740-1742.

Chen JC, Ramirez VD. 1988. In vivo dopaminergic activity from nucleus accumbens, substantia nigra and ventral tegmental area in the freely moving rat: basal neurochemical output and prolactin effect. Neuroendocrinology 48: 329-335.

Chiodo LA, Caggiula AR, 1980. Alterations in basal firing rate and autoreceptor sensitivity of dopamine neurons in the substantia nigra following acute and extended exposure to estrogen. Eur J Pharmacol 67(1): 165-166.

Clapp CL, Torner L, Butierrez-Ospina G, Alcantara E, Lopez-Gomez FJ, et al. 1994. The prolactin gene is expressed in the hypothalamic neurohypophyseal system and the protein is processed into a 14-kDa fragment with activity like 16-kDa prolactin. Proc Natl Acad Sci USA 91: 10384-10388.

Clarke DL, Linzer DI. 1993. Changes in prolactin receptor expression during pregnancy in the mouse ovary. Endocrinology 133: 224-232.

Coirini H, Schumacher M, Flanagan LM, McEwen BS. 1991. Transport of estrogen-induced oxytocin receptors in the ventromedial hypothalamus. J Neurosci 11: 3317-3324.

Connor JR, Davis HN. 1980a. Postpartum estrous in Norway rats: I Behavior. Biol Reprod 23: 994-999.

Connor JR, Davis HN. 1980b. Postpartum estrous in Norway rats II Physiology. Biol Reprod 23: 1000-1006.

Cornish JL, Duffy P, Kalivas PW. 1999. A role for nucleus accumbens glutamate transmission in the relapse to cocaine-seeking behavior. Neuroscience 93: 1359-1367.

Consiglio AR, Lucion AB, 1996. Lesion of hypothalamic paraventricular nucleus and maternal aggressive behavior in female rats. Physiol Behav 59(4–5): 591-596.

Corodimas KP, Morrell JI. 1990. Estradiol-sensitive neurons in the preoptic area are afferent to the medial basal hypothalamus. Exp Brain Res 80: 381-386.

Corodimas KP, Rosenblatt JS, Canfield ME, Morrell JI. 1993. Neurons in the lateral subdivision of the habenular complex mediate the hormonal onset of maternal behavior in rats. Behav Neurosci 107: 827-843.

Corodimas KP, Rosenblatt JS, Morrell JI. 1992. The habenular complex mediates hormonal stimulation of maternal behavior in rats. Behav Neurosci 106: 853-865.

Creutz LM, Kritzer MF, 2002. Estrogen receptor-beta immunoreactivity in the midbrain of adult rats: regional, subregional, and cellular localization in the A10, A9, and A8 dopamine cell groups. J Comp Neurol 446(3): 288-300.

Crowley WR, O'Donohue TL, Jacobowitz DM. 1978a. Changes in catecholamine content in discrete brain nuclei during the estrous cycle of the rat. Brain Res 147(2): 315-326.

Crowley WR, O'Donohue TL, Wachslicht H, Jacobowitz DM. 1978b. Effects of estrogen and progesterone on plasma gonadotropins and on catecholamine levels and turnover in discrete brain regions of ovariectomized rats. Brain Res 154: 345-357.

Cummings AM, Laws SC. 1981. Endocrine control of implantation and delayed implantation in rats and mice. J Reprod Fertil Supp 29: 97-109.

Curran-Rauhut MA, Petersen SL, 2002. The distribution of progestin receptor mRNA in rat brainstem. Brain Res Gene Expr Patterns 1(3–4): 151-157.

De Kloet ER, Voorhuis DA, Boschma Y, Elands J. 1986. Estradiol modulates denstiry of putative "oxytocin receptors" in discrete rat brain regions. Neuroendcrinology 44: 415-421.

Del Cerro MC, Izquierdo MA, Collado P, Segovia S, Guillamon A. 1991. Bilateral lesions of the bed nucleus of the accessory olfactory tract facilitate maternal behavior in virgin female rats. Physiol Behav 50: 67-71.

Devito WJ, Avakian C, Stone S, Ace CI. 1992. Estardiol increases prolactin synthesis and prolactin messenger ribonucleic acid in selected brain regions in the hypohyposectomized female rat. Endocrinology 131: 2154-2160.

Di Paolo T, Bedard PJ, Dupont A, Poyet P, Labrie F. 1982a. Effects of estradiol on intact and denervated striatal dopamine receptors and on dopamine levels: a biochemical and behavioral study. Can J Physiol Pharmacol 60(3): 350-357.

Di Paolo T, Dupont A, Daigle M, 1982b. Effect of chronic estradiol treatment on dopamine concentrations in discrete brain nuclei of hypophysectomized female rats. Neurosci Lett 32(3): 295-300.

Di Paolo T, Falardeau P, Morissette M. 1988. Striatal D-2 dopamine agonist binding sites fluctuate during the rat estrous cycle. Life Sci 43: 665-672.

Disshon KA, Boja JW, Dluzen DE. 1998. Inhibition of striatal dopamine transporter activity by 17 beta estradiol. Eur J Pharmacol 345: 207-211.

Dluzen DE, Ramirez VD. 1984. Bimodal effect of progesterone on in vitro dopamine function of the rat corpus striatum. Neuroendocrinology 39: 149-155.

Dluzen DE, Ramirez VD. 1987. Intermittent infusion of progesterone potentiates where continuous infusion reduces amphetamine-stimulated dopamine release from ovariectomized estrogen-primed rat striatal fragments superfused in vitro. Brain Res 406: 1-9.

Dluzen DE, Ramirez VD. 1989a. Progesterone effects upon dopamine release from the corpus striatum of female rats. I. Evidence for interneuronal control. Brain Res 476: 332-337.

Dluzen DE, Ramirez VD. 1989b. Progesterone effects upon dopamine release from the corpus striatum of female rats. II. Evidence for a membrane site of action and the role of albumen. Brain Res 476: 338-344.

Dluzen DE, Ramirez VD. 1990. In vivo changes in responsiveness of the caudate nucleus to L-dopa infusion as a function of the estrous cycle. Brain Res 536: 163-168.

Doerr HK, Siegel HI, Rosenblatt JS. 1981. Effects of progesterone withdrawal and estrogen on maternal behavior in nulliparious rats. Behav Neural Biol 32: 35-44

DonCarlos LL, Greene GL, Morrell JI. 1989. Estrogen plus progesterone increases progestin receptor immunoreactivity in the brain of ovariectomized guinea pigs. Neuroendocrinology 50: 613-623.

Drago F, Bohus B, Canonico PL, Scapagnini U. 1981. Prolactin induces grooming in the rat: possible involvement of nigrostriatal dopaminergic system. Pharmacol Biochem Behav 15(1): 61-63.

Drago F, Caldwell JD, Pedersen CA, Continella G, Scapagnini U, et al. 1986. Dopamine neurotransmission in the nucleus accumbens may be involved in oxytocin-enhanced grooming behavior of the rat. Pharmacol Biochem Behav 24: 1185-1188.

Dupont A, Di Paolo T, Gagne B, Barden N. 1981. Effects of chronic estrogen treatment on dopamine concentrations and turnover in discrete brain nuclei of ovariectomized rats. Neurosci Lett 22(1): 69-74.

Eichenbaum H, Cohen NJ. 2001. From conditioning to conscious recollection: memory systems of the brain. Working memory and the prefrontal cortex. New York: Oxford University Press; pp. 471-506.

Emanuele NV, Jurgens JK, Halloran MM, Tentler JJ, Lawrence AM, et al. 1992. The rat prolactin gene is expressed in brain tissue: detection of normal and alternatively spliced prolactin messenger RNA. Mol Endocrinol 6: 35-42.

Euvard C, Oberlander C, Bossier JR. 1980. Antidopaminergic effect of estrogens at the striatal level. J Pharmacol Exp Ther 214: 179-185.

Everitt BJ, Wolf ME. 2002. Psychomotor stimulant addiction: a neural systems perspective. J Neurosci 22: 3312-3320.

Fahrbach SE, Morrell JI, Pfaff DW. 1985. Role of oxytocin in the onset of estrogen-facilitated maternal behavior.

Oxytocin:clinical and laboratory studies. Amico JA, Robinson AG, editors. Amsterdam: Elsevier; pp. 372-388.

Fahrbach SE, Morrell JI, Pfaff DW. 1986. Identification of medial preoptic neurons that concentrate estradiol and project to the midbrain in the rat. J Comp Neurol 247: 364-382.

Fahrbach SE, Pfaff DW. 1982. Hormonal and neural mechanisms underlying maternal behavior in the rat. The physiological mechanisms of motivation. Pfaff DW, editor. New York: Springer-Verlag; pp. 253-285.

Fahrbach SE, Pfaff DW. 1986. Effect of preoptic region implants of dilute estradiol on the maternal behavior of ovariectomized, nulliparous rats. Horm Behav 20: 354-363.

Febo M, Gonzalez-Rodriguez LA, Capo-Ramos DE, Gonzalez-Segarra NY, Segarra AC. 2003. Estrogen-dependent alterations in D2/D3-induced G protein activation in cocaine-sensitized female rats. J Neurochem 86(2): 405-412.

Felton TM, Linton L, Rosenblatt JS, Morrell JI. 1999. First and Second order afferents of the lateral hebenula relevant for maternal behavior. Neuro report 10: 883-887.

Felton TM, Linton L, Rosenblatt JS, Morrell JI. 1998. Intact neurons of the lateral habenular nucleus are necessary for the nonhormonal, pup-mediated display of maternal behavior in sensitized virgin female rats. Behav Neurosci 112(6): 1458-1465.

Fernandez-Ruiz JJ, Hernandez ML, de Miguel R, Ramos JA. 1991. Nigrostriatal and mesolimbic dopaminergic activities were modified throughout the ovarian cycle of female rats. J Neural Transm Gen Sect 85(3): 223-229.

Ferris CF, Kulkarni P, Sullivan JM, Harder JA, Messenger TL, et al. 2005. Pup suckling is more rewarding than cocaine: evidence from functional magnetic resonance imaging and three-dimensional computational analysis. J Neurosci 25: 149-156.

Fields JZ, Gordon JH. 1982. Estrogen inhibits the dopaminergic supersensitivity induced by neuroleptics. Life Sci 30: 229-234.

Fisher AE. 1956. Maternal and sexual behavior induced by intracranial chemical stimulation. Science 124: 228-229.

Flannelly KJ, Kemble ED, Blanchard DC, Blanchard RJ. 1986. Effects of septal-forebrain lesions on maternal aggression and maternal care. Behav Neural Biol 45(1): 17-30.

Fleischer S, Slotnick BM. 1978. Disruption of maternal behavior in rats with lesions of the septal area. Physiol Behav 21(2): 189-200.

Fleming AS. 1976. Control of food intake in the lactating rat: role of suckling and hormones. Physiol Behav 17: 841-848.

Fleming AS, Cheung U, Myhal N, Kessler Z. 1989. Effects of maternal hormones on 'timidity' and attraction to pup-related odors in female rats. Physiol Behav 46(3): 449-453.

Fleming AS, Korsmit M, Deller M. 1994. Rat pups are potent reinforcers to the maternal animal: effects of experience, partiy, hormones, and dopamine function. Psychobiology 22: 44-53.

Fleming AS, Luebke C. 1981. Timidity prevents the virgin female rat from being a good mother: emotionality differences between nulliparous and parturient females. Physiol Behav 27(5): 863-868.

Fleming AS, Rosenblatt JS. 1974a. Olfactory regulation of maternal behavior in rats. I. effects of olfactory bulb removal in experienced and inexperienced lactating and cycling females. J Comp Physiol Psychol 86(2): 221-232.

Fleming AS, Rosenblatt JS. 1974b. Olfactory regulation of maternal behavior in rats. II. Effects of peripherally induced anosmia and lesions of the lateral olfactory tract in pup-induced virgins. J Comp Physiol Psychol 86(2): 233-246.

Fleming AS, Walsh C. 1994. Neuropsychology of maternal behavior in the rat: c-fos expression during mother litter interactions. Psychoneuroendocrinology 19: 429-443.

Fleming AS, Vaccarino F, Luebke C. 1980. Amygdaloid inhibition of maternal behavior in the nulliparous female rat. Physiol Behav 25(5): 731-743.

Fliestra RJ, Voogt JL. 1997. Lactogenic hormones of the placenta and pituitary inhibit suckling-induced prolactin (PRL) release but not the ante-partum PRL surge. Proc Soc Exp Biol Med 214: 258-264.

Francis DD, Champagne FC, Meaney MJ. 2000. Variations in maternal behaviour are associated with differences in oxytocin receptor levels in the rat. J Neuroendocrinol 12: 1145-1148.

Friedman MI, Bruno JP, Alberts JR. 1981. Physiological and behavior consequences in rats of water recycling during lactation. J Comp Physiol Psychol 95: 26-35.

Franklin TR, Druhan JP. 2000a. Expression of fos-related antigens in the nucleus accumbens and associated regions following exposure to a cocaine-paired environment. Eur J Neurosci 12: 2097-2106.

Franklin TR, Druhan JP. 2000b. Involvement of the nucleus accumbens and medial prefrontal cortex in the expression of conditioned hyperactivity to a cocaine associated environment in rats. Neuropsychopharmacology 23: 633-644.

Freeman ME. 1988. The ovarian cycle in the rat. The physiology of reproduction, vol 1. Knobil E, Neill JD, editors. New York: Raven Press; pp. 1893-1928.

Frye CA. 2001. The role of neurosteroids and non-genomic effects of progestins and androgens in mediating sexual receptivity of rodents. Brain Res Rev 37: 201-222.

Fuchs RA, Evans KA, Parker MP, See RE. 2004. Differential involvement of orbitofrontal cortex subregions in conditioned cue-induced and cocaine-primed reinstatement of cocaine seeking in rats. J Neurosci 24: 6600-6610.

Gaffori O, Le Moal M. 1979. Disruption of maternal behavior and appearance of cannibalism after ventral mesencephalic tegmentum lesions. Physiol Behav 23: 317-323.

Gammie SC, Nelson RJ. 1999. Maternal aggression is reduced in neuronal nitric oxide synthase-deficient mice. J Neurosci 19: 8027-8035.

Gandelman R, Zarrow MX, Denenberg VH, 1970. Maternal behavior: differences between mother and virgin mice as a function of the testing procedure. Dev Psychobiol 3(3): 207-214.

Gandelman R, Zarrow MX, Denenberg VH. 1972. Reproductive and maternal performance in the mouse following removal of the olfactory bulbs. J Reprod Fertil 28(3): 453-456.

Garland HO, Atherton JC, Baylis C, Morgan MRA, Milne CM. 1987. Hormonal profiles for progesterone, oestradiol, prolactin, plasma renin activity, aldosterone and corticosterone during pregnancy and pseudopregnancy in two strains of rat: correlation with renal studies. J Endocrinol 113: 435-444.

Gibori G, Chatterton RT, Chien JL. 1979. Ovarian and serum concentrations of androgen throughout pregnancy in the rat. Biol Reprod 21: 53-56.

Gibori G, Khan I, Warshaw ML, McLean MP, Puryear TK, et al. 1988. Placental-derived regulators and the complex control of luteal cell function. Recent Prog Horm Res 44: 377-425.

Gibori G, Sridaran R. 1981. Sites of androgen and estradiol production in the second half of pregnancy in the rat. Biol Reprod 24: 249-256.

Giordano AL, Johnson AE, Rosenblatt JS. 1990. Haloperidol-induced disruption of retrieval behavior and reversal with apomorphine in lactating rats. Physiol Behav 48(1): 211-214.

Giordano AL, Siegel HI, Rosenblatt JS. 1991. Nuclear estrogen receptor binding in microdissected brain regions of female rats during pregnancy: implications for maternal and sexual behavior. Physiol Behav 50(6): 1263-1267.

Gingrich JA, Hen R. 2000. Commentary The broken mouse: the role of development, plasticity, and environment in the interpretation of phenotypic changes in knockout mice. Curr Opin Neurobiol 10: 146-152.

Gonzalez-Mariscal G, Jimenez P, Beyer C, Rosenblatt JS. 2003. Androgens stimulate specific aspects of maternal nest-building and reduce food intake in rabbits. Horm Behav 43: 312-317.

Gordon JH, Nance DM, Wallis CJ, Gorski RA. 1977. Effects of estrogen on dopamine turnover, glutamic acid decarboxylase activity and lordosis behavior in septal lesioned female rats. Brain Res Bull 2(5): 341-346.

Grattan DR. 2001. The actions of prolactin in the brain during pregnancy and lactation. The maternal brain neurobiological and neuroendocrine adaptation and disorders in pregnancy and postpartum. Progress in Brain Research, vol 133. Russell JA, Douglas AJ, Windle RJ, Ingram CD, editors. Amsterdam, New York, Oxford: Elsevier; pp. 153-173.

Greco B, Lubbers LS, Blaustein JD. 2003. Estrogen receptor beta messenger ribonucleic acid expression in the forebrain of proestrus, pregnant, and lactating female rats. Endocrinology 144: 1869-1875.

Groenewegen HJ, Wright CI, Beijer AV. 1996. The nucleus accumbens: gateway for limbic structures to reach the motor system? Prog Brain Res 107: 485-511.

Groenewegen HJ, Uylings HB. 2000. The prefrontal cortex and the integration of sensory, limbic and autonomic information. Prog Brain Res 126: 3-28.

Groenewegen HJ, van den Heuvel OA, Cath DC, Voorn P, Veltman DJ. 2003. Does an imbalance between the dorsal and ventral striatopallidal systems play in Tourette's syndrome? A neuronal circuit approach. Brain Dev 25(Suppl 1): S3-S14.

Grota LJ, Eik-Nes KB. 1967. Plasma progesterone concentrations during pregnancy and lactation in the rat. J Reprod Fert 13: 83-91.

Gualillo O, Cominos J, Blanco M, Garcia-Caballero T, Kojima M, et al. 2001. Grehlin, a novel placental-derived hormone. Endocrinology 142: 788-794.

Gunnersen JM, Crawford RJ, Tregear GW. 1995. Expression of the relaxin gene in rat tissues. Mol Cell Endocrinol 110: 55-64.

Hansen S. 1994. Maternal behavior of female rats with 6-OHDA lesions in the ventral striatum: characterization of the pup retrieval deficit. Physiol Behav 55(4): 615-620.

Hansen S, Bergvall AH, Nyiredi S. 1993. Interaction with pups enhances dopamine release in the ventral striatum of maternal rats: a microdialysis study. Pharmacol Biochem Behav 45(3): 673-676.

Hansen S, Harthon C, Wallin E, Lofberg L, Svensson K. 1991a. The effects of 6-OHDA-induced dopamine depletions in the ventral or dorsal striatum on maternal and sexual behavior in the female rat. Pharmacol Biochem Behav 39(1): 71-77.

Hansen S, Harthon C, Wallin E, Lofberg L, Svensson K. 1991b. Mesotelencephalic dopamine system and reproductive behavior in the female rat: effects of ventral tegmental 6-hydroxydopamine lesions on maternal and sexual responsiveness. Behav Neurosci 105(4): 588-598.

Hard E, Hansen S. 1985. Reduced fearfulness in the lactating rat. Physiol Behav 35(4): 641-643.

Hauser H, Gandelman R. 1985. Lever pressing for pups: evidence for hormonal influence upon maternal behavior of mice. Horm Behav 19(4): 454-468.

Heap RB, Flint APF. 1986. Pregnancy. Hormonal control of reproduction (2nd edition). Austin CR, Short RV, editors. Cambridge England: Cambridge University Press; pp. 153-194.

Heimer L, Zahm DS, Alheid GF. 1995. Basal ganglia. The rat nervous system. Paxinos G, editor. San Diego, New York: Academic Press; pp. 579-629.

Heine PA, Di S, Ross LR, Anderson LL, Jacobson CD. 1997. Relaxin-induced expression of Fos in the forebrain of the late pregnant rat. Neuroendocrinology 66: 38-46.

Hernandez ML, Fernandez-Ruiz JJ, Navarro M, de Miguel R, Cebeira M, et al. 1994. Modifications of mesolimbic and nigrostriatal dopaminergic activities after intracerebroventricular administration of prolactin. J Neural Transm Gen Sect 96(1): 63-79.

Hernandez ML, Fernandez-Ruiz JJ, de Miguel R, Ramos JA. 1991. Time-dependent effects of ovarian steroids on tyrosine hydroxylase activity in the limbic forebrain of ♀ rats. Journal of Neural Transmission 83: 77-84.

Hiroi N, White NM. 1991. The lateral nucleus of the amygdala mediates expression of the amphetamine-produced conditioned place preference. J Neurosci 11: 2107-2116.

Hollaway WR, Dollinger MJ, Denenberg VH. 1980. Parturition in the rat: description and assessment. Maternal infleunces and early behavior. Bell RW, Smotherman WP, editors. New York: SP Med. Sci. Books; pp. 1-26.

Hornsby DJ, Wilson BC, Summerlee AJ. 2001. Relaxin and drinking in pregnant rats. Prog Brain Res 133: 229-240.

Hruska RE, Nowak MW. 1988. Estrogen treatment increases the density of D1 dopamine receptors in the rat striatum. Brain Res 442: 349-350.

Hruska RE, Pitman KT. 1982. Distribution and localization of estrogen-sensitive dopamine receptors in the rat brain. J Neurochem 39(5): 1418-1423.

Hu M, Becker JB. 2003. Effects of sex and estrogen on behavioral sensitization to cocaine in rats. J Neurosci 23(2): 693-699.

Hull EM, Lorrain DS, Du J, Matuszewich L, Lumley LA, et al. 1999. Hormone-neurotransmitter interactions in the control of sexual behavior. Behav Brain Res 105: 105-116.

Ikemoto S, Panksepp J. 1999. The role of nucleus accumbens dopamine in motivated behavior: a unifying interpretation with special reference to reward-seeking. Brain Res Rev 31: 6-41.

Insel TR. 1986. Postpartum increases in brain oxytocin binding. Neuroendocrinology 44(4): 515-518.

Insel TR. 1990. Regional changes in brain oxytocin receptor post-partm: time-course and relationship to maternal behaviour. J Neuroendocrinol 2: 539-545.

Insel TR, Harbaugh CR. 1989. Lesions of the hypothalamic paraventricular nucleus disrupt the initiation of maternal behavior. Physiol Behav 45(5): 1033-1041.

Isaac WL, Nonneman AJ, Neisewander J, Landers T, Bardo MT. 1989. Prefrontal cortex lesions differentially disrupt cocaine-reinforced conditioned place preference but not conditioned taste aversion. Behav Neurosci 103: 345-355.

Jacobson CD, Terkel J, Gorski RA, Sawyer CH. 1980. Effects of small medial preoptic area lesions on maternal behavior: retrieving and nest building in the rat. Brain Res 194(2): 471-478.

Jirik-Babb P, Manaker S, Tucker AM, Hofer MA. 1984. The role of the accessory and main olfactory systems in maternal behavior of the primiparous rat. Behav Neural Biol 40: 170-178.

Jones HM, Vernon MW, Rush ME. 1987. Androgenic modulation of periovulatory follicle-stimulating hormone release in the rat. Biol Reprod 37: 268-276.

Joyce JN, Van Hartesveldt C. 1984. Estradiol application to one striatum produces postural deviation to systemic apomorphine. Pharmacol Biochem Behav 20: 575-581.

Kalinichev M, Rosenblatt JS, Nakabeppu Y, Morrell JI. 2000. Induction of c-fos-like and fosB-like immunoreactivity reveals forebrain neuronal populations involved differentially in pup-mediated maternal behavior in juvenile and adult rats. J Comp Neurol 416(1): 45-78.

Kalivas PW, Barnes CD. 1994. Limbic motor circuits and neuropsychiatry. Boca Raton: CRC Press.

Kalivas PW, Nakamura M. 1999. Neural systems for behavioral activation and reward. Curr Opin Neurobiol 9: 223-227.

Kaltwasser MT, Crawley JN. 1987. Oxytocin and cholecysokinin induce grooming behavior in the ventral tegmentum of the rat. Brain Res 426: 1-7.

Kanyicska B, Nguyen DN, Arey BJ, Freeman ME. 1990. Hysterectomy-induced alterations in prolactin secretion of lactating rats. Endocrinology 126: 2021-2026.

Kazandjian A, Spyraki C, Papadopoulou Z, Sfikakis A, Varonos DD. 1988. Behavioural and biochemical effects of haloperidol during the oestrous cycle of the rat. Neuropharmacology 27(1): 73-78.

Keating RJ, Tcholakian RK. 1979. In vivo patterns of circulating steroid in adult male rats. I. Variation in testosterone during twenty four and forty eight hour standard and reverse light dark cycles. Endocrinology 104(1): 184-193.

Kelley AE. 2004. Ventral striatal control of appetitive motivation: role in ingestive behavior and reward-related learning. Neurosci Biobehav Rev 27: 765-776.

Kelley AE, Berridge KC. 2002. The neuroscience of natural rewards: relevance to addictive drugs. J Neurosci 22(9): 3306-3311.

Kelly PA, Binart N, Lucas B, Bouchard B, Goffin V. 2001. Implications of multiple phenotypes observed in prolactin receptor knockout mice. Front Neuroendocrinol 22(2): 140-145.

Kenyon P, Cronin P, Keeble S. 1983. Role of the infraorbital nerve in retrieving behavior in lactating rats. Behav Neurosci 97: 255-269.

Kilpatrick IC, Starr MS, Fletcher A, James TA, MacLeod NK. 1980. Evidence for a GABAergic nigrostriatal pathway in the rat. I. Behavioural and biochemical studies. Exp Brain Res 40: 45-54.

Kirkby RJ. 1967. Caudate nucleus lesions and maternal behavior in the rat. Psychonomic Science 9: 601-602.

Kolunie JM, Stern JM. 1995. Maternal aggression in rats: effects of olfactory bulbectomy, ZnSO4-induced anosmia, and vomeronasal organ removal. Horm Behav 29(4): 492-518.

Koob GF, Vaccarino FJ, Amaltric M, Bloom FE. 1987. Positive reinforcement properties of drugs: search for neural substrates. Brain reward systems and abuse. Engel J, Oreland L, editors. New York: Raven Press; pp. 35-50.

Kremarik P, Fruend-Mercier MJ, Stoeckel ME. 1995. Estrogen-sensitive oxytocin binding sites and differentially regulation by progesterone in the telencephalon and the hypothalamus of the rat. J Neuroendocrinol 281-289.

Kritzer MF. 1997. Selective colocalization of immunoreactivity for intracellular gonadal hormone receptors and tyrosine hydroxylase in the ventral tegmental area, substantia nigra, and retrorubral fields in the rat. J Comp Neurol 379(2): 247-260.

Kubota Y, Inagaki S, Kito S, Wu JY. 1987. Dopaminergic axons directly make synapses with GABAergic neurons in the rat neostriatum. Brain Res 406: 147-156.

Lambert RC, Moos FC, Ingram CD, Wakerley JB, Kremarik P, et al. 1993. Electrical activity of neurons in the ventrolateral septum and bed nuclei of the stria terminalis in suckled rats: statistical analysis gives evidence for sensitivity to oxytocin and for relation to the milk-ejection reflex. Neuroscience 54(2): 361-376.

Lammers CH, D'Souza U, Qin ZH, Lee SH, Yajima S, Mouradian MM. 1999. Regulation of striatal dopamine receptors by estrogen. Synapse 34(3): 222-227.

Laping NJ, Ramirez VD. 1988. Prolactin-induced yawning behavior requires an intact nigro-striatal dopamine system. Pharmacol Biochem Behav 29: 59-62.

Leckman JF, Herman AE. 2002. Maternal behavior and developmental psychopathology Biol Psychiatry 51: 27-43.

Lee A, Clancy S, Fleming AS. 2000. Mother rats bar-press for pups: effects of lesions of the mpoa and limbic sites on maternal behavior and operant responding for pup-reinforcement. Behav Brain Res 108(2): 215-231.

Lee A, Li M, Watchus J, Fleming AS. 1999. Neuroanatomical basis of maternal memory in postpartum rats: selective role for the nucleus accumbens. Behav Neurosci 113: 523-538.

Lee CK, Kang HS, Lee BJ, Kang HM, Choi WS, et al. 1999. Effects of dopamine and melatonin on the regulation of PIT-1 isotype placental growth hormone and lactogen gene expression in the rat placenta. Mol Cells 9: 646-651.

Leon M, Woodside B. 1983. Energetic limits on reproduction: maternal food intake. Physiol Behav 30(6): 945-957.

Levesque D, Gagne B, Barden N, Di Paolo T. 1992. Chronic estradiol treatment increases anterior pituitary but not striatal D2 dopamine receptor mRNA levels in rats. Neurosci Lett 140: 5-8.

Li Y, Hu X-T, Berney TG, Vartanian AJ, Stine CD, et al. 1999. Both glutamate receptor antagonists and prefrontal cortex lesions prevent induction of cocaine sensitization and associated neuroadaptions. Synapse 34: 169-180.

Lindblom C, Sodersten P, Eneroth P. 1985. Effects of pup sucking behavior on inhibition of sexual behavior and ovulatory secretion of LH in lactating rats. J Endocrinol 104: 419-425.

Linden A, Uvnas-Moberg K, Eneroth P, Sodersten P. 1989. Stimulation of maternal behaviour in rats with chlocystokinin octapeptide. J Neuroendocrinol 1: 389-392.

Lisciotto CA, Morrell JI. 1993. Circulating gonadal steroid hormones regulate estrogen receptor mRNA in the male rat forebrain. Mol Brain Res 20: 79-90.

Lofstrom A. 1977. Catecholamine turnover alterations in discrete areas of the median eminence of the 4- and 5-day cyclic rat. Brain Res 120(1): 113-131.

Lofström A, Eneroth P, Gustaffson JA, Skett P. 1977. Effects of estradiol benzoate on catecholamine levels and turnover in discrete areas of the median eminence and the limbic forebrain, and on serum luteinizing hormone, follicle stimulating hormone and prolactin concentrations in the ovariectomized female rat. Endocrinology 101: 1559-1569.

Lonstein JS. 2002. Effects of dopamine receptor antagonism with haloperidol on nuturing behavior in the biparental prairie vole. Pharmacol Biochem Behav 74: 11-19.

Lonstein JS. 2005. Reduced anxiety in postpartum rats requires recent physical contact with pups, but is independent of suckling or peripheral sources of hormones. Hormones and Behavior, 47(3): 241-255.

Lonstein JS, Blaustein JD. 2004. Immunocytochemical investigation of nuclear progestin receptor expression within dopaminergic neurones of the female rat brain. J Neuroendocrinol 16: 534-543.

Lonstein JS, Dominguez JM, Putnam SK, DeVries GJ, Hull EM. 2003. Intracellular preoptic and striatal monoamines in pregnant and lactating rats: possible role in maternal behavior. Brain Res 970: 149-158.

Lonstein JS, Fleming AS. 2001. Parental behavior in rats and mice. Current protocols in neuroscience. New York: Wiley and Sons; pp. 8.15.1-8.15.26.

Lonstein JS, Gammie SC. 2002. Sensory, hormonal, and neural control of maternal aggression in laboratory rodents. Neurosci Biobehav Rev 26(8): 869-888.

Lorberbaum JP, Newman JD, Dubno JR, Horwitz AR, Nahas Z, et al. 1999. Feasibility of using fMRI to study mothers responding to infant cries. Depress Anxiety 10: 99-104.

Lonstein JS, Simmons DA, Stern JM. 1998. Functions of the caudal periqueductal gray in lactating rats: kyphosis, lordosis, maternal aggression and fearfulness. Behav Neurosci 112: 1502-1518.

Lonstein JS, Stern JM. 1997a. Role of the midbrain periaqueductal gray in maternal nurturance and aggression: c-fos and electrolytic lesion studies in lactating rats. J Neurosci 17(9): 3364-3378.

Lonstein JS, Stern JM. 1997b. Somatosensory contributions to c-fos activation within the caudal periaqueductal gray of lactating rats: effects of perioral, rooting, and suckling stimuli from pups. Horm Behav 32: 155-166.

Lonstein JS, Stern JM. 1998. Site and behavioral specificity of periaqueductal gray lesions on postpartum sexual, maternal, and aggressive behaviors in rats. Brain Res 804: 21-35.

Lonstein JS, Simmons DA, Swann JM, Stern JM. 1998. Forebrain expression of c-fos due to active maternal behaviour in lactating rats. Neuroscience 82: 267-281.

Magnusson JE, Fisher K. 2000. The involvement of dopamine in nociception: the role of D(1) and D(2) receptors in the dorsolateral striatum. Brain Res 855(2): 260-266.

Magnusson JE, Fleming AS. 1995. Rat pups are reinforcing to the maternal rat: role of sensory cues. Psychobiology 23: 69-75.

Mann PE, Bridges RS. 2001. Lactogenic hormone regulation of maternal behavior. The maternal brain neurobiological and neuroendocrine adaptation and disorders in pregnancy and postpartum. Progress in Brain Research, vol 133. Russell JA, Douglas AJ, Windle RJ, Ingram CD, editors. Amsterdam, New York, Oxford: Elsevier; pp. 252-262.

Mann PE, Bridges RS. 2002. Prolactin receptor gene expression in the forebrain of pregnant and lactating rats. Mol Brain Res 105: 136-145.

Mann PE, Felicio LF, Bridges RS. 1995. Investigation into the role of cholecystokinin (CCK) in the induction and maintenance of maternal behavior in rats. Horm Behav 29: 392-406.

Mantalenakis SJ, Ketchel MM. 1966. Frequency and extent of delayed implantation in lactating rats and mice. J Reprod Fertil 12: 391-394.

Marinari KT, Moltz H. 1978. Serum prolactin levels and vaginal cyclicity in concaveated and lactating female rats. Physiol Behav 21: 525-528.

Mattheij JAM, Swarts HJM, Verstijnen CPHJ. 1984. The response of plasma prolactin to suckling during normal and prolonged lactation in the rat. Horm Res 20: 261-268.

Mattheij JAM, Swarts HJM, van Mourik S. 1985. Plasma prolactin in rat during suckling without prior separation from pups. Acta Endocrinol 108(4): 468-474.

Matthews J, Gustafsson JA. 2003. Estrogen signaling: a subtle balance between ER alpha and ER beta. Mol Interv 3: 281-292.

Matthews JC, Beveridge MJ, Malandro MS, Rothstein JD, Campbell-Thompson M, et al. 1998. Activity and protein localization of multiple glutamate transporters in gestation day 14 vs 20 rat placenta. Am J Physiol 274: C603-C614.

Matthews-Felton T, Corodimas KP, Rosenblatt JS, Morrell JI. 1995. Lateral habenula neurons are necessary for the hormonal onset of maternal behavior and for the display of postpartum estrus in naturally parturient female rats. Behav Neurosci 109(6): 1172-1188.

Matthews-Felton T, Linton LN, Rosenblatt JS, Morrell JI. 1999. Estrogen implants in the lateral habenular nucleus does not stimulate the onset of maternal behavior in female rats. Horm Behav 35: 71-80.

Mattson BJ. 2002. Competing appetitive processes in the lactating, maternal rodent: natural and pharmacological stimuli. PhD thesis Rutgers, Newark NJ: The State University of New Jersey; pp. 1-211.

Mattson BJ, Morrell JI. 2005. Preference for cocaine- versus pup-associated cues differentially engages neurons expressing either Fos or CART in lactating, maternal rat. Neuroscience 135: 315-328.

Mattson BJ, Williams SE, Rosenblatt JS, Morrell JI. 2001. Comparison of two positive reinforcing stimuli: pups and cocaine throughout the postpartum period. Behav Neurosci 115: 683-694.

Mattson BJ, Williams SE, Rosenblatt JS, Morrell JI. 2003. Preferences for cocaine- or pup-asssociated chambers differentiates otherwise behaviorally identical postpartum maternal rats. Psychopharmacology 167: 1-8.

Mayer AD, Rosenblatt JS, 1980. Hormonal interaction with stimulus and situational factors in the initiation of maternal behavior in nonpregnant rats. J Comp Physiol Psychol 94(6): 1040-1059.

McDermott JL. 1993. Effects of estrogen upon dopamine release from the corpus striatum of young and aged female rats. Brain Res 606: 118-125.

McDermott JL, Anderson LI, Dluzen DE. 1997. Interactive effects of tamoxifen and estrogen upon the nigrostriatal dopaminergic system. Neuroendocrinology 66(3): 181-187.

McFarland K, Lapish CC, Kalivas PW. 2003. Prefrontal glutamate release into the core of the nucleus accumbens mediates cocaine-induced reinstatement of drug-seeking behavior. J Neurosci 23: 3531-3537.

McGinnis MY, Gordon JH, Gorski RA. 1980. Time course and localization of the effects of estrogen on glutamic acid decarboxylase activity. J Neurochem 34(4): 785-792.

McKinley MJ, Burns P, Colvill LM, Oldfield BJ, Wade JD, et al. 1997. Distribution of Fos immunoreactivity in the lamina terminalis and hypothalamus induced by centrally administered relaxin in conscious rats. J Neuroendocrinol 9: 431-437.

Mermelstein PG, Becker JB, Surmeier DJ. 1996. Estradiol reduces calcium currents in rat neostriatal neurons via a membrane receptor. J Neurosci 16: 595-604.

Miceli MO, Fleming AS, Malsbury CW. 1983. Disruption of maternal behaviour in virgin and postparturient rats following sagittal plane knife cuts in the preoptic area-hypothalamus. Behav Brain Res 9(3): 337-360.

Miranda-Paiva CM, Ribeiro-Barbosa ER, Canteras NS, Felicio LF. 2003. A role for the periaqueductal grey in opioidergic inhibition of maternal behavior. Eur J Neurosci 18: 667-674.

Mittler T, Cho J, Peoples LL, West MO. 1994. Representtion of the body in the lateral striatum of the freely moving rat: single neurons related to licking. Exp Brain Res 98: 163-167.

Moltz H, Geller D, Levin R. 1967. Maternal behavior in totally mammectomized rats. J Comp Physiol Psychol 64: 225-229.

Moltz H, Levin R, Leon M. 1969a. Prolactin in the postpartum rat: synthesis and release in the absence of suckling stimulation. Science 163: 1083-1084.

Moltz H, Levin R, Leon M. 1969b. Differential effect of progesterone on the maternal behavior of primiparous and multiparous rats. J Comp Physiol Psychol 67: 36-40.

Moltz H, Lubin M, Leon M, Numan M. 1970. Hormonal induction of maternal behavior in the ovariectomized nulliparous rat. Physiol Behav 5: 1373-1377.

Moltz H, Robbins D, Parks M. 1966. Caesarean delivery and maternal behavior of primiparous and multiparous rats. J Comp Physiol Psychol 61: 455-460.

Mora S, Dussaubat N, Diaz-Veliz G. 1996. Effects of the estrous cycle and ovarian hormones on behavior indices of anxiety in female rats. Psychoneuroendocrinology 21: 609-620.

Morgan HD, Watchus JA, Milgram NW, Fleming AS. 1999. The long lasting effects of electrical simulation of the medial preoptic area and medial amygdala on maternal behavior in female rats. Behav Brain Res 99(1): 61-73.

Morissette M, Biron D, DiPaolo T. 1990a. Effect of estradiol and progesterone on rat striatl dopamine uptake sites. Brain Res Bull 25: 419-422.

Morissette M, DiPaolo T. 1993. Effect of chronic estradiol and progesterone treatments of ovariectomized rats on brain dopamine uptake sites. J Neurochem 60(5): 1876-1883.

Morissette M, Levesque D, Belanger A, DiPaolo T. 1990b. A physiological dose of estradiol with progesterone affects striatum biogenic amines. Can J Physiol Pharmacol 68: 1520-1526.

Morrell JI, Wagner CK, Malik KF, Lisciotto CA. 1995. Estradiol receptor RNA: neuroanatomy and regulation in three physiological models. Neurobiological effects of sex steroid hormones. Micevych PE, Hammer RP, editors. New York NY: Cambridge University Press; pp. 57-84.

Moss FA. 1924. Study of animal drives. J Exp Psychol 3: 165-185.

Mufson EJ, Cai WJ, Jaffar S, Chen E, Stebbins G, et al. 1999. Estrogen receptor immunoreactivity within subregions of the rat forebrain: neuronal distribution and association with perikarya containing choline acetyltranferase. Brain Res 849: 253-274.

Mustafa A, Nyberg F, Bogdanovic N, Islam A, Suliman I, et al. 1995. Prolactin binding sites in rat brain and liver: effects of long-term ovariectomy and ovarian steroids. Neurosci Lett. Nov 24; 200(3): 179-182.

Nicoletti F, Patti F, Ferrara N, Canonico PL, Giammona G, et al. 1982. Comparative effects of estrogens and prolactin on nigral and striatal GAD activity. Brain Res 232(1): 238-241.

Nicoletti F, Speciale C, Sortino MA, Panetta MS, Di Giorgio RM, et al. 1985. Estrogen effects on nigral glutamic acid decarboxylase activity: a possible role for catecholestrogen. Eur J Pharmacol 115(2–3). 297-300.

Neisewander JL, Baker DA, Fuchs RA, Tran-Nguyen LT, Palmer A, et al. 2000. Fos protein expression and cocaine-seeking behavior in rats after exposure to a cocaine self-administration environment. J Neurosci 20: 798-805.

Nissen HW. 1930. A study of maternal behavior in the white rat by means of the obstruction method. J Genetic Psychol 37: 377-393.

Niswender GD, Juengel JL, McGuire WJ, Belfiore CJ, Wiltbank MC. 1994. Luteal function: the estrous cycle and early pregnancy. Biol Reprod 50: 239-247.

Novakova V, Sterc J, Kuchar S, Mozes S, 1993. Maternal behaviour in septal rat females. Physiol Res 42(5): 351-360.

Numan M. 1978. Progesterone inhibition of maternal behavior in the rat. Horm Behav 11: 209-231.

Numan M. 1988. Maternal behavior. The physiologiy of reproductive behavior. Knobil E, Neill J, editors. New York: Raven Press, Ltd.

Numan M. 1994. Maternal behavior. The physiology of reproduction. Knobil E, Neil JD, editors. New York: Raven Press; pp. 223-283.

Numan M. 1990. Long-term effects of preoptic area knife cuts on the maternal behavior of postpartum rats. Behav Neural Biol 53(2): 284-290.

Numan M, Numan MJ. 1996. A lesion and neuroanatomical tract-tracing analysis of the role of the bed nucleus of the stria terminalis in retrieval behavior and other aspects of maternal responsiveness in rats. Dev Psychobiol 29(1): 23-51.

Numan M, Callahan EC. 1980. The connections of the medial preoptic region and maternal behavior in the rat. Physiol Behav 25(5): 653-665.

Numan M, Corodimas KP. 1985. The effects of paraventricular hypothalamic lesions on maternal behavior in rats. Physiol Behav 35(3): 417-425.

Numan M, Corodimas KP, Numan MJ, Factor EM, Piers WD. 1988. Axon-sparing lesions of the preoptic region and substantia innominata disrupt maternal behavior in rats. Behav Neurosci 102(3): 381-396.

Numan M, Insel TR. 2003. The Neurobiology of Parental Behavior. New York: Springer-Verlag.

Numan M, McSparren J, Numan MJ. 1990. Dorsolateral connections of the medial preoptic area and maternal behavior in rats. Behav Neurosci 104: 964-979.

Numan M, Nagle DS. 1983. Preoptic area and substantia nigra interact in the control of maternal behavior in the rat. Behav Neurosci 97(1): 120-139.

Numan M, Numan MJ. 1994. Expression of Fos-like immunoreactivity in the preoptic area of maternally behaving virgin and postpartum rats. Behav Neurosci 108(2): 379-394.

Numan M, Numan MJ. 1995. Importance of pup-related sensory inputs and maternal performance for the expression of Fos-like immunoreactivity in the preoptic area and ventral bed nucleus of the stria terminalis of postpartum rats. Behav Neurosci 109(1): 135-149.

Numan M, Numan MJ. 1997. Projection sites of medial preoptic area and ventral bed nucleus of the stria terminalis neurons that express Fos during maternal behavior in female rats. J Neuroendocrinol 9(5): 369-384.

Numan M, Numan MJ, English JB. 1993. Excitotoxic amino acid injections into the medial amygdala facilitate maternal behavior in virgin female rats. Horm Behav 27: 56-81.

Numan M, Rosenblatt JS, Komisaruk BR. 1977. Medial preoptic area and onset of maternal behavior in the rat. J Comp Physiol Psychol 91(1): 146-164.

Numan M, Roach JK, Del Cerro MC, Guillamon A, Segovia S, et al. 1999. Expression of intracellular progesterone receptors in rat brain during different reproductive states, and involvement in maternal behavior. Brain Res 830(2): 358-371.

Numan M, Smith HG. 1984. Maternal behavior in rats: evidence for the involvment of preoptic projections to the ventral tegmental area. Behav Neurosci 98: 712-727.

Ogawa S, Eng V, Taylor J, Lubahn DB, Korach KS, et al. 1998. Roles of estrogen receptor-α gene expression in reproduction-related behaviors in female mice. Endocrinology 139: 5070-5081.

Olazabal DE, Ferreira A. 1997. Maternal behavior in rats with kainic acid-induced lesions of the hypothalamic paraventricular nucleus. Physiol Behav 61(5): 779-784.

Olazabal DE, Abercrombie E, Rosenblatt JS, Morrell JI. 2004. The content of dopamine, serotonin, and their metabolites in the neural circuit that mediates maternal behavior in juvenile and adult rats. Brain Res Bull 63: 259-268.

Olazabal DE, Kalinichev M, Morrell JI, Rosenblatt JS. 2002. MPOA Cytotoxic lesions and maternal behavior in the rat: Effects of midpubertal lesions on maternal behavior and the role of ovarian hormones in the maturation of MPOA control of maternal behavior. Horm Behav 41: 126-138.

Osheroff P, Phillips HS. 1991. Autoradiographic localization of relaxin binding sites in the rat brain. Proc Natl Acad Sci 88: 6413-6417.

Ouhtit A, Morel G, Kelly PA. 1993. Visualization of gene expression of short and long forms of prolactin receptor in rat reproductive tissues. Biol Reprod 49: 528-536.

Paladini CA, Celada P, Tepper JM. 1999. Striatal, pallidal, and pars reticulata evoked inhibition of nigrostriatal dopaminergic neurons is mediated by $GABA_A$ receptors in vivo. Neuroscience 89: 799-812.

Pare D, Collins DR. 2000. Neuronal correlates of fear in the lateral amygdala: multiple extracellular recordings in conscious cats. J Neurosci 20: 2701-2710.

Park WK, Bari AA, Jey AR, Anderson SM, Spealman RD, et al. 2002. Cocaine administered into the medial prefrontal cortex reinstates cocaine-seeking behavior by increasing AMPA receptor-mediated glutamate transmission in the nucleus accumbens. J Neurosci 22: 2916-2925.

Parkinson JA, Crofts HS, McGuigan M, Tomic DL, Everitt BJ, et al. 2001. The role of the primate amygdala in conditioned reinforcement. J Neurosci 21: 7770-7780.

Parsons B, Rainbow TC, MacLusky NJ, McEwen BS. 1982. Progestin receptor levels in rat hypothalamus and limbic nuclei. J Neurosci 2: 1446-1452.

Pasqualini C, Olivier V, Guibert B, Frain O, Leviel V. 1995. Acute stimulatory effect of estradiol on striatal dopamine synthesis. J Neurochem 65(4): 1651-1657.

Patchev VK, Schlosser SF, Hassan AH, Almeida OF. 1993. Oxytocin binding sites in rat limbic and hypothalamic structures: site-specific modulation by adrenal and gonadal steroids. Neuroscience 57: 537-543.

Pedersen CA. 1997. Oxytocin control of maternal behavior. Regulation by sex steroids and offspring stimuli. Ann NY Acad Sci 807: 126-145.

Pedersen CA, Ascher JA, Monroe YL, Prange AJ. 1982. Oxytocin induces maternal behavior in virgin female rats. Science 216: 648-650.

Pedersen CA, Boccia ML. 2003. Oxytocin antagonism alters rat dams' oral grooming and upright posturing over pups. Physiol Behav 80: 233-241.

Pedersen CA, Caldwell JD, Walker C, Ayers G, Mason GA. 1994. Oxytocin activates the postpartum onset of rat maternal behavior in the ventral tegmental and medial preoptic areas. Behav Neurosci 108(6): 1163-1171.

Pennartz CMA, Groenewegen HJ, Lopes Da Silva FH. 1994. The nucleus accumbens as a complex of functionally distinct neuronal ensembles: an integration of behavioural, electronphysiological and anatomical data. Prog Neurobiol 42: 719-761.

Pfaff D, Keiner M. 1973. Atlas of estradiol-concentrating cells in the central nervous system of the female rat. J Comp Neurol 151(2): 121-158.

Pi XJ, Grattan DR. 1999a. Increased expression of both short and long forms of prolactin receptor mRNA in hypothalamic nuclei of lactating rats. J Mol Endocrinol 23: 13-22.

Pi XJ, Grattan DR. 1999b. Expression of prolactin receptor mRNA is increased in the preoptic area of lactating rats. Endocrine 11: 91-98.

Pi XJ, Grattan DR. 1999c. Increased prolactin receptor immunoreactivity in the hypothalamus of lactating rats. J Neuroendocrinol 11(9): 693-705.

Pi X, Voogt JL. 2000. Effect of suckling on prolactin receptor immunoreactivity in the hypothalamus of the rat. Neuroendocrinology 71(5): 308-317.

Pi X, Voogt JS. 2001. Mechanisms for suckling-induced changes in expression of prolactin receptor in the hypothalamus of the lactating rat. Brain Res 891: 197-205.

Pi X, Zhang B, Li J, Voogt JL. 2003. Promoter usage and estrogen regulation of prolactin receptor gene in the brain of the female rat. Neuroendocrinology 77: 187-197.

Piccirillo M, Alpert JE, Cohen DJ, Shaywitz BA. 1980. Effects of 6-hydroxydopamine and amphetamine on rat mothering behavior and offspring development. Pharmacol Biochem Behav 13(3): 391-395.

Pifl C, Giros B, Caron MG. 1993. Dopamine transporter expression confers cytoxicity to low doses of the parkinsonism-inducing neurotoxin 1-methyl-4-phenylpyridinum. J Neurosci 12(10): 4246-4253.

Pryce CR, Dobeli M, Martin RD. 1993. Effects of sex steroids on maternal motivation in the common marmoset (*Callithrix jacchus*): development and application of an operant system with maternal reinforcement. J Comp Psychol 107(1): 99-115.

Quinones-Jenab V, Jenab S, Ogawa S, Adan RA, Burbach JP, et al. 1997. Effects of estrogen on oxytocin receptor messenger ribonucleic acid expression in the uterus, pituitary and forebrain of the female rat. Neuroendocrinology 65: 9-17.

Ramirez VD, Zheng J. 1996. Membrane sex-steroid receptors in the brain. Front Neuroendocrinol 17: 402-439.

Rees SL, Panesar S, Steiner M, Fleming AS. 2004. The effects of adrenalectomy and corticosterone replacement on maternal behavior in the postpartum rat. Horm Behav 46: 411-419.

Rehavi M, Goldin M, Roz N, Weizman A. 1998. Regulation of rat brain vesicular monoamine transporter by chronic treatment with ovarian hormones. Brain Res Mol Brain Res 57(1): 31-37.

Reid MS, O'Connor WT, Herrera-Marschitz M, Ungerstedt U. 1990. The effects of intranigral GABA and dynorpin A injections on striatal dopamine and GABA release: evidence that dopamine provides inhibitory regulation of striatal GABA neurons via D2 receptors. Brain Res 519: 255-260.

Reisbick S, Rosenblatt JS, Mayer AD. 1975. Decline of maternal behavior in the virgin and lactating rat. J Comp Physiol Psychol 89(7): 722-732.

Robbins TW, Everitt BJ. 1999. Motivation and Reward. Fundamental Neuroscience. Zigmond M, et al. editors. Academic Press; New York Chp 48 pp. 1245-1278.

Robertson MC, Gillespie B, Friesen HG. 1982. Characterization of the two forms of rat placental lactogens (rPL): rPL-I and rPL-II. Endocrinology 111: 1862-1866.

Rodriguez M, Gonzalez-Hernandez T. 1999. Electrophysiological and morphological evidence for a GABAergic nigrostrial pathway. J Neurosci 19: 4682-4894.

Roky R, Paut-Pagano L, Goffin V, Kitahama K, Valatx JL, et al. 1996. Distribution of prolactin receptors in the rat forebrain. Immunohistochemical study. Neuroendocrinology 63: 422-429.

Rosenberg P, Halaris A, Moltz H. 1977. Effects of central norepinephrine depletion on the initiation and maintenance of maternal behavior in the rat. Pharmacol Biochem Behav 6(1): 21-24.

Rosenblatt JS. 1967. Non-hormonal basis of maternal behavior in the rat. Science 156: 1512-1514.

Rosenblatt JS, Siegel HI, Mayer AD. 1979. Progress in the Study of maternal behavior in the rat: Hormonal, Nonhormonal, Sensory, and Developmental Aspects. Advances in the Study of Behavior, vol 10. Rosenblatt JS, Hinde RA, Beer C, Busnel M-C, editors. NY London: Academic Press; pp. 225-311.

Rosenblatt JS. 1995. Hormonal basis of parenting. Handbook of Parenting volume 2 Biology and Ecology of Parenting. Bornstein MH, editor. New Jersey, Hove UK: Lawrence Erlbaum Associates, Publishers Mahwah.

Rosenblatt JS, Ceus K. 1998. Estrogen implants in the medial preoptic area stimulate maternal behavior in male rats. Horm Behav 33: 23-30.

Rosenblatt JS, Olufowobi A, Siegel HI. 1998. Effects of pregnancy hormones on maternal responsiveness, responsiveness to estrogen stimulation of maternal behavior, and lordosis response to estrogen stimulation. Horm Behav 33: 104-114.

Rosenblatt JS, Siegel HI, 1975. Hysterectomy-induced maternal behavior during pregnancy in the rat. J Comp Physiol Psychol 89(7): 685-700.

Rosenblatt JS, Wagner CK, Morrell JI. 1994. Hormonal priming and triggering of maternal behavior in the rat with special reference to the relations between estrogen receptor binding and ER mRNA in specific brain regions. Psychoneuroendocrinology 19: 543-552.

Rosenkranz JA, Grace AA. 2001. Dopamine attenuates prefrontal cortical suppression of sensory inputs to the basolateral amygdala of rats. J Neurosci 21: 4090-4103.

Roth LL, Rosenblatt JS. 1968. Mammary glands of pregnant rats: development stimulated by licking. Science 151: 1403-1404.

Roy EJ, Buyer DR, Licari VA. 1990. Estradiol in the striatum: effects on behavior and dopamine receptors but no evidence for membrane steroid receptors. Brain Res Bull 25: 221-227.

Rubin BS, Bridges RS. 1984. Disruption of ongoing maternal responsiveness in rats by central administration of morphine sulfate. Brain Res 307(1–2): 91-97.

Rush ME, Blake CA. 1982. Serum testosterone concentrations during the 4-day cycle in normal and adrenalectomized rats. Proc Soc Exp Biol Med 169: 216-221.

Russell JA, Leng G, Douglas AJ. 2003. The magnocellular oxytocin system, the fount of maternity: adaptations in pregnancy. Front Neuroendoncrinol 24: 27-61.

Russo SJ, Festa ED, Fabian SJ, Gazi FM, Kraish M, et al. 2003. Gonadal hormones differentially modulate cocaine-induced conditioned place preference in male and female rats. Neuroscience 120(2): 523-33.

Sakaguchi K, Tanaka M, Ohkubo T, Doh-ura K, Fujikawa T, et al. 1996. Induction of brain prolactin receptor long-form mRNA expression and maternal behavior in pup-contracted male rats: promotion by prolactin administration and suppression by female contact. Neuroendocrinology 63: 559-568.

Sakamoto Y, Suga S, Sakuma Y. 1993. Estrogen-sensitive neurons in the female rat ventral tegmental area: a dual route for the hormone action. J Neurophysiol 70(4): 1469-1475.

Salzberg HC, Lonstein JS, Stern JM. 2002. $GABA_A$ receptor regulation of kyphotic nursing and female sexual behavior in the caudal ventrolateral periaqueductal gray of postpartum rats. Neuroscience 114: 675-687.

Sanyal MKJ. 1978. Secretion of progesterone during gestation in the rat. Endocrinology 79: 179-190.

Saper C. 2000. Hypothalamic connections with the cerebral cortex. Prog Brain Res 126: 40-48.

Sar M. 1988. Distribution of progestin-concentrating cells in rat brain: colocalization of [3H]ORG.2058, a synthetic progestin, and antibodies to tyrosine hydroxylase in hypotyhalamus by combined autoradiography and immunocytochemistry. Endocrinology 123: 1110-1118.

Schlinger BA, Arnold AP. 1993. Estrogen síntesis in vivo in the adult zebra finch: additional evidence that circulating estrogens can originate in the brain. Endo 133: 2610-2616.

Schneider JS, Stone MK, Wynne-Edwards KE, Horton TH, Lydon J, et al. 2003. Progesterone receptors mediate male aggression toward infants. PNS 100: 2951-2956.

Schroeder BE, Binzak JM, Kelley AM. 1999. A common profile of prefrontal cortical activation following exposure to nicotine or chocolate-associated contextual cues. Neuroscience 105: 535-545.

Schulz H, Kovacs GL, Telegdy G. 1979. Action of posterior pituitary neuropeptides on the nigrostriatal dopaminergic system. Eur J Pharmacol 57: 185-190.

Schumacher M, Coirini H, Frankfurt M, McEwen BS. 1989. Localized actions of progesterone in the hypothalamus involve oxytocin. Proc Natl Acad Sci USA 86: 6798-6801.

Schwartz E, Rowe FA. 1976. Olfactory bulbectomy: influences on maternal behavior in primiparous and multiparous rats. Physiol Behav 17: 879-883.

Sell SL, Scalzitti JM, Thomas ML, Cunningham KA. 2000. Influence of ovarian hormones and estrous cycle on the behavioral response to cocaine in female rats. J Pharm Exp Ther 293: 879-886.

Shaikh AA. 1971. Estrone and estradiol levels in the ovarian venous blood from rats during estrus cycle and pregnancy. Biol Reprod 5: 297-307.

Shearman LP, McReynolds AM, Zhou FC, Meyer JS. 1998. Relationship between 125 RTI-55-labeled cocaine binding sites and the serotonin transporter in rat placenta. Am J Physiol 275: 1621-1629.

Sheehan TP, Cirrito J, Numan MJ, Numan M. 2000. Using c-Fos immunocytochemistry to identify forebrain regions that may inhibit maternal behavior in rats. Behav Neurosci 114(2): 337-352.

Sheehan TP, Numan M. 1997. Microinjection of the tachykinin neuropeptide K into the ventromedial hypothalamus disrupts the hormonal onset of maternal behavior in female rats. J Neuroendocrinol 9(9): 677-687.

Sheehan T, Paul M, Amaral E, Numan MJ, Numan M. 2001. Evidence that the medial amygdala projects to the anterior/ventromedial hypothalamic nuclei to inhibit maternal behavior in rats. Neuroscience. 106(2): 341-356.

Sherwood OD. 1988. Relaxin. The physiology of reproduction, vol 2. Knobil E, Neill JD, editors. New York: Raven Press.

Sherwood OD, Crnekovic VE, Gordon WL, Rutherford JE. 1980. Radioimmunoassay of relaxin throughout pregnancy and during parturition in the rat. Endocrinology 107: 691-698.

Shimizu H, Bray GA 1993. Effects of castration, estrogen replacement and estrus cycle on monoamine metabolism in the nucleus accumbens, measured by microdialysis. Brain Res 621(2): 200-206.

Shingo T, Gregg C, Enwere E, Fujikawa H, Hassam R, et al. 2003. Pregnancy-stimulated neurogenesis in the adult female forebrain mediated by prolactin. Science 299: 117-120.

Siegel HI, Rosenblatt JS. 1978. Duration of estrogen stimulation and progesterone inhibition of maternal behavior in pregnancy-terminated rats. Horm Behav 11: 12-19.

Siegel HI, Rosenblatt JS. 1975a. Estrogen-induced maternal behavior in hysterectomized-ovariectomized virgin rats. Physiol Behav 14: 465-471.

Siegel HI, Rosenblatt JS. 1975b. Hormonal basis of hysterectomy-induced maternal behavior during pregnancy in the rat. Horm Behav 6: 211-222.

Silva MR, Bernardi MM, Cruz-Casallas PE, Felicio LF. 2003. Pimozide injections into the nucleus accumbens disrupt maternal behaviour in lactating rats. Pharmacol Toxicol 93: 42-47.

Silva MR, Bernardi MM, Felicio LF. 2001. Effects of dopamine receptor antagonists on ongoing maternal behavior in rats. Pharmacol Biochem Behav 68(3): 461-468.

Simerly RB, Swanson LW. 1986. The organization of neural inputs to the medial preoptic nucleus of the rat. J Comp Neurol 246: 312-342.

Simerly RB, Swanson LW. 1988. Projections of the medial preoptic nucleus: a *Phaseolus vulgaris* leucoagglutinin anterograde tract tracting study in the rat. J Comp Neurol 270: 209-242.

Simmons R. 1924. The relative effectiveness of certain incentives in animal learning. Comp Psychol Monog 2: 79.

Sinnamon HM. 1993. Preoptic and hypothalamic neurons and the initiation of locomotion in the anesthetized rats. Prog Neurobiol 41: 323-344.

Smith JT, Waddell BJ. 2002. Leptin receptor expression in the rat placenta: changes in ob-ra, ob-rb, and ob-re with gestational age and suppression of glucocorticoids. Biol Reprod 67(4): 1204-1210.

Smith MO, Holland RC. 1975. Effects of lesions of the nucleus accumbens on lactation and postpartum behavior. Physiol Psychol 3: 331-336.

Smith MS, Freeman ME, Neill JD. 1975. The control of progesterone secretion during the estrous cycle and early pseudopregnancy in the rat: prolactin, gonadotrophin, and steroid levels associated with rescue of the corpus luteum of pseudopregnancy. Endocrinology 96: 219-226.

Smith MS, Neill JD. 1977. Inhibition of gonadotrophin secretion during lactation in the rat: relative contribution of suckling and ovarian steroids. Biol Reprod 17: 255-261.

Smith KS, Mattson BJ, Morrell JI. 2002. The MPOA is engaged in the motivation to perform maternal behavior. Abstract, Society for Behavioral Neuroendocrinology.

Sorenson CA, Gordon M. 1975. Effects of 6-hydroxydopamine on shock-elicited aggression, emotionality and maternal behavior in female rats. Pharmacol Biochem Behav 3(3): 331-335.

Spanagel R, Weiss F. 1999. The dopamine hypothesis of reward: past and current status. Trends Neurosci 22: 521-527.

Spielewoy C, Roubert C, Hamon M, Nosten-Bertrand M, Betancur C, et al. 2000. Behavioural disturbances associated with hyperdopaminergia in dopamine-transporter knockout mice. Behav Pharmacol 11(3–4): 279-290.

Stack EC, Balakrishnan R, Numan MJ, Numan M. 2002. A functional neuroanatomical investigation of the role of the medial preoptic area in neural circuits regulating maternal behavior. Behav Brain Res 131: 17-36.

Stack EC, Numan M. 2000. The temporal course of expression of c-Fos and Fos B within the medial preoptic area and other brain regions of postpartum female rats during prolonged mother-young interactions. Behav Neurosci 114: 609-622.

Stack EC, Numan MJ, Numan M. 2001. The effects of bilateral lesions of the shell of the nucleus accumbens on maternal behavior in the rat. Soc Neurosci Abst 857.1.

Steele MK, Rowland D, Moltz H. 1979. Initiation of maternal behavior in the rat: possible involvement of limbic norepinephrine. Pharmacol Biochem Behav 11(2): 123-130.

Stern JM. 1989. Maternal behavior: sensory, hormonal, and neural determinants. Psychoneuroendocrinology. Brush FR, Levine S, editors. San Diego: Academic Press; pp. 102-226.

Stern JM. 1996. Somatosensation and maternal care in Norway rats. Parental Care - Evolution, Mechanisms, and Adaptive Significance. Advances in the Study of Behavior, vol. 25. Rosenblatt JS, Snowdon CT, editors. New York: Academic Press.

Stern JM. 1991. Nursing posture is elicited rapidly in maternally naive, haloperidol-treated female and male rats in response to ventral trunk stimulation from active pups. Horm Behav 25(4): 504-517.

Stern JM, Johnson SK. 1989. Perioral somatosensory determinants of nursing behavior in Norway rats (Rattus norvegicus). J Comp Psychol 103(3): 269-280.

Stern JM, Keer SE. 1999. Maternal motivation of lactating rats is disrupted by low dosages of haloperidol. Behav Brain Res 99(2): 231-239.

Stern JM, Keer SE. 2002. Acute hunger of pups elicits increased kyphotic nursing and shorter intervals between nursing bouts: implications for changes in nursing with time postpartum. J Comp Psychol 116: 83-92.

Stern JM, Mackinnon DA. 1976. Postpartum, hormonal, and nonhormonal induction of maternal behavior in rats: effects on T-maze retrieval of pups. Horm Behav 7(3): 305-316.

Stern JM, Mackinnon DA. 1978. Sensory regulation of maternal behavior in rats: effects of pup age. Dev Psychobiol 11(6): 579-586.

Stern JM, Protomastro M. 2000. Effects of low dosages of apomorphine on maternal responsiveness in lactating rats. Pharmacol Biochem Behav 66: 353-359.

Stern JM, Taylor LA. 1991. Haloperidol inhibits maternal retrieval and licking, but enhances nursing behavior and litter weight gains in lactating rats. J Neuroendocrinol 3: 591-596.

Stivers JA, Kaltwasser MT, Hill PS, Hruby VJ, Crawley JN. 1988. Ventral tegmental oxytocin induces grooming. Peptides 9: 223-231.

Sugiyama T, Minoura H, Kawabe N, Tanaka M, Nakashima K, 1994. Preferential expression of long form prolactin receptor mRNA in the rat brain during the oestrous cycle, pregnancy and lactation: hormones involved in its gene expression. J Endocrinol 141(2): 325-333.

Sugiyama T, Minoura H, Toyoda N, Sakaguchi K, Tanaka M, et al. 1996. Pup contact induces the expression of long form prolactin receptor mRNA in the brain of female rats: effects of ovariectomy and hypophysectomy on receptor gene expression. J Endocrinol 149(2): 335-340.

Summerlee AJ, Ramsey DG, Poterski RS. 1998. Neutralization of relaxin within the brain affects the timing of birth in rats. Endocrinology 139: 479-484.

Swanson LW. 1988–1989. The neural basis of motivated behavior. Acta Morphologica Neerlando-Scandinavica 26: 165-176.

Swanson LW. 1998. Brain maps: structure of the rat brain, second edition. Amsterdam: Elsevier.

Swanson LW, Mogenson GJ, Simerly RB, Wu M. 1987. Anatomical and electrophysiological evidence for a projection from the medial preoptic area to the 'mesencephalic and subthalamic locomotor regions' in the rat. Brain Res 405(1): 108-122.

Takeo T, Sakuma Y. 1995. Diametrically opposite effects of estrogen on the excitability of female rats medial and lateral preoptic neurons with axons to the midbrain locomotor region. Neurosci Res 22: 73-80.

Talamantes F, Ogren L. 1988. The placenta as an endocrine organ: Polypeptides. The physiology of reproduction, vol 2. Knobil E, Neill JD, editors. New York: Raven Press; pp. 2093-2144.

Tansey EM, Aubuthnott GW, Fink G, Whale D. 1983. Oestradiol-17 beta increases the firing rate of antidromically indentified neurones of the rat neostriatum. Neuroendocrinology 37: 106-110.

Taya K, Greenwald GS. 1982. Peripheral blood and ovarian levels of sex steroid in the lactating rat. Endocrinology Japan 29: 453-459.

Taya K, Sasamoto S. 1981. Changes in FSH, LH and prolactin secretion and ovarian follicular development during lactation in the rat. Endocrinology Japan 28: 187-196.

Tepper JM, Martin LP, Anderson DR. 1995. $GABA_A$ receptor-mediated inhibition of rat substantia nigra dopaminergic neurons by pars reticulata projection neurons. J Neurosci 15: 3092-3103.

Terkel J, Bridges RS, Sawyer CH. 1979. Effects of transecting lateral neural connections of the medial preoptic area on maternal behavior in the rat: nest building, pup retrieval and prolactin secretion. Brain Res 169: 369-380.

Terkel J, Rosenblatt JS. 1972. Humoral factors underlying maternal behavior at parturition: cross transfusions between freely moving rats. J Comp Physiol Psychol 80: 355-371.

Thoman EB, Levine S. 1970. Effects of adrenalectomy on maternal behavior in rats. Dev Psychobiol 3: 237-244.

Thoman EB, Sproul M, Seeler B, Levine S. 1970. Influence of adrenalectomy in female rats on reproductive processes including effects of the foetus and offspring. J Endocrinol 46: 297-303.

Thompson TL. 1999. Attenuation of dopamine uptake in vivo following priming with estradiol benzoate. Brain Res 834 (1–2): 164-167.

Thompson TL, Moss RL. 1994. Estrogen regulation of dopamine release in the nucleus accumbens: genomic- and non-genomic-mediated effects. J Neurochem 62(5): 1750-1756.

Thompson TL, Moss RL. 1997. Modulation of mesolimbic dopaminergic activity over the rat estrous cycle. Neurosci Lett 229(3): 145-148.

Thompson TL, Moore CC, Smith B. 2000. Estrogen priming modualtes autoreceptor-mediated potentiation of dopamine uptake. European Journal of Pharmacology;, 401: 357-363.

Tonegawa S, Nakazawa K, Wilson MA. 2003. Genetic neuroscience of mammalian learning and memory. Philos Trans R Society Lond B Biol Sci 358: 787-795.

Torner L, Toschi N, Nava G, Clapp C, Neumann ID. 2002. Increased hypothalamic expression of prolactin in lactation: involvement in behavioural and neuroendocrine stress responses. Eur J Neurosci 15(8): 1381-1389.

Tribollet E, Audigier S, Dubois-Dauphin M, Dreifuss JJ. 1990. Gonadal steroid regulation oxytocin receptors but not vasopressin receptors in the brain of male and female rats. An autoradiographical study. Brain Res 511: 129-140.

Tucker AM. 1988. Lactation and its hormonal control. The physiology of reproduction, vol 2. Knobil E, Neill JD, editors. New York: Raven Press; pp. 2235-2258.

Tyler JL, Gordon JH, Gorski RA. 1979. Effects of olfactory bulbectomy and estrogen on tyrosine hydroxylase and glutamic acid decarboxylase in the nigrostriatal and mesolimbic dopamine systems of adult female rats. Pharmacol Biochem Behav 11(5): 549-552.

Tzschentke TM. 1998. Measuring reward with the conditioned place preference paradigm: a comprehensive review of drug effects, recent progress and new issues. Prog Neurobiol 56: 613-672.

Van Der Shooter P, Lankhorst RR, De Roo JA, de Greef WJ. 1978. Suckling stimulus, lactation and suppression of ovulation in the rat. Endocrinology 103: 949-956.

Van Hemel SB. 1973. Pup retrieving as a reinforcer in nulliparous mice. J Exper Anal Behav 19: 233-238.

Vernotica EM, Rosenblatt JS, Morrell JI. 1999. Microinfusion of cocaine into the medial preoptic area or nucleus accumbens transiently impairs maternal behavior in the rat. Behav Neurosci 113(2): 377-390.

Voloschin LM, Tramezzani JH. 1984. Relationship of prolactin release in lactating rats to milk ejection, sleep state, and ultrasonic vocalizations by the pups. Endocrinology 114: 618-623.

Voorn P, Vanderschuren LJMJ, Groenewegen HJ, Robbins TW, Pennartz CMA. 2004 Putting a spin on the dorsal–ventral divide of the striatum. Trends Neurosci 27: 468-474.

Wagner CK, Morrell JI. 1995. In situ analysis of estrogen receptor mRNA expression in the brain of female rats during pregnancy. Brain Res Mol Brain Res 33(1): 127-135.

Wagner CK, Morrell JI. 1996. Levels of estrogen receptor immunoreactivity are altered in behaviorally-relevant brain regions in female rats during pregnancy. Brain Res Mol Brain Res 42(2): 328-336.

Wagner CK, Silverman AJ, Morrell JI. 1998. Evidence for estrogen receptor in cell nuclei and axon terminals within the lateral habenula of the rat: regulation during pregnancy. J Comp Neurol 392: 330-342.

Walsh CJ, Fleming AS, Lee A, Magnusson JE. 1996. The effects of olfactory and somatosensory desensitization on Fos-like immunoreactivity in the brains of pup- exposed postpartum rats. Behav Neurosci 110(1): 134-153.

Walsh RJ, Slaby FJ, Posner BI. 1987. A receptor-mediated mechanism for the transport of prolactin from blood to cerebrospinal fluid. Endocrinology 120: 1846-1850.

Wansaw MP, Williams S, Rosenblatt JS, Morrell JI. 2002. Cocaine and pup induced conditioned place preference in the lactating maternal rat. Abstract. Orlando, Florida: Society for Neuroscience.

Wansaw MP, Reiss J, Morrell JI. 2003a. Varying the time dams are deprived of pups has minimal effects on preferences for pup-associated cues in the early, and substantial effects in the late postpartum period. Abstract. New Orleans, Louisiana: Society for Neuroscience.

Wansaw MP, Reiss J, Morrell JI. 2003b. Dynamic changes in motivation to seek either natural or pharmacological rewards occur as the postpartum period progresses in the rat. NIDA Symposium Abstract. New Orleans Louisiana: Society for Neuroscience.

Wayner MJ, Barone FC, Scharoun SL, Guevara-Aguilar R, Aguilar-Baturoni HU. 1983. Limbic connections to the lateral preoptic area: a horseradish peroxidase study in the rat. Neurosci Biobehav Rev 7: 375-384.

Wellman J, Carr D, Graham A, Jones H, Human JL, et al. 1997. Preoptic area infusions of morphine disrupt – and naloxone restores – parental-like behavior in juvenile rats. Brain Res Bull 44(2): 183-191.

Wilsoncroft WE. 1969. Babies by bar-press: maternal behavior in the rat. Behav Res Meth Instrum 1: 229-230.

Wise RA. 1998. Drug-activation of brain reward pathways. Drug Alcohol Depend 51: 13-22.

Woodside B, Abizaid A, Walker CD. 2000. Changes in leptin levels during lactation: implications for lactational hyperphagia and anovulation. Horm Behav 37: 353-365.

Woodside B, Popeski N. 1999. The contribution of changes in milk delivery to the prolongation of lactational infertility induced by food restriction or increased litter size. Physiol Behav 65: 711-715.

Ying S-Y, Gove S, Fang VS, Greep RO. 1973. Ovulation in the postpartum rat. Endocrinology 92: 108-116.

Young LJ, Muns S, Wang Z, Insel TR. 1997. Changes in oxytocin receptor mRNA in rat brain during pregnancy and the effects of estrogen and interleukin-6. J Neuroendocrinol 9(11): 859-865.

Xiao L, Becker JB. 1998. Effects of estrogen agonists on amphetamine-stimulated striatal dopamine release. Synapse 29: 379-391.

Xiao L, Jackson LR, Becker JB. 2003. The effect of estradiol in the striatum is blocked by ICI 182,780 but not tamoxifen: pharmacological and behavioral evidence. Neuroendocrinology 77: 239-245.

Zaborszky L, Alheld GF, Beinfeld MC, Eiden LE, Heimer L, et al. 1985. Cholecystokinin innervation of the ventral striatum: a morphological and radioimmunological study. Neuroscience 14: 427-453.

Zahm DS, Brog JS. 1992. On the significance of subterritories in the "accumbens" part of the rat ventral striatum. Neuroscience 50: 751-767.

Zhou W, Cunningham KA, Thomas ML. 2002. Estrogen regulation of gene expression in the brain: a possible mechanism altering the response to psychostimulants in female rats. Brain Res Mol Brain Res 100(1–2): 75-83.

Zsarnovszky A, Scalise TJ, Horvath TL, Naftolin F. 2000. Estrogen effects on tyrosine hydroxylase-immunoreactive cells in the ventral mesencephalon of the female rat: further evidence for the two cell hypothesis of dopamine function. Brain Res 868(2): 363-366.

6 Neuroendocrinology, Neurochemistry, and Molecular Neurobiology of Affiliative Behavior

E. A. D. Hammock · L. J. Young

Abstract: This chapter reviews current understanding and hypotheses in social behavior from molecular mechanisms of social feeding in *Caenorhabditis elegans* to social cognition in humans. The influential roles of steroids and neurohormones on the neural circuitry of social behavior are emphasized.

1 Introduction

Affiliation is the basis of all social units, from membership in a colony, troupe, or hive, to the selective attachments found in parent–offspring bonds, pair bonds, or other nonfamilial bonds. The complexity of the expression of affiliative behavior ranges broadly from aggregation behavior in *Caenorhabditis elegans* to the organization of human societies. The most basic definition of affiliation includes any prosocial behavior that brings two or more individuals into close proximity. Affiliation involves a complex set of processes including sensory detection of conspecifics, recognition of familiar individuals, and motivation to initiate proximity or contact. In addition, incompatible processes such as aversion or aggression must be suppressed. Here, we discuss the neuroendocrinological, molecular neurobiological, and neurochemical mechanisms underlying various components of affiliative behavior. We draw from studies on a variety of species to highlight unifying themes in the regulation of affiliative behavior.

2 A Brief Historical Context for the Study of Affiliation

Studies of the biological basis of affiliation began in the 1920s, with a focused interest on maternal behavior. Affiliation research dramatically increased throughout the 1990s (Levine et al., 1997), with a broadened emphasis on other forms of attachment in addition to the mother–infant bond, such as pair bonding and colony membership.

Historically, there have been two main approaches to investigating the mechanisms underlying affiliative behavior: behavioral endocrinology and functional neuroanatomy. Both approaches have been crucial in formulating our current understanding of the biological mechanisms of affiliation. Behavioral endocrinologists investigate the powerful relationship between changes in hormone concentrations and changes in behavior and manipulate hormone concentrations to determine cause and effect relationships. One of the earliest examples of this involved a study of prolactin-induced maternal behavior in birds and rats (Riddle, 1935a, b). In contrast, functional neuroanatomists investigate the role of different brain areas in the expression of behavior. A classic example of this approach is Michael Numan's (Numan et al., 1977) influential work demonstrating that knife cuts disrupting the medial preoptic area (MPOA) function block maternal behavior.

Within the past 30 years, these two approaches have merged to create a more unified field of Behavioral Neuroscience, which encompasses a wide range of topics including genetics, molecular and cell biology, neuroanatomy, and neuroendocrinology. Today, researchers in this area seek to understand the integration of hormonal states and neural circuits in regulating complex behaviors in variable genetic and environmental backgrounds. An example integrating these approaches involved injecting prolactin into the MPOA and observing increases in maternal behavior in steroid-primed rats (Bridges and Freemark, 1995). The advent of new experimental tools has allowed the behavioral neuroscientist to further explore the molecular neurobiology of affiliation. These tools include transgenic animals (knockouts), viral vector gene transfer, antisense, and microarray technology. With the rapid development of bioinformatics, behavioral neuroscience is branching out and seeking companionship in behavioral genetics (fruit fly, *C. elegans*, etc.) and neurogenomics. Interactions among researchers in these fields hold great promise for elucidating neurobiological mechanisms of affiliative behavior.

3 The Point of Social Behavior

A unifying theme across species that exhibit social behavior is that these species do so in response to stress (Wilson, 2000; Sokolowski, 2002). The simplest prosocial behavior is that of the single-celled prokaryotes like the bacterium *Myxococcus xanthus*, and the social amoeba, *Dictyostelium discoideum*. Both these species aggregate in response to starvation stress to form fruiting bodies. In both cases, only some of the original cells go on to reproduce after this event, meaning that some cells sacrifice their genes for the "greater good" (species survival as opposed to individual survival, i.e., kin selection). Although there are cheaters in this scenario (Strassmann et al., 2000; Velicer et al., 2000), the cost of social cooperation might provide the benefit of increased fitness to genetically similar "cooperative" individuals. Regardless of an organism's awareness of it, prosocial behavior has a point: it is an evolutionary survival strategy.

While it may come across as a debatable model system for behavioral neuroscience, the single-celled organism – with no nerve cell, much less an entire nervous system – serves as an excellent reminder of the molecular nature of all behaviors. From the simplest "social" aggregation of single cells via sticky surface proteins (Rainey and Rainey, 2003) to the complex angst of the human teenager in the throes of love, all social behavior, no matter how complex, is a result of specific molecular interactions in the context of cells and systems.

4 Invertebrate Model Systems Informing the Molecular Biology of Social Behavior

4.1 The Roundworm

The roundworm, *C. elegans*, has become an informative model of the molecular neurobiology of social behavior. Its genes are easily manipulated and it has a very well-defined neuroanatomy, with precisely 302 neurons of known developmental origin and connectivity. It has become a model for studies of locomotion, sensory systems, survival, reproduction, learning and memory, and drug-related and most recently, social behaviors (Rankin, 2002).

4.1.1 Natural Variation in Social Behavior

In natural populations of *C. elegans*, individuals display one of these two responses to a food stimulus: solitary or social feeding. De Bono and Bargmann (1998) traced the variation in this behavior to a single gene that varies across strains of worms: *npr-1*. This gene encodes a G-protein coupled neuropeptide Y receptor-like protein (NPR-1). Interestingly, the mammalian homologue, the neuropeptide Y receptor, is also involved in feeding behavior. The variation in the *npr-1* gene across strains with different behavior is a single base-pair modification that results in a single amino acid change in the NPR-1 protein. The NPR-1 protein of all the solitary worm strains (five strains) contains a valine at amino acid number 215, whereas all the social worms (11 strains) have a phenylalanine at that position. This strong association of the genotype with the behavioral phenotype suggested that the genetic variation might actually cause the behavioral differences across strains. To test this idea, de Bono and Bargmann (1998) set out to manipulate the *npr-1* gene to see if genetic manipulation at this specific locus could cause a behavioral change. First, in social worms they changed the gene to code for a valine (solitary allele) instead of a phenylalanine (social allele). The behavior of these social worms was transformed into solitary-like behavior with this genetic change, demonstrating that these allelic differences between the strains significantly contribute to the strain-typical behavior. Furthermore, when the entire gene was removed in a solitary strain, the strain became social (de Bono and Bargmann, 1998). These results indicate that the normal role of NPR-1 (i.e., the solitary allele) is to inhibit aggregation behavior. The single amino acid change between solitary and social strains is thought to alter the efficiency of the biochemical interaction of NPR-1 with its appropriate G-proteins. In the solitary strains, with a valine at position 215, the interaction of NPR-1 with its G-proteins is very efficient. This efficient interaction serves to inhibit aggregation behavior. In the social strains, the efficacy is reduced and they have less inhibition of social behavior and are therefore more social.

4.1.2 Neural Circuitry of *C. elegans* Social Behavior

As mentioned above, *C. elegans* is a fantastic model system, because its neuroanatomy is known. Because of this, it is possible to deduce which neurons contribute to social behavior by identifying which neurons contain the NPR-1 protein. NPR-1 is expressed in neurons in the head, ventral nerve chord, and preanal ganglion of *C. elegans*, as measured by an NPR-1-GFP fusion protein (de Bono and Bargmann, 1998). Three specific neurons, URX, AQR, and PQR, are particularly interesting, because they border the body cavity (the pseudocoelom) of the worm (Coates and de Bono, 2002). In *npr-1* mutant worms, which have a social feeding phenotype, selective restoration of functional NPR-1 to just these three neurons suppressed the social feeding phenotype. Selectively inhibiting these three pairs of neurons also results in the suppression of *npr-1* mutation-induced aggregation behavior. Thus, NPR-1 in these three neurons alone sufficiently antagonizes aggregation behavior. Likewise, without NPR-1, these three neurons promote aggregation behavior. Therefore, there must be proteins in these neurons whose activity results in increased cellular output to drive aggregation behavior. These three neurons express gene products that promote aggregation behavior: TAX-2 and TAX-4. These two genes encode two subunits of a cGMP-gated cation channel (Coates and de Bono, 2002). In these neurons, the activity of the TAX-2/TAX-4 cation channel increases aggregation behavior. This cation channel is inhibited by the activity of NPR-1. The authors hypothesize that these neurons, containing molecules that antagonize each other's actions, modulate the animal's response to a signal from the environment. One such signal may be oxygen.

C. elegans prefer a specific range of oxygen levels. Too little oxygen prevents proper cell functioning and too much oxygen results in oxidative stress and damage to the organism. Behaviorally, worms aggregate more in higher levels of oxygen, perhaps to reduce levels of oxygen in their microenvironment. The same three neurons described earlier also contain oxygen sensors that double as producers of cyclic GMP: GCY-35 and GCY-36 (Cheung et al., 2004; Gray et al., 2004). Therefore, in high levels of oxygen, GCY-35/GCY-36 binds oxygen and increases production of cGMP. These increased levels of cGMP stimulate the TAX-2/TAX-4 cGMP-gated ion channel in URX, AQR, and PQR neurons. The activity of these neurons increases aggregation behavior.

Taken together, these data suggest a model for converging genetic and environmental regulation of aggregating behavior (➲ *Figure 6-1*). At the genetic level, variation at the *npr-1* locus contributes to the probability of aggregation: in solitary strains, the normal function of NPR-1 in URX, AQR, and PQR neurons is to inhibit the function of the TAX-2/TAX-4 cGMP-gated ion channel and thereby inhibit aggregation behavior. In contrast, this seemingly genetically determined behavior can be influenced by environmental conditions. When elevated oxygen levels cause more oxygen to bind to GCY-35 or GCY-36, more cGMP, which stimulates the activity of the cGMP-gated ion channel, is produced, thus promoting aggregation behavior. Coates and De Bono (2002) suggest that the interplay between these components (as well as other as yet unidentified components) contributes to the *probability* of social aggregating behavior. Just as we expect in vertebrates, the neurons of this invertebrate compute an output based on multiple converging inputs. This computation includes the integration of a genetic predisposition for a behavior with changing environmental conditions. Additionally, this is a very lucid example of how prosocial behavior reduces exposure to "stress," which in this case is oxygen free-radical toxicity.

The fruit fly, *Drosophila melanogaster*, has also been used as a model to understand the molecular biology of social behaviors. Interestingly, the neuropeptide Y system regulates aggregation behavior in flies too. *Drosophila* larvae show an interesting social burrowing behavior. They aggregate and dig into the culture medium together, and prepare to pupate. Cooperative aggregation behavior is beneficial to the animals presumably because it allows for more effective burrowing into the substrate to be used for pupation. The onset of this behavior is correlated with dNPF expression levels (Wu et al., 2003), which is the ligand for the *Drosophila* neuropeptide Y receptor. Experimental downregulation of dNPF or its receptor causes aggregation in young larvae, while upregulation of either results in a prolonged solitary feeding stage, clearly demonstrating a role for dNPF in social feeding in *Drosophila* (Wu et al., 2003). It is interesting that this neuropeptide system is used in the regulation of social feeding behavior in both *C. elegans* and *Drosophila* (Sokolowski, 2003).

4.1.3 Chemodetection ("olfaction") is a Key Component of *C. elegans* Social Behavior

The "decision" to aggregate is an important one and should require high levels of regulation at various levels. The sensory detection of appropriate environmental stimuli is one such important level of

Figure 6-1

Genetic and environmental regulation of aggregation behavior in *C. elegans*. Social feeding in *C. elegans* involves the activity of many components. Genetic predisposition for social or solitary feeding and environmental influence of oxygen levels coverge in URX, AQR, and PQR neurons, where increased activity of the cGMP-gated TAX-2/TAX-4 ion channel promotes aggregation behavior. Environmentally, the activity of this channel can be enhanced by the soluble oxygen sensors GCY-35 and GCY-36 by oxygen-induced upregulation of cGMP production. In contrast, the activity of TAX-2/TAX-4 (and thus social feeding) can be inhibited by the activity of NPR-1. Genetic variants of NPR-1 differ in their ability to downregulate the TAX-2/TAX-4 ion channel, leading to heritable variation in social behavior. (F = phenylalanine, V = valine)

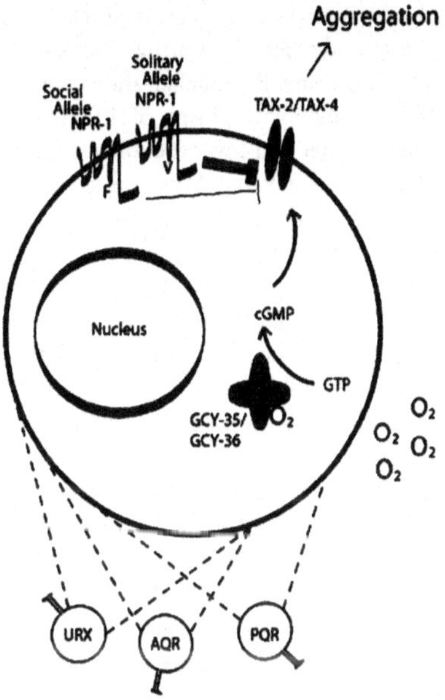

regulation. In addition to oxygen levels, *C. elegans* use food cues and social odor cues associated with population density to influence the probability of aggregation. In the laboratory, population density is directly correlated with levels of the constitutively secreted dauer pheromone on the plate of worms. This pheromone inhibits the expression of a gene called *daf-7* in ASI chemosensory neurons (Ren et al., 1996; Schackwitz et al., 1996). This gene encodes a TGFβ-like neuromodulator, which normally inhibits aggregation, among other things. When worms have a mutated *daf-7* gene, they aggregate. Environmental influences of population density and food availability on aggregation behavior are thought to be regulated in part by modulating expression levels of this *daf-7* gene in chemosensory neurons. High population density decreases *daf-7* expression and removes the inhibition of aggregation, while high food abundance increases *daf-7* expression and increases the inhibition of aggregation (de Bono, 2003).

In addition to the *daf-7* gene in ASI chemosensory neurons, there are other genes in other sensory neurons that participate in the regulation of aggregation behavior. Because *C. elegans* is such a well-defined experimental system, De Bono and colleagues (2002) were able to ask if discrete changes in sensory systems modulate aggregation behavior in social strains. Social behavior in social strains could be eliminated by selective laser ablation of two sensory neurons: ASH and ADL. In addition to neuronal lesions, social behavior was also eliminated by mutations in genes encoding cation channel subunits involved in chemosensation: *osm-9* and *ocr-2*. Social behavior was rescued in these mutants by selective expression of the *osm-9* or *ocr-2* genes in ASH and ADL neurons. Additionally, genes involved in olfaction, *odr-4* and *odr-8*, when mutated, disrupted the ability of *npr-1* mutants to aggregate in response to food. The product of *odr-8* is currently unknown. However, *odr-4* is exceptionally interesting because its gene product is thought to traffic odorant receptors to the distal cilia of ADL neurons. Therefore, a mutation in *odr-4* potentially disrupts the function of several other gene products in olfactory pathways. It is possible that these receptors on these neurons detect noxious and odorant stimuli given off by the food (*C. elegans* eats *Escherichia coli*, which is occasionally toxic) and that this information is used to encourage the individuals to aggregate as a survival strategy (de Bono et al., 2002). These genetic studies clearly demonstrate the role of specific gene products in specific neurons for aggregation behavior to occur.

4.2 The Honeybee

4.2.1 Regulation of Division of Labor for Caregiving Behavior

The honeybee, *Apis mellifera*, is an asocial insect with a clear age-related division of labor among workers. Young adult bees take care of the young, and older adults bring food back to the hive. To begin to dissect the molecular players in this behavioral dichotomy, Ben-Shahar and coworkers (2002) drew a comparison of this behavioral dichotomy to the behavioral polymorphisms of sitter and rover flies. There are two foraging behavioral isoforms in adult *Drosophila*, sitter and rover (Sokolowski, 1980). Rovers wander much farther in search of food than do sitters. This behavioral variability maps to variation in a particular gene, *dg2* (also known as *for*), which codes for a protein kinase G (PKG) (Osborne et al., 1997). Sitter flies express lower levels of *dg2* mRNA, and lower PKG activity than do rover flies. Ben-Shahar and coworkers hypothesized that young sitter-like bees would have low PKG activity, and older rover-like bees would have higher PKG activity. The honeybee gene, *Amfor*, is orthologous to the *for* (*dg-2*) gene from flies described earlier. *Amfor* mRNA was indeed found to be upregulated in the adult foragers, and there was a fourfold increase in PKG activity in foragers relative to nurse bees (Ben-Shahar et al., 2002). To rule out whether this change in *Amfor* expression and PKG activity was just due to the confounding nature of age (foragers are older) or really related to the behavior, Ben-Shahar and coworkers (2002) set up a new hive containing only young nurse bees. This resulted in the generation of a few precocious foragers. These young adults, who would normally be caring for the brood, were now out foraging for food. As expected, these young foragers also showed changes in *Amfor* mRNA levels and PKG activity, demonstrating that the changes are related to the behavior and not due to age. Additionally, pharmacological upregulation of PKG activity resulted in the increased likelihood of precocious foraging in a dose-dependent manner. It seems that low levels of *Amfor* and related PKG activity are required for bees to remain as social caregivers. Parenting social behavior, presented here in honeybees, is likely the output of a system (the bee) computing important social inputs. In particular, the transition to foraging is heavily influenced by social signals, such as the number of foragers already in the colony, and by pheromones from the queen and the brood (Robinson, 2002).

4.2.2 Frontiers in "Sociogenomics"

An exciting postscript to the story above involves the use of microarrays to compare the gene expression profiles of forager versus nurse bees. Recently, Whitfield and coworkers (2003) used microarrays composed of bee brain expressed sequence tags (est) to determine the "molecular signature" of the changes in gene expression associated with the two behaviors. Not only did they confirm the upregulation of *Amfor* mRNA in the forager bee as described above, but they also observed changes in the expression of many other genes (Whitfield et al., 2003). Based on the differences in the gene expression profiles, they identified two distinct patterns of gene expression that were associated with either nurse or forager behavior. Because the transition to foraging involves many aspects of social behavior, like communicative dancing and detection of social stimuli, some of the changes in gene expression may be involved in pathways that mediate those behaviors. This kind of study opens the door for exciting revelations of the molecular machinery underlying the regulation of social behavior.

4.3 Conclusions from the Molecular Dissection of Social Behavior in Invertebrates

Some may argue against the value of a discussion of invertebrate social behavior in the context of social affiliation, but it is important to remember that because of the advanced molecular and genetic techniques available in invertebrate models, these models provide very keen insight into the mechanisms underlying more complex affiliative behaviors in other animals. The important lessons learned from molecular genetic studies of social feeding in *C. elegans* and division of labor in the honeybee are fourfold.

First, natural variation in behavior (e.g., social versus solitary response to food; nurse versus forager division of labor) can be used to uncover molecular substrates of behavior. Social behavior is a rapidly evolving trait. Natural selection acts on genes and it is not surprising that heritable variation in social behavior can be traced to individual variation in single gene locus. It is important to keep in mind that *variation* in behavior can sometimes be traced to variation in a single gene. However, this does not mean that a behavior is a result of a single gene. This brings us to the next concept.

Second is the concept of locus heterogeneity. This means that the gene "locus" for a behavior is spread across the genome of the organism and is distributed across many genes rather than in one particular gene. In *C. elegans*, several different mutations resulted in the same behavioral phenotype. In the microarray study in honeybees, many gene products were associated with the nurses or foragers. In sum, there is no single gene for social behavior; rather, there are many genes whose products contribute to a highly regulated system.

Third, to engage in social behavior, individuals must be able to detect conspecifics. In the worm, social detection occurs through chemosensory cells containing gene products that respond to chemical social cues in the environment. In honeybees, individuals must gauge their potential role in the hive by taking in information about what other individuals are doing. This too requires detection, and perhaps discrimination, of conspecifics.

Finally, behavior is multisystemic in that the animal uses multiple sensory inputs, cellular memory, and state information to compute the appropriate behavior. Just as there is no single gene for a complex behavior, there is no single circuit. The behavior itself is composed of the probabilistic output of multiple systems, including motor systems and sensory feedback. Even in organisms as relatively simple as *C. elegans*, there are multiple pathways to influence social behavior. Each level of regulation contributes to the probability of occurrence of a particular behavior. The overall output of such a probabilistic system influences both the quantity and quality of behavior.

These important points apply not only to *C. elegans* and honeybees or just to invertebrates, but to all organisms. Although not quite as elegantly dissected as in invertebrates, we shall see more examples of these points in various vertebrates in the next sections.

5 Vertebrate Model Systems Informing the Neurobiology of Social Behavior

5.1 Rats and Mice

Most of what is known about affiliative behavior in rats and mice comes from a long history of studies of maternal behavior and social recognition memory in rats. Maternal behavior is a classic affiliative behavior because it brings a mother in close physical contact with her offspring. (A thorough presentation of maternal behavior can be found in the chapter by Lonstein and Morrell, in this volume.) Additionally, social recognition memory is a meaningful prerequisite to selective attachment behavior. Historically, mice were rarely studied for understanding social behavior; however, the transgenic mouse has changed that. Rats and mice are considered here together to illustrate important concepts in mammalian social behavior. In particular, we discuss the roles of gonadal hormones, and neurohormones like prolactin, oxytocin, and vasopressin.

5.1.1 Maternal Behavior

It is important to note that mice and rats display a not-so-subtle difference in their maternal behavioral phenotype. Female mice of some laboratory strains are spontaneously maternal, whereas adult female rats display an initial aversion to pups and overcome this through pregnancy and parturition, or long periods of repeated exposure to pups (Rosenblatt, 1967).

Originally, the main thrust of research on maternal behavior was to understand the role of pregnancy hormones in regulating the onset of maternal behavior. Estradiol, progesterone, and prolactin levels change during pregnancy. Specifically, 17β-estradiol and prolactin levels rise, while progesterone levels fall toward the end of pregnancy. In addition to preparing the uterus and mammary tissue, these changes appear to be important for initiating the appropriate events in the brain to prepare the animal for maternal behavior (Rosenblatt and Siegel, 1975; Bridges et al., 1977), although these hormones are not required for the maintenance of maternal behavior (Rosenblatt et al., 1988). This indicates that changes in the levels of these hormones somehow result in long-lasting changes in the brain structures that underlie these behaviors.

Which brain areas might be regulated by these circulating hormones? The MPOA has one of the highest concentrations of estradiol-binding sites in the brain (Pfaff and Keiner, 1973), and is therefore a likely target for the effects of pregnancy-induced changes in estradiol levels. Michael Numan's classic work, using lesion studies (knife lesions, radiofrequency lesions, neurochemical lesions, electrical lesions, etc), clearly identifies the MPOA as a critical brain structure in the onset and maintenance of maternal behavior (Numan, 1994). Implants of estradiol benzoate directly into the MPOA facilitate the onset of maternal behavior (Numan et al., 1977). Estrogen receptor-alpha knockout mice have poor maternal behavior (Ogawa et al., 1998), suggesting that this is likely to be the substrate with which estradiol interacts in the MPOA. The estrogen receptor alpha is a nuclear hormone receptor that, when bound to estrogens, interacts with regulatory regions of genes to change the transcriptional profile of the MPOA. Additionally, mice lacking the immediate early gene FosB have major defects in nurturing (Brown et al., 1996). It is likely that FosB contributes to the organization of transcriptional changes in the MPOA that must occur around the time of parturition for nurturing behavior (Lin et al., 1998). The hormone-induced changes in the MPOA may be either self-perpetuating or maintained by some other brain activity, potentially by activation of brain circuitry during interactions with offspring (Stern and Lonstein, 2001).

Estradiol-induced changes in the MPOA are not the only variable contributing to the onset of maternal behavior. Prolactin in the MPOA also facilitates maternal behavior, and prolactin receptor levels rise in the MPOA as parturition approaches (Bridges and Ronsheim, 1990; Bridges et al., 1990). Prolactin receptor knockout mice have pup retrieval and nesting deficits (Lucas et al., 1998). It is therefore apparent that the activities of the MPOA, by the actions of many gene products, affect maternal behavior.

The MPOA is heavily interconnected with many brain regions that are also involved in the regulation of maternal behavior. Sensory input from the main and accessory olfactory bulbs reaches the MPOA by way of the medial amygdala (MeA) and the bed nucleus of the stria terminalis (BST). Lesions of the vomeronasal organ in naive female rats can actually facilitate the onset of maternal behavior, suggesting that the accessory olfactory bulbs normally inhibit the onset of maternal behavior (Fleming et al., 1979). Lesions of the olfactory bulb and the bed nucleus of the accessory olfactory tract also facilitate maternal behavior in naive females (Del Cerro et al., 1991). Male rats are never naturally paternal and they have larger olfactory bulbs than females. Lesions of the olfactory bulb in males also permit parental behavior. These data fit with the idea that in naive rats, the olfactory bulb serves to inhibit parental care. Lesions of the olfactory nuclei, the amygdala, or the BST will result in the onset of parental behavior (Fleming et al., 1979, 1980; Del Cerro et al., 1991). It appears that the inhibitory nature of the olfactory pathway to the MPOA changes during pregnancy so that the odor of rat pups becomes attractive to lactating rats (Smotherman et al., 1974). Recently, it has been demonstrated that prolactin induces neurogenesis in the subventricular zone, which gives rise to new neurons in the olfactory bulb (Shingo et al., 2003). This could be a potential mechanism contributing to the pregnancy-induced onset of maternal behavior by affecting the olfactory processing of pup odor cues.

Oxytocin is a very important hormone during labor in mammals. Oxytocin promotes uterine contractions and regulates milk ejection. The source of oxytocin for these events is the magnocellular neurons in the paraventricular and supraoptic nuclei of the hypothalamus, which release their oxytocin content into the neurohypophysis, or posterior pituitary. Oxytocin is also produced in the parvocellular neurons of the hypothalamus. Hypothalamic neurons send projections to forebrain regions like the hippocampus, amygdala, striatum, hypothalamus, and to mid- and hindbrain nuclei, such as the locus coeruleus and nucleus of the tractus solitarius, as well as the spinal cord (Sofroniew, 1983). PVN lesions result in a near-complete loss of the brain oxytocinergic system (De Vries and Buijs, 1983) and a delay in the onset of maternal behavior in naive animals (Insel and Harbaugh, 1989), but have no effect in animals that are already maternal (Numan and Corodimas, 1985).

Oxytocin from the PVN may act on oxytocin receptors (OTR) throughout the brain to promote maternal behavior. Oxytocin delivered into the ventricles in female rats enhances maternal behavior (Pedersen and Prange, 1979; Pedersen et al., 1982; Fahrbach et al., 1984) and reduces infanticidal behavior in female house mice (McCarthy, 1990). In contrast, OTR antagonists delivered to the ventricles delay the onset of maternal behavior in hormone-primed females (Fahrbach et al., 1985; van Leengoed et al., 1987). Finally, OTR antagonists injected directly into the MPOA and ventral tegmental area (VTA) inhibit maternal behavior (Pedersen et al., 1994). The above data indicate that oxytocin has a very clear role in maternal behavior. However, there has been some controversy surrounding the findings of the role of oxytocin in maternal behavior. Oxytocin does not always facilitate maternal behavior. It appears that the effects of manipulation of oxytocin systems are influenced by many factors such as the species or even the strain of the animal, housing conditions, and other experimental manipulations (Bolwerk and Swanson, 1984; Fahrbach et al., 1986). These kinds of contrasting data highlight the importance of other factors in determining how hormones can affect behavior and they also emphasize the complexity of biological systems.

The receptor for oxytocin (OTR) has been localized in the rat brain by mRNA in situ hybridization, immunocytochemisty, and receptor autoradiography (Brinton et al., 1984; van Leeuwen et al., 1985; Freund-Mercier et al., 1987, 1988a, b; Elands et al., 1988; Yoshimura et al., 1993; Adan et al., 1995). Oxytocin and OTR levels are regulated by estradiol and progesterone, followed by progesterone withdrawal in the hypothalamus and several areas throughout the brain (Bale and Dorsa, 1995; Amico et al., 1997). In particular, levels of OTR in the hypothalamus and MPOA are heavily influenced by estradiol. During pregnancy, when 17β-estradiol levels rise, there is an increase in the number of OTR in these brain areas (Young et al., 1997b). Treatment with estradiol also results in increased levels of OTR by estrogen receptor alpha (Breton and Zingg, 1997; Young et al., 1998). It seems that the hormones of pregnancy assist the brain in preparing to interpret a bolus of oxytocin that is released during pregnancy by increasing the ability to detect oxytocin.

Because OTRs are located in the olfactory bulb, amygdala, and BST, oxytocin might be acting on the aforementioned olfactory pathway to release the olfactory-mediated inhibition of the MPOA. Oxytocin delivered to the olfactory bulbs of parturient rats facilitates the onset of maternal behavior, whereas the application of an OTR antagonist inhibits the onset of maternal behavior (Yu et al., 1996a, b).

Curiously, oxytocin knockout mice have deficits in lactation, but initially appeared to have normal maternal behavior (Nishimori et al., 1996). However, with more sensitive and quantitative measures of maternal behavior, a deficit was detected in the oxytocin knockout mouse. These mice show a 22% reduction in the amount of pup licking in inexperienced females (Pedersen et al., 2004). In recent studies, the OTR knockout mouse has shown a maternal behavior deficit. Interestingly, naive receptor knockout mice show deficits in licking behavior similar to the oxytocin knockout, as well as deficits in other behaviors such as huddling with the pups [Takayanagi et al.(submitted); Olazabal et al., 2004]. Perhaps in mice, oxytocin promotes nurturing in both naive and experienced females and in rats, oxytocin is important in mediating the *switch* from pup avoidance to maternal behavior. In sum, the effect of oxytocin on maternal behavior in rats is quite dramatic, but less dramatic in laboratory strains of mice. Curiously however, the *peg3* knockout mouse, with a reduction in the numbers of oxytocin neurons in the PVN, exhibits almost no maternal behavior (nest building, pup retrieval, crouching) (Li et al., 1999). In wild-type animals, PEG3 is present in brain areas involved in maternal behavior such as the MPOA, BNST, PVN, and MeA. *Peg3* knockouts also have a lactation deficit. These data suggest that PEG3 activity may be upstream of oxytocin and many other gene products in the maternal behavior neural circuitry and that the activity of PEG3 may contribute to the organization of the behavior.

Individual differences in OTR levels have been associated with individual differences in maternal care (Francis et al., 2000; Champagne et al., 2001). In particular, rat dams that show high levels of licking and grooming of their pups have higher levels of OTR binding in the BST, the central amygdala, and the MPOA. Intriguingly, dams that show high levels of licking and grooming behavior rear female pups that grow up to have higher levels of OTR in the central nucleus of the amygdala and the BST (Francis et al., 2002).

It appears that good mothering results in good mothering in subsequent generations (Francis et al., 1999). Perhaps this is due in part to behavioral transmission of brain OTR levels.

It is clear that the MPOA is the site of converging action of multiple hormones (e.g., estradiol, prolactin, and oxytocin) and sensory input for maternal behavior. Once the MPOA has integrated that information, what does it do with it? The MPOA projects to numerous areas, most notably to reward circuitry such as the VTA. The VTA sends dopaminergic projections to ventral forebrain areas such as the nucleus accumbens (NAcc). Chemical lesions of dopaminergic projections of the VTA prevent the onset of maternal behavior (Hansen et al., 1991). It is thought that this circuitry underlies the motivational state or drive for maternal behavior (see Lonstein and Morrell's chapter for a detailed description).

This section on rodent maternal behavior was presented from the perspective of the mother, with no consideration for the behavior or the physiology of the pup. This was done only for clarity of discussion. In fact, pups do contribute to the mother–infant interactions in very proactive ways and many of the same hormones regulate pup behavior. For example, oxytocin knockout mouse pups show a deficit in separation-induced vocalization (Winslow et al., 2000), and pharmacological manipulation of central oxytocin systems alters isolation-induced calling in rat pups (Insel and Winslow, 1991). Recently, Moles and colleagues (2004) demonstrated that pups lacking the mu-opioid receptor gene do not cry in response to removal from their mother and show no preference for mother-associated cues. Perhaps normal pups have an activated endogenous opioid system during contact with the mother, but when removed from the mother, they suffer the equivalent of a painful opiate withdrawal (Nelson and Panksepp, 1998). The mu-opioid knockout mouse pup would not detect the rewarding state, nor the withdrawal state, and therefore would not cry in pain upon separation (Moles et al., 2004). Pup–mother interactions are critical for the success of the pups. In addition, it is very clear that a pup's early experience can dramatically affect behavioral phenotype later in life by epigenomic mechanisms (Francis et al., 1999; Weaver et al., 2004).

The main impact of the studies of maternal care on our understanding of social bonding has been to realize that steroid and neurohormones prime specific brain regions and organize behavioral efforts. Maternal behavior is very complex and involves many brain areas and gene products. There are many knockout lines of mice with maternal behavioral deficits, reminding us that there is not "a gene" for maternal behavior; rather there are many genes operating within a system that each contribute to the regulation of behavior. Other knockouts with maternal behavior deficits are reviewed in Leckman and Herman (2002).

5.1.2 Social Recognition in Mice and Rats

Social recognition is a prerequisite for more meaningful social encounters. Social recognition requires a complex set of processes: social approach and investigation, sensory processing, and learning and memory. Social recognition in rodents can be measured by allowing an animal to investigate a novel animal for a brief period followed by reexposure to the same or a novel stimulus animal. With each exposure, the investigator records the duration of anogenital and perioral sniffing by the test animal. With repeated presentations of the same stimulus animal, the duration of sniffing behavior by the test animal decreases. This decrease in the amount of investigation by the test animal is interpreted as social recognition. If the stimulus animal is novel, then the test animal reverts to a thorough anogenital/perioral investigation behavior. In this assay, social recognition memory was determined to be around 2 h or less for rats and mice, but can be prolonged or truncated with various behavioral and pharmacological manipulations.

For individuals to recognize each other, they must produce some unique identifying characteristic(s) that are stable over time. Rodents use olfaction as a primary sensory modality and therefore use unique odor cues to distinguish among individuals. There are two classes of polymorphic proteins found in urine that may provide a signal for individuality, or "odortype": the major urinary proteins (MUPs) and major histocompatability complex genes (MHC). The MUPs are highly polymorphic across individuals and stable over time, suggesting that they are good candidates for individual markers (Beynon and Hurst, 2003). In addition to MUPs, each individual has a set of highly variable MHC. These genes code for proteins that present both endogenous and exogenous protein fragments to the immune system (Germain, 1994). This mechanism teaches the immune system to recognize "self" versus "not-self." MHC gene products are expressed throughout the body, and are present in both the urine and in olfactory sensory neurons (Brennan, 2004). The exact molecular mechanisms of how these MHC genes contribute to individual recognition are still unknown. Although MHC molecules are polymorphic, their molecular signature changes with experimental parameters, making them less than ideal candidates to represent the odortype of an individual (Beynon and Hurst, 2003). Behaviorally, the role of MHC genes is clear. Individuals tend to choose mates with less similar MHC genes and show increased cooperative and altruistic behavior toward individuals with more similar MHC genes (Potts et al., 1991; Manning et al., 1992; Yamazaki et al., 2000). These preferences appear to be learned and not innate, because cross-fostering of animals disrupts disassortative mating and preferences for cooperation with siblings and parents (Yamazaki et al., 1988; Penn and Potts, 1998). It seems that the MHC genes play a role in reducing inbreeding while increasing kin selection behaviors. Both MUPs and MHC genes will need to be studied further to gain a clearer picture of the mechanism of odortype diversity.

While the exact mechanism of odortype diversity and constancy is still unclear, it is clear that this information is processed in olfactory pathways in the brain. In rodents, much of the brain effort underlying social recognition memory involves the main and accessory olfactory bulbs and their projections. Social recognition memory mediated by these pathways is facilitated by the neurohormones oxytocin and vasopressin.

As described earlier, oxytocin plays a role in mother–infant interactions. Oxytocin also plays a role in social recognition. Oxytocin infused into the olfactory bulbs of rats minutes before behavioral testing prolongs the apparent duration of a social recognition response (Dluzen et al., 1998b), and this effect appears to be mediated through norepinephrine. 6-OHDA lesions of norepinephrine cells in the bulb prevent the prolonged social recognition response to oxytocin (Dluzen et al., 1998a), whereas stimulation of alpha-2 noradrenergic receptors with clonidine or blockade of norepinephrine reuptake by nisoxetine increases the social recognition response, even without oxytocin coadministration (Dluzen et al., 2000; Shang and Dluzen, 2001). Also, oxytocin treatment results in increased norepinephrine release as measured by microdialysis of the bulb (Dluzen et al., 2000). The emerging model is that oxytocin potentiates the release of norepinephrine in the olfactory bulb, and this increased norepinephrine stimulates alpha-2 noradrenergic receptors, which inhibit local inhibitory granule cells. This inhibition of granule cells *disinhibits* the output cells of the olfactory bulb to increase the social-recognition response. This model also applies to the mechanism underlying the Bruce effect, which is the phenomenon that female mice will resorb developing embryos when exposed to odors of a novel male. This phenomenon requires the discrimination of males and is blocked by 6-OHDA-induced lesions of noradrenergic inputs to the olfactory bulb (Rosser and Keverne, 1985).

Oxytocin appears to act in other brain areas in addition to the olfactory bulb to regulate social-recognition behavior. Oxytocin in the LS and MPOA of male rats can prolong social-recognition (Popik and van Ree, 1991, 1992). Male oxytocin knockout mice have an impaired ability to remember a mouse they just met in the kind of social recognition test described above (Ferguson et al., 2000). This deficit is reversed by infusion of oxytocin into the MeA before the learning trial, and in wild-type animals, the deficit is mimicked with an OTR antagonist into the MeA (Ferguson et al., 2001). The timing of treatment with oxytocin and the receptor antagonist indicates that oxytocin plays a role in the early stages of social recognition memory formation rather than in the subsequent expression of social recognition behavior. Similarly, estrogen receptor-alpha and -beta knockout mice also show social recognition deficits (Choleris et al., 2003). The female oxytocin knockout mouse has deficits in discriminating between healthy and parasitized male odors, but no impairment in tasks that require the discrimination of nonsocial odors (Kavaliers et al., 2003). Additionally, estrogen receptor-alpha and -beta knockout mice also have deficits in the ability to discriminate, remember, and avoid parasitized male odors (Kavaliers et al., 2004a). In a forced choice test, wild-type mice choose urine odors derived from other wild-type mice compared with urine odors from the oxytocin knockout and estrogen receptor-beta knockout mice (Kavaliers et al., 2004b). In contrast, the same wild-type mice chose odors from the estrogen receptor-alpha knockout mouse compared with wild type (Kavaliers et al., 2004b). This suggests that not only do these knockout animals have decreased social discrimination, but they also possibly produce different odor cues themselves.

Estradiol improves social memory in female rats (Hlinak, 1993), and oxytocin in the PVN and OTRs in the amygdala are regulated by estradiol through estrogen receptor-beta and -alpha, respectively (Young et al., 1998; Patisaul et al., 2003). Because all three knockouts (oxytocin, estrogen receptor-alpha, and -beta) have social-recognition deficits, Choleris and coworkers proposed a "four gene micronet" to synthesize the results from the knockout studies. The four genes encode oxytocin, OTR, estrogen receptor-alpha, and estrogen receptor-beta. In this "micronet" (➋ *Figure 6-2*), social odor cues are detected by the main and accessory olfactory bulbs, which project to the OTR-dense amygdala. The amygdala receives oxytocinergic input from the hypothalamus. The amount of OTRs in the amygdala is determined in part by the circulating estrogens acting at the estrogen receptor-alpha. Likewise, the amount of oxytocin in the hypothalamus is determined in part by the circulating estrogens acting through the estrogen receptor-beta. In sum, the circulating estrogens may set the gain on a social recognition network by upregulating oxytocin and its receptor via estrogen receptors. In some yet undetermined way, the output of the amygdala contributes to social recognition behavior. While this is a simple model with just four genes in two brain areas, it is a decent framework to begin to understand the neural correlates of social recognition.

Figure 6-2

Social recognition is an important component of affiliative behavior. Social recognition can be examined by presenting a test animal with a series of novel and familiar stimulus animals. The test animal is first presented with a novel animal (trial 1) and spends roughly 80% of the 60-s trial investigating this novel individual. In the next three trials, the test animal is presented with the same stimulus animal. In these trials, the test subject spends less time investigating the stimulus animal. This decrease in investigation in trials 2, 3, and 4 represents social recognition. In a fifth trial, the test subject is presented with a novel stimulus animal and resumes full investigation. Knockout mice for oxytocin (OT), estrogen receptor-alpha (ER-alpha), and -beta (ER-beta) all have deficits in social recognition. This is a complex process and involves many gene products acting within a system. Choleris and colleagues have proposed a "four gene micronet," which includes the genes encoding OT, ER-alpha, ER-beta, and the oxytocin receptor (OTR), acting within the hypothalamus and amygdala. Social odor cues are processed in the main and accessory olfactory bulbs, which project to the amygdala, the output of which modulates social recognition. Oxytocinergic projections from the hypothalamus modulate the activity of the amygdala. The activity of this micronet is upregulated by circulating estrogens acting at ER-alpha and ER-beta to increase levels of OTR and OT, respectively. This figure was adapted from Choleris et al. (2004)

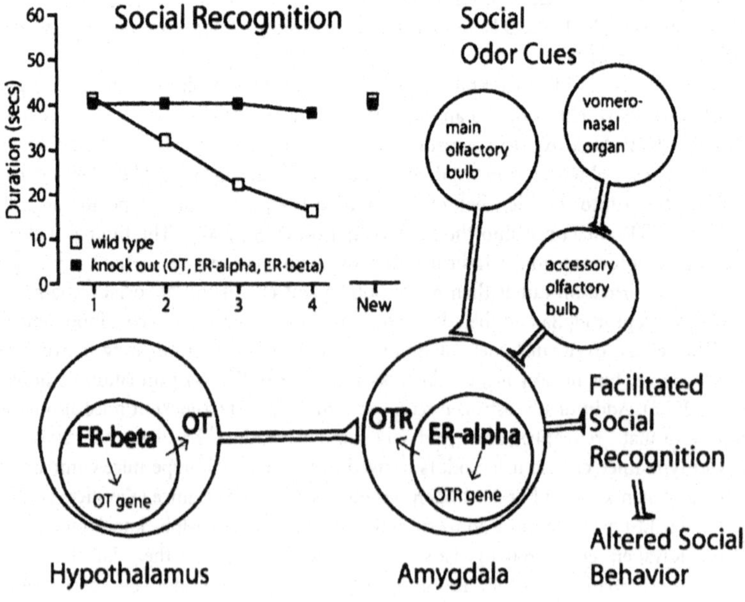

Vasopressin is a nine amino acid peptide hormone that is very closely related to oxytocin. Like oxytocin, vasopressin is also involved in social recognition. Central administration of vasopressin prolongs the duration of social recognition memory in rats (Le Moal et al., 1987). Vasopressin applied specifically to the LS also prolongs the social recognition response, while a vasopressin 1a receptor (V1aR) antagonist applied into the LS of rats inhibits social recognition (Dantzer et al., 1988; Popik and van Ree, 1991). The Brattleboro rat, which has a natural mutation in its vasopressin gene, has a social recognition deficit, and application of vasopressin directly into the LS of these rats restores social recognition (Engelmann and Landgraf, 1994). Downregulation of the V1aR in the dorsal LS of normal rats with an antisense oligonucleotide to V1aR inhibits social recognition behavior (Landgraf et al., 1995). Additionally, increasing V1aR density in the LS of normal rats by viral vector gene transfer of V1aR enhances the durability of social recognition memories (Landgraf et al., 2003). V1aR knockout mice also have a social recognition deficit (Bielsky et al., 2004a) and this deficit is corrected with replacement of the receptor by viral vector gene transfer into the LS (Bielsky et al., 2004b). The vasopressin 1b receptor (V1bR) also appears to be capable of influencing social recognition. The V1bR knockout also has a modest impairment in social recognition, although the overwhelming phenotype is a reduction in aggression (Wersinger et al., 2002). Before the social recognition processing occurs in the LS, however, the olfactory input is first processed in the olfactory bulbs. In fact, direct application of vasopressin to the olfactory bulbs prolongs the expression of the social recognition response (Dluzen et al., 1998b), and as with oxytocin, the treatment effect may depend on intact noradrenergic signaling in the olfactory bulb (Dluzen et al., 1998a).

It is clear from studies in rats and mice that both oxytocin and vasopressin play a role in social recognition behavior. These neuropeptides and their receptors can modulate this behavior in several brain areas, including the olfactory bulbs, the MeA, and the LS. It is not surprising that this complex behavior, which includes sensory processing, learning, and memory as well as motivation and motor activity, is influenced by such a broad network of brain areas and molecular mediators.

The need to develop rapid attachments to the caregiver is very strong in altricial species. In rodent pups, the ability to associate maternal care with the mother happens very quickly. Maternal licking and grooming of the anogenital region of the pup is a major aspect of rodent maternal care. Maternal licking can be experimentally simulated by stroking the pup with a paintbrush. Pups that have been stroked with a paintbrush and simultaneously exposed to an odor like peppermint will show a learned preference for peppermint odor (Sullivan et al., 1989). Therefore, maternal licking enhances olfactory learning in the pup. Rat pups have increased noradrenergic neurotransmission from the locus coeruleus into the olfactory bulb relative to adults (Sullivan et al., 2000; Sullivan, 2003). This increase in noradrenergic neurotransmission is present during a critical window of olfactory learning and coincides with increased olfactory stimulation representing the mother's odor in the olfactory bulb of the pup. This change in neural activity and modulation by ascending noradrenergic activity results in plasticity in the pup olfactory bulb and therefore learning of the mother's odor (Sullivan, 2003). This change in the bulb not only results in the learning of the mother's odor, but also a preference for her odor. In fact, this learning of and subsequent preference for an odor can be reproduced with exogenous application of norepinephrine into the bulb or stimulation of the locus coeruleus (Sullivan et al., 2000). Oxytocin and vasopressin may play a role in social recognition in pups as well. Normal rat pups learn to associate a nonsocial odor (such as lemon scent) with the odor of their mother and will subsequently show a preference for the lemon scent even in the absence of their mother. In contrast, preweanling vasopressin-deficient Brattleboro rats do not make the association of mother odor cues with a nonsocial odor, or at least they do not show a preference for the lemon scent after repeated mother-odor pairings (Nelson and Panksepp, 1998). Likewise, in normal preweanling rats, OTR antagonist administration i.c.v. before the learning trial inhibits the association of mother odor with a nonsocial odor (Nelson and Panksepp, 1996). Perhaps in rat pups, oxytocin and/or vasopressin potentiates the release of norepinephrine to modulate social memory as oxytocin does in adult rats (Dluzen et al., 2000). Therefore, an emerging model of social olfactory learning in the rodent pup includes increased noradrenergic drive to the olfactory bulb, which is potentiated by oxytocin and/or vasopressin released with social stimuli. This increased noradrenergic drive probably reduces the inhibitory tone on the olfactory bulb, allowing increased activity of the main output cells of the bulb. This increased excitation in the output cells of the bulb probably results in plasticity in both the bulb and its downstream projections. Individual differences in developmental plasticity of the olfactory bulb may affect adult sociobehavioral phenotypes by differential tuning of the system to attend differently to social odors. There is a very similar norepinephrine mechanism acting in the olfactory bulb of the ewe when she learns her lamb odor, which we shall see later.

In rats and mice, it is very clear that appropriate social behavior requires an intact sensory modality (olfaction) for social recognition, a way to assess the valency of the olfactory input (as in the amygdala), and regulation of the motivation and reward circuitry (ventral forebrain). This circuitry is influenced by steroid and neuropeptide hormones, endogenous opioids, and classical neurotransmitters like norepinephrine. This is clearly a very simplified model as there are many other brain regions containing many hormones and gene products that contribute to the regulation of the many aspects of complex vertebrate affiliative behavior.

5.2 Maternal Bonding in Sheep

Sheep have become a very useful model of attachment and bonding, because like rats they also show strong maternal behavior after parturition. However, in contrast to rats and mice, sheep show *selective* maternal behavior. Maternal rats and mice take care of all infants, but the ewe cares only for her own lamb with which she bonds just after delivery. Many of the same molecular correlates for maternal behavior and social recognition in rats and mice have emerged in sheep.

The changes in hormones during pregnancy and the actual event of delivery are required for the onset of maternal behavior in ewes. Hormone priming with 17β-estradiol and progesterone followed by vaginocervical stimulation induces maternal behavior in virgin ewes (Keverne et al., 1983; Yeo and Keverne, 1986, Kendrick et al., 1991). Vaginocervical stimulation results in increased oxytocin release, measured in the CSF and in the brain (Kendrick et al., 1986; Kendrick et al., 1988a, b). Additionally, oxytocin levels rise in the MPOA, BST, and olfactory bulbs during parturition (Kendrick et al., 1988b, 1992b). The ascending sensory input from the vagina appears to be important for the release of oxytocin in the brain because peridural-anesthetized females require exogenous application of oxytocin to exhibit appropriate maternal behavior (Levy et al., 1992). Therefore, the effects of vaginocervical stimulation can be replicated by intracranial delivery of oxytocin. In fact, as in rats, maternal behavior can be induced by oxytocin infusion into the ventricles in steroid-primed ewes (Kendrick et al., 1987).

As in rats, estradiol and progesterone levels change throughout pregnancy and at parturition, and they alter gene expression in the brain. These changes in gene expression prepare the brain to interpret signals at parturition (such as the flood of oxytocin) to promote/allow the onset of maternal behavior. 17β-estradiol and progesterone have been shown to increase levels of oxytocin, opioids, corticotrophin-releasing hormone, OTR, and prolactin receptors (Schumacher et al., 1992, 1993; Broad et al., 1993a, b, 1995, 1999). These gene expression changes probably also act to increase the salience of lamb-related stimuli for long-term maintenance of maternal behavior. In summary, there are very clear similarities in neural mechanisms of maternal behavior across rats, mice, and sheep.

As described above, sheep come with an added feature in that they also display selective attachment. Ewes become maternal immediately after delivery; however, within a few hours they become selectively maternal and will only accept their own lamb to nurse. This behavior is thought to be mediated in part by reorganization of the olfactory bulb (Kendrick et al., 1992a). Nulliparous ewes find odors associated with a placenta repulsive (Levy et al., 1983). However, with parturition, ewes find these odors extremely attractive. Some important changes must happen in the ewe's brain to change the perception of a stimulus from repulsive to attractive, and to learn that the attractive quality belongs to her lamb but not to others.

One site of learning in the brain for this dramatic behavioral switch appears to be the olfactory bulb. Vaginocervical stimulation like parturition results in changes in various neurotransmitters in the sheep olfactory bulb. Oxytocin, noradrenaline, acetylcholine, glutamate, and GABA levels rise following parturition or vaginocervical stimulation (Kendrick et al., 1988a, b; Keverne et al., 1993; Levy et al., 1993). Oxytocin appears to modulate the release of the other neurotransmitters, suggesting that the release of oxytocin *organizes* the changes in the olfactory bulb after parturition or vaginocervical stimulation

(Levy et al., 1995). The cells in the olfactory bulb become very highly tuned to detect lamb odors and specifically to discriminate among lambs. After giving birth, the output of the olfactory bulb changes in response to lamb odors. Mitral cells are the output cells of the olfactory bulb. The number of mitral cells that respond to lamb odors increases greatly after parturition, and these increases in activity are associated with increased cholinergic and noradrenergic neurotransmitter release (Kendrick et al., 1992a). The majority of these mitral cells respond to any lamb odor, while a subset of the mitral cells respond strongly and selectively to the specific lamb's odors. Associated with this selective responding is an increase in the number of mitral cell to granule cell synapses and an increase in glutamate and GABA signaling in response to one's own lamb (Kendrick et al., 1992a). Granule cells are inhibitory interneurons that form dendrodendritic synapses with the mitral cells. It seems that the selective increase in glutamate signaling is due to nitric oxide acting as a retrograde messenger within the olfactory bulb. Specifically, glutamate binds to NMDA receptors on both mitral and granule cells and this stimulates nitric oxide synthase. This enzyme catalyzes the production of nitric oxide, which diffuses throughout the bulb. Within the mitral cells, the increased nitric oxide levels then increase glutamate signaling via a cGMP-dependent mechanism (Kendrick et al., 1997b). Inhibitors of the nitric oxide synthase enzyme inhibit olfactory learning as do lesions of more upstream components of this process such as noradrenergic projections to the bulbs or application of beta-noradrenergic antagonists (➋ *Figure 6-3*; Kendrick et al., 1997a).

To summarize, in sheep, sensory information about vaginocervical stimulation ascends through the spinal cord, stimulating A1, A2, and A6 hindbrain noradrenergic cell groups. These cells project to the PVN that has been primed with the hormones of pregnancy to rapidly release large amounts of oxytocin into the posterior pituitary and throughout the brain. Several OTR-expressing areas [MPOA, VTA, BNST, MeA, and olfactory bulb (Broad et al., 1999)] respond to the increased levels of oxytocin and permit or promote maternal behavior. In particular, changes in the olfactory bulb represent the associative learning of the specific odor of the lamb with the events surrounding birth. These changes result in long-lasting memory of and selective attachment to the lamb.

5.3 Pair Bonding in Voles

5.3.1 Monogamous Behavior

Voles are a diverse genus of rodents and vole species display a wide array of social behaviors, from asocial to prosocial and promiscuous to monogamous (Carter et al., 1995a). Comparative analyses of vole species with divergent social structures have been used to unveil the neural mechanisms of social behavior. Studies of voles for their species diversity in social behaviors began in the field (Getz et al., 1981). In the 1980s voles were brought into the lab and comparative analyses of the brains were begun to gain insight into the underlying neural mechanisms of the species-typical behavior (Getz et al., 1981). Monogamy is defined here as a long-term selective association with a partner (Carter et al., 1995a). Prairie voles form lasting attachments with a partner throughout breeding and nonbreeding seasons, and males contribute to care of the young. Additionally, prairie voles show selective aggression toward novel conspecifics after becoming sexually experienced (Winslow et al., 1993; Insel et al., 1995). This definition of monogamy does not preclude sexual promiscuity.

In the field, monogamy is observed when animals are routinely trapped together and share nest sites. In the laboratory, monogamy is assessed by measuring several social behaviors: partner preference, biparental care of offspring, and selective aggression toward unfamiliar intruders. Monogamous-typical behaviors include high levels of affiliation with juveniles, high levels of paternal care, high levels of selective aggression toward a novel conspecific after sexual experience, and a persistent preference for a familiar partner animal (Carter et al., 1995a). Also, pups of monogamous prairie voles, but not promiscuous montane voles, show a robust stress response to maternal separation with increased vocalization and increased serum corticosterone levels (Shapiro and Insel, 1990).

In addition to the work on selective bonding in sheep, much of the laboratory work that has been done to understand the molecular neurobiology of selective social attachment has used the prairie vole pair bond

Figure 6-3
The sheep olfactory bulb undergoes changes at parturition to learn the odor of the lamb. Odors are detected by olfactory receptor neurons and this information is transferred to the mitral cells of the olfactory bulb. These cells make dendro-dendritic contacts with several other cell types. In particular, the mitral to granule cell synapses appear to be the major site of plasticity in the bulb. Immediately after birth, glutamate levels rise throughout the bulb due to the increased inhibition of inhibitory neurons in the bulb. Glutamate release from the mitral cell stimulates the postsynaptic granule cell to produce the retrograde messenger, nitric oxide (NO). This NO diffuses back to the presynaptic mitral cell and further potentiates the release of glutamate and strengthens the synapse. This plasticity is most robust at mitral-granule cell synapses that have received stimulation from the olfactory bulb (i.e., neurons whose activity represents the odor of the lamb). This figure is taken from Kendrick and coworkers (1997), with permission. The contribution of oxytocin, which serves to facilitate olfactory memory formation in the bulb by potentiating the release of neurotransmitters is not shown here. The sub-cellular location of the OTR is currently unknown

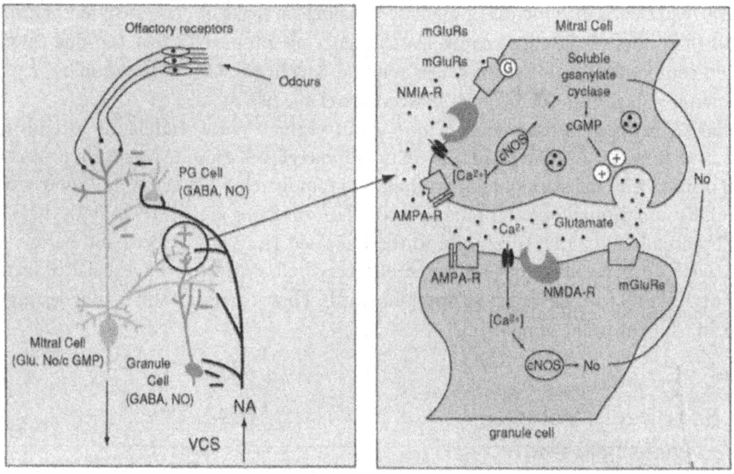

as a model. To establish a pair bond in the lab, sexually naive animals are paired for a cohabitation period (24 h of cohabitation with mating will produce reliable partner preferences in at least 80% of the animals). After 24 h of cohabitation, the animals are tested in a 3-h "partner-preference test." The testing chamber consists of three cages connected by Plexiglas tubes (➋ *Figure 6-4a*). The test subject has access to all three cages. The test subject's "partner" (from the cohabitation period) is tethered in one of the cages; a "stranger" animal is tethered in a second cage; and the third cage is considered "neutral" (it does not contain any animal). Both males and females can be tested in this apparatus. A "pair bond" or "partner preference" is operationally defined by the amount of time the test animal spends with the partner compared with the time it spends with the stranger. The animals are considered "pair bonded" when the test animal spends twice as much time with the partner than with a stranger. Monogamous animals like prairie and pine voles show pair-bond formation by this definition with prolonged cohabitation, which is facilitated by mating. In contrast, promiscuous meadow and montane voles do not show a preference for either animal and actually spend more time alone in the neutral cage than monogamous voles (➋ *Figure 6-4b*), although meadow voles have been observed to form pair bonds in certain circumstances (Parker, 2001). Parameters of the cohabitation period can be altered to increase or decrease the frequency of pair-bond formation in a given sample of prairie voles. For both males and females, increased cohabitation lengths can increase the frequency of pair-bond formation, even in the absence of mating. Decreasing the length of cohabitation and preventing mating during the cohabitation period reduce the frequency of pair-bond formation. It is important to note that the specific conditions ensuring robust partner-preference formation do vary from laboratory to laboratory, indicating that there are high levels of environmental regulation of social bonding.

Figure 6-4

The pair-bonding component of monogamous social behavior is tested in the laboratory using the partner-preference test. (a) A three-chambered apparatus is used for this test. The partner is tethered in one cage, while a stranger is tethered in another cage. The test animal has access to these two stimulus animals and a neutral cage via Plexiglas tubes. The amount of time spent in side-by-side contact with either the stranger or the partner can be quantified. Additionally, the amount of time the animal spends in each of the three cages can also be quantified. Prairie voles, but not montane voles, spend significantly more time in the cage with their partner than with the stranger (b; adapted from Insel et al., 1995). These species differences in behavior are correlated with species differences in oxytocin receptor (OTR) and vasopressin 1a receptor (V1aR) distribution in the reward areas of the brain. Specifically, prairie voles (c) have more OTR than montane voles (d) in the nucleus accumbens. Additionally, prairie voles (e) have more V1aR than montane voles (f) in the VP

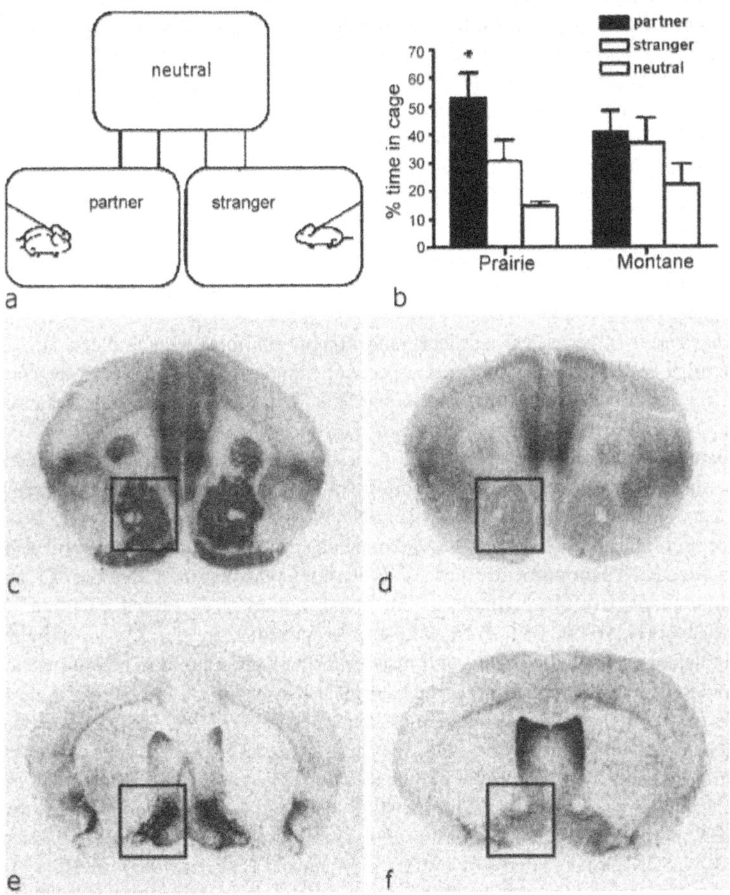

5.3.2 Oxytocin

As just described, in both male and female prairie voles, copulation and cohabitation are involved in the natural development of pair bonds and subsequent selective aggression toward strange intruders (Carter et al., 1988; Winslow et al., 1993). What occurs in the brains of prairie voles as two individuals spend time together and/or mate? As described previously, oxytocin plays a very important role in other social behaviors such as social recognition, maternal behavior, and maternal–infant bonding. Oxytocin injection into the brain increases social contact in female prairie voles (Witt et al., 1990), male rats (Witt et al., 1992), and male squirrel monkeys (Winslow and Insel, 1992). Because of the abundance of data demonstrating an important role for oxytocin in other prosocial behaviors, oxytocin was hypothesized to play an important role in social bonding in voles. Indeed, central infusion of oxytocin facilitates partner-preference formation in female prairie voles, and OTR antagonists inhibit partner-preference formation even if the females have not mated with their partner (Williams et al., 1994; Insel and Hulihan, 1995). Injections of oxytocin facilitate partner-preference formation in ovariectomized female prairie voles (Williams et al., 1992a, 1994), demonstrating that oxytocin is sufficient and that gonadal hormones are not necessary.

Where in the female brain does oxytocin exert its effects on partner-preference behavior? There are drastic species differences in OTR-distribution patterns across monogamous and nonmonogamous vole species (Insel and Shapiro, 1992). Some regions of the brain like the caudate and the NAcc have large numbers of receptors in monogamous species compared with nonmonogamous species (➊ *Figure 6-4c, d*). It has been hypothesized that the species differences in receptor-distribution patterns are responsible for the species differences in response to central administration of oxytocin. OTR antagonists applied directly to the NAcc or prelimbic cortex of female prairie voles inhibit partner-preference formation (Young et al., 2001). The mechanism by which species differences in OTR are generated is still unclear, although the observed species differences are most likely not posttranslational (Young et al., 1996). It is possible that the species differences in the regulatory regions of OTR genes guide the neuroanatomical distributions of the receptor (Young et al., 1996).

5.3.3 Vasopressin

In male prairie voles, it is the related neurohormone, vasopressin, that seems to play a larger role in the regulation of pair-bond formation and other aspects of monogamy like paternal care. Vasopressin is involved in many social behaviors in males including flank marking and aggression in hamsters (Albers and Bamshad, 1998) and social recognition in rats and mice (Bielsky and Young, 2004). Vasopressin-distribution patterns are highly conserved across species (Wang et al., 1996). Several brain areas contain either vasopressinergic cells (cells expressing vasopressin mRNA and vasopressin immunoreactivity) and/or vasopressin immunoreactive fibers (Bamshad et al., 1993; Wang et al., 1994a, 1996; Wang, 1995; Lim et al., 2004a). Vasopressin immunoreactive and mRNA-containing cells are found in the hypothalamus (suprachiasmatic nucleus, SCN; paraventricular nucleus, PVN; supraoptic nucleus, SON), the BST, and the MeA. Vasopressin immunoreactive fibers appear in the lateral septum (LS), ventral pallidum (VP), lateral habenular nucleus (LH), MPOA, BST, PVN, and MeA. Vasopressin expression levels in extrahypothalamic areas of the brain are sexually dimorphic, with males showing higher numbers of vasopressin-expressing cells in BST and more fibers in the LS and lateral habenula (Bamshad et al., 1993; Wang et al., 1994, 1996). Vasopressin levels are regulated by gonadal steroids (Wang and De Vries, 1993), which may coordinate the behavior of the animal with reproductive capacity (Carter et al., 1995a). Specifically, when sexually naive males are castrated, the levels of vasopressin immunoreactivity are reduced in cells and fibers of brain areas thought to be involved in paternal care: cells of the BST and MeA, and fibers within the LS, which receives input from BST and MeA (Wang and De Vries, 1993). Vasopressin levels in hypothalamic regions of the brain (PVN, SCN, SON) are not altered by castration (Wang and De Vries, 1993). Infusion of vasopressin into the LS of castrated males increases pup-directed behaviors and antagonist treatment decreases pup-directed behaviors (Wang et al., 1994a). Vasopressin immunoreactivity is altered in male prairie voles after cohabitation and mating and pup exposure (Bamshad et al., 1993, 1994). Vasopressin immunoreactivity decreases in the LS with sexual experience and the onset of parental care, while vasopressin mRNA levels increase in the BST (Bamshad et al., 1994; Wang et al., 1994b). The BST projects to the LS, so the changes in vasopressin are probably a result of vasopressin release in the LS and a compensatory increase in vasopressin production. The changes in vasopressin levels with the onset of paternal care could be a result of the ingestion of hypertonic urine during anogenital licking and grooming of the pups (Bamshad et al., 1993; Wang et al., 1994a). In contrast to the findings presented above, castrated sexually naive prairie voles, with no vasopressin immunoreactivity in the LS, show only mild deficits in spontaneous paternal behavior (Lonstein and De Vries, 1999). This indicates that vasopressin may not be required for the expression of paternal care, although it does not preclude a role for its influence.

As mentioned above, pair-bond formation relies on prolonged cohabitation with a partner facilitated by concomitant mating. Vasopressin is thought to be released in the brain during the cohabitation due to both mating and social contact. Injections of vasopressin directly into the brain facilitate partner-preference formation (without mating) and receptor antagonists block partner-preference formation in male prairie voles (Winslow et al., 1993). Considering that vasopressin distribution is so highly conserved across species, it is interesting that it has such dramatic differences in behavioral effects when injected into the brain (Young et al., 1997a). The brain-distribution pattern for the vasopressin 1a receptor (V1aR) is not conserved even across closely related species like prairie and montane voles (➋ *Figure 6-4e, f*) . Like the OTR, species differences in V1aR-distribution patterns have been hypothesized to contribute to the species differences in social structure (Insel et al., 1994; Young et al., 1997a). In particular, the V1aR levels are higher in the VP, the central nucleus of the amygdala, cingulate cortex, and the laterodorsal thalamus in monogamous prairie voles than in promiscuous montane voles (Insel et al., 1994). Within the prairie vole species, there do not appear to be any sex differences in the distribution patterns of V1aR (Insel et al., 1994; Phelps and Young, 2003; Lim et al., 2004a). To test the idea that species-specific V1aR-distribution patterns have important behavioral consequences, site-specific modulation of V1aR has been used. For example, V1a receptor blockers prevent partner-preference formation when applied directly to the VP at doses that are ineffective when delivered into the lateral ventricles (Lim and Young, 2004). A viral vector containing the prairie vole *v1ar* with a neuron-specific enolase promoter has been used to alter V1aR levels in specific brain regions. With such a tool, it is possible to test the hypotheses about the molecular and neural substrates of pair bonding. Increasing V1aR in the VP of monogamous prairie voles using this V1aR-expressing virus facilitates partner-preference formation (Pitkow et al., 2001) in the absence of mating.

Clearly, the species differences in V1aR-distribution patterns have important consequences for behavior. The differences could be due to differences in receptor pharmacology across the two species or the differences in gene regulation. It appears that the latter is the case. The two species are highly homologous in the coding region for the *v1ar* gene (Young et al., 1999) and consequently there are no differences in the receptor pharmacology across the two species (Insel et al., 1994). Additionally, the species differences in receptor binding levels are apparent at the perinatal period, indicating that environmental effects on regulation are less likely than genetic effects (Wang et al., 1997). Furthermore, the species differences in distribution are not only apparent at the level of receptor binding, but also at the mRNA level, suggesting species differences in gene regulation rather than in posttranslational processing (Young et al., 1997a). In the regulatory region of the *v1ar* gene, there is a striking species difference at about 660-bp upstream of the transcription start site (➋ *Figure 6-5a*). In monogamous prairie and pine voles, there is a 500-bp repetitive expansion at this locus, which is only about 50-bp long in the promiscuous meadow and montane voles (Young et al., 1999). It is possible that this microsatellite modifies gene expression patterns by changing the promoter structure of the *v1ar* gene across monogamous and promiscuous species (Young et al., 1999; Hammock and Young, 2002). A transgenic mouse for the prairie vole *v1ar* gene, including this microsatellite, has a receptor-distribution pattern that is more like that of a prairie vole than like a wild-type mouse (➋ *Figure 6-5b, c, d*), suggesting that sequences in the prairie vole *v1ar* gene contribute to species-specific distribution patterns (Young et al., 1999; Hammock and Young, 2002). In cell culture, this prairie-vole-specific microsatellite modulates gene expression. Specifically, deleting the microsatellite results in an increase in the activity of a reporter gene in some, but not all, of the cell lines that were tested (Hammock and Young, 2004). This indicates that the microsatellite acts in a cell-type-dependent manner to regulate gene expression. We would expect that such a regulatory mechanism is also functioning in the brain, because not all brain areas show species differences in receptor-binding levels. In addition, when the prairie vole microsatellite was replaced with the montane vole microsatellite, the short montane vole microsatellite also increased reporter gene activity relative to the long prairie microsatellite (➋ *Figure 6-5e*), demonstrating that species differences in microsatellite length affect gene regulation (Hammock and Young, 2004). It is plausible that the species differences in gene structure lead to changes in gene expression patterns, which ultimately have behavioral consequences.

Figure 6-5
Species differences in V1aR distribution patterns are associated with a functional microsatellite polymorphism in the 5′ regulatory region of the *v1ar* gene. The *v1ar* gene in montane and prairie voles is highly homologous in coding regions (a; hatched boxes), but not in the 5′ regulatory region. Specifically, there is an expanded repetitive microsatellite in the prairie vole *v1ar* 5′ region but not in the montane vole *v1ar* gene (shaded box). Additionally, the *v1ar* gene appears to have undergone a gene duplication event for this locus and therefore this species has two copies of the *v1ar* gene. One copy, however, contains a mutation in the coding region that would result in a truncated form of the receptor. In support of the role of the microsatellite in the regulatory control of the distribution of V1aR, transgenic mice (d) for the prairie vole *v1ar* gene (including the microsatellite) have V1aR distribution patterns that resemble prairie voles (b) more than wild-type mice (c). In vitro, species-specific lengths of the microsatellite modify gene expression (e), demonstrating that it is possible that this rapidly evolving sequence can regulate gene expression in the brain. (Figures adapted from Young et al., 1999 and Hammock and Young, 2004)

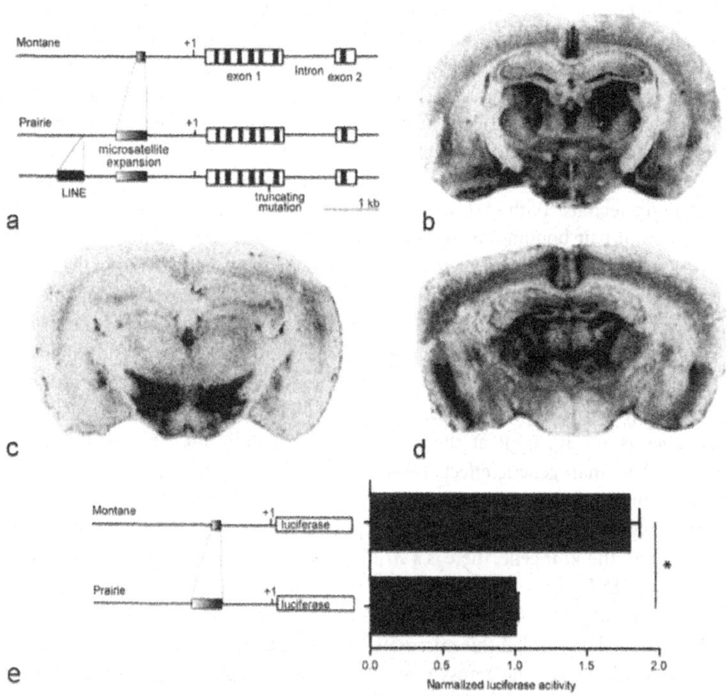

5.3.4 Neural Circuitry of Pair Bonding

If the main difference in the brain contributing to the observed species differences in pair-bond formation in male voles were the amount of V1aR in the VP, then one would predict that changes in V1aR levels in the VP would modulate behavior. Lim and colleagues (2004) tested this idea by injecting promiscuous meadow voles with a V1aR-expressing viral vector directly into the VP. After injection and recovery, the animals were tested for their ability to form pair bonds in the partner-preference test. All the meadow voles that were tested had been transformed by the addition of V1aR in the VP: they all preferred to be in side-by-side contact with their partner. It was very intriguing to see that changes in the levels of one protein in the brain can dramatically alter species-typical behavior.

Because both the NAcc and VP are key relay nuclei in the brain circuits involved in reward, it has been hypothesized that both oxytocin and vasopressin may be facilitating affiliation and social attachment in monogamous species by modulating these reward pathways (Young et al., 2001; Young and Wang, 2004; Lim et al., 2004b). The NAcc plays a major role in the regulation of pair-bond formation in female prairie voles. After mating, dopamine turnover increases in the NAcc (Gingrich et al., 2000). Dopamine D2 receptor antagonists in the NAcc block partner-preference formation and agonists facilitate female partner-preference formation (Gingrich et al., 2000). Concurrent activation of D2 dopamine receptors and OTRs in the NAcc is necessary for partner-preference formation in female prairie voles (Liu and Wang, 2003). Additionally, dopamine in the NAcc is necessary for partner-preference formation in the male prairie vole. Similar to female prairie voles, in male prairie voles dopamine turnover increases in the NAcc with mating (Aragona et al., 2003a). Dopamine D2 antagonists in the NAcc in males block partner-preference formation while D2 receptor agonists facilitate partner-preference formation (Aragona et al., 2003a). The VP and the NAcc are heavily interconnected. Perhaps the activation of OTR in the NAcc of females and V1aR in the VP of males, along with concurrent activation of dopamine receptors in the reward circuitry in both males and females, provides sex-specific, but parallel mechanisms for pair-bond formation in this species (➲ *Figure 6-6*; (Lim et al., 2004a, b)).

Figure 6-6

Proposed neural circuitry of pair-bond formation in male and female prairie voles. The olfactory signature of a particular individual is combined with sex-induced activity of the reward circuitry. Presumably, future exposure to that individual could continue to activate endogenous reward pathways even in the absence of mating. Individuals therefore associate specific individuals with reward and then prefer that particular individual for their acquired rewarding properties. (Adapted from Lim et al., 2004)

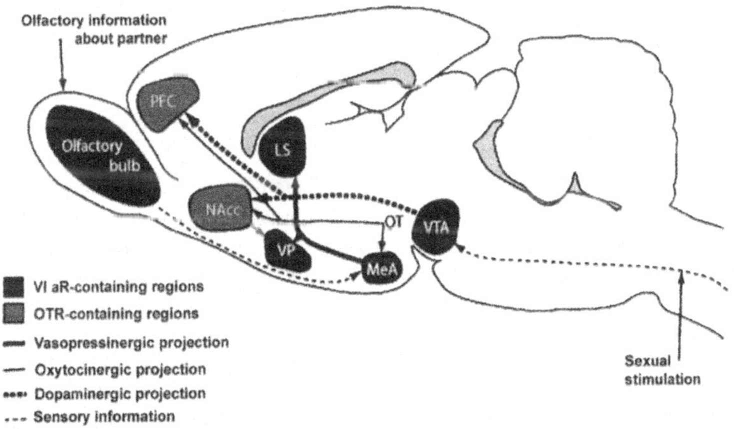

It appears that the species-specific distribution of V1aR and OTR in this reward circuitry is a defining difference in pair-bonding behavior between monogamous and promiscuous vole species. In the meadow vole experiment by Lim and colleagues described earlier, the effect of V1aR application to the VP was reversed when those same animals were treated with D2 receptor antagonists. This demonstrates that in these meadow voles, the DA reward circuitry was playing a critical role in the formation of the pair bonds. Therefore, it does seem likely that the addition of the V1aR in the VP, by changes in the 5′ regulatory region of the *v1ar* gene, was permissive for the natural selection of monogamous behavior in the evolutionary history of prairie voles.

It is easy to see how a mechanism based in reward circuitry could allow for the formation of pair bonds. However, if the concurrent activation of D2 receptors and OTR or V1aR in the reward circuitry is part of the mechanism for pair-bond formation, and animals engage in extrapair copulations, what prevents the formation of pair bonds with all individuals? What stabilizes the monogamous social structure, as it is known that even socially monogamous animals like prairie voles do engage in extrapair copulations?

It appears that dopamine D1 receptors in the NAcc may play a role. D1 receptor activation actually inhibits pair-bond formation. In the transition from "not bonded" to "pair bonded," D1 receptor levels rise in the NAcc. It is possible that simple increases in the levels of the D1 receptor antagonize the future development of multiple pair bonds (Aragona et al., 2003b).

For voles to be able to form enduring social bonds, several events have to take place. The animals must first be able to detect each other and discriminate among individuals to make the association with the partner and "reward." The ability to detect each other and discriminate among individuals starts with chemosensory detection by the olfactory bulb. Lesions of the main and accessory olfactory bulb inhibit partner-preference formation in female prairie voles (Williams et al., 1992b; Curtis et al., 2001). In male prairie voles, lesions of the olfactory bulb result in decreased social behavior (Kirkpatrick et al., 1994). With an intact olfactory system, the social information probably passes through to the amygdala, where saliency may be ascribed (Meredith and Westberry, 2004). Social exposure to a male and subsequent parturition in females alter neurogenesis in the amygdala and subventricular zone, which sends new neurons to the olfactory bulb (Fowler et al., 2002). This neurogenesis may influence social recognition by increasing the discriminatory capacity of the olfactory bulb–amygdala social discrimination pathway. Perhaps in voles, as in mice (Shingo et al., 2003), this neurogenesis is a result of elevated prolactin levels associated with mating and pregnancy. The duration of "social memory" of prairie voles is between 1 and 2 weeks (Carter et al., 1995a), whereas the social memory of rats and mice is less than 2 h (Dantzer et al., 1987). However, when group-housed, mice display drastic improvements (days) in the duration of social memory (Kogan et al., 2000). Interestingly, while there are clear data that suggest a role for vasopressin and oxytocin in social recognition memory in rats and mice, the role of these neurohormones in the requisite social recognition component of pair-bonding behavior in voles is not known. Due to the prolonged duration of social memory in monogamous voles, there are perhaps additional neural mechanisms underlying social recognition memory compared with rats and mice.

5.3.5 Bonding and Stress

There is evidence suggesting that adrenal hormones and the hypothalamic-pituitary-adrenal (HPA) or "stress" axis play a role in the social behavior of voles. Presentation of a novel intruder to a sexually experienced vole, but not a naive animal results in an increase in serum corticosterone, and the removal of a mate results in an increase in corticosterone levels, while the subsequent reunion with a mate decreases corticosterone levels (Carter et al., 1995a, b). The direction of change in stress hormone levels indicates that familiarity among prairie voles buffers the activity of the HPA axis.

Prairie voles are sexually dimorphic in their response to stress and stress hormones. Stress antagonizes partner-preference formation in females and promotes partner-preference formation in males. A forced swim stressor inhibits partner-preference formation in females and facilitates partner-preference formation in males (De Vries et al., 1996). Adrenalectomy (and therefore elimination of corticosterone) inhibits partner preference in males (De Vries et al., 1996); however, the same treatment facilitates partner preference in females (De Vries et al., 1995). Furthermore, intraperitoneal injections of 20 to 200 μg corticosterone facilitates partner-preference formation in male prairie voles after just 6 h of cohabitation with a female (De Vries et al., 1996). When very low doses of corticotropin releasing factor (CRF) are injected directly into the brain, male partner-preference formation is facilitated with just 3 h of cohabitation (De Vries et al., 2002). On the basis of previous findings with species differences in behavioral response to oxytocin and vasopressin, one would hypothesize that brain distribution patterns of receptors for CRF would also vary across behavioral phenotypes. Specifically, we would predict that CRF receptor-distribution patterns might vary at least across sex within the prairie vole species and possibly across vole species with divergent social structures. These hypotheses are currently being investigated.

Interestingly, both maternal behavior and pair bonding involve modifications to olfactory centers in the brain and are heavily influenced by the reward circuitry, although by different mechanisms. Additionally, it is possible that both maternal behavior and pair bonding benefit the animal by reducing lifetime exposure to stress hormones via an oxytocinergic mechanism (De Vries et al., 2003). Other hormones, neuropeptides, receptors, and brain areas, in addition to those described here, are likely to be important for social bonding.

5.4 Nonhuman Primates

5.4.1 Stress Reactivity and Social Behavior

Much of the work on the physical correlates of affiliation in nonhuman primates has been on the effects of social interaction on the activity of the HPA-axis. Like prairie voles, monogamous titi monkeys form strong partner preferences and display increased aggression toward novel strangers (Mendoza and Mason, 1997). Upon separation, males and females show behavioral and endocrine signs of distress. Specifically, titi monkeys show an increased cortisol response to separation (Mendoza and Mason, 1986). This cortisol response is similar to that seen in prairie voles when pair-bonded individuals are separated from each other. Interestingly, infants of titi monkeys show higher cortisol response when separated from their father than from their mother (Hoffman et al., 1995). Additionally, pairs of titi monkeys have reduced cortisol response to novelty than do individuals (Hennessy et al., 1995), suggesting that the presence of the mate serves as a buffer to stress.

In contrast, squirrel monkeys are not monogamous and do not have different cortisol responses to novelty when tested alone or with a cagemate (Hennessy et al., 1995). Even though they do not form adult heterosexual monogamy-like attachments, squirrel monkeys do show maternal–infant bonds and adult friendships. Social separation in mother–infant pairs or like-sex adult friendships results in increased cortisol production in squirrel monkeys (Lyons et al., 1995). Even though these monkeys have reduced adrenocorticotropic hormone (ACTH) concentrations, the adrenal cortex appears to be more sensitive to ACTH and this results in increased serum cortisol levels following a social separation (Levine et al., 1997).

Interestingly, levels of the brain stress neurohormone, CRF were observed to be correlated with social and emotional reactivity behavioral traits in two species of macaques (Rosenblum et al., 2002). Bonnet macaques are gregarious animals with relatively low stress reactivity, which spend a lot of time in social huddles composed of both relatives and nonkin alike. In contrast, pigtail macaques are emotionally volatile and tend to spend more time in isolation, and when they do interact with others it is usually only with their kin. The highly social, temperamentally stable bonnet macaques had low levels of CRF in their spinal fluid compared with the high levels seen in the asocial, volatile pigtail macaques. Furthermore, the social bonnet macaques had high levels of oxytocin in their spinal fluid compared with the asocial pigtail macaques. Across individual members of each species, the levels of oxytocin were modestly positively correlated with CRF in the bonnet macaques and strongly negatively correlated in the pigtail macaques. The strong correlations between these two neurohormones within individuals suggest some underlying shared mechanism. The mechanisms underlying the observed biobehavioral correlation across these two macaque species are currently unknown. It is certainly intriguing that oxytocin has again emerged as a correlate of social bonding.

5.4.1 Endocrine and Socioenvironmental Influences on Social Behavior

In promiscuous matrilineal rhesus monkeys, female–female grooming maintains normal female troupe relationships, except during the periovulatory period. During ovulation and related mating, females appear to either increase aggressive or submissive behaviors toward other females based on social rank. In contrast, affiliative grooming behavior by a female toward a male occurs only during this periovulatory portion of the female's cycle, when estradiol levels peak. In this matrilineal society, unsolicited mating by a male could be disastrous for the male. It is thought that the transient increase in female–male social grooming helps to maintain a short-term relationship to indicate to the male that mounting behavior is appropriate (Wallen and Tannenbaum, 1997). If a rise in estradiol levels permits the inclusion of males in the matrilineal social structure, then groups composed of females with no chance of increased estradiol levels (i.e., ovariectomized females) should be less likely to include males in their group. Interestingly, a group of adult ovariectomized females did not integrate a group of young male rhesus monkeys even after 2.5 years of cohousing. However, when those females were given periovulatory doses of estradiol, they immediately integrated the males into the colony, as assessed by grooming of the males. This integration of males into the female group lasted long after the hormone treatment ended, but only for the males who mated with the females (Tannenbaum and Wallen, 1997). Perhaps in rhesus monkeys, as in rats, mice and sheep, estradiol exerts its effects on social behavior by regulating prosocial brain oxytocin systems. Estradiol might also influence oxytocin systems important for maternal care. In pigtail macaques, individual differences in late pregnancy estradiol/progesterone ratios were highly positively correlated with the frequency of infant contact (Maestripieri and Zehr, 1998). In addition, ovariectomized females engaged in infant contact less often than ovariectomized females receiving estradiol replacement or intact females (Maestripieri and Zehr, 1998). As in rats, these differences in maternal behavior probably have effects on subsequent adult phenotypes of their offspring. As an extreme example of variation in maternal care, rhesus monkeys that were reared in a nursery instead of by their mother, showed reduced social reciprocity and increased agonistic behavior as adults. These changes were associated with reduced levels of oxytocin in the cerebrospinal fluid (Winslow et al., 2003).

5.4.2 Prolactin and Paternal Care in Monogamous Primates

In several New World primate species, males contribute to the care of the offspring. The male assists by carrying the infants and sharing his food with his young. A large body of research on monogamous biparental birds and rodents demonstrates that in males, prolactin levels contribute to or are associated with paternal care (Schradin and Anzenberger, 1999; Ziegler, 2000). Interestingly, in monogamous biparental monkeys, plasma prolactin levels in males correlate with levels of paternal care, suggesting that prolactin may play a role in primate paternal care as well (Schradin et al., 2003). Signals from pregnant females and/or contact with offspring may increase plasma prolactin levels to further promote parenting behavior in males (Ziegler, 2000). Prolactin may influence paternal care in human males as well. In pregnant couples, men had increased plasma prolactin levels toward the end of their partner's pregnancy and there was a positive relationship between plasma prolactin levels and couvade (pregnancy-like) symptoms in the men (Storey et al., 2000). While there is mounting evidence that prolactin is involved in paternal care in biparental species, the mechanisms are unknown (Schradin and Anzenberger, 1999).

5.5 Humans

Our understanding of the molecular neurobiology, neurochemistry, and neuroendocrinology of social behavior in humans is very limited. Studies in humans are limited to genetics, blood and CSF measures, relatively low-resolution neuroimaging, and postmortem analyses. As we gather information from experimentally tractable model systems such as maternal–infant bonding in rodents and sheep and pair bonding in voles, we can begin to piece together hypotheses about what mechanisms the human brain might use to achieve complex social behavior.

5.5.1 Oxytocin and Vasopressin

In studies of other organisms, it is clear that oxytocin and vasopressin systems are involved in social behavior. Studies in humans suggest that these same molecules may play a role in our own social behavior. For example, oxytocin is released in women during childbirth and subsequent lactation and nipple stimulation (Sogolow, 1966; Christensson et al., 1989), which is temporally appropriate for it to serve a role in mother–infant bonding. Vasopressin levels in serum are elevated in men during sexual arousal (Murphy et al., 1987), and oxytocin levels in serum are elevated in both males and females at orgasm

(Carmichael et al., 1987; Murphy et al., 1987). These changes in neurohormone levels occur at the appropriate time to permit vasopressin and oxytocin to play a role in social bond formation in humans in a similar manner to their roles in monogamous voles.

A recent study used intranasal application of oxytocin via a nasal spray to assess a role for oxytocin in social behavior. Four groups of men were tested in a social stress paradigm (they had to give an unprepared speech and do mental arithmetic in front of strangers). One group received a placebo nasal spray, the next group received nasal spray containing oxytocin, the third group received placebo but were accompanied by their best friend, and the fourth group received oxytocin and were also accompanied by their best friend. After giving the speech, subjects in all four groups donated serum for measurement of cortisol. The group without extra oxytocin or their best friend reported the most stress while in the stress task and had the highest levels of cortisol. In contrast, the participants who received oxytocin and had their best friend had the lowest self-report of stress and the lowest cortisol measures. The two groups with either oxytocin or their best friend, but not both, had intermediate levels of stress and cortisol levels. This experiment effectively demonstrates the role of both a social buffer on stress (the presence of the best friend) and the role of oxytocin in stress buffering (Heinrichs et al., 2003).

Social contact in rodents produces endocrine changes that are consistent with increased parasympathetic activity conducive to stress reduction (Uvnas-Moberg, 1997). In humans, social contact such as a Swedish massage results in modest increases in serum oxytocin (Turner et al., 1999, see also Wikstrom et al., 2003). Such increased levels of oxytocin could mediate the calming effects of social contact and perhaps facilitate social memory. Additionally, cultural traditions that employ some sort of touch during social introductions such as the western handshake or cheek-to-cheek touch may be involved in reducing stress associated with introduction of novel individuals and/or facilitate social recognition memory. If so, it is plausible that this stress reduction and/or facilitation of social memory may occur through oxytocin or vasopressin mechanisms.

In support of the idea that oxytocin and vasopressin are involved in human social behavior, children with autism have lower serum levels of oxytocin (Modahl et al., 1998) and higher oxytocin prohormone/oxytocin ratios (Green et al., 2001), suggesting deficits in oxytocin metabolism. In the prairie vole, it has been demonstrated that a 5′ regulatory microsatellite in the vasopressin 1a receptor gene modifies expression levels of the gene and can potentially direct regional expression (Hammock and Young, 2004), which has important behavioral consequences (Lim et al., 2004c). In the human vasopressin 1a receptor gene, there is also a polymorphic microsatellite that could potentially contribute to gene expression differences across individuals (Thibonnier, 2000), and also potentially have behavioral consequences (Hammock and Young, 2002). There are two reports indicating that particular alleles of this 5′ regulatory microsatellite may be associated with autism (Kim et al., 2002; Wassink et al., 2004).

5.5.2 Brain Imaging

Some of the neural correlates of human social behavior have been identified in imaging studies. Imaging technology, such as MRI, fMRI, and PET, allows visualization and hypothesis testing of the human "social brain." These techniques can measure structural changes in the brain in patients with social deficits in comparison with healthy controls. Additionally, changes in activity in the brain associated with performance in sociobehavioral tasks can also be observed.

Abnormal social behavior in humans helps to provide insight into how the normal brain functions. There are various disease states in humans that result in impaired social behavior; these include Rett's syndrome, Asperger's syndrome, schizophrenia, and autism as mentioned above. Rett's syndrome occurs primarily in females and is caused by a mutation in the MecP2 gene (Amir et al., 1999). The causes of the remaining disorders are still unknown. It is possible that because these disorders are so heterogeneous there are multiple contributing factors across or even within individuals. One emerging mechanism of social behavior dysfunction, at least in schizophrenia and autism, is that of defects in visual processing of faces. Social behavior in rodents and sheep is very olfactory-driven, as that is the main sensory modality used to discriminate individuals. In contrast, humans are a visual species, and we use visual information about faces to aid in distinguishing among individuals. The cognitive processing required to discriminate individuals appears to be mediated in part by the fusiform gyrus in the temporal lobe (Kanwisher et al., 1997; Gauthier et al., 2000). Autistic patients have an increased volume of this brain area (Frith, 2003), while schizophrenics have a decrease in volume in this brain area (Onitsuka et al., 2003) compared with healthy controls. In functional imaging studies, autistic patients show a lack of activation of this brain area when presented with face stimuli (Schultz et al., 2000). Perhaps these differences in size of this area result in (or from) the visual processing deficit in these disorders.

Which brain areas are activated while attributing "sociality" to abstract objects, and are these areas part of the human "social behavior circuit"? To answer this question, Castelli and coworkers used a PET study in which subjects viewed abstract objects that were behaving "socially" or "mechanically" or "randomly" (Castelli et al., 2002). Interestingly, regions that have been implicated across vertebrate species as playing a role in social behavior were significantly more activated when the abstract objects were behaving "socially." These areas include the lateral fusiform gyrus, the superior temporal sulcus, the ventromedial prefrontal cortex, and the amygdala.

There are some interesting structural changes associated with deficits in social behavior in humans. A single individual, patient S.M., with Urbach-Wiethe disease, which produces bilateral amygdaloid damage, was reported in 1994 to be unable to recognize the facial emotion that represents fear (Adolphs et al., 1994). With further characterization, this deficit was detected as specific to recognition of the facial expression of fear and not in understanding the idea of fear (Adolphs et al., 1995). In a follow-up study, several other individuals with bilateral damage to the amygdala tended to rate pictures of unfamiliar faces as "more trustworthy" and "approachable" than normal subjects who would give negative ratings to the same faces (Adolphs et al., 1998). These results are interesting because the results from nonhuman primate studies suggest that the amygdala serves as a filter, or gate, regulating the saliency of social information, with most emphasis on fear-eliciting social stimuli (Amaral, 2003). In a human imaging study, the amygdala was robustly activated when parents heard the sounds of infants crying compared with nonparents (Seifritz et al., 2003). Amygdaloid abnormalities have been implicated in autism, but the results are equivocal (Palmen and Van Engeland, 2004; Palmen et al., 2004). Most recently, the deficit in patient S.M. has been attributed to a lack of attention to the eyes when looking at pictures of faces. When she was reminded to look at the eyes, she easily recognized the facial expression of fear (Adolphs et al., 2005).

In other animals, it is clear that social attachment requires utilization of the brain circuitry underlying motivation and reward, such as the MPOA, the VTA, and the ventral striatum. Neuroimaging studies in humans have recently provided evidence that brain activity in these brain regions may also play a role in human bonding. Using healthy adults as volunteers, Bartels and Zeki showed brain activation in ventral forebrain and midbrain reward circuitry when males and females looked at pictures of their partners (Bartels and Zeki, 2000). In this study, increased brain activity was observed in the anterior cingulate cortex, the medial insula, and the ventral forebrain, including the caudate and putamen, the hypothalamus, and the midbrain substantia nigra/VTA. In particular, the activation of reward areas such as the substantia nigra/VTA point to convergent mechanisms of social bonding in mammals. During this task, several brain areas were less active including the posterior cingulate gyrus and the amygdala. In a follow-up study Bartels and Zeki compared the neural correlates of maternal and romantic "love" (Bartels and Zeki, 2004). Again using healthy volunteers, they first had mothers look at photos of their children and separate photos of other kids matched for age and familiarity. When looking at the pictures of their own children, the mothers had significantly increased brain activation in the anterior cingulate cortex, middle insula, striatum, periaqueductal gray, and substantia nigra. Additionally, the amygdala was less activated when looking at pictures of their own versus other children. This deactivation of the amygdala occurs in both maternal and romantic love. This fits with the idea that the role of the output of the amygdala is to signal "fear" to other brain areas: it would be contradictory to activate fear circuitry when looking at one's child or partner. Additionally, when Bartels and Zeki compared the activations in both maternal and romantic "love," they observed consistent activations in brain regions that were previously reported to contain OTRs and vasopressin receptors, such as the substantia nigra/VTA (Loup et al., 1989, 1991). This suggests that the oxytocin/vasopressin systems observed in the reward pathways in monogamous voles may play a similar role in social bonding in humans (➋ *Figure 6-7*).

Figure 6-7

There are both common and distinct neural correlates in humans underlying maternal love and romantic attachment. Images here show neural activation in the brain in the "maternal" condition (mothers looking at their own versus a familiar but unrelated child) and in the "romantic" condition (women viewing pictures of a loved partner versus three friends). All the labeled areas in this figure have been previously reported to contain vasopressin receptors or oxytocin receptors. Some of the brain areas that show altered activity in these conditions overlap with such brain areas. (a) sagittal, (b–c) coronal, (d) transverse. Abbreviations: C, caudate nucleus; GP, glubus pallidus; hi, hippocampus; hTh, hypothalamus; P, putamen; PAG, periaqueductal (central) gray; M, nucleus of Meynert; rf, retrorubal fields/intralaminar/subthalamic nuclei; SN, substantia nigra; Tha, lateral thalamus; VTA, ventral tegmental area. (This figure was taken from Bartels and Zeki 2004, with permission)

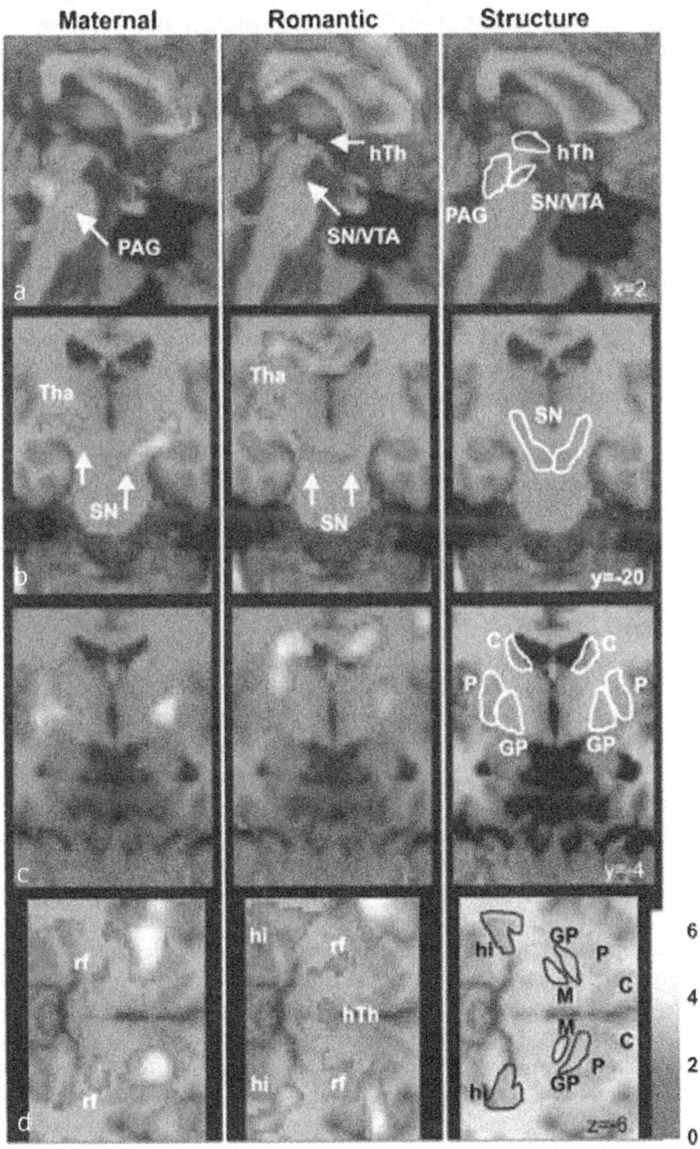

If similar mechanisms of pair bonding occur in humans as in voles, then mating should involve the activation of the reward circuitry. Holstege and colleagues (2003) measured changes in blood flow during ejaculation using PET imaging. Heterosexual adult males were placed in the PET scanner and stimulated by their partners. Among other brain areas with increased blood flow, the VTA showed increased blood flow during ejaculation, which is similar to brain activation during a heroin rush (Holstege et al., 2003).

5.5.3 Pain and Social Cognition

Because social behavior is so important to the survival of prosocial species, it is important for an individual to perceive changes in social context. Activation of pain systems may be one way to motivate an individual to avoid or rectify certain social situations that have negative implications (Panksepp, 2003). An individual with an activated pain system should try to remedy the situation to reduce the perceived pain. Social behavior may have evolved to use pain-perception mechanisms to regulate behavior. Eisenberger and coworkers recently demonstrated that perceived social exclusion is associated with changes in brain activity involved in the perception of physical pain, such as the anterior cingulate cortex and the right ventral prefrontal cortex. Using fMRI, they had subjects engage in a virtual reality ball tossing game, where the ball was tossed back and forth from the subject and two other "players." The other "players" were actually part of a preprogrammed computer game. Part of the way through the ball tossing game, the subject in the scanner was excluded from the game. After the scan, the subjects reported that they had felt excluded and they were distressed. This measure of distress was significantly correlated with activation in the anterior cingulate cortex, which in turn was strongly negatively correlated with activity in the ventral prefrontal cortex (Eisenberger et al., 2003). These brain areas have been shown to be involved in the perception of physical pain. Interestingly, lesions of the anterior cingulate in female hamsters result in a decrease in pup retrieval behaviors, and the same lesions in squirrel monkeys result in a decrease in isolation calls. The anterior cingulate is activated in human mothers listening to infant cries (Murphy et al., 1981; MacLean and Newman, 1988; Lorberbaum et al., 1999). Perhaps across species, activation of the anterior cingulate results from cries of the infant or feelings of social isolation, which may be perceived as physically painful and could produce a motivational state to remedy the situation.

5.5.4 Foundations

While we still know relatively little about the neurochemical, neuroendocrine, and molecular neurobiological mechanisms of human affiliation, studies in simpler model systems have yielded promising insights. ➋ *Figure 6-8* briefly summarizes human social behavior from a systems perspective. As in rodents, we must detect and discriminate among individuals. Visual pathways primarily serve this process, whereas in rodents these processes are primarily olfactory. Next, the identification of specific individuals must be accompanied by a signal of valency. We tend to have strong positive or negative feelings about others and current data suggest that the amygdala plays some role in this process. Before the initiation of prosocial behavior, we must be motivated to act, and our levels of motivation are influenced by the activity of the ventral forebrain reward circuitry. Our genes and hormonal milieu permit the development of such a system as well as act within the system to produce behavior. Likewise, the environment in which humans develop is critical for the development of appropriate social behavior. Early experience serves to tune the genetically and hormonally wired system and influence our choices. Extreme social deprivation early in life has profound effects on adult social behavior. For example, children reared in extremely deprived environments (e.g., the infamous orphanages of Romania) often display autistic-like behavior in addition to dramatic deficits in growth (Carlson and Earls, 1997). Other studies have indicated that social touch in humans is important for infant growth. Massage of premature infants can result in rapid increases in weight gain and reduction in hospital stays (Field, 2002). These studies indicate that early experience with social interaction in humans

Figure 6-8
Affiliative behavior requires detection and discrimination of social stimuli through highly developed sensory systems. In humans, visual processing of faces is an important component of social recognition, while in rodents, social recognition is primarily an olfactory process. The affective value of the social stimulus may be assessed or ascribed in the amygdala. After such information processing, motivational and reward systems must be engaged to prompt the animal to act. This is a very simplified view in an attempt to summarize common themes that have emerged from vertebrate models of affiliation

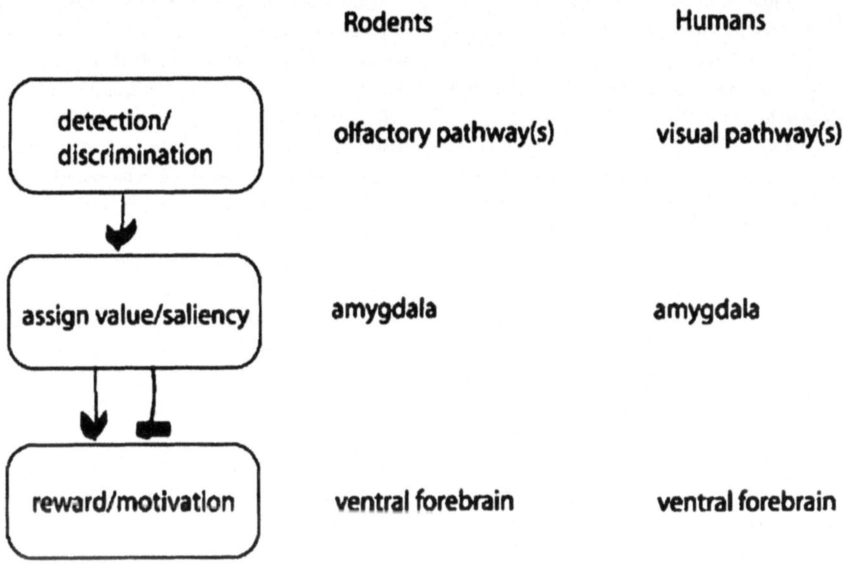

has important consequences not only for the appropriate development of social cognition, but also for the basic mechanisms of growth and survival. These data underscore the importance of basic science research into the underlying mechanisms of social behavior in animals, because it pertains to both the psychological and physical well-being of our own exceedingly social species.

6 Conclusions

Social behavior is highly variable across species. This fact allows for informative comparative studies. An ongoing comparative approach of natural variation has revealed mechanisms of social behavior at the genetic, molecular, cellular, and systems levels of understanding. Social behavior is multisystemic. Proper social behavior requires integration of internal states with environmental stimuli. In vertebrate species, adrenal and gonadal hormones serve to organize behavior, as do central neurotransmitter systems such as prolactin, oxytocin, and vasopressin. Similar humoral mechanisms function in invertebrates as well. Studies in invertebrates have certainly aided our understanding of social behavior. We can apply the mechanistic principles obtained from simpler systems to generate hypotheses and guide our understanding of the biological substrates of more complex social behavior.

There are numerous remaining questions from these cross-species studies of social behavior. Of course, many of the species-specific circuit and molecular details remain to be elucidated. There are currently more questions than answers, especially of our own species. However, the once seemingly impossible task of elucidating mechanisms of complex social behavior is now a tractable topic of investigation with appropriate model systems and new techniques.

References

Adan RA, Van Leeuwen FW, Sonnemans MA, Brouns M, Hoffman G, et al. 1995. Rat oxytocin receptor in brain, pituitary, mammary gland, and uterus: partial sequence and immunocytochemical localization. Endocrinology 136: 4022-4028.

Adolphs R, Gosselin F, Buchanan TW, Tranel D, Schyns P, et al. 2005. A mechanism for impaired fear recognition after amygdala damage. Nature 433: 68-72.

Adolphs R, Tranel D, Damasio AR. 1998. The human amygdala in social judgment. Nature 393: 470-474.

Adolphs R, Tranel D, Damasio H, Damasio A. 1994. Impaired recognition of emotion in facial expressions following bilateral damage to the human amygdala. Nature 372: 669-672.

Adolphs R, Tranel D, Damasio H, Damasio AR. 1995. Fear and the human amygdala. J Neurosci 15: 5879-5891.

Albers HE, Bamshad M. 1998. Role of vasopressin and oxytocin in the control of social behavior in Syrian hamsters (*Mesocricetus auratus*). Prog Brain Res 119: 395-408.

Amaral DG. 2003. The amygdala, social behavior, and danger detection. Ann NY Acad Sci 1000: 337-347.

Amico JA, Thomas A, Hollingshead DJ. 1997. The duration of estradiol and progesterone exposure prior to progesterone withdrawal regulates oxytocin mRNA levels in the paraventricular nucleus of the rat. Endocr Res 23: 141-156.

Amir RE, Veyver Van den IB, Wan M, Tran CQ, Francke U, et al. 1999. Rett syndrome is caused by mutations in X-linked MECP2, encoding methyl-CpG-binding protein 2. Nat Genet 23: 185-188.

Aragona BJ, Liu Y, Curtis JT, Stephan FK, Wang Z. 2003a. A critical role for nucleus accumbens dopamine in partner-preference formation in male prairie voles. J Neurosci 23: 3483-3490.

Aragona BJ, Liu Y, Yu YJ, Damron A, Perlman G, et al. 2003b. Opposite modulation of social attachment by D1- and D2-type dopamine receptor activation in nucleus accumbens shell. Horm Behav 44: 37.

Bale TL, Dorsa DM. 1995. Sex differences in and effects of estrogen on oxytocin receptor messenger ribonucleic acid expression in the ventromedial hypothalamus. Endocrinology 136: 27-32.

Bamshad M, Novak MA, De Vries GJ. 1993. Sex and species differences in the vasopressin innervation of sexually naive and parental prairie voles, *Microtus ochrogaster* and meadow voles, *Microtus pennsylvanicus*. J Neuroendocrinol 5: 247-255.

Bamshad M, Novak MA, de Vries GJ. 1994. Cohabitation alters vasopressin innervation and paternal behavior in prairie voles (*Microtus ochrogaster*). Physiol Behav 56: 751-758.

Bartels A, Zeki S. 2000. The neural basis of romantic love. Neuroreport 11: 3829-3834.

Bartels A, Zeki S. 2004. The neural correlates of maternal and romantic love. Neuroimage 21: 1155-1166.

Ben-Shahar Y, Robichon A, Sokolowski MB, Robinson GE. 2002. Influence of gene action across different time scales on behavior. Science 296: 741-744.

Beynon RJ, Hurst JL. 2003. Multiple roles of major urinary proteins in the house mouse, *Mus domesticus*. Biochem Soc Trans 31: 142-146.

Bielsky IF, Hu SB, Ren X, Terwilliger EF, Young LJ. 2004b. V1a vasopressin receptor expression in the septum is necessary and sufficient for social recognition in mice: a knockout and gene replacement study. Program No. 552.13. In: 2004 Abstract Viewer/Itinerary Planner.: Washington, DC: Society for Neuroscience, 2004. Online.

Bielsky IF, Hu SB, Szegda KL, Westphal H, Young LJ. 2004a. Profound impairment in social recognition and reduction in anxiety-like behavior in vasopressin V1a receptor knockout mice. Neuropsychopharmacology 29: 483-493.

Bielsky IF, Young LJ. 2004. Oxytocin, vasopressin, and social recognition in mammals. Peptides 25: 1565-1574.

Bolwerk EL, Swanson HH. 1984. Does oxytocin play a role in the onset of maternal behaviour in the rat? J Endocrinol 101: 353-357.

Brennan PA. 2004. The nose knows who's who: chemosensory individuality and mate recognition in mice. Horm Behav 46: 231-240.

Breton C, Zingg HH. 1997. Expression and region-specific regulation of the oxytocin receptor gene in rat brain. Endocrinology 138: 1857-1862.

Bridges RS, Feder HH, Rosenblatt JS. 1977. Induction of maternal behaviors in primigravid rats by ovariectomy, hysterectomy, or ovariectomy plus hysterectomy: effect of length of gestation. Horm Behav 9: 156-169.

Bridges RS, Freemark MS. 1995. Human placental lactogen infusions into the medial preoptic area stimulate maternal behavior in steroid-primed, nulliparous female rats. Horm Behav 29: 216-226.

Bridges RS, Numan M, Ronsheim PM, Mann PE, Lupini CE. 1990. Central prolactin infusions stimulate maternal behavior in steroid-treated, nulliparous female rats. Proc Natl Acad Sci USA 87: 8003-8007.

Bridges RS, Ronsheim PM. 1990. Prolactin (PRL) regulation of maternal behavior in rats: bromocriptine treatment delays and PRL promotes the rapid onset of behavior. Endocrinology 126: 837-848.

Brinton RE, Wamsley JK, Gee KW, Wan YP, Yamamura HI. 1984. [3H]oxytocin binding sites in the rat brain demonstrated by quantitative light microscopic autoradiography. Eur J Pharmacol 102: 365-367.

Broad KD, Kendrick KM, Sirinathsinghji DJ, Keverne EB. 1993a. Changes in oxytocin immunoreactivity and mRNA expression in the sheep brain during pregnancy, parturition and lactation and in response to oestrogen and progesterone. J Neuroendocrinol 5: 435-444.

Broad KD, Kendrick KM, Sirinathsinghji DJ, Keverne EB. 1993b. Changes in pro-opiomelanocortin and preproenkephalin mRNA levels in the ovine brain during pregnancy, parturition and lactation and in response to oestrogen and progesterone. J Neuroendocrinol 5: 711-719.

Broad KD, Keverne EB, Kendrick KM. 1995. Corticotrophin releasing factor mRNA expression in the sheep brain during pregnancy, parturition and lactation and following exogenous progesterone and oestrogen treatment. Brain Res Mol Brain Res 29: 310-316.

Broad KD, Levy F, Evans G, Kimura T, Keverne EB, et al. 1999. Previous maternal experience potentiates the effect of parturition on oxytocin receptor mRNA expression in the paraventricular nucleus. Eur J Neurosci 11: 3725-3737.

Brown JR, Ye H, Bronson RT, Dikkes P, Greenberg ME. 1996. A defect in nurturing in mice lacking the immediate early gene fosB. Cell 86: 297-309.

Carlson M, Earls F. 1997. Psychological and neuroendocrinological sequelae of early social deprivation in institutionalized children in Romania. Ann NY Acad Sci 807: 419-428.

Carmichael MS, Humbert R, Dixen J, Palmisano G, Greenleaf W, et al. 1987. Plasma oxytocin increases in the human sexual response. J Clin Endocrinol Metab 64: 27-31.

Carter CS, De Vries AC, Getz LL. 1995a. Physiological substrates of mammalian monogamy: the prairie vole model. Neurosci Biobehav Rev 19: 303-314.

Carter CS, De Vries AC, Taymans SE, Roberts RL, Williams JR, et al. 1995b. Adrenocorticoid hormones and the development and expression of mammalian monogamy. Ann NY Acad Sci 771: 82-91.

Carter CS, Witt DM, Thompson EG, Carlstead K. 1988. Effects of hormonal, sexual, and social history on mating and pair bonding in prairie voles. Physiol Behav 44: 691-697.

Castelli F, Frith C, Happe F, Frith U. 2002. Autism, Asperger syndrome and brain mechanisms for the attribution of mental states to animated shapes. Brain 125: 1839-1849.

Champagne F, Diorio J, Sharma S, Meaney MJ. 2001. Naturally occurring variations in maternal behavior in the rat are associated with differences in estrogen-inducible central oxytocin receptors. Proc Natl Acad Sci USA 98: 12736-12741.

Cheung BH, Arellano-Carbajal F, Rybicki I, de Bono M. 2004. Soluble guanylate cyclases act in neurons exposed to the body fluid to promote *C. elegans* aggregation behavior. Curr Biol 14: 1105-1111.

Choleris E, Gustafsson JA, Korach KS, Muglia LJ, Pfaff DW, et al. 2003. An estrogen-dependent four-gene micronet regulating social recognition: a study with oxytocin and estrogen receptor-alpha and -beta knockout mice. Proc Natl Acad Sci USA 100: 6192-6197.

Choleris E, Kavaliers M, Pfaff DW. 2004. Functional genomics of social recognition. J Neuroendocrinol 16: 383-389.

Christensson K, Nilsson BA, Stock S, Matthiesen AS, Uvnas-Moberg K. 1989. Effect of nipple stimulation on uterine activity and on plasma levels of oxytocin in full term, healthy, pregnant women. Acta Obstet Gynecol Scand 68: 205-210.

Coates JC, de Bono M. 2002. Antagonistic pathways in neurons exposed to body fluid regulate social feeding in *Caenorhabditis elegans*. Nature 419: 925-929.

Curtis JT, Liu Y, Wang Z. 2001. Lesions of the vomeronasal organ disrupt mating-induced pair bonding in female prairie voles (*Microtus ochrogaster*). Brain Res 901: 167-174.

Dantzer R, Bluthe RM, Koob GF, Le Moal M. 1987. Modulation of social memory in male rats by neurohypophyseal peptides. Psychopharmacology (Berl) 91: 363-368.

Dantzer R, Koob GF, Bluthe RM, Le Moal M. 1988) Septal vasopressin modulates social memory in male rats. Brain Res 457: 143-147.

de Bono M. 2003. Molecular approaches to aggregation behavior and social attachment. J Neurobiol 54: 78-92.

de Bono M, Bargmann CI. 1998. Natural variation in a neuropeptide Y receptor homolog modifies social behavior and food response in *C. elegans*. Cell 94: 679-689.

de Bono M, Tobin DM, Davis MW, Avery L, Bargmann CI. 2002. Social feeding in *Caenorhabditis elegans* is induced by neurons that detect aversive stimuli. Nature 419: 899-903.

De Vries GJ, Buijs RM. 1983. The origin of the vasopressinergic and oxytocinergic innervation of the rat brain with special reference to the lateral septum. Brain Res 273: 307-317.

Del Cerro MC, Izquierdo MA, Collado P, Segovia S, Guillamon A. 1991. Bilateral lesions of the bed nucleus of the accessory olfactory tract facilitate maternal behavior in virgin female rats. Physiol Behav 50: 67-71.

De Vries AC, De Vries MB, Taymans S, Carter CS. 1995. Modulation of pair bonding in female prairie voles. *Microtus ochrogaster*) by corticosterone. Proc Natl Acad Sci USA 92: 7744-7748.

De Vries AC, De Vries MB, Taymans SE, Carter CS. 1996. The effects of stress on social preferences are sexually dimorphic in prairie voles. Proc Natl Acad Sci USA 93: 11980-11984.

De Vries AC, Glasper ER, Detillion CE. 2003. Social modulation of stress responses. Physiol Behav 79: 399-407.

De Vries AC, Guptaa T, Cardillo S, Cho M, Carter CS. 2002. Corticotropin-releasing factor induces social preferences in male prairie voles. Psychoneuroendocrinology 27: 705-714.

Dluzen DE, Muraoka S, Engelmann M, Ebner K, Landgraf R. 2000. Oxytocin induces preservation of social recognition in male rats by activating alpha-adrenoceptors of the olfactory bulb. Eur J Neurosci 12: 760-766.

Dluzen DE, Muraoka S, Engelmann M, Landgraf R. 1998b. The effects of infusion of arginine vasopressin, oxytocin, or their antagonists into the olfactory bulb upon social recognition responses in male rats. Peptides 19: 999-1005.

Dluzen DE, Muraoka S, Landgraf R. 1998a. Olfactory bulb norepinephrine depletion abolishes vasopressin and oxytocin preservation of social recognition responses in rats. Neurosci Lett 254: 161-164.

Eisenberger NI, Lieberman MD, Williams KD. 2003. Does rejection hurt? An FMRI study of social exclusion. Science 302: 290-292.

Elands J, Beetsma A, Barberis C, de Kloet ER. 1988. Topography of the oxytocin receptor system in rat brain: an autoradiographical study with a selective radioiodinated oxytocin antagonist. J Chem Neuroanat 1: 293-302.

Engelmann M, Landgraf R. 1994. Microdialysis administration of vasopressin into the septum improves social recognition in Brattleboro rats. Physiol Behav 55: 145-149.

Fahrbach SE, Morrell JI, Pfaff DW. 1984. Oxytocin induction of short-latency maternal behavior in nulliparous, estrogen-primed female rats. Horm Behav 18: 267-286.

Fahrbach SE, Morrell JI, Pfaff DW. 1985. Possible role for endogenous oxytocin in estrogen-facilitated maternal behavior in rats. Neuroendocrinology 40: 526-532.

Fahrbach SE, Morrell JI, Pfaff DW. 1986. Effect of varying the duration of pre-test cage habituation on oxytocin induction of short-latency maternal behavior. Physiol Behav 37: 135-139.

Ferguson JN, Aldag JM, Insel TR, Young LJ. 2001. Oxytocin in the medial amygdala is essential for social recognition in the mouse. J Neurosci 21: 8278-8285.

Ferguson JN, Young LJ, Hearn EF, Matzuk MM, Insel TR, et al. 2000. Social amnesia in mice lacking the oxytocin gene. Nat Genet 25: 284-288.

Field T. 2002. Preterm infant massage therapy studies: an American approach. Semin Neonatol 7: 487-494.

Fleming A, Vaccarino F, Tambosso L, Chee P. 1979. Vomeronasal and olfactory system modulation of maternal behavior in the rat. Science 203: 372-374.

Fleming AS, Vaccarino F, Luebke C. 1980. Amygdaloid inhibition of maternal behavior in the nulliparous female rat. Physiol Behav 25: 731-743.

Fowler CD, Liu Y, Ouimet C, Wang Z. 2002. The effects of social environment on adult neurogenesis in the female prairie vole. J Neurobiol 51: 115-128.

Francis D, Diorio J, Liu D, Meaney MJ. 1999. Nongenomic transmission across generations of maternal behavior and stress responses in the rat. Science 286: 1155-1158.

Francis DD, Champagne FC, Meaney MJ. 2000. Variations in maternal behaviour are associated with differences in oxytocin receptor levels in the rat. J Neuroendocrinol 12: 1145-1148.

Francis DD, Young LJ, Meaney MJ, Insel TR. 2002. Naturally occurring differences in maternal care are associated with the expression of oxytocin and vasopressin (V1a) receptors: gender differences. J Neuroendocrinol 14: 349-353.

Freund-Mercier MJ, Dietl MM, Stoeckel ME, Palacios JM, Richard P. 1988b. Quantitative autoradiographic mapping of neurohypophysial hormone binding sites in the rat forebrain and pituitary gland–II. Comparative study on the Long-Evans and Brattleboro strains. Neuroscience 26: 273-281.

Freund-Mercier MJ, Stoeckel ME, Dietl MM, Palacios JM, Richard P. 1988a. Quantitative autoradiographic mapping of neurohypophysial hormone binding sites in the rat forebrain and pituitary gland–I. Characterization of different types of binding sites and their distribution in the Long-Evans strain. Neuroscience 26: 261-272.

Freund-Mercier MJ, Stoeckel ME, Palacios JM, Pazos A, Reichhart JM, et al. 1987. Pharmacological characteristics and anatomical distribution of [3H]oxytocin-binding sites in the Wistar rat brain studied by autoradiography. Neuroscience 20: 599-614.

Frith C. 2003. What do imaging studies tell us about the neural basis of autism? Novartis Found Symp 251: 149-166; discussion 166–176, 281–197.

Gauthier I, Tarr MJ, Moylan J, Skudlarski P, Gore JC, et al. 2000. The fusiform "face area" is part of a network that processes faces at the individual level. J Cogn Neurosci 12: 495-504.

Germain RN. 1994. MHC-dependent antigen processing and peptide presentation: providing ligands for T lymphocyte activation. Cell 76: 287-299.

Getz LL, Carter CS, Gavish L. 1981. The mating system of the prairie vole, Microtus ochrogaster: field and laboratory evidence for pair-bonding. Behav Ecol Sociobiol 8: 189-194.

Gingrich B, Liu Y, Cascio C, Wang Z, Insel TR. 2000. Dopamine D2 receptors in the nucleus accumbens are important for social attachment in female prairie voles (*Microtus ochrogaster*). Behav Neurosci 114: 173-183.

Gray JM, Karow DS, Lu H, Chang AJ, Chang JS, et al. 2004. Oxygen sensation and social feeding mediated by a *C.elegans* guanylate cyclase homologue. Nature 430: 317-322.

Green L, Fein D, Modahl C, Feinstein C, Waterhouse L, et al. 2001. Oxytocin and autistic disorder: alterations in peptide forms. Biol Psychiatry 50: 609-613.

Hammock EA, Young LJ. 2002. Variation in the vasopressin V1a receptor promoter and expression: implications for inter- and intraspecific variation in social behaviour. Eur J Neurosci 16: 399-402.

Hammock EA, Young LJ. 2004. Functional microsatellite polymorphism associated with divergent social structure in vole species. Mol Biol Evol 21: 1057-1063.

Hansen S, Harthon C, Wallin E, Lofberg L, Svensson K. 1991. The effects of 6-OHDA-induced dopamine depletions in the ventral or dorsal striatum on maternal and sexual behavior in the female rat. Pharmacol Biochem Behav 39: 71-77.

Heinrichs M, Baumgartner T, Kirschbaum C, Ehlert U. 2003. Social support and oxytocin interact to suppress cortisol and subjective responses to psychosocial stress. Biol Psychiatry 54: 1389-1398.

Hennessy MB, Mendoza SP, Mason WA, Moberg GP. 1995. Endocrine sensitivity to novelty in squirrel monkeys and titi monkeys: species differences in characteristic modes of responding to the environment. Physiol Behav 57: 331-338.

Hlinak Z. 1993. Social recognition in ovariectomized and estradiol-treated female rats. Horm Behav 27: 159-166.

Hoffman KA, Mendoza SP, Hennessy MB, Mason WA. 1995. Responses of infant titi monkeys, *Callicebus moloch*, to removal of one or both parents: evidence for paternal attachment. Dev Psychobiol 28: 399-407.

Holstege G, Georgiadis JR, Paans AMJ, Meiners LC, Graaf van der FHCE, et al. 2003. Brain Activation during Human Male Ejaculation. J Neurosci 23: 9185-9193.

Insel TR, Harbaugh CR. 1989. Lesions of the hypothalamic paraventricular nucleus disrupt the initiation of maternal behavior. Physiol Behav 45: 1033-1041.

Insel TR, Hulihan TJ. 1995. A gender-specific mechanism for pair bonding: oxytocin and partner preference formation in monogamous voles. Behav Neurosci 109: 782-789.

Insel TR, Shapiro LE. 1992. Oxytocin receptor distribution reflects social organization in monogamous and polygamous voles. Proc Natl Acad Sci USA 89: 5981-5985.

Insel TR, Wang ZX, Ferris CF. 1994. Patterns of brain vasopressin receptor distribution associated with social organization in microtine rodents. J Neurosci 14: 5381-5392.

Insel TR, Winslow JT. 1991. Central administration of oxytocin modulates the infant rat's response to social isolation. Eur J Pharmacol 203: 149-152.

Insel TR, Preston S, Winslow JT. 1995. Mating in the monogamous male: behavioral consequences. Physiol Behav 57: 615-627.

Kanwisher N, McDermott J, Chun MM. 1997. The fusiform face area: a module in human extrastriate cortex specialized for face perception. J Neurosci 17: 4302-4311.

Kavaliers M, Agmo A, Choleris E, Gustafsson JA, Korach KS, et al. 2004b. Oxytocin and estrogen receptor alpha and beta knockout mice provide discriminably different odor cues in behavioral assays. Genes Brain Behav 3:189-195.

Kavaliers M, Choleris E, Agmo A, Pfaff DW. 2004a. Olfactory-mediated parasite recognition and avoidance: linking genes to behavior. Horm Behav 46: 272-283.

Kavaliers M, Colwell DD, Choleris E, Agmo A, Muglia LJ, et al. 2003. Impaired discrimination of and aversion to parasitized male odors by female oxytocin knockout mice. Genes Brain Behav 2: 220-230.

Kendrick KM, Da Costa AP, Broad KD, Ohkura S, Guevara R, et al. 1997a. Neural control of maternal behaviour and olfactory recognition of offspring. Brain Res Bull 44: 383-395.

Kendrick KM, Guevara-Guzman R, Zorrilla J, Hinton MR, Broad KD, et al. 1997b. Formation of olfactory memories mediated by nitric oxide. Nature 388: 670-674.

Kendrick KM, Keverne EB, Baldwin BA. 1987. Intracerebroventricular oxytocin stimulates maternal behaviour in the sheep. Neuroendocrinology 46: 56-61.

Kendrick KM, Keverne EB, Baldwin BA, Sharman DF. 1986. Cerebrospinal fluid levels of acetylcholinesterase, monoamines and oxytocin during labour, parturition, vaginocervical stimulation, lamb separation and suckling in sheep. Neuroendocrinology 44: 149-156.

Kendrick KM, Keverne EB, Chapman C, Baldwin BA. 1988a. Microdialysis measurement of oxytocin, aspartate, gamma-aminobutyric acid and glutamate release from the olfactory bulb of the sheep during vaginocervical stimulation. Brain Res 442: 171-174.

Kendrick KM, Keverne EB, Chapman C, Baldwin BA. 1988b. Intracranial dialysis measurement of oxytocin, monoamine and uric acid release from the olfactory bulb and substantia nigra of sheep during parturition, suckling, separation from lambs and eating. Brain Res 439: 1-10.

Kendrick KM, Keverne EB, Hinton MR, Goode JA. 1992b. Oxytocin, amino acid and monoamine release in the region of the medial preoptic area and bed nucleus of the stria terminalis of the sheep during parturition and suckling. Brain Res 569: 199-209.

Kendrick KM, Levy F, Keverne EB. 1992a. Changes in the sensory processing of olfactory signals induced by birth in sleep. Science 256: 833-836.

Keverne EB, Levy F, Guevara-Guzman R, Kendrick KM. 1993. Influence of birth and maternal experience on olfactory bulb neurotransmitter release. Neuroscience 56: 557-565.

Kim SJ, Young LJ, Gonen D, Veenstra-Vander Weele J, Courchesne R, et al. 2002. Transmission disequilibrium testing of arginine vasopressin receptor 1A (AVPR1A) polymorphisms in autism. Mol Psychiatry 7: 503-507.

Kirkpatrick B, Williams JR, Slotnick BM, Carter CS. 1994. Olfactory bulbectomy decreases social behavior in male prairie voles (*M. ochrogaster*). Physiol Behav 55: 885-889.

Kogan JH, Frankland PW, Silva AJ. 2000. Long-term memory underlying hippocampus-dependent social recognition in mice. Hippocampus 10: 47-56.

Landgraf R, Frank E, Aldag JM, Neumann ID, Sharer CA, et al. 2003. Viral vector-mediated gene transfer of the vole V1a vasopressin receptor in the rat septum: improved social discrimination and active social behaviour. Eur J Neurosci 18: 403-411.

Landgraf R, Gerstberger R, Montkowski A, Probst JC, Wotjak CT, et al. 1995. V1 vasopressin receptor antisense oligodeoxynucleotide into septum reduces vasopressin binding, social discrimination abilities, and anxiety-related behavior in rats. J Neurosci 15: 4250-4258.

Le Moal M, Dantzer R, Michaud B, Koob GF. 1987. Centrally injected arginine vasopressin (AVP) facilitates social memory in rats. Neurosci Lett 77: 353-359.

Leckman JF, Herman AE. 2002. Maternal behavior and developmental psychopathology. Biol Psychiatry 51: 27-43.

Levine S, Lyons DM, Schatzberg AF. 1997. Psychobiological consequences of social relationships. Ann NY Acad Sci 807: 210-218.

Levy F, Guevara-Guzman R, Hinton MR, Kendrick KM, Keverne EB. 1993. Effects of parturition and maternal experience on noradrenaline and acetylcholine release in the olfactory bulb of sheep. Behav Neurosci 107: 662-668.

Levy F, Kendrick KM, Goode JA, Guevara-Guzman R, Keverne EB. 1995. Oxytocin and vasopressin release in the olfactory bulb of parturient ewes: changes with maternal experience and effects on acetylcholine, gamma-aminobutyric acid, glutamate and noradrenaline release. Brain Res 669: 197-206.

Levy F, Kendrick KM, Keverne EB, Piketty V, Poindron P. 1992. Intracerebral oxytocin is important for the onset of maternal behavior in inexperienced ewes delivered under peridural anesthesia. Behav Neurosci 106: 427-432.

Levy F, Poindron P, Le Neindre P. 1983. Attraction and repulsion by amniotic fluids and their olfactory control in the ewe around parturition. Physiol Behav 31: 687-692.

Li L, Keverne EB, Aparicio SA, Ishino F, Barton SC, et al. 1999. Regulation of maternal behavior and offspring growth by paternally expressed Peg3. Science 284: 330-333.

Lim MM, Hammock EA, Young LJ. 2004b. The role of vasopressin in the genetic and neural regulation of monogamy. J Neuroendocrinol 16: 325-332.

Lim MM, Murphy AZ, Young LJ. 2004a. Ventral striatopallidal oxytocin and vasopressin V1a receptors in the monogamous prairie vole (*Microtus ochrogaster*). J Comp Neurol 468: 555-570.

Lim MM, Wang Z, Olazabal DE, Ren X, Terwilliger EF, et al. 2004c. Enhanced partner preference in a promiscuous species by manipulating the expression of a single gene. Nature 429: 754-757.

Lim MM, Young LJ. 2004. Vasopressin-dependent neural circuits underlying pair bond formation in the monogamous prairie vole. Neuroscience 125: 35-45.

Lin SH, Miyata S, Weng W, Matsunaga W, Ichikawa J, et al. 1998. Comparison of the expression of two immediate early gene proteins, FosB and Fos in the rat preoptic area, hypothalamus and brainstem during pregnancy, parturition and lactation. Neurosci Res 32: 333-341.

Liu Y, Wang ZX. 2003. Nucleus accumbens oxytocin and dopamine interact to regulate pair bond formation in female prairie voles. Neuroscience 121: 537-544.

Lonstein JS, De Vries GJ. 1999. Sex differences in the parental behaviour of adult virgin prairie voles: independence from gonadal hormones and vasopressin. J Neuroendocrinol 11: 441-449.

Lorberbaum JP, Newman JD, Dubno JR, Horwitz AR, Nahas Z, et al. 1999 Feasibility of using fMRI to study mothers responding to infant cries. Depress Anxiety 10: 99-104.

Loup F, Tribollet E, Dubois-Dauphin M, Dreifuss JJ. 1991. Localization of high-affinity binding sites for oxytocin and vasopressin in the human brain. An autoradiographic study. Brain Res 555: 220-232.

Loup F, Tribollet E, Dubois-Dauphin M, Pizzolato G, Dreifuss JJ. 1989. Localization of oxytocin binding sites in the human brainstem and upper spinal cord: an autoradiographic study. Brain Res 500: 223-230.

Lucas BK, Ormandy CJ, Binart N, Bridges RS, Kelly PA. 1998. Null mutation of the prolactin receptor gene produces a defect in maternal behavior. Endocrinology 139: 4102-4107.

Lyons DM, Ha CM, Levine S. 1995. Social effects and circadian rhythms in squirrel monkey pituitary-adrenal activity. Horm Behav 29: 177-190.

MacLean PD, Newman JD. 1988. Role of midline frontolimbic cortex in production of the isolation call of squirrel monkeys. Brain Res 450: 111-123.

Maestripieri D, Zehr JL. 1998. Maternal responsiveness increases during pregnancy and after estrogen treatment in macaques. Horm Behav 34: 223-230.

Manning CJ, Wakeland EK, Potts WK. 1992. Communal nesting patterns in mice implicate MHC genes in kin recognition. Nature 360: 581-583.

McCarthy MM. 1990. Oxytocin inhibits infanticide in female house mice (*Mus domesticus*). Horm Behav 24: 365-375.

Mendoza SP, Mason WA. 1986. Contrasting responses to intruders and to involuntary separation by monogamous and polygynous New World monkeys. Physiol Behav 38: 795-801.

Mendoza SP, Mason WA. 1997. Attachment relationships in New World primates. Ann N Y Acad Sci 807: 203-209.

Meredith M, Westberry JM. 2004. Distinctive responses in the medial amygdala to same-species and different-species pheromones. J Neurosci 24: 5719-5725.

Modahl C, Green L, Fein D, Morris M, Waterhouse L, et al. 1998. Plasma oxytocin levels in autistic children. Biol Psychiatry 43: 270-277.

Moles A, Kieffer BL, D'Amato FR. 2004. Deficit in attachment behavior in mice lacking the mu-opioid receptor gene. Science 304: 1983-1986.

Murphy MR, MacLean PD, Hamilton SC. 1981. Species-typical behavior of hamsters deprived from birth of the neocortex. Science 213: 459-461.

Murphy MR, Seckl JR, Burton S, Checkley SA, Lightman SL. 1987. Changes in oxytocin and vasopressin secretion during sexual activity in men. J Clin Endocrinol Metab 65: 738-741.

Nelson E, Panksepp J. 1996. Oxytocin mediates acquisition of maternally associated odor preferences in preweanling rat pups. Behav Neurosci 110: 583-592.

Nelson EE, Panksepp J. 1998. Brain substrates of infant-mother attachment: contributions of opioids, oxytocin, and norepinephrine. Neurosci Biobehav Rev 22: 437-452.

Nishimori K, Young LJ, Guo Q, Wang Z, Insel TR, et al. 1996. Oxytocin is required for nursing but is not essential for parturition or reproductive behavior. Proc Natl Acad Sci USA 93: 11699-11704.

Numan M. 1994. A neural circuitry analysis of maternal behavior in the rat. Acta Paediatr Suppl 397: 19-28.

Numan M, Corodimas KP. 1985. The effects of paraventricular hypothalamic lesions on maternal behavior in rats. Physiol Behav 35: 417-425.

Numan M, Rosenblatt JS, Komisaruk BR. 1977. Medial preoptic area and onset of maternal behavior in the rat. J Comp Physiol Psychol 91: 146-164.

Ogawa S, Eng V, Taylor J, Lubahn DB, Korach KS, et al. 1998. Roles of estrogen receptor-alpha gene expression in reproduction-related behaviors in female mice. Endocrinology 139: 5070-5081.

Olazabal DE, Ross HE, Nishimori K, Young LJ. 2004. Female oxytocin receptor knockout mice show a deficit in 'spontaneous' maternal behavior. Program No. 437.6. In: 2004 Abstract Viewer/Itinerary Planner: Washington, DC: Society for Neuroscience, 2004. Online.

Onitsuka T, Shenton ME, Kasai K, Nestor PG, Toner SK, et al. 2003. Fusiform gyrus volume reduction and facial recognition in chronic schizophrenia. Arch Gen Psychiatry 60: 349-355.

Osborne KA, Robichon A, Burgess E, Butland S, Shaw RA, et al. 1997. Natural behavior polymorphism due to a cGMP-dependent protein kinase of *Drosophila*. Science 277: 834-836.

Palmen SJ, Van Engeland H. 2004. Review on structural neuroimaging findings in autism. J Neural Transm 111: 903-929.

Palmen SJ, Van Engeland H, Hof PR, Schmitz C. 2004. Neuropathological findings in autism. Brain Dec 2004; 127(pt 12): 2572-83.

Panksepp J. 2003. Neuroscience. Feeling the pain of social loss. Science 302: 237-239.

Parker KJ, Phillips, KM, Lee TM. 2001. Development of selective partner preferences in captive male and female meadow voles, *Microtus pennsylvanicus*. Animal Behaviour 61: 1217-1226.

Patisaul HB, Scordalakes EM, Young LJ, Rissman EF. 2003. Oxytocin, but not oxytocin receptor, is regulated by oestrogen receptor beta in the female mouse hypothalamus. J Neuroendocrinol 15: 787-793.

Pedersen CA, Ascher JA, Monroe YL, Prange AJ Jr. 1982. Oxytocin induces maternal behavior in virgin female rats. Science 216: 648-650.

Pedersen CA, Caldwell JD, Walker C, Ayers G, Mason GA. 1994. Oxytocin activates the postpartum onset of rat maternal behavior in the ventral tegmental and medial preoptic areas. Behav Neurosci 108: 1163-1171.

Pedersen CA, Prange AJ Jr. 1979. Induction of maternal behavior in virgin rats after intracerebroventricular administration of oxytocin. Proc Natl Acad Sci USA 76: 6661-6665.

Pedersen CA,Vadlamudi S,Boccia ML,Amico JA.2004. Quantitative measurements reveal pup-licking deficits in the maternal behavior of nulliparous oxytocin knockout mice. Program No. 440.2. In: 2004 Abstract Viewer/Itinerary Planner.: Washington, DC: Society for Neuroscience, 2004. Online.

Penn D, Potts W. 1998. MHC-disassortative mating preferences reversed by cross-fostering. Proc R Soc Lond B Biol Sci 265: 1299-1306.

Pfaff D, Keiner M. 1973. Atlas of estradiol-concentrating cells in the central nervous system of the female rat. J Comp Neurol 151: 121-158.

Phelps SM, Young LJ. 2003. Extraordinary diversity in vasopressin (V1a) receptor distributions among wild prairie voles (*Microtus ochrogaster*): patterns of variation and covariation. J Comp Neurol 466: 564-576.

Pitkow LJ, Sharer CA, Ren X, Insel TR, Terwilliger EF, et al. 2001. Facilitation of affiliation and pair-bond formation by vasopressin receptor gene transfer into the ventral forebrain of a monogamous vole. J Neurosci 21: 7392-7396.

Popik P, van Ree JM. 1991. Oxytocin but not vasopressin facilitates social recognition following injection into the medial preoptic area of the rat brain. Eur Neuropsychopharmacol 1: 555-560.

Popik P, Van Ree JM. 1992. Long-term facilitation of social recognition in rats by vasopressin related peptides: a structure-activity study. Life Sci 50: 567-572.

Potts WK, Manning CJ, Wakeland EK. 1991. Mating patterns in seminatural populations of mice influenced by MHC genotype. Nature 352: 619-621.

Rainey PB, Rainey K. 2003. Evolution of cooperation and conflict in experimental bacterial populations. Nature 425: 72-74.

Rankin CH. 2002. From gene to identified neuron to behaviour in *Caenorhabditis elegans*. Nat Rev Genet 3: 622-630.

Ren P, Lim CS, Johnsen R, Albert PS, Pilgrim D, et al. 1996. Control of *C. elegans* larval development by neuronal expression of a TGF-beta homolog. Science 274: 1389-1391.

Riddle O, Bates RW, Lahr EL. 1935a. Maternal behavior induced in virgin rats by prolactin. Proc Soc Exp Biol Med 32: 730-734.

Riddle O, Bates RW, Lahr EL. 1935b. Prolactin induces broodiness in fowl. Am J Physiol 111: 352-360.

Robinson GE. 2002. Genomics and integrative analyses of division of labor in honeybee colonies. The American Naturalist 160: s160-s172.

Rosenblatt JS. 1967. Nonhormonal basis of maternal behavior in the rat. Science 156: 1512-1514.

Rosenblatt JS, Mayer AD, Giordano AL. 1988. Hormonal basis during pregnancy for the onset of maternal behavior in the rat. Psychoneuroendocrinology 13: 29-46.

Rosenblatt JS, Siegel HI. 1975. Hysterectomy-induced maternal behavior during pregnancy in the rat. J Comp Physiol Psychol 89: 685-700.

Rosenblum LA, Smith EL, Altemus M, Scharf BA, Owens MJ, et al. 2002. Differing concentrations of corticotropin-releasing factor and oxytocin in the cerebrospinal fluid of bonnet and pigtail macaques. Psychoneuroendocrinology 27: 651-660.

Rosser AE, Keverne EB. 1985. The importance of central noradrenergic neurones in the formation of an olfactory memory in the prevention of pregnancy block. Neuroscience 15: 1141-1147.

Schackwitz WS, Inoue T, Thomas JH. 1996. Chemosensory neurons function in parallel to mediate a pheromone response in *C. elegans*. Neuron 17: 719-728.

Schradin C, Anzenberger G. 1999. Prolactin, the Hormone of Paternity. News Physiol Sci 14: 223-231.

Schradin C, Reeder DM, Mendoza SP, Anzenberger G. 2003. Prolactin and paternal care: comparison of three species of monogamous new world monkeys (*Callicebus cupreus, Callithrix jacchus*, and *Callimico goeldii*). J Comp Psychol 117: 166-175.

Schultz RT, Gauthier I, Klin A, Fulbright RK, Anderson AW, et al. 2000. Abnormal ventral temporal cortical activity during face discrimination among individuals with autism and Asperger syndrome. Arch Gen Psychiatry 57: 331-340.

Schumacher M, Coirini H, Flanagan LM, Frankfurt M, Pfaff DW, et al. 1992. Ovarian steroid modulation of oxytocin receptor binding in the ventromedial hypothalamus. Ann NY Acad Sci 652: 374-386.

Schumacher M, Coirini H, Johnson AE, Flanagan LM, Frankfurt M, et al. 1993. The oxytocin receptor: a target for steroid hormones. Regul Pept 45: 115-119.

Seifritz E, Esposito F, Neuhoff JG, Luthi A, Mustovic H, et al. 2003. Differential sex-independent amygdala response to infant crying and laughing in parents versus nonparents. Biol Psychiatry 54: 1367-1375.

Shang Y, Dluzen DE. 2001. Nisoxetine infusion into the olfactory bulb enhances the capacity for male rats to identify conspecifics. Neuroscience 104: 957-964.

Shapiro LE, Insel TR. 1990. Infant's response to social separation reflects adult differences in affiliative behavior: a comparative developmental study in prairie and montane voles. Dev Psychobiol 23: 375-393.

Shingo T, Gregg C, Enwere E, Fujikawa H, Hassam R, et al. 2003. Pregnancy-stimulated neurogenesis in the adult female forebrain mediated by prolactin. Science 299: 117-120.

Smotherman WP, Bell RW, Starzec J, Elias J, Zachman TA. 1974. Maternal responses to infant vocalizations and olfactory cues in rats and mice. Behav Biol 12: 55-66.

Sofroniew MV. 1983. Vasopressin and Oxytocin in the Mammalian Brain and Spinal-Cord. Trends in Neurosciences 6: 467-472.

Sogolow SR. 1966. An historical review of the use of oxytocin prior to delivery. Obstet Gynecol Surv 21: 155-172.

Sokolowski MB. 1980. Foraging strategies of *Drosophila melanogaster*: a chromosomal analysis. Behav Genet 10: 291-302.

Sokolowski MB. 2002. Neurobiology: social eating for stress. Nature 419: 893-894.

Sokolowski MB. 2003. NPY and the regulation of behavioral development. Neuron 39: 6-8.

Stern JM, Lonstein JS. 2001. Neural mediation of nursing and related maternal behaviors. Prog Brain Res 133: 263-278.

Storey AE, Walsh CJ, Quinton RL, Wynne-Edwards KE. 2000. Hormonal correlates of paternal responsiveness in new and expectant fathers. 21: 79–95.

Strassmann JE, Zhu Y, Queller DC. 2000. Altruism and social cheating in the social amoeba Dictyostelium discoideum. Nature 408: 965-967.

Sullivan R, Wilson D, Leon M. 1989. Norepinephrine and learning-induced plasticity in infant rat olfactory system. J Neurosci 9: 3998-4006.

Sullivan RM. 2003. Developing a sense of safety: the neurobiology of neonatal attachment. Ann NY Acad Sci 1008: 122-131.

Sullivan RM, Stackenwalt G, Nasr F, Lemon C, Wilson DA. 2000. Association of an odor with activation of olfactory bulb noradrenergic [beta]-receptors or locus Coeruleus stimulation is sufficient to produce learned approach responses to that odor in neonatal rats. Behavioral Neuroscience 114: 957-962.

Takayanagi Y, Yoshida M, Bielsky IF, Ross HE, Kawamata M, Onaka T, Yanagisawa T, Kimura T, Matzuk MM, Young LJ, Nishimori K. Pervasive social deficits, but normal parturition, in oxytocin receptor-deficient mice. Proc Natl Acad Sci USA (2005) Nov 102(44): 16096-101.

Tannenbaum PL, Wallen K. 1997. Sexually initiated affiliation facilitates rhesus monkey integration. Ann NY Acad Sci 807: 578-582.

Thibonnier M, Graves MK, Wagner MS, Chatelain N, Soubrier F, et al. 2000. Study of V(1)-vascular vasopressin receptor gene microsatellite polymorphisms in human essential hypertension. J Mol Cell Cardiol 32: 557-564.

Turner RA, Altemus M, Enos T, Cooper B, McGuinness T. 1999. Preliminary research on plasma oxytocin in normal cycling women: investigating emotion and interpersonal distress. Psychiatry 62: 97-113.

Uvnas-Moberg K. 1997. Physiological and endocrine effects of social contact. Ann NY Acad Sci 807: 146-163.

van Leengoed E, Kerker E, Swanson HH. 1987. Inhibition of post-partum maternal behaviour in the rat by injecting an oxytocin antagonist into the cerebral ventricles. J Endocrinol 112: 275-282.

van Leeuwen FW, van Heerikhuize J, Meulen van der G, Wolters P. 1985. Light microscopic autoradiographic localization of [3H]oxytocin binding sites in the rat brain, pituitary and mammary gland. Brain Res 359: 320-325.

Velicer GJ, Kroos L, Lenski RE. 2000. Developmental cheating in the social bacterium *Myxococcus xanthus*. Nature 404: 598-601.

Wallen K, Tannenbaum PL. 1997. Hormonal modulation of sexual behavior and affiliation in rhesus monkeys. Ann NY Acad Sci 807: 185-202.

Wang Z. 1995. Species differences in the vasopressin-immunoreactive pathways in the bed nucleus of the stria terminalis and medial amygdaloid nucleus in prairie voles (*Microtus ochrogaster*) and meadow voles (*Microtus pennsylvanicus*). Behav Neurosci 109: 305-311.

Wang Z, De Vries GJ. 1993. Testosterone effects on paternal behavior and vasopressin immunoreactive projections in prairie voles (*Microtus ochrogaster*). Brain Res 631: 156-160.

Wang Z, Ferris CF, De Vries GJ. 1994a. Role of septal vasopressin innervation in paternal behavior in prairie voles (*Microtus ochrogaster*). Proc Natl Acad Sci USA 91: 400-404.

Wang Z, Smith W, Major DE, De Vries GJ. 1994b. Sex and species differences in the effects of cohabitation on vasopressin messenger RNA expression in the bed nucleus of the stria terminalis in prairie voles (*Microtus ochrogaster*) and meadow voles (*Microtus pennsylvanicus*). Brain Res 650: 212-218.

Wang Z, Young LJ, Liu Y, Insel TR. 1997. Species differences in vasopressin receptor binding are evident early in development: comparative anatomic studies in prairie and montane voles. J Comp Neurol 378: 535-546.

Wang Z, Zhou L, Hulihan TJ, Insel TR. 1996. Immunoreactivity of central vasopressin and oxytocin pathways in microtine rodents: a quantitative comparative study. J Comp Neurol 366: 726-737.

Wassink TH, Piven J, Vieland VJ, Pietila J, Goedken RJ, et al. 2004. Examination of AVPR1a as an autism susceptibility gene. Mol Psychiatry 9(10): 968-72.

Weaver IC, Cervoni N, Champagne FA, D'Alessio AC, Sharma S, et al. 2004. Epigenetic programming by maternal behavior. Nat Neurosci 7: 847-854.

Wersinger SR, Ginns EI, O'Carroll AM, Lolait SJ, Young WS 3rd. 2002. Vasopressin V1b receptor knockout reduces aggressive behavior in male mice. Mol Psychiatry 7: 975-984.

Whitfield CW, Cziko AM, Robinson GE. 2003. Gene expression profiles in the brain predict behavior in individual honey bees. Science 302: 296-299.

Wikstrom S, Gunnarsson T, Nordin C.. 2003. Tactile stimulus and neurohormonal response: a pilot study. Int J Neurosci 113: 787-793.

Williams JR, Carter CS, Insel T. 1992a. Partner preference development in female prairie voles is facilitated by mating or the central infusion of oxytocin. Ann NY Acad Sci 652: 487-489.

Williams JR, Insel TR, Harbaugh CR, Carter CS. 1994. Oxytocin administered centrally facilitates formation of a partner preference in female prairie voles (*Microtus ochrogaster*). J Neuroendocrinol 6: 247-250.

Williams JR, Slotnick BM, Kirkpatrick BW, Carter CS. 1992b. Olfactory bulb removal affects partner preference development and estrus induction in female prairie voles. Physiol Behav 52: 635-639.

Wilson EO. 2000 Sociobiology: the new synthesis; 25th anniversary edition. In. Cambridge, Mass: Belknap Press of Harvard University Press.

Winslow JT, Hastings N, Carter CS, Harbaugh CR, Insel TR. 1993. A role for central vasopressin in pair bonding in monogamous prairie voles. Nature 365: 545-548.

Winslow JT, Hearn EF, Ferguson J, Young LJ, Matzuk MM, et al. 2000. Infant vocalization, adult aggression, and fear behavior of an oxytocin null mutant mouse. Horm Behav 37: 145-155.

Winslow JT, Insel TR. 1992. Social and environmental determinants of centrally administered oxytocin effects on male squirrel monkey behavior. Ann NY Acad Sci 652: 452-455.

Winslow JT, Noble PL, Lyons CK, Sterk SM, Insel TR. 2003. Rearing effects on cerebrospinal fluid oxytocin concentration and social buffering in rhesus monkeys. Neuropsychopharmacology 28: 910-918.

Witt DM, Carter CS, Walton DM. 1990. Central and peripheral effects of oxytocin administration in prairie voles (*Microtus ochrogaster*). Pharmacol Biochem Behav 37: 63-69.

Witt DM, Winslow JT, Insel TR. 1992. Enhanced social interactions in rats following chronic, centrally infused oxytocin. Pharmacol Biochem Behav 43: 855-861.

Wu Q, Wen T, Lee G, Park JH, Cai HN, et al. 2003. Developmental control of foraging and social behavior by the *Drosophila* neuropeptide Y-like system. Neuron 39: 147-161.

Yamazaki K, Beauchamp GK, Curran M, Bard J, Boyse EA. 2000. Parent-progeny recognition as a function of MHC odortype identity. Proc Natl Acad Sci USA 97: 10500-10502.

Yamazaki K, Beauchamp GK, Kupniewski D, Bard J, Thomas L, et al. 1988. Familial imprinting determines H-2 selective mating preferences. Science 240: 1331-1332.

Yoshimura R, Kiyama H, Kimura T, Araki T, Maeno H, et al. 1993. Localization of oxytocin receptor messenger ribonucleic acid in the rat brain. Endocrinology 133: 1239-1246.

Young LJ, Huot B, Nilsen R, Wang Z, Insel TR. 1996. Species differences in central oxytocin receptor gene expression: comparative analysis of promoter sequences. J Neuroendocrinol 8: 777-783.

Young LJ, Lim MM, Gingrich B, Insel TR. 2001. Cellular mechanisms of social attachment. Horm Behav 40: 133-138.

Young LJ, Muns S, Wang Z, Insel TR. 1997b. Changes in oxytocin receptor mRNA in rat brain during pregnancy and the effects of estrogen and interleukin-6. J Neuroendocrinol 9: 859-865.

Young LJ, Nilsen R, Waymire KG, Mac Gregor GR, Insel TR. 1999. Increased affiliative response to vasopressin in mice expressing the V1a receptor from a monogamous vole. Nature 400: 766-768.

Young LJ, Wang Z. 2004. The neurobiology of pair bonding. Nat Neurosci 7: 1048-1054.

Young LJ, Wang Z, Donaldson R, Rissman EF. 1998. Estrogen receptor alpha is essential for induction of oxytocin receptor by estrogen. Neuroreport 9: 933-936.

Young LJ, Winslow JT, Nilsen R, Insel TR. 1997a. Species differences in V1a receptor gene expression in monogamous and nonmonogamous voles: behavioral consequences. Behav Neurosci 111: 599-605.

Yu GZ, Kaba H, Okutani F, Takahashi S, Higuchi T. 1996a. The olfactory bulb: a critical site of action for oxytocin in the induction of maternal behaviour in the rat. Neuroscience 72: 1083-1088.

Yu GZ, Kaba H, Okutani F, Takahashi S, Higuchi T. et al. 1996b. The action of oxytocin originating in the hypothalamic paraventricular nucleus on mitral and granule cells in the rat main olfactory bulb. Neuroscience 72: 1073-1082.

Ziegler TE. 2000. Hormones associated with nonmaternal infant care: a review of mammalian and avian studies. Folia Primatol (Basel) 71: 6-21.

7 Neurochemistry and Molecular Neurobiology of Aggressive Behavior

K. A. Miczek · S. P. Faccidomo · E. W. Fish · J. F. DeBold

Abstract: The molecular events through which social experiences shape future aggressive acts are beginning to be deciphered. The brain serotonin system is by far the major focus of neurobiological inquiries into the mechanisms mediating certain types of aggressive behavior. Its significant role in impulsive, hostile, antisocial and intensely violent outbursts has been supported by several findings including: deficient levels of 5-HT and its acid metabolite; polymorphisms in the genes coding for its metabolic enzymes; mutations in the genes for some of its receptor and transporter molecules; and altered responses to pharmacological challenges of 5-HT agonists. The reciprocal interactions of serotonin with other amines, acids, peptides and steroids provide further evidence that this indolamine mediates impulsive aggressive behavior. Afferent and efferent projections from dopamine and norepinephrine cells to serotonin cells as well as serotonergic presynaptic modulation of catecholaminergic transmission are the basis for serotonin's impact on sensory, integrative motivational and motor facets of aggressive acts. Similarly, serotonin influences the actions of oxytocin and vasopressin as well as corticotrophin releasing factor and opioid peptides in their ultimate effect on aggressive behavior. New tools are required to determine how experience-dependent phasic changes ("state") in serotonergic terminals in prefrontal cortex, ventral striatum, central and basolateral amygdala, tegmental area and raphé nuclei are superimposed on the serotonin tone in violence-prone individuals ("trait"). Among the urgent research needs are the development of laboratory model systems that capture the transition from the species-typical patterns of dominance or territorial aggression to escalated forms of aggression, since the latter is of primary clinical relevance. The emerging neurobiological research areas focus on: the allosteric modulation of the $GABA_A$ receptor and its subunit composition as a site for the escalation of aggressive behavior; the triggering of aggressive acts by glutamate action, presumably via NMDA receptors, in the lateral hypothalamic area; the predisposing role of specific polymorphisms in enzymes and transporter molecules that are critical for the inactivation of monoamines; and the dissociation between catecholaminergic activities in anticipation of intense aggressive encounters vs. the recovery from such interactions. Important targets for intervention are the long-term neuroadaptive changes in DA, opioid peptides and glutamate in the VTA as they are engendered by repeated aggressive experiences, both in the perpetrator and in the victim of aggression.

1 Opening Vignettes: Lessons from Lobsters and Birds

Integrating and evaluating the current research on the neurobiological mechanisms of aggressive and violent behavior will develop an agenda for urgent research needs. Although the WHO "World report on violence and health" (2002) ignores this topic entirely, understanding the molecular events through which social experiences shape future aggressive acts may bridge the rift between the social and biological science approaches to aggressive behavior—a problem for the public health and the criminal justice systems.

Understanding the chemical characteristics of neural circuits that mediate different kinds of aggressive behavior began in the 1960s by focusing on the prominent role of brain serotonin, and this effort continues even today. The initial findings correlated low levels of this indolamine in the brainstem of isolated mice with their propensity to fight (Giacalone et al., 1968). A decade later, this observation was complemented by the measurements of low levels of the major serotonin metabolite 5-hydroxyindolacetic acid (5-HIAA) in the cerebrospinal fluid of violence-prone marine soldiers (Brown et al., 1979). This finding is often, but not always, replicated in various species, including humans; and it is the basis for linking the serotonin deficiency to individuals prone to engage in the cluster of impulsive, antisocial, hostile, intense violent outbursts relative to instrumental, calculating aggressive acts (Stoff and Vitiello, 1996; Vitiello and Stoff, 1997).

The anatomy of the serotonin cell bodies in the midline raphe region with their efferents and afferents are remarkably similar throughout phylogeny (Parent et al., 1984). Fundamental insights for serotonin into aggressive behavior emerged from measurements in crustaceans that confronted each other. These studies showed that serotonin could exert multiple roles—one role at the neuromuscular junction and another in the central nerve cord (Livingstone et al., 1980; Edwards and Kravitz, 1997; Kravitz and Huber, 2003). Contrary to the serotonin deficiency in many aggressive vertebrates, lobster or crayfish that assume an aggressive stance, threaten, and attack an opponent, are characterized by high levels of serotonin in the central cord. Injections of serotonin and indirect serotonergic agonists into a lobster at rest trigger aggressive displays (Kravitz and Huber, 2003), and these receptor subtype–specific serotonin effects in crayfish depend on the individual's prior experience as subordinate or dominant (Yeh et al., 1996). The studies in invertebrates led to the proposal that serotonin in the ventral cord exerts a gain-setting role in the neural circuit, which integrates aggression-provoking sensory information with signals that initiate the motor acts for aggressive displays (Kravitz, 1988). This integrative function of serotonin at the neural level of crustaceans is instructive, and may serve as model for conceptualizing how serotonin in corticomeso-limbic circuits operate in the mammalian nervous system, albeit in a functionally opposite manner.

Linking the direct measurement of neurotransmitters such as serotonin and its metabolites with behavioral events remains a traditional approach to the study of the neurochemistry of aggression, but is clearly limited by the *correlational* nature of the evidence. A considerably more cogent strategy is based on the endocrine research paradigm that was introduced by Arnold Berthold (1849). When the glandular source of the endogenous candidate substance was removed (i.e., the roosters were castrated), aggressive displays declined. Upon replacement of the glandular material, aggressive behavior recovered (Berthold, 1849). The depletion and repletion effects for testosterone have been replicated in animal species ranging from lizards, snakes, and fish to birds, for whom testosterone is obligatory in order to engage in aggressive behavior (Crews and Moore, 1986; Wingfield et al., 1987). By contrast, in mammalian species, experience can *modulate* the obligatory role of androgens on aggressive behavior, sometimes attenuating the effects of castration (Christie and Barfield, 1979; DeBold and Miczek, 1981). The endocrine research paradigm of depletion and repletion has become a model for studying the canonical neurotransmitters, such as the monoamines, and their role in aggressive behavior (e.g., Kantak et al., 1981). More recently, this logic has been applied with molecular biology tools to the study of aggression. Once aggressive behavior is established, the expression of a gene of interest can be suppressed with the so-called knock-down techniques and subsequently reexpressed with the tetracycline-gene regulation system (Chen et al., 1997).

Pharmacological interventions offer a further important approach to the neurobiological study of aggressive behavior. Drugs that selectively modulate discrete elements of a neurochemical system can reveal a neurotransmitter's specific function in the initiation, execution, and termination of aggressive behavior. For example, the simple functional label of serotonergic inhibition of aggression has yielded to a more intricate perspective, which is prompted by the knowledge of an array of receptor subtypes and transporter molecules for serotonin and discrete anatomical 5-HT pathways (Hoyer et al., 2002). Activation of the 5-HT$_{1A}$ and 5-HT$_{1B}$ receptor subtypes, for example, produces divergent effects in various behavioral and physiological activities such as movements, alimentary behavior, and sexual libido, and it is possible to extrapolate such opposing roles also to aggressive behavior (Jenck et al., 1989; Dulawa et al., 2000; Olivier, 2004; Faccidomo et al., 2005).

2 Behavioral Foundations

Aggression encompasses a range of diverse behavior patterns, each with unique distal and proximal antecedents and various functions. Aggression is by no means unitary in its origins, motivations, expressions, or functions, implying diverse neurobiological mechanisms. From a clinical perspective, it has been useful to distinguish between kinds of aggression, such as the impulsive–reactive–hostile–affective, as opposed to the controlled–proactive–instrumental–predatory subtype of human aggression (Vitiello and Stoff, 1997). Behavioral biologists focus on types of aggressive behavior that have evolved as social and reproductive adaptations. In socially cohesive species, such as primates and rats, males establish and maintain dominance hierarchies, whereas dispersed species, such as mice, mark, patrol, and defend territories with the overarching objective of securing the resources to reproduce successfully and transmit genes. Similarly, females suppress the reproductive activities of rivals and defend their offspring (Huntingford and Turner, 1987). Experimental models of aggressive behavior that escalate the intensity and frequency beyond the species-typical norms are particularly relevant to understanding the types of human aggression that are of clinical concern.

2.1 Species-Typical Aggressive Behavior: Fighting as an Adaptation

2.1.1 Dominance Aggression

In socially organized and cohesive species, dominance hierarchies are established and maintained by aggressive acts, postures, and displays that result in submissive, defensive, and evasive responses by lower-ranking individuals. Facial grimaces, body postures, and mounting are prominently displayed by both old world and new world primates that are occasionally used in neurochemical studies of aggressive behavior (Mehlman et al., 1994; Higley et al., 1996; Fairbanks et al., 1999; Bennett et al., 2002). Intense, frequent, and potentially injurious aggression characterizes the early phase of forming a dominance hierarchy in rhesus monkeys (Bernstein et al., 1974; Bernstein, 1981) and in rats (Steiniger, 1950), particularly in the presence of females. Later on, dominance is maintained by threat displays. These displays are part of an elaborate repertoire of acts and postures that have evolved in each species to communicate social signals and to engage in aggressive confrontations, often referred to as agonistic behavior (Scott and Fredericson, 1951). This behavioral repertoire of aggressive acts, postural, and facial displays, develops during rough-and-tumble play fighting in the prepubertal period (Pellis and Pellis, 1987). In many established groups of primates, dominance is determined early in life by the social rank of the mother (Walters and Seyfarth, 1986). In contrast to the prevailing ethological thesis that dominance aggression consists chiefly of ritualized displays (Eibl-Eibesfeldt, 1961; Lorenz, 1966; Goodall, 1986), rare, but memorable instances of injurious aggression have been documented in chimpanzees that track down neighboring groups for the purpose of deadly attacks (Nishida et al., 1985; Wrangham, 1999).

The functional significance of dominance aggression is expressed in its operational definition of gaining ready access to essential resources such as preferred foods, protected sleeping places, and high-status mates. From a sociobiological perspective, the transmission of genetic information into the next generation is the essential definition of dominance (Wilson, 1974). Yet, the significance of lethal raiding parties in hominids remains a challenging topic, as it focuses on relatively rare behavioral events that are associated with intense autonomic excitement as well as careful planning and coordination.

2.1.2 Territorial Aggression

Much neurochemical work on aggressive behavior is performed in murine species that are characterized by territorial aggression (Garattini et al., 1967; Henley et al., 1973; Saudou et al., 1994b). Adult breeding male mice patrol, mark, and defend their territory that becomes the site of the deme (breeding unit, "Grossfamilie") when they attract fertile females and exclude other breeding males (Eibl-Eibesfeldt, 1950; Crowcroft and Rowe, 1963; Reimer and Petras, 1967; Cairns and Nakelski, 1971; Poole and Morgan, 1976). The term territorial aggression is applied to the behavior of a male in control of an "exclusive territory" or of one male who dominates others ("dominance territory"). Even under laboratory conditions, nearly all offspring are sired by dominant males, whereas subordinate males are rarely seen mating and are the targets of frequent attacks in this pugnacious species (Mackintosh, 1970). A breeding male resident mouse attacks an adult male who intrudes into the territory rapidly, vehemently, and frequently. Under controlled laboratory conditions, this type of confrontation is referred to as *resident–intruder aggression* (Crawley et al., 1975; Miczek and O'Donnell, 1978) (*Figure 7-1*).

Figure 7-1

Mouse agonistic behavior. Behaviors of resident and intruder mice engaged in an aggressive confrontation: (a) the resident leaps and bites the intruder as the intruder attempts to escape; (b) the resident (*right*) threatens as the intruder (*left*) holds a defensive upright posture; (c) the resident investigates the intruder's anogenital region; (d) the resident pursues the fleeing intruder; (e) both resident and intruder engage in a mutual upright defensive posture. Copyright 1978 by Springer-Verlag. Reprinted with permission from Miczek and O'Donnell (1978)

Figure 7-1 (continued)

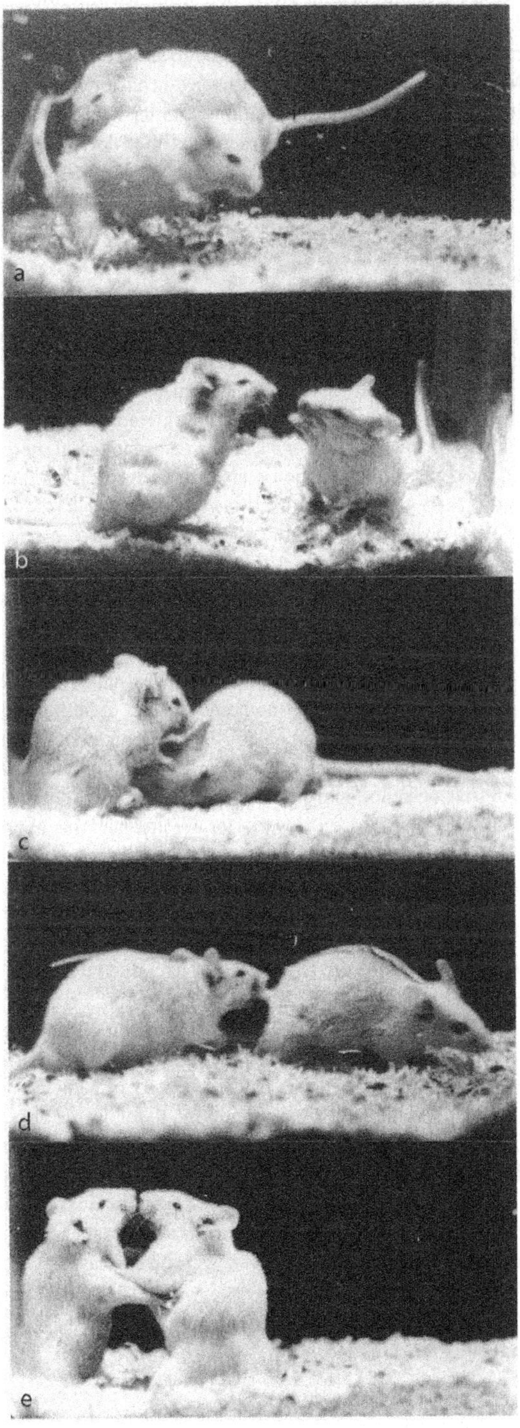

Laboratory mice are housed most often in groups, preventing the formation of demes, and under these conditions territorial aggression is seen only in some proportion of outbred mice. Most inbred mice, particularly those used as background strains for transgenic mice, exhibit rarely species-normative social and aggressive behavior (Silver, 1995; Parmigiani et al., 1999; Miczek et al., 2001). Social experiences during development are required for most male *Mus musculus* to engage in aggressive behavior during adulthood (Cairns and Nakelski, 1971). In addition to the aggressive confrontations with territorial intruders, outbred laboratory mice such as Swiss mice that are housed in large same-sex groups often develop despotic relationships, with one mouse dominating all other mice (Uhrich, 1938; Poole and Morgan, 1973; Brain, 1975).

2.1.3 Female Aggression

Rivalries among breeding females and protection of the offspring from harm are the key determinants of aggressive behavior in females of most mammalian species (Hurst, 1987). In the presence of dominant females, the endocrine cyclicity and fecundity of low-ranking females are suppressed, as illustrated in several primate species such as marmosets and baboons (Smuts, 1986). Harassing attacks by dominant females interrupt opportunistic matings by low-ranking females. The neurobiological mechanisms for these types of female–female rivalry in the context of mate choice have not been studied so far.

Nearly all neurobiological research on aggressive behavior in females has focused on postpartum or maternal aggression in rodents (Lonstein and Gammie, 2002). In fact, female mice, rats, voles, and hamsters fight more near their nest site, possibly a parallel to the male territorial defense. Exposure to pups can promote aggression in females, even in virgins, given the appropriate endocrine conditions. In most rodents, aggressive behavior becomes more frequent, shortly before gestation, reaches a peak during the first week postpartum, and declines thereafter (Noirot et al., 1975; Erskine et al., 1980; Mann and Svare, 1982). The behavioral repertoire for maternal aggression shares many similarities with male aggression, but importantly includes frequent bites directed at the snout and the face of both male and female opponents that are defensive in character (Haney et al., 1989).

Even when not lactating and without maternal experience, female rodents and primates can exhibit substantial amount of aggressive behavior, and females compete for reproductive opportunities (Smuts, 1986; Palanza et al., 2005). In mice and rats, even ovariectomized females that are housed with a male will attack intruders (DeBold and Miczek, 1981; Zitzman et al., 2004). The degree to which the neurobiological basis of aggressive behavior by nonlactating and maternal females overlap, remains to be studied.

2.2 Escalated Aggressive Behavior: Models for Disorders

Aggressive behavior in animals becomes increasingly relevant to clinical concerns in humans when it exceeds the species-typical pattern (Miczek et al., 2004a; de Almeida et al., 2005). The ethical dilemma of investigating escalated forms of aggressive behavior is the necessity of allowing potentially harmful and injurious behavior to proceed, and at the same time, attempting to follow the principle of harm avoidance as much as possible (Miczek, 2001). Aside from illegal breeding practices in fowl, dogs, and cattle, mice have been bred for aggressive behavior under controlled laboratory conditions (Lagerspetz, 1964). These genetic selection studies are particularly relevant for questions addressing individual differences in aggressive trait characteristics. In addition, provoking an individual with frustrative experiences or with social instigations examines the resilience and the vulnerability to engage in aggressive behavior.

2.2.1 Bred for Aggression

Domestication of feral animals amply demonstrates the success with selective breeding for nonaggressive behavioral characteristics (Blanchard et al., 1998; Koolhaas et al., 1999). Conversely, neurobiological studies have taken advantage of genetic selection for aggressive behavioral phenotypes, by comparing lines of rodents and canine species that express high levels of aggressive behavior relative to low levels (Weerts et al., 1992; Gariepy et al., 1998; Van Der Vegt et al., 2001; Badino et al., 2004).

Systematic efforts in selective breeding for aggression in mice have demonstrated that lines of highly aggressive and nonaggressive males diverge promptly and stabilize at high and low levels after ca. 15 generations ("top–down genetics") (Lagerspetz, 1964; van Oortmerssen and Bakker, 1981; Sandnabba, 1986). Individual differences in aggressive behavior have been characterized in strains or breeds of animals that show extreme expressions of the behavior, and this strategy has been exploited in the study of various neurochemical markers in postmortem assays of neural tissue.

An alternative approach focuses on a single candidate gene for the pathogenesis of escalated aggressive behavior ("bottom–up genetics"), and it has become feasible by techniques that enable the deletion or overexpression of a specific gene. This approach has pointed to the influence of several genes on most chromosomes, ranging from the 5-HT_{1B} receptors, nNOS, MAO_A, preproenkephalin, adenosine A_{2a} receptor, neural cell adhesion molecule, estrogen receptor beta, and adrenergic alpha2 receptor to oxytocin (Miczek et al., 2001). Concordant with the polygenic influences on aggressive behavior, mouse mutants have demonstrated that several gene products are important in the development of escalated aggressive behavior (➋ *Figure 7-2*).

The genetics of aggressive behavior may be advanced by the recently sequenced DNA of the fruit fly (*Drosophila melanogaster*) and by the detailed analysis of its aggressive behavior (Chen et al., 2002). The techniques for generating genetic mutants in fruit flies are fully established, and it should be possible to identify all the relevant genes that are critical for escalated aggressive behavior in Drosophila.

Figure 7-2

Gene deletions and mouse aggression. Gene deletions affecting aggressive phenotypes in mice. This figure portrays representations of 21 chromosomes (*vertical bars*) and the approximate location of genes (*filled and open horizontal bars*) that have been knocked out in mice and engendered heightened- (*filled bars, upward arrow*) or suppressed- (*open bars, downward arrow*) levels of aggression. The proteins targeted by the gene deletion are indicated in abbreviated text. Deletions that have not affected aggressive behavior are not shown. Abbreviations and references: 5-HT_{1B}, serotonin receptor 1B (Saudou et al., 1994b); A1AR, adenosine receptor A1 (Saudou et al., 1994a); A_{2a}, adenosine receptor 2A (Ledent et al., 1997); AchE, acetylcholinesterase (Duysen et al., 2002); αCaMK II, alpha-calcium-calmodulin kinase II (Chen et al., 1994); Ar, aromatase (Matsumoto et al., 2003); AVPR3, arginine vasopressin 1B receptor (Wersinger et al., 2002); β2m, beta2-microglobulin (Loconto et al., 2003); CB1, cannabinoid receptor 1 (Martin et al., 2002); CNGA2, cyclic nucleotide gated channel alpha 2 (Mandiyan et al., 2005);COMT, catechol-*O*-methyltransferase (Gogos et al., 1998); DAT, dopamine transporter (Rodriguiz et al., 2004); DBH, dopamine beta hydroxylase (Marino et al., 2005); eNOS, endothelial nitric oxide synthase (Demas et al., 1999); ENK, enkephalin (Konig et al., 1996); EP1, prostaglandin E receptor 1 (Matsuoka et al., 2005); ERα, estrogen receptor alpha (Ogawa et al., 1998); ERβ, estrogen receptor beta (Ogawa et al., 1999); GAD_{65}, glutamic acid decarboxylase (65 amino acids) (Stork et al., 2000); GDIA, guanosine diphosphate (GDP) dissociation inhibitor 1 (D'Adamo et al., 2002); H1, histamine receptor 1 (Yanai et al., 1998); MAO_A, monoamine oxidase A (Cases et al., 1995); MC5R, melanocortin-5 receptor (Morgan and Cone, 2006); MTL1, nuclear receptor subfamily 2, group E, member 1 (aka "FIERCE") (Young et al., 2002a); Naα_{2C}, adrenergic alpha receptor 2C (Sallinen et al., 1998); NCAM, neural cell adhesion molecule (Stork et al., 1997); NEP, neutral endopeptidase (Fischer et al., 2000); NET, norepinephrine transporter (Haller et al., 2002); NK-1r, neurokinin receptor 1 (De Felipe et al., 1998); nNOS, neuronal nitric oxide synthase (Nelson et al., 1995); OT, oxytocin (Winslow et al., 2000); PAC1, adenylate cyclase activating polypeptide 1 receptor 1 (Nicot et al., 2004); PET-1, ETS (E26 transformation specific) domain transcription factor (Hendricks et al., 2003); RGS2, regulator of G protein signaling (Oliveira-dos-Santos et al., 2000); SERT, serotonin transporter (Holmes et al., 2002); Trp2, transient receptor potential family 2 (Stowers et al., 2002); V1$R_{a/b}$, a cluster of vomeronasal receptor genes located on chromosome six, V1ra_{1-9} and V1$rb_{1-4,\ 7-9}$ (Del Punta et al., 2002);Y1, neuropeptide Y receptor 1 (Karl et al., 2004). As of March, 2006

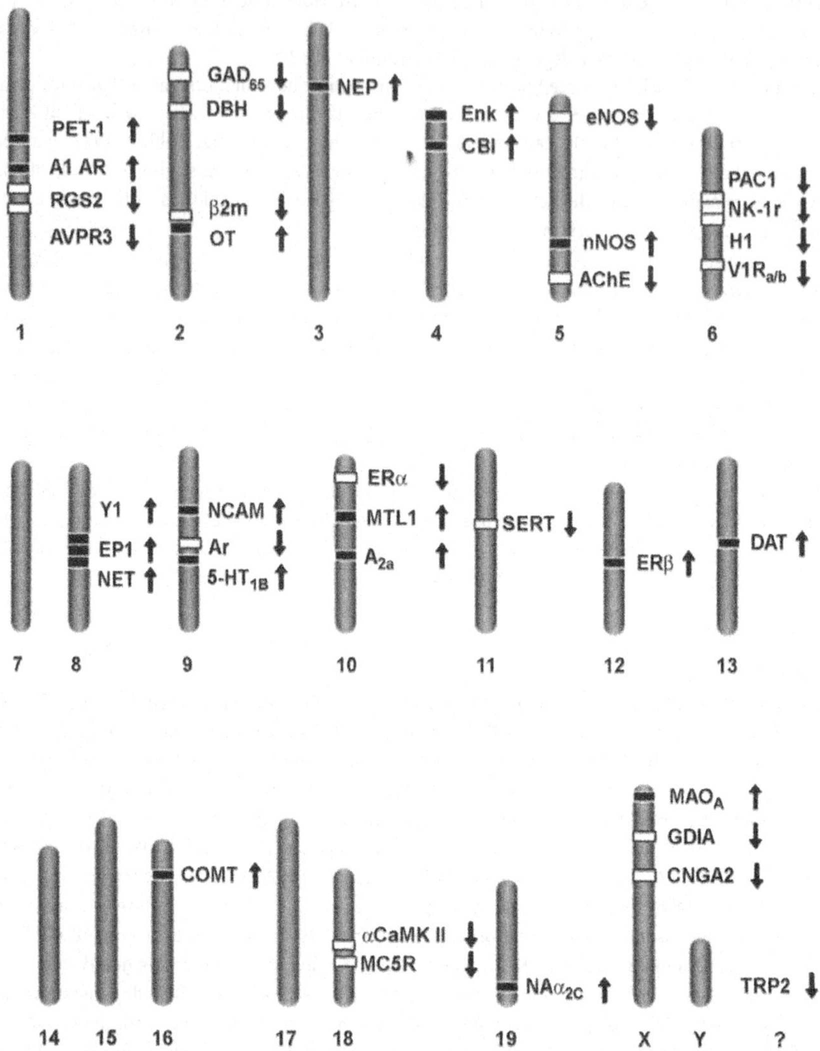

2.2.2 Frustration Aggression

By far, one of the generally most effective ways to escalate aggressive behavior is the exposure to frustrative experiences (Dollard et al., 1939). When a scheduled reward is suddenly omitted, the probability of heightened aggressive behavior increases greatly in various animal species, including humans. Violent youths characteristically show a low-frustration threshold (Calkins and Johnson, 1998; Nock and Kazdin, 2002). Several experimental protocols in avian species, rodents, and primates have captured this phenomenon (Azrin et al., 1966; Cherek and Heistad, 1971). The basic principle of omitting a scheduled reward as a trigger for escalated aggressive behavior can be readily implemented in laboratory mice (Fish et al., 1999; de Almeida and Miczek, 2002) (➋ *Figure 7-3*).

In spite of the ubiquitous nature of frustration-escalated aggressive behavior, neurobiological studies have been limited to pharmacological manipulations involving therapeutic agents and drugs of abuse

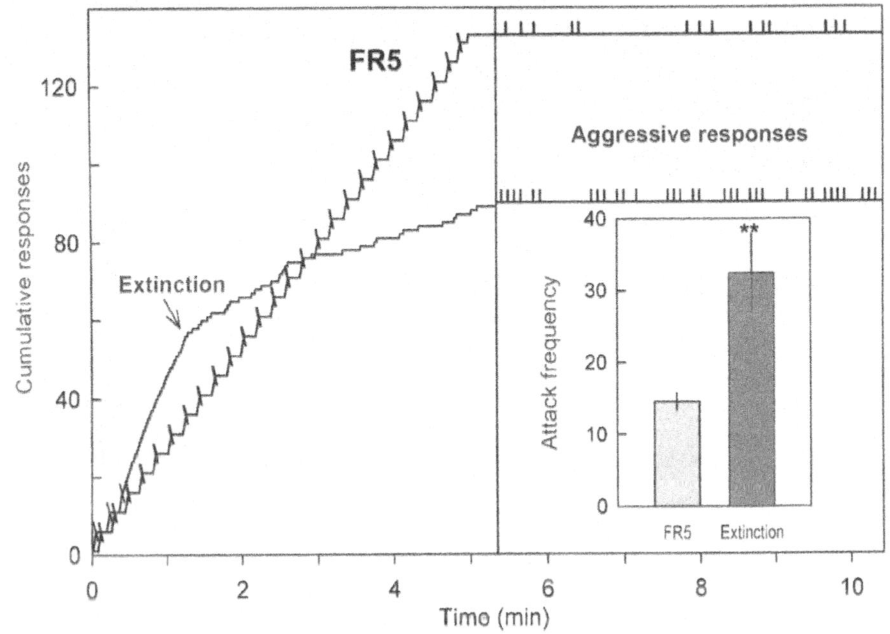

Figure 7-3
"Frustration-induced" aggression in a resident male mouse after omission of scheduled reinforcement. (*Left panel*) During a 5 min session, the mouse responds on a fixed ratio 5 schedule of positive reinforcement for the delivery of a sucrose solution (indicated by a slash) and then confronts an intruder. On extinction sessions, only three reinforcers are available. (*Right panel*) Aggressive responses (indicated by vertical deflections in the time line) are heightened after extinction. (*Inset*) The mean frequency of attack bites (±SEM, vertical lines) in the reinforced (FR5, light gray bar) and extinguished (dark gray bar) conditions, **denotes $p < 0.01$. Copyright 2002 by Springer-Verlag. Reprinted with permission from Miczek et al. (2002)

(Cherek and Pietras, 2003). So far, the neural mechanisms through which a frustrative experience triggers aggressive outbursts remain to be delineated; similarly, the trait characteristics of resilience to frustration experience or, operationally, resistance to extinction, have not been characterized in terms of neurobiological mechanisms.

2.2.3 Instigation or Social Provocation

Closely related to the escalated aggression resulting from frustrative experiences are the provocations by a potential opponent. In experimental laboratory preparations with research animals, a resident male fish, mouse, rat, or hamster is exposed to a breeding male, which is placed into the resident's home cage and protected by a screen for a brief period (Heiligenberg, 1974; Potegal and Tenbrink, 1984; Kudryavtseva, 1991; Potegal, 1991; Fish et al., 1999). The presence of a breeding male provokes the resident into intense and frequent aggressive behavior, when given the opportunity. This provocation is hypothesized to induce specifically "aggressive arousal" or "attack readiness," presumably due to the olfactory, visual, and auditory cues emanating from the intruder.

So far, escalated aggression due to social instigation has begun to be characterized in terms of GABAergic and serotonergic activity (Fish et al., 1999; de Almeida and Miczek, 2002), and the neurochemical basis for the postulated aggressive arousal remains to be identified (➋ *Figure 7-4*).

Figure 7-4

Social instigation of aggression. Heightened aggression by social instigation. For 5 min, a resident male mouse is exposed to an intruder male that is protected by a screen through which olfactory and visual cues are still available. After an interval, the resident attacks an unprotected intruder with greater frequency. Bars represent the median attack bites and vertical lines represent the interquartile range after control (light gray) and instigated (dark gray) conditions

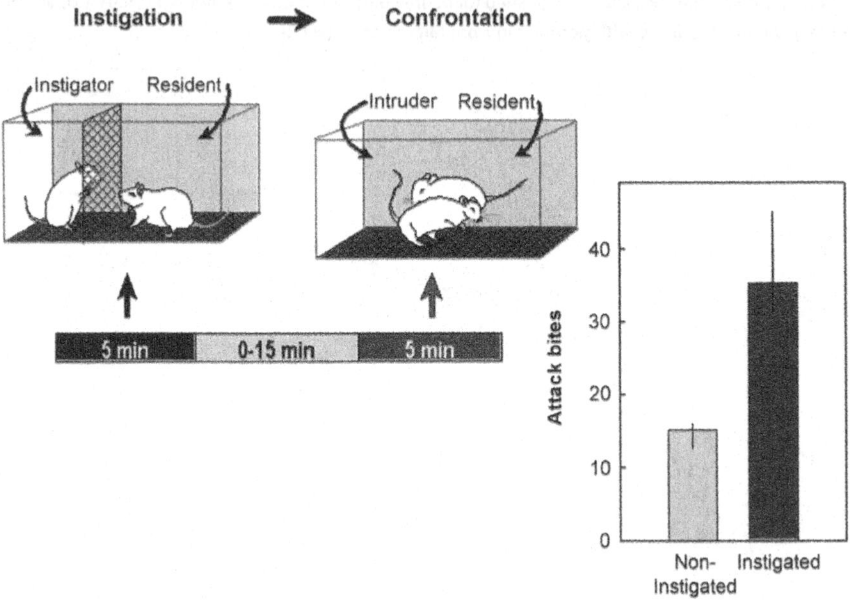

As discussed earlier, aggression is behaviorally rich and functionally complex. An equally rich and complex neurochemistry underlies its expression. Situations and stimuli that provoke specific forms of aggression activate shared and unique neural mechanisms to coordinate how an individual initiates, performs, responds to, and recovers from these acts. In some instances, such as maternal aggression, these neural mechanisms are well characterized and distinguished from those involved in other forms of aggression. In others, such as escalated aggression, the distinctive features are currently being advanced. Across all forms of aggression, a prominent role exists for the neurotransmitter 5-HT, and this is highlighted in the following sections. However, 5-HT depends upon its interaction with amino acids, other monoamines, neuropeptides, and steroids (Miczek et al., 2002).

3 Neurotransmitters and Aggression

3.1 Excitatory and Inhibitory Amino Acids: A Delicate Balance

Most synaptic transmission in the CNS relies on excitatory and inhibitory amino acids, and their role in the mechanisms mediating aggressive behavior has been delineated with increasing success. Based on their cellular mode of action, the earliest hypothesis simply extrapolated the excitatory effects of glutamate and the inhibitory effects of GABA to the level of aggressive behavior (Mandel et al., 1979). This approach was elaborated by the "limbic dyscontrol" hypothesis of aggression, which portrays the expression of aggressive outbursts as a result of an imbalance between glutamate excitation and GABA inhibition (Monroe, 1978).

Clinically, the imbalance between excitatory and inhibitory neurotransmission is evident in patients with temporal lobe epilepsy, who engage in explosive aggressive acts or "interictal rage" (Mirsky and Harman, 1974; Bear and Fedio, 1977; Monroe, 1978; Weiger and Bear, 1988; Siegel and Mirsky, 1994). However, it is not known if the seizures themselves are the causal mechanisms for the aggressive behavior (Fenwick, 1989; Marsh and Krauss, 2000). Animal models of seizure disorders, such as in cats or rats exposed to repeated intermittent stimulation with electric pulses in the amygdala or hippocampus (i.e., kindling) or administration with a neurotoxin like trimethyltin, also provide evidence for increased emotional reactivity and defensive aggressive responses (Pinel et al., 1977; Adamec, 1990; Kalynchuk et al., 1992). The enhanced neural activation due to repetitive stimulation seems to exaggerate the responses to already threatening stimuli, rather than causing indiscriminate or stereotypical aggressive behavior. Thus, under conditions of heightened excitation, an animal may engage in defensive biting or attack, an otherwise mild stimulus.

Support for the glutamate–GABA imbalance hypothesis derives from the initial measurements of higher levels of glutamate and lower levels of GABA in whole brain, hypothalamus, amygdala, frontal cortex, and olfactory bulbs of aggressive relative to nonaggressive animals (Agrawal et al., 1967; Early and Leonard, 1980; Simler et al., 1982; Clement et al., 1987; Munoz-Blanco and Porras, 1987). Moreover, pharmacologically decreasing glutamate or increasing GABA levels inhibits various kinds of aggressive behavior (Puglisi-Allegra, 1980; Puglisi-Allegra et al., 1981; Lumley et al., 2004). Similar antiaggressive effects occur in rats and mice that are treated with direct agonists to $GABA_A$ receptors such as muscimol (Haug et al., 1980; Puglisi-Allegra, 1980; Mandel et al., 1981; Molina et al., 1986). Clinically, anticonvulsant medications, such as valproate, carbemazepine, and phenytoin, restore the balance between glutamate and GABA, and these compounds are also efficacious antiaggressive treatments (Barratt, 1993; Pabis and Stanislav, 1996; Lindenmayer and Kotsaftis, 2000; Stanford et al., 2001).

However, there are some notable exceptions to the imbalance hypothesis. Deleting the gene that codes for the GAD_{65} isoform of the synthetic enzyme for GABA, reduces brain levels of GABA and decreases aggression (Stork et al., 2000), and higher levels of GAD_{65} are found in the brains of hamsters induced to high levels of aggression by prior exposure to cocaine (Ricci et al., 2005). These findings demonstrate that there are instances when GABA may facilitate, rather than inhibit, aggressive behavior.

3.1.1 Critical Neuroanatomy

Preclinical studies have begun to identify the neural sites of action for glutamate and GABA within the pathways that regulate aggression. Although several brain regions are known to critically modulate aggressive and defensive behavior (Chi and Flynn, 1971), a region of the hypothalamus, referred to as the hypothalamic attack area (HAA), is thought to be of particular importance for the control of offensive attack behavior (Siegel et al., 1999) (➋ *Figure 7-5*).

Electrical stimulation of the HAA elicits promptly the complete coordinated pattern of offensive attacks toward a conspecific in rats and cats, provided an opponent is present. More medial periventricular regions, extending from the septal area to the medial hypothalamus and periaqueductal gray area, have been identified to control the expression of defensive behavior (Bandler et al., 1985; Siegel et al., 1999). The HAA is not localized to a conventional anatomical nucleus, but consists of a zone that extends from the anterior hypothalamus (AH) through the lateral and parts of the ventromedial hypothalamus (Hrabovszky et al., 2005). Its neurons appear to be predominantly glutamatergic, because they express more mRNA for the type 2 vesicular glutamate transporter rather than for the enzyme glutamic acid decarboxylase $(GAD)_{67}$ (Hrabovszky et al., 2005). This cytochemical difference suggests the presence of excitatory efferents from the HAA that may be inhibited by local GABAergic interneurons (Gregg and Siegel, 2001). Further support for tonic inhibition of offensive aggressive acts by hypothalamic GABAergic interneurons derives from the effects of microinjection of GABA antagonists or, alternatively, glutamate agonists into the HAA (Hrabovszky et al., 2005) and/or GABA antagonists (Adams et al., 1993; Roeling et al., 1993).

In addition to its excitatory role in the HAA and in the amygdaloid and hippocampal seizure foci, glutamate is a key transmitter in the neural processing of olfactory cues that provoke or inhibit aggression. The projections from the olfactory bulb, via the lateral olfactory tract to the amygdala, are glutamatergic

Figure 7-5
Hypothalamic attack area. Three-dimensional portrayal of hypothalamic areas that elicit offensive aggressive behavior when electrically stimulated. The hypothalamic attack area is shown in red (or dark gray) and consists of portions of the anterior hypothalamus, ventromedial hypothalamus, the subfornical hypothalamus, and an area between the ventromedial hypothalamus and the medial forebrain bundle. For orientation, the ventrolateral hypothalamus (VMH), paraventricular nucleus (PVN), fornix (F) and third ventricle (IIIV) are also illustrated (PVN) and rostral to caudal and medial to lateral coordinants are given relative to bregma and midline. Copyright 2005 by Elsevier Ltd. Reprinted with permission from Hrabovsky et al. (2005)

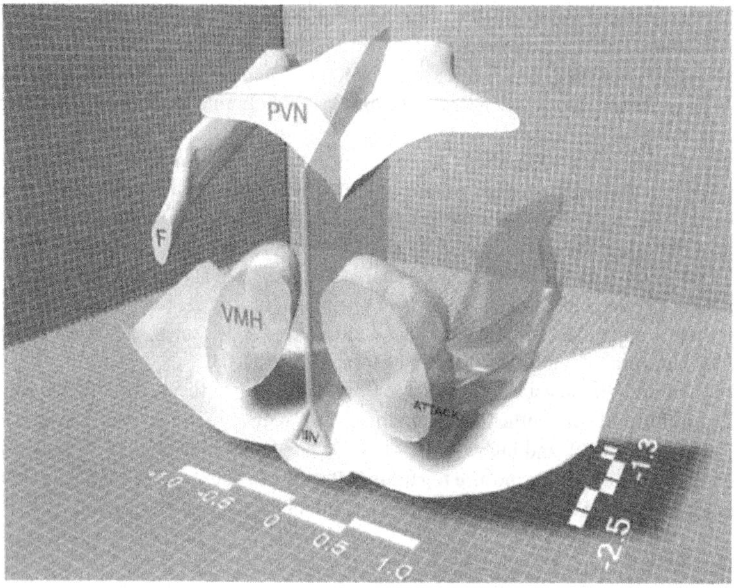

and are essential for the display of sexual, maternal, and aggressive behavior (Gandelman et al., 1971; Denenberg et al., 1973; Larsson, 1975; Kendrick et al., 1997). Considerable evidence documents glutamatergic projections from the medial prefrontal cortex to the amygdala (Gerfen and Wilson, 1996; Rosenkranz and Grace, 2002), and this pathway is activated by various types of stressors (McFarland et al., 2004). Within the mPFC of maternal sheep, glutamate activity increases following exposure to novel stimuli that prompt the display of aggressive behavior (Broad et al., 2002b), and it would be useful to directly measure glutamate in the initiation and the termination of aggressive confrontations.

There is some support for the early hypothesis that aggressive behaviors are influenced by the balance between the excitatory actions of glutamate and the inhibitory actions of GABA. However, this attractively simple hypothetical framework needs now to accommodate the diverse range of receptor subtypes that are the targets for these ubiquitous transmitters. Stimulation, blockade, and modulation of some of the glutamate and GABA receptor subtypes, particularly $GABA_A$ receptors, have begun to reveal very significant roles in different kinds of aggressive behavior.

3.1.2 Glutamate Receptors

Most pharmacological studies on glutamate and aggression have relied on rather nonspecific agents that primarily target the NMDA receptors and do not differentiate between receptor subtypes (Broad et al., 2002a). Meanwhile, it is clear that ionotropic (iGluR) and metabotropic (mGluR) receptors mediate the neurochemical and behavioral effects of glutamate. The NMDA, AMPA, and kainate glutamate receptors are ligand-gated ion channels that permit the flow of Na^+ and Ca^{2+} and typically depolarize the membrane. Additionally, the group I (mGluR1 and mGluR5) and group II (mGluR2 and mGluR3) receptors increase the activity of phospholipase C (PLC), whereas the group III (mGluR4, mGluR6–8) decreases the activity of adenylate cyclase. The emergence of novel and more selective drugs for each of these receptor subtypes holds promise for elucidating their specific aggression-modulating effects (Kew and Kemp, 2005).

3.1.2.1 NMDA Receptors Classic antagonists of the NMDA receptors, including phencyclidine (PCP) and dizocilpine (MK-801) that block the ion channel in the receptor, have inconsistent effects on aggressive behavior that are accompanied by motor side-effects including incoordination. In some individuals, in certain test conditions (as discussed later), these compounds increase aggressive behavior (Rewerski et al., 1971; Krsiak, 1974; Burkhalter and Balster, 1979; Musty and Consroe, 1982; Wilmot et al., 1987; McAllister, 1990), while in others, they are suppressive and sedative (Tyler and Miczek, 1982; Miczek and Haney, 1994; Lang et al., 1995; Belozertseva and Bespalov, 1999). Behaviorally nonspecific effects of these high-affinity antagonists are not surprising, given their dissociative and hallucinogenic effects. The more recently developed low-affinity channel blockers, such as memantine and MRZ 2/579, or partial agonists to the glycine binding site on the receptor, such as HA-966 (Kew and Kemp, 2005), appear to have fewer adverse side effects (Rogawski, 2000; Kemp and McKernan, 2002), and could be useful pharmacotherapeutic options for aggressive disorders due to their primary action on escalated, rather than on basal excitation. Neither memantine nor the related compound MRZ 2/579 altered basal levels of aggression or the latency to attack (Belozertseva and Bespalov, 1999; Minkeviciene et al., 2004), but they specifically reduced the intense aggression that followed withdrawal from morphine (Sukhotina and Bespalov, 2000).

Two key determinants for the disparate effects of NMDA receptor manipulations across studies on aggression are the individual's prior history with aggressive behavior and the intensity of the encounter. NMDA receptor antagonists render inexperienced animals that fight at low rates more aggressive, while experienced, intensely aggressive animals become calmer—an example of the law of initial value (Wilder, 1958). This pattern raises the interesting possibility that repeated aggressive experiences alter the regulation of NMDA receptor expression along the neural pathway that regulates the behavior. As aggression becomes more likely and more intense, the role of glutamate becomes more significant.

Changes in glutamate neurotransmission and signaling have long been identified as significant targets for repeated exposures to salient stimuli, and such changes are strongly related to learning and memory and adaptation to stressors (Holscher, 1997; Braunewell and Manahan-Vaughan, 2001). Following aggressive experience, there are also changes in glutamate (Broad et al., 2002b), and this effect needs to be further characterized. Changes in the NMDA receptor system occur in repeatedly defeated rats and mice, as evidenced by the prevention of the sensitized responses to psychomotor stimulants by protective treatments with NMDA receptor antagonists (Covington and Miczek, 2002; Yap et al., 2005). Exposure to repeated anabolic steroids, which can facilitate aggressive behaviors in rats (e.g., McGinnis, 2004), decreases the expression of mRNA for the NR2A subunits in the hypothalamus and hippocampus, as well as the NR2B subunit in the hypothalamus (LeGreves et al., 1997), but in lizards (*Anolis carolinensis*) these proteins upregulate following establishment of dominant-subordinate status (Summers et al., 2005). These changes in the NR2B receptor due to aggressive experiences should prompt direct manipulations with subunit-selective antagonists, such as ifenprodil, in an effort to alter the long-term consequences of aggressive experience. It can be hypothesized that the most important role of NMDA receptors in aggression is to determine whether an individual will attack or submit in future encounters.

3.1.2.2 AMPA and Kainate Receptors Regarding aggressive behaviors, AMPA and kainate receptors have been studied much less than the NMDA receptors. This has been due, in part, to the lack of selective pharmacological agents for these receptors. Genetic studies have so far identified subunits of the AMPA receptor as important for aggressive behavior. A QTL study by Brodkin et al. (2002) identified the AMPA3 subunit gene as a candidate related to attack behavior, whereas mice without the gene for the AMPA2 subunit show very little offensive attacks (Vekovischeva et al., 2004). At this early stage, it is evident that certain AMPA receptor subtypes can significantly modulate aggressive behavior. It is likely that other AMPA subunits contribute to the neural mechanisms of aggression.

3.1.3 GABA$_A$ Receptors

Direct stimulation and positive modulation of GABA$_A$ receptors have a range of effects on aggressive behavior, from inhibition to escalation (Miczek et al., 2003). These receptors are composed of five transmembrane proteins, encoded by at least 17 separate genes, forming a Cl^- channel that when opened typically promotes the inward flow of Cl^- to hyperpolarize the surrounding membrane (Mohler et al., 1995). Their sensitivity is determined in part by the combination of individual proteins to the receptor complex, comprising alpha, beta, gamma, and delta subunits in mammalian brain neurons (review refs Mehta and Ticku, 1999). Drugs such as benzodiazepines, barbiturates, alcohol, and certain neurosteroids allosterically modulate the GABA$_A$ receptor to potentiate the actions of GABA and to increase Cl^- flow.

3.1.3.1 Positive Allosteric Modulation These positive modulators are used clinically to reduce anxiety, seizures, muscle tension, and sleep disorders (Shader and Greenblatt, 1993). Alcohol has many of these same behavioral effects, and though this ubiquitous agent affects a variety of ionotropic receptors, its positive modulation of the GABA$_A$ receptor complex is particularly important for these acute effects (Liljequist and Engel, 1982; Vanover et al., 1999a, b). When used clinically, the rapid administration of GABA$_A$ receptor positive modulators at substantial doses effectively reduces aggressive outbursts (Dietch and Jennings, 1988). However, early astute clinical observations noted patients who become more aggressive, particularly after benzodiazepine treatments (DiMascio, 1973). In laboratory animals, heightened aggression can occur not only after benzodiazepines, but also after the neurosteroid allopregnanolone or the synthetic steroid alphaxalone (Fish et al., 2001). These reactions have been considered "paradoxical," because they contradict the primarily calming effects of these drugs.

Several factors interact to determine how benzodiazepines affect aggressive behavior. These include the drug itself, the dose, the context of administration, and the individual's history (Gardos et al., 1968; Bond et al., 1995). In humans, increased aggression has been reported after treatment with the benzodiazepines diazepam, chlordiazepoxide, lorazepam, flunitrazepam, and alprazolam (DiMascio et al., 1969; Brown, 1978; Bond et al., 1995; Weisman et al., 1998; Daderman and Lidberg, 1999), particularly if these individuals have a history of previous violent behavior. Chlordiazepoxide, diazepam, and midazolam have all been shown to increase dominance, territorial, and maternal aggressive behavior in mice, rats, pigs, and monkeys (Cole and Wolf, 1970; Miczek, 1974; Arnone and Dantzer, 1980; Miczek and O'Donnell, 1980b; Rodgers and Waters, 1985; Mos and Olivier, 1989; Weerts and Miczek, 1996; Ferrari et al., 1997; Gourley et al., 2005). This pro-aggressive effect is reversible by antagonists at the benzodiazepine binding site on the GABA$_A$ receptor complex, indicating that this site is critical for heightened aggression (e.g., Olivier et al., 1991), and the search for the relevant subunit composition of the GABA$_A$ receptor for the aggression-modulating effects has begun (Gourley et al., 2005). However, regardless of an individual's past violent behavior, other benzodiazepines, such as oxazepam, clorazepate, and zolpidem, either reduce aggressive behavior or have no effect at all (DiMascio, 1973; Bond and Lader, 1988; Pabis and Stanislav, 1996; Weisman et al., 1998; Martin-Lopez and Navarro, 2002; de Almeida et al., 2004). One hypothesis for these differential effects of benzodiazepines suggests that compounds with higher intrinsic activity at the GABA$_A$ receptor are more likely to reduce aggressive behavior (Ducic et al., 1993; Gourley et al., 2005).

Another important determinant of the pro-aggressive effects is the dose of the GABA$_A$ positive modulator. All GABA$_A$ receptor positive modulators that heighten aggression do so at low to moderate doses and are sedative at higher doses, which at least partially accounts for their antiaggressive effects. It is very likely that doses along the ascending limb of the bidirectional dose–effect curve recruit different mechanisms than those along the descending limb (➋ *Figure 7-6*).

Context and prior experiences with aggressive behavior are two further determinants of the effects on aggression that result from stimulation of GABA$_A$ receptors. When benzodiazepines are given to individuals without an identified pathology or are used in a clinical setting, they typically decrease aggressive and

Figure 7-6
GABA$_A$ positive allosteric modulators and aggression. Biphasic effects of GABA$_A$ receptor positive modulators on aggression in rats (*top panel*) and mice (*bottom panel*). Low doses of alcohol (*filled hexagons*), the benzodiazepines diazepam (*filled diamonds*, rats only) and midazolam (*filled circles*) and the neurosteroid allopregnanolone (*filled triangles*, mice only) increase the mean (±SEM, *vertical lines*) number of attack bites, expressed as a percentage of vehicle control, while higher doses decrease this measure of aggression. Triazolam (*filled upward triangles*) increases attack bites in rats but not mice. No increase in aggression was seen after treatment with zolpidem, the α_1 receptor selective agonist, (*filled downward triangles*, mice only). The *dotted line* represents the baseline at 100%

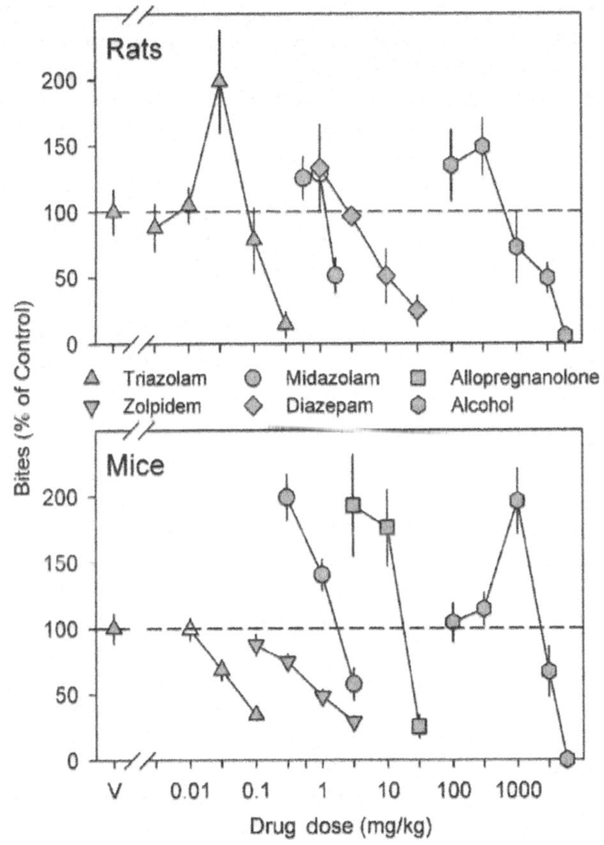

hostile behavior (Cherek et al., 1990; Cherek et al., 1991; Dorevitch et al., 1999; Pietras et al., 2005). However, the same compounds can increase aggression when given to or used by individuals who are prone to violence (DiMascio et al., 1969; Brown, 1978; Daderman and Lidberg, 1999). For example, flunitrazepam can be used to calm violent patients in emergency rooms, but when used repeatedly in recreational settings, the same compound has been associated with sexual and aggressive assaults (Daderman and Lidberg, 1999; Dorevitch et al., 1999). The importance of experience and context to GABA$_A$ positive modulator effects on aggression is also evident in mice. Chlordiazepoxide increased aggressive behavior in those animals with previous aggressive experience but decreased aggressive behaviors of mice without previous aggressive experience (Rodgers and Waters, 1985; Ferrari et al., 1997). Similarly, allopregnanolone increased aggression when given to resident male mice with repeated aggressive experiences (Fish et al., 2001), but decreased aggression when given to socially isolated male mice without a history of aggression (Guidotti et al., 2001; Pinna et al., 2003). Moreover, chlordiazepoxide and midazolam exert larger aggression-heightening effects when they are tested under conditions that engender low, rather than high, levels of aggressive behavior (Miczek and O'Donnell, 1980a, b; Fish et al., 2005). The molecular basis for these powerful effects of experience with aggressive behavior apparently includes changes in the expression of $GABA_A$ receptors and in particular some of the yet to be determined subunits.

An individual's prior exposure to drugs may also affect its response to positive modulators of the $GABA_A$ receptor. Alcohol's effect on aggression varies substantially across individuals; a small percentage becomes consistently more aggressive while moderately intoxicated, but most are not affected (e.g., Miczek et al., 1998a). However, mice that have received prior daily alcohol, but not morphine, administrations may "sensitize" to the aggression-heightening effects of alcohol and allopregnanolone (Fish et al., 2002; Miczek et al., 2004b). The ongoing challenge is to understand how an individual's prior history and current level of aggression modify the $GABA_A$ receptor complexes that are necessary for these actions of $GABA_A$-positive modulators.

Such divergent effects of $GABA_A$ receptor positive modulators may be the result of alterations in the subunit composition of the $GABA_A$ receptor complex. The receptor's subunit composition has been proven important for the binding of a drug as well as for particular drug effects. For example, in order for a benzodiazepine to bind, the $GABA_A$ receptor must contain a γ_2 and an α_1, α_2, α_3, or α_5 subunit (Pritchett et al., 1989; Rudolph et al., 1999). The sedative effects of benzodiazepines appear to result from actions at the α_1, rather than the α_2 and α_3 subunits, and the latter appear necessary for antianxiety effects (Low et al., 2000; Rudolph et al., 2001; Rowlett et al., 2005). The functional importance of the α_1 subunit protein has begun to be investigated for its role in aggressive behavior, primarily due to the availability of α_1-preferring agonists, such as zolpidem, and antagonists such as β-CCt and 3-PBC (Cox et al., 1995; Huang et al., 2000). Antagonist effects suggest that the α_1 subunit is important in the aggression-heightening effects of midazolam and alcohol (de Almeida et al., 2004; Gourley et al., 2005). Furthermore, these data suggest a structural dissociation between the anxiety-attenuating and aggression-heightening effects of $GABA_A$ receptor positive modulators.

3.2 Monoamines: Anticipation, Permission, Execution, and Pattern

The canonical amines, norepinephrine (NE), dopamine (DA), and serotonin, are the traditional targets for pharmacotherapeutic interventions of pathological aggression during the past decades. Next to the androgens, they have been studied most extensively for their role in the neurobiology of different kinds of aggressive behavior. The early emphasis on catecholamines in the "fight–flight" syndrome (Cannon, 1929) has shifted to recognizing the importance of serotonin, and this indolamine has emerged as a major research focus. Molecular neurobiological techniques have greatly advanced the study of serotonin and identified several components in the life cycle and receptor subtype diversity of this system that are related to aggression. Although a role for brain serotonin is repeatedly demonstrated for certain types of hostile–impulsive–violent behavior, the precise mechanism through which 5-HT controls these types of aggression is yet to be convincingly demonstrated. One complication to demonstrating mechanisms is the broad distribution of 5-HT throughout the brain and the diversity of the receptor subtypes.

3.2.1 Norepinephrine

Confrontations between individuals are characterized by a high degree of arousal and likewise, increased central and peripheral NE activity. Fighting is one of many behavioral events associated with noradrenergic activation, and this finding has been confirmed in various animal species and contexts (Welch and Welch, 1965; Bell and Hepper, 1987; Summers and Greenberg, 1994; Sgoifo et al., 1996; Gerra et al., 1997). Elevations in the activity of noradrenergic cell bodies in the locus coeruleus and in cortical NE occur when events are important and command an individual's attention; they are not specific to agonistic interactions. In fact, merely observing a fight is sufficient to elevate NE levels in the cortex of mice (Hendley et al., 1973). Perhaps because of the importance of NE to general arousal, a consistent relationship between aggressive behavior and NE is yet to be established. Nonetheless, treatments using β-blockers, such as propranolol, can be highly effective for reducing aggressive outbursts (Elliott, 1977; Yudofsky et al., 1981; Sorgi et al., 1986; Ratey and Gordon, 1993).

3.2.1.1 Levels The correlations between aggressive behavior and levels of NE or its metabolite 3-methoxy-4-hydroxyphenylethylene glycol (MHPG) in the brain or CSF are not as strong or as replicable across studies as those for 5-HT. Some have found positive correlations (Brown et al., 1979; Higley et al., 1992; Traskman-Bendz et al., 1992; Placidi et al., 2001; Kaplan et al., 2002; van der Vegt et al., 2003), while others find inverse (Bernard, 1975; Linnoila et al., 1983) or no correlations (Brown et al., 1982; Reisner et al., 1996; Höglund et al., 2000; Kim et al., 2000). The positive correlations are suggestive for a link between brain NE and aggressive behavior, but genetic manipulations in NE levels offer stronger support.

So far, three genetic manipulations of the NE system have resulted in altered levels of NE and altered aggressive behavior. The effects of deleting the enzyme catechol-*O*-methyltransferase (COMT) are discussed later in the dopamine section, because this mutation does not alter NE levels. Elevations in NE levels have occurred in mice lacking the gene for the norepinephrine transporter (NET) or for the catabolic enzyme monoamine oxidase A (MAO-A) (Cases et al., 1995; Haller et al., 2002). Mice from the 129SvJ × C57BL/6J strain exhibit few retaliatory attacks toward a larger, aggressive resident mouse (Haller et al., 2002). However, those that are missing the NET are more likely to attack the aggressor during the first, but not later, encounters. In agreement with the findings of heightened aggression in male members of a Dutch family with a point mutation in the *MAO-A* gene (Brunner et al., 1993), C3H/HeJ mice without this gene attack intruders faster than do C3H wild-type mice and have significantly more wounding when living in a cage with other males (Cases et al., 1995). Deletion of MAO-A also alters 5-HT levels, so the importance of NE relative to 5-HT to the aggressive phenotype is debatable. NE remains elevated in the MAO-A KO mice throughout adulthood, whereas 5-HT deficiencies are largest during early development, suggesting a more likely acute contribution of NE (Cases et al., 1995; Shih et al., 1999a). Genetically reducing NE in 129/SvEV × C57Bl6/J mice by knocking out the gene coding for dopamine β-hydroxylase produces nonaggressive mice (Marino et al., 2005). Interestingly, the effects of these gene manipulations of the NE system seem to be most significant on the initiation of attack.

Antidepressant treatments that increase extracellular NE by inhibiting the NET or MAO-A have been studied extensively for their effects on aggressive behavior. Under different testing conditions, the aggressive behavior of mice, rats, cats, and monkeys is altered by tricyclics like desipramine, imipramine, and amitriptyline, but there is a significant amount of variability across studies (for review see Miczek et al., 1994a). Many studies have found inhibitory effects on aggressive behavior (e.g., Cook and Wiedley, 1960; Krsiak, 1979; Tobe et al., 1981; Yoshimura, 1984), but there are also examples when low doses of desipramine and imipramine can increase aggressive behavior, such as when isolated mice confront each other (Matsumoto et al., 1991; Cai et al., 1993; Cutler et al., 1997) or when rats respond defensively after receiving footshock (Allikmets and Lapin, 1967; Burov, 1975). NE microinjections into the medial hypothalamic area, which regulates "affective defense" in cats, facilitated this behavioral display (Barrett et al., 1987; Barrett et al., 1990). Conversely, decreasing levels of NE can attenuate offensive aggressive patterns (Crawley and Contrera, 1976) and facilitate some forms of defensive aggression (Thoa et al., 1972).

3.2.1.2 α and β Receptors The actions of NE on aggressive behavior depend upon which receptor is targeted, although limited receptor selectivity of agonists and antagonists prompt cautionary interpretation of these findings (Bell and Hepper, 1987; Haller et al., 1998). The antiaggressive effect of β blockers in human and nonhuman subjects suggests that NE facilitates intensely arousing, hostile aggressive behavior via action at this receptor. However, this calming effect may be accompanied by significant slowing of motor activity, as evident by the sedative effects of β receptor blockade (Hegstrand and Eichelman, 1983; Matsumoto et al., 1991; Bell and Hobson, 1993; Gao and Cutler, 1993). In addition, propranolol, pindolol, and similar β blockers bind to 5-HT_{1A} receptors and antagonize the antiaggressive effects of 8-OH-DPAT (Sanchez et al., 1996). Modulation of 5-HT neurotransmission could be a further mechanism through which β blockers affect aggression.

Both agonists and antagonists at α_2 noradrenergic receptors produce antiaggressive effects, but these effects on aggressive behavior are embedded in very different behavioral and physiological effects of these compounds (Haller and Kruk, 2003)—a constellation of effects that is also apparent with agonists and antagonists at D2 DA receptors. Aggressive mice treated with α_2 receptor antagonists, such as yohimbine, respond with increased defensive, rather than offensive, behaviors (Kemble et al., 1991; Haller et al., 1996b), which may reflect the anxiogenic effects of these compounds (Charney et al., 1983). In rats, there are biphasic effects of α_2 receptor antagonists; low doses increase and high doses decrease aggression, even when the rats are highly aggressive (Haller, 1995). Modulation of aggressive behavior in mice after gene deletion and over-expression provides additional evidence for an inhibitory role of the α_2 receptor subtype. Isolated mice lacking the gene for the α_{2c} receptor attacked an intruder faster than the wild-type mice, whereas mice over-expressing this receptor had the opposite behavioral phenotype (Sallinen et al., 1998). The major effect of this receptor was on the initiation of the attack, because once the fight began, the mice attacked equally.

It is evident that noradrenergic activation is a necessary prerequisite for and consequence of aggressive behavior. One key function of cortical NE is its role in attention to salient events, and aggressive confrontations certainly fulfill this definition. Other functions of locus coeruleus activation during stressful and anxiogenic experiences appear pertinent for the increased defensive reactions after NE agonist treatment. NE can regulate aggressive behavior by acting upon aggression-facilitating β receptors or aggression-inhibiting α adrenoreceptors. However, basal NE levels do not consistently differentiate individuals with a history of past aggression, from those with a history of nonaggressive behavior. This suggests that the largest effects of NE may occur at the time a fight is initiated, possibly by determining whether an individual will fight or flee. Previous experience with victory or defeat is likely to be an important factor in this outcome, as locus coeruleus activation is heavily influenced by past social experiences (Kollack-Walker et al., 1997).

3.2.2 Dopamine

Three sources of information implicate dopaminergic synaptic activity in the neurobiological mechanisms of aggressive behavior, particularly escalated forms of aggression, in both animals and humans. First, studies of the neural mechanisms via which neuroleptic drugs achieve their pharmacotherapeutic effects in violent patients, point to the DA D2 receptor family as a target for effective antiaggressive interventions (Itil and Wadud, 1975; Humble and Berk, 2003). Second, investigations of preclinical model systems, mainly in rodents and cats, highlight the obligatory role of intact DA activity as a prerequisite for initiating and executing aggressive and defensive behavior (Miczek et al., 1994b; Miczek et al., 2002). Thirdly, intense violence has been linked to illicit drugs, but this link refers primarily to trafficking and the violent drug trade (e.g., National Drug Intelligence Center. 2005), and is only very indirectly related to the action of cocaine and amphetamines on brain DA. Careful examination of the data does not support a direct causative link between these drugs and violence in people (Parker and Auerhahn, 1998; Hoaken and Stewart, 2003). Controlled laboratory studies, in which graded doses of amphetamines are administered to human volunteers, have found moderately disruptive effects on aggressive behavior (Cherek et al., 1986; Beezley et al., 1987; Cherek and Steinberg, 1987). These data from experimental studies contrast with the clinical case studies of a select group of addicts, who take large amounts of amphetamines or its congeners and are documented to engage in heinous violent acts (Ellinwood Jr, 1971).

3.2.2.1 Amphetamines and Apomorphine Diverse findings in various animal species and contexts prompt a careful examination of the exact experimental details and behavioral measures when developing an emerging role for brain DA systems in aggressive and defensive behavior. Low doses of indirect and nonselective agonists of DA, such as amphetamines or apomorphine, can increase aggressive behavior of isolated mice or provoked rats, whereas higher doses of amphetamine increase defensive reactions to electric shock pulses or to attacks by an aggressive opponent (Senault, 1968, 1971; Crowley, 1972; Hasselager et al., 1972; Miczek, 1974; Puech et al., 1974; Miczek and O'Donnell, 1978; Ray et al., 1983;

Miczek and Haney, 1994). The hyper-defensive behavior in amphetamine-treated animals may be due to the general stimulus reactivity rather than a specific effect on defense. Ever since the introduction of amphetamines, the effects of enhancing performance in fatigued individuals have been exploited, and this phenomenon can be reproduced under controlled laboratory conditions during extended fights in mice (Winslow and Miczek, 1983). In monkeys and mice with extensive experience in aggressive confrontations, amphetamines disrupt the intricate pattern of aggressive and social interactions (Hodge and Butcher, 1975; Miczek and O'Donnell, 1978; Miczek and Yoshimura, 1982; Field and Pellis, 1994). Finally, chronic treatment with such drugs may have different effects than acute or subchronic exposure (Sokolov et al., 2004). These pharmacological data suggest a role of DA in neurobiological mechanisms of aggressive behavior at multiple levels—to enhance low-level aggressive tendencies, to disorganize species-typical ritualized aggressive displays, to intensify stereotypical aggressive acts, and to prolong aggressive activities. These changes in frequency, duration, intensity, and patterning of aggressive behavior as a result of dopaminergic stimulation parallel those seen on many other motivated behaviors such as feeding, drinking, sexual behavior, or drug taking.

3.2.2.2 Antipsychotics Treatment with antipsychotic drugs that function as DA receptor antagonists continue to be commonly used to reduce aggressive outbursts in schizophrenic patients (Citrome and Volavka, 1997a, b). In addition, such pharmacotherapies can be effective in the treatment of impulsive aggression in conditions like borderline personality disorder (Friedel, 2004). Initially, the usually employed treatments were classic antipsychotic drugs like haloperidol, which are primarily DA D2 antagonists. However, more recently, the atypical neuroleptics that have pronounced effects on other neurotransmitter receptors have been found to be effective at reducing aggressive outbursts with a superior profile of side effects. For example, risperidone, olanzapine, and clozapine have all been reported to be more effective than haloperidol in reducing aggressive behavior in schizophrenic patients (Volavka et al., 2004; Bitter et al., 2005). These atypical antipsychotics drugs are now being used in various psychiatric conditions for reducing aggression and agitation (Bhana et al., 2001; Kennedy et al., 2001; Aman et al., 2004; Ballard and Waite, 2006). In clinical practice, the atypical antipsychotic drugs that are increasingly used for control of aggressiveness are less specific for DA receptors. These so-called atypical compounds have additional effects on neurotransmitter receptors other than DA, including those for serotonin. In addition, one of the populations in which atypical antipsychotics have been reported to be effective are children and adolescents (Findling et al., 2005; Patel et al., 2005). Using an entirely different approach to manipulating DA activity, children and adolescents receive indirect DA agonists (i.e., amphetamine, methylphenidate) to reduce aggressive outbursts associated with ADHD and some other conditions (Connor and Steingard, 1996; Connor et al., 2002; Hazell and Stuart, 2003).

3.2.2.3 Genetic Association and Manipulation Links between polymorphisms in the dopamine transporter (*DAT1*) gene and increased aggressive behavior have been reported for several patients groups (Young et al., 2002b; Chen et al., 2005). In addition, differences in striatal *DAT* distribution or heterogeneity have been reported when comparing violent and nonviolent alcoholics (Tiihonen et al., 1995; Kuikka et al., 1998). Disruption of the *DAT1* gene in mice also leads to an increase in aggressive behavior toward conspecifics (Rodriguiz et al., 2004). This pattern of results could be interpreted to suggest that increased DA in synapses contributes to the escalation of aggression.

The role of catecholamines like DA are implicated in aggressive traits by research on the two primary catabolic enzymes involved in the clearance of catecholamines, COMT and MAO-A. Genetic disruption of either COMT or *MAO-A* genes using the gene deletion ("knock-out") methodology has been found to increase aggressive behavior in male mice (Cases et al., 1995; Gogos et al., 1998). These findings are particularly interesting, because a proclivity to increased aggressive behavior has been reported in patients with a point mutation that reduces MAO-A activity (Brunner et al., 1993). Furthermore, a specific polymorphism in the gene for COMT is also associated with increased aggressive behavior and violence in schizophrenic men (Strous et al., 1997; Lachman and Papolos, 1998; Kotler et al., 1999). However, these enzymes are involved in the degradation of multiple neurotransmitters and thus do not specifically implicate DA in the behavioral outcomes. Furthermore, interpretation of differences in behavior associated with genetic mutations or disruptions must take into account that the individual's development may have been altered in such a way that there are other physiological differences that are a consequence of compensations to altered levels of neurotransmitters during development. Nonetheless, this association of increased aggressive behavior in mice and men with reduced COMT or MAO-A is intriguing and needs to be delineated in terms of anatomical and developmental specificity.

3.2.2.4 D1, D2, and D3 Receptors Preclinical studies using various experimental protocols in animal species have afforded the opportunity to manipulate and measure DA and its receptor systems to assess their role in aggressive behavior. In rodents, drugs that act either as DA D1 or as DA D2 antagonists reduce male aggressive behavior (Janssen et al., 1960; Rolinski, 1975; Krsiak et al., 1981; Olivier and van Dalen, 1982; McMillen et al., 1989; Tidey and Miczek, 1992a; Aguilar et al., 1994; Miczek et al., 1994b; Rodriguez-Arias et al., 1998, 1999b; Bondar and Kudryavtseva, 2005). However, the pharmacological characterization of the DA D1 and DA D2 receptors in terms of their role in aggressive behavior presents a pattern that requires careful and detailed analysis. Both DA D1 and DA D2 agonists and antagonists can decrease aggressive behavior, but the antiaggressive effects of these compounds are embedded in distinctively different profiles of effects. While particularly the DA D2 antagonists slow motor routines and reduce locomotor activity (Fowler and Liou, 1998), the antiaggressive effects appear to be part of the sedative effects of these drugs. By contrast, DA D2 agonists disrupt and fragment the behavioral sequences required for species-typical aggressive behavior. When animals are undergoing withdrawal from opiates, their sensitivity to DA D2 agonists greatly increases (Gianutsos and Lal, 1978). The aggression-heightening effects of indirect and direct DA agonists are readily seen in opiate-withdrawing mice (Kantak and Miczek, 1988; Tidey and Miczek, 1992b). A promising novel target among the DA receptor subtypes is the DA D3 receptor; agonists such as 7-OH-DPAT or PD128907 reduce aggressive behavior without concurrent sedative effects (Gendreau et al., 2000), but the DA D3 antagonist U-99194A also exerts similar antiaggressive effects (Rodriguez-Arias et al., 1999a). As more selectively acting compounds become available and more types of aggressive behavior are studied, the respective role of DA D1, D2, and D3 receptors in reducing aggression will be more adequately characterized.

3.2.2.5 Dopamine Levels The actions of specific DA receptor agonists and antagonists depend on the ongoing up or downregulation of receptors, processes that are regulated by impulse flow in DA pathways that is engendered by aggressive behavior itself. Early postmortem tissue assays pointed already to increased DA levels in the frontal cortex and ventral striatum after the display of aggressive and defensive behavior in mice or rats (Mos and Van Valkenburg, 1979; Haney et al., 1990; Puglisi-Allegra and Cabib, 1990). In the meantime, in vivo microdialysis has provided real-time measurements of DA in corticolimbic structures as aggressive behavior is initiated, executed, terminated, and undergoes recovery. Rises in extracellular DA, presumably reflecting release, in nucleus accumbens and prefrontal cortex are evident in both the offensively aggressive rat and in the defensive–submissive opponent (Tidey and Miczek, 1996; Miczek et al., 1999; Van Erp and Miczek, 2000) (➲ *Figure 7-7*).

With increasing temporal resolution, it may become possible to identify a more precise synchrony between the increase in DA and specific aggressive or defensive acts. The DA release in nucleus accumbens is particularly noteworthy in rats that are targets of threats and react defensively, suggesting a role for accumbal DA that is separate from the often-reiterated interpretation of a reward signal.

After repeated aggressive episodes in regular intervals, neuroadaptive changes in accumbal DA become evident. In anticipation of the very next aggressive confrontation, the extracellular DA in nucleus accumbens rises even before the specific time of the fight, and the heightened DA persists, even in the absence of the accustomed confrontation (Ferrari et al., 2003) (➲ *Figure 7-8*).

These findings suggest that the release of monoamines can be conditioned to prepare the individual for action in a salient event. Heightened corticolimbic DA appears to characterize the individual who is reacting to an aggressive confrontation as well as anticipating such an event, and these different roles may become dissociated with more precise anatomical and temporal analyses.

Figure 7-7
Dopamine and serotonin during aggression. Measurements of extracellular dopamine and serotonin via in vivo microdialysis in resident male rats before, during, and after a confrontation with an intruder. (a) In the nucleus accumbens (*top panel*), dopamine levels (gray circles) rise and remain elevated after the confrontation, while serotonin levels (black diamonds) do not significantly change. (b) In the prefrontal cortex (*bottom panel*), dopamine levels rise after the confrontation, while serotonin decline and remain lower after the confrontation. Samples were collected every 10 min and levels are expressed as mean (±SEM, vertical lines) percent of baseline. Baseline was measured for 50 min before the fight. The vertical light gray bar indicates the occurrence of the 10-min fight. * and ** represent significance from baseline (dashed line) at the $p < 0.05$ and $p < 0.01$ levels, respectively. Copyright 2000 by the Society for Neuroscience. Reprinted with permission from Van Erp and Miczek (2000)

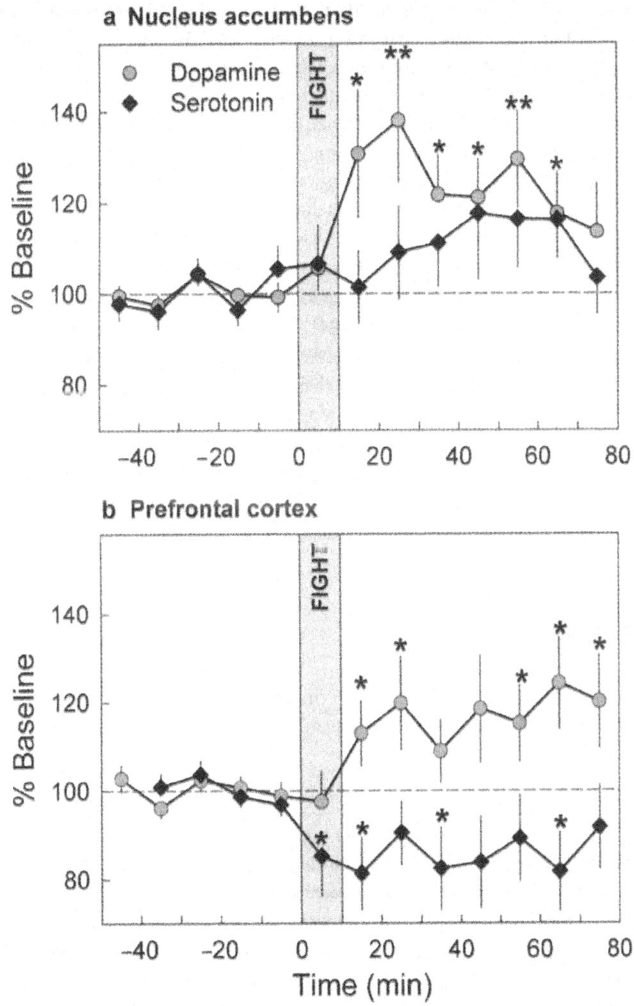

3.3 Serotonin: Inhibitory Control versus Impulse

Brain 5-hydroxytryptamine (5-HT, serotonin) has been the focus of neurobiological research on aggression and violence more than any other neurotransmitter. Ever since the discovery of this indolamine in the mammalian brain (Twarog et al., 1953), a tropotrophic action was proposed as its all-encompassing function (Brodie and Shore, 1957), including also its calming effects on aggression.

3.3.1 CSF Metabolite

The initial evidence relied on postmortem assays of brainstem tissue from previously isolated Swiss albino mice. These animals became aggressive when confronting other isolated mice and also showed lower concentrations of 5-HT and 5-HIAA in the lower portion of the brain, including the brainstem (Giacalone et al., 1968). The first neurochemical data from humans were derived from marine soldiers who were dismissed from the service on account of their life history of repeated aggressive behavior and who had reduced CSF levels of 5-HIAA (negative correlation between CSF 5-HIAA and life history of aggression of $r = -0.78$) (Brown et al., 1979). The inverse correlation between CSF 5-HIAA and trait measures of impulsive aggression has been confirmed subsequently often, particularly in psychopathic deviants (Brown et al., 1982), in impulsive violent offenders (Linnoila et al., 1983), in type II alcoholic violent arsonists (Virkkunen et al., 1987, 1989, 1994), and in behaviorally disruptive youngsters (Kruesi et al., 1990). On the other hand, in several studies, this inverse correlation was not seen, possibly due to the intensity and nature of the aggressive behavior (Gardner et al., 1990; Castellanos et al., 1994; Moller et al., 1996; Coccaro et al., 1997a). In nonhuman primate studies, a similar discrepancy remains unresolved. Significant inverse correlations between CSF 5-HIAA and an individual's prior history of impulsive aggressive behavior are evident in samples of rhesus and pigtail macaques and vervet monkeys (Higley et al., 1992; Westergaard et al., 1999; Manuck et al., 2003; Fairbanks et al., 2004), but not in talapoin monkeys and squirrel monkeys (Yodyingyuad et al., 1985; Miczek and Donat, 1989). The significance of CSF 5-HIAA measures is compromised by the considerable uncertainty as to their precise anatomical origin. CSF samples, whether obtained from the lumbar region or the foramen magnum, reflect the activity of several brain regions, particularly those close to the ventricles, and 5-HT exerts different functions in these cell groups than in circuits that are more distant from the ventricles. Nonetheless, it has been proposed that lumbar samples of CSF 5-HIAA correspond to serotonergic activity in the prefrontal cortex (Doudet et al., 1995), a structure that is significant for behavioral inhibition and impulsive behavior (Raine et al., 1998; Best et al., 2002).

3.3.2 Pharmaco-Challenge

A second approach to define a serotonergic marker in aggressive individuals relies on challenging brain serotonin, either with an agonist such as D,L-fenfluramine or buspirone, or alternatively, a brief exposure

Figure 7-8

Anticipatory changes in nucleus accumbens serotonin and dopamine and heart rate in male rats. The *filled circles* represent the mean (±SEM, vertical lines) measurements from rats that fought an intruder at a specific time (vertical gray bar) everyday over the initial 10 days. The open circles represent measurements from rats without the history of daily aggression and that were entrained to the onset of the dark cycle. Measurements were taken on the 11th day, when no aggressive confrontation occurred. In the *top panel,* dopamine levels (expressed as a percent of baseline) rise in the 10 min before- and in the *middle panel* serotonin levels (expressed as a percent of baseline) decline after the regularly scheduled aggression test. No such changes occur in rats without the prior history of aggression. In the *bottom panel,* the peak heart rate is higher in the rats anticipating the aggression test (*thick lines*) than in those anticipating the light change (*thin lines*). The dark portion of the light–dark cycle is indicated in the horizontal black bar. The *denotes significance ($p < 0.05$) from the light entrained group. Copyright 2003 by Federation of European Neuroscience Societies. Reprinted with permission from Ferrari et al. (2003)

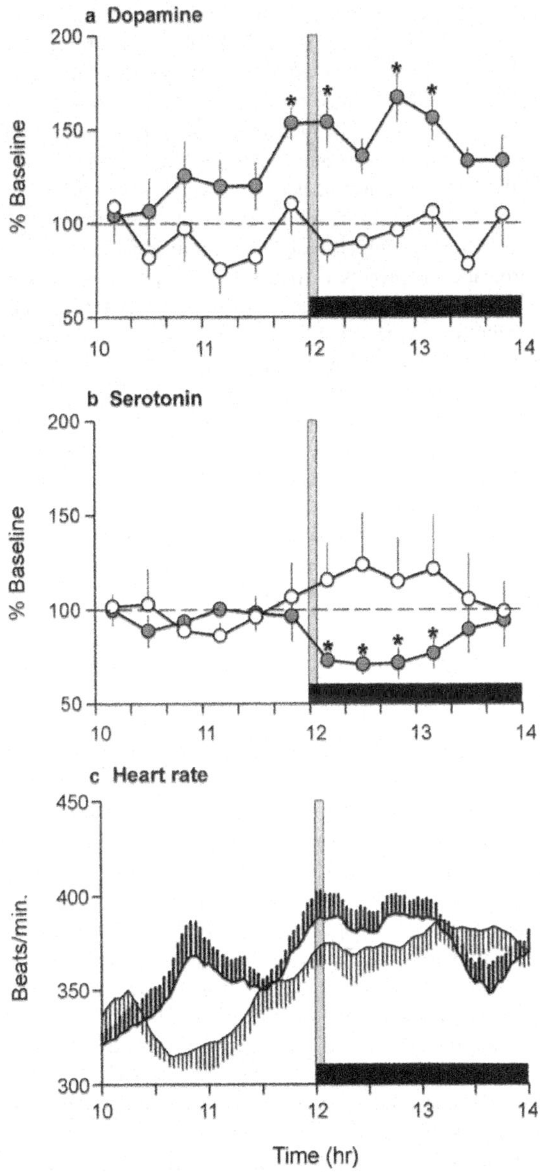

to tryptophan-depleted diet. Initial studies in patients with personality and mood disorders, and also in patients with a history of suicide attempts, demonstrated a blunted plasma prolactin response to a D,L-fenfluramine challenge, relative to patients without those disorders (Coccaro et al., 1997a, b, 1989). This finding has been replicated in violent abstinent alcoholics (Moss et al., 1990; Handelsman et al., 1996), sociopathic violent offenders (O'Keane et al., 1992), personality-disordered patients, and nonpatients (New et al., 1997; Manuck et al., 1998). But, similar to the CSF 5-HIAA studies, there are also notable failures to replicate with blunted prolactin responses to fenfluramine that comprise studies in children and adolescents with attention-deficit and disruptive behavioral problems (Halperin et al., 1994; Pine et al., 1997) and in cocaine addicts (Fishbein et al., 1989; Handelsman et al., 1998). At present, at least two serotonin receptor subtypes, 5-HT$_{1A}$ and 5-HT$_{2C}$, have been proposed as the eventual sites of action, through which indirect or partial serotonin agonists produce the blunted prolactin response in aggressive individuals (Cherek and Lane, 1999; Cherek et al., 1999). Presumably, such an action occurs through either raphé projections to PVN neurons that release oxytocin or vasopressin (VA) which affect prolactin, or via raphé projections to GABAergic neurons in the mediobasal hypothalamus which then inhibit the DA neurons (Mirkes and Bethea, 2001; Emiliano and Fudge, 2004). It will be important to link this mechanism of action with the characteristic corticolimbic activity of impulsive aggressive individuals.

When the essential aromatic acid precursor L-tryptophan is omitted acutely from the diet in humans, an immediate depletion of plasma tryptophan and presumably brain serotonin results (Delgado et al., 1990). In nonpatient research volunteers, acute tryptophan depletion resulted in increased aggressive responding in experimental situations that provoke the subjects by a fictitious opponent and that allow the subject to inflict harm by administering an electrical shock or to take monetary rewards away (Cleare and Bond, 1995; Pihl et al., 1995; Moeller et al., 1996). Conversely, under similar conditions, buspirone and eltoprazine, drugs with partial agonist actions at 5-HT$_{1A}$ receptors, decrease aggressive responding in the point subtraction aggression paradigm (Cherek et al., 1995, 1999). Earlier studies in laboratory mice had provided evidence that tryptophan-free diet resulted in increased aggressive behavior (Kantak et al., 1980), and dietary tryptophan supplements decreased murine aggressive behavior (Lasley and Thurmond, 1985). Overall, these studies on changes in serotonin via dietary manipulations support the serotonin deficiency hypothesis of increased aggressive behavior.

3.3.3 In vivo Microdialysis

Considerably more detailed insights into the dynamic serotonin fluctuations associated with aggressive behavior begin to emerge with in vivo microdialysis sampling. Instead of relying on a single sample of CSF, a peripheral marker, or an endocrine response to a single pharmaco-challenge, in vivo microdialysis offers the opportunity to monitor 5-HT activity in a specific brain region while the experimental animal rests, initiates, executes and terminates aggressive acts and postures, and recovers from an aggressive confrontation. It is noteworthy that during the initiation phase of an aggressive encounter, cortical 5-HT of a resident rat confronting an intruder remains unchanged, whereas a decline in 5-HT in prefrontal cortex, but not in nucleus accumbens, is measured during the actual performance of aggressive behavior and during the termination and recovery (Van Erp and Miczek, 2000). These 5-HT changes in the termination phase of aggressive behavior can be dissociated from the vigorous motor activity that characterizes fighting. Resident rats that have confronted an intruder regularly at a specific time of the day, show a decrease in 5-HT in nucleus accumbens on a day when the fight was omitted (Ferrari et al., 2003). These data suggest that the anticipation of a fight and its termination is sufficient to lower 5-HT in the nucleus accumbens. It will be essential to develop in vivo methodologies that provide higher temporal resolution, so that aggressive acts and samples of serotonergic activity can be better synchronized.

3.3.4 5-HT Receptors

The identification of the genetic basis and molecular features of several receptor families for 5-HT and the 5-HT transporter (SERT) has emerged as the most significant accomplishment during the past decade (Hoyer et al., 2002). This information prompts questions as to the functional significance of these receptors, including their role in modulating aggressive behavior. So far, only the first two families of serotonin receptors, belonging to the G-protein coupled superfamily, have been studied with regard to their role in aggressive behavior.

3.3.4.1 Serotonin$_1$ Family The 5-HT$_1$ family comprises 5-HT$_{1A}$, 5-HT$_{1B}$, and 5-HT$_{1D}$ receptors (Barnes and Sharp, 1999; Hoyer et al., 2002), and it has been implicated in the modulation of various behaviors ranging from sex, feeding, drug self-administration, sleep, anxiety, depression, and also aggression (Olivier et al., 1995; Lee and Simansky, 1997; Ahlenius and Larsson, 1998; Hillegaart and Ahlenius, 1998; Parsons et al., 1998; Sari, 2004). A high concentration of 5-HT$_{1A}$ receptors is located in the raphé nuclei and provides negative feedback via somatodendritic synapses (Zifa and Fillion, 1992). A second population of 5-HT$_{1A}$ receptors is found as postsynaptic heteroreceptors in the terminal regions of serotonergic projections. Two major populations of 5-HT$_{1B}$ receptors can be distinguished; one serves as presynaptic autoreceptors on axon terminals and a second one as postsynaptic heteroreceptors in nonserotonergic neurons in terminal areas (Zifa and Fillion, 1992). There are also 5-HT$_{1B}$ receptors found in the raphé nuclei acting as autoreceptors (Hoyer et al., 1985, 1986; Sari et al., 1999; Sari, 2004). Application of 5-HT$_{1A}$ agonists onto midbrain raphé tissue slices decreases the firing rate of these 5-HT neurons (Sinton and Fallon, 1988a, b; Sprouse and Aghajanian, 1988; Sprouse, 1991). In contrast, 5-HT$_{1B}$ agonists have either no effect or potentiate the firing rate of raphé neurons (Sprouse and Aghajanian, 1987; Evrard et al., 1999). The contrasting effects of 5-HT$_{1A}$ and 5-HT$_{1B}$ agonists on raphé electrophysiology may reflect activation of somatodendritic versus heteroreceptors.

Behaviorally, systemic administration of 5-HT$_1$ agonists dose dependently decreases species-typical levels of aggression in mice and rats (Olivier and Mos, 1986, 1992; Muehlenkamp et al., 1995; Sanchez and Meier, 1997; Miczek et al., 1998b; de Boer et al., 1999; Fish et al., 1999; de Boer et al., 2000; McKenzie-Quirk et al., 2005). In the dose range that produces effective reductions in aggressive behavior, most of these ligands also sedate, slow motor routines, or can induce stereotyped movements, indicating a behaviorally nonspecific effect of these drugs on aggressive behavior (Miczek et al., 1998b; de Boer and Koolhaas, 2005; McKenzie-Quirk et al., 2005). Alnespirone, S15535, and some other compounds in development that act presumably on subpopulations of 5-HT$_{1A}$ receptors differ from the prototypic 8-OH-DPAT in that they reduce specifically aggressive behavior in feral rats (de Boer et al., 1999, 2000; de Boer and Koolhaas, 2005). The reversal by WAY 100635 points to the 5-HT$_{1A}$ receptor as the site of action for the antiaggressive effects of these compounds. Genetic studies, however, generate evidence that is difficult to reconcile with the pharmacological data. Mutant mice without the gene for the 5-HT$_{1A}$ receptor protein actually exhibit less aggressive behavior (Zhuang et al., 1999). In mice that have been selectively bred for attacking rapidly, the mRNA for 5-HT$_{1A}$ receptors and the binding to this receptor are significantly higher in the hippocampus and prefrontal cortex (Korte et al., 1996). When challenged with the 5-HT$_{1A}$ receptor agonist alnespirone, short attack latency mice and feral aggressive rats were significantly more sensitive to the hypothermic effects, suggesting a heightened, rather than a deficient, 5-HT$_{1A}$ receptor activity in highly aggressive individuals (Van Der Vegt et al., 2001).

The distinctive feature of 5-HT$_{1B}$ receptor agonists is that, in most situations, they appear to specifically reduce aggressive behavior without sedation (Olivier et al., 1990; Fish et al., 1999; de Almeida et al., 2001; de Almeida and Miczek, 2002; de Boer and Koolhaas, 2005) (➋ *Figure 7-9*).

Ligands such as CP-94,253 and CP-93,129, that selectively bind to the 5-HT$_{1B}$ receptor subtype (Macor et al., 1990; Koe et al., 1992), proved to be promising pharmacological tools with antiaggressive effects in mice and rats. Microinjections of CP-94,253 into the ventral orbitofrontal cortex effectively reduced aggressive behavior in mice (de Almeida et al., 2006).

The precise mechanisms and sites of action for the antiaggressive effects by 5-HT$_{1A}$ and 5-HT$_{1B}$ receptor agonists remain unclear. Microdialysis measurements indicate that systemic administration of 5-HT$_{1A}$ agonists decreases extracellular levels of 5-HT in the striatum, hippocampus, frontal cortex, nucleus accumbens, and medial septum, although the magnitude of the decrease varies between brain regions (Hjörth and Sharp, 1991; Kreiss and Lucki, 1994; Gobert et al., 1998; Dekeyne et al., 2000). Similarly, systemic administration of 5-HT$_{1B}$ receptor agonists decreases extracellular levels of 5-HT in the striatum, hippocampus, and prefrontal cortex (Knobelman et al., 2000a, b; Johnson et al., 2001; De Groote et al., 2003). These data would point to an inhibition of 5-HT release by 5-HT$_{1A}$ and 5-HT$_{1B}$ receptor agonists, presumably via stimulation of somatodendritic receptors at raphé neurons (Blier et al., 1998; Adell et al., 2002). The decrease in extracellular levels of corticolimbic 5-HT on the one hand, and

■ Figure 7-9

Serotonin$_{1B}$ receptor agonists and aggression. Specific reduction of aggressive behavior by 5-HT$_{1B}$ receptor agonists. Panels a–c portray the effects of the varying doses of 5-HT$_{1B}$ agonists, eltoprazine, CGS-12066B, and CP-93,129 on aggressive and nonaggressive behaviors in male wild-type Groningen rats confronting an intruder. Panel d portrays the effect of the 5-HT$_{1B/1D}$ receptor antagonist GR-127935. Copyright 2005 by Elsevier B.V. Reprinted with permission from de Boer and Koolhaas (2005)

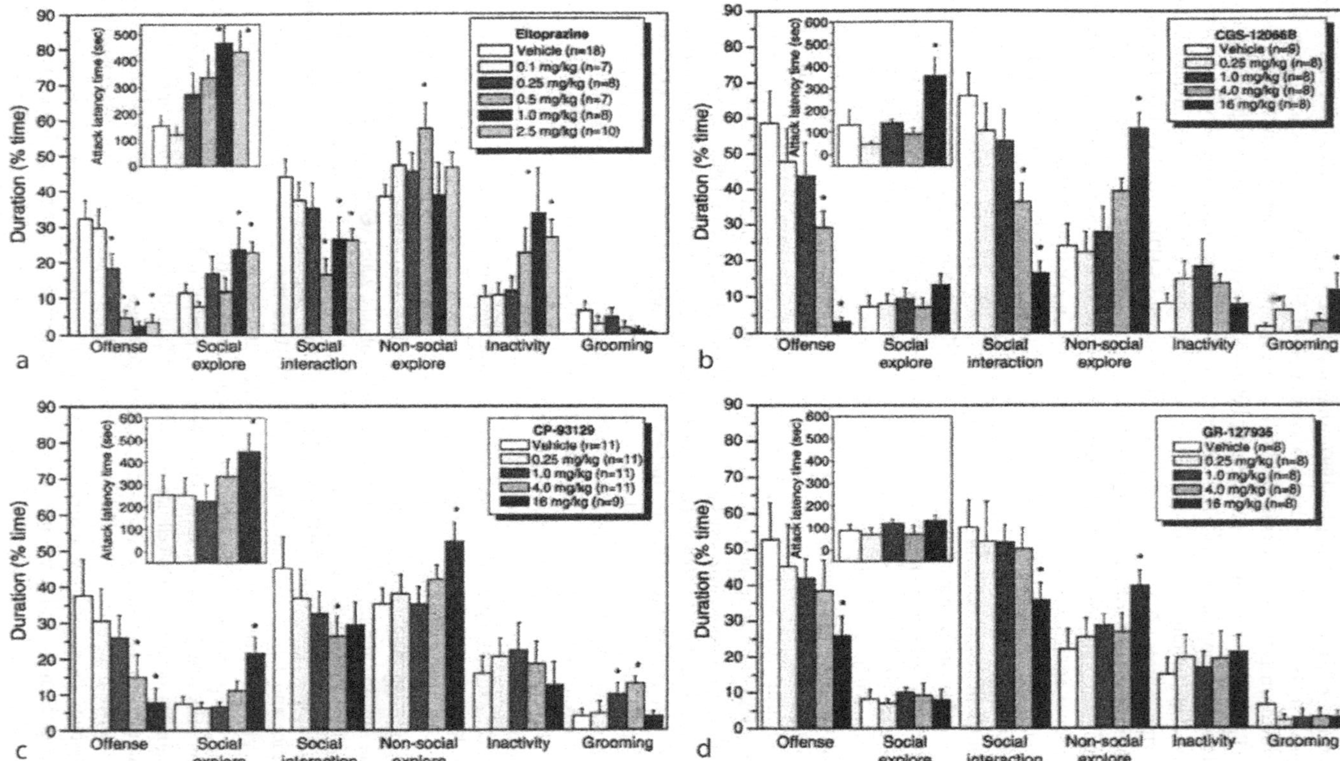

the antiaggressive effects on the other, represent a significant challenge to the serotonin deficiency hypothesis of aggression.

A second mechanism for the antiaggressive effects of 5-HT_{1A} and 5-HT_{1B} receptor agonists focuses on the postsynaptic heteroreceptors (Olivier and Van Oorschot, 2005). For example, when presynaptic 5-HT neurons are destroyed by the neurotoxin 5,7-DHT, the antiaggressive effects of 8-OH-DPAT, eltoprazine, and zolmitriptan remain unaffected, suggesting a postsynaptic site of action (Sijbesma et al., 1991; de Almeida et al., 2001). If in fact, low-5-HT levels are inversely related to intense, impulsive aggressive behavior, then differential sensitivity due to presynaptic gating via 5-HT_{1B} receptors would be predicted. Yet, 5-HT_{1B} receptor agonists reduce aggressive behavior independent from the basal level; species-typical normative as well as escalated types of aggressive behavior due to social instigation or consumption of low alcohol doses is effectively reduced by 5-HT_{1B} receptor agonists (Mos et al., 1992; Fish et al., 1999; de Almeida and Miczek, 2002). These observations can be interpreted in support of a postsynaptic site of action for the antiaggressive effects of 5-HT_{1B} receptor agonists.

So far, the contribution of gene deletion studies of the 5-HT_{1A} and 5-HT_{1B} receptors has been obscured by the influence of the background strain of mice in which null mutations were implemented. It is difficult to interpret the level of aggressive behavior in mice in which the gene for the 5-HT_{1B} receptor was deleted, when in fact the level of aggressive behavior in the wild-type 129 strain mice is species-atypically absent, possibly influenced by abnormalities in brain structures (Saudou et al., 1994b; Bouwknecht et al., 2001; Wahlsten et al., 2001). Moreover, pharmacological antagonists of 5-HT_{1B} receptors do not alter different types of aggressive behavior, which has been interpreted to reflect the low serotonergic tone at this receptor (Olivier and Van Oorschot, 2005).

Some data begin to emerge that link allelic variation in the polymorphic gene for the 5-HT_{1B} receptor to type II antisocial alcoholism (Lappalainen et al., 1998), repeated suicidal behavior (New et al., 2001), and attention deficits and behavior disorders in children. There is no differentiation between pre- and postsynaptic 5-HT_{1B} receptors in terms of allele frequency, and the functional significance of these polymorphisms remains to be determined.

3.3.4.2 Serotonin$_2$ Family The 5-HT_2 receptor family became relevant to the neurobiology of aggression, with the development of neuroleptics that achieve their effects via antagonism of 5-HT_{2A} receptors. These compounds proved effective in the clinical management of aggressive behavior in children, adolescents, middle-aged, and elderly patients with diagnoses ranging from schizophrenia, dementia, depression, and posttraumatic stress disorder (Czobor et al., 1995; Buckley et al., 1997; De Deyn et al., 1999; Keck Jr et al., 2000; Buitelaar et al., 2001; Zarcone et al., 2001; McCracken et al., 2002; Swanson et al., 2004; Bitter et al., 2005; Kratochvil et al., 2005; LeBlanc et al., 2005; Nolan et al., 2005; Volavka et al., 2005). As discussed with neuroleptic drugs acting on DA D2 receptors, the lack of behavioral specificity of 5-HT_{2A} agonists in reducing aggressive behavior is a critical issue. Preclinical studies in isolated male mice highlight that the antiaggressive effects of risperidone and similarly acting drugs can be achieved at doses that also inhibit motor activity (Rodriguez-Arias et al., 1998), and similar results are seen with ketanserin in MAO-A-deficient mice (Shih et al., 1999b).

Microinjections of DOI, a substituted phenylisopropylamine that acts as an agonist at both 5-HT_{2A} and HT_{2C} receptors, into the periaqueductal gray of cats enhance the defensive hissing that was evoked by hypothalamic stimulation (Shaikh et al., 1997). This effect is most likely related to the anxiety-like behavior that results from activation of these receptors (Lucki and Wieland, 1990; Nogueira and Graeff, 1995). In experimental models of murine aggression, agonists at 5-HT_{2A} and 5-HT_{2C} receptors decrease aggressive behavior concurrently with other motor activities (Sanchez et al., 1993; Bonson et al., 1994; de Almeida and Lucion, 1994; Baxter et al., 1995; Muehlenkamp et al., 1995; Olivier et al., 1995). A more satisfactory characterization of 5-HT_2 receptor subtypes regarding their role in aggressive behavior will have to await the development of more specific molecular tools for targeting these receptors.

3.3.4.3 Serotonin Transporter In support of the 5-HT deficiency hypothesis, restoring levels of 5-HT represents an effective means to calm and sedate aggressive individuals. Indeed, drugs that block the reuptake mechanism for 5-HT, the 5-HT transporter (SERT), reduce aggressive outbursts, particularly when administered over extended periods of time (Fava and Rosenbaum, 1993; Knutson et al., 1998; Swann, 2003; New et al., 2004). Although there are occasional reports of increased aggressivity and suicidal tendencies among those receiving selective serotonin reuptake inhibitors (SSRIs) (Troisi et al., 1995; Spigset, 1999), these compounds are beneficial for most patients, as documented by large meta-analyses of the accumulated clinical studies (Heiligenstein et al., 1993; Walsh and Dinan, 2001). The reports of increased aggressive behavior should not be dismissed, however, because they reveal the existence of individual differences in the response to SSRIs. These differences may be related to the etiology of the aggressive behavior (i.e., whether it is associated with symptoms of an underlying disorder such as depression or with a personality disorder such as an antisocial personality) or to the individual's genotype and unique expression profile of the SERT. As modification of 5-HT levels continues to be a focus of pharmacotherapies, understanding the nature of these individual differences may lead to more precisely targeted and safer treatments.

The importance of the SERT to aggressive behavior is also evident in studies on animal aggression. Acutely, SSRIs reduce aggression in species as diverse as lobsters, dogs, monkeys, hamsters, and mice (Olivier et al., 1989; Ferris and Delville, 1994; Dodman et al., 1996; Huber et al., 1997; Pinna et al., 2003; Sanchez et al., 2003). The few long-term studies that have administered SSRIs repeatedly point to important treatment conditions and prevailing levels of aggressive behavior as critical determinants of outcome. Laboratory rats that typically do not engage in fighting have been shown to become moderately more aggressive after long-term SSRIs (Mitchell, 2005). However, in mice, consuming alcohol and fighting at high levels of repeated citalopram administration reduces aggression (Caldwell and Miczek, 2005); similar findings characterize prairie voles (Villalba et al., 1997). Apparently concordant with the 5-HT deficiency hypothesis, the antiaggressive effects of SSRIs can be interpreted to result from the elevations in extracellular 5-HT, but this interpretation does not differentiate between the acute and the chronic effects. It is also possible that the effects of SSRIs increase the $GABA_A$ receptor positive modulator allopregnanolone (Pinna et al., 2003). This mechanism is interesting, because allopregnanolone can inhibit or activate aggression, depending on the individual's prior experience (see GABA section), and could contribute to the occasional increases in aggression during the clinical use of SSRIs.

The genes coding for the SERT have also been related to aggressive behavior, as demonstrated by studies in mutant mice ("knockout") and associations with polymorphisms. Creation of mice in which the *SERT* gene was deleted results in a long-lasting reduction of aggressive behavior, increased 5-HT synthesis, and reduced 5-HT stores (Holmes et al., 2002; Kim et al., 2005). SERT KO mice also show elevated anxiety-like behavior (Holmes et al., 2003a, b), which may also contribute to the reduced aggressive behavior. A further hypothesis postulates that the altered aggressive behavior results from long-lasting developmental consequences of *SERT* gene deletion. 5-HT is essential for neural development (Lauder, 1993; Azmitia, 1999), and sustained 5-HT elevations during early life following repeated exposure to an SSRI generate a more anxious-like phenotype into adulthood (Ansorge et al., 2004). Whether the altered aggressive phenotype is the result of ongoing or past changes in 5-HT neurotransmission remains an unresolved but urgent issue.

Human and nonhuman primate studies have focused on polymorphisms in the *SERT* gene that confer variation in SERT expression and 5-HT uptake. The 5-HT transporter linked polymorphic region (5-HTTLPR) has been associated with several behavioral and personality traits related to aggression and other affective disorders (Lesch et al., 1996; Lesch, 2005), but there is controversy regarding the particular allelic variation. Many studies find a stronger association between the short allele, which renders lower expression of the transporter, and affective disturbances (e.g., Lesch et al., 1996; Hallikainen et al., 1999; Gorwood et al., 2000; Greenberg et al., 2000; Haberstick et al., 2006). However, the long allele has also been associated with aggressive feelings and prior aggressive acts (Twitchell et al., 2001; Zalsman et al., 2001), not only in children but also alcoholics (Parsian and Cloninger, 2001) and patients with dementia (Sukonick et al., 2001; Sweet et al., 2001). The apparent disagreement between these findings does not discount the importance of this association between the 5-HTTLPR and aggression. The contribution of the 5-HTTLPR to the variability in aggressive personality traits is relatively small and subject to obligatory interactions between this gene, other genes, and the environmental triggers (Bouchard, 1994; Lesch and Mossner, 1998). Nonetheless, a thorough understanding of how SERT expression influences aggressive behavior and extracellular concentrations of 5-HT in specific brain regions could prompt a revision of the 5-HT deficiency hypothesis.

3.4 Dynamic Interactions: Reciprocal Neurochemistry Coordinates Behavior

One persistent puzzle with brain serotonin is how so very little of this amine can affect so very much. The consistent link of brain 5-HT to aggressive behavior, although often seemingly contradictory, is based on action of this indolamine at distinctively different levels of the neuroaxis modulating sensory, motivational, and performance aspects of aggression. The 5-HT system is anatomically organized to both directly and indirectly affect several facets of aggression. First, serotonergic projections arising from the dorsal, median, and linear raphé nuclei are widely distributed throughout the mid- and forebrain. The areas most heavily influenced by 5-HT include the striatum, hippocampus, hypothalamus, thalamus, amygdala, septum, frontal cortex, substantia nigra, ventral tegmental area, locus coeruleus, and the periaqueductal gray area. By acting directly within these pathways, 5-HT can influence aggression in several ways, from the perception of triggering stimuli, which involves the olfactory bulb, the amygdala, and frontal cortex, to the intricately patterned motor output necessary for aggressive acts and postures, which involves the substantia nigra, striatum, thalamus, motor cortex, and the descending motor neurons. When these regions are interconnected, 5-HT may further affect their functions by altering the activity of projection neurons. Serotonergic activity in the prefrontal cortical terminals is of particular relevance in the initiation and termination of aggressive acts as part of the inhibitory control of behavior. To illustrate the diversity in 5-HT's neural control over aggression behavior, the following section focuses on 5-HT's interactions with other neurochemical systems in the brainstem, hypothalamus, and cortex (➋ *Figure 7-10*).

3.4.1 Mesolimbic Dopamine

Serotonergic neurons from the dorsal raphé (DR) nucleus project directly to the VTA and substantia nigra (Herve et al., 1987; Corvaja et al., 1993; Gervais and Rouillard 2000). Serotonergic 5-HT_{1A} and 5-HT_{2A} receptors can also be found in the VTA and substantia nigra (Wright et al., 1995), and colocalization of 5-HT_{2A} receptors on TH positive neurons has been described (Doherty and Pickel, 2000). Pharmacological methods have demonstrated that this serotonergic input can effectively modulate mesencephalic DA activity, which in turn alters telencephalic DA release (Schmidt and Fadayel, 1995; Pehek et al., 2001; Nocjar et al., 2002). This neuroanatomical and neuropharmacological evidence suggests a basis for serotonergic modulation of DA-mediated motivational and motoric processes of aggressive behavior.

In the ventral striatum, the nucleus accumbens septi is an area that is critical for the motivation of many goal-directed behaviors (Robbins and Everitt, 1996; Wise, 2005), including aggressive behavior (Miczek et al., 1994b) and impulsive choices (Cardinal et al., 2001). Serotonergic, dopaminergic, and noradrenergic projections overlap in the nucleus accumbens in both rats and humans (Gaspar et al., 1985; Delfs et al., 1998; Tong et al., 2006). The demonstration of increases in accumbal DA in advance of the regularly scheduled aggressive confrontations provides evidence of a potentially important role of this amine in the anticipation, preparation, and prediction for a salient event (Ferrari et al., 2003). The anatomical data indicate how this catecholaminergic activity in aggressive individuals is modulated by serotonergic input.

3.4.2 Hypothalamic Vasopressin

Neuroanatomical and neurophysiolgical studies established hypothalamic regions to be essential for the performance of attack behavior, identifying HAAs in rats and cats (Flynn, 1967; Halasz et al., 2002). 5-HT neurons project into this area and regulate the release and action of VP, a neuropeptide associated with high levels of aggression in several animal species (Compaan et al., 1992; Winslow et al., 1993; Ferris, 1996; Haller et al., 1996a; Ferris et al., 1997; Bester-Meredith and Marler, 2001), through actions at 5-HT_{1A} and 5-HT_{1B} receptors (Ferris et al., 1997, 1999). A reduction in VP levels in the AH may also contribute to the antiaggressive effects of fluoxetine (Altemus et al., 1992; Ferris, 1996).

A possible interaction between 5-HT and VP on aggression is an opposing influence on the same neuron. VP and 5-HT innervation are proximally oriented in the AH (Ferris et al., 1997). If 5-HT and VP

Figure 7-10
Serotonin interactions and facets of aggression. Serotonergic innervation of brain structures involved in various aspects of aggressive behavior. The brain in sagittal view is presented schematically with raphé nuclei shown in dark gray oblong shapes and major ascending and descending serotonergic projections shown in black arrow paths. Regions that are prominently involved in aggression are highlighted in text boxes. Each text box indicates neurochemicals with which serotonin interacts and suggests a functional outcome of these interactions. For simplicity, several projection regions are not highlighted, such as the hippocampus, thalamus, septum, and nucleus accumbens

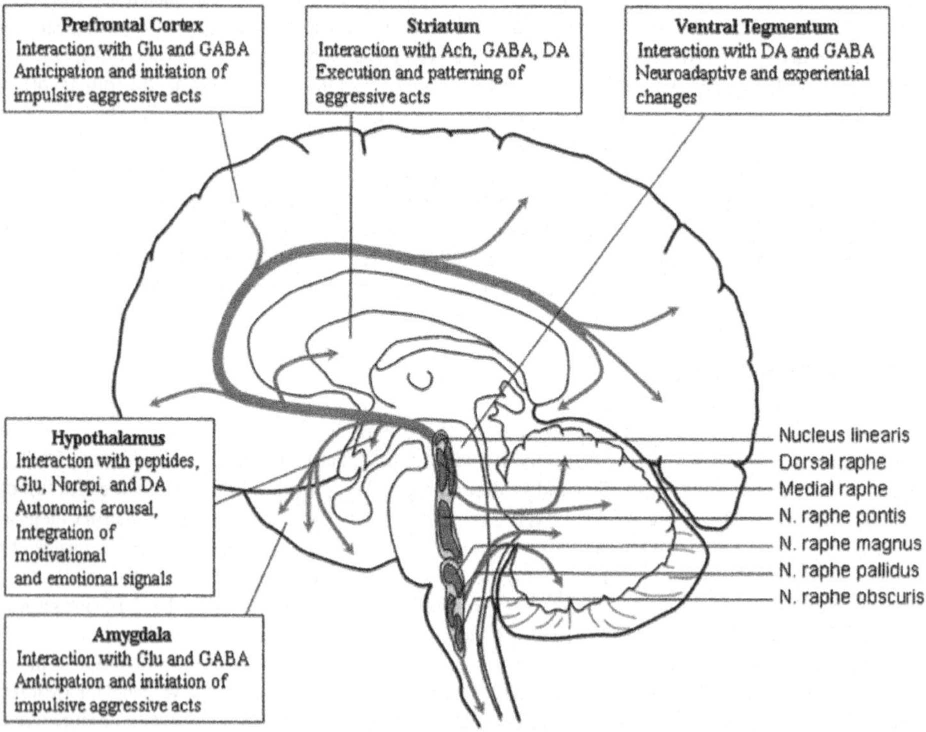

affect the same neuron, it is likely to be glutamatergic, given the high expression of vesicular glutamate transzrter in the HAA (Hrabovszky et al., 2005) and the importance of glutamate stimulation of attack (Siegel et al., 1999). A second and not exclusive interaction is that 5-HT affects the soma of VP neurons at other locations, such as the medial supraoptic nucleus and nucleus circularis to alter release (Ferris et al., 1997). The change in VP levels after fluoxetine may be anatomically specific to the AH, because fluoxetine has been shown to increase peripheral VP levels (Gibbs and Vale, 1983). Moreover, administration of 5-HT_1 and 5-HT_2 agonists did not affect VP mRNA levels in the SON (Jorgensen et al., 2003), though this does not preclude an effect on release.

3.4.3 Cortical Glutamate and GABA

A further region where 5-HT can act to regulate aggressive behavior is the PFC, particularly the medial PFC. In human studies, this brain area is consistently recognized as being important to several psychiatric diseases with aggressive symptoms, as well as being involved in impulsive and aggressive behaviors (Raine et al., 1998, 2000; Best et al., 2002). The PFC coordinates behavioral and autonomic responses to complex environmental stimuli, and has been implicated in sensory motor gating, attention and cognitive processes, motivation, motor control, as well as the hormonal and cardiovascular responses to stressors (Verberne and Owens, 1998). All of these processes are involved in various components of aggressive behavior, such as the motivation to fight, the initiation and maintenance of fighting, and the termination and recovery from fighting.

Anatomically, the PFC is well situated for these diverse functions and to influence aggressive behavior. It projects to motivational and motor regions such as the dorsal and the ventral striatum, ventral tegmental area, and spinal cord, sensory regions such as the mediodorsal thalamus, and affective regions including the basolateral amygdala, lateral hypothalamus, septum, PAG, DR, and nucleus of the tractus solitarius (Verberne and Owens, 1998; Carr and Sesack, 2000; Grace and Rosenkranz, 2002; Gabbott et al., 2005). Several of these regions directly influence aggressive behavior (Siegel et al., 1999).

In the PFC, 5-HT modulates the release of other neurotransmitters, particularly GABA and glutamate. The 5-HT projection to the PFC incorporates several receptors, including the SERT, 5-HT$_{1A}$, 5-HT$_{1B}$, and 5-HT$_2$ receptors. 5-HT can act at presynaptic heteroreceptors or postsynaptic receptors to alter cellular activity and extracellular GABA and glutamate concentrations, particularly in cortical layer V, and lead to further changes in any of the PFC projection regions. The PFC's inhibition of activity and 5-HT release in the DR (Celada et al., 2001) or the activity of the BLA (Rosenkranz and Grace, 2002) are two examples of how 5-HT's initial actions in the PFC can stretch throughout the neural circuitry regulating aggression.

The effect of 5-HT on GABA and glutamate concentrations is differentially regulated by 5-HT$_{1A}$ and 5-HT$_2$ receptors. Stimulation of 5-HT$_2$ receptors increases levels of GABA and glutamate by depolarizing GABA interneurons and by acting at the terminals of glutamatergic afferents from the thalamus (Marek and Aghajanian, 1998; Lambe et al., 2000) in layer V of the cortex (Lambe et al., 2000). 5-HT$_{1A}$ receptors are located postsynaptically on pyramidal and GABA interneurons and can be colocalized with 5-HT$_{2A}$ receptors (Amargos-Bosch et al., 2004; Santana et al., 2004). Stimulation of 5-HT$_{1A}$ receptors in the PFC hyperpolarizes neurons and can reduce glutamate and 5-HT release in this region (Calcagno et al., 2006).

3.4.4 Dorsal and Median Raphé Auto- and Heteroregulation

The dorsal and median raphé nuclei are important sites in which 5-HT and other neurochemicals can interact to control aggressive behavior. It is in this region that the activity of 5-HT can be modulated by several distal projection neurons as well as local interneurons and its own release through somatodendritic autoreceptors. 5-HT can be influenced by the catecholamines through NE fibers from the locus coeruleus and DA from the SN and VTA. The function of NE is to increase 5-HT cell firing (Baraban and Aghajanian, 1980) and release, when measured in the raphé (Adell and Artigas, 1999) or PFC (Linner et al., 2004). This effect most likely involves the α1 receptor (Svensson et al., 1975), because the antagonist prazosin inhibits NE-stimulated release (Adell and Artigas, 1999). α2 receptors in the DR and MR appear to have the opposite effect by inhibiting 5-HT release (Clement et al., 1992), and antagonists to these receptors facilitate 5-HT release (de Boer et al., 1996). However, the actions of DA on the DR 5-HT activity are less clearly defined (Adell et al., 2002). GABA can decrease 5-HT through neurons projecting from the PAG and hypothalamus as well as local interneurons (Tao et al., 1996; Abellan et al., 2000b; Tao and Auerbach, 2002). This inhibition is caused by GABA$_A$ receptors located on 5-HT cell bodies (Gao et al., 1993), rather than GABA$_B$ receptors that stimulate 5-HT activity (Abellan et al., 2000a). An opposing excitation can arise from the mPFC through NMDA, AMPA, and kainate receptors (Tao and Auerbach, 2000). Changes in 5-HT may represent a common path through which these neurotransmitters can influence aggression. The ultimate expression of a neurotransmitter's effect on aggression may depend upon its actions at 5-HT cells in the raphé. Modulation of 5-HT may also be a mechanism through which genetic manipulations affect aggression.

3.4.5 Indirect Alterations by Gene Manipulation

Serotonergic function has been indirectly altered in several knockout mice that have unusual aggressive phenotypes, such as the MAO-A, NAα2C, NCAM, H1 receptor, α-CaMKII, Pet-1, and nNOS knockouts

(Miczek et al., 2001). For example, the high levels of aggressive behavior by mutant mice lacking the gene for the neuronal isoform of nitric oxide synthase (nNOS$^{-/-}$) occur with impaired 5-HT turnover and reduced sensitivity of 5-HT$_{1A}$ and 5-HT$_{1B}$ receptor-mediated feedback control (Chiavegatto et al., 2001; Chiavegatto and Nelson, 2003). The lack of nNOS apparently causes dysfunctions in 5-HT metabolism in cortex, hypothalamus, midbrain, and cerebellum, while the endothelial isoform of NOS appears to facilitate 5-HT turnover. NO may modulate the receptor function for the 5-HT$_{1A}$ and 5-HT$_{1B}$ subtypes in these areas, rendering them less effective for mediating antiaggressive effects.

4 Conclusions

Aggressive behavior is the expression of several neurochemical interactions, each of which regulates the display of different behavioral elements. There is a consistently demonstrated significant role of 5-HT synthesis, release, and interaction with receptor and uptake sites, particularly in the affective dimension of impulsive, hostile, reactive, and violent behavior. The instrumental dimension, which requires proactive calculation, planning, premeditation, and memory, is based on a considerably more distributed neural circuitry. The mesocorticolimbic DA system in interaction with 5-HT, excitatory and inhibitory amino acids, monoamines, and peptides is critical for the rewarding aspects of aggressive behavior and may form the basis of recidivist acts of aggression.

A research strategy on aggressive behavior needs to pursue 5-HT as a potential target for pharmacotherapies and/or a mechanism underlying aggressive phenotypes. Given the limited available tools, stimulation of 5-HT$_{1A}$ and 5-HT$_{1B}$ receptors are promising targets for effective antiaggressive treatments, with the latter showing considerably more behavioral specificity than stimulation or blockade of other 5-HT receptor subtypes. We discuss evidence that emphasizes the reciprocal interactions between 5-HT, monoamines, and peptides.

It will be useful to learn how 5-HT interacts with DA, opioid peptides, and glutamate in the VTA for the long-term neuroadaptive changes that are engendered by repeated aggressive experiences. The interactions between 5-HT, glutamate, and GABA in the prefrontal cortex need to be defined systematically to understand their roles in the initiation and the inhibition of impulsive aggressive acts. In addition to the cortical influences on limbic and hypothalamic regions with significance for aggressive behavior, amygdaloid modulation of these subcortical areas requires neurochemical analysis that is related to aggressive outcomes. The promising information on CRF, GABA, and glutamate in these amygdaloid connections with hypothalamic and brainstem structures in situations of intense emotional behavior (Davis and Whalen, 2001; Pare et al., 2004) should prompt a close examination of these candidate mechanisms in escalated types of aggressive behavior.

The research agenda for molecular neurochemical studies needs to define the cascade of events that convey vulnerability and resilience to aggressive behavior. Early life experiences during critical developmental periods trigger gene expression that appears relevant to adult aggressive behavior, and the recent studies on early childhood maltreatment and adult changes in 5-HT metabolism represent promising examples of this interaction (Caspi et al., 2002). Individuals displaying hypoarousal and low glucocorticoid activation may be particularly prone to engage in intensely violent acts (Raine, 1996; Haller and Kruk, 2006). Further advances in in vivo methodologies will create the opportunity to study momentary neurochemical changes that match the time scale of the sudden behavioral outbursts more adequately than the current microdialysis approach. For example, it will be important to learn to which degree dopaminergic signals in the prefrontal cortex correspond to uncontrollable aggressive impulses devoid of serotonergic inhibition. Moreover, rapid measurements could resolve the issue of cause and consequence. Are altered 5-HT release and impulse flow necessary to alter aggressive behavior, or are they consequences of altered aggressive behavior? These need not be mutually exclusive possibilities but the distinction may be important for diagnosis and treatment, as well as for understanding mechanism. Additionally, molecular neurochemical studies may identify novel components of the 5-HT system that specifically regulate the threshold to trigger aggressive behavior but do not disrupt 5-HT's coordination of ongoing neural activity.

Acknowledgments

The authors express their gratitude to J. Thomas Sopko for his assistance with manuscript preparation and to NIAAA and NIDA grants (AA013983, DA02632) for financial support.

References

Abellan MT, Martin-Ruiz R, Artigas F. 2000b. Local modulation of the 5-HT release in the dorsal striatum of the rat: an in vivo microdialysis study. Eur Neuropsychopharmacol 10: 455-462.

Abellan MT, Jolas T, Aghajanian GK, Artigas F. 2000a. Dual control of dorsal raphe serotonergic neurons by GABA(B) receptors. Electrophysiological and microdialysis studies. Synapse 36: 21-34.

Adamec RE. 1990. Does kindling model anything clinically relevant? Biol Psychiatry 27: 249-279.

Adams DB, Boudreau W, Cowan CW, Kokonowski C, Oberteuffer K, et al. 1993. Offense produced by chemical stimulation of the anterior hypothalamus of the rat. Physiol Behav 53: 1127-1132.

Adell A, Artigas F. 1999. Regulation of the release of 5-hydroxytryptamine in the median raphe nucleus of the rat by catecholaminergic afferents. Eur J Neurosci 11: 2305-2311.

Adell A, Celada P, Abellan MT, Artigas F. 2002. Origin and functional role of the extracellular serotonin in the midbrain raphe nuclei. Brain Res Rev 39: 154-180.

Agrawal HC, Fox MW, Himwich WA. 1967. Neurochemical and behavioral effects of isolation-rearing in the dog. Life Sci 6: 71-78.

Aguilar MA, Minarro J, Perez-Iranzo N, Simon VM. 1994. Behavioral profile of raclopride in agonistic encounters between male mice. Pharmacol Biochem Behav 47: 753-756.

Ahlenius S, Larsson K. 1998. Evidence for an involvement of 5-HT_{1B} receptors in the inhibition of male rat ejaculatory behavior produced by 5-HTP. Psychopharmacology 137: 374-382.

Allikmets LK, Lapin IP. 1967. Influence of lesions of the amygdaloid complex on behaviour and on effects of antidepressants in rats. Int J Neuropharmacol 6: 99-108.

Altemus M, Cizza G, Gold PW. 1992. Chronic fluoxetine treatment reduces hypothalamic vasopressin secretion in vitro. Brain Res 593: 311-313.

Aman MG, Binder C, Turgay A. 2004. Risperidone effects in the presence/absence of psychostimulant medicine in children with ADHD, other disruptive behavior disorders, and subaverage IQ. J Child Adol Psychopharmacol 14: 243-254.

Amargos-Bosch M, Bortolozzi A, Puig MV, Serrats J, Adell A, et al. 2004. Co-expression and in vivo interaction of serotonin$_{1A}$ and serotonin$_{2A}$ receptors in pyramidal neurons of prefrontal cortex. Cereb Cortex 14: 281-299.

Ansorge MS, Zhou M, Lira A, Hen R, Gingrich JA. 2004. Early-life blockade of the 5-HT transporter alters emotional behavior in adult mice. Science 306: 879-881.

Arnone M, Dantzer R. 1980. Effects of diazepam on extinction induced aggression in pigs. Pharmacol Biochem Behav 13: 27-30.

Azmitia EC. 1999. Serotonin neurons, neuroplasticity, and homeostasis of neural tissue. Neuropsychopharmacology 21: 33-45.

Azrin NH, Hutchinson RR, Hake DF. 1966. Extinction-induced aggression. J Exp Anal Behav 9: 191-204.

Badino P, Odore R, Osella MC, Bergamasco L, Francone P, et al. 2004. Modifications of serotonergic and adrenergic receptor concentrations in the brain of aggressive *Canis familiaris*. Comp Biochem Physiol Mol Int Physiol 139: 343-350.

Ballard C, Waite J. 2006. The effectiveness of atypical antipsychotics for the treatment of aggression and psychosis in Alzheimer's disease. Cochrane Database Syst Rev.

Bandler R, Depaulis A, Vergnes M. 1985. Identification of midbrain neurons mediating defensive behaviour in the rat by microinjections of excitatory amino acids. Behav Brain Res 15: 107-119.

Baraban JM, Aghajanian GK. 1980. Suppression of firing activity of 5-HT neurons in the dorsal raphe by alpha-adrenoceptor antagonists. Neuropharmacology 19: 355-363.

Barnes NM, Sharp T. 1999. A review of central 5-HT receptors and their function. Neuropharmacology 38: 1083-1152.

Barratt ES. 1993. The use of anticonvulsants in aggression and violence. Psychopharmacol Bull 29: 75-81.

Barrett JA, Edinger H, Siegel A. 1990. Intrahypothalamic injections of norepinephrine facilitate feline affective aggression via α2-adrenoceptors. Brain Res 525: 285-293.

Barrett JA, Shaikh MB, Edinger H, Siegel A. 1987. The effects of intrahypothalamic injections of norepinephrine upon affective defense behavior in the cat. Brain Res 426: 381-384.

Baxter G, Kennett G, Blaney F, Blackburn T. 1995. 5-HT2 receptor subtypes: a family re-united? Trends Pharmacol Sci 16: 105-110.

Bear DM, Fedio P. 1977. Quantitative analysis of interictal behavior in temporal lobe epilepsy. Arch Neurol 34: 454-467.

Beezley DA, Gantner AB, Bailey DS,Taylor SP. 1987. Amphetamines and human physical aggression. J Res Pers 21: 52–60.

Bell R, Hepper PG. 1987. Catecholamines and aggression in animals. Behav Brain Res 23: 1-21.

Bell R, Hobson H. 1993. Effects of (-)-pindolol and SDZ 216–525 on social and agonistic behavior in mice. Pharmacol Biochem Behav 46: 873-880.

Belozertseva IV, Bespalov AY. 1999. Effects of NMDA receptor channel blockade on aggression in isolated male mice. Aggress Behav 25: 381-396.

Bennett AJ, Lesch KP, Heils A, Long JC, Lorenz JG, et al. 2002. Early experience and serotonin transporter gene variation interact to influence primate CNS function. Mol Psychiatry 7: 118-122.

Bernard BK. 1975. Aggression and the brain monoamines: what are the answers, but of more importance, what are the questions? Aminergic hypotheses of behavior: reality or cliche? (NIDA research monograph series 3). Bernard BK, editor. Rockville, MD: National Institute on Drug Abuse; pp. 71-84.

Bernstein IS. 1981. Dominance: the baby and the bathwater. Behav Brain Sci 4: 419-457.

Bernstein IS, Gordon TP, Rose RM. 1974. Aggression and social controls in rhesus monkey (*Macaca mulatta*) groups revealed in group formation studies. Fol Primatol 21: 81-107.

Berthold AA. 1849. Transplantation der Hoden. Archiv fuer Anatomie, Physiologie und wissenschaftliche Medicin 16: 42-46.

Best M, Williams JM, Coccaro EF. 2002. Evidence for a dysfunctional prefrontal circuit in patients with an impulsive aggressive disorder. Proc Natl Acad Sci USA 99: 8448-8453.

Bester-Meredith JK, Marler CA. 2001. Vasopressin and aggression in cross-fostered California mice (*Peromyscus californicus*) and white-footed mice (*Peromyscus leucopus*). Horm Behav 40: 51-64.

Bhana N, Foster RH, Olney R, Plosker GL. 2001. Olanzapine: an updated review of its use in the management of schizophrenia. Drugs 61: 111-161.

Bitter I, Czobor P, Dossenbach M, Volavka J. 2005. Effectiveness of clozapine, olanzapine, quetiapine, risperidone, and haloperidol monotherapy in reducing hostile and aggressive behavior in outpatients treated for schizophrenia: a prospective naturalistic study (IC-SOHO). Eur Psychiatry 20: 403-408.

Blanchard RJ, Hebert MA, Ferrari P, Palanza P, Figueira R, et al. 1998. Defensive behaviors in wild and laboratory (Swiss) mice: the mouse defense test battery. Physiol Behav 65: 201-209.

Blier P, Pineyro G, El Mansari M, Bergeron R, de Montigny C. 1998. Role of somatodendritic 5-HT autoreceptors in modulating 5-HT neurotransmission. Ann N Y Acad Sci 861: 204-216.

Bond A, Lader M. 1988. Differential effects of oxazepam and lorazepam on aggressive responding. Psychopharmacology 95: 369-373.

Bond AJ, Curran HV, Bruce MS, O'Sullivan G, Shine P. 1995. Behavioural aggression in panic disorder after 8 weeks' treatment with alprazolam. J Affect Disord 35: 117-123.

Bondar NP, Kudryavtseva NN. 2005. Repeated aggressive experience and efficacy of anxiolytics. Eur Neuropsychopharmacol 15: S142.

Bonson KR, Johnson RG, Fiorella D, Rabin RA, Winter JC. 1994. Serotonergic control of androgen-induced dominance. Pharmacol Biochem Behav 49: 313-322.

Bouchard TJJ. 1994. Genes, environment, and personality. Science 264: 1700-1701.

Bouwknecht JA, HijzenTH, van der GJ, Maes RA, Hen R, et al. 2001. Absence of 5-HT(1B) receptors is associated with impaired impulse control in male 5-HT(1B) knockout mice. Biol Psychiatry 49: 557-568.

Brain PF. 1975. What does individual housing mean to a mouse? Life Sci 16: 187-200.

Braunewell KH, Manahan-Vaughan D. 2001. Long-term depression: a cellular basis for learning? Rev Neurosci 12: 121-140.

Broad KD, Hinton MR, Keverne EB, Kendrick KM. 2002a. Involvement of the medial prefrontal cortex in mediating behavioural responses to odour cues rather than olfactory recognition memory. Neuroscience 114: 715-729.

Broad KD, Hinton MR, Keverne EB, Kendrick KM. 2002b. Involvement of the medial prefrontal cortex in mediating behavioural responses to odour cues rather than olfactory recognition memory. Neuroscience 114: 715-729.

Brodie BB, Shore PA. 1957. A concept for a role of serotonin and norepinephrine as chemical mediators in the brain. Analgesia 66: 631-642.

Brodkin ES, Goforth SA, Keene AH, Fossella JA, Silver LM. 2002. Identification of quantitative trait Loci that affect aggressive behavior in mice. J Neurosci 22: 1165-1170.

Brown CR. 1978. The use of benzodiazepines in prison populations. J Clin Psychiatry 38: 219-222.

Brown GL, Goodwin FK, Ballenger JC, Goyer PF, Major LF. 1979. Aggression in humans correlates with cerebrospinal fluid amine metabolites. Psychiatry Res 1: 131-139.

Brown GL, Ebert MH, Goyer PF, Jimerson DC, Klein WJ, et al. 1982. Aggression, suicide, and serotonin-relationships to CSF amine metabolites. Am J Psychiatry 139: 741-746.

Brunner HG, Nelen M, Breakefield XO, Ropers HH, van Oost BA. 1993. Abnormal behavior associated with a point mutation in the structural gene for monoamine oxidase A. Science 262: 578-580.

Buckley PF, Ibrahim ZY, Singer B, Orr B, Donenwirth K, et al. 1997. Aggression and schizophrenia: efficacy of risperidone. J Am Acad Psychiatry Law 25: 173-181.

Buitelaar JK, van der Gaag RJ, Cohen-Kettenis P, Melman TM. 2001. A randomized controlled trial of risperidone in the treatment of aggression in hospitalized adolescents with subaverage cognitive abilities. J Clin Psychiatry 62: 239-248.

Burkhalter JE, Balster RL. 1979. The effects of phencyclidine on isolation-induced aggression in mice. Psychol Rep 45: 571-576.

Burov YV. 1975. The influence of psychotropic drugs upon emotions. CNS Behav Pharmacol 3: 197-205.

Cai B, Matsumoto K, Ohta H, Watanabe H. 1993. Biphasic effects of typical antidepressants and mianserin, an atypical antidepressant, on aggressive behavior in socially isolated mice. Pharmacol Biochem Behav 44: 519-525.

Cairns RB, Nakelski JS. 1971. On fighting in mice: ontogenetic and experiential determinants. J Comp Physiol Psychol 74: 354-364.

Calcagno E, Carli M, Invernizzi RW. 2006. The 5-HT(1A) receptor agonist 8-OH-DPAT prevents prefrontocortical glutamate and serotonin release in response to blockade of cortical NMDA receptors. J Neurochem 96: 853-860.

Caldwell EE, Miczek KA. 2005. Effects of chronic citalopram treatment on alcohol-heightened aggression in mice Program No. 76.11. 2005 *Abstract Viewer/Itinerary Planner.* Washington, DC: Society for Neuroscience, 2005. Online.

Calkins SD, Johnson MC. 1998. Toddler regulation of distress to frustrating events: temperamental and maternal correlates. Infant Behav Dev 21: 379-395.

Cannon WB. 1929. Organization of physiological homeostasis. Physiol Rev 9: 399-431.

Cardinal RN, Pennicott DR, Sugathapala CL, Robbins TW, et al. 2001. Impulsive choice induced in rats by lesions of the nucleus accumbens core. Science 292: 2499-2501.

Carr DB, Sesack SR. 2000. Projections from the rat prefrontal cortex to the ventral tegmental area: target specificity in the synaptic associations with mesoaccumbens and mesocortical neurons. J Neurosci 20: 3864-3873.

Cases O, Seif I, Grimsby J, Gaspar P, Chen K, et al. 1995. Aggressive behavior and altered amounts of brain serotonin and norepinephrine in mice lacking MAOA. Science 268: 1763-1766.

Caspi A, McClay J, Moffitt TE, Mill J, Martin J, et al. 2002. Role of genotype in the cycle of violence in maltreated children. Science 297: 851-854.

Castellanos FX, Elia J, Kruesi MJP, Gulotta CS, Mefford IN, et al. 1994. Cerebrospinal-fluid monoamine metabolites in boys with attention-deficit hyperactivity disorder. Psychiatry Res 52: 305-316.

Celada P, Puig MV, Casanovas JM, Guillazo G, Artigas F. 2001. Control of dorsal raphe serotonergic neurons by the medial prefrontal cortex: involvement of serotonin-1A, GABA(A), and glutamate receptors. J Neurosci 21: 9917-9929.

Charney DS, Heninger GR, Redmond DE. 1983. Yohimbine induced anxiety and increased noradrenergic function in humans—effects of diazepam and clonidine. Life Sci 33: 19-29.

Chen C, Rainnie DG, Greene RW, Tonegawa S. 1994. Abnormal fear response and aggressive behavior in mutant mice deficient for alpha-calcium-calmodulin kinase II. Science 265: 291-294.

Chen JS, Kelz MB, Hope BT, Nakabeppu Y, Nestler EJ. 1997. Chronic Fos-related antigens: stable variants of delta FosB induced in brain by chronic treatments. J Neurosci 17: 4933-4941.

Chen S, Lee AY, Bowens NM, Huber R, Kravitz EA. 2002. Fighting fruit flies: a model system for the study of aggression. Proc Natl Acad Sci USA 99: 5664-5668.

Chen TJH, Blum K, Mathews D, Fisher L, Schnautz N, et al. 2005. Are dopaminergic genes involved in a predisposition to pathological aggression? Hypothesizing the importance of "super normal controls" in psychiatricgenetic research of complex behavioral disorders. Med Hypoth 65: 703-707.

Cherek DR, Heistad GT. 1971. Fixed-interval induced aggression. Psychon Sci 25: 7-8.

Cherek DR, Lane SD. 1999. Effects of *d,l* fenfluramine on aggressive and impusive responding in adult males with a history of conduct disorder. Psychopharmacology 146: 473-481.

Cherek DR, Moeller FG, Khan-Dawood F, Swann A, Lane SD. 1999. Prolactin response to buspirone was reduced in violent compared to nonviolent parolees. Psychopharmacology 142: 144-148.

Cherek DR, Pietras CJ. 2003. Human aggression: biological correlates and environmental influences. Disorders of brain and mind. Ron MA, Robbins TW, editors. Cambridge: Cambridge University Press; pp. 375-399.

Cherek DR, Spiga R, Creson DL. 1995. Acute effects of eltoprazine on aggressive and point-maintained responding of normal male participants: phase I study. Exp Clin Psychopharmacol 3: 287-293.

Cherek DR, Spiga R, Roache JD, Cowan KA. 1991. Effects of triazolam on human aggressive, escape and point-maintained responding. Pharmacol Biochem Behav 40: 835-839.

Cherek DR, Steinberg JL. 1987. Effects of drugs on human aggressive behavior. Advances in human psychopharmacology.

Burrows GD, editor. Greenwich, CN: JAI Press; pp. 239-290.

Cherek DR, Steinberg JL, Kelly TH, Robinson DE. 1986. Effects of D-amphetamine on human aggressive behavior. Psychopharmacology 88: 381-386.

Cherek DR, Steinberg JL, Kelly TH, Robinson DE, Spiga R. 1990. Effects of acute administration of diazepam and D-amphetamine on aggressive and escape responding of normal male subjects. Psychopharmacology 100: 173-181.

Chi CC, Flynn JP. 1971. Neuroanatomic projections related to biting attack elicited from hypothalamus in cats. Brain Res 35: 49-66.

Chiavegatto S, Nelson RJ. 2003. Interaction of nitric oxide and serotonin in aggressive behavior. Horm Behav 44: 233-241.

Chiavegatto S, Dawson VL, Mamounas LA, Koliatsos VE, Dawson TM, et al. 2001. Brain serotonin dysfunction accounts for aggression in male mice lacking neuronal nitric oxide synthase. Proc Natl Acad Sci USA 98: 1277-1281.

Christie MH, Barfield RJ. 1979. Effects of castration and home cage residency on aggressive behavior in rats. Horm Behav 13: 85-91.

Citrome L, Volavka J. 1997a. Psychopharmacology of violence: part I: assessment and acute treatment. Psychiatr Ann 27: 691-695.

Citrome L, Volavka J. 1997b. Psychopharmacology of violence: part II: beyond the acute episode. Psychiatr Ann 10: 696-703.

Cleare AJ, Bond AJ. 1995. The effect of tryptophan depletion and enhancement on subjective and behavioural aggression in normal male subjects. Psychopharmacology 118: 72-81.

Clement HW, Gemsa D, Wesemann W. 1992. The effect of adrenergic drugs on serotonin metabolism in the nucleus raphe dorsalis of the rat, studied by in vivo voltammetry. Eur J Pharmacol 217: 43-48.

Clement J, Simler S, Ciesielski L, Mandel P, Cabib S, et al. 1987. Age-dependent changes of brain GABA levels, turnover rates and shock-induced aggressive behavior in inbred strains of mice. Pharmacol Biochem Behav 26: 83-88.

Coccaro EF, Kavoussi RJ, Cooper TB, Hauger RL. 1997a. Central serotonin activity and aggression: inverse relationship with prolactin response to D-Fenfluramine, but not CSF 5-HIAA concentration, in human subjects. Am J Psychiatry 154: 1430-1435.

Coccaro EF, Kavoussi RJ, Trestman RL, Gabriel SM, Cooper TB, et al. 1997b. Serotonin function in human subjects: intercorrelations among central 5-HT indices and aggressiveness. Psychiatry Res 73: 1-14.

Coccaro EF, Siever LJ, Klar HM, Maurer G, Cochrane K, et al. 1989. Serotonergic studies in patients with affective and personality disorders. Arch Gen Psychiatry 46: 587-599.

Cole HF, Wolf HH. 1970. Laboratory evaluation of aggressive behavior of the grasshopper mouse (*Onychomys*). J Pharm Sci 59: 969-971.

Compaan JC, Koolhaas JM, Buijs RM, Pool CW, de Ruiter AJH, et al. 1992. Vasopressin and the individual differentiation in aggression in male house mice. Oxytocin in maternal, sexual, and social behaviors. Pedersen CA, editor. New York: New York Academy of Sciences; pp. 458-459.

Connor DF, Steingard RJ. 1996. A clinical approach to the pharmacotherapy of aggression in children and adolescents. Understanding Aggressive Behavior in Children. Ann N Y Acad Sci 794: 290-307.

Connor DF, Glatt SJ, Lopez ID, Jackson D, Melloni RH. 2002. Psychopharmacology and aggression. I: a meta-analysis of stimulant effects on overt/covert aggression-related behaviors in ADHD. J Am Acad Child Adoles Psychiatry 41: 253-261.

Cook L, Wiedley E. 1960. Effects of a series of psychopharmacological agents on isolation induced attack behavior in mice. Fed Proceed 19: 22.

Corvaja N, Doucet G, Bolam JP. 1993. Ultrastructure and synaptic targets of the raphe-nigral projection in the rat. Neuroscience 55: 417-427.

Covington HE, Miczek KA. 2002. NMDA receptor antagonists block behavioral sensitization, but not enhanced cocaine self-administration after social defeat stress. Abstr Soc Neurosci 28: 897.11.

Cox ED, Hagen TJ, McKernan RM, Cook JM. 1995. BZ[1] receptor subtype specific ligands. Synthesis and biological properties of betaCCt, a BZ[1] receptor subtype specific antagonist. Med Chem Res 5: 710-718.

Crawley JN, Contrera JF. 1976. Intraventricular 6-hydroxydopamine lowers isolation-induced fighting behavior in male mice. Pharmacol Biochem Behav 4: 381-384.

Crawley JN, Schleidt WM, Contrera JF. 1975. Does social environment decrease propensity to fight in male mice? Behav Biol 15: 73-83.

Crews D, Moore MC. 1986. Evolution of mechanisms controlling mating behavior. Science 231: 121-125.

Crowcroft P, Rowe FP. 1963. Social organization and territorial behaviour in the wild house mouse. Proc Zool Soc London 140: 517-531.

Crowley TJ. 1972. Dose-dependent facilitation or supression of rat fighting by methamphetamine, phenobarbital, or imipramine. Psychopharmacologia 27: 213-222.

Cutler MG, Rodgers RJ, Jackson JE. 1997. Behavioural effects in mice of subchronic buspirone, ondansetron, and tianeptine. I. Social interactions. Pharmacol Biochem Behav 56: 287-293.

Czobor P, Volavka J, Meibach RC. 1995. Effect of risperidone on hostility in schizophrenia. J Clin Psychopharmacol 15: 243-249.

D'Adamo P, Welzl H, Papadimitriou S, Raffaele dB, Tiveron C, et al. 2002. Deletion of the mental retardation gene Gdi1 impairs associative memory and alters social behavior in mice. Hum Mol Genet 11: 2567-2580.

Daderman AM, Lidberg L. 1999. Flunitrazepam (Rohypnol) abuse in combination with alcohol causes premeditated, grievous violence in male juvenile offenders. J Am Acad Psychiatry Law 27: 83-99.

Davis M, Whalen PJ. 2001. The amygdala: vigilance and emotion. Mol Psychiatry 6: 13-34.

de Almeida RM, Rosa MM, Santos DM, Saft DM, Benini Q, et al. 2006. 5-HT(1B) receptors, ventral orbitofrontal cortex, and aggressive behavior in mice. Psychopharmacology (Berl) 185: 441-450.

de Almeida RMM, Lucion A. 1994. Effects of intracerebroventricular administration of 5-HT receptor agonists on the maternal aggression of rats. Eur J Neurosci 264: 445-448.

de Almeida RMM, Miczek KA. 2002. Aggression escalated by social instigation or by discontinuation of reinforcement ("frustration") in mice: inhibition by anpirtoline—a 5-HT1B receptor agonist. Neuropsychopharmacology 27: 171-181.

de Almeida RMM, Ferrari PF, Parmigiani S, Miczek KA. 2005. Escalated aggressive behavior: dopamine, serotonin and GABA. Eur J Pharmacol 526: 51-64.

de Almeida RMM, Nikulina EM, Faccidomo S, Fish EW, Miczek KA. 2001. Zolmitriptan-a 5-HT1B/D agonist, alcohol, and aggression in mice. Psychopharmacology (Berl) 157: 131-141.

de Almeida RMM, Rowlett JK, Cook JM, Yin W, Miczek KA. 2004. GABAA/alpha1 receptor agonists and antagonists: effects on species-typical and heightened aggressive behavior after alcohol self-administration in mice. Psychopharmacology (Berl) 172: 255-263.

de Boer SF, Koolhaas JM. 2005. 5-HT1A and 5-HT1B receptor agonists and aggression: a pharmacological challenge of the serotonin deficiency hypothesis. Eur J Pharmacol 526: 125-139.

de Boer SF, Lesourd M, Mocaer E, Koolhaas JM. 1999. Selective antiaggressive effects of alnespirone in resident-intruder test are mediated via (5-hydroxytryptamine)1A receptors: a comparative pharmacological study with 8-hydroxy-2-dipropylaminotetralin, ipsapirone, buspirone, eltoprazine, and WAY-100635. J Pharmacol Exp Ther 288: 1125-1133.

de Boer SF, Lesourd M, Mocaer E, Koolhaas JM. 2000. Somatodendritic 5-HT(1A) autoreceptors mediate the anti-aggressive actions of 5-HT(1A) receptor agonists in rats: an ethopharmacological study with S-15535, alnespirone, and WAY-100635. Neuropsychopharmacology 23: 20-33.

de Boer TH, Nefkens F, van Helvoirt A, van Delft AM. 1996. Differences in modulation of noradrenergic and serotonergic transmission by the alpha-2 adrenoceptor antagonists, mirtazapine, mianserin and idazoxan. J Pharmacol Exp Ther. 277: 852-860.

De Bold JF, Miczek KA. 1981. Sexual dimorphism in the hormonal control of aggressive behavior of rats. Pharmacol Biochem Behav 14 Suppl 1: 89-93.

De Deyn PP, Rabheru K, Rasmussen A, Bocksberger JP, Dautzenberg PL, et al. 1999. A randomized trial of risperidone, placebo, and haloperidol for behavioral symptoms of dementia. Neurology 53: 946-955.

De Felipe C, Herrero JF, O'Brien JA, Palmer JA, Doyle CA, et al. 1998. Altered nocieption, analgesia and aggression in mice lacking the receptor for substance P. Nature 392: 394-397.

De Groote L, Olivier B, Westenberg HG. 2003. Role of 5-HT1B receptors in the regulation of extracellular serotonin and dopamine in the dorsal striatum of mice. Eur J Pharmacol 476: 71-77.

Dekeyne A, Brocco M, Adhumeau A, Gobert A, Millan MJ. 2000. The selective serotonin $(5\text{-HT})_{1A}$ receptor ligand, S15535, suppresses dialysate levels of 5-HT in teh dorsal hippocampus of freely-moving rats and displays anxiolytic properties in the social interaction and Vogel paradigms: a comparison to other anxiolytic agents. Psychopharmacology 152: 55-66.

Del Punta K, Leinders-Zufall T, Rodriguez I, Jukam D, Wysocki CJ, et al. 2002. Deficient pheromone responses in mice lacking a cluster of vomeronasal receptor genes. Nature 419: 70-74.

Delfs JM, Zhu Y, Druhan JP, Aston-Jones G.S. 1998. Origin of noradrenergic afferents to the shell subregion of the nucleus accumbens: anterograde and retrograde tract-tracing studies in the rat. Brain Res 806: 127-140.

Delgado PL, Charney DS, Price LH, Aghajanian GK, Landis H, et al. 1990. Serotonin function and the mechanism of antidepressant action—reversal of antidepressant-induced remission by rapid depletion of plasma tryptophan. Arch Gen Psychiatry 47: 411-418.

Demas GE, Kriegsfeld LJ, Blackshaw S, Huang P, Gammie SC, et al. 1999. Elimination of aggressive behavior in male mice lacking endothelial nitric oxide synthase. J Neurosci 19: 1-5.

Denenberg VH, Gaulin-Kremer E, Gandelman R, Zarrow MX. 1973. The development of standard stimulus animals for mouse (*Mus musculus*) aggression testing by means of olfactory bulbectomy. Anim Behav 21: 590-598.

Dietch JT, Jennings RK. 1988. Aggressive dyscontrol in patients treated with benzodiazepines. J Clin Psychiatry 49: 184-188.

Di Mascio A. 1973. The effects of benzodiazepines on aggression: reduced or increased? Psychopharmacologia 30: 95-102.

Di Mascio A, Shader RI, Harmatz J. 1969. Psychotropic drugs and induced hostility. Psychosomatics 10: 46-47.

Dodman NH, Donnelly R, Shuster L, Mertens P, Rand W, et al. 1996. Use of fluoxetine to treat dominance aggression in dogs. J Am Vet Med Assoc 209: 1585-1587.

Doherty MD, Pickel VM. 2000. Ultrastructural localization of the serotonin 2A receptor in dopaminergic neurons in the ventral tegmental area. Brain Res 864: 176-185.

Dollard J, Doob L, Miller N, Mowrer O, Sears R. 1939. Frustration and aggression. New Haven: Yale University Press.

Dorevitch A, Katz N, Zemishlany Z, Aizenberg D, Weizman A. 1999. Intramuscular flunitrazepam versus intramuscular haloperidol in the emergency treatment of aggressive psychotic behavior. Am J Psychiatry 156: 142-144.

Doudet D, Hommer D, Higley JD, Andreason PJ, Moneman R, et al. 1995. Cerebral glucose metabolism, CSF 5-HIAA levels, and aggressive behavior in rhesus monkeys. Am J Psychiatry 152: 1782-1787.

Ducic I, Puia G, Vicini S, Costa E. 1993. Triazolam is more efficacious than diazepam in a broad spectrum of recombinant GABAA receptors. Eur J Pharmacol 244: 29-35.

Dulawa SC, Gross C, Stark KL, Hen R, Geyer MA. 2000. Knockout mice reveal opposite roles for serotonin 1A and 1B receptors in prepulse inhibition. Neuropsychopharmacology 22: 650-659.

Duysen EG, Stribley JA, Fry DL, Hinrichs SH, Lockridge O. 2002. Rescue of the acetylcholinesterase knockout mouse by feeding a liquid diet; phenotype of the adult acetylcholinesterase deficient mouse. Brain Res Dev Brain Res 137: 43-54.

Early CJ, Leonard BE. 1980. The effect of testosterone and cyproterone acetate on the concentration of γ-aminobutyric acid in brain areas of aggressive and non-aggressive mice. Pharmacol Biochem Behav 12: 189-193.

Edwards DH, Kravitz EA. 1997. Serotonin, social status and aggression. Curr Opin Neurobiol 7: 812-819.

Eibl-Eibesfeldt I. 1950. Beiträge zur Biologie der Haus- und der Ährenmaus nebst einigen Beobachtungen an anderen Nagern. Z Tierpsychol 7: 558-587.

Eibl-Eibesfeldt I. 1961. The fighting behavior of animals. Sci Am 203: 112-120.

Ellinwood EH Jr. 1971. Assault and homicide associated with amhetamine abuse. Am J Psychiatry 127: 90-95.

Elliott FA. 1977. Propanolol for the control of belligerant behavior following acute brain damage. Ann Neurol 1: 489-491.

Emiliano AB, Fudge JL. 2004. From galactorrhea to osteopenia: rethinking serotonin-prolactin interactions. Neuropsychopharmacology 29: 833-846.

Erskine MS, Barfield RJ, Goldman BD. 1980. Postpartum aggression in rats: II. Dependence on maternal sensitivity to young and effects of experience with pregnancy and parturition. J Comp Physiol Psychol 94: 495-505.

Evrard A, Laporte AM, Chastanet M, Hen R, Hamon M, et al. 1999. 5-HT1A and 5-HT1B receptors control the firing of serotoninergic neurons in the dorsal raphe nucleus of the mouse: studies in 5-HT1B knock-out mice. Eur J Neurosci 11: 3823-3831.

Faccidomo S, Bannai M, van Trigt RL, De Bold JF, Miczek KA. 2005. Alcohol-heightened aggression and corticolimbic 5-HT in mice: infusion and reverse microdialysis of 5-HT1B agonists into the infralimbic and orbitofrontal cortex. Abstr Soc Neurosci.

Fairbanks LA, Fontenot MB, Phillips- Conroy JE, Jolly CJ, Kaplan JR, et al. 1999. CSF monoamines, age and impulsivity in wild grivet monkeys (*Cercopithecus aethiops aethiops*). Brain Behav Evol 53: 305-312.

Fairbanks LA, Jorgensen MJ, Huff A, Blau K, Hung YY, et al. 2004. Adolescent impulsivity predicts adult dominance attainment in male. Am J Primatol 64: 1-17.

Fava M, Rosenbaum JF. 1993. Psychopharmacology of pathologic aggression. Harvard Rev Psychiatry 1: 244-246.

Fenwick P. 1989. The nature and management of aggression in epilepsy. J Neuropsychiatry Clin Neurosci 1: 418-425.

Ferrari PF, Parmigiani S, Rodgers RJ, Palanza P. 1997. Differential effects of chlordiazepoxide on aggressive behavior in male mice: the influence of social factors. Psychopharmacology 134: 258-265.

Ferrari PF, Van Erp AMM, Tornatzky W, Miczek KA. 2003. Accumbal dopamine and serotonin in anticipation of the next aggressive episode in rats. Eur J Neurosci 17: 371-378.

Ferris CF. 1996. Serotonin diminishes aggression by suppressing the activity of the vasopressin system. Understanding aggressive behavior in children. Ferris CF, editor. New York: The New York Academy of Sciences, New York; pp. 98-103.

Ferris CF, Delville Y. 1994. Vasopressin and serotonin interactions in the control of agonistic behavior. Psychoneuroendocrinology 19: 593-601.

Ferris CF, Stolberg T, Delville Y. 1999. Serotonin regulation of aggressive behavior in male golden hamsters (*Mesocricetus auratus*). Behav Neurosci 113: 804-815.

Ferris CF, Melloni RH, Koppel G, Perry KW, Fuller RW, et al. 1997. Vasopressin/serotonin interactions in the anterior hypothalamus control aggressive behavior in golden hamsters. J Neurosci 17: 4331-4340.

Field EF, Pellis SM. 1994. Differential effects of amphetamine on the attack and defense components of play fighting in rats. Physiol Behav 56(2): 325-330.

Findling RL, Steiner H, Weller EB. 2005. Use of antipsychotics in children and adolescents. J Clin Psychiatry 66: 29-40.

Fischer HS, Zernig G, Schuligoi R, Miczek KA, Hauser KF, et al. 2000. Alterations within the endogenous opioid system in mice with targeted deletion of the neutral endopeptidase ("enkephalinase") gene. Regul Pept 96: 53-58.

Fish EW, De Bold JF, Miczek KA. 2002. Repeated alcohol: behavioral sensitization and alcohol-heightened aggression in mice. Psychopharmacology (Berl) 160: 39-48.

Fish EW, De Bold JF, Miczek KA. 2005. Escalated aggression as a reward: corticosterone and GABAA receptor positive modulators in mice. Psychopharmacology 182: 116-127.

Fish EW, Faccidomo S, Miczek KA. 1999. Aggression heightened by alcohol or social instigation in mice: reduction by the 5-HT1B receptor agonist CP-94,253. Psychopharmacology 146: 391-399.

Fish EW, Faccidomo S, De Bold JF, Miczek KA. 2001. Alcohol, allopregnanolone and aggression in mice. Psychopharmacology 153: 473-483.

Fishbein DH, Lozovsky D, Jaffe JH. 1989. Impulsivity, aggression, and neuroendocrine responses to serotonergic stimulation in substance abusers. Biol Psychiatry 25: 1049-1066.

Flynn JP. 1967. The neural basis of aggression in cats. Neurophysiology and emotion. Glass DC, editor. New York: Rockefeller University Press; pp. 40-60.

Fowler SC, Liou JR. 1998. Haloperidol, raclopride, and eticlopride induce microcatalepsy during operant performance in rats, but clozapine and SCH 23390 do not. Psychopharmacology 140: 81-90.

Friedel RO. 2004. Dopamine dysfunction in borderline personality disorder: a hypothesis. Neuropsychopharmacology 29: 1029-1039.

Gabbott PL, Warner TA, Jays PR, Salway P, Busby SJ. 2005. Prefrontal cortex in the rat: projections to subcortical autonomic, motor, and limbic centers. J Comp Neurol 492: 145-177.

Gandelman R, Zarrow MX, Denenberg VH, Myers M. 1971. Olfactory bulb removal eliminates maternal behavior in the mouse. Science 171: 210-211.

Gao B, Cutler MG. 1993. Buspirone increases social envestigation in pair-housed male mice—comparison with the effects of chlordiazepoxide. Neuropharmacology 32: 429-437.

Gao B, Fritschy JM, Benke D, Mohler H. 1993. Neuron-specific expression of GABAA receptor subtypes: differential association of the a1 and a3 subunits with serotonergic and GABAergic neurons. Neuroscience 54: 881-892.

Garattini S, Giacalone E, Valzelli L. 1967. Isolation, aggressiveness and brain 5-hydroxytryptamine turnover. J Pharm Pharmacol 19: 338-339.

Gardner DL, Lucas PB, Cowdry RW. 1990. Csf metabolites in borderline personality-disorder compared with normal controls. Biol Psychiatry 28: 247-254.

Gardos G, Di Mascio A, Salzman C, Shader RI. 1968. Differential actions of chlordiazepoxide and oxazepam on hostility. Arch Gen Psychiatry 18: 757-760.

Gariepy JL, Gendreau PL, Cairns RB, Lewis MH. 1998. D1 dopamine receptors and the reversal of isolation-induced behaviors in mice. Behav Brain Res 95: 103-111.

Gaspar P, Berger B, Alvarez C, Vigny A, Henry JP. 1985. Catecholaminergic innervation of the septal area in man—immunocytochemical study using Th and Dbh antibodies. J Comp Neurol 241: 12-33.

Gendreau PL, Petitto JM, Petrova A, Gariepy J, Lewis MH. 2000. D(3) and D(2) dopamine receptor agonists differentially modulate isolation-induced social-emotional reactivity in mice. Behav Brain Res 114: 107-117.

Gerfen CR, Wilson CJ. 1996. The basal ganglia. Handbook of chemical neuroanatomy, integrated systems of the CNS, part III. Bjorklund A, Hokfelt T, Swanson LW, editors. Amsterdam: Elsevier Science; pp. 369-466.

Gerra G, Zaimovic A, Avanzini P, Chittolini B, Giucastro G, et al. 1997. Neurotransmitter-neuroendocrine responses to experimentally induced aggression in humans: influence of personality variable. Psychiatry Res 66: 33-43.

Gervais J, Rouillard C. 2000. Dorsal raphe stimulation differentially modulates dopaminergic neurons in the ventral tegmental area and substantia nigra. Synapse 35: 281-291.

Giacalone E, Tansella M, Valzelli L, Garattini S. 1968. Brain serotonin metabolism in isolated aggressive mice. Biochem Pharmacol 17: 1315-1327.

Gianutsos G, Lal H. 1978. Psychopharmacology of aggression (Modern problems of pharmacopsychiatry, Vol. 13). Narcotic analgesics and aggression. Valzelli L, editor. New York: S. Karger; pp. 114-138.

Gibbs DM, Vale W. 1983. Effect of the serotonin reuptake inhibitor fluoxetine on corticotropin-releasing factor and vasopressin secretion into hypophysial portal blood. Brain Res 280: 176-179.

Gobert A, Rivet JM, Audinot V, Newman-Tancredi A, Cistarelli L, et al. 1998. Simultaneous quantification of serotonin, dopamine and noradrenaline levels in single frontal cortex dialysates of freely-moving rats reveals a complex pattern of reciprocal auto- and heteroreceptor-mediated control of release. Neuroscience 84: 413-429.

Gogos JA, Morgan M, Luine V, Santha M, Ogawa S, et al. 1998. Catechol-O-methyltransferase-deficient mice exhibit sexually dimorphic changes in catecholamine levels and behavior. Proc Natl Acad Sci USA 95: 9991-9996.

Goodall J. 1986. The chimpanzees of Gombe: patterns of behavior. Belknap Press of Harvard University Press. Cambridge, Massachusetts:

Gorwood P, Batel P, Ades J, Hamon M, Boni C. 2000. Serotonin transporter gene polymorphisms, alcoholism, and suicidal behavior. Biol Psychiatry 48: 259-264.

Gourley SL, De Bold JF, Yin W, Cook J, Miczek KA. 2005. Benzodiazepines and heightened aggressive behavior in rats: reduction by GABA(A)/alpha(1) receptor antagonists. Psychopharmacology (Berl) 178: 232-240.

Grace AA, Rosenkranz JA. 2002. Regulation of conditioned responses of basolateral amygdala neurons. Physiol Behav 77: 489-493.

Greenberg BD, Li Q, Lucas FR, Hu S, Sirota LA, et al. 2000. Association between the serotonin transporter promoter polymorphism and personality traits in a primarily female population sample. Am J Med Genet 96: 202-216.

Gregg TR, Siegel A. 2001. Brain structures and neurotransmitters regulating aggression in cats: implications for human aggression. Prog Neuropsychopharmacol Biol Psychiatry 25: 91-140.

Guidotti A, Dong E, Matsumoto K, Pinna G, Rasmusson AM, et al. 2001. The socially-isolated mouse: a model to study the putative role of allopregnanolone and 5 alpha-dihydroprogesterone in psychiatric disorders. Brain Res Rev 37: 110-115.

Haberstick BC, Smolen A, Hewitt JK. 2006. Family-based association test of the 5HTTLPR and aggressive behavior in a general population sample of children. Biol Psychiatry 59 (9): 836-843.

Halasz J, Liposits Z, Meelis W, Kruk MR, Haller J. 2002. Hypothalamic attack area-mediated activation of the forebrain in aggression. Neuroreport 13: 1267-1270.

Haller J. 1995. Alpha-2 adrenoceptor blockade and the response to intruder aggression in Long-Evans rats. Physiol Behav 58(1): 101-106.

Haller J, Kruk MR. 2003. Neuroendocrine stress responses and aggression. Neurobiology of aggression: understanding and preventing violence. Mattson MP, editor. Totowa, NJ: Humana Press; pp. 93-118.

Haller J, Kruk MR. 2006. Normal and abnormal aggression: human disorders and novel laboratory models. Neurosci Biobehav Rev 30: 292-303.

Haller J, Makara GB, Kovacs JL. 1996b. The effect of alpha-3 adrenoceptor blockers on aggressive behavior in mice: implications for the actions of adrenoceptor agents. Psychopharmacology 126: 345-350.

Haller J, Makara GB, Kruk MR. 1998. Catecholaminergic involvement in the control of aggression: hormones, the peripheral sympathetic, and central noradrenergic systems. Neurosci Biobehav Rev 22: 85-97.

Haller J, Makara GB, Barna I, Kovacs K, Nagy J, et al. 1996a. Compression of the pituitary stalk elicits chronic increases in CSF vasopressin, oxytocin as well as in social investigation and aggressiveness. J Neuroendocrinol 8: 361-365.

Haller J, Bakos N, Rodriguiz RM, Caron MG, Wetsel WC, et al. 2002. Behavioral responses to social stress in noradrenaline transporter knockout mice: effects on social behavior and depression. Brain Res Bull 58: 279-284.

Hallikainen T, Saito T, Lachman HM, Volavka J, Pohjalainen T, et al. 1999. Association between low activity serotonin transporter promoter genotype and early onset alcoholism with habitual impulsive violent behavior. Mol Psychiatry 4: 385-388.

Halperin JM, Sharma V, Siever LJ, Schwartz ST, Matier K, et al. 1994. Serotonergic function in aggressive and nonaggressive boys with attention deficit hyperactivity disorder. Am J Psychiatry 151: 243-248.

Handelsman L, Holloway K, Kahn RS, Sturiano C, Rinaldi PJ, et al. 1996. Hostility is associated with a low prolactin response to meta-chlorophenylpiperazine in abstinent alcoholics. Alcohol Clin Exp Res 20: 824-829.

Handelsman L, Kahn RS, Sturiano C, Rinaldi PJ, Gabriel S, et al. 1998. Hostility is associated with a heightened prolactin response to meta-chlorophenylpiperazine in abstinent cocaine addicts. Psychiatry Res 80: 1-12.

Haney M, De Bold JF, Miczek KA. 1989. Maternal aggression in mice and rats towards male and female conspecifics. Aggress Behav 15: 443-453.

Haney M, Noda K, Kream R, Miczek KA. 1990. Regional serotonin and dopamine activity: sensitivity to amphetamine and aggressive behavior in mice. Aggress Behav 16: 259-270.

Hasselager E, Rolinski Z, Randrup A. 1972. Specific antagonism by dopamine inhibitors of items of amphetamine induced aggressive behaviour. Psychopharmacologia 24: 485-495.

Haug M, Simler S, Kim L, Mandel P. 1980. Studies on the involvement of GABA in the aggression directed by groups of intact or gonadectomized male and female mice towards lactating intruders. Pharmacol Biochem Behav 12: 189-193.

Hazell PL, Stuart JE. 2003. A randomized controlled trial of clonidine added to psychostimulant medication for hyperactive and aggressive children. J Am Acad Child Adoles Psychiatry 42: 886-894.

Hegstrand LR, Eichelman B. 1983. Increased shock-induced fighting with supersensitive beta-adrenergic receptors. Pharmacol Biochem Behav 19: 313-320.

Heiligenberg W. 1974. Processes governing behavioral states of readiness. Adv Stud Behav 5: 173-200.

Heiligenstein JH, Beasley CM Jr, Potvin JH. 1993. Fluoxetine not associated with increased aggression in controlled clinical trials. Int Clin Psychopharmacol 8: 277-280.

Hendley ED, Moisset B, Welch BL. 1973. Catecholamine uptake in cerebral cortex: adaptive change induced by fighting. Science 180: 1050-1052.

Hendricks TJ, Fyodorov DV, Wegman LJ, Lelutiu NB, Pehek EA, et al. 2003. Pet-1 ETS gene plays a critical role in 5-HT neuron development and is required for normal anxiety-like and aggressive behavior. Neuron 37: 233-247.

Henley ED, Moisset B, Welch BL. 1973. Catecholamine uptake in cerebral cortex: adaptive change induced by fighting. Science 180: 1050-1052.

Herve D, Pickel VM, Joh TH, Beaudet A. 1987. Serotonin axon terminals in the ventral tegmental area of the rat—fine-structure and synaptic input to dopaminergic-neurons. Brain Res 435: 71-83.

Higley JD, Mehlman PT, Taub DM, Higley SB, Suomi SJ, et al. 1992. Cerebrospinal fluid monoamine and adrenal correlates of aggression in free-ranging rhesus monkeys. Arch Gen Psychiatry 49: 436-441.

Higley JD, King ST, Hasert MF, Champoux M, Suomi SJ, et al. 1996. Stability of interindividual differences in serotonin function and its relationship to severe aggression and competent social behavior in rhesus macaque females. Neuropsychopharmacology 14: 67-76.

Hillegaart V, Ahlenius S. 1998. Facilitation and inhibition of male rat ejaculatory behaviour by the respective 5-HT1A and 5-HT1B receptor agonists 8-OH-DPAT and anpirtoline, as evidenced by use of the corresponding new and selective receptor antagonists NAD-299 and NAS-181. Brit J Pharmacol 125: 1733-1743.

Hjörth S, Sharp T. 1991. Effect of the 5-HT_{1A} receptor agonist 8-OH-DPAT on the release of 5-HT in dorsal and median raphe-innervated rat brain regions as measured by in vivo microdialysis. Life Sci 48: 1779-1786.

Hoaken PNS, Stewart SH. 2003. Drugs of abuse and the elicitation of human aggressive behavior. Addict Behav 28: 1533-1554.

Hodge GK, Butcher LL. 1975. Catecholamine correlates of isolation-induced aggression in mice. Eur J Pharmacol 31: 81-93.

Höglund E, Balm PH, Winberg S. 2000. Skin darkening, a potential social signal in subordinate arctic charr (*Salvelinus alpinus*): the regulatory role of brain monoamines and pro-opiomelanocortin-derived peptides. J Exp Biol 203 (Pt11): 1711–1721.

Holmes A, Murphy DL, Crawley JN. 2002. Reduced aggression in mice lacking the serotonin transporter. Psychopharmacology 161: 160-167.

Holmes A, Li Q, Murphy DL, Gold E, Crawley JN. 2003a. Abnormal anxiety-related behaviour in serotonin transporter null mutant mice: the influence of genetic background. Genes Brain Behav 2: 365-380.

Holmes A, Yang RJ, Lesch KP, Crawley JN, Murphy DL. 2003b. Mice lacking the serotonin transporter exhibit 5-HT1A receptor-mediated abnormalities in tests for anxiety-like behavior. Neuropsychopharmacology 28: 2077-2088.

Holscher C. 1997. Long-term potentiation: a good model for learning and memory? Prog Neuropsychopharmacol Biol Psychiatry 21: 47-68.

Hoyer D, Engel G, Kalkman HO. 1985. Characterization of the 5-HT1B recognition site in rat brain: binding studies with (-)[125I]iodocyanopindolol. Eur J Pharmacol 118: 1-12.

Hoyer D, Hannon JP, Martin GR. 2002. Molecular, pharmacological and functional diversity of 5-HT receptors. Pharmacol Biochem Behav 71: 533-554.

Hoyer D, Pazos A, Probst A, Palacios JM. 1986. Serotonin receptors in the human brain. I. Characterization and autoradiographic localization of 5-HT1A recognition sites. Apparent absence of 5-HT1B recognition sites. Brain Res 376: 85-96.

Hrabovszky E, Halasz J, Meelis W, Kruk MR, Liposits Z, et al. 2005. Neurochemical characterization of hypothalamic neurons involved in attack behavior: glutamatergic dominance and co-expression of thyrotropin-releasing hormone in a subset of glutamatergic neurons. Neuroscience 133: 657-666.

Huang Q, He XH, Ma CR, Liu RY, Yu S, et al. 2000. Pharmacophore/receptor models for GABAA/BzR subtypes (alpha 1 beta 3 gamma 2, alpha 5 beta 3 gamma 2, and alpha 6 beta 3 gamma 2) via a comprehensive ligand-mapping approach. J Med Chem 43: 71-95.

Huber R, Smith K, Delago A, Isaksson K, Kravitz EA. 1997. Serotonin and aggressive motivation in crustaceans: Altering the decision to retreat. Proc Natl Acad Sci USA 94: 5939-5942.

Humble F, Berk M. 2003. Pharmacological management of aggression and violence. Hum Psychopharmacol 18: 423-436.

Huntingford FA, Turner AK. 1987. Animal conflict. Chapman and Hall. London:

Hurst JL. 1987. Behavioral variation in wild house mice *Mus domesticus* Rutty: a quantitative assessment of female social organization. Anim Behav 35: 1846-1857.

Itil TM, Wadud A. 1975. Treatment of human aggression with major tranquilizers, antidepressants and newer psychotropic drugs. J Nerv Ment Dis 160: 83-99.

Janssen PAJ, Jageneau AH, Niemegeers JE. 1960. Effects of various drugs on isolation-induced fighting behavior of male mice. J Pharmacol Exp Ther 129: 471-475.

Jenck F, Broekkamp CLE, Van Delft AML. 1989. Effects of serotonin receptor antagonists on PAG stimulation induced aversion: different contributions of 5HT1, 5HT2 and 5HT3 receptors. Psychopharmacology 97: 489-495.

Johnson DE, Rollema H, Schmidt AW, McHarg AD. 2001. Serotonergic effects and extracellular brain levels of eletriptan, zolmitriptan and sumatriptan in rat brain. Eur J Pharmacol 425: 203-210.

Jorgensen H, Riis M, Knigge U, Kjaer A, Warberg J. 2003. Serotonin receptors involved in vasopressin and oxytocin secretion. J Neuroendocrinol 15: 242-249.

Kalynchuk LE, Dalal S, Mana MJ, Pinel JP. 1992. Nifedipine blocks the development of tolerance to the anticonvulsant effects of ethanol. Ann N Y Acad Sci 654: 459-460.

Kantak KM, Miczek KA. 1988. Social, motor, and autonomic signs of morphine withdrawal: differential sensitivities to catecholaminergic drugs in mice. Psychopharmacology 96: 468-476.

Kantak KM, Hegstrand LR, Eichelman B. 1980. Dietary tryptophan modulation and aggressive behavior in mice. Pharmacol Biochem Behav 12: 675-679.

Kantak KM, Hegstrand LR, Eichelman B. 1981. Dietary tryptophan reversal of septal lesion and 5,7-DHT lesion elicited shock-induced fighting. Pharmacol Biochem Behav 15: 343-350.

Kaplan JR, Manuck SB, Fontenot MB, Mann JJ. 2002. Central nervous system monoamine correlates of social dominance in cynomolgus monkeys (*Macaca fascicularis*). Neuropsychopharmacology 26: 431-443.

Karl T, Lin S, Schwarzer C, Sainsbury A, Couzens M, et al. 2004. Y1 receptors regulate aggressive behavior by modulating serotonin pathways. Proc Natl Acad Sci USA 101: 12742-12747.

Keck PE Jr, Strakowski SM, McElroy SL. 2000. The efficacy of atypical antipsychotics in the treatment of depressive symptoms, hostility, and suicidality in patients with schizophrenia. J Clin Psychiatry 61 Suppl 3: 4-9.

Kemble ED, Behrens M, Rawleigh JM, Gibson BM. 1991. Effects of yohimbine on isolation-induced aggression, social attraction, and conspecific odor preference in mice. Pharmacol Biochem Behav 40: 781-785.

Kemp JA, McKernan RM. 2002. NMDA receptor pathways as drug targets. Nat Neurosci 5: 1039-1042.

Kendrick KM, Da Costa APC, Broad KD, Ohkura S, Guevara R, et al. 1997. Neural control of maternal behaviour and olfactory recognition of offspring. Brain Res Bull 44: 383-395.

Kennedy JS, Bymaster FP, Schuh L, Calligaro DO, Nomikos G, et al. 2001. A current review of olanzapine's safety in the geriatric patient: from pre-clinical pharmacology to clinical data. Int J Geriatr Psychiatry 16: S33-S61.

Kew JNC, Kemp JA. 2005. Ionotropic and metabotropic glutamate receptor structure and pharmacology (vol 179, pg 4, 2005). Psychopharmacology (Berl) 182: 320.

Kim DH, Jung JS, Yan JJ, Suh HW, Son BK, et al. 2000. Increased plasma corticosterone, aggressiveness and brain monoamine changes induced by central injection of pertussis toxin. Eur J Pharmacol 409: 67-72.

Kim DK, Tolliver TJ, Huang SJ, Martin BJ, Andrews AM, et al. 2005. Altered serotonin synthesis, turnover and dynamic regulation in multiple brain regions of mice lacking the serotonin transporter. Neuropharmacology 49: 798-810.

Knobelman DA, Kung HF, Lucki I. 2000a. Regulation of extracellular concentrations of 5-hydroxytryptamine (5-HT) in mouse striatum by 5-HT(1A) and 5-HT(1B) receptors. J Pharmacol Exp Ther 292: 1111-1117.

Knobelman DA, Kung HF, Lucki I. 2000b. Regulation of extracellular concentrations of 5-hydroxytryptamine (5-HT) in mouse striatum by 5-HT(1A) and 5-HT(1B) receptors. J Pharmacol Exp Ther 292: 1111-1117.

Knutson B, Wolkowitz OM, Cole SW, Chan T, Moore EA, et al. 1998. Selective alteration of personality and social behavior by serotonergic intervention. Am J Psychiatry 155: 373-379.

Koe BK, Nielsen JA, Macor JE, Heym J. 1992. Biochemical and behavioral studies of the 5-HT_{1B} receptor agonist, CP-94, 253. Drug Dev Res 26: 241-250.

Kollack-Walker S, Watson SJ, Akil H. 1997. Social stress in hamsters: defeat activates specific neurocircuits within the brain. J Neurosci 17: 8842-8855.

Konig M, Zimmer AM, Steiner H, Holmes PV, Crawley JN, et al. 1996. Pain responses, anxiety and aggression in mice deficient in pre-proenkephalin. Nature 383: 535-538.

Koolhaas JM, Korte SM, de Boer SF, Van Der Vegt BJ, Van Reenen CG, et al. 1999. Coping styles in animals: current status in behavior and stress-physiology. Neurosci Biobehav Rev 23: 925-935.

Korte SM, Meijer OC, De Kloet ER, Buwalda B, Keijser J, et al. 1996. Enhanced 5-HT1A receptor expression in forebrain regions of aggressive house mice. Brain Res 736: 338-343.

Kotler M, Barak P, Cohen H, Averbuch IE, Grinshpoon A, et al. 1999. Homicidal behavior in schizophrenia associated with a genetic polymorphism determining low catechol O-methyltransferase (COMT) activity. Am J Med Genet 88: 628-633.

Kratochvil CJ, Findling RL, McDougle CJ, Scahill L, Hamarman S. 2005. Pharmacological management of agitation and aggression in an adolescent with autism. J Am Acad Child Adoles Psychiatry 44: 829-832.

Kravitz EA. 1988. Hormonal control of behavior: amines and the biasing of behavioral output in lobsters. Science 241: 1775-1781.

Kravitz EA, Huber R. 2003. Aggression in invertebrates. Curr Opin Neurobio 13: 736-743.

Kreiss DS, Lucki I. 1994. Differential regulation of serotonin (5-HT) release in the striatum and hippocampus by 5-HT_{1A} autoreceptors of the dorsal and median raphe nuclei. J Pharmacol Exp Ther 269: 1268-1279.

Krsiak M. 1974. Behavioral changes and aggressivity evoked by drugs in mice. Res Commun Chem Pathol Pharmacol 7: 237-257.

Krsiak M. 1979. Effects of drugs on behaviour of aggressive mice. Brit J Pharmacol 65: 525-533.

Krsiak M, Sulcova A, Tomasikova Z, Dlohozkova N, Kosar E, et al. 1981. Drug effects on attack, defense and escape in mice. Pharmacol Biochem Behav 14: 47-52.

Kruesi MJP, Rapoport JL, Hamburger S, Hibbs E, Potter WZ, et al. 1990. Cerebrospinal fluid monoamine metabolites, aggression, and impulsivity in disruptive behavior disorders of children and adolescents. Arch Gen Psychiatry 47: 419-426.

Kudryavtseva NN. 1991. A sensory contact model for the study of aggressive and submissive behavior in male mice. Aggress Behav 17: 285-291.

Kuikka JT, Tiihonen J, Bergstrom KA, Karhu J, Rasanen P, 1998. Abnormal structure of human striatal dopamine re-uptake sites in habitually violent alcoholic offenders: a fractal analysis. 253: 195–197.

Lachman HM, Papolos DF. 1998. Chromosome 22q11 deletions and aggressive behaviour. Brit J Psychiatry 172: 540.

Lagerspetz K. 1964. Studies on the aggressive behaviour of mice. Ann Acad Sci Fenn 131: 1-131.

Lambe EK, Goldman- Rakic PS, Aghajanian GK. 2000. Serotonin induces EPSCs preferentially in layer V pyramidal neurons of the frontal cortex in the rat. Cereb Cortex 10: 974-980.

Lang A, Harro J, Soosaar A, Koks S, Volke V, et al. 1995. Role of N-methyl-D-aspartic acid and cholecystokinin receptors in apomorphine-induced aggressive behaviour in rats. Naunyn Schmiedebergs Arch Pharmacol 351: 363-370.

Lappalainen J, Long JC, Eggert M, Ozaki N, Robin RW, et al. 1998. Linkage of antisocial alcoholism to the serotonin 5-HT_{1B} receptor gene in 2 populations. Arch Gen Psychiatry 55: 989-994.

Larsson K. 1975. Sexual impairment of inexperienced male rats following pre- and postpuberal olfactory bulbectomy. Physiol Behav 14: 195-199.

Lasley SM, Thurmond JB. 1985. Interaction of dietary tryptophan and social isolation on territorial aggression, motor activity, and neurochemistry in mice. Psychopharmacology 87: 313-321.

Lauder JM. 1993. Neurotransmitters as growth regulatory signals—role of receptors and 2nd messengers. Trends Neurosci 16: 233-240.

Le Blanc JC, Binder CE, Armenteros JL, Aman MG, Wang JS, et al. 2005. Risperidone reduces aggression in boys with a disruptive behaviour disorder and below average intelligence quotient: analysis of two placebo-controlled randomized trials. Int Clin Psychopharmacol 20: 275-283.

Ledent C, Vaugeois JM, Schiffmann SN, Pedrazzini T, El Yacoubi M, et al. 1997. Aggressiveness, hypoalgesia and high blood pressure in mice lacking the adenosine A(2a) receptor. Nature 388: 674-678.

Lee M, Simansky K. 1997. CP-94,253: a selective serotonin 5-HT1B agonist that promotes satiety. Psychopharmacology 131: 264-270.

Le Greves P, Huang W, Johansson P, Thornwall M, Zhou Q, et al. 1997. Effects of an anabolic-androgenic steroid on the regulation of the NMDA receptor NR1, NR2A and NR2B subunit mRNAs in brain regions of the male rat. Neurosci Lett 226: 61-64.

Lesch KP. 2005. Alcohol dependence and gene × environment interaction in emotion regulation: is serotonin the link? Eur J Pharmacol 526: 113-124.

Lesch KP, Mossner R. 1998. Genetically driven variation in serotonin uptake: is there a link to affective spectrum, neurodevelopmental, and neurodegenerative disorders? Biol Psychiatry 44: 179-192.

Lesch KP, Bengel D, Heils A, Sabol SZ, Greenberg BD, et al. 1996. Association of anxiety-related traits with a polymorphism in the serotonin transporter gene regulatory region. Science 264: 1537-1551.

Liljequist S, Engel J. 1982. Effects of GABAergic agonists and antagonists on various ethanol-induced behavioral changes. Psychopharmacology 78: 71-75.

Lindenmayer JP, Kotsaftis A. 2000. Use of sodium valproate in violent and aggressive behaviors: a critical review. J Clin Psychiatry 61: 123-128.

Linner L, Wiker C, Arborelius L, Schalling M, Svensson TH. 2004. Selective noradrenaline reuptake inhibition enhances serotonergic neuronal activity and transmitter release in the rat forebrain. J Neural Transm 111: 127-139.

Linnoila M, Virkkunen M, Scheinin M, Nuutila A, Rimon R, et al. 1983. Low cerebrospinal fluid 5-hydroxyindoleacetic acid concentration differentiates impulsive from nonimpulsive violent behavior. Life Sci 33: 2609-2614.

Livingstone MS, Harris-Warrick RM, Kravitz EA. 1980. Serotonin and octopamine produce opposite postures in lobsters. Science 208: 76-79.

Loconto J, Papes F, Chang E, Stowers L, Jones EP, et al. 2003. Functional expression of murine V2R pheromone receptors involves selective association with the M10 and M1 families of MHC class Ib molecules. Cell 112: 607-618.

Lonstein JS, Gammie SC. 2002. Sensory, hormonal, and neural control of maternal aggression in laboratory rodents. Neurosci Biobehav Rev 26: 869-888.

Lorenz K. 1966. On aggression. London: Methuen.

Low K, Crestani F, Keist R, Benke D, Brunig I, et al. 2000. Molecular and neuronal substrate for the selective attenuation of anxiety. Science 290: 131-134.

Lucki I, Wieland S. 1990. 5-Hydroxytryptamine-1A receptors and behavioral responses. Neuropsychopharmacology 3: 481-493.

Lumley LA, Robison CL, Slusher BS, Wozniak K, Dawood M, et al. 2004. Reduced isolation-induced aggressiveness in mice following NAALADase inhibition. Psychopharmacology (Berl) 171: 375-381.

Mackintosh JH. 1970. Territory formation by laboratory mice. Anim Behav 18: 177-183.

Macor JE, Burkhart CA, Heym JH, Ives JL, Lebel LA, et al. 1990. 3-(1,2,5,6-Tetrahydropyrid-4-Yl)pyrrolo[3,2-B]pyrid-5-one—a potent and selective serotonin (5-Ht1B) agonist and rotationally restricted phenolic analog of 5-methoxy-3-(1,2,5,6-tetrahydropyrid-4-Yl)indole. J Med Chem 33: 2087-2093.

Mandel P, Mack G, Kempf E. 1979. Molecular basis of some models of aggressive behavior. Psychopharmacology of aggression. Sandler M, editor. New York: Raven Press; pp. 95-110.

Mandel P, Ciesielski L, Maitre M, Simler S, Kempf E, et al. 1981. Inhibitory amino acids, aggressiveness, and convulsions. Amino acid neurotransmitters. De Feudis FV, editor. New York: Raven Press; pp. 1-9.

Mandiyan VS, Coats JK, Shah NM. 2005. Deficits in sexual and aggressive behaviors in Cnga2 mutant mice. Nat Neurosci 8: 1660-1662.

Mann MA, Svare B. 1982. Factors influencing pregnancy-induced aggression in mice. Behav Neural Biol 36: 242-258.

Manuck SB, Kaplan JR, Rymeski BA, Fairbanks LA, Wilson ME. 2003. Approach to a social stranger is associated with low central nervous system serotonergic responsivity in female cynomolgus monkeys (*Macaca fascicularis*). Am J Primatol 61: 187-194.

Manuck SB, Flory JD, McCaffery JM, Matthews KA, Mann JJ, et al. 1998. Aggression, impulsivity, and central nervous system serotonergic responsivity in a nonpatient sample. Neuropsychopharmacology 19: 287-299.

Marek GJ, Aghajanian GK. 1998. 5-Hydroxytryptamine-induced excitatory postsynaptic currents in neocortical layer V pyramidal cells: suppression by mu-opiate receptor activation. Neuroscience 86: 485-497.

Marino MD, Bourdelat-Parks BN, Liles LC, Weinshenker D. 2005. Genetic reduction of noradrenergic function alters social memory and reduces aggression in mice. Behav Brain Res 161: 197-203.

Marsh L, Krauss GL. 2000. Aggression and violence in patients with epilepsy. Epilepsy Behav 1: 160-168.

Martin M, Ledent C, Parmentier M, Maldonado R, Valverde O. 2002. Involvement of CB1 cannabinoid receptors in emotional behaviour. Psychopharmacology (Berl) 159: 379-387.

Martin-Lopez M, Navarro JF. 2002. Antiaggressive effects of zolpidem and zopiclone in agonistic encounters between male mice. Aggress Behav 28: 416-425.

Matsumoto K, Cai B, Satoh T, Ohta H, Watanabe H. 1991. Desipramine enhances isolation-induced aggression behavior in mice. Pharmacol Biochem Behav 39: 167-170.

Matsumoto T, Honda S, Harada N. 2003. Neurological effects of aromatase deficiency in the mouse. J Steroid Biochem Mol Biol 86: 357-365.

Matsuoka Y, Furuyashiki T, Yamada K, Nagai T, Bito H, et al. 2005. Prostaglandin E receptor EP1 controls impulsive behavior under stress. Proc Natl Acad Sci USA 102: 16066-16071.

McAllister KH. 1990. Ethological analysis of the effects of MK-801 upon aggressive male mice: similarity to chlordiazepoxide. Pharmacol Biochem Behav 37: 101-106.

McCracken JT, McGough J, Shah B, Cronin P, Hong D, et al. 2002. Risperidone in children with autism and serious behavioral problems. New Eng J Med 347: 314-321.

McFarland K, Davidge SB, Lapish CC, Kalivas PW. 2004. Limbic and motor circuitry underlying footshock-induced reinstatement of cocaine-seeking behavior. J Neurosci 24: 1551-1560.

McGinnis MY. 2004. Anabolic androgenic steroids and aggression—studies using animal models. Youth Violence: Scientific Approaches to Prevention. Ann N Y Acad Sci 1036: 399-415.

McKenzie-Quirk SD, Girasa KA, Allan AM, Miczek KA. 2005. 5-HT3 receptors, alcohol and aggressive behavior in mice. Behav Pharmacol 16: 163-170.

McMillen BA, Da Vanzo EA, Song AH, Scott SM, Rodriguez ME. 1989. Effects of classical and atypical antipsychotic drugs on isolation-induced aggression in male mice. Eur J Pharmacol 160: 149-153.

Mehlman PT, Higley JD, Faucher I, Lilly AA, Taub DM, et al. 1994. Low CSF 5-HIAA concentrations and severe aggression and impaired impulse control in nonhuman primates. Am J Psychiatry 151: 1485-1491.

Mehta AK, Ticku MK. 1999. An update on $GABA_A$ receptors. Brain Res Rev 29: 196-217.

Miczek KA. 1974. Intraspecies aggression in rats: effects of D-amphetamine and chlordiazepoxide. Psychopharmacologia 39: 275-301.

Miczek KA. 2001. Research on animal aggression: emerging successes for understanding determinants of human violence. Animal research and human health:advancing human welfare through behavioral science. Carrol ME, Overmier JB, editors. Washington, D.C: American Psychological Association.

Miczek KA, Donat P. 1989. Brain 5-HT systems and inhibition of aggressive behavior. Behavioural pharmacology of 5-HT. Bevan P, Cools AR, Archer T, editors. Hillsdale, NJ: Lawrence Erlbaum Associates; pp. 117-144.

Miczek KA, Haney M. 1994. Psychomotor stimulant effects of D-amphetamine, MDMA and PCP: Aggressive and schedule-controlled behavior in mice. Psychopharmacology 115: 358-365.

Miczek KA, O'Donnell JM. 1978. Intruder-evoked aggression in isolated and nonisolated mice: effects of psychomotor stimulants and L-dopa. Psychopharmacology 57: 47-55.

Miczek KA, O'Donnell JM. 1980a. Alcohol and chlordiazepoxide increase suppressed aggression in mice. Psychopharmacology 69: 39-44.

Miczek KA, O'Donnell JM. 1980b. Alcohol and chlordiazepoxide increase suppressed aggression in mice. Psychopharmacology 69: 39-44.

Miczek KA, Yoshimura H. 1982. Disruption of primate social behavior by D-amphetamine and cocaine: differential antagonism by antipsychotics. Psychopharmacology 76: 163-171.

Miczek KA, Fish EW, De Bold JF. 2003. Neurosteroids, GABAA receptors, and escalated aggressive behavior. Horm Behav 44: 242-257.

Miczek KA, Hussain S, Faccidomo S. 1998b. Alcohol-heightened aggression in mice: attenuation by 5-HT1A receptor agonists. Psychopharmacology 139: 160-168.

Miczek KA, Barros HM, Sakoda L, Weerts EM. 1998a. Alcohol and heightened aggression in individual mice. Alcohol Clin Exp Res 22: 1698-1705.

Miczek KA, Fish EW, De Bold JF, de Almeida RMM. 2002. Social and neural determinants of aggressive behavior: pharmacotherapeutic targets at serotonin, dopamine and γ-aminobutyric acid systems. Psychopharmacology 163: 434-458.

Miczek KA, Maxson SC, Fish EW, Faccidomo S. 2001. Aggressive behavioral phenotypes in mice. Behav Brain Res 125: 167-181.

Miczek KA, Weerts EM, Haney M, Tidey J. 1994b. Neurobiological mechanisms controlling aggression: preclinical developments for pharmacotherapeutic interventions. Neurosci Biobehav Rev 18: 97-110.

Miczek KA, Fish EW, de Almeida RMM, Faccidomo S, DeBold JF. 2004b. Role of alcohol consumption in escalation to violence. Ann N Y Acad Sci 1036: 278-289.

Miczek KA, Haney M, Tidey J, Vivian J, Weerts E. 1994a. Neurochemistry and pharmacotherapeutic management of violence and aggression. Understanding and preventing violence: biobehavioral influences on violence. Reiss AJ, Miczek KA, Roth JA, editors. Washington, DC: National Academy Press; pp. 244-514.

Miczek KA, Mutschler NH, Van Erp AMM, Blank AD, McInerney SC. 1999. D-Amphetamine "cue" generalizes to social defeat stress: sensitization and role of accumbens dopamine. Psychopharmacology 147: 190-199.

Miczek KA, Faccidomo S, de Almeida RMM, Bannai M, Fish EW, et al. 2004a. Escalated aggressive behavior: new pharmacotherapeutic approaches and opportunities. Ann N Y Acad Sci 1036: 336-355.

Minkeviciene R, Banerjee P, Tanila H. 2004. Memantine improves spatial learning in a transgenic mouse model of Alzheimer's disease. J Pharmacol Exp Ther 311: 677-682.

Mirkes SJ, Bethea CL. 2001. Oestrogen, progesterone and serotonin converge on GABAergic neurones in the monkey hypothalamus. J Neuroendocrinol 13: 182-192.

Mirsky AF, Harman N. 1974. On aggressive behavior and brain disease-some questions and possible relationships derived from the study of men and monkeys. The neuropsychology of aggression. Whalen RF, editor. New York: Plenum Press; pp. 185-210.

Mitchell PJ. 2005. Antidepressant treatment and rodent aggressive behaviour. Eur J Pharmacol 526: 147-162.

Moeller FG, Dougherty DM, Swann AC, Collins D, Davis CM, et al. 1996. Tryptophan depletion and aggressive responding in healthy males. Psychopharmacology 126: 97-103.

Mohler H, Benke D, Benson J, Luscher B, Fritschy JM. 1995. GABA(A)-receptor subtypes in vivo: cellular localization, pharmacology and regulation. GABA(A) receptors and anxiety. Biggio G, editor. 227 East Washington Sq Philadelphia PA 19106: Raven Press; pp. 41-56.

Molina V, Ciesielsski L, Gobaille S, Mandel P. 1986. Effects of the potentiation of the GABAergic neurotransmission in the olfactory bulbs on mouse-killing behavior. Pharmacol Biochem Behav 24: 657-664.

Moller SE, Mortensen EL, Breum L, Alling C, Larsen OG, et al. 1996. Aggression and personality: association with amino acids and monoamine metabolites. Psychol Med 26: 323-331.

Monroe RR. 1978. Brain dysfunction in aggressive criminals. DC Heath and Company. Lexington:

Morgan C, Cone RD. 2006. Melanocortin-5 receptor deficiency in mice blocks a novel pathway influencing pheromone-induced aggression. Behav Genet: 1–10.

Mos J, Olivier B. 1989. Quantitative and comparative analyses of pro-aggressive actions of benzodiazepines in maternal aggression of rats. Psychopharmacology 97: 152-153.

Mos J, Van Valkenburg CFM. 1979. Specific effect on social stress and aggression on regional dopamine metabolism in rat brain. Neurosci Lett 15: 325-327.

Mos J, Olivier B, Poth M, Aken H. 1992. The effects of intraventricular administration of eltoprazine, 1-(3-trifluoromethylphenyl) piperazine hydrochloride and 8-OH-DPAT on resident intruder aggression in the rat. Eur J Pharmacol 212: 295-298.

Moss HB, Yao JK, Panzak GL. 1990. Serotonergic responsivity and behavioral dimensions in antisocial personality disorder with substance abuse. Biol Psychiatry 28: 325-338.

Muehlenkamp F, Lucion A, Vogel WH. 1995. Effects of selective serotenergic agonists on aggressive behavior in rats. Pharmacol Biochem Behav 50: 671-674.

Munoz-Blanco J, Porras CA. 1987. Changes in neurotransmitter amino acids content in several CNS areas from aggressive and non-aggressive bull strains. Physiol Behav 39: 453-457.

Musty RE, Consroe PF. 1982. Phencyclidine produces aggressive behavior in rapid eye movement sleep-deprived rats. Life Sci 30: 1733-1738.

National Drug Intelligence Center. 2005. National Drug Threat Assessment.

Nelson RJ, Demas GE, Huang PL, Fishman MC, Dawson VL, et al. 1995. Behavioural abnormalities in male mice lacking neuronal nitric oxide synthase. Nature 378: 383-386.

New AS, Trestman RL, Mitropoulou V, Benishay DS, Coccaro E, et al. 1997. Serotonergic function and self-injurious behavior in personality disorder patients. Psychiatry Res 69: 17-26.

New AS, Gelernter J, Goodman M, Mitropoulou V, Koenigsberg H, et al. 2001. Suicide, impulsive aggression, and HTR1B genotype. Biol Psychiatry 50: 62-65.

New AS, Buchsbaum MS, Hazlett EA, Goodman M, Koenigsberg HW, et al. 2004. Fluoxetine increases relative metabolic rate in prefrontal cortex in impulsive aggression. Psychopharmacology 176: 451-458.

Nicot A, Otto T, Brabet P, Dicicco-Bloom EM. 2004. Altered social behavior in pituitary adenylate cyclase-activating polypeptide type I receptor-deficient mice. J Neurosci 24: 8786-8795.

Nishida T, Haraiwa-Hasegawa M, Takahata Y. 1985. Group extinction and female transfer in wild chimpanzees in the Mahale National Park, Tanzania. Zeitschrift für Tierpsychologie 67: 284-301.

Nocjar C, Roth BL, Pehek EA. 2002. Localization of 5-HT2A receptors on dopamine cells in subnuclei of the midbrain A10 cell group. Neuroscience 111: 163-176.

Nock MK, Kazdin AE. 2002. Parent-directed physical aggression by clinic-referred youths. J Clin Child Adolesc Psychol 31: 193-205.

Nogueira RL, Graeff FG. 1995. Role of 5HT receptor subtypes in the modulation of dorsal periaqueductal gray generated aversion. Pharmacol Biochem Behav 52: 1-6.

Noirot E, Goyens J, Buhot MC. 1975. Aggressive behavior of pregnant mice towards males. Horm Behav 6: 9-17.

Nolan KA, Volavka J, Czobor P, Sheitman B, Lindenmayer JP, et al. 2005. Aggression and psychopathology in treatment-resistant inpatients with schizophrenia and schizoaffective disorder. J Psychiatr Res 39: 109-115.

Ogawa S, Chan J, Chester AE, Gustafsson JA, Korach KS, et al. 1999. Survival of reproductive behaviors in estrogen receptor beta gene-deficient (beta ERKO) male and female mice. Proc Natl Acad Sci USA 96: 12887-12892.

Ogawa S,Washburn TF, Taylor J, Lubahn DB, Korach KS, et al. 1998. Modifications of testosterone-dependent behaviors by estrogen receptor-à gene disruption in male mice. Endocrinology 139: 5058-5069.

O'Keane V, Moloney E, O'Neill H, O'Connor A, Smith C, et al. 1992. Blunted prolactin responses to D-fenfluramine in sociopathy—evidence for subsensitivity of central serotonergic function. Brit J Psychiatry 160: 643-646.

Oliveira-dos-Santos AJ, Matsumoto G, Snow BE, Bai D, Houston FP, et al. 2000. Regulation of T cell activation, anxiety, and male aggression by RGS2. Proc Natl Acad Sci USA 97: 12272-12277.

Olivier B. 2004. Serotonin and aggression. Ann N Y Acad Sci 1036: 382-392.

Olivier B, Mos J. 1986. Serenics and aggression. Stress Med 2: 197-209.

Olivier B, Mos J. 1992. Rodent models of aggressive behavior and serotonergic drugs. Prog Neuropsychopharmacol Biol Psychiatry 16: 847-870.

Olivier B, van Dalen D. 1982. Social behaviour in rats and mice: an ethologically based model for differentiating psychoactive drugs. Aggress Behav 8: 163-168.

Olivier B, Van Oorschot R. 2005. 5-HT1B receptors and aggression: a review. Eur J Pharmacol 526: 207-217.

Olivier B, Mos J, Miczek KA. 1991. Ethopharmacological studies of anxiolytics and aggression. Eur Neuropsychopharmacol 1: 97-100.

Olivier B, Mos J, Rasmussen D. 1990. Behavioural pharmacology of the serenic, eltoprazine. Rev Drug Metab Drug Interact 8: 31-83.

Olivier B, Mos J, Van der Heyden J, Hartog J. 1989. Serotonergic modulation of social interactions in isolated male mice. Psychopharmacology 97: 154-156.

Olivier B, Mos J, Van Oorschot R, Hen R. 1995. Serotonin receptors and animal models of aggressive behavior. Pharmacopsychiatry 28: 80-90.

Pabis DJ, Stanislav SW. 1996. Pharmacotherapy of aggressive behavior. Ann Pharmacother 30: 278-287.

Palanza P, Della Seta D, Ferrari PF, Parmigiani S. 2005. Female competition in wild house mice depends upon timing of female/male settlement and kinship between females. Anim Behav 69: 1259-1271.

Pare D, Quirk GJ, Le Doux JE. 2004. New vistas on amygdala networks in conditioned fear. J Neurophysiol 92: 1-9.

Parent A, Poitras D, Dube L. 1984. Comparative anatomy of central monoaminergic systems. Handbook of chemical neuroanatomy. Bjorklund A, editor. Vol. 2 classical transmitters in the CNS. Part I. Elsevier Science Publishers B.V., Amsterdam, pp. 409-439.

Parker RN, Auerhahn K. 1998. Alcohol, drugs, and violence. Ann Rev Sociol 24: 291-311.

Parmigiani S, Palanza P, Rogers J, Ferrari PF. 1999. Selection, evolution of behavior and animal models in behavioral neuroscience. Neurosci Biobehav Rev 23: 957-969.

Parsian A, Cloninger CR. 2001. Serotonergic pathway genes and subtypes of alcoholism: association studies. Psychiatric Genetics 11: 89-94.

Parsons LH, Weiss F, Koob GF. 1998. Serotonin(1B) receptor stimulation enhances cocaine reinforcement. J Neurosci 18: 10078-10089.

Patel NC, Crismon ML, Hoagwood K, Jensen PS. 2005. Unanswered questions regarding atypical antipsychotic use in aggressive children and adolescents. J Child Adol Psychopharmacol 15: 270-284.

Pehek EA, McFarlane HG, Maguschak K, Price B, Pluto CP. 2001. M100,907, a selective 5-HT2A antagonist, attenuates dopamine release in the rat medial prefrontal cortex. Brain Res 888: 51-59.

Pellis SM, Pellis VC. 1987. Play-fighting differs from serious fighting in both target of attack and tactics of fighting in the laboratory rat *Rattus norvegicus*. Aggress Behav 13: 227-242.

Pietras CJ, Lieving LM, Cherek DR, Lane SD, Tcheremissine OV, et al. 2005. Acute effects of lorazepam on laboratory measures of aggressive and escape responses of adult male parolees. Behav Pharmacol 16: 243-251.

Pihl RO, Young SN, Harden P, Plotnick S, Chamberlain B, 1995. Acute effect of altered tryptophan levels and alcohol on aggression in normal human males. Psychopharmacology 119: 353-360.

Pine DS, Coplan JD, Wasserman GA, Miller LS, Fried JE, et al. 1997. Neuroendocrine response to fenfluramine challenge in boys—associations with aggressive behavior and adverse rearing. Arch Gen Psychiatry 54: 839-846.

Pinel JPJ, Treit D, Rovner LI. 1977. Temporal lobe aggression in rats. Science 197: 1088-1089.

Pinna G, Dong E, Matsumoto K, Costa E, Guidotti A. 2003. In socially isolated mice, the reversal of brain allopregnanolone down-regulation mediates the anti-aggressive action of fluoxetine. Proc Natl Acad Sci USA 100: 2035-2040.

Placidi GPA, Oquendo MA, Malone KM, Huang YY, Ellis SP, et al. 2001. Aggressivity, suicide attempts, and depression: relationship to cerebrospinal fluid monoamine metabolite levels. Biol Psychiatry 50: 783-791.

Poole TB, Morgan DR. 1973. Differences in aggressive behavior between male mice (*Mus musculus* L.) in colonies of different sizes. Anim Behav 21: 788-795.

Poole TB, Morgan HDR. 1976. Social and territorial behavior of laboratory mice (*Mus-musculus* L) in small complex areas. Anim Behav 24: 476-478.

Potegal M. 1991. Attack priming and satiation in female golden hamsters: tests of some alternatives to the aggression arousal interpretation. Aggress Behav 17: 327-335.

Potegal M, Tenbrink L. 1984. Behavior of attack-primed and attack-satiated female golden hamsters (*Mesocricetus auratus*). J Comp Psychol 98: 66-75.

Pritchett DB, Sontheimer H, Shivers BD, Ymer S, Kettenmann H, et al. 1989. Importance of a novel GABAA receptor subunit for benzodiazepine pharmacology. Nature 338: 582-585.

Puech AJ, Simon P, Chermat R, Boisseir JR. 1974. Profil neuropsychopharmacologique de l'apomorphine. J Pharmacol 5: 241-254.

Puglisi-Allegra S. 1980. Effects of sodium *n*-dipropylacetate, muscimol hydrobromide and (R,S) nipecotic acid amide on isolation-induced aggressive behavior. Psychopharmacology 70: 287-290.

Puglisi-Allegra S, Cabib S. 1990. Effects of defeat experiences on dopamine metabolism in different brain areas of the mouse. Aggress Behav 16: 271-284.

Puglisi-Allegra S, Simler S, Kempf E, Mandel P. 1981. Involvement of the GABAergic system on shock-induced aggressive behavior in two strains of mice. Pharmacol Biochem Behav 14, S1: 13-18.

Raine A. 1996. Autonomic nervous system factors underlying disinhibited, antisocial, and violent behavior. Biosocial perspectives and treatment implications. Ann N Y Acad Sci 794: 46-59.

Raine A, Lencz T, Bihrle S, La Casse L, Colletti P. 2000. Reduced prefrontal gray matter volume and reduced autonomic activity in antisocial personality disorder. Arch Gen Psychiatry 57: 119-127.

Raine A, Meloy JR, Bihrle S, Stoddard J, La Casse L, et al. 1998. Reduced prefrontal and increased subcortical brain functioning assessed using positron emission tomography in predatory and affective murderers. Behav Sci Law 16: 319-332.

Ratey JJ, Gordon A. 1993. The psychopharmacology of aggression: toward a new day. Psychopharmacol Bull 29: 65-73.

Ray A, Sharma KK, Alkondon M, Sen P. 1983. Possible interrelationship between the biogenic amines involved in the modulation of footshock aggression in rats. Arch Int Pharmacodyn Ther 265: 36-41.

Reimer JD, Petras ML. 1967. Breeding structure of the house mouse, *Mus musculus*, in a population cage. J Mammal 48: 88-99.

Reisner IR, Mann JJ, Stanley M, Huang YY, Houpt KA. 1996. Comparison of cerebrospinal fluid monoamine metabolite levels in dominant-aggressive and non-aggressive dogs. Brain Res 714: 57-64.

Rewerski W, Kostowski W, Piechocki T, Rylski M. 1971. The effects of some hallucinogens on aggressiveness of mice and rats. I. Pharmacology 5: 314-320.

Ricci LA, Grimes JM, Knyshevski I, Melloni RH. 2005. Repeated cocaine exposure during adolescence alters glutamic acid decarboxylase-65 (GAD(65)) immunoreactivity in hamster brain: correlation with offensive aggression. Brain Res 1035: 131-138.

Robbins TW, Everitt BJ. 1996. Neurobehavioural mechanisms of reward and motivation. Curr Opin Neurobiol 6: 228-236.

Rodgers RJ, Waters AJ. 1985. Benzodiazepines and their antagonists: a pharmacoethological analysis with particular reference to effects on "aggression." Neurosci Biobehav Rev 9: 21-35.

Rodriguez-Arias M, Felip CM, Broseta I, Minarro J. 1999a. The dopamine D3 antagonist U-99194A maleate increases social behaviors of isolation-induced aggressive male mice. Psychopharmacology 144: 90-94.

Rodriguez-Arias M, Pinazo J, Minarro J, Stinus L. 1999b. Effects of SCH 23390, raclopride, and haloperidol on morphine withdrawal-induced aggression in male mice. Pharmacol Biochem Behav 64: 123-130.

Rodriguez-Arias M, Minarro J, Aguilar MA, Pinazo J, Simon VM. 1998. Effects of risperidone and SCH 23390 on isolation-induced aggression in male mice. Eur Neuropsychopharmacol 8: 95-103.

Rodriguiz RM, Chu R, Caron MG, Wetsel WC. 2004. Aberrant responses in social interaction of dopamine transporter knockout mice. Behav Brain Res 148: 185-198.

Roeling TAP, Kruk MR, Schuurmans R, Veening JG. 1993. Behavioural responses of bicucculline methiodide injections into the ventral hypothalamus of freely moving, socially interacting rats. Brain Res 615: 121-127.

Rogawski MA. 2000. Low affinity channel blocking (uncompetitive) NMDA receptor antagonists as therapeutic agents—toward an understanding of their favorable tolerability. Amino Acids 19: 133-149.

Rolinski Z. 1975. Pharmacological studies on isolation-induced aggressiveness in mice in relation to biogenic amines. Pol J Pharmacol Pharm 27: 37-44.

Rosenkranz JA, Grace AA. 2002. Cellular mechanisms of infralimbic and prelimbic prefrontal cortical inhibition and dopaminergic modulation of basolateral amygdala neurons in vivo. J Neurosci 22: 324-337.

Rowlett JK, Platt DM, Lelas S, Atack JR, Dawson GR. 2005. Different GABA(A) receptor subtypes mediate the anxiolytic, abuse-related, and motor effects of benzodiazepine-like drugs in primates. Proc Natl Acad Sci USA 102: 915-920.

Rudolph U, Crestani F, Mohler H. 2001. GABA(A) receptor subtypes: dissecting their pharmacological functions. Trends Pharmacol Sci 22: 188-194.

Rudolph U, Crestani F, Benke D, Brünig I, Benson JA, et al. 1999. Benzodiazepine actions mediated by specific gamma-aminobutyric acid-A receptor subtypes. Nature 401: 796-800.

Sallinen J, Haapalinna A, Viitamaa T, Kobilka BK, Scheinin M. 1998. Adrenergic alpha2c receptors modulate the acoustic startle reflex, prepulse inhibition, and aggression in mice. J Neurosci 18: 3035-3042.

Sanchez C, Meier E. 1997. Behavioral profiles of SSRIs in animal models of depression, anxiety and aggression. Psychopharmacology 129: 197-205.

Sanchez C, Arnt J, Moltzen EK. 1996. The antiaggressive potency of (-)-penbutolol involves both 5- HT1A and 5- HT1B receptors and beta-adrenoceptors. Eur J Pharmacol 297: 1-8.

Sanchez C, Arnt J, Hyttel J, Moltzen EK. 1993. The role of serotonergic mechanisms in inhibition of isolation-induced aggression in male mice. Psychopharmacology 110: 53-59.

Sanchez C, Bergqvist PB, Brennum LT, Gupta S, Hogg S, et al. 2003. Escitalopram, the S-(+)-enantiomer of citalopram, is a selective serotonin reuptake inhibitor with potent effects in animal models predictive of antidepressant and anxiolytic activities. Psychopharmacology (Berl) 167: 353-362.

Sandnabba NK. 1986. Effects of selective breeding for high and low aggressiveness and of fighting experience on odor discrimination in mice. Aggress Behav 12: 359-366.

Santana N, Bortolozzi A, Serrats J, Mengod G, Artigas F. 2004. Expression of serotonin1A and serotonin2A receptors in pyramidal and GABAergic neurons of the rat prefrontal cortex. Cereb Cortex 14: 1100-1109.

Sari Y. 2004. Serotonin1B receptors: from protein to physiological function and behavior. Neurosci Biobehav Rev 28: 565-582.

Sari Y, Miquel MC, Brisorgueil MJ, Ruiz G, Doucet E, et al. 1999. Cellular and subcellular localization of 5-Hydroxytryptamine$_{1B}$ receptors in the rat central nervous system: immunocytochemical, autoradiographic and lesion studies. Neuroscience 88: 899-915.

Saudou F, Amara DA, Dierich A, Lemeur M, Ramboz S, 1994b. Enhanced aggressive behavior in mice lacking 5-HT$_{1B}$ receptor. Science 265: 1875-1878.

Saudou F, Amara DA, Dierich A, Lemeur M, Ramboz S, et al. 1994a. Enhanced aggressive behavior in mice lacking 5-HT$_{1B}$ receptor. Science 265: 1875-1878.

Schmidt CJ, Fadayel GM. 1995. The selective 5-Ht2A receptor antagonist, Mdl-100,907, increases dopamine efflux in the prefrontal cortex of the rat. Eur J Pharmacol 273: 273-279.

Scott JP, Fredericson E. 1951. The causes of fighting in mice and rats. Physiol Zool 24: 273-309.

Senault B. 1968. Syndrome agressif induit par l'apomorphine chez le Rat. J Physiol 60: 543-544.

Senault B. 1971. Influence de l'isolement sur le comportement d'agressivit, intrasp,cifique induit par l'apomorphine chez le rat. Psychopharmacologia 20: 389-394.

Sgoifo A, deBoer SF, Haller J, Koolhaas JM. 1996. Individual differences in plasma catecholamine and corticosterone stress responses of wild-type rats: relationship with aggression. Physiol Behav 60(6): 1403-1407.

Shader RI, Greenblatt DJ. 1993. Drug-therapy—use of benzodiazepines in anxiety disorders. New Eng J Med 328: 1398-1405.

Shaikh MB, De Lanerolle NC, Siegel A. 1997. Serotonin 5-HT1A and 5-HT2/1C receptors in the midbrain periaqueductal gray differentially modulate defensive rage behavior elicited from the medial hypothalamus of the cat. Brain Res 765: 198-207.

Shih JC, Chen K, Ridd MJ. 1999a. Monoamine oxidase: from genes to behavior. Ann Rev Neurosci 22: 197-217.

Shih JC, Ridd MJ, Chen K, Meehan WP, Kung MP, et al. 1999b. Ketanserin and tetrabenazine aggression in mice lacking monoamine oxidase A Brain Res 835: 104-112.

Siegel A, Mirsky AF. 1994. The neurobiology of violence and aggression. Reiss Jr, AJ Understanding and preventing violence. Biobehavioral influences. National Academy Press; Washington, DC:pp. 59-172. editor.

Siegel A, Roeling TAP, Gregg TR, Kruk MR. 1999. Neuropharmacology of brain-stimulation-evoked aggression. Neurosci Biobehav Rev 23: 359-389.

Sijbesma H, Schipper J, De Kloet ER, Mos J, Van Aken H, et al. 1991. Postsynaptic 5-HT_1 receptors and offensive aggression in rats: a combined behavioural and autoradiographic study with eltoprazine. Pharmacol Biochem Behav 38: 447-458.

Silver LM. 1995. Mouse genetics. Concepts and Applications.

Simler S, Puglisi-Allegra S, Mandel P. 1982. Gamma-aminobutyric acid in brain areas of isolated aggressive or non-aggressive inbred strains of mice. Pharmacol Biochem Behav 16: 57-61.

Sinton CM, Fallon SL. 1988b. Electrophysiological evidence for a functional differentiation between subtypes of the 5-HT1 receptor. Eur J Pharmacol 157: 173-181.

Sinton CM, Fallon SL. 1988a. Electrophysiological evidence for a functional differentiation between subtypes of the 5-HT1 receptor. Eur J Pharmacol 157: 173-181.

Smuts BB. 1986. Sexual competition and mate choice. Primate societies. Smuts BB, Cheney DL, Seyfarth RM, Wrangham RW, Struhsaker TT, editors. Chicago: University of Chicago Press; pp. 385-399.

Sokolov BP, Schindler CW, Cadet JL. 2004. Chronic methamphetamine increases fighting in mice. Pharmacol Biochem Behav 77: 319-326.

Sorgi PJ, Ratey JJ, Polakoff S. 1986. Beta-adrenergic blockers for the control of aggressive behaviors in patients with chronic schizophrenia. Am J Psychiatry 143: 775-776.

Spigset O. 1999. Adverse reactions of selective serotonin re-uptake inhibitors—Reports from a spontaneous reporting system. Drug Safety 20: 277-287.

Sprouse JS. 1991. Inhibition of dorsal raphe cell firing by MDL 73005EF, a novel 5-HT1A receptor ligand. Eur J Pharmacol 201: 163-169.

Sprouse JS, Aghajanian GK. 1987. Electrophysiological responses of serotoninergic dorsal raphe neurons to 5-HT1A and 5-HT1B agonists. Synapse 1: 3-9.

Sprouse JS, Aghajanian GK. 1988. Responses of hippocampal pyramidal cells to putative serotonin 5-HT1A and 5-HT1B agonist: a comparative study with dorsal raphe neurons. Neuropharmacology 27: 707-715.

Stanford MS, Houston RJ, Mathias CW, Greve KW, Villemarette-Pittman NR, et al. 2001. A double-blind placebo-controlled crossover study of phenytoin in individuals with impulsive aggression. Psychiatry Res 103: 193-203.

Steiniger F. 1950. Beitrag zur Soziologie und sonstigen Biologie der Wanderratte. Zeitschrift für Tierpsychologie 7: 356-379.

Stoff DM, Vitiello B. 1996. Role of serotonin in aggression of children and adolesecents: biochemical and phamacological studies. Aggression and violence: Genetic, neurobiological and biosocial perspectives. Stoff DM, editor. Mahwah, NJ: Lawrence Erlbaum Associates; pp. 101-123.

Stork O, Welzl H, Cremer H, Schachner M. 1997. Increased intermale aggression and neuroendocrine response in mice deficient for the neural cell adhesion molecule (NCAM). Eur J Neurosci 9: 1117-1125.

Stork O, Ji FY, Kaneko K, Stork S, Yoshinobu Y, et al. 2000. Postnatal development of a GABA deficit and disturbance of neural functions in mice lacking GAD65. Brain Res 865: 45-58.

Stowers L, Holy TE, Meister M, Dulac C, Koentges G. 2002. Loss of sex discrimination and male–male aggression in mice deficient for TRP2. Science 295: 1493-1500.

Strous RD, Bark N, Parsia SS, Volavka J, Lachman HM. 1997. Analysis of a functional catechol-O-methyltransferase gene polymorphism in schizophrenia: evidence for association with aggressive and antisocial behavior. Psychiatry Res 69: 71-77.

Sukhotina IA, Bespalov AY. 2000. Effects of the NMDA receptor channel blockers memantine and MRZ 2/579 on morphine withdrawal-facilitated aggression in mice. Psychopharmacology 149: 345-350.

Sukonick DL, Pollock BG, Sweet RA, Mulsant BH, Rosen J, et al. 2001. The 5-HTTPR*S/*L polymorphism and aggressive behavior in Alzheimer disease. Arch Neurol 58: 1425-1428.

Summers CH, Greenberg N. 1994. Somatic correlates of adrenergic activity during aggression in the lizard, *Anolis carolinensis*. Horm Behav 28: 29-40.

Summers CH, Forster GL, Korzan WJ, Watt MJ, Larson ET, et al. 2005. Dynamics and mechanics of social rank reversal. J Comp Physiol A Neuroethol Sens Neural Behav Physiol 191: 241-252.

Svensson TH, Bunney BS, Aghajanian GK. 1975. Inhibition of both noradrenergic and serotonergic neurons in brain by the alpha-adrenergic agonist clonidine. Brain Res 92: 291-306.

Swann AC. 2003. Neuroreceptor mechanisms of aggression and its treatment. J Clin Psychiatry 64: 26-35.

Swanson JW, Swartz MS, Elbogen EB, Van Dorn RA. 2004. Reducing violence risk in persons with schizophrenia: olanzapine versus risperidone. J Clin Psychiatry 65: 1666-1673.

Sweet RA, Pollock BG, Sukonick DL, Mulsant BH, Rosen J, et al. 2001. The 5-HTTPR polymorphism confers liability to a combined phenotype of psychotic and aggressive behavior in Alzheimer disease. Int Psychogeriatrics 13: 401-409.

Tao R, Auerbach SB. 2000. Regulation of serotonin release by GABA and excitatory amino acids. J Psychopharmacol 14: 100-113.

Tao R, Auerbach SB. 2002. GABAergic and glutamatergic afferents in the dorsal raphe nucleus mediate morphine-induced increases in serotonin efflux in the rat central nervous system. J Pharmacol Exp Ther 303: 704-710.

Tao R, Ma Z, Auerbach SB. 1996. Differential regulation of 5-hydroxytryptamine release by GABAA and GABAB receptors in midbrain raphe nuclei and forebrain of rats. Br J Pharmacol 119: 1375-1384.

Thoa NB, Eichelman B, Richardson JS, Jacobowitz D. 1972. 6-Hydroxydopa depletion of brain norepinephrine and the facilitation of aggressive behavior. Science 178: 75-77.

Tidey JW, Miczek KA. 1992a. Effects of SKF38393 and quinpirole on patterns of aggressive, motor and schedule-controlled behaviors in mice. Behav Pharmacol 3: 553-565.

Tidey JW, Miczek KA. 1992b. Morphine withdrawal aggression: modification with D1 and D2 receptor agonists. Psychopharmacology 108: 177-184.

Tidey JW, Miczek KA. 1996. Social defeat stress selectively alters mesocorticolimbic dopamine release: an in vivo microdialysis study. Brain Res 721: 140-149.

Tiihonen J, Kuikka J, Bergstrom K, Hakola P, Karhu J, et al. 1995. Altered striatal dopamine re-uptake site densities in habitually violent and non-violent alcoholics. Nat Med 1: 654-657.

Tobe A, Yoshida Y, Ikoma H, Tonomura S, Kikumoto R. 1981. Pharmacological evaluation of 2-(4-methylaminobutoxy) diphenylmethane hydrochloride (MCI-2016), a new psychotropic drug with antidepressant activity. Arzneimittelforschung/Drug Research 31: 1278-1285.

Tong J, Hornykiewicz O, Kish SJ. 2006. Identification of a noradrenaline-rich subdivision of the human nucleus accumbens. J Neurochem 96: 349-354.

Traskman-Bendz L, Alling C, Oreland L, Regnell G, Vinge E, et al. 1992. Prediction of suicidal behavior from biologic tests. J Clin Psychopharmacol 12: 21S-26S.

Troisi A, Vicario E, Nuccetelli F, Ciani N, Pasini A. 1995. Effects of fluoxetine on aggressive behavior of adult inpatients with mental retardation and epilepsy. Pharmacopsychiatry 28: 1-4.

Twarog BM, Page IH, Bailey H. 1953. Serotonin content of some mammalian tissues and urine and a method for its determination. Am J Physiol 175: 157-161.

Twitchell GR, Hanna GL, Cook EH, Stoltenberg SF, Fitzgerald HE, et al. 2001. Serotonin transporter promoter polymorphism genotype is associated with behavioral disinhibition and negative affect in children of alcoholics. Alcohol Clin Exp Res 25: 953-959.

Tyler CB, Miczek KA. 1982. Effects of phencyclidine on aggressive behavior in mice. Pharmacol Biochem Behav 17: 503-510.

Uhrich J. 1938. The social hierarchy in albino mice. J Comp Psychol 25: 373-413.

van der Vegt BJ, Lieuwes N, Cremers TIFH, de Boer SF, Koolhaas JM. 2003. Cerebrospinal fluid monoamine and metabolite concentrations and aggression in rats. Horm Behav 44: 199-208.

Van Der Vegt BJ, de Boer SF, Buwalda B, de Ruiter AJ, de Jong JG, et al. 2001. Enhanced sensitivity of postsynaptic serotonin-1A receptors in rats and mice with high trait aggression. Physiol Behav 74: 205-211.

Van Erp AMM, Miczek KA. 2000. Aggressive behavior, increased accumbal dopamine, and decreased cortical serotonin in rats. J Neurosci 20: 9320-9325.

van Oortmerssen GA, Bakker TCM. 1981. Artificial selection for short and long attack latencies in wild *Mus musculus domesticus*. Behav Genet 11: 115-126.

VanoverKE, SurukiM, RobledoS, HuberM, WielandS, et al. 1999a. Positive allosteric modulators of the $GABA_A$ receptor: differential interacton of benzodiazepines and neuroactive steroids with ethanol. Psychopharmacology 141: 77-82.

Vanover KE, Suruki M, Robledo S, Huber M, Wieland S, et al. 1999b. Positive allosteric modulators of the $GABA_A$ receptor: differential interacton of benzodiazepines and neuroactive steroids with ethanol. Psychopharmacology 141: 77-82.

Vekovischeva OY, Aitta-aho T, Echenko O, Kankaanpaa A, Seppala T, et al. 2004. Reduced aggression in AMPA-type glutamate receptor GluR-A subunit-deficient mice. Genes Brain Behav 3: 253-265.

Verberne AJ, Owens NC. 1998. Cortical modulation of the cardiovascular system. Prog Neurobiol 54: 149-168.

Villalba C, Boyle PA, Caliguri EJ, De Vries GJ. 1997. Effects of the selective serotonin reuptake inhibitor fluoxetine on social behaviors in male and female prairie voles (*Microtus ochrogaster*). Horm Behav 32: 184-191.

Virkkunen M, De Jong J, Bartko J, Linnoila M. 1989. Psychobiological concomitants of history of suicide attempts among violent offenders and impulsive fire setters. Arch Gen Psychiatry 46: 604-606.

Virkkunen M, Nuutila A, Goodwin FK, Linnoila M. 1987. Cerebrospinal-fluid monoamine metabolite levels in male arsonists. Arch Gen Psychiatry 44: 241-247.

Virkkunen M, Rawlings R, Tokola R, Poland RE, Guidotti A, et al. 1994. Csf biochemistries, glucose-metabolism, and diurnal activity rhythms in alcoholic, violent offenders, fire setters, and healthy-volunteers. Arch Gen Psychiatry 51: 20-27.

Vitiello B, Stoff DM. 1997. Subtypes of aggression and their relevance to child psychiatry. J Am Acad Child Adolesc Psychiatry 36: 307-315.

Volavka J, Czobor P, Nolan K, Sheitman B, Lindenmayer JP, et al. 2004. Overt aggression and psychotic symptoms in patients with schizophrenia treated with clozapine, olanzapine, risperidone, or haloperidol. J Clin Psychopharmacol 24: 225-228.

Volavka J, Czobor P, Citrome L, McQuade RD, Carson WH, et al. 2005. Efficacy of aripiprazole against hostility in schizophrenia and schizoaffective disorder: data from 5 double-blind studies. J Clin Psychiatry 66: 1362-1366.

Wahlsten D, Crabbe JC, Dudek BC. 2001. Behavioural testing of standard inbred and $5HT_{1B}$ knockout mice: implications of absent corpus callosum. Behav Brain Res 125: 23-32.

Walsh MT, Dinan TG. 2001. Selective serotonin reuptake inhibitors and violence: a review of the available evidence. Acta Psychiatr Scand 104: 84-91.

Walters JR, Seyfarth RM. 1986. Conflict and cooperation. Primate societies. Smuts BB, Cheney DL, Seyfarth RM, Wrangham RW, Struhsaker TT, editors. Chicago: The University of Chicago Press; pp. 306-317.

Weerts EM, Miczek KA. 1996. Primate vocalizations during social separation and aggression: effects of alcohol and benzodiazepines. Psychopharmacology 127: 255-264.

Weerts EM, Miller LG, Hood KE, Miczek KA. 1992. Increased $GABA_A$-dependent chloride uptake in mice selectively bred for low aggressive behavior. Psychopharmacology 108: 196-204.

Weiger WA, Bear DM. 1988. An approach to the neurology of aggression. J Psychiatr Res 22: 85-98.

Weisman AM, Berman ME, Taylor SP. 1998. Effects of clorazepate, diazepam, and oxazepam on a laboratory measurement of aggression in men. Int Clin Psychopharmacol 13: 183-188.

Welch BL, Welch AS. 1965. Effect of grouping on the level of brain norepinephrine in white swiss mice. Life Sci 4: 1011-1018.

Wersinger SR, Ginns EI, O'Carroll AM, Lolait SJ, Young WS III. 2002. Vasopressin V1b receptor knockout reduces aggressive behavior in male mice. Mol Psychiatry 7: 975-984.

Westergaard GC, Suomi SJ, Higley JD, Mehlman PT. 1999. CSF 5-HIAA and aggression in female macaque monkeys: species and interindividual differences. Psychopharmacology 146: 440-446.

Wilder J. 1958. Modern psychophysiology and the law of initial value. Am J Psychother 12: 199-221.

Wilmot CA, Vander Wende C, Spoerlein MT. 1987. The effects of phencyclidine on fighting in differentially housed mice. Pharmacol Biochem Behav 28: 341-346.

Wilson EO. 1974. Sociobiology. Cambridge: The Belknap Press.

WingfieldJC, Ball GF, Dufty AM Jr, Hegner RE, Ramenofsky M. 1987. Testosterone and aggression in birds. Am Sci 75: 602-608.

Winslow JT, Miczek KA. 1983. Habituation of aggression in mice: pharmacological evidence of catecholaminergic and serotonergic mediation. Psychopharmacology 81: 286-291.

Winslow JT, Hastings N, Carter CS, Harbaugh CR, Insel TR. 1993. A role for central vasopressin in pair bonding in monogamous prairie voles. Nature 365: 545-548.

Winslow JT, Hearn EF, Ferguson J, Young LJ, Matzuk MM, et al. 2000. Infant vocalization, adult aggression, and fear behavior of an oxytocin null mutant mouse. Horm Behav 37: 145-155.

Wise RA. 2005. Forebrain substrates of reward and motivation. J Comp Neurol 493: 115-121.

Wrangham RW. 1999. Evolution of coalitionary killing. Am J Phys Anthropol Suppl 29: 1-30.

Wright DE, Seroogy KB, Lundgren KH, Davis BM, Jennes L. 1995. Comparative localization of serotonin1A, 1C, and 2 receptor subtype mRNAs in rat brain. J Comp Neurol 351: 357-373.

Yanai K, Son LZ, Endou M, Sakurai E, Nakagawasai O, et al. 1998. Behavioral characterization and amounts of brain monoamines and their metabolites in mice lacking histamine H_1 receptors. Neuroscience 87: 479-487.

Yap JJ, Covington HE III, Gale MC, Datta R, Miczek KA. 2005. Behavioral sensitization due to social defeat stress in mice: antagonism at mGluR5 and NMDA receptors. Psychopharmacology (Berl) 179: 230-239.

Yeh SR, Fricke RA, Edwards DH. 1996. The effect of social experience on serotonergic modulation of the escape circuit of crayfish. Science 271: 366-369.

Yodyingyuad U, de la Riva C, Abbott DH, Herbert J, Keverne EB. 1985. Relationship between dominance hierarchy, cerebrospinal fluid levels of amine transmitter metabolites (5-hydroxyindole acetic acid and homovanillic acid) and plasma cortisol in monkeys. Neuroscience 16: 851-858.

Yoshimura H. 1984. Pharmaco-ethological analysis of agonistic behavior between resident and intruder mice: effects of psychotropic drugs. Folia Pharmacol Jpn 84: 221-228.

Young KA, Berry ML, Mahaffey CL, Saionz JR, Hawes NL, et al. 2002a. Fierce: a new mouse deletion of Nr2e1; violent behaviour and ocular abnormalities are background-dependent. Behav Brain Res 132: 145-158.

Young SE, Smolen A, Corley RP, Krauter KS, DeFries JC, et al. 2002b. Dopamine transporter polymorphism associated with externalizing behavior problems in children. Am J Med Genet 114: 144-149.

Yudofsky S, Williams D, Gorman J. 1981. Propranolol in the treatment of rage and violent behavior in patients with chronic brain syndromes. Am J Psychiatry 138: 218-220.

Zalsman G, Frisch A, Bromberg M, Gelernter J, Michaelovsky E, et al. 2001. Family-based association study of serotonin transporter promoter in suicidal adolescents: no association with suicidality but possible role in violence traits. Am J Med Genet 105: 239-245.

Zarcone JR, Hellings JA, Crandall K, Reese RM, Marquis J, et al. 2001. Effects of risperidone on aberrant behavior of persons with developmental disabilities: I. A double-blind crossover study using multiple measures. Am J Ment Retard 106: 525-538.

Zhuang X, Gross C, Santarelli L, Compan V, Trillat AC, et al. 1999. Altered emotional states in knockout mice lacking 5-HT_{1A} or 5-HT_{1B} receptors. Neuropsychopharmacology 21: S52-S60.

Zifa E, Fillion G. 1992. 5-Hydroxytryptamine receptors. Pharmacol Rev 44: 401-457.

Zitzman DL, De Bold JF, Rocha B, Miczek KA. 2004. Effects of estradiol on aggressive behavior in ovariectomized female mice. Abstr Soc Neurosci.

8 Novel Mechanisms Underlying Neuroendocrine Regulation of Aggression: A Synthesis of Rodent, Avian, and Primate Studies

G. E. Demas · M. A. Cooper · H. E. Albers · K. K. Soma

Abstract: An import role for gonadal steroid hormones, particularly testosterone, in mediating aggressive behavior is well-established across vertebrate taxa. Due to the emphasis placed on testosterone much less is know regarding the potential role of other steroid and peptide hormones in the regulation of aggression. The idea that any single hormone mediates aggression, however, is overly simplistic. In fact, research over the last two decades or more have has suggested that aggression in a complex behavior that is regulated by a wide range of hormones in addition to testosterone. This chapter reviews several novel neuroendocrine mechanisms that have been recently identified as being implicated in the regulation of aggressive behavior. Specific focus is placed on studies conducted in rodents, birds and primates, as the majority of research in this area has focused on these groups of animals.

List of Abbreviations: ACTH, adrenocorticotropic hormone; ADHD, attention deficit hyperactivity disorder; AE, androstenedione; ALLO, allopregnanolone; AR, androgen receptor; Aromatase, cytochrome P450 aromatase; AVP, arginine vasopressin; AVT, arginine vasotocin; BERKO, estrogen receptor β knock-out; BNST, bed nucleus of the stria terminalis; CNS, central nervous system; CORT, cortisol; 5α-DHT, 5α-dihydrotestosterone; DHEA-S, dehydroepiandrosterone sulfate; DHEA, dehydroepiandrosterone; E_1, estrone; E_2, 17β-estradiol; ER, estrogen receptor; ERKO, estrogen receptor knock-out; GABA, gamma-aminobutyric acid; 17β-HSD, 17β-hydroxysteroid dehydrogenase; 3β-HSD, 3β-hydroxysteroid dehydrogenase/isomerase; HP, hippocampus; HPA, hypothalamo-pituitary-adrenal; LD, light:dark; LH, luteinizing hormone; MPOA, medial preoptic area; N, nidopallium; NCM, caudomedial nidopallium; P450c17, cytochrome P450 17α-hydroxylase/C17,20 lyase; P450scc, cytochrome P450 side chain cleavage; POA, preoptic areaPREG, pregnenolone; PROG, progesterone; RA, robust nucleus of the arcopallium; STI, simulated territorial intrusion; T, testosterone; Tn, nucleus taeniae of the amygdala; TP, testosterone propionate; VMN, ventromedial nucleus

1 Introduction

► "It is the habit of every aggressor to claim that it is acting on the defensive" (Jawaharlal Nehru)

Among the wide array of social behaviors displayed by organisms, undoubtedly one of the most important and well studied is aggression. Aggression is a highly complex behavior displayed by virtually all living organisms, and serves a wide range of adaptive functions. The possibility for aggressive behavior exists whenever the interests of two or more individuals are in conflict, typically involving limited resources (e.g., food, territories, and mates). In nature, social interactions dictate which animals gain access to a limited resource and which ones do not. In many cases, a submissive posture displayed by one animal avoids the necessity of physical aggression. Additionally, animals may engage in threat displays or ritualized combat in which dominance is determined in the absence of physical harm. If such displays are ineffective, however, physical aggression can result. In some cases a material goal for aggression cannot be identified, and animals appear to fight over dominance status (cf. Mason, 1993). Aggression is an essential part of the socialization process, as mothers and other adults use aggression to modify inappropriate behavior in developing animals.

Despite its importance, aggression is a notoriously nebulous concept that has been defined and categorized in a multitude of ways over the years. Aggression has traditionally been defined as overt behavior with the intention of inflicting physical damage upon another individual or "goal entity" (Moyer, 1971). One of the most commonly employed classifications of aggression was described by Moyer (1971), who divided aggression into specific subtypes based on differences in social conditions in which the behavior was elicited. Among these forms of aggression were predatory aggression, intermale aggression, fear-induced aggression, irritable aggression, maternal aggression, territorial defense, and instrumental aggression. One of the primary tenets of Moyer's classification system was that, although these different forms of aggression share behavioral features, the environmental factors eliciting these responses and the biological substrates underlying their manifestation differ markedly. More recently, a simplified classification of aggression has been suggested (Blanchard and Blanchard, 1988) in which aggression is divided into *offensive* and *defensive* aggression. Offensive aggression refers to behaviors used in attack, whereas defensive behaviors do not involve an active approach to the opponent; rather, they serve as a defense against an attack. This latter classification system provides a useful framework with which to identify and describe aggressive behavior across many species.

Aggressive behavior has received extensive study under a wide range of environmental settings and experimental conditions; the experimental models used, the types of aggression measured, and the species tested, however, can vary considerably from study to study. Thus, it is often difficult to compare results across a range of studies. Although a relatively large number of experimental paradigms have been developed to test aggression (e.g., electric shock or brain lesion-induced aggression, conditioned aggression), one of the most prevalent models of assessing offensive aggression in rodents has been the resident–intruder model. This test is intended to simulate rodent territorial aggression and involves introducing a group-housed, nonexperimental "intruder" (typically younger and smaller than the experimental animal) into the home cage of an experimental animal, and the amount and duration of aggressive behavior (e.g., chases, attacks, bites) are subsequently recorded in a timed test. Another less-commonly employed but useful model is aggression in a neutral arena, which involves placing two animals in a novel "neutral" cage and recording the amount of aggression directed toward each animal. In addition, this latter model has the added benefit of assessing the formation of dominance relationships among animals because "territories" have not been established at the time of testing. Finally, it is important to consider the time of testing (i.e., day versus night), as most rodents are nocturnal and display significantly more behavior during night; many primates in contrast are diurnal. Thus, behavioral testing is more appropriately assessed during the day.

Much of the early research on the neuroendocrine mechanisms of aggression focused on the role of gonadal steroid hormones, and predominantly testosterone (T), as the primary factor regulating aggression (➋ *Figure 8-1*). In fact, most behaviorally oriented textbooks discuss the role of T in aggression almost exclusively, with much less emphasis placed on the role of other factors (e.g., neuropeptides, neurotransmitters) in mediating this behavior. One of the goals of this chapter is to suggest that the idea that a single hormone or neurotransmitter mediates aggression is overly simplistic. For example, recent evidence indicates that T can be converted to 17β-estradiol within the brain and that estrogens may mediate aggressive behavior, at least in some species and contexts (Soma et al., 2000a, b). Alternatively, the so-called

Figure 8-1
Simplified diagram of sex steroid synthesis. ***Steroids*****: PREG = pregnenolone; PROG = progesterone; DHEA = dehydroepiandrosterone; AE = androstenedione; T = testosterone; E_1 = estrone; E_2 = 17β-estradiol.** ***Enzymes*****:** ***P450scc*** **= Cytochrome P450 side chain cleavage;** ***P450c17*** **= Cytochrome P450 17α-hydroxylase/C17,20 lyase;** ***3β-HSD*** **= 3β-hydroxysteroid dehydrogenase/isomerase;** ***17β-HSD*** **= 17β-hydroxysteroid dehydrogenase;** ***Aromatase*** **= Cytochrome P450 aromatase**

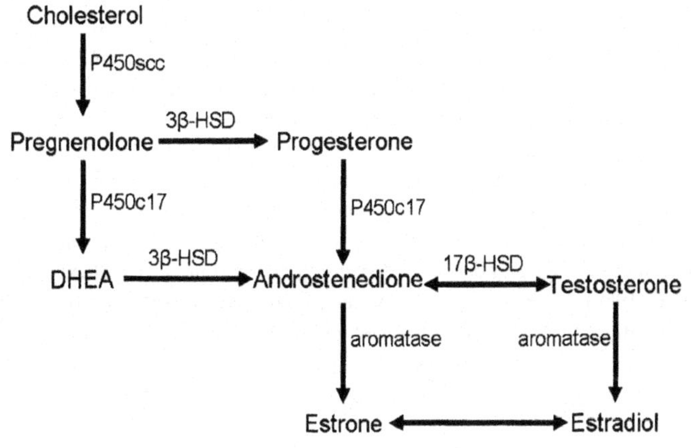

"weak" androgens such as dehydroepiandrosterone (DHEA) may be produced in extragonadal tissue (e.g., adrenals) or *de novo* within the brain. In fact, it has been suggested that both adrenal and neurally derived androgens, the latter called neurosteroids (Baulieu, 1991), may play an important role in regulating aggressive responses (Simon, 2002). Collectively, these important findings have helped elucidate several possible pathways by which androgens can act on the brain, either directly or indirectly, to affect aggression (➊ *Figure 8-2*).

In addition to androgens, a wide range of neuroendocrine factors have now been established as playing a major role in aggressive behaviors. We will review important classical findings, as well as more recent discoveries for some of the key factors; however, an encyclopedic review of all the neuroendocrine factors regulating aggression is beyond the scope of this chapter. For a more in-depth discussion of specific neuroendocrine factors and their effects on aggression, we refer the reader to some excellent recent reviews (Ferris and Delville, 1994; Albers and Bamshad, 1998; Ferguson et al., 2002; Chiavegatto and Nelson, 2003).

Another important goal of this chapter is to present a comparative approach to the study of aggression. Historically, much of the early work on the physiology of aggression placed considerable emphasis on rodent models (e.g., Edwards et al., 1969; Brain and Poole, 1974). In fact, much of this research has focused on highly domesticated species (e.g., inbred strains of rats and mice). Although these models have provided important insights into the mechanisms of aggression, and continue to do so (especially in the light of recent advances in molecular and genetic techniques), recent evidence suggests that specific mechanisms mediating aggression can differ markedly across taxa (reviewed in Wingfield et al., 1997). Thus, an important aim of this chapter is to integrate research findings from several different taxa with the goal of elucidating common themes and noteworthy differences, with respect to the neuroendocrine regulation of aggression. With this goal in mind, we have chosen to focus on three groups of animals – rodents, birds, and primates—where considerable data from both laboratory and field settings are available. The application of a comparative approach to the study of aggression will allow for the development of a more comprehensive understanding of the factors mediating social behavior.

2 Neuroendocrine Regulation of Aggression in Rodents

Much of the early work on the neuroendocrine regulation of aggression focused on the role of gonadal steroids in mediating intermale aggression and a majority of this work was conducted using laboratory-bred strains of rats and mice. Although space constraints do not permit an exhaustive discussion of this literature, a review of the major findings in this area will serve as a foundation for interpreting more recent findings, as well as exploring alternative physiological mechanisms regulating aggression in rodents.

2.1 Androgens: Organizational Versus Activational Effects

The link between T and aggression in rodents was first reported by Uhrich, who demonstrated that castration of postpubertal male rats decreased aggression, and that T replacement substantially reversed this effect (Uhrich, 1938). Since this initial finding, literally thousands of papers have been published examining the relationship between androgens and aggression (Beeman, 1947; Bevan et al., 1958, 1960; Yen et al., 1962; Sigg et al., 1969; Suchowsky et al., 1969; Brain and Haug, 1992). An important distinction must be made, however, between *organizational* and *activational* effects of androgens on aggression. Ever since the initial findings of Young and colleagues studying the effects of steroids on reproductive behaviors in guinea pigs, a well-established tenet of behavioral endocrinology is that steroid hormones can exert differential effects on physiology and behavior, depending on the time course of exposure during development (Phoenix et al., 1959). For example, for some behavioral responses (e.g., reproduction), androgens are required during a critical or "sensitive" period of perinatal development to permanently organize the neuroendocrine substrates required for adequate display of appropriate responses during adulthood. In many cases, the presence of androgens during adulthood is also required to activate the appropriate responses.

Figure 8-2
Steroid hormones can reach the brain to affect aggression via several possible mechanisms. (a) Gonadal testosterone can act directly on the brain; (b) gonadal T can be converted locally to estradiol; (c) adrenal DHEA can be converted locally to T or E_2; (d) neurosteroids can be produced locally in the absence of gonadal and adrenal steroid production

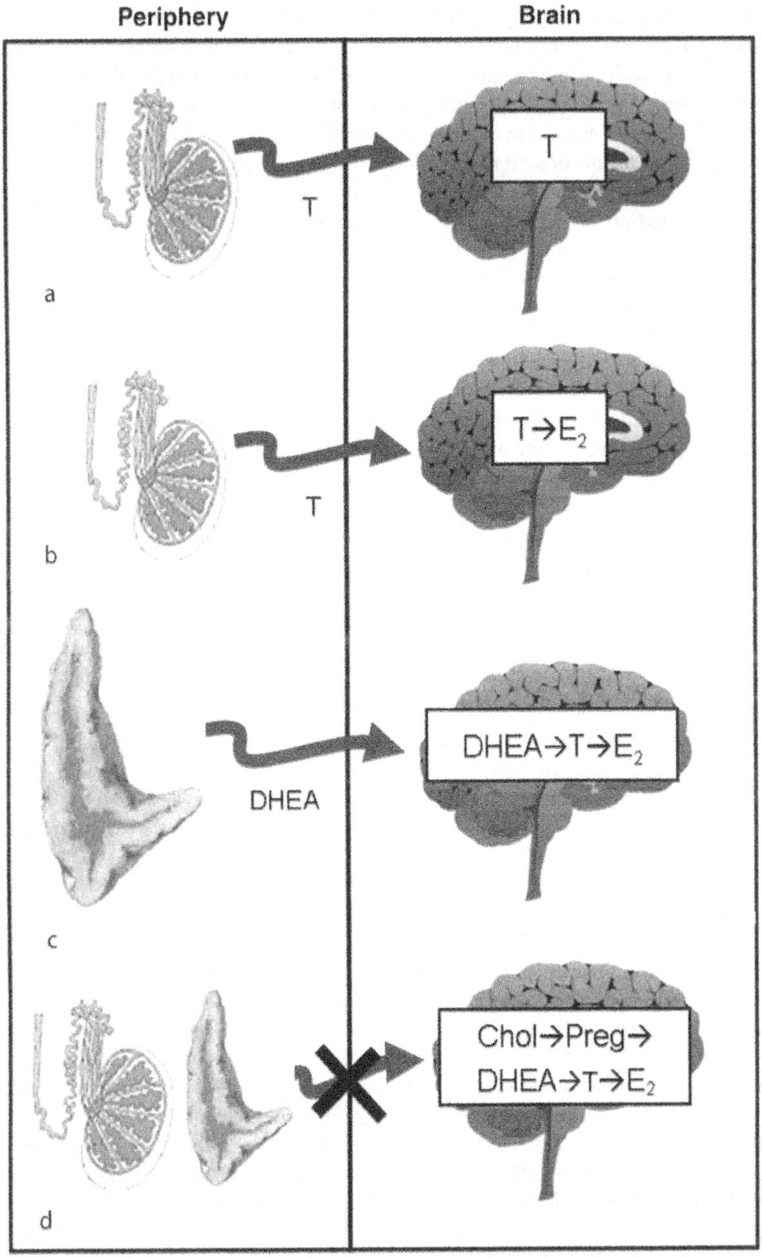

A large number of studies have demonstrated both organizational and activational effects of androgens on aggression in rodents. For example, castration of male mice on the day of birth eliminates aggression during adulthood, despite the presence or absence of exogenous androgens at the time of testing (Bronson and Desjardins, 1969). However, treatment of castrated pups with exogenous T immediately following surgery will result in normal levels of male aggression in adult animals, provided that androgens are given at the time of behavioral testing. Interestingly, the level of aggression displayed by adult female mice can be increased to that seen in male mice by treatment with exogenous androgens early during development, again provided that injections of hormone are also administered at the time of testing, consistent with an organizational effect of androgens on aggression (Edwards, 1968; Bronson and Desjardins, 1970). Male-like aggression in females treated with T prepubertally is not seen if T is not administered in adulthood, suggesting that activation of aggression is androgen-dependent.

The sensitive period for the organizational effects of androgens appears to be relatively long, although exogenous treatment with T is most effective when administered close to the time of birth in mice. For example, a single injection of T in castrated mice is sufficient to organize adult aggression, but this treatment is more effective on Day 0 than on Day 10 after birth (Edwards, 1969; Bronson and Desjardins, 1970). Furthermore, single injections of T become ineffective between 12–24 days of postnatal age, although more prolonged treatment (e.g., 20 days) can effectively organize adult aggression as late as 30 days postnatal (Edwards, 1970).

The widely accepted relationship between the presence of the testes (and normal circulating T concentrations) and aggression, however, is based primarily on studies of male–male aggression in a limited number of highly domesticated species such as laboratory rats (*Rattus norvegicus*) and house mice (*Mus musculus*) typically maintained in 12:12 light—dark (LD) cycles and tested during the light phase. In these species, removal of the gonads results in substantial decreases in circulating T and subsequently, reduced aggression (e.g., Edwards, 1969, 1970). When males of several nondomesticated species are examined, however, exceptions to the relationship of low circulating T concentrations and reduced aggression begin to emerge (Mathewson, 1961; Tiefer, 1970; Whitsett, 1975; Caldwell et al., 1984; Demas et al., 1999; Wiley and Goldizen, 2003; Gottreich et al., 2001). For example, several species of animals, including Mongolian gerbils (*Meriones unguiculatus*), prairie voles (*Microtus ochrogaster*), Syrian hamsters (*Mesocricetus auratus*), blind mole rats (*Spalax Ehrenberg*), saddle-back tamarins (*Saguinus fuscicollis*), European starlings (*Sturnus vulgaris*), and red-sided garter snakes (*Thamnophis sirtalis*) do not display decreased aggression after castration (Davis, 1957; Mathewson, 1961; Tiefer, 1970; Christianson et al., 1972; Epple, 1978; Dewsbury, 1991; Demas et al., 1999b). However, even in laboratory species, T does not stimulate aggression under all conditions. For example, in rats, T appears to be most effective in stimulating aggression during competitive experiences, but not during other social encounters (Albert et al., 1992).

2.2 Androgens and Seasonal Aggression

Several studies have examined the role of T in aggression *indirectly* by manipulation of the photoperiod. Many nontropical rodent species are seasonal breeders, maintaining reproductive function during summer and curtailing breeding during the winter. Ambient day length (photoperiod) is the proximal environmental cue used by individuals within these species to coordinate their reproduction to the appropriate season (Goldman, 2001). For example, reproductive function (and high levels of circulating T) is maintained during long "summer-like" days (e.g., >12.5 h of light/day) whereas reproductive regression, including virtual collapse of the gonads and marked decreases in T occur during the short "winter-like" days (i.e., <12.5 h of light/day) (Goldman, 2001). Interestingly, in male Syrian hamsters maintaining animals in short days actually *increases* resident–intruder aggression compared with long-day hamsters (Garrett and Campbell, 1980). Specifically, adult male Syrian hamsters housed in short days for nine weeks display approximately twice the amount of aggression in a resident–intruder test compared with long-day controls when tested 4 h before dark, despite gonadal regression (Garrett and Campbell, 1980). After prolonged maintenance in short days (>15 weeks) hamsters typically undergo spontaneous gonadal recrudescence (i.e., increased testicular mass and circulating T), despite continued maintenance in short days. The short-day increases in aggressive behavior largely disappear in animals undergoing spontaneous recrudescence, returning to long-day levels of aggression by 21 weeks (Garrett and Campbell, 1980). More recently, short-day increases in aggression in male Syrian hamsters have been confirmed (Jasnow et al., 2002; Caldwell and Albers, 2004). For example, Syrian hamsters housed in short days (LD 10:14) for 10 weeks displayed a significantly greater number of attacks and a longer duration of attacks than did long-day hamsters when tested using a resident–intruder test (Jasnow et al., 2002). Furthermore, timed daily melatonin injections mimicking short-day patterns of the hormone in long-day, pineal-intact animals will produce short-day-like increases in aggression. Because these injections occurred for only 10 days, the gonadal mass and circulating levels of T are unaffected, supporting the idea that photoperiodic changes in aggression are not mediated by changes in gonadal steroids in this species (Jasnow et al., 2002). In contrast, these results suggest that levels of aggressive behavior are mediated by changes in the pattern of melatonin secretion.

Photoperiodic changes in aggression have been demonstrated in females of at least one species, Syrian hamsters (Fleming et al., 1988; Badura and Nunez, 1989). Female hamsters were housed in long (LD 14:10) or short days (LD 6:18) for 12 weeks and then both offensive and defensive aggression were tested (Fleming et al., 1988). Female hamsters maintained in short days displayed significantly less defensive aggression compared to long-day animals and thus had a higher ratio of offensive to defensive aggression than long-day animals (Fleming et al., 1988).

In virtually all mammals, photoperiodic responses are mediated by changes in the pineal indolamine melatonin. Melatonin is secreted in abundance during darkness, whereas daylight inhibits pineal melatonin secretion (Goldman, 2001). Thus, changes in ambient day length result in changes in the pattern of secretion of melatonin. In this manner, it is the precise pattern of melatonin secretion, and not the amount of hormone per se, that provides the biochemical "code" for day length (Goldman, 2001).

Pinealectomy, which eliminates melatonin secretion and renders animals physiologically "blind" to day length, prevents the short-day increase in aggression in female Syrian hamsters, whereas treatment of long-day hamsters with exogenous short-day-like melatonin increases aggression in female Syrian hamsters (Fleming et al., 1988). Ovariectomy, in contrast, has no effect of aggression. This finding suggests that photoperiodic changes in aggression are independent of changes in gonadal steroids in female Syrian hamsters (Fleming et al., 1988). A subsequent study in female Syrian hamsters confirmed these findings and provided further support for a role of pineal melatonin in mediating photoperiod changes in aggression. Specifically, a higher percentage of female hamsters housed in short days (LD 6:18) showed aggressive behavior compared with long day housed (LD 16:8) hamsters (Badura and Nunez, 1989). Consistent with previous findings, short-day aggression was attenuated by pinealectomy, but treatment with exogenous estradiol (E_2) (alone or in combination with progesterone) had no effect on aggression. These results support the hypothesis that photoperiodic changes in aggression are mediated by pineal melatonin, but independent of gonadal steroids, at least in female Syrian hamsters.

In Syrian hamsters, unlike most rodent species, females are more aggressive than males (Marques and Valenstein, 1977; Ciacco et al., 1979). Few studies have examined the role of photoperiod on male aggression in rodents displaying typical male-dominant aggression. Unlike Syrian hamsters, male Siberian hamsters (*Phodopus sungorus*) display significantly more aggression than females. It has recently been demonstrated that short-day male Siberian hamsters are significantly more aggressive than long-day animals (Jasnow et al., 2000; Jasnow et al., 2002), consistent with previous studies in Syrian hamsters. Specifically, male Siberian hamsters housed in short days (LD 8:16) for 10 weeks display a greater number of attacks during a resident–intruder test and have a lower latency to initial attack, relative to long-day (LD 16:8) animals. As previously reported for many rodent species, prolonged maintenance on short days (i.e., 20 weeks) resulted in spontaneous reproductive recrudescence in which the gonads, and thus T, returned to normal long-day levels (Jasnow et al., 2000). Interestingly, gonadally recrudesced hamsters displayed less aggression than gonadally regressed animals even though both groups experienced the same photoperiod and melatonin signal; levels of aggression in recrudesced hamsters were generally indistinguishable from long-day hamsters (Jasnow et al., 2000). These results support previous findings in male Syrian hamsters (Garrett and Campbell, 1980). When short-day Siberian hamsters were implanted with Silastic capsules containing T (to achieve long-day-like levels), aggression actually *decreased* compared with short-day control animals (Jasnow et al., 2000), suggesting that short-day increases in aggression may be *inversely* related to serum T concentrations.

Despite growing evidence that short-day increases in aggression are independent of (or inversely related to) circulating levels of T, much less is known about the precise neuroendocrine mechanisms underlying seasonal aggression in rodents. As previously described, several studies have implicated changes in the pineal hormone melatonin in mediating short-day aggression. More recent research in male Siberian hamsters (Demas et al., 2004) confirms previous findings that treatment of long-day animals with short-day-like levels of melatonin mimics photoperiodic changes in aggression; long-day hamsters given daily timed injections of melatonin 2 h before lights-out to mimic short-day levels of the hormone displayed elevated aggression in a resident–intruder test compared with control animals. As with previous studies, these results were not likely due to changes in gonadal steroids, as serum T was unaffected by this injection protocol.

The effects of melatonin on aggression in rodents may be due to direct actions of this hormone on neural substrates mediating aggression (e.g., hypothalamus, limbic system). Alternatively, melatonin-induced aggression may be indirectly due to changes in hypothalamo–pituitary–adrenal (HPA) activity, as adrenal hormones have been implicated in aggressive behavior (Haller and Kruk, 2003). In support of the latter hypothesis, changes in both the size and function of the adrenal gland are associated with changes in aggression (Paterson and Vickers, 1981). In addition, male house mice housed in a LD 12:12 photoperiod and treated with melatonin display increased territorial aggression but decreased adrenal masses compared to saline-treated animals (Paterson and Vickers, 1981). The increases in aggression displayed by melatonin-treated animals, however, can be blocked by adrenalectomy (Paterson and Vickers, 1981). Experimental reductions in both adrenomedullary catecholamines, as well as adrenocortical glucocorticoids, are associated with decreased aggression in rodents (Patterson and Vickers, 1981; Haller and Kruk, 2003) and reductions in glucocorticoids via pharmacological blockade of adrenocorticotropic hormone (ACTH) release can attenuate melatonin-induced increases in aggression in mice (Patterson and Vickers, 1981). Thus, exogenous melatonin, despite reducing adrenal mass, appears to increase aggression by stimulating adrenocortical steroid release. These results are particularly intriguing given that house mice have traditionally been assumed to be photoperiodically nonresponsive (Nelson, 1990).

More recently, research has implicated changes in adrenocortical hormones in mediating melatonin- and possibly short-day-induced aggression in Siberian hamsters. As described previously, long-day hamsters treated with short-day-like levels of melatonin displayed increased aggression, comparable to levels seen in short-day animals (Demas et al., 2004). Interestingly, melatonin-induced aggression could be blocked by bilateral adrenalectomy, consistent with previous results in house mice (Paterson and Vickers, 1981). Adrenal demedullation, which eliminates adrenal catechloamines (i.e., epinephrine) but leaves adrenocortical steroid release (i.e., cortisol, DHEA) intact, had no effect on melatonin-induced aggression (Demas et al., 2004). Collectively, these results support the hypothesis that the effects of exogenous melatonin on aggression are mediated by the effects of this hormone on adrenocortical steroids. However, it is currently not known which class of steroid hormones may mediate this effect, as adrenal androgens (e.g., DHEA) and glucocorticoids (e.g., cortisol) have both been implicated in aggression in rodents (Schlegel et al., 1985; Haller and Kruk, 2003). In laboratory rats and mice, corticosterone (CORT) is the predominant adrenal glucocorticoid, and these species secrete little to no adrenal DHEA. In contrast, in hamsters, as in humans, cortisol is the primary adrenal glucocorticoid, and both hamsters and humans secrete measurable amounts of DHEA and its sulfated form, DHEA-S (Pieper et al., 2000; Mellon and Vaudry, 2001).

Recent evidence in avian species suggests that aggression in the nonbreeding season (i.e., winter) may be mediated by changes in DHEA (Soma et al., 2000; see ➋ Section 4). Although similar evidence suggesting a role for DHEA in mediating photoperiodic changes in aggression in rodents is lacking, studies in mice suggest that exogenous melatonin can stimulate DHEA production from cultured adrenal glands (Haus et al., 1996). Behavior was not examined in this study; however, these results are consistent with the hypothesis that short-day increases in melatonin may increase adrenal production of DHEA and thus affect aggression.

Several field studies published to date support the laboratory data discussed above suggesting differential dependence on gonadal steroids in animals during breeding season versus when they are not breeding. Specifically, male rat-like hamsters (*Cricetus triton*) in the field display elevated aggression during the winter nonbreeding season, despite low levels of plasma T (Zhang et al., 2001). Seasonal changes in aggression appear independent of seasonal changes in circulating T in wild wood rats (*Neotoma fuscipes*) (Caldwell et al., 1984). Pronounced seasonal changes in aggression are seen in male wood rats, with high levels during midbreeding season and low levels during the nonbreeding season. Despite differences in circulating T levels at these two time points, castration has no affect on aggression, suggesting an independence of seasonal aggression from circulating levels of T (Caldwell et al., 1984). More recently, seasonal changes in aggressive encounters have been examined in free-living arctic ground squirrels (Buck and Barnes, 2003). The effects of challenges by conspecific males on circulating T levels varied seasonally, with challenges by male intruders eliciting significant increases in circulating T during the spring breeding season, but similar challenges failed to trigger increase in androgen at the end of the summer after the breeding season. These results suggest that androgens play a more important role during the breeding season than during the nonbreeding season (i.e., late summer). Collectively, these studies fail to support the simple notion that all forms of aggression are mediated by circulating T by providing salient examples of T-*independent* aggression, at least with respect to circulating levels of the hormone. Unlike other forms of aggression, however, very little is known regarding the neuroendocrine mechanisms underlying seasonal changes in aggression in mammals.

2.3 Estrogens and Aggression

The specific role of estrogens in the regulation of aggression is complicated by the fact that both estrogens and androgens (if aromatized to estrogens) can act on estrogen receptors (ER) within the brain. Thus, many of the effects of androgens on aggression traditionally ascribed to actions via androgen receptors (AR) may in fact be due to activation of ER following aromatization. In addition, unlike the AR, two isoforms of ER, ERα, and ERβ, have been identified, and these receptors can exert differential effects on rodent aggression (Nomura et al., 2002). Much of the experimental evidence implicating ER in the regulation of intermale aggression has been determined using genetic knockout (KO) techniques in inbred male mice. For example, initial studies using ERKO mice that lack functional ERα demonstrated that intermale aggression was markedly reduced and male-typical offensive attacks were rarely displayed by ERKO mice (Ogawa et al., 1997). Moreover, both castrated wild-type (WT) and ERKO mice displayed comparably low levels of aggression; T replacement increased aggression in castrated WT males, but had no effect on ERKO aggression (Ogawa et al, 1998).

Unlike studies using ERKO mice, male mice that lack functional ERβ (BERKO) display relatively normal or even slightly elevated levels of aggressive behavior compared with WT males (Ogawa et al., 1999). Male aromatase KO mice that lack a functional aromatase enzyme necessary for the conversion of androgens to estrogens also display marked reductions in aggression (Matsumoto et al., 2003). This finding, coupled with the data from ERKO studies, suggests that the effects of androgens on intermale aggression may be due, at least in part, to conversion of these hormones to estrogens and the subsequent action of estrogens on ER within the brain. Furthermore, these results support the idea that ERα is the active form of ER for regulating neuroendocrine effects on aggression (Ogawa, 2003), although more research is needed. These results suggest that ERβ may provide a "brake" on aggression.

2.4 Glucocorticoids and the Development of Aggression

Although many pre-pubertal rodents display some form of "play aggression" prior to the development of fully functioning testes (Delville et al., 2003), fighting among males typically begins at the onset of puberty. Postpubertal aggression is mediated by the surge in circulating steroids that can "activate" aggressive behavior. Much of the work in this area has been performed in Syrian hamsters. For example, repeated social stress in male Syrian hamsters inhibits aggression and increases submissive behaviors in adulthood. Acute social defeat in adult male Syrian hamsters increases subsequent submissive behavior and produces a complete loss of normal territorial aggression (Huhman et al., 2003). This change in agonistic behavior, however, appears mediated by the CNS and not by glucocorticoid feedback (Cooper and Huhman, 2003). Interestingly, recent evidence suggests that social stress during puberty enhances offensive aggression, most likely via the actions of glucocorticoids on the organization of aggression in this species (although a potential role for the adrenal androgen DHEA has not been examined to date). Syrian hamsters initiate aggression on approximately postnatal Day 20, engaging in considerable flank marking (a stereotyped form of communicative behavior employing odors) and play fighting (Goldman and Swanson, 1975; Ferris et al., 1996). Play fighting occurs as soon as coordinated movement is possible, and juvenile hamsters engage in more attacks during agonistic behavior than adult hamsters (Pellis and Pellis, 1998; Wommack et al., 2003). The number of attacks increases over development, reaching peak levels at approximately postnatal Day 35, and subsequently decreases until stable levels are reached at ~45 days postnatal (reviewed in Delville, 2003).

It is important to note that the development of aggression in Syrian hamsters occurs when juvenile hamsters are attempting to establish independence from other territorial males and thus, are potentially exposed to social stressors from their dominant conspecifics (Delville, 2003). Recent research has examined the role of the stress response, and particularly the activation of the HPA axis and subsequent release of glucocorticoids, in the development of agonistic behavior in juvenile Syrian hamsters (Delville et al., 1998; Wommack et al., 2003). Specifically, juvenile hamsters were placed daily in the home cages of an aggressive adult hamster for short periods from weaning to midpuberty, whereas control animals were moved to an empty cage. Initially, both groups of animals displayed significantly increased levels of circulating cortisol. By two weeks, however, only hamsters exposed to males maintained elevated cortisol levels. Unlike previous studies in rats (Blanchard et al., 1995), chronic social stress did not result in altered body mass or lower circulating T concentrations. Social stress did not affect the frequency of attacks against a smaller intruder during early puberty (Wommack et al., 2003); however, at midpuberty, socially stressed hamsters displayed significantly more attacks than control animals (Wommack et al., 2003). Socially subjugated animals also performed adult-like attacks at an earlier developmental age, while decreasing the amount of play-fighting (Wommack et al., 2003). More recent data provide support for the idea that social stress during development may enhance offensive aggression in Syrian hamsters. Specifically, hamsters were exposed to social subjugation or isolation (control) from postnatal day 28 until midpuberty (Wommack et al., 2004). Hamsters were then assessed under basal conditions or after social defeat during early puberty, midpuberty or early adulthood. Socially stressed animals had lower post-defeat cortisol levels than control males during midpuberty, as well as complete inhibition of olfactory investigation of aggressive adults (Wommack et al., 2004). Collectively, these results suggest an important role for social stress, and subsequent HPA activation, in the development of age-appropriate levels of aggression.

2.5 Neurosteroids and Aggression

The concept of "neurosteroids" was first introduced by Baulieu to describe the high levels of the androgen DHEA, and its sulfated form DHEA-S, seen in rat brain even after castration and adrenalectomy (Baulieu, 1981; Corpechot et al., 1981). It is now established that DHEA, among other steroid hormones [e.g., allopregnanolone (ALLO)], can be synthesized *de novo* within the central nervous system and can act locally on specific neural substrates to regulate behavior (Simon, 2002). For example, a recent study has demonstrated that intermale aggression is associated with changes in brain neurosteroid synthesis (Pinna et al., 2005). Specifically, administration of testosterone propionate (TP) to male mice decreased brain ALLO content by ~40% and was correlated with increased aggression. Increasing brain ALLO levels pharmacologically attenuated aggressive behavior in these mice. It remains unclear, however, how neurosteroids might be synthesized in adult rodent brain, given that some of the relevant synthetic enzymes (e.g., P450c17) are apparently nondetectable in the brains of many rodents (Mellon and Vaudry, 2001).

DHEA has also been implicated in the regulation of aggression in rodents. For example, DHEA has been shown to be a potent inhibitor of female-typical aggression (Bayart et al., 1986; Haug et al., 1989, 1992; Young et al., 1996; Perché et al., 2000). Attack behavior in ovariectomized females, intact females, or castrated male mice toward lactating females can be significantly reduced following treatment with exogenous DHEA (Bayart et al., 1989; Brain and Haug, 1992). Because this effect requires prolonged (~15 days) DHEA treatment, it has been suggested that the results are consistent with a genomic mechanism of action (Lu et al., 2003). More recently, it has been shown that prenatal T treatment enhances DHEA-induced inhibition of aggression in female offspring (Perché et al., 2001). It is important to note, however, that this testing paradigm (i.e., aggression toward a lactating intruder) is less commonly employed than more traditional models of aggression (e.g., intermale aggression); thus, the role of DHEA in mediating other forms of aggression has received very little experimental attention. One recent finding, however, suggests that increases in aggressive behavior can be induced by a single injection of DHEA-S immediately prior to behavioral testing in mice tested in a neutral arena (Nicolas et al., 2001). Furthermore, treatment with COUMATE, a drug that inhibits the steroid sulfatase enzyme involved in converting DHEA-S to DHEA, also increased aggression in male mice (Nicolas et al., 2001). Despite these intriguing results, more studies are needed to fully elucidate the role of DHEA and DHEA-S in intermale aggression.

The effects of DHEA on aggression have led to an attempt at determining the physiological mechanisms underlying this behavioral response, with much of the work focused on the interaction of this steroid with gamma aminobuteric acid (GABA) neurotransmission (reviewed in Simon, 2002). It has been demonstrated that DHEA alters brain levels of pregnenolone sulfate (PREG-S), a neurosteroid that inhibits GABAergic actions via the $GABA_A$ receptor. Specifically, DHEA-induced changes in GABA activity appear responsible for the effects of DHEA on aggression. These results are consistent with previous findings demonstrating an inhibitory effect of GABA on aggression (Miczek et al., 1994, 1997). Although the precise mechanisms of action are still unknown, DHEA can modulate the actions of both GABA and glutamatergic NMDA receptors (Labrie, 1998; Mellon and Vaudry, 2001). Alternatively, DHEA may act by further conversion in the brain to T and E_2 via 3β-HSD and the aromatase enzyme (Soma et al., 2004).

More recent findings suggest that an additional mechanism for the effects of DHEA on aggression may exist (Simon, 2002). Although is has long been assumed that DHEA is incapable of directly interacting with androgen receptors (AR), recent data suggest that DHEA upregulates AR in mouse brains (Lu et al., 2003). For example, because intact and gonadectomized, T-treated males do not display aggression toward females and prolonged treatment with DHEA is necessary to elicit antiaggressive effects, it has been speculated that androgens such as DHEA play an inhibitory role, presumably through a genomic action by binding directly on AR and altering AR transcription and translation (Simon, 2002). For example, it has recently been demonstrated that DHEA can compete for recombinant AR binding, up-regulate neural AR protein levels in mouse brains and immortalized GT1-7 hypothalamic cells, and induce transcription through AR in CV-1 cells, suggesting direct actions of DHEA on AR (Simon, 2002). Although this idea is intriguing, the affinity of DHEA for AR is low relative to T, so high local levels of DHEA would be necessary for this mechanism to be plausible.

2.6 Arginine Vasopressin and Dominant/Subordinate Relationships

Arginine vasopressin (AVP) when released as a neurohormone from the posterior pituitary is involved in the control of several important homeostatic functions, including the regulation of cardiovascular responses and fluid balance. However, there are also neural circuits in the brain that contain AVP and that have important effects on social behavior, including aggression (Ferris and Delville, 1994; Albers and Bamshad, 1998; Albers et al., 2002). One of the first demonstrations of a role for AVP in agonistic behavior occurred accidentally when it was found that injections of AVP into the anterior hypothalamus (AH) elicited flank marking in Syrian hamsters (Ferris et al., 1984). Since this initial discovery, a number of studies have implicated AVP in the regulation of aggressive behavior and dominance status (Ferris and Delville, 1994; Albers and Bamshad, 1998, Albers et al., 2002). For example, AVP released within the medial preoptic area (MPOA)-AH appears to be critical for the communication of dominance status by flank marking in Syrian hamsters (Ferris et al., 1986). AVP antagonists injected into dominant hamsters inhibit the high levels of flank marking observed when these hamsters are placed with their subordinate partners. In response to this inhibition of flank marking in the dominant hamsters, subordinate hamsters significantly increase their levels of flank marking. Injections of AVP into the MPOA-AH of subordinate hamsters can also increase flank marking. Although injections of AVP and AVP antagonists can significantly modulate flank marking in both dominant and subordinate hamsters, these injections do not appear to modify the basic dominant/subordinate relationship (Ferris et al., 1986).

Studies on both Syrian hamsters and laboratory rats also suggest a role for AVP in the regulation of offensive aggression, most typically studied using the resident–intruder model. For example, injections of an AVP receptor antagonist into the AH trigger a dose-dependent decrease in offensive aggression of a resident male hamster toward an intruder (Potegal and Ferris, 1990). Similar results were found when animals were tested using a neutral arena. Injection of AVP into the AH increases aggression; however, the concentration of AVP administered must be below the levels that induce flank marking (Ferris et al., 1997; Caldwell and Albers, 2004). In addition, it appears that AVP enhances aggression only in hamsters that have been exposed to social conditions that increase their potential for aggression (see Caldwell and Albers, 2004). There is also evidence that AVP injected into the ventrolateral hypothalamus can facilitate offensive aggression (Ferris and Delville, 1994).

Aggressive behavior plays a critical role in the establishment of dominant/subordinate relationships in rodents. For example, when two male rodents that have not experienced one another previously are placed in a neutral arena, an initial period of high aggression is followed by a decrease in aggression once one male has established dominance. Injection of a specific AVP V_{1a} receptor antagonist into hamsters that have not previously interacted prevents the formation of a dominant/subordinate relationship (reviewed in Albers and Bamshad, 1998), suggesting that the period of aggression or flank marking is necessary for the establishment of social hierarchies. In contrast, injections of either AVP or AVP antagonists in hamsters with established hierarchies can affect the amount of flank marking displayed by both dominant and submissive hamsters. For example, injection of AVP into the MPOA-AH of subordinate hamsters increases flank marking in the presence of the dominant animal. Conversely, injections of an AVP antagonist into a dominant hamster reduces flank marking, despite the presence of the subordinate hamster (reviewed in Albers and Bamshad, 1998). The development of stable dominant/subordinate relationships appears to result in marked changes in the AVP system within the CNS. For example, subordinate hamsters in stable dominance relationships have reduced levels of AVP-ir and fewer AVP-ir fibers in the AH, but not in other brain regions, relative to dominant animals (reviewed in Albers and Bamshad, 1998). Lastly, knock-out mice that lack the V_{1b} AVP receptor display a marked reduction in aggression (Wersinger et al., 2002).

In addition to mediating aggression in adult animals, AVP has been implicated in the *development* of aggression in several rodent species. For example, adult sexually naïve male prairie voles (*Microtus ochrogaster*) are typically not aggressive; however, significant aggression is seen toward strange males within 24 h of mating. Furthermore, icv injections of AVP can induce intermale aggression in sexually naïve prairie voles, mimicking the effects of social experience (Stribley and Carter, 1999). Recently, it has been shown that developmental exposure of sexually naïve male prairie voles to AVP can elicit long-lasting changes in social behavior, with sexual naïve males exhibiting levels of aggression comparable to that displayed by male postmating, suggesting an important role for early AVP in the development of aggression in this species (Stribley and Carter, 1999). The monogamous California mouse (*Peromyscus californicus*) is typically more aggressive than its polygynous relative, the white-footed mouse (*Peromyscus leucopus*). Cross-fostering pups from these two species can reverse the species-specific bias toward aggression, with California mouse pups becoming less aggressive in a resident–intruder test and white-footed mouse pups becoming more aggressive in a neutral arena test (Bester-Meredith and Marler, 2001). Interestingly, the decrease in aggression seen in California mice was associated with decreases in AVP-immunoreactive in the bed nucleus of the stria teminalis (BnST), supraoptic nucleus, and medial amygdala (Bester-Meredith and Marler, 2001). In Syrian hamsters, repeated androgenic steroid treatment during adolescence increases hypothalamic V_{1a} AVP receptor staining and facilitates offensive aggression (DeLeon et al., 2002). Collectively, these studies suggest that in addition to regulating adult aggression, AVP plays an important role in the development of agonistic behavior in a range of rodent species.

3 Neuroendocrine Regulation of Aggression in Birds

Studies in birds have contributed greatly to our current understanding of the neuroendocrine regulation of aggression (Harding and Follett, 1979; Wingfield et al., 1987; Konishi et al., 1989; Schlinger and Callard, 1990; Goodson et al., 2005). Indeed, the first study of hormones and aggressive behavior was conducted in birds, as described below. More recently, field studies of wild songbirds, such as sparrows, have been invaluable for revealing hormone–behavior relationships across species, seasons, and habitats, as well as the molecular mechanisms underlying territorial aggression. Free-living animals are often more aggressive and have higher circulating T levels than captive animals (Wingfield, 1984a, b; Schwabl and Kriner, 1991; Smulders et al., 2000). In addition, testing conditions in the laboratory can have large, unexpected effects on behavior, particularly when animals are forced to interact in small spaces without the opportunity to flee. For these reasons, field studies can be an important complement to laboratory experiments. Such approaches in birds have produced novel insights into the social regulation of T levels, the expression of aggression during the nonbreeding season, the "costs" of elevated T levels, and species-specific effects of hormones (Goodson et al., 2005).

3.1 Aggressive Behavior

Our knowledge of the natural history, ethology, social systems, and seasonal reproductive patterns of birds is by far greater than for other classes of vertebrates. There are approximately 9,000 species of birds, living in habitats as diverse as deserts, arctic tundra, and tropical rainforest (DeVoogd et al., 1993; Brenowitz, 1997). A total of 5,300 species belong to the order Passeriformes ("perching birds"), which is composed of the oscine suborder ("true songbirds," 4,000 species) and the suboscine suborder (1,300 species; e.g., antbirds, flycatchers). There is a rich diversity of social systems from colonial to strictly territorial species, permitting comparative studies of aggression and the underlying physiological mechanisms (Soma and Wingfield, 1999; Goodson et al., 2005). In addition, many birds are terrestrial and diurnal, making their natural behavior relatively easy to observe in the field. Moreover, some species are highly territorial and show little spatial movement, facilitating long-term behavioral studies of individual animals.

For example, the ethologist Nice (1943) carried out groundbreaking studies of song sparrows (*Melospiza melodia*), a common North American songbird, by marking animals with unique combinations of color bands and conducting detailed behavioral observations, sometimes following individuals for several years. Nice (1943) provides descriptions of territorial aggression, including postures and vocalizations. For example, song sparrows use songs, specific threat postures, feather "puffing," wing "waving," and actual physical contact during territorial conflicts. Such encounters can go on for hours or days. Nice (1943) also described seasonal patterns of territorial behavior and the development of such behaviors in juveniles. Moreover, she was able to incorporate these behavioral observations into the larger context, including weather and population ecology. This foundation provides significant advantages when studying the hormonal and neural mechanisms underlying aggressive behavior in song sparrows (see later). Similar data are available for many other species, including European robins (*Erithacus rubecula*), European starlings (*Sturnus vulgaris*), and red-winged blackbirds (*Agealius phoenecius*), in which the neuroendocrinology of aggression has been investigated (Beletsky et al., 1990; Kriner and Schwabl, 1991; Schwabl and Kriner, 1991; Pinxten et al., 2002).

3.2 The Role of Testicular Hormones: Berthold's Capons

The study of hormones and aggression, and of hormones in general, can be traced back to the work of Berthold (1849). Berthold removed the testes of immature male chickens (the castrated animals are called capons) and found a decrease in some secondary sex characteristics and male-typical behaviors. Specifically, capons did not crow, did not try to mate with females, and did not fight aggressively with other males. Importantly, transplantation of testes into castrated animals restored male sexual and aggressive behavior. When a single autotransplanted testis was replaced after all original vascular and neural connections to the testis were cut, these animals showed normal male aggressive behavior. Further, when a single heterotransplanted testis was replaced, these animals also showed typical male fighting. Additionally, when the testis from one heterotransplanted bird was removed, the animal behaved as a capon. Upon dissection, Berthold found that the transplanted testes in the abdomen had established new vascular connections. From these results, Berthold concluded that the testes release a substance into the blood that affects behavior and the body in general. Thus, from its beginning, the study of hormones and aggression was focused on gonadal secretions, primarily T. This conceptual framework has greatly influenced subsequent studies up to the present day. It is clear that T secreted by the testes is important for the expression of aggressive behavior. However, an *exclusive* focus on the testes and circulating levels of T is too simplistic and can often be misleading.

In seasonally breeding avian species, the gonads grow (recrudesce) markedly prior to the breeding season and regress following the termination of breeding. In parallel, circulating sex steroids fluctuate, generally with high levels during the breeding season and basal levels during other times of the annual cycle (e.g., molt, nonbreeding season) (Farner and Wingfield, 1980; Wingfield and Farner, 1993). In many species, territorial behavior shows a similar seasonal pattern: animals are territorial in spring, but gregarious and form flocks in autumn and winter (Wingfield et al., 1997, 2001). Note, however, that there are exceptions to this general pattern, and such exceptions can yield novel insights into the neuroendocrinology of aggression (see ➋ Section 3.6 and also ➋ Section 2.2).

3.3 Field Endocrinology and the Challenge Hypothesis

Ethologists generally focus on the behavior of free-living birds, whereas early behavioral endocrinologists focused on domesticated species, such as chickens. These two fields were combined in a series of elegant studies of wild songbirds by Wingfield and colleagues (Wingfield et al., 1987, 1990). Previous laboratory studies had demonstrated that long day lengths stimulate testis growth and increase plasma T levels in birds, but field studies were important for identifying the role of social interactions. In field studies, the temporal pattern of T titers differed from lab studies, and the absolute values of T concentrations were higher in wild-caught animals (Wingfield, 1984a, b). These two discrepancies suggested that environmental information in addition to day length was critical for T secretion.

As mentioned previously, song sparrows are common North American songbirds. In a migratory population, plasma T levels are very high in free-living males when they first arrive in early spring and are aggressively establishing breeding territories (Wingfield, 1984a, b). After pairing with females, plasma T levels in males decline, even though day lengths are increasing at this time. These data raised the hypothesis that aggressive male–male interactions stimulate T secretion, which may be important during periods of intense fighting or challenge. This hypothesis was tested experimentally in both the field and laboratory. In the field, wild territorial males were given either long-term subcutaneous T implants or empty implants in spring (Wingfield, 1984a, b). T-treated males were more aggressive, as measured by simulated territorial intrusions (STI) using taped song playback. Hormone levels were measured in the untreated *neighbors* of implanted animals. The immediate neighbors of T-treated males had higher circulating T levels than the immediate neighbors of control subjects. The hormone levels of distant neighbors were not affected. These results suggest that cues from aggressive males can stimulate T secretion (Wingfield, 1984a, b). In a separate field experiment, free-living males were exposed to a simulated territorial intrusion (STI) using both song playback and a live caged decoy for up to 120 min (Wingfield, 1985). The combined visual and auditory cues provoked a robust aggressive response. In addition, plasma T levels were significantly higher in males exposed to STIs than in controls. Further field experiments demonstrated that rapid changes in plasma T can occur within only 10 min of STI exposure (Wingfield and Wada, 1989).

In one laboratory study, the timecourse and specificity of STI response were investigated (Wingfield and Wada, 1989). Breeding male song sparrows were captured in the field and brought into captivity. After interaction with a novel male conspecific, plasma T titers were elevated. Interaction with a control stimulus (i.e., heterospecific male) had no effect. Lab experiments also assessed the relative importance of auditory and visual cues using taped song playback only or a devocalized male only. Exposure to the devocalized male significantly increased plasma T, whereas there was only a trend for the song playback to do so (Wingfield and Wada, 1989).

These results and studies in many other bird species led to the formulation of the "challenge hypothesis" (Wingfield et al., 1990). This hypothesis states that plasma T levels and aggression are positively correlated during periods of social instability, or challenge, such as establishment of territorial boundaries or attempts at territory takeover. During such periods, aggressive interactions are more frequent, leading to high T concentrations. In contrast, during times of social stability, status or boundaries are maintained by social inertia, and plasma T levels are lower. Under such stable conditions, aggression and plasma T may not be correlated. This idea has been extremely influential and has been examined in a variety of birds, fish, lizards, rodents, and primates.

It may be advantageous to limit high circulating T levels to periods of instability during the breeding season because high T levels incur "costs" (Ketterson et al., 1992; Ketterson and Nolan, 1999). For example, systemic T treatment can increase the metabolic rate (Wikelski et al., 1999) and decrease body mass and fat stores in birds (Wingfield, 1984a, b; Ketterson et al., 1991). In addition, T treatment suppresses the immune function in some, but not all, species of birds (Casto et al., 2001; Owen-Ashley et al., 2004). These costs of high circulating T are likely to be most evident in the field, where food is not typically available ab libitum and parasites are often present. Moreover, several field studies have shown that T treatment can reduce male parental care (Hegner and Wingfield, 1987; Ketterson et al., 1992). In many bird species, biparental care is important, or even obligatory, for offspring survival. In such species, there should be strong evolutionary selection for low T levels during the parental phase of the breeding season (Van Roo, 2004). The relative importance of the costs of T will likely change across different stages of the life history cycle (e.g., breeding versus nonbreeding season) and across different habitats (e.g., temperate zone versus tropics) (Levin and Wingfield, 1992; Goymann et al., 2004).

3.4 Possible Neural Sites of Action of Testosterone

Where does T act in the avian brain to modulate aggression? In an important early study, Barfield et al. (1978) used steroid autoradiography and injected radiolabeled T into adult (3–4 months old) male chickens. The subjects were castrated at 5 weeks of age to remove gonadal T that would compete with radiolabeled T for receptors (Barfield et al., 1978). This method does not distinguish between direct uptake of T and uptake after local metabolism of T to 5α-DHT, E_2 or other steroids. Other caveats are that castration can affect the expression of hormone receptors, and castration will not reduce the competition of possible nongonadal androgens with radiolabeled T for receptors. Nevertheless, the results demonstrated that radioactivity accumulated in the POA (➋ *Figure 8-3*), several regions of the hypothalamus, nucleus taeniae (Tn, homologous to the mammalian medial amygdala), lateral septum, and midbrain (including the central gray) (Barfield et al., 1978). This pattern was similar to that of other vertebrates and coincided with regions known to be important for reproduction. Several of these interconnected regions are considered part of a general "social behavior network" in vertebrates (Wood, 1997; Goodson et al., 2005).

More recent studies have examined the localization of androgen (AR) and ER in the avian brain using immunocytochemistry and in situ hybridization (Gahr et al., 1993; Soma et al., 1999; Fusani et al., 2000; Kim et al., 2004). In addition to the steroid autoradiography results, recent studies demonstrate that there is robust expression of AR and ERα in the BnST, a region involved in regulating aggression (Saldanha and Coomaralingam, 2005). There is less information regarding the expression of ERβ in the avian brain, but ERβ mRNA is present in the avian amygdala and POA (Bernard et al., 1999). Currently, there is no clear evidence for membrane AR or ER in avian brain, and this is a key topic for future research.

It is important to note that songbirds, unlike nonsongbird species such as chickens, have additional sites of abundant AR and ER expression in the song control system, which controls the production of birdsong in aggressive and sexual contexts (Arnold et al., 1976; Brenowitz et al., 1997; Schlinger, 1997; Tramontin and Brenowitz, 2000). The song control system is a circuit comprising interconnected, discrete regions that

Figure 8-3
Rostral (a) to caudal (e) distribution of brain aromatase mRNA from autoradiographic film from a wild male song sparrow in spring. (a) Expression in the medial magnocellular nucleus of the anterior nidopallium (mMAN). (b) Expression in the hyperpallium (H) near the ventricles (v), the nidopallium (N), and preoptic area (POA). (c) Expression in the bed nucleus of the stria terminalis (BNST), POA, and ventromedial nulceus (VMN). (d) Expression in caudomedial nidopallium (NCM), nucleus taeniae of the amygdala (Tn), and hippocampus (HP). A few cells in caudomedial HVC expressed aromatase. (e) Expression in the caudal nidopallium, but not in the robust nucleus of the arcopallium (RA)

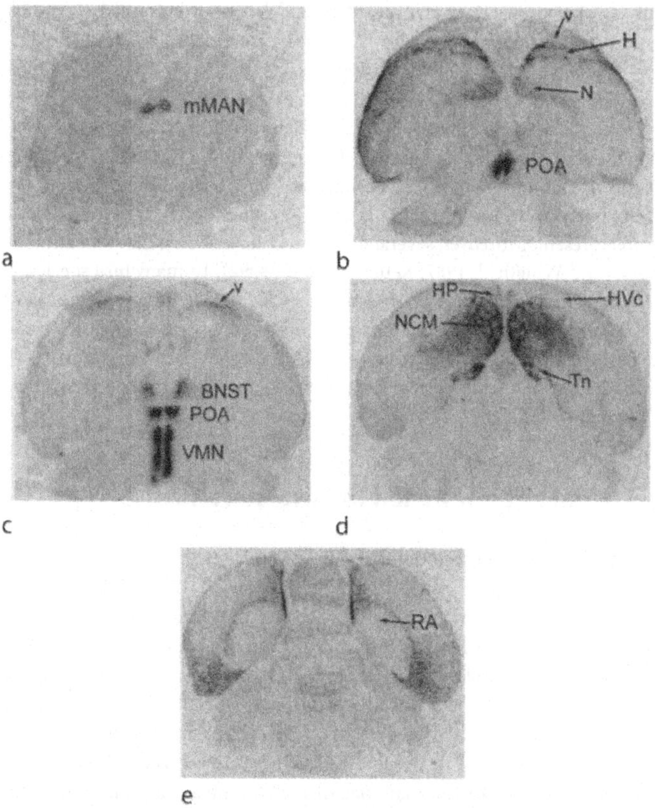

controls the muscles of the syrinx (the vocal production organ in birds) (Nottebohm et al., 1976; Brenowitz et al., 1997). Several telencephalic song nuclei, such as HVC, robust nucleus of the arcopallium (RA), and Area X of the striatum, contain AR (Arnold et al., 1976; Harding et al., 1984; Soma et al., 1999; Kim et al., 2004) (➋ *Figure 8-3*). HVC also contains ERα (Walters and Harding, 1988; Gahr et al., 1993; Perlman and Arnold, 2003; Saldanha and Coomaralingam, 2005). No song nuclei are known to express ERβ (Bernard et al., 1999).

In addition to localizing hormone receptors, a complementary approach is to assess the effects of local, intracerebral hormone implants in castrated animals. Small TP implants into the POA-AH do not activate aggression in chickens, but are effective in ring doves (Barfield, 1971; Phillips and Barfield, 1977). TP implants near Tn or the lateral septum do not increase agonistic behavior in chickens or ring doves (Barfield, 1971). It remains possible, however, that *multiple* sites must be activated by T to elicit aggression. Administration of hormone receptor antagonists locally in intact animals can be used to test this idea.

3.5 Brain Aromatase and Aggression

In many cases, the effects of plasma T are mediated by local metabolism within the brain. For example, T can be metabolized to 17β–E_2 by the enzyme P450 aromatase or to 5α-DHT by 5α-reductase (Schlinger and Callard, 1987; Fusani et al., 2000; Balthazart et al., 2003). Importantly, aromatase is expressed at high levels in the avian brain, facilitating its measurement in birds relative to mammals. 5β-reductase metabolizes T to 5β-DHT, an apparently inactive steroid, and brain 5β-reductase is present at very high levels in birds, but not rats and mice (Schlinger and Callard, 1987).

A series of laboratory studies with breeding Japanese quail (*Coturnix coturnix japponica*) illustrates the key role of brain aromatase in the expression of male aggressive behavior (Schlinger and Callard, 1990). To assess aggressive behavior, males were allowed to view a stimulus bird (i.e., nonaggressive female) through a glass partition (Schlinger and Callard, 1989). The number of pecks directed toward the glass partition and locomotor activity were recorded for 2 min. Males were killed either 90 s or 24 h after behavioral tests to measure aromatase, 5α-reductase, and 5β-reductase activities in several brain regions. Enzyme activities were assessed by measuring the metabolism of radiolabeled androgen (androstenedione, AE) by tissue homogenates in vitro. Aggressive behavior was positively correlated with aromatase activity in the hypothalamus but not in other brain regions (Schlinger and Callard, 1989; Schlinger and Arnold, 1992). Aggressive behavior was not correlated with 5α- or 5β-reductase or circulating T or estrogen levels. In a separate study, aggression was correlated with nuclear ("occupied") ER in the hypothalamic-POA (Schlinger and Callard, 1989). Moreover, in breeding males, experimental treatment with an aromatase inhibitor or ER antagonist decreased aggression, but a 5α-reductase inhibitor and AR antagonist had no effect (Schlinger and Callard, 1990). Taken together, these data argue that T affects aggression in quail via the aromatase pathway. In addition, there is evidence for a negative correlation between aggression and 5β-reductase activity in quail (Delville et al., 1984).

Generally similar results were obtained in studies of zebra finches (*Taeniopygia guttata*) and red-winged blackbirds. Captive reproductively active male zebra finches were castrated and treated with AE (androstenedione; an aromatizable androgen), AE + an aromatase inhibitor (ATD), or AE + ATD + E_2 (Walters and Harding, 1988). Although levels of aggression in this social species were generally low, only the AE and AE + ATD + E_2 groups displayed aggression, suggesting that the aromatase inhibitor reduced aggression (Walters and Harding, 1988). In captive breeding male red-winged blackbirds, castration greatly reduced aggression. Treatment of castrated animals with aromatizable androgens (T or AE) or E_2 + 5α-DHT in combination restored aggression (Harding et al., 1988). Treatments with E_2 alone or 5α-DHT alone were not effective, suggesting that the presence of both metabolites was necessary for the full expression of aggression (Harding et al., 1988). In a separate study of captive red-winged blackbirds, treatment with an AR antagonist (flutamide) decreased dominance and aggression, but not singing (Searcy and Wingfield, 1980). In a field experiment, blackbirds treated with flutamide and an aromatase inhibitor in combination lost parts of their territories (Beletsky et al., 1990). These data suggest that both estrogenic and androgenic metabolites of T are important for aggression in some avian species.

Field studies have examined seasonal changes in brain aromatase and/or 5α- and 5β-reductase in wild birds. In pied flycatchers (*Ficedula hypoleuca*) in Sweden, during early summer, males switch from defending a territory to helping with parental care. At the same time, plasma T levels decrease, and aromatase activity and immunoreactive cells in the diencephalon decline (Foidart et al., 1998). In another study of this species, aggression was measured using a simulated territorial intrusion, or STI; males were captured immediately afterwards; and brain tissue was collected (Silverin et al., 2004). Aromatase activity was measured by the release of tritiated water during aromatization of [1β-^{3}H]-AE, which is a sensitive technique but one that does not assess 5α- or 5β-reductase. Aggression was not correlated with plasma T or 5α-DHT. Aggression was, however, positively correlated with aromatase activity in the POA-anterior diencephalon, but not in the posterior diencephalon or telencephalon (Silverin et al., 2004). Specific telencephalic regions were not examined (e.g., Tn, BnST, lateral septum). It is possible that the STI rapidly upregulates aromatase activity, and this is a topic for future studies.

In Lapland longspurs (*Calcarius lapponicus*) in arctic Alaska, during the brief breeding season, males rapidly court females, then aggressively guard mates, and finally switch to parental care. Aggression is only robustly expressed during the mate-guarding phase. In this species, plasma T is correlated with singing behavior but not territorial aggression, and T treatment increases song but not aggression (Hunt et al., 1995, 1997). Aromatase, 5α-, and 5β-reductase activities were measured in seven brain regions during all three phases of the breeding season (Soma et al., 1999). Aromatase activity in the anterior diencephalon generally matched temporal changes in aggressive behavior. Changes in 5β-reductase did not explain the pattern of aggressive behavior (Soma et al., 1999). Taken together, these data suggest a significant role of aromatase in the anterior diencephalon/POA in the control of aggressive behavior. In addition, studies in song sparrows suggest the presence of aromatase in Tn and perhaps BnST is involved in nonbreeding territoriality (see Section 3.6).

3.6 Territorial Aggression During the Nonbreeding Season: Beyond Berthold

During the autumn and winter, many birds are in nonbreeding condition, which is generally characterized by regressed gonads and secondary sex characteristics (e.g., cloacal protuberance, wattles) and basal plasma sex steroid levels. Typically, nonbreeding birds abandon exclusive territories in favor of flocks. Some species, however, aggressively defend territories during the nonbreeding season, even if plasma T is basal at this time (Wingfield et al., 1997; Soma and Wingfield, 1999). The physiological regulation of breeding territoriality has been the focus of numerous studies (Wingfield et al., 1987), but the proximate mechanisms underlying nonbreeding territoriality have been enigmatic.

Winter territoriality may be dissociated from plasma T because of the costs of circulating T. In seasonally breeding birds, sex steroids in the general circulation may have particularly high costs during the nonbreeding season. The nonbreeding season may last the majority of the year and can be a difficult time because of low ambient temperatures and reduced food supply (Soma and Wingfield, 1999). Moreover, diurnal birds do not forage during the long nights of winter. Many costs of T during the nonbreeding season may be energetic in nature. For example, T treatment can increase the basal metabolic rate (Wikelski et al., 1999). T treatment also decreases fat stores (Owen-Ashley et al., 2004), which are important for surviving snow and ice storms. T treatment stimulates secondary sex characteristics (Soma et al., 2002), which require energy to grow and maintain. Other costs of T are not directly related to energetic constraints. Importantly, T can suppress the immune system of birds (Casto et al., 2001); T also inhibits molt (Schleussner and Gwinner, 1989). Finally, T treatment may stimulate reproductive behavior, which is inappropriate during the winter (Logan and Carlin, 1991). These various effects may explain why T treatment of wild birds reduces overwinter survival in some species (Dufty, 1989; Moss et al., 1994; Ketterson et al., 1996).

In Washington State, song sparrows (*Melospiza melodia morphna*) are sedentary. Males are territorial during the spring (breeding season), when plasma T is high (Wingfield and Hahn, 1994; Wingfield and Soma, 2002). After breeding, the animals molt their feathers (August–September), and during the molt, plasma T levels are basal, gonads are regressed, and aggression is greatly reduced. Following completion of the molt, there is a resurgence of territorial aggression in the autumn (nonbreeding season). Males in autumn and spring respond similarly during the STI, but autumn birds are less "persistent" after the intrusion has ended. After the decoy is removed, territorial behavior is extinguished more rapidly in autumn (Wingfield, 1994). The social context of territoriality also changes seasonally. Spring territories are defended by a breeding pair, but autumn territories can be defended by individuals, pairs, or larger groups (Wingfield and Monk, 1992; Wingfield, 1994). Plasma T is nondetectable during the autumn, and the testes and cloacal protuberance are fully regressed (Wingfield and Hahn, 1994; Soma and Wingfield, 1999). While agonistic interactions between males increase plasma T in spring (Section 3.3; Wingfield and Hahn, 1994), they do not affect circulating T in autumn (Wingfield and Hahn, 1994; Soma and Wingfield, 2001). Plasma 17β-E_2, 5α-DHT, AE, and estrone concentrations are also basal in males during the nonbreeding season (Soma and Wingfield, 1999). Moreover, castration does not decrease aggressive behavior in autumn (Wingfield, 1994). Thus, aggression may be independent of T in autumn.

The hypothesis that nonbreeding season aggression is independent of T was tested in three field experiments by treating wild song sparrows with pharmacological inhibitors of aromatase, with or without an AR antagonist. A combined treatment of an aromatase inhibitor (ATD) and AR antagonist (flutamide) decreases nonbreeding aggression in free-living males after 30 days, but not after just 7 days (Soma et al., 1999). Fadrozole, a more potent and specific aromatase inhibitor than ATD, strongly decreases nonbreeding aggression within 10 days. The effects of fadrozole are rescued by E_2 replacement (Soma et al., 2000). Moreover, fadrozole did not affect body condition or plasma CORT levels, indicating that the animals were not affected in a nonspecific manner (Soma et al., 2000). Additional studies suggest that fadrozole can reduce some aspects of autumnal aggression within only 24 h (Soma et al., 2000). These data indicate that sex steroids, particularly estrogens, are important for the expression of aggressive behavior in the nonbreeding season, even though plasma sex steroids are nondetectable and castration has no effect.

In addition, studies have also examined regional and seasonal differences in song sparrow brain aromatase using biochemical and molecular techniques. In situ hybridization reveals that aromatase mRNA is highly expressed in the POA, ventromedial nucleus of the hypothalamus, Tn (avian medial amygdala), BnST, caudomedial nidopallium (NCM, implicated in song perception), and medial magnocellular nucleus of the anterior nidopallium (MMAN, a song nucleus) (Soma et al., 2003). In addition, brain regions were dissected and androgen-metabolizing enzyme activities were measured in several brain regions during spring, molt, and autumn. Aromatase activity in the ventromedial telencephalon (includes Tn) is specifically reduced during molt, matching seasonal changes in aggression (Soma et al., 2003). Aromatase activity in the diencephalon, however, is high only during spring. 5β-reductase is not elevated during molt and thus cannot explain the low aggression during molt. These results suggest that changes in aromatase activity may regulate seasonal changes in aggression in birds.

Despite the evidence implicating aromatization of androgens, the precise source of androgen substrate for brain aromatase in the nonbreeding season remains unknown, because plasma levels of aromatizable androgens (e.g., T and AE) are basal at this time. One endocrine candidate, DHEA, is considered an "inert" androgen precursor and does not bind to AR or ER with high affinity (Kroboth et al., 1999). DHEA, however, can be converted into active sex steroids within tissues that express the appropriate enzymes (Labrie et al., 2001). Interestingly, plasma levels of DHEA are detectable and elevated (several-fold higher than plasma T) in nonbreeding song sparrows (Soma and Wingfield, 2001). Circulating DHEA may originate from the adrenals or regressed testes in autumn (Soma and Wingfield, 2001). As described for rodents, another, as yet untested, possibility is that the brain synthesizes DHEA de novo from cholesterol in the autumn. Across the annual cycle, circulating DHEA levels are specifically reduced during the molt, the one life-history stage when song sparrows show reduced aggressiveness (Soma and Wingfield, 2001). In a separate experiment, treatment of wild nonbreeding males with high physiological levels of DHEA increased territorial singing behavior (but not other territorial behaviors) and the size of the song-control nucleus HVC (Soma et al., 2002). DHEA treatment did not, however, stimulate the growth of a peripheral secondary sex characteristic (cloacal protuberance). More recent results suggest that DHEA treatment, unlike T, does not inhibit the immune system of nonbreeding song sparrows (Owen-Ashley et al., 2004).

Next, DHEA metabolism was examined in song sparrow and zebra finch brain. The in vitro assay measures the conversion of tritiated DHEA to AE and estrogens by the sequential activities of 3β-HSD and aromatase (Soma et al., 2002; Soma, 2004; Soma et al., 2004). The song sparrow brain can indeed convert DHEA to androgens and estrogens, with the highest levels of 3β-HSD activity in the diencephalon and telencephalon (K. Soma, D. Wacker, J. Wingfield, B. Schlinger, unpublished results). Interestingly, seasonal studies demonstrate that neural DHEA metabolism is highest in the nonbreeding season (unpublished results). These data support the novel hypothesis that in nonbreeding song sparrows, adrenal and/or gonadal DHEA synthesis is coupled with neural DHEA metabolism to supply sex steroids to brain circuits controlling aggression. This mechanism would reduce the exposure of peripheral tissues to T, which is deleterious in winter. Thus, nonbreeding song sparrows may circumvent many costs of circulating T. These studies on song sparrows provide a broad picture of hormone–behavior relationships in this species, integrating field studies of behavior, endocrinology, and ecological constraints with biochemical and molecular mechanisms (Wingfield and Soma, 2002).

Similar results have been obtained in field and laboratory studies of tropical birds. Several avian species that breed in the tropics defend territories year-round and have very low levels of circulating sex steroids throughout the year (Levin and Wingfield, 1992; Goymann et al., 2004). One example is the spotted antbird

(*Hylophylax n. naevioides*) in Panama. In this species, both sexes sing and aggressively defend territories year-round in the rainforest (Wikelski et al., 2000). Wild spotted antbirds have basal plasma T and E_2 levels, even during the breeding season, except for transient increases during territorial encounters (Wikelski et al., 1999). In a laboratory experiment, a combined treatment of aromatase inhibitor and AR antagonist decreased male aggressive vocalizations (songs and "snarls") in the breeding season, even though plasma T was basal (Hau et al., 2000). In the nonbreeding season, both sexes show high levels of aggression, primarily to intruders of the same sex (Hau et al., 2004). Both males and females have regressed gonads and low levels of plasma sex steroids in the nonbreeding season (Hau et al., 2004). However, plasma levels of DHEA are elevated in both sexes during the nonbreeding season, and in males, plasma DHEA levels are positively correlated with aggressive vocalizations and/or the duration of territorial intrusions (Hau et al., 2004). These data suggest that DHEA regulates aggressive behavior, particularly vocal behavior, in both males and females of this species. In tropical species facing a large number of parasites, the immune function may be particularly important (Martin et al., 2004), thus favoring the evolution of endocrine mechanisms which avoid the immunosuppressive effects of circulating T.

Taken together, such studies suggest that nonbreeding aggression is not regulated by gonadal T in some species. Instead, circulating DHEA, from the adrenal glands or regressed gonads, may be converted to active androgens and estrogens locally within the brain. It is also possible that DHEA and/or sex steroids are synthesized de novo from cholesterol by the brain (Corpechot et al., 1981). These alternate mechanisms have important consequences for interpreting data on peripheral levels of T and the effects of castration.

Such mechanisms are likely to be of general importance in birds and other vertebrates. Winter territoriality is not uncommon among birds. For example, excellent field studies have documented nonbreeding territories in mockingbirds (Logan and Wingfield, 1990), willow tits (Silverin et al., 1984), European robins (Kriner and Schwabl, 1991; Schwabl, 1992), and European stonechats (Gwinner et al., 1994), to name just a few cases. In these species, robust territorial aggression is expressed despite basal circulating T levels during the autumn and winter. In European robins, an AR antagonist (flutamide) decreases aggression in spring but not in winter (Schwabl and Kriner, 1991). In European stonechats, a combined treatment of flutamide and an aromatase inhibitor (ATD) for 7 days does not decrease aggression in winter (Canoine and Gwinner, 2002). It would be useful to examine the effects of a more potent aromatase inhibitor, such as fadrozole, in this species. Red grouse in Scotland also defend territories in autumn, but in this species, there is a small peak in plasma T during autumn (Mougeot et al., 2005). Thus, the regulation of territorial behavior outside the breeding season is not identical in all species. In addition, birds that do not defend a winter territory may nonetheless display aggressive behavior within dominance hierarchies in winter flocks, for example, European starlings and dark-eyed juncos (Ketterson, 1979; Pinxten et al., 2000; Pinxten et al., 2002; Pinxten et al., 2003). The endocrine mechanisms regulating dominance in flocks are largely unknown.

3.7 Testosterone and Aggression in Juvenile Birds

Several interesting studies have documented the presence of steroids in egg yolk and consequences for behavioral development. For example, in an important study, it was demonstrated that yolk has significant quantities of AE, 5α-DHT, and T in the eggs of captive canaries (*Serinus canaria*) (Schwabl, 1993). In addition, the concentration of T depended on the order in which the eggs were laid. The T content increased with laying order (i.e., the first egg in a clutch had the lowest T levels). Moreover, the social rank of siblings was correlated with the T content in the yolk. These and other data suggest that steroid exposure in ovo can affect subsequent behavior (Whittingham and Schwabl, 2002). Manipulation of steroids in yolk is easier than manipulations of pregnant mammals, making birds excellent animal models for studying the organizational effects of steroids.

In addition to maternal steroids from yolk, young chicks can synthesize steroids independently. For example, black-headed gull (*Larus ridibundus*) chicks defend territories from conspecifics and transiently increase plasma T levels (Ros et al., 2002). An extreme form of aggression in juvenile birds is siblicide. Lethal attacks against siblings have been reported in several avian species. For example, in a seabird such as the Nazca booby (*Sula nebouxii*), each clutch has 2 eggs that hatch 4–7 days apart (Ferree et al., 2004). If both eggs hatch, the first chick attacks the second chick and aggressively pushes it out of the nest. The second chick then dies of starvation or predation. A recent study indicates that plasma T levels are generally basal in chicks (~0.04 ng/ml), but transiently elevated during aggressive interactions with siblings (up to 0.15 ng/ml) (Ferree et al., 2004). This was true for both first and second chicks. However, even these peaks of plasma T are far lower than typical levels in many breeding adult birds. Plasma DHEA levels were much higher (0.5–4.0 ng/ml) but did not differ among groups. In general, little is known regarding the control of aggression in developing birds, and this remains an excellent topic for future studies.

3.8 Corticosterone and Aggression

In birds, the predominant circulating glucocorticoid is CORT. In song sparrows, aggressive interactions do not appear to increase plasma CORT levels (Wingfield, 1985). However, treatment of wild males with exogenous CORT during the breeding season affected territorial aggression (Wingfield and Silverin, 1986). After 18–24 h of treatment, the majority of CORT-treated males failed to respond to an STI, and the ones that responded did so weakly (Wingfield and Silverin, 1986). Surprisingly, CORT treatment did not affect luteinizing hormone (LH) levels and only decreased plasma T levels slightly.

A similar study was conducted on tree sparrows (*Spizella arborea*) breeding in arctic Alaska (Astheimer et al., 2000). Interestingly, CORT treatment had no effect on aggressive behavior or reproductive hormones in tree sparrows. It is possible that ecological constraints, such as a short breeding season, reduce the sensitivity of neural circuits to CORT, allowing animals to successfully breed in spite of environmental stressors. This could be achieved, for example, by modulating corticosterone binding globulin (CBG) or glucocorticoid receptor levels (Wingfield and Sapolsky, 2003).

3.9 Arginine Vasotocin and Aggression

Arginine vasotocin (AVT) is a neuropeptide homologous to AVP in mammals. Classically, AVT/AVP neurons in the hypothalamus have been shown to play an important role in water balance. More recently, AVT/AVP neurons in the extended amygdala (medial amygdala and BnST) have been implicated in the regulation of social behavior (Goodson and Bass, 2001). These neurons project to many regions, including the lateral septum, nucleus accumbens, and periaqueductal gray (Goodson and Bass, 2001). In particular, the lateral septum receives AVT input from the BnST and contains high levels of AVT/AVP receptors.

In songbirds, a series of studies by Goodson and colleagues have examined the behavioral effects of septal AVT infusions in different species. Interestingly, intraseptal AVT administration had opposite effects in species with different social systems. In a colonial species, the zebra finch, AVT infusions facilitated aggressive behavior (Goodson and Adkins-Regan, 1999). In contrast, in territorial species, the violet-eared waxbill (*Uraeginthus granatina*) and field sparrow (*Spizella pusilla*), AVT infusions inhibited overt physical aggression (Goodson, 1998a, b). Zebra finches and violet-eared waxbills are both estrildid finches and share critical features of breeding ecology, except for social organization. Thus, this species comparison is less likely to be confounded by phylogeny or other ecological variables.

Song sparrows have also been a useful model system for understanding the role of AVT in territorial aggression. In one experiment, immediate early gene (ZENK) responses were assessed following STI and/or nonsocial stress (restraint) in breeding male song sparrows under semi-natural conditions. Exposure to STI and/or restraint significantly increased ZENK protein in the lateral septum, whereas the medial BnST showed a highly selective response to STI. Interestingly, infusion of an AVT/AVP antagonist (Manning compound, specific to V_1 receptors) into the lateral ventricle abolished these ZENK responses, and influenced ZENK-ir in the medial BnST only after STI, but not after restraint (Goodson and Evans, 2004). These data suggest that AVT acts within the BnST to specifically modulate neural responses to social stimuli. In a second experiment, the role of DHEA in nonbreeding song sparrows was examined. Captive male song sparrows were exposed to an STI and then perfused 90 min later. Plasma DHEA levels were *negatively* correlated with immediate early gene expression in medial BnST (J. Goodson, K. Soma, unpublished results). In addition, plasma DHEA levels were *positively* correlated with AVT cell counts in the medial BnST, which may indicate decreased transport and release of AVT (unpublished results). Although speculative, one possible interpretation of these data is that DHEA acts within the medial BnST to decrease the transport and release of AVT in the lateral septum. Because intraseptal AVT inhibits aggressive behavior in territorial species, DHEA would be reducing a "brake" on aggression.

4 Neuroendocrine Regulation of Aggression in Primates

Aggression has been a central topic of primate research for decades. A great deal of research indicates the importance of environmental and social influences on the development and expression of primate aggressive behavior. Dominance is certainly the most well-known social influence on aggression, and dominance hierarchies are considered a major organizing principle in many primate societies. Knowing an animal's dominance rank provides valuable information about that individual's aggressive tendencies. In primates, high-ranking males are more likely to initiate and win aggressive encounters, but they are not necessarily the most aggressive animals in the group (Bernstein et al., 1983). Furthermore, unlike in rodents, in primates individual aggressive power does not necessarily correlate with dominance rank. For instance, the best fighters do not inevitably achieve the highest dominance rank because alliances are critical for acquiring and maintaining high dominance rank (Chapais, 1992). Young primates must learn the proper settings for expressing aggression and the social skills required for recruiting allies. In sum, achieving high dominance rank in primate species probably depends as much on inhibiting aggression as it does on expressing aggression.

Although the study of the hormonal mechanisms of aggression has some experimental attention, much less is known than in the case of rodents. Most recent studies on the neuroendocrine regulation of primate aggression have used fecal or urinary assays to measure androgens. Although the advent of these assays has stimulated new research questions in primate socioendocrinology, their limitations need to be recognized. The different steroids and steroid metabolites present in urine and feces are species-specific, and the levels that are excreted frequently represent a fraction of plasma concentrations. The extended time tag for fecal excretion probably favors the use of fecal samples, and to a lesser extent urinary samples, for assaying hormonal baselines; however, this characteristic makes fecal steroids a less-sensitive measure of hormonal responses to a single behavioral event. Also, most immunoassay kits used to detect steroids were developed for use with human serum or plasma, and may not have been tested for cross-reactivity with metabolites that may be present in urine or fecal samples. High cross-reactivity between excreted metabolites can limit conclusions regarding the origin of a steroid and its mechanism of action. In addition, plasma, urinary, and fecal assays are not recommended for investigating a steroid's mechanism of action due to the importance of local metabolism. Although most primate studies report little cross-reactivity between steroids and related metabolites, some studies have reported considerable cross-reactivity. For instance, 46% cross-reactivity between T and DHT was reported in a study using urine samples in chimpanzees (*Pan troglodytes*) (Muller and Wrangham, 2004), which makes conclusions regarding specific hormone–behavior relationships difficult. Finally, most research on the neuroendocrinology of primate aggression has been conducted with Old World monkeys such as macaques, baboons, and vervets. Apes, prosimians, and New World monkeys share some characteristics with Old World monkeys, but unfortunately much less is known about the neuroendocrine mechanisms regulating aggression in these primates. Thus, our review is biased toward Old World monkeys, although we include examples from other primate taxa, including humans, when possible.

4.1 Androgens and Aggression

In general, high rates of aggression tend to be positively correlated with elevated androgen concentrations in nonhuman primates. This correlation is observed in seasonal changes in androgens and rates of aggression, sex differences in aggression, and increased aggression at puberty. Many primates are seasonal breeders and these species have provided much of the information available on correlations between androgens and aggression. Seasonality in testicular activity, with circulating T levels peaking during the mating season, has been reported in many primate species, including squirrel monkeys (*Saimiri sciureus*: Wiebe et al., 1988), Hanuman langurs (*Presbytis entellus*: Lohiya et al., 1998), ringtailed lemurs (*Lemur catta*: Cavigelli and Pereira, 2000), mandrills (*Mandrillus sphinx*: Setchell and Dixon, 2001), and several species of macaques (*Macaca mulatta*: Gordon et al., 1976; *M. radiata*: Glick, 1979; *M. fuscata*: Barrett et al., 2002). Both environmental and social cues may elicit the onset of increased T production. Some studies have shown that environmental cues, such as photoperiod, are sufficient to produce seasonal fluctuations in T secretion (Rostal et al., 1986; Perret, 1992; Herndon et al., 1996). In contrast, other studies have found that exposure to females is necessary to maintain a seasonal pattern of T secretion (Gordon et al., 1978; Schiml et al., 1996), and that exposure to sexually active females outside the mating season can elicit increases in T (Ruiz de Elvira et al., 1982). Several studies have shown that rates of aggression in males tend to parallel the rise in T during the mating season (Gordon et al., 1976; Rostal et al., 1986; Cavigelli and Pereira, 2000). In addition, seasonal changes in T have been associated with aggression in females. In ringtailed lemurs, aggressive conflicts between females were found to increase during the mating season and correspond to a seasonal increase in fecal androgen concentrations (von Engelhardt et al., 2000). However, in this study individual rates of aggression did not correlate with androgen levels. It is interesting to note that T fluctuations do not always peak during the mating season. For instance, male muriqui monkeys (*Brachyteles arachnoides*) do not compete aggressively for mates, and fecal T concentrations do not increase during the mating season (Strier et al., 1999).

Although exceptions exist, the majority of research suggests that androgens do not necessarily increase aggression in nonhuman primates. For instance, in Japanese macaques seasonal increases in serum T and 5α-DHT levels were found to precede seasonal increases in aggression by 1–2 months, which was interpreted to mean that neither hormone affects aggression in a simple causal way (Rostal et al., 1986). Also, pharmacologically elevating T does not reliably increase rates of aggression. In rhesus monkeys, twice-weekly injections of human chorionic gonadotropin stimulated T production but had no consistent effect on rates of aggression (Gordon et al., 1979). Rather, increased circulating T levels were associated with an intensification of existing behavior, which did not disrupt group stability. On the other hand, Rejeski et al. (1990) showed evidence of a causal link between anabolic steroids and aggression in male long-tailed macaques (*M. fascicularis*). They found that injection of TP increased aggression in dominants, although it also increased submission in subordinates. When male marmosets (*Callithrix jacchus*) were castrated as neonates and tested as adults, they were found to display high rates of aggression with female partners and low rates with male partners (Dixson, 1993a). This pattern of aggression is opposite to that expressed by intact male marmosets. Further, the effects of neonatal castration on aggressive behavior were reversed by TP treatment in adulthood (Dixson, 1993b). Finally, several studies have found that castration of adult males has little effect on their aggressive behavior, suggesting that T is not required to maintain aggression in adults (Wilson and Vessey, 1968; Epple, 1978; Dixson, 1993c).

A common finding is that winning increases T, while losing decreases T. Among macaque males, animals that lost social conflicts showed a reduction in plasma T, especially if the defeat resulted in a fall in dominance rank (Rose et al., 1972; Bernstein et al., 1979). Similarly, on days when vervet (*Cercopithecus aethiops*) males fought, plasma T levels were higher in the winners than in the losers (Steklis et al., 1985). Also, several studies have shown that during periods of group formation plasma T concentrations increase in future dominants and decline in future subordinates, but that plasma T levels prior to group formation do not predict dominance rank (Rose et al., 1975; Mendoza et al., 1979; Keverne et al., 1982). In sum, these studies suggest that changes in T may be a consequence, rather than a cause, of aggression. In humans, physical aggression among males also seems unrelated to plasma T concentrations. Rather, T appears associated with the achievement and maintenance of high social status (Mazur and Booth, 1998). For instance, winning at tennis (Booth et al., 1989), chess (Mazur et al., 1992), or soccer (Neave and Wolfson, 2003) has been associated with transient increases in circulating T. Similarly, the rise in T during puberty has been related to nonaggressive symptoms of conduct disorder in boys with deviant peers, whereas it has been related to leadership in boys without deviant peers (Rowe et al., 2004).

Correlations between androgens and dominance rank appear to be mediated by the relationship between rank and aggression. This was demonstrated by the pioneering work of Sapolsky (1983; reviewed in 1993) on olive baboons (*Papio anubis*). This work demonstrated that during periods of instability in the dominance hierarchy, dominant males were more aggressive and had greater plasma T concentrations than subordinates; in contrast, no correlations existed during stable periods. Since then, numerous studies have found that during stable periods, individual differences in rates of aggression and dominance rank do not correlate with T concentrations (e.g., Eaton and Resko, 1974; Steklis et al., 1986; Nieuwenhuijsen et al., 1987; van Schaik et al., 1991), whereas such correlations have been found during periods with unstable hierarchies or other social challenges (e.g., Wicklings and Dixson, 1992; Brockman et al., 1998). In chimpanzees, however, dominance rank has been correlated with afternoon (but not morning) urinary T concentrations during a period with a stable dominance hierarchy (Muller and Wrangham, 2004). A similar correlation has been found in chimpanzees by measuring fecal T during stable conditions (Muehlenbein et al., 2004). As the authors suggested, these correlations may be due to high-ranking males being more aggressive than low-ranking males regardless of hierarchy stability.

4.2 The Challenge Hypothesis

Recent studies on the neuroendocrinology of primate aggression have addressed the challenge hypothesis, which states that T levels are related to the degree of reproductive competition rather than mating activity (Wingfield et al., 1990; ➋ section 3.3). As in birds, T levels in primates appear more closely associated with competition than with reproduction. In primates, T facilitates spermatogenesis and sexual behavior but has little effect above minimum threshold levels (Weinbauer et al., 1988; Michael and Zumpe, 1993). In contrast, high rates of aggression have been associated with maximal plasma T levels (Rose et al., 1971). A similar relationship has been found between rates of aggression and T levels measured in CSF (Higley et al., 1996c). Only mild forms of aggression, however, correlate with increased T, whereas severe aggression (i.e., contact aggression such as biting) occurs throughout the year and tends to be independent of plasma and CSF T levels (Rose et al., 1978; Bernstein, 1983; Higley et al., 1996c). In Japanese macaques fecal T was found to correlate with noncontact aggression during the mating season, but no correlation was found with contact aggression (Barrett et al., 2002). Also, correlations between aggression and T levels predominately occur in competitive situations such as rank instability (Sapolsky, 1983; Higley et al., 1996c; Brockman et al., 2001) and mate competition (Cavigelli and Pereira, 2000; Barrett et al., 2002; Muller and Wrangham, 2004). Interspecies comparisons are also consistent with the challenge hypothesis. In multimale groups, where male reproductive competition is expected to be more frequent, breeding season concentrations of serum T are higher than those in unimale groups, where male challenges over mates occur more sporadically (Whitten, 2000).

Some primate studies have found mixed support for the challenge hypothesis, while others have suggested it needs a slight modification when applied to primates. In capuchin monkeys (*Cebus apella*), which show little direct mate competition among males, fecal T concentrations increased during a period of synchronized female sexual activity, but rates of male aggression did not correspond to the peak in T (Lynch et al., 2002). In male redfronted lemurs (*Eulemur fulvus*), fecal T concentrations tracked the increase in aggression observed during the mating season, but the highest T levels were recorded during the birth season (Ostner et al., 2002). These studies demonstrate that T levels can show a mating season peak without overt reproductive competition, and that T levels can be associated with social challenges outside a reproductive context. The risk of infanticide in some primate species may produce more predictable challenges, and may require males to elevate T and be prepared to aggressively defend infants against intruders during relatively well-defined periods of high risk (Ostner et al., 2002; Whitten et al., 2004). As previously mentioned, male chimpanzees respond to competition for estrous females with elevated urinary T (Muller and Wrangham, 2004), although urinary and fecal T concentrations remain positively correlated with dominance rank during periods of relative social stability (Muller and Wrangham, 2004; Muehlenbein et al., 2004). The fission–fusion society of chimpanzees, in which individuals regularly form small subgroups for foraging but occasionally join together in larger groups, creates unpredictable social challenges for males, and thus may require dominant males to maintain relatively high levels of circulating T to remain ready for potential confrontations with other males (Muller and Wrangham, 2004;

Muehlenbein et al., 2004). In addition, Muehlenbein et al. (2004) suggested that male chimpanzees may selectively lower T levels during infection to prevent the immunosuppressive effects of high circulating T.

4.3 Estrogens and Aggression

Although estrogens contribute to the expression of aggression in nonhuman primates, the available research is mixed. Some studies suggest that E_2 facilitates aggression in females; other studies suggest that estrogens inhibit aggression, and still others have found no effect. In rhesus macaques, females were found to display increased rates of noncontact forms of aggression (e.g., threats) during the middle of the menstrual cycle, when E2 peaks (Walker et al., 1983). Similarly, in ovariectomized rhesus, E_2 replacement has been shown to increase rates of noncontact aggression compared with nonhormone treated ovariectomized controls (Michael and Zumpe, 1993). In contrast, in long-tailed macaques ovariectomy without hormone replacement was shown to increase aggression, including biting, compared to sham-operated controls (Stavisky et al., 1999). A similar effect on aggression has been reported in a prosimian primate, the greater galago (*Galago crassicaudatus*). In this case, ovariectomy increased aggression and E_2 replacement decreased it (Dixson, 1978). Finally, a study on long-tailed macaques found no effect of ovariectomy on rates of aggression (Shively et al., 1986).

Clearly, the effect of estrogens on aggression in female nonhuman primates is equivocal. One possible reason for the variable results is the variety of social conditions used for behavioral testing such as male–female pairs, small same-sex groups, or large mixed-sex colonies. These different social conditions may also influence the type of aggressive behavior displayed such as contact and noncontact forms of aggression. Further, it should be noted that the reduction in E_2 produced by ovariectomy does not affect only aggression, but it can also decrease affiliation (Shively et al., 1986; Stavisky et al., 1999). Interestingly, recent data suggest that aggression in males may also be modulated by an estrogen-dependent pathway. In male long-tailed macaques, diets high in isoflavones, which are partial agonists of ERβ, have been associated with increased rates of aggression (Simon et al., 2004). Although aromatase inhibitors have been shown to decrease sexual behavior in male macaques (Zumpe et al., 1996), their effect of aggression has not been investigated. Indeed, the local metabolism of T into E_2 is an excellent area for future primate research.

Some studies also suggest that progesterone increases aggression in nonhuman primates. In vervet monkeys (*Cercopithecus aethiops*), aggression was shown to peak during the late luteal phase of the menstrual cycle, when progesterone levels are high (Rapkin et al., 1995). Also, medroxyprogesterone acetate (MPA), a synthetic progestin, has been shown to increase rates of aggression in female macaques (Linn and Steklis, 1990; Pazol et al., 2004). However, which steroid receptor mediates the effect of MPA is unclear.

4.4 Adrenal DHEA

Studies from rodents and birds suggest that androgens such as DHEA may regulate aggression in situations where aggression seems otherwise T-independent. Some studies in primates also suggest that aggressive behavior can be unrelated to fecal T levels (Lynch et al., 2002; Ostner et al., 2002). Although primate adrenal glands are known to secrete high levels of DHEA and DHEA-S (Rehman and Carr, 2004), the role of adrenal androgens in primate aggression is largely unknown. In some primate studies, measures of androgens may be confounded by adrenal DHEA and DHEA-S, causing difficulty in interpretation. For instance, Mohle et al. (2002) showed that in macaques metabolites of T and DHEA may cross-react with T antibodies in urine and fecal assays (Mohle et al., 2002). Thus, when a correlation is found between aggression and fecal or urinary androgens, it may be difficult to conclude with certainty the origin of the steroid. In addition, a lack of a correlation between aggression and excreted T does not necessarily mean that aggression is androgen-independent, as other androgens such as DHEA may regulate behavior.

One study has assessed circulating DHEA-S levels in a population of wild baboons (Sapolsky et al., 1993). They found that DHEA-S concentrations were high in both male and female baboons and showed marked age-related decreases in both sexes; however, circulating levels of DHEA-S were not compared with aggression. Adrenal androgens appear to play a role in human aggression, as indicated by studies on conduct disorder, which is a term used to describe a collection of symptoms including aggression to people or animals, destruction of property, theft, and serious violations of rules. Prepubertal boys with conduct disorder were found to have higher levels of plasma DHEA-S, but not T, than normal control boys (van Goozen et al., 1998). Also, DHEA-S concentrations were correlated with the intensity of aggression as rated by parents and teachers. In another study, plasma DHEA-S concentrations were found to be higher in boys with conduct disorder than in boys with attention-deficit/hyperactivity disorder (ADHD) or normal controls (van Goozen et al., 2000). These studies suggest that adrenal androgens play an important role in the onset of aggression in adolescent boys. Adrenal androgens may also contribute to the regulation of aggression in human females. Adolescent and adult females with congenital adrenal hyperplasia, who were exposed to high levels of adrenal androgens in the prenatal and early postnatal periods, were found to have greater self-reported aggression ratings than were control females (Berenbaum and Resnick, 1997).

5 Conclusions

In rodents, birds, and primates, rates of aggression in males are often associated with circulating levels of T. This relationship has been extensively demonstrated in rodents, predominantly rats and mice, in which castration has been shown to reduce aggression and exogenous T treatment restores aggressive behavior (Edwards, 1969, 1970; Barfield, 1971). In birds, a positive correlation between plasma T and rates of aggression has been repeatedly documented during periods of social challenge, and particularly during mate competition (Wingfield et al., 1987). Likewise, recent data from primates has largely supported the challenge hypothesis (Cavigelli and Pereira, 2000; Barrett et al., 2002; Muller and Wrangham, 2004). These findings in primates build on previous work demonstrating a positive relationship between aggression and circulating T during periods of social instability (Sapolsky, 1983). It is important to remember, however, that in many cases, aggression is independent or even inversely related to circulating levels of T.

Seasonal aggression provides a valuable paradigm with which to study the neuroendocrine mechanisms of aggression. In many species (e.g., song sparrows, Siberian and Syrian hamsters, capuchin monkeys), high rates of aggression in males occur outside the period of high plasma T levels (Garrett and Campbell, 1980; Soma et al., 2000; Lynch et al., 2002, respectively). The neuroendocrine mechanisms regulating aggression under these conditions are poorly understood, but recent data provide some interesting possibilities. In birds, the metabolism of T to E_2 may mediate aggression during the mating season. During the nonmating season when circulating T is basal, steroid metabolism may still mediate aggression, albeit via a different endocrine mechanism (Soma et al., 2000). Despite low levels of circulating T, the androgen precursor DHEA remains elevated in blood, and conversion of DHEA to E_2 within the brain appears to play an important role in nonbreeding season aggression in some species (Soma et al., 2000; Soma and Wingfield, 2001).

In rodents, the aromatization of androgens also appears to regulate, at least in part, intermale aggression. Although seasonal changes in rodent aggression are less well-studied than in birds, recent data suggest that melatonin may mediate the increased aggression observed in hamsters during winter-like photoperiods (Jasnow et al., 2002; Demas et al., 2004). It seems likely that melatonin might regulate aggression under these conditions by affecting adrenal steroids such as cortisol and DHEA (Haus et al., 1996).

In nonhuman primates, little is known about the regulation of aggression when plasma T levels are low. In prepubertal boys, plasma DHEA-S levels have been associated with conduct disorder, and in particular with the aggressive symptoms of conduct disorder (van Goozen et al., 1998). Human and nonhuman primates secrete relatively high levels of DHEA and DHEA-S from the adrenal glands, but the role of DHEA in mediating nonhuman primate aggression has received little attention and is an exciting area for future research. An exclusive focus on T levels limits our understanding of the neuroendocrine regulation of aggression and strengthens the mistaken perception that aggression is simply a function of T. The idea that circulating T is not always associated with physical aggression, and instead is more often associated with competition in specific situations, has been extended to humans (see Mazur and Booth, 1998). It seems likely that continued research on the role of steroid hormones in the regulation of aggression in animals will improve our understanding of aggression and violence in humans.

In addition to steroid hormones, a large number of peptides and neurotransmitters have been implicated in the regulation of aggression, including the neuropeptides oxytocin and vasopressin, as well as serotonin and nitric oxide among other factors. A brief discussion of only some of these factors could be presented here. It is important to keep in mind that, although each of these factors is typically studied in isolation, no single factor can sufficiently explain the wide range of aggressive behavior displayed by humans and nonhuman animals. Rather, a comprehensive examination of the multiple factors that regulate aggression and how these factors interact is needed to provide a complete understanding of the neuroendocrine regulation of aggression.

Acknowledgments

The authors would like to thank Dr. James Goodson, Dr. Michaela Hau, Melissa Scotti, and Amy Newman for helpful comments on the manuscript. This work was supported in part by NIH award NS10596, an HHMI Capstone award (G.E.D), NIH F32 award MH072085 (M.A.C), NIH award MH062641, and the STC program of the NSF (H.E.A) and NSERC, Canadian Institutes of Health Research, Michael Smith Foundation for Health Research, and Canada Foundation for Innovation (K.K.S).

References

Albers HE, Bamshad M. 1998. Role of vasopressin and oxytocin in the control of social behavior in Syrian hamsters (*Mesocricetus auratus*). Prog Brain Res 119: 395-408.

Albers HE, Huhman KL, Meisel RL. 2002. Hormonal basis of social conflict and communication. Hormones, brain and behavior. Pfaff DW, Arnold AP, Etgen AM, Fahrbach SE, Rubin RT, editors. New York: Academic; pp. 393-433

Albert DJ, Jonik RH, Walsh ML. 1992. Hormone dependent aggression in male and female rats: Experimental, hormonal, and neural foundations. Neurosci Biobehav Rev 16: 177-192.

Arnold AP, Nottebohm F, Pfaff DW. 1976. Hormone concentrating cells in vocal control and other areas of brain of zebra finch (*Poephila guttata*). J Comp Neurol 165: 487-511.

Astheimer LB, Buttemer WA, Wingfield JC. 2000. Corticosterone treatment has no effect on reproductive hormones or aggressive behavior in free-living male tree sparrows, *Spizella arborea*. Horm Behav 37: 31-39.

Badura LL, Nunez AA. 1989. Photoperiodic modulation of sexual and aggressive behavior in female golden hamsters (*Mesocricetus auratus*): Role of the pineal gland. Horm Behav 23: 27-42.

Balthazart J, Baillien M, Charlier TD, Cornil CA, Ball GF. 2003. Multiple mechanisms control brain aromatase activity at the genomic and nongenomic level. J Steroid Biochem Molec Biol 86: 367-379.

Barfield RJ. 1971. Activation of sexual and aggressive behavior by androgen implanted into male ring dove brain. Endocrinology 89: 1470-1477.

Barfield RJ, Ronay G, Pfaff DW. 1978. Autoradiographic localization of androgen-concentrating cells in brain of male domestic fowl. Neuroendocrinology 26: 297-311.

Barrett GM, Shimizu K, Bardi M, Asaba S, Mori A. 2002. Endocrine correlates of rank, reproduction, and female-directed aggression in male Japanese macaques (*Macaca fuscata*). Horm Behav 42: 85-96.

Baulieu EE. 1981. Steroid hormone regulation of the brain. Fuxe K, Gustafsson JA, editors. Oxford: Pergamon Press; pp. 3-14.

Baulieu EE. 1991. Neurosteroids: A new function in the brain. Biol Cell 71: 3-10.

Bayart F, Spetz JP, Cittone L, Haug M. 1989. The role of gender and hormonal state on aggression during encounters between residential and intruder mice. Med Sci Res 17: 517-519.

Beeman EA. 1947. The effect of male hormone on aggressive behavior in mice. Physiol Zool 20: 373-404.

Beletsky LD, Orians GH, Wingfield JC. 1990. Effects of exogenous androgen and antiandrogen on territorial and nonterritorial red-winged blackbirds. Ethology 85: 58-72.

Berenbaum SA, Resnick SM. 1997. Early androgen effects on aggression in children and adults with congenital adrenal hyperplasia. Psychoneuroendocrinology 22: 505-15.

Bernard DJ, Bentley GE, Balthazart J, Turek FW, Ball GF. 1999. Androgen receptor, estrogen receptor α, and estrogen receptor β show distinct patterns of expression in forebrain song control nuclei of European starlings. Endocrinology 140: 4633-4643.

Bernstein I, Williams L, Ramsay M. 1983. The expression of aggression in Old World monkeys. Int J Primatol 4: 113-125.

Berthold AA. 1849. Transplantation of testes. Trans. By DP Quiring. Bull History Med, 1944 16: 399-401.

Bester-Meredith JK, Marler CA. 2001. Vasopressin and aggression in cross-fostered California mice (*Peromyscus californicus*) and white-footed mice (*Peromyscus leucopus*). Horm Behav 40: 51-64.

Bevan JM, Bevan W, Williams BF. 1958. Spontaneous aggressiveness in young castrate C3H mice treated with three dose levels of testosterone. Physiol Zool 31: 284-288.

Bevan WD, Daves WF, Levy FW. 1960. The relation of castration androgen therapy, and pre-test fighting experience to competitive aggression in male C57BL/10 mice. Anim Behav 8: 6-12.

Blanchard DC, Blanchard RJ. 1988. Ethoexperimental approaches to the biology of emotion. Annu Rev Psychol 39: 43-68.

Blanchard DC, Spencer RL, Weiss SM, Blanchard RJ, McEwen B, et al. 1995. Visible burrow system as a model of chronic social stress: Behavioral and neuroendocrine correlates. Psychoneuroendocrinology 20: 117-134.

Booth A, Shelley G, Mazur A, Tharp G, Kittok R. 1989. Testosterone and winning and losing in human competition. Horm Behav 23: 556-571.

Bottriech A, Zuir I, Hammel I, Terkel J. 2001. Non-involvement of testosterone in aggressive defense behavior in the male blind mole rat *Spalax ehrenbergi*. Agg Behav 27: 64-72.

Brain PF. 1972. Mammalian behavior and the adrenal cortex. A review. Behav Biol 7: 453-477.

Brain PF. 1975. What does individual housing mean to a mouse? Life Sci 16: 187-200.

Brain PF, Poole AE. 1974. Some studies on the use of 'standard opponents' in intermale aggression testing in TT albino mice. Behaviour 50: 100-110.

Brain PF, Haug M. 1992. Hormonal and neurochemical correlates of various forms of animal "aggression". Psychoneuroendocrinology 17: 537-551.

Brenowitz EA. 1997. Comparative approaches to the avian song system. J Neurobiol 33: 517-531.

Brenowitz EA, Margoliash D, Nordeen KW. 1997. An introduction to birdsong and the avian song system. J Neurobiol 33: 495-500.

Brockman DK, Whitten PL, Richard AF, Schneider A. 1998. Reproduction in free-ranging male Propithecus verreauzi: The hormonal correlates of mating and aggression. Am J Phys Anthropol 105: 137-151.

Brockman DK, Whitten PL, Richard AF, Benander B. 2001. Birth season testosterone levels in male Verreaux's sifaka, Propithecus verreauxi: Insights into socio-demographic factors mediating seasonal testicular function. Behav Ecol Sociobiol 49: 117-127.

Bronson FH, Desjardins C. 1969. Aggressive behavior and seminal vesicle function in mice: Differential sensitivity to androgen given neonatally. Endocrinology 85: 971-974.

Bronson FH, Desjardins C. 1970. Neonatal androgen administration and adult aggressiveness in female mice. Gen Comp Endocrinol 15: 320-325.

Buck CL, Barnes BM. 2003. Androgen in free-living arctic ground squirrels: Seasonal changes and influence of staged male-male aggressive encounters. Horm Behav 43: 318-326.

Caldwell HK, Albers HE. 2004. Effect of photoperiod on vasopressin-induced aggression in Syrian hamsters. Horm Behav 46: 444-449.

Caldwell GS, Glickman SE, Smith ER. 1984. Seasonal aggression independent of seasonal testosterone in wood rats. Proc Natl Acad Sci USA 81: 5255-5257.

Canoine V, Gwinner E. 2002. Seasonal differences in the hormonal control of territorial aggression in free-living European stonechats. Horm Behav 41: 1-8.

Casto JM, Nolan V, Ketterson ED. 2001. Steroid hormones and immune function: Experimental studies in wild and captive dark-eyed juncos (*Junco hyemalis*). Am Nat 157: 408-420.

Cavigelli SA, Pereira ME. 2000. Mating season aggression and fecal testosterone levels in male ring-tailed lemurs (*Lemur catta*). Horm Behav 37: 246-255.

Chapais B. 1992. The role of alliances in the social inheritance of rank among female primates. Coalitions and alliances in humans and other animals. Harcourt AH, de Waal FBM, editors. Oxford: Oxford University Press; pp. 29-60.

Chiavegatto S, Nelson RJ. 2003. Interaction of nitric oxide and serotonin in aggressive behavior. Horm Behav 44: 233-241.

Christianson T, Wallen K, Brown B, Glickman SE. 1972. Effects of castration, blindness, and anosmia on social activity in the male Mongolian gerbil (*Meriones unguiculatus*). Physiol Behav 10: 989-994.

Ciacco LA, Lisk RD, Rueter LA. 1979. Prelordotic behavior in the hamster: A hormonally modulated transition from aggression to sexual receptivity. J Comp Physiol Psychol 93: 771-780.

Cooper MA, Huhman KL. 2003. Corticotropin-releasing hormone, and not glucocorticoids, modulated the acquisition of conditioned defeat in male Syrian hamsters. San Diego, CA: Society for Neuroscience.

Corpechot C, Robel P, Axelson M, Sjovall J, Baulieu EE. 1981. Characterization and measurement of dehydroepiandrosterone sulfate in rat brain. Proc Natl Acad Sci 78: 4704-4707.

Davis DE. 1957. Aggressive behavior in castrated starlings. Science 130: 253.

Delville Y, Hendrick JC, Sulon J, Balthazart J. 1984. Testosterone metabolism and testosterone-dependent characteristics in Japanese quail. Physiol Behav 33: 817-823.

Demas GE, Moffatt CA, Drazen DL, Nelson RJ. 1999b. Castration does not inhibit aggressive behavior in adult male prairie voles. Physiol Behav 6: 59-61.

Demas GE, Polacek KM, Durazzo A, Jasnow AM. 2004. Adrenal hormones mediate melatonin-induced increases in aggression in male Siberian hamsters (*Phodopus sungorus*). Horm Behav 46: 582-591.

Dewsbury DA. 1991. Individual attributes generate contrasting degrees of sociality in voles. Social systems and population cycles in voles. Tamarkin RH, Ostfeld RS, Pugh SR, Bujalska G, editors. Boston: Birkhäuser Verlag; pp. 1-9.

DeLeon KR, Grimes JM, Melloni RH Jr. 2002. Repeated anabolic-androgenic steroid treatment during adolescence increases vasopressin V(1a) receptor binding in Syrian hamsters: Correlation with offensive aggression. Horm Behav 42: 182-191.

Delville Y, David JT, Taravosh-Lahn K, Wommack JC. 2003. Stress and the development of agonistic behavior in golden hamsters. Horm Behav 44: 263-270.

DeVoogd TJ, Krebs JR, Healy SD, Purvis A. 1993. Relations between song repertoire size and the volume of brain nuclei related to song: Comparative evolutionary analyses amongst oscine birds. Proc Roy Soc London B 254: 75-82.

Dixson AF. 1978. Effects of ovariectomy and estradiol replacement therapy upon the sexual and aggressive behavior of the greater galago (*Galago crassicaudatus crassicaudatus*). Horm Behav 10: 61-70.

Dixson AF. 1993a. Observations on effects of neonatal castration upon sexual and aggressive behavior in the male common marmoset (*Callithrix jacchus*). Am J Primatol 31: 1-10.

Dixson AF. 1993b. Effects of testosterone propionate upon the sexual and aggressive behavior of adult male marmosets (*Callithrix jacchus*) castrated as neonates. Horm Behav 27: 216-230.

Dixson AF. 1993c. Sexual and aggressive behavior of adult male marmosets (*Callithrix jacchus*) castrated neonatally, prepubertally, or in adulthood. Physiol Behav 54: 301-307.

Dufty AM. 1989. Testosterone and survival: A cost of aggressiveness. Horm Behav 23: 185-193.

Eaton GG, Resko JA. 1974. Plasma testosterone and male dominance in a Japanese macaque (*Macaca fuscata*) troop compared with repeated measures of testosterone in laboratory males. Horm Behav 5: 251-259.

Edwards DA. 1968. Mice: Fighting by neonatally androgenized females. Science 161: 1027-1028.

Edwards DA. 1969. Early androgen stimulation and aggressive behavior in male and female mice. Physiol Behav 4: 333-338.

Edwards DA. 1970. Post-neonatal androgenization and adult aggressive behavior in female mice. Physiol Behav 5: 465-446.

Epple G. 1978. Lack of effects of castration on scent marking, displays, and aggression in a South American primate (*Saguinus fuscicollis*). Horm Behav 11: 139-150.

Farner DS, Wingfield JC. 1980. Reproductive endocrinology of birds. Annu Rev Physiol 42: 457-472.

Ferguson JN, Young LJ, Insel TR. 2002. The neuroendocrine basis of social recognition. Front Neuroendocrinol 23: 200-224.

Ferree ED, Wikelski MC, Anderson DJ. 2004. Hormonal correlates of siblicide in Nazca boobies: Support for the challenge hypothesis. Horm Behav 46: 655-662.

Ferris CF, Albers HE, Wesolowski SM, Goldman BD, Luman SE. 1984. Vasopressin injected into the hypothalamus triggers a stereotypic behavior in golden hamsters. Science 224: 521-523.

Ferris CF, Meenan DM, Axelson JF, Albers HE. 1986. A vasopressin antagonist can reverse dominant/subordinate behavior in hamsters. Physiol Behav 38: 135-138.

Ferris CF, Delville Y. 1994. Vasopressin and serotonin interactions in the control of agonistic behavior. Psychoneuroendocrinology 19: 593-601.

Ferris CF, Delville Y, Brewer JA, Mansour K, Yules B, Melloni RH Jr. 1996. Vasopressin and developmental onset of flank marking behavior in golden hamsters. J Neurobiol 30: 192-204.

Ferris CF, Melloni RH, Koppel G, Perry KW, Fuller RW, et al. 1997. Vasopressin/serotonin interactions in the anterior hypothalamus control aggressive behavior in Syrian hamsters. J Neurosci 17: 4331-4340.

Fleming AS, Phillips A, Rydall A, Levesque L. 1988. Effects of photoperiod, the pineal gland and the gonads on agonistic behavior in female golden hamsters (*Mesocricetus auratus*). Physiol Behav 44: 227-234.

Fusani L, Van't Hof T, Hutchison JB, Gahr M. 2000. Seasonal expression of androgen receptors, estrogen receptors, and aromatase in the canary brain in relation to circulating androgens and estrogens. J Neurobiol 43: 254-268.

Gahr M, Guttinger HR, Kroodsma DE. 1993. Estrogen receptors in the avian brain: Survey reveals general distribution and forebrain areas unique to songbirds. J Comp Neurol 327: 112-122.

Garrett JW, Campbell CS. 1980. Changes in social behavior of the male golden hamster accompanying photoperiodic changes in reproduction. Horm Behav 14: 303-319.

Glick BB. 1979. Testicular size, testosterone level, and body weight in male *Macaca radiata*. Folia Primatol 32: 268-289.

Goldman L, Swanson HH. 1975. Developmental changes in pre-adult behavior in confined colonies of golden hamsters. Dev Psychobiol 8: 137-150.

Goldman BD. 2001. Mammalian photoperiodic system: Formal properties and neuroendocrine mechanisms of photoperiodic time measurement. J Biol Rhythms 16: 283-301.

Goodson JL. 1998a. Territorial aggression and dawn song are modulated by septal vasotocin and vasoactive intestinal polypeptide in male field sparrows (*Spizella pusilla*). Horm Behav 34: 67-77.

Goodson JL. 1998b. Vasotocin and vasoactive intestinal polypeptide modulate aggression in a territorial songbird, the violet-eared waxbill (Estrildidae: Uraeginthus granatina). Gen Comp Endocrinol 111: 233-244.

Goodson JL, Adkins-Regan E. 1999. Effect of intraseptal vasotocin and vasoactive intestinal polypeptide infusions on courtship song and aggression in the male zebra finch (*Taeniopygia guttata*). J Neuroendocrinol 11: 19-25.

Goodson JL, Bass AH. 2001. Social behavior functions and related anatomical characteristics of vasotocin/vasopressin systems in vertebrates. Brain Res Rev 35: 246-265.

Goodson JL, Evans AK. 2004. Neural responses to territorial challenge and nonsocial stress in male song sparrows: Segregation, integration, and modulation by a vasopressin V1 antagonist. Horm Behav 46: 371-381.

Goodson JL, Evans AK, Lindberg L, Allen CD. 2005. Neuro-evolutionary patterning of sociality. Proc Roy Soc B 272: 227-235.

Goodson JL, Saldanha CJ, Hahn TP, Soma KK. 2005. Recent advances in behavioral neuroendocrinology: Insights from studies on birds. Horm Behav 48(4): 461-473.

Gordon TP, Rose RM, Bernstein IS. 1976. Seasonal rhythm in plasma testosterone levels in the rhesus monkey (*Macaca mulatta*): A three year study. Horm Behav 7: 229-243.

Gordon TP, Bernstein IS, Rose RM. 1978. Social and seasonal influences on testosterone secretion in the male rhesus monkey. Physiol Behav 21: 623-627.

Gordon TP, Rose RM, Grady CL, Bernstein IS. 1979. Effects of increased testosterone secretion on the behavior of adult male rhesus living in a social group. Folia Primatol 32: 149-160.

Gottreich A, Zuri I, Hammel I, Terkel J. 2001. Noninvolvement of testosterone in aggressive defense behavior in the male blind mole rat Spalax ehrenbergi. Agg Behav 27: 64-72.

Goymann W, Moore IT, Scheuerlein A, Hirschenhauser K, Grafen A, et al. 2004. Testosterone in tropical birds: Effects of environmental and social factors. Am Nat 164: 327-334.

Gwinner E, Rodl T, Schwabl H. 1994. Pair territoriality of wintering stonechats: Behavior, function and hormones. Behav Ecol Sociobiol 34: 321-327.

Haller J, Kruk MR. 2003. Neuroendocrine stress responses and aggression. Mattson MP, editor. Neurobiology of Aggression. NJ: Humana Press; Totawa, pp. 93-118.

Harding CF, Follett BK. 1979. Hormone changes triggered by aggression in a natural population of blackbirds. Science 203: 918-920.

Harding CF, Walters MJ, Collado D, Sheridan K. 1988. Hormonal specificity and activation of social behavior in male red-winged blackbirds. Horm Behav 22: 402-418.

Harding CF, Walters MJ, Parsons B. 1984. Androgen receptor levels in hypothalamic and vocal control nuclei in the male zebra finch. Brain Res 106: 333-339.

Hau M, Wikelski M, Soma KK, Wingfield JC. 2000. Testosterone and year-round territorial aggression in a tropical bird. Gen Comp Endocrinol 117: 20-33.

Hau M, Stoddard ST, Soma KK. 2004. Territorial aggression and hormones during the non-breeding season in a tropical bird. Horm Behav 45: 40-49.

Haug M, Johnson F, Brain PF. 1992. Biological correlates of attack on lactating intruders by female mice: A topical review. Bjorkvist K, Niemala P. (eds). Of mice and women In Aspects of female Aggression Academic Press: San Diego, pp 381-393.

Haus E, Nicolau GY, Ghinea E, Dumitriu L, Petrescu E, et al. 1996. Stimulation of the secretion of dehydroepiandrosterone by melatonin in mouse adrenals in vitro. Life Sci 58: PL263-PL267.

Hegner RE, Wingfield JC. 1987. Effects of experimental manipulation of testosterone levels on parental investment and breeding success in male house sparrows. Auk 104: 462-469.

Herndon JG, Bein ML, Nordmeyer DL, Turner JJ. 1996. Seasonal testicular function in male rhesus monkeys. Horm Behav 30: 266-271.

Higley JD, Mehlman PT, Poland RE, Taub DM, Vickers J, et al. 1996c. CSF testosterone and 5-HIAA correlate with different types of aggressive behaviors. Biol Psychiat 40: 1067-1082.

Huhman KL, Solomon MB, Janicki M, Harmon AC, Lin SM, Israel JE, Jasnow AM. 2003. Conditioned defeat in male and female Syrian hamsters. Horm Behav 44: 293-299.

Hunt K, Wingfield JC, Astheimer LB, Buttemer WA, Hahn TP. 1995. Temporal patterns of territorial behavior and circulating testosterone in the lapland longspur and other Arctic passerines. Am Zool 35: 274-284.

Hunt KE, Hahn TP, Wingfield JC. 1997. Testosterone implants increase song but not aggression in male Lapland longspurs. Anim Behav 54: 1177-1192.

Jasnow AM, Huhman KL, Bartness TJ, Demas GE. 2000. Short-day increases in aggression are inversely related to circulating testosterone concentrations in male Siberian hamsters (*Phodopus sungorus*). Horm Behav 38: 102-110.

Jasnow AM, Huhman KL, Bartness TJ, Demas GE. 2002. Short days and exogenous melatonin increase aggression in male Syrian hamsters (*Mesocricetus auratus*). Horm Behav 42: 13-20.

Ketterson ED. 1979. Aggressive behavior in wintering dark-eyed juncos: Determinants of dominance and their possible relation to geographic variation in sex ratio. Wilson Bulletin 91: 371-383.

Ketterson ED, Nolan V. 1999. Adaptation, exaptation, and constraint: A hormonal perspective. Am Nat 154: S4-S25.

Ketterson ED, Nolan V, Cawthorn MJ, Parker PG, Ziegenfus C. 1996. Phenotypic engineering: Using hormones to explore the mechanistic and functional bases of phenotypic variation in nature. Ibis 138: 70-86.

Ketterson ED, Nolan V, Wolf L, Ziegenfus C. 1992. Testosterone and avian life histories: Effects of experimentally elevated testosterone on behavior and correlates of fitness in the dark-eyed junco (*Junco hyemalis*). Am Nat 140: 980-999.

Ketterson ED, Nolan V, Wolf L, Ziegenfus C, Dufty AM, et al. 1991. Testosterone and avian life histories: The effect of experimentally elevated testosterone on corticosterone and body mass in dark-eyed juncos. Horm Behav 25: 489-503.

Keverne EB, Meller RE, Eberhart A. 1982. Dominance and subordination: Concepts and physiological states. Advanced views in primate biology. Chiarelli AB, Corruccini RS, editors. New York: Springer-Verlag; pp. 81-94.

Kim YH, Perlman WR, Arnold AP. 2004. Expression of androgen receptor mRNA in zebra finch song system: Developmental regulation by estrogen. J Comp Neurol 469: 535-547.

Konishi M, Emlen ST, Ricklefs RE, Wingfield JC. 1989. Contributions of bird studies to biology. Science 246: 465-472.

Kriner E, Schwabl H. 1991. Control of winter song and territorial aggression of female robins (*Erithacus rubecula*) by testosterone. Ethology 87: 37-44.

Kroboth PD, Salek FS, Pittenger AL, Fabian TJ, Frye RF. 1999. DHEA and DHEA-S: A review. J Clin Pharmacol 39: 327-348.

Labrie F, Belanger A, Luu-The V, Labrie C, Simard J, Cusan L, Gomez JL, Candas B. 1998. DHEA and the intracrine formation of androgens and estrogens in peripheral target tissues: Its role during aging. Steroids. 63: 322-328.

Labrie F, Luu-The V, Labrie C, Simard J. 2001. DHEA and its transformation into androgens and estrogens in peripheral target tissues: Intracrinology. Front Neuroendocrinol 22: 185-212.

Levin RN, Wingfield JC. 1992. The hormonal control of territorial aggression in tropical birds. Ornis Scand 23: 284-291.

Linn FS, Steklis HD. 1990. The effects of depo-medroxyprogesterone acetate (DMPA) on copulation-related and agonistic behaviors in an island colony of stumptail macaques (*Macaca arctoides*). Physiol Behav 47: 403-408.

Lohiya NK, Sharma RS, Manivannan B, Kumar TCA. 1998. Reproductive exocrine and endocrine profiles and their seasonality in male langur monkeys (*Presbytis entellus entellus*). J Med Primatol 27: 15-20.

Logan CA, Carlin CA. 1991. Testosterone stimulates reproductive behavior during autumn in mockingbirds (*Mimus polyglottos*). Horm Behav 25: 229-241.

Logan CA, Wingfield JC. 1990. Autumnal territorial aggression is independent of plasma testosterone in mockingbirds. Horm Behav 24: 568-581.

Lu SF, Mo Q, Hum S, Garippam C, Simon NG. 2003. Dehydroepiandrosterone upregulates neural androgen receptor level and transcriptional activity. J Neurobiol 57: 163-171.

Lynch JW, Ziegler TE, Strier KB. 2002. Individual and seasonal variation in fecal testosterone and cortisol levels of wild male tufted capuchin monkeys, *Cebus apella nigritus*. Horm Behav 41: 275-287.

Marques DM, Valenstein ES. 1977. Individual differences in aggressiveness of female hamsters: Response to intact and castrated males and to females. Anim Behav 25: 131-139.

Martin LB, Pless M, Svoboda J, Wikelski M. 2004. Immune activity in temperate and tropical house sparrows: A common-garden experiment. Ecology 85: 2323-2331.

Mason WA. 1993. The nature of social conflict: A psycho-ethological perspective. Mason WA, Mendoza SP (eds). Primate Social Conflict. Albany: SUNY Press; pp. 13-47.

Matsumoto T, Honda S, Harada N. 2003. Neurological effects of aromatase deficiency in the mouse. J Steroid Biochem Mol Biol 86: 357-365.

Mathewson SF. 1961. Gonadotrophic hormones affect aggressive behavior in starlings. Science 134: 1522-1523.

Mazur A, Booth A. 1998. Testosterone and dominance in men. Behav Brain Sci 21: 353-397.

Mazur A, Booth A, Dabbs JM. 1992. Testosterone and chess competition. Soc Psychol Q 55: 70-77.

Mellon SH, Vaudry H. 2001. Biosynthesis of neurosteroids and regulation of their synthesis. Int Rev Neurobiol 46: 33-78.

Mendoza SP, Coe CL, Lowe EL, Levine S. 1979. The physiological response to group formation in adult male squirrel monkeys. Psychoneuroendocrinology 3: 221-229.

Michael RP, Zumpe D. 1993. A review of hormonal factors influencing the sexual and aggressive behavior of macaques. Am J Primatol 30: 213-241.

Miczek KA, DeBold JF, van Erp AM. 1994. Neuropharmacological characteristics of individual differences in alcohol effects on aggression in rodents and primates. Behav Pharmacol 5: 407-421.

Mohle U, Heistermann M, Palme R, Hodges JK. 2002. Characterization of urinary and fecal metabolites of testosterone and their measurement for assessing gonadal endocrine function in male nonhuman primates. Gen Comp Endocrinol 129: 135-145.

Moss R, Parr R, Lambin X. 1994. Effects of testosterone on breeding density, breeding success and survival of red grouse. Proc Roy Soc London B 258: 175-180.

Mougeot F, Dawson A, Redpath SM, Leckie F. 2005. Testosterone and autumn territorial behavior in male red grouse, *Lagopus lagopus scoticus*. Horm Behav 47: 576-584.

Moyer KE. The Physiology of Hostility. Chicago: Markham; 1971.

Muehlenbein MP, Watts DP, Whitten PL. 2004. Dominance rank and fecal testosterone levels in adult male chimpanzees (*Pan troglodytes schweinfurthii*) at Ngogo, Kibale National Park, Uganda. Am J Primatol 64: 71-82.

Muller MN, Wrangham RW. 2004. Dominance, aggression and testosterone in wild chimpanzees: A test of the 'challenge hypothesis'. Anim Behav 67: 113-123.

Neave N, Wolfson S. 2003. Testosterone, territoriality, and the 'home advantage'. Physiol Behav 78: 269-275.

Nelson, RJ. 1990. Photoperiodic responsiveness in house mice. Physiol Behav 48: 403-408.

Nicolas LB, Pinoteau W, Papot S, Routierm S, Guillaumetm G, et al. 2001. Aggressive behavior induced by the steroid sulfatase inhibitor COUMATE and by DHEAS in CBA/H mice. Brain Res 922: 216-222.

Nieuwenhuijsen K, de Neef KJ, van der Werff ten Bosch JJSAK. 1987. Testosterone, testis size, seasonality, and behavior in group-living stumptail macaques (*Macaca arctoides*). Horm Behav 21: 153-169.

Nomura M, Durbak L, Chan J, Smithies O, Gustafsson JA, Korach KS, Pfaff DW, Ogawa S. 2002. Genotype/age interactions on aggressive behavior in gonadally intact estrogen receptor beta knockout (betaERKO) male mice. Horm Behav 41: 288-296.

Nottebohm F, Stokes TM, Leonard CM. 1976. Central control of song in the canary, *Serinus canarius*. J Comp Neurol 165: 457-486.

Ogawa S, Lubahnm DB, Korach KS, Pfaff DW. 1997. Behavioral effects of estrogen receptor gene disruption in male mice. Proc Natl Acad Sci USA 94: 1476-1481.

Ogawa S, Washburn TF, Taylor J, Lubahn DB, Korach KS, et al. 1998. Modifications of testosterone-dependent behaviors by estrogen receptor-alpha gene disruption in male mice. Endocrinology 139: 5058-5069.

Ogawa S, Chanm J, Chesterm AE, Gustafssonm JA, Korach KS, et al. 1999. Survival of reproductive behaviors in estrogen receptor beta gene-deficient (betaERKO) male and female mice. Proc Natl Acad Sci USA 96: 12887-12892.

Ogawa S. 2003. Differential roles of estrogen receptor alpha and beta in the regulation of neuroendocrine functions and socio-sexual behaviors. J Nippon Med Sch 70: 530.

Ostner T, Kappeler PM, Heistermann M. 2002. Seasonal variation and social correlates of androgen excretion in male redfronted lemurs (*Eulemur fulvus rufus*). Behav Ecol Sociobiol 52: 485-495.

Owen-Ashley NT, Hasselquist D, Wingfield JC. 2004. Androgens and the immunocompetence handicap hypothesis: Unraveling direct and indirect pathways of immunosuppression in song sparrows. Am Nat 164: 490-505.

Paterson AT, Vickers C. 1981. Melatonin and the adrenal cortex: Relationship to territorial aggression in mice. Physiol Behav 27: 983-987.

Pazol K, Wilson ME, Wallen K. 2004. Medroxyprogesterone acetate antagonizes the effects of estrogen treatment on social and sexual behavior in female macaques. J Clin Endocrinol Metabol 89: 2998-3006.

Pellis SM, Pellis VC. 1998. Play fighting of rats in comparative perspective: A schema for neurobehavioral analyses. Neurosci Biobehav Rev 23: 87-101.

Perché F, Young J, Robel P, Simon NG, Haug M. 2000. Prenatal testosterone treatment potentates the aggression-suppressive effect of the neurosteroid dehydroandrosterone in female mice. Agg Behav 27: 130-138.

Perlman WR, Arnold AP. 2003. Expression of estrogen receptor and aromatase mRNAs in embryonic and posthatch zebra finch brain. J Neurobiol 55: 204-219.

Perret M. 1992. Environmental and social determinants of sexual function in the male lesser mouse lemur (*Microcebus murinus*). Folia Primatol 59: 1-25.

Phillips RE, Barfield RJ. 1977. Effects of testosterone implants in midbrain vocal areas of capons. Brain Res 122: 378-381.

Phoenix CH, Goy RW, Gerall AA, Young WC. 1959. Organizing action of prenatally administered testosterone propionate on the tissues mediating mating behavior in the female guinea pig. Endocrinology 65: 369-382.

Pieper DR, Lobocki CA. 2000. Characterization of serum dehydroepiandrosterone secretion in golden hamsters. Proc Soc Exp Biol Med 224: 278-284.

Pinna G, Costa E, Guidotti A. 2005. Changes in brain testosterone and allopregnanolone biosynthesis elicit aggressive behavior. Proc Natl Acad Sci USA. 102: 2135-2140.

Pinxten R, De Ridder E, Balthazart J, Berghman L, Eens M. 2000. The effect of castration on aggression in the nonbreeding season is age-dependent in male European starlings. Behaviour 137: 647-661.

Pinxten R, De Ridder E, Balthazart J, Eens M. 2002. Context-dependent effects of castration and testosterone treatment on song in male European starlings. Horm Behav 42: 307-318.

Pinxten R, De Ridder E, De Cock M, Eens M. 2003. Castration does not decrease nonreproductive aggression in yearling male European starlings (*Sturnus vulgaris*). Horm Behav 43: 394-401.

Potegal M, Ferris CF. 1990. Intraspecific aggression in male hamsters is inhibited by intrahypothalamic VP receptor antagonist. Aggressive Behav 15: 311-320.

Rapkin AJ, Pollack DB, Raleigh MJ, Stone B, McGuire MT. 1995. Menstrual cycle and social behavior in vervet monkeys. Psychoneuroendocrinology 20: 289-297.

Rehman KS, Carr BR. 2004. Sex differences in adrenal androgens. Semin Reprod Med 22: 349-360.

Rejeski WJ, Gregg E, Kaplan JR, Manuck SB. 1990. Anabolic androgenic steroids: Effects on social behavior and baseline heart rate. Health Psychol 9: 774-791.

Ros AFH, Dieleman SJ, Groothuis TGG. 2002. Social stimuli, testosterone, and aggression in gull chicks: Support for the challenge hypothesis. Horm Behav 41: 334-342.

Rose RM, Bernstein IS, Gordon TP. 1975. Consequences of social conflict on plasma testosterone levels in rhesus monkeys. Psychosom Med 37: 50-61.

Rose RM, Holaday JW, Bernstein IS. 1971. Plasma testosterone, dominance rank and aggressive behavior in male rhesus monkeys. Nature 231: 366-368.

Rose RM, Gordon TP, Bernstein IS. 1972. Plasma testosterone levels in the male rhesus: Influences of sexual and social stimuli. Science 178: 643-645.

Rose RM, Gordon TP, Bernstein IS. 1978. Diurnal variation in plasma testosterone and cortisol in rhesus monkeys living in social groups. J Endocrinol 76: 67-74.

Rostal DC, Glick BB, Eaton GC, Resko JA. 1986. Seasonality of adult male Japanese macaques (*Macaca fuscata*): Androgens and behavior in a confined troop. Horm Behav 20: 452-462.

Rowe R, Maughan B, Worthman CM, Costello EJ, Angold A. 2004. Testosterone, antisocial behavior, and social dominance in boys: Pubertal development and biosocial interaction. Biol Psychiat 55: 546-552.

Ruiz de Elvira MC, Herndon JG, Wilson ME. 1982. Influence of estrogen-treated females on sexual behavior and male testosterone levels of a social group of rhesus monkeys during the nonbreeding season. Biol Reprod 26: 825-834.

Saldanha CJ, Coomaralingam L. 2005. Overlap and co-expression of estrogen synthetic and responsive neurons in the songbird brain: A double-label immunocytochemical study. Gen Comp Endocrinol 141: 66-75.

Sapolsky RM. 1983. Endocrine aspects of social instability in the olive baboon (*Papio anubis*). Am J Primatol 5: 365-379.

Sapolsky RM, Vogelman JH, Orentreich N, Altmann J. 1993. Senescent decline in serum dehydroepiandrosterone sulfate concentrations in a population of wild baboons. J Gerontol 48: B196-B200.

Schiml PA, Mendoza SP, Saltzman W, Lyons DM, Mason WA. 1996. Seasonality in squirrel monkeys (*Saimiri sciureus*), social facilitation by females. Physiol Behav 60: 1105-1113.

Schlegel ML, Spetz JF, Robel P, Haug M. 1985. Studies on the effects of dehydroepiandrosterone and its metabolites on attack by castrated mice on lactating intruders. Physiol Behav 34: 867-870.

Schleussner G, Gwinner E. 1989. Photoperiodic time measurement during the termination of photorefractoriness in the starling (*Sturnus vulgaris L.*). Gen Comp Endocrinol 75: 54-61.

Schlinger BA. 1997. Sex steroids and their actions on the birdsong system. J Neurobiology 33: 619-631.

Schlinger BA, Arnold AP. 1992. Circulating estrogens in a male songbird originate in the brain. Proc Natl Acad Sci 89: 7650-7653.

Schlinger BA, Callard GV. 1987. A comparison of aromatase, 5α-, and 5β-reductase activities in the brain and pituitary of male and female quail. J Exp Zool 242: 171-180.

Schlinger BA, Callard GV. 1989a. Aromatase activity in quail brain: Correlation with aggressiveness. Endocrinology 124: 437-443.

Schlinger BA, Callard GV. 1989b. Estrogen receptors in quail brain. A functional relationship to aromatase and aggressiveness. Biol Reprod 40: 268-275.

Schlinger BA, Callard GV. 1990a. Aggressive behavior in birds: An experimental model for studies of brain-steroid interactions. Comp Biochem Physiol A 97: 307-316.

Schlinger BA, Callard GV. 1990b. Aromatization mediates aggressive behavior in quail. Gen Comp Endocrinol 79: 39-53.

Schwabl H. 1992. Winter and breeding territorial behavior and levels of reproductive hormones of migratory European robins. Ornis Scand 23: 271-276.

Schwabl H. 1993. Yolk is a source of maternal testosterone for developing birds. Proc Natl Acad Sci 90: 11446-11450.

Schwabl H, Kriner E. 1991. Territorial aggression and song of male European robins (*Erithacus rubecula*) in autumn and spring-effects of antiandrogen treatment. Horm Behav 25: 180-194.

Searcy WA, Wingfield JC. 1980. The effects of androgen and antiandrogen on dominance and aggressiveness in male red-winged blackbirds. Horm Behav 14: 126-135.

Setchell JM, Dixson AF. 2001. Circannual changes in the secondary sexual adornments of semifree-ranging male and female mandrills (*Mandrillus sphinx*). Am J Primatol 53: 109-121.

Shively CA, Kaplan JR, Adams MR. 1986. Effects of ovariectomy, social instability and social status on female *Macaca fascicularis* social behavior. Physiol Behav 36: 1147-1153.

Sigg EB. 1969. Relationship of aggressive behavior to adrenal and gonadal function in male mice. Aggressive behavior (monograph). Garattini S, Sigg EB, editors. Amsterdam: Excerpta Medica.

Silverin B, Baillien M, Balthazart J. 2004. Territorial aggression, circulating levels of testosterone, and brain aromatase activity in free-living pied flycatchers. Horm Behav 45: 225-234.

Silverin B, Viebke PA, Westin J. 1984. Plasma levels of luteinizing hormone and steroid hormones in free-living winter groups of willow tits (*Parus montanus*). Horm Behav 18: 367-379.

Simon NG. 2002. Hormonal processes in the development and expression of aggressive behavior. In: Pfaff DW, Arnold AP, Etgen AM, Fahrbach SE, Rubin RT, (eds). Hormones, Brain and behavior, vol. 1., New York: Academic Press. pp. 339-392.

Simon NG, Kaplan JR, Hu S, Register TC, Adams MR. 2004. Increased aggressive behavior and decreased affiliative behavior in adult male monkeys after long-term consumption of diets rich in soy protein and isoflavones. Horm Behav 45: 278-284.

Scott JP, Fredericson E. 1951. The causes of fighting in mice and rats. Physiol Zool 24: 273.

Smulders TV, Casto JM, Nolan V Jr, Ketterson ED, DeVoogd TJ. 2000. Effects of captivity and testosterone on the volumes of four brain regions in the dark-eyed junco (*Junco hyemalis*). J Neurobiol 43: 244-253.

Soma KK. 2004. Steroid regulation of territorial behavior in wild songbirds. Horm Behav 46: 135.

Soma KK, Alday NA, Hau M, Schlinger BA. 2004. Dehydroepiandrosterone metabolism by 3β-hydroxysteroid dehydrogenase/δ5-δ4 isomerase in adult zebra finch brain: Sex difference and rapid effect of stress. Endocrinology 145: 1668-1677.

Soma KK, Alday NA, BA. Schlinger 2002. 3β-HSD and aromatase in songbird brain: DHEA metabolism, aggression, and song. Society for Neuroscience Abstracts.

Soma KK, Bindra RK, Gee J, Wingfield JC, Schlinger BA. 1999. Androgen-metabolizing enzymes show region-specific changes across the breeding season in the brain of a wild songbird. J Neurobiol 41: 176-188.

Soma KK, Hartman VN, Wingfield JC, Brenowitz EA. 1999. Seasonal changes in androgen receptor immunoreactivity in the song nucleus HVc of a wild bird. J Comp Neurol 409: 224-236.

Soma KK, Schlinger BA, Wingfield JC, Saldanha CJ. 2003. Brain aromatase, 5α-reductase and 5β-reductase change seasonally in wild male song sparrows: Relationship to sexual and aggressive behaviors. J Neurobiol 56: 209-221.

Soma KK, Sullivan K, Wingfield J. 1999. Combined aromatase inhibitor and antiandrogen treatment decreases territorial aggression in a wild songbird during the nonbreeding season. Gen Comp Endocrinol 115: 442-453.

Soma KK, Sullivan KA, Tramontin AD, Saldanha CJ, Schlinger BA, et al. 2000a. Acute and chronic effects of an aromatase inhibitor on territorial aggression in breeding and nonbreeding male song sparrows. J Comp Physiol A 186: 759-769.

Soma KK, Tramontin AD, Wingfield JC. 2000b. Oestrogen regulates male aggression in the non-breeding season. Proc Roy Soc London B 267: 1089-1096.

Soma KK, Wingfield JC. 1999. Endocrinology of aggression in the nonbreeding season. 22nd International Ornithological Congress. Adams N, Slotow R, editors. Durban, South Africa: University of Natal. pp. 1606-1620.

Soma KK, Wingfield JC. 2001. Dehydroepiandrosterone in songbird plasma: Seasonal regulation and relationship to territorial aggression. Gen Comp Endocrinol 123: 144-155.

Soma KK, Wissman AM, Brenowitz EA, Wingfield JC. 2002. Dehydroepiandrosterone (DHEA) increases territorial song and the size of an associated brain region in a male songbird. Horm Behav 41: 203-212.

Stavisky RC, Register TC, Watson SL, Weaver DS, Kaplan JR. 1999. Behavioral responses to ovariectomy and chronic anabolic steroid treatment in female cynomolgus macaques. Physiol Behav 66: 95-100.

Steklis HD, Brammer GL, Raleigh MJ, McGuire MT. 1985. Serum testosterone, male dominance, and aggression in captive groups of vervet monkeys (*Cercopithecus aethiops sabaeus*). Horm Behav 19: 154-163.

Steklis HD, Raleigh MJ, Kling AS, Tachiki K. 1986. Biochemical and hormonal correlates of dominance and social behavior in all-male groups of squirrel monkeys (*Saimiri sciureus*). Am J Primatol 11: 133-145.

Stribley JM, Carter CS. 1999. Developmental exposure to vasopressin increases aggression in adult prairie voles. Proc Natl Acad Sci USA 196: 12601-12604.

Strier KB, Ziegler TE, Wittwer DJ. 1999. Seasonal and social correlates of fecal testosterone and cortisol levels in wild male muriquis (*Brachyteles arachnoides*). Horm Behav 35: 125-134.

Suchowsky GK, Pegrassi L, Bonsignori A. 1969. The effect of steroids on aggressive behavior in isolated male mice. In: Garattini S, Sigg EB (eds). Aggressive Behaviour. Proc. Int. Symp. on Biology of Aggressive Behaviour. Amsterdam: Exerpta Medica Foundation, 164-171.

Suchowsky GK, Pegrassi L, Bonsignori. 1969. The effect of steroids on aggressive behaviour in isolated male mice.

Aggressive behaviour. Garattini S, Sigg EB, editors. New York: Wiley; pp. 164-171.

Tiefer L. 1970. Gonadal hormones and mating behavior in the adult golden hamster. Horm Behav 1: 189-202.

Tramontin AD, Brenowitz EA. 2000. Seasonal plasticity in the adult brain. Trends Neurosci 23: 251-258.

Uhrich J. 1938. The social hierarchy in albino mice. J Comp Pscyhol 25: 373.

van Goozen SHM, Matthys W, Cohen-Kettenis PT, Thijssen JHH, van Engeland H. 1998. Adrenal androgens and aggression in conduct disorder prepubertal boys and normal controls. Biol Psychiat 43: 156-158.

van Goozen SH, van den Ban E, Matthys W, Cohen-Kettenis PT, Thijssen JH, et al. 2000. Increased adrenal androgen functioning in children with oppositional defiant disorder: A comparison with psychiatric and normal controls. J Am Acad Child Adolesc Psychiat 39: 1446-1451.

Van Roo BL. 2004. Exogenous testosterone inhibits several forms of male parental behavior and stimulates song in a monogamous songbird: The blue-headed vireo (*Vireo solitarius*). Horm Behav 46: 678-683.

van Schaik CP, van Noordwijk MA, van Bragt T, Blankenstein MA. 1991. A pilot study of the social correlates of levels of urinary cortisol, prolactin, and testosterone in wild long-tailed macaques (*Macaca fascicularis*). Primates 32: 345-356.

von Engelhardt N, Kappeler PM, Heistermann M. 2000. Androgen levels and female social dominance in *Lemur catta*. Proc Roy Soc London B 267: 1533-1539.

Walker ML, Wilson ME, Gordon TP. 1983. Female rhesus monkey aggression during the menstrual cycle. Anim Behav 31: 1047-1054.

Walters MJ, Harding CF. 1988. The effects of an aromatization inhibitor on the reproductive behavior of male zebra finches. Horm Behav 22: 207-218.

Walters MJ, McEwen BS, Harding CF. 1988. Estrogen receptor levels in hypothalamic and vocal control nuclei in the male zebra finch. Brain Res 459: 37-43.

Weinbauer GF, Gockeler E, Nieschlag E. 1988. Testosterone prevents complete suppression of spermatogenesis in the gonadotropin-releasing hormone antagonist-treated non-human primate (*Macaca fascicularis*). J Clin Endocrinol Metabol 67: 284-290.

Whitsett JM. 1975. The development of aggressive and marking behavior in intact and castrated male hamsters. Horm Behav 6: 47-57.

Whitten PL. 2000. Evolutionary endocrinology of the cercopithecoids. Old World monkeys. Whitehead PF, Jolly CJ, editors. Cambridge: Cambridge University Press; pp. 269-297.

Whitten PL, Turner TR. 2004. Male residence and the patterning of serum testosterone in vervet monkeys (*Cercopithecus aethiops*). Behav Ecol Sociobiol 56: 565-578.

Wickings EJ, Dixson AF. 1992. Testicular function, secondary sexual development, and social-status in male mandrills (*Mandrillus sphinx*). Physiol Behav 52: 909-916.

Wiebe RH, Williams LE, Abee CR, Yeoman RR, Diamond EJ. 1988. Seasonal changes in serum dehydroepiandrosterone, androstenedione, and testosterone levels in the squirrel monkey (*Saimiri boliviensis boliviensis*). Am J Primatol 14: 285-291.

Wikelski M, Hau M, Wingfield JC. 1999. Social instability increases plasma testosterone in a year- round territorial neotropical bird. Proc Roy Soc London B 266: 551-556.

Wikelski M, Hau M, Wingfield JC. 2000. Seasonality of reproduction in a neotropical rain forest bird. Ecology 81: 2458-2472.

Wikelski M, Lynn S, Breuner C, Wingfield JC, Kenagy GJ. 1999. Energy metabolism, testosterone and corticosterone in white-crowned sparrows. J Comp Physiol A 185: 463-470.

Wiley CJ, Goldizen AW. 2003. Testosterone is correlated with courtship but not aggression in the tropical buff-banded rail, *Gallirallus philippensis*. Horm Behav 43: 554-560.

Wilson AP, Vessey SH. 1968. Behavior of free-ranging castrated rhesus monkeys. Folia Primatologica 9: 1-14.

Wingfield JC. 1984a. Environmental and endocrine control of reproduction in the song sparrow, *Melospiza melodia*. Vol. I. Temporal organization of the breeding cycle. Gen Comp Endocrinol 56: 406-416.

Wingfield JC. 1984b. Environmental and endocrine control of reproduction in the song sparrow, *Melospiza melodia*. Vol. II. Agonistic interactions as environmental information stimulating secretion of testosterone. Gen Comp Endocrinol 56: 417-424.

Wingfield JC. 1985. Short-term changes in plasma levels of hormones during establishment and defense of a breeding territory in male song sparrows, *Melospiza melodia*. Horm Behav 19: 174-187.

Wingfield JC. 1994. Control of territorial aggression in a changing environment. Psychoneuroendocrinology 19: 709-721.

Wingfield JC, Hegner RE, Dufty AM Jr, Ball GF. 1990. The "challenge hypothesis": Theoretical implications for patterns of testosterone secretion, mating systems, and breeding strategies. Am Nat 136: 829-846.

Wingfield JC, Hahn T. 1994. Testosterone and territorial behaviour in sedentary and migratory sparrows. Anim Behav 47: 77-89.

Wingfield JC, Jacobs J, Hillgarth N. 1997. Ecological constraints and the evolution of hormone-behavior interrelationships. Ann NY Acad Sci 807: 22-41.

Wingfield JC. 1994. Regulation of territorial behavior in the sedentary song sparrow, *Melospiza melodia morphna*. Horm Behav 28: 1-15.

Wingfield JC, Ball GF, Dufty AM, Hegner RE, Ramenofsky M. 1987. Testosterone and aggression in birds. Am Sci 75: 602-608.

Wingfield JC, Farner DS. 1993. Endocrinology of reproduction in wild species. Avian biology. Farner DS, King J, Parkes K, editors. San Diego, CA: Academic; pp. 163-327.

Wingfield JC, Hegner RE, Dufty AM, Ball GF. 1990. The challenge hypothesis: Theoretical implications for patterns of testosterone secretion, mating systems, and breeding strategies. Am Nat 136: 829-846.

Wingfield JC, Jacobs J, Hillgarth N. 1997. Ecological constraints and the evolution of hormone-behavior interrelationships. Integrative neurobiology of affiliation. Carter CS, Lederhendler II, Kirkpatrick B, editors. New York: New York Academy of Sciences; pp. 22-41.

Wingfield JC, Lynn SE, Soma KK. 2001. Avoiding the 'costs' of testosterone: Ecological bases of hormone-behavior interactions. Brain Behav Evol 57: 239-251.

Wingfield JC, Monk D. 1992. Control and context of year-round territorial aggression in the nonmigratory song sparrow, *Zonotrichia melodia morphna*. Ornis Scand 23: 298-303.

Wingfield JC, Sapolsky RM. 2003. Reproduction and resistance to stress: When and how. J Neuroendocrinol 15: 711-724.

Wingfield JC, Silverin B. 1986. Effects of corticosterone on territorial behavior of free-living male song sparrows, *Melospiza melodia*. Horm Behav 20: 405-417.

Wingfield JC, Soma KK. 2002. Spring and autumn territoriality in song sparrows: Same behavior, different mechanisms? Integr Comp Biol 42: 11-20.

Wingfield JC, Wada M. 1989. Changes in plasma levels of testosterone during male-male interactions in the song sparrow, *Melospiza melodia*: Time course and specificity of response. J Comp Physiol A 166: 189-194.

Whittingham LA, Schwabl H. 2002. Maternal testosterone in tree swallow eggs varies with female aggression. Anim Behav 63: 63-67.

Wommack JC, Salinas A, Melloni RH Jr, Delville Y. 2004. Behavioural and neuroendocrine adaptations to repeated stress during puberty in male golden hamsters. J Neuroendocrinol 16: 767-775.

Wommack JC, Taravosh-Lahn K, David JT, Delville Y. 2003. Repeated exposure to social stress alters the development of agonistic behavior in male golden hamsters. Horm Behav 43: 229-236.

Wood RI. 1997. Thinking about networks in the control of male hamster sexual behavior. Horm Behav 32: 40-45.

Yen, HCY, Day DA, Sigg EB. 1962. Influence of endocrine factors on development of fighting behavior in rodents. Pharmacologist 4: 173.

Young J, Corpechot C, Perche F, Eychenne B, Haug M, Baulieu EE, Robel P. 1996. Neurosteroids in the mouse brain: Behavioral and pharmacological effects of a 3β hydroxysteroid dehydrogenase inhibitor. Steroids, 61: 144-149.

Zhang J-X, Zhang Z-B, Wang Z-W. 2001. Seasonal changes in and effects of familiarity on agonistic behaviors of rat-like hamsters (*Cricetulus triton*). Ecol Res 16: 309-317.

Zumpe D, Clancy AN, Bonsall RW, Michael RP. 1996. Behavioral responses to Depo-Provera, Fadrozole, and estradiol in castrated, testosterone-treated cynomolgus monkeys (*Macaca fascicularis*): The involvement of progestin receptors. Physiol Behav 60: 531-540.

9 Molecular Neurobiology of Bird Song

D. F. Clayton

Abstract: Songbirds (oscines of Order Passeriformes) are emerging as a major focus of research in neurobiology. This review considers five broad research topics that are proving amenable to molecular biological analysis in songbirds: (1) molecular anatomy of a functioning neural system (the song system); (2) regulation of learning and memory formation; (3) brain circuit development, including roles for sex steroids and ongoing neurogenesis; (4) integration of social and environmental signals; (5) species diversity and evolution. Results from ~400 references are considered, illustrating the enormous breadth and vitality of bird song molecular neurobiology.

List of Abbreviations: AFP, anterior forebrain pathway; BAC, bacterial artificial chromosome; BDNF, brain-derived neurotrophic factor; BOS, bird's own song; CMM, caudomedial mesopallium; CREB, cyclic AMP response element binding protein; ERK, extracellular signal-regulated kinase; EST, expressed sequence tag; FOXP2, forkhead box P2; HVC, (proper name for nidopallial song nucleus previously misnamed Hyperstriatum Ventrale pars caudalis and commonly referred to as High Vocal Center, see Reiner et al., 2004b); IEG, immediate early gene; IGF-II, Insulin-related growth factor II; LMAN, lateral magnocelluar nucleus of the anterior nidopallium; LTD, long-term depression; LTP, long-term potentiation; NCM, caudomedial nidopallium; NMDA, N-methyl-D-aspartic acid; RA, robust nucleus of the arcopallium; SNAg, song system nuclear antigen; ZRALDH, zebra finch retinaldehyde dehydrogenase

1 Introduction: Why This—A Molecular Neurobiology of Bird Song?

Songbirds have fascinated humans probably since history began, and are still worshiped as gods by the aborigines of Australia (Zann, 1996). However, only in the last few decades songbirds have become common subjects of formal scientific investigation. Interest in songbirds is now skyrocketing. The annual rate of new publications is now approaching 400 per year (PubMed), up tenfold from a decade ago. More than 10 million dollars in research grants from the NIH and more than 50 major labs around the world are involved in work that contributes to songbird neurobiology. Songbird research (Dingemanse et al., 2004) was the focus of a recent editorial in the New York Times (Editorial, 2005). An opinion piece in Nat Rev Neurosci (Jarvis et al., 2005) on a seemingly arcane topic—a revised nomenclature for avian neuroanatomy deriving primarily from songbird work—was treated as a major news story in the popular press. In October, 2005, the National Human Genome Research Institute targeted a songbird, the zebra finch, for whole-genome sequencing (http://www.genome.gov/11007951).

Why this sudden interest? And why should we advance a molecular neurobiology of songbirds? Birds do have a natural appeal, and they have contributed to many areas of scientific research (Konishi et al., 1989). In the last 20 years, the questions and tools of molecular biology have increasingly been brought to bear on songbird research, as in every other area of biology. But beyond these general factors are five specific reasons why the "molecular neurobiology of bird song" is emerging as a discrete field of scientific study.

1.1 Reason #1: The Song-Control System

Songbirds communicate through learned vocalizations, an ability shared only by humans and a small number of other species (Doupe and Kuhl, 1999; Jarvis, 2004). The neural system that controls vocal learning and production in songbirds lies within the basic framework of sensory, motor, and modulatory pathways conserved in all vertebrates (Reiner et al., 2004b; Jarvis et al., 2005). Yet it also sits apart, comprising a set of unique, interconnected "song nuclei" that can often be seen by the naked eye in the unstained brain and are readily highlighted by distinctive patterns of gene expression (➋ *Section 2*). The motor components of the system appear to be dedicated to a single function: learning and production of species-specific vocalizations. The ready ability to identify the functional components of the system, to manipulate them, and to study their function in a seemingly simple yet "higher" integrative behavior has established the song system as a superb model for fundamental neuroscience.

Definition of the song-control system began in 1976, with this simple declaration: "We have traced central nervous pathways controlling bird song in the canary using a combination of behavioral and anatomical techniques" (Nottebohm et al., 1976). In that report, a large superficial nucleus in the forebrain [proper name for nucleus commonly known as High Vocal Center] was identified by a lesion as essential for singing behavior. Two other connected nuclei (RA, Area X) were identified by degenerating fibers following the lesion (➋ *Figure 9-1a*) .

In the years since Nottebohm's seminal 1976 publication, any illusions that the song-control system could be defined by merely three nuclei have been dashed by the steady identification of other interconnected, functionally related components (➋ *Figure 9-1b*). Although the system as we now understand it appears complex, it can be broken down into three functional subsystems, as indicated by shadings and dashed lines in ➋ *Figure 9-1b*. Elements associated primarily with motor control are shown in gray. On the left edge of the figure two of the nuclei identified in (Nottebohm et al., 1976) HVC and RA can be found. They constitute the essential telencephalic output pathway for motor control of singing. HVC sends a major projection to RA, and in this review we shall refer frequently to the HVC–RA pathway.

HVC also sends a separate projection to a striatal nucleus, Area X. This forms the beginning of the second subsystem, known as the "anterior forebrain pathway" (AFP). The AFP, which is indicated in the figure by the dashed line, proceeds through a striatal > thalamic > cortical loop and ultimately converges again onto the motor output pathway at RA via a projection from nucleus LMAN. The AFP is more clearly involved in learning how to sing than in song performance, as we shall see later.

The third subsystem to consider in song communication is the auditory forebrain, which forms the primary representations of song patterns heard. The auditory forebrain has a complex internal organization that is still being sorted out. Auditory afferents from the thalamus enter the forebrain primarily at an area known as Field L2a, but molecular analyses have focused mainly on two reciprocally connected stations further upstream, representing the caudomedial portions of nidopallium and mesopallium, respectively. In this review, we shall refer to these areas, considered functionally analogous to secondary auditory cortex, by their anatomical abbreviations NCM and CMM.

1.1.1 A Note on Nomenclature

In 2002, a consortium of avian brain scientists began working on new nomenclature to resolve explicit conflicts between mammalian and avian designations, while preserving as much as possible the familiar acronyms for major avian brain centers like the song-control nuclei. The results of their deliberations were published (Reiner et al., 2004b) and have been generally adopted in the songbird research community. In this article, we shall exclusively adhere to the new nomenclature and shall generally rely on acronyms, where those are preserved between the old and the new. The interested or confused reader should consult the following for more guidance: Perkel (2004); Reiner et al. (2004a, b); Jarvis et al. (2005).

1.2 Reason #2: Rich Opportunities to Study Song Learning

Song learning is central to the social life of a songbird, and in more ways than one. In fact, the term encompasses multiple discrete situations, processes, and phases. Most common is *song-recognition learning*. Birds of both sexes and all ages learn to recognize songs and discriminate individuals on the basis of those songs. Song recognition is used in territoriality (in some species) and probably colony organization (in others); in courtship; in maintenance of pair bonds; and perhaps in other social situations (Zann, 1996). Song-recognition learning can occur quickly, within a day at most and probably within minutes. Behavioral assays have been developed to demonstrate this learning, and molecular correlates of song-recognition learning have been well established (➋ *Section 3.1*).

A more specialized aspect is *song-production learning*, or learning to sing. In many species, only males actually learn to sing, even though both sexes produce vocalizations. Song-production learning itself can be broken into two separate phases (Thorpe, 1958; Konishi, 1965; Nottebohm, 1968; Immelmann, 1969; Marler and Peters, 1977; Marler and Waser, 1977; Marler, 1997). First is a "sensitive period" during which an initial auditory memory is formed of the song of a tutor, typically the bird's father. This is followed by a "sensorimotor" period during which the bird attempts to replicate the model song on his own through a

Figure 9-1

Two views of the song system. a: From the original description of elements in the central nervous system responsible for control of song production (Nottebohm et al., 1976). Reprinted with permission. b: After a recent review of elements now associated with song perception and performance (Theunissen et al., 2004). Brain regions in this panel are identified using the new revised nomenclature (Reiner et al., 2004b). Areas in gray represent components of the primary motor control pathway, the output of which proceeds from RA "to motor organs" (via the nucleus of the 12th tracheosyringeal nerve, as shown in Panel a). Unfilled areas represent areas associated primarily with auditory processing. The dotted lines (with arrowheads) connect the nuclei of the Anterior Forebrain Pathway (AFP)

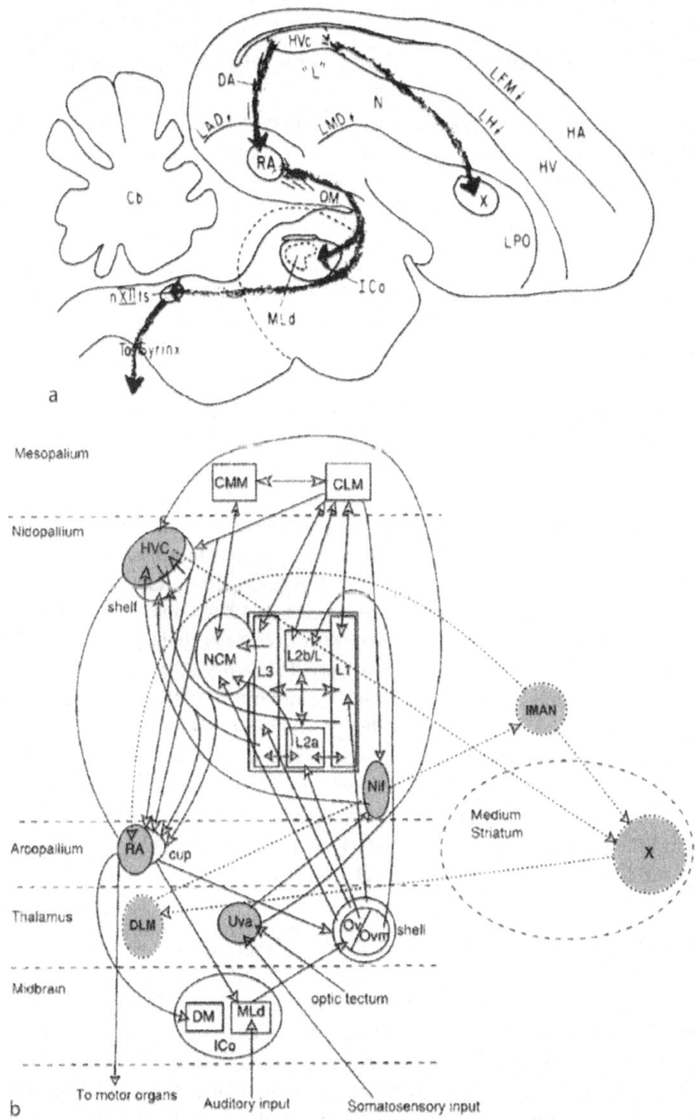

process of trial and error. These two periods may be separated in time by months as in the case of the swamp sparrow (Marler and Peters, 1982) or nightingale (Geberzahn and Hultsch, 2003), or they may closely overlap as in the case of the zebra finch (➲ *Section 3.2*). In many species, including the zebra finch, this entire process occurs only once during a restricted period in juvenile development; with sexual maturity the song "crystallizes" and is retained with little change throughout the rest of the bird's life (Price, 1979; Marler and Peters, 1987).

As models for analysis of learning, both song recognition and song production share several advantages. Both are forms of natural learning; they occur spontaneously and are deeply important to the individual's social life, survival, and reproduction. The primary sensory stimuli – songs – can be easily manipulated (e.g., through sound recording, editing, and playback). The learned responses are easily recorded and quantified.

1.3 Reason #3: Circuit Formation, Brain, Sex, and Neurogenesis

The song-control system, especially of the zebra finch, has emerged as an exceptional model for the study of a range of fundamental issues in developmental neurobiology. There are several reasons for this. The heart of the circuit, the projection from HVC to RA, develops long after the rest of the brain is anatomically mature, making it easy to study against a relatively stable developmental background (Konishi and Akutagawa, 1985; Konishi and Akutagawa, 1988; Mooney and Rao, 1994). The HVC–RA pathway is never even completed in the female zebra finch, reflecting the sharp sex difference in singing behavior in this species and setting the stage for fundamental analysis of the signals and mechanisms associated with regulated circuit formation (Holloway and Clayton, 2001). There are also major differences between the sexes in the volume of the song-control nuclei (Wade and Arnold, 2004).

The song-control nuclei are very responsive to sex steroids (Ball and Balthazart, this volume). Almost half the cells in HVC and LMAN of adult males contain androgen receptor proteins, which are also present in RA (Arnold and Saltiel, 1979; Arnold, 1980; Nordeen et al., 1986; Bottjer, 1987; Nordeen et al., 1987a; Balthazart et al., 1992; Nastiuk and Clayton, 1995). Both projections from HVC (to RA and Area X) include a substantial proportion of androgen-binding neurons (Sohrabji et al., 1989; see also Gahr et al., 1996; Gahr and Metzdorf, 1997; Bernard et al., 1999; Gahr and Metzdorf, 1999; Metzdorf et al., 1999; Maney et al., 2001; Perlman et al., 2003; Kim et al., 2004). HVC also contains estrogen receptors in neurons that project to Area X but not in those projecting to RA (Nordeen et al., 1987b; Gahr et al., 1993; Johnson and Bottjer, 1995; Gahr, 1996; Jacobs et al., 1996; Gahr and Metzdorf, 1997; Bernard et al., 1999; Gahr and Metzdorf, 1999; Metzdorf et al., 1999; Perlman and Arnold, 2003). In females, exposure to estrogen in the first weeks of life increases the size of the song nuclei, stimulates the formation of synapses between HVC and RA, and instills the capacity for song production (Gurney and Konishi, 1980; Simpson and Vicario, 1991b; Simpson and Vicario, 1991a).

Study of how the song nuclei develop and can change has already led to two unexpected discoveries that have potentially wide ramifications. First was the discovery that new neurons are continually being born in the adult avian brain (Goldman et al., 1992). This came at a time when the dogma was that neurogenesis was limited to early development in the vertebrate, and it reawakened interest in the neurogenic potential of other animals, including humans (Nottebohm, 2004). Second was the discovery that "gonadal" (sex) steroid synthesis can occur within the brain itself (➲ *Section 4*).

1.4 Reason #4: Social and Environmental Integration

With their abundant presence in nature, diversity of niches, and many different adaptations and social structures, songbirds are a natural focus for research in behavioral ecology and environmental influences on brain and behavior. Singing is associated with reproduction, and so not surprisingly, there are profound effects of season and photoperiod on form and function in the song system (Schwabl, 1996; Bernard and Ball, 1997; Kirn and Schwabl, 1997; Smith et al., 1997; Brenowitz et al., 1998; Gulledge and Deviche, 1998; Whaling et al., 1998; Tramontin et al., 1999; Reeves et al., 2003; Singh et al., 2003; Bhardwaj and Anushi, 2004; Perfito et al., 2004). These themes are reviewed in detail elsewhere (Ball et al., 2004; Ball and Balthazart, this volume) and discussed briefly later in this article (➲ *Section 5*).

1.5 Reason #5: Perspectives on Evolution and Diversity

Almost half of the more than 9,000 avian species now on Earth are songbirds, oscines of the order Passeriformes (Barker et al., 2004). This extraordinary diversification began ~65 million years ago, after divergence from nonsongbirds such as the chicken, quail, and pigeon. Thus, even the most distantly related oscines are closer to one another in evolution than, for example, many mammals. This has two consequences for the molecular neurobiologist. First, it offers an incredible range of species, habitats, and adaptation to study. Second, the relatively high relatedness may facilitate ultimate comparative genomic analyses aimed at identifying key genetic events associated with species differences; for example, in singing behavior and learning ability.

For the moment, this research potential is largely untapped. Roughly half of all song-related publications listed in PubMed focus on a single species, the zebra finch (Williams, 2004). There are several good reasons why the zebra finch has come to be favored, especially for laboratory research. Unlike the large majority of oscines, zebra finches breed well in captivity (given a warm and humid environment), and have long been a favorite of bird fanciers. Unlike seasonal species like the domesticated canary, the zebra finch breeds year-round and is comparatively insensitive to potentially confounding photoperiod effects. It has a relatively short generation time, reaching sexual maturity 3–4 months after hatching. It is relatively small (~10 g) and easy to rear. The bulk of the research reviewed here focuses on zebra finch studies.

1.6 Major Questions for Songbird Molecular Neurobiology

Each of the points above defines a focal point for research into associated molecular mechanisms. The rest of this review focuses on these five questions:

1. What molecular properties (e.g., specific gene expression) characterize the song-control system and contribute to its unique organization?
2. What molecular events accompany or determine the observable learning events (song recognition, sensory template formation, sensorimotor learning, song crystallization)?
3. What molecular events determine sex-specific developmental decisions, including the relative production and survival of new neurons?
4. What molecular processes allow external factors (social context, photoperiod) to modulate song learning and singing behavior?
5. What molecular (genetic) variations underlie species differences?

2 Molecular Characteristics of the Song System

2.1 Molecular Anatomy

The song system is unique to songbirds and anatomically distinct – are there also unique or distinguishing molecular properties? The songbird brain probably expresses at least 30,000 different gene products (Clayton and Huecas, 1990), and a definitive answer to the question awaits a thorough characterization of at least a modest fraction of these (until early 2004, sequences for only ~35 had appeared in Genbank). With the rapid development of new technologies in genomics and proteomics and their application to songbird research (reviewed in Clayton, 2004b), that day may not be so far in the future.

For now, however, we must focus on the much smaller set of gene products that have already been described in some part or all of the song system [➋ *Table 9-1*; for additional summary information and discussion, see also Clayton (1997, 2004a)]. A review of these references suggests that each song nucleus generally tends to express the gene products that are characteristic of the surrounding brain region in which it sits. No single marker has yet been found that is expressed in all song nuclei and only in song nuclei. Nevertheless, each song nucleus can be frequently distinguished from the surrounding brain by higher or lower expression of specific gene products.

Table 9-1
Described molecular components of the song system

Familiar name	Key references
Cell-surface signals and receptors	
Adrenergic receptor (Alpha-2)	Bernard and Ball, 1995; Casto and Ball, 1996
BDNF	Akutagawa and Konishi, 1998; Dittrich et al., 1999; Rasika et al., 1999; Li et al., 2000
Cannabinoid receptor	Soderstrom and Johnson, 2000, 2001, 2003; Whitney et al., 2003
Cholinergic (Muscarinic) receptor	Ryan and Arnold, 1981; Watson et al., 1988; Ball et al., 1990
Cholinergic (Nicotinic) receptor	Salgado-Commissariat et al., 2004
Dopamine receptor	Casto and Ball, 1994; Ball et al., 1995
Dopamine-beta-Hydroxylase	Mello et al., 1998
GABA synthesis (GAD-65)	Pinaud et al., 2004
Glutamate receptor, other subunits	Wada et al., 2004
Glutamate receptor (NMDA)	Many studies, reviewed: Nordeen and Nordeen, 2004
Glutamate receptor NR2A	Heinrich et al., 2002
Glutamate receptor NR2B	Basham et al., 1999; Singh et al., 2000; Heinrich et al., 2003; Singh et al., 2003
IGF I	Jiang et al., 1998
IGF II	Holzenberger et al., 1997
Melatonin receptor	Schneider, 1995; Gahr and Kosar, 1996; Whitfieldrucker and Cassone, 1996
Met-enkephalin	Durand et al., 1998
NGF	Contreras and Wade, 1999; Fiore et al., 1999
Nitric Oxide (NADH Diaphorase)	Wallhausser-Franke et al., 1995a, b
Reelin	Absil et al., 2003
TrkB	Rasika et al., 1999; Wade, 2000
Tyrosine Hydroxylase	Bottjer and Hewer, 1992; Soha et al., 1996
Intracellular signal transduction	
Canarigranin, HAT-14	Siepka et al., 1994; Clayton, 1997
CaM Kinase II	Singh et al., 2005
ERK, MAPK	Cheng and Clayton, 2004
MEK-1, MAPKK, HAT-5	George et al., 1994
n-chimaerin, HAT-2	George and Clayton, 1992
Neurocalcin	Veney et al., 2003
Parvalbumin	Braun et al., 1985, 1991
Protein Kinase C (PKC)	Sakaguchi and Yamaguchi, 1997; Watanabe et al., 2002; Yoshida et al., 2003
Steroid synthesis	
3-beta-HSD	Vanson et al., 1996; Cam and Schlinger, 1998; Soma et al., 2004
Aromatase	Shen et al., 1994, 1995; Saldanha and Schlinger, 1997; Saldanha et al., 1998
CYP17	Schlinger et al., 1999; London et al., 2003
retinaldehyde dehydrogenase	Denisenko-Nehrbass et al., 2000

Table 9-1 (continued)

Familiar name	Key references
Steroid receptor	
Androgen receptor	Nastiuk and Clayton, 1994, 1995; Gahr and Wild, 1997; Bernard et al., 1999; Metzdorf et al., 1999; Perlman et al., 2003; Kim et al., 2004
Estrogen receptor-alpha	Jacobs et al., 1996
Estrogen receptor- beta	Bernard et al., 1999; Perlman and Arnold, 2003
Structural (misc.	
Brain Lipid Binding Protein	Rousselot et al., 1997
Myelin components	Herrmann and Bischof, 1986; Kafitz et al., 1992; Campagnoni et al., 1994; Clayton, 1997; Siepka, 1997
Neurofilament (medium MW)	Siepka and Clayton, 1995; Clayton, 1997; Siepka, 1997
Synaptosomal	
GAP-43	Sakaguchi and Saito, 1996; Clayton, 1997
SNAP-25	Voigt et al., 2004
Synaptoporin	Voigt et al., 2004
Synelfin, alpha-synuclein, HAT-3	George et al., 1995; Jin and Clayton, 1995; Hartman et al., 2001
Transcription coactivator/modulator	
CBP	Auger et al., 2002
SRC-1	Charlier et al., 2003
Transcription (DNA-binding factor	
c-fos	Kimpo and Doupe, 1997; Bailey et al., 2002
c-jun	Nastiuk et al., 1994
c-myc	Collum et al., 1991
CREB	Sakaguchi et al., 1999
FoxP2, FoxP1	Haesler et al., 2004; Teramitsu et al., 2004
n-myc	Collum et al., 1991
zenk	Mello et al., 1992 and other references in *Sections 2* and *3*
Unknown Function	
SNAg	Akutagawa and Konishi, 2001

To illustrate, consider the example in *Figure 9-2.* This shows the level of expression in four song nuclei of a protein initially called synelfin in the songbird (George et al., 1995). Synelfin is recognized as the ortholog of alpha-synuclein, a presynaptic protein associated with neurodegenerative disease in humans (George, 2001). Synelfin is a striking marker of these four nuclei by its relative *absence* compared with the surrounding brain. This absence is not constitutive as synelfin mRNA is transiently expressed in juvenile LMAN, with the accumulation of the protein in axonal terminals in RA up to the age of ~45 days (Jin and Clayton, 1997b). In sparrows, more subtle changes in gene expression also occur in Area X in response to testosterone treatment (Hartman et al., 2001). The protein is believed to have some role in presynaptic plasticity, but the functional significance of its early expression and then virtual removal from the song system remains to be established.

A complementary example is seen in the expression of the gene encoding retinaldehyde dehydrogenase (zRalDH), which is responsible for the synthesis of retinoic acid (Denisenko-Nehrbass et al., 2000). Very high expression was observed rather specifically in the HVC neurons that project to Area X. Developmental expression was observed in RA, peaking at day 38 and declining by day 60. Pharmacological interference with retinoid synthesis in HVC beginning at day 30–35 disrupted normal song crystallization (Denisenko-Nehrbass et al., 2000).

Figure 9-2

Synelfin (alpha-synuclein) immunoreactivity demarcates song nuclei (from George et al., 1995). Panels show immunostaining of parasagittal canary brain sections using monoclonal antibody H3-C, revealed by immunofluorescence (where light indicates presence and dark indicates absence of the synelfin protein). a: Song nuclei LMAN (*upper arrow*) and Area X (*lower arrow*) are revealed by relative absence of immunoreactive synelfin. b: Song nuclei HVC (*upper arrow*) and RA (*lower arrow*) are revealed by relative absence of immunoreactive synelfin; note also the evident fibers (decreased staining) extending from HVC to RA. c: Higher power view of the boundary between interior of RA (*lower right half*) and the surrounding arcopallium (*upper left half*), as revealed by synelfin immunoreactivity. In all panels, dorsal is up and rostral is to the left. Bar = 1 mm (panel a), 1 mm (panel b) and 20 μm (panel c). Reprinted with permission

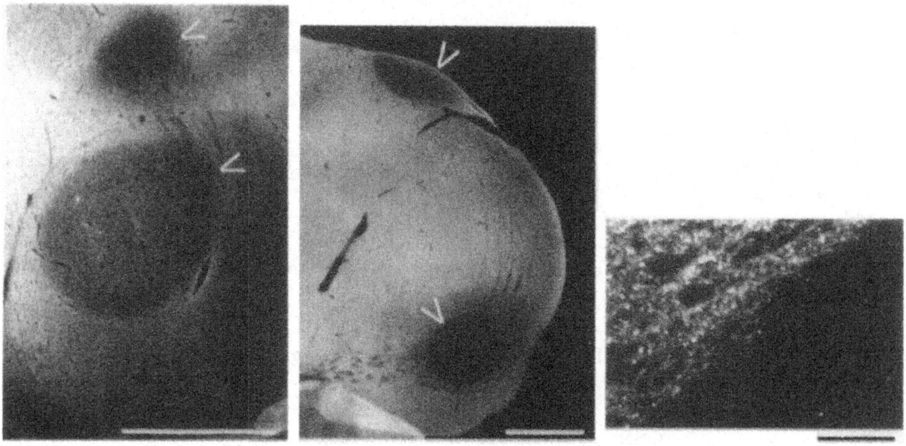

A third example is provided by a monoclonal antibody epitope referred to as song system nuclear antigen (SNAg) . It reacts rather specifically with some of the neuronal cell nuclei in several of the song nuclei, but apparently only in the zebra finch and other members of the Estrilidine family of songbirds. Its expression increases between days 20 and 40 and only in males or in females masculinized by early estrogen treatment (Akutagawa and Konishi, 2001). The antigen has not been identified.

These three examples show that molecules that show striking differential expression in the song system can be readily identified. However, as in most of the examples found so far, each pattern is distinct. Synelfin/alpha-synuclein is selectively absent from song nuclei, except early in the critical period. SNAg is selectively present in a different subset of song nuclei, but only in males and perhaps only in one family of songbirds. ZRalDH is selectively present in another subset of song nuclei (especially HVC) and developmentally regulated in RA. The example of zRalDH in particular illustrates the power of an experimental approach that uses differential gene expression within the song system to identify regulated molecules; follows with the characterization of the predicted proteins; and finally employs localized pharmacological interference to move from molecular anatomy to physiological function.

2.2 Functional Anatomy Revealed by Gene Activation

2.2.1 *ZENK* and Other Immediate Early Genes

Most of the gene products listed in *Table 9-1* display relatively stable patterns of expression in the adult brain. Notable exceptions to this are found in the list of transcription factors. Three of the genes in this category (*c-fos, c-jun, zenk*) are well-established examples of "immediate early genes" (IEGs). IEGs show rapid changes in transcription in response to cell stimulation and activity (Lanahan and Worley, 1998).

A pulse of IEG expression has been likened to a "genomic action potential," and may be a means by which neurons integrate information over periods of minutes to hours and modulate long-term information storage (Clayton, 2000). Links between IEG expression and song memory formation are considered further in ➋ *Section 3* (see also Ribeiro and Mello, 2000). Apart from their functional role in memory formation, IEGs have proven very useful in identifying sites in the brain that are physiologically active during specific behaviors or in response to specific sensory stimuli (Mello, 2002). In songbirds, this was first demonstrated in studies of the gene known as *zenk* (Mello et al., 1992), an acronym for *zif-268, egr-1, ngfi-a, krox-24*—all different names assigned to the same gene in mammals (Milbrandt, 1987; Christy et al., 1988; Lemaire et al., 1988; Sukhatme et al., 1988).

2.2.2 IEG Activation by the Sound of Song

The sound of birdsong triggers a sharp pulse of *zenk* expression in various parts of the auditory system of songbirds (Mello et al., 1992; Mello and Clayton, 1994). This can be observed in awake, unrestrained birds, even outdoors in natural populations (Jarvis et al., 1997). Changes in gene expression can be easily measured in histological sections of brain tissue, using in situ hybridization (to measure the mRNA) or immunocytochemistry (to measure the protein). In unstimulated birds, few cells in the auditory forebrain show much evidence of *zenk* mRNA, but the fraction of *zenk*-expressing cells increases to 30–90% after the bird hears a complex conspecific song stimulus (Mello et al., 1992). From the results of double-label analyses, the population of responding cells includes both GABAergic and non-GABAergic neurons (Pinaud et al., 2004). The mRNA peaks in cells about 30 min after the onset of stimulus (Mello and Clayton, 1994; Mello et al., 1995) and the protein follows, peaking after 1–2 h (Mello and Ribeiro, 1998). As little as a single presentation of a 2-s song will elicit a detectable *zenk* response, developing on this same slow timecourse (Kruse et al., 2000).

Zenk gene induction by song was among the first demonstrations of gene activation in the brain driven by a natural behavioral stimulus, and the discovery prompted a significant shift in thinking and research about song representation systems in the songbird brain. Before this, the focus of song perceptual research had been almost exclusively on the distinctive song-control nuclei that are now appreciated more for their role in male song production. Mello et al. (1992) showed that the sound of song playbacks had no detectable effect on the *zenk* expression within the song-control nuclei. Instead, the responsive regions defined an interconnected network distinct from but communicating with the classical song-control nuclei (➋ *Figure 9-3*). The most intense responses were observed in NCM and CMM, and these areas have become the focus of ongoing research into the mechanisms of song learning (➋ *Section 3*).

Within NCM and CMM, variations have been observed in the regional density of *zenk*-expressing cells depending on the stimulus, and these observations may give clues to the internal functional organization of the auditory representation system. For example, in canaries, distinct bands of cells show relative preferences for different frequencies and acoustic features in natural vocalizations (Ribeiro et al., 1998). In female starlings, the ventral portion of NCM specifically shows heightened responsiveness to long male songs, which the birds also prefer by behavioral criteria (Gentner et al., 2001). However, it would be premature to draw broad generalizations about the functional organization of NCM/CMM based on the present results, as different patterns have been observed in different species and with various stimuli. For example, in a study of chickadees, the dorsal portion of NCM and not the ventral portion showed greater response specificity (Phillmore et al., 2003). In house finches (Hernandez and MacDougall-Shackleton, 2004) and white-crowned sparrows (Maney et al., 2003), CMM showed more specificity than did NCM for conspecific song over heterospecific song. In female zebra finches, the opposite result was observed, with conspecific specificity more evident in NCM than in CMM (Bailey and Wade, 2003).

2.2.3 IEG Activation in Other Contexts

Activation of IEGs has also been used to identify and map patterns of brain activation associated with other behavioral or perceptual contexts, including song production and sexual imprinting. The act of singing

Figure 9-3
Sites of *zenk* response to song stimulation (after Mello et al., 2004). Gray areas represent regions where the *zenk* gene is induced in response to conspecific song presentation; thick solid arrows represent auditory pathways and broken arrows represent connections of the motor pathway for song production. Thin solid arrows represent minor or to-be-confirmed projections. Abbreviations: CMM and CLM, medial and lateral portions of the caudomedial mesopallium; DM, dorsomedial nucleus of the intercollicular complex; L1, L2a, L2b, and L3, subdivisions of field L; midb, midbrain; MLd, nucleus mesencephalicus lateralis, pars dorsalis; NCM, caudomedial nidopallium; NIf, nucleus interfacialis; Ov, nucleus ovoidalis; CSt, caudal striatum; RA, nucleus robustus arcopaliallis; telen, telencephalon, tha, thalamus

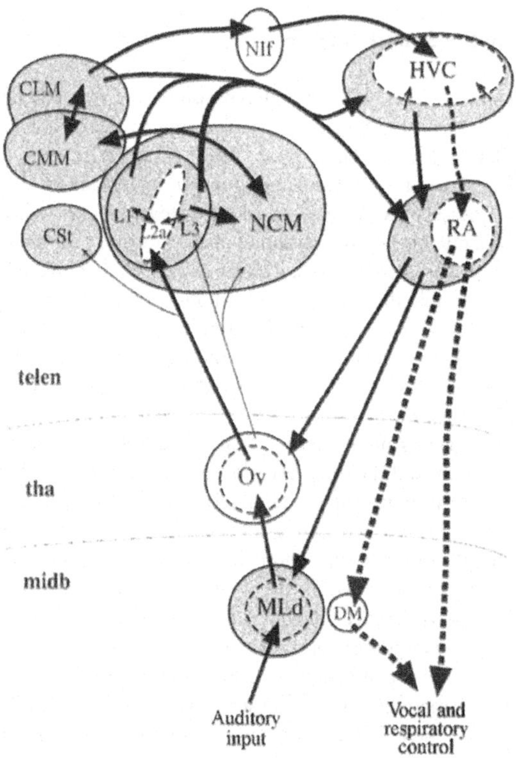

induces IEG expression in the primary song-control nuclei, including HVC, RA, LMAN, and Area X (Jarvis and Nottebohm, 1997; Kimpo and Doupe, 1997; Jin and Clayton, 1997a). Of great interest, the pattern of expression varies with developmental and contextual factors, which again may provide clues about the internal functional organization of the system. Within RA, the *zenk* response varies with developmental age in the zebra finch, becoming more restricted with the progress of the juvenile song-learning period (Jin and Clayton, 1997a). In adults, the response is much more widespread throughout the song system when the bird is singing an undirected song instead of presenting a song to a female (Jarvis et al., 1998). The significance of these last two examples is not known, but in both cases one notes a correlation of broader *zenk* activation with periods of more acute song rehearsal.

2.2.4 IEG Activation in the Songbird Hippocampus

In other studies, *zenk* and *c-fos* have been used to measure lateralized patterns of activation in the hippocampus and the visual system associated with sexual imprinting and courtship in zebra finches

(Sadananda and Bischof, 2002; Lieshoff et al., 2004; Sadananda and Bischof, 2004) and also in the hippocampus associated with food-hoarding behavior in the chickadee (Smulders and DeVoogd, 2000). Interestingly, IEG responses to song stimulation have generally not been observed in the hippocampus, with two exceptions: the experiments of Bailey and Wade on female zebra finches (Bailey et al., 2002; Bailey and Wade, 2003) and in one experiment by Kruse et al. in which the song was specifically paired with a novel visual stimulus (Kruse et al., 2004). Taken together, these few observations suggest that IEG responses in the songbird hippocampus may involve or require some multimodal or visual associations, even if an auditory stimulus is the immediate trigger.

3 Learning Mechanisms

What molecular events accompany or determine observable events in song learning? As introduced earlier (Section 8.1.2), there are at least two forms of song learning to study, song recognition and song production. Fascination with the cultural passage of song from generation to generation motivated early studies of songbird learning (Thorpe, 1958; Immelmann, 1959; Marler, 1976). A distinctive advantage in the study of song-production learning is the ability to observe a clear behavioral endpoint—the song produced by the adult bird. A disadvantage (for the molecular biologist) is the slow timecourse of its development. In the zebra finch, it can take as much as two months from the time of tutor exposure before the final song has crystallized. At the molecular level, this may mean that molecular differences associated with the learning process itself will be hard to isolate. The result, a crystallized song performance, may be the accretion of many small, transient learning events; each daily rehearsal may mark a separate short pulse of synaptic plasticity and reinforcement.

Another challenge for mechanistic analysis of song-production learning is the entanglement of multiple stages and processes. Consider this example: a young bird is exposed to two different tutors early in juvenile life. When he reaches adulthood, the song he produces is more like one tutor than the other, but a perfect copy of neither. What mechanisms led to this result? It could be that he established a more accurate and lasting auditory memory of the tutor he copied more thoroughly. It could be that the auditory memories were equivalent, but other factors led him to reproduce one over the other—maybe he simply liked one tutor better. Why was his song copy imperfect? Maybe some elements of the tutor's song were simply too hard for him to copy, or maybe the influence of the second tutor or other distractions biased his earliest attempts and the variations quickly became engrained with rehearsal. In other words, song-production learning is an integration of multiple processes—sensory template formation, motor learning, feedback-mediated sensory-motor integration, selective attention—which may be distributed in multiple neuroanatomical loci and play themselves out slowly over weeks and months.

Song-recognition learning occurs on a much quicker timescale, and may be a more tractable problem for molecular and biochemical analysis. Even so, if behavioral measures are used to score learning, again there are confounding results (Cahill et al., 2001). Song responses may be affected not just by learning and familiarity, but also by complex factors such as motivation and attentional state. The discovery of sharp changes in gene expression in the auditory system of songbirds hearing song playbacks (➋ *Section 2.2.2*) ignited interest in recognition learning partly because it opened the door to direct observation of neurophysiological "learning," at least for the moment bypassing some of the ambiguities of behavioral analysis. Of course, the ultimate goal is to link behavior and neurophysiological function. Below we begin with a review of the behavioral and molecular biology of song recognition, and then return to the more complex topic of song production learning.

3.1 Song-Recognition Learning

3.1.1 Field Observations

Songbirds use song (and more broadly, vocal communication) for many purposes. At least among the species most commonly studied, there is clear evidence that songbirds learn to recognize and discriminate between individual birds by the songs they sing. Females recognize the song of their mates (Miller, 1979a; Lind et al., 1996) and fathers (Miller, 1979b) and respond more to strangers' songs if they sound more like their fathers' (O'Loghlen and Beecher, 1999). Male song sparrows, a territorial species, not only recognize their neighbors but come to know their individual repertoires and display their knowledge in a form of countersinging called "repertoire matching." The countersinging bird responds with a different song from the one presented, but one that is also contained in the repertoire shared between the two individuals (Beecher et al., 1996; Beecher et al., 2000). Birds also learn to associate a particular song with a particular context, and modify their territorial responses to the song if the context of its presentation changes (Brooks and Falls, 1975; Falls, 1982; Stoddard, 1996).

Zebra finches are an interesting case because they are colonial, and there is no known territorial function for song in zebra finches. Yet, as Zann summarizes in his thorough review (Zann, 1996), song is used by zebra finches on a daily basis to recognize mates, kin, and potentially other members of the colony. Mate recognition is important for maintenance of the male–female pair bond, especially during the looser bonding period necessary for egg incubation and brooding. Fledglings recognize their parents by their vocalizations, and preferentially associate with their siblings (Burley et al., 1990). Although the evidence is anecdotal, there is reason to think that song may be used to maintain the dynamic organization of foraging groups, within which preferential mate/kin associations are locally maintained. In addition, song may be used to learn about new arrivals in the colony (a daily occurrence in the wild, due to emigration and "extended local excursions"). Theoretically, the structure of a song may convey not only the identity of the singer, but also socially relevant information such as the singer's age and geographic origin. Individual nests within a socially cohesive colony can lie as far away as a kilometer from the center, and yet the birds commute throughout the colony and congregate and assort daily, for flock foraging and other social activities (Zann, 1996).

3.1.2 Laboratory Studies of Song-Recognition Learning

Two threads run through the literature on song-recognition learning. One emphasizes the effects of early juvenile experience on later preferences; the other emphasizes day-to-day learning in adulthood. In both cases, it is worth keeping in mind that a bird's behavioral response to a song (a "preference" or "recognition") may depend greatly on the context in which the song is heard, including the hormonal status of the listener. A bird may apply different criteria depending on whether the song is a stimulus for mating, affiliation, or competition.

For studying juvenile preference learning, a common approach has been to focus on females, who respond to preferred songs with sexual displays that are easy to quantify (in this case, the preference index is specifically related to mating). Female zebra finches will perform more sexual displays for songs they were tutored on early in life, even preferring songs of another subspecies if that is what they were trained on (Clayton, 1990). Riebel (2000) tested whether early exposure to a tape-recorded song was sufficient for inducing a stable preference in female zebra finches, using an operant task in which pecking of response keys led to song playback. Tutored females repeatedly preferred the song they were tutored on. In a related study (Riebel et al., 2002), both males and females showed clear and equivalent preferences for their father's song in an operant training assay (in which song playback itself was the reward for key pecking). The amount of variability was as great between siblings as between nest groups, suggesting that song preferences developed independent of inherited factors or quality of father's song. It is notable that song recognition in the zebra finch does not show a strong sex bias despite the essentially absolute sex difference in song-production ability in this species. Interestingly, although effects of early experience on song preferences have been demonstrated in some other species [e.g., canaries (Nagle and Kreutzer, 1997) and white-crowned sparrows (Casey and Baker, 1992; Macdougall-Shackleton et al., 2001)], not all species show evidence of early preference learning. Examples where preferences (e.g., for local dialects) appear to be independent of early experience include the house finch (Hernandez and MacDougall-Shackleton, 2004), the swamp sparrow (Balaban, 1988), and chaffinch (Riebel and Slater, 1998).

In addition to preferences based on juvenile experience, adult zebra finches continue to form new memories and associations based on the songs they hear later in life. In response to operant conditioning, naïve adult zebra finches learn to perform a song discrimination task over a period of 1–8 days (Cynx and Nottebohm, 1992). However, once the birds learn the task (i.e., to respond to the right song to get a food reward), they can learn to discriminate songs within just a few hours (Shy et al., 1986; Weary, 1989; Beecher et al., 1994; Jarvis et al., 1995; Braaten, 2000; Benney and Braaten, 2000). Thus, adult zebra finches apparently have the capacity to form memories and associations linked to specific songs they hear within a single day.

The observation of rapid induction and stimulus-specific habituation of gene responses to song playbacks also implies the formation of stable song-specific memories in a matter of hours (➋ *Section 3.1.2*). A simple "listening" assay was developed as a way to measure behavioral correlates of genomic habituation, without introducing any explicit reinforcement or operant training requirement (Stripling et al., 2003). Zebra finches show visible evidence of attending to a song playback, when song playback first begins. They perch in one place, stop vocalizing, and orient toward the speaker from which the song is coming. The first time a song is played, this "listening behavior" may be sustained for 10–20 min or even longer. After the bird has been exposed to a series of playbacks of the same song, however, subsequent presentations of the same song elicit a much shorter listening response. The term "habituation" has been used to describe the shortening of the listening response after a test song has been repeated. The development of behavioral habituation correlates with habituation of IEG responses to the song, as considered in the next section.

3.1.3 Role of Gene Regulation in Song-Recognition Learning

Thus song-recognition learning has been observed on two different time scales: across juvenile development, and in the course of a single day in adult life. IEG responses have been implicated in both, but in contrasting ways. Put simply, experiments on learned preferences emphasize the use of IEG activation as an indicator of past learning completed. Experiments on adult song habituation emphasize IEG expression as a measure of new learning in progress. The goal of this section is to synthesize the diverse observations (➋ *Table 9-2*) into a coherent mechanistic account of the function of IEG activities associated with learning and memory.

Workers focused on early preference learning have tended to regard the adult IEG response as an analog or correlate of song preference—which is to say, past learning or "memory strength" [e.g., Bolhuis et al., 2000; Bolhuis et al., 2001; see also Ball and Balthazart (2001), for a variant of this idea, emphasizing the relationship to "sign" stimuli]. The "memory strength" interpretation rests on two assumptions: (1) that prior learning leads to increased activity in the system that represents the learned object or association; (2) that the amount of IEG expression is proportional to the amount of that representational activity. Neither of these points is particularly well established on its own. Indeed, some evidence in other systems indicates that learning leads to a more efficient representation that requires less metabolic activity (Chialvo and Bak, 1999), and there are a number of examples where IEG expression clearly diverges from cellular metabolic or synaptic activity (Clayton, 2000).

Nevertheless, the "memory strength" argument has been propelled by observations such as those of Bolhuis and colleagues, who found a correlation between the magnitude of gene response to tutor song playbacks and the fractional representation of tutor song elements in an adult zebra finch's own song (Bolhuis et al., 2000; Bolhuis et al., 2001; Terpstra et al., 2004). When the bird's own song (BOS) was used as the stimulus, there was no significant correlation between the magnitude of each bird's response and the similarity of the BOS to the original tutor song. Thus, the birds who copied the tutor song most closely showed selectively greater IEG responses to the tutor song than did the other birds.

Bolhuis et al. concluded that tutor song is better represented in the birds with the higher responses, and the magnitude of the response reflects this degree of representation. An equally plausible possibility not considered is that the birds who originally copied the tutor song did so better because they liked it (or the

Table 9-2
Two views of the link between IEG expression and learning

Reference	Species	"Learning completed"		"Learning in progress"	
		Correlation with behavioral response preference	Effect of prior "critical period" exposure	Effect of immediate contextual novelty	Correlation with new behavioral learning
Mello et al., 1992	zebra finch, canary	Con > hetero			
Bailey et al., 2002	zebra finch	Con > hetero			
Hernandez and MacDougall-Shackleton, 2004	house finch	Con > hetero	suppressing		
Maney et al., 2003	white-crowned sparrow	Local > foreign	enhancing		
Gentner et al., 2001	starling	Sexual prefs			
Phillmore et al., 2003	black capped chickadee	Song > call	no effect		
Eda-Fujiwara et al., 2003	budgerigar	Complex > simple			
Jin and Clayton, 1997a	zebra finch		enhancing		
Bolhuis et al., 2000	zebra finch		enhancing		
Bolhuis et al., 2001	zebra finch		enhancing		
Terpstra et al., 2004	zebra finch		enhancing		
Sockman et al., 2002	starling		enhancing		
Jarvis et al., 1995	canary			enhancing	yes
Mello et al., 1995	zebra finch			enhancing	Stripling et al., 2003
Kruse et al., 2004	zebra finch			enhancing	

The table shows how reports of IEG responses to song stimuli have emphasized two different functional interpretations. Some reports have emphasized relationships to innate or learned preferences ("learning completed"). Others have emphasized the recent novelty of the stimulus and the context in which it is experienced ("learning in progress"). See text for further discussion and reconciliation of these interpretations.

tutor) more, and this relative preference is reflected in greater "strength of activation" to that stimulus in adulthood. In other words, some birds invest the tutor's song with greater significance, and this may be the true proximal correlate of IEG response magnitude.

Workers focused on acquisition of new song memories in adulthood have regarded the IEG response as an indicator of active, current learning ("learning state"). This interpretation, which is developed in detail elsewhere (Clayton, 2000), is rooted in a major tenet of learning and memory research, that a wave of new gene expression is necessary to consolidate labile short-term memories into stable long-term memories (McGaugh, 2000; Kandel, 2001; Sweatt, 2003). In this view, induction of IEGs by song stimulation is an essential part of the brain's mechanism for storing new information—either memories of the song itself, or new associations *about* that song and the context in which it was heard.

An immediate impetus for extending this view to songbird research is the phenomenon of song habituation (➋ *Figure 9-4*). When one song is played repeatedly over several hours or days, *zenk* mRNA levels in NCM/CMM return to the baseline measured in unstimulated controls, and subsequent presentations of that song no longer elicit any increase in *zenk*. However, a new round of *zenk* expression occurs when the bird is presented with a different song (Mello et al., 1995). This result was extended by Kruse et al. (2004) to show that the habituated song can reinduce a *zenk* response if the context of its presentation is changed (position in space, association with novel visual elements). This suggests that IEG expression is driven not so much by the novelty of the song itself as it is by the novelty of the context in which the song is heard. The study of Jarvis et al. (1995) also attempted to place the *zenk* response in the context of associational learning and generated a similar observation: a *zenk* response to habituated song can be restored by pairing the habituated song with an aversive shock, even though shock alone has no detectable effect on *zenk* levels in NCM/CMM.

Electrophysiological responses in NCM (measured as spike frequency) also undergo changes described as "habituation" when a song is repeated (Chew et al., 1995, 1996; Stripling et al., 1997). The habituation is highly selective for the repeated song. It takes approximately one half hour from stimulus onset before spike habituation is stable (Stripling et al., 1997), and the habituation is not sustained if protein synthesis inhibitors are injected at the recording site (Chew et al., 1995). Taken together, these observations are consistent with a requirement for new protein synthesis (IEG expression) in the consolidation of synaptic change in NCM/CMM. Importantly, these changes may underlie formation of memories or associations *about* the song, and not necessarily *of* the song's acoustic pattern (Jarvis et al., 1995; Kruse et al., 2004).

How then to reconcile these two interpretations, "memory strength" versus "learning state?" If IEG induction is all about the formation of new memories, why does it appear to be so clearly correlated with past learning and behavioral preference (➋ *Table 9-2*)? Subtly, the two perspectives can be resolved by recognizing that preferred stimuli are also the ones that will trigger the most attention and provide the most salient motivation for new learning and associations. The magnitude of the IEG response intrinsically integrates past experience, current context, and any innate perceptual filtering. The outcome of this integration can be thought of as an indicator of *significance*. Significance is highest if perceptual filtering indicates that the stimulus is conspecific (Mello et al., 1992); if the stimulus has been associated with significant others (such as Father or Tutor) earlier in life (Bolhuis et al., 2001); and if the stimulus is immediately associated with a strongly aversive (Jarvis et al., 1995) or an unexpectedly novel context (Kruse et al., 2004). Thus, both interpretations capture part of the dynamic reality: IEG expression has a function in new memory formation, but its appearance is also correlated with old memory strength because old memories are the scaffolding upon which new learning is constructed.

3.1.4 Other Molecular Mechanisms in Song-Recognition Learning

If IEG expression is part of the mechanism underlying song-recognition learning, how does it carry out this function? In answer to this fascinating question, at this point one can offer only broad hypothetical models (Clayton, 2000) and a few specific observations. The most familiar IEGs, including *zenk*, encode DNA-binding transcription factors (Lanahan and Worley, 1998). These are presumed to have their effects by activating or repressing other genes, and indeed a number of potential target genes under apparent regulation by the *zenk* protein have been identified, including genes that encode synaptic proteins [see Knapska and Kaczmarek (2004) for a recent review]. However, not all IEGs are transcription factors, and some may have immediate effects on neuronal form and growth. For example, the gene for the growth factor BDNF responds as an IEG and is induced in HVC by singing activity (Li et al., 2000). Understanding how IEG activity might affect memory formation requires information about two points that for the moment lie out of reach: a functional "proteomic" analysis of all the regulated proteins and how they interact (e.g., O'Donovan et al., 1999); and more insight into the precise synaptic structure of song representations and associations within the song system.

What are the proximal cellular signals that determine whether a given pattern of activity induces an IEG response? Regulation of common IEGs (and specifically *zenk*) in other systems has been extensively

Figure 9-4

Song-specific habituation of the *zenk* response in NCM/CMM (reprinted with permission from Mello et al., 1995). Zebra finches were presented with a repeated song stimulus for a 2.5 h training period, immediately followed by a second test stimulus for another 30 min. The combined training-test protocol for each experimental group is indicated (first stimulus/second stimulus). S1 and S2 represent the specific songs of two different individuals from another aviary, and 0 represents silence. a: Representative in situ hybridization autoradiograms of parasagittal brain sections (200 μm from midline). The diagram on the upper right indicates the visible structures: Cb, cerebellum; H, hyperpallium; Hp, hippocampus; LPO, lobus parolfactorius (now referred to as medial striatum); NCM, caudomedial nidopallium (also including a portion of CMM, caudomedial mesopallium, not labeled). Scale bar, 3 mm. b: Quantitative analysis of *zenk* hybridization signal in NCM. Each dot represents *zenk* mRNA abundance in NCM in one animal, and the bars represent means for each experimental group

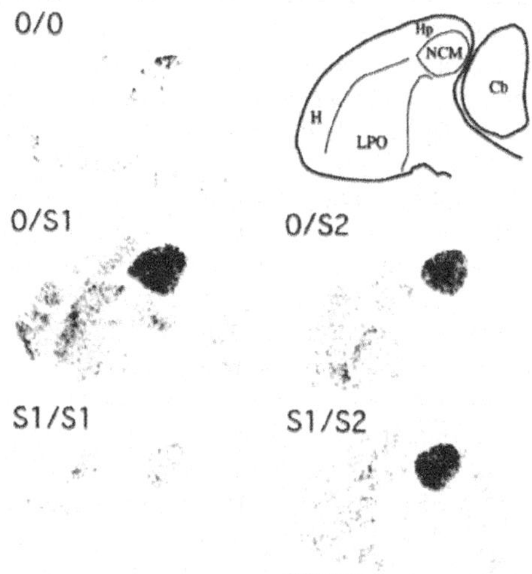

a

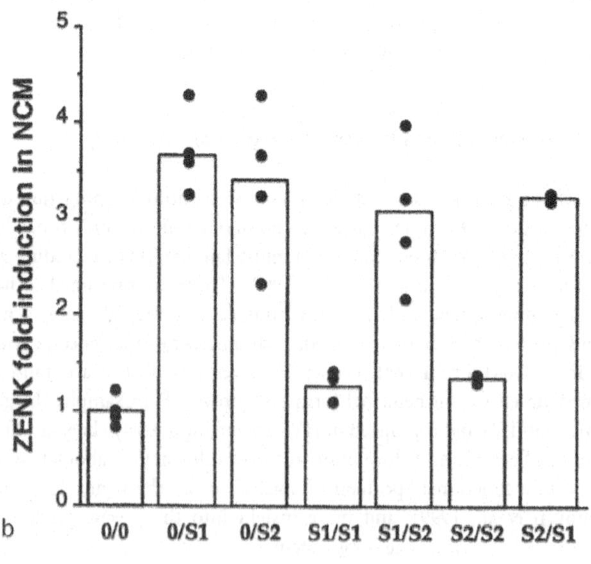

b

reviewed elsewhere (Lanahan and Worley, 1998; Davis et al., 2003; Knapska and Kaczmarek, 2004). The proximal cause of *zenk* gene transcription in the rodent brain appears to be the phosphorylation of constitutively expressed transcription factor proteins, which bind to specific recognition sites in the gene's promoter; the critical transcription factors include Elk-1 and CREB. Elk-1 and CREB phosphorylations are caused by ERK (MAP Kinase), and a complex network of signaling pathways couples the stimulation of various cell-surface receptors (e.g., for neurotransmitters) to functional activation of ERK.

In songbirds, a critical role for ERK in *zenk* gene expression has also been demonstrated recently (Cheng and Clayton, 2004). Song presentations resulted in a very rapid pulse of ERK activation measured by immunoblotting using an antibody specific for the activated form of the protein. Pharmacological inhibition of ERK activation in NCM/CMM blocked the *zenk* response to song. Interestingly, the activation of ERK was transient, reaching a sharp peak in less than 2 min from song onset. ERK activation also showed stimulus-specific habituation, analogous to that described earlier for both *zenk* expression and electrophysiological spike rate. Other evidence suggests that ERK activation by song depends on stimulation of the noradrenaline/α_1-receptor (Ribeiro and Mello, 2000; Cheng, 2003). These results suggest that the *zenk* gene response to a song can be influenced by diffuse neuromodulator systems, thus defining a pathway by which information about context and significance might be integrated with the primary song representations generated in the activity of neurons in NCM/CMM (Clayton, 2000).

Evidence for CREB activation following song stimulation has also been presented, but the evidence is difficult to interpret (Sakaguchi et al., 1999). The authors reported that conspecific song presentation selectively induced an increase in pCREB in adult male zebra finches after 30 min of stimulation. Surprisingly, this effect was reported for HVC and not NCM/CMM. The authors did not offer any comment or show any data on CREB or pCREB in NCM/CMM, but the absence of evident labeling in the shelf and nidopallium surrounding HVC in the photomicrographs suggests a lack of any effect. In HVC, they localized CREB to the cells that project to Area X, but did not observe any increase in pCREB in Area X itself. Nor did they observe any increase in pCREB in HVC associated with singing activity. These results thus contradict expectations on a number of fronts. There is scant evidence that zebra finches retain significant sensory plasticity related to song production in adulthood (see ➲ *Section 3.2.4*). The sound of song has not been shown to induce other plasticity-associated activities in HVC (e.g., *zenk*, *c-jun*, *c-fos*, phospho-ERK). Conversely, singing (but not listening) has been shown to induce *zenk* and *c-fos* in the HVC of adult singers, which may be interpreted in the context of ongoing engrainment of motor patterns (Lombardino and Nottebohm, 2000). Given the small group sizes ($n = 3–5$) and the unspecified criteria for judging cells as immunopositive, replication of this intriguing observation seems especially important.

Another rather surprising enzyme activity has been observed in NCM/CMM following novel song presentation: song presentation has been observed to trigger an increase in caspase-3 enzyme activity, specifically in or near post-synaptic densities (Huesmann and Clayton, 2002; Huesmann and Clayton, submitted). The activity peaks about 10 min after song onset; it shows song-specific habituation much as observed for *zenk*, ERK, and electrophysiological spike rate; and pharmacological inhibition of it appears to block *zenk* habituation. Caspase-3 activation is commonly thought of as the terminal step in the commitment of cells to apoptotic death, but there is no evidence that song presentation triggers apoptosis. The phenomenon is currently under investigation as a potential mechanism for learning-related synaptic modification.

3.2 Song-Production Learning

A male zebra finch's song performance is refined through practice over a critical period spanning the second and third months of life after hatching, and auditory feedback throughout is required for normal song development. He begins to practice singing at (30–40 days of age and develops a stable song pattern by 65 days (Arnold, 1975; Bottjer et al., 1984). During this time, he continues to be receptive to influences from additional tutors (Slater et al., 1991), but once adulthood is reached (approximately day 90), his song is fully crystallized and normally will not change significantly for the rest of his life (Arnold, 1975; Price, 1979).

A deep understanding of song-production learning will require knowledge of how specific song patterns are represented in the neurophysiological activity of the song system. It seems very likely that there are multiple representations (e.g., tutor song, maps for sensorimotor matching, crystallized motor performance), perhaps distributed and woven throughout the system and interacting differently as the animal passes through different states of arousal and maturation. The greatest progress has been made in working out the respective roles of spike activity in HVC and RA in sequencing the elements of a mature vocal performance (Margoliash, 1997; Fee et al., 2004; Rose et al., 2004; Leonardo and Fee, 2005). Mechanisms for representing an auditory song memory are much less clear. Until recently, research was heavily focused on examination of auditory responses in HVC and the AFP in anesthetized birds. Neurons that responded selectively to playback of the BOS were identified. Methods for analyzing responses in awake birds have improved; however, the interpretation of BOS responses in the motor system has become cloudy (Dave et al., 1998; Hessler and Doupe, 1999a; Konishi, 2004; Leonardo, 2004), and more researchers are turning their attention to examination of the auditory regions such as NCM/CMM (for further considerations of song selectivity, see Theunissen et al., 2004).

Regarding the learning process itself, neurobiologists have been compelled by two major questions about song development: how is the system structured so that vocal performance will gradually evolve to match a template acquired from early auditory experience? And what determines the developmental boundaries of the learning process? There are reasonable ideas about how these feats might be accomplished (e.g., Adret, 2004; Bottjer, 2004; Brainard, 2004; Deregnaucourt et al., 2004; Doupe et al., 2004; Mooney, 2004), but there is little conclusive evidence about the critical molecular processes or control mechanisms. Rather, the field has seen an accumulation of bits of evidence for various factors or processes, each of which may have at least a contributing role in some aspect of song learning.

3.2.1 Boundary of the Sensitive Period

A young male zebra finch will sing the song of his father if exposed to him only until day 35, the approximate age of independence from his parents (Böhner, 1990; Funabiki and Konishi, 2003). Thus, a complete and intact sensory memory must already be formed by then. However, a young bird is still sensitive or receptive to further experience until ~day 65. If raised in the presence of his father but then isolated at day 35 and exposed sequentially to other tutors, he will copy elements of new tutor songs he hears before ~day 65 (Eales, 1985; Slater et al., 1991). This suggests that juvenile finches have the capacity to update or overwrite their initial song models during the period between days 35 and 65 (Slater et al., 1991). In general, social interactions are clearly very significant in determining whether or not a bird will copy a particular song (Clayton, 1987a, b).

Receptivity to new song models may be extended into adulthood under conditions of social isolation. Birds raised by females only, in isolation from their father or other potential adult tutors, will still copy large amounts of the father's song if first exposed to him as late as day 65 (Eales, 1987). In a more extreme case, birds raised by their mother alone and then kept in sound-proof chambers until day 120 developed very abnormal and variable songs that contained an unusually large number of elements, suggesting that crystallization had not occurred. When later exposed to a tutor (beginning at day 120) they showed some learning from the tutor (Slater et al., 1993). Similar results were obtained in another study, which also found that birds would copy elements of tutor songs presented in adulthood after they had been maintained in visual and social (but not necessarily auditory) isolation from other birds (Morrison and Nottebohm, 1993). It is not known whether conditions of extreme social impoverishment such as these show their effects by modulating normal development (i.e., "extend the critical period") or whether they force adaptations that may have no direct relationship to the mechanisms that constrain learning under normal circumstances.

3.2.2 Sensorimotor Learning

The process by which a young male zebra finch gradually refines his own stereotyped vocal performance requires auditory feedback, physical completion of the primary control pathway (HVC–RA), and an intact

AFP. Active song-like vocalizations around the 30th day after hatching correspond to the arrival in RA of axonal fibers from HVC (Konishi and Akutagawa, 1985; Mooney and Rao, 1994; Holloway and Clayton, 2001). Microstimulation of HVC and RA can elicit vocalizations (Vu et al., 1994; Vicario and Simpson, 1995), and electrophysiological analyses have correlated spike activity of individual RA neurons with specific elements of a bird's song (Yu and Margoliash, 1996). HVC neurons appear to fire sequentially in short bursts to form an explicit representation of the temporal organization of the song (Hahnloser et al., 2002). Thus, the primary excitatory drive for song production is provided by inputs from HVC, and much of the mystery of song-production learning may be transposed into questions of mechanism for adjusting or maintaining specific HVC–RA synaptic relationships.

An intact AFP is also required for normal song development (Bottjer et al., 1984; Scharff and Nottebohm, 1991), and its precise functional role is a subject of intense current investigation (Adret, 2004; Bottjer, 2004; Brainard, 2004; Deregnaucourt et al., 2004; Doupe et al., 2004; Mooney, 2004; Kao et al., 2005). The AFP can be considered a specialization within the general "cortical > striatal > thalamic > cortical" organization of processing in the vertebrate brain, important for skill and sequence learning (reviewed in Perkel, 2004). The AFP converges onto the HVC–RA pathways via an NMDA-receptor-mediated projection from LMAN onto RA neurons; these terminals are present and functional by day 25, before the arrival in RA of terminals from HVC (Mooney, 1992). Abarbanel et al. (2004) constructed a biophysical model based on observed timing delays in inputs from AFP versus HVC onto RA, and concluded that subtle variations in these timing relationships could readily shift the balance between synaptic potentiation and depression via NMDA receptors in RA.

Two examples of developmentally restricted, spike timing-dependent synaptic plasticity in the AFP have been reported (based on observations made in anesthetized animals or brain slices). Boettiger and Doupe (2001) observed NMDA-receptor-dependent long-term potentiation (LTP) of synapses from recurrent collaterals in LMAN, and long-term depression (LTD) of out-of-phase thalamic afferents. Both of these phenomena could be elicited at day 20 but not at day 60, thus corresponding to the early sensory period for song-model acquisition. This suggests the possibility that tutor song exposure might refine a representation in LMAN, which may then be used as the template for a subsequent error correction phase of sensorimotor learning. Working even further up the AFP, Ding and Perkel (2004) observed NMDA- and dopamine D1-receptor-dependent LTP in Area X, but only after day 47. This might provide a mechanism for refining, reinforcing, or maintaining the bird's own vocal performance.

A recent investigation examined the effects of microstimulation of LMAN on song performance in adult zebra finches, where the stimulus was applied only during a specific song syllable by use of a computer-controlled system for syllable detection and stimulus generation (Kao et al., 2005). Intriguingly, this resulted in acute changes in the amplitude and fundamental frequency of the song element being performed concurrently (or more precisely, within 35–70 ms of microstimulation). This suggests that activity in the AFP may contribute directly to song performance, even in adult birds. Consistent with this, a correlation was measured between variability in the multiunit firing pattern of LMAN (higher in undirected than directed song) and variability of the fundamental frequency of concurrent song elements (also higher when undirected than when directed). The authors point out that the effect of AFP activity could be instructive (by directly influencing the structure of specific song elements), and yet also supportive of a selection-based learning mechanism, in which variability in AFP activity generates variability in song performance—a necessary prerequisite for trial-and-error learning. For the molecular neurobiologist, it is especially intriguing to note that undirected song also triggers a much larger IEG response in the AFP than does directed song (Jarvis et al., 1998). This provides another, yet rather novel, association of IEG expression with neuroplasticity and learning; in this case, the plasticity is a context-dependent increase in variability of spike rate, which may provide the substrate for long-term learning by trial, error, and selection.

3.2.3 Do Songbirds Dream Song?

For most models of song learning that involve error correction at the level of HVC and RA, confounding the problem is the lack of evidence that neurons in these nuclei respond to the natural auditory feedback generated during the bird's own singing (McCasland and Konishi, 1981; Konishi, 2004). Intriguingly, however, individual neurons in RA respond to BOS playbacks when the bird is asleep with a firing pattern that is similar to the pattern they generate during wakeful singing (Dave and Margoliash, 2000). Moreover, spontaneous activity of these neurons during sleep matches their sensorimotor activity. This raises the possibility that mechanisms of critical aspects of sensorimotor learning, integration, and template-matching may occur "offline," during sleep [see Margoliash (2002), for detailed discussion].

Further support for this has recently come from a careful daily analysis of song ontogeny, which showed a fascinating contrast: song performance deteriorated after overnight sleep during periods of learning, and yet the birds showing more post-sleep deterioration ended up producing better imitations of the tutor song at the end of the learning period (Deregnaucourt et al., 2005). Do birds need to sleep, dream, and forget—the better to learn?

In humans and in other species, sleep is critical for memory consolidation (Walker et al., 2003; Walker and Stickgold, 2004) and is associated with major changes in gene expression (Cirelli and Tononi, 2000b; Cirelli et al., 2004). Some learning-associated IEGs are suppressed during sleep (Cirelli and Tononi, 2000a; Cirelli and Tononi, 2000c). Yet *zenk* itself may be activated specifically during REM sleep after periods of waking stimulation (Ribeiro et al., 1999). The role of sleep in the song-learning process is an important emerging area of research, and analysis of state-dependent changes in gene expression may be especially critical.

3.2.4 Song Crystallization

A striking feature of the mature song performance, especially in zebra finches, is its stereotypy and resistance to change. The process of song crystallization may be delayed by social isolation or auditory interference, but once a bird is given the opportunity to hear his own song in a normal social context, song moves inevitably over about a month toward a highly stereotyped performance (Funabiki and Konishi, 2003). Song crystallization may also be induced prematurely by androgen treatment (Marler et al., 1988; Korsia and Bottjer, 1991). Once crystallized, learned song is a tenacious motor memory. Changes in song pattern may not emerge for many weeks (if at all) even in birds deafened with no auditory feedback (Nordeen and Nordeen, 1992), or with active perturbation of auditory feedback (Leonardo and Konishi, 1999), or after sudden perturbation of song performance via peripheral nerve injury (Williams and McKibben, 1992).

To what degree is crystallization an active process or just the passive endpoint of some developmental trajectory? This distinction potentially has significant implications for the underlying molecular and cellular biology of neural plasticity. For example, singing activity triggers *zenk* gene expression in song nuclei, even in adult zebra finches past the age of crystallization (Jarvis et al., 1998). How can the stability of song performance be reconciled with the active recruitment of a molecular mechanism associated with plastic change (Clayton, 2000)?

One possibility is that the "plasticity" signals in the adult song system simply have the effect of reinforcing existing physiological organization, in the absence of any mismatch between template memory and current performance. What happens then if mismatch is introduced, e.g., by deafening? Brainard and Doupe (2000, 2001) found that changes in song induced by deafening could be suppressed by lesioning LMAN, the output nucleus of the AFP. Thus, the AFP is necessary not only for initial song learning, but also for song plasticity induced in adulthood by altered sensory feedback. This does not clarify whether song crystallization is truly a default state, which may be overridden by instructive signals from the AFP only in cases of sensorimotor mismatch; or whether the AFP continually provides some reinforcing signal, which only has demonstrable influence in the absence of auditory feedback [but see Kao et al. (2005)].

In a technical tour de force, Leonardo (2004) recorded from single neurons in LMAN of awake unanaesthetized adult zebra finches, and used a computer-controlled system to perturb auditory feedback during singing. The goal was to test whether spike activity in LMAN is sensitive to auditory perturbations, as one would predict from a model in which the AFP performs a comparison of auditory feedback to an auditory memory and delivers an error correction signal. Contradicting such a model, spike activity in

LMAN was insensitive to auditory perturbations. One interpretation of this result is that, at least in the adult zebra finch, the AFP provides an "efference copy" of the song, i.e., a sequence that predicts the motor patterns associated with the desired output (Troyer and Doupe, 2000a, b). A caveat is that in this study, activity was only recorded during directed song, when the hypothesized sensorimotor error correction process may be taken offline (Jarvis et al., 1998; Hessler and Doupe, 1999b). Clearly, more work needs to be done to clarify the physiological role of the AFP in song system function—but this stands as an exceptionally promising system that may eventually illuminate the relationships between neural systems design, physiological function, molecular regulation, and behavioral endpoint.

3.2.5 Gene Expression Across the Critical Period

As the previous sections should demonstrate, song-production learning is a complex affair. Moreover, it is occurring simultaneously with the primary formation of the central HVC–RA projection in the telencephalon, and in young animals as they progress from parental dependency to independence and sexual maturity. So perhaps it comes as no surprise that the list of regulated genes, proteins, and signals in the developing song system is long. As reviewed elsewhere (Clayton, 1997, 2004a), developmental changes have been described in the expression of at least 20 different gene products in the song system. A partial list appears here:

1. Androgen receptor (Arnold and Saltiel, 1979; Bottjer, 1987; Nordeen et al., 1987a; Perlman et al., 2003; Kim et al., 2004)
2. Aromatase (Perlman and Arnold, 2003)
3. BDNF (Akutagawa and Konishi, 1998), [but see Dittrich et al. (1999), Johnson et al. (2000)]
4. CaM Kinase II phosphorylation (Singh et al., 2005)
5. Catecholaminergic innervation (Sakaguchi and Saito, 1989; Bottjer, 1993; Soha et al., 1996)
6. Cholinergic innervation (Sakaguchi and Saito, 1989; Sakaguchi and Saito, 1991)
7. Cyp17 (London et al., 2003)
8. Estrogen receptor (Gahr and Konishi, 1988; Konishi and Akutagawa, 1988; Jacobs et al., 1999; Perlman and Arnold, 2003)
9. Metabolic activity measured by cytochrome oxidase (Adret and Margoliash, 2002)
10. NMDA receptors in song nuclei (Aamodt et al., 1995; Aamodt et al., 1996; Basham et al., 1996; Livingston and Mooney, 1997; Basham et al., 1999; White et al., 1999; Livingston et al., 2000; Singh et al., 2000; Heinrich et al., 2002; Heinrich et al., 2003; Singh et al., 2003; Scott et al., 2004)
11. Retinaldehyde dehydrogenase (Denisenko-Nehrbass et al., 2000)
12. Synelfin/alpha-synuclein mRNA in LMAN cell bodies and protein in presynaptic terminals onto RA (Jin and Clayton, 1995; Jin and Clayton, 1997b)
13. Trk B (Wade, 2000)
14. Zenk mRNA levels in both NCM/CMM and in RA (Jin and Clayton, 1997a)

Attempts to establish causal links between any of these changes and steps in song production learning have for the most part been unsastisfying (Clayton, 1997, 2004a). The subject of the most intensive study has been the NMDA receptor, reviewed in detail in Nordeen and Nordeen (2004). But despite its apparent developmental regulation in the song system, and the obvious potential that might have to enable or suppress song learning, the evidence does not support the hypothesis that NMDA gene regulation sets the boundaries of song learning in the zebra finch critical period (Nordeen and Nordeen, 2004).

3.3 Critical Issues for the Molecular Biology of Song Learning

The great limitation of molecular-level analyses of song learning so far is that they have been almost entirely correlational. Some aspect of learning occurs (song-response habituation, singing crystallization) and changes in gene expression are observed to occur in some temporal relationship. Are those changes a cause or an effect? In model systems with a strong basis in experimental genetics (i.e., drosophila, mouse), tools are available to produce transgenic animals in which specific gene regulation is altered. Methods may eventually be refined for efficient production of transgenic zebra finches, but a more direct strategy may be to exploit the high degree of anatomical definition in the song system for cannula-directed delivery of gene expression vectors and interfering agents (antisense oligonucleotides, RNAi). Intracerebral delivery of steroids, peptides, and other pharmacological agents has been widely effective in songbird research (Grisham et al., 1994; Basham et al., 1996; Johnson et al., 1997; Maney et al., 1997; Goodson, 1998a, b; Maney and Wingfield, 1998; Nordeen et al., 1998; Goodson and Adkins-Regan, 1999; Rasika et al., 1999; Brenowitz and Lent, 2002; Alvarez-Borda et al., 2004; Goodson et al., 2004).

4 Song Pathway Development: Sex Differences, Sex Hormones, and Neuronal Replacement

4.1 Sex Differences in Song-Control System

In the brain of zebra finches and many other species, the song nuclei differ greatly in volume between the sexes (Nottebohm and Arnold, 1976; Arnold and Saltiel, 1979; Gahr et al., 1993; Arnold et al., 1996; Brenowitz et al., 1996; Gahr et al., 1998). Not only are the nuclei larger in males than in females, but the primary motor output pathway (HVC–RA) does not develop in the female zebra finch (Konishi and Akutagawa, 1985; Mooney and Rao, 1994; Holloway and Clayton, 2001). Recent studies have defied expectations that these sex differences could be explained simply by the differential production of gonadal steroids in the sexes acting on a common brain substrate—the situation as it is understood in mammals (Schlinger et al., 2001; Wade and Arnold, 2004). Instead, it now appears that the genotype of the zebra finch brain itself has a significant role in determining whether the song system develops in masculine or feminine form.

The most striking evidence for this comes from an analysis of a rare spontaneous occurrence of a gynandromorphic zebra finch, in which the right half of the brain was genetically male and the left half genetically female (Agate et al., 2003). The neural song circuit on the right-hand side had a more masculine phenotype than that on the left, despite the fact both sides were exposed to a common gonadal hormone environment. An implication of this and other work [reviewed in Arnold (2004), Wade and Arnold (2004)] is that the sex-linked genes act locally within the nervous system to influence the way specific circuits develop. Clearly, one would like to identify these genes. An important advance that may help is the completion of the first draft genomic sequence of the chicken (Hillier et al., 2004), and the planned production of a zebra finch genomic sequence and physical map in the near future (http://www.genome.gov/11007951).

4.2 Sex Hormone Production in the Brain

Ironically, the mechanism by which sex-linked genes act directly within the brain may be to influence the production of androgen and estrogens—but by the brain itself and not from gonadal sources. The surprising capacity of the zebra finch brain for producing sex steroids was first indicated by studies showing high activities in the brain of aromatase, the enzyme responsible for producing estradiol-17-beta from testosterone (Schlinger and Arnold, 1991; Schlinger et al., 1994; Shen et al., 1994). Further studies have identified in the brain all the enzymes necessary to synthesize estrogens from cholesterol (Schlinger et al., 1995; Wade et al., 1995; Vanson et al., 1996; Freking et al., 1998; Schlinger et al., 1999; London et al., 2003; Soma et al., 2004).

A direct demonstration of functional brain estrogen production in song-circuit development was obtained from a study of cultured tissue slices prepared from juvenile zebra finches at 25 days of age, just before HVC axon begin their extension toward RA 2–3 mm below (Holloway and Clayton, 2001). After

1 week in culture, slices from both sexes formed a projection that reached the boundaries of RA. After 2–3 weeks the projection had entered RA in the male slices, but appeared to avoid the interior of the nucleus in the female slices. When both sexes were cultured together, both developed a male-like projection. Pharmacological inhibition of aromatase or the estrogen receptor blocked masculine development in male and cocultured female slices. Estradiol-17-beta was detected in the slice culture medium even after 3 weeks, and the levels were significantly higher in the male cultures than the females during the first culture week.

Key questions now are to identify more precisely the actual sites of steroid synthesis and action in the song system. Possibilities to consider include the cellular targets within RA; the tract leading from HVC to RA; existing neurons within HVC (and their associated glia) that are mounting the projections; and neuronal precursors arising from the ventricular zone that may be stimulated or recruited by local steroids (see next section). It should be possible to resolve these issues through a combination of localizing the synthetic enzymes (e.g., London et al., 2003) and microapplication of agonists and antagonists within the cultured slice or brain of the intact animal. Results of two studies indirectly favor the hypothesis of actions at the level of HVC of the associated neurogenic ventricular zone. First, BDNF is increased within RA-projecting cells in HVC of males relative to females, by day 30–35; levels are low in RA itself (Dittrich et al., 1999). BDNF levels in HVC were prematurely stimulated by estrogen (estradiol-17-beta) treatment, suggesting a mechanism by which estrogens acting in HVC might promote recruitment and survival of new neurons or stimulate axonal outgrowth. Second, a study of the effects of intracerebral testosterone implants in the white-crowned sparrow observed growth in ipsilateral HVC, RA, and Area X when the implants were placed near HVC but not RA (Brenowitz and Lent, 2002). It would be interesting to see the results of similar analyses using estradiol in the young zebra finch.

The significance of sex steroids produced locally in the brain may not be limited just to development. Estrogens have neuroprotective effects (Wise, 2002). Injury in the zebra finch brain results in an acute localized increase in aromatase activity, the enzyme that produces estradiol from testosterone, and the effect of this appears to be neuroprotective (Peterson et al., 2001; Wynne and Saldanha, 2004; Peterson et al., 2004). Aromatase is coexpressed with NMDA receptors in some neurons, and estradiol increases somal size and innervation of these neurons and the frequency of their presynaptic NMDA receptors (Saldanha et al., 2004).

4.3 Regulation of Neuronal Birth, Death, and Replacement

New neurons are continually added to various parts of the adult avian forebrain. This was first discovered through a study of seasonal changes in the volume of song nuclei in canaries (Goldman and Nottebohm, 1983; Nottebohm, 2004). Neurons are born at sites along the lateral ventricles and migrate to destinations that include HVC, Area X, and NCM, where they may be incorporated to replace dying neurons of the same type (Scharff et al., 2000). Song nucleus HVC sits adjacent to the neurogenic ventricular zone and incorporates a particularly large number of neurons. As much as 50% of the RA-projecting neurons in HVC can be replaced in a 6-month period (Kirn and Nottebohm, 1993).

Nottebohm hypothesized that the continual replacement of neurons could provide a mechanism for new learning (Nottebohm, 1989; Nottebohm, 2002). Many studies have now been performed in various songbird species to assess this idea. In a detailed and thoughtful review, Wilbrecht and Kirn (2004) concluded that the linkage to learning per se was rather indirect: "song instability, rather than song learning, may be a better correlate of new HVC neuron recruitment levels. . .new neurons may provide the raw variability in song upon which a new song can be sculpted" (p. 670).

Regardless of their precise relationship to song learning, the phenomena of adult neurogenesis are of considerable interest for insights they may offer into neuronal development and for the potential application of these insights into strategies for brain repair. A number of influences on neuronal recruitment and survival have been identified. One is the target site for neuronal incorporation, as neurons are recruited into HVC but not LMAN in the nidopallium. They are recruited into Area X, but not RA (Alvarez-Buylla and Kirn, 1997). In HVC, they replace RA projecting neurons, but not X-projecting neurons (Alvarez-Buylla et al., 1988; Nordeen and Nordeen, 1988).

Another factor may be "access" or "space," as periods of increased neuronal incorporation in the canary appear to follow seasonal periods of increased cell death (Kirn et al., 1994), and targeted ablation of neurons in HVC results in compensatory replacement (Scharff et al., 2000). These effects may be mediated by local production of neurotrophic factors. For example, IGF-I is produced by radial cells (Jiang et al., 1998), which are the probable stem cells for adult neurogenesis (Alvarez-Buylla et al., 1990; Goldman et al., 1996; Alvarez-Buylla and Temple, 1998). In HVC, X-projecting cells produce IGF-II and the protein is accumulated by RA-projecting cells, the population of which is turning over (Holzenberger et al., 1997). Neuronal recruitment and survival can be influenced by testosterone and estradiol (Rasika et al., 1994; Hidalgo et al., 1995), use and activity (Alvarez-Borda and Nottebohm, 2002; Lipkind et al., 2002; Wilbrecht et al., 2002a,b), and photoperiod (Kirn and Schwabl, 1997). BDNF enhances survival of a subset of young neurons (Alvarez-Borda et al., 2004), and it appears to mediate some of the effects listed earlier as its production is boosted by testosterone (Rasika et al., 1999), estradiol (Dittrich et al., 1999), and singing activity (Li et al., 2000).

5 Social and Environmental Integration

One of the most compelling aspects of songbird model is the clear influence of an array of modulatory factors on singing behavior, song learning, and associated brain anatomy. These factors can reveal how underlying cellular and molecular processes are integrated and embedded in the world of natural experience. Differential gene expression is likely to play a central role in these processes. Gene regulation works to establish the receptor systems and circuits that mediate integrative processes, and the genome itself plays a dynamic role in the responses to social and environmental stimuli, as the following examples show.

5.1 Influence of Social Factors on Song-System Molecular Biology

Social factors have an obvious role in song learning, as a young bird must interact with one or more tutors to acquire a learned song model (Clayton, 1987a; Slater et al., 1988). A male zebra finch modulates its vocal response to the calls of its female partner based on the presence and mating status of other birds in the "audience" (Vignal et al., 2004). In cowbirds, social transmission of courtship behaviors has been described (Freeberg, 2004).

Three examples show that social factors can modulate gene expression and cellular metabolism within the songbird brain. The act of singing induces *zenk* gene expression in the male zebra finch song-control system—but the anatomical pattern in which the gene response occurs differs depending upon the audience. Intriguingly, and perhaps counterintuitively, gene activation is more intense and widespread when the bird is singing by himself (undirected song) than when he is presenting essentially the same vocal performance to a female (directed song). The significance and mechanistic basis for this is yet unknown, although a favorite speculation is that undirected singing represents the explicit rehearsal of singing, during which genome-dependent processes of plasticity are brought online (Jarvis et al., 1998). Enforced isolation and restraint of a zebra finch change the selectivity of the *zenk* response to sounds (conspecific versus heterospecific songs or noise). Selectivity for conspecific song is greatest when the bird is free to move and has experienced shorter periods of social isolation (Park and Clayton, 2002). As a third example, social environment influences the recruitment and survival of new neurons in song processing centers in the adult zebra finch brain. Birds placed in a large heterosexual group for 40 days had more new neurons than birds kept singly or as male–female pairs (Lipkind et al., 2002).

5.2 Seasonal and Photoperiod Effects

The discovery of robust adult avian neurogenesis followed on the heels of a previous discovery of "a brain for all seasons: cyclical anatomical changes in song-control nuclei of the canary brain" (Nottebohm, 1981). Further analyses eventually showed photoperiod effects on neuronal death and replacement in canaries, independent of chronic changes in gonadal steroid hormone levels (Kirn and Schwabl, 1997; Huang et al., 1998). Photoperiod effects at the level of *zenk* gene responsiveness to preferred songs have also been demonstrated (Sockman et al., 2002). The extensive literature on seasonal and photoperiod effects in the song-control nuclei is reviewed in detail elsewhere (Tramontin and Brenowitz, 2000; Ball et al., 2004; Ball and Balthazart, this volume; Brenowitz, 2004).

5.3 Linking Environment to Genome

The mechanisms by which social and environmental factors have both immediate and long-term effects on brain gene activity and neuronal survival are not understood. There are quite a few possibilities, however, four of which are considered briefly here [see also Ball et al. (2004), Ball and Balthazart (this volume)].

5.3.1 Dopamine and Other Catecholamines

Catecholamine cell groups project to forebrain song-control nuclei [reviewed in Ball et al. (2003)]. Simulated territorial intrusion activates immediate early gene expression (*c-fos*) in two of these groups in the brainstem, AVT, and midbrain central gray (Maney and Ball, 2003). Lesioning of noradrenergic inputs selectively suppresses courtship singing (Barclay et al., 1996). Given the evidence that dopamine exerts presynaptic depression on glutamatergic inputs to neurons in Area X (Ding et al., 2003), these observations suggest mechanisms for how the social context of singing (directed versus undirected song) alters the *zenk* gene response in basal ganglia component of the AFP, Area X (Jarvis et al., 1998).

5.3.2 Adrenocorticoids

Zebra finches maintain lifelong pair bonds, and both members of a pair can recognize their mates' vocalizations (Miller, 1979a; Vignal et al., 2004). A study of the effects of mate separation showed behavioral alterations and rises in plasma corticosterone; these effects were reversed upon reunion of mates but not upon exposure to other same-sex individuals or opposite-sex partners (Remage-Healey et al., 2003). Thus, a circulating hormonal signal is modulated by social perceptions.

5.3.3 Gonadal Steroids

Testosterone may be a transducer for aspects of natural experience, such as agonist interactions. For example, in a study of free-living male song sparrows, birds with increased exposure to testosterone-treated males (who behaved more aggressively) had higher circulating testosterone levels themselves (Wingfield, 1984). Testosterone in turn can have a variety of neural effects, both direct and indirect, including suppression of song plasticity (Marler et al., 1988), modulation of NMDA receptors (White et al., 1999), regulation of catecholamine levels (Ball et al., 2003), and elevation of BDNF in HVC (Rasika et al., 1999).

5.3.4 Endocannabinoids

Cannabinoids, the active constituents of marijuana, have diverse effects on learning and memory, perception, and feeding behavior. They act through abundant receptors in the brain, which are responsive to endogenous cannabinoids, lipids related to arachidonic acid. The normal function of the endocannabinoid system is not well understood. Soderstrom and colleagues turned to the zebra finch model to gain some insight, and reported that cannabinoid treatment alters song learning,but not production of song after it has been learned (Soderstrom and Johnson, 2003; Soderstrom and Tian, 2004). Birds treated during juvenile learning learned fewer notes, and their song was less stereotyped yet also tended to contain more repetitive note production. An intriguing integrative role for endocannabinoids, penetrating to the level of gene response modulation, is suggested by the observation that cannabinoid treatment reduces the *zenk* response to song in NCM—yet it has no effect on the responses in primary auditory Fields L1 and L3. Moreover, agonist treatment during 2 days of training exposures apparently blocked habituation of the *zenk* response (when tested on the third day in the absence of the drug) (Whitney et al., 2003). Furthermore, endocannabinoid levels are elevated in response to periods of limited food availability, and limited food availability also suppresses the *zenk* response (much as does cannabinoid treatment). This suggests a functional integration of hunger and food availability with the *zenk*-mediated tendency to form auditory memories and associations, all linked through an endocannabinoid mechanism (Soderstrom et al., 2004).

6 Diversity and Evolution

Songbird research has tended to focus on just a handful of species, and the zebra finch in particular. Zebra finches are especially attractive for laboratory studies because they are easy to breed in captivity, and the sharp dichotomy in behavior and brain anatomy of males and females immediately suggests incisive experimental comparisons. However, the new tools of molecular genomics make it easier to move between and compare species at a deep level (Clayton, 2004a), and songbirds as a group offer exciting opportunities for comparative analyses.

The species diversity of Order Passeriformes is comparable to that of living mammals (Barker et al., 2004). The diversification of oscines began only about 65 million years ago, and much of it occurred in the last 10–20 million years (➋ *Figure 9-5*). Efforts to map out the evolutionary pathways and relationships are increasingly directed to the study of specific nuclear gene sequences, including some genes of special interest to neurophysiology, e.g., *zenk* (Long and Salbaum, 1998; Chubb, 2004a, b), *c-myc* (Ericson and Johansson, 2003), and *rag-1* (Ericson and Johansson, 2003; Barker et al., 2004).

At the moment, the primary focus is on refining the avian evolutionary tree as a basis for consideration of biogeography, history, and broad mechanisms of adaptive evolution (Grant and Grant, 1997). As this research progresses, however, it may become feasible to mine the phylogenetic tree for specific events in gene evolution that correlate with emergence of specific phenotypic characteristics, such as the ability to learn and modify song. There are already a few examples that illustrate this line of research.

6.1 Aromatase and Estrogen Receptor

Although songbirds (oscines) predominate in the Order Passeriformes, there are also suboscines in the order that did not evolve a song system and do not show evident song-learning ability. Saldanha et al. (2000) asked whether one such species, the golden-collared manakin, expresses aromatase (estrogen synthase) in the telencephalon at high levels like the zebra finch (oscine passerine) or at low levels like the quail (nonpasserine). Restriction of aromatase to oscines would support the hypothesis that evolution of the song-learning ability occurred in concert with evolution of brain estrogen production. The results, however, show high aromatase in several telencephalic loci of the manakin brain (despite absence of detectable song nuclei). Thus, forebrain aromatase may have emerged in passerines before the divergence of oscines.

Figure 9-5

Avian phylogeny emphasizing passerines, including both oscines (songbirds) and suboscines (no vocal learning) (reprinted with permission from Barker et al., 2004). Relationships are based on analysis of combined *rag-1* and *-2* sequences (146 taxa, 4,126 aligned nucleotide positions). The topology presented is the best ML estimate, rooted using Gallus and Coracias. Multiple exemplars of genera and certain higher taxa have been collapsed for clarity (number of species indicated in parentheses after genus or group). See original reference for further description (Barker et al., 2004)

Figure 9-5 (continued)

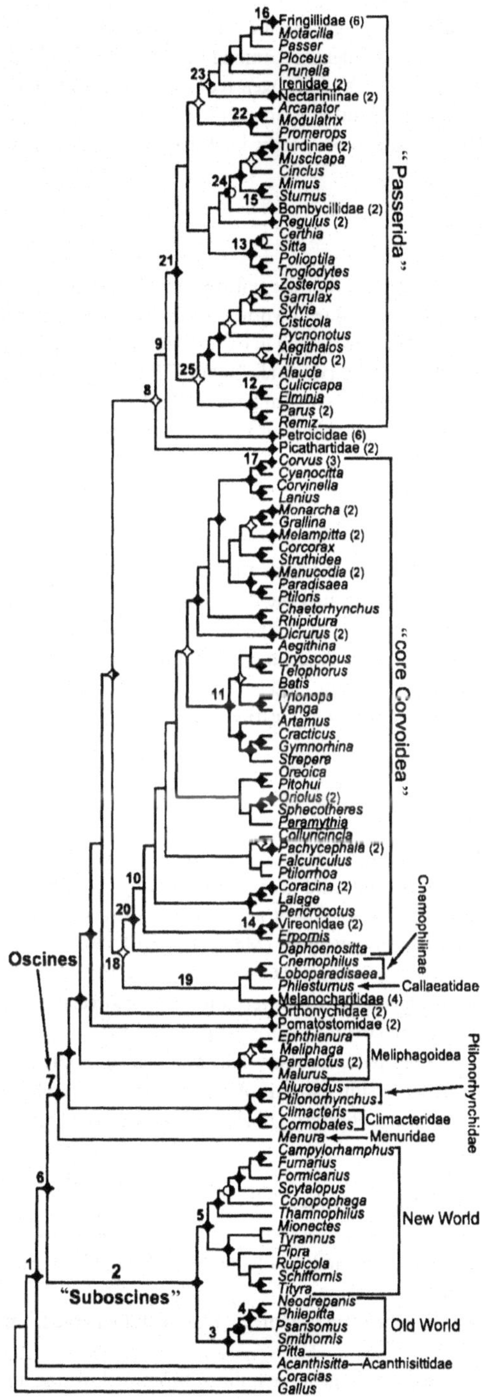

Brain distribution of estrogen receptor itself was examined in 26 different avian species, including 3 suboscines (nonsongbirds in Passeriformes) and nine species outside Passeriformes (Gahr et al., 1993). Immunoreactivity in song nuclei (or the general areas in which song nuclei are embedded) was observed only in the songbirds. There were also differences among the songbird species. For example, adult males of the Estrildinae had much lower numbers of estrogen receptor immunoreactive cells than did adult males of the Fringillidae, Paridae, Sturnidae, and Ploceinae. It may be interesting to note that the SNAg antigen, discussed in ➋ *Section 2.1*, is reported to be specific for Estrildinae (Akutagawa and Konishi, 2001).

6.2 *ZENK* Expression

Although oscines are the most common, they are not the only avian order in which vocal learning ability evolved. Analysis of *zenk* gene activation during vocal behavior revealed the presence of seven functionally homologous forebrain areas active during singing in oscines and the other avian vocal learners, parrots, and hummingbirds (Jarvis and Mello, 2000; Jarvis et al., 2000). These three groups have been thought to have evolved vocal learning independently, and the similarities in patterns of vocal activation in the brain may indicate specific constraints on the organization of vocal learning systems.

6.3 FoxP2

FoxP2 is a transcription factor, mutations in which are genetically linked to abnormalities in human speech production. Two research groups investigated the possibility that FoxP2 might serve a convergent role in songbird vocal learning as well (Haesler et al., 2004; Teramitsu et al., 2004). Although some evidence was found for regulated expression in Area X during developmental (zebra finch) and seasonal (canary) plasticity (Haesler et al., 2004), expression patterns in other nonvocal-learning species undermine any simple specific association of FoxP2 with vocal learning ability.

6.4 Beyond the Song

Although the focus of this review (and a lot of research) has been on phenomena related to song communication, the taxonomic richness and accessibility of songbirds have made them favored subjects for a variety of other research endeavors. Drent and colleagues are reshaping the study of genetics and personality through consideration of songbirds (Marchetti and Drent, 2000; Carere et al., 2001; Dingemanse et al., 2003; Drent et al., 2003; Dingemanse et al., 2004; van Oers et al., 2004a, b, c). Comparison of mainland versus island white-winged fairy wrens recently led to appreciation of the divergent evolution of melanin production and the melanocortin receptor locus (Doucet et al., 2004). And nearly 200 years after inspiring Charles Darwin with their patterns of diversity and relatedness, the songbird species of the genus *Geospiza* revealed variations in expression patterns of a growth factor, Bmp4, that may account for some of the morphological variations Darwin observed (Abzhanov et al., 2004).

7 Critical Issues for the Future

We have reviewed five broad areas of research involving the biology of bird song, which define fundamental issues of molecular neurobiology:

1. How does a neural "system" (the song system) map onto and emerge from the underlying code of the genome?
2. Within such a system, how does the genome limit or promote learning and behavioral plasticity?

3. How do genes regulate the formation of specific neural circuits (e.g., sex-specific song-control pathways) and the turnover of neurons within those circuits?
4. How is all this molecular biology coordinated and integrated to result in adaptive responses to social and environmental exigencies?
5. How did the sophisticated and adaptive neurobiology of learned vocal communication evolve, across the thousands of different songbird species extant on planet Earth?

Each of the studies reviewed in this article focused on one or at most a handful of gene products. However, there is no reason to think that any of the questions given earlier can be answered by consideration of a single gene product. Indeed, a paradigm shift is underway in how molecular biological research is framed and conducted. With the tools of modern genomics, literally thousands of gene products can be studied simultaneously (e.g., high throughput sequencing, microarray hybridizations, mass spectrometry-based protein analysis). This not only increases the rate of discovery, but it also makes possible a new kind of observation, of multigenic interactions and gene networks.

7.1 Gene Network Analysis

For the songbird community, the infrastructure for this new kind of research is now being developed. In the US, there are two organized efforts to produce large systematic collections of mRNA sequences expressed in the zebra finch brain ("Expressed Sequence Tags," ESTs). EST collections are produced using robotics to partially determine the sequences of randomly selected cDNA clones prepared from the tissue of interest. The term "tag" is used because the sequences are typically incomplete, but are large enough to be useful for identifying and measuring the gene products. ESTs can be used to assess whether a particular protein sequence has been conserved in songbirds, and whether detection reagents (e.g., antibodies) developed in other species should crossreact with songbird material. The ESTs themselves can be used to generate nucleic acid probes for measuring the specific mRNAs from which they were derived. Moreover, the ESTs can be used as a basis for microarray development, which allows measurement of literally thousands of mRNA simultaneously in a single experiment (below).

As of the time of writing this article, more than 60,000 zebra finch brain ESTs have been sequenced by the two groups in the US, and these define more than 22,000 unique gene products (http://titan.biotec.uiuc.edu/songbird/). The efforts now underway in the songbird research community should lead to the initial identification in the near future of ESTs representing most of the 30,000 genes expressed in the songbird brain (Clayton, 2004a). Information about these efforts as well as progress toward whole-genome sequencing is available at: http://www.songbirdgenome.org.

EST sequencing provides information about the sequences of mRNAs, but not about the DNA sequences associated with the corresponding gene in the chromosome. To obtain the complete genomic DNA sequence for a given gene, other approaches are needed. The complete sequence may not be important for investigators only interested in knowing which proteins are expressed in their tissue of interest, or using microarrays to measure the amount of mRNA produced by various genes. However, to understand the mechanisms that regulate the expression of any gene, one must eventually consider the structure of its genomic DNA and the sequence of the gene's promoter. This sort of analysis in songbirds has recently been facilitated by the construction of a zebra finch brain Bacterial Artificial Chromosome (BAC) library. BAC clones are large fragments of foreign chromosomal DNA maintained in a relatively stable form by bacteria. Additional information about this resource may be found at this URL: www.genome.arizona.edu. The recent availability of the draft sequence of the chicken genome (Hillier et al., 2004) should facilitate the analysis of the other avian genomes, by providing an evolutionarily "recent" high-resolution template for physical mapping and functional annotation.

Another major tool under development for bird song molecular neurobiology is the DNA microarray, which allows measurement of expression levels for many genes simultaneously. The rich literature on this topic includes early examples (Schena et al., 1995; DeRisi et al., 1997), detailed methodological reviews

(Churchill, 2002; Nadon and Shoemaker, 2002; Yang and Speed, 2002; Simon and Dobbin, 2003), and an initial proof-of-concept study of microarray development for songbird research (Wade et al., 2004). Using robotics, thousands of different DNA probes are arrayed on a single support, the size of a standard microscope slide. RNAs extracted from tissues of interest are then labeled with fluorescent dyes and hybridized to the DNA probes on the glass slide, and the intensity of binding monitored by computerized optical methods.

For songbird researchers, microarrays will have at least two broad uses: functional diagnostics and gene discovery. In functional diagnostics, the investigator uses patterns of gene expression known to be associated with a particular cell type or physiological state to assess the status of a biological sample. For example, a songbird researcher might want to monitor how gene profiles associated with neurogenesis or synaptic plasticity change across seasons or during the juvenile song acquisition period. In gene discovery, an investigator compares tissues that differ in a specific functional property of interest and seeks to discover genes whose expression is correlated with that functional difference.

7.2 Targeted Gene Manipulation

A technical frontier with a longer horizon in songbird research is the development of genetic methods for direct manipulation of the genes—either within the germline (whole organism) or within specific targeted regions of the developing or adult brain. These methods include organized breeding programs to isolate distinct strains for conventional genetic mapping, and gene transfer methods for targeted introduction of specific engineered gene sequences into the organism. Potential methods of gene transfer applicable to songbird research include variations on the "transgenic" approach (whereby a new gene sequence is stably integrated into the chromosomes of germ cells followed by breeding and rearing), and the "expression vector" approach (whereby a DNA or RNA sequence is delivered directly into targeted brain cells in the juvenile or adult organism). Such methods are central to research with mice, fruit flies, and several other common model organisms in biological science, but researchers are just beginning to explore their application in songbirds (see also ➋ *Section 3.3*).

7.3 Conclusion

The widespread utility and appeal of songbirds for biological research justifies a significant and ongoing investment in the development of new tools, technologies, and experimental strategies for songbird molecular genetic analysis. In bird song, nature offers us an extraordinary model for appreciating the deep relationships of genes, brains, and behavior.

Acknowledgments

The author thanks the many colleagues who have inspired, instructed, and challenged him over the years—and especially the dozen or so graduate students who have shared his path of discovery. This work was supported by NIH grants RO1 NS045264 and RO1 MH52086.

References

Aamodt SM, Nordeen EJ, Nordeen KW. 1995. Early isolation from conspecific song does not affect the normal developmental decline of N-methyl-D-aspartate receptor binding in an avian song nucleus. J Neurobiol 27: 76-84.

Aamodt SM, Nordeen EJ, Nordeen KW. 1996. Blockade of NMDA receptors during song model exposure impairs song development in juvenile zebra finches. Neurobiol Learn Mem 65: 91-98.

Abarbanel HD, Gibb L, Mindlin GB, Rabinovich MI, Talathi S. 2004. Spike timing and synaptic plasticity in the premotor pathway of birdsong. Biol Cybernet 91: 159-167.

Absil P, Pinxten R, Balthazart J, Eens M. 2003. Effects of testosterone on reelin expression in the brain of male European starlings. Cell Tissue Res 312: 81-93.

Abzhanov A, Protas M, Grant BR, Grant PR, Tabin CJ. 2004. Bmp4 and morphological variation of beaks in Darwin's finches. Science 305: 1462-1465.

Adret P. 2004. In search of the song template. Ann NY Acad Sci 1016: 303-324.

Adret P, Margoliash D. 2002. Metabolic and neural activity in the song system nucleus robustus archistriatalis: effect of age and gender. J Comp Neurol 454: 409-423.

Agate RJ, Grisham W, Wade J, Mann S, Wingfield J, et al. 2003. Neural, not gonadal, origin of brain sex differences in a gynandromorphic finch. Proc Natl Acad Sci USA 100: 4873-4878.

Akutagawa E, Konishi M. 1998. Transient expression and transport of brain-derived neurotrophic factor. Proc Natl Acad Sci USA 95: 11429-11434.

Akutagawa E, Konishi M. 2001. A monoclonal antibody specific to a song system nuclear antigen in estrildine finches. Neuron 31: 545-556.

Alvarez-Borda B, Nottebohm F. 2002. Gonads and singing play separate, additive roles in new neuron recruitment in adult canary brain. J Neurosci 22: 8684-8690.

Alvarez-Borda B, Haripal B, Nottebohm F. 2004. Timing of brain-derived neurotrophic factor exposure affects life expectancy of new neurons. Proc Natl Acad Sci USA 101: 3957-3961.

Alvarez-Buylla A, Kirn JR. 1997. Birth, migration, incorporation, and death of vocal control neurons in adult songbirds. J Neurobiol 33: 585-601.

Alvarez-Buylla A, Temple S. 1998. Stem cells in the developing and adult nervous system. J Neurobiol 36: 105-110.

Alvarez-Buylla A, Theelen M, Nottebohm F. 1988. Birth of projection neurons in the higher vocal center of the canary forebrain before, during, and after song learning. Proc Natl Acad Sci USA 85: 8722-8726.

Alvarez-Buylla A, Theelen M, Nottebohm F. 1990. Proliferation "hot spots" in adult avian ventricular zone reveal radial cell division. Neuron 5: 101-109.

Arnold AP. 1975. The effects of castration on song development in zebra finches (*Poephila guttata*). J Exp Zool 191: 261-278.

Arnold AP. 1980. Quantitative analysis of sex differences in hormone accumulation in the zebra finch brain: methodological and theoretical issues. J Comp Neurol 189: 421-436.

Arnold AP. 2004. Sex chromosomes and brain gender. Nat Rev Neurosci 5: 701-708.

Arnold AP, Saltiel A. 1979. Sexual difference in pattern of hormone accumulation in the brain of a songbird. Science 205: 702-705.

Arnold AP, Wade J, Grisham W, Jacobs EC, Campagnoni AT. 1996. Sexual differentiation of the brain in songbirds. Dev Neurosci 18: 124-136.

Auger CJ, Bentley GE, Auger AP, Ramamurthy M, Ball GF. 2002. Expression of cAMP response element binding protein-binding protein in the song control system and hypothalamus of adult European starlings (*Sturnus vulgaris*). J Neuroendocrinol 14: 805-813.

Bailey DJ, Wade J. 2003. Differential expression of the immediate early genes *fos* and *zenk* following auditory stimulation in the juvenile male and female zebra finch. Brain Res Mol Brain Res 116: 147-154.

Bailey DJ, Rosebush JC, Wade J. 2002. The hippocampus and caudomedial neostriatum show selective responsiveness to conspecific song in the female zebra finch. J Neurobiol 52: 43-51.

Balaban E. 1988. Cultural and genetic variation in swamp sparrows (*Melospiza georgiana*). II. Behavioral salience of geographic song variants. Behaviour 105: 292-321.

Ball GF, Balthazart J. 2001. Ethological concepts revisited: immediate early gene induction in response to sexual stimuli in birds. Brain Behav Evol 57: 252-270.

Ball GF, Casto JM, Balthazart J. 1995. Autoradiographic localization of D1-like dopamine receptors in the forebrain of male and female Japanese quail and their relationship with immunoreactive tyrosine hydroxylase. J Chem Neuroanat 9: 121-133.

Ball GF, Nock B, Wingfield JC, McEwen BS, Balthazart J. 1990. Muscarinic cholinergic receptors in the songbird and quail brain: a quantitative autoradiographic study. J Comp Neurol 298: 431-442.

Ball GF, Castelino CB, Maney DL, Appeltants D, Balthazart J. 2003. The activation of birdsong by testosterone: multiple sites of action and role of ascending catecholamine projections. Ann NY Acad Sci 1007: 211-231.

Ball GF, Auger CJ, Bernard DJ, Charlier TD, Sartor JJ, et al. 2004. Seasonal plasticity in the song control system: multiple brain sites of steroid hormone action and the importance of variation in song behavior. Ann NY Acad Sci 1016: 586-610.

Balthazart J, Foidart A, Wilson EM, Ball GF. 1992. Immunocytochemical localization of androgen receptors in the male songbird and quail brain. J Comp Neurol 317: 407-420.

Barclay SR, Harding CF, Waterman SA. 1996. Central DSP-4 treatment decreases norepinephrine levels and courtship behavior in male zebra finches. Pharmacol Biochem Behav 53: 213-220.

Barker FK, Cibois A, Schikler P, Feinstein J, Cracraft J. 2004. Phylogeny and diversification of the largest avian radiation. Proc Natl Acad Sci USA 101: 11040-11045.

Basham ME, Nordeen EJ, Nordeen KW. 1996. Blockade of NMDA receptors in the anterior forebrain impairs sensory acquisition in the zebra finch (*Poephila guttata*). Neurobiol Learn Mem 66: 295-304.

Basham ME, Sohrabji F, Singh TD, Nordeen EJ, Nordeen KW. 1999. Developmental regulation of NMDA receptor 2B subunit mRNA and ifenprodil binding in the zebra finch anterior forebrain. J Neurobiol 39: 155-167.

Beecher MD, Campbell SE, Burt JM. 1994. Song perception in the song sparrow: birds classify by song type but not by singer. Anim Behav 47: 1343-1351.

Beecher MD, Stoddard PK, Campbell SE, Horning CL. 1996. Repertoire matching between neighbouring song sparrows. Anim Behav 51: 917-923.

Beecher MD, Campbell SE, Burt JM, Hill CE, Nordby JC. 2000. Song-type matching between neighbouring song sparrows. Anim Behav 59: 21-27.

Benney KS, Braaten RF. 2000. Auditory scene analysis in estrilid finches (*Taeniopygia guttata* and *Lonchura striata domestica*): a species advantage for detection of conspecific song. J Comp Psychol 114: 174-182.

Bernard DJ, Ball GF. 1995. Two histological markers reveal a similar photoperiodic difference in the volume of the high vocal center in male European starlings. J Comp Neurol 360: 726-734.

Bernard DJ, Ball GF. 1997. Photoperiodic condition modulates the effects of testosterone on song control nuclei volumes in male European starlings. Gen Comp Endocrinol 105: 276-283.

Bernard DJ, Bentley GE, Balthazart J, Turek FW, Ball GF. 1999. Androgen receptor, estrogen receptor alpha, and estrogen receptor beta show distinct patterns of expression in forebrain song control nuclei of European starlings. Endocrinol 140: 4633-4643.

Bhardwaj SK, Anushi N. 2004. The effect of duration and time of food availability on the photoperiodic response in the male house sparrow, *Passer domesticus*. Reprod Nutr Dev 44: 29-35.

Boettiger CA, Doupe AJ. 2001. Developmentally restricted synaptic plasticity in a songbird nucleus required for song learning. Neuron 31: 809-818.

Böhner J. 1990. Early acquisition of song in the zebra finch, *Taeniopygia guttata*. Anim Behav 39: 369-374.

Bolhuis JJ, Zijlstra GG, den Boer-Visser AM, Van Der Zee EA. 2000. Localized neuronal activation in the zebra finch brain is related to the strength of song learning. Proc Natl Acad Sci USA 97: 2282-2285.

Bolhuis JJ, Hetebrij E, Den Boer-Visser AM, De Groot JH, Zijlstra GG. 2001. Localized immediate early gene expression related to the strength of song learning in socially reared zebra finches. Eur J Neurosci 13: 2165-2170.

Bottjer SW. 1987. Ontogenetic changes in the pattern of androgen accumulation in song-control nuclei of male zebra finches. J Neurobiol 18: 125-139.

Bottjer SW. 1993. The distribution of tyrosine hydroxylase immunoreactivity in the brains of male and female zebra finches. J Neurobiol 24: 51-69.

Bottjer SW. 2004. Developmental regulation of basal ganglia circuitry during the sensitive period for vocal learning in songbirds. Ann NY Acad Sci 1016: 395-415.

Bottjer SW, Hewer SJ. 1992. Castration and antisteroid treatment impair vocal learning in male zebra finches. J Neurobiol 23: 337-353.

Bottjer SW, Miesner EA, Arnold AP. 1984. Forebrain lesions disrupt development but not maintenance of song in passerine birds. Science 224: 901-903.

Braaten RF. 2000. Multiple levels of representation of song by European starlings (*Sturnus vulgaris*): open-ended categorization of starling song types and differential forgetting of song categories and exemplars. J Comp Psychol 114: 61-72.

Brainard MS. 2004. Contributions of the anterior forebrain pathway to vocal plasticity. Ann NY Acad Sci 1016: 377-394.

Brainard MS, Doupe AJ. 2000. Interruption of a basal ganglia–forebrain circuit prevents plasticity of learned vocalizations. Nature 404: 762-766.

Brainard MS, Doupe AJ. 2001. Postlearning consolidation of birdsong: stabilizing effects of age and anterior forebrain lesions. J Neurosci 21: 2501-2517.

Braun K, Scheich H, Heizmann CW, Hunziker W. 1991. Parvalbumin and calbindin-D28K immunoreactivity as developmental markers of auditory and vocal motor nuclei of the zebra finch. Neuroscience 40: 853-869.

Braun K, Scheich H, Schachner M, Heizmann CW. 1985. Distribution of parvalbumin, cytochrome oxidase activity and 14C-2-deoxyglucose uptake in the brain of the zebra finch. I. Auditory and vocal motor systems. Cell Tissue Res 240: 101-115.

Brenowitz EA. 2004. Plasticity of the adult avian song control system. Ann NY Acad Sci 1016: 560-585.

Brenowitz EA, Lent K. 2002. Act locally and think globally: intracerebral testosterone implants induce seasonal-like growth of adult avian song control circuits. Proc Natl Acad Sci USA 99: 12421-12426.

Brenowitz EA, Arnold AP, Loesche P. 1996. Steroid accumulation in song nuclei of a sexually dimorphic duetting bird, the rufous and white wren. J Neurobiol 31: 235-244.

Brenowitz EA, Baptista LF, Lent K, Wingfield JC. 1998. Seasonal plasticity of the song control system in wild nuttalls white-crowned sparrows. J Neurobiol 34: 69-82.

Brooks RJ, Falls JB. 1975. Individual recognition by song in white-throated sparrows. 1. Discrimination of songs of neighbors and strangers. Can J Zool 53: 879-888.

Burley N, Minor C, Strachan C. 1990. Social preferences of zebra finches for siblings, cousins and non-kin. Anim Behav 39: 775-784.

Cahill L, McGaugh JL, Weinberger NM. 2001. The neurobiology of learning and memory: some reminders to remember. Trends Neurosci 24: 578-581.

Cam V, Schlinger BA. 1998. Activities of aromatase and 3-beta-hydroxysteroid dehydrogenase delta(4)-delta(5) isomerase in whole organ cultures of tissues from developing zebra finches. Hormones Behav 33: 31-39.

Campagnoni CW, Kampf K, Mason B, Handley VW, Campagnoni AT. 1994. Isolation and characterization of a cDNA encoding the zebra finch myelin proteolipid protein. Neurochem Res 19: 1061-1065.

Carere C, Welink D, Drent PJ, Koolhaas JM, Groothuis TG. 2001. Effect of social defeat in a territorial bird (*Parus major*) selected for different coping styles. Physiol Behav 73: 427-433.

Casey RM, Baker MC. 1992. Early social tutoring influences female sexual response in white-crowned sparrows. Anim Behav 44: 983-986.

Casto JM, Ball GF. 1994. Characterization and localization of D1 dopamine receptors in the sexually dimorphic vocal control nucleus, Area X, and the basal ganglia of European starlings. J Neurobiol 25: 767-780.

Casto JM, Ball GF. 1996. Early administration of 17-beta-estradiol partially masculinizes song control regions and alpha(2)-adrenergic receptor distribution in European starlings (*Sturnus vulgaris*). Hormones Behav 30: 387-406.

Charlier TD, Balthazart J, Ball GF. 2003. Sex differences in the distribution of the steroid receptor coactivator SRC-1 in the song control nuclei of male and female canaries. Brain Res 959: 263-274.

Cheng H-Y. 2003. The ups and downs of MAP kinase ERK: modulation of molecular responses to song in the zebra finch auditory forebrain, Doctoral Thesis. University of Illinois: Urbana, IL.

Cheng HY, Clayton DF. 2004. Activation and habituation of extracellular signal-regulated kinase phosphorylation in zebra finch auditory forebrain during song presentation. J Neurosci 24: 7503-7513.

Chew SJ, Vicario DS, Nottebohm F. 1996. A large-capacity memory system that recognizes calls and songs of individual birds. Proc Natl Acad Sci USA 93: 1950-1955.

Chew SJ, Mello C, Nottebohm F, Jarvis E, Vicario DS. 1995. Decrements in auditory responses to a repeated conspecific song are long-lasting and require two periods of protein synthesis in the songbird forebrain. Proc Natl Acad Sci USA 92: 3406-3410.

Chialvo DR, Bak P. 1999. Learning from mistakes. Neuroscience 90: 1137-1148.

Christy B, Lau L, Nathans, D. 1988. A gene activated in mouse 3T3 cells by serum factors encodes a protein with "zinc finger" sequences. Proc Natl Acad Sci USA 85: 7857-7861.

Chubb AL. 2004a. New nuclear evidence for the oldest divergence among neognath birds: the phylogenetic utility of *zenk* (i). Mol Phylogenet Evol 30: 140-151.

Chubb AL. 2004b. Nuclear corroboration of DNA–DNA hybridization in deep phylogenies of hummingbirds, swifts, and passerines: the phylogenetic utility of *zenk* (ii). Mol Phylogenet Evol 30: 128-139.

Churchill GA. 2002. Fundamentals of experimental design for cDNA microarrays. Nat Genet 32(Suppl):490-495.

Cirelli C, Tononi G. 2000a. Differential expression of plasticity-related genes in waking and sleep and their regulation by the noradrenergic system. J Neurosci 20: 9187-9194.

Cirelli C, Tononi G. 2000b. Gene expression in the brain across the sleep-waking cycle. Brain Res 885: 303-321.

Cirelli C, Tononi G. 2000c. On the functional significance of *c-fos* induction during the sleep-waking cycle. Sleep 23: 453-469.

Cirelli C, Gutierrez CM, Tononi G. 2004. Extensive and divergent effects of sleep and wakefulness on brain gene expression. Neuron 41: 35-43.

Clayton DF. 1997. Role of gene regulation in song circuit development and song learning. J Neurobiol 33: 549-571.

Clayton DF. 2000. The genomic action potential. Neurobiol Learn Mem 74: 185-216.

Clayton DF. 2004a. Songbird genomics: methods, mechanisms, opportunities and pitfalls. Behav Neurobiol Birdsong, Zeigler PH, Marler P, editors. New York: New York Academy of Sciences; pp. 45-60.

Clayton DF. 2004b. Songbird genomics: methods, mechanisms, opportunities, and pitfalls. Annals of New York Academy of Sciences. Zeigler PH, Marler P, editors; pp. 45–60.

Clayton DF, Huecas M. 1990. Forebrain-enriched RNAs of the canary: a population analysis using hybridization kinetics. Mol Brain Res 7: 23-30.

Clayton NS. 1987a. Song tutor choice in zebra finches. Anim Behav 35: 714-721.

Clayton NS. 1987b. Song learning in cross-fostered zebra finches: a re-examination of the sensitive phase. Behaviour 102: 67-81.

Clayton NS. 1990. Subspecies recognition and song learning in zebra finches. Anim Behav 40: 1009-1017.

Collum RG, Clayton DF, Alt FW. 1991. Structure and expression of canary *myc* family genes. Mol Cell Biol 11: 1770-1776.

Contreras ML, Wade J. 1999. Interactions between nerve growth factor binding and estradiol in early development of the zebra finch telencephalon. J Neurobiol 40: 149-157.

Cynx J, Nottebohm F. 1992. Role of gender, season and familiarity in discrimination of conspecific song by zebra finches (*Taeniopygia guttata*). Proc Natl Acad Sci USA 89: 1368-1371.

Dave AS, Margoliash D. 2000. Song replay during sleep and computational rules for sensorimotor vocal learning. Science 290: 812-816.

Dave AS, Yu AC, Margoliash D. 1998. Behavioral state modulation of auditory activity in a vocal motor system. Science 282: 250-254.

Davis S, Bozon B, Laroche S. 2003. How necessary is the activation of the immediate early gene zif268 in synaptic plasticity and learning? Behav Brain Res 142: 17-30.

Denisenko-Nehrbass NI, Jarvis E, Scharff C, Nottebohm F, Mello CV. 2000. Site-specific retinoic acid production in the brain of adult songbirds. Neuron 27: 359-370.

Deregnaucourt S, Mitra PP, Feher O, Pytte C, Tchernichovski O. 2005. How sleep affects the developmental learning of bird song. Nature 433: 710-716.

Deregnaucourt S, Mitra PP, Feher O, Maul KK, Lints TJ, et al. 2004. Song development: in search of the error-signal. Ann NY Acad Sci 1016: 364-376.

DeRisi JL, Iyer VR, Brown PO. 1997. Exploring the metabolic and genetic control of gene expression on a genomic scale. Science 278: 680-686.

Ding L, Perkel DJ. 2004. Long-term potentiation in an avian basal ganglia nucleus essential for vocal learning. J Neurosci 24: 488-494.

Ding L, Perkel DJ, Farries MA. 2003. Presynaptic depression of glutamatergic synaptic transmission by D1-like dopamine receptor activation in the avian basal ganglia. J Neurosci 23: 6086-6095.

Dingemanse NJ, Both C, Drent PJ, Tinbergen JM. 2004. Fitness consequences of avian personalities in a fluctuating environment. Proc R Soc Lond B Biol Sci 271: 847-852.

Dingemanse NJ, Both C, van Noordwijk AJ, Rutten AL, Drent PJ. 2003. Natal dispersal and personalities in great tits (*Parus major*). Proc R Soc Lond B Biol Sci 270: 741-747.

Dittrich F, Feng Y, Metzdorf R, Gahr M. 1999. Estrogen-inducible, sex-specific expression of brain-derived neurotrophic factor mRNA in a forebrain song control nucleus of the juvenile zebra finch. Proc Natl Acad Sci USA 96: 8241-8246.

Doucet SM, Shawkey MD, Rathburn MK, Mays HL Jr, Montgomerie R. 2004. Concordant evolution of plumage colour, feather microstructure and a melanocortin receptor gene between mainland and island populations of a fairy-wren. Proc R Soc Lond B Biol Sci 271: 1663-1670.

Doupe AJ, Kuhl PK. 1999. Birdsong and human speech: Common themes and mechanisms. Annu Rev Neurosci 22: 567-631.

Doupe AJ, Solis MM, Kimpo R, Boettiger CA. 2004. Cellular, circuit, and synaptic mechanisms in song learning. Ann NY Acad Sci 1016: 495-523.

Drent PJ, van Oers K, van Noordwijk AJ. 2003. Realized heritability of personalities in the great tit (*Parus major*). Proc R Soc Lond B Biol Sci 270: 45-51.

Durand SE, Liang WR, Brauth SE. 1998. Methionine enkephalin immunoreactivity in the brain of the budgerigar (*Melopsittacus undulatus*)—similarities and differences with respect to oscine songbirds. J Comp Neurol 393: 145-168.

Eales LA. 1985. Song learning in zebra finches: some effects of song model availability on what is learnt and when. Anim Behav 33: 1293-1300.

Eales LA. 1987. Song learning in female-raised zebra finches: another look at the sensitive phase. Anim Behav 35: 1356-1365.

Eda-Fujiwara H, Satoh R, Bolhuis JJ, Kimura T. 2003. Neuronal activation in female budgerigars is localized and related to male song complexity. Eur J Neurosci 17: 149-154.

Editorial. 2005. My little chickadee. New York: The New York Times.

Ericson PG, Johansson US. 2003. Phylogeny of Passerida (Aves: Passeriformes) based on nuclear and mitochondrial sequence data. Mol Phylogenet Evol 29: 126-138.

Falls JB. 1982. Individual recognition by sounds in birds. Acoustic communication in birds. Kroodsma DE, Miller DH, Ouellet H, editors. New York: Academic Press; pp. 237-278.

Fee MS, Kozhevnikov AA, Hahnloser RH. 2004. Neural mechanisms of vocal sequence generation in the songbird. Ann NY Acad Sci 1016: 153-170.

Fiore M, Clayton NS, Pistillo L, Angelucci F, Alleva E, et al. 1999. Song behavior, NGF level and NPY distribution in the brain of adult male zebra finches. Behav Brain Res 101: 85-92.

Freeberg TM. 2004. Social transmission of courtship behavior and mating preferences in brown-headed cowbirds, *Molothrus ater.* Learn Behav 32: 122-130.

Freking F, Ramachandran B, Schlinger BA. 1998. Regulation of aromatase, 5-alpha- and 5-beta-reductase in primary cell cultures of developing zebra finch telencephalon. J Neurobiol 36: 30-40.

Funabiki Y, Konishi M. 2003. Long memory in song learning by zebra finches. J Neurosci 23: 6928-6935.

Gahr M. 1996. Developmental changes in the distribution of oestrogen receptor mRNA expressing cells in the forebrain of female, male and masculinized female zebra finches. Neuroreport 7: 2469-2473.

Gahr M, Konishi M. 1988. Developmental changes in estrogen-sensitive neurons in the forebrain of the zebra finch. Proc Natl Acad Sci USA 85: 7380-7383.

Gahr M, Kosar E. 1996. Identification, distribution and developmental changes of a melatonin-binding site in the song control system of the zebra finch. J Comp Neurol 367: 308-318.

Gahr M, Metzdorf R. 1997. Distribution and dynamics in the expression of androgen and estrogen receptors in vocal control systems of songbirds. Brain Res Bull 44: 509-517.

Gahr M, Metzdorf R. 1999. The sexually dimorphic expression of androgen receptors in the song nucleus hyperstriatalis ventrale pars caudale of the zebra finch develops independently of gonadal steroids. J Neurosci 19: 2628-2636.

Gahr M, Wild JM. 1997. Localization of androgen receptor mRNA-containing cells in avian respiratory-vocal nuclei—an in situ hybridization study. J Neurobiol 33: 865-876.

Gahr M, Guttinger H-R, Kroodsma D. 1993. Estrogen receptors in the avian brain: survey reveals general distribution and forebrain areas unique to songbirds. J Comp Neurol 327: 112-122.

Gahr M, Metzdorf R, Aschenbrenner S. 1996. The ontogeny of the canary HVC revealed by the expression of androgen and oestrogen receptors. Neuroreport 8: 311-315.

Gahr M, Sonnenschein E, Wickler W. 1998. Sex difference in the size of the neural song control regions in a dueting songbird with similar song repertoire size of males and females. J Neurosci 18: 1124-1131.

Geberzahn N, Hultsch H. 2003. Long-time storage of song types in birds: evidence from interactive playbacks. Proc R Soc Lond B Biol Sci 270: 1085-1090.

Gentner TQ, Hulse SH, Duffy D, Ball GF. 2001. Response biases in auditory forebrain regions of female songbirds following exposure to sexually relevant variation in male song. J Neurobiol 46: 48-58.

George JG. 2001. The synucleins. Genome Biol 3 (Reviews 3002.3001-3002.3006).

George JM, Clayton DF. 1992. Differential regulation in the avian song control circuit of an mRNA predicting a highly conserved protein related to protein kinase C and the *bcr* oncogene. Mol Brain Res 12: 323-329.

George JM, Jin H, Woods WS, Clayton DF. 1994. MAP kinase kinase (MEK-1) is enriched in radial cell processes in zebra finch brain. Soc Neurosci Abstr 20: 1437.

George JM, Jin H, Woods WS, Clayton DF. 1995. Characterization of a novel protein regulated during the critical period for song learning in the zebra finch. Neuron 15: 361-372.

Goldman SA, Nottebohm F. 1983. Neuronal production, migration and differentiation in a vocal control nucleus of the adult female canary brain. Proc Natl Acad Sci USA 80: 2390-2394.

Goldman SA, Zaremba A, Niedzwiecki D. 1992. In vitro neurogenesis by neuronal precursor cells derived from the adult songbird brain. J Neurosci 12: 2532-2541.

Goldman SA, Zukhar A, Barami K, Mikawa T, Niedzwiecki D. 1996. Ependymal/subependymal zone cells of postnatal and adult songbird brain generate both neurons and nonneuronal siblings in vitro and in vivo. J Neurobiol 30: 505-520.

Goodson JL. 1998a. Territorial aggression and dawn song are modulated by septal vasotocin. Hormones Behav 34: 67-77.

Goodson JL. 1998b. Vasotocin and vasoactive intestinal polypeptide modulate aggression in a territorial songbird, the violet-eared waxbill (estrildidae, *Uraeginthus granatina*). Gen Comp Endocrinol 111: 233-244.

Goodson JL, Adkins-Regan E. 1999. Effect of intraseptal vasotocin and vasoactive intestinal polypeptide infusions on courtship song and aggression in the male zebra finch (*Taeniopygia guttata*). J Neuroendocrinol 11: 19-25.

Goodson JL, Lindberg L, Johnson P. 2004. Effects of central vasotocin and mesotocin manipulations on social behavior in male and female zebra finches. Hormones Behav 45: 136-143.

Grant PR, Grant BR. 1997. Genetics and the origin of bird species. Proc Natl Acad Sci USA 94: 7768-7775.

Grisham W, Mathews GA, Arnold AP. 1994. Local intracerebral implants of estrogen masculinize some aspects of the zebra finch song system. J Neurobiol 25: 185-196.

Gulledge CC, Deviche P. 1998. Photoperiod and testosterone independently affect vocal control region volumes in adolescent male songbirds. J Neurobiol 36: 550-558.

Gurney M, Konishi M. 1980. Hormone induced sexual differentiation of brain and behavior in zebra finches. Science 208: 1380-1382.

Haesler S, Wada K, Nshdejan A, Morrisey EE, Lints T, et al. 2004. FoxP2 expression in avian vocal learners and non-learners. J Neurosci 24: 3164-3175.

Hahnloser RH, Kozhevnikov AA, Fee MS. 2002. An ultra-sparse code underlies the generation of neural sequences in a songbird. Nature 419: 65-70.

Hartman VN, Miller MA, Clayton DF, Liu W-C, Kroodsma DE, et al. 2001. Testosterone regulates alpha-synuclein mRNA in the avian song system. Neuroreport 12: 943-946.

Heinrich JE, Singh TD, Nordeen KW, Nordeen EJ. 2003. NR2B downregulation in a forebrain region required for avian vocal learning is not sufficient to close the sensitive period for song learning. Neurobiol Learn Mem 79: 99-108.

Heinrich JE, Singh TD, Sohrabji F, Nordeen KW, Nordeen EJ. 2002. Developmental and hormonal regulation of NR2A mRNA in forebrain regions controlling avian vocal learning. J Neurobiol 51: 149-159.

Hernandez AM, MacDougall-Shackleton SA. 2004. Effects of early song experience on song preferences and song control and auditory brain regions in female house finches (*Carpodacus mexicanus*). J Neurobiol 59: 247-258.

Herrmann K, Bischof H-J. 1986. Delayed development of song control nuclei in the zebra finch is related to behavioral development. J Comp Neurol 245: 167-175.

Hessler NA, Doupe AJ. 1999a. Singing-related neural activity in a dorsal forebrain–basal ganglia circuit of adult zebra finches. J Neurosci 19: 10461-10481.

Hessler NA, Doupe AJ. 1999b. Social context modulates singing-related neural activity in the songbird forebrain. Nat Neurosci 2: 209-211.

Hidalgo A, Barami K, Iversen K, Goldman SA. 1995. Estrogens and non-estrogenic ovarian influences combine to promote the recruitment and decrease the turnover of new neurons in the adult female canary brain. J Neurobiol 27: 470-487.

Hillier LW, Miller W, Birney E, Warren W, Hardison RC, et al. 2004. Sequence and comparative analysis of the chicken genome provide unique perspectives on vertebrate evolution. Nature 432: 695-716.

Holloway CC, Clayton DF. 2001. Estrogen synthesis in the male brain triggers development of the zebra finch song control circuit in vitro. Nat Neurosci 4: 170-175.

Holzenberger M, Jarvis ED, Chong C, Grossman M, Nottebohm F, et al. 1997. Selective expression of insulin-like growth factor II in the songbird brain. J Neurosci 17: 6974-6987.

Huang LY, Devries GJ, Bittman EL. 1998. Photoperiod regulates neuronal bromodeoxyuridine labeling in the brain. J Neurobiol 36: 410-420.

Huesmann G, Clayton DF. 2002. Is it memory or is it death? Rapid non-lethal caspase-3 activation during acoustic memory trace formation. Soc Neurosci Abstr 28: #382.383.

Immelmann K. 1959. Experimentelle Untersuchungen über die biologische Bedeutung artsspezifischer Merkmale beim Zebrafinken (*Taeniopygia guttata* Gould). Zoologische Jahrbücher Abteilung für Systematik Okologie und Geographie der Tiere 86: 437-592.

Immelmann K. 1969. Song development in the zebra finch and other estrilid finches. Bird vocalizations. Hinde RA, editor. Cambridge, UK: Cambridge University Press; pp. 61-74

Jacobs EC, Arnold AP, Campagnoni AT. 1996. Zebra finch estrogen receptor cDNA—cloning and mRNA expression. J Steroid Biochem Mol Biol 59: 135-145.

Jacobs EC, Arnold AP, Campagnoni AT. 1999. Developmental regulation of the distribution of aromatase- and estrogen-receptor- mRNA-expressing cells in the zebra finch brain. Dev Neurosci 21: 453-472.

Jarvis ED. 2004. Learned birdsong and the neurobiology of human language. Ann NY Acad Sci 1016: 749-777.

Jarvis ED, Mello CV. 2000. Molecular mapping of brain areas involved in parrot vocal communication. J Comp Neurol 419: 1-31.

Jarvis ED, Nottebohm F. 1997. Motor-driven gene expression. Proc Natl Acad Sci USA 94: 4097-4102.

Jarvis ED, Mello CV, Nottebohm F. 1995. Associative learning and stimulus novelty influence the song-induced expression of an immediate early gene in the canary forebrain. Learn Mem 2: 62-80.

Jarvis ED, Schwabl H, Ribeiro S, Mello CV. 1997. Brain gene regulation by territorial singing behavior in freely ranging songbirds. Neuroreport 8: 2073-2077.

Jarvis ED, Scharff C, Grossman MR, Ramos JA, Nottebohm F. 1998. For whom the bird sings—context-dependent gene expression. Neuron 21: 775-788.

Jarvis ED, Ribeiro S, da Silva ML, Ventura D, Vielliard J, et al. 2000. Behaviourally driven gene expression reveals song nuclei in hummingbird brain. Nature 406: 628-632.

Jarvis ED, Gunturkun O, Bruce L, Csillag A, Karten H, et al. 2005. Opinion: avian brains and a new understanding of vertebrate brain evolution. Nat Rev Neurosci 6: 151-159.

Jiang JW, McMurtry J, Niedzwiecki D, Goldman SA. 1998. Insulin-like growth factor-1 is a radial cell-associated neurotrophin that promotes neuronal recruitment from the adult songbird ependyma/subependyma. J Neurobiol 36: 1-15.

Jin H, Clayton DF. 1995. Synelfin mRNA expression is differentially regulated in nucleus LMAN of zebra finches raised under acoustic isolation. Soc Neurosci Abstr 21: 960.

Jin H, Clayton DF. 1997a. Localized changes in immediate-early gene regulation during sensory and motor learning in zebra finches. Neuron 19: 1049-1059.

Jin H, Clayton DF. 1997b. Synelfin regulation during the critical period for song learning in normal and isolated juvenile zebra finches. Neurobiol Learn Mem 68: 271-284.

Johnson F, Bottjer SW. 1995. Differential estrogen accumulation among populations of projection neurons in the higher vocal center of male canaries. J Neurobiol 26: 87-108.

Johnson F, Norstrom E, Soderstrom K. 2000. Increased expression of endogenous biotin, but not BDNF, in telencephalic song regions during zebra finch vocal learning. Brain Res Dev Brain Res 120: 113-123.

Johnson F, Hohmann SE, Distefano PS, Bottjer SW. 1997. Neurotrophins suppress apoptosis induced by deafferentation of an avian motor-cortical region. J Neurosci 17: 2101-2111.

Kafitz KW, Herth G, Bartsch U, Guttinger HR, Schachner M. 1992. Application of testosterone accelerates oligodendrocyte maturation in brains of zebra finches. Neuroreport 3: 315-318.

Kandel ER. 2001. The molecular biology of memory storage: a dialogue between genes and synapses. Science 294: 1030-1038.

Kao MH, Doupe AJ, Brainard MS. 2005. Contributions of an avian basal ganglia–forebrain circuit to real-time modulation of song. Nature 433: 638-643.

Kim YH, Perlman WR, Arnold AP. 2004. Expression of androgen receptor mRNA in zebra finch song system: developmental regulation by estrogen. J Comp Neurol 469: 535-547.

Kimpo RR, Doupe AJ. 1997. Fos is induced by singing in distinct neuronal populations in a motor network. Neuron 18: 315-325.

Kirn JR, Nottebohm F. 1993. Direct evidence for loss and replacement of projection neurons in adult canary brain. J Neurosci 13: 1654-1663.

Kirn JR, Schwabl H. 1997. Photoperiod regulation of neuron death in the adult canary. J Neurobiol 33: 223-231.

Kirn JR, O'Loughlin B, Kasparian S, Nottebohm F. 1994. Cell death and neuronal recruitment in the high vocal center of adult male canaries are temporally related to changes in song. Proc Natl Acad Sci USA 91: 7844-7848.

Knapska E, Kaczmarek L. 2004. A gene for neuronal plasticity in the mammalian brain: Zif268/Egr-1/NGFI-A/Krox-24/TIS8/ZENK? Prog Neurobiol 74: 183-211.

Konishi M. 1965. The role of auditory feedback in the control of vocalizations in the white-crowned sparrow. Z Tierpsychol 22: 770-783.

Konishi M. 2004. The role of auditory feedback in birdsong. Ann NY Acad Sci 1016: 463-475.

Konishi M, Akutagawa E. 1985. Neuronal growth, atrophy and death in a sexually dimorphic song nucleus in the zebra finch. Nature 315: 145-147.

Konishi M, Akutagawa E. 1988. A critical period for estrogen action on neurons of the song control system in the zebra finch. Proc Natl Acad Sci USA 85: 7006-7007.

Konishi M, Emlen ST, Ricklefs RE, Wingfield JC. 1989. Contributions of bird studies to biology. Science 246: 465-472.

Korsia S, Bottjer S. 1991. Chronic testosterone treatment impairs vocal learning in male zebra finches during a restricted period of development. J Neurosci 11: 2362-2371.

Kruse AA, Stripling R, Clayton DF. 2000. Minimal experience required for immediate-early gene induction in zebra finch neostriatum. Neurobiol Learn Mem 74: 179-184.

Kruse AA, Stripling R, Clayton DF. 2004. Context-specific habituation of the *zenk* gene response to song in adult zebra finches. Neurobiol Learn Mem 82: 99-108.

Lanahan A, Worley P. 1998. Immediate-early genes and synaptic function. Neurobiol Learn Mem 70: 37-43.

Lemaire P, Revelant O, Bravo R, Charnay P. 1988. Two mouse genes encoding potential transcription factors with identical DNA-binding domains are activated by growth factors in cultured cells. Proc Natl Acad Sci USA 85: 4691-4695.

Leonardo A. 2004. Experimental test of the birdsong error-correction model. Proc Natl Acad Sci USA 101: 16935-16940.

Leonardo A, Fee MS. 2005. Ensemble coding of vocal control in birdsong. J Neurosci 25: 652-661.

Leonardo A, Konishi M. 1999. Decrystallization of adult birdsong by perturbation of auditory feedback. Nature 399: 466-470.

Li XC, Jarvis ED, Alvarez-Borda B, Lim DA, Nottebohm F. 2000. A relationship between behavior, neurotrophin expression, and new neuron survival. Proc Natl Acad Sci USA 97: 8584-8589.

Lieshoff C, Grosse-Ophoff J, Bischof HJ. 2004. Sexual imprinting leads to lateralized and non-lateralized expression of the immediate early gene *zenk* in the zebra finch brain. Behav Brain Res 148: 145-155.

Lind H, Dabelsteen T, McGregor PK. 1996. Female great tits can identify mates by song. Anim Behav 52: 667-671.

Lipkind D, Nottebohm F, Rado R, Barnea A. 2002. Social change affects the survival of new neurons in the forebrain of adult songbirds. Behav Brain Res 133: 31-43.

Livingston FS, Mooney R. 1997. Development of intrinsic and synaptic properties in a forebrain nucleus essential to avian song learning. J Neurosci 17: 8997-9009.

Livingston FS, White SA, Mooney R. 2000. Slow NMDA–EPSCs at synapses critical for song development are not required for song learning in zebra finches. Nat Neurosci 3: 482-488.

Lombardino AJ, Nottebohm F. 2000. Age at deafening affects the stability of learned song in adult male zebra finches. J Neurosci 20: 5054-5064.

London SE, Boulter J, Schlinger BA. 2003. Cloning of the zebra finch androgen synthetic enzyme CYP17: a study of its neural expression throughout posthatch development. J Comp Neurol 467: 496-508.

Long KD, Salbaum JM. 1998. Evolutionary conservation of the immediate-early gene *zenk*. Mol Biol Evol 15: 284-292.

Macdougall-Shackleton SA, Macdougall-Shackleton EA, Hahn TP. 2001. Physiological and behavioural responses of female mountain white-crowned sparrows to natal- and foreign-dialect songs. Can J Zool 79: 325-333.

Maney DL, Ball GF. 2003. Fos-like immunoreactivity in catecholaminergic brain nuclei after territorial behavior in free-living song sparrows. J Neurobiol 56: 163-170.

Maney DL, Wingfield JC. 1998. Central opioid control of feeding behavior in the white-crowned sparrow, *Zonotrichia leucophrys* gambelii. Hormones Behav 33: 16-22.

Maney DL, Bernard DJ, Ball GF. 2001. Gonadal steroid receptor mRNA in catecholaminergic nuclei of the canary brainstem. Neurosci Lett 311: 189-192.

Maney DL, Goode CT, Wingfield JC. 1997. Intraventricular infusion of arginine vasotocin induces singing in a female songbird. J Neuroendocrinol 9: 487-491.

Maney DL, MacDougall-Shackleton EA, MacDougall-Shackleton SA, Ball GF, Hahn TP. 2003. Immediate early gene response to hearing song correlates with receptive behavior and depends on dialect in a female songbird.

J Comp Physiol A Neuroethol Sens Neural Behav Physiol 189: 667-674.

Marchetti C, Drent PJ. 2000. Individual differences in the use of social information in foraging by captive great tits. Anim Behav 60: 131-140.

Margoliash D. 1997. Functional organization of forebrain pathways for song production and perception. J Neurobiol 33: 671-693.

Margoliash D. 2002. Evaluating theories of bird song learning: implications for future directions. J Comp Physiol A Neuroethol Sens Neural Behav Physiol 188: 851-866.

Marler P. 1976. An ethological theory of the origin of vocal learning. Ann NY Acad Sci 280: 386-395.

Marler P. 1997. Three models of song learning—evidence from behavior. J Neurobiol 33: 501-516.

Marler P, Peters S. 1977. Selective vocal learning in a sparrow. Science 198: 519-521.

Marler P, Peters S. 1982. Long-term storage of learned bird songs prior to production. Anim Behav 30: 479-482.

Marler P, Peters S. 1987. A sensitive period for song acquisition in the song sparrow, *Melospiza melodia:* a case of age-limited learning. Ethology 76: 89-100.

Marler P, Waser M. 1977. The role of auditory feedback in canary song development. J Comp Physiol Psychol 91: 8-16.

Marler P, Peters S, Ball GF, Dufty AM Jr, Wingfield JC. 1988. The role of sex steroids in the acquisition and production of birdsong. Nature 336: 770-772.

McCasland JS, Konishi M. 1981. Interaction between auditory and motor activities in an avian song control nucleus. Proc Natl Acad Sci USA 78: 7815-7819.

McGaugh JL. 2000. Memory—a century of consolidation. Science 287: 248-251.

Mello CV. 2002. Mapping vocal communication pathways in birds with inducible gene expression. J Comp Physiol A Neuroethol Sens Neural Behav Physiol 188: 943-959.

Mello CV, Clayton DF. 1994. Song-induced *zenk* gene expression in auditory pathways of songbird brain and its relation to the song control system. J Neurosci 14: 6652-6666.

Mello CV, Ribeiro S. 1998. Zenk protein regulation by song in the brain of songbirds. J Comp Neurol 393: 426-438.

Mello CV, Nottebohm F, Clayton DF. 1995. Repeated exposure to one song leads to a rapid and persistent decline in an immediate early gene's response to that song in zebra finch telencephalon. J Neurosci 15: 6919-6925.

Mello CV, Pinaud R, Ribeiro S. 1998. Noradrenergic system of the zebra finch brain: immunocytochemical study of dopamine-beta-hydroxylase. J Comp Neurol 400: 207-228.

Mello CV, Velho TA, Pinaud R. 2004. Song-induced gene expression: a window on song auditory processing and perception. Ann NY Acad Sci 1016: 263-281.

Mello CV, Vicario DS, Clayton DF. 1992. Song presentation induces gene expression in the songbird forebrain. Proc Natl Acad Sci USA 89: 6818-6822.

Metzdorf R, Gahr M, Fusani L. 1999. Distribution of aromatase, estrogen receptor, and androgen receptor mRNA in the forebrain of songbirds and nonsongbirds. J Comp Neurol 407: 115-129.

Milbrandt J. 1987. A nerve growth factor-induced gene encodes a possible transcriptional regulatory factor. Science 238: 797-799.

Miller DB. 1979a. The acoustic basis of mate recognition by female zebra finches (*Taeniopygia guttata*). Anim Behav 27: 376-380.

Miller DB. 1979b. Long-term recognition of father's song by female zebra finches. Nature 280: 389-391.

Mooney R. 1992. Synaptic basis for developmental plasticity in a birdsong nucleus. J Neurosci 12: 2464-2477.

Mooney R. 2004. Synaptic mechanisms for auditory–vocal integration and the correction of vocal errors. Ann NY Acad Sci 1016: 476-494.

Mooney R, Rao M. 1994. Waiting periods versus early innervation: the development of axonal connections in the zebra finch song system. J Neurosci 14: 6532-6543.

Morrison R, Nottebohm F. 1993. Role of a telencephalic nucleus in the delayed song learning of socially isolated zebra finches. J Neurobiol 24: 1045-1064.

Nadon R, Shoemaker J. 2002. Statistical issues with microarrays: processing and analysis. Trends Genet 18: 265-271.

Nagle L, Kreutzer ML. 1997. Song tutoring influences female song preferences in domesticated canaries. Behaviour 134: 89-104.

Nastiuk KL, Clayton DF. 1994. Seasonal and tissue-specific regulation of the canary androgen receptor mRNA. Endocrinol 134: 640-649.

Nastiuk KL, Clayton DF. 1995. The canary androgen receptor mRNA is localized in the song control nuclei of the brain and is rapidly regulated by testosterone. J Neurobiol 26: 213-224.

Nastiuk KL, Mello CV, George JM, Clayton DF. 1994. Immediate-early gene responses in the avian song control system: cloning and expression analysis of the canary c-jun cDNA. Mol Brain Res 27: 299-309.

Nordeen KW, Nordeen EJ. 1988. Projection neurons within a vocal motor pathway are born during song learning in zebra finches. Nature 334: 149-151.

Nordeen KW, Nordeen EJ. 1992. Auditory feedback is necessary for the maintenance of stereotyped song in adult zebra finches. Behav Neural Biol 57: 58-66.

Nordeen KW, Nordeen EJ. 2004. Synaptic and molecular mechanisms regulating plasticity during early learning. Ann NY Acad Sci 1016: 416-437.

Nordeen KW, Nordeen EJ, Arnold AP. 1986. Estrogen establishes sex differences in androgen accumulation in zebra finch brain. J Neurosci 6: 734-738.

Nordeen EJ, Nordeen KW, Arnold AP. 1987a. Sexual differentiation of androgen accumulation within the zebra finch brain through selective cell loss and addition. J Comp Neurol 259: 393-399.

Nordeen KW, Nordeen EJ, Arnold AP. 1987b. Estrogen accumulation in zebra finch song control nuclei: implications for sexual differentiation and adult activation of song behavior. J Neurobiol 18: 569-582.

Nordeen EJ, Voelkel L, Nordeen KW. 1998. Fibroblast growth factor-2 stimulates cell proliferation and decreases sexually dimorphic cell death in an avian song control nucleus. J Neurobiol 37: 573-581.

Nottebohm F. 1968. Auditory experience and song development in the Chaffinch, *Fringilla coelebs*. Ibis 110: 549-568.

Nottebohm F. 1981. A brain for all seasons: cyclical anatomical changes in song control nuclei of the canary brain. Science 214: 1368-1370.

Nottebohm F. 1989. From birdsong to neurogenesis. Sci Am 260: 74-79.

Nottebohm F. 2002. Why are some neurons replaced in adult brain? J Neurosci 22: 624-628.

Nottebohm F. 2004. The road we travelled: discovery, choreography, and significance of brain replaceable neurons. Ann NY Acad Sci 1016: 628-658.

Nottebohm F, Arnold A. 1976. Sexual dimorphism in vocal control areas of the songbird brain. Science 194: 211-213.

Nottebohm F, Stokes T, Leonard CM. 1976. Central control of song in the canary. J Comp Neurol 165: 457-486.

O'Donovan KJ, Tourtellotte WG, Milbrandt J, Baraban JM. 1999. The EGR family of transcription-regulatory factors: progress at the interface of molecular and systems neuroscience. Trends Neurosci 22: 167-173.

O'Loghlen AL, Beecher MD. 1999. Mate, neighbour and stranger songs: a female song sparrow perspective. Anim Behav 58: 13-20.

Park KH, Clayton DF. 2002. Influence of restraint and acute isolation on the selectivity of the adult zebra finch *zenk* gene response to acoustic stimuli. Behav Brain Res 136: 185-191.

Perfito N, Tramontin AD, Meddle S, Sharp P, Afik D, et al. 2004. Reproductive development according to elevation in a seasonally breeding male songbird. Oecologia 140: 201-210.

Perkel DJ. 2004. Origin of the anterior forebrain pathway. Ann NY Acad Sci 1016: 736-748.

Perlman WR, Arnold AP. 2003. Expression of estrogen receptor and aromatase mRNAs in embryonic and posthatch zebra finch brain. J Neurobiol 55: 204-219.

Perlman WR, Ramachandran B, Arnold AP. 2003. Expression of androgen receptor mRNA in the late embryonic and early posthatch zebra finch brain. J Comp Neurol 455: 513-530.

Peterson RS, Saldanha CJ, Schlinger BA. 2001. Rapid upregulation of aromatase mRNA and protein following neural injury in the zebra finch (*Taeniopygia guttata*). J Neuroendocrinol 13: 317-323.

Peterson RS, Lee DW, Fernando G, Schlinger BA. 2004. Radial glia express aromatase in the injured zebra finch brain. J Comp Neurol 475: 261-269.

Phillmore LS, Bloomfield LL, Weisman RG. 2003. Effects of songs and calls on *zenk* expression in the auditory telencephalon of field- and isolate-reared black capped chickadees. Behav Brain Res 147: 125-134.

Pinaud R, Velho TA, Jeong JK, Tremere LA, Leao RM, von Gersdorff H, et al. 2004. GABAergic neurons participate in the brain's response to birdsong auditory stimulation. Eur J Neurosci 20: 1318-1330.

Price P. 1979. Developmental determinants of structure in zebra finch song. J Comp Physiol Psychol 93: 260-277.

Rasika S, Alvarez-Buylla A, Nottebohm F. 1999. BDNF mediates the effects of testosterone on the survival of new neurons in an adult brain. Neuron 22: 53-62.

Rasika S, Nottebohm F, Alvarez-Buylla A. 1994. Testosterone increases the recruitment and/or survival of new high vocal center neurons in adult female canaries. Proc Natl Acad Sci USA 91: 7854-7858.

Reeves BJ, Beecher MD, Brenowitz EA. 2003. Seasonal changes in avian song control circuits do not cause seasonal changes in song discrimination in song sparrows. J Neurobiol 57: 119-129.

Reiner A, Perkel DJ, Mello CV, Jarvis ED. 2004a. Songbirds and the revised avian brain nomenclature. Ann NY Acad Sci 1016: 77-108.

Reiner A, Perkel DJ, Bruce L, Butler A, et al. 2004b. Revised nomenclature for avian telencephalon and some related brainstem nuclei. J Comp Neurol 473: 377-414.

Remage-Healey L, Adkins-Regan E, Romero LM. 2003. Behavioral and adrenocortical responses to mate separation and reunion in the zebra finch. Hormones Behav 43: 108-114.

Ribeiro S, Mello CV. 2000. Gene expression and synaptic plasticity in the auditory forebrain of songbirds. Learn Mem 7: 235-243.

Ribeiro S, Cecchi GA, Magnasco MO, Mello CV. 1998. Toward a song code—evidence for a syllabic representation in the canary brain. Neuron 21: 359-371.

Ribeiro S, Goyal V, Mello CV, Pavlides C. 1999. Brain gene expression during REM sleep depends on prior waking experience. Learn Mem 6: 500-508.

Riebel K. 2000. Early exposure leads to repeatable preferences for male song in female zebra finches. Proc R Soc Lond B Biol Sci 267: 2553-2558.

Riebel K, Slater PJB. 1998. Testing female chaffinch song preferences by operant conditioning. Anim Behav 56: 1443-1453.

Riebel K, Smallegange IM, Terpstra NJ, Bolhuis JJ. 2002. Sexual equality in zebra finch song preference: evidence for a dissociation between song recognition and production learning. Proc R Soc Lond B Biol Sci 269: 729-733.

Rose GJ, Goller F, Gritton HJ, Plamondon SL, Baugh AT, et al. 2004. Species-typical songs in white-crowned sparrows tutored with only phrase pairs. Nature 432: 753-758.

Rousselot P, Heintz N, Nottebohm F. 1997. Expression of brain lipid binding protein in the brain of the adult canary and its implications for adult neurogenesis. J Comp Neurol 385: 415-426.

Ryan S, Arnold A. 1981. Evidence for cholinergic participation in the control of birdsong: acetylcholinesterase distribution and muscarinic receptor autoradiography in the zebra finch brain. J Comp Neurol 202: 211-219.

Sadananda M, Bischof HJ. 2002. Enhanced fos expression in the zebra finch (*Taeniopygia guttata*) brain following first courtship. J Comp Neurol 448: 150-164.

Sadananda M, Bischof HJ. 2004. c-fos is induced in the hippocampus during consolidation of sexual imprinting in the zebra finch (*Taeniopygia guttata*). Hippocampus 14: 19-27.

Sakaguchi H, Saito N. 1989. The acetylcholine and catecholamine contents in song control nuclei of zebra finch during song ontogeny. Dev Brain Res 47: 313-317.

Sakaguchi H, Saito N. 1991. Developmental change of cholinergic activity in the forebrain of the zebra finch during song learning. Dev Brain Res 62: 223-228.

Sakaguchi H, Saito N. 1996. Developmental changes in axon terminals visualized by immunofluorescence for the growth-associated protein, GAP-43, in the robust nucleus of the archistriatum of the zebra finch. Dev Brain Res 95: 245-251.

Sakaguchi H, Yamaguchi A. 1997. Early song-deprivation affects the expression of protein kinase c in the song control nuclei of the zebra finch during a sensitive period of song learning. Neuroreport 8: 2645-2650.

Sakaguchi H, Wada K, Maekawa M, Watsuji T, Hagiwara M. 1999. Song-induced phosphorylation of cAMP response element-binding protein in the songbird brain. J Neurosci 19: 3973-3981.

Saldanha CJ, Schlinger BA. 1997. Estrogen synthesis and secretion in the brown-headed cowbird (*Molothrus ater*). Gen Comp Endocrinol 105: 390-401.

Saldanha CJ, Popper P, Micevych PE, Schlinger BA. 1998. The passerine hippocampus is a site of high aromatase: inter- and intraspecies comparisons. Hormones Behav 34: 85-97.

Saldanha CJ, Schlinger BA, Micevych PE, Horvath TL. 2004. Presynaptic N-methyl-D-aspartate receptor expression is increased by estrogen in an aromatase-rich area of the songbird hippocampus. J Comp Neurol 469: 522-534.

Saldanha CJ, Schultz JD, London SE, Schlinger BA. 2000. Telencephalic aromatase but not a song circuit in a sub-oscine passerine, the golden collared manakin (*Manacus vitellinus*). Brain Behav Evol 56: 29-37.

Salgado-Commissariat D, Rosenfield DB, Helekar SA. 2004. Nicotine-mediated plasticity in robust nucleus of the archistriatum of the adult zebra finch. Brain Res 1018: 97-105.

Scharff C, Nottebohm F. 1991. A comparative study of the behavioral deficits following lesions of various parts of the zebra finch song system: implications for vocal learning. J Neurosci 11: 2896-2913.

Scharff C, Kirn JR, Grossman M, Macklis JD, Nottebohm F. 2000. Targeted neuronal death affects neuronal replacement and vocal behavior in adult songbirds. Neuron 25: 481-492.

Schena M, Shalon D, Davis RW, Brown PO. 1995. Quantitative monitoring of gene expression patterns with a complementary DNA microarray. Science 270: 467-470.

Schlinger BA, Arnold AP. 1991. Brain is the major site of estrogen synthesis in a male songbird. Proc Natl Acad Sci USA 88: 4191-4194.

Schlinger BA, Soma KK, London SE. 2001. Neurosteroids and brain sexual differentiation. Trends Neurosci 24: 429-431.

Schlinger BA, Amurumarjee S, Campagnoni AT, Arnold AP. 1995. 5-beta-reductase and other androgen-metabolizing enzymes in primary cultures of developing zebra finch telencephalon. J Neuroendocrinol 7: 187-192.

Schlinger BA, Lane NI, Grisham W, Thompson L. 1999. Androgen synthesis in a songbird: a study of Cyp17 (17 alpha-hydroxylase/C17, 20-lyase) activity in the zebra finch. Gen Comp Endocrinol 113: 46-58.

Schlinger BA, Amurumarjee S, Shen P, Campagnoni AT, Arnold AP. 1994. Neuronal and non-neuronal aromatase in primary cultures of developing zebra finch telencephalon. J Neurosci 14: 7541-7552.

Schneider T. 1995. Distribution of 2-[125I]Iodomelatonin binding sites in the brain of the pied flycatcher (*Ficedula hypoleuca*) and the zebra finch (*Taeniopygia guttata*). J Exp. Biol. 198: 1943-1949.

Schwabl H. 1996. Environment modifies the testosterone levels of a female bird and its eggs. J Exp Zool 276: 157-163.

Scott LL, Singh TD, Nordeen EJ, Nordeen KW. 2004. Developmental patterns of NMDAR expression within the song system do not recur during adult vocal plasticity in zebra finches. J Neurobiol 58: 442-454.

Shen P, Schlinger BA, Campagnoni AT, Arnold AP. 1995. An atlas of aromatase mRNA expression in the zebra finch brain. J Comp Neurol 360: 172-184.

Shen P, Campagnoni CW, Kampf K, Schlinger BA, Arnold AP, et al. 1994. Isolation and characterization of a zebra finch aromatase cDNAa—in situ hybridization reveals high aromatase expression in brain. Mol Brain Res 24: 227-237.

Shy E, McGregor PK, Krebs JR. 1986. Discrimination of song types by great tits. Behav Processes 13: 1-12.

Siepka SM. 1997. Molecular analysis of neural structures during song learning in zebra finch, Doctoral Dissertation. University of Illinois: Urbana, IL.

Siepka S, Clayton DF. 1995. Changes in neurofilament protein immunoreactivity during development of the song control circuit in zebra finches. Soc Neurosci Abstr 21: 39.

Siepka SM, George JM, Jin H, Clayton DF. 1994. The HAT-14 mRNA is enriched in several avian song control nuclei and encodes an RC3/neurogranin-related protein. Soc Neurosci Abstr 20: 1436.

Simon RM, Dobbin K. 2003. Experimental design of DNA microarray experiments. Biotechniques Suppl: 16-21.

Simpson HB, Vicario DS. 1991a. Early estrogen treatment alone causes female zebra finches to produce learned, male-like vocalizations. J Neurobiol 22: 755-776.

Simpson HB, Vicario DS. 1991b. Early estrogen treatment of female zebra finches masculinizes the brain pathway for learned vocalizations. J Neurobiol 22: 777-793.

Singh TD, Nordeen EJ, Nordeen KW. 2005. Song tutoring triggers CaMKII phosphorylation within a specialized portion of the avian basal ganglia. J Neurobiol 65: 179-191.

Singh TD, Basham ME, Nordeen EJ, Nordeen KW. 2000. Early sensory and hormonal experience modulate age-related changes in NR2B mRNA within a forebrain region controlling avian vocal learning. J Neurobiol 44: 82-94.

Singh TD, Heinrich JE, Wissman AM, Brenowitz EA, Nordeen EJ, et al. 2003. Seasonal regulation of NMDA receptor NR2B mRNA in the adult canary song system. J Neurobiol 54: 593-603.

Slater PJB, Eales LA, Clayton NS. 1988. Song learning in zebra finches (*Taeniopygia guttata*): Progress and prospects. Adv Study Behav 18: 1-34.

Slater PJB, Jones A, Ten Cate C. 1993. Can lack of experience delay the end of the sensitive phase for song learning? Netherlands J Zool 43: 80-90.

Slater PJB, Richards C, Mann NI. 1991. Song learning in zebra finches exposed to a series of tutors during the sensitive phase. Ethology 88: 163-171.

Smith GT, Brenowitz EA, Wingfield JC. 1997. Roles of photoperiod and testosterone in seasonal plasticity of the avian song control system. J Neurobiol 32: 426-442.

Smulders TV, DeVoogd TJ. 2000. Expression of immediate early genes in the hippocampal formation of the black-capped chickadee (*Poecile atricapillus*) during a food-hoarding task. Behav Brain Res 114: 39-49.

Sockman KW, Gentner TQ, Ball GF. 2002. Recent experience modulates forebrain gene-expression in response to mate-choice cues in European starlings. Proc R Soc Lond B Biol Sci 269: 2479-2485.

Soderstrom K, Johnson F. 2000. CB1 cannabinoid receptor expression in brain regions associated with zebra finch song control. Brain Res 857: 151-157.

Soderstrom K, Johnson F. 2001. Zebra finch CB1 cannabinoid receptor: pharmacology and in vivo and in vitro effects of activation. J Pharmacol Exp Ther 297: 189-197.

Soderstrom K, Johnson F. 2003. Cannabinoid exposure alters learning of zebra finch vocal patterns. Brain Res Dev Brain Res 142: 215-217.

Soderstrom K, Tian Q. 2004. Distinct periods of cannabinoid sensitivity during zebra finch vocal development. Brain Res Dev Brain Res 153: 225-232.

Soderstrom K, Tian Q, Valenti M, Di Marzo V. 2004. Endocannabinoids link feeding state and auditory perception-related gene expression. J Neurosci 24: 10013-10021.

Soha JA, Shimizu T, Doupe AJ. 1996. Development of the catecholaminergic innervation of the song system of the male zebra finch. J Neurobiol 29: 473-489

Sohrabji F, Nordeen KW, Nordeen EJ. 1989. Projections of androgen-accumulating neurons in a nucleus controlling avian song. Brain Res 488: 253-259.

Soma KK, Alday NA, Hau M, Schlinger BA. 2004. Dehydroepiandrosterone metabolism by 3beta-hydroxysteroid dehydrogenase/Delta5-Delta4 isomerase in adult zebra finch brain: sex difference and rapid effect of stress. Endocrinol 145: 1668-1677.

Stoddard PK. 1996. Vocal recognition of neighbors by territorial passerines. Ecology and evolution of acoustic communication in birds. Kroodsma DE, Miller EH, editors. Ithaca, NY: Cornell University Press; pp. 356-374.

Stripling R, Volman S, Clayton D. 1997. Response modulation in the zebra finch caudal neostriatum: relationship to nuclear gene regulation. J Neurosci 17: 3883-3893.

Stripling R, Milewski L, Kruse AA, Clayton DF. 2003. Rapidly learned song-discrimination without behavioral reinforcement in adult male zebra finches (*Taeniopygia guttata*). Neurobiol Learn Mem 79: 41-50.

Sukhatme VP, Cao X, Chang LC, Tsai-Morris CH, Stamenkovitch D, et al. 1988. A zinc finger-encoding gene coregulated with c-fos during growth and differentiation, and after cellular depolarization. Cell 53: 37-43.

Sweatt JD. 2003. Mechanisms of memory. New York: Academic Press; pp. 1-350.

Teramitsu I, Kudo LC, London SE, Geschwind DH, White SA. 2004. Parallel FoxP1 and FoxP2 expression in songbird and human brain predicts functional interaction. J Neurosci 24: 3152-3163.

Terpstra NJ, Bolhuis JJ, den Boer-Visser AM. 2004. An analysis of the neural representation of birdsong memory. J Neurosci 24: 4971-4977.

Theunissen FE, Amin N, Shaevitz SS, Woolley SM, Fremouw T, et al. 2004. Song selectivity in the song system and in the auditory forebrain. Ann NY Acad Sci 1016: 222-245.

Thorpe WH. 1958. The learning of song patterns by birds, with especial reference to the song of the chaffinch, *Fringilla coelebs.* Ibis 100: 535-570.

Tramontin AD, Brenowitz EA. 2000. Seasonal plasticity in the adult brain. Trends Neurosci 23: 251-258.

Tramontin AD, Wingfield JC, Brenowitz PA. 1999. Contributions of social cues and photoperiod to seasonal plasticity in the adult avian song control system. J Neurosci 19: 476-483.

Troyer TW, Doupe AJ. 2000a. An associational model of birdsong sensorimotor learning I. Efference copy and the learning of song syllables. J Neurophysiol 84: 1204-1223.

Troyer TW, Doupe AJ. 2000b. An associational model of birdsong sensorimotor learning II. Temporal hierarchies and the learning of song sequence. J Neurophysiol 84: 1224-1239.

van Oers K, de Jong G, Drent PJ, Noordwijk van AJ. 2004a. A genetic analysis of avian personality traits: correlated, response to artificial selection. Behav Genet 34: 611-619.

van Oers K, Drent PJ, de Goede P, Noordwijk van AJ. 2004b. Realized heritability and repeatability of risk-taking behaviour in relation to avian personalities. Proc R Soc Lond B Biol Sci 271: 65-73.

van Oers K, Drent PJ, de Jong G, Noordwijk van AJ. 2004c. Additive and nonadditive genetic variation in avian personality traits. Heredity 93: 496-503.

Vanson A, Arnold AP, Schlinger BA. 1996. 3-beta-hydroxysteroid dehydrogenase/isomerase and aromatase activity in primary cultures of developing zebra finch telencephalon—dehydroepiandrosterone as substrate for synthesis of androstenedione and estrogens. Gen Comp Endocrinol 102: 342-350.

Veney SL, Peabody C, Smith GW, Wade J. 2003. Sexually dimorphic neurocalcin expression in the developing zebra finch telencephalon. J Neurobiol 56: 372-386.

Vicario DS, Simpson HB. 1995. Electrical stimulation in forebrain nuclei elicits learned vocal patterns in songbirds. J Neurophysiol 73: 2602-2607.

Vignal C, Mathevon N, Mottin S. 2004. Audience drives male songbird response to partner's voice. Nature 430: 448-451.

Voigt C, Metzdorf R, Gahr M. 2004. Differential expression pattern and steroid hormone sensitivity of SNAP-25 and synaptoporin mRNA in the telencephalic song control nucleus HVC of the zebra finch. J Comp Neurol 475: 83-94.

Vu ET, Mazurek ME, Kuo Y-C. 1994. Identification of a forebrain motor programming network for the learned song of zebra finches. J Neurosci 14: 6924-6934.

Wada K, Sakaguchi H, Jarvis ED, Hagiwara M. 2004. Differential expression of glutamate receptors in avian neural pathways for learned vocalization. J Comp Neurol 476: 44-64.

Wade J. 2000. TrkB-like immunoreactivity in the song system of developing zebra finches. J Chem Neuroanat 19: 33-39.

Wade J, Arnold AP. 2004. Sexual differentiation of the zebra finch song system. Ann NY Acad Sci 1016: 540-559.

Wade J, Schlinger BA, Arnold AP. 1995. Aromatase and 5 beta-reductase activity in cultures of developing zebra finch brain: an investigation of sex and regional differences. J Neurobiol 27: 240-251.

Wade J, Peabody C, Coussens P, Tempelman RJ, Clayton DF, et al. 2004. A cDNA microarray from the telencephalon of juvenile male and female zebra finches. J Neurosci Methods 138: 199-206.

Walker MP, Stickgold R. 2004. Sleep-dependent learning and memory consolidation. Neuron 44: 121-133.

Walker MP, Brakefield T, Hobson JA, Stickgold R. 2003. Dissociable stages of human memory consolidation and reconsolidation. Nature 425: 616-620.

Wallhausser-Franke E, Collins CE, Devoogd TJ. 1995a. Developmental changes in the distribution of NADPH-diaphorase-containing neurons in telencephalic nuclei of the zebra finch song system. J Comp Neurol 356: 345-354.

Wallhausser-Franke E, Nixdorf-Bergweiler BE, Devoogd TJ. 1995b. Song isolation is associated with maintaining high spine frequencies on zebra finch iman neurons. Neurobiol Learn Mem 64: 25-35.

Watanabe A, Kimura T, Sakaguchi H. 2002. Expression of protein kinase C in song control nuclei of deafened adult male Bengalese finches. Neuroreport 13: 127-132.

Watson JT, Adkins-Regan E, Whiting P, Lindsotrm JM, Podleski TR. 1988. Autoradiographic localization of nicotinic acetylcholine receptors in the brain of the zebra finch (*Poephila guttata*). J Comp Neurol 274: 255-264.

Weary DM. 1989. Categorical perception of bird song: how do great tits (*Parus major*) perceive temporal variation in their song? J Comp Psychol 103: 320-325.

Whaling CS, Soha JA, Nelson DA, Lasley B, Marler P. 1998. Photoperiod and tutor access affect the process of vocal learning. Anim Behav 56: 1075-1082.

White SA, Livingston FS, Mooney R. 1999. Androgens modulate NMDA receptor-mediated EPSCs in the zebra finch song system. J Neurophysiol 82: 2221-2234.

Whitfieldrucker MG, Cassone VM. 1996. Melatonin binding in the house sparrow song control system—sexual dimorphism and the effect of photoperiod. Hormones Behav 30: 528-537.

Whitney O, Soderstrom K, Johnson F. 2003. CB1 cannabinoid receptor activation inhibits a neural correlate of song recognition in an auditory/perceptual region of the zebra finch telencephalon. J Neurobiol 56: 266-274.

Wilbrecht L, Kirn JR. 2004. Neuron addition and loss in the song system: regulation and function. Ann NY Acad Sci 1016: 659-683.

Wilbrecht L, Crionas A, Nottebohm F. 2002a. Experience affects recruitment of new neurons but not adult neuron number. J Neurosci 22: 825-831.

Wilbrecht L, Petersen T, Nottebohm F. 2002b. Bilateral LMAN lesions cancel differences in HVC neuronal recruitment induced by unilateral syringeal denervation. Lateral magnocellular nucleus of the anterior neostriatum. J Comp Physiol A Neuroethol Sens Neural Behav Physiol 188: 909-915.

Williams H. 2004. Birdsong and singing behavior. Ann NY Acad Sci 1016: 1-30.

Williams H, McKibben JR. 1992. Changes in stereotyped central motor patterns controlling vocalization are induced by peripheral nerve injury. Behav Neural Biol 57: 67-78.

Wingfield J. 1984. Environmental and endocrine control of reproduction in the song sparrow, *Melospiza melodia* 2. Agonistic interactions as environmental information stimulating secretion of testosterone. Gen Comp Endocrinol 56: 417-424.

Wise PM. 2002. Estrogens and neuroprotection. Trends Endocrinol Metab 13: 229-230.

Wynne RD, Saldanha CJ. 2004. Glial aromatization decreases neural injury in the zebra finch (*Taeniopygia guttata*): influence on apoptosis. J Neuroendocrinol 16: 676-683.

Yang YH, Speed T. 2002. Design issues for cDNA microarray experiments. Nat Rev Genet 3: 579-588.

Yoshida Y, Yamada T, Sakaguchi H. 2003. Activation of protein kinase C by the error signal from a basal ganglia–forebrain circuit in the zebra finch song control nuclei. Neuroreport 14: 645-649.

Yu AC, Margoliash D. 1996. Temporal hierarchical control of singing in birds. Science 273: 1871-1875.

Zann RA. 1996. The zebra finch: a synthesis of field and laboratory studies. Oxford: Oxford University Press; pp. 1-335.

10 The Neuroendocrinology and Neurochemistry of Birdsong

G. F. Ball · J. Balthazart

Abstract: Birdsong is a complex learned motor skill that is used in the context of territory defense and mate choice. Songbirds have evolved a specialized neural circuit that controls the learning, production, and perception of song. Studies of the hormonal regulation of neurotransmitter systems in this specialized circuit provide an opportunity for neuroscientists to investigate the cellular neurochemistry of complex behaviors. In this review, we review the hormonal regulation of seasonal neuroplasticity in this circuit with special emphasis on the action of androgens including their estrogenic and androgenic metabolites. One theme that emerges is that androgens can induce seasonal changes in the morphology of the song circuit but that the ability of testosterone to be metabolized into androgenic or estrogenic metabolites also changes seasonally. The basic chemical neuroanatomy of the song system is reviewed and possible sites for the modulation of these transmitter systems by androgens are identified.

Lsit of Abbreviations: A, arcopallium; α_2, α_2- adrenergic receptors; Ach, acetylcholine; Achesterase, acetylcholine esterase; AF, aspiny fast-firing neurons; ant.Di, anterior diencephalon; AP5, 2-amino-5-phosphonovaleric acid; APH, area parahippocampalis; AR, androgen receptors; β1, Beta1 adrenergic receptors; β2, Beta2 adrenergic receptors; BDNF, brain-derived neurotrophic factor; BST, bed nucleus of the stria terminalis; BSTM, bed nucleus of the stria terminalis, medial part; CA, commissura anterior; Cereb, Cerebellum; ChAT, Choline acetyl transferase; DA, Dopamine; DBC, decussatio brachiorum conjunctivorum; DHEA, dehydroepiandrosterone; 5α-DHT, 5α-dihydrotestosterone; 5β-DHT, 5β-dihydrotestosterone; DLM, dorsolateral thalamic nucleus; DM, dorsomedial portion of the nucleus intercollicularis; DSD, decussatio supraoptica dorsalis; DSP-4, N-[2-chloroethyl]-N-ethyl-2-bromobenzylamine; E, Entopallium; ERα, Estrogen Receptors of the Alpha Subtype; ERβ, Estrogen Receptors of the Beta Subtype; FLM, fasciculus longitudinalis medialis; GABA, γ-amino butyric acid; GAD, glutamic acid decarboxylase; GCt, substantia grisea centralis (=mesencephalic central gray); GLv, nucleus geniculatus lateralis pars ventralis; GnRH, gonadotropin-releasing hormone; GP, globus pallidus; HA, accessory part of the hyperpallium; ^{3}H, tritium; HVC, Used as the proper name for the nucleus; ICC, immunocytochemistry; ISH, *in situ* hybridization; ICo, nucleus Intercollicularis; Km, estimated affinity; LMAN, lateral magnocellular nucleus of the anterior nidopallium; LSt, lateral striatum; LTP, Long Term Potentiation; M, mesopallium; MLd, nucleus mesencephalicus lateralis, pars dorsalis; Musc, Muscarinic Cholinergic Receptors; mRNA, messenger ribonucleic acid; N, nidopallium; NC, caudal nidopallium; NE, Norepinephrine; Nico, Nictotinic cholinergic receptors; Nif, nucleus interfacialis (=interface); NMDA, N-methyl-D-aspartic acid; OV, nucleus ovoidalis; PMH, Posterior nucleus of the medial hypothalamus; nXIIts, tracheosyringeal part of the nucleus of the XIIth cranial nerve; O.L, Optic Lobes; POA, Preoptic region; POM, medial preoptic nucleus; post.Di, posterior diencephalon; PVN, paraventricular nucleus; RA, robust nucleus of the arcopallium; Ram, nucleus retroambigualis; Rt, nucleus rotundus; rVRG, rostral Ventral Respiratory Group; SN , spiny neurons; T, Testosterone; Tel, Telencephalon; TH, tyrosine hydroxylase; TnA, nucleus taeniae of the amygdala; TSM, septopallial mesencephalic tract; Tu, nucleus tuberis; VIP, vasoactive intestinal polypeptide; Vmax, maximum velocity; VMN, Ventromedial nucleus of the hypothalamus; VTA, Ventral tegmental area; VT, vasotocin; X, area X of the medial striatum; ZENK, zif-268, egr-1, NGFI-A, or Krox-24

1 Birdsong and the Song System

1.1 Why is Song Interesting from a Neuroendocrine/Neurochemical Perspective?

Among vertebrate species, neurochemistry has been studied in most detail in mammalian species, especially rats and mice. The initial rationale for the overwhelming focus on these species is that they are convenient to study in the laboratory and by elucidating principles of neurochemical organization and function in these species, one can glean a general understanding of neurochemistry applicable to all mammals including humans. With the recent development of sophisticated genetic tools, various strains of the house mouse (*Mus musculus*) have emerged as an especially attractive group for investigation. Comparative neurochemistry is a relatively underdeveloped field. One obvious reason for this underdevelopment is that a reductionist approach to structure and function of neural circuits requires that one builds on previous discoveries, and replicating basic investigations in a variety of species is generally considered to be time consuming and wasteful. Therefore, so-called alternative species for investigation at this level of analysis should be selected with care. Why then would the neurochemistry and neuroendocrinology of a particular behavior (song) produced by a particular taxon of birds (songbirds) be of general interest to neuroscientists? First, it should be observed that there is a long tradition in neurobiology of studying a particular species if it facilitates our understanding of a fundamental principle of nervous system functioning. For example, the basic properties of synaptic transmission were first explicated in the giant axons of squid (Hodgkin, 1976). Kandel and colleagues initiated studies of the mollusc *Aplysia californica* in order to understand synaptic changes associated with learning (Kandel, 1995). Songbirds are a taxon that possesses a neural circuit that controls a learned complex species-typical vocalization, song, and many properties of this circuit are of general interest to neuroscientists (Arnold, 1990; Brenowitz et al., 1997; Doupe and Kuhl, 1999; Nottebohm, 2004). In this review, we first concisely describe these properties and then outline the scope of coverage of the paper.

1.2 Brief Description of Songbirds, Song, and the Song System

The term "songbird" defines an order in the avian class. Songbirds are by far the largest avian order (Passeriformes), which contains over half of the living bird species whose number exceeds 9000 (Sibley and Monroe, 1990). All songbirds have a repertoire of up to 20 or so distinct vocal sounds that they use for communication about danger, food, sex, or group movements, and for other purposes (Nottebohm, 1975; Marler, 2004). In general, the most complex vocalization produced from a prominent perch in a stereotypic fashion is referred to as the bird's "song." The main functions that have been ascribed to song behavior are territory defense (or spacing behavior) and mate attraction, as opposed to calls, which are involved in such functions as signaling danger or food and maintaining flock cohesion (Catchpole and Slater, 1995). Songs, especially among songbird species in the temperate zone, are usually produced by males (Nottebohm, 1975). Among tropical species, females as well as males often sing (Morton, 1996). In some species, males and females sing a co-ordinated song that is known as a duet (Morton, 1996).

Unlike most calls, songs are learned: they develop abnormally if a young male is reared out of the hearing range of the sounds of adults. A common consequence of this dependence on learning among oscine songbirds (i.e., members of the large suborder of Passeriformes that exhibit vocal learning) is the emergence of local song dialects, varying on much the same geographic scale as dialects in human speech. Songbirds are unique among animals in the many analogies that can be struck between song learning and the acquisition of human speech (Ball and Hulse, 1998; Doupe and Kuhl, 1999). Avian vocal development provides one of the few tractable animal models for studying the behavioral, hormonal, and neural bases of vocal plasticity (Ball and Hulse, 1998). No nonhuman primate is known that depends on learning for the development of its natural vocal repertoire though learning does influence usage (Seyfarth and Cheney, 1997). Other than humans, cetaceans and perhaps some bats are the only mammals that appear to rely on learning for vocal development (Tyack and Sayigh, 1997). The avian groups known to have learned songs include hummingbirds, parrots, and all oscine songbirds (Jarvis, 2004).

Songbirds have evolved, in association with their very sophisticated vocal abilities, a suite of neural specializations that include an interconnected circuit of telencephalic, diencephalic, mesencephalic, and myencephalic nuclei that regulate the learning, production, and perception of song (Brenowitz et al., 1997). This circuit can be divided into two main parts: a caudal motor pathway and a more rostral forebrain pathway mostly involved in song learning and auditory feedback needed to maintain adult song (See ➋ *Figure 10-1*). The caudal pathway consists of nucleus HVC (used as a proper name; sometimes referred to as the high vocal center), the robust nucleus of the arcopallium (RA) of the telencephalon and the intercollicular complex (ICo) in the mesencephalon (see Reiner et al., 2004 for the recently revised nomenclature of the avian brain).

HVC (originally misnamed as hyperstriatum ventrale, pars caudale see Brenowitz et al., 1997; Reiner et al., 2004) is a telencephalic nucleus that appears to be unique to species in the songbird suborder

Figure 10-1
Generalized view of the songbird vocal control system. Several of these telencephalic nuclei including HVC, RA, and lMAN represent neural specializations for vocal learning that are observed only in songbird species. One forebrain pathway HVC to RA to ICo and nXIIts is involved in song production. Another pathway HVC to X to lMAN to RA is involved in song learning during ontogeny and song maintenance in adulthood. The POM a key area in the control of male sexual behaviors is also illustrated. See text for more detail

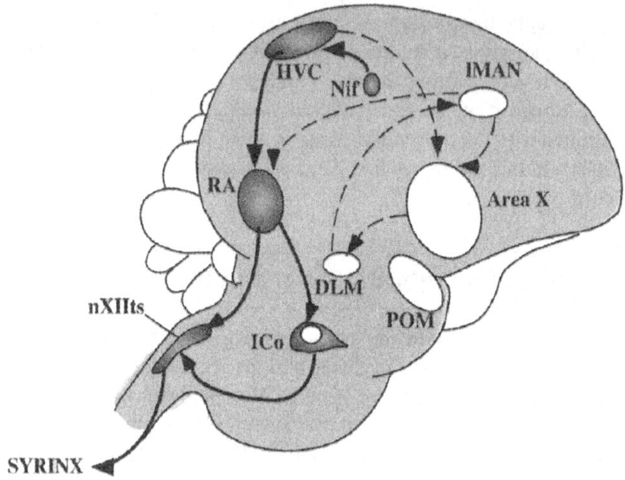

(Nottebohm, 1980; Kroodsma and Konishi, 1991; Ball, 1994; Brenowitz, 1997). It is a key part of the song system involved in the learning, production, and perception of song (Nottebohm, 1980, 1993; Brenowitz, 1997; Wild, 2004). HVC projects to the nucleus robustus arcopallialis (RA) that in turn projects to nucleus intercollicularis (ICo), in particular the dorsomedial portion of this complex (DM). Both RA and DM project to several medullary components of this circuit including the tracheosyringeal part of the nucleus of the XIIth cranial nerve (nXIIts) that innervates the vocal production organ the syrinx as well as to nucleus retroambigualis (RAm) and the rostral ventral respiratory group of neurons (rVRG) that co-ordinate respiratory activity with song production (Wild, 1994; 2004).

This pathway (HVC → RA → ICo → nXIIts) is involved in the motor production of song: lesions to nuclei in the pathway block song production (Nottebohm et al., 1976; Simpson and Vicario, 1990), and both immediate early gene induction studies (Jarvis and Nottebohm, 1997; Kimpo and Doupe, 1997) and electrophysiological studies (Yu and Margoliash, 1996; Leonardo and Fee, 2005) indicate that neurons in these nuclei are active in association with song production.

The second major pathway is an anterior forebrain pathway that includes HVC, area X of the medial striatum (homolog of the caudate/putamen) and the lateral magnocellular nucleus of the anterior nidopallium (LMAN), all of these are in the telencephalon. Another nucleus in this circuit is in the diencephalon, the medial part of the dorsolateral thalamic nucleus (DLM). This pathway is organized as follows HVC → X → DLM → lMAN → RA (➋ *Figure 10-1*; see Doupe et al., 2005 for a review). LMAN also projects to area X.

Thus, there are two projection pathways that go from HVC to RA. The caudal pathway, described above, is clearly essential for song production based on lesion studies (Nottebohm et al., 1976; Simpson and Vicario, 1990), immediate early gene induction (Jarvis and Nottebohm, 1997; Kimpo and Doupe, 1997), and electrophysiological recordings (Yu and Margoliash, 1996; Leonardo and Fee, 2005). The anterior forebrain pathway that consists of a series of more indirect projections between HVC and RA plays a role in song learning (see Bottjer and Johnson, 1997; Doupe et al., 2005, for reviews) and in the maintenance of stereotypic adult song (Benton et al., 1998), but lesions to nuclei within this pathway do not block adult song production (Bottjer et al., 1984; Sohrabji et al., 1990; Scharff and Nottebohm, 1991). Nuclei such as area X exhibit immediate early gene induction in singing male zebra finches that is context dependent (Jarvis et al., 1998). Males singing song that is directed at a female exhibit almost no induction of the immediate early gene ZENK in area X whereas ZENK induction is high in males singing in isolation or in the presence of other males (Jarvis et al., 1998). Similar findings have been reported for electrophysiological activity, in which multiunit activity in area X is much lower when the birds are engaging in song directed at a female than when they are singing in social isolation (Hessler and Doupe, 1999a, b). These findings suggest that this anterior forebrain circuit may modulate song output in adulthood, but may not be directly responsible for the motor output. Currently, one favored hypothesis is that the anterior forebrain pathway conveys an error correction signal between the learned auditory memory that the birds are trying to produce and the ongoing vocal output providing the feedback necessary to maintain high-quality learned song (Brainard, 2004). Blocking auditory feedback via deafening (Nordeen and Nordeen, 1992) or distorting it via a delayed feedback procedure (Leonardo and Konishi, 1999) results in a decline in the quality of song production due to an interruption in ongoing auditory feedback. Two studies by Brainard and colleagues, in particular, support the hypothesis that the anterior forebrain pathway mediates the corrective effects of auditory feedback that maintain stereotypic high-quality ongoing song production (Brainard and Doupe, 2000; Kao et al., 2005). In one case, it was shown that subsequent lesions to a nucleus in the anterior forebrain pathway essentially reverse the negative effects of deafening (Brainard and Doupe, 2000). This study suggests that if the active error correction signal provided by the anterior forebrain pathway is removed, the song does not deteriorate. In another study, the song-induced microstimulation of lMAN, the key output nucleus of the anterior forebrain pathway, was found to disrupt the ongoing song again, suggesting that this pathway mediates the ongoing auditory feedback needed to maintain stereotypic song behavior (Kao et al., 2005). Overall, this work supports the notion that one key function of the anterior forebrain pathway is to maintain ongoing song via an error correction process.

1.3 Strategy and Scope of Chapter

Song behavior and the neural circuit regulating song are profoundly regulated by steroid hormones, in particular androgens and estrogens (Schlinger, 1997; Ball and Balthazart, 2002; Ball et al., 2002; Brenowitz, 2004). Therefore, a major component of this review will be to document hormone effects on song behavior, the chemical neuroanatomy of how such hormones can regulate this behavior, and the regulation of hormone receptors and metabolizing enzymes that are involved in this regulation. We also describe the organization of neurotransmitter systems relevant to song and propose a theory concerning the role played by these different systems in song regulation. We also consider how hormones regulate these transmitter systems. One of the most interesting features of the song system is the pronounced degree to which adult neuroplasticity is exhibited within key forebrain nuclei in the song system such as HVC, RA, and area X (Ball et al., 2002; Brenowitz, 2004). This neuroplasticity can be observed at a rather gross histological level by comparing changes in nuclear volume as defined in Nissl-stained material in birds collected during different seasons and/or in different hormonal conditions. The cellular basis of these changes in volume involve changes in cell number in some nuclei such as HVC as well as changes in cell size and shape as seems to be the case in RA (Brenowitz, 2004). Thus, the major theme of the chapter is how one can discern general principles about how steroid hormones can induce changes in neurochemistry and brain morphology in adult animals.

2 Neuroendocrinology of the Song System

2.1 Hormone Effects on Song Behavior

The link between sex steroid hormone action and song behavior was first made based on the field studies correlating seasonal changes in gonadal size and other aspects of endocrine physiology with changes in song behavior. These descriptive studies have been reviewed in some detail (Ball, 1999; Tramontin and

Brenowitz, 2000) and only major features of these findings will be presented here. First, many male temperate-zone songbirds sing at high rates in the spring as compared to other seasons (e.g Cox, 1944; Slagsvold, 1977; see Catchpole and Slater, 1995 for a review). In these species, seasonal differences in male song are positively correlated with dramatic seasonal increases and decreases in aspects of reproductive physiology such as gonadal size and plasma hormone concentrations (Wingfield and Farner, 1993). However, among these temperate-zone birds there is interspecific variability in the degree to which maximal rates of singing are observed outside the breeding period. For example, robins (*Erithacus rubecula*) living in northern Europe sing at relatively high rates throughout the year (Hoelzel, 1986) only pausing in July (Cox, 1944) whereas most other songbirds living in the same region do not. Birds that sing outside the breeding season do exhibit seasonal cycles in gonad size and endocrine secretions that are similar to other temperate-zone species. Studies of selected species that exhibit territorial song production in the autumn such as song sparrows (*Melospiza melodia*) in the western USA, mockingbirds, and the European robin clearly suggest that song behavior in the fall can be elicited by the appropriate stimulus in the absence of substantial concentrations of testosterone (T) of gonadal origin (Logan and Wingfield, 1990; Schwabl and Kriner, 1991; Wingfield, 1994; Wingfield and Hahn, 1994). Interestingly, in the case of the song sparrow there is evidence that this autumnal singing does involve estrogen acting in the brain even though the gonad is inactive (Soma et al., 2000). The source of this neuroactive estrogen in this case does not appear to be from T of gonadal origin that is then locally metabolized (Soma et al., 2000). It could be derived from the neural aromatization of a substrate, (potentially dehydroepiandrosterone, DHEA) produced by the adrenals or synthesized de novo from cholesterol in the brain (i.e., a neurosteroid, Soma and Wingfield, 2001; Soma et al., 2002, 2004). Overall, these data indicate that there is not necessarily a tight correlation between endocrine activity and song rate in all temperate-zone species.

However, seasonal changes in reproductive physiology in these species may relate to changes in other aspects of song such as repertoire size or stereotypy. For example, seasonal changes in song repertoires have been observed in European starlings (*Sturnus vulgaris*, Eens, 1997) and canaries (Nottebohm et al., 1986), where the number of song types and other measures of song complexity may change. Additionally, seasonal changes in other measures of song such as stereotypy have been described in white-crowned sparrows (*Zonotrichia leucophrys*, Smith et al., 1995) and song sparrows (*Melospiza melodia*, Smith et al., 1997a).

A careful consideration of behavioral data from temperate-zone songbirds suggests that although song output is positively correlated with various measures of reproductive physiology, including hormone concentration in the plasma, there is not necessarily a strong causal relationship between the two as is the case for sex steroids and certain reproductive behaviors such as lordosis in rats (Pfaff et al., 1994), the bow coo display in male ring doves (Lehrman, 1965), or male-typical copulatory behaviors in Japanese quail (Balthazart et al., 2004). Experimental studies on the effects of exogenous hormone administration or castration with hormone replacement on song have been performed on a relatively small number of species but these studies confirm this view of the relationship between steroids and song behavior. Administering exogenous T can clearly increase song rate (e.g., Nowicki and Ball, 1989; Hunt et al., 1997). Several independent studies of zebra finches have shown that castration greatly reduces but does not eliminate male-typical song (e.g., Arnold, 1975; Harding et al., 1983), whereas in red-winged blackbirds castration was reported to eliminate adult song production (Harding et al., 1988). In the case of song sparrows in the western USA, castrated males were able to maintain fall territories and sang at high rates in response to territorial challenge in a manner that was indistinguishable from intact controls (Wingfield, 1994). The hormonal control of song behavior therefore appears to be a clear case of a hormone-enhanced, rather than hormone-dependent behavior. Species-typical stimulus factors (the presence of a conspecific male and/or female as well as a nest site or a favorable environment) promote song production in males. The presence of gonadal steroids in the plasma can increase the probability and intensity of these behavioral responses to the appropriate stimulus but this presence is not essential for behavioral activation (Wingfield, 1994). Therefore, one should not be surprised by the reports of substantial song production being observed in association with low steroid hormone concentrations in some cases. The stimulus factors releasing song can be so strong in some cases that high gonadal steroid concentrations are not necessary for song production to be observed.

In some cases, T effects on song may be limited to certain social contexts. For example, castrated European starlings continue to express high basal rates of singing but fail to exhibit an increase in singing rate when presented with a female (Pinxten et al., 2002). A female-induced increase in singing rate was, however, observed in castrates treated with T (Pinxten et al., 2002). Breeding season song in starlings usually declines after mating. If male starlings are treated with T during the incubation period, there is a robust increase in the song rate while a similar treatment during the nestling feeding period has little effect (De Ridder et al., 2000). The authors interpret the function of these findings as follows. In the population they studied in Belgium, during the incubation period there are still a large number of reproductively active females for the males to direct their song while by the time the nestling feeding stage starts there are few receptive females available (De Ridder et al., 2000). Thus, T was only effective in inducing an increase in song in breeding starlings when females available for mating were present. Finally, comparing the effects of a female on male song rate in starlings in the spring when T concentrations are high and the fall when they are low reveals an enhancing effect of a female only in the spring (Riters et al., 2000). These data are all consistent with the notion that in starlings, T is effective in enhancing the song produced in response to the presence of a female.

Although T does not appear to be necessary for the initiation of song production in all cases, it does appear to influence aspects of song quality such as stereotypy. For example, castration prevents the onset of crystallized (i.e., stereotyped) adult song in 1-year old song and swamp sparrows singing for the first time in the spring (Marler et al., 1988). Upon receiving T, the song rapidly crystallizes (Marler et al., 1988). As mentioned previously, in both white-crowned sparrows and song sparrows, fall song in the presence of low concentrations of T is less stereotypic than spring song produced in the presence of high concentrations of testosterone (Smith et al., 1995, 1997a). Overall, the data indicate that song can be produced (though at a low rate) in the presence of low concentrations of T but that, the stereotypic quality of the song is also regulated by the presence of T.

Finally, some steroid hormone replacement studies indicate that both androgenic and estrogenic metabolites of T are needed to fully restore high rates of singing (e.g., Harding et al., 1983, 1988). In zebra finches, it has been suggested that estrogenic metabolites selectively promote female-directed song (Walters et al., 1991). Similarly, in canaries there are data suggesting that estradiol selectively activates syllables that are particularly attractive to females while other aspects of song are activated by androgenic metabolites (Rybak and Gahr, 2004). These studies indicates that there may be selective actions of the two primary metabolites of testosterone on song behavior.

2.2 Localization of Sex Steroid Hormone Receptors in the Song System and Other Areas Relevant to Reproductive Behavior

Singing activity is enhanced by T in many songbird species, and the song control circuit is a neural specialzation that controls song production. Therefore, some of the first questions asked by behavioral neuroendocrinologists concerned whether T is able to act directly on song control nuclei to activate song and whether T acts via androgenic or estrogenic metabolites at these brain sites. Initial autoradiographic studies using [^{3}H] T revealed uptake in several of the forebrain song control nuclei (Arnold et al., 1976; Arnold and Saltiel, 1979). Subsequent studies using immunocytochemistry (ICC) to identify the receptor protein or in situ hybridization (ISH) to localize the corresponding messenger RNA have confirmed the presence of steroid-sensitive neurons in these nuclei. These studies have also allowed for the specification as to whether the sites accumulating radioactivity that had been detected by in vivo autoradiography with [^{3}H] T as the ligand correspond to androgen (AR) or estrogen receptors (ER). This presence of brain areas that highly express AR in discrete nuclei in the telencephalon represents a neural specialization that distinguishes songbirds from other vertebrates (Kelley and Pfaff, 1978; Morrell and Pfaff, 1978). In this section, we will review briefly what is known about the distribution of AR and ER in the song control system (see ➋ *Figure 10-2* for a summary).

Androgen Receptors. In general, the distribution of AR in birds is restricted to the septal-preoptic area, and to the various nuclei in the hypothalamus and in the midbrain (for review, Balthazart, 1983; Ball, 1990; Brenowitz, 1991; Ball and Balthazart, 2002) as is commonly observed in all vertebrate classes (Morrell et al.,

Figure 10-2

Schematic illustration the distribution of androgen receptors (AR), estrogen receptors alpha (ERα), and estrogen receptors beta (ERβ) in the song control nuclei and in the medial preoptic nucleus (POM) that also participates to the activation of singing behavior. The nuclei of the forebrain pathway involved in song production (HVC, RA, ICo, and nXIIts) are connected by black arrows. Nuclei of the anterior forebrain pathway involved in song learning during ontogeny and song maintenance in adulthood (HVC, area X, DLM, lMAN, and RA) are connected by dashed arrows. The presence of crosshatches in or near a given nucleus means that the receptor is expressed in the brain area. In the case of ERα in HVC there is species diversity in this expression. See text for more detail

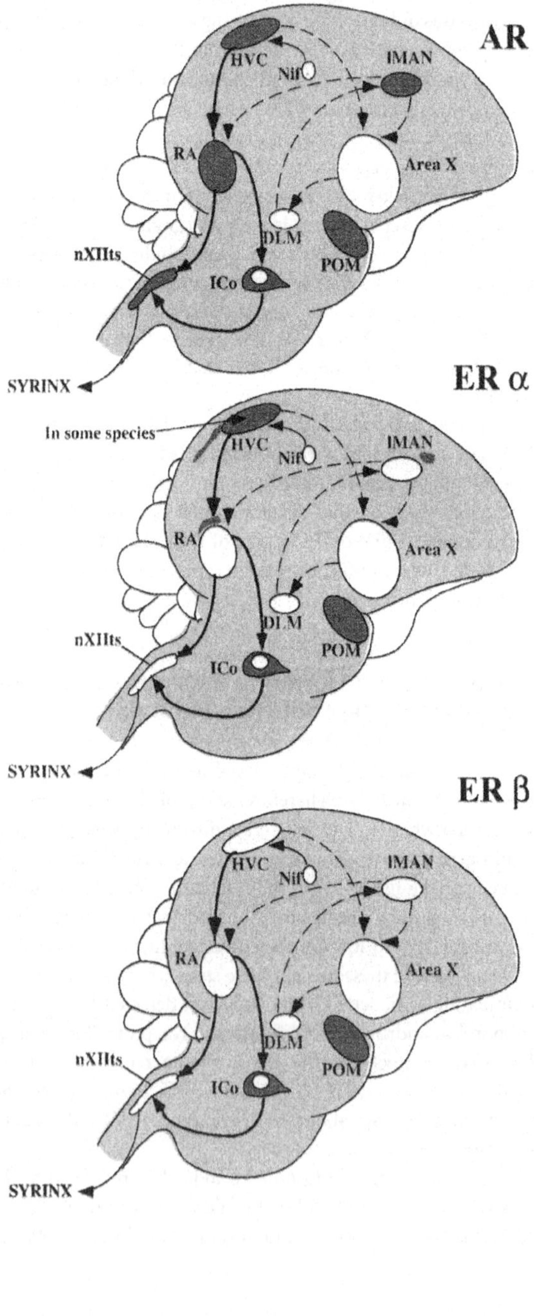

1975; Pfaff, 1976; Stumpf and Sar, 1978). In addition to these receptors in brain areas common to all avian (and vertebrate) species, songbirds possess androgen-sensitive brain areas that are part of the song control system (Arnold et al., 1976; Nottebohm, 1980; Konishi, 1985; Nottebohm et al., 1990). HVC, RA, and MAN contain AR based on both autoradiographic (Zigmond et al., 1973; Arnold et al., 1976; Lücke and Haase, 1980; Zigmond et al., 1980; Gahr, 1990; for review: Ball, 1990; Brenowitz, 1991; Ball and Balthazart, 2002) and binding assay (Harding et al., 1984) methods (see *Figure 10-2* top panel).

Additional autoradiographic studies using the nonaromatizable androgen [^{3}H] dihydrotestosterone as a ligand (Arnold and Saltiel, 1979; Nordeen et al., 1987a; Sohrabji et al., 1989) have confirmed that these telencephalic-binding sites, identified by T autoradiography, are specific for androgens (not estrogens derived by aromatization). The presence of AR in these nuclei has also been confirmed by ICC (Balthazart et al., 1992b; Smith et al., 1996; Soma et al., 1999b). The AR has also been cloned and sequenced in canaries (Nastiuk and Clayton, 1994; Gahr and Metzdorf, 1997) and starlings (Bernard et al., 1999) and probes based on these studies have been used to localize the AR mRNA in the canary (Nastiuk and Clayton, 1995; Metzdorf et al., 1999; Fusani et al., 2000) and starling brain (Bernard et al., 1999; Maney et al., 2001). These ISH studies have largely confirmed the AR distribution identified by in vivo autoradiography or ICC. Interestingly, however, AR have also been identified by ISH within the area X of adult starlings (Bernard et al., 1999) and this presence was recently confirmed in the developing zebra finch brain (Perlman et al., 2003).

Weakly labeled AR-ir cells can also be observed in a position adjacent to the canary RA, in a "hook-like" structure that runs laterally from RA and then turns in a ventral direction (Balthazart et al., 1992b). This area had been previously identified in zebra finches as a region receiving projections from the lateral MAN (Bottjer et al., 1989) and labeled by the a high density of α_2 adrenergic receptors in starlings (Ball, 1990; 1994). This area should thus be considered as part of the song system. Several medullary nuclei that receive inputs from RA, such as the nucleus hypoglossus pars tracheosyringealis (syringeal motonucleus or nXIIts), the nucleus retroambigualis (RAm), and the rostroventral respiratory group (rVRG) (Wild, 1997; Wild et al., 1997; Reinke and Wild, 1998) are also defined by the presence of AR in songbirds (Gahr and Wild, 1997). However, these areas appear to be devoid of these receptors in most other birds (Gahr, 2000; or at least there is no report on the topic), with the exception of the Anna's hummingbird (*Calypte anna*), a species that belongs to another bird family where song learning has been identified. Caution should be exercised when making definitive conclusions about species variation in the absence of many studies. For example, it has been reported that AR are present in the syringeal motor nucleus of chickens, though they are absent in quail (Shaw and Kennedy, 2002). These data raise questions concerning the general nature of the presence of these receptors in the nXIIts of songbirds as opposed to other avian species and studies in additional species should be undertaken to understand the rules governing expression of AR in this motor nucleus.

Estrogen Receptors of the Alpha Subtype (ERα). Estrogen receptors, as was the case for AR, were originally studied in songbirds by in vivo autoradiography. In the late 1990s, a new ER was cloned in mammals (Kuiper et al., 1996; Mosselman et al., 1996; Tremblay et al., 1997) and later identified in avian species (Lakaye et al., 1998; Bernard et al., 1999; Foidart et al., 1999). This second ER was called estrogen receptor beta (ERβ) to distinguish it from the previously described receptor it was then renamed as ERα. We first review here, work that was performed before the second receptor was known and we now think primarily concern the first identified form of receptor (ERα). The distribution of ERβ will be considered subsequently.

The distribution of ER without regard to subtype was investigated in several avian species, including songbirds, by a variety of methods (in vivo autoradiography, in vitro-binding, ICC and, in a few cases, ISH). Similar to what has been observed for AR, ER binding sites appear to be restricted to the hypothalamic and limbic structures, and to the mesencephalic intercollicular nucleus in nonsongbirds and songbirds (e.g., Martinez-Vargas et al., 1975; 1976; Kim et al., 1978; Watson and Adkins-Regan, 1989), but additional binding sites are found in telencephalic song nuclei of oscines (e.g., Nordeen et al., 1987b).

Immunocytochemistry has also been used to analyze the distribution of ERα in zebra finches and canaries first (Gahr et al., 1987) and then in many avian species ($n = 26$) belonging to a large number of avian orders (Gahr et al., 1993). The distribution of ERα immunoreactive cells in the diencephalic and limbic structures is very similar across all species. All songbirds, however, also display significant numbers of ERα–expressing cells in three structures of the nonlimbic forebrain: the caudal nidopallium including in some species HVC, the dorso-rostral area surrounding RA, and an area in the rostral forebrain, dorsal to the lamina mesopallialis and rostral to the nucleus MAN (➋ *Figure 10-2* middle panel). ERα was also cloned in zebra finches and canaries and the distribution of the corresponding mRNA was analyzed in the brain of these species by ISH. These studies confirmed with a few exceptions the results that were previously established by autoradiographic and ICC methods (e.g., presence of ERα in the nido- and arcopallial areas adjacent to HVC and RA, see Jacobs et al., 1996). One noticeable difference between canaries and zebra finches should be highlighted here, however, ERα mRNA and protein are present within HVC in canaries but not in zebra finches where they are located in the surrounding nidopallium but not specifically in HVC itself (Gahr and Metzdorf, 1997; Metzdorf et al., 1999; Fusani et al., 2000).

*Estrogen Receptors of the Beta Subtype (ERβ).*The identification of a second type of ER with a distinct neuroanatomical distribution (Kuiper et al., 1996; 1998) suggested new ways in which estrogens could act in the brain. Several species, for example, do not express ERα in HVC (e.g., the zebra finch or starling) and the discovery of the new receptor type raised the possibility that estrogens could act in this nucleus via binding to ERβ if it were expressed in this nucleus. ERβ was cloned and sequenced in Japanese quail (Foidart et al., 1999; Lakaye et al., 1998) and then European starlings (Bernard et al., 1999), and a specific riboprobe was used to analyze the neuroanatomical distribution of the mRNA in the starling brain with a special focus on the brain areas containing the song control nuclei (Bernard et al., 1999). ERβ mRNA was not expressed at detectable levels in any forebrain song control nucleus (➋ *Figure 10-2* bottom panel). A low level of expression was observed in the caudomedial nidopallium close to but not within HVC, with a pattern of distribution distinct from ERα. As was observed in a nonsongbird species, the Japanese quail (Foidart et al., 1999), a high level of expression was observed in starlings in the medial preoptic nucleus (POM), bed nucleus of the stria terminalis (BST), and in the nucleus taeniae of the amygdala (TnA), a number of brain areas that are implicated in the mediation of sexual behavior (Panzica et al., 1996; Balthazart et al., 1998; Thompson et al., 1998; Absil et al., 2002). These data therefore provide no evidence that ERβ directly mediates the control of singing in the telencephalic song control nuclei.

2.3 Localization of Steroid Hormone Metabolizing Enzymes in the Song System

As discussed previously, estrogenic and androgenic stimulation is usually required for the optimal activation of singing behavior (Harding et al., 1983, 1988; Walters et al., 1991). It is therefore important to investigate the neural sites where testosterone can be metabolized into behaviorally active steroids and to analyze the control of these enzymatic activities.

In the avian brain, three enzymes catalyze the transformation of T into behaviorally relevant metabolites (Balthazart, 1989; Hutchison, 1991): aromatase leads to the production of estrogens, 5α-reductase produces 5α-dihydrotestosterone (5α-DHT), and 5β-reductase leads to the formation of the behaviorally inactive 5β-dihydrotestosterone (5β-DHT) which apparently represents an inactivation shunt for T (Hutchison and Steimer, 1981; Steimer and Hutchison, 1981). These transformations are thermodynamically irreversible, at least in physiological conditions.

The presence of these three enzymes has been confirmed in the brain of zebra finches and starlings during in vitro experiments (Vockel et al., 1990b; Schlinger and Arnold, 1991, 1992a; Riters et al., 2001). The presence of an active aromatase and 5β-reductase has also been identified in zebra finches in vivo (Schlinger and Arnold, 1992a). With only a few exceptions (Hutchison et al., 1986; Schlinger and Arnold, 1992a), all available data concerning T metabolism in the avian brain are derived from in vitro studies carried out on brain homogenates. These studies demonstrate the presence of the enzymes in the central nervous system but the measures of activity obtained in such studies may ignore some regulatory mechanisms that could play a key role in the live animal. Some caution should therefore be exercised when using these assay data to interpret physiological phenomena.

Interestingly, there is an inverse relationship between the estimated affinity (Km) and maximum velocity (Vmax) of the three enzymes that irreversibly metabolize T (see Vockel et al., 1988; for data in zebra finches). Enzymes that produce biologically active metabolites such as aromatase and 5α-reductase have a high affinity (small Km) for T and therefore a privileged access to the substrate (T), but they have a limited maximum velocity and cannot produce very large amounts of these biologically active substances (E2 and 5α-DHT). Conversely, the enzyme that produces biologically inactive steroids, such as the 5β-reduced androgens does not have a priority of access to the substrate but can metabolize large amounts of the substrate (high Vmax). In physiological conditions where a relatively low concentration of T is available, the steroid should thus be preferentially transformed into estrogens by aromatase, but if higher amounts of substrate are present (e.g. after the injection of exogenous T), a larger proportion could be inactivated into 5β-reduced steroids that have no hormonal activity by themselves.

The activity of aromatase, 5α-, and 5β-reductase was originally measured by radioenzyme assays in zebra finch brain nuclei that had been dissected by the Palkovits punch technique (Vockel et al., 1990a, b). The 5α-reductase activity was more or less evenly distributed in all brain nuclei that were considered with the noticeable exception of RA, where enzyme activity was twice as high as in all other regions (➋ *Figure 10-3a*). High levels of 5β-reductase activity were found throughout the brain (note the very different scale on the graph) but especially in MAN, HVC, and in the medial part of the bed nucleus stria terminalis (BST). These two reductases were even present in two steroid-insensitive regions that had been included in the study as controls, the entopallium (formerly, ectostriatum; see Reiner et al., 2004) and the nucleus rotundus.

High levels of aromatase were observed in the diencephalic and limbic nuclei that are well known to express this enzyme in mammals and other avian species, namely the nucleus preopticus anterioris (POA), the nucleus periventricularis magnocellularis (PVN), the nucleus medialis hypothalami posterioris (PMH, named more accurately as the ventromedial nucleus of the hypothalamus, VMN), BST, and TnA. Interestingly, these studies also reported the presence of significant levels of aromatase activity in telencephalic song control nuclei such as the area X, MAN, HVC, and RA. They also indicated that the area parahippocampalis (APH) that had originally been selected as a control region was characterized by a surprisingly high level of enzymatic activity, greater in fact than in any of the diencephalic or limbic nuclei that had been studied (Vockel et al., 1990b). Additional studies confirmed the presence of a very high aromatase activity in the dorsal telencephalon and demonstrated that the enzyme is actually present in the entire roof of the forebrain in the nidopallium (formerly neostriatum), hippocampal, and parahippocampal region. This high level of telencephalic aromatase activity was independently confirmed (Schlinger and Arnold, 1991, 1992a) and a series of elegant studies based on the injection of radioactive androgens in the brain or periphery combined with blood collection in the jugulars or carotids has provided experimental evidence indicating that brain aromatase substantially contributes to the circulating levels of estrogens in the Zebra Finch (Schlinger and Arnold, 1992a, 1993).

More recently, the distributions of the aromatase protein or mRNA have been described at a cellular level of resolution in a number of songbird species (e.g., zebra finch, canary, pied flycatcher) by ICC or ISH (Balthazart et al., 1990; Shen et al., 1994, 1995; Balthazart et al., 1996; Foidart et al., 1998; Metzdorf et al., 1999; Saldanha et al., 2000). All studies confirmed the presence of aromatase and its mRNA in the diencephalic and limbic regions such as the POA, PVN, VMN, BST, and TnA that are implicated in the control of reproduction where a high level of enzyme activity had been measured. Unexpectedly, however, all telencephalic song control nuclei were found to contain only minimal, often nondetectable, concentrations of the protein or its mRNA. The presence of large amounts of aromatase (protein and mRNA) in the nidopallium has been confirmed and its distribution was characterized. At the level of HVC, or in slightly more rostral parts of the telencephalon, aromatase and its mRNA are present in broad areas of the medial nidopallium adjacent to the lateral ventricles. At more caudal levels, the cluster of aromatase-expressing cells extends more medially as a band coursing along the dorsal edge of the lamina arcopallialis dorsalis; they are also present in the ventrolateral aspects of the nidopallium (See ➋ *Figure 10-3b* for an illustration of this distribution in male pied flycatchers, similar data were obtained in the other investigated species). Based on these detailed anatomical data, it is likely that the aromatase activity that had been detected in HVC and RA by radioenzyme assays performed on nuclei microdissected by the Palkovits punch technique originated from a contamination by nidopallial or arcopallial tissue immediately surrounding these nuclei. A number of aromatase-immunoreactive neuronal processes originating in the adjacent nidopallium (Saldanha et al., 2000) or in other aromatase-immunoreactive neurons (neurons in Nif express aromatase mRNA and project to HVC; Saldanha et al., 2000) seem to enter HVC where they could release locally produced estrogens. These immunopositive fibers may also partially explain the significant level of aromatase activity detected by radioenzyme assays in this song control nucleus.

Interestingly, enzyme assays performed on cell fractions separated by differential centrifugation of the quail preoptic area (Schlinger and Callard, 1989) or the telencephalon of zebra finches (Schlinger and Arnold, 1992b) indicate that aromatase activity is found not only in the microsomal but also in synaptosomal (pinched-out nerve terminals) subfractions. The presence of aromatase-immunoreactive material has furthermore been confirmed in cell processes of the quail brain that have an axonal morphology (Evrard et al., 2004; Foidart et al., 1994) and, at the electron microscope level, it was confirmed that aromatase-immunoreactive material is present in high density at the level of presynaptic boutons (Naftolin et al., 1996; see Peterson et al., 2005 also for recent data in songbirds). This local production of estrogens in the close vicinity of synapses may provide steroid signals that could affect brain activity independent of the binding to nuclear receptors acting as transcription factors (i.e., by nongenomic mechanisms). It is known that estrogens are able to modify the electrical activity of the neuronal membranes within seconds after their application, which essentially rules out the possibility of a classical steroid receptor-mediated effect on protein synthesis (for additional discussion, see Blaustein and Olster, 1989; Schumacher, 1990; McEwen, 1994; Ramirez et al., 1996). Other effects of estrogens on brain functioning that appear to be independent of their binding to intracellular receptors have also been described (e.g., Thompson and Moss, 1994; Pasqualini et al., 1995; Mermelstein et al., 1996). Whether the presence of aromatase activity in brain areas such as part of the nidopallium that are apparently devoid of intracellular ER relates to these membrane effects of estrogens is unknown at present.

Most of the studies described so far on androgen metabolizing enzymes used laboratory-reared, domesticated species (e.g., Japanese quail, ring dove, zebra finch) and their relevance for free-living species was difficult to establish. One unusual feature of songbirds that was suggested by this work is the presence of surprisingly high levels of aromatase activity in the dorsal telencephalon as compared to limbic and

Figure 10-3

Distribution of testosterone metabolizing enzymes in the brain of songbirds. (a) Formation of estradiol (aromatization), 5α-dihydrotestosterone (5α-reduction) and 5β-dihydrotestosterone (5β-reduction) by different brain areas of adult male zebra finches dissected by the Palkovits punch technique. Data represent the mean ± SE of 6–11 replicates in each case. Different scales have been used to represent the activity of the three enzymes. X: area X of the medial striatum (formerly lobus parolfactorius); MAN: nucleus of the anterior nidopallium; HVC: used a a proper name (formerly high vocal center); RA: robust nucleus of the arcopallium; ICo: nucleus intercollicularis; POA: nucleus preopticus anterioris; PVN: nucleus periventricularis magnocellularis; PMH: nucleus medialis hypothalami posterioris (or VMN); E: entopallium; Rt: nucleus rotundus; BSTM: nucleus stria terminalis pars medialis; TnA: nucleus taeniae of the amygdala; APH: area parahippocampalis. Based on data from Vockel et al., (1990b). (b) Schematic representation of the distribution of aromatase immunoreactive cells in the male pied flycatcher brain. Successive rostral to caudal levels are shown from left to right. Darkly labeled cells are indicated by dots in the diencephalon. In the telencephalon, the distribution of cells showing a weaker immunoreactivity is indicated by the shaded areas. A group of slightly darker cells is indicated by a darker shading in the ventromedial corner of the neostriatum and archistriatum Abbreviations: A: arcopalium; APH: area parahippocampalis; BSTM: nucleus stria terminalis pars medialis; Cb: cerebellum; CA: commissura anterior; DBC: decussatio brachiorum conjunctivorum; DSD: decussatio supraoptica dorsalis; E: Entopallium; FLM: fasciculus longitudinalis medialis; GCt: substantia grisea centralis; GLv: nucleus geniculatus lateralis pars ventralis; GP: globus pallidus, formerly paleostriatum primitivum; HA: accessory part of the hyperpallium; Hp: hippocampus; HVC: song control nucleus formerly known as high vocal center; ICo: nucleus intercollicularis; LSt: lateral striatum formerly paleostriatum augmentatum; M: mesopallium (formerly hyperstriatum ventrale); MLd: nucleus mesencephalicus lateralis pars dorsalis; N: nidopallium; NC: caudal nidopallium; OV: nucleus ovoidalis; POM: nucleus

preopticus medialis; Rt: nucleus rotundus; TnA: nucleus taeniae of the amygdala; TSM: septopallial mesencephalic tract; Tu: nucleus tuberis; VMN: nucleus ventromedialis hypothalami. Modified from data in Foidart et al., (1998). (c) Comparison of the aromatase activity in the telencephalon and diencephalon of five species of songbirds and in quail or doves. Each data point represents the mean ± SE of 5–7 determinations except for doves where $n = 2$. Data are expressed as enzyme activity per hour per mg protein (left) or for the entire brain region considered (right). Modified from the data in Silverin et al. (2000)

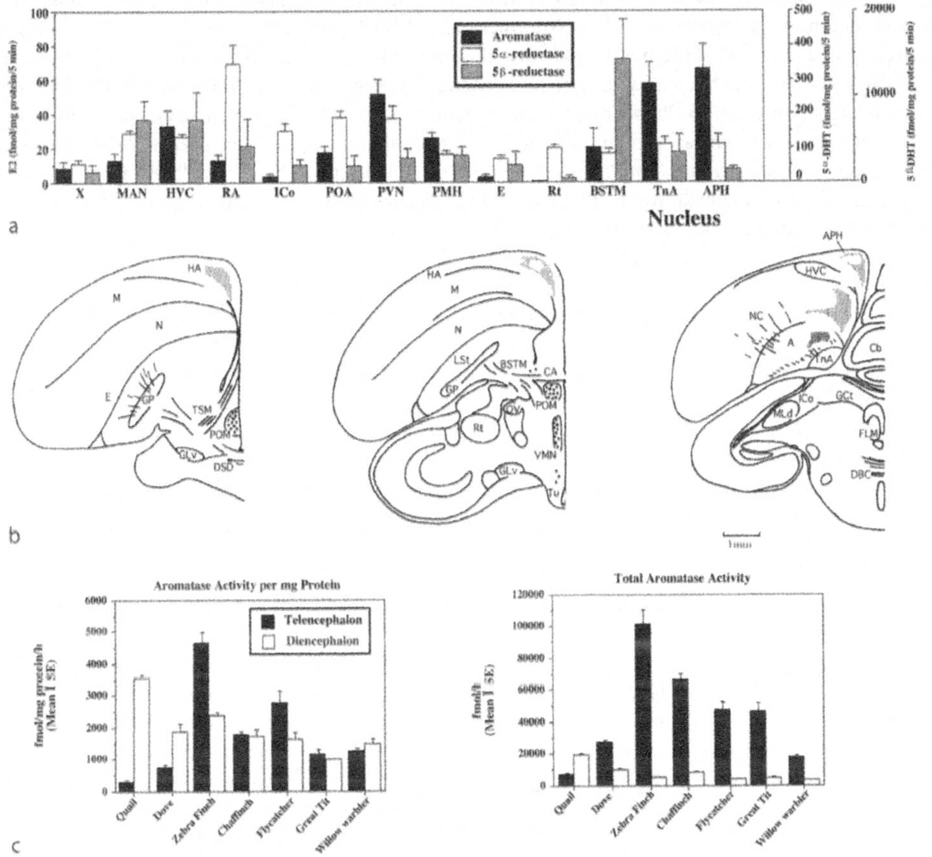

hypothalamic areas. Although, it was quite tempting to generalize these features of the zebra finch aromatase to the entire songbirds order (passeriformes), data supporting this generalization were missing with the exception of limited studies devoted to a few species such as the brown-headed cowbird (*Molothrus ater*), white-crowned sparrow, or house sparrow (*Passer domesticus*) (Schlinger et al., 1992; Schlinger, 1996; Saldanha and Schlinger, 1997). Considering that zebra finches studied in the laboratory are derived from semidomesticated stocks and they represent a non- (or only marginally) photoperiodic species that inhabits the Australian deserts (see Bentley et al., 2000), their use as a model species for the songbird group that contains many photoperiodic species originating from the temperate zone appears questionable. We investigated whether the high level of telencephalic aromatase that had been detected in zebra finches would also be found in a variety of European songbird species caught in the wild during the breeding season. Within the same experiment, aromatase activity was thus measured in the brain of five songbirds species from different families including chaffinches (*Fringilla coelebs*, Fringillidae), pied flycatchers (*Ficedula hypoleuca*, Muscicapidae), great tits (*Parus major*, Paridae), willow warblers (*Phylloscopus trochilus*,

Sylviidae), and zebra finches (*Taeniopygia guttata*, Estrildidae) and in two nonsongbird species, the Japanese quail (*Coturnix japonica*, Phasianidae) and ring dove (*Streptopelia roseogrisea* [= *risoria*]).

Aromatase activity was measured in triplicate in five different brain regions: the telencephalon, diencephalon, optic lobes, cerebellum, and brain stem. The enzyme activity, expressed per mg protein, was larger in the telencephalon than in the diencephalon in zebra finches but not in quail and doves (see *Figure 10-3c*, left panel). The same situation as in zebra finches was observed in the flycatcher but not in the other songbird species that were considered (chaffinch, tit, warbler), therefore questioning the generalization of this pattern to the passeriformes order. It is important to note that the telencephalon mass is at least one order of magnitude larger than the diencephalon. When aromatase activity was therefore calculated for the entire brain area, the activity difference between the telencephalon and the diencephalon became more prominent (5–20 fold excess) in all songbird species but not in quail and dove (see *Figure 10-3c*, right panel). Aromatase activity was low in all other brain regions (see Silverin et al., 2000 for detail). These data thus support the notion that songbirds are characterized by a high level of telencephalic aromatase activity by comparison with other avian groups. The difference is, however, not as dramatic as previously assumed on the basis of the zebra finch data.

Aromatase activity is present in the telencephalon of all species studied so far including nonsongbirds. The evolutionary significance of the high or low ratio between telencephalic and diencephalic aromatase activities remains unclear at present and will probably be impossible to specify, until the enzyme has been studied in a larger number of species originating from a broad range of families and displaying a variety of ecological life styles. The role of this high level of estrogen production in broad areas of the telencephalon that often do not express nuclear ER also remains to be identified. The presence of aromatase at the presynaptic level suggests that a paracrine mode of action is possible (see (Ball and Balthazart, 2002) for a broader discussion of this topic). The demonstration that infusion of radioactive androstenedione in the zebra finch telencephalon results in increased levels of radioactive estrogens in the jugular blood (Schlinger and Arnold, 1992a, 1993) even suggests an endocrine action for brain aromatase.

2.4 Localization of Receptors for Other Hormones in the Song Control System

Although it has been hypothesized for many years that pineal melatonin exerts important effects on brain function, it was not until the 1980s that autoradiographic procedures were developed that allowed for the reliable visualization of melatonin receptors. In birds, the surprise that emerged based on initial studies in chicks is that binding sites for iodomelatonin are widely distributed throughout the visual system involving both thalamic nuclei such as the lateral geniculate nucleus, and the optic tectum, as well as telencephalic structures such as the entopallium (Rivkees et al., 1989). Subsequently, detailed autoradiographic studies were carried out in zebra finches (Gahr and Kosar, 1996), house sparrows (Whitfield-Rucker and Cassone, 1996), and European starlings (Bentley and Ball, 2001). The striking finding for songbirds is that the song nucleus HVC exhibits high receptor density in all species investigated, whereas RA also exhibits significant receptor binding though at lower densities than HVC. Area X is more variable, but this is probably due to the fact that melatonin receptors in this nucleus exhibit a marked seasonal regulation in both starlings and house sparrows (Whitfield-Rucker and Cassone, 1996; Bentley and Ball, 2001). In starlings, it is clear that receptor densities are high when birds are photosensitive and photorefractory but not when photostimulated, and this regulation is independent of gonadal steroid hormones (Bentley and Ball, 2001).

2.5 Hormonally Regulated Adult Neuroplasticity in the Song Control System: Seasonal and Sex Differences

Along with the seasonal changes in the gonads and T described above, one can observe seasonal changes in the volumes of several song control nuclei and aspects of song behavior. These correlated changes could reflect direct actions of steroids on the song control nuclei and on song expression, but alternative mechanisms have also been suggested.

Since the discovery of seasonal anatomical changes in the song control system of canaries by Fernando Nottebohm (1981), studies in several songbird species have revealed strong positive relationships between the volumes of song nuclei, typically HVC, RA, and area X, and both seasonal fluctuations in T and aspects of song production (e.g., Nottebohm et al., 1986, 1987; Smith et al., 1997a; Riters et al., 2000; Tramontin and Brenowitz, 2000). The primary environmental factor influencing the volumes of song nuclei is day length. In the laboratory, male songbirds exposed to artificial photoperiods typical of spring have significantly larger song control nuclei and elevated T concentrations than males housed under fall-like photoperiod conditions (Bernard and Ball, 1995). Similarly, in males captured from the field, seasonal peaks in T, associated with long day lengths in spring in most temperate-zone songbirds, correspond to increases in the volumes of song control nuclei (e.g., Smith, 1996; Smith et al., 1997a, b; Brenowitz et al., 1998; Soma et al., 1999b). A recent study of wild song sparrows carefully correlated the late winter and spring onset of testis growth and other measures of reproductive development with the growth of the song system in wild sparrows and found that increases in HVC volume clearly preceded the growth and development of the reproductive system (Tramontin et al., 2001). Similar results have also been reported for blue tits (*Parus caeruleus*) in Corsica (Caro et al., 2005). Thus, the maximal concentrations of T associated with breeding are not needed for the growth of HVC to be observed (Tramontin et al., 2001; Caro et al., 2005).

In addition to day length, several additional environmental factors interact to influence the volumes of song control nuclei including social factors (Tramontin and Brenowitz, 2000; Tramontin et al., 2001) and interactions between T concentrations and day length (Bernard and Ball, 1997). Not all song control nuclei show the same pattern of volume change (e.g., Smith et al., 1997a), but in general HVC changes are followed by changes in RA and area X depending on the species and whether the bird is studied in captivity or captured from the wild (see Tramontin et al., 2000 for general discussion of this sequence). Compared to laboratory-housed males, males captured in the field are exposed to a much fuller array of environmental stimuli. This likely accounts for differences sometimes observed in the volumes of song control nuclei in males studied in the laboratory compared with males captured in the field. For example, in starlings exposed to a spring-like photoperiod, only the volume of HVC was enlarged relative to males under fall-like photoperiod conditions (Bernard and Ball, 1995, 1997; Bentley and Ball, 2000). In contrast, males captured in the field and housed in outdoor aviaries HVC, RA, and area X displayed significant seasonal changes in volume (Riters et al., 2002).

The changes in the volumes of song control nuclei reflect changes in the number of neurons, the size of the neurons, or changes in dendritic fields within the nuclei. In RA, seasonal changes in volume do not appear to reflect an increase or decrease in neuron number, but they reflect changes in the size of the neurons, spacing between neurons, and changes in dendritic arborization (Smith et al., 1997a; Tramontin et al., 1998). Similary, studies of the cellular basis of seasonal changes in the volume of area X in song sparrows indicate that the 75% variation in volume between spring and fall birds is due to changes in neuronal size and neuronal density (Thompson and Brenowitz, 2005). In contrast to RA and area X, new neurons are incorporated into HVC seasonally (Nottebohm, 1989; Brenowitz et al., 1991; Goldman et al., 1992; Smith et al., 1997c). Given that HVC tends to be at its largest extent when T concentrations are highest (e.g., Kirn et al., 1994; Riters et al., 2002) though see (Tramontin et al., 2001), and cell death in HVC is associated with low concentrations of T (Kirn et al., 1994; Nottebohm, 1987; Nottebohm et al., 1987), it was suggested that these cellular changes are associated with changes in song behaviors (Nottebohm et al., 1987; Kirn et al., 1994). It was also hypothesized that T regulates the incorporation of new neurons into HVC, possibly by influencing neurotrophic factors (Rasika et al., 1999). There is now direct evidence that this is the case (Alvarez-Borda and Nottebohm, 2002). Furthermore manipulations of the neurotrophin, brain-derived neurotrophic factor (BDNF) induce song variability in adult zebra finches suggesting that T exerts morphological and behavioral effrects via actions on BDNF in HVC (Rasika et al., 1999; Kittelberger and Mooney, 2005).

Although seasonal changes in T and the volumes of song nuclei have been reported for several songbird species, the precise behavioral function of seasonal changes in the song control system is still a matter of debate. Based on work in canaries, the original hypothesis was that seasonal changes in the volumes of song nuclei reflected seasonal changes in song learning (e.g., Nottebohm et al., 1986). Male canaries learn new songs each year outside of the breeding season. During song learning, the songs that are produced are variable and the volumes of song control nuclei are small. However, with the increase in T concentrations during the breeding season, song becomes much more stable and stereotyped, and the volumes of HVC and RA increase 99% and 76%, respectively (Nottebohm, 1981), suggesting that volume increases play a role in the storage of a learned song. However, more recent studies have called this possibility into question (Brenowitz et al., 1995; Bernard et al., 1997; Smith et al., 1997b). For example, in rufous-sided towhees seasonal changes in the volumes of song control nuclei are observed in a species that does not learn to produce new songs in adulthood (Brenowitz et al., 1991). An alternative to the hypothesis that changes in nuclei relate to song learning is that seasonal changes in song nuclei are related to seasonal changes in aspects of song performance such as the amount of song produced or changes in song structure (Ball, 1999; Tramontin and Brenowitz, 2000). As in canaries, in song sparrows the largest volumes of HVC and RA were observed at times of the year when male song is extremely stereotyped, whereas smaller volumes were associated with times during which song was more variable (Smith et al., 1997b). Given that song sparrows do not learn new songs each year, seasonal changes in the volume of nuclei do not seem to relate to changes in song learning, but seem instead to relate to changes in song structure. Like canaries and song sparrows, male starlings sing at relatively high rates for most of the year (Feare, 1984). Unlike song sparrows, starlings appear to be able to learn new songs throughout their lives (Chaiken et al., 1994) and unlike canaries song learning in starlings does not appear to be restricted to a particular season (Böhner et al., 1990). However, in starlings the volumes of HVC, RA, and area X change considerably across an annual cycle (Riters et al., 2002), again suggesting that changes in volume do not correspond well to song learning. It is unknown whether starlings display seasonal changes in stereotypy; however, starlings do sing a longer song bout in spring than in fall indicating that some structural changes do occur seasonally in this species (Riters et al., 2000). Furthermore, during spring a strong relationship was identified in male starlings between the volumes of HVC and RA and song rate (Bernard et al., 1996). Together these data build a strong case for the idea that changes in the volumes of song control nuclei underlie seasonal changes in aspects of song production such as stereotypy or other aspects of song structure.

2.6 Seasonal Dynamics in the Steroid Metabolizing Enzymes

Aspects of the hormonal mechanisms regulating song and song system exhibit variation as a function of sex, seasonal, or hormonal condition besides the morphology of the song control system. For example, biochemical studies carried out on microdissected brain regions demonstrated that the activity of the T-metabolizing enzymes in zebra finches is markedly affected by the age and sex of the birds as well as by their endocrine condition (i.e., castrates treated or not treated with exogenous T (Balthazart et al., 1986; Vockel et al., 1988, 1990a, b). The most dramatic changes observed after treatment with exogenous T concerned aromatase activity and fewer changes in the activity of the two reductases (5-alpha and 5-beta) were detected. In the hypothalamic and preoptic nuclei of zebra finches, aromatase activity was, as in other vertebrates, much higher in males than in females and this sex difference appeared to result mainly if not exclusively from the presence of higher T levels in males. Gonadectomy reduced the enzyme activity to basal levels and treatment with T usually induced high levels of enzymatic activity in both sexes (Vockel et al., 1990a). This pattern is very similar to what had been previously described for many hypothalamic structures in other avian species such as the quail and dove (Balthazart, 1989; Ball and Balthazart, 2002 for reviews) and in mammals (Roselli et al., 1985, 1987; Roselli and Resko, 1989; Roselli, 1991).

A different pattern emerged however in telencephalic samples. In areas such as the dorsal nidopallium or the area parahipocampalis, no clear sex difference in enzyme activity could be detected in sexually mature, gonadally intact birds. Furthermore, gonadectomy associated or not with a treatment with exogenous T, did not seem to have much of an effect on enzyme activity (Vockel et al., 1990a). The data therefore suggested that telencephalic aromatase is not sexually differentiated and not regulated by T. A lack of a sex difference in the telencephalic aromatase activity was also reported independently in young zebra finches (Schlinger and Arnold, 1992b). This differential regulation of aromatase in telencephalic and limbic areas suggested that different mechanisms regulate aromatase expression in these two parts of the brain, and more recent studies characterizing the promoters of the aromatase gene actually provide some support for this idea (Ramachandran et al., 1999). As already discussed previously, zebra finches are not a typical photoperiodic species so that extrapolation of the control mechanisms observed in this species to other songbirds that live in the temperate or boreal zones may be hazardous.

Seasonally breeding species show marked seasonal changes in testis mass and in plasma T levels. These variations are paralleled by pronounced behavioral modifications. The high level of aggressiveness that characterizes the establishment of the territory is generally lost during the period of incubation and feeding of nestlings (Silverin, 1993). Both aggressive and reproductive male behaviors are activated by T in many vertebrate species (Silver et al., 1979; Balthazart, 1983; Wingfield, 1985) and the central aromatization of T appears to be a critical step in the activation of male copulatory behavior and also possibly of parts of the aggressive repertoire (Balthazart, 1989; Balthazart et al., 2004). Little data are available concerning the neuroanatomical distribution and functional regulation of the aromatase enzyme in the brain of free-living species. In the recent years, a few studies have begun to characterize the seasonal variations in the activity of T-metabolizing enzymes in free-living songbirds. This will be illustrated here by studies carried out in three wild species.

2.6.1 Lapland Longspur

The Lapland longspur (*Calcarius lapponicus*) is an Arctic-breeding songbird that exhibits rapid behavioral changes during a short breeding season. Changes in plasma T in the spring are correlated in this species with singing activity but not with territorial aggression. Furthermore, T treatment increases fall song but not aggression in this species. Soma and collaborators (1999a) asked whether localized or temporal differences in the activity of T-metabolizing enzymes could explain these differential relationships between hormones and behavior in this species. The activity of aromatase and of the two reductases was measured in seven brain regions at three different stages of the breeding season (display, mate guarding, and incubation periods). Significant changes in aromatase and 5α-reductase activities were found in several brain regions but the activity of the 5β-reductase changed only in one area, the caudal telencephalon. The most dramatic changes that were observed concerned aromatase activities in the rostral and caudal hypothalamus. In particular, a very pronounced drop of aromatase activity was observed during incubation in the rostral hypothalamus corresponding to the period when the plasma levels of T sharply decrease. This might explain why exogenous testosterone does not activate aggression at that time. These data also provided a first indication that the production of behaviorally relevant metabolites of T changes seasonally in one species of free-living bird.

2.6.2 Pied Flycatcher

Seasonal changes in aromatase activity were also studied during the spring breeding season in male pied flycatchers *Ficedula hypoleuca*, a migratory species common in northern Europe. The pied flycatcher is a polyterritorial, hole-nesting bird that breeds preferentially in deciduous forests (Silverin, 1983a, b). Male pied flycatchers arrive in the breeding area and occupy territories before the females arrive. When a female settles in a territory, she starts nest building more or less immediately. Within just a day or two, the male is found to have drastically increased plasma T titers that remain elevated during the early part of the nest-building period (Silverin and Wingfield, 1982). Egg laying is followed by an early and extremely rapid gonadal regression starting during the later part of the incubation period so that males have basal T levels at the time of egg hatching (Silverin, 1975; Silverin and Wingfield, 1982; Silverin, 1983b).

Males were captured at two different stages of the breeding cycle: in late May/early June when they defend a secondary territory and in late June, when they had returned to their primary territory and were helping females in feeding the nestlings. The existence of a statistically significant drop in plasma T levels and in testicular mass was confirmed between these two stages. Aromatase activity was then measured by the radioenzyme assay in six brain regions: the telencephalon (Tel.), the anterior diencephalon (ant.Di., corresponding to the preoptic area and rostral hypothalamus), the posterior diencephalon (post.Di., including the ventromedial nucleus of the hypothalamus and the tuberal region), the optic lobes (O.L.), the cerebellum (Cereb.), and the brain stem (B.S.); see Foidart et al., (1998) for a full description of these results.

As already reported in other species, the highest level of aromatase activity was observed in the telencephalon where enzyme activity clearly exceeded the activity detected in any other brain area including the anterior or posterior parts of the diencephalon (see ➊ *Figure 10-4a*). Low to nondetectable levels of enzyme activity were observed in the other brain regions. Aromatase activity per mg protein was approximately twice higher in the telencephalon than in the diencephalon, a difference that, given the large size of the telencephalon, translates into a 20-fold excess when total activities are computed for the entire area.

There was a very marked change in aromatase activity in some of the brain areas (➊ *Figure 10-4a*); the activity was significantly higher in the territorial males than in males feeding nestlings in the anterior and posterior diencephalon. This difference was not present in the other brain regions including the telencephalon. These data therefore confirm the existence of seasonal changes in aromatase activity in parallel with changes in plasma T in some brain areas but not in others.

2.6.3 European Starlings

European starlings are widely distributed seasonal breeders that also show marked seasonal fluctuations in circulating T concentrations and reproductive behaviors (➊ *Figure 10-4b*). In the spring, T and its metabolites act within portions of the diencephalon to regulate the pituitary-gonadal axis and activate courtship, including singing and copulation. Song in male starlings is critical for mate attraction during the breeding season (Eens, 1997); it is regulated by steroid-sensitive nuclei in the telencephalon and diencephalon (Ball et al., 2002). Outside the breeding season, T is low to undetectable in the plasma but males continue to sing at high rates (Eens, 1997). Singing outside of the breeding season might thus not be T-dependent or alternatively, singing when T is low might continue to be regulated by T due to an increased sensitivity of the brain to the action of the steroid. This increase in sensitivity could be mediated by changes in intracellular T metabolism leading to increased production of active or decreased production of inactive metabolites.

This hypothesis was tested by measuring the activity of T-metabolizing enzymes (aromatase, 5α-reductase, and 5β-reductase) in the diencephalon and telencephalon in every month throughout an annual cycle (Riters et al., 2001). In the anterior and posterior diencephalon, the highest aromatase actvity was observed in spring when this region is critical for courtship and copulation (➊ *Figure 10-4c–e*). A different pattern of seasonal changes was observed in the telencephalon, where aromatase activity was highest and 5β-reductase activity was lowest throughout the winter months well prior to the reproductive season. This enzymatic balance presumably maximizes T-activity within this region during the winter (high production of the active metabolite, estradiol, and decreased production of the behaviorally inactive 5β-reduced metabolites). Although these data do not indicate whether the metabolic changes occur specifically within song nuclei, these findings are consistent with the idea that singing in male starlings outside the breeding season may be regulated by steroids despite the presence of low serum T concentrations, because brain sensitivity to these low levels is maximized by the pattern of T-metabolizing enzymatic activities.

In conclusion, it can be confidently stated at this point that T metabolism is exquisitely regulated during the annual cycle in the brain of songbirds and that these temporal changes present a high degree of neuroanatomical specificity. This makes it possible, in theory, to modulate specifically the action of T on a variety of behaviors. There are still many questions about the exact contribution provided by these metabolic changes to the control of the seasonal cycle in the behavior of songbirds.

3 Neurochemistry of the Song System

The chemical phenotype of the connections of the various song control nuclei is, as one would expect, based on the basic vertebrate pattern of brain organization and chemical neuroanatomy (Deutch and Roth, 2003). That is to say that amino acid transmitters mediate most connections among the song nuclei. Extrinsic

■ Figure 10-4

Seasonal changes in the activity of steroid metabolizing enzymes in the brain of songbirds. (a) Mean levels of aromatase activity (fmoles per mg protein per hour) observed in six parts of male flycatcher (*Ficedula hypoleuca*) brains that were either defending a secondary territory or feeding nestlings. * = $p < 0.05$; * = $p < 0.01$ by comparison with territory birds. Modified from the data in Foidart et al. (1998). (b) Changes in serum concentrations of T (a) and in the activity of aromatase (b), 5α-reductase (c), and 5β-reductase (d) in the telencephalon, anterior diencephalon, and posterior diencephalon of male starlings sacrificed each month from September to August. Data are mean ± SEM. Modified from the data in Riters et al. (2001)

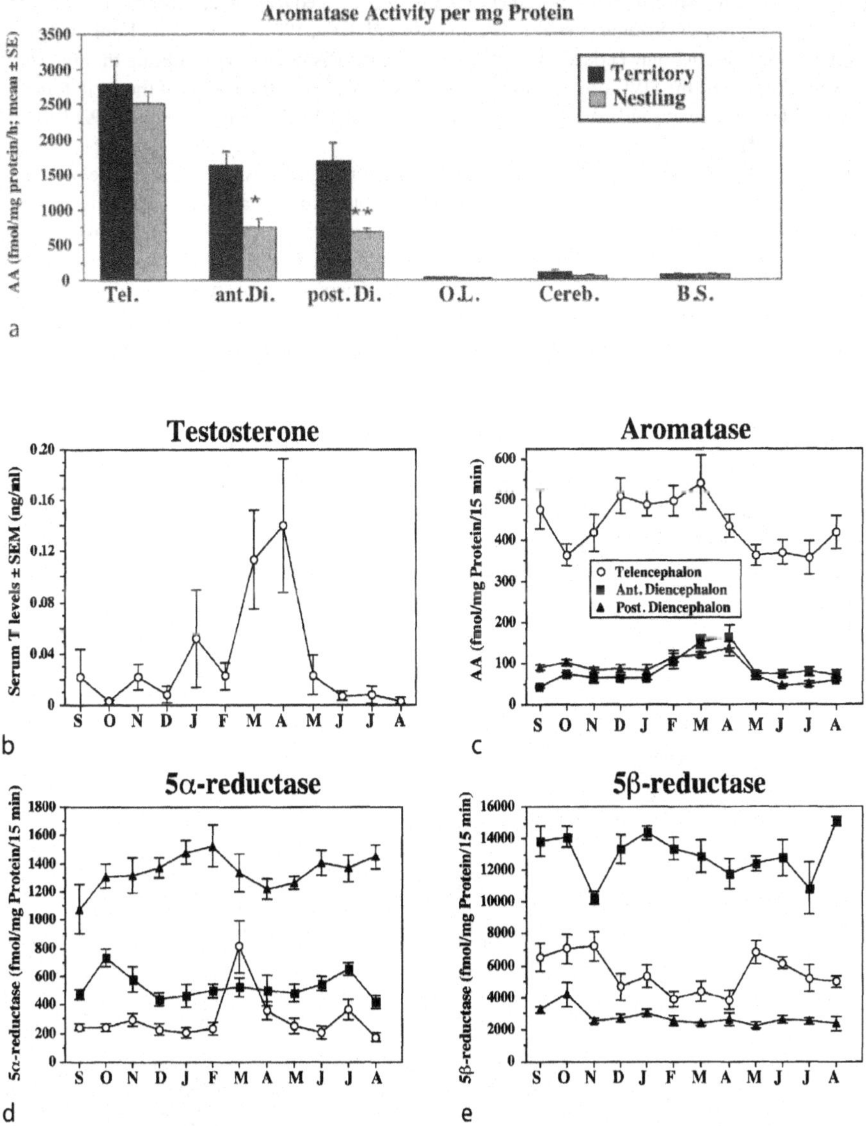

projections from aceytlcholine and catecholamines to these nuclei modulate this ongoing amino acid transmission. In this section, we therefore start with a discussion of the projections mediated by excitatory and inhibitory amino acid transmitters.

3.1 Amino Acid Transmitters

Most of our knowledge about the transmitters mediating synaptic transmission among the key forebrain song control nuclei are based on in vitro studies of brain slices derived from male zebra finches that contain subsets of the song nuclei (Sakaguchi et al., 1987; Kubota and Saito, 1991; Mooney and Konishi, 1991; Luo and Perkel, 1999a; Farries and Perkel, 2002). For example, studies of this type indicate that the synaptic contacts resulting from the HVC projections to area X and RA involve the release of the excitatory amino acid transmitters, glutamate and aspartate (Mooney and Konishi, 1991; Sakaguchi et al., 1992; Farries and Perkel, 2002; see *Figure 10-5*).

Similarly, synapses associated with the projection from lMAN to RA involve the release of glutamate and aspartate (Sakaguchi et al., 1992). The appropriate receptors for glutamatergic transmission are also present. The majority (21 out of 26) of the known amino acid transmitter receptor subunits have been subcloned in zebra finches and their distributions are mapped based on ISH histochemistry procedures that localize the mRNA on brain slices (Wada et al., 2004). These receptor subtypes exhibit a complex distribution in the forebrain song control system of male zebra finches with some subtypes more enriched in the song control areas than surrounding structures, whereas others are expressed at a lower level

Figure 10-5

Schematic presentation of the excitatory and inhibitory amino acid transmissions in the song system. Glutamatergic non-NMDA- and NMDA-mediated transmissions and GABAergic transmission are indicated by different symbols and associated with plus and minus signs for excitatory and inhibitory inputs, respectively. Two types of neurons present in area X use GABAergic transmission, the spiny neurons (SN), and the aspiny fast-firing (AF) neurons

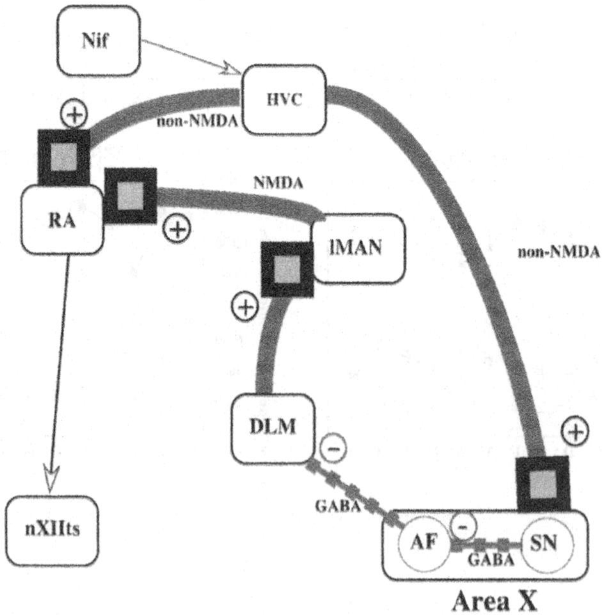

(Wada et al., 2004). These studies of the mRNA expression reveal patterns similar to what has been described based on autoradiographic studies localizing specific glutamatergic-binding sites. This is the case for HVC in zebra finches which is outlined from the surrounding neostriatum by a lower density of NMDA glutamatergic receptors (Aamodt et al., 1992, 1995), based on autoradiographic labeling of receptors.

In vitro studies with specific receptor blockers have revealed diversity in the receptor types that mediate neurotransmission among the major projections of the song nuclei. For example, the projection from HVC to RA involves binding to receptors of the non-NMDA ionotropic subtype while the lMAN to RA projection involves those of the NMDA subtype. The HVC to area X excitatory synapse on spiny-type neurons in X is also mediated by non-NMDA ionotropic glutamatergic receptors (Ding et al., 2003; see *Figure 10-5*).

In contrast to these excitatory synapses, the projection from area X to DLM appears to use the major inhibitory amino acid transmitter, γ-amino butyric acid (GABA) (Luo and Perkel, 1999a, b). GABAergic cells have been localized in male and female zebra finches based on an immunohistochemical study of GABA (Grisham and Arnold, 1994). As would be expected for such a ubiquitous transmitter, GABA immunoreactive cells were found in all the song control nuclei studied, including HVC, area X, RA, MAN, and ICo. Most of these labeled cells are in all probability local interneurons. However, in the case of cells that project from area X to DLM there is now a convincing amount of evidence demonstrating that these are also GABAergic. Cell bodies staining positive for the GABA synthesizing enzyme glutamic acid decarboxylase (GAD) are present in X but only terminals are observed in the target nucleus DLM (Luo and Perkel, 1999b). These terminals in DLM are greatly reduced when X is subjected to excitotoxic lesions (Luo and Perkel, 1999b). In vitro studies of brain slices containing the X to DLM projection found strong inhibitory postsynaptic potentials resulting from the activation of afferents from X that were unchanged in the presence of glutamate antagonists but blocked by the $GABA_A$ receptor antagonist bicuculline methiodide (Luo and Perkel, 1999a). This inhibitory projection raised questions about the processing of auditory information in the anterior forebrain pathway that includes X, given that auditory input derived from HVC is measurable upstream in this pathway in X and downstream in lMAN (Kimpo et al., 2003).

How can this auditory coding be maintained when the projections through the pathway move from excitatory to inhibitory and then back to excitatory? Detailed studies by Farries and Perkel (2002) of the physiology of different cell types within X have led to the proposal that X has properties characteristic of the mammalian striatum as well as the globus pallidus. Farries and Perkel propose that there are two inhibitory synapses between HVC and DLM outputs that ultimately result in an excitatory signal from HVC to DLM. HVC stimulates spiny neurons in X, which would then inhibit the long projecting aspiny fast-firing cells (AF), reducing their spontaneous activity which would in turn transiently reduce tonic inhibition in DLM cells allowing them to fire (Farries and Perkel, 2002). In mammals, the spiny neurons are characteristic of the striatum but the long projecting AF neurons are more characteristic of the globus pallidus. Other chemical neuroanatomical studies are consistent with the idea that although area X is essentially striatal in nature some cell groups do have pallidal characteristics (Carrillo and Doupe, 2004). In any case, this hypothesis requires further testing to be confirmed in detail (Perkel, 2004), but there is no question that a complex inhibitory projection is present from HVC to X to DLM.

The situation for interneurons within HVC is complex and a clear picture is just emerging. Although these interneurons clearly use amino acid transmitters, their transmitter phenotype does not distinguish different cell types. Rather the identity of calcium-binding proteins expressed by the interneurons seems to be a key variable (Wild et al., 2005). Double and triple-label studies using antibodies against paralbumin, calbindin, and calretinin indicate that long projecting cells to area X or HVC in male zebra finches tend not to express calcium-binding proteins (Wild et al., 2005). Some interneurons tend to express two or three calcium-binding proteins and these cells tend to have a fast-spiking physiological profile while other interneurons that tend not to express any calcium-binding protein tend to have a slow-spiking profile (Wild et al., 2005). The story emerging from this work and other physiological studies is that HVC contains a microcircuit of X and RA projecting cells whose activity is co-ordinated by interneurons so that motor output to RA and some sort of efferent copy to the X and the anterior forebrain circuit can be co-ordinated (Mooney and Prather, 2005).

3.2 Localization of Catecholamines and Acetylcholine

3.2.1 Acetylcholine

Another feature of the neurochemistry of the song system is that amine transmitters such as acetylcholine and the catecholamines are synthesized in brain areas outside the song system per se, but these areas send long distance projections that modulate ongoing neural activity. This pattern is again in common with the basic vertebrate pattern of brain organization. In the case of acetylcholine, there is intrinsic synthesis in some areas such as area X of the medial striatum and in other cases such as HVC, RA, and lMAN acetylcholine is synthesized outside the nucleus in brain areas that project to these song control nuclei (Li and Sakaguchi, 1997; *Figure 10-6*).

Figure 10-6
Schematic presentation of cholinergic innervation and neurotransmission in the song system. Although most song control nuclei contain acetylcholine (ACh), acetylcholine esterase (AChesterase), and the corresponding muscarinic (Musc) and nicotinic (Nico) receptors, evidence for ACh synthesis (i.e. presence of Choline acetyl transferase, ChAT) is only available for area X. Large and small symbols for receptors indicate that a receptor density is respectively higher or lower in the nucleus than in the surrounding tissue. Tract-tracing studies have indicated that the Ach is present in HVC, and RA is produced in the anterior and posterior parts of the ventral pallidum (VP), respectively

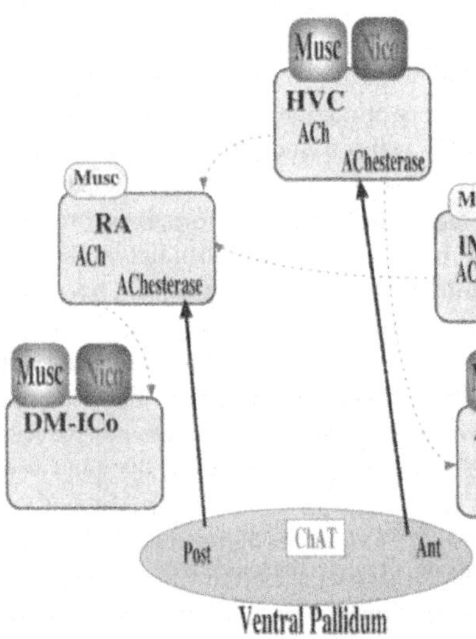

Arnold and collaborators originally demonstrated that several song control nuclei, including HVC, RA, lMAN, and X can be identified in zebra finch brain sections by the presence of markers of cholinergic transmission such as the enzyme involved in transmitter degradation, acetylcholinesterase, or the acetylcholine receptors of the muscarinic subtype (Ryan and Arnold, 1981). Subsequent autoradiographic work using a ligand that is not subtype selective identified muscarinic receptors in the song nuclei of European starlings (Ball et al., 1990; Bernard et al., 1993; Ball et al., 1994) and various nicotinic receptor receptor subtypes in similar brain area of male zebra finches (Watson et al., 1988). Again, as was observed in the case of glutamate receptors, the pattern of receptor density was such that in some cases such as RA the density appeared lower than the surrounding arcopallium, whereas in other cases such as area X the density of muscarinic receptors was higher than the surrounding medial striatum (Ryan and Arnold, 1981; Ball et al., 1990, 1994). Subsequent studies of the synthesizing enzyme choline acetyltransferase and the degrading enzyme acetylcholinesterase indicated that most of the song nuclei receive cholinergic input, but clear evidence for intrinsic synthesis was only apparent in X (Zuschratter and Scheich, 1990). Tract tracing studies combined with the ICC localization of choline acetyltransferase identified the ventral pallidum (VP) as the source of cholinergic input to HVC and RA, but could not identify the source of the innervation of lMAN (Li and Sakaguchi, 1997). VP appears homologous to the mammalian basal forebrain cholinergic system that includes the nucleus Basalis of Meynert (Li and Sakaguchi, 1997).

3.2.2 Catecholamines

A prominent catecholaminergic innervation of the song control system has been demonstrated in several songbird species, most notably zebra finches by the presence of fibers immunoreactive for the catecholamine synthesizing enzymes tyrosine hydroxylase (TH; Bottjer, 1993; Soha et al., 1996) and/or dopamine beta-hydroxylase (Mello et al., 1998). Similarly, ICC studies of TH in male canaries have found that the boundaries of HVC and RA can be defined in a manner in agreement with Nissl-defined boundaries by a higher density of fibers immunoreactive for tyrosine hydroxylase compared to the surrounding neostriatum (Appeltants et al., 2001).

High concentrations of NE and DA have accordingly been measured in the song control nuclei of zebra finches (Barclay and Harding, 1988; 1990; Sakaguchi and Saito, 1989) (● *Figure 10-7*). Autoradiographic studies have revealed binding sites for [^{3}H] RX821002, a ligand specific for α_2-adrenergic receptors, to be enriched in forebrain song control nuclei of this species (Riters and Ball, 2002). In European starlings, high densities of noradrenergic receptors of the α_2 and $\beta1/\beta2$ subtypes are also present in several song control nuclei (HVC, RA, and area X; Ball, 1994; Ball et al., 1994; Riters et al., 2002) and high densities of dopamine

■ Figure 10-7

Schematic presentation of catecholaminergic innervation and neurotransmission in the song system. The relative concentration of dopamine (DA) and norepinephrine (NE) in each nucleus is illustrated by the >, <, or = signs. The presence of dopaminergic (D1) or noradrenergic (α1, α 2, β1, and β2) receptors is also indicated for each nucleus. Large and small symbols indicate a receptor density that is respectively higher or lower in the nucleus than in the surrounding tissue (based mainly on the data in Ball, 1990, 1994; Barclay and Harding, 1988; 1990; Casto and Ball, 1994; Riters et al., 2002)

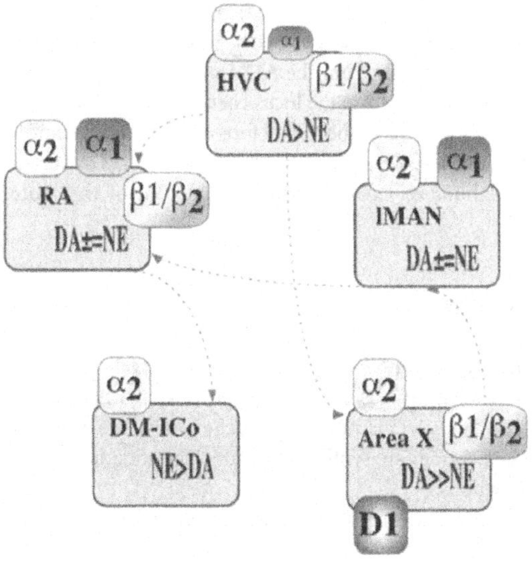

D1 receptors have been described in area X of the medial striatum in the same species (Casto and Ball, 1994). In these cases, boundaries of these song control nuclei can be defined based on the high receptor densities compared to the surrounding brain area that are consistent with Nissl-defined boundaries.

The cells that synthesize catecholamines and give rise to the catecholaminergic innervation of the song control nuclei have been investigated by tract-tracing studies in canaries (➋ *Figure 10-8*). HVC receives dopaminergic inputs mainly from the mesencephalic central gray (GCt; group A11 in the nomenclature of

◻ Figure 10-8

Generalized view of the songbird vocal control system schematically illustrating the catecholaminergic innervation of the song control nuclei (based on the data in Lewis et al., 1981; Appeltants et al., 2000, 2002a)

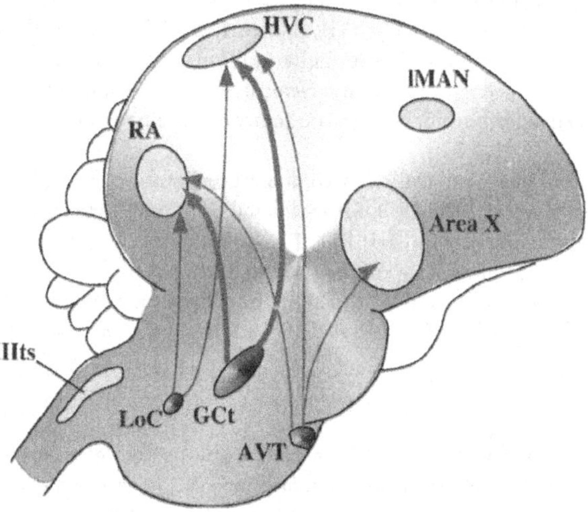

Dahlström and Fuxe 1964; Reiner et al., 1994, for a review of this nomenclature use in birds) and to a lesser extent from the ventral tegmental area (VTA; formally, area ventralis of Tsai; group A10) and noradrenergic inputs from the complex of the locus coeruleus (Loc; group A6) (Appeltants et al., 2000). Similarly, dopaminergic inputs to RA come mostly from the A11 (mesencephalic central gray) and A10 (VTA) cell groups, and noradrenergic inputs come from the locus coeruleus and subcoeruleus (A6) (Appeltants et al., 2002a). In contrast, area X, a key nucleus of the rostral forebrain pathway controlling song learning, but not song production, seems to receive CA inputs exclusively from VTA (Lewis et al., 1981). These data thus suggest that different populations of CA cell groups could control the motor and anterior forebrain pathways.

3.2.3 Neuropeptide Systems

Of the many neuropeptides that are known to be active as neurotransmitters or neuromodulators in the brain, only a subset has been investigated in the song control system. In most cases, these neuropeptides have been localized to the song nuclei via immunohistochemistry, but we know little about the receptors they employ or whether the peptides are synthesized within the song nuclei or by cells outside the nuclei that project to these brain regions.

The neuropeptide methionine (Met) enkephalin was shown to label several song nuclei in zebra finches and two other songbird species, the song sparrow and the European starling (Ryan et al., 1981; Ball et al., 1988). Vasoactive intestinal polypeptide (VIP) has also been found to be widely localized in forebrain song control nuclei including HVC and lMAN (Ball et al., 1988) (Bottjer and Alexander, 1995; Ball and collaborators (1988) similarly showed that several neuropeptides (Met-enkephalin, cholecystokinin, substance P) are distributed in a unique manner in the song control nuclei of two songbird species, the song sparrow and the European starling. Neuropeptides such as VIP or Met-enkephalin are also specific markers for HVC and RA in zebra finches, and the volumes defined by these peptidergic innervations match closely the volumes defined by more traditional histological methods (Ball et al., 1995a).

Neurochemistry has also been used to demonstrate the existence of a sex difference in the nucleus interfacialis (Nif) that cannot be observed by traditional histological methods. Nif specifically projects to HVC and is considered as part of the motor circuit of the song control system. Nif was originally identified by retrograde tract tracing from HVC (Nottebohm et al., 1982), but it is not readily visible in Nissl stained sections. It is however characterized by an accumulation of Met-enkephalin and VIP-immunoreactive fibers and the volume occupied by these immunoreactive fibers is larger in males than in females (Ball et al., 1995b).

Several of these neurochemical markers of song control nuclei are neuropeptides (e.g., VIP, met-enkephalin) that are not contained in neurons located within HVC or RA but are present in fibers originating from neurons, whose location has not been identified. Given that the song nuclei defined by these fibers are different in males and females, these projections and the cell groups from which they originate may also be sexually dimorphic. Studies combining retrograde tract tracing with ICC should be carried out to identify these sexually dimorphic cell groups.

The distribution of vasotocin-immunoreactive (VT-ir) fibers has been studied in detail by ICC in the brain of the canary (Kiss et al., 1987; Voorhuis et al., 1990), zebra finch (Voorhuis and de Kloet, 1992) and Junco (*Junco hyemalis*; Panzica et al., 1999). VT-ir fiber endings are observed in broad areas of the telencephalon, but in particular a large supply of VT-ir fibers is present around RA (Kiss et al., 1987; Panzica et al., 1999) where they correspond to the presence of vasotocin-binding sites (Voorhuis et al., 1990). The mesencephalic song control nucleus ICo is also labeled by VT-ir fibers in both oscines and nonoscine birds.

3.3 Functional Studies of Neurotransmitter Systems

There is still much to be learned about the functional interactions among the transmitter systems that mediate neural transmission in the song control circuit. We highlight some major observations in this section. As would be expected based on the wealth of data in other systems, excitatory amino acid transmission acting via NMDA receptors is involved in the song learning process (Nordeen and Nordeen, 2004). For example, administering the NMDA receptor anatagonist, AP5 to nucleus lMAN just before a song tutoring session will significantly attenuate the number of songs reproduced by the pupil (Basham et al., 2004). Treatment on alternate days in which tutoring does not occur or infusing the compound to nonsong control regions has no effect (Basham et al., 2004). There are developmental changes in NMDA receptor subunits that correlate with the opening and closing of the sensitive period for song learning (Nordeen and Nordeen, 2004). However, experimental manipulations (such as early social deprivation or premature testosterone treatment) can modify the timing of the ability to learn song but they do not change the developmental progression of NMDA receptor subunits in parallel so these changes cannot in and of themselves explain the closing of the sensitive period for song learning (Nordeen and Nordeen, 2004). Consistent with the notion that excitatory amino acid transmission plays a potentially important role in the song learning process is the description of NMDA-dependent long term potentiation (LTP) in nucleus area X both in juvenile and in adult zebra finches (Ding and Perkel, 2004).

Specific modulatory roles for ascending amine transmitter systems on ongoing amino acid transmission have also been described. We will focus here on two specific examples, one involving cholinergic mechanisms and the other involving the noradrenergic system. Several forebrain nuclei in the song control circuit including HVC exhibit remarkable patterns of auditory responsivity. In anesthetized birds, cells in song nuclei in the anterior forebrain pathway are tuned to fire at high rates only to sounds of the bird's own song (Margoliash and Konishi, 1985; see Theunissen et al., 2004 for a review). These responses, though, are gated while birds are awake; they may or may not exhibit the responses depending on the particular behavioral conditions (Schmidt and Konishi, 1999). While they are asleep they always exhibit such responses (Dave et al., 1998). This gating of auditory responsiveness to the bird's own song in HVC appears to be modulated by basal forebrain cholinergic projections to HVC (Shea and Margoliash, 2003). The administration of muscarinic cholinergic agonists to HVC was found to block the expression of neural responses to sounds from the bird's own song and this could be reversed by the administration of antagonists (Shea and Margoliash, 2003). Similarly, stimulation of the VP, the assumed source of the relevant cholinergic input also blocked physiological responses to these sounds (Shea and Margoliash, 2003).

Ascending noradrenergic projections to the song system have been implicated in the regulation of social modulation of gene expression in the song control circuit (Castelino and Ball, 2005). The act of singing drives the expression of an immediate early gene ZENK (acronym derived from homologues in other species; Mello et al., 1992) in several forebrain nuclei of the song control system in male zebra finches (Jarvis and Nottebohm, 1997). Zebra finch song is produced in different social contexts that result in dramatically different degrees of ZENK expression in song control nuclei (Jarvis et al., 1998). Motor-driven ZENK expression levels in song control nuclei are high as a result of solo song and much lower after male or female-directed song (Jarvis et al., 1998). This effect is most apparent in area X of the medial striatum where there is a qualitative shift in the pattern of gene expression from the solo/undirected context (very high expression) to the directed context (much lower expression; Jarvis et al., 1998). Systemic administration of the noradrenergic neurotoxin DSP-4 was found to result in a significant increase in the number of ZENK expressing cells in area X of male finches engaged in directed song (Castelino and Ball, 2005). The increase in ZENK expression after the administration of a neurotoxin that depletes noradrenergic concentrations in the song system supports the hypothesis that the noradrenergic system plays a key role in downregulating ZENK expression during directed song (Castelino and Ball, 2005).

Some of the ideas expressed in this section will be further developed in the next section, where steroid–neurotransmitter interactions are discussed.

4 Hormone–Neurotransmitter Interactions

4.1 Hormone Effects on Neurotransmitter Systems

As mentioned above, in songbirds, catecholamines and their synthetic enzymes are present at high concentrations in brain nuclei that control song (Barclay and Harding, 1988; Sakaguchi and Saito, 1989; Barclay and Harding, 1990; Bottjer, 1993; Soha et al., 1996; Mello et al., 1998; Appeltants et al., 2001). The boundaries of several song control nuclei can be defined by high densities of α_2-adrenergic (Ball, 1994; Ball et al., 1994) or dopaminergic (Casto and Ball, 1994) receptors.

Recent work has identified two brain sites outside the song control system proper where steroids could act to modify song production. The POM that regulates appetitive aspects of male sexual behavior in a variety of species (Panzica et al., 1996; Balthazart et al., 1998), including starlings (Riters and Ball, 1999; Riters et al., 2000), and the catecholaminergic nuclei of the mesencephalon and pons that send afferent projections to the song control nuclei. The anatomical connections between these brain areas and the neural circuit that controls singing are still poorly understood. No direct projection between the POM and the song control nuclei has been described to this date. Given the large number of tract-tracing studies that were previously carried out, it appears probable that such connections should have been detected if they exist. Available evidence suggests however one indirect link between these structures. It has been recently shown that in Japanese quail, the aromatase-immunoreactive cells of the POM massively project to the mesencephalic central gray (Absil et al., 2001). In canaries, dopaminergic cells of the central gray are known to be the major source of catecholaminergic inputs to HVC (Appeltants et al., 2000) and RA (Appeltants et al., 2002a). Recent work in starlings indicates that projections from the POM to central gray are also present similarly to what has been described in quail (Riters and Alger, 2004). These projections could therefore represent the anatomical substrate that co-ordinates the actions of POM and of song control nuclei on song production.

Interestingly, the prominent catecholaminergic innervation of the song control nuclei is modulated by steroids. This is demonstrated by ICC studies of tyrosine hydroxylase in canaries (including sex differences detected in HVC and RA; Appeltants et al., 2001; increases in TH immunoreactivity following treatment with exogenous testosterone; Appeltants et al., 2003), by the presence of a steroid modulation of NE and DA baseline levels and turnover (Barclay and Harding, 1988, 1990), and by the observation that the density of α_2 noradrenergic receptors in HVC and RA varies seasonally in starlings (see later).

This steroid modulation of catecholaminergic inputs represents one possible way that steroids may influence song learning (Marler et al., 1988; Soha et al., 1996) and production (Harding et al., 1983; Bottjer and Johnson, 1997; Schlinger, 1997). Indeed functional interactions between steroids and catecholamines have been reported in many vertebrate species, and catecholamines are involved in the control of a variety of steroid-regulated reproductive behaviors (Balthazart and Ball, 1989; Blaustein and Olster, 1989; Barclay et al., 1991; Blaustein et al., 1995; Barclay et al., 1996). Thus, it has been speculated that steroids may influence song control nuclei and singing activity at least in part by a direct action at the level of catecholaminergic structures. However, the roles of NE and DA and of their specific receptors located within the song system are not well understood.

The specific function of the catecholaminergic afferent inputs to the song system has not been studied in great detail. In zebra finches, systemic injections of DSP-4, a specific noradrenergic neurotoxin that preferentially destroys catecholaminergic innervation of telencephalic regions (Fritschy and Grzanna, 1989; Zaczek et al., 1990) decreases male courtship behavior including female-directed singing (Barclay et al., 1991, 1996). This effect appears to result from attention deficits rather than from impairments of the motor aspects of song because in DSP-4-treated birds the latency to initiate singing was increased, but once song begun it was quite normal.

In female canaries, the pharmacological depletion of central norepinephrine levels obtained by injection of DSP4 decreases the incidence of copulation solicitation displays that are produced in response to the playback of sexually stimulating male songs (Appeltants et al., 2002b). This effect of DSP4 was observed in conditions when the sexually stimulating songs were partly masked by nonstimulating wild canary songs or heterospecific songs, but not when they were masked by simpler signals such as white noise. These data therefore suggest that central noradrenergic inputs (presumably to auditory areas such as Field L or to HVC) modulate sexual responsiveness of female canaries by affecting the auditory processing of the relevant information contained in sexually stimulating male songs.

Dave and colleagues (1998) also demonstrated that NE acting in HVC regulates in a gate-like fashion the auditory responsiveness of efferent neurons to RA. Studies of the modulation of the neural activity in the anterior forebrain pathway [area X, dorsolateral nucleus of the anterior thalamus (DLM) and nucleus magnocellularis of the anterior neostriatum (lMAN)] as a function of social context have also implicated ascending catecholaminergic projections (Jarvis et al., 1998; Hessler and Doupe, 1999b) as being responsible for the contextual regulation.

These studies suggest that catecholamines have an important role in the control of certain aspects of song behavior but there are still major lacunae in our knowledge about the chemical neuroanatomy of this system in songbirds.

One possibility is that NE is involved in differences in song expressed by males within and outside the context of reproduction. In a reproductive context, male song directed toward females is critical for mate attraction (Catchpole and Slater, 1995). Outside the context of reproduction, males often sing less and do not direct song toward females, or use song for purposes of dominance or territory maintenance (Wiley et al., 1993; Catchpole and Slater, 1995). NE levels within some song nuclei are positively correlated with courtship singing in male zebra finches (Barclay et al., 1991), and selective neurotoxic lesions of noradrenergic neurons produce specific behavioral deficits in male courtship song displays (Barclay et al., 1991; 1996), indicating a role for NE in song sung within a reproductive context. Song behavior is related to the activity of testosterone (T) and its metabolites, presumably acting within the song control system (Arnold, 1981). Estrogenic T metabolites, known specifically to increase female-directed courtship song in male zebra finches (Walters et al., 1991), increase NE turnover in the song system (Barclay and Harding, 1988; 1990), suggesting that the role of NE in song relates to T activity.

In contrast to zebra finches, a species that does not restrict breeding to a particular season, many temperate-zone songbirds display seasonal variation in T concentrations, song behavior, and reproductive activities (Wingfield and Farner, 1993; Ball, 1999). Changes in T are associated with changes in the context in which males will sing. During both the breeding and nonbreeding seasons, when T is high and low respectively, song can be observed during intermale aggression and in males singing alone; however, only during the breeding season male song is observed to play a direct role in mate attraction (Wingfield and Farner, 1993; Catchpole and Slater, 1995; Eens, 1997; Ball, 1999).

The studies in zebra finches discussed above suggest that NE plays a role in the control of courtship song and that T (or its metabolite, estradiol) mediates this relationship. If this relationship exists, then in seasonally breeding birds, differences might be expected in NE or NE receptors in birds during the breeding season as compared with outside the breeding season. In agreement with this idea, T regulates concentration of NE and the density of its receptors in specific brain areas of at least one seasonally breeding bird (a nonsongbird, the Japanese quail; Balthazart and Ball, 1989; Balthazart et al., 1992a). These data suggest specifically that seasonal changes in T might be associated with variation in noradrenergic transmission within the brains of seasonally breeding birds. Several of these catecholaminergic cell groups express nuclear receptors for sex steroids (Maney et al., 2001) and could therefore represent alternative sites where steroids could act in the brain to control aspects of singing behavior.

4.1.1 Seasonal Changes in Alpha-Two Adrenergic Receptor Densities

In songbirds also, T can regulate the activity of catecholamines through the regulation of receptors. Consistent with this idea are studies of T and α_2-adrenergic receptors in songbirds. In seasonally breeding male starlings, seasonal changes in the density of α_2-adrenergic receptors were identified within HVC and RA (Riters et al., 2002), that correspond with changes in courtship song. These changes mirrored both seasonal changes in T concentrations and changes in HVC and RA volume. T-dependent seasonal decreases in NE within the song system might result in an upregulation of α_2-adrenergic receptors in fall, and a downregulation of receptors in spring, a pattern that would be consistent with the seasonal changes in density observed in starlings.

α_2-adrenergic receptors were least dense in spring when song is important for a male to attract a female. In contrast, during molt, fall, and winter nuclei were smallest, and α_2-adrenergic receptors were densest when males were likely not singing (during molt; Eens, 1997), or when song is not directly sung to attract mates for immediate reproduction (Feare, 1984; Eens, 1997). These data suggest that changes in α_2-adrenergic receptors might reflect differential NE involvement in female-directed spring song compared with undirected fall song, with NE possibly regulating the attention a male directs toward a potential mate during female-directed song, as proposed previously in zebra finches (Barclay et al., 1996).

A general concept emerging in the study of hormones and behavior is that the steroids may regulate brain morphology in areas important for reproductive behaviors not only by acting directly on cells in the brain area of interest, but also by acting indirectly (i.e., transsynaptically) via cells at a distant brain site that then projects to the site of interest and modifies neural activity (Beyer, 1987; Balthazart and Ball, 1995; 1998). This mode of action is well illustrated by studies of sex steroid feedback on the gonadotropin-releasing hormone system (GnRH). Sex steroids are known to exert negative feedback on GnRH immunoreactive cells but these neurons do not contain detectable levels of receptors for sex steroid hormones. Rather, the effects of steroids on these neurons are mediated by changes in the activity of peptidergic (e.g., opioids) and catecholaminergic (e.g., norepinephrine) inputs to the GnRH cells (Barraclough and Wise, 1982; Weiner et al., 1988; Herbison, 1998; Kelly and Ronnekleiv, 2002).

The presence of AR and/or ERα in the neural network mediating the learning, perception, and production of song in oscine songbirds has often been taken as evidence indicating that steroids control singing behavior by binding directly to these receptors. However, steroid receptors are also found in the catecholaminergic cell groups (Maney et al., 2001) that are known to project to the song control nuclei such as the locus coeruleus, the substantia nigra, and the ventral tegmental area (Lewis et al., 1981; Burd et al., 1986; Appeltants et al., 2000, 2002a) so that steroids could also affect song by modulating these catecholaminergic inputs. There would therefore be a direct action of steroids in the telencephalic song control nuclei and in an indirect action through the modulation of catecholaminergic inputs.

These indirect effects of steroids could themselves reflect two different modes of actions. Steroids are known to modify the release and thus turnover of NE and DA in song control nuclei. This possibly results from a direct nongenomic action at the level of the catecholaminergic terminals. Alternatively, steroid hormones could modify the expression of catecholamine synthesizing enzymes by acting genomically on catecholaminergic perikarya. This mode of action is mainly suggested by the observation that the treatment of castrated male canaries with testosterone increases the density of tyrosine hydroxylase immunoreactive fibers in HVC and RA (Appeltants et al., 2003). This change in immunoreactivity is likely to reflect changes in concentration of the enzyme that most likely are caused by increases in transcription of the corresponding protein. This effect thus has to take place at the level of the catecholaminergic perikarya that are at the origin of the tyrosine hydroxylase immunoreactive fibers, which innervate the song control nuclei. The presence of steroid receptors in the nuclei that contain these perikarya (Maney et al., 2001) clearly supports this mode of indirect action for steroids.

Taken together, the available data suggest that seasonal variation and/or steroid-induced changes in catecholaminergic afferent input into the song control system are the important factors regulating changes in the morphology of the vocal control system and in vocal learning or production.

5 Conclusion

In this review, we have tried to highlight the many facets of the neurochemistry and the neuroendocrinology of song and the song control circuit. We have focused on the role of steroid hormones in regulating forebrain nuclei in the song control circuit involved in the learning and production of song. This system is one of the few clear examples in the vertebrate brain where steroid hormones have well-documented and dramatic effects on the morphology and physiology of pallial structures. Steroids undoubtedly have important effects on pallial structures in other vertebrate taxa, but it may well be the case that fundamental insights will be gleaned in species exhibiting the most obvious effects such as songbirds.

Acknowledgments

Research in the authors' laboratory and preparation of this review were supported by grants from the NIH/NINDS (R01 NS 35467) to G.F.B., and by a grant from the Belgian FRFC (2.4562.05) to J.B.

References

Aamodt SM, Kozlowski MR, Nordeen EJ, Nordeen KW. 1992. Distribution and developmental change in [3H]MK-801 binding within zebra finch song nuclei. J Neurobiol 23: 997-1005.

Aamodt SM, Nordeen EJ, Nordeen KW. 1995. Early isolation from conspecific song does not affect the normal developmental decline of N-methyl-D-aspartate receptor binding in an avian song nucleus. J Neurobiol 27: 76-84.

Absil P, Braquenier JB, Balthazart J, Ball GF. 2002. Effects of lesions of nucleus taeniae on appetitive and consummatory aspects of male sexual behavior in Japanese quail. Brain Behav Evol 60(1): 13-35.

Absil P, Riters LV, Balthazart J. 2001. Preoptic aromatase cells project to the mesencephalic central gray in the male Japanese quail (*Coturnix japonica*). Horm Behav 40(3): 369-383.

Alvarez-Borda B, Nottebohm F. 2002. Gonads and singing play separate, additive roles in new neuron recruitment in adult canary brain. J Neurosci 19: 8684-8690.

Appeltants D, Absil P, Balthazart J, Ball GF. 2000. Identification of the origin of catecholaminergic inputs to HVc in canaries by retrograde tract tracing combined with tyrosine hydroxylase immunocytochemistry. J Chem Neuroanat 18: 117-133.

Appeltants D, Ball GF, Balthazart J. 2001. The distribution of tyrosine hydroxylase in the canary brain: demonstration of a specific and sexually dimorphic catecholaminergic innervation of the telencephalic song control nuclei. Cell Tissue Res 304(2): 237-259.

Appeltants D, Ball GF, Balthazart J. 2002a. The origin of catecholaminergic inputs to the song control nucleus RA in canaries. Neuroreport 13: 649-653.

Appeltants D, Ball GF, Balthazart J. 2003. Song activation by testosterone is associated with an increased catecholaminergic innervation of the song control system in female canaries. Neuroscience 121(3): 801-814.

Appeltants D, Del Negro C, Balthazart J. 2002b. Noradrenergic control of auditory information processing in female canaries. Behav Brain Res 133(2): 221-235.

Arnold AP. 1975. The effects of castration and androgen replacement on song courtship, and aggression in zebra finches. J Exp Zool 191: 309-326.

Arnold AP. 1981. Logical levels of steroid hormone action in the control of vertebrate behavior. Amer Zool 21: 233-242.

Arnold AP. 1990. The passerine bird song system as a model in neuroendocrine research. J Exp Zool 4(Suppl): 22-30.

Arnold AP, Nottebohm F, Pfaff DW. 1976. Hormone concentrating cells in vocal control areas of the brain of the zebra finch (*Poephila guttata*). J Comp Neurol 165: 487-512.

Arnold AP, Saltiel A. 1979. Sexual difference in pattern of hormone accumulation in the brain of a song bird. Science 205: 702-705.

Ball GF. 1990. Chemical neuroanatomical studies of the steroid-sensitive songbird vocal control system: a comparative approach. Hormones, brain and behaviour in vertebrates. 1. Sexual differentiation, neuroanatomical aspects, neurotransmitters and neuropeptides. Balthazart J, editor. Comp Physiol Basel: Karger; 8: 148-167.

Ball GF. 1994. Neurochemical specializations associated with vocal learning and production in songbirds and budgerigars. Brain Behav Evol 44: 234-246.

Ball GF. 1999. Neuroendocrine basis of seasonal changes in vocal behavior among songbirds. The design of animal communication. Hauser M, Konishi M, editors. Cambridge, MA: MIT Press; pp. 213-253.

Ball GF, Absil P, Balthazart J. 1995a. Assessment of volumetric sex differences in the song control nuclei HVC and RA in zebra finches by immunocytochemistry for methionine enkephalin and vasoactive intestinal polypeptide. Brain Res 699: 83-96.

Ball GF, Absil P, Balthazart J. 1995b. Peptidergic delineations of nucleus interface reveal a sex difference in volume. Neuroreport 6: 957-960.

Ball GF, Balthazart J. 2002. Neuroendocrine mechanisms regulating reproductve cycles and reproductive behavior in birds. hormones, brain and behavior, Vol. 2. Pfaff DW, Arnold AP, Etgen AM, Fahrbach SE, Rubin RT, editors. San Diego, CA: Academic; pp. 649-798.

Ball GF, Casto JM, Bernard DJ. 1994. Sex differences in the volume of avian song control nuclei: comparative studies and the issue of brain nucleus delineation. Psychoneuroendocrinology 19: 485-504.

Ball GF, Faris PL, Hartman BK, Wingfield JC. 1988. Immunohistochemical localization of neuropeptides in the vocal control regions of two songbird species. J Comp Neurol 268: 171-180.

Ball GF, Hulse SH. 1998. Bird song. Am Psychol 53: 37-58.

Ball GF, Nock B, Wingfield JC, McEwen BS, Balthazart J. 1990. Muscarinic cholinergic receptors in the songbird and quail brain: a quantitative autoradiographic study. J Comp Neurol 298: 431-442.

Ball GF, Riters LV, Balthazart J. 2002. Neuroendocrinology of song behavior and avian brain plasticity: multiple sites of action of sex steroid hormones. Front Neuroendocrinol 23: 137-178.

Balthazart J. 1983. Hormonal correlates of behavior. Avian biology Vol VII. Farner DS, King JR, Parkes KC, editors. New York: Academic; pp. 221-365.

Balthazart J. 1989. Steroid metabolism and the activation of social behavior. Advances in comparative and environmental physiology, Vol 3. Balthazart J, editor. Berlin: Springer; pp. 105-159.

Balthazart J, Absil P, Foidart A, Houbart M, Harada N, et al. 1996. Distribution of aromatase-immunoreactive cells in the forebrain of zebra finches (*Taeniopygia guttata*): implications for the neural action of steroids and nuclear definition in the avian hypothalamus. J Neurobiol 31: 129-148.

Balthazart J, Absil P, Gérard M, Appeltants D, Ball GF. 1998. Appetitive and consummatory male sexual behavior in Japanese quail are differentially regulated by subregions of the preoptic medial nucleus. J Neurosci 18: 6512-6527.

Balthazart J, Baillien M, Cornil CA, Ball GF. 2004. Preoptic aromatase modulates male sexual behavior: slow and fast mechanisms of action. Physiol Behav 83: 247-270.

Balthazart J, Ball GF. 1989. Effects of the noradrenergic neurotoxin DSP-4 on luteinizing hormone levels, catecholamine concentrations, alpha2-adrenergic receptor binding, and aromatase activity in the brain of the Japanese quail. Brain Res 492: 163-175.

Balthazart J, Ball GF. 1995. Sexual differentiation of brain and behavior in birds. Trends Endocrinol Metab 6: 21-29.

Balthazart J, Ball GF. 1998. New insights into the regulation and function of brain estrogen synthase (aromatase). TINS 21: 243-249.

Balthazart J, Foidart A, Sante P, Hendrick JC. 1992a. Effects of alpha-methyl-para-tyrosine on monoamine levels in the Japanese quail: sex differences and testosterone effects. Brain Res Bull 28: 275-288.

Balthazart J, Foidart A, Surlemont C, Vockel A, Harada N. 1990. Distribution of aromatase in the brain of the Japanese quail, ring dove, and zebra finch: an immunocytochemical study. J Comp Neurol 301: 276-288.

Balthazart J, Foidart A, Wilson EM, Ball GF. 1992b. Immunocytochemical localization of androgen receptors in the male songbird and quail brain. J Comp Neurol 317: 407-420.

Balthazart J, Schumacher M, Pröve E. 1986. Brain testosterone metabolism during ontogeny in the Zebra finch. Brain Res 378: 240-256.

Barclay SR, Harding CF. 1988. Androstenedione modulation of monoamine levels and turnover in hypothalamic and vocal control nuclei in the male zebra finch: Steroid effects on brain monoamines. Brain Res 459: 333-343.

Barclay SR, Harding CF. 1990. Differential modulation of monoamine levels and turnover rates by estrogen and/or androgen in hypothalamic and vocal control nuclei of male zebra finches. Brain Res 523: 251-262.

Barclay SR, Harding CF, Waterman SA. 1991. Correlations between catecholamine levels and sexual behavior in male zebra finches. Pharmacol Biochem Behav 41: 195-201.

Barclay SR, Harding CF, Waterman SA. 1996. Central DSP-4 treatment decreases norepinephrine levels and courtship behavior in male zebra finches. Pharmacol Biochem Behav 53: 213-220.

Barraclough CA, Wise PM. 1982. The role of catecholamines in the regulation of pituitary luteinizing hormone and follicle stimulating-hormone secretion. Endocr Rev 3: 91-119.

Basham ME, Nordeen EJ, Nordeen KW. 2004. Blockade of NMDA receptors in the anteriori forebrain imparis sensory acquisition in the zebra finch (*Poephila guttata*). Neurobiol Learn Mem 66: 295-304.

Bentley GE, Ball GF. 2000. Photoperiod-dependent and -independent regulation of melatonin receptors in the forebrain of songbirds. J Neuroendocrinol 12: 745-752.

Bentley GE, Ball GF. 2001. Steroid-melatonin interactions and seasonal regulation of the song control system. Avian endocrinology. Dawson A, Chaturvedi CM, editors. New Delhi, India: Narosa Publishing House.

Bentley GE, Spar BD, MacDougall-Shackleton SA, Hahn TP, Ball GF. 2000. Photoperiodic regulation of the reproductive axis in male zebra finches, Taeniopygia guttata. Gen Comp Endocrinol 117: 449-455.

Benton S, Nelson DA, Marler P, DeVoogd TJ. 1998. Anterior forebrain pathway is needed for stable song expression in adult male white-crowned sparrows (*Zonotrichia leucophrys*). Behav Brain Res 96: 135-150.

Bernard DJ, Ball GF. 1995. Two histological markers reveal a similar photoperiodic difference in the volume of the high vocal center in male European starlings. J Comp Neurol 360: 726-734.

Bernard DJ, Ball GF. 1997. Photoperiodic condition modulates the effects of testosterone on song control nuclei volumes in male European starlings. Gen Comp Endocrinol 105: 276-283.

Bernard DJ, Bentley GE, Balthazart J, Turek FW, Ball GF. 1999. Androgen receptor, estrogen receptor alpha, and estrogen receptor beta show distinct patterns of expression in forebrain song control nuclei of European starlings. Endocrinology 140: 4633-4643.

Bernard DJ, Casto JM, Ball GF. 1993. Sexual dimorphism in the volume of song control nuclei in European starlings: assessment by a Nissl stain and autoradiography for muscarinic cholinergic receptors. J Comp Neurol 334: 559-570.

Bernard DJ, Eens M, Ball GF. 1996. Age- and behavior-related variation in volumes of song control nuclei in male European starlings. J Neurobiol 30: 329-339.

Bernard DJ, Wilson FE, Ball GF. 1997. Testis-dependent and -independent effects of photoperiod on volumes of song control nuclei in American tree sparrows (*Spizella arborea*). Brain Res 760: 163-169.

Beyer C. 1987. Sex steroids and afferent input: their roles in brain sexual differentiation. Ann Rev Physiol 49: 349-364.

Blaustein JD, Olster DH. 1989. Gonadal steroid hormone receptors and social behaviors. Advances in comparative and environmental physiology, Vol 3. Balthazart J, editor. Berlin: Springer; pp. 31-104.

Blaustein JD, Tetel MJ, Meredith JM. 1995. Neurobiological regulation of hormonal response by progestin and estrogen receptors. Neurobiological effects of sex steroid hormones. Micevych PE, Hammer RPJ, editors. Cambridge: Cambridge University Press; pp. 324-349.

Böhner J, Chaiken M, Ball GF, Marler P. 1990. Song acquisition in photosensitive and photorefractory male European starlings. Horm Behav 24: 582-594.

Bottjer SW. 1993. The distribution of tyrosine hydroxylase immunoreactivity in the brains of male and female zebra finches. J Neurobiol 24: 51-69.

Bottjer SW, Alexander G. 1995. Localization of met-enkephalin and vasoactive intestinal polypeptide in the brains of male zebra finches. Brain Behav Evol 45: 153-177.

Bottjer SW, Halsema KA, Brown SA, Miesner EA. 1989. Axonal connections of a forebrain nucleus involved with vocal learning in zebra finches. J Comp Neurol 279: 312-326.

Bottjer SW, Johnson F. 1997. Circuits, hormones, and learning: Vocal behavior in songbirds. J Neurobiol 33: 602-618.

Bottjer SW, Miesner EA, Arnold AP. 1984. Forebrain lesions disrupt development but not maintenance of song in passerine birds. Science 224: 901-903.

Brainard MS. 2004. Contributions of the anterior forebrain pathway to vocal plasticity. Behavioral neurobiology of birdsong, Vol. 1016. Zeigler HP, Marler P, editors. New York: Annals of the New York Academy of Sciences. pp. 377-394.

Brainard MS, Doupe AJ. 2000. Interruption of a basal ganglia-forebrain circuit prevents plasticity of learned vocalizations. Nature 404.

Brenowitz E, Lent K, Kroodsma DE. 1995. Brain space for learned song in birds develops independently of song learning. J Neurosci 15: 6281

Brenowitz EA. 1991. Evolution of the vocal control system in the avian brain. Sem Neurosci 3: 399-407.

Brenowitz EA. 1997. Comparative approaches to the avian song system. J Neurobiol 33: 517-531.

Brenowitz EA. 2004. Plasticity of the adult avian song control system. Behavioral neurobiology of birdsong, Zeigler HP, Marler P, editors. Ann NY Acad Sci 1016: 560-585.

Brenowitz EA, Baptista LF, Lent K, Wingfield JC. 1998. Seasonal plasticity of the song control system in wild Nuttall's white-crowned sparrows. J Neurobiol 34: 69-82.

Brenowitz EA, Margoliash D, Nordeen KW. 1997. An introduction to birdsong and the avian song system. J Neurobiol 33: 495-500.

Brenowitz EA, Nalls B, Wingfield JC, Kroodsma DE. 1991. Seasonal changes in avian song nuclei without seasonal changes in song repertoire. J Neurosci 11: 1367-1374.

Burd GD, Paton H, Hemmings H, Heintz J, Nottebohm F. 1986. Dopamine innervation of newly generated neurons in the forebrain of adult canaries. J Histochem Cytochem 34: 109.

Caro SP, Lambrechts MM, Balthazart J. 2005. Early seasonal development of brain song control nuclei in male blue tits. Neurosci Lett 386: 139-144.

Carrillo GD, Doupe AJ. 2004. Is the songbird area X striatal, pallidal or both? An anatomical study. J Comp Neurol 473: 415-437.

Castelino CB, Ball GF. 2005. A role for norepinephrine in the regulation of context-dependent ZENK expression in male zebra finches, Taeniopygia guttat. Eur J Neurosci 21: 1962-1972.

Casto JM, Ball GF. 1994. Characterization and localization of D1 dopamine receptors in the sexually dimorphic vocal control nucleus, area X, and the basal ganglia of European starlings. J Neurobiol 25: 767-780.

Catchpole CK, Slater PJB. 1995. Bird song: biological themes and variations. 1st edn. Cambridge/New York: Cambridge University Press.

Chaiken M, Bohner J, Marler P. 1994. Repertoire turnover and the timing of song acquisition in European starlings. Behaviour 128: 25-39.

Cox PR. 1944. A statistical investigation into bird-song. Br Birds 38.

Dahlström A, Fuxe K. 1964. Evidence for the existence of monoamine-containing neurons in the central nervous system. I. Demonstration of monoamines in the cell bodies of brain stem neurones. Acta Physiol Scand 62(Suppl. 232): 1-55.

Dave AS, Yu AC, Margoliash D. 1998. Behavioral state modulation of auditory activity in a vocal motor system. Science 282: 2250-2254.

De Ridder E, Pinxten R, Eens M. 2000. Experimental evidence of a testosterone-induced shift from paternal to mating behaviour in a facultatively polygynous songbird. Behav Ecol Sociobiol 49: 24-30.

Deutch AY, Roth RH. 2003. Neurotransmitters. Fundamental neuroscience. Squire LR, Bloom FE, Mcconnell SK, Roberts JL, Spitzer NC, et al. editors. San Diego, CA: Academic; pp. 163-196.

Ding L, Perkel DJ. 2004. Long-term potentiation in an avian basal ganglia nucleus essential for vocal learning. J Neurosci 24: 488-494.

Ding L, Perkel DJ, Farries MA. 2003. Presynaptic depression of glutamatergic synaptic transmission by D1-like dopamine receptor activation in the avian basal ganglia. J Neurosci 23(14): 6086-6095.

Doupe AJ, Kuhl PK. 1999. Birdsong and human speech: common themes and mechanisms. Annu Rev Neurosci 22: 567-631.

Doupe AJ, Perkel DJ, Reiner A, Stern EA. 2005. Birdbrains could teach basal ganglia research a new song. Trends Neurosci 28(7): 353-363.

Eens M. 1997. Understanding the complex song of the European starling: an integrated ethological approach. Adv Study Anim Behav 26: 355-434.

Evrard HC, Harada N, Balthazart J. 2004. Immunocytochemical localization of aromatase in sensory and integrating nuclei of the hindbrain in Japanese quail (*Coturnix japonica*). J Comp Neurol 473(2): 194-212.

Farries MA, Perkel DJ. 2002. A telencephalic nucleus essential for song learning contains neurons with physiological characteristics of both striatum and globus pallidus. J Neurosci 22: 3776-3787.

Feare CJ. 1984. The starling. Oxford: Oxford University Press.

Foidart A, Harada N, Balthazart J. 1994. Effects of steroidal and non steroidal aromatase inhibitors on sexual behavior and aromatase-immunoreactive cells and fibers in the quail brain. Brain Res 657: 105-123.

Foidart A, Lakaye B, Grisar T, Ball GF, Balthazart J. 1999. Estrogen receptor-beta in quail: cloning, tissue expression and neuroanatomical distribution. J Neurobiol 40: 327-342.

Foidart A, Silverin B, Baillien M, Harada N, Balthazart J. 1998. Neuroanatomical distribution and variations across the reproductive cycle of aromatase activity and aromatase-immunoreactive cells in the pied flycatcher (*Ficedula hypoleuca*). Horm Behav 33: 180-196.

Fritschy J-M, Grzanna R. 1989. Immunohistochemical analysis of the neurotoxic effects of DSP-4 identifies two populations of noradrenergic axon terminals. Neurosci 30: 181-197.

Fusani L, Van't Hof T, Hutchison JB, Gahr M. 2000. Seasonal expression of androgen receptors, estrogen receptors, and aromatase in the canary brain in relation to circulating androgens and estrogens. J Neurobiol 43: 254-268.

Gahr M. 1990. Localization of androgen receptors and estrogen receptors in the same cells of the songbird brain. Proc Natl Acad Sci USA 87: 9445-9448.

Gahr M. 2000. Neural song control system of hummingbirds: comparison to swifts, vocal learning (songbirds) and nonlearning (suboscines) passerines, and vocal learning (budgerigars) and nonlearning (dove, owl, gull, quail, chicken) nonpasserines. J Comp Neurol 426: 182-196.

Gahr M, Flügge G, Güttinger HR. 1987. Immunocytochemical localization of estrogen binding neurons in the songbird brain. Brain Res 402: 173-177.

Gahr M, Güttinger H-R, Kroodsma DE. 1993. Estrogen receptors in the avian brain: survey reveals general distribution and forebrain areas unique to songbirds. J Comp Neurol 327: 112-122.

Gahr M, Kosar E. 1996. Identification, distribution, and developmental changes of a melatonin binding site in the song control system of the zebra finch. J Comp Neurol 367: 308-318.

Gahr M, Metzdorf R. 1997. Distribution and dynamics in the expression of androgen and estrogen receptors in vocal control systems of songbirds. Brain Res Bull 44: 509-517.

Gahr M, Wild JM. 1997. Localization of androgen receptor mRNA-containing cells in avian respiratory-vocal nuclei: an in situ hybridization study. J Neurobiol 33: 865-876.

Goldman SA, Zaremba A, Niedzwiecki D. 1992. In vitro neurogenesis by neuronal precursor cells derived from the adult songbird brain. J Neurosci 12: 2532-2541.

Grisham W, Arnold AP. 1994. Distribution of GABA-like immunoreactivity in the song system of the zebra finch. Brain Res 651: 115-122.

Harding CF, Sheridan K, Walters MJ. 1983. Hormonal specificity and activation of sexual behavior in male zebra finches. Horm Behav 17: 111-133.

Harding CF, Walters MJ, Collado D, Sheridan K. 1988. Hormonal specificity and activation of social behavior in male red-winged blackbirds. Horm Behav 22: 402-418.

Harding CF, Walters MJ, Parsons B. 1984. Androgen receptor levels in hypothalamic and vocal control nuclei in the male zebra finch. Brain Res 306: 333-339.

Herbison AE. 1998. Multimodal influence of estrogen upon gonadotropin-releasing hormone neurons. Endo Rev 19(3): 302-330.

Hessler NA, Doupe AJ. 1999a. Singing-related neural activity in a dorsal forebrain-basal ganglia circuit of adult zebra finches. J Neurosci 19: 10461-10481.

Hessler NA, Doupe AJ. 1999b. Social context modulates singing-related neural activity in the songbird forebrain. Nature Neurosci 2: 209-211.

Hodgkin A. 1976. Chance and design in electrophysiology: an informal account of certain experiments on nerves carred out between 1942 and 1952. J Physiol (Lond.) 263: 1-21.

Hoelzel AR. 1986. Song characteristics and response to playback of male and female Robins *Erithacus rubecula*. Ibis 128: 115-127.

Hunt KE, Hahn TP, Wingfield JC. 1997. Testosterone implants increase song but not aggression in male Lapland longspurs. Anim Behav 54: 1177-1192.

Hutchison JB. 1991. Hormonal control of behaviour: steroid action in the brain. Curr Opin Neurobiol 1: 562-571.

Hutchison JB, Steimer T. 1981. Brain 5β-reductase. A correlate of behavioral sensitivity to androgen. Science 213: 244-246.

Hutchison JB, Steimer T, Hutchison RE. 1986. Formation of behaviorally active estrogen in the dove brain: induction of preoptic aromatase by intracranial testosterone. Neuroendocrinology 43: 416-427.

Jacobs EC, Arnold AP, Campagnoni AT. 1996. Zebra finch estrogen receptor cDNA: cloning and mRNA expression. J Steroid Biochem Mol Biol 59: 135-145.

Jarvis ED. 2004. Learned birdsong and the neurobiology of human language. Behavioral neurobiology of birdsong, Zeigler HP, Marler P, editors. Ann NY Acad Sci 1016: 749-777.

Jarvis ED, Nottebohm F. 1997. Motor-driven gene expression. Proc Natl Acad Sci USA 94: 4097-4102.

Jarvis ED, Scharff C, Grossman MR, Ramos JA, Nottebohm F. 1998. For whom the bird sings: context-dependent gene expression. Neuron 21: 775-788.

Kandel ER. 1995. Cellular mechanisms of learning and memory. Essentials of neural science and behavior. Kandel ER, Schwartz JH, Jessell TM, editors. East Norwalk, Conn.: Appleton & Lange; pp. 667-694.

Kao MH, Doupe AJ, Brainard MS. 2005. Contributions of an avian basal ganglia-forebrain circuit to real-time modulation of song. Nature 433: 638-643.

Kelley DB, Pfaff DW. 1978. Generalizations from comparative studies on neuroanatomical and endocrine mechanisms of sexual behaviour. Biological determinants of sexual behaviour. Hutchison JB, editor. Chichester: Wiley; pp. 225-254.

Kelly MJ, Ronnekleiv OK. 2002. Rapid membrane effects of estrogen in the central nervous system. Hormones, brain and behavior, Vol. 3. Pfaff DW, Arnold AP, Etgen AM,

Fahrbach SE, Rubin RT, editors. San Diego: Academic; pp. 361-380.

Kim YS, Stumpf WE, Sar M, Martinez-Vargas MC. 1978. Estrogen and androgen target cells in the brain of fishes, reptiles, and birds: phylogeny and ontogeny. Amer Zool 18: 425-433.

Kimpo RR, Doupe AJ. 1997. FOS is induced by singing in distinct neuronal populations in a motor network. Neuron 18: 315-325.

Kimpo RR, Theunissen FE, Doupe AJ. 2003. Propagation of correlated activity through multiple stages of a neural circuit. J Neurosci 23: 5750-5761.

Kirn J, O'Loughlin B, Kasparian S, Nottebohm F. 1994. Cell death and neuronal recruitment in the high vocal center of adult male canaries are temporally related to changes in song. Proc Natl Acad Sci USA 91: 7844-7848.

Kiss JZ, Voorhuis TAM, van Eekelen JAM, de Kloet ER, de Wied D. 1987. Organization of vasotocin-immunoreactive cells and fibers in the canary brain. J Comp Neurol 263: 347-364.

Kittelberger JM, Mooney R. 2005. Acute injections of brain-derived neurotrophic factor in a vocal premotor nculeus reversibly disrupt adult birdsong stability and trigger syllable deletion. J Neurobiol 62: 406-424.

Konishi M. 1985. Birdsong: from behavior to neuron. Ann Rev Neurosci 8: 125-170.

Kroodsma D, Konishi M. 1991. A suboscine bird (Eastern phoebe, Sayornis phoebe) develops normal song without auditory feedback. Anim Behav 42: 477-487.

Kubota M, Saito N. 1991. Sodium- and calcium-dependent conductances of neurones in the zebra finch hyperstriatum ventrale pars caudale in vitro. J Physiol (Lond.) 440: 131-142.

Kuiper GGJM, Enmark E, Pelto-Huikko M, Nilsson S, Gustafsson JÅ. 1996. Cloning of a novel estrogen receptor expressed in rat prostate and ovary. Proc Natl Acad Sci USA 93: 5925-5930.

Kuiper GGJM, Shughrue PJ, Merchenthaler I, Gustafsson J-Å. 1998. The estrogen receptor β subtype: a novel mediator of estrogen action in neuroendocrine systems. Front Neuroendocrinol 19: 253-286.

Lakaye B, Foidart A, Grisar T, Balthazart J. 1998. Partial cloning and distribution of estrogen receptor beta in the avian brain. Neuroreport 9: 2743-2748.

Lehrman DS. 1965. Interaction between internal and external environments in the regulation of the reproductive cycle of the ring dove. Sex and behavior. Beach FA, editor. New York: Wiley; pp. 355-380.

Leonardo A, Fee MS. 2005. Ensemble coding of vocal control in birdsong. J Neuroscience 25(3): 652-661.

Leonardo A, Konishi M. 1999. Decrystallization of adult birdsong by perturbation of auditory feedback. Nature 399: 466-470.

Lewis JW, Ryan SM, Arnold AP, Butcher LL. 1981. Evidence for catecholamine projection to area X in the zebra finch. J Comp Neurol 196: 347-354.

Li R, Sakaguchi H. 1997. Cholinergic innervation of the song control nuclei by the ventral paleostriatum in the zebra finch: a double-labeling study with retrograde fluorescent tracers and choline acetyltransferase immunohistochemistry. Brain Res 763: 239-246.

Logan CA, Wingfield JC. 1990. Autumnal territorial aggression is independent of plasma testosterone in mockingbirds. Horm Behav 24: 568-581.

Lücke J, Haase E. 1980. Autoradiographische Untersuchungen am Gehirn von Bergfinken (*Fringilla montifringilla* L.) nach injektion von 3H-Testosteron. J Hirnforsch 21: 369-380.

Luo MM, Perkel DJ. 1999a. A GABAergic, strongly inhibitory projection to a thalamic nucleus in the zebra finch song system. J Neurosci 19: 6700-6711.

Luo MM, Perkel DJ. 1999b. Long-range GABAergic projection in a circuit essential for vocal learning. J Comp Neurol 403: 68-84.

Maney DL, Bernard DJ, Ball GF. 2001. Gonadal steroid receptor mRNA in catecholaminergic nuclei of the canary brainstem. Neurosci Lett 311: 189-192.

Margoliash D, Konishi M. 1985. Auditory representation of autogenous song in the song system of white-crowned sparrows. Proc Natl Acad Sci USA 82: 5997-6000.

Marler P. 2004. Bird calls: their potential for behavioral neurobiology. Ann NY Acad Sci 1016: 31-44.

Marler P, Peters S, Ball GF, Dufty AM, Wingfield JC. 1988. The role of sex steroids in the acquisition and production of birdsong. Nature 336: 770-772.

Martinez-Vargas MC, Stumpf WE, Sar M. 1975. Estrogen localization in the dove brain. Phylogenetic considerations and implications for nomenclature. Anatomical neuroendocrinology. Basel: Karger; pp. 166-175.

Martinez-Vargas MC, Stumpf WE, Sar M. 1976. Anatomical distribution of estrogen target cells in the avian CNS: a comparison with the mammalian CNS. J Comp Neurol 167: 83-104.

McEwen BS. 1994. Steroid hormone actions on the brain: when is the genome involved? Horm Behav 28: 396-405.

Mello CV, Pinaud R, Ribeiro S. 1998. Noradrenergic system of the zebra finch brain: immunocytochemical study of dopamine-beta-hydroxylase. J Comp Neurol 400: 207-228.

Mello CV, Vicario DS, Clayton DF. 1992. Song presentation induces gene expression in the songbird forebrain. Proc Natl Acad Sci USA 89: 6818-6822.

Mermelstein PG, Becker JB, Surmeier DJ. 1996. Estradiol reduces calcium currents in rat neostriatal neurons via a membrane receptor. J Neurosci 16: 595-604.

Metzdorf R, Gahr M, Fusani L. 1999. Distribution of aromatase, estrogen receptor, and androgen receptor mRNA in the forebrain of songbirds and nonsongbirds. J Comp Neurol 407: 115-129.

Mooney R, Konishi M. 1991. Two distinct inputs to an avian song nucleus activate different glutamate receptor subtypes on individual neurons. Proc Natl Acad Sci USA 88: 4075-4079.

Mooney R, Prather JF. 2005. The HVC microcircuit: The synaptic basis for interactions between song motor and vocal plasticity pathways. J Neurosci 25: 1952-1964.

Morrell JI, Kelley DB, Pfaff DW. 1975. Sex steroid binding in the brain of vertebrates. Brain-endocrine interactions II. Knigge KM, Scott DE, Kobayashi H, Miura S, Ishii S, editors. Basel: Karger; pp. 230-256.

Morrell JI, Pfaff DW. 1978. A neuroendocrine approach to brain function: localization of sex steroid concentrating cells in vertebrate brains. Amer Zool 18: 447-460.

Morton ES. 1996. A comparison of vocal behavior among tropical and temperate passerine birds. Ecology and evolution of acoustic communication in birds. Kroodsma DE, Miller EH, editors. Ithaca, NY: Cornell University Press; pp. 258-268.

Mosselman S, Polman J, Dijkema R. 1996. ER beta: identification and characterization of a novel human estrogen receptor. FEBS Lett 392: 49-53.

Naftolin F, Horvath TL, Jakab RL, Leranth C, Harada N, et al. 1996. Aromatase immunoreactivity in axon terminals of the vertebrate brain -- An immunocytochemical study on quail, rat, monkey and human tissues. Neuroendocrinology 63: 149-155.

Nastiuk KL, Clayton DF. 1994. Seasonal and tissue-specific regulation of canary androgen receptor messenger ribonucleic acid. Endocrinology 134: 640-649.

Nastiuk KL, Clayton DF. 1995. The canary androgen receptor mRNA is localized in the song control nuclei of the brain and is rapidly regulated by testosterone. J Neurobiol 26: 213-224.

Nordeen EJ, Nordeen KW, Arnold AP. 1987a. Sexual differentiation of androgen accumulation within the zebra finch brain through selective cell loss and addition. J Comp Neurol 259: 393-399.

Nordeen KW, Nordeen EJ. 1992. Auditory feedback is necessary for the maintenance of stereotyped song in adult zebra finches. Behav Neural Biol 59: 58-66.

Nordeen KW, Nordeen EJ. 2004. Synaptic and molecuarl mechanisms regulating plasticity during early learning. Behavioral neurobiology of birdsong, Zeigler HP, Marler P, editors. Ann NY Acad Sci 1016: 416-437.

Nordeen KW, Nordeen EJ, Arnold AP. 1987b. Estrogen accumulation in zebra finch song control nuclei: implications for sexual differentiation and adult activation of song behavior. J Neurobiol 18: 569-582.

Nottebohm F. 1975. Vocal behavior in birds. Avian biology, Vol. 5. Farner DS, King JR, editors. New York: Academic; pp. 287-332.

Nottebohm F. 1980. Brain pathways for vocal learning in birds: A review of the first 10 years. Progress in Psychobiology and Physiological Psychology, Vol 9. Sprague JM, Epstein AN, editors. New York: Academic; pp. 85-214.

Nottebohm F. 1981. A brain for all seasons: cyclical anatomical changes in song-control nuclei of the canary brain. Science 214: 1368-1370.

Nottebohm F. 1987. Plasticity in the adult avian central nervous system: possible relation between hormones, learning, and brain repair. Handbook of physiology-the nervous system. pp. 85-108.

Nottebohm F. 1989. From bird song to neurogenesis. Sci Am 260: 74-79.

Nottebohm F. 1993. The search for neural mechanisms that define the sensitive period for song learning in birds. Neth J Zool 43: 193-234.

Nottebohm F. 2004. The road we travelled: discovery, choreography, and significance of brain replaceable neurons. Behavioral neurobiology of birdsong, Vol. 1016. Zeigler HP, Marler P, editors. pp. 628-658.

Nottebohm F, Alvarez-Buylla A, Cynx J, Kirn J, Ling CY, et al. 1990. Song learning in birds: the relation between perception and production. Philos Trans R Soc Lond (Biol) 329: 115-124.

Nottebohm F, Kelley DB, Paton JA. 1982. Connections of vocal control nuclei in the canary telencephalon. J Comp Neurol 207: 344-357.

Nottebohm F, Nottebohm ME, Crane L. 1986. Developmental and seasonal changes in canary song and their relation to changes in the anatomy of song-control nuclei. Behav Neural Biol 46: 445-471.

Nottebohm F, Nottebohm ME, Crane LA, Wingfield JC. 1987. Seasonal changes in gonadal hormone levels of adult male canaries and their relation to song. Behav Neural Biol 47: 197-211.

Nottebohm F, Stokes TM, Leonard CM. 1976. Central control of song in the canary, Serinus canarius. J Comp Neurol 165: 457-486.

Nowicki S, Ball GF. 1989. Testosterone induction of song in photosensitive and photorefractory male sparrows. Horm Behav 23: 514-525.

Panzica GC, Plumari L, García-Ojeda E, Deviche P. 1999. Central vasotocin-immunoreactive system in a male passerine bird (*Junco hyemalis*). J Comp Neurol 409: 105-117.

Panzica GC, Viglietti-Panzica C, Balthazart J. 1996. The sexually dimorphic medial preoptic nucleus of quail: a key brain area mediating steroid action on male sexual behavior. Front Neuroendocrinol 17: 51-125.

Pasqualini C, Olivier V, Guibert B, Frain O, Leviel V. 1995. Acute stimulatory effect of estradiol on striatal dopamine synthesis. J Neurochem 65: 1651-1657.

Perkel DJ. 2004. Origin of the anterior forebrain pathway. Behavioral Neurobiology of Birdsong, Zeigler HP, Marler P, editors. Ann NY Acad Sci 1016: 736-748.

Perlman WR, Ramachandran B, Arnold AP. 2003. Expression of androgen receptor mRNA in the late embryonic and early posthatch zebra finch brain. J Comp Neurol 455(4): 513-530.

Peterson RS, Yarram L, Schlinger BA, Saldanha CJ. 2005. Aromatase is pre-synaptic and sexually dimorphic in the adult zebra finch brain. Proc Biol Sci 272: 2089-2096.

Pfaff DW. 1976. The neuroanatomy of sex hormone receptors in the vertebrate brain. Neuroendocrine regulation of fertility. Anand Kumar TC, editor. Basel: Karger; pp. 30-45.

Pfaff DW, Schwartz-Giblin S, McCarthy MM, Kow LM. 1994. Cellular and molecular mechanisms of female reproductive behaviors. The physiology of reproduction. Knobil E, Neill JD, editors. New York: Raven Press; pp. 107-220.

Pinxten R, De Ridder E, Balthazart J, Eens M. 2002. Context-dependent effects of castration and testosterone treatment on song in male European starlings. Horm Behav 42: 307-318.

Ramachandran B, Schlinger BA, Arnold AP, Campagnoni AT. 1999. Zebra finch aromatase gene expression is regulated in the brain through an alternate promoter. Gene 240: 209-216.

Ramirez VD, Zheng JB, Siddique KM. 1996. Membrane receptors for estrogen, progesterone, and testosterone in the rat brain: Fantasy or reality. Cell Mol Neurobiol 16: 175-198.

Rasika S, Alvarez-Buylla A, Nottebohm F. 1999. BDNF mediates the effects of testosterone on the survival of new neurons in an adult brain. Neuron 22: 53-62.

Reiner A, Karle E, Jerson KD, Medina L. 1994. Catecholaminergic perikarya and fibers in the avian nervous system. Phylogeny and development of catecholamine systems in the CNS of vertebrates. Smeets WJAJ, Reiner A, editors. Cambridge UK: Cambridge University Press; pp. 135-181.

Reiner A, Perkel DJ, Bruce LL, Butler AB, Csillag A, et al. 2004. Revised nomenclature for avian telencephalon and some related brainstem nuclei. J Comp Neurol 473(3): 377-414.

Reinke H, Wild JM. 1998. Identification and connections of inspiratory premotor neurons in songbirds and budgerigar. J Comp Neurol 391: 147-163.

Riters LV, Alger SJ. 2004. Neuroanatomical evidence for indirect connections between the medial preoptic nucleus and the song control system: possible neural substrates for sexually motivated song. Cell Tissue Res 316(1): 35-44.

Riters LV, Baillien M, Eens M, Pinxten R, Foidart A, et al. 2001. Seasonal variation in androgen-metabolizing enzymes in the diencephalon and telencephalon of the male European starling (*Sturnus vulgaris*). J Neuroendocrinol 13(11): 985-997.

Riters LV, Ball GF. 1999. Lesions to the medial preoptic area affect singing in the male European starling (*Sturnus vulgaris*). Horm Behav 36: 276-286.

Riters LV, Ball GF. 2002. Sex differences in the densities of alpha2-adrenergic receptors in the song control system, but not the medial preoptic nucleus in zebra finches. J Chem Neuroanat 23(4): 269-277.

Riters LV, Eens M, Pinxten R, Ball GF. 2002. Seasonal changes in the densities of $\alpha 2$-noradrenergic receptors are inversely related to changes in testosterone and the volumes of song control nuclei in male European starlings. J Comp Neurol 444: 63-74.

Riters LV, Eens M, Pinxten R, Duffy DL, Balthazart J, et al. 2000. Seasonal changes in courtship song and the medial preoptic area in male European starlings (*Sturnus vulgaris*). Horm Behav 38: 250-261.

Rivkees SA, Cassone VM, Weaver DR, Reppert SM. 1989. Melatonin receptors in chick brain: Characterization and localization. Endocrinology 125: 363-368.

Roselli CE. 1991. Synergistic induction of aromatase activity in the rat brain by estradiol and 5alpha-dihydrotestosterone. Neuroendocrinol 53: 79-84.

Roselli CE, Horton LE, Resko JA. 1985. Distribution and regulation of aromatase activity in the rat hypothalamus and limbic system. Endocrinology 117: 2471-2477.

Roselli CE, Horton LE, Resko JA. 1987. Time-course and steroid specificity of aromatase induction in rat hypothalamus-preoptic area. Biol Reprod 37: 628-633.

Roselli CE, Resko JA. 1989. Testosterone regulates aromatase activity in discrete brain areas of male rhesus macaques. Biol Reprod 40: 929-934.

Ryan S, Arnold AP. 1981. Evidence for cholinergic participation in the control of bird song: acetylcholinesterase distribution and muscarinic receptor autoradiography in the zebra finch brain. J Comp Neurol 202: 211-219.

Ryan SM, Arnold AP, Elde RP. 1981. Enkephalin-like immunoreactivity in vocal control regions of the zebra finch brain. Brain Res 229: 236-240.

Rybak F, Gahr M. 2004. Modulation by steroid hormones of a "sexy" acoustic signal in an oscine species, the common canary, Serinus canaria. An Acad Bras Cienc 76(2): 365-367.

Sakaguchi H, Asano M, Yamamoto K, Saito N. 1987. Release of endogenous γ-aminobutyric acid from vocalization nucleus, the robust nucleus of the archistriatum of zebra finch in vitro. Brain Res 410: 380-384.

Sakaguchi H, Kubota M, Saito N. 1992. In vitro release of glutamate and aspartate from zebra finch song control nuclei. Exp Brain Res 88: 560-562.

Sakaguchi H, Saito N. 1989. The acetylcholine and catecholamine contents in song control nuclei of zebra finch during song ontogeny. Dev Brain Res 47: 313-317.

Saldanha CJ, Schlinger BA. 1997. Estrogen synthesis and secretion in the brown-headed cowbird (*Molothrus ater*). Gen Comp Endocrinol 105: 390-401.

Saldanha CJ, Tuerk MJ, Kim YH, Fernandes AO, Arnold AP, et al. 2000. Distribution and regulation of telencephalic aromatase expression in the zebra finch revealed with a specific antibody. J Comp Neurol 423: 619-630.

Scharff C, Nottebohm F. 1991. A comparative study of the behavioral deficits following lesions of various parts of the zebra finch song system: Implications for vocal learning. J Neurosci 11: 2896-2913.

Schlinger BA. 1996. Evolution of steroid-metabolizing enzyme expression in the avian brain. It J Anat Embryol 101 (Suppl. 1): 67-68.

Schlinger BA. 1997. Sex steroids and their actions on the birdsong system. J Neurobiol 33: 619-631.

Schlinger BA, Arnold AP. 1991. Brain is the major site of estrogen synthesis in a male songbird. Proc Natl Acad Sci USA 88: 4191-4194.

Schlinger BA, Arnold AP. 1992a. Circulating estrogens in a male songbird originate in the brain. Proc Natl Acad Sci USA 89: 7650-7653.

Schlinger BA, Arnold AP. 1992b. Plasma sex steroids and tissue aromatization in hatchling zebra finches: Implications for the sexual differentiation of singing behavior. Endocrinology 130: 289-299.

Schlinger BA, Arnold AP. 1993. Estrogen synthesis in vivo in the adult zebra finch: Additional evidence that circulating estrogens can originate in brain. Endocrinology 133: 2610-2616.

Schlinger BA, Callard GV. 1989. Localization of aromatase in synaptosomal and microsomal subfractions of quail (*Coturnix coturnix japonica*) brain. Neuroendocrinol 49: 434-441.

Schlinger BA, Slotow RH, Arnold AP. 1992. Plasma estrogens and brain aromatase in winter white-crowned sparrows. Ornis Scand 23: 292-297.

Schmidt MF, Konishi M. 1999. Gating of auditory responses in the vocal control system of awake songbirds. Nat Neurosci 1: 513-518.

Schumacher M. 1990. Rapid membrane effects of steroid hormones: an emerging concept in neuroendocrinology. TINS 13: 359-362.

Schwabl H, Kriner E. 1991. Territorial aggression and song of male European robins (Erithacus rubecula) in autumn and spring: Effects of antiandrogen treatment. Horm Behav 25: 180-194.

Seyfarth RM, Cheney DL. 1997. Some general features of vocal development in nonhuman primates. Social influences on vocal development. Snowden CT, Hausberger M, editors. Cambridge, UK: Cambridge; pp. 249-273.

Shaw BK, Kennedy GG. 2002. Evidence for species differences in the pattern of androgen receptor distribution in relation to species differences in an androgen-dependent behavior. J Neurobiol 52: 203-220.

Shea SD, Margoliash D. 2003. Basal forebrain chlinergic modulation of auditory activity in the zebra finch song system. Neuron 40: 1213-1226.

Shen P, Campagnoni CW, Kampf K, Schlinger BA, Arnold AP, et al. 1994. Isolation and characterization of a zebra finch aromatase cDNA: In situ hybridization reveals high aromatase expression in brain. Mol Brain Res 24: 227-237.

Shen P, Schlinger BA, Campagnoni AT, Arnold AP. 1995. An atlas of aromatase mRNA expression in the zebra finch brain. J Comp Neurol 360: 172-184.

Sibley CG, Monroe JBL. 1990. Distribution and taxonomy of the birds of the world. New Haven, CT: Yale University Press.

Silver R, O'Connell M, Saad R. 1979. The effects of androgens on the behavior of birds. Endocrine control of sex behavior. Beyer C, editor. New York: Raven Press; pp. 223-279.

Silverin B. 1975. Reproductive organs and breeding behaviour of the male pied flycatcher Ficedula hypoleuca (Pallas). Ornis Scand 6: 15-26.

Silverin B. 1983a. Population endocrinology and gonadal activities of the male pied flycatcher (Ficedula hypoleuca) Avian endocrinology: environmental and ecological perspectives. Japan Sci Soc. Mikami S, Homma K, Wada M, editors. Tokyo/Berlin: Press/Springer-Verlag; pp. 289-305.

Silverin B. 1983b. Population endocrinology of the female pied flycatcher, Ficedula hypoleuca. Hormones and Behaviour in Higher Vertebrates. Balthazart J, Pröve E, Gilles R, editors. Berlin: Springer Verlag; pp. 388-397.

Silverin B. 1993. Territorial aggressiveness and its relation to the endocrine system in the pied flycatcher. Gen Comp Endocrinol 89: 206-213.

Silverin B, Baillien M, Foidart A, Balthazart J. 2000. Distribution of aromatase activity in the brain and peripheral tissues of passerine and nonpasserine avian species. Gen Comp Endocrinol 117: 34-53.

Silverin B, Wingfield JC. 1982. Patterns of breeding behaviour and plasma levels of hormones in a free-living population of pied flycatchers Ficedula hypoleuca. J Zool Lond 198: 117-129.

Simpson HB, Vicario DS. 1990. Brain pathways for learned and unlearned vocalizations differ in zebra finches. J Neurosci 10: 1541-1556.

Slagsvold T. 1977. Bird song activity in relation to breeding cycle, spring weather, and environmental physiology. Ornis Scand 8: 197-222.

Smith GT, Brenowitz EA, Beecher MD, Wingfield JC. 1997a. Seasonal changes in testosterone, neural attributes of song control nuclei, and song structure in wild songbirds. J Neurosci 17: 6001-6010.

Smith GT, Brenowitz EA, Prins GS. 1996. Use of PG-21 immunocytochemistry to detect androgen receptors in the songbird brain. J Histochem Cytochem 44: 1075-1080.

Smith GT, Brenowitz EA, Wingfield JC. 1997b. Roles of photoperiod and testosterone in seasonal plasticity of the avian song control system. J Neurobiol 32: 426-442.

Smith GT, Brenowitz EA, Wingfield JC. 1997c. Seasonal changes in the size of the avian song control nucleus HVC defined by multiple histological markers. J Comp Neurol 381: 253-261.

Smith GT, Brenowitz EA, Wingfield JC, Baptista LF. 1995. Seasonal changes in song nuclei and song behavior in Gambels white-crowned sparrows. J Neurobiol 28: 114-125.

Smith TR. 1996. Seasonal plasticity in the song nuclei of wild rufous-side towhees. Brain Res 734: 79-85.

Soha JA, Shimizu T, Doupe AJ. 1996. Development of the catecholaminergic innervation of the song system of the male zebra finch. J Neurobiol 29: 473-489.

Sohrabji F, Nordeen EJ, Nordeen KW. 1990. Selective impairment of song learning following lesions of a forebrain nucleus in juvenile zebra finches. Behav Neural Biol 53: 51-63.

Sohrabji F, Nordeen KW, Nordeen EJ. 1989. Projections of androgen-accumulating neurons in a nucleus controlling avian song. Brain Res 488: 253-259.

Soma KK, Alday NA, Hau M, Schlinger BA. 2004. Dehydroepiandrosterone metabolism by 3beta-hydroxysteroid dehydrogenase/Delta5-Delta4 isomerase in adult zebra finch brain: Sex difference and rapid effect of stress. Endocrinology 145(4): 1668-1677.

Soma KK, Bindra RK, Gee J, Wingfield JC, Schlinger BA. 1999a. Androgen-metabolizing enzymes show region-specific changes across the breeding season in the brain of a wild songbird. J Neurobiol 41: 176-188.

Soma KK, Hartman VN, Wingfield JC, Brenowitz EA. 1999b. Seasonal changes in androgen receptor immunoreactivity in the song nucleus HVc of a wild bird. J Comp Neurol 409: 224-236.

Soma KK, Tramontin AD, Wingfield JC. 2000. Oestrogen regulates male aggression in the non-breeding season. Proc R Soc Lond (Biol) 267: 1089-1096.

Soma KK, Wingfield JC. 2001. Dehydroepiandrosterone in songbird plasma: Seasonal regulation and relationship to territorial aggression. Gen Comp Endocrinol 123(2): 144-155.

Soma KK, Wissman AM, Brenowitz EA, Wingfield JC. 2002. Dehydroepiandrosterone (DHEA) increases territorial song and the size of an associated brain region in a male songbird. Horm Behav 41(2): 203-212.

Steimer T, Hutchison JB. 1981. Metabolic control of the behavioral action of androgens in the dove brain: testosterone inactivation by 5β-reduction. Brain Res 209: 189-204.

Stumpf WE, Sar M. 1978. Anatomical distribution of estrogen, androgen, progestin, corticoid and thyroid hormone target sites in the brain of mammals: phylogeny and ontogeny. Amer Zool 18: 435-445.

Theunissen FE, Amina N, Shaevitz SS, Woolley SMN, Fremouw T, et al. 2004. Song selectivity in the song system and in the auditory forebrain. Behavioral neurobiology of birdsong, Zeigler HP, Marler P, editors. Ann NY Acad Sci 1016: 222-245.

Thompson CK, Brenowitz EA. 2005. Seasonal changes in neuron size and spacing but not neuronal recruitment in a basal ganglia nucleus in the avian song control system. J Comp Neurol 481: 276-283.

Thompson RR, Goodson JL, Ruscio MG, Adkins-Regan E. 1998. Role of the archistriatal nucleus taeniae in the sexual behavior of male Japanese quail (*Coturnix japonica*): a comparison of function with the medial nucleus of the amygdala in mammals. Brain Behav Evol 51: 215-229.

Thompson TL, Moss RL. 1994. Estrogen regulation of dopamine release in the nucleus accumbens: genomic- and nongenomic-mediated effects. J Neurochem 62: 1750-1756.

Tramontin AD, Brenowitz EA. 2000. Seasonal plasticity in the adult brain. TINS 23: 251-258.

Tramontin AD, Hartman VN, Brenowitz EA. 2000. Breeding conditions induce rapid and sequential growth in adult avian song control circuits: a model of seasonal plasticity in the brain. J Neurosci 20: 854-861.

Tramontin AD, Perfito N, Wingfield JC, Brenowitz EA. 2001. Seasonal growth of song control nuclei precedes seasonal reproductive development in wild adult song sparrows. Gen Comp Endocrinol 122(1): 1-9.

Tramontin AD, Smith GT, Breuner CW, Brenowitz EA. 1998. Seasonal plasticity and sexual dimorphism in the avian song control system: Stereological measurement of neuron density and number. J Comp Neurol 396: 186-192.

Tremblay GB, Tremblay A, Copeland NG, Gilbert DJ, Jenkins NA, et al. 1997. Cloning, chromosomal localization, and functional analysis of the murine estrogen receptor beta. Mol Endocrinol 11: 353-365.

Tyack PL, Sayigh LS. 1997. Vocal learning in cetaceans. Social influences on vocal development. Snowden CT, Hausberger M, editors. Cambridge, UK: Cambrige University Press; pp. 208-233.

Vockel A, Pröve E, Balthazart J. 1988. Changes in the activity of testosterone-metabolizing enzymes in the brain of male and female zebra finches during the post-hatching period. Brain Res 463: 330-340.

Vockel A, Pröve E, Balthazart J. 1990a. Effects of castration and testosterone treatment on the activity of testosterone-metabolizing enzymes in the brain of male and female zebra finches. J Neurobiol 21: 808-825.

Vockel A, Pröve E, Balthazart J. 1990b. Sex- and age-related differences in the activity of testosterone-metabolizing enzymes in microdissected nuclei of the zebra finch brain. Brain Res 511: 291-302.

Voorhuis TAM, de Kloet ER. 1992. Immunoreactive vasotocin in the zebra finch brain (*Taeniopygia guttata*). Dev Brain Res 69: 1-10.

Voorhuis TAM, de Kloet ER, de Wied D. 1990. The vasotocin system in the canary brain. Hormones, brain and behaviour in vertebrates, Vol. 1. Sexual differentiation, neuroanatomical aspects, neurotransmitters and neuropeptides. Balthazart J, editor. Basel, NY: Karger; pp. 168-179.

Voorhuis TAM, Elands JPM, de Kloet ER. 1990. Vasotocin target sites in the capsular region surrounding the nucleus robustus archistriatalis of the canary brain. J Neuroendocrinol 2: 653-657.

Wada K, Sakaguchi H, Jarvis ED, Hagiwara M. 2004. Differential expression of glutamate receptors in avian neural pathways for learned vocalization. J Comp Neurol 476(1): 44-64.

Walters MJ, Collado D, Harding CF. 1991. Oestrogenic modulation of singing in male zebra finches: Differential effects on directed and undirected songs. Anim Behav 42: 445-452.

Watson JT, Adkins-Regan E. 1989. Neuroanatomical localization of sex steroid-concentrating cells in the Japanese quail (*Coturnix japonica*): Autoradiography with [3H]-testosterone, [3H]-estradiol, and [3H]-dihydrotestosterone. Neuroendocrinol 49: 51-64.

Watson JT, Adkins-Regan E, Whiting P, Lindstrom JM, Podleski TR. 1988. Autoradiographic localization of nicotinic acetylcholine receptors in the brain of teh zebra finch (*Poephila guttata*). J Comp Neurol 274: 255-264.

Weiner RI, Findell PR, Kordon C. 1988. Role of classic and peptide neuromediators in the neuroendocrine regulation of LH and prolactin. The physiology of reproduction. Knobil E, Neill J, et al. editors. New York: Raven Press; pp. 1235-1281.

Whitfield-Rucker MG, Cassone VM. 1996. Melatonin binding in the house sparrow song control system: Sexual dimorphism and the effect of photoperiod. Horm Behav 30: 528-537.

Wild JM. 1994. The auditory-vocal-respiratory axis in birds. Brain Behav Evol 44: 192-209.

Wild JM. 1997. Neural pathways for the control of birdsong production. J Neurobiol 33: 653-670.

Wild JM. 2004. Functional neuroanatomy of the sensorimotor control of singing. Ann NY Acad Sci 1016: 438-462.

Wild JM, Li DF, Eagleton C. 1997. Projections of the dorsomedial nucleus of the intercollicular complex (DM) in relation to respiratory-vocal nuclei in the brainstem of pigeon (*Columba livia*) and zebra finch (*Taeniopygia guttata*). J Comp Neurol 377: 392-413.

Wild JM, Williams MN, Howie GJ, Mooney R. 2005. Calcium-binding proteins define interneurons in HVC of the zebra finch (*Taeniophygia guttata*). J Comp Neurol 483: 76-90.

Wiley RH, Piper WH, Archawaranon M, Thompson EW. 1993. Singing in relation to social dominance and testosterone in white-throated sparrows. Behaviour 127: 175-190.

Wingfield JC. 1985. Environmental and endocrine control of territorial behavior in birds. The endocrine system and the environment. Follett BK, Ishii S, Chandola A, editors. Tokyo/Berlin: Japan Scientific Society Press/Springer-Verlag; pp. 265-277.

Wingfield JC. 1994. Regulation of territorial behavior in the sedentary song sparrow, Melospiza melodia morphna. Horm Behav 28: 1-15.

Wingfield JC, Farner DS. 1993. Endocrinology of reproduction in wild species. Avian biology, vol. 9. Farner DS, King J, Parkes KC, editors. New York: Academic; pp. 163-327.

Wingfield JC, Hahn TP. 1994. Testosterone and territorial behaviour in sedentary and migratory sparrows. Anim Behav 47: 77-89.

Yu AC, Margoliash D. 1996. Temporal hierarchical control of singing in birds. Science 273: 1871-1875.

Zaczek R, Fritschy J-M, Culp S, De Souza EB, Grzanna R. 1990. Differential effects of DSP-4 on noradrenaline axons in cerebral cortex and hypothalamus may reflect heterogeneity of noradrenaline uptake sites. Brain Res 522: 308-314.

Zigmond RE, Detrick RA, Pfaff DW. 1980. An autoradiographic study of the localization of androgen concentrating cells in the chaffinch. Brain Res 182: 369-381.

Zigmond RE, Nottebohm F, Pfaff DW. 1973. Androgen-concentrating cells in the midbrain of a songbird. Science 179: 1005-1007.

Zuschratter W, Scheich H. 1990. Distribution of choline acteylytransferase and acetylcholinesterase in the vocal motor system of zebra finches. Brain Res 513: 193-201.

11 Sex Differences in Neurotransmitters and Behavior: Development

M. M. McCarthy, PhD

Abstract: The ability of gonadal steroids to impact on the developing brain and organize permanent sex differences during a perinatal sensitive period has been established for over 50 years. What has been lacking is an understanding of the mechanistic basis of this process, termed sexual differentiation of the brain. Recent advances have identified target-derived growth factors to direct neuronal circuits and an unexpected role for prostaglandins in synaptic patterning. The advent of transgenic mice with null mutations for estrogen receptor isoforms, aromatase and the androgen receptor have confirmed and extended our understanding of steroid action in the developing brain. Differential cell death regulated by steroids in males versus females is a major determinant of volumetric sex differences in particular regions. Sex differences in reproductively relevant brain areas, predominantly in the diencephalon, are far greater in magnitude and more readily generalize across species than those related to cognition or emotionality. Nonetheless, the potential heuristic value of studying sex differences, and its potential importance to the etiology of gender-biased neurological disorders and diseases of mental health, have not been fully realized.

List of Abbreviations: AVPv, anteroventral periventricular nucleus; BNST, bed nucleus of the stria terminalis; CAH, congental adrenal hyperplasia; COX-1,2, cyxlooxegenase 1,2; FSH, follicle stimulating hormone; GABA, gamma-amino butyric acid; GAD 65,67, glutamic acid decarboxlase 65,67; GAP-43, growth associated protein 43; INAH, interstitial nucleus of the anterior hypothalamus; LH, lutenizing hormone; LHRH, lutenizing hormone releasing hormone; LTD, long term depression; LTP, long term potentiation; MRI, magnetic resonance imaging; PGE2, prostaglandin E2; PN, postnatal; POA, preoptic area; PVN, paraventricular nucleus; SDN, sexually dimorphic nucleus; SNB, spinal nucleus of the bulbocavernosus

1 Introduction

The study of sex differences in the brain is a topic of general interest, but opinion on *why* it is of interest varies considerably. Some think it is interesting to study sex differences in the brain because of the importance to fundamental reproductive physiology, which begins with the LHRH neuron, continues with sexual behavior, and ends with parental care and in mammals' lactation. Others find it fascinating because all the known sex differences in behavior and physiology are not directly related to the need to procreate but impact on fitness, survival, or just the desire to have fun. These would be the gender differences observed in aggression, learning strategies, food preferences, stress responsiveness, and novelty seeking. There are also profound sex differences in the relative risk of developing specific neurological disorders or diseases related to mental health, something that touches almost everyone's life directly or indirectly at some time. Still others think studying the mechanistic basis for sex differences in the brain is an advantageous window to normal developmental processes caused by a readily available and controllable variable, namely, gonadal steroids, which can be used to trigger differentiation of the brain in a particular direction. And finally, many are fascinated by all these aspects of sex differences in the brain and are enamored of meeting the challenge of integrating levels of analysis that range between biopsychology to physiology, endocrinology, and neuroscience.

2 Historical Overview

The modern concept of sexual differentiation of brain and behavior is generally tied to a single seminal paper by Phoenix, Goy, Gerall, and Young (Phoenix et al., 1959), which postulated the organizational/activational hypothesis of steroid effects on the brain. All four authors went on to pursue highly distinguished careers in behavioral neuroendocrinology, and their work continues to stand as the framework on which most hypotheses are hung for analyses, even if for purposes of refuting their original hypothesis. It is important, however, to acknowledge the tremendously rich history on which Phoenix et al. were hanging their own hypothesis. Even a cursory reading of the book, *Hormones and Behavior* published by Frank A. Beach in 1948 reveals a strikingly sophisticated view of an already mature field. Subject matter ranges from hormonal effects in males and females on fish, amphibia, reptiles, birds, and mammals, bisexuality of mating behavior, emotion, conditioning, and other types of learning, morphology, metabolism, metamorphosis, and molting to difficulties of experimentation and interpretation. References dating back to the 1920s and 1930s are common, including one titled "The female sex cycle as a factor in learning in the rat" published in 1926 (Ball, 1926), which sounds like it could have been published in yesterday's issue of *Brain Research.*

Returning to the work of Phoenix et al.—female guinea pigs were treated with testosterone during pregnancy, and the female offspring born to these mothers did not act like females as adults, meaning there was no evidence of an estrous cycle (the rodent equivalent of a menstrual cycle), and the females did not display sexual receptivity when in the presence of a male suitor. Even more striking, if provided with testosterone as adults, these genetic females behaved exactly as genetic males. Similar studies were conducted on the rat by Charles Barraclough and colleagues, with the distinction that in these experiments the females were not treated with testosterone until they were born, and the focus was the impact on reproductive physiology rather than behavior (Gorski and Barraclough, 1963). Two important things emerged. One was that the rat was a better model than the guinea pig, which is a highly precocial species, unlike humans. The second was that a graduate student involved in these early findings named Roger Gorski decided to pursue the neuroanatomical correlates of this early hormone action and was the first to report a major sexual dimorphism in the mammalian brain in 1978 (Gorski et al., 1978). In the interest of historical fairness, however, he was not the first person to think of looking for sex differences in the mammalian brain. Studies conducted 10 to 15 years earlier by Don Pfaff (Pfaff, 1966) and the team of Raisman and Field (Raisman and Field, 1971) had reported small but significant differences in the microneuroarchitecture of the rat brain, differences that required an electron microscope to be detected. In retrospect, these researchers can appreciate that they might have missed the forest for the trees in their search for sex differences in the brain.

No one expected there to be major neuroanatomical differences—it had been assumed if there were sex differences in the brain at all, they would be subtle and exacting. The discarding of this dogma came from an unlikely source, the canary. The complex and delightful song of male canaries is familiar to many. Reasoning that such a highly refined behavior must have a neuroanatomical underpinning, Fernando Nottebohm and his student at the time, Arthur Arnold, set about discovering the song control nuclei in the canary brain and reported in 1976 that these nuclei were either wholly absent or greatly diminished in female canaries (Nottebohm and Arnold, 1976), which are subdued both in plumage and in song. By analogy, Roger Gorski decided it would be a good idea to step back and take a more distant look at the rat brain, particularly in those regions controlling sexual behavior, and thus was discovered the sexually dimorphic nucleus (SDN) of the preoptic area. Many other brain regions have since been documented to be larger in one sex than in the other, although how this occurs mechanistically remains a mystery and what it means functionally is equally unclear. Within about a decade of the discovery of volumetric sex differences in the brain were a series of studies describing remarkable differences in synaptic patterning in discrete regions of the male versus female brain (Matsumoto and Arai, 1980). Although much harder to study, these synaptic patterning differences are much easier to relate to function.

The 1990s was an era of intense interest in sex differences in the human brain and how these might relate to sexual preference (LeVay, 1991; Breedlove, 1994). Emphasis on the complicated question of human sexual preference distracted attention away from the basic question of sex differences and in many ways undermined the field by infusing it with political agendas. What this might ultimately have cost will never be known. This brings us to the present day and the advent of transgenic mice, gene arrays, explant cultures, and other modern tools for exploring the multitude of ways in which a male and female can differ. These tools are reenergizing the field and paradigm shifts, such as the notion of a genetic contribution to brain sex differences (Arnold et al., 2004) and differential de novo steroidogenesis in select brain regions (Prange-Kiel et al., 2003; Amateau et al., 2004) promise to shed light on this fascinating topic.

2.1 Organizational/Activational Hypothesis

Returning to the organizational/activational hypothesis of hormone action, it can be said that it is in essence a framework for the concept that early hormone action exerts permanent (organizational) effects on the neural architecture that are then acted upon in the adult. In most species the early exposure occurs perinatally, meaning around birth. In rats and mice gestation lasts 20–22 days with litter size being on the order of 6–12. During late gestation there is copious production of testicular androgens leading to a hormonal surge selectively in males. A second surge occurs at birth, such that males experience two- to threefold more androgen exposure than females. In rodents these hormonal surges are the result of a complex interplay between active steroidogenesis, probably from the adrenals and gonads, and changes in metabolism that lead to a buildup of steroids (Baum et al., 1988). In primates, there is an activation of the hypothalamic–pituitary–gonadal axis during fetal development in males that must then be actively suppressed postnatally until puberty (Plant, 2001). Regardless of differences in etiology, in both species the outcome is the same—a sexually dimorphic exposure to gonadal steroids during a restricted developmental period. The principles of the organizational/activational hormone effects are illustrated in ➋ *Figure 11-1.*

Figure 11-1

The organizational/activational construct for sexual differentiation is based on the principle of an early sensitive period that is operationally defined as beginning with the onset of gonadal secretion of steroid hormones in the male, and as the termination being that point at which the female becomes unresponsive to the impact of exogenously administered steroid hormone. The actions of gonadal steroids during this sensitive period are considered organizational, and generally permanent, and this differentiated neural substrate is then acted upon (activational) by the appropriate gonadal steroids in adulthood to produce stereotypic sexual behavior or physiology. Different physiological and behavioral endpoints vary in the precise timing of the sensitive period although most differentiation is complete by the end of the first postnatal week of life

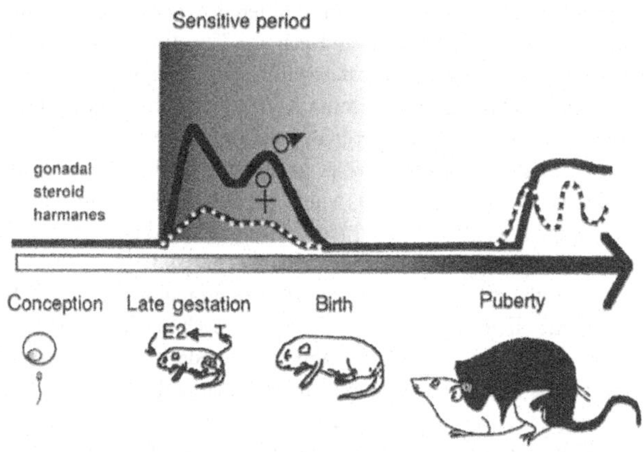

2.2 The Sensitive Period

Sensitive periods are a general phenomenon in developmental neurobiology, referring to an opportunistic window during which events must occur or forever be precluded. An illustration of this, familiar to many, is the lazy eye. If the good eye is patched when an individual is still a child so that the lazy eye has to work harder, the condition is corrected. If, however, the good eye is not patched until adulthood, the lazy eye cannot be coaxed into compensating. This example actually involves the musculature controlling eye movement but the same is true for the brain's detection of visual stimuli. If kittens are raised in an environment that consists only of horizontal bars they will be forever incapable of "seeing" vertical bars and vice versa (Hubel, 1967). There are many other examples of sensitive periods that occur at different times during development and involve both intrinsic and extrinsic stimuli (Kameyama, 1991; Morreale de Escobar, 2001; Lewis, 2004). In the case of sex differences in the brain, the stimulus is intrinsic, being hormones. The sensitive period is operationally defined by the onset when hormonal secretion begins and the termination being that point beyond which extrinsic hormone is ineffective at exerting an organizational effect. Exogenous administration of testosterone to females has revealed different timing for the termination of the sensitive period for distinct endpoints. The two major endpoints are physiology and behavior, the latter being the control of gonadotropin secretion from the anterior pituitary and the former being largely reproductive behavior. The sensitive period for determining whether lutenizing hormone (LH) secretion from the anterior pituitary is responsive to both positive and negative feedback (females) or only negative feedback (males) is generally postnatal, whereas the control of sexual behavior has two overlapping time periods, one for masculinization of behavior and another for elimination of female behavior, also known as defeminization. Sensitive periods have also been explored for morphometric endpoints. The ability of exogenous steroid to increase the volume of the SDN ends precipitously at postnatal day 6 (PN6) (Rhees et al., 1990). The corpus callosum, the major fiber track between the two cerebral hemispheres, also exhibits sexually dimorphic characteristics, and these appear to be established much later than most other endpoints, occurring around PN 10–14 (Fitch et al., 1990), a time at which gonadal steroids are not generally considered sexually dimorphic, although truth be told, this has not been carefully investigated.

2.2.1 Is There a Sensitive Period in Humans?

In humans, the sensitive period for sex determination (i.e., development of the gonads and accessory duct systems) occurs relatively early, within the first trimester. This is followed by a period of gonadal activation during the second trimester, leading to elevated androgen in males. At birth, there is a second surge in androgen secretion in males in response to a release of LH from the newborn pituitary (Corbier et al., 1990, 1992). As a result, newborn males have two- to threefold higher levels of testosterone than females for the first day or two of life. Androgen levels then drop and remain relatively low in both sexes until they begin to diverge again at about 12 years of age, at which point males again have two- to threefold higher levels, as opposed to the fivefold higher levels that characterize the adult (Granger et al., 1999). This is in part due to a drop in androgen production or increased aromatization by the female adrenal postpuberty. Because of the elevated androgens in males shortly after sex determination in the first trimester, this period is often sited as the time when sexual differentiation of the brain occurs but is based on no definitive evidence other than that this is when androgens are elevated. In the rodent, we have been able to clearly define the end of the sensitive period as refractoriness to the effects of exogenous hormone administered to females. In humans, we have no parallel data, leaving the end of the sensitive period for determination of sex differences in human brain an open question.

2.3 Organizational Hypothesis—Revised

Whereas the organizational/activational hypothesis remains a viable and useful framework for empirical investigation of the etiology of sex differences, it also is not an absolute, there being many and varied exceptions. This includes the venerable SDN. The initial determination of the sensitive period was based on the ability of exogenous steroid administration to females to increase SDN volume and this was found to drop rather precipitously at PN6. Further work, however, established that removal of the gonads in males as late as PN29 reduced SDN volume, indicating a need for prolonged exposure to some gonadal factor for final crystallization of SDN size (Davis et al., 1995). This finding illustrates a more general point, that all organization does not occur within the first few days of life. In fact there appears to be a second organizational period associated with puberty in which many of the neural circuits regulating adult behaviors are "finished" or fully organized. This principle is illustrated by the simple observation that administration of gonadal hormones to animals prior to puberty does not induce the normal complement of adult behavior, and conversely, depriving an animal of steroids during puberty permanently changes sensitivity to subsequent steroid exposure (see for review, Sisk and Foster, 2004). Lastly, other brain regions remain sensitive to gonadal steroids throughout life. The posterodorsal subdivision of the medial amygdala receives olfactory input and is related to male arousal. It is larger in males than females, and if adult males are deprived of testosterone, it shrinks to female size and reduces the response of males to female olfactory cues (Cooke et al., 1999). The cellular mechanisms regulating organization events associated with puberty and morphological endpoints that remain sensitive to steroids in adulthood remain relatively unexplored.

2.4 Sex Differences Versus Hormonally Modulated Endpoints

Related to a revised organizational hypothesis is the need for clarification of what constitutes a sex difference. Many behavioral and physiological endpoints are modulated by circulating steroid hormones in the adult animal. This includes the obvious, such as reproductive behavior, but also includes responses such as anxiety, learning and memory, or aggression. In the case of reproductive behavior, there is both a sex difference and a hormonal modulation. Only masculinized brains will respond to testosterone to increase male sexual behavior, and the more testosterone the more robust the behavior (within limits). For learning and memory, there is a mean level of responding in males when quantified by latency in a maze or threshold for induction of long-term potentiation (LTP), and the response in females is different from males but the mean is higher than males under one hormonal condition and lower than males under another hormonal condition (Warren et al., 1995; Cordoba Montoya and Carrer, 1997; Ito et al., 1999). This is not a sex difference, it is a hormonally modulated response. The distinction is more than just semantics. Understanding whether something is a genuine sex difference or a hormonally modulated response speaks to mechanism (Becker et al., 2005).

2.5 The Aromatization Hypothesis

When exploring the specificity of hormonal effects on the developing brain, researchers were quick to notice that compared to testosterone, estradiol was as effective, and often even more effective, at inducing masculinization. This seemed counterintuitive given that estradiol is equated with femininity, and in fact estradiol must be suppressed for proper development of the testis, whereas testosterone is the quintessential male hormone. The conundrum was clarified when several additional pieces were added to the puzzle with the use of the rodent model. First is that fetal and newborn rodents have high circulating levels of a steroid-binding globulin called α-fetoprotein, which has a strong affinity for estradiol. This helps to sequester the estradiol of maternal origin in the bloodstream of the fetus, effectively preventing its entry into neurons and other cells. Testosterone, however, is not bound by α-fetoprotein and moves freely into neurons. The second piece of the puzzle is that aromatase, the p450scc enzyme that converts testosterone to estradiol, is found at very high levels in some regions of the developing brain. Neurons of the hypothalamus, preoptic area, and hippocampus exhibit more aromatase activity during the perinatal sensitive period than at any other time in life (Maclusky et al., 1994). About one week after birth, levels will drop precipitously and for the rest of life the majority of estradiol synthesis will be in the periphery, mostly the ovary, although neurons continue to locally produce estradiol from testosterone throughout life. A third piece of the puzzle is the distribution of estrogen receptors, which is broader and for the most part denser than that of androgen receptors. The ability to locally synthesize estradiol within neurons allows for fine-tuning of steroid hormone differentiation within discrete subregions of the brain. Again, a lesson should be taken from research on songbirds in which the brain has been found capable of synthesizing estradiol de novo; this possibility has now been confirmed in mammalian brains. The adult male hippocampus makes small but detectable quantities of estradiol de novo from estradiol and the rate of synthesis is regulated by neuronal activity as transduced by NMDA receptor activation and calcium influx (Hojo et al., 2003). The female hippocampus also appears capable of making its own estradiol particularly early in development (Amateau et al., 2004), which may provide this telencephalic brain structure with necessary trophic effects provided by this powerful steroid.

The role of aromatization in primates appears to be much less than that for rodents. Instead, it is the direct action of androgens on neurons that differentiates the brain into a masculine phenotype (Wallen and

Baum, 2002). There also appears to be far less of a role for early postnatal hormone exposure in primate models than in rodents, but both this and the potential for an important involvement of estradiol in primate brain differentiation remain to be definitively established. Regardless of the specifics of the hormone and the timing of its action, it is clear that gonadal steroids exert profound and permanent effects on the primate brain that are analogous to those seen in rodents.

3 Morphometric Sex Differences in the Brain

Sex differences in morphometry can be divided into two broad categorizations based on level of analysis: (1) differences in the size of a brain region or structure and (2) differences in the shape and connectivity of individual neurons and glia. The majority of attention has been focused on differences in size of readily identifiable structures such as fiber tracks or particular nuclei. This is in part because of ease of measure. Volumetric measures can be made on fixed tissue sections, including that from postmortem human brain, and overall size can be extrapolated from a few sections. Other advantages of size analysis include the ability to measure the size of a nucleus based on neurochemical identity, such as staining for calbindin or vasopressin, or it can be done just based on a Nissl stain. Lastly, the use of MRI to measure the size of structures in the living human brain has provided considerable insight into the dynamics of size changes in relation to growth. The latter illustrates an important principle in any studies of brain morphometry, that it is often very dynamic. Size of a particular structure may be dimorphic between the sexes at one point in life but not another, which raises the interesting question of what the functional significance of size is, and how such dimorphisms are regulated.

By synaptic patterning we mean the frequency and density of synaptic contacts within a particular brain region. Not all synapses are made equal and they come in three basic flavors. One is the axosomatic synapse, referring to an axon terminating on the soma (cell body) of another neuron. Second is the axodendritic synapse, when an axon terminates on the dendrite of a target neuron, and the last is a specialization of this known as the axodendritic spine synapse. Spines are small protuberances on dendrites that are the site of predominantly excitatory synapses and changes in spine density are a hallmark of synaptic plasticity. We used to think of the brain as a relatively hard wired structure; once connections were made they were permanent. This is still true to some degree, as discussed in the previous section on sensitive periods. However, it is now apparent that the brain is also constantly revising some proportion of its synaptic connections, strengthening some and dismantling others. These changes can underlie learning and memory, sensitization or changes associated with stress and disease. Hormones too can exert profound effects on synaptic patterning in the adult brain, a topic covered in other chapters in this text. These effects are transient, and portions of the female brain may remodel with each successive estrous cycle (in rodents, and perhaps with each menstrual cycle in women, although currently unknown). It is now apparent that hormones sculpt the neuroarchitecture of the developing brain and establish permanent sex differences in synaptic patterning. For instance, the density of axodendritic spine synapses is twice as great in one hypothalamic region (the arcuate nucleus) in females as males whereas the opposite is true in another hypothalamic area (the preoptic area), with males having double the number of dendritic synapses as females. Surprisingly, it appears that a specialized type of glia, known as astrocytes, are playing a central role in the establishment of sexually dimorphic synaptic patterning (McCarthy et al., 2002). What makes it surprising is that this cell type is not generally thought of as a primary target for steroid action, in part because there is little evidence that it possesses receptors for estradiol.

3.1 Morphometric Sex Differences Correlate with Sex Differences in Physiology and Behavior

As noted earlier, we have made substantial progress at cataloging the many morphometric changes induced in the developing brain in response to gonadal steroid exposure during the perinatal sensitive period. These studies are frequently conducted by anatomists using techniques ranging from histochemistry to electron microscopy. Another cadre of researchers focuses on sex differences in behavior and how they are modulated by steroids developmentally and in adulthood. A particularly powerful approach is a combined morphometric and physiological/behavioral assay analysis but the restriction to only correlative statements remains. Nonetheless, there has been sufficient advance to allow strong inference regarding how changes in brain morphology might determine physiological and behavioral outcomes in some instances. Therefore, rather than separately reviewing the current state of the art on each, a synthesis of the neuroanatomy, physiology, and behavior may provide an integrated view of sexual differentiation of the brain, the cause and the consequences.

4 Sexual Differentiation of Sex Behavior

Studies attempting to elucidate the mechanistic basis of sex differences in the brain benefit from a robust and reliable endpoint, namely, sexual behavior in the albino laboratory rat. Whether an adult rat responds to sexual advances from another adult rat with a male response (mounting, intromitting, and ejaculating) or a female response (lordosis) is entirely dependent on two variables: (1) the hormonal milieu during a perinatal sensitive window that must match (2) the hormonal milieu of the adult. Male sexual behavior is opportunistic and readily expressed whenever a receptive female is present. Female sexual behavior is physiologically constrained to be expressed only in proximity to ovulation. A fascinating but mechanistically unexplained distinction in male and female sexual behavior is the impact of experience. Males improve with practice and if testosterone is removed (via castration), will continue to exhibit high levels of copulatory behavior that only gradually extinguishes over a period of months. Females, by contrast, get it right the first time and every time, but, they will only exhibit lordosis under the proper hormonal umbrella. If steroids are eliminated, so is the behavior. In terms of ultimate causation, the adaptive basis for the distinction is obvious; there is no benefit and possible cost to females mating outside the window of opportunity for conception, whereas males maximize fitness by never missing an opportunity to share the wealth. Proximately, meaning at the mechanistic level, this dichotomy in the nature of the plasticity attendant to both behaviors suggests distinct and separate neuronal circuitries, be they physically or merely functionally so.

4.1 Masculinization, Defeminization, and Feminization

A useful framework for investigating mechanistic questions of sexual differentiation is the operationally defined and distinct processes of masculinization, feminization, and defeminization. **Masculinization** refers to an active developmental process initiated by gonadal steroids during the perinatal sensitive period followed by expression of normal male copulatory behavior in adulthood. **Feminization** is essentially what happens in the absence of masculinization, meaning it is the default pathway leading to expression of lordosis under the proper hormonal conditions in adulthood. **Defeminization** is distinct from but normally occurs in tandem with masculinization and refers to the process whereby the ability to express female sexual behavior is lost (see ❯ *Figure 11-2*). Defeminization appears to be exclusively regulated by estradiol, whereas masculinization involves both estrogens and androgens (Whalen and Edwards, 1967; Vreeburg et al., 1977; Auger et al., 2002).

The neural circuitry controlling male sexual behavior has been mapped by lesion studies and expression of the immediate early gene, c-fos. Olfactory input via the main and accessory olfactory bulbs is transmitted to the subnuclei of the amygdala and converges on the preoptic area (POA). From there the information is sent down to the midbrain and critical spinal cord nuclei for the control of the penis and suppression of micturation (Wood, 1997; Coolen et al., 1997; Murphy and Hoffman, 1998). The POA serves as a critical nodal point for the integration of sensory cues and is a principal site for steroid hormone action. Developmentally, the POA is exquisitely sensitive to estradiol, which is locally aromatized from testicular testosterone. Males have two- to threefold more estradiol in the POA than females over the first few days of life (Rhoda et al., 1984; Amateau et al., 2004), and the actions of this steroid masculinize the neuronal

Figure 11-2

The tripartite nature of sexual differentiation of reproductive behavior has been defined as Masculinization, the active induction of neural substrates that mediate male sexual behavior that is triggered by the gonadal steroid testosterone and its metabolite, estradiol; Feminization, the default pathway that leads to a neural substrate conducive to female reproductive behavior; and Defeminization, which is the active removal of the default pathway and is also triggered by gonadal steroids during the sensitive period for sexual differentiation

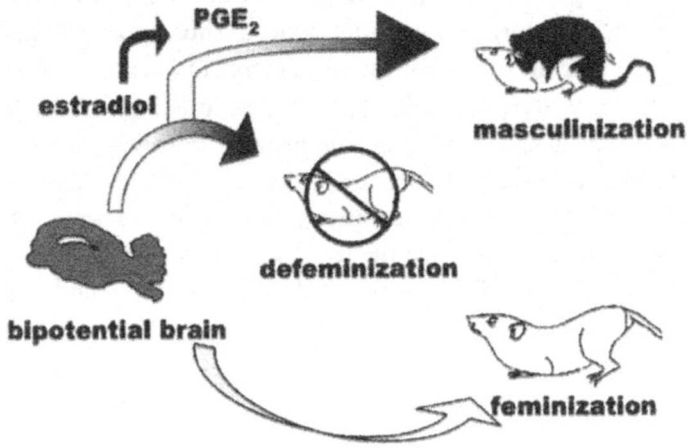

substrate. As a result, the POA is a prominently sexually dimorphic brain region at both the volumetric and cellular levels. The SDN of the POA is 5–7 times larger in males, and there is a greater density of dendritic spine synapses on POA neurons in males (Raisman and Field, 1973; Amateau and McCarthy, 2002a). A growth associated protein (GAP-43) selectively concentrated in axonal growth cones during development is significantly higher in newborn male POA than female (Shughrue and Dorsa, 1994), and astrocytes of the male POA have more and longer processes that branch more frequently than those of females (Amateau and McCarthy, 2002b). All of these morphometric sex differences are dependent upon perinatal estradiol. Exogenous administration of estradiol increases the number of dendritic spines on the POA neurons of females to a level equal to that of males, and differentiates astrocytes and increases the expression of GAP-43. Together it can be assumed that these differences in morphometry converge in a coherent manner that then provides the neural substrate mediating the expression of sex-typic reproductive behavior. However, this still does not tell us how the morphometric differences were established, and what the precise action of estradiol is. Surprisingly, the effects of estradiol on dendritic spines are mimicked by exogenous administration of the prostaglandin (PGE2) and prevented by indomethacin, an inhibitor of prostaglandin synthesis (Amateau and McCarthy, 2002a, 2004), raising the interesting possibility that a prostaglandin is a major determinant of sex differences in synaptic patterning in the POA.

4.2 Prostaglandin E2 Mediates Behavioral Masculinization of Sex Behavior

Prostaglandins are synthesized from the precursor, arachidonic acid, by the rate limiting enzyme cyclooxegenase (COX), which comes in three forms. COX-1 is constitutively active whereas COX-2 and -3 are inducible forms. COX-3 is particularly enriched in brain (Shaftel et al., 2003). Arachidonic acid is converted to the prostanoid prostaglandin H2 (PGH2), and then prostaglandin G2 (PGG2), before diverging into multiple endpoints as a function of specific synthetic enzymes leading to the synthesis of eight distinct prostanoids. The synthesis of PGE2 requires PGE synthase, which also has a constitutively active and an inducible form. The inducible form, mPGE2, is yoked to COX-2 such that stimuli that induce an increase in COX-2 inevitably lead to an increase in PGE2. In the developing POA, estradiol acts to increase both the mRNA and protein for COX-2, having no effect on PGE synthase, nonetheless, estradiol treatment results in a sevenfold increase in POA PGE2 levels, without altering the other prostanoids (Amateau and McCarthy, 2002a).

This still leaves the question of how a prostaglandin could increase the density of dendritic spines on POA neurons unanswered. The importance of astrocytes to neuronal functioning is continuing to emerge. Recent evidence from other brain regions reveals that glutamate is released by astrocytes in a calcium-dependent manner, and that this glutamate impacts on neuronal function. A prominent trigger of astrocytic glutamate release is PGE2 (Sanzgiri et al., 1999). A major regulator of dendritic spines is glutamate, and in the developing POA the effects of either estradiol or PGE2 on dendritic spines can be blocked with the specific glutamate receptor antagonist, NBQX, which is specific to the AMPA form of the glutamate receptor. Antagonists to the NMDA form were without effect. Thus, in an attempt to connect the dots, we have developed a working model in which testosterone enters POA neurons where it is locally converted to estradiol. Estradiol binds to its receptor and induces transcription of the COX-2 gene. Induction of COX-2 leads to increased PGE2 production within the neuron, from which it is released to act on the adjacent astrocyte. The effects of PGE2 on the astrocytes are twofold. One is to contribute to the morphometric differentiation of the astrocytes; the other we speculate is to induce the release of glutamate, which in turn acts back on the neuron to evoke the formation of dendritic spines, thereby establishing a sexually dimorphic synaptic pattern (see ➋ *Figure 11-3*). This is a working model and not all components

◘ Figure 11-3

A working model proposes that within the preoptic area, estradiol is derived by local aromatization from its precursor, testosterone, (1) and induces synthesis of COX-2 through conventional genomic means (2). COX-2 is the inducible form of cyclooxygenase and is the rate-limiting enzyme in all prostaglandin production but is preferentially yoked with the production of PGE2 (3). There is no definitive evidence that both COX-2 and PGE2-synthase are exclusively in neurons but we speculate such. We further speculate that PGE2 is released by neurons to act on the EP2 and EP3 receptors that are localized to astrocytes (4), inducing the release of glutamate (5), which in turn acts in part via AMPA-receptors (6) to induce the formation of dendritic spines on nearby neurons (7). The increased dendritic spines are presumed to represent a parallel increase in the density of excitatory synapses that will ultimately determine the expression of male sexual behavior in the adult

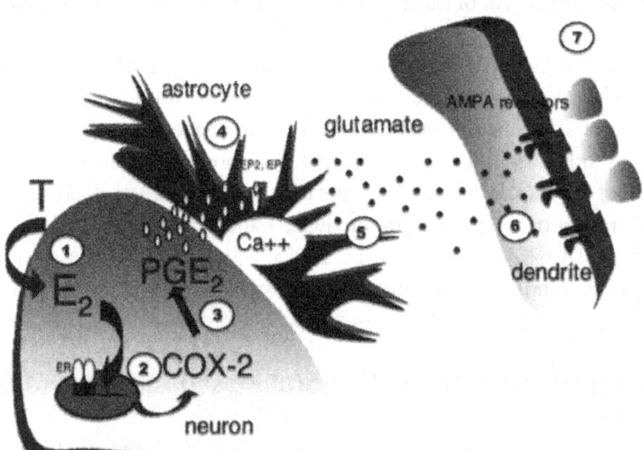

are equally firm, but that there is an interaction between neurons and glia in the establishment of synaptic sex differences seems clear. This principle is generalizable to other sexually dimorphic brain regions (see later).

But again, critical questions remain. What do changes in dendritic spines on POA neurons have to do with behavior? One way to find out is to treat newborn females with a dose of PGE2 known to induce dendritic spines to the same level as seen in males, raise them to adulthood, remove the ovaries, and provide them with testosterone before testing them for male sexual behavior. Alternatively, males deprived of their

PGE2 by neonatal treatment with a COX-2 inhibitor could also be tested for ability to display male sexual behavior. We conducted these experiments and the results were conclusive—females neonatally treated with PGE2 exhibited adult levels of male sexual behavior identical to that seen in control males, whereas males developmentally deprived of PGE2 were duds, exhibiting no sexual interest in the stimulus females across three 30-min trials. At first pass, it is hard to imagine that a brief disruption in PGE2 synthesis developmentally could have such profound and apparently permanent effects. Equally implausible is the observation that two injections of PGE2, one on the day of birth and one the day after, are sufficient to fully masculinize the female brain in regard to sexual behavior. However, when we examined the brain for prostaglandin mediated changes in dendritic spines, we found that the effects that are manifested within the first two days of life are still apparent at 21 days and, even more impressively, at 90 days of age (Amateau and McCarthy, 2004). This provides us with a striking correlation between changes in dendritic spine synapses and adult behavior, but it is nonetheless, a correlation. Many neonatal manipulations will impair adult male sexual performing, but to our knowledge there are no previous reports in which a female brain is masculinized without steroid exposure. This is not just an interesting anomaly; the ability to masculinize the brain without steroids allows for investigation of questions not previously possible. For instance, can an animal be masculinized but not defeminized? Also, is a male-sized SDN a requisite for male sexual behavior? Furthermore we can ask, are the effects of PGE2 specific to male sexual behavior, or do they also effect maternal behavior, anxiety, olfaction, or any of a host of other relevant behaviors?

4.3 Other Factors Contribute to Masculinization and Defeminization

Given that steroid receptors are primarily transcription factors, an obvious and fundamental question is what genes are being activated as a result of hormone action during development? Based on the changes in both the mRNA and protein levels of COX-2 and the importance of prostaglanin production, we can say with some confidence that COX-2 is likely to be a direct product of gene induction by estradiol. However, it is certainly not the only gene product likely to be regulated by direct transcriptional regulation via estrogen receptors during the perinatal sensitive period. Progesterone receptors are induced by estradiol and are higher in males (Quadros et al., 2002) but this is not unexpected given that induction of progesterone receptors is a well-established response to estradiol in the adult brain. Sex differences in synaptic patterning and size of specific projections might be explained in part by GAP-43, a growth-associated protein abundant in axons but not dendrites and elevated in the developing brain during the period of neurite outgrowth, extension, and synaptogenesis. During the sensitive period for sexual differentiation, males have markedly higher levels of GAP-43 in the POA (Shughrue and Dorsa, 1994). Growth factors have also been implicated in the establishment of sex differences in the brain, but not strongly so. Infusion of antibodies against NGF blocked the defeminization of female sexual behavior but did not alter control of gonadotropin release (Kornack et al., 1991; Ricceri et al., 1997), further supporting the notion that these are physiological processes undergoing distinct differentiation. Dopamine and serotonin are both prominently involved in male sexual performance in the adult animal, in both the appetitive and consummatory components. The possibility that dopamine is a causal agent establishing sex differences has been proposed, as it can either enhance or inhibit neurite outgrowth depending on the receptor composition of a particular neuron (Hull et al., 1998, 1999). There is empirical evidence for both also contributing to normal male brain development, but there is little known mechanistically as to precisely how either of these neurotransmitters mediates masculinization.

The use of subtractive hybridization of cDNA isolated from newborn male and female rat hypothalamus identified granulin (epithelin) as a gene expressed at higher levels in males. A curious and interesting aspect of this finding is that just prior to and at birth, males and females have equally high levels of granulin, which then drop in females, resulting in a marked sex difference by postnatal day 10. Functional significance of this difference was demonstrated with the use of antisense oligonucleotides to reduce granulin protein during the sensitive period, resulting in compromised adult male sexual behavior (Suzuki et al., 2000; Suzuki and Nishiahara, 2002). The opposite, inducing granulin production in females to masculinize the brain, has not yet been achieved leaving the precise role of this protein unknown.

Now that we have some insight into the cellular mechanism of masculinization, and we know that it is not involved in defeminization or feminization, the next step will be to identify the mechanisms mediating these other developmental phenomena. In the case of defeminization we are aided in part by knowing the hormonal trigger, estradiol, leaving feminization as perhaps the most intractable of the three processes. To our knowledge, there have been no clear examples of disruption of feminization with the exception of studies suggesting a role for estradiol in normal feminization (McCarthy et al., 1993; Bakker et al., 2002). Thus, one of the greatest challenges ahead is determining how the neural circuitries controlling female sexual behavior are meant to develop unless coopted by male gonadal steroids and defeminized.

4.4 The Sexually Dimorphic Nucleus of the Preoptic Area Does Not Correlate with Masculinization of Behavior

In the rat, the SDN of the POA is a major brain region regulating expression of male sexual behavior, but there is no unique role elucidated for the SDN. The SDN-POA is an area that is several fold larger in males than females (Gorski et al., 1980) and is entirely a product of differential estradiol exposure. Morphologically, it is a collection of large dense and darkly staining cells embedded within the medial preoptic nucleus. Other than their size and propensity for taking up Nissl stain, there is little to distinguish these cells from the surrounding ones. A notable exception is high levels of the calcium-binding protein, calbindin (Sickel and McCarthy, 2000). Neurons of the surrounding POA seem entirely devoid of this particular cellular marker. Neurons of the SDN contain estrogen receptors, express aromatase, and are largely GABAergic (Sagrillo and Selmanoff, 1997). Projections to and from the SDN are profuse, and given its location in the heart of the POA, it seems as natural as a major nodal site for the regulation of male sexual behavior. Analogous nuclei have been identified in a variety of species including guinea pigs, gerbils, sheep, nonhuman primates, and humans (Bleier et al., 1982; Swaab and Fliers, 1985), but with the exception of the gerbil in which the SDN regulates scent marking (Heeb and Yahr, 1996), there remains a lack of understanding of precisely why this collection of cells exists. Conflicting reports indicated the SDN of male rats was not critical for control of sex behavior (Anderson et al., 1986; Houtsmuller et al., 1994). Masculinization induced by exogenous administration of PGE2 to females allowed for the expression of male sex behavior in females that were not concurrently defeminized (because there was no steroid provided) and these masculinized females had a female-sized SDN (Amateau and McCarthy, 2004), supporting speculation that the SDN may relate more to defeminization than masculinization. Recent data suggest a different role for the SDN, specifically, the preference for an estrous female over a male (Houtsmuller et al., 1994; Woodson and Gorski, 2000; Roselli et al., 2002). Although an interesting idea, this too remains in the realm of correlation, as there is no clear way to manipulate the volume of the SDN without also manipulating gonadal steroids during the perinatal sensitive period. Understanding how the SDN becomes dimorphic may provide the needed insight into definitively establishing its function(s).

4.4.1 The SDN-POA Illustrates General Principles of Morphometric Sex Differences

So what do we know about how the dimorphism of the SDN is established? Actually, a fair amount. We know that it is not due to sex differences in the number of neurons born or their migration into a particular location, although these two variables may play minor roles (Jacobson and Gorski, 1981). We also know that the sexual dimorphism is not evident until as late as postnatal day 5, long after the neonatal surge in testosterone in males, and is not fully established until at least a week after birth (Sickel and McCarthy, 2000). It appears a single variable will ultimately determine the volume of the SDN, that is, how many neurons succumb to naturally occurring cell death or apoptosis (Arai et al., 1996; Davis et al., 1996). The single variable determining whether the cells live or die is estradiol. If there is enough, the neurons survive, if there is not, they die. Once outside the sensitive period, no amount of estradiol will restore SDN volume, but as to what estradiol is doing, dead neurons tell no tales.

Thus, it appears we have a nucleus of unknown function and unknown origin, yet the SDN remains a topic of considerable intrigue and serves as a valuable example of many principles of neuroanatomical sex differences. First, variance in size does not necessarily correspond to variance in function. Second, sex differences in the size of a region or particular nucleus appear to occur via the common route of differential apoptosis, either induced by or protected from the actions of gonadal steroids. Put quite simply, males and females start out with the same number of neurons in a particular area (meaning there is no sex difference in neurogenesis or migration) and then neurons begin to die in one sex either as a result of insufficient hormone exposure or in response to hormone exposure. This scenario has been established for the SDN, a motor nucleus in the spinal cord controlling the penis called the SNB, a subdivision of the bed nucleus of the stria terminalis (BNSTc) and the hippocampus, all of which are larger in males, and for a nucleus in the hypothalamus that controls the LH-surge, the anteroventral periventricular nuclues (AVPv), and the visual cortex, both of which are larger in females. Exactly how steroids are dictating whether neurons will live or die is unknown and complicated by the fact that the peak of cell death (or survival) appears to be several days after differential hormone exposure. Intriguingly, for the BNST and AVPv, brain regions in which estradiol has opposite effects on cell death, the mechanism appears to involve the cell death gene BAX in both cases (Forger et al., 2004). That these two interconnected cell groups showing opposite patterns of sexual differentiation in response to estradiol are determined by the same gene highlights how much we have to learn about the cellular and molecular mechanisms of sexual differentiation. And if it were not already confusing and confounding enough, in the same brain, BAX does not regulate the sex difference in the number of dopaminergic neurons found in the AVPv, suggesting further heterogeneity of mechanism (Forger et al., 2004).

A third general principle is that no one sexually dimorphic structure stands alone. Just as the SDN is embedded in the medial POA, the mPOA is also intricately and reciprocally connected with other brain structures that are either themselves dimorphic or constitute part of a neural network controlling sexually dimorphic physiology and behavior. Virtually all components of this network have a high concentration of neurons expressing receptors for estrogen and/or androgens. This principle critically relates to a fourth, identifying the mechanism by which steroids modulate neuroanatomical sex differences is complicated by the potential of steroid effects that act directly on the cells of interest as well as indirectly via hormonal modulation of afferent input. It is further complicated by the fact that parameters affecting cell death appear to be intermingled with factors determining neurochemical phenotype, making for a complex mosaic of effects that remains poorly understood.

5 Sexual Differentiation of Gonadotropin Secretion

The gonadotrophs of the anterior pituitary are the site of synthesis of the gonadotropins, LH and FSH (follicle stimulating hormone), two essential ingredients of the hypothalamic–pituitary–gonadal axis. Both are released from the gonadotrophs directly into the bloodstream and travel to their target organs, the gonads, where they promote steroidogenesis and gamete production and/or maturation, respectively. In males of nonseasonally breeding species, such as man and his commensal rodents, there is a constant and pressing need for both steroidogenesis and gamete production and both are maintained by regular pulses (hourly or less) of LH and FSH into the blood stream. The pulses of LH and FSH are in turn maintained at constant levels via negative feedback exerted by the gonadal steroids they are designed to produce, creating a closed loop system that maintains homeostasis. In females, by contrast, the gametes are sequestered in the ovary and only a select few are allowed to mature with each reproductive cycle. The maturation process is regulated by an intricate balance between LH and FSH and steroid production, and this culminates in the process of ovulation whereby the ova are released into the body to be made available for fertilization. The primary event responsible for ovulation is a massive surge in release of LH, which will act on the mature follicles to soften the stigma and promote expulsion of the ova along with its attendant radiating crown of granulosa cells. The surge of LH release from the gonadotrophs is the direct result of a surge in release of LHRH from the neurosecreteory neurons of the preoptic area and mediobasal hypothalamus that project to the median eminence and thereby gain access to the primary portal plexus leading to the anterior pituitary.

Achieving sufficient release of LHRH requires synchronized bursting of the neurons and this is achieved via mechanisms that remain poorly understood but clearly involve rising estradiol on a background of previously low levels of estradiol. The combination of synchronized LHRH neuronal firing, increased LH synthesis, and greater sensitivity of the gonadotrophs to LHRH is referred to as "positive feedback" effects of estradiol and is required for ovulation to occur. A masculinized brain does not respond to rising estradiol with positive feedback and the loss of this ability is determined by gonadal steroids during the perinatal sensitive period. Newborn females treated exogenously with testosterone, or its aromatized product, estradiol, will be forever sterile due to the inability to generate an LH surge and thereby ovulate. If females are given low doses of masculinizing hormone, or the exposure to steroid occurs late in the sensitive period, the adult female will show some fertility shortly after puberty but then will display rapid reproductive aging and enters into a period known as "persistent estrus" characterized by chronically elevated estradiol due to the absence of ovulation and therefore, there is no formation of a corpora lutea and progesterone production. Alternatively, restricted prenatal androgen exposure is sufficient to completely prevent LH surges in adulthood and this appears to be, at least in part, because of a lack of induction of progesterone receptors by estradiol in the preoptic area (Foecking et al., 2005).

Although the activity of the LHRH neuron is the central mediator of LH release, afferent input to these neurons plays an important modulatory role, particularly in controlling the precise temporal aspects of positive feedback. Local GABAergic neurons are particularly important and their escape from tonic inhibition prior to the LH surge is a prerequisite for the surge to occur. GABA neurons in the vicinity of LHRH neurons possess estrogen receptors of both subtypes and increasing estradiol is associated with decreased levels of the GABA synthetic enzymes, GAD-65 and GAD-67 (Grattan et al., 1996). A potential target for differentiation of the effects of estradiol on gonadotropin neurons is these local inhibitory interneurons. Use of microdialysis to accurately characterize the dynamic nature of GABA release in relation to the timing of the LH surge indicates that the profile of release is markedly sexually dimorphic in the presence of equivalent hormonal milieus as adults (Tin-Tin-Win-Shwe et al., 2004). This could reflect changes in the morphometry of the GABA neurons, changes in the synthesis of GABA, or changes in the mechanics of release. Regardless of this, the findings highlight the potential for a particular neurochemical cellular phenotype to be permanently differentiated in development to mediate adult reproductive responses.

In addition to the local control, important regulatory components of the LHRH cell bodies originate from projections outside the preoptic area. Some afferent inputs are derived from nearby neighborhoods, such as those originating in the AVPv, which is within the hypothalamus/preoptic continuum, whereas others are as distant as the brain stem (see Simerly, 2002 for review). There is also evidence of an additional modulatory role for the arcuate nucleus, through which the LHRH axons course on their way to the median eminence.

5.1 A Neuroanatomical Circuit for Sex Differences in Positive Feedback

Given that investigation of the cellular mechanisms mediating positive feedback effects of estradiol on LHRH release remains an enigmatic and highly controversial area, it is not surprising that relatively few inroads have been made into understanding how this sexually dimorphic response is established. The greatest traction has been achieved in the elegant work of Simerly and others on two related fronts. One is that the AVPv (also called the PePOA) exhibits an unusual sexual dimorphism in that it is larger in females than males (Simerly, 1998). Interestingly, this sex difference is entirely due to estradiol, just like the SDN-POA, and it is also entirely because of differential, naturally occurring cell death or apoptosis, just like the SDN-POA. Only in the AVPv the effect is in the opposite direction; estradiol actually increases cell death (Murakami and Arai, 1989). The remarkable opposing actions of estradiol on cell death in two diencephalic nuclei that are in close proximity to each other have never been explained. It is also unclear how the volume of the AVPv relates to its role in control of the LH surge. Neurons of this region project directly to LHRH neurons and a strong correlation has been found between levels of c-fos expression in neurons of the AVPv and LHRH neurons at the time of the LH surge (Le et al., 1999), suggesting that excitation in the AVPv is directly related to excitation of LHRH neurons. This raises the question of what regulates excitation of AVPv neurons and relates to the second major avenue through which insight is being gained into the sexual differentiation of gonadotropin secretion. Considerable focus has been on either the volume of a brain region (usually a function of the number of neurons) or the morphometry of individual neurons. Equally important, however, is how many neurons project to a particular region, in other words, the size of an afferent input. One of the most, if not the most, sexually dimorphic projection in the brain is that from the principal nucleus of the bed nucleus of the stria terminalis (pBNST) to the AVPv (Hutton et al., 1998), being almost ten times larger in males than females.

An essential first step in determining how the sex dimorphism in the projection from the pBNST to the AVPv is established was achieved with the use of a combined explant culture system. This provided the critical advantage of being able to ask whether a pBNST from a male neonate would innervate an AVPv derived from a female and vice versa. Surprisingly, the AVPv of males (which is smaller) releases a target derived factor that promotes innervation by pBNST neurons, regardless of whether they are from males or females. Treating the AVPv in culture with estradiol achieves the same end (Ibanez et al., 2001). To date, this is the only known example of steroid induction of a target-derived diffusible factor that regulates innervation. The neurochemical nature of the innervation provides additional critical information as it appears to be largely GABAergic (Polston et al., 2004). Thus, a working model can be constructed in which a tenfold greater inhibitory GABAergic projection for the pBNST to the twofold to threefold smaller AVPv of males results in tonic inhibition of the AVPv and provides a physiological explanation for the failure of males to exhibit an LH surge in response to estradiol treatment in adulthood (Simerly, 2002). This is a beautiful example of a functional circuit directly related to sexually dimorphic reproductive function. This simplistic view, however, requires reexamining and refinement in light of recent evidence that AVPv neurons responsive to estradiol are actually of a dual phenotype, both GABAergic and glutamatergic, with the relative impact of each varying as a result of differing estradiol levels and photoperiodic timing of the LH surge (Ottem et al., 2004).

5.2 Are Sex Differences in Astrocyte Morphology Relevant to Positive Feedback?

As mentioned earlier, there is additional modulation of the LH surge by regulation of the LHRH neurons at the level of the arcuate/median eminence. Astrocytes are a subtype of glia, and can be further divided into reactive and protoplasmic astrocytes, as well as tanycytes, a specialized astrocyte found only in the arcuate/ median eminence. Protoplasmic astrocytes can be identified by their expression of glial fibrillary acidic protein (GFAP), which when visualized by immunocyochemistry, reveals long sinuous and multiple branching processes. These nonneuronal cells have long been implicated in the process of synaptic plasticity in neuroendocrine brain regions that integrate physiologic stimuli from the body. Garcia-Segura, Naftolin, and colleagues (Garcia-Segura et al., 1996; Garcia-Segura et al., 2000) have shown a similar involvement of astrocytes in synaptic plasticity in the arcuate nucleus with the exception that the dominant stimulus is estradiol, which drives changes in astrocyte morphology over the course of the estrous cycle. The surface area covered by astrocytic processes is dramatically increased on the afternoon of proestrus and correlates with a reduction in GABAergic synapses, leading to the postulate that disinhibition in the arcuate plays a positive role in the induction of the LH surge (Olmos et al., 1989; Parducz et al., 1993; Horvath et al., 1997). There is also a role for tanycytes, the end-feet of which appear to surround the terminals of LHRH neurons as they enter into the median eminence in proximity to the primary portal plexus. Steroid hormone treatments that induce an LH surge result in a retraction of this glial ensheathment of the LHRH nerve terminals, thereby increasing availability of the releasing hormone to the vasculature (King and Rubin, 1994). Whether developmentally established sex difference in the responsiveness of either protoplasmic astrocytes or tanycytes to gonadal steroids in adulthood contributes to sexual differentiation of gonadotropin secretion, remains unknown.

Noting the marked remodeling of arcuate astrocytes that occurs in the adult female brain in response to estradiol exposure observed by Garcia-Segura and colleagues, the potential for a similar effect was investigated in the developing brain. A sexually dimorphic morphology of arcuate astrocytes is established as early as the day of birth and maintained throughout life. Males have longer and more frequently branching astrocytes with an overall stellate appearance than the predominantly bipolar and simple morphology of astrocytes in the arcuate nucleus of immature females. This sex difference appears to be entirely a function of testicularly derived testosterone being aromatized to estradiol in the neonatal male brain, as astrocytes of the female can be induced to differentiate to a point that is indistinguishable from that of males by exogenous testosterone or estradiol, but not dihydrotestoterone (Mong et al., 1996). A plausible scenario for estradiol regulation of astrocytes would be via direct binding to ER in the astrocytes and regulation of a target structural gene, such as GFAP. However, as with most things, when it comes to steroids and reproduction, the cellular mechanisms were anything but obvious. Efforts to find ER in astrocytes of the arcuate nucleus produced only negative results (Mong and McCarthy, 1999), but this can only be taken as absence of evidence rather than evidence of absence. However, it was further demonstrated that the amino acid transmitter, GABA, is a critical mediator of estradiol-induced astrocyte differentiation. Blocking GABA synthesis for the first two days of life with antisense oligonucleotides to the rate limiting enzymes GAD-65 and GAD-67 prevented the differentiation of astrocytes, which occurs in males or androgenized females. More important, administration of the GABA-A receptor agonist, muscimol, early in life, induced differentiation of astrocytes in females that was of the same magnitude as that normally occurring in males (Mong et al., 2002). Astrocytes in vitro have been shown to possess GABA-A receptors (Fraser et al., 1994) and to change morphology in response to receptor activation (Matsutani and Yamamoto, 1997). Given that GAD is synthesized exclusively in neurons, this suggests that GABA is released by neurons to act on nearby astrocytes to alter their morphology. Thus, it seems that astrocytes are not the primary site of hormone action, but instead they are responsive to hormonally induced signals originating in the neurons.

Interest in glial–neuronal interactions has increased dramatically in the last half decade. That glia would have a sexually dimorphic morphology permanently established during the neonatal sensitive period seemed surprising at first, but when considered in the context of how intimate the relationship between neurons and glia is, and how neurons exhibit lifelong differences in morphology, it becomes apparent that what would be truly surprising is if astrocytes were NOT sexually dimorphic. An important question is the relationship between astrocytic and neuronal morphology. In the arcuate nucleus, increased astrocyte complexity is inversely correlated with the density of dendritic spine synapses on the nearby neurons (Mong et al., 1999). The mechanistic basis for this inverse relationship is not known but has been speculated to be a result of physical inhibition of spine formation. In other words, the astrocytes act like a blanket over the neurons, suppressing the sprouting of spines. There is no evidence either in support of or against this theory, but it seemed plausible enough until other brain regions were investigated.

The discovery of a sex difference in astrocyte morphology in the arcuate nucleus prompted a similar analysis in the preoptic area. Here too, astrocytes of males have more, longer, and increasingly bifurcated processes than females and as with the arcuate, this is a function of estradiol action (Amateau and McCarthy, 2002b). Parallel analysis of neuronal morphology, however, demonstrated that the relationship between glia and neurons was exactly the opposite to that seen in the arcuate; more complex astrocytes were positively correlated with increased density of dendritic spine synapses (Amateau and McCarthy, 2002a). Another notable difference was the mechanism regulating astrocyte morphology. Instead of GABA, morphology appears to be determined at least in part by PGE2, the same prostaglandin shown to increase dendritic spines and masculinize the POA in regard to male sex behavior (McCarthy et al., 2002). Once again, the same steroid hormone estradiol regulates markedly different pathways to achieve the same end, more complex astrocytes. Whether or not estradiol acts directly on astrocytes and if astrocytes in the POA possess estrogen receptors, is currently unknown.

6 Sexual Differentiation of Cognition

Cognition, also referred to as learning and memory, is a value-laden metric even when assessed with apparently objective criteria. Rats and mice have been frequently used to demonstrate sex differences in cognition, and underlying morphometric differences in the hippocampus or cortex are often cited as further evidence for differential learning between the sexes. It is presumed that this is based in early organizational events similar to that observed for reproductive endpoints. But is this assumption warranted? The hippocampus is vitally important to learning and memory, and there is no argument that despite having little or no role in the control of reproductive function, it is a sensitive target for gonadal steroids. Considerable attention has been paid to estradiol modulation of synaptic plasticity in the adult (Desmond and William, 1998; Woolley, 1999) and the effects of hormonal treatment on cognitive functioning (Kampen and Sherwin, 1994; Sherwin, 1997; Daniel et al., 1999; Daniel and Dohanich, 2001; Sandstrom and Williams, 2001; Williams, 2002; Luine et al., 2003; Li et al., 2004). Interest in estradiol action on the mature and aging hippocampus has been further increased by controversy surrounding the use of hormone replacement therapy in women. There is irrefutable evidence that estradiol potently enhances synaptic efficacy in the hippocampus by increasing the density and functionality of dendritic spine synapses in the female rat (Woolley et al., 1997; Rudick and Woolley, 2003). Behavioral tests designed to temporally correlate physiological changes with spatial learning have provided equally strong evidence of enhanced cognitive ability in response to estradiol action in the hippocampus (Sandstrom and Williams, 2001). The considerable body of work on estradiol and the hippocampus leads toward the logical assumption that there are substantial sex differences in the hippocampus as well. Indeed, the effects of estradiol on spatial learning tasks in rats and verbal recall tasks in humans are often interpreted as the basis for sex differences in cognitive abilities. However, when questions are specifically framed to ask are there *sex differences* in the hippocampus and hippocampally regulated functions, the evidence is often in favor of either no sex difference, very subtle sex differences, or differences in which hormonal modulation in the female results in a mean response that varies but stays within the range of the male response.

6.1 What is the Evidence of Sex Differences in the Hippocampus?

Volumetrically, males have a larger hippocampus than females (Madeira and Lieberman, 1995; Nunez et al., 2000), with the CA1 region being 16% larger in males in one study (Isgor and Sengelaub, 1998). A sex difference is evident within the first week of life, although it is observed inconsistently and is small in magnitude (~10%) (Hilton et al., 2003; Nunez et al., 2003). In mice, males have more granule cells in the dentate gyrus than females, but this was observed only in three of six strains examined (Wimer and Wimer, 1985). It is important to compare these small and unreliable sex differences with the robust volumetric sex differences observed in subnuclei of the reproductively important POA where differences are in the range of 300–700% (Gorski et al., 1980; Simerly, 2000, 2002).

When examined at the level of physiology, estradiol enhances LTP in ovariectomized females (Cordoba Montoya and Carrer, 1997) and in naturally cycling females (Warren et al., 1995), and this is proposed as the physiological basis for increased learning efficacy following estradiol exposure. Unfortunately, few studies have compared LTP induction across the sexes and in the one that has, the magnitude of LTP induced in females on the afternoon of proestrus (high estradiol) was the same as that seen in males during the morning (Warren et al., 1995). There was also no sex difference in the input–output curves generated by evoked excitatory field potentials. Hippocampal-dependent fear conditioning, another form of learning, was similarly examined for sex differences using cycling females and intact males. In this instance, males and estrous females performed similarly, whereas females in proestrus showed less spatial–contextual conditioning than both males and estrous females (Markus and Zecevic, 1997). In contrast to the well-established ability of estradiol to enhance LTP in CA1 induced by Schaffer collateral stimulation, it reduces LTP at perforant path synapses, an effect believed to be functionally related to the impaired contextual (hippocampal-dependent) fear conditioning also observed in estrogen-replaced ovariectomized rats (Gupta et al., 2001). This was again one of the few studies that compared sexes and interestingly a different pattern appears in that intact males and ovariectomized females show identical fear conditioning but now both differ significantly from estradiol-treated females. Induction of LTD at CA3-CA1 synapses is also enhanced by estradiol treatment of females. A low-frequency (2 Hz), asynchronous conditioning stimulation protocol does not produce LTD in slices prepared from ovariectomized females but does so readily in slices from estradiol replaced females (Zamani et al., 2000). Unfortunately, there was no sex comparison in this study, so it is unknown if the threshold for LTD induction varies between males and females. Changes in the density of dendritic spines on CA1 pyramidal neurons are correlated with LTP and again, there is a sex difference but the direction of the sex difference varies with the phase of the female estrous cycle, being higher in males than in the proestrous females, but lower than males in diestrous females (Shors et al., 2001). Thus, in regard to the cognitively relevant endpoints of LTP, dendritic spine density, and fear conditioning, it appears there is a hormone-dependent modulation of the response in females, but this response varies around the mean in males and therefore does not constitute a sex difference per se.

At the behavioral level, males learn spatial tasks like the radial arm maze or Morris water maze faster than females, but with practice they both perform the task equally well and empirical evidence indicates no difference in memory capacity between males and females. Thus, sex differences are limited to acquisition of the task, not steady-state performance. In fact, when the cues used for acquisition are altered, females outperform males. Systematic control of environmental variables reveals that males use the geometrical cues, such as the shape of the room or location of large objects as guides, whereas females attend to local small-scale details (Williams, 2002). Thus, both sexes learn: they just use different strategies to do it. Even more important, if males and females are familiarized with the spatial task in advance, resulting in a reduction in the natural tendency of females to show greater thigmotaxis than males, females actually outperform males on finding the hidden platform in the water maze. Moreover, there is no sex difference during the probe trial in which the platform is removed (Perrot-Sinal, 1996). These studies illustrate the potential for nonlearning-related parameters, such as response to novelty, to skew results to appear as if there are sex differences in spatial ability when in fact there are not.

Learning can also be assessed with conditioning paradigms such as contextual fear and passive avoidance. Males reach criteria for acquisition in both of these tasks faster than females. The phosphorylation, and thereby activation of the transcription factor CREB, is often used as a marker for neuronal activation and as an indicator of the active gene transcription required for consolidation of long-term memory. The number of neurons expressing pCREB increased in the CA1 of male rats after fear conditioning, but did not change in females (Kudo et al., 2004). Removal of gonadal steroids in males did not alter their performance, hinting at the possibility that this sex difference is determined organizationally, but by no means establishing it.

Sex differences in spatial navigation tasks have been examined within the framework of organizational versus organizational hormone actions. All the traditional strictures apply in that removal of the male testis early in life feminizes learning in adulthood, and treating newborn females with testosterone or estradiol masculinizes their responses as adults (Williams et al., 1990; Williams and Meck, 1991; Roof, 1993). Thus, as with the majority of other sex differences in the brain, estradiol has been implicated as the masculinizing factor. Females treated with estradiol for the first 10 days of life show masculine learning patterns as adults, and males castrated neonatally are predictably female-like in their responses (Williams et al., 1990). However, note that females were not treated with an aromatase inhibitor and removing the testis in males would remove both androgens and estrogens. In a separate study, the effects of hormonal manipulation during late gestation included treatment with the nonaromatizable androgen, dihydrotestosterone, and the antiandrogen flutamide. Animals were tested on a water maze as adults and at the behavioral level the findings are unambiguously interpretable as an *androgen*-mediated effect on spatial learning, with there being no effect of exogenously administered estradiol (Isgor and Sengelaub, 1998). This study is not without its caveats, but raises the interesting possibility that androgens, not estrogens, mediate sex differences in spatial ability. Interestingly, these same authors performed a morphometric analysis of adult brains from neonatally treated animals and found androgen effects on neuron number in CA3 that were consistent with enhanced spatial performance, but in CA1 cell number was increased by estradiol. We have also noted an effect of increased cell number by postnatal estradiol administration to females that is selective to the CA1 field (Hilton et al., 2003), suggesting there may be considerable regional heterogeneity of steroid action and/or availability within the developing hippocampus. Estradiol effects restricted to the CA1 field may be the functional basis for the simplifying of associational–perceptual processes that improve acquisition of spatial tasks as proposed by Williams and Meck (1991).

An additional approach for assessing the role of estradiol on hippocampal function is to knock it out. This has been achieved with transgenic technology in the form of the aromatase deficient mouse (ArKO). Male ArKO mice have severely disrupted male sexual behavior (Honda et al., 1998), consistent with the effects of perinatal aromatase inhibition in the rat. However, both males and females are equally impaired in a spatial reference test, the Y-maze (Martin et al., 2003). Adult female mice lacking functional ERα (ERαKO) show retention deficits on a hippocampal dependent inhibitory avoidance task, a form of trace conditioning (Fugger, 2000).

So are there sex differences in the hippocampus? Yes, there clearly are, but they are of a much smaller magnitude than generally assumed and perhaps more important, they may have little if anything to do with dichotomous developmental exposure to estradiol.

7 Sexual Differentiation of Stress and Anxiety

As central as the hippocampus is to learning and memory, it is equally important to the negative feedback control of the hypothalamic–pituitary–adrenal axis or stress response system. Pyramidal neurons express high levels of glucocorticoid receptors, Type I and Type II, and projections via the subiculum regulate the output of the paraventricular nucleus (PVN), which projects directly to the anterior pituitary to regulate ACTH release into circulation. Increased circulating levels of glucocorticoids act on the hippocampus in a negative feedback loop to decrease the output of the PVN and return the system to homeostasis (see for review Van de Kar and Blair, 1999). Dysfunction of this feedback loop is observed in a significant portion of patients suffering debilitating depression (Wong and al., 2000) and reduced hippocampal volume is associated with a number of neuropsychiatric disorders including post-traumatic stress syndrome and schizophrenia (Nopoulos et al., 1997).

There are sex differences in stress response in adults, with females exhibiting a more robust and longer-lasting increase in glucocorticoids and behavioral impairment following exposure to a stressor (Shors et al., 2001; Bale et al., 2002), and this sex difference appears to be organizationally determined (McCormick et al., 1998). Sex differences in the hippocampus have been speculated to underlie the variance observed in adult stress response, but this connection has not been definitively established. The stress axis is highly complex with multiple components and involves several extra-hippocampal sites of negative feedback including the highly sexually dimorphic POA (Viau and Meaney, 1996). Experimental designs previously employed would preclude definitively concluding that any sex differences observed in stress responding are the result of organizational steroid influences on the hippocampus.

Learning and memory and stress intersect in the form of the Yerkes–Dodson Law, which states that learning is enhanced by stress at one level, but impaired if that stress is increased or unaffected if the stress is decreased (see for review de Kloet et al., 1999; Williams, 2002). This centuries old tenet has been recast recently in light of sex differences. Exposure to acute stress in male rats enhances learning, but has the opposite effect in females, impairing their performance in an eye-blink conditioned reflex experiment. Treatment of neonatal females with testosterone results in a male-like performance as adults, whereas castration of newborn males leads to a female-like response. Antagonizing the androgen receptor during late gestation blocks the enhanced learning induced by stress in adult males (Shors and Miesegaes, 2002), suggesting these effects are mediated by androgens, not estrogens.

These brief reviews of sex differences in cognition and stress response are not meant to provide a comprehensive overview of these highly complex and multivariable systems. Instead, the intent is to highlight that when considering sex differences in these systems, there is a need for caution in evoking similar terminology or in use of assumptions regarding sex differences in brain regions of physiological mechanisms that are directly relevant to reproduction. A complete understanding of the ways in which the sexes do NOT differ is just as important as the ways in which they do.

8 Evidence for Sexual Differentiation of the Human Brain

Interest in sex differences of the human brain predates all of the animal literature and can reasonably be assumed to have emerged shortly after the realization that the brain is in fact the site of control of thought, movement, emotion, and much physiology. As with rodents, male human brains are larger than females, and as with rodents, there are many subregions that are larger in males than females. However, unlike the research on animal models, studies in humans are fraught with numerous technical as well as political pitfalls. Technically, the greatest challenge had been the predominant source of the tissue, cadavers. Differences in time from death to collection and fixation combined with variance in sex, age, ethnicity, and life history combine to make for a heterogeneous sample under the best of circumstances. However, the recent extensive use of magnetic resonance imaging (MRI) highlights an unexpected advantage of studies on humans, namely, the ability to precisely quantify large regions in regard to gray versus white matter and to prospectively study the same individual over time (Giedd, 2004). The use of functional MRI has also allowed for assessing the relative activity of a particular brain region in men versus women and revealed sex differences in uniquely human aspects of speech (Shaywitz et al., 1995).

The use of MRI has also confirmed and extended a fundamental principle in the study of human brain sex differences, the importance of developmental age. The size of particular brain regions is amazingly dynamic and comparisons between the sexes must incorporate a longitudinal view to be truly interpretable. Early studies of a human analog of the SDN found a hypothalamic nucleus that is larger in males than females. A more detailed analysis across the lifespan revealed that the degree of difference varies with age, not manifesting until after about 10 years of age, becoming maximal in young adulthood and gradually declining in both sex difference and in size in older adults (Swaab and Fliers, 1985). In some ways, this even more neatly parallels the rat in that the sex difference in SDN volume is not apparent until 5 days of age and is maximal just postpuberty. However, enthusiasm must be tempered by the fact that the sexual activity and life experiences of humans are also going to be dramatically different between 10 years of age and young adulthood and it is plausible that experience, rather than sex (meaning gender in this case) has a major impact on the size of a particular brain region. Other investigators have examined the human hypothalamus and seen a more complex picture. Rather than a single SDN, they have characterized a series of hypothalamic nuclei referred to as the interstitial nuclei of the anterior hypothalamus (INAH) and numbered 1 through 4 (INAH-1,-2,-3,-4) with INAH-1 being the original SDN. Working with Roger Gorski, Laura Allen found no sex difference in INAH-1 but instead reported that both INAH-2 and -3 were larger in males (Allen et al., 1989).

The technical limitations inherent in use of tissue from cadavers may have been a major contributing factor to inconsistencies in reports of sex differences on cortical volume, size of the cerebellum, and in particular, the size and shape of the corpus callosum, the major fiber tract between the two cerebral hemispheres. Like the hypothalamus, all these studies suffered the same limitations of being conducted on postmortem tissue. The increasing use of MRI to quantify the volume of brain regions in living subjects has helped to clarify some of these discrepancies in that the approach offers the great advantage of allowing for a precise matching of the age, socioeconomic status, and ethnicity of the male and female subjects through the use of recruitment. But like all advances, it also has shortcomings. Although the resolution of MRI scanning has increased tremendously in the last decade, the precision of the images is still far from that which is obtained on fixed tissue, and questions such as cell density or cell type cannot be answered with this approach. Smaller brain structures, such as the INAHs, cannot yet be visualized by MRI.

One of the great advantages of MRI is the ability to distinguish white from gray matter and thereby gain insight into the rate of myelination of particular regions and fiber tracts. Human brain development proceeds in the same way as other mammals in that myelination is a relatively late process that occurs in tandem with the pruning of excessive synapses. Relative changes in white versus gray matter can be taken as an index of the processes of myelination and synaptic pruning, respectively. In a cross-sectional study, changes in white versus gray matter were examined in boys and girls ranging in age from 7 to 17. Both sexes exhibit an increase in white matter and decrease in gray matter with age, but the relative changes in males were significantly greater, having over a threefold loss in gray matter volume and more than twofold increase in white matter than females (De Bellis et al., 2001). Females also showed a significant age-related change in gray and white matter, but the change appeared to be occurring at a different rate compared to males. This echoes the general principle of brain development in males and females elucidated earlier; the same things happen, they just happen in different degrees and/or at different times.

Establishing cellular mechanisms that determine sex differences in human neuroanatomy or behavior can be considered challenging at best. The use of strong inference based on studies done on primates and so-called natural experiments on humans is currently the most promising approach. Studies on primates indicate that the overall process of sexual differentiation in regard to reproductive physiology and behavior is a hormonally driven process analogous to that in rodents, but differs in some important details. One is that it appears that the sensitive period for sexual differentiation of nonhuman primates is entirely prenatal, with little or no impact of exogenous hormone administration once birth has occurred. The second is that in contrast to rodents in which estradiol is the masculinizing hormone, in nonhuman primates it appears that androgens are the causative agent (see for review Wallen and Baum, 2002). Interestingly, in primates α-fetoprotein binds both androgens and estrogens, as opposed to only estrogens in rodents. Presumably, the male primate fetal testis makes sufficiently high quantities of androgens to overwhelm the absorptive barrier of α-fetoprotein, although a role for adrenal androgens cannot be ruled out. A potential role for estrogens in primate brain sexual differentiation, of either males or females, also cannot be ruled out, but precisely what that role might be is unclear at this time.

The most commonly explored naturally occurring experiment for human brain sexual differentiation is girls exposed to high levels of androgens early in gestation because of a genetic defect of steroidogenisis in the adrenal that results in excessive androgen production. This is referred to as congenital adrenal hyperplasia (CAH) and is most frequently due to defective 11-β-dehydroxylase in the glucocorticoid synthetic pathway. Affected girls are born with ambiguous genitalia that are partially masculinized. The hormonal imbalance is corrected at birth and the physical abnormalities are sometimes, but not always, corrected surgically within the first few years of life. These girls have been the subject of intense study in a number of different cultures and this topic has been extensively reviewed (Berenbaum, 1999; Hines, 2002, 2004). A sweeping generalization that can be made about almost every endpoint that has been assessed is that CAH girls fall somewhere between unaffected male and female siblings on measures of masculinization. This is true whether the endpoint is rough-and-tumble play, sexual attraction to girls, spatial learning, or expression through spontaneous drawing (Iijima et al., 2001). The caveat that these girls might be raised differently than unaffected girls can never be fully dismissed, but on balance the evidence suggests that prenatal androgens influence brain development in a manner that is analogous to, but perhaps not equivalent in magnitude to, the permanent organizational effects exerted by estradiol in the perinatal rodent brain.

9 Why is It Important to Study Sexual Differentiation of the Brain?

Across the spectra of neurological and mental health disorders and diseases, an individual's gender is a major predictive factor of relative risk. Interestingly, the risk clusters according to the developmental timing of onset. Males are at considerably higher risk for disorders with very early onset, such as autism, dyslexia, stuttering, attention deficit, hyperactivity disorder, early onset of schizophrenia, and Tourette's syndrome. By contrast, females are at higher risk of postpubertal adult onset of illnesses such as major depressive disorder, general anxiety disorder, obsessive/compulsive disorder, and anorexia/bulimia. Many factors are likely to contribute to this difference, including cultural and environmental variables, but endemic differences in the brain are a critically important yet poorly understood contributor to the etiology of sex differences in mental health and disease (Crick and Zahn-Waxler, 2003). That architectural and neurochemical sex differences in the brain are permanently established early in development by the steroid hormone milieu, has been known for close to 50 years. The cellular mechanisms of how neuronal sex differences are established has been woefully underinvestigated, despite its clear utility for understanding basic brain development and potential as an entry point into the origins of mental illness and neurological disease.

10 Conclusions and Future Directions

Sex differences in brain and behavior will no doubt continue to fascinate the researcher and lay person alike. It is a question that impacts on medicine, education, the criminal justice system, and even religion. From the point of view of biologists, it is a lucrative entry point into a multitude of systems endpoints ranging from signal transduction to complex social behaviors. Given all this interest, do we know as much as we should? The phenomenon has been described for close to 50 years, and we do know a great deal about where and when sex differences in the brain arise, but as to how, we know very little. When phenomena are poorly understood at the mechanistic level, it is often because the subject is either exceedingly complex, or not of much interest. In the case of sexual differentiation, the endpoints are robust, the effects of hormones are reliable, and the outcomes are highly relevant, but in terms of complexity it probably trumps all others due to the convergence of biological (genetic and hormonal), social, environmental, and cultural influences in both the quantifiable outcomes and the interpretation of those outcomes. The National Institute of Mental Health, National Institute of Neurological Disorders and Diseases, and the National Institute of Drug Abuse have all identified sex differences in the brain as an underinvestigated topic and one targeted for increased research emphasis (http://www.nimh.nih.gov/strategic/strategicplanmenu.cfm). Gender is a major predictor of relative risk of a range of neurological disorders and diseases of mental health, response to injury, and predilection to drug addiction. Moreover, only by understanding how the brain differs between men and women can we also understand how it does not. The similarities are as important as the differences.

References

Allen LS, Hines M, Shryne JE, Gorski RA. 1989. Two sexually dimorphic cell groups in the human brain. J Neurosci 9: 497-506.

Amateau SK, McCarthy MM. 2002a. A novel mechanism of dendritic spine plasticity involving estradiol induction of prostglandin-E2. J Neurosci 22: 8586-8596.

Amateau SK, McCarthy MM. 2002b. Sexual differentiation of astrocyte morphology in the developing rat preoptic area. J Neuroendo 14: 904-910.

Amateau SK, McCarthy MM. 2004. Induction of PGE(2) by estradiol mediates developmental masculinization of sex behavior. Nat Neurosci 7: 643-650.

Amateau SK, Alt JJ, Stamps CL, McCarthy MM. 2004. Brain estradiol content in newborn rats: Sex differences, regional heterogeneity, and possible de novo synthesis by the female telencephalon. Endocrinology 145: 2906-2917.

Anderson RH, Fleming DE, Rhees RW, Kinghorn E. 1986. Relationships between sexual activity, plasma testosterone, and the volume of the sexually dimorphic nucleus of the preoptic area in prenatally stressed and non-stressed rats. Brain Res 370: 1-10.

Arai Y, Sekine Y, Murakami S. 1996. Estrogen and apoptosis in the developing sexually dimorphic preoptic area in female rats. Neurosci Res 25: 403-407.

Arnold AP, Xu J, Grisham W, Chen X, Kim YH, Itoh Y. 2004. Minireview: Sex chromosomes and brain sexual differentiation. Endocrinology 145: 1057-1062.

Auger AP, Tetel MJ, McCarthy MM. 2002. Steroid receptor coactivator-1 (SRC-1) mediates the development of sex-specific brain morphology and behavior. Proc Natl Acad Sci 97: 7551-7555.

Bakker J, Honda SNH, Balthazart J. 2002. The aromatase knock-out mouse provides new evidence that estradiol is required during development in the female for the expression of sociosexual behaviors in adulthood. J Neurosci 22: 9104-9112.

Bale TL, Picetti R, Contarino A, Koob GF, Vale WW. 2002. Mice deficient for both corticotrophin-releasing factor receptor 1 (CRFR1) and CRFR2 have an impaired stress response and display sexually dichotomous anxiety-like behavior. J Neurosci 22: 193-199.

Ball J. 1926. The female sex cycle as a factor in learning in the rat. Am J Physiol 78: 533-536.

Baum MJ, Brand T, Ooms MP, Vreeburg JT, Slob AK. 1988. Immediate postnatal rise in whole body androgen content in male rats: Correlation with increased testicular content and reduced body clearance of testosterone. Bio Reprod 38: 980-986.

Becker JB, Arnold AP, Berkley KJ, Blaustein JD, Eckel LA, et al. 2005. Strategies and methods for research on sex differences in brain and behavior. Endocrinology 146: 1650-1673.

Berenbaum SA. 1999. Effects of early androgens on sex-typed activities and interests in adolescents with congenital adrenal hyperplasia. Horm Behav 35: 102-110.

Bleier R, Byne W, Sigglekow I. 1982. Cytoarchitectonic sexual dimorphism of the medial preoptic area and anterior hypothalamic area in guinea pig, rat, hamster and mouse. J Comp Neurol 275: 725-730.

Breedlove SM. 1994. Sexual differentiation of the human nervous system. Annu Rev Psychol 45: 389-418.

Cooke BM, Breedlove SM, Jordan CL. 1999. A brain sexual dimorphism controlled by adult circulating androgens. PNAS 96: 7538-7540.

Coolen LM, Olivier B, Peters HJ, Veening JG. 1997. Demonstration of ejaculation-induced neural activity in the male rat brain using 5-HT1A agonist 8-OH-DPAT. Physiol Behav 62: 881-891.

Corbier P, Dehennin L, Castanier M, Mebaza A, Edwards DA, et al. 1990. Sex differences in serum luteinizing hormone and testosterone in the human neonate during the first few hours after birth. J Clin Endocrinol Metab 71: 1344-1348.

Corbier P, Edwards DA, Roffi J. 1992. The neonatal testosterone surge: A comparative study. Arch Int Physiol Biochem Biophys 100: 127-131.

Cordoba Montoya DA, Carrer HF. 1997. Estrogen facilitates induction of long term potentiation in the hippocampus of awake rats. Brain Res 778: 430-438.

Crick NR, Zahn-Waxler C. 2003. The development of psychopathology in females and males: Current progress and future challenges. Dev Psychopathol 15: 719-742.

Daniel JM, Dohanich GP. 2001. Acetylcholine mediates the estrogen-induced increase in NMDA receptor binding in CA1 of the hippocampus and the associated improvement in working memory. J Neurosci 21: 6949-6956.

Daniel JM, Roberts SL, Dohanich GP. 1999. Effects of ovarian hormones and environment on radial maze and water maze performance of female rats. Physiol Behav 66: 11-20.

Davis EC, Shryne JE, Gorski RA. 1995. A revised critical period for the sexual differentiation of the sexually dimorphic nucleus of the preoptic area in the rat. Neuroendocrinology 62: 579-585.

Davis EC, Popper P, Gorski RA. 1996. The role of apoptosis in sexual differentiation of the rat sexually dimorphic nucleus of the preoptic area. Brain Res 734: 10-18.

De Bellis MD, Keshavan MS, Beers SR, Hall J, Frustaci K, et al. 2001. Sex differences in brain maturation during childhood and adolescence. Cereb Cortex 11: 552-557.

de Kloet ER, Oitzl MS, Joels M. 1999. Stress and cognition: Are corticosteroids good or bad guys? TINS 22: 422-426.

Desmond NL, William B. 1998. Free postsynaptic densities in the hippocampus of the female rat. NeuroReport 9: 1975-1979.

Fitch RH, Berrebi AS, Cowell PE, Schrott LM, Denenberg VH. 1990. Corpus Callosum: Effects of neonatal hormones on sexual dimorphism in the rat. Brain Res 515: 111-116.

Foecking EM, Szabo M, Schawartz NB, Levine JE. 2005. Neuroendocrine consequences of prenatal androgen exposure in the female rat: Absence of luteinizing hormone surges, suppression of progesterone receptor gene expression, and acceleration of the gonadotropin-releasing hormone pulse generator. Biol Reprod epub.

Forger NG, Rosen GJ, Waters EM, Jacob D, Simerly RB, et al. 2004. Deletion of Bax eliminates sex differences in the mouse forebrain. PNAS 101: 13666-13671.

Fraser DD, Mudrick-Donnon LA, Mac Vicar BA. 1994. Astrocytic GABA receptors. Glia 11: 83-93.

Fugger HN, Foster, Thomas C, Gustafsson, Jan-Åke, Rissman, Emilie F. 2000. Novel effects of estradiol and estrogen receptor α and ß on cognitive function. Brain Res 883: 258-264.

Garcia-Segura LM, Chowen JA, Naftolin F. 1996. Endocrine glia: Roles of glial cells in the brain actions of steroid and thyroid hormones and in the regulation of hormone secretion. Front Neuroendocrinol 17: 180-211.

Garcia-Segura LM, Cardona-Gomez GP, Trejo JL, Fernandez-Galaz MC, Chowen JA. 2000. Glial cells are involved in organizational and activational effects of sex hormones in the brain. Sexual Differentiation of the Brain. Matsumoto A, editor. Boca Raton: CRC Press; pp. 83-93.

Giedd JN. 2004. Structural magnetic resonance imaging of the adolescent brain. Ann NY Acad Sci 1021: 77-85.

Gorski RA, Barraclough CA. 1963. Effects of low dosages of androgen on the differentiation of hypothalamic regulatory control of ovulation in the rat. Endocrinology 73: 210-216.

Gorski RA, Gordon JH, Shryne JE, Southam AM. 1978. Evidence for a morphological sex difference within the medial preoptic area of the rat brain. Brain Res 148: 333-346.

Gorski RA, Harlan RE, Jacobson CD, Shryne JE, Southam AM. 1980. Evidence for the existence of a sexually dimorphic nucleus in the preoptic area of the rat. J Comp Neurol 193: 529-539.

Granger DA, Schwartz EB, Booth A, Arentz M. 1999. Salivary testosterone determination in studies of child health and development. Horm Behav 35: 18-27.

Grattan DR, Rocca MS, Strauss KI, Sagrillo CA, Selmanoff M, et al. 1996. GABAergic neuronal activity and mRNA levels for both forms of glutamic acid decarboxylase (GAD65 and GAD67) are reduced in the diagonal band of Broca during the afternoon of proestrous. Brain Res 733: 46-55.

Gupta RR, Sen S, Diepenhorst LL, Rudick CN, Maren S. 2001. Estrogen modulates sexually dimorphic contextual fear conditioning and hippocampal long-term potentiation (LTP) in rats. Brain Res 888: 356-365.

Heeb MM, Yahr P. 1996. c-fos immunoreactivity in the sexually dimorphic area of the hypothalamus and related brain regions of male gerbils after exposure to sex-related stimuli or performance of specific sexual behaviors. Neuroscience 72: 1049-1071.

Hilton GD, Nunez JL, McCarthy MM. 2003. Sex differences in response to kainic acid and estradiol in the hippocampus of newborn rats. Neuroscience 117: 383-391.

Hines M. 2002. Sexual differentiation of human brain and behavior. Hormones, Brain and Behavior. Pfaff D, editor. London, UK: Academic; pp. 425-462.

Hines M. 2004. Psychosexual development in individuals who have female pseudohermaphroditism. Child Adolesc Psychiatr 13: 641-656.

Hojo Y, Hattori TA, Enami T, Furukawa A, Suzuki K, Ishii HT, Mukai H, Morrison JH, Janssen WG, Kominami S,

Harada N, Kimoto T, Kawato S. 2003. Adult male rat hippocampus synthesizes estradiol from pregnenolone by cytochromes P45017alpha and P450 aromatase localized in neurons. Proc Natl Sci USA 101(3): 865-870.

Honda S, Harada N, Ito S, Takagi Y, Maeda S. 1998. Disruption of sexual behavior in male aromatase-deficient mice lacking exons 1 and 2 of the cyp 19 gene. Biochem Biophys Res Commun 252: 445-449.

Horvath TL, Garcia-Segura LM, Naftolin F. 1997. Lack of gonadotrophin-positive feedback in the male rat is associated with lack of estrogen-induced synaptic plasticity in the arcuate nucleus. Neuroendocrinology 65: 136-140.

Houtsmuller EJ, Brand T, De Jonge FH, Joosten RN, van den Poll N, et al. 1994. SDN-POA volume, sexual behavior, and partner preference of male rats affected by perinatal treatment with ATD. Physiol Behav 56: 535-541.

Hubel DH. 1967. Eleventh Bowditch Lecture. Effects of distortion of sensory input on the visual system of kittens. Physiologist 10: 17-45.

Hull EM, Lorrain DS, Du J, Matuszewich L, Bitran D, et al. 1998. Organizational and activational effects of dopamine on male sexual behavior. Males, females and behavior: Toward biological understanding. Ellis L, Ebertz L, editors. Greenwood; pp. 79-96.

Hull EM, Lorrain DS, Du J, Matuszewich L, Lumley LA, et al. 1999. Hormone–neurotransmitter interactions in the control of sexual behavior. Behav Brain Res 105: 105-116.

Hutton LA, Gu G, Simerly RB. 1998. Development of a sexually dimorphic projection from the bed nuclei of the stria terminalis to the anteroventral periventricular nucleus in the rat. J Neurosci 18: 3003-3013.

Ibanez MA, Gu G, Simerly RB. 2001. Target-dependent sexual differentiation of a limbic–hypothalamic neural pathway. J Neurosci 21: 5652-5659.

Iijima M, Arisaka O, Minamoto F, Arai Y. 2001. Sex differences in children's free drawings: A study on girls with congenital adrenal hyperplasia. Horm Behav 40: 99-104.

Isgor C, Sengelaub DR. 1998. Prenatal gonadal steroids affect adult spatial behavior, CA1 and CA3 pyramidal cell morphology in rats. Horm Behav 34: 183-198.

Ito K, Skinkle KL, Hicks TP. 1999. Age-dependent, steroid-specific effects of oestrogen on long-term potentiation in rat hippocampal slices. J Physiol (Lond) 515: 209-220.

Jacobson CD, Gorski RA. 1981. Neurogenesis of the sexually dimorphic nucleus of the preoptic area in the rat. J Comp Neurol 196: 519-529.

Kameyama K. 1991. Sensitive phases for teratogen-induced developmental defects in the brain. J Toxicol Sci 16: 25-35.

Kampen DL, Sherwin BB. 1994. Estrogen use and verbal memory in healthy postmenopausal women. Obstet Gynecol 83: 979-983.

King JC, Rubin BS. 1994. Dynamic changes in LHRH neurovascular terminals with various endocrine conditions in adults. Horm Behav 28: 349-356.

Kornack DR, Lu B, Black IB. 1991. Sexually dimorphic expression of the NGF receptor gene in the developing rat brain. Brain Res 542: 171-174.

Kudo K, Qiao CX, Kanba S, Arita J. 2004. A selective increase in phosphorylation of cyclic AMP response element-binding protein in hippocampal CA1 region of male, but not female, rats following contextual fear and passive avoidance conditioning. Brain Res 1024: 233-243.

Le WW, Berghorn KA, Rassnick S, Hoffman GE. 1999. Periventricular preoptic area neurons coactivated with luteinizing hormone (LH)-releasing hormone (LHRH) neurons at the time of the LH surge are LHRH afferents. Endocrinology 140: 510-519.

LeVay S. 1991. A difference in hypothalamic structure between heterosexual and homosexual men. Science 9: 497-506.

Lewis MH. 2004. Environmental complexity and central nervous system development and function. Ment Retard Dev Disabil Res Rev 10: 91-95.

Li C, Brake WG, Romeo RD, Dunlop JC, Gordon M, et al. 2004. Estrogen alters hippocampal dendritic spine share and enhances synaptic protein immunoreactivity and spatial memory in female mice. Proc Natl Acad Sci 101: 2185-2190.

Luine VN, Jacome LF, Maclusky NJ. 2003. Rapid enhancement of visual and place memory by estrogens in rats. Endocrinology 144: 2836-2844.

Maclusky NJ, Walters MJ, Clark AS, Toran-Allerand CD. 1994. Aromatase in the cerebral cortex, hippocampus, and midbrain: Ontogeny and developmental implications. Mol Cell Neurosci 5: 691-698.

Madeira MD, Lieberman AR. 1995. Sexual dimorphism in the mammalian limbic system. Prog Neurobiol 45: 275-333.

Markus EJ, Zecevic M. 1997. Sex differences and estrous cycle changes in hippocampus-dependent fear conditioning. Psychobiology 25: 246-252.

Martin S, Jones M, Simpson EM, VDB. 2003. Impaired spatial reference memory in aromatase deficient (ArKO) mice. Neuroreport 14: 1979-1982.

Matsumoto A, Arai Y. 1980. Sexual dimorphism in "wiring pattern" in the hypothalamic arcuate nucleus and its modification by neonatal hormonal environment. Brain Res 19: 238-242.

Matsutani S, Yamamoto N. 1997. Neuronal regulation of astrocyte morphology in vitro is mediated by GABAergic signaling. Glia 20: 1-9.

McCarthy MM, Schlenker EH, Pfaff DW. 1993. Enduring consequences of neonatal treatment with antisense oligodeoxynucleotides to estrogen receptor mRNA on sexual differentiation of rat brain. Endocrinology 133: 433-439.

McCarthy MM, Amateau SK, Mong JA. 2002. Steroid modulation of astrocytes in the neonatal brain: Implications for adult reproductive function. Bio Reprod 67: 691-698.

McCormick CM, Furey BF, Child M, Sawyer MJ, Donohue SM. 1998. Neonatal sex hormones have "organization" effects on the hypothalamic–pituitary–adrenal axis of male rats. Dev Brain Res 105: 295-307.

Mong JA, McCarthy MM. 1999. Steroid-induced developmental plasticity in hypothalamic astrocytes: Implications for synaptic patterning. J Neurobiol 40: 602-619.

Mong JA, Glaser E, McCarthy MM. 1999. Gonadal steroids promote glial differentiation and alter neuronal morphology in the developing hypothalamus in a regionally specific manner. J Neurosci 19: 1464-1472.

Mong JA, Nunez JL, McCarthy MM. 2002. GABA mediates steroid-induced astrocyte differentiation in the neonatal rat hypothalamus. J Neuroendocrinol 14: 1-16.

Mong JA, Kurzweil RL, Davis AM, Rocca MS, McCarthy MM. 1996. Evidence for sexual differentiation of glia in rat brain. Horm Behav 30: 553-562.

Morreale de Escobar G. 2001. The role of thyroid hormone in fetal neurodevelopment. J Pediatr Endocrinol Metab 14: 1453-1462.

Murakami S, Arai Y. 1989. Neuronal death in the developing sexually dimorphic periventricular nucleus of the preoptic area in the female rat: Effect of neonatal androgen treatment. Neurosci Lett 102: 185-190.

Nopoulos P, Flaum M, Andreasen NC. 1997. Sex differences in brain morphology in Schizophrenia. Am J Psychiatry 154: 1648-1654.

Nottebohm F, Arnold AP. 1976. Sexual dimorphism in vocal control areas of the songbird brain. Science 194: 211-213.

Nunez JL, Koss WA, Juraska JM. 2000. Hippocampal anatomy and water maze performance are affected by neonatal cryoanesthesia in rats of both sexes. Horm Behav 37: 169-178.

Nunez JL, Alt J, McCarthy MM. 2003. A new model for prenatal brain damage: I. GABAA receptor activation induces cell death in developing rat hippocampus. Exp Neurol 181: 258-269.

Olmos G, Naftolin F, Peres J, Tranque PA, Garcia-Segura LM. 1989. Synaptic remodeling in the rat arcuate nucleus during the estrous cycle. Neuroscience 32: 663-667.

Ottem EN, Godwon JG, Krishnan S, Petersen SL. 2004. Dual-phenotype GABA/glutamate neurons in adult preoptic area: Sexual dimorphism and function. J Neurosci 24: 8097-8105.

Parducz A, Perez J, Garcia-Segura LM. 1993. Estradiol induces plasticity of GABAergic synapses in the hypothalamus. Neuroscience 53: 395-401.

Perrot-Sinal TS. 1996. Sex differences in performance in the Morris water maze and the effects of initial nonstationary hidden platform training. Behav Neurosci 110: 1309-1320.

Pfaff DW. 1966. Morphological changes in the brains of adult male rats after neonatal castration. J Endocrinol 36: 415-416.

Phoenix CH, Goy RW, Gerall AA, Young WC. 1959. Organizing action of prenatally administered testosterone propionate on the tissues mediating mating behavior in the female guinea pig. Endocrinology 65: 369-382.

Plant TM. 2001. Neurobiological bases underlying the control of onset of puberty in the rhesus monkey: A representative higher primate. Front Neuroendocrinol 22: 107-139.

Polston EK, Gu G, Simerly RB. 2004. Neurons in the principal nucleus of the bed nuclei of the stria terminalis provide a sexually dimorphic GABAergic input to the anteroventral periventricular nucleus of the hypothalamus. Neuroscience 123: 793-803.

Prange-Kiel J, Wehrenberg U, Jarry H, Rune GM. 2003. Para/autocrine regulation of estrogen receptors in hippocampal neurons. Hippocampus 13: 226-234.

Quadros PS, Goldstein AY, De Vries GJ, Wagner CK. 2002. Regulation of sex differences in progesterone receptor expression in the medial preoptic nucleus of postnatal rats. J Neuroendocrinol 14: 761-767.

Raisman G, Field PM. 1971. Sexual dimorphism in the preoptic area of the rat. Science 173: 731-733.

Raisman G, Field PM. 1973. Sexual dimorphism in the neuropil of the preoptic area of the rat and its dependence on neonatal androgen. Brain Res 54: 1-29.

Rhees RW, Shryne JE, Gorski RA. 1990. Termination of the hormone-sensitive period for differentiation of the sexually dimorphic nucleus of the preoptic area in male and female rats. Dev Brain Res 52: 17-23.

Rhoda J, Corbier P. Roffi J. 1984. Gonadal steroid concentrations in serum and hypothalamus of the rat at birth: Aromatization of testosterone to 17β-estradiol. Endocrinology 114: 1754-1760.

Ricceri L, Ewusi A, Calamandrei G, Berger-Sweeney J. 1997. Sexually dimorphic effects of anti-NGF treatment in neonatal rats. Dev Brain Res 101: 273-276.

Roof RL. 1993. Neonatal exogenous testosterone modifies sex difference in radial arm and Morris water maze performance in prepubescent and adult rats. Behav Brain Res 53: 1-10.

Roselli CE, Resko JA, Stormshak F. 2002. Hormonal influences on sexual partner preference in rams. Arch Sex Behav 31: 43-49.

Rudick CN, Woolley C. 2003. Selective estrogen receptor modulators regulate phasic activation of hippocampal CA1 pyramidal cells by estrogen. Endocrinology 144: 179-187.

Sagrillo CA, Selmanoff M. 1997. Castration decreases single cell levels of mRNA encoding glutamic acid decarboxylase in the rostral hypothalamus. J Neuroendocrinol 9: 699-706.

Sandstrom N, Williams CL. 2001. Memory retention is modulated by acute estradiol and progesterone replacement. Behav Neurosci 115: 384-393.

Sanzgiri RP, Araque A, Haydon PG. 1999. Prostaglandin E(2) stimulates glutamate receptor-dependent astrocyte neuromodulation in cultured hippocampal cells. J Neurobiol 41(2): 221-229.

Shaftel SS, Olschowka JA, Hurley SD, Moore AH, O'Banion Mk. 2003. COX-3: A splice variant of cyclooxygenase-1 in mouse neural tissue cells. Mol Brain Res 119: 213-215.

Shaywitz BA, Shaywitz SE, Pugh KR, Constable RT, Sckudlarski P, et al. 1995. Sex differences in the functional organization of the brain for language. Nature 373: 607-608.

Sherwin BB. 1997. Estrogen effects on cognition in menopausal women. Neurology 48: S21-S26.

Shors TJ, Miesegaes G. 2002. Testosterone in utero and at birth dictates how stressful experience will affect learning in adulthood. Proc Natl Acad Sci 99: 13955-13960.

Shors TJ, Chua C, Falduto J. 2001. Sex differences and opposite effects of stress on dendritic spine density in the male versus female hippocampus. J Neurosci 21: 6292-6297.

Shughrue PJ, Dorsa DM. 1994. The ontogeny of GAP-43 (Neuromodulin) mRNA in postnatal rat brain: Evidence for a sex dimorphism. J Comp Neurol 340: 174-184.

Sickel MJ, McCarthy MM. 2000. Calbindin D28-K immunoreactivity is a marker for a subdivision of the sexual dimorphic nucleus of the preoptic area in the rat: Developmental profile and gonadal steroid modulation. J Neurobiol 12: 397-402.

Simerly RB. 1998. Organization and regulation of sexually dimorphic neuroendocrine pathways. Behav Brain Res 92: 195-203.

Simerly RB. 2000. Development of sexually dimorphic forebrain pathways. Sexual differentiation of the brain. Matsumoto A, editor. Boca Raton, FL: CRC Press; pp. 175-202.

Simerly RB. 2002. Wired for reproduction: Organization and development of sexually dimorphic circuits in the mammalian forebrain. Annu Rev Neurosci 25: 507-536.

Sisk CL, Foster DL. 2004. The neural basis of puberty and adolescence. Nat Neurosci 7: 1040-1047.

Suzuki M, Nishiahara M. 2002. Granulin precursor gene: A sex steroid-inducible gene involved in sexual differentiation of the rat brain. Mol Genet Metab 75: 31-37.

Suzuki M, Bannai M, Matsumuro M, Furuhata Y, Ikemura R, et al. 2000. Suppression of copulatory behavior by intracerebroventricular infusion of antisense oligodeoxynucleotide of granulin in neonatal male rats. Physiol Behav 68: 707-713.

Swaab DF, Fliers E. 1985. A sexually dimorphic nucleus in the human brain. Science 228: 1112-1115.

Tin-Tin-Win-Shwe D, Mitsushima D, Shinohara K, Kimura F. 2004. Sexual dimorphism of GABA release in the medial preoptic area and luteinizing hormone release in gonadectomized estrogen-primed rats. Neuroscience 127: 243-250.

Van de Kar L, Blair ML. 1999. Forebrain pathways mediating stress-induced hormone secretion. Front Neuroendocrinol 20: 1-48.

Viau V, Meaney MJ. 1996. The inhibitory effect of testosterone on hypothalamic–pituitary–adrenal responses to stress is mediated by the medial preoptic area. J Neurosci 16: 1866-1876.

Vreeburg JT, van der Vaart PD, van der Shoot P. 1977. Prevention of central defeminization but not masculinization in male rats by inhibition neonatally of oestrogen biosynthesis. J Endocrinol 74: 375-382.

Wallen K, Baum MJ. 2002. Masculinization and defeminization in altricial and precocial mammals: Comparative aspects of steroid hormone action. Hormones brain and behavior. Pfaff D, editor. London, UK: Academic; pp. 385-424.

Warren SG, Humphreys AG, Juraska JM, Greenough WT. 1995. LTP varies across the estrous cycle: Enhanced synaptic plasticity in proestrus rats. Brain Res 703: 26-30.

Whalen R, Edwards D. 1967. Hormonal determinants of the development of masculine and feminine behavior in male and female rats. Anat Rec 157: 173-180.

Williams CL. 2002. Hormones and cognition in nonhuman animals. Behavioral Endocrinology, 2nd Edition. al. JBBe, editor. Cambridge MA: MIT Press; pp. 527-578.

Williams CL, Meck WH. 1991. The organizational effects of gonadal steroids on sexually dimorphic spatial ability. Psychoneuroendocrinology 16: 155-176.

Williams CL, Barnett AM, Meck WH. 1990. Organizational effects of early gonadal secretions on sexual differentiation in spatial memory. Behav Neurosci 104: 84-97.

Wimer RE, Wimer C. 1985. Three sex dimorphims in the granule cell layer of the hippocampus in house mice. Brain Res 328: 105-109.

Wong M-L, al. e. 2000. Pronounced and sustained central hypernoradrenergic function in major depression with melancholic features: Relation to hypercortisolism and corticotropin-releasing hormone. Proc Natl Acad Sci 97: 325-330.

Wood RI. 1997. Thinking about networks in the control of male hamster sexual behavior. Horm Behav 32: 40-45.

Woodson JC, Gorski RA. 2000. Structural sex differences in the mammalian brain: Reconsidering the male/female dichotomy. Sexual differentiation of the brain. Matsumoto A, editor. Boca Raton, FL: CRC Press; pp. 229-255.

Woolley CS. 1999. Effects of estrogen in the CNS. Cur Op Neurobiol 9: 349-354.

Woolley CS, Weiland NG, McEwen BS, Schwartzkroin PA. 1997. Estradiol increases the sensitivity of hippocampal CA1 pyramidal cells to NMDA receptor-mediated synaptic input: Correlation with dendritic spine density. J Neurosci 17: 1848-1859.

Zamani MR, Desmond NL, Levy WB. 2000. Estradiol modulates long-term synaptic depression in female rat hippocampus. J Neurophysiol 84: 1800-1808.

12 Sex Differences in Neurotransmitters Systems; Vasopressin as an Example

G. J. de Vries

1 Introduction

Current research on sex differences in the brain stems from two relatively independent lines of research. One focused on sex differences in gonadotropic hormone release, the other on hormonal control of sexual behavior. The first line traces back to Harris and Jacobsohn's demonstration that pituitary transplants placed directly under the hypothalamus are able to drive ovarian cycles in female rats irrespective of whether they came from male or female donors. This dispelled an earlier claim that the pituitary was responsible for sex differences in gonadotropic hormone release and pointed at the brain instead (Harris and Jacobsohn, 1952). Barraclough and his collaborators subsequently showed that injecting testosterone propionate shortly after birth renders female rats permanently infertile (Barraclough and Leathem, 1954; Barraclough, 1961). After showing that the anterior preoptic area was responsible for cyclicity in gonadotropic hormone release, Barraclough and Gorski (1961) suggested that in the presence of androgens the anterior preoptic loses its ability to sustain this cyclicity. Around the same time, Phoenix and colleagues (1959) showed that testosterone propionate injections into pregnant guinea pigs increased the display of male sexual behavior in female offspring while it had the opposite effects on female sexual behavior. This work led to the realization that gonadal hormones influence sexually dimorphic behaviors in two fundamentally different ways. Early in life, they direct the development of neural circuitry that generates male- or female-typical functions and behaviors in adulthood. These developmental effects are permanent and called organizational. For example, testosterone exposure during development increases the likelihood that animals will show male sexual behavior as adults. However, to show male sexual behavior, animals have to be exposed to testosterone in adulthood as well. This adult effect is transient and therefore called activational (Becker et al., 2005).

Phoenix and colleagues did not expect that the developmental effects would cause as dramatic changes in morphology as those observed in reproductive organs. They suggested that early androgen exposure causes "a more subtle change reflected in function rather than in visible structure." The first reports of sex differences hinted that differences were indeed subtle. For example, in 1960, Kato detected higher serotonin levels in female than in male rat brains. A little while later, Pfaff (1966) showed that neonatal castration of rats permanently changed the size of nucleoli in hypothalamic neurons. In 1970, McEwen and his colleagues showed that neonatal steroid treatment changed testosterone and estradiol uptake in rat brains (McEwen and Pfaff, 1970; McEwen et al., 1970). And one year after that, Raisman and Field (1971) reported the first sex difference in neural connectivity: male rats showed more synapses from non-strial origin on dendritic shafts and fewer on dendritic spines than did females. Although the functional significance of this finding is still completely unclear, it was a milestone because it showed that neonatal manipulations of gonadal hormone levels could reverse morphological sex differences levels (Raisman and Field, 1971, 1973). This mapped very well onto the ideas of Phoenix et al. (1959) on the organizational effects of gonadal hormones.

After these admittedly subtle findings, Nottebohm and Arnold (1976) reported some very conspicuous differences in the size of song control nuclei in the brain of zebra finches and canaries. These differences reinforced the notion that studying sex differences in brain structure could narrow the gap between structure and function, because the correlation between structure and function was very striking in birds. Males, which sing, had much larger song control nuclei than did females, which do not sing. Soon thereafter, equally impressive differences were found in gross anatomy of mammalian brains. Arguably, the most famous of these was found in the sexually dimorphic nucleus of the medial preoptic area (SDN) in the rat, which was discovered by Gorski and his colleagues (1978). This nucleus is five times larger in males than in females, which is remarkable considering that many researchers must have overlooked this difference, which is visible to the naked eye in Nissl-stained sections.

Currently, hundreds if not thousands of sex differences in neural structure have been found almost anywhere in the brain (see for example, Juraska, 1991; Segovia and Guillamon, 1993; Kawata, 1995; Madeira and Lieberman, 1995; Tobet and Hanna, 1997; Cooke et al., 1998; Breedlove et al., 1999; Segovia et al., 1999; Morris et al., 2004). There are also a number of excellent reviews on development and function of sex differences in other neurotransmitter systems (see for example, Fink, 1998; McCarthy et al., 2002, 2005).

This review describes how focusing on the neurotransmitter components of sexually dimorphic structures helps in studying the development and function of sex differences in the brain.

2 Function of Sex Differences in the Brain

Studying sex differences in the brain holds a number of promises. First, studying sex differences in hormonally manipulated animals offers a unique perspective for studying how neural systems develop to generate sex-specific behaviors. Second, sex differences allow one to study how differences in brain structure translate into differences in function. There are sound clinical reasons for pursuing these questions as solving them may explain why so many behavioral and neurological disorders show marked sex differences in incidence and severity (Swaab et al., 2003). Nature, however, has been reluctant to deliver on these promises. Although for many sex differences, we know which hormones control sexual differentiation and when they do that, we barely know how. Only a smattering of molecular and cellular processes that mediate hormonal effects on sexual differentiation has been identified (see e.g., Auger et al., 2000; Amateau and McCarthy, 2004; Forger et al., 2004; Forger, 2006). In addition, except for a handful of sexually dimorphic cell groups, most of which control specific sexually dimorphic muscle systems (Kelley, 1988; Breedlove, 1992), the functional significance of practically all sex differences in the central nervous system is unknown (De Vries and Boyle, 1998). The complexity of the connections of the areas where these differences are found and the technical difficulties of manipulating specifically the sexually dimorphic elements in these areas form significant roadblocks.

One way around these roadblocks may be to focus on the neurotransmitter systems that make up these areas. That approach may reveal whether sexual differentiation selectively affects particular cell systems, which will help in revealing the cellular processes underlying differentiation. Focusing on neurotransmitter systems also helps in tracing the anatomical connections of sexually dimorphic areas, and therefore in assessing the impact of a particular dimorphism on other brain areas. Finally, knowing which neurotransmitter systems are involved will allow manipulations such as injecting receptor agonists and antagonists to test the function of specific sexually dimorphic components (De Vries, 1990). This chapter illustrates how focusing on neurotransmitter systems throws light on the function of sex differences in the brain by focusing on the sexually dimorphic vasopressin projections of the bed nucleus of the stria terminalis (BST) and the medial amygdaloid nucleus (MeA). This is one of the most consistently found sex differences among vertebrates and probably one of the best understood (De Vries and Panzica, 2006).

3 Sex Differences in Vasopressin Projections in the Brain

3.1 Bed Nucleus of the Stria Terminalis and Amygdala

The BST and MEA are telencephalic structures that show a high degree of sexual dimorphism and that have been implicated in the regulation of autonomic as well as reproductive functions, for example, gonadotropin release and male sexual behavior (Harris and Sachs, 1975; Emery and Sachs, 1976; Beltramino and Taleisnik, 1978, 1980, 1985; Valcourt and Sachs, 1979). In rats as well as mice, subregions of the BST, such as its principal nucleus, are bigger and contain more neurons in males than in females (Del Abril et al., 1987; Guillamon et al., 1988; Hines et al., 1992; Forger et al., 2004). The MeA shows similar differences (Mizukami et al., 1983; Hines et al., 1992; Cooke et al., 1999). There are also differences in the number of synapses in the MeA (Nishizuka and Arai, 1983) and in specific neurotransmitters other than vasopressin. For example, the posterodorsal area of the MeA and the principal nucleus of the BST contain more cholecystokinin-immunoreactive cells and receive denser substance P-immunoreactive innervation in males than in females (Malsbury and McKay, 1987, 1989, 1994; Micevych et al., 1988a, b).

In rats, the MeA and BST share similar patterns of projections to the hypothalamus (Simerly, 1990; Canteras et al., 1995; Gu et al., 2001). The principal nucleus of the BST, which shows the most striking sex differences in volume of all subregions of the BST (Hines et al., 1992; Forger et al., 2004), sends dense projections to nuclei in the periventricular zone of the hypothalamus, especially the anteroventral periventricular nucleus, preoptic periventricular nucleus, and the arcuate nucleus of the hypothalamus, each of which regulates hormonal secretion from the anterior pituitary (Simerly, 1995). However, the MeA and BST also innervate the medial preoptic and ventromedial nuclei of the hypothalamus, which are more commonly associated with behavior. Because the MeA and BST relay multimodal sensory information to the hypothalamus (Alheid et al., 1995), Simerly has proposed that the larger size of subnuclei of MeA and BST in males suggest that the degree of convergence of sensory information from these limbic regions onto neurons in the hypothalamus may differ as well (Simerly, 2002). The vasopressin projections from the MeA and BST, which are the focus of this chapter, project primarily to areas outside the hypothalamus that are much less intimately linked to reproductive functions (De Vries et al., 1985).

3.2 Nature of Sex Difference

We found the sex differences in these vasopressin projections by chance, before any other sex difference had been detected in a specific neurotransmitter system by anatomical means, and before we knew that these projections came from the BST and MeA. We noted an enormous variability in the density of vasopressin innervation of the lateral septum and lateral habenular nucleus in rats of twelve days and older while we were studying the development of what we thought were the projections of the suprachiasmatic nucleus. Separating subjects by sex revealed that vasopressin-immunoreactive (-ir) projections are much denser in males than in females from the twelfth postnatal day onwards (De Vries et al., 1981). At the time we found this sex difference, the suprachiasmatic and paraventricular nuclei were seen as the most important sources of vasopressin innervation of the brain (De Vries and Buijs, 1983). There are, however, several other areas outside the hypothalamus that contain vasopressin-expressing cells that do not stain as readily for vasopressin as do cells in the suprachiasmatic and paraventricular nucleus. The biggest cohorts of these cells were discovered in the BST and the MeA after pretreating animals with the axonal transport blocker colchicine (Caffe and Van Leeuwen, 1983; Van Leeuwen and Caffe, 1983). By mapping how lesions of the different possible sources of vasopressin innervation affected vasopressin fiber staining throughout the brain and by retrograde tracing studies that focused on the lateral septum, we showed that the BST and MeA provide a large share of the central vasopressin projections, especially to limbic structures such as the lateral septum, the lateral habenular nucleus, and several midbrain structures such as midbrain central gray, dorsal raphe nucleus, pontine peripeduncular nucleus and the locus coeruleus (De Vries and Buijs, 1983; Caffe et al., 1987). The sex differences in, and steroid responsiveness of, vasopressin fiber staining matched our anatomical studies remarkably well and therefore strengthened our conclusions: vasopressin cells and projections from the paraventricular and suprachiasmatic nucleus did not show obvious sex differences and steroid effects whereas those derived from the BST and MeA did (De Vries et al., 1985; De Vries and Al-Shamma, 1990) (➋ *Figure 12-1*).

Corresponding to the sex difference in the density of vasopressin innervation, the BST of male rats has about two to three times more vasopressin-ir cells (Van Leeuwen et al., 1985; De Vries and Al-Shamma, 1990; Wang et al., 1993) and cells that can be labeled for vasopressin mRNA than does the female BST (Miller et al., 1989b; De Vries et al., 1994). Although a similar trend was found in the number of vasopressin-ir cells in the MeA in two studies (De Vries and Al Shamma, 1990; Wang et al., 1993), these differences were not significant, presumably because vasopressin staining of MeA cells was more variable than that of BST cells. A birth-date labeling study that used large numbers of male and female rats did find significantly more vasopressin-ir cells in the MeA, however (Al Shamma and De Vries, 1996). In addition, in situ hybridization studies consistently labeled two times more MeA cells for vasopressin mRNA in males than in females (Szot and Dorsa, 1993; Wang and De Vries, 1995). Although vasopressin-ir cells are present in the posterodorsal area of the MeA and the principal nucleus of the BST, the subregions with the most

Figure 12-1
Sex differences in vasopressin innervation. Dark-field microphotographs of vasopressin-immunoreactive fibers in a male- (*right panels*) and female rat (*left panels*) in the lateral septum (*LS; top panels*), which receives its innervation from the BST and MeA, and the periventricular nucleus of the hypothalamus (*Pe; bottom panels*), which receives suprachiasmatic nucleus innervation. Note that fiber density is sexually dimorphic in the LS (*arrows*) but not in the Pe; * lateral ventricle in *top panels* and third ventricle in *bottom panel*

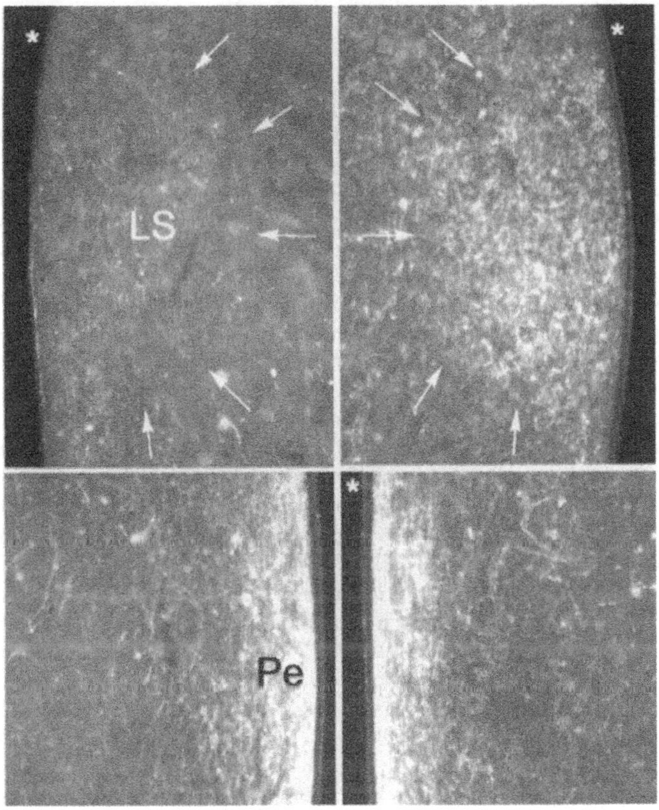

conspicuous size differences (Hines et al., 1992; Cooke et al., 1999; Forger et al., 2004), in the rat BST, most vasopressin cells are found lateral and ventral to these areas (Caffé and Van Leeuwen, 1983; Van Leeuwen and Caffé, 1983).

Although global sex differences in cell number and density of the projections have not been found in other vasopressin-ir projections, some of them are partly sexually dimorphic. For example, male gerbils have much denser vasopressin-ir projections to the sexually dimorphic area of the preoptic/anterior hypothalamic area (SDA) than do females, but projections to the periventricular nucleus of the hypothalamus do not differ (Crenshaw et al., 1992); both projections most likely come from the same source, the suprachiasmatic nucleus (Crenshaw and De Vries, 1991). The global differences in BST and MeA projections may be related to differences in the number of vasopressin-ir cells in the BST and MeA, whereas the partial sex differences in suprachiasmatic projections may be related to sex differences in the targets they innervate; in this case, to a presence of a sex difference in the SDA, an area homologous to the medial preoptic nucleus in rats (De Vries et al., 1988), and the absence of a similar difference in the periventricular nucleus (Commins and Yahr, 1984).

3.3 Causes of Sex Difference

Sex differences in vasopressin innervation of the brain depend on organizational and activational effects of gonadal hormones and possibly also on the compositions of the sex chromosomes, independently of gonadal hormones.

3.3.1 Activational Effects

3.3.1.1 Effects of Gonadectomy In rats, the vasopressin projections of the BST and MeA are exquisitely responsive to gonadal steroids. Gonadectomy eliminates vasopressin mRNA and peptide expression in BST and MeA cells, but treatment with gonadal steroids reverses these changes (De Vries et al., 1984, 1985; Van Leeuwen et al., 1985; Miller et al., 1989a). These changes are relatively slow. In males, vasopressin immunoreactivity disappears gradually from BST and MeA projections in about two to three months (De Vries et al., 1984); in mice, an equally slow decline was found (Mayes et al., 1988). Biosynthesis of vasopressin probably declines much faster because already one day after castration, vasopressin mRNA levels are significantly lowered, while 1 week later only one tenth of labeled cells remains (Miller et al., 1992). Vasopressin peptide apparently hangs around in terminals for several weeks after vasopressin biosynthesis has been shut down. This discrepancy in the rate of decline of vasopressin peptide and message suggests that castration acutely reduces vasopressin release. The lingering peptide levels suggest that BST projections maintain the capacity to influence brain functions and behavior by releasing vasopressin.

Although no studies have addressed directly whether gonadectomy reduces vasopressin release in BST and MeA projection, indirect evidence suggests that it does. If rats are primed with an intracerebroventricular (icv) vasopressin injection, a second vasopressin injection given two days later will cause motor disturbances (Poulin and Pittman, 1991). Injections of hypertonic saline will likewise sensitize rats to the motor effects of icv vasopressin injections, presumably because hypertonic saline boosts septal vasopressin release (Demotes-Mainard et al., 1986; Landgraf et al., 1988). Once rats have been castrated, icv vasopressin injections still sensitize rats to the motor effects of subsequent vasopressin injections, but hypertonic saline does not, suggesting that castration has blocked septal vasopressin release (Poulin and Pittman, 1991).

Although the sex differences in, and hormonal effects upon, vasopressin-ir projections likely influence vasopressin release in rats, they do not change receptor density or sensitivity. Long-term castration does not affect vasopressin binding in rats (Tribollet et al., 1988), nor does it affect number and affinity of vasopressin receptors, vasopressin-stimulated phospho-inositol hydrolysis in septal tissue, or the ability of vasopressin to sensitize septal tissue to the motor effects of a subsequent vasopressin injection (Poulin and Pittman, 1991). This is unexpected, because changes in other peptide systems typically change receptor function (Catt et al., 1979). In BST and MeA projections, however, steroids appear to influence vasopressin function at the level of production and release rather than at the receptive site.

3.3.1.2 Estrogen Versus Androgen Effects Testosterone influences vasopressin production by androgen as well as by estrogen receptor-mediated mechanisms. In castrated male rats, estradiol—an estrogenic metabolite of testosterone (Naftolin et al., 1975)–partially restores vasopressin immunostaining in castrated male rats, while dihydrotestosterone--a non-aromatizable, androgenic metabolite of testosterone (Lieberburg and McEwen, 1975)–cannot by itself restore vasopressin immunostaining. However, in combination with estradiol, dihydrotestosterone enhances vasopressin immunostaining (De Vries et al., 1986). These same steroids influence vasopressin mRNA production in much the same way (De Vries et al., 1994; Wang and De Vries, 1995). Estrogen receptors are also more important for vasopressin expression in mice, as deletions of the aromatase and estrogen receptor alpha in knockout strains of mice dramatically reduce vasopressin expression in this system (Plumari et al., 2002; Scordalakes and Rissman, 2004). Since virtually all vasopressin cells in the BST and MeA in rats are immunoreactive for estrogen as well as androgen receptors (Axelson and Van Leeuwen, 1990; Zhou et al., 1994), androgen and estrogen may influence vasopressin production by directly acting on vasopressin-ir cells.

Because there are no apparent androgen- and estrogen-responsive elements in the promoter region of the vasopressin gene (Young, 1992; Adan and Burbach, 1992), it is not clear whether androgens and estrogens influence the transcription of vasopressin messenger RNA by acting directly on the promoter. Experiments with luciferase reporter cell systems suggest that there is an estrogen response element about 4 kb upstream from the vasopressin gene, which may confer estrogen sensitivity to vasopressin gene expression in vivo (Shapiro et al., 2000). There is some doubt, however, whether steroids influence vasopressin mRNA transcription at all, since a nuclear run-on assay of BST tissue did not show any effect of castration on vasopressin gene transcription. Since the same study showed that castration decreased, while testosterone increased the length of the poly (A) tail of vasopressin mRNA–which may enhance its stability–, gonadal hormones may influence vasopressin mRNA levels at a posttranscriptional level (Carter and Murphy, 1993).

Androgens and estrogens may influence the vasopressin production indirectly, by influencing each other's effectiveness. Estrogens might, for example, increase the responsiveness of individual BST and MeA cells to dihydrotestosterone treatment by preventing the metabolic inactivation of dihydrotestosterone in the brain (cf. Södersten, 1980). Estrogens might also increase the effectiveness of androgen receptors of vasopressin-producing cells by altering the duration of androgen receptor occupation, as has been observed in the preoptic area of male rats (Roselli and Fasasi, 1992).

3.3.1.3 Role of Circulating Hormones Given the hormone responsiveness of BST and MeA projections, sex differences in circulating gonadal hormones can certainly contribute to differences in vasopressin expression. Although females have relatively high levels of estradiol, the level at which they activate androgen receptors is probably lower. In addition, their higher progesterone levels may further depress vasopressin expression because, in male rats, progesterone inhibits vasopressin expression (Auger and Vanzo, 2006). This may be a direct effect, because practically all vasopressin neurons in the BST and MeA express progesterone receptor (Auger and De Vries, 2002). It is not known, however, whether progesterone has the same effects in females as it does in males.

Sex differences in gonadal hormone levels cannot explain all sex differences in vasopressin cells and projections that are observed in intact animals (De Vries et al., 1981; Miller et al., 1989b). Gonadectomized male and female rats that are treated with similar levels of testosterone still show differences in cell number and density of the projections (De Vries and Al-Shamma, 1990; Wang et al., 1993; De Vries et al., 1994). Circulating hormones may, however, be the main factor in species such as prairie voles, where differences in vasopressin-ir fiber density (Bamshad et al., 1993) or vasopressin mRNA expression (Wang et al., 1994a) are extreme. These differences are still impressive, but much smaller if adult male and female voles are treated with similar levels of testosterone (Lonstein et al., 2005).

Differences in testosterone metabolism can contribute to such residual sex differences in testosterone-treated rats because the male BST has higher levels of aromatase, the enzyme that catalyzes the aromatization of testosterone into estradiol, than does the female BST (Roselli, 1991). Testosterone may, therefore, stimulate BST cells more effectively in male than in female rats. However, a sex difference in aromatase levels cannot be the only factor, because there are still sex differences in vasopressin mRNA levels if rats are treated with a combination of estradiol and dihydrotestosterone (De Vries et al., 1994). In males, more BST cells produce vasopressin mRNA in response to estrogen stimulation. In addition, these cells show more labeling per cell, suggesting that in addition to a difference in the number of cells that express vasopressin, there are also sex differences in estrogen responsiveness of individual cells.

Even more so than differences in estrogen responsiveness, however, sex differences in androgen responsiveness contribute to sex differences in vasopressin-producing cells in the BST. Dihydrotestosterone, when given in combination with estradiol, significantly increased the number of vasopressin mRNA labeled cells in males but not in females (De Vries et al., 1994).

Since it is not known at which level androgens and estrogens act on vasopressin synthesis, it is not yet clear which factor contributes to sex differences in androgen and estrogen responsiveness of vasopressin cells. One factor may be steroid receptors levels of individual cells. Although no sex differences have been found in estrogen receptors (Brown et al., 1992), the number of androgen receptors associated with the cell nuclear fraction is higher in the BST of males than of females (Roselli, 1991). Such a sex difference might explain why dihydrotestosterone increases the number of vasopressin mRNA labeled cells in males but not in females.

3.3.2 Organizational Effects

Research on sexual differentiation of these projections in mammals has been done predominantly in rats and mice. When we first tested whether gonadal hormones influence sexual differentiation of vasopressin-ir projections (De Vries et al., 1983), we did not know yet that gonadal hormones influenced vasopressin expression in adulthood. We manipulated rats hormonally in the first three weeks after birth and measured vasopressin innervation at postnatal day 28, at which age the sex difference in vasopressin innervation was extreme, with females showing virtually no vasopressin innervation and males dense fiber networks in the lateral septum (De Vries et al., 1981). We found that gonadal hormones influenced these projections in the first as well as in the third postnatal week (De Vries et al., 1983), much later than what could be expected based on the critical periods within which gonadal hormones influence the development of other sexually dimorphic systems (Gorski 1984; Yahr, 1988; Forger, 2001). In retrospect, that study did not answer whether perinatal levels of gonadal hormones influence sexual differentiation of vasopressin-ir projections, because subjects were put to death within four weeks after hormonal manipulation, which is considerably sooner than what is needed for vasopressin to disappear from the fibers after gonadectomy (De Vries et al., 1984). We readdressed the same question by comparing the effects of neonatal manipulations on vasopressin expression in three-month-old rats all of which had been treated with testosterone for four weeks before they were put to death. We first showed that male rats that were castrated as adults had more vasopressin-ir cells in the BST and a higher density of vasopressin-ir fibers in the lateral septum than did neonatally castrated male rats. The latter had cell numbers and fiber densities equivalent to those of female rats, whether ovariectomized at birth or as adults. This suggested that testicular secretions after birth masculinize vasopressin-ir projections of the BST. A second experiment showed that males castrated at birth or at postnatal day 7 had fewer vasopressin-ir cells in the BST and MeA and a lower vasopressin-ir fiber density in the lateral septum than did male rats castrated at postnatal day 21 or as adults, suggesting that testicular secretions masculinize vasopressin-ir projections around day 7. Treatment of neonatally gonadectomized male and female rats with testosterone propionate at postnatal day 7 confirmed this, because it increased vasopressin-ir fiber density in the lateral septum. It also fully restored the number of vasopressin-ir cells in the BST of neonatally castrated males but not of females (Wang et al., 1993). This discrepancy may be due to prenatal differences in gonadal hormones. Another possibility is that sex chromosomal complement (XX vs XY) makes a difference independently of gonadal hormone levels (see ➋ section 12.6.3).

3.4 Development of Cellular Phenotype

3.4.1 Ontogeny

To understand the mechanisms by which steroids influence the development of vasopressin expression, one would like to study the early development of these cells. This is made difficult by the late onset of vasopressin expression in the BST and MeA. In males, vasopressin mRNA becomes first detectable at postnatal days three and five, and in females at postnatal days 21 and 35, respectively (Szot and Dorsa, 1993). This makes it nearly impossible to recognize future vasopressin cells during the period that gonadal steroids influence their sexual differentiation. There may be indirect ways to distinguish vasopressin cells from surrounding cells in the BST. Using the thymidine analog bromo-2-deoxy-5-uridine (BrdU) as a birth marker, we determined that most vasopressin-ir cells are born on embryonic day 12 and 13 (counting the day that sperm plugs were found as embryonic day 1), which places them in the earliest cohort of cells that survive to form the adult forebrain (Al-Shamma and De Vries, 1996). This was a surprise, because most cells in the BST and MeA are born on embryonic days 14–16 (Bayer, 1980; Bayer, 1987). This difference in cell birth may be exploited to study the sexual differentiation of the vasopressin-ir cells by following the development of cells labeled by BrdU injections on embryonic days 12 and 13. Alternatively, future vasopressin cells may be labeled by staining them for peptides that are co-expressed such as galanin (see below).

3.4.2 Differentiation of Cellular Phenotype

Although differences in the level of vasopressin mRNA expression per cell may contribute to the differences in the density of vasopressin projections, differences in number of vasopressin cells are probably the main factor. In theory, two fundamentally different types of processes could cause a difference in vasopressin cell number: first, processes that determine absolute cell number, such as cell birth, cell death, or cell migration, and second, processes that influence the phenotype of existing cells. The early birth of vasopressin cells eliminates differential cell birth and migration as likely factors, because the cells are born at least a week before hormones trigger their sexual differentiation (Wang et al., 1993; Al-Shamma and De Vries, 1996). Differential cell death can probably also be ruled out. Recently, we compared wild-type mice and mice with a null mutation in the gene coding for the cell death factor, Bax. This mutation blocks most neuronal cell death and eliminates sex differences in cell number of several brain areas (Forger et al., 2004). This mutation increased the number of vasopressin cells in males as well as females, but kept the sex difference intact (De Vries et al., unpublished). This suggests that developmentally programmed cell death determines the final number of potential vasopressin cells in males and females but not the differences between them.

All evidence points at testosterone stimulating an already existing group of cells to express vasopressin. After discovering that practically all vasopressin cells in the BST co-express galanin, but not all galanin cells vasopressin, and that males have more vasopressin cells than do females but similar numbers of galanin cells, Planas et al. (1995) proposed that differences in vasopressin expression depend on the percentage of galanin cells that co-express vasopressin. During development, higher levels of testosterone may stimulate more galanin neurons to co-express vasopressin. In support of this idea, vasopressin and galanin neurons in the BST and MeA show the same unusual birth profile with both days of birth earlier than most surrounding cells (Han and De Vries, 1999). As in mice, BST cells produce galanin well before birth (Tobet et al., 1999), galanin may be a good marker for the cells in which sexual differentiation of vasopressin production takes place.

It is currently unknown whether gonadal steroids target BST and MeA cells directly or act via other cells to differentiate vasopressin expression. Androgen- and estrogen receptors appear to be involved, as estradiol and dihydrotestosterone can masculinize vasopressin expression (Han and De Vries, 2003). Although vasopressin cells in the areas express estrogen receptor alpha, androgen, and progesterone receptors in adult rats (Axelson and Van Leeuwen, 1992; Zhou et al., 1994; Auger and De Vries, 2002), it is unknown whether the same cells do this during sexual differentiation. It is unknown, however, whether vasopressin cells also express gonadal steroid receptors during sexual differentiation.

4 Comparative Aspects

The sex difference in, and hormone responsiveness of, vasopressin innervation are conserved features. These features have been found in various rodent species, i.e., European (Buijs et al., 1986) and Djungarian hamsters (Bittman et al., 1991), Mongolian gerbils (Crenshaw et al., 1992), prairie and meadow voles (Bamshad et al., 1993), mice (Mayes et al., 1988), and in other mammalian species, e.g., ferrets (De Vries and Baum, unpublished). Similar dimorphism and hormonal responsiveness are present in homologous vasotocin-immunoreactive projections in nonmammalian vertebrates, i.e., in amphibians: rough-skinned newts (Moore, 1992) and bullfrogs (Boyd et al., 1992); in reptiles: the lizards *Gekko gecko* (Stoll and Voorn, 1985) and *anolis carolinensis* (Propper et al., 1992), the turtle *Pseudemys scripta elegans* and the snake *Python regius* (Smeets et al., 1990); and in birds: Japanese quails (Viglietti-Panzica et al., 1992), and canaries (Voorhuis et al., 1988). More complete listings can be found in Goodson and Bass (2001) and De Vries and Panzica (2006). Not all species with homologous vasotocin-immunoreactive or vasopressin-immunoreactive projections show conspicuous sex differences, for example, rainbow trouts (Van den Dungen et al., 1982), the amphibians *Rana ridibunda, Xenopus-laevis, and Pleurodeles waltii* (Gonzales and Smeets, 1992a, b), and guinea pigs (Dubois-Dauphin et al., 1989). Since differences were not quantified, it is not known whether sex differences are non-existent or just subtle in these species. The BST of humans and the BST and MeA of macaques have vasopressin-ir cells, and although neither species have shown significant innervation of the lateral septum, other areas that receive hormone-responsive vasopressin innervation in rats (i.e., the ventral tegmental area and midbrain central gray of macaques, and the locus coeruleus of humans and macaques) did show such projections, although without notable sex differences (Fliers et al., 1986; Caffe et al., 1989). It is not known whether vasopressin projections in these species are steroid responsiveness either. However, the marmoset, another primate, has more vasopressin-ir cells in the BST of males than of females (Wang et al., 1997). Finally, Syrian hamsters stand apart in that no trace of vasopressin projections of the BST and MeA can be found (Dubois-Dauphin et al., 1990; Albers et al., 1991; Ferris et al., 1995; Miller et al., 1999). Curiously, the lateral septum of Syrian hamster has vasopressin binding sites (Ferris et al., 1993) and is sensitive to the behavioral effects of vasopressin (Irvin et al., 1990). These variations in vasopressin/vasotocin projections may be exploited to reveal the processes underlying their sexual differentiation and identify their function (see e.g., Goodson and Bass, 2001).

5 Function of Sex Differences in Vasopressin Expression

5.1 Function of Vasopressin Innervation in the Brain

The widespread occurrence of sex differences in vasopressin and vasotocin projections suggests that these differences serve a function important enough to conserve it through evolution. Identifying this function is intimately connected to determining the role of vasopressin and vasotocin in the brain. This review will focus mainly on vasopressin. For excellent reviews on vasotocin, see, for example, Goodson and Bass, 2001; Panzica et al., 2001; Rose and Moore, 2002.

Revealing what functions the sexually dimorphic vasopressin projections of the BST and MeA may serve, is helped by vasopressin being one of the first peptides found to influence behavior and, in fact, among the first to be called a neuropeptide (De Wied, 1969). A great many studies address vasopressin effects on centrally regulated functions and behaviors, such as learning and memory (De Wied, 1969), cardiovascular functions (Versteeg et al., 1983), thermoregulation (Kasting, 1989), and motor behaviors (Kasting et al., 1980). Many of these studies involved injecting vasopressin or its analogs into the ventricles or into areas that do not necessarily receive vasopressin-ir input. Since there are several different vasopressin systems (De Vries and Miller, 1998), the results of such studies cannot always be related to any specific vasopressin system. More recent studies, done with the anatomy of the central vasopressin pathways in mind, have indicated a number of functions in which specifically BST and MA projections may be involved. For example, the vasopressin innervation of the septal area has been implicated in thermoregulation, osmoregulation, social recognition memory, motor disturbances, and aggressive behavior (see below). Competition studies using nonradioactive vasopressin analogs (Dorsa et al., 1988), electrophysiological studies (Raggenbass et al., 1987), in situ hybridization studies (Ostrowski et al., 1992), and functional studies (see below) indicate that vasopressin receptors in this area resemble vasopressor receptors (V_{1A} receptors), and not the antidiuretic receptor (the V_2 receptor) or the vasopressin-oxytocin receptor which is found in the hippocampus (Audigier and Barberis, 1985). Given the sexually dimorphic nature and extreme hormone responsiveness of vasopressin projections, one is tempted to conclude that this system is involved in sexually dimorphic functions or at least in functions that are influenced by gonadal steroids, such as reproductive behaviors and the regulation of gonadal hormone release. Lesion studies have indeed implicated the BST and MeA in male sexual behavior, the regulation of gonadal hormone release, and the control of parental behavior (Harris and Sachs, 1975; Emery and Sachs, 1976; Beltramino and Taleisnik, 1978, 1980, 1985; Valcourt and Sachs, 1979; Fleming et al., 1980; Numan and Numan, 1996), but it is unknown specifically which BST and MeA projections contribute to these functions. In fact, research of the past fifteen years suggests that the sex differences and hormone responsiveness of vasopressin projections are somewhat of a red herring. Some of these reproductive functions are discussed first, followed by functions that have been directly related to vasopressin projections of the BST. This is done against the backdrop of sex differences in vasopressin projections.

5.1.1 Sexually Dimorphic Behaviors

5.1.1.1 Female sexual behavior Vasopressin projections of the BST and MeA may control female sexual behavior. Lesion and stimulation studies suggest that areas innervated by these projections influence female sexual behavior. The lateral septum, for example, has an inhibitory influence on the lordosis response (Nance et al., 1974; Zasorin et al., 1975). Its vasopressin innervation may play a role in this, because intraventricular injections of vasopressin stimulate female sexual behavior whereas similar injections of vasopressin antagonist inhibit it (Södersten et al., 1985). Similar effects have recently been obtained by Pedersen and Boccia (2006). Intraventricular injections, however, may interact with any vasopressin system in the brain. In fact, circadian fluctuations in the effects of these injections and of vasopressin levels in the brain made us suggest that suprachiasmatic, and not BST and MeA, projections formed the substrate for vasopressin's effects on female sexual behavior (Södersten et al., 1985). A direct link between the vasopressin innervation of the lateral septum and female sexual behavior has never been tested. Nevertheless, if septal vasopressin inhibits female sexual behavior, then the higher levels of vasopressin in males versus females correlate well with males being far less likely to show female sexual behavior than are females.

5.1.1.2 Male sexual behavior Remarkable similarities in hormone effects on vasopressin expression and behavior in rats make it particularly tempting to associate the vasopressin projections of the BST and MeA with male sexual behavior. Castration causes an equally gradual decline in male sexual behavior as it does in vasopressin content of the BST and MeA projections (Davidson, 1966; De Vries et al., 1984). Similarly, testosterone metabolites stimulate male sexual behavior the same way as vasopressin expression, with estradiol being much more effective than 5α-dihydrotestosterone (Baum and Vreeburg, 1973; Larsson et al., 1973; De Vries et al., 1986; De Vries et al., 1994; Wang and De Vries, 1995). Although injections of vasopressin directly into the lateral septum do not affect male sexual behavior (Koolhaas et al., 1991), the effects of vasopressin or its antagonists on male sexual behavior in other areas that receive input from the BST and MeA have not been studied. However, treating rats with a centrally acting vasopressin analogue reversed the decline of male sexual behavior following castration (Bohus, 1977), which supports the idea that central vasopressin pathways stimulate male sexual behavior. Comparative research suggests, however, that vasopressin's role may be more complicated because intracerebroventricular injections of vasotocin into Japanese quail, which show similar sex differences in vasotocin innervation as do rats in vasopressin innervation, inhibit certain aspects of male sexual behavior (Castagna et al., 1998; Panzica et al., 2002). One could argue that if vasopressin projections are an integral part of the circuitry that controls male sexual behavior, whether in an activating or inhibitory role, it still makes sense that males have a denser innervation than do females. This idea, however, seems less compelling than the one proposed earlier for a role of the sex difference in vasopressin innervation in female sexual behavior. However, at the very least, it suggests that correlations should be treated with caution if one wants to link sex differences in brain structure with sex differences in brain function.

5.1.1.3 Aggressive Behavior Septal vasopressin has also been implicated in other sexually dimorphic behaviors that are influenced by gonadal steroids, for example, intermale aggression. Just as the vasopressin content of the BST and MeA projections, intermale aggression declines gradually after castration (DeBold and Miczek, 1984). Furthermore, injections of vasopressin into the lateral septum or the MeA reverse this decline (Koolhaas et al., 1990, 1991). In fact, in all vertebrate classes, vasopressin and its non-mammalian homologue, vasotocin, have been linked to aggressive behavior and related behaviors, such as social dominance (Ferris et al., 1986; Ferris and Potegal, 1988; Winslow et al., 1993; Marler et al., 1995; Delville et al., 1996; Chu et al., 1998; Goodson, 1998a, b; Semsar et al., 1998, 2001; Goodson and Adkins-Regan, 1999; Bester-Meredith and Marler, 2001; Tito et al., 1999; Lema and Nevitt, 2004; Thompson et al., 2004; Goodson et al., 2004b; Maney et al., 2005), including in humans (Coccaro et al., 1998; Thompson et al., 2004). As with female sexual behavior, the higher levels of vasopressin/vasotocin in males correlate well with the higher levels of resident-intruder aggression displayed by males versus females. Interestingly, in hyenas, in which females are more aggressive and dominant over males (Matthews, 1939; Hamilton, 1986), males have either the same or a lower vasopressin fiber density in the septum and other BST and MeA projection sites (Rosen et al., 2001). And perhaps it is no coincidence either that in Syrian hamsters, which as mentioned earlier lack the vasopressin projections of the BST and MeA entirely; females are as aggressive as males and tend to dominate males of similar body weight (Payne and Swanson, 1970; Marques and Valenstein 1977; Huhman et al., 2003; Taravosh-Lahn and Delville, 2004).

5.1.2 Nondimorphic Functions

5.1.2.1 Social Behaviors Unlike sexual and aggressive behavior, several other functions are influenced by septal vasopressin for which there is no clear relationship with the sex differences in vasopressin innervation (De Vries, 1988). For example, in male rats, social recognition memory can be boosted or inhibited by injections of vasopressin or a V_{1A} antagonist, respectively, into the lateral septum (Dantzer et al., 1988). In females, which do not differ from males in social recognition memory, injections of a V_{1A} antagonist do not inhibit social recognition memory (Bluthe and Dantzer, 1990). In this case, one cannot correlate the sex difference in vasopressin innervation with a functional sex difference. Rather, these data suggest that the neurochemical underpinnings of social recognition memory differ between the sexes.

5.1.2.2 Homeostatic Functions When we first discovered the sex difference in vasopressin projections, septal vasopressin had only been linked to functions that did not differ dramatically between sexes, for example, fever control (Cooper et al., 1979). In fact, this is a very well established function for vasopressin because the fever-reducing effects of vasopressin and fever-enhancing effects of $V1_a$ receptor antagonists injected specifically into the ventral septal area have been demonstrated repeatedly for a variety of mammals (Kasting, 1989). Although the source of vasopressin innervation in the ventral septal area has not been traced directly, it almost certainly comes from the BST. This innervation disappears after lesioning the BST but not after lesioning other sites that produce vasopressin (De Vries and Buijs, 1983). It is also denser in males than in females (De Vries and Al-Shamma, 1990) and loses its vasopressin content after castration (De Vries et al., 1985). In addition, electrical stimulation of the BST influences electrical activity of ventral septal neurons similarly as does exogenous application of vasopressin (Disturnal et al., 1985a, b). Finally, electrical stimulation of the BST reduces pyrogen-induced fever, presumably by enhancing vasopressin release because V_{1A} receptor antagonist injected into the ventral septal area can block this effect (Naylor et al., 1988).

The steroid-responsiveness of the vasopressin innervation of the ventral septal area has been exploited to demonstrate that endogenous vasopressin release reduces fever. Castration, which presumably reduces vasopressin release in the lateral septum, lengthens pyrogen-induced fever (Pittman et al., 1988). Another study considered how castration affected the ability of interleukin-1 to induce not only fever but also "sickness behavior" (increased sleepiness, lethargy, reduced social activities, reduced food intake), and showed that vasopressin treatment can reverse this castration effect. Taking social investigation of juvenile conspecifics as an index of sickness behavior, these studies showed that castrated rats were more sensitive to the fever-inducing and behavioral effects of interleukin-1. In addition, vasopressin more effectively attenuated the effects of interleukin 1 in castrated- than in intact rats and, conversely, the V_{1A} receptor antagonist potentiated the effects of interleukin 1 in intact- but not in castrated rats (Dantzer et al., 1991; Bluthe and Dantzer, 1992).

In European hamsters, which unlike Syrian hamsters show vasopressin projections from the BST (Buijs et al., 1986), vasopressin release in the lateral septum may prevent hypothermia. In this species, the density of the vasopressin-ir fiber plexus in the lateral septum is high during the summer, when testosterone levels are high, and low during fall and winter, when testosterone levels are low and the animals hibernate (Buijs et al., 1986). Chronic infusions of vasopressin into the lateral septum prevented the bouts of hypothermia that are associated with hibernation (Hermes et al., 1989). In addition, silastic implants with testosterone given at the beginning of the hibernation season kept the density of vasopressin-ir innervation of the lateral septum high and prevented hibernation, possibly by sustaining the release of vasopressin in the lateral septum. Together, these studies show a good correlation between levels of vasopressin expression and temperature regulation that may be related to levels of circulating gonadal hormones. Sex differences in vasopressin expression and temperature regulation do not show such a correlation.

6 Sex Differences in the Neurochemical Underpinnings of Functions

When we realized after finding the sex difference in vasopressin innervation that vasopressin had only been implicated in functions that do not differ clearly by sex, we suggested that there may be compensatory mechanisms, such as sex differences in other transmitter systems, that may keep such functions similar (De Vries et al., 1981). I soon forgot our own suggestion and, when reviewing sex differences in neurotransmitter systems, I mainly linked differences in neurotransmitter systems to differences in overt functions and behaviors (see for example, De Vries, 1984, 1990). Our work on voles forced me to revisit the idea that sex differences in neurotransmitter systems may just as well be linked to behaviors that do not differ ostensibly by sex.

Such differences may be adaptive if sex differences in circulating gonadal hormones would otherwise cause undesirable sex differences in overt physiology or behavior. For example, the wide-spread distribution of gonadal steroid receptors in multimodal areas such as the lateral septum and the CA1 region of the hippocampus, which are involved in many functions that are not, or only marginally, sexually dimorphic, may require compensatory sex differences to keep such functions similar. Because these areas are constantly exposed to circulating gonadal steroids levels that differ by sex, their level of activation may differ as well. Although differences in activation may be expected for sexually dimorphic functions controlled by such areas, they are more difficult to explain for functions that do not show clear sex differences. For example, practically every neuron in the lateral septum expresses androgen receptors (Sar and Stumpf, 1975; Simerly et al., 1990; Zhou and De Vries, 1994). The higher levels of circulating androgen in males versus females may therefore cause chronic differences in androgen receptor activation in practically every lateral septal neuron. Such a difference might make sense for the inhibitory actions of the septum on female sexual behavior (Nance et al., 1974; Zasorin et al., 1975; De Vries, 1990), but is more difficult to understand for functions that are not or only marginally sexually dimorphic, such as thermoregulation. Perhaps additional sex differences such as the denser innervation of vasopressin in males compensate for differential androgenic effects, thereby preventing undesirable sex differences in some functions, while generating them in others.

6.1 Parental Behavior

Sex differences in neurochemistry may also compensate for changes in gonadal hormones that occur in one sex, but not in the other. Parental behavior in species where males and females provide parental care, such as prairie voles (Lonstein and De Vries, 1999a), may serve as an example. In rodents, hormonal changes associated with pregnancy increase parental responsiveness so that once pups are born females are fully maternal (Rosenblatt et al., 1988; Bridges, 1990). In voles, the stakes are even higher as females also have to undergo parturition before they are fully parental. Females whose pups were delivered by Caesarian section do not show increased parental responsiveness (Hayes et al., 2005). In males, paternal behavior must clearly be induced through different mechanisms. This may imply that neural circuits involved in parental behavior must differ to guarantee a similar behavioral output between males and females.

The sexually dimorphic vasopressin innervation may play such a role in prairie voles. In this species, sexually naive males are commonly spontaneously parental, whereas sexually naive females are not (Lonstein and De Vries, 1999b). In males, mating activated vasopressin mRNA production in BST and MeA cells (Wang et al., 1994a), while decreasing vasopressin content of BST and MeA projections to the lateral septum

(Bamshad et al., 1994). These changes suggest that mating increases vasopressin release in the lateral septum. Recently, Lim and Young (2004) provided additional evidence for mating-induced increase of vasopressin from BST projections in males, by showing that boosting V_{1A} receptor expression increases mating-induced Fos expression in the ventral pallidum, another target of BST projections. The changes in vasopressin neurotransmission in males may very well explain the increase in paternal responsiveness seen in males after mating (Bamshad et al., 1994), because vasopressin injections into the lateral septum stimulated paternal responsiveness in virgin voles, whereas injections of a V_{1A} receptor antagonist inhibited it (Wang et al., 1994b). Vasopressin does not appear to play a role of significance in females, because vasopressin expression is nearly absent in the BST and does not change after mating (Bamshad et al., 1993; Wang et al., 1994a). In this case, the sex difference in vasopressin innervation may allow fathers to show similar responses to pups as do mothers.

Interestingly, this does not necessarily mean that vasopressin is essential for paternal behavior under all hormonal conditions. Castration, which eliminates vasopressin expression in BST and MeA projections in voles as it does in rats (Wang and De Vries, 1993), does not eliminate parental behavior in males (Lonstein and De Vries, 1999b). The same is true for social recognition memory, which does not change dramatically after long-term castration but can no longer be blocked by vasopressin antagonists (Bluthe et al., 1990). This suggests that the neurochemical underpinnings of some behaviors may vary between different hormonal states the same way as they do between males and females.

6.2 Pairbonding

Mating-induced changes in vasopressin release may affect other social behaviors in voles as well (Young and Wang, 2004). Following mating, prairie voles form stable pairbonds and display increased aggression towards unfamiliar conspecifics and increased paternal responsiveness (Getz et al., 1981). In males, these changes could be blocked by intracerebroventricular injections of a V_{1A} antagonist (Winslow et al., 1993) suggesting that endogenous vasopressin release drives these changes. Interestingly, pairbonding may also be regulated differently in males and females. Whereas intracerebroventricular injections of vasopressin stimulate pairbonding, oxytocin injections do not. In female prairie voles, however, the reverse is true (Insel and Hulihan, 1995). In addition, adrenalectomy prevents pairbonding in males, but facilitates it in females; in both cases corticosteroid injections counter the effects of adrenalectomy (DeVries et al., 1996). These sex-dependent neuropeptide and corticosteroid effects may very well reflect differences in the neurochemical underpinnings of pairbonding in voles.

Between the extreme possibilities of functions that do not differ by sex, but the underlying neurochemistry does (such as parental behavior in voles), or functions that are sexually dimorphic thanks to sex differences in underlying neurochemistry (such as aggressive behavior), there are probably many intermediate situations. For example, fear and anxiety-related behaviors, which vary by sex, but not extremely so (Palanza, 2002; Toufexis et al., 2006), show dramatic differences in the extent to which they are controlled by vasopressin. For example, male vasopressin V_{1A} receptor knockout mice show a reduction in anxiety-like behavior (Bielsky et al., 2004), but their female counterparts are unaffected (Bielsky et al., 2005), presumably because a loss of V_{1A} receptors has less of an impact in animals that had little vasopressin to begin with, but perhaps also because the neural circuitry underlying anxiety-like behaviors differs in males and females.

6.3 Sex Differences as Compensation for Sex Chromosomal Effects

A relatively new idea is that sex chromosomal complement (XX versus XY) directly affects sex differences in the brain (Arnold, 1996, 2004). Such direct effects of sex chromosomal complement can be detected with a model system that can distinguish between differences caused by sex chromosomal complement (XX vs. XY) or different gonads (testes vs. ovaries) (De Vries et al., 2002). In this model, female mice with an XX genotype were crossed with males with an XY^- Sry genotype. These XY^- *Sry* mice lack the Sry gene on the

Y chromosome, which normally directs the growth of the testis (Koopman et al., 1991). Nevertheless, they develop a male phenotype, because they have an *Sry* transgene inserted on an autosomal chromosome. This cross generates XX and XY⁻ mice of either sex depending on whether they inherited the *Sry* transgene. Comparing XX and XY⁻ mice within sex (defined on basis of type gonad) revealed a number of differences that may be caused directly by differences in sex chromosomal complement, including differences in vasopressin innervation of the lateral septum, which was denser in XY⁻ than in XX animals irrespective of gonadal sex (De Vries et al., 2002; Gatewood et al., 2006). As these mice still showed respectable differences between gonadal males and females irrespective of sex chromosomal complement, hormones as well as sex chromosomal complement appear to control sexual differentiation of vasopressin expression in mice. Interestingly, females, but not males, showed an effect of sex chromosomes on parental and aggressive behavior (Gatewood et al., 2006), both of which are modulated by vasopressin (see Sections 12.5.1.1.3 and 12.6.1).

If sex chromosomal complement, like sex steroids, can induce sex differences in neural structure, there is no reason why such differences may also cause as well as prevent sex differences in behavior. Although the idea that sex differences at the molecular and cellular level, whether caused by sex chromosomal complement or hormones, may generate as well as prevent sex differences in behavior may sound unorthodox, similar ideas are well accepted in other fields of biology. In fact, sex differences in mechanisms underlying any physiological or behavioral endpoint may be the norm, not the exception. This is definitely the case at the chromosomal level. In mammals, every cell treats X chromosomes differently, depending on whether it is found in a male or female body. In females, cells show random X inactivation, the silencing of most genes on one of the two X chromosomes; in males, cells do not (Lyon, 1999). This X inactivation is generally seen as a compensatory mechanism, meant to ensure that X chromosomal genes, many of which serve basic household functions, are expressed at roughly the same rate in males and females (Lyon, 1999). The energy spent on inactivating one of the two X chromosomes is, presumably, a cost of the evolution of two sexes (Charlesworth, 1996). Clearly, differential expression of sex chromosomal genes is essential for generating a male and female genotype. But for all we know, much of this expression is restricted to only a few tissues at rather restricted periods of our life. In mice, for example, expression of the Sry gene is needed for only half a day (embryonic day 11–11.5), in only one cell-type (cells that will become Sertoli cells), to set in motion a cascade of events that generates the male phenotype (Burgoyne et al., 1988; Lovell-Badge et al., 1995). X inactivation, however, has to take place in practically all cells of the female body, throughout life, in any tissue, sexually dimorphic or not (Lyon, 1999). Similar compensatory processes, meant to counter undesired side effects of sexual differentiation, may take place repeatedly, in developing as well as in adult animals, from the molecular to the macroscopic level.

6.4 Sex Difference in Other Neurotransmitter Systems

6.4.1 Cholecystokinin

Brains may be exactly the place to look for such compensatory processes, in particular because like sex chromosomes, neural circuits often serve more than one function, some dimorphic, others not. Generating a sex difference in any such circuit, for example, to generate reproductive behavior is therefore likely to cause sex differences in other functions served by the same circuits. If such sex differences are maladaptive, evolution will favor the development of compensatory mechanisms. It is, therefore, unlikely that the sexually dimorphic vasopressin innervation of the brain would be the only system that may cause sex differences in some functions but prevent them in others. Cholecystokinin neurotransmission may provide another example. Cholecystokinin binding in the ventromedial hypothalamic nucleus is higher in males than in females (Micevych and Ulibarri, 1992), which fits with a potential role of cholecystokinin in female sexual behavior (Micevych and Sinchak, this volume). However, cholecystokinin has also been implicated in food intake, which does not show nearly as dramatic a sex difference as does female sexual behavior (Willis et al., 1986; Schick et al., 1994). The challenge is to understand how sexually dimorphic systems such as the cholecystokinin system can be involved in diverse functions–some sexually dimorphic, others not, and how such sexually dimorphic systems can prevent sex differences in some functions, whereas causing them in others. Solving this problem will require a much better understanding of the various components that contribute to the control of specific functions. How daunting such a task is, is illustrated in the chapter, in this volume, by Micevych and Sinchak, who review the various transmitter systems that have been implicated in female sexual behavior.

6.4.2 Sry and Dopamine

Dopamine cells in the substantia nigra offer another interesting example. Male mice have about 10% to 20% more dopamine cells in their substantia nigra than do females (Dewing et al., 2006). Although sex differences in dopamine neurotransmission in the striatum have been linked to sex differences in certain motor behaviors (Becker, 1999), it is not clear whether this particular sex difference contributes to those differences. Behind this relatively moderate sex difference, however, hides a rather impressive sex difference in how sex chromosomal genes affect these cells. In males, mesencephalic dopamine cells express the *Sry* gene (Dewing et al., 2006). Treating males with *Sry* antisense oligonucleotides blocks the production of *Sry* protein as well as lowers tyrosine hydroxylase levels, the rate-limiting enzyme in the production of dopamine, and causes motor defects. In females, the same treatment has no effect. This suggests that *Sry* is necessary to normalize the motor function. Therefore, this may be an example where a difference in sex chromosomal gene expression compensates for a sex difference in this system, possibly caused by differential exposure to gonadal hormones (Dewing et al., 2006).

6.4.3 Sex Chromosomal Factors Other than Sry and Dopamine

Sry may not be the only sex chromosomal gene that acts on dopamine cells. Cultures of mesencephalic cells harvested on embryonic day 14 differ in the number of cells that will express tyrosine hydroxylase (TH) depending on the sex of origin (Reisert and Pilgrim, 1991; Sibug et al., 1996). As these cells were harvested after the onset of gonadal differentiation including the development of the testosterone-producing fetal Leydig cells (Haider, 2004), one cannot exclude hormones as a factor. However, by using the same cross of XX with XY$^-$ *Sry* mice described earlier, Carruth et al. (2002) showed that sex chromosomal complement rather than previous exposure to gonadal hormones determines the level of TH expression in vitro. Cultures from XY$^-$ mice developed more TH neurons than did XX cultures regardless of the sex of origin. As *Sry* gene expression was similar between XX and XY$^-$ mice (either absent in gonadal females or present in gonadal males), these differences must be caused by other factors, for example, expression of Y chromosomal genes other than *Sry*, dosage differences of X chromosomal genes, or differential imprinting of maternal and paternal X chromosomes (Arnold, 2004).

Interestingly, mesencephalic cells do not show a sex difference during development in vivo (Lieb et al., 1996). One possible explanation is that, in vivo, the internal milieu prevents XX and XY specific gene expression from inducing sex differences in mesencephalic cells in vivo. A much more provocative idea is that XX and XY specific gene expression occurs in vivo, perhaps to compensate for effects of factors derived from other brain regions that would have caused undesirable sexual differentiation if left unchecked. Ideas like these are perfectly testable. For example, one could identify the elements on X and Y chromosomes that cause the differences in TH expression in vitro. Equating the influence of these elements in XX and XY animals should eliminate the difference in vitro and induce a difference in vivo. In fact, such experiments have been done, not necessarily with testing this idea in mind. For example, Bluthe and Dantzer's (1990) demonstration that vasopressin V1a receptor antagonists blocked social recognition memory in males, but not in females, could be regarded as a test of the idea that the sexually dimorphic vasopressin innervation keeps social recognition memory similar. This manipulation caused a sex difference, where there was not one before. In fact, sexually dimorphic involvement of neurotransmitter systems in functions that are similar between sexes may be suspected in all cases where drug manipulations causes sex differences that were not present in untreated animals (Villalba et al., 1997).

7 Sex Differences in Human Brains

Like animal brains, the human brain is sexually dimorphic. There are, for example, volume differences in several brain structures, including the anterior commissure, the left planum temporale, and several nuclei in the hypothalamus and BST (Swaab and Fliers, 1985; Allen et al., 1989, 1991; Allen and Gorski, 1990, 1991; Hofman and Swaab, 1991; Zhou et al., 1995; Wisniewski, 1998). In addition, several neurotransmitters differ in levels, metabolism, or distribution; for example, serotonin (Arato et al., 1991; Biver et al., 1996; Nishizawa et al., 1997), dopamine (Pohjalainen et al., 1998; Kaasinen et al., 2001), vasoactive intestinal polypeptide, and somatostatin (Zhou et al., 1995; Kruiver et al., 2000; Chung et al., 2002). Although sex differences in human brains are often associated with differences in behavior (for example, cortical difference with differences in cognitive abilities and hypothalamic differences with differences in sexual behavior and sexual orientation), these sex differences may also normalize behavior in men and women.

In fact, the human literature provides ample evidence that male and female brains may use different strategies and, therefore, different neural systems to perform the same tasks. For example, strokes can have different outcomes in men and women, even if they occur in the same regions (Inglis et al., 1982; Kimura et al., 1985; Hier et al., 1994; Roof and Hall, 2000), and functional imaging studies suggest that men and women use telencephalic regions differently even for functions that do not differ by sex (Shaywitz et al., 1995; Jaeger et al., 1998; Derbyshire et al., 2002; Grabowski et al., 2003; Cahill et al., 2004; Piefke et al., 2005). Ironically, the prediction that the neurochemical underpinnings of nondimorphic behaviors may also differ in humans is tested each time, when medical practitioners prescribe psychoactive drugs in similar dosages to men and women to treat their behavioral disorders (Hrdina, 2000). Even though these disorders may present similar symptoms, many of the drugs used to treat them act on neurotransmitter systems that are sexually dimorphic, such as the central serotonin innervation (Arato et al., 1991; Biver et al., 1996; Nishizawa et al., 1997). For that reason alone, the effectiveness of these treatments may vary by sex. Unfortunately, few researchers take note of the results of these unintended experiments (Hrdina, 2000). Experiments designed to test the activity of psychotropic drugs on physiological parameters such as the activity of the hypothalamo-pituitary adrenal axis, however, show that drugs can have different effects in men and women (Rhodes and Rubin, 1999). This will be an even bigger issue in developing drug interventions geared to interact with systems, such as the central vasopressin innervation, that show in their central fuctions.

Acknowledgment

This review was written while the author was funded by NIMH grant RO1 MH47538 and KO2 MH01497.

References

Adan RAH, Burbach JPH. 1992. Regulation of vasopressin and oxytocin gene expression by estrogen and thyroid hormone. Prog Brain Res 92: 127-136.

Albers HE, Rowland CM, Ferris CF. 1991. Arginine-vasopressin immunoreactivity is not altered by photoperiod or gonadal hormones in the Syrian hamster (*Mesocricetus auratus*). Brain Res 539: 137-142.

Alheid GF, De Olmos JS, Beltramino CA. 1995. Amygdala and extended amygdala. The Rat Nervous System. Second Edition, Paxinos G. editor. New York: Academic Press; pp. 495-578.

Allen LS, Gorski RA. 1991. Sexual dimorphism of the anterior commissure and massa intermedia of the human brain. J Comp Neurol 312: 97-104.

Allen LS, Gorski RA. 1990. Sex difference in the bed nucleus of the stria terminalis of the human brain. J Comp Neurol 302: 697-706.

Allen LS, Hines M, Shryne JE, Gorski RA. 1989. Two sexually dimorphic cell groups in the human brain. J Neurosci 9: 497-506.

Allen LS, Richey MF, Chai YM, Gorski RA. 1991. Sex differences in the corpus callosum of the living human being. J Neurosci 11: 933-942.

Al-Shamma HA, De Vries GJ. 1996. Neurogenesis of the sexually dimorphic VP cells of the bed nucleus of the stria terminalis and amygdala of rats. J Neurobiol 29: 91-98.

Amateau SK, McCarthy MM. 2004. Induction of PGE2 by estradiol mediates developmental masculinization of sex behavior. Nat Neurosci 7: 643-650.

Arato M, Frecska E, Tekes K, Mac Crimmon DJ. 1991. Serotonergic interhemispheric asymmetry: gender difference in the orbital cortex. Acta Psychiatr Scand 84: 110-111.

Arnold AP. 1996. Genetically triggered sexual differentiation of brain and behavior. Horm Behav 30: 495-505.

Arnold AP. 2004. Sex chromosomes and brain gender. Nat Rev Neurosci 5: 701-708.

Audigier S, Barberis C. 1985. Pharmacological characterization of two specific binding sites for neurohypophyseal hormones in hippocampal synaptic plasma membranes of the rat. EMBO J 4: 1407-1412.

Auger AP, Tetel MJ, McCarthy MM. 2000. Steroid receptor coactivator-1 (SRC-1) mediates the development of sex-specific brain morphology and behavior. Proc Natl Acad Sci USA 97: 7551-7555.

Auger CA, De Vries GJ. 2002. Progestin receptor immunoreactivity within steroid-responsive vasopressin-immunoreactive cells in the male and female rat brain. J Neuroend 14: 561-567.

Auger CJ, Vanzo RJ. 2006. Progesterone treatment of adult male rats suppresses arginine vasopressin expression in the bed nucleus of the stria terminalis and the centromedial amygdala. J Neuroendocrinol 18: 187-194.

Axelson JF, Van Leeuwen FW. 1990. Differential localization of estrogen receptors in various vasopressin synthesizing nuclei of the rat brain. J Neuroendocrinol 2: 209-216.

Bamshad M, Novak MA, De Vries GJ. 1993. Sex and species differences in the vasopressin innervation of sexually naive and parental prairie voles, *Microtus ochrogaster* and meadow voles, *Microtus pennsylvanicus.* J Neuroendocrinol 5: 247-255.

Bamshad M, Novak MA, De Vries GJ. 1994. Cohabitation alters vasopressin innervation and paternal behavior in Prairie voles, *Microtus ochrogaster.* Physiol Behav 56: 751-758.

Barraclough CA. 1961. Production of anovulatory, sterile rats by single injections of testosterone propionate. Endocrinology 68: 62-67.

Barraclough CA, Gorski RA. 1961. Evidence that the hypothalamus is responsible for androgen- induced sterility in the female rat. Endocrinology 68: 68-79.

Barraclough CA, Leathem JH. 1954. Infertility induced in mice by a single injection of testosterone propionate. Proc Soc Exp Biol Med 85: 673-674.

Baum MJ, Vreeburg JTM. 1973. Copulation in castrated male rats following combined treatment with estradiol and dihydrotestosterone. Science 182: 293-285.

Bayer SA. 1980. Quantitative [3H]thymidine radiographic analysis of neurogenesis in the rat amygdala. J Comp Neurol 194: 845-875.

Bayer SA. 1987. Neurogenetic and morphogenetic heterogeneity in the bed nucleus of the stria terminalis. J Comp Neurol 265: 47-64.

Becker JB, Arnold AP, Berkley KJ, Blaustein JD, Eckel LA, et al. 2005. Strategies and methods for research on sex differences in brain and behavior. Endocrinology 146: 1650-1673.

Becker JB. 1999. Gender differences in dopaminergic function in striatum and nucleus accumbens. Pharmacol Biochem Behav 64: 803-812.

Beltramino C, Taleisnik S. 1978. Facilitatory and inhibitory effects of electrochemical stimulation of the amygdala on the release of luteinizing hormone. Brain Res 144: 95-107.

Beltramino C, Taleisnik S. 1980. Dual action of electrochemical stimulation of the bed nucleus of the stria terminalis on the release of LH. Neuroendocrinology 30: 238-242.

Beltramino C, Taleisnik S. 1985. Ventral premammillary nuclei mediate pheromonal-induced LH release stimuli in the rat. Neuroendocrinology 41: 119-124.

Bester-Meredith JK, Marler CA. 2001. Vasopressin and aggression in cross-fostered California mice (*Peromyscus californicus*) and white-footed mice (*Peromyscus leucopus*). Horm Behav 40: 51-64.

Bielsky IF, Hu SB, Young LJ. 2005. Sexual dimorphism in the vasopressin system: lack of an altered behavioral phenotype in female V1a receptor knockout mice. Behav Brain Res 164: 132-136.

Bielsky IF, Hu S-B, Szegda KL, Westphal H, Young LJ. 2004. Profound impairment in social recognition and reduction in anxiety-like behavior in vasopressin V1a receptor knockout mice. Neuropsychopharmacology 29: 483-493.

Bittman EL, Bartness TJ, Goldman BD, De Vries GJ. 1991. Suprachiasmatic and paraventricular control of photoperiodism in Siberian hamsters. Amer J Physiol 260: R90-R101.

Biver F, Lotstra F, Monclus M, Wikler D, Damhaut P, et al. 1996. Sex difference in 5HT2 receptor in the living human brain. Neurosci Lett 204: 25-28.

Bluthe R-M, Dantzer R. 1990. Social recognition does not involve vasopressinergic neurotransmission in female rats. Brain Res 535: 301-304.

Bluthe R-M, Dantzer R. 1992. Chronic intracerebral infusions of vasopressin and vasopressin antagonists modulate behavioral effects of interleukin-1 in rat. Brain Res Bull 29: 897-900.

Bluthe R-M, Schoenen J, Dantzer R. 1990. Androgen-dependent vasopressinergic neurons are involved in social recognition in rats. Brain Res 519: 150-157.

Bohus B. 1977. Postcastration masculine behavior in the rats: the role of hypothalamo hypophyseal peptides. Exp Brain Res 28: R8.

Boyd SK, Tyler CJ, De Vries GJ. 1992. Sexual dimorphism in the vasotocin system of the bullfrog (*Rana catesbeiana*). J Comp Neurol 325: 313-325.

Breedlove SM. 1992. Sexual dimorphism in the vertebrate nervous system. J Neurosci 12: 4133-4142.

Breedlove SM, Cooke BM, Jordan CL. 1999. The orthodox view of brain sexual differentiation. Brain Behav Evol 54: 8-14.

Bridges RS. 1990. Endocrine regulation of parental behavior in rodents. Mammalian Parenting. Biochemical Neurobiological, and Behavioral Determinants. Krasgenor NA, Bridges RS, editors. New York: Oxford University Press; pp. 93 117.

Brown TJ, Naftolin F, MacLusky NJ. 1992. Sex differences in estrogen receptor binding in the rat hypothalamus: effects of subsaturating pulses of estradiol. Brain Res 578: 129-134.

Buijs RM, Pevet P, Masson-Pevet M, Pool CW, De Vries GJ, et al. 1986. Seasonal variation in vasopressin innervation in the brain of the European hamster (*Cricetus cricetus*) Brain Res 371: 193-196.

Burgoyne PS, Buehr M, Koopman P, Rossant J, McLaren A. 1988. Cell-autonomous action of the testis-determining gene: Sertoli cells are exclusively XY in XX----XY chimaeric mouse testes. Development 102: 443-450.

Caffé AR, Van Leeuwen FW. 1983. Vasopressin immunoreactive cells in the dorsomedial hypothalamic region, medial amygdaloid nucleus and locus coeruleus of the rat. Cell Tiss Res 233: 23-33.

Caffé AR, Van Leeuwen FW, Luiten PGM. 1987. Vasopressin cells in the medial amygdala of the rat project to the lateral septum and ventral hippocampus. J Comp Neurol 261: 237-252.

Caffé AR, Van Ryen PC, Woude Van der TP, Van Leeuwen FW. 1989. Vasopressin and oxytocin systems in the brain and upper spinal cord of *Macaca fascicularis*. J Comp Neurol 287: 302-325.

Cahill L, Uncapher M, Kilpatrick L, Alkire MT, Turner J. 2004. Sex-related hemispheric lateralization of amygdala function in emotionally influenced memory: An fMRI investigation Learning Memory 11: 261-266.

Canteras NS, Simerly RB, Swanson LW. 1995. Organization of projections from the medial nucleus of the amygdala: a PHAL study in the rat. J Comp Neurol 360: 213-245.

Carruth LL, Reisert I, Arnold AP. 2002. Sex chromosome genes directly affect brain sexual differentiation. Nat Neurosci 5: 933-934.

Carter DA, Murphy D. 1993. Regulation of vasopressin (VP) gene expression in the bed nucleus of the stria terminalis: gonadal steroid-dependent changes in VP mRNA accumulation are associated with alternations in mRNA poly (A) tail length but are independent of the rate of VP gene transcription. J Neuroendocrinol 5: 509-515.

Castagna C, Absil P, Foidart A, Balthazart J. 1998. Systemic and intracerebroventricular injections of vasotocin inhibit appetitive and consummatory components of male sexual behavior in Japanese quail. Behav Neurosci 112: 233-250.

Catt KJ, Harwood JP, Aguilera G, Dufau ML. 1979. Hormonal regulation of peptide receptors and target cell responses. Nature 280: 109-116.

Charlesworth B. 1996. The evolution of chromosomal sex determination and dosage compensation. Curr Biol 6: 149-162.

Chu J, Marler CA, Wilczynski W. 1998. The effects of arginine vasotocin on the calling behavior of male cricket frogs in changing social contexts. Horm Behav 34: 248-261.

Chung WC, De Vries GJ, Swaab DF. 2002. Sexual differentiation of the bed nucleus of the stria terminalis in humans may extend into adulthood. J Neurosci 22: 1027-1033.

Coccaro EF, Kavoussi RJ, Hauger RL, Cooper TB, Ferris CF. 1998. Cerebrospinal fluid vasopressin levels: correlates with aggression and serotonin function in personality-disordered subjects. Arch Gen Psychiatry 55: 708-714.

Commins D, Yahr P. 1984. Adult testosterone levels influence the morphology of a sexually dimorphic area in the mongolian gerbil brain. J Comp Neurol 224: 132-140.

Compaan JC, Buijs RM, Pool CW, De Ruiter AJ, Koolhaas JM. 1993. Differential lateral septal vasopressin innervation in aggressive and nonaggressive male mice. Brain Res Bull 30: 1-6.

Cooke B, Hegstrom CD, Villeneuve LS, Breedlove SM. 1998. Sexual differentiation of the vertebrate brain: principles and mechanisms. Front Neuroendocrinol 19: 323-362.

Cooke BM, Tabibnia G, Breedlove SM. 1999. A brain sexual dimorphism controlled by adult circulating androgens. Proc Natl Acad Sci USA 96: 7538-7540.

Cooper KE, Kasting NW, Lederis K, Veale WL. 1979. Evidence supporting a role for endogenous vasopressin in natural suppression of fever in the sheep. J Physiol 295: 33-45.

Crenshaw BL, De Vries GJ. 1991. Vasopressin innervation of the medial preoptic area in gerbils originates in the suprachiasmatic nucleus. Soc Neurosci Abs 17: 1231.

Crenshaw BL, De Vries GJ, Yahr PI. 1992. Vasopressin innervation of sexually dimorphic structures of the gerbil forebrain under various hormonal conditions. J Comp Neurol 322: 589-598.

Dantzer R, Bluthe R-M, Kelley KW. 1991. Androgen dependent vasopressinergic neurotransmission attenuates interleukin 1 induced sickness behavior. Brain Res 557: 115-120.

Dantzer R, Koob GF, Bluthé RM, Le Moal M. 1988. Septal vasopressin modulates social memory in male rats. Brain Res 457: 143-147.

Davidson JM. 1966. Characteristics of sex behaviour in male rats following castration. Animal Behav 14: 266-272.

De Vries GJ, Al-Shamma HS. 1990. Sex differences in hormonal responses of vasopressin pathways in the rat brain. J Neurobiol 21: 686-693.

De Vries GJ, Boyle PA. 1998. Double duty for sex differences in the brain. Behav Brain Res 92: 205-213.

De Vries GJ, Buijs RM. 1983. The origin of vasopressinergic and oxytocinergic innervation of the rat brain with special reference to the lateral septum. Brain Res 273: 307-317.

De Vries GJ, Miller MA. 1998. Anatomy and function of extrahypothalamic vasopressin systems in the brain. Prog Brain Res 119: 3-20.

De Vries GJ, Panzica GC. 2006. Sexual differentiation of central vasopressin and vasotocin systems in vertebrates: different mechanisms, similar endpoints. Neuroscience 138: 947-955.

De Vries GJ, Best W, Sluiter AA. 1983. The influence of androgens on the development of a sex difference in the vasopressinergic innervation of the rat lateral septum. Dev Brain Res 8: 377-380.

De Vries GJ, Buijs RM, Sluiter AA. 1984. Gonadal hormone actions on the morphology of the vasopressinergic innervation of the adult rat brain. Brain Res 298: 141-145.

De Vries GJ, Buijs RM, Swaab DF. 1981. Ontogeny of the vasopressinergic neurons of the suprachiasmatic nucleus and their extrahypothalamic projections in the rat brain presence of a sex difference in the lateral septum. Brain Res 218: 67-78.

De Vries GJ, Buijs RM, Van Leeuwen FW. 1984. Sex differences in vasopressin and other neurotransmitter systems in the brain. Prog Brain Res 61: 185-203.

De Vries GJ, Duetz W, Buijs RM, Van Heerikhuize J, Vreeburg JTM. 1986. Effects of androgens and estrogens on the vasopressin and oxytocin innervation of the adult rat brain. Brain Res 399: 296-302.

De Vries GJ, Gonzales CL, Yahr P. 1988. Afferent connections of the sexually dimorphic area of the hypothalamus of male and female gerbils. J Comp Neurol 271: 91-105.

De Vries GJ, Rissman EF, Simerly RB, Yang LY, Scordalakes EM, et al. 2002. A model system for study of sex chromosome effects on sexually dimorphic neural and behavioral traits. J Neurosci 22: 9005-9014.

De Vries GJ, Wang ZX, Bullock NA, Numan S. 1994. Sex differences in the effects of testosterone and its metabolites on vasopressin messenger RNA levels in the bed nucleus of the stria terminalis of rats. J Neurosci 14: 1789-1794.

De Vries GJ. 1990. Sex differences in neurotransmitter systems. J Neuroend 2: 1-13.

De Wied D. 1969. Effects of peptide hormones on behavior. Frontiers in Neuroendocrinology. Ganong WF, Martini L, editors. London: Oxford University Press; pp. 97-140.

DeBold JE, Miczek KA. 1984. Aggression persist after gonadectomy in female rats. Hormones Behav 18: 177-190.

Del Abril A, Segovia S, Guillamon A. 1987. The bed nucleus of the stria terminalis in the rat: regional sex differences controlled by gonadal steroids early after birth. Dev Brain Res 32: 295-300.

Delville Y, Mansour KM, Ferris CF. 1996. Testosterone facilitates aggression by modulating vasopressin receptors in the hypothalamus. Physiol Behav 60: 25-29.

Demotes Mainard J, Chauveau J, Rodriguez R, Vincent JD, Poulain DA. 1986. Septal release of vasopressin in response to osmotic, hypovolemic and electrical stimulation in rats. Brain Res 381: 314.

Derbyshire SWG, Nichols TE, Firestone L, Townsend DW, Jones AKP. 2002. Gender differences in patterns of cerebral activation during equal experience of painful laser stimulation. J Pain 3: 401-411.

DeVries GJ, Buijs RM, Van Leeuwen FW, Caffé AR, Swaab DF. 1985. The vasopressinergic innervation of the brain in normal and castrated rats. J Comp Neurol 233: 236-254.

DeVries AC, DeVries MB, Taymans SE, Carter CS. 1996. The effects of stress on social preferences are sexually dimorphic in prairie voles. Proc Natl Acad Sci USA 93: 11980-11984.

Dewing P, Chiang CW, Sinchak K, Sim H, Fernagut PO, et al. 2006. Direct regulation of adult brain function by the male-specific factor SRY. Curr Biol 16: 415-420.

Disturnal JE, Veale WL, Pittman QJ. 1985b. Modulation by arginine vasopressin of glutamate excitation in the ventral septal area of the rat brain. Can J Phsyiol Pharmacol 65: 30-35.

Disturnal JE, Veale WL, Pittman QJ. 1985a. Electrophysiological analysis of potential arginine vasopressin projections to the ventral septal area of the rat. Brain Res 342: 162-167.

Dorsa DM, Brot MD, Shewey LM, Meyers KM, Szot P, et al. 1988. Interaction of a vasopressin antagonist with vasopressin receptors in the septum of the rat brain. Synapse 2: 205-211.

Dubois Dauphin M, Tribollet E, Dreifuss JJ. 1989. Distribution of neurohypophysial peptides in the guinea pig brain.

An immunocytochemical study of the vasopressin related glycopeptide. Brain Res 496: 45-65.

Emery DE, Sachs BD. 1976. Copulatory behavior in male rats with lesions in the bed nucleus of the stria terminalis. Physiol Behav 17: 803-806.

Ferris CF, Potegal M. 1988. Vasopressin receptor blockade in the anterior hypothalamus suppresses aggression in hamsters. Physiol Behav 44: 235-239.

Ferris CF, Delville Y, Grzonka Z, Luber-Narod, Insel TR. 1993. An iodinated vasopressin (V1) antagonist blocks flank marking and selectively labels neural binding sites in golden hamsters. Physiol Behav 54: 737-747.

Ferris CF, Delville Y, Miller MA, Dorsa DM, De Vries GJ. 1995. Distribution of small vasopressinergic neurons in golden hamsters. J Comp Neurol 360: 589-598.

Ferris CF, Meenan DM, Axelson JF, Albers HE. 1986. A vasopressin antagonist can reverse dominant/subordinate behavior in hamsters. Physiol Behav 38: 135-138.

Fink G, Sumner BE, McQueen JK, Wilson H, Rosie R. 1998. Sex steroid control of mood, mental state and memory. Clin Exp Pharmacol Physiol 25: 764-775.

Fleming AS, Vaccarino F, Luebke C. 1980. Amygdaloid inhibition of maternal behavior in the nulliparous female rat. Physiol Behav 25: 731-743.

Fliers E, Guldenaar SEF, van der Wal N, Swaab DF. 1986. Extrahypothalamic vasopressin and oxytocin in the human brain; presence of vasopressin cells in the bed nucleus of the stria terminalis. Brain Res 375: 363-367.

Forger NG, Rosen GJ, Waters EM, Jacob D, Simerly RB, et al. 2004. Deletion of Bax eliminates sex differences in the mouse forebrain. Proc Natl Acad Sci USA 101: 13666-13671.

Forger NG. 2001. The development of sex differences in the nervous system. The Handbook of Behavioral Neurobiology, Volume 13: Developmental Psycholobiology. Blass E, editor. New York: Plenum. pp. 153-208.

Forger NG. 2006. Cell death and sexual differentiation of the nervous system. Neuroscience 138: 929-936.

Gatewood JD, Wills A, Shetty S, Xu J, Arnold AP, et al. Sex chromosome complement and gonadal sex influence aggressive and parental behaviors in mice. J Neurosci 26: 2335-2342.

Getz LL, Carter CS, Gavish L. 1981. The mating system of prairie vole, *Microtus ochrogaster*: Field and laboratory evidence for pair bonding. Behav Ecol Sociobiol 8: 189-194.

Gonzalez A, Smeets WJAJ. 1992b. Comparative analysis of the vasotocinergic and mesotocinergic cells and fibers in the brain of two amphibians, the Anuran *Rana ridibunda* and the Urodele *Pleurodeles waltlii*. J Comp Neurol 315: 53-73.

Gonzalez A, Smeets WJAJ. 1992a. Distribution of vasotocin- and mesotocin-like immunoreactivities in the brain of South African clawed frog *Xenopus-laevis* J Chem Neuroanat 5: 465-479.

Goodson JL, Adkins-Regan E. 1999. Effect of intraseptal vasotocin and vasoactive intestinal polypeptide infusions on courtship song and aggression in the male zebra finch (*Taeniopygia guttata*). J Neuroendocrinol 11: 19-25.

Goodson JL, Bass AH. 2001. Social behavior functions and related anatomical characteristics of vasotocin/vasopressin systems in vertebrates. Brain Res Brain Res Rev 35: 246-265.

Goodson JL, Lindberg L, Johnson P. 2004b. Effects of central vasotocin and mesotocin manipulations on social behavior in male and female zebra finches. Horm Behav 45: 136-143.

Goodson JL. 1998a. Territorial aggression and dawn song are modulated by septal vasotocin and vasoactive intestinal polypeptide in male field sparrows (*Spizella pusilla*). Horm Behav 34: 67-77.

Goodson JL. 1998b. Vasotocin and vasoactive intestinal polypeptide modulate aggression in a territorial songbird, the violet-eared waxbill (Estrildidae: *Uraeginthus granatina*). Gen Comp Endocrinol 111: 233-244.

Gorski RA. 1984. Critical role for the medial preoptic area in the sexual differentiation of the brain. Prog Brain Res 61: 129-146.

Gorski RA, Gordon JH, Shryne JE, Southam AM. 1978. Evidence for a morphological sex difference within the medial preoptic area of the rat brain. Brain Res 148: 333-346.

Grabowski TJ, Damasio H, Eichhorn GR, Tranel D. 2003. Effects of gender on blood flow correlates of naming concrete entities. Neuroimage 20: 940-954.

Gu G, Cornea A, Simerly RB. 2001. Sexual differentiation of projections from the principal nucleus of the bed nuclei of the stria terminalis. J Comp Neurol 460: 542-562.

Guillamon A, Segovia S, Del Abril A. 1988. Early effects of gonadal steroids on the neuron number in the medial posterior and the lateral divisions of the bed nucleus of the stria terminalis in the rat. Dev Brain Res 44: 281-290.

Haider SG. 2004. Cell biology of Leydig cells in the testis. Int Rev Cytol 233: 181-241.

Hamilton WJ, Tilson RL, Frank LG. 1986. Sexual monomorphism in spotted hyenas, *Crocuta crocuta*. Ethology 71: 63-73.

Han TM, De Vries GJ. 1999. Neurogenesis of galanin cells in the bed nucleus of the stria terminalis and centromedial amygdala in rats: a model for sexual differentiation of neuronal phenotype. J Neurobiol 38: 491-498.

Han TM, De Vries GJ. 2003. Organizational effects of testosterone, estradiol, and dihydrotestosterone on vasopressin mRNA expression in the bed nucleus of the stria terminalis. J Neurobiol 54: 502-510.

Harris GW, Jacobsohn D. 1952. Functional grafts of the anterior pituitary gland. Proc R Soc Lond B Biol Sci 139: 263-276.

Harris VH, Sachs BD. 1975. Copulatory behavior in male rats following amygdaloid lesions. Brain Res 86: 514-518.

Hayes UL, Tam JB, De Vries GJ. 2005. Role of labor in the induction of maternal behavior in female prairie voles. Soc Neurosci Abs 436: 8.

Hermes MLHJ, Buijs RM, Masson-Pevet M, Van der Woude TP, Pevet P. 1989. Central vasopressin infusion prevents hibernation in the European hamster (*Cricetus cricetus*). Proc Natl Acad Sci USA 86: 6408-6411.

Hier DB, Yoon WB, Mohr JP, Price TR, Wolf PA. 1994. Gender and aphasia in the stroke data-bank. Brain Lang 47: 155-167.

Hines M, Allen LS, Gorski RA. 1992. Sex differences in subregions of the medial nucleus of the amygdala and the bed nucleus of the stria terminalis of the rat. Brain Res 579: 321-326.

Hofman MA, Swaab DF. 1991. Sexual dimorphism of the human brain: myth and reality. Exp Clin Endocrinol 98: 161-170.

Hrdina P. 2000. Sex-related differences: do they matter? J Psych Neurosci 25: 319-320.

Huhman KL, Solomon MB, Janicki M, Harmon AC, Lin SM, et al. 2003. Conditioned defeat in male and female Syrian hamsters. Horm Behav 44: 293-299.

Inglis J, Ruckman M, Lawson JS, MacLean AW, Monga TN. 1982. Sex differences in the cognitive effects of unilateral brain damage. Cortex 18: 257-275.

Insel TR, Hulihan TJ. 1995. A gender-specific mechanism for pair bonding: oxytocin and partner preference formation in monogamous voles. Behav Neurosci 109: 782-789.

Irvin RW, Szot P, Dorsa DM, Potegal M, Ferris CF. 1990. Vasopressin in the septal area of the golden hamster controls scent marking and grooming. Physiol Behav 48: 693-699.

Jaeger JJ, Lockwood AH, Van Valin RD, Kemmerer DL, Murphy BW, et al. 1998. Sex differences in brain regions activated by grammatical and reading tasks. Neuroreport 9: 2803-2807.

Juraska JM. 1991. Sex differences in "cognitive" regions of the rat brain. Psychoneuroendocrinology 16: 105-109.

Kaasinen V, Nagren K, Hietala J, Farde L, Rinne JO. 2001. Sex Differences in Extrastriatal Dopamine D(2)-Like Receptors in the Human Brain. Am J Psychiatry 158: 308-311.

Kasting NW, Veale WL, Cooper KE. 1980. Convulsive and hypothermic effects of vasopressin in the brain of the rat. Can J Physiol Pharmacol 58: 316-319.

Kasting NW. 1989. Criteria for establishing a physiological role for brain peptides a case in point: the role of vasopressin in thermoregulation during fever and antipyresis. Brain Res Rev 14: 143-153.

Kato R. 1960. Serotonin content of rat brain in relation to sex and age. J Neurochem 5: 202.

Kawata M. 1995. Roles of steroid hormones and their receptors in structural organization in the nervous system. Neurosci Res 24: 1-46.

Kelley DB. 1988. Sexually dimorphic behavior. Annu Rev Neurosci 11: 225-251.

Kimura D, Harshman RA. 1984. Sex differences in brain organization for verbal and non-verbal functions. Prog Brain Res 61: 423-441.

Koolhaas JM, Moor E, Hiemstra Y, Bohus B. 1991. The testosterone dependent vasopressinergic neurons in the medial amygdala and lateral septum: involvement in social behaviour of male rats. Vasopressin, Colloque INSERM, Vol. 208. Jard S, Jamison R, editors. London: John Libbey Eurotext Ltd.; pp. 213-219.

Koolhaas JM, Van den Brink THC, Roozendaal B, Boorsma F. 1990. Medial amygdala and aggressive behavior: interaction between teststerone and vasopressin. Aggressive Behav 16: 223-229.

Koopman P, Gubbay J, Vivian N, Goodfellow P, Lovell-Badge R. 1991. Male development of chromosomally female mice transgenic for *Sry*. Nature 351: 117-121.

Kruijver FP, Zhou JN, Pool CW, Hofman MA, Gooren LJ, et al. 2000. Male-to-female transsexuals have female neuron numbers in a limbic nucleus. J Clin Endocrinol Metab 85: 2034-2041.

Landgraf R, Neumann I, Schwarzberg H. 1988. Central and peripheral release of vasopressin and oxytocin in the conscious rat after osmotic stimulation. Brain Res 457: 219-225.

Larsson K, Södersten P, Beyer C. 1973. Sexual behavior in male rats treated with estrogen in combination with dihydrotestosterone. Horm Behav 4: 289-299.

Lee JH, Jordan CL, Arnold AP. 1989. Critical period for androgenic regulation of soma size of sexually dimorphic motoneurons in rat lumbar spinal cord. Neurosci Lett 98: 79-84.

Lema SC, Nevitt GA. 2004. Exogenous vasotocin alters aggression during agonistic exchanges in male Amargosa River pupfish (*Cyprinodon nevadensis amargosae*). Horm Behav 46: 628-637.

Lieb K, Andersen C, Lazarov N, Zienecker R, Urban I, et al. 1996. Pre- and postnatal development of dopaminergic neuron numbers in the male and female mouse midbrain. Dev Brain Res 94: 37-43.

Lieberburg I, McEwen BS. 1975. Brain cell nuclear retention of testosterone metabolites 5α dihydrotestosterone and estradiol 17 in adult rat. Endocrinology 100: 588-597.

Lim MM, Young LJ. 2004. Vasopressin-dependent neural circuits underlying pair bond formation in the monogamous prairie vole. Neuroscience 125: 35-45.

Lonstein JS, Rood BD, De Vries GJ. 2005. Unexpected effects of perinatal gonadal hormone manipulations on sexual differentiation of the extrahypothalamic arginine-vasopressin system in prairie voles. Endocrinology 146: 1559-1567.

Lonstein JS, De Vries GJ. 1999a. Comparison of the parental behavior of pair-bonded female and male prairie voles (*Microtus ochrogaster*). Physiol Behav 66: 33-40.

Lonstein JS, De Vries GJ. 1999b. Sex differences in the parental behaviour of adult virgin prairie voles: independence from gonadal hormones and vasopressin. J Neuroendocrinol 11: 441-449.

Lovell-Badge R, Hacker A. 1995. The molecular genetics of Sry and its role in mammalian sex determination. Philos Trans R Soc Lond B Biol Sci 350: 205-214.

Lyon MF. 1999. X-chromosome inactivation. Curr Biol 9: R235-R237.

Madeira MD, Lieberman AR. 1995. Sexual dimorphism in the mammalian limbic system. Prog Neurobiol 45: 275-333.

Malsbury CW, McKay K. 1987. A sex difference in the pattern of substance P like immunoreactivity in the bed nucleus of the stria terminalis. Brain Res 420: 365-370.

Malsbury CW, McKay K. 1989. Sex difference in the substance P immunoreactive innervation of the medial nucleus of the amygdala. Brain Res Bull 23: 561-567.

Malsbury CW, McKay K. 1994. Neurotrophic effects of testosterone on the medial nucleus of the amygdala in adult male rats. J Neuroendocrinol 6: 57-69.

Maney DL, Erwin KL, Goode CT. 2005. Neuroendocrine correlates of behavioral polymorphism in white-throated sparrows. Horm Behav 48: 196-206.

Marler CA, Chu J, Wilczynski W. 1995. Arginine vasotocin injection increases probability of calling in cricket frogs, but causes call changes characteristic of less aggressive males. Horm Behav 29: 554-570.

Marques DM, Valenstein ES. 1977. Individual differences in aggressiveness of female hamsters: response to intact and castrated males and females. Anim Behav 25: 131-139.

Matthews LH. 1939. Reproduction in the spotted hyena (*Crocuta crocuta*, Erxleben) Philos Trans R Soc London Ser B 230: 1-78.

Mayes CR, Watts AG, McQueen JK, Fink G, Charlton HM. 1988. Gonadal steroids influence neurophysin II distribution in the forebrain of normal and mutant mice. Neuroscience 25: 1013-1022.

McCarthy MM, Konkle AT. 2005. When is a sex difference not a sex difference? Front Neuroendocrinol 26: 85-102.

McCarthy MM, Auger AP, Perrot-Sinal TS. 2002. Getting excited about GABA and sex differences in the brain. Trends Neurosci 25: 307-312.

McEwen BS, Pfaff DW. 1970. Factors influencing sex hormone uptake by rat brain regions. I. Effects of neonatal treatment, hypophysectomy, and competing steroid on estradiol uptake. Brain Res 21: 1-16.

McEwen BS, Pfaff DW, Zigmond RE. 1970. Factors influencing sex hormone uptake by rat brain regions. II. Effects of neonatal treatment and hypophysectomy on testosterone uptake. Brain Res 21: 17-28.

Micevych P, Akesson T, Elde R. 1988b. Distribution of cholecystokinin-immunoreactive cell bodies in the male and female rat: II. Bed nucleus of the stria terminalis and amygdala. J Comp Neurol 269: 381-391.

Micevych P, Ulibarri C. 1992. Development of the limbic-hypothalamic cholecystokinin circuit: a model of sexual differentiation. Dev Neurosci 14: 11-34.

Micevych PE, Matt DW, Go VL. 1988a. Concentrations of cholecystokinin, substance P, and bombesin in discrete regions of male and female rat brain: sex differences and estrogen effects. Exp Neurol 100: 416-425.

Miller MA, De Vries GJ, Al Shamma HA, Dorsa DM. 1992. Rate of decline of vasopressin immunoreactivity and messenger RNA levels in the bed nucleus of the stria terminalis. J Neurosci 12: 2881-2887.

Miller MA, Ferris CF, Kolb PE. 1999. Absence of vasopressin expression by galanin neurons in the golden hamster: implications for species differences in extrahypothalamic vasopressin pathways. Brain Res Mol Brain Res 67: 28-35.

Miller MA, Urban JA, Dorsa DM. 1989a. Steroid dependency of vasopressin neurons in the bed nucleus of the stria terminalis by in situ hybridization. Endocrinology 125: 2335-2340.

Miller MA, Vician L, Clifton DK, Dorsa DM. 1989b. Sex differences in vasopressin neurons in the bed nucleus of the stria terminalis by in situ hybridization. Peptides 10: 615-619.

Mizukami S, Nishizyka M, Arai Y. 1983. Sexual difference in nuclear volume and its ontogeny in the rat amygdala. Exp Neurol 79: 569-575.

Moore FL. 1992. Evolutionary precedents for behavioral actions of oxytocin and vasopressin. Ann NY Acad Sci 652: 156-165.

Morris JA, Jordan CL, Breedlove SM. 2004. Sexual differentiation of the vertebrate nervous system. Nat Neurosci 7: 1034-1039.

Naftolin F, Ryan KJ, Davies IJ, Reddy VV, Flores F, et al. 1975. The formation of estrogens by central neuroendocrine tissue. Rec Progr Horm Res 31: 295-316.

Nance DM, Shryne J, Gorski RA. 1974. Septal lesions: effects on lordosis behavior and pattern of gonadotropin release. Horm Behav 5: 73-81.

Naylor AM, Pittman QJ, Veale WL. 1988. Stimulation of vasopressin release in the ventral septum of the rat brain suppresses prostaglandin E1 fever. J Physiol Lond 399: 177-189.

Nephew BC, Aaron RS, Romero LM. 2005. Effects of arginine vasotocin (AVT) on the behavioral, cardiovascular, and corticosterone responses of starlings (*Sturnus vulgaris*) to crowding. Horm Behav 47: 280-289.

Nishizawa S, Benkelfat C, Young SN, Leyton M, Mzengeza S, et al. 1997. Differences between males and females in rates of serotonin synthesis in human brain. Proc Natl Acad Sci USA 94: 5308-5313.

Nishizuka M, Arai Y. 1983. Regional difference in sexually dimorphic synaptic organization of the medial amygdala. Exp Brain Res 49: 462-465.

Nottebohm F, Arnold AP. 1976. Sexual dimorphism in vocal control areas of the songbird brain. Science 194: 211-213.

Numan M, Numan M. 1996. A lesion and neuroanatomical tract-tracing analysis of the role of the bed nucleus of the stria terminalis in retrieval behavior and other aspects of maternal responsiveness in rats. Dev Psychobiol 29: 23-51.

Ostrowski NL, Lolait SJ, Bradley DJ, O'Carroll A-M, Brownstein MJ, et al. 1992. Distribution of V1a and V2 vasopressin receptor messenger ribonucleic acids in rat liver, kidney, pituitary and brain. Endocrinology 131: 533-535.

Palanza P. 2005. Animal models of anxiety and depression: how are females different? Neurosci Biobehav Rev 25: 219-233.

Panzica GC, Aste N, Castagna C, Viglietti-Panzica C, Balthazart J. 2001. Steroid-induced plasticity in the sexually dimorphic vasotocinergic innervation of the avian brain: behavioral implications. Brain Res Rev 37: 178-200.

Panzica GC, Balthazart J, Pessatti M, Viglietti-Panzica C. 2002. The parvocellular vasotocin system of Japanese quail: a developmental and adult model for the study of influences of gonadal hormones on sexually differentiated and behaviorally relevant neural circuits. Environ Health Perspect 110, Suppl 3: 423-428.

Payne AP, Swanson HH. 1970. Agonistic behaviour between pairs of hamsters of the same and opposite sex in a neutral observation area. Behaviour 36: 260-269.

Pedersen CA, Boccia ML. 2006. Vasopressin interactions with oxytocin in the control of female sexual behavior. Epub ahead of print.

Pfaff DW. 1966. Morphological changes in the brains of adult male rats after neonatal castration. J Endocrinol 36: 415-416.

Phoenix CH, Goy RW, Gerall AA, Young WC. 1959. Organizing action of prenatally administered testosterone propionate on the tissues mediating mating behavior in the female guinea pig. Endocrinology 65: 369-382.

Piefke M, Weiss PH, Markowitsch HJ, Fink GR. 2005. Gender differences in the functional neuroanatomy of emotional episodic autobiographical memory. Hum Brain Mapp 24: 313-324.

Pittman QJ, Malkinson TJ, Kasting NW, Veale WL. 1988. Enhanced fever following castration: possible involvement of brain arginine vasopressin. Am J Physiol 254: R513-R517.

Planas B, Kolb PE, Raskind MA, Miller MA. 1995. Sex difference in coexpression by galanin neurons accounts for sexual dimorphism of vasopressin in the bed nucleus of the stria terminalis. Endocrinology 136: 727-733.

Plumari L, Viglietti-Panzica C, Allieri F, Honda S, Harada N, et al. 2002. Changes in the arginine-vasopressin immunoreactive systems in male mice lacking a functional aromatase gene. J Neuroendocrinol 14: 971-978.

Pohjalainen T, Rinne JO, Nagren K, Syvalahti E, Hietala J. 1998. Sex differences in the striatal dopamine D2 receptor binding characteristics in vivo. Am J Psychiatry 155: 768-773.

Poulin P, Pittman QJ. 1991. Septal arginine vasopressin (AVP) receptor regulation in rats depleted of septal AVP following long term castration. J Neurosci 11: 1531-1539.

Propper CR, Jones RE, Lopez KH. 1992. Distribution of arginine vasotocin in the brain of the lizard Anolis carolinensis. Cell Tiss Res 267: 391-398.

Raggenbass M, Tribollet E, Dreiffuss JJ. 1987. Electrophysiological and autoradiographical evidence of V1 vasopressin receptors in the lateral septum of the rat brain. Proc Natl Acad Sci USA 84: 7778-7783.

Raisman G, Field PM. 1971. Sexual dimorphism in the preoptic area of the rat. Science 173: 731-733.

Raisman G, Field PM. 1973. Sexual dimorphism in the neuropil of the preoptic area of the rat and its dependence on neonatal androgen. Brain Res 54: 1-29.

Reisert I, Pilgrim C. 1991. Sexual differentiation of monoaminergic neurons–genetic or epigenetic? Trends Neurosci 14: 468-473.

Rhodes ME, Rubin RT. 1999. Functional sex differences ('sexual diergism') of central nervous system cholinergic systems, vasopressin, and hypothalamic-pituitary-adrenal axis activity in mammals: a selective review. Brain Res Rev 30: 135-152.

Roof RL, Hall ED. 2000. Gender differences in acute CNS trauma and stroke: neuroprotective effects of estrogen and progesterone. J Neurotrauma 17: 367-388.

Rose JD, Moore FL. 2002. Behavioral neuroendocrinology of vasotocin and vasopressin and the sensorimotor processing hypothesis. Front Neuroendocrinol 23: 317-341.

Roselli CE, Fasasi TA. 1992. Estradiol increases the duration of nuclear androgen receptor occupation in the preoptic area of the male rat treated with dihydrotestosterone. J Steroid Biochem Mol Biol 42: 161-168.

Roselli CE. 1991. Sex differences in androgen receptors and aromatase activity in microdissected regions of the rat brain. Endocrinology 128: 1310-1316.

Rosen GJ, De Vries GJ, Villalba C, Weldele ML, Place NJ, 2006. The distribution of vasopressin in the forebrain of spotted hyenas. J Comp Neurol 498: 80-92.

Rosenblatt JS, Mayer AD, Giordano AL. 1988. Hormonal basis during pregnancy for the onset of maternal behavior in the rat. Psychoneuroend 13: 29-46.

Schick RR, Schusdziarra V, Yaksh TL, Go VL. 1994. Brain regions this is what I want it to read. Ann NY Acad Sci 713: 242-254.

Scordalakes EM, Rissman EF. 2004. Aggression and arginine vasopressin immunoreactivity regulation by androgen receptor and estrogen receptor alpha. Genes Brain Behav 3: 20-26.

Segovia S, Guillamon A. 1993. Sexual dimorphism in the vomeronasal pathway and sex differences in reproductive behaviors. Brain Res Brain Res Rev 18: 51-74.

Segovia S, Guillamon A, del Cerro MC, Ortega E, Perez-Laso C, et al. 1999. The development of brain sex differences: a multisignaling process. Behav Brain Res 105: 69-80.

Semsar K, Kandel FL, Godwin J. 2001. Manipulations of the AVT system shift social status and related courtship and aggressive behavior in the bluehead wrasse. Horm Behav 40: 21-31.

Semsar K, Klomberg KF, Marler C. 1998. Arginine vasotocin increases calling-site acquisition by nonresident male grey treefrogs. Anim Behav 56: 983-987.

Shapiro RA, Xu C, Dorsa DM. 2000. Differential transcriptional regulation of rat vasopressin gene expression by estrogen receptor alpha and beta. Endocrinology 141: 4056-4064.

Shaywitz BA, Shaywitz SE, Pugh KR, Constable RT, Skudlarski P, et al. 1995. Sex differences in the functional organization of the brain for language. Nature 373: 607-609.

Sibug R, Kuppers E, Beyer C, Maxson SC, Pilgrim C, et al. 1996. Genotype-dependent sex differentiation of dopaminergic neurons in primary cultures of embryonic mouse brain. Dev Brain Res 93: 136-142.

Simerly RB. 2002. Wired for reproduction: organization and development of sexually dimorphic circuits in the mammalian forebrain. Annu Rev Neurosci 25: 507-536.

Simerly RB. 1990. Hormonal control of neuropeptide gene expression in sexually dimorphic olfactory pathways. Trends Neurosci 13: 104-110.

Simerly RB. 1995. Anatomical substrates of hypothalamic integration. The Rat Nervous System, Second Edition. Paxinos G, editor. San Francisco: Academic Press, Inc.; pp. 353-376.

Simerly RB, Chang C, Muramatsu M, Swanson LW. 1990. Distribution of androgen and estrogen receptor mRNA-containing cells in the rat brain: an in situ hybridization study. J Comp Neurol 294: 76-95.

Smeets WJAJ, Sevensma JJ, Jonker AJ. 1990. Comparative analysis of vasotocin like immunoreactivity in brain of the turtle Pseudemys scripta elegans and the snake Python regius. Brain Beh Evol 35: 65-84.

Södersten P, De Vries GJ, Buijs RM, Melin P. 1985. A daily rhythm in behavioural vasopressin sensitivity and brain vasopressin concentrations. Neurosci Lett 58: 37-41.

Södersten P. 1980. A way in which oestradiol might play a role in the sexual behaviour of male rats. Horm Behav 14: 271-274.

Stoll CJ, Voorn P. 1985. The distribution of hypothalamic and extrahypothalamic vasotocinergic cells and fibers in the brain of a lizard, Gekko gecko: presence of a sex difference. J Comp Neurol 239: 193-204.

Swaab DF, Chung WC, Kruijver FP, Hofman MA, Hestiantoro A. 2003. Sex differences in the hypothalamus in the different stages of human life. Neurobiol Aging 24 Suppl 1: S1-S19.

Swaab DF, Fliers E. 1985. A sexually dimorphic nucleus in the human brain. Science 228: 1112-1115.

Szot P, Dorsa DM. 1993. Differential timing and sexual dimorphism in the expression of the vasopressin gene in the developing rat brain. Dev Brain Res 73: 177-183.

Taravosh-Lahn K, Delville Y. 2004. Aggressive behavior in female golden hamsters: development and the effect of repeated social stress. Horm Behav 46: 428-435.

Thompson R, Gupta S, Miller K, Mills S, Orr S. 2004. The effects of vasopressin on human facial responses related to social communication. Psychoneuroendocrinol 29: 35-48.

Tito MB, Hoover MA, Mingo AM, Boyd SK. 1999. Vasotocin maintains multiple call types in the gray treefrog, Hyla versicolor. Horm Behav 36: 166-175.

Tobet SA, Hanna IK. 1997. Ontogeny of sex differences in the mammalian hypothalamus and preoptic area. Cell Mol Neurobiol 17: 565-601.

Toufexis DJ, Myers KM, Davis M. 2006. The effect of gonadal hormones and gender on anxiety and emotional learning in the rat. Horm Behav, In press.

Tribollet E, Audigier S, Dubois-Dauphin M, Dreiffus JJ. 1988. Gonadal steroids regulate oxytocin receptors but not vasopressin receptors in the brain of male and female rats. An autoradiographic study. Brain Res 511: 129-140.

Valcourt RJ, Sachs BD. 1979. Penile reflexes and copulatory behavior in male rats following lesions in the bed nucleus of the stria terminalis. Brain Res Bull 4: 131-133.

Van den Dungen HM, Buijs RM, Pool CW, Terlou M. 1982. The distribution of vasotocin and isotocin in the brain of the rainbow trout. J Comp Neurol 212: 146-157.

Van Leeuwen FW, Caffe R. 1983. Vasopressin immunoreactive cell bodies in the bed nucleus of the stria terminalis of the rat. Cell Tiss Res 228: 525-534.

Van Leeuwen FW, Caffé AR, De Vries GJ. 1985. Vasopressin cells in the bed nucleus of the stria terminalis of the rat: sex differences and the influence of androgens. Brain Res 325: 391-394.

Versteeg CAM, Cransberg K, de Jong W, Bohus B. 1983. Reduction of a centrally induced pressor response by neurohypophyseal peptides: The involvement of lower brainstem mechanisms. Eur J Pharmacol 94: 133-140.

Viglietti-Panzica C, Anselmetti GC, Balthazart J, Aste N, Panzica GC. 1992. Vasotocinergic innervation of the septal region in Japanese quail: sexual differences and the influence of testosterone. Cell Tiss Res 267: 261-265.

Villalba C, Boyle PA, De Vries GJ. 1997. Effects of the selective serotonin reuptake inhibitor, fluoxetine, on social behaviors in male and female prairie voles (*Microtus ochrogaster*). Horm Behav 32: 184-191.

Voorhuis TAM, Kiss JZ, De Kloet ER, De Wied D. 1988. Testosterone-sensitive vasotocin-immunoreactive cells and fibers in the canary brain. Brain Res 442: 139-146.

Wang ZX, De Vries GJ. 1993. Testosterone effects on paternal behavior and vasopressin immunoreactive projections in prairie voles (*Microtus ochrogaster*). Brain Res 631: 156-160.

Wang ZX, De Vries GJ. 1995. Androgen and estrogen effects on vasopressin messenger RNA expression in the medial amygdaloid nucleus in male and female rats. J Neuroendocrinol 7: 827-831.

Wang ZX, Bullock NA, De Vries GJ. 1993. Sexual differentiation of vasopressin projections of the bed nucleus of the stria terminalis and medial amygdaloid nucleus in rats. Endocrinology 132: 2299-2306.

Wang ZX, Ferris CF, De Vries GJ. 1994b. The role of septal vasopressin innervation in paternal behavior in prairie voles (*Microtus ochrogaster*). Proc Natl Acad Sci USA 91: 400-404.

Wang ZX, Moody K, Newman JD, Insel TR. 1997. Vasopressin and oxytocin immunoreactive neurons and fibers in the forebrain of male and female common marmosets (*Callithrix jacchus*). Synapse 27: 14-25.

Wang ZX, Smith W, Major DE, De Vries GJ. 1994a. Sex and species differences in the effects of cohabitation on vasopressin messenger RNA expression in the bed nucleus of the stria terminalis in prairie voles (*Microtus ochrogaster*) and meadow voles (*Microtus pennsylvanicus*). Brain Res 650: 212-218.

Willis GL, Hansky J, Smith GC. 1986. Central and peripheral proglumide administration and cholecystokinin-induced satiety. Regul Pept 15: 87-98.

Winslow JT, Hastings N, Carter CS, Harbaugh CR, Insel TR. 1993. A role for central vasopressin in pair bonding in monogamous prairie voles. Nature 365: 545-547.

Winslow JT, Hastings N, Carter CS, Harbaugh CR, Insel TR. 1993. A role for central vasopressin in pair bonding in monogamous prairie voles. Nature 365: 545-548.

Wisniewski AB. 1998. Sexually-dimorphic patterns of cortical asymmetry, and the role for sex steroid hormones in determining cortical patterns of lateralization. Psychoneuroendocrinology 23: 519-547.

Yahr P. 1988. Sexual differentiation of behavior in the context of developmental psychobiology. Handbook of Behavioral Neurobiology. Blass EM, editor. New York: Plenum Press; pp. 197-243.

Young LJ, Wang Z. 2004. The neurobiology of pair bonding. Nat Neurosci 7: 1048-1054.

Young WS III. 1992. Expression of the oxytocin and vasopressin genes. J Neuroend 4: 527-540.

Zasorin NL, Malsbury CW, Pfaff DW. 1975. Suppression of lordosis in the hormone primed female hamster by electrical stimulation of the septal area. Physiol Behav 14: 595-599.

Zhou JN, Hofman MA, Gooren LJ, Swaab DF. 1995. A sex difference in the human brain and its relation to transsexuality. Nature 378: 68-70.

Zhou L, Blaustein JD, De Vries GJ. 1994. Distribution of androgen receptor immunoreactivity in vasopressin-immunoreactive and oxytocin-immunoreactive neurons in the male rat brain. Endocrinol 134: 2622-2627.

13 Neurochemical Systems Regulating the Hypothalamo–Pituitary–Adrenocortical Axis

J. P. Herman · H. F. Figueiredo · N. K. Mueller · M. M. Ostrander · R. Zhang · M. Tauchi · D. C. Choi · A. R. Furay · N. K. Evanson · E. B. Nelson · Y. M. Ulrich-Lai

Abstract: Control of glucocorticoid secretion is essential for the health and survival of all vertebrate organisms. Hyper- and hypo-secretion of glucocorticoids are associated with disease processes, underlying the importance of maintaining normal daily glucocorticoid rhythms and generating appropriate glucocorticoid responses to stress. This chapter reviews the principle neurochemical mechanisms that operate in the CNS to regulate excitation and inhibition of the hypothalam-pituitary-adrenocortical (HPA) axis. Daily glucocorticoid secretion is controlled by monoamine and GABAergic circuitry, likely relayed through the suprachiasmatic nucleus. Glucocorticoids appear to play a role in circadian inhibition, exerted via the mineralocorticoid receptor. Neurochemical activation of the HPA axis is highly dependent on modality and intensity. Notably, brainstem norepinephrine/epinephrine neurons are selectively involved in HPA axis activation by systemic stressors. Activation of the HPA axis by psychogenic stressors is intensity-dependent, with peptidergic (vasopressin, Orphanin FQ) and glutamatergic systems playing a role in responses to mild, but not intense stressors. Responses to intense psychogenic stressors appear to involve serotonergic and peptidergic systems (e.g., brainstem glucagon-like peptide 1). Inhibition of the HPA axis is accomplished by GABAergic signals and glucocorticoid feedback, the latter of which is controlled by combined actions at glucocorticoid and mineralocorticoid receptors. The neurochemical systems underlying chronic stress-induced changes in HPA function remain to be elucidated. Overall, the data to date identify numerous candidate neurochemical systems capable of modulating HPA axis activity. Selective targeting of these systems may prove useful for treatment of HPA axis-related disease states.

List of Abbreviations: 5,7-DHT, 5,7-dihydroxytryptamine; 5-HT, 5-hydroxytryptamine (serotonin); 6-OHDA, 6-hydroxydopamine; ACTH, adrenocorticotropic hormone; ADM, adrenomedullin; ANP, atrial natriuretic peptide; AVP, arginine vasopressin; BMI, bicuculline methiodide; CART, cocaine and amphetamine-regulated transcript; CB1, cannabinoid receptor 1; CBG, corticosteroid binding globin; CCK, cholecystokinin; CGRP, calcitonin gene-related peptide; CNP, C-type natriuretic peptide; CNS, central nervous system; CO, carbon monoxide; CRH, corticotropin-releasing hormone; CRH R1, CRH receptor 1; DHT, dihydrotestosterone; DYN, dynorphin; E, epinephrine; ENK, enkephalin; ER, estrogen receptor; GABA, gamma-aminobutyric acid; GAD, glutamic acid decarboxylase; GALP, galanin-like peptide; GLP-1, glucagon-like petide-1; GR, glucocorticoid receptor; HA, histamine; HPA, hypothalamo-pituitary-adrenocortical axis; i.c.v., Intracerebroventricular; i.v., Intravenous; IL-1β, interleukin 1-beta; LC, locus coeruleus; MCH, melanin concentrating hormone; MCR, melanocortin receptor; MR, mineralocorticoid receptor; MSH, melanocyte stimulating hormone; N/OFQ, Nociceptin/orphanin FQ; NE, norepinephrine; NK, neurokinin; NMDA, N-methyl D-aspartate; NO, nitric oxide; NPB, neuropeptide B; NPW, neuropeptide W; NPY, neuropeptide Y; NTS, nucleus of the solitary tract; PVN, paraventricular nucleus; SCN, suprachiasmatic nucleus; SP, substance P; VIP, vasoactive intestinal polypeptide; VNAB, ventral noradrenergic bundle

1 Introduction

1.1 Hypothalamo-Pituitary-Adrenocortical Axis

The hypothalamo–pituitary–adrenocortical (HPA) axis is a key regulator of systemic homeostasis. Stimulation of this axis culminates in the secretion of glucocorticoid hormones, which act on multiple peripheral effectors as well as the brain to promote adaptation to adversity. In most contexts, the primary function of glucocorticoids is to redistribute energy resources. Contemporary theory holds that this redistribution process is involved in restoration or defense of homeostasis following challenge. For example, glycogenolytic effects of glucocorticoids promote mobilization of glucose in the wake of energy expenditure, whereas antiinflammatory actions of glucocorticoids are thought to "brake" ongoing immune responses (Munck et al., 1984).

However, it is important to emphasize that release of glucocorticoids also occurs in anticipation of adverse events. There are numerous environmental factors that will trigger glucocorticoid release in the absence of a frank physical stimulus. For example, rodents exhibit marked hormonal responses to open, illuminated spaces (Emmert and Herman, 1999), presumably to reduce risk of predation, as well as to predator odors (Morrow et al., 2000), or the physical presence of predators (Figueiredo et al., 2003). Furthermore, there is ample evidence that secretion of glucocorticoids can be paired to previously innocuous stimuli, i.e., the organism can "learn" to produce a glucocorticoid response. The relevance of the anticipatory response is likely to be hinged on the predicted occurrence of energy expenditure; the process is in motion before the survival challenge (prolonged flight, fight, and injury) is encountered.

The characteristics of stimulus-induced glucocorticoid release are tightly linked to the concept of "stress." The definition of "stress" has been through numerous revisions since the original coining of the term by Selye (1956); within the context of our work, stress can be viewed as "a real or perceived threat to homeostasis." The response to stress takes many forms, of which glucocorticoid secretion is one. Notably, "stressors" typically engender a very rapid and short-lived sympathetic response and a rise in plasma norepinephrine (sympathetic nerves) and epinephrine (adrenal medulla), comprising the so-called 'sympathoadrenomedullary' response. In addition, stressors may invoke behavioral responses that may or may not occur in concert with physiological changes. Unlike the physiological responses, these behavioral responses are quite dependent on internally or externally perceived cues. Thus, the glucocorticoid response is one component of the integrated set of events responsible for adapting to "stress" – and should not be confused with being synonymous with the stress response.

The multifaceted array of stimuli that provoke stress responses can be coalesced into generalized categories that have some construct validity. Stress responses initiated by the types of stimuli relying on physical perturbations (e.g., hypoglycemia, inflammation, blood loss, and pain) are direct "reactions" to these stimuli. These types of stimuli are currently categorized as "systemic," "interoceptive," or "physiological" stressors. Stress reactions initiated by innate or learned cues occur before a physical perturbation is present; i.e., they "anticipate" the physical stimulus. In these cases, the "stimuli" for the stress response are generated within the brain as a result of the interpretation of physically innocuous sensory information. This subtype of stressor has been categorized as "psychogenic," "neurogenic," "processive," "exteroceptive," or "emotional" stressors. The latter category of stressor triggers "innate" response patterns, programmed into the brain over the course of evolution, and/or "learned" responses, i.e., association of sensory stimuli with danger or adversity. This hypothetical classification of stimuli – which we will refer to as "systemic" or "psychogenic" stress (Herman et al., 2003) – has heuristic value in understanding the neurochemistry of stress (at least as concerns the HPA axis) and will be a recurring theme of this chapter.

The hypothalamic paraventricular nucleus (PVN) is the initiation point of neuronally-mediated stress responses (Antoni, 1986; Whitnall, 1993) (➋ *Figure 13-1*). Activation of the HPA axis is mediated by a discrete set of neurosecretory neurons localized in the medial parvocellular zone of this nucleus. These neurons project to blood vessels in the external lamina of the median eminence, where they release numerous ACTH "secretagogues," the most important of which are corticotropin-releasing hormone (CRH) and arginine vasopressin (AVP) (Antoni, 1986; Whitnall, 1993). Importantly, these neurons also express glucocorticoid receptors (GR) (Uht et al., 1988) and may serve as a site of "negative feedback" regulation of the HPA axis by glucocorticoids.

Generally, dendrites of medial parvocellular PVN neurons do not extend beyond the PVN proper (van den Pol, 1982; Rho and Swanson, 1989). Thus, processing of afferent information largely occurs within the nucleus, either by direct synaptic innervation or interactions among neighboring cell populations (e.g., magnocellular neurons or glial cells). This point is important, as numerous regions, including limbic structures such as the hippocampus, medial amygdala, and lateral septum (Oldfield and Silverman, 1985; Oldfield et al., 1985; Canteras and Swanson, 1992; Cullinan et al., 1993; Canteras et al., 1995), send projections to the neuropil immediately outside the PVN. The absence of dendritic extensions from PVN neurons into this "peri-PVN zone" makes it unlikely that these limbic regions directly influence the PVN.

The PVN is one of the most highly vascularized regions in the brain (McGinty et al., 1983). Thus, the PVN, including its parvocellular component, has direct access to signaling molecules that can traverse the blood–brain barrier. In addition, there is some evidence suggesting that the PVN vessels themselves can produce signaling molecules, such as prostaglandins, nitric oxide (NO), or carbon monoxide (CO)

Figure 13-1

Organization of the HPA axis. Corticotropin releasing hormone (CRH) neurons located within the medial parvocellular region (mp) of the hypothalamic paraventricular nucleus (PVN) drive pituitary corticotrophs via the portal vasculature, stimulating ACTH. ACTH, in turn, mediates the synthesis and release of corticosteroids from the adrenal glands

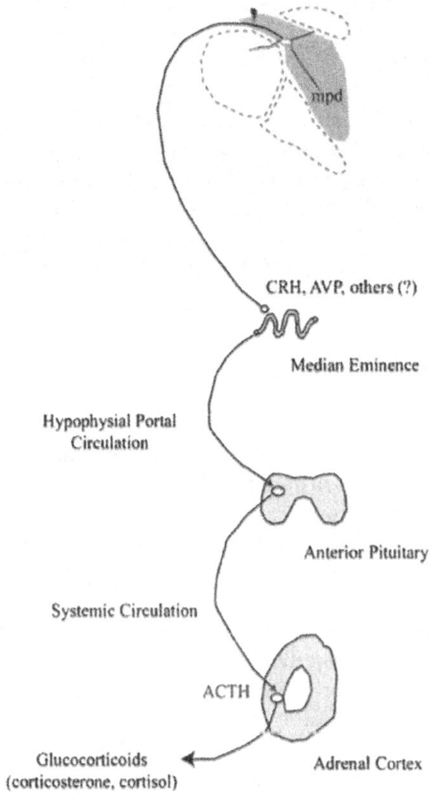

(Vincent et al., 1994; Rivest, 2001; Rivier, 2002), following appropriate stimulation (e.g., cytokine binding). The vascularity also assures parvocellular PVN neurons access to glucocorticoids, which distribute freely across the blood–brain barrier.

1.2 Normal Domains of HPA Axis Function

1.2.1 Basal Corticosteroid Secretion

It is well known that the HPA axis is driven by cues related to "stressors"; however, it is important to note that there is a powerful circadian rhythm of glucocorticoid secretion. In most organisms, glucocorticoids peak immediately before the onset of waking (Dallman et al., 1987, for review). The levels attained during the diurnal peak are not trivial; typically, in rats, circadian peak levels of corticosterone are roughly half the secreted product generated during a robust stress response. This contrasts markedly with extremely low levels seen during the nadir, which can approach the lower limits of detection in some assays. The net impact of the rhythm is a roughly 40- to 50-fold fluctuation in corticosterone levels every day. Daily corticosterone variations have a potentially profound impact on brain (and body) glucocorticoid signaling across the day; low levels seen during the nadir are only sufficient to bind the high-affinity mineralocorticoid receptor (MR), whereas peak levels can occupy the majority of GRs in addition to MRs (see De Kloet et al., 1998, for review). Emerging evidence ascribes different functions to MR, GR, and possibly, MR–GR heterodimers (e.g., Trapp et al., 1994), indicating that the level of glucocorticoids may indeed dictate how glucocorticoid signals are interpreted in the brain and periphery.

1.2.2 Acute Stress Responses

Glucocorticoid secretion is generally part of the acute response of the organism to stress. It is often erroneously referred to as part of the "fight or flight" response; in fact, the actions of glucocorticoids, being mostly genomic in nature, are generally delayed, and are thus more relevant to survival well after an initial, sympathetically-mediated "fight or flight" reaction has occurred. Indeed, it is thought by some that the glucocorticoid response is primarily geared to restoration of homeostasis (Munck et al., 1984). The net result of glucocorticoid release is best thought of as a redistribution of resources, e.g., mobilizing energy reserves, modulating immune function, and/or inhibiting energy-intensive systems not required for survival.

1.2.3 Glucocorticoid Negative Feedback

Glucocorticoids released following stress have negative feedback actions on the HPA axis. Feedback effects of glucocorticoids happen in three different time frames: rapid effects occur within minutes, intermediate feedback over hours, and delayed feedback over days (Keller-Wood and Dallman, 1984). Rapid or "fast," feedback is rate-sensitive (occurs when corticosteroid levels are rising) and is exerted at the level of ACTH secretion, perhaps via the PVN; these rapid effects are almost certainly nongenomic and thus unlikely to act through the "traditional" GR. Intermediate feedback effects operate on the time frame of genomic responses, but typically do not involve changes in pituitary proopiomelanocortin (ACTH precursor) synthesis or storage; rather, the history of glucocorticoid release is able to modify later secretory patterns, perhaps via neuronal mechanisms. The last realm of feedback is mediated by direct changes in transcription and protein synthesis so as to downregulate the capacity of the pituitary to make and secrete ACTH (Keller-Wood and Dallman, 1984).

Neuronal pathways mediating glucocorticoid negative feedback are yet to be defined. The GR is abundantly expressed in the pituitary, PVN, and numerous brain regions implicated in HPA axis regulation (Aronsson et al., 1988; Arriza et al., 1988; Ahima and Harlan, 1990; Herman, 1993), and studies have implicated some or all of these loci in feedback processes (Herman et al., 2003). Given that glucocorticoids are capable of rapid inhibition of PVN activity and presynaptic glutamate release (Saphier, 1987; Di et al., 2003), it is likely that fast feedback is mediated to some degree at the PVN. The proopiomelanocortin gene has a known negative glucocorticoid response element that affords direct transcriptional repression by glucocorticoids (Drouin et al., 1989), which may well account for delayed feedback effects. The intermediate realm of feedback is currently under some debate; it appears to occur at the level of the brain (Levin et al., 1988), but the exact location is poorly understood. Strong cases can be made for the PVN itself, the hippocampus, and the medial prefrontal cortex as potential feedback loci; however, in most cases the data in the literature are mixed (see Herman et al., 2003, for review) and it is likely that feedback is mediated by distributed rather than dedicated circuits.

1.2.4 Chronic Stress

Chronic drive of the HPA axis with prolonged stress precipitates long-term changes in both central and peripheral effector systems. These include adrenal hypertrophy, associated with prolonged exposure of the adrenal cortex to ACTH, and thymic atrophy, a probable result of chronic glucocorticoid elevation (Prewitt and Herman, 1997). Chronic glucocorticoid release consequent to stress may also have an impact on other endocrine systems [e.g., inhibition of testosterone release (Tamashiro et al., 2005)]. Glucocortioids are directly linked to chronic stress-related hippocampal damage and cognitive deficits (McEwen and Sapolsky, 1995; Sapolsky, 2000) and may also be related to the decreased food intake (Dallman et al., 2003), behavioral anhedonia (Willner et al., 1992), and altered cardiovascular tone (Grippo et al., 2002) characteristic of prolonged stress exposure.

The overall impact of a given chronic stress regimen on HPA axis function is highly dependent on design. Most chronic stress designs create long-term increases in basal corticosterone release (Ottenweller et al., 1992; Gomez et al., 1996; Prewitt and Herman, 1997). This enhanced release can be observed 16–24 h after the last stress exposure, indicating that the HPA system has undergone long-term upregulation. Enhanced basal release of ACTH may or may not accompany the corticosterone changes, suggesting that in some cases sensitivity of the adrenal may be affected by chronic stress. At the level of the pituitary, chronic stress generally enhances proopiomelanocortin mRNA and pituitary ACTH stores (Shiomi et al., 1986); in contrast, CRH-R1 binding is decreased (Anderson et al., 1993; Fuchs and Flugge, 1995), suggesting a downregulatory response to enhanced CRH secretion. In the hypothalamus, chronic stress generally increases PVN CRH and AVP mRNA expression (DeGoeij et al., 1991, 1992b; Imaki et al., 1991; Herman et al., 1995; Prewitt and Herman, 1997), PVN and median eminence CRH peptide levels (Chappell et al., 1986), and the extent of CRH/AVP colocalization in the median eminence (Whitnall, 1989), all of which are consistent with enhanced PVN activity and secretagogue release. Enhanced PVN CRH expression is negatively correlated with PVN glucocorticioid receptor levels (Herman et al., 1995; Makino et al., 1995; Prewitt and Herman, 1997), implying reduced hypothalamic glucocorticoid feedback efficacy following chronic stress. Chronic stress also causes changes in upstream stress regulatory circuits; for example, chronic regimens can reduce hippocampal expression of glucocorticoid and/or MRs (Sapolsky et al., 1984; Herman et al., 1995; Gomez et al., 1996; Prewitt and Herman, 1997), and increase expression of tyrosine hydroxylase in the locus coeruleus (LC) (Mamalaki et al., 1993). The overall connection between these mRNA changes in stress regulatory regions and the stress-induced drive of the HPA axis remains to be determined.

Repeated exposure to some stressors (restraint, cold) can cause habituation of HPA axis responses (Akana et al., 1992; Cole et al., 2000). For example, the corticosterone response to a fifth daily session of restraint is roughly half the magnitude of the first (Cole et al., 2000). It is important to note that the animal still responds to an individual stressor and is therefore still subject to the cumulative impact of multiple stress exposures. The habituation response is stressor-specific; for example, HPA axis responses to immobilization do not show decrements with repeated exposure (Dobrakovova et al., 1982). Failure to habituate is likely associated with stressor intensity (e.g., movement of the limbs is completely prohibited in immobilization, whereas limited movement is allowed in restraint).

Finally, chronic stress can also result in HPA axis facilitation or sensitization, whereby responses to novel stressors are enhanced following repeated homotypic stress exposure (Akana et al., 1992; Bhatnagar and Dallman, 1998). In facilitation paradigms, responses are either the same as naïve controls (i.e., blockade of habituation) or substantially higher. Facilitation appears to be mediated within the CNS (see later).

Chronic stress results in numerous physiological and behavioral changes that are similar to human disease or aging pathology. For example, chronic stress-induced adrenal hypertrophy, weight loss, anhedonia, and dexamethasone feedback resistance are highly reminiscent of melancholic depression (Willner et al., 1992), a disease with clear links to glucocorticoid hypersecretion and stress. In addition, chronic stress can cause memory impairment and cellular atrophy in the hippocampus, both of which are seen in rodent (and human) aging models (McEwen and Sapolsky, 1995). Thus, chronic drive of the HPA axis represents a significant health risk factor for the organism.

2 Neurochemistry of HPA Axis Regulation: Overview

The neurochemistry of HPA axis regulation has been studied extensively and has yielded a voluminous (if not confusing) literature. In this chapter, we focus on understanding how various neurotransmitter, neuropeptide, and steroid systems affect the central activation of the HPA axis, with a particular focus on the PVN. This limits the interpretation of our review in two important ways, and this should be kept in mind as one proceeds. First, this chapter focuses on central actions of neurochemical systems on the HPA axis, centered on the PVN CRH neuron. Thus, we will not discuss results of systemic injections of neuromodulatory agonists/antagonists to any substantial extent, unless these injections are made in conjunction with some central treatment or measure. The reader should note that multiple systems have the capacity to act at the pituitary and/or adrenal to directly modulate ACTH and corticosteroid release, respectively. While these actions are not discussed here, it is important to consider that direct pituitary or adrenal regulation of HPA axis output offers major modulatory mechanisms that should not be underestimated. Second, the roles of particular neurochemical species are discussed in the context of the domains of HPA axis function: basal (circadian), acute stress, and chronic stress, under the assumption that the given neurochemical systems may affect different aspects of HPA axis integration. Importantly, a large number of neuropeptides activate the HPA axis after central injection; however, few of these species have been tested for necessity (usually due to lack of specific receptor antagonists) and most have not been placed in the context of diurnal regulation. Thus, we discuss these as possible candidate regulators, but do not incorporate them into our general neurochemical scheme for HPA integration. Finally, the "other" stress responses noted above (autonomic, behavioral) will have input into HPA axis activity. For example, autonomic activity during stress can have an impact on PVN activation by multiple central and peripheral mechanisms, and indeed, it is fair to say that a substantial portion of the HPA axis stress response may be a reaction to sympathoadrenomedullary activation (Ulrich-Lai and Engeland, 2002). In addition, an organism's behavior may also affect HPA outcome. For example, avoiding an anxiety-provoking situation (e.g., open arms of an elevated plus-maze in rodents) may effectively decrease the magnitude of the HPA response by minimizing exposure to the "danger" of falling. Thus, by showing greater "anxiety," the organism can conceivably diminish homeostatic threat and the attendant glucocorticoid response. Thus, while we do not address sympathoadrenal and behavioral stress systems directly, it is important to acknowledge their possible roles in "tuning" the HPA axis.

3 Glucocorticoids and Feedback

Glucocorticoids have powerful inhibitory actions on the HPA axis. Genomic actions of glucocorticoids are mediated by two types of GRs: type I, or MRs and type II, or GRs. Differential ligand affinities [the K_D for MR is 0.5–1.0 nM, GR is 2.5–5 nM (Reul and de Kloet, 1985)], brain region specificity (McEwen et al., 1968), and receptor binding mechanisms (Rogerson et al., 2004) likely dictate the role of the two receptors in basal HPA axis regulation, circadian drive, and the HPA axis response to stress. Under conditions of low glucocorticoid secretion (e.g., circadian secretory nadir), both MR and GR are in the cytosol (van Eekelen et al., 1987), encased in chaperone complexes [consisting of heat shock proteins 90, 70, 30, heat shock cognate 70, p23, and an immunophilin moiety FKBP51 or 52 (Fuller et al., 2000)]. These stabilize tertiary structure by encouraging proper protein folding and conformation, maintaining the receptor in a high affinity state for glucocorticoid binding. Upon ligand binding, a nuclear localization signal is exposed and receptors translocate to the nuclear compartment, where they enhance or repress transcription by (1) binding as homodimers (or perhaps heterodimers) on DNA response elements (Trapp et al., 1994) or (2) altering transcriptional effects of other nuclear factors by way of protein–protein interactions (Gottlicher et al., 1998).

The two adrenocorticosteroid receptors are differentially distributed in the brain; MR is primarily located in limbic system brain regions (i.e., hippocampus and septum), whereas GR is widely distributed in the CNS, with high levels of expression in the hippocampus, the cortex, the amygdala, and in CRH neurons of the PVN (Herman, 1993). Thus, both receptor species are expressed in regions implicated in inhibition of the HPA axis. Expression patterns are both autonomous (e.g., MR in lateral septum, GR in PVN) and overlapping (e.g., hippocampus), affording the ability of the two receptors to work individually or in concert. Colocalization of MR and GR is evident in neurons of the hippocampus, allowing for possible interaction between the two receptors via heterodimerization, a process that has been demonstrated in vitro (Herman et al., 1989; Van Eekelen and De Kloet, 1992; van Steensel et al., 1996; Han et al., 2005). As would be predicted by the differential affinities of these receptors for ligand, approximately 70% of MR is available to bind glucocorticoid at the circadian nadir; little unbound MR is available during the circadian rise in corticosterone secretion. Availability of the GR varies substantially across the diurnal cycle; binding capacity is much greater at the circadian low (85–100% unbound), decreasing to 60–75% around the circadian corticosteroid peak (Spencer et al., 1993).

Administration of spironolactone, a potent MR antagonist, increases morning ACTH release in non-adrenalectomized (Spencer et al., 1998) and corticosterone replaced (20% and 50%, respectively) adrenalectomized rats. Higher levels of corticosterone replacement (50%) are necessary for PM inhibition of ACTH, suggesting a role for GR in this process; however, inhibition of MR prohibits inhibitory actions of corticosterone on the PM rise (Bradbury et al., 1994). In intact rats, combined treatment with GR and MR antagonists increases circadian peak corticosterone secretion, consistent with the notion that the two receptors work together to control peak daily secretory rhythms (Spencer et al., 1998). Together, these data suggest that both receptors are involved in basal HPA axis inhibition. However, central injection studies indicate that i.c.v. MR antagonist treatment increases resting morning corticosterone levels in the rat, whereas GR antagonist is without effect, suggesting that binding to the MR is important in the control of basal HPA activity (Ratka et al., 1989; van Haarst et al., 1997), at least at the level of the CNS. Experiments in humans indicate that blockade of the MR increases basal cortisol and ACTH levels (Deuschle et al., 1998).

Glucocorticoid inhibition of acute HPA axis stress responses may also be mediated by the combined actions of GR and MR. Systemic injection of either a GR (RU45055) or MR (RU28318) antagonist does not affect corticosterone secretion driven by acute restraint; however, blockade of both receptors increases the corticosteroid responses to novelty, consistent with a role for both receptors in acute stress feedback (Spencer et al., 1998). Notably, blockade of GR but not MR produces long-term elevation of corticosterone observed following an acute stress, suggesting a role for GR in terminating corticosteroid secretory responses (Moldow et al., 2005). Results following central injections of adrenocorticosteroid receptor antagonists generally support the combined-action model. Central administration of the GR antagonist RU486 enhances the magnitude of the corticosteroid response to novelty, whereas both MR and GR antagonists prolong the magnitude of the HPA response to this stimulus, consistent with a role for both receptors in stress control (Ratka et al., 1989). However, it is important to note that central administration of corticosterone itself can enhance HPA axis activity in adrenalectomized rats, increasing ACTH responses to stress and PVN CRH mRNA levels (Laugero et al., 2002).

The primary brain regions responsible for GR and/or MR control of the HPA axis are currently being established. Previous studies suggest that limbic structures, such as the hippocampus and prefrontal cortex, are involved in corticosteroid-dependent termination of stress responses; however, there is considerable conflicting data on this topic (Jacobson and Sapolsky, 1991; Herman et al., 2003). Recent studies employing a conditional forebrain GR knockout mouse model, which has a selective deletion of GR in the hippocampus and cortex, reveal elevated corticosterone and ACTH at both the trough and nadir of the circadian rhythm. Basal PM glucocorticoid secretion is resistant to dexamethasone suppression (Boyle et al., 2005). Together, these data implicate forebrain GRs in inhibition of basal HPA tone. This work is supported by data from transgenic mouse models that overexpress GR (reduced basal secretion) or an antisense GR construct (increased basal secretion) in neurons (Kellendonk et al., 2002).

Chronic exposure to homotypic stressors often promotes habituation of corticosterone (Keim and Sigg, 1976) and ACTH (Campmany et al., 1996; Gadek-Michalska and Bugajski, 2003) responses to stress. Acute treatment with MR or combined MR–GR antagonists prevents habituation, whereas GR antagonist alone is ineffective (Cole et al., 2000). These data indicate that the MR may be selectively involved in HPA axis habituation.

Finally, it is important to consider the role of glucocorticoids in fast feedback inhibition of the HPA axis. Glucocorticoids inhibit PVN neurons within seconds to minutes of exposure (Saphier, 1989) and it is well established that systemic administration of glucocorticoids immediately prior to stress can effectively block ACTH release (Keller-Wood and Dallman, 1984). Genomic actions of corticosteroids are unlikely to account for such rapid inhibition of the HPA axis. Thus, numerous investigators propose the existence of a membrane GR that can bind corticosteroids and effect immediate physiological changes (Orchinik et al., 1991; Evans et al., 2000; Di et al., 2003). A receptor with appropriate characteristics has been detected in amphibian (Orchinik et al., 1991), but is yet to be definitively identified and cloned. Recent studies suggest that the fast feedback effects of corticosteroids on PVN neurons are mediated through inhibition of presynaptic glutamate release, by way of a G-protein coupled receptor activating synthesis of endocannabinoids (acting as retrograde messengers) (Di et al., 2003).

4 Monoamine Neurotransmission

4.1 Norepinephrine/Epinephrine

Brainstem norepinephrine (NE)/epinephrine (E) systems provide direct innervation to the medial parvocellular PVN and play a major role in HPA axis regulation. The A2/C2 (nucleus of the solitary tract) and A1/C1 (ventrolateral medulla) cell groups provide the primary NE/E innervation to the PVN, with the former comprising the major input to the hypophysiotropic, CRH-containing medial parvocellular region (see *Figure 13-2*). Axons of these cell groups ascend in the ventral noradrenergic-ascending bundle

Figure 13-2
Innervation of the PVN by norepinephrine/epinephrine and serotonin terminals. Corticotropin releasing hormone (CRH) neurons densely populate the medial parvocellular (mp) subdivision of the paraventricular nucleus (a). The mp PVN region is richly invested by fibers positive for dopamine beta-hydroxylase, an enzyme, specific for norepinephrine and epinephrine neurons (b). In contrast, the mp PVN contains a relatively small number of serotonin (5-HT) positive fibers, with most terminal localized outside the PVN proper (c)

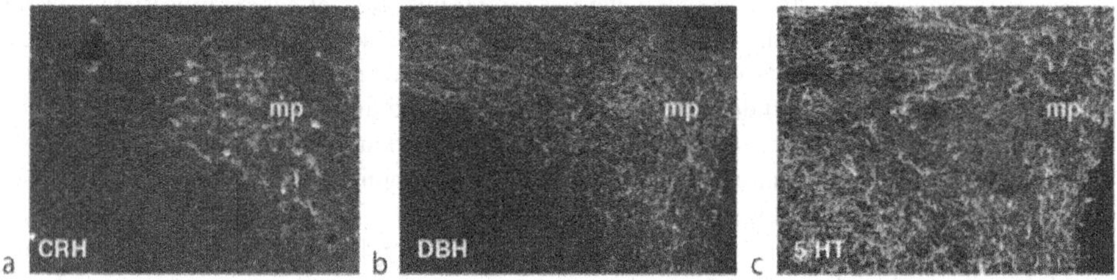

(VNAB) (Fuxe, 1965; Swanson et al., 1981; Cunningham and Sawchenko, 1988) to form synaptic connections with CRH containing neurons in the PVN (Liposits et al., 1986). The LC also contributes to HPA axis regulation; however, this region does not directly innervate the medial parvocellular PVN (Cunningham and Sawchenko, 1988). Thus, the effects of the LC on PVN neurons are likely transsynaptic in nature, acting through other brain regions known to be involved in the regulation of the HPA axis (e.g., prefrontal cortex, amygdala, and hippocampus) (Herman and Cullinan, 1997). Consistent with the importance of NE/E signaling in PVN integration, parvocellular PVN neurons express α_{1A}, α_{1B}, α_{1D}, and α_2-adrenoceptors, with clear evidence for colocalization of α_{1B} (and to a lesser extent α_{1D}) in CRH neurons (Cummings and Seybold, 1988; Williams and Morilak, 1997; Day et al., 1997; Sands and Morilak, 1999).

In the PVN, extracellular NE levels and α2-NE receptor binding show diurnal patterns that parallel plasma corticosterone levels (Jhanwar-Uniyal et al., 1986; Stanley et al., 1989). Deprivation of NE following destruction of the VNAB with 6-hydroxydopamine (6-OHDA) reduced the amplitude of diurnal patterns of ACTH and corticosterone secretion (Szafarczyk et al., 1985), consistent with α1 and/or α2 regulation of basal secretory patterns. Firing rates of LC neurons also show a diurnal pattern (Aston-Jones and Bloom, 1981a) that regulates state-dependent gene expressions in many brain regions (Cirelli and Tononi, 2004). As the LC does not have direct projections to the PVN, it may affect circadian activities of the HPA axis indirectly through the suprachiasmatic nucleus, the main regulatory site of the biological clock in the brain (Cagampang et al., 1994; Sage et al., 2001). Taken together, the brainstem NE system plays a significant role in regulating the circadian rhythm of HPA axis activity.

Norepinephrine appears to play a major stimulatory role in HPA axis activation. Stimulation of the A1, A2, LC, or VNAB increases the firing rate of PVN neurons (Day et al., 1985; Saphier, 1989; Saphier and Feldman, 1989). Electrical stimulation of the VNAB activates the HPA axis, indicated by an increase of CRH release into the hypophysial-portal system (Plotsky, 1987). In addition, local administration of NE rapidly activates CRH and AVP gene transcription (Itoi et al., 1999) and c-Fos expression in PVN neurons (Cole and Sawchenko, 2002), consistent with direct action at the level of CRH neurons. The effects of central NE–E administration or VNAB stimulation on portal CRH release or ACTH/corticosterone secretion can be blocked by either α1 (Plotsky, 1987; Szafarczyk et al., 1987) or α2 (Daniels et al., 1993) adrenoreceptor antagonists.

A variety of systemic as well as psychogenic stressors activate NE cells in the brainstem (A1/C1 and A2/C2 regions) (Li et al., 1996; Dayas et al., 2001b). Release of NE in the hypothalamus or PVN can be observed following exposure to different stressors, including inescapable shock (Anisman and Sklar, 1979), immobilization (Yokoo et al., 1990a; Pacak et al., 1993), conditioned fear (Yokoo et al., 1990b; Cullinan et al., 1993), ether inhalation (Szafarczyk et al., 1985), and formalin injection (Palkovits et al., 1999). Lesion studies further corroborate NE involvement in acute stress responses; for example, HPA axis responses elicited by various stressors are attenuated by brainstem hemisections (Pacak et al., 1996) and 6-OHDA lesions of PVN terminals (Gibson et al., 1986) or the VNAB (Gaillet et al., 1993).

Despite the ability of a variety of stressors to elicit PVN NE secretion, accumulating evidence indicates that ascending NE/E systems may be preferentially involved in responses to systemic stress (Herman and Cullinan, 1997). 6-OHDA lesions of the VNAB or the PVN markedly attenuate ACTH and/or corticosterone responses to systemic stressors such as ether (Szafarczyk et al., 1985; Gaillet et al., 1991; Herman et al., 2002), interleukin-1 (Chuluyan et al., 1992), or hypoglycemia (Ritter et al., 2003). In contrast, corticosterone responses to psychogenic stressors such as restraint stress (Gaillet et al., 1991; Chuluyan et al., 1992; Herman et al., 2002), immobilization (Pacak et al., 1993), and forced swim (Ritter et al., 2003) are not attenuated, suggesting that effects of NE/E are modality-specific. Notably, brainstem hemisections that markedly decrease PVN NE release, NE turnover, and CRH mRNA expression do not affect corticosterone responses to immobilization (Pacak et al., 1993, 1996), suggesting that the HPA axis response to this stimulus is independent of PVN NE, and is possibly driven by forebrain or midbrain circuitry. Some studies also suggest that NE/E lesions may dissociate ACTH and corticosterone responses to stress; although VNAB lesions did not attenuate corticosterone responses to restraint stress, ACTH responses were significantly blunted (Gaillet et al., 1991). VNAB lesions seem to have more marked effects on stress-activated ACTH release than corticosterone release following systemic stressors such as ether (Szafarczyk et al., 1985). Thus, it is plausible that NE/E lesions may also be affecting the HPA axis at the level of the adrenal.

Noradrenergic neurons originating in the LC appear to have indirect effects on HPA axis activity. Various stressors activate NE cells at the LC (Aston-Jones and Bloom, 1981b; Abercrombie and Jacobs, 1987; Grant et al., 1988). Despite the lack of a direct connection between the LC and medial parvocellular PVN, lesions of the LC attenuate ACTH and corticosterone responses to restraint stress (Ziegler et al., 1999), implying that the LC transsynaptically activates the HPA axis. Indeed, following stress, an increase in NE level is observed in many regions of the brain receiving projections from LC, including the amygdala, the hippocampus, and the medial frontal cortex (Abercrombie et al., 1988; Tanaka et al., 1991; Cenci et al., 1992), all of which are implicated in stress regulation (Herman et al., 2003).

Chronic stress appears to sensitize electrophysiological responses of LC NE neurons (Jedema and Grace, 2003). The LC NE system also exhibits chronic stress facilitation, whereby both LC c-Fos induction and forebrain NE release to a novel, heterotypic stressor are markedly potentiated (Nisenbaum et al., 1991; Gresch et al., 1994; Jedema et al., 1999). In general, NE turnover is increased in chronically stressed animals (Thierry et al., 1968; Irwin et al., 1986; Nisenbaum et al., 1991), implying an ongoing increase in NE release with repeated stimulation. These data strongly suggest that the LC participates in adaptations to chronic stress; however, the role of the LC in chronic drive of the HPA axis is less clear. Lesions of the LC do not alter elevated PVN CRH mRNA, adrenal hypertrophy, thymic atrophy, or corticosterone hypersecretion seen following exposure to a chronic variable stress regimen (Ziegler et al., 1999), suggesting that the influence of the LC on long-term stress integration is not pronounced.

4.2 Serotonin

The HPA axis is regulated by ascending input from the 5-hydroxytryptamine (5-HT; serotonin) containing neurons in the raphe nuclei (see Fuller, 1992; Van de Kar, 1997, for review). Dorsal raphe neurons provide a limited direct innervation of CRH neurons in the PVN (➋ *Figure 13-2*) (Sawchenko and Swanson, 1985; Assenmacher et al., 1987; Kitazawa et al., 1987; Liposits et al., 1987) that may underlie 5-HT actions. However, it should be noted that serotonin richly innervates numerous CNS structures involved in stress regulation, as well as the immediate surround of the PVN (Sawchenko et al., 1983). Thus, 5-HT effects on the HPA axis may be either direct or transsynaptic.

Serotonin seems to be particularly important in the regulation of HPA axis rhythmicity. Lesions of the raphe nuclei (Vermes et al., 1974; Lissak et al., 1975) or pharmacological depletion of brain 5-HT by the 5-HT synthesis inhibitor, p-chlorophenylalanine, disrupt the circadian corticosterone rhythm (Krieger and Rizzo, 1969; Scapagnini et al., 1971; Vernikos-Danellis et al., 1973). Injection of the serotonin neurotoxin 5,7-dihydroxytryptamine (5,7-DHT) into the suprachiasmatic nucleus led to flattening of circadian corticosterone fluctuations 2–7 weeks later (Williams et al., 1983; Banky et al., 1986). Studies in genetically modified mice also support a role for 5-HT in basal HPA integration; for example, 5-HT transporter knockout mice show decreased basal plasma corticosterone levels (Lanfumey et al., 2000), and basal AVP (but not CRH) mRNA is reduced in the PVN of 5-HT$_{3A}$ receptor knockout mice (Bhatnagar et al., 2004).

There is also substantial evidence for a role of 5-HT in HPA activation. 5-HT stimulates CRH secretion from isolated rat hypothalamic explants (Calogero et al., 1989a) and there is strong evidence that in vivo administration of 5-HT precursors (5-HTP, L5-HTP), releasers (p-chloroamphetamine, fenfluramine), and reuptake inhibitors (fluoxetine, citalopram) activate the HPA axis (Fuller and Snoddy, 1990; Spinedi and Gaillard, 1991; Fuller, 1992; Fuller et al., 1996; Moncek et al., 2003) and increase PVN CRH mRNA (Jorgensen et al., 2002) in the rat. However, it is important to note that other studies suggest that the role of 5-HT may be more complex. For example, administration of the 5 HT-1A agonists 8-OH DPAT or ipsapirone in conjunction with stress augments stress-induced HPA activation (Korte et al., 1992; Cassano and D'Mello A, 2001). In addition, blockade of 5-HT reuptake by fluoxetine does not potentiate forced swim or hypoglycemia-induced increases in plasma corticosterone secretion (Fuller and Snoddy, 1977).

Specific 5-HT receptor subtypes appear to be involved in the stimulatory effects of 5-HT on the HPA axis. Systemic administration of 5-HT$_{1A}$, 5-HT$_{1A/7}$, or 5-HT$_{2C}$ receptor-selective agonists induce the release of ACTH and corticosterone in vivo (Koenig et al., 1987; Lorens and Van de Kar, 1987; Gilbert et al., 1988; Przegalinski et al., 1989; Van de Kar et al., 1989; Alper, 1990; Owens et al., 1991; Li et al., 1992, 1994b; Matheson et al., 1997a, b; Mikkelsen et al., 2004). In addition, systemic administration of 5-HT$_{1A/7}$ or 5-HT$_{2C}$ receptor-selective agonists cause c-Fos and CRH mRNA induction in the PVN (Jorgensen et al., 2002; Mikkelsen et al., 2004). Furthermore, direct application of 5-HT$_{1A}$, 5-HT$_{1A/7}$, or 5-HT$_{2A/2C}$ receptor agonists into the hypothalamus leads to the release of CRH and plasma ACTH (Haleem et al., 1989; Di Sciullo et al., 1990; Pan and Gilbert, 1992; Rittenhouse et al., 1994; Bluet Pajot et al., 1995). These data indicate that 5-HT$_{1A}$ and 5-HT$_2$ receptors appear to be important mediators for 5-HT control of the HPA axis at the level of the hypothalamus.

Subtraction experiments aimed at reducing 5-HT levels have led to more mixed results. Pharmacological studies indicate that blockade of 5-HT$_{1A}$ receptors reduces HPA axis responses to both restraint and immune challenge (Jorgensen et al., 1998); however, blockade does not prohibit HPA axis responses triggered by ultrasonic vocalizations (Groenink et al., 1996), suggesting that serotonergic actions may be associated with stressor modality or intensity. Posterior hypothalamic deafferentation or injections of the serotonergic neurotoxin 5,7-DHT into the raphe or PVN inhibit the effect of several stressors on plasma corticosterone secretion (Feldman et al., 1984, 1987, 1990; Feldman and Weidenfeld, 1995), again consistent with the removal of excitatory input. However, electrolytic lesions of the midbrain raphe nuclei cause facilitation of stress-induced plasma corticosterone secretion (Cassano and D'Mello A, 2001). Other studies indicate that destruction of 5-HT terminals by i.c.v. injection of 5,6-dihydroxytryptamine decreases plasma corticosterone secretion after immobilization stress (Richardson, 1984). Notably, stress-induced secretion of plasma ACTH and corticosterone is decreased following injections of 5,7-DHT or ketanserin into the amygdala (Feldman and Weidenfeld, 1998; Feldman et al., 1998), suggesting that the inhibitory effects of 5-HT may be transsynaptic in nature.

Serotonin receptor or 5-HT transporter knockout mice display altered HPA axis responses to acute stress. Serotonin transporter knockout mice have elevated plasma ACTH secretion in response to saline injection (Li et al., 2004), but do not hypersecrete ACTH in response to injection of a 5-HT_2 receptor agonist (Li et al., 2003a) or following immobilization stress (Tjurmina et al., 2002). Plasma corticosterone concentrations are elevated by the 5-HT_{1A} receptor agonist flesinoxan in wild-type mice but not in 5-HT_{1A} receptor knockout mice (Pattij et al., 2001). Additionally, 5-HT_{1A} receptor knockout mice show decreased corticosterone secretion after open field exposure (Gross et al., 2000), whereas 5-HT_{3A} receptor knockout mice show reduced plasma ACTH secretion following restraint or lipopolysaccharide administration (Bhatnagar et al., 2004), suggesting both receptors may be involved in HPA stimulation. However, aged 5-HT_{2C} receptor knockout mice show increased ACTH and corticosterone secretion after restraint stress (Chou-Green et al., 2003), arguing for functional differences among the 5-HT receptors in HPA regulation. Thus, it is likely that the effects of 5-HT on the HPA system are determined by the actions of distinct receptor subunit expression patterns in stress-integrative brain regions.

Serotonin is also involved in the control of HPA axis responses to chronic stress. For example, systemic injection of 8-OH-DPAT decreased chronic mild stress-induced ACTH secretion (Grippo et al., 2004). Increasing serotonin availability through inhibition of reuptake appears to attenuate HPA axis responses to chronic stress; for example, long-term pretreatment with imipramine blocks repeated immobilization-induced increases in PVN CRH mRNA levels and glutamic acid decarboxylase 65/67 (GABA synthesizing enzyme) mRNA in the BST (Butterweck et al., 2001). CRH mRNA upregulation in the PVN during exposure to chronic variable stress is attenuated by concurrent administration of venlafaxine (a 5-HT/NE reuptake inhibitor) (Stout et al., 2002). Central serotonin reduction by an i.c.v. injection of 5,7-DHT had no effect on the progressive reduction in corticosterone secretion with increasing number of tests (Chung et al., 1999), indicating that serotonin depletion is not able to disrupt HPA axis habituation to this stressor.

4.3 Dopamine

The neurotransmitter dopamine appears to play an excitatory role in the regulation of both basal and acute responses of the HPA axis. Dopamine depletion of the medial prefrontal cortex, dorsal striatum, and ventral striatum by 6-OHDA lesion of the ventral tegmental area attenuates basal as well as stress-induced plasma corticosterone secretion (Casolini et al., 1993). Systemic or i.c.v. administration of D1 and D2 dopamine receptor agonist elevates serum ACTH (Borowsky and Kuhn, 1992), and infusion of the selective dopamine uptake inhibitor GBR12909 into the third ventricle or PVN increases serum ACTH levels (Borowsky and Kuhn, 1993). Dopamine may modulate HPA axis activity at extrahypothalamic sites, as infusions of a mixture of D1 and D2 dopamine receptor agonists (SKF 38393 and quinpirole) into the medial prefrontal cortex, dorsal striatum, and ventral striatum elevate plasma corticosterone (Ikemoto and Goeders, 1998). Dopamine may also regulate HPA axis responses to stressors differently depending on stressor type; D1 and D2 dopamine receptor antagonists attenuate plasma ACTH responses to systemic stressor (IL1-β), but not mild psychogenic stressors (air puff) (Spencer et al., 2004).

4.4 Histamine

Histamine (HA) is a monoamine transmitter that is synthesized in a limited population of neurons in the tuberomammillary nucleus of the hypothalamus, thought to be important in neuronal control of sleep–wakefulness and arousal (Siegel, 2004). A role for HA in the circadian rhythm of the HPA axis is supported by evidence that central depletion of HA attenuates the diurnal variation of corticosterone and ACTH secretion without affecting the mean basal levels of these hormones (Itowi et al., 1989). However, treatment with the H_1 antagonist mepyramine did not affect the basal rates of CRH, AVP, or OT mRNA expression, while H_2-receptor blockade using cimetidine had only a mild inhibitory effect (Kjaer et al., 1998). Moreover, inhibition of HA synthesis and release using either α-fluromethylhistamine (FHA) or treatment with an H_3-receptor agonist did not affect basal secretion of ACTH (Soe-Jensen et al., 1993). Thus, HA may play a role in rhythmic activity of the HPA, but may not be directly responsible for inhibition or activation of CRH neurons.

Neuronal HA may play a more important role in regulating the HPA axis response to stress (Kjaer, 1996). Central injections of HA elevate plasma levels of ACTH and beta-endorphin in a dose-dependent manner (Kjaer et al., 1992a). This effect appears to be mediated by the HA type 1 and type 2 (H_1 and H_2) receptors, as central injections of H_1 and H_2-receptor agonists produce similar effects on pituitary stress hormones (Kjaer et al., 1992b). Evidence suggests that the HA-induced activation of the HPA axis is in part due to increased CRH secretion from the PVN. Specifically, intracerebroventricular (i.c.v.) administration of HA, HA_1-, and HA_2-receptor agonists has been shown to increase CRH mRNA (Kjaer et al., 1998) and c-Fos expression (Kjaer et al., 1994) in CRH containing neurons in the PVN. Moreover, these agents have been shown to increase CRH release into the portal venous blood with an associated three to fourfold increase in plasma ACTH levels (Kjaer et al.,1992a). This effect on ACTH secretion was diminished by pretreatment with anti-CRH antibodies (Kjaer et al., 1992a). A role for vasopressin in the stress- and HA-induced activation of the HPA axis has also been described (Kjaer et al., 1993), possibly via a mediating action produced by binding to V_{1a} receptors located on CRH neurons (Kjaer, 1996), as well as a permissive effect on CRH-induced ACTH secretion by the anterior pituitary of stress hormones (Kjaer et al., 1993).

Evidence also supports a role for HA in the stress-induced activation of the HPA axis, as pretreatment with H_1- and H_2-receptor antagonists (Knigge et al., 1989), as well as bilateral posterior hypothalamic lesion and treatment with agonists of the HA type 3 (H_3) presynaptic autoreceptor (Knigge et al., 1999), attenuate secretion of pituitary hormones in response to stress.

Although differences in brain HA turnover in the diencephalon (Ito et al., 1999) and brain H3 receptor density (Endou et al., 2001) have been reported in rats exposed to chronic versus acute stress, the role of HA in the regulation of the HPA response to chronic stress is not yet defined.

5 Acetylcholine and Amino Acid Neurotransmission

5.1 Acetylcholine

The HPA axis is also regulated by acetylcholine. Acetylcholine is a potent activator of CRH release in vitro, effects that appear to be mediated by either muscarinic or nicotinic receptors (c.f., Hillhouse et al., 1975; Jones et al., 1976; Tsagarakis et al., 1988; Calogero et al., 1989b). In vivo stimulatory effects of acetylcholine have been verified by central injection studies, e.g., central application of nicotine activates the HPA axis (Matta et al., 1987). Direct injections of a nicotinic antagonist into the PVN were not sufficient to block effects of nicotine on HPA activation, suggesting that CRH neurons are not directly activated by nicotinic receptors (Matta et al., 1987). In contrast, injections of acetylcholine directly into the region of the PVN increase ACTH release in a dose-dependent manner; these effects appear to be mediated by muscarinic receptors (Ohmori et al., 1995). Subsequent work indicates that nicotinic effects are likely mediated through ascending catecholaminergic inputs from the brainstem (Fu et al., 1997). Thus, central acetylcholine activation of the HPA axis appears to have a PVN component, mediated by muscarinic receptors, and a brainstem component, mediated by nicotinic receptors. Interestingly, the magnitude of responses to peripheral muscarinic and nicotinic stimulation differs in males and females, suggesting the possibility for sex differences in central responsiveness to acetylcholine or cholinergic agents (Rhodes et al., 2001). This regulatory scheme predicts a multifaceted role for acetylcholine in HPA axis regulation.

Studies using nicotinic receptor antibodies report decreased ACTH and corticosterone release following ether stress (Weidenfeld et al., 1983) or acoustic (but not photic or sciatic) stimulation (Weidenfeld et al.,

1983, 1988), suggesting a role for the nicotinic cholinergic receptor system in HPA activation following acute systemic or psychogenic stress. Notably, chronic homecage crowding reduces ACTH and corticosterone responses to central muscarnic stimulation (Bugajski, 1999) and peripheral nicotine (Bugajski et al., 2002), suggesting a possible role for cholinergic receptor downregulation in chronic stress-induced changes in HPA axis function.

5.2 Gamma-Aminobutyric Acid

GABA is a major inhibitory modulator of HPA axis activity. Electron microscopic studies have documented that half of all synapses in the PVN are GABAergic (Decavel and Van Den Pol, 1990), suggesting that GABA plays a major regulatory role in HPA axis activation. In support of this hypothesis, microinjection of the GABA-A receptor antagonist bicuculline methiodide (BMI) into the PVN is sufficient to cause corticosterone secretion and induction of c-Fos expression in parvocellular PVN neurons (Cole and Sawchenko, 2002; Bartanusz et al., 2004). These results indicate that GABA provides tonic inhibition under basal HPA axis conditions. In addition, BMI injections increase AVP heteronuclear RNA in the medial parvocellular PVN (Kovacs et al., 2004), suggesting that the expression of this HPA "amplification factor" is also under tonic inhibition by GABA. In addition, BMI induce transcription of CRH and AVP in hypothalamic slices (Bali and Kovacs, 2003), confirming GABAergic inhibition of secretagogue biosynthesis. Ibotenate lesions of PVN-projecting cell populations in the posterior bed nucleus of the stria terminalis increase CRH mRNA in the medial parvocellular PVN under resting conditions (Herman et al., 1994). Given that this region sends a predominantly GABAergic projection to the PVN (Cullinan et al., 1993), these data suggest that GABA innervation may be necessary for maintenance of basal HPA tone.

Pharmacological analyses confirm that GABA is a major inhibitor of stress responses. Corticosterone and ACTH secretion are attenuated by i.c.v. injection of a GABA-A agonist (muscimol) (Makara and Stark, 1974) as well as by local injection of benzodiazepines (Imaki et al., 1995). Microinjection of the GABA-A agonist muscimol directly into the PVN inhibits stress-induced corticosterone release and medial parvocellular PVN c-Fos induction, whereas infusion of muscimol into the PVN by microdialysis decreased PVN c-Fos activation in response to ether stress (Kovacs et al., 2004). These data support the hypothesis that GABA directly inhibits stress-induced activation of CRH neurons. Cell culture and electrophysiological studies confirm direct GABAergic effects; for example, direct application of muscimol to parvocellular PVN neurons inhibits miniature excitatory postsynaptic currents (Boudaba et al., 1996).

Sources of GABAergic innervation to the PVN are fairly well characterized. GABA input emanates from several major sources, including the peri-PVN region, dorsomedial hypothalamic nucleus, medial preoptic nucleus/area, and bed nucleus of the stria terminalis, among others. Thus, it is not possible to identify a single source of GABA innervation to the PVN, indicating that inhibition of the PVN may reflect the net activity of several GABA circuits. However, it is important to note that many of these GABA systems are activated by stressful stimuli; for example, swim stress activates glutamic acid decarboxylase mRNA-expressing (i.e., GABAergic) neurons in the peri-PVN region, dorsomedial hypothalamus, preoptic area, and bed nucleus of the stria terminalis (Cullinan et al., 1995, 1996; Campeau and Watson, 1997), and glutamic acid decarboxylase mRNA expression is elevated in these same PVN-projecting regions following restraint (Bowers et al., 1998).

GABA systems may also play a role in chronic stress-induced HPA axis hyperactivity. Elevated glutamic acid decarboxylase 65 mRNA expression is observed in numerous PVN-projecting GABA cell populations following chronic variable stress (Bowers et al., 1998), consistent with chronic drive of these cell populations. Importantly, chronic stress decreases expression of key GABA-A receptor subunits in the parvocellular PVN (Cullinan and Wolfe, 2000), suggesting a reduced net impact of GABA on CRH neurons following prolonged activity. Consistent with this hypothesis, electrophysiological analyses indicate that GABAergic inhibitory postsynaptic currents are reduced following chronic stress (Joels et al., 2003; Verkuyl et al., 2004), likely due to a reduced number of GABAergic synaptic contacts (Verkuyl et al., 2004) resulting in a disinhibition of HPA axis activity.

5.3 Glutamate

Glutamatergic terminals directly innervate CRH neurons (see Figure 13-3) and are sufficient to induce activation of the HPA axis. Infusion of glutamate into the third ventricle increases the secretion of corticosterone in anesthetized rats (Makara and Stark, 1975). Similarly, injection of glutamate directly

Figure 13-3

Confocal micrographs demonstrating appositions between glutamatergic boutons (green) and CRH neurons (red). This is a three-dimensional rendering of a glutamate (vesicular glutamate transporter 2(VGLUT2)-positive)-innervated CRH neuron, showing the shape of the CRH immunoreactive soma (a), and the patterning of boutons around the cell (b). The overlay (c) demonstrates close apposition of VGLUT2 boutons to the CRH neuron, including overlap of immunoreactivity (yellow regions), and occupancy of apparent depression in the cell soma. The results strongly suggest innervation of CRH neurons by VGLUT2-expressing cell populations. Reprinted with permission from Ziegler, Cullinan and Herman, J. Comp. Neurol. 484 (2005)

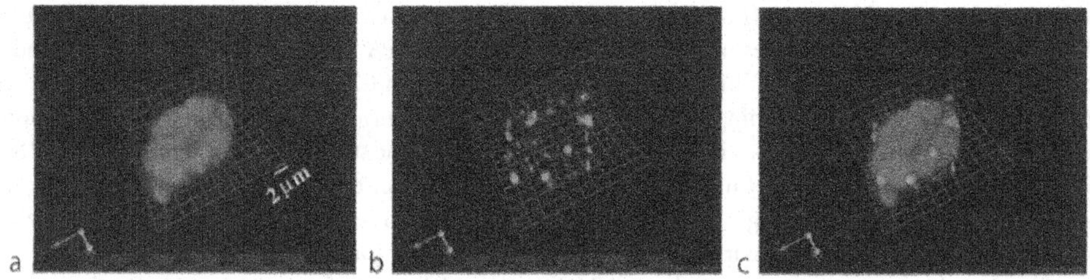

into the PVN increases the secretion of ACTH, depletes the median eminence of CRH, and elevates basal corticosterone in anesthetized rats (Darlington et al., 1989). These acute effects of glutamate may be mediated by NMDA receptors (Feldman and Weidenfeld, 1997). Interestingly, peripheral administration of MK-801, an NMDA receptor antagonist, induces ACTH (Jezova et al., 1995) and corticosterone secretion (Pechnick et al., 1989) in unstressed animals, but blunts the ACTH response to stress (Jezova et al., 1995). These effects were probably due to central action of the drug, because peripheral administration of an NMDA antagonist that does not cross the blood–brain barrier does not affect basal ACTH levels. NonNMDA receptors may also play a role in acute effects of glutamate on the HPA axis, as administration of DNQX, a nonNMDA antagonist, into the lateral ventricles stimulates corticosterone and ACTH secretion (Tokarev and Jezova, 1997).

Peripheral administration of NMDA and kainate, like their antagonists, stimulates ACTH and corticosterone responses in rats (Zelena et al., 2005). Surgical lesions of the PVN prevent the ACTH and corticosterone responses to NMDA and kainate, suggesting that they may be due to actions on the PVN or higher centers (Zelena et al., 2005). In support of this idea, the ACTH response induced by NMDA and kainate receptor agonists is blocked by passive immunization with anti-CRH antibodies (Chautard et al., 1993). It is curious to note that both antagonists and agonists of these ionotropic glutamate receptors seem to have qualitatively the same effect on acute HPA output. A possible explanation of this phenomenon may be that the drugs are working on different brain regions. For example, kynurenic acid, an ionotropic glutamate receptor antagonist, decreases the corticosterone response to acute restraint stress when injected into the PVN, but increases the corticosterone response when injected dorsal to the PVN (Ziegler and Herman, 2000). The latter data indicate that whereas glutamate can directly activate PVN neurons, neighboring GABA neurons are also receptive to glutamate and may confer inhibition of CRH neurons. This supposition is confirmed by recent studies indicating that application of glutamate to an area dorsal and lateral to the PVN proper selectively activates glutamic acid decarboxylase-expressing neurons in the peri-PVN zone (Cole and Sawchenko, 2002). Thus, the common effects of peripheral injections of agonists/antagonists may be related to differential activation of circuits relevant to HPA excitation and inhibition. Alternatively, an optimal "glutamate tone" may be required for control of the HPA axis, whereby too little or too much glutamate signaling results in HPA activation.

In addition to ionotropic glutamate receptors, metabotropic glutamate receptors may also be involved in regulating the HPA response to stress. Acute systemic administration of an mGluR2/3 antagonist induces corticosterone secretion in mice (Scaccianoce et al., 2003). The effect of mGluR2/3 antagonist on corticosterone release is presumably mediated at the hypothalamus, as this drug is sufficient to induce CRH release from isolated hypothalami in vitro. However, neither mGluR2/3 agonists nor antagonists significantly alter corticosterone levels in vivo in the rat (Johnson et al., 2001). Acute dosing with group III agonists (mGluR4, 6, 7, and 8) stimulate a corticosterone response (Johnson et al., 2001). Interestingly, both agonists and antagonists to group I receptors (mGluR1 and mGluR5) increase serum corticosterone (Johnson et al., 2001), similar to results discussed already for ionotropic glutamate receptors. Thus, it is apparent that the different classes of metabotropic glutamate receptors have quite different profiles of HPA axis regulation.

Chronic administration of the NMDA antagonist felbamate increases basal blood levels of corticosterone in mice, but attenuates the corticosterone response to chronic social defeat stress (Pistovcakova et al., 2005). In contrast, chronic blockade of both NMDA and AMPA receptors attenuates the response to acute immobilization stress, but not to footshock or ether stress (Zelena et al., 1999). Chronic blockade of NMDA or AMPA receptors does not affect the HPA response to any of these stressors. The data suggest a possible role for glutamate in long-term regulation of the HPA axis. In support of this hypothesis, chronic mild stress alters the expression of both ionotropic and metabotropic glutamate receptors; for example chronic mild stress increases expression of mGluR5 in the CA1 region of the hippocampus, whereas expression is decreased in the CA3 region (Wieronska et al., 2001). Recent data from our group indicate that a 2-week chronic variable stress regimen increases GluR5 mRNA expression in the medial parvocellular PVN, consistent with enhanced glutamate signaling through nonNMDA channels. Interestingly, chronic stress reduces NMDA-R2B mRNA expression, while NMDA-R1 and NMDA-R2A subunit mRNAs are not affected (Ziegler et al., 2005). Receptor complexes containing the NMDA-R2B subunit show less calcium permeability than those containing NMDA-R2A (Loftis and Janowsky, 2003); thus, reduction in expression of this subunit at the expense of others also predicts enhanced glutamate excitability.

6 Neuropeptide Systems

6.1 Opioid Peptides

Opioid peptides are known to be important in pain processing in the CNS (Akil et al., 1984). These peptides are richly expressed in numerous PVN-projecting regions and the known inhibitory effects of opioids on synaptic transmission predict an inhibitory role in HPA regulation. In humans, this appears to be the case, as systemic administration of opiate agonists inhibits the HPA axis (Grossman et al., 1982), whereas opiate antagonists are stimulatory (Volavka et al., 1979; Morley et al., 1980; Judd et al., 1981; Grossman et al., 1982; Cohen et al., 1985). However, this does not appear to be the case in rodents, as administration of either opiate receptor agonists (Pfeiffer et al., 1985; Pechnick et al., 1985a, b; Iyengar et al., 1986, 1987; Nikolarakis et al., 1987; Gunion et al., 1991; Eisenberg, 1993) or antagonists (Eisenberg, 1984; Pechnick et al., 1985a, b; Nikolarakis et al., 1987) stimulates the HPA axis. As was the case with glutamate, it is possible that an optimal "opioid tone" needs to be present to permit normal HPA axis function.

6.1.1 β-Endorphin

β-endorphin has an excitatory role in HPA axis regulation. While there is little information regarding the role of opioid peptides on basal/circadian regulation of the HPA axis, several studies indicate that i.c.v. administration of β-Endorphin potently stimulates ACTH (Yamauchi et al., 1997) and corticosterone (Haracz et al., 1981; Iyengar et al., 1987; Gunion et al., 1991) release in a dose-dependent manner that is reversible by systemic naloxone (Haracz et al., 1981; Iyengar et al., 1987; Yamauchi et al., 1997), suggesting that centrally located opioid receptors are responsible for modulation of HPA axis activity. Systemic administration of CRH antiserum blocks β-Endorphin-mediated increases in plasma ACTH and restraint stress-induced ACTH is inhibited by immunoneutralization of β-endorphin (Yamauchi et al., 1997). While some have argued that β-endorphin stimulates the HPA axis through an action at the μ opioid receptor (Gunion et al., 1991), other studies indicate that β-endorphin activates the HPA axis in a μ and κ opioid receptor-independent manner (Iyengar et al., 1986), and thus the mechanism is currently unclear.

In vitro analyses suggest that β-endorphin may also have inhibitory effects on the HPA axis. β-endorphin has been reported to inhibit release of CRH into the hypophysial-portal circulation in vivo (Plotsky, 1986) and from the hypothalamus in vitro (Buckingham, 1986; Plotsky, 1986; Yajima et al., 1986). Central administration of β-endorphin also attenuates CRH release in the hypophysial-portal circulation during nitroprusside-induced hypertension (Plotsky, 1986), suggesting an inhibitory role of this peptide in stimulating CRH secretion.

6.1.2 Enkephalin

As with the other opioid peptides, little is known about basal/circadian influences of enkephalin peptides on HPA axis function and regulation. Administration of i.c.v. met- or leu-enkephalin increases corticosterone secretion in rats, an effect that is attenuated by pretreatment with naloxone (Gadek-Michalska and Bugajski, 1996). In addition, administration of met-enkephalin i.c.v. potentiates restraint stress-induced plasma corticosterone secretion in rats (Gadek-Michalska et al., 1997) and ether stress-induced plasma corticosterone secretion in mice (Gibson et al., 1979, 1980). However, leu-enkephalin attenuates ether stress-induced plasma corticosterone in mice (Gibson et al., 1980), suggesting the possibility of peptide-specific actions. Met- and leu-enkephalin are reported to increase (Buckingham, 1982) or decrease (Yajima et al., 1986) in vitro CRH release from rat hypothalamic explants in a naloxone-reversible manner. Thus, the majority of studies suggest that enkephalins (particularly met-enkephalin, the major product of the proenkephalin precursor) appear to have an excitatory influence on the HPA axis; however, there are conflicting data in the literature and thus the overall contribution of these peptides remains to be definitively determined.

6.1.3 Dynorphin

Similar to other opioid peptides, dynorphin also appears to mediate an excitatory influence on the HPA axis in vivo, as i.c.v. injections potently stimulate corticosterone secretion (Iyengar et al., 1987; Gunion et al., 1991). It is unclear whether the dynorphin effects on the HPA axis are mediated through opioid receptors, as some groups have reported reversibility with naloxone (Iyengar et al., 1987) while others have not (Gunion et al., 1991). However, like β-endorphin, dynorphin inhibits basal secretion of CRH into the hypophysial-portal circulation (Plotsky, 1986) and basal in vitro release from the hypothalamus (Buckingham, 1986; Plotsky, 1986; Yajima et al., 1986). Furthermore, nitroprusside-hypotension-induced CRH release is decreased by prior i.c.v. administration of dynorphin (Plotsky, 1986). Thus, the impact of dynorphin appears to depend on the experimental preparation employed.

6.1.4 Nociceptin/Orphanin FQ

The role of nociceptin (N)/orphanin FQ (OFQ) in the regulation of the HPA axis is equivocal. Disruption of the N/OFQ gene in mice produces elevated morning plasma corticosterone levels (Koster et al., 1999), while i.c.v. administration of antisense oligonucleotides directed against the N/OFQ receptor decreases AM and PM plasma corticosterone levels in rats (Blakley et al., 2004). Administration of i.c.v. N/OFQ elevates circulating plasma ACTH and corticosterone in a dose-dependent manner, and enhances hormonal responses to mild psychogenic stressors (elevated plus-maze or open field exposure) (Devine et al., 2001; Fernandez et al., 2004), but not to more intense stressors (Devine et al., 2001). However, N/OFQ-deficient mice exhibit elevated plasma corticosterone levels following exposure to elevated plus-maze relative to wild-type littermates (Koster et al., 1999).

6.1.5 Endomorphins

Endomorphins are endogenous opioid peptides with high affinity and selectivity for the μ opioid receptors. I.c.v. administration of endomorphin-1 or endomorphin-2 does not affect plasma corticosterone, CRH, and AVP mRNA expression in the PVN or proopiomelanocortin mRNA expression in the anterior pituitary (Coventry et al., 2001), suggesting that these endogenous opioids do not play an important role in the regulation of the HPA axis.

6.2 CRH and Related Peptides

A considerable body of evidence ascribes a role for CRH in neurotransmission. In particular, central CRH systems are involved in CNS regulation of fear, anxiety, and anorexia (Heinrichs and Richard, 1999; Takahashi, 2001). In addition, several additional CRH-like compounds (urocortins I-III) have been discovered, all of which have actions on the known CRH receptors (CRH-R1 and CRH-R2).

6.2.1 Corticotropin-Releasing Hormone

Notably, the principal receptor for CRH itself is CRH-R1, which is the primary transducer of CRH action at the pituitary. The presence of CRH-R1 receptors in the pituitary largely negates the value of peripheral antagonist treatments in understanding central CRH action and as a consequence, there are relatively few studies that address central CRH actions on the HPA axis.

The role of CRH in promoting ACTH release is well established. However, it is clear that central CRH systems project to PVN neurons (Champagne et al., 1998) and CRH-synapses can be observed in apposition to CRH neurons (Silverman et al., 1989). In addition, CRH-R1 receptors are expressed in the medial parvocellular PVN following stressful stimuli (Imaki et al., 1996), whereas intermediate levels of CRH-R2 are expressed in the magnocellular PVN (Chalmers et al., 1995); these receptors provide a possible substrate for direct or local actions of CRH and related peptides on HPA axis activity.

Studies assessing the role of central CRH in HPA function point toward an important role of synaptic CRH in stress responses. Central injection of CRH elicits the release of ACTH and corticosterone (as well as NE) (Brown et al., 1982; Donald et al., 1983; Kalin et al., 1983) and induces c-Fos and CRH mRNA expression in parvocellular PVN neurons, proving that CRH itself is sufficient to activate the HPA axis, and may exert positive feedback on its own synthesis. Central (i.c.v.) injection of alpha-helical CRH attenuates ACTH responses to a conditioned fear stimulus (Nijsen et al., 2000) and reduces corticosterone (but not ACTH) responses to immune stimulation (Kaneta and Kusnecov, 2005). Central alpha-helical CRH also inhibits PVN induction of c-Fos following restraint stress (Imaki et al., 2001). However, alpha-helical CRH was not sufficient to block ACTH or corticosterone release in a different conditioned fear test (defensive prod burying task) (Korte et al., 1994), suggesting that the role of central CRH in HPA axis function may be related to stressor intensity.

As noted above, the locus of central CRH action on the HPA axis is yet to be determined. While CRH synapses can be localized to the PVN, very few CRH-R1 or R2 receptors are localized to PVN neurons under unstimulated conditions (Potter et al., 1994). Recent evidence suggests that CRH has pronounced physiological and behavioral effects at the level of the LC (Valentino and Van Bockstaele, 2001) and raphe nuclei (Roche et al., 2003). Thus, it is possible that the impact of CRH on the HPA may be transsynaptic.

Central CRH systems may also play a role in development and/or maintenance of chronic stress-induced HPA axis activation. Chronic delivery of CRH into the ventricular system results in marked anorexia, decreased thymus weight, and increased adrenal weight, similar to that seen following chronic stress. Importantly, these effects are not due to leakage of CRH into the portal circulation. The central drive of the HPA by CRH may be drived from any of several sources, including the bed nucleus of the stria terminalis or central amygdaloid nucleus. Importantly, CRH mRNA is upregulated in the former structures following pharmacological doses of glucocorticoids or severe stress, suggesting a possible role for these CRH systems in chronic stress pathologies (Makino et al., 1994a, b; Makino et al., 1999).

6.2.2 Urocortin (I)

Urocortins are homologous to CRH and show a particularly high specificity for CRH-R2. In rats, intravenous (i.v.) administration of urocortin I increases ACTH, corticosterone secretion, and pituitary proopiomelanocortin mRNA, while decreasing CRH-R1 mRNA in the pituitary. However, urocortin effects are almost entirely attributable to the affinity of urocortin for CRH-R1 (Asaba et al., 1998) and it is unlikely that urocortin is a major endogenous regulator of pituitary ACTH secretion (Masuzawa et al., 1999; Turnbull et al., 1999). This conclusion is supported by data from urocortin knockout mice indicating normal basal and stress-stimulated plasma corticosterone levels (Vetter et al., 2002; Wang et al., 2002).

6.2.3 Urocortin II and III

Urocortin II (Reyes et al., 2001) and III (Hsu and Hsueh, 2001) are more recent additions to the CRH-urocortin peptide family. They appear to be exclusive CRH-R2 ligands (Dautzenberg and Hauger, 2002, review; Venihaki et al., 2004).

While Urocortin II and III are localized to regions of brain believed to be involved in stress regulation, available evidence does not indicate an important role for these peptides in HPA axis regulation. Administration of mouse urocortin II and human urocortin III had no effect on ACTH release in mice (Pelleymounter et al., 2004). Mouse urocortin III does not exert any effect on the HPA axis stress response, but stimulates behaviors associated with reduced anxiety (Venihaki et al., 2004). In combination with the selective specificity for CRH-R2, it is possible that these peptides play a role in behavioral but not endocrine aspects of stress responsiveness (Pelleymounter et al., 2004).

6.3 Neurohypophysial Peptides

The neurohypophysial peptides vasopressin and oxytocin are both localized in magnocellular neurons of the PVN, from whence they influence water balance, blood pressure, parturition, and lactation (Hayward, 1975). Both peptides can directly influence ACTH release; in the case of vasopressin, this is largely accomplished via colocalization and corelease with CRH (Whitnall, 1993). A stimulatory influence of oxytocin on ACTH release has been repeatedly documented (Plotsky et al., 1985; Sapolsky et al., 1989; Dohanics et al., 1991); however, oxytocin is not colocalized with CRH in parvocellular neurons, suggesting a contribution of the magnocellular system in HPA axis regulation. The exact nature of magnocellular regulation of ACTH release remains somewhat vague; there is limited evidence for release of neurohypophysial peptides from axons of passage in the internal zone of the median eminence (Holmes et al., 1986), and some suggest that peptides can be diverted into the anterior pituitary portal system by retrograde blood flow. Whatever the mechanism, it is clear that there is an opportunity for the magnocellular system to contribute to HPA axis integration.

6.3.1 Arginine Vasopressin

As noted, the actions of vasopressin on CRH-induced ACTH release are well documented. However, vasopressin also modulates HPA function within the hypothalamus itself. Most notably, microinjections of vasopressin into suprachiasmatic terminal fields in the subPVN region or dorsomedial hypothalamus attenuate elevated corticosteroid levels seen following suprachiasmatic lesions, whereas injections of a vasopressin type 1 receptor antagonist increase HPA axis activity in intact animals (Kalsbeek et al., 1992). Circadian studies indicate that vasopressin release and gene transcription are negatively correlated with circulating corticosterone (Kalsbeek et al., 1996a; Yambe et al., 2002). Together, the data strongly implicate vasopressin in inhibition of basal corticosterone release during the nadir of the circadian cycle. Lesions of the suprachiasmatic nucleus are known to increase corticosterone release following mild stress (Buijs et al., 1993); thus, vasopressin may also play a role in the inhibition of stress responses.

6.3.2 Oxytocin

Oxytocin also modulates ACTH release within the CNS. Intracerebroventricular injections of an oxytocin antagonist attenuate both basal and stress-induced ACTH release (elevated platform or swim) (Neumann et al., 2000a, b). Microdialysis studies indicate that stressful stimuli release oxytocin into the PVN (Nishioka et al., 1998; Neumann et al., 2000b) and retrodialysis of an oxytocin antagonist decreases basal and stress-induced ACTH release (Neumann et al., 2000b), implying an action of local oxytocin on CRH neurons. The local influence of oxytocin may emanate from either axon collaterals of neurohypophysial/preautonomic oxytocin neurons or from local dendritic release of peptide (Ludwig, 1998).

6.4 HPA-Excitatory Neuropeptides

6.4.1 Angiotensin II

Central angiotensin II pathways are implicated in fluid balance and blood pressure regulation (Fitzsimons, 1998). Angiotensin II is richly expressed in the subfornical organ, which has direct projections to the parvocellular PVN (Lind et al., 1984). Central administration of angiotensin II results in a dose-dependent increase in CRH in the portal plasma that is reversed by the angiotensin II receptor antagonist, saralasin (Plotsky et al., 1988). Stimulation of the subfornical organ results in saralasin-reversible activation of CRH release, verifying the physiological relevance of this pathway (Plotsky et al., 1988). However, the nature of angiotensin II modulation of HPA axis function in intact animals is currently unclear; blockade of type I angiotensin receptors attenuates HPA axis responses to isolation stress, but does not affect responses to immobilization.

6.4.2 Cholecystokinin

Cholecystokinin (CCK) is a gut–brain peptide that plays a role in satiety (Smith and Gibbs, 1994). Systemic administration of CCK-A and CCK-B receptor antagonists at the peak of the circadian rhythm does not affect plasma ACTH or corticosterone levels and does not activate c Fos expression in neurons of the parvocellular PVN (Ruiz-Gayo et al., 2000; Cano et al., 2003). These data suggest that CCK signaling is not involved in regulating the basal activity of the HPA axis.

CCK does not regulate HPA axis activity in response to most types of acute stress. For example, systemic pretreatment with a CCK-A receptor antagonist does not affect the plasma ACTH or corticosterone response to acute forced swim stress in rats (Hernando et al., 1996). Systemic pretreatment with CCK-A or CCK-B receptor antagonists does not affect the plasma ACTH and corticosterone response to systemic IL-1β (Day and Akil, 1999). Similarly, intraparaventricular thalamic nucleus infusions of CCK-B receptor antagonist do not affect the HPA axis response to restraint stress (Bhatnagar et al., 2000). However, CCK signaling does appear to activate the HPA axis in response to acute metabolic stress. Systemic pretreatment with CCK-A but not CCK-B receptor antagonist prevents increases in plasma ACTH and corticosterone in response to acute fasting during the dark period of the circadian rhythm (Ruiz-Gayo et al., 2000). Both CCK-A and CCK-B receptor antagonists attenuate c-Fos expression induced in parvocellular PVN neurons in response to acute fasting (Cano et al., 2003). Moreover, administration of a CCK-A receptor antagonist blocks increases in c-Fos expression in parvocellular PVN in response to 2-deoxyglucose treatment (Cano et al., 2003). Collectively, these data demonstrate that CCK signaling activates the HPA axis during acute metabolic stress.

Central CCK systems may also play a role in modulating HPA axis responses to repeated or prolonged stress. Infusion of CCK-B receptor antagonist into the paraventricular thalamic nucleus increases the plasma ACTH response to novel restraint stress in chronic cold stressed rats, but not euthermic controls, suggesting that during chronic stress CCK-ergic inputs to the paraventricular thalamus are activated and act downstream on the PVN to limit chronic stress-induced facilitation of the HPA axis response to novel stress (Bhatnagar et al., 2000).

6.4.3 Glucagon-Like Peptide-1

Glucagon-like peptide-1 (GLP-1) is an insulinotropic gastrointestinal peptide that is expressed in the central nervous system, where it exerts anorectic actions (Kinzig et al., 2002). GLP-1 is expressed exclusively in the nucleus of the solitary tract (NTS) and ventrolateral medulla (Drucker, 1990; Larsen et al., 1997; Merchenthaler et al., 1999). Notably, despite the rich presence of NE/E neurons in this region, colocalization of catecholaminergic markers and GLP-1 is minimal, indicating that these are mutually exclusive neuronal populations. Nonetheless, like the NE/E systems, GLP-1 containing neuronal fibers heavily innervate the parvocellular PVN (Drucker, 1990; Shughrue et al., 1996; Merchenthaler et al., 1999) and form direct synaptic contacts with CRH-immunoreactive cell bodies (Sarkar et al., 2003).

Several studies implicate GLP-1 in HPA axis regulation. I.c.v. infusions of GLP-1 increase plasma ACTH and/or corticosterone (Larsen et al., 1997; Kinzig et al., 2003), whereas pretreatment with a GLP-1 antagonist attenuates ACTH/corticosterone responses to lithium chloride and elevated platform exposure (Kinzig et al., 2003). Lithium chloride induces c-Fos expression in GLP-1 cells in the NTS (Rinaman, 1999), further supporting a role for endogenous GLP-1 in stress integration. These data suggest that GLP-1 is involved in responses to both psychogenic and systemic stressors. However, GLP-1 receptor knockout mice show an inconsistent phenotype; for example, adrenal weights are reduced, hypothalamic CRH mRNA expression is not altered, and responses to a mild stressor are slightly increased (MacLusky et al., 2000). In chicks, central administration of GLP-1 does not trigger plasma corticosterone secretion (Furuse et al., 1997). Thus, while the results suggest a role for GLP-1 in stress integration in rats, the receptor mechanism and species specificity are yet to be established.

6.4.4 Neuropeptide Y

Central neuropeptide Y (NPY) neurons are implicated in the regulation of energy balance (Woods et al., 1998). Immunoneutralization with NPY antiserum (i.c.v.) does not affect nonstress plasma ACTH and cortisol levels in dogs (Inui et al., 1990), suggesting that NPY signaling does not regulate basal HPA axis activity.

Central administration of NPY (i.c.v. and intraPVN) to rats increases circulating ACTH and corticosterone levels (Harfstrand et al., 1987; Wahlestedt et al., 1987; Leibowitz et al., 1988; Hanson and Dallman, 1995; Sainsbury et al., 1996) and activates c-Fos expression in parvocellular PVN neurons (Li et al., 1994a; Lambert et al., 1995; Xu et al., 1995). Similarly, i.c.v. infusion of NPY activates the HPA axis neuroendocrine response in dogs (Inoue et al., 1989) and sheep (Brooks et al., 1994; Liu et al., 1994). Treatment of hypothalamic explants with NPY in vitro evokes the release of CRH (Tsagarakis et al., 1989; Blasquez et al., 1995). These results demonstrate that NPY has the ability to activate the HPA axis centrally.

Administration of NPY during acute stress facilitates the HPA axis response. For example, central administration (intraPVN or intracisternal) of NPY to rats during anesthesia/surgical stress amplifies plasma ACTH and corticosterone levels and increases CRH protein and mRNA in the hypothalamus (Haas and George, 1987; Wahlestedt et al., 1987; Suda et al., 1993). Furthermore, immunoneutralization with NPY antiserum (i.c.v.) decreases the plasma ACTH and cortisol response to acute hypoglycemia stress in dogs (Inui et al., 1990). Collectively, these data suggest that NPY acts during acute stress to enhance the HPA axis response.

Chronic continuous i.c.v. infusion of NPY produces an initial increase in plasma ACTH and corticosterone that largely normalizes by day 4 of treatment, and results in thymic atrophy (Akana et al., 1996; Sainsbury et al., 1997; Baran et al., 2002). The normalization of basal HPA axis activity during chronic NPY infusion is likely due to glucocorticoid negative feedback as subsequent challenge with acute restraint or cold stress is unable to activate the HPA axis (Akana et al., 1996). Moreover, chronic i.c.v. NPY infusion in rats with limited negative feedback due to clamped levels of corticosterone results in increased morning plasma ACTH before stress, and a facilitated plasma ACTH response to restraint (Akana et al., 1996). These studies demonstrate the ability of NPY to provide prolonged activation of the HPA axis in the absence of negative feedback. Moreover, the data suggest that NPY treatment is not sufficient to mimic chronic stress-induced facilitation of the HPA axis response to novel acute stress.

6.4.5 Neurotensin

Neurotensin appears to activate the HPA axis under both basal and stimulated conditions. While central administration of the neurotensin receptor antagonist SR48692 to rats had no effect on morning ACTH and corticosterone, it was able to attenuate the evening rise in ACTH (but not corticosterone) (Rowe et al., 1997), suggesting the involvement of neurotensin in the evening rise in HPA activity. Importantly, chronic local injection of SR48692 into the PVN region attenuates corticosterone and ACTH response to novelty and restraint (Nicot et al., 1997; Rowe et al., 1997), consistent with a role for neurotensin in mediating HPA responses to psychogenic stressors. Neurotensin antagonist injections also reduce PVN CRH mRNA, which may contribute to the blockade of HPA responsiveness (Nicot et al., 1997). Central administration of neurotensin 1–13 or 8–13 is sufficient to induce the release of ACTH and corticosterone in a CRH-dependent manner (Gudelsky et al., 1989; Rowe et al., 1995), further consistent with a role in HPA excitation.

6.4.6 Vasoactive Intestinal Polypeptide

Vasoactive intestinal polypeptide (VIP) is expressed in the SCN and displays a circadian biosynthetic rhythm that is opposite that of SCN AVP [increased at the end of the light phase (Yang et al., 1993)]. Unlike vasopressin, microinjection into the PVN region stimulates ACTH and corticosterone secretion (Alexander and Sander, 1994). In combination, these data have led some to suggest that SCN VIP may play a role in the circadian rise in HPA axis secretory activity (Kalsbeek et al., 1996b).

6.5 HPA-Inhibitory Neuropeptides

6.5.1 Atrial Natriuretic Peptide

Atrial natriuretic peptide (ANP) is involved in the regulation of plasma sodium levels and cardiac function (McGrath et al., 2005). There are three known members of the natriuretic peptide family; of these, ANP and C-type natriuretic peptide (CNP) are expressed in the brain (ironically, the third, "brain natriuretic peptide" is not) (Langub et al., 1995; Ryan and Gundlach, 1995). These peptides are somewhat unique, in that they signal through natriuretic peptide A and B receptors that function as guanylate cyclases (NPR-A and NPR-B). ANP appears to play an inhibitory role in central HPA integration. ANP blocks CRH release in vitro (Ibanez-Santos et al., 1990), while i.c.v. application of ANP antiserum attenuates ether-stimulated corticosteroid secretion in rats (Fink et al., 1991). In humans, intranasal application of ANP inhibits ACTH and cortisol responses to insulin-induced hypoglycemia (Perras et al., 2004).

6.5.2 Galanin

The role of galanin in basal HPA axis regulation is unclear. Transgenic mice with either an upregulation of brain galanin or a knockout of galanin have normal serum corticosterone levels (Hohmann et al., 2003). However, in rats, i.c.v. immunoneutralization with galanin antiserum in the morning increases plasma ACTH levels, indicating that galanin normally acts centrally to suppress HPA axis activity at the diurnal nadir (Hooi et al., 1990). Infusion of M40, a galanin antagonist, into the bed nucleus of the stria terminalis does not affect morning nonstress levels of ACTH, suggesting that the suppressive action of galanin at the diurnal nadir occurs at brain sites other than the bed nucleus of the stria terminalis (Khoshbouei et al., 2002). Infusion of galanin into the PVN of rats in the predark period (near the peak of the circadian rhythm) decreases plasma corticosterone, whereas galanin infusions in the prelight period (near the nadir of the circadian rhythm) have no effect on plasma corticosterone, indicating that the exogenous galanin inhibits HPA axis activity and that this activity is linked to the circadian cycle (Tempel and Leibowitz, 1990).

Similarly, acute or chronic i.c.v. infusions of galanin in rats and mice do not affect corticosterone levels near the nadir of the diurnal rhythm (Melander et al., 1987; Smith et al., 1994; Hohmann et al., 2003). The lack of effect of central galanin infusion on plasma corticosterone at the nadir of the rhythm may indicate that galanin suppression of the HPA axis is already maximal, so the addition of exogenous galanin is ineffective.

The ability of acute central infusion of galanin to decrease nonstress plasma corticosterone depends on the time of administration (see above). In contrast, acute central infusion of galanin decreases the HPA axis neuroendocrine response after stress regardless of the time of day. Galanin administration (i.c.v.) to anesthetized rats attenuates surgery-induced increases in plasma corticosterone (Balment and al Barazanji, 1992). IntraPVN infusion of galanin prevents the small increases in plasma ACTH that are observed in saline-infused controls and decreases the plasma ACTH response to ether stress (Hooi et al., 1990). These data suggest that galanin can act in the PVN to suppress HPA axis activation during acute stress. In contrast, infusion of M40, a galanin receptor antagonist, into the bed nucleus of the stria terminalis immediately before immobilization stress decreases the stress-evoked plasma ACTH response, demonstrating that galanin signaling in this region normally acts to facilitate the HPA axis response to acute stress (Khoshbouei et al., 2002). These data suggest that galanin acts differentially at multiple brain sites to modulate HPA axis responses to acute stress.

Galanin-like peptide (GALP), a neuropeptide with sequence homology to galanin, has recently been characterized. Amino acids 9–21 of GALP are identical to the biologically active N-terminal (amino acids 1–13) of galanin. Central administration (i.c.v.) of GALP increases plasma ACTH in a dose-dependent manner, in contrast to the observation of reduced plasma ACTH following galanin infusion (Onaka et al., 2005). GALP has a higher affinity for galanin receptor 1 (GALR1) than for galanin receptor 2 (GALR2), whereas galanin has an equal affinity for the two receptors, suggesting that the opposing effects of these neuropeptides may be due to differential activation of galanin receptor subtypes.

6.5.3 Melanin-Concentrating Hormone

Central administration of melanin-concentrating hormone (MCH) in nonstressed rats produces contradictory results. Some investigators report that i.c.v. and intraPVN MCH infusion in male rats near the nadir of the circadian rhythm activates the HPA axis, increasing both plasma ACTH and corticosterone (Jezova et al., 1992; Kennedy et al., 2003) and augments the number of c-Fos positive neurons in the PVN (Hervieu, 2003). In support of these findings, MCH increases CRH release from hypothalamic explants in vitro (Kennedy et al., 2003). However, others report no effect of MCH administered near the nadir of the diurnal rhythm and suppression of plasma ACTH is observed when MCH is administered near the peak of the circadian rhythm (Nahon, 1994; Bluet-Pajot et al., 1995). Similarly, i.c.v. MCH does not affect plasma corticosterone levels at the nadir of the diurnal rhythm in female rats (Tsukamura et al., 2000). Hypothalamic MCH mRNA displays a diurnal variation that peaks near the end of the lights-on period (Nahon, 1994), suggesting that the ability of exogenous MCH to regulate nonstress HPA axis activity may depend in part on the amount of endogenous MCH present at a particular point in the diurnal cycle.

During acute stress, infusion (i.c.v.) of MCH attenuates HPA axis activation in rats. For example, MCH prevents increases in plasma ACTH and corticosterone after mild handling stress (Ludwig et al., 1998) and blocks the ACTH response to ether stress (Nahon, 1994; Bluet-Pajot et al., 1995). Collectively these results suggest that MCH has a stimulatory effect on the HPA axis during times of low HPA axis activity and an inhibitory effect on the HPA axis during times of high HPA axis activity.

6.5.4 Somatostatin

Central immunoneutralization of somatostatin via i.c.v. or intrahippocampal (CA3 and DG) infusion of antiserum in rats does not affect plasma corticosterone levels near the peak of the circadian rhythm, but prevents dexamethasone-induced suppression of plasma corticosterone (Ferrara et al., 1991). These data suggest that somatostatin does not modulate the circadian rise of plasma corticosterone, but may mediate glucocorticoid negative feedback, thereby affecting basal HPA axis tone.

Acute central administration (i.c.v.) of somatostatin analogs does not affect plasma ACTH in sheep (Wang et al., 1987) or cortisol in dogs (Miles et al., 1994), whereas chronic intermittent infusion (i.c.v.) of somatostatin analogs decreases plasma ACTH in rats (Starcevic et al., 2000). Moreover, acute central (i.c.v.) infusion of somatostatin analogs attenuates the HPA axis response to acute stress, including (1) inhibition of ether and tail-suspension stress-induced increases in plasma ACTH in rats (Brown et al., 1984); (2) prevention of carbachol-induced increases in plasma cortisol in dogs (Miles et al., 1994); and (3) abolishment of hemorrhage-induced increases in plasma ACTH in sheep (Wang et al., 1987). Lastly, immunoneutralization (i.v.) of somatostatin during anesthesia or handling stress augments the stress-induced increase in plasma corticosterone in cockerels (Cheung et al., 1988). Collectively, these data demonstrate that the somatostatin analogs act to attenuate HPA axis responses during acute stress.

6.5.5 Substance P

Central administration (i.c.v.) of nonspecific neurokinin receptor (NK-R) and specific neurokinin-1 receptor (NK-1R) antagonists to nonstressed rats increases plasma ACTH, plasma corticosterone, and CRH mRNA expression in the medial parvocellular PVN (Larsen et al., 1993; Jessop et al., 2000), suggesting that substance P (SP) exerts tonic inhibitory control over basal HPA axis activity.

Substance P decreases potassium-stimulated release of CRH from in vitro preparations of rat hypothalamus (Faria et al., 1991). I.c.v. or intraPVN infusions decrease plasma ACTH and corticosterone in nonstressed rats (Unger et al., 1988; Saphier et al., 1994; Culman et al., 1995). Central infusion (i.c.v.) of SP also decreases plasma ACTH in rats during stress, including surgical anesthesia (Chowdrey et al., 1990). These data show that SP can act centrally to inhibit HPA axis activity. Intracerebroventricular infusion of an NK-1R specific antagonist prolongs the plasma ACTH and corticosterone response to restraint stress and increases CRH mRNA expression in parvocellular PVN neurons of stressed rats (Jessop et al., 2000). Moreover, systemic treatment with NK-R antagonist increases the plasma ACTH and corticosterone responses to acute ether and cold stress (Malendowicz et al., 1996). These studies demonstrate that SP normally acts to limit HPA axis activation during acute stress.

Substance P signaling may also attenuate HPA axis responses during chronic stress. Systemic treatment with an NK-1R antagonist augments increases in plasma ACTH, plasma corticosterone, and CRH mRNA expression in the PVN during adjuvant-induced arthritis (Chowdrey et al., 1995). Intracerebroventricular infusion of NK-R antagonist similarly enhances plasma ACTH, plasma corticosterone, and CRH mRNA expression in the medial parvocellular PVN of rats given chronic saline treatment (Larsen et al., 1993). These data suggest that SP also acts to reduce HPA axis activation during chronic inflammatory and osmotic stressors.

6.6 Putative Excitatory Neuropeptides

In addition to the neuropeptide systems noted above, there are several neuropeptides that are sufficient to activate the HPA axis upon central injections under unstimulated conditions. As it is not yet clear whether these peptides are necessary for induction of HPA axis responses to stress/circadian drive, it is too early to determine whether they are physiological activators of the HPA axis. These neuropeptidergic systems are noted in *Table 13-1*.

6.6.1 Calcitonin Gene-Related Peptide/Adrenomedullin

Central application (i.c.v.) of calcitonin gene-related peptide (CGRP) to rats increases plasma corticosterone and this effect is blocked by pretreatment with CRH antiserum (Kovacs et al., 1995). IntraPVN

Table 13-1
Neuropeptides Sufficient to Elicit HPA Axis Activation Under Unstimulated Conditions

Neuropeptide	Corticosteroid	ACTH	CRH release	PVN c-FOS	Other
Adrenomedullin		*	*		
AGRP		*	*		
CGRP	*	*	*		
CART	*	*	*	*	*a
CNP		*	*		
Ghrelin	*	*	*	*	
α-MSH[†]	*	*	*	*	*a,b
Neuropeptide B	*	*	*		
Neuropeptide W	*		*		*c
Orexins (A and B)	*(A>B)	*(A>B)	*(A)	*(A)	*c,d

a, induces CREB phosphorylation in medial parvocellular PVN neurons; b, increases CRH hnRNA expression in PVN; c, depolarizes PVN neurons; d, increases CRH and/or AVP mRNA in PVN. Abbreviations: AGRP, agouti-related peptide; CGRP, calcitonin gene-related peptide; CART, cocaine-amphetamine regulated transcript; CNP, C-type natriuretic peptide; α-MSH, alpha melanocyte stimulating hormone

[†]α-MSH may also inhibit HPA axis responses to stress: see text

infusion of CGRP increases plasma ACTH and corticosterone (Dhillo et al., 2003). In addition, CGRP treatment of hypothalamic explants in vitro stimulates the release of CRH and AVP (Dhillo et al., 2003). Adrenomedullin (ADM), another member of the calcitonin peptide family, has homology to CGRP and has similar biological activity (Wimalawansa, 1996). Like CGRP, central administration (i.c.v.) of ADM activates the HPA axis, including increases in plasma ACTH and induction of c-Fos in CRH neurons in the medial parvocellular PVN (Shan and Krukoff, 2001; Ji et al., 2004). Collectively, these data demonstrate that central CGRP and ADM can activate the HPA axis, and suggest that CGRP and ADM may be similarly involved in HPA axis activation in response to stress.

6.6.2 Cocaine- and Amphetamine-Regulated Transcript

CRH-positive neurons in the PVN are densely innervated by cocaine- and amphetamine-regulated transcript (CART)-positive nerve fibers (Vrang et al., 2000; Sarkar et al., 2004). Central administration of CART (i.c.v.) induces c-Fos expression and the phosphorylation of CREB in CRH-positive neurons in the medial parvocellular portion of the PVN in rats (Vrang et al., 1999, 2000; Sarkar et al., 2004). CART stimulates the release of CRH from medial basal hypothalamic explants that include the PVN (Stanley et al., 2001). Moreover, i.c.v. administration of CART evokes increases in plasma ACTH and corticosterone (Vrang et al., 2000; Stanley et al., 2001; Baranowska et al., 2004), and intraPVN injection of CART increases plasma ACTH in rats (Stanley et al., 2001). Together, these data demonstrate the ability of central CART to activate the HPA axis and suggest that CART may be similarly involved in mediating the stress response of the HPA axis stress response.

6.6.3 C-type Natriuretic Peptide

Unlike ANP, i.c.v. injection of CNP stimulates ACTH release in sheep (Charles et al., 1992) and decreases hypothalamic CRH levels in the rat (Gardi et al., 1997), suggesting a stimulatory role for this peptide in HPA axis regulation.

6.6.4 Ghrelin

Ghrelin is a recently discovered gut peptide that is expressed in the brain and has a role in energy balance (Horvath et al., 2001). Central application (i.c.v.) of ghrelin to rodents increases plasma ACTH and corticosterone (Wren et al., 2002; Bjursell et al., 2005). Moreover, intraPVN infusion of ghrelin increases the number of c-Fos-positive neurons in the PVN (Olszewski et al., 2003), and application of ghrelin to rat hypothalamic explants evokes the release of CRH and AVP (Wren et al., 2002; Mozid et al., 2003). These data suggest that ghrelin may act centrally to activate the HPA axis and modulate its response to stress.

6.6.5 Melanocortins

Melanocortins are thought to play a major role in central regulation of food intake and body weight (Schwartz et al., 2000). Transgenic mice with disruptions of the melanocortin signaling system, including melanocortin receptor 4 (MC4R) knockout mice (Huszar et al., 1997), and agouti and AGRP overexpressing mice (Graham et al., 1997; Harris et al., 2001) have normal nonstress plasma corticosterone levels. Moreover, acute and chronic central administration (i.c.v.) of MCR nonselective or of MC4R selective antagonists in rats does not affect plasma corticosterone levels (Von Frijtag et al., 1998; Baran et al., 2002; Lu et al., 2003; Cragnolini et al., 2004). Similarly, i.c.v. infusion of [Nle^4,D-Phe^7]-α-MSH (NDP-MSH; a long-acting α-MSH analog) in the morning in rhesus monkeys does not affect plasma cortisol levels (Xiao et al., 2003). These data suggest that the melanocortin system is not involved in regulating the nonstress activity of the HPA axis. However, humans with a heterozygous missense mutation in MC4R (a G to A substitution at codon 103) have higher nonstress salivary cortisol levels in the morning (near the peak of the circadian rhythm) that decrease to normal in the evening, relative to G/G homozygotes, suggesting that alterations in MC4R can affect the basal circadian rhythm of HPA axis activity in humans (Rosmond et al., 2001).

Acute central administration (i.c.v.) of nonselective melanocortin agonists, including α-MSH, $ACTH_{1-24}$, and MTII, increases plasma ACTH and corticosterone in rats (Wiegant et al., 1979; Ludwig et al., 1998; Von Frijtag et al., 1998; Lu et al., 2003; Cragnolini et al., 2004), whereas administration of a selective MC3R agonist does not (Von Frijtag et al., 1998). Activation of the HPA axis by nonselective melanocortin agonists is blocked by cotreatment with MC4R selective antagonists or with MCR nonselective antagonists, suggesting that central activation occurs via actions at MC4R (Von Frijtag et al., 1998; Lu et al., 2003). Central administration (i.c.v.) of α-MSH also modulates PVN activity, increasing phosphorylation of CREB (Sarkar et al., 2002), and induction CRH gene transcription (Lu et al., 2003) in the medial parvocellular PVN. Infusion of NDP-MSH into the PVN increases plasma ACTH and corticosterone, and treatment of PVN-containing hypothalamic explants with α-MSH increases the release of CRH and AVP (Dhillo et al., 2002). Thus, central treatment with melanocortin agonists activates the HPA axis, probably via actions at MC4R.

Conversely, melanocortin signaling appears to attenuate HPA axis responses during acute stress. Transgenic mice overexpressing ectopic agouti protein (an endogenous antagonist of MCR) show increased plasma corticosterone responses to restraint stress (Harris et al., 2001; Karkaeva et al., 2005). In addition, Cragnolini et al. (2004) demonstrated that while i.c.v. infusion of α-MSH evokes increases in plasma corticosterone in rats, infusion of either γ-MSH (a selective MC3R agonist) or lower doses of α-MSH inhibits IL-1β-induced corticosterone release; this inhibition by α-MSH was prevented by cotreatment with MC4R selective or MCR nonselective receptor antagonists (Cragnolini et al., 2004). Similarly, central infusion (i.c.v.) of AGRP, a melanocortin receptor antagonist, in rhesus monkeys increased the plasma ACTH response to IL-1β (Xiao et al., 2003), suggesting that endogenous melanocortin agonists normally act to suppress the HPA axis response to this stressor. Moreover, i.c.v. infusion of AGRP in rhesus monkeys (Xiao et al., 2003) and intraPVN infusion of AGRP in rats (Dhillo et al., 2002) increases plasma ACTH, and treatment of PVN-containing rat hypothalamic explants with AGRP increases the release of CRH and AVP (Dhillo et al., 2002). These data suggest that administration of an exogenous melanocortin receptor antagonist may act to block endogenous melanocortin receptor signaling, thereby activating the HPA axis. This indicates that endogenous melanocortin signaling normally acts to inhibit the HPA axis or alternatively AGRP may be acting via a currently unidentified melanocortin receptor and/or brain site to activate the HPA axis.

6.6.6 Neuropeptide B

Central treatment (i.c.v.) with Neuropeptide B (NPB) increases plasma ACTH and corticosterone and this hormonal release is attenuated by pretreatment with anti-CRH antiserum (Samson et al., 2004). These data demonstrate that central NPB is capable of activating the HPA axis and suggest that NPB may be similarly involved in HPA axis activation in response to stress.

6.6.7 Neuropeptide W (NPW)

Central administration (i.c.v.) of NPW increases plasma corticosterone (Baker et al., 2003; Taylor et al., 2005). Systemic pretreatment with CRH receptor antagonist attenuates the plasma corticosterone response to centrally applied NPW (Taylor et al., 2005). Moreover, application of NPW to hypothalamic slice preparations depolarizes parvocellular PVN neurons (Taylor et al., 2005). These data demonstrate that NPW can act centrally to activate the HPA axis, possibly via direct actions on hypophysiotropic CRH neurons.

6.6.8 Orexins/Hypocretins

The orexins (hypocretins) are hypothalamic peptides believed to be involved in the regulation of sleep and wakefulness (Siegel, 2004). Central administration (i.c.v.) of orexin A and orexin B increases plasma ACTH and corticosterone in rats; however, the response to orexin A is more robust than the response to orexin B (Hagan et al., 1999; Ida et al., 2000; Jaszberenyi et al., 2000; Kuru et al., 2000; Al-Barazanji et al., 2001; Jones et al., 2001; Russell et al., 2001; Samson et al., 2002; Brunton and Russell, 2003). Moreover, i.c.v. orexin A activates c-Fos mRNA and protein expression (Date et al., 1999; Edwards et al., 1999; Kuru et al., 2000; Sakamoto et al., 2004), and increases CRH and AVP mRNA expression (Al-Barazanji et al., 2001; Brunton and Russell, 2003) in parvocellular cells of the PVN. In vitro application of orexin A depolarizes PVN parvocellular neurons in hypothalamic slice preparations (Samson et al., 2002) and stimulates the release of CRH from hypothalamic explants (Russell et al., 2001). Collectively, these data suggest that orexins may act directly or indirectly on PVN hypophysiotropic neurons to activate the HPA axis.

7 Local Transmission

7.1 Nitric Oxide

Neurons of the PVN express high levels of neuronal nitric oxide synthase and thus have the capacity to signal through generation of NO. A role for this gaseous transmitter in HPA axis integration has been documented (Rivier, 2001). Central administration of the nitric oxide synthesis inhibitor L-NAME attenuates ACTH release following footshock stress (Kim and Rivier, 2000), whereas i.c.v. administration of the NO donor SIN-1 elicits ACTH release in unstressed rats (Seo et al., 2002). Interestingly, the effects of i.c.v SIN-1 can be blocked by immunoneutralization of CRH and AVP (Seo et al., 2002), further suggesting that the central excitatory effects of NO are mediated through release of one or both secretagogues. The fact that SIN-1 induces transcription of the immediate early gene NGFI-B in the medial parvocellular region of the PVN is consistent with this hypothesis (Seo et al., 2002).

7.2 Carbon Monoxide

CO is generated in the CNS by the action of hemoxygenases, which can be localized to neurons of the PVN (Vincent et al., 1994). Intracerebroventricular injections of the hemoxygenase inhibitor protoporhyrin decrease ACTH release, PVN NGFI-B induction, and PVN CRH heteronuclear RNA levels following stress (Turnbull et al., 1998; Kim and Rivier, 2000), all suggestive of an excitatory role for endogenous CO in HPA axis excitation. It should be noted that CO inhibition also caused a small increase in blood pressure, which may or may not be of relevance to the observed diminution of the HPA axis response.

7.3 Endocannabinoids

Endocannabinoids function as retrograde messengers in the brain. Recent data indicate that endocannabinoids act within the PVN to inhibit presynaptic glutamate release and thereby inhibit parvocellular neurons (Di et al., 2003). Endocannabinoid inhibition is triggered by glucocorticoids, suggesting an involvement in fast feedback inhibition of the HPA axis (Di et al., 2003). However, more global manipulations have contrasting effects; systemic administration of the cannabinoid receptor-1 (CB1) antagonist SR141716 inhibits corticosterone release and PVN c-Fos activation following restraint. Central or peripheral injections of the endogenous endocannabinoid anandamide increase PVN c-Fos expression, and elicit ACTH and corticosterone release (Weidenfeld et al., 1994; Wenger et al., 1997). It should be noted that CB1 receptors are widespread in the CNS; thus, it is possible that global cannabinoid manipulations affect upstream circuitry that results in HPA axis activation.

8 Gonadal Steroids

Effects of estrogens, progestins, and androgens on the HPA axis are evident in multiple steroid manipulation models and mediate, at least in part, marked sex differences in HPA axis regulation. In rodents, it is well known that females display higher basal and stress-related HPA activity than males (Kitay, 1961; Critchlow et al., 1963; Le Mevel et al., 1979; Viau and Meaney, 1991; Carey et al., 1995; Figueiredo et al., 2002). Importantly, the enhanced glucocorticoid or pituitary responses to acute stress observed in female rats are particularly pronounced during the proestrous phase of the estrous cycle (Ogle and Kitay, 1977; Viau and Meaney, 1991) and are abolished by ovariectomy (Critchlow et al., 1963; Seale et al., 2004a), highlighting the influence of ovarian steroids on the HPA axis. Females are known to secrete higher levels of corticosteroid-binding globulin (CBG), which can reduce free corticosterone levels. However, it is apparent that CBG differences are not sufficient to negate stress-induced corticosterone surges (McCormick et al., 1995), indicating that females remain hyperesponsive to stress.

8.1 Estrogens/Progestins

The finding that ovariectomy attenuates corticosterone hypersecretion in female rats suggests a stimulatory influence of ovarian steroids on the HPA axis. Indeed, estradiol administration to ovariectomized females enhances basal ACTH secretion (Carey et al., 1995; Viau and Meaney, 1991) and reestablishes the ultradian pattern of corticosterone secretion (Seale et al., 2004b). However, the role of estrogens on stress responsiveness appears to be more complex than previously thought. While early investigations using supraphysiological doses or prolonged (21 days) estradiol regimens demonstrated stimulation of ACTH and corticosterone responses to a variety of stressors (e.g., restraint, footshock, and novelty) (Viau and Meaney, 1991; Burgess and Handa, 1992; Carey et al., 1995), emerging studies using physiological doses or shorter (7 days) estradiol treatments indicate that this steroid inhibits ACTH response (Redei et al., 1994; Young, 1995; Dayas et al., 2000) and decreases or has no effect on corticosterone responses to acute stressors, including restraint or footshock (Redei et al., 1994; Young, 1995).

The mechanisms responsible for estradiol effects on the HPA axis remain poorly understood. While it is clear that estrogens can act at multiple levels of the HPA axis, there is considerable controversy among these studies. For example, while some studies indicate that estradiol stimulates CRH mRNA expression in the PVN (Patchev and Almeida, 1996; Li et al., 2003b), similar to its role in vitro (Vamvakopoulos and Chrousos, 1993), others show that estradiol either decreases PVN CRH peptide/ mRNA (Haas and George, 1988; Paulmyer-Lacroix et al., 1996) or has no effect on its transcription (Redei et al., 1994). Here, it is important to note that estrogen receptor ERα expression is lacking in the PVN (Shughrue et al., 1997; Laflamme et al., 1998) and while ERβ is richly expressed in the magnocellular oxytocin and AVP neurons of the PVN, its colocalization with CRH PVN neurons, if any, is very limited (Alves et al., 1998; Hrabovszky et al., 1998; Isgor et al., 2003). These findings suggest that estrogenic effects on CRH-secreting parvocellular neurons occur most likely via nongenomic mechanisms or via transsynaptic inputs, or both.

The role of progesterone in HPA axis activity has not been sufficiently addressed. Although the study by Viau and Meaney (1991) indicated that progesterone inhibits the facilitatory effects of estradiol on ACTH responses to stress, other studies do not find a role for this progestin on the basal or stress-related HPA axis activity (Carey et al., 1995; Young, 1995).

Although relevant to human health, the role of ovarian steroids on chronic stress responses has received little attention. High doses of estrogens potentiated plasma corticosterone secretion and AVP mRNA expression in the PVN of ovariectomized female rats submitted to repeated-restraint stress (Lunga and Herbert, 2004). Interestingly, while estradiol decreased basal CRH mRNA in the PVN of ovariectomized rats compared vehicle-treated animals, it had no effect on that of acute or repeated stressed rats (Lunga and Herbert, 2004), suggesting the presence of additional mechanisms in the stress response of females. Future studies probing the precise involvement of estrogens on chronic stress will be valuable for the understanding and treatment of stress-related pathologies (e.g., major depression and anxiety) that affect women at a higher rate than men.

Overall, a consensus has not yet emerged regarding the impact of estrogens or progestins on CRH neurons. The dissociation between central, estradiol-induced downregulation of CRH, reduced ACTH release under physiological replacement regimens, and possible stimulatory effects at the adrenal suggest that female gonadal steroids influence all aspects of the HPA system.

8.2 Androgens

Several lines of evidence demonstrate that testosterone inhibits basal and stress-induced HPA axis activity in male rats (for review, see Viau, 2002). Under basal conditions, gonadectomized male rats have higher plasma ACTH and plasma corticosterone levels than intact controls, as well as increases in corticosterone pulsatility (Seale et al., 2004a). Similarly, gonadectomized males show higher corticosterone and ACTH responses to numerous stressors, including footshock, novel open field, restraint, noise, or lipopolysaccharide administration than gonad-intact males (Handa et al., 1994; Viau and Meaney, 1996; Seale et al., 2004b). The facilitatory effect of gonadectomizing on HPA axis responses to acute stress is reversed with testosterone or with the nonaromatizable androgen dihydrotestosterone proprionate (DHT) (Handa et al., 1994; Viau and Meaney, 1996; Seale et al., 2004b), indicating that the inhibitory effect of androgens on the HPA axis most likely occurs via an androgen receptor-mediated mechanism.

Gonadectomy enhances AVP, CRH, and GR mRNA expression at the level of the PVN under resting conditions in male rats. Importantly, Viau et al. (1999) demonstrated that, under basal conditions, the activity of the HPA axis is regulated by corticosterone-dependent effects on CRH and testosterone-dependent effects on ACTH in the PVN. For example, gonadectomy blocked the enhanced AVP mRNA, but not CRH mRNA expression in the PVN following adrenalectomy in male rats (Viau et al., 1999). On the other hand, stress-induced PVN activity is regulated by corticosterone and testosterone interactions (Viau, 2002).

Although androgen receptor immunoreactivity is present in the oxytocin- and vasopressin-containing neurons of the medial parvocellular PVN, it is absent in other parts of the PVN including the CRH-containing parvocellular neurons (Zhou et al., 1994; Martini et al., 1996). In addition, very little, if any,

ACTH-secreting cells of the anterior pituitary gland express androgen receptors (Thieulant and Duval, 1985), suggesting that testosterone acts at or above the level of the PVN. Additionally, it is important to bear in mind that testosterone can be metabolized in the brain to DHT and estradiol (Martini et al., 1996; Celotti et al., 1997) and thus may act through estrogen receptors as well.

There is strong evidence to support the idea that testosterone can act at extraPVN sites to modulate HPA axis activity. For example, crystalline testosterone or corticosterone implants in the medial preoptic area, a brain region rich in GR and androgen receptor, reduce plasma ACTH and corticosterone responses to restraint, and decrease AVP but not CRH in the median eminence (Viau and Meaney, 1996). In addition, lesions of the medial preoptic area abolished the inhibitory effects of testosterone on ACTH and corticosterone responses to restraint stress (Viau and Meaney, 1996). Furthermore, systemic testosterone stimulated CRH and AVP mRNA expression in the anterior fusiform and posterior bed nucleus of the stria terminalis, respectively, and enhanced AVP message in the medial amygdala in a glucocorticoid-independent manner (Viau et al., 2001).

It is well established that increased plasma testosterone correlates with dominant behavior and consequently, reduced testosterone levels are related to subordinate behavior. For example, in the visible burrow system, a model of rodent social stress (Blanchard et al., 1995), subordinate male rats are characterized by severe weight loss, decreased plasma testosterone levels, and increased basal levels of corticosterone compared with dominant males (Monder et al., 1994). However, the neural mechanisms relating plasma testosterone levels with social stress remain poorly understood. In one study, a subgroup of subordinate males had impaired corticosterone response to restraint stress and display a lower number of CRH-containing neurons in the PVN compared with dominant males or responsive subordinates (Albeck et al., 1997). Interestingly, AVP mRNA levels did not change with behavioral rank of these animals. In another study (DeGoeij et al., 1992a), subordinate male rats displayed lower testosterone levels and increased AVP-ir but not CRH-ir in the median eminence. As in the case of estrogens and progestins, more studies are needed to unravel the role of testosterone in chronically stressed males.

9 Discussion

The reports summarized above highlight the multiplicity of neurochemical systems involved in HPA axis integration. General features that emerge from this dataset suggest that, while complex, several candidate neurochemical systems can be implicated in various aspects of HPA integration. In addition, it is clear that the involvement of a given neurochemical species in HPA axis regulation is multifaceted; involvement in one aspect of HPA axis regulation does not necessarily imply a role in other functional domains. Regulation of particular aspects of HPA axis function by the various neurochemical systems can be summarized briefly as follows:

9.1 Basal Regulation of the HPA Axis

Studies involving neurochemical regulation of basal HPA axis (➋ *Figure 13-4*) tone address three major features: circadian secretion across the diurnal cycle, activation of corticosteroid secretion prior to the circadian peak, and inhibition of corticosterone secretion at the trough of the rhythm. Lesion studies provide solid evidence that monoamines are involved in the maintenance of diurnal rhythms. In particular, serotonin has been implicated as a major factor in this process, likely working indirectly through the suprachiasmatic nucleus. The effects of NE on rhythmicity appear to be mediated by the ventral noradrenergic bundle, suggesting that effects can be direct (via PVN) or indirect (via other hypothalamic structures).

Circumstantial evidence implicates vasoactive intestinal peptide circadian drive of the HPA axis. Effects of VIP would likely be mediated by neurons in the suprachiasmatic nucleus. Neurotensin may also play a role in this process, as antagonism of neurotensin receptors blocks the evening rise in ACTH in the rat; however, no effects on evening corticosterone secretion were seen, making the interpretation of these data somewhat unclear.

Figure 13-4

Neurochemical systems mediating basal tone of the HPA axis. Basal function of the HPA axis is characterized by a marked circadian rhythm in glucocorticoid and to a lesser extent, ACTH release. Circadian secretory rhythms appear to be mediated by noradrenergic and serotonergic systems. Orphanin FQ has also been implicated in this process. GABAergic, galaninergic and vasopressinergic (AVP) systems are associated with generation of the circadian trough in glucocorticoid secretion; one or more of these systems may be communicating corticosteroid feedback information driven by the mineralocorticoid receptor. In addition, dopamine and opioid peptides are implicated in basal HPA axis excitation, whereas substance P may be inhibitory; the role of these systems in circadian rhythms has not been definitively established

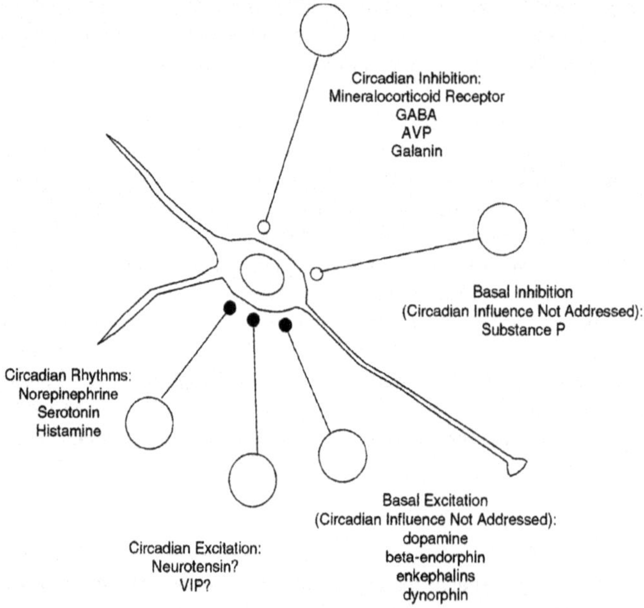

Several factors are implicated in basal HPA inhibition, determined from studies indicating that damage or receptor blockade can result in elevated basal corticosterone at the circadian nadir. Not surprisingly, GABA, a primary inhibitor of the HPA axis, is prominent among these. A series of studies by Buijs' group indicate that AVP is an inhibitor of circadian secretion in the morning (Kalsbeek et al., 1992, 1996a, b); interestingly, AVP actions appear to involve the subPVN region and dorsomedial hypothalamus, both GABA-rich regions. Thus, it is plausible that AVP actually exerts its effect by way of PVN-projecting GABAergic neurons. Substance P also plays a major role in basal inhibition of parvocellular CRH neurons, likely through projections from regions such as the bed nucleus of the stria terminalis, dorsomedial nucleus, and lateral hypothalamus (Bittencourt et al., 1991); again, these are known to be major sources of GABAergic innervation to the PVN. Finally, a number of studies implicate the MR as a steroidal mediator of basal inhibition; the site at which glucocorticoid binding to MR regulates basal tone is yet to be established.

9.2 Acute Stress Excitation

The above reports indicate that two major issues need be considered in defining factors important in HPA axis activation: modality and intensity (*Figure 13-5*). The former concept fits into current notions dividing stressors into "psychogenic" and "systemic" categories (Herman and Cullinan, 1997; Sawchenko et al., 2000; Dayas et al., 2001a; Herman et al., 2003), while the latter indicates that the "psychogenic" category should be subdivided according to intensity.

Figure 13-5
Neurochemical systems mediating excitation and inhibition of the acute stress response. Different neurochemical systems appear to be responsible for stressor modality-specific excitation and inhibition of the HPA axis. Norepinphrine (NE) and perhaps epinephrine (E) appear to be selectively involved in HPA axis responses to systemic stressors. There is also some evidence to suggest a role for serotonin (5-HT) in this process. Neuropeptide Y (NPY), glucagon-like peptide 1 (GLP-1), nitric oxide (NO), carbon monoxide (CO), and dopamine also mediate systemic stress responses, while the influence of CCK seems to be limited to metabolic stimuli. Serotonin, NO, CO, and GLP-1 are also involved in communicating responses to psychogenic stressors of various intensity and may be ubiquitous initiators of HPA to all stress modalities. Glutamate, AVP, orphanin FQ, and angiotensin II appear to be selectively involved in activation of the HPA axis by "mild" stressors. Stress inhibition is mediated by glucocorticoid information from MR and GR-sensing cells and by GABAergic, galaninergic, and melanin concentrating hormone systems. The latter systems inhibit both systemic and psychogenic stress responses. Systemic stress responses are also mediated by opioid peptides [beta-endorphin (beta-End), enkephalins (Enk) and dynorphin (Dyn)] and substance P

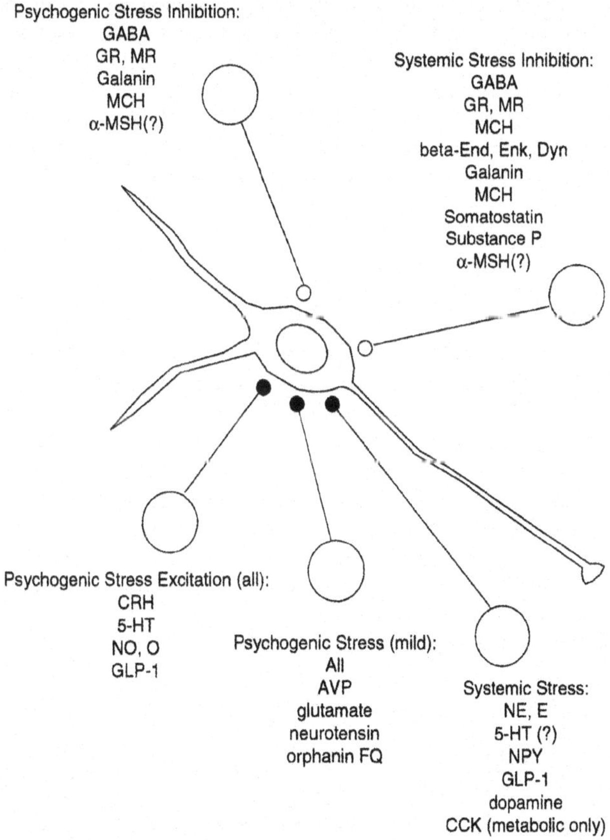

9.2.1 Psychogenic Stress (Mild)

Several reports have emerged that indicate that some neurochemical systems appear important in activating responses to mild stress (e.g., novelty), but have no influence on more "severe" stressors (e.g., restraint, footshock). Systems falling into this category include glutamate, AVP, OFQ, and angiotensin II. These systems may thus be involved in setting the threshold of psychogenic stress responses.

Interpretation of the case for glutamate requires some caveats. The data supporting selective involvement of glutamate in mild stress employ i.c.v. administration of antagonists. Glutamate is a fairly ubiquitous transmitter and is present in both excitatory and inhibitory stress circuits in the brain. In addition, data from our group have shown that corticosterone responses to restraint are blocked by intraPVN injection of the pan-ionotropic glutamate receptor antagonist kynurenic acid, whereas injections immediately outside the PVN exacerbate stress responses (Ziegler and Herman, 2000). Thus, the ability of global delivery of antagonist to modify the activity of both positive and negative inputs indicates that high-resolution methods are required to fully define the role of glutamate in stress integration.

9.2.2 Psychogenic Stress (General)

Several different systems are implicated in psychogenic stress activation: serotonin, NO, CO, prostaglandins, GLP-1, and perhaps, NE originating in the LC. In the case of GLP-1, excitation is likely derived from direct PVN projections from the NTS. Serotonin is a bit harder to pin down, as both the PVN and the GABA-rich PVN surroundings are in receipt of serotonin afferents; wherever the locus of action, it is clear that local serotonin release is involved in HPA axis activation. Interestingly, the gaseous transmitters are likely generated in the PVN region itself, emanating either from the parvocellular neurons themselves, from neighboring magnocellular neurons or even from blood vessels. These data suggest a "feed-forward" mechanism whereby these factors can reinforce an initiated response. Finally, there are some data implicating the LC in activation of the HPA axis; this is likely mediated through extraPVN NE actions on other forebrain stress circuits.

9.2.3 Systemic Stress

The most notable difference between systemic and the aforementioned psychogenic class of stressor is the prominent involvement of NTS or ventral medullary NE and/or E in the former. Indeed, there is convincing evidence that lesions that severely compromise PVN NE levels have little effect on psychogenic stress responses (Pacak et al., 1993). Thus, it is likely that the ascending NE systems from the NTS and perhaps the ventrolateral medulla, are differentially sensitive to stimuli that signal internally driven homeostatic disruption (e.g., blood loss, respiratory distress, infection). This contrasts with the neighboring, yet distinct GLP-1 system, which is involved in both psychogenic and systemic stress integration.

NPY also plays a role in HPA axis responses to systemic stress. The source of PVN NPY appears to be divided between the arcuate nucleus and NTS. Given that NPY is colocalized with NE in the NTS (Hokfelt et al., 1987), it is tempting to speculate that the excitatory effects of NPY are mediated through this circuit.

As was the case with psychogenic stress, the gaseous transmitters NO and CO are stress-excitatory. Given the sensitivity of these systems to cytokines, these systems may be important in activation of the HPA by inflammatory stimuli.

9.3 Acute Stress Inhibition

In contrast with activational circuitry, differential inhibition of psychogenic versus systemic responses has not been extensively assessed. Thus, the designation in ❸ *Figure 13-3* indicates the contexts in which the agents have been studied, rather than reflecting direct comparison among stressors (❸ *Figure 13-6*).

In cases where both classes of stressors have been studied, it is apparent that, not surprisingly, GABA inhibits HPA axis responses to both modalities. These data are in keeping with the preponderance of GABA innervation to the PVN (Decavel and Van Den Pol, 1990), the presence of ample GABA-A receptors in CRH neurons (Cullinan, 2000), and the documented direct influence of GABA on the PVN (Cullinan, 1998; Cole and Sawchenko, 2002). The rich network of systems supplying GABAergic terminals to the PVN affords inhibition at multiple levels through multiple pathways.

Figure 13-6
Neurochemical systems mediating chronic stress responses. Systems responsible for drive of the HPA axis by chronic stress are currently unknown; ascending NE from the locus coeruleus system and central CRH neurons may play a role in this process. The process of chronic stress facilitation appears to involve serotonergic and CCK-ergic systems, whereas habituation is subject to control by MR. The neurochemical phenotype responsible for MR inhibition has yet to be determined. Chronic stress-induced hyperfunction of CRH neurons may be mediated by alterations in the PVN itself; both CRH and AVP are increased in CRH neurons, and GR mRNA is decreased, suggesting possible reduction in feedback. In addition, expression of GABA-A receptor subunit expression is reduced in the medial parvocellular PVN, consistent with reduced GABA signaling, whereas glutamate receptor subunit mRNA expression is altered in such a manner as to predict enhanced excitability (GluR5 is increased, NMDAR2B is decreased relative to NMDAR2A)

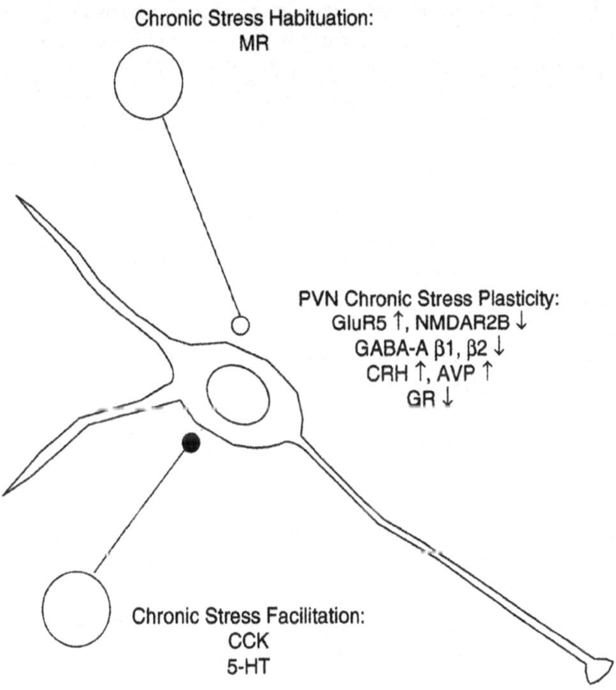

Like GABA, the neuropeptides galanin, MCH, and SP inhibit responses to both psychogenic and systemic stressors. Galanin is present in numerous circuits innervating the PVN, notably including the dorsomedial hypothalamus and the PVN itself, whereas MCH is present in nearby dorsal hypothalamic regions. Both areas project directly to the PVN and are rich in GABA neurons. In addition, PVN-projecting SP neurons are localized in the bed nucleus of the stria terminalis, dorsomedial hypothalamus, and lateral hypothalamus, all of which are rich in GABAergic neurons. Together, these data raise the possibility that these peptides may be released in conjunction with GABA.

There is some evidence to support an inhibitory effect of testosterone on HPA axis stress responses, elaborated through the medial preoptic area (Viau and Meaney, 1996). Once again, these effects may be mediated by PVN-projecting GABAergic systems in this region.

Finally, it is clear that glucocorticoids play multiple inhibitory roles in HPA integration. First, fast feedback effects are apparent at the level of the PVN, mediated by an as yet to be determined receptor. This receptor has been demonstrated in vitro; this hypothesized mechanism is yet to be tested in vivo. Second, central injections of the GR and MR antagonists prolong HPA axis responses to novelty and GR blockade enhances peak responsiveness, indicating that both nuclear receptors act to inhibit HPA responses, probably through transcriptional mechanisms. Finally, a wealth of systemic injection studies using agonists and antagonists indicate glucocorticoid inhibition of the HPA axis; however, it is important to consider that peripheral injections do not readily distinguish central from pituitary actions of glucocorticoids. In this regard, it is important to note that recent studies suggest that central corticosterone injections can actually increase ACTH release (Tanimura and Watts, 1998; Laugero et al., 2002), suggesting that glucocorticoids can have central facilitatory actions on the axis.

9.4 Chronic Stress Regulation

Despite a wealth of data indicating that chronic stress elevates biosynthetic and probably secretory activity of hypophysiotrophic PVN neurons, the precise neurochemical systems mediating this action remain to be determined. Circumstantial evidence implicates GABAergic and glutamatergic signaling in this process, as PVN GABA-A and nonNMDA receptor subunit expression are altered in a way that predicts enhanced excitability of CRH neurons (Cullinan, 1998; Ziegler et al., 2005). Downregulation of PVN or upstream limbic GRs may also be involved in enhanced HPA activation, through the release of glucocorticoid negative feedback.

Repeated exposure to a given stressor can result in marked habituation of the HPA axis response. This habituation appears to be heavily influenced by glucocorticoids acting through the MR. Notably, chronic stress exposure also produces HPA axis facilitation, whereby the HPA axis response to a novel stressor is enhanced. Facilitation occurs regardless of whether the chronic stress protocol involves repeated exposure to a single stress (c.f., Akana et al., 1992) or random exposure to a variety of different stressors (chronic variable stress) (Herman et al, unpublished observations). The process of facilitation appears to be under the control of CCK circuits at the level of the paraventricular thalamus (Bhatnagar and Dallman, 1998).

9.5 Other HPA Modulators

There are considerable data implicating various neurochemical systems in HPA regulation by virtue of the ability of i.c.v. injection to activate corticosterone or ACTH release and/or CRH neurons. Many of these have not been examined extensively with respect to effects on basal drive, acute stress, or chronic stress, and thus the overall relevance remains to be determined. It is important to realize that all these are important candidates that may be worthy of future attention.

In addition, it is also clear that several neuropeptide systems have mixed effects on HPA axis function. For example, α-MSH has opposite effects on basal (excitatory) versus stress-induced (inhibitory) HPA activity. Similarly, opioid peptides activate the HPA axis under basal conditions, but can have both stimulatory and inhibitory effects on stress-induced secretion. Thus, the impact of these two neuropeptidergic systems on HPA axis function appears to be context-dependent.

10 Summary

Overall, it is apparent that multiple neurochemical systems combine to regulate basal HPA axis tone, activation and inhibition. The neurochemical regulatory data add weight to connectional data linking the suprachiasmatic nucleus to the control of circadian corticosteroid rhythms; the brainstem to the control of systemic stress processing; and the limbic system to the control of psychogenic responses. Riding atop these generalized pathways are modality and intensity characteristics that are likely communicated differentially by various sensory and associational circuits, and summated at the PVN from the array of input from multiple neurochemical sources. Neurochemical control of chronic stress remains unresolved; it is likely that glucocorticoid-related information is a determining factor in the seemingly counterintuitive ability to maintain or even facilitate HPA axis function in a high-stress environment, but the pathways to this end are far from clear. Given that chronic stress likely lies at the center of human stress-related dysfunction, an understanding of the mechanism of maladaptive HPA axis responses in chronic stress will be important in the development of new treatments and strategies for intervention in affective disease states, psychosis, and drug addiction.

Acknowledgments

This work was supported by research grants MH049698 (JPH), MH069680 (JPH), MH069725 (JPH), AG012962 (JPH), MH067705 (EBN), MH065770 (NKM), DA016466 (MMO), and DK059803 (YMU).

References

Abercrombie ED, Jacobs BL. 1987. Single-unit response of noradrenergic neurons in the locus coeruleus of freely moving cats. I. Acutely presented stressful and nonstressful stimuli. J Neurosci 7(9): 2837-2843.

Abercrombie ED, Keller RW Jr, Zigmond MJ. 1988. Characterization of hippocampal norepinephrine release as measured by microdialysis perfusion: Pharmacological and behavioral studies. Neuroscience 27(3): 897-904.

Ahima RS, Harlan RE. 1990. Charting of type II glucocorticoid receptor-like immunoreactivity in the rat central nervous system. Neuroscience 39: 579-604.

Akana SF, Dallman MF, Bradbury MJ, Scribner KA, Strack AM, et al. 1992. Feedback and facilitation in the adrenocortical system: Unmasking facilitation by partial inhibition of the glucocorticoid response to prior stress. Endocrinology 131(1): 57-68.

Akana SF, Hanson ES, Horsley CJ, Strack AM, Bhatnagar S, et al. 1996. Clamped corticosterone (B) reveals the effect of endogenous B on both facilitated responsivity to acute restraint and metabolic responses to chronic stress. Stress 1(1): 33-49.

Akil H, Watson SJ, Young E, Lewis ME, Khachaturian H, et al. 1984. Endogenous opioids: Biology and function. Annu Rev Neurosci 7: 223-255.

Al-Barazanji KA, Wilson S, Baker J, Jessop DS, Harbuz MS. 2001. Central orexin-A activates hypothalamic-pituitary-adrenal axis and stimulates hypothalamic corticotropin releasing factor and arginine vasopressin neurones in conscious rats. J Neuroendocrinol 13(5): 421-424.

Albeck DS, McKittrick CR, Blanchard DC, Blanchard RJ, Nikulina J, et al. 1997. Chronic social stress alters levels of corticotropin-releasing factor and arginine vasopressin mRNA in rat brain. J Neurosci 17(12): 4895-4903.

Alexander LD, Sander LD. 1994. Vasoactive intestinal peptide stimulates ACTH and corticosterone release after injection into the PVN. Regul Pept 51(3): 221-227.

Alper RH. 1990. Evidence for central and peripheral serotonergic control of corticosterone secretion in the conscious rat. Neuroendocrinology 51(3): 255-260.

Alves SE, Lopez V, McEwen BS, Weiland NG. 1998. Differential colocalization of estrogen receptor beta (ERbeta) with oxytocin and vasopressin in the paraventricular and supraoptic nuclei of the female rat brain: An immunocytochemical study. Proc Natl Acad Sci USA 95(6): 3281-3286.

Anderson SM, Kant GJ, De Souza EB. 1993. Effects of chronic stress on anterior pituitary and brain corticotropin-releasing factor receptors. Pharmacol Biochem Behav 44(4): 755-761.

Anisman H, Sklar LS. 1979. Catecholamine depletion in mice upon reexposure to stress: Mediation of the escape deficits produced by inescapable shock. J Comp Physiol Psychol 93(4): 610-625.

Antoni FA. 1986. Hypothalamic control of adrenocorticotropin secretion: Advances since the discovery of 41-residue corticotropin-releasing factor. Endocr Rev 7(4): 351-378.

Aronsson M, Fuxe K, Dong Y, Agnati LF, Okret S, et al. 1988. Localization of glucocorticoid receptor mRNA in the male rat brain by in situ hybridization. Proc Natl Acad Sci USA 85: 9331-9335.

Arriza JL, Simerly RB, Swanson LW, Evans RM. 1988. Neuronal mineralocorticoid receptor as a mediator of glucocorticoid response. Neuron 1: 887-900.

Asaba K, Makino S, Hashimoto K. 1998. Effect of urocortin on ACTH secretion from rat anterior pituitary in vitro and in vivo: Comparison with corticotropin-releasing hormone. Brain Res 806(1): 95-103.

Assenmacher I, Szafarczyk A, Alonso G, Ixart G, Barbanel G. 1987. Physiology of neural pathways affecting CRH secretion. Ann NY Acad Sci 512: 149-161.

Aston-Jones G, Bloom FE. 1981a. Activity of norepinephrine-containing locus coeruleus neurons in behaving rats anticipates fluctuations in the sleep-waking cycle. J Neurosci 1(8): 876-886.

Aston-Jones G, Bloom FE. 1981b. Nonrepinephrine-containing locus coeruleus neurons in behaving rats exhibit pronounced responses to non-noxious environmental stimuli. J Neurosci 1(8): 887-900.

Baker JR, Cardinal K, Bober C, Taylor MM, Samson WK. 2003. Neuropeptide W acts in brain to control prolactin, corticosterone, and growth hormone release. Endocrinology 144(7): 2816-2821.

Bali B, Kovacs KJ. 2003. GABAergic control of neuropeptide gene expression in parvocellular neurons of the hypothalamic paraventricular nucleus. Eur J Neurosci 18(6): 1518-1526.

Balment RJ, al Barazanji K. 1992. Renal, cardiovascular and endocrine effects of centrally administered galanin in the anaesthetised rat. Regul Pept 38(1): 71-77.

Banky Z, Halasz B, Nagy G. 1986. Circadian corticosterone rhythm did not develop in rats seven weeks after destruction with 5,7-dihydroxytryptamine of the serotoninergic nerve terminals in the suprachiasmatic nucleus at the age of 16 days. Brain Res 369(1–2): 119-124.

Baran K, Preston E, Wilks D, Cooney GJ, Kraegen EW, et al. 2002. Chronic central melanocortin-4 receptor antagonism and central neuropeptide-Y infusion in rats produce increased adiposity by divergent pathways. Diabetes 51(1): 152-158.

Baranowska B, Wolinska-Witort E, Martynska L, Chmielowska M, Baranowska-Bik A. 2004. Effects of cocaine-amphetamine regulated transcript (CART) on hormone release. Regul Pept 122(2): 55-59.

Bartanusz V, Muller D, Gaillard RC, Streit P, Vutskits L, et al. 2004. Local gamma-aminobutyric acid and glutamate circuit control of hypophyseotrophic corticotropin-releasing factor neuron activity in the paraventricular nucleus of the hypothalamus. Eur J Neurosci 19(3): 777-782.

Bhatnagar S, Dallman M. 1998. Neuroanatomical basis for facilitation of hypothalamic-pituitary-adrenal responses to a novel stressor after chronic stress. Neuroscience 84(4): 1025-1039.

Bhatnagar S, Sun LM, Raber J, Maren S, Julius D, et al. 2004. Changes in anxiety-related behaviors and hypothalamic-pituitary-adrenal activity in mice lacking the 5-HT-3A receptor. Physiol Behav 81(4): 545-555.

Bhatnagar S, Viau V, Chu A, Soriano L, Meijer OC, et al. 2000. A cholecystokinin-mediated pathway to the paraventricular thalamus is recruited in chronically stressed rats and regulates hypothalamic-pituitary-adrenal function. J Neurosci 20(14): 5564-5573.

Bittencourt JC, Benoit R, Sawchenko PE. 1991. Distribution and origins of substance P-immunoreactive projections to the paraventricular and supraoptic nuclei: Partial overlap with ascending catecholaminergic projections. J Chem Neuroanat 4(1): 63-78.

Bjursell M, Egecioglu E, Gerdin AK, Svensson L, Oscarsson J, et al. 2005. Importance of melanin-concentrating hormone receptor for the acute effects of ghrelin. Biochem Biophys Res Commun 326(4): 759-765.

Blakley GG, Pohorecky LA, Benjamin D. 2004. Behavioral and endocrine changes following antisense oligonucleotide-induced reduction in the rat NOP receptor. Psychopharmacology (Berl) 171(4): 421-428.

Blanchard DC, Spencer RL, Weiss SM, Blanchard RJ, McEwen B, et al. 1995. Visible burrow system as a model of chronic social stress: Behavioral and neuroendocrine correlates. Psychoneuroendocrinology 20(2): 117-134.

Blasquez C, Jegou S, Friard O, Tonon MC, Fournier A, et al. 1995. Effect of centrally administered neuropeptide Y on hypothalamic and hypophyseal proopiomelanocortin-derived peptides in the rat. Neuroscience 68(1): 221-227.

Bluet Pajot MT, Mounier F, di Sciullo A, Schmidt B, Kordon C. 1995. Differential sites of action of 8OHDPAT, a 5HT1A agonist, on ACTH and PRL secretion in the rat. Neuroendocrinology 61(2): 159-166.

Bluet-Pajot MT, Presse F, Voko Z, Hoeger C, Mounier F, et al. 1995. Neuropeptide-E-I antagonizes the action of melanin-concentrating hormone on stress-induced release of adrenocorticotropin in the rat. J Neuroendocrinol 7(4): 297-303.

Borowsky B, Kuhn CM. 1992. D1 and D2 dopamine receptors stimulate hypothalamo-pituitary-adrenal activity in rats. Neuropharmacology 31(7): 671-678.

Borowsky B, Kuhn CM. 1993. GBR12909 stimulates hypothalamo-pituitary-adrenal activity by inhibition of uptake at hypothalamic dopamine neurons. Brain Res 613(2): 251-258.

Boudaba C, Szabo K, Tasker JG. 1996. Physiological mapping of local inhibitory inputs to the hypothalamic paraventricular nucleus. J Neurosci 16(22): 7151-7160.

Bowers G, Cullinan WE, Herman JP. 1998. Region-specific regulation of glutamic acid decarboxylase (GAD) mRNA expression in central stress circuits. J Neurosci 18(15): 5938-5947.

Boyle MP, Brewer JA, Funatsu M, Wozniak DF, Tsien JZ, et al. 2005. Acquired deficit of forebrain glucocorticoid receptor produces depression-like changes in adrenal axis regulation and behavior. Proc Natl Acad Sci USA 102(2): 473-478.

Bradbury MJ, Akana SF, Dallman MF. 1994. Roles of type I and II corticosteroid receptors in regulation of basal activity in the hypothalamo-pituitary-adrenal axis during the diurnal trough and the peak: Evidence for a nonadditive effect of combined receptor occupation. Endocrinology 134(3): 1286-1296.

Brooks AN, Howe DC, Porter DW, Naylor AM. 1994. Neuropeptide-Y stimulates pituitary-adrenal activity in fetal and adult sheep. J Neuroendocrinol 6(2): 161-166.

Brown MR, Fisher LA, Spiess J, Rivier C, Rivier J, Vale W. et al. 1982. Corticotropinreleasing factor: Actions on the sympathetic nervous system and metabolism. Endocrinology 111: 928-931.

Brown MR, Rivier C, Vale W. 1984. Central nervous system regulation of adrenocorticotropin secretion: Role of somatostatins. Endocrinology 114(5): 1546-1549.

Brunton PJ, Russell JA. 2003. Hypothalamic-pituitary-adrenal responses to centrally administered orexin-A are suppressed in pregnant rats. J Neuroendocrinol 15(7): 633-637.

Buckingham JC. 1982. Secretion of corticotrophin and its hypothalamic releasing factor in response to morphine and opioid peptides. Neuroendocrinology 35(2): 111-116.

Buckingham JC. 1986. Stimulation and inhibition of corticotrophin releasing factor secretion by beta endorphin. Neuroendocrinology 42(2): 148-152.

Bugajski J. 1999. Social stress adapts signaling pathways involved in stimulation of the hypothalamic-pituitary-adrenal axis. J Physiol Pharmacol 50(3): 367-379.

Bugajski J, Gadek-Michalska A, Bugajski AJ. 2002. Involvement of prostaglandins in the nicotine-induced pituitary-adrenocortical response during social stress. J Physiol Pharmacol 53(4 Pt 2): 847-857.

Buijs RM, Kalsbeek A, Woude van der TP, van Heerikhuize JJ, Shinn S. 1993. Suprachiasmatic nucleus lesion increases corticosterone secretion. Am J Physiol 264(6 Pt 2): R1186-R1192.

Burgess LH, Handa RJ. 1992. Chronic estrogen-induced alterations in adrenocorticotropin and corticosterone secretion and glucocorticoid receptor-mediated functions in female rats. Endocrinology 131(3): 1261-1269.

Butterweck V, Winterhoff H, Herkenham M. 2001. St John's wort, hypericin, and imipramine: A comparative analysis of mRNA levels in brain areas involved in HPA axis control following short-term and long-term administration in normal and stressed rats. Mol Psychiatry 6(5): 547-564.

Cagampang FR, Okamura H, Inouye S. 1994. Circadian rhythms of norepinephrine in the rat suprachiasmatic nucleus. Neurosci Lett 173(1–2): 185-188.

Calogero AE, Bernardini R, Margioris AN, Bagdy G, Gallucci WT, et al. 1989a. Effects of serotonergic agonists and antagonists on corticotropin-releasing hormone secretion by explanted rat hypothalami. Peptides 10(1): 189-200.

Calogero AE, Kamilaris TC, Gomez MT, Johnson EO, Tartaglia ME, et al. 1989b. The muscarinic cholinergic agonist arecoline stimulates the rat hypothalamic-pituitary-adrenal axis through a centrally-mediated corticotropin-releasing hormone-dependent mechanism. Endocrinology 125(5): 2445-2453.

Campeau S, Watson SJ. 1997. Neuroendocrine and behavioral responses and brain pattern of c-fos induction associated with audiogenic stress. J Neuroendocrinol 9(8): 577-588.

Campmany L, Pol O, Armario A. 1996. The effects of two chronic intermittent stressors on brain monoamines. Pharmacol Biochem Behav 53(3): 517-523.

Cano V, Ezquerra L, Ramos MP, Ruiz-Gayo M. 2003. Characterization of the role of endogenous cholecystokinin on the activity of the paraventricular nucleus of the hypothalamus in rats. Br J Pharmacol 140(5): 964-970.

Canteras NS, Simerly RB, Swanson LW. 1995. Organization of projections from the medial nucleus of the amygdala: A PHAL study in the rat. J Comp Neurol 360(2): 213-245.

Canteras NS, Swanson LW. 1992. Projections of the ventral subiculum to the amygdala, septum, and hypothalamus: A PHAL anterograde tract-tracing study in the rat. J Comp Neurol 324(2): 180-194.

Carey MP, Deterd CH, de Koning J, Helmerhorst F, de Kloet ER. 1995. The influence of ovarian steroids on hypothalamic-pituitary-adrenal regulation in the female rat. J Endocrinol 144(2): 311-321.

Casolini P, Kabbaj M, Leprat F, Piazza PV, Rouge-Pont F, et al. 1993. Basal and stress-induced corticosterone secretion is decreased by lesion of mesencephalic dopaminergic neurons. Brain Res 622(1–2): 311-314.

Cassano WJ Jr, D'Mello AP. 2001. Acute stress-induced facilitation of the hypothalamic-pituitary-adrenal axis: Evidence for the roles of stressor duration and serotonin. Neuroendocrinology 74(3): 167-177.

Celotti F, Negri-Cesi P, Poletti A. 1997. Steroid metabolism in the mammalian brain: 5alpha-reduction and aromatization. Brain Res Bull 44(4): 365-375.

Cenci MA, Kalen P, Mandel RJ, Bjorklund A. 1992. Regional differences in the regulation of dopamine and noradrenaline release in medial frontal cortex, nucleus accumbens and caudate-putamen: A microdialysis study in the rat. Brain Res 581(2): 217-228.

Chalmers DT, Lovenberg TW, De Souza EB. 1995. Localization of novel corticotropin-releasing factor receptor (CRF2) mRNA expression to specific subcortical nuclei in rat brain: Comparison with CRF1 receptor mRNA expression. J Neurosci 15(10): 6340-6350.

Champagne D, Beaulieu J, Drolet G. 1998. CRFergic innervation of the paraventricular nucleus of the rat hypothalamus: A tract-tracing study. J Neuroendocrinol 10(2): 119-131.

Chappell PB, Smith MA, Kilts CD, Bissette G, Ritchie J, et al. 1986. Alterations in corticotropin-releasing factor-like immunoreactivity in discrete brain regions after acute and chronic stress. J Neurosci 6: 2908-2914.

Charles CJ, Richards AM, Espiner EA. 1992. Central C-type natriuretic peptide but not atrial natriuretic factor lowers blood pressure and adrenocortical secretion in normal conscious sheep. Endocrinology 131: 1721-1726.

Chautard T, Boudouresque F, Guillaume V, Oliver C. 1993. Effect of excitatory amino acid on the hypothalamo-pituitary-adrenal axis in the rat during the stress-hyporesponsive period. Neuroendocrinology 57(1): 70-78.

Cheung A, Harvey S, Hall TR, Lam SK, Spencer GS. 1988. Effects of passive immunization with antisomatostatin serum on plasma corticosterone concentrations in young domestic cockerels. J Endocrinol 116(2): 179-183.

Chou-Green JM, Holscher TD, Dallman MF, Akana SF. 2003. Repeated stress in young and old 5-HT(2C) receptor knockout mice. Physiol Behav 79(2): 217-226.

Chowdrey HS, Jessop DS, Lightman SL. 1990. Substance P stimulates arginine vasopressin and inhibits adrenocorticotropin release in vivo in the rat. Neuroendocrinology 52(1): 90-93.

Chowdrey HS, Larsen PJ, Harbuz MS, Lightman SL, Jessop DS. 1995. Endogenous substance P inhibits the expression of corticotropin-releasing hormone during a chronic inflammatory stress. Life Sci 57(22): 2021-2029.

Chuluyan HE, Saphier D, Rohn WM, Dunn AJ. 1992. Noradrenergic innervation of the hypothalamus participates in adrenocortical responses to interleukin-1. Neuroendocrinology 56(1): 106-111.

Chung KK, Martinez M, Herbert J. 1999. Central serotonin depletion modulates the behavioural, endocrine and physiological responses to repeated social stress and subsequent c-fos expression in the brains of male rats. Neuroscience 92 (2): 613-625.

Cirelli C, Tononi G. 2004. Locus ceruleus control of state-dependent gene expression. J Neurosci 24(23): 5410-5419.

Cohen MR, Cohen RM, Pickar D, Kreger D, McLellan C, et al. 1985. Hormonal effects of high dose naloxone in humans. Neuropeptides 6(4): 373-380.

Cole MA, Kalman BA, Pace TW, Topczewski F, Lowrey MJ, et al. 2000. Selective blockade of the mineralocorticoid receptor impairs hypothalamic-pituitary-adrenal axis expression of habituation. J Neuroendocrinol 12(10): 1034-1042.

Cole RL, Sawchenko PE. 2002. Neurotransmitter regulation of cellular activation and neuropeptide gene expression in the paraventricular nucleus of the hypothalamus. J Neurosci 22 (3): 959-969.

Coventry TL, Jessop DS, Finn DP, Crabb MD, Kinoshita H, et al. 2001. Endomorphins and activation of the hypothalamo-pituitary-adrenal axis. J Endocrinol 169(1): 185-193.

Cragnolini AB, Perello M, Schioth HB, Scimonelli TN. 2004. alpha-MSH and gamma-MSH inhibit IL-1beta induced activation of the hypothalamic-pituitary-adrenal axis through central melanocortin receptors. Regul Pept 122(3): 185-190.

Critchlow V, Liebelt RA, Bar-Sela M, Mountcastle W, Lipscomb HS. 1963. Sex difference in resting pituitary-adrenal function in the rat. Am J Physiol 205(5): 807-815.

Cullinan WE. 1998. Evidence for a PVN site of action of GABA in the regulatory control of the rat stress axis. Physiologist (Suppl): 41353.

Cullinan WE. 2000. GABA(A) receptor subunit expression within hypophysiotropic CRH neurons: A dual hybridization histochemical study. J Comp Neurol 419(3): 344-351.

Cullinan WE, Helmreich DL, Watson SJ. 1996. Fos expression in forebrain afferents to the hypothalamic paraventricular nucleus following swim stress. J Comp Neurol 368(1): 88-99.

Cullinan WE, Herman JP, Battaglia DF, Akil H, Watson SJ. 1995. Pattern and time course of immediate early gene expression in rat brain following acute stress. Neuroscience 64(2): 477-505.

Cullinan WE, Herman JP, Watson SJ. 1993. Ventral subicular interaction with the hypothalamic paraventricular nucleus: Evidence for a relay in the bed nucleus of the stria terminalis. J Comp Neurol 332: 1-20.

Cullinan WE, Wolfe TJ. 2000. Chronic stress regulates levels of mRNA transcripts encoding beta subunits of the GABA(A) receptor in the rat stress axis. Brain Res 887(1): 118-124.

Culman J, Itoi K, Unger T. 1995. Hypothalamic tachykinins. Mediators of stress responses? Ann NY Acad Sci 771: 204-218.

Cummings S, Seybold V. 1988. Relationship of alpha-1- and alpha-2-adrenergic-binding sites to regions of the paraventricular nucleus of the hypothalamus containing corticotropin-releasing factor and vasopressin neurons. Neuroendocrinology 47(6): 523-532.

Cunningham ET Jr, Sawchenko PE. 1988. Anatomical specificity of noradrenergic inputs to the paraventricular and supraoptic nuclei of the rat hypothalamus. J Comp Neurol 274(1): 60-76.

Dallman MF, Akana SF, Cascio CS, Darlington DN, Jacobson L, et al. 1987. Regulation of ACTH secretion: Variations on a theme of B. Recent Prog Horm Res 43(113): 113-173.

Dallman MF, Pecoraro N, Akana SF, La Fleur SE, Gomez F, et al. 2003. Chronic stress and obesity: A new view of "comfort food". Proc Natl Acad Sci USA 100(20): 11696-11701.

Daniels WM, Jaffer A, Russell VA, Taljaard JJ. 1993. Alpha 2- and beta-adrenergic stimulation of corticosterone secretion in rats. Neurochem Res 18(2): 159-164.

Darlington DN, Miyamoto M, Keil LC, Dallman MF. 1989. Paraventricular stimulation with glutamate elicits bradycardia and pituitary responses. Am J Physiol 256(1 Pt 2): R112-R119.

Date Y, Ueta Y, Yamashita H, Yamaguchi H, Matsukura S, et al. 1999. Orexins, orexigenic hypothalamic peptides, interact with autonomic, neuroendocrine and neuroregulatory systems. Proc Natl Acad Sci USA 96(2): 748-753.

Dautzenberg F, Hauger RL. 2002. The CRF peptide family and their receptors: Yet more partners discovered. Trends Pharmacol Sci 23: 71-77.

Day HE, Akil H. 1999. Evidence that cholecystokinin receptors are not involved in the hypothalamic-pituitary-adrenal response to intraperitoneal administration of interleukin-1beta. J Neuroendocrinol 11(7): 561-568.

Day HE, Campeau S, Watson SJ Jr, Akil H. 1997. Distribution of alpha 1a-, alpha 1b- and alpha 1d-adrenergic receptor mRNA in the rat brain and spinal cord. J Chem Neuroanat 13(2): 115-139.

Day TA, Ferguson AV, Renaud LP. 1985. Noradrenergic afferents facilitate the activity of tuberoinfundibular neurons of the hypothalamic paraventricular nucleus. Neuroendocrinology 41(1): 17-22.

Dayas CV, Buller KM, Crane JW, Xu Y, Day TA. 2001a. Stressor categorization: Acute physical and psychological stressors elicit distinctive recruitment patterns in the amygdala and in medullary noradrenergic cell groups. Eur J Neurosci 14(7): 1143-1152.

Dayas CV, Buller KM, Day TA. 2001b. Medullary neurones regulate hypothalamic corticotropin-releasing factor cell responses to an emotional stressor. Neuroscience 105(3): 707-719.

Dayas CV, Xu Y, Buller KM, Day TA. 2000. Effects of chronic oestrogen replacement on stress-induced activation of hypothalamic-pituitary-adrenal axis control pathways. J Neuroendocrinol 12(8): 784-794.

De Kloet ER, Vreugdenhil E, Oitzl MS, Joels M. 1998. Brain corticosteroid receptor balance in health and disease. Endocr Rev 19(3): 269-301.

Decavel C, Pol VanDen AN. 1990. GABA: A dominant neurotransmitter in the hypothalamus. J Comp Neurol 302: 1019-1037.

DeGoeij DC, Binnekade R, Tilders FJ. 1992a. Chronic stress enhances vasopressin but not corticotropin-releasing factor secretion during hypoglycemia. Am J Physiol 263: E394-399.

DeGoeij DC, Jezova D, Tilders FJ. 1992b. Repeated stress enhances vasopressin synthesis in CRF neurons in the paraventricular nucleus. Brain Res 577: 165-168.

De Goeij DC, Kvetnansky R, Whitnall MH, Jezova D, Berkenbosch F, et al. 1991. Repeated stress-induced activation of corticotropin-releasing factor neurons enhances vasopressin stores and colocalization with corticotropin-releasing factor in the median eminence of rats. Neuroendocrinology 53(2): 150-159.

Deuschle M, Weber B, Colla M, Muller M, Kniest A, et al. 1998. Mineralocorticoid receptor also modulates basal activity of hypothalamuspituitaryadrenocortical system in humans. Neuroendocrinology 68(5): 355-360.

Devine DP, Watson SJ, Akil H. 2001. Nociceptin/orphanin FQ regulates neuroendocrine function of the limbic-hypothalamic-pituitary-adrenal axis. Neuroscience 102(3): 541-553.

Dhillo WS, Small CJ, Jethwa PH, Russell SH, Gardiner JV, et al. 2003. Paraventricular nucleus administration of calcitonin gene-related peptide inhibits food intake and stimulates the hypothalamo-pituitary-adrenal axis. Endocrinology 144(4): 1420-1425.

Dhillo WS, Small CJ, Seal LJ, Kim MS, Stanley SA, et al. 2002. The hypothalamic melanocortin system stimulates the hypothalamo-pituitary-adrenal axis in vitro and in vivo in male rats. Neuroendocrinology 75(4): 209-216.

Di S, Malcher-Lopes R, Halmos KC, Tasker JG. 2003. Nongenomic glucocorticoid inhibition via endocannabinoid release in the hypothalamus: A fast feedback mechanism. J Neurosci 23(12): 4850-4857.

Di Sciullo A, Bluet-Pajot MT, Mounier F, Oliver C, Schmidt B, et al. 1990. Changes in anterior pituitary hormone levels after serotonin 1A receptor stimulation. Endocrinology 127 (2): 567-572.

Dobrakovova M, Kvetnansky R, Torda T, Murgas K. 1982. Changes of plasma and adrenal catecholamines and corticosterone in stressed rats with septal lesions. Physiol Behav 29: 41-45.

Dohanics J, Hoffman GE, Verbalis JG. 1991. Hyponatremia-induced inhibition of magnocellular neurons causes stressor-selective impairment of stimulated adrenocorticotropin secretion in rats. Endocrinology 128(1): 331-340.

Donald RA, Redekopp C, Cameron V, Nicholls MG, Bolton J, Livesey J, Espiner EA, River J, Vale W. 1983. The hormonal actions of corticotropin-releasing factor in sheep: Effect of intravenous and intracerebroventricular injection. Endocrinology **113**: 866-870.

Drouin J, Sun YL, Nemer M. 1989. Glucocorticoid repression of proopiomelanocortin gene transcription. J Steroid Biochem 34: 63-69.

Drucker DJ. 1990. Glucagon and the glucagon-like peptides. Pancreas 5(4): 484-488.

Edwards CM, Abusnana S, Sunter D, Murphy KG, Ghatei MA, et al. 1999. The effect of the orexins on food intake: Comparison with neuropeptide Y, melanin-concentrating hormone and galanin. J Endocrinol 160(3): R7-R13.

Eisenberg RM. 1984. Effects of naltrexone on plasma corticosterone in opiate-naive rats: A central action. Life Sci 34(12): 1185-1191.

Eisenberg RM. 1993. DAMGO stimulates the hypothalamo-pituitaryadrenal axis through a mu-2 opioid receptor. J Pharmacol Exp Ther 266(2): 985-991.

Emmert MH, Herman JP. 1999. Differential forebrain c-fos mRNA induction by ether inhalation and novelty: Evidence for distinctive stress pathways. Brain Res 845(1): 60-67.

Endou M, Yanai K, Sakurai E, Fukudo S, Hongo M, et al. 2001. Food-deprived activity stress decreased the activity of the histaminergic neuron system in rats. Brain Res 891 (1–2): 32-41.

Evans SJ, Murray TF, Moore FL. 2000. Partial purification and biochemical characterization of a membrane glucocorticoid receptor from an amphibian brain. J Steroid Biochem Mol Biol 72(5): 209-221.

Faria M, Navarra P, Tsagarakis S, Besser GM, Grossman AB. 1991. Inhibition of CRH-41 release by substance P, but not substance K, from the rat hypothalamus in vitro. Brain Res 538(1): 76-78.

Feldman S, Conforti N, Melamed E. 1987. Paraventricular nucleus serotonin mediates neurally stimulated adrenocortical secretion. Brain Res Bull 18(2): 165-168.

Feldman S, Conforti N, Saphier D. 1990. Adrenal responses and neurotransmitters in posterior hypothalamic deafferentation. Brain Res Bull 25(1): 75-78.

Feldman S, Melamed E, Conforti N, Weidenfeld J. 1984. Effect of central serotonin depletion on adrenocortical responses to neural stimuli. Exp Neurol 85(3): 661-666.

Feldman S, Newman ME, Gur E, Weidenfeld J. 1998. Role of serotonin in the amygdala in hypothalamopituitaryadrenocortical responses. Neuroreport 9(9): 2007-2009.

Feldman S, Weidenfeld J. 1995. Posterior hypothalamic deafferentation or 5,7-dihydroxytryptamine inhibit corticotropin-releasing hormone, ACTH and corticosterone responses following photic stimulation. Neurosci Lett 198(2): 143-145.

Feldman S, Weidenfeld J. 1997. Hypothalamic mechanisms mediating glutamate effects on the hypothalamo-pituitary-adrenocortical axis. J Neural Transm 104(6–7): 633-642.

Feldman S, Weidenfeld J. 1998. The excitatory effects of the amygdala on hypothalamo-pituitary-adrenocortical responses are mediated by hypothalamic norepinephrine, serotonin and CRF-41. Brain Res Bull 45(4): 389-393.

Fernandez F, Misilmeri MA, Felger JC, Devine DP. 2004. Nociceptin/orphanin FQ increases anxiety-related behavior and circulating levels of corticosterone during neophobic tests of anxiety. Neuropsychopharmacology 29(1): 59-71.

Ferrara C, Cocchi D, Muller EE. 1991. Somatostatin in the hippocampus mediates dexamethasone-induced suppression of corticosterone secretion in the rat. Neuroendocrinology 53(4): 428-431.

Figueiredo HF, Bodie BL, Tauchi M, Dolgas CM, Herman JP. 2003. Stress integration after acute and chronic predator stress: Differential activation of central stress circuitry and sensitization of the hypothalamo-pituitary-adrenocortical axis. Endocrinology 144(12): 5249-5258.

Figueiredo HF, Dolgas CM, Herman JP. 2002. Stress activation of cortex and hippocampus is modulated by sex and stage of estrus. Endocrinology 143(7): 2534-2540.

Fink G, Dow RC, Casley D, Johnston CI, Lim AT, et al. 1991. Atrial natriuretic peptide is a physiological inhibitor of ACTH release: Evidence from immunoneutralization in vivo. J Endocrinol 131(3): R9-R12.

Fitzsimons JT. 1998. Angiotensin, thirst, and sodium appetite. Physiol Rev 78(3): 583-686.

Fu Y, Matta SG, Valentine JD, Sharp BM. 1997. Adrenocorticotropin response and nicotine-induced norepinephrine secretion in the rat paraventricular nucleus are mediated through brainstem receptors. Endocrinology 138(5): 1935-1943.

Fuchs E, Flugge G. 1995. Modulation of binding sites for corticotropin-releasing hormone by chronic psychosocial stress. Psychoneuroendocrinology 20(1): 33-51.

Fuller PJ, Lim-Tio SS, Brennan FE. 2000. Specificity in mineralocorticoid versus glucocorticoid action. Kidney Int 57 (4): 1256-1264.

Fuller RW. 1992. The involvement of serotonin in regulation of pituitary-adrenocortical function. Front Neuroendocrinol 13(3): 250-270.

Fuller RW, Perry KW, Hemrick-Luecke SK, Engleman E. 1996. Serum corticosterone increases reflect enhanced uptake inhibitor-induced elevation of extracellular 5-hydroxytryptamine in rat hypothalamus. J Pharm Pharmacol 48(1): 68-70.

Fuller RW, Snoddy HD. 1977. Elevation of plasma corticosterone by swim stress and insulin-induced hypoglycemia in control and fluoxetine-pretreated rats. Endocr Res Commun 4(1): 11-23.

Fuller RW, Snoddy HD. 1990. Serotonin receptor subtypes involved in the elevation of serum corticosterone concentration in rats by direct- and indirect-acting serotonin agonists. Neuroendocrinology 52(2): 206-211.

Furuse M, Matsumoto M, Saito N, Sugahara K, Hasegawa S. 1997. The central corticotropin-releasing factor and glucagon-like peptide-1 in food intake of the neonatal chick. Eur J Pharmacol 339(2–3): 211-214.

Fuxe K. 1965. Evidence for the existence of monoamine neurons in the central nervous system. IV. Distribution of monoamine nerve terminals in the central nervous system. Acta Physiol Scand 64(Suppl 247): 37-85.

Gadek-Michalska A, Bugajski J. 1996. Stimulatory effect of intracerebroventricular met- and leu-enkephalin on corticosterone secretion in rats. Folia Med Cracov 37(3–4): 3-12.

Gadek-Michalska A, Bugajski J. 2003. Repeated handling, restraint, or chronic crowding impair the hypothalamic-pituitary-adrenocortical response to acute restraint stress. J Physiol Pharmacol 54(3): 449-459.

Gadek-Michalska A, Cetera B, Bugajski J. 1997. Corticosterone response induced by intracerebroventricular administration of met-enkephalin and naloxone in rats under stress. Folia Med Cracov 38(3–4): 17-26.

Gaillet S, Alonso G, Le Borgne R, Barbanel G, Malaval F, et al. 1993. Effects of discrete lesions in the ventral noradrenergic ascending bundle on the corticotropic stress response depend on the site of the lesion and on the plasma levels of adrenal steroids. Neuroendocrinology 58(4): 408-419.

Gaillet S, Lachuer J, Malaval F, Assenmacher I, Szafarczyk A. 1991. The involvement of noradrenergic ascending pathways in the stress-induced activation of ACTH and corticosterone secretions is dependent on the nature of stressors. Exp Brain Res 87(1): 173-180.

Gardi J, Biro E, Vecsernyes M, Julesz J, Nyari T, et al. 1997. The effects of brain and C-type natriuretic peptides on corticotropin-releasing factor in brain of rats. Life Sci 60(23): 2111-2117.

Gibson A, Ginsburg M, Hall M, Hart SL. 1979. The effect of intracerebroventricular administration of methionine-enkephalin on the stress-induced secretion of corticosterone in mice. Br J Pharmacol 66(2): 164-166.

Gibson A, Hart SL, Patel S. 1986. Effects of 6-hydroxydopamine-induced lesions of the paraventricular nucleus, and of prazosin, on the corticosterone response to restraint in rats. Neuropharmacology 25(3): 257-260.

Gibson A, Hart SL, Shabib A. 1980. Leucine-enkephalin and methionine-enkephalin produce opposing effects on plasma corticosterone levels in ether-stressed mice. Br J Pharmacol 70(4): 509-511.

Gilbert F, Brazell C, Tricklebank MD, Stahl SM. 1988. Activation of the 5-HT1A receptor subtype increases rat plasma ACTH concentration. Eur J Pharmacol 147(3): 431-439.

Gomez F, Lahmame A, de Kloet ER, Armario A. 1996. Hypothalamic-pituitary-adrenal response to chronic stress in five inbred rat strains: Differential responses are mainly located at the adrenocortical level. Neuroendocrinology 63(4): 327-337.

Gottlicher M, Heck S, Herrlich P. 1998. Transcriptional cross-talk, the second mode of steroid hormone receptor action. J Mol Med 76(7): 480-489.

Graham M, Shutter JR, Sarmiento U, Sarosi I, Stark KL. 1997. Overexpression of Agrt leads to obesity in transgenic mice. Nat Genet 17(3): 273-274.

Grant SJ, Aston-Jones G, Redmond DE Jr. 1988. Responses of primate locus coeruleus neurons to simple and complex sensory stimuli. Brain Res Bull 21(3): 401-410.

Gresch PJ, Sved AF, Zigmond MJ, Finlay JM. 1994. Stress-induced sensitization of dopamine and norepinephrine efflux in medial prefrontal cortex of the rat. J Neurochem 63(2): 575-583.

Grippo AJ, Moffitt JA, Johnson AK. 2002. Cardiovascular alterations and autonomic imbalance in an experimental model of depression. Am J Physiol Regul Integr Comp Physiol 282(5): R1333-R1341.

Grippo AJ, Sullivan NR, Damjanoska KJ, Crane JW, Carrasco GA, et al. 2004. Chronic mild stress induces behavioral and physiological changes, and may alter serotonin 1A receptor function, in male and cycling female rats. Psychopharmacology (Berl) 179(4): 769-780.

Groenink L, Mos J, Gugten Van der J, Olivier B. 1996. The 5-HT1A receptor is not involved in emotional stress-induced rises in stress hormones. Pharmacol Biochem Behav 55(2): 303-308.

Gross C, Santarelli L, Brunner D, Zhuang X, Hen R. 2000. Altered fear circuits in 5-HT(1A) receptor KO mice. Biol Psychiatry 48(12): 1157-1163.

Grossman A, Gaillard RC, McCartney P, Rees LH, Besser GM. 1982. Opiate modulation of the pituitaryadrenal axis: Effects of stress and circadian rhythm. Clin Endocrinol (Oxf) 17(3): 279-286.

Gudelsky GA, Berry SA, Meltzer HY. 1989. Neurotensin activates tuberoinfundibular dopamine neurons and increases serum corticosterone concentrations in the rat. Neuroendocrinology 49(6): 604-609.

Gunion MW, Rosenthal MJ, Morley JE, Miller S, Zib B, et al. 1991. mu-receptor mediates elevated glucose and corticosterone after third ventricle injection of opioid peptides. Am J Physiol 261(1 Pt 2): R70-R81.

Haas DA, George SR. 1987. Neuropeptide Y administration acutely increases hypothalamic corticotropin-releasing factor immunoreactivity: Lack of effect in other rat brain regions. Life Sci 41(25): 2725-2731.

Haas DA, George SR. 1988. Gonadal regulation of corticotropin-releasing factor immunoreactivity in hypothalamus. Brain Res Bull 20(3): 361-367.

Hagan JJ, Leslie RA, Patel S, Evans ML, Wattam TA, et al. 1999. Orexin A activates locus coeruleus cell firing and increases arousal in the rat. Proc Natl Acad Sci USA 96(19): 10911-10916.

Haleem DJ, Kennett GA, Whitton PS, Curzon G. 1989. 8-OH-DPAT increases corticosterone but not other 5-HT1A receptor-dependent responses more in females. Eur J Pharmacol 164(3): 435-443.

Han F, Ozawa H, Matsuda K, Nishi M, Kawata M. 2005. Colocalization of mineralocorticoid receptor and glucocorticoid receptor in the hippocampus and hypothalamus. Neurosci Res 51(4): 371-381.

Handa RJ, Nunley KM, Lorens SA, Louie JP, McGivern RF, et al. 1994. Androgen regulation of adrenocorticotropin and corticosterone secretion in the male rat following novelty and foot shock stressors. Physiol Behav 55(1): 117-124.

Hanson ES, Dallman MF. 1995. Neuropeptide Y (NPY) may integrate responses of hypothalamic feeding systems and the hypothalamopituitaryadrenal axis. J Neuroendocrinol 7 (4): 273-279.

Haracz JL, Bloom AS, Wang RI, Tseng LF. 1981. Effect of intraventricular beta-endorphin and morphine on hypothalamic-pituitary-adrenal activity and the release of pituitary beta-endorphin. Neuroendocrinology 33(3): 170-175.

Harfstrand A, Eneroth P, Agnati L, Fuxe K. 1987. Further studies on the effects of central administration of neuropeptide Y on neuroendocrine function in the male rat: Relationship to hypothalamic catecholamines. Regul Pept 17(3): 167-179.

Harris RB, Zhou J, Shi M, Redmann S, Mynatt RL, et al. 2001. Overexpression of agouti protein and stress responsiveness in mice. Physiol Behav 73(4): 599-608.

Hayward JN. 1975. Neural control of the posterior pituitary. Annu Rev Physiol 37: 191-210.

Heinrichs SC, Richard D. 1999. The role of corticotropin-releasing factor and urocortin in the modulation of ingestive behavior. Neuropeptides 33(5): 350-359.

Herman JP. 1993. Regulation of adrenocorticosteroid receptor mRNA expression in the central nervous system. Cell Mol Neurobiol 13: 349-372.

Herman JP, Adams D, Prewitt CM. 1995. Regulatory changes in neuroendocrine stress-integrative circuitry produced by a variable stress paradigm. Neuroendocrinology 61: 180-190.

Herman JP, Cullinan WE. 1997. Neurocircuitry of stress: Central control of the hypothalamo-pituitary-adrenocortical axis. Trends Neurosci 20(2): 78-84.

Herman JP, Cullinan WE, Watson SJ. 1994. Involvement of the bed nucleus of the stria terminalis in tonic regulation of paraventricular hypothalamic CRH and AVP mRNA expression. J Neuroendocrinol 6: 433-442.

Herman JP, Figueiredo H, Mueller NK, Ulrich-Lai Y, Ostrander MM, et al. 2003. Central mechanisms of stress integration: Hierarchical circuitry controlling hypothalamo-pituitary-adrenocortical responsiveness. Front Neuroendocrinol 24(3): 151-180.

Herman JP, McCreary BJ, Bettenhausen K, Ziegler DR. 2002. Neurocircuit activation of the hypothalamo-pituitary-adrenocortical axis: Roles for ascending norepinephrine systems. McCarty R, Aguilera G, Sabban E, Kvetnansky R, Stress: Neural, endocrine and molecular studies. Taylor and Francis; London and New York: pp. 19-24.

Herman JP, Patel PD, Akil H, Watson SJ. 1989. Localization and regulation of glucocorticoid and mineralocorticoid receptor messenger RNAs in the hippocampal formation of the rat. Mol Endocrinol 3(11): 1886-1894.

Hernando F, Fuentes JA, Ruiz-Gayo M. 1996. Impairment of stress adaptive behaviours in rats by the CCKA receptor antagonist, devazepide. Br J Pharmacol 118(2): 400-406.

Hervieu G. 2003. Melanin-concentrating hormone functions in the nervous system: Food intake and stress. Expert Opin Ther Targets 7(4): 495-511.

Hillhouse EW, Burden J, Jones MT. 1975. The effect of various putative neurotransmitters on the release of corticotrophin releasing hormone from the hypothalamus of the rat in vitro. I. The effect of acetylcholine and noradrenaline. Neuroendocrinology 17(1): 1-11.

Hohmann JG, Krasnow SM, Teklemichael DN, Clifton DK, Wynick D, et al. 2003. Neuroendocrine profiles in galanin-overexpressing and knockout mice. Neuroendocrinology 77(6): 354-366.

Hokfelt T, Millhorn D, Seroogy K, Tsuruo Y, Ceccatelli S, et al. 1987. Coexistence of peptides with classical transmitters. Experientia 43: 768-780.

Holmes MC, Antoni FA, Aguilera G, Catt KJ. 1986. Magnocellular axons in passage through the median eminence release vasopressin. Nature 319: 326-329.

Hooi SC, Maiter DM, Martin JB, Koenig JI. 1990. Galaninergic mechanisms are involved in the regulation of corticotropin and thyrotropin secretion in the rat. Endocrinology 127(5): 2281-2289.

Horvath TL, Diano S, Sotonyi P, Heiman M, Tschop M. 2001. Minireview: Ghrelin and the regulation of energy balance – a hypothalamic perspective. Endocrinology 142(10): 4163-4169.

Hrabovszky E, Kallo I, Hajszan T, Shughrue PJ, Merchenthaler I, et al. 1998. Expression of estrogen receptor-beta messenger ribonucleic acid in oxytocin and vasopressin neurons of the rat supraoptic and paraventricular nuclei. Endocrinology 139(5): 2600-2604.

Hsu SY, Hsueh AJ. 2001. Human stresscopin and stresscopin-related peptide are selective ligands for the type 2 corticotropin-releasing hormone receptor. Nat Med 7(5): 605-611.

Huszar D, Lynch CA, Fairchild-Huntress V, Dunmore JH, Fang Q, et al. 1997. Targeted disruption of the melanocortin-4 receptor results in obesity in mice. Cell 88(1): 131-141.

Ibanez-Santos J, Tsagarakis S, Rees LH, Besser GM, Grossman A. 1990. Atrial natriuretic peptides inhibit the release of corticotrophin-releasing factor-41 from the rat hypothalamus in vitro. J Endocrinol 126(2): 223-228.

Ida T, Nakahara K, Murakami T, Hanada R, Nakazato M, et al. 2000. Possible involvement of orexin in the stress reaction in rats. Biochem Biophys Res Commun 270(1): 318-323.

Ikemoto S, Goeders NE. 1998. Microinjections of dopamine agonists and cocaine elevate plasma corticosterone: Dissociation effects among the ventral and dorsal striatum and medial prefrontal cortex. Brain Res 814(1–2): 171-178.

Imaki T, Katsumata H, Miyata M, Naruse M, Imaki J, et al. 2001. Expression of corticotropin-releasing hormone type 1 receptor in paraventricular nucleus after acute stress. Neuroendocrinology 73(5): 293-301.

Imaki T, Nahan JL, Rivier C, Sawchenko PE, Vale W. 1991. Differential regulation of corticotropin-releasing factor mRNA in rat brain regions by glucocorticoids and stress. J Neurosci 11(3): 585-599.

Imaki T, Naruse M, Harada S, Chikada N, Imaki J, et al. 1996. Corticotropin-releasing factor up-regulates its own receptor mRNA in the paraventricular nucleus of the hypothalamus. Brain Res Mol Brain Res 38(1): 166-170.

Imaki T, Wang XQ, Shibasaki T, Harada S, Chikada N, et al. 1995. Chlordiazepoxide attenuates stress-induced activation of neurons, corticotropin-releasing factor (CRF) gene transcription and CRF biosynthesis in the paraventricular nucleus (PVN). Brain Res Mol Brain Res 32 (2): 261-270.

Inoue T, Inui A, Okita M, Sakatani N, Oya M, et al. 1989. Effect of neuropeptide Y on the hypothalamic-pituitary-adrenal axis in the dog. Life Sci 44(15): 1043-1051.

Inui A, Inoue T, Nakajima M, Okita M, Sakatani N, et al. 1990. Brain neuropeptide Y in the control of adrenocorticotropic hormone secretion in the dog. Brain Res 510(2): 211-215.

Irwin J, Ahluwalia P, Anisman H. 1986. Sensitization of norepinephrine activity following acute and chronic footshock. Brain Res 379(1): 98-103.

Isgor C, Shieh KR, Akil H, Watson SJ. 2003. Colocalization of estrogen beta receptor messenger RNA with orphanin FQ, vasopressin and oxytocin in the rat hypothalamic paraventricular and supraoptic nuclei. Anat Embryol (Berl) 206(6): 461-469.

Ito C, Shen H, Toyota H, Kubota Y, Sakurai E, et al. 1999. Effects of the acute and chronic restraint stresses on the central histaminergic neuron system of Fischer rat. Neurosci Lett 262(2): 143-145.

Itoi K, Helmreich DL, Lopez-Figueroa MO, Watson SJ. 1999. Differential regulation of corticotropin-releasing hormone and vasopressin gene transcription in the hypothalamus by norepinephrine. J Neurosci 19(13): 5464-5472.

Itowi N, Yamatodani A, Cacabelos R, Goto M, Wada H. 1989. Effect of histamine depletion on circadian variations of corticotropin and corticosterone in rats. Neuroendocrinology 50(2): 187-192.

Iyengar S, Kim HS, Wood PL. 1986. Kappa opiate agonists modulate the hypothalamic-pituitary-adrenocortical axis in the rat. J Pharmacol Exp Ther 238(2): 429-436.

Iyengar S, Kim HS, Wood PL. 1987. Mu-, delta-, kappa- and epsilon-opioid receptor modulation of the hypothalamic-pituitary-adrenocortical (HPA) axis: Subchronic tolerance studies of endogenous opioid peptides. Brain Res 435(1–2): 220-226.

Jacobson L, Sapolsky RM. 1991. The role of the hippocampus in feedback regulation of the hypothalamo-pituitary-adrenocortical axis. Endocrine Rev 12: 118-134.

Jaszberenyi M, Bujdoso E, Pataki I, Telegdy G. 2000. Effects of orexins on the hypothalamic-pituitary-adrenal system. J Neuroendocrinol 12(12): 1174-1178.

Jedema HP, Grace AA. 2003. Chronic exposure to cold stress alters electrophysiological properties of locus coeruleus neurons recorded in vitro. Neuropsychopharmacology 28 (1): 63-72.

Jedema HP, Sved AF, Zigmond MJ, Finlay JM. 1999. Sensitization of norepinephrine release in medial prefrontal cortex: Effect of different chronic stress protocols. Brain Res 830(2): 211-217.

Jessop DS, Renshaw D, Larsen PJ, Chowdrey HS, Harbuz MS. 2000. Substance P is involved in terminating the hypothalamo-pituitary-adrenal axis response to acute stress through centrally located neurokinin-1 receptors. Stress 3(3): 209-220.

Jezova D, Bartanusz V, Westergren I, Johansson BB, Rivier J, et al. 1992. Rat melanin-concentrating hormone stimulates adrenocorticotropin secretion: Evidence for a site of action in brain regions protected by the blood-brain barrier. Endocrinology 130(2): 1024-1029.

Jezova D, Tokarev D, Rusnak M. 1995. Endogenous excitatory amino acids are involved in stress-induced adrenocorticotropin and catecholamine release. Neuroendocrinology 62(4): 326-332.

Jhanwar-Uniyal M, Roland CR, Leibowitz SF. 1986. Diurnal rhythm of alpha 2-noradrenergic receptors in the paraventricular nucleus and other brain areas: Relation to circulating corticosterone and feeding behavior. Life Sci 38(5): 473-482.

Ji SM, Wang ZM, Li XP, He RR. 2004. Intracerebroventricular administration of adrenomedullin increases the expression of c-fos and activates nitric oxide-producing neurons in rat cardiovascular related brain nuclei. Sheng Li Xue Bao 56(3): 328-334.

Joels M, Verkuyl JM, Van Riel E. 2003. Hippocampal and hypothalamic function after chronic stress. Ann NY Acad Sci 1007: 367-378.

Johnson MP, Kelly G, Chamberlain M. 2001. Changes in rat serum corticosterone after treatment with metabotropic glutamate receptor agonists or antagonists. J Neuroendocrinol 13(8): 670-677.

Jones DN, Gartlon J, Parker F, Taylor SG, Routledge C, et al. 2001. Effects of centrally administered orexin-B and orexin-A: A role for orexin-1 receptors in orexin-B-induced hyperactivity. Psychopharmacology (Berl) 153(2): 210-218.

Jones MT, Hillhouse EW, Burden J. 1976. Effect on various putative neurotransmitters on the secretion of corticotrophin-releasing hormone from the rat hypothalamus in vitro-a model of the neurotransmitters involved. J Endocrinol 69(1): 1-10.

Jorgensen H, Knigge U, Kjaer A, Moller M, Warberg J. 2002. Serotonergic stimulation of corticotropin-releasing hormone and pro-opiomelanocortin gene expression. J Neuroendocrinol 14(10): 788-795.

Jorgensen H, Knigge U, Kjaer A, Vadsholt T, Warberg J. 1998. Serotonergic involvement in stress-induced ACTH release. Brain Res 811(1–2): 10-20.

Judd LL, Janowsky DS, Zettner A, Huey LY, Takahashi KI. 1981. Effects of naloxone-HCl on cortisol levels in patients with affective disorder and normal controls. Psychiatry Res 4(3): 277-283.

Kalin NH, Shelton SE, Kraemer GW, McKinney WT. 1983. Corticotropin-releasing factor administered intraventricularly to rhesus monkeys. Peptides 4: 217-220.

Kalsbeek A, Buijs RM, van Heerikhuize JJ, Arts M, Woude van der TP. 1992. Vasopressin-containing neurons of the suprachiasmatic nuclei inhibit corticosterone release. Brain Res 580(1–2): 62-67.

Kalsbeek A, Vliet van der J, Buijs RM. 1996a. Decrease of endogenous vasopressin release necessary for expression of the circadian rise in plasma corticosterone: A reverse microdialysis study. J Neuroendocrinol 8(4): 299-307.

Kalsbeek A, van Heerikhuize JJ, Wortel J, Buijs RM. 1996b. A diurnal rhythm of stimulatory input to the hypothalamo-pituitary-adrenal system as revealed by timed intrahypothalamic administration of the vasopressin V1 antagonist. J Neurosci 16(17): 5555-5565.

Kaneta T, Kusnecov AW. 2005. The role of central corticotropin-releasing hormone in the anorexic and endocrine effects of the bacterial T cell superantigen, Staphylococcal enterotoxin A. Brain Behav Immun 19(2): 138-146.

Karkaeva NR, Bazhan NM, Yakovleva TV, Makarova EN. 2005. Function of the hypothalamo-hypophyseal-adrenal system in mice with ectopic hyperproduction of the agouti protein. Neurosci Behav Physiol 35(2): 187-191.

Keim KL, Sigg EB. 1976. Physiological and biochemical concomitants of restraint stress in rats. Pharmacol Biochem Behav 4(3): 289-297.

Kellendonk C, Gass P, Kretz O, Schutz G, Tronche F. 2002. Corticosteroid receptors in the brain: Gene targeting studies. Brain Res Bull 57(1): 73-83.

Keller-Wood M, Dallman MF. 1984. Corticosteroid inhibition of ACTH secretion. Endocrine Rev. 5(1): 1-24.

Kennedy AR, Todd JF, Dhillo WS, Seal LJ, Ghatei MA, et al. 2003. Effect of direct injection of melanin-concentrating hormone into the paraventricular nucleus: Further evidence for a stimulatory role in the adrenal axis via SLC-1. J Neuroendocrinol 15(3): 268-272.

Khoshbouei H, Cecchi M, Morilak DA. 2002. Modulatory effects of galanin in the lateral bed nucleus of the stria terminalis on behavioral and neuroendocrine responses to acute stress. Neuropsychopharmacology 27(1): 25-34.

Kim CK, Rivier CL. 2000. Nitric oxide and carbon monoxide have a stimulatory role in the hypothalamic-pituitary-adrenal response to physico-emotional stressors in rats. Endocrinology 141(6): 2244-2253.

Kinzig KP, D'Alessio DA, Herman JP, Sakai RR, Vahl TP, et al. 2003. CNS glucagon-like peptide-1 receptors mediate endocrine and anxiety responses to interoceptive and psychogenic stressors. J Neurosci 23(15): 6163-6170.

Kinzig KP, D'Alessio DA, Seeley RJ. 2002. The diverse roles of specific GLP-1 receptors in the control of food intake and the response to visceral illness. J Neurosci 22(23): 10470-10476.

Kitay JI. 1961. Sex differences in adrenal cortical secretion in the rat. Endocrinology 68: 818-824.

Kitazawa S, Shioda S, Nakai Y. 1987. Catecholaminergic innervation of neurons containing corticotropin-releasing factor in the paraventricular nucleus of the rat hypothalamus. Acta Anat (Basel) 129(4): 337-343.

Kjaer A. 1996. Neurohypophysial peptides. Histaminergic regulation and function in adenohypophysial secretion. Dan Med Bull 43(5): 391-406.

Kjaer A, Knigge U, Bach FW, Warberg J. 1992a. Histamine- and stress-induced secretion of ACTH and beta-endorphin: Involvement of corticotropin-releasing hormone and vasopressin. Neuroendocrinology 56(3): 419-428.

Kjaer A, Knigge U, Bach FW, Warberg J. 1993. Permissive, mediating and potentiating effects of vasopressin in the ACTH and beta-endorphin response to histamine and restraint stress. Neuroendocrinology 58(5): 588-596.

Kjaer A, Knigge U, Plotsky PM, BachFW, Warberg J. 1992b. Histamine H1 and H2 receptor activation stimulates ACTH and beta-endorphin secretion by increasing corticotropin-releasing hormone in the hypophyseal portal blood. Neuroendocrinology 56(6): 851-855.

Kjaer A, Larsen PJ, Knigge U, Jorgensen H, Warberg J. 1998. Neuronal histamine and expression of corticotropin-releasing hormone, vasopressin and oxytocin in the hypothalamus: Relative importance of H1 and H2 receptors. Eur J Endocrinol 139(2): 238-243.

Kjaer A, Larsen PJ, Knigge U, Moller M, Warberg J. 1994. Histamine stimulates c-fos expression in hypothalamic vasopressin-, oxytocin-, and corticotropin-releasing hormone-containing neurons. Endocrinology 134(1): 482-491.

Knigge U, Matzen S, Bach FW, Bang P, Warberg J. 1989. Involvement of histaminergic neurons in the stress-induced release of pro-opiomelanocortin-derived peptides in rats. Acta Endocrinol (Copenh) 120(4): 533-539.

Knigge U, Soe-Jensen P, Jorgensen H, Kjaer A, Moller M, et al. 1999. Stress-induced release of anterior pituitary hormones: Effect of H3 receptor-mediated inhibition of histaminergic activity or posterior hypothalamic lesion. Neuroendocrinology 69(1): 44-53.

Koenig JI, Gudelsky GA, Meltzer HY. 1987. Stimulation of corticosterone and beta-endorphin secretion in the rat by selective 5-HT receptor subtype activation. Eur J Pharmacol 137(1): 1-8.

Korte SM, Bouws GA, Bohus B. 1992. Adrenal hormones in rats before and after stress-experience: Effects of ipsapirone. Physiol Behav 51(6): 1129-1133.

Korte SM, Korte-Bouws GA, Bohus B, Koob GF. 1994. Effect of corticotropin-releasing factor antagonist on behavioral and neuroendocrine responses during exposure to defensive burying paradigm in rats. Physiol Behav 56(1): 115-120.

Koster A, Montkowski A, Schulz S, Stube EM, Knaudt K, et al. 1999. Targeted disruption of the orphanin FQ/nociceptin gene increases stress susceptibility and impairs stress adaptation in mice. Proc Natl Acad Sci USA 96(18): 10444-10449.

Kovacs A, Biro E, Szeleczky I, Telegdy G. 1995. Role of endogenous CRF in the mediation of neuroendocrine and behavioral responses to calcitonin gene-related peptide in rats. Neuroendocrinology 62(4): 418-424.

Kovacs KJ, Miklos IH, Bali B. 2004. GABAergic mechanisms constraining the activity of the hypothalamo-pituitary-adrenocortical axis. Ann NY Acad Sci 1018: 466-476.

Krieger DT, Rizzo F. 1969. Serotonin mediation of circadian periodicity of plasma 17-hydroxycorticosteroids. Am J Physiol 217(6): 1703-1707.

Kuru M, Ueta Y, Serino R, Nakazato M, Yamamoto Y, et al. 2000. Centrally administered orexin/hypocretin activates HPA axis in rats. Neuroreport 11(9): 1977-1980.

Laflamme N, Nappi RE, Drolet G, Labrie C, Rivest S. 1998. Expression and neuropeptidergic characterization of estrogen receptors (ERalpha and ERbeta) throughout the rat brain: Anatomical evidence of distinct roles of each subtype. J Neurobiol 36(3): 357-378.

Lambert PD, Phillips PJ, Wilding JP, Bloom SR, Herbert J. 1995. c-fos expression in the paraventricular nucleus of the hypothalamus following intracerebroventricular infusions of neuropeptide Y. Brain Res 670(1): 59-65.

Lanfumey L, Mannoury La Cour C, Froger N, Hamon M. 2000. 5-HT-HPA interactions in two models of transgenic mice relevant to major depression. Neurochem Res 25 (9–10): 1199-1206.

Langub MC Jr, Watson RE Jr, Herman JP. 1995. Distribution of atrial natriuretic peptide and C-type natriuretic peptide precursor mRNAs in the male rat brain. J Comp Neurol 356: 183-199.

Larsen PJ, Jessop D, Patel H, Lightman SL, Chowdrey HS. 1993. Substance P inhibits the release of anterior pituitary adrenocorticotrophin via a central mechanism involving corticotrophin-releasing factor-containing neurons in the hypothalamic paraventricular nucleus. J Neuroendocrinol 5(1): 99-105.

Larsen PJ, Tang-Christensen M, Jessop DS. 1997. Central administration of glucagon-like peptide-1 activates hypothalamic neuroendocrine neurons in the rat. Endocrinology 138(10): 4445-4455.

Laugero KD, Gomez F, Manalo S, Dallman MF. 2002. Corticosterone infused intracerebroventricularly inhibits energy storage and stimulates the hypothalamo-pituitary axis in adrenalectomized rats drinking sucrose. Endocrinology 143(12): 4552-4562.

Le Mevel JC, Abitbol S, Beraud G, Maniey J. 1979. Temporal changes in plasma adrenocorticotropin concentration after repeated neurotropic stress in male and female rats. Endocrinology 105(3): 812-817.

Leibowitz SF, Sladek C, Spencer L, Tempel D. 1988. Neuropeptide Y, epinephrine and norepinephrine in the paraventricular nucleus: Stimulation of feeding and the release of corticosterone, vasopressin and glucose. Brain Res Bull 21(6): 905-912.

Levin N, Shinsako J, Dallman MF. 1988. Corticosterone acts on the brain to inhibit adrenalectomy-induced adrenocorticotropin secretion. Endocrinology 122(2): 694-701.

Li BH, Xu B, Rowland NE, Kalra SP. 1994a. c-fos expression in the rat brain following central administration of neuropeptide Y and effects of food consumption. Brain Res 665(2): 277-284.

Li HY, Ericsson A, Sawchenko PE. 1996. Distinct mechanisms underlie activation of hypothalamic neurosecretory neurons and their medullary catecholaminergic afferents in categorically different stress paradigms. Proc Natl Acad Sci USA 93(6): 2359-2364.

Li Q, Brownfield MS, Levy AD, Battaglia G, Cabrera TM, et al. 1994b. Attenuation of hormone responses to the 5-HT1A agonist ipsapirone by long-term treatment with fluoxetine, but not desipramine, in male rats. Biol Psychiatry 36(5): 300-308.

Li Q, Holmes A, Ma L, Van de Kar LD, Garcia F, et al. 2004. Medial hypothalamic 5-hydroxytryptamine (5-HT)1A receptors regulate neuroendocrine responses to stress and exploratory locomotor activity: Application of recombinant adenovirus containing 5-HT1A sequences. J Neurosci 24(48): 10868-10877.

Li Q, Rittenhouse PA, Levy AD, Alvarez Sanz MC, Van de Kar LD. 1992. Neuroendocrine responses to the serotonin2 agonist DOI are differentially modified by three 5-HT1A agonists. Neuropharmacology 31(10): 983-989.

Li Q, Wichems CH, Ma L, Van de Kar LD, Garcia F, et al. 2003a. Brain region-specific alterations of 5-HT2A and 5-HT2C receptors in serotonin transporter knockout mice. J Neurochem 84(6): 1256-1265.

Li XF, Mitchell JC, Wood S, Coen CW, Lightman SL, et al. 2003b. The effect of oestradiol and progesterone on hypoglycaemic stress-induced suppression of pulsatile luteinizing hormone release and on corticotropin-releasing hormone mRNA expression in the rat. J Neuroendocrinol 15(5): 468-476.

Lind RW, Swanson LW, Ganten D. 1984. Angiotensin II immunoreactive pathways in the central nervous system of the rat: Evidence for a projection from the subfornical organ to the paraventricular nucleus of the hypothalamus. Clin Exp Hypertens A 6: 1915-1920.

Liposits Z, Phelix C, Paull WK. 1987. Synaptic interaction of serotonergic axons and corticotropin releasing factor (CRF) synthesizing neurons in the hypothalamic paraventricular nucleus of the rat. A light and electron microscopic immunocytochemical study. Histochemistry 86(6): 541-549.

Liposits Z, Sherman D, Phelix C, Paull WK. 1986. A combined light and electron microscopic immunocytochemical method for the simultaneous localization of multiple tissue antigens. Tyrosine hydroxylase immunoreactive innervation of corticotropin releasing factor synthesizing neurons in the paraventricular nucleus of the rat. Histochemistry 85(2): 95-106.

Lissak K, Vermes I, Telegdy G. 1975. Effect of midbrain raphe lesion on diurnal and stress-induced changes in serotonin content of discrete regions of the limbic system and adrenal function in the rat. Prog Brain Res 42: 327-328.

Liu JP, Clarke IJ, Funder JW, Engler D. 1994. Studies of the secretion of corticotropin-releasing factor and arginine vasopressin into the hypophysial-portal circulation of the conscious sheep. II. The central noradrenergic and neuropeptide Y pathways cause immediate and prolonged hypothalamic-pituitary-adrenal activation. Potential involvement in the pseudo-Cushing's syndrome of endogenous depression and anorexia nervosa. J Clin Invest 93(4): 1439-1450.

Loftis JM, Janowsky A. 2003. The N-methyl-D-aspartate receptor subunit NR2B: Localization, functional properties, regulation, and clinical implications. Pharmacol Ther 97(1): 55-85.

Lorens SA, Kar Van de LD. 1987. Differential effects of serotonin (5-HT1A and 5-HT2) agonists and antagonists on renin and corticosterone secretion. Neuroendocrinology 45(4): 305-310.

Lu XY, Barsh GS, Akil H, Watson SJ. 2003. Interaction between alpha-melanocyte-stimulating hormone and corticotropin-releasing hormone in the regulation of feeding and hypothalamopituitaryadrenal responses. J Neurosci 23(21): 7863-7872.

Ludwig DS, Mountjoy KG, Tatro JB, Gillette JA, Frederich RC, et al. 1998. Melanin-concentrating hormone: A functional melanocortin antagonist in the hypothalamus. Am J Physiol 274(4 Pt 1): E627-E633.

Ludwig M. 1998. Dendritic release of vasopressin and oxytocin. J Neuroendocrinol 10(12): 881-895.

Lunga P, Herbert J. 2004. 17Beta-oestradiol modulates glucocorticoid, neural and behavioural adaptations to repeated restraint stress in female rats. J Neuroendocrinol 16(9): 776-785.

Mac Lusky NJ, Cook S, Scrocchi L, Shin J, Kim J, et al. 2000. Neuroendocrine function and response to stress in mice with complete disruption of glucagon-like peptide-1 receptor signaling. Endocrinology 141(2): 752-762.

Makara GB, Stark E. 1974. Effects of gamma-aminobutyric acid (GABA) and GABA antagonist drugs on ACTH release. Neuroendocrinology 16(3–4): 178-190.

Makara GB, Stark E. 1975. Effect of intraventricular glutamate on ACTH release. Neuroendocrinology 18(2): 213-216.

Makino S, Gold PW, Schulkin J. 1994a. Corticosterone effects on corticotropin-releasing hormone mRNA in the central nucleus of the amygdala and the parvocellular region of the paraventricular nucleus of the hypothalamus. Brain Res 640: 105-112.

Makino S, Gold PW, Schulkin J. 1994b. Effects of corticosterone effects on corticotropin-releasing hormone mRNA and content in the bed nucleus of the stria terminalis; comparison with the effects in the central nucleus of the amygdala and the paraventricular nucleus of the hypothalamus. Brain Res 657: 141-149.

Makino S, Shibasaki T, Yamauchi N, Nishioka T, Mimoto T, Wakabayashi I, Gold PW, Hashimoto K. 1999. Psychological stress increased corticotropin-releasing hormone mRNA and content in the central nucleus of the amygadala but not in the hypothalamic paraventricular nucleus in the rat. Brain Res 850: 136-143.

Makino S, Smith MA, Gold PW. 1995. Increased expression of corticotropin-releasing hormone and vasopressin messenger ribonucleic acid (mRNA) in the hypothalamic paraventricular nucleus during repeated stress: Association with reduction in glucocorticoid receptor mRNA levels. Endocrinology 136: 3299-3309.

Malendowicz LK, Andreis PG, Nussdorfer GG, Markowska A. 1996. The possible role of endogenous substance P in the modulation of the response of rat pituitaryadrenal axis to stresses. Endocr Res 22(3): 311-318.

Mamalaki E, Kvetnansky R, Brady LS, Gold PW, Herkenham M. 1993. Repeated immobilization stress alters tyrosine hydroxylase, corticotropin-releasing hormone and corticosteroid receptor ribonucleic acid levels in rat brain. J Neuroendocrinol 4(6): 689-699.

Martini L, Celotti F, Melcangi RC. 1996. Testosterone and progesterone metabolism in the central nervous system: Cellular localization and mechanism of control of the enzymes involved. Cell Mol Neurobiol 16(3): 271-282.

Masuzawa M, Oki Y, Ozawa M, Watanabe F, Yoshimi Y. 1999. Corticotropin-releasing factor but not urocortin is involved in adrenalectomy-induced adrenocorticotropin release. J Neuroendocrinol 11: 71-74.

Matheson GK, Knowles A, Gage D, Michel C, Guthrie D, et al. 1997a. Modification of hypothalamicpituitaryadrenocortical activity by serotonergic agents in the rat. Pharmacology 55(2): 59-65.

Matheson GK, Knowles A, Guthrie D, Gage D, Weinzapfel D, et al. 1997b. Actions of serotonergic agents on hypothalamic-pituitary-adrenal axis activity in the rat. Gen Pharmacol 29(5): 823-828.

Matta SG, Beyer HS, McAllen KM, Sharp BM. 1987. Nicotine elevates rat plasma ACTH by a central mechanism. J Pharmacol Exp Ther 243(1): 217-226.

McCormick CM, Smythe JW, Sharma S, Meaney MJ. 1995. Sex-specific effects of prenatal stress on hypothalamic-pituitary-adrenal responses to stress and brain glucocorticoid receptor density in adult rats. Brain Res Dev Brain Res 84(1): 55-61.

McEwen BS, Sapolsky RM. 1995. Stress and cognitive function. Curr Opin Neurobiol 5(2): 205-216.

McEwen BS, Weiss JM, Schwartz LS. 1968. Selective retention of corticosterone by limbic structures in rat brain. Nature 220(170): 911-912.

McGinty JF, Koda LY, Bloom FE. 1983. A combined vascular-catecholamine fluorescence method reveals the relative vascularity of rat locus coeruleus and the paraventricular and supraoptic nuclei of the hypothalamus. Neurosci Lett 36(2): 117-123.

McGrath MF, Kuroski de Bold ML, de Bold AJ. 2005. The endocrine function of the heart. Trends Endocrinol Metab 16(10): 469-477.

Melander T, Fuxe K, Harfstrand A, Eneroth P, Hokfelt T. 1987. Effects of intraventricular injections of galanin on neuroendocrine functions in the male rat. Possible involvement of hypothalamic catecholamine neuronal systems. Acta Physiol Scand 131(1): 25-32.

Merchenthaler I, Lane M, Shughrue P. 1999. Distribution of pre-pro-glucagon and glucagon-like peptide-1 receptor messenger RNAs in the rat central nervous system. J Comp Neurol 403(2): 261-280.

Mikkelsen JD, Hay-Schmidt A, Kiss A. 2004. Serotonergic stimulation of the rat hypothalamo-pituitary-adrenal axis: Interaction between 5-HT1A and 5-HT2A receptors. Ann NY Acad Sci 1018: 65-70.

Miles PD, Yamatani K, Brown MR, Lickley HL, Vranic M. 1994. Intracerebroventricular administration of somatostatin octapeptide counteracts the hormonal and metabolic responses to stress in normal and diabetic dogs. Metabolism 43(9): 1134-1143.

Moldow RL, Beck KD, Weaver S, Servatius RJ. 2005. Blockage of glucocorticoid, but not mineralocorticoid receptors prevents the persistent increase in circulating basal corticosterone concentrations following stress in the rat. Neurosci Lett 374(1): 25-28.

Moncek F, Duncko R, Jezova D. 2003. Repeated citalopram treatment but not stress exposure attenuates hypothalamic-pituitary-adrenocortical axis response to acute citalopram injection. Life Sci 72(12): 1353-1365.

Monder C, Sakai RR, Miroff Y, Blanchard DC, Blanchard RJ. 1994. Reciprocal changes in plasma corticosterone and testosterone in stressed male rats maintained in a visible burrow system: Evidence for a mediating role of testicular 11 beta-hydroxysteroid dehydrogenase. Endocrinology 134 (3): 1193-1198.

Morley JE, Baranetsky NG, Wingert TD, Carlson HE, Hershman JM, et al. 1980. Endocrine effects of naloxone-induced opiate receptor blockade. J Clin Endocrinol Metab 50(2): 251-257.

Morrow BA, Redmond AJ, Roth RH, Elsworth JD. 2000. The predator odor, TMT, displays a unique, stress-like pattern of dopaminergic and endocrinological activation in the rat. Brain Res 864(1): 146-151.

Mozid AM, Tringali G, Forsling ML, Hendricks MS, Ajodha S, et al. 2003. Ghrelin is released from rat hypothalamic explants and stimulates corticotrophin-releasing hormone and arginine-vasopressin. Horm Metab Res 35(8): 455-459.

Munck A, Guyre PM, Holbrook NJ. 1984. Physiological functions of glucocorticoids in stress and their relations to pharmacological actions. Endocrine Rev 5: 25-44.

Nahon JL. 1994. The melanin-concentrating hormone: From the peptide to the gene. Crit Rev Neurobiol 8(4): 221-262.

Neumann ID, Torner L, Wigger A. 2000a. Brain oxytocin: Differential inhibition of neuroendocrine stress responses and anxiety-related behaviour in virgin, pregnant and lactating rats. Neuroscience 95(2): 567-575.

Neumann ID, Wigger A, Torner L, Holsboer F, Landgraf R. 2000b. Brain oxytocin inhibits basal and stress-induced activity of the hypothalamo-pituitary-adrenal axis in male and female rats: Partial action within the paraventricular nucleus. J Neuroendocrinol 12(3): 235-243.

Nicot A, Rowe WB, De Kloet ER, Betancur C, Jessop DS, et al. 1997. Endogenous neurotensin regulates hypothalamic-pituitary-adrenal axis activity and peptidergic neurons in the rat hypothalamic paraventricular nucleus. J Neuroendocrinol 9(4): 263-269.

Nijsen MJ, Croiset G, Diamant M, Stam R, Kamphuis PJ, et al. 2000. Endogenous corticotropin-releasing hormone inhibits conditioned-fear-induced vagal activation in the rat. Eur J Pharmacol 389(1): 89-98.

Nikolarakis K, Pfeiffer A, Stalla GK, Herz A. 1987. The role of CRF in the release of ACTH by opiate agonists and antagonists in rats. Brain Res 421(1–2): 373-376.

Nisenbaum LK, Zigmond MJ, Sved AF, Abercrombie ED. 1991. Prior exposure to chronic stress results in enhanced synthesis and release of hippocampal norepinephrine in response to a novel stressor. J Neurosci 11(5): 1478-1484.

Nishioka T, Anselmo-Franci JA, Li P, Callahan MF, Morris M. 1998. Stress increases oxytocin release within the hypothalamic paraventricular nucleus. Brain Res 781(1–2): 56-60.

Ogle TF, Kitay JI. 1977. Ovarian and adrenal steroids during pregnancy and the oestrous cycle in the rat. J Endocrinol 74(1): 89-98.

Ohmori N, Itoi K, Tozawa F, Sakai Y, Sakai K, et al. 1995. Effect of acetylcholine on corticotropin-releasing factor gene expression in the hypothalamic paraventricular nucleus of conscious rats. Endocrinology 136(11): 4858-4863.

Oldfield BJ, Hou-Yu A, Silverman AJ. 1985. A combined electron microscopic HRP and immunocytochemical study of the limbic projections to rat hypothalamic nuclei containing vasopressin and oxytocin neurons. J Comp Neurol 231(2): 221-231.

Oldfield BJ, Silverman AJ. 1985. A light microscopic HRP study of limbic projections to the vasopressin-containing nuclear groups of the hypothalamus. Brain Res Bull 14(2): 143-157.

Olszewski PK, Grace MK, Billington CJ, Levine AS. 2003. Hypothalamic paraventricular injections of ghrelin: Effect on feeding and c-Fos immunoreactivity. Peptides 24(6): 919-923.

Onaka T, Kuramochi M, Saito J, Ueta Y, Yada T. 2005. Galanin-like peptide stimulates vasopressin, oxytocin and adrenocorticotropic hormone release in rats. Neuroreport 16(3): 243-247.

Orchinik M, Murray TF, Moore FL. 1991. A corticosteroid receptor in neuronal membranes. Science 252(5014): 1848-1851.

Ottenweller JE, Servatius RJ, Tapp WN, Drastal SD, Bergen MT, et al. 1992. A chronic stress state in rats: Effects of repeated stress on basal corticosterone and behavior. Physiol Behav 51(4): 689-698.

Owens MJ, Knight DL, Ritchie JC, Nemeroff CB. 1991. The 5-hydroxytryptamine2 agonist, (+)-1-(2,5-dimethoxy-4-bromophenyl)-2-aminopropane stimulates the hypothalamic-pituitary-adrenal (HPA) axis. I. Acute effects on HPA axis activity and corticotropin-releasing factor-containing neurons in the rat brain. J Pharmacol Exp Ther 256(2): 787-794.

Pacak K, Palkovits M, Kvetnansky R, Kopin IJ, Goldstein DS. 1993. Stress-induced norepinephrine release in the paraventricular nucleus of rats with brainstem hemisections: A microdialysis study. Neuroendocrinology 58(2): 196-201.

Pacak K, Palkovits M, Makino S, Kopin IJ, Goldstein DS. 1996. Brainstem hemisection decreases corticotropin-releasing hormone mRNA in the paraventricular nucleus but not in the central amygdaloid nucleus. J Neuroendocrinol 8(7): 543-551.

Palkovits M, Baffi JS, Pacak K. 1999. The role of ascending neuronal pathways in stress-induced release of noradrenaline in the hypothalamic paraventricular nucleus of rats. J Neuroendocrinol 11: 529-539.

Pan L, Gilbert F. 1992. Activation of 5-HT1A receptor subtype in the paraventricular nuclei of the hypothalamus induces CRH and ACTH release in the rat. Neuroendocrinology 56(6): 797-802.

Patchev VK, Almeida OF. 1996. Gonadal steroids exert facilitating and "buffering" effects on glucocorticoid-mediated transcriptional regulation of corticotropin-releasing hormone and corticosteroid receptor genes in rat brain. J Neurosci 16(21): 7077-7084.

Pattij T, Hijzen TH, Groenink L, Oosting RS, van der Gugten J, et al. 2001. Stress-induced hyperthermia in the 5-HT(1A) receptor knockout mouse is normal. Biol Psychiatry 49(7): 569-574.

Paulmyer-Lacroix O, Hery M, Pugeat M, Grino M. 1996. The modulatory role of estrogens on corticotropin-releasing factor gene expression in the hypothalamic paraventricular nucleus of ovariectomized rats: Role of the adrenal gland. J Neuroendocrinol 8(7): 515-519.

Pechnick R, George R, Poland RE. 1985a. Identification of multiple opiate receptors through neuroendocrine responses. I. Effects of agonists. J Pharmacol Exp Ther 232(1): 163-169.

Pechnick R, George R, Poland RE. 1985b. Identification of multiple opiate receptors through neuroendocrine responses. II. Antagonism of mu, kappa and sigma agonists by naloxone and WIN 44,441–3. J Pharmacol Exp Ther 232 (1): 170-177.

Pechnick RN, George R, Poland RE. 1989. Characterization of the effects of the acute and repeated administration of MK-801 on the release of adrenocorticotropin, corticosterone and prolactin in the rat. Eur J Pharmacol 164(2): 257-263.

Pelleymounter M, Joppa M, Foster AC. 2004. Behavioral and neuroendocrine effects of the selective CRF2 receptor agonist urocortin II and urocortin III. Peptides 25: 659-666.

Perras B, Schultes B, Behn B, Dodt C, Born J, et al. 2004. Intranasal atrial natriuretic peptide acts as central nervous inhibitor of the hypothalamo-pituitary-adrenal stress system in humans. J Clin Endocrinol Metab 89(9): 4642-4648.

Pfeiffer A, Herz A, Loriaux DL, Pfeiffer DG. 1985. Central kappa- and mu-opiate receptors mediate ACTH-release in rats. Endocrinology 116(6): 2688-2690.

Pistovcakova J, Makatsori A, Sulcova A, Jezova D. 2005. Felbamate reduces hormone release and locomotor hypoactivity induced by repeated stress of social defeat in mice. Eur Neuropsychopharmacol 15(2): 153-158.

Plotsky PM. 1986. Opioid inhibition of immunoreactive corticotropin-releasing factor secretion into the hypophysial-portal circulation of rats. Regul Pept 16(3–4): 235-242.

Plotsky PM. 1987. Facilitation of immunoreactive corticotropin-releasing factor secretion into the hypophysial-portal circulation after activation of catecholaminergic pathways or central norepinephrine injection. Endocrinology 121(3): 924-930.

Plotsky PM, Bruhn TO, Vale W. 1985. Evidence for multifactor regulation of the adrenocorticotropin secretory response to hemodynamic stimuli. Endocrinology 116(2): 633-639.

Plotsky PM, Sutton SW, Bruhn TO, Ferguson AV. 1988. Analysis of the role of angiotensin II in mediation of adrenocorticotropin secretion. Endocrinology 122: 538-545.

Potter E, Sutton S, Donaldson C, Chen R, Perrin M, et al. 1994. Distribution of corticotropin-releasing factor receptor mRNA expression in the rat brain and pituitary. Proc Natl Acad Sci USA 91(19): 8777-8781.

Prewitt CM, Herman JP. 1997. Hypothalamo-pituitary-adrenocortical regulation following lesions of the central nucleus of the amygdala. Stress 1(4): 263-280.

Przegalinski E, Budziszewska B, Warchol-Kania A, Blaszczynska E. 1989. Stimulation of corticosterone secretion by the selective 5-HT1A receptor agonist 8-hydroxy-2-(di-n-propylamino)tetralin (8-OH-DPAT) in the rat. Pharmacol Biochem Behav 33(2): 329-334.

Ratka A, Sutanto W, Bloemers M, de Kloet ER. 1989. On the role of brain mineralocorticoid (type I) and glucocorticoid (type II) receptors in neuroendocrine regulation. Neuroendocrinology 50(2): 117-123.

Redei E, Li L, Halasz I, McGivern RF, Aird F. 1994. Fast glucocorticoid feedback inhibition of ACTH secretion in the ovariectomized rat: Effect of chronic estrogen and progesterone. Neuroendocrinology 60(2): 113-123.

Reul JM, de Kloet ER. 1985. Two receptor systems for corticosterone in rat brain: Microdistribution and differential occupation. Endocrinology 117(6): 2505-2511.

Reyes TM, Lewis K, Perrin MH, Kunitake KS, Vaughan J, et al. 2001. Urocortin II: A member of the corticotropin-releasing factor (CRF) neuropeptide family that is selectively bound by type 2 CRF receptors. Proc Natl Acad Sci USA 98: 2843-2848.

Rho JH, Swanson LW. 1989. A morphometric analysis of functionally defined subpopulations of neurons in the paraventricular nucleus of the rat with observations on the effects of colchicine. J Neurosci 9(4): 1375-1388.

Rhodes ME, O'Toole SM, Czambel RK, Rubin RT. 2001. Male-female differences in rat hypothalamic-pituitary-adrenal axis responses to nicotine stimulation. Brain Res Bull 54(6): 681-688.

Richardson JS. 1984. Brain part monoamines in the neuroendocrine mechanisms activated by immobilization stress in the rat. Int J Neurosci 23(1): 57-67.

Rinaman L. 1999. Interoceptive stress activates glucagon-like peptide-1 neurons that project to the hypothalamus. Am J Physiol 277(2 Pt 2): R582-R590.

Rittenhouse PA, Bakkum EA, Levy AD, Li Q, Carnes M, et al. 1994. Evidence that ACTH secretion is regulated by serotonin2A/2C (5-HT2A/2C) receptors. J Pharmacol Exp Ther 271(3): 1647-1655.

Ritter S, Watts AG, Dinh TT, Sanchez-Watts G, Pedrow C. 2003. Immunotoxin lesion of hypothalamically projecting norepinephrine and epinephrine neurons differentially affects circadian and stressor-stimulated corticosterone secretion. Endocrinology 144(4): 1357-1367.

Rivest S. 2001. How circulating cytokines trigger the neural circuits that control the hypothalamic-pituitary-adrenal axis. Psychoneuroendocrinology 26(8): 761-788.

Rivier C. 2001. Role of gaseous neurotransmitters in the hypothalamic-pituitary-adrenal axis. Ann NY Acad Sci 933: 254-264.

Rivier C. 2002. Role of nitric oxide and carbon monoxide in modulating the activity of the rodent hypothalamic-pituitary-adrenal axis. Front Horm Res 29: 15-49.

Roche M, Commons KG, Peoples A, Valentino RJ. 2003. Circuitry underlying regulation of the serotonergic system by swim stress. J Neurosci 23(3): 970-977.

Rogerson FM, Brennan FE, Fuller PJ. 2004. Mineralocorticoid receptor binding, structure and function. Mol Cell Endocrinol 217(1–2): 203-212.

Rosmond R, Chagnon M, Bouchard C, Bjorntorp P. 2001. A missense mutation in the human melanocortin-4 receptor gene in relation to abdominal obesity and salivary cortisol. Diabetologia 44(10): 1335-1338.

Rowe W, Viau V, Meaney MJ, Quirion R. 1995. Stimulation of CRH-mediated ACTH secretion by central administration of neurotensin: Evidence for the participation of the paraventricular nucleus. J Neuroendocrinol 7(2): 109-117.

Rowe WB, Nicot A, Sharma S, Gully D, Walker CD, et al. 1997. Central administration of the neurotensin receptor antagonist, SR48692, modulates diurnal and stress-related hypothalamic-pituitary-adrenal activity. Neuroendocrinology 66(2): 75-85.

Ruiz-Gayo M, Garrido MM, Fuentes JA. 2000. Inhibition of the hypothalamic-pituitary-adrenal axis in food-deprived rats by a CCK-A receptor antagonist. Br J Pharmacol 129(5): 839-842.

Russell SH, Small CJ, Dakin CL, Abbott CR, Morgan DG, et al. 2001. The central effects of orexin-A in the hypothalamic-pituitary-adrenal axis in vivo and in vitro in male rats. J Neuroendocrinol 13(6): 561-566.

Ryan MC, Gundlach AL. 1995. Anatomical localisation of preproatrial natriuretic peptide in the rat brain by in situ hybridization histochemistry: Novel identification in olfactory regions. J Comp Neurol 356: 168-182.

Sage D, Maurel D, Bosler O. 2001. Involvement of the suprachiasmatic nucleus in diurnal ACTH and corticosterone responsiveness to stress. Am J Physiol Endocrinol Metab 280(2): E260-E269.

Sainsbury A, Rohner-Jeanrenaud F, Cusin I, Zakrzewska KE, Halban PA, et al. 1997. Chronic central neuropeptide Y infusion in normal rats: Status of the hypothalamopituitaryadrenal axis, and vagal mediation of hyperinsulinaemia. Diabetologia 40(11): 1269-1277.

Sainsbury A, Rohner-Jeanrenaud F, Grouzmann E, Jeanrenaud B. 1996. Acute intracerebroventricular administration of neuropeptide Y stimulates corticosterone output and feeding but not insulin output in normal rats. Neuroendocrinology 63(4): 318-326.

Sakamoto F, Yamada S, Ueta Y. 2004. Centrally administered orexin-A activates corticotropin-releasing factor-containing neurons in the hypothalamic paraventricular nucleus and central amygdaloid nucleus of rats: Possible involvement of central orexins on stress-activated central CRF neurons. Regul Pept 118(3): 183-191.

Samson WK, Baker JR, Samson CK, Samson HW, Taylor MM. 2004. Central neuropeptide B administration activates stress hormone secretion and stimulates feeding in male rats. J Neuroendocrinol 16(10): 842-849.

Samson WK, Taylor MM, Follwell M, Ferguson AV. 2002. Orexin actions in hypothalamic paraventricular nucleus: Physiological consequences and cellular correlates. Regul Pept 104(1–3): 97-103.

Sands SA, Morilak DA. 1999. Expression of alpha1D adrenergic receptor messenger RNA in oxytocin- and corticotropin-releasing hormone-synthesizing neurons in the rat paraventricular nucleus. Neuroscience 91(2): 639-649.

Saphier D. 1987. Cortisol alters firing rate and synaptic responses of limbic forebrain units. Brain Res Bull 19(5): 519-524.

Saphier D. 1989. Catecholaminergic projections to tuberoinfundibular neurones of the paraventricular nucleus: I. Effects of stimulation of A1, A2, A6 and C2 cell groups. Brain Res Bull 23(6): 389-395.

Saphier D, Feldman S. 1989. Catecholaminergic projections to tuberoinfundibular neurones of the paraventricular nucleus: II. Effects of stimulation of the ventral noradrenergic ascending bundle: Evidence for cotransmission. Brain Res Bull 23(6): 397-404.

Saphier D, Welch JE, Farrar GE, Nguyen NQ, Aguado F, et al. 1994. Interactions between serotonin, thyrotropin-releasing hormone, and substance P in the CNS regulation of adrenocortical secretion. Psychoneuroendocrinology 19(8): 779-797.

Sapolsky RM. 2000. Glucocorticoids and hippocampal atrophy in neuropsychiatric disorders. Arch Gen Psychiatry 57(10): 925-935.

Sapolsky RM, Armanini MP, Sutton SW, Plotsky PM. 1989. Elevation of hypophysial portal concentrations of adrenocorticotropin secretagogues after fornix transection. Endocrinology 125(6): 2881-2887.

Sapolsky RM, Krey LC, McEwen BS. 1984. Stress down-regulates corticosterone receptors in a site-specific manner in the brain. Endocrinology 114(1): 287-292.

Sarkar S, Fekete C, Legradi G, Lechan RM. 2003. Glucagon like peptide-1 (7–36) amide (GLP-1) nerve terminals densely innervate corticotropin-releasing hormone neurons in the hypothalamic paraventricular nucleus. Brain Res 985(2): 163-168.

Sarkar S, Legradi G, Lechan RM. 2002. Intracerebroventricular administration of alpha-melanocyte stimulating hormone increases phosphorylation of CREB in TRH- and CRH-producing neurons of the hypothalamic paraventricular nucleus. Brain Res 945(1): 50-59.

Sarkar S, Wittmann G, Fekete C, Lechan RM. 2004. Central administration of cocaine- and amphetamine-regulated transcript increases phosphorylation of cAMP response element binding protein in corticotropin-releasing hormone-producing neurons but not in prothyrotropin-releasing hormone-producing neurons in the hypothalamic paraventricular nucleus. Brain Res 999(2): 181-192.

Sawchenko PE, Li HY, Ericsson A. 2000. Circuits and mechanisms governing hypothalamic responses to stress: A tale of two paradigms. Prog Brain Res 122: 61-78.

Sawchenko PE, Swanson LW. 1985. Localization, colocalization, and plasticity of corticotropin-releasing factor immunoreactivity in rat brain. Fed Proc 44(1 Pt 2): 221-227.

Sawchenko PE, Swanson LW, Steinbusch HW, Verhofstad AA. 1983. The distribution and cells of origin of serotonergic inputs to the paraventricular and supraoptic nuclei of the rat. Brain Res 277(2): 355-360.

Scaccianoce S, Matrisciano F, Del Bianco P, Caricasole A, Di Giorgi Gerevini V et al. 2003. Endogenous activation of group-II metabotropic glutamate receptors inhibits the hypothalamic-pituitary-adrenocortical axis. Neuropharmacology 44(5): 555-561.

Scapagnini U, Moberg GP, Van Loon GR, De Groot J, Ganong WF. 1971. Relation of brain 5-hydroxytryptamine content to the diurnal variation in plasma corticosterone in the rat. Neuroendocrinology 7(2): 90-96.

Schwartz MW, Woods SA, Porte D Jr, Seeley RJ, Baskin DG. 2000. Central nervous system control of food intake. Nature 404(6778): 661-671.

Seale JV, Wood SA, Atkinson HC, Bate E, Lightman SL, et al. 2004a. Gonadectomy reverses the sexually diergic patterns of circadian and stress-induced hypothalamic-pituitary-adrenal axis activity in male and female rats. J Neuroendocrinol 16(6): 516-524.

Seale JV, Wood SA, Atkinson HC, Harbuz MS, Lightman SL. 2004b. Gonadal steroid replacement reverses gonadectomy-induced changes in the corticosterone pulse profile and stress-induced hypothalamic-pituitary-adrenal axis activity of male and female rats. J Neuroendocrinol 16(12): 989-998.

Selye H. 1956. The stress of life. McGraw-Hill; New York: pp. 324.

Seo DO, Lee S, Rivier C. 2002. Comparison between the influence of the intravenous and intracerebroventricular injection of a nitric oxide donor on adrenocorticotropic hormone release and hypothalamic neuronal activity. J Neuroendocrinol 14(7): 568-573.

Shan J, Krukoff TL. 2001. Intracerebroventricular adrenomedullin stimulates the hypothalamic-pituitary-adrenal axis, the sympathetic nervous system and production of hypothalamic nitric oxide. J Neuroendocrinol 13(11): 975-984.

Shiomi H, Watson SJ, Kelsey JE, Akil H. 1986. Pretranslational and posttranslational mechanisms for regulating beta-endorphin-adrenocorticotropin of the anterior pituitary lobe. Endocrinology 119(4): 1793-1799.

Shughrue PJ, Lane MV, Merchenthaler I. 1996. Glucagon-like peptide-1 receptor (GLP1-R) mRNA in the rat hypothalamus. Endocrinology 137(11): 5159-5162.

Shughrue PJ, Lane MV, Merchenthaler I. 1997. Comparative distribution of estrogen receptor-alpha and -beta mRNA in the rat central nervous system. J Comp Neurol 388(4): 507-525.

Siegel JM. 2004. The neurotransmitters of sleep. J Clin Psychiatry 65(Suppl 16): 4-7.

Silverman AJ, Hou-Yu A, Chen WP. 1989. Corticotropin-releasing factor synapses within the paraventricular nucleus of the hypothalamus. Neuroendocrinology 49(3): 291-299.

Smith BK, York DA, Bray GA. 1994. Chronic cerebroventricular galanin does not induce sustained hyperphagia or obesity. Peptides 15(7): 1267-1272.

Smith GP, Gibbs J. 1994. Satiating effect of cholecystokinin. Ann NY Acad Sci 713: 236-241.

Soe-Jensen P, Knigge U, Garbarg M, Kjaer A, Rouleau A, et al. 1993. Responses of anterior pituitary hormones and hypothalamic histamine to blockade of histamine synthesis and to selective activation or inactivation of presynaptic histamine H3 receptors in stressed rats. Neuroendocrinology 57(3): 532-540.

Spencer RL, Kim PJ, Kalman BA, Cole MA. 1998. Evidence for mineralocorticoid receptor facilitation of glucocorticoid receptor-dependent regulation of hypothalamic-pituitary-adrenal axis activity. Endocrinology 139(6): 2718-2726.

Spencer RL, Miller AH, Moday H, Stein M, McEwen BS. 1993. Diurnal differences in basal and acute stress levels of type I and type II adrenal steroid receptor activation in neural and immune tissues. Endocrinology 133(5): 1941-1950.

Spencer SJ, Ebner K, Day TA. 2004. Differential involvement of rat medial prefrontal cortex dopamine receptors in modulation of hypothalamic-pituitary-adrenal axis responses to different stressors. Eur J Neurosci 20(4): 1008-1016.

Spinedi E, Gaillard RC. 1991. Stimulation of the hypothalamo-pituitary-adrenocortical axis by the central serotonergic pathway: Involvement of endogenous corticotropin-releasing hormone but not vasopressin. J Endocrinol Invest 14(7): 551-557.

Stanley BG, Schwartz DH, Hernandez L, Hoebel BG, Leibowitz SF. 1989. Patterns of extracellular norepinephrine in the paraventricular hypothalamus: Relationship to circadian rhythm and deprivation-induced eating behavior. Life Sci 45(4): 275-282.

Stanley SA, Small CJ, Murphy KG, Rayes E, Abbott CR, et al. 2001. Actions of cocaine- and amphetamine-regulated transcript (CART) peptide on regulation of appetite and hypothalamo-pituitary axes in vitro and in vivo in male rats. Brain Res 893(1–2): 186-194.

Starcevic V, Milosevic V, Brkic B, Severs WB. 2000. Effects of centrally applied somatostatin on pituitary adrenocorticotropes in female rats. Pharmacology 60(4): 203-207.

Stout SC, Owens MJ, Nemeroff CB. 2002. Regulation of corticotropin-releasing factor neuronal systems and hypothalamic-pituitary-adrenal axis activity by stress and chronic antidepressant treatment. J Pharmacol Exp Ther 300(3): 1085-1092.

Suda T, Tozawa F, Iwai I, Sato Y, Sumitomo T, et al. 1993. Neuropeptide Y increases the corticotropin-releasing factor messenger ribonucleic acid level in the rat hypothalamus. Brain Res Mol Brain Res 18(4): 311-315.

Swanson LW, Sawchenko PE, Berod A, Hartman BK, Helle KB, et al. 1981. An immunohistochemical study of the organization of catecholaminergic cells and terminal fields in the paraventricular and supraoptic nuclei of the hypothalamus. J Comp Neurol 196(2): 271-285.

Szafarczyk A, Alonso G, Ixart G, Malaval F, Assenmacher I. 1985. Diurnal-stimulated and stress-induced ACTH release in rats is mediated by ventral noradrenergic bundle. Am J Physiol 249(2 Pt 1): E219-E226.

Szafarczyk A, Malaval F, Laurent A, Gibaud R, Assenmacher I. 1987. Further evidence for a central stimulatory action of catecholamines on adrenocorticotropin release in the rat. Endocrinology 121(3): 883-892.

Takahashi LK. 2001. Role of CRF(1) and CRF(2) receptors in fear and anxiety. Neurosci Biobehav Rev 25(7–8): 627-636.

Tamashiro KL, Nguyen MM, Sakai RR. 2005. Social stress: From rodents to primates. Front Neuroendocrinol 26(1): 27-40.

Tanaka T, Yokoo H, Mizoguchi K, Yoshida M, Tsuda A, et al. 1991. Noradrenaline release in the rat amygdala is increased by stress: Studies with intracerebral microdialysis. Brain Res 544(1): 174-176.

Tanimura SM, Watts AG. 1998. Corticosterone can facilitate as well as inhibit corticotropin-releasing hormone gene expression in the rat hypothalamic paraventricular nucleus. Endocrinology 139(9): 3830-3836.

Taylor MM, Yuill EA, Baker JR, Ferri CC, Ferguson AV, et al. 2005. Actions of neuropeptide W in paraventricular hypothalamus: Implications for the control of stress hormone secretion. Am J Physiol Regul Integr Comp Physiol 288(1): R270-R275.

Tempel DL, Leibowitz SF. 1990. Galanin inhibits insulin and corticosterone release after injection into the PVN. Brain Res 536(1–2): 353-357.

Thierry AM, Javoy F, Glowinski J, Kety SS. 1968. Effects of stress on the metabolism of norepinephrine, dopamine and serotonin in the central nervous system of the rat. I. Modifications of norepinephrine turnover. J Pharmacol Exp Ther 163(1): 163-171.

Thieulant ML, Duval J. 1985. Differential distribution of androgen and estrogen receptors in rat pituitary cell populations separated by centrifugal elutriation. Endocrinology 116(4): 1299-1303.

Tjurmina OA, Armando I, Saavedra JM, Goldstein DS, Murphy DL. 2002. Exaggerated adrenomedullary response to immobilization in mice with targeted disruption of the serotonin transporter gene. Endocrinology 143(12): 4520-4526.

Tokarev D, Jezova D. 1997. Effect of central administration of the non-NMDA receptor antagonist DNQX on ACTH and corticosterone release before and during immobilization stress. Methods Find Exp Clin Pharmacol 19(5): 323-328.

Trapp T, Rupprecht R, Castern M, Reul JHM, Holsboer F. 1994. Heterodimerization between mineralocorticoid and glucocorticoid receptor: A new principle of glucocorticoid action in the CNS. Neuron 13: 1457-1462.

Tsagarakis S, Holly JM, Rees LH, Besser GM, Grossman A. 1988. Acetylcholine and norepinephrine stimulate the release of corticotropin-releasing factor-41 from the rat hypothalamus in vitro. Endocrinology 123(4): 1962-1969.

Tsagarakis S, Rees LH, Besser GM, Grossman A. 1989. Neuropeptide-Y stimulates CRF-41 release from rat hypothalami in vitro. Brain Res 502(1): 167-170.

Tsukamura H, Thompson RC, Tsukahara S, Ohkura S, Maekawa F, et al. 2000. Intracerebroventricular administration of melanin-concentrating hormone suppresses pulsatile luteinizing hormone release in the female rat. J Neuroendocrinol 12(6): 529-534.

Turnbull A, Vaughan J, Rivier JE, Vale WW, Rivier C. 1999. Urocortin is not a significant regulator of intermittent electrofootshock-induced adrenocorticotropin secretion in the intact male rat. Endocrinology 140: 71-78.

Turnbull AV, Kim CK, Lee S, Rivier CL. 1998. Influence of carbon monoxide, and its interaction with nitric oxide, on the adrenocorticotropin hormone response of the normal rat to a physico-emotional stress. J Neuroendocrinol 10(10): 793-802.

Uht RM, McKelvy JF, Harrison RW, Bohn MC. 1988. Demonstration of glucocorticoid receptor-like immunoreactivity in glucocorticoid-sensitive vasopressin and corticotropin-releasing factor neurons in the hypothalamic paraventricular nucleus. J Neurosci Res 19(4): 405-411.

Ulrich-Lai YM, Engeland WC. 2002. Adrenal splanchnic innervation modulates adrenal cortical responses to dehydration stress in rats. Neuroendocrinology 76(2): 79-92.

Unger T, Carolus S, Demmert G, Ganten D, Lang RE, et al. 1988. Substance P induces a cardiovascular defense reaction in the rat: Pharmacological characterization. Circ Res 63 (4): 812-820.

Valentino RJ, Van Bockstaele E. 2001. Opposing regulation of the locus coeruleus by corticotropin-releasing factor and opioids. Potential for reciprocal interactions between stress and opioid sensitivity. Psychopharmacology (Berl) 158(4): 331-342.

Vamvakopoulos NC, Chrousos GP. 1993. Evidence of direct estrogenic regulation of human corticotropin-releasing hormone gene expression. Potential implications for the sexual dimophism of the stress response and immune/ inflammatory reaction. J Clin Invest 92(4): 1896-1902.

Kar Van de LD. 1997. 5-HT receptors involved in the regulation of hormone secretion. Baumgarten HG, Gothert M, Handbook of experimental pharmacology. Springer-Verlag; Berlin: pp. 537-562.

Kar Van de LD, Carnes M, Maslowski RJ, Bonadonna AM, Rittenhouse PA, et al. 1989. Neuroendocrine evidence for denervation supersensitivity of serotonin receptors: Effects of the 5-HT agonist RU 24969 on corticotropin, corticosterone, prolactin and renin secretion. J Pharmacol Exp Ther 251(2): 428-434.

Pol van den AN. 1982. The magnocellular and parvocellular paraventricular nucleus of rat: Intrinsic organization. J Comp Neurol 206(4): 317-345.

Van Eekelen JA, De Kloet ER. 1992. Co-localization of brain corticosteroid receptors in the rat hippocampus. Prog Histochem Cytochem 26(1–4): 250-258.

van Eekelen JA, Kiss JZ, Westphal HM, De Kloet ER. 1987. Immunocytochemical study on the intracellular localization of the type 2 glucocorticoid receptor in the rat brain. Brain Res 436(1): 120-128.

van Haarst AD, Oitzl MS, De Kloet ER. 1997. Facilitation of feedback inhibition through blockade of glucocorticoid receptors in the hippocampus. Neurochem Res 22(11): 1323-1328.

van Steensel B, van Binnendijk EP, Hornsby CD, van der Voort HT, Krozowski ZS, et al. 1996. Partial colocalization of glucocorticoid and mineralocorticoid receptors in discrete compartments in nuclei of rat hippocampus neurons. J Cell Sci 109 (Pt 4): 787-792.

Venihaki M, Sakihara S, Subramanian S, Dikkes P, Weninger SC, et al. 2004. Urocortin III, a brain neuropeptide of the corticotropin-releasing hormone family: Modulation by stress and attenuation of some anxiety-like behaviors. J Neuroendocrinol 16: 411-422.

Verkuyl JM, Hemby SE, Joels M. 2004. Chronic stress attenuates GABAergic inhibition and alters gene expression of parvocellular neurons in rat hypothalamus. Eur J Neurosci 20(6): 1665-1673.

Vermes I, Telegdy G, Lissak K. 1974. Effect of midbrain raphe lesion on diurnal and stress-induced changes in serotonin content of discrete regions of the limbic system and in adrenal function in the rat. Acta Physiol Acad Sci Hung 45(3–4): 217-224.

Vernikos-Danellis J, Berger P, Barchas JD. 1973. Brain serotonin and pituitaryadrenal function. Prog Brain Res 39: 301-310.

Vetter DE, Li C, Zhao L, Contarino A, Liberman MC, et al. 2002. Urocortin-deficient mice show hearing impairment and increased anxiety-like behavior. Nat Genet 31(4): 363-369.

Viau V. 2002. Functional cross-talk between the hypothalamic-pituitary-gonadal and -adrenal axes. J Neuroendocrinol 14(6): 506-513.

Viau V, Chu A, Soriano L, Dallman MF. 1999. Independent and overlapping effects of corticosterone and testosterone on corticotropin-releasing hormone and arginine vasopressin mRNA expression in the paraventricular nucleus of the hypothalamus and stress-induced adrenocorticotropic hormone release. J Neurosci 19(15): 6684-6693.

Viau V, Meaney MJ. 1991. Variations in the hypothalamic-pituitary-adrenal response to stress during the estrous cycle in the rat. Endocrinology 129(5): 2503-2511.

Viau V, Meaney MJ. 1996. The inhibitory effect of testosterone on hypothalamic-pituitary-adrenal responses to stress is mediated by the medial preoptic area. J Neurosci 16(5): 1866-1876.

Viau V, Soriano L, Dallman MF. 2001. Androgens alter corticotropin releasing hormone and arginine vasopressin mRNA within forebrain sites known to regulate activity in the hypothalamic-pituitary-adrenal axis. J Neuroendocrinol 13(5): 442-452.

Vincent SR, Das S, Maines MD. 1994. Brain heme oxygenase isoenzymes and nitric oxide synthase are co-localized in select neurons. Neuroscience 63(1): 223-231.

Volavka J, Cho D, Mallya A, Bauman J. 1979. Naloxone increases ACTH and cortisol levels in man. N Engl J Med 300(18): 1056-1057.

Von Frijtag JC, Croiset G, Gispen WH, Adan RA, Wiegant VM. 1998. The role of central melanocortin receptors in the activation of the hypothalamus-pituitary-adrenal-axis and the induction of excessive grooming. Br J Pharmacol 123 (8): 1503-1508.

Vrang N, Larsen PJ, Kristensen P, Tang-Christensen M. 2000. Central administration of cocaine-amphetamine-regulated transcript activates hypothalamic neuroendocrine neurons in the rat. Endocrinology 141(2): 794-801.

Vrang N, Tang-Christensen M, Larsen PJ, Kristensen P. 1999. Recombinant CART peptide induces c-fos expression in central areas involved in control of feeding behaviour. Brain Res 818(2): 499-509.

Wahlestedt C, Skagerberg G, Ekman R, Heilig M, Sundler F, et al. 1987. Neuropeptide Y (NPY) in the area of the hypothalamic paraventricular nucleus activates the pituitary-adrenocortical axis in the rat. Brain Res 417(1): 33-38.

Wang X, Su H, Copenhagen LD, Vaishnav S, Pieri F, et al. 2002. Urocortin-deficient mice display normal stress-induced anxiety behavior and autonomic control but an impaired acoustic startle response. Mol Cell Biol 22(18): 6605-6610.

Wang XM, Tresham JJ, Coghlan JP, Scoggins BA. 1987. Intracerebroventricular infusion of a cyclic hexapeptide analogue of somatostatin inhibits hemorrhage-induced ACTH release. Neuroendocrinology 45(4): 325-327.

Weidenfeld J, Brenner T, Conforti N, Bodoff M, Feldman S. 1988. Differential adrenocortical responses to neural stimuli following intracerebroventricular injection of nicotinic acetylcholine receptor antibodies in rats. Exp Neurol 100(3): 578-582.

Weidenfeld J, Feldman S, Mechoulam R. 1994. Effect of the brain constituent anandamide, a cannabinoid receptor agonist, on the hypothalamo-pituitary-adrenal axis in the rat. Neuroendocrinology 59(2): 110-112.

Weidenfeld J, Siegel R, Conforti N, Mizrachi R, Brenner T. 1983. Effect of intracerebroventricular injection of nicotinic acetylcholine receptor antibodies on ACTH, corticosterone and prolactin secretion in the male rat. Brain Res 265(1): 152-156.

Wenger T, Jamali KA, Juaneda C, Leonardelli J, Tramu G. 1997. Arachidonyl ethanolamide (anandamide) activates the parvocellular part of hypothalamic paraventricular nucleus. Biochem Biophys Res Commun 237(3): 724-728.

Whitnall MH. 1989. Stress selectively activates the vasopressin-containing subset of corticotropin-releasing hormone neurons. Neuroendocrinology 50(6): 702-707.

Whitnall MH. 1993. Regulation of the hypothalamic corticotropin-releasing hormone neurosecretory system. Prog Neurobiol 40: 573-629.

Wiegant VM, Jolles J, Colbern DL, Zimmermann E, Gispen WH. 1979. Intracerebroventricular ACTH activates the pituitary–adrenal system: Dissociation from a behavioral response. Life Sci 25(21): 1791-1796.

Wieronska JM, Branski P, Szewczyk B, Palucha A, Papp M, et al. 2001. Changes in the expression of metabotropic glutamate receptor 5 (mGluR5) in the rat hippocampus in an animal model of depression. Pol J Pharmacol 53(6): 659-662.

Williams AM, Morilak DA. 1997. alpha1B adrenoceptors in rat paraventricular nucleus overlap with, but do not mediate, the induction of c-fos expression by osmotic or restraint stress. Neuroscience 76(3): 901-913.

Williams JH, Miall-Allen VM, Klinowski M, Azmitia EC. 1983. Effects of microinjections of 5,7-dihydroxytryptamine in the suprachiasmatic nuclei of the rat on serotonin reuptake and the circadian variation of corticosterone levels. Neuroendocrinology 36(6): 431-435.

Willner P, Muscat R, Papp M. 1992. Chronic mild stress-induced anhedonia: A realistic animal model of depression. Neurosci Biobehav Rev 16(4): 525-534.

Wimalawansa SJ. 1996. Calcitonin gene-related peptide and its receptors: Molecular genetics, physiology, pathophysiology, and therapeutic potentials. Endocr Rev 17(5): 533-585.

Woods SC, Seeley RJ, Porte D Jr, Schwartz MW. 1998. Signals that regulate food intake and energy homeostasis. Science 280(5368): 1378-1383.

Wren AM, Small CJ, Fribbens CV, Neary NM, Ward HL, et al. 2002. The hypothalamic mechanisms of the hypophysiotropic action of ghrelin. Neuroendocrinology 76(5): 316-324.

Xiao E, Xia-Zhang L, Vulliemoz NR, Ferin M, Wardlaw SL. 2003. Agouti-related protein stimulates the hypothalamic-pituitary-adrenal (HPA) axis and enhances the HPA response to interleukin-1 in the primate. Endocrinology 144(5): 1736-1741.

Xu B, Li BH, Rowland NE, Kalra SP. 1995. Neuropeptide Y injection into the fourth cerebroventricle stimulates c-Fos expression in the paraventricular nucleus and other nuclei in the forebrain: Effect of food consumption. Brain Res 698(1–2): 227-231.

Yajima F, Suda T, Tomori N, Sumitomo T, Nakagami Y, et al. 1986. Effects of opioid peptides on immunoreactive corticotropin-releasing factor release from the rat hypothalamus in vitro. Life Sci 39(2): 181-186.

Yamauchi N, Shibasaki T, Wakabayashi I, Demura H. 1997. Brain beta-endorphin and other opioids are involved in restraint stress-induced stimulation of the hypothalamic-pituitary-adrenal axis, the sympathetic nervous system, and the adrenal medulla in the rat. Brain Res 777(1–2): 140-146.

Yambe Y, Arima H, Kakiya S, Murase T, Oiso Y. 2002. Diurnal changes in arginine vasopressin gene transcription in the rat suprachiasmatic nucleus. Brain Res Mol Brain Res 104(2): 132-136.

Yang J, Cagampang FR, Nakayama Y, Inouye SI. 1993. Vasoactive intestinal polypeptide precursor mRNA exhibits diurnal variation in the rat suprachiasmatic nuclei. Brain Res Mol Brain Res 20(3): 259-262.

Yokoo H, Tanaka M, Tanaka T, Tsuda A. 1990a. Stress-induced increase in noradrenaline release in the rat hypothalamus assessed by intracranial microdialysis. Experientia 46(3): 290-292.

Yokoo H, Tanaka M, Yoshida M, Tsuda A, Tanaka T, et al. 1990b. Direct evidence of conditioned fear-elicited enhancement of noradrenaline release in the rat hypothalamus assessed by intracranial microdialysis. Brain Res 536(1–2): 305-308.

Young EA. 1995. The role of gonadal steroids in hypothalamic-pituitary-adrenal axis regulation. Crit Rev Neurobiol 9(4): 371-381.

Zelena D, Makara GB, Jezova D. 1999. Simultaneous blockade of two glutamate receptor subtypes (NMDA and AMPA) results in stressor-specific inhibition of prolactin and corticotropin release. Neuroendocrinology 69 (5): 316-323.

Zelena D, Mergl Z, Makara GB. 2005. Glutamate agonists activate the hypothalamic-pituitary-adrenal axis through hypothalamic paraventricular nucleus but not through vasopressinerg neurons. Brain Res 1031(2): 185-193.

Zhou L, Blaustein JD, De Vries GJ. 1994. Distribution of androgen receptor immunoreactivity in vasopressin- and oxytocin-immunoreactive neurons in the male rat brain. Endocrinology 134(6): 2622-2627.

Ziegler DR, Cass WA, Herman JP. 1999. Excitatory influence of the locus coeruleus in hypothalamic-pituitary-adrenocortical axis responses to stress. J Neuroendocrinol 11(5): 361-369.

Ziegler DR, Cullinan WE, Herman JP. 2005. Organization and regulation of paraventricular nucleus glutamate signaling systems: N-methyl-D-aspartate receptors. J Comp Neurol 484(1): 43-56.

Ziegler DR, Herman JP. 2000. Local integration of glutamate signaling in the hypothalamic paraventricular region: Regulation of glucocorticoid stress responses. Endocrinology 141(12): 4801-4804.

14 Neuroendocrinology of Stress

B. S. McEwen · S. Chattarji

Abstract: The response to stress involves the active release of hormones and other mediators that produce adaptation in the aftermath of acute stress (allostasis) and yet can lead to pathophysiology when the same mediators are not shut off or are dysregulated over weeks and months (allostatic load). The brain is the key organ of stress since it interprets what is potentially threatening and therefore stressful and it determines the behavioral and physiological responses of the individual. Behavioral responses include fighting or fleeing, eating too much of the wrong things, smoking, drinking and losing sleep at night. Behavioral responses may also include health promoting behaviors such as regular moderate exercise. Hence lifestyle and behavior are key contributors or "being stressed out" and hence to allostatic load. The brain also is a target of stress and remodeling of dendrites and formation or down-regulation of synapses is a key aspect. In hippocampus and prefrontal cortex, repeated stress causes dendrites to shrink and spine synapses to decrease in number whereas in basolateral amygdala stress causes increased dendritic branching and synapse density. Mediators of structural remodeling include not only adrenal steroids but endogenous neurotransmitters, neurotrophins, the extracellular protease, tissue plasminogen activator, and cell surface adhesion molecules. Remodeling of neural circuitry under repeated stress leads to impaired memory, increased anxiety and aggression and impaired attention. Changes of this type may occur in depression and anxiety disorders.

List of Abbreviations: ADX, adrenalectomy; EAA, excitatory amino acids; LTP, long term potentiation; LTD, long term depression; PBP, primed burst potentiation; GRE, glucocorticoid response element; NMDA, N-methyl-D-aspartate; HPA, hypothalamo-pituitary-adrenal; CRF, corticotrophin releasing factor; AVP, arginine vasopressin; Acg, anterior cingulated; PL, prelimbic; IL, infralimbic; CE, central amygdala; B, basal amygdala; AB, anterior basal amygdala

1 Introduction

We all think we know what stress is, but it is still difficult to define it. For the purposes of this chapter, we define stress as a "threat, real or implied, to homeostasis," and we define stress response as involving both behavioral (e.g., "fight or flight") and physiological (i.e., neuroendocrine, autonomic, and immune) responses. In reality, stress is a word we give to the challenges to an organism of many kinds, whether physical (temperature, physical exertion, trauma) or psychological (surprise of the unexpected, threat of danger, defeat). But in fact, the physiological systems that respond to physical and psychological stressors are called into play during virtually every moment of the day, and are closely linked to biological rhythms of sleep and waking, and to the day—night cycle and patterns of feeding, and physical activity. As a result, we introduce two terms, allostasis and allostatic load, which focus attention on the physiological systems that respond to stressors with two, somewhat paradoxical, outcomes, namely, they help an organism to adapt in the immediate aftermath of a challenge but are also involved in damage and pathophysiology when they are overused or dysregulated.

Besides physiology of adaptation, the word stress often implies psychological factors such as frustration, exhaustion, fear, anger, and hostility (as in being "stressed out"), which are products of the activity of the nervous system. Consequently, we discuss brain areas such as the amygdala, hippocampus, and prefrontal cortex, which are involved in processing emotions and emotional memories, and which also regulate output of the physiological systems of allostasis and allostatic load. We discuss how these brain regions are altered by acute and chronic stress and how these changes, in turn, contribute to physiological and behavioral responses to stressful events, culminating in successful adaption, on the one hand, and mood and anxiety disorders, on the other hand, that have consequences for physical as well as mental well-being. The purpose of this chapter is, therefore, to integrate the behavioral and physiological aspects of allostasis and allostatic load into a seamless interaction between the nervous system and the neuroendocrine, autonomic and immune mediators that operate throughout the whole body.

2 The Response to Stress: Neuroendocrinology and Neurochemistry

2.1 Allostasis and Allostatic Load

When a stressful event occurs, the initial response of the brain, body, and behavior is a protective one, and hormones, cytokines, and other mediators, such as the neurotransmitters, are used to survive and to adapt to the challenge. Many functions are enhanced by acute stress, including aspects of memory, immune function, and metabolism. However, repeated stressful experiences can have deleterious effects, in part because the very same mechanisms that help protect in the short run are now either mismanaged and/or overused (McEwen, 1998). And, over weeks, months, and years, the dysregulation and overactivity of these systems can promote changes that appear to be deleterious. The following examples illustrate the paradox of protection versus damage.

For the *immune system*, acute stress promotes immune function by enhancing movement of immune cells to places in the body where they are needed to defend against a pathogen; yet, chronic stress suppresses immune function and uses the same hormonal mediators to suppress immune function (Dhabhar and McEwen, 1999). Thus, animals that evade a predator but are injured have the acute enhancement of immune function to help heal wounds and fight infections.

For the *cardiovascular system*, we see a similar paradox. Getting out of bed in the morning requires an increase in blood pressure and a reapportioning of blood flow to the head so we can stand up and not faint (Sterling and Eyer, 1988). Blood pressure rises and falls during the day as physical and emotional demands change, providing adequate blood flow as needed. Yet, repeatedly elevated blood pressure promotes generation of atherosclerotic plaques, particularly when combined with a supply of cholesterol, lipids, and oxygen-free radicals that damage the coronary artery walls (Manuck et al., 1995). Beta adrenergic receptor blockers are known to inhibit this cascade of events and to slow down the atherosclerosis that is accelerated in dominant male cynomologus monkeys exposed to an unstable dominance hierarchy (Manuck et al., 1991).

For *metabolism*, the paradox is also evident. Glucocorticoids, so named because of their ability to promote conversion of protein and lipids to usable carbohydrates, serve the body well in the short run, by replenishing energy reserves after a period of activity, like running away from a predator. Glucocorticoids also act on the brain to increase appetite for food, and to increase locomotor activity and food seeking behavior (Leibowitz and Hoebel, 1997), thus regulating behaviors that control energy input and expenditure. This is very useful in manual labor or in playing active sports, but it is not beneficial when we grab a pizza and a beer while watching television or writing a paper, particularly when these activities may also be generating psychological stress—e.g., watching distressing news or worrying about getting the paper done in time. Comfort foods appear to reduce anxiety by reducing activity of the hypothalamic-pituitary-adrenal (HPA) axis and subsequently reducing activity of the anxiogenic corticotropin-releasing hormone (CRH) system of the amygdala (Dallman, 2003). Inactivity and lack of energy expenditure creates a situation in which chronically elevated glucocorticoids that may result from either poor sleep, ongoing stress, or as side effects of excessive intake of "comfort food" impedes the action of insulin to promote glucose uptake. One of the results of this interaction is that insulin levels increase and, together, insulin and glucocorticoid elevations promote the deposition of body fat, and this combination of hormones also promotes the formation of atherosclerotic plaques in the coronary arteries (Brindley and Rolland, 1989). Thus, whether psychological stress, sleep deprivation, or a rich diet is increasing the levels of glucocorticoids, the consequences in terms of allostatic load are the same—insulin resistance and increased risk for cardiovascular disease. Thus, catecholamines and a combination of glucocorticoids and insulin can have dangerous effects on the body, besides their important short-term adaptive roles (Brindley and Rolland, 1989).

Finally, *in brain*, the normal activity of stress hormones and excitatory amino acid (EAA) neurotransmitters promotes memory and synaptic plasticity; yet somewhat higher, acute levels of glucocorticoids can inhibit synaptic plasticity and promote long-term depression (LTD), as is described later in this article. See ➋ *Figure 14-1*, which depicts this relationship as an inverted U. Moreover, as also depicted in ➋ *Figure 14-1*,

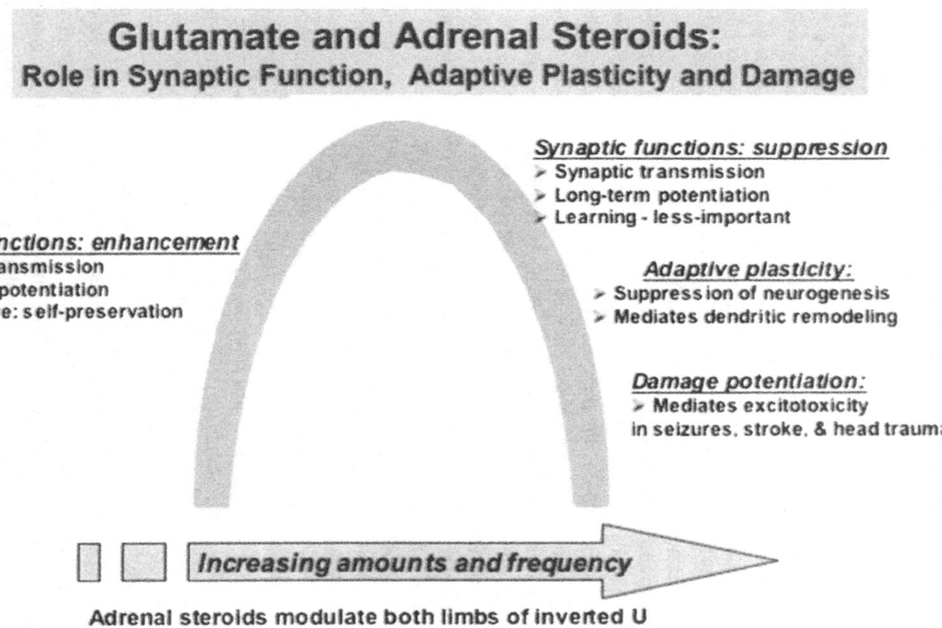

Figure 14-1

Adrenal steroids and the excitatory amino acid (EAA) neurotransmitter, glutamate, play a synergistic role in synaptic functions, adaptive plasticity, and damage in the hippocampus and possibly in other brain regions. The intensity and duration of the activation of the HPA axis and the EAA system determines which of the outcomes will occur. The downward limb of the inverted U includes 3 levels of effect, beginning with acute and transient effects on synaptic function and then leading to chronic effects that produce adaptive plasticity. Finally, overstimulation of the HPA axis and EAA system in head trauma, seizures, and ischemia leads to permanent damage in which EAA and glucocorticoids also participate

prolonged overactivity of stress hormones in the blood and EAA in the brain suppresses neurogenesis in dentate gyrus (DG) and causes debranching of dendrites in hippocampus and medial prefrontal cortex, whereas chronic stress causes neurons in amygdala to show dendritic growth (McEwen, 1999), as we summarize later in this chapter (Sousa et al., 2000; Wellman, 2001; Vyas et al., 2002). Furthermore, the inverted U depicted in Figure 14-1 has a more sinister aspect, in that extreme overactivity of stress hormones and EAA in head trauma, seizures, and stroke causes permanent damage and cell loss in the hippocampus (Sapolsky, 1992).

The hippocampus contains receptors for adrenal steroids, which regulate excitability and morphological changes. Along with many other brain regions, the amygdala and the prefrontal cortex also contain adrenal steroid receptors, which influence function in this structure as well, along with other tissue mediators such as extracellular proteases, neurotrophins, and neurotransmitters, as are summarized in Section 3. However, because adrenal steroids have led the way in the understanding of stress effects on the brain and body, we emphasize their role in both allostasis and allostatic load, recognizing that they are but one of a system of interacting mediators.

2.2 Adrenal Steroids and Their Receptors and Effects

2.2.1 Adrenal Steroid Receptors and Effects on Excitability and Neurochemistry

Adrenal steroids modulate the excitability of hippocampal neurons, as illustrated by the phenomenon of long-term potentiation (LTP) (Bliss and Lomo, 1973). A single burst of high-frequency stimulation to hippocampal afferents immediately alters the responsiveness of neurons to subsequent acute stimulation, an effect lasting from over many hours to days. A number of recent studies on the hippocampal CA1 field and the dentate gyrus have demonstrated that acute stress and acute glucocorticoid elevation produce an impairment in LTP or its close relative, primed-burst potentiation, PBP (Diamond et al., 1992, 1994; Pavlides et al. 1993). There is a U-shaped dose response curve, with low levels of corticosterone facilitating PBP and high levels inhibiting PBP in the CA1 region (Diamond et al., 1992). See *Figure 14-1*. In both the dentate gyrus and CA1 and CA3 fields, LTP can be modulated rapidly (within 1 h) and biphasically by adrenal steroids acting, respectively, via Type I and Type II receptors (Pavlides et al., 1995a, b, 1996). Moreover, in awake, freely moving adrenalectomized (ADX) rats, the enhancement of LTP by the Type I receptor agonist, aldosterone, lasts for at least 24 h, at which time it is still markedly higher than that found in ADX rats given only vehicle treatment before LTP induction (Pavlides et al., 1994).

Regarding how these biphasic effects come about, it is very likely that principal neurons in dentate gyrus and Ammon's horn contain both types of receptors, considering the distribution of mRNA, immunocytochemical reactivity, and binding for Type I and Type II adrenal steroid receptor subtypes (Herman et al. 1989; van Steensel, 1995). In studies on pyramidal neurons of the CA1 region, adrenal steroids have been shown to act via Type I and Type II adrenal steroid receptors to maintain and modulate excitability of hippocampal neurons (Beck et al., 1994; Joels and De Kloet, 1994; Birnstiel and Beck, 1995). Type I receptor activation in hippocampus from ADX rats is associated with reduced calcium currents through voltage-gated channels, reduced responses to serotonin via 5HT-1A receptors and to carbachol via muscarinic receptors and stable responses to synaptic inputs involving excitatory and inhibitory amino acids (Hesen and Joels, 1996a, b). Additional activation of Type II receptors causes increased calcium currents and enhanced responses to EAA, serotonin and carbachol (Joels and De Kloet, 1994; Joels, 1997), and very high levels of Type II receptor activation markedly increases calcium currents (Kerr and Campbell, 1992), and also leads to increased NMDA receptor expression on hippocampal neurons (Weiland et al., 1995). Acute stress also increases NMDA R1 mRNA but decreases AMPA receptor subunit A mRNA levels without affecting mRNA levels of subunits B and C (Bartanusz et al., 1995). Kainate receptor mRNA levels were also affected by acute corticosterone treatment, with low dose occupancy of Type I receptors increasing mRNA levels for kainate receptor 1 and 2 and also for the GluR7 subunit of the AMPA receptor and high dose occupancy of Type II as well as Type I receptors reversing this effect (Joels et al., 1996). Besides these effects, specific Type I and Type II agonists given to bilaterally ADX rats produce a variety of other effects on various aspects of gene expression in hippocampus associated with neurotransmission (Lupien and McEwen, 1997).

What is surprising about these nonoverlapping actions of Type I and Type II receptors on gene products in hippocampus is that they defy the classical model of adrenal steroid receptor action via a common glucocorticoid response element (GRE) (Evans and Arriza, 1989) and point to a different and possibly more complex mode of mineralocorticoid and glucocorticoid regulation of gene expression (Miner and Yamamoto, 1991; Reichardt and Schutz, 1998). It should be noted that the regulation of preprotachykinin A gene mRNA levels in rat forebrain regions does follow the predictions of the classic GRE in that both Type I and Type II agonist treatments of ADX rats elevate mRNA levels for this neuropeptide (Pompei et al., 1995).

The fact that Type I and Type II adrenal steroid receptor activation has been characterized separately in ADX rats using selective agonists raises the question of what happens when both receptors are activated simultaneously over the physiological range of corticosterone. Under such conditions it is necessary to consider Type I/Type II receptor heterodimers (Trapp et al., 1994), and those of other comodulators of steroid-regulated gene expression such as immediate early genes (see later). This is important, because we have noted that hippocampal neurons in Ammon's horn and dentate gyrus express both types of receptors, and it remains to be seen how different the consequences of combined receptor activation are from the information summarized using selective Type I and Type II receptor agonists and antagonists.

Another issue to be resolved is the extent to which pyramidal neurons and granule neurons show similar molecular responses to adrenal steroids in affecting their excitability. There is also the unanswered question whether neurons in other brain regions, such as amygdala and prefrontal cortex (discussed later) also respond in the same ways to adrenal steroids. From the standpoint of LTP, the biphasic actions of adrenal steroids in dentate gyrus and Ammon's horn occur in pathways using NMDA receptors and not in the mossy fiber non-NMDA pathway (Pavlides and McEwen, 1999). The question whether there is commonality of adrenal steroid effects on NMDA receptor expression, on calcium channel activity, and on the sensitivity to carbachol and to serotonin via 5HT1A receptors (discussed earlier) is therefore an important issue when looking at different brain regions.

2.2.2 Adrenal Steroid Metabolism as a Control Point for Glucocorticoid Actions

Glucocorticoids such as cortisol and corticosterone are converted to the 11-dehydro form by 11-hydroxysteroid dehydrogenase-2 and the products, cortisone and 11-dehydrocorticosterone, are converted back to the parent glucocorticoids by 11-hydroxysteroid dehydrogenase-1 (Funder et al., 1988; Seckl and Walker, 2001). Each enzyme has differential expression in various tissues. 11-HSD-1 is very prominent in the liver and overexpression of this enzyme in liver leads to obesity (Masuzaki et al., 2001). 11-HSD-2 is expressed in kidney collecting tubules and inactivates cortisol and corticosterone to prevent continual sodium reabsorbtion and hypertension, which occurs after blocking 11-HSD-2 by glycyrrhetinic acid, a constituent of licorice (Funder et al., 1988). 11-HSD-2 expression is low in brain whereas 11-HSD-1 expression is measurable in the brain (Sandeep et al., 2004). Knockout of 11-HSD-1 in mice results in a slower rate of brain aging with respect to memory tasks that are dependent on the hippocampus (Yau et al., 2001). Blockade of 11-HSD-1 activity for months in elderly human subjects improves cognitive function and the effect is particularly evident in individuals with Type 2 diabetes. It is interesting to note that poor glucose tolerance is associated with impaired declarative memory and hippocampal volume reduction associated with "mild cognitive impairment" (Convit et al., 2003). Moreover, Type 2 diabetes is associated with increased risk for Alzheimer's disease (Arvanitakis et al., 2004). In this regard, the hippocampus is sensitive to the effects of Type 1 diabetes and expresses two insulin sensitive glucose transporters, Glut 4 and Glut 8, for which the translocation to membranes is regulated by circulating insulin and glucose (McEwen and Reagan, 2004). Insulin is reported to enhance cognitive function in cognitively impaired elderly subjects (Craft et al., 1995), and IGF-1 is able to increase neurogenesis in the dentate gyrus and mediates exercise-induced increases of cell proliferation in this brain region (Aberg et al., 2000; Carro et al., 2000). Thus, the hippocampus is influenced by insulin related molecules and by glucocorticoids.

2.3 Allostasis and Mechanisms for Neuroendocrine Adaptation of HPA Axis

The HPA axis is controlled via releasing factors secreted into the portal blood supply in the median eminence of the hypothalamus. Corticotrophin releasing hormone (CRH) is the principal factor and is produced by neurons in the paraventricular nucleus of the hypothalamus. Vasopressin also plays a regulatory role and is more important for some stressors than others as a stimulus of ACTH release. Glucocorticoid negative feedback takes two forms, rate sensitive control that is related to rising and falling levels of glucocorticoids and level sensitive feedback that is more tonic and sustained. Whereas rate sensitive feedback may involve rapid nongenomic actions of glucocorticoids, level sensitive feedback is classically viewed as a direct action of glucocorticoids on glucocorticoid receptors in the PVN. Yet, recent evidence challenges this mechanism in that, after adrenalectomy, increased ACTH levels are reduced not only by glucocorticoids but also by metabolizable sugars.

The hypothalamic regulation of CRH and ACTH release is influenced by pathways from many forebrain structures (➊ *Figure 14-2*). The hippocampus and amygdala are key limbic brain structures that process experiences by interfacing with lower vegetative brain areas, such as hypothalamus, and higher cortical centers. They also help to interpret, on the basis of current and past experiences, whether an event is threatening or otherwise stressful and help to determine the behavioral, neuroendocrine, and autonomic responses. See ➊ *Figure 14-3*. The amygdala is an essential neural component in the memory of fearful and emotionally laden events, whereas the hippocampus is concerned with determining the context in which

Figure 14-2
The HPA axis is regulated by excitatory and inhibitory neural inputs to the bed nucleus of the stria terminalis (BNST). The BNST relays information to the paraventricular nucleus of the hypothalamus (PVN), which, in turn, regulates ACTH secretion from the anterior pituitary via corticotropin releasing factor (CRF) and arginine vasopressin (AVP). The hippocampus, through the ventral subiculum, and the medial prefrontal cortex (mPFC) are involved in shutting off the HPA stress response via the BNST and the preoptic region of the hypothalamus (POA) (Herman and Cullinan, 1997). The brainstem catecholaminergic system has a direct stimulatory role in the PVN. Glucocorticoid negative feedback operates at multiple levels and via both rapid and delayed actions (Keller-Wood and Dallman, 1984); in the amygdala, there is positive feedback, whereas in the PVN and anterior pituitary there is negative feedback. Metabolic hormones also play a role in negative feedback processes (Pecoraro et al., 2004)

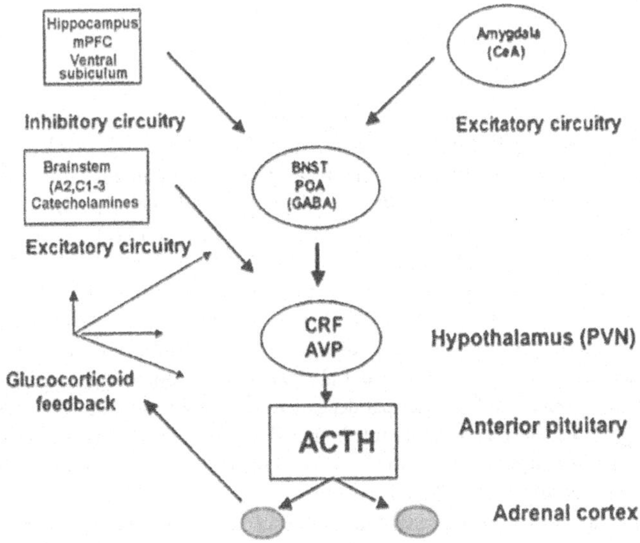

Figure 14-3
Chronic stress increases fear and anxiety while impairing spatial memory. The amygdala, which plays a pivotal role in fear and anxiety, shows growth of dendrites and spine synapses with chronic stress that is mediated, in part, by CRF and tissue plasminogen activator (tPA). After chronic stress, the hippocampus, which is important for spatial memory, shows decreased neural activity and suppression of dentate gyrus neurogenesis and dendritic remodeling; there may also be alterations in spine synapse number and distribution. Mediators of these changes include glutamate and glucocorticoids along with neurotrophins such as BDNF, NMDA receptors, and calcium fluxes, CREB and tau phosphorylation. After chronic stress, the medial prefrontal cortex, which is involved in extinction of fear memory, as well as working memory and executive function, shows dendritic remodeling and loss of spine synapses. Although not yet tested, it is possible that this results in impairment of fear extinction and working memory and executive function

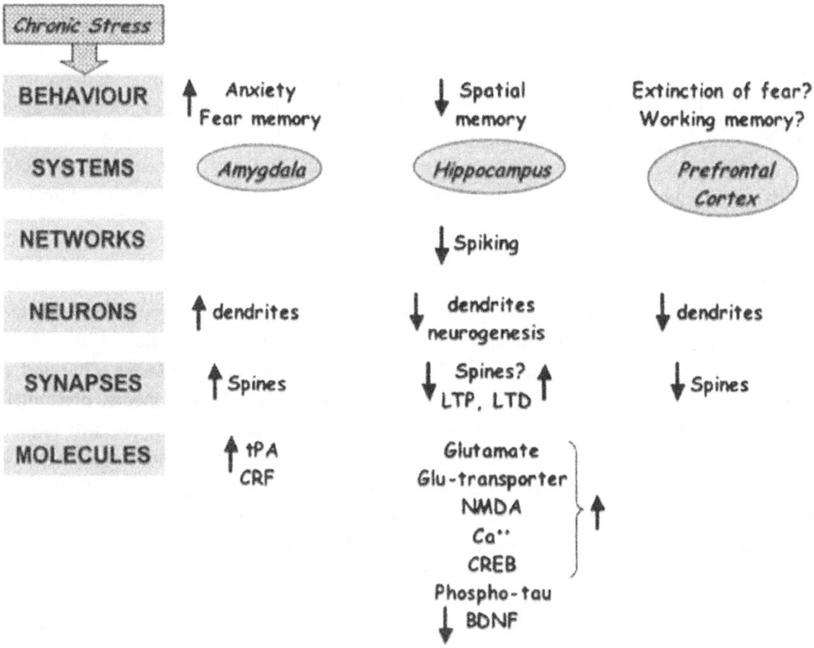

such events take place as well as other aspects of episodic and declarative memory (Squire and Zola-Morgan, 1991; Eichenbaum and Otto, 1992; Phillips and LeDoux, 1992). Whereas lesions of the central or lateral amygdala will abolish conditioning of the freezing response of an animal to a tone paired with a shock, a hippocampal lesion has no such effect; on the other hand, the hippocampal lesion will abolish conditioning of the freezing response to the "context", i.e., to the environment of a particular conditioning chamber (Phillips and LeDoux, 1992).

The amygdala and hippocampus are also linked to each other anatomically and functionally (Knigge, 1961; Pitkanen et al., 2000; Petrovich et al., 2001). For example, lesions of the basolateral amygdaloid nucleus reduce long-term potentiation in the dentate gyrus and stimulation of this nucleus facilitates dentate gyrus long-term potentiation (Ikegaya et al., 1994, 1995). Moreover, amygdala lesions block the inhibitory effects of stressors on hippocampal-dependent memory tasks (Kim et al., 2001). The hippocampus and amygdala also have a regulatory role in the HPA axis activity, with the hippocampus in general being inhibitory and the amygdala acting as a facilitator of the HPA stress response (Knigge, 1961; McEwen, 1977; Jacobson and Sapolsky, 1991; Herman et al. 1996).

However, this statement oversimplifies a great deal of complexity. For example, within the hippocampus, certain sites respond to electrical stimulation by increasing HPA activity (Dunn and Orr, 1984), even though, as noted earlier, the overall role of the hippocampus appears to be to shut off HPA activity. Moreover, other brain areas are involved, e.g., a brain lesion and steroid implant study indicate that parts of the medial prefrontal cortex (mPFC; see *Figure 14-3*) are a site of glucocorticoid negative feedback and play a role in containing the HPA response to psychological (e.g., restraint) stress, but not to ether stress (Diorio et al., 1993). Yet, the mPFC role is not simple either, since another report indicates that other parts of the mPFC participate in the stimulation of HPA activity (Sullivan and Gratton, 1999).

As noted, glucocorticoid implants into the mPFC reduce the magnitude of the HPA response to stress (Diorio et al., 1993) pointing to the important topic of steroid feedback in the control of HPA activity. It is important to note that the HPA axis is dynamically regulated, and steroid feedback operates at several levels in relation to neural control of the turning on and shutting off of the stress response (Akana et al., 1988; Jacobson et al., 1988). Besides rate sensitive and level sensitive feedback, delayed feedback may be viewed as both a thermostat (steroid elevation turning down ACTH release) and a modulation of neural activity impinging upon the CRF and AVP neurons PVN. The glucocorticoid-modulated input can be inhibitory (perhaps via the GABA system) as well as excitatory (Herman et al., 1996).

The demonstration that constant steroid feedback via corticosterone pellets implanted into ADX rats normalizes ACTH levels but allows for sustained ACTH secretion after stress highlights the importance of neural control in the shutoff of the HPA stress response (Akana et al., 1988; Jacobson et al., 1988). In the same study, diurnal exposure to CORT in the drinking water also normalized ACTH levels in ADX rats but allowed for a more rapid termination of the HPA stress response, even when no steroid was present. This further highlights the importance of understanding the role of adrenal steroids in priming neural mechanisms that subserve shutoff of the HPA axis (Akana et al., 1988; Jacobson et al., 1988).

Because of these two interrelated roles of the hippocampus, that is, a role in aspects of memory (*Figure 14-3*) and regulation of HPA activity (*Figure 14-2*), impairment of hippocampal function through changes in either excitability, reversible plasticity, or permanent damage may be expected to have two effects. The first is to impair hippocampal involvement in episodic, declarative, contextual, and spatial memory; impairments of these functions are likely to debilitate an individual's ability to process information in new situations and to make decisions about how to deal with new challenges. The second effect is to impair the hippocampal role in regulating HPA activity, particularly the shutoff of the stress response, leading to elevated HPA activity and further exacerbating the actions of adrenal steroids in the long-term effects of repeated stress. This concept, first called the "glucocorticoid cascade hypothesis" of hippocampal aging (Sapolsky et al., 1986), stands at the center of the notion of "allostasis" and "allostatic load," described earlier in this chapter. Yet, in looking into the roles of adrenal steroids and other modulators in the effects of stress on the hippocampus, it is important not only to recognize the roles of other brain regions such as the amygdala and prefrontal cortex, but also to consider the adaptive role played by the acute secretion of adrenal steroids in the response to acute stressors.

3 The Effects of Stress: Behavior, Cells, Synapses, and Molecules

Over the past decade, there is growing appreciation of the fact that, in addition to affecting endocrinologic and physiologic aspects of body and brain function, stress also has a profound impact on neural plasticity at multiple levels of neural organization. These effects on plasticity mechanisms are manifested as behavioral effects at one end of the spectrum and down to the level of cells, synapses, and molecules at the other (*Figure 14-3*). Moreover, the rich database of hippocampal synaptic plasticity has provided a very useful foundation to gain mechanistic insights into the cellular and molecular effects of stress in excitatory neurons. Although much of the early evidence on stress-induced plasticity emerged from studies of the hippocampus, more recent work has taken into account the important role played by other areas, such as the amygdala and prefrontal cortex (*Figure 14-3* and *14-4*). Findings from these studies have given rise to an intellectual framework that enables us to investigate molecular and biochemical underpinnings of neuronal plasticity mechanisms, and how these mechanisms are adversely affected by stress and emotional disorders. A combination of behavioral, morphological, electrophysiological, pharmacological,

Figure 14-4
Remodeling of neurons occurs in amygdala, hippocampus, and medial prefrontal cortex of rats after chronic stress. Repeated restraint or immobilization stress causes reduction in dendritic length and branch points of CA3 pyramidal neurons in the hippocampus; CA1 pyramidal cells do not exhibit the same degree of atrophy. Repeated stress also impairs neurogenesis in the dentate gyrus (DG). In striking contrast to the hippocampal effects, chronic stress increases dendritic arborization in spiny pyramidal and stellate neurons, but not in bipolar neurons. Chronic stress also causes dendritic hypertrophy in bipolar neurons of the bed nucleus of stria terminalis (BNST), but not central nucleus (CeA). Further, molecular markers related to structural remodeling, such as tPA and GAP-43, are up-regulated in the medial (MeA) and central (CeA) nuclei by repeated stress. Repeated restraint stress also leads to dendritic atrophy in pyramidal neurons of the anterior cingulated region (ACg) of the medial prefrontal cortex

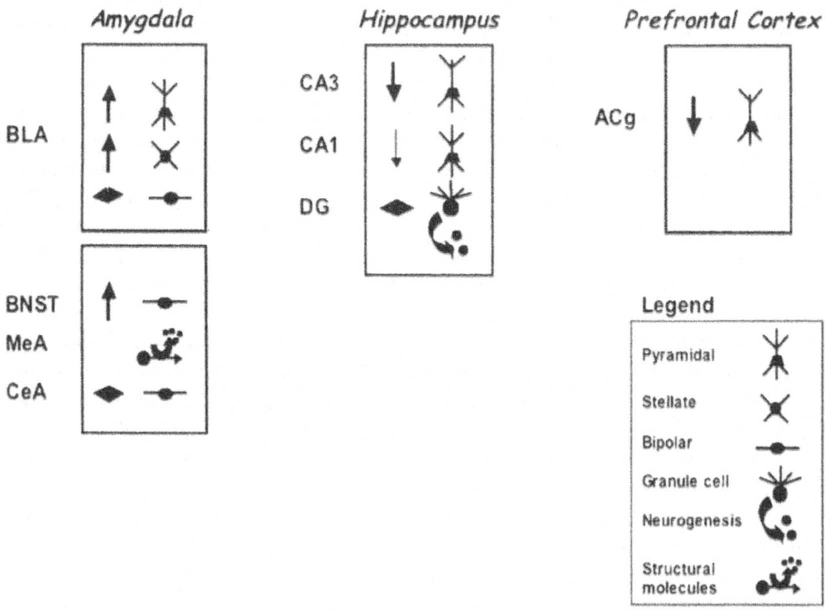

and molecular techniques have been used to explain how stress elicits *global* changes in a brain's output, on the basis of *local* changes in individual neurons and their networks.

3.1 Behavioral Adaptations to Stressors

We begin with the facilitative effects of adrenal steroids on adaptive behaviors, i.e., how they enhance allostasis that leads to behavioral adaptation. The amygdala and hippocampus are both involved in contextual fear conditioning and in passive avoidance learning. In fear conditioning, glucocorticoids enhance learned fear (Corodimas et al., 1994). Moreover, they play an important role in forming the memory of context in contextual fear conditioning (*Figure 14-3*) but not of the actual effect of footshock in rats that are already familiar with the context where the shock is administered (Pugh et al., 1997a, b). This suggests that the hippocampal role in contextual fear conditioning is enhanced by moderate levels of glucocorticoids, but the fear conditioning is either not so dependent on glucocorticoids or is so strong that glucocorticoid influences are hard to demonstrate. Yet, there is evidence for an influence of glucocorticoids on the flow of information within the amygdala.

Glucocorticoids potentiate serotonin inhibition of the processing of excitatory input to the lateral amygdala from the thalamus, suggesting that there is a mechanism for containing or limiting the sensory input that is important for fear conditioning (Stutzmann et al., 1998). Thus, adrenal steroids may regulate the nature of the signals that reach the amygdala and allow for greater discrimination of the most salient cues for learning.

Moreover, in passive avoidance, both catecholamines and glucocorticoids play a role in facilitating the learning (Cahill et al., 1994; Roozendaal, 2000). Catecholamines work outside of the blood–brain barrier and their effects can be blocked by beta adrenergic blocking agents that do not cross the blood–brain barrier (Cahill et al., 1994). Glucocorticoids enter the brain, and local implants of exogenous corticosterone into hippocampus, amygdala, and nucleus tractus solitarii (NTS) are all able to enhance passive avoidance learning (Roozendaal, 2000).

Adrenal steroids also play a supporting role in the learning of a spatial navigation task in mice (Oitzl et al., 2001). Adrenalectomy impairs the acquisition of the memory of hidden platform location in the Morris water maze, and glucocorticoid administration restores the normal learning curve; however, in mice in which the glucocorticoid receptor was deleted and replaced with a GR that lacks the DNA binding domain, glucocorticoids have no effect to improve task acquisition (Oitzl et al., 2001). This finding illustrates a role for glucocorticoid receptors acting upon the genome in a task that is known to depend on the hippocampus. Interestingly, other actions of glucocorticoids via glucocorticoid receptors are known to involve the protein–protein interactions that are not prevented in mice carrying the GR defective in the DNA binding domain (Reichardt and Schutz, 1998).

Other evidence for glucocorticoid actions supports an inverted U-shaped dose-response curve in which low to moderate levels of adrenal steroids enhance acquisition of tasks that involve the hippocampus, whereas high levels of glucocorticoids disrupt task acquisition (Diamond et al., 1992; Pugh et al., 1997b; Diamond et al., 1999; Conrad et al., 1999a). And, indeed, elevated levels of glucocorticoids, along with overactivity of EAA suppress neurogenesis, promote dendritic remodeling, and even higher activity of glucocorticoids and excitatory amino acids promote damage in strokes, seizures, and head trauma (➋ *Figure 14-1*). Adrenal steroids have biphasic effects upon excitability of hippocampal neurons that may underlie their biphasic actions on memory and recall (Diamond et al., 1992; Pavlides et al., 1994, 1995a, b; Pavlides and McEwen, 1999).

Whereas some studies have reported spatial memory deficits following stress (Luine et al., 1994) or chronic corticosterone treatment (Bodnoff et al., 1995), other studies (Conrad et al., 1999b) suggest that stress might also impair memory through nonhippocampal mechanisms, such as enhanced emotionality. Furthermore, stress facilitates classical eye-blink conditioning (Shors et al., 1992) and this facilitation requires activation of NMDA receptors in the basolateral amygdala (BLA) (Shors and Mathew, 1998). Corticosterone injections have also been shown to potentiate fear conditioning (Corodimas et al., 1994).

3.2 Stress and Dendritic Remodeling: Hippocampus and Beyond

Prolonged and severe stress leads to behavioral abnormalities manifested as cognitive impairments as well as affective disorders. Investigations into cellular mechanisms underlying stress-induced cognitive deficits have focused primarily on the hippocampus, a structure involved in forming declarative/episodic/spatial memories (O'Keefe and Nadel, 1978; Squire and Zola-Morgan, 1991; McEwen, 1999; Martin et al., 2000), not only because of its susceptibility to stress-related damage but also because of its negative feedback regulation (➋ *Figure 14-2*) of the stress response via the HPA axis (Herman et al., 1989; Jacobson and Sapolsky, 1991; Sapolsky et al., 1991; Herman and Cullinan, 1997). Earlier studies on how stress and stress hormones affect the rat hippocampus revealed that 21 days (6 h per day) of repeated restraint stress produces significant dendritic remodeling in CA3 pyramidal neurons (➋ *Figure 14-3* and ➋ *Figure 14-4*) (Watanabe et al., 1992; Magarinos and McEwen, 1995a, b; Sousa et al., 2000). This dendritic remodeling is characterized by a reversible shortening and debranching of apical dendrites (Conrad et al., 1999b) and is mediated by mechanisms involving high levels of glucocorticoid secretion, glutamate, and serotonin (Magarinos and McEwen, 1995a, b). These findings, in turn, have contributed to rodent models of stress-induced neuronal atrophy that may provide one potential explanation for the hippocampal shrinkage associated with post-traumatic stress disorder, recurrent depressive illness, and Cushing's disease (Starkman et al., 1992; Bremner et al., 1995; Sheline et al., 1996; Bremner et al., 1997; Starkman et al., 1999, 2003).

Although hippocampal plasticity may mediate cognitive aspects of behavioral impairments caused by severe stress, changes in the amygdala are more likely to contribute to the affective aspects of stress disorders. In addition to a critical role for the amygdala in fear and anxiety (Davis, 1992; LeDoux, 1994; Davis et al., 1994), there is growing evidence for striking differences between the hippocampus and amygdala with respect to stress. This contrast between the hippocampus and amygdala is manifested in two ways—how these two structures affect the stress response and how they, in turn, are affected by stress. First, anatomical studies indicate that limbic inputs impinging on the paraventricular nucleus (PVN) of the hypothalamus and hypothalamic GABA-ergic neurons can be either excitatory from the hippocampus and thereby enhancing GABA-ergic tone, or inhibitory from the amygdala and thereby reducing GABA-ergic tone (Herman et al., 1989; Jacobson and Sapolsky, 1991; Sapolsky et al., 1991; Pitkanen and Amaral, 1994; Herman and Cullinan, 1997). This in turn implies that whereas enhanced hippocampal input would suppress the HPA axis, enhanced amygdaloid input could have the opposite effect on HPA activity. Thus, one potential difference between the hippocampus and amygdala with respect to the neural circuitry underlying stress comes from their disparate roles in the regulation of the HPA axis (➋ *Figure 14-2*).

Evidence for a second important difference comes from behavioral studies demonstrating how chronic stress affects hippocampal- or amygdala-dependent learning and memory (➋ *Figure 14-3*). In rodents, severe stress facilitates fear and anxiety-like behavior (Conrad et al., 1999b; McGaugh and Roozendaal, 2002; Korte and De Boer, 2003; Vyas et al., 2003a, b), but impairs spatial learning (Luine et al., 1994). Whereas repeated stress that produces dendritic remodeling in the CA3 region impairs hippocampal-dependent learning (Conrad et al., 1996), the basolateral amygdala has been shown to be essential for stress-induced facilitation of aversive learning (Liang et al., 1994; Shors and Mathew, 1998). Further, there are reports indicating that the debilitating effects of chronic stress on anxiety and fear can be long-lasting and persist well beyond the actual period of exposure to stress (Adamec et al., 1999; Conrad et al., 1999b). These observations are particularly striking in comparison to the mixed results from studies on the effects of stress on hippocampus-dependent behavior. Some of these reports indicate that hippocampal spatial memory deficits are reversible and temporally limited (Luine et al., 1994; McEwen, 1997). Some findings also suggest that the capacity for stress to impair memory is influenced by the brain memory system (hippocampal versus nonhippocampal) involved in solving the task (Kim et al., 2001). Therefore, the same behavioral perturbation, in the form of chronic stress, elicits contrasting functional outputs from the hippocampus and amygdala that appear to have divergent properties in terms of their temporal persistence after the termination of stress.

One possible explanation for these conflicting results involving hippocampal spatial learning comes from a study reporting that stress-induced dendritic atrophy of CA3 neurons is reversible within 7–10 days after the termination of the 21 days of restraint stress paradigm (Luine et al., 1994; Conrad et al., 1999b). The divergent nature of the behavioral outputs of the hippocampus and amygdala in response to stress is further highlighted in an interesting pharmacological study by Conrad et al. (Conrad et al., 1999b). This study reports that although tianeptine, an atypical antidepressant, prevented CA3 atrophy, it failed to prevent enhanced fear and anxiety after repeated restraint stress. The particularly relevant finding of this study was that repeated restraint stress facilitated fear conditioning to both context and tone independently of causing hippocampal CA3 dendritic atrophy. This led the authors to conclude that chronic stress may have a powerful effect on the amygdala, which could override any influence of the hippocampus. Taken together, these findings raise the possibility that cellular substrates for stress-induced changes in emotional behavior may reside in the amygdala and recent reports, reviewed in the next section, have provided new evidence in this direction.

3.3 Stress and Structural Plasticity in the Amygdala: Implications for Anxiety-Like Behavior

Unlike the hippocampus, relatively little is known about how stress affects the amygdala and the nature of its role in the stress response. One recent study (Vyas et al., 2002) used the same experimental strategy that previously elucidated properties of stress-induced dendritic remodeling in hippocampal CA3 neurons. In this study, the effects of two different models of chronic stress on hippocampal and amygdaloid neuronal morphology in rats were examined. In agreement with previous reports, chronic immobilization stress (2 h per day for 10 days) induced dendritic atrophy and debranching in CA3 pyramidal neurons of the hippocampus. In striking contrast, principal neurons in the basolateral complex of the amygdala (BLA) exhibited *enhanced* dendritic arborization in response to the same chronic stress. This stress-induced enhancement in dendritic arborization was restricted only to BLA pyramidal and stellate neurons, which are presumably excitatory projection neurons (McDonald, 1982, 1992) (➋ *Figure 14-4*). Moreover, these patterns varied depending on the type of chronic stress used. Thus, chronic unpredictable stress, which was relatively less effective in remodeling CA3 pyramidal neurons, caused atrophy in bipolar/bitufted neurons without affecting spiny pyramidal and stellate cells of the BLA. The efficacy of chronic immobilization, and not unpredictable stress, in eliciting dendritic hypertrophy in BLA was also shown to be relevant in terms of its anxiogenic properties. Thus, only chronic immobilization stress caused a significant increase in anxiety-like behavior as manifested by a reduction in open-arm exploration and risk-assessment behavior in the elevated plus maze (Vyas and Chatterji, 2004). Interestingly, even after 21 days of stress-free recovery, animals exposed to immobilization stress continued to exhibit enhanced anxiety (Vyas and Chatterji, 2004; Vyas et al., 2004). At the cellular level, chronic immobilization stress (CIS)-induced dendritic remodeling of spiny BLA pyramidal neurons is also as persistent as enhanced anxiety after 21 days of recovery. Moreover, BLA hypertrophy, elicited by chronic immobilization stress, is distinct from hippocampal CA3 atrophy, which is reversible within the same period of stress-free recovery (Vyas and Chatterji, 2004). These findings on persistent dendritic remodeling in the amygdala, in addition to highlighting important differences with hippocampal structural plasticity, raise the possibility that certain forms of chronic stress, by affecting specific neuronal elements in the amygdala, may lead to the long-lasting behavioral manifestations of enhanced emotionality.

Although plasticity mechanisms in the BLA, which play a key role in the acquisition and consolidation of fear memories, have received considerable attention, recent reports also point to the importance of noncortical amygdalar nuclei that influence its output. For example, work by Davis and colleagues (Davis and Shi, 1999; Davis, 2000), using the startle reflex, indicates that the bed nucleus of stria terminalis (BNST) in the so-called extended amygdala, may be involved in processing signals more akin to cue-nonspecific fear or anxiety, whereas the central nucleus of the amygdala (CeA) is more involved in cue-specific fear. Interestingly, chronic stress also enhances dendritic arborization in BNST neurons (Vyas et al., 2003a, b). Further evidence for stress-induced plasticity in amygdalar nuclei, which receive information from the BLA, comes from a recent study using a knockout mouse model (Pawlak et al., 2003). This study identified the serine protease tissue–plasminogen activator (tPA) in the amygdala, but not the hippocampus, as a critical component in the sequence of molecular events linking repeated restraint stress-induced amygdalar plasticity with the development of anxiety-like behavior.

These findings also point to novel aspects of stress-induced plasticity, in the temporal and spatial domain, which will require further investigation. First, unlike stress-induced dendritic hypertrophy in the BLA, stress-induced up-regulation of tPA was restricted to the medial (MeA) and central (CeA) nuclei of the amygdala. This highlights the need to examine subtle region-specific, possibly even cell-specific, effects within the amygdalar network. Second, the temporal features of these effects are equally important. For example, an acute episode of restraint stress triggered a cascade of molecular signals related to tPA within the first few hours following exposure to stress, and all of these preceded the eventual development of anxiety in the animal. In other words, there may be considerable differences between the spatio-temporal dynamics of stress-induced changes in molecular signals, neuronal plasticity, and ultimately, behavioral output. Further, an acute episode of stress may elicit plasticity that is considerably different in terms of its spatio-temporal features from that induced by repeated episodes of the same stressor over time.

In summary, recent evidence on stress-induced plasticity in various nuclei of the amygdala expand the scope for studying effects of stress on emotional behavior. In particular, a growing body of findings indicates that plasticity mechanisms triggered as a result of stress-induced amygdala activity can strengthen the excitatory drive within the BLA and thereby influence subsequent information processing by its downstream targets, such as the MeA, CeA, and BNST. This, in turn, suggests that chronic stress could lead to an imbalance in HPA axis function through a gradual loss of hippocampal inhibitory control as well as gain in excitatory control exerted by the amygdala. Taken together, these studies on stress-induced plasticity in the various nuclei of the amygdala may provide putative cellular substrates for exploring psychiatric disorders that are characterized by diminished cognitive capabilities and abnormally high fear response.

3.4 Stress and Structural Plasticity in the Prefrontal Cortex

The medial prefrontal cortex (mPFC) and prelimbic cortex show structural remodeling after repeated stress and repeated corticosterone treatment, which is similar to that found in the hippocampus, namely a shortening of dendritic length and a simplification of dendritic branching in layer II/III neurons (➋ *Figure 14-4*) (Wellman, 2001; Wellman, 2004; Radley et al., 2004) There is also a reduction after 21days of chronic restraint stress in spine density on the apical dendrites of layer II/III neurons (Radley et al., 2004). The net result of these changes is estimated to be a reduction of 40% in synaptic input to the mPFC. Neurons in this region receive input from the amygdala and may be responding to increased output from the amygdala during stress (Kim et al., 2001; Diamond et al., 2004). It is not yet known if the infralimbic cortex of the mPFC also shows remodeling after 21 days of chronic restraint stress. This would be important because of the involvement of this region in extinction of fear conditioning (Santini et al., 2004). A prediction of these structural changes is that prefrontal cortical functions in working memory, executive function, and memory extinction would be impaired by chronic stress.

3.5 Cellular and Synaptic Correlates for Stress-Induced Plasticity in the Hippocampus

Investigations into cellular mechanisms underlying stress-induced impairments have focused largely on two common metrics of hippocampal plasticity—one structural, and the other electrophysiological. The morphological changes triggered by stress have been described earlier. In this section, we describe some data obtained from the second approach that has been guided by the hypothesis that stress, possibly by disrupting synaptic transmission and plasticity mechanisms, leads to deficits in hippocampal memory.

A variety of rodent models of stress have been shown to impair hippocampal LTP (McEwen, 1999; Kim and Diamond, 2002). Evidence for the involvement of excitatory amino acids and NMDA receptors, in hippocampal dendritic remodeling (Magarinos and McEwen, 1995b; McEwen, 1999), provided early insights into the underlying synaptic signaling mechanisms. A series of recent studies using whole-cell recordings in CA3 pyramidal cells indicate that repeated restraint stress increased the amplitude and deactivation time-constant of the NMDA current, without affecting the AMPA/kainate receptor mediated currents at commissural–associational inputs (Kole et al., 2002). This stress-induced enhancement in NMDA-receptor signaling, however, is a condition that is likely to facilitate, and not impair LTP induction. Moreover, intracellular calcium, a key determinant of hippocampal LTP, is also enhanced by glucocorticoids (Kerr and Campbell, 1992; Joels and Vreugdenhil, 1998). These findings, therefore, are difficult to reconcile with the significant body of evidence for stress-induced impairment of LTP.

But there is also evidence that not all forms of LTP are equally susceptible to suppression following exposure to stress or corticosterone (Alfarez et al., 2002). It should also be noted that a majority of reports on stress-induced impairment of LTP are not based on studies in area CA3 of the hippocampus, which undergoes the most significant atrophy among all hippocampal regions. The same group that earlier reported stress-induced increases in NMDA currents at CA3 commissural–associational synapses (Kole et al., 2002), has recently demonstrated that brief social stress prevented induction of LTP by low-frequency tetanic stimuli, even after a 21-day time delay (Kole et al., 2004a). Interestingly, repetition of the same stress did not result in a greater suppression of LTP, but a reversal of synaptic potentials resembling LTD, at commissural–associational synapses (Kole et al., 2004b).

The apparent contradiction between stress-induced impairment of LTP and increase in NMDA-receptor signaling may be explained by the possibility that enhanced NMDA currents give rise to a form of metaplasticity that favors a saturation of LTP (Kim and Yoon, 1998). This saturation of LTP, in turn, raises the threshold for subsequent LTP induction, which is manifested either as a reduction in the magnitude of LTP elicited, or a complete prevention of further LTP, or, in extreme cases, a depression of synaptic strength in the form of LTD. Such a scenario is based on the Bienenstock–Cooper–Munro (BCM) theory of synaptic plasticity, which proposes that the threshold for triggering LTP or LTD, itself, is modifiable and depends on the previous history of activity of the synaptic inputs (Bienenstock et al., 1982; Abraham and Bear, 1996).

In other words, the same tetanic stimuli that elicit robust LTP in an "unstressed" or "naïve" synapse, can fail to do so in a "stressed" synapse that has already strengthened or saturated due to stress-induced increase in NMDA currents or calcium influx, thereby reaching a state in which further LTP is not favored. This framework, rooted in the concepts of BCM theory and stress-induced metaplasticity (Bienenstock et al., 1982; Abraham and Bear, 1996; Kim and Yoon, 1998), provides a useful basis for future studies that will require careful examination of the various forms of hippocampal synaptic plasticity, that depend on both NMDA-receptor dependent and independent mechanisms (Johnston et al., 1992).

The traditional approach to examining the functional consequences of stress-induced dendritic remodeling has been driven by the logic that a reduction in postsynaptic dendritic surface will adversely affect the availability of synaptic inputs, and thereby synaptic plasticity. This framework, which focuses primarily on synaptic transmission and plasticity, overlooks the important point that changes in dendritic architecture can also affect intrinsic excitability properties of neurons in ways that do not necessarily have to be mediated directly through modulation of synaptic transmission. Theoretical studies point to a significant role for dendritic morphology in modulating various facets of neuronal excitability including patterns of action potential firing (Mainen and Sejnowski, 1996; Krichmar et al. 2002; van Ooyen et al., 2002; Narayanan et al., 2004), and propagation of action potentials (Vetter, 2001).

Such modulation of intrinsic excitability gains particular significance in light of the strong excitatory recurrent connections through associational inputs to the *stratum radiatum* region of CA3, which also exhibits the most robust stress-induced dendritic remodeling among all subregions of the hippocampus. Interestingly, in vivo recordings in area CA3 indicate that, in addition to suppressing LTP, repeated stress also leads to epileptic afterdischarges and shifts in current sources and sinks along apical dendrites (Pavlides et al., 2002). Thus, the cumulative effects of altered synaptic and neuronal excitability, propagated through recurrent loops, may affect the balance of excitation and inhibition in the hippocampal CA3 network. Such a shift in the overall balance of excitation and inhibition, due to repeated and severe stress, could contribute to excitotoxicity and eventual neurodegeneration in the hippocampus (Orchinik et al., 2001).

This view raises another fundamental question that is at the core of the large body of experimental evidence for stress-induced dendritic remodeling: Is this simply a reflection of damage caused by repeated stress or a manifestation of adaptive plasticity that attempts to counter stress-induced damage? Although a comprehensive answer to this complex question still eludes us, there are several interesting findings that may provide an explanation. First, stress-induced CA3 atrophy is reversible, which is indicative of homeostatic mechanisms that enable the neuron to adapt to the levels and duration of stressful conditions. For example, in hibernation, there is rapid remodeling of dendrites and synaptic connections in the hippocampus that appears to reduce the vulnerability of the DG-CA3 region to permanent damage when energy supplies are low (Popov and Bocharova, 1992; Popov et al., 1992; Arendt et al., 2003).

Second, modulation of synaptic transmission and plasticity caused by stress could represent a second category of plasticity, some of which could contribute to functional impairment in learning and memory. Such deficits in synaptic plasticity and behavioral output, however, could still lie within a range that is reversible depending on the severity and duration of the stressor. Third, mechanisms of intrinsic plasticity, as described earlier, could also provide an important means of homeostasis that is yet to be studied in detail. One recent report (Kole et al., 2004b), using the tree shrew, suggests that chronic stress can indeed trigger homeostatic maintenance in excitability of CA3 pyramidal neurons. In principle, a reduction in dendritic arbors would predict an increase in input resistance, and consequently enhanced rates of action potential firing in CA3 neurons (Narayanan et al., 2004). Surprisingly, results from whole-cell recordings used in this study indicate that chronic stress actually reduced the input resistance by 20–25% (Kole et al., 2004b). Moreover, all active membrane properties, including depolarization-induced action potential kinetics and complex spiking patterns, did not change after chronic stress. Taken together, these data show that despite dendritic remodeling and enhanced cortisol release, the efficacy of somatic excitability of CA3 pyramidal neurons is largely preserved after long-term stress. This is a striking example of how intrinsic mechanisms of homeostatic plasticity can coexist with and counter structural remodeling caused by repeated stress.

In summary, a growing body of experimental evidence indicates that chronic exposure to stress can activate a complex repertoire of plasticity mechanisms in the hippocampus. Based on the considerable evidence pointing to factors that are common to both stress-induced damage and synaptic plasticity mechanisms subserving normal hippocampal function, the challenge now is to gain a mechanistic understanding of how these factors may contribute to degeneration, as well as adaptive changes, caused by stress.

3.6 Cellular and Synaptic Correlates for Stress-Induced Plasticity in the Amygdala

Findings on the cellular and molecular underpinnings of plasticity mechanisms elicited by stress in the hippocampus, as outlined earlier, will also provide a useful framework for future investigations of such mechanisms in the amygdala. Although at present little is known about such plasticity mechanisms in the amygdala, some recent evidence provides interesting insights into putative mechanisms that may play a critical role in the amygdala. In one study (Stutzmann et al., 1998), aimed at investigating the in vivo modulation of sensory transmission in the lateral amygdala (LA) by corticosterone, iontophoretically applied serotonin (5-HT) inhibited both synaptically and glutamate-evoked action potentials in a majority of neurons examined. However, after adrenalectomy (ADX), which eliminates endogenous corticosterone, 5-HT no longer inhibited evoked activity in the LA. High, but not low, doses of corticosterone administrated to ADX animals reinstated the inhibition of excitatory transmission of 5-HT. Furthermore, immunocytochemical labeling of the glucocorticoid receptor in the intact rat demonstrated nuclear staining throughout several amygdala regions, including the LA. In contrast, no nuclear labeling was visible after ADX. These data suggest that the ability of 5-HT to modulate glutamatergic activity in thalamic inputs to the LA depends on the levels of corticosterone present. However, they do not necessarily mean that genomic actions of glucocorticoids are involved, as the effects described by Stutzman et al. (1998) are rapid and as nongenomic actions of glucocorticoids are known in brain (Makara and Haller, 2001).

Moreover, these findings also point to the potential importance of both serotonergic and inhibitory modulation of amygdalar activity during a stressful experience. Interestingly, the importance of the serotonergic system in the amygdala is also evident from recent studies on the modulation of fear (Burghardt et al., 2004) and stress-related amygdalar plasticity with tianeptine, an atypical antidepressant that is believed to act on the 5-HT reuptake mechanism (reviewed in McEwen and Chattarji, 2004).

A key role for inhibition comes from another recent report that examined how synaptic plasticity in the amygdala, induced by corticotrophin releasing factor (CRF), may translate stress into anxiety-like behavior (Rainnie et al., 2004). In this study, non-anxiety-inducing doses of a potent CRF type 1 and 2 receptor agonist, urocortin, was infused locally into the BLA of rats. After five daily injections of urocortin, the animals developed anxiety-like behavior that persisted for more than 30 days. Importantly, whole-cell patch-clamp recordings from BLA neurons of these anxious animals revealed a pronounced reduction in both spontaneous and stimulation-evoked inhibitory postsynaptic potentials (IPSPs), leading to a hyper-excitability of the BLA network. Moreover, this form of urocortin-induced plasticity appears to be dependent on NMDA receptors and subsequent calcium–calmodulin-dependent protein kinase II (CaMKII) activation.

Preliminary studies (Bennur and Chattarji, 2004) also show, using whole-cell recordings in amygdala slices, that chronic immobilization stress leads to a reduction in inhibition along with enhanced NMDA receptor currents in principal neurons of the LA. Thus, future studies will have to focus on the causal links between stress-induced plasticity mechanisms that involve modulation of glutamatergic, serotonergic, and inhibitory transmission in the amygdala and how these collectively give rise to stress disorders at the behavioral level in which CRF may play an important role. In this connection, recent evidence from the mouse indicates that the ability of acute stress to increase anxiety and promote plasticity within the medial and central regions of the amygdala via a tissue plasminogen activator-dependent mechanism

(Pawlak et al., 2003), as discussed earlier, is dependent on CRF-induced activation of tPA activity (Matys et al., 2004).

4 Future Directions

Finally, although much of the earlier discussion treats stress-induced plasticity in the hippocampus, amygdala, and prefrontal cortex as separate or independent processes, there is considerable physiological and anatomical evidence (e.g., Petrovich et al., 2001) pointing to interactions among these brain areas that could have implications for how stress responses are integrated across structures at the systems level. We have already referred to the extensive connectivity among these three brain regions (➋ *Figure 14-5*). A key

Figure 14-5

The amygdala is reciprocally connected to the medial prefrontal cortex and sends projections to Ammon's horn of the hippocampus [depicted as an unfolded flat map, (Petrovich et al., 2001)] but not to the dentate gyrus (Petrovich et al., 2001). It is possible that stress-induced adaptive plasticity in both hippocampus and medial prefrontal cortex depends on activity in amygdala

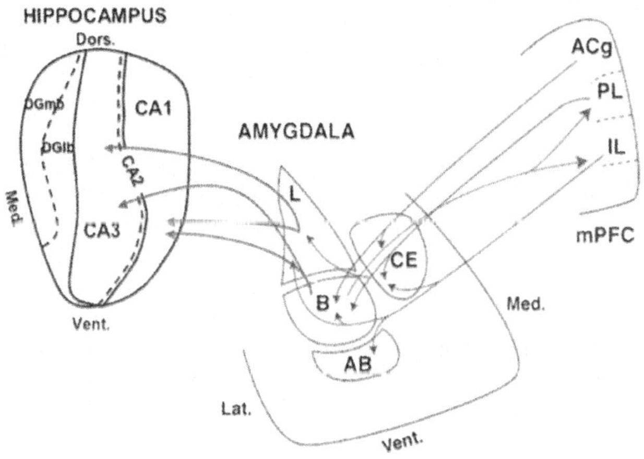

question is whether activity in the amygdala may play a key role in driving changes in hippocampus and prefrontal cortex. This possibility is suggested by two observations already described in this article. First, as reviewed earlier, there is the finding that acute stress causes rapid changes in amygdala structural plasticity in both rats and mice that occur before the cumulative effects of repeated stress can be found in the prefrontal cortex and hippocampus. Second, in human depressive illness, hyperactivity in the amygdala is reported in the first episode of depression (Drevets, 2000), and there are several reports on increased amygdala volume (Frodl et al., 2002, 2003), whereas hippocampal shrinkage is reported as a function of the duration of symptoms of depression (Sheline et al., 1999; MacQueen et al., 2003) and also in post-traumatic stress disorder (Bremner et al., 1995).

Furthermore, there is accumulating evidence for a pivotal role for the amygdala in modulating synaptic plasticity in the hippocampus (Diamond et al., 1994; Kim and Diamond, 2002). For example, pharmacological inactivation of NMDA receptors in the BLA impairs LTP in the dentate gyrus (Abe, 2001). High-frequency stimulation of the BLA has also been shown to facilitate LTP in dentate gyrus (Abe, 2001). Activation of the amygdala, as well as exposure to stress, elicits a biphasic modulation of hippocampal LTP, which is manifested first as an immediate excitatory effect, followed by an inhibitory effect that is more long-lasting (Akirav and Richter-Levin, 1999). Since hippocampal responses are altered following lesions or pharmacological inactivation of the amygdala, one experimental approach to studying interactions among these three brain areas is to lesion or reversibly inactivate key efferent pathways from the amygdala while applying stressors that normally cause structural plasticity. The use of viruses to introduce genes that alter responses in amygdala neurons should also be considered.

5 Conclusion

Future studies will have to focus on integrating current and new findings in two broad directions. First, within each of the three brain areas that constitute the key components of the neural circuitry of stress (hippocampus, amygdala, prefrontal cortex), an integration across multiple levels of neural organization has to be achieved—from genetic and molecular mechanisms at one end of the spectrum, to systems and behavioral level analyses at the other. Second, integration across the three structures will also be critically important in elucidating their roles in the emotional and cognitive aspects of stress disorders.

Acknowledgments

Research support has come from the National Institute of Mental Health Grants MH 41256 and MH58911. S.C. is supported by an International Senior Research Fellowship from The Wellcome Trust.

References

Abe K. 2001. Modulation of hippocampal long-term potentiation by the amygdala: a synaptic mechanism linking emotion and memory. Jpn J Pharmacol 86: 18-22.

Aberg MA, Aberg ND, Hedbacker H, Oscarsson J, Eriksson PS. 2000. Peripheral infusion of IGF-1 selectively induces neurogenesis in the adult rat hippocampus. J Neurosci 20: 2896-2903.

Abraham WC, Bear MF. 1996. Metaplasticity: the plasticity of synaptic plasticity. Trends Neurosci 19: 126-130.

Adamec RE, Burton P, Shallow T, Budgell J. 1999. NMDA receptors mediate lasting increases in anxiety-like behavior produced by the stress of predator exposure—implications for anxiety associated with posttraumatic stress disorder. Physiol Behav 65: 723-737.

Akana S, Jacobson L, Cascio C, Shinsako J, Dallman M. 1988. Constant corticosterone replacement normalizes basal adrenocorticotropin (ACTH) but permits sustained ACTH hypersecretion after stress in adrenalectomized rats. Endocrinology 122: 1337-1342.

Akirav I, Richter-Levin G. 1999. Biphasic modulation of hippocampal plasticity by behavioral stress and basolateral amygdala stimulation in the rat. J Neurosci 19: 10530-10535.

Alfarez DN, Wiegert O, Joels M, Krugers HJ. 2002. Corticosterone and stress reduce synaptic potentiation in mouse hippocampal slices with mild stimulation. Neurosci 115: 1119-1126.

Arendt T, Stieler J, Strijkstra AM, Hut RA, Rudiger J, et al. 2003. Reversible paired helical filament-like phosphorylation of tau is an adaptive process associated with neuronal plasticity in hibernating animals. J Neurosci 23: 6972-6981.

Arvanitakis Z, Wilson RS, Bienias JL, Evans DA, Bennett DA. 2004. Diabetes mellitus and risk of Alzheimer disease and decline in cognitive function. Arch Neurol 61: 661-666.

Bartanusz V, Aubry JM, Pagliusi S, Jezova D, Baffi J. 1995. Stress-induced changes in messenger RNA levels of N-methyl-D-aspartate and AMPA receptor subunits in selected regions of the rat hippocampus and hypothalamus. Neuroscience 66: 247-252.

Beck SG, List TJ, Choi KC. 1994. Long- and short-term administration of corticosterone alters CA1 hippocampal neuronal properties. Neuroendocrinology 60: 261-272.

Bennur S, Chattarji S. 2004. Effects of chronic stress on intrinsic and synaptic plasticity in principal neurons of the basolateral amygdala. Soc for Neurosci Abst 30: 511.2.

Bienenstock EL, Cooper LN, Munro PW. 1982. Theory for the development of neuron selectivity: orientation specificity and binocular interaction in visual cortex. J Neurosci 2: 32-48.

Birnstiel, S. Beck SG. 1995. Modulation of the 5-hydroxytryptamine4 receptor-mediated response by short-term and long-term administration of corticosterone in rat CA1 hippocampal pyramidal neurons. J Pharmacol Exp Ther 273: 1132-1138.

Bliss TVP, Lomo T. 1973. Long-lasting potentiation of synaptic transmission in the dentate area of the anaesthetized rabbit following stimulation of the perforant path. J Physiol 232: 331-356.

Bodnoff SR, Humphreys AG, Lehman JC, Diamond DM, Rose GM, et al. 1995. Enduring effects of chronic corticosterone treatment on spatial learning, synaptic plasticity, and hippocampal neuropathology in young and mid-aged rats. J Neurosci 15: 61-69.

Bremner JD, Randall P, Scott TM, Bronen RA, Seibyl JP, et al. 1995. MRI-based measurement of hippocampal volume in patients with combat-related posttraumatic stress disorder. Am J Psychiatry 152: 973-981.

Bremner, JD, Randall P, Vermetten E, Staib L, Bronen RA, et al. 1997. Magnetic resonance imaging-based measurement of hippocampal volume in posttraumatic stress disorder related to childhood physical and sexual abuse–a preliminary report. Biol Psychiatry 41: 23-32.

Brindley D, Rolland Y. 1989. Possible connections between stress, diabetes, obesity, hypertension and altered lipoprotein metabolism that may result in atherosclerosis. Clin Sci 77: 453-461.

Burghardt NS, Sullivan GM, McEwen BS, Gorman JM, Le Doux JE. 2004. The selective reuptake inhibitor citalopram increases fear and acute treatment but reduces fear with chronic treatment: a comparison with tianeptine. Biol Psychiatry 55: 1171-1178.

Cahill L, Prins B, Weber M, McGaugh JL. 1994. Beta-adrenergic activation and memory for emotional events. Nature 371: 702-704.

Carro E, Nunez A, Busiguina S, Aleman I. 2000. Circulating insulin-like growth factor I mediates effects of exercise on the brain. J Neurosci 20: 2926-2933.

Conrad CD, Galea LAM, Kuroda Y, McEwen BS. 1996. Chronic stress impairs rat spatial memory on the Y-maze and this effect is blocked by tianeptine pre-treatment. Behav Neurosci 110: 1321-1334.

Conrad CD, Lupien SJ, McEwen BS. 1999a. Support for a bimodal role for type II adrenal steroid receptors in spatial memory. Neurobiol LearnMem 72: 39-46.

Conrad CD, Magarinos AM, LeDoux JE, McEwen BS. 1999b. Repeated restraint stress facilitates fear conditioning independently of causing hippocampal CA3 dendritic atrophy. Behav Neurosci 113: 902-913.

Convit A, Wolf OT, Tarshish C, Leon MJ. 2003. Reduced glucose tolerance is associated with poor memory performance and hippocampal atrophy among normal elderly. Proc Natl Acad Sci USA 100: 2019-2022.

Corodimas KP, Le Doux JE, Gold PW, Schulkin J. 1994. Corticosterone potentiation of learned fear. Annals of the New York Acad Sci 746: 392

Craft S, Newcomer JW, Kanne S, Jack S, Cryer P, et al. 1995. Memory improvement following induced hyperinsulinemia in Alzheimer's disease. Neurobiol Aging 17: 123-130.

Dallman MF. 2003. Chronic stress and obesity: a new view of 'comfort food'. Proc Natl Acad Sci USA 100: 11696-11701.

Davis M. 1992. The role of the amygdala in fear and anxiety. Annu Rev Neurosci 15: 353-375.

Davis M. 2000. The role of the amygdala in conditioned and unconditioned fear and anxiety. The amygdala: a functional analysis. Anonymous: Oxford University Press; pp. 213-287.

Davis M, Rainnie D, Cassell M. 1994. Neurotransmission in the rat amygdala related to fear and anxiety. Trends Neurosci 17: 208.

Davis M, Shi C. 1999. The extended amygdala: are the central nucleus of the amygdala and the bed nucleus of the stria terminalis differentially involved in fear versus anxiety? Ann NY Acad Sci 877: 281-291.

Dhabhar F, McEwen B. 1999. Enhancing versus suppressive effects of stress hormones on skin immune function. Proc Natl Acad Sci USA 96: 1059-1064.

Diamond DM, Bennett MC, Fleshner M, Rose GM. 1992. Inverted-U relationship between the level of peripheral corticosterone and the magnitude of hippocampal primed burst potentiation. Hippocampus 2: 421-430.

Diamond DM, Fleshner M, Rose GM. 1994. Psychological stress repeatedly blocks hippocampal primed burst potentiation in behaving rats. Behav Brain Res 62: 1-9.

Diamond DM, Park CR, Heman KL, Rose GM. 1999. Exposing rats to a predator impairs spatial working memory in the radial arm water maze. Hippocampus 9: 542-552.

Diamond DM, Park CR, Woodson JC. 2004. Stress generates emotional memories and retrograde amnesia by inducing an endogenous form of hippocampal LTP. Hippocampus 14: 281-291.

Diorio D, Viau V, Meaney MJ. 1993. The role of the medial prefrontal cortex (cingulate gyrus) in the regulation of hypothalamic-pituitary-adrenal responses to stress. J Neurosci 13: 3839-3847.

Drevets WC. 2000. Neuroimaging studies of mood disorders. Biol Psychiatry 48: 813-829.

Dunn J, Orr S. 1984. Differential plasma corticosterone responses to hippocampal stimulation. Exp Brain Res 54: 1-6.

Eichenbaum H, Otto T. 1992. The hippocampus - what does it do? Behav Neural Biol 57: 2-36.

Evans RM, Arriza, JL. 1989, A molecular framework for the actions of glucocorticoid hormones in the nervous system. Neuron 2: 1105-1112.

Frodl T, Meisenzahl E, Zetzsche T, Bottlender R, Born C, et al. 2002. Enlargement of the amygdala in patients with a first episode of major depression. Biol Psychiatry 51: 708-714.

Frodl T, Meisenzahl EM, Zetzsche T, Born C, Jager M, et al. 2003. Larger amygdala volumes in first depressive episode as compared to recurrent major depression and healthy control subjects. Biol Psychiatry 53: 338-344.

Funder J, Pearce P, Smith R, Smith A. 1988. Mineralocorticoid action: target tissue specificity is enzyme, not receptor, mediated. Science 242: 583-585.

Herman J, Schafer M, Young E, Thompson R, Douglass J, et al. 1989. Evidence for hippocampal regulation of neuroendocrine neurons of hypothalamo-pituitary-adrenocortical axis. J Neurosci 9: 3072-3082.

Herman JP, Cullinan, WE. 1997. Neurocircuitry of stress: central control of the hypothalamo-pituitary-adrenocortical axis. Trends Neurosci 20: 78-84.

Herman JP, Patel PD, Akil H, Watson SJ. 1989. Localization and regulation of glucocorticoid and mineralocorticoid receptor messenger RNAs in the hippocampal formation of the rat. Mol Endocrinol 3: 1886-1894.

Herman JP, Prewitt CMF, Cullinan WE. 1996. Neuronal circuit regulation of the hypothalamo-pituitary-adrenocortical stress axis. Crit RevNeurobiol 10: 371-394.

Hesen W, Joels, M. 1996a. Carbachol responsiveness of rat CA1 hippocampal neurons in vitro: modulation by corticosterone and stress. Stress 1: 63-72.

Hesen W, Joels M. 1996b. Modulation of $5HT_{1A}$ responsiveness in CA1 pyramidal neurons by *in vivo* activation of corticosteroid receptors. J Neuroendocrinol 8: 433-438.

Ikegaya Y, Saito H, Abe K. 1994. Attenuated hippocampal long-term potentiation in basolateral amygdala-lesioned rats. Brain Res 656: 157-164.

Ikegaya Y, Saito H, Abe K. 1995. High-frequency stimulation of the basolateral amygdala facilitates the induction of long-term potentiation in the dentate gyrus in vivo. Neurosci Res 22: 203-207.

Jacobson L, Akana SF, Cascio CS, Shinsako J, Dallman MF. 1988. Circadian Variations in Plasma Corticosterone Permit Normal Termination of Adrenocorticotropin Responses to Stress. Endocrinology 122: 1343-1348.

Jacobson L, Sapolsky R. 1991. The role of the hippocampus in feedback regulation of the hypothalamic-pituitary-adrenocortical axis. Endocr Rev 12: 118-134.

Joels M. 1997. Steroid hormones and excitability in the mammalian brain. Front Neuroendocrinol 18: 2-48.

Joels M, Bosma A, Hendriksen H, Diegenbach P, Kamphuis W. 1996. Corticosteroid actions on the expression of kainate receptor subunit mRNAs in rat hippocampus. Mol Brain Res 37: 15-20.

Joels M, De Kloet ER. 1994. Mineralocorticoid and glucocorticoid receptors in the brain. Implications for ion permeability and transmitter systems. Progr Neurobiol 43: 1-36.

Joels M, Vreugdenhil E. 1998. Corticosteroids in the brain. Mol Neurobiol 17: 87-198.

Johnston D, Williams S, Jaffe D, Gray R. 1992. NMDA-receptor-independent long-term potentiation. Ann Rev Physiol 54: 489-505.

Keller-Wood M, Dallman M. 1984. Corticosteroid inhibition of ACTH secretion. Endocr Rev 5: 1-24.

Kerr D, Campbell L. 1992. Hippocampal glucocorticoid receptor activation enhances voltage-dependent ca2+ conductances: relevance to brain aging. Proc Natl Acad Sci USA 89: 8527-8531.

Kim JJ, Diamond DM. 2002. The stressed hippocampus, synaptic plasticity and lost memories. Nat Rev Neurosci 3: 453-462.

Kim JJ, Lee HJ, Han J-S, Packard MG. 2001. Amygdala is critical for stress-induced modulation of hippocampal long-term potentiation and learning. J Neurosci 21: 5222-5228.

Kim JJ, Yoon KS. 1998. Stress: metaplastic effects in the hippocampus. TINS 21: 505-509.

Knigge K. 1961. Adrenocortical response to stress in rats with lesions in hippocampus and amygdala. Proc Soc Exp Biol Med 180: 18-20.

Kole MHP, Costoli T, Koolhaas JM, Fuchs E. 2004a. Bidirectional shift in the cornu ammons 3 pyramidal dendritic organization following brief stress. Neuroscience 125: 337-347.

Kole MHP, Czeh B, Fuchs E. 2004b. Homeostatic maintenance in excitability of tree shrew hippocampal CA3 pyramidal neurons after chronic stress. Hippocampus 14: 742-751.

Kole MHP, Swan L, Fuchs E. 2002. The antidepressant tianeptine persistently modulates glutamate receptor currents of the hippocampal CA3 commissural-associational synapse in chronically stressed rats. Eur J Neurosci 16: 807-816.

Korte M, De Boer SF. 2003. A robust animal model of state anxiety: fear-potentiated behaviour in the elevated plus-maze. Eur J Pharmacol 463: 163-175.

Krichmar JL, Nasuto SJ, Scorcioni R, Washington SD, Ascoli GA. 2002. Effects of dendritic morphology on CA3 pyramidal cell electrophysiology. Brain Res 941: 11-28.

Le Doux JE. 1994. The amygdala: contributions to fear and stress. Sem Neurosci 6: 231-237.

Leibowitz SF, Hoebel BG. 1997. Behavioral neuroscience of obesity. Handbook of obesity. Bray GA, Bouchard C, James WPT, editors. New York: Marcel Dekker; pp. 313-358.

Liang KC, Hon W, Davis M. 1994. Pre-and posttraining infusion of N-methyl-D-aspartate receptor antagonists into the amygdala impair memory in an inhibitory avoidance task. Behav Neurosci 108: 241-253.

Luine V, Villegas M, Martinez C, McEwen BS. 1994. Repeated stress causes reversible impairments of spatial memory performance. Brain Res 639: 167-170.

Lupien SJ, McEwen BS. 1997. The acute effects of corticosteroids on cognition: integration of animal and human model studies. Brain Res Rev 24: 1-27.

MacQueen GM, Campbell S, McEwen BS, Macdonald K, Amano S, et al. 2003. Course of illness, hippocampal function, and hippocampal volume in major depression. Proc Natl Acad Sci USA 100: 1387-1392.

Magarinos AM, McEwen BS. 1995a. Stress-induced atrophy of apical dendrites of hippocampal CA3c neurons: comparison of stressors. Neuroscience 69: 83-88.

Magarinos AM, McEwen BS. 1995b. Stress-induced atrophy of apical dendrites of hippocampal CA3c neurons: involvement of glucocorticoid secretion and excitatory amino acid receptors. Neuroscience 69: 89-98.

Mainen Z, Sejnowski T. 1996. Influence of dendritic structure on firing pattern in model neocortical neurons. Nature 382: 363-366.

Makara GB, Haller J. 2001. Non-genomic effects of glucocorticoids in the neural system. Evidence, mechanisms and implications. Prog Neurobiol 65: 367-390.

Manuck SB, Kaplan JR, Adams MR, Clarkson TB. 1995. Studies of psychosocial influences on coronary artery atherosclerosis in cynomolgus monkeys. Health Psychol 7: 113-124.

Manuck SB, Kaplan JR, Muldoon MF, Adams MR, Clarkson TB. 1991. The behavioral exacerbation of atherosclerosis and its inhibition by propranolol. Stress, coping and disease. McCabe PM, Schneiderman N, Field TM, Skyler JS, editors. Hove and London: Lawrence Erlbaum Associates; pp. 51-72.

Martin SJ, Grimwood PD, Morris RGM. 2000. Synaptic plasticity and memory: an evaluation of the hypothesis. Annu Rev Neurosci 23: 649-711.

Masuzaki H, Paterson J, Shinyama H, Morton NM, Mullins JJ, et al. 2001. A transgenic model of visceral obesity and the metabolic syndrome. Science 294: 2166-2170.

Matys T, Pawlak R, Matys E, Pavlides C, McEwen BS, et al. 2004. Tissue plasminogen activator promotes the effects of corticotropin releasing factor on the amygdala and anxiety-like behavior. Proc Natl Acad Sci USA 101: 16345-16350.

McDonald AJ. 1982. Neurons of the lateral and basolateral amygdaloid nuclei: a Golgi study in the rat. J Comp Neurol 212: 293-312.

McDonald AJ. 1992. Cell types and intrinsic connections of the amygdala. The amygdala: neurobiological aspects of emotion, memory, and mental dysfunction. Aggleton JP, editor. Wiley-Liss Inc NY; pp. 67-96.

McEwen BS. 1977. Adrenal steroid feedback on neuroendocrine tissues. Ann NY Acad Sci 297: 568-579.

McEwen BS. 1997. Possible mechanisms for atrophy of the human hippocampus. Mol Psychiatry 2: 255-262.

McEwen BS. 1998. Protective and damaging effects of stress mediators. New England J Med 338: 171-179.

McEwen BS. 1999. Stress and hippocampal plasticity. Annu Rev Neurosci 22: 105-122.

McEwen BS, Chattarji S. 2004. Molecular mechanisms of neuroplasticity and pharmacological implications: the example of tianeptine. Eur Neuropsychopharmacol 14: 5497-5502.

McEwen BS, Reagan LP. 2004. Glucose transporter expression in the central nervous system: relationship to synaptic function. Eur J Pharmacol 490: 13-24.

McGaugh JL, Roozendaal B. 2002. Role of adrenal stress hormones in forming lasting memories in the brain. Curr Opin Neurobiol 12: 205-210.

Miner JN, Yamamoto KR. 1991. Regulatory crosstalk at composite response elements. Trends Biochem Sci 16: 423-426.

Narayanan R, Narayan A, Chattarji S. 2004. A probabilistic framework for region-specific remodeling of dendrites in three-dimensional neuronal reconstructions. Neural Comput 17: 1-22.

O'Keefe J, Nadel L. 1978. The hippocampus as a cognitive map. Oxford: Oxford University Press.

Oitzl MS, Reichardt HM, Joels M, de Kloet ER. 2001. Point mutation in the mouse glucocorticoid receptor preventing DNA binding impairs spatial memory. Proc Natl Acad Sci USA 98: 12790-12795.

Orchinik M, Carroll SS, Li Y-H, McEwen BS, Weiland NG. 2001. Heterogeneity of hippocampal $GABA_A$ receptors: regulation by corticosterone. J Neurosci 21: 330-339.

Pavlides C, Kimura A, Magarinos AM, McEwen BS. 1994. Type I adrenal steroid receptors prolong hippocampal long-term potentiation. Neuroreport 5: 2673-2677.

Pavlides C, Kimura A, Magarinos AM, McEwen BS. 1995. Hippocampal homosynaptic long-term depression/depotentiation induced by adrenal steroids. Neuroscience 68: 379-385.

Pavlides C, McEwen BS. 1999. Effects of mineralocorticoid and glucocorticoid receptors on long-term potentiation in the CA3 hippocampal field. Brain Res 851: 204-214.

Pavlides C, Nivon LG, McEwen BS. 2002. Effects of chronic stress on hippocampal long-term potentiation. Hippocampus 12: 245-257.

Pavlides C, Ogawa S, Kimura A, McEwen B. 1996. Role of adrenal steroid mineralocorticoid and glucocorticoid receptors in long-term potentiation in the CA1 field of hippocampal slices. Brain Res 738: 229-235.

Pavlides C, Watanabe Y, Magarinos AM, McEwen BS. 1995. Opposing role of adrenal steroid Type I and Type II receptors in hippocampal long-term potentiation. Neuroscience 68: 387-394.

Pavlides C, Watanabe Y, McEwen BS. 1993. Effects of glucocorticoids on hippocampal long-term potentiation. Hippocampus 3: 183-192.

Pawlak R, Magarinos AM, Melchor J, McEwen B, Strickland S. 2003. Tissue plasminogen activator in the amygdala is critical for stress-induced anxiety-like behavior. Nature Neurosci 6: 168-174.

Pecoraro N, Reyes F, Gomez F, Bhargava A, Dallman MF. 2004. Chronic stress promotes palatable feeding, which reduces signs of stress: feedforward and feedback effects of chronic stress. Endocrinology 145: 3754-3762.

Petrovich GD, Canteras NS, Swanson LW. 2001. Combinatorial amygdalar inputs to hippocampal domains and hypothalamic behavior systems. Brain Res Rev 38: 247-289.

Phillips RG, Le Doux JE. 1992. Differential contribution of amygdala and hippocampus to cued and contextual fear conditioning. Behav Neurosci 106: 274-285.

Pitkanen A, Amaral DG. 1994. The distribution of GABAergic cells, fibers, and terminals in the monkey amygdaloid complex: an immunohistochemical and in situ hybridization study. J Neurosci 14: 2200-2224.

Pitkanen A, Pikkarainen M, Nurminen N, Ylinen A. 2000. Reciprocal connections between the amygdala and the hippocampal formation, perirhinal cortex, and postrhinal cortex in rat. Ann NY Acad Sci 911: 369-391.

Pompei P, Riftina F, McEwen BS. 1995. Effect of adrenal steroids on preproneurokinin-A gene expression in discrete regions of the rat brain. Mol Brain Res 33: 209-216.

Popov VI, Bocharova LS. 1992. Hibernation-induced structural changes in synaptic contacts between mossy fibres and hippocampal pyramidal neurons. Neuroscience 48: 53-62.

Popov VI, Bocharova LS, Bragin AG. 1992. Repeated changes of dendritic morphology in the hippocampus of ground squirrels in the course of hibernation. Neuroscience 48: 45-51.

Pugh CR, Fleshner M, Rudy JW. 1997a. Type II glucocorticoid receptor antagonists impair contextual but not auditory-cue fear conditioning in juvenile rats. Neurobiol Learn Mem 67: 75-79.

Pugh CR, Tremblay D, Fleshner M, Rudy JW. 1997b. A selective role for corticosterone in contextual-fear conditioning. Behav Neurosci 111: 503-511.

Radley JJ, Rocher AB, Miller M, Janssen WGM, Liston C, Hof PR, et al. 2006. Chronic behavioral stress induces apical dendritic spine loss in pyramidal neurons of the rat medial prefrontal cortex. Repeated stress induces dendritic spine loss in the rat medial prefrontal cortex. Cereb Cortex 16: 313-320.

Radley JJ, Sisti HM, Hao J, Rocher AB, McCall T, et al. 2004. Chronic behavioral stress induces apical dendritic reorganization in pyramidal neurons of the medial prefrontal cortex. Neuroscience 125: 1-6.

Rainnie DG, Bergeron R, Sajdyk TJ, Patil M, Gehlert DR, Shekhar A. 2004. Corticotrophin releasing factor-induced synaptic plasticity in the amygdala translates stress into emotional disorders. J Neurosci 24: 3471-3479.

Reichardt HM, Schutz G. 1998. Glucocorticoid signalling—multiple variations of a common theme. Mol Cell Endocrinol 146: 1-6.

Roozendaal B. 2000. Glucocorticoids and the regulation of memory consolidation. Psychoneuroendocrinology 25: 213-238.

Sandeep TC, Yau JLW, MacLullich AMJ, Noble J, Deary IJ, et al. 2004. 11β-Hydroxysteroid dehydrogenase inhibition improves cognitive function in healthy elderly men and type 2 diabetics. Proc Natl Acad Sci USA 101: 6734-6739.

Santini E, Ge H, Ren K, Pena de Ortiz S, Quirk GJ. 2004. Consolidation of fear extinction requires protein synthesis in the medial prefrontal cortex. J Neurosci 24: 5704-5710.

Sapolsky R. 1992. Stress, the aging brain and the mechanisms of neuron death. Cambridge: MIT Press; pp. 1-423.

Sapolsky RM, Krey LC, McEwen BS. 1986. The neuroendocrinology of stress and aging: the glucocorticoid cascade hypothesis. Endocr Rev 7: 284-301.

Sapolsky RM, Zola-Morgan S, Squire LR. 1991. Inhibition of glucocorticoid secretion by the hippocampal formation in the primate. J Neurosci 11: 3695-3704.

Seckl JR, Walker BR. 2001. Minireview: 11β -Hydroxysteroid dehydrogenase type 1—a tissue-specific amplifier of glucocorticoid action. Endocrinology 142: 1371-1376.

Sheline YI, Sanghavi M, Mintun MA, Gado MH. 1999. Depression duration but not age predicts hippocampal volume loss in medically healthy women with recurrent major depression. J Neurosci 19: 5034-5043.

Sheline YI, Wang PW, Gado MH, Csernansky JC, Vannier MW. 1996. Hippocampal atrophy in recurrent major depression. Proc Natl Acad Sci USA 93: 3908-3913.

Shors TJ, Mathew PR. 1998. NMDA receptor antagonism in the lateral–basolateral but not central nucleus of the amygdala prevents the induction of facilitated learning in response to stress. Learn Mem 5: 220-230.

Shors TJ, Weiss C, Thompson RF. 1992. Stress-induced facilitation of classical conditioning. Science 257: 537-539.

Sousa N, Lukoyanov NV, Madeira MD, Almeida OFX, Paula- Barbosa MM. 2000. Reorganization of the morphology of hippocampal neurites and synapses after stress-induced damage correlates with behavioral improvement. Neuroscience 97: 253-266.

Squire LR, Zola-Morgan S. 1991. The medial temporal lob memory system. Science 253: 1380-1386.

Starkman MN, Gebarski SS, Berent S, Schteingart DE. 1992. Hippocampal formation volume, memory dysfunction, and cortisol levels in partiens with Cushing's syndrome. Biol Psychiatry 32: 756-765.

Starkman MN, Giordani B, Gebarski SS, Schteingart DE. 2003. Improvement in learning associated with increase in hippocampal formation volume. Biol Psychiat 53: 233-238.

Starkman MN, Giordani B, Gebrski SS, Berent S, Schork MA, et al. 1999. Decrease in cortisol reverses human hippocampal atrophy following treatment of Cushing's disease. Biol Psychiat 46: 1595-1602.

Sterling P, Eyer J. 1988. Allostasis: a new paradigm to explain arousal pathology. Handbook of life stress, cognition and health. Fisher S, Reason J, editors. New York: John Wiley & Sons; pp. 629-649.

Stutzmann GE, McEwen BS, Le Doux JE. 1998. Serotonin modulation of sensory inputs to the lateral amygdala: dependency on corticosterone. J Neuroscience 18: 9529-9538.

Sullivan RM, Gratton A. 1999. Lateralized effects of medial prefrontal cortex lesions on neuroendocrine and autonomic stress responses in rats. J Neurosci 19: 2834-2840.

Trapp T, Rupprecht R, Castren M, Reul JMHM, Holsboer F. 1994. Heterodimerization between mineralocorticoid and glucocorticoid receptor: a new principle of glucocorticoid action in the CNS. Neuron 13: 1457-1462.

van Ooyen A, Duijnhouwer J, Remme MW, van Pelt J. 2002. The effect of dendritic topology on ring patterns in model neurons. Network: Computation in Neural Systems 13: 311-325.

van Steensel B. 1995. Steroid receptors and nuclear structure. PhD Thesis. Amsterdam: University of Amsterdam; pp. 1-94.

Vetter M. 2001. A turn of the helix: preventing the glial fate. Neuron 29: 559-562.

Vyas A, Bernal S, Chattarji S. 2003a. Effects of chronic stress on dendritic arborization in the central and extended amygdala. Brain Res 965: 290-294.

Vyas A, Chattarji S. 2004. Modulation of different states of anxiety-like behavior by chronic stress. Behav Neurosci 118: 1450-1454.

Vyas A, Mitra R, Chattarji S. 2003b. Enhanced anxiety and hypertrophy in basolateral amygdala neurons following chronic stress in rats. Ann NY Acad Sci 985: 554-555.

Vyas A, Mitra R, Rao BSS, Chattarji S. 2002. Chronic stress induces contrasting patterns of dendritic remodeling in hippocampal and amygdaloid neurons. J Neurosci 22: 6810-6818.

Vyas A, Pillai AG, Chattarji S. 2004. Recovery after chronic stress fails to reverse amygdaloid neuronal hypertrophy and enhanced anxiety-like behavior. Neuroscience 128: 667-673.

Watanabe Y, Gould E, McEwen BS. 1992. Stress induces atrophy of apical dendrites of hippocampus CA3 pyramidal neurons. Brain Res 588: 341-344.

Weiland NG, Orchinik M, McEwen BS. 1995. Corticosterone regulates mRNA levels of specific subunits of the NMDA receptor in the hippocampus but not in cortex of rats. Abstracts, Soc Neurosci 21: #207.12-pp. 502.

Wellman CL. 2001. Dendritic reorganization in pyramidal neurons in medial prefrontal cortex after chronic corticosterone administration. J Neurobiol 49: 245-253.

Wellman CL. 2004. Chronic stress alters dendritic morphology in rat medial prefrontal cortex. J Neurobiol 60: 236-248.

Yau JLW, Noble J, Kenyon CJ, Hibberd C, Kotelevtsev Y, et al. 2001. Lack of tissue glucocorticoid reactivation in 11β-hydroxysteroid dehydrogenase type 1 knockout mice ameliorates age-related learning impairments. Proc Natl Acad Sci USA 98: 4716-4721.

15 Maternal Programming of Glucocorticoid Receptor Expression and HPA Responses to Stress Through DNA Methylation in the Rat

M. J. Meaney · I. C. G. Weaver · T. Wu · I. Hellstrom · J. Diorio · Moshe Szyf

Abstract: Increased levels of pup licking/grooming and arched-back nursing by rat mothers over the first week of life alter the epigenome at a glucocorticoid receptor gene promoter in the hippocampus of the offspring. Differences in the DNA methylation pattern between the offspring of high- and low-licking/grooming—arched-back mothers emerge over the first week of life. They are reversed with cross-fostering, persist into adulthood, and are associated with altered histone acetylation and transcription factor (NGFI-A) binding to the glucocorticoid receptor promoter. Central infusion of the adult offspring with the histone deacetylase (HDAC) inhibitor trichostatin A removes the previously defined epigenomic group differences in histone acetylation, DNA methylation, NGFI-A binding, glucocorticoid receptor expression, and hypothalamic–pituitary–adrenal responses to stress. This suggests a causal relation between the epigenomic state, glucocorticoid receptor expression, and the effects of maternal care on stress responses in the offspring. These findings demonstrate that an epigenomic state of a gene can be established through a behavioral mode of programming, and that in spite of the inherent stability of this epigenomic mark, it is dynamic and potentially reversible.

1 Introduction

Epidemiological studies reveal the importance of family function and early life events as predictors of health in adulthood (Repetti et al., 2002). Such studies show that the quality of family life influences the development of individual differences in vulnerability to illness throughout life. As adults, victims of childhood physical or sexual abuse are at considerably greater risk for mental illness, as well as for obesity, diabetes, and heart disease (e.g., Bifulco et al., 1991; Brown and Anderson, 1991; McCauley et al., 1991; Felitti et al., 1998; Heim et al., 2002; Newport et al., 2002). Persistent emotional neglect, family conflict, and conditions of harsh, inconsistent discipline all serve to compromise growth (e.g., Montgomery et al., 1997), intellectual development (Ammerman et al., 1986; Trickett and McBride-Chang, 1995), and to increase the risk for adult obesity (Lissau and Sorensen, 1994), depression, and anxiety disorders (Holmes and Robins, 1987, 1988; Gottman, 1998). Cold, distant parent–child relationships are associated with a significantly increased risk of chronic illness in later life (e.g., Parker, 1981; Canetti et al., 1997; Russak and Schwartz, 1997).

"Stress diathesis" models have emerged as explanations for the relationship between the quality of early life and health in adulthood. These models suggest that adversity in early life alters the development of neural and endocrine systems in a manner that predisposes individuals to disease in adulthood. The relation between the quality of the early environment and health in adulthood appears to be mediated by parental influences on the development of neural systems that underlie the expression of behavioral and endocrine responses to stress (Seckl and Meaney, 1993; Nemeroff, 1996; Sroufe, 1997; Francis and Meaney, 1999; Heim et al., 2000; Repetti et al., 2002). Adversity or decreased quality of parental investment increases the magnitude of emotional, autonomic, and hypothalamic–pituitary–adrenal (HPA) responses to stress in adulthood. These models are constructed on two principal assumptions. First, that prolonged activation of neural and hormonal responses to stress can promote illness, and second that early environmental events influence the development of these responses. There is strong evidence in favor of both ideas. In humans, forms of parenting that enhance the risk of chronic illness in the offspring increase endocrine and autonomic responses to stress in adulthood (De Bellis et al., 1994; Heim et al., 2000). There is considerable evidence for comparable effects in primates (Higley et al., 1991; Suomi, 1997; Bennett et al., 2002) and rodents (Meaney, 2001). Moreover, prolonged exposure to elevated levels of stress hormones, including corticotropin-releasing factor (CRF), catecholamines, most notably norepinepherine, and glucocorticoids promote the development of a diverse range of high-risk conditions, such as visceral obesity, hypertension, and insulin intolerance, or overt pathology, including diabetes, depression, drug addiction, and multiple forms of coronary heart disease. The clinical risks associated with prolonged activation of the HPA and autonomic systems are a logical consequence of the otherwise adaptive stress response. In response to neural signals associated with the stressor, there is an increased release of glucocorticoids from the adrenal gland and catecholamines, particularly norepinephrine from the sympathetic system. The combined actions of these hormones increase the availability of energy substrates, such as those derived from lipids and glucose metabolism, to maintain the normal cellular output and organ efficiency. These actions protect against catastrophes such as hypotensive shock. These hormones, along with the central CRF and catecholamines, act on multiple brain regions to increase vigilance and fear, and enhance avoidance learning and fear conditioning, which reduces the chances of further encounters with the offending conditions.

Support for the basic elements stress diathesis models appears compelling. Adversity during perinatal life alters development in a manner that seems likely to promote vulnerability, especially for stress-related diseases. Diathesis describes the interaction between development, including the potential influence of genetic factors, and the prevailing level of stress in predicting health outcomes. Such models have considerable appeal, and could potentially identify both the origins and the nature of vulnerability derived from either epigenetic influences, such as early family life, or genomic variations (e.g., Bennet et al., 2002; Caspi et al., 2003).

2 The Development of Individual Differences in Stress Responses

2.1 Handling Studies

In the late 1950s and early 1960s, the pages of *Science* and *Nature* were frequently dedicated to articles reporting the effects of postnatal handling (also known as infantile stimulation) on the development of responses to adverse stimuli, or stressors (Levine, 1957, 1962). The handling paradigm involves a brief (i.e., ~15 min) separation of the pups from the dam. This period falls well within the range of normal mother–pup separations that fall between nursing bouts, and does not constitute any major deprivation of

Figure 15-1
The nexus of the hypothalamic–pituitary–adrenal axis are the corticotropin-releasing factor (CRF) neurons of the paraventricular nucleus of the hypothalamus. CRF is released into the portal system of the anterior pituitary stimulating the synthesis and release of adrenocorticotropin (ACTH), which then stimulates adrenal glucocorticoid release. Glucocorticoids act on mineralocorticoid and glucocorticoid receptors in multiple brain regions, including the hippocampus, to inhibit the synthesis and release of CRF (i.e., glucocroticoid negative feedback)

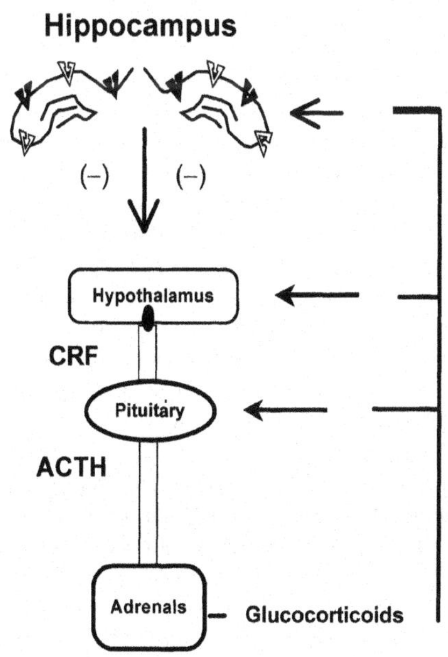

parental care. In infant rats and mice, handling during infancy decreases the magnitude of both behavioral and HPA responses to stress in adulthood. These findings demonstrate that the early environment influences the development of even rudimentary defensive responses to threat.

Levine and others suggested that the effects of handling are actually mediated by changes in maternal care. Indeed, handling increases the licking/grooming of pups by the mother (Lee and Williams, 1975; Liu et al., 1997). Subsequent studies support the maternal-mediation hypothesis. One approach was to examine the consequences of naturally occurring variations in maternal licking/grooming. These studies indicate that the adult offspring of high-licking/grooming (LG) mothers resembled postnatally handled animals on measures of behavioral and endocrine responses to stress, while those of low-LG mothers were comparable to nonhandled animals. Cross-fostering studies, where pups born to high-LG mothers were fostered at birth to low-LG mothers (and vice versa), suggest a direct relationship between maternal care and the postnatal development of individual differences in behavioral and HPA responses to stress (Francis et al., 1999; Caldji et al., 2003). Finally, these studies suggest that variations within a normal range of parental care can dramatically alter development. As in humans, parental care need not include forms of overt abuse or extreme neglect to influence the development of the offspring. In large measure, this is likely because natural selection has shaped offspring to respond to subtle variations in parental behaviors as a forecast of the environmental conditions they will ultimately face following independence from the parent (Hinde, 1986).

2.2 Maternal Care in the Rat: Behavioral and HPA Responses to Stress

Central CRF systems furnish the critical signal for the activation of behavioral, emotional, autonomic, and endocrine responses to stressors. There are two major CRF pathways regulating the expression of these stress responses. First, a CRF pathway from the parvocellular regions of the paraventricular nucleus of the hypothalamus to the hypophysial-portal system of the anterior pituitary, which serves as the principal mechanism for the transduction of a neural signal into a pituitary–adrenal response (Rivier and Plotsky, 1986; Plotsky, 1991; Antoni, 1993; Whitnall, 1993). In responses to stressors, CRF, as well as cosecretagogues, such as arginine and vasopressin, are released from hypothalamic neurons into the portal blood supply of the anterior pituitary where they stimulate the synthesis and release of adrenocorticotropin hormone (ACTH). Pituitary ACTH, in turn, causes the release of glucocorticoids from the adrenal gland. CRF synthesis and release are subsequently inhibited through a glucocorticoid negative-feedback system mediated by both mineralocorticoid and glucocorticoid receptors in a number of brain regions including and perhaps especially in the hippocampus (De Kloet et al., 1998).

CRF neurons in the central nucleus of the amygdala project directly to the locus coeruleus and increase the firing rate of locus coeruleus neurons, resulting in increased noradrenaline release in the vast terminal fields of this ascending noradrenergic system. Thus, i.c.v infusion of CRF increases extracellular noradrenaline levels (Emoto et al., 1993; Lavicky and Dunn, 1993; Page and Valentino, 1994; Valentino et al., 1998). The amygdaloid CRF projection to the locus coeruleus (Moga and Gray, 1989; Koegler-Muly et al., 1993; Gray and Bingaman, 1996; Van Bockstaele et al., 1996; Valentino et al., 1998) is also critical for the expression of behavioral responses to stress (Butler et al., 1990; Liang et al., 1992; Swiergiel et al., 1993; Koob et al., 1994; Schulkin et al., 1994; Stenzel-Poore et al., 1994; Bakshi et al., 2000; Davis and Whalen, 2001). Hence, the CRF neurons in the hypothalamus and the central nucleus of the amygdala serve as important mediators of both behavioral and endocrine responses to stress.

We examine the relation between maternal care and the development of behavioral and endocrine responses to stress using a rather simple model of naturally occurring variations in maternal behavior over the first 8 days after birth (Champagne et al., 2003). We characterize individual differences in maternal behavior through direct observation of mother–pup interactions in normally reared animals. These observations reveal considerable variation in two forms of maternal behavior—LG of pups and arched-back nursing (ABN). Licking/grooming includes both body as well as anogenital licking. ABN, also referred to as "crouching," is characterized by a dam nursing her pups with her back conspicuously arched and legs splayed outward (Stern, 1997). Although common, it is not the only posture from which dams nurse. A blanket posture represents a more relaxed version of the arched-back position where the mother is almost lying on the suckling pups. As can be imagined, it provides substantially less opportunity for movement by the pups such as nipple switching. Dams also nurse from their sides and often move from one posture to another over the course of a nursing bout. Interestingly, the frequency of LG and ABN is correlated across animals and thus we are able to define mothers according to both behaviors; high- or low-LG-ABN mothers. For the sake of most of the studies described here, high- and low-LG-ABN mothers are females whose scores on both measures were ±1 SD above (high) or below (low) the mean for their cohort. Importantly, high- and low-LG-ABN mothers do not differ in the amount of contact time with pups. Differences in the frequency of LG or ABN do not occur simply as a function of time in contact with pups. High and low-LG-ABN mothers raise a comparable number of pups to weaning, and there are no differences in the weaning weights of the pups, which suggest an adequate level of maternal care across the groups. These findings also suggest that we are examining the consequences of variations in maternal care that occur within a normal range. Indeed, the frequency of both pup LG and ABN is normally distributed across large populations of lactating female rats (Champagne et al., 2003).

The critical question concerns the potential consequences of these differences in maternal behavior for the development of behavioral and neuroendocrine responses to stress. By comparison with the adult offspring of low-LG-ABN mothers, as adults, the offspring of high-LG-ABN mothers show reduced plasma ACTH and corticosterone responses to acute stress (Liu et al., 1997; Weaver et al., 2004). Circulating glucocorticoids act at glucocorticoid and mineralocorticoid receptor sites in corticolimbic structures, such as the hippocampus, to regulate HPA activity. Such feedback effects commonly target CRF synthesis and release at the level of the hypothalamus. The high-LG-ABN offspring showed significantly increased hippocampal glucocorticoid receptor mRNA expression, enhanced glucocorticoid negative feedback sensitivity, and decreased hypothalamic CRF mRNA levels. Moreover, (Liu et al., 1997) found that the magnitude of the corticosterone response to acute stress was significantly correlated with the frequency of both maternal LG ($r = 61$) and ABN ($r = 0.64$) during the first week of life, as was the level of hippocampal glucocorticoid receptor mRNA and hypothalamic CRF mRNA expression (all $r > 0.70$).

The offspring of the high- and low-LG-ABN mothers also differ in behavioral responses to stress (Caldji et al., 1998a,b; Francis et al., 1999; Menard et al., 2004). As adults, the offspring of the high LG-ABN show decreased startle responses, increased open-field exploration, and shorter latencies to eat food provided in a novel environment. The offspring of low-LG-ABN mothers also show greater burying in the defensive burying paradigm (Menard et al., 2004), which involves an *active* response to a threat. The offspring of the high-LG-ABN mothers exhibit decreased CRF receptor levels in the locus coeruleus, and increased GABAA/benzodiazepine receptor levels in the basolateral and central nucleus of the amygdala, as well as in the locus coeruleus (Caldji et al., 1998a,b, 2003) and decreased CRF mRNA expression in the CnAmy. Benzodiazepine agonists suppress CRF expression in the amygdala (Owens et al., 1991). Predictably, stress-induced increases in hypothalamic levels of noradrenaline that are normally stimulated by CRF were significantly higher in the offspring of the low-LG-ABN offspring.

Maternal care during the first week of life is associated with stable individual differences in GABAA receptor subunit expression in the brain regions that regulate stress reactivity. The adult offspring of high-LG-ABN mothers show significantly higher levels of GABAA/benzodiazepine receptor binding in the basolateral and central nuclei of the amygdala as well as the locus coeruleus. These findings provide a mechanism for increased GABAergic inhibition of amygdala–locus coeruleus activity. Importantly, maternal care also affects the behavioral sensitivity to acute benzodiazepine administration (Fries et al., in press). A series of in situ hybridization studies (Caldji et al., 2003) illustrate the molecular mechanism for these differences in receptor binding and suggest that variations in maternal care might actually permanently alter the subunit composition of the GABAA receptor complex in the offspring. The offspring of the high-LG-ABN mothers show increased levels of the mRNAs for the $\gamma 1$ and $\gamma 2$ subunits, and contribute to the formation of a functional benzodiazepine binding site. Such differences are not unique to the γ subunits. Levels of mRNA for the $\alpha 1$ subunit of the GABAA/benzodiazepine receptor complex are significantly higher in the amygdala and locus coeruleus of high compared with Low-LG-ABN offspring. The $\alpha 1$ subunit appears to confer higher affinity for GABA, providing the most efficient form of the GABAA receptor complex through increased receptor affinity for GABA. The adult offspring of the low-LG-ABN mothers actually show increased expression of the mRNAs for the $\alpha 3$ and $\alpha 4$ subunits in the amygdala and the locus coeruleus. Interestingly, GABAA/CBZ receptors composed of the α3 and α4 subunits show a reduced affinity for GABA compared with the α1 subunit. Moreover, the α4 subunit does not contribute to the formation of a benzodiazepine receptor site. These differences in subunit expression are tissue specific; no such differences are apparent in the hippocampus, hypothalamus, or cortex. Thus, differences in GABAA/benzodiazepine receptor binding are not simply due to a deficit in subunit expression in the offspring of the low-LG-ABN mothers, but of an apparently "active" attempt to maintain a specific GABAA/benzodiazepine receptor profile in selected brain regions.

Together with the findings from earlier handling studies, the results of these studies suggest that the behavior of the mother toward her offspring can "program" behavioral and neuroendocrine responses to stress in adulthood. These effects are associated with sustained changes in the expression of genes in brain regions that mediate responses to stress and form the basis for stable individual differences in stress reactivity. These findings provide a potential mechanism for the influence of parental care on vulnerability/resistance to stress-induced illness over the lifespan.

2.3 Cross-fostering Studies: Evidence for Direct Maternal Effects

Individual differences in behavioral and neuroendocrine responses to stress in the rat are associated with naturally occurring variations in maternal care. Such effects might serve as a possible mechanism by which selected traits might be transmitted from one generation to another. Indeed, low-LG-ABN mothers are more fearful in response to stress than are high-LG-ABN dams (Francis et al., 2000). Individual differences in stress reactivity are apparently transmitted across generations: Fearful mothers beget more stress reactive offspring. The obvious question is whether the transmission of these traits occurs only as a function of genomic-based inheritance. If this is the case, then the differences in maternal behavior may simply be an epiphenomenon, and not causally related to the development of individual differences in stress responses. The issue is not one of inheritance, but of the mode of inheritance.

The results of recent studies provide evidence for a nongenomic transmission of individual differences in stress reactivity and maternal behavior (Francis et al., 1999). One study involved a reciprocal cross-fostering of the offspring of low- and high-LG-ABN mothers. The primary concern here was that the wholesale fostering of litters between mothers is known to affect maternal behavior (Maccari et al., 1995). To avert this problem and maintain the original character of the host litter, no more than 2 of 12 pups were fostered into or from any one litter (McCarty and Lee, 1996). The critical groups of interest are the biological offspring of low-LG-ABN mothers fostered onto high-LG-ABN dams, and vice versa. The limited cross-fostering design did not result in any effect on group differences in maternal behavior. Hence, the frequency of pup licking/grooming and ABN across all groups of high-LG-ABN mothers was significantly higher than that for any of the low-LG-ABN dams regardless of litter composition.

The results of the behavioral studies are consistent with the idea that variations in maternal care are causally related to individual differences in the behavior of the offspring. The biological offspring of low-LG-ABN dams reared by high-LG-ABN mothers were significantly less fearful under conditions of novelty than were the offspring reared by low LG-ABN mothers, including the biological offspring of high-LG-ABN mothers (Francis et al., 1999). Subsequent studies reveal similar findings for hippocampal glucocorticoid receptor expression and for the differences in both the α1 and γ2 GABAA receptor subunit expression in the amygdala (Caldji et al., 2003). These findings suggest that individual differences in patterns of gene expression and behavior can be directly linked to maternal care over the first week of life.

3 Environmental Programming of Glucocorticoid Receptor Expression

3.1 Molecular Basis for the Effect of Maternal Care on HPA Responses to Stress

Postnatal handling, which increases maternal licking/grooming, and increased levels of licking/grooming increase 5-HT turnover in the hippocampus in day 6 rat pups. Interestingly, postnatal handling results in specific increases in 5-HT in the hippocampus and prefrontal cortex, where glucocorticoid receptor expression is increased. Serotonin levels in the hypothalamus, septum, and amygdala are unaffected and glucocorticoid receptor levels in these regions are not altered by handling.

In vitro, the treatment of primary hippocampal cell cultures with 5-HT increases glucocorticoid receptor expression and this effect is mediated by 5-HT_7 receptor activation (Mitchell et al., 1990a,b, 1992; Laplante et al., 2002). The 5-HT_7 receptor, positively coupled to cAMP and glucocorticoid receptor expression in cultured hippocampal neurons, is significantly increased after treatment with 8-bromo cAMP or with various doses of the specific 5-HT_7 receptor agonist, 3-(2-Aminoethyl)-1H-indole-5-carboxamide maleate (5-carboxamidotryptamine; 5-CT) for 4 days. The effect of 5-CT on glucocorticoid receptor expression is blocked by methiothepin. Likewise, 5-CT produces a significant increase in cAMP levels and the effect is blocked by methiothepin. Pindolol, which binds to the 5-HT_{1A} but not the 5-HT_7 receptor, has little effect [also see Mitchell et al. (1992)]. These results further implicate the 5-HT_7 receptor. The increase in glucocorticoid receptor expression is also mimicked with 5-methoxytryptamine (5-MeOT), an effect blocked with methiothepin as well as H8, an inhibitor of PKA (cyclic nucleotide-dependent protein kinase). Over the course of these studies, we found that other serotonergic agonists could partially mimic the 5-HT effect on GR levels; however, this was the first evidence that a more selective serotonergic agonist, 5-CT, could fully mimic the 5-HT effect. Moreover, across all studies, the magnitude of the serotonergic effect on cAMP concentrations is highly correlated ($r = 0.97$) with that on glucocorticoid receptor expression. This observation is consistent with the idea that the effect of 5-HT of glucocorticoid receptor expression in hippocampal neurons is mediated by a 5-HT_7 receptor via the activation of cAMP.

Significantly, the increase in glucocorticoid receptor binding capacity following 5-HT treatment persists following 5-HT removal from the medium; for as long as the cultures can be maintained, there is a sustained increase in glucocorticoid receptor levels as long as 50 days beyond the removal of 5-HT from the medium. Thus, 5-HT can act directly on hippocampal neurons to increase glucocorticoid receptor expression, and the effect of 5-HT on glucocorticoid receptor density observed in hippocampal culture cells mimics the long-term effects of early environmental events.

Activation of cAMP pathways can regulate gene transcription through effects on a number of transcription factors, including of course, the cAMP-response element binding protein (CREB) via an enhanced phosphorylation of CREB. pCREB regulates gene transcription through pathways that involve the transcriptional cofactor, CREB-binding protein (CBP). Primary hippocampal cell cultures treated with 8-bromo cAMP, 5-CT, or 5-HT show a significant increase in CBP expression.

The 5-HT_7 receptor is positively coupled to adenylyl cyclase. In vivo, both handling and increased maternal licking/grooming result in an increased level of hippocampal cAMP concentrations and the activation of PKA over the first week of postnatal life (Mitchell et al., 1992). Activation of PKA results in the tissue-specific induction of a number of transcription factors. The day 6 offspring of high-LG mothers or pups of the same age exposed to handling show increased hippocampal expression of NGFI-A (also known as, zif-268, krox-24, egr-1, zenk, etc). In vitro, 5-HT increases NGFI-A expression in cultured hippocampal neurons and the effect of 5-HT on glucocorticoid receptor expression in hippocampal cultures is completely blocked by concurrent treatment with an oligonucleotide antisense directed at the NGFI-A mRNA.

These studies suggest that maternal licking/grooming results in an increased expression of NGFI-A, which in turn might then regulate glucocorticoid receptor expression. Other rodent models examining environmental regulation of hippocampal glucocorticoid receptor expression also suggest a correspondence between NGFI-A levels and glucocorticoid receptor expression (Mohammed et al., 1993; Andrews et al., 2004). In each case, increased levels of NGFI-A are associated with enhanced glucocorticoid receptor expression. However, the critical site for glucocorticoid receptor regulation remains to be defined.

3.2 Glucocorticoid Receptor Gene Regulation

We assumed that a potential target for regulation is the promoter region of the glucocorticoid receptor gene. We (McCormick et al., 2000) identified and characterized several new glucocorticoid receptor mRNAs cloned from rat hippocampus. They encode a common protein, but differ in their 5′-leader sequences presumably because of alternative splicing of potentially several different sequences from the 5′ noncoding exon 1 region of the glucocorticoid receptor gene. The alternate exon 1 sequences are unlikely to alter the amino acid sequence of the glucocorticoid receptor protein; an in-frame stop codon is present immediately 5′ to the translation initiation site in exon 2, common to all the mRNA variants. Of the ten alternate exon 1 sequences identified by 5′-RACE, four correspond to alternative exon 1 sequences previously identified in mouse, exons 1_1, 1_5, 1_9, and 1_{10} (Strähle et al., 1992; Chen et al., 1999). There is a consensus 5′ splice site immediately downstream of each of these exon 1 sequences. Thus, each alternative exon 1 is spliced onto the first coding exon to create diverse glucocorticoid receptor mRNAs. Most alternative exons are located in a 3-kb CpG island upstream of exon 2 that exhibits substantial promoter activity in transfected cells. Ribonuclease protection assays demonstrate significant levels of six alternative exon 1 sequences in vivo in the rat, with differential expression in the liver, hippocampus, and thymus presumably reflecting tissue-specific differences in promoter activity. Hippocampal RNA contains significant levels of the exon 1_7-containing glucocorticoid receptor mRNA variants expressed at undetectable levels in the liver and thymus.

Figure 15-2

The 5′ noncoding variable exon 1 region of the hippocampal glucocorticoid receptor gene contains multiple alternate exon 1 sequences (McCormick et al., 2000), four of which correspond to alternative exon 1 sequences previously identified in mouse, exons 1_1, 1_5, 1_9 and 1_{10}. Transfection studies show that the activity of individual constructs fused to a luciferase reporter in different cell types is similar with one notable exception; the exon 1_7 promoter sequence has the highest activity of any single promoter construct (McCormick et al., 2000)

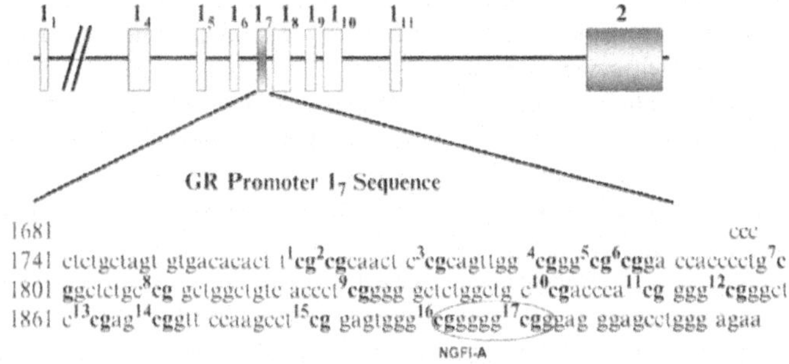

In transient transfection experiments, a construct encoding the whole CpG island of the glucocorticoid receptor gene, including eight of the alternate exon 1 sequences and the splice acceptor site within the intron 5′ of exon 2, fused to a luciferase reporter gene within exon 2, exhibited substantial promoter activity in all cell lines tested. This activity results from transcripts originating at any point within the CpG island that are spliced from an appropriate donor site onto the splice acceptor site 5′ to exon 2 and, we presume, represents the sum of the activity of individual promoters on the genomic DNA fragment. Promoter activity is also associated with particular regions of the CpG island, where the fusion to luciferase was made within specific exon 1 sequences. In these cases, no splice acceptor site is available within the luciferase gene, and a transcriptional fusion was generated between the specific exon 1 and the luciferase reporter; luciferase activity therefore reflects transcription through the specific exon 1. Relative activity of these constructs in different cell types is similar with one notable exception, the exon 1_7 promoter sequence. The exon 1_7, fused to luciferase within exon 1_7, has the highest activity of any single promoter construct in B103 and C6 cells, both of which are CNS derived. The activity of this construct is lower in hepatic cells. In vivo, glucocorticoid receptor mRNA transcripts containing exon 1_7 are present at significant levels in the hippocampus, but absent in the liver, suggesting that factors present in the cells of CNS origin are responsible for transcription initiation at the promoter upstream of exon 1_7 in rat hippocampus.

Interestingly, exon 1_7 usage is altered by postnatal handling, which increases glucocorticoid receptor expression in the hippocampus. Handling selectively elevated glucocorticoid receptor mRNA containing exon 1_7; there is, for example, no effect on exon 1_{10} (McCormick et al., 2000). Predictably, maternal care also affected the expression of glucocorticoid receptor splice variants: Variants containing the exon 1_7 sequence were significantly increased in the adult offspring of high-LG-ABN mothers.

3.3 NGFI-A Regulation of Glucocorticoid Receptor Gene Expression

The CpG island of the glucocorticoid receptor gene contains GC boxes (GCGGGGGCG), which form the core consensus site for NGFI-A. There is indeed a sequence exactly matching the consensus-binding site for the family of zinc finger proteins that includes NGFI-A (Crosby et al., 1991) within exon 1_7. Thus, increases in NGFI-A induced by maternal licking/grooming could increase transcription from a promoter adjacent to exon 1_7, leading to increased glucocorticoid receptor mRNA. In previous studies, we found that handling increased the binding of both NGFI-A to a promoter sequence for the human glucocorticoid receptor containing consensus sequences for both transcription factors. Since neonatal handling increases maternal licking/grooming, these finding suggest that naturally occurring variations in maternal behavior might regulate glucocorticoid receptor expression in neonatal offspring through a 5-HT-induced increase in NGFI-A expression, and the subsequent binding of NGFI-A to the exon 1_7 promoter. Recent findings support this idea, including studies using the chromatin immunoprecipitation (ChIP) assay in which the in vivo formation of protein–DNA complexes is examined using cross-linking with paraformaldehyde perfusion and subsequently precipitated from soluble hippocampal samples using specific antibodies. Protein binding, defined by the specificity of the antibody, to specific DNA sequences is then quantified following PCR amplification with targeted primers and Southern blotting. ChIP analysis of hippocampal samples from postnatal day 6 pups reveals dramatically increased NGFI-A binding to the exon 1_7 promoter in the offspring of high- compared with low-LG-ABN mothers. These findings confirm that maternal care regulates the binding of NGFI-A to the exon 1_7 promoter sequence in pups.

More recent studies confirm the transactivational effect of NGFI-A at the exon 1_7 sequence. We used a cotransfection model with HEK cells (intentionally aiming as far from the neural target as possible) with an NGFI-A expression vector and an exon 1_7-luciferase construct. Cotransfection of the NGFI-A vector and the exon1_7-luciferase construct resulted in a robust increase in luciferase activity, reflecting NGFI-A-induced activation of transcription through the exon 1_7 promoter. Recall that an NGFI-A antisense completely blocks the effects of 5-HT on glucocorticoid receptor expression in hippocampal cell cultures.

These findings suggest that NGFI-A might increase glucocorticoid receptor expression in hippocampal neurons, and provide a mechanism for the effect of maternal care over the first week of life. However, while there are striking differences in NGFI-A expression in the offspring of high- and low-LG mothers at day 6 of postnatal life, hippocampal NGFI-A expression in adulthood is unaffected by maternal care. There is no difference in hippocampal NGFI-A expression in the adult offspring of high- and low-LG-ABN dams. We are thus left with the defining question of early experience studies: How are the effects of early life events sustained into adulthood?

4 The Epigenome and Epigenetic Programming of Stress Responses

4.1 DNA Methylation and Chromatin Structure

While we are all familiar with linear models of DNA and protein–DNA interactions, such models ignore the fact that most of the DNA is tightly packed into nucleosomes that involve a close relationship between DNA wrapped around a core of histone proteins. The conformation or structure of the histone–DNA configuration regulates gene expression. The relation between DNA and histone is maintained, in part, by electrostatic bonds occurring between the positively charged histones and the negatively charged DNA. This chromatin structure commonly precludes transcription factor binding to DNA and underscores the importance of enzymes that modify histone–DNA interactions. One class of such proteins, histone acetyltransferase (HAT; Roth et al., 2001), catalyzes the acetylation of selected amino acids on the protruding histone tails, most commonly histone 3 (H3) or H4. Positively charged amino acids such as lysine and arginine are common targets for acetylation. Histone acetylation modifies the histone–DNA relation. For example, acetylation of the lysine (K) residue on H3 serves to neutralize the positively charged histone, opening the histone–DNA relationship, and facilitating transcription factor binding to DNA. Thus, H3-K9 acetylation serves as a marker of active gene transcription. Many known transcriptional cofactors, such as CBP, are HATs. Histone acetylation is dynamic and is regulated by HDACs, which serve to block histone acetylation and suppress gene expression. Thus, chromatin structure can be viewed as dynamic and clearly subject to modification through intracellular signals that trigger either HATs or HDACs downstream (Bird and Wolffe, 1999; Bird, 2001; Li, 2003). Indeed, in addition to CBP, several known transcriptional cofactors act as histone acetylases. The study of histone acetylation provides a remarkable advance in our understanding of the dynamic and complex regulation of gene expression. Likewise, histone modifications involving methylation, phosphorylation, ribosylation, and ubiquitination can all modify chromatin structure and thus gene expression.

4.2 The Chemistry of DNA Methylation

In addition to chromatin, which is associated with DNA, the DNA molecule is itself chemically modified by methyl residues at the 5′ position of the cytosine rings in the sequence CG in vertebrates (Razin and Riggs, 1980; Wolffe et al., 1999). What distinguishes DNA methylation in vertebrate genomes is the fact that not all CGs are methylated in any given cell type (Razin and Szyf, 1984). Different CGs are methylated in different cell types, generating cell-type-specific patterns of methylation. Thus, the DNA methylation pattern confers upon a genome itself its cell type identity. Since DNA methylation is part of the chemical structure of the DNA, it remains long after all other proteins and epigenomic marks are degraded, and thus it has extremely important diagnostic potential (Shi et al., 2003). It was originally believed that the DNA methylation pattern is established during development, and is then maintained faithfully throughout life by the maintenance DNA methyltransferase (Razin and Szyf, 1984; Reik et al., 2000). The DNA methylation reaction was believed to be irreversible; thus the only way methyl residues could be lost was believed to be through replication in the absence of DNA methyltransferase by passive demethylation (Razin and Riggs, 1980; Wolffe et al., 1999). This mechanism is not applicable to postmitotic tissue such as neurons in the brain. However, our recent data and the data reviewed here support an alternative model that the DNA methylation pattern is dynamic and is an equilibrium of methylation and demethylation reaction (Szyf, 2001). We propose that DNA methylation is a reversible signal like any other biological signal and could therefore potentially change in response to environmental and physiological signals. The notion that DNA methylation is reversible in postmitotic cells has immense implications on our understanding of the potential role of DNA methylation in marking gene expression in the brain.

The hallmark of DNA methylation patterns is the correlation between chromatin and the DNA methylation pattern. Active chromatin is usually associated with unmethylated DNA while inactive chromatin is associated with methylated DNA (Razin and Cedar, 1977; Kadonaga, 1998; Bird and Wolffe, 1999). This tight linkage between DNA methylation and chromatin structure has important implications on our understanding of the function of DNA methylation as well as the processes responsible for generating, maintaining, and altering DNA methylation patterns under physiological and pathological conditions. It was originally believed that DNA methylation precedes and is dominant over chromatin structure changes (e.g., Hashimshony et al., 2003). Methylation was believed to be generated independently of chromatin structure. Methylated DNA attracts methylated DNA binding proteins that recruit repressor complexes containing HDACs, which results in an inactive chromatin (Jones et al., 1998; Nan et al., 1998). The model positioning DNA methylation as driving chromatin inactivation is widespread and profoundly influences our understanding of how altered DNA methylation is involved in cancer. Nevertheless, at present there are data suggesting that the state of chromatin structure can also determine DNA methylation and that chromatin can affect DNA methylation in both directions triggering either de novo DNA methylation or demethylation (Cervoni and Szyf, 2001). These data revise the classic model of a DNA methylation pattern, which is predetermined during development and is then maintained through life, and gives a more dynamic view of the DNA methylation pattern as an interface between the dynamic environment and the static genome. Thus, although DNA methylation is an extremely stable signal, it can be altered later in life when there is a sufficiently stable and consistent signal to activate the chromatin. Transient changes in chromatin structure are not accompanied by changes in DNA methylation. The DNA methylation pattern is proposed to guard the epigenome from random noise and protect it from drifting (Szyf, 2001). This relation between the chromatin state and DNA methylation forms a molecular link through which environmental signals might alter DNA methylation in specific genes in postmitotic neurons. Behavioral signals trigger cellular signaling pathways whose downstream consequence is activation of *trans*-acting factors. These trans-acting factors recruit HAT to the target gene resulting in increased histone acetylation, chromatin opening, and increased accessibility to DNA demethylases. Since methylation of cytosine is an extremely stable chemical bond on DNA, this modification will remain stable for years. For methylation signals to serve as stable marks, they should not be responsive to chromatin noise or short-term signals. The mechanism proposed here also allows for a reversal of the methylation mark by a similar intense change in chromatin structure later in life (Szyf, 2001). This model has important implications on our understanding of how behavioral signals can stably alter glucocorticoid gene expression and how these stable marks in the genome are potentially reversible later in life by specific interventions in spite of their remarkable stability.

DNA methylation marks genes for silencing by a number of mechanisms. The first mechanism is indirect and links DNA methylation to inactive chromatin structure. A region of methylated DNA juxtaposed to regulatory regions of genes attracts different members of a family of methylated DNA binding proteins. The better-studied member of this class is MeCP2, which recruits HDACs (Jones et al., 1998; Nan et al., 1998) and histone methyltransferases (Fuks et al., 2003) to methylated genes (Razin, 1998; Bird and Wolffe, 1999). This results in a modification of chromatin around the gene precipitating an inactive chromatin structure. A different mechanism, which is relevant to our discussion here, involves direct interference of a specific methylated CG residing within a *cis*-recognition sequence for a transcription factor with the interaction of transcription factors, such as the inhibition of binding of cMyc to its recognition element when it is methylated. Essentially, the methylated C serves as a mutation of the recognition element. A third mechanism involves a combination of binding a methylated DNA binding protein and inhibition of activity of a transcription factor (Kudo, 1998). While the first mechanism is dependent on the general density of methyl cytosines within the region associated with a gene rather than methylation of a specific CG, the second mechanism requires a discrete methylation event and is relevant to the mechanism proposed here.

4.3 Maternal Care and DNA Methylation

The important consideration is the stability of cytosine methylation, which is preserved by covalent carbon–carbon bonds and could therefore as a long-term memory of early life behavioral signaling lead to chromatin activation of the glucocorticoid receptor promoter in the offspring of high LG-ABN and lack of chromatin activation in low LG-ABN. DNA methylation changes might provide a mechanism for environmental programming effects occurring in early development. Glucocorticoid receptor gene expression is increased throughout the hippocampus in the adult offspring of high- compared with low-LG-ABN mothers (Liu et al., 1997). The exon 1_7 glucocorticoid receptor promoter sequence appears to be significantly more active in the adult offspring of high compared with low-LG-ABN mothers and was therefore the focus of initial studies of possible maternal effects on DNA methylation. To test the hypothesis that maternal care alters the DNA methylation mark of the glucocorticoid receptor promoter, Weaver et al. (2004) examined the level of methylation across the entire exon 1_7 glucocorticoid receptor promoter sequence in the hippocampus using the sodium bisulfite mapping technique in the adult offspring of high- and low-LG-ABN mothers. Sodium bisulfite treatment of DNA samples converts nonmethylated cytosines to uracils, which are then detected as thymidine on subsequent sequencing gels (Frommer et al., 1992). Methylated cytosines are unaffected by sodium bisulfite and the differences in methylation status are thus apparent and easily quantifiable on sequencing gels. We found significantly greater methylation of the exon 1_7 glucocorticoid receptor promoter sequence in the offspring of the low-LG-ABN mothers. These findings are consistent with the hypothesis that maternal effects alter DNA methylation patterns in the offspring.

To determine whether DNA methylation of specific target sites on the glucocorticoid receptor promoter changes in response to maternal care, we mapped the differences in methylation using the NaBis mapping technique, focusing on a region around the NGFI-A consensus sequence within the exon 1_7 promoter. The results reveal significant differences in the methylation of specific regions of the exon 1_7 glucocorticoid receptor promoter sequence. Notably, the cytosine within the 5′ CpG dinucleotide of the NGFI-A consensus sequence is always methylated in the offspring of low-LG-ABN mothers, and rarely methylated in the offspring of high-LG-ABN dams. This is consistent with site-specific DNA methylation silencing of the glucocorticoid receptor promoter.

To directly examine a causal relation between maternal behavior and DNA methylation changes within the exon 1_7 glucocorticoid receptor promoter, Weaver and colleagues (2004) performed an adoption study. As per this study, the biological offspring of high- or low-LG-ABN mothers were cross-fostered to either high or low dams within 12 h of birth (as described earlier; see Francis et al., 1999; Caldji et al., 2003). These studies could rule out either a purely genetic or a prenatal basis for the variation in DNA methylation in the offspring of high-LG-ABN versus low-LG-ABN. Cross-fostering the biological offspring of high- or low-LG-ABN mothers produced a pattern of exon 1_7 glucocorticoid receptor promoter methylation associated with the rearing mother. The cross-fostering procedure reverses the difference in methylation at specific cytosines. The cytosine within the 5′ CpG dinucleotide of the NGFI-A consensus sequence is hypomethylated following cross-fostering of the offspring of low-LG-ABN to high-LG-ABN dams, with no effect at the cytosine within the 3′ CpG dinucleotide. Thus, the pattern of methylation of the cytosine within the 5′ CpG dinucleotide of the NGFI-A consensus sequence within the exon 1_7 glucocorticoid receptor promoter of the biological offspring of low-LG-ABN mothers cross-fostered to high-LG-ABN dams is indistinguishable from that of the biological offspring of high-LG-ABN mothers. The reverse is true for the offspring of high-LG-ABN mothers fostered to low-LG-ABN dams. Interestingly, cross-fostering does not have the same effect on the CpG methylation status on each individual dinucleotide in the exon 1_7 sequence. For example, the CpG dinucleotide of the AP-1 consensus sequence within the exon 1_7 glucocorticoid receptor promoter of the biological offspring of high-LG-ABN mothers cross-fostered to low-LG-ABN dams remains hypomethylated. However, the CpG dinucleotide of the AP-1 consensus sequence within the exon 1_7 glucocorticoid receptor promoter of the biological offspring of low-LG-ABN mothers cross-fostered to high-LG-ABN dams remains hypermethylated. The molecular basis for such selectivity is unknown. These findings suggest that variations in maternal care alter the methylation status within specific sites of the exon 1_7 promoter of the glucocorticoid receptor gene and represent the first demonstration of a DNA methylation pattern established through a behavioral mode of programming. Parental imprinting (Sapienza, 1990), a well-established paradigm of inheritance of an epigenomic mark, requires germ-line transmission (Li, 2003).

4.4 Site-Specific Methylation of the 5′ CpG Dinucleotide of the NGFI-A Response Element Blocks Transcription Factor Binding

The obvious question concerns the functional importance of such differences in methylation. DNA methylation affects gene expression either by attracting methylated DNA binding proteins to a densely methylated region of a gene or by site-specific interference with the binding of a transcription factor to its recognition element (Razin, 1998; Bird and Wolffe, 1999). Data showing site-specific demethylation of the cytosine within the 5′ CpG dinucleotide of the NGFI-A response element is consistent with the hypothesis that methylation at this site interferes with the binding of NGFI-A protein to its binding site. To address this question, we determined the in vitro binding of increasing concentrations of purified recombinant NGFI-A protein (Milbrandt et al., 1987) to its response element under different states of methylation using the electrophilic mobility shift assay (EMSA) technique with four ^{32}P-labeled synthetic oligonucleotide sequences bearing the NGFI-A binding site. This binding site was either (a) nonmethylated, (b) methylated in the 3′ CpG site, (c) methylated in the 5′ CpG site, (d) methylated in both sites, or (e) mutated at the two CpGs with an adenosine replacing the cytosines. NGFI-A formed a protein–DNA complex with the nonmethylated oligonucleotide, while the protein was unable to form a complex with either a fully methylated sequence or a sequence methylated at the 5′ CpG site. NGFI-A binding to its response element was only slightly reduced with the sequence methylated at the 3′ CpG site. The effect of selective cytosine methylation on NGFI-A binding was further confirmed by competition experiments. NGFI-A recombinant protein was incubated with a labeled, nonmethylated oligonucleotide in the presence of increasing concentrations of nonlabeled oligonucleotides containing the NGFI-A consensus sequence that were either 3′ CpG methylated, 5′ CpG methylated, methylated at both sites, or mutated at the two CpGs with an adenosine replacing the cytosines. The nonmethylated oligonucleotide completely eliminates the formation labeled oligonucleotide protein–DNA complex, while the mutated oligonucleotide is unable to compete away the labeled oligonucleotide protein–DNA complex. Neither the oligonucleotide methylated in both the 3′ and 5′ CpGs nor the 5′ CpG methylated oligonucleotide were able to compete. Importantly, the 3′ CpG methylated oligonucleotide, which mimics the sequence observed in the offspring of high-LG-ABN mothers, exhibited substantial competition, suggesting that binding activity is retained despite the methylation at this site. The results indicate that while methylation of the cytosine within the 5′ CpG dinucleotide reduces NGFI-A protein binding to the same extent as methylation in both CpG sites, methylation of the cytosine within the 3′ CpG dinucleotide only partially reduces NGFI-A protein binding. These data support the hypothesis that methylation of the cytosine within the 5′ CpG dinucleotide in the NGFI-A response element of the exon 1_7 glucocorticoid receptor promoter region in the offspring of low LG-ABN mothers inhibits NGFI-A protein binding.

This is an important finding for our understanding of the processes by which maternal care programs hippocampal glucocorticoid receptor expression and thus HPA responses to stress. Although there are substantial differences in NGFI-A expression between the offspring of high- and low-LG-ABN mothers in early postnatal life, no such differences are apparent in adulthood. Our hypothesis is that the cytosine methylation in the binding site for NGFI-A interferes with NGFI-A binding to the glucocorticoid receptor exon 1_7 promoter. Therefore, it was predicted that the lower cytosine methylation in the adult offspring of high- compared with low-LG-ABN mothers would result in greater NGFI-A binding to the exon 1_7 promoter. This prediction was confirmed using a chromatin-immunoprecipitation (ChIP) assay examining in vivo formation of protein–DNA complexes in hippocampal tissue from adult animals (Weaver et al., 2004). Animals were perfused with paraformaldehyde to fix protein–DNA complexes at the time of sacrifice. NGFI-A-bound DNA complexes were then immunoprecipitated using a selective antibody. The protein–DNA complexes were uncross-linked, and the precipitated genomic DNA was subjected to PCR amplification with primers specific for the exon 1_7 glucocorticoid receptor promoter sequence. The results indicated a threefold greater binding of NGFI-A protein to the hippocampal exon 1_7 glucocorticoid receptor promoter in the adult offspring of high- than in low-LG-ABN mothers. Using the same tissue samples and an antibody against the acetylated form of H3, we also found dramatically increased acetylated H3 association with the exon 1_7 glucocorticoid receptor promoter in the offspring of the high-LG-ABN mothers. As described earlier, histone acetylation is associated with active states of gene expression. These findings are therefore consistent with the idea of increased NGFI-A binding to the exon 1_7 promoter and increased transcriptional activation.

We confirmed that DNA methylation inhibits the ability of NGFI-A to activate the exon 1_7 promoter in isolation from other potential differences between adult offspring of high- and low-LG-ABN dams using a transient cotransfection assay in human HEK 293 cells. These cells are not of hippocampal origin and thus allow us to measure the transcriptional consequences of the interaction of NGFIA with either a methylated or nonmethylated version of the glucocorticoid receptor exon 1_7 promoter per se. Although transfection of HEK cells containing an exon 1_7-luciferase reporter construct with an NGFI-A expression vector significantly increases luciferase activity, this effect is dramatically reduced if the CpG dinucleotides within the exon 1_7 sequence are methylated. Moreover, the effect of NGFI-A on transcription through an exon

1_7-luciferase reporter construct was almost completely abolished with a point mutation at the 5′ cytosine (a cytosine to adenosine mutation). Taken together, these findings suggest that an "epimutation" at a single cytosine within the NGFI-A consensus sequence alters the binding of NGFI-A and might therefore explain the sustained effect of maternal care on hippocampal glucocorticoid receptor expression and HPA responses to stress.

4.5 How Does Maternal Care Alter Cytosine Methylation?

Maternal behavior could either inhibit de novo methylation or stimulate demethylation. To address this question, we performed a developmental study of the methylation pattern of glucocorticoid receptor exon 1_7 promoter from embryonic day 20 to day 90. High- and low-LG-ABN mothers differ in the frequency of pup licking/grooming and ABN only during the first week of life. More important, this period corresponds to the appearance of the difference in DNA methylation in the offspring in studies using sodium bisulfite mapping to precisely map the methylation status of the cytosines within the exon 1_7 glucocorticoid receptor promoter over multiple developmental time points. This analysis demonstrates that just before birth, on embryonic day 20, the entire exon 1_7 region is unmethylated in both groups. Strikingly, 1 day following birth (postnatal day 1) the exon 1_7 glucocorticoid receptor promoter is de novo methylated in both groups. The 5′ and 3′ CpG sites of the exon 1_7 glucocorticoid receptor NGFI-A response element in the offspring of both high- and low-LG-ABN mothers, which exhibit differential methylation later in life, are de novo methylated to the same extent. These data show that both the basal state of methylation and the first wave of de novo methylation after birth occur similarly in both groups. Whereas it is generally accepted that DNA methylation patterns are formed prenatally and that de novo methylation occurs early in development, there is at least one documented example of postnatal de novo methylation of the HoxA5 and HoxB5 genes (Hershko et al., 2003). Since similar analyses are not documented for other genes, it is unknown yet whether changes in methylation are common around birth or whether they are unique to this glucocorticoid receptor promoter. One aspect of these findings that is important is that of the complete absence of cytosine methylation on embryonic day 20. As the majority of the pyramidal cells of Ammon's Horn are born between embryonic days 16 and 20, it seems unlikely that methylation patterns, at least on the exon 1_7 promoter of the glucocorticoid receptor, are generated at the time of DNA replication and cell division, as would normally be the case with imprinted genes.

The differences in the status of methylation of the exon 1_7 glucocorticoid receptor developed between the two groups emerge between postnatal days 1 and 6, which is precisely the period when differences in the maternal behavior of high- and low-LG-ABN dams are apparent. There are no differences in maternal licking/grooming between high- and low-LG-ABN mothers beyond day 8 (Caldji et al., 1998a,b; Champagne et al., 2003). By postnatal day 6, the 5′ CpG dinucleotide of the NGFI-A response element is demethylated in the high, but not in the low-LG-ABN group. These findings are consistent with data from the cross-fostering experiment, which illustrates that the differences between the two groups developed following birth in response to maternal behavior. The group difference in CpG dinucleotide methylation then remains consistent through to adulthood. Our findings suggest that the group difference in DNA methylation occurs as a function of maternal behavior over the first week of life. The results of earlier studies indicated that the first week of postnatal life is indeed a "critical period" for the effects of early experience on hippocampal glucocorticoid receptor expression.

4.6 Reversal of the Maternal Effect on Glucocorticoid Receptor Expression and HPA Responses to Stress

These findings suggest that maternal behavior produces an active demethylation process at selected and presumably targeted sites. The resulting demethylation of the 5′ CpG dinucleotide within the NGFI-A response element of the exon 1_7 promoter enhances NGFI-A binding, increasing glucocorticoid receptor gene transcription and HPA responses to stress.

These findings raise the question of how maternal high LG-ABN might activate a demethylation of the glucocorticoid receptor exon 1_7 promoter. As discussed earlier, a testable working hypothesis is that high LG-ABN leads to activation of NGFI-A as a downstream effector of activation of a 5-HT signaling through increased cAMP and PKA. Increased NGFI-A increases the frequency of occupancy of the GR exon 1_7 promoter. NGFI-A interaction with the GR exon 1_7 promoter leads to increased acetylation and increased accessibility of the GR exon 1_7 promoter to demethylase resulting in DNA demethylation. This hypothesis predicts that pharmacological activation of chromatin using HDAC inhibitors should result in the activation of GR exon 1_7 promoter demethylation. However, the question is whether this reversibility will be limited to early life exclusively or whether it is possible to reverse these marks later in life as well, if the appropriate signals to activate the chromatin structure are applied or by a pharmacological activation of chromatin structure.

Our hypothesis is that the DNA methylation is a steady state of DNA methylation and demethylation whose direction is determined by the state of chromatin structure (Szyf, 2001). This hypothesis predicts that both DNMTs and demethylases are present in adult neurons and that if the chromatin state is altered by either persistent physiological or pharmacological signals one should be able to change the state of methylation of a gene in postmitotic tissue such as adult hippocampal neurons. We previously established that pharmacological activation of chromatin structure by HDAC inhibitors could trigger replication-independent active demethylation of DNA (Cervoni and Szyf, 2001). The hypothesis was tested and showed that the demethylation of the GR exon 1_7 promoter is driven by histone acetylation and could be activated in adult neurons as well. This understanding leads to an obvious prediction: HDAC inhibition should reverse the effects of cytosine methylation on NGFI-A binding to the exon 1_7 promoter, GR expression, and HPA responses to stress. We tested this idea with central infusion of adult offspring of high- or low-LG-ABN mothers with the HDAC inhibitor, trichostatin A (TSA), for four consecutive days. As expected, ChIP assays revealed that TSA infusion significantly increased the level of acetylated H3 at the exon 1_7 site (i.e., HDAC inhibition resulted in increased histone acetylation) in the offspring of low-LG-ABN mothers to levels comparable with those observed in the offspring of high LG-ABN mothers. The increased histone acetylation is associated and enhanced with NGFI-A binding to the exon 1_7 promoter sequence and completely eliminates the effect of maternal care. As expected, enhanced NGFI-A binding to the exon 1_7 promoter increased hippocampal glucocorticoid receptor expression. Hippocampal glucocorticoid receptor expression in the TSA-treated adult offspring of low-LG-ABN mothers is indistinguishable from that of the high LG-ABN groups. Most important, TSA infusion also eliminates the effect of maternal care on HPA responses to stress. During and following exposure to acute stress, plasma corticosterone levels in TSA-treated offspring of low-LG-ABN mothers are indistinguishable from those of TSA- or vehicle-treated high-LG-ABN mothers. There was no effect of TSA on any measure in the offspring of high-LG-ABN animals. This is understandable because under normal circumstances there is considerable H3 acetylation and NGFI-A binding at the exon 1_7 sequence in these animals.

These findings have important implications on our understanding of the mechanisms linking early maternal behavior and stable changes in behavior later in adulthood, as well as on our understanding of the mechanisms responsible for maintaining the DNA methylation pattern in adult postmitotic tissues. First, our data support the idea that demethylation is driven by activation of chromatin and that HDAC inhibitors produce demethylation even in nondividing cells (i.e., in a replication-independent manner). Second, our data are consistent with the hypothesis that the demethylation of GR exon 1_7 in offspring of high-LG-ABN rats early after birth is driven by increased histone acetylation as discussed earlier. Third, these data provide the first evidence that molecular mechanisms that underlie the effects of early life-experience neural function are potentially reversible in adulthood. This consideration is of obvious social and therapeutic implications. Fourth, these data provide the first in vivo evidence for our hypothesis that the DNA methylation pattern is dynamic even in postmitotic tissues and that its steady state is maintained by the state of chromatin acetylation (Szyf, 2001). Fifth, our data provide a general mechanism for how external signals could change the DNA methylation pattern and thus the chemistry of the genome itself even during adulthood.

4.7 Dissection of the Molecular Mechanisms Linking Maternal Behavior and Active Demethylation of GR exon 1_7 Promoter in the Hippocampus

The data discussed above support the hypothesis that histone acetylation could produce active demethylation of the GR exon 1_7 promoter, yet several questions remain unanswered. How, for example, is histone acetylation targeted to the exon 1_7 promoter as a consequence of maternal behavior? We propose that maternal behavior stimulates 5-HT, which stimulates NGFI-A and that NGFI-A then targets HATS and eventually demethylases to the glucocorticoid receptor exon 1_7 promoter. To dissect the different molecular components of this hypothesis, we took advantage of both hippocampal primary neuronal cell cultures as well as nonneuronal cell lines. The two systems have different strengths and could be used to test different components of the model. First, we tested the hypothesis that 5-HT acts through cAMP to produce hypomethylation. Hippocampal cell cultures treated with either 5-HT or 8-bromo-cAMP, a stable cAMP analog, show increased glucocorticoid receptor expression following 4 days of treatment. Treatment of hippocampal cells in culture with 5-HT also results in the hypomethylation of the 5′ CpG dinucleotide of the NGFI-A consensus sequence within the exon 1_7 promoter of the glucocorticoid receptor gene, with no effect at the 3′ site. Treatment with 8-bromo-cAMP produces an even more pronounced effect on cytosine methylation at the 5′ CpG site. In both studies, cultures maintained under control conditions show complete methylation of both the 5′ and 3′ CpG sites of the NGFI-A consensus sequence. Bromodeoxyuridine labeling, which marks newly generated cells, reveals little or no cell replication in the cultures at the time of 5-HT treatment. These findings reinforce the idea that the alterations in cytosine methylation occur independently of cell replication and in response to intracellular signals associated with variations in maternal care. These cells establish that 5-HT signaling induced by maternal care triggers replication-independent changes in methylation of GR exon 1_7 promoter through an increase in cAMP. Since increased cAMP activates NGFI-A, it seems that ectopic expression of NGFI-A can target the demethylation process to the GR exon 1_7 promoter. However, this would not explain the selective effect at the 5′ cytosine.

To test the hypothesis that NGFI-A targets demethylation to GR exon 1_7 promoter, nonneuronal cell line HEK 293 is resorted to. Here, we can isolate the direct effect of NGFI-A from other neuron-specific events that might confound the interpretation of data from the hippocampal cultures. By comparing the fate of a transiently transfected methylated GR exon 1_7 promoter–luciferase vector in the presence and absence of NGFI-A, we could better determine the specific effects of NGFI-A on demethylation. Since the nonintegrated plasmid does not bear an origin of replication and does not replicate in HEK 293 cells, the assay measures the effects of NGFI-A on active replication-independent demethylation. Whereas in vitro methylated glucocorticoid receptor exon 1_7 promoter–luciferase vector remains methylated in HEK 293 cells, coexpression of NGFI-A results in active demethylation of a significant fraction of the transfected plasmids. To demonstrate that this DNA demethylation requires direct contact between NGFI-A and its recognition element, site-directed mutagenesis of the two CpGs included in the NGFI-A recognition element was performed. Our preliminary results suggest that these manipulations abolished the ability of NGFI-A to activate and demethylate the glucocorticoid receptor exon 1_7 promoter (Weaver et al., unpublished data, 2005). These experiments provide a molecular mechanism on how demethylation is triggered to specific sequences. The outstanding question is to determine how NGFI-A triggers demethylation on binding to specific sequences. One possibility is that NGFI-A directly recruits a demethylase to the gene or that as proposed before it recruits a HAT that increases acetylation, thus increasing the accessibility to demethylase as proposed earlier (Cervoni and Szyf, 2001).

4.8 The Putative Demethylase

The results of the studies with hippocampal cell cultures, the HEK 293 cells, and the developmental timecourse study suggest a process of active demethylation. In the developmental studies, the 5′ CpG site is initially methylated to the same extent in the offspring of high- and low-LG-ABN mothers. Over the course of the first week following birth, the methylation mark is functionally removed from the 5′ site in the offspring of high-LG-ABN mothers. Prevailing models of DNA methylation propose that the methylation pattern of newly synthesized DNA is exclusively determined by the methylation pattern of the parental strand, and thus thought to be preserved in somatic cells. This model cannot explain the data described here nor recent data clearly demonstrating demethylation in response to changes in chromatin structure, such as those induced by TSA (Cervoni and Szyf, 2001). In response to such findings, Szyf and colleagues (Bhattacharya et al., 1999) proposed that DNA methylation is enzymatically reversible and that DNA methylation is dynamic in fully differentiated cells. This idea remains controversial. Active demethylation was nevertheless clearly demonstrated early in embryogenesis and the parental genome undergoes replication-independent, active demethylation hours after fertilization, well before the initiation of replication. Demethylation at very early stages in development has been relatively accepted, but the possibility of postnatal demethylation, and especially in fully differentiated somatic cells, has been hotly disputed. However, active replication demethylation was demonstrated in EBV-infected B cells and more recently, it was repeatedly demonstrated in HEK 293 cells. The HEK 293 transient transfection provided direct evidence that active replication-independent demethylation takes place in somatic cells and that it is dependent on the chromatin state. Fully in vitro methylated plasmid is transiently transfected into the cells. The plasmid, which does not bear an origin of replication, does not replicate in these cells and this has been validated using a DpnI restriction analysis (Cervoni and Szyf, 2001). Upon treatment of the cells with TSA, the plasmid undergoes complete demethylation that could only be accomplished by a processive demethylase. A recent paper on altered expression of IL-2 expression T lymphocytes following activation clearly implicates an active process of demethylation in a normal nontransformed somatic cell. Our results (Weaver et al., 2004) are the first demonstration of active demethylation in postmitotic neurons in vivo.

Earlier studies from Szyf's laboratory extracted active DNA demethylase activity from a human lung cancer cell line and identified a protein with demethylase activity (Bhattacharya et al., 1999), which was cloned concurrently by Bird's group and named MBD2 (Hendrich and Bird, 1998). Interestingly, the protein, MBD2, was found by Bird's group and others to also associate with a chromatin-remodeling complex containing HDAC, which is involved in silencing of gene expression through the recruitment of a repressor complex. The assignment of a demethylase (dMTase) function to a protein that was independently discovered as a recruiter of repressor complexes triggered the expected controversy in the field and there are reports that MBD2 failed to produce demethylase activity. However, the observation that MBD2/dMTase expression produces the demethylation of some but not all promoters in a dose- and time-dependent manner has been confirmed (e.g., Cervoni et al., 2001; Detich et al., 2002). Clearly, the contextual factors that determine MBD2 demethylase activity remain to be fully explained. Interestingly, MBD2 increased gene expression in those instances where promoter demethylation occurred, suggesting that not all promoters respond in the same orderly manner. Indeed, the same is true for DNA methylation, which impedes the DNA binding of most, but not all transcription factors; SP1 binds to methylated DNA. Antisense knock down of MBD2 resulted in inhibition of active demethylation induced by valproate and caused hypermethylation and silencing of the prometastatic gene uPA in metastatic breast cancer cells. Another group reported that ectopic expression of MBD2/dMTase in a hepatocyte cell line caused demethylation and activation of the hexokinase type 2 gene (Goel et al., 2003). Additional support for the demethylase activity of MBD2/dMTase emerges from the finding that expression of MBD2/dMTase is correlated with demethylation within the promoters of *C-ERBB-2* and *SURVIVIN* genes in ovarian cancers (Hattori et al., 2001a,b) and hypomethylated *cMyc* in gastric cancer (Fang et al., 2004). In addition, the Drosophila homolog of MBD2, dMBD2/3, formed foci that associated with DNA at the cellular blastoderm stage, concurrent with the activation of the embryonic genome, and also associated with the active Y chromosome (Marhold et al., 2002).

To test the hypothesis that MBD2 is associated with maternally induced demethylation, we performed an in situ hybridization assay with probes for the mRNAs for a number of methylated binding proteins at day 6 postpartum. Our analysis revealed that MBD2/dMTase expression is elevated in the hippocampus at this point in time in offspring of high- versus low-LG-ABN mothers. A ChIP analysis with an antiMBD2/dMTase antibody demonstrates significantly increased binding of MBD2/dMTase to the exon 1_7 GR promoter in day 6 offspring of high- versus low-LG-ABN mothers. We also found increased NGFI-A binding to the same sequence in day 6 offspring of high-LG-ABN offspring. We then performed a bisulfite mapping of the state of methylation of the exon 1_7 glucocorticoid receptor promoter bound to MBD2 and precipitated in the ChIP assay with antiMBD2 antibody. If MBD2 is the demethylase involved in this process or if it is part of the demethylase complex, then MBD2-bound exon 1_7 sequences at day 6 should be found in the process of demethylation. Indeed, most of the MBD2-bound DNA was unmethylated or partially unmethylated.

In summary, our findings suggest that shortly after birth there is a wave of de novo methylation that results in the methylation of both CpG sites within the NGFI-A consensus sequence. Such events would impede the binding of NGFI-A to the exon 1_7 promoter. However, in the offspring of the high-LG-ABN mothers, NGFI-A expression is increased to the point where binding occurs despite the "low-affinity" status of the binding site. The binding of NGFI-A is associated with histone acetylation and the subsequent availability of the site to demethylase. In support of this idea, the treatment of the adult offspring of the low-LG-ABN mothers with TSA increases H3 acetylation and NGFI-A binding (see earlier) and results in the demethylation of the 5′CpG site of the NGFI-A consensus sequence (Weaver et al., 2004). Although this model remains speculative (and controversial) at this time, these findings do suggest that modifications to the DNA methylation status in fully differentiated cells are clearly possible and pharmacologically reversible, an idea that holds considerable potential therapeutic implications.

5 Experience-Dependent Chromatin Plasticity? Environmental Variability Meets Epigenomic Predictability

The defining question of early experience studies concerns the mechanism by which environmental effects occurring in early development are "biologically embedded" and thus sustained into adulthood (i.e., so-called "environmental programming" effects). The offspring of high-LG-ABN mothers exhibit increased hippocampal glucocorticoid receptor expression from the exon 1_7 promoter and dampened HPA responses to stress that persist into adulthood. We propose that the differential epigenomic status of the exon 1_7 glucocorticoid receptor promoter in the offspring of high-LG-ABN mothers serves as a mechanism for this maternal effect. It is important to note that these findings are restricted to the study of a single promoter of but one gene in one region of the brain. The degree to which such results might generalize to other instances of environmental programming remains to be determined. Moreover, further studies are required to determine how maternal behavior alters the epigenomic status of the exon 1_7 glucocorticoid receptor promoter. The developmental timecourse study is critical. Recall that the 5′ CpG dinucleotide of the NGFI-A consensus sequence of the exon 1_7 promoter is methylated to the same elevated level in the newborn offspring of high- and low-LG-ABN mothers. It is only over the first week of life that the difference emerges, with the decline in the methylation of the 5′ CpG site in the offspring of high but not low-LG-ABN mothers. Note that no such demethylation occurs at the neighboring 3′ CpG site. The impressive selectivity suggests a demethylation process that is targeted in some manner. It is critical to define the processes by which such apparently active demethylation might occur. Regardless of these yet unanswered questions, these findings provide the first evidence that maternal behavior stably alters the epigenome of the offspring, providing a mechanism for the long-term effects of early experience on gene expression in the adult. These studies offer an opportunity to clearly define the nature of gene–environment interactions during development and how such effects result in the sustained "environmental programming" of gene expression and function over the life span. Finally, it is important to note that maternal effects on the expression of defensive responses, such as increased HPA activity, are a common theme in biology such that the magnitude of the maternal influence on the development of HPA and behavioral responses to stress in the rat should not be surprising. Maternal effects on defensive responses to threat are apparent in plants, insects, and reptiles. Such effects commonly follow from the exposure of the mother to the same or similar forms of threat and may represent examples where the environmental experience of the mother is translated through an epigenetic mechanism of inheritance into phenotypic variation in the offspring. Indeed, maternal effects could result in the transmission of adaptive responses across generations (Meaney, 2001). Epigenomic modifications of targeted regulatory sequences in response to even reasonably subtle variations in environmental conditions might serve as a major source of epigenetic variation in gene expression and function and ultimately as a process mediating such maternal effects. We propose that epigenomic changes serve as an intermediate process that imprints dynamic environmental experiences on the fixed genome resulting in stable alterations in phenotype. Such variations may then serve as a source of individual differences in vulnerability/resistance to chronic illness over the life span depending on the conditions of the prevailing environment. It is a question of fit.

References

Ammerman RT, Cassisi JE, Hersen M, van Hasselt VB. 1986. Consequences of physical abuse and neglect in children. Clin Psychol Rev 6: 291-310.

Andrews MH, Kostaki A, Setiawan E, McCabe L, Owen D, et al. 2004. Developmental regulation of the 5-HT7 serotonin receptor and transcription factor NGFI-A in the fetal guinea-pig limbic system: Influence of GCs. J Physiol 555: 659-670.

Antoni FA. 1993. Vasopressinergic control of pituitary adrenocorticotropin secretion comes of age. Front Neuroendocrinol 14: 76-122.

Bakshi VP, Shelton SE, Kalin NH. 2000. Neurobiological correlates of defensive behaviors. Prog Brain Res 122: 105-115.

Bennett AJ, Lesch KP, Heils A, Long JC, Lorenz JG, et al. 2002. Early experience and serotonin transporter gene variation interact to influence primate CNS function. Mol Psychiatr 7: 118-122.

Bhattacharya SK, Ramchandani S, Cervoni N, Szyf, M. 1999. A mammalian protein with specific demethylase activity for mCpG DNA. Nature 397: 579-583.

Bifulco A, Brown GW, Adler Z. 1991. Early sexual abuse and clinical depression in adult life. Br J Psychiatr 159: 115-122.

Bird AP. 2001. Methylation talk between histones and DNA. Science 294: 2113-2115.

Bird A, Wolffe AP. 1999. Methylation-induced repression—Belts, Braces, and Chromatin. Cell 99: 451-454.

Brinton RD, Yamazaki R, Gonzalez CM, O'Neill K, Schreiber SS. 1998. Vasopressin-induction of the immediate early gene, NGFI-A, in cultured hippocampal glial cells. Brain Res Mol Brain Res 57: 73-85.

Brown GR, Anderson B. 1991. Psychiatric morbidity in adult inpatients with childhood histories of sexual and physical abuse. Am J Psychiatr 148: 55-61.

Butler PD, Weiss JM, Stout JC, Nemeroff CB. 1990. Corticotropin-releasing factor produces fear-enhancing and behavioural activating effects following infusion into the locus coeruleus. J Neurosci 10: 176-183.

Caldji C, et al. 1998a. Maternal care during infancy regulates the development of neural systems mediating the expression of fearfulness in the rat. Proc Natl Acad Sci USA 95: 5335-5340.

Caldji C, Tannenbaum B, Sharma S, Francis D, Plotsky PM, et al. 1998b. Maternal care during infancy regulates the development of neural systems mediating the expression of behavioral fearfulness in adulthood in the rat. Proc Natl Acad Sci 95: 5335-5340.

Caldji C, Diorio J, Meaney MJ. 2003. Variations in maternal care alter GABAA receptor subunit expression in brain regions associated with fear. Neuropsychopharmacology 28: 150-159.

Canetti L, Bachar E, Galili-Weisstub E, De-Nour AK, Shalev AY. 1997. Parental bonding and mental health in adolesence. Adolesence 32: 381-394.

Caspi A, Sugden K, Moffitt TE, Taylor A, Caraig IW, et al. 2003. Influence of life stress on depression: Moderation by a polymorphism in the 5-HTT gene. Science 301: 386-390.

Cervoni N, Szyf M. 2001. Demethylase is directed by histone acetylation. J Biol Chem 276: 40778-40787.

Champagne FA, Francis DD, Mar A, Meaney MJ. 2003. Naturally-occurring variations in maternal care in the rat as a mediating influence for the effects of environment on the development of individual differences in stress reactivity. Physiol Behav 79: 359-371.

Chen FH, Watson CS, Gametchu B. 1999. Multiple glucocorticoid receptor transcripts in membrane glucocorticoid receptor-enriched S-49 mouse lymphoma cells. J Cell Biochem 74: 418-429.

Cintra A, et al. 1994a. Glial and neuronal glucocorticoid receptor immunoreactive cell populations in developing, adult, and aging brain. Ann NY Acad Sci 746: 42-61; [discussion pp. 61–63].

Cintra A, et al. 1994b. Mapping and computer assisted morphometry and microdensitometry of glucocorticoid receptor immunoreactive neurons and glial cells in the rat central nervous system. Neuroscience 62: 843-897.

Clark SJ, Harrison J, Paul CL, Frommer M. 1991. High sensitivity mapping of methylated cytosines. Nucleic Acids Res 22: 2990-2997.

Comb M, Goodman HM. 1990. CpG methylation inhibits proenkephalin gene expression and binding of the transcription factor AP-2. Nucleic Acids Res 18: 3975-3982.

Crosby SD, Puetz JJ, Simburger KS, Fahrner TJ, Milbrandt J. 1991. The early response gene NGFI-C encodes a zinc finger transcriptional activator and is a member of the GCGGGGGCG (GSG) element-binding protein family. Mol Cell Biol 11: 3835-3841.

Davis M, Whalen PJ. 2001. The amygdala: Vigilance and emotion. Mol Psychiatry 6: 13-34.

De Bellis MD, Chrousos GP, Dorn LD, Burke L, Helmers K, et al. 1994. Hypothalamic–pituitary–adrenal dysregulation in sexually abused girls. J Clin Endocrinol Metab 78: 249-255.

De Kloet ER, Vregdenhil E, Oitzl MS, Joels M. 1998. Brain corticosteroid receptor balance in health and disease. Endo Rev 19: 269-301.

Detich N, Theberge J, Szyf M. 2002. Promoter-specific activation and demethylation by MBD2/demethylase. J Biol Chem 277: 35791-35794.

Emoto H, Yokoo H, Yoshida M, Tanaka M. 1993. Corticotropin-releasing factor enhances noradrenaline release in the rat hypothalamus assessed by intracerebral microdialysis. Brain Res 601: 286-288.

Fang JY, Cheng ZH, Chen YX, Lu R, Yang L, et al. 2004. Expression of Dnmt1, demethylase, MeCP2 and methylation of tumor-related genes in human gastric cancer. World J Gastroenterol 10: 3394-3398.

Felitti VJ, Anda RF, Nordenberg D, Williamson DF, Spitz AM, et al. 1998. Relationship of childhood abuse and household dysfunction to many of the leading causes of death in adults. Am J Prevent Med 14: 245-258.

Fleming AS, O'Day DH, Kraemer GW. 1999. Neurobiology of mother–infant interactions: Experience and central nervous system plasticity across development and generations. Neurosci Biobehav Rev 23: 673-685.

Francis DD, Meaney MJ. 1999. Maternal care and the development of stress responses. Curr Opin Neurobiol 9: 128-134.

Francis DD, Champagne F, Meaney MJ. 2000. Variations in maternal behaviour are associated with differences in oxytocin receptor levels in the rat. J Neuroendocrinol 12: 1145-1148.

Francis DD, Diorio J, Liu D, Meaney MJ. 1999. Nongenomic transmission across generations in maternal behavior and stress responses in the rat. Science 286: 1155-1158.

Frommer M, et al. 1992. A genomic sequencing protocol that yields a positive display of 5-methylcytosine residues in individual DNA strands. Proc Natl Acad Sci USA 89: 1827-1831.

Fuks F, Hurd PJ, Wolf D, Nan X, Bird AP, et al. 2003. The methyl-CpG-binding protein MeCP2 links DNA methylation to histone methylation. J Biol Chem 278: 4035-4040.

Goel A, Saroj P, Mathupala P, Pedersen PL. 2003. Glucose metabolism in cancer. Evidence that demethylation events play a role in activating type II hexokinase gene expression. J Biol Chem 278: 15333-15340.

Gottman JM. 1998. Psychology and the study of marital processes. Ann Rev Psychol 49: 169-197.

Gray TS, Bingaman EW. 1996. The amygdala: Corticotropin-releasing factor, steroids, and stress. Crit Rev Neurobiol 10: 155-168.

Hashimshony T, Zhang J, Keshet I, Bustin M, Cedar H. 2003. The role of DNA methylation in setting up chromatin structure during development. Nature Genet 34: 187-192.

Hattori M, Sakamoto H, Yamamoto T. 2001a. DNA demethylase expression correlates with lung resistance protein expression in common epithelial ovarian cancers. J Int Med Res 29: 204-213.

Hattori M, Sakamoto H, Satoh K, Yamamoto T. 2001b. DNA demethylase is expressed in ovarian cancers and the expression correlates with demethylation of CpG sites in the promoter region of c-erbB-2 and survivin genes. Cancer Lett 169: 155-164.

Heim C, Newport DJ, Heit S, Graham YP, Wilcox M, et al. 2000. Pituitary–adrenal and autonomic responses to stress in women after sexual and physical abuse in childhood. JAMA 284: 592-597.

Heim C, Newport DJ, Wagner D, Wilcox MM, Miller AH, et al. 2002. The role of early adverse experience and adulthood stress in the prediction of neuroendocrine stress reactivity in women: A multiple regression analysis. Depress Anxiety 15: 117-125.

Hendrich B, Bird AP. 1998. Identification and characterization of a family of mammalian methyl-CpG binding proteins. Mol Cell Biol 18: 6538-6547.

Hershko AY, Kafri T, Fainsod A, Razin A. 2003. Methylation of HoxA5 and HoxB5 and its relevance to expression during mouse development. Gene 302: 65-72.

Higley JD, Hasert MF, Suomi SJ, Linnoila M. 1991. Nonhuman primate model of alcohol abuse: Effects of early experience, personality and stress on alcohol consumption. Proc Natl Acad Sci USA 88: 7261-7265.

Hinde RA. 1986. Some implications of evolutionary theory and comparative data for the study of human prosocial and aggressive behaviour. Development of anti-social and prosocial behaviour. Olweus D, Block J, Radke-Yarrow M, editors. Orlando: Academic Press; pp. 13-32.

Holmes SJ, Robins LN. 1987. The influence of childhood disciplinary experience on the development of alcoholism and depression. J Child Psychol Psychiatr 28: 399-415.

Holmes S, Robins LN. 1988. The role of parental disciplinary practices in the development of depression and alcoholism. Psychiatry 51: 24-36.

Jones PL, Veenstra GJ, Wade PA, Vermaak D, Kass SU, et al. 1998. Methylated DNA and MeCP2 recruit histone deacetylase to repress transcription. Nat Genet 19: 187-191.

Kadonaga JT. 1998. Eukaryotic transcription: An interlaced network of transcription factors and chromatin-modifying machines. Cell 92: 307-313.

Keshet I, Yisraeli J, Cedar H. 1985. Effect of regional DNA methylation on gene expression. Proc Natl Acad Sci USA 82: 2560-2564.

Koegler-Muly SM, Owens MJ, Ervin GN, Kilts CD, Nemeroff CB. 1993. Potential corticotropin-releasing factor pathways in the rat brain as determined by bilateral electrolytic lesions of the central amygdaloid nucleus and paraventricular nucleus of the hypothalamus. J Neuroendocrinol 5: 95-98.

Koob GF, Heinrichs SC, Menzaghi F, Pich EM, Britton KT. 1994. Corticotropin-releasing factor, stress and behavior. Sem Neurosci 6: 221-229.

Kudo S. 1998. Methyl-CpG-binding protein MeCP2 represses SP1-activated transcription of the human leukosialin gene when the promoter is methylated. Mol Cell Biol 18: 5492-5499.

Laplante P, Diorio J, Meaney MJ. 2002. Serotonin regulates hippocampal glucocorticoid receptor expression via a 5-HT7 receptor. Brain Res Dev Brain Res 139: 199-203.

Lavicky J, Dunn AJ. 1993. Corticotropin-releasing factor stimulates catecholamine release in hypothalamus and prefrontal cortex in freely moving rats as assessed by microdialysis. J Neurochem 60: 602-612.

Lee MHS, Williams DI. 1975. Long term changes in nest condition and pup grouping following handling of rat litters. Dev Psychobiol 8: 91-95.

Levine S. 1957. Infantile experience and resistance to physiological stress. Science 126: 405-406.

Levine S. 1962. Plasma-free corticosteroid response to electric shock in rats stimulated in infancy. Science 135: 795-796.

Levine S. 1994. The ontogeny of the hypothalamic–pituitary–adrenal axis. The influence of maternal factors. Ann NY Acad Sci 746, 275-88; Discussion 289–293.

Li E. 2003. Chromatin modification and epigenetic reprogramming in mammalian development. Nat Rev Genet 3: 662-673.

Liang KC, Melia KR, Miserendino MJ, Falls WA, Campeau S, et al. 1992. Corticotropin releasing factor: Long-lasting facilitation of the acoustic startle reflex. J Neurosci 12: 2303-2312.

Lissau I, Sorensen TIA. 1994. Parental neglect during childhood and increased risk of obesity in young adulthood. Lancet 343: 324-327.

Liu D, Tannenbaum B, Caldji C, Francis D, Freedman A, et al. 1997. Maternal care, hippocampal glucocorticoid receptor gene expression and hypothalamic–pituitary–adrenal responses to stress. Science 77: 1659-1662.

Maccari S, Piazza PV, Kabbaj M, Barbazanges A, Simon H, et al. 1995. Adoption reverses the long-term impairments in glucocorticoid feedback induced by prenatal stress. J Neurosci 15: 110-116.

Marhold J, Zbylut M, Lankenau DH, Li M, Gerlich D, et al. 2002. Stage-specific chromosomal association of Drosophila dMBD2/3 during genome activation. Chromosoma 111: 13-21.

McCarty R, Lee JH. 1996. Maternal influences on adult blood pressure of SHRs: A single pup cross-fostering study. Physiol Behav 59: 71-75.

McCauley J, Kern DE, Kolodner K, Dill L, Schroeder AF, et al. 1991. Clinical characteristics of women with a history of childhood abuse: Unhealed wounds. JAMA 277: 1362-1368.

McCormick JA, Lyons V, Jacobson MD, Noble J, Diorio J, et al. 2000. 5′-heterogeneity of glucocorticoid receptor messenger RNA is tissue specific: Differential regulation of variant transcripts by early-life events. Mol Endocrinol 14: 506-517.

Meaney MJ. 2001. The development of individual differences in behavioral and endocrine responses to stress. Ann Rev Neurosci 24: 1161-1192.

Meaney MJ, Diorio J, Francis D, La Rocque S, O'Donnell D, et al. 1994. Environmental regulation of the development of glucocorticoid receptor systems in the rat forebrain: The role of serotonin. Ann NY Acad Sci 746: 260-275.

Meaney MJ, Aitken DH, Viau V, Sharma S, Sarrieau, A. 1989 Neonatal handling alters adrenocortical negative feedback sensitivity and hippocampal type II glucocorticoid receptor binding in the rat. Neuroendocrinology 50: 597-604.

Menard J, Champagne D, Meaney MJ. 2004. Maternal care alters behavioral and neural activity patterns in the defensive burying paradigm. Neurosci 129: 297-308.

Milbrandt J. 1987. A nerve growth factor-induced gene encodes a possible transcriptional regulatory factor. Science 238: 797-799.

Mitchell JB, Rowe W, Boksa P, Meaney MJ, 1990b. Serotonin regulates type II, corticosteroid receptor binding in cultured hippocampal cells. J Neurosci 10: 1745-1752.

Mitchell JB, Betito K, Rowe W, Boksa P, Meaney MJ. 1992. Serotonergic regulation of type II corticosteroid receptor binding in cultured hippocampal cells: The role of serotonin-induced increases in cAMP levels. Neurosci 48: 631-639.

Mitchell JB, Iny LJ, Meaney MJ. 1990a. The role of serotonin in the development and environmental regulation of hippocampal type II corticosteroid receptors. Dev Brain Res 55: 231-235.

Moga MM, Gray TS. 1989. Evidence for corticotropin-releasing factor, neurotensin, and somatostatin in the neural pathway from the central nucleus of the amygdala to the parabrachial nucleus. J Comp Neurol 241: 275-284.

Mohammed AH, Henricksson BG, Söderström S, Ebendal T, Olsson T, et al. 1993. Environmental influences on the central nervous system and their implications for the aging rat. Behav Brain Res 57: 183-191.

Montgomery SM, Bartley MJ, Wilkinson RG. 1997 Family conflict and slow growth. Arch Dis Child 77: 326-330.

Myers MM, Brunelli SA, Shair HN, Squire JM, Hofer MA. 1989. Relationships between maternal behavior of SHR and WKY dams and adult blood pressures of cross-fostered F1 pups. Dev Psychobiol 22: 55-67.

Nan X, Ng HH, Johnson CA, Laherty CD, Turner BM, et al. 1998. Transcriptional repression by the methyl-CpG-binding protein MeCP2 involves a histone deacetylase complex. Nature 393: 386-389.

Nemeroff CB. 1996. The corticotropin-releasing factor (CRF) hypothesis of depression: New findings and new directions. Mol Psychiatr 1: 336-342.

Newport DJ, Stowe ZN, Nemeroff CB. 2002. Parental depression: Animal models of an adverse life event. Am J Psychiatr 159: 1265-1283.

Owens MJ, Vargas MA, Knight DL, Nemeroff CB. 1991. The effects of alprazoplam on corticotropin-releasing factor neurons in the rat brain: Acute time course, chronic treatment and abrupt withdrawal. J Pharmacol Exp Ther 258: 349-356.

Page ME, Valentino RJ. 1994. Locus coeruleus activation by physiological challenges. Brain Res Bull 35: 557-560.

Parker G. 1981. Parental representations of patients with anxiety neurosis. Acta Psychiatr Scand 63: 33-36.

Plotsky PM. 1991. Pathways to the secretion of adrenocorticotropin: A view from the portal. J Neuroendocrinol 3: 1-9.

Razin A, Cedar H. 1977. Distribution of 5-methylcytosine in chromatin. Proc Natl Acad Sci USA 74: 2725-2728.

Razin A. 1998. CpG methylation, chromatin structure and gene silencing—a three-way connection. EMBO J 17: 4905-4908.

Razin A, Riggs AD. 1980 DNA methylation and gene function. Science 210: 604-610.

Razin A, Szyf M. 1984 DNA methylation patterns. Formation and function. Biochim Biophys Acta 782: 331-342.

Reik W, Dean W, Walter J. 2000. Epigenetic reprogramming in mammalian development. Science 293: 1089-1093.

Repetti RL, Taylor SE, Seeman TE. 2002. Risky families: Family social environments and the mental and physical health of offspring. Psychol Bull 128: 330-366.

Rivier C, Plotsky PM. 1986. Mediation by corticotropin-releasing factor of adenohypophysial hormone secretion. Ann Rev Physiol 48: 475-489.

Rossiter MC. 1999. Maternal effects as adaptations. Mousseau TA, Fox CW, editors. London: Oxford University Press.

Roth SY, Denu JM, Allis CD. 2001. Histone acetyltransferases. Ann Rev Biochem 70: 81-120.

Russak LG, Schwartz GE. 1997. Feelings of parental care predict health status in midlife: A 35 year follow-up of the Harvard Mastery of Stress Study. J Behav Med 20: 1-11.

Sapienza C. 1990. Parental imprinting of genes. Sci Am 263: 52-60.

Schulkin J, McEwen BS, Gold PW. 1994 Allostasis, the amygdala and anticipatory angst. Neurosci Biobehav Rev 18: 385-396.

Seckl JR, Meaney MJ. 1993. Early life events and later development of ischaemic heart disease. Lancet 342: 1236.

Shi H, Yan PS, Chen CM, Rahmatpanah F, Lofton-Day C, et al., Expressed CpG island sequence tag microarray for dual screening of DNA hypermethylation and gene silencing in cancer cells. Cancer Res 62: 3214–3220.

Sroufe LA. 1997. Psychopathology as an outcome of development. Dev Psychopathol 9: 251-268.

Stenzel-Poore MP, Heinrichs SC, Rivest S, Knob GF, Vale WW. 1994. Overproduction of corticotropin-releasing factor in transgenic mice: A genetic model of anxiogenic behavior. J Neurosci 14: 2579-2584.

Stern JM. 1997. Offspring-induced nurturance: Animal–human parallels. Dev Psychobiol 31: 19-37.

Strähle U, Schmidt A, Kelsey G, Stewart AF, Cole TJ, et al. 1992. At least three promoters direct expression of the mouse glucocorticoid receptor gene. Proc Natl Acad Sci USA 89: 6731-6735.

Strobl JS. 1990. A role for DNA methylation in vertebrate gene expression. Mol Endocrinol 4: 181-183.

Suomi SJ. 1997. Early determinants of behaviour: Evidence from primate studies. Br Med Bull 53: 170-184.

Swiergiel AH. Takahashi LK, Kahn NH. 1993. Attenuation of stress-induced by antagonism of corticotropin-releasing factor receptors in the central amygdala in the rat. Brain Res 623: 229-234.

Szyf M. 2001. Towards a pharmacology of DNA methylation. Trends Pharmacol Sci 22: 350-354.

Trickett PK, McBride-Chang C. 1995. The developmental impact of different forms of child abuse and neglect. Dev Rev 15: 311-337.

Valentino RJ, Curtis AL, Page ME, Pavcovich LA, Florin-Lechner SM. 1998. Activation of the locus coeruleus brain noradrenergic system during stress: Circuitry, consequences, and regulation. Adv Pharmacol 42: 781-784.

Van Bockstaele EJ, Colago EE, Valentino RJ. 1996. Corticotropin-releasing factor-containing axon terminals synapse onto catecholamine dendrites and may presynaptically modulate other afferents in the rostral pole of the nucleus locus coeruleus in the rat brain. J Comp Neurol 364: 523-534.

Weaver ICG, La Plante P, Weaver S, Parent A, Sharma S, et al. 2001. Early environmental regulation of hippocampal glucocorticoid receptor gene expression: characterization of intracellular mediators and potential genomic target sites. Mol Cell Endocrinol 185: 205-218.

Weaver ICG, Cervoni N, D'Alessio, AC, Champagne FA, Seckl JR, et al. 2004. Epigenetic programming through maternal behavior. Nat Neurosci 7: 847-854.

Whitnall MH. 1993. Regulation of the hypothalamic corticotropin-releasing hormone neurosecretory system. Prog Neurobiol 40: 573-629.

Wolffe AP, Jones PL, Wade PA. 1999. DNA demethylation. Proc Natl Acad Sci USA 96: 5894-5896.

16 Energy Balance and Feeding

S. C. Benoit · D. J. Clegg · S. C. Woods

List of Abbreviations: AgRP, agouti-related peptide; ARC, arcuate; CCK, cholecystokinin; CHO, Chinese hamster ovary; CNS, central nervous system; ER, estrogen receptor; GC, glucocorticoid; HSPGs, heparan sulfate proteoglycans; IRS, insulin receptor substrate; IRs, insulin receptors; JAK, Janus Kinases; LH, lateral hypothalamic; LTP, long-term potentiation; MAPK, mitogen activated protein kinase; MC, melanocortin; NEAT, nonexercise activity thermogenesis; NPY, neuropeptide Y; POMC, pro-opiomelanocortin; PVN, paraventricularRQ, respiratory quotient; SOCS, suppressors of cytokine signaling proteins; SPA, spontaneous physical activity; STAT, signal transducers and activators of transcription; WAT, white adipose tissue

1 Introduction

1.1 Body Weight Is a Regulated Variable

Body weight (or more accurately body adiposity) is a regulated variable. To maintain body adiposity over time, caloric intake must equal caloric expenditure. Such an intricate process relies on the interactions of a number of physiological systems. As an example, there is a negative feedback regulator system that is composed of signals derived from adipose tissue, which inform the central nervous system (CNS) about the status of peripheral energy balance. Signals from adipose tissue or peripheral fat stores comprise one side of the hypothesized feedback loop. On the receiving side of this regulatory system are one or more central effector systems that translate adipose-store information into appropriate subsequent behavior. When the system detects low adipose stores, food intake is increased whereas energy expenditure is decreased in an attempt to maintain body adiposity and body weight. This negative feedback loop also functions in the presence of high adipose stores, such that intake is reduced while energy expenditure is increased, again in an attempt to maintain body weight.

1.1.1 Leptin

There are at least two peripherally derived hormones that provide key afferent information to the CNS for body weight regulation. Leptin, a peptide hormone secreted from adipocytes in proportion to fat mass, has received tremendous attention during the last several years. Considerable evidence has been derived that implicates leptin as one of the body's adiposity signals (Elmquist et al., 1999; Ahima et al., 2000; Schwartz et al., 2000; Cone et al., 2001). Leptin levels in the blood correlate directly with body fat, and peripheral or central administration of leptin reduces food intake and increases energy expenditure.

More important, leptin levels are better correlated with subcutaneous fat than with visceral fat in humans, such that the reliability of leptin as an adiposity signal varies with the distribution of body fat. As is described later, there is a sexual dimorphism with respect to how body fat is distributed. Males tend to have more body fat located in the visceral adipose depot, whereas females tend to have more fat in the subcutaneous depot. Because females tend to have more subcutaneous fat than males, on the average, leptin is therefore a better correlate of total adiposity in females than in males (Havel et al., 1996b). Further, when energy balance is suddenly changed (for example, if an individual has fasted for a day), plasma leptin levels decrease far more than body adiposity in the short term (Ahren et al., 1997; Buchanan et al., 1998; Havel, 1999). Hence, although much has been written about leptin as an adiposity signal, it is not ideal in and of itself, suggesting that other signals may exist. One candidate is the pancreatic hormone, insulin.

1.1.2 Insulin

Insulin is well known for its role in regulating glucose homeostasis, however, an oft underrepresented role for insulin is as an adiposity signal. Plasma insulin levels also directly correlate with adiposity, and whereas leptin is a better correlate of subcutaneous adiposity, insulin correlates better with visceral adiposity (Bjorntorp, 1991, 1992, 1997a; Woods et al., 1998). Moreover, when energy balance changes, there are changes in plasma insulin that closely follow changes in homeostasis (Schwartz et al., 1992a). Therefore, both leptin and insulin can be considered as adiposity signals, each indicating something different to the brain; insulin is a correlate of visceral adiposity and leptin is a correlate of subcutaneous adiposity and together or separately, they are markers of changes of metabolic status.

2 Peripheral Signals of Energy Balance

2.1 Leptin

Leptin is a 146-amino acid peptide that is expressed primarily in adipose tissue. Obesity is associated with increased leptin synthesis and secretion, whereas fasting and weight loss are associated with decreased leptin synthesis and secretion (Zhang et al., 1994; Maffei et al., 1995; Ahima et al., 1996; Campfield et al., 1996; Considine et al., 1996; Dagogo-Jack et al., 1996; Dua et al., 1996; Havel et al., 1996b; Kolaczynski et al., 1996; Matson et al., 1996; White and Tartaglia, 1996; Woods et al., 1997; Brunner and Levens, 1998; Buchanan et al., 1998; Clement et al., 1998; Keim et al., 1998; Banks et al., 1999; Baskin et al., 1999a, b; Seeley et al., 1999a, b; Trayhurn et al., 1999; van Dijk et al., 1999; Ahima et al., 2000; Jeanrenaud and Rohner-Jeanrenaud, 2000). Leptin influences many physiological processes, including lipid and glucose metabolism, thermogenesis, reproduction, and many endocrine functions (Zhang et al., 1994; Halaas et al., 1995; Maffei et al., 1995; Friedman, 1997; Kamohara et al., 1997; Friedman, 1998; van Dijk et al., 1999). Compelling evidence suggests that a primary site of leptin action is in the CNS. As body fat increases, more leptin is secreted and enters the brain (Caro et al., 1996; Schwartz et al., 1996a). Leptin acts upon its receptors in the brain to cause a net catabolic response including reduced caloric intake and loss of body weight.

2.1.1 Leptin Receptor Activation: STAT3

Leptin receptors belong to the class I cytokine receptor family (Nakashima et al., 1997), which acts through Janus Kinases (JAK) and signal transducers and activators of transcription (STAT) (Baumann et al., 1996; Ghilardi et al., 1996; White et al., 1997). Several isoforms of the leptin receptor can be produced by alternative splicing, resulting in receptors with the same extracellular domain but different intracellular domains (Tartaglia et al., 1995; Chua et al., 1996; Cioffi et al., 1996; Lee et al., 1996; White and Tartaglia, 1996; White et al., 1997). Isoforms with longer intracellular domains (i.e., with more amino acids) interact with more varied intracellular signaling systems. When the longest and most reactive isoform, called Ob-Rb, binds with a ligand, JAKs become autophosphorylated and catalyze the tyrosine phosphorylation of several STAT proteins. The activated STATs in turn dimerize and translocate to the nucleus, where specific gene responses are elicited (Vaisse et al., 1996; McCowen et al., 1998). Only STAT3 has been found to be activated by leptin in vivo (Vaisse et al., 1996), and this occurs in several hypothalamic areas including the paraventricular (PVN), arcuate (ARC), periventricular, and lateral hypothalamic (LH) nuclei (Bjorbaek et al., 1998; Elmquist et al., 1998a, b; McCowen et al., 1998). A ratio of phosphorylated to nonphosphorylated STAT3 is used as a marker of Ob-Rb activation.

2.1.2 Leptin Activation of SOCS-3

Within the hypothalamus, leptin is thought to stimulate some cell types and to inhibit others. c-Fos is a widely used marker of neuronal activation, and leptin administration elicits c-fos expression in certain hypothalamic neuronal populations (van Dijk et al., 1996). On the inhibitory side, Bjorbaek et al. (1998, 1999a, b) demonstrated that leptin administration rapidly induces the expression of a member of the family of suppressors of cytokine signaling proteins (SOCS)-3 (Hirano et al., 1997; Starr et al., 1997). SOCS-3 is an intracellular protein induced by activation of cytokine receptors such as the long form of the leptin receptor (Tartaglia et al., 1995; Bjorbaek et al., 1998, 1999a, b). SOCS-3 attenuates leptin signaling by blocking phosphorylation of the receptor and downstream STAT proteins involved in signal transduction (Starr et al., 1997; Nicholson and Hilton, 1998). In Chinese hamster ovary (CHO) cells transfected with Ob-Rb, expression of SOCS-3 inhibits leptin-induced tyrosine phosphorylation of JAK2 proteins, suggesting that SOCS-3 is a leptin-regulated inhibitor of proximal leptin signaling in vivo (Bjorbaek et al., 1999a, b). The point is that when leptin interacts with its receptor, it initiates several intracellular processes (e.g., JAK–STAT pathways) that contribute to decreased food intake and related responses; and it simultaneously activates other processes (e.g., SOCS-3) that turn off the JAK–STAT activation, thus limiting its time of action. Very little SOCS-3 mRNA is present in the unstimulated rat brain, but systemic leptin administration rapidly induces expression of the SOCS-3 gene in regions of the rat brain that express Ob-Rb mRNA (Elmquist et al., 1998a, b; Elias et al., 1999). Thus, excessive SOCS-3 activity in leptin responsive neurons is a potential mechanism for leptin resistance (Bjorbaek et al., 1999a, b).

2.1.3 Leptin/Insulin Convergence

Leptin functionally enhances or "sensitizes" some actions of insulin. The underlying molecular mechanisms for the insulin-sensitizing effects of leptin are unclear, and results obtained from studies are conflicting regarding the effect of leptin on insulin-stimulated signal transduction. Whereas the long form of the leptin receptor has the capacity to activate the JAK/STAT3 (Vaisse et al., 1996; Tartaglia, 1997) and mitogen activated protein kinase (MAPK) pathways, leptin is also able to stimulate tyrosine phosphorylation of insulin receptor substrate (IRS-1) (Vaisse et al., 1996), and to increase transcription of fos, jun (Cohen et al., 1996). Some early data suggested that leptin can impair the early steps of insulin signaling including autophosphorylation of the insulin receptor and tyrosine phosphorylation of IRS-1 in certain cell types such as rat-1 fibroblasts and hepatocytes (Cohen et al., 1996). Other studies demonstrated that leptin can mimic effects of insulin such as stimulation of tyrosine phosphorylation of IRS-1 in 293 cells and stimulation of glucose transport and glycogen synthesis in C2C12 myotubes (Berti et al., 1997). The latter may be mediated by stimulation of PI 3-kinase, although unlike insulin, this does not involve IRS-1 (Kellerer et al., 1997). Even under conditions in which leptin decreased tyrosine phosphorylation of IRS-1 in hepatocytes, IRS-1-associated-PI3 kinase activity increased (Cohen et al., 1996) indicating possible crosstalk of the insulin and leptin signaling pathways, but not a complete convergence.

2.2 Insulin

Considerable evidence implicates insulin as an adiposity signal and a key regulator of food intake. Kennedy (1953) hypothesized that adipose tissue produced a hormone that acts as a negative feedback signal for the control of adiposity, and one early suggestion was that this signal is insulin (Woods and Porte, 1976, Baskin et al., 1987; Woods et al., 1996). Data supporting insulin's role as an adiposity signal have been collected over the past two decades and include studies using several species and techniques (Woods et al., 1985, 1990; Schwartz et al., 1994). To summarize that literature, plasma insulin levels correlate directly with body weight and in particular with body adiposity (Bagdade et al., 1967; Polonsky et al., 1988); like leptin, insulin is secreted in direct proportion to fat mass; obese animals and humans have higher basal insulin levels, and they secrete more insulin in response to a meal than lean individuals (Bagdade et al., 1967; Woods et al., 1974). During positive energy balance, insulin increases, whereas insulin decreases during negative energy balance. Insulin resistance in obese animals and humans is characterized by the requirement of greater insulin secretion to maintain euglycemia. Type 2 diabetes mellitus is characterized by insulin resistance, and this is often associated with elevated levels of plasma glucose and obesity, and specifically with visceral adiposity (Bjorntorp, 1997a).

2.2.1 Peripheral Effects

The second line of evidence consistent with insulin being a prime candidate as an adiposity signal is that insulin maps onto acute changes of energy metabolism and adiposity more accurately and more rapidly than its adipocyte cousin, leptin. Insulin secretion tracks changes of energy balance on the order of minutes to hours, as opposed to days as is the case for leptin, and these changes are always in direct proportion to the size of the adipose mass (Polonsky et al., 1988). Additionally, insulin directly regulates glucose and lipid utilization and storage. In the absence of insulin, or during times of insulin resistance and therefore insufficiency of insulin action, insulin sensitive tissues cannot take up glucose, and therefore glucose accumulates in the blood. During these times, adipocytes cannot take up glucose and store fat. When more insulin is secreted to regulate glucose during insulin resistance, the excess insulin causes increased accumulation of fat in adipocytes and leads to obesity. Reductions of insulin sensitivity are associated with both obesity and hyperglycemia, leading ultimately to diabetes.

2.2.2 Central Effects

In the brain, insulin receptors (IRs) and insulin receptor mRNA are found in regions involved in the regulation of food intake and body weight, especially in the hypothalamic ARC nucleus (Corp et al., 1986; Baskin et al., 1990; Schwartz et al., 1992a, b). Direct injections of insulin into the third ventricle (i3vt), which has direct access to the ARC, decrease expression of the anabolic orexigenic effector peptide, neuropeptide Y (NPY) in the ARC (Sipols et al., 1995). ARC NPY fibers project to the paraventricular hypothalamic nucleus (PVN), and intra-third-cerebroventricle (i3vt) insulin causes increased expression of corticotrophin-releasing hormone mRNA in the PVN (Sipols et al., 1995; Schwartz et al., 1996b). These results demonstrate that the insulin signal is tied to the changes in food intake associated with hypothalamic neuropeptides (e.g., Benoit et al., 2002).

Insulin enters the brain via a saturable transport process that moves insulin from the plasma into brain interstitial fluid (Hachiya et al., 1988; Schwartz et al., 1990; Baura et al., 1993; Banks et al., 1997; Woods et al., 2003). Fluctuations of plasma insulin directly affect the rate of entry of insulin to CNS, although higher levels of insulin exceed the saturation point (Baura et al., 1992, 1993). Once saturation has occurred, the entry of insulin to brain remains relatively constant (Schwartz et al., 1992a, 1994). Disruptions in insulin transport have been associated with dysregulation of body weight and obesity.

2.2.3 Central Insulin and Food Intake

Administration of exogenous insulin either as a single injection or over repeated injections into the brain reduces food intake, consistent with its being an adiposity signal (Woods et al., 1979; Brief and Davis, 1984; Plata-Salaman et al., 1986; Chavez et al., 1995a; Woods et al., 1996; Air et al., 2002). Insulin reduces food intake in part by modulating the body's response to short-term signals that terminate meals. Central administration of insulin potentiates the anorexic effects of peripherally administered cholecystokinin (CCK) (Riedy et al., 1995) and centrally administered CRF (Richardson et al., 2000). Additionally, peripherally administered insulin, in amounts that do not cause hypoglycemia, also decreases food intake (Brief and Davis, 1984; McGowan et al., 1990; Chavez et al., 1995b). Consistent with these data, administration of antibodies to insulin directly into the brain increases food intake and body weight (Strubbe and Mein, 1977; McGowan et al., 1992), as does selective absence of neuronal insulin receptors (Bruning et al., 2000) or compromise of the ability to synthesize insulin receptors locally in the hypothalamus (Obici et al., 2002). It is important to note that there is no evidence that alterations in food intake after administration of insulin, either systemic or central, produce aversive consequences (Chavez et al., 1995b). That is, administration of exogenous insulin does not appear to make animals ill. Collectively, these data suggest that insulin is an adiposity signal that provides information to the CNS on the state of energy balance.

2.3 Ghrelin

Recently, the 28-amino acid gastric hormone, ghrelin, has been linked with meal initiation. Ghrelin is also expressed in pancreas, duodenum, and hypothalamus (Kojima et al., 1999), and ghrelin and its receptors are also located in the hypothalamus. Ghrelin administration stimulates food intake and increases body fat mass (Tschöp et al., 2000). The ghrelin receptor, also known as the growth hormone secretagogue receptor (GHS-R1a) (Smith et al., 1999), is localized in specific neurons of the arcuate nucleus that coexpress the orexigenic neuropeptides NPY and AGRP (e.g., Cowley et al., 2003). Endogenous ghrelin is increased during fasting and prior to meal initiation and decreased immediately following nutrient intake in rodents and humans (Tschöp et al., 2000, 2001). Based on these findings, ghrelin has been proposed to be an endogenous meal initiation factor (Cummings et al., 2001). Whereas numerous reports have focused on the orexigenic effects of ghrelin (Horvath et al., 2001; Inui, 2001; Cummings and Shannon, 2003), some have also investigated its effects on activity or energy expenditure (Tschöp et al., 2000; Asakawa et al., 2001; Muccioli et al., 2002), demonstrating that ghrelin can significantly attenuate energy expenditure.

2.4 Dysregulation by Dietary-Induced Obesity

When leptin or insulin is administered into the brain of experimental animals, there is a selective reduction of body fat, with lean body mass being spared (Chen et al., 1996). Consistent with this, when insulin is administered into the brain, there is a reduction of the respiratory quotient (RQ) suggesting that the body is oxidizing relatively more fat (Park et al., 1992). These observations suggest that one action of these adipose signals within the brain is to reduce body fat, and a corollary of this is that fat intake would be expected to be reduced as well. Consistent with this, when insulin is administered i3vt, fat intake is selectively reduced (Chavez et al., 1996). Hence, it is reasonable to hypothesize that leptin and insulin, acting in the brain, reduce body fat by increasing lipid mobilization and oxidation and simultaneously by reducing the consumption of dietary fat.

The epidemiological data that increasing dietary fat accelerates the development of obesity are quite compelling and have been summarized in several reviews (Gibney, 1995; Bray and Popkin, 1998; Samaras et al., 1998). Additionally, studies on animals provide strong corroborative evidence; i.e., across numerous experiments, diets, and species, the conclusion that increased consumption of high-fat (HF) diets leads to increased body fat is inescapable (Bray et al., 1990; Hill et al., 1992; Warwick and Schiffman, 1992; Warwick, 1996; Golay and Bobbioni, 1997; West and York, 1998). More important, there are strong genetic influences that dictate whether or not a given individual will be prone or resistant to becoming obese when exposed to an HF diet (Levin and Routh, 1996; West, 1996; Leibel et al., 1997; Levin et al., 1997; Reed et al., 1997; Bray and Popkin, 1998; West and York, 1998). As Bray and Popkin point out, an HF diet can be viewed as the environmental agent that acts on a susceptible host animal to produce the noninfectious disease, obesity.

Experiments in which animals were rendered obese and then placed on a low(er) fat (LF) diet have been somewhat equivocal, with some reporting loss of body weight (Bernstein et al., 1975; Hill et al., 1992) and others no weight reduction (Faust et al., 1978; Rolls et al., 1980; Harris et al., 1986; Uhley and Jen, 1989; Hill et al., 1992). One important parameter is evidently the age at which the obesity is initially induced. Younger rats (as well as older rats made obese by maintenance on an HF diet for longer intervals) may increase their number of adipocytes (Lemonnier, 1972; Faust et al., 1978; Hill et al., 1992). When subsequently placed on an LF diet, such animals tend not to lose weight (or body fat). However, if the number of adipocytes does not increase, obese animals placed on an LF diet lose weight to the level of rats never made obese at all (Hill et al., 1992). Without assessing fat cell number, we have observed that adult rats given an HF diet and held at an obese weight for a prolonged interval lose weight to control levels when returned to an LF diet (Bernstein et al., 1975). One conclusion that has been reached from this literature is that it is easier to induce obesity in a lean individual with an HF diet than it is to induce leanness in an obese individual with an LF diet (Hill et al., 1992; Bray and Popkin, 1998).

3 Central Effectors

Within the CNS, in order to regulate food intake and body weight effectively, diverse signals need to interact in meaningful ways and to engage neurochemical systems that influence energy intake and energy expenditure. The best known of these CNS systems are in the ventral hypothalamus, and they can be roughly partitioned into those whose activity reduces body fat (catabolic effector systems) and those whose activity increases body fat (anabolic effector systems). Anabolic effectors elicit increased food intake, decreased energy expenditure, and consequently increased stored energy in the form of adipose tissue. They are hypothesized to become more active when energy stores are low as indicated by reduced levels of insulin and leptin (i.e., when the body is in negative energy balance). Catabolic effectors do just the opposite. Activated by positive energy balance, they decrease food intake, increase energy expenditure, and result in decreased adipose tissue mass. A critical aspect of this negative feedback model is that hormones responsive to the level of adiposity inhibit anabolic pathways while activating catabolic pathways, and it is the balance between these two pathways that ultimately determines the animal's ingestive behavior and defended level of adiposity.

The catabolic and anabolic effector systems are in actuality a series of discrete neurotransmitter systems and axonal pathways in the brain, and many of the key details of this overall schema have emerged in the last few years. Although receptors for leptin and insulin are located throughout the CNS, both are concentrated in the ARC in the ventral hypothalamus. Hence, ARC neurons are sensitive to these hormones and consequently to the amount of adipose tissue in the body.

3.1 Neuropeptide-Y

The best-described anabolic effector peptide in the brain is NPY. Although NPY mRNA and peptide are distributed widely throughout the CNS, NPY-containing cell bodies in the ARC are especially important in the control of energy homeostasis (Schwartz et al., 1992b). Although these ARC NPYergic neurons directly influence several areas of the brain, major projections are to the nearby PVN and the lateral hypothalamic area (LHA). ARC NPYergic neurons respond to negative energy balance (e.g., to food deprivation) by synthesizing more NPY mRNA, and they consequently release more NPY in the PVN (Kalra et al., 1991; Schwartz et al., 1992b) and presumably the LHA as well. More important, animals in negative energy balance have low levels of adiposity hormones, resulting in elevated NPY mRNA in the ARC. Local replacement of either insulin or leptin in the vicinity of the ARC normalizes the elevated NPY mRNA in the ARC in fasted animals (Schwartz et al., 1992b; Sipols et al., 1992). Hence, the activity of these ARC NPY neurons is under the direct influence of at least two adiposity signals.

3.2 Melanocortins

The hypothalamic melanocortin (MC) system is the best-known central effector system controlling food intake and energy balance. The evidence for this is multifold: First is the phenotype of the agouti mouse. This yellow mouse has an autosomally transmitted trait resulting from deletion of DNA 3' of the coding region of the agouti gene, which results in ectopic expression of the agouti protein (Bultman et al., 1992). In melanocytes, this inappropriate expression results in antagonism of alpha-melanocyte-stimulating-hormone (a-MSH) signaling. The resulting phenotype is a yellow coat. However, the agouti mouse is obese as well as yellow. This observation led to the hypothesis that ectopic agouti protein antagonizes central MC receptors involved in food intake. Indeed, though melanocortins have been known to influence food intake since the 1980s, it was only during the last decade that their receptors were cloned and localized to the hypothalamus (Mountjoy et al., 1992). As predicted, agouti is an antagonist of these receptors (Lu et al., 1994). More important, an endogenous agouti-related peptide (AgRP) was subsequently found to be expressed in the ARC and to project to the sites of the hypothalamic MC receptors (Fan et al., 1997).

Consistent with the hypothesis that the MC system mediates the effects of adipose hormones is the finding that expression of the MC gene products is regulated by energy balance. During periods of negative energy balance (and consequently low adipose hormones), expression of AgRP mRNA is increased whereas expression of the melanocortin precursor molecule, pro-opiomelanocortin (POMC), is decreased (Mizuno et al., 1999). During positive energy balance (and high levels of adipose hormones) on the other hand, expression of POMC mRNA is increased and AgRP is decreased. Further, POMC-containing neurons also have receptors for the adipocyte hormone, leptin (Cheung et al., 1997; Mountjoy and Wong, 1997; Seeley et al., 1997). These findings suggest that the hypothalamic MC system is a likely central target of adipose signals and a mediator of their effects on food intake.

The brain expresses two types of MC receptors, MC3R and MC4R (Mountjoy et al., 1994). Distribution of MC3R is limited to areas of the hypothalamus, whereas MC4R are located throughout the brain. Most data suggest that MC4R are critical for the MC involvement in food intake. Indeed, an important piece of evidence linking MCs to food intake also supports the hypothesis that MC4R is the critical receptor. Mice with targeted deletion of the MC4R gene are phenotypically similar to the yellow (agouti) mouse (Huszar et al., 1997). Although not yellow, they exhibit profound obesity as well as hyperinsulinemia. As predicted, nonselective MC3/4R agonists (e.g., MTII) do not reduce food intake in MC4R-/- mice (Marsh et al., 1999). Although the issue is complicated by findings that MC3R-/- also have increased body fat but not increased food intake (Chen et al., 2000), such findings support the hypothesis that the MC system plays an important role in the control of energy balance.

Additional evidence suggesting a role of MCs in the control of food intake comes from experimental administration of both naturally occurring and synthetic peptides. Ii3vt administration of a-MSH decreases food intake (Tsujii and Bray, 1989; McMinn et al., 2000), as does i3vt administration of synthetic agonists, including MTII and Ro27-3225 (Fan et al., 1997; Thiele et al., 1998; Benoit et al., 2000). Conversely, administration of MC receptor antagonists, such as AgRP or the synthetic agonist, SHU-9119, elicits long-lasting increase in food intake (Fan et al., 1997; Ollmann et al., 1997; Rossi et al., 1998; Hagan et al., 1999; 2000). These data are consistent with the hypothesized role of the MC system in the control of food intake. Further, the presence of both an endogenous agonist and antagonist of the same receptor make this system a prime candidate for the translation of negative feedback signals from adipose tissue.

4 Sex Difference in Body Weight Regulation

4.1 Estrogens

Although estrogens are mainly produced in the ovaries and testes, a significant proportion is synthesized in adipose tissue, from the precursor steroids, dehydroepiandrosterone and androstenedione (Horton and Tait, 1966; Grodin et al., 1973; Corbould et al., 1998). The local concentration of estrogen within adipose tissue is largely determined by the expression of the enzyme, p450 aromatase, which is responsible for the conversion of androstenedione to estrone and T to estrogen (Horton and Tait, 1966; Corbould et al., 1998). Sex hormone- and fat depot-specific differences in the activity of aromatase, and therefore local estrogen biosynthesis, may underpin depot- and sex-specific patterns of fat distribution. The aromatase P450 gene contains a glucocorticoid (GC) response element, indicating the potential for GC regulation of P450 aromatase (Zhao et al., 1995). Furthermore, cortisol (the human GC) in women, but not men, increases aromatase activity in subcutaneous fat (McTernan et al., 2002). The observation that individuals of both sexes with natural mutations of the gene encoding aromatase develop visceral obesity associated with insulin resistance, hypercholesterolemia, and increased triglyceride levels (Conte et al., 1994; Morishima et al., 1995; Carani et al., 1997) suggests an important role of estrogens in limiting visceral fat and thus determining fat distribution. Estrogen, therefore, may have hormonal and/or paracrine actions in fat depots as adipocytes also express estrogen receptors.

The two isoforms of the estrogen receptor, ERα and ERβ, are the products of separate genes, and both are expressed in the brain and adipose tissue (Lieberburg and McEwen, 1975; McEwen 1975; McEwen et al., 1975a, b; Maclusky et al., 1976; Gray and Wade, 1981; Korach et al., 1996; Heine et al., 2000).

Comparison of the ligand-binding specificity and tissue distribution of ER α and β suggests that there are specific actions of estradiol that can be attributed to one receptor but not the other (Krege et al., 1998; Schomberg et al., 1999; Vidal et al., 2000). The role of each in mediating the attenuating effects of estrogens on food intake, body weight, and body fat distribution has not been conclusively determined. When estrogen acts on its receptors, it increases PI3K activation (Simoncini and Genazzani, 2003; Simoncini et al., 2003) and stimulates its downstream signaling cascade analogously to what occurs following insulin administration and binding to the insulin receptor (Niswender et al., 2003). Mice with targeted deletion of the ERα (but not the ER β) have increased total white adipose tissue (WAT), increased insulin resistance, and impaired glucose tolerance in both sexes (Heine et al., 2000; Ohlsson et al., 2000; Vidal et al., 2000; Geary et al., 2001). Humans lacking either ERα or aromatase also have increased insulin resistance (Morishima et al., 1995; Simpson et al., 2002). Novel compounds are now available that have various estrogenic and antiestrogenic activities on specific estrogen receptors that could be used to test the specific roles of ERα and/or ERβ.

In sum, higher estrogen levels are associated with a pattern of relatively more subcutaneous and less visceral fat, and these in turn are inversely correlated with the risk of metabolic complications (elevated visceral fat is highly correlated with disease). Estrogen receptors in adipose tissue itself and/or in brain are likely to mediate these effects, potentially through activation of PI3K.

4.2 Androgens

In obese men, there is a reciprocal relation between adipose mass and androgens. Specifically, total and free testosterone are both decreased in proportion to the degree of obesity (Zumoff et al., 1990; Isidori et al., 1999) and aging (Hartman et al., 2001). Testosterone levels within the normal physiologic range promote insulin sensitivity (Phillips, 1977; Marin et al., 1992b) and fat deposition in the visceral depot (Marin, 1995; Bhasin et al., 1996; Bjorntorp, 1997); however, decreased testosterone results in insulin insensitivity (Laughlin and Barrett-Connor, 2000; Tsai et al., 2000), and an even greater increase of visceral adiposity (Marin and Arver, 1998). Visceral fat as well as the brain contains androgen receptors, implying that androgen action in either or both locations could contribute to fat distribution.

4.3 Gonadal Regulation of Leptin

Leptin levels are higher in females, even before puberty, compared with males, and this is independent of differences in body composition (Frederich et al., 1995; Ostlund et al., 1996; Rosenbaum et al., 1996; Kennedy et al., 1997; Saad et al., 1997; Demerath et al., 1999; Clegg et al., 2003). After puberty, estrogen and testosterone modulate leptin synthesis and secretion directly, apparently via sex steroid receptor-dependent transcriptional mechanisms (Machinal et al., 1999). In rats, ovariectomy increases leptin levels acutely, and this is reversed by administration of estradiol (Shimizu et al., 1997; Machinal et al., 1999). Leptin secretion rate and leptin mRNA expression are two to three times higher in subcutaneous than visceral fat (Montague et al., 1997; Casabiell et al., 1998).

Leptin levels are inversely correlated with testosterone (Vettor et al., 1997; Wabitsch et al., 1997; Luukkaa et al., 1998; Isidori et al., 1999; Kristensen et al., 1999) and exposure of human fat cells to testosterone or dihydrotestosterone inhibits leptin expression (Wabitsch et al., 1997). In aging and obese men, there is increased aromatase activity and conversion of androgens to estrogen, resulting in increased plasma leptin (Zumoff et al., 1990; Jockenhovel et al., 1997; Morley and Perry, 2000). Testosterone replacement normalizes elevated serum leptin levels in hypogonadal men and in castrated male rats. In hyperandrogenic women, testosterone causes accumulation of visceral adipose tissue. Subcutaneous adipose tissue, on the other hand, has few androgen receptors and estrogen down-regulates the androgen receptors found there (Bjorntorp, 1997b). Estrogen may therefore be protective against androgen effects on female body fat distribution. When estrogen levels become sufficiently low, visceral fat accumulation increases in females.

4.4 Body Fat Distribution

Body fat distribution is an important correlate of metabolic disorders (Kannel, 1985; Larsson, 1991). Excess fat in the central region of the body (visceral, "android," or male-pattern obesity) (Wajchenberg, 2000) is correlated with increased risk and mortality from disorders including diabetes mellitus, hyperlipidemia, hypertension, and atherosclerosis. In contrast, subcutaneous ("gynoid," or female-pattern) fat distribution is poorly correlated with risk for these metabolic diseases (Lapidus et al., 1984; Ohlson et al., 1985; Donahue and Abbott, 1987; Donahue et al., 1987; Bjorntorp, 1996).

4.4.1 Visceral Fat

Visceral fat is characterized by having relatively more capillaries and efferent sympathetic axons per unit volume than subcutaneous white fat, and it drains into the hepatic portal vein. Visceral fat has four times more GC receptors than subcutaneous fat and it is therefore more sensitive to the fat-accumulating effects of circulating GCs and triglycerides (Pedersen et al., 1994). Visceral fat tissue responds to circulating GCs by further increasing its size. Large central fat deposits may contribute to disease in part by releasing free fatty acids into the portal circulation, which can promote other risk factors, such as synthesis of cholesterol and insulin resistance (Bjorntorp, 1997a). Consistent with this, visceral fat is relatively insensitive to insulin (Bolinder et al., 1983a, b; Holmang and Bjorntorp, 1992; Marin et al., 1992b) and insulin action is markedly impaired in individuals with visceral obesity (O'Shaughnessy et al., 1995; Carey et al., 1996). Weight loss is characterized by an initial reduction in visceral fat (as opposed to subcutaneous fat), in part because visceral adipocytes are more metabolically active (Micheli et al., 1969; Bjorntorp, 1992b, c, d, e; Marin et al., 1993). Surgical removal of intra-abdominal fat in rodents prevents the onset of age-dependent insulin resistance and glucose intolerance (Gabriely et al., 2002). Surgical removal of subcutaneous fat tissue of equal weight has no appreciable impact on the same parameters (Gabriely et al., 2002).

Gonadal steroids have been proposed as regulators of the distribution of fat into various depots (Elbers et al., 1999a, b). Visceral fat increases with androgen levels (as well as with age and GC levels) in both genders (Enzi et al., 1986; Bjorntorp, 1992c, e; Bouchard et al., 1993). Postmenopausal women develop increased visceral adiposity, and those who receive estrogen replacement therapy do not (Haarbo et al., 1991a, b; Gambacciani et al., 1997), suggesting a role of estrogen in limiting visceral fat mass (Bjorkelund et al., 1996; Colombel and Charbonnel, 1997). Hence, both androgens and estrogens have been postulated to influence visceral fat.

4.4.2 Subcutaneous Fat

In contrast, relatively little is known of the control of subcutaneous fat. These adipocytes are scattered within and beneath a broad area of skin, are relatively poorly innervated and vascularized, and have a larger average diameter than visceral adipocytes (Wajchenberg, 2000). Because subcutaneous fat has historically been more difficult to measure precisely, its parameters have often been inferred by subtracting visceral fat from total body fat (Wajchenberg, 2000).

5 Central Integration

Leptin and insulin fill distinct niches in the endocrine system. Although leptin has been implicated in several systemic processes, such as angiogenesis (Schwartz et al., 1996a), the primary role of leptin appears to be as a negative feedback adiposity signal that acts in the brain to suppress food intake and net catabolic effector (Porte et al., 1998; Woods et al., 1998; Woods et al., 2003). Consistent with this, animals lacking leptin or functional leptin receptors are grossly obese. Insulin, in contrast, has a primary action in the periphery to regulate blood glucose and stimulate glucose uptake by most tissues. Analogous to leptin, however, deficits in insulin signaling are also associated with hyperphagia in humans, and animals that lack normal insulin signaling in the brain are also obese (Tartaglia et al., 1995; Porte et al., 1998; Woods et al., 1998; Woods et al., 2003).

5.1 Adiposity Signals: Colocalization

The potential for redundancy between leptin and insulin has been highlighted by several recent studies in which leptin and insulin have been found to share both intracellular and neuronal signaling pathways. Although the melanocortin system has long been thought to mediate the central actions of leptin, recent studies, in which insulin significantly stimulated POMC expression in fasted rats and insulin-induced hypophagia was blocked by a nonspecific MC receptor antagonist, (Fan et al., 1997; Ollmann et al., 1997; Seeley et al., 1997; Rossi et al., 1998; Hagan et al., 1999, 2000) strongly support a role for the MC system in the regulation of energy balance by insulin as well. Furthermore, phosphatidylinositol-3-OH kinase (PI(3) K), an enzyme which is an intracellular mediator of insulin signaling (Niswender et al., 2003), appears to play a crucial role in the leptin-induced anorexia signal transduction pathway as well (Niswender et al., 2003). Although these data are consistent with the concept that leptin and insulin share such pathways, they also suggest that over time, this redundancy dissipates and their pathways diverge.

5.2 Plasticity

Recent reports have identified critical roles for mechanisms of plasticity in the control of food intake and body weight, specifically related to the function of long-term adiposity signals. Two important studies have demonstrated that exogenous administration of leptin results in increased axonal outgrowth in mouse pups lacking the leptin gene (obo) (Bouret et al., 2004), to a similar extent seen in normal, wild type mice. Additionally, leptin appears to regulate synaptic structure in hypothalamic tissue of adult rodents (Pinto et al., 2004). Previous work had already suggested a role for leptin in synaptic plasticity and neuronal outgrowth (Shanley et al., 2001; Wayner et al., 2004). For example, ob/ob mice exhibit impaired hippocampal long-term potentiation (LTP) and mice that lack leptin signaling are deficient in some memory tasks (Li et al., 2002; Ohta et al., 2003). However, the recent demonstrations of leptin action on plasticity in hypothalamic tissues suggests that the regulation of energy homeostasis systems may involve cellular mechanisms that have most often been relegated to CNS structures involved in learning. These data suggest that the energy balance regulatory system is linked to systems of structural plasticity. In fact, this may be one mechanism whereby the body resets the adipostatic set-point, as occurs in obesity. We have recently uncovered a role in energy balance for some genes normally associated with "plasticity," known as syndecans.

5.3 Plasticity in Energy Balance

Syndecans are a family of highly abundant cell-surface heparan sulfate proteoglycans (HSPGs) (proteins with covalently attached, highly acidic sugar chains), which are unique in their ability to bind extracellular peptides such as hormones and growth factors. They act as coreceptors by modulating interactions of peptide ligands with their activity-generating receptors. In mammals, the syndecans are composed of four transmembrane HSPGs and members of the glypican family of glycerophosphoinositol-linked HSPGs. Together, these account for nearly all HSPGs that are unbiquitously expressed at cell surfaces (Bernfield et al., 1999; Park et al., 2000). Syndecan family members are found on virtually every cell type (Bernfield et al., 1999) but are differentially expressed in a tissue-specific manner. Syndecan-1 is found predominately on epithelial cells whereas syndecan-3 is found primarily on neural crest-derived cells and neurons. They are induced during development and injury and in response to a wide spectrum of physiological stimuli (Lauri et al., 1998; Bernfield et al., 1999; Hsueh et al., 1999).

The discovery of an important energy balance function of syndecan-3 paralleled the discovery of AgRP's action during the course of studies into the obesity of the agouti mouse. The association of an obese phenotype with a peripheral physiological system led to the hypothesis that a related central system should exist. During the course of studies on a transgene of syndecan-1, Reizes et al. (2001) inferred and subsequently confirmed that endogenous CNS syndecan-3 plays an important role in the regulation of food intake and body weight.

In those studies, Reizes induced overexpression of syndecan-1. For reasons that are still not clear, this did not result in ubiquitous expression, but rather yielded high levels of syndecan-1 expression in specific areas of the brain where it would normally not be found, and in particular, the syndecan-1 transgene was highly expressed in hypothalamus. Syndecan-1 transgenic mice have severe maturity-onset obesity and type-II diabetes. The phenotype of this obesity closely resembles that of previously characterized mice with disruptions of the MC-signaling pathway including the agouti lethal-yellow, AgRP-overexpressers, and MC4-receptor knockout mice (Huszar et al., 1997; Ollmann et al., 1997). Additionally, syndecan-1 was found to potentiate the obesity of the lethal yellow (Ay/a) mice, and to potentiate the activity of AgRP and agouti signaling protein in cell culture preparations. Finally, it was also discovered that syndecan-1 only promoted obesity when it was localized to the cell surface. Mice that constitutively shed syndecan-1 from the cell membrane, because the membrane-binding region of the gene had been deleted, had a complete reversal of the obese phenotype (Reizes et al., 2001). These findings were consistent with the hypothesis that the syndecan-1 transgene acts as a coreceptor for AgRP on MC3 and MC4 receptor-containing neurons. This, in turn, led to the hypothesis that syndecan-3, expressed in hypothalamic tissues, is a coreceptor for endogenous AgRP antagonism.

6 Behavioral Outcomes

There is a long history of research on the behavioral mechanisms that influence food intake. We have previously described several of the possible "psychological" or "motivational" factors that can play important roles in the regulation of food intake and meal size—factors that may contribute to the development of obesity (Woods 1991; Woods and Strubbe, 1994). At the simplest level of analysis, "behavioral outcomes" can include changes in activity or energy expenditure that obviously impact on energy balance and food intake. At a more sophisticated level, these processes might also include psychological mechanisms that underlie changes in intake or food-seeking behaviors. We address these briefly here but recommend the work by Stricker and Woods (2004) for a more extensive discussion.

6.1 Activity and Energy Expenditure

Body fat reflects the difference between energy intake and energy expenditure over time (Ravussin and Gautier, 1999; Ravussin, 2005). Thus, understanding the regulation of energy balance can be partitioned into understanding the factors that regulate caloric intake and factors that regulate caloric expenditure (Schwartz et al., 2003; Seeley and Woods, 2003). Although much is being learned about the mechanisms of caloric intake, less attention has been paid to the control of energy expenditure (e.g., physical activity or exercise) (Levine, 2003) or how dietary fats might affect caloric expenditure.

A compelling case can be made that obesity is associated with lowered rates of energy expenditure (Ravussin and Bogardus, 1992). For example, previous research indicates a strong link between time spent watching television or working at a computer with obesity in children and adults (Arluk et al., 2003; Georgiades et al., 2003; Giammattei et al., 2003). Particularly relevant to human obesity is the phenomenon of nonconscious, nonexercise activity thermogenesis (NEAT), or spontaneous physical activity (SPA). NEAT is defined as the energy expended for everything we do that is not sleeping, eating, or sports-like exercise (Levine, 2002, 2003; Levine et al., 2002). Even trivial physical activities can increase metabolic rate substantially and it is the cumulative impact of a multitude of such exothermic actions that culminate in an individual's daily NEAT (Snitker et al., 2001; Ravussin, 2005).

6.2 Reward and Hedonics

One obvious mechanism by which different dietary fats might give rise to different amounts of caloric intake is the concept of "palatability" or hedonics. It is well known that certain tastes and combinations of macronutrients elicit greatly increased food intake in rats and other species. Generally, this has been described using concepts like hedonic reward, pleasure, and other psychological factors. Although there is a rich and instructive historical debate on the functions and utilities of these conceptual frameworks, one can conceptualize such issues simply: "Do humans and other animals consume more of a given food because they like it more than other food stuffs?" Although seemingly simplistic, it is perhaps one of the most important questions that we can ask with regard to body weight and obesity in humans. If increased intake and resultant changes in body weight are due solely to differences in palatability or hedonic value, then it makes little sense to ask whether HF diets first impair physiological systems (e.g., insulin signaling) to cause obesity.

The first line of evidence supporting the role of palatability in the control of intake comes from experiments that examine learning about flavors. Numerous experiments indicate that the acceptance or avoidance of flavors can be manipulated through conditioning procedures (Capaldi et al., 1983; Sclafani and Nissenbaum, 1988; Elizalde and Sclafani, 1990; Capaldi, 1991; Sclafani, 1991). For example, rats will increase their intake of a less preferred flavor solution (such as the bitter taste of quinine) if its presentation has historically been followed by another, preferred flavor solution (Breslin et al., 1990). In other kinds of experiments, delivery of a flavor that is normally preferred can support the learning of a conditioned place preference in rats (Delamater et al., 2000). In these paradigms, rats are given an equal amount of experience in two novel contexts, each of which has a drinking tube. In one context, the drinking tube delivers water or a nonpreferred solution. In the other, the tube delivers a highly preferred sucrose solution. At a later test, rats are given a choice between the two contexts but without access to the drinking tubes. The common finding is that they spend more time in the context where they had received the sucrose solution. These data indicate that flavor itself can be a powerful stimulus capable of supporting learning.

Dianne Figlewicz-Lattemann, has recently obtained evidence that leptin reduces the reinforcement potency of sweet tastes. In that experiment, food-deprived rats were given access to sucrose in a conditioned place preference paradigm after receiving leptin or saline. As occurs when rats are trained just after eating and are sated, leptin attenuated subsequent preference for the chamber that contained sucrose. More important, the amount of sucrose consumed during training was extremely small (1.5 kcal), strongly implicating taste rather than calories as the key aspect of the stimulus in the paradigm (Figlewicz et al., 2001).

7 Concluding Remarks

The overview presented here is only a small fraction of the vast literature on the complex regulation of food intake, body weight, and energy expenditure. The goal of this chapter has been to provide a concise sample of several critical variables and hypotheses regarding this important system. However, as research continues in the topics of food intake and obesity, the field will undoubtedly change and adopt new as-yet-unexpected ideas about the control of body weight.

References

Ahima RS, Prabakaran D, Mantzoros C, Qu K, Lowell B, et al. 1996. Role of leptin in the neuroendocrine response to fasting. Nature 382: 250-252.

Ahima RS, Saper CB, Flier JS, Elmquist JK. 2000. Leptin regulation of neuroendocrine systems. Front Neuroendocrinol 21: 263-307.

Ahren B, Mansson S, Gingerich RL, Havel PJ. 1997. Regulation of plasma leptin in mice: Influence of age, high-fat diet and fasting. Am J Physiol 273: R113-R120.

Air EL, Benoit SC, Blake Smith KA, Clegg DJ, Woods SC. 2002. Acute third ventricular administration of insulin decreases food intake in two paradigms. Pharmacol Biochem Behav 72(1–2): 423-429.

Arluk SL, Branch JD, Swain DP, Dowling EA. 2003. Childhood obesity's relationship to time spent in sedentary behavior. Mil Med 168(7): 583-586.

Asakawa A, Inui A, Kaga I, Yuzuriha H, Nagata T, et al. 2001. A role of ghrelin in neuroendocrine and behavioral responses to stress in mice. Neuroendocrinology 74: 143-147.

Bagdade JD, Bierman EL, Porte D Jr. 1967. The significance of basal insulin levels in the evaluation of the insulin response to glucose in diabetic and nondiabetic subjects. J Clin Invest 46: 1549-1557.

Banks WA, DiPalma CR, Farrell CL. 1999. Impaired transport of leptin across the blood–brain barrier in obesity. Peptides 20: 1341-1345.

Banks WA, Jaspan JB, Huang W, Kastin AJ. 1997. Transport of insulin across the blood–brain barrier: Saturability at euglycemic doses of insulin. Peptides 18: 1423-1429.

Baskin DG, Figlewicz-Lattemann D, Seeley RJ, Woods SC, Porte D Jr., et al. 1999. Insulin and leptin: Dual adiposity signals to the brain for the regulation of food intake and body weight. Brain Res 848: 114-123.

Baskin DG, Figlewicz DP, Woods SC, D Porte Jr., Dorsa DM. 1987. Insulin in the brain. Annu Rev Physiol 49: 335-347.

Baskin DG, Hahn TM, Schwartz MW. 1999. Leptin sensitive neurons in the hypothalamus. Horm Metab Res 31(5): 345-350.

Baskin DG, Marks JL, Schwartz MW, Figewicz DP, Woods SC, et al. 1990. Insulin and insulin receptors in the brain in relation to food intake and body weight. Endocrine and Nutritional Control of Basic Biological Functions, Lehnert H, Murison R, Weiner H, Hellhammer D, Beyer J. Stuttgart: Hogrefe & Huber: 202-222.

Baumann H, Morella KK, White DW, Dembski M, Bailon PS, et al. 1996. The full-length leptin receptor has signaling capabilities of interleukin 6-type cytokine receptors. Proc Natl Acad Sci USA 93: 8374-8378.

Baura G, Foster D, D Porte Jr., Kahn SE, Bergman RN, et al. 1993. Saturable transport of insulin from plasma into the central nervous system of dogs in vivo: a mechanism for regulated insulin delivery to the brain. J Clin Invest 92: 1824-1830.

Baura GD, Kahn SE, Taborsky GJ Jr., Porte D Jr., Foster D, et al. 1992. Evidence for saturable transport of insulin from plasma into cerebrospinal fluid in vivo. Clin Res 40: 55A.

Benoit SC, Air EL, Coolen LM, Strauss R, Jackman A, et al. 2002. The catabolic action of insulin in the brain is mediated by melanocortins. J Neurosci 22(20): 9048-9052.

Benoit SC, Schwartz MW, Lachey JL, Hagan MM, Rushing PA, et al. 2000. A novel selective melanocortin-4 receptor agonist reduces food intake in rats and mice without producing aversive consequences. J Neurosci 20(9): 3442-3448.

Bernfield M, Gotte M, Park PW, Reizes O, Fitzgerald ML, et al. 1999. Functions of cell surface heparan sulfate proteoglycans. Annu Rev Biochem 68: 729-777.

Bernstein IL, Lotter EC, Kulkosky PJ. 1975. Effect of force-feeding upon basal insulin levels in rats. Proc Soc Exp Biol Med 150: 546-548.

Berti L, Kellerer M, Capp E, Haring HU. 1997. Leptin stimulates glucose transport and glycogen synthesis in C2C12 myotubes: evidence for a P13-kinase mediated effect. Diabetologia 40(5): 606-609.

Bhasin S, Storer TW, Berman N, Callegari C, Clevenger B, et al. 1996. The effects of supraphysiologic doses of testosterone on muscle size and strength in normal men. N Engl J Med 335(1): 1-7.

Bjorbaek C, El-Haschimi K, Frantz JD, Flier JS. 1999. The role of SOCS-3 in leptin signaling and leptin resistance. J Biol Chem 274(42): 30059-30065.

Bjorbaek C, Elmquist JK, El- Haschimi K, Kelly J, Ahima RS, et al. 1999. Activation of SOCS-3 messenger ribonucleic acid in the hypothalamus by ciliary neurotrophic factor. Endocrinology 140(5): 2035-2043.

Bjorbaek C, Elmquist JK, Michl P, Ahima RS, van Bueren A, et al. 1998. Expression of leptin receptor isoforms in rat brain microvessels. Endocrinology 139: 3485-3491.

Bjorkelund C, Lissner L, Andersson S, Lapidus L, Bengtsson C. 1996. Reproductive history in relation to relative weight and fat distribution. Int J Obes Relat Metab Disord 20(3): 213-219.

Bjorntorp P. 1991. Metabolic implications of body fat distribution. Diabetes Care 14(12): 1132-1143.

Bjorntorp P. 1992a. Abdominal fat distribution and disease: An overview of epidemiological data. Ann Med 24(1): 15-18.

Bjorntorp P. 1992b. Abdominal fat distribution and the metabolic syndrome. J Cardiovasc Pharmacol 20 (Suppl. 8): S26-S28.

Bjorntorp P. 1992c. Abdominal obesity and the metabolic syndrome. Ann Med 24(6): 465-468.

Bjorntorp P. 1992d. Hormonal effects on fat distribution and its relationship to health risk factors. Acta Paediatr (Suppl. 383): 59–60; discussion 61.

Bjorntorp P. 1992e. Metabolic abnormalities in visceral obesity. Ann Med 24(1): 3-5.

Bjorntorp P. 1996. The android woman–a risky condition. J Intern Med 239(2): 105-10.

Bjorntorp P. 1997. Body fat distribution, insulin resistance, and metabolic diseases. Nutrition 13: 795-803.

Bjorntorp P. 1997. Hormonal control of regional fat distribution. Hum Reprod (Suppl. 1): 12 21-25.

Bolinder J, Engfeldt P, Ostman J, Arner P. 1983. Site differences in insulin receptor binding and insulin action in subcutaneous fat of obese females. J Clin Endocrinol Metab 57(3): 455-461.

Bolinder J, Ostman J, Arner P. 1983. Influence of aging on insulin receptor binding and metabolic effects of insulin on human adipose tissue. Diabetes 32(10): 959-964.

Bouchard C, Despres JP, Mauriege P. 1993. Genetic and nongenetic determinants of regional fat distribution. Endocr Rev 14(1): 72-93.

Bouret SG, Draper SJ, Simerly RB. 2004. Trophic action of leptin on hypothalamic neurons that regulate feeding. Science 304(5667): 108-110.

Bray GA, Popkin BM. 1998. Dietary fat does affect obesity. Am J Clin Nutr 68: 1157-1173.

Bray GA, Fisler J, York DA. 1990. Neuroendocrine control of the development of obesity: Understanding gained from studies of experimental animal models. Front Neuroendocrinol 11: 128-181.

Breslin PAS, Davidson TL, Grill HJ. 1990. Conditioned reversal of reactions to normally avoided tastes. Physiol Behav 47: 535-538.

Brief DJ, Davis JD. 1984. Reduction of food intake and body weight by chronic intraventricular insulin infusion. Brain Res Bull 12: 571-575.

Bruning JC, Gautam D, Burks DJ, Gillette J, Schubert M, et al. 2000. Role of brain insulin receptor in control of body weight and reproduction. Science 289(5487): 2122-2125.

Brunner L, Levens N. 1998. The regulatory role of leptin in food intake. Curr Opin Clin Nutr Metab Care 1(6): 565-571.

Buchanan C, Mahesh V, Zamorano P, Brann D. 1998. Central nervous system effects of leptin. Trends Endocrinol Metabol 9(4): 146-150.

Bultman S, Michaud E, Woychik R. 1992. Molecular characterization of the mouse agouti locus. Cell 71(Dec 24): 1195-1204.

Campfield LA, Smith FJ, Burn P. 1996. The OB protein (leptin) pathway–a link between adipose tissue mass and central neural networks. Horm Metab Res 28: 619-632.

Capaldi ED. 1991. Hunger and the learning of flavor preferences. Bolles RC, editors. et al. The hedonics of taste. Lawrence Erlbaum Associates, Inc., Hillsdale, NJ, USA; pp. 127-142.

Capaldi ED, Myers DE, Campbell DH, Sheffer JD. 1983. Conditioned flavor preferences based on hunger level during original flavor exposure. Learn Behav 11(1): 107-115.

Carani C, Qin K, Simoni M, Faustini-Fustini M, Serpente S, et al. 1997. Effect of testosterone and estradiol in a man with aromatase deficiency. N Engl J Med 337(2): 91-95.

Carey DG, Jenkins AB, Campbell LV, Freund J, Chisholm DJ. 1996. Abdominal fat and insulin resistance in normal and overweight women: Direct measurements reveal a strong relationship in subjects at both low and high risk of NIDDM. Diabetes 45(5): 633-638.

Caro JF, Kolaczynski JW, Nyce MR, Ohannesian JP, Opentanova I, et al. 1996. Decreased cerebrospinal-fluid/serum leptin ratio in obesity: a possible mechanism for leptin resistance. Lancet 348: 159-161.

Casabiell X, Pineiro V, Peino R, Lage M, Camina J, et al. 1998. Gender differences in both spontaneous and stimulated leptin secretion by human omental adipose tissue in vitro: Dexamethasone and estradiol stimulate leptin release in women, but not in men. J Clin Endocrinol Metab 83(6): 2149-2155.

Chavez M, Kaiyala K, Madden LJ, Schwartz MW, Woods SC. 1995. Intraventricular insulin and the level of maintained body weight in rats. Behav Neurosci 109: 528-531.

Chavez M, Riedy CA, van Dijk G, Woods SC. 1996. Central insulin and macronutrient intake in the rat. Am J Physiol 271: R727-R731.

Chavez M, Seeley RJ, Woods SC. 1995. A comparison between the effects of intraventricular insulin and intraperitoneal LiCl on three measures sensitive to emetic agents. Behav Neurosci 109: 547-550.

Chen AS, Marsh DJ, et al. 2000. Inactivation of the mouse melanocortin-3 receptor results in increased fat mass and reduced lean body mass. Nat Genet 26: 97-102.

Chen G, Koyama K, Yuan X, Lee Y, Zhou YT, et al. 1996. Disappearance of body fat in normal rats induced by adenovirus-mediated leptin gene therapy. Proc Natl Acad Sci, USA 93: 14795-14799.

Cheung CC, Clifton DK, Steiner RA. 1997. Proopiomelanocortin neurons are direct targets for leptin in the hypothalamus. Endocrinology 138: 4489-4492.

Chua SC, Chung WK, Wu-Peng XS, Zhang Y, Liu S, et al. 1996. Phenotypes of mouse diabetes and rat fatty due to mutations in the OB (Leptin) receptor. Science 271: 994-996.

Cioffi JA, Shafer AW, Zupancic TJ, Smith- Gbur J, Mikhail A, et al. 1996. Novel B219/OB receptor isoforms: Possible role of leptin in hematopoiesis and reproduction. Nat Med 2(5): 585-589.

Clegg DJ, Benoit SC, Fisher ME, Barrera JG, Seeley RJ, et al. 2003. Sex hormones determine body fat distribution and sensitivity to adiposity signals. Appetite 40: 324.

Clement K, Vaisse C, Lahlou N, Cabrol S, Pelloux V, et al. 1998. A mutation in the human leptin receptor gene causes obesity and pituitary dysfunction (see comments). Nature 392(6674): 398-401.

Cohen B, Novick D, Rubinstein M. 1996. Modulation of insulin activities by leptin. Science 274(5290): 1185-1188.

Colombel A, Charbonnel B. 1997. Weight gain and cardiovascular risk factors in the post-menopausal women. Hum Reprod 12(Suppl. 1):134-145.

Cone RD, Cowley MA, Butler AA, Fan W, Marks DL, et al. 2001. The arcuate nucleus as a conduit for diverse signals relevant to energy homeostasis. Int J Obes Relat Metab Disord 25 (Suppl. 5): S63-S67.

Considine RV, Sinha MK, Heiman ML, Kriaucinas A, Stephens IW, et al. 1996. Serum immunoreactive-leptin concentrations in normal-weight and obese humans. N Engl J Med 334(5): 292-295.

Conte FA, Grumbach MM, Ito Y, Fisher CR, Simpson ER. 1994. A syndrome of female pseudohermaphrodism, hypergonadotropic hypogonadism, and multicystic ovaries associated with missense mutations in the gene encoding aromatase (P450arom). J Clin Endocrinol Metab 78(6): 1287-1292.

Corbould AM, Judd SJ, Rodgers RJ. 1998. Expression of types 1, 2, and 3 17β-hydroxysteroid dehydrogenase in subcutaneous abdominal and intraabdominal adipose tissue of women. J Clin Endocrinol Metab 83(1): 187-194.

Corp ES, Woods SC, D Porte Jr, Dorsa DM, Figlewicz DP, et al. 1986. Localization of ^{125}I-insulin binding sites in the rat hypothalamus by quantitative autoradiography. Neurosci Lett 70: 17-22.

Cowley MA, Smith RG, Diano S, Tschop M, Pronchuk N, et al. 2003. The distribution and mechanism of action of ghrelin in the CNS demonstrates a novel hypothalamic circuit regulating energy homeostasis. Neuron 37(4): 649-661.

Cummings DE, Purnell JQ, Frayo RS, Schmidova K, Wisse BE, et al. 2001. A preprandial rise in plasma ghrelin levels suggests a role in meal initiation in humans. Diabetes 50 (8): 1714-1719.

Cummings DE, Shannon MH. 2003. Roles for ghrelin in the regulation of appetite and body weight. Arch Surg 138: 389-396.

Dagogo-Jack S, Fanelli C, Paramore D, Landt M. 1996. Plasma leptin and insulin relationships in obese and nonobese humans. Diabetes 45: 695-698.

Delamater A, Sclafani A, Bodnar R. 2000. Pharmacology of sucrose-reinforced place-preference conditioning: Effects of naltrexone. Pharmacol Biochem Behav 65(Apr): 697-704.

Demerath EW, Towne B, Wisemandle W, Blangero J, Chumlea WC, et al. 1999. Serum leptin concentration, body composition, and gonadal hormones during puberty. Int J Obes Relat Metab Disord 23(7): 678-685.

Donahue RP, Abbott RD. 1987. Central obesity and coronary heart disease in men. Lancet 2(8569): 1215.

Donahue RP, Orchard TJ, Becker DJ, Kuller LH, Drash AL. 1987. Sex differences in the coronary heart disease risk profile: a possible role for insulin. The Beaver County Study. Am J Epidemiol 125(4): 650-657.

Dua A, Hennes MI, Hoffman RG, Maas DL, Krakower GR, et al. 1996. Leptin: A significant indicator of total body fat but not of visceral fat and insulin insensitivity in African-American women. Diabetes 45: 1635-1637.

Elbers JM, Asscheman H, Seidell JC, Gooren LJ. 1999. Effects of sex steroid hormones on regional fat depots as assessed by magnetic resonance imaging in transsexuals. Am J Physiol 276(2 Pt 1): E317-E325.

Elbers JM, de Roo GW, Popp-Snijders C, Nicolaas-Merkus A, Westerveen E, et al. 1999. Effects of administration of 17β-oestradiol on serum leptin levels in healthy postmenopausal women. Clin Endocrinol (Oxf) 51(4): 449-454.

Elias CF, Aschkenasi C, Lee C, Kelly J, Ahima RS, et al. 1999. Leptin differentially regulates NPY and POMC neurons projecting to the lateral hypothalamic area. Neuron 23(4): 775-786.

Elizalde G, Sclafani A. 1990. Fat appetite in rats: Flavor preferences conditioned by nutritive and non-nutritive oil emulsions. Appetite 15(Dec): 189-197.

Elmquist JK, Ahima RS, Elias CF, Flier JS, Saper CB. 1998. Leptin activates distinct projections from the dorsomedial and ventromedial hypothalamic nuclei. Proc Natl Acad Sci USA 95: 741-746.

Elmquist JK, Bjorbaek C, Ahima RS, Flier JS, Saper CB. 1998. Distributions of leptin receptor mRNA isoforms in the rat brain. J Comp Neurol 395(4): 535-547.

Elmquist JK, Elias CF, Saper CB. 1999. From lesions to leptin: Hypothalamic control of food intake and body weight. Neuron 22: 221-232.

Enzi G, Gasparo M, Biondetti PR, Fiore D, Semisa M, et al. 1986. Subcutaneous and visceral fat distribution according to sex, age, and overweight, evaluated by computed tomography. Am J Clin Nutr 44(6): 739-746.

Fan W, Boston B, Kesterson R, Hruby V, Cone R. 1997. Role of melanocortinergic neurons in feeding and the agouti obesity syndrome. Nature 385(Jan 9): 165-168.

Faust IM, Johnson PR, Stern JS, Hirsch J. 1978. Diet-induced adipocyte number increase in adult rats: A new model of obesity. Am J Physiol 235: E279-E286.

Figlewicz DP, Higgins MS, Ng- Evans SB, Havel PJ. 2001. Leptin reverses sucrose-conditioned place preference in food-restricted rats. Physiol Behav 73(1–2): 229-234.

Frederich RC, Lollmann B, Hamann A, Napolitano- Rosen A, Kahn BB, et al. 1995. Expression of OB mRNA and its encoded protein in rodents. J Clin Invest 96: 1658-1663.

Friedman MI. 1997. An energy sensor for control of energy intake. Proc Nutr Soc 56(1A): 41-50.

Friedman MI. 1998. Fuel partitioning and food intake. Am J Clin Nutr 67 (Suppl. 3): 513S-518S.

Gabriely I, Ma XH, Yang XM, Atzmon G, Rajala MW, et al. 2002. Removal of visceral fat prevents insulin resistance and glucose intolerance of aging: an adipokine-mediated process? Diabetes 51(10): 2951-2958.

Gambacciani M, Ciaponi M, Cappagli B, Piaggesi L, De Simone L, et al. 1997. Body weight, body fat distribution, and hormonal replacement therapy in early postmenopausal women. J Clin Endocrinol Metab 82(2): 414-417.

Geary N, Asarian L, Korach KS, Pfaff DW, Ogawa S. 2001. Deficits in E2-dependent control of feeding, weight gain, and cholecystokinin satiation in ER-α null mice. Endocrinology 142(11): 4751-4757.

Georgiades E, Reilly JJ, Stathopoulou E, Livingston AM, Pitsiladis YP. 2003. BMI distribution changes in adolescent British girls. Arch Dis Child 88(11): 978-979.

Ghilardi N, Ziegler S, Wiestner A, Stoffel R, Heim MH, et al. 1996. Defective STAT signaling by the leptin receptor in diabetic mice. Proc Natl Acad Sci USA 93(13): 6231-6235.

Giammattei J, Blix G, Marshak HH, Wollitzer AO, Pettitt DJ. 2003. Television watching and soft drink consumption: Associations with obesity in 11- to 13-year-old schoolchildren. Arch Pediatr Adolesc Med 157(9): 882-886.

Gibney MJ. 1995. Epidemiology of obesity in relation to nutrient intake. Int J Obes 19 (Suppl. 5): S1-S3.

Golay A, Bobbioni E. 1997. The role of dietary fat in obesity. Int J Obes Relat Metab Disord 21 (Suppl. 3): S2-S11.

Gray JM, Wade GN. 1981. Food intake, body weight, and adiposity in female rats: Actions and interactions of progestins and antiestrogens. Am J Physiol 240(5): E474-E481.

Grodin JM, Siiteri PK, Mac Donald PC. 1973. Source of estrogen production in postmenopausal women. J Clin Endocrinol Metab 36(2): 207-214.

Haarbo J, Hansen BF, Christiansen C. 1991. Hormone replacement therapy prevents coronary artery disease in ovariectomized cholesterol-fed rabbits. APMIS 99(8): 721-727.

Haarbo J, Marslew U, Gotfredsen A, Christiansen C. 1991. Postmenopausal hormone replacement therapy prevents central distribution of body fat after menopause. Metabolism 40(12): 1323-1326.

Hachiya HL, Halban PA, King GL. 1988. Intracellular pathways of insulin transport across vascular endothelial cells. Am J Physiol 255: C459-C464.

Hagan M, Rushing P, Schwartz M, Yagaloff K, Burn P, et al. 1999. Role of the CNS melanocortin system in the response to overfeeding. J Neurosci 19(Mar 15): 2362-2367.

Hagan MM, Rushing PA, Pritchard LM, Schwartz MW, Strack AM, et al. 2000. Long-term orexigenic effects of AgRP-(83-132) involve mechanisms other than melanocortin receptor blockade. Am J Physiol 279: R47-R52.

Halaas JL, Gajiwala KS, Maffel M, Cohen SL, Chait BI, et al. 1995. Weight-reducing effects of the plasma protein encoded by the obese gene. Science 269: 543-546.

Harris RBS, Kasser TR, Martin RJ. 1986. Dynamics of recovery of body composition after overfeeding, food restriction or starvation of mature female rats. J Nutr 116: 2536-2546.

Hartman TJ, Dorgan JF, Woodson K, Virtamo J, Tangrea JA, et al. 2001. Effects of long-term α-tocopherol supplementation on serum hormones in older men. Prostate 46(1): 33-38.

Havel PJ. 1999. Mechanisms regulating leptin production: Implications for control of energy balance. Am J Clin Nutr 70: 305-306.

Havel PJ, Kasim-Karakas S, Dubuc GR, Mueller W, Phinney SD. 1996. Gender differences in plasma leptin concentrations. Nat Med 2(9): 949-950.

Havel PJ, Kasim Karakas S, Mueller W, Johnson PR, Gingerich RL, et al. 1996. Relationship of plasma leptin to plasma insulin and adiposity in normal weight and overweight women: Effects of dietary fat content and sustained weight loss. J Clin Endocrinol Metab 81: 4406-4413.

Heine PA, Taylor JA, Iwamoto GA, Lubahn DB, Cooke PS. 2000. Increased adipose tissue in male and female estrogen receptor α knockout mice. Proc Natl Acad Sci USA 97(23): 12729-12734.

Hill JO, Lin D, Yakubu F, Peters JC. 1992. Development of dietary obesity in rats: influence of amount and composition of dietary fat. Int J Obes 16: 321-333.

Hirano T, Nakajima K, Hibi M. 1997. Signaling mechanisms through gp130: a model of the cytokine system. Cytokine Growth Factor Rev 8(4): 241-252.

Holmang A, Bjorntorp P. 1992. The effects of testosterone on insulin sensitivity in male rats. Acta Physiol Scand 146(4): 505-510.

Horton R, Tait JF. 1966. Androstenedione production and interconversion rates measured in peripheral blood and studies on the possible site of its conversion to testosterone. J Clin Invest 45(3): 301-313.

Horvath TL, Diano S, Sotonyi P, Heiman M, Ischop M. 2001. Minireview: Ghrelin and the regulation of energy balance–a hypothalamic perspective. Endocrinology 142(10): 4163-4169.

Hsueh YP, Sheng M. 1999. Regulated expression and subcellular localization of syndecan heparan sulfate proteoglycans and the syndecan-binding protein CASK/LIN-2 during rat brain development. J Neurosci 19(17): 7415-7425.

Huszar D, Lynch C, Victoria F-H, Dunmore J, Fang Q, et al. 1997. Targeted disruption of the melanocortin-4 receptor results in obesity in mice. Cell 88(Jan 10): 131-141.

Inui A. 2001. Ghrelin: an orexigenic and somatotrophic signal from the stomach. Nat Rev Neurosci 2: 551-560.

Isidori AM, Caprio M, Strollo F, Moretti C, Frajese G, et al. 1999. Leptin and androgens in male obesity: Evidence for leptin contribution to reduced androgen levels. J Clin Endocrinol Metab 84(10): 3673-3680.

Jeanrenaud B, Rohner-Jeanrenaud F. 2000. CNS–periphery relationships and body weight homeostasis: Influence of the glucocorticoid status. Int J Obes Relat Metab Disord 24 (Suppl. 2): S74-S76.

Jockenhovel F, Blum WF, Vogel E, Englaro P, Muller-Wieland D, et al. 1997. Testosterone substitution normalizes elevated serum leptin levels in hypogonadal men. J Clin Endocrinol Metab 82(8): 2510-2513.

Kalra SP, Dube MG, Sahu A, Phelps CP, Kalra P. 1991 Neuropeptide Y secretion increases in the paraventricular nucleus in association with increased appetite for food. Proc Natl Acad Sci USA 88: 10931-10935.

Kamohara S, Burcelin R, Halaas JL, Friedman JM, Charron MJ. 1997. Acute stimulation of glucose metabolism in mice by leptin treatment. Nature 389: 374-377.

Kannel WB. 1985. Lipids, diabetes, and coronary heart disease: insights from the Framingham Study. Am Heart J 110 (5): 1100-1107.

Keim NL, Stern JS, Havel PJ. 1998. Relation between circulating leptin concentrations and appetite during a prolonged, moderate energy deficit in women. Am J Clin Nutr 68: 794-801.

Kellerer M, Koch M, Metzinger E, Mushack J, Capp E, et al. 1997. Leptin activates PI-3 kinase in C2C12 myotubes via janus kinase-2 (JAK-2) and insulin receptor substrate-2 (IRS-2) dependent pathways. Diabetologia 40(11): 1358-1362.

Kennedy A, Gettys TW, Watson P, Wallace P, Ganaway E, et al. 1997. The metabolic significance of leptin in humans: gender-based differences in relationship to adiposity, insulin sensitivity, and energy expenditure. J Clin Endocrinol Metab 82(4): 1293-1300.

Kennedy GC. 1953. The role of depot fat in the hypothalamic control of food intake in the rat. Proc R Soc Lond (Biol) 140: 579-592.

Kojima M, Hosoda H, Date Y, Nakazato M, Matsuo H, et al. 1999. Ghrelin is a growth-hormone-releasing acylated peptide from stomach. Nature 402(6762): 656-660.

Kolaczynski JW, Considine RV, Ohannesian J, Marco C, Opentanova I, et al. 1996. Responses of leptin to short-term fasting and refeeding in humans: A link with ketogenesis but not ketones themselves. Diabetes 45: 1511-1515.

Korach KS, Couse JF, Curtis SW, Washburn TF, Lindzey J, et al. 1996. Estrogen receptor gene disruption: Molecular characterization and experimental and clinical phenotypes. Recent Prog Horm Res 51:159-86; discussion 186–8.

Krege JH, Hodgin JB, Couse JF, Enmark E, Warner M, et al. 1998. Generation and reproductive phenotypes of mice lacking estrogen receptor β. Proc Natl Acad Sci USA 95 (26): 15677-15682.

Kristensen K, Pedersen SB, Richelsen B. 1999. Regulation of leptin by steroid hormones in rat adipose tissue. Biochem Biophys Res Commun 259(3): 624-630.

Lapidus L, Bengtsson C, Larsson B, Pennert K, Rybo E, et al. 1984. Distribution of adipose tissue and risk of cardiovascular disease and death: a 12-year follow up of participants in the population study of women in Gothenburg, Sweden. BMJ 289: 1257-1261.

Larsson B. 1991. Obesity, fat distribution and cardiovascular disease. Int J Obes 15 (Suppl. 2): 53-57.

Laughlin GA, Barrett- Connor E. 2000. Sexual dimorphism in the influence of advanced aging on adrenal hormone levels: the Rancho Bernardo Study. J Clin Endocrinol Metab 85 (10): 3561-3568.

Lauri SE, Rauvala H, Kaila K, Taira I. 1998. Effect of heparin-binding growth-associated molecule (HB-GAM) on synaptic transmission and early LIP in rat hippocampal slices. Eur J Neurosci 10(1): 188-194.

Lee G, Proenca R, Montez JM, Carroll KM, Darvishzadeh JG, et al. 1996. Abnormal splicing of the leptin receptor in diabetic mice. Nature 379: 632-635.

Leibel RL, Chung WK, Chua SC Jr. 1997. The molecular genetics of rodent single gene obesities. J Biol Chem 272 (51): 31937-31940.

Lemonnier D. 1972. Effect of age, sex and site on the cellularity of adipose tissue in mice and rats rendered obese by a high-fat diet. J Clin Invest 51: 2907-2915.

Levin BE, Dunn-Meynell AA, Balkan B, Keesey RE. 1997. Selective breeding for diet-induced obesity and resistance in Sprague-Dawley rats. Am J Physiol 273: R725-R730.

Levin BE, Routh VH. 1996. Role of the brain in energy balance and obesity. Am J Physiol 271: R491-R500.

Levine J, Peters J, Saris W, Hill J. 2002. Impact of physical activity on the emerging crisis of obesity in Asia. Asia Pac J Clin Nutr 11 (Suppl. 8): S710-S713.

Levine JA. 2002. Non-exercise activity thermogenesis (NEAI). Best Pract Res Clin Endocrinol Metab 16(4): 679-702.

Levine JA. 2003. Non-exercise activity thermogenesis. Proc Nutr Soc 62(3): 667-679.

Li XL, Aou S, Oomura Y, Hori N, Fukunaga K, et al. 2002 Impairment of long-term potentiation and spatial memory in leptin receptor-deficient rodents. Neuroscience 113(3): 607-615.

Lieberburg I, McEwen BS. 1975. Estradiol-17β: a metabolite of testosterone recovered in cell nuclei from limbic areas of adult male rat brains. Brain Res 91(1): 171-174.

Lu D, Willard D, Patel I, Kadwell S, Overton L, et al. 1994. Agouti protein is an antagonist of the melanocyte-stimulating-hormone receptor. Nature 371 (Oct 27): 799-802.

Luukkaa V, Pesonen U, Huhtaniemi I, Lehtonen A, Tilvis R, et al. 1998. Inverse correlation between serum testosterone and leptin in men. J Clin Endocrinol Metab 83(9): 3243-3246.

Machinal F, Dieudonne MN, Leneveu MC, Pecquery R, Giudicelli Y. 1999. In vivo and in vitro ob gene expression and leptin secretion in rat adipocytes: Evidence for a regional specific regulation by sex steroid hormones. Endocrinology 140(4): 1567-1574.

Maclusky NJ, Chaptal C, Lieberburg I, McEwen BS. 1976. Properties and subcellular inter-relationships of presumptive estrogen receptor macromolecules in the brains of neonatal and prepubertal female rats. Brain Res 114(1): 158-165.

Maffei M, Halaas J, Rayussin E, Pratley RE, Lee GM, et al. 1995. Leptin levels in human and rodent: Measurement of plasma leptin and *ob* RNA in obese and weight-reduced subjects. Nat Med 1(11): 1155-1161.

Marin BV, Tschann JM, Gomez CA, Kegeles SM. 1993. Acculturation and gender differences in sexual attitudes and behaviors: Hispanic vs non-Hispanic white unmarried adults. Am J Public Health 83(12): 1759-1761.

Marin P. 1995. Testosterone and regional fat distribution. Obes Res 3 (Suppl 4): 609S-612S.

Marin P, Andersson B, Ottosson M, Olbe L, Chowdhury B, et al. 1992. The morphology and metabolism of intraabdominal adipose tissue in men. Metabolism 41(11): 1242-1248.

Marin P, Arver S. 1998. Androgens and abdominal obesity. Bailliere's Clin Endocrinol Metab 12(3): 441-451.

Marin P, Holmang S, Jonsson L, Sjostrom L, Kvist H, et al. 1992. The effects of testosterone treatment on body composition and metabolism in middle-aged obese men. Int J Obes Relat Metab Disord 16(12): 991-997.

Marsh D, Hollopeter G, Huszar D, Laufer R, Yagaloff K, et al. 1999. Response of melanocortin-4 receptor-deficient mice to anorectic and orexigenic peptides. Nat Genet 21(Jan): 119-122.

Matson CA, Wiater MF, Weigle DS. 1996. Leptin and the regulation of body adiposity: A critical review. Diabetes Rev 4(4): 488-508.

McCowen KC, Chow JC, Smith RJ. 1998. Leptin signaling in the hypothalamus of normal rats in vivo. Endocrinology 139(11): 4442-4447.

McEwen BS. 1975. Hormones, receptors and brain function. Adv Pathobiol(1): 56–69.

McEwen BS, Pfaff DW, Chaptal C, Luine VN. 1975. Brain cell nuclear retention of [^{3}H]estradiol doses able to promote lordosis: Temporal and regional aspects. Brain Res 86(1): 155-161.

McEwen BS, Plapinger L, Chaptal C, Gerlach J, Wallach G. 1975. Role of fetoneonatal estrogen binding proteins in the associations of estrogen with neonatal brain cell nuclear receptors. Brain Res 96(2): 400-406.

McGowan MK, Andrews KM, Grossman SP. 1992. Chronic intrahyphothalamic infusions of insulin or insulin antibodies alter body weight and food intake in the rat. Physiol Behav 51: 753-766.

McGowan MK, Andrews KM, Kelly J, Grossman SP. 1990. Effects of chronic intrahypothalamic infusion of insulin on food intake and diurnal meal patterning in the rat. Behav Neurosci 104: 373-385.

McMinn JE, Wilkinson CW, Havel PJ, Woods SC, Schwartz MW. 2000. Effect of intracerebroventricular α-MSH on food intake, adiposity, c-Fos induction, and neuropeptide expression. Am J Physiol 279: R695-R703.

McTernan PG, Anderson LA, Anwar AJ, Eggo MC, Crocker J, et al. 2002. Glucocorticoid regulation of p450 aromatase activity in human adipose tissue: Gender and site differences. J Clin Endocrinol Metab 87(3): 1327-1336.

Micheli H, Carlson LA, Hallberg D. 1969. Comparison of lipolysis in human subcutaneous and omental adipose tissue with regard to effects of noradrenaline, theophylline, prostaglandin E1 and age. Acta Chir Scand 135(8): 663-670.

Mizuno TM, Makimura H, Silverstein J, Roberts JL, Lopingco T, et al. 1999. Fasting regulates hypothalamic neuropeptide Y, agouti-related peptide, and proopiomelanocortin in diabetic mice independent of changes in leptin or insulin. Endocrinology 140(10): 4551-4557.

Montague CT, Farooqi IS, Whitehead JP, Soos MA, Rau H, et al. 1997. Congenital leptin deficiency is associated with severe early-onset obesity in humans. Nature 387: 903-908.

Morishima A, Grumbach MM, Simpson ER, Fisher C, Qin K. 1995. Aromatase deficiency in male and female siblings caused by a novel mutation and the physiological role of estrogens. J Clin Endocrinol Metab 80(12): 3689-3698.

Morley JE, Perry HM, 3rd. 2000. Androgen deficiency in aging men: Role of testosterone replacement therapy. J Lab Clin Med 135(5): 370-378.

Mountjoy K, Mortrud M, Low M, Simerly R, Cone R. 1994. Localization of the melanocortin-4 receptor (MC4-R) in neuroendocrine and autonomic control circuits in the brain. Mol Endocrinol 8(Oct): 1298-1308.

Mountjoy K, Robbins L, Mortrud M, Cone R. 1992. The cloning of a family of genes that encode the melanocortin receptors. Science 257(Aug 28): 1248-1251.

Mountjoy K, Wong J. 1997. Obesity, diabetes and functions for proopiomelanocortin-derived peptides. Mol Cell Endocrinol 128(Apr 4): 171-177.

Muccioli G, Tschop M, Papotti M, Deghenghi R, Heiman M, et al. 2002. Neuroendocrine and peripheral activities of ghrelin: Implications in metabolism and obesity. Eur J Pharmacol 440(2–3): 235-54.

Nakashima K, Narazaki M, Taga T. 1997. Leptin receptor (OB-R) oligomerizes with itself but not with its closely related cytokine signal transducer gp130. FEBS Lett 403(1): 79-82.

Nicholson SE, Hilton DJ. 1998. The SOCS proteins: a new family of negative regulators of signal transduction. J Leukoc Biol 63(6): 665-668.

Niswender KD, Morrison CD, Clegg DJ, Olson R, Baskin DG, et al. 2003. Insulin activation of phosphatidylinositol-3-kinase in the hypothalamic arcuate nucleus: a key mediator of insulin-induced anorexia. Diabetes 52(2): 227-231.

Niswender KD, Schwartz MW. 2003. Insulin and leptin revisited: Adiposity signals with overlapping physiological and intracellular signaling capabilities. Front Neuroendocrinol 24: 1-10.

O'Shaughnessy IM, Myers TJ, Stepniakowski K, Nazzaro P, Kelly TM, et al. 1995. Glucose metabolism in abdominally obese hypertensive and normotensive subjects. Hypertension 26(1): 186-192.

Obici S, Zhang BB, Karkanias G, Rossetti L. 2002. Hypothalamic insulin signaling is required for inhibition of glucose production. Nat Med 8(12): 1376-1382.

Ohlson LO, Larsson B, Svardsudd K, Welin L, Eriksson H, et al. 1985. The influence of body fat distribution on the incidence of diabetes mellitus 13.5 years of follow-up of the participants in the study of men born in 1913. Diabetes 34 (10): 1055-1058.

Ohlsson C, Hellberg N, Parini P, Vidal O, Bohlooly M, et al. 2000. Obesity and disturbed lipoprotein profile in estrogen receptor-α-deficient male mice. Biochem Biophys Res Commun 278(3): 640-645.

Ohta R, Shigemura N, Sasamoto K, Koyano K, Ninomiya Y. 2003. Conditioned taste aversion learning in leptin-receptor-deficient db/db mice. Neurobiol Learn Mem 80(2): 105-112.

Ollmann M, Wilson B, Yang Y, Kerns J, Chen Y, et al. 1997. Antagonism of central melanocortin receptors in vitro and in vivo by agouti-related protein. Science 278135-138. (Oct 3):

Ostlund RE Jr., Yang JW, Klein S, Gingerich R. 1996. Relation between plasma leptin concentration and body fat, gender, diet, age, and metabolic covariates. J Clin Endocrinol Metab 81(11): 3909-3913.

Park C, Chavez M, Woods SC. 1992. IVT insulin decreases respiratory quotient in rats. Soc Neurosci Abstr 18: 939 (abstr).

Park PW, Reizes O, Bernfield M. 2000. Cell surface heparan sulfate proteoglycans: selective regulators of ligand–receptor encounters. J Biol Chem 275(39): 29923-29926.

Pedersen SB, Jonler M, Richelsen B. 1994. Differences in glucocorticoid receptors and lipoprotein lipase activity in human adipose tissue. J Clin Endocrinol Metab 78(6): 1354-1359.

Phillips GB. 1977. Relationship between serum sex hormones and glucose, insulin and lipid abnormalities in men with myocardial infarction. Proc Natl Acad Sci USA 74 (4): 1729-1733.

Pinto S, Roseberry AG, Liu H, Diano S, Shanabrough M, et al. 2004. Rapid rewiring of arcuate nucleus feeding circuits by leptin. Science 304(5667): 110-115.

Plata- Salaman CR, Oomura Y, Shimizu N. 1986. Dependence of food intake on acute and chronic ventricular administration of insulin. Physiol Behav 37: 717-734.

Polonsky KS, Given E, Carter V. 1988. Twenty-four-hour profiles and pulsatile patterns of insulin secretion in normal and obese subjects. J Clin Invest 81: 442-448.

Porte DJ, Seeley RJ, Woods SC, Baskin DG, Figlewicz DP, et al. 1998. Obesity, diabetes and the central nervous system. Diabetologia 41: 863-881.

Ravussin E. 2005. Physiology. A NEAT way to control weight? Science 307(5709): 530-531.

Ravussin E, Bogardus C. 1992. A brief overview of human energy metabolism and its relationship to essential obesity. Am J Clin Nutr 55: 242S-245S.

Ravussin E, Gautier JF. 1999. Metabolic predictors of weight gain. Int J Obes Relat Metab Disord 23 (Suppl. 1): 37-41.

Reed DR, Bachmanov AA, Beauchamp GK, Tordoff MG, Price RA. 1997. Heritable variation in food preferences and their contribution to obesity. Behav Genet 27: 373-387.

Reizes O, Lincecum J, Wang Z, Goldberger O, Huang L, et al. 2001. Transgenic expression of syndecan-1 uncovers a physiological control of feeding behavior by syndecan-3. Cell 106(1): 105-116.

Richardson RD, Boswell T, Woods SC, Wingfield JC. 2000. Intracerebroventricular corticotropin-releasing factor decreases food intake in white-crowned sparrows. Physiol Behav 71(1–2): 213-216.

Riedy CA, Chavez M, Figlewicz DP, Woods SC. 1995. Central insulin enhances sensitivity to cholecystokinin. Physiol Behav 58: 755-760.

Rolls BJ, Rowe EA, Turner RC. 1980. Persistent obesity in rats following a period of consumption of mixed, high energy diet. J Physiol (London) 298: 415-427.

Rosenbaum M, Nicolson M, Hirsch J, Heymsfield SB, Gallagher D, et al. 1996. Effects of gender, body composition, and menopause on plasma concentrations of leptin. J Clin Endocrinol Metab 81: 3424-3427.

Rossi M, Kim M, Morgan D, Small C, Edwards C, et al. 1998. A C-terminal fragment of Agouti-related protein increases feeding and antagonizes the effect of α-melanocyte stimulating hormone in vivo. Endocrinology 139(Oct): 4428-4431.

Saad MF, Damani S, Gingerich RL, Riad-Gabriel MG, Khan A, et al. 1997. Sexual dimorphism in plasma leptin concentration. J Clin Endocrinol Metab 82(2): 579-584.

Samaras K, Kelly PJ, Chiano MN, Arden N, Spector TD, et al. 1998. Genes versus environment. Diabetes Care 21: 2069-2076.

Schomberg DW, Couse JF, Mukherjee A, Lubahn DB, Sar M, et al. 1999. Targeted disruption of the estrogen receptor-α gene in female mice: characterization of ovarian responses and phenotype in the adult. Endocrinology 140(6): 2733-2744.

Schwartz MW, Figlewicz DP, Baskin DG, Woods SC, Porte D Jr. 1992. Insulin in the brain: a hormonal regulator of energy balance. Endocr Rev 13: 387-414.

Schwartz MW, Figlewicz DP, Baskin DG, Woods SC, Porte D Jr. 1994. Insulin and the central regulation of energy balance: Update 1994. Endocrine Monographs Reviews 2: 109-113.

Schwartz MW, Peskind E, Raskind M, Nicolson M, Moore J, et al. 1996. Cerebrospinal fluid leptin levels: Relationship to plasma levels and to adiposity in humans. Nat Med 2(5): 589-593.

Schwartz MW, Seeley RJ, Campfield LA, Burn P, Baskin DG. 1996. Identification of hypothalamic targets of leptin action. J Clin Invest 98: 1101-1106.

Schwartz MW, Sipols AJ, Kahn SE, Lattemann DP, Taborsky GJ Jr., et al. 1990. Kinetics and specificity of insulin uptake from plasma into cerebrospinal fluid. Am J Physiol 259: E378-E383.

Schwartz MW, Sipols AJ, Marks JL, Sanacora G, White JD, et al. 1992. Inhibition of hypothalamic neuropeptide Y gene expression by insulin. Endocrinology 130: 3608-3616.

Schwartz MW, Woods SC, Porte DJ, Seeley RJ, Baskin DG. 2000. Central nervous system control of food intake. Nature 404: 661-671.

Schwartz MW, Woods SC, Seeley RJ, Barsh GS, Baskin DG, et al. 2003. Is the energy homeostasis system inherently biased toward weight gain? Diabetes 52: 232-238.

Sclafani A. 1991. Conditioned food preferences and appetite. Appetite 17(Aug): 71-72.

Sclafani A, Nissenbaum J. 1988. Robust conditioned flavor preference produced by intragastric starch infusions in rats. Am J Physiol 255(Oct): R672-R675.

Seeley R, Schwartz M. 1999. Neuroendocrine regulation of food intake. Acta Paediatr Supplement 88(Feb): 58-61.

Seeley R, Yagaloff K, Fisher S, Burn P, Thiele T, et al. 1997. Melanocortin receptors in leptin effects. Nature 390 (Nov 27): 349.

Seeley RJ, Woods SC. 1999. Leptin actions in the hypothalamus. Regulation of pituitary hormone secretion. In: P. Beck-Peccoz (ed) Regulation of Pituitry Hormone Secretion. Bioscientifica, Bristol UK.

Seeley RJ, Woods SC. 2003. Monitoring of stored and available fuel by the CNS: Implications for obesity. Nat Rev Neurosci 4(11): 901-909.

Shanley LJ, Irving AJ, Harvey J. 2001. Leptin enhances NMDA receptor function and modulates hippocampal synaptic plasticity. J Neurosci 21(24): RC186.

Shimizu H, Shimomura Y, Nakanishi Y, Futawatari T, Ohtani K, et al. 1997. Estrogen increases in vivo leptin production in rats and human subjects. J Endocrinol 154(2): 285-292.

Simoncini T, Genazzani AR. 2003. Non-genomic actions of sex steroid hormones. Eur J Endocrinol 148(3): 281-292.

Simoncini T, Rabkin E, Liao JK. 2003. Molecular basis of cell membrane estrogen receptor interaction with phosphatidylinositol-3-kinase in endothelial cells. Arterioscler Thromb Vasc Biol 23(2): 198-203.

Simpson ER, Clyne C, Rubin G, Boon WC, Robertson K, et al. 2002. Aromatase–a brief overview. Annu Rev Physiol 64: 93-127.

Sipols AJ, Baskin DG, Schwartz MW. 1995. Effect of intracerebroventricular insulin infusion on diabetic hyperphagia and hypothalamic neuropeptide gene expression. Diabetes 44: 147-151.

Sipols AJ, Brief DJ, Ginter KL, Saghafi S, Woods SC. 1992. Neuropeptide Y paradoxically increases food intake yet causes conditioned flavor aversions. Physiol Behav 51: 1257-1260.

Smith RG, Palyha OC, Feighner SD, Tan CP, McKee KK, et al. 1999. Growth hormone releasing substances: Types and their receptors. Horm Res 51 (Suppl. 3): 1-8.

Snitker S, Tataranni PA, Ravussin E. 2001. Spontaneous physical activity in a respiratory chamber is correlated to habitual physical activity. Int J Obes Relat Metab Disord 25(10): 1481-1486.

Starr R, Willson TA, Viney EM, Murray LJ, Rayner JR, et al. 1997. A family of cytokine-inducible inhibitors of signalling. Nature 387(6636): 917-921.

Stricker EM, Woods SC. 2004. Neurobiology of Food and Fluid Intake (Handbooks of Behavioral Neurobiology). USA, Plenum.

Strubbe JH, Mein CG. 1977. Increased feeding in response to bilateral injection of insulin antibodies in the VMH. Physiol Behav 19: 309-313.

Tartaglia LA. 1997. The leptin receptor. J Biol Chem 272: 6093-6096.

Tartaglia LA, Dembski M, Weng X, Deng N, Culpepper J, et al. 1995. Identification and expression cloning of a leptin receptor, OB-R. Cell 83: 1263-1271.

Thiele I, van DG, Yagaloff K, Fisher S, Schwartz M, et al. 1998. Central infusion of melanocortin agonist MTII in rats: Assessment of c-Fos expression and taste aversion. Am J Physiol 274(Jan): R248-R254.

Trayhurn P, Hoggard N, Mercer JG, Rayner DV. 1999. Leptin: Fundamental aspects. Int J Obes Relat Metab Disord 23 (Suppl. 1): 22-28.

Tsai EC, Boyko EJ, Leonetti DL, Fujimoto WY. 2000. Low serum testosterone level as a predictor of increased visceral fat in Japanese-American men. Int J Obes Relat Metab Disord 24(4): 485-491.

Tschöp M, Smiley DL, Heiman ML. 2000. Ghrelin induces adiposity in rodents. Nature 407(6806): 908-913.

Tschöp M, Weyer C, Tataranni PA, Devanarayan V, Ravussin E, et al. 2001. Circulating ghrelin levels are decreased in human obesity. Diabetes 50(4): 707-709.

Tsujii S, Bray GA. 1989. Acetylation alters the feeding response to MSH and β-endorphin. Brain Res Bull 23: 165-169.

Uhley VE, Jen KLC. 1989. Changes in feeding efficiency and carcass composition in rats on repeated high-fat feedings. Int J Obes 13: 849-856.

Vaisse C, Halaas JL, Horvath CM, JE Darnell Jr., Stoffel M, et al. 1996. Leptin activation of Stat3 in the hypothalamus of wild-type and ob/ob mice but not db/db mice. Nat Genet 14(1): 95-97.

van Dijk G, Seeley RJ, Thiele TE, Friedman MI, Ji H, et al. 1999. Metabolic, gastrointestinal, and CNS neuropeptide effects of brain leptin administration in the rat. Am J Physiol 276: R1425-R1433.

van Dijk G, Thiele TE, Donahey JCK, Campfield LA, Smith FJ, et al. 1996. Central infusion of leptin and GLP-1 (7-36) amide differentially stimulate c-Fos-like immunoreactivity in the rat brain. Am J Physiol 271: R1096-R1100.

Vettor R, De Pergola G, Pagano C, Englaro P, Laudadio E, et al. 1997. Gender differences in serum leptin in obese people: Relationships with testosterone, body fat distribution and insulin sensitivity. Eur J Clin Invest 27(12): 1016-1024.

Vidal O, Lindberg MK, Hollberg K, Baylink DJ, Andersson G, et al. 2000. Estrogen receptor specificity in the regulation of skeletal growth and maturation in male mice. Proc Natl Acad Sci USA 97(10): 5474-5479.

Wabitsch M, Blum WF, Muche R, Braun M, Hube F, et al. 1997. Contribution of androgens to the gender difference in leptin production in obese children and adolescents. J Clin Invest 100(4): 808-813.

Wajchenberg BL. 2000. Subcutaneous and visceral adipose tissue: Their relation to the metabolic syndrome. Endocr Rev 21(6): 697-738.

Warwick ZS. 1996. Probing the causes of high-fat diet hyperphagia: A mechanistic and behavioral dissection. Neurosci Biobehav Rev 16: 585-596.

Warwick ZS, Schiffman SS. 1992. Role of dietary fat in calorie intake and weight gain. Neurosci Biobehav Rev 16: 585-596.

Wayner MJ, Armstrong DL, Phelix CF, Oomura Y. 2004. Orexin-A (Hypocretin-1) and leptin enhance LTP in the dentate gyrus of rats in vivo. Peptides 25(6): 991-996.

West DB. 1996. Genetics of obesity in humans and animal models. Endocrinol Metab Clin North Am 25: 801-813.

West DB, York B. 1998. Dietary fat, genetic predisposition, and obesity: Lessons from animal models. Am J Clin Nutr 67 (Suppl. 3): 505S-512S.

White DW, Kuropatwinski KK, Devos R, Baumann H, Tartaglia LA. 1997. Leptin receptor (OB-R) signaling. Cytoplasmic domain mutational analysis and evidence for receptor homo-oligomerization. J Biol Chem 272: 4065-4071.

White DW, Tartaglia LA. 1996. Leptin and OB-R: Body weight regulation by a cytokine receptor. Cytokine Growth Factor Rev 7: 303-309.

Woods SC. 1991. The eating paradox: How we tolerate food. Psychol Rev 98: 488-505.

Woods SC, Chavez M, Park CR, Riedy C, Kaiyala K, et al. 1996. The evaluation of insulin as a metabolic signal controlling behavior via the brain. Neurosci Biobehav Rev 20: 139-144.

Woods SC, Decke E, Vasselli JR. 1974. Metabolic hormones and regulation of body weight. Psychol Rev 81(1): 26-43.

Woods SC, Figlewicz DP, Schwartz MW, Porte D Jr. 1990. A re-assessment of the regulation of adiposity and appetite by the brain insulin system. Int J Obesity 14(Suppl. 3): 69-76.

Woods SC, Lotter EC, McKay LD, Porte D Jr. 1979. Chronic intracerebroventricular infusion of insulin reduces food intake and body weight of baboons. Nature 282: 503-505.

Woods SC, Porte D Jr. 1976. Insulin and the set-point regulation of body weight. In: Novin D, Bray GA, Wyrwichka W (eds) Hunger: Basic Mechanisms and Clinical Implications: Raven Press, New York, p 273–280.

Woods SC, Porte DJ, Bobbioni E, Ionescu E, Sauter JF, et al. 1985. Insulin: its relationship to the central nervous system and to the control of food intake and body weight. Am J Clin Nutr 42: 1063-1071.

Woods SC, Seeley RJ, Baskin DG, Schwartz MW. 2003. Insulin and the blood–brain barrier. Curr Pharm Des 9: 795-800.

Woods SC, Seeley RJ, Porte DJ, Schwartz MW. 1998. Signals that regulate food intake and energy homeostasis. Science 280: 1378-1383.

Woods SC, Seeley RJ, Schwartz MW. 1997. Leptin as an adiposity signal. Endocrinol Metab 4: 77-79.

Woods SC, Strubbe JH. 1994. The psychobiology of meals. Psychonomic Bulletin and Review 1: 141-155.

Zhang Y, Proenca R, Maffei M, Barone M, Leopold L, et al. 1994. Positional cloning of the mouse obese gene and its human homologue. Nature 372: 425-432.

Zhao Y, Nichols JE, Bulun SE, Mendelson CR, Simpson ER. 1995. Aromatase P450 gene expression in human adipose tissue. Role of a Jak/STAT pathway in regulation of the adipose-specific promoter. J Biol Chem 270(27): 16449-16457.

Zumoff B, Strain GW, Miller LK, Rosner W, Senie R, et al. 1990. Plasma free and non-sex-hormone-binding-globulin-bound testosterone are decreased in obese men in proportion to their degree of obesity. J Clin Endocrinol Metab 71(4): 929-931.

17 The Neuroendocrinology, Neurochemistry and Molecular Biology of Thirst and Salt Appetite

A. K. Johnson · R. L. Thunhorst

List of Abbreviations: 11β-HSD, 11β-hydroxysteroid dehydrogenase; 5-HT, serotonin; 8-OH-DPAT, 8-hydroxy-2-(di-*n*-propylamine) tetralin; ACE, angiotensin converting enzyme; ACTH, adrenocorticotropic hormone; ADH, antidiuretic hormone; AdMnSOD, adenoviral-vector encoding human mitochondrial superoxide dismutase; AGT, angiotensinogen; ANG, angiotensin; ANP, atrial natriuretic peptide; AP, area postrema; AT_1, angiotensin Type 1; AV3V, anteroventral third ventricle; CAP, captopril; CeA, amygdala; CNS, central nervous system; CSF, cerebrospinal fluid; CVOs, circumventricular organs; DOCA, deoxy-corticosterone; DOI, 2,5-dimethoxy-4-iodoamphetamine bromide; ECF, extracellular fluid compartment; FURO, furosemide; i.p., intraperitoneal; ICF, intracellular fluid compartment; icv, intracerebroventricular; iv, intravenously; LPBN, lateral parabrachial nucleus; MePO, median preoptic nucleus; Methy, methyseride; NE, Norepinephrine; NTS, nucleus of the solitary tract; OT, oxytocin; OVLT, organum vasculosum of the lamina terminalis; PVN, paraventricular nucleus; RAS, renin-angiotensin system; s.c., subcutaneous; SAD, sinoaortic baroreceptor denervation; SCVOs, sensory circumventricular organs; SFO, subfornical organ; SON, supraoptic nucleus; TRPV1, transient receptor potential channel, vanilloid subfamily Type 1; TRPV4, transient receptor potential channel, vanilloid subfamily Type 4; *trpv4-/-*, *trpv4* null or knockout; VP, Vasopressin; VR-OAC, vanilloid receptor-related osmotically activated channel

1 Introduction

1.1 Homeostatic Reflexes and Behaviors Maintain Body Fluid Balance

The maintenance of the internal milieu requires coordinated efforts of autonomic and endocrine reflexes and behaviors. Terrestrial animals constantly lose water and sodium to the environment. Thus, the volumes and constituents of the various body fluids are always in flux. Renal losses of sodium and water are minimized by autonomic and endocrine responses [e.g., activation of the sympathetic nervous system; release of aldosterone and the antidiuretic hormone, vasopressin (VP)]. Behavioral responses include the seeking out and ingestion of water and salty substances. The consumption of water and sodium is required to replenish the body with these substances. The motivational states that drive animals to find and consume water and salty substances are *thirst* and *salt appetite* and are defined operationally by measuring their consumption under specified experimental conditions.

Homeostatic reflexes and behaviors operate with different latencies. Sympathetic reflexes begin within moments after an imbalance of body fluids is detected. Endocrine effects are felt within minutes. Behavioral responses take somewhat longer. Thus, reflex mechanisms minimize homeostatic disturbances and allow time for behaviors to be mobilized. Coordination of these reflex and behavioral responses requires the integrative action of the central nervous system (CNS) which continuously monitors body fluid status. The reflexes and behaviors that maintain body fluid homeostasis often employ common mechanisms. They are activated by similar afferent signals to the brain and use many of the same CNS structures and neurochemical systems to accomplish their complementary roles. For example, the hormone angiotensin (ANG) II stimulates sympathetic outflow, VP release, and water and sodium ingestion through overlapping neural structures and pathways. This review summarizes the major neural and humoral signaling mechanisms that apprise the CNS of body fluid status and the key brain systems (structures and neurochemicals) that generate thirst and salt appetite.

1.2 Determinants of the Volume and Distribution of Body Fluids

Approximately 60% of the weight of the body is water. Almost all of this water is contained in the intracellular and the extracellular fluid compartments with the amount inside the cells being approximately twice that outside. The solute composition of these fluid spaces is markedly different, and the maintenance of a 2:1 ratio of intra- to extracellular water is mostly the result of the ratio of sodium to water (sodium concentration) in the extracellular fluid. Because the cell membrane is effectively impermeable to sodium, this monovalent ion generates an osmotic gradient, which is responsible for the movement of water between the fluid compartments.

Under normal conditions, water and sodium are gradually lost from the body (see *Figure 17-1*), and exposure to moderate physiological challenges such as periodic exercise, increases in ambient temperature, lactation and gestation will accelerate such losses. The rates of water and sodium loss to the environment can be quite different from each other. Hence, the net effect of ongoing depletion of the substances may result in a profile of loss that produces an internal state that is hypotonic, isotonic, or hypertonic. The consequence is that water will not necessarily remain distributed in a constant 2:1 proportion between the intra- and extracellular fluid compartments.

Figure 17-1

Schematic of changes in body fluid distribution and composition as a result of manipulations that produce negative and positive water and sodium balances. Above each diagram is a description of the body fluid status and an example of a manipulation used to induce it. To the right of the diagram is an indication of the control actions elicited attempting to compensate for or correct the disturbance. ECF, extracellular fluid compartment; ICF, intracellular fluid compartment; thirst = drinking water; sodium appetite = salt appetite = ingestion of salty substances, notably hypertonic NaCl solutions. Reprinted from Johnson (1990); originally adapted from Andersson (1971)

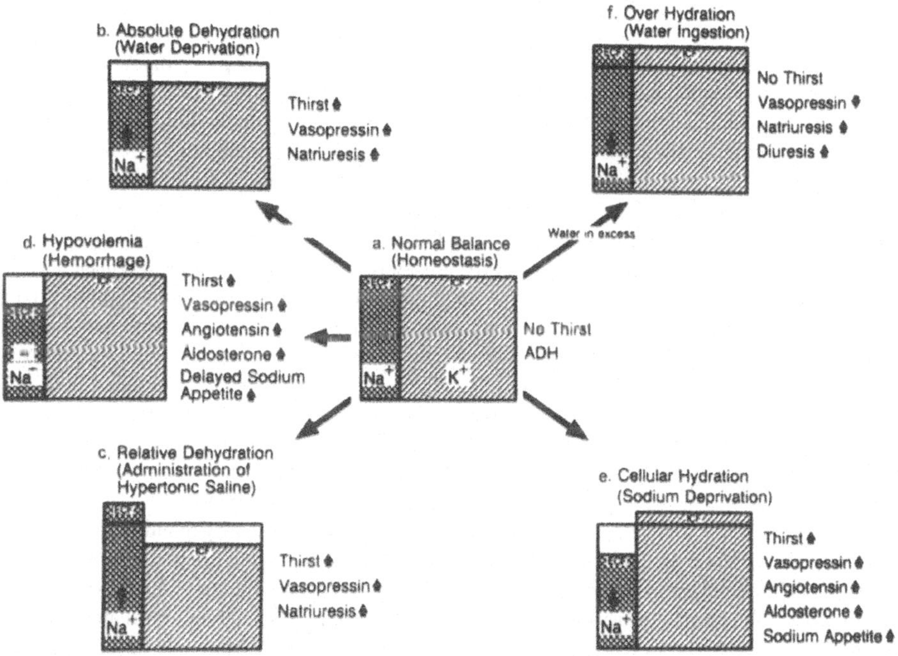

The extracellular fluid compartment is composed of two subcompartments—the cell-free fluid within the cardiovascular system (i.e., plasma) and the fluid comprising the interstitial space (i.e., that surrounding cells). The movement of water and sodium between the vasculature and interstitial spaces is virtually unimpeded so that a reduction or expansion in one of these two subcompartments is rapidly accompanied by a proportional change in volume in the other. The relative changes in amount of fluid in the extracellular space can be easily quantified by measuring changes in hematocrit or in plasma protein concentration (Stricker, 1966).

In the face of dehydration of either the cellular or the extracellular fluid compartments, physiological mechanisms are activated to minimize the rate of water loss from the body, and behaviors are mobilized to find and drink water (i.e., thirst). In the case of extracellular depletion, additional responses are initiated to retain and to seek out and consume sodium (i.e., salt appetite). The kidneys are major targets of the physiological controls of the rates of water and sodium losses from the body. Both the sympathetic component of the autonomic nervous system and various hormones act on the kidneys to modulate the rates of sodium and water excretion (Samson, 2004). Increased sympathetic neural tone exerted on the proximal renal tubules increases the reabsorption of sodium and water to reduce the rate of loss of these substances (DiBona, 1982). Similarly, several hormonal agents each act on one or more segments of the renal tubule or collecting duct system to either increase or decrease the rates of sodium and water reabsorption. Although the kidneys represent major targets for the control of water and sodium excretion, it is worth noting that a degree of physiological control is also exerted on the sweat and salivary glands to modulate rates of loss.

Sympathetic and hormonal responses can temporarily stem the insult of severe dehydrating challenges by slowing the concomitant loss of water and sodium through the volume and composition of urine and to a lesser extent by reducing salivary secretion and sweat rates. However, it is very important to appreciate that under normal conditions the only way that the state of homeostasis can be restored in the face of water or sodium deficit is by arousal of behaviors consistent with finding and consuming sufficient amounts of water and sodium.

1.3 A Brief History of the Study of Thirst and Salt Appetite

A fair starting point for the scientific analysis of thirst is Haller's (1747) summary of what was then known about the topic, and his restatement of the age-old "dry mouth" notion, namely, that thirst arises from sensations in the mouth and throat when they become dry (Grossman, 1967; Fitzsimons, 1973). This idea was largely dispelled by the work that followed. One classic refutation was provided by the great physiologist Bernard (1856) who prepared dogs and horses with esophageal fistulas so that water imbibed by the thirsty animals passed out of a hole in the throat and did not reach the stomach and intestines for absorption. The animals drank to exhaustion apparently without relief although their mouths and throats were continuously bathed by water. By the start of the last century, thirst was largely viewed as a kind of general sensation arising from a lack of water in the blood. Mayer (1900) found increases in the osmotic pressure of blood from dogs deprived of water for several days. He thought this thickening of the blood accounted for thirst. He provided a key metric of dehydration by determining the osmotic pressure of plasma. Measuring the osmotic pressure presented a quantifiable index of a physical change that accompanies thirst. Wettendorf (1901) was impressed by how modest the change in blood osmolality is in the early stages of water deprivation. He believed that water is drawn from the tissues into the systemic circulation keeping the blood changes small, and that tissue dehydration, not increased osmotic pressure *per se*, was the stimulus for thirst.

With his influential Croonian lecture, W.B. Cannon (1918) defended the "dry mouth" theory. Cannon argued that thirst comes from local tissue desiccation, specifically of the mouth. He posited that these local conditions can be caused by the loss of water from the whole body, but that thirst does not arise from nerve endings all over the body that project to the brain. Thirst, he claimed, results from nerves innervating the mouth region that sense the local desiccation occurring there. By force of his stature in the field, Cannon's defense of the dry mouth theory unfortunately steered the course of research largely into investigations of the relationship of salivary flow and thirst.

Fundamental advances were delayed until the 1930s when seminal work was performed. Two kinds of dehydration were described—one that accompanies loss of water and the other that accompanies loss of sodium (Kerpel-Fronius, 1935). Darrow and Yannet (1935) detailed the changes in distribution of body fluids on addition and subtraction of sodium. Using dogs, they showed that sodium administered into the peritoneal cavity expanded plasma volume and contracted cell volume. They remarked that this made the dogs thirsty, but they did not measure the amount of water drunk. Upon the loss of sodium from the body, they showed contraction of plasma volume and expansion of cell volume. Their dogs were not thirsty after sodium loss, even though the mucus membranes of the mouth indicated considerable dehydration.

Gilman's (1937) work refocused interest on the cellular dehydration theory of thirst. Gilman infused dogs intravenously (iv) with equiosmolar concentrations of urea, which readily passes through the cell membrane, and of sodium, which does not. The infusions of sodium caused twice as much drinking as the infusions of urea. Gilman concluded that sodium was a more effective stimulus by virtue of being excluded from the cells and thus causing the cells to shrink as water was withdrawn from them. After Gilman, the focus of much research was on the location of the cells responsible for detecting hydrational status. In the course of studying VP release, Verney (1947) developed the concept of *osmoreceptors*—cells activated when they shrink from dehydration (see ➲ *Section 2.2.1*). Furthermore, Verney localized osmoreceptors to the base of the brain (e.g., hypothalamus) by comparing infusions of urea and sodium into the carotid arteries and other vessels on the release of VP. Infusions of sodium into the carotid arteries, which supply the hypothalamus, worked best. Wolf (1950) applied the idea of osmoreceptors to the sensing of dehydration that triggers thirst.

By mid-century, it was clear that an exclusive cellular theory of thirst did not satisfactorily explain all drinking. For example, animals deprived of water drink more than animals administered enough hypertonic saline to cause equivalent increases in osmolality. Furthermore, animals depleted of sodium eventually develop large increases in water drinking even though the cells are expanded. Wolf (1958) suggested that osmoreceptors responded not only to shrinkage but also to expansion so that any *change* in cell size is an adequate stimulus for thirst. However, Adolph and group (1954) explicitly argued the case for multiple factors in thirst. Fitzsimons (1961) effectively settled the issue in favor of Adolph and colleagues (1954) by showing that extracellular fluid depletion, specifically hypovolemia, causes thirst. Fitzsimons' strategy was to cause an effective loss of extracellular fluid by injecting animals with isotonic hyperoncotic colloid[1], which causes a localized edema by sequestering isotonic fluid into the region surrounding the injection site. This reduces plasma volume, and since the sequestered fluid is iso-osmotic, there is no effect on the volume of the intracellular compartment. This selective depletion of extracellular fluid volume causes animals to drink.

Fitzsimons' (1966, 1969b; Fitzsimons and Simons, 1969) subsequent discovery that a renal factor (i.e., renin acting through ANG II) partly mediates hypovolemic thirst was a major advance in the endocrinology of drinking (see ➲ *Section 2.2.2.1*). The notion that cellular and extracellular dehydrations are independent causes of thirst was formally stated as the *double depletion hypothesis* of thirst by both Epstein (1973) and Fitzsimons (1973). Cellular and extracellular factors are, by and large, additive in their effects on thirst (see ➲ *Section 2.4.1*).

With refinement of stereotaxic technique and its routine use in the 1950s and 1960s came rapid advances in understanding of the role of the brain (i.e., mostly the hypothalamus) in thirst and salt appetite. Electrolytic lesions of the hypothalamus profoundly alter water intake (Stevenson et al., 1950) and sodium ingestion (Wolf, 1964a). Electrical stimulation using electrodes implanted into the hypothalamus causes drinking in conscious animals (Andersson and McCann, 1955). Andersson (1952) injected concentrated saline solutions into the hypothalamus through a chronically implanted cannula and elicited water drinking in water-replete goats. Andersson's studies convinced him that the sought after osmoreceptor for thirst was really a *sodium receptor* (see ➲ *Section 2.2.1*). Grossman's (1960, 1962) ability to selectively stimulate feeding or drinking by administering adrenergic or cholinergic agents, respectively, through the same cannula implanted into the lateral hypothalamus (see ➲ *Section 4.1.2*) provided an example of chemical coding of behavior in the brain, and ushered in the neurochemistry and neuropharmacology of thirst.

Major work in salt appetite began in the 1930s. Richter (1936) showed the profound sodium chloride intake of adrenalectomized rats caused by uncontrollable urinary loss of sodium following the removal of the source of the mineralocorticoid hormone, aldosterone. Robust salt appetite is also produced by excess mineralocorticoids (Rice and Richter, 1943). By the 1960s, it was clear that salt appetite is produced under many of the same conditions as extracellular thirst, including sodium loss by glucose dialysis (Falk, 1961), hypovolemia through colloid administration (Stricker and Wolf, 1966) and procedures that generally

---

[1] A large molecular weight substance that remains in a depot in the intersitial subcompartment but that does not enter the systemic circulation or the cellular compartment.

increase renin secretion. This suggests that salt appetite is under many of the same controls as extracellular thirst. Spurred by the discovery of an intrinsic renin-angiotensin system (RAS) in the brain (Ganten et al., 1971a; Ganten et al., 1971b; see *Section 4.1.1*), it was shown that ANG II stimulated salt appetite upon central injection (Buggy and Fisher, 1974; Chiaraviglio, 1976). Although the doses employed in these initial studies were rather high, the dose could be lowered considerably by pretreating animals with systemic mineralocorticoids (Fluharty and Epstein, 1983). This work suggested the *synergy hypothesis* of salt appetite whereby high levels of cerebral ANG and systemic mineralocorticoids can independently stimulate salt appetite but together act synergistically (see *Section 2.4.2*).

The last 30 years or so have been notable for the diversity of work in the areas of thirst and salt appetite. The circumventricular organs (CVOs) of the brain have become regarded as key areas for sensing humoral factors because of their unique absence of a blood–brain barrier (Johnson and Gross, 1993). The CVOs are established as areas housing osmoreceptors and receptors for ANG II. The list of hormones affecting thirst and salt appetite continues to grow. There are new hormones (e.g., relaxin) and neurotransmitters under investigation as are new roles for old hormones (glucocorticoids) and neuroactive agents. The interactions among extracellular factors (i.e., pressure, volume, ANG) that produce thirst and salt appetite have become clearer. Research into the cellular basis of the *synergy hypothesis* is proceeding as are alternative accounts of the underlying mechanisms for synergy. Major advances have been made into understanding the mechanisms of satiation of thirst and salt appetite, and newer avenues of investigation include examination of the molecular bases of these states and behaviors.

1.4 Experimental Strategies and Animal Models

A variety of techniques has been used to produce deficits of body water or electrolytes in order to investigate compensatory physiological and behavioral responses. The ability to produce selective deficits in the cellular and extracellular fluid compartments has greatly facilitated the study of the underlying mechanisms of thirst and salt appetite.

Water depletion normally results from a simple lack of water to drink. It produces deficits in both the cellular and the extracellular compartments. Water is lost first from the extracellular compartment, through skin, lungs, and urine. This increases osmotic pressure of the extracellular fluid, which dehydrates cells, ultimately resulting in diminished volumes of both compartments (Marriott, 1950; Strauss, 1957). Water depletion is produced experimentally by depriving people or animals of water. Thirst is the earliest symptom, and it is usually progressive (Dill, 1938). However, a small but reliable salt appetite also occurs as a consequence of the extracellular fluid deficits (Weisinger et al., 1985a, b; Sato et al., 1996).

A relatively pure sodium depletion can occur in people who lose substantial sodium by sweating, alimentary secretions (e.g., diarrhea), or urine (Addison's disease) provided they are liberally drinking water. Sodium deficiency is produced experimentally in animals by dialysis against glucose (Gilman, 1934; Darrow and Yannet, 1935) or in humans by extensive sweating (McCance, 1936a, b; Takamata et al., 1994) or aspiration of alimentary secretions (Nadal et al., 1941) with water as the only drinking fluid. With loss of the main extracellular electrolyte—sodium—osmotic pressure falls and water shifts into the cells. This reduces extracellular fluid volume and may result in a fall in blood pressure (Gilman, 1934). Over time, the kidneys excrete water in an effort to achieve isotonicity.

Dialysis against glucose involves injections of a volume of isotonic (5%) glucose into the peritoneal space (Gilman, 1934; Darrow and Yannet, 1935, 1936). Over a few hours, there is almost complete equilibration and exchange of extracellular sodium with glucose. Subsequent removal of an equal volume of peritoneal fluid removes large amounts of sodium without a net loss of body water. The decreased concentration of extracellular sodium results in a shift of water into cells and a decrease in extracellular fluid volume. In the early use of this technique, investigators reported both diminished (Darrow and Yannet, 1935, 1936; Huang, 1955; Semple, 1952) and increased (Cizek et al., 1951; Semple, 1952; Huang, 1955) water drinking. Holmes (1960) suggested two phases to the drinking response, an acute phase with absence of thirst and a chronic phase with thirst present. Acutely, animals are described as "languid" (Darrow and Yannet, 1935). The cells are greatly expanded at this time and water drinking is almost universally found to be diminished. Later, increased water intake is usually observed, although with considerable variability among animals. Water drinking is also observed in dogs depleted of sodium by a combination of sucrose diuresis and low-sodium diet (Holmes and Cizek, 1951) and by maintenance on sodium-deficient diets for a sufficient time (Kerpel-Fronius, 1935; Swanson et al., 1935; Radford Jr., 1959).

In human studies involving specific sodium loss, there is a reported lack of thirst (Nadal et al., 1941) or a sensation somehow distinct from thirst (McCance, 1936a, b). People may drink water, but it is not satisfying. Ingestion of sodium provides immediate relief. Sodium deficiency in human subjects produced by exercise-induced sweating followed by rehydration with water produces enhanced palatability of saline solutions (Takamata et al., 1994).

Further insight into the question of whether sodium deficiency produces thirst or salt appetite can be gained from the studies of Falk (1961, 1965, 1966). He produced sodium deficiency in rats by intraperitoneal dialysis. When water and concentrated saline were presented together so that animals have a choice of fluids to drink, the greatest increase in drinking is of the saline solution (Falk, 1965). This is appropriate as the animals have preferentially lost sodium through dialysis. This supports the idea that sodium loss does not generate true thirst even though there may be increased water intake involved.

Selective depletion of the cellular compartment is achieved by adding electrolyte to the extracellular space. This happens whenever one eats salty foods. As noted earlier, Darrow and Yannet (1935) injected sodium into the peritoneal cavity of experimental animals and observed a decrease in cell volume and increase in plasma and blood volume indicating an osmotic shift of water from the cells into the extracellular space. They reported that their animals were thirsty. Gilman (1937) later showed that the osmotic shift of water out of cells is a potent stimulus for thirst.

Selective depletion of the extracellular compartment produces hypovolemia which causes not only thirst but also salt appetite (Stricker and Wolf, 1966). After the induction of hypovolemia, animals typically ingest water for a few hours before consuming salt solutions. If only water is available for drinking, hypovolemic rats stop drinking before blood volume is restored (Stricker, 1969; Stricker and Jalowiec, 1970) because the ingested water excessively dilutes body fluids. However, by ingesting salt solutions extracellular tonicity is maintained, and animals drink sufficient amounts to fully repair blood volume deficits (Stricker and Jalowiec, 1970; Stricker et al., 1992).

Hemorrhage is a specific loss of extracellular fluid that does not affect cell volume. Although hemorrhage has been historically thought to be a potent stimulus for thirst (Starling, 1909; Cannon, 1918; Gregersen, 1941), it actually seems to be unreliable in inducing thirst. Holmes and Montgomery (1953) report that blood donors do not get thirsty. They were also unable to elicit thirst in dogs via hemorrhage (Holmes and Montgomery, 1951). Nevertheless, Fitzsimons (1961, 1969a; Fitzsimons and Oatley, 1968) found thirst after hemorrhage in rats. However it has been pointed out that the degree of hemorrhage needed to produce thirst may be debilitating because of the loss of red blood cells (i.e., acute anemia) which may account for the unreliable drinking found in earlier studies (Fitzsimons, 1972).

Procedures that interfere with the circulation can produce or mimic aspects of extracellular fluid depletion. Fitzsimons has shown that ligating the abdominal vena cava reduces venous return to the heart, and sometimes reduces blood pressure. Drinking occurs within a couple of hours (Fitzsimons, 1964, 1969b; Fitzsimons and Moore-Gillon, 1980; Fitzsimons and Elfont, 1982).

2 The Sensory Modalities that Signal the CNS about the Status of Body Water and Sodium

2.1 Overview of Chemical and Mechanical Afferent Signaling Mechanisms

There are both chemo- and mechanoreceptors that are responsive to stimuli that are indices of body water and sodium status. These receptors are located in the systemic viscera, particularly the vasculature, kidney, and heart and in the brain. Many of the receptors and their afferents are comparable to those informational inputs critical for regulating cardiovascular functions (e.g., heart rate and blood pressure).

2.2 Humoral Mediators of Thirst and Salt Appetite

Increased osmolality of blood perfusing the brain induces thirst, releases VP, and increases sympathetic nerve activity (*Figure 17-2*). The immediate consequence of increased brain osmolalities is an acute increase in blood pressure. If the stimulus is sustained, an expansion of body fluids will occur as a result of water retention and drinking. This pattern of physiological and behavioral responses is consistent with defending a deficit and restoring body water. Decreased intracellular fluid is believed to be sensed by osmoreceptors. The term *osmoreceptor* was used to describe a hypothetical sensory cell type that transduces hyperosmolarity of extracellular fluids into neural output. The term was coined nearly 60 years ago by Verney (1947). In turn, Wolf (1958) applied the same term and concepts to account for the sensory mechanism responsible for the induction of thirst and drinking produced by administration of hypertonic solutions.

Figure 17-2
Water intake during infusion of hypertonic solutions in rats (2 ml/20 min, either split between both carotid arteries or into the vena cava). The sensors for the drinking responses are cephalic because intracarotid infusions of hyperosmotic substances worked at doses that were ineffective intravenously. The sensors are located outside the blood–brain barrier because dehydration of cells located inside the blood–brain barrier with infusions of glycerol did not work (note: glycerol passes into cells, but does not cross the blood–brain barrier). The sensors are osmosensitive rather than strictly sodium-sensitive because hypertonic mannitol as well as hypertonic saline was effective. *Indicates difference between intracarotid and intravenous infusion (n = 8) From Schoorlemmer et al. (2000)

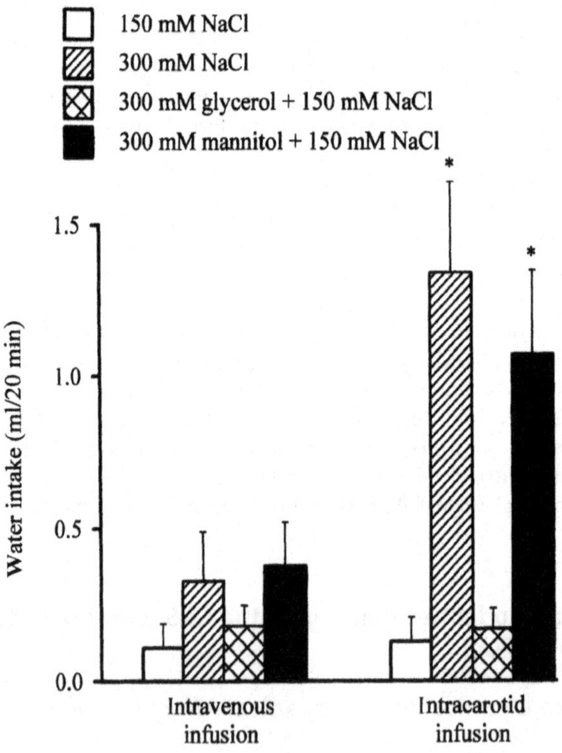

A second humoral signal that evokes reflexes and behavioral responses consistent with restoration and expansion of body fluids is the peptide ANG II. Systemic administration of exogenous ANG II elevates arterial blood pressure, releases VP and aldosterone, and increases the ingestion of water and salt. With the exception of aldosterone release, these responses can also be elicited by intracranial injections of ANG II at doses lower than those that are effective peripherally (see ➋ *Section 4.1.1*).

ANG is generated in the systemic circulation by the action of renin, which is synthesized and released from the kidney to act on a circulating substrate, angiotensinogen (AGT). The product of this enzymatic reaction is ANG I, a decapeptide which in turn is converted into ANG II by angiotensin converting enzyme (ACE) that removes two N-terminal amino acids from ANG I. ANG II is the primary effector peptide of the renin-angiotensin metabolic cascade. Although it is the most thoroughly studied, the blood-borne (also referred to as the classic or renal) RAS is not the only RAS. Beginning about 35 years ago, investigators began to find evidence for *de novo* synthesis of components of the RAS in various tissues including brain (Ganten et al., 1971a, b; see ➋ *Section 4.1.1*).

Although there are many regions in the CNS that are sensitive to osmotic changes and to ANG II, there are three specific structures lying outside the blood–brain barrier that contain receptive elements that respond to one or both of these humoral stimuli. Because of their sensitivity to blood-borne factors, these structures have been referred to collectively as *sensory circumventricular organs* (SCVOs) (Johnson and Gross, 1993; McKinley et al., 2003). Specifically the SCVOs are the subfornical organ (SFO), the organum vasculosum of the lamina terminalis (OVLT) and the area postrema (AP).

In recent years, considerable attention has been paid to mineralocorticoid receptors located in the brain and their contribution to sodium appetite (Epstein, 1982). When given in pharmacological doses (Wolf, 1964b) or in more physiological concentrations in conjunction with ANG II (Fluharty and Epstein, 1983; Fluharty and Manaker, 1983), mineralocorticoid agonists (desoxycorticosterone; aldosterone) generate the ingestion of salty solutions.

2.2.1 Osmo- and Sodium Receptors

The nature of osmoreceptors and whether there are sensory cells detecting the concentration of extracellular sodium which are distinct from osmoreceptors are the major questions in this field that have not been definitively addressed. Beginning in the 1960s Andersson and colleagues studying the goat challenged the concept of a cerebral osmoreceptor and instead proposed that the critical sensory cells are actually sodium receptors (Andersson et al., 1967; Olsson, 1969; Andersson, 1971). Their reasoning was in part based on evolving knowledge about the blood–brain barrier and the fact that Verney (1947) had based the idea of osmoreceptors on studies where hyperosmotic solutions were administered on the blood (i.e., body) side of the blood–brain barrier. Under such circumstances, it was proposed that water moved out of the brain, crossing the blood–brain barrier at a faster rate than the extracellular fluid solutes. The net effect was to increase the extracellular sodium concentration in the brain. As a test of this hypothesis, Andersson and coworkers (1967; Olsson, 1969) infused hypertonic solutions of glucose or of sucrose through the cerebral ventricular system and saw no effects on thirst or antidiuresis (i.e., VP release). Both drinking and antidiuresis were observed when an equivalent osmotic solution of hypertonic saline was delivered by the same central route. McKinley and colleagues (1974) reasoned that perhaps the sugar solutions used by the Andersson laboratory diluted CNS sodium and nonspecifically disrupted brain function. In experiments where sucrose was added to normal artificial cerebrospinal fluid (CSF), McKinley and coworkers (1974) demonstrated that such hypertonic solutions readily elicited drinking. Interesting, however, is the fact that the artificial CSF made hypertonic with sucrose is not as effective a dipsogen as infusate made equally hypertonic by adding NaCl to artificial CSF. Consequently, these investigators have speculated that there might be both osmo- and sodium receptors in the brain (McKinley et al., 1974, 1978).

There is evidence that osmo- and sodium receptors are located both in the periphery and in the CNS. One region of the body where cells are likely to be exposed to frequent and large changes in osmolality and sodium concentration is the splanchnic circulation, particularly the hepatoportal region. Assaults on fluid homeostasis are likely to occur with each meal or bout of drinking. Haberich (1968) demonstrated that water infused into the portal vein was more effective in producing diuresis than when infused systemically. Changes in vagal afferent nerve activity are produced by increasing and decreasing the osmolarity of injections into the portal vein (Niijima, 1969; Andrews and Orbach, 1974; Adachi et al., 1976; Rogers et al., 1979). The same or similar receptor systems are also likely to contribute to the control of thirst and salt appetite. Blake and Lin (1978) found that the volume of isotonic saline consumed after water deprivation was reduced by hepatic portal vein infusions of saline and that this attenuating effect was abolished by right vagotomy. Portal vein infusions of saline have also been shown to reduce salt appetite in sodium-depleted rats (Tordoff et al., 1986, 1987). Kraly and coworkers (1995) demonstrated that thirst can be induced by intragastric delivery of hypertonic saline at a dose that it does not raise systemic osmolality, and that subdiaphragmatic vagotomy attenuated this drinking response.

Although systemic osmo- and sodium receptors may function as early sentinels and act in a feed-forward manner to initiate corrective actions, there are other osmo- and sodium receptor systems located near or within the tissues of the CNS. As previously noted, it was the work and thinking of Verney (1947), Andersson (1971; Andersson et al., 1967), and Wolf (1958) that directed attention to the importance of cerebral osmo- and sodium sensors in the control of body fluid and cardiovascular homeostasis. There are many early extracellular recording and lesion studies implicating several brain regions as being sensitive to systemically or centrally administered hypertonic solutions. For example, Silverstone and group (Clemente et al., 1957) demonstrated that systemically injected hypertonic saline increased firing of neurons in the vicinity of the AP, the most caudal of the brain's SCVOs. Similarly, Malmo and Mundl (1975) recorded osmotically activated units in the lateral preoptic area, which had previously been implicated in the control of osmotically induced thirst (Blass and Epstein, 1971; Peck and Novin, 1971) in lesion studies. Lesion and *in situ* extracellular recording studies may provide supporting evidence to implicate a region as osmo- or sodium-receptive. However, neither one of these methods is sufficient to unequivocally fulfill the criteria that a given cell type in a brain region is intrinsically responsive to extracellular osmolality or sodium concentrations. To accomplish this, it is necessary to examine responsiveness of cells when they are synaptically isolated.

From their studies on VP release, Jewell and Verney (1957) suggested that the hypothalamic cells that synthesize VP may be osmoreceptors. Mason (1980) found osmosensitive cells in the supraoptic nucleus (SON) by employing *in vitro* intracellular recording in hypothalamic slices. In more recent studies, Bourque and his colleagues found that changes in the volume of SON neurons result in modulation of mechanosensitive gadolinium-sensitive 33 pS stretch-inactivated cation channels (Bourque et al., 2002; Oliet and Bourque, 1993).

Although the VP-synthesizing cells of the SON have provided an excellent model for studying osmo- and sodium sensitivity (Oliet and Bourque, 1993; Bourque et al., 2002), it is likely that these neurons require additional input from other osmosensitive regions to effect VP release in response to physiological changes (Honda et al., 1990; Bourque and Richard, 2001). Also because the principle target of magnocellular neurons is the posterior pituitary, it is not likely that they make an important direct contribution to the neural circuitry involved in the generation of thirst or salt appetite. The periventricular region surrounding the anteroventral third ventricle (AV3V) (see Johnson and Thunhorst, 1997 for review) is another diencephalic region that has been implicated in sensing changes in osmolality or extracellular sodium concentration and providing information pertinent to thirst, release of posterior pituitary peptides and activity of the sympathetic nervous system. The importance of the AV3V region in the sensing and processing of osmotic and sodium concentration information was first appreciated in the mid 1970s from studies which found that rats with AV3V lesions had global defects in their capacity to activate many of the physiological systems that maintain body fluid and cardiovascular homeostasis (see ➋ *Section 3.1.1.1* and ➋ *Table 17-1*). One specific structure within the AV3V that is implicated in osmoreception is the OVLT. Richard and Bourque (1995) studied the relationship of OVLT with SON neurons in an explant preparation that allows selective manipulation of OVLT osmolality. Osmotic stimulation of the OVLT increased the firing rate of SON magnocellular neurons. Further evidence indicates that this excitatory drive is mediated through the release of glutamate from OVLT originating neurons that terminate on SON magnocellular neurons. (see Bourque and Richard, 2001 for review).

2.2.1.1 Transient Receptor Potential Ion Channels Recent studies employing genomic manipulations provide further insight into the molecular nature of the osmo- and sodium-sensing process. Unique

Table 17-1
Acute and Chronic Effects of AV3V* Lesions in the Rat

Acute Effects
Adipsia
Impaired secretion of vasopressin
Severe weight loss and if untreated, debilitation and death due to dehydration
Chronic Effects
Reductions in body weight
Recovery of ad libitum drinking
Impaired drinking responses to experimental challenges
• Attenuated water deprivation-induced drinking
• Abolished cellular dehydration-induced drinking
• Extracellular thirst
– Abolished angiotensin II-induced drinking
– Attenuated isoproterenol-induced drinking
– Attenuated caval ligation-induced drinking
– Attenuated polyethylene glycol treatment-induced drinking
Altered ad libitum sodium intake
Impaired sodium depletion-induced salt appetite
Impaired vasopressin secretion to hypertonic saline and ANG II
Impaired natriuresis
Impaired pressor responses to hypertonic saline and ANG II
Protection against most forms of experimentally induced hypertension
Hypernatremia
Impaired behavioral responses to defend against hyperthermia induced by heating
Impaired hemodynamic responses to defend against hyperthermia induced by heating

*AV3V is the periventricular tissue surrounding the anteroventral third cerebral ventricle

proteins associated with ion channels and transporters have attracted particular attention. Several groups recognized at about the same time that products of genes encoding members of transient receptor potential ion channels may be stretch-sensitive and activated by osmotic stimuli (Suzuki et al., 1999; Liedtke et al., 2000; Strotmann et al., 2000). Liedtke and colleagues (2000) used a candidate gene approach and found a vertebrate homolog of an osmotically gated ion channel related to the OSM-9 gene of the worm, *Caenorhabditis elegans*, and to the vanilloid receptor. They identified a novel protein in human, rat, mouse, and chicken, which has characteristics of a nonselective cation channel (initially called the vanilloid receptor related osmotically activated channel; VR-OAC) which is now called the transient receptor potential channel, vanilloid subfamily Type 4 (TRPV4). Using *in situ* hybridization, TRPV4 mRNA has been detected in neurons of the OVLT, SFO, and median preoptic nucleus (MePO) as well as in the ependymal cells of the choroid plexus in the lateral ventricles (Liedtke et al., 2000).

Employing strategies to disrupt the TRPV4 gene, two groups have studied osmotic responses in *trpv4* null (*trpv4*-/-) mice. Mizuno and coworkers (2003) did not find effects on water intake or serum osmolality in the *trpv4*-/- mouse under conditions of ad libitum access to food and water. In contrast, Liedtke and Friedman (2003) found that *trpv4*-/- mice drink less water and become more hyperosmotic than wild-type littermates when both were given access to water but no food. In addition, under conditions of chronic infusion of a synthetic antidiuretic to induce hyponatremia, the plasma of *trpv4*-/- mice was more hypotonic, and the animals showed less suppression of drinking than wild-type controls. Currently, it appears that there may be a small role for the TRPV4-related cation channel in facilitating and inhibiting thirst under hyper and hypoosmolar conditions, respectively. However, it would appear that this may be only one molecular component of an osmosensor (Liedtke and Friedman, 2003) or that it is redundant with other osmo- and sodium receptors.

Recently, Bourque's laboratory (Naeini et al., 2006) has provided data indicating that an N-terminal variant of the transient receptor potential channel, vanilloid subfamily Type 1 (TRPV1) channel is required for osmosensory transduction in supraoptic magnocellular neurons. These cells express a form of TRPV1 that antibodies directed toward the N-terminal of the protein will bind. However, antibodies directed toward the C-terminal do not react with the same type of supraoptic cell. These researchers have also demonstrated that *trpv1* knockout mice are hyperosmotic under basal conditions and that magnocellular neurons from these animals do not show increased membrane conductance in response to hyperosmotic stimulation.

2.2.1.2 Na_x Channel The Na_x channel is a membrane-associated protein that has come under scrutiny for the sensing of extracellular sodium status. Previously called the Na_v 2.3- (human), NaG/SCL11- (rat) and Na_v 2.3-channel (mouse), the Na_x channel is a member of a subfamily of voltage-gated sodium channels. Using Na_x knockout mice and wild-type rats, Noda and collaborators (Watanabe et al., 2000; Hiyama et al., 2002, 2004) and Mouginot and colleagues (Grob et al., 2004), respectively, have conducted both behavioral and electrophysiological studies on the Na_x channel. Na_x null mice drink excess amounts of hypertonic sodium chloride solution after dehydration by water deprivation (Watanabe et al., 2000), and populations of Na_x neurons have been identified in the SFO, OVLT, MePO, and periventricular nuclei (Watanabe et al., 2000; Grob et al., 2004).

Intracerebroventricular (icv) infusion of hypertonic saline into water deprived wild-type mice induces water intake and diminishes saline consumption, but this latter attenuated response to salt is not observed in Na_x knockout animals (Hiyama et al., 2004). However, the attenuated response to the salt solution is restored in Na_x knockout mice by transducing the SFO with Na_x cDNA (Hiyama et al., 2004). Taken together, these results indicate that structures located along the lamina terminalis house neurons that are sodium-sensitive and endowed with Na_x channels. These cells are likely to contribute to the behavioral control of hydromineral balance and cardiovascular homeostasis.

2.2.2 Systemic Peptides and Steroids

2.2.2.1 The Systemic Renin-Angiotensin System Seminal work of Fitzsimons (1969b) implicated circulating ANG II as a major stimulus for thirst. He discovered that arising from simulated hypovolemia (i.e., reduced venous return to the heart) thirst is greatly diminished by nephrectomy. He further demonstrated that renal extracts contain a thirst factor, which is identical to renin (Fitzsimons, 1969b). In general, treatments that activate the RAS (e.g., hypotension, hypovolemia, and vasodilatory drugs) induce drinking (see Fitzsimons, 1972, 1979, 1998 for reviews). Fitzsimons and Simons (1969) subsequently showed that high doses of ANG II are dipsogenic upon iv infusion. Hsiao and colleagues (1977) were able to reduce the dose that reliably induced drinking in rats by refining the drinking protocol. Mann and coworkers (1980) established that the rates of iv ANG II used by Hsiao and group (1977) produced circulating plasma levels of ANG II in the high physiological range (i.e., approximately 200 pmol/ml; *Figure 17-3*). The relatively large doses required for the dipsogenic response to ANG II has produced several challenges to accepting its physiological role in drinking (Abraham et al., 1975; Stricker, 1977, 1978; Atkinson et al., 1979; Anke et al., 1988; Pawloski and Fink, 1990). An alternative possibility has been that circulating ANG II is merely permissive of drinking during thirst-inducing procedures (e.g., hypotension, hypovolemia) by maintaining blood pressure at levels compatible with behavioral competence. Some of these concerns have been resolved by work examining the critical role of arterial pressure in the dipsogenic potency of ANG II (Evered, 1992; Johnson and Thunhorst, 1995; see *Section 2.4.3*).

It has been difficult to establish the role of the endocrine RAS in salt appetite (Fitzsimons, 1979) because of competing explanations for observed results. Surgical procedures that greatly affect sodium intake, presumably by altering the activity of the RAS (i.e., nephrectomy, adrenalectomy, renal artery constriction or ureteric ligation), do more than just produce a change in renin secretion and production of ANG II. As examples, nephrectomy may abolish salt appetite by removing the source of renin or by causing anuria (Fitzsimons and Stricker, 1971; Stricker et al., 1979). Salt appetite caused by the massive loss of sodium

Figure 17-3
Summary indicating plasma levels of angiotensin (ANG) II obtained following various challenges to activate the endogenous renin-angiotensin system. These are presented on the regression line of the relationship between infused ANG II dose (abscissa) and circulating level of octapeptide (ordinate). Intact-25 pmol/kg/min and Intact-100 pmol/kg/min indicate levels of circulating ANG II after 60-min infusions of the indicated dose into unanesthetized rats. Lower cross-hatched areas represent the range of control values of plasma ANG II. Upper cross-hatched area represents approximate ANG II plasma level at dipsogenic threshold. The threshold at which ANG II induces drinking with iv infusion was established by Hsiao et al. (1977). Abbreviations: DI, diabetes insipidus; RH, renal hypertension; PEG, polyethylene glycol; ISOP, isoproterenol. Based on the data from Johnson et al. (1981) and Mann et al. (1980); figure from Johnson and Thunhorst (1995, 1997)

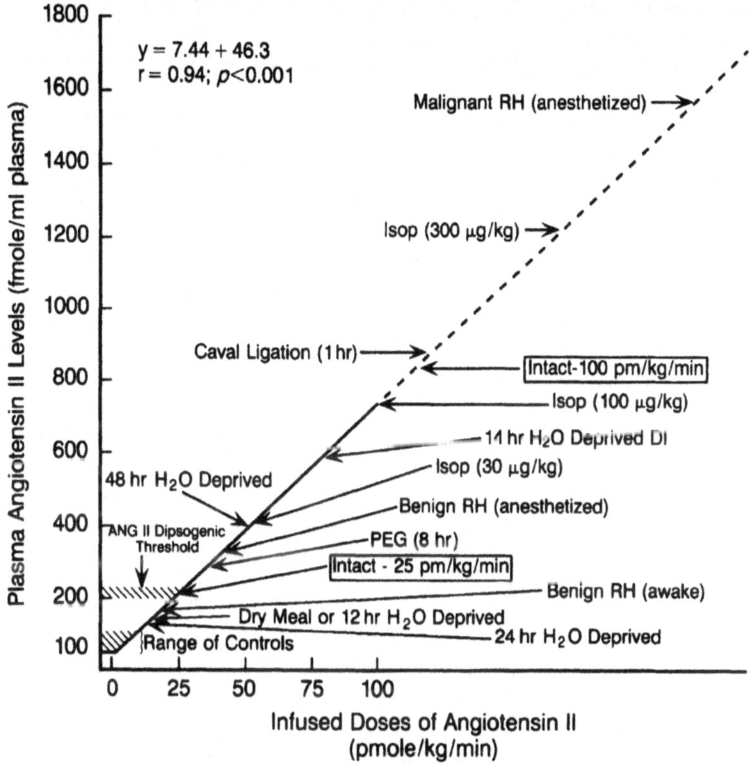

following adrenalectomy may be due to increased renin secretion and formation of ANG II or to changes in salivary sodium content (Fregly and Rowland, 1985) or sodium concentration at brain sodium receptors (Weisinger et al., 1985a, b). Salt appetite caused by clipping of the renal arteries may be a pathological response to excessive renin secretion or because of sodium loss produced by a pressor natriuresis (Mohring et al., 1975). Ureteric ligation suppresses salt appetite in the face of markedly increased renin secretion due to altered renal function or to unidentified inhibitory factors (Fitzsimons and Stricker, 1971; Michell, 1995). Pharmacological interventions have their own problems. Intravenous infusions of ANG II do not consistently stimulate salt appetite in sodium-replete animals (Fitzsimons and Wirth, 1978; Findlay and Epstein, 1980; Tarjan et al., 1988; Sakai et al., 1990) other than sheep (Weisinger et al., 1986), and the effect has been attributed to pressure natriuresis and subsequent negative sodium balance (Weisinger et al., 1986; Tarjan et al., 1988; Yang and Epstein, 1991). Pharmacological blockade of the renal RAS with an ANG II receptor antagonist does not prevent salt appetite of sodium-depleted rats (Sakai et al., 1990). Studies that used systemically administered ACE inhibitor to prevent salt appetite implicated an inhibition of the brain

RAS rather than the peripheral RAS (Moe et al., 1984; Sakai et al., 1990) because of the high doses that were employed.

The brain RAS was further implicated by studies showing that centrally administered ANG II stimulates salt appetite without prior natriuresis or negative sodium balance (Buggy and Fisher, 1974; Chiaraviglio, 1976; Avrith and Fitzsimons, 1980; Bryant et al., 1980; Fluharty and Epstein, 1983; Fluharty and Manaker, 1983). Furthermore, depletion-induced salt appetite is easily inhibited by the central administration of ACE inhibitors and ANG II receptor antagonists (Buggy and Jonklaas, 1984; Moe et al., 1984; Sakai et al., 1986; Weiss et al., 1986). For these reasons, the natriorexigenic role of circulating ANG II has been minimized. According to the *synergy hypothesis* of salt appetite, ANG II derived from renal renin release serves only to stimulate the secretion of aldosterone following sodium depletion, and it is brain-derived ANG II that synergizes with aldosterone causing salt appetite through combined activity within the CNS (Epstein, 1982; Sakai et al., 1990; Yang and Epstein, 1991).

Weisinger and colleagues showed that ANG II stimulates salt appetite upon iv administration to sodium-deficient animals, including cow (Blair-West et al., 1988), sheep (Weisinger et al., 1987), rabbit (Tarjan et al., 1993), and mouse (Denton et al., 1990). The sodium-deficient animals were given enough ACE inhibitor to block the formation of endogenous ANG II peripherally and presumably in the brain. This total-body ACE blockade reduced or abolished the salt appetite response in each case. Subsequent systemic administration of ANG II stimulated salt appetite in the blocked animals, or prevented the reduction in salt appetite from occurring if administered during the ACE blockade. Since ANG II does not cross the blood–brain barrier, the results clearly show that the infused ANG II is acting on sites accessible from the blood. Others have shown that iv infusion of ANG II stimulates salt appetite in sodium-deficient rats (Thunhorst and Fitts, 1994; Fitts and Thunhorst, 1996; Schoorlemmer et al., 2001). Importantly, these later studies show that selective blockade of the systemic RAS is sufficient to prevent sodium hyphen depletion-induced salt appetite. In these studies, the rats were given doses of ACE inhibitor that prevented formation of ANG II in the circulation but not in the brain (Thunhorst and Fitts, 1994; Fitts and Thunhorst, 1996). Selective blockade of the endocrine RAS completely prevented salt appetite in response to sodium depletion despite a functioning brain RAS. Subsequent iv infusions of ANG II restored the salt appetite response within minutes. Similar results were obtained in experiments using sodium-depleted, adrenalectomized rats that completely lack endogenous mineralocorticoids (Schoorlemmer et al., 2001; ➋ *Figure 17-4*). The results point to a direct stimulatory role for the endocrine RAS in producing the salt appetite of sodium-depleted animals.

2.2.2.2 Aldosterone Aldosterone is essential for the retention of sodium by the kidney and is thus critical for normal body fluid balance. Mineralocorticoids are increased during periods of sodium deficiency, including hypovolemia (Stricker, 1983). Administration of exogenous mineralocorticoids stimulates salt appetite in both intact and adrenalectomized rats (Wolf, 1965; Fregly and Waters, 1966). The appetite depends in part on the central actions of mineralocorticoids as blockade of central aldosterone receptors (Sakai et al., 1986) or the reduction of their expression in the brain (Sakai et al., 1996, 2000) diminishes the salt appetite of sodium deficiency. However, aldosterone is not strictly essential for salt appetite because adrenalectomized rats have a pronounced salt appetite (Richter, 1936) that can be regarded as a behavioral assay of a successful adrenalectomy. Thus, the relationship between mineralocorticoid levels and magnitude of the salt appetite response is described by a U-shaped function (Fregly and Waters, 1966). Salt appetite is pronounced in the absence of mineralocorticoids, for example, after adrenalectomy when the response is driven by a severe renal loss of sodium from the body. Sodium intake is low in intact animals expressing low, basal levels of mineralocorticoids, and in adrenalectomized animals infused with low levels of exogenous mineralocorticoids that permit renal sodium retention. Salt appetite is again pronounced in either intact or adrenalectomized animals administered mineralocorticoids in excess of levels needed for sodium retention where the appetite is driven by direct actions on the brain (Fregly and Waters, 1966).

There are two types of adrenal steroid receptors, the Type 1 (mineralocorticoid) receptor and Type II (glucocorticoid) receptor (de Kloet et al., 1998). Interestingly, glucocorticoids normally occupy both subtypes in the brain (de Kloet et al., 1998). Specificity to these receptors is conferred by the enzyme 11β-hydroxysteroid dehydrogenase (11β-HSD), which converts glucocorticoids to inactive molecules thus

Figure 17-4

Top, intake of water and saline by adrenalectomized rats after sodium depletion. Either captopril or vehicle was infused intravenously ($n = 5$). Selective blockade of the systemic renin-angiotensin system nearly abolished the salt appetite response. Bottom, intake of saline by adrenalectomized rats after sodium depletion. Either isotonic saline ($n = 5$), ANG II ($n = 5$), or antidiuretic hormone (ADH; $n = 4$) was infused intravenously in a captopril (2.5 mg/h, iv) solution. ANG II restored the salt appetite response of captopril-blocked rats, but equipressor doses of ADH did not. *Indicates significant difference, $p < 0.05$. Adapted from Schoorlemmer et al. (2001)

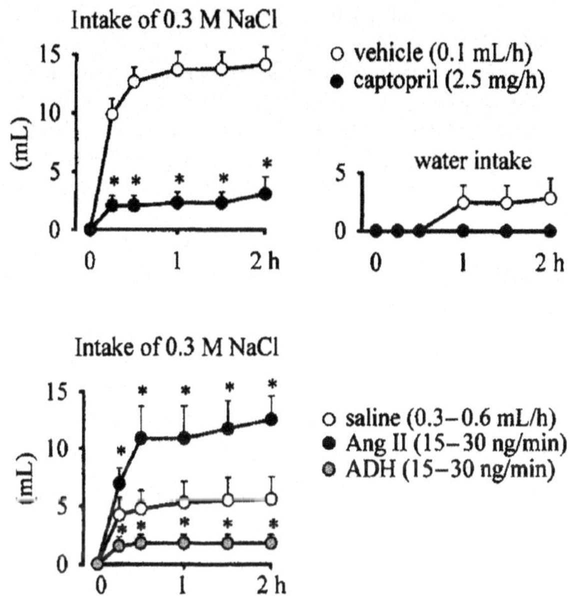

making the Type 1 receptor available to mineralocorticoids. That is, 11β-HSD protects Type 1 receptors from occupancy by glucocorticoids. At high levels, endogenous glucocorticoids may saturate the 11β-HSD enzyme (Ferrari, 2003). The 11β-HSD is found in the brain predominantly in the anterior hypothalamus and CVOs (de Kloet et al., 1998).

Recently, Loewy and colleagues argued that aldosterone-sensitive neurons should possess three characteristics: (1) receptors for mineralocorticoids so that aldosterone can bind there, (2) 11β-HSD to convert glucocorticoids and allow aldosterone access to mineralocorticoid receptors, and (3) accessibility to aldosterone, which according to the authors does not readily cross the blood–brain barrier (Geerling et al., 2006a). In a survey of the brain (Geerling et al., 2006b), only the nucleus of the solitary tract (NTS) satisfied all criteria (i.e., cells possessing both mineralocorticoid receptors and 11β-HSD, and a location adjacent to the AP, which lacks a blood–brain barrier). Reportedly, the ventromedial hypothalamic nucleus and the medial vestibular nucleus possess 11β-HSD, but the amygdala does not (Geerling et al., 2006a). Several salt appetite inducing procedures activate NTS neurons that have 11β-HSD, and these cells are inactivated rapidly upon ingestion of sodium (Geerling et al., 2006b). It is not clear how this recent work accounts for studies implicating forebrain areas such as the amygdala in salt appetite (e.g., Sakai et al., 2000).

There are noted species differences in that mineralocorticoids are relatively ineffective in producing salt appetite in rabbit and sheep (Denton, 1982), or in dogs, gerbils and hamsters (Rowland and Fregly, 1988b). The phenomenon is controversial in mice (Rowland and Fregly, 1988a, b; Blair-West et al., 1995).

2.2.2.3 Relaxin Relaxin is a hormone secreted by the corpus luteum during pregnancy. It is dipsogenic upon iv (Sinnayah et al., 1999) and icv (Summerlee and Robertson, 1995; Thornton and Fitzsimons, 1995) administration. Drinking in response to icv relaxin is dose-related, and is greater during the dark part of the light/dark cycle (Summerlee et al., 1998). Drinking in response to iv relaxin is also dose-related and sensitive to the presence of ANG II, as a nondipsogenic dose of ANG II triples the water drinking response to iv relaxin (Sinnayah et al., 1999). The effects of relaxin are mediated in part by a central angiotensinergic pathway as central administration of an ANG II receptor antagonist reduces drinking to systemic relaxin (Sinnayah et al., 1999). Although icv administration of relaxin causes water drinking, it does not stimulate sodium consumption (Thornton and Fitzsimons, 1995). However, antagonism of serotonin (5-HT) receptors in the lateral parabrachial nucleus (LPBN) stimulates robust sodium consumption in response to icv relaxin (Menani et al., 2004). The salt appetite response to relaxin that is observed after 5-HT receptor blockade is completely abolished by central injections of losartan, an ANG Type 1 (AT_1) receptor antagonist (Menani et al., 2004). This is further evidence that a central angiotensinergic mechanism at least partly mediates the central responses to relaxin. The icv administration of relaxin causes c-*fos* expression in brain areas associated with drinking and body fluid regulation, including the SFO, OVLT, MePO of the lamina terminalis, the SON, and the hypothalamic paraventricular nucleus (PVN; McKinley et al., 1997). Circulating relaxin activates neurons of the SFO and OVLT that directly project to the SON and PVN (Sunn et al., 2001).

2.2.2.4 Atrial Natriuretic Peptide (ANP) It has long been suspected that there should be hormonal mechanisms that cause sodium and water loss in states of volume overload (Smith, 1957). Although many substances have been proposed as primary natriuretic or diuretic agents, it was not until 1981 when de Bold and coworkers (1981) made the seminal finding that infusions of extracts of atrial tissue produced natriuresis in the rat. This led to a rapid application of biochemical techniques to identify and sequence the peptide and clone the genes of a family of natriuretic peptides that are present not only in the heart but also in other tissues including the brain (see Levin et al., 1998 for review). Both increased systemic osmolality and hypervolemia elevate circulating ANP. Systemic administration of ANP attenuates osmotically induced thirst in humans (Burrell et al., 1991). Icv injections of ANP attenuate both thirst (Antunes-Rodrigues et al., 1985) and sodium appetite (Fitts et al., 1985a).

2.3 Vascular Baroreceptors

2.3.1 Overview

The receptors identified as most important for sensing changes linked to blood volume (and hence extracellular fluid volume) and blood pressure are mechanoreceptors. These sensors are composed of free nerve endings, which are insinuated into the walls of the great veins, the heart and large arteries, and they are depolarized as these hollow structures are stretched. Depolarization generates increased nerve traffic that is proportional to the distention of the vessel or organ. The great veins and the atria are readily distensible, and the walls of these structures expand and contract as more or less blood volume fills the capacitance portion of the circulation. This yields good coupling between blood and extracellular volume and nerve traffic generated from receptors on the low-pressure side of the circulation. These sensors are referred to as low-pressure receptors (or unfortunately using a somewhat confusing and ambiguous nomenclature often as cardiopulmonary receptors).

A direct relationship between blood volume and baroreceptor activity does not exist for the pressure receptors located on the arterial side of the circulation. The arterial (or high pressure) baroreceptors are located in the aortic arch and carotid sinuses, and the afferent nerve activity from these receptors reflects changes in the arterial pressure wave throughout each cardiac cycle. Since cardiac output and arterial pressure are maintained relatively constant in the face of significant changes in blood volume, for example, under conditions of stage I shock (i.e., compensated shock), arterial baroreceptors do not always provide good information to the CNS about extracellular fluid status. However, when hypovolemia becomes sufficiently severe, cardiac output cannot be maintained and blood pressure falls. This subsequent decrease in arterial pressure can then provide information that hypovolemia is quite profound.

Input from low-pressure (i.e., blood volume) receptors and the high-pressure receptors in the aortic arch is conducted into the brain through afferents running in the vagus nerve (X cranial nerve). Arterial pressure receptors in the carotid sinuses project their information centrally by way of the glossopharyngeal (IXth cranial) nerve. The NTS is the terminal site for both of these the visceral afferent nerves (Norgren, 1981).

2.3.2 Low-Pressure Receptors

The role of vascular baroreceptors in the control of drinking is demonstrated by experimentally "loading" and "unloading" the receptors. Loading and unloading of the receptors is accomplished by procedures that produce or mimic increases or decreases in blood volume or blood pressure. For example, unloading of low-pressure receptors is produced by the constriction of a cuff placed around the vena cava (Thrasher et al., 1982a, b; Quillen Jr. et al., 1988) or by the inflation of a balloon surgically implanted inside the vena cava (Fitzsimons and Moore-Gillon, 1980). Both procedures impede blood flow through the vena cava, thereby reducing venous return to the heart, unloading of the low-pressure receptors and sometimes unloading of arterial baroreceptors as well (i.e., if the reduction in cardiac return is sufficient). These means of unloading low-pressure receptors are associated with drinking in dogs. Baroreceptors are involved in this drinking because surgical transection of the nerves supplying either the low-pressure or the sinoaortic baroreceptors (Quillen Jr. et al., 1990) reduces the drinking response. The drinking is completely abolished by simultaneous denervation of both sets of baroreceptors (Quillen Jr. et al., 1990). Drinking by hypovolemic sheep is reduced by crushing the low-pressure receptors located in the left atrial appendage (Zimmerman et al., 1981). Thus, in dog and sheep, vascular baroreceptors may provide neural afferent signals that stimulate thirst upon volume loss. In rat, a portion of hypovolemic thirst typically survives nephrectomy and the consequent loss of the source of renin (Fitzsimons, 1961). By inference, the remaining thirst is probably mediated neurally by baroreceptor input.

Loading of low-pressure baroreceptors can be mimicked by inflation of an intravascular balloon placed at the junction of the left pulmonary vein and the left atrium. Loading of low-pressure baroreceptors reduces water drinking in dogs in response to drug treatment with isoproterenol, iv infusions of hypertonic saline, and water deprivation (Moore-Gillon and Fitzsimons, 1982). Temporary vagal blockade of these baroreceptor signals increases water intake in response to reduced venous return (Fitzsimons and Moore-Gillon, 1980). These results suggest that low-pressure baroreceptors provide signals during volume expansion that inhibit further water intake.

Elegant work by Thrasher and colleagues (1999) nicely demonstrates baroreceptor control of drinking in dogs. Thrasher and group (1999) specifically increased either right or left atrial pressure during hypotension and measured drinking. Only increased left atrial pressure reduced (i.e., abolished) drinking in response to hypotension. This accords well with their work showing that increased left, but not right, atrial pressure inhibits hypotension-induced VP secretion (Andersen et al., 1994, 1995). In rats, it is the right atrium that provides inhibitory signals for thirst (Kaufman, 1984). Species differences in baroreceptor control of drinking are also underscored by Schreihofer's (1999) findings in rats that NTS lesions, which interrupt afferent input from both high- and low-pressure receptors, had no effect on water intake in response to hypovolemia. However, Quillen and coworkers (1990) found that simultaneous removal of high- and low-pressure receptor inputs by nerve transection totally abolished hypovolemic drinking in dogs. These different results may indicate that hypovolemic drinking is largely neurally mediated in dogs and hormonally mediated (i.e., ANG II) in rats. This suggestion must be tempered by the caveat that the techniques for abolishing baroreceptor input in these studies are not comparable.

The evidence for a neural contribution from baroreceptors in controlling salt appetite is less developed than for thirst. A problem with accepting a role for baroreceptors in the initiation of salt appetite is the time it takes for hypovolemia to stimulate salt appetite (Fitzsimons, 1979; Stricker, 1991). Hypovolemic treatments typically require several hours to stimulate salt intake, a delay that seems at odds with a neurally mediated response. Stricker's work suggests that the delay between the onset of hypovolemia and salt appetite is an artifact of maintaining rats on standard laboratory diet with its high sodium content (Stricker, 1981; Stricker, 1983). In effect, standard diet permits animals to acquire a reserve of body sodium that buffers the extracellular depleting effects of hypovolemia. Rats maintained on a sodium-deficient diet for as little as 2–4 days before study develop salt appetite with the onset of hypovolemia. Rats maintained on sodium deficient diets have reduced plasma volume and sodium concentrations (Stricker, 1981; Stricker et al., 1987) and increased levels of plasma renin activity and aldosterone (Stricker et al., 1979) compared with rats maintained on standard diet. Consequently, hypovolemic treatments cause greater reductions in plasma volume, larger increases in plasma renin activity (Stricker et al., 1979) and aldosterone (Stricker et al., 1979), and blunted secretion of oxytocin (OT; Stricker et al., 1994; see *Section 4.2.1*) in rats maintained on sodium deficient diet than those maintained on standard diet. In addition, rats maintained on sodium-deficient diet have difficulty maintaining arterial pressure during hypovolemia (Stricker et al., 1979, 1987). Coupled with the greater reductions in plasma volume, the difficulty maintaining arterial pressure provides a basis by which arterial and cardiopulmonary baroreceptors may contribute important neural inputs under these conditions. The ability of atrial distension to inhibit salt appetite (Toth et al., 1987) may be circumstantial evidence that baroreceptors mediate the inhibition although it cannot be ruled out that ANP (see *Section 2.2.2.4*) released during atrial distension accounts for the effect.

2.3.3 Arterial Baroreceptors

Loading and unloading of arterial (i.e., high-pressure) baroreceptors are typically accomplished by infusions of vasoconstrictor or vasodilator drugs that raise or lower blood pressure, respectively. Increases in arterial pressure inhibit osmotic thirst (Stocker et al., 2001), hypovolemic thirst (Stocker et al., 2001), salt appetite (Thunhorst and Johnson, 1994a), and drinking in response to ANG II administered peripherally (Robinson and Evered, 1987; Evered et al., 1988; Evered, 1992) or into the brain (Thunhorst et al., 1993). Removal of arterial baroreceptor input to the brain by sinoaortic baroreceptor denervation (SAD) eliminates the inhibition of thirst to ANG II administered peripherally (Stocker et al., 2002) but not centrally (Thunhorst et al., 1993). Denervation of the arterial baroreceptors reduces salt appetite in sodium-depleted rats by half (Thunhorst et al., 1994). This may be due to the absence of baroreceptor signals, altered hormonal responses (e.g., reduced renin secretion), or both. Rats with SAD have reduced salt intake over a range of NaCl concentrations at or above isotonic levels (Rocha et al., 1997). The reduced ingestion is specific for salt because SAD does not change ingestion of water, sucrose, or dilute saline solutions (Rocha et al., 1997). When offered only 2% NaCl to drink, SAD rats consume less than controls (Rocha et al., 1993) but have exaggerated release of VP and OT (Rocha et al., 1993). This led to the idea that exaggerated OT secretion upon initial ingestion of concentrated saline might inhibit further ingestion in SAD rats (Thunhorst et al., 1994). However, OT antisense injected directly into the PVN region decreased, rather than increased, the saline consumption of SAD rats when compared with controls (Morris et al., 1995).

Lesions of the NTS remove neural input from both arterial and cardiopulmonary baroreceptors (Schreihofer et al., 1999). Rats with NTS lesions have increased thirst and salt appetite responses to hypovolemia under some conditions. Rats with NTS lesions drink exactly as much as controls during hypovolemia when water or isotonic saline are available (Schreihofer et al., 1999). However, when a choice of water and concentrated saline is provided, rats with NTS lesions drink more of both in response to hypovolemia (Schreihofer et al., 1999). Rats with NTS lesions drink greater amounts of water, with reduced latency, in response to iv infusions of ANG II (Schreihofer et al., 2000). Increased drinking in response to ANG II is also observed in dogs lacking all arterial and cardiopulmonary input after surgical denervation (Klingbeil et al., 1991) and in rats with lesions of the LPBN (Ohman and Johnson, 1986).

2.4 Interactions Between Systems

2.4.1 The Additivity of Cellular and Extracellular Thirsts

Thirst arising from simple water deprivation, probably the most common and naturally occurring situation producing thirst, involves loss of water from both fluid compartments. A thorough analysis of thirst following water deprivation requires understanding how stimuli arising from the two major fluid compartments act simultaneously. It is possible that when multiple stimuli for thirst are present, drinking could result solely from the effects of the more intense stimulus. This is not the case, as Oatley (1964) showed in rat that the combination of hemorrhage and injection of hypertonic saline produces more drinking than either stimulus alone. Alternatively, the effects of multiple stimuli can sum, multiply, or interact in some other fashion. For example, Corbit (1968) demonstrated that cellular dehydration (i.e., i.p. hypertonic saline) and hypovolemia (i.e., hyperoncotic colloid) are independent and additive in their effects on thirst and do not interact in a multiplicative way. Fitzsimons and Oatley (1968) found that both cellular dehydration (i.e., iv hypertonic saline) and hypovolemia (i.e., hemorrhage) caused additional drinking in water-deprived animals. Again, the amount of water drunk after the combination of the osmotic or hypovolemic stimulus with water deprivation was simply the sum of the amounts drunk after each stimuli (i.e., it was additive). Furthermore, there was no interaction effect of combining the stimuli.

A variety of combinations of thirst stimuli involving intracellular and extracellular deficits have additive effects on drinking. These include hypertonic saline and hemorrhage (Oatley, 1964), water deprivation and hypertonic saline or hemorrhage (Fitzsimons and Oatley, 1968), hypertonic saline and hyperoncotic colloid (Corbit, 1968; Stricker, 1969; Blass and Fitzsimons, 1970), caval ligation and water deprivation or hypertonic saline or hypertonic sucrose (Fitzsimons, 1969a), and ANG and caval ligation (Fitzsimons and Simons, 1969) or hypertonic saline (Fitzsimons and Simons, 1969; Hsiao and Epstein, 1973). Severs and coworkers (1974) found that the amount of water drunk after icv administration of ANG II was additive with both cellular (i.e., s.c. hypertonic saline) and extracellular (i.e., polyethylene glycol) dehydration-induced drinking. In this case, a central stimulus (icv ANG) summed with both types of systemic stimuli (osmotic or volumetric). In all cases, drinking produced by one stimulus simply sums with the drinking produced by another stimulus. Additivity of thirst stimuli has been shown in gerbil (Vanderweele, 1974; Wright et al., 1987), degu and hamster (Wright et al., 1987) using hypertonic saline and hyperoncotic colloid.

In an interesting study on sham drinking in rats with gastric fistulas, Salisbury and Rowland (1990) found the usual additivity of drinking to combined cellular (i.e., hypertonic saline) and extracellular (i.e., s.c. colloid) dehydration with the fistula closed. They also found typical sham drinking responses to the individual stimuli with the fistula open (i.e., doubling of the drinking response). However, sham drinking in response to the combination of cellular and extracellular dehydration was no greater than the amount drunk with the fistula closed. That is, there was no additional drinking under sham conditions when the stimuli were combined. This contrasted with sham drinking in response to simple water deprivation, which was three times greater than with the fistula closed. It is not clear why two conditions involving combined cellular and extracellular deficits (i.e., water deprivation vs. combined colloid and hypertonic saline) should produce disparate sham drinking responses.

2.4.2 The Synergy Hypothesis

The capacity of ANG II and adrenal steroids to stimulate salt appetite led to the suggestion that they may act cooperatively or synergistically to promote salt appetite when levels of both are elevated during hypovolemia or sodium depletion (Epstein, 1982, 1984; Fluharty and Epstein, 1983; Sakai et al., 1986). As originally formulated, the *synergy hypothesis* holds that mineralocorticoids of peripheral origin (i.e., the adrenal glands) act together with ANG of central origin (i.e., the brain) to promote salt appetite. The combined effect is described as synergistic because the resulting sodium ingestion is greater than the sum of the intakes produced by the hormones separately across a range of doses. Experiments showed that subthreshold doses of systemically administered deoxycorticosterone (DOCA) or aldosterone, and centrally administered ANG II readily caused saline intake when administered together (Fluharty and Epstein, 1983; Sakai, 1986). Converging evidence showed that blockade of either ANG II or aldosterone receptors alone only partially prevented salt appetite in response to sodium depletion, but simultaneous blockade abolished it (Sakai et al., 1986). The inability to prevent depletion-induced salt appetite by peripheral infusion of ANG II antagonists (Sakai et al., 1990) led to the conclusion that ANG II of central origin, not renal origin, is responsible for the synergistic response. However, this conclusion leaves no stated mechanism to increase brain ANG II levels following sodium depletion. A possible solution to this need is the finding that circulating ANG II is a potent stimulus of salt appetite in sodium-depleted animals (see Section 2.2.2.1), and if connected in series with the brain RAS, could serve as the stimulus for increased levels of brain ANG II that are vital for the synergistic response.

Since its formulation, much work on the *synergy hypothesis* has concerned cellular mechanisms by which mineralocorticoids and ANG II interact in the brain. For example, elevated mineralocorticoid levels increase ANG II receptor binding in brain tissue, specific brain nuclei, and neuronal cultures (Wilson et al., 1986; King et al., 1988; Sumners and Fregly, 1989; De Nicola et al., 1993), probably through both genomic (Fluharty and Sakai, 1995; Daniels and Fluharty, 2004) and nongenomic (Sakai et al., 2000) mechanisms. An intriguing aspect of the work regarding the proposed cellular basis of the ANG II/aldosterone synergy is the prominent role given to CVOs as the anatomical locus for the interactions (Daniels and Fluharty, 2004).

An alternative account for the synergistic behavioral response upon dual administration of mineralocorticoids and ANG II relies on the ability of mineralocorticoids to suppress secretion of OT, an inhibitor of sodium consumption (Stricker et al., 1987, 1994). In this scenario, circulating mineralocorticoids create the conditions suitable for initiating a salt appetite response by suppressing OT activity in the brain. This removes a brake on sodium consumption hence permitting greater responding to ANG II.

A synergy of ANG II and aldosterone has been found in pigeon (Massi and Epstein, 1990) but does not appear to hold for sheep and cow (Weisinger et al., 1996).

2.4.3 Interactions Between Arterial Blood Pressure and Angiotensin

Considerable work has clarified the interactions between hormonal and baroreceptor signals in the control of body fluid regulation (Stricker, 1991; Evered, 1992; Thunhorst and Johnson, 1993, 1994b; Johnson and Thunhorst, 1995). Fitzsimons and Simons (1969) first established that ANG II is dipsogenic upon iv administration in rats. The dipsogenic nature of ANG II has subsequently been confirmed in several species (Cooling and Day, 1975; Fitzsimons, 1975; Trippodo et al., 1976; Schwob and Johnson, 1977; Simpson et al., 1978), but the often weak nature of the dipsogenic response argues against an important role for ANG II in drinking behavior (Stricker, 1977, 1978; Anke et al., 1988). A resolution of the problem comes with the understanding of the confounding effects of acute hypertension on drinking behavior (Robinson and Evered, 1987; Evered et al., 1988; Evered, 1992). ANG II causes striking increases of arterial pressure in fluid-replete animals (Evered, 1992). Under natural conditions of dehydration involving hypovolemia, endogenous ANG II serves to maintain arterial pressure only at approximately normotensive levels, not at hypertensive levels. Animals drink more in response to iv infusion of ANG II if the arterial pressure response is blunted (Robinson and Evered, 1987; Evered et al., 1988).

Low-pressure receptors and arterial baroreceptors mediate the pressor-induced inhibition of water drinking in response to ANG II. In dogs, surgical removal of both sets of these baroreceptors permits greater drinking during iv ANG II (Klingbeil et al., 1991). Initial studies in rats did not find that removal of the arterial baroreceptors by themselves affected drinking in response to systemically administered ANG (Rettig and Johnson, 1986; Kadekaro et al., 1989). However, recent work by Stocker and coworkers (2002) clearly shows that removal of arterial baroreceptor input (i.e., SAD) permits greater drinking in response to high doses of iv infused ANG II. The different results may be due to the degree of completeness of the surgical denervations or the routes and doses of ANG II. Therefore, it seems that sensing increased arterial pressure inhibits drinking caused by ANG II.

Increased arterial pressure inhibits drinking when ANG II is administered centrally (Johnson et al., 1977; Harland et al., 1988; Kucharczyk, 1988). Modest reductions in arterial pressure are associated with a doubling of the dipsogenic response to centrally administered ANG II (Thunhorst and Johnson, 1993). However, the ability of arterial pressure to affect the dipsogenic response of centrally administered ANG II is not influenced by surgical denervation of the arterial baroreceptors (Thunhorst et al., 1993). The possibility remains that other systemic baroreceptors (e.g., renal baroreceptors; Stella and Zanchetti, 1991; Thunhorst et al., 1996) mediate the influences of arterial pressure on drinking in response to centrally administered ANG II.

Less thoroughly studied are the interactions between hormonal and baroreceptor signals controlling salt appetite. Although aldosterone and ANG II stimulate salt appetite, neither is essential (Stricker, 1981; Fregly and Rowland, 1985), so other mechanisms may be involved including cerebral sodium sensors (Weisinger et al., 1985a, b) and baroreceptors. The relatively small salt appetite response following prolonged iv infusions of ANG II into fluid-replete animals may be due to pressure-mediated inhibition of the response. For example, iv infusions of ANG II that elevate arterial pressure above baseline slightly reduce salt appetite of sodium-deficient sheep (Bott et al., 1967) and rats (Fitts and Thunhorst, 1996). Hypotension accelerates the appearance of salt appetite in hypovolemic rats following diuretic treatment (Thunhorst et al., 1999). The onset of salt appetite is also hastened by administering a low dose of ACE inhibitor to animals made hypovolemic by polyethylene glycol (Stricker, 1983) or diuretic treatment (Fitts and Masson, 1989; Masson and Fitts, 1989; Thunhorst and Johnson, 1994a). In these hypovolemic situations, peripheral ACE blockade abruptly reduces arterial pressure (Gardiner and Bennett, 1986; Thunhorst and Johnson, 1994a) which aids the salt appetite response (Thunhorst and Johnson, 1994a). When arterial pressure is prevented from falling under these conditions, for example by iv infusion of the pressure agent, phenylephrine, the salt appetite response is blunted whereas the water drinking response is not (Thunhorst and Johnson, 1994a). It is not known if hypotension aids salt appetite by providing neural input to the CNS or by increasing renin secretion. Increased arterial pressure may inhibit salt appetite under some circumstances. The few experiments examining salt intake in animals lacking baroreceptors suggest that the salt appetite response is severely diminished (Thunhorst et al., 1994; Morris et al., 1995; Rocha et al., 1997).

2.4.4 Interactions Between Vascular Volume and Angiotensin

Prevailing levels of extracellular fluid volume, as with arterial pressure, modify the dipsogenic and natriorexigenic responses to ANG II. Experiments in which endogenous ANG II levels are increased by various means in volume- and sodium-replete animals fail to show convincing salt appetite responses (e.g., Bott et al., 1967; Fitzsimons and Wirth, 1978). It is only when ANG II is administered in sufficiently high doses to cause pressure-induced natriuresis and hypovolemia that salt appetite is observed, and usually only after considerable delay (Fitzsimons, 1980; Findlay and Epstein, 1980). On the other hand, systemic infusions of ANG II readily elicit water drinking and sodium ingestion from animals that are water and/or sodium-deplete (Weisinger et al., 1996). This has been explicitly demonstrated in experiments using rats that were sodium- and volume-deplete following diuretic treatment (Fitts and Thunhorst, 1996) or adrenalectomy (Schoorlemmer et al., 2001). In both experiments, the typical salt appetite responses following hypovolemia were prevented by interrupting the endogenous formation of ANG II in the peripheral circulation with doses of ACE inhibitor. In both experiments, the hypovolemic animals began ingesting sodium within minutes when supplied with exogenous ANG II through iv infusion. Therefore, the ability of circulating ANG II to stimulate water and sodium ingestion is modified by extracellular volume.

Other work suggests that volume and pressure status modifies salt appetite and thirst, in response to central administration of ANG II (Thunhorst and Johnson, 2001). In a series of tests, icv infusions of ANG II were given to rats that were either normotensive or hypotensive, and to rats there were either fluid-replete or fluid-deplete. Similar to a previous report (Thunhorst and Johnson, 1993), water-drinking responses to icv infusions of ANG II were greater under hypotensive conditions than normotensive conditions, even though the intakes were accompanied by substantial urinary retention of water and sodium. Saline drinking in response to the infusion was not affected by hypotension. Likewise, saline drinking in response to icv infusions of ANG II was not affected by hypovolemia following diuresis. However, conditions that simultaneously produced hypotension and hypovolemia increased both water and saline drinking. Therefore, while reductions in arterial pressure are sufficient to increase water-drinking responses to centrally administered ANG II (Harland et al., 1988; Thunhorst et al., 1993; Thunhorst and Johnson, 1994b), reductions in both arterial blood pressure and volume are required to increase saline-drinking responses to centrally administered ANG II (Thunhorst and Johnson, 2001).

Some factors that change with the levels of extracellular fluid volume and could interact with ANG II include levels of circulating aldosterone and ANP and also neural signals from baroreceptors (Kaufman, 1984; Toth et al., 1987). Systemic mineralocorticoids potentiate the salt appetite response to centrally administered ANG II (Fluharty and Epstein, 1983), either by acting synergistically with ANG II in the brain (Fluharty and Epstein, 1983) or by suppressing secretion of OT (Stricker and Verbalis, 1990). Reductions in volume and pressure could remove residual inhibitory signals arising from baroreceptors (Toth et al., 1987) or circulating ANP that may arise when icv ANG II is infused in fluid-replete, normotensive animals.

2.4.5 The Effects of Glucocorticoids on Mineralocorticoid-Induced Salt Appetite

The hormones of stress, including glucocorticoids, participate in water and sodium homeostasis. Many situations involving stress or disease (e.g., hemorrhage, heat stress, diarrhea) are accompanied by loss of sodium from the body and increased secretion of hormones from the hypothalamus–pituitary–adrenal axis, including adrenocorticotropic hormone (ACTH). ACTH stimulates salt appetite in rabbits (Tarjan and Denton, 1991), rats (Weisinger et al., 1978), mice (Denton et al., 1999), and sheep (Weisinger et al., 1980), and water intake in mice (Blair-West et al., 1996). The hypothalamic releasing hormone for ACTH, corticotrophin releasing factor, stimulates salt appetite in rabbits (Tarjan and Denton, 1991; Tarjan et al., 1991) and mice (Denton et al., 1999) upon central, but not systemic administration (Denton et al., 1999) and thus probably acts by releasing ACTH. ACTH stimulates salt appetite when given systemically (Weisinger et al., 1978, 1980; Tarjan and Denton, 1991; Denton et al., 1999), but not centrally (Tarjan and Denton, 1991; Denton et al., 1999). A "cocktail" of adrenal hormones, including aldosterone, cortisol, and corticosterone, elicits robust salt appetite (Blaine et al., 1975; Weisinger et al., 1980). Collectively, these works suggest that the stress hormones stimulate salt appetite, and adrenal steroids, likely a combination of glucocorticoids and mineralocorticoids, ultimately mediate the salt appetite response. Animals with access to sodium solutions during ACTH treatment ingest sodium before natriuresis, and therefore not secondarily to sodium loss (Weisinger et al., 1978, 1980). However, animals without a source of sodium in addition to their food during ACTH treatment excrete sodium in excess of what they obtain from their food (Tarjan et al., 1991). The salt appetite response to stress hormones may be beneficial in preventing electrolyte imbalances.

Hypovolemic treatments that stimulate thirst and salt appetite are associated with increased levels of the glucocorticoid, corticosterone, in addition to aldosterone and ANG (Stricker et al., 1979; Rowland and Morian, 1999). Pharmacological doses of glucocorticoids increase water-drinking and pressor responses induced by ANG II (Ganesan and Sumners, 1989; Sumners et al., 1991; Scheuer and Bechtold, 2001) and both water drinking and sodium ingestion in mineralocorticoid-treated animals (Wolf, 1965; Ma et al., 1993; Zhang et al., 1993; Fluharty and Sakai, 1995; Shelat et al., 1999a, b; Fluharty, 2002). It is postulated that the powerful influences of glucocorticoids on responses to mineralocorticoids reflects synergy between these classes of steroid hormones at the cellular level (Fluharty and Sakai, 1995; Daniels and Fluharty, 2004). Thus, new versions of the *synergy hypothesis* include prominent roles for glucocorticoids and their receptors (Fluharty and Sakai, 1995; Fluharty, 2002). In the brain, glucocorticoids increase ANG II receptor binding and amplify ANG II-related intracellular signaling processes (Sumners et al., 1991; Shelat et al., 1998, 1999a, b; Fluharty, 2002; Daniels and Fluharty, 2004) and also increase Type 1 mineralocorticoid receptor binding (Fluharty and Sakai, 1995).

Glucocorticoids have powerful effects on urinary excretion that offer another mechanism for their effects on thirst and salt appetite. Basal levels of glucocorticoids are required for normal renal function (Ganong, 2005). Glucocorticoids at physiological and pharmacological levels increase glomerular filtration rate in animals and humans (Baylis et al., 1990), and they increase urine volume (Okuno et al., 1981; Suzuki et al., 1982; Yoshida et al., 1988; Baylis et al., 1990), urinary sodium excretion (Okuno et al., 1981; Suzuki et al., 1982; Yoshida et al., 1988; Baylis et al., 1990), and potassium excretion (Baylis et al., 1990). Forty years ago, Wolf (1965) speculated that glucocorticoids, namely corticosterone, potentiate sodium ingestion during mineralocorticoid treatment by stimulating glomerular filtration rate and essentially countering the sodium retaining effects of mineralocorticoids. By facilitating the excretion of sodium, corticosterone could allow sodium to be ingested at a higher rate (Wolf, 1965).

3 Central Nervous System (CNS) Integration of Signals for Thirst and Salt Appetite

3.1 Primary Brain Structures and Systems Implicated in Thirst and Salt Appetite

A variety of physiological and neuroanatomical mapping techniques, such as ablation, electrical stimulation, metabolic activation (e.g., 2-deoxyglucose), and expression of immediate early genes (e.g., c-*fos*) has established a collection of brain structures that form a neural network that integrates information related to body fluid balance. Some structures and pathways are particularly important. Processing of visceral information within this network or *visceral neuraxis* is broadly distributed and provides redundancy so that insult to one part may not be catastrophic. This network may exhibit plasticity. For example, the central resetting of baroreceptor reflexes is well known. Likewise, the chronically enhanced sodium preference and salt appetite that result from a single episode of sodium deficiency (Falk, 1966; Sakai et al., 1987) probably involve some form of neural plasticity. Several forebrain structures, including those of the lamina terminalis and the amygdala, have demonstrated roles in facilitating thirst and sodium appetite. A hindbrain system including the AP, the NTS, and the parabrachial nucleus (Shapiro and Miselis, 1985; Cunningham Jr. et al., 1994) appears to inhibit these behavioral responses (Johnson and Edwards, 1990; Johnson and Thunhorst, 1997).

3.1.1 Forebrain

3.1.1.1 Lamina Terminalis During embryonic development, the rostral end of the neural tube closes to form the lamina terminalis. In mature vertebrates, this tissue becomes the anterior wall of the third cerebral ventricle which is associated with and houses three midline nuclei—the SFO, the MePO, and the OVLT. The SFO and OVLT, respectively, are situated dorsally and ventrally along the lamina terminalis and the MePO resides between the two. Morphological studies led to speculation that the SFO is somehow involved in sodium and water homeostasis (Diereckx, 1963; Palkovits, 1966). Simpson and Routtenberg (1973) subsequently showed that the SFO is one site in the brain that is responsible for the dipsogenic action of ANG II. They demonstrated that very low doses of ANG II injected into the SFO of rats caused drinking. Furthermore, lesions of the SFO eliminate drinking to iv administration of ANG II in rat (Simpson et al., 1978), dog (Thrasher et al., 1982a, b), and quail (Takei, 1977) but not in the sheep (McKinley et al., 1986).

The ventral area of the lamina terminalis is also important for the dipsogenic response to ANG II. Ablation of tissue surrounding the anteroventral portion of the third (cerebral) ventricle (AV3V) abolishes drinking to both systemically and centrally administered ANG II (Buggy and Johnson, 1977, 1978; Johnson and Buggy, 1978, 1997). This ablation, termed the AV3V lesion, destroys the OVLT, the ventral part of the MePO, the periventricular preoptic nuclei, and the periventricular nuclei of the anterior hypothalamus. The AV3V lesion produces a host of deficits in responding to body fluid challenges (➋ *Table 17-1*). The totality of these effects is not localizable to a single structure within the area but is obtained only from the destruction of the whole AV3V region. Reflex and behavioral responses to humoral factors, including ANG II and increased plasma osmolality are greatly affected by AV3V lesions. Animals with AV3V lesions will evidence some drinking to a hypovolemic challenge as long as it is sustained and there is still presumably significant input from vascular baroreceptors (Buggy and Johnson, 1977). However, the drinking to hypovolemia is impaired (Lind and Johnson, 1983), probably from failure to respond to the ANG II component (Johnson et al., 1981). Selective lesions of the OVLT attenuate drinking to infusions of ANG II and to increased osmolality (Thrasher and Keil, 1987). Several reviews of the numerous aspects of the effects of AV3V lesions on body fluid and cardiovascular regulation are available (Brody and Johnson, 1980; Johnson, 1985a, b; Johnson and Edwards, 1990; Johnson et al., 1996; Johnson and Thunhorst, 1997).

In elegant studies using systemic and icv administration of osmotic solutions, two groups of investigators, McKinley and colleagues (1978) in sheep and Thrasher and coworkers (1980) in dog, made a strong case that cephalic osmoreceptors are probably located in CVOs. Further *in vivo* studies employing tissue ablation and site-specific administration of hypertonic solutions provide evidence that the structures of the lamina terminalis contain osmo- or sodium receptors responsible for drinking, antidiuretic responses, and pressor responses to hypertonic stimuli (Buggy and Johnson, 1977; Johnson and Buggy, 1978; Johnson et al., 1978; Buggy et al., 1979; Hosutt et al., 1981; Lind et al., 1984) and *in vitro* methods have lent additional support (Bourque et al., 1994). Individual or combined lesion studies of the nuclear groups of the lamina terminalis show that all parts contribute to drinking and VP secretion in response to osmotic stimuli (McKinley et al., 1982). However, ablation of nearly the entire lamina terminalis is required to abolish these responses (McKinley et al., 1999).

The participation of the SFO and the AV3V region in salt appetite is logical based on their known roles in mediating the effects of ANG II (for a fuller account see Johnson and Thunhorst, 1997). The relative importance of the SFO for salt appetite depends on the strength of the response, and duration over which the response is produced and measured. Ablation of the SFO almost completely eliminates salt appetite that is rapidly developing and measured over a few hours after sodium depletion (Thunhorst et al., 1999) or iv infusion of ANG II (Morris et al., 2002) and reduces by half the salt appetite response produced by 24 h of sodium depletion (Thunhorst et al., 1990; Weisinger et al., 1990). On the other hand, SFO ablation or interruption of its major efferent pathway to the rest of the brain does not attenuate total saline consumption over an extended 48-h period following sodium depletion (Schulkin et al., 1983) or the robust salt appetite of adrenalectomy (Wilson et al., 2002). Longer testing periods may allow for compensatory or redundant mechanisms to contribute to the salt appetite response in animals lacking an SFO. Lesions of the SFO prevent the permanent elevation in daily saline intakes that are sometimes observed after episodes of sodium depletion (Ruhf et al., 2001).

Lesions of the AV3V region also diminish depletion-induced salt appetite (De Luca Jr. et al., 1992). Lesions that encompass the OVLT and the adjacent third of the ventral MePO reduce sodium ingestion in response to renin-dependent forms of salt appetite but increase sodium intake elicited by DOCA (Fitts et al., 1990; Fitts, 1991). Selective ablations of the OVLT reduce the salt intake by one-half after sodium depletion (Chiaraviglio and Perez Guaita, 1984) and iv infusion of ANG II (Morris et al., 2002). In heroic work, it was found that simultaneous lesions of the SFO and OVLT more effectively reduce salt appetite in response to sodium depletion than either lesion alone (Fitts et al., 2004). Taken together, the functional studies illustrate the redundancy and complex unity of function (Johnson and Wilkin, 1987; Lind, 1988; Johnson, 1990) among neural groups along the lamina terminalis.

Discrete injections or infusions of ANG II into the OVLT and adjacent MePO stimulate ingestion of both water and saline (Fitts and Masson, 1990). However, injections of ANG II into the SFO produce water drinking (Simpson and Routtenberg, 1973; Simpson et al., 1978) but do not usually support saline drinking (Fitts and Masson, 1990; Colombari et al., 1996). It is suggested that ANG II may act at the SFO to engage mechanisms that inhibit salt appetite, for example OT secretion (Ferguson and Renaud, 1986; Ferguson and Kasting, 1987; Stricker et al., 1987, 1994; Ferguson and Kasting, 1988; Stricker and Verbalis, 1990; Bartanusz and Jezova, 1994). In light of this, it is interesting that injections of ANG II into the SFO cause substantial saline drinking provided that the postulated inhibitory system of the LPBN is first blocked by the administration of 5-HT antagonist (Colombari et al., 1996; see ➋ *Section 3.1.2.2*).

3.1.1.2 Amygdala Subdivisions of the amygdala are implicated in salt appetite (Johnson et al., 1999). Ablation of the basolateral amygdala diminishes salt appetite induced by DOCA treatment (Nachman and Ashe, 1974). Ablation of the medial region of the amygdala diminishes salt appetite produced by the administration of mineralocorticoids but not by sodium depletion (Nitabach et al., 1989; Zhang et al., 1993). This selective impairment is not from disrupting pathways to the ventral forebrain through the stria terminalis (Black et al., 1992). Although lesions of the medial amygdala do not affect renin-dependent salt appetite, injections of tachykinins into the medial amygdala inhibit renin-dependent forms of salt appetite (Massi et al., 1990). Ablation of the central nucleus of the amygdala (CeA) greatly diminishes the salt appetite response to DOCA, sodium depletion, s.c. administration of yohimbine and icv ANG II, and abolishes ad libitum sodium intake (Galaverna et al., 1992; Zardetto-Smith et al., 1994). The general lack of salt intake by rats with lesions of the CeA may not result from disordered gustatory function (Galaverna et al., 1993). Rats with CeA lesions appropriately discriminate between different tastes based on their internal states, yet do not consume sodium (Seeley et al., 1993). However, it is also reported that lesions of the CeA increase the aversive value of taste stimuli (Touzani et al., 1997).

The amygdala and bed nucleus of the stria terminalis are anatomically related structures that are likely to have similar functions (Johnston, 1923; Alheid et al., 1985; de Olmos et al., 1985; Price et al., 1987). They are reciprocally connected with several nuclei involved in cardiovascular control and fluid balance (Sofroniew, 1983; Holstege et al., 1985; Moga et al., 1989). Electrolytic ablation of either of them significantly reduces sodium intake following systemic administration of yohimbine or sodium depletion (Zardetto-Smith et al., 1993; Johnson et al., 1999). However such lesions do not affect water intake to s.c. ANG II or to s.c. hypertonic saline, indicating specificity of the lesion for salt appetite.

3.1.2 Hindbrain

The hindbrain has been studied for its role in reflex control of the circulation and fluid balance at least since Claude Bernard noted that puncturing the floor of the fourth cerebral ventricle increases urinary excretion (Cushny, 1926). The role of this region in the behavioral control of fluid balance has been studied for only about 30 years.

3.1.2.1 The Area Postrema (AP) and Nucleus of the Solitary Tract (NTS) The AP is found in the fourth ventricle at the caudal aspect of the rhomboid fossa. It was the first CVO shown to be a central target for circulating ANG II (Joy and Lowe, 1970; Ferrario et al., 1972; Ueda et al., 1972) in its capacity to reduce the gain of cardiovascular baroreceptor reflexes in response to ANG II. The AP is small and difficult to manipulate. Most functional studies of the AP involve ablation, and the lesions typically encroach on adjacent NTS tissue or otherwise cause degeneration of NTS neuropil because of the rich neural interconnectivity. Thus, AP lesions are sometimes denoted as AP/mNTS lesions. Contreras and Stetson (1981) first demonstrated effects of AP lesions on water and sodium ingestion. They described increased ad libitum consumption of NaCl solutions over a range of concentrations. These intakes can be phenomenal. AP/mNTS ablated rats in sodium balance consume the equivalent of 25% of their extracellular sodium in only 3 h after first being given access to hypertonic saline (Edwards et al., 1993). The increased sodium intake may be secondary to increased sodium excretion observed after AP ablation (Hyde and Miselis, 1984), but AP lesions sometimes have minimal effects on sodium excretion (Contreras and Stetson, 1981; Watson, 1985).

Lesions of AP/mNTS also produce a general increase in water intake in response to extracellular dehydration and isoproterenol compared with intact controls (Edwards and Ritter, 1982), but they do not increase water intake in response to intracellular dehydration (Edwards and Ritter, 1982) or to centrally administered carbachol (Ohman and Johnson, 1989). Therefore, the AP/mNTS may be a part of a neural system that inhibits water and sodium ingestion and may be more important in regulating extracellular than intracellular volume.

3.1.2.2 Parabrachial Nucleus The NTS and AP both send projections to the LPBN (Loewy and Burton, 1978; Norgren, 1978; Ricardo and Koh, 1978; van der Kooy and Koda, 1983; Shapiro and Miselis, 1985; Cunningham Jr. et al., 1994) including a substantial ascending serotonergic pathway from the AP (Lança and van der Kooy, 1985). In turn, the LPBN sends projections to forebrain areas involved in fluid balance such as the PVN, the CeA, and MePO (Ciriello et al., 1984; Fulwiler and Saper, 1984; Jhamandas et al., 1992; Krukoff et al., 1993; Jhamandas et al., 1996). Initially, it was thought that lesions of the LPBN would diminish thirst by interrupting ascending vascular baroreceptor information from the NTS. However, such electrolytic or chemical lesions had the opposite effect and greatly increased water drinking in response to extracellular fluid challenges (Ohman and Johnson, 1986, 1989; Edwards and Johnson, 1991; Menani et al., 1995) but not to intracellular-related fluid challenges (Ohman and Johnson, 1986, 1989; Edwards and Johnson, 1991). These results are similar to those observed after lesions of the AP and suggest that the LPBN and AP/mNTS comprise a hindbrain system that functions to limit extracellular volume expansion

(Johnson and Edwards, 1990, 1991; Johnson and Thunhorst, 1997). Support for this idea is the finding that right atrial stretch no longer inhibits some forms of extracellular thirst after lesions of the LPBN (Ohman and Johnson, 1995). In contrast to their similar effects on thirst, lesions of the LPBN and AP/mNTS have different effects on salt appetite. As noted, AP/mNTS lesions increase salt appetite under ad libitum conditions and in response to experimental challenges (Contreras and Stetson, 1981; Kosten et al., 1983; Hyde and Miselis, 1984; Watson, 1985; Edwards et al., 1993); however, LPBN lesions do not (unpublished observations).

Pharmacological manipulations of serotonergic receptors located within the LPBN have profound effects on both thirst and salt appetite. Discrete injections of 5-HT or certain selective 5-HT agonists into the LPBN suppress water intake whereas similar injections of the nonselective 5-HT receptor antagonist methysergide greatly increase water intake (Menani and Johnson, 1995). In addition, injections of 5-HT antagonist into the LPBN produce some of the largest intakes of hypertonic saline solution yet observed (Colombari et al., 1996; Menani et al., 1996, 1998b; *Figure 17-5*). This is not a nonspecific increase in behavior as NaCl intake increases preferentially to water when both are available for drinking (Menani et al., 1998a, b, 2000, 2002, 2004). In addition, ingestion of palatable sucrose solutions is not affected by similar

Figure 17-5

(a) Cumulative 0.3M NaCl intakes, (b) Individual 0.3 M NaCl intakes, (c) Cumulative water intakes, (d) Individual water intakes induced by subcutaneous furosemide (FURO; 10 mg/kg) + captopril (CAP; 5/mg/kg) after previous bilateral injections of vehicle or methysergide (4 μg/200 ml) into the lateral parabrachial nucleus (LPBN). (a, c) Values are means ± SE. *Significantly different from vehicle pretreatment tested by *t*-tests ($p < 0.05$). $n = 8$ rats. From Menani et al. (1996)

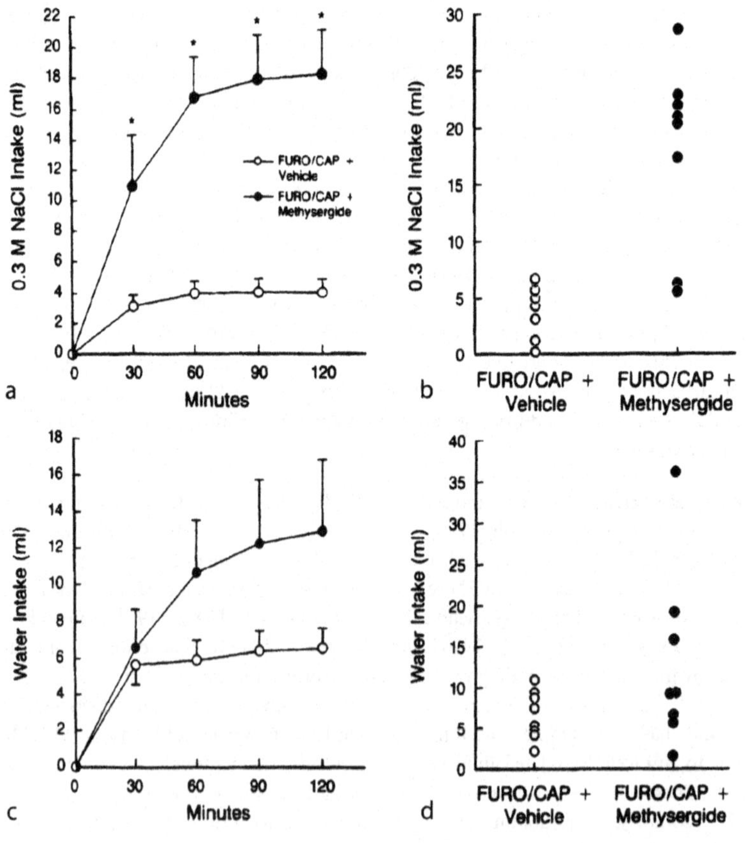

treatment (Menani et al., 1998b). The serotonergic innervation of the LPBN thus appears to limit both thirst and salt appetite responses.

Antagonism of 5-HT receptors in the LPBN increases salt intake under most conditions. It increases salt appetite in both renin-dependent and non renin-dependent models (Menani et al., 1998a, b; De Gobbi et al., 2000), and it greatly exaggerates salt appetite in response to stimuli known to produce salt intake (e.g., sodium depletion; hypovolemia; ANG II; DOCA; Menani and Johnson, 1995; Menani et al., 1998a, b; De Gobbi et al., 2000). 5-HT receptor antagonism initiates salt appetite under conditions when salt appetite is not normally found (i.e., in the first few hours after diuretic treatment; after injections of ANG II into the SFO; Colombari et al., 1996; Menani et al., 2000; ➋ *Figure 17-6*). Finally, it increases salt appetite in response to stimuli that normally inhibit salt appetite (i.e., icv carbachol and relaxin; intragastric loads of hypertonic saline; after s.c. isoproterenol treatment; Menani et al., 2000, 2002, 2004; De Luca Jr. et al., 2003). Yet, there is no effect of 5-HT blockade of the LPBN on the unstimulated animal (i.e., without some dipsogenic or natriorexigenic stimulus also present).

Injections of a 5-$HT_{2A/2C}$ receptor agonist, 2,5-dimethoxy-4-iodoamphetamine bromide (DOI), into the LPBN reduce saline and water intake after sodium depletion (Menani et al., 1996). In contrast,

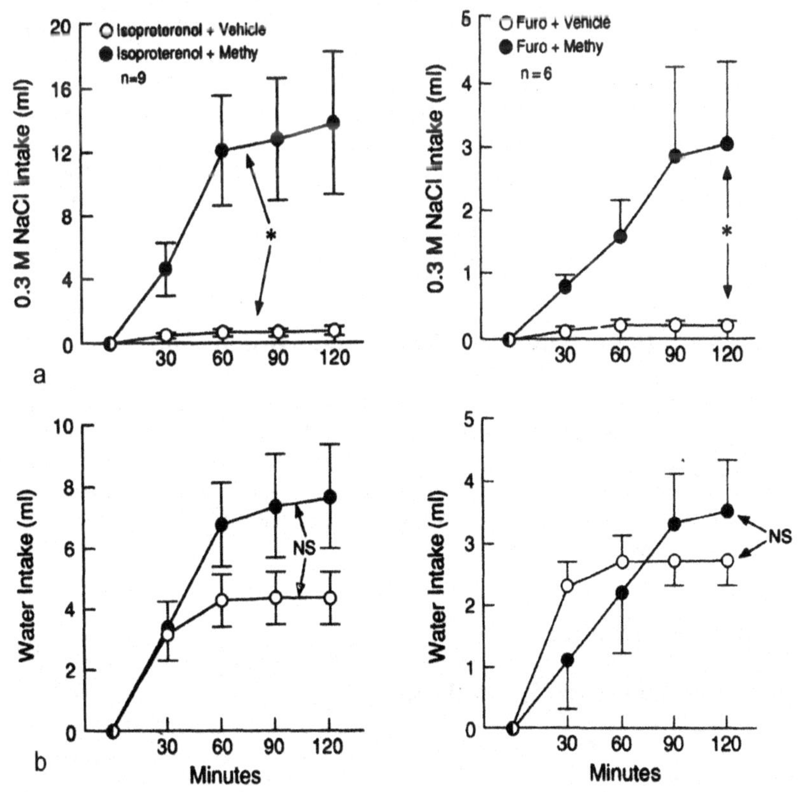

■ Figure 17-6

The induction of salt appetite by lateral parabrachial nuclei (LPBN) injections of the nonselective 5-HT receptor antagonist, methysergide (Methy) in rats pretreated with dipsogenic stimuli (either the β-adrenoreceptor agonist, isoproterenol = left-hand panels or the diuretic, furosemide = right-hand panels) that generate virtually no hypertonic saline intake when administered by themselves. The panels show cumulative intakes of 0.3 M NaCl or water after the respective systemic dipsogenic treatments followed by either LPBN vehicle or LPBN Methy treatments. Methy treatments significantly increased the intake of 0.3 M NaCl (*$p < 0.05$), but not of water. (NS = nonsignificant; $p > 0.05$). From Menani et al. (2000)

LPBN injections of the 5-HT_{1A} receptor agonist, 8-hydroxy-2-(di-*n*-propylamine) tetralin (8-OH-DPAT) enhanced saline and water intake of sodium-deficient rats (De Gobbi et al., 2005).

Evidence from functional neuroanatomical studies combining immunocytochemistry for FOS with immunocytochemistry for 5-HT (Franchini et al., 2002) indicate that serotonergic neurons in the rostral portions of the dorsal and medial raphe groups are active when rats are in a sodium-replete state. Activity in these cells is reduced in sodium deficient rats and returns as depleted animals restore their sodium deficit. In contrast, 5-HT neurons in the AP show no response to sodium depletion but increase their activity upon ingesting sodium. Taken together, these results seem to indicate that neurons in the caudal 5-HT cell group (i.e., AP region) are phasically activated to acutely inhibit excess salt appetite, whereas cells in the anterior groups exert a tonic inhibitory influence on this ingestive behavior.

Other neurotransmitters/neuromodulators that act within the LPBN and also affect thirst and salt appetite include cholecystokinin (Menani and Johnson, 1998; De Gobbi et al., 2001), α_2-adrenergic receptor agonist (Andrade et al., 2004), GABA (Callera et al., 2005) and corticotrophin-releasing hormone (De Castro e Silva et al., 2006). Injections of cholecystokinin into the LPBN inhibit thirst and salt appetite responses to ANG II and hypovolemia (Menani and Johnson, 1998), whereas the antagonists increase these behaviors. It is interesting that the activation of α_2-adrenergic receptors in the LPBN increases water and NaCl intake after hypovolemic treatment (Andrade et al., 2004) because similar activation of α_2-adrenergic receptors in the forebrain inhibits water and sodium intake (Fregly et al., 1981; Menani et al., 1999). Stimulation of $GABA_A$ receptors in the LPBN produces a robust salt appetite response in unstimulated, euvolemic animals (i.e., in animals not otherwise hydrationally challenged; Callera et al., 2005).

Other novel findings are that the antagonism of 5-HT receptors in the LPBN causes salt appetite in response to ANG II administered into the SFO (Colombari et al., 1996), which normally produces only water intake. In addition, antagonism of AT_1 receptors in the SFO prevents the increased, or additional, salt appetite response caused by antagonism of 5-HT receptors in the LPBN during renin-dependent models of salt appetite (Menani et al., 1998a).

Taken together, results from studying the actions of several classes of neurally active agents in the LPBN indicate that many neurotransmitter systems exert an inhibitory action in this pontine structure to keep excessive sodium intake in check. Under basal conditions sodium appetite is inhibited through activity in the LPBN (probably GABAergic). When animals are predisposed to consume water or saline solutions, the amounts are modulated by actions of neurotransmitter release into the LPBN. Removal of these inhibitory actions is tantamount to fixing the switch for sodium intake to the "on" position. Understanding the mechanisms of what leads to excessive activation of sodium appetite promises to be important for gaining insight into why humans consume excess dietary sodium chloride, a predilection that has significant negative health-related consequences (MacGregor and de Wardener, 1998).

4 Major Brain Neurotransmitter/Neuromodulator Systems

Several brain neurohumoral and neurochemical systems have been implicated as major factors in the control of thirst and salt appetite. These systems often exert their actions at many sites throughout the neuraxis. As a generalization, most of these can be classified as either facilitory or inhibitory to thirst or sodium intake. In this section, several key, centrally acting agents that have been identified in the control of hydromineral intake will be categorized and discussed based on their capacity to promote or retard the intake of water or sodium solutions.

4.1 Facilitory Mechanisms

4.1.1 Brain Renin-Angiotensin System (RAS)

Shortly after Fitzsimons (1969b) implicated the renal (or systemic) RAS in thirst, he along with his coworkers (Epstein et al., 1970) demonstrated that ANG II in doses on the order of 5 to 4000 pmoles produced avid drinking when injected into several limbic (e.g., septal area) and basal forebrain (e.g., preoptic and anterior hypothalamic nuclei) tissue sites. At the time of its publication, the Epstein and group (1970) study seemed somewhat problematic or paradoxical because the physiological source of ANG II that acted on the brain was believed to be from the systemic circulation. It was difficult to conceive that the peptide ANG II, which has a molecular weight of slightly more than 1000 Daltons, could pass the blood–brain barrier to act at central sites such as the preoptic area.

Subsequently, it was discovered that few, if any, of the initially implicated tissue sites were in fact sensitive to ANG II (Johnson and Epstein, 1975). Most of the injections in the initial studies (Epstein et al., 1970) probably leaked or diffused into the cerebral ventricles (Johnson and Epstein, 1975). Actually, the forebrain sites most responsive to ANG II are located in structures surrounding the brain ventricular system and include areas that lie both inside and outside of the blood–brain barrier. Simpson and colleagues (Simpson and Routtenberg, 1973; Simpson et al., 1978) implicated the SFO in ANG-induced drinking in the rat. In addition to the SFO, the OVLT may sense ANG II and contribute to drinking in dog and sheep (Thrasher et al., 1982a, b; McKinley et al., 1986). The MePO is ANG sensitive and located within the blood–brain barrier. It is not accessed by circulating ANG, but this structure can be readily reached by CSF-borne peptide (Johnson and Epstein, 1975).

At the time of increasing functional evidence for the central actions of ANG II (e.g., Epstein et al., 1970; Ferrario et al., 1972), two groups (Fischer-Ferraro et al., 1971; Ganten et al., 1971a, b) provided evidence that the components of the RAS were located in the brain. This gave rise to the concept of tissue RASs in which it was proposed that the agents found in this biochemical cascade were generated within specific structures, such as brain (Ganten et al., 1971a, b). Although this concept was initially received with skepticism from some investigators, the body of biochemical evidence generated over the past 35 years argues strongly for a brain RAS. Although the systemic RAS and the brain RAS may be biochemically independent, it has been proposed that the two systems are functionally coupled (Johnson, 1985a, b; Johnson and Wilkin, 1987). For example, in the case of thirst it has been hypothesized that in the state of hypovolemia circulating ANG II acts on the SFO in the mode of a hormone which in turn leads to the activation of angiotensinergic neurons that project from the SFO into areas like the MePO where synaptically released ANG II acts like a neurotransmitter (Johnson, 1985a, b; Johnson and Wilkin, 1987).

In comparison to the systemic RAS, the physiology and biochemistry of the brain RAS has proved to be a formidable scientific challenge. For the systemic RAS, the cellular and organ sources of renin, AGT, and ACE are distinct and spatially widely separated. In the brain, multiple cell types (glia vs. neurons) lie in close proximity with one another and may generate the same RAS constituent. For example, there has been a major discord regarding the cellular localization of AGT. Although there is reasonable agreement among laboratories as to the structures where AGT mRNA and protein are present, there is major disagreement as to whether it is produced only in astrocytes or also in neurons and whether different pools have different functions. These factors make the study of regulation and physiology of the brain RAS extremely complex. Recently, knockout and transgenic rodent models have provided a means to begin to dissect the complexities of the brain RAS. Such physiological genomic approaches to studying the brain RAS have often included measures of ad libitum and induced salt and water intake in their characterization of the role of this system in body fluid and cardiovascular regulation.

Early transgenic RAS models developed in mouse (Davisson et al., 1998) and rat (Mullins et al., 1990) introduced genes from other species (human AGT and human renin into mouse[2]; mouse Ren-2 into rat). Such rats and mice show increased activity of RASs at various tissue sites, including brain. The transgenic animals also have increased circulating ANG II and are hypertensive. Functional studies indicate that a central action of ANG II makes a major contribution to the elevated blood pressure of the double transgenic mice (Davisson et al., 1998). Transgenic rats with an additional Ren-2 mouse gene show increased ad libitum food and water intake, but no change in the food to water ratio (Szczepanska-Sadowska et al., 2003).

There is a unique species specificity of renin-AGT interaction. Human renin does not cleave mouse AGT and vice versa. To enhance the formation of ANG, it is necessary to cross a mouse carrying the human renin gene with a mouse carrying the human AGT gene (i.e., produce a double transgenic model).

Recently, more informative transgenic models have been developed using techniques to affect specific brain cell types for overexpression of components of the RAS. This has been achieved by generating transgenic mice with either glia- or neuronal-targeted human renin or human AGT (Morimoto et al., 2001, 2002). The synapsin-1 promoter was used to drive human renin (Morimoto et al., 2002) and human AGT in neurons and the glial fibrillary acidic protein (Morimoto et al., 2001) drives the expression of human renin and AGT in astrocytes. In these models, expression of the human transgenes is restricted to the predicted brain cell types with minimal systemic expression. Cross-breeding various combinations of these models so that ANG was likely to be generated resulted in at best only modest increases in arterial blood pressure (Morimoto et al., 2001, 2002). However, several of the crosses resulted in increased saline intake (Morimoto et al., 2001, 2002).

Genetically manipulated mouse models are also being used to address the role of the ANG receptor subtypes in thirst and fluid homeostasis. Rodents have three ANG receptor subtypes, specifically AT_{1a}, AT_{1b}, and AT_2. Knockout models implicate all three in thirst. Using AT_{1a} null mice and AT_{1b} null mice, Davisson and colleagues (2000) found the pressor response to centrally injected ANG II to be dependent on AT_{1a} receptors, but the drinking response to be mediated by AT_{1b} receptors. Li and group (2003) on the other hand found evidence for AT_2 and AT_{1b} receptors in mediating ANG-induced drinking in their mice knockout models. The AT_2 null mouse has an impaired drinking response to water deprivation (Hein et al., 1995). It is not clear why different groups reach different conclusions about the roles of the ANG receptor subtypes in thirst. Li and group (2003) have suggested that different methods for assessing drinking may contribute to the discrepancies. Another variable not considered by investigators is the effect of manipulating receptors that simultaneously affect drinking and blood pressure since blood pressure interacts with the drinking response (see *Section 2.4.3*). Careful behavioral analysis and the control of blood pressure as a variable will be needed to clarify many of these issues.

Recently developed molecular techniques for the analysis of ANG actions in the brain have also been applied to the analysis of intracellular second messenger signaling. In studies from Davisson's laboratory (Zimmerman et al., 2002), a novel ANG II signaling mechanism in brain has been demonstrated to involve reactive oxygen species. In these studies, mice were injected icv with an adenoviral-vector encoding human mitochondrial superoxide dismutase (AdMnSOD) or control vectors. Evidence of human mitochondrial superoxide dismutase was observed in periventricular tissue of the AdMnSOD animals a few days after treatment. When challenged with central injections of ANG II, the AdMnSOD-treated mice showed attenuated pressor and heart rate responses. In addition, central ANG II-induced drinking was significantly attenuated in the AdMnSOD mice.

4.1.2 Acetylcholine

Probably the first reported neurotransmitter to be injected into the brain and induce copious water intake was acetylcholine. The classic studies by Grossman (1960, 1962) demonstrated that acetylcholine or the cholinergic receptor agonist, carbachol, applied in the lateral hypothalamus produced water intake in rats. The phenomena were specific in that the cholinergic stimulation did not produce food intake, but norepinephrine (NE) delivered at the same site did induce feeding. Further characterization of this form of chemically induced drinking indicated that the intake is a result of action on muscarinic receptors (Stein and Seifter, 1962). Studies by Fisher and Coury (1962) demonstrated that crystalline carbachol applied to several sites throughout the forebrain generated drinking. These researchers (Fisher and Coury, 1962) proposed a cholinergic thirst circuit distributed throughout the limbic system. There is some evidence to suggest that cholinergic thirst in the rat mediates cellular dehydration-induced drinking as opposed to extracellular thirst (Fisher, 1973). Interestingly, central administration of carbachol strongly inhibits experimentally induced salt appetite (Fitts et al., 1985b).

In spite of the remarkable water intake induced in rats by cholinergic stimulation, and its specificity, little additional insight into the physiological significance of cholinergic thirst has been gained over the past 40 plus years. Unlike thirst in response to centrally applied ANG II which has been demonstrated in a wide number of vertebrate species (see Schwob and Johnson, 1977 for a summary of species), a reliable vigorous drinking response induced by muscarinic agonists is limited mainly to the rat (Note: there is a small variable response in dog; Ramsay and Reid, 1975). Specification of the central site(s) of action of the cholinergic agonists that induce drinking remains problematic. The speculation that crystalline carbachol when applied to limbic tissue sites may leak into the brain ventricles and act at the SFO (Routtenberg and Simpson, 1971) has not been fully addressed. While being a sensitive site for cholinergic actions, the SFO does not appear to be necessary for the cholinergic drinking response. Ablation of the structure (Buggy and Fisher, 1976) or cutting its major efferent pathway (Masson and Fitts, 1989) has no effect on cholinergic-induced drinking.

4.1.3 Norepinephrine (NE)

Among the widely distributed neurotransmitter systems in the brain are the biogenic amines, NE, 5-HT, and dopamine. These systems consist of a relatively few cell bodies located in discrete clusters throughout the brain stem and have axons, which diverge to innervate nearly all portions of the brain. Numerous early studies with neurotoxins indicate that wholesale depletion of brain dopamine and NE renders rats adipsic and aphagic. Many of these effects on ingestive behaviors are nonspecific and are a result of the destruction of ascending nigrostriatal dopamine projections.

Using discrete methods to administer the neurotoxin 6-hydroxydopamine (McRae-Degueurce et al., 1986; Bellin et al., 1987a, b, 1988; Cunningham and Johnson, 1989), it has been possible to demonstrate that integrity of the NE innervation of the AV3V is necessary for normal dipsogenic responses to ANG II. Such a NE terminal region is not necessary for drinking induced by carbachol or by hypertonic saline. Replacing locally depleted NE with exogenous icv infusions or transplantation of neonatal NE synthesizing cell bodies into the AV3V (McRae-Degueurce et al., 1986; Cunningham and Johnson, 1989) restores drinking in impaired rats.

The AV3V region receives ascending NE innervation from the A1/C1 region of the ventrolateral medulla and the A2/C2 region of the dorsomedial medulla (Saper et al., 1983). NE pathways are activated by various stressors including hypovolemic/hypotensive challenges. Turnover of NE in the region of the MePO is increased by hypovolemic treatment that would normally induce drinking (Wilkin et al., 1987). More recently, Miyakubo and colleagues (2003) used microdialysis to show that hypovolemic rats have increased extracellular NE in the MePO before they drink. The elevated NE then decreases as drinking proceeds. Taken together, such data have been used to hypothesize that circulating ANG II acts on the SFO to activate a descending angiotensinergic pathway to the MePO, where the peptide interacts with extracellular NE released from ascending pathways carrying input from systemic baroreceptors (see Johnson and Wilkin, 1987; Johnson and Thunhorst, 1997 for review and references).

4.2 Inhibitory Mechanisms

4.2.1 Vasopressin (VP) and Oxytocin (OT)

The two long recognized and extensively studied hormones of the posterior pituitary are VP and OT. Each of these peptides is synthesized in their own unique type of magnocellular neuron in the PVN and SON and is transported to the posterior pituitary for release into the systemic circulation. Classically, VP was recognized for its antidiuretic action and as a pressor agent. OT was also noted for its effects on contractile tissues and implicated in uterine contraction and milk ejection. In addition to the hypothalamo-neurohypophyseal systems for VP and OT, there are multiple groups of VP and OT cell bodies and dispersed projections throughout the brain (Buijs, 1978).

In light of the classic functional roles of VP and OT, it seems highly plausible that VP in particular should be implicated in behaviors related to fluid balance. There is evidence in the dog that centrally administered VP induces drinking (Szczepanska-Sadowska et al., 1982). Although numerous investigators have injected VP centrally in rats in anticipation of effects on drinking, the most common behavior observed is rotation of the body around its longitudinal axis (barrel rotations). Such a dramatic motor response greatly complicates the study of ingestive behaviors. Recent work using another approach employing central administration of OT and VP receptor antagonists suggests that VP in intrinsic brain pathways may play a facilitory role in saline intake in response to sodium depletion (Flynn et al., 2002). In light of the VP-induced motor complications, new strategies will be required to fully explore the role of VP in thirst and salt appetite.

In comparison to VP there is a much larger body of evidence implicating OT in the behavioral control of fluid balance, particularly in the central inhibition of salt appetite. Initial observations that rats maintained on a sodium deficient diet have blunted OT secretion in response to hypovolemia suggested a role of this peptide in fluid homeostasis (Stricker et al., 1987, 1994). Because systemic administration of OT has no apparent inhibitory effect on salt or water intake, (Stricker and Verbalis, 1987) attention has been directed toward central OT actions and pathways. In this light, circulating OT might serve as an index of the activity in central OT pathways. Several findings support the idea that central OT mechanisms act to limit salt intake. For example, treatments (e.g., hyperosmolality, naloxone, nausea) that stimulate OT secretion are associated with reduced salt appetite produced by hypovolemia. In addition, treatments suppressing OT release (e.g., sodium deprivation, alcohol intake) increase sodium intake (Stricker and Verbalis, 1987; Stricker et al., 1987; Blackburn et al., 1992a, b, 1994). Salt appetite induced by centrally administered ANG II is enhanced by central application of an OT receptor antagonist or after destruction of central OT neurons with a selective neurotoxin (Blackburn et al., 1992a, b, 1995). It is especially noteworthy that the concept of having an OT brake on salt intake is appealing from the standpoint of biological consistency in that OT also induces natriuresis at physiological levels in the rat (Verbalis et al., 1991).

4.2.2 Tachykinins

The neuropeptides of the tachykinin family are coded by two different genes. The preprotachykinin A gene encodes substance P, neurokinin A, neuropeptide γ, and neurokinin K; the preprotachykinin B gene codes for neurokinin B (Krause et al., 1987; MacDonald et al., 1988). Tachykinin NK1, NK2, and NK3 receptor subtypes preferentially bind substance P, NKA, and NKB, respectively (Maggi et al., 1987). Icv injections of NK3 receptor antagonists suppress both ad libitum and induced salt appetite (Massi et al., 1988a, b; Smith and Flynn, 1994). When animals are in a state of sodium deficiency, gene expression for NKB is reduced relative to the sodium-replete condition (Pompei et al., 1997). Taken together, there is a substantial body of data emerging to indicate that central NKB exerts an inhibitory effect on salt intake.

5 Summary and Conclusions

Thirst and salt appetite are appetitive states that have been viewed as generalized sensations, instinctive behaviors, drives, and negative affective conditions. They are states that give rise to behaviors that are necessary to maintain and correct homeostatic disruptions in fluid balance and distribution. Of these two appetitive conditions, thirst is the most common and easily recognized by humans. Nevertheless, salt appetite will develop in individuals as a sodium-deficient state becomes more protracted and can be defined by an increase in the palatability of concentrated salt solutions.

Significant progress has been made in the past half century by pursuing what has been in effect a sensory analysis of thirst and salt appetite. As contrasted with senses like vision or audition, there are no specific sensory organs that are devoted exclusively to sensing thirst or salt appetite. These states must be generated from available humoral and mechanical stimuli that are associated with water and sodium perturbations and cardiovascular changes (➋ *Figure 17-7*).

Water loss from either the cellular or the extracellular compartment gives rise to thirst (i.e., the *double depletion hypothesis)* whereas, loss of isotonic fluid from the extracellular space (i.e., hypovolemia) produces a hunger for salt. Humoral and physical stimuli associated with cellular and extracellular depletions serve to

Figure 17-7

Schematic depicting the nature of neural and hormonal inputs into the brain and the central neural pathways that mediate sensory integration of signals for the generation of drinking (thirst) and sodium ingestion (salt appetite). Both inhibitory and excitatory inputs from the periphery derive from arterial and cardiopulmonary baroreceptors and probably other visceral receptors (e.g., gastric, hepatic, portal, or renal). Information carried in afferent nerves projects mainly to the nucleus of the tractus solitarius (NTS). Angiotensin (ANG) acts in the form of ANG II on angiotensin type 1 (AT_1) receptors in the subfornical organ (SFO). Information reflecting input to the SFO is carried in *descending* pathways, some of which are likely to use ANG in the mode of a neurotransmitter, to forebrain structures such as those in the anteroventral third ventricle (AV3V). *Ascending* information to the forebrain is carried in from noradrenergic cell groups (A) in the hindbrain, which are activated by arterial and cardiopulmonary receptor input under conditions of hypotension and/or hypovolemia. ANG and noradrenergic (NE) inputs act synergistically in the forebrain nuclei. A hindbrain inhibitory pathway originating in the area postrema (AP) and medial NTS ascends to the lateral parabrachial nucleus (LPBN). This projection uses serotonin (5-HT), as one of several neurotransmitters, to guard against excessive sodium and water intake thereby limiting excessive expansion of extracellular fluid volume. Inhibitory input is likely to ascend the neuraxis either directly or indirectly to interact with the forebrain structures within the visceral neuraxis devoted to regulating body fluid and cardiovascular homeostasis. From Johnson and Thunhorst (1997)

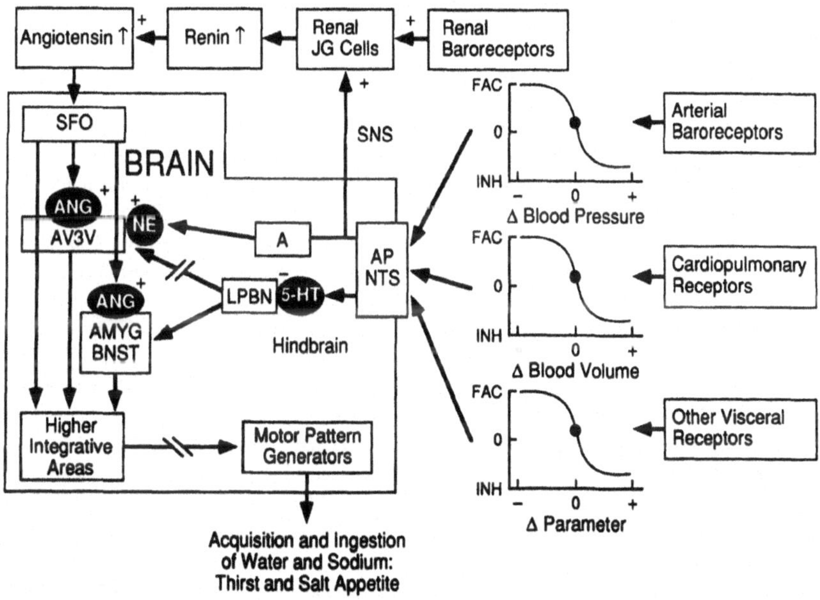

inform the brain about the status of fluid balance and distribution. Cellular dehydration is thought to be sensed by hypothesized *osmoreceptors* (and probably sodium receptors) located in the systemic viscera and in the brain. The most extensively studied osmoreceptors are in the brain. The SCVOs located along the lamina terminalis (i.e., the SFO and OVLT) are major candidate regions for containing osmosensitive cells that trigger cellular dehydration thirst. Hypovolemia, which generates extracellular thirst, is signaled by ANG II and afferent input from systemic baroreceptors located on both the high- and low-pressure sides of the circulation and possibly in the kidney. The same baroreceptors and ANG II also seem to contribute to salt appetite. Given the longer latency for the onset of the sodium seeking, salt appetite may have a higher threshold for its activation or require the addition or removal of other factors (e.g., ↑ aldosterone or ↓ OT= facilitation) for its expression.

Humoral input reaching the structures of the laminal terminalis and ascending visceral input entering the brain stem through the NTS project information about the status of body fluids into a neural network often referred to as the *visceral neuraxis.* Key structures in this circuitry include the NTS, parabrachial nucleus, PVN, SON, SFO, MePO, septal area, and components of the extended amygdala. In this system, various neuroactive agents have both excitatory (e.g., ANG II, acetylcholine, NE, glutamate) and inhibitory (e.g., tachykinins, OT, 5-HT, ANP, GABA) influences on thirst and salt appetite. To produce thirst and salt appetite, information is integrated and processed in the brain. In other words, the states of thirst and hunger for salt, as with other sensations and perceptions, are the result of sensing by receptors and then synthesis of the state or perceived condition by the brain. Finally, to understand the complex sensory biology and information processing involved in thirst and salt appetite, the field of research is constantly striving to apply the evolving concepts and methods of contemporary neuroscience.

Acknowledgments

Support for research from the authors' laboratory reported in this manuscript and for preparation of this review was provided by grants from the National Heart, Lung, and Blood Institute HL14388 and HL57472, National Institute of Diabetes and Digestive and Kidney Diseases DK66086, National Institute on Aging AG25465, and National Institute of Mental Health MH59239.

References

Abraham SF, Baker RM, Blaine EH, Denton DA, McKinley MJ. 1975. Water drinking induced in sheep by angiotensin–a physiological or pharmacological effect? J Comp Physiol Psychol 88: 503-518.

Adachi A, Niijima A, Jacobs HL. 1976. An hepatic osmoreceptor mechanism in the rat: electrophysiological and behavioral studies. Am J Physiol 231: 1043-1049.

Adolph EF, Barker JP, Hoy PA. 1954. Multiple factors in thirst. Am J Physiol 178: 538-562.

Alheid GF, de Olmos JS, Beltramino CA. 1985. Amygdala and extended amygdala. The Rat Nervous System, 2nd edn. editor. Paxinos G, New York: Academic Press; pp. 495-578.

Andersen JL, Andersen LJ, Thrasher TN, Keil LC, Ramsay DJ. 1994. Left heart and arterial baroreceptors interact in control of plasma vasopressin, renin, and cortisol in awake dogs. Am J Physiol Regul Integr Comp Physiol 266: R879-R888.

Andersen LJ, Andersen JL, Thrasher TN, Keil LC, Ramsay DJ. 1995. Effect of loading right atrial and ventricular receptors on stimulated AVP, ACTH, and renin secretion in awake dogs. Am J Physiol Regul Integr Comp Physiol 268: R1069-R1077.

Andersson B. 1952. Polydipsia caused by intrahypothalamic injections of hypertonic NaCl solutions. Experientia 8: 157-158.

Andersson B. 1971. Thirst--and brain control of water balance. Am Sci 59: 408-415.

Andersson B, McCann SM. 1955. A further study of polydipsia evoked by hypothalamic stimulation in the goat. Acta Physiol Scand 33: 333-346.

Andersson B, Olsson K, Warner RG. 1967. Dissimilarities between the central control of thirst and the release of antidiuretic hormone (ADH). Acta Physiol Scand 71: 57-64.

Andrade CA, Barbosa SP, De Luca LA Jr, Menani JV. 2004. Activation of alpha2-adrenergic receptors into the lateral parabrachial nucleus enhances NaCl intake in rats. Neuroscience 129: 25-34.

Andrews WH, Orbach J. 1974. Sodium receptors activating some nerves of perfused rabbit livers. Am J Physiol 227: 1273-1275.

Anke J, Van Eekelen M, Phillips MI. 1988. Plasma angiotensin II levels at moment of drinking during angiotensin II intravenous infusion. Am J Physiol Regul Integr Comp Physiol 255: R500-R506.

Antunes-Rodrigues J, McCann SM, Rogers LC, Samson WK. 1985. Atrial natriuretic factor inhibits dehydration- and angiotensin II-induced water intake in the conscious, unrestrained rat. Proc Natl Acad Sci USA 82: 8720-8723.

Atkinson J, Kaesermann HP, Lambelet J, Peters G, Peters-Haefeli L. 1979. The role of circulating renin in drinking in response to isoprenaline. J Physiol (Lond) 291: 61-73.

Avrith DB, Fitzsimons JT. 1980. Increased sodium appetite in the rat induced by intracranial administration of components of the renin-angiotensin system. J Physiol (Lond) 301: 349-364.

Bartanusz V, Jezova D. 1994. Angiotensin II induces reduced oxytocin but normal corticotropin release in rats with lesions of the subfornical organ. Fundam Clin Pharmacol 8: 539-545.

Baylis C, Handa RK, Sorkin M. 1990. Glucocorticoids and control of glomerular filtration rate. Semin Nephrol 10: 320-329.

Bellin SI, Bhatnagar RK, Johnson AK. 1987a. Periventricular noradrenergic systems are critical for angiotensin-induced drinking and blood pressure responses. Brain Res 403: 105-112.

Bellin SI, Landas SK, Johnson AK. 1987b. Localized injections of 6-hydroxydopamine into lamina terminalis-associated structures: effects on experimentally induced drinking and pressor responses. Brain Res 416: 75-83.

Bellin SI, Landas SK, Johnson AK. 1988. Selective catecholamine depletion of structures along the ventral lamina terminalis: effects on experimentally-induced drinking and pressor responses. Brain Res 456: 9-16.

Bernard C. 1856. Leçons de physiologie expérimentale appliquée à la médecine faites au Collège de France. Cours du Semestre d'été. Henri Lefèvre, editor. Paris: Baillière; pp. 49-52.

Black RM, Weingarten HP, Epstein AN, Maki R, Schulkin J. 1992. Transection of the stria terminalis without damage to the medial amygdala does not alter behavioural sodium regulation in rats. Acta Neurobiol Exp (Warsz) 52: 9-15.

Blackburn RE, Stricker EM, Verbalis JG. 1992a. Central oxytocin mediates inhibition of sodium appetite by naloxone in hypovolemic rats. Neuroendocrinology 56: 255-263.

Blackburn RE, Stricker EM, Verbalis JG. 1994. Acute effects of ethanol on ingestive behavior in rats. Alcohol Clin Exp Res 18: 924-930.

Blackburn RE, Demko AD, Hoffman GE, Stricker EM, Verbalis JG. 1992b. Central oxytocin inhibition of angiotensin-induced salt appetite in rats. Am J Physiol Regul Integr Comp Physiol 263: R1347-R1353.

Blackburn RE, Samson WK, Fulton RJ, Stricker EM, Verbalis JG. 1995. Central oxytocin and ANP receptors mediate osmotic inhibition of salt appetite in rats. Am J Physiol Regul Integr Comp Physiol 269: R245-R251.

Blaine EH, Covelli MD, Denton DA, Nelson JF, Shulkes AA. 1975. The role of ACTH and adrenal glucocorticoids in the salt appetite of wild rabbits (*Oryctolagus cuniculus* (L)). Endocrinology 97: 793-801.

Blair-West JR, Denton DA, McBurnie MI, Weisinger RS. 1996. The effect of adrenocorticotrophic hormone on water intake in mice. Physiol Behav 60: 1053-1056.

Blair-West JR, Denton DA, McKinley MJ, Weisinger RS. 1988. Angiotensin-related sodium appetite and thirst in cattle. Am J Physiol Regul Integr Comp Physiol 255: R205-R211.

Blair-West JR, Denton DA, McBurnie M, Tarjan E, Weisinger RS. 1995. Influence of adrenal steroid hormones on sodium appetite of Balb/c mice. Appetite 24: 11-24.

Blake WD, Lin KK. 1978. Hepatic portal vein infusion of glucose and sodium solutions on the control of saline drinking in the rat. J Physiol (Lond) 274: 129-139.

Blass EM, Epstein AN. 1971. A lateral preoptic osmosensitive zone for thirst in the rat. J Comp Physiol Psychol 76: 378-394.

Blass EM, Fitzsimons JT. 1970. Additivity of effect and interaction of a cellular and an extracellular stimulus of drinking. J Comp Physiol Psychol 70: 200-205.

Bott E, Denton DA, Weller S. 1967. The effect of angiotensin II infusion, renal hypertension and nephrectomy on salt appetite of sodium-deficient sheep. Aust J Exp Biol Med Sci 45: 595-612.

Bourque CW, Richard D. 2001. Axonal projections from the organum vasculosum lamina terminalis to the supraoptic nucleus: functional analysis and presynaptic modulation. Clin Exp Pharmacol Physiol 28: 570-574.

Bourque CW, Oliet SH, Richard D. 1994. Osmoreceptors, osmoreception, and osmoregulation. Front Neuroendocrinol 15: 231-274.

Bourque CW, Voisin DL, Chakfe Y. 2002. Stretch-inactivated cation channels: cellular targets for modulation of osmosensitivity in supraoptic neurons. Prog Brain Res 139: 85-94.

Brody MJ, Johnson AK. 1980. Role of the anteroventral third ventricle region in fluid and electrolyte balance, arterial pressure regulation, and hypertension. Frontiers in Neuroendocrinology, Vol. 6. Martini L, Ganong WF, editors. New York: Raven Press; pp. 249-292.

Bryant RW, Epstein AN, Fitzsimons JT, Fluharty SJ. 1980. Arousal of a specific and persistent sodium appetite in the rat with continuous intracerebroventricular infusion of angiotensin II. J Physiol (Lond) 301: 365-382.

Buggy J, Fisher AE. 1974. Evidence for a dual central role for angiotensin in water and sodium intake. Nature 250: 733-735.

Buggy J, Fisher AE. 1976. Anteroventral third ventricle site of action for angiotensin induced thirst. Pharmacol Biochem Behav 4: 651-660.

Buggy J, Johnson AK. 1977. Preoptic-hypothalamic periventricular lesions: thirst deficits and hypernatremia. Am J Physiol Regul Integr Comp Physiol 233: R44-R52.

Buggy J, Johnson AK. 1978. Angiotensin-induced thirst: effects of third ventricle obstruction and periventricular ablation. Brain Res 149: 117-128.

Buggy J, Jonklaas J. 1984. Sodium appetite decreased by central angiotensin blockade. Physiol Behav 32: 737-742.

Buggy J, Hoffman WE, Phillips MI, Fisher AE, Johnson AK. 1979. Osmosensitivity of rat third ventricle and interactions with angiotensin. Am J Physiol Regul Integr Comp Physiol 236: R75-R82.

Buijs RM. 1978. Intra- and extrahypothalamic vasopressin and oxytocin pathways in the rat. Pathways to the limbic system, medulla oblongata and spinal cord. Cell Tissue Res 192: 423-435.

Burrell LM, Lambert HJ, Baylis PH. 1991. Effect of atrial natriuretic peptide on thirst and arginine vasopressin release in humans. Am J Physiol Regul Integr Comp Physiol 260: R475-R479.

Callera JC, Oliveira LB, Barbosa SP, Colombari DS, De Luca LA Jr. et al. 2005. GABA(A) receptor activation in the lateral parabrachial nucleus induces water and hypertonic NaCl intake. Neuroscience 134: 725-735.

Cannon WB. 1918. The physiological basis of thirst. Proc Roy Soc London B 90: 283-301.

Chiaraviglio E. 1976. Effect of renin-angiotensin system on sodium intake. J Physiol (Lond) 255: 57-66.

Chiaraviglio E, Perez Guaita MF. 1984. Anterior third ventricle (A3V) lesions and homeostasis regulation. J Physiol (Paris) 79: 446-452.

Ciriello J, Lawrence D, Pittman QJ. 1984. Electrophysiological identification of neurons in the parabrachial nucleus projecting directly to the hypothalamus in the rat. Brain Res 322: 388-392.

Cizek LJ, Semple RE, Huang KC, Gregersen MI. 1951. Effect of extracellular electrolyte depletion on water intake in dogs. Am J Physiol 164: 415-422.

Clemente CD, Sutin J, Silverstone JT. 1957. Changes in electrical activity of the medulla on the intravenous injection of hypertonic solutions. Am J Physiol 188: 193-198.

Colombari DS, Menani JV, Johnson AK. 1996. Forebrain angiotensin type 1 receptors and parabrachial serotonin in the control of NaCl and water intake. Am J Physiol Regul Integr Comp Physiol 271: R1470-R1476.

Contreras RJ, Stetson PW. 1981. Changes in salt intake lesions of the area postrema and the nucleus of the solitary tract in rats. Brain Res 211: 355-366.

Cooling MJ, Day MD. 1975. Angiotensin-induced drinking in the cat. Control of Mechanisms of Drinking. Peters G, Fitzsimons JT, Peters-Haefeli L, editors. New York: Springer-Verlag; pp. 132-137.

Corbit JD. 1968. Cellular dehydration and hypovolaemia are additive in producing thirst. Nature 218: 886-887.

Cunningham ET Jr, Miselis RR, Sawchenko PE. 1994. The relationship of efferent projections from the area postrema to vagal motor and brain stem catecholamine-containing cell groups: an axonal transport and immunohistochemical study in the rat. Neuroscience 58: 635-648.

Cunningham JT, Johnson AK. 1989. Decreased norepinephrine in the ventral lamina terminalis region is associated with angiotensin II drinking response deficits following local 6-hydroxydopamine injections. Brain Res 480: 65-71.

Cushny AR. 1926. The Secretion of the Urine. London: Longmans Green; pp. 135-137.

Daniels D, Fluharty SJ. 2004. Salt appetite: a neurohormonal viewpoint. Physiol Behav 81: 319-337.

Darrow DC, Yannet H. 1935. The changes in the distribution of body water accompanying increase and decrease in extracellular electrolyte. J Clin Invest 14: 266-275.

Darrow DC, Yannet H. 1936. Metabolic studies of the changes in body electrolyte and distribution of body water induced experimentally by deficit of extracellular electrolyte. J Clin Invest 15: 419-427.

Davisson RL, Oliverio MI, Coffman TM, Sigmund CD. 2000. Divergent functions of angiotensin II receptor isoforms in the brain. J Clin Invest 106: 103-106.

Davisson RL, Yang G, Beltz TG, Cassell MD, Johnson AK, et al. 1998. The brain renin-angiotensin system contributes to the hypertension in mice containing both the human renin and human angiotensinogen transgenes. Circ Res 83: 1047-1058.

de Bold AJ, Borenstein HB, Veress AT, Sonnenberg H. 1981. A rapid and potent natriuretic response to intravenous injection of atrial myocardial extract in rats. Life Sci 28: 89-94.

De Castro e Silva E, Fregoneze JB, Johnson AK. 2006. Corticotropin-releasing hormone in the lateral parabrachial nucleus inhibits sodium appetite in rats. Am J Physiol Regul Integr Comp Physiol 290: R1136-R1141.

De Gobbi JIF, De Luca LA Jr, Menani JV. 2000. Serotonergic mechanisms of the lateral parabrachial nucleus on DOCA-induced sodium intake. Brain Res 880: 131-138.

De Gobbi JIF, De Luca LA Jr, Johnson AK, Menani JV. 2001. Interaction of serotonin and cholecystokinin in the lateral parabrachial nucleus to control sodium intake. Am J Physiol Regul Integr Comp Physiol 280: R1301-R1307.

De Gobbi JIF, Barbosa SP, De Luca LA Jr, Thunhorst RL, Johnson AK, et al. 2005. Activation of serotonergic 5-HT (1A) receptors in the lateral parabrachial nucleus increases NaCl intake. Brain Res 1066: 1-9.

de Kloet ER, Vreugdenhil E, Oitzl MS, Joels M. 1998. Brain corticosteroid receptor balance in health and disease. Endocr Rev 19: 269-301.

De Luca LA Jr, Barbosa SP, Menani JV. 2003. Brain serotonin blockade and paradoxical salt intake in rats. Neuroscience 121: 1055-1061.

De Luca LA Jr, Galaverna O, Schulkin J, Yao SZ, Epstein AN. 1992. The anteroventral wall of the third ventricle and the angiotensinergic component of need-induced sodium intake in the rat. Brain Res Bull 28: 73-87.

De Nicola AF, Seltzer A, Tsutsumi K, Saavedra JM. 1993. Effects of deoxycorticosterone acetate (DOCA) and aldosterone on Sar1-angiotensin II binding and angiotensin-converting enzyme binding sites in brain. Cell Mol Neurobiol 13: 529-539.

de Olmos J, Alheid GF, Beltramino CA. 1985. Amygdala. The Rat Nervous System, Volume 1, Forebrain and Midbrain. Paxinos G, editor. Orlando: Academic Press; pp. 223-334.

Denton D. 1982. The Hunger for Salt. New York: Springer-Verlag.

Denton DA, Blair-West JR, McBurnie M, Osborne PG, Tarjan E, et al. 1990. Angiotensin and salt appetite of BALB/c mice. Am J Physiol Regul Integr Comp Physiol 259: R729-R735.

Denton DA, Blair-West JR, McBurnie MI, Miller JA, Weisinger RS, et al. 1999. Effect of adrenocorticotrophic hormone on sodium appetite in mice. Am J Physiol Regul Integr Comp Physiol 277: R1033-R1040.

DiBona GF. 1982. The functions of the renal nerves. Rev Physiol Biochem Pharmacol 94: 75-181.

Dierckx K. 1963. The subfornical organ, a specialized osmoreceptor. Naturwissenschaften 50: 163-164.

Dill DB. 1938. Life, Heat, and Altitude. Cambridge: Harvard University Press.

Edwards GL, Johnson AK. 1991. Enhanced drinking after excitotoxic lesions of the parabrachial nucleus in the rat. Am J Physiol Regul Integr Comp Physiol 261: R1039-R1044.

Edwards GL, Ritter RC. 1982. Area postrema lesions increase drinking to angiotensin and extracellular dehydration. Physiol Behav 29: 943-947.

Edwards GL, Beltz TG, Power JD, Johnson AK. 1993. Rapid-onset "need-free" sodium appetite after lesions of the dorsomedial medulla. Am J Physiol Regul Integr Comp Physiol 264: R1242-R1247.

Epstein AN. 1973. Epilogue: retrospect and prognosis. The Neuropsychology of Thirst: New Findings and Advances in Concepts. Epstein AN, Kissileff HR, Stellar E, editors. Washington DC: VH Winston; pp. 315-332.

Epstein AN. 1982. Mineralocorticoids and cerebral angiotensin may act together to produce sodium appetite. Peptides 3: 493-494.

Epstein AN. 1984. The dependence of the salt appetite of the rat on the hormonal consequences of sodium deficiency. J Physiol (Paris) 79: 496-498.

Epstein AN, Fitzsimons JT, Rolls BJ. 1970. Drinking induced by injection of angiotensin into the brain of the rat. J Physiol (Lond) 210: 457-474.

Evered MD. 1992. Investigating the role of angiotensin II in thirst: interactions between arterial pressure and the control of drinking. Can J Physiol Pharmacol 70: 791-797.

Evered MD, Robinson MM, Rose PA. 1988. Effect of arterial pressure on drinking and urinary responses to angiotensin II. Am J Physiol Regul Integr Comp Physiol 254: R69-R74.

Falk JL. 1961. The behavioral regulation of water-electrolyte balance. Nebraska Symposium on Motivation. Jones MR, editor. Lincoln: University of Nebraska Press; pp. 1-37.

Falk JL. 1965. Water intake and NaCl appetite in sodium depletion. Psychol Rev 16: 315-325.

Falk JL. 1966. Serial sodium depletion and NaCl solution. Physiol Behav 1: 75-77.

Ferguson AV, Kasting NW. 1987. Activation of subfornical organ efferents stimulates oxytocin secretion in the rat. Regul Pept 18: 93-100.

Ferguson AV, Kasting NW. 1988. Angiotensin acts at the subfornical organ to increase plasma oxytocin concentrations in the rat. Regul Pept 23: 343-352.

Ferguson AV, Renaud LP. 1986. Systemic angiotensin acts at subfornical organ to facilitate activity of neurohypophysial neurons. Am J Physiol Regul Integr Comp Physiol 251: R712-R717.

Ferrari P. 2003. Cortisol and the renal handling of electrolytes: role in glucocorticoid-induced hypertension and bone disease. Baillieres Best Pract Res Clin Endocrinol Metab 17: 575-589.

Ferrario CM, Gildenberg PL, McCubbin JW. 1972. Cardiovascular effects of angiotensin mediated by the central nervous system. Circ Res 30: 257-262.

Findlay AL, Epstein AN. 1980. Increased sodium intake is somehow induced in rats by intravenous angiotensin II. Horm Behav 14: 86-92.

Fischer-Ferraro C, Nahmod VE, Goldstein DJ, Finkielman S. 1971. Angiotensin and renin in rat and dog brain. J Exp Med 133: 353-361.

Fisher AE. 1973. Relationships between cholinergic and other dipsogens in the central mediation of thirst. The Neuropsychology of Thirst: New Findings and Advances in Concepts. Epstein AN, Kissileff HR, Stellar E, editors. Washington DC: VH Winston & Sons; pp. 243-278.

Fisher AE, Coury JN. 1962. Cholinergic tracing of a central neural circuit underlying the thirst drive. Science 138: 691-693.

Fitts DA. 1991. Effects of lesions of the ventral ventral median preoptic nucleus or subfornical organ on drinking and salt appetite after deoxycorticosterone acetate or yohimbine. Behav Neurosci 105: 721-726.

Fitts DA, Freece JA, Van Bebber JE, Zierath DK, Bassett JE. 2004. Effects of forebrain circumventricular organ ablation on drinking or salt appetite after sodium depletion or hypernatremia. Am J Physiol Regul Integr Comp Physiol 287: R1325-R1334.

Fitts DA, Masson DB. 1989. Forebrain sites of action for drinking and salt appetite to angiotensin or captopril. Behav Neurosci 103: 865-872.

Fitts DA, Masson DB. 1990. Preoptic angiotensin and salt appetite. Behav Neurosci 104: 643-650.

Fitts DA, Thunhorst RL. 1996. Rapid elicitation of salt appetite by an intravenous infusion of angiotensin II in rats. Am J Physiol Regul Integr Comp Physiol 270: R1092-R1098.

Fitts DA, Thunhorst RL, Simpson JB. 1985a. Diuresis and reduction of salt appetite by lateral ventricular infusions of atriopeptin II. Brain Res 348: 118-124.

Fitts DA, Thunhorst RL, Simpson JB. 1985b. Modulation of salt appetite by lateral ventricular infusions of angiotensin II and carbachol during sodium depletion. Brain Res 346: 273-280.

Fitts DA, Tjepkes DS, Bright RO. 1990. Salt appetite and lesions of the ventral part of the ventral median preoptic nucleus. Behav Neurosci 104: 818-827.

Fitzsimons JT. 1961. Drinking by rats depleted of body fluid without increase in osmotic pressure. J Physiol (Lond) 159: 297-309.

Fitzsimons JT. 1964. Drinking caused by constriction of the inferior vena cava in the rat. Nature 204: 479-480.

Fitzsimons JT. 1966. Hypovolaemic drinking and renin. J Physiol (Lond) 186: 130P-131P.

Fitzsimons JT. 1969a. Effect of nephrectomy on the additivity of certain stimuli of drinking in the rat. J Comp Physiol Psychol 68: 308-314.

Fitzsimons JT. 1969b. The role of a renal thirst factor in drinking induced by extracellular stimuli. J Physiol (Lond) 201: 349-368.

Fitzsimons JT. 1972. Thirst. Physiol Rev 52: 468-561.

Fitzsimons JT. 1973. Some historical perspectives in the physiology of thirst. The Neuropsychology of Thirst: New Findings and Advances in Concepts. Epstein AN, Kissileff HR, Stellar E, editors. Washington DC: VH Winston; pp. 3-33.

Fitzsimons JT. 1975. The renin-angiotensin system and drinking behavior. Prog Brain Res 42: 215-233.

Fitzsimons JT. 1979. The Physiology of Thirst and Sodium Appetite. Cambridge: Cambridge University Press.

Fitzsimons JT. 1980. Angiotensin and other peptides in the control of water and sodium intake. Proc R Soc Lond B Biol Sci 210: 165-182.

Fitzsimons JT. 1998. Angiotensin, thirst, and sodium appetite. Physiol Rev 78: 583-686.

Fitzsimons JT, Elfont RM. 1982. Angiotensin does contribute to drinking induced by caval ligation in rat. Am J Physiol Regul Integr Comp Physiol 243: R558-R562.

Fitzsimons JT, Moore-Gillon MJ. 1980. Drinking and antidiuresis in response to reductions in venous return in the dog: neural and endocrine mechanisms. J Physiol (Lond) 308: 403-416.

Fitzsimons JT, Oatley K. 1968. Additivity of stimuli for drinking in rats. J Comp Physiol Psychol 66: 450-455.

Fitzsimons JT, Simons BJ. 1969. The effect on drinking in the rat of intravenous infusion of angiotensin, given alone or in combination with other stimuli of thirst. J Physiol (Lond) 203: 45-57.

Fitzsimons JT, Stricker EM. 1971. Sodium appetite and the renin-angiotensin system. Nat New Biol 231: 58-60.

Fitzsimons JT, Wirth JB. 1978. The renin-angiotensin system and sodium appetite. J Physiol (Lond) 274: 63-80.

Fluharty SJ. 2002. Neuroendocrinology of body fluid homeostasis. Hormones, Brain, and Behavior. Pfaff DW, Arnold A, Etgen A, Fahrbach S, Rubin R, editors. Boston: Academic Press; pp. 525-570.

Fluharty SJ, Epstein AN. 1983. Sodium appetite elicited by intracerebroventricular infusion of angiotensin II in the rat: II. Synergistic interaction with systemic mineralocorticoids. Behav Neurosci 97: 746-758.

Fluharty SJ, Manaker S. 1983. Sodium appetite elicited by intracerebroventricular infusion of angiotensin II in the rat: I. Relation to urinary sodium excretion. Behav Neurosci 97: 738-745.

Fluharty SJ, Sakai RR. 1995. Behavioral and cellular analysis of adrenal steroid and angiotensin interactions mediating salt appetite. Progress in Psychobiology and Physiological Psychology. Fluharty SJ, Morrison AR, Sprague JM, Stellar E, editors. San Diego: Academic Press; pp. 177-212.

Flynn FW, Kirchner TR, Clinton ME. 2002. Brain vasopressin and sodium appetite. Am J Physiol Regul Integr Comp Physiol 282: R1236-R1244.

Franchini LF, Johnson AK, de Olmos J, Vivas L. 2002. Sodium appetite and Fos activation in serotonergic neurons. Am J Physiol Regul Integr Comp Physiol 282: R235-R243.

Fregly MJ, Rowland NE. 1985. Role of renin-angiotensin-aldosterone system in NaCl appetite of rats. Am J Physiol Regul Integr Comp Physiol 248: R1-R11.

Fregly MJ, Waters IW. 1966. Effect of mineralocorticoids on spontaneous sodium chloride appetite of adrenalectomized rats. Physiol Behav 1: 65-74.

Fregly MJ, Kelleher DL, Greenleaf JE. 1981. Antidipsogenic effect of clonidine on angiotensin II-, hypertonic saline-, pilocarpine- and dehydration-induced water intakes. Brain Res Bull 7: 661-664.

Fulwiler CE, Saper CB. 1984. Subnuclear organization of the efferent connections of the parabrachial nucleus in the rat. Brain Res 319: 229-259.

Galaverna O, De Luca LA Jr, Schulkin J, Yao SZ, Epstein AN. 1992. Deficits in NaCl ingestion after damage to the central nucleus of the amygdala in the rat. Brain Res Bull 28: 89-98.

Galaverna OG, Seeley RJ, Berridge KC, Grill HJ, Epstein AN, et al. 1993. Lesions of the central nucleus of the amygdala. I: Effects on taste reactivity, taste aversion learning and sodium appetite. Behav Brain Res 59: 11-17.

Ganesan R, Sumners C. 1989. Glucocorticoids potentiate the dipsogenic action of angiotensin II. Brain Res 499: 121-130.

Ganong WF. 2005. Review of Medical Physiology. New York: McGraw-Hill.

Ganten D, Marquez-Julio A, Granger P, Hayduk K, Karsunky KP, et al. 1971a. Renin in dog brain. Am J Physiol Regul Integr Comp Physiol 221: 1733-1737.

Ganten D, Minnich JL, Granger P, Hayduk K, Brecht HM, et al. 1971b. Angiotensin-forming enzyme in brain tissue. Science 173: 64-65.

Gardiner SM, Bennett T. 1986. Pressor contributions from angiotensin and vasopressin after polyethylene glycol. Am J Physiol Regul Integr Comp Physiol 251: R769-R774.

Geerling JC, Kawata M, Loewy AD. 2006b. Aldosterone-sensitive neurons in the rat central nervous system. J Comp Neurol 494: 515-527.

Geerling JC, Engeland WC, Kawata M, Loewy AD. 2006a. Aldosterone target neurons in the nucleus tractus solitarius drive sodium appetite. J Neurosci 26: 411-417.

Gilman A. 1934. Experimental sodium loss analogous to adrenal insufficiency: the resulting water shift and sensitivity to hemorrhage. Am J Physiol 108: 662-669.

Gilman A. 1937. The relation between blood osmotic pressure, fluid distribution and voluntary water intake. Am J Physiol 120: 323-328.

Gregersen MI. 1941. Total water balance: thirst, dehydration and water intoxication. MacLeod's Physiology in Modern Medicine. Bard P, editor. St. Louis: CV Mosby Co; pp. 1072-1083.

Grob M, Drolet G, Mouginot D. 2004. Specific Na^+ sensors are functionally expressed in a neuronal population of the median preoptic nucleus of the rat. J Neurosci 24: 3974-3984.

Grossman SP. 1960. Eating or drinking elicited by direct adrenergic or cholinergic stimulation of hypothalamus. Science 132: 301-302.

Grossman SP. 1962. Direct adrenergic and cholinergic stimulation of hypothalamic mechanisms. Am J Physiol 202: 872-882.

Grossman SP. 1967. Physiological mechanisms of thirst. A Textbook of Physiological Psychology. Grossman SP, editor. New York: John Wiley & Sons, Inc; pp. 395-445.

Haberich FJ. 1968. Osmoreception in the portal circulation. Fed Proc 27: 1137-1141.

Haller, Albrecht von. 1747. Primae lineae physiologiae in usum praelectionem academicarum Göttingen.

Harland D, Gardiner SM, Bennett T. 1988. Cardiovascular and dipsogenic effects of angiotensin II administered i.c.v. in Long-Evans and Brattleboro rats. Brain Res 455: 58-64.

Hein L, Barsh GS, Pratt RE, Dzau VJ, Kobilka BK. 1995. Behavioural and cardiovascular effects of disrupting the angiotensin II type-2 receptor in mice. Nature 377: 744-747.

Hiyama TY, Watanabe E, Okado H, Noda M. 2004. The subfornical organ is the primary locus of sodium-level sensing by Na(x) sodium channels for the control of salt-intake behavior. J Neurosci 24: 9276-9281.

Hiyama TY, Watanabe E, Ono K, Inenaga K, Tamkun MM, et al. 2002. Na(x) channel involved in CNS sodium-level sensing. Nat Neurosci 5: 511-512.

Holmes JH. 1960. Mechanisms producing the hyponatremic syndrome. Edema, Mechanisms and Management. Moyer JH, Fuchs M, editors. Philadelphia: WB Saunders Co; pp. 166-171.

Holmes JH, Cizek LJ. 1951. Observations on sodium chloride depletion in the dog. Am J Physiol 164: 407-414.

Holmes JH, Montgomery AV. 1951. Observations on relation of hemorrhage to thirst. Am J Physiol 167: 796.

Holmes JH, Montgomery AV. 1953. Thirst as a symptom. Am J Med Sci 225: 281-286.

Holstege G, Meiners L, Tan K. 1985. Projections of the bed nucleus of the stria terminalis to the mesencephalon, pons, and medulla oblongata in the cat. Exp Brain Res 58: 379-391.

Honda K, Negoro H, Dyball RE, Higuchi T, Takano S. 1990. The osmoreceptor complex in the rat: evidence for interactions between the supraoptic and other diencephalic nuclei. J Physiol (Lond) 431: 225-241.

Hosutt JA, Rowland N, Stricker EM. 1981. Impaired drinking responses of rats with lesions on the subfornical organ. J Comp Physiol Psychol 95: 104-113.

Hsiao S, Epstein AN. 1973. Additivity of dipsogens: angiotensin plus cell dehydration. Fed Proc 32: 384.

Hsiao S, Epstein AN, Camardo JS. 1977. The dipsogenic potency of peripheral angiotensin II. Horm Behav 8: 129-140.

Huang KC. 1955. Effect of salt depletion and fasting on water exchange in the rabbit. Am J Physiol 181: 609-615.

Hyde TM, Miselis RR. 1984. Area postrema and adjacent nucleus of the solitary tract in water and sodium balance. Am J Physiol Regul Integr Comp Physiol 247: R173-R182.

Jewell PA, Verney EB. 1957. An experimental attempt to determine the site of the neurohypophysial osmoreceptors in the dog. Philos Trans R Soc Lond B 240: 197-324.

Jhamandas JH, Harris KH, Petrov T, Krukoff TL. 1992. Characterization of the parabrachial input to the hypothalamic paraventricular nucleus in the rat. J Neuroendocrinol 4: 461-471.

Jhamandas JH, Petrov T, Harris KH, Vu T, Krukoff TL. 1996. Parabrachial nucleus projection to the amygdala in the rat: electrophysiological and anatomical observations. Brain Res Bull 39: 115-126.

Johnson AK. 1985a. Role of the periventricular tissue surrounding the anteroventral third ventricle (AV3V) in the regulation of body fluid homeostasis. Vasopressin. Schrier RW, editor. New York: Raven Press; pp. 319-331.

Johnson AK. 1985b. The periventricular anteroventral third ventricle (AV3V): its relationship with the subfornical organ and neural systems involved in maintaining body fluid homeostasis. Brain Res Bull 15: 595-601.

Johnson AK. 1990. Brain mechanisms in the control of body fluid homeostasis. Perspectives in Exercise Science and Sports Medicine: Fluid Homeostasis During Exercise. Gisolfi CV, Lamb DR, editors. Carmel: Benchmark Press; pp. 347-419.

Johnson AK, Buggy J. 1978. Periventricular preoptic-hypothalamus is vital for thirst and normal water economy. Am J Physiol Regul Integr Comp Physiol 234: R122-R129.

Johnson AK, Buggy J. 1977. A critical analysis of the site of action for the diposgenic effect of angiotensin II. Central Actions of Angiotensin and Related Hormones. Buckley JP, Ferrario CM, Lokhandwala MF, editors. New York: Pergamon Press; pp. 357-386.

Johnson AK, Cunningham JT, Thunhorst RL. 1996. Integrative role of the lamina terminalis in the regulation of cardiovascular and body fluid homeostasis. Clin Exp Pharmacol Physiol 23: 183-191.

Johnson AK, de Olmos J, Pastuskovas CV, Zardetto-Smith AM, Vivas L. 1999. The extended amygdala and salt appetite. Ann NY Acad Sci 877: 258-280.

Johnson AK, Edwards GL. 1990. The neuroendocrinology of thirst: afferent signaling and mechanisms of central integration. Current Topics in Neuroendocrinology, Vol. 10. Pfaff DW, Ganten D, editors. Heidelberg: Springer-Verlag; pp. 149-190.

Johnson AK, Edwards GL. 1991. Central projections of osmotic and hypovolaemic signals in homeostatic thirst. Thirst. Ramsay DJ, Booth DA, editors. London: Springer-Verlag; pp. 149-175.

Johnson AK, Epstein AN. 1975. The cerebral ventricles as the avenue for the dipsogenic action of intracranial angiotensin. Brain Res 86: 399-418.

Johnson AK, Gross PM. 1993. Sensory circumventricular organs and brain homeostatic pathways. FASEB J 7: 678-686.

Johnson AK, Hoffman WE, Buggy J. 1978. Attenuated pressor responses to intracranially injected stimuli and altered antidiuretic activity following preoptic-hypothalamic periventricular ablation. Brain Res 157: 161-166.

Johnson AK, Mann JF, Rascher W, Johnson JK, Ganten D. 1981. Plasma angiotensin II concentrations and experimentally induced thirst. Am J Physiol Regul Integr Comp Physiol 240: R229-R234.

Johnson AK, Phillips MI, Mohring J, Ganten D. 1977. Angiotensin-induced drinking in rats wth hereditary hypothalamic diabetes insipidus. Neurosci Lett 4: 327-330.

Johnson AK, Thunhorst RL. 1995. Sensory mechanisms in the behavioral control of body fluid balance: thirst and salt appetite. Prog Psychobiol Physiol Psychol 16: 145-176.

Johnson AK, Thunhorst RL. 1997. The neuroendocrinology of thirst and salt appetite: visceral sensory signals and mechanisms of central integration. Front Neuroendocrinol 18: 292-353.

Johnson AK, Wilkin LD. 1987. The lamina terminalis. Circumventricular Organs and Body Fluids. Gross P, editor. Boca Raton: CRC Press; pp. 125-141.

Johnston JB. 1923. Further contributions to the study of the evolution of the forebrain. J Comp Neurol 35: 337-481.

Joy MD, Lowe RD. 1970. Evidence that the area postrema mediates the central cardiovascular response to angiotensin II. Nature 228: 1303-1304.

Kadekaro M, Creel M, Terrell ML, Lekan HA, Gary HE Jr, et al. 1989. Effects of sinoaortic denervation on glucose utilization in the subfornical organ and pituitary neural lobe during administration of angiotensin II. Peptides 10: 103-108.

Kaufman S. 1984. Role of right atrial receptors in the control of drinking in the rat. J Physiol (Lond) 349: 389-396.

Kerpel-Fronius E. 1935. über die Beziehungen zwischen Salz- and Wasserhaushalt bei experimentellen Wasserverlusten. Z Kinderheilkd 57: 489-504.

King SJ, Harding JW, Moe KE. 1988. Elevated salt appetite and brain binding of angiotensin II in mineralocorticoid-treated rats. Brain Res 448: 140-149.

Klingbeil CK, Brooks VL, Quillen EW Jr, Reid IA. 1991. Effect of baroreceptor denervation on stimulation of drinking by angiotensin II in conscious dogs. Am J Physiol 260: E333-E337.

Kosten T, Contreras RJ, Stetson PW, Ernest MJ. 1983. Enhanced saline intake and decreased heart rate after area postrema ablations in rat. Physiol Behav 31: 777-785.

Kraly FS, Kim YM, Dunham LM, Tribuzio RA. 1995. Drinking after intragastric NaCl without increase in systemic plasma osmolality in rats. Am J Physiol Regul Integr Comp Physiol 269: R1085-R1092.

Krause JE, Chirgwin JM, Carter MS, Xu ZS, Hershey AD. 1987. Three rat preprotachykinin mRNAs encode the neuropeptides substance P and neurokinin A. Proc Natl Acad Sci USA 84: 881-885.

Krukoff TL, Harris KH, Jhamandas JH. 1993. Efferent projections from the parabrachial nucleus demonstrated with the anterograde tracer Phaseolus vulgaris leucoagglutinin. Brain Res Bull 30: 163-172.

Kucharczyk J. 1988. Inhibition of angiotensin-induced water intake following hexamethonium pretreatment in the dog. Eur J Pharmacol 148: 213-219.

Lança AJ, van der Kooy D. 1985. A serotonin-containing pathway from the area postrema to the parabrachial nucleus in the rat. Neuroscience 14: 1117-1126.

Levin ER, Gardner DG, Samson WK. 1998. Natriuretic peptides. N Engl J Med 339: 321-328.

Li Z, Iwai M, Wu L, Shiuchi T, Jinno T, et al. 2003. Role of AT2 receptor in the brain in regulation of blood pressure and water intake. Am J Physiol Heart Circ Physiol 284: H116-H121.

Liedtke W, Friedman JM. 2003. Abnormal osmotic regulation in trpv4-/- mice. Proc Natl Acad Sci USA 100: 13698-13703.

Liedtke W, Choe Y, Marti-Renom MA, Bell AM, Denis CS, et al. 2000. Vanilloid receptor-related osmotically activated channel (VR-OAC), a candidate vertebrate osmoreceptor. Cell 103: 525-535.

Lind RW. 1988. Angiotensin and the lamina terminalis: illustrations of a complex unity. Clin Exp Hypertens A 10 Suppl 1: 79-105.

Lind RW, Johnson AK. 1983. A further characterization of the effects of AV3V lesions on ingestive behavior. Am J Physiol Regul Integr Comp Physiol 245: R83-R90.

Lind RW, Thunhorst RL, Johnson AK. 1984. The subfornical organ and the integration of multiple factors in thirst. Physiol Behav 32: 69-74.

Loewy AD, Burton H. 1978. Nuclei of the solitary tract: efferent projections to the lower brain stem and spinal cord of the cat. J Comp Neurol 181: 421-449.

Ma LY, McEwen BS, Sakai RR, Schulkin J. 1993. Glucocorticoids facilitate mineralocorticoid-induced sodium intake in the rat. Horm Behav 27: 240-250.

MacDonald MR, McCourt DW, Krause JE, MacDonald MR, McCourt DW, et al. 1988. Posttranslational processing of alpha-, beta-, and gamma-preprotachykinins. Cell-free translation and early posttranslational processing events. J Biol Chem 263: 15176-15183.

MacGregor GA, de Wardener HE. 1998. Salt, Diet and Health. Cambridge: Cambridge University Press.

Maggi CA, Giuliani S, Santicioli P, Regoli D, Meli A. 1987. Peripheral effects of neurokinins: functional evidence for the existence of multiple receptors. J Auton Pharmacol 7: 11-32.

Malmo RB, Mundl WJ. 1975. Osmosensitive neurons in the rat's preoptic area: medial-lateral comparison. J Comp Physiol Psychol 88: 161-175.

Mann JF, Johnson AK, Ganten D. 1980. Plasma angiotensin II: dipsogenic levels and angiotensin-generating capacity of renin. Am J Physiol Regul Integr Comp Physiol 238: R372-R377.

Marriott HL. 1950. Water and Salt Depletion. Springfield: Charles C. Thomas.

Masson DB, Fitts DA. 1989. Subfornical organ connectivity and drinking to captopril or carbachol in rats. Behav Neurosci 103: 873-880.

Mason WT. 1980. Supraoptic neurones of rat hypothalamus are osmosensitive. Nature 287: 154-157.

Massi M, Epstein AN. 1990. Angiotensin/aldosterone synergy governs the salt appetite of the pigeon. Appetite 14: 181-192.

Massi M, Gentili L, Perfumi M, de Caro G, Schulkin J. 1990. Inhibition of salt appetite in the rat following injection of tachykinins into the medial amygdala. Brain Res 513: 1-7.

Massi M, Perfumi M, de Caro G, Epstein AN. 1988a. Inhibitory effect of kassinin on salt intake induced by different natriorexigenic treatments in the rat. Brain Res 440: 232-242.

Massi M, Polidori C, Gentili L, Perfumi M, de Caro G, et al. 1988b. The tachykinin NH2-senktide, a selective neurokinin B receptor agonist, is a very potent inhibitor of salt appetite in the rat. Neurosci Lett 92: 341-346.

Mayer A. 1900. Variations de la tension osmotique du sang chez les animaux privés de liquides. CR Seances Soc Biol Fil 52: 153-155.

McCance RA. 1936a. Experimental sodium chloride deficiency in man. Proc R Soc Lond B Biol Sci 119: 245-268.

McCance RA. 1936b. Medical problems in mineral metabolism. Lancet 227: 823-830.

McKinley MJ, Blaine EH, Denton DA. 1974. Brain osmoreceptors, cerebrospinal fluid electrolyte composition and thirst. Brain Res 70: 532-537.

McKinley MJ, Denton DA, Weisinger RS. 1978. Sensors for antidiuresis and thirst--osmoreceptors or CSF sodium detectors? Brain Res 141: 89-103.

McKinley MJ, Denton DA, Park RG, Weisinger RS. 1986. Ablation of subfornical organ does not prevent angiotensin-induced water drinking in sheep. Am J Physiol Regul Integr Comp Physiol 250: R1052-R1059.

McKinley MJ, Mathai ML, Pennington G, Rundgren M, Vivas L. 1999. Effect of individual or combined ablation of the nuclear groups of the lamina terminalis on water drinking in sheep. Am J Physiol Regul Integr Comp Physiol 276: R673-R683.

McKinley MJ, Burns P, Colvill LM, Oldfield BJ, Wade JD, et al. 1997. Distribution of Fos immunoreactivity in the lamina terminalis and hypothalamus induced by centrally administered relaxin in conscious rats. J Neuroendocrinol 9: 431-437.

McKinley MJ, Denton DA, Leksell LG, Mouw DR, Scoggins BA, et al. 1982. Osmoregulatory thirst in sheep is disrupted by ablation of the anterior wall of the optic recess. Brain Res 236: 210-215.

McKinley MJ, McAllen RM, Davern P, Giles ME, Penschow J, et al. 2003. The sensory circumventricular organs of the mammalian brain. Adv Anat Embryol Cell Biol 172: 1-127.

McRae-Degueurce A, Bellin SI, Landas SK, Johnson AK. 1986. Fetal noradrenergic transplants into amine-depleted basal forebrain nuclei restore drinking to angiotensin. Brain Res 374: 162-166.

Menani JV, Johnson AK. 1995. Lateral parabrachial serotonergic mechanisms: angiotensin-induced pressor and drinking responses. Am J Physiol Regul Integr Comp Physiol 269: R1044-R1049.

Menani JV, Johnson AK. 1998. Cholecystokinin actions in the parabrachial nucleus: effects on thirst and salt appetite. Am J Physiol Regul Integr Comp Physiol 275: R1431-R1437.

Menani JV, Beltz TG, Johnson AK. 1995. Effects of lidocaine injections into the lateral parabrachial nucleus on dipsogenic and pressor responses to central angiotensin II in rats. Brain Res 695: 250-252.

Menani JV, De Luca LA Jr, Johnson AK. 1998b. Lateral parabrachial nucleus serotonergic mechanisms and salt appetite induced by sodium depletion. Am J Physiol Regul Integr Comp Physiol 274: R555-R560.

Menani JV, Thunhorst RL, Johnson AK. 1996. Lateral parabrachial nucleus and serotonergic mechanisms in the control of salt appetite in rats. Am J Physiol Regul Integr Comp Physiol 270: R162-R168.

Menani JV, De Luca LA Jr, Thunhorst RL, Johnson AK. 2000. Hindbrain serotonin and the rapid induction of sodium appetite. Am J Physiol Regul Integr Comp Physiol 279: R126-R131.

Menani JV, Barbosa SP, De Luca LA Jr, De Gobbi JI, Johnson AK. 2002. Serotonergic mechanisms of the lateral parabrachial nucleus and cholinergic-induced sodium appetite. Am J Physiol Regul Integr Comp Physiol 282: R837-R841.

Menani JV, Barbosa SP, McKinley MJ, Wade JD, De Luca LA Jr. 2004. Serotonergic mechanism of the lateral parabrachial nucleus and relaxin-induced sodium intake. Brain Res 1030: 74-80.

Menani JV, Colombari DS, Beltz TG, Thunhorst RL, Johnson AK. 1998a. Salt appetite: interaction of forebrain angiotensinergic and hindbrain serotonergic mechanisms. Brain Res 801: 29-35.

Menani JV, Sato MA, Haikel L, Vieira AA, de Andrade CA, et al. 1999. Central moxonidine on water and NaCl intake. Brain Res Bull 49: 273-279.

Michell AR. 1995. The Clinical Biology of Sodium: The Physiology and Pathophysiology of Sodium in Mammals. New York: Elsevier Science.

Miyakubo H, Yamamoto K, Hatakenaka S, Hayashi Y, Tanaka J. 2003. Drinking decreases the noradrenaline release in the median preoptic area caused by hypovolemia in the rat. Behav Brain Res 145: 1-5.

Mizuno A, Matsumoto N, Imai M, Suzuki M. 2003. Impaired osmotic sensation in mice lacking TRPV4. Am J Physiol Cell Physiol 285: C96-C101.

Moe KE, Weiss ML, Epstein AN. 1984. Sodium appetite during captopril blockade of endogenous angiotensin II formation. Am J Physiol Regul Integr Comp Physiol 247: R356-R365.

Moga MM, Saper CB, Gray TS. 1989. Bed nucleus of the stria terminalis: cytoarchitecture, immunohistochemistry, and projection to the parabrachial nucleus in the rat. J Comp Neurol 283: 315-332.

Mohring J, Petri M, Mohring B. 1975. Salt appetite during the early phase of renal hypertension in rats. Pflugers Archiv - Eur J Physiol 356: 153-158.

Moore-Gillon MJ, Fitzsimons JT. 1982. Pulmonary vein-atrial junction stretch receptors and the inhibition of drinking. Am J Physiol Regul Integr Comp Physiol 242: R452-R457.

Morimoto S, Cassell MD, Sigmund CD. 2002. Neuron-specific expression of human angiotensinogen in brain causes increased salt appetite. Physiol Genomics 9: 113-120.

Morimoto S, Cassell MD, Beltz TG, Johnson AK, Davisson RL, et al. 2001. Elevated blood pressure in transgenic mice with brain-specific expression of human angiotensinogen driven by the glial fibrillary acidic protein promoter. Circ Res 89: 365-372.

Morris M, Li P, Barrett C, Callahan MF. 1995. Oxytocin antisense reduces salt intake in the baroreceptor-denervated rat. Regul Pept 59: 261-266.

Morris MJ, Wilson WL, Starbuck EM, Fitts DA. 2002. Forebrain circumventricular organs mediate salt appetite induced by intravenous angiotensin II in rats. Brain Res 949: 42-50.

Mullins JJ, Peters J, Ganten D. 1990. Fulminant hypertension in transgenic rats harbouring the mouse Ren-2 gene. Nature 344: 541-544.

Nachman M, Ashe JH. 1974. Effects of basolateral amygdala lesions on neophobia, learned taste aversions, and sodium appetite in rats. J Comp Physiol Psychol 87: 622-643.

Nadal JW, Pedersen S, Maddock WG. 1941. A comparison between dehydration from salt loss and from water deprivation. J Clin Invest 20: 691-703.

Naeini RS, Witty MF, Seguela P, Bourque CW. 2006. An N-terminal variant of Trpv1 channel is required for osmosensory transduction. Nat Neurosci 9: 93-98.

Niijima A. 1969. Afferent discharges from osmoreceptors in the liver of the guinea pig. Science 166: 1519-1520.

Nitabach MN, Schulkin J, Epstein AN. 1989. The medial amygdala is part of a mineralocorticoid-sensitive circuit controlling NaCl intake in the rat. Behav Brain Res 35: 127-134.

Norgren R. 1978. Projections from the nucleus of the solitary tract in the rat. Neuroscience 3: 207-218.

Norgren R. 1981. The central organization of the gustatory and visceral afferent systems in the nucleus of the solitary tract. Brain Mechanisms of Sensation. Katsuki Y, Norgren R, Sato M, editors. New York: Wiley & Sons; pp. 143-160.

Oatley K. 1964. Changes in blood volume and osmotic pressure in the production of thirst. Nature 202: 1341-1342.

Ohman LE, Johnson AK. 1989. Brain stem mechanisms and the inhibition of angiotensin-induced drinking. Am J Physiol Regul Integr Comp Physiol 256: R264-R269.

Ohman LE, Johnson AK. 1986. Lesions in lateral parabrachial nucleus enhance drinking to angiotensin II and isoproterenol. Am J Physiol Regul Integr Comp Physiol 251: R504-R509.

Ohman LE, Johnson AK. 1995. Role of lateral parabrachial nucleus in the inhibition of water intake produced by right atrial stretch. Brain Res 695: 275-278.

Okuno T, Suzuki H, Saruta T. 1981. Dexamethasone hypertension in rats. Clin Exp Hypertens 3: 1075-1086.

Oliet SH, Bourque CW. 1993. Mechanosensitive channels transduce osmosensitivity in supraoptic neurons. Nature 364: 341-343.

Olsson K. 1969. Studies on central regulation of secretion of antidiuretic hormone (ADH) in the goat. Acta Physiol Scand 77: 465-474.

Palkovits M. 1966. The role of the subfornical organ in the salt and water balance. Naturwissenschaften 53: 336.

Pawloski CM, Fink GD. 1990. Circulating angiotensin II and drinking behavior in rats. Am J Physiol Regul Integr Comp Physiol 259: R531-R538.

Peck JW, Novin D. 1971. Evidence that osmoreceptors mediating drinking in rabbits are in the lateral preoptic area. J Comp Physiol Psychol 74: 134-147.

Pompei P, Lucas LR, Angeletti S, Massi M, McEwen BS. 1997. In situ hybridization analysis of preprotachykinin-A and -B mRNA levels in short-term sodium depletion. Brain Res Mol Brain Res 49: 149-156.

Price JL, Russchen FT, Amaral DG. 1987. The limbic region. II: The amygdaloid complex. Handbook of Chemical Neuroanatomy. Björklund A, Hokfelt T, Swanson LW, editors. New York: Elsevier. pp. 279-388.

Quillen EW Jr, Keil LC, Reid IA. 1990. Effects of baroreceptor denervation on endocrine and drinking responses to caval constriction in dogs. Am J Physiol Regul Integr Comp Physiol 259: R618-R626.

Quillen EW Jr, Reid IA, Keil LC. 1988. Cardiac and arterial baroreceptor influences on plasma vasopressin and drinking. Vasopressin: Cellular and Integrative Functions. Cowley Jr, AW Liard J-F, Ausiello DA, editors. New York: Raven Press; pp. 405-411.

Radford EP Jr. 1959. Factors modifying water metabolism in rats fed dry diets. Am J Physiol 196: 1098-1108.

Ramsay DJ, Reid IA. 1975. Some central mechanisms of thirst in the dog. J Physiol (Lond) 253: 517-525.

Rettig R, Johnson AK. 1986. Aortic baroreceptor deafferentation diminishes saline-induced drinking in rats. Brain Res 370: 29-37.

Ricardo JA, Koh ET. 1978. Anatomical evidence of direct projections from the nucleus of the solitary tract to the hypothalamus, amygdala, and other forebrain structures in the rat. Brain Res 153: 1-26.

Rice KK, Richter CP. 1943. Increased sodium chloride and water intake of normal rats treated with desoxycorticosterone acetate. Endocrinology 33: 106-115.

Richard D, Bourque CW. 1995. Synaptic control of rat supraoptic neurones during osmotic stimulation of the organum vasculosum lamina terminalis in vitro. J Physiol (Lond) 489: 567-577.

Richter CP. 1936. Increased salt appetite in adrenalectomized rats. Am J Physiol 115: 155-161.

Robinson MM, Evered MD. 1987. Pressor action of intravenous angiotensin II reduces drinking response in rats. Am J Physiol Regul Integr Comp Physiol 252: R754-R759.

Rocha MJ, Callahan MF, Morris M. 1997. Baroreceptor regulation of salt intake. Brain Res Bull 42: 147-151.

Rocha MJ, Callahan MF, Sundberg DK, Morris M. 1993. Sinoaortic denervation alters the molecular and endocrine responses to salt loading. Neuroendocrinology 57: 729-737.

Rogers RC, Novin D, Butcher LL. 1979. Electrophysiological and neuroanatomical studies of hepatic portal osmo- and sodium-receptive afferent projections within the brain. J Auton Nerv Syst 1: 183-202.

Routtenberg A, Simpson JB. 1971. Carbachol-induced drinking at ventricular and subfornical organ sites of application. Life Sci I 10: 481-490.

Rowland NE, Fregly MJ. 1988a. Characteristics of thirst and sodium appetite in mice (*Mus musculus*). Behav Neurosci 102: 969-974.

Rowland NE, Fregly MJ. 1988b. Sodium appetite: species and strain differences and role of renin-angiotensin-aldosterone system. Appetite 11: 143-178.

Rowland NE, Morian KR. 1999. Roles of aldosterone and angiotensin in maturation of sodium appetite in furosemide-treated rats. Am J Physiol Regul Integr Comp Physiol 276: R1453-R1460.

Ruhf AA, Starbuck EM, Fitts DA. 2001. Effects of SFO lesions on salt appetite during multiple sodium depletions. Physiol Behav 74: 629-636.

Sakai RR. 1986. The hormones of renal sodium conservation act synergistically to arouse a sodium appetite in the rat. The Physiology of Thirst and Sodium Appetite. de Caro G, Epstein AN, Massi M, editors. New York: Plenum Press; pp. 425-430.

Sakai RR, Chow SY, Epstein AN. 1990. Peripheral angiotensin II is not the cause of sodium appetite in the rat. Appetite 15: 161-170.

Sakai RR, Nicolaidis S, Epstein AN. 1986. Salt appetite is suppressed by interference with angiotensin II and aldosterone. Am J Physiol Regul Integr Comp Physiol 251: R762-R768.

Sakai RR, Fine WB, Epstein AN, Frankmann SP. 1987. Salt appetite is enhanced by one prior episode of sodium depletion in the rat. Behav Neurosci 101: 724-731.

Sakai RR, Ma LY, Zhang DM, McEwen BS, Fluharty SJ. 1996. Intracerebral administration of mineralocorticoid receptor antisense oligonucleotides attenuate adrenal steroid-induced salt appetite in rats. Neuroendocrinology 64: 425-429.

Sakai RR, McEwen BS, Fluharty SJ, Ma LY. 2000. The amygdala: site of genomic and nongenomic arousal of aldosterone-induced sodium intake. Kidney Int 57: 1337-1345.

Salisbury JJ, Rowland NE. 1990. Sham drinking in rats: osmotic and volumetric manipulations. Physiol Behav 47: 625-630.

Samson WK. 2004. The posterior pituitary and water metabolism. Textbook of Endocrine Physiology. Griffin JE, Ojeda SR, editors. New York: Oxford University Press.

Saper CB, Reis DJ, Joh T. 1983. Medullary catecholamine inputs to the anteroventral third ventricular cardiovascular regulatory region in the rat. Neurosci Lett 42: 285-291.

Sato MA, Yada MM, De Luca LA Jr. 1996. Antagonism of the renin-angiotensin system and water deprivation-induced NaCl intake in rats. Physiol Behav 60: 1099-1104.

Scheuer DA, Bechtold AG. 2001. Glucocorticoids potentiate central actions of angiotensin to increase arterial pressure. Am J Physiol Regul Integr Comp Physiol 280: R1719-R1726.

Schoorlemmer GHM, Johnson AK, Thunhorst RL. 2000. Effect of hyperosmotic solutions on salt excretion and thirst in rats. Am J Physiol Regul Integr Comp Physiol 278: R917-R923.

Schoorlemmer GHM, Johnson AK, Thunhorst RL. 2001. Circulating angiotensin II mediates sodium appetite in adrenalectomized rats. Am J Physiol Regul Integr Comp Physiol 281: R723-R729.

Schreihofer AM, Stricker EM, Sved AF. 2000. Nucleus of the solitary tract lesions enhance drinking, but not vasopressin release, induced by angiotensin. Am J Physiol Regul Integr Comp Physiol 279: R239-R247.

Schreihofer AM, Anderson BK, Schiltz JC, Xu L, Sved AF, et al. 1999. Thirst and salt appetite elicited by hypovolemia in rats with chronic lesions of the nucleus of the solitary tract. Am J Physiol Regul Integr Comp Physiol 276: R251-R258.

Schulkin J, Eng R, Miselis RR. 1983. The effects of disconnecting the subfornical organ on behavioral and physiological responses to alterations of body sodium. Brain Res 263: 351-355.

Schwob JE, Johnson AK. 1977. Angiotensin-induced dipsogenesis in domestic fowl (Gallus gallus). J Comp Physiol Psychol 91: 182-188.

Seeley RJ, Galaverna O, Schulkin J, Epstein AN, Grill HJ. 1993. Lesions of the central nucleus of the amygdala. II: Effects on intraoral NaCl intake. Behav Brain Res 59: 19-25.

Semple RE. 1952. Compensatory changes in the rat following removal of electrolytes by intraperitoneal dialysis. Am J Physiol 168: 55-65.

Severs WB, Summy-Long J, Daniels-Severs A. 1974. Angiotensin interaction with thirst mechanisms. Am J Physiol 226: 340-344.

Shapiro RE, Miselis RR. 1985. The central neural connections of the area postrema of the rat. J Comp Neurol 234: 344-364.

Shelat SG, Flanagan-Cato LM, Fluharty SJ. 1999a. Glucocorticoid and mineralocorticoid regulation of angiotensin II type 1 receptor binding and inositol triphosphate formation in WB cells. J Endocrinol 162: 381-391.

Shelat SG, Fluharty SJ, Flanagan-Cato LM. 1998. Adrenal steroid regulation of central angiotensin II receptor subtypes and oxytocin receptors in rat brain. Brain Res 807: 135-146.

Shelat SG, King JL, Flanagan-Cato LM, Fluharty SJ. 1999b. Mineralocorticoids and glucocorticoids cooperatively increase salt intake and angiotensin II receptor binding in rat brain. Neuroendocrinology 69: 339-351.

Simpson JB, Routtenberg A. 1973. Subfornical organ: site of drinking elicitation by angiotensin II. Science 181: 1172-1175.

Simpson JB, Epstein AN, Camardo JS Jr. 1978. Localization of receptors for the dipsogenic action of angiotensin II in the subfornical organ of rat. J Comp Physiol Psychol 92: 581-601.

Sinnayah P, Burns P, Wade JD, Weisinger RS, McKinley MJ. 1999. Water drinking in rats resulting from intravenous relaxin and its modification by other dipsogenic factors. Endocrinology 140: 5082-5086.

Smith HW. 1957. Salt and water volume receptors: an exercise in physiologic apologetics. Am J Med 23: 623-652.

Smith ME, Flynn FW. 1994. Tachykinin NK3 receptor agonist blocks sodium deficiency-induced shift in taste reactivity. Brain Res 665: 123-126.

Sofroniew MV. 1983. Direct reciprocal connections between the bed nucleus of the stria terminalis and dorsomedial medulla oblongata: evidence from immunohistochemical detection of tracer proteins. J Comp Neurol 213: 399-405.

Starling EH. 1909. The Fluids of the Body. Chicago: WT Keener & Co.

Stein L, Seifter J. 1962. Muscarinic synapses in the hypothalamus. Am J Physiol 202: 751-756.

Stella A, Zanchetti A. 1991. Functional role of renal afferents. Physiol Rev 71: 659-682.

Stevenson JA, Welt LG, Orloff J. 1950. Abnormalities of water and electrolyte metabolism in rats with hypothalamic lesions. Am J Physiol 161: 35-39.

Stocker SD, Stricker EM, Sved AF. 2001. Acute hypertension inhibits thirst stimulated by ANG II, hyperosmolality, or hypovolemia in rats. Am J Physiol Regul Integr Comp Physiol 280: R214-R224.

Stocker SD, Stricker EM, Sved AF. 2002. Arterial baroreceptors mediate the inhibitory effect of acute increases in arterial blood pressure on thirst. Am J Physiol Regul Integr Comp Physiol 282: R1718-R1729.

Strauss MB. 1957. Body Water in Man. The Acquisition and Maintenance of the Body Fluids. Boston: Little, Brown & Co.

Stricker EM. 1966. Extracellular fluid volume and thirst. Am J Physiol 211: 232-238.

Stricker EM. 1969. Osmoregulation and volume regulation in rats: inhibition of hypovolemic thirst by water. Am J Physiol 217: 98-105.

Stricker EM. 1977. The renin-angiotensin system and thirst: a reevaluation. II. Drinking elicited in rats by caval ligation or isoproterenol. J Comp Physiol Psychol 91: 1220-1231.

Stricker EM. 1978. The renin-angiotensin system and thirst: some unanswered questions. Fed Proc 37: 2704-2710.

Stricker EM. 1981. Thirst and sodium appetite after colloid treatment in rats. J Comp Physiol Psychol 95: 1-25.

Stricker EM. 1983. Thirst and sodium appetite after colloid treatment in rats: role of the renin-angiotensin-aldosterone system. Behav Neurosci 97: 725-737.

Stricker EM. 1991. Central control of water and sodium chloride intake in rats during hypovolaemia. Thirst: Physiological and Psychological Aspects. Ramsay DJ, Booth DA, editors. New York: Springer-Verlag; pp. 194-206.

Stricker EM, Jalowiec JE. 1970. Restoration of intravascular fluid volume following acute hypovolemia in rats. Am J Physiol 218: 191-196.

Stricker EM, Verbalis JG. 1987. Central inhibitory control of sodium appetite in rats: correlation with pituitary oxytocin secretion. Behav Neurosci 101: 560-567.

Stricker EM, Verbalis JG. 1990. Sodium appetite. Handbook of Behavioral Neurobiology, Vol. 10, Neurobiology of Food and Fluid Intake. Stricker EM, editor. New York: Plenum Press; pp. 387-419.

Stricker EM, Wolf G. 1966. Blood volume and tonicity in relation to sodium appetite. J Comp Physiol Psychol 62: 275-279.

Stricker EM, Gannon KS, Smith JC. 1992. Thirst and salt appetite induced by hypovolemia in rats: analysis of drinking behavior. Physiol Behav 51: 27-37.

Stricker EM, Hosutt JA, Verbalis JG. 1987. Neurohypophyseal secretion in hypovolemic rats: inverse relation to sodium appetite. Am J Physiol Regul Integr Comp Physiol 252: R889-R896.

Stricker EM, Schreihofer AM, Verbalis JG. 1994. Sodium deprivation blunts hypovolemia-induced pituitary secretion of vasopressin and oxytocin in rats. Am J Physiol Regul Integr Comp Physiol 267: R1336-R1341.

Stricker EM, Vagnucci AH, McDonald RH Jr, Leenen FH. 1979. Renin and aldosterone secretions during hypovolemia in rats: relation to NaCl intake. Am J Physiol Regul Integr Comp Physiol 237: R45-R51.

Strotmann R, Harteneck C, Nunnenmacher K, Schultz G, Plant TD. 2000. OTRPC4, a nonselective cation channel that confers sensitivity to extracellular osmolarity. Nat Cell Biol 2: 695-702.

Summerlee AJ, Robertson GL. 1995. Central administration of porcine relaxin stimulates drinking behaviour in rats: an effect mediated by central angiotensin II. Endocr J 3: 377-387.

Summerlee AJ, Hornsby DJ, Ramsey DG. 1998. The dipsogenic effects of rat relaxin: The effect of photoperiod and the potential role of relaxin on drinking in pregnancy. Endocrinology 139: 2322-2328.

Sumners C, Fregly MJ. 1989. Modulation of angiotensin II binding sites in neuronal cultures by mineralocorticoids. Am J Physiol 256: C121-C129.

Sumners C, Gault TR, Fregly MJ. 1991. Potentiation of angiotensin II-induced drinking by glucocorticoids is a specific glucocorticoid type II receptor (GR)-mediated event. Brain Res 552: 283-290.

Sunn N, McKinley MJ, Oldfield BJ. 2001. Identification of efferent neural pathways from the lamina terminalis activated by blood-borne relaxin. J Neuroendocrinol 13: 432-437.

Suzuki H, Handa M, Kondo K, Saruta T. 1982. Role of renin-angiotensin system in glucocorticoid hypertension in rats. Am J Physiol 243: E48-E51.

Suzuki M, Sato J, Kutsuwada K, Ooki G, Imai M. 1999. Cloning of a stretch-inhibitable nonselective cation channel. J Biol Chem 274: 6330-6335.

Swanson PP, Timson GH, Frazier E. 1935. Some observations on the physiological adjustment of the albino rat to a diet poor in salts when edestin is the source of dietary protein. J Biol Chem 109: 729-737.

Szczepanska-Sadowska E, Paczwa P, Dobruch J. 2003. Enhanced food and water intake in renin transgenic rats. J Physiol Pharmacol 54: 81-88.

Szczepanska-Sadowska E, Sobocinska J, Sadowski B. 1982. Central dipsogenic effect of vasopressin. Am J Physiol Regul Integr Comp Physiol 242: R372-R379.

Takamata A, Mack GW, Gillen CM, Nadel ER. 1994. Sodium appetite, thirst, and body fluid regulation in humans during rehydration without sodium replacement. Am J Physiol Regul Integr Comp Physiol 266: R1493-R1502.

Takei Y. 1977. The role of the subfornical organ in drinking induced by angiotensin in the Japanese quail, *Coturnix coturnix japonica*. Cell Tissue Res 185: 175-181.

Tarjan E, Denton DA. 1991. Sodium/water intake of rabbits following administration of hormones of stress. Brain Res Bull 26: 133-136.

Tarjan E, Denton DA, Weisinger RS. 1991. Corticotropin-releasing factor enhances sodium and water intake/excretion in rabbits. Brain Res 542: 219-224.

Tarjan E, Denton DA, McBurnie MI, Weisinger RS. 1988. Water and sodium intake of wild and New Zealand Rabbits following angiotensin. Peptides 9: 677-679.

Tarjan E, Ferraro T, May C, Weisinger RS. 1993. Converting enzyme inhibition in rabbits: effects on sodium and water intake/excretion and blood pressure. Physiol Behav 53: 291-299.

Thornton SM, Fitzsimons JT. 1995. The effects of centrally administered porcine relaxin on drinking behaviour in male and female rats. J Neuroendocrinol 7: 165-169.

Thrasher TN, Keil LC. 1987. Regulation of drinking and vasopressin secretion: role of organum vasculosum laminae terminalis. Am J Physiol Regul Integr Comp Physiol 253: R108-R120.

Thrasher TN, Keil LC, Ramsay DJ. 1982a. Hemodynamic, hormonal, and drinking responses to reduced venous return in the dog. Am J Physiol Regul Integr Comp Physiol 243: R354-R362.

Thrasher TN, Keenan CR, Ramsay DJ. 1999. Cardiovascular afferent signals and drinking in response to hypotension in dogs. Am J Physiol Regul Integr Comp Physiol 277: R795-R801.

Thrasher TN, Simpson JB, Ramsay DJ. 1982b. Lesions of the subfornical organ block angiotensin-induced drinking in the dog. Neuroendocrinology 35: 68-72.

Thrasher TN, Brown CJ, Keil LC, Ramsay DJ. 1980. Thirst and vasopressin release in the dog: an osmoreceptor or sodium receptor mechanism? Am J Physiol Regul Integr Comp Physiol 238: R333-R339.

Thunhorst RL, Fitts DA. 1994. Peripheral angiotensin causes salt appetite in rats. Am J Physiol Regul Integr Comp Physiol 267: R171-R177.

Thunhorst RL, Johnson AK. 1993. Effects of arterial pressure on drinking and urinary responses to intracerebroventricular angiotensin II. Am J Physiol Regul Integr Comp Physiol 264: R211-R217.

Thunhorst RL, Johnson AK. 1994a. Renin-angiotensin, arterial blood pressure, and salt appetite in rats. Am J Physiol Regul Integr Comp Physiol 266: R458-R465.

Thunhorst RL, Johnson AK. 1994b. The role of arterial pressure and arterial baroreceptors in the modulation of the drinking response to centrally-administered angiotensin II. Integrative and Cellular Aspects of Autonomic Functions: Temperature and Osmoregulation. Pleschka R, Gerstberger R, editors. Paris: John Libbey Eurotext; pp. 397-405.

Thunhorst RL, Johnson AK. 2001. Effects of hypotension and fluid depletion on central angiotensin-induced thirst and salt appetite. Am J Physiol Regul Integr Comp Physiol 281: R1726-R1733.

Thunhorst RL, Beltz TG, Johnson AK. 1999. Effects of subfornical organ lesions on acutely induced thirst and salt appetite. Am J Physiol Regul Integr Comp Physiol 277: R56-R65.

Thunhorst RL, Ehrlich KJ, Simpson JB. 1990. Subfornical organ participates in salt appetite. Behav Neurosci 104: 637-642.

Thunhorst RL, Kirby RF, Johnson AK. 1996. Role of renal nerves in sodium depletion-induced salt appetite. Am J Physiol Regul Integr Comp Physiol 271: R806-R812.

Thunhorst RL, Lewis SJ, Johnson AK. 1993. Role of arterial baroreceptor input on thirst and urinary responses to intracerebroventricular angiotensin II. Am J Physiol Regul Integr Comp Physiol 265: R591-R595.

Thunhorst RL, Lewis SJ, Johnson AK. 1994. Effects of sinoaortic baroreceptor denervation on depletion-induced salt appetite. Am J Physiol 267: R1043-R1049.

Tordoff MG, Schulkin J, Friedman MI. 1986. Hepatic contribution to satiation of salt appetite in rats. Am J Physiol Regul Integr Comp Physiol 251: R1095-R1102.

Tordoff MG, Schulkin J, Friedman MI. 1987. Further evidence for hepatic control of salt intake in rats. Am J Physiol Regul Integr Comp Physiol 253: R444-R449.

Toth E, Stelfox J, Kaufman S. 1987. Cardiac control of salt appetite. Am J Physiol Regul Integr Comp Physiol 252: R925-R929.

Touzani K, Taghzouti K, Velley L. 1997. Increase of the aversive value of taste stimuli following ibotenic acid lesion of the central amygdaloid nucleus in the rat. Behav Brain Res 88: 133-142.

Trippodo NC, McCaa RE, Guyton AC. 1976. Effect of prolonged angiotensin II infusion on thirst. Am J Physiol Regul Integr Comp Physiol 230: 1063-1066.

Ueda H, Katayama S, Kato R. 1972. Area postrema – angiotensin-sensitive site in brain. Control of Renin Secretion. Assaykeen TA, editor. New York: Plenum Press; pp. 109-116.

van der Kooy D, Koda LY. 1983. Organization of the projections of a circumventricular organ: the area postrema in the rat. J Comp Neurol 219: 328-338.

Vanderweele DA. 1974. The effects of taste adulteration, hypertonic and hyperoncotic solutions on water ingestion in the gerbil. Anim Learn Behav 2: 309-312.

Verbalis JG, Mangione MP, Stricker EM. 1991. Oxytocin produces natriuresis in rats at physiological plasma concentrations. Endocrinology 128: 1317-1322.

Verney EB. 1947. The antidiuretic hormone and the factor which determines its release. Proc R Soc Lond B Biol Sci 135: 27-106.

Watanabe E, Fujikawa A, Matsunaga H, Yasoshima Y, Sako N, et al. 2000. Nav2/NaG channel is involved in control of salt-intake behavior in the CNS. J Neurosci 20: 7743-7751.

Watson WE. 1985. The effect of removing area postrema on the sodium and potassium balances and consumptions in the rat. Brain Res 359: 224-232.

Weisinger RS, Denton DA, McKinley MJ, Nelson JF. 1978. ACTH induced sodium appetite in the rat. Pharmacol Biochem Behav 8: 339-342.

Weisinger RS, Denton DA, McKinley MJ, Nelson JF. 1985a. Dehydration-induced sodium appetite in rats. Physiol Behav 34: 45-50.

Weisinger RS, Denton DA, McKinley MJ, Muller AF, Tarjan E. 1985b. Cerebrospinal fluid sodium concentration and salt appetite. Brain Res 326: 95-105.

Weisinger RS, Denton DA, McKinley MJ, Muller AF, Tarjan E. 1986. Angiotensin and Na appetite of sheep. Am J Physiol Regul Integr Comp Physiol 251: R690-R699.

Weisinger RS, Blair-West JR, Burns P, Denton DA, McKinley MJ, et al. 1996. The role of angiotensin II in ingestive behaviour: a brief review of angiotensin II, thirst and Na appetite. Regul Pept 66: 73-81.

Weisinger RS, Coghlan JP, Denton DA, Fan JS, Hatzikostas S, et al. 1980. ACTH-elicited sodium appetite in sheep. Am J Physiol 239: E45-E50.

Weisinger RS, Denton DA, Di Nicolantonio R, Hards DK, McKinley MJ, et al. 1990. Subfornical organ lesion decreases sodium appetite in the sodium-depleted rat. Brain Res 526: 23-30.

Weisinger RS, Denton DA, Di Nicolantonio R, McKinley MJ, Muller AF, et al. 1987. Role of angiotensin in sodium appetite of sodium-deplete sheep. Am J Physiol Regul Integr Comp Physiol 253: R482-R488.

Weiss ML, Moe KE, Epstein AN. 1986. Interference with central actions of angiotensin II suppresses sodium appetite. Am J Physiol Regul Integr Comp Physiol 250: R250-R259.

Wettendorff H. 1901. Modifications du sang sous l'influencxe de la privation d'eau: contribution á l'étude de la soif. Travaux du Laboratoire de Physiologie, Instituts Soplvay 4: 353-484.

Wilkin LD, Patel KP, Schmid PG, Johnson AK. 1987. Increased norepinephrine turnover in the median preoptic nucleus following reduced extracellular fluid volume. Brain Res 423: 369-372.

Wilson KM, Sumners C, Hathaway S, Fregly MJ. 1986. Mineralocorticoids modulate central angiotensin II receptors in rats. Brain Res 382: 87-96.

Wilson WL, Starbuck EM, Fitts DA. 2002. Salt appetite of adrenalectomized rats after a lesion of the SFO. Behav Brain Res 136: 449-453.

Wolf AV. 1950. Osmometric analysis of thirst in man and dog. Am J Physiol 161: 75-86.

Wolf AV. 1958. Thirst: Physiology of the Urge to Drink and Problems of Water Lack. Springfield: Charles C. Thomas.

Wolf G. 1964a. Effect of dorsolateral hypothalamic lesions on sodium appetite elicited by desoxycorticosterone and by acute hyponatremia. J Comp Physiol Psychol 58: 396-402.

Wolf G. 1964b. Sodium appetite elicited by aldosterone. Psychonomic Science I: 211-212.

Wolf G. 1965. Effect of deoxycorticosterone on sodium appetite of intact and adrenalectomized rats. Am J Physiol 208: 1281-1285.

Wright JW, Morseth SL, Fairley PC, Petersen EP, Harding JW. 1987. Angiotensin's contribution to dipsogenic additivity in several rodent species. Behav Neurosci 101: 361-370.

Yang ZF, Epstein AN. 1991. Blood-borne and cerebral angiotensin and the genesis of salt intake. Horm Behav 25: 461-476.

Yoshida S, Yamada K, Tamura Y, Yoshida S. 1988. Role of the renal nerve in glucocorticoid hypertension. Endocrinol Jpn 35: 877-884.

Zardetto-Smith AM, Beltz TG, Johnson AK. 1994. Role of the central nucleus of the amygdala and bed nucleus of the stria terminalis in experimentally-induced salt appetite. Brain Res 645: 123-134.

Zardetto-Smith AM, Thunhorst RL, Cicha MZ, Johnson AK. 1993. Afferent signaling and forebrain mechanisms in the behavioral control of extracellular fluid volume. Ann NY Acad Sci 689: 161-176.

Zhang DM, Epstein AN, Schulkin J. 1993. Medial region of the amygdala: involvement in adrenal-steroid-induced salt appetite. Brain Res 600: 20-26.

Zimmerman MB, Blaine EH, Stricker EM. 1981. Water intake in hypovolemic sheep: effects of crushing the left atrial appendage. Science 211: 489-491.

Zimmerman MC, Lazartigues E, Lang JA, Sinnayah P, Ahmad IM, et al. 2002. Superoxide mediates the actions of angiotensin II in the central nervous system. Circ Res 91: 1038-1045.

18 Neurochemistry/ Neuropharmacology of Fear and Fear Conditioning

G. E. Schafe · J. E. LeDoux

Abstract: Over the last 20 years, research on emotion has enjoyed a resurgence in the neurosciences. Nowhere has this been more evident than in the fear system of the brain. In this chapter, we discuss the brain's "fear learning" system from a neurochemical and pharmacological perspective. We focus on the neurochemical and biochemical events that underlie fear "memory consolidation" or the processes whereby short-term fear memories are gradually transformed into stable, long-term fear memories.

List of Abbreviations: AKAP, A-kinase anchoring protein; APV, D-2-amino-5-phosphonovaleric acid; BPAP, back-propagating action potential; CaMKII, Ca^{2+}/calmodulin-dependent protein kinase II; CR, conditioned response; CRE, cAMP-response element; CREB, cAMP-response-element binding protein; CS, conditioned stimulus; DAG, diacylglycerol; E-LTP, early-LTP; ERK, extracellular signal-regulated kinase; IP_3, inositol 1,4,5-trisphosphate; LA, lateral amygdala; L-LTP, late-LTP; LTM, long-term memory; LTP, long-term potentiation; L-VGCC, L-type voltage-gated calcium channel; MAPK, mitogen-activated protein kinase; MEK, mitogen-activated protein kinase kinase; mGluR, metabotropic glutamate receptor; MPEP, 2-methyl-6-(phenylethynyl)-pyridine; NMDAR, N-methyl-D-aspartate receptor; PKA, protein kinase A; PKC, protein kinase C; PSD, postsynaptic density; STM, short-term memory; US, unconditioned stimulus

1 Introduction

In classical or Pavlovian *fear conditioning*, an animal learns to respond defensively to an emotionally neutral stimulus (the conditioned stimulus; CS) that acquires emotional significance after being associated with a noxious stimulus (the unconditioned stimulus; US). In the laboratory, this type of associative learning has been most widely studied in rodents, where a discrete cue (such as a tone, light, or odor; CS) is paired with a brief electric shock to the feet (US). Before conditioning, the CS does not elicit fearful or defensive behavior. After as little as one CS–US pairing, however, the animal begins to exhibit a range of conditioned responses (CRs), both to the tone CS and to the context in which conditioning occurs (e.g., the conditioning chamber). In rats, these CRs include "freezing" or immobility (the rat's species-typical behavioral response to a threatening stimulus), autonomic and endocrine adjustments (such as changes in heart rate and blood pressure, defecation, and increased levels of circulating stress hormones), and other changes such as the potentiation of reflexes like the acoustic startle response (Blanchard and Blanchard, 1969; Kapp et al., 1979; LeDoux et al., 1988; Roozendaal et al., 1991; Davis, 1997) (*Figure 18-1*).

Figure 18-1
Pavlovian fear conditioning. *Top*: Fear conditioning involves the presentation of an initially innocuous stimulus, such as a tone (conditioned stimulus; CS), that is paired or associated with a noxious stimulus, such as a brief electric shock to the feet (unconditioned stimulus; US). *Bottom*: Before conditioning, the CS elicits little response from the animal. After conditioning, the CS elicits a wide range of behavioral and physiological responses that are characteristically elicited by naturally aversive or threatening stimuli

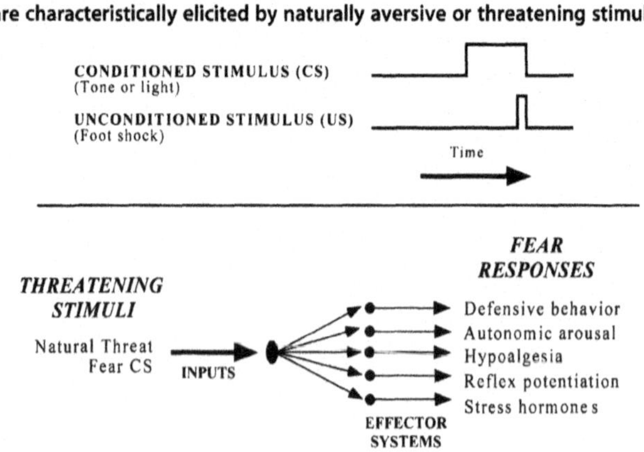

1.1 The Amygdala and Fear Conditioning

The neurobiological substrates of fear conditioning have been studied most extensively using the "auditory fear conditioning" paradigm, where an animal learns to fear a tone (CS) that is paired with footshock (US). In this review, we therefore emphasize the findings from the auditory conditioning literature, although similar mechanisms have also been proposed for conditioning to visual and contextual stimuli (Davis, 1992, 1997; Maren, 2001).

Auditory fear conditioning involves transmission of auditory CS and somatosensory US information to the lateral amygdala (LA), an area in the temporal lobe that lesion and functional inactivation studies have shown to be critical for the learning (LeDoux et al., 1990; Helmstetter and Bellgowan, 1994; Campeau and Davis, 1995; Muller et al., 1997; Wilensky et al., 2000). Anatomical tract tracing studies have shown that cells in the LA receive direct glutamatergic projections from areas of the auditory thalamus and cortex, specifically from the medial division of the *medial geniculate body* and the posterior intralaminar nucleus (MGm/PIN), and area TE3, respectively (LeDoux et al., 1985; LeDoux and Farb, 1991; Bordi and LeDoux, 1992; Romanski and LeDoux, 1993; McDonald, 1998; Doron and LeDoux, 1999). Electrophysiological evidence has indicated that inputs from each of these auditory areas converge onto single neurons in the LA (Li et al., 1996). Further, cells in the same or nearby regions of the thalamus are responsive to the footshock US (Bordi and LeDoux, 1994), as are cells in the LA (Romanski et al., 1993) (*Figure 18-2*).

Figure 18-2
Anatomy of the fear system. (a) Auditory fear conditioning involves the transmission of CS sensory information from areas of the auditory thalamus and cortex to the lateral amygdala (LA), where it can converge with incoming somatosensory information from the footshock US. It is in the LA that alterations in synaptic transmission are thought to encode key aspects of the learning. (b) During fear expression, the LA engages the central nucleus of the amygdala (CE), which projects widely to many areas of the forebrain and brainstem that control the expression of fear CRs, including freezing, HPA axis activation, and alterations in cardiovascular activity

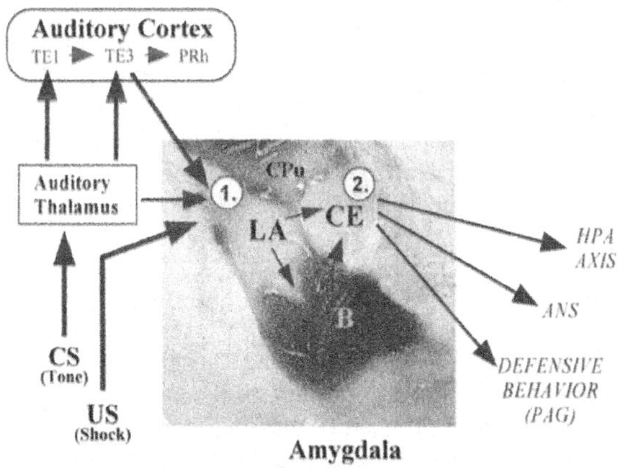

The convergence of auditory and somatosensory information onto individual LA neurons is thought to be critical for the "plasticity" underlying fear conditioning (LeDoux, 2000; Maren, 2001; Blair et al., 2001). In support of this view, individual cells in the LA alter their electrophysiological response properties when CS and US are paired during fear conditioning. Specifically, LA neurons that are initially weakly responsive to auditory input respond vigorously to the same input after fear conditioning (Quirk et al., 1995, 1997; Maren, 2000; Repa et al., 2001; Blair et al., 2003). These changes in the responsiveness of LA cells that occur as the result of training have contributed to the view that neural plasticity in the LA encodes key aspects of fear learning and memory storage (Fanselow and LeDoux, 1999; Maren, 2001; Blair et al., 2001; Schafe et al., 2001).

In the following sections, we discuss the neurochemical and biochemical mechanisms that likely underlie plasticity and memory formation at LA synapses. We begin with a discussion of long-term potentiation (LTP), as it has been proposed that this type of synaptic plasticity is the most likely type of mechanism that underlies fear memory formation (Maren, 1999; Blair et al., 2001; Schafe et al., 2001).

2 Neurochemical Mechanisms of Fear Memory Formation: The Importance of LTP

The change in the responsiveness of LA cells during fear conditioning suggests that alterations in excitatory synaptic transmission in the LA might be critical for fear conditioning. Accordingly, many of the recent studies that have examined the neurochemical basis of fear conditioning have drawn on a larger literature that has focused on the biochemical events that underlie LTP, an activity-dependent form of synaptic plasticity that was initially discovered in the hippocampus (Bliss and Lømo, 1973). There are several good reasons behind this strategy, including the fact that LTP has also been demonstrated in thalamic and cortical auditory input pathways to the LA (Chapman et al., 1990; Clugnet and LeDoux, 1990; Rogan and LeDoux, 1995; Huang and Kandel, 1998; Weisskopf et al., 1999; Weisskopf and LeDoux, 1999). Further, auditory fear conditioning itself has been shown to lead to electrophysiological changes in the LA that resemble artificial LTP induction (McKernan and Shinnick-Gallagher, 1997; Rogan et al., 1997). Collectively, these findings provide strong support for the hypothesis that an LTP-like process in the LA may underlie fear conditioning. This, in turn, suggests that fear memory acquisition and consolidation may share a common neurochemical substrate with LTP.

There are several pharmacologically distinct forms of LTP, most of which have been identified in the hippocampus. One form critically involves the N-methyl-D-aspartate receptor (NMDAR), which is normally blocked by Mg^{2+} but which can be opened following sufficient depolarization of the postsynaptic cell during LTP induction (Malenka and Nicoll, 1993). The other, less widely studied form involves the L-type voltage-gated calcium channel (L-VGCC). Other forms require a combination of both NMDARs and L-VGCCs (Grover and Teyler, 1990; Cavus and Teyler, 1996). Importantly, both NMDAR and L-VGCC-mediated forms of LTP have been discovered in the LA (Bauer et al., 2002). The hallmark of each form of LTP is the entry of Ca^{2+} into the postsynaptic spine, which initiates a biochemical cascade of events that leads to strengthening of the synapse. Some of these biochemical cascades lead to a transient change in synaptic strength known as "early-LTP" (E-LTP) that is independent of de novo RNA and protein synthesis. Other intracellular cascades lead to a more permanent alteration in cell excitability known as "late-LTP" (L-LTP). Unlike E-LTP, L-LTP requires de novo RNA and protein synthesis (Nguyen et al., 1994; Nguyen and Kandel, 1996).

2.1 Early-LTP Versus Late-LTP

Both LTP induction and E-LTP are known to require NMDAR-mediated activation of protein kinase signaling pathways (Malenka and Nicoll, 1993; Soderling and Derkach, 2000). At least two protein kinase signaling pathways have been implicated in E-LTP: the Ca^{2+}/calmodulin-dependent protein kinase II (CaMKII) and protein kinase C (PKC). Both the alpha isoform of CaMKII (αCaMKII) and PKC are known to undergo a rapid, Ca^{2+}-induced "autophosphorylation," a state in which these enzymes can remain active in the absence of further Ca^{2+} entry for a period of time after LTP induction (Sweatt et al., 1998; Sweatt, 1999; Soderling and Derkach, 2000). In this state, both αCaMKII and PKC are thought to translocate to the postsynaptic density (PSD) where they can target and transiently influence the conductances of NMDARs and AMPARs (Lee et al., 2000, 2003; Soderling and Derkach, 2000). The transient enhancement of AMPA receptor conductance is one mechanism by which E-LTP is thought to be maintained in the absence of new protein synthesis (Sweatt, 1999; Soderling and Derkach, 2000).

While E-LTP is a short-lived phenomenon, L-LTP lasts many hours, possibly days to weeks (or longer). This persistence of L-LTP is thought to be due, in part, to structural changes at the synapse itself (Milner et al., 1998). Accordingly, L-LTP requires new RNA and protein synthesis and the activation of different classes of protein kinase signaling pathways. These include protein kinase A (PKA) and the mitogen-activated protein kinase (MAPK) (Nguyen and Kandel, 1996; English and Sweatt, 1997; Huang et al., 2000). Once activated by stimulation that promotes L-LTP, both PKA and MAPK are thought to translocate to the cell nucleus and engage activators of transcription, including the cAMP-response-element binding protein (CREB) and cAMP-response element (CRE)-mediated gene expression (Impey et al., 1996, 1998). It is the activation of CREB and CRE-mediated genes that ultimately leads to the protein and RNA synthesis-dependent functional and/or structural changes that are thought to underlie L-LTP (Frank and Greenberg, 1994; Yin and Tully, 1996; Silva et al., 1998; Stevens, 1998; Holt and Maren, 1999).

In the subsequent sections, we will explore how the biochemical mechanisms of E-LTP and L-LTP are related to memory formation in the LA. As shown later, many of these have been implicated in short-term memory (STM) and long-term memory (LTM) of fear conditioning, respectively.

3 Neurochemical Mechanisms Underlying Acquisition and STM Formation of Fear Conditioning in the Amygdala

Like E-LTP, STM is a short-lasting form of memory that does not require new protein or RNA synthesis (Milner et al., 1998). While no consistent time frame of STM has been defined in the literature, it is generally tested shortly after training, usually within 1 h. Further, deficits in STM formation are typically assumed to reflect deficits in memory acquisition, although it should be emphasized that acquisition and STM formation are likely subserved by distinct molecular mechanisms (Rodrigues et al., 2004). In this section, we examine how glutamatergic transmission, αCaMKII, and PKC might contribute to fear memory acquisition and STM formation in the LA.

3.1 NMDA Receptors

Inspired by the hippocampal LTP literature, early pharmacological studies of Pavlovian fear conditioning examined the role of the NMDA receptor in the amygdala in fear acquisition. These initial experiments consistently showed that blockade of NMDARs in the LA impaired fear conditioning (Miserendino et al., 1990; Kim et al., 1991; Campeau et al., 1992). Later reports, however, indicated that NMDAR blockade in the amygdala also impaired the expression of previously acquired fear responses (Maren et al., 1996). These findings are consistent with electrophysiological evidence showing that NMDA receptors are involved, at least in part, in routine synaptic transmission in the LA (Weisskopf and LeDoux, 1999). As such, it has been difficult to conclude unambiguously that NMDARs are required for fear acquisition independently of a role in routine synaptic transmission.

The role of NMDARs in fear conditioning has recently been revisited by examining the effects of selective blockade of the NR2B subunit of the NMDA receptor in the LA. NMDARs are heteromeric complexes composed of several subunits, including the NR1 subunit, which is essential for channel function, as well as a range of NR2 subunits that regulate channel function (Monyer et al., 1992; Nakanishi, 1992). In vitro studies have shown that the NR1–NR2B complex exhibits longer EPSPs than the NR1–NR2A complex (Monyer et al., 1992). This characteristic of NR2B containing NMDARs is thought to provide a longer time window for coincidence detection, which is thought to be especially important during synaptic plasticity (Tsien, 2000). Indeed, recent molecular genetic studies have implicated the NR2B subunit in both synaptic plasticity and memory formation; overexpression of NR2B in the forebrain of mice results in enhanced LTP and memory formation for a variety of tasks, including fear conditioning (Tang et al., 1999).

Most previous studies that examined the role of NMDARs in the amygdala in fear conditioning have used the NMDAR antagonist D-2-amino-5-phosphonovaleric acid (APV), which blocks the entire NMDA receptor complex (Miserendino et al., 1990; Kim et al., 1991; Campeau et al., 1992). In contrast, a recent study used ifenprodil, a drug which selectively blocks the NR2B subunit of the NMDA receptor (Rodrigues et al., 2001). Pretraining intraamygdala infusions of ifenprodil dose-dependently impaired formation of both STM and LTM of fear conditioning (Rodrigues et al., 2001); that is, memory was impaired both at 1 and 24 h after infusion and training (➋ *Figure 18-3a* and *b*). In contrast, infusions of ifenprodil before testing at either time point had no effect on fear expression. These results suggest that ifenprodil lacks the nonspecific effects on routine transmission that are characteristic of the more global NMDAR antagonist

APV. In support of this hypothesis, bath application of ifenprodil to amygdala slices also impairs LTP at thalamic inputs to LA neurons, but has no effect on routine synaptic transmission (Bauer et al., 2002; ➋ *Figure 18-3c*). These results are also consistent with those of a recent study that examined the effects of APV on acquisition of fear potentiated startle (Walker and Davis, 2000), showing that APV can, under certain circumstances, have selective effects on plasticity. Collectively, findings suggest that the NMDA receptor in the amygdala plays an essential role in both the acquisition and STM of conditioned fear.

3.2 Metabotropic Glutamate Receptors

In contrast to the ionophoric NMDAR, the metabotropic glutamate receptor (mGluR) is a G-protein linked receptor that, when bound to glutamate, can modulate synaptic transmission via linkage to a variety of intracellular cascades (Nakanishi, 1992; Pin and Duvoisin, 1995; Anwyl, 1999). The Group I mGluRs, which consist of the subtypes mGluR1 and mGluR5, appear to be particularly important for synaptic plasticity (Huber et al., 1998; Balschun et al., 1999; Kleppisch et al., 1999). Transgenic mice lacking mGluR5 display a complete loss of the NMDAR-mediated component of LTP in the CA1 region of the hippocampus (Lu et al., 1997; Jia et al., 1998, 2001) and show impairments in NMDAR-dependent hippocampal memory tasks, including spatial learning in the Morris water maze (Lu et al., 1997).

Several recent studies have examined the role of mGluRs in fear conditioning. Transgenic mice lacking mGluR5 are impaired in fear conditioning tasks (Lu et al., 1997), as rats are injected systemically with the selective mGluR5 antagonist 2-methyl-6-(phenylethynyl)-pyridine (MPEP) before fear conditioning (Fendt and Schmid, 2002). In a recent study, Rodrigues et al., showed that mGluR5 was localized in LA spines postsynaptic to auditory thalamic inputs and required for synaptic plasticity at thalamic inputs to LA neurons (Rodrigues et al., 2002). Further, in behavioral experiments, intraamygdala infusion of MPEP before fear conditioning impaired formation of both STM and LTM of fear conditioning (Rodrigues et al., 2002; ➋ *Figure 18-3d* and *e*), and also impaired LTP at thalamic inputs to the LA (➋ *Figure 18-3f*). Similar to the results with the NR2B antagonist ifenprodil, infusion of MPEP prior to training blocked both STM and LTM, while infusion immediately before testing at either time point had no effect.

These findings suggest that mGluRs, and in particular mGluR5, are required for fear conditioning and STM formation in the amygdala. Future experiments, however, will be required to understand the exact mechanisms by which mGluRs contribute to fear conditioning. One attractive hypothesis is that activation of mGluR5 in the amygdala recruits the PKC signaling pathway. Both mGluR1 and mGluR5, for example, are positively coupled to phospholipase C, activation of which leads to the hydrolysis of phosphatidylinositol 4,5-biphosphate into inositol 1,4,5-trisphosphate (IP_3) and diacylglycerol (DAG), two effector proteins that are directly upstream from PKC. As discussed earlier, in its active form PKC can modulate the

Figure 18-3
Neurochemical mechanisms of fear acquisition and STM Formation in the LA. Both STM (a) and LTM (b) of auditory fear conditioning is dose-dependently impaired by intra-LA infusions of ifenprodil, a selective NR2B antagonist (A–B adapted, with permission, from Rodrigues et al., 2001, Copyright by the Society for Neuroscience). (c). LTP at thalamic inputs to the LA is also impaired by ifenprodil (adapted, with permission, from Bauer et al., 2002, Copyright by the Society for Neuroscience). (d–e) Both STM and LTM of auditory fear conditioning are dose-dependently impaired by intra-LA infusions of MPEP, a selective mGluR5 antagonist. (f) LTP at thalamic inputs to the LA is also impaired by MPEP (d–f adapted, with permission, from Rodrigues et al., 2002, *Copyright Society for Neuroscience*). (g) Fear conditioning results in an increase in autophosphorylated αCaMKII in LA spines. Here, rats were conditioned and activated αCaMKII was detected in LA spines using EM and an antibody against autophosphorylated αCaMKII at Thr^{286}. The image on the left shows a labeled spine (LSp) in the LA that contains numerous αCaMKII-immunogold labeled particles. The graph on the right shows that paired, but not unpaired, training leads to significant elevations in αCaMKII-labeled particles in LA spines ($p<0.05$). (h) Both STM and LTM of auditory fear conditioning are impaired after intra-LA infusion of KN-62, a CaMKII antagonist. (i) LTP at thalamic inputs to the LA is also impaired by KN-62 (G–I adapted, with permission, from Rodrigues et al., 2004, Copyright by the Society for Neuroscience)

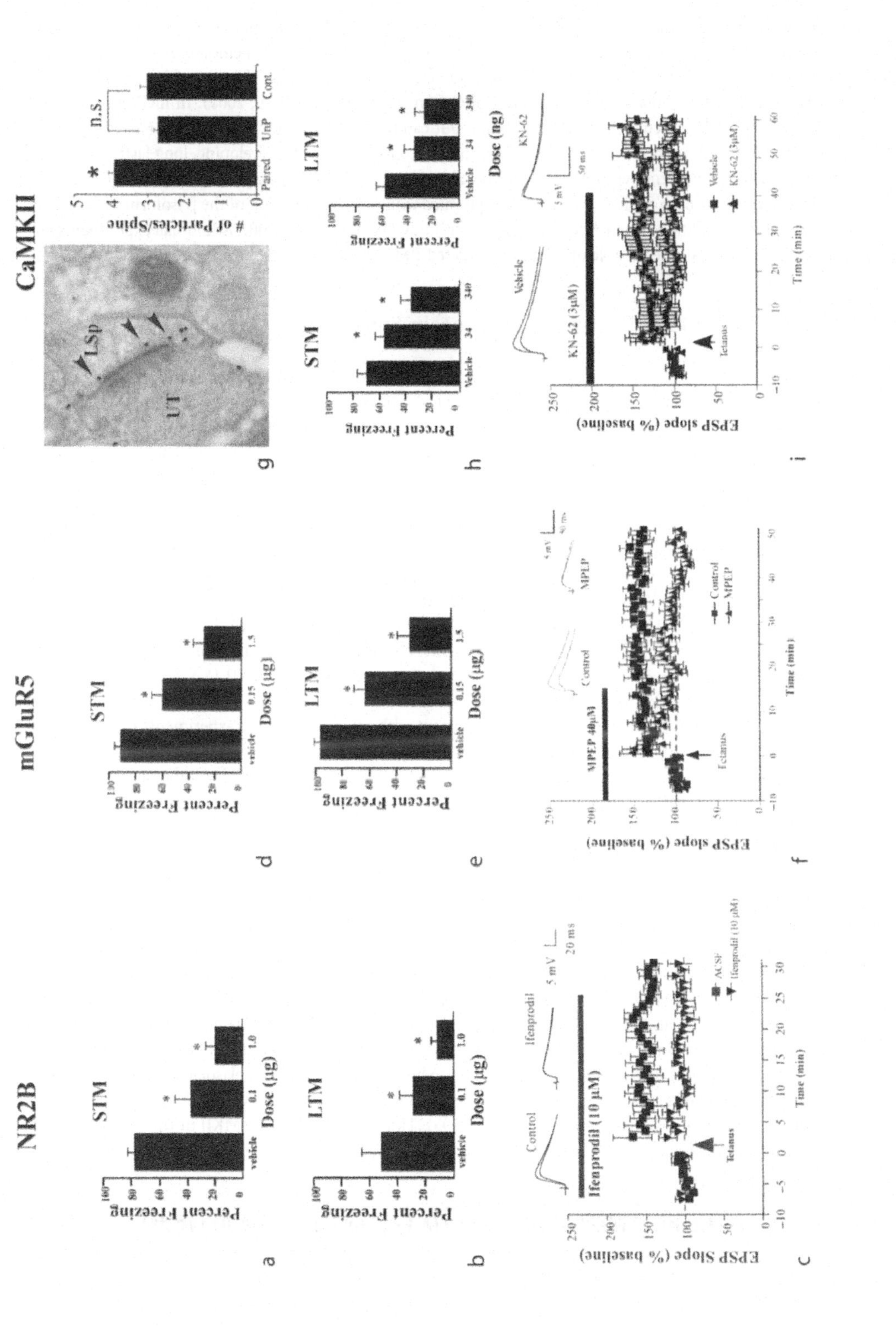
NR2B
mGluR5
CaMKII
STM
LTM
Percent Freezing
Dose (µg)
Dose (ng)
Ifenprodil (10 µM)
MPEP 40µM
KN-62 (3µM)
Tetanus
EPSP Slope (% baseline)
Time (min)
# of Particles/Spine
n.s.
Paired
UnP
Cont.
a
b
c
d
e
f
g
h
i

conductances of a number of synaptic proteins, including NMDARs and AMPA receptors, leading to short-term alterations in synaptic transmission, which could be important for memory formation. PKC, for example, is known to target two phosphorylation sites on the C-terminal domain of NR2B (Liao et al., 2001), which can modulate NMDAR conductance (Anwyl, 1999; Liao et al., 2001). In mGluR5 knockout mice, LTP of NMDAR currents in CA1 is absent, but can be rescued by activators of PKC (Jia et al., 1998). Further, an mGluR5 agonist (CHPG) has been reported to induce a slowly developing, long-lasting potentiation of NMDAR currents via PKC (Doherty et al., 1997). The role of PKC in fear acquisition and STM formation has not been explicitly tested, although mice with a specific deletion of the β isoform of PKC have impaired fear conditioning when tested 24 h after training (Weeber et al., 2000). Additional experiments will be necessary to examine the role of mGluR5-mediated PKC activation in the LA in fear conditioning.

3.3 Ca^{2+}/Calmodulin-Dependent Protein Kinase

CaMKII has been widely implicated in synaptic plasticity and memory formation (Fukunaga and Miyamoto, 1999; Fink and Meyer, 2002; Lisman et al., 2002). CaMKII consists of a family of isoforms, with the α and β isoforms being the most prominent (Bennett et al., 1983; Miller and Kennedy, 1985). αCaMKII, in particular, has been implicated as a major contributor to synaptic plasticity and memory formation. Transgenic mice with a deletion of the αCaMKII gene display deficits in hippocampal LTP and hippocampal-dependent spatial memory (Silva et al., 1992a, 1992b). Similarly, pharmacological inhibition of CaMKII blocks the induction of LTP in the hippocampal area CA1 (Ito et al., 1991; Stanton and Gage, 1996), and impairs hippocampal-dependent learning and memory (Tan and Liang, 1996).

αCaMKII has also been implicated in fear conditioning. αCaMKII is robustly expressed in LA pyramidal neurons (McDonald), where it coexists with NR2B in LA spines postsynaptic to terminals that originate in the auditory thalamus (Rodrigues et al., 2004). Fear conditioning has recently been shown to lead to increases in the autophosphorylated form of αCaMKII at Thr^{286} in spines of LA neurons (➋ *Figure 18-3g*). Further, intraamygdala infusion or bath application of an inhibitor of CaMKII (KN-62) impairs acquisition and STM formation of fear conditioning and LTP at thalamic inputs to LA neurons, respectively (➋ *Figure 18-3h* and *i*). This latter finding is consistent with molecular genetic experiments indicating that induced overexpression of active αCaMKII by a transgene that replaces Thr^{286} with an aspartate residue in the amygdala and striatum results in a reversible deficit in fear conditioning (Mayford et al., 1996).

3.4 AMPA Receptor Regulation

One way that αCaMKII might promote STM formation is by targeting and altering the conductance of glutamate receptors at the PSD (Shen and Meyer, 1999; Shen et al., 2000). Autophosphorylated αCaMKII, for example, is known to target and phosphorylate the GluR1 subunit of the AMPA receptor at Ser^{831} (Barria et al., 1997; Lee et al., 2000). Phosphorylation of AMPARs at Ser^{831} has been shown to enhance their conductance (Barria et al., 1997; Benke et al., 1998), and promote their mobilization and insertion into synapses (Hayashi et al., 2000; Passafaro et al., 2001). Upregulation of AMPAR conductance and/or number is in fact one mechanism believed to underlie the temporary persistence of memory (Malinow, 2003). Phosphorylation of GluR1 has recently been shown to be required for synaptic plasticity in the hippocampus and the retention of spatial memory (Lee et al., 2003). Further, fear conditioning leads to the trafficking of AMPA receptors to LA synapses (Rumpel et al., 2005). Thus, activation of αCaMKII during fear acquisition may regulate AMPARs at LA synapses and thereby contribute to the formation and maintenance of STM.

4 Neurochemical Mechanisms of LTM Formation in the Amygdala

As its name implies, LTM is a long-lasting phenomenon that can last many hours, days, weeks, or even years (Milner et al., 1998). Accordingly, LTM is typically tested at longer intervals after training, usually starting at

24 h. Further, like L-LTP, LTM requires the activation of PKA and MAPK and is critically dependent on de novo RNA and protein synthesis (Davis and Squire, 1984; Milner et al., 1998). In this section, we discuss in detail how each of these processes has been linked to LTM formation in the LA. We begin with a discussion of L-type VGCCs, as recent work has suggested that these channels play an essential role in promoting LTM formation in the LA.

4.1 L-type VGCCs

Recent experiments have shown that LTP at thalamic input synapses to the LA is, under certain conditions, L-VGCC dependent, and NMDAR independent (Weisskopf et al., 1999). These experiments used a pairing protocol in which subthreshold presynaptic stimulation of auditory afferents were paired with brief postsynaptic depolarizations (Johnston et al., 1999; Magee and Johnston, 1997; Markram et al., 1997). In this protocol, back-propagating action potentials (BPAPs) originating in the soma are thought to invade the dendrites and interact with EPSPs leading to Ca^{2+} influx through VGCCs (Magee and Johnston, 1997; Johnston et al., 1999; Stuart and Hausser, 2001). Accordingly, LTP induced by pairing in the thalamic pathway is blocked by application of the L-VGCC blockers *nifedipine* or *verapamil* (Weisskopf et al., 1999; Bauer et al., 2002).

Until recently, the contribution of L-VGCCs to fear conditioning had not been established. Bauer et al., however, examined the effect of intraamygdala infusion of the L-VGCC blocker verapamil on the acquisition and consolidation of auditory fear conditioning (Bauer et al., 2002). The findings revealed that blockade of L-VGCCs before conditioning selectively impaired LTM formation of fear conditioning at 24 h after training; acquisition and STM, assessed at 1 h, was left intact. These findings, together with those of studies that examined the role of NMDAR function in fear conditioning discussed earlier, suggest that there are two sources of Ca^{2+} in the LA that are critical for fear memory formation. One, mediated by NMDARs, appears to be selectively involved in fear acquisition and STM formation of fear conditioning (Walker and Davis, 2000; Rodrigues et al., 2001). The second, mediated by L-VGCCs, is selectively involved in LTM formation. While the effects of L-VGCC blockade are not apparent in fear conditioning for many hours after training, it is important to note that this is likely due to interference with a process that is set in motion at the time of CS–US pairing and fear acquisition. Consistent with that notion, recent reports have demonstrated that L-VGCCs play a selective role in signaling to the nucleus and initiating CRE-mediated transcription, which is known to be required for long-term synaptic plasticity and memory formation (Dolmetsch et al., 2001). Additional experiments will be necessary to determine the contribution of L-VGCCs to the activation of protein kinases and CRE-driven gene expression in the LA following fear conditioning.

4.2 Protein Kinase A and Mitogen-Activated Protein Kinase

The rise in intracellular Ca^{2+} during fear acquisition (CS–US pairing) leads, either directly or indirectly, to the activation of protein kinase second messenger pathways that are necessary for long-term synaptic plasticity and memory. The role of two of these in the LA has been studied fairly extensively in relation to fear conditioning: PKA and MAPK. There has been a great deal of recent interest in both PKA and ERK/MAPK, in part because they have been shown to be essential for the late phase of multiple forms of synaptic plasticity and memory. This has included fear conditioning, where mice that overexpress an inhibitory form of PKA, R(AB), exhibit impaired L-LTP in hippocampal area CA1 and selective deficits in LTM, but not STM, of contextual fear conditioning (Abel et al., 1997). Similarly, mice that lack *Ras*, an upstream regulator of ERK/MAP kinase, have impaired memory consolidation of auditory and contextual fear conditioning, as well as impaired amygdala LTP (Brambilla et al., 1997).

Recent pharmacological experiments have examined the role of PKA and ERK/MAPK in amygdala LTP and in fear conditioning. Huang et al., showed that bath application of inhibitors of PKA or ERK/MAPK to amygdala slices impairs L-LTP at thalamic and cortical inputs to the LA, but has no effect on E-LTP (Huang et al., 2000; ➋ *Figure 18-4a–c*). Consistent with those findings, infusion of a PKA inhibitor or of a peptide that blocks the association of PKA with the A-kinase anchoring protein (AKAP) in the LA impairs LTM, but not STM of fear conditioning (Schafe and LeDoux, 2000; Moita et al., 2002; ➋ *Figure 18-4h*). Further, fear conditioning results in a transient activation of ERK/MAPK in the LA (➋ *Figure 18-4d–f*), and infusion of an ERK/MAPK inhibitor into the LA before fear conditioning impairs memory consolidation; that is, rats have intact STM and impaired LTM (Schafe et al., 2000; ➋ *Figure 18-4g*). Collectively, these findings support the hypothesis that both PKA and ERK/MAPK contribute to fear memory formation by engaging cellular processes, possibly those in the nucleus, that are necessary for long-term synaptic plasticity and memory formation.

4.3 Transcriptional Regulation and Macromolecular Synthesis

Both L-LTP in the LA (Huang and Kandel, 1998; Huang et al., 2000) and LTM of fear conditioning (Bailey et al., 1999; Schafe and LeDoux, 2000) are known to require new RNA and protein synthesis in the LA (➋ *Figure 18-4i*). The requirement for de novo RNA synthesis is particularly important, because it suggests that a nuclear event is required for both L-LTP and the transition between short- and long-term memory formation. Several transcription factors have been implicated in long-term synaptic plasticity and in memory formation, but CREB is perhaps the best studied. CREB is a family of transcription factors consisting of several functionally distinct isoforms. Some, known as activator isoforms, bind to DNA at CRE promoter regions and promote transcription. Others, known as repressor isoforms, compete with the binding of activator isoforms to DNA (Bartsch et al., 1995; Abel et al., 1998; Silva et al., 1998). CREB is an attractive candidate molecule for memory consolidation because it has direct interaction with the transcriptional machinery and also contains phosphorylation sites for the major protein kinase signaling pathways that are known to be involved in memory formation, including PKA, ERK/MAPK, and CaMKII (Silva et al., 1998).

The first evidence to suggest that CREB might be involved in memory consolidation came from a study employing a Pavlovian conditioning task in *Drosophila*. Overexpression of a dominant negative (repressor) isoform of CREB in flies impaired LTM formation in a conditioned odor aversion task (Yin et al., 1994).

◘ Figure 18-4

Neurochemical mechanisms of LTM formation in the LA. (a–c) Long-term potentiation at thalamic inputs to the LA is impaired by inhibitors of MAPK, PKA, and protein synthesis, respectively (adapted, with permission, from Huang et al., 2000, Copyright by the Society for Neuroscience). (d) Fear conditioning leads to an increase in phosphorylated ERK1 and ERK2 at t = 60 min after training. In these experiments, rats were trained, sacrificed at different time points after conditioning, and LA homogenates were probed with antibodies that recognize phosphorylated ERK/MAPK. ERK1 (black bars) and ERK2 (gray bars) are the two isoforms of ERK/MAPK recognized by the anti-phospho-ERK antibody ($p < 0.05$). (e) The increase in activated ERK/MAPK is not accounted for by a change in the amount of total (unphosphorylated) ERK/MAPK. (f) Immuocytochemical localization of phosphorylated ERK/MAPK in the LA after fear conditioning. The image shows ERK-labeled cells in three different regions of the LA (dorsal, LAd; ventromedial, LAvm; and ventrolateral, LAvl), with most of the label concentrated in the ventral portions of the LAd, and throughout the LAvm and LAvl. Activated ERK/MAPK is also highly expressed in the nearby central nucleus (CE) and the amygdala-striatal transition zone (AST). d–f adapted from Schafe et al. (2000), copyright by the Society for Neuroscience. (g–i) LTM, but not STM, in the LA requires MAPK, PKA, and protein synthesis. In these studies, rats received intraamygdala infusions of U0126 (an MEK inhibitor, which is an upstream regulator of ERK/MAPK activation; G), Rp-cAMPS (a PKA inhibitor; H), or anisomycin (a protein synthesis inhibitor; I) at or around the time of training and were assayed for both short-term memory (1–4 h later) and long-term memory (24 h later) of auditory fear conditioning. In each figure, vehicle-treated rats are represented by the white bars, while drug-treated animals are represented by the black bars. *$p<0.05$ relative to vehicle controls. g adapted from Schafe et al. (2000a), copyright by the Society for Neuroscience. H–I adapted from Schafe et al. (2000b), copyright by the Society for Neuroscience

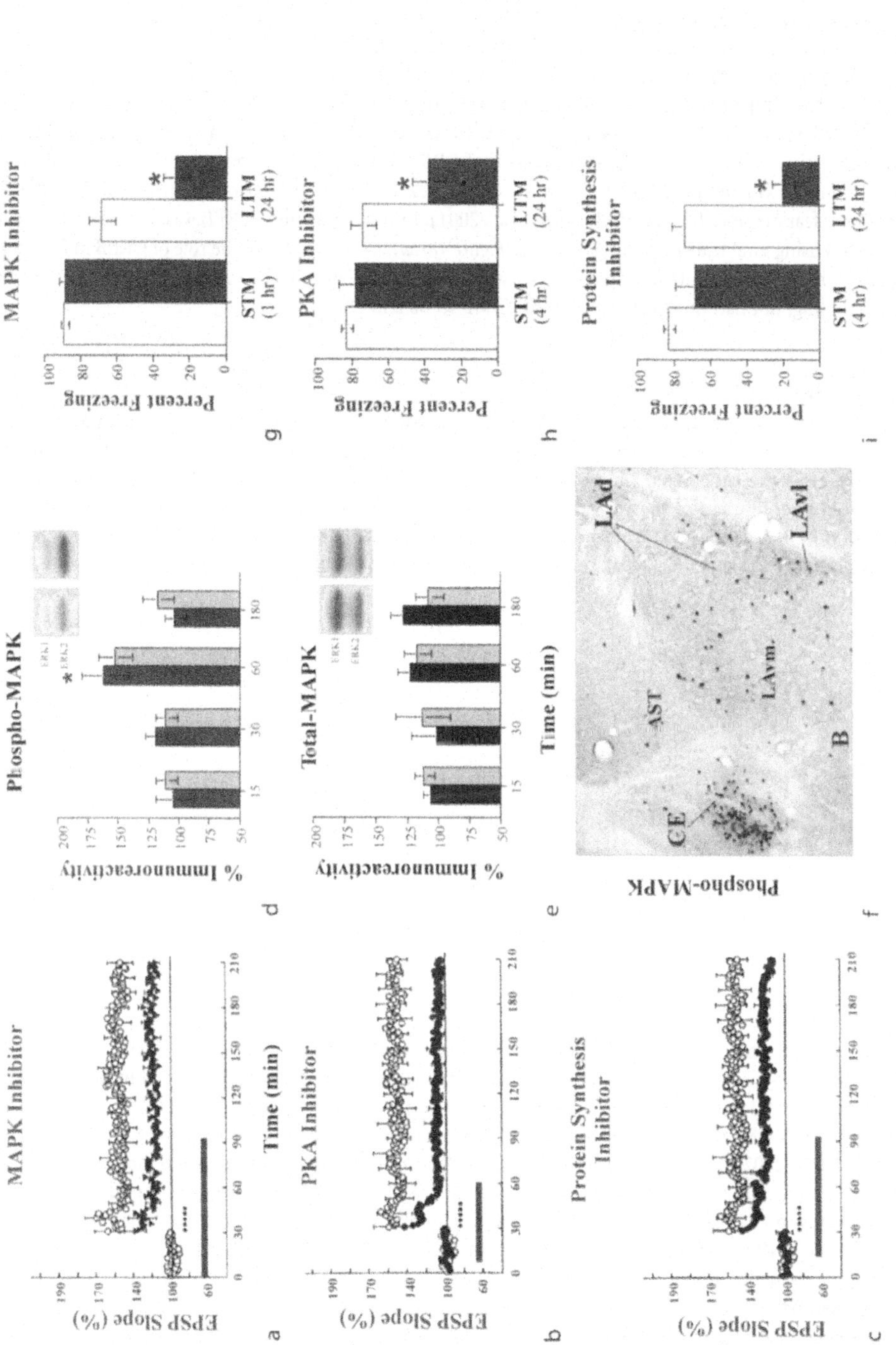
MAPK Inhibitor
PKA Inhibitor
Protein Synthesis Inhibitor
EPSP Slope (%)
Time (min)
a
b
c
Phospho-MAPK
Total-MAPK
% Immunoreactivity
ERK1
ERK2
d
e
LAd
AST
LAvm
LAvl
CE
B
f
Percent Freezing
STM (1 hr)
LTM (24 hr)
STM (4 hr)
g
h
i

Conversely, overexpression of an activator isoform of CREB facilitated LTM; that is, behavioral training that would normally produce only STM was effective at producing LTM (Yin et al., 1995).

CREB has also been implicated in fear conditioning. Mice lacking two critical isoforms of CREB, the α and δ, have impaired hippocampal L-LTP and memory consolidation for auditory and contextual fear conditioning; that is, LTM is impaired, while STM is intact (Bourtchuladze et al., 1994). Further, induced overexpression of a dominant negative isoform of CREB in the forebrain impairs LTM formation of fear conditioning (Kida et al., 2002). Conversely, overexpression of the transcription factor CREB in the LA facilitates fear memory formation (Josselyn et al., 2001). In this latter study, CREB was overexpressed locally in the LA using viral transfection methods (➤ *Figure 18-5a*). Consistent with the role of CREB in long-term synaptic plasticity and memory formation, overexpression of CREB in the LA facilitated the LTM of fear conditioning, but had no effect on STM (➤ *Figure 18-5b* and *c*).

Figure 18-5

CREB and fear memory formation. Overexpression of CREB in the amygdala via viral transfection facilitates consolidation of fear conditioning. (a) Example of viral transfection in the LA using a control virus that overexpresses the LacZ gene, which encodes the protein β-galactosidase. (b) LTM of fear conditioning is enhanced after infusion of the virus that encodes a wild-type CREB (HSV-CREB), but not by infusion of the control virus (HSV-LacZ). (c) STM of fear conditioning is not affected by infusion of HSV-CREB into the amygdala. Adapted from Josselyn et al. (2001), copyright by the Society for Neuroscience

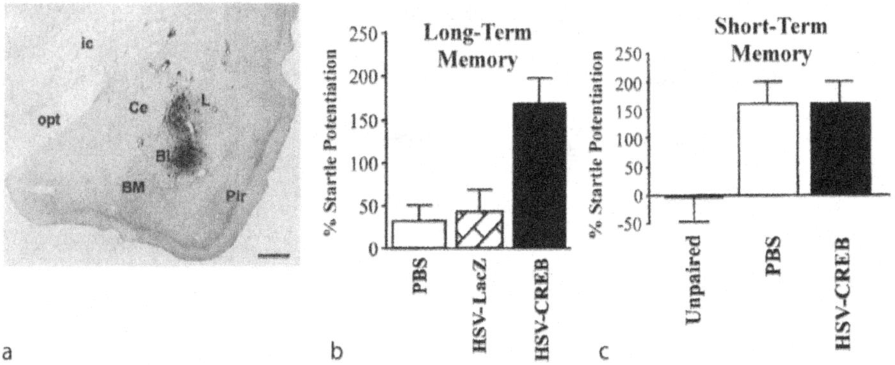

While CRE-mediated transcription clearly supports the development of long-term plasticity and memory, the downstream targets of CREB have remained largely unknown. However, a number of studies have shown that fear conditioning induces the expression of both immediate-early (Beck and Fibiger, 1995; Rosen et al., 1998; Malkani and Rosen, 2000; Scicli et al., 2004) and downstream genes (Stork et al., 2001; Ressler et al., 2002; Rattiner et al., 2004;) in the LA. While the specific contributions of many of these genes to fear conditioning is still unclear, it is widely believed that learning-induced gene expression ultimately contributes to changes in cell (especially synaptic) structure that stabilizes memory (Bailey and Kandel, 1993; Woolf, 1998; Rampon et al., 2000; Sweatt, 2004), presumably by altering the actin cytoskeleton underlying synaptic organization (van Rossum and Hanisch, 1999; Matus, 2000; Kasai et al., 2003). Such changes in synaptic structure have been well documented in invertebrates, where stimulation that promotes long-term synaptic plasticity has been shown to lead to an increase in new synaptic contacts (Bailey et al., 1992, 1994; Bailey and Kandel, 1993). Further, both learning and LTP result in a number of structural changes in the hippocampus and cortex, including an increase in spine head volume and widening and shortening of the spine neck (Van Harreveld and Fifkova, 1975; Fifkova and Van Harreveld, 1977; Fifkova and Anderson, 1981), spine perforation (Toni et al., 1999), and an increase in the total number of spines (Engert and Bonhoeffer, 1999; Leuner et al., 2003).

While there is little evidence that changes in synaptic structure occur in the LA following fear conditioning, recent studies have shown that fear conditioning leads to the transcription of genes involved in cytoskeletal remodeling, including the CRE-mediated gene NF-1 (Ressler et al., 2002). Further, interference with molecular pathways known to be involved in structural plasticity during early development, such as the Rho-GAP signaling pathway, disrupts memory formation (Lamprecht et al., 2002). Additional studies will be required to examine whether fear conditioning results in structural changes at LA synapses and what mechanisms might underlie these dynamic changes.

5 A Model of Fear Memory Acquisition and Consolidation

In summary, the converging evidence from a number of recent studies supports a model of fear conditioning in which CS and US inputs converge onto individual LA neurons and initiate changes in synaptic function and/or structure (Blair et al., 2001; ➲ *Figure 18-6*). The convergence of CS and US inputs onto LA

Figure 18-6

A model of fear memory consolidation in the amygdala. See text for details. (a) Acquisition and STM formation of fear conditioning requires events at the postsynaptic density, including activation of NMDARs, mGluR5, αCaMKII, and, possibly, PKC. Both αCaMKII and PKC may contribute to STM by influencing the conductance of NMDARs and AMPARs. (b) LTM formation of fear conditioning requires the activation of L-type VGCCs, PKA, and ERK/MAPK. Both PKA and MAPK are thought to translocate to the nucleus to influence gene expression. (c) CREB and CRE-mediated transcription are both required for LTM of fear conditioning. (d) The translation of CRE-mediated genes into proteins may lead to structural changes at LA spines that contribute to the permanence of LTM formation

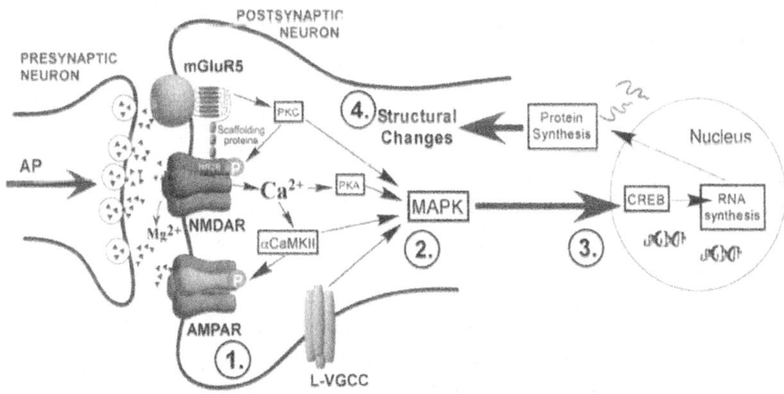

principal cells during training leads to Ca^{2+} influx through both NMDARs (Miserendino et al., 1990; Kim et al., 1991; Campeau et al., 1992; Walker and Davis, 2000; Rodrigues et al., 2001) and also L-VGCCs (Bauer et al., 2002). Both NMDARs and L-VGCCs are likely opened by the strong depolarization induced by the footshock US by BPAPs that invade the dendrites of LA neurons (Magee and Johnston, 1997; Blair et al., 2001). The NMDAR-mediated increase in intracellular Ca^{2+}, together with mGluR5 (Rodrigues et al., 2002), leads to the activation of a variety of local protein kinases at the PSD, including αCaMKII (Rodrigues et al., 2004) and likely PKC, that promote STM formation by targeting and modulating the conductance of NMDARs and AMPARs at LA synapses (Barria et al., 1997; Benke et al., 1998). The combined entry of Ca^{2+} through both NMDARs and L-VGCCs, however, leads to the activation of PKA and ERK/MAPK (Schafe et al., 2000; Schafe and LeDoux, 2000), which appear to be exclusively involved in the formation of LTM, possibly via translocation to the cell nucleus and activation of transcription factors such as CREB (Josselyn et al., 2001). The activation of CREB by PKA and ERK/MAPK promotes CRE-mediated gene transcription

(Bailey et al., 1999; Ressler et al., 2002) and the synthesis of new proteins (Schafe and LeDoux, 2000), which likely promotes LTM formation by leading to alterations in the structure of LA synapses (Lamprecht et al., 2002; Ressler et al., 2002).

6 Conclusions

In summary, in this review we have emphasized some key synaptic events and downstream cellular cascades that are responsible for the acquisition and consolidation of fear conditioning in the LA. These findings provide a foundation for the continued study of the neural basis of emotional learning and memory at the cellular level, and also for bridging the gap between studies of memory formation and synaptic plasticity in the mammalian brain.

References

Abel T, Martin KC, Bartsch D, Kandel ER. 1998. Memory suppressor genes: Inhibitory constraints on the storage of long-term memory. Science 279: 338-341.

Abel T, Nguyen PV, Barad M, Deuel TA, Kandel ER, et al. 1997. Genetic demonstration of a role for PKA in the late phase of LTP and in hippocampus-based long-term memory. Cell 88: 615-626.

Anwyl R. 1999. Metabotropic glutamate receptors: Electrophysiological properties and role in plasticity. Brain Res Brain Res Rev 29: 83-120.

Bailey CH, Alberini C, Ghirardi M, Kandel ER. 1994. Molecular and structural changes underlying long-term memory storage in Aplysia. Adv Second Messenger Phosphoprotein Res 29: 529-544.

Bailey CH, Kandel ER. 1993. Structural changes accompanying memory storage. Annu Rev Physiol 55: 397-426.

Bailey CH, Montarolo P, Chen M, Kandel ER, Schacher S. 1992. Inhibitors of protein and RNA synthesis block structural changes that accompany long-term heterosynaptic plasticity in Aplysia. Neuron 9: 749-758.

Bailey DJ, Kim JJ, Sun W, Thompson RF, Helmstetter FJ. 1999. Acquisition of fear conditioning in rats requires the synthesis of mRNA in the amygdala. Behav Neurosci 113: 276-282.

Balschun D, Manahan-Vaughan D, Wagner T, Behnisch T, Reymann KG, et al. 1999. A specific role for group I mGluRs in hippocampal LTP and hippocampus-dependent spatial learning. Learn Mem 6: 138-152.

Barria A, Muller D, Derkach V, Griffith LC, Soderling TR. 1997. Regulatory phosphorylation of AMPA-type glutamate receptors by CaM-KII during long-term potentiation. Science 276: 2042-2045.

Bartsch D, Ghirardi M, Skehel PA, Karl KA, Herder SP, et al. 1995. Aplysia CREB2 represses long-term facilitation: Relief of repression converts transient facilitation into long-term functional and structural change. Cell 83: 979-992.

Bauer EP, Schafe GE, LeDoux JE. 2002. NMDA receptors and L-type voltage-gated calcium channels contribute to long-term potentiation and different components of fear memory formation in the lateral amygdala. J Neurosci 22: 5239-5249.

Beck CH, Fibiger HC. 1995. Conditioned fear-induced changes in behavior and in the expression of the immediate early gene c-fos: With and without diazepam pretreatment. J Neurosci 15: 709-720.

Benke TA, Luthi A, Isaac JT, Collingridge GL. 1998. Modulation of AMPA receptor unitary conductance by synaptic activity. Nature 393: 793-797.

Bennett MK, Erondu NE, Kennedy MB. 1983. Purification and characterization of a calmodulin-dependent protein kinase that is highly concentrated in brain. J Biol Chem 258: 12735-12744.

Blair HT, Schafe GE, Bauer EP, Rodrigues SM, LeDoux JE. 2001. Synaptic plasticity in the lateral amygdala: A cellular hypothesis of fear conditioning. Learning & Memory 8(5): 229-242.

Blair HT, Tinkelman A, Moita MA, LeDoux JE. 2003. Associative plasticity in neurons of the lateral amygdala during auditory fear conditioning. Ann N Y Acad Sci 985: 485-487.

Blanchard RJ, Blanchard DC. 1969. Crouching as an index of fear. J Comp Physiol Psychol 67: 370-375.

Bliss TV, Lømo T. 1973. Long-lasting potentiation of synaptic transmission in the dentate area of the anaesthetized rabbit following stimulation of the perforant path. J Physiol (Lond) 232: 331-356.

Bordi F, LeDoux J. 1992. Sensory tuning beyond the sensory system: An initial analysis of auditory response properties of neurons in the lateral amygdaloid nucleus and overlying areas of the striatum. J Neurosci 12: 2493-2503.

Bordi F, LeDoux JE. 1994. Response properties of single units in areas of rat auditory thalamus that project to the amygdala. II.

Cells receiving convergent auditory and somatosensory inputs and cells antidromically activated by amygdala stimulation. Exp Brain Res 98: 275-286.

Bourtchuladze R, Frenguelli B, Blendy J, Cioffi D, Schutz G, et al. 1994. Deficient long-term memory in mice with a targeted mutation of the cAMP-responsive element-binding protein. Cell 79: 59-68.

Brambilla R, Gnesutta N, Minichiello L, White G, Roylance AJ, et al. 1997. A role for the Ras signalling pathway in synaptic transmission and long-term memory. Nature 390: 281-286.

Campeau S, Davis M. 1995. Involvement of the central nucleus and basolateral complex of the amygdala in fear conditioning measured with fear-potentiated startle in rats trained concurrently with auditory and visual conditioned stimuli. J Neurosci 15: 2301-2311.

Campeau S, Miserendino MJ, Davis M. 1992. Intra-amygdala infusion of the N-methyl-D-aspartate receptor antagonist AP5 blocks acquisition but not expression of fear-potentiated startle to an auditory conditioned stimulus. Behav Neurosci 106: 569-574.

Cavus I, Teyler T. 1996. Two forms of long-term potentiation in area CA1 activate different signal transduction cascades. J Neurophysiol 76: 3038-3047.

Chapman PF, Kairiss EW, Keenan CL, Brown TH. 1990. Long-term synaptic potentiation in the amygdala. Synapse 6: 271-278.

Clugnet MC, LeDoux JE. 1990. Synaptic plasticity in fear conditioning circuits: Induction of LTP in the lateral nucleus of the amygdala by stimulation of the medial geniculate body. J Neurosci 10: 2818-2824.

Davis HP, Squire LR. 1984. Protein synthesis and memory: A review. Psychol Bull 96: 518-559.

Davis M. 1992. The role of the amygdala in fear-potentiated startle: Implications for animal models of anxiety. Trends Pharmacol Sci 13: 35-41.

Davis M. 1997. Neurobiology of fear responses: The role of the amygdala. J Neuropsychiatry Clin Neurosci 9: 382-402.

Doherty AJ, Palmer MJ, Henley JM, Collingridge GL, Jane DE. 1997. (RS)-2-chloro-5-hydroxyphenylglycine (CHPG) activates mGlu5, but no mGlu1, receptors expressed in CHO cells and potentiates NMDA responses in the hippocampus. Neuropharmacology 36: 265-267.

Dolmetsch RE, Pajvani U, Fife K, Spotts JM, Greenberg ME. 2001. Signaling to the nucleus by an L-type calcium channel-calmodulin complex through the MAP kinase pathway. Science 294: 333-339.

Doron NN, LeDoux JE. 1999. Organization of projections to the lateral amygdala from auditory and visual areas of the thalamus in the rat [published erratum appears in J Comp Neurol 2000 Feb 14;417(3):385–6]. J Comp Neurol 412: 383-409.

Engert F, Bonhoeffer T. 1999. Dendritic spine changes associated with hippocampal long-term synaptic plasticity. Nature 399: 66-70.

English JD, Sweatt JD. 1997. A requirement for the mitogen-activated protein kinase cascade in hippocampal long term potentiation. J Biol Chem 272: 19103-19106.

Fanselow MS, LeDoux JE. 1999. Why we think plasticity underlying Pavlovian fear conditioning occurs in the basolateral amygdala. Neuron 23: 229-232.

Fendt M, Schmid S. 2002. Metabotropic glutamate receptors are involved in amygdaloid plasticity. Eur J Neurosci 15: 1535-1541.

Fifkova E, Anderson CL. 1981. Stimulation-induced changes in dimensions of stalks of dendritic spines in the dentate molecular layer. Exp Neurol 74: 621-627.

Fifkova E, Van Harreveld A. 1977. Long-lasting morphological changes in dendritic spines of dentate granular cells following stimulation of the entorhinal area. J Neurocytol 6: 211-230.

Fink CC, Meyer T. 2002. Molecular mechanisms of CaMKII activation in neuronal plasticity. Curr Opin Neurobiol 12: 293-299.

Frank DA, Greenberg ME. 1994. CREB: A mediator of long-term memory from mollusks to mammals. Cell 79: 5-8.

Fukunaga K, Miyamoto E. 1999. Current studies on a working model of CaM kinase II in hippocampal long-term potentiation and memory. Jpn J Pharmacol 79: 7-15.

Grover LM, Teyler TJ. 1990. Two components of long-term potentiation induced by different patterns of afferent activation. Nature 347: 477-479.

Hayashi Y, Shi SH, Esteban JA, Piccini A, Poncer JC, et al. 2000. Driving AMPA receptors into synapses by LTP and CaMKII: Requirement for GluR1 and PDZ domain interaction. Science 287: 2262-2267.

Helmstetter FJ, Bellgowan PS. 1994. Effects of muscimol applied to the basolateral amygdala on acquisition and expression of contextual fear conditioning in rats. Behav Neurosci 108: 1005-1009.

Holt W, Maren S. 1999. Muscimol inactivation of the dorsal hippocampus impairs contextual retrieval of fear memory. J Neurosci 19: 9054-9062.

Huang YY, Kandel ER. 1998. Postsynaptic induction and PKA-dependent expression of LTP in the lateral amygdala. Neuron 21: 169-178.

Huang YY, Martin KC, Kandel ER. 2000. Both protein kinase A and mitogen-activated protein kinase are required in the amygdala for the macromolecular synthesis-dependent late phase of long-term potentiation. J Neurosci 20: 6317-6325.

Huber KM, Sawtell NB, Bear MF. 1998. Effects of the metabotropic glutamate receptor antagonist MCPG on phosphoinositide turnover and synaptic plasticity in visual cortex. J Neurosci 18: 1-9.

Impey S, Mark M, Villacres EC, Poser S, Chavkin C, et al. 1996. Induction of CRE-mediated gene expression by stimuli that generate long-lasting LTP in area CA1 of the hippocampus. Neuron 16: 973-982.

Impey S, Obrietan K, Wong ST, Poser S, Yano S, et al. 1998. Cross talk between ERK and PKA is required for Ca2+ stimulation of CREB-dependent transcription and ERK nuclear translocation. Neuron 21: 869-883.

Ito I, Hidaka H, Sugiyama H. 1991. Effects of KN-62, a specific inhibitor of calcium/calmodulin-dependent protein kinase II, on long-term potentiation in the rat hippocampus. Neurosci Lett 121: 119-121.

Jia Z, Lu Y, Henderson J, Taverna F, Romano C, et al. 1998. Selective abolition of the NMDA component of long-term potentiation in mice lacking mGluR5. Learn Mem 5: 331-343.

Jia Z, Lu YM, Agopyan N, Roder J. 2001. Gene targeting reveals a role for the glutamate receptors mGluR5 and GluR2 in learning and memory. Physiol Behav 73: 793-802.

Johnston D, Hoffman DA, Colbert CM, Magee JC. 1999. Regulation of back-propagating action potentials in hippocampal neurons. Curr Opin Neurobiol 9: 288-292.

Josselyn SA, Shi C, Carlezon WA Jr, Neve RL, Nestler EJ, et al. 2001. Long-term memory is facilitated by cAMP response element-binding protein overexpression in the Amygdala. J Neurosci 21: 2404-2412.

Kapp BS, Frysinger RC, Gallagher M, Haselton JR. 1979. Amygdala central nucleus lesions: Effect on heart rate conditioning in the rabbit. Physiol Behav 23: 1109-1117.

Kasai H, Matsuzaki M, Noguchi J, Yasumatsu N, Nakahara H. 2003. Structure-stability-function relationships of dendritic spines. Trends Neurosci 26: 360-368.

Kida S, Josselyn SA, de Ortiz SP, Kogan JH, Chevere I, et al. 2002. CREB required for the stability of new and reactivated fear memories. Nat Neurosci 5: 348-355.

Kim JJ, DeCola JP, Landeira-Fernandez J, Fanselow MS. 1991. N-methyl-D-aspartate receptor antagonist APV blocks acquisition but not expression of fear conditioning. Behav Neurosci 105: 126-133.

Kleppisch T, Pfeifer A, Klatt P, Ruth P, Montkowski A, et al. 1999. Long-term potentiation in the hippocampal CA1 region of mice lacking cGMP-dependent kinases is normal and susceptible to inhibition of nitric oxide synthase. J Neurosci 19: 48-55.

Lamprecht R, Farb CR, LeDoux JE. 2002. Fear memory formation involves p190 RhoGAP and ROCK proteins through a GRB2-mediated complex. Neuron 36: 727-738.

LeDoux JE. 2000. Emotion circuits in the brain. Annu Rev Neurosci 23: 155-184.

LeDoux JE, Cicchetti P, Xagoraris A, Romanski LM. 1990. The lateral amygdaloid nucleus: Sensory interface of the amygdala in fear conditioning. J Neurosci 10: 1062-1069.

LeDoux JE, Farb CR. 1991. Neurons of the acoustic thalamus that project to the amygdala contain glutamate. Neurosci Lett 134: 145-149.

LeDoux JE, Iwata J, Cicchetti P, Reis DJ. 1988. Different projections of the central amygdaloid nucleus mediate autonomic and behavioral correlates of conditioned fear. J Neurosci 8: 2517-2529.

LeDoux JE, Ruggiero DA, Reis DJ. 1985. Projections to the subcortical forebrain from anatomically defined regions of the medial geniculate body in the rat. J Comp Neurol 242: 182-213.

Lee HK, Barbarosie M, Kameyama K, Bear MF, Huganir RL. 2000. Regulation of distinct AMPA receptor phosphorylation sites during bidirectional synaptic plasticity. Nature 405: 955-959.

Lee HK, Takamiya K, Han JS, Man H, Kim CH, et al. 2003. Phosphorylation of the AMPA receptor GluR1 subunit is required for synaptic plasticity and retention of spatial memory. Cell 112: 631-643.

Leuner B, Falduto J, Shors TJ. 2003. Associative memory formation increases the observation of dendritic spines in the hippocampus. J Neurosci 23: 659-665.

Li XF, Stutzmann GE, LeDoux JE. 1996. Convergent but temporally separated inputs to lateral amygdala neurons from the auditory thalamus and auditory cortex use different postsynaptic receptors: In vivo intracellular and extracellular recordings in fear conditioning pathways. Learn Mem 3: 229-242.

Liao GY, Wagner DA, Hsu MH, Leonard JP. 2001. Evidence for direct protein kinase-C mediated modulation of N-methyl-D-aspartate receptor current. Mol Pharmacol 59: 960-964.

Lisman J, Schulman H, Cline H. 2002. The molecular basis of CaMKII function in synaptic and behavioural memory. Nat Rev Neurosci 3: 175-190.

Lu YM, Jia Z, Janus C, Henderson JT, Gerlai R, et al. 1997. Mice lacking metabotropic glutamate receptor 5 show impaired learning and reduced CA1 long-term potentiation (LTP) but normal CA3 LTP. J Neurosci 17: 5196-5205.

Magee JC, Johnston D. 1997. A synaptically controlled, associative signal for Hebbian plasticity in hippocampal neurons. Science 275: 209-213.

Malenka RC, Nicoll RA. 1993. NMDA-receptor-dependent synaptic plasticity: Multiple forms and mechanisms. Trends Neurosci 16: 521-527.

Malinow R. 2003. AMPA receptor trafficking and long-term potentiation. Philos Trans R Soc Lond B Biol Sci 358: 707-714.

Malkani S, Rosen JB. 2000. Specific induction of early growth response gene 1 in the lateral nucleus of the amygdala following contextual fear conditioning in rats. Neuroscience 97: 693-702.

Maren S. 1999. Long-term potentiation in the amygdala: A mechanism for emotional learning and memory. Trends Neurosci 22: 561-567.

Maren S. 2000. Auditory fear conditioning increases CS-elicited spike firing in lateral amygdala neurons even after extensive overtraining. Eur J Neurosci 12: 4047-4054.

Maren S. 2001. Neurobiology of Pavlovian fear conditioning. Annu Rev Neurosci 24: 897-931.

Maren S, Aharonov G, Stote DL, Fanselow MS. 1996. N-methyl-D-aspartate receptors in the basolateral amygdala are required for both acquisition and expression of conditional fear in rats. Behav Neurosci 110: 1365-1374.

Markram H, Lubke J, Frotscher M, Sakmann B. 1997. Regulation of synaptic efficacy by coincidence of postsynaptic APs and EPSPs. Science 275: 213-215.

Matus A. 2000. Actin-based plasticity in dendritic spines. Science 290: 754-758.

Mayford M, Bach ME, Huang YY, Wang L, Hawkins RD, et al. 1996. Control of memory formation through regulated expression of a CaMKII transgene. Science 274: 1678-1683.

McDonald AJ. 1998. Cortical pathways to the mammalian amygdala. Prog Neurobiol 55: 257-332.

McKernan MG, Shinnick-Gallagher P. 1997. Fear conditioning induces a lasting potentiation of synaptic currents in vitro. Nature 390: 607-611.

Miller SG, Kennedy MB. 1985. Distinct forebrain and cerebellar isozymes of type II Ca2+/calmodulin-dependent protein kinase associate differently with the postsynaptic density fraction. J Biol Chem 260: 9039-9046.

Milner B, Squire LR, Kandel ER. 1998. Cognitive neuroscience and the study of memory. Neuron 20: 445-468.

Miserendino MJ, Sananes CB, Melia KR, Davis M. 1990. Blocking of acquisition but not expression of conditioned fear-potentiated startle by NMDA antagonists in the amygdala. Nature 345: 716-718.

Moita MA, Lamprecht R, Nader K, LeDoux JE. 2002. A-kinase anchoring proteins in amygdala are involved in auditory fear memory. Nat Neurosci 5: 837-838.

Monyer H, Sprengel R, Schoepfer R, Herb A, Higuchi M, et al. 1992. Heteromeric NMDA receptors: Molecular and functional distinction of subtypes. Science 256: 1217-1221.

Muller J, Corodimas KP, Fridel Z, LeDoux JE. 1997. Functional inactivation of the lateral and basal nuclei of the amygdala by muscimol infusion prevents fear conditioning to an explicit conditioned stimulus and to contextual stimuli. Behav Neurosci 111: 683-691.

Nakanishi S. 1992. Molecular diversity of glutamate receptors and implications for brain function. Science 258: 597-603.

Nguyen PV, Abel T, Kandel ER. 1994. Requirement of a critical period of transcription for induction of a late phase of LTP. Science 265: 1104-1107.

Nguyen PV, Kandel ER. 1996. A macromolecular synthesis-dependent late phase of long-term potentiation requiring cAMP in the medial perforant pathway of rat hippocampal slices. J Neurosci 16: 3189-3198.

Passafaro M, Piech V, Sheng M. 2001. Subunit-specific temporal and spatial patterns of AMPA receptor exocytosis in hippocampal neurons. Nat Neurosci 4: 917-926.

Pin JP, Duvoisin R. 1995. The metabotropic glutamate receptors: Structure and functions. Neuropharmacology 34: 1-26.

Quirk GJ, Armony JL, LeDoux JE. 1997. Fear conditioning enhances different temporal components of tone-evoked spike trains in auditory cortex and lateral amygdala. Neuron 19: 613-624.

Quirk GJ, Repa C, LeDoux JE. 1995. Fear conditioning enhances short-latency auditory responses of lateral amygdala neurons: Parallel recordings in the freely behaving rat. Neuron 15: 1029-1039.

Rampon C, Tang YP, Goodhouse J, Shimizu E, Kyin M, et al. 2000. Enrichment induces structural changes and recovery from nonspatial memory deficits in CA1 NMDAR1-knockout mice. Nat Neurosci 3: 238-244.

Rattiner LM, Davis M, French CT, Ressler KJ. 2004. Brain-derived neurotrophic factor and tyrosine kinase receptor B involvement in amygdala-dependent fear conditioning. J Neurosci 24: 4796-4806.

Repa JC, Muller J, Apergis J, Desrochers TM, Zhou Y, et al. 2001. Two different lateral amygdala cell populations contribute to the initiation and storage of memory. Nat Neurosci 4: 724-731.

Ressler KJ, Paschall G, Zhou XL, Davis M. 2002. Regulation of synaptic plasticity genes during consolidation of fear conditioning. J Neurosci 22: 7892-7902.

Rodrigues SM, Bauer EP, Farb CR, Schafe GE, LeDoux JE. 2002. The group I metabotropic glutamate receptor mGluR5 is required for fear memory formation and long-term potentiation in the lateral amygdala. J Neurosci 22: 5219-5229.

Rodrigues SM, Farb CR, Bauer EP, LeDoux JE, Schafe GE. 2004. Pavlovian fear conditioning regulates Thr286 autophosphorylation of Ca2+/calmodulin-dependent protein kinase II at lateral amygdala synapses. J Neurosci 24: 3281-3288.

Rodrigues SM, Schafe GE, LeDoux JE. 2001. Intraamygdala blockade of the NR2B subunit of the NMDA receptor disrupts the acquisition but not the expression of fear conditioning. J Neuroscience 21(17): 6889-6896.

Rogan MT, LeDoux JE. 1995. LTP is accompanied by commensurate enhancement of auditory-evoked responses in a fear conditioning circuit. Neuron 15: 127-136.

Rogan MT, Staubli UV, LeDoux JE. 1997. Fear conditioning induces associative long-term potentiation in the amygdala. Nature 390: 604-607.

Romanski LM, Clugnet MC, Bordi F, LeDoux JE. 1993. Somatosensory and auditory convergence in the lateral nucleus of the amygdala. Behav Neurosci 107: 444-450.

Romanski LM, LeDoux JE. 1993. Information cascade from primary auditory cortex to the amygdala: Corticocortical and corticoamygdaloid projections of temporal cortex in the rat. Cereb Cortex 3: 515-532.

Roozendaal B, Koolhaas JM, Bohus B. 1991. Attenuated cardiovascular, neuroendocrine, and behavioral responses after a single footshock in central amygdaloid lesioned male rats. Physiol Behav 50: 771-775.

Rosen JB, Fanselow MS, Young SL, Sitcoske M, Maren S. 1998. Immediate-early gene expression in the amygdala following footshock stress and contextual fear conditioning. Brain Res 796: 132-142.

Rumpel S, LeDoux J, Zador A, Malinow R. 2005. Postsynaptic receptor trafficking underlying a form of associative learining. Science 308: 83-88.

Schafe GE, Atkins CM, Swank MW, Bauer EP, Sweatt JD, et al. 2000. Activation of ERK/MAP kinase in the amygdala is required for memory consolidation of pavlovian fear conditioning. J Neurosci 20: 8177-8187.

Schafe GE, LeDoux JE. 2000. Memory consolidation of auditory Pavlovian fear conditioning requires protein synthesis and protein kinase A in the amygdala. J Neurosci 20: RC96.

Schafe GE, Nader K, Blair HT, LeDoux JE. 2001. Memory consolidation of Pavlovian fear conditioning: A cellular and molecular perspective. Trends Neurosci 24: 540-546.

Scicli AP, Petrovich GD, Swanson LW, Thompson RF. 2004. Contextual fear conditioning is associated with lateralized expression of the immediate early gene c-fos in the central and basolateral amygdalar nuclei. Behav Neurosci 118: 5-14.

Shen K, Meyer T. 1999. Dynamic control of CaMKII translocation and localization in hippocampal neurons by NMDA receptor stimulation. Science 284: 162-166.

Shen K, Teruel MN, Connor JH, Shenolikar S, Meyer T. 2000. Molecular memory by reversible translocation of calcium/calmodulin-dependent protein kinase II. Nat Neurosci 3: 881-886.

Silva AJ, Kogan JH, Frankland PW, Kida S. 1998. CREB and memory. Annu Rev Neurosci 21: 127-148.

Silva AJ, Paylor R, Wehner JM, Tonegawa S. 1992a. Impaired spatial learning in alpha-calcium-calmodulin kinase II mutant mice. Science 257: 206-211.

Silva AJ, Stevens CF, Tonegawa S, Wang Y. 1992b. Deficient hippocampal long-term potentiation in alpha-calcium- calmodulin kinase II mutant mice. Science 257: 201-206.

Soderling TR, Derkach VA. 2000. Postsynaptic protein phosphorylation and LTP. Trends Neurosci 23: 75-80.

Stanton PK, Gage AT. 1996. Distinct synaptic loci of Ca2+/calmodulin-dependent protein kinase II necessary for long-term potentiation and depression. J Neurophysiol 76: 2097-2101.

Stevens CF. 1998. A million dollar question: Does LTP = memory? Neuron 20: 1-2.

Stork O, Stork S, Pape HC, Obata K. 2001. Identification of genes expressed in the amygdala during the formation of fear memory. Learn Mem 8: 209-219.

Stuart GJ, Hausser M. 2001. Dendritic coincidence detection of EPSPs and action potentials. Nat Neurosci 4: 63-71.

Sweatt JD. 1999. Toward a molecular explanation for long-term potentiation. Learn Mem 6: 399-416.

Sweatt JD. 2004. Mitogen-activated protein kinases in synaptic plasticity and memory. Curr Opin Neurobiol 14: 311-317.

Sweatt JD, Atkins CM, Johnson J, English JD, Roberson ED, et al. 1998. Protected-site phosphorylation of protein kinase C in hippocampal long- term potentiation. J Neurochem 71: 1075-1085.

Tan SE, Liang KC. 1996. Spatial learning alters hippocampal calcium/calmodulin-dependent protein kinase II activity in rats. Brain Res 711: 234-240.

Tang YP, Shimizu E, Dube GR, Rampon C, Kerchner GA, et al. 1999. Genetic enhancement of learning and memory in mice. Nature 401: 63-69.

Toni N, Buchs PA, Nikonenko I, Bron CR, Muller D. 1999. LTP promotes formation of multiple spine synapses between a single axon terminal and a dendrite. Nature 402: 421-425.

Tsien JZ. 2000. Building a brainier mouse. Sci Am 282: 62-68.

Van Harreveld A, Fifkova E. 1975. Swelling of dendritic spines in the fascia dentata after stimulation of the perforant fibers as a mechanism of post-tetanic potentiation. Exp Neurol 49: 736-749.

van Rossum D, Hanisch UK. 1999. Cytoskeletal dynamics in dendritic spines: Direct modulation by glutamate receptors? Trends Neurosci 22: 290-295.

Walker DL, Davis M. 2000. Involvement of NMDA receptors within the amygdala in short- versus long-term memory for fear conditioning as assessed with fear-potentiated startle. [In Process Citation]. Behav Neurosci 114: 1019-1033.

Weeber EJ, Atkins CM, Selcher JC, Varga AW, Mirnikjoo B, et al. 2000. A role for the beta isoform of protein kinase C in fear conditioning. J Neurosci 20: 5906-5914.

Weisskopf MG, Bauer EP, LeDoux JE. 1999. L-type voltage-gated calcium channels mediate NMDA-independent associative long-term potentiation at thalamic input synapses to the amygdala. J Neurosci 19: 10512-10519.

Weisskopf MG, LeDoux JE. 1999. Distinct populations of NMDA receptors at subcortical and cortical inputs to principal cells of the lateral amygdala. J Neurophysiol 81: 930-934.

Wilensky AE, Schafe GE, LeDoux JE. 2000. The amygdala modulates memory consolidation of fear-motivated inhibitory avoidance learning but not classical fear conditioning. J Neurosci 20: 7059-7066.

Woolf NJ. 1998. A structural basis for memory storage in mammals. Prog Neurobiol 55: 59-77.

Yin JC, Del Vecchio M, Zhou H, Tully T. 1995. CREB as a memory modulator: Induced expression of a dCREB2 activator isoform enhances long-term memory in Drosophila. Cell 81: 107-115.

Yin JC, Tully T. 1996. CREB and the formation of long-term memory. Curr Opin Neurobiol 6: 264-268.

Yin JC, Wallach JS, Del Vecchio M, Wilder EL, Zhou H, et al. 1994. Induction of a dominant negative CREB transgene specifically blocks long-term memory in Drosophila. Cell 79: 49-58.

19 Neurochemistry and Molecular Neurobiology of Memory

P. Dash · A.N. Moore

"Is it the blood, or air, or fire by which we think? Or is it none of these, and does the brain furnish the perceptions of hearing and sight and smell, and do memory and opinion arise from these, and does knowledge come from memory and opinion when they have attained fixity?" – Socrates (469–399 B.C.).

Abstract: The quest to understand the process of forming and storing memories has engaged scientists for hundreds of years. Recent advances in biology have led to the identification of some of the cellular and molecular mechanisms associated with these processes. Animal studies using both invertebrates and vertebrates have revealed that the mechanisms of learning and memory at the molecular level have been conserved throughout evolution, and that more complex forms of learning have been built on these evolutionarily conserved mechanisms. In this chapter, we discuss how behavioral training alters neuronal activity and neurotransmitter release leading to changes in intracellular kinase activity, protein phosphorylation, protein synthesis, and gene expression. We describe how the various phases of memory: working memory (lasting for seconds), short-term memory (lasting minutes), intermediate-term memory (lasting for hours), and long-term memory (lasting days to a lifetime) are established by the above intracellular changes. We present theories on how permanent memory storage occurs, how memories are recalled, and how they are extinguished.

List of Abbreviations: 5-HT, serotonin; AMPA, alpha-amino-3-hydroxy-5-methyl-4-isoxazolepropionic acid; C/EBP, CCAAT/enhancer-binding protein; CA, cornus ammonis; CaMK, calcium/calmodulin-dependent protein kinase; cAMP, cyclic adenosine monophosphate; CPEB, cytoplasmic polyadenylation element binding protein; CRE, cAMP response element; CREB, calcium/cAMP response element binding protein; CS, conditioned stimulus; DBH, dopamine β-hydroxylase; DG, dentate gyrus; EPSP, excitatory postsynaptic potential; ERK, extracellular signal-regulated protein kinase; GAP43, growth associated protein 43; GTP, Guanine triphosphate; I-1, inhibitor protein-1; IkB, Inhibitory kappa B; IL, infralimbic cortex; IP3, inositol trisphosphate; L-DOPA, 3,4-dihydroxy-L-phenylalanine; L-DOPS, L-Threo-3,4-dihydroxyphenylserine; LTD, long-term depression; LTP, long-term potentiation; MAP-2, microtubule associated protein 2; MAPK, mitogen-activated protein kinase; MRI, magnetic resonance imaging; MTT, Multiple Trace Theory; NFkB, nuclear factor kappaB; NMDA, N-methyl-D-aspartate; NMR, nictating membrane response; NO, nitric oxide; p90rsk, p90 ribosomal S6 kinase; PFC, prefrontal cortex; PKA, cAMP-dependent protein kinase A; PKC, protein kinase C; PKM, protein kinase M; PP-1, protein phosphatase I; Ser, serine; SRF, serum response factor; SRF, serum response factor; Thr, threonine; Tyr, tyrosine; US, unconditioned stimulus; VTA, ventral tegmental area

1 Introduction

Learning is the process of acquiring knowledge, while memory results from the process by which the knowledge is consolidated, stored, and later retrieved. The ability of the brain to store information about each individual's current and past experiences in a usable form – a "memory" – which can be retrieved at a later point, allows for continuity and cohesion between successive events in one's life, giving rise to a sense of self. We are who we are, because of what we learn and what we remember. As a result of acquiring and retaining knowledge, we change our behavior. When learning and memory processes are impaired, as seen in conditions such as Alzheimer's disease or following injury to the brain, a person's behavior is altered and a sense of self is often lost.

One of the major goals of learning and memory research is to understand how acquiring knowledge changes the brain and how these changes alter subsequent behavior. Over the past two decades, a great deal of progress has been made in identifying some of the molecular mechanisms underlying these processes. In this section, we discuss the cellular and molecular mechanisms that contribute to both explicit (declarative) and implicit (nondeclarative) learning and memory. Electrophysiological recordings, pharmacological and molecular biological agents, and genetic manipulation in experimental animals have been highly successful in elucidating the cellular and molecular mechanisms underlying both explicit and implicit memory formation and storage. We illustrate that the mechanisms of learning and memory at the molecular level have been conserved throughout evolution and more complex forms of learning have been built on these evolutionarily conserved mechanisms.

2 Modulation of Neuronal Communication Underlies Memory Formation

In 1890, the American psychologist William James posed the following question in his analysis of the phenomenon of memory: “How does a man come, after having the thought of A, to have the thought of B the next moment? Or how does he come to think of A and B always together?” (James, 1890). To answer his question on how the brain forms associations between two events, James opines “When two elementary brain processes have been active together or in immediate succession, one of them on recurring, tends to propagate its excitement into the other.” Although the interconnection between two memories appeared evident, the way in which this relationship developed was not clear. The synaptic mechanism for memory was proposed by Tanzi and later by Ramon y Cajal to suggest that changes in the strength of synaptic connections between neurons underlie information storage (Tanzi, 1893; Ramon y Cajal, 1911). This proposal was extended by Donald Hebb who postulated that “when an axon of cell A is near enough to excite a cell B and repeatedly or persistently takes part in firing it, some growth process or metabolic change takes place in one or both cells such that A’s efficiency, as one of the cells firing B, is increased” (Hebb, 1949). This increase in efficiency is thought to be a neural correlate for memory storage. Hebb’s postulate can be better appreciated with the following example. Pavlov’s famous experiment with dogs showed that repeated pairing of the sound of a bell with the presentation of food lead to an association between the bell and food presentation (➋ *Figure 19-1*) (Pavlov, 1927). For simplicity, assume that the sound of the bell is processed by neuron A and saliva production is caused by neuron B. Prior to the pairing, the sound of the bell (i.e., neuron A) is unable to activate neuron B and saliva production. However, the ringing bell caused neuron A to fire an action potential and presenting the food caused neuron B to fire (which results in saliva production). Repeated pairing of sound with food caused both neurons A and B to fire together. Repeated firing of the two neurons as a result of training increases the efficiency of neuron A to fire B to such an extent that activation of neuron A (as result of the bell) causes the firing of neuron B and saliva production without the presentation of food.

Several plausible mechanisms (e.g., growth of new synaptic connections, increased excitability, synaptic modification) can increase the efficiency of neuron A to cause neuron B to fire. For example, the formation of additional synapses onto neuron B would cause more transmitter release in response to the sound of the bell, and increase the efficiency of A to generate action potentials in B. As long as the synaptic strength between A and B remains high, the sound of the bell can bring back the memory of food. Thus, the cellular and molecular mechanisms that give rise to changes in synaptic strength are thought to be the mechanisms of memory storage. Within the last 15 years, neuroscience research has demonstrated that a number of biochemical activities take place within neurons (both in A and B from the above example) that allow for these changes and are required for memory storage.

Since the original discoveries made by Bliss and Lomo (1973) in the hippocampus, potentiation of synaptic transmission has been observed in several brain regions. Several experiments have established the Hebbian nature of some forms of synaptic plasticity, although non-Hebbian forms also exist. It has been proposed that the molecular mechanisms of memory storage may share similarities with those that give rise to long-term potentiation (LTP) and long-term depression (LTD).

2.1 Glutamate Receptors

Glutamate is the most prominent neurotransmitter in the brain and has been intimately linked to LTP and LTD. Approximately half of the synapses (and nearly all excitatory neurons) use glutamate as a neurotranmitter. When this transmitter is relased in response to neuronal activity (such as during learning), it binds to specific glutamate receptors. These receptors can be divided into two classes: the ionotropic glutamate receptors and the metabotropic glutamate receptors.

Figure 19-1
Enhanced neuronal communication following associative learning. Prior to conditioning, neuron A responds to the sound of the bell but lacks the strength to cause neuron B to fire *(top left)*. Neuron B triggers saliva production in response to food presentation *(top right)*. During conditioning in which the sound of the bell and food reward are paired *(bottom left)*, neuron A and neuron B fire together. This pairing of the neuronal activities of neuron A and neuron B increases the synaptic strength (and/or excitability) between neurons A and B as result of formation of additional synaptic connections. Following the conditioning, the sound of the bell alone is now sufficient to cause neuron B to fire as a result of more transmitter release (and/or more number of action potentials) from these newly formed synaptic sites, triggering saliva production

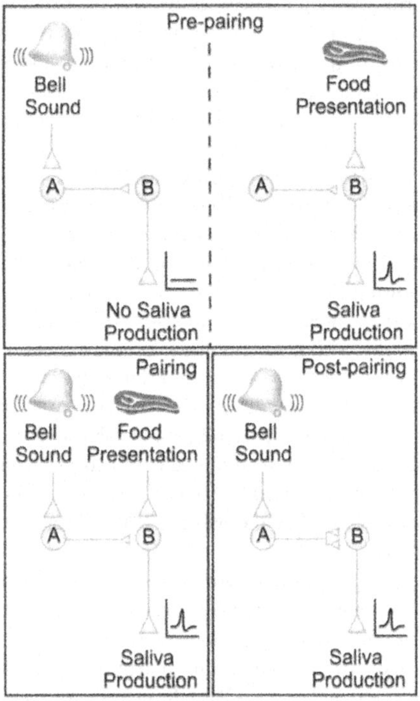

Three subtypes of ionotropic receptors, originally classified on the basis of the action of pharamacological agonist, have been identified: AMPA, kinate and NMDA, which are abundantly expressed in the brain (Nicoll et al., 1990; Hollmann and Heinemann, 1994). AMPA and kinate receptors are sometimes referred to as non-NMDA type receptors. The binding of glutamate to these ionotropic receptors results in the opening of ion channels, which allows the movement of ions (Na^+ and Ca^{2+} in, and K^+ out) across the membrane. Functional AMPA and kinate receptors are made up of various combinations of subunits (GluR1, GluR2, GluR3, and GluR4 for AMPA; GluR5, GluR6, GLuR7, KA1, and KA2 for kianate) giving rise to a large number of receptors with varied ionic conductance. There are at least five forms of NMDA receptor subunits (NMDAR1, NMDAR2A, NMDAR2B, NMDAR2C, and NMDAR2D). NMDAR1 is essential for the formation of a functional receptor, since the subunits NMDAR2A-D do not form channels when expressed singly or in combination. Coexpression of NMDAR1 with other subunits forms channels with varying ionic conductances. These receptors have two interesting properties: (1) high permeability to Ca^{2+} and (2) opening that is dependent on the presence of glutamate and membrane voltage. At resting membrane potential, these receptors are blocked due to the binding of a Mg^{2+} ion. Mg^{2+} is expelled from the receptor when the neuron is sufficiently depolarized (as a result of the opening of AMPA receptors) allowing the flow of ions. These excitatory ionotropic receptors are essential for communication between neurons and are consequently required for activity-dependent changes in neuronal morphology, the presumed cellular basis of storage of memories. This supposition has since been supported by numerous studies showing that blocking glutamate receptors in structures involved in memory functions results in anterograde amnesia.

There are three classes of metabotropic glutamate receptors (Class I: mGluR1, mGluR5; Class II: mGluR2, mGluR3; Class III: mGluR4, mGluR6, mGluR7, mGluR8). Activation of these receptors leads to increases in second messenger molecules (such as cAMP and calcium) and can lead to inhibition of postsynatic Na^+ and Ca^{2+} channels. Unlike ionotropic glutamate receptors, activation of metabotropic receptors can either increase or decrease excitability of postsynaptic neurons. Thus, their role in learning and memory is complex.

2.2 Hebbian LTP Requires Glutamate Receptors and Activation of Intracellular Kinases

LTP is defined as an enduring increase (greater than one hour) in synaptic efficacy that results from high-frequency stimulation of an afferent pathway. LTP exhibits three phases, (1) short: dependent on protein phosphorylation, (2) intermediate: dependent on protein synthesis, and (3) long: requiring gene expression. Molecular investigation into LTP mechanisms started with the demonstration that an antagonist for NMDA subtype of glutamate receptors blocks LTP in the hippocampal CA1 subfield (Nicoll and Malenka, 1995). At the resting potential of a neuron (approximately –70 mV), magnesium ions block the entry of calcium through the NMDA cation channel. When the neuron depolarizes and glutamate is bound, however, the magnesium ions are released from the receptor allowing the influx of calcium from the extracellular space. Rises in intracellular calcium concentrations are obligatory for all types of LTP. In addition to NMDA subtype glutamate receptors, AMPA receptors play a prominent role. These receptors are expressed throughout the brain and mediate fast excitatory synaptic transmission. Activation of AMPA receptors allows positively charged sodium ions to enter cells through its pore. This results in depolarization of the postsynaptic neuron and generation of a fast excitatory postsynaptic potential (EPSP). Several studies have shown that AMPA receptors participate in LTP in two prominent ways during induction and maintenance. First, during LTP induction, postsynaptic AMPA receptors are activated, depolarizing the neurons. This depolarization allows NMDA receptors to conduct Ca^{2+}. Following LTP induction, the number of AMPA receptors on postsynaptic neurons is gradually increased (Liao et al., 1995). This results in a larger EPSP or potentiation in response to presynaptic stimulation.

Interestingly, it was observed that in certain synapses where AMPA receptors were absent, LTP induction resulted in the insertion of these receptors into the synaptic site. This process was found to be dependent on NMDA receptor activation and calcium influx. This key observation led to the silent synapse theory in which certain synapses were silent because they lacked AMPA receptors, although they have NMDA receptors (Liao et al., 1995). During LTP induction, intracellular calcium reaches a level capable of stimulating several calcium-sensitive enzymes including calcium/calmodulin-dependent protein kinase II (CaMKII). CaMKII is one of the most abundant proteins in the brain, accounting for approximately 2% of all proteins. Once activated by the calcium-calmodulin complex, CaMKII phosphorylates the GluR1 subunit of the AMPA receptors on Ser^{831} (Barria et al., 1997). Phosphorylation of AMPA receptors by CaMKII is thought to facilitate insertion of these receptors into the silent synapse allowing the potentiated state to persist. Conversely, inhibition of CaMKII leads to rapidly decaying LTP.

In addition to CaMKII, increased intracellular calcium can activate other signaling molecules including protein kinase C (PKC), adenylyl cyclase, and mitogen-activated protein kinase/extracellular signal-regulated protein kinase (MAPK/ERK). Stimulation of adenylyl cyclase activity by calcium leads to increased synthesis of cAMP and activation of cAMP-dependent protein kinase A (PKA). Initial studies using pharmacological inhibitors and activators of the cAMP/PKA system led to the observation that this pathway is required for the long-term maintenance of the potentiated state. In neuronal cells, enhanced activity of the cAMP/PKA pathway can activate b-Raf (via GEF Rap1) resulting in the phosphorylation and activation of the ERK cascade (Weeber and Sweatt, 2002). Activated ERK translocates into the cell nucleus where it phosphorylates two important transcription factors: calcium/cAMP response element binding protein (CREB) and serum response factor (SRF). Phosphorylation of these factors stimulates transcription of selective genes that carry the binding sites for these factors in their promoter regions. These gene products are thought to elicit growth processes in the neurons (as initially proposed by Ramon y Cajal) to maintain the potentiated state for long periods of time. As discussed below, these molecules have been shown to be required for memory formation in animals.

2.3 LTD Requires Calcium-Dependent Intracellular Signaling

In addition to potentiation of synapses, information in the brain can also be stored (or erased) by a decrease in synaptic communication between neurons, referred to as long-term depression (LTD). LTD can be induced at the Schaffer collaterals-CA1 synapses of the hippocampus by low frequency stimulation (Mulkey and Malenka, 1992; Dudek and Bear, 1993). Similar to LTP, LTD depends on AMPA and NMDA receptor functions, as well as increased intracellular calcium. However, the low frequency stimulation used to induce LTD causes only modest increases in intracellular calcium resulting in the activation of the calcium/calmodulin-dependent protein phosphatase calcineurin (also known as PP2B), but not CaMKII, which requires higher concentrations of calcium for activation. These studies show that while LTP involves protein phosphorylation, LTD is caused by dephosphorylation of specific substrate proteins. Among the proteins that are regulated by calcineurin, dephosphorylation of inhibitor protein-1 [I-1; a protein which blocks the activity of protein phosphatase I (PP-1)] has been shown to be critical. Dephosphorylated I-1 loses its ability to inhibit PP-1 allowing PP-1 to dephosphorylate substrate proteins including AMPA receptors. Dephosphorylated AMPA receptors are removed from the synaptic sites via endocytosis, thereby depressing synaptic communication.

LTD in the cerebellum has been proposed to be a mechanism for motor learning and memory. A type of associative learning called eyeblink response (or nictating membrane response, NMR) has been used to explore some of the cellular and molecular mechanisms required for motor learning and memory. In this task, a tone is used as the conditioned response (CS) and an air puff is used as the unconditioned response (US). The CS information is conveyed to the cerebellar cortex (and the deep nuclei of the cerebellum) by the mossy fibers that originate in the pontine nuclei. The mossy fibers form synapses with the granule cells of the cerebellum whose axons give rise to the parallel fibers. The US information arrives at the cerebellar cortex (and the deep nuclei) via the climbing fibers that originate in the inferior olivary nucleus. Purkinje cells send their dendrites to the molecular layer of the cerebellar cortex and form synapses with the parallel and climbing fibers. Thus, the information carried by the CS and the US converge on the dendrites of the Purkinje cells (CS and US also converge at deep nuclei). The axons of the Purkinje cells make inhibitory synapses on neurons located in the deep nuclei. Thus, stimulation of parallel fibers in the cerebellum gives rise to an EPSP in the Purkinje cell, which inhibits the activity of the deep nuclei.

When stimulation of the climbing fibers is paired with stimulation of the parallel fibers (such as when the CS and US are co-presented), the size of the EPSP in the Purkinje cell in response to parallel fiber stimulation is reduced. Thus, learning decreases the ability of Purkinje cells to inhibit neurons in the deep nuclei. This allows for the development of an eyeblink response to the CS. LTD at parallel fibers-Purkinje synapses has been shown to require increases in intracellular calcium (via ionotropic and metabotropic glutamate receptors as well as via voltage-dependent calcium channels) and subsequent activation of PKC. Although several hypotheses have been proposed as to how PKC-mediated phosphorylation causes LTD, the exact mechanism has not been established.

3 Implicit and Explicit Forms of Memory

Different types of memory can be classified into two broad categories: implicit memory and explicit memory (Squire, 1987). Implicit memory (also called nondeclarative memory or procedural memory) does not require conscious recollection and includes simple associative and nonassociative forms of memory such as classical conditioning, sensitization, and habituation. Examples of implicit memory include riding a bicycle, driving a car, or Pavlovian conditioning. In mammals, implicit memory involves many brain structures including the cerebellum and basal ganglia. Explicit memory (also called declarative memory) involves conscious recollection of information about people, places, and things. Examples of this type of memory are remembering the words to a song, the names of streets in your neighborhood or describing the appearance of a friend. In mammals, explicit memory depends on the integrity of the temporal lobe including the hippocampus, entorhinal, perirhinal, and parahippocampal cortices.

3.1 Implicit Memory

3.1.1 Habituation and Sensitization

Invertebrate model systems have contributed significantly to the understanding of cellular and molecular mechanisms of simple forms of implicit memory. Studies in *Aplysia* by Eric Kandel, James Schwartz, Vincent Castelluci, Tom Carew, John Byrne and others on habituation and sensitization have identified several molecular steps required for memory formation and storage (Kandel and Schwartz, 1982; Byrne and Kandel, 1996). Many of these mechanisms appear to be conserved all the way to humans. Implicit memories have three forms: short-, intermediate-, and long-term, all of which are associated with the facilitation of neuronal communication with distinct molecular mechanisms underlying each (Milner et al., 1998).

One behavioral response in particular, the gill withdrawal reflex in *Aplysia*, has been well characterized (Pinsker et al., 1970). A particularly well studied component of the gill withdrawal reflex is the monosynaptic connection between the sensory neurons that innervate the siphon skin and the motor neurons that innervate the gill (➋ *Figure 19-2*). When a tactile stimulus is presented to the siphon, the siphon sensory neurons at first produce an EPSP in the motor neurons they innervate. This causes a brisk contraction of the gill (and siphon). With repeated stimulation, a progressive decrease in the amplitude of the postsynaptic

Figure 19-2

Siphon-gill component of the withdrawal reflex in *Aplysia*, which has been used to examine the cellular and molecular mechanisms underlying sensitization, habituation and classical conditioning. The siphon sensory neurons transmit the tactile information from the siphon skin to interneurons and gill motor neurons that cause gill contraction. With repeated tactile stimulation of the siphon, the gill contraction habituates due to less transmitter release from the sensory neurons. Noxious stimulation of the tail excites a different group of facilitatory neurons that release modulatory neurotransmitters onto siphon sensory neuron synaptic sites. This increases the strength between siphon sensory neurons and gill motor neurons, causing a stronger gill contraction in response to subsequent siphon stimulation. Thus, the response is sensitized

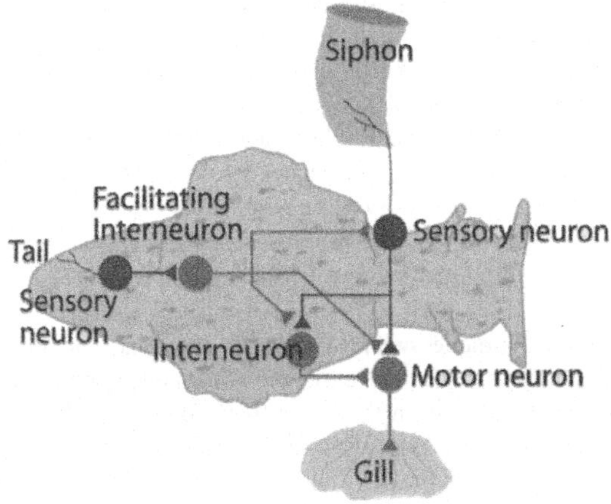

potential produced by the sensory neurons onto gill motor neurons occurs causing the reflex to habituate. This decrease in synaptic effectiveness results from a decrease in the amount of transmitter released from the terminals of sensory neurons onto the gill motor neuron.

In contrast to habituation, when a sensitizing stimulus (e.g., mild electric shock) is applied to the tail or the head region, the gill withdrawal reflex in response to a tactile stimulus is enhanced. The sensitizing stimulus activates facilitating interneurons that act on siphon sensory neurons to enhance transmitter release onto gill (and siphon) motor neurons. This causes an increase in the amplitude of the postsynaptic potential and a stronger withdrawal of the gill and siphon in response to siphon sensory neuron activity. One group of facilitating interneurons uses serotonin (5-HT) as a neurotransmitter to stimulate the sensory neurons. Other groups of facilitatory interneurons use other neurotransmitters, such as small cardioactive peptide. As discussed in *Section 4*, the action of these neurotransmitters changes synaptic communication (or synaptic strength) by stimulating distinct molecular cascades, giving rise to the different forms (short-, intermediate-, and long-term) of memory.

3.1.2 Classical Conditioning: Invertebrates

In classical conditioning (or associative learning), animals learn to associate the occurrence of two events. In other words, animals demonstrate an instinctive response (e.g., salivation) when presented with a stimulus (e.g., tone presentation) that has been paired with a second stimulus (e.g., food). As mentioned earlier, the behavioral aspects of classical conditioning were first described by Pavlov, and thus it is often referred to as Pavlovian conditioning. An understanding of the molecular events underlying classical conditioning has since emerged from studies carried out in *Aplysia, Hermissenda,* and the fruit fly *Drosophila* (Benzer, 1971; Tully, 1984; Byrne, 1987). The gill- (and siphon-) withdrawal reflex of *Aplysia* exhibits classical conditioning when a tactile stimulus [e.g., a gentle touch, the conditioned stimulus (CS)] to the siphon is paired with an electrical shock to the tail [the unconditioned stimulus (US)]. After repeated pairing of the tactile stimulus to the siphon with the tail shock (separated by about 0.5 second), the animal exhibits a robust gill contraction to the previously innocuous gill touch in anticipation of tail shock.

Both pre- and postsynaptic molecular changes contribute to classical conditioning. Presynaptically, the CS (gentle touch of the siphon) causes firing of action potentials in the siphon sensory neurons causing entry of calcium into the cell. Calcium in complex with calmodulin stimulates the activity of calcium-sensitive adenylyl cyclase, the enzyme that synthesizes cAMP. The US (tail shock) activates facilitatory interneurons resulting in release of serotonin onto the siphon sensory neuron terminals. This is thought to activate G-protein-coupled 5-HT receptors and the GTP-binding protein $G_{\alpha s}$, which also stimulates the adenylyl cyclase activity leading to cAMP production and PKA activation. Therefore, following the paired activity of the gill sensory neurons with the facilitatory interneuron, both Gαs activation and calcium influx can lead to robust PKA activation. This allows for a larger glutamate release from the siphon sensory neurons in response to a subsequent action potential. Glutamate release onto the gill motor neuron results in gill contraction. Postsynaptically, glutamate binds to AMPA receptors resulting in depolarization. Depolarization in conjunction with binding of glutamate to NMDA receptors activates them, allowing calcium entry into the motor neuron. In addition to enhancing the firing of motor neurons, which would result in robust gill contraction, postsynaptic calcium also generates a retrograde signal, possibly nitric oxide (NO), which acts upon the presynaptic terminal to stimulate transmitter release further, thereby establishing a communication pathway between the presynaptic and postsynaptic changes (*Figure 19-3*).

While *Aplysia* has provided a model system with many advantages for studying memory formation, they are not readily amenable to many of the advanced molecular and genetic manipulations as other model systems such as the fruit fly *Drosophila.* In flies and other arthropods, olfactory learning and memory involves the function of a pair of structures called mushroom bodies. Since the genetics of *Drosophila* is known in great detail, manipulation of genes within the mushroom bodies has allowed the roles of these

Figure 19-3
Molecular mechanism underlying classical conditioning in *Aplysia*. (1) The conditioned stimulus (CS, siphon touch) triggers action potentials in the presynaptic sensory neurons causing opening of voltage-sensitive calcium channels and calcium entry. (2) The unconditioned stimulus (US, tail shock) activates facilitatory interneurons that release serotonin onto the sensory neuron. (3) Both calcium- and serotonin-induced events combine to robustly stimulate calcium-sensitive adenylyl cyclase and cAMP production. The binding of cAMP to the regulatory subunit of PKA (R) releases the catalytic subunit (C) and activates the kinase. Activated PKA leads to the closure of potassium channels and increased Ca^{2+} influx allowing subsequent firing of the sensory neuron to release more glutamate. (4) Increased synaptic glutamate depolarizes the postsynaptic terminal, and together with glutamate, activates NMDA receptors allowing the influx of calcium. (5) Increased postsynaptic calcium enhances the activity of several biochemical cascades (e.g., NO production) further enhancing neuronal communication between the pre- and postsynaptic neurons

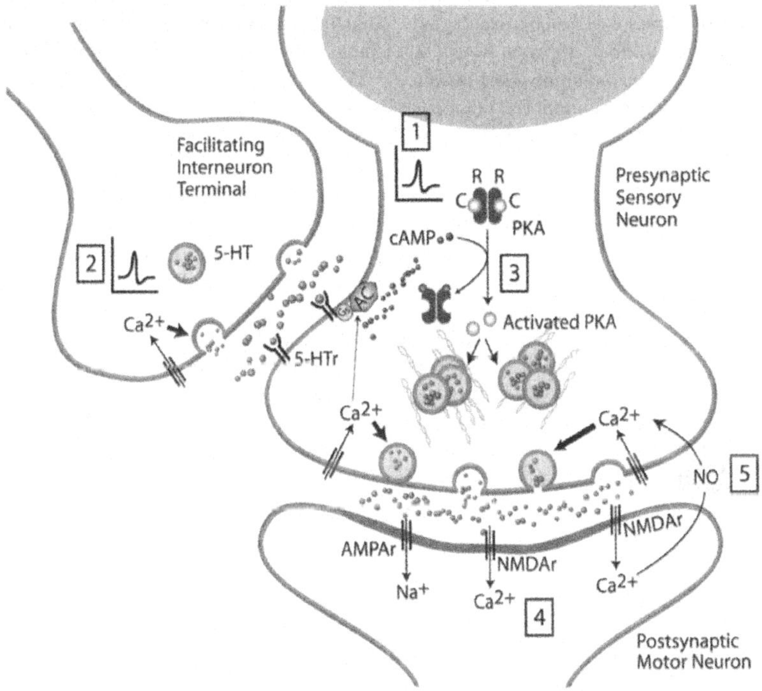

genes to be investigated. As a result of training, flies can learn to avoid an odor (CS) that has been associated with a foot shock (US). Using this paradigm, several mutant strains of flies which failed to make the required association were isolated, including *dunce*, *rutabaga*, *amnesiac*, and PKA-R1 (Dudai et al., 1976; Tully and Quinn, 1985; Han et al., 1996). Subsequent cloning of the mutant genes and measurement of enzymatic activity of the mutant proteins showed that *dunce* lacks a cyclic nucleotide phosphodiesterase, *rutabaga* is defective in calcium-sensitive adenylyl cyclase, *amnesiac* lacks peptide that stimulate adenylyl cyclase, and PKA-R1 is deficient in PKA activity. Analysis of *dunce* and *rutabaga* mutants showed that although cAMP is required for classical conditioning, either an excess (in *dunce*) or deficiency (in *rutabaga*) in cAMP could impair classical conditioning. These and other studies have shown that learning and memory, in general, are linked to the underlying neurochemical changes in an "inverted-U" manner and that a narrow range of biochemical activity is required for proper information storage.

3.1.3 Classical Conditioning in Mammals: Fear Conditioning

Memories for fearful or emotional events are robust and can be formed after only one exposure. For example, when rats are exposed to a tone (the CS) and receive a mild footshock (the US) in a training chamber, fear-related responses such as freezing (lack of movement) and increased heart rate are elicited. Following the training, if the animal is put back in the training chamber (contextual memory) or is presented the tone in an otherwise nonthreatening environment (cued memory), it will freeze and its heart rate will increase.

A great deal of evidence, from both pharmacological and lesion studies indicates that the expression of fear (both in response to contextual and salient cues) is dependent on the function of the amygdala (Cahill et al., 1995; LeDoux, 2000). Recently, studies have shown that the amygdala is not only required for the expression of fear, but that this structure also forms and stores the memory for the CS–US association. For example, intraamygdalal infusion of an acetylcholine receptor antagonist decreases fear memory, while agonists increase it (Power et al., 2003). If a tone is used as the CS, the perception of the tone is carried to the lateral amygdala via auditory afferents. When this afferent stimulation is paired with a footshock and neuronal response is tested in the amygdala, enhanced synaptic strength can be observed in these neurons in response to just the auditory stimuli. Development of this association in the lateral amygdala coincides with the formation of freezing behavior to the CS. In addition to its role in modulating memory formation, the amygdala also appears to be required for the recall of fear (or emotional) memory.

Trace conditioning, a form of classical conditioning in which a time gap separates the occurrence of the CS and the US, is thought to be a form of explicit memory, and various types of studies have implicated the hippocampus in both trace eyeblink and trace fear conditioning. For example, bilateral lesions of the hippocampus markedly impair trace learning and memory, but do not significantly impair these processes when a delay conditioning paradigm is used (McEchron et al., 1998). Recently, studies indicate that in addition to the hippocampus, the prefrontal cortex (PFC; medial prefrontal in rats) is critical for the formation of the CS–US association in trace conditioning. Furthermore, the memory for this association is also stored, at least in part, within this structure (Runyan et al., 2004). As discussed later, the neurotransmitter dopamine plays a key role in the functions of the PFC, with both excessive and insufficient levels of this neurotransmitter being linked to prefrontal dysfunction. Consistent with the key role of this neurotransmitter, blockade of dopamine D1 receptor activity within the PFC impairs long-term memory for the CS–US association following trace fear conditioning (Runyan and Dash, 2004b).

3.2 Explicit Memory

3.2.1 The Medial Temporal Lobe and Explicit Memory

The participation of the medial temporal lobe in explicit memory formation and consolidation was first demonstrated in the patient H.M. by William Scoville and Brenda Milner (Scoville and Milner, 2000). When H.M. was a young boy, he suffered a traumatic brain injury (TBI) as result of a bicycle accident. He subsequently developed posttraumatic epilepsy, which was found to be resistant to pharmacological treatment. As a final effort to lessen the seizures, the neurosurgeon, Scoville, surgically removed parts of the medial temporal lobe from both sides of H.M.'s brain. The extent of medial temporal lobe lesions in H.M. has been determined using MRI and included the hippocampus, amygdala, and overlaying parahippocampal gyrus. Although the surgery lessened the seizure episodes, it left H.M. with profound anterograde memory (new memories) deficits. His first neurological evaluation, performed nearly 2 years after the surgery, revealed that his intelligence, motivation, personality, abstract thinking, and perception were intact. However, he is unable to acquire new long-term memories, although remote memories for events in his childhood remain intact. Because of his severe impairment in forming new memories and the fact that his remote memories remain intact, H.M. has a false sense of self, thinking he is younger than he actually is, and being unable to recognize a current picture of himself. H.M. does not recognize people, including Dr. Milner with whom he has met repeatedly over a period of 40 years, requiring her to reintroduce herself with every meeting. He displays deficits with spatial navigation, frequently becoming lost in the neighborhood he moved into following his surgery. However, H.M. is able to form new nondeclarative (implicit) memories even though his ability to form new declarative memories is severely impaired. For example, he was taught to draw by looking at his hand in a mirror, a procedural learning task that requires a great deal of practice. Although he has become quite adept at performing this task, he does not remember ever being taught to do it, nor does he remember the countless times he practiced the task. The case study of H.M., combined with subsequent studies of other patients with temporal lobe damage, has revealed that this area and the hippocampus in particular are important for the establishment of new explicit memories.

3.2.2 The Hippocampus and Spatial Memory

Experiments using animals, especially rats and mice, have been invaluable in exploring the neurochemical basis of explicit memory formation. One type of explicit memory in particular, spatial memory (memory of one's environment used for spatial navigation), has provided us considerable insight into the molecular mechanisms of memory storage. Using intrahippocampal recordings, John O'Keefe and John Dostrovsky demonstrated that groups of hippocampal neurons fire when a rat is in a specific location within a room (O'Keefe and Dostrovsky, 1971). For example, neuron A will fire when the animal is in the northwest corner of a room, whereas neuron B may fire when the rat is along the west wall. Interestingly, if the visual cues in the room are moved, the neurons will continue to fire in relation to the new cue positions, indicating that allocentric (outside self) visuospatial information is important for the activity of these neurons. Because of their location-specific firing, these neurons are referred to as "place" cells. Based on these findings, O'Keefe and Lynn Nadel proposed that in a new environment, stable place cell firing patterns develop which allow the animal to form a spatial map that can be used to navigate in space (O'Keefe and Nadel, 1978). Consistent with this, if a rat is allowed to freely explore a room and form stable place fields, then is blindfolded, the place cells will continue to fire based on where the animal "thinks" it is within the room. From these and other observations, behavioral tasks have been designed to specifically test the involvement of, and neuronal changes within, the hippocampus during spatial learning and memory (➔ *Figure 19-4*).

One such task, the Morris water maze (named for its inventor, Richard Morris), was designed to test the influence of temporal lobe lesions on spatial memory (Morris et al., 1986). In this task, the goal for the animal is to locate an escape platform submerged under opaque water using only extra maze cues located throughout the room. With repeated training, the animal's latency to locate the platform decreases. This decrease in latency is used as an indicator of memory acquisition. Memory retention can be tested by removing the platform and measuring the time spent in the area in which the platform was located during training. Similar to the deficits observed in H.M., Morris and colleagues demonstrated that in the absence of the hippocampus, rats are incapable of learning or remembering the location of the hidden platform, demonstrating profound spatial deficits. Specifically, it was determined that the dorsal hippocampus in particular was important for spatial learning. In these studies, Morris demonstrated that even 25% of the hippocampus could support learning, if the residual tissue was located at the dorsal pole (Moser et al., 1995).

The radial arm maze, developed by David Olton, has also been used to provide evidence for a role of the hippocampus in spatial memory storage (Walker and Olton, 1979). In this task, eight passageways or "arms" radiate out from a central hub. Visual cues placed around the interior of the central hub or outside the maze, are used by the animals to identify the individual arms. Once placed in the maze, animals will explore the arms searching for food pellets placed at their ends. With repeated training, rats learn to use the visual cues to determine which arms have already been explored, successfully retrieving all food pellets after entering each arm only once. Hippocampally lesioned animals, by comparison, learn to search the arms as do normal animals, but fail to remember which arms have already been searched resulting in repeated entries into some arms while leaving others unexplored. These data suggest that although procedural learning and memory remains intact following hippocampal lesions (allowing the animal to perform the task), spatial memory is impaired requiring that the animal use a trial-and-error approach to finding the food pellets.

Figure 19-4
Spatial tasks can be used to examine explicit memory formation and storage in rodents. (a) The Morris water maze requires that the animal learn the position of a submerged platform using only allocentric (outside self) cues spread throughout the room. A computer-based tracking system allows for the interrogation of the search strategy and swimming path utilized by the animal to navigate to the platform from a distance. (b) The radial-arm maze tests spatial memory by allowing animals to use visual cues to identify those arms, which were baited with food during training. Correct and incorrect entries are counted and used as a measure of spatial learning and memory

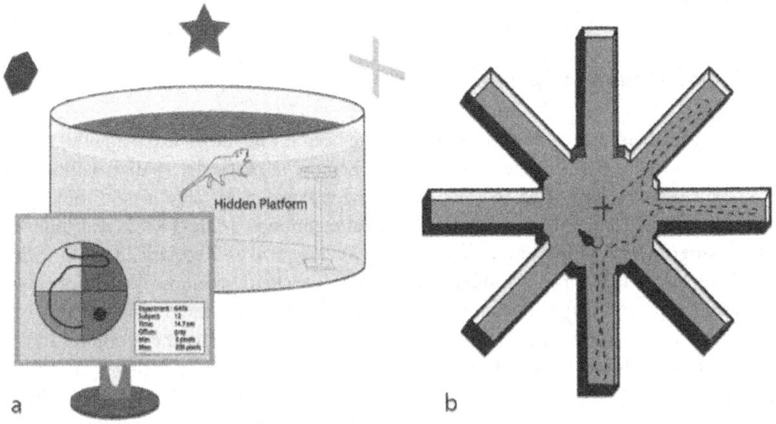

3.2.3 Neuroanatomy of the Hippocampal Formation

Explicit memory formation in the hippocampus begins with the flow of highly processed sensory information into the temporal lobe from cortical association areas such as the prefrontal and parieto-occipital cortices. Input first reaches the parahippocampal and perirhinal cortices, then the entorhinal cortex (*Figure 19-5*). The entorhinal cortex sends information to the hippocampus via a bundle of axons called the perforant path, which originate in the superficial cell layers II and III. These axons form synapses with the neurons of the dentate gyrus (DG), also called granule neurons because of their granular appearance under the microscope. As mentioned earlier, in 1966 Terje Lomo observed that following a period of conditioning stimulation, a single test stimulation of the perforant path caused a potentiated response in the DG. Later, Bliss and Lomo reported that trains of high frequency stimulation of the perforant path caused a sustained potentiation of synaptic transmission in the granule cells of DG (Bliss and Lomo, 1973). The granule neurons of the DG send their axons (called the mossy fibers) to the pyramidal neurons of the CA3 subfield, which in turn, send part of their projections to the CA1 subfield in a bundle called the Schaffer collateral. Information from the hippocampus returns to the deep layer (layer V) of entorhinal cortex, then to the parahippocampal and perirhinal cortices and finally back to the associational cortices. The hippocampus also has direct connections with neocortical areas such as the PFC. In addition, efferent and afferent connections via the fornix link the hippocampus to the mammillary bodies of the hypothalamus (postcommissural fornix) and to the septal nuclei (precommissural fornix).

The fornix, which projects to the septal and the mammillary nuclei, is a structure that is very important for hippocampal function. Substantial input from the septal nucleus and the nucleus basalis of Meynert (also called the diagonal band of Broca) supplies the hippocampus with cholinergic input via the fornix, which is essential to hippocampal function. Therefore, damage to the fornix results in memory deficits similar to those seen with hippocampal damage. Alzheimer's disease, which is characterized by a progressive memory loss, partially results from the loss of cholinergic input to the hippocampal formation as well as other brain regions.

Figure 19-5
a. Explicit or spatial information enters the hippocampus from various association cortices such as prefrontal and parieto-occipital cortices. From there, information enters the parahippocampal and perirhinal cortices, then the entorhinal cortex (EC) and finally enters the hippocampus where it is thought to be stored temporarily. The hippocampus also sends projections to and receives projections from mammillary and septal nuclei via the fornix. b. Information from the entorhinal cortex travels via the perforant path into the hippocampus. Granule cells in the dentate gyrus (DG synapse with the perforant path, while their axons (the mossy fibers) project to the pyramidal neurons of the CA3 subfield. Output from the CA3 pyramidal neurons travels to the CA1 subfield via the Schaffer collaterals. Among other areas, information leaving the hippocampus is returned to the entorhinal cortex. The EC also sends information directly to the dorsal CA1 region thereby bypassing the trisynaptic circuit (not shown)

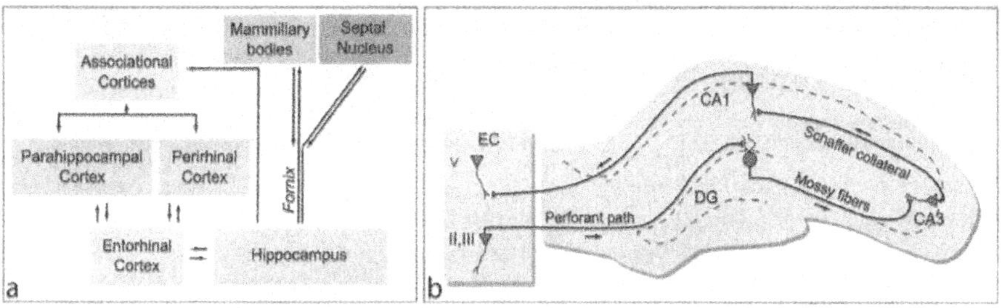

3.2.4 The Entorhinal Cortex and Memory

As shown in *Figure 19-5*, the hippocampus receives information from and sends information to the neocortex via the entorhinal cortex. It is well accepted that the entorhinal–hippocampal circuit is obligatory for performance in spatial memory, as disruption of this structure results in an interruption in the flow of information into and out of the hippocampus. However, the role of the entorhinal cortex in memory storage has only recently been explored. Studies using noninvasive approaches that selectively impair plasticity within the entorhinal cortex indicate that this structure not only stores long-term memory, but that this information is not redundant with that stored within the hippocampus (Hebert and Dash, 2002, 2004). For example, blockade of plasticity in the hippocampus results in a loss of precise location information, whereas inhibition of information storage in the entorhinal cortex results in a loss of broad location information. These influences can be observed during memory testing of rats following Morris water maze training. In the absence of hippocampal memory storage, rats will enter the quadrant of the maze, which previously contained the platform, but will not perform a localized search. Rather, the rats will return to the tank wall to reorient themselves before reentering the target quadrant, only to once again abandon the search. In contrast, animals lacking entorhinal information storage aimlessly swim within the tank. If, by chance, the animal swims into the immediate proximity of the platform, it begins a localized search. Taken together, these results suggest that information stored in these two structures may allow for navigation from a distance (entorhinal) to a precise (hippocampus) unmarked goal.

4 Each Phase of Memory Storage Requires Distinct Molecular Events

Based on the duration, memory is thought to have four phases: working (lasting seconds), short- (lasting minutes to hours), intermediate- (lasting several hours), and long-term memory (lasting days to a lifetime). Engagement of different phases of memory depends on the amount and pattern of training as well as on the learning paradigm. Recent experiments are beginning to address if the different phases of memory are interdependent or if they can develop independently of one other. For example, can long-term memory be formed in absence of short-term memory? As we will see, distinct molecular mechanisms underlie these phases of memory (➋ *Figure 19-6*) and in some situations, memory phases can develop independently.

Figure 19-6
Different phases of memory allow for the storage of information ranging from seconds to years. Working memory (lasting for up to 20 seconds), short-term memory (lasting for minutes), intermediate-term memory (lasting for hours) and long-term memory (lasting for hours to days) can be distinguished not only by their temporal profiles, but also by the underlying molecular mechanisms. Long-term memories can become permanent through a process called consolidation. Although it was thought that the memory phases are interdependent, giving rise to a temporal continuum of memory, recent experiments suggest that the different phases of memory may develop independently

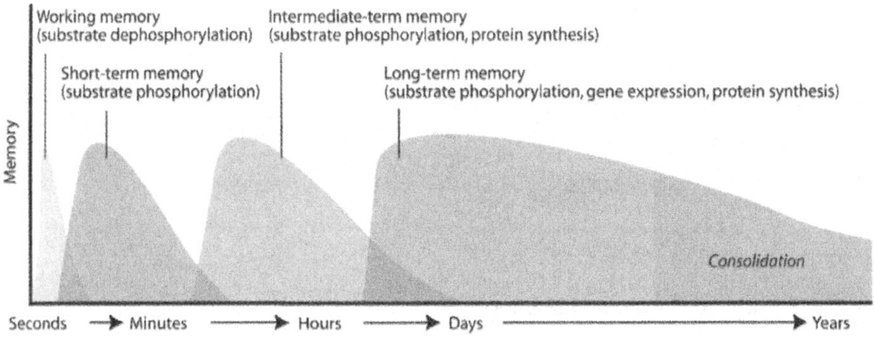

4.1 Working Memory Requires Persistent Neuronal Activity and Calcium-Sensitive Phosphatase Activity

Working memory is defined as the process of actively maintaining and integrating information for a relatively short period of time for directing goal-oriented action (Fuster, 1973). A common example is remembering a phone number long enough to dial the number. Once the number is dialed, the memory for the number is forgotten. As discussed in ➋ *Section 6*, working memory is also necessary for recreating or retrieving previously stored memory. It is thought that reverberating neuronal activity maintains working memory. Electrophysiological recordings and functional imaging studies have indicated the involvement of the dorsolateral PFC in monkeys and humans in working memory (➋ *Figure 19-7*). In rats, the medial part of the prefrontal cortex (mPFC) is considered to be the homologue of the primate dorsolateral PFC. Several tasks, such as the delay match-to-sample and delay non-match-to-sample, have been developed to evaluate working memory function in animals. In these task, animals are required to remember a cue for a period of seconds (the delay period) in order to perform an appropriate response. For example, in a delay match-to-sample task, a monkey is shown an object (e.g. a blue ball) that is then hidden from view for a period of 5 seconds. After the delay, the subject is shown the same blue ball along with another, novel object (e.g. a red cube). The animal must correctly identify the blue ball as the familiar object in order to receive a food reward. In a delay non-match-to-sample task, the monkey must identify the novel object (e.g. the red cube in the above example) as the non-matching object. 'Delay cells' within the PFC of monkeys and rats have been shown to remain active during working memory tasks, suggesting that these neurons may transiently store information (Fuster and Alexander, 1971). Appreciation of the importance of the PFC, and delay cells in particular, has facilitated the design of experiments to test the molecular mechanisms underlying working memory.

The work of Brozoski et al. was the first to demonstrate that dopamine plays a critical role in the modulation of the working memory function of the PFC (Brozoski et al., 1979). In this experiment,

Figure 19-7

a. Color-shaded regions represent approximate anatomical location of dorsolateral prefrontal cortex (Brodmann areas 10 and 46) involved in working memory. b. Location of the rodent medial prefrontal cortex (PL+IL) in a representation of a coronal brain section. AC: anterior cingulate, cc: corpus callossum, PL: prelimbic cortex, IL: infralimbic cortex

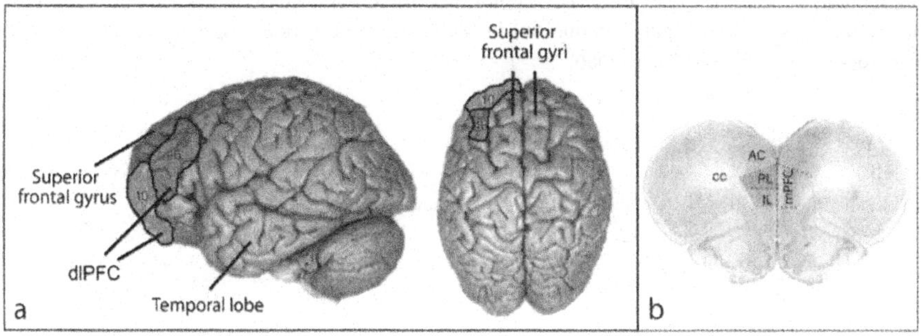

depletion of dopamine in the PFC of rhesus monkeys produced impairments in working memory performance nearly as severe as that caused by surgical ablation of the same area. Furthermore, this behavioral deficit could be pharmacologically reversed with L-DOPA or apomorphine, a mixed dopamine D1/D2 receptor agonist. The involvement of dopamine is further supported by electrophysiological recording in monkeys performing working memory tasks in which iontophoresis (a method of administering a compound using an electric current) of dopamine onto dorsolateral prefrontal neurons enhanced memory-related firing during the delay period (Williams and Goldman Rakic, 1995). Although D1 receptor stimulation appears to be integral for proper working memory, overstimulation can impair working memory function. It has been recognized for more than 20 years that exposure to even mild stress greatly increases dopamine (as well as norepinephrine) levels in the PFC, and that these changes in neurotransmitter levels are associated with impaired working memory function. In contrast, exposure to mild stress has little effect or actually improves performance of tasks dependent on the inferior temporal cortex or cerebellum. These and other studies indicate that a limited range of dopamine levels in the PFC is optimal for working memory function. Both excessive and insufficient dopamine receptor stimulation can impair working memory performance giving rise to an inverted "U-shaped" dependency of working memory on dopamine levels (➋ *Figure 19-8*).

Dopaminergic projections to the PFC originate predominately from the ventral tegmental area (VTA), and are often referred to as mesoprefrontal dopamine projections. Dopamine receptors are metabotropic, G-protein coupled receptors that induce their effects on neuronal activity through the stimulation or inhibition of the activity of intracellular enzymes. There are two different types or classes of dopamine-binding receptors: D1-type receptors comprised of D1- and D5- receptor isoforms (which can increase cAMP levels) and D2-type receptors comprised of D2-, D3-, and D4- receptor isoforms (which decrease cAMP levels). As indicated in ➋ *Figure 19-8*, working memory is dependent on, and is highly sensitive to, D1 receptor activity. Stimulation of D1 receptors increases cAMP formation and the activity of cAMP-dependent PKA. Although many of the actions of dopamine are thought to be mediated by the PKA cascade, it has been shown that inhibition of PKA activity in the PFC does not interfere with working memory. In contrast, stimulation of PKA activity by intraprefrontal infusion of Sp-cAMPS, a membrane permeable cAMP analogue that is resistant to breakdown by phosphodiesterases, impairs working memory in normal animals (Taylor et al., 1999). This suggests that although PKA activity is not required for the formation of working memory, excess PKA activity actually interferes with the process. Interestingly, the aging-associated decline in working memory can be blocked by the cAMP analogue Rp-cAMP, an inhibitor of PKA, suggesting that an increase in prefrontal PKA activity as a result of aging may underlie some of the working memory problems observed in this population.

Figure 19-8
An inverted "U"-shaped function for the role of PFC dopamine/D1 receptors in working memory. Physiological and behavioral studies indicate that normal working memory functions only within a limited range of dopamine D1 receptor activation. When prefrontal dopamine is below the optimal range (as in aging and Parkinson's disease) or above the optimal range (as occurs in stressful situations and amphetamine psychosis), working memory is impaired. Increasing D1 receptor activation (agonism) in low dopamine conditions, and decreasing D1 receptor activation (antagonism) in high dopamine conditions, improve working memory. Figure adapted from Goldman-Rakic (2000)

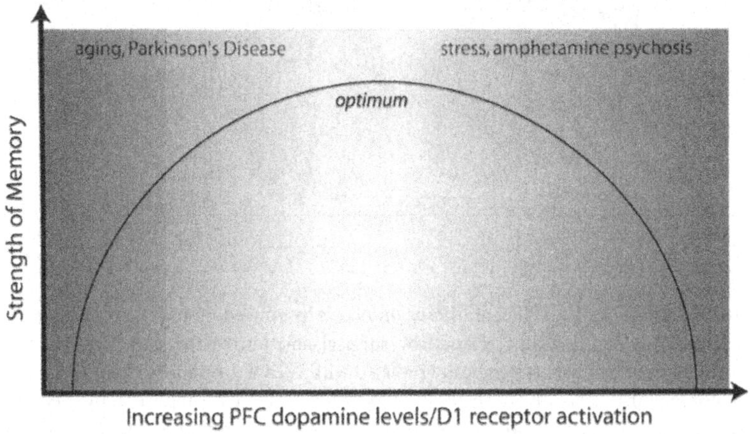

In addition to stimulating cAMP signaling, D1 receptors can increase phospholipase C activity via the $G_{q/11}$ proteins (*Figure 19-9*). Phospholipase C cleaves membrane phosphatidylinositol phospholipids to generate diacylglycerol and inositol trisphosphate (IP_3). IP_3 causes calcium release from intracellular stores by binding to its receptor located on endoplasmic reticulum. Increased calcium release leads to the sequential activation of calcium/calmodulin-dependent phosphatase calcineurin, calcium/

Figure 19-9
Dopamine D1 receptor activation can increase intracellular levels of cAMP and calcium. Binding of dopamine to D1 receptors activates GTP-binding proteins G_s and $G_{q/11}$ which, in turn, stimulate the activities of adenylyl cyclase (AC) and phospholipase C (PLC), respectively. Adenylyl cyclase converts ATP to cAMP, which stimulates protein kinase A (PKA). PLC cleaves the membrane lipid phosphoinositide bisphosphate (PIP_2) to generate inositol trisphosphate (IP_3) and diacylglycerol (DAG). IP3 binds to its receptor on the endoplasmic reticulum causing the release of calcium, which stimulates calcium-sensitive enzymes such as calcineurin, calcium/calmodulin-dependent protein kinase and protein kinase C

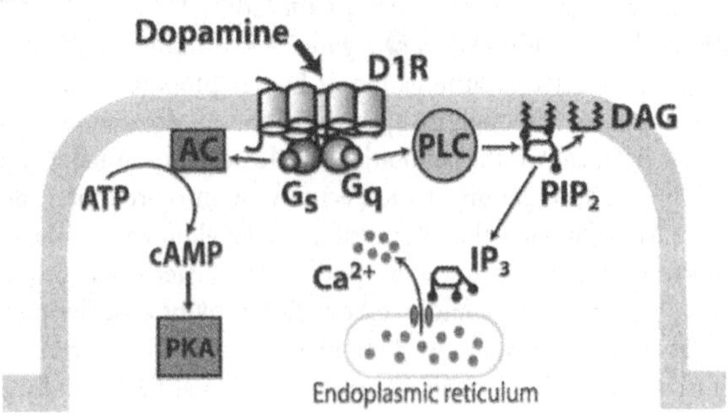

calmodulin-dependent protein kinase (CaMK) and calcium/diacylglycerol-dependent PKC as calcium accumulates. Recent biochemical studies show that while calcineurin is required for working memory, activation of PKC and CaMK (similar to that observed for PKA activation) exert a negative influence (Runyan et al., 2005). It has been proposed that following small increases in intracellular calcium, the calcium/calmodulin-activated phosphatase calcineurin is activated, leading to the dephosphorylation of specific intracellular substrates. During working memory, activation of calcineurin may lead to the dephosphorylation of ion channels in the plasma membrane of prefrontal neurons. This dephosphorylation changes the biophysical properties of the neurons leading to enhanced activity during the delay period, thereby actively maintaining working memory. As calcium levels continue to rise as a result of neuronal activity and dopamine levels, PKC and CaMK (via calmodulin) become activated. Enhanced activity of these protein kinases increases the phosphorylation of channels (possibly those previously dephosphorylated by calcineurin), thereby decreasing or blocking delay period activity and working memory. As stated above, exposure to even mild stress has been shown to increase dopamine (and norepinephrine) levels resulting in impaired working memory. Thus, mild stress may impair working memory by increasing intracellular calcium to a level sufficient to activate calcium-dependent protein kinase activity, impeding the cell's ability to dephosphorylate substrates required to maintain delay activity. Consistent with this possibility, it has been shown that stress-induced working memory impairment is due to, in part, over activation of PKC (Birnbaum et al., 2004).

4.2 Short-Term Memory Requires Kinase Activation and Phosphorylation of Preexisting Proteins but not Protein Synthesis or Gene Expression

A learning event leading to short-term memory formation (lasting minutes) causes the release of glutamate and other neurotransmitters. These neurotransmitters act upon their receptors located on postsynaptic neurons resulting in their depolarization and increase in the levels of intracellular calcium (which is approximately 100 nM in resting neurons). Increases in the levels of intracellular calcium can directly (CaMKII, PKC) and indirectly (PKA) stimulate the activity of intracellular second messenger cascades. For example, short-term sensitization of the gill-withdrawal reflex in *Aplysia* results in the release of serotonin (5-HT) from presynaptic interneurons onto the siphon sensory neurons. Binding of released 5-HT with its receptor activates the enzyme adenylyl cyclase resulting in increased cAMP levels. Among other things, this rise in cAMP leads to activation of PKA, which phosphorylates specific potassium channels (S- or serotonin-sensitive potassium channel) (Shuster et al., 1986). Phosphorylation of this channel causes prolonged channel closure, thus preventing the efflux of potassium to the extracellular space and slowing repolarization. The increase in intracellular potassium causes broadening of the action potential (since the neuron cannot quickly repolarize) allowing for more Ca^{2+} entry and enhancing transmitter release (a presynaptic mechanism for short-term memory). Moreover, the closing of the potassium channel increases the input resistance of the membrane because fewer channels are open at rest, thereby increasing neuronal excitability (i.e., more action potentials in response to the same level of depolarization) and transmitter release. As a result of these transient changes, the same stimulus can, for a short time, elicit greater transmitter release and a larger response. As long as the channels remain phosphorylated (which is on the order of minutes), short-term memory persists. Eventually, the channels are dephosphorylated by protein phosphatases, returning the neuron to a lower state of activity allowing the memory to decay. Thus, the underlying molecular mechanisms required for the formation of working memory, and those required for short-term memory, appear to be antagonistic. Recently, short-term memory mechanisms that are independent of the phosphorylation of potassium channels and the duration of action potential have also been identified (Sugita et al., 1997).

The biochemical mechanisms involved in explicit learning and memory were first demonstrated in rats performing the Morris water maze task. Richard Morris and colleagues demonstrated that rats injected with AP5, a NMDA-selective antagonist, have impaired spatial memory formation (Morris et al., 1986). Subsequently, it was shown that mice lacking the NMDAR1 subunit (also called NR1) of the NMDA receptor (resulting in impaired calcium conductance) have similar impairments (Sakimura et al., 1995).

Interestingly, overexpression of the NMDAR2B subunit (also called NR2B) of the NMDA receptor, which allows more calcium entry, enhances memory formation. As described above, calcium influx activates intracellular second messenger kinases. Studies in rats and mice have shown that inhibition of CaMKII, PKC, or PKA by pharmacological inhibitors or genetic techniques impairs short-term spatial- and fear memory, indicating that increased protein phosphorylation is required for short-term memory. For example, genetically modified mice lacking the alpha subunit of CaMKII, as well as those with a constitutively active form of the enzyme (caused by a substitution of the autophosphorylation site Thr^{286} with an Asp), have impaired spatial memory formation. In these mice, hippocampal place cells are unstable, suggesting that the animal cannot form a spatial map. Consistent with a requirement for substrate phosphorylation in short-term memory, inhibition of the calcium-sensitive protein phosphatase calcineurin leads to short-term memory enhancement (Malleret et al., 2001). These findings are consistent with those observed in *Aplysia* and other model systems, demonstrating that the molecular mechanisms underlying short-term memory have been conserved throughout evolution.

The PKC family of protein kinases is subdivided into three types according to their mechanism of activation: conventional (α, β(1), β(2), and γ), novel (δ, ε, η, θ and μ), and atypical (ζ, λ and τ). Conventional PKC isozymes are activated in response to increases in intracellular calcium and binding to diacylglycerol as a result of phospholipase C activity. Novel PKCs require diacylglycerol, but not calcium. Atypical isoforms are dependent on neither calcium nor diacylglycerol for activation. Structurally, PKC isoforms are single polypeptides consisting of two domains: (a) the N-terminal regulatory domain contains binding sites for cofactors and a pseudosubstrate (a pseudosubstrate has a consensus sequence for phosphorylation without a Serine/Threonine) that inhibits the kinase activity and (b) the C-terminal catalytic domain. The role of PKC activity in memory storage has been addressed both pharmacologically as well as with knockout mice. For example, administration of phorbol esters, which can activate novel and conventional isoforms of PKC, have been found to elicit synaptic enhancement which is indistinguishable from LTP (Malenka et al., 1986). These effects, as well as LTP itself, could be blocked by the PKC inhibitors polymyxin B and K-252b (Reymann et al., 1988a, b). Once activated, PKC translocates from the cytosol to the cell membrane where it is thought to phosphorylate several plasticity-related proteins including the presynaptic protein growth associated protein 43 (GAP43) and the postsynaptic calmodulin reservoir protein neurogranin. Although the involvement of PKC activity in neuronal plasticity has been well described, the roles of the individual isozymes have yet to be fully explored.

Although the mechanisms by which each of these kinases enhances short-term explicit memory has not been fully delineated, activation of silent synapses as result of AMPA receptor insertion (a postsynaptic mechanism for short-term memory) or inactivation of potassium channels are possibilities. Phosphorylation of AMPA receptors by protein kinases is a required step for their insertion at the postsynaptic site. Both PKC and CaMKII can phosphorylate GluR1 subunit of AMPA receptors at the same site at Ser^{831} (Barria et al., 1997). In addition, prolonged activation of PKC via the calcium-stimulated neutral protease calpain has also been implicated in certain types of short-term memory. In response to increases in intracellular calcium, calpain activity is increased resulting in the cleavage of specific substrate proteins such as PKC. Multiple PKC isozymes can be cleaved by calpain (removing the N-terminal regulatory domain) to generate constitutively active kinases (referred to as PKM) that can continue to phosphorylate substrate proteins even after calcium concentrations have returned to normal levels. In addition, PKM can diffuse away from the plasma membrane, freeing it to possibly interact with and phosphorylate targets distinct from membrane bound PKC isoforms. Although the specific substrates of PKM are still being identified, this may represent one mechanism by which protein phosphorylation is maintained, giving rise to short-term memory formation.

Although short-term memory requires protein phosphorylation, it is independent of new protein synthesis and gene expression. As discussed in the following sections, longer-lasting forms of memory result from neuronal changes, which persist beyond the temporal limitations of protein phosphorylation. While activation of protein kinases is still critical for these forms of memory, long-lasting neuronal changes require the synthesis of new proteins.

4.3 Intermediate-Term Memory Requires Protein Synthesis from Preexisting mRNA

Memory lasting for hours has been classified as intermediate-term memory. Unlike short- and long-term memory, the molecular mechanisms contributing to intermediate-term memory are only recently beginning to be examined. Studies performed in *Aplysia* indicate that a correlate of intermediate-term memory (synaptic facilitation) lasts for over 2 h, while short-term facilitation lasts less than 30 min (Sutton et al., 2004). Mechanistically, intermediate-term memory differs from short-term memory in its requirement for protein synthesis. Inhibition of protein synthesis by drugs such as anisomycin or gelonin blocks intermediate-term memory without interfering with short-term memory. Interestingly, protein synthesis at synaptic sites, but not in the soma, is required for intermediate-term memory formation suggesting a synaptic mechanism of induction (Sherff and Carew, 2004). Messenger RNAs for several molecules implicated in memory formation (e.g., CaMK II, Fragil X, nur77, PKMξ) and structural proteins [e.g., tubulin and microtubule associated protein 2 (MAP-2)] are found within the dendrites and perhaps at the synapse. How can a learning event activate synaptic protein synthesis? One model that can be envisioned is based on recent findings by Kandel and colleagues in *Aplysia*. These researchers have shown that the cytoplasmic polyadenylation element binding protein (CPEB) and its mRNA are present in the sensory-motor neuron terminals involved in the gill-withdrawal reflex (Si et al., 2003). CPEB is capable of increasing protein synthesis by elongating the poly A tails of mRNA. The release of neurotransmitters (serotonin in *Aplysia*) at specific synapses as a result of learning can activate CPEB (possibly via phosphorylation-induced conformational change, protein aggregation and/or enhanced CPEB synthesis), which would lead to enhanced synaptic protein synthesis required for intermediate term memory. An interesting aspect of CPEB is its ability to remain active via oliogomer formation and propagation of activity to newly synthesized CPEB via a prion-like mechanism. Such a self-propagating activity could be used to maintain intermediate-term memory.

The continuous activity of a specific isoform of PKC (PKMξ) has also been shown to maintain LTP in the rat hippocampus and associative memory in *Drosophila* (Drier et al., 2002). PKMξ is related to PKCξ, but it lacks the regulatory N-terminal domain making it autonomously active (Hernandez et al., 2003). The absence of the N-terminal domain was originally thought to be due to proteolytic cleavage of PKCξ, but recent evidence has demonstrated that PKMξ is translated directly from its own mRNA. PKMξ mRNA is localized at dendritic sites where it is normally repressed by BC-1, a nontranslatable mRNA that blocks the translation initiation complex. In response to learning, BC-1 repression is released allowing PKMξ mRNA to be translated into protein (➋ *Figure 19-10*). Once translated, PKMξ phosphorylates dendritic receptors and channels which can maintain memory for hours. PKMξ activity has been shown to be not only necessary, but also sufficient for memory maintenance.

The duration of these protein synthesis-dependent persistent kinase activities determines the duration of intermediate-term memory. While intermediate-term memory requires local protein synthesis, its formation is independent of gene transcription. As seen below, the inhibition of transcription selectively impairs long-term memory without interfering with either short- or intermediate-term memory.

4.4 Long-Term Memory Requires Kinase Activation, Gene Expression, and Protein Synthesis

Similar to the shorter forms of memory, long-term memory formation begins with stimulation of neurotransmitter receptors as well as neurotrophic factor receptors during learning. Activation of NMDA receptors in conjunction with cell depolarization during learning causes a rise in calcium levels inside neurons. Elevations in intracellular calcium can activate several enzyme pathways including PKA (via adenylyl cyclase and increased cAMP), PKC, CaMK, and ERK cascades, each of which have been implicated in the formation of long-term implicit and explicit memories. Several studies demonstrated that, similar to

Figure 19-10

Translation of preexisting mRNA is required for intermediate-term memory. PKMξ mRNA is found in the dendrites of neurons where it is maintained in an untranslated state by its interaction with BC-1 (not shown). In response to learning, BC-1 repression is released, allowing PKMξ to be translated into protein. PKMξ, being constitutively active, can maintain the phosphorylation of ion channels and/or receptors for hours allowing memory to persist beyond the temporal limits of short-term memory

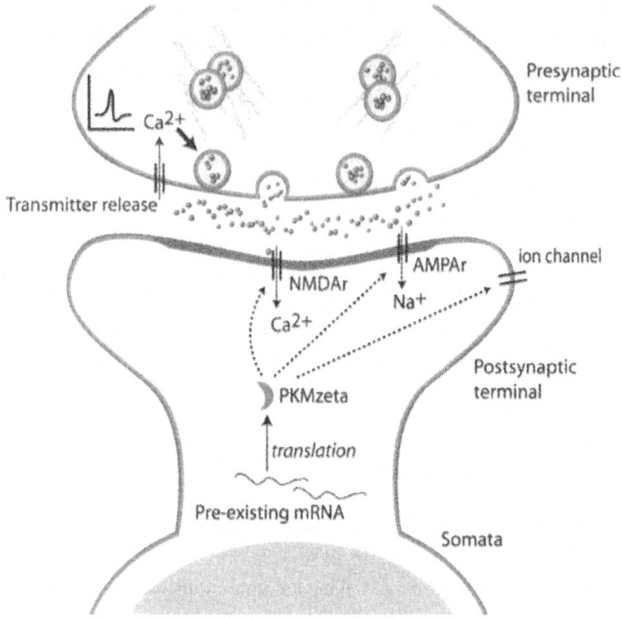

other forms of memory, long-term memory requires an optimal cAMP level. Consistent with this, decreased expression of G_i (a GTP-binding protein that inhibits adenylyl cyclase activity) in the rodent hippocampus results in a substantial increase in intracellular cAMP and impaired long-term spatial memory. In contrast, administration of rolipram, a cyclic nucleotide phosphodiesterase inhibitor, modestly increases cAMP levels and enhances long-term spatial memory.

As in short-term facilitation, long-term facilitation in *Aplysia* requires active PKA. As indicated before, binding of cAMP to PKA causes dissociation of the regulatory subunits from the catalytic subunit, and this catalytic subunit is responsible for phosphorylation of potassium channels. A protease (possibly proteosome or one of its subunits) that is synthesized as a result of behavioral training degrades the regulatory subunits leaving the catalytic subunit free (Chain et al., 1999). Inhibition of protein synthesis blocks this persistent kinase activity and memory. This results in persistent PKA activity that is independent of its requirement for cAMP, and thus receptor activation. The involvement of the cAMP/PKA cascade in memory formation in a wide variety of species including *Aplysia*, *Drosophila*, rats, and mice indicates an evolutionarily conserved mechanism for memory formation.

Increases in cAMP/PKA activity, as well as increases in intracellular calcium and neurotrophic factor signaling, can activate the ERK cascade. ERK is activated via dual phosphorylation on Thr and Tyr by the kinase MEK (MAP/ERK kinase). Using phosphorylation-specific antibodies for ERK, it has been demonstrated that training of rodents in the Morris water maze task increases ERK phosphorylation in hippocampal neurons. Furthermore, when ERK activation is inhibited in the hippocampus or the entorhinal cortex (the major input/output structure for the hippocampus, *Figure 19-5*) by infusion of pharmacological inhibitors of MEK, long-term memory is impaired. However, short-term memory remains intact. Two isoforms of CaMK, CaMK II (primarily cytosolic), and CaMKIV (primarily nuclear) have also been demonstrated to be involved in memory. As discussed earlier, CaMKII activity appears to be necessary for the formation (learning) of spatial memory. In contrast, CaMKIV mutant mice show impairments in long-term spatial memory retention, and hippocampal late-LTP without having any appreciable differences in basic synaptic function or early-LTP (Kang et al., 2001). These effects of CaMKIV appear to be due to its ability to stimulate the expression of genes required for long-lasting changes in neuronal function.

Activation of the above mentioned protein kinases leads to long-term memory formation through their ability to phosphorylate and activate transcription factors, especially the cyclic AMP/calcium-response element binding protein (CREB) (➊ *Figure 19-11*). The CREB protein was originally identified as the transcription factor that induces the expression of the *somatostatin* gene in response to increases in

Figure 19-11

A molecular model for the formation of long-term memory. Training-induced release of neurotransmitters (glutamate) and neurotrophic factors (BDNF) activates cell surface receptors and increases intracellular concentrations of second messengers (e.g., Ca^{2+} and cAMP). Second messenger-stimulated kinases phosphorylate the transcription factors CREB and SRF, as well as other transcription factors such as NFkB and C/EBP (not shown). Phosphorylated CREB protein increases the transcription of genes carrying the CRE (cAMP response element) sequence in their promotors. Some of these gene products activate late effector genes whose protein products are needed for synapse formation (or elimination) and/or enhanced excitability, transmitter release and receptor levels. These modifications result in the long-lasting changes in neuronal communication that underlie long-term memory storage

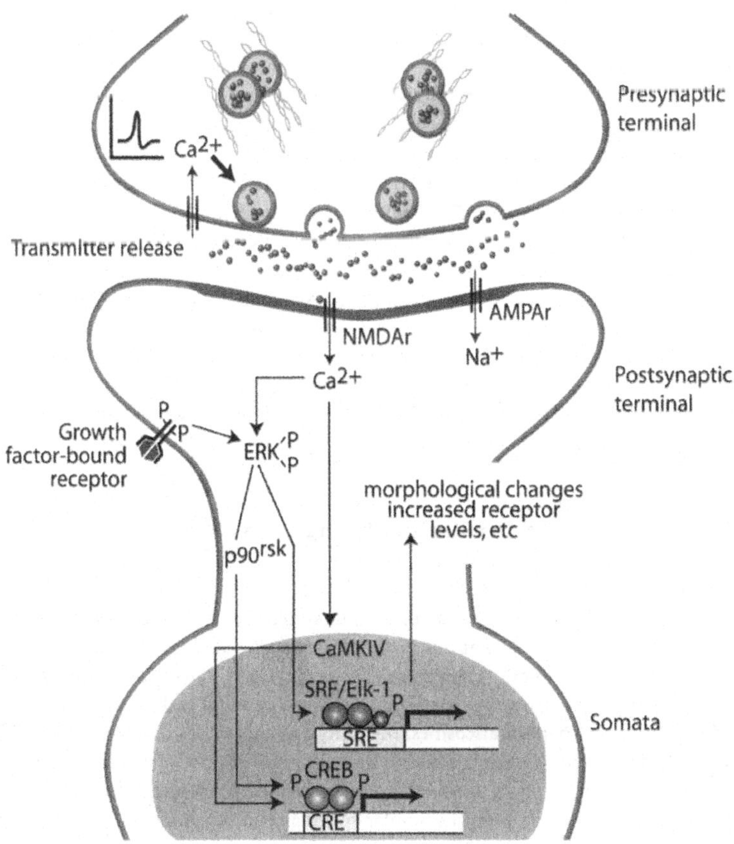

intracellular cAMP. CREB binds to a specific enhancer sequence called the CRE element (TGACGTCA) that is located in the promoter region of *somatostatin* and several other genes whose transcription is regulated by cAMP. A sustained increase in cAMP leads to the translocation of the catalytic subunit of PKA into the nucleus where it phosphorylates Ser^{133} of CREB. Phosphorylation of CREB on Ser^{133} recruits transcriptional coactivators to the promoter complex and enhances mRNA transcription (Yamamoto et al., 1988). CREB has also been shown to increase the expression of CRE-sequence containing genes in response to increased intracellular calcium levels, or activation of ERK. It is thought that CaMKIV, which is localized in the cell nucleus, phosphorylates CREB on Ser^{133} in response to sustained calcium level increases, whereas ERK can indirectly lead to CREB activation via $p90^{rsk}$. Thus, CREB can act as a convergence site for several protein kinases implicated in memory formation.

The first study demonstrating the involvement of CREB in long-term memory was carried out in *Aplysia* (Dash et al., 1990). Long-term facilitation was found to require CREB mediated gene expression because microinjection of CRE decoy oligonucleotides into the nucleus of *Aplysia* sensory neurons blocked this phenomenon. This effect was specific to long-term facilitation, as CRE oligonucleotide injection did not affect the short-term facilitation caused by brief 5-HT application. Following these discoveries in *Aplysia*, it was demonstrated that mutation of the *Drosophila* CREB gene (*dCREB2*) impairs long-term memory for classical conditioning (Yin et al., 1994). Interestingly, overexpression of CREB may enhance long-term memory in this task. In addition, genetically altered mice carrying targeted disruptions of selective CREB isoforms also show impaired spatial memory. Since these isoforms of CREB were missing throughout development of the animal, one could argue that observed deficit is due to some subtle developmental abnormalities (Bourtchuladze et al., 1994). This potential problem was eliminated by decreasing CREB protein levels in adult animals by injecting antisense oligonucleotides directly into the hippocampi. Similar to the impairments seen in CREB null mice, animals receiving antisense oligonucleotides displayed long-term spatial memory deficits.

Since these initial experiments which demonstrated a requirement for CREB in long-term memory storage, several other transcription factors have been implicated including NFkB, SRF, zif268, c-Fos, c-Jun, Jun-B, and C/EBP (➋ *Figure 19-11*). For example, intrahippocampal infusion of double-stranded oligonucleotide decoys containing the binding sites for either nuclear factor kappaB (NFkB) or serum response factor (SRF) impair long-term spatial memory (Dash et al., 2004). NFkB, when bound to its regulatory protein IkB (inhibitor of NFkB), is normally cytosolic. Upon appropriate stimulation, IkB is phosphorylated and quickly degraded, unmasking the nuclear localization sequence in NFkB allowing it to translocate to the nucleus and initiate gene expression (Mercurio et al., 1997). As this phosphorylation of IkB can be blocked by inhibitors of the ERK cascade and ERK has been shown to phosphorylate the SRF/Elk-1 complex leading to enhanced subunit interaction and DNA binding (Gille et al., 1995), this suggests that these transcription factors may mediate some of the memory-related gene expression changes initiated by activation of the ERK cascade. Consistent with this, sequestration of SRF resulted in memory deficits, which appear to result from a loss of precise platform location information, effects consistent with those observed following inhibition of the ERK cascade in the hippocampus (Hebert and Dash, 2004).

4.4.1 Morphological Changes and Storage of Long-Term Memory

Ramon y Cajal was the first to postulate that morphological changes at the synapse may underlie lasting memories (Ramon y Cajal, 1911). Later, studies in both vertebrates and invertebrates showed that long-lasting morphological changes in the neurons involved in learning are likely to constitute the underlying cellular mechanism for long-term memory storage (Bailey and Chen, 1989). Long-term behavioral sensitization and habituation in *Aplysia* involves significant structural changes in the sensory neurons. It has been shown that the number of synaptic varicosities, the number and size of active zones, and the number of synaptic vesicles all change following long-term training (Nazif et al., 1991). When compared to untrained control animals, these morphological features increased in long-term sensitized animals, whereas they decreased in long-term habituated animals. Similar to long-term sensitization training, 5-HT or cAMP treatment of *Aplysia* sensory neurons in culture also produce morphological changes that have been shown to be dependent on protein and RNA synthesis, possibly as a result of CREB-mediated gene expression.

Examples of morphological changes associated with memory formation have also been observed following LTP induction in the CA1 and DG subfields of the hippocampus and in intact, behaving animals. As discussed above, several aspects of synaptic structure appear to change with experience. The most consistent of these structural alterations is an alteration in the number and/or pattern of synaptic connections. The formation of additional synaptic connections as a result of training is thought to increase the synaptic efficiency between two neurons by allowing more neurotransmitter release in response to presynaptic stimulation (*Figure 19-1*). As the addition of new synaptic connections is likely to be longer lasting than increases in transmitter release caused by the phosphorylation of ion channels and receptors, changes in the morphology of neurons allow for longer lasting forms of memory.

4.4.2 Synaptic Tagging

Long-lasting forms of synaptic plasticity (or synaptic efficiency) have been shown to require gene expression and protein synthesis. Since genes are transcribed at the nucleus in the cell soma, the signals generated locally at a synaptic site must be propagated to the nucleus, and the gene products, in turn, transported from the nucleus to the synapse. As the newly synthesized proteins are transported to all synapses, this raises the question of how gene products can be targeted to the specific synapse(s) where modification is needed. Frey and Morris showed that once long-term plasticity has been induced in one pathway of a hippocampal slice, this long-term change can be "captured" by a second pathway with a stimulus that would normally produce only short-term changes (Frey and Morris, 1997). Using a single bifurcated *Aplysia* sensory neuron that forms synapses with two motor neurons, Martin and colleagues independently demonstrated the synaptic "capture" phenomenon which led to the formulation of the "synaptic tagging" hypothesis (Martin et al., 1997). This hypothesis proposes that following the induction of long-lasting plasticity at a specific synapse (called the tagged synapse), gene products are synthesized and are delivered to all the synapses of a neuron, but are incorporated into only the synapse that has been "tagged" by previous synaptic activity. In addition, since gene products are transported to all the synapses of a neuron, a weak stimulation at a previously unstimulated synapse can utilize the gene products that are already present to induce long-lasting plasticity at this second synapse. Recently, Kandel and colleagues have proposed a molecular mechanism for synaptic tagging (Si et al., 2003). In this model, CPEB (the self-propagating protein suggested to play a role in intermediate-term memory, *Section 4.3*) marks the active synapse until new gene products arrive for the cell soma.

5 The Consolidation of Memories

Long-term storage of explicit memories requires a process of consolidation, a concept and term that was first introduced by Muller and Pilzecker in 1888, when they proposed that learning does not immediately induce permanent memories but that this requires a period of time for memory to be "consolidirung" (consolidated). During this period of time, memories remain vulnerable to disruption. Memory consolidation can therefore be defined as the process by which information becomes resistant to disruption within the brain, though it is not known exactly why a consolidation period is necessary. This requirement is illustrated by the retrograde amnesia observed following traumatic brain injury. These patients demonstrated temporally-graded retrograde amnesia in that older memories were found to be spared while more recent memories were lost.

Although it is clear that the medial temporal lobe plays an integral role in memory formation, there has been considerable discussion about what function the medial temporal lobe performs that is necessary for the consolidation of explicit memory. From this debate, two conclusions can be made about the medial temporal lobe's involvement in memory consolidation: (1) it is required for the formation of usable memory, and (2) it supports explicit memory for a period of time during which the memory is more susceptible to disruption. However, it is thought that permanent memory storage occurs outside the medial temporal lobe, within the higher association areas of the cerebral cortex. The medial temporal lobe is connected to many cortical areas, including the higher association areas, via the perirhinal and enthorhinal cortices as well as the formix and anterior commissure. It has been suggested that this connectivity would allow for a time-limited involvement of the medial temporal lobe in the storage and/or retrieval of memory representations within these cortical loci. However, it has also been proposed that certain types of memory pertaining to contextual and specific spatial representations are permanently stored within the hippocampus.

5.1 Consolidation Models

5.1.1 The Standard Model

In 1900, Muller and Pilzecker proposed that the formation of permanent memory takes time, and that during this period memory remains vulnerable to disruption. As indicated in ➲ *Section 3.2*, the patient H.M. showed profound amnesia for recent explicit memories with earlier memories remaining intact. From these observations, the Standard Model for memory consolidation was developed in order to explain the occurrence of progressively worse amnesia for more recent memories compared to remote memories (i.e., temporally graded retrograde amnesia) following hippocampal damage. This model posits that memory is initially stored in the hippocampus using sparse, nonoverlapping representations. Over time, the hippocampus induces a distributed representation of the memory in the neocortex, after which the hippocampus is no longer required (Marr, 1971). The time course for the development of this distributed neocortical representation is the key determinant for the duration of systems consolidation.

In 1985, Teyler and DiScenna proposed a variation of the Standard Model that stated that LTP in the hippocampus initially stores an index of the neocortical areas activated by the experience (Teyler and DiScenna, 1985). During memory retrieval, this hippocampal memory storage is necessary for reactivating the unique pattern of cortical areas involved in the original experiential event in the same spatio-temporal sequence. Over time, repeated hippocampal reactivation of a set of neocortical areas composing an experiential event is proposed to be involved in establishing a cortically based memory trace that is capable of hippocampus-independent memory retrieval.

5.1.2 The Multiple Trace Theory

The Multiple Trace Theory (MTT) was proposed to explain observations of flat retrograde amnesia (the loss of both recent and remote memories equally) for episodic memory (memory of a personally experienced event) following hippocampal damage (Nadel et al., 2000). The MTT incorporates a key feature of the Standard Model in the idea that the hippocampal complex (hippocampus proper, subiculum, entorhinal and perirhinal cortices, and parahippocampus) rapidly encodes all information. However, the MTT states that only nonepisodic memories (memories for facts and concepts, also called semantic memory) become hippocampus-independent over time and that episodic memory remains bound to the hippocampus. According to this theory, each reactivation of a memory results in the creation of a new hippocampally-stored memory trace. Over time, as more traces for a memory are created, the memory encompasses an increasingly larger hippocampal area. Therefore, a partial hippocampal lesion would not result in amnesia for older memories, but would give rise to a pronounced amnesia for recent memories. Complete hippocampal lesions were proposed to result in flat retrograde amnesia for hippocampal-dependent memories. However, animal studies have revealed that complete removal of the hippocampus in rats can result in a temporally graded amnesia for spatial and other hippocampal-dependent memories. Consistent with these experimental findings, it has been reported that patients with severe damage to the hippocampal region demonstrate a temporally graded retrograde amnesia for episodic- and semantic-, as well as spatial- and non-spatial memory.

5.1.3 The "C" Theory

There is increasing evidence that long-term memory-associated molecular changes occur in both the hippocampus and the neocortex as a direct result of learning. As outlined in the previous sections, some of these early changes have been shown to be required for long-term memory. These data are not accounted for by the previous consolidation theories that maintain that initial memory storage takes place solely within the hippocampus. In addition, these previous theories do not account for the apparent longer consolidation periods in species with physically larger brains. To account for these observations, the C theory was proposed (Dash et al., 2004). This theory postulates that the initial learning event triggers changes in gene expression in neurons within both the hippocampal and neocortical areas (and possibly subcortical areas), resulting in memory storage in these structures after learning. During recall, the hippocampus serves as an initial memory coordinator for the retrieval of the components of memories stored in these neocortical and subcortical areas. This is in contrast to the Standard Model (➋ *Section 5.1.1*) which proposed that the hippocampus itself induces memory storage in the cortical areas.

The period in which the hippocampus is required as the coordinator is the time required for consolidation. As mentioned earlier, the length of this period seems to correlate with the size of the brain. The C theory proposes that, in addition to the hippocampus, the learning event also alters gene expression in the soma of other neurons, which connect the cortical and/or subcortical areas where components of memory are stored. However, until the gene products arrive at the synapses (which are likely to be tagged) of these cells to cause the required morphological changes, the hippocampus continues to coordinate retrieval and possibly maintain the synaptic tags. As the length of time required for the transport of proteins would be dependent on the length of the projections, larger brain sizes would require more time to develop this hippocampus-independent ability for memory retrieval.

6 The Retrieval of Memories

Explicit memory retrieval involves the conscious recollection of information acquired from past learning experiences. It is thought that information is retrieved by and held in mind through working memory in response to appropriate retrieval cues. Therefore, explicit memory retrieval involves prefrontal cortical regions. During memory retrieval, working memory assembles stored representations within cortical networks. In humans, memory encoding involves the left PFC more than the right, while the retrieval of stored memory activates the right prefrontal more than the left. This observation has led to the proposal that the lateralization of frontal lobe function in memory encoding and retrieval is due to the stronger involvement of the right hemisphere in emotional- and imagery-based autobiographical episodes.

There are some data supporting the view that memory loss results from a deficit in retrieval processes, rather than a loss of memory itself. For instance, patients with memory disorders show a benefit from recognition and contextual cues, suggesting that damage to the hippocampus, rather than resulting in the loss of memory per se, results in the inability to retrieve recent memories. These findings are consistent with the C theory since memory is initially formed not only in the hippocampus but also in the neocortex. Since the hippocampus initially coordinates retrieval of memory, damage to this structure would result in impaired retrieval of information stored in neocortical sites. Although the neurochemical basis of memory retrieval is not well understood, recent studies employing gene deletion techniques have implicated norepinephrine/epinephrine in the retrieval of spatial and contextual memory. For example, deletion of the gene encoding dopamine β-hydroxylase (DBH), the enzyme responsible for norepinephrine/epinephrine synthesis, results in impaired memory recall (Murchison et al., 2004). Interestingly, if a precursor for norepinephrine (L-DOPS), which can be converted to norepinephrine in the absence of DBH activity, is injected into the hippocampus of *dbh* null mice prior to memory testing, their memory recall is normal. Pharmacological studies corroborate the role of norepinephrine in memory recall and further extend these findings to implicate the β1-adrenergic receptor in the retrieval process.

Recent studies also indicate that memory retrieval is not a passive process. These experiments have demonstrated that when a memory is reactivated (remembered), strong or robust memories can be temporarily suppressed if protein synthesis is impaired (Nader et al., 2000; Runyan and Dash, 2004a). For example, the association of a tone with a footshock in a trace conditioning paradigm (when the CS and US are separated by a time gap) requires hippocampal function and memory storage. As expected, if protein synthesis is impaired by anisomycin infusion immediately after training, cue-elicited long-term memory is impaired. If the memory is allowed to form, anisomycin infusion no longer has any effect. However, if a reminder (e.g., tone presentation) is given, then anisomycin is infused, and the original memory is lost. This effect appears to be related to memory strength, as weaker memories are not influenced by anisomycin infusion (Eisenberg et al., 2003). Although the reason for this effect is not known, this could arise from a weakening of the ability of the hippocampus to coordinate memory retrieval.

6.1 Forgetting and Memory Extinction

Forgetting is a normal process and several psychological theories have been proposed as to its cause (Mansuy, 2005). Dominant amongst these is the idea that memories are most often forgotten when they encounter interference during the consolidation period. Thus, older memories (those that have been consolidated) are predicted to decay at a slower rate than newer memories of the same synaptic weight. This interference could be the result of experiences, which are similar to the initial learning event, sometimes referred to as cue overload. However, interference cannot explain all forms of forgetting, especially those related to forced forgetting. This type of memory loss is most often observed in persons following a traumatic event such as sexual abuse.

Pathological conditions such as stroke, trauma, vitamin B1 deficiency (Korsakoff's syndrome), and following electroconvulsive therapy to treat depression, all result in forgetting (often referred to as amnesia). For example, electroconvulsive therapy is thought to interfere with gene expression and protein synthesis, impairing the consolidation of new memories while leaving older memories intact. Similarly, mild brain trauma can cause both limited retrograde (impairment of memory recall for events prior to the condition) and anterograde (impairment of new memory formation following the pathological condition) amnesias in the absence of overt cell loss, possibly by altering neuronal function via aberrant expression of genes and proteins.

Unlike forgetting, memory extinction is new learning and not an erasure of the initial memory. For example, if a CS is repeatedly presented in the absence of the original US, the conditioned response will be lessened. This reduction in response is thought to occur as a result of competing memory representations. Imaging studies in humans, and lesion studies in animals, have implicated the PFC in memory extinction. For example, patients with posttraumatic stress disorder (an anxiety disorder in which extreme distress occurs after a person is reminded of the event by a personal "trigger") have depressed ventral PFC activity (Bremner, 1999). Regardless of new experience, these patients are incapable of learning that the "trigger" does not predict a new traumatic event. Similarly, rats with mPFC lesions can acquire fear in a delayed tone-shock training paradigm, but require many more unpaired tone presentations to extinguish the fear than control animals (Morgan et al., 1993). In a recent study, Milad and Quirk demonstrated that tone (CS) paired with brief electrical stimulation of the infralimbic cortex (IL) (part of the mPFC, ➊ *Figure 19-7*) in a delayed conditioning paradigm caused enhanced memory extinction (Milad and Quirk, 2002). These and other studies show that PFC activity is required for memory extinction and that a loss of prefrontal function results in attenuated extinction. Since extinction is thought to be new learning, the process is likely to require some of the same molecular mechanisms necessary for memory formation. Consistent with this, inhibition of either protein synthesis or the ERK cascade within the mPFC impairs extinction memory (Hugues et al., 2004; Santini et al., 2004). When animals are tested a day or two following an extinction session, no reduction in the robustness of memory is observed.

7 Summary

Memory at the cellular level is stored via altered communication between neurons as a result of synaptic modulation and/or changes in excitability. Both explicit- (also called declarative memory requiring concious recollection) and implicit (also called nondeclarative memory) memory encoding utilize similar molecular mechanisms. Working memory, defined as the online holding of information for guiding behavior, is required for higher cognitive functions and is dependent on the function of the PFC. Both reverberating circuits within the PFC and intracellular protein dephosphorylation/phosphorylation, modulated by the action of dopamine and other neurotransmitters, are required for normal working memory. Short-term memory, lasting for minutes-to-hours, requires the phosphorylation of preexisting proteins including cell surface ion channels, and may involve insertion of AMPA receptors into the synapse resulting in enhanced synaptic communication. Intermediate-term memory, lasting for minute-to-hours requires the synthesis of proteins from preexisting mRNA, which may lead to prolonged kinase activation and protein phosphorylation. Long-term memory, lasting for days to a lifetime, is dependent on gene expression and protein synthesis giving rise to long-lasting morphological changes. This form of memory is dependent on the activation of specific transcription factors, of which CREB has been the best characterized. Explicit memory storage is initally dependent on the function of the hippocampus, a structure in the medial temporal lobe. These memories are made permanent by a process called consolidation, for which numerous theories have been proposed. The "C" theory posits that the molecular events underlying long-term memory storage are initiated in the hippocampus and neocortex (and possibly in subcortical structures) with similar time courses and that the period of hippocampal dependency is determined by the time required to transport the newly synthesized proteins to the appropriate cortical synapses. Explicit memory retrieval, like memory encoding, involves the PFC and working memory. During retrieval, working memory processes activate stored representations allowing for behavioral action, and if required, further modification of the memory. Memories can be actively suppressed by a process of extinction, in which new, conflicting memories compete. The molecular mechanisms underlying extinction appear to be similar to those involved in the original learning experience.

Acknowledgments

The authors would like to thank Drs. John H. Byrne and John Redell, as well as Diasinou Fiorovante, Evangelos Antzoulatos, and the members of the Dash laboratory for their critical readings of this material. The work in the authors' laboratory is supported by grants from NIH, NS049160, MH072933 and the Mission Connect/TIRR consortium.

References

Bailey CH, Chen M. 1989. Time course of structural changes at identified sensory neuron synapses during long-term sensitization in Aplysia. J Neurosci 9: 1774-1780.

Barria A, Derkach V, Soderling T. 1997. Identification of the Ca2+/calmodulin-dependent protein kinase II regulatory phosphorylation site in the alpha-amino-3-hydroxyl-5-methyl-4-isoxazole-propionate-type glutamate receptor. J Biol Chem 272: 32727-32730.

Benzer S. 1971. From the gene to behavior. JAMA 218: 1015-1022.

Birnbaum SG, Yuan PX, Wang M, Vijayraghavan S, Bloom AK, et al. 2004. Protein kinase C overactivity impairs prefrontal cortical regulation of working memory. Science 306: 882-884.

Bliss TV, Lomo T. 1973. Long-lasting potentiation of synaptic transmission in the dentate area of the anaesthetized rabbit following stimulation of the perforant path. J Physiol 232: 331-356.

Bourtchuladze R, Frenguelli B, Blendy J, Cioffi D, Schutz G, et al. 1994. Deficient long-term memory in mice with a targeted mutation of the cAMP-responsive element-binding protein. Cell 79: 59-68.

Bremner JD. 1999. Alterations in brain structure and function associated with post-traumatic stress disorder. Semin Clin Neuropsychiatry 4: 249-255.

Brozoski TJ, Brown RM, Rosvold HE, Goldman PS. 1979. Cognitive deficit caused by regional depletion of dopamine in prefrontal cortex of rhesus monkey. Science 205: 929-932.

Byrne JH. 1987. Cellular analysis of associative learning. Physiol Rev 67: 329-439.

Byrne JH, Kandel ER. 1996. Presynaptic facilitation revisited: state and time dependence. J Neurosci 16: 425-435.

Cahill L, Babinsky R, Markowitsch HJ, McGaugh JL. 1995. The amygdala and emotional memory [letter]. Nature 377: 295-296.

Chain DG, Casadio A, Schacher S, Hegde AN, Valbrun M, et al. 1999. Mechanisms for generating the autonomous cAMP-dependent protein kinase required for long-term facilitation in Aplysia. Neuron 22: 147-156.

Dash PK, Hebert AE, Runyan JD. 2004. A unified theory for systems and cellular memory consolidation. Brain Res Brain Res Rev 45: 30-37.

Dash PK, Hochner B, Kandel ER. 1990. Injection of the cAMP-responsive element into the nucleus of Aplysia sensory neurons blocks long-term facilitation. Nature 345: 718-721.

Dash PK, Orsi SA, Moore AN. 2004. Sequestration of serum response factor in the hippocampus impairs long-term spatial memory. J Neurochem 93: 269-278.

Drier EA, Tello MK, Cowan M, Wu P, Blace N, et al. 2002. Memory enhancement and formation by atypical PKM activity in Drosophila melanogaster. Nat Neurosci 5: 316-324.

Dudai Y, Jan YN, Byers D, Quinn WG, Benzer S. 1976. dunce, a mutant of Drosophila deficient in learning. Proc Natl Acad Sci USA 73: 1684-1688.

Dudek SM, Bear MF. 1993. Bidirectional long-term modification of synaptic effectiveness in the adult and immature hippocampus. J Neurosci 13: 2910-2918.

Eisenberg M, Kobilo T, Berman DE, Dudai Y. 2003. Stability of retrieved memory: inverse correlation with trace dominance. Science 301: 1102-1104.

Frey U, Morris RG. 1997. Synaptic tagging and long-term potentiation. Nature 385: 533-536.

Fuster JM. 1973. Unit activity in prefrontal cortex during delayed-response performance: neuronal correlates of transient memory. J Neurophysiol 36: 61-78.

Fuster JM, Alexander GE. 1971. Neuron activity related to short-term memory. Science 173: 652-654.

Gille H, Kortenjann M, Thomae O, Moomaw C, Slaughter C, et al. 1995. ERK phosphorylation potentiates Elk-1-mediated ternary complex formation and transactivation. EMBO J 14: 951-962.

Han PL, Meller V, Davis RL. 1996. The Drosophila brain revisited by enhancer detection. J Neurobiol 31: 88-102.

Hebb DO. 1949. The organization of behavior; a neuropsychological theory. New York: Wiley.

Hebert AE, Dash PK. 2002. Extracellular signal-regulated kinase activity in the entorhinal cortex is necessary for long-term spatial memory. Learn Mem 9: 156-166.

Hebert AE, Dash PK. 2004. Nonredundant roles for hippocampal and entorhinal cortical plasticity in spatial memory storage. Pharmacol Biochem Behav 79: 143-153.

Hernandez AI, Blace N, Crary JF, Serrano PA, Leitges M, et al. 2003. Protein kinase M zeta synthesis from a brain mRNA encoding an independent protein kinase C zeta catalytic domain. Implications for the molecular mechanism of memory. J Biol Chem 278: 40305-40316.

Hollmann M, Heinemann S. 1994. Cloned glutamate receptors. Annu Rev Neurosci 17: 31-108.

Hugues S, Deschaux O, Garcia R. 2004. Postextinction infusion of a mitogen-activated protein kinase inhibitor into the medial prefrontal cortex impairs memory of the extinction of conditioned fear. Learn Mem 11: 540-543.

James W. 1890. "The principals of psychology". Vols I-III. New York: Henry Holt and Company. Unaltered reproduction, New York: Dover, 1950.

Kandel ER, Schwartz JH. 1982. Molecular biology of learning: modulation of transmitter release. Science 218: 433-443.

Kang H, Sun LD, Atkins CM, Soderling TR, Wilson MA, et al. 2001. An important role of neural activity-dependent CaMKIV signaling in the consolidation of long-term memory. Cell 106: 771-783.

Le Doux JE. 2000. Emotion circuits in the brain. Annu Rev Neurosci 23: 155-184.

Liao D, Hessler NA, Malinow R. 1995. Activation of postsynaptically silent synapses during pairing-induced LTP in CA1 region of hippocampal slice. Nature 375: 400-404.

Malenka RC, Madison DV, Nicoll RA. 1986. Potentiation of synaptic transmission in the hippocampus by phorbol esters. Nature 321: 175-177.

Malleret G, Haditsch U, Genoux D, Jones MW, Bliss TV, et al. 2001. Inducible and reversible enhancement of learning, memory, and long-term potentiation by genetic inhibition of calcineurin. Cell 104: 675-686.

Mansuy IM. 2005. Forgetting : theories and potential mechanisms. Med Sci (Paris) 21: 83-88.

Marr D. 1971. Simple memory: a theory for archicortex. Philos Trans R Soc Lond B Biol Sci 262: 23-81.

Martin KC, Casadio A, Zhu H, Rose JC, Chen M, et al. 1997. Synapse-specific, long-term facilitation of aplysia sensory to motor synapses: a function for local protein synthesis in memory storage. Cell 91: 927-938.

McEchron MD, Bouwmeester H, Tseng W, Weiss C, Disterhoft JF. 1998. Hippocampectomy disrupts auditory trace fear conditioning and contextual fear conditioning in the rat. Hippocampus 8: 638-646.

Mercurio F, Zhu H, Murray BW, Shevchenko A, Bennett BL, et al. 1997. IKK-1 and IKK-2: cytokine-activated IkappaB kinases essential for NF-kappaB activation. Science 278: 860-866.

Milad MR, Quirk GJ. 2002. Neurons in medial prefrontal cortex signal memory for fear extinction. Nature 420: 70-74.

Milner B, Squire LR, Kandel ER. 1998. Cognitive neuroscience and the study of memory. Neuron 20: 445-468.

Morgan MA, Romanski LM, Le Doux JE. 1993. Extinction of emotional learning: contribution of medial prefrontal cortex. Neurosci Lett 163: 109-113.

Morris RG, Anderson E, Lynch GS, Baudry M. 1986. Selective impairment of learning and blockade of long-term potentiation by an N-methyl-D-aspartate receptor antagonist, AP5. Nature 319: 774-776.

Moser MB, Moser EI, Forrest E, Andersen P, Morris RG. 1995. Spatial learning with a minislab in the dorsal hippocampus. Proc Natl Acad Sci USA 92: 9697-9701.

Mulkey RM, Malenka RC. 1992. Mechanisms underlying induction of homosynaptic long-term depression in area CA1 of the hippocampus. Neuron 9: 967-975.

Murchison CF, Zhang XY, Zhang WP, Ouyang M, Lee A, et al. 2004. A distinct role for norepinephrine in memory retrieval. Cell 117: 131-143.

Nadel L, Samsonovich A, Ryan L, Moscovitch M. 2000. Multiple trace theory of human memory: computational, neuroimaging, and neuropsychological results. Hippocampus 10: 352-368.

Nader K, Schafe GE, Le Doux JE. 2000. Fear memories require protein synthesis in the amygdala for reconsolidation after retrieval. Nature 406: 722-726.

Nazif FA, Byrne JH, Cleary LJ. 1991. cAMP induces long-term morphological changes in sensory neurons of Aplysia. Brain Res 539: 324-327.

Nicoll RA, Malenka RC. 1995. Contrasting properties of two forms of long-term potentiation in the hippocampus. Nature 377: 115-118.

Nicoll RA, Malenka RC, Kauer JA. 1990. Functional comparison of neurotransmitter receptor subtypes in mammalian central nervous system. Physiol Rev 70: 513-565.

O'Keefe J, Dostrovsky J. 1971. The hippocampus as a spatial map. Preliminary evidence from unit activity in the freely-moving rat. Brain Res 34: 171-175.

O'Keefe JA, Nadel L. 1978. The Hippocampus as a Cognitive Map. London: Oxford University Press.

Pavlov IP. 1927. Conditioned reflexes: an investigation of the physiological processes of the cerebral cortex. London: Oxford UP.

Pinsker H, Kupfermann I, Castellucci V, Kandel E. 1970. Habituation and dishabituation of the gill-withdrawal reflex in Aplysia. Science 167: 1740-1742.

Power AE, Vazdarjanova A, McGaugh JL. 2003. Muscarinic cholinergic influences in memory consolidation. Neurobiol Learn Mem 80: 178-193.

Ramon y Cajal S. 1911. Histologie du Systeme Nerveux de l'Homme et des Vertebres. L Azoulay, trans. Paris: Maloine. Translated to English (N Swanson and LW Swanson) New York: Oxford University Press, 1995.

Reymann KG, Frey U, Jork R, Matthies H. 1988a. Polymyxin B, an inhibitor of protein kinase C, prevents the maintenance of synaptic long-term potentiation in hippocampal CA1 neurons. Brain Res 440: 305-314.

Reymann KG, Schulzeck K, Kase H, Matthies H. 1988b. Phorbol ester-induced hippocampal long-term potentiation is counteracted by inhibitors of protein kinase C. Exp Brain Res 71: 227-230.

Runyan JD, Dash PK. 2004a. Inhibition of hippocampal protein synthesis following recall disrupts expression of episodic-like memory in trace conditioning. Hippocampus.

Runyan JD, Dash PK. 2004b. Intra-medial prefrontal administration of SCH-23390 attenuates Erk phosphorylation and long-term memory for trace fear conditioning in rats. Neurobiol Learn Mem 65-70.

Runyan JD, Moore AN, Dash PK. 2004. A role for prefrontal cortex in memory storage for trace fear conditioning. J Neurosci 24: 1288-1295.

Runyan JD, Moore AN, Dash PK. 2005. A role for prefrontal calcium-sensitive protein phosphatase and kinase activities in working memory. Learning & Memory 12: 103-110.

Sakimura K, Kutsuwada T, Ito I, Manabe T, Takayama C, et al. 1995. Reduced hippocampal LTP and spatial learning in mice lacking NMDA receptor epsilon 1 subunit. Nature 373: 151-155.

Santini E, Ge H, Ren K, Pena dO, Quirk GJ. 2004. Consolidation of fear extinction requires protein synthesis in the medial prefrontal cortex. J Neurosci 24: 5704-5710.

Scoville WB, Milner B. 2000. Loss of recent memory after bilateral hippocampal lesions. 1957. J Neuropsychiatry Clin Neurosci 12: 103-113.

Sherff CM, Carew TJ. 2004. Parallel somatic and synaptic processing in the induction of intermediate-term and long-term synaptic facilitation in Aplysia. Proc Natl Acad Sci USA 101: 7463-7468.

Shuster MJ, Camardo JS, Siegelbaum SA, Kandel ER. 1986. Modulation of the 'S' K+ channel by cAMP-dependent protein phosphorylation in cell-free membrane patches. Prog Brain Res 69: 119-132.

Si K, Giustetto M, Etkin A, Hsu R, Janisiewicz AM, et al. 2003. A neuronal isoform of CPEB regulates local protein synthesis and stabilizes synapse-specific long-term facilitation in aplysia. Cell 115: 893-904.

Squire LR. 1987. Memory and Brain. New York: Oxford University Press.

Sugita S, Baxter DA, Byrne JH. 1997. Differential effects of 4-aminopyridine, serotonin, and phorbol esters on facilitation of sensorimotor connections in Aplysia. J Neurophysiol 77: 177-185.

Sutton MA, Bagnall MW, Sharma SK, Shobe J, Carew TJ. 2004. Intermediate-term memory for site-specific sensitization in aplysia is maintained by persistent activation of protein kinase C. J Neurosci 24: 3600-3609.

Tanzi E. 1893. I fatti e la induzioni nell'odierna istologia del sistema nervoso. Riv. Sper. Med. Leg. Alienazioni Met. Soc Ital Psichiatira 19: 419-472.

Taylor JR, Birnbaum S, Ubriani R, Arnsten AF. 1999. Activation of cAMP-dependent protein kinase A in prefrontal cortex impairs working memory performance. J Neurosci 19: RC23.

Teyler TJ, DiScenna P. 1985. The role of hippocampus in memory: a hypothesis. Neurosci Biobehav Rev 9: 377-389.

Tully T. 1984. Drosophila learning: behavior and biochemistry. Behav Genet 14: 527-557.

Tully T, Quinn WG. 1985. Classical conditioning and retention in normal and mutant Drosophila melanogaster. J Comp Physiol [A] 157: 263-277.

Walker JA, Olton DS. 1979. Spatial memory deficit following fimbria-fornix lesions: independent of time for stimulus processing. Physiol Behav 23: 11-15.

Weeber EJ, Sweatt JD. 2002. Molecular neurobiology of human cognition. Neuron 33: 845-848.

Williams GV, Goldman-Rakic PS. 1995. Modulation of memory fields by dopamine D1 receptors in prefrontal cortex. Nature 376: 572-575.

Yamamoto KK, Gonzalez GA, Biggs WH, Montminy MR. 1988. Phosphorylation-induced binding and transcriptional efficacy of nuclear factor CREB. Nature 334: 494-498.

Yin JC, Wallach JS, Del Vecchio M, Wilder EL, Zhou H, et al. 1994. Induction of a dominant negative CREB transgene specifically blocks long-term memory in Drosophila. Cell 79: 49-58.

20 Neurochemistry and Molecular Neurobiology of Reward

J. B. Becker · R. L. Meisel

Abstract: The reward system consists of multiple interactive neural systems. This chapter provides an overview of the neurotransmitters and brain regions involved with reward. The neural systems important for reward include brain areas involved in learning that a stimulus is rewarding or associated with reward and separate neural systems that mediate "wanting" and "liking" for a stimulus. The ascending dopamine systems are thought to be necessary for "wanting," whereas "liking" is mediated by GABAergic and opioid neurons in the nucleus accumbens shell in association with the ventral pallidum and parabrachial nucleus. The specific systems and molecular mechanisms mediating food reward versus the rewarding aspects of drugs of abuse are discussed. Sex differences in the reward system are highlighted.

1 Introduction

Some behaviors are exhibited because they result in the consummation of a primary reinforcer such as food, drink, or sex. Other behaviors are learned through association with primary reinforcers or conditioned reinforcers. The neural system that receives and evaluates the rewarding properties of stimuli has been known for some time as the reward system. In this chapter, the reward system is described as multiple interactive neural systems. The goal of this chapter is to provide an overview of the neurotransmitters and brain regions involved with reward, and to discuss sex differences in the reward system.

Historically, the ascending dopamine projection from the midbrain to the nucleus accumbens (NAcc) has been viewed as coding some aspect(s) of reward, or being the currency of reward itself in the brain (Wise, 1978; Bassareo et al., 2002; Di Chiara, 2002; Schultz, 2002; Wise, 2004). While this dopamine projection plays a role in some aspects of reward, it has been clear for some time that many other neurotransmitter systems are involved. In addition, neural and behavioral adaptations occur as an unconditioned stimulus acquires rewarding properties, or as compulsive drug-taking behavior develops in an addicted animal. In other words, neural plasticity is an important part of the neurochemistry and molecular biology of reward. Consequently, the study of reward is described in the language and based on the principles of learning theory (e.g., Di Chiara, 1998; Berridge and Robinson, 1998b; Kelley and Berridge, 2002; Salamone and Correa, 2002; Berridge and Robinson, 2003; Cardinal and Everitt, 2004; Schultz, 2004). A thorough discussion of the many views of reward from a learning theory perspective is beyond the scope of the current review; the reader is referred to recent reviews (Salamone and Correa, 2002; Berridge and Robinson, 2003). The interpretation of the role different neurotransmitters play in reward, however, does require an understanding of how various psychological constructs are being defined. Therefore, a brief description is provided.

1.1 History of the Study of "Reward"

The study of the neurobiology of reward began more than 50 years ago when Olds and Milner (1954) reported that rats would press a bar to receive electrical stimulation in certain areas of the brain. In this initial study, animals were reported to make more than 700 bar presses/h, but many self-stimulation studies since then have reported more than 100 responses/min (e.g., Wise, 2004). Subsequent "intracranial self-stimulation" studies singled out the medial forebrain bundle as being crucial for positive reinforcement from electric shock (Olds and Olds, 1969). From these studies, it was concluded that there were "rewarding" areas in the brain that were referred to as "pleasure centers" of the brain. The concept of a "reward system" in the brain followed from these important observations.

During this same time period, Teitelbaum and Epstein (1962) demonstrated that electrolytic lesions of the lateral hypothalamus caused aphagia, and adipsia. This phenomenon was called the "lateral hypothalamic syndrome," and animals were shown to exhibit extreme sensory neglect, which contributed to the failure to eat or drink during early stages of recovery (Levitt and Teitelbaum, 1975). It was the work of Ungerstedt, which first demonstrated that the neurotransmitter dopamine was playing an important role for both intracranial self-stimulation and the lateral hypothalamic syndrome.

In his research using the selective catecholamine neurotoxin 6-hydroxydopamine, Ungerstedt showed that when dopamine neurons in the midbrain are neurochemically destroyed, the result is severe akinesia, aphagia, and adipsia (Ungerstedt, 1971). A few years earlier, investigators in the same laboratory in Sweden, had shown that catecholamine neurons from the midbrain project via the medial forebrain bundle through the lateral hypothalamus to the striatum, NAcc, amygdala, and medial prefrontal cortex (Anden et al., 1966). Thus, most of the deficits found in the lateral hypothalamic syndrome were caused by the destruction of axons from dopamine neurons in the midbrain that project through the lateral hypothalamus on their way to their forebrain targets.

As the pieces of the puzzle began falling into place, it was generally accepted that the reinforcing properties of intracranial self-stimulation, drugs of abuse, and natural rewards were mediated by the activation of the ascending dopamine pathway, and in the absence of dopamine these things were not rewarding. As Wise (1984) summarized, "The notion that cocaine and amphetamine can centrally activate natural reward mechanisms and that they owe their reinforcing action to such activation ... is consistent with the fact that these agents have the traditional properties of conventional reinforcers ... In both cases the central reinforcer can be more powerful than more "natural" reinforcers, even in cases of acute need; access to psychomotor stimulant reinforcement ... or to electrical brain stimulation ...can cause self starvation to the point of severe weight loss."

Nevertheless, from early in the discussion it has been clear that dopamine is not the only mediator of reward. Going back to self-stimulation as a tool to uncover the neurobiological mechanisms of reward, there are some areas of the brain where animals will respond for self-stimulation, yet discrete dopamine lesions at these sites do not block responding (for a discussion see Wise, 1978). In a later review of the role of dopamine in reward, Wise and Rompre (1989 p. 220) concluded, "While the evidence is strong that dopamine plays some fundamental and special role in the rewarding effects of brain stimulation, psychomotor stimulants, opiates, and food, the exact nature of that role is not clear. One thing is clear: Dopamine is not the only reward transmitter, and dopaminergic neurons are not the final common path for all rewards" (Wise and Rompre, 1989). Therefore, a discussion of the neurochemistry and molecular biology of reward must involve a discussion of many neural systems.

1.2 Definitions

Before we go much farther, it is necessary to define some of the terms that are used in the study of reward.

1.2.1 Reward Versus Reinforcement

In the psychological literature "reward" is used as a noun to refer to an object or incentive that an animal or individual will work to receive. A natural reward is an unconditioned stimulus that has some intrinsic value. Rewarding is used to refer to the properties attributed to a stimulus that gives value to an animal or individual. A reinforcer is any natural reward or conditioned stimulus that when associated with a behavior, increases the probability that the given behavior will be repeated. When Wise (1984) postulated that drugs of abuse like amphetamine and cocaine "activate natural reward mechanisms, and that they owe their reinforcing action to such activation," he was referring to activation of neural systems that respond to natural rewards.

Berridge and Robinson (2003) argue that to understand the neurobiology of reward one must begin to dissect or parse reward into its psychological components. They identify these components as: (1) learning; (2) affect or what they refer to as "liking"; and (3) motivation or "wanting". In other words, an individual *learns* that a stimulus is rewarding due to its *hedonic or affective characteristics* and *how much these characteristics are valued or wanted* by the individual. For some stimuli, the first time an animal encounters the item it is rewarded (e.g., the intrinsic value of a good tasting food item is immediately recognized by an animal or individual). For the example of a good tasting food item, the animal learns that something that looks like the item tastes good. The visual stimuli associated with the item (color, shape, size, etc.) are not rewarding until the animal has tasted the food. The neuroanatomical and neurochemical systems that respond to and adapt with changes in reinforcement history are discussed in more detail below.

1.2.2 The Role of Learning and Memory in Reward

With all rewards, an animal must learn whether something is rewarding. Food preferences are learned by eating the food, or through interaction with an animal that has eaten a novel food (Galef and Whiskin, 2003). Sexual experience results in changes in motivational and consummatory aspects of sexual behavior in male rats (Pfaus et al., 2001) and female rats (Jenkins and Becker, 2003a, b). Other stimuli acquire rewarding properties by association with external or internal primary reinforcers.

Are there stimuli that are so powerful that they are immediately approached and sought after even in animals that have never had contact with the item before? There is one example of this that comes to mind: pheromones. Pheromones produced by the female silk moth will attract male silk moths from miles away. In some mammals, there are clearly powerful effects of pheromones on mating behavior. The male hamster will not mate without an intact vomeronasal organ, demonstrating the important role of chemosensory input for behavior in the hamster. There is still a role for learning in the effects of pheromones in the male hamster, as the neural systems activated by female hamster pheromones are different in sexually experienced and naïve male hamsters (Westberry and Meredith, 2003). In addition, only experienced male hamsters can be trained to bar press for access to female hamster vaginal secretions (Coppola and O'Connell, 1988). Therefore, in mammals, even with highly potent natural rewards, some experience with the stimulus is important for it to acquire rewarding properties.

As summarized by Robinson and Berridge, "Individuals are guided to incentive stimuli by the influence of Pavlovian stimulus–stimulus (S–S) associations on motivational systems, which are psychologically separable from the symbolic cognitive systems that mediate conscious desire, declarative expectancies of reward, and act-outcome representations. Prefrontal and other cortical areas primarily mediate cognitive forms of desire and act-outcome representations, whereas NAcc-related circuitry (especially dopamine-related systems) plays a more important role in Pavlovian-guided attributions of incentive salience" (Robinson and Berridge, 2003). The study of the neurochemistry and molecular biology of reward involves the integration of neural systems that mediate learning and memory, with the neural systems that attribute value to a stimulus.

As indicated in the quotation above, the neural systems that are implicated in the neurobiological changes associated with learning include the orbitofrontal cortex, cingulate cortex, and the amygdala (➋ *Figure 20-1*). These brain regions are activated by the rewarding stimuli (for review see Berridge and Robinson, 2003) and have been shown to exhibit electrophysiological and morphological neuroadaptations following treatment with drugs such as cocaine, amphetamine, or morphine (Cador et al., 1989; Everitt et al., 1989; Robinson and Kolb, 1997, 1999; Robinson et al., 2001; Volkow et al., 2002; Carelli et al., 2003). Although other brain regions are involved in learning associated with reward, these areas of the brain are most consistently implicated. It should also be noted that these brain regions may also be important for the interpretation of the hedonic and motivational value of stimuli.

1.2.3 Hedonic Value Versus Motivational Value ("Liking" Versus "Wanting")

It is generally assumed that the reason we work for natural rewards is that these rewards have intrinsic hedonic value. We like the way something tastes or how it makes us feel and that is the basis for the rewarding value of natural rewards. This is not always true, of course. Sometimes we eat food because we are hungry, even though we do not really like the taste. Sometimes we drink because we are thirsty, not for the hedonic value of water. As we shall see, with sexual behavior in the female rat (for example), the act of engaging in sex may be rewarding only under certain conditions. Both the internal state of the animal (hunger, thirst, hormone primed to be sexually receptive, etc.) and the environmental conditions in which an event occurs can affect the hedonic value of a natural reward or conditioned stimulus (Rolls, 2004).

Figure 20-1

Simplified schematic of the neural systems important for reward. Of course, no brain region is involved in only one aspect of reward. Cortical areas and the amygdala are also implicated in "wanting." Habit learning is likely to take place in the dorsal striatum and NAcc. There are also serotonin connections from the raphe nucleus to the shell of the NAcc and acetylcholine projections from the pedunculo pontine nucleus to the VTA involved in the regulation of reward. Red arrows: dopamine projections; Green arrows: GABA projections; Blue arrows: glutamate Projections; Black arrows: unknown or multiple; Blue boxes: important for the learning components of reward; Purple boxes: important for "liking"; Black boxes: important for "wanting." NAcc, nucleus accumbens; PBN, parabrachial nucleus

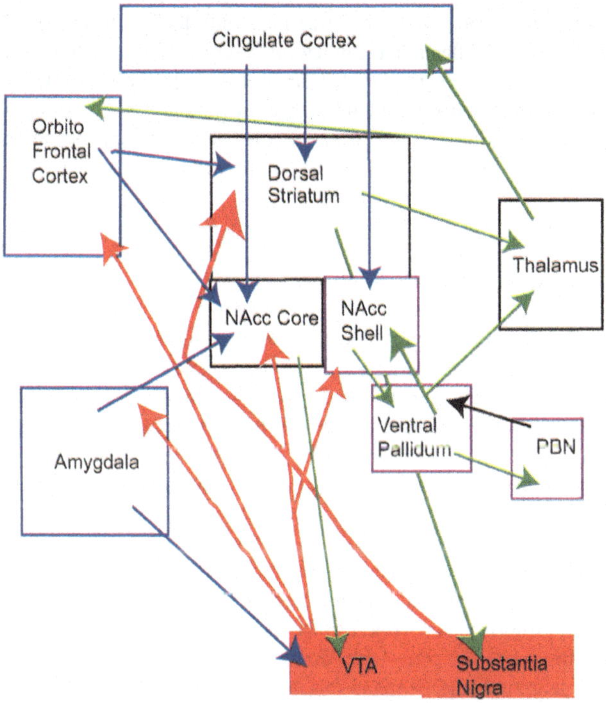

For the reasons discussed above, and others, the hedonic value of a stimulus is not the only property of a stimulus that can make it rewarding. Berridge and Robinson have termed the other component of reward as "wanting" or "incentive salience" (Robinson and Berridge, 1993; Berridge and Robinson, 2003). "Incentive salience is a motivational, rather than an affective, component of reward. Its attribution transforms mere sensory information about rewards and their cues (sights, sounds, and smells) into attractive, desired, riveting incentives. The sight of food, drugs, or other incentives is merely a sensory configuration of shape and color that is not intrinsically motivating. Attribution of incentive salience to a percept or other representation is what is suggested here to make it a "wanted" target of "motivation" (Berridge and Robinson, 2003).

Berridge and Robinson have written extensively in support of separate neural systems for "wanting" or incentive salience and "liking" or intrinsic hedonic value (Robinson and Berridge, 1993; Berridge and Robinson, 1995; Berridge, 1996; Berridge and Robinson, 1998a, b; Robinson and Berridge, 2000, 2001; Berridge and Robinson, 2003; Robinson and Berridge, 2003). They propose that the ascending dopamine system mediates incentive salience or "wanting" for a reward, whereas "liking" is mediated primarily by the actions of endogenous opioids and gamma-aminobutyric acid (GABA) acting in the shell of the NAcc, the parabrachial nucleus (PBN), and ventral pallidum (Berridge and Winkielman, 2003; Robinson and Berridge, 2003). Their argument is based on evidence that rats will continue to exhibit and learn new hedonic responses after dopamine depletion; therefore, dopamine is not necessary for hedonic attributes of a stimulus.

Changes in dopamine activity in NAcc, however, can change the incentive salience attributed to a cue for a reward, so that animals will work harder for a reward if dopamine activity is enhanced. This is true even when conditions are such that there can be no contribution from the hedonic value of the reward. (Berridge and Winkielman, 2003; Robinson and Berridge, 2003). These data and results from laboratories with alternative views will be discussed in more detail later as the neurochemistry of the reward system is discussed.

In addition to separate neural systems for "wanting" and "liking," there is evidence that separate neuronal populations respond to different types of reward. Neurons in NAcc have been shown to selectively respond to cocaine versus water or food reinforcers (Carelli et al., 2000; Carelli, 2004). Neurons in NAcc and amygdala show adaptations as a cue or stimulus changes in value through learning (Carelli et al., 2003; Peters et al., 2003; Carelli, 2004). Furthermore, stimuli acquire value through association with both positive and negative outcomes, and the NAcc is involved in the acquisition of both types of information (Schoenbaum and Setlow, 2003; Setlow et al., 2003).

1.2.4 The Neurobiology of the Reward System

The areas of the brain that are thought to be important for the neurobiology of motivation are illustrated in ➋ *Figure 20-1*. While no diagram can do justice to the complexity and subtleties of the relations among these brain regions, this figure will allow the reader to follow the discussion. At the core of the reward system are the ascending dopamine systems that project from the substantia nigra to the dorsal striatum and from the ventral tegmental area (VTA) to the NAcc, amygdala, and frontal cortex. As initially hypothesized by Robinson and Berridge (2003), these dopamine systems are necessary for incentive salience or wanting.

The NAcc is composed of the core and shell, which differ in their afferent input and efferent projections. The hippocampus projects to both core and shell, with the dorsal subiculum projecting to the core and ventral subiculum projecting to the shell. Prelimbic prefrontal cortex projects to the core of the NAcc whereas infralimbic and piriform cortex project to the shell (Brog et al., 1993). Specific subcompartments of the amygdala also project to the core versus shell (Wright et al., 1996). Both core and shell receive input from dopamine neurons in the VTA, and this input is topographically organized. The output from the NAcc core connects to the ventral pallidum, subthalamic nucleus, and substantia nigra, while the shell projects more to the subcortical limbic regions including the VTA, ventral medial pallidum, and lateral hypothalamus.

The NAcc core appears to be important for cocaine-seeking behavior that is acquired using conditioned reinforcers, but lesions of the core did not affect acquisition of responding for cocaine under continuous reinforcement conditions (Ito et al., 2004). On the other hand, if one looks at what happens in the NAcc after repeated exposure to cocaine, there is increased dendritic branching in the core and the shell of the NAcc only in animals that exhibited behavioral sensitization. Animals that were exposed to the same amount of cocaine, but did not develop behavioral sensitization had increased dendritic branching only in the shell of the NAcc (Li et al., 2004). Thus, the core of the NAcc exhibits structural plasticity under conditions where behavioral sensitization occurs.

The NAcc shell in association with the ventral pallidum and PBN is thought to be important for the hedonic value of stimuli or "liking" (Berridge and Robinson, 2003). Selective excitotoxic lesions of the NAcc shell do not impair drug self-administration or the acquisition of cocaine-seeking behavior, but do attenuate the psychostimulant effects of cocaine (Ito et al., 2004).

Information from the core and the shell of the NAcc is integrated at the level of the thalamus. Cortical areas (amygdala, and orbitofrontal, perhaps the cingulate cortex) are important for learning the association between a CS and reward as well as for mediating changes in the incentive salience of stimuli.

1.3 Summary

The study of reward has been focused on the ascending dopamine systems for many years, but it is now clear that complex relations among multiple neurotransmitter systems mediate reward. Neural systems important for reward include brain areas that are important for learning whether a stimulus is rewarding or it is associated with a rewarding stimulus (e.g., prefrontal cortex and NAcc). In addition, it is thought that there are separate neural systems that mediate "wanting" and "liking" for a stimulus. The ascending dopamine systems are thought to be necessary for "wanting" but not liking. It is proposed that liking is mediated by the NAcc shell in association with ventral pallidum and PBN.

2 The Neurochemistry of Reward

2.1 Food Reward

2.1.1 Dopamine

Most discussions of the neurochemistry of feeding behavior and food reward begin with a discussion of the role of dopamine in feeding behavior. This idea follows from the work of Teitelbaum (discussed earlier), who showed that lesions of the ascending dopamine pathway in the lateral hypothalamus resulted in profound aphagia and adipsia. What Teitelbaum also showed, however, is that when rats with lateral hypothalamic lesions were fed a highly palatable food they could recover from these lesions, and would eventually eat standard laboratory rat chow without a restoration of dopamine function. Therefore, while dopamine clearly plays a role in feeding behavior, there are other neurotransmitters involved.

Evidence supporting a role for dopamine in feeding behavior includes studies investigating the rewarding effects of sham-fed sucrose. Dopamine increases in dialysate from the NAcc in response to sucrose, and previous experience with sucrose enhances this effect (Smith, 2004). Dopamine antagonists decrease sucrose intake as a function of the concentration of sucrose, suggesting that dopamine in the NAcc is important for the rewarding aspects of sucrose. Depletion of NAcc dopamine with 6-hydroxydopamine, however, does not affect the hedonic response to sucrose in the taste reactivity test (Berridge et al., 1989). Furthermore, Wyvell and Berridge (2001) demonstrated that sensitization of animals with repeated amphetamine did not increase the hedonic impact of sucrose, but did increase responding for sucrose in response to a previously paired conditioned stimulus. These results taken together suggest that sensitization of the ascending dopamine system with amphetamine increases the incentive salience or "wanting" for sucrose.

It may seem counterintuitive, but other evidence in support of a role for dopamine in feeding behavior comes from the widespread use of amphetamine for the treatment of obesity. Amphetamine stimulates dopamine release and blocks dopamine reuptake. So it seems that an excess of dopamine decreases feeding behavior, just the opposite of what happens when endogenous dopamine increases. This indicates that there is an inverted U-shaped curve relationship between dopamine release and feeding behavior, with dopamine release facilitating feeding behavior at physiological levels and inhibiting feeding at pharmacological levels (➋ *Figure 20-2*). In support of this idea, mice with reduced expression of the dopamine transporter, which elevates intracellular dopamine, are hyperphagic and exhibit behaviors consistent with an increase in "wanting" for sucrose, without an increase in "liking" for sucrose (Pecina et al., 2003). On the other hand, in mice that cannot make norepinephrine or epinephrine due to the lack of dopamine-β-hydroxylase, amphetamine reduces food intake equally well as in wild-type mice (Cannon et al., 2004). This effect is not blocked by the dopamine antagonist haloperidol, even though locomotor behavior was suppressed. These results indicate that there had not been compensation during development by D2 dopamine receptors mediating the hypophagia induced by amphetamine, and that hypophagia was not a secondary consequence of enhanced locomotor activity preventing feeding behavior (Cannon et al., 2004). In contrast, mice lacking dopamine are resistant to the hypophagic effects of amphetamine and will actually increase food intake after amphetamine (these animals require supplementation with l-dopa for normal food intake; Cannon et al., 2004).

Figure 20-2

Hypothetical relationship between dopamine release and food consumption. As the concentration of dopamine in the synapse increases, there is initially an increase in food consumption. After reaching the asymptote, additional increases in dopamine result in decreased food consumption

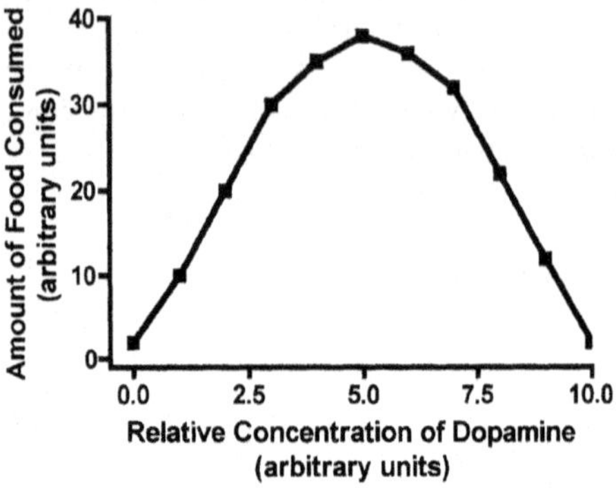

Using extracellular recording from dopamine neurons in the substantia nigra and VTA of awake monkeys, Schultz has shown that dopamine neurons initially fire in response to delivery of a juice reward. If the reward is paired with a predictive stimulus (CS) dopamine neurons come to fire with the CS (Schultz and Romo, 1990; Schultz, 1992; Schultz et al., 1992, 1993; Schultz, 1997; Fiorillo et al., 2003; Tobler et al., 2003). Dopamine neurons are also activated by rewards that are considered more valuable than would have been predicted by the CS, uninfluenced by rewards that occur as predicted, and depressed by rewards that are worse than predicted (Schultz, 2004). Schultz has suggested that the dopamine signal is important for risk taking behavior, rather than being related to the type of reward (such as food). As uncertainty in reward value increases, there is a greater increase in dopamine neuron firing (Fiorillo et al., 2003). These data can be summarized as indicating that dopamine neurons fire in response to a learned (or unlearned) reward or CS and code information about reward value and reward predictability.

If dopamine activity is reduced with a low dose of a dopamine receptor antagonist or depletion of dopamine in the NAcc, food intake and affective responses to food tastes are not affected, but behaviors emitted to gain access to food are suppressed (Berridge, 1996). Similarly, infusion of selective D1 or D2 dopamine receptor antagonists into the NAcc core or shell of food-restricted rats has no effect on food intake, but will suppress locomotor activity usually associated with hunger (Kelley, 2004). Kelley (2004) hypothesizes that dopamine plays a broad role in adaptive motor behavior and learning, rather than a specific role in food intake. Salamone (Salamone and Correa, 2002) finds that NAcc dopamine depletions decrease an animal's likelihood of engaging in instrumental responding that has a high cost-benefit ratio. He proposes that NAcc dopamine is important for maintaining effort in instrumental responding over time in the absence of primary reinforcement (Salamone and Correa, 2002). This interpretation is not inconsistent with dopamine coding reward value or "wanting."

2.1.2 GABA and Glutamate

Research on the roles of GABA, glutamate and the endogenous opiates in feeding behavior has been reviewed recently, so the reader is referred to that excellent article (Kelley, 2004). The Kelley lab has shown that neurons in the NAcc shell are important for modulating feeding behavior. Antagonists of glutamatergic AMPA receptors (but not NMDA antagonists) in the shell induce the immediate onset of feeding behavior. Inhibition of the NAcc shell with the GABA-A agonist muscimol or the GABA-B agonist baclofen also induces feeding (Kelley, 2004). These data suggest that corticolimbic glutamatergic input to the NAcc shell is involved in inhibition of feeding, and that GABA released with feeding is important for self-inhibition of competing behaviors, and enhanced feeding behavior.

2.1.3 Endogenous Opiates

Opioid agonists stimulate food intake and antagonists depress feeding behavior. It has been suggested that opioids are important for taste palatability or how much the animals like a food substance (Kelley et al., 2002). Opioid stimulation of the NAcc induces feeding behavior. Conversely, regular feeding of animals with a highly palatable food can result in down regulation of striatal enkephalin gene expression (Kelley, 2004).

When areas of the NAcc that are important for feeding behavior were mapped, the caudal region of the shell was found to contain the opioid neural circuits where morphine activation is sufficient to increase responding on the taste reactivity test, which indicated enhanced "liking" for food (Berridge, 2003). As reviewed by Berridge (2003), opiate drugs will also enhance how much an animal eats and rats will work for morphine infusion into the NAcc. So, opioid systems in the NAcc are involved in both "liking" and "wanting" for food reward.

2.2 Drugs of Abuse

Drugs of abuse have been used extensively in research to tap into the reward system(s). This is not to say that drugs of abuse activate the exact same neural systems that are activated by natural rewards. As White (1996) points out, drugs such as amphetamine and morphine, when given systemically, can simultaneously cause a conditioned place preference and a conditioned taste aversion. These simultaneous effects are thought to be due to the actions of these drugs at multiple regions in the brain. For example, conditioned place preference for amphetamine requires the lateral nucleus of the amygdala, the NAcc, and the tegmental pedunculo pontine nucleus. In contrast, conditioned taste aversions can be induced by amphetamine in the area postrema and blocked by lesions of the basolateral amygdala (White, 1996). So, the use of drugs of abuse to investigate the neural mechanisms of reward can be informative, but it is important to keep in mind that these drugs are having many other actions in the brain as well.

2.2.1 Dopamine

Psychomotor Stimulants. During microdialysis in the NAcc, dopamine increases in both the shell and core during intravenous cocaine self-administration. Noncontingent CS presentation led to increased dopamine only in the NAcc core (Ito et al., 2000). Using a second-order schedule of reinforcement, dopamine in the dorsal striatum is also increased (Ito et al., 2002). These authors suggest that dopamine in the NAcc core may be involved in the acquisition of the CS–cocaine association, but that once the association is established, other neural systems, including dopamine in the dorsal striatum take precedence (Ito et al., 2002).

When dopamine antagonists are given to animals that are lever pressing for brain self-stimulation, or other rewards, responding declines over time (Wise, 2004). The animals initiate responding normally and then the behavior drops off. A sensory challenge can reinstate responding, suggesting that the decline in behavior is not due to the effect of the drug on motor activity. Wise (2004) concludes, "most normally rewarding stimuli fail to serve as effective reinforcers in dopamine-compromised animals.

Psychomotor stimulant drugs, such as cocaine, amphetamine, and methylphenidate act by blocking dopamine reuptake at the dopamine transporter and thereby increasing the amount of dopamine available at postsynaptic receptors (Schenk, 2002). Amphetamine acts at the dopamine transporter to increase dopamine release (Fischer and Cho, 1976). Amphetamine is also active at norepinephrine and serotonin transporters, although within the striatum, it is the action of amphetamine at the dopamine transporter that accounts for most of the behavioral effects of amphetamine (Heikkila et al., 1975; Raiteri et al., 1975).

Amphetamine also redistributes dopamine from within vesicles to the cytosol where it promotes reverse transport out of the neuron through the mechanism of exchange diffusion (Sulzer et al., 1995). This is an active, energy-dependent process that derives its energy from the effect of amphetamine as a weak base that results in a pH gradient that promotes reverse transport (Sulzer et al., 1993), and from the polarization of the neuronal membrane: transporter activity increases with increased membrane polarization (Kanner, 1983).

Cocaine has equivalent affinities at the dopamine transporter, norepinephrine transporter, and serotonin transporter, but it is the binding potency at the dopamine transporter that is correlated with the effects of cocaine in tests of reward (Kuhar et al., 1988). The hypothesis that the monoamine transporters were involved in the effect of cocaine lead to the cloning of these transporter genes and the development of animals with overexpression, underexpression, or knockout of this gene. As discussed by Uhl et al. (2002), the subsequent experimental use of these animals demonstrated that the dopamine transporter is necessary for cocaine-induced locomotor behavior above an already elevated baseline of locomotor activity. On the other hand, dopamine transporter knockout mice express a normal conditioned place preference for cocaine and will learn to self-administer cocaine (Sora et al., 1998; Uhl et al., 2002).

Interestingly, double knockouts of the dopamine transporter and the serotonin transporter do not show a conditioned place preference for cocaine, whereas double knockouts of the serotonin transporter and the norepinephrine transporter have an enhanced conditioned place preference for cocaine. The results with the various monoamine transporter knockouts illustrate that cocaine can produce its rewarding effects by acting at multiple sites, and through multiple mechanisms. These results also suggest that there are both rewarding and aversive aspects of cocaine's action being mediated by the coordinated actions of blocking different transporters. Animals with double knockouts of the dopamine transporter and the serotonin transporter have reward deficits, while double knockouts of the serotonin transporter and the norepinephrine transporter do not exhibit conditioned aversion with cocaine (Uhl et al., 2002). Of course, experiments with animals that have been genetically altered since birth must be interpreted cautiously as there may be considerable adaptation during development as a result of the altered gene expression. It will be important to determine if the areas of the brain identified as important for conditioned place preference and conditioned taste aversion in knockout animals also mediate the rewarding and aversive effects of cocaine in the wild-type mice.

Opiates. Considerable evidence supports a role for dopamine in the reinforcing value of opiates. In animals depletion of dopamine in the VTA or NAcc, the acquisition or maintenance of heroin or morphine reinforcement are abolished, as are conditioned place preferences for these drugs; D1 dopamine receptor antagonists also decrease the rewarding value of opiates (Shippenberg and Herz, 1987, 1988; Shippenberg et al., 1993; Xi and Stein, 2002). Studies using in vivo microdialysis have shown that morphine induces a significant increase in extracellular dopamine in the NAcc (Leone et al., 1991; Pothos et al., 1991). The exact role of dopamine in opiate-induced reward remains a topic of discussion, but it seems clear that dopaminergic mechanisms are involved to some degree.

Nicotine. Nicotine is thought to increase dopamine release by acting at presynaptic dopamine receptors to increase the rapidly releasable pool of dopamine (Turner, 2004). In striatal slices from guinea pig brains, nicotine enhances dopamine release during phasic, but not tonic activity (Rice and Cragg, 2004). In corticostriatal slices from mouse brain, nicotine depresses dopamine released by stimuli inducing tonic firing, but increases dopamine released from stimuli emulating phasic firing (Zhang and Sulzer, 2004). These results suggest that nicotine enhances incentive salience during periods of dopaminergic activation.

2.2.2 Glutamate

The presence of glutamate afferents to the VTA and NAcc suggests an interaction between dopamine and glutamate in the mediation of the effects of both natural rewards and drugs of abuse. The effects of glutamate are mediated by receptor subtypes from both metabotropic and ionotropic receptor classes (➋ *Table 20-1*). The significance of these receptor classes is that the ionotropic receptors can modulate rapid excitatory neurophysiological responses (Chuhma et al., 2004) as well as mediate plastic changes in synaptic contacts (Carroll and Zukin, 2002). The metabotropic receptors are linked to classic intracellular signaling pathways that can regulate protein phosphorylation and transcription, which may also lead to long-term plasticity (Schoepp et al., 1999).

Table 20-1
Organization of Glutamate Receptor Subtypes

Receptor class	Receptor subtype	Receptor subunits	Signaling
Ionotropic[1]	AMPA Receptors	GluR1-4	$Na^+/K^+/(Ca^{2+})$ influx[2]
	NMDA Receptors	NR1	$Ca^{2+}/Na^+/K^+$ influx
		NR2A-D	$Ca^{2+}/Na^+/K^+$ influx
		NR3A	$Ca^{2+}/Na^+/K^+$ influx
Metabotropic	Group I	mGluR1/5	Gq; IP_3 activation
	Group II	mGluR2/3	Gi/o; Adenylyl cyclase inhibition
	Group III	mGluR4/6-8	Gi/o; Adenylyl cyclase inhibition

[1]Kainate receptors are not included here as their role in reward-related processes is uncertain
[2]Permeability to Ca^{2+} is inhibited if the R2 subunit is present

There are three groups of metabotropic glutamate receptors (Groups I–III) containing eight receptor subtypes. Group I receptors (mGluR1/5) are Gq coupled, whereas group II receptors (mGluR2/3) and group III receptors (mGluR4/6–8) are Gi/o coupled. The ionotropic glutamate receptors are referred to by the ligands they preferentially bind (NMDA versus AMPA), giving rise to a series of NMDA (NR1-3) and AMPA receptor (GluR1-4) subtypes based on the subunits that comprise the receptors (*Table 20-1*). The combination of inputs from different forebrain regions and multiple receptor types offers a remarkably complicated array of means through which glutamate can impact reward processes. Still there are some clear patterns in the mechanisms through which glutamate can mediate reward-related processes.

Psychomotor Stimulants. In addition to the regulation of dopamine neurotransmission, drugs of abuse produce changes in synaptic levels of glutamate. For example, acute treatment with amphetamine increases extracellular levels of glutamate in the VTA and NAcc (Xue et al., 1996; Wolf, 1998; Wolf et al., 2000). The effects of amphetamine on glutamate release, at least in the VTA, depend on efferents from the prefrontal cortex (Li et al., 1999). Repeated amphetamine administration does not increase glutamate release; however, stimulated glutamate release is enhanced with sensitization (Wolf, 1998; Giorgetti et al., 2001).

Less is known about the effects of cocaine on glutamate release. Chronic cocaine elevates glutamate levels in the VTA (Kalivas and Duffy, 1998), but not in the NAcc (Robinson et al., 1997). Oddly, aspartate concentrations, which can be used as an indication of glutamate release, did increase in the NAcc following chronic cocaine (Robinson et al., 1997). Another potential (though certainly indirect) measure of glutamate neurotransmission is the abundance of glutamate in presynaptic nerve terminals. Glutamate immunoreactivity was measured in the NAcc and VTA in response to either acute or chronic cocaine administration (Meshul et al., 1998; Kozell and Meshul, 2001, 2003, 2004). Acute cocaine decreased glutamate staining in terminals in both the shell and core of the NAcc, an effect that persisted in the core, but not in the shell. This effect was not found in glutamate terminals in either the VTA or prefrontal cortex. Chronic cocaine increased glutamate immunoreactivity in the VTA, and the effect was magnified in behaviorally sensitized animals. After withdrawal, however, VTA glutamate was increased in the nonsensitized cocaine-treated animals. Despite the minimal evidence for a link between cocaine and glutamate release, these results are consistent with the findings from experiments with amphetamine. At the same time, it does not appear that presynaptic glutamate content is related to the development of sensitization following cocaine administration (Kozell and Meshul, 2004).

One way that drugs such as amphetamine or cocaine may impact glutamate neurotransmission is by acting at metabotropic or ionotropic glutamate receptors. In one electrophysiological study, VTA metabotropic glutamate receptor postsynaptic inhibitory potentials were dramatically inhibited by amphetamine, whereas ionotropic receptor-mediated postsynaptic currents were not affected (Paladini et al., 2001). When the relative contributions of currents derived from AMPA and NMDA receptors were parsed; however, amphetamine produced an increase in the ratio of AMPA:NMDA currents in the VTA relative to saline controls (Faleiro et al., 2004).

Chronic treatment with cocaine increased both GluR1 and NR1 (but not GluR2/3) content in the VTA one day after the last cocaine injection (Churchill et al., 1999). This effect was restricted to behaviorally sensitized animals, and was gone 3 weeks later. In the NAcc there was no difference between cocaine-sensitized animals in GluR1 or NR1 subunits one day after treatment termination. Three weeks later, however, GluR1 levels increased in sensitized rats, with GluR2/3 and NR1 subunits still unchanged. The lack of effect of chronic cocaine on NR1/2A-2C receptor levels immediately after treatment has also been seen by other investigators (Yamaguchi et al., 2002).

Withdrawal from amphetamine produced a complex pattern of changes in AMPA receptor subunit expression that depended on the period of withdrawal and the neural region (Lu and Wolf, 1999). First, no changes in GluR2/3, GluR2/4, or GluR4 subunit expression were seen at any time point. In the NAcc, there was no change in GluR1 nor in GluR2 expression after 3 days of withdrawal, but at 14 days of withdrawal, GluR2 levels decreased in both the NAcc shell and core, but GluR1 levels decreased only in the NAcc shell. In the prefrontal cortex, only GluR1 expression was affected, with a decrease in subunit levels at 3 days of withdrawal, normalizing by 14 days of withdrawal.

Paradoxically, NMDA and AMPA agonists and antagonists both increase locomotor activity (Kelley and Throne, 1992; Burns et al., 1994; Danysz et al., 1994; Leriche et al., 2003; David et al., 2004), possibly through interactions with dopamine D2 and/or D3 receptors (Choi et al., 2000; Leriche et al., 2003; David et al., 2004). Amphetamine directly infused into the core of the NAcc also increases locomotor activity, but these effects of amphetamine are blocked by intraNAcc pretreatment with either an NMDA receptor agonist or an antagonist (Kelley and Throne, 1992; David et al., 2004). Pretreatment with an AMPA receptor agonist or antagonist produced a similar attenuation of amphetamine's effects (Vanover, 1998; David et al., 2004). This overall pattern of results is seen with both cocaine and amphetamine treatment (Pulvirenti et al., 1994).

One possible resolution to this apparent paradox is that glutamate receptors are found both on dopaminergic presynaptic terminals and on postsynaptic dopaminergic target neurons (David et al., 2004). Another variable often overlooked in these paradigms is the environment in which the drugs are given. One notable study tested the ability of NMDA receptor antagonists to counteract the sensitizing effects of repeated amphetamine treatments (Battisti et al., 2000). In this case, the antagonists were effective when the animal was tested in the same environment in which the amphetamine was administered, but not when the animals were tested in a novel environment. If locomotor activity is dependent on dopamine neurotransmission, then the integrated outcome of modulating both dopamine release and postsynaptic responsiveness in a context-sensitive manner could differentially regulate locomotor activity.

Similar to the studies on ionotropic glutamate receptors, the locomotor stimulating effects of pharmacological agents targeting the metabotropic glutamate receptors have been evaluated. Injections of group II mGluR antagonists either systemically or in the NAcc will antagonize the increases in activity produced by amphetamine administration (Cartmell et al., 2000; Kim et al., 2000; Breysse et al., 2002; David and Abraini, 2003). The effect of group I antagonists on amphetamine or cocaine stimulated locomotor behavior are similar to that of antagonizing Group II receptors. Group I antagonists block the effects of amphetamine or cocaine on activity in some studies (David and Abraini, 2003; Herzig and Schmidt, 2004), but not in another (Kim et al., 2000).

Though locomotor activity typically is a property associated with the rewarding effects of drugs of abuse, group I antagonists affect reward differently for individual drugs. Cocaine, amphetamine, and morphine-related reward processes are inconsistently disrupted by group I antagonists (Darracq et al., 2001; Popik and Wrobel, 2002; McGeehan and Olive, 2003; Herzig and Schmidt, 2004; Kim et al., 2005). Another way to look at this question is to examine expression of receptors following drug sensitization (Mao and Wang, 2001). Of group I receptors, a sensitizing amphetamine regimen increases mGluR1 expression in the NAcc, an effect that disappears during withdrawal. In contrast, amphetamine experience decreased mGluR5 expression in the NAcc, and this effect persisted during a period of drug withdrawal. Why mGluR5 receptor expression is elevated in the NAcc during withdrawal from a sensitizing cocaine regimen (Ghasemzadeh et al., 1999) is unclear. Thus, the effects of sensitization on mGluRs depend on the particular drug administered as well as the pattern of administration. Consequently, it seems unlikely that activation of metabotropic glutamate receptors forms a common mechanism for the behavioral consequences of drug exposure.

Opiates. Acute morphine treatment had a bidirectional effect on ionotropic glutamate receptor subunits in the core of the NAcc (Jacobs et al., 2005). Three days after injection there was a decrease in expression of NR1, NR2B/2C, and GluR1-4 subunits as measured by the real time PCR. By 21 days there appeared to be a compensatory rebound as expressions of NR1, NR2A-D, NR3, and GluR1-4 were significantly elevated compared with saline-treated controls. It is clear that these drugs produce a complicated pattern of changes in glutamate ionotropic receptors, a pattern that may depend on time after drug administration and brain region. Further, the degree to which fluctuations in receptor subunits contribute to active or silent synapses needs to be evaluated (Nakayama et al., 2005). Still, the suggestion that increased expression of GluR1 mediates sensitization to repeated drug exposure (Carlezon and Nestler, 2002) warrants further study.

Reward Learning. Besides the glutamate modulation of the sensitization of locomotor responses to drugs of abuse, the role of glutamate in more conventional learning paradigms has been assessed. In a conditioned place preference paradigm, inhibition of glutamate release through systemic treatment with riluzole prevented conditioned place preference to either morphine or amphetamine (Tzschentke and Schmidt, 1998). Further, NMDA antagonists blocked either the acquisition or reinstatement of conditioned morphine place preference (Harris and Aston-Jones, 2003; Ribeiro Do Couto et al., 2005). A role for specific subunits of the NMDA receptor was evaluated by intracranial passive immunization to the NR1, NR2A, or NR2B subunits (Narita et al., 2000). The effectiveness of inactivation of each of the subunits was demonstrated by increased NMDA-induced seizure threshold. Subunit specificity for conditioned place preference for morphine was demonstrated in that inactivation of the NR2B subunit eliminated conditioned place preference, whereas inactivation of the other subunits was without effect. Similar effects of AMPA and NMDA antagonists were seen with cocaine-induced place preference, though the effectiveness of these agents was limited to infusions in the VTA (Harris and Aston-Jones, 2003), a site specificity also noted for morphine-induced place preference (Harris et al., 2004). Oddly, systemic administration of MS-153, a reported activator of the glutamate transporter, also antagonized conditioned place preference to cocaine, morphine, or methamphetamine (Nakagawa et al., 2005). A simple explanation for this discrepancy awaits further study.

Instrumental learning tasks appear to be facilitated by glutamate. Spatial memory as measured in a radial arm maze is disrupted by treatment with NMDA or AMPA antagonists (Klein et al., 2004) and enhanced by drugs that prolong AMPA receptor mediated currents (Davis et al., 1997). NMDA antagonists will interfere with instrumental responses for food or response to novel objects when infused into the NAcc or related structures (Maldonado-Irizarry and Kelley, 1994; Andrzejewski et al., 2004; Kelley, 2004). Likewise, GluR2 subunit knockout mice a disruption in instrumental responding to a food reward (Mead and Stephens, 2003). Glutamate activity does not seem to be sufficient in this regard, instead requiring an interaction between dopamine and glutamate receptor activation (Smith-Roe and Kelley, 2000). One means through which glutamate affects instrumental responding is through a discrimination of responses directed at cues related to the value of the reward. In particular, animals treated with NMDA or AMPA antagonists respond strongly to cues associated with reward, but lose the ability to selectively respond to the cue signaling the reward of greater value (Giertler et al., 2005).

2.2.3 GABA

GABA is a ubiquitous neurotransmitter that is present in both interneurons and in long-projecting neurons in the mesolimbic circuitry (Van Bockstaele and Pickel, 1995). Besides the potential role of these neuronal types in the mediation of reward, GABA has two classes of receptors, the ionotropic $GABA_A$ and metabotropic $GABA_B$ receptors (Matsumoto, 1989; Bowery, 1993). Both receptor subtypes are coexpressed through most regions of the brain (Bowery, 1993), though different patterns of localization of the receptor subtypes are found due to circuitry and cellular localization. $GABA_A$ and $GABA_B$ receptors occur pre and postsynaptically, with the presynaptic receptors acting as autoreceptors and modulating the release of other neurotransmitters (Bowery, 1993). The implication here is that the functional consequences of GABA neurotransmission will depend on receptor subtype and location (e.g., Akiyama et al., 2004).

In general, GABA neurotransmission is thought to be inhibitory to the rewarding effects of drugs of abuse or natural stimuli (Ikemoto and Wise, 2004). Consequently, one therapeutic approach taken to antagonize the rewarding effects of drugs is the administration of gamma vinylGABA (GVG). GVG is an irreversible inhibitor of GABA transaminase (an enzyme that degrades GABA). Treatment with GVG results in prolonged elevation of synaptic GABA. Rats sensitized to repeated cocaine administration have reduced GABA release (e.g., Jung et al., 1999). Further, overexpression of the GABA transporter leads to a reduction in reward efficacy of morphine (Hu et al., 2003). Therefore, sustained elevations of GABA antagonize reward processes.

Several examples illustrate this conclusion. GVG dose-dependently depressed cocaine self-administration when given within 3 h of access to cocaine (Kushner et al., 1999). The GVG was without effect if administered the previous day. Also, drugs like cocaine reduce the threshold for intracranial brain stimulation, and GVG given in conjunction with cocaine restores the brain stimulation threshold (Kushner et al., 1997). An interesting outcome of conditioned place preference studies is that GVG antagonizes conditioned place preference to cocaine, but it is without effect on conditioned place preference to a food reward (Dewey et al., 1998). On the other hand, GVG does antagonize bar pressing for food (Kushner et al., 1999). In fact, as discussed above, GABA neurotransmission plays a very different role in food-related reward (Berridge, 1996; Kelley and Berridge, 2002; Berridge and Robinson, 2003).

The generalization of the role of GABA in reward processes is complicated by the fact that there are regional differences in the responsiveness to GABAergic agents both between brain regions and within brain regions (Berridge, 2003). In addition, at the cellular level there are bidirectional effects of both $GABA_A$ and $GABA_B$ receptor stimulation (Cruz et al., 2004; Ikemoto and Wise, 2004; Laviolette and van der Kooy, 2004). For $GABA_A$ receptor mediated processes, it seems that the effects on reward are a function of both dopamine-dependent and dopamine-independent processes, which may mediate both excitatory and inhibitory signaling mechanisms (Laviolette and van der Kooy, 2001, 2004).

In the case of $GABA_B$ signaling, it appears that GABA could differentially alter cellular activity based on the efficacy of the coupling of receptors to functional ion channels. Dopamine and GABA neurons in the VTA differ in the subunit composition of G-protein-gated inwardly rectifying potassium channels, with a much lower efficiency of coupling in dopamine versus GABA neurons (Cruz et al., 2004). Therefore, the passage of information through dopaminergic or GABAergic neurons depends on the concentrations of extracellular GABA and the GABA receptor subtype. Coupling of $GABA_B$ receptors to G-proteins could be one of the mechanisms mediating these differences in signaling efficacy. The reported effects of chronic amphetamine on $GABA_B$ coupling to G-proteins make this a likely candidate for plastic changes underlying reward-related events (Zhang et al., 2000).

Another link to $GABA_B$ receptors in reward is the observation that the effects of GVG on reward-related processes are antagonized by a $GABA_B$ antagonist (Ashby et al., 1999). Baclofen, a commonly used $GABA_B$ receptor agonist, prevents the elevation in extracellular dopamine (a hallmark effect of drugs of abuse) following nicotine, cocaine, or morphine treatment (Fadda et al., 2003; Jayaram and Steketee, 2004). When placed directly in the VTA, baclofen prevents the locomotor sensitizing effects of repeated morphine treatment (Leite-Morris et al., 2004) and self-administration of cocaine (Brebner et al., 2000). Besides diminishing responsiveness to primary rewards, baclofen also prevents responsiveness to drug-related cues (Hotenspiller and Wolf, 2003; Kaplan et al., 2003).

Collectively, these results contribute to the view that drugs of abuse may modulate GABA responsiveness in a way that influences both dopamine and glutamate release (Giorgetti et al., 2002). For example, systemic or local administration of heroin or morphine inhibits neuronal activity in the NAcc where the majority of neurons are GABAergic, and administration of drugs that activate GABA-B receptors blocks heroin self-administration in a dose-dependent manner (Xi and Stein, 1998a, 2002). Furthermore, sensitization with cocaine results in an increase in the basal extracellular concentrations of GABA and desensitization of GABA-B receptors (Xi et al., 2003). One possibility is that the output of medium spiny GABA projection neurons mediate the rewarding effects of drugs of abuse, as inhibition of GABA activity by dopamine or directly by some drugs of abuse is necessary for drugs to be rewarding (Xi and Stein, 1998a, 2002).

2.2.4 Opioids

In the central nervous system (CNS) there are three classes of opiate receptors: μ, σ and κ. Morphine and heroin act primarily at μ-opiate receptors, and animals will self-administer selective μ agonists. Endogenously, endorphins and enkephalins act at μ or σ receptors. On the other hand, κ agonists tend to lack reinforcing effects or produce conditioned place aversions, and the endogenous ligand is dynorphin (Xi and Stein, 2002). Reward mediated by μ-opioid receptors is hypothesized to be mediated primarily via actions on dopamine, GABA, and perhaps glutamate neurons. Activation of opioid receptors in the VTA inhibits GABA release, and as a result of the release of inhibition on VTA dopamine neurons there is an increase in dopamine release in the NAcc (Leone et al., 1991). Opioid receptors on GABA neurons in the NAcc can similarly inhibit GABA release and enhance dopamine release (Xi and Stein, 1998b, 2002). So, activation of opioid receptors by morphine or heroin is hypothesized to produce rewarding effects through secondary actions on other neurotransmitter systems.

2.3 Summary

Table 20-2 presents a simplified summary of the role of different neurotransmitters in food reward and drug abuse. It should be noted that the relationship between relative dopamine concentrations and consumption of food or drugs of abuse is the same. On the other hand, both glutamate and GABA have different effects on feeding behavior versus intake of drugs of abuse. This suggests that glutamate and GABA are indirectly modulating reward behavior, and different factors contribute to the effects of these neurotransmitters on feeding versus responding for drugs of abuse.

Table 20-2

Summary of neurotransmitter effects on responding to natural and drug-related rewards. Indications of changes in a particular category reflect how increased activity in that neurotransmitter system affects an animal's response for either a natural reward (e.g. food) or for drugs (e.g. cocaine)

Neurotransmitter	Natural reward (Food intake)	Drug-related reward
Dopamine	↑ Responding[1]	↑ Responding[1]
Glutamate	↓ Responding[2]	↑ Responding
GABA	↑ Responding	↓ Responding
Opiates	↑ Responding	↑ Responding[3]

[1]Bell-shaped dose response curve: initially there is increased food intake or operant responding for drugs of abuse as dopamine concentrations increase, then decreased food intake and decreased self-administration as dopamine concentrations increase further

[2]However, glutamate *increases* cue salience

[3]Effects are mediated through other neurotransmitters

3 Cellular and Molecular Mechanisms of Reward

Inherent in reward processes, whether through natural behaviors, laboratory paradigms, or pharmacological agents, are variations in behavioral expression. The assumption is that the underlying behavioral plasticity is a corresponding neural plasticity and that the artificially induced neural plasticity is mediated by neural processes used in behavioral plasticity. A prominent example of the later is hippocampal long-term potentiation studied as a mechanistic basis for learning. Of significance here is that long-term potentiation can also be demonstrated among neurons in the mesolimbic system and that there are common molecular mechanisms for long-term potentiation in the hippocampal and mesolimbic systems.

In this section, we review some of the cellular and molecular adaptations thought to mediate plasticity in reward systems.

3.1 Interaction Between Drugs of Abuse and Natural Rewards

An ongoing assumption is that drugs of abuse activate neural pathways underlying natural rewards. Both types of processes seemingly activate common neural pathways and utilize similar cellular mechanisms (Nestler, 2002). Nonetheless, this does not automatically mean that the underlying mechanisms are isomorphic to drugs of abuse and natural rewards. One way in which this isomorphism was demonstrated was through the parity of experience in one modality with consequent adaptations in the behavioral responses in the other. This approach began with a rather clever experiment by Mitchell and Stewart (1990). In this study, male rats were given a sensitizing regimen of morphine in either their home cage or in a novel arena. Saline was administered in a complementary manner. The rats were tested for copulatory behavior in either their home cage or in the previously unfamiliar arena. All rats received equal morphine exposure. Copulatory parameters were enhanced only in the arena (either the home cage or the unfamiliar cage) in which the males had previously received morphine, compared to copulation in the arena in which the animals had received saline. Following this study, others showed that prior exposure to sensitizing amphetamine treatments also had a positive impact on copulation in males (Fiorino and Phillips, 1999a, b). In further support of these findings, Barr et al. (1999) found an inhibition of copulation when male rats were tested during withdrawal from a sensitizing amphetamine regimen.

It may not be surprising that potent pharmacological agents like morphine and amphetamine could alter the nervous system in a way that affects the display of a natural behavior, such as sexual behavior. Perhaps more dramatic are studies showing that prior experience with a natural behavior can sensitize the animal's behavioral response to an initial drug exposure. In one paradigm, animals experience an aggressive encounter in which they are defeated. This stressful social experience enhanced the behavioral responses to amphetamine or cocaine in animals that were previously drug-naïve (Marrow et al., 1999; Miczek, 2004; Miczek et al., 2004; de Jong et al., 2005). Both social defeat and aggression induce a stress response. Furthermore, stress can enhance the behavioral response to drugs of abuse. It is possible, therefore, that the initial response to amphetamine or cocaine is altered by the stress hormones and the behaviors per se are not directly involved (Miczek et al., 2004). Results from similar studies with natural reinforcers support the interpretation that such behavioral experiences do sensitize neural systems underlying responsiveness to drugs (Bradley and Meisel, 2001).

3.2 Alterations in Dendritic Structure

One consequence of experience with both drugs of abuse and natural rewards is changes in dendritic structure, particularly in the NAcc and prefrontal cortex. The plasticity of dendritic structure and particularly that of dendritic spines is linked to the dynamic regulation of intrinsic cytoskeleton proteins, whose expression is a function of synaptic activity (Carlisle and Kennedy, 2005). The drugs of abuse, amphetamine (Robinson and Kolb, 1999; Robinson et al., 2002; Li et al., 2003), cocaine (Robinson and Kolb, 1999; Robinson et al., 2002; Norrholm et al., 2003; Li et al., 2004), morphine (Robinson and Kolb, 1999; Robinson et al., 2002), or nicotine (Brown et al., 2000; Brown et al., 2001) all affect dendritic length and/or spine density compared to that of controls, with some differences in the pattern of dendritic changes for individual drug treatments (➋ *Figure 20-3*).

Chronic treatment (3–5 weeks) with either amphetamine or cocaine increased both total dendritic length and dendritic spine densities in medium spiny neurons of the NAcc and caudate, as well as in the apical dendrites of layer III pyramidal neurons of the prefrontal cortex (Robinson and Kolb, 2004). For the NAcc, these changes are seen in both the shell and core subregions. With the exception of Li et al. (2003), all spine density measurements were taken from the distal dendritic branch. Li et al. (2003) confirmed that the changes in spine density indeed occur only in the distal branches and not in the more proximal dendritic

Figure 20-3

An illustration of the effects of sexual experience on dendritic structure. Sexual experience in female hamsters produced region specific effects on spine density in the medial prefrontal cortex, NAcc significantly decreased spine density (left panel) in the terminal branches of the apical dendrites of prefrontal cortex pyramidal neurons, while increasing spine density on terminal branches of accumbens medium spiny neurons. Sexual experience had no effect on caudate medium spiny neurons

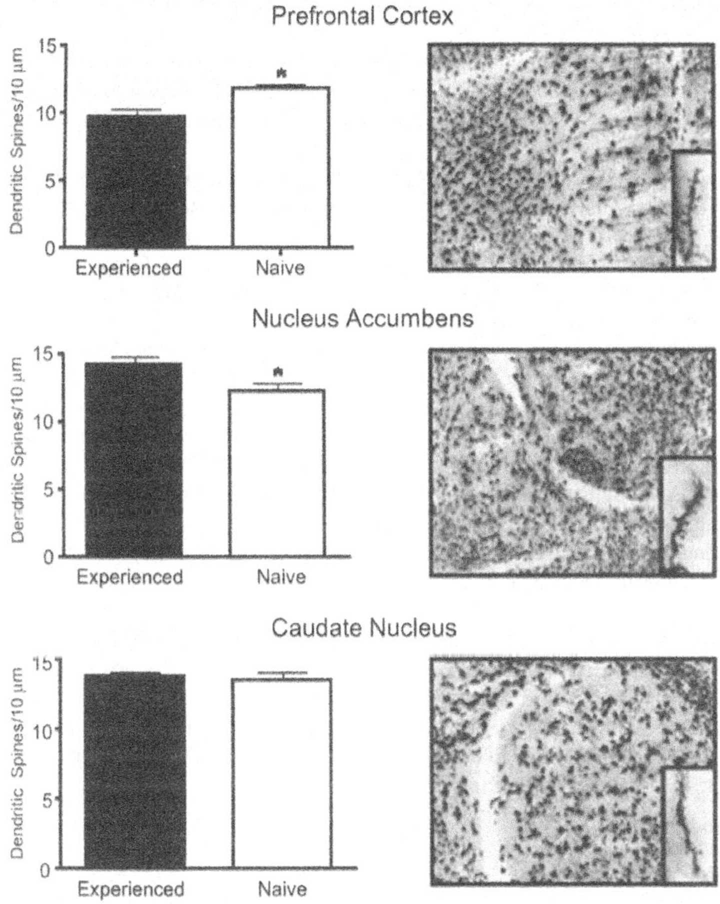

branches of the medium spiny neurons of the NAcc and caudate. Robinson et al. (2001) performed an interesting twist on this approach using animals self-administering cocaine. In this paradigm animals pressed a bar to receive intravenous cocaine for a 7 week period. The same pattern of dendritic changes was observed in the NAcc, indicating that the morphological changes could be produced when the self-administered drugs are related to specific effects of cocaine in the brain. Supporting this conclusion was the observation that additional rats bar pressing for food pellets over this same time period had no change in dendritic measurements (Robinson et al., 2001). On the other hand, only changes in the NAcc core are specific to animals that had shown behavioral sensitization (Li et al., 2004). In this experiment, animals were treated with cocaine in a novel environment or their home cage. Animals that received cocaine in the home cage did not sensitize, while animals treated with cocaine in the novel environment did. Both groups showed increased branching in the NAcc shell, while only the sensitized group had increased branching in the NAcc core (Li et al., 2004). These findings indicate that cocaine induces increased branching in the NAcc shell, independent of behavioral changes, whereas dendritic branching changes in the NAcc core are related to behavioral sensitization.

Nicotine injections over a period of 5 weeks produced a somewhat different pattern of dendritic response (Brown and Kolb, 2001). For the NAcc, the pattern was the same as that of amphetamine and cocaine. That is, there was an increase in both dendritic length and spine density. In the pyramidal cells of the prefrontal cortex, spine density increased in both the basilar and apical dendrites. On the other hand, the dendritic length increased in the basilar, but not in the apical dendrites of the prefrontal cortex neurons.

In contrast to drugs having stimulant properties, chronic morphine treatment reduced dendritic length and spine density in neurons of both the NAcc and prefrontal cortex (Robinson and Kolb, 1999; Robinson et al., 2002). Like the effects seen with cocaine, it did not matter whether the morphine was injected or self-administered by the rat.

One interesting feature of these experiments is the persistence of the drug effects on dendritic morphology. Though a time course of dendritic changes has not been established (after all these are heroic anatomical studies), the delay between drug administration and sacrifice has been as short as 2 days (Norrholm et al., 2003) and as long as 3.5 months later (Li et al., 2003). Consequently, the structural changes appear relatively soon after at least the cessation of drug administration, but are retained even in the absence of further drug administration.

The absence of an effect of bar pressing for food on dendritic morphology would suggest that these structural changes are a consequence of the pharmacological actions of the drugs rather than an inherent plasticity of these neurons and activation of afferent pathways. Nevertheless, several recent studies indicate otherwise. Induction of a salt appetite through diuretic treatments followed by consumption of saline led to an increase in dendritic length in medium spiny neurons of the shell of the NAcc (Roitman et al., 2002). There was no effect on spine density, though as the authors note the static spine density coupled with an increase in dendritic length would yield an increase in total numbers of spines. Fiorino and Kolb (2003) found that dendritic length increased in both apical and basilar dendrites of prefrontal cortex pyramidal neurons following sexual experience in male rats. In contrast, there was a decrease in dendritic length in medium spiny neurons of the NAcc. Spine densities were not reported in this study. In a similar study in female hamsters (Mullins et al., 2004), spine density was increased in the NAcc, but decreased in apical dendrites of prefrontal cortical neurons following sexual experience. There was no difference between experienced and naive hamsters in spine density in medium spiny neurons of the caudate nucleus. Total dendritic length was not measured. One interesting speculation derived from this work is that comparisons between the drug effects and those following natural behaviors may offer insight into the pathways and neurotransmitters recruited by different behaviors.

3.3 Synaptic Mechanisms

Considerable research effort has been aimed at discovering molecular mechanisms mediating drug-induced sensitization and natural rewards. Based on our foregoing discussions, we can look for clues to these mechanisms from observations of behavioral sensitization, structural (particularly dendritic spine) alterations, and dopamine/glutamate interactions related to both repeated drug treatments and experience with natural rewards. The predominant approach has been that of hypothesis testing, in which individual endpoints are measured in response to particular treatment conditions. Recently, a discovery approach has been taken using DNA microarrays to broadly screen differences in gene expression across experimental conditions (Gonzalez-Nicolini and McGinty, 2002; Backes and Hemby, 2003; McClung and Nestler, 2003; Sokolov et al., 2003; Yuferov et al., 2003; Bradley et al., 2005). In many cases, we can identify the "usual suspects" in these cellular and molecular mechanisms, though there are some unsuspected signaling pathways at work here as well. Because the goal of the microarray analyses is to generate "hypotheses," the value of this approach will only be known following future experimental tests of these hypotheses. For this review, we will focus on the results of the conventional approach.

Both presynaptic and postsynaptic mechanisms of dopamine neurotransmission have been explored, with the obvious candidates emerging. Increases in extracellular dopamine following repeated experience with drugs or natural behaviors are certainly related to neuronal firing rate (Floresco et al., 2003). Still, the sensitized and cross-sensitized responses to drugs and natural behaviors suggest that these augmented synaptic levels of dopamine include changes in dopamine autoreceptor and transporter function. To be fair, for both functions contradictions exist in whether repeated drug treatments yield prolonged effects. Some of these contradictions have been resolved based on methodological differences in route or timing of drug administration (Henry and White, 1991; King et al., 1994; Henry et al., 2002). Regarding autoreceptor function, the ability of autoreceptor agonists to inhibit dopamine release is reduced in animals sensitized to cocaine or amphetamine (Ackerman and White, 1990; Pierce and Kalivas, 1995; Pierce et al., 1995; Jones et al., 1996; Chen et al., 1999). The reduction in autoreceptor responsiveness is a product of both receptor sensitivity and overactivity of pertussis toxin-sensitive effector mechanisms (e.g., Bates et al., 1991). In fact, infusions of pertussis toxin into the NAcc sensitize the locomotor response to an acute cocaine challenge (Hummel and Unterwald, 2003).

Parallel to autoreceptor changes, effects of drugs on dopamine transporters could also augment synaptic concentrations of dopamine. Repeated administration of methamphetamine, amphetamine, or cocaine yields consistent decreases in NAcc uptake of exogenously applied dopamine, mRNA for the dopamine transporter, increased transporter turnover, or decreased transporter binding sites (Cass et al., 1993; Masserano et al., 1994; Boulay et al., 1996; Hitry et al., 1996; Letchworth et al., 1997, 2001).

On the postsynaptic side, chronic drug administration has been reported to produce transient increases in dopamine D1 receptor binding (Alburges et al., 1993; Bonhomme et al., 1995), long-term increases in D1 binding (Unterwald et al., 1996), decreases in D1 receptor binding (DeMontis et al., 1993), or no effect on D1 binding (Nestby et al., 1997). A similar lack of congruence is seen in dopamine D3 receptor regulation (Chiang et al., 2003; Le Foll et al., 2003).

In addition to dopamine receptor resistance to repeated drug treatments, G-protein levels are not altered by a sensitizing drug regimen. Perrine et al. (2005) administered cocaine to rats for 14 days, after which the rats displayed sensitized locomotor responses. Although there was some pattern of alterations in mRNA expression, protein levels of Gαs, Gαolf, Gαo, or Gαi1 were unchanged in either the NAcc, prefrontal cortex, or striatum after 1, 3, or 14 days of cocaine exposure. In contrast, repeated treatments with drugs (Roseboom et al., 1990; Van Vliet et al., 1993; Schoffelmeer et al., 1996; Unterwald et al., 1996) or sexual experience (Bradley et al., 2004) enhance cAMP accumulation through dopamine D1 receptor stimulation. These latter findings suggest that although levels of dopamine D1 receptors and their cognate G-proteins are not affected by drug administration or experience with a natural reward, the coupling of the receptor/G-protein complex to downstream signaling pathways, for example through RGS protein inhibition (Rahman et al., 2003), is more efficacious. In support of this interpretation are the findings that (1) intracerebral administration of forskolin, a direct activator of adenylyl cyclase, facilitates locomotor sensitization to cocaine (Schroeder et al., 2004) and (2) cAMP-dependent protein kinase A inhibition blocks conditioned place preference to amphetamine (Beninger et al., 2003).

Besides dopamine signaling, NMDA receptor activation will lead to a biochemical cascade that can affect activity of dopamine neurons. The two most notable consequences of calcium entry through NMDA receptors are the activation of the calcium/calmodulin system along with the activation of the MAP kinase pathway (Adams et al., 2000; Licata et al., 2000). Increased intracellular calcium levels through NMDA receptors, or for that matter through L-type calcium channels increase calmodulin activity, which in turn activates calcium/calmodulin protein kinase (CaMKII). CaMKII thereby phosphorylates CREB, which leads to an increase in gene expression. Calmodulin will also activate raf, which forms an entry point into the MAP kinase system. Further, the MAP kinase pathway is activated through phospholipase C mediated pathways as well as binding to cell-surface growth factor receptors such as receptor tyrosine kinase.

What makes this system so notable with respect to reward-related processes is that cAMP accumulation through G-protein coupled receptors (e.g., dopamine D1 receptors) also provides an entry into the MAP kinase pathway. This interaction becomes more salient with the demonstration of a reciprocal regulation of the phosphorylation of DARPP-32 through dopamine and NMDA receptor activation (Valjent et al., 2005).

Collectively, this system is key to the sensitizing effects of drugs of abuse and natural rewards. A host of manipulations, including stimulation of L-type calcium channels, administration of brain-derived neurotropic factor (an endogenous ligand for receptor tyrosine kinase), elimination of DARPP, blockade of protein kinase C, or direct manipulations of the MAP kinase pathway illustrate the importance of these signaling systems in reward (e.g., Horger et al., 1999; Licata et al., 2000; Zachariou et al., 2002; Gerdjikov et al., 2004).

Not clear from the foregoing discussion is how these transient events lead to long-term changes in neural and behavioral plasticity. One proposed resolution to this question was the discovery of the transcription factor deltaFosB (Kelz et al., 1999; McClung and Nestler, 2003). Because deltaFosB expression is stable over long periods of time, repeated neuronal stimulation produces an incremental accumulation of this protein. Elevated deltaFosB increases the sensitivity of the cells to further stimulation (e.g., by cocaine) meaning a higher level of behavioral output. As would be expected from this hypothesis, overexpression of deltaFosB yields a sensitized response to cocaine (Colby et al., 2003). Rounding things out, NMDA receptor activity can regulate the protein kinase Cdk5 (Wei et al., 2005). Cdk5 regulates NMDA receptor activity, calmodulin, protein kinase A, and DARPP-32 dephosphorylation (Bibb, 2003). Cdk5 may link these processes to the formation of dendritic spines, serving a common function in the biochemical and structural plasticity of dopamine responsive neurons.

3.4 Summary

An individual's experience with both drugs and natural rewards produces alterations in behavior that reflect changes in underlying neural systems. The commonality of these neural alterations is predicted by the reciprocal effects of drugs and natural experiences on an individual's future reaction to drugs or behavioral expression. Indeed, these experiences are represented by a morphological and biochemical plasticity that encompasses several brain regions, several neurotransmitter systems, and interacting signaling pathways. At the same time, there are important differences in the cellular and molecular responses to drug exposure and one's ongoing natural behaviors. One goal of future research will be to disentangle the complexity of these experiential effects to distinguish those neural responses forming the basis for behavioral flexibility from the pathological corruption of these pathways leading to addiction.

4 Sex Differences in Reward

Very little research has investigated sex differences in reward systems, except for some recent work investigating sex differences in drug abuse. There is some research on sexual motivation in male and female rats, but studies that directly investigate differences between males and females are rare. Nevertheless, as will be discussed below, there are sex differences in and hormonal influences on drug taking behavior, the behavioral response to psychomotor stimulants, sexual motivation, and feeding behavior.

The differences between males and females in the modulation of natural rewards are in part due to the modulation of these behaviors by ovarian hormones. Estradiol decreases feeding behavior and in combination with progesterone activates sexual behavior. In other words, there is day-to-day variation in the value of natural rewards in females. As a consequence of the ovarian cycle, the reward system in females is rarely in a state that is maximally activated by stimuli. In fact, the rewarding value of many stimuli is in a state of constant variation, due to hormonal changes of the estrous cycle. In males, enhanced testosterone secretion that occurs in response to some behaviors may be rewarding, in addition to the rewarding value of the stimulus. As a consequence, sex differences in steroid secretion may affect the neurochemical tone in reward pathways differentially between males and females. Similarly, the salience of external stimuli (e.g., food or conspecifics) may fluctuate as a function of hormonal state. These variations in activity of reward pathways would be expected to impact the individual's response to pharmacological agents that similarly activate these neural pathways. And in fact this is the case. If the goal of motivated behaviors is the activation of internal reward processes rather than the homeostatic biological mechanisms that are necessary for survival and reproduction, these behaviors serve a homeostatic function. Evolutionary pressures led to a situation that optimized the link between reward and homeostasis.

Regulation of behaviors optimizes activation of reward pathways with reproduction, regulation of metabolic fuels, electrolyte balance, and the secondary consequences of the rewarding value of behavioral activity. Again, this would suggest that there are both points of convergence and divergence between the sexes in the responsiveness of these reward pathways. Thus, in females when drugs of abuse drive neural activity in the reward system day after day, the system shows greater changes than seen in males, because the effect of the drugs is more dramatic. It is not just the magnitude of the drug effect, but that the drug effect occurs repeatedly without waxing and waning. The ways in which males and females differ, as well as the dramatic influences of ovarian hormones on these behaviors, may provide us with new paradigms to tease apart the neurochemical and molecular mechanisms mediating reward.

4.1 Sexual Behavior

Sexual behavior has both appetitive (motivational) and consummatory components. This has been demonstrated most elegantly by experiments from the Everitt laboratory (Everitt and Stacey, 1987; Everitt, 1990). In experiments using a second-order schedule of reinforcement, Everitt and collaborators demonstrated that male rats would bar press for access to a sexually receptive female rat. They went on to show that bar pressing for access to the female rat (i.e., sexual motivation) was abolished by lesions of the basolateral amygdala. In contrast, lesions of the preoptic area had no effect on operant responding for access to the female, but when the female rat was available, the male does not copulate with her (Everitt and Stacey, 1987; Everitt, 1990). These experiments clearly demonstrated that there are distinct neural substrates necessary for sexual motivation versus sexual ability. Furthermore, when amphetamine was delivered to the NAcc of the male rat with basolateral amygdala lesions, bar pressing for access to the female was reinstated. (Everitt, 1990). Since in the rat there are projections from the basolateral amygdala to the NAcc, dopamine in the NAcc was implicated in sexual motivation.

Subsequently, a number of investigators have demonstrated that dopamine increases in the NAcc of male rats in anticipation of gaining access to a sexually receptive female rat. (Pfaus et al., 1990; Pleim et al., 1990; Pfaus and Phillips, 1991; Damsma et al., 1992). Furthermore, both male and female rats exhibit an increase in extracellular concentrations of dopamine in the NAcc during sexual behavior (Pfaus et al., 1990; Mermelstein and Becker, 1995; Pfaus et al., 1995).

Sexual behavior in the female rat is unique in that it is not rewarding to the female under traditional laboratory testing conditions (Oldenberger et al., 1992; Paredes and Alonso, 1997; Paredes and Vazquez, 1999; Martinez and Paredes, 2001; Jenkins and Becker, 2003b). In the laboratory, copulation has historically been studied under conditions where contacts are initiated by the male who engages in a series of mounts and intromissions that ultimately lead to an ejaculation (Bermant, 1961; Adler, 1969). Under these conditions, the male controls the rate of copulation and he will intromit with a female at a relatively rapid rate.

On the other hand, if the female is given a place where she can escape from the male, she will establish longer latencies between sexual contacts (Adler, 1969, 1978; McClintock, 1984; Erskine et al., 1989). In the wild, rats engage in mating in groups of several females and males, and males are able to achieve relatively rapid intromissions with different females. The females, in turn, achieve their preferred rate of copulation by withdrawing from the male and then returning after the preferred interval (McClintock, 1984). This behavior of the females, of withdrawing and then returning to the male after copulatory contacts is known as pacing behavior (Erskine et al., 1989). Pacing behavior allows the female to control both the rate and duration of the copulatory bout. Importantly, sexual behavior is rewarding to the female rat when she achieves her preferred rate of copulation (Paredes and Alonso, 1997; Paredes and Vazquez, 1999; Martinez and Paredes, 2001; Jenkins and Becker, 2003b), whether or not she is actively pacing the rate of copulation (Jenkins and Becker, 2003a).

In support for a role for dopamine in sexual reward, female rats only exhibit an increase in NAcc dopamine if they can achieve a preferred rate of intromissions. This can be accomplished by either the female actively controlling or "pacing" the rate of copulation or if the experimenter removes the male and then returns it to the female's chamber at appropriate intervals during copulation (Mermelstein and Becker, 1995; Becker et al., 2001). Female hamsters also exhibit an increase of dopamine in dialysate during sexual behavior (Meisel et al., 1993). Achieving the preferred rate of copulation is important for the female rat to optimize the rate of vaginocervical stimulation received from a male which activates a neuroendocrine reflex that is necessary for pregnancy to occur (Adler, 1969; McClintock and Adler, 1977; Adler, 1978; McClintock, 1984; Erskine, 1989).

It should be noted that females engaged in sexual behavior at their preferred pacing interval had greater increases in dopamine in dialysate from the NAcc than did animals in which the male rat was removed and then returned to the female's chamber either too rapidly or much later (Becker et al., 2001). Furthermore, dopamine increases in the NAcc occurred prior to coital stimulation when intromissions are received at the female's preferred pacing interval, but not when coital stimulation occurred under other conditions (Jenkins and Becker, 2003a). Thus, increases in NAcc dopamine are not induced by coital stimulation or escape from/removal of the male rat. Instead, dopamine increases occur in anticipation of coital stimulation that occurs at a specific interval. These data support the hypothesis that dopamine increases in the NAcc signal the impending receipt of coital stimulation at the female's preferred pacing interval.

Is the increase in dopamine in the NAcc necessary for a female to find sexual behavior rewarding? In hamsters, pretreatment of females with D2-dopamine receptor antagonist raclopride blocked conditioned place preferences for the place in which mating occurred (Meisel, 1996). On the other hand, Paredes found that pretreatment with the dopamine antagonists flupentixol or raclopride did not block conditioned place preference induced by paced mating in female rats (Garcia-Horsman and Paredes, 2004), while naloxone treatment prevented establishment of conditioned place preference induced by paced mating (Paredes and Martinez, 2001). These results suggest that activation of D2 dopamine receptors is not necessary for sexual behavior to be rewarding, but μ-opioid receptor mediated activation, with the downstream effects on GABA, dopamine, and glutamate neurotransmission is important for sexual motivation in the female rat. The study by Garcia-Horseman and Paredes (2004) used relatively low dose of raclopride, so it is possible that in the reward circuit this dose was not sufficient to completely block the effect of dopamine. Alternatively, if sexual reward can be dissociated into "wanting" and "liking," it is possible that the way the conditioned place preference test is conducted (animals are placed into the test apparatus for training immediately after receiving an ejaculation) is particularly sensitive for "liking." This interpretation would be consistent with the findings, as dopamine is thought to be important for "wanting," while GABA and opioid systems are important for "liking."

Sexual behavior of the female rat requires new considerations and interpretations of the role of dopamine in reward. Female rats find sexual behavior rewarding and have increased NAcc dopamine when they engage in sex at their preferred interval (Mermelstein and Becker, 1995; Paredes and Alonso, 1997; Becker et al., 2001; Martinez and Paredes, 2001). Sex that is rewarding has been shown to be associated with the triggering of a neuroendocrine reflex necessary for pregnancy (Adler, 1974; Gilman et al., 1979; Erskine et al., 1989). One possibility is that the changes in dopamine observed here represent a coupling of the sexual interaction and its physiological consequences, both of which may be necessary for sexual behavior to be rewarding. In other words, increases in dopamine predict the receipt of coital stimulation, but only when the coital stimulation occurs at such a rate that triggers the neuroendocrine reflex necessary for successful pregnancy to occur. Coital stimulation is known to induce the release of oxytocin in rats and other species (Flanagan et al., 1993). The neuroendocrine reflex that is activated in the female rat results in the release of prolactin (Erskine, 1995). Since oxytocin is thought to induce the release of prolactin, one possibility is that the activation of the intrahypothalamic neurons necessary for coital-induced release of oxytocin also enhances dopamine release in the NAcc.

4.2 Drug Abuse

The evidence for sex differences in drug abuse and drug taking behavior has been the subject of recent reviews (Lynch et al., 2002; Carroll et al., 2004; Roth et al., 2004), so this review will be brief. Cocaine abuse by women has increased in the last decade so that of the 1.8 million Americans who use cocaine, approximately 30% are now female (Wetherington and Roman, 1995). According to a recent report, 9% of women aged 12 and above have used cocaine. The only illicit drug used more by women is marijuana (28% have used marijuana) (Kandel et al., 1995). Among women who have used cocaine, prevalence of lifetime dependence for cocaine is 14.9 ± 2.0% (mean ± S.D.). This is in contrast to the case with alcohol where 79% have used it, but only 9.2 ± 0.8% have developed lifetime dependence (Kandel et al., 1995). The use of all illicit drugs has been increasing among women in the past decade, and stimulant drug use and dependence among women, in particular, is a growing public health concern (Wetherington and Roman, 1995; Lynch et al., 2002; Carroll et al., 2004).

Psychomotor Stimulants. Sex differences in the pattern of cocaine abuse and behavioral responses to cocaine indicate that the pattern of cocaine use and the onset of addiction to cocaine are more rapid in women than men (Carroll et al., 2004). Women begin using cocaine and enter treatment at earlier ages than men (Griffin et al., 1989; Mendelson et al., 1991) and have more severe cocaine use at intake than men (Kosten et al., 1993). Furthermore, cocaine cues induce more drug craving in female than male addicts (Robbins et al., 1999). Collectively, these results suggest that women may be more sensitive to the addictive properties of cocaine than men.

Basic research on the role of sex and ovarian hormones in the neurochemical and behavioral responses to acute and repeated exposure to drugs of abuse also finds sex differences. In order to investigate sex differences, the issues involved are complex (Becker et al., 2005). First, adult males and females do not differ solely in their genetic composition, but also in the hormones secreted by their gonads and the patterns of hormone secretion (i.e., females exhibit a cyclic release of estradiol and progesterone, while males exhibit a tonic release of testosterone). These patterns of hormone release are a consequence of hormone exposure during sexual differentiation of the brain. This means that there are three different ways that males and females can be different: (1) sex differences independent of the milieu of gonadal hormones in the adult, due to developmental sex differences in the brain, which are driven either by secretions from the fetal gonads or due to cell-autonomous sexual differentiation; (2) sex differences in effects of gonadal hormones in the adult acting on a sexually dimorphic brain; and (3) sex differences due to the different hormones produced by males and females.

The acute behavioral response to psychomotor stimulants that rodents exhibit can reflect both sex differences and can be modulated by gonadal hormones in males and females. Furthermore, the effect of repeated exposure to psychomotor stimulants results in behavioral sensitization. Behavioral sensitization can be different in males and females, and can also be differentially affected by gonadal steroid hormones. In humans, these factors are intermingled because chronic cocaine use can disrupt and even cause cessation of a woman's menstrual cycle (Mello, 1995). In such women, estradiol may play a role in acquisition of drug taking behaviors, but not in maintenance of these behaviors (because in women with amenorrhea the serum concentrations of estradiol are extremely low).

Research on rodents and humans indicate that the behavioral effects of drugs of abuse and the psychomotor stimulants, in particular, are both sexually dimorphic and modulated by the gonadal steroid hormones (e.g., Gordon, 1980; Hruska and Silbergeld, 1980; Becker and Ramirez, 1981; Di Paolo et al., 1981; Joyce et al., 1982; Dluzen and Ramirez, 1984; Becker and Beer, 1986; Di Paolo et al., 1986; Hruska, 1988; Van Hartesveldt et al., 1989; Dluzen and Ramirez, 1990; Bazzett and Becker, 1994; Lynch et al., 2002; Sell et al., 2002; Carroll et al., 2004). If one considers sensitization of amphetamine or cocaine-induced psychomotor behavior to be the absolute increase in the behavioral response exhibited when two tests are compared, females exhibit more robust sensitization than do intact males (Robinson et al., 1982; Robinson, 1984; Camp and Robinson, 1988a, b; van Haaren and Meyer, 1991; Forgie and Stewart, 1994).

The preponderance of evidence indicates that ovariectomy (OVX) of female rats does not affect the induction or expression of sensitization to amphetamine (Robinson et al., 1982; Robinson, 1984; Camp and Robinson, 1988a, b; Forgie and Stewart, 1994). Yet in some studies, OVX females do not exhibit significant sensitization of cocaine-induced locomotor activity when intact females or gonadal hormone treated females do (van Haaren and Meyer, 1991; Sircar and Kim, 1999). Furthermore, estradiol treatments in OVX rats enhance sensitization of locomotor activity induced by amphetamine or cocaine

(Peris et al., 1991; Forgie and Stewart, 1994). Thus, it is possible that variability in the behavior of intact female rats across the estrous cycle may obscure the effects of ovarian hormones on the induction and/or expression of sensitized psychomotor behaviors in experiments where intact and OVX rats are compared. Effects of ovarian hormones are likely to be seen more clearly when females are OVX and treated with estradiol.

Female rats have also been shown to acquire cocaine self-administration more rapidly than males (Lynch and Carroll, 1999; Carroll et al., 2002; Roth et al., 2004), and cocaine self-administration in female rodents varies across the estrous cycle (Lynch et al., 2000). Female rats will work harder for cocaine (by making more lever presses for an infusion of cocaine) during the estrous phase of the cycle than during other phases of the cycle, and females work harder than male rats (Roberts et al., 1989). In contrast, sucrose self-administration does not vary across the estrous cycle (Hecht et al., 1999).

Estradiol administration to OVX females affects many psychostimulant drug-induced behaviors, including self-administration (Verimer et al., 1981; Peris et al., 1991; Morissette and Di Paolo, 1993; Thompson and Moss, 1994; Grimm and See, 1997; Becker, 1999; Sircar and Kim, 1999; Quinones-Jenab et al., 2000; Freeman et al., 2001). For example, Hu et al. (2004) found that in OVX female rats, exogenous estradiol treatment alone was sufficient to facilitate acquisition of cocaine self-administration. Estradiol-facilitated cocaine self-administration has also been found in other studies (Roberts et al., 1989; Freeman et al., 2001).

In contrast to estradiol, the subjective effects of psychomotor stimulant drugs are negatively correlated with salivary progesterone levels (White, 2002). Furthermore, in rodents, progesterone inhibits cocaine-mediated behaviors, such as estradiol-enhanced locomotor activity and sensitization of cocaine-induced stereotyped behavior, compared to OVX females treated with estradiol. Recently, it was reported that concurrent administration of progesterone with estradiol counteracts the effect of estradiol on acquisition of cocaine self-administration behavior (Jackson et al., 2005). Therefore, over the course of the estrous cycle, and perhaps across the menstrual cycle, there are peaks and valleys during which females are more or less susceptible to the reinforcing properties of cocaine.

Castration (CAST) of males has been reported to enhance sensitization of amphetamine- or cocaine-induced psychomotor behavior (e.g., Robinson, 1984; Camp and Robinson, 1988a, b), although this result has not been found consistently (van Haaren and Meyer, 1991; Forgie and Stewart, 1994). It has been hypothesized that if CAST enhances the induction and/or expression of behavioral sensitization, testosterone treatment should reverse this effect. This is not the case, however, as testosterone treatment has not been found to affect behavioral sensitization in CAST males (Forgie and Stewart, 1994). Furthermore, there is no effect of CAST on acquisition of cocaine self-administration behavior and a dose of estradiol that enhances self-administration in female rats has no effect on cocaine self-administration behavior in male rats. Thus, the effects of estradiol on the acquisition of cocaine self-administration are sexually dimorphic. There are also sex differences in the maintenance of cocaine self-administration, and in the reinstatement of responding after abstinence (Roth et al., 2004). It should be noted that sex differences are most robust at lower doses of cocaine, and at higher doses of cocaine, differences between males and females are less evident.

Opiates, Nicotine, and Alcohol. There are also sex differences in humans and laboratory animals in acquisition, maintenance, and relapse seen with other drugs of abuse. This literature has been reviewed extensively, so the review here will be brief. The reader is referred to recent reviews for additional information (Lynch et al., 2002; Roth et al., 2004).

In laboratory animals, not all studies find a sex difference in self-administration of opiates (heroin, morphine, and fentanyl). When there is a sex difference, females tend to acquire self-administration more rapidly and take more drugs during the maintenance phase (see *Table 20-1* in Roth et al., 2004). In humans, however, there is no sex difference found in the pattern of opiate use (Lynch et al., 2002). Female rats also acquire nicotine self-administration more rapidly than males, and will work harder to receive nicotine than males (reviewed in Lynch et al., 2002; Roth et al., 2004). Women report shorter intervals between cigarettes, and find it more difficult to quit smoking than men (reviewed in Lynch et al., 2002; Roth et al., 2004).

Fewer women than men abuse alcohol (7–12% versus 20%), yet the frequency that young women are becoming intoxicated on alcohol is rising, and the medical consequences of chronic alcohol consumption are more severe for women than for men (Almeida et al., 1998; Devaud et al., 2003). It has also been suggested that women become addicted to alcohol more rapidly than do men (Lynch et al., 2002). In laboratory animals, there are sex differences in the development of and recovery from ethanol dependence. Female rats have decreased seizure threshold following withdrawal and have a more rapid return to the control level of seizure susceptibility (Devaud and Chadda, 2001). Devaud and colleagues have found that ethanol administration affects $GABA_A$ and NMDA receptor subtypes differently in male and female rat brain, and the direction of the sex difference varies among brain regions (Devaud and Alele, 2004). Whether the sex differences in the effects of alcohol and $GABA_A$ and NMDA receptors mediate sex differences in the long-term consequences of alcohol dependence is not known. Further research in the clinical setting and the laboratory is needed to clarify the causes and extend our understanding of the nature of the sex differences.

4.3 Summary

It is clear from this brief discussion that there are sex differences in the reward system. Studies of the response to cocaine in gonadectomized male and female rats suggest that there is an underlying sex difference due to sexually dimorphic development of the brain that, in part, mediates the sex difference in reward behavior. In addition, there are effects of gonadal hormones that modulate the reward system. In particular, estradiol enhances the rewarding value of drugs, whereas progesterone counteracts the effect of estradiol.

5 General Principles and Significance

The principles governing the motivation to engage in behaviors motivated by natural rewards are the same as those governing disorders such as drug abuse and other pathological conditions. In general, activation of dopamine release in the NAcc is an internal cue that tells an animal that something desired is available (or is soon to be available). We conclude, as others have, that this dopamine signal is associated with enhanced incentive salience or "wanting" of the desired stimulus. Glutamate and GABA neurons modulate dopamine release, providing input related to context and motivational state. GABA and endogenous opioids also provide information about the hedonic value of the stimulus or "liking."

Sex differences in motivation and motivational systems may provide key insights into the fundamental processes mediating the switch from natural rewards to pathological behaviors, such as drug addiction. We argue that knowledge of sex differences in reward processes is essential for the development of appropriate therapeutic approaches. A rational therapeutic treatment must accommodate differences between males and females.

Acknowledgments

This research was supported by grants from the USPHS to JBB (DA12677 & NS48141) and to RLM (DA 13680).

References

Ackerman JM, White FJ. 1990. A10 somatodendritic dopamine autoreceptor sensitivity following withdrawal from repeated cocaine treatment. Neurosci Lett 117: 181-187.

Adams JP, Roberson ED, English JD, Selcher JC, Sweatt JD. 2000. MAPK regulation of gene expression in the central nervous system. Acta Neurobiol Exp 60: 377-394.

Adler NT. 1969. Effects of the male's copulatory behavior on successful pregnancy of the female rat. J Comp Physiol Psychol 69: 613-622.

Adler NT. 1974. The behavioral control of reproductive physiology. Adv Behav Biol 11: 259-286.

Adler NT. 1978. On the mechanisms of sexual behavior and their evolutionary constraints. Hutchison JB, Biological determinants of sexual behavior. New York: Wiley; pp. 657-694.

Akiyama G, Ikeda H, Matsuzaki S, Sato M, Moribe S, et al. 2004. GABAA and GABAB receptors in the nucleus accumbens shell differentially modulate dopamine and acetylcholine receptor-mediated turning behaviour. Neuropharmacology 46: 1082-1088.

Alburges ME, Narang N, Wamsley JK. 1993. Alterations in the dopaminergic receptor system after chronic administration of cocaine. Synapse 14: 314-323.

Almeida OF, Shoaib M, Deicke J, Fischer D, Darwish MH, et al. 1998. Gender differences in ethanol preference and ingestion in rats. The role of the gonadal steroid environment. J Clin Invest 101: 2677-85.

Anden NE, Dahlstrom A, Fuxe K, Larsson K, Olson L, et al. 1966. Ascending monoamine neurons to the telencephalon and diencephalon. Act Physiol Scand 67: 313-323.

Andrzejewski ME, Sadeghian K, Kelley AE. 2004. Central amygdalar and dorsal striatal NMDA receptor involvement in instrumental learning and spontaneous behavior. Behav Neurosci 118: 715-729.

Ashby CR Jr, Rohatgi R, Ngosuwan J, Borda T, Gerasimov MR, et al. 1999. Implication of the GABA(B) receptor in gamma vinyl-GABA's inhibition of cocaine-induced increases in nucleus accumbens dopamine. Synapse 31: 151-153.

Backes E, Hemby SE. 2003. Discrete cell gene profiling of ventral tegmental dopamine neurons after acute and chronic cocaine self-administration. J Pharmacol Exp Ther 307: 450-459.

Barr AM, Fiorino DF, Phillips AG. 1999. Effects of withdrawal from an escalating dose schedule of d-amphetamine on sexual behavior in the male rat. Pharmacol Biochem Behav 64: 597-604.

Bassareo V, De Luca M, Di Chiara G. 2002. Differential expression of motivational stimulus properties by doamine in the nucleus accumbens shell versus core and prefrontal cortex. J Neurosci 22: 4709-4719.

Bates MD, Senogles SE, Bunzow JR, Liggett SB, Civelli O, et al. 1991. Regulation of responsiveness at D2 dopamine receptors by receptor desensitization and adenylyl cyclase sensitization. Mol Pharmacol 39: 55-63.

Battisti JJ, Uretsky NJ, Wallace LJ. 2000. NMDA glutamate receptor role in the development of context-dependent and independent sensitization of the induction of stereotypy by amphetamine or apomorphine. Behav Brain Res 114: 167-174.

Bazzett TJ, Becker JB. 1994. Sex differences in the rapid and acute effects of estrogen on striatal D2 dopamine receptor binding. Brain Res 637: 163-172.

Becker JB. 1999. Gender differences in dopaminergic function in striatum and nucleus accumbens. Pharmacol Biochem Behav 64: 803-812.

Becker JB, Arnold AP, Berkley KJ, Blaustein JD, Eckel LA, et al. 2005. Strategies and methods for research on sex differences in brain and behavior. Endocrinology 146: 1650-1673.

Becker JB, Beer ME. 1986. The influence of estrogen on nigrostriatal dopamine activity: Behavioral and neurochemical evidence for both pre- and postsynaptic components. Behav Brain Res 19: 27-33.

Becker JB, Ramirez VD. 1981. Sex differences in the amphetamine stimulated release of catecholamines from rat striatal tissue in vitro. Brain Res 204: 361-372.

Becker JB, Rudick CN, Jenkins WJ. 2001. The role of dopamine in the nucleus accumbens and striatum during sexual behavior in the female rat. J Neurosci 21: 3236-3241.

Beninger RJ, Nakonechny PL, Savina I. 2003. cAMP-dependent protein kinase and reward-related learning: Intra-accumbens Rp-cAMPS blocks amphetamine-produced place conditioning in rats. Psychopharmacology 170: 23-32.

Bermant G. 1961. Response latencies of female rats during sexual intercourse. Science 133: 1771-1773.

Berridge KC.1996. Food reward: Brain substrates of wanting and liking. Neurosci Biobehav Rev 20: 1-25.

Berridge KC. 2003. Pleasures of the brain. Brain Cogn 52: 106-128.

Berridge KC, Robinson TE. 1995. The mind of an addicted brain: Neural sensitization of wanting versus liking. Curr Direct Psychol Sci 4: 71-76.

Berridge KC, Robinson TE. 1998a. What is the role of dopamine in reward: Hedonic impact, reward learning, or incentive salience? Brain Res Rev 28: 309-369.

Berridge KC, Robinson TE. 1998b. The role of dopamine in reward: Hedonics, learning and incentive salience after 6-hyproxydopamine lesions. Brain Res Rev 28: 309-369.

Berridge KC, Robinson TE. 2003. Parsing reward. Trends Neurosci 26: 507-515.

Berridge KC, Venier IL, Robinson TE. 1989. Taste reactivity analysis of 6-hydroxydopamine-induced aphagia: Implications for arousal and anhedonia hypotheses of dopamine function. Behav Neurosci 103: 36-45.

Berridge KC, Winkielman P. 2003. What is an unconscious emotion? (The case for unconscious "liking"). Cognition and Emotion 17: 181-211.

Bibb JA. 2003. Role of Cdk5 in neuronal signaling, plasticity, and drug abuse. Neurosignals 12: 191-199.

Bonhomme N, Cador M, Stinus L, Le Moal M, Spampinato U. 1995. Short and long-term changes in dopamine and serotonin receptor binding sites in amphetamine-sensitized rats: A quantitative autoradiographic study. Brain Res 675: 215-223.

Boulay D, Duterte-Boucher D, Leroux-Nicollet I, Naudon L, Costentin J. 1996. Locomotor sensitization and decrease in [3H]mazindol binding to the dopamine transporter in the nucleus accumbens are delayed after chronic treatments by GBR12783 or cocaine. J Pharmacol Exp Ther 278: 330-337.

Bowery NG. 1993. GABAB receptor pharmacology. Annu Rev Pharmacol Toxicol 33: 109-147.

Bradley KC, Meisel RL. 2001. Sexual behavior induction of c-Fos in the nucleus accumbens and amphetamine-stimulated locomotor activity are sensitized by previous sexual experience in female Syrian hamsters. J Neurosci 21: 2123-2130.

Bradley KC, Mullins AJ, Meisel RL, Watts VJ. 2004. Sexual experience alters D1 receptor-mediated cyclic AMP production in the nucleus accumbens of female Syrian hamsters. Synapse 53: 20-27.

Brebner K, Phelan R, Roberts DC. 2000. Intra-VTA baclofen attenuates cocaine self-administration on a progressive ratio schedule of reinforcement. Pharmacol Biochem Behav 66: 857-862.

Breysse N, Baunez C, Spooren W, Gasparini F, Amalric M. 2002. Chronic but not acute treatment with a metabotropic glutamate 5 receptor antagonist reverses the akinetic deficits in a rat model of parkinsonism. J Neurosci 22: 5669-5678.

Brog JS, Salyapongse A, Deutch AY, Zahm DS. 1993. The patterns of afferent innervation of the core and shell in the "accumbens" part of the rat ventral striatum: Immunohistochemical detection of retrogradely transported fluorogold. J Comp Neurol 338: 255-278.

Brown RW, Gonzalez CL, Kolb B. 2000. Nicotine improves Morris water task performance in rats given medial frontal cortex lesions. Pharmacol Biochem Behav 67: 473-478.

Brown RW, Gonzalez CL, Whishaw IQ, Kolb B. 2001. Nicotine improvement of Morris water task performance after fimbria-fornix lesion is blocked by mecamylamine. Behav Brain Res 119: 185-192.

Brown RW, Kolb B. 2001. Nicotine sensitization increases dendritic length and spine density in the nucleus accumbens and cingulate cortex. Brain Res 899: 94-100.

Burns LH, Everitt BJ, Kelley AE, Robbins TW. 1994. Glutamate-dopamine interactions in the ventral striatum: Role in locomotor activity and responding with conditioned reinforcement. Psychopharmacology 115: 516-528.

Cador M, Robbins TW, Everitt BJ. 1989. Involvement of the amygdala in stimulus-reward associations: Interaction with the ventral striatum. Neuroscience 30: 77-86.

Camp DM, Robinson TE. 1988a. Susceptibility to sensitization. II. The influence of gonadal hormones on enduring changes in brain monoamines and behavior produced by the repeated administration of D-amphetamine or restraint stress. Behav Brain Res 30: 69-88.

Camp DM, Robinson TE. 1988b. Susceptibility to sensitization. I. Sex differences in the enduring effects of chronic D-amphetamine treatment on locomotion, stereotyped behavior and brain monoamines. Behav Brain Res 30: 55-68.

Cannon CM, Abdallah L, Tecott LH, During MJ, Palmiter RD. 2004. Dysregulation of striatal dopamine signaling by amphetamine inhibits feeding by hungry mice. Neuron 44: 509-520.

Cardinal RN, Everitt BJ. 2004. Neural and psychological mechanisms underlying appetitive learning: Links to drug addiction. Curr Opin Neurobiol 14: 156-162.

Carelli RM. 2004. Nucleus accumbens cell firing and rapid dopamine signaling during goal-directed behaviors in rats. Neuropharmacology 47: 180-189.

Carelli RM, Ijames SG, Crumling AJ. 2000. Evidence that separate neural circuits in the nucleus accumbens encode cocaine versus "natural" (water and food) reward. J Neurosci 20: 4255-4266.

Carelli RM, Williams JG, Hollander JA. 2003. Basolateral amygdala neurons encode cocaine self-administration and cocaine-associated cues. J Neurosci 23: 8204-8211.

Carlezon WA Jr, Nestler EJ. 2002. Elevated levels of GluR1 in the midbrain: A trigger for sensitization to drugs of abuse? Trends Neurosci 25: 610-615.

Carlisle HJ, Kennedy MB. 2005. Spine architecture and synaptic plasticity. Trends Neurosci 28: 182-187.

Carroll ME, Lynch W, Roth M, Morgan A, Cosgrove K. 2004. Sex and estrogen influence drug abuse. Trends Pharmacol Sci 25: 273-279.

Carroll ME, Morgan AD, Lynch WJ, Campbell UC, Dess NK. 2002. Intravenous cocaine and heroin self-administration in rats selectively bred for differential saccharin intake: Phenotype and sex differences. Psychopharmacology 161: 304-313

Carroll R, Zukin R. 2002. NMDA-receptor trafficking and targeting: Implications for synaptic transmission and plasticity. Trends Neurosci 25: 571-577.

Cartmell J, Monn JA, Schoepp DD. 2000. The mGlu(2/3) receptor agonist LY379268 selectively blocks amphetamine ambulations and rearing. Eur J Pharmacol 400: 221-224.

Cass WA, Gerhardt GA, Gillespie K, Curella P, Mayfield RD et al. 1993. Reduced clearance of exogenous dopamine in rat nucleus accumbens, but not in dorsal striatum, following cocaine challenge in rats withdrawn from repeated cocaine administration. J Neurochem 61: 273-283.

Chen JC, Su HJ, Huang LI, Hsieh MM. 1999. Reductions in binding and functions of D2 dopamine receptors in the rat ventral striatum during amphetamine sensitization. Life Sci 64: 343-354.

Chiang YC, Chen PC, Chen JC. 2003. D(3) dopamine receptors are down-regulated in amphetamine sensitized rats and their putative antagonists modulate the locomotor sensitization to amphetamine. Brain Res 972: 159-167.

Choi KH, Zarandi B, Todd KG, Biondo AM, Greenshaw AJ. 2000. Effects of AMPA/kainate receptor blockade on responses to dopamine receptor agonists in the core and shell of the rat nucleus accumbens. Psychopharmacology 150: 102-111.

Chuhma N, Zhang H, Masson J, Zhuang X, Sulzer D, et al. 2004. Dopamine neurons mediate a fast excitatory signal via their glutamatergic synapses. J Neurosci 24: 972-981.

Churchill L, Swanson CJ, Urbina M, Kalivas PW. 1999. Repeated cocaine alters glutamate receptor subunit levels in the nucleus accumbens and ventral tegmental area of rats that develop behavioral sensitization. J Neurochem 72: 2397-2403.

Colby CR, Whisler K, Steffen C, Nestler EJ, Self DW. 2003. Striatal cell type-specific overexpression of ΔFosB enhances incentive for cocaine. J Neurosci 23: 2488-2493.

Coppola D, O'Connell R. 1988. Are pheromones their own reward? Physiol Behav 44: 811-816.

Cruz HG, Ivanova T, Lunn ML, Stoffel M, Slesinger PA, et al. 2004. Bi-directional effects of GABA(B) receptor agonists on the mesolimbic dopamine system. Nature Neurosci 7: 153-159.

Damsma G, Pfaus JG, Wenkstern D, Phillips AG, Fibiger HC. 1992. Sexual behavior increases dopamine transmission in the nucleus accumbens and striatum of male rats: Comparison with novelty and locomotion. Behav Neurosci 106: 181-191.

Danysz W, Essmann U, Bresink I, Wilke R. 1994. Glutamate antagonists have different effects on spontaneous locomotor activity in rats. Pharmacol Biochem Behav 48: 111-118.

Darracq L, Drouin C, Blanc G, Glowinski J, Tassin JP. 2001. Stimulation of metabotropic but not ionotropic glutamatergic receptors in the nucleus accumbens is required for the D-amphetamine-induced release of functional dopamine. Neuroscience 103: 395-403.

David HN, Abraini JH. 2003. Blockade of the locomotor stimulant effects of amphetamine by group I, group II, and group III metabotropic glutamate receptor ligands in the rat nucleus accumbens: Possible interactions with dopamine receptors. Neuropharmacology 44: 717-727.

David HN, Sissaoui K, Abraini JH. 2004. Modulation of the locomotor responses induced by D1-like and D2-like dopamine receptor agonists and D-amphetamine by NMDA and non-NMDA glutamate receptor agonists and antagonists in the core of the rat nucleus accumbens. Neuropharmacology 46: 179-191.

Davis CM, Moskovitz B, Nguyen MA, Tran BB, Arai A, et al. 1997. A profile of the behavioral changes produced by facilitation of AMPA-type glutamate receptors. Psychopharmacology 133: 161-167.

de Jong JG, Wasilewski M, Vegt van der BJ, Buwalda B, Koolhaas JM. 2005. A single social defeat induces short-lasting behavioral sensitization to amphetamine. Physiol Behav 83: 805-811.

DeMontis MG, Gambarana C, Gessa GL, Meloni D, Tagliamonte A, Stefanini E. 1993. Reduced [3H]SCH 23390 binding and DA-sensitive adenylyl cyclase in the limbic system of ethanol-preferring rats. Alcohol 28: 387-400.

Devaud LL, Alele P. 2004. Differential effects of chronic ethanol administration and withdrawal on gamma-aminobutyric acid type A and NMDA receptor subunit proteins in male and female rat brain. Alcohol Clin Exp Res 28: 957-65.

Devaud LL, Alele P, Ritu C. 2003. Sex differences in the central nervous system actions of ethanol. Crit Rev Neurobiol 15: 41-59.

Devaud LL, Chadda R. 2001. Sex differences in rats in the development of and recovery from ethanol dependence assessed by changes in seizure susceptibility. Alcohol Clin Exp Res 25: 1689-96.

Dewey SL, Morgan AE, Ashby CR Jr, Horan B, Kushner SA, et al. 1998. A novel strategy for the treatment of cocaine addiction. Synapse 30: 119-129.

Di Chiara G. 1998. A motivational learning hypothesis of the role of mesolimbic dopamine in compulsive drug use. J Psychopharmacol 12: 54-67.

Di Chiara G. 2002. Nucleus accumbens shell and core dopamine: Differential role in behavior and addiction. Behav Brain Res 137: 75-114.

Di Paolo T, Levesque D, Daigle M. 1986. A physiological dose of progesterone affects rat striatum biogenic amine metabolism. Eur J Pharmacol 125: 11-16.

Di Paolo T, Poyet P, Labrie F. 1981. Effect of chronic estradiol and haloperidol treatment on striatal dopamine receptors. Eur J Pharmacol 73: 105-106.

Dluzen DE, Ramirez VD. 1984. Bimodal effect of progesterone on in vitro dopamine function of the rat corpus striatum. Neuroendocrinol 39: 149-155.

Dluzen DE, Ramirez VD. 1990. In vitro progesterone modulation of amphetamine-stimulated dopamine release from the corpus striatum of ovariectomized estrogen-treated female rats: Response characteristics. Brain Res 517: 117-122.

Erskine MS. 1989. Solicitation behavior in the estrous female rat: A review. Horm Behav 23: 473-502.

Erskine MS. 1995. Prolactin release after mating and genitosensory stimulation in females. Endocr Rev 16: 508-528.

Erskine MS, Kornberg E, Cherry JA. 1989. Paced copulation in rats: Effects of intromission frequency and duration on luteal activation and estrus length. Physiol Behav 45: 33-39.

Everitt BJ. 1990. Sexual motivation: A neural and behavioural analysis of the mechanisms underlying appetitive and copulatory responses of male rats. Neurosci Biobehav Rev 14: 217-232.

Everitt BJ, Cador M, Robbins TW. 1989. Interactions between the amygdala and ventral striatum in stimulus-reward associations: Studies using a second-order schedule of sexual reinforcement. Neuroscience 30: 63-75.

Everitt BJ, Stacey P. 1987. Studies of instrumental behaviour with sexual reinforcement in male rats (Rattus norvegicus): II Effects of preoptic area lesions, castration and testosterone. J Comp Psychol 101: 407-419.

Fadda P, Scherma M, Fresu A, Collu M, Fratta W. 2003. Baclofen antagonizes nicotine-, cocaine-, and morphine-induced dopamine release in the nucleus accumbens of rat. Synapse 50: 1-6.

Faleiro L, Jones S, Kauer J. 2004. Rapid synaptic plasticity of glutamatergic synapses on dopamine neurons in the ventral tegmental area in response to acute amphetamine injection. Neuropsychopharmacology 29: 2115-2125.

Fiorillo C, Tobler P, Schultz W. 2003. Discrete coding of reward probability and uncertainty by dopamine neurons. Science 299: 1898-1902.

Fiorino DF, Kolb BS. 2003. Sexual experience leads to long-lasting morphological changes in male rat prefrontal cortex, pareital cortex, and nucleus accumbens neurons. Society for neuroscience. New Orleans, LA: 2003 Abstract Viewer and Itinerary Planner. Washington, DC.

Fiorino DF, Phillips AG. 1999a. Facilitation of sexual behavior in male rats following d-amphetamine-induced behavioral sensitization. Psychopharmacology 142: 200-208.

Fiorino DF, Phillips AG. 1999b. Facilitation of sexual behavior and enhanced dopamine efflux in the nucleus accumbens of male rats after D-amphetamine-induced behavioral sensitization. J Neurosci 19: 456-463.

Fischer JF, Cho AK. 1976. Properties of dopamine efflux from rat striatal tissue caused by amphetamine and p-hydroxyamphetamine. Proc West Pharmacol Soc 19: 179-182.

Flanagan LM, Pfaus JG, Pfaff DW, McEwen BS. 1993. Induction of FOS immunoreactivity in oxytocin neurons after sexual activity in female rats. Neuroendocrinology 58: 352-358.

Floresco SB, West AR, Ash B, Moore H, Grace AA. 2003. Afferent modulation of dopamine neuron firing differentially regulates tonic and phasic dopamine transmission. Nat Neurosci 6: 968-973.

Forgie ML, Stewart J. 1994. Sex differenc in amphetamine-induced locomotor activity in adult rats: Role of tetosterone exposure in the neonatal period. Pharmacol Biochem Behav 46: 637-645.

Freeman WM, Brebner K, Lynch WJ, Robertson DJ, Roberts DC, et al. 2001. Cocaine-responsive gene expression changes in rat hippocampus. Neuroscience 108: 371-380.

Galef BJ, Whiskin E. 2003. Socially transmitted food preferences can be used to study long-term memory in rats. Learn Behav 31: 160-164.

Garcia-Horsman P, Paredes RG. 2004. Dopamine antagonists do not block conditioned place preference induced by paced mating behavior in female rats. Behav Neurosci 118: 356-364.

Gerdjikov TV, Ross GM, Beninger RJ. 2004. Place preference induced by nucleus accumbens amphetamine is impaired by antagonists of ERK or p38 MAP kinases in rats. Behav Neurosci 118: 740-750.

Ghasemzadeh MB, Nelson LC, Lu XY, Kalivas PW. 1999. Neuroadaptations in ionotropic and metabotropic glutamate receptor mRNA produced by cocaine treatment. J Neurochem 72: 157-165.

Giertler C, Bohn I, Hauber W. 2005. Involvement of NMDA and AMPA/KA receptors in the nucleus accumbens core in instrumental learning guided by reward-predictive cues. Eur J Neurosci 21: 1689-1702.

Gilman DP, Mercer LF, Hitt JC. 1979. Influence of female copulatory behavor on the induction of pseudopregnancy in the female rat. Physiol Behav 22: 675-678.

Giorgetti M, Hotsenpiller G, Froestl W, Wolf ME. 2002. In vivo modulation of ventral tegmental area dopamine and glutamate efflux by local GABA(B) receptors is altered after repeated amphetamine treatment. Neuroscience 109: 585-595.

Giorgetti M, Hotsenpiller G, Ward P, Teppen T, Wolf ME. 2001. Amphetamine-induced plasticity of AMPA receptors in the ventral tegmental area: Effects on extracellular levels of dopamine and glutamate in freely moving rats. J Neurosci 21: 6362-6369.

Gonzalez-Nicolini V, McGinty JF. 2002. Gene expression profile from the striatum of amphetamine-treated rats: A cDNA array and in situ hybridization histochemical study. Brain Res Gene Expr Patterns 1: 193-198.

Gordon JH. 1980. Modulation of apomorphine-induced stereotypy by estrogen: Time course and dose response. Brain Res Bull 5: 679-682.

Griffin ML, Weiss RD, Lange U. 1989. A comparison of male and female cocaine abuse. Arch Gen Psychiatry 46: 122-126.

Grimm JW, See RE. 1997. Cocaine self-administration in ovariectomized rats is predicted by response to novelty, attenuated by 17-beta estradiol, and associated with abnormal vaginal cytology. Physiol Behav 61: 755-761.

Harris GC, Aston-Jones G. 2003. Critical role for ventral tegmental glutamate in preference for a cocaine-conditioned environment. Neuropsychopharmacology 28: 73-76.

Harris GC, Wimmer M, Byrne R, Aston-Jones G. 2004. Glutamate-associated plasticity in the ventral tegmental area is necessary for conditioning environmental stimuli with morphine. Neuroscience 129: 841-847.

Hecht GS, Spear NE, Spear LP. 1999. Changes in progressive ratio responding for intravenous cocaine throughout the reproductive process in female rats. Dev Psychobiol 35: 136-145.

Heikkila RE, Orlansky H, Mytilineou C, Cohen G. 1975. Amphetamine: Evaluation of d- and l-isomers as releasing agents and uptake inhibitors for 3H-dopamine and 3H-norepinephrine in slices of rat neostriatum and cerebral cortex. J Pharmacol Exp Ther 194: 47-56.

Henry DJ, White FJ. 1991. Repeated cocaine administration causes persistent enhancement of D1 dopamine receptor sensitivity within the rat nucleus accumbens. J Pharmacol Exp Ther 258: 882-890.

Henry SA, Lehmann-Masten V, Gasparini F, Geyer MA, Markou A. 2002. The mGluR5 antagonist MPEP, but not the mGluR2/3 agonist LY314582, augments PCP effects on prepulse inhibition and locomotor activity. Neuropharmacology 43: 1199-1209.

Herzig V, Schmidt WJ. 2004. Effects of MPEP on locomotion, sensitization and conditioned reward induced by cocaine or morphine. Neuropharmacology 47: 973-984.

Hitry A Little KY, Ellenwood EH Jr. 1996. Effect of cocaine on dopamine transporter receptors depends on routes of chronic cocaine administration. Neuropsychopharmacology 14: 205-210.

Horger BA, Iyasere CA, Berhow MT, Messer CJ, Nestler EJ, et al. 1999. Enhancement of locomotor activity and conditioned reward to cocaine by brain-derived neurotropic factor. J Neurosci 19: 4110-4122.

Hotenspiller G, Wolf ME. 2003. Baclofen attenuates conditioned locomotion to cues associated with cocaine administration and stabilizes extracellular glutamate levels in rat nucleus accumbens. Neuroscience 118: 123-134.

Hruska RE. 1988. 17βEstradiol regulation of DA receptor interactions with G-proteins. Soc Neurosci Abstr 14: 454.

Hruska RE, Silbergeld EK. 1980. Increased dopamine receptor sensitivity after estrogen treatment using the rat rotation model. Science 208: 1466-1468.

Hu JH, Yang N, Ma YH, Zhou XG, Zhang XY, et al. 2003. Decrease of morphine-induced reward effects and withdrawal symptoms in mice overexpressing gamma-aminobutyric acid transporter I. J Neurosci Res 74: 614-621.

Hu M, Crombag HS, Robinson TE, Becker JB. 2004. Biological basis of sex differences in the propensity to self-administer cocaine. Neuropsychopharmacology 29: 81-85.

Hummel M, Unterwald EM. 2003. Intra-accumbens pertussis toxin sensitizes rats to the locomotor activating effects of a single cocaine challenge. Brain Res 965: 100-107.

Ikemoto S, Wise RA. 2004. Mapping of chemical trigger zones for reward. Neuropharmacology 1: 190-201.

Ito R, Dalley JW, Howes SR, Robbins TW, Everitt BJ. 2000. Dissociation in conditioned dopamine release in the nucleus accumbens core and shell in response to cocaine cures and during cocaine-seeking behavior in rats. J Neurosci 20: 7489-7495.

Ito R, Dalley JW, Robbins TW, Everitt BJ. 2002. Dopamine release in the dorsal striatum during cocaine-seeking behavior under the control of a drug-associated cue. J Neurosci 22: 6247-6253.

Ito R, Robbins TW, Everitt BJ. 2004. Differential control over coaine-seeking behavior by nucleus accumbens core and shell. Nat Neurosci 7: 389-397.

Jackson LR, Robinson TE, Becker JB. 2005. Sex differences and hormonal influences on acquisition of cocaine self-administration in rats. Neuropsychopharmacology 31: 129-138.

Jacobs E, Wardeh G, Smit A, Schoffelmeer A. 2005. Morphine causes a delayed increase in glutamate receptor functioning in the nucleus accumbens core. Eur J Pharmacol 511: 27-30.

Jayaram P, Steketee JD. 2004. Effects of repeated cocaine on medial prefrontal cortical GABAB receptor modulation of neurotransmission in the mesocorticolimbic dopamine system. J Neurochem 90: 839-847.

Jenkins WJ, Becker JB. 2003a. Dynamic increases in dopamine during paced copulation in the female rat. Eur J Neurosci 18: 1997-2001.

Jenkins WJ, Becker JB. 2003b. Female rats develop conditioned place preferences for sex at their preferred interval. Horm Behav 43: 503-507.

Jones SR, Lee TH, Wightman RM, Ellinwood EH. 1996. Effects of intermittent and continuous cocaine administration on dopamine release and uptake regulation in the striatum: In vitro voltammetric assessment. Psychopharmacology (Berl) 126: 331-338.

Joyce JN, Smith RL, Van Hartesveldt C. 1982. Estradiol suppresses then enhances intracaudate dopamine-induced contralateral deviation. Eur J Pharmacol 81: 117-122.

Jung BJ, Dawson R Jr, Sealey SA Peris J. 1999. Endogenous GABA release is reduced in the striatum of cocaine-sensitized rats. Synapse 34: 103-110.

Kalivas PW, Duffy P. 1998. Repeated cocaine administration alters extracellular glutamate in the ventral tegmental area. J Neurochem 70: 1497-1502.

Kandel DB, Warner MPP, Kessler RC. 1995. The epidemiology of substance abuse and dependence among women. Drug addiction research and the health of women. Wetherington CL, Roman AR, editors. Rockville, MD: US Department of Health and Human Services; pp. 105-130.

Kanner BI. 1983. Bioenergetics of neurotransmitter transport. Biochim Biophys Acta 726: 293-316.

Kaplan GB, Leite-Morris KA, Joshi M, Shoeb MH, Carey RJ. 2003. Baclofen inhibits opiate-induced conditioned place preference and associated induction of Fos in cortical and limbic regions. Brain Res 987: 122-125.

Kelley AE. 2004. Ventral striatal control of appetitive motivation: Role in ingestive behavior and reward-related learning. Neurosci Biobehav Rev 27: 765-776.

Kelley AE, Bakshi VP, Haber SN, Steininger TL, Will MJ, et al. 2002. Opioid modulation of taste hedonics within the ventral striatum. Physiol Behav 76: 365-377.

Kelley AE, Berridge KC. 2002. The neuroscience of natural rewards: Relevance to addictive drugs. J Neurosci 22: 3306-3311.

Kelley AE, Throne LC. 1992. NMDA receptors mediate the behavioral effects of amphetamine infused into the nucleus accumbens. Brain Res Bull 29: 247-254.

Kelz MB, Chen J, Carlezon WA Jr, Whisler K, Gilden L, et al. 1999. Expression of the transcription factor ΔFosB in the brain controls sensitivity to cocaine. Nature 401: 272-276.

Kim JH, Austin JD, Tanabe L, Creekmore E, Vezina P. 2005. Activation of group II mGlu receptors blocks the enhanced drug taking induced by previous exposure to amphetamine. Eur J Neurosci 21: 295-300.

Kim JH, Beeler JA, Vezina P. 2000. Group II, but not group I, metabotropic glutamate receptors in the rat nucleus accumbens contribute to amphetamine-induced locomotion. Neuropharmacology 39: 1692-1699.

King GR, Ellinwood EH Jr, Silvia C, Joyner CM, Xue Z, et al. 1994. Withdrawal from continuous or intermittent cocaine administration: Changes in D2 receptor function. J Pharmacol Exp Ther 269: 743-749.

Klein S, Hadamitzky M, Koch M, Schwabe K. 2004. Role of glutamate receptors in nucleus accumbens core and shell in spatial behaviour of rats. Neuroscience 128: 229-238.

Kosten TA, Gawin FH, Kosten TR, Rounsaville BJ. 1993. Gender differeces in cocaine use and treatment response. J Subst Abuse Treat 10: 63-66.

Kozell B, Meshul K. 2003. Alterations in nerve terminal glutamate immunoreactivity in the nucleus accumbens and ventral tegmental area following single and repeated doses of cocaine. Psychopharmacology 165: 337-345.

Kozell LB, Meshul CK. 2001. The effects of acute or repeated cocaine administration on nerve terminal glutamate within the rat mesolimbic system. Neuroscience 106: 15-25.

Kozell LB, Meshul CK. 2004. Nerve terminal glutamate immunoreactivity in the rat nucleus accumbens and ventral tegmental area after a short withdrawal from cocaine. Synapse 51: 224-232.

Kuhar MJ, Ritz MC, Sharkey J. 1988. Cocaine receptors on dopamine transporters mediate cocaine-reinforced beahvior. NIDA Res Monogr 88: 14-22.

Kushner SA, Dewey SL, Kornetsky C. 1997. Gamma-vinyl GABA attenuates cocaine-induced lowering of brain stimulation reward thresholds. Psychopharmacology 133: 383-388.

Kushner SA, Dewey SL, Kornetsky C. 1999. The irreversible gamma-aminobutyric acid (GABA) transaminase inhibitor gamma-vinyl-GABA blocks cocaine self-administration in rats. J Pharmacol Exp Ther 290: 797-802.

Laviolette SR, Kooy van der D. 2001. GABA(A) receptors in the ventral tegmental area control bidirectional reward signalling between dopaminergic and non-dopaminergic neural motivational systems. Eur J Neurosci 13: 1009-1015.

Laviolette SR, Kooy van der D. 2004. GABAA receptors signal bidirectional reward transmission from the ventral tegmental area to the tegmental pedunculopontine nucleus as a function of opiate state. Eur J Neurosci 20: 2179-2187.

Le Foll B, Diaz J, Sokoloff P. 2003. Increased dopamine D3 receptor expression accompanying behavioral sensitization to nicotine in rats. Synapse 47: 176-183.

Leite-Morris KA, Fukudome EY, Shoeb MH, Kaplan GB. 2004. GABA(B) receptor activation in the ventral tegmental area inhibits the acquisition and expression of opiate-induced motor sensitization. J Pharmacol Exp Ther 308: 667-678.

Leone P, Popcock D, Wise RA. 1991. Morphine-dopamine interaction: Ventral tegmental morphine increases nucleus accumbens dopamine release. Pharmacol Biochem Behav 39: 469-472.

Leriche L, Schwartz JC, Sokoloff P. 2003. The dopamine D3 receptor mediates locomotor hyperactivity induced by NMDA receptor blockade. Neuropharmacology 45: 174-181.

Letchworth SR, Daunais JB, Hedgecock AA, Porrino LJ. 1997. Effects of chronic cocaine administration on dopamine transporter mRNA and protein in the rat. Brain Res 750: 214-222.

Letchworth SR, Nader MA, Smith HR, Friedman DP, Porrino LJ. 2001. Progression of changes in dopamine transporter binding site density as a result of cocaine self-administration in rhesus monkeys. J Neurosci 21: 2799-2807.

Levitt D, Teitelbaum P. 1975. Somnolence, akinesia, and sensory activation of motivated behavior in the lateral hypothalamic syndrome. Proc Natl Acad Sci USA 72: 2819-2823.

Li Y, Acerbo MJ, Robinson TE. 2004 The induction of behavioural sensitization is associated with cocaine-induced structural plasticity in the core (but not shell) of the nucleus accumbens. Eur J Neurosci 20: 1647-1654.

Li Y, Hu XT, Berney TG, Vartanian AJ, Stine CD, et al. 1999. Both glutamate receptor antagonists and prefrontal cortex lesions prevent induction of cocaine sensitization and associated neuroadaptations. Synapse 34: 169-180.

Li Y, Kolb B, Robinson TE. 2003. The location of persistent amphetamine-induced changes in the density of dendritic spines on medium spiny neurons in the nucleus accumbens and caudate-putamen. Neuropsychopharmacology 28: 1082-1085.

Licata SC, Freeman AY, Pierce-Bancroft AF, Pierce RC. 2000. Repeated stimulation of L-type calcium channesl in the rat ventral tremental area mimics the initiation of behavioral sensitization to cocaine. Psychopharmacology 152: 110-118.

Lu W, Wolf ME. 1999. Repeated amphetamine administration alters AMPA receptor subunit expression in rat nucleus accumbens and medial prefrontal cortex. Synapse 32: 119-131.

Lynch WJ, Carroll ME. 1999. Sex differences in the acquisition of intravenously self-administered cocaine and heroin in rats. Psychopharmacology 144: 77-82.

Lynch WJ, Roth ME, Carroll ME. 2002. Biological basis of sex differences in drug abuse: Preclinical and clinical studies. Psychopharmacology 164: 121-137.

Maldonado-Irizarry CS, Kelley AE. 1994. Differential behavioral effects following microinjection of an NMDA antagonist into nucleus accumbens subregions. Psychopharmacology 116: 65-72.

Mao L, Wang JQ. 2001. Differentially altered mGluR1 and mGluR5 mRNA expression in rat caudate nucleus and nucleus accumbens in the development and expression of behavioral sensitization to repeated amphetamine administration. Synapse 41: 230-240.

Marrow LP, Overton PG, Brain PF, Clark D. 1999. Encounters with aggressive conspecifics enhance the locomotor-activating effects of cocaine in the rat. Addiction Biology 4: 437-441.

Martinez I, Paredes RG. 2001. Only self-paced mating is rewarding in rats of both sexes. Horm Behav 40: 510-517.

Masserano JM, Venable D, Wyatt RJ. 1994. Effects of chronic cocaine administration on [3H]dopamine uptake in the nucleus accumbens, striatum and frontal cortex of rats. J Pharmacol Exp Ther 270: 133-141.

Matsumoto RR. 1989. GABA receptors: Are cellular differences reflected in function? Brain Res Brain Res Rev 14: 203-225.

McClintock MK. 1984. Group mating in the domestic rat as context for sexual selection: Consequences for the analysis of sexual behavior and neuroendocrine responses. Adv Study of Behav 14: 1-50.

McClintock MK, Adler NT. 1977. The role of the female during copulation in wild and domestic norway rats. Behav 67: 67-96.

McClung CA, Nestler EJ. 2003. Regulation of gene expression and cocaine reward by CREB and δFosB. Nat Neurosci 6: 1208-1215.

McGeehan AJ, Olive MF. 2003. The mGluR5 antagonist MPEP reduces the conditioned rewarding effects of cocaine but not other drugs of abuse. Synapse 47: 240-242.

Mead AN, Stephens DN. 2003. Involvement of AMPA receptor GluR2 subunits in stimulus-reward learning: Evidence from glutamate receptor gria2 knock-out mice. J Neurosci 23: 9500-9507.

Meisel RL, Camp DM, Robinson TE. 1993. A microdialysis study of ventral striatal dopamine during sexual behaivor in female Syrian hamsters. Behav Brain Res 55: 151-157.

Meisel RL, Joppa MA, Rowe RK. 1996. Dopamine receptor antagonists attenuate conditioned place preference following sexual behavior in female Syrian hamsters. Eur J Pharmacol 309: 21-4.

Mello NK. 1995. Cocaine abuse and reproductive function in women. Drug addiction research and the health of women. Wetherington CL, Roman AR, editors. Rockville, MD: US Department of Health and Human Services.

Mendelson JH, Weiss R, Griffin M, Mirin SM, Teoh SK, et al. 1991. Some special considerations for treatment of drug abuse and dependence in women. NIDA Res Monogr 106: 313-327.

Mermelstein PG, Becker JB. 1995. Increased extracellular dopamine in the nucleus accumbens and striatum of the female rat during paced copulatory behavior. Behav Neurosci 109: 354-365.

Meshul CK, Noguchi K, Emre N, Ellison G. 1998. Cocaine-induced changes in glutamate and GABA immunolabeling within rat habenula and nucleus accumbens. Synapse 30: 211-220.

Miczek KA. 2004. Animal models of social stress and drug abuse. Biol Psychiatry 55: 193S-193S.

Miczek KA, Covington HE, Nikulina EA, Hammer RP. 2004. Aggression and defeat: Persistent effects on cocaine self-administration and gene expression in peptidergic and aminergic mesocorticolimbic circuits. Neurosci Biobehav Rev 27: 787-802.

Mitchell JB, Stewart J. 1990. Facilitation of sexual behaviors in the male rat associated with intra-VTA injections of opiates. Pharmacol Biochem Behav 35: 643-650.

Morissette M, Di Paolo T. 1993. Effect of chronic estradiol and progesterone treatments of ovariectomized rats on brain dopamine uptake sites. J Neurochem 60: 1876-1883.

Mullins AJ, Sengelaub DR, Meisel RL. 2004. Effects of sexual experience in female hamsters on MAP kinase signaling and dendritic morphology. In: Society for Neuroscience. San Diego: 2004 Abstract Viewer and Itinerary Planner. Washington, DC.

Nakagawa T, Fujio M, Ozawa T, Minami M, Satoh M. 2005. Effect of MS-153, a glutamate transporter activator, on the conditioned rewarding effects of morphine, methamphetamine and cocaine in mice. Behav Brain Res 156: 233-239.

Nakayama K, Kiyosu K, Taguchi T. 2005. Diminished neuronal activity increases neuron-neuron connectivity underlying silent synapse formation and the rapid conversion of silent to functional synapses. J Neurosci 25: 4040-4051.

Narita M, Aoki T, Suzuki T. 2000. Molecular evidence for the involvement of NR2B subunit containing N-methyl-D-aspartate receptors in the development of morphine-induced place preference. Neuroscience 101: 601-606.

Nestby P, Schotte A, Janssen PF, Tjon GH, Vanderschuren LJ, et al. 1997. Striatal dopamine receptors in rats displaying long-term behavioural sensitization to morphine. Synapse 27: 262-265.

Nestler EJ. 2002. Common molecular and cellular substrates of addiction and memory. Neurobiol Learn Mem 78: 637-647.

Norrholm SD, Bobb JA, Nestler EJ, Ouimet CC, Taylor JR, et al. 2003. Cocaine-induced proliferation of dendritic spines in nucleus accumbens is dependent on the activity of cyclin-dependent kinase-5. Neuroscience 116: 19-22.

Oldenberger WP, Everitt BJ, de Jong FH. 1992. Conditioned place preference induced by sexual interaction in female rats. Horm Behav 26: 214-228.

Olds J, Milner P. 1954. Positive reinforcement produced by electrical stimulationof septal area and other areas of the brain. J Comp Physiol Psychol 47: 419-427.

Olds M, Olds J. 1969. Effects of lesions in the medial forebrain bundle on self-stimulation behavior. Am J Physiol 217: 1253-1264.

Paladini CA, Fiorillo CD, Morikawa H, Williams JT. 2001. Amphetamine selectively blocks inhibitory glutamate transmission in dopamine neurons. Nat Neurosci 4: 275-281.

Paredes RG, Alonso A. 1997. Sexual behavior regulated (paced) by the female induces conditioned place preference. Behav Neurosci 111: 123-128.

Paredes RG, Martinez I. 2001. Naloxone blocks place preference conditioning after paced mating in female rats. Behav Neurosci 115: 1363-1367.

Paredes RG, Vazquez B. 1999. What do female rats like about sex? Paced mating. Behav Brain Res 105: 117-127.

Pecina S, Cagniard B, Berridge K, Aldridge J, Zhuang X. 2003. Hyperdopaminergic mutant mice have a higher "wanting: But not "liking" for sweet rewards. J Neurosci 23: 9395-9402.

Peris J, Decambre N, Coleman-Hardee M, Simpkins J. 1991. Estradiol enhances behavioral sensitization to cocaine and amphetamine-stimulated [3H]dopamine release. Brain Res 566: 255-264.

Perrine SA, Schroeder JA, Unterwald EM. 2005. Behavioral sensitization to binge-pattern cocaine administration is not associated with changes in protein levels of four major G-proteins. Brain Res Mol Brain Res 133: 224-232.

Peters YM, O'Donnell P, Carelli RM. 2003. Prefrontal cortical responses during maintenance vs extinction of goal directed behavior for water reinforcement in rats. Behav Pharmacol 14: S74.

Pfaus JG, Damsma G, Nomikos GG, Wenkstern DG, Blaha CD, et al. 1990. Sexual behavior enhances central dopamine transmission in the male rat. Brain Res 530: 345-348.

Pfaus JG, Damsma G, Wenkstern D, Fibiger HC. 1995. Sexual activity increases dopamine transmission in the nucleus accumbens and striatum of female rats. Brain Res 693: 21-30.

Pfaus JG, Kippin TE, Centeno S. 2001. Conditioning and sexual behavior: A review. Horm Behav 40: 291-321.

Pfaus JG, Phillips AG. 1991. Role of dopamine in anticipatory and consummatory aspects of sexual behavior in the male rat. Behav Neurosci 105: 727-743.

Pierce RC, Duffy P, Kalivas PW. 1995. Sensitization to cocaine and dopamine autoreceptor subsensitivity in the nucleus accumbens. Synapse 20: 33-36.

Pierce RC, Kalivas PW. 1995. Amphetamine produces sensitized increases in locomotion and extracellular dopamine preferentially in the nucleus accumbens shell of rats administered repeated cocaine. J Pharmacol Exp Ther 275: 1019-1029.

Pleim ET, Matochik JA, Barfield RJ, Auerbach SB. 1990. Correlation of dopamine release in the nucleus accumbens with masculine sexual behavior in rats. Brain Res 524: 160-163.

Popik P, Wrobel M. 2002. Morphine conditioned reward is inhibited by MPEP, the mGluR5 antagonist. Neuropharmacology 43: 1210-1217.

Pothos E, Rada P, Mark GP, Hoebel BG. 1991. Dopamine microdialysis in the nucleus accumbens during acute and chronic morphine, naloxone-precipitated withdrawal and clonidine treatment. Brain Res 566: 348-350.

Pulvirenti L, Berrier R, Kreifeldt M, Koob GF. 1994. Modulation of locomotor activity by NMDA receptors in the nucleus accumbens core and shell regions of the rat. Brain Res 664: 231-236.

Quinones-Jenab V, Perrotti LI, Mc Monagle J, Ho A, Kreek MJ. 2000. Ovarian hormone replacement affects cocaine-induced behaviors in ovariectomized female rats. Pharmacol Biochem Behav 67: 417-422.

Rahman Z, Schwarz J, Gold SJ, Zachariou V, Wein MN, et al. 2003. RGS9 modulates dopamine signaling in the basal ganglia. Neuron 38: 941-952.

Raiteri M, Bertollini A, Angelini F, Levi G. 1975. d-Amphetamine as a releaser or reuptake inhibitor of biogenic amines in synaptosomes. Eur J Pharmacol 34: 189-195.

Ribeiro Do Couto B, Aguilar MA, Manzanedo C, Rodriguez-Arias M, Minarro J. 2005. NMDA glutamate but not dopamine antagonists blocks drug-induced reinstatement of morphine place preference. Brain Res Bull 64: 493-503.

Rice M, Cragg J. 2004. Nicotine amplifies reward-related dopamine signals in striatum. Nat Neurosci 7: 583-584.

Robbins SJ, Ehrman RN, Childress AR, O'Brien CP. 1999. Comparing levels of cocaine cue reactivity in male and female outpatients. Drug Alcohol Depend 53: 223-230.

Roberts DCS, Bennett SAL, Vickers GJ. 1989. The estrous cycle affects cocaine self-administration on a progressive ratio schedule in rats. Psychopharmacology 98: 408-411.

Robinson S, Kunko P, Smith J, Wallace M, Mo Q, et al. 1997. Extracellular aspartate concentration increases in nucleus accumbens after cocaine sensitization. Eur J Pharmacol 319: 31-36.

Robinson TE. 1984. Behavioral sensitization: Characterization of enduring changes in rotational behavior produced by intermittent injections of amphetamine in male and female rats. Psychopharmacology (Berl) 84: 466-475.

Robinson TE, Becker JB, Presty SK. 1982. Long-term facilitation of amphetamine-induced rotational behavior and striatal dopamine release produced by a single exposure to amphetamine: Sex differences. Brain Res 253: 231-241.

Robinson TE, Berridge KC. 1993. The neural basis of drug craving: An incentive-sensitization theory of addiction. Brain Res Rev 18: 247-291.

Robinson TE, Berridge KC. 2000. The psychology and neurobiology of addiction: An incentive-sensitization view. Addiction 95: S91-S117.

Robinson TE, Berridge KC. 2001. Incentive-sensitization and addiction. Addiction 96: 103-114.

Robinson TE, Berridge KC. 2003. Addiction. Annu Rev Psychol 54: 25-53.

Robinson TE, Kolb B. 2004. Structural plasticity associated with exposure to drugs of abuse. Neuropharmacology 47: 33-46.

Robinson TE, Gorny G, Mitton E, Kolb B. 2001. Cocaine self-administration alters the morphology of dendrites and dendritic spines in the nucleus accumbens and neocortex. Synapse 39: 257-266.

Robinson TE, Gorny G, Savage VR, Kolb B. 2002. Widespread but regionally specific effects of experimenter- versus self-administered morphine on dendritic spines in the nucleus accumbens, hippocampus, and neocortex of adult rats. Synapse 46: 271-279.

Robinson TE, Kolb B. 1997. Persistent structural modifications in nucleus accumbens and prefrontal cortex neurons produced by previous experience with amphetamine. J Neurosci 17: 8491-8497.

Robinson TE, Kolb B. 1999. Morphine alters the structure of neurons in the nucleus accumbens and neocortex of rats. Synapse 33: 160-162.

Roitman MF, Na E, Anderson G, Jones TA, Bernstein IL. 2002. Induction of a salt appetite alters dendritic morphology in nucleus accumbens and sensitizes rats to amphetamine. J Neurosci 22: RC225.

Rolls ET. 2004. Smell, taste, texture, and temperature multimodal representations in the brain, and their relevance to the control of appetite. Nutr Rev 62: S193-S204.

Roseboom PH, Hewlett GH, Gnegy ME. 1990. Repeated amphetamine administration alters the interaction between D1-stimulated adenylyl cyclase activity and calmodulin in rat striatum. J Pharmacol Exp Ther 255: 197-203.

Roth M, Cosgrove K, Carroll M. 2004. Sex differences in the vulnerability to drug abuse: A review of preclinical studies. Neurosci Biobehav Rev 28: 533-546.

Salamone JD, Correa M. 2002. Motivational views of reinforcement: Implications for understanding the behavioral functions of nucleus accumbens dopamine. Behav Brain Res 137: 3-25.

Schenk JO. 2002. The functioning neuronal transporter for dopamine: Kinetic mechanisms and effects of amphetamines, cocaine and methylphenidate. Prog Drug Res 59: 111-131.

Schoenbaum G, Setlow B. 2003. Lesions of nucleus accumbens disrupt learning about aversive outcomes. J Neurosci 23: 9833-9841.

Schoepp DD, Jane DE, Monn JA. 1999. Pharmacological agents acting at subtypes of metabotropic glutamate receptors. Neuropharmacology 38: 1431-1476.

Schoffelmeer AN, Voorn P, Jonker AJ, Wardeh G, Nestby P, et al. 1996. Morphine-induced increase in D-1 receptor regulated signal transduction in rat striatal neurons and its facilitation by glucocorticoid receptor activation: Possible role in behavioral sensitization. Neurochem Res 21: 1417-1423.

Schroeder JA, Hummel M, Unterwald EM. 2004. Repeated intracerebroventricular forskolin administration enhances behavioral sensitization to cocaine. Behav Brain Res 153: 255-260.

Schultz W. 1992. Activity of dopamine neurons in the behaving primate. Sem Neurosci 4: 129-138.

Schultz W. 1997. Dopamine neurons and their role in reward mechanisms. Curr Opin Neurobiol 7: 191-197.

Schultz W. 2002. Getting formal with dopamine and reward. Neuron 36: 241-263.

Schultz W. 2004. Neural coding of basic reward terms of animal learning theory, game theory, microeconomics andbehavioral ecology. Curr Opin Neurobiol 14: 139-147.

Schultz W, Apicella P, Ljungberg T, Romo R, Scarnati E. 1993. Reward-related activity in the monkey striatum and substantia nigra. Prog Brain Res 99: 227-235.

Schultz W, Apicella P, Scarnati E, Ljungberg T. 1992. Neuronal activity in monkey ventral striatum related to the expectation of reward. J Neurosci 12: 4595-4610.

Schultz W, Romo R. 1990. Dopamine neurons of the monkey midbrain: Contingencies of responses to stimuli eliciting immediate behavioral reactions. J Neurophysiol 63: 607-624.

Sell SL, Thomas ML, Cunningham KA. 2002. Influence of estrous cycle and estradiol on behavioral sensitization to cocaine in female rats. Drug Alcohol Depend 67: 281-290.

Setlow B, Schoenbaum G, Gallagher M. 2003. Neural encoding in ventral striatum during olfactory discrimination learning. Neuron 38: 625-636.

Shippenberg TS, Bals-Kubik R, Herz A. 1993. Examination of the neurochemical substrates mediating the motivational effects of opioids: Role of the mesolimbic dopamine system and D-1 vs. D-2 dopamine receptors. J Pharmacol Exp Ther 265: 53-59.

Shippenberg TS, Herz A. 1987. Place preference conditioning reveals the involvement of D1-dopamine receptors in the motivational properties of mu- and kappa-opioid agonists. Brain Res 436: 169-172.

Shippenberg TS, Herz A. 1988. Motivational effects of opioids: Influence of D-1 versus D-2 receptor antagonists. Eur J Pharmacol 151: 233-242.

Sircar R, Kim D. 1999. Female gonadal hormones differentially modulate cocaine-induced behavioral sensitization in Fischer, Lewis and Sprague-Dawley rats. J Pharmacol Exp Ther 289: 54-65.

Smith GP. 2004. Accumbens dopamine mediates the rewarding effect of orosensory stimulation by sucrose. Appetite 43: 11-13.

Smith-Roe SL, Kelley AE. 2000. Coincident activation of NMDA and dopamine D1 receptors within the nucleus accumbens core is required for appetitive instrumental learning. J Neurosci 20: 7737-7742.

Sokolov BP, Polesskaya OO, Uhl GR. 2003. Mouse brain gene expression changes after acute and chronic amphetamine. J Neurochem 84: 244-252.

Sora I, Wichems C, Takahashi N, Li XF, Zeng Z, Revay R, et al. 1998. Cocaine reward models: Conditioned place preference can be established in dopamine- and serotonin-transporter knockout mice. Proc Natl Acad Sci USA 95: 7699-7704.

Sulzer D, Chen T, Lau Y, Kristensen H, Rayport S, et al. 1995. Amphetamine redistributes dopamine from synaptic vesicles to the cytosol and promotes reverse transport. J Neurosci 15: 4102-4108.

Sulzer D, Maidment NT, Rayport S. 1993. Amphetamine and other weak bases act to promote reverse transport of dopamine in ventral midbrain neurons. J Neurochem 60: 527-535.

Teitelbaum P, Epstein AN. 1962. The lateral hypothalamic syndrome: Recovery of feeding and drinking after lateral hypothalamic lesions. Psych Rev 69: 74-90.

Thompson TL, Moss RL. 1994. Estrogen regulation of dopamine release in the nucleus accumbens: Genomic- and nongenomic-mediated effects. J Neurochem 62: 1750-1756.

Tobler P, Dickinson A, Schultz W. 2003. Coding of predicted reward omission by dopamine neurons in a conditioned inhibition paradigm. J Neurosci 23: 10402-10410.

Turner TJ. 2004. Nicotine enhancement of dopamine release by a calcium-dependent increase in the size of the readily releasable pool of synaptic vesicles. J Neurosci 24: 11328-11326.

Tzschentke TM, Schmidt WJ. 1998. Blockade of morphine- and amphetamine-induced conditioned place preference in the rat by riluzole. Neurosci Lett 242: 114-116.

Uhl GR, Hall FS, Sora I. 2002. Cocaine, reward, movement and monoamine transporters. Mol Psychiatry 7: 21-26.

Ungerstedt U. 1971. Adipsia and aphagia after 6-hydroxydopamine induced degeneration of the nigro-striatal dopamine system. Acta Physiol Scand Suppl 367: 95-122.

Unterwald EM, Fillmore J, Kreek MJ. 1996. Chronic repeated cocaine administration increases dopamine D1 receptor-mediated signal transduction. Eur J Pharmacol 318: 31-35.

Unterwald EM, Ho A, Rubenfeld JM, Kreek MJ. 1994. Time course of the development of behavioral sensitization and dopamine receptor up-regulation during binge cocaine administration. J Pharmacol Exp Ther 270: 1387-1396.

Valjent E, Pascoli V, Svenningsson P, Paul S, Enslen H, et al. 2005. Regulation of a protein phosphatase cascade allows convergent dopamine and glutamate signals to activate ERK in the striatum. Proc Natl Acad Sci USA 102: 491-496.

Van Bockstaele EJ, Pickel VM. 1995. GABA-containing neurons in the ventral tegmental area project to the nucleus accumbens in rat brain. Brain Res 682: 215-221.

van Haaren F, Meyer M. 1991. Sex differences in the locomotor activity after acute and chronic cocaine administration. Pharmacol Biochem Behav 39: 923-927.

Van Hartesveldt C, Cottrell GA, Meyer ME. 1989. Effects of intrastriatal hormones on the dorsal immobility response in male rats. Pharmacol Biochem Behav 35: 307-310.

Van Vliet BJ, Van Rijswijk AL, Wardeh G, Mulder AH, Schoffelmeer AN. 1993. Adaptive changes in the number of Gs- and Gi-proteins underlie adenylyl cyclase sensitization in morphine-treated rat striatal neurons. Eur J Pharmacol 245: 23-29.

Vanover KE. 1998. Effects of AMPA receptor antagonists on dopamine-mediated behaviors in mice. Psychopharmacology 136: 123-131.

Verimer T, Arneric SP, Long JP, Walsh BJ, Abou Zeit-Har MS 1981. Effects of ovariectomy, castration, and chronic lithium chloride treatment on stereotyped behavior in rats. Psychopharmacology 75: 273-276.

Volkow ND, Fowler JS, Wang GJ, Goldstein RZ. 2002. Role of dopamine, the frontal cortex and memory circuits in drug addiction: Insight from imaging studies. Neurobiol Learn Mem 78: 610-624.

Wei FY, Tomizawa K, Ohshima T, Asada A, Saito T, et al. 2005. Control of cyclin-dependent kinase 5 (Cdk5) activity by glutamatergic regualtion of p35 stability. J Neurochem 93: 502-512.

Westberry J, Meredith M. 2003. The influence of chemosensory input and gonadotropin releasing hormone on mating behavior circuits in male hamsters. Brain Res 974: 1-16.

Wetherington CL, Roman AR, editors. 1995. Drug addiction research and the health of women. Rockville, MD: US Department of Health and Human Services.

White FJ. 2002. A behavioral/systems approach to the neuroscience of drug addiction. J Neurosci 22: 3303-3305.

White NM. 1996. Addictive drugs as reinforcers: Multiple partial actions on memory systems. Addiction 91: 921-949.

Wise RA. 1978. Catecholamine theories of reward: A critical review. Brain Res 152: 215-247.

Wise RA. 1984. Neural mechanisms of the reinforcing action of cocaine. NIDA Res Monogr 50: 15-33.

Wise RA. 2004. Dopamine, learning and motivation. Nat Rev Neurosci. 5: 483-494.

Wise RA, Rompre P. 1989. Brain dopamine and reward. Annu Rev Psychol 40: 191-225.

Wolf ME. 1998. The role of excitatory amino acids in behavioral sensitization to psychomotor stimulants. Prog Neurobiol 54: 679-720.

Wolf ME, Xue CJ, Li Y, Wavak D. 2000. Amphetamine increases glutamate efflux in the rat ventral tegmental area by a mechanism involving glutamate transporters and reactive oxygen species. J Neurochem 75: 1634-1644.

Wright C, Beijer A, Groenewegen HJ. 1996. Basal amygdaloid complex afferent to the rat nucleus accumbens are compartmentally organized. J Neurosci 16: 1877-1893.

Wyvell CL, Berridge KC. 2001. Incentive sensitization by previous amphetamine exposure: Increased cue-triggered "wanting" for sucrose reward. J Neurosci 21: 7831-7840.

Xi ZX, Stein EA. 1998a. Nucleus accumbens dopamine release modulation by mesolimbic GABA(A) receptors – an in vivo electrochemical study. Brain Res 798: 156-165.

Xi ZX, Stein EA. 1998b. Nucleus accumbens dopamine release modulation by mesolimbic GABAA receptors – an in vivo electrochemical study. Brain Res 798: 156-165.

Xi ZX, Stein EA. 2002. GABAergic mechanisms of opiate reinforcement. Alcohol Alcohol 37: 485-494.

Xi ZX, Ramamoorthy S, Shen H, Lake R, Samuvel DJ, et al. 2003. GABA transmission in the nucleus accumbens is altered after withdrawal from repeated cocaine. J Neurosci 23: 3498-3505.

Xue CJ, Ng JP, Li Y, Wolf ME. 1996. Acute and repeated systemic amphetamine administration: Effects on extracellular glutamate, aspartate, and serine levels in rat ventral tegmental area and nucleus accumbens. J Neurochem 67: 352-363.

Yamaguchi M, Suzuki T, Abe S, Hori T, Kurita H, et al. 2002. Repeated cocaine administration differentially affects NMDA receptor subunit (NR1, NR2A-C) mRNAs in rat brain. Synapse 46: 157-169.

Yuferov V, Kroslak T, Laforge KS, Zhou Y, Ho A, et al. 2003. Differential gene expression in the rat caudate putamen after "binge" cocaine administration: Advantage of triplicate microarray analysis. Synapse 48: 157-169.

Zachariou V, Benoit-Marand M, Allen PB, Ingrassia P, Feinberg AA, et al. 2002. Reduction of cocaine place preference in mice lacking the protein phosphatase 1 inhibitors DARPP 32 or Inhibitor 1. Biol Psychiatry 51: 612-620.

Zhang H, Sulzer D. 2004. Frequency-dependent modulation of dopamine release by nicotine. Nat Neurosci 7: 581-582.

Zhang K, Tarazi FI, Campbell A, Baldessarini RJ. 2000. GABA(B) receptors: Altered coupling to G-proteins in rats sensitized to amphetamine. Neuroscience 101: 5-10.

21 Neuroendocrinology of Memory and Cognitive Function

V. N. Luine

Abstract: Cognitive function – the ability to learn, retain, and recall information – changes over the lifespan and shows some sexual dimorphism. These patterns of cognitive ability appear to depend, in part, on the changing/different mileu of circulating steroid hormones. This review examines the extent to which steroid hormones, gonadal, and adrenal, influence cognitive function. First, emerging studies indicate that estrogens (directly secreted or derived from androgens) and glucocorticoids (cortisol in humans and corticosterone in most rodents) exert programming effects on cognition during the perinatal period which are subsequently expressed at adulthood. At adulthood, hormones also promote or impair cognition, depending on the specific hormone, the sex of the subject, and the situation/function examined. Finally, steroids appear to be important contributors to age-related losses in memory function. Hormonal actions on cognition, like their well-described effects on homeostatic function, rely on specific genomic activations of neurons via morphological and neurochemical mechanisms in brain areas underlying memory function. Newer evidence also suggests that estrogens rapidly activate signal transduction pathways via membrane receptors to enhance memory. Using classic, posttraining paradigms for assessing learning and memory, these rapid effects of estrogen on neural function have been shown to reflect enhancements in the consolidation of memory. While it is clear that hormonal effects on cognition are not generally large, this review posits that hormonal influences are nonetheless pervasive throughout the lifespan and provide fundamental, important regulation over higher order neural function, i.e., the intellectual/cognitive realms of life.

List of Abbreviations: BDNF, Brain Derived Neurotrophic Factor; CEE, Conjugated equine estrogens; CNS, Central Nervous System; CREB, cAMP Response Element-binding protein; DES, Diethystilbestrol; EB, Estradiol Benzoate; ERα, Estrogen receptor α; ERβ, Estrogen receptor β; HPA, Hypothalamic-Pituitary-Adrenal (Axis); HPG, Hypothalamic-Pituitary-Gonadal (Axis); IGF2, Insulin-like Growth Factor 2; MAP Kinase, Mitogen-activated Protein Kinase; MPA, Medroxyprogesterone; NGF, Nerve Growth Factor; Ovx, Ovariectomized; PS, Prenatal Stress; RAM, Radial arm maze; SC, Subcutaneous; SERM, Selective Estrogen Receptor Modulator; SGRM, Selective Glucocorticoid Receptor Modulator; SRM, Selective Receptor Modulator; Sts, Stress

1 Introduction

The emergence and progress of neuroendocrine research during the latter half of the 20th century demonstrated without question that numerous hormones act within specific sites of the mammalian brain to regulate homeostatic functions critical for life. Previously, the pituitary was thought to be the site for regulation of hormonal secretions and actions. Thus, it was surprising when endocrine studies demonstrated that hypothalamic and preoptic area sites were responsible for hormonal regulation of ovulation and sexual behavior (Everett, 1965; Gorski, 1971). Currently, endocrine glands that synthesize centrally acting hormones include – but are not limited to – the gonads, adrenal gland, thyroid gland, gut, and pancreas (van Wimersma Greidanus and Rigter, 1989; Nelson, 2005). Thus, hormones acting in the brain regulate such diverse functions as reproduction and growth, and hormones also influence ingestion and the sleep–wake cycle (Nelson, 2005). Traditionally, this regulation has involved hormonal action at brain sites and systems that are not involved in complex, higher order brain function, for example at sites in hypothalamic, limbic, and spinal cord areas.

Current neuroendocrinological research, driven by advances in molecular biology and technology, suggests that hormones also impact higher order brain functions such as mood and psychiatric symptoms (Ostlund et al., 2003; Payne, 2003), cognitive function (Luine and Harding, 1994; Dohanich, 2002), and bonding/affiliation (Carter et al., 1997). These effects are mediated by hormonal actions at additional sites and/or neural systems in cerebral cortex, basal forebrain, hippocampus and striatum which are responsible for complex, higher order neural function. Thus, hormones may not only regulate homeostatic/housekeeping functions of the central nervous system (CNS), but they may also influence the intellectual/cognitive realms of life.

This review will examine and discuss evidence for steroid hormone – adrenal and gonadal – influences on cognitive processes. The primary focus is on basic research in rodents but pertinent, relevant data in primates/humans, will also be provided in order to highlight the relevance of the animal data. Hormonal influences on cognition will be examined over the lifespan as it is important to note that hormones exert potent influences during critical developmental periods before birth, during the neonatal period, at puberty and adulthood, and also impact the aging process. While hormonal influences on cognition at each period of development are not generally large, the accumulation of effects across the life span are particularly evident in old age, especially in females who experience the greatest age-dependent changes in circulating hormones. *Figure 21-1* highlights sex differences and the effects of stress and aging on performance of the spatial memory task, radial arm maze, in rats (see Section 4.3 for specific information on the radial arm maze task).

Figure 21-1

Sex, stress, and aging affect radial arm maze performance in rats. Performance of the spatial memory task, radial arm maze, is shown in male and female rats of various ages and after chronic stress. Bars are the average ± SEM of the choice where the subject made the first mistake (larger numbers indicate better performance). Young males and females (2–3 months) served as controls or received daily restraint for 21 days. Old rats were 22–24 months old. Chronic stress impairs young male performance but enhances female performance. Both aged males and females are impaired as compared to young controls, and aged females show the worst performance of all groups. Data are combined from a number of studies and are for illustrative purposes only as no statistical tests were performed to test group differences. See the original studies for details and statistical analyses. Data from Luine and Rodriguez (1994), Luine et al. (1994), Luine and Hearns (1990), Bowman et al. (2001)

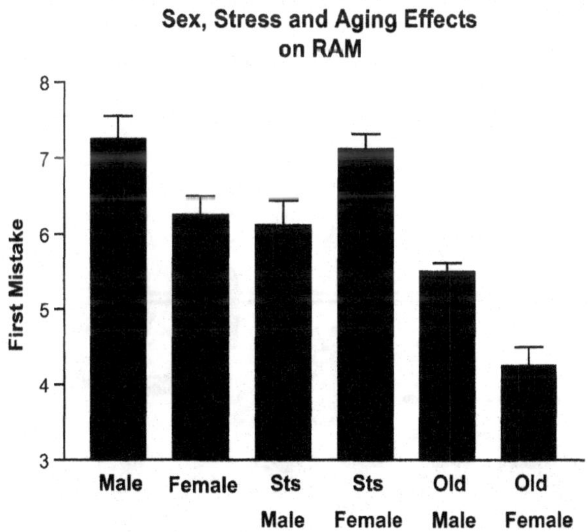

2 The Endocrine Brain – Mechanisms and Sites for Steroid Hormone Modulation of Neural Function

2.1 Hormonal Activation Through Intracellular Receptors

Receptors for the gonadal hormone estradiol were first identified and characterized from the uterus approximately 40 years ago and subsequently identified in specific brain sites (see Jensen and Jacobson, 1962 for review). Recently, a second type of estrogen receptor has been identified and receptors binding estradiol are currently designated as estrogen receptor alpha (ERα) and estrogen receptor beta (ERβ) (Kuiper et al., 1997). Both receptors are ligand-dependent transcription factors and through interactions at specific sites on DNA, initiate a cascade of intracellular reactions culminating in unique physiological responses within estrogen target tissues like uterus, breast, osteoclasts, and neurons of the brain (➲ *Figure 21-2*). ➲ *Figure 21-2* also illustrates that recent studies show hormonal activations by estrogens through membrane receptors. This mechanism is discussed further in ➲ Section 2.4. See McEwen and Alves (1999) for a detailed description of genomic mechanisms for estrogen action in the CNS.

Receptors for glucocorticoids (cortisol in humans and corticosterone in most rodents) and mineralocorticoids (aldosterone) as well as other gonadal hormones like progesterone and testosterone were identified around the same time as the estrogen receptor, and all are now known to be part of the steroid super family of receptors which also includes vitamin D (Tsai and O'Malley, 1994; Smith and

Figure 21-2
Diagram of possible mechanisms by which estradiol may act in neurons to alter function and thereby enhance memory. (a) Genomic pathway through classical cytosolic/nuclear receptors acting as nuclear transcriptional factors. This pathway would be present in only the soma of neurons. (b) Nongenomic pathway mediating rapid effects via signal transduction mechanisms through unusual membrane receptor (B2) or classical cytosolic/nuclear receptor spanning through plasma membrane (B1) or through multiproteic complex associated to the inner part of plasma membrane (B3). Nongenomic pathways could be present in the soma, in dendritic spines, or in nerve terminals. E_2: estrogens; iER: classical cytosolic/nuclear ER; mER: membrane ER; Gp: G protein; iNOS: inducible nitric oxide synthase; NO: nitric oxide; TK: tyrosine kinases; AC: adenylate cyclase; SH2: Src homology domain; PLC: phospholipase C; PKA: protein kinase A; ERK1/2: ERKs 1/2; DAG: diacyl glycerol; IP3: inositol 3-phosphate; PKC: protein kinase C; cAMP: cyclic adenosin monophosphate; PIP2: phosphatidyl inositol bisphosphate. Reprinted from Luconi M, Forti G, Baldi E. 2002. Genomic and nongenomic effects of estrogens: molecular mechanisms of action and clinical implications for male reproduction. *J Steroid Biochem Molec Biol* 80, with permission from Elsevier

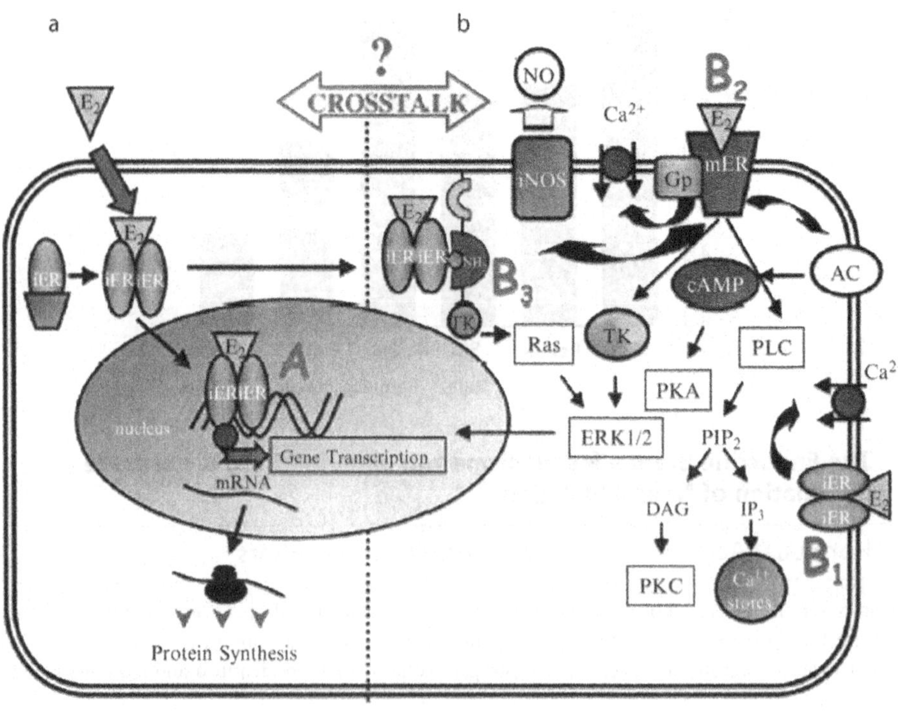

O'Malley, 2004). Like estrogen receptors, adrenal steroid receptors have a unique distribution within the brain and act as ligand-dependent transcription factors to alter neural function (See deKloet et al., 1994 for details). When steroids are bound to their respective receptors, they initiate new protein synthesis through interactions with specific sites on DNA and in this way influence function in their target neurons (➋ *Figure 21-2*).

2.2 Brain Areas Involved in Steroid Effects on Cognition

Brain areas which mediate cognition in animals and humans include, but are not limited to, frontal cortex, basal forebrain, hippocampus, amygdala, and striatum (see ➋ Section 3.1 for details) and in these areas, estradiol, testosterone, progesterone, and adrenal hormones exert wide ranging effects (see below). It is noteworthy, in this context, that ERβ is found in several cortical areas, including frontal cortex, and the hippocampus (Mitra et al., 2003). In contrast, ERα is present at low levels in these areas (Shughrue et al., 1997). Thus, current theories hypothesize an important role for ERβ in cognitive function (Rissman et al., 2002), a hypothesis requiring further testing and confirmation (see ➋ Section 4.3).

The foundation for glucocorticoid effects on learning and memory was laid with the remarkable demonstration in the early 1970's that ^{3}H-corticosterone was avidly deposited within pyramidal neurons of the hippocampus when it was administered to adrenalectomized rats (McEwen et al., 1968). Subsequently, receptors for glucocorticoids have been measured in areas subserving both the stress response and cognitive function (van Steensel et al., 1996; Helm et al., 2002); thus providing a strong context for stress influences on cognition. Influences of estrogen on cognition stemmed from the demonstration that cognitive loss in Alzheimer's disease derived from loss of cholinergic neurons and that estrogen activated cholinergic neurons in this circuit through genomic activation by estrogen receptors (Luine et al., 1975, 1994; Luine, 1985; Wickelgren, 1997). Current studies have amplified these observations of steroid hormone influences on cognitive function (Gibbs and Aggarwal, 1998).

2.3 Morphological and Neurochemical Effects of Steroids

At the morphological level, steroids regulate the structure of dendrites, spines, and synapses in brain regions critical for memory. For example, estradiol treatment to ovariectomized (Ovx) rodents increases the number of dendritic spines in CA1 hippocampal pyramidal neurons (Gould et al., 1990; Woolley and McEwen, 1992; Li et al., 2004), and estradiol and testosterone both increase the density of spine synapses in CA1 pyramidal neurons (Leranth et al., 2000; MacLusky et al., 2005). Cortical neurons and terminals have received less investigation, but Kritzer and Kohama, (1998) and Markham and Juraska (2002) found that hormone induced plasticity is present in these areas as well. Recent studies have confirmed estrogen-dependent influences on spines, synapses, or neural markers in nonhuman primates (Leranth et al., 2002; Hao et al., 2003; Tang et al., 2004; Voytko and Tinkler, 2004). Moreover, estradiol administration is also associated with increased neural cell birth in the dentate gyrus of the hippocampus (Tanapat et al., 1999).

Stress hormones also regulate morphological structure in CNS; however, these effects are generally opposite to gonadal hormone effects. Chronic restraint stress or corticosteroid administration is associated with a pruning of the apical dendritic arbor of CA3 pyramidal neurons as well as layer III frontal cortex pyramidal neurons (Watanabe et al., 1992; McEwen, 2001; Wellman, 2001). In cases of severe and prolonged stress, neuronal loss in the hippocampus has been reported (Uno et al., 1989); however, this effect awaits further validation.

Using neurochemical techniques, estrogens, progestins, and glucocorticoids have been shown to affect activity in a variety of neuronal systems: the basal forebrain–hippocampal cholinergic system, serotonergic cell bodies and terminals, monoaminergic terminals and δ-amino-butyric acid (GABA) containing interneurons and projecting systems (see McEwen and Alves, 1999; Bowman et al., 2003 for reviews). The synthesis and release of a number of trophic factors and/or their receptors are also affected by gonadal hormones in sites subserving memory and include BDNF, NGF, IGFII, and others (Gibbs and Aggarwal, 1998; Gibbs, 1999). Thus, direct genomic activation of diverse neural circuits by hormones has been demonstrated.

In summary, steroid hormones exert a wide range of effects on neural function and these effects may in turn influence cognitive function. Genomic actions at the cell nucleus by the hormones result in long-lasting, sustained effects on neural function. Thus, it is likely that genomic changes subserve developmental hormone effects, which exert life-long actions in programming sex-dependent functions and sex-dependent differences in functions. In adults, genomic effects also likely underlie changes over the menstrual and estrous cycles, and during aging. Alterations in neural function during aging may also derive from sustained down regulation of genomic function by age-related decreases in circulating gonadal hormones. This notion is expanded and elaborated upon in the following sections.

2.4 Hormonal Activations Through Membrane Receptors

Recently, estrogen receptors have been identified in extranuclear sites, primarily the membrane, in peripheral tissues like the breasts and epithelial cells, and in the CNS (Levin, 2002; Toran-Allerand et al., 2002). See *Figure 21-2*. During development, nonnuclear receptors appear in areas of the cerebral cortex of mice (Toran-Allerand et al., 2002). These receptors, in contrast to classically identified ERα and β receptors, mediate rapid hormonal effects (sec to min) through signal transduction pathways (see Levin, 2002 for review and details). While the extranuclear presence of receptors in neurons remains to be validated, several studies indicate putative estrogen receptors (whether they are ERα or β, or another different receptor remains to be determined) in membranes of cell bodies, axons, spines, presynaptic terminals, and near postsynaptic neurotransmitter receptors (*Figure 21-2*; Blaustein, 1992; Milner et al., 2001; Towart et al., 2003). Thus, the possibility is raised that estradiol, or other steroids, could exert rapid changes in cognitive and other functions through these receptors.

As the demonstration of nonnuclear steroid receptors is recent, their role in brain function is currently unclear. Changes in several signal transduction pathways, for example, MAP kinase (mitogen-activated protein kinase), CREB (cAMP response element-binding protein), and several immediate, early genes have been shown within minutes following estrogen in vitro and in vivo (Toran-Allerand etal., 1999, 2002; Bi et al., 2000, 2001; Singh, 2001;Wade et al., 2001; Znamensky et al., 2003). See *Figure 21-2* for schematic of the relationship between these pathways. Interestingly, estrogen-dependent enhancements in recognition memory have been demonstrated within 4h, an effect which may be extragenomic (Luine et al., 2003; see Section 4.5). The membrane receptors may also cross-talk with the genomic receptors to amplify or modify hormonal effects (See *Figure 21-2*). However, it remains unclear, at this time, to what extent nongenomic mechanisms of hormone action contribute to neural function and enhancement of cognition.

3 The Cognitive Brain – Mediation of Learning and Memory

3.1 Sites for Learning and Memory

Cognition represents a complex, multidimensional set of intellectual functions whose component processes are subserved by specific, yet interrelated, brain sites. In this context, no generally accepted framework for the acquisition, consolidation, and retention of memory has been agreed upon and a number of neural substrates of memory, "memory molecules," have been delineated, but which may be most important and how the molecules are coordinated is still generally unclear (but see Bailey et al., 2004; Ge et al., 2004). However, some agreement as to brain areas/systems necessary for cognition exists, and several different theories link these critical brain sites together in diverse ways. Thus, as a detailed description of the neural, chemical, or anatomical bases for cognitive function is still not available, neuroendocrinologists are unable to delineate how hormones might alter memory. Nonetheless, a brief synopsis of currently held notions of cognition and how hormones might intervene to alter cognitive function follows.

At a minimum, brain areas incorporated into theories of cognition function include the hippocampus, dorsal striatum (basal ganglia), amygdala, and several cortical areas (Ennaceur and Aggleton, 1994; Packard et al., 1994; Ennaceur et al., 1997). Most researchers have hypothesized multiple "memory systems" which operate in parallel to process information; however, the exact number varies. White and McDonald (2002) recently reviewed evidence for three systems relying on "three central structures," the hippocampus, the matrix compartment of the dorsal striatum (caudate putamen), and the amygdala (*Figure 21-3*). All

Figure 21-3
The concept of parallel processing of information as reviewed by White and McDonald (2002). Three systems, which have as primary structures the hippocampus, amygdala, and dorsal striatum/caudate putamen and which may receive input from an alternate system (cortex, lateral septum/mammillary bodies, and thalamus), function independently to facilitate learning. Each system receives input which leads to learning and memory formation, but each system processes that information differently (processing style). These systems influence behavior/memory in either a cooperative or competitive manner, but all systems lead to storage of information (memory) that influences future processing of similar information

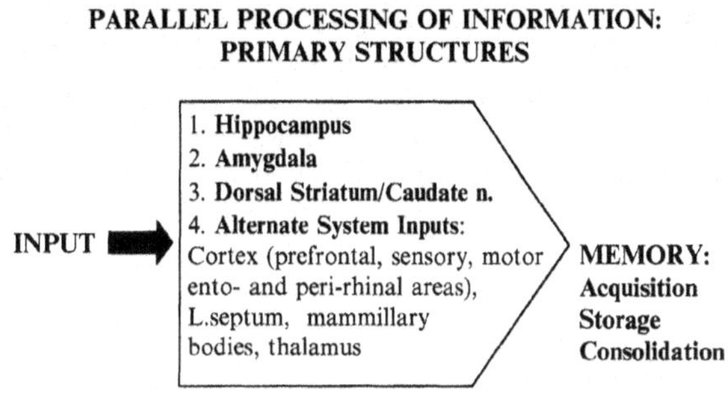

systems access the same information at acquisition, but each is specialized to take in and process a different kind of relationship between stimuli, responses, and reinforcers which flow through it. In contrast, Steckler et al. (1998) posited two parallel systems for recognition memory in rats. Network I is essential for the processing of nonspatial/item recognition memory processes and incorporates the cortical association areas, the rhinal cortices, the mediodorsal thalamic nucleus, and prefrontal cortical areas. Network II comprises the hippocampus, mammillary bodies, anterior thalamic nuclei and medial prefrontal areas, and is suggested to be pivotal for the processing of spatial recognition memory.

3.2 Mechanisms for Learning and Memory

Current hypotheses of cognitive function are congruent with the concept of memory consolidation as originally advanced at the beginning of the 20th century by the laboratory of Muller and Pilzecher (see McGaugh, 2000 for discussion), which hypothesized that newly acquired information or memories (short-term memory) were consolidated into long-term memory. The exact nature of this process still remains unclear, but see discussion below. McGaugh and coworkers have highlighted the critical role of the amygdala in the consolidation, but not necessarily the storage, of all forms of memory in multiple brain areas (McGaugh, 1989; Packard et al., 1994; Roozendaal et al., 2004).

In relation to the action of steroids within current tenets of memory function, it is notable that nuclear or other receptors for steroids are present in the four key brain's areas underlying cognitive function (described above). However, the strongest association of steroids with retention of memory is found with spatial memory functions of the hippocampus. This relationship is present for gonadal hormones (see Section 4) and for adrenal hormones (see Section 5).

Progress on the molecular/neurochemical basis of memory is spearheaded by elegant studies of the Kandel (Bailey et al., 2004) and Tully (Ge et al., 2004) groups. While details of these studies are beyond the scope of this review, several signal transduction pathways which may form the basis of short-term and long-term memory have been identified in aplysia, the fruit fly, and in some cases, rats or mice. Integral to this process is CREB and other signal transduction pathway intermediaries like MAP kinase and the immediate early genes cJun and cFos (See Figure 21-2). Recent studies in brain show that estrogen affects these signal molecules (Bi et al., 2000, 2001; Znamensky et al., 2003; Akama and McEwen, 2003) and coupled with the identification of estrogen receptors outside of the nucleus in cortex, striatal, and hippocampal areas, it can be hypothesized that gonadal hormones may influence consolidation of memory through these signal transduction pathways (See Figure 21-2).

In summary, a strong basis for hormones influencing cognition can be advanced as the structures critical for memory express receptors which bind and mediate steroid hormone action (McEwen, 2001; de Kloet, 2004). The looming challenge is for neuroendocrinologists and cognitive scientists to collaborate in designing experiments to test critical hypotheses: a new field of cognitive neuroendocrinology?

4 The Sexual Brain – Gonadal Hormone Influences on Cognition

Gonadal hormones initiate reproduction and attendant sexual behaviors in males and females through central regulation of the hypothalamic/pituitary/gonadal axis. Neuroendocrinological research in the past 50 years has documented the roles and neural sites of estrogens, progestins, and androgens in the activation and maintenance of these events (the sexual brain). The breadth and depth of this research has established that hormonal regulation of these reproductive functions resides within the CNS (see Micevych and Hammer, 1995).

4.1 Gonadal Hormone Activation of Cognition

Recent studies have extended observations on hormonal effects in brain and suggest that gonadal hormones also modulate neural areas responsible for complex, higher order function (Luine, 1994; Luine, 1997; McEwen et al., 1995; McEwen and Alves, 1999; Dohanich, 2002). While this hypothesis has not received widespread investigation, a body of evidence is building. Studies demonstrating gonadal hormone modulation of cognitive processes are highlighted in the following sections.

4.2. Sex/Gender Differences in Cognition

In adults, an initial observation suggesting gonadal hormone influences on cognitive function was that males and females differ in performance of some tasks/tests requiring memory. For example, men outperform women in tasks requiring spatial contexts (Hampson, 1990). This pattern of male gender superiority is also seen in rodents where males show better performance on radial arm maze (Williams et al., 1990; Luine and Rodriguez, 1994) and object placement (Beck and Luine, 2002; Bisagno et al., 2003). See *Figures 21-1* and *21-4* for illustration of radial arm maze performance in male and female, rats and mice. Women are superior to men in various aspects of verbal memory (Hampson, 1990; Nelson, 2005); however, such tasks cannot be assessed in rodents, and nonhuman primates have not been investigated.

Sex differences in memory tasks are not large (usually less than a standard deviation unit), but these differences suggest that gonadal hormones, estrogens, androgens, or both, influence function since different circulating levels of these hormones are found in the sexes (It remains possible that genetic differences, presence or absence of specific genes on the Y chromosome which only males possess, may also be responsible/contribute. Little research has addressed this difficult question (but see Arnold, 2004). An additional, important question which has not been widely investigated in relation to these sex differences is whether the differences are due to gonadal hormones acting during development (organizing/programming effect) to establish long-lasting neuroanatomical, chemical, and/or functional sex differences (Matsumoto, 1999). Developmental/organizational effects of gonadal hormones on cognitive function are detailed next in Section 6. Alternatively, sex differences in cognitive function may arise from differences in circulating hormones that act on neurons to influence their function (activational effect) (McEwen, 2001). Moreover, sex differences in cognitive function may depend on both organizational and activational effects of hormones, and genes on the Y chromosome may also contribute. The extent and

Figure 21-4
Male rodents perform the radial arm maze task more accurately than females. Performance is shown in young male and female mice (*top* panel) and rats (*bottom* panel). In mice, females make more errors to complete the tasks than males. In rats, the females make their first error on earlier choices than males. Rat data from Luine and Rodriguez (1994). Mice data is from Kneavel, Christakos and Luine, unpublished. Student's *t*-test value is $^{**}p < 0.01$

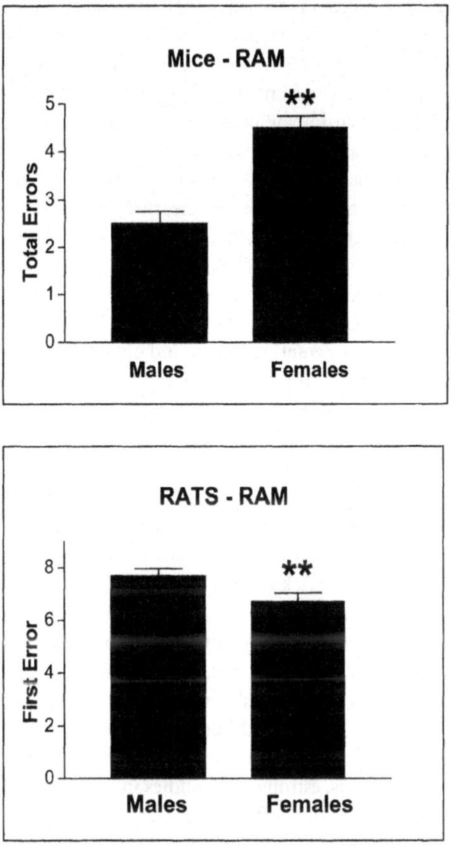

importance of gonadal hormone influences on cognition, i.e., learning and memory, remains somewhat controversial in both animals and humans, but a growing literature providing support for this idea is reviewed below.

4.3 Estrogens Enhance Spatial Memory Tasks in Rodents

In rats and mice, a variety of memory tasks relying on different reinforcements or contingencies (for example, the aversive stimulation of electric shock), measuring various types of memory (for example, spatial or visual memory) and which assess acquisition (learning) or memory (long- or short-term, working versus reference) or both, have been investigated in relation to possible gonadal hormone influences on their performance. In Dohanich (2002), a comprehensive list of these tasks, their primary properties in relation to learning and memory, and the effects of gonadal hormones can be found.

Spatial memory is the most widely assessed form of memory in rodents, and a variety of tasks have been developed for its measurement. These are strongly hippocampal-dependent tasks that rely on the innate ability of rodents to know and defend a territory by utilization of salient environmental landmarks to establish a cognitive map which resides within hippocampal neurons or networks (Mumby et al., 2002).

In the widely applied radial arm maze, subjects receive a food reward at the end of arms of the maze (mazes include from 8–17 arms). This task takes advantage of the natural foraging strategy of rats, and the ability to complete the task without reentering arms depends on building a cognitive map of the cues around the maze, and remembering the arms entered using the cues. The Morris water maze uses a similar spatial context, but subjects learn the location of one, nonvisible platform from which to escape onto from a rat-sized, swimming pool. Several studies show acute or chronic estradiol treatment to Ovx rats or mice enhances, during acquisition, the performance of the radial arm maze and Morris water maze task (Luine and Rodriguez, 1994; Sandstrom and Williams, 2001; Daniel et al., 1997; Luine et al., 1998; Daniel and Dohanich 2001). When subjects have acquired the contingency rules of the specific task, often referred to as reference memory, estradiol appears to cease enhancing performance (Williams et al., 1994; Luine et al., 1998; Fader et al., 1999) and may even impair performance of the radial arm and Morris water maze, although the data are inconclusive and interactive effects with progesterone have not been properly controlled (see Dohanich, 2002 for discussion). When performance of the radial arm maze, Morris water maze and recognition memory tasks are configured to measure working memory (to use and remember within trial information) or short-term memory, then estradiol administration to Ovx subjects usually enhances these spatial memory tasks. ➋ *Figure 21-5* shows that estradiol enhances performance of the spatial memory task, object placement, in Ovx mice when the hormone is given 24 h before testing. Recognition memory tasks such as object placement rely on the observation that rodents seek novelty. When a delay is interspersed between presentation of new and old objects (object recognition) or objects in new or old locations (object placement), memory can be assessed by determining whether the subject spends more time exploring the new object or location than the old (Ennaceur and Aggleton, 1994; Ennaceur et al., 1997).

Results from a recent study provide a provocative interpretation for the lack of estrogen enhancement in several Morris water maze studies (Frick et al., 2004). Stress, engendered by forced swimming in the face of drowning, may interfere with possible mechanisms underlying estrogen effects, i.e., estrogen-dependent increases in spine density in hippocampal CA1 pyramidal neurons may be blocked in estrogen-treated subjects which swam in the water maze (Frick et al., 2004; see ➋ Section 5 for further discussion of stress effects on memory). Nonetheless, when spatial memory tasks are applied to gonadally intact rats experiencing regular estrous cycles, a less clear picture of estrogenic influences also emerges. For acquisition parameters, proestrous rats are often impaired (Frye, 1995; Warren and Jaraska, 1997; Bowman et al., 2001), a result which is inconsistent with estrogen's enhancement of spatial memory in Ovx rats because proestrous rats have high estradiol levels. It should be noted that gonadally intact rats experience daily, hourly and moment to moment changes in gonadotropins, gonadal hormones and adrenal hormones depending on the estrous cycle day. Thus, estrous cycle studies in relation to cognitive function are difficult to implement and complex to interpret. In addition, estrogen levels are high only on early proestrus and decline concomitant with a rise in progestins. Thus, time of memory assessments on proestrus may be critical. Other studies fail to show performance changes over estrous cycle days (Berry et al., 1997; Stackman et al., 1997), but these studies assessed reference memory, a paradigm where estrogen administration in Ovx models also appears generally ineffective (see Dohanich, 2002 for discussion). Recent studies also indicate that strategy selection or memory system utilized (➋ *Figure 21-3*) may change depending on the presence or absence of specific hormones and thereby impact performance (Korol, 2004; Daniel and Lee, 2004). Clearly, further studies of spatial memory in intact females are necessary and should include measurements of acquisition, reference and working memory components as well as several time points during each estrous cycle day in tasks which minimize stress to the subjects and allow assessment of performance style.

4.4 Estrogens Enhance Performance of Specific Memory Tasks in Humans

In humans, spatial memory shows both sexual dimorphisms (see above) and sensitivity to hormonal fluctuations. Young, premenopausal women who have undergone ovarian removal or treatments to lower estradiol because of gynecological problems, show a decline in some aspects of memory, an effect which is prevented by estradiol replacement (Sherwin, 2003). Likewise, some studies in women treated with antiestrogens (tamoxifen or raloxifen) for estrogen-sensitive breast cancer show some cognitive/memory

Figure 21-5
Estradiol enhances object recognition and placement in ovariectomized mice. Ovariectomized mice received sesame oil or estradiol benzoate (5 μg SC) 24 h before testing. Groups consisted of 10 mice. The delay between sample trial (T1) and the recognition trial (T2) was 1 h. *Top* panel is performance of object recognition and *bottom* panel is object placement. Data analyzed by ANOVA (object × treatment) and differences between old and new location tested within groups by paired *t*-test, *$p < 0.05$. Luine, Gordon and McEwen, unpublished. See Li et al. (2004) for other published studies in mice

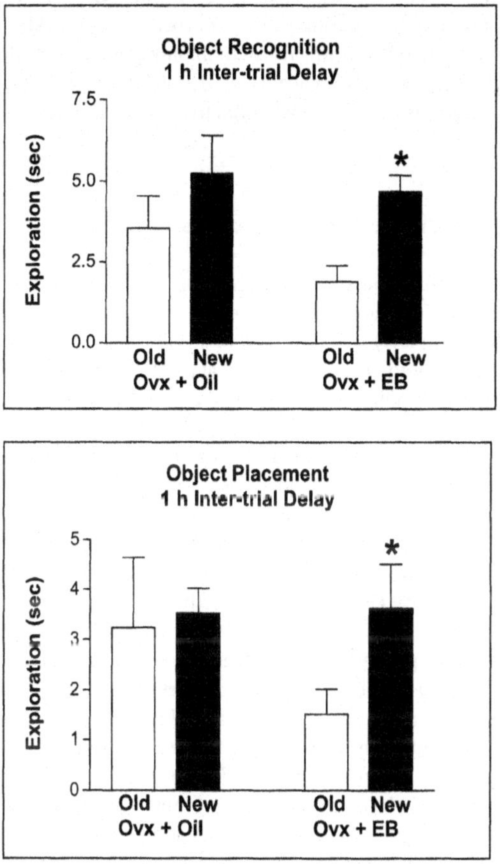

losses as well as postmenopausal women who have little circulating estrogen (Sherwin, 2002; Nelson et al., 2002; Maki and Resnik, 2001). These results suggest that circulating estradiol may promote memory function in adult women. Yet, like rodents, examination of cognitive function over the menstrual cycle gives a conflicting array of information. Tests where women excel compared to men were enhanced during the period when estrogens are high while tasks where men are better than women were worse under high hormonal conditions (Kimura, 1992). The same considerations and caveats for examining memory over the estrous cycle appear germane to the menstrual cycle.

4.5 Estrogens Enhance Nonspatial Memory Tasks in Rodents

While spatial memory tasks provide a convenient measure for memory function and a sensitive read-out for hippocampal function in relation to memory, spatial tasks do not necessarily assess higher order memory or executive function which rely on cortical sites (Ennaceur and Aggleton, 1994; Ennaceur et al., 1997, Mumby et al., 2002). Other memory tasks or tests like visual and verbal memory require complex integration of hippocampal circuits with cortical areas that receive and store sensory information and a knowledge base. Thus, effects of gonadal hormones on associative learning (conditioning, for example, as in active or passive avoidance or eye blink) or on object recognition may also be informative and relate more closely to primate cognitive processes. Avoidance conditioning and T-maze performance is sensitive to circulating estrogen levels in rats and mice (Dohanich et al., 1994; Singh et al., 1994; Gibbs, 2000; Rhodes and Frye, 2004). Recognition memory, in which subjects are tested for the ability to discriminate between familiar and unfamiliar objects or objects in familiar or unfamiliar locations, is also enhanced by administration of estradiol or other estrogens to Ovx rats (Luine et al., 2003) or mice (➋ *Figure 21-5*; Li et al., 2004). These tasks share the common property of assessing short-term memory or working memory. This aspect of memory is in contrast to most spatial tasks which are commonly configured to measure reference memory or acquisition. Interestingly, Sandstrom and Williams (2001, 2004) reported that performance of the spatial memory task, Morris water maze, is enhanced by estrogen treatment when the task is configured to measure working memory.

4.6 Posttraining Activation of Memory by Estrogens

Consistent with the demonstration of putative, nonnuclear steroid receptors in brain (➋ *Figure 21-2*), rapid enhancements in performance of memory tasks by estradiol treated rats have been recently demonstrated. Recognition memory (object or place) was enhanced 4.5 h following estradiol or diethylstilbestrol (DES, synthetic estrogen) administration to Ovx rats (Luine et al., 2003). Behavioral effects were consistent with the stereochemistry of estrogen binding at membrane sites, 17α-estradiol as well as 17β-estradiol, were effective (Toran-Allerand et al., 1999, 2002; Wade et al., 2001). In contrast, genomic responses to estradiol are activated only by 17β isomers, and 17α estrogens have extremely low potencies (Kuiper et al., 1997). Supporting these novel behavioral findings, Rhodes and Frye (2004) reported that inhibitory avoidance is also enhanced by estradiol within 4h after treatment.

The short latency of above responses to estrogens allows for investigating whether enhanced performance of memory tasks is due to hormonal actions on performance/psychological parameters like motivation, arousal, and explorative effects which indirectly enhance performance or whether hormones act directly on mnemonic mechanisms (see Packard, 1998; McGaugh et al., 1989; McGaugh, 2000). Such experiments utilize a posttraining paradigm for assessment of memory function, i.e., drugs or hormones are given following the rats' learning of the location of a platform in the water maze or after subjects have explored objects in object recognition and placement tasks. This paradigm is based on the idea that after a training or sample trial, new information requires consolidation and that drugs or hormones could influence memory storage processes during the period immediately training. This hypothesis was confirmed, beginning in the 1960s, by McGaugh (2000), who found that immediate posttraining injections of specific drugs could facilitate memory in rats. An important additional finding was that the effectiveness of posttraining treatments was time-limited; only treatments given 1–2 h after training/sampling trials were effective, and treatments given later were not effective. Moreover, enhancements after immediate, but not delayed, posttrial injections rule out the possibility that enhancing effects derive from nonmnemonic effects (Packard et al., 1994; Packard, 1998).

Evidence that estrogens act on consolidation of memory has been provided in four memory tasks – water maze, inhibitory avoidance, object recognition, and object placement. For object recognition and placement, immediate, but not delayed, subcutaneous (SC) injections of 17β-estradiol, diethylstilbestrol (DES), or 16α-iodoestradiol after the sample trial enhanced discrimination 4 h later in Ovx rats (➋ *Figure 21-6*; Luine et al., 2003). In the water maze, subcutaneous or intrahippocampal estrogens given immediately following, but not 2 h after, the training trials enhanced the ability of Ovx or castrated rats to locate the hidden platform (Packard and Theather, 1997a, b). Likewise, estradiol administration to Ovx rats immediately, but not 1, 2, or 3h posttraining, increased crossover latencies as compared to vehicle for inhibitory avoidance (Rhodes and Frye, 2004). Thus, the temporal relationship between hormonal application and performance enhancement is consistent with augmentation of mnemonic processes. It is interesting that an increase in spine synapse density in hippocampal CA1 pyramidal neurons has also

Figure 21-6
Diethylstilbestrol (DES) enhances object recognition when given before or immediately after the sample trial (posttrain paradigm). The time spent exploring the old object (closed bars) and the new object (open bars) in the recognition/retention trial (T2) is shown. Ovariectomized subjects received vehicle (control) or DES (15 μg/kg) 30 min before T1 (pre T1), immediately following T1 (DES post T1) or 1 h post T1 (DES del post T1). (Controls were collapsed for pre and post T1). Only DES given before T1 or immediately post T1 enhanced recognition. The intertrial delay was 4 h. Data analyzed by ANOVA (object × treatment) and differences between time spent at old and new location tested within groups by paired *t*-test, **$p < 0.01$. This result shows that an estrogen enhances mnemonic function. Data from Luine et al. (2003)

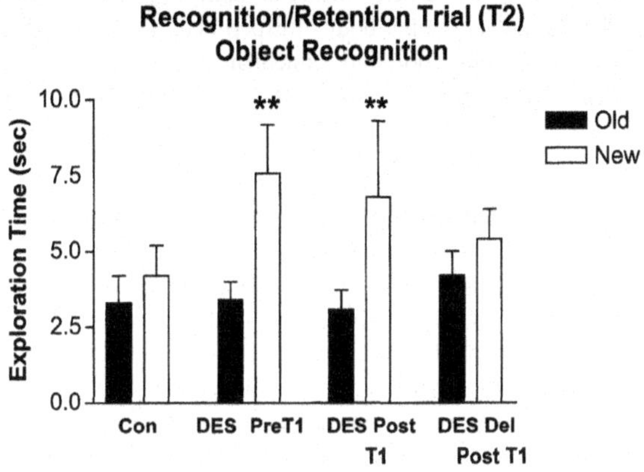

been reported 30 min following injection of estradiol; both 17α and β-estradiol are effective, but the α form is more potent (MacLusky et al., 2005). Increased synapse density may provide the neural substrate underlying the mnemonic enhancements as both processes share similar dose–response curves and ligand specificities. Consistent with this hypothesis, results by Li et al. (2004) show that an increase in mushroom type spines in the hippocampal CA1 region of mice by estradiol occurs with the same dose regimen and time course as an enhancement in object placement performance.

While these memory data and possible activation of signal transduction mechanisms for estradiol action provide a new avenue for hormonal effects, possible relationships to cognition in humans are totally unknown. However, the considerable differences in steroid efficacies and target tissue responses between nuclear and nonnuclear receptor mediated processes may be valuable for designing new programs of hormone replacement following menopause or in aged subjects (see the aged brain, ➋ Section 7).

4.7 Other Gonadal Hormones and Memory

It is notable that few studies have examined whether progestins, the other major gonadal hormone secreted by the ovaries, influence cognitive function. In rodents, progesterone reaches high levels on proestrus, and in primates, it remains elevated for most of the luteal portion of the menstrual cycle (Becker et al., 2005). Progestins can act as classical progestational agents through binding to nuclear progestin receptors to initiate transcription and as neurosteroids through local metabolism. Thus, progestins provide a rich context for possible effects on cognitive function. Due to the paucity of studies, the role of this hormone in learning and memory remains unknown, but, like its effects in other hormone target tissues and neurons, progestins could synergize with or oppose estrogenic effects as well as exert progestational effects. While androgens have clear growth (and cancer) promoting properties in peripheral target tissues, a pivotal role in regulation of reproductive function, and influence aggression, previous studies have shown little influence on cognitive function. However, recent studies have shown that testosterone enhances performance of some cognitive tasks in rats (Roof and Heavens, 1992; Kritzer et al., 2001; Daniel et al., 2003) and men (Yaffe et al.,

2002; Edinger et al., 2004). Clearly, androgens, like progestins, require further investigation in relation to cognition.

4.8 Conclusion: Gonadal Hormones Influence Cognition

Overall, compelling support for the view that gonadal hormones influence cognitive function in adult animals and humans is now extant: foremostly, female rodents and women exhibit a variety of substantial differences from male rodents and men in aspects of cognitive function. Moreover, gonadal hormones, the most documented being estradiol, enhance performance of many memory tasks and mnemonic processes. On the other hand, results from a substantial body of experiments do not support this view or suggest alternative effects. The continued application and refinement of neuroendocrine research strategies will ultimately resolve these differing views concerning the extent to which gonadal hormones influence cognition.

5 The Coping Brain – Adrenal Hormone Influences on Cognition

Over the lifespan, animals as well as humans are presented situations requiring rapid and judicious responses. For animals, an example is a deer leaping and sprinting to escape from predators. Humans, at this juncture in our evolution, not only face similar life threatening situations like war, natural disasters, accidents, and abuse, but also psychologically stressful situations at the workplace and home. Does the release of corticosterone (most rodents) or cortisol (primates) from the adrenals assist, impede, or not affect the performance of daily tasks and/or remembrance of salient events? From the commonly heard phrases "pumped up, burned out, or stressed out" it might be inferred that stress, the hallmark of which is high circulating levels of adrenal steroids, affects people's functioning or at least their view of their own functionality. What is the evidence for glucocorticoid influences on adult cognitive function?

5.1 Acute Stress Effects

The response to stress is characterized by rapid and large outpourings of hypothalamic, pituitary, and adrenal hormones, which provide a positive impact on brain and body for coping with acute stressors and which attempt to restore homeostasis (the fight or flight response, the details of which have been widely described previously; See also *Table 21-1*). There does not appear to be a general pattern in cognitive

Table 21-1
Effects of stress on physiological function

Short period – adaptive	Prolonged period – maladaptive
↑ Energy and Oxygen Use	↑ Myopathy, Fatigue, Ulcers
↑ Analgesia	↓ Immune Function with ↑ in Disease and Cancer
↓ Digestion and growth	Hypertension, Impotence
↑ or ↓ Cognition	↓ Cognition
Males	
Adaptive → ———— Time Course ————	Maladaptive →
Females	
Adaptive → ———————— Time Course ————————	Maladaptive →

responses to acute stress although many responses are enhanced. Acute foot shock enhances several conditioning paradigms in male rats (Shors et al., 1992; Shors, 2001) but impairs recognition (Baker and Kim, 2002) and spatial memory (Diamond et al., 1999). These effects are rapid in onset and may last up to a few days. Responses to acute stress in female rodents are less investigated, but results suggest sex differences in the response. Acute stress impairs eye blink conditioning in females while it enhances such conditioning in males (Wood and Shors, 1998), and performance of some spatial memory tasks is enhanced in female rats but impaired in male rats following acute stress (Conrad et al., 2004).

5.2 Chronic Stress Effects

When stress continues for an extended period (chronic stress), hormonal-dependent, adaptive responses become maladaptive, and subjects suffer negative consequences from the extended secretion of adrenal steroids (see ➲ *Table 21-1* and Nelson, 2005). In an effort to determine possible consequences of chronic stress on cognition, various stressors (restraint, immobilization, noise, lights, cold water) were applied to rats and mice over days and weeks, and effects on performance of cognitive tasks determined. Performance of several spatial memory tasks, radial arm maze (➲ *Figure 21-1* and Luine et al., 1994), Y-maze (Conrad et al., 1996), object placement (Beck and Luine, 1999, 2002) and Morris water maze (Park et al., 2001; Kitraki et al., 2004) is impaired in male rats following these chronic stress regimens. Some nonspatial tasks including recognition memory are also impaired following daily chronic restraint stress (Beck and Luine, 1999, 2002 and see ➲ *Figure 21-7*). Chronic stress effects are not large and appear reversible. That is, cognitive function returns to prestress or control levels in a few days to a few weeks.

5.3 Sex Differences in Chronic Stress Effects on Cognition

Not surprisingly, less information is available for females, rodent or human, but chronic stress appears to affect this sex quite differently than males. Daily restraint to female rats for the same period as males, 21 days, results in enhanced female performance of a number of spatial memory tasks: object placement (Beck and Luine, 2002), Y-maze (Conrad et al., 2003), radial arm maze (Bowman et al., 2001, 2002 and see ➲ *Figure 21-1*) and Morris water maze (Kitraki et al., 2004). This change in performance following stress is opposite to that in males receiving the same stress. In addition, no effects of chronic stress were noted on object recognition in females (Beck and Luine, 2002; Bisagno et al., 2004), whereas males were impaired on this task (Beck and Luine, 1999, 2002). See ➲ *Figure 21-7* which shows effects of chronic stress on the performance of male and female rats in the object recognition task. If female rats are stressed for a longer period (28 days) or Ovx, then enhanced performance is no longer present following stress (Bowman et al., 2001). Interestingly, when males receive a shorter stress, 13 days of daily restraint, radial arm maze performance is enhanced (Luine et al., 1996). These differences in cognitive responding between the sexes may be due to a right shift in the female curve of the continuum from adaptive to maladaptive effects of stress (➲ *Table 21-1*; Bowman et al., 2003). This shift may result from sex differences in processing by adrenal steroid receptors (De kloet et al., 1999; de Kloet, 2004). The shift may also result from the presence in females of higher levels of circulating estrogens which may provide neuroprotection through antioxidant or antiapoptotic effects and/or upregulation of trophic factors (Garcia-Segura et al., 2001; Wise et al., 2001) as well as through direct activation of the brain areas mediating cognition (see ➲ Section 4).

5.4 Morphological and Neurochemical Effects of Stress

The paradigm of 21 days of daily restraint stress is associated with some sexually dimorphic morphological effects in the hippocampus. In males, stress or corticosterone administration is associated with decreased dendritic branches and length in CA3 pyramidal neurons (Watanabe et al., 1992; Coburn-Litvak et al., 2004) and frontal cortex (Wellman, 2001). These hippocampal changes are not present in stressed females

Figure 21-7

Chronic restraint stress impairs object recognition in male, but not female, rats. Young male (*top* panel) and female (*bottom* panel) rats received daily restraint stress for 6 h/day for 21 days or only handling (control). The time spent exploring the old (solid bars) and new (open bars) objects in the recognition/retention trial is shown. The intertrial delay between sample trial (T1) and the recognition/retention trial was 4 h. Data were analyzed by two-way ANOVA (object × treatment) and differences in time spent between old and new object tested within groups by paired *t*-test, $*p < 0.05$. Stress impaired male, but not female, performance on the task. Data from Beck and Luine (2002)

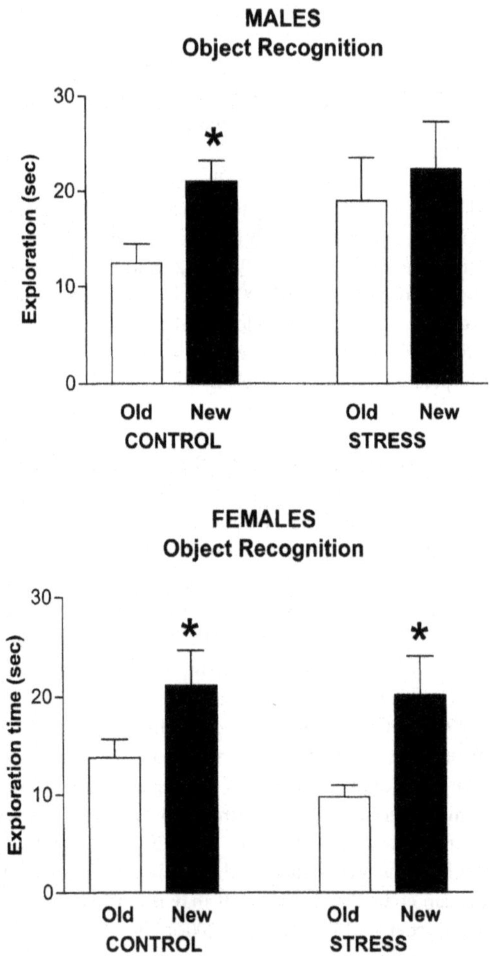

(Galea et al., 1997). Acute stress also has opposite effects in males and females on spine density in CA1 (Shors et al., 2001). In addition, receptors for glucocorticoids are increased in males and decreased in females following chronic stressors (Kitraki et al., 2004; de Kloet, 2004). Monoamines are also altered in the CA3 subfield of the hippocampus: levels of 5-HT and NE were increased in females, but not males, whereas males showed increased levels of GABA (Luine, 2002). Thus, stress appears to affect the male and female hippocampus differently in terms of behavioral output and neuroanatomical and neurochemical properties.

5.5 Stress and Human Cognition

While less is known concerning effects of stress on humans, acute stress can exert either impairing or enhancing effects on memory depending on the temporal context, the relationship of the stress to the task, and the sex of the subject (Kudielka et al., 1998; Wolf, 2003). The Trier social stress test causes acute stress and assessment of memory immediately following this paradigm using word recall found no significant stress effect, but male cortisol levels were correlated with performance; thus, the subjects with highest increases in cortisol had lowest performance (Wolf et al., 2001). No such correlation between cortisol levels and performance was found in women. Processing of emotionally influenced memory storage may also differ between the sexes but follow a different pattern than non-emotion based memories (Cahill, 2001).

While difficult to study in humans, chronic stressors such as stays in boot camps, catastrophic illnesses, or death of spouses result in some form of impaired cognitive function; Posttraumatic stress disease sufferers or those who have experienced extended, high secretions of cortisol show similar deficits (Seeman et al., 1997; McEwen, 2000; Wolf, 2003).

5.6 Conclusion: Adrenal Hormones Influence Cognition

Stress, given both acutely and chronically, appears to affect cognitive function in rodents and in humans. Surprisingly, qualitative and quantitative sex differences in the response appear prominent even though few studies have assessed cognition in females. Since stress has increased for all segments of society, the long-term effects of stress on cognition should become all the more apparent in this new century.

6 The Fetal/Neonatal Brain – Programming of Adult Cognitive Function

Previous discussions of steroid hormone effects on cognition in humans and rodents have highlighted sex differences in performance of memory tasks and sex differences in the effects of both gonadal and adrenal hormones (see ➋ Sections 4 and 5, above). The perinatal developmental period (pre and postnatal) is critical for establishing patterns of brain function and sex differences in these functions. Research beginning in the 1950s established that sex differences in reproductive function (both behavior and gonadotrophin secretion) depend on the presence of gonadal hormones in order to establish a male or female adult pattern of function (Matsumoto, 1999). Hormonal secretions from the testes of fetal and newborn males act within the brain to establish phenotypic male behaviors and a tonic pattern of HPG axis hormonal secretions. On the other hand, phenotypic female patterns – behavior and cyclic patterns of HPG axis hormonal secretions – are inherent in the brain and expressed without secretions from the ovary.

6.1 Perinatal Gonadal Hormone Influences

Perinatal events such as secretion or lack of secretion, of gonadal hormones also program sex specific aspects of the HPA axis and its response to stress in adulthood (Patchev et al., 1995; McCormick et al., 1998; Patchev and Almeida, 1998). For example, males and females show differences in circulating levels of adrenal steroids and their receptors in the brain (Handa et al., 1994; MacLusky et al., 1996; Bowman et al., 2004). Surprisingly, few studies have addressed whether sexual differences in cognition are dependent on perinatal hormonal secretions from the gonads, the adrenals, or both. However, Williams et al. (1990) found that sex differences in radial arm maze performance (males are better than females) were abolished if males were castrated immediately after birth or if females were given estradiol immediately following birth. These results suggest that cognitive sex differences may be programmed similarly to reproductive behaviors, that is, through aromatization of testosterone to estrogen within the brains of males.

6.2 Perinatal Glucocorticoid Influences

During development, as well as in adulthood, the HPG and HPA axes interact to influence brain function. For example, stressing pregnant dams causes a premature cessation and decline in testicular secretions in male fetuses which is accompanied by incomplete masculinization and some feminization of sexual behaviors (Reznekov et al., 2001). Likewise, the HPA axis responds to the effects of early hormonal exposure to adrenal steroids (McCormick et al., 1998; Bowman et al., 2004), and the ability to successfully respond to stress at adulthood, that is, maintain glucocorticoids at a low level, appears to depend on perinatal stress exposure.

Recent studies show a similar pattern of prenatal stress effects on cognitive function as on sexual behavior: stressed males' spatial memory performance is lower than control males' performance (Szuran et al., 1994, 2000; Bowman et al., 2004) and stressed females' performance is improved, and not different, from control male performance (Bowman et al., 2004). Interestingly, prenatally stressed females appear masculinized in other brain properties in addition to spatial memory: anxiety and basal and stress induced corticosterone levels (Bowman et al., 2004). Taken together these results suggest that androgens, as well as corticosterone, may be released from the adrenal gland following stress and alter neural function in females (Frye and Wawrzykci, 2003; Bowman et al., 2004). Prenatal stress is also associated with alterations in monoamines and metabolites at adulthood in areas of the brain associated with cognition, frontal cortex and hippocampus (Bowman et al., 2004; See ➲ *Figure 21-8*). Thus, the prenatal hormonal milieu presumably exerts powerful organizing effects on development of cognitive function. While these hormonally driven, prenatal effects leave a permanent imprint through adulthood on neural and behavioral function, they have received little study.

7 The Aged Brain – Cognitive Loss and Steroid Hormones

With advancing age, most aspects of cognitive function suffer some degree of deterioration, ranging from modest declines in healthy individuals to devastating losses in people with neurodegenerative diseases associated with aging like Alzheimer's and Parkinson's (McEwen et al., 1995; Fillit and Luine, 1997). See also ➲ *Figure 21-1* which illustrates age-related memory declines in radial arm maze performance in male and female rats.

7.1 Gonadal Hormones and Cognition During Aging

Ascribing a specific role for hormones or other parameters to age-related cognitive loss is difficult because all factors are confounded by the aging process itself. Nonetheless, it is notable that in postmenopausal women, gonadal hormone levels plummet to near undetectable levels, and testosterone levels in men, starting at approximately 50 years, decline 30–60% (Yaffe et al., 2000, 2002; Sherwin, 2002). Thus, hormones that tend to promote cognitive function (gonadal) are lower in aged individuals.

In relation to neurodegenerative disease, some evidence suggests that estradiol replacement postmenopause, is associated with either a delay in the onset or a prevention of Alzheimer's disease (Tang et al., 1996; Slooter et al., 1999). Postulated neuroprotective, antioxidant, or antiaging effects of estrogen in brain may be responsible for this effect (Wise et al., 2001; Garcia-Segura et al., 2001).

In healthy, aged women from the Baltimore Longitudinal Study of Aging, gonadal hormone replacement was associated with better maintenance of many aspects of cognitive function (see Maki and Resnick, 2001, for example), a result also shown in several other studies (Gibbs and Aggarwal, 1998; Yaffe et al., 2000). However, results of the recent Women's Health Initiative (WHI) trial supported by NIH and other trials have not replicated these results (Zec and Trevidi, 2002). In addition, hormonal replacement with either conjugated equine estrogens (CEE) alone (Anderson et al., 2004) or in combination with medroxyprogesterone (MPA) (Wassertheil-Smoller et al., 2003) showed some unexpected, deleterious effects in subjects. CEE was associated with a small increase in stroke while CEE+MPA subjects show an increased incidence of breast cancer and cardiovascular complications. Endocrinologists have pointed out that this population of women was generally older (67% were age 60–79) and heavier (69% were overweight or

Figure 21-8

Prenatal stress is associated with altered expression at adulthood of frontal cortex monoaminergic activity. Data are expressed as pg/μg protein (mean ± SEM). Group differences are identified by main effect of sex (a) and the sex X prenatal treatment interaction (x). Females had higher levels of NE ($p < 0.006$) and 5HT ($p < 0.05$) than males, regardless of prenatal treatment. 5HIAA levels were influenced both by sex (higher in females, $p < 0.001$) and the sex X treatment interaction (lowest in PS males, $p < 0.02$). Dopamine activity, as indexed by HVA ($p < 0.03$) and DOPAC ($p < 0.03$) metabolite levels, was influenced by the sex X treatment interaction, with lowest levels observed in PS males and highest levels in PS females. From Bowman et al. (2004). Sexually dimorphic effects of prenatal stress on cognition, hormonal responses, and central neurotransmitters. *Endocrinology* 145: 3778–3787. Copyright 2004, The Endocrine Society

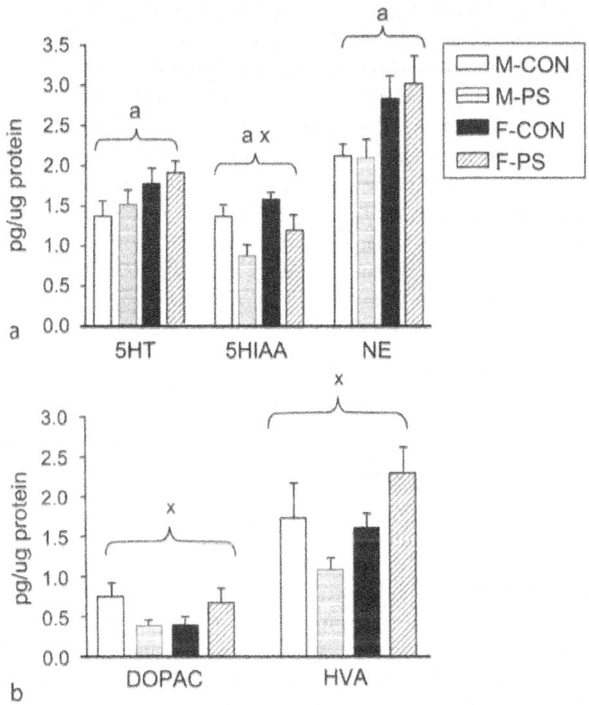

obese) than populations in other epidemiological studies like the Baltimore study. In addition, 74% had never used hormone replacement and were thus estrogen-deficient for more than a decade. These risk factors may have contributed to the untoward cardiovascular and other effects. The use in this trial of widely prescribed replacement drugs (CEE and MPA), not estradiol or progesterone, may have also been problematic (See Turgeon et al., 2004 for a discussion of these studies). Thus, the WHI study has discouraged use of estrogens in postmenopausal women. The implementation of better hormone formulations, different modes of administration, and initiation of therapy at the menopause may provide a better hormonal replacement outcome, one consistent with positive effects of estrogen in animal models (see above) and in premenopausal women (Sherwin, 2003).

7.2 Adrenal Hormones and Cognition During Aging

Compounding the decline of gonadal hormones, the level of basal, circulating glucocorticoids tends to increase with aging (McEwen et al., 1995), and high levels of these steroids, for extended periods in rodents and humans, are associated with poorer cognitive performance (See Section 5). Aged individuals therefore possess a hormonal mileu, high glucocorticoids and low gonadal hormones, not optimal for cognition. Thus, the ability to maintain cortisol at a low level has been linked to maintenance of cognitive function and "successful" aging (Lupien et al., 1999), and treatments or behaviors which maintain low cortisol levels may be beneficial to cognitive function and a slowing of aging. However, some recent studies have not supported the idea that high cortisol is linked to poor cognitive performance in aged individuals (See Wolf, 2003 for discussion).

7.3 Hormonal Therapy for Age-Related Memory Loss

Since hormones influence neurons over the entire life cycle, the use of hormonal derivatives or "designer" hormones to promote memory in aged individuals would appear to offer advantages. The goal would be to eliminate the cancer promoting and cardiovascular damage that estrogens may exert but still take advantage of growth promoting and neuroprotective effects. The development of such selective receptor modulators (SRMs; Smith and O'Malley, 2004), especially selective estrogen receptor modulators (SERMs) is promising in this regard, and raloxifene (Evista), the second generation of such drugs has proven effective for osteoporosis without unwanted effects in the uterus or breast, but raloxifene unfortunately appears largely inactive in neural tissue (Tierney and Luine, 1997). Estrogens that act through membrane receptors offer promise as well because 17α-estrogens are more potent than the 17β-estrogens, and 17α-estrogens show little cancer promoting properties (See Luine et al., 2003; MacLusky et al., 2005 for more discussion). These estrogens may offer new routes for treatment of memory loss and neurodegenerative disease.

The blocking of deleterious effects from elevated glucocorticoids might also be advantageous in aged populations. To date, little progress on identifying selective glucocorticoid receptor modulators (SGRMs) or related antagonists has been made. Interestingly, gene therapy has recently shown promise in an animal model. A viral amplicon, containing a chimeric glucocorticoid receptor ligand-binding domain fused to the estrogen receptor DNA binding domain, was placed in the rat hippocampus. The size of a lesion after neurological damage was reduced by 63% (Kaufer et al., 2004) and the viral vector attenuated acute stress-dependent impairments in object recognition memory (Ferguson et al., 2004). Thus, neuroprotective effects of estrogen appear to have mitigated deleterious effects of glucocorticoids.

8 Concluding Remarks

Steroid hormones influence cognitive function in animal models and humans over the entire life span. Some hormonal influences on memory appear to be mediated directly through actions on neurons and circuits subserving memory and on memory consolidation itself while other mnemonic changes may be indirect and result from hormonal influences on psychological/performance parameters. Clearly, further research on how cognition is altered by hormones is critical. Increased knowledge of the substrates and the circuits responsible for acquiring and maintaining memories would provide neuroendocrinologists the information upon which to develop novel, hormonally based therapies (SRM drugs-selective receptor modulator drugs) or specific hormone–gene targets to enhance cognitive function. Hormonal or genetic based interventions could promote recovery from strokes or injury and also intervene in preventing or attenuating age- and neurodegenerative disease-related memory losses.

Acknowledgments

Research in the author's laboratory is supported by NIH grants RISE, GM60665; SCORE, GM60654; MIDARP, DA12136, and RCMI, RR03037. Advice and discussions with Gary Dohanich in the preparation of this manuscript is greatly appreciated. Dedicated work by former and current students Kevin Beck, Rachel Bowman, Deveroux Ferguson, Luis Jacome, Govini Mohan, Marisa Gordon, and Shannon Richards are gratefully acknowledged.

References

Akama KT, McEwen BS. 2003. Estrogen stimulates postsynaptic density-95 rapid protein synthesis via the Akt/protein kinase B pathway. J Neurosci 23: 2333-2339.

Anderson GL, Limacher M, Assaf AR, Bassford T, Beresford SA, et al. 2004. Effects of conjugated equine estrogen in postmenopausal women with hysterectomy: the Women's Health Initiative randomized controlled trial. JAMA 291: 1701-1712.

Arnold AP. 2004. Sex chromosomes and brain gender. Nat Rev Neurosci 5: 701-708.

Bailey CH, Kandel ER, Si, K. 2004. The persistence of long-term memory: a molecular approach to self-sustaining changes in learning-induced synaptic growth. Neuron 44: 49-57.

Baker KB, Kim JJ. 2002. Effects of stress and hippocampal NMDA receptor antagonism on recognition memory in rats. Learn Mem 9: 58-65.

Beck KD, Luine VN. 1999. Food deprivation modulates chronic stress effects on object recognition in male rats: role of monoamines and amino acids. Brain Res 830: 56-71.

Becker JB, Arnold AP, Berkley KJ, Blaustein JD, Eckel LA, 2005. Strategies and methods for research on sex differences in brain and behavior. Endocrinology 2004 Dec 23 (Epub ahead of print).

Berry B, McMahan R, Gallagher M. 1997. Spatial learning and memory at defined points of the estrous cycle: effects on performance of a hippocampal-depedent task. Behav Neurosci 11: 267-274.

Bi R, Broutman G, Foy MR, Thompson RF, Baudry M. 2000. The tyrosine kinase and mitogen-activated protein kinase pathways mediate multiple effects of estrogen in hippocampus. Proc Natl Acad Sci USA 97: 3602-3607.

Bi R, Foy MR, Vouimba RM, Thompson RF, Baudry M. 2001. Cyclic changes in estradiol regulate synaptic plasticity through theMAPkinase pathway. Proc Natl Acad Sci USA 98: 13391-13395.

Bisagno V, Ferguson D, Luine VN. 2003. Chronic d-amphetamine induces sexually dimorphic effects on locomotion, recognition memory and brain monoamines. Pharmac Biochem Behav 74: 859-867.

Bisagno V, Grill CA, Piroli GG, Giraldo P, McEwen BS, et al. 2004. Chronic stress alters the behavioral and neural effects of amphetamine. Pharmac Biochem Behav 78: 541-550.

Blaustein JD. 1992. Cytoplasmic estrogen receptors in rat brain: immunocytochemical evidence using three antibodies with distinct epitopes. Endocrinology 131: 1336-1342.

Bowman RE, Beck KD, Luine VN. 2003. Chronic stress effects on memory: sex differences in performance and monoamines. Horm Behav 43: 48-59.

Bowman RE, Ferguson D, Luine VN. 2002. Effects of chronic restraint stress and estradiol on open field activity, spatial memory, and monoaminergic neurotransmitters in ovariectomized rats. Neuroscience 113: 401-410.

Bowman RE, MacLusky NJ, Sarmiento Y, Frankfurt M, Gordon M, et al. 2004. Sexually dimorphic effects of prenatal stress on cognition, hormonal responses and central neurotransmitters. Endocrinology 145: 3778-3787.

Bowman RE, Zrull MC, Luine VN. 2001. Chronic restraint stress enhances radial arm maze performance in female rats. Brain Res 904, 279-289.

Cahill L, Haier RJ, White NS, Fallon J, Kilpatrick L, et al. 2001. Sex-related difference in amygdala activity during emotionally influenced memory storage. Neurobiol Learn Mem 75: 1-9.

Carter CS, Lederhendler II, Kirkpatrick B. 1997. Integrative neurobiologoy of affiliation. Vol. 807. New York: New York Academy of Sciences.

Coburn-Litvak PS, Tata DA, Gorby HE, McCloskey DP, Richardson G, et al. 2004. Chronic corticosterone affects brain weight, and mitochondrial, but not glial volume fraction in hippocampal area CA3. Neuroscience 124: 429-438.

Conrad CD, Galea LA, Kuroda Y, McEwen BS. 1996. Chronic stress impairs rat spatial memory on the Y maze, and this effect is blocked by tianeptine pretreatment. Behav Neurosci 110: 1321-1334.

Conrad CD, Grote KA, Hobbs RJ, Ferayorni A. 2003. Sex differences in spatial and non-spatial Y-maze performance after chronic stress. Neurobiol Learn Mem 79: 32-40.

Conrad CD, Jackson JL, Wieczorek L, Baran SE, Harman JS, et al. 2004. Acute restraint stress impairs spatial memory in male but not female rats: influence of estrous cycle. Pharmac Biochem Behav 78: 569-579.

Daniel JM, Fader AJ, Spencer A, Dohanich GP. 1997. Estrogen enhances performance of female rats during acquisition of a radial arm maze. Horm Behav 32: 217-225.

Daniel JM, Dohanich GP. 2001. Acetylcholine mediates the estrogen-induced increase in NMDA receptor binding in CA1 of the hippocampus and the associated improvement in working memory. J Neurosci 21: 6949-6956.

Daniel JM, Lee DC. 2004. Estrogen replacement in ovariectomized rats affects strategy selection in the Morris water maze. Neurobiol Learn Mem 82: 142-19l.

Daniel JM, Winsauer PJ, Moerschbaecher JM. 2003. Castration in rats impairs performance during acquisition of a working memory task and exacerbates deficits in working memory produced by scopolamine and mecamylamine. Psychopharm 170: 294-300.

de Kloet ER. 2004. Hormones and the stressed brain. Ann NY Acad Sci 1018: 1-15.

deKloet ER, Azmitia EC, Landfield PW. (editors) 1994. Brain corticosteroid receptors: studies on the mechanism, function and neurotoxicity of corticosteroid action, Ann NY Acad Sci 746.

De Kloet ER, Oitzl MS, Joels M. 1999. Stress and cognition: are corticostereoids good or bad guys? Trends Neurosci 22: 422-426.

Diamond DM, Par CR, Herman KL, Rose GM. 1999. Exposing rats to a predator impairs spatial working memory in the radial arm water maze. Hippocampus 9: 942-952.

Dohanich GP. 2002. Gonadal steroids, learning and memory. Hormones, brain and behavior. San Diego: Academic Press; pp. 265-327.

Dohanich GP, Fader AJ, Javorsky DJ. 1994. Estrogen and estrogen–progesterone treatments counteract the effect of scopolamine on reinforced T-maze alternation in female rats. Behav Neurosci 108: 988-992.

Edinger KL, Lee B, Frye CA. 2004. Mnemonic effects of testosterone and its 5α-reduced metabolites in the conditioned fear and inhibitory avoidance tasks. Pharmac Biochem Behav 78: 559-568.

Ennaceur A, Aggleton JP. 1994. Spontaneous recognition of object configurationsin rats: effects of fornix lesions. Exp Brain Res 100: 85-92.

Ennaceur A, Neave N, Aggleton JP. 1997. Spontaneous object recognition and object location memory in rats: the effects of lesions in the cingulate cortices, the medial prefrontal cortex, the cingulum bundle and the fornix. Exp Brain Res 113: 509-519.

Everett JW. 1965. Ovulation in rats from preoptic stimulation through platinum electrodes. Important of duration and spread of stimuli. Endocrinology 76: 1195.

Fader AJ, Johnson PEM, Dohanich GP. 1999. Estrogen improves working but not reference memory and prevents amnestic effects of scopolamine on a radial-arm maze. Pharmacl Biochem Behav 62: 711-717.

Ferguson D, Dumas T, Sapolsky RM. 2004. Gene therapy to spare cognition from the disruptive effects of stress. Program No. 770.10. Abstract Viewer/Itinerary Planner. Washington, DC: Society for Neuroscience.

Fillit H, Luine V. 1997. The neurobiology of gonadal hormones and cognitive decline in late life. Maturitas 26: 159-164.

Frick KM, Fernandez SM, Bennett JC, Prange-Kiel J, Maclusky NJ, et al. 2004. Behavioral training interferes with the ability of gonadal hormones to increase CAI spine synapse density in ovariectomized female rats. European J Neurosci 19: 1-7.

Frye CA. 1995. Estrus-associated decrements in a water maze task are limited to acquisition. Physiol Behav 57: 5-14.

Frye CA, Wawrzycki J. 2003. Effect of prenatal stress and gonadal hormone condition on depressive behaviors of female and male rats. Horm Behav. 44: 319-26.

Galea LA, McEwen BS, Tanapat P, Deak T, Spencer RL, et al. 1997. Sex differences in dendritic atrophy of CA3 pyramidal neurons in response to chronic restraint stress. Neuroscience 81: 689-697.

Garcia-Segura LM, Azcortia I, Don Carlos LC. 2001. Neuroprotection by estrogen. Prog Neurobiol 63: 29-60.

Ge X, Hannan F, Xie Z, Feng C, Tully T, et al. 2004. Notch signaling in Drosophila long-term memory formation. Proc Natl Acad Sci USA 101: 10172-10176.

Gibbs RB. 1999. Treatment with estrogen and progestereone affects relative levels of brain-derived neurotrophic factor mRNA and protein in different regions of the adult rat brain. Brain Res 844: 20-27.

Gibbs RB. 2000. Basal forebrain cholinergic neurons are necessary for estrogen to enhance acquisition of a delayed matching-to-position T-maze task. Horm Behav 42: 245-257.

Gibbs RB, Aggarwal P. 1998. Estrogen and basal forebrain cholinergic neurons: implications for brain aging and Alzheimer's disease-related cognitive decline. Horm Behav 334: 98-111.

Gorski RA. 1971. Gonadal hormones and the perinatal development of neuroendocrine function. Frontiers in Neuroendocrinology. Martini L, Ganong WF. (editors), New York: Oxford Press; p. 237.

Gould E, Woolley CS, Frankfurt M, McEwen BS. 1990. Gonadal steroids regulate dendritic spine density in hippocampal pyramidal cells in adulthood. J Neurosci 10: 1286-1291.

Hampson E. 1990. Estrogen-related variations in human spatial and ariticulatory-motor skills. Psychoneuroendo 15: 97-111.

Handa RJ, Burgess LH, Kerr JE, O'Keefe JA. 1994. Gonadal steroid hormone receptors and sex differences in the hypothalamo-pituitary-adrenal axis. Horm Behav 28: 464-476.

Hao J, Janssen WG, Tang Y, Roberts JA, McKay H, et al. 2003. Estrogen increases the number of spinophilin-immunoreactive spines in the hippocampus of young and aged female rhesus monkeys. J Comp Neurol 465: 540-550.

Helm KA, Hand JS, Gallagher M. 2002. Effects of cholinergic lesions produced by infusions of 192 IgG-saporin on glucocorticoid receptor mRNA expression in hippocampus and medial prefrontal cortex of the rat. Neuroscience 115: 765-774.

Jensen E, Jacobson H. 1962. Basic guides to the mechanism of esrogen action. Rec Progr Horm Res 18: 387-408.

Kaufer D, Ogle WO, Pincus ZS, Clark KL, Nicholas AC, et al. 2004. Restructuring the neuronal stress response with anti-glucocorticoid gene delivery. Nat Neurosci 7: 899-900.

Kimura D. 1992. Sex differences in the brain. Sci Am September: 119–125.

Kitraki E, Kremmyda O, Youlatos D, Alexis MN, Kittas C. 2004. Gender-dependent alterations in corticosteroid receptor status and spatial performance following 21 days of restrain stress. Neuroscience 125: 47-55.

Korol DL. 2004. Role of estrogen in balancing contributions from multiple memory systems. Neurobiol Learn Mem 82: 309-323.

Kritzer MF, Kohama SG. 1998. Ovarian hormones influence morphology, distribution and density of tyrosine hydroxylase immunoreactive axons in the dorsolater prefrontal cortex of adult rhesus monkeys. J Comp Neurol 395: 1-170.

Kritzer MF, McLaughlin PJ, Smirlis T, Robinson JK. 2001. Gonadectomy impairs T-maze acquisition in adult male rats. Horm Behav 39: 167-174.

Kudielka BM, Hellhammer J, Hellhammer DH, Wolf OT, Pirke KM, et al. 1998. Sex differences in endocrine and psychological responses to psychosocial stress in healthy elderly subjects and the impact of a 2-week dehydroepiandrosterone treatment. J Clin Endocrinol Metab 83: 1756-1761.

Kuiper GG, Carlsson B, Grandien K, Enmark E, Haggblad J, et al. 1997. Comparison of the ligand binding specificity and transcript tissue distribution of estrogen receptors α and β. Endocrinology 138: 863-870.

Leranth C, Hajszan T, MacLusky NJ. 2004. Androgens increase spine synapse density in the CA1 hippocampal subfield of ovariectomized female rats. J Neurosci 24: 495-499.

Leranth C, Shanabrough M, Horvath TL. 2000. Hormonal regulation of hippocampal spine synapse density involves subcortical mediation. Neuroscience 101: 349-356.

Leranth C, Shanabrough M, Redmond DE. Jr. 2002. Gonadal hormones are responsible for maintaining the integrity of spine synapses in the CA1 hippocampal subfield of female nonhuman primates. J Comp Neurol 447: 34-42.

Levin ER. 2002. Cellular functions of plasma membrane estrogen receptors. Steroids 67: 471-475.

Li C, Brake WG, Romeo RD, Dunlop JC, Gordon M, et al. 2004. Estrogen alters hippocampal dendritic spine shape and enhances synaptic protein immunoreactivity and spatial memory in female mice. Proc Natl Acad Sci USA 101: 2185-2190.

Luconi M, Forti G, Baldi E. 2002. Genomic and nongenomic effects of estrogens: molecular mechanisms of action and clinical implications for male reproduction. J Steroid Biochem Molec Biol 80: 369-381.

Luine V, Harding CF. (Editors) 1994. Hormonal restructuring of the adult brain: Basic and clinical perspectives. Ann NY Acad Sci 743.

Luine V, Hearns M. 1990. Spatial memory deficits in aged rats: Contributions of the cholinergic system assessed by ChAT. Brain Res 523: 321-324.

Luine V, Martinez C, Villegas M, Magarinos AM, McEwen B. 1996. Restraint stress reversibly enhances spatial memory performance, Physiology and Behavior 59:27-32.

Luine V, Richards ST, Wu VY, Beck K. 1998. Estradiol enhances learning and memory in a spatial memory task and effects levels of monoaminergic neurotransmitters. Horm Behav 34: 149-162.

Luine V, Rodriguez M. 1994. Effects of estradiol on radial arm maze performance of young and aged rats. Behav Neural Biol 62: 230-236.

Luine V, Villegas M, Martinez C, McEwen BS. 1994. Repeated stress causes reversible impairments of spatial memory performance. Brain Res 639: 167-170.

Luine VN. 1985. Estradiol increases choline acetyltransferase activity in specific basal forebrain nuclei and projection areas of female rats. Exp Neurol 89: 484-490.

Luine VN. 1994. Steroid hormone influences on spatial memory. Ann NY Acad Sci 743: 201-211.

Luine VN. 1997. Steroid hormone modulation of hippocampal dependent spatial memory. Stress 2: 21-36.

Luine VN. 2002. Sex differences in chronic stress effects on memory in rats. Stress 5: 205-216.

Luine VN, Jacome LF, Maclusky NJ. 2003. Rapid enhancement of visual and place memory by estrogens in rats. Endocrinology 144: 2836-2844.

Luine VN, Khylchevskaya RI, McEwen BS. 1975. Effects of gonadal steroids on activities of Monoamine oxidase and Choline acetylase in rat brain. Brain Res 86: 293-306.

Lupien SJ, Nair NP, Briere S, Maheu F, Tu MT, et al. 1999. Increased cortisol levels and impaired cognition in human aging: implication for depression and dementia in later life. Rev Neurosci 10: 117-139.

MacLusky N, Luine VN, Hajszan T, Leranth C. 2005. The 17 α and β isomers of estradiol both induce rapid spine synapse formation in the CA1 hippocampal subfield of ovariectomized female rats. Endocrinology 146: 287-293.

MacLusky NJ, Yuan H, Elliott J, Brown TJ. 1996. Sex differences in corticosteroid binding in the rat brain: an in vitro autoradiographic study. Brain Res 708: 71-81.

Maki PM, Resnick SM. 2001. Effects of estrogen on patterns of brain activity at rest and during cognitive activity: a review of neuroimaging studies. Neuroimage 14: 789-801.

Markham JA, Juraska JM. 2002. Aging and sex influence the anatomy of rat anterior cingulated cortex. Neurobiol Aging 23: 579-588.

Matsumoto A (Ed). 1999 Sexual Differentiation of the Brain. CRC Press. See this multi-author book for in depth understanding of sexual differentiation of the brain.

Micevych PE, Hammer RP Jr. (Editors) 1995. Neurobiological Effects of Sex Steroid Hormones. Cambridge University Press. See this multi-author book for articles on all aspects of gonadal steroid regulation of sexual behavior.

McCormick CM, Furey BF, Child M, Sawyer MJ, Donohue SM. 1998. Neonatal sex hormones have 'organizational' effects on the hypothalamic-pituitary-adrenal axis of male rats. Brain Res Dev 105: 295-307.

McEwen BS. 2000. The neurobiology of stress: from serendipity to clinical relevance. Brain Res 886: 172-189.

McEwen BS. 2001. Invited review: estrogens effects on the brain: multiple sites and molecular mechanisms. J Appl Physiol 91: 2785-2801.

McEwen BS, Alves SE. 1999. Estrogen actions in the central nervous system. Endocr Rev 20: 279-307.

McEwen BS, Gould E, Orchinik M, Weiland NG, Woolley CS. 1995. Oestrogens and the structural and functional plasticity of neurons: implications for memory, ageing and neurodegenerative processes. Ciba Found Symp 191: 52-73.

McEwen BS, Weiss JM, Schwartz LS. 1968. Selective retention of corticostereone by limbic structures in rat brain. Nature 22: 911-912.

McGaugh JL. 1989. Disassociating learning and performance: drug and hormone enhancement of memory storage. Brain Res Bull 23: 339-345.

McGaugh JL. 2000. Memory-a century of consolidation 2000. Science 287: 248-251.

Micevych PE, Hammer RP Jr. (Editors) 1995. Neurobiological effects of sex steroid hormones. Cambridge University Press. See this multi-author book for articles on all aspects of gonadal steroid regulation of sexal behavior.

Milner TA, McEwen BS, Hayashi S, Li CJ, Reagan LP, et al. 2001. Ultrastructural evidence that hippocampal α estrogen receptors are located at extranuclear sites. J Comp Neurol 429: 355-371.

Mitra SW, Hoskin E, Yudkovitz J, Pear L, Wilkinson HA, et al. 2003. Immunolocalization of estrogen receptor beta in the mouse brain: comparison with estrogen receptor alpha. Endocrinology 144: 2055-2067.

Mumby DG, Gaskin S, Glenn MJ, Schramek TE, Lehmann H. 2002. Hippocampal damage and exploratory preferences in rats: memory for objects, places, and contexts. Learn Mem 9: 49-57.

Nelson HD, Humphrey LL, Nygren P, Teutsch SM, Allan JD. 2002. Postmenopausal hormone replacement therapy: scientific review. JAMA 288: 872-881.

Nelson RJ. 2005. An introduction to behavioral endocrinology, Sinauer Assoc., 3rd Edition,Textbook provides excellent background material on Endocrinology. For this review see faceplates for listing of vertebrate steroid hormones; Chapter 11 on Stress, p. 557; Chapter 12 on Learning and Memory, p. 593 and Chapter 4 Sex Differences in Behavior: Animal Models and Humans, p. 183.

Ostlund H, Keller E. Hurd YL. 2003. Estrogen receptor gene expression in relation to neuropsychiatric disorders. Ann NY Acad Sci 1007: 54-63.

Packard MG. 1998. Posttraining estrogen and memory modulation. Horm Behav 34: 126-139.

Packard MG, Cahill L, McGaugh JL. 1994. Amygdala modulation of hippocampal-dependent and caudate nucleus-dependent memory processes. Proc Natl Acad Sci USA 91: 8477-8481

Packard MG, Teather, LA. 1997a. Intra-hippocampal estradiol infusion enhances memory in ovariectomized rats. Neuroreport 8: 3009-3013.

Packard MG, Teather LA. 1997b. Posttraining estradiol injections enhance memory in ovariectomized rats: cholinergic blockade and synergism. Neurobiol Learn Mem 68: 172-188.

Park CR, Campbell AM, Diamond DAM. 2001. Chronic psychosocial stresss impairs learning and memory and increases sensitiveity to yohimbine in rats. Biol Psychiatry 50: 994-1004.

Patchev VK, Almeida OF. 1998. Gender specificity in the neural regulation of the response to stress: new leads from classical paradigms. Mol Neurobiol 16: 63-77.

Patchev VK, Hayashi S, Orikasa C, Almeida OF. 1995. Implications of estrogen-dependent brain organization for gender differences in hypothalamo-pituitary-adrenal regulation. FASEB J 9: 419-423.

Payne JL. 2003. The role of estrogen in mood disorders in women. Int Rev Psyciatry 15: 280-290.

Reznikov AG, Nosenko ND, Tarasenko LV, Sinitsyn PV, Polyakova LI. 2001. Early and long-term neuroendocrine effects of prenatal stress in male and female rats. Neurosci Behav Physiol 31:1-5.

Rhodes ME, Frye CA. 2004. Estrogen has mnemonic-enhancing effects in the inhibitory avoidance task. Pharmacol Biochem Behav 78: 551-558.

Rissman EF, Heck AL, Leonard JE, Shupnik MA, Gustafsson JA. 2002. Disruption of estrogen receptor β gene impairs spatial learning in female mice. Proc Natl Acad Sci USA 99: 3996-4001.

Roof RL, Havens MD. 1992. Testosterone improves maze performance and induces development of a male hippocampus in females. Brain Res 572: 310-313.

Roozendaal B, McReynolds JM, McGaugh JL. 2004. The basolateral amygdala interacts with the medial prefrontal cortex in regulating glucocorticoid effects on working memory impairment. J Neurosci 24: 1385-1392.

Sandstrom NJ, Williams CL. 2001. Memory retention is modulated by acute estradiol and progesterone replacement. Behav Neurosci 115: 384-393.

Sandstrom NJ, Williams CL. 2004. Spatial memory retention is enhanced by acute and continuous estradiol replacement. Horm Behav 45: 128-135.

Seeman TE, McEwen BS, Singer BH. 1997. Increase in urinary cortisol excretion and memory declinesL: MacArthur studies of successful aging. J Clin Endo Metab 82: 2458-2465.

Sherwin BB. 2002. Estrogen and cognitive aging in women. Trends Pharmacol Sci 23: 527-534.

Sherwin BB. 2003. Estrogen and cognitive function in women. Endocrin Reviews 24: 133-151.

Shors TJ. 2001. Actue stress rapidly and persistently enhances memory formation in the male rat. Neurobiol Learn Mem 75: 1929.

Shors TJ, Chua C, Falduto J. 2001. Sex differences and opposite effects of stress on dendritic spine density in the male versus female hippocampus. J Neurosci 21: 6292-6297.

Shors TJ, Weiss, Thompson RF. 1992. Stress-induced facilitation of classical conditioning. Science 257: 537-539.

Shughrue PJ, Lane MV, Merchenthaler I. 1997. Comparative distribution of estrogen receptor-α and -β in the rat central nervous system. J Comp Neurol 388: 507-525.

Shughrue PJ, Lane MV, Merchenthaler I. 1999. Biologically active estrogen receptor-β: evidence from *in vivo* autoradiographic studies with estrogen receptor α-knockout mice. Endocrinology 140: 2613-2620.

Singh M. 2001. Ovarian hormones elicit phosphorylation of Akt and extracellular- signal regulated kinase in explants of the cerebral cortex. Endocrine 14: 407-415.

Singh M, Meyer EM, Millard WJ, Simpkins JW. 1994. Ovarian steroid deprivation results in a reversible learning impairment and compromised cholinergic function in female Sprague-Dawley rats. Brain Res 644: 305-312.

Slooter AJ, Bronzova J, Witteman JC, Van Broeckhoven C, Hofman A, et al. 1999. Estrogen use and early onset Alzheimer's disease: a population- based study. J Neurol Neurosurg Psychiatry 67: 779-781.

Smith CL, O'Malley. 2004. Coregulator function: a key to understanding tissue specificity of selective receptor modulators. Endo Rev 25: 45-71.

Stackman WR, Blasberg ME, Langan CJ, Clark AS. 1997. Stability of spatial working memory across the estrous cycle of Long-Evens rats. Neurobiol Learn Mem 67: 167-171.

Steckler T, Drinkenburgh WHIM, Sahgal A, Aggleton JPP. 1998. Recognition memory in rats--II. Neuroanatomical substrates. Prog Neurobiol 54: 313-332.

Szuran TF, Pliska V, Pokorny J, Welzl H. 2000. Prenatal stress in rats: effects on plasma corticosterone, hippocampal glucocorticoid receptors, and maze performance. Physiol Behav 71:353-362.

Tang Y, Janssen W, Hao J, Roberts JA, McKay H, et al. 2004. Estrogen replacement increase spinophilin-immunoreactive spine number in the prefrontal cortex of female rhesus monkeys. Cerebral Cortex 14: 215-223.

Tanapat P, Hastings NB, Reeves AJ, Gould E. 1999. Estrogen stimulates a transient increase in the number of new neurons in the dentate gyrus of the adult female rat. J Neurosci 19: 5792-5801.

Tierney MC, Luine VN. 1997. New Concepts in Hormone Replacement: Selective Estrogen Receptor Modulators (SERMS) IV. Effects of estrogens and antiestrogens on the CNS. Supplement to J Can Soc Of Gyn and Obst 19: 46-56.

Toran-Allerand CD, Guan X, MacLusky NJ, Horvath TL, Diano S, Singh M, Connolly ES, Nethrapalli TS, Tinnikov AA, 2002. ER-X: a novel, plasma membrane-associated, putative estrogen receptor that is regulated during development and after ischemic brain injury. J Neurosci 22: 8391-8401.

Toran-Allerand CD, Singh M, Setalo G Jr. 1999. Novel mechanisms of estrogen action in the brain: new players in an old story. Front Neuroendocrinol 20: 97-121.

Towart LA, Alves SE, Znamensky V, Hayashi S, McEwen BS, et al. 2003. Subcellular relationships between cholinergic terminals and estrogen receptor α in the dorsal hippocampus. J Comp Neurol 463: 390-401.

Tsai M-J, O'Malley BW. 1994. Molecular mechanisms of action of steroid/thyroid receptor superfamily members. Ann Rev Biochem 63: 451-486.

Turgeon JL, McDonnell DP, Martin KA, Wise PM. 2004. Hormone therapy: Physiological complexity belies therapeutic simplicity. Science 304: 1269-1273.

Uno H, Tarara R, Else JG, Suleman MA, Sapolsky RM. 1989. Hippocampal damage associated with prolonged and fatal stress in primates. J Neurosci 9: 1705-1711.

van Steensel B, van Binnendijk EP, Hornsby CDE. 1996. Partial colocalization of glucocorticoid and mineralocorticoids receptors in discrete compartments in nuclei of rat hippocampus neurons. J Cell Sci 109: 787-792.

van Wimersma Greidanus TB, Rigter H. 1989. Hormonal regulation of learning performance. Psychoendocrinology, Brush, FB, Levine S. (editors), New York: Academic Press, Inc; pp. 271-304.

Voytko ML, Tinkler GP. 2004. Cognitive function and its neural mechanisms in nonhuman primate modesl of aging, Alzheimer disease and menopause Front Biosci 9: 1899-1914.

Wade CB, Robinson S, Shapiro RA, Dorsa DM. 2001. Estrogen receptor (ER) α and ERβ exhibit unique pharmacologic properties when coupled to activation of the mitogen-activated protein kinase pathway. Endocrinology 142: 2336-2342.

Warren SG, Juraska JM. 1997. Spatial and nonspatial learning across the rat estrous cycle. Behav Neurosci 111: 259-266.

Wassertheil-Smoller S, Hendrix SL, Limacher M, Heiss G, Kooperberg C, et al. 2003. Effects of estrogen plus progestin on stroke in postmenopausal women: the Women's Health Initiative: a randomized trial. JAMA 289: 2673-2684.

Watanabe Y, Gould E, McEwen BS. 1992. Stress induces atrophy of apical dendrites of hippocampal CA3 pyramidal neurons. Brain Res 588: 341-345.

Wellman CL. 2001. Dendritic reorganization in pyramidal neurons in medial prefrontal cortex after chronic corticosterone administration. J Neurobiol 49: 245-253.

White NM, McDonald RB. 2002. Theoretical Review: Multiple Parallel memory systems in the brain of the rat. Neurobiol Learn & Mem 77: 125-184.

Wickelgren, I. 1997. Research News-Estrogen stakes claim to cognition, Science 276: 675-678.

Williams CL, Barnett AM, Meck WH. 1990. Organizational effects of early gonadal secretions on sexual differentiation in spatial memory. Behav Neurosci 104: 84-97.

Wise PM, Dubal DB, Wilson ME, Rau SW, Bottner M. 2001. Minireview: neuroprotective effects of estrogen—New insights into mechanisms of action. Endocrinology 142: 969-973.

Wolf OT. 2003. HPA axis and memory. Best Pract Res Clin Endo And Metab 17: 287-299.

Wolf OT, Schommer NC, Hellhammer DH, McEwen BS. Kirschbaum C. 2001. The relationship between stress induced cortisol levels and memory differs between men and women. Psychoneuroendocrinology 26: 711-720.

Wood GE, Shors TJ. 1998. Stress facilitates classical conditioning in males, but impairs classical conditioning in females through activational effects of ovarian hormones. Proc Natl Acad Sci USA 95: 4066-4071.

Woolley CS, McEwen BS. 1992. Estradiol mediates fluctuation in hippocampal synapse density during the estrous cycle in the adult rat. J Neurosci 12: 2549-2554.

Yaffe K, Lui LY, Grady D, Cauley J, Kramer J, et al. 2000. Cognitive decline in women in relation to non-protein-bound oestradiol concentrations. Lancet 356: 708-712.

Yaffe K, Lui LY, Zmuda J, Cauley J. 2002. Sex hormones and cognitive functionin older men. J Am Geriatr Soc 50: 707-712.

Zec RF, Trivedi MA. 2002. Effects of hormone replacement therapy on cognitive aging and dementia risk in postmenopausal women: a review of ongoing large-scale, long-term clinical trials. Climacteric 5: 122-134.

Znamensky V, Akama KT, McEwen BS, Milner TA. 2003. Estrogen levels regulate the subcellular distribution of phosphorylated Akt in hippocampal CA1 dendrites. J Neurosci 23: 2340-2347.

22 The Mammalian Circadian System: from Genes to Behavior

B. H. Miller · E. L. McDearmon · J. S. Takahashi

List of Abbreviations: ANS, Autonomic nervous system; CT, circadian time; CRE, cyclic-AMP response element; CREB, CRE-binding protein; FSH, follicle stimulating hormone; GRP, gastrin-related peptide; GnRH, gonadotropin-releasing hormone; G-protein, guanine nucleotide binding protein; HPA, hypothalamic-pituitary-adrenal; HPG, hypothalamic-pituitary-gonadal; IGL, intergeniculate leaflet; LD, light:dark; LH, luteinizing hormone; MEF, mouse embryonic fibroblast; MUA, multiunit electrical activity; NPY, neuropeptide Y; PACAP, pituitary adenylate cyclase activating polypeptide; POA, preoptic area; RGC, retinal ganglion cell; RHT, retinohypothalamic tract; RORE, retinoic acid related orphan receptor response element; 5-HT, serotonin; SCN, suprachiasmatic nucleus; VIP, vasoactive intestinal peptide; AVP, vasopressin; TTX, tetrodotoxin; ZT, Zeitgeber time

1 Introduction

Virtually all organisms studied to date, from bacteria to humans, possess an internal clock that has an intrinsic period of approximately 24 h. This endogenous pacemaker drives circadian rhythms ("circa" = about, "dies" = a day) of myriad biological processes, from cell metabolism to behavioral state. Although the anatomical location and genetic components of the pacemaker differ among species, all circadian systems can be described by a three-component model that includes: (1) the pacemaker, a molecular clock that has a period of approximately 24 h and that continues to oscillate even in the prolonged absence of any timing cues, (2) an input pathway that conveys environmental time cues to the pacemaker, and (3) output pathways by which the pacemaker can influence biochemistry and behavior (Takahashi et al., 2001). This review focuses on mammalian circadian rhythms, and covers the anatomical and molecular basis for the master pacemaker, the neurochemical processes responsible for pacemaker entrainment, and the diverse systems involved in pacemaker output.

2 Historical Background

2.1 Discovery and Properties of the Circadian System

The rotation of the earth and the accompanying cycle of day and night produce a clear and dependable external time cue, known as a "zeitgeber" ("time giver"). Changes in light intensity are particularly important for photosynthetic organisms, and the ability to predict energy (light) availability is an advantageous adaptation that enables organisms to coordinate photosynthesis and nitrogen fixation to the appropriate time of day. Accordingly, circadian rhythms were first observed and studied in plants. It was long assumed that daily variations such as leaf movement and petal opening occurred in response to external stimuli, but in the early 1700s Jean Jacques d'Ortous de Mairan conducted an insightful experiment demonstrating the existence of an endogenous timekeeper. de Mairan placed heliotrope plants, which open and close their leaves in a circadian manner, in total darkness, and documented a persistence in the rhythm of leaf movement despite the absence of photic timing cues (de Mairan, 1729). Later, Carolus Linnaeus, the creator of the modern taxonomical system, took advantage of the circadian regulation of petal opening to create a "circadian garden," with each hour represented by a flower that opened at a particular time of day (von Marilaun, 1895). Since that time, endogenous circadian rhythms have been identified in a wide variety of organisms, allowing for the prediction and testing of hypotheses related to the properties of circadian systems.

Primary to the function of endogenous circadian rhythms is the ability to be entrained, or synchronized to the phase and periodicity of the external world (Pittendrigh, 1981). In fact, because endogenous rhythms have a period of nearly, but not exactly, 24 hours, a periodic external signal (zeitgeber) is required to adjust the period and phase of the internal rhythm to that of the environment. For entrainment to occur, the endogenous system must be able to adopt and maintain a stable phase angle relationship to the zeitgeber; if the zeitgeber shifts, the endogenous system must also shift until it reaches the same phase relationship with the external signal as before (for an in-depth review, see Daan and Aschoff, 2001). A stable phase relationship occurs when the period of the endogenous oscillator closely matches the period of the environment, and includes a mechanism for phase adjustment. The maintenance of a stable phase relationship allows for predictive homeostasis: the endogenous system gains the ability to anticipate changes in the environment and to adjust behavior and physiology accordingly. An important property of entrainment is that the response of the internal system to the zeitgeber depends on the phase at which the zeitgeber occurs. This effect, which is represented as a phase response curve, can be defined by administering brief entraining stimuli at different phases under otherwise constant conditions and measuring the resulting phase changes (Daan and Aschoff, 2001). In nocturnal rodents, for example, a brief light pulse administered during the early subjective night results in a phase delay in activity rhythms, whereas a light pulse administered during the subjective night results in a phase advance (Pittendrigh, 1981). For the majority of organisms, the dominant zeitgeber is the light–dark (LD) cycle; however, in the absence of a LD cycle, other environmental cues, such as cycles in temperature, feeding, or social activity, can entrain endogenous rhythms (Gwinner, 1966; Engelmann et al., 1974; Edgar and Dement, 1991). Even in the presence of a LD cycle, nonphotic cues may help to stabilize entrainment to the environment (Amir and Stewart, 1996).

The methods used to study mammalian circadian rhythms have remained essentially unchanged since rigorous study began in the early 1950s. Model organisms—generally rats, hamsters, or mice—are placed within an environment where the LD conditions can be regulated, and the properties of circadian rhythms are monitored by observing one or more circadian-regulated activities, such as wheel running, drinking, or feeding. These activities represent the output of the internal pacemaker, and, as such, can be used to measure the properties of the endogenous clock. In some cases, animals are placed under constant conditions (constant dark, DD, or constant light, LL) to observe rhythms in the absence of entraining cues; the length of the cycle of rhythms observed under DD reflects the "true," or steady state, period (τ) of the internal pacemaker, and animals under constant conditions are said to be "free running." In the majority of experiments, animals are maintained on a cycle of 12 h light, 12 h dark (LD 12:12); where, by convention, lights-on is known as ZT0 and lights-off as ZT12. Under constant conditions, circadian time (CT) 0 to CT12 is the subjective day, and CT12-24 is the subjective night.

In the 1920s, Richter documented endogenous circadian rhythms in the rat by measuring locomotor and drinking activity, and showed that activity rhythms could be entrained either by light or by feeding (Richter, 1922, 1967). Richter found that rats, which are nocturnal, restrict the majority of both drinking and locomotor activities to the actual or subjective night. Johnson documented similar circadian, environmentally entrained activity in mice (Johnson, 1926). Later, Aschoff and Wever extended Richter's findings to humans, demonstrating the presence of endogenous circadian rhythms in humans by isolating subjects in an underground bunker and observing the persistence of circadianly organized activity in the absence of any entraining signals (Aschoff, 1965; Wever, 1979).

2.2 Identification of the Suprachiasmatic Nucleus

Although multiple examples of mammalian circadian rhythms—including the sleep–wake cycle, locomotor activity, and endocrine rhythms—had been identified in the mid-1900s, little was known about how such rhythms were controlled. Rhythms were remarkably persistent, continuing after such insults as anesthesia and seizures, and were not perturbed by the removal of candidate pacemakers such as the adrenal and pineal glands. The sheer number of physiological systems showing circadian rhythmicity suggested that either there was a single master pacemaker that drove and coordinated slave oscillators throughout the body, or that peripheral systems contained individual clocks that were phase-coordinated by cross talk between systems (Pittendrigh, 1960). Then, in the 1970s, anatomical tracing of the retinohypothalamic tract (RHT), the source of photic input to the circadian system, led two separate groups of researchers to identify a hypothalamic area known as the suprachiasmatic nuclei (SCN) as the region responsible for all measurable rhythms, including activity, sleep, and drinking (Moore and Eichler, 1972; Stephan and Zucker, 1972; Menaker et al., 1978). Stephan and Zucker electrically lesioned one or both SCN nuclei and found that bilateral SCN lesions resulted in arrhythmicity, whereas unilateral SCN lesions resulted only in a reduction in the amplitude of circadian drinking rhythms (Stephan and Zucker, 1972). Working independently of Stephan and Zucker, Moore and Eichler found that bilateral SCN lesions resulted in the loss of circadian corticosterone rhythms in rats (Moore and Eichler, 1972).

The SCN quickly became the focus of intense interest among circadian biologists, and further evidence of the role of the SCN as master pacemaker was soon reported. Inouye and Kawamura demonstrated that the SCN pacemaker functioned autonomously by using knife-cuts to isolate SCN tissue *in vivo* and subsequently measure multiunit electrical activity (MUA) within the SCN and other brains regions (Inouye and Kawamura, 1979). In SCN-intact animals, the amplitude of electrical activity in the brain had been shown to have a clear circadian rhythm, with firing rates peaking during the day in the SCN and during the night in other brain tissues. In SCN-isolated animals, electrical activity continued in a rhythmic manner in the SCN but became arrhythmic in non-SCN areas, indicating that the SCN contains a self-sustained pacemaker, and that neural outputs from the SCN are responsible for driving the circadian organization of electrical activity in other parts of the brain. These results supported those from experiments conducted by Schwartz and Gainer, who showed that the uptake of 14-C-deoxyglucose is rhythmic in the SCN, with peak uptake occurring during the day, but that uptake in other brains regions is nonrhythmic (Schwartz and Gainer, 1977).

2.3 Use of Mutant Animal Models to Study the Circadian System

The identification of a region as a pacemaker requires not only that the putative pacemaker shows autonomous rhythms, but that it can also drive and sustain rhythms throughout the body. In a series of papers using a circadian mutant hamster, Ralph and Menaker showed both that the SCN was the master pacemaker and that the period length of circadian rhythms in mammals was genetically determined. During the observation of circadian behavior in individual animals from a particular shipment of hamsters, Ralph and Menaker identified a single hamster that had a shortened period of free-running wheel-running activity: the normal period for hamsters in DD is approximately 24.1 h, but the free-running period of the mutant hamster was only 22 h (Ralph and Menaker, 1988). Subsequent breeding experiments showed that this mutation, which was named "*tau*," was autosomal and acted in a semidominant fashion, such that the homozygous offspring of the heterozygous founder had a free-running period of 20 h. Ralph and Menaker then showed that they could make a genetic *tau* mutant run with a normal 24-h period by transplanting the SCN from a wild-type embryo into the third ventricle of an SCN-lesioned mutant (Ralph et al., 1990). The converse was also true: a genetically wild-type animal could be made to show the circadian phenotype of a *tau* mutant if the wild-type animal was SCN-lesioned and then implanted with the SCN from a mutant hamster.

The genetic control of the circadian timing signal thus established, the hunt was on for genes involved in the mammalian pacemaker. Although advances in the identification of genes regulating circadian rhythms had been made in nonmammalian models (Sargent and Woodward, 1969; Feldman and Hoyle, 1973), the first breakthrough in mammals did not occur until the mid-1990s, when Vitaterna and colleagues identified a mouse with abnormal circadian activity during a screen for ENU-induced circadian mutants (Vitaterna et al., 1994). The founder mouse had a free-running period almost an hour longer than normal, and, like the *tau* mutant, the responsible mutation turned out to be semidominant: mice homozygous for the subsequently identified *Clock* gene mutation had a free-running period of almost 4 h longer than normal, and often became arrhythmic after about 2 weeks in constant darkness (Vitaterna et al., 1994) (➲ *Figure 22-1*). The identification of the gene responsible for the *Clock* mutant phenotype was crucial, as it provided an entrée into the genetic underpinnings of the mammalian pacemaker (Antoch et al., 1997; King et al., 1997). Other core circadian genes were subsequently identified, and it was ultimately shown that neurons in the SCN of all mammalian species studied thus far contain a cell-autonomous molecular feedback loop composed of positive and negative elements that cycle with a rhythm of approximately 24 h. The molecular mechanism of circadian clocks is discussed in detail later in this chapter. Briefly, CLOCK and its binding partner BMAL1 form the positive limb of the loop by driving the transcription of a number of genes, including *Period1*, *Period2*, *Cryptochrome1*, and *Cryptochrome2*, which form the negative limb of the loop by

Figure 22-1
Circadian Wheel-Running Activity in Wild-Type and Clock Mutant Mice. Activity records of wheel-running behavior from wildtype, *Clock/+*, and *Clock/Clock* mice in 12 h lights:12 h dark (LD) and constant darkness (DD). Records are double-plotted. Courtesy of K. Shimomura

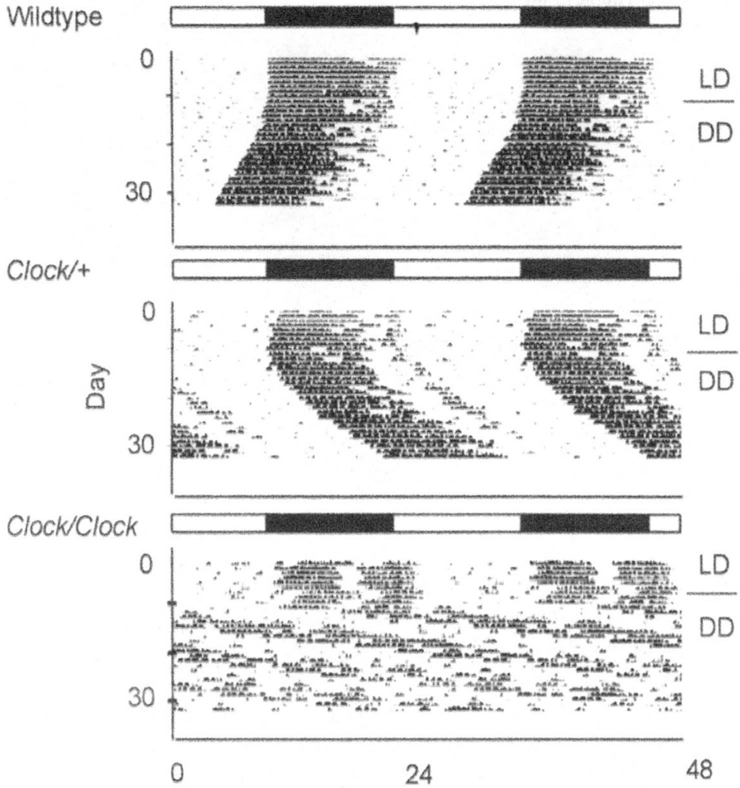

inhibiting their own transcription. Turnover of the negative limb proteins allows CLOCK:BMAL1-induced transcription to begin anew (Lowrey and Takahashi, 2000). The biochemical components of the clock have since been shown to be rhythmically expressed in many other tissues, including liver and pituitary, but the SCN remain the only tissue capable of phase-coordinating the circadian activity of oscillators throughout the body (Brown and Schibler, 1999; King and Takahashi, 2000; Lowrey and Takahashi, 2004; Yoo et al., 2004).

3 Structure and Function of the SCN

3.1 Rhythmic Properties of the SCN

The SCN are bilateral nuclei, each made up of approximately 10,000 neurons, that reside in the anterior hypothalamus immediately superior to the optic chiasm and lateral to the third ventricle (➋ *Figure 22-2*). As described earlier, previous studies showed that neurons in the SCN exhibit rhythmic patterns in glucose utilization and electrical activity, with both measures peaking during the day. These electrical and metabolic rhythms persist under constant conditions, and, *in vivo*, are phased similarly in both diurnal and nocturnal

Figure 22-2

Anatomical Location of the Suprachiasmatic Nuclei. The SCN are bilateral nuclei located at the base of the anterior hypothalamus, superior to the optic chiasm and lateral to the third ventricle. This coronal section from an adult mouse brain has been hybridized with an oligonucleotide probe for *Period2*, a core clock gene that is highly expressed and rhythmic in the SCN

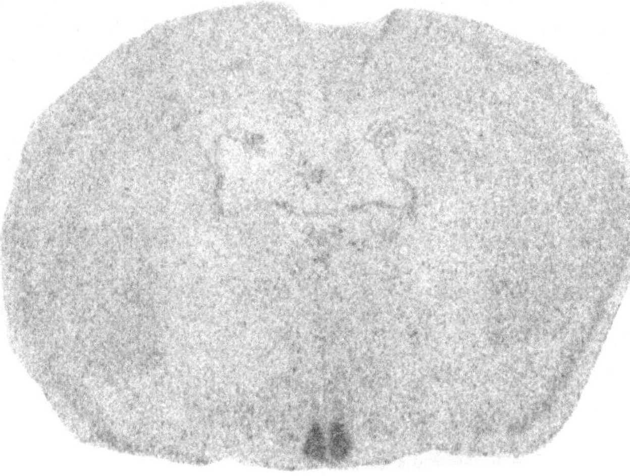

animals. Some of the first studies of electrical activity measured output from multiple neurons; more recently, several groups have recorded circadian rhythms in electrical activity, calcium concentration, and membrane conductance in isolated SCN neurons, indicating that these rhythms are an intrinsic property of individual cells and are not dependent on functional interneuronal connections (Welsh et al., 1995; Gillette et al., 1995b).

MUA measurements from SCN slices have shown that the rhythm of spontaneous neuronal activity has a periodicity of approximately 24 h (Bos and Mirmiran, 1990). Even when SCN slices are cultured for many weeks, rhythmic spontaneous activity is stably maintained, despite the absence of humoral and endogenous time cues normally present *in vivo*. Application of agents such as tetrodotoxin (TTX), a sodium channel blocker, inhibits neuronal activity and, if administered *in vivo*, rhythmic locomotor activity, but spontaneous firing reappears after washout with the phase unchanged (Schwartz et al., 1987). This indicates that neuronal activity is not required for continued function of the pacemaker within individual cells, although more recent studies have shown that disrupting electrical activity does reduce the amplitude of oscillation (Yamaguchi et al., 2003). *In vivo* studies using electrodes implanted in the brains of freely moving hamsters support the *in vitro* findings, showing that MUA in the SCN peaks during the day, and, moreover, has a period that matches the period of free-running locomotor activity. Yamazaki and colleagues measured MUA from wild-type and *tau* mutant hamsters, and found that, though the period of MUA was 24 h in wild-types, the period of MUA in *tau* mutants was only about 20 h, similar to the period of activity measured by wheel running (Yamazaki et al., 1998).

Other measures of neuronal properties also show circadian rhythmicity in SCN slices. In addition to glucose metabolism and electrical activity, intracellular calcium levels peak during the day, although calcium rhythms occur only in the cytoplasm and not in the nucleus. Using optical imaging of the Ca^{+2}-sensitive dye Fura2-AM in acute slices, Colwell determined that calcium rhythms are present in SCN sections removed from animals maintained in DD, suggesting that changes in rhythmic calcium concentration are regulated by the internal pacemaker, rather than dependent on external cues (Colwell, 2000a). More recently, Ikeda and colleagues transfected organotypic SCN cultures with a calcium-sensitive fluorescent protein and determined that the length of the Ca^{+2} period matches that of the MUA, but is phase-advanced in comparison to the MUA (Ikeda et al., 2003). Unlike rhythmic MUA, which can be blocked by TTX, Ca^{+2}

fluctuations continue in the absence of electrical activity. Thus, Ca^{+2} rhythms are driven by the intracellular clock, rather than by neuronal activity (Shibata and Moore, 1993; Ikeda et al., 2003).

3.2 SCN Neurons as Cell-Autonomous Pacemakers

A major advance occurred when individual SCN neurons were found to have many of the same rhythmic properties as SCN slices. Studies of spontaneous activity, membrane resistance, and, more recently, circadian gene expression in individual neurons indicate that between 75% and 99% of SCN neurons possess oscillatory characteristics (de Jeu et al., 1988; Yamaguchi et al., 2003). In 1995, Welsh and colleagues measured rhythmic spontaneous activity from individual SCN neurons cultured on fixed microelectrode arrays and found that single cell rhythmicity can be stably maintained for weeks (Welsh et al., 1995). Furthermore, as had been shown in whole slices, application of TTX for up to 2.5 days inhibited neuronal activity, but did not stop each neuron's internal clock. Surprisingly, the activity of individual neurons within the same culture was not necessarily in phase, a finding that has since been replicated many times (Gillette et al., 1995a). Both acute and long-term measurements of individual SCN neurons have shown that there is a wide distribution of phase and period between different neurons from the same SCN. In measurements from dissociated cells in culture, the intrinsic periods among individual neurons may differ by up to 8 h, whereas measurements from cells in organotypic SCN cultures show a period range of up to 5 h suggesting that intact neuronal connections, though not necessary for the generation of rhythmicity, do contribute to phase-coupling (Herzog et al., 1997, 2004; Honma et al., 1998). Despite the range of periods within single cells, the mean period for the entire SCN is equivalent to the species-specific period in locomotor activity (Liu et al., 1997; Herzog et al., 1998).

To understand how multiple oscillators with different periods can produce a behavior with a single stable ~24-h period, circadian biologists have used theoretical models of oscillators. In these models, each neuron represents a single oscillator with an individual period and phase, and synchronous output is achieved through chemical and/or electrical coupling among neurons. In support of this hypothesis, several groups have shown that the period of locomotor activity represents the mean period of all neurons sampled from a single SCN (Liu et al., 1997; Yamazaki et al., 1998). For example, in the *tau* mutant hamster, the range of periods expressed by individual neurons depends on genotype. Liu and colleagues measured the period of electrical activity from individual SCN neurons cultured from wild-type hamsters and from hamsters homozygous for the *tau* mutation and found that periods in wild-type neurons ranged from 20 to 25 h with a mean of ~24 h, whereas periods in *tau* mutant neurons ranged from 16.5 to 21.5 h with a mean of ~20 hours (Liu et al., 1997). Thus, the mean neuronal activity period length for each genotype is equal to the period of free-running locomotor activity, as wild-type hamsters have a period of 24 h and *tau* mutant homozygotes have a wheel-running period of approximately 20 h. Yamazaki and colleagues later confirmed these findings by measuring the period of electrical activity in SCNs of freely moving wild-type and *tau* mutant hamsters (Yamazaki et al., 1998).

The production of mice with targeted knockout or induced mutations of specific core clock genes has provided approaches to investigate the underpinnings of rhythmic neuronal activity. Using chimeric mice made of cells from wild-type and *Clock/Clock* mice, Low-Zeddies and Takahashi showed a direct correlation between the period of circadian wheel-running activity and the percent of SCN cells carrying the *Clock* mutation: the greater the percent of mutant cells in the SCN, the longer the behavioral period (Low-Zeddies and Takahashi, 2001). Herzog and colleagues, also using *Clock* mutant mice, found that neurons isolated from heterozygous mice, which have a free-running activity period of approximately 24.7 h, show a lengthened period of spontaneous activity, whereas neurons from homozygous mutants, which become arrhythmic under constant conditions, have arrhythmic spontaneous activity (Herzog et al., 1998). Similarly, the deletion of the *Cryptochrome 1* and *Cryptochrome 2* genes, part of the negative limb of the core pacemaker, results in arrhythmicity of both locomotor and neuronal activity under constant conditions (Albus et al., 2002).

A relatively new method of studying the characteristics of individual oscillatory cells within the SCN has arisen from the development of transgenic mouse and rat strains that produce a molecular reporter, such as luciferase, under the control of rhythmic promoters from the *Per1* and *Per2* genes (Yamazaki et al., 2000; Yamaguchi et al., 2003; Yoo et al., 2004). Studies measuring rhythms in luciferase luminescence under the control of the *Per1* promoter have confirmed that between 75% and 100% of SCN cells have intrinsic oscillators, and that these oscillators exhibit a range of periods and peak times (Quintero et al., 2003; Yamaguchi et al., 2003). In the transgenic models, mean bioluminescence peaks during the day, similar to the peak in electrical activity, and the mean period for individual neurons is the same length as the period of locomotor activity. Recently, one group used neurons containing the *Per1*-luciferase transgene to confirm and extend previous electrophysiological findings regarding the lack of an effect of inhibition of electrical activity on the endogenous clock. Yamaguchi and colleagues, using bioluminescence imaging of SCN slice cultures, found that the application of TTX does not prevent rhythmic luciferase expression in individual neurons (Yamaguchi et al., 2003). However, TTX application does damp mean circadian luciferase expression across an entire slice. By analyzing luciferase expression at the single-cell level, it was found that that TTX-induced damping resulted from a combination of the desynchronization of individual neurons and a reduction, but not elimination, of luciferase amplitude within individual cells. This indicates that, while neuronal activity is not mandatory for intrinsic expression of the molecular pacemaker, it is important for the synchronization of individual cells. These results also suggest that TTX-sensitive sodium channels are somehow important to the coupling of neuronal activity that is required to generate a stable, coherent output from the SCN. Gap junctions have also been implicated in neuronal coupling in the SCN, but the molecular mechanism underlying activity coupling remains relatively unknown (Colwell, 2000b).

3.3 Shell versus Core: the Division of Labor Within the SCN

Although rhythmic neurons are distributed throughout the SCN, functional and anatomical studies suggest that the nuclei contain two major divisions known as the shell and the core (Leak and Moore, 2001). Although there are some species-specific differences, the shell and core are organized in a similar manner across mammals (Moore et al., 2002). The core is immediately superior to the optic chiasm, and it receives the majority of afferents from the visual system. Studies have shown direct connections between retinal ganglion cells (RGC) and neurons within the core; this pathway is known as the RHT, and is the primary pathway for phase-resetting photic stimuli (Moore and Lenn, 1972) (*Figure 22-3*). The core also receives photic input via an indirect pathway involving the intergeniculate leaflet (IGL) (Moore and Card, 1994).

Figure 22-3

Major Efferent and Afferent SCN projections. Model depicts a sagittal cross section of mouse brain. Light is detected by the retinal neurons at the back of the eyes, which communicate to the suprachiasmatic nuclei (SCN) directly via the retinal hypothalamic tract (RHT) and indirectly via the intergeniculate leaflet (IGL). As the master coordinator of circadian timing, the SCN communicates via neuronal projections to important regulatory regions of the brain such as the mPOA, the dorsomedial hypothalamus (DMH), the paraventricular nucleus (PVN) and the paraventricular nucleus of the thalamus (PVT). The SCN may also communicate with the subparaventricular zone (sPVz) via secreted factors (dashed line)

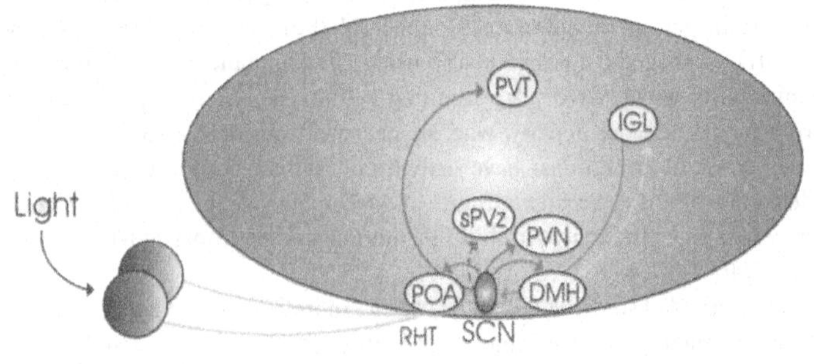

In addition to receiving RHT input, a defining characteristic of the core is its high level of vasoactive intestinal peptide (VIP) expression: 26% of SCN neurons express VIP, and these neurons reside almost exclusively in the core. Core neurons also preferentially express neurotensin and gastrin-related peptide (GRP) (Moore et al., 2002). Neurons within the core densely innervate each other and the neurons in the shell, but account for only a minority of projections from the SCN to other brain regions (Watts et al., 1987). Recent studies using several different retrograde tracers indicate that the only regions innervated by the core are the subparaventricular zone of the hypothalamus, thalamic nuclei, and the peri-suprachiasmatic area immediately adjacent to the SCN (Leak and Moore, 2001).

An abundance of evidence suggests that the core is responsible for responding to and integrating photic cues. Electron microscopic studies have shown that RHT terminals make direct contact with the VIP- and GRP-expressing core neurons (Ibata et al., 1989; Tanaka et al., 1997), and that these same neurons respond to light pulses by upregulating expression of the immediate early genes (IEGs) *c-fos*, *egr3*, and *junB* (Romijn et al., 1996; Castel et al., 1997; Beaule and Amir, 1999). Furthermore, light-induced expression of the clock gene *Per1* occurs mainly within core neurons, particularly those expressing GRP and/or VIP (Yan et al., 1999; Kulhman et al., 2003). In hamsters, light-induced *Per1* and *Per2* expression is further restricted to a small population of calbindin-positive neurons within the core that only express the *Per* genes after light exposure at night (Silver et al., 1996b).

The shell, which, in most mammals, fully encapsulates the core, receives the majority of its inputs from the core. Shell neurons make few projections back to the core, but are responsible for the majority of SCN efferents (Swanson and Cowan, 1975). Retrograde tracing studies have shown that shell neurons make monosynaptic connections to the medial preoptic area of the hypothalamus (mPOA), the dorsomedial (DMH) and ventromedial (VMH) hypothalamus, the paraventricular thalamic nuclei (PVT), the lateral septum, and the bed nucleus of stria terminalis (BNST), among others (Leak and Moore, 2001). Multi-synaptic efferents connect shell neurons to endocrine organs such as the pituitary and pineal glands, and to the autonomic nervous system (ANS). Shell neurons are characterized by high expression of vasopressin (AVP) and calretinin (Moore et al., 2002). AVP is of particular interest, as its expression in the SCN is directly controlled by core clock genes and RNA levels are rhythmic even under constant conditions (Lee et al., 2001). Thus, AVP is a likely candidate for relaying timing signals from the SCN to other brain regions (Kalsbeek and Buijs, 2002).

The pattern of neural connections into, out of, and within the SCN suggests a relatively simple subdivision of labor in which the core receives photic input—the major phase resetting environmental cue—and integrates this input via activation of IEGs and their downstream genes, and regulation of *Per1* and *Per2* expressions and protein turnover (Ginty et al., 1993; Okamura et al., 1999; Yamazaki et al., 2000; Kornhauser et al., 2002). The core then conveys phase-resetting information to the shell, which is responsible for the majority of SCN efferents. SCN output, both humoral and neuronal, phase-coordinates circadian activity throughout the brain and periphery.

4 Photic and Nonphotic Input to the SCN

4.1 Overlapping Roles of Rods, Cones, and Retinal Ganglion Cells

In mammals, entrainment to an environmental LD cycle sets the phase of behavioral activity and rest cycles, fluctuations in body temperature and hormone levels (e.g., melatonin), and circadian gene expression patterns. The entrainment of mammalian circadian rhythms has been shown to require photoreceptive cells within the eye, as enucleation (removal of the eyes) prevents entrainment to a LD cycle as well as attenuates the response to light in constant darkness (Nelson and Zucker, 1981; Foster et al., 1991). The retina acts as the light-sensitive region of the eye and is composed of multiple neuronal cell layers that capture light and transmit image-forming and nonimage-forming signals to other areas of the brain. The photoreceptor cells are primarily responsible for mammalian visual perception and are organized into two major classes: rods, which are responsible for vision in dim light, and cones, which mediate color vision. Due to differential expression of the opsin photopigments, rod and cone cells are receptive to distinct wavelengths of light.

In rod cells, rhodopsin predominates and is maximally stimulated by light of approximately 500 nm wavelength. Cone cells are classified based on their distinct action spectra peaks. Humans have three distinct classes of cones, UV or short wave (S), middle wave (M) and long wave (LW), whereas rodents have only M- and S-type cones (Neitz and Neitz, 2001). Wavelengths of 360 nm (blue/UV) and 510 nm (green) light maximally stimulate the S and M cone opsins, respectively (Jacobs, 1993; Foster, 1998). Studies in rodents have shown that phase shifting of wheel-running activity is most sensitive to light near 500 nm wavelength, so it is plausible that classic photoreceptor cells may also communicate light input for circadian entrainment purposes (Takahashi et al., 1984; Provencio and Foster, 1995).

Spontaneous mutants and genetically altered rodent models have proven to be useful in determining how distinct retinal cells are involved in circadian photic entrainment. Mice that are homozygous for the *rd/rd* mutation, which results in gradual rod and cone degeneration, retain the ability to entrain to an LD cycle and to respond to night-time light with melatonin suppression (Foster et al., 1991; Provencio et al., 1994; Lucas et al., 1999). Another transgenic mouse line, *rdta*, which displays complete loss of rods and most cones by 28 days of age, is also able to entrain to an LD cycle (McCall et al., 1996; Lupi et al., 1999). However, these mice exhibit some circadian abnormalities, such as a slightly shorter free-running period in DD and an increased sensitivity to phase-shifting light pulses, suggesting that rod/cone input may play a role during the development of circadian rhythms (McCall et al., 1996). Because some cone cells remain in the *rd/rd* and *rdta* mutants, it was long thought that these photoreceptors were responsible for photic input to the SCN. However, in 1999, two groups working independently crossed the *rd/rd* or *rdta* mutants to a strain of mice lacking cones (*cl*) to produce mice that completely lacked rods and cones (Freedman et al., 1999; Lucas et al., 1999). The offspring retained the ability to entrain to an LD cycle and phase shift both melatonin and wheel-running activity in response to a light pulse, indicating that neither rods nor cones are necessary for photic regulation of circadian rhythms.

Recently, another class of retinal neurons known as RGCs has been identified as a mediator of photic stimulation and communication to the SCN. Berson and colleagues found that certain RGCs are intrinsically light-sensitive: neither blocking neuron–neuron communication from rods or cones, nor isolating individual rat RGCs prevented the cells from firing in response to light stimulation (Berson et al., 2002). In addition, RGCs are not as sensitive to light levels as rods and cones: they are slow to react to increased light intensity, but then respond continuously without adapting over a relatively long period of time (seconds to minutes). These results correlate with the predicted characteristics of the mammalian circadian photoreceptive system based on the findings that hamsters required a relatively high threshold of light intensity and long duration of light exposure to maximally phase-shift wheel-running activity (Takahashi et al., 1984). As further evidence for the role of RGCs in circadian entrainment, Wee and colleagues characterized circadian activity in a mouse line rendered deficient in the majority of RGCs by the disruption of *Math5*, a gene necessary for RGC differentiation. $Math5^{-/-}$ mice do not have an observable RHT and are unable to entrain to a LD cycle (Brown et al., 2001; Wang et al., 2001; Wee et al., 2002; Brzezinski et al., 2005). These studies provide strong evidence that RGCs are the retinal cells primarily responsible for mediating photic input to the circadian system.

4.2 Biochemical Effectors of Photic Input: Opsins and Cryptochromes

4.2.1 Opsins

The molecular basis for mammalian light detection has been extensively studied in rodent rods and cones. Mammalian photoreceptor opsins are transmembrane receptors that contain a lysine residue in the third transmembrane domain, which binds the light-sensitive chromophore, 11-cis-retinal (vitamin A), through a Schiff-base linkage. Once exposed to light, the chromophore is converted to all-trans-retinal, which induces a conformational change in the opsin protein. Intracellularly, photoreceptor opsins are coupled to guanine nucleotide-binding proteins (G-proteins) by their third cytoplasmic loop. The light-activated conformational change in the opsin activates the bound G-protein and subsequent signaling pathways required for visual perception. Rod cells, which represent 97% of mouse photoreceptors (Young, 1985), are dependent on the retinal pigment epithelium (RPE) layer to isomerize the chromophore back to 11-cis-retinal. The chromophore is then transported back to the photoreceptor cells where it reassociates with the photo-bleached opsins and awaits reactivation by light.

In 1998, Provencio and colleagues discovered a novel opsin in the photosensitive dermal melanophores of *Xenopus laevis*, which they named melanopsin (Provencio et al., 1998). Two years later, they described the identification of melanopsin in human retina, and localized the mRNA to the amacrine and ganglion cell layers in both mouse and monkey retina (Provencio et al., 2000). Phylogenetic comparisons revealed that both amphibian and mammalian melanopsin were more similar to invertebrate opsins than to vertebrate isoforms at the amino acid level: mammalian opsins contain a lysine residue in the third transmembrane domain responsible for chromophore binding, but melanopsin sequences contain an aromatic residue substitution (tyrosine). Therefore, melanopsin, like invertebrate opsins, may be able to retain the covalently linked chromophore, allowing for repeated interconversion between the two states and eliminating the dependency on the RPE for chromophore supply. Recent findings have shown that exogenously expressed melanopsin does indeed exhibit both inherent photosensitivity and photoisomerase activity (Melyan et al., 2005; Panda et al., 2005). These features make melanopsin a likely candidate for a circadian photoreceptor, as melanopsin-containing cells can autonomously function in widely dispersed regions.

In 2002, a series of elegant studies were published characterizing both melanopsin and the RGCs that express the photopigment. Provencio and colleagues (2002) identified what they termed an 'expansive photoreceptive net' in the mouse inner retina by immunolabeling flat mount sections with antiserum against melanopsin. The melanopsin-labeled cells were present in the RGC layer, and dendrites from the cells extended to different regions of the retinal inner plexiform layer, creating an extensive network capable of receiving photic stimulation. Gooley and colleagues (2001) used retrograde tracing to show that, in rats, the majority of melanopsin-containing RGCs project to the SCN through the RHT, and the majority of RGCs that project to the SCN contain melanopsin. These studies indicate that melanopsin-containing RGCs have a wide dendritic network for efficient photic capture, can fire in response to sustained increases in light, and project to the SCN—necessary characteristics for mediators of circadian entrainment. However, subsequent knockout experiments showed that mice with a null mutation of the melanopsin gene (*Opn4*) had a subtle defect in light-induced phase shifts, but were capable of entraining to an LD cycle and had a normal free-running activity period (Panda et al., 2002b; Ruby et al., 2002).

The results obtained with *Opn4* knockout mice indicated that compensation by the rods, cones, and/or another photoreceptor could occur in the absence of melanopsin. Hattar and colleagues (2003) definitively showed that RGCs, rods, and cones are the only cells required for normal circadian entrainment. The authors first examined the action spectrum for phase-shifting circadian wheel-running activity in rodless, coneless mice (*rd/rd cl*), and observed a maximal effect of light of approximately 481 nm wavelength, which corresponds to the wavelength required for maximal stimulation of endogenous melanopsin-expressing RGCs in the rat, as well as exogenous melanopsin-expressing *Xenopus* oocytes (Berson et al., 2002; Panda et al., 2005). Interestingly, mice that are rendered functionally deficient in all three major photoreceptor pathways ($Gnat1^{-/-}$ $Cnga3^{-/-}$ $Opn4^{-/-}$) could not entrain to an LD cycle, although they exhibited a normal free-running period. Thus, circadian entrainment is mediated by the three major photoreceptive systems in the mammal: rods, cones, and melanopsin-containing RGCs. Recent findings have begun to elucidate the intracellular biochemical pathways through which melanopsin may function, as Panda and colleagues (2005) have shown that the $G\alpha_q/G\alpha_i$ family of G-proteins and the transient receptor potential (TRP) subfamily C (TRPC) channel are involved in light-dependent activation of melanopsin-expressing *Xenopus* oocyte membrane currents.

4.2.2 Cryptochromes

In plants, blue-light photoreceptors known as cryptochromes (*Crys*) transduce sunlight, which is required for a variety of functions such as growth and reproduction. Cryptochromes are distinct from opsin photopigments in that they do not use vitamin A as a chromophore. Instead, vitamin-B based FAD and folate are the bound cofactors required for light sensitivity (Kavakli and Sancar, 2002). The discovery of mammalian *Cry* homologs *Cry1* and *Cry2* and their expression in retinal cell layers, including the ganglion cells, made these proteins potential candidates for the circadian photoreceptors (Miyamoto and Sancar, 1998). In addition, a study from Sancar's group showed that vitamin-A dependent opsins may not be required for circadian entrainment, as mice that were fed a vitamin-A-deficient diet and genetically lacked the serum retinol transport protein could still respond to light by upregulating *Per2* in the SCN (Thompson et al., 2001). However, mammalian *Crys* exhibit maximal light absorption at 420 nm wavelength, shorter than that required for maximal circadian response (~500 nm) (Hsu et al., 1996). Furthermore, mice that carried null mutations of both *Cry* genes (*Cry1*$^{-/-}$, *Cry2*$^{-/-}$) respond to an LD cycle (van der Horst et al., 1999; Vitaterna et al., 1999). However, Selby and colleagues crossed *Cry* double knockout mice with *rd/rd* mice and found that triple mutant mice (*Cry1*$^{-/-}$, *Cry2*$^{-/-}$, *rd/rd*) were unable to entrain to a LD cycle, suggesting that the cryptochromes are also involved in circadian entrainment (Selby et al., 2000). As both *Cry* genes and melanopsin are expressed in the RGC layer, it is possible that they are both involved in stimulation of these neurons and may act in concert. The ability of these retinal cells to compensate for the loss of one or even two of the retinal photoreceptive systems indicates the crucial role of environmental input to the circadian system.

4.3 Neurotransmitters and Intracellular Pathways

As previously described, RGCs project to neurons in the core region of the SCN through the RHT. Upon light stimulation, RGCs fire and release neurotransmitters—primarily glutamate and pituitary adenylate cyclase activating polypeptide (PACAP)—that can act on the SCN. Both glutamate and PACAP have been observed in RHT neurons that terminate in the SCN, and receptors for both these neurotransmitters (NMDA and non-NMDA type glutamate receptors, PAC1, and VPAC2) are expressed in the SCN (Cagampang et al., 1998a, b; Hannibal, 2002). Studies examining the role that these neurotransmitters play in transmission of photic signal have relied primarily on the phase-shifting effects of light pulses on animals housed in DD. Pulses of light given to nocturnal rodents during the early subjective night, or active phase, result in a delay in the circadian phase, whereas light pulses given during late subjective night result in phase advances (Daan, 1977). Light administered during the subjective day usually does not alter phase. Phase shifting can be detected by changes in the time of wheel-running activity onset, the electrical firing rate of explanted SCN tissue, and/or expression of the circadian clock genes *Per1* and *Per2* in the SCN.

In vitro, application of glutamate or NMDA to SCN slice preparations induces a phase shift in the electrical firing rate of SCN neurons (Ding et al., 1994; Shibata et al., 1994). *In vivo*, light pulses or injection of NMDA near the SCN produce similar phase shifts in wheel-running behavior, whereas injection of glutamate receptor antagonists can block the phase-shifting effects of light on wheel-running behavior (Colwell et al., 1990, 1991; Colwell and Menaker, 1992; Mintz et al., 1999). Evidence also points to a role for PACAP in light-induced phase shifts, as PACAP immunoreactivity in the SCN is almost completely lost following enucleation in hamsters (Hannibal et al., 1998). Microinjection of PACAP into the SCN produces phase shifts in both wheel-running activity and neuronal firing rates, and PACAP antagonists attenuate light-induced phase shifts (Harrington et al., 1999; Piggins et al., 2001; Minami et al., 2002; Bergstrom et al., 2003). Recently, Colwell and colleagues found that PACAP-deficient mice show attenuated phase shifts in wheel-running activity in response to light (Colwell et al., 2004). Therefore, both glutamate and PACAP appear to communicate the photic signal directly to the SCN neurons.

PACAP and glutamate may also modulate one another's ability to evoke a postsynaptic response. Glutamate normally induces a phase advance in SCN slice firing rate when administered at late subjective night, and a phase delay when administered at early subjective night. However, Chen and colleagues showed that treatment of SCN slices with PACAP blocked the normal glutamate-induced phase advance at CT19, but potentiated the glutamate-induced phase delay at CT14 (Chen et al., 1999). Defining the light-induced regulation and interactions of the two major neurotransmitters of the RHT may help explain why light has differential phase-shifting properties at distinct points throughout the circadian cycle. The characterization of this modulation, especially with respect to time of day, will further our understanding of the complex mechanisms that regulate phase-shifting input to the SCN.

RHT input to SCN neurons results in the activation of several intracellular signaling pathways, including the calcium/calmodulin kinase (CAMK), ERK/MAP kinase, and protein kinase A (PKA) pathways. All these kinases can activate the cAMP response element (CRE) CRE-binding protein (CREB) transcriptional pathway, which regulates many cellular processes. Obrietan and colleagues used transgenic mice expressing a CRE-β-galactosidase construct to show that both reporter gene activity and endogenous CREB phosphorylation exhibit a circadian rhythm in the SCN (Obrietan et al., 1999). It has since been shown that three amino acids, serines 133, 142, and 143, on CREB are important for activation by light and glutamate (Ginty et al., 1993; Gau et al., 2002; Kornhauser et al., 2002), as mutation of Serine 142 results in an attenuation of light-induced phase shifts of wheel-running activity and *Per1* gene expression (Gau et al., 2002).

Phosphorylated CREB also regulates the expression of IEGs such as *c-fos* in response to light stimulation. *C-fos, Per1,* and *Per2* all contain CRE sites in their promoter regions, and injection of an oligodeoxynucleotide CRE decoy into the SCN inhibits light or glutamate-induced phase shifts of wheel-running activity and *Per1* gene expression. This suggests that CRE/CREB-mediated transcriptional activation plays an important role in photic entrainment (Ginty et al., 1993; Tracnickova et al., 2002; Tischkau et al., 2003). Expression of *Per1* and *Per2* mRNA, which is normally low during the night, is rapidly induced by light exposure at night/subjective night, but is not induced by light exposure during the subjective day (Shearman et al., 1997). Thus, the regulation of the *Period* genes may be key to behavioral phase-shifting in response to light.

4.4 Nonphotic Input Modulates Circadian Rhythms

In addition to light, external cues such as activity/exercise and social behavior can modulate SCN-dependent circadian rhythms (Edgar and Dement, 1991). Anatomical and pharmacological studies have shown that nonphotic input to the SCN is communicated by the IGL and raphe nuclei, and is mediated primarily by the neurotransmitters neuropeptide Y (NPY) and serotonin (5-HT), respectively (Yannielli and Harrington, 2004). Application of either NPY or 5-HT during the subjective day results in phase advances of circadian rhythms of wheel-running activity and SCN gene expression (Yannielli and Harrington, 2004). However, these neurotransmitters mediate their effects through separate molecular pathways: NPY acts through the NPY Y1, Y2 and Y5 receptors to activate PKC and to elicit phase shifts, whereas the 5-HT family of receptors mediates serotonin-induced phase shifts through PKA and potassium channel activation (Yannielli and Harrington, 2004).

Nonphotic input to the SCN not only has a direct influence on circadian rhythms, but can also modulate the phase-shifting effects of light, and vice versa. Both NPY and serotonin can inhibit light-induced phase shifts during subjective night, with serotonin likely acting through 5HT1A/7 and 5HT1B receptors (Pickard et al., 1996; Weber et al., 1998; Yannielli and Harrington, 2004). Interestingly, though the 5HT1A/7 receptors are located on SCN neurons, the 5HT1B receptors are localized to RHT axons. Activation of these presynaptic receptors decreases glutamatergic excitatory postsynaptic currents (EPSCs) in the SCN, suggesting that serotonin modulates light-induced phase shifts by regulating both the stimulatory effects of the RHT and the cellular response of SCN neurons (Pickard et al., 1999; Smith et al., 2001). NPY has been shown to inhibit both *in vivo* light responses and NMDA-induced phase advances of spontaneous firing rate in SCN slice preparations (Yannielli and Harrington, 2000, 2001a, b). Y5 receptor antagonists block the inhibitory effects of NPY on light-induced wheel-running phase shifts, in addition to potentiating light-induced phase advances in the absence of exogenous NPY (Yannielli et al., 2004; Yannielli and Harrington, 2001a, b). This suggests that the NPY pathway normally acts to restrict the shifting effects of light during subjective night, perhaps as a way to fine-tune the effects of phase-shifting stimuli.

A recent study has uncovered a mechanism coupling the photic and nonphotic pathways in the SCN. *Dexras1*, a guanine nucleotide exchange factor, shows a circadian pattern of expression in the SCN, with mRNA levels peaking in early subjective night, and DEXRAS1 guanine nucleotide exchange activity increases in response to NMDA stimulation of cortical neurons (Fang et al., 2000; Panda et al., 2002a;

Takahashi et al., 2003). Cheng and colleagues found that mice lacking the *dexras1* gene exhibit attenuated phase delays and MAPK pathway component phosphorylation in response to light pulses, and also exhibit dramatically increased phase advances when placed in a novel environment (Cheng et al., 2004). SCN preparations from *dexras1*$^{-/-}$ mice exhibited an increased sensitivity to NPY blockade of NMDA-induced firing rate phase shifts compared with sections from wild-type mice, implying a role for *dexras1* in nonphotic input (Cheng et al., 2004). These results provide evidence for a common molecular pathway that can receive and respond to photic and nonphotic stimuli to appropriately adjust an organism's circadian rhythm to environmental cues.

5 The Molecular Clock

Many of the advances in our understanding of the molecular pacemaker have come either from the identification of mammalian homologs of circadian genes in other organisms, or from the induced or targeted disruption of individual clock genes (Lowrey and Takahashi, 2004). Beginning with the identification of the *Clock* gene in 1997, the number of confirmed mammalian clock genes has rapidly risen to include six core genes necessary for generating rhythmicity and a number of other genes that modify pacemaker characteristics (Takahashi, 2004; Lowrey and Takahashi, 2004). In the present chapter, we define the core clock genes as those necessary for pacemaker function, mutation or deletion of which results in an extreme phenotype such as molecular and behavioral arrhythmicity. In contrast, clock-modifying genes are not essential for the generation of rhythmicity, but do play a role in regulating the properties of circadian rhythms (Takahashi, 2004).

At its most basic level, the molecular circadian clock present within each SCN neuron is composed of a transcription–translation loop: two proteins, CLOCK and BMAL1 (also known as MOP3) induce the transcription of four other genes, *Per1*, *Per2*, *Cry1*, and *Cry2*, which, after translation, dimerize, re-enter the nucleus, and inhibit their own transcription (King and Takahashi, 2000) (➋ *Figure 22-4*). Following turnover of the inhibitory proteins, the transcription–translation cycle resumes. This molecular cycle occurs over a period of approximately 24 h, and continues indefinitely with a stable period in the absence of any input.

5.1 *Clock* and *Bmal1/Mop3*

As previously described, the *Clock* mutant mouse was discovered in 1994 during a screen for ENU-induced mutations that resulted in abnormal circadian phenotypes (Vitaterna et al., 1994). Following the identification and characterization of the mutant phenotype, Vitaterna and colleagues determined that the responsible mutation mapped to a single locus on mouse chromosome 5. King and colleagues (1997) used positional cloning to characterize and map the *Clock* gene. The *Clock* gene spans almost 100 kb and has at least two splice variants, both of which have 23 exons (King et al., 1997). The resulting protein is a member of the basic helix-loop-helix (bHLH) PAS (Period-Arnt-Sim) transcription factor family, and shares high homology with bHLH-PAS member NPAS2 and lower homology with other family members such as ARNT, HIFα, and SIM1. CLOCK has a bHLH domain, important for DNA binding, PAS-A and PAS-B domains, which are involved in protein–protein dimerization, and a C-terminus glutamine-rich region associated with transactivation of the transcription initiation complex. It is within this glutamine-rich region that the single base-pair mutation responsible for the *Clock* mutant phenotype occurs: an A–>T transversion at the exon–intron junction 3′ to exon 19 results in a 51-amino acid internal deletion (King et al., 1997). The *Clock* mutation has a dominant negative functional effect: the mutant CLOCK protein is able to bind to other transcription factors and to DNA, but is unable to initiate transcription, resulting in neuronal and behavioral arrhythmicity under constant conditions (King et al., 1997; Gekakis et al., 1998; King and Takahashi, 2000).

Bmal1, another bHLH-PAS family member, was discovered when Weitz and Takahashi used a yeast two-hybrid screen to identify CLOCK-binding partners (Gekakis et al., 1998). Both *Clock* and *Bmal1* are highly

Figure 22-4

The Circadian Transcription–Translation Feedback Loop. The proteins CLOCK, BMAL1/MOP3, PERIOD1 and 2, and CRYPTOCHROME1 and 2 make up the core of the molecular circadian pacemaker. CLOCK and BMAL1 dimerize and activate the transcription of the *Period* and *Cryptochrome* genes, the protein products of which inhibit their own transcription following homo- and heterodimerization. There are a number of other proteins, including CK1ε and REV-ERB α, that can regulate the speed at which the pacemaker runs by regulating transcription or protein turnover of the core clock genes. Reprinted by permission of Annual Review of Genetics and Human Genomics

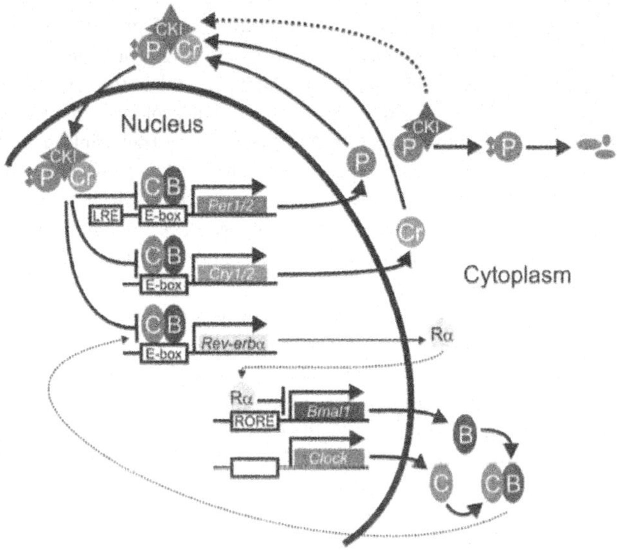

expressed in the SCN; although *Clock* expression is not rhythmic, *Bmal1* expression is regulated in a circadian manner with peak expression during the subjective night (CT15-CT18) (Shearman et al., 2000b). BMAL1 and CLOCK heterodimerize and, like other bHLH-PAS proteins, bind to an E-box element on genomic DNA. This E-box sequence, CACGTGA, is involved in the induction of gene transcription by bHLH-PAS transcription factors, and E-boxes have since been identified in the promoter regions of many circadian genes (Hogenesch et al., 1998; Panda et al., 2002a). Mice with a targeted disruption of *Bmal1* lack both rhythmic *Per1* and *Per2* mRNA expression in the SCN and rhythmic locomotor activity in constant darkness (Bunger et al., 2000).

5.2 The *Period* and *Cryptochrome* Paralogs

The CLOCK:BMAL1 heterodimer induces coordinated transcription of *Per1*, *Per2*, *Cry1*, and *Cry2*, which have been shown, through the use of knockout mice, to be the essential components of the core circadian mechanism. The *Period* paralogs and the *Cryptochrome* paralogs are homologous to genes known to be essential to circadian rhythms in other organisms, including *Drosophila*. The *Per* genes, members of the PAS family, are highly enriched in the SCN, and are rhythmically expressed with peak *Per1* mRNA expression around CT6, peak *Per2* mRNA expression at ZT9, and peak protein expression of both approximately 3–6 h phase-delayed to mRNA expression (Zylka et al., 1998; Zheng et al., 2001). The PER proteins have been shown to homodimerize, and to heterodimerize with each other and with both CRY1 and CRY2 (Griffin et al., 1999; Bae et al., 2001; Zheng et al., 2001). Dimerization occurs in the cytoplasm and results in the exposure of a nuclear localization sequence (NLS) and subsequent nuclear entry, where the dimers inhibit CLOCK:BMAL1-induced gene transcription (Kume et al., 1999).

Mice with a targeted deletion of either the *Per1* gene or the *Per2* gene display a free-running period approximately 1 h shorter than normal, and in the case of *Per2* show a loss of rhythmicity during extended periods in DD (Zheng et al., 1999, 2001; Bae et al., 2001). *Per1/Per2* double knockouts exhibit locomotor arrhythmicity immediately upon introduction to DD, indicating that while the molecular pacemaker continues to function—albeit somewhat impaired—with one normal *Period* gene, the loss of both *Period* genes completely disrupts pacemaker function (Bae et al., 2001; Zheng et al., 2001). However, there are important phenotypic differences between *Per1* and *Per2* knockouts: *Per1* deletion has little effect on *Per2*, *Cry1*, or *Bmal1* mRNA expression, but reduces the level of CLOCK protein expression, whereas *Per2* deletion results in a reduction in the amplitude of both *Per1* and *Per2* gene expressions (Bae et al., 2001). Furthermore, *Per2* knockout mice have a shorter free-running period than the *Per1* knockout mice, and the majority of *Per2* knockout mice lose circadian rhythmicity with prolonged time in constant darkness (Zheng et al., 1999). It has been proposed that *Per2* regulates *Bmal1* transcription, which might explain why *Per2* knockout mice have a more severe phenotype than *Per1* knockouts (Shearman et al., 2000a). Microarray analysis also indicates that *Per1* and *Per2* have differential effects on clock-controlled genes (Zheng et al., 2001). Unlike *Per1* and *Per2*, a third paralog, *Per3*, does not appear to be necessary for circadian behavioral rhythms, as *mPer3* knockout mice show only a slightly shortened period and no change in rhythmic clock gene expression (Shearman et al., 2000a). PER3 does not appear to interact with the CRY proteins, perhaps explaining why it cannot compensate for the loss of both *Per1* and *Per2* (Lee et al., 2004).

Like the *Period* genes, *Cryptochromes 1* and *2* are orthologs of core circadian genes in *Drosophila* (Kume et al., 1999). CRY1 and CRY2 strongly inhibit CLOCK:BMAL1-induced transcription, and have been shown to dimerize with each other and with the PER proteins (Kume et al., 1999). *Cry1* and *Cry2* are rhythmically expressed in the SCN, with mRNA expression peaking toward the end of the subjective day, and are unaffected by nighttime light exposure (Okamura et al., 1999; van der Horst et al., 1999). CRY1 and CRY2 are present in both the retina and the SCN, and are likely involved in the photic entrainment pathway.

Cry1 and *Cry2* loss-of-function mutations have opposite effects on circadian period: the mutation of *Cry1* results in an ~1-h shortening of the free-running period, whereas the mutation of *Cry2* results in an ~1-h lengthening of the period in mice (van der Horst et al., 1999; Thresher et al., 1998). This suggests that *Cry1* and *Cry2* have overlapping functions as core clock components, but opposite functions in affecting period length. As might be predicted from their behavioral effects, *Cry1* and *Cry2* differentially regulate *Period* gene expression. Deletion of *Cry1* results in an earlier peak of *Per1* and *Per2* expression in the SCN, as would be expected from the shortened period of locomotor activity, and may blunt light-induced *Per1* expression at night (Okamura et al., 1999) (Vitaterna et al., 1999). Deletion of *Cry2* results in a delayed peak of *Period* gene expression, but has no effect on light-induced *Per1* expression (van der Horst et al., 1999). Similar to *Per1/Per2* double knockouts, *Cry1/Cry2* double knockouts become arrhythmic in DD. *Per1* and *Per2* levels were found to be elevated and arrhythmic in *Cry1/Cry2* knockouts; further investigation has shown that the CRY proteins are potent inhibitors of the CLOCK:BMAL1 complex via protein:protein interactions (Okamura et al., 1999; Vitaterna et al., 1999).

5.3 *Rev-erbα*, *Timeless*, and Other Modifiers of the Circadian Pacemaker

The mutation of *Clock*, *Bmal1*, both *Per1* and *Per2*, or both *Cryptochrome* genes results in behavioral arrhythmicity in the absence of entraining cues. However, there are a number of other proteins that are not essential for sustaining rhythmicity, but do affect the length or properties of the molecular clock. These genes include enzymes such as *Casein Kinase 1ε*, which regulates PER accumulation by phosphorylation, and transcription factors such as *Rev-erbα*, which regulates the transcription of *Bmal1* (Takahashi, 2004; Lowrey and Takahashi, 2004). Lowrey and colleagues (2000) used positional syntenic cloning to identify a mutation in the *Casein kinase 1ε* gene as the cause of the shortened period length in *tau* mutant hamsters. *CK1ε* is a serine/threonine protein kinase that normally phosphorylates cytoplasmic PER1 and PER2, resulting in PER degradation. The mutant version of *CK1ε* fails to effectively phosphorylate the PERIOD proteins, allowing PER to accumulate more rapidly than normal (Lee et al., 2001). This effectively shortens the molecular cycle, resulting in the shortened activity period observed in *tau* mutants.

The transcription factor *Rev-erbα* has recently been shown to couple the positive and negative limbs of the core pacemaker by regulating a component of the positive limb. *Rev-erbα* mRNA is highly rhythmic in the SCN, and expression levels peak early in the subjective day (Ueda et al., 2002). Expression is induced by the CLOCK:BMAL1 complex binding to an E-box sequence on the *Rev-erbα* promoter, but, unlike the *Period* and *Cryptochrome* genes, *Rev-erbα* feeds back to inhibit transcription of *Bmal1* and other genes that have retinoic acid related orphan receptor response elements (ROREs) within their promoters (Ueda et al., 2002; Sato et al., 2004). Preitner and colleagues (2002) determined that the *Bmal1* promoter contains two ROREs that bind *Rev-erbα*. In addition to being regulated by CLOCK:BMAL1, expression of *Rev-erbα* is negatively regulated by PER2, and levels of *Rev-erbα* mRNA are high and nonrhythmic in *Per2* knockout mice. Because of its role in repression of *Bmal1* expression, *Bmal1* levels are constitutively high in *Rev-erbα* knockout mice. However, locomotor activity is only slightly shorter in the knockouts, indicating that rhythmic expression of *Bmal1* is not essential for proper pacemaker function (Preitner et al., 2002).

The role of a third candidate gene, *Timeless*, is still under investigation. *Timeless* is a core component of the *Drosophila* molecular clock, and flies carrying a null mutation of timeless (tim^{01}) are arrhythmic, but the function of the mammalian homolog has been difficult to ascertain because deletion of *Tim* results in embryonic lethality (Gotter et al., 2000). However, Barnes and colleagues have recently used *in vitro* knockdown of *Tim* to suggest a circadian function for the gene (Hastings et al., 1999; Barnes et al., 2003). Application of antisense oligodeoxynucleotides against *Tim* to SCN slices results in the disruption of rhythmic VIP expression and phase-shifts spontaneous neuronal activity when applied during the subjective day (Barnes et al., 2003; Tischkau et al., 2003). Barnes and colleagues suggest that previous attempts to study *Tim* have been confounded by the abundance of a short, nonrhythmic isoform of the protein, and that it is the less-abundant full-length isoform that is transcribed rhythmically and contributes to normal *Period* and *Cryptochrome* expression.

In addition to *CK1k1ε*, *Rev-erbα*, and *Timeless*, there are several other genes that may modify period length but whose functions have not been studied as extensively. These include: (a) albumin D-element binding protein (DBP), a member of the PAR family of basic leucine (bZip) transcription factors, expression of which is induced by the CLOCK:BMAL complex (Ripperger et al., 2000; Yamaguchi et al., 2000b); (b) *Dec1* and *Dec2*, bHLH transcription factors that are rhythmically expressed in the SCN and inhibit CLOCK:BMAL1-induced transcription of *Per1* by competing for E-box binding and/or directly binding to BMAL1 (Honma et al., 2002); (c) *Math5*, a helix-loop-helix protein that is involved in RHT development and the deletion of which results in both the inability to entrain to an LD cycle and an ~1-h lengthening of the free-running period (Wee et al., 2002); and (d) *Rab3a*, a ras-associated protein that regulates synaptic vesicle trafficking and mutation of which results in a shortened free-running period (Kapfhamer et al., 2002). Also of interest is *Npas2*, the bHLH-PAS family member most similar to *Clock* (King et al., 1997). Although NPAS2 does not appear to play a role in pacemaker activity in the SCN, it is expressed throughout the forebrain and forms a transcription initiation complex with BMAL1, the DNA-binding activity of which is regulated by redox state (Rutter et al., 2001). NPAS2 may therefore be involved in the entrainment of the pacemaker by food, linking cellular redox state to the circadian pacemaker.

In summary, the core clock genes are the basis for the cell-autonomous molecular clock observed in SCN neurons. CLOCK, BMAL, the PERIODs, and the CRYPTOCHROMEs form an autoregulatory loop that functions with or without external cues. However, the time required for the loop to run full-circle can be regulated by a number of other proteins that act as transcription factors or phosphatases. As the CLOCK: BMAL complex drives transcription of core component genes, it also induces or inhibits transcription of numerous other genes, some of which are themselves transcription factors and some of which may be directly involved in circadian phenotypes observed at the cellular level, such as neuropeptide expression and membrane conductance (Panda et al., 2002a).

5.4 Transcriptional and Functional Regulation of Clock Genes I: Transcription

Circadian expression of genes can be controlled at one or more of three levels: (1) at the level of gene expression, as in the induction or inhibition of transcription, (2) at the level of protein function through mechanisms such as phosphorylation and dimerization that regulate protein activity or cellular localization, or (3) at the level of protein degradation, with rhythmic proteins targeting other proteins for removal. Of these three mechanisms, the best studied is transcriptional control, and it also appears to be the most important means of regulating circadian gene/protein expression.

The primary mechanism of circadian control over gene expression occurs via CLOCK:BMAL1 binding at specific E-box elements located within the regulatory regions of a number of clock-controlled genes. E-boxes are hexameric DNA sequences, CANNTG, recognized by all bHLH transcription factors; the sequence preferentially recognized by the CLOCK:BMAL complex is (G/T)G(A/G)A*CACGTG*ACCC (Hogenesch et al., 1998; Panda et al., 2002a). The primacy of the E-box to circadian transcription was first recognized in flies, when it was shown that an E-box-containing 69-bp region of the *per* promoter was enough to induce robust circadian transcription of a reporter gene (Hao et al., 1997); indeed, a multimer of the E-box plus the 6-bp immediately 5′ and 3′ could also produce circadian transcription (Darlington et al., 2000). In mammals, *Per1* has 5 circadian E-boxes in the 6.8 kb upstream to the transcription start site. These E-boxes are necessary for circadian transcription and act in an additive manner: mutation of all but one upstream E-box results in greatly reduced, but still rhythmic, transcription, whereas mutation of all five results in low, nonrhythmic transcription of a reporter gene (Hida et al., 2000; Yamaguchi et al., 2000a).

A number of other clock-controlled genes also contain E-boxes within their promoter regions. By using microarrays to analyze gene expression in the SCN and liver over 48 h, Panda and colleagues identified 127 cycling transcripts, 9 of which contained a CACGTGA sequence within the 10 kb upstream to the transcription start site (Panda et al., 2002a). However, a number of clock-controlled genes, such as *Per2*, do not contain a canonical E-box within the promoter, suggesting that there are other, as-yet unidentified sequences involved in circadian transcription, and that there are sequences other than the canonical CACGTGA sequence that are necessary for regulating CLOCK:BMAL1 binding. Indeed, a variant of the circadian E-box has recently been identified by Yoo and colleagues, who found that the majority of the circadian transcription of *Per2* can be accounted for by a noncanonical CACGTT sequence 20 bp upstream of the *Per2* transcription start site (Yoo et al., 2005). Chromatin immunoprecipitation shows that this E2 sequence, like the canonical E-box, constitutively binds the CLOCK:BMAL1 dimer *in vivo*, and a 210-bp E2-containing enhancer sequence is sufficient for driving *in vivo* luciferase rhythms in transgenic mice (Yoo et al., 2005). Further investigation may show that the E2 element is as important as the canonical E-box for regulating genome-wide circadian transcription.

E-box sequences are relatively common throughout the genome, yet only a subset of genes with E-box-containing promoters is transcribed rhythmically. To identify non-E-box regions involved in circadian transcription, Munoz and colleagues compared the promoter regions of two genes, *vasopressin* (AVP) and *cyclin-b* (CYC) (Munoz et al., 2002). Both promoters contain a single CACGTG sequence, but expression of AVP in the SCN is rhythmic and controlled by CLOCK:BMAL, whereas expression of CYC is not rhythmic (Hwang et al., 1995; Jin et al., 1999). In addition to showing that CYC transcription is not affected by CLOCK:BMAL, suggesting that there are sequences adjacent to the E-box that increase or decrease a promoter's affinity for CLOCK:BMAL, Munoz found that the AVP promoter has lower affinity than the CYC promoter for the universal E-box factor USF. This leads to the hypothesis that nonrhythmic E-box containing genes combine a low affinity for CLOCK:BMAL1 with a high affinity for constitutively expressed transcription factors, whereas rhythmic genes have a lower affinity for constitutive transcription factors and a high affinity for CLOCK:BMAL1 (Munoz et al., 2002). Additional regulation of circadian transcription likely occurs via both spatial control of transcription factor expression and concentration-dependent competition by other bHLH transcription factors.

Although Munoz identified several E-box-like sequences in the AVP promoter as possible enhancers for CLOCK:BMAL1 binding, there are still no non-E-box sequences known to confer transcriptional rhythmicity. However, the cAMP-responsive element (CRE) has been identified as important for light-dependent induction of the *Period* genes, and, as described earlier, the RORE plays a role in the regulation of rhythmic *Bmal1* expression (Tracnickova et al., 2002; Tischkau et al., 2003). The ROREs in the *Bmal1* promoter appear to have a bidirectional effect on *Bmal1* expression depending on which transcription factors are present: whereas *Rev-Erb* α inhibits *Bmal1* transcription via RORE binding, the orphan nuclear receptor RORα induces *Bmal1* transcription (Preitner et al., 2002; Ueda et al., 2002; Najima et al., 2004; Sato et al., 2004). Recently, Ueda and colleagues classified circadian transcriptional regulators into three groups: those that act via E/E1-boxes, those that act on ROR elements, and those that act on DBP/E4BP4 elements (Ueda et al., 2005). They showed that the rhythmicity, amplitude, and phase of circadian gene expression can be fine-tuned by the combination of one of more of these elements regulating the expression of transcriptional activators and repressors. Given that microarray studies have indicated that a large number of transcription factors exhibit rhythmic expression, it is possible that the CLOCK:BMAL complex only directly regulates rhythmic expression of genes with E-box-like promoter sequences, and that the rhythmic expression of all other clock controlled genes is dependent on the rhythmic expression of transcription factors, such as *Dbp*, downstream of CLOCK:BMAL transcriptional activity (Panda et al., 2002a; Hayes et al., 2005).

5.5 Transcriptional and Functional Regulation of Clock Genes II: Posttranslational Regulation

The presence and activity of core clock genes are regulated at the protein level by several mechanisms, including the timing of translation, spatial localization, phosphorylation, and degradation. Although transcriptional regulation allows for gross control of the availability of clock genes, transcript processing also appears to be regulated. For example, peak *Per* gene expression occurs around CT6 in the SCN, and PER and CRY protein levels peak at approximately CT12 (Kume et al., 1999; Field et al., 2000; Shearman et al., 2000b; Maywood et al., 2003). Thus, the negative loop component protein levels are phase-delayed with respect to mRNA expression by about 3–6 hours. Shearman and colleagues (2000b) suggested that *Bmal1* had a similar translational delay of 4–6 h, which would result in peak BMAL1 protein at the approximate time of *Per* gene induction. However, a later study using mouse liver extracts found very little delay between the peak time of *Bmal1* mRNA and protein expression, suggesting that as-yet-unknown factors may regulate the rate at which core clock genes are translated, affecting the bioavailability of the resulting proteins for dimerization (Lee et al., 2001).

Following translation, clock proteins must dimerize and re-enter the nucleus to regulate transcription of other genes. Nuclear entry is associated with protein phosphorylation, which may regulate both spatial localization and degradation. Lee and colleagues (2001), using mouse liver extracts to characterize the levels and localization of core circadian clock proteins, observed that the molecular weights of several core clock proteins showed a circadian rhythm. From CT15-03, higher molecular weight forms of PER1 and PER2 appeared, peaked, and then gradually diminished. BMAL1 and CLOCK also exhibited fluctuations in molecular weight over the 24-h period. By treating proteins immunoprecipitated from liver extracts with protein phosphatase, the authors determined that the changes in molecular weight were due to phosphorylation of these clock components. Lee and colleagues (2001) also showed that phosphorylation was associated with nuclear accumulation, as the phosphorylated forms of core clock proteins predominated in the nuclear fractions. Little phosphorylated CLOCK or BMAL1 was observed in the cytoplasm, suggesting that phosphorylation either results in, or occurs immediately after, nuclear entry.

Currently, two isoforms of casein kinase 1 (CK1ε and CK1δ) are the only kinases specifically known to play a role in the mammalian circadian clock. The mutation in the *tau* hamster results in an arginine-to-cysteine amino acid substitution in the substrate-binding region of the CK1ε protein, resulting in reduced CK1ε activity and decreased phosphorylation of the PER proteins (Lowrey et al., 2000). A later human genetic study correlated a mutation in *Per2* with familial advanced sleep phase syndrome (FASPS), in which affected individuals awake very early in the morning (Toh et al., 2001). The mutation was mapped to Serine 662, an acceptor site for CK1ε phosphorylation. Thus, the interaction of PER2 and CK1ε appears to be an evolutionarily conserved mechanism required for proper circadian function. In addition, CK1δ, a homolog of CK1ε, is expressed in the SCN and has been shown to associate with and phosphorylate mPER1 and mPER2 (Vielhaber et al., 2000; Camacho et al., 2001; Ishida et al., 2001). More recently, a missense mutation (T44A) in the gene encoding CK1δ was also shown to result in FASPS in humans (Xu et al., 2005). The region downstream of the CK1ε-binding site in the PER1 protein contains various acceptor residues which, when phosphorylated, can mask the nuclear localization sequence and prevent the protein from entering the nucleus (Vielhaber et al., 2000). Retention of PER proteins in the cytoplasm makes them more susceptible to degradation; therefore, mutations that reduce PER phosphorylation appear to shorten the circadian cycle by increasing the rate of PER dimerization and nuclear accumulation.

Several other core clock proteins, including BMAL1, CRY1, and CRY2, may also be regulated by phosphorylation, possibly by components of the MAPK pathway (Sanada et al., 2002, 2004). Because the MAPK pathway is thought to play a role in communicating photic input to the SCN, it is possible that photic control of circadian rhythms is partially mediated by clock protein phosphorylation. Future studies will undoubtedly add to the list of protein kinases that modify and regulate known clock components.

Phosphatase-mediated removal of phosphate groups also appears to be involved in clock component regulation. Butcher and colleagues showed that the MAPK phosphatases MKP-1 and -2 are expressed in mouse SCN and acutely regulate light-induced MAPK activity; similar results were obtained by Hoyashi and colleagues, who used chick pineal gland as a model (Hayashi et al., 2001; Butcher et al., 2003). In *Drosophila*, protein phosphatase 2A has been implicated in directly regulating dPER by stabilizing the protein in central clock neurons (Sathyanarayanan et al., 2004). The modification and regulation of circadian clock proteins by the addition or removal of phosphate groups is therefore crucial to the timing and plasticity of circadian function.

A third mechanism of control at the protein level involves regulation of the formation of heterodimer complexes by the limitation of component availability. Because *Bmal1* mRNA levels fluctuate over a 24-h period while *Clock* mRNA levels remain constant, it had been proposed that BMAL1 is the rate-limiting factor in the positive arm of the molecular clock feedback loop (Shearman et al., 2000b). However, overexpression of a *Clock*-containing bacterial artificial chromosome (BAC) in transgenic mice resulted in a shortened free-running period, suggesting that under normal conditions, CLOCK is either not in excess of BMAL1 or not completely functionally available (Antoch et al., 1997). It has also been shown that CLOCK nuclear localization is dependent on BMAL1, as no CLOCK immunoreactivity is detected in nuclear fractions from *Bmal1*-deficient mouse liver tissue (Kondratov et al., 2003). If these results are representative of all tissues, the *Bmal1* knockout mouse model may also be considered functionally deficient in CLOCK, due to the lack of CLOCK nuclear translocation. Regarding the negative limb components, CRY proteins have been shown to be in molar excess of PER proteins and are present in the cytoplasm throughout the day in liver extracts, leading Lee and colleagues to suggest that that PER proteins are the limiting factors in the negative limb of the feedback loop (Lee et al., 2001). In support of this, although CRY is present in the cytoplasm of *Per* double knockouts, very little CRY protein can be detected in nuclear extracts from livers of *Per1/2* double knockouts (Lee et al., 2001).

A final mechanism of protein regulation is degradation or turnover. Based on *in vitro* studies, it is clear that phosphorylation of circadian clock proteins precedes and most likely triggers the degradation process. The exact mechanism for this is still unknown, but two recent studies provide evidence that the 26S ubiquitin-mediated proteosome pathway is involved in the regulation of clock gene activity (Yagita et al., 2002; Kondratov et al., 2003). The 26S pathway of degradation begins with the addition of multiple ubiquitin polypeptide units onto lysine residues of the substrate protein by a family of proteins known as ubiquitin ligases (E1, E2, and E3). Once polyubiquitylated, the protein is targeted to the 26S-proteasome, where it is proteolytically cleaved and degraded. Yagita and colleagues found that mouse embryonic fibroblast (MEF) cells from *Cry1/2* knockout mice had very little PER2 in the nucleus, but when the cells were treated with a specific inhibitor of 26S proteosome-mediated protein degradation (MG132), a dramatic increase in nuclear and cytoplasmic PER2 was observed (Yagita et al., 2002). Additionally, coexpression of *Cry1* prevented ubiquitylation of PER2 in COS7 cells in a dose-dependent manner, and the presence of PER2 reduced ubiquitylation of CRY1. CLOCK and BMAL1 levels may also be regulated by ubiquitylation: Antoch's group has shown that phosphorylation of BMAL1 is dependent on CLOCK: BMAL1 dimerization, and that this dimerization triggers rapid degradation of the CLOCK protein (Kondratov et al., 2003). Furthermore, inhibition of the 26S pathway in HEK 293 cells results in CLOCK nuclear accumulation, indicating that CLOCK is normally degraded via a proteosome-dependent mechanism after it has heterodimerized with BMAL1 and translocated to the nucleus. Interestingly, multiple members of the ubiquitin-mediated degradation pathway (ubiquitin-conjugating enzymes, ubiquitin-like proteins, and proteosome subunits) exhibit circadian regulation of mRNA expression, suggesting that the degradation of circadian clock proteins (as well as other proteins) may also be circadianly regulated (Panda et al., 2002a).

Increasing efforts in the circadian field are being directed at understanding how fluctuating protein levels, protein modification, and core clock component localization regulate cellular pathways and subsequent tissue physiology. In addition to regulating the transcription of genes involved in cellular output pathways, the core clock proteins may directly interact with ion channel proteins or other proteins involved in regulating neuronal activity. By elucidating the spatial and temporal regulation of protein function, we will increase our understanding of the control over circadian physiology and behavior of an organism.

6 Neuronal and Humoral Output from the SCN

The fundamental function of the SCN is to transmit time-of-day information to the rest of the body. For this to occur, the molecular pacemaker within each SCN neuron must be translated into an output signal—such as a change in neuronal excitability or neurotransmitter production—that can be communicated to other brain regions that directly or indirectly regulate activity at the systems level. Molecularly, this requires a change in the expression, localization, or stability of proteins involved in neuronal communication. Recent gene expression studies have identified a number of candidates that are rhythmically expressed and are known to be involved in pathways associated with neural output. These genes include neurotransmitter receptors, such as the vasopressin receptor (*Avpr1a*), which regulates neuron response to incoming signals, ion channel proteins, such as the large conductance Ca-channel potassium channel (*mSlo*), which affects neuronal excitability, and neuropeptides and secretory factors, such as secretogranin III (*Scg3*) and vasopressin (*Avp*), which affect the communication of neurons with one another (Panda et al., 2002a). These changes occur at the single-cell level, but through coupling of individual SCN neurons, produce time-of-day dependent changes in the properties of the SCN as a whole. The SCN effects changes in the periphery via neuronal and humoral outputs to other regions of the brain, such as the hypothalamus and the pineal gland, that then regulate physiological and behavioral processes.

6.1 Hypothalamic Projections

The most abundant group of SCN efferents are the projections to other nuclei in the hypothalamus. Kalsbeek and Buijs have grouped SCN targets in the hypothalamus into three categories: endocrine neurons such as GnRH neurons, which directly control pituitary function; interneurons, which likely act as integrators of circadian and noncircadian information and regulate endocrine neuron function; and autonomic neurons, which regulate the function of the peripheral nervous system (Kalsbeek and Buijs, 2002). SCN communication has been shown to occur through both direct neuronal connections and humoral mechanisms: when a transplanted SCN is placed inside a membrane that allows for the release of diffusible factors but prevents the establishment of synaptic connections, certain rhythms, such as locomotor activity, are restored, but many endocrine rhythms are not (Lehman et al., 1987; Silver et al., 1996a; Meyer-Bernstein et al., 1999).

As described earlier, SCN efferents generally express AVP or VIP, and arise primarily from the ventrolateral shell. These neuropeptides are thought to be the primary messengers of the timing signal, and changes in AVP and VIP signaling have been shown to be important in the regulation of endocrine function (Kalsbeek and Buijs, 2002). However, circadian wheel-running behavior appears to be regulated differently from endocrine function, and two other molecules, TGF-α and prokineticin-2 (PK2), have been identified as involved in controlling circadian locomotor activity (Kramer et al., 2001; Cheng et al., 2002).

The following section addresses the molecular mechanisms of circadian regulation of three specific processes: the hypothalamic-pituitary-gonadal axis (HPG) in the female, the hypothalamic-pituitary-adrenal (HPA) axis, and locomotor activity.

6.2 The Hypothalamic-Pituitary-Gonadal (HPG) Axis

Although reproduction in the male is generally not under circadian control—except for seasonally regulated breeders such as hamsters—the female reproductive axis is under SCN control in all mammals, and the components of the axis are well understood. The gatekeepers of reproductive physiology are gonadotropin-releasing hormone (GnRH) neurons, which are diffusely distributed throughout the mPOA of the anterior hypothalamus (Wray and Hoffman, 1986). GnRH neurons act as integrating stations, receiving input from the SCN (signaling time of day), from the ovaries (signaling follicular readiness via estradiol and progesterone production), and from fat pads (signaling nutrient availability) (Evans et al., 1994; Hamann and Matthaei, 1996; Magni et al., 1999). On the afternoon of proestrus, which occurs every 4–5 days in mice and rats, incoming signals coordinate GnRH neuronal activity, inducing a bolus of GnRH release into the portal hypophysial circulatory system. GnRH stimulates the pituitary to release the gonadotropins luteinizing hormone (LH) and follicle-stimulating hormone (FSH), which induce ovulation and follicular recruitment. The ovaries produce the hormones estradiol and progesterone, which feed back to regulate GnRH release via steroid receptors on interneurons and on GnRH neurons themselves (Peterson et al., 2003).

The role of the SCN in controlling GnRH neuron activity was first suggested by SCN ablation studies that resulted in vaginal acyclicity and cessation of reproductive function and later confirmed by tract-tracing studies showing direct projections of SCN neurons to GnRH neurons (Brown-Grant and Raisman, 1977; Van der Beek et al., 1997; Wiegand et al., 1980). The GnRH surge occurs only on the afternoon of proestrus, around the light–dark transition, and if release is inhibited by phenobarbital administration on the morning of proestrus, it will occur exactly 24 h later (Everett and Sawyer, 1950). It has been suggested that the SCN transmits a daily signal to GnRH neurons that permits GnRH release when other factors, such as estradiol and progesterone levels, are also permissive (Levine et al., 1991, 1995).

The neuropeptides AVP and VIP have both been implicated in the circadian control of LH release (Kalsbeek and Buijs, 2002). Electron microscopy and immunohistochemical studies have shown that VIP-containing SCN projections directly contact GnRH neurons; the population of GnRH neurons receiving VIP input is the same population that is preferentially activated (*c-fos* expressing) on the afternoon of proestrus (Van der Beek et al., 1994; Smith et al., 2000). In middle-aged female rats, VIP-mediated transmission is lower, possibly contributing to the altered LH surge profile characteristic of early reproductive senescence (Krajnak et al., 2001). AVP expression is circadianly controlled, elevated in the hypothalamus immediately before GnRH release, and capable of inducing an LH surge in SCN-lesioned rats (Gillette and Reppert, 1987; Jin et al., 1999; Palm et al., 1999). Furthermore, estrous cyclicity is abnormal in the Brattleboro rat, a strain that does not synthesize AVP (Boer et al., 1981). The *vasopressin* promoter contains an E-box element with a CLOCK-binding sequence (CACGTG), and *vasopressin* is transcribed in the same phase as other CLOCK-controlled genes such as *Per1* and *Per2.* In male *Clock* mutant mice, circadian transcription of *vasopressin* in the SCN is damped, as is transcription of the *Per1* and *Per2* genes (Jin et al., 1999). Because AVP secretion is stimulatory to GnRH release, we have recently hypothesized that reduced SCN-derived AVP output from the SCN may account for the lack of a coordinated proestrus LH surge in *Clock* mutant mice (Miller et al., 2004).

Certain aspects of pregnancy are also under circadian control. Following ovulation, the corpus luteum (CL) of the ovulated oocyte begins producing progesterone. The CL will quickly recede unless mating occurs. The mating stimulus triggers twice-daily circadian-regulated prolactin release that maintains the CL for approximately 10 days following copulation (Barkley et al., 1978). Maintenance of the progesterone-producing CL is an absolute requirement of pregnancy: if prolactin secretion is interrupted during pregnancy, progesterone levels drop and the developing fetuses are reabsorbed (Smith, 1980). Several studies have linked the control of postcopulatory prolactin release to vasopressinergic transmission from the SCN, presumably via contacts to dopaminergic neurons that regulate prolactin release from the pituitary, and we have recently described mid-gestational pregnancy failure in *Clock/Clock* females that is likely due to abnormal prolactin release during early- and mid-pregnancy (Palm et al., 2001; Miller et al., 2004). Finally, parturition is temporally restricted to the nighttime/early morning in rodents, although the interplay of maternal and fetal circadian rhythms that control the timing of birth is poorly understood (Rowland et al., 1991; Muglia, 2000).

6.3 The Hypothalamic-Pituitary-Adrenal (HPA) Axis

The adrenal cortex regulates organismal response to stress by releasing steroid hormones and catecholamines, and adrenal gland function is controlled both by the hypothalamic-pituitary axis and by the ANS. Under normal conditions, the hypothalamus is the primary modulator of adrenal function, resulting in basal corticosterone release that is low during the behaviorally inactive period and rises before the onset of activity. Under stressful conditions, the sympathetic nervous system releases norepinephrine, which stimulates the adrenal glands to produce epinephrine.

The role of AVP in SCN regulation of adrenal function can be surmised both by the daily timing of AVP and corticosterone release, and by changes in adrenal function following infusion of a vasopressin receptor antagonist into the dorsomedial hypothalamus (DMH) (Inouye, 1996). As described earlier, AVP is released rhythmically into the hypothalamus and CSF by neurons projecting from the SCN. In rats, AVP release is highest during the early day (ZT2-4), when corticosterone levels are at a nadir, and as AVP release decreases later in the day, corticosterone levels rise (de Boer and Van der Gugten, 1987; Gillette and Reppert, 1987). Tract-tracing and electron microscopy studies have shown that AVP-containing neurons project from the SCN to the paraventricular nucleus (PVN) and the dorsomedial hypothalamus (DMH), the location of the corticotropin-releasing hormone (CRH) neurons that stimulate pituitary release of corticotropin (ACTH) (Kalsbeek and Buijs, 2002). Although there are few direct AVP-CRH synapses, SCN fibers densely innervate interneurons that likely regulate CRH neuron activity (Kalsbeek et al., 1992). Functional studies have shown that, in SCN-lesioned rats, AVP infusion into the PVN results in rapid inhibition of corticosterone levels, likely via actions on GABA-ergic PVN neurons (Kalsbeek et al., 1992; Hermes et al., 2000). The effects of a vasopressin receptor antagonist vary with time of day: in rats, infusion during the morning results in a rapid increase in plasma corticosterone, whereas treatment during the early evening or night has no effect on corticosterone release (Kalsbeek et al., 1996).

It has recently been shown that the SCN can also regulate adrenal activity by way of a multisynaptic autonomic pathway, suggesting that autonomic output from the SCN modulates adrenal responsiveness to ACTH. Retrograde tracing following injection of pseudorabies virus into the liver or spinal cord shows that SCN neurons make third-order synaptic connections with both sympathetic and parasympathetic autonomic neurons (Buijs et al., 1999, 2003). SCN projections to preautonomic neurons in the PVN are separated into parasympathetic and sympathetic branches, but there appears to be no regional separation within the SCN, and SCN projections to both branches are primarily vasopressinergic, raising the question of how a single signal can regulate opposing outputs. Buijs and colleagues (2003) suggest that vasopressinergic outputs to the PVN may have different cotransmitters, such as GABA or glutamate, depending on which branch of the ANS is innervated. If this is the case, the SCN may regulate ANS control of adrenergic function by simultaneously stimulating the parasympathetic nervous system and inhibiting the sympathetic nervous systems.

6.4 Locomotor Activity

Although the measurement of wheel-running activity has been the gold standard for insight into the circadian pacemaker for almost a century, we have only recently begun to understand the molecular mechanisms behind circadian control of locomotor activity. SCN transplant studies involving encapsulated SCN explants indicated that the factor(s)controlling activity had to be diffusible, as transplants could restore locomotor rhythms without establishing neural connections, but the first advance in molecular knowledge occurred only when a Harvard group conducted a systematic screen to look at the behavioral effects of factors secreted from the SCN (Silver et al., 1996a; Kramer et al., 2001). Kramer and colleagues performed a yeast-secretion trap on a hamster SCN cDNA library to identify previously unknown secreted factors, and these factors were then infused into the third ventricle of freely moving hamsters though wheel-running activity was monitored. Kramer was looking for locomotor inhibitory and excitatory factors that functioned downstream of the molecular pacemaker, specifically affecting wheel running without resetting the underlying circadian clock. This screen identified the candidate peptide transforming growth factor-alpha (TGFα), which, upon application, completely blocked wheel-running activity for the duration of the infusion. When TGFα application was halted, wheel-running activity reappeared with the predicted phase and period, indicating that the cessation of circadian locomotor activity was not due to an effect on the circadian pacemaker (Kramer et al., 2001).

There are several characteristics that make TGFα a prime candidate for transmission of a locomotor control signal. First, TGFα is expressed in the SCN in a rhythmic manner, with peak expression during the inactive phase and trough expression during the active phase, as would be expected of a locomotor inhibitory factor. Second, the TGFα receptor, epidermal growth factor receptor (EGFR), is highly expressed in the subparaventricular zone (SPZ), a target area for SCN projection neurons that is spatially close enough to the SCN to receive diffusible factors, and lesions of which disrupt locomotor activity (Watts et al., 1987; Lu et al., 2001). Finally, mice with the *waved-2* mutation, which decreases the activity of the EGFR by approximately 90%, exhibit excessive daytime activity but have a normal activity period length (Kramer et al., 2001).

More recently, a second candidate molecule has been identified (Cheng et al., 2002). Prokineticin 2 (PK2) is a cysteine-rich protein that is expressed rhythmically in the SCN in a manner similar to TGFα and the *Period* genes, with peak expression during the inactive period and trough expression during the active period. The PK2 promoter contains 4 E-boxes and a CRE, the CLOCK:BMAL1 complex induces PK2 transcription, and expression is negatively regulated by PERIOD and CRYPTOCHROME proteins. The PK2 receptor, a G-protein coupled receptor, is expressed in several regions that receive SCN projections, including the paraventricular thalamic nucleus (PVT), the PVN, and the DMH, although there are no PK2 receptors in the SPZ, the target for TGFα activity. Finally, administration of PK2 into the third ventricle in rats shows that PK2 administration at CT14, during the active period, suppresses wheel-running activity (Cheng et al., 2002). It is interesting to note that both candidate molecules for the control of the SCN over locomotor activity are inhibitory, suggesting that activity is the basal state and the function of the SCN is to restrict activity to an appropriate time period.

Two other molecules, DBP and histidine decarboxylase (HDC), have been shown to affect circadian locomotor activity. Both mice with a targeted deletion of HDC and mice lacking DBP, a PAR leucine transcription factor that is rhythmically expressed in the SCN, display significantly reduced locomotor activity (Lopez-Molina et al., 1997; Abe et al., 2004). However, both mutants also display altered free-running periods, suggesting that these proteins are either involved in regulating the core molecular clock, or are part of a feedback mechanism from behavior to the SCN. Finally, Shimomura and colleagues have exploited the differences in circadian behavior among inbred mouse strains to identify candidate genes involved in circadian phenotype, including activity amplitude. This genome-wide approach involving quantitative trait locus analysis (QTL) of circadian phenotypes in C57BL/6J x BALBc/J F2 hybrids has identified two loci, *Activity-1* and *Activity-2*, linked to changes in circadian activity level, although the associated genes have yet to be cloned (Shimomura et al., 2001).

6.5 Peripheral Oscillators

In additional to regulating physiology and behavior via endocrine and neural output, the SCN also modulates physiology by phase-coordinating oscillators that are present in peripheral cells. Until recently, it was commonly believed that the oscillator in the SCN and the oscillators in peripheral tissues functioned in a master–slave relationship, with the autonomous pacemaker in the SCN driving peripheral oscillators.

In the absence of the SCN, peripheral oscillators would quickly run down. However, new data indicate that this model is incorrect, and that peripheral oscillators are in fact autonomous, continuing to oscillate even without SCN input (Yoo et al., 2004). Therefore, the function of the SCN appears to be to phase-coordinate oscillators in peripheral tissues.

Following the cloning and identification of the core mammalian clock components, it was found that genes such as *Per1* and *Bmal1* were expressed almost ubiquitously throughout the organism, and that expression was rhythmic in almost every tissue, including liver, lung, muscle, and kidney (Balsolobre et al., 1998; Reppert and Weaver, 2001). Balsolobre and colleagues (1998) showed that a serum shock could induce rhythmic transcription of *Per1* and *Per2* in cultured Rat-1 fibroblasts, and that such oscillations could be sustained for approximately 3 days with a stable period of 22.5 h. Later, Yagita and colleagues (2001) found that serum shock of NIH3T3 fibroblasts induced robust rhythmic expression of *Bmal1*, the *Period* genes, and *Cry1*, with the *Bmal1* and *Per* expression levels antiphase to each other and *Cry1* peak expression delayed relative to peak *Per1* expression by 4–8 h, as is observed in SCN neurons. The period of clock gene expression in peripheral tissues is regulated by genotype: MEFs from *Cry1* knockout mice, which have a shortened free-running period of locomotor activity, show short periodicity of *Dbp* expression, whereas MEFs from *Cry2* knockout mice, which have a lengthened activity period, have a lengthened periodicity of *Dbp* expression. MEFs cultured from *Cry1/Cry2* double knockout mice, which are arrhythmic under constant conditions, showed constitutively high, nonrhythmic *Dbp* expression (Yagita et al., 2001).

The SCN primarily regulates peripheral oscillators via endocrine output. Balsolobre and colleagues (2000) showed that the glucocorticoid hormone analog dexamethasone could induce rhythmic *Period, Cry,* and *Dbp* expression in Rat-1 fibroblasts. Furthermore, mice injected with dexamethasone showed transient *Per1* induction in the liver in a CT-dependent manner: injections at ZT14 and ZT21, after normal peak *Per1* expression, resulted in phase delays of *Per1* and *Dbp* expression, whereas injections at ZT1, before the normal *Per1* peak, phase-advanced *Per1* and *Dbp* expression (Balsolobre et al., 2000). Importantly, the effects of dexamethasone on clock gene expression in the liver occurred in the absence of any effects on *Per1* expression in the SCN, indicating that glucocorticoids function downstream of the SCN to regulate peripheral clocks.

Although the SCN is the primary phase-coordinator of peripheral rhythms, recent studies have shown that peripheral oscillators can be uncoupled from the SCN by the application of stimuli, such as restricted feeding, that bypass the SCN (Damiola et al., 2000; Stokkan et al., 2001). Damiola and colleagues (2000) maintained mice on a normal LD12:12 cycle, but restricted food availability either to the dark or to the light phase. *Period, Cry1,* and *Rev-Erbα* gene expression in livers from mice fed at night showed a normal rhythm and phase angle of expression, with peak peripheral clock gene expression phase delayed by ~8 h relative to peak gene expression in the SCN. However, clock gene expression in livers from mice restricted to day-time feeding was 8–12 h out of phase with that of from night-time fed mice, although there was no effect of day-time feeding on the phase of rhythmic gene expression in the SCN (Damiola et al., 2000). Restricted feeding appears to phase shift peripheral tissues via global circulating signals, as tissues not directly involved in food response, including heart and lung, also show phase shifts in clock gene expression following restricted feeding. However, the nature of the entraining signal is unknown: although glucocorticoids can phase shift peripheral rhythms, Stokkan and colleagues (2001) have suggested that, although the pattern of corticosterone secretion changes during restricted feeding, the new peak of corticosterone occurs at a time when peripheral tissues are refractory to the phase-shifting effects of glucocorticoids.

The ability of peripheral clocks to be uncoupled from SCN regulation reflects the physiological role of peripheral oscillators. The SCN restricts certain behaviors, such as locomotor activity and reproductive receptivity, to the night, when nocturnal rodents can best avoid prey and are most likely to have conspecific interactions. However, peripheral clocks in organs such as the liver and pancreas, both involved in feeding, must be able to respond whenever food is available. Peripheral circadian gene expression plays a crucial role in many biochemical pathways, as microarray experiments have shown that up to 10% of the genes expressed in all organs studied shows rhythmic transcription profiles (Akhtar et al., 2002; Panda et al., 2002a; Storch et al., 2002). In keeping with the different functions that individual organs perform, comparisons of rhythmic gene expression between the SCN, heart, and liver show that the majority of genes that are rhythmic in one tissue are not rhythmic—and sometimes not even expressed—in the other tissues. Out of approximately 650 cycling transcripts in mouse SCN and liver, Panda and colleagues found that only 28 genes were rhythmic in both tissues, including genes like *Per2* that were already known to be involved in the clock mechanism (Panda et al., 2002a). Although the function of some of the 28 genes is unknown, it is likely that all are involved either in the regulation of the core molecular clock or in the basic circadian cellular outputs.

Observation of peripheral molecular clocks at the level of the individual cell has shown that rhythms can persist for weeks or months in the absence of the SCN, although individual cells may drift out of phase with each other. Yoo and colleagues (2004) developed a reporter mouse that carries the luciferase gene fused to the 3′ end of the *Per2* gene, so that luciferase is under the same transcriptional and posttranscriptional regulation as *Per2*. Unlike earlier work that involved transgenes using the *Per1* promoter fused to the *luciferase* gene, all of which had shown that rhythms in explanted peripheral tissues damped after several days, Yoo found that all explanted tissues, including pituitary, lung, lung, kidney, and tail, showed robust rhythms that persisted for up to 20 days. Furthermore, peripheral tissues removed from behaviorally arrhythmic SCN-lesioned animals showed rhythmic oscillations, although there were significant differences in tissue phase within and among animals (Yoo et al., 2004). Thus, the Yoo study suggests that it is time to revise our understanding of the role the SCN plays: rather than driving peripheral rhythms, SCN output is responsible for phase-coordinating the oscillators within peripheral cells and tissue.

7 Conclusion

The tissue-specific regulation of circadian gene expression indicates that many specialized processes are under circadian control; presence of cell-autonomous oscillators in almost every cell type studied and the extent of circadian regulation of the SCN and peripheral transcriptomes suggest that the ability to time-regulate physiological functions is extremely beneficial evolutionarily. Under normal circumstances, the SCN is entrained by a predictable LD cycle, and subsequently phase-coordinates peripheral oscillators to organize behavior and physiology appropriately. In nocturnal rodents, physiological processes such as metabolism and ovulation are restricted to the dark period, when feeding and mating are most likely to occur, and behaviors such as locomotor activity—important in food gathering—and estrus behaviors, crucial for mating, are also restricted to the night, when the risk of predation is lower. However, in situations when nonphotic zeitgebers are out of phase with photic cues, peripheral tissues are able to respond by functioning independently of the SCN.

Although humans have escaped from the confines of the LD cycle imposed by the earth's rotation, we are still subject to the regulation of physiology by the SCN. This regulation is normally only obvious to us as the sleep–wake cycle, but also becomes apparent following circadian disruptions such as shift-work and long-distance travel, which have been linked to gastric disruptions, disrupted reproductive function, mood disorders, and even a heightened risk of cancer (Hastings et al., 2003). However, based on the genome-wide circadian regulation of gene expression observed in mice, it is likely that circadian rhythms in humans have a multitude of more subtle effects that may turn out to be important. A new field, chronotherapeutics, recognizes that there are CT-dependent changes in both disease processes and the ability to metabolize therapeutic drugs; if this field is successful, we may be able to make circadian gene regulation work in our favor even while defying the natural LD cycle.

References

Abe H, Honma S, Ohtsu H, Honma K. 2004. Circadian rhythms in behavior and clock gene expressions in the brain of mice lacking histidine decarboxylase. Mol Brain Res 124: 178-187.

Akhtar RA, Reddy AB, Maywood ES, Clayton JD, King VM, et al. 2002. Circadian cycling of the mouse liver transcriptome, as revealed by cDNA microarray, is driven by the suprachiasmatic nucleus. Curr Biol 12: 540-550.

Albus H, Bonnefont X, Chaves I, Yasui A, Doczy J, et al. 2002. *Cryptochrome*-deficient mice lack circadian electrical activity in the suprachiasmatic nuclei. Curr Biol 12: 1130-1133.

Amir S, Stewart JM. 1996. Resetting of the circadian clock by a conditioned stimulus. Nature 379: 542-545.

Antoch MP, Song EJ, Chang A-M, Vitaterna MH, Zhao Y, et al. 1997. Functional identification of the mouse circadian *Clock* gene by transgenic BAC rescue. Cell 89: 655-667.

Aschoff J. 1965. Circadian rhythms in man. Science 148: 1427-1432.

Bae K, Xiaowei J, Maywood ES, Hastings MH, Reppert SM, et al. 2001. Differential functions of *mPer1*, *mPer2*, and *mPer3* in the SCN circadian clock. Neuron 30: 525-536.

Balsolobre A, Brown SA, Marcacci L, Tronche F, Kellendonk C, et al. 2000. Resetting of circadian time in peripheral tissues by glucocorticoid signaling. Science 289: 2344-2347.

Balsolobre A, Damiola F, Schibler U. 1998. A serum shock induces circadian gene expression in mammalian tissue culture cells. Cell 93: 929-937.

Barkley M, Bradford G, Geschwind I. 1978. The pattern of plasma prolactin concentration during the first half of mouse gestation. Biol Reprod 19: 291-296.

Barnes JW, Tischkau SA, Barnes JA, Mitchell JW, Burgoon PW, et al. 2003. Requirement of mammalian *Timeless* for circadian rhythmicity. Science 302: 439-442.

Beaule C, Amir S. 1999. Photic entrainment and induction of immediate-early genes within the rat circadian system. Brain Res 821: 95-100.

Bergstrom AL, Hannibal J, Hindersson P, Fahrenkrug J. 2003. Light-induced phase shift in the Syrian hamster (*Mesocricetus auratus*) is attenuated by the PACAP receptor antagonist PACAP6-38 or PACAP immunoneutralization. Eur J Neurosci 18: 2552-2562.

Berson DM, Dunn FA, Takao M. 2002. Phototransduction by retinal ganglion cells that set the circadian clock. Science 295: 1070-1073.

Boer K, Boer GJ, Swaab DF. 1981. Reproduction in Brattleboro rats with diabetes insipidus. J Reprod Fertil 61: 273-280.

Bos NP, Mirmiran M. 1990. Circadian rhythms in spontaneous neuronal discharges of the cultured suprachiasmatic nucleus. Brain Res 511: 158-162.

Brown NL, Patel S, Brzezinski J, Glaser T. 2001. Math5 is required for retinal ganglion cell and optic nerve formation. Development 128: 2497-2508.

Brown SA, Schibler U. 1999. The ins and outs of circadian timekeeping. Curr Opin Gen Dev 9: 588-594.

Brown- Grant K, Raisman G. 1977. Abnormalities in reproductive function associated with the destruction of the suprachiasmatic nuclei in female rats. Proc R Soc Lond B Biol Sci 198: 279-296.

Brzezinski JAt, Brown NL, Tanikawa A, Bush RA, Sieving PA, et al. 2005. Loss of circadian photoentrainment and abnormal retinal electrophysiology in Math5 mutant mice. Invest Ophthalmol Vis Sci 46: 2540-2551.

Buijs RM, la Fleur SE, Wortel J, Van Heyningen C, Zuiddam L, et al. 2003. The suprachiasmatic nucleus balances sympathetic and parasympathetic output to peripheral organs through separate preautonomic neurons. J Comp Neurol 464: 36-48.

Buijs RM, Wortel J, Van Heerikhuize JJ, Feenstra MGP, Ter Horst GJ, et al. 1999. Anatomical and functional demonstration of a multisynaptic suprachiasmatic nucleus adrenal (cortex) pathway. Eur J Neurosci 11: 1535-1544.

Bunger MK, Wilsbacher LD, Moran SM, Clendenin C, Radcliffe LA, et al. 2000. *Mop3* is an essential component of the master circadian pacemaker in mammals. Cell 103: 1009-1017.

Butcher GQ, Lee B, Obrietan K. 2003. Temporal regulation of light-induced extracellular signal-regulated kinase activation in the suprachiasmatic nucleus. J Neurophysiol 90: 3854-3863.

Cagampang FR, Piggins HD, Sheward WJ, Harmar AJ, Coen CW. 1998a. Circadian changes in PACAP type 1 (PAC1) receptor mRNA in the rat suprachiasmatic and supraoptic nuclei. Brain Res 813: 218-222.

Cagampang FR, Sheward WJ, Harmar AJ, Piggins HD, Coen CW. 1998b. Circadian changes in the expression of vasoactive intestinal peptide 2 receptor mRNA in the rat suprachiasmatic nuclei. Brain Res Mol Brain Res 54: 108-112.

Camacho F, Cilio M, Guo Y, Virshup DM, Patel K, et al. 2001. Human casein kinase Idelta phosphorylation of human circadian clock proteins period 1 and 2. FEBS Lett 489: 159-165.

Castel M, Belenky M, Cohen S, Wagner S, Schwartz WJ. 1997. Light-induced c-Fos expression in the mouse suprachiasmatic nucleus: immunoelectron microscopy reveals co-localization in multiple cell types. Eur J Neurosci 9: 1950- 1960.

Chen D, Buchanan GF, Ding JM, Hannibal J, Gillette MU. 1999. Pituitary adenylyl cyclase-activating peptide: a pivotal modulator of glutamatergic regulation of the suprachiasmatic circadian clock. Proc Natl Acad Sci USA 96: 13468-13473.

Cheng HY, Obrietan K, Cain SW, Lee BY, Agostino PV, et al. 2004. Dexras1 potentiates photic and suppresses nonphotic responses of the circadian clock. Neuron 43: 715-728.

Cheng MY, Bullock CM, Li C, Lee AG, Bermak JC, et al. 2002. Prokineticin 2 transmits the behavioural circadian rhythm of the suprachiasmatic nucleus. Nature 417: 405-410.

Colwell C. 2000a. Circadian modulation of calcium levels in cells in the suprachiasmatic nucleus. Eur J Neurosci 12: 571-576.

Colwell CS. 2000b. Rhythmic coupling among cells in the suprachiasmatic nucleus. J Neurobiol 43: 379-388.

Colwell CS, Menaker M. 1992. NMDA as well as non-NMDA receptor antagonists can prevent the phase-shifting effects of light on the circadian system of the golden hamster. J Biol Rhythms 7: 125-136.

Colwell CS, Foster RG, Menaker M. 1991. NMDA receptor antagonists block the effects of light on circadian behavior in the mouse. Brain Res 554: 105-110.

Colwell CS, Ralph MR, Menaker M. 1990. Do NMDA receptors mediate the effects of light on circadian behavior? Brain Res 523: 117-120.

Colwell CS, Michel S, Itri J, Rodriguez W, Tam J, et al. 2004. Selective deficits in the circadian light response in mice lacking PACAP. Am J Physiol Regul Integr Comp Physiol 287, R1194-1201.

Daan S. 1977. Tonic and phasic effects of light in the entrainment of circadian rhythms. Ann NY Acad Sci 290: 51-59.

Daan S, Aschoff J. 2001. The entrainment of circadian systems., In Handbook of Behavioral Neurobiology: Circadian Clocks, Takahashi JS, Turek FW, Moore RY, editors. New York: Kluwer/Plenum.

Damiola F, Minh NL, Preitner N, Kornmann B, Fleury- Olela F, et al. 2000. Restricted feeding uncouples circadian oscillators in peripheral tissues from the central pacemaker in the suprachiasmatic nucleus. Genes Dev 14: 2950-2961.

Darlington TK, Lyons LC, Hardin PE, Kay SA. 2000. The period E-box is sufficient to drive circadian oscillation of transcription *in vivo.* J Biol Rhythms 15: 462-471.

de Boer SF, Van der Gugten J. 1987. Daily variations in plasma noradrenaline, adrenaline, and corticosterone concentrations in rats. Physiol Behav 40: 323-328.

de Jeu M, Hermes M, Pennartz C. 1988. Circadian modulation of membrane properties in slices of rat suprachiasmatic nucleus. Neuroreport 9: 3725-3729.

de Mairan J-J. 1729. Observation botanique. Histoir de l'Academie Royale des Science, 35–36.

Ding JM, Chen D, Weber ET, Faiman LE, Rea MA, et al. 1994. Resetting the biological clock: mediation of nocturnal circadian shifts by glutamate and NO. Science 266: 1713-1717.

Edgar DM, Dement WC. 1991. Regularly scheduled voluntary exercise synchronizes the mouse circadian clock. Am J Physiol 261: R928-933.

Engelmann W, Eger I, Johnsson A,Karlsson HG. 1974. Effects of temperature pulses on the petal rhythm of Kalanchoe: An experimental and theoretical study. Int J Chronobiol, 29: 347–358

Evans NP, Dahl GE, Glover BH, Karsch FJ. 1994. Central regulation of pulsatile gonadotropin-releasing hormone (GnRH) secretion by estradiol during the period leading up to the preovulatory GnRH surge in the ewe. Endocrinology 134: 1806-1811.

Everett JW, Sawyer CH. 1950. A 24 h periodicity in the LH-release apparatus of female rats, disclosed by barbituate sedation. Endocrinology 46: 196-216.

Fang M, Jaffrey SR, Sawa A, Ye K, Luo X, et al. 2000. Dexras1: a G protein specifically coupled to neuronal nitric oxide synthase via CAPON. Neuron 28: 183-193.

Feldman JF, Hoyle MN. 1973. Isolation of circadian clock mutants of Neurospora crassa. Genetics 75: 605-613.

Field M, Maywood E, O'Brien J, Weaver D, Reppert S, et al. 2000. Analysis of clock proteins in mouse SCN demonstrates phylogenetic divergence of the circadian clockwork and resetting mechanisms. Neuron 25: 437-447.

Foster RG. 1998. Shedding light on the biological clock. Neuron 20: 829-832.

Foster RG, Provencio I, Hudson D, Fiske S, De Grip W, et al. 1991. Circadian photoreception in the retinally degenerate mouse (rd/rd). J Comp Physiol A 169: 39-50.

Freedman MS, Lucas RJ, Soni B, von Schantz M, Munoz M, et al. 1999. Regulation of mammalian circadian behavior by non-rod, non-cone, ocular photoreceptors. Science 284: 502-504.

Gau D, Lemberger T, von Gall C, Kretz O, Le Minh N, et al. 2002. Phosphorylation of CREB Ser142 regulates light-induced phase shifts of the circadian clock. Neuron 34: 245-253.

Gekakis N, Staknis D, Nguyen HB, Davis FC, Wilsbacher LD, et al. 1998. Role of the CLOCK protein in the mammalian circadian mechanism. Science 280: 1564-1569.

Gillette M, Reppert S. 1987. The hypothalamic suprachiasmatic nuclei: Circadian patterns of vasopressin secretion and neuronal activity *in vitro.* Brain Res Bull 19: 135-139.

Gillette MU, Medanic M, McArthur AJ, Liu C, Ding JM, et al. 1995a. Intrinsic neuronal rhythms in the suprachiasmatic nuclei and their adjustment. Ciba Found Symp 183: 134-144; discussion 144–153.

Gillette MU, Medanic M, McArthur AJ, Liu C, Ding JM, et al. 1995b. Intrinsic neuronal rhythms in the suprachiasmatic nuclei and their adjustment, In Circadian Clocks and their Adjustment. Wiley and Sons, pp. 134–153.

Ginty DD, Kornhauser JM, Thompson MA, Bading H, Mayo KE, et al. 1993. Regulation of CREB phosphorylation in the suprachiasmatic nucleus by light and a circadian clock. Science 260: 238-241.

Gooley JJ, Lu J, Chou TC, Scammell TE, Saper CB. 2001. Melanopsin in cells of origin of the retinohypothalamic tract. Nat Neurosci 4: 1165

Gotter AL, Manganaro T, Weaver DR, Kolakowski LF, Possidente B, et al. 2000. A time-less function of mouse *Timeless.* Nat Neurosci 3: 755-756.

Griffin EA, Staknis D, Weitz CJ. 1999. Light-independent role of CRY1 and CRY2 in the mammalian circadian clock. Science 286: 768-771.

Gwinner E. 1966. Entrainment of a circadian rhythm in birds by species-specific song cycles (Aves, Fringillidae; *Cardeulis spinus, Serinus serinus*). Experientia 22: 1-3.

Hamann A, Matthaei S. 1996. Regulation of energy balance by leptin. Exp Clin Endocrinol Diabetes 104: 293-300.

Hannibal J. 2002. Neurotransmitters of the retino-hypothalamic tract. Cell Tissue Res 309: 73-88.

Hannibal J, Ding JM, Chen D, Fahrenkrug J, Larsen PJ, et al. 1998. Pituitary adenylate cyclase activating peptide (PACAP) in the retinohypothalamic tract: a daytime regulator of the biological clock. Ann NY Acad Sci 865: 197-206.

Hao H, Allen DL, Hardin PE. 1997. A circadian enhancer mediates PER-dependent mRNA cycling in *Drosophila melanogaster.* Mol Cell Biol 17: 3687-3693.

Harrington ME, Hoque S, Hall A, Golombek D, Biello S. 1999. Pituitary adenylate cyclase activating peptide phase shifts circadian rhythms in a manner similar to light. J Neurosci 19: 6637-6642.

Hastings MH, Reddy AB, Maywood ES. 2003. A clockwork web: circadian timing in brain and periphery, in health and disease. Nat Rev Neurosci 4: 649-661.

Hastings MH, Field MD, Maywood ES, Weaver DR, Reppert SM. 1999. Differential regulation of mPER1 and mTIM proteins in the mouse suprachiasmatic nuclei: new insights into a core clock mechanism. J Neurosci 19(12):RC11.

Hattar S, Lucas RJ, Mrosovsky N, Thompson S, Douglas RH, et al. 2003. Melanopsin and rod-cone photoreceptive systems account for all major accessory visual functions in mice. Nature 424: 76-81.

Hayashi Y, Sanada K, Fukada Y. 2001. Circadian and photic regulation of MAP kinase by Ras- and protein phosphatase-dependent pathways in the chick pineal gland. FEBS Lett 491: 71-75.

Hayes KR, Baggs JE, Hogenesch JB. 2005. Circadian clocks are seeing the systems biology light. Genome Biol 6: 219

Hermes ML, Ruijter JM, Klop A, Buijs RM, Renaud LP. 2000. Vasopressin increases GABAergic inhibition of rat hypothalamic paraventricular nucleus neurons *in vitro.* J Neurophysiol 83: 705-711.

Herzog E, Geusz M, Khalsa S, Straume M, Block G. 1997. Circadian rhythms in mouse suprachiasmatic nucleus explants on multimicroelectrode plates. Brain Res 757: 285-290.

Herzog ED, Takahashi JS, Block GD. 1998. Clock controls circadian period in the isolated suprachiasmatic nucleus neurons. Nat Neurosci 1: 708-713.

Herzog ED, Aton SJ, Numano R, Sakaki Y, Tei H. 2004. Temporal precision in the mammalian circadian system: a reliable clock from less reliable neurons. J Biol Rhythms 19: 35-46.

Hida A, Koike N, Hirose M, Hattori M, Sakaki Y, et al. 2000. The human and mouse *Period1* genes: five well-conserved E-boxes additively contribute to the enhancement of *mPer1* transcription. Genomics 65: 224-233.

Hogenesch JB, Gu YZ, Jain S, Bradfield CA. 1998. The basic-helix-loop-helix-PAS orphan MOP3 forms transcriptionally active complexes with circadian and hypoxia factors. Proc Natl Acad Sci USA 95: 5474-5479.

Honma S, Shirakawa T, Katsuno Y, Namihira M, Honma K. 1998. Circadian periods of single suprachiasmatic neurons in rats. Neurosci Lett 250: 157-160.

Honma S, Kawamoto T, Takagi Y, Fujimoto K, Sato F, et al. 2002. *Dec1* and *Dec2* are regulators of the mammalian molecular clock. Nature 419: 841-844.

Hsu DS, Zhao X, Zhao S, Kazantsev A, Wang RP, et al. 1996. Putative human blue-light photoreceptors hCRY1 and hCRY2 are flavoproteins. Biochemistry 35: 13871-13877.

Hwang A, Maity A, McKenna WG, Muschel RJ. 1995. Cell cycle-dependent regulation of the cyclin B1 promoter. J Biol Chem 270: 28419-28424.

Ibata Y, Takahashi Y, Okamura H, Kawakami F, Terubayashi H, et al. 1989. Vasoactive intestinal peptide (VIP)-like immunoreactive neurons located in the rat suprachiasmatic nucleus receive a direct retinal projection. Neurosci Lett 97: 1-5.

Ikeda M, Sugiyama T, Wallace CS, Gompf HS, Yoshioka T, et al. 2003. Circadian dynamics of cytosolic and nuclear Ca+2 in single suprachiasmatic nucleus neurons. Neuron 38: 253-263.

Inouye S-IT. 1996. Circadian rhythms of neuropeptides in the suprachiasmatic nucleus. Progress in Brain Research, Buijs R, Kalsbeck A, Romijn H, Pennartz C, Mirmiran M, editors. Elsevier Science.

Inouye ST, Kawamura H. 1979. Persistance of circadian rhythmicity in a mammalian hypothalamic "island" containing the suprachiasmatic nucleus. Proc Natl Acad Sci USA 76: 5962-5966.

Ishida Y, Yagita K, Fukuyama T, Nishimura M, Nagano M, et al. 2001. Constitutive expression and delayed light response of casein kinase Iepsilon and Idelta mRNAs in the mouse suprachiasmatic nucleus. J Neurosci Res 64: 612-616.

Jacobs GH. 1993. The distribution and nature of colour vision among the mammals. Biol Rev Camb Philos Soc 68: 413-471.

Jin X, Shearman LP, Weaver DR, Zylka MJ, De Vries GJ, et al. 1999. A molecular mechanism regulating rhythmic output from the suprachiasmatic circadian clock. Cell 96: 57-68.

Johnson MS. 1926. Activity and distribution of certain wild mice in relation to biotic communities. J Mammal 7: 245-275.

Kalsbeek A, Buijs RM. 2002. Output pathways of the mammalian suprachiasmatic nucleus: coding circadian time by transmitter selection and specific targeting. Cell Tissue Res 309: 109-118.

Kalsbeek A, Buijs RM, van Heerikhuize JJ, Arts M, van der Woude TP. 1992. Vasopressin-containing neurons of the suprachiasmatic nuclei inhibit corticosterone release. Brain Res 580: 62-67.

Kalsbeek A, van Heerikhuize JJ, Wortel J, Buijs RM. 1996. A diurnal rhythm of stimulatory input to the hypothalamo-pituitary-adrenal system as revealed by timed intrahypothalamic administration of the vasopressin V1 antagonist. J Neurosci 16: 5555-5565.

Kapfhamer D, Valladares O, Sun Y, Nolan PM, Rux JJ, et al. 2002. Mutations in *Rab3a* alter circadian period and homeostatic response to sleep loss in the mouse. Nat Genet 32: 290-295.

Kavakli IH, Sancar A. 2002. Circadian photoreception in humans and mice. Mol Interv 2: 484-492.

King DP, Takahashi JS. 2000. Molecular genetics of circadian rhythms in mammals. Annu Rev Neurosci 23: 713-742.

King DP, Zhao Y, Sangoram AM, Wilsbacher LD, Tanaka M, et al. 1997. Positional cloning of the mouse circadian *Clock* gene. Cell 89: 641-653.

Kondratov RV, Chernov MV, Kondratova AA, Gorbacheva VY, Gudkov AV, et al. 2003. BMAL1-dependent circadian oscillation of nuclear CLOCK: posttranslational events induced by dimerization of transcriptional activators of the mammalian clock system. Genes Dev 17: 1921-1932.

Kornhauser JM, Cowan CW, Shaywitz AJ, Dolmetsch RE, Griffith EC, et al. 2002. CREB transcriptional activity in neurons is regulated by multiple, calcium-specific phosphorylation events. Neuron 34: 221-233.

Krajnak K, Rosewell KL, Wise PM. 2001. Fos-induction in gonadotropin-releasing hormone neurons receiving vasoactive intestinal polypeptide innervation is reduced in middle-aged female rats. Biol Reprod 64: 1160-1164.

Kramer A, Yang F-C, Snodgrass P, Li X, Scammell TE, et al. 2001. Regulation of daily locomotor activity and sleep by hypothalamic EGF receptor signaling. Science 294: 2511-2515.

Kulhman SJ, Silver R, Le Sauter J, Bult- Ito A, McMahon DG. 2003. Phase resetting light pulses induce *Per1* and persistant spike activity in a subpopulation of biological clock neurons. J Neurosci 23: 1441-1450.

Kume K, Zylka MJ, Sriram S, Shearman LP, Weaver DR, et al. 1999. mCRY1 and mCRY2 are essential components of the negative limb of the circadian click feedback loop. Cell 98: 193-205.

Leak RK, Moore RY. 2001. Topographic organization of suprachiasmatic nucleus projection neurons. J Comp Neurol 433: 312-334.

Lee C, Weaver DR, Reppert SM. 2004. Direct association between mouse PERIOD and CKIepsilon is critical for a functioning circadian clock. Mol Cell Biol 24: 584-594.

Lee CC, Etchegaray J-P, Cagampang FRA, Loudon AS, Reppert SM. 2001. Posttranslational mechanisms regulate the mammalian circadian clock. Cell 107: 855-867.

Lehman MN, Silver R, Gladstone WR, Kahn RM, Gibson M, et al. 1987. Circadian rhythmicity restored by neural transplant. Immunocytochemical characterization of the graft and its integration with the host brain. J Neurosci 7: 1626-1638.

Levine JE, Bauer-Dantoin AC, Besecke LM, Conaghan LA, Legan SJ, et al. 1991. Neuroendocrine regulation of the luteineizing hormone-releasing hormone pulse generator in the rat. Rec Prog Horm Res 47: 97-153.

Levine JE, Chappell PE, Besecke LM, Bauer-Dantoin AC, Wolfe AM, et al. 1995. Amplitude and frequency modulation of pulsatile luteinizing hormone-releasing hormone release. Cell Mol Neurobiol 15: 117-139.

Liu C, Weaver DR, Strogatz SH, Reppert SM. 1997. Cellular construction of a circadian clock: period determination in the suprachiasmatic nuclei. Cell 91: 855-860.

Lopez- Molina L, Conquet F, Dubois- Dauphin M, Schibler U. 1997. The DBP gene is expressed according to a circadian rhythm in the suprachiasmatic nucleus and influences circadian behavior. EMBO J 16: 6762-6771.

Low- Zeddies SS, Takahashi JS. 2001. Chimera analysis of the *Clock* mutation in mice shows that complex cellular integration determines circadian behavior. Cell 105: 25-42.

Lowrey PL, Takahashi JS. 2000. Genetics of the mammalian circadian system: Photic entrainment, circadian pacemaker mechanisms, and postranslational regulation. Annu Rev Genet 34: 533-562.

Lowrey PL, Takahashi JS. 2004. Mammalian circadian biology: elucidating genome-wide levels of temporal organization. Annu Rev Genomics Hum Genet 5: 407-441.

Lowrey PL, Shimomura K, Antoch MP, Yamazaki S, Zemenides PD, et al. 2000. Positional syntenic cloning and functional cloning characterization of the mammalian circadian mutation *tau*. Science 288: 483-491.

Lu J, Zhang YH, Chou TC, Gaus SE, Elmquist JK, et al. 2001. Contrasting effects of ibotenate lesions of the paraventricular nucleus and subparaventricular nucleus on sleep-wake cycle and temperature regulation. J Neurosci 21: 4864-4874.

Lucas RJ, Freedman MS, Munoz M, Garcia- Fernandez JM, Foster RG. 1999. Regulation of the mammalian pineal by non-rod, non-cone, ocular photoreceptors. Science 284: 505-507.

Lupi D, Cooper HM, Froehlich A, Standford L, McCall MA, et al. 1999. Transgenic ablation of rod photoreceptors alters the circadian phenotype of mice. Neuroscience 89: 363-374.

Magni P, Vettor R, Pagano C, Calcagno A, Beretta E, et al. 1999. Expression of a leptin receptor in immortalized gonadotropin-releasing hormone-secreting neurons. Endocrinology 140: 1581-1585.

Maywood ES, O'Brien JA, Hastings MH. 2003. Expression of mCLOCK and other circadian clock-relevant proteins in the mouse suprachiasmatic nuclei. J Neuroendocrinol 15: 329-334.

McCall MA, Gregg RG, Merriman K, Goto Y, Peachey NS, et al. 1996. Morphological and physiological consequences of the selective elimination of rod photoreceptors in transgenic mice. Exp Eye Res 63: 35-50.

Melyan Z, Tarttelin EE, Bellingham J, Lucas RJ, Hankins MW. 2005. Addition of human melanopsin renders mammalian cells photoresponsive. Nature 433: 741-745.

Menaker M, Takahashi J, Eskin A. 1978. The physiology of circadian pacemakers. Ann Rev Physiol 40: 501-526.

Meyer-Bernstein EL, Jetton AE, Markuns JF, Matsumoto S-I, Lehman MN, et al. 1999. Effects of suprachiasmatic transplants on circadian rhythms of neuroendocrine function in golden hamsters. Endocrinology 140: 207-218.

Miller BH, Olson SL, Turek FW, Levine JE, Horton TH, et al. 2004. Circadian *Clock* mutation disrupts estrous cyclicity and maintenance of pregnancy. Curr Biol 14: 1367-1373.

Minami Y, Furuno K, Akiyama M, Moriya T, Shibata S. 2002. Pituitary adenylate cyclase-activating polypeptide produces a phase shift associated with induction of mPer expression in the mouse suprachiasmatic nucleus. Neuroscience 113: 37-45.

Mintz EM, Marvel CL, Gillespie CF, Price KM, Albers HE. 1999. Activation of NMDA receptors in the suprachiasmatic nucleus produces light-like phase shifts of the circadian clock *in vivo*. J Neurosci 19: 5124-5130.

Miyamoto Y, Sancar A. 1998. Vitamin B2-based blue-light photoreceptors in the retinohypothalamic tract as the photoactive pigments for setting the circadian clock in mammals. Proc Natl Acad Sci USA 95: 6097-6102.

Moore RY, Card JP. 1994. Intergeniculate leaflet: an anatomically and functionally distinct subdivision of the lateral geniculate complex. J Comp Neurol 344: 403-430.

Moore RY, Eichler VB. 1972. Loss of a circadian adrenal corticosterone rhythm following suprachiasmatic lesions in the rat. Brain Res 42: 201-206.

Moore RY, Lenn NJ. 1972. A retinohypothalamic projection in the rat. J Comp Neurol 146: 1-14.

Moore RY, Speh JC, Leak RK. 2002. Suprachiasmatic nucleus organization. Cell Tissue Res 309: 89-98.

Muglia LJ. 2000. Genetic analysis of fetal development and parturition control in the mouse. Ped Res 47: 437-443.

Munoz E, Brewer M, Baler R. 2002. Circadian transcription: Thinking outside the E-box. J Biol Chem 277: 36009-36017.

Najima Y, Ikeda M, Kimura T, Honma S, Ohmiya Y, et al. 2004. Bidirectional role of orphan nuclear receptor ROR-alpha in clock gene transcriptions demonstrated by a novel reporter assay system. FEBS Lett 565: 122-126.

Neitz M, Neitz J. 2001. The uncommon retina of the common house mouse. Trends Neurosci 24: 248-250.

Nelson R, Zucker I. 1981. Absence of extraocular photoreception in diurnal and nocturnal rodents exposed to direct sunlight. Comp Biochem Physiol 69: 145-148.

Obrietan K, Impey S, Smith D, Athos J, Storm DR. 1999. Circadian regulation of cAMP response element-mediated gene expression in the suprachiasmatic nuclei. J Biol Chem 274: 17748-17756.

Okamura H, Miyake S, Sumi Y, Yamaguchi S, Yasui A, et al. 1999. Photic induction of *mPer1* and *mPer2* in *Cry*-deficient mice lacking a biological clock. Science 286: 2531-2534.

Palm IF, Van der Beek EM, Wiegant VM, Buijs RM, Kalsbeek A. 1999. Vasopressin induces a luteinizing hormone surge in ovariectomized, estradiol-treated rats with lesions of the suprachiasmatic nucleus. Neuroscience 93: 659-666.

Palm IF, Van der Beek EM, Swarts HJM, Van der Vliet J, Wiegant VM, et al. 2001. Control of the estradiol-induced prolactin surge by the suprachiasmatic nucleus. Endocrinology 142: 2296-2302.

Panda S, Antoch MP, Miller BH, Su AI, Schook AB, et al. 2002a. Coordinated transcription of key pathways in the mouse by the circadian clock. Cell 109: 307-320.

Panda S, Nayak SK, Campo B, Walker JR, Hogenesch JB, et al. 2005. Illumination of the melanopsin signaling pathway. Science 307: 600-604.

Panda S, Sato TK, Castrucci AM, Rollag MD, DeGrip WJ, et al. 2002b. Melanopsin (Opn4) requirement for normal light-induced circadian phase shifting. Science 298: 2213-2216.

Peterson SL, Ottem EN, Carpenter CD. 2003. Direct and indirect regulation of gonadotropin-releasing hormone neurons by estradiol. Biol Reprod 69: 1771-1778.

Pickard GE, Weber ET, Scott PA, Riberdy AF, Rea MA. 1996. 5HT1B receptor agonists inhibit light-induced phase shifts of behavioral circadian rhythms and expression of the immediate-early gene c-fos in the suprachiasmatic nucleus. J Neurosci 16: 8208-8220.

Pickard GE, Smith BN, Belenky M, Rea MA, Dudek FE, et al. 1999. 5-HT1B receptor-mediated presynaptic inhibition of retinal input to the suprachiasmatic nucleus. J Neurosci 19: 4034-4045.

Piggins HD, Marchant EG, Goguen D, Rusak B. 2001. Phase-shifting effects of pituitary adenylate cyclase activating polypeptide on hamster wheel-running rhythms. Neurosci Lett 305: 25-28.

Pittendrigh CS. 1960. Circadian rhythms and the circadian organization of living systems. Cold Spring Harbor Symp Quant Biol 25: 159-182.

Pittendrigh CS. 1981. Circadian systems: entrainment, In Handbook of Behavioral Biology: Biological Rhythms, Aschoff J, editor. New York: Plenum Press; pp. 95-124.

Preitner N, Damiola F, Molina LL, Zakany J, Duboule D, et al. 2002. The orphan nuclear receptor REV-ERBalpha controls circadian transcription within the positive limb of the mammalian circadian oscillator. Cell 110: 251-260.

Provencio I, Foster RG. 1995. Circadian rhythms in mice can be regulated by photoreceptors with cone-like characteristics. Brain Res 694: 183-190.

Provencio I, Rollag MD, Castrucci AM. 2002. Photoreceptive net in the mammalian retina. This mesh of cells may explain how some blind mice can still tell day from night. Nature 415: 493.

Provencio I, Jiang G, De Grip WJ, Hayes WP, Rollag MD. 1998. Melanopsin: An opsin in melanophores, brain, and eye. Proc Natl Acad Sci USA 95: 340-345.

Provencio I, Wong S, Lederman AB, Argamaso SM, Foster RG. 1994. Visual and circadian responses to light in aged retinally degenerate mice. Vision Res 34: 1799-1806.

Provencio I, Rodriguez IR, Jiang G, Hayes WP, Moreira EF, et al. 2000. A novel human opsin in the inner retina. J Neurosci 20: 600-605.

Quintero JE, Kulhman SJ, McMahon DG. 2003. The biological clock nucleus: a multiphasic oscillator network regulated by light. J Neurosci 23: 8070-8076.

Ralph MR, Menaker M. 1988. A mutation of the circadian system in golden hamsters. Science 241: 1225-1227.

Ralph MR, Foster RG, Davis FC, Menaker M. 1990. Transplanted suprachiasmatic nucleus determines circadian period. Science 247: 975-978.

Reppert SM, Weaver DR. 2001. Molecular analysis of mammalian circadian rhythms. Annu Rev Physiol 63: 647-676.

Richter CP. 1922. A behavioristic study of activity in the rat. Comp Psych Monographs 1: 1-55.

Richter CP. 1967. Sleep and activity: their relation to the 24-hour clock. Proc Assoc Res Nerv Ment Dis 45: 8-27.

Ripperger JA, Shearman LP, Reppert SM, Schibler U. 2000. CLOCK, an essential pacemaker component, controls expression of the circadian transcription factor DBP. Gen Dev 14: 679-689.

Romijn HJ, Sluiter AA, Pool CW, Wortel J, Buijs RM. 1996. Differences in the colocalization between Fos and PHI, GRP, VIP, and VP in neurons of the rat suprachiasmatic nucleus after a light stimulus during the phase delay versus the phase advance period of the night. J Comp Neurol 372: 1-8.

Rowland DL, Wagonblast AL, Dylestra TA. 1991. Timing of parturition in the rat: an analysis of successive births. Chronobiologia 18: 31-38.

Ruby NF, Brennan TJ, Xie X, Cao V, Franken P, et al. 2002. Role of melanopsin in circadian responses to light. Science 298: 2211-2213.

Rutter J, Reick M, Wu LC, McKnight SL. 2001. Regulation of clock and NPAS2 DNA binding by the redox state of NAD cofactors. Science 293: 510-514.

Sanada K, Okano T, Fukada Y. 2002. Mitogen-activated protein kinase phosphorylates and negatively regulates basic helix-loop-helix-PAS transcription factor BMAL1. J Biol Chem 277: 267-271.

Sanada K, Harada Y, Sakai M, Todo T, Fukada Y. 2004. Serine phosphorylation of mCRY1 and mCRY2 by mitogen-activated protein kinase. Genes Cells 9: 697-708.

Sargent ML, Woodward DO. 1969. Genetic determinants of circadian rhythmicity in Neurospora. J Bacteriol 97: 861-866.

Sathyanarayanan S, Zheng X, Xiao R, Sehgal A. 2004. Posttranslational regulation of Drosophila PERIOD protein by protein phosphatase 2A. Cell 116: 603-615.

Sato TK, Panda S, Miraglia LJ, Reyes TM, Rudic RD, et al. 2004. A functional genomics strategy reveals Rora as a component of the mammalian circadian clock. Neuron 43: 527-537.

Schwartz WJ, Gainer H. 1977. Suprachiasmatic nucleus: use of 14C-labeled deoxyglucose uptake as a functional marker. Science 197: 1089-1091.

Schwartz WJ, Gross RA, Morton MT. 1987. The suprachiasmatic nuclei contain a tetrodotoxin-resistant circadian pacemaker. Proc Natl Acad Sci USA 84: 1694-1698.

Selby CP, Thompson C, Schmitz TM, Van Gelder RN, Sancar A. 2000. Functional redundancy of cryptochromes and classical photoreceptors for nonvisual ocular photoreception in mice. Proc Natl Acad Sci USA 97: 14697-14702.

Shearman LP, Jin X, Lee CC, Reppert SM, Weaver DR. 2000a. Targeted disruption of the *mPer3* gene: subtle effects on circadian clock function. Mol Cell Biol 20: 6269-6275.

Shearman LP, Zylka MJ, Weaver DR, Kolakowski LF, Reppert SM. 1997. Two *period* homologs: circadian and photic regulation in the suprachiasmatic nucleus. Neuron 19: 1261-1269.

Shearman LP, Sriram S, Weaver DR, Maywood ES, Chaves I, et al. 2000b. Interacting molecular loops in the mammalian circadian clock. Science 288: 1013-1018.

Shibata S, Moore RY. 1993. Tetrodotoxin does not affect circadian rhythms in neuronal activity and metabolism in rodent suprachiasmatic nucleus *in vitro*. Brain Res 606: 259-266.

Shibata S, Watanabe A, Hamada T, Ono M, Watanabe S. 1994. N-methyl-D-aspartate induces phase shifts in circadian rhythm of neuronal activity of rat SCN *in vitro*. Am J Physiol 267, R360-364.

Shimomura K, Low- Zeddies SS, King DP, Steeves TDL, Whiteley A, et al. 2001. Genome-wide epistatic interaction analysis reveals complex determinants of circadian behavior in mice. Gen Res 11: 959-980.

Silver R, LeSauter J, Tresco PA, Lehman MN. 1996a. A diffusable coupling signal from the transplanted suprachiasmatic nucleus controlling circadian locomotor rhythms. Nature 382: 810-813.

Silver R, Romero MT, Besmer HR, Leak RK, Nunez JM, et al. 1996b. Calbindin-D28K cells in the hamster SCN express light-induced c-fos. Neuroreport 7: 1224-1228.

Smith BN, Sollars PJ, Dudek FE, Pickard GE. 2001. Serotonergic modulation of retinal input to the mouse suprachiasmatic nucleus mediated by 5-HT1B and 5-HT7 receptors. J Biol Rhythms 16: 25-38.

Smith MS. 1980. Role of prolactin in mammalian reproduction, Reproductive Physiology III, Greep RO, editor. Baltimore: University Park Press; pp. 249-276.

Smith MJ, Jennes L, Wise PM. 2000. Localization of the VIP2 receptor protein on GnRH neurons in the female rat. Endocrinology 141: 4317-4320.

Stephan FK, Zucker I. 1972. Circadian rhythms in drinking behavior and locomotor activity of rats are eliminated by hypothalamic lesions. Proc Natl Acad Sci USA 69: 1583-1586.

Stokkan K-A, Yamazaki S, Tei H, Sakaki Y, Menaker M. 2001. Entrainment of the circadian clock in the liver by feeding. Science 291: 490-493.

Storch KF, Lipna O, Leykin I, Viswanathan N, Davis FC, Wang WH, Weitz CJ. 2002. Extensive and divergent gene expression in liver and heart. Nature 417:78-83.

Swanson LW, Cowan WM. 1975. The efferent connections of the suprachiasmatic nucleus of the hypothalamus. J Comp Neurol 160 1-12.

Takahashi H, Umeda N, Tsutsumi Y, Fukumura R, Ohkaze H, et al. 2003. Mouse dexamethasone-induced RAS protein 1 gene is expressed in a circadian rhythmic manner in the suprachiasmatic nucleus. Brain Res Mol Brain Res 110: 1-6.

Takahashi JS. 2004. Finding new clock components: Past and future. J Biol Rhythms 19: 339-347.

Takahashi JS, Turek FW, Moore RY, editors. 2001. Circadian Clocks, Kluwer Academic/Plenum Publishers.

Takahashi JS, De Coursey PJ, Bauman L, Menaker M. 1984. Spectral sensitivity of a novel photoreceptive system mediating entrainment of mammalian circadian rhythms. Nature 308: 186-188.

Tanaka M, Hayashi S, Tamada Y, Ikeda M, Hisa Y, et al. 1997. Direct retinal projections to GRP neurons in the suprachiasmatic nucleus of the rat. Neuroreport 8: 2187-2191.

Thompson CL, Blaner WS, Van Gelder RN, Lai K, Quadro L, et al. 2001. Preservation of light signaling to the suprachiasmatic nucleus in vitamin A-deficient mice. Proc Natl Acad Sci USA 98: 11708-11713.

Thresher RJ, Vitaterna MH, Miyamoto Y, Kazantsev A, Hsu DS, et al. 1998. Role of mouse cryptochrome blue-light photoreceptor in circadian photoresponses. Science 282: 1490-1494.

Tischkau SA, Mitchell JW, Tyan S-H, Buchanan GF, Gillette MU. 2003. Ca^{+2}/cAMP response element-binding protein (CREB)-dependent activation of *Per1* is required for light-induced signaling in the suprachiasmatic nucleus circadian clock. J Biol Chem 278: 718-723.

Toh KL, Jones CR, He Y, Eide EJ, Hinz WA, et al. 2001. An hPer2 phosphorylation site mutation in familial advanced sleep phase syndrome. Science 291: 1040-1043.

Tracnickova Z, Cermakian N, Reppert SM, Sassone-Corsi P. 2002. Bimodal regulation of *mPeriod* promoters by CREB-dependent signaling and CLOCK/BMAL1 activity. Proc Natl Acad Sci 99: 7728-7733.

Ueda HR, Chen W, Adachi A, Wakamatsu H, Hayashi S, et al. 2002. A transcription factor response element for gene expression during circadian night. Nature 418: 534-539.

Ueda HR, Hayashi S, Chen W, Sano M, Machida M, et al. 2005. System-level identification of transcriptional circuits underlying mammalian circadian clocks. Nat Genet 37: 187-192.

Van der Beek EM, Horvath TL, Wiegant VM, Van den Hurk R, Buijs RM. 1997. Evidence for a direct neuronal pathway from the suprachiasmatic nucleus to the gonadotropin-releasing hormone system: combined tracing and light and electron microscopic immunocytochemical studies. J Comp Neurol 384: 569-579.

Van der Beek EM, van Oudheusden HJC, Buijs RM, van der Hurk HA, van der Honk R, et al. 1994. Preferential induction of c-fos immunoreactivity in vasoactive intestinal polypeptide-innervated gonadotropin-releasing hormone neurons during a steroid-induced luteinizing hormone surge in the female rat. Endocrinology 134: 2637-2644.

van der Horst GTJ, Muijtjens M, Kobayashi K, Takano R, Kanno S, et al. 1999. Mammalian *Cry1* and *Cry2* are essential for maintenance of circadian rhythms. Nature 398: 627-630.

Vielhaber E, Eide E, Rivers A, Gao ZH, Virshup DM. 2000. Nuclear entry of the circadian regulator mPER1 is controlled by mammalian casein kinase I epsilon. Mol Cell Biol 20: 4888-4899.

Vitaterna MH, King DP, Chang A, Kornhauser JM, Lowrey PL, et al. 1994. Mutagenesis and mapping of a mouse gene, *Clock*, essential for circadian behavior. Science 264: 719-725.

Vitaterna MH, Selby CP, Todo T, Niwa H, Thompson C, et al. 1999. Differential regulation of mammalian period genes and circadian rhythmicity by cryptochromes 1 and 2. Proc Natl Acad Sci USA 96: 12114-12119.

von Marilaun AK. 1895. The Natural History of Plants, Vol. 2, New York: Henry Holt.

Wang SW, Kim BS, Ding K, Wang H, Sun D, et al. 2001. Requirement for math5 in the development of retinal ganglion cells. Genes Dev 15: 24-29.

Watts AG, Swanson LW, Sanchez-Watts G. 1987. Efferent projections of the suprachiasmatic nucleus: I. Studies using anterograde transport of Phaseolus vulgaris leucoagglutinin in the rat. J Comp Neurol 258: 204-229.

Weber ET, Gannon RL, Rea MA. 1998. Local administration of serotonin agonists blocks light-induced phase advances of the circadian activity rhythm in the hamster. J Biol Rhythms 13: 209-218.

Wee R, Castrucci AM, Provencio I, Gan L, Van Gelder RN. 2002. Loss of photic entrainment and altered free-running circadian rhythms in math5-/- mice. J Neurosci 22: 10427-10433.

Welsh DK, Logothetis DE, Meister M, Reppert SM. 1995. Individual neurons dissociated from rat suprachiasmatic nucleus express independently phased circadian firing rhythms. Neuron 14: 697-706.

Wever RA. 1979. The Circadian System of Man: Results of Experiments Under Temporal Isolation, New York: Springer-Verlag.

Wiegand SJ, Terasawa E, Bridson WE, Goy RW. 1980. Effects of discrete lesions of preoptic and suprachiasmatic structures in the female rat. Alterations in the feedback regulation of gonadotropin secretion. Neuroendocrinology 31: 147-157.

Wray S, Hoffman GE. 1986. A developmental study of the quantitative distribution of LHRH neurons within the central nervous system of postnatal male and female rats. J Comp Neurol 252: 522-531.

Xu Y, Padiath QS, Shapiro RE, Jones CR, Wu SC, et al. 2005. Functional consequences of a CKIdelta mutation causing familial advanced sleep phase syndrome. Nature 434: 640-644.

Yagita K, Tamanini F, van der Horst GTJ, Okamura H. 2001. Molecular mechanisms of the biological clock in cultured fibroblasts. Science 292: 278-281.

Yagita K, Tamanini F, Yasuda M, Hoeijmakers JH, van der Horst GT, et al. 2002. Nucleocytoplasmic shuttling and mCRY-dependent inhibition of ubiquitylation of the mPER2 clock protein. EMBO J 21: 1301-1314.

Yamaguchi S, Isejima H, Matsuo T, Okura R, Yagita K, et al. 2003. Synchronization of cellular clocks in the suprachiasmatic nucleus. Science 302: 1408-1412.

Yamaguchi S, Mitsui S, Miyake S, Yan L, Onishi H, et al. 2000a. The 5′ upstream region of *mPer1* gene contains two promoters and is responsible for circadian oscillation. Curr Biol 10: 873-876.

Yamaguchi S, Mitsui S, Yan L, Yagita K, Miyake S, et al. 2000b. Role of DBP in the circadian oscillatory mechanism. Mol Cell Biol 20: 4773-4781.

Yamazaki S, Kerbeshian MC, Hocker CG, Block GD, Menaker M. 1998. Rhythmic properties of the hamster suprachiasmatic nucleus *in vivo*. J Neurosci 18: 10709-10723.

Yamazaki S, Numano R, Abe M, Hida A, Takahashi R, et al. 2000. Resetting central and peripheral circadian oscillators in transgenic rats. Science 288: 682-685.

Yan L, Takekida S, Shigeyoshi Y, Okamura H. 1999. *Per1* and *Per2* gene expression in the rat suprachiasmatic nucleus: circadian profile and the compartment-specific response to light. Neuroscience 94: 141-150.

Yannielli P, Harrington ME. 2004. Let there be "more" light: enhancement of light actions on the circadian system through non-photic pathways. Prog Neurobiol 74: 59-76.

Yannielli PC, Brewer JM, Harrington ME. 2004. Blockade of the NPY Y5 receptor potentiates circadian responses to light: complementary *in vivo* and *in vitro* studies. Eur J Neurosci 19: 891-897.

Yannielli PC, Harrington ME. 2000. Neuropeptide Y applied *in vitro* can block the phase shifts induced by light *in vivo*. Neuroreport 11: 1587-1591.

Yannielli PC, Harrington ME. 2001a. Neuropeptide Y in the mammalian circadian system: effects on light-induced circadian responses. Peptides 22: 547-556.

Yannielli PC, Harrington ME. 2001b. The neuropeptide Y Y5 receptor mediates the blockade of "photic-like" NMDA-induced phase shifts in the golden hamster. J Neurosci 21: 5367-5373.

YooS-H, Yamazaki S, Lowrey PL, Shimomura K, Ko CH, et al. 2004. PERIOD2:LUCIFERASE real-time reporting of circadian dynamics reveals persistent circadian oscillations in mouse peripheral tissues. Proc Natl Acad Sci USA 101: 5339-5346.

Yoo SH, Ko CH, Lowrey PL, Buhr ED, Song EJ, et al. 2005. A noncanonical e-box enhancer drivers mouse *Period2* circadian oscillations *in vivo*. Proc Natl Acad Sci USA 102: 2608-2613.

Young RW. 1985. Cell differentiation in the retina of the mouse. Anat Rec 212: 199-205.

Zheng B, Albrecht U, Kaasik K, Sage M, Lu W, et al. 2001. Nonredundant roles of the *mPer1* and *mPer2* genes in the mammalian circadian clock. Cell 105: 683-694.

Zheng B, Larkin DW, Albrecht U, Sun ZS, Sage M, et al. 1999. The *mPer2* gene encodes a functional component of the mammalian circadian clock. Nature 400: 169-173.

Zylka MJ, Shearman LP, Weaver DR, Reppert SM. 1998. Three period homologs in mammals: differential light responses in the suprachiasmatic circadian clocks and oscillating transcripts outside of brain. Neuron 20: 1103-1110.

23 Neuroendocrinology of Behavioral Rhythms

T. M. Lee · L. Smale

1 Introduction

From ancient times, humans have appreciated the role of specific organs in regulating behavior and physiological function, particularly the reproductive organs. The practice of castration of animals and men has been used to control behavior and fertility since earliest written records exist. In this chapter, we briefly review some history of how we came to appreciate the influence of hormones on behavior, the control of many of those hormonal systems by the brain, and how the exquisite timing of many behavioral events involves an interaction between the circadian mechanism (described in Chapter 22) and the neural control of hormones.

1.1 Historical Background on Neuroendocrine Relationship to Behavior

The earliest record of manipulation of behavior by manipulating hormones is described in ancient texts such as the Bible and writings of Aristotle (384–322 BCE). Aristotle compared the effects of castration of birds and of mammals (including humans). He noted that removal of the testes resulted in reduction of male typical traits such as the crowing of the rooster, the deep voice of the human male, the aggressive behavior of male ungulates during the breeding season, and the sexual interest in most males. Understanding the role of the testis in production of male specific traits led humans to consume testes, or eventually inject testicular extracts, in an attempt to regain and maintain masculine traits. Similar treatments were used on animals. These attempts to make use of the masculinizing, androgenic substances from the testis to improve strength, sexual prowess and fertility are little different from the use of androgenic compounds today.

The first successful experimental manipulation of an endocrine gland that resulted in behavioral changes was reported in 1849 by A.A. Berthold. He systematically gonadectomized roosters and noted the loss of crowing behavior, loss of size and color of the comb and wattle, and loss of male-male aggression and sexual interest. When he put a single testis back into the caponized rooster and restored these traits to normal, he concluded that the testis must secrete a substance into the blood that maintains male behavior and secondary sex characteristics.

Around the same time, other scientists began to demonstrate that a number of physiological problems were related to atrophy of various glands and recovery could occur by replacement with extracts from the glands. For example, Thomas Addison (1855) described a human condition (now called Addison's disease) associated with deterioration of the adrenal cortex, and Murray (1891) prepared emulsions of sheep thyroid glands that helped patients suffering from hypothyroidism (Turner and Bhagnara, 1971).

The birth of the field of Endocrinology occurred around 1902–1905 (Bayliss and Starling, 1902) with studies demonstrating the secretion of pancreatic juices into the gut without assistance of the nervous system. Over the next 30 years, researchers discovered most of the glands that produce the key hormones that are recognized today as modulating growth, metabolism, energy storage and release, and reproduction in both sexes. The field boomed from the 1940s as new hormones were isolated and their biochemical and physiological properties were described. During this same period of time, the role of the hypothalamus in controlling the release of hormones was coming to be appreciated. Harris (1937) demonstrated that electrical stimulation of the hypothalamus could induce ovulation in the rabbit. From the 1930s to early 1950s researchers tried to determine the relationship between the brain and pituitary in controlling testes and ovaries. By 1932, Moore and Price proposed a model of pituitary control of the gonads with gonadal steroids negatively regulating pituitary secretions to maintain a balance in the male and cyclicity in the female. They conveniently ignored the evidence of neural mediation of gonadal function from copulation-induced ovulation and the effects of other external factors, such as season of the year, on gonadal function. Their model was soon called into question by studies that moved the pituitaries to other locations in the body, demonstrated that they were still healthy, but no longer stimulated gonadal function (Hohlweg and Junkmann, 1932). G. W. Harris continued this line of research, theorizing that hypothalamic neurons must produce secretions that entered the hypophyseal portal vasculature to regulate the release of pituitary hormones. Eventually, he and his colleagues demonstrated the existence of what we now know as gonadotropin-releasing hormone (GnRH) and its control over pituitary LH and FSH release, which in turn controlled ovarian function. In a 1955 monograph, Harris summarized the results of more than 20 years of research demonstrating that the hypothalamus, pituitary, and gonads each produced their own secretions which influenced the activity of the tissues above them, such that pituitary hormones had a negative effect on GnRH release from the hypothalamus, and ovarian or testicular hormones had a negative effect on production of GnRH, LH and FSH.

The marvelous experiments of Harris quickly led other researchers to demonstrate similar relationships. Schally (1968) demonstrated with in vitro systems the release of corticotrophin releasing hormone (CRH) that controlled pituitary adrenal corticotrophic hormone (ACTH) release, and ultimately glucocorticoid (cortisol or corticosterone, CORT) release from the adrenal. While Harris was widely recognized for his contribution in proving the existence of the hypothalamic control over pituitary hormone release, it was ultimately Andrew Schally who received the Nobel Prize in 1977 for Physiology and Medicine for his work on these principles; Harris simply did not live long enough (died in Nov, 1971). In his acceptance speech and later writings, Schally frequently cited the path-breaking work of Harris, and many felt that Harris would also have received the prize had he lived.

The final leap forward in our understanding of the role of hormones and their neural controls opened in 1959 with the publication by Phoenix et al. (1959) demonstrating that gonadal steroids during embryonic life not only organized the secondary genital tissues, but also organized the central nervous system. They found that female guinea pigs exposed to androgens during a specific period of maximal sensitivity lost the ability to express estrous behavior. Research during the 1960s with altricial, muroid rodents (rat, mouse, hamster) demonstrated that similar organizing effects could occur postnatally as well as prenatally. By the early 1970s, researchers (Beach, 1971; Goy and Goldfoot, 1973) were introducing and clarifying the concepts of organizational and activational effects of steroid hormones, as well as critical (or sensitive) periods for masculinization and defeminization of behavior, and showing that those principles generalized to a wide variety of species. Books began to appear in the early 1980s that summarized the rapidly advancing field of Neuroendocrinology and its relationship to behavior (Goy and McEwen, 1980; Pfaff, 1980; Adler, 1981; Adler, et al., 1985). This relationship is now well established and represented in the second edition of two popular text books by Becker et al. (2002) and Nelson (2000).

The recognition that biological rhythms influence the neurohormonal control of behavior and physiology is, at once, quite old and relatively new. As mentioned earlier, from most ancient times humans have recognized the seasonal changes in behavior and reproductive activity of the animals and plants around them. The manipulations of testes and androgens, described earlier, made it clear that seasonal control of reproduction came about by altering the secretions from the gonads. However, it is only in the last 50 years, and primarily since the early 1970s when radioimmunoassays were developed that allowed the measurement of blood concentrations of hormones, that the circadian nature of these secretions and their influence on behavior has been clear (Moore-Ede et al., 1982).

1.2 Goal of this Chapter

The goal of this chapter is to describe the relationship between the circadian mechanism and neuroendocrine axes. Hormones such as insulin, glucagon, and gastrin are important for physiologic functions that affect behavior; their secretion is not primarily controlled by the central nervous system. This chapter focuses on a few hormones whose release is directly controlled by the brain and which have large effects on behavior. We will examine how the circadian rhythms generated by the suprachiasmatic nucleus (SCN) are transmitted to endocrine tissues so as to organize the temporal patterns of release of several hormones. We begin with melatonin, released by the pineal gland, which is under direct neural control of the SCN. We then examine the role the SCN plays in regulating release of hypothalamic factors that control the pituitary, and ultimately gonadal and adrenal tissues. This is not meant to be an exhaustive review of all neuroendocrine mechanisms that are influenced by the SCN. Instead, we look at exemplars of direct and indirect control of circadian hormone secretion and its impact on behavior.

2 Circadian Rhythms and Neuroendocrine Control

2.1 Overview of Circadian Control of Hormone Release

Dorothy T. Krieger edited the first book that seriously examined the range of circadian rhythms exhibited by the endocrine system (1979). This came after a period of rapid development of radioimmunoassays (RIAs) used by researchers such as Weitzman and co-workers (1966, 1971, 1974, 1976; Hellman et al., 1970) to quantify the episodic release of hormones, such as cortisol. In Krieger's book, a chapter by one of the founders of the field of chronobiology, Jurgen Aschoff (1979), describes the daily pattern of release of cortisol, growth hormone, aldosterone, prolactin, testosterone, thyrotropin, LH and FSH in humans. The examples are illustrative of one of the key features of the circadian system: while it regulates many physiological and behavioral processes very precisely, the rhythms have a great variety of relationships to one another and the LD cycle. For example, in the diurnal human, cortisol, aldosterone, prolactin, and testosterone peak late at night or at dawn. In contrast, growth hormone, thyrotropin, melatonin and LH are elevated during the dark, with the first three peaking near the onset of sleep. Only melatonin has the same phase relationship to the LD cycle in essentially all vertebrates, rising during the night and falling during the day. The remaining hormone patterns vary between species, often as a function of whether they are diurnal or nocturnal.

To determine whether any daily rhythm is circadian, that is, endogenously produced with an approximately 24 h period, the animal must be housed in conditions without information about time of day (i.e., constant conditions such as constant darkness, DD, or constant light, LL). If the daily rhythm persists, then it can be described as truly circadian. Some daily hormone rhythms are not directly coupled to the SCN, but are rather driven by other systems; for example, growth hormone is only released during slow wave sleep. If sleep is prevented, the daily rhythm of growth hormone disappears. However, many of the hormones with a major influence on behavior are circadian.

Once a rhythm is determined to be circadian, the next obvious question to answer is what environmental cues are synchronizing, or entraining, the circadian rhythm to a 24 h period. Not surprisingly, given that the endogenous period is close to 24 h, the same as that of the earth's rotation, the 24 h light: dark (LD) cycle provides the most potent source of entraining signal ("zeitgeber" from the German for "time giver"). Other non-photic signals, such as regular food intake, social interactions, and exercise can also produce entraining effects on circadian rhythms, but light is by far the dominant signal. Chapter 21 describes, in great depth, how circadian rhythms are generated and how they are controlled by light. For our purposes, it is important to note that some tissues outside the SCN also have endogenous circadian rhythms. In mammals, these oscillating tissues (including the adrenal cortex, for example) are considered to be slave oscillators since, under stable environmental conditions, the SCN indirectly controls their daily activity. In mammals, the SCN receives direct retinal input through the retinohypothalamic tract that maintains entrainment of the endogenously rhythmic cells with the LD cycle. Interestingly, in nonmammalian species, even among non-mammalian vertebrates, it is common for more than one tissue to have its own photoreceptors that can entrain the circadian activity of that tissue independently of all other tissues. One such well-described tissue is the pineal gland, which produces the daily rhythm of melatonin. However, in mammals, the pineal, as well as all other endocrine systems with circadian release patterns, is under direct or indirect control of the SCN.

The importance of the SCN for maintaining internal synchronization of endocrine and other physiological processes is best understood when we examine the consequences of internal desynchronization in humans who make transmeridian trips (jetlag; rapid shift in the phase relationship to the LD cycle) or have jobs with rotating shift schedules (Klein et al., 1972; Tapp and Natelson, 1989). Jet lag can cause myriad physical, emotional, and psychiatric problems in humans (Winget et al., 1984; Cho et al., 2000; Cho, 2001; Katz et al., 2001). Some of the symptoms of jetlag, such as ulcers, depression, and emotional distress, are also commonly associated with stress and an increase in cortisol secretion. In women, the timing of reproductive events, including the preovulatory LH surge, is linked to the LD cycle (Kapen et al., 1973; Ewards, 1981; Seibel et al., 1982; Testart et al., 1982; Hoff et al., 1983). Frequent disruptions of circadian rhythms in women are correlated with disruptions in reproductive function. Female shift workers have a higher risk of having low birth weight babies, spontaneous abortion, and subfecundity, and can experience irregular menstrual cycles accompanied by changes in secretion patterns of ovarian and pituitary hormones (Scott, 2000; Knutson, 2003; Lohstron et al., 2003). Female flight attendants have similar negative reproductive outcomes. Additionally, sleep disturbance, a common side effect of disrupted circadian rhythms, in late pregnancy is associated with increased labor duration and increased likelihood of medical intervention (cesarean section) (Lee and Gay, 2004). The problems with labor are most likely due to a disruption in the nightly rise in oxytocin release that prepares the uterus for labor in the last weeks of pregnancy (Hirst et al., 1993).

As mentioned earlier, the circadian control of hormone release can occur in one of several ways. In the following sections, we discuss the direct neural pathway between the SCN and the pineal gland of mammals that controls the nightly release of melatonin. Other circadian hormone release patterns are modulated more indirectly. The circadian control of ovulation and estrous behavior in female rodents, and perhaps some other female mammals, occurs because time of day information from the SCN is integrated with other important information about the internal state of the female (e.g., energy balance, age, recent pregnancy or parturition) and the state of the environment (e.g., season of the year, presence of appropriate mate) at the GnRH cells in the hypothalamus. In some cases, part of the integration includes endogenous circadian oscillations within a part of the complex axis, as occurs with the adrenal cortex and perhaps GnRH cells.

2.2 Neural Control of Melatonin Release

One of the most pervasive sets of behavioral changes are those associated with preparation for the winter in temperate zone species. Among vertebrates, these changes include migration, hibernation, molting of outer coverings, changes in color, cessation of reproduction, decreases in aggression, and alterations in aggregation patterns. Increasing or decreasing the amount of light each day (photoperiod) can induce estrus in the non-breeding season (ferrets—Bissonette, 1932; field voles—Baker and Ranson, 1932; sheep—Yeates, 1949), as well as a host of other changes. Rusak and Morin (1976) demonstrated that SCN lesions prevent short day lengths from causing regression of the testis. Rusak and Morin concluded that the SCN controlled seasonal changes by altering the circadian release of melatonin because of two earlier findings: (1) Axelrod (1974) had demonstrated that nerves from the superior cervical ganglion (SCG) synapse near blood vessels around the pineal gland, and (2) Moore and Klein (1974) found that the circadian rhythm of the enzyme necessary for production of melatonin (pineal N-acetyltransferase) was abolished by lesions of the SCN.

Subsequently, Turek and Campbell (1979) demonstrated that administration of melatonin to a hamster when it was housed in long day lengths could, like a shortening of the photoperiod, cause testicular regression. In the early 1980s, anatomical work by Klein et al. (1983) and endocrine work by Goldman et al. (1981) demonstrated that the SCN sends efferents to the paraventricular nucleus (PVN), which projects to the SCG, which synapses on blood vessels near the pineal gland, and that manipulation of day length directly affects melatonin release. Interruption of this circuitous pathway at any location blocks the daily rhythm of melatonin release. Goldman et al. (1984, 1991; Elliott et al., 1989) demonstrated that the duration of the melatonin signal, whether exogenously produced during the day or naturally produced during the night, determines whether the animal interprets the photoperiod as being long or short. Virtually all seasonal changes in behavior and physiology have subsequently been connected to the circadian pattern of melatonin release. Even circannual rhythms are synchronized to a 12 month period by seasonal changes in the circadian pattern of melatonin (Woodfill et al., 1994; Hiebert et al., 2000).

Further evidence of the circadian nature of melatonin release is the persistence of the rhythm in animals living in DD (Illnerova and Vanecek, 1983; Sun et al., 2002, 2003). Because of the minimal number of synapses between the SCN and the pineal gland, the melatonin rhythm has frequently been used as a marker of SCN function. Earlier it was thought that shifts in the LD cycle "instantaneously" altered the function of SCN (Moore-Ede et al., 1981), however melatonin rhythms very clearly do not recover within the

Figure 23-1

Pathway by which the SCN controls the production and secretion of pineal melatonin. Axons from the eye, via the retinohypothalamic tract (RHT), synapse on ventral–lateral cells of the suprachiasmatic nucleus (SCN). The SCN projects to the paraventricular nucleus of the hypothalamus (PVN), which in turn projects to the intermediolateral column (IML) of the spinal cord. The IML projects to neurons in the superior cervical ganglion (SCG) of the sympathetic nervous system, which ultimately project to the pineal gland where it regulates the synthesis of melatonin

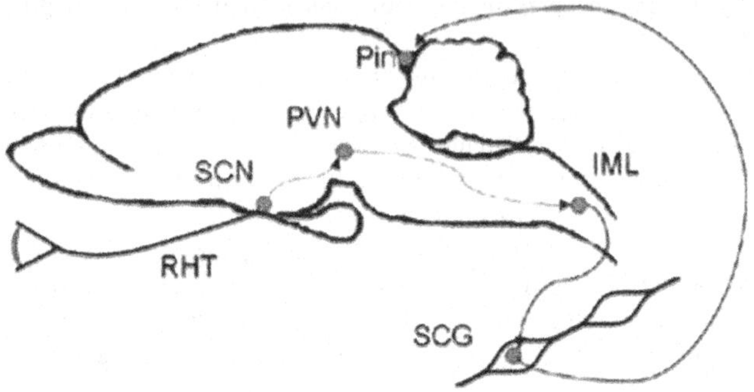

1–2 days expected based on the SCN models (Illnerova and Vanecek, 1983; Humlova and Illnerova, 1992). Recent reports analyzing the response of melatonin rhythms (Liu and Borjigin, 2005) and *per* gene transcription in the SCN (Nagano et al., 2003) to phase shifts of the LD cycle reinforce how tightly melatonin is driven by the SCN, and that both take a number of days to recover from a shift.

Because melatonin is thought to be a completely passive system, driven only by the neural signals from the SCN, one would expect to find that the SCN and melatonin require approximately the same amount of time to recover from a shift. Liu and Borjigin (2005) used in vivo microdialysis with a 10 min sampling schedule to detect changes in pineal melatonin release after phase delays of 3, 6, or 12 h. Their data demonstrate that recovery of melatonin entrainment, like the SCN (Nagano et al., 2003) takes several days to achieve. The process occurred in three steps. First, there was a rapid shift of the melatonin onset at lights-out within 1–2 cycles, much as the ventral–lateral SCN shifted (Nagano et al., 2003). Secondly, the fall in melatonin at the end of the dark phase shifted rapidly for 3–4 days, overshooting the correct phase so that melatonin was still elevated during the early lights-on phase. And lastly, a period of several days passed (increasing with increasing size of the shift), when, both, the rising and the falling phases of the melatonin rhythm make small adjustments to reach a steady state. The initial large shift that brought the melatonin rhythm close to entrainment after 1–3 days correlates with the initial rapid shift of the ventral–lateral SCN, while the slow final resolution of the entrained phase correlates with the slower dorsal–medial SCN recovery time (Nagano et al., 2003). It is interesting that in each case, the melatonin rhythm lags behind that of the SCN by 1–2 cycles. The *rPer1* rhythm from the shifting dorsal–medial SCN does not appear to overshoot the entrained phase, as the melatonin rhythm does. Since the in situ study of the SCN did not collect data from animals for every day during the recovery period after the shift, and the intervals of collection were 2 h, it remains possible that the dorsal–medial SCN also phase shifts too far initially, and then returns to the correct phase. The precision of the melatonin rhythm most likely provides a more fine-tuned analysis of the time it takes for the SCN to recover completely from a shift in the LD cycle than does the *rPer1*data (Nagano et al., 2003; Liu and Borjigin, 2005).

2.3 Effects of Melatonin on the SCN

The interaction between the SCN and the pineal is not as unidirectional as the data mentioned earlier would suggest. Although early studies demonstrated that pinealectomy did not alter entrained or free-running rhythms (Quay, 1968), later work demonstrated that pinealectomy of rats and hamsters enhanced the rate of recovery from a shift in the LD cycle (Quay, 1970; Finkelstein et al., 1978). Furthermore, pinealectomized rats housed in LL quickly produced only ultradian (no circadian) rhythms (Cassone, 1992), and pinealectomy enhanced splitting of activity rhythms in hamsters housed in LL (Aguilar-Roblero and Vega-Gonzalez, 1993). Redman et al. (1983) demonstrated that daily injections of melatonin at the same time of day could entrain circadian rhythms in animals free-running in DD. Cassone et al. (1986) then demonstrated that the effect was apparent with physiological levels of melatonin and that saline injections at the same time of day had no effect on the free-running rhythms of the animals.

The mechanism underlying these effects is still unknown, although melatonin receptors are present in the SCN of some mammals, and the numbers of receptors varies across the day (Gauer et al., 1993; 1994a, b; Weaver et al., 1993). Two species with reduced or undetectable numbers of melatonin receptors in the SCN (adult Syrian hamsters and mink) also do not entrain to injections of melatonin (Hastings et al., 1992; Bonnefond et al., 1993). Thus melatonin receptors in the SCN may be necessary for entrainment feedback effects.

To summarize, the SCN controls the circadian pattern of melatonin secretion in mammals by a four synapse pathway. Melatonin has multiple effects on physiology and behavior, and therefore represents one indirect pathway through which the SCN can influence physiology and behavior. The SCN is also the target through which melatonin can entrain circadian rhythms. It seems likely that the rates and patterns of phase recovery for the clock rhythms in the dorsal–medial SCN and for the rhythms in melatonin secretion from the pineal gland following a shift in the LD cycle is a consequence of interactions between these neural and endocrine structures, respectively.

3 Circadian Mechanisms Regulating the LH Surge and Estrous Behavior

Successful reproduction requires coordinated timing of male and female behavior so that sperm arrives soon after the egg is expelled from the follicle and is ready for fertilization. For males, the temporal coordination is relatively simple. During times of post-pubertal life, when females are available to be inseminated, they produce relatively continuous secretion of high levels of one primary hormone (testosterone), and a steady supply of sperm. Such a male is in a position to take advantage of opportunities to mate whenever they arise. For a female, the process is far more complex. She must develop mechanisms that will couple the occurrence of her sexual behavior to the LH surge at a time when the follicles are ready for ovulation, and the uterus must be prepared to receive and nurture an embryo. The nature of the mechanisms coordinating these events varies considerably across species. For example, ovulation can be spontaneous or induced, and it may or may not be tightly regulated by a circadian time-keeping system. Here we focus on the role of endogenous circadian timekeeping mechanisms responsible for the temporal coordination of estrus-related events in mammals. The system is bidirectional, in that it involves not only circadian mechanisms that regulate hormone secretion, but also influences of those hormones on the circadian system. We focus first, and primarily, on the role of the circadian system in the regulation of neuroendocrine events associated with estrus, and then turn briefly to consider how ovarian hormones can feed back to influence circadian systems.

The two initial events that the circadian system helps to coordinate are the occurrence of the surge of secretion of LH from the anterior pituitary gland that triggers ovulation, and mating behavior that makes possible the fertilization of the eggs that have been released by that ovulation. Both of these events require a period of exposure to steroid hormones secreted by the ovary. These secretions, over time, prepare the brain so as to lead females to seek out and solicit reproductive males, and respond to male advances with receptive behavior. Steroid hormones also prepare the brain to emit the signals that ultimately lead to ovulation. Specifically, ovulation is triggered by a surge in the release of LH from the anterior pituitary, which is promoted by the release of GnRH from the hypothalamus into the median eminence. This basic system can be considered a part of the positive limb of the hypothalamic–pituitary–gonadal (HPG) axis. The negative side of this axis involves a system of feedback through which the pituitary hormones [luteinizing hormone (LH) and follicle stimulating hormone (FSH)] inhibit the hypothalamic production of GnRH, and the ovarian hormones (progesterone-P and estradiol-E) inhibit both the pituitary and the hypothalamus. The last basic element of this system key to understanding ovulation is a fascinating switch from a negative to a positive response by the GnRH neurons to rising estradiol concentrations as the follicles develop. That is, just before the GnRH and LH surge occurs, the GnRH neurons change, such that they respond to estradiol by increasing rather than decreasing the release of GnRH. This switch from a negative to a positive feedback effect of estradiol is a phenomenon that has puzzled neuroendocrinologists for years and, more recently, has drawn the attention of mathematicians who have begun to model the process. However, we bypass that mystery here and focus, instead, directly on the role of the circadian timekeeping system in the regulation of the events leading to ovulation and mating behavior. These events are primed by steroid hormones, but they are also influenced by, and in some species depend upon, signals from the circadian time-keeping system. Indeed, disruption of genes intrinsic to the molecular clock can lead to reductions in fertility (Chappell et al., 2003).

To understand just how a signal from the circadian system might regulate estrus, it is important to first consider the nature of the GnRH cells that the SCN signals must ultimately reach, as these cells represent a final common pathway upon which many kinds of signals converge to regulate reproductive physiology and behavior. The GnRH neurons are large bipolar cells that are few in number and diffusely distributed in a network that crosses traditional cytoarchitectonic boundaries (Silverman et al., 1994). The rostral extent of their distribution begins in the septum and diagonal band of Broca, and they become most concentrated caudal to that in the medial regions of the preoptic area of the hypothalamus (POA). In some species, such as rats and hamsters, they continue caudally to the end of the optic chiasm, while in others, such as guinea pigs and rhesus monkeys, they continue through the arcuate nucleus (Silverman et al., 1994). The diffuse nature of their distribution has made it difficult to identify the afferents to the GnRH cells. One approach that has been taken is to examine details of synapses on their cell bodies, which has revealed the nature of the peptides and transmitters impinging on them. However, probably the best characterized afferents of the GnRH neurons are those through which the circadian information reaches them.

Direct evidence that GnRH cells are rhythmic has been obtained through studies of Fos expression, and in situ hybridization and direct measurements of GnRH. Fos expression within GnRH neurons peaks late in the light phase on the day of proestrus; this peak coincides with the LH surge in rats (Lee et al., 1990; van der Beek et al., 1994) and guinea pigs (King et al., 1998), but occurs immediately after the surge in hamsters (Doan and Urbanski, 1994). A peak at the time of the LH surge in Fos expression within GnRH neurons has also been seen in the diurnal grass rat during a post partum estrus, approximately 12 h out of phase with its occurrence in the nocturnal rodents (McElhinny et al., 1999). The rhythm in Fos expression in GnRH neurons appears to be endogenous in both nocturnal lab rats and diurnal grass rats (Mahoney et al., 2004).

3.1 The Circadian System is Involved in the Triggering of the LH Surge

A clear role for the circadian timekeeping in the direct regulation of the LH surge has been established in closely related murid rodents, with short cycles, that ovulate spontaneously but do not have a spontaneous luteal phase (rats, hamsters, and mice). The development of our understanding of this fascinating interplay between steroids and circadian signals has its origins in a 1950 paper by Everett and Sawyer (1950). These investigators injected intact female rats with pentobarbital at different times on the day of proestrous, and found that there was a critical phase of the daily rhythm during which the drugs could block ovulation, and the delay was not simply for the duration of the anesthesia, but rather, it lasted for 24 h (reviewed in Rusak and Zucker, 1979b). This study provided evidence that there is a critical phase of the daily rhythm during which a daily neural signal triggers ovulation. The same method has been used to establish that the postpartum LH surge of rats is also triggered by a precisely timed daily signal (Fox and Smith, 1984). This conclusion has been bolstered by many kinds of experimental evidence. For example, Legan and Karsch (1975) implanted ovariectomized (OVX) rats with silastic capsules and found that LH surges occurred every day at approximately the same time, 3–4 h before lights-out; estrogen-treated OVX mice are similarly only able to surge around lights off, regardless of when the steroid stimulation occurs

(Bronson and Vom Saal, 1979). The postpartum LH surge of rats also occurs 3–4 h before lights-off and is also gated by a time-keeping system such that appropriately timed injections of pentobarbital lead to a 24 h delay (Fox and Smith, 1984).

The question of whether the phase of the daily signal is determined by cues from the environment or from an endogenous time-keeping system was addressed by Alleva et al. (1971), who found that when hamsters were kept in constant light, they had free-running "quadradian" rhythms in the occurrence of a variety of indices of the estrous cycle including gonadotropin release, with periods that were four times those of their circadian rhythms. Since these early studies, compelling evidence has accumulated from a variety of very different studies. For example, when the period of the circadian rhythm of hamsters was lengthened by administering deuterium oxide (heavy water), the period of the estrous cycle increased proportionally in hamsters (Fitzgerald and Zucker, 1976). When hamsters were maintained in LL and their wheel-running rhythms split into two bouts in antiphase relationships, they developed two surges per day, approximately 12 h apart (Swann and Turek, 1985). When rats were exposed to LL for long periods, their behavioral rhythms became disorganized, and under these conditions the animals ceased having regular estrous cycles and became induced ovulators (Brown-Grant et al., 1973). While the role of the circadian system in the LH surge has been well established in some rodents, this system has less direct control over the surges, and may be less essential, in some other species, including sheep and rhesus monkeys, and diurnal species with a spontaneous LH surge and a spontaneous luteal phase (Jackson et al., 1975; Tersawa et al., 1984).

3.2 The SCN

Even before it was clearly established that the murid rodent's LH surge is regulated by a circadian "gate," and before the relationship between the SCN and circadian rhythms had been clearly established, Critchlow (1963) (reviewed in Rusak and Zucker, 1979) proposed that a retinohypothalamic tract projecting to the SCN mediates the entrainment of the estrous cycle by the LD cycle. This conclusion was based on the finding that SCN lesions led to constant estrus and acyclicity in females. Since that time, many lesion studies have led to the same basic conclusion, although the details of the nature of the effects of the lesions have varied (Stetson and Watson-Whitmyre, 1976; Gray et al., 1978b; Wiegand et al., 1980; Kawakami and Arita, 1981; Meyer-Bernstein et al., 1999b). Differences in the size of lesions and the degree to which they altered projections to the anteroventral preoptic area (AVPV), an area slightly rostral to the SCN, may account for the discrepancy. When Wiegand et al. (1980) produced tiny lesions in different regions along the rostral–caudal axis from the SCN to the AVPV (which they refer to as the medial preoptic nucleus), they found that the LH surge of rats was blocked if the lesion destroyed either the SCN or the AVPV. However, even as the SCN lesions disrupted ovulation in response to E alone, small LH surges did occur in response to E plus P injections. Animals were unable to surge in response to any hormone treatment when the AVPV was destroyed. This work led to the notion that under intact conditions a "surge generator" in the AVPV integrates circadian signals from the SCN with information on levels of steroid hormones to determine whether a surge will occur.

As noted previously, the circadian system plays less of a role in the regulation and generation of the surge of some species, and in these cases the SCN, not unexpectedly, is not as critical. In both guinea pigs and rhesus monkeys, knife cuts that severed the SCN from the caudal population of GnRH neurons did not interfere with the ability to produce an LH surge (Butler and Donovan, 1971; Krey et al., 1975). It is also the case that in some species, including humans, females may typically produce LH surges at a specific time of day, even if it can be generated at any time. In these cases, it is not yet known whether the normal circadian timing of ovulation is important to fertility.

It should also be noted that the region immediately dorsal to the SCN, the lower subparaventricular zone (LSPV), may contribute to the temporal organization of reproductive events. The LSPV receives a massive input from the SCN and projects to many of the same targets as the SCN (Watts and Swanson, 1987; Morin et al., 1994), including cells that contain GnRH and ER (De La Iglesia et al., 1995; Watson et al., 1995; van der Beek et al., 1997), and is rhythmic (Schwartz et al., 2004), and in OVX lab rats, lesions of the LSPV result in an increase in tonic levels of LH (Doecke et al., 1982).

3.3 Signals from the SCN to the GnRH Neurons

More recent studies of murid rodents have been aimed at examining which specific pathways emanating from the SCN communicate temporal information to GnRH neurons. At a very general level, the SCN can get its message out via traditional neural efferents and also via an as yet unidentified diffusible substance (Silver et al., 1996). The possibility that a diffusible signal might serve to communicate temporal information from the SCN to its targets has been tested most carefully in hamsters. This notion was first suggested by the finding that although SCN lesions abolished wheel running rhythms, core knife cuts that severed all fibers into and out of the SCN did not (Silver and LeSauter, 1995). The hypothesis was further supported by the finding that behavioral rhythms could be restored in SCN-lesioned animals by transplants of fetal SCNs into the third ventricle of lesioned animals even when the transplanted tissue was encapsulated in a material through which neural processes could not pass (Silver et al., 1996). It thus seems that, although traditional efferents may contribute to behavioral rhythms, the rhythms do not depend on them. It should be noted, however, that this model has not been tested in other species and may not apply to rats (Nunez and Casati, 1979; Nunez et al., 1985).

In the case of neuroendocrine rhythms, the situation appears to be different. Here, the data suggest that a diffusible signal is not sufficient for the generation of the rhythms. The issue has been addressed in a number of ways. In rats, knife cuts that severed the lateral, dorsal, and caudal connections of the SCN eliminated estrous cycles, suggesting that neither a diffusible signal nor the rostral pathway is sufficient (Nunez and Casati, 1979). More recently, Myer-Bernstein et al. (1999) used SCN transplants in hamsters to address the question of which rhythms can and cannot be restored in SCN-lesioned animals. These

Figure 23-2
Proposed direct pathway by which the SCN may regulate the timing of GnRH (gonadotropin-releasing hormone) release into the portal vasculature around the pituitary. GnRH then stimulates the surge of luteinizing hormone (LH) that causes ovulation

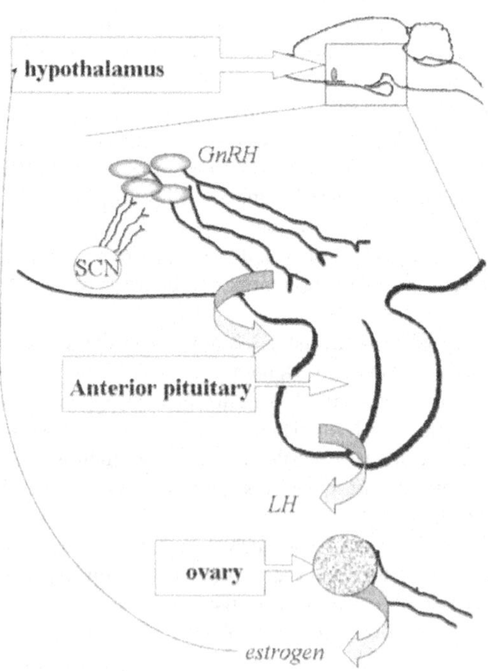

investigators found that transplants that restored wheel-running rhythms in female hamsters did not bring back estrous cycles or the ability of OVX females to surge in response to E. This study also found that rhythms in pineal, gonadal, and adrenal function of males could not be restored by the grafts. These results suggest that, although a diffusible factor might contribute to neuroendocrine rhythms, these rhythms are dependent on neural pathways connecting the SCN to its various targets.

A recent study by de la Iglesia et al. (2003) provides further compelling evidence that a neural rather than a humoral signal is central to circadian regulation of the GnRH system. This study examined patterns of laterality in GnRH neuron function of hamsters with split activity rhythms. As noted previously, splitting involves the dissociation of both a single daily activity bout and a single LH surge into two that are in an antiphase relationship, with smaller peaks in activity and LH, respectively, approximately 12 h apart (Swann and Turek, 1985). In these animals, the two bilaterally paired SCN also exhibited antiphase relationships with respect to their rhythms in PER, one of the proteins intrinsic to the circadian clock. In split hamsters, rhythms in FOS within GnRH neurons on the two sides of the brain were also in an antiphase relationship relative to each other, further indicating that axonal outputs of the SCN, which are primarily ipsilateral, rather than a diffusible factor, trigger the activation of the GnRH neurons (de la Iglesia et al., 2003).

Axons emanating from the SCN could theoretically influence the GnRH neurons via indirect pathways, perhaps via ER-containing cells in the AVPV, or through direct projections. Tract-tracing studies have revealed the existence of a direct pathway in rats (van der Beek et al., 1997a), hamsters (De La Iglesia et al., 1995), and Nile grass rats (unpublished data). In hamsters and grass rats, these projections have only been examined at the light level, but in rats they have been clearly verified with electron microscopy. Estimates of the numbers of GnRH neurons receiving SCN input in these species are 11–19% for hamsters and lab rats, and 60% for grass rats. The majority of GnRH cells contacted by SCN efferents in these species are located in the region of the organum vasculosum of the lamina terminalis (OVLT) and the POA of lab rats and grass rats, and in the diagonal band of Broca (DBB) and POA of hamsters (de la Iglesia et al., 1995; van der Beek et al., 1997).

Recent data from Kallo et al. (2001) raise the possibility that GnRH neurons receiving direct input from the SCN could themselves represent one site at which circadian signals converge with feedback signals from ovarian steroids to provoke an ovulatory surge in LH. This is based on their observation that ER-beta is ordinarily present in over 50–65% of GnRH cells in the POA of lab rats, but that when they are exposed to high levels of E, as occurs prior to the surge, this percentage drops to 26–46% (depending on the time and dose of E administration), potentially leading to a decrease in negative feedback effects of E. If these cells are among those receiving input directly from the SCN, this could heighten their response to a circadian signal, enabling a surge of GnRH release into the hypophyseal portal system to occur.

3.4 VIP and VP

The question of which SCN cells and transmitters convey temporal signals to GnRH neurons has received considerable attention in recent years. Two of the best understood signaling peptides found in SCN efferents are vasopressin (VP) and vasoactive intestinal polypeptide (VIP) (Moore et al., 2002). These peptides are present in a large proportion of SCN cells and their axons, and they appear to play important roles in the regulation of rhythms in glucocorticoids, feeding and drinking, as well as reproduction (Salisbury et al., 1980; Buijs et al., 2003). These populations of cells are different with respect to their inputs and outputs as well as some of their functional characteristics. Within the hypothalamus, VIP cell bodies, which are present in every species that has been examined, are concentrated in the "core" region of the SCN, where they receive direct input from the retina but exhibit rhythms in the absence of an LD cycle. Cells containing VP are located in the "shell" of the SCN which is different from the core with respect to its inputs and its intrinsic connections as well as its rhythms. Both populations of cells appear to be involved in communicating from the SCN to GnRH neurons. Somewhat conflicting evidence exists, however, regarding whether these peptides have inhibitory or excitatory effects on the ovulatory surge in GnRH and, subsequently, LH (Vijayan et al., 1979; Stobie and Weick, 1989; Weick and Stobie, 1992). The discrepancies are most likely due to variations in methodology with respect to the route of administration of the various agents, the specific agents administered, and the doses and animal models used (van der Beek, 1996; Palm et al., 1999b). Here we highlight some of the data bearing on the respective roles of VIP and VP, and then consider some of the attempts to develop models that may account for them.

Evidence of a role for VIP as an efferent signal emanating from the SCN comes from a variety of studies, both correlational and experimental. In female lab rats, hamsters, and grass rats, contacts between VIP fibers and GnRH cells have been seen using light level microscopes (Van Der Beek et al., 1998; Kriegsfeld et al., 2002; Mahoney and Smale, 2005). In lab rats, SCN lesions eliminated most of these contacts (Van der Beek et al., 1993), and the presence of synapses between VIP terminals and GnRH neurons has been verified with electron microscopy (van der Beek et al., 1997). Triple immunocytochemical labeling has also revealed that approximately 50% of GnRH cells that expressed Fos at the time of the surge are contacted by VIP fibers (van der Beek et al., 1994). These appositions are more numerous in females, which produce a surge at regular intervals, than in males, which are unable to generate one (Horvath et al., 1998b). A role for VIP is also indicated by the presence of VIP2 receptors on GnRH neurons, both in vivo and in vitro (Smith et al., 2000; Olcese et al., 2003).

The issues become somewhat less clear and the data less consistent when one considers effects of experimental manipulations intended to block or increase availability of VIP. Although VIP administered into the ventricles of lab rats has been reported to stimulate the release of LH in some conditions (Vijayan et al., 1979), other studies have found that administration of VIP inhibits LH (Alexander et al., 1985; Stobie and Weick, 1989; Weick and Stobie, 1992). Experimental disruption of VIP, however, delayed and attenuated the surge in two studies. In one, the production of VIP was disrupted by antisense administered into the SCN (Harney et al., 1996), and in the other, anti-VIP antisera was delivered into the lateral ventricle just before the expected onset of the surge (van der Beek, 1999). Several potentially confounding variables may account for the discrepancies, such as differences in the timing of experimental manipulations of the system (discussed in van der Beek et al., 1999).

VP cell bodies are found in a number of brain regions, including the PVN, making interpretations regarding their relationship to the circadian system more problematic. A role for VP as a potential efferent signal of the SCN was first suggested by the discovery of rhythms in its release into the cerebral spinal fluid that peak during the day in both nocturnal and diurnal mammals; these rhythms were abolished by SCN lesions (Reppert et al., 1987). More recently, rhythms in VP release within the SCN of rats have been documented via microdialysis (Kalsbeek et al., 1995). These VP rhythms also peak during the day, during the critical period 2 h before the LH surge, when it can be blocked by administration of barbiturates (see above).

VP fibers make contacts with GnRH neurons in the hypothalamus of female lab rats, hamsters, and grass rats, as well as primates (Thind et al., 1991; Horvath et al., 1998; Abizaid et al., 2004). In lab rats, there is a sex difference in the number of VP contacts with GnRH neurons in the diagonal band of Broca such that males actually have more than females (Palm, 2001), and the number of contacts is reduced by lesions of the SCN, suggesting that at least some of this input comes from cells within the SCN (Palm et al., 1999a).

As with VIP, some studies in which VP has been administered to laboratory rodents suggest an inhibitory role, and others suggest an excitatory one (reviewed in Palm et al., 1999a). The most recent work used reverse microdialysis to deliver relatively minute quantities of VP specifically into the POA; this generated an LH surge in SCN-lesioned rats (Palm et al., 2001a). This work also found that the effect of VP occurred when its administration began at ZT 8 but not when it began at ZT 0 or ZT 4. Further evidence that VP may stimulate the release of GnRH comes from the finding that administering it into the POA induced GnRH release (Funabashi et al., 2000).

At least two studies have directly compared VP and VIP with respect to their potential roles in the regulation of the surge. In one, Krajnak et al. (1998) evaluated mRNA rhythms in the SCN of male and female rats. The rising phases of the rhythm in VIP mRNA occurred 8–12 h earlier in female than male rats, whereas VP mRNA rhythms were the same. This pattern is consistent with the possibility that a relatively early rise in production, and perhaps release, of VIP may be important for the initiation of the LH surge. The potential roles of VIP and VP have also been investigated in vitro. Funabashi et al. (2000) found that when preoptic area tissue and the SCN were cultured together, GnRH, VP, and VIP were all released rhythmically. Most importantly, the GnRH rhythms were tightly coupled to those of VP but not VIP. They also found that administration of VP stimulates the release of GnRH from cultured cells in a dose-dependent manner.

As one can see from the above considerations of the respective roles of VIP and VP, the data are somewhat conflicting. Several explanations have been offered. On the basis of the sex difference in rhythms in its mRNA, Krajnak et al. (1998) suggested that VIP plays an important role in the timing of the surge. However, although the presence of the sex differences may suggest a role for VIP, the absence of a sex difference in rhythms in VP mRNA does not argue against the possibility that VP plays an important role. Palm et al. (1999) attempted to account for the data by suggesting that VP plays an important role in the initiation of the surge, but that it can only do this during a brief window of time determined by the release of VIP. This hypothesis is difficult to reconcile with the fact that in vitro rhythms in VP release are tightly coupled to those of GnRH, which occurs at all phases of the rhythms in VIP (Funabashi et al., 2000). Funabashi et al. (2000) suggest that the data are consistent with a model, whereby VP mediates the effect of the clock on the LH surge and any effect of VIP is through its actions within the SCN, where it modulates entrainment. This model offers no explanation for the function of the VIP contacts on GnRH cells, or for the dramatic sex difference in VIP mRNA rhythms within the SCN, which is not paralleled by a sex difference in entrainment. Clearly, although these SCN efferents are likely to play a role in the generation of the LH surge, the exact nature of that role remains to be elucidated.

3.5 Molecular Oscillators Within GnRH Neurons

A fascinating new part of the story of circadian regulation of GnRH comes from recent discoveries of molecular oscillators in brain regions beyond the SCN (Reppert and Weaver, 2002). For example, Abe et al. (2002) found in vitro rhythms in *Per1* that persisted for approximately 3–5 cycles in 11 isolated brain regions. There are now several reports of in vivo rhythms of PER proteins in several brain regions outside the SCN in male rats, including the bed nucleus of the stria terminalis (BNST) and the centromedial and basolateral amygdala (Amir et al., 2004; Waddington Lamont et al., 2005). This work was conducted with adult males, but there is also evidence of molecular oscillators within regions of the brain beyond the SCN in female rats, specifically within neuroendocrine cells (Gillespie et al., 2003; Kriegsfeld et al., 2003; Sellix et al., 2006). The suggestion that oscillations in the expression of clock genes occur within GnRH cells has come from in vitro studies of GT1–7 cell lines (Chappell et al., 2003; Gillespie et al., 2003; Olcese et al., 2003). Gillespi et al. (2003) demonstrated mRNA rhythms for several clock genes in this cell line after the cells had been synchronized with serum shock; the phase relationships among these rhythms were similar to those seen in cells within the SCN of intact animals. The GnRH gene was expressed rhythmically in these cells, and experimentally altering patterns of expression of two of the clock genes led to changes in the pattern of GnRH secretion from these cells (Chappell et al., 2003). These data raise the tantalizing possibility that GnRH neurons represent "slave" oscillators entrained by the SCN, perhaps allowing for greater flexibility in the phase of rhythms in GnRH expression (de la Iglesia, 2006). This might also account for the reduced fertility apparent in mice with a mutation in *Cry*, an important element of the clock (Dolatshad et al., 2004; Kenneway et al., 2006).

3.6 Circadian Influences on GnRH Neurons via ER-Containing Cells

Another way in which the circadian system could influence estrus-related events is via the production of rhythms in responsiveness to ovarian hormones, as estradiol and progesterone have a major influence on both the LH surge and sex behavior in rodents. Here we focus on estrogen receptors (ERs), as more is known about the input of ER-containing neurons to the SCN. Numbers of ERs have, in fact, been reported to change across the day. An early indication that this might be the case came from examination of cytosolic binding to E in tissue containing the hypothalamus, POA, and amygdala from OVX female rats put to death across the day (Roy and Wilson, 1981). This binding underwent 24 hour rhythms with a peak just before lights-out and a trough in the middle of the dark period, at ZT 18. The amplitude of the rhythm was relatively low, with peak values approximately 30% higher than trough values. A similar rhythm has been reported in OVX hamsters (O'Connor et al., 1985). The functional significance of these rhythms is unclear, as the receptors are highest after the period during which behavioral and neuroendocrine functions are maximally sensitive to circulating E.

More recent work has focused on where in the hypothalamus ovarian hormones might interact with circadian signals to promote the surge. Much of this work has focused on the AVPV, which contains a high concentration of ERs and PRs and, as noted earlier, is essential for the generation of the surge (Tersawa et al., 1980; Tersawa et al., 1984). Watson and Langub (1992) discovered that VP fibers form synapses on ER-immunocytochemically labeled cells in the AVPV, raising the possibility that VP cells within the SCN project to ER neurons in that region. Watson et al. (1995) then directly addressed the question of whether the SCN projects to ER cells in the AVPV by delivering an anterograde tracer into the SCN of lab rats and then staining the tissue for visualization of ERs. Labeled fibers were observed in the AVPV, where they formed synapses with cells containing ERs; injections of the tracer that were restricted to the LSPV also revealed such synapses. Evidence that the SCN projects to ER-containing neurons has also been obtained in hamsters where, using light-microscopy, contacts were seen on 8–30% of ER-containing cells (De La Iglesia et al., 1995). These data suggest that AVPV cells that contain ERs and receive input from the SCN may be a site at which rising levels of E converge with the circadian signal that triggers the LH surge. The resulting signal could be carried from the AVPV through its direct projections to GnRH neurons (Gu and Simerly, 1997). Some of the effects of VIP or VP or both on the GnRH and LH surge, described above, might occur through projections from the SCN to the AVPV (Chappell, 2005).

Recent data have pointed to the SCN as another possible site for the integration of circadian and steroid signals. This possibility is suggested by evidence that ER-beta is present in the SCN, where rhythms in its mRNA with a sharp peak at ZT 4 have been documented (Wilson et al., 2002). This peak could theoretically contribute to a rhythmic gating of the GnRH/LH surge. This could happen, for example, if there were a time lag between the elevated ER mRNA and elevated protein in the SCN, and a further lag between the rise in ER-beta protein and the time at which the binding of E to ER promotes a functional change in the cell. Such a sequence of steroid-induced events might then facilitate the production of the circadian signal thought to be emitted around this time.

It is clear that a great deal has been learned over the years about the influence of the circadian system on the LH surge. However, it should also be clear from the above discussion that the full story has not yet come together, that many interesting questions about these mechanisms remain, and that there is still a great deal of room for speculation.

3.7 Rhythms in Sex Behavior

An integration of signals from the circadian system and steroid hormones is also important for another critical estrus-related event: sex behavior. The mechanisms exquisitely timing the surge would be irrelevant if that surge were not tightly coupled to female sexual behavior. Not surprisingly, therefore, animals do exhibit rhythms in their copulatory behavior, with diurnal rodents typically mating during daylight hours and nocturnal rodents displaying mating behavior throughout the dark period (reviewed in Mahoney and Smale, 2005b). In lab rats, estrous behavior continues to occur during subjective night even when animals are housed in constant lighting conditions, and these rhythms are abolished by lesions of the SCN (Gray et al., 1978a; Hansen et al., 1978a; Hansen, 1979; Hansen et al., 1979; Eskes, 1984; Meyer-Bernstein et al., 1999a).

Daily rhythms in female sexual behaviors appear to be due to temporal patterns of change in steroid hormone secretion, as well as changes in responsiveness to these hormones (Hansen et al., 1978b). Ovariectomized lab rats implanted with estradiol capsules and tested for sexual behavior three days later had peak lordosis quotients (LQ, position of arched back and deflected tail by the female that allows male to successfully mate), indicative of heightened sexual behavior, during the dark, and the lowest LQs during the light portion of the LD cycle (Hansen et al., 1978b; Hansen et al., 1979). Not all studies of lab rats have found daily rhythms in sexual behavior, perhaps because different strains of rats have been used in these studies (Erskine et al., 1980). In diurnal female grass rats primed with E and P at different times of day, rates of sexual behavior three hours after hormone injections were highest one hour before the lights came on, as is the case for intact females in estrus (Mahoney and Smale, 2005b). Nocturnal and diurnal female rodents thus appear to have rhythms in responsiveness to steroid hormones that are 12 h out of phase with one another. The mechanisms responsible for that difference are unknown, but it is tempting to speculate that they involve differences within ER-containing neurons that receive input from the SCN.

This possibility is related to a more general question that emerges from consideration of whether efferent signals from the SCN that regulate behavior are conveyed through a process of diffusion, or whether they are carried along more traditional axonal pathways (Silver et al., 1996). Transplant studies (above) have provided evidence that a diffusible signal is essential and sufficient for the generation of behavioral rhythms. However, those studies have focused on patterns of general activity, feeding or drinking, rather than reflexive behaviors expressed in more specialized circumstances. In the case of lordosis, the fact that the reflex is modulated by estradiol, and that the SCN projects onto ER-containing cells in regions involved in its regulation, suggests the possibility that axonal outputs, rather than diffusible signals, may be key.

3.8 Ovarian Steroids Affect the Circadian System

Until now, we have been considering the ways in which the circadian system can influence reproductive processes and behavior. It is also important to appreciate how the neuroendocrine system can also alter circadian timekeeping systems. Gonadal hormones act early in development to promote sex differences in circadian systems, as well as again in adulthood to modulate these systems, across the estrous cycle as well as over the course of pregnancy and lactation (e.g., Rosenwasser et al., 1987). Here we discuss effects of ovarian secretions associated with the estrous cycle as an example of how hormones may feed back to the circadian system. Direct and indirect evidence for effects of E and P on circadian rhythms has been obtained from a range of species, including carnivores and primates, but the best understood species are rodents, reviewed here.

An initial appreciation of how ovarian hormones can influence the phase and period of circadian systems began with observations by Morin et al. (1977) of changes in the onset of activity that correlated with the estrous cycle of female hamsters kept in a light–dark cycle. A "scalloping" effect was seen whereby the onset of daily wheel running occurred earlier on days 3 and 4 of the estrous cycle than on days 1 and 2, when animals were kept in LL as well as in an LD cycle. Evidence that rising levels of E were responsible for the advance was provided by the observation that in OVX females, subcutaneous implants of estradiol advanced the phase angle of entrainment and shortened the period of the rhythm in free-running conditions; the changes in phase may have been due to the changes in the period. Estradiol implants have also been reported to help maintain synchrony of rhythms of female hamsters kept in constant conditions that commonly lead to dissociated activity rhythms such as those seen in the case of splitting (described above; Morin, 1980; Morin and Cummings, 1982). Administration of E also reduces variability in the onset of activity of free-running hamsters, an effect that is counteracted by P (Takahashi and Menaker, 1980). Very similar effects have been seen in female rats (Albers et al., 1981), although they are strain- and housing specific. It has been suggested that E may shorten the period of the activity rhythm and trigger the LH surge by acting on common circadian mechanisms (Fitzgerald and Zucker, 1976). The function of the circadian regulation of responsiveness to E may be to speed up the clock in a manner that helps ensure that a female is awake and active when she needs to find a mate (Takahashi and Menaker, 1980). It could be quite costly for a short-lived rodent to miss out on the opportunity to mate on the day of estrus, as she may then have to wait 4–5 more days, which is not an insignificant portion of the reproductive life of such rodents.

Another species in which dramatic effects of hormones on rhythms have been seen is the *Octodon degus* (degu), a diurnal, hystricomorph rodent from South America. These females have a 21 day estrous cycle which, unlike those of rats and hamsters, has a spontaneous luteal phase. Pronounced changes in patterns of daily rhythms and in their phase angles of entrainment to an LD cycle are associated with the estrous cycle, and rates at which the rhythms can adjust to phase shifts in a LD cycle are modulated by ovarian hormones (Labyak and Lee, 1995). As with hamsters, the cyclic effects of E and P on entrained phase disappear with ovariectomy, and phase is altered by replacement hormones. However, unlike the hamster, it does not appear that either hormone is directly altering the underlying period of the circadian mechanism (Labyak and Lee, 1995; Jechura and Lee, 2004). Interestingly, in both hamsters and degus, the sex difference in circadian period is determined by exposure to postnatal estrogen. This occurs perinatally in the hamster (Albers et al. 1981; Davis et al. 1983), but long after puberty in the degu (Hummer et al., in press).

Over the years, since the effects of steroid hormones on behavioral and physiological rhythms were discovered, it has become clear that these hormones have myriad activational effects on a variety of features of the SCN, such as on the presence of gap junctions and associated proteins (Shinohara et al., 2003). Most recently, it has been shown that E influences the molecular clock within the SCN, as well as in the uterus. Nakamura et al. (2005) found that E treatment led to an advance in the peak and a decrease in the amplitude of the *Per2* rhythm, but no change in *Per1*, within the SCN of OVX rats. Interestingly, E induced a biphasic rhythm in *Per1* and *Per 2* within the uterus of these animals, significantly altering phase relationships between rhythms in the SCN and uterus. Effects of E on the clock within the SCN could occur indirectly through the ER-alpha containing cells afferent to the SCN, noted above (De La Iglesia et al., 1995), or directly through ER-beta containing cells within the SCN (Wilson et al., 2002).

4 Circadian Rhythms in Neuroendocrine Dopamine Neurons and Prolactin

Prolactin (PRL) is an anterior pituitary hormone that plays a critical role in a far-ranging suite of behavioral and physiological functions, from the promotion of "water drive" in amphibians and crop sack development in birds to a variety of osmoregulatory functions in all vertebrate classes. In mammals, its role has been best studied in the context of female reproduction, where, as the name implies, one of its functions is to promote lactation. The secretion of prolactin from the anterior pituitary is regulated by the release of both inhibitory and stimulatory factors. Stimulatory factors may include oxytocin, thyrotropin releasing hormone, VIP and VP. The latter two factors may exert their effects on PRL secretion via their release into the hypothalamus, where they may act as neuromodulators, as well as through their direct effects on the pituitary gland. The primary inhibitory factor is dopamine (DA), whose synthesis is regulated by the enzyme tyrosine hydroxylase (TH). DA from TH-producing cells reaches the anterior pituitary directly through axons projecting from the hypothalamus to the median eminence, as well as indirectly through its release into the intermediate and posterior lobes of the pituitary; it travels from these lobes to the anterior pituitary through short portal blood vessels.

Rhythms in PRL secretion are evident in females under a variety of reproductive conditions, as reviewed by Freeman et al. (2000). In pregnant and pseudopregnant animals, the rhythms vary somewhat across species. In rats, a bimodal pattern of prolactin secretion is evident for the first ten days of pregnancy, until the placenta matures and begins to secrete its own luteotrophic hormone, at which time the prolactin secretion, initially necessary for maintenance of pregnancy, drops precipitously, and then remains low until the end of pregnancy. A circadian pattern of PRL release is less clear during lactation, when the stimulation a mother receives from her nursing offspring becomes the primary determinant of the pattern of secretion of the hormone. When animals are not pregnant or lactating, prolactin secretion varies systematically as a function of the stage of the estrous cycle and the time of day. Here we focus on the circadian control of the peri-ovulatory PRL surge in the nonpregnant female rat, the best studied model for the circadian regulation of this hormone.

In rats, the preovulatory surge of prolactin parallels that of LH, and is similarly regulated by a circadian signal from the SCN in the context of rising levels of E. As is true of the LH surge, the PRL surge is blocked by lesions of the SCN (Kawakami et al., 1980), which may influence circadian timing of release through VIP and VP signals. As with LH, the PRL surge requires high levels of E, whose effects appear to be mediated primarily by ER-alpha- and ER-beta-containing cells located in the POA. In rats, P enhances the magnitude of the PRL surge as it does for LH. However, the mechanisms controlling the PRL surge are quite different from those controlling the LH surge in some important ways. The primary hypothalamic factor influencing PRL release is DA, which has an inhibitory rather than a stimulatory effect on its secretion, and DA release is reduced rather than increased at the time of ovulation. The circadian signals involved must therefore, directly or indirectly, inhibit the release of DA from the hypothalamus.

Studies of the circadian regulation of prolactin by DA have focused on three primary populations of TH neurons, whose roles are described in Moore and Lookingland (1995). These include TH-producing neuroendocrine cells referred to as the periventricular hypophyseal DA (PHDA) neurons. These cells represent a subset of the A14 group of TH neurons and project to the intermediate lobe of the pituitary where they play a role in the regulation of melanocyte stimulating hormone (MSH). A second group of neuroendocrine DA cells are the tuberohypophyseal TH neurons (THDA neurons), a subset of A12 neurons; these cells are located in the anterior part of the arcuate and the periventricular region. THDA neurons are thought to project to the posterior neural lobe of the pituitary. A third group of neuroendocrine DA cells are referred to as the tuberoinfundibular neurons (TIDA neurons), and represents another subset of the A12 neurons. These are located in the dorsomedial arcuate, caudal to the THDA neurons. These cells release DA into the median eminence (ME) and play a major role in the regulation of prolactin secretion from the anterior pituitary. Research on the question of what might generate a circadian rhythm in prolactin secretion has focused on these three populations of TH neurons.

Evidence that neuroendocrine TH neurons are rhythmic comes from examination of patterns of DA turnover in the pituitary of OVX animals, as well as expression of immediate early genes in TH cells. DA turnover in all three lobes of the pituitary was rhythmic in animals kept in LD cycles as well as in DD, though the patterns were somewhat different in these two conditions (Sellix and Freeman, 2003). Immunohistochemical studies of the immediate early gene FRA have enabled investigators to determine which TH neurons change around the time of the PRL surge (Hoffman et al., 1994; Lerant and Freeman, 1997). Hoffman et al. (1994) first reported a decline in FRA expression within TIDA neurons across the day of proestrous, suggesting that a decrease in their activity precedes the PRL surge. More detailed analyses showed a decrease in FRA within all three groups of neuroendocrine DA cells (THDA and PHDA, as well as TIDA neurons) in both OVX females and intact females on the day of proestrus (Lerant and Freeman, 1997). Interestingly, the decrease occurred in OVX females, despite the fact that PRL secretion did not change over the course of the day in these animals. Lerant and Freidman (1997) suggest that the apparent paradox may reflect increased responsiveness of the pituitary to inhibitory factors, such as DA, over the course of the day, or a change in stimulatory signals that act in parallel with the inhibitory neuroendocrine DA neurons (NDA neurons) or both. Another possibility is that a dissociation develops between mechanisms responsible for an increase in FRA and a change in DA secreted by these cells.

As is the case with GnRH neurons, tract-tracing studies are consistent with the hypothesis that rhythms in NDA neurons may be regulated by both direct and indirect input from the SCN. Horvath (1997) provided evidence for direct projections by showing contacts between fibers labeled with an anterograde tracer subsequent to its injection into the SCN, and cell bodies labeled by a retrograde tracer injected into the peritoneal cavity; (the tracer is picked up by terminals in the pituitary and carried back to neuroendocrine cells in the hypothalamus). Indirect input from the SCN to TH neurons was suggested by the observation that it projects to hypothalamic regions afferent to these neurons, including those in the POA (Watts et al., 1987). As with the circadian regulation of GnRH cells, VP and VIP appear to be important signals communicating temporal information along pathways from the SCN, directly or indirectly or both, to the NDA neurons.

Several lines of evidence suggest that the SCN may convey some temporal information directly to NDA neurons through the release of VIP. VIP fibers form synapses on NDA neurons, which express a VIP2 receptor (Gerhold et al., 2001). Interestingly, the number of contacts between VIP fibers and NDA neurons was elevated by treatment with E and P (Gerhold et al., 2001). VIP may also influence these neurons indirectly, as injections of VIP into the preoptic area stimulate PRL release (Akema et al., 1988). Blocking the signal by administering VIP antisera into the medial preoptic area may also block the normal rise of PRL (van der Beek et al., 1999; but see Harney et al., 1996). Two studies have been conducted using delivery of VIP antisense into the SCN to evaluate its role in the regulation of the PRL surge, and have lead to somewhat conflicting results. In the first, which looked at PRL as the endpoint, the rhythm was not affected (Harney et al., 1996), and in the second, the expression of FRA within NDA neurons was significantly decreased at ZT 13 but not at ZT 1 (Gerhold et al., 2002). These authors did not mention the earlier antisense study, and the reasons for the apparent discrepancy are unclear. Gerhaold et al. (2002) suggest that VIP, coming from the SCN, may help shape PRL rhythms by inhibiting NDA neurons on the evening of proestrus, leading to decreased release of DA and, consequently, increased secretion of prolactin.

A role for VP release from SCN terminals in the regulation of the PRL surge has recently been suggested by Palm et al. (2001). These investigators found that delivery of VP into the POA by reverse microdialysis on the afternoon of proestrus, when VP release normally declines, blocked the PRL surge in OVX female rats treated with E (Palm et al., 2001b). These investigators suggest that, under intact conditions, the drop in secretion of VP from SCN terminals in the POA may help to promote the rise in PRL at that time. It is not yet clear how VP might act under physiological conditions, as attempts to influence PRL release by administration of one VP receptor antagonist, V1aR, were unsuccessful. Nevertheless, when taken together, it appears that an increase in release of both VP and VIP from SCN terminals into the POA and onto NDA neurons may contribute to the circadian regulation of the periovulatory surge in PRL secretion by triggering a decrease in DA secretion. This is especially interesting, as the same SCN outputs appear to trigger an increase in release of GnRH at this time.

Another important part of the mechanism generating circadian rhythms of PRL release may involve rhythmic mechanisms intrinsic to DA neurons. One of the more exciting recent developments relating to rhythms in prolactin is the discovery in female rats, of rhythms in PER1 and PER2 in several populations of cells that contain TH. In the PHDA neurons, Sellix et al. (2006) found a PER1 rhythm with a peak that occurred at ZT 12, and a PER2 rhythm that peaked between ZT 6 and ZT 12. In the THDA neurons, Sellix et al. (2006) found a PER2 rhythm in cells with a peak between ZT 6 and ZT 12; PER1 was present in this group of cells, but was not rhythmic in them. In the TIDA neurons, which are most important for the regulation of prolactin secretion, PER1 peaked at ZT18, and PER2 was highest at ZT 6 and 12 (Sellix et al., 2006). Another study of clock mechanisms in TH neurons, combined TIDA and THDA cells for analysis of PER1 protein and found significantly higher levels in female mice euthanized at ZT 10 than at ZT 22 (Kriegsfeld et al., 2003).

5 Circadian Rhythms in the Hypothalamic–Pituitary–Adrenal Axis

Activation of glucocorticoids by stressful stimuli is an adaptive mechanism that serves many functions (Sapolsky et al. 2000). The underlying circadian pattern of the hypothalamic–pituitary–adrenal (HPA) activity keeps this system in tune with the environment. Chronically elevated glucocorticoid levels, however, can have detrimental effects on an organism and change its response to novel stressors. Species typically produce predominantly either corticosterone or cortisol (CORT) from the adrenal cortex, although a few produce both. Interestingly, chronically elevated CORT levels and circadian or sleep disturbances tend to coexist in a myriad of human diseases. The following section outlines the basic features of the HPA axis, anatomy and interactions with the circadian axis, which when disrupted correlates with many stress-related disorders. Chapters 13, 14, and 15 examine the stress axis in much greater detail. The focus of this section is on its relationship with the circadian axis.

In response to stressful or challenging stimuli (henceforth referred to as "stressors"), the activity in the sympathoadrenal system (SAS) and HPA-axis is increased. The SAS releases norepinephrine and epinephrine at synaptic terminals and into the bloodstream, respectively, and is largely responsible for the initial, so-called "fight or flight" response to a stressor. The HPA-axis is also activated, and has the potential to lead to more long-lasting effects. Chronic exposure to the same stressor can result in an inhibited CORT and ACTH response (e.g., habituation) compared to those of stress-naive animals

Figure 23-3
The mechanism by which the daily circadian release of glucocorticoids is controlled by the SCN. The SCN projects to the PVN which releases corticotrophin-releasing hormone (CRH) into the portal vasculature to cause the release of adrenocorticotropic hormone (ACTH). ACTH, in turn, controls the release of corticosterone/cortisol from the adrenal cortex. The glucocorticoids have a negative feedback effect at the pituitary and PVN, as well as effecting other brain areas not shown in the diagram

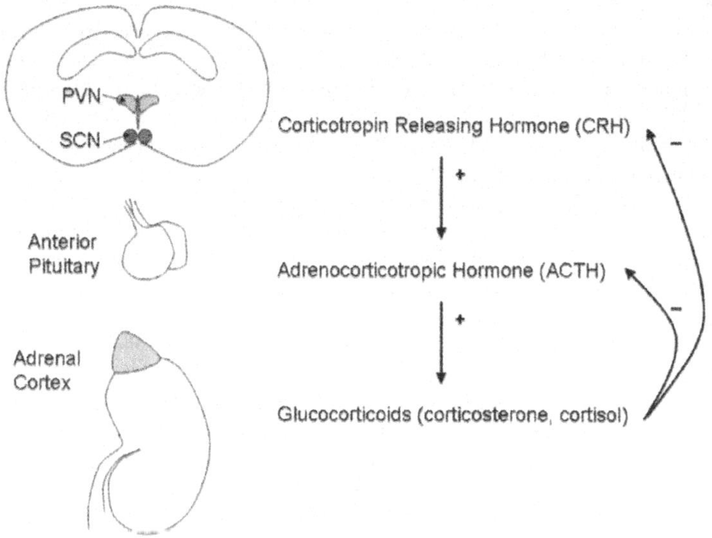

(Dobrakovova et al., 1993; Campmany et al., 1996; Bhatnagar et al., 2002; Bhatnagar et al., 2004; although see Pitman et al., 1988). Habituation following chronic stress, however, is coupled with an enhanced response (facilitation) to novel heterotypic stressors (Bhatnagar and Meaney, 1995; Bhatnagar and Dallman, 1998; Johnson et al., 2002); a more robust CORT and ACTH response is observed following acute presentation of the novel stimulus (Scribner et al., 1991; Bhatnagar and Dallman, 1998). The sympathoadrenal system also appears to be modulated by chronic stress (Dobrakovova et al., 1993; Terrazzino et al., 1995).

Activation of the HPA-axis by a stressor results in release of CRH from the PVN of the hypothalamus, and CRH then stimulates the pituitary to secrete adrenocorticotropic hormone (ACTH). Hypothalamic AVP release tends to enhance this response (Antoni, 1993) while oxytocin tends to suppress it (Swaab et al., 2005). ACTH then stimulates the adrenal cortex to release glucocorticoids. Under normal conditions, the resulting increase in CORT then feeds back at the level of the hypothalamus and pituitary to down-regulate production and release of CRH and ACTH, thus ultimately slowing its own production (Ziegler and Herman, 2002).

Glucocorticoids exert their effects by binding to mineralocorticoid (MR, or Type 1) and glucocorticoid (GR, or Type 2) receptors (De Kloet et al., 1998). Low, basal levels of CORT result in only activation of MR, for which CORT has a higher affinity, to regulate metabolism. The heightened levels of CORT, induced by stress, tend to result in activation of GR, allowing an organism to respond to the stressor by mobilizing glucose and suppressing energetically costly functions that are not of immediate importance, such as immune function and some reproductive functions (Sapolsky et al., 2000; Charmandari et al., 2004). When the HPA axis has been activated repeatedly for prolonged periods of time, a suite of serious health problems can develop. In addition, further glucocorticoid release can become resistant to suppression via negative feedback.

5.1 Anatomy of the HPA Axis

The PVN receives direct, excitatory input from the brainstem (McKellar and Loewy, 1981; Cunningham and Sawchenko, 1988). These projections are catecholaminergic in nature, with norepinephrine stimulating the hypophysiotrophic cells of the PVN (Itoi et al., 1994; Itoi et al., 1998). This allows for a fast HPA response to so called "systemic stressors," challenges that present an immediate, physical threat to homeostasis.

The PVN also receives indirect input from a number of limbic system structures, including the hippocampus and amygdala (Herman, et al., 2003). These limbic inputs are thought to be involved in activation of "processive" stressors (Herman and Cullinan, 1997). These are the "psychological" stressors that are learned through experience, such as fears and anxieties related to potential threats. The dominating presence of GRs in the hippocampus (Han et al., 2005) makes that brain area a likely player in the process of glucocorticoid negative feedback to the PVN.

Inputs from hypothalamic areas, including the BNST (Cullinan et al., 1993, 1996), medial preoptic area (MPOA) (Cullinan et al., 1996), subparaventricular zone (Roland and Sawchenko, 1993), and SCN (Watts et al., 1987) are also present in the PVN. The BNST both excites and inhibits the HPA axis via GABA synapses in the PVN (Herman and Cullinan, 1997). The BNST also acts as a relay station to the PVN (Herman et al., 2005), receiving innervation from limbic and brainstem nuclei that influence PVN activity (Weller and Smith, 1982; Cullinan et al., 1993). The MPOA, providing inhibitory input to the PVN (Cullinan et al., 1996; Viau and Meaney, 1996), is a likely point of interface for the HPA and HPG axes, thus allowing for integration of the reproductive and stress systems. The subparaventricular zone contributes GABAergic inhibition to the PVN (Roland and Sawchenko, 1993) and, being a recipient of input from the SCN (Abrahamson and Moore, 2001), is likely to play a role in circadian regulation of the HPA axis. The SCN innervates a number of other hypothalamic nuclei that also receive input from the PVN (Cullinan et al., 1996; Abrahamson and Moore, 2001) and also projects directly to the parvocellular PVN (Boudaba et al., 1996; Abrahamson and Moore, 2001).

The thalamus also provides important input to the PVN. Sensory information is relayed through thalamic nuclei to provide stress-related information to the PVN. Through projections to the PVN, the posterior paraventricular nucleus of the thalamus (PVTh) plays a role in HPA response to chronic stress (Bhatnagar and Dallman, 1998), and provides another mechanism of circadian control of the HPA axis.

5.2 Daily Rhythms in the Adrenal Axis

The HPA axis provides an excellent example of the hierarchical control of circadian rhythms (e.g., Aschoff, 1981; Moore-Ede et al., 1982; Panda and Hogenesch, 2004). On the one hand, the adrenal gland appears to generate its own rhythms that are then entrained by the SCN. The rhythm in adrenal secretion of CORT appears to act, in turn, as a zeitgeber for other peripheral organs, such as the liver, without having a feedback effect on the SCN (Balsabore et al., 2000; LeMinh et al., 2001; Gachon et al., 2004).

5.2.1 The Circadian Adrenal Gland

The adrenal gland was the first tissue outside the SCN in mammals that was reported to have the ability to oscillate independently. The earliest direct evidence that the adrenal cortex produced endogenous circadian oscillations was provided by adrenal glands cultured in vitro for several days. These glands maintained a nearly 24 h period of CORT release into the culture medium for several days (Andrews and Folk, 1964; Andrews, 1968; Shiotsuka et al., 1974). Subsequently, Moore and Eichler (1972) demonstrated a dramatic dampening of the circadian pattern after SCN lesions, suggesting that, at least in vivo, the adrenal circadian rhythm requires an input of circadian information from the SCN. However, as Davidson et al. (2005) point out, these early conclusions of dependence upon the brain for rhythmicity may have been erroneous, since they were based on the averaged data from groups of animals whose adrenal glands could have produced free-running rhythms following SCN lesions. Moore and Eichler (1972) also collected blood samples at only four times each day, so that even the individual rhythms of the animals might have looked dampened if not collected before a free-running rhythm emerged.

Recent studies using mouse knockins of luciferase attached to the clock genes *mPer1* or *mPer2*, have demonstrated that many peripheral tissues have strong circadian oscillations in vivo and in vitro (Yamazaki et al., 2000; Yoo, et al., 2004; Ishida et al., 2005). The protein produced by the gene construct can be visualized by the bioluminescence of the luciferase when exposed to an appropriate light source. Yamazaki et al. (2000) found that the oscillations of mPer1:luciferase in peripheral tissues (liver, lung, and skeletal muscle) dampened after only 2–7 days in culture while the SCN continued to oscillate for 32 days. The same laboratory subsequently found that the peripheral oscillations last as long as do the SCN oscillations when a mPer2:luciferase construct is monitored rather then Per1 (Yoo et al., 2004). It is unclear why the clock gene used should make such a difference, but the only laboratory to recently report on adrenal clock gene activity used the mPer1:luciferase construct (Ishida et al., 2005). They found that light was able to induce *mPer1* and *mPer2* only in the cortex of the adrenal gland. When animals were housed in DD, the circadian rhythm of Per1 bioluminescence persisted. Not surprisingly, when the animals had their SCNs lesioned, light no longer elicited a rapid increase in Per. However, they did not examine whether SCN-lesioned animals continued to express a daily bioluminescence rhythm of Per in the adrenal cortex (or CORT secretion) in individual animals. Thus, although the technology is available to answer the question of whether the adrenal cortex oscillates endogenously, the current data leave that question open for future research.

It seems likely that the adrenal gland is capable of producing endogenous oscillations, as many of the tissues outside the brain that express clock genes are able to continue oscillating for an extended time. Those tissues now include cornea, pituitary, liver, kidney, and lung (Yoo et al., 2004). Additionally, CORT rhythms seem capable of re-setting the period of clock genes in several peripheral tissues acting through GR, including liver, kidney, and heart (Balsalobre et al., 2000; LeMinh et al., 2001). Thus, whether the adrenal is capable of oscillating independently is still an open question that deserves attention.

5.2.2 Circadian Control of HPA Axis

The HPA axis is tightly regulated by the circadian system under stable entrained conditions. There is an underlying rhythm to the production of CRH, ACTH, and CORT (Ixart et al., 1977; Kaneko et al., 1981; Atkinson and Waddell, 1997), and this rhythm differs between chronotypes. Nocturnal species typically reach peak glucocorticoid levels just before lights-off (Ixart et al., 1977; D'Agostino et al., 1982; Muglia et al., 1997); diurnal animals' glucocorticoid levels peak around the time of lights-on. This rhythm of glucocorticoid secretion is controlled by the rhythm of ACTH (review by Panda and Hogenesch, 2004). In nocturnal rodents, vasopressin release from the SCN appears to exert inhibitory control upon the HPA-axis, which could reflect part of a mechanism by which the SCN entrains oscillators in the adrenal gland that generate rhythms in CORT secretion (Kalsbeek et al., 1992).

The adrenal gland also appears to receive information from the SCN via sympathetic brainstem nuclei (Buijs et al., 1999). This raises the possibility that the SCN might influence CORT secretion in a manner that is independent of the HPA axis. While the basic rhythm in CORT secretion appears to be driven by the pattern of ACTH release, there may also be a contribution from the sympathetic nervous system (England and Arnhold, 2005).

5.2.3 Circadian Disruptions Elicit Stress Responses

Alterations of the LD cycle, typical of transmeridian travel or shift-work, can result in increased glucocorticoid secretion in a number of species (Cho et al., 2000; Leproult et al., 2001; Sei et al., 2003; Ishida et al., 2005). In lab rats, light can also suppress adrenal function more directly, in a manner independent of the circadian system (Buijs et al., 1999). In contrast to findings from mice, in which a light pulse stimulated an increase in CORT secretions (Ishida et al., 2005), Buijs et al. (1999) found a suppression of CORT in rats when light was presented close to the onset of darkness (ZT14). Interestingly, both sets of investigators attributed the change in glucocorticoid levels not to HPA activation, but to sympathetic innervation of the adrenal by the SCN (Buijs et al., 1999; Ishida et al., 2005).

Data from human studies provide support for an interaction between disruption of the circadian system and alterations in HPA functioning. Both, long-term transmeridian travel (Cho et al., 2000) and exposure to a light pulse (Leproult et al., 2001), result in elevated CORT levels. However, in one study, a phase-shift failed to produce statistically significant increases in mean CORT levels, although a trend towards such, as well as an abbreviated quiescent period, was observed (Caufriez et al., 2002). Additionally, desynchrony of the rhythms of ACTH and cortisol following a shift in the LD cycle (Desir et al., 1981) further suggests that circadian manipulations can have an effect on the human HPA axis.

The rodent literature points to a link between glucocorticoid activation and recovery time following a photic phase-shift. The time it takes female rats to reentrain following a shift in the LD cycle is positively correlated with their CORT reactivity under entrained conditions (Weibel et al., 2002). This suggests that HPA activation resulting from the phase-shift could be contributing to interindividual variability in reentrainment rate. This hypothesis is further supported by the finding that adrenalectomy can accelerate reentrainment (Sage et al., 2004).

5.3 Glucocorticoids Affect Circadian Rhythm Expression

There is a large body of evidence supporting a role for glucocorticoids in the timing of circadian systems. Glucocorticoids can induce mouse *Per1* expression in a number of peripheral tissues (Balsalobre et al., 2000; Yamamoto et al., 2005), but not the SCN (Balsalobre et al., 2000). Although it has been thought that the SCN is insensitive because of a lack of GR (Rosenfeld et al., 1988, 1993), we have recently labeled cells with GR in both the ventral–lateral and dorsal–medial SCN of rats and found that a light pulse during the night elicits c-FOS expression within those GR-containing cells (Lee and Mohawk, unpublished data). In addition, there is induction of cFOS in the SCN following glucocorticoid administration (Briski et al., 1997). CRH-deficient mice lack the normal rhythm of μ-opioid receptor expression in the brain, and this rhythm can be reinstated by administration of exogenous CORT (Yoshida et al., 2005). Together, these data suggest that the circadian rhythm of CORT may act as a zeitgeber for both peripheral tissues and for some CNS structures, and may influence SCN function.

The effects of stress on a stable (entrained or free-running) circadian system do not appear to be robust, as it does not result in altered free-running period (Barrington et al., 1993; Meerlo et al., 1997). However, stress can have a reliable effect on another important parameter of a circadian rhythm, its amplitude. Decreases in the amplitude of body temperature rhythms in rats have been seen following both social stress (Meerlo et al., 1997) and footshock (Kant et al., 1991). Chronic, intermittent cold stress can also decrease the amplitude of body temperature rhythms in rats (Bhatnagar and Dallman, 1999). Surgical stress increases rats' heart rates and the amplitude of their body temperature rhythms (Harper et al., 1996). There is also minimal evidence of an ability of stress and glucocorticoids to affect phase angle of entrainment. Injection of dexamethasone, a glucocorticoid agonist, has been shown to have some effect on the entrained phase angle in free-running rats; however, dexamethasone was only effective at high doses (Horseman and Ehret, 1982). Restraint stress does not have an effect on the phase angle of entrainment of locomotor activity rhythms in rats (Barrington et al., 1993) or mice (Yamamoto et al., 2005), but has been shown to cause phase delays in rhythms of free-running hamsters (Van Reeth et al., 1991; Rosenwasser and Dwyer, 2002).

5.4 Glucocorticoids Alter Recovery from Phase Shifts of the Light Cycle

In contrast to the minimal effect of CORT on entrained circadian rhythms, it appears to have a major impact on the rate of recovery from shifts in the LD cycle. We found that recovery of locomotor activity synchronization is altered by restraint stress in the diurnal rodent *Octodon degus* (degu) and the nocturnal rat. Sixty minutes after the new lights-on, animals underwent 60 min of restraint stress. The number of days it took each animal to reentrain its activity rhythms to the new LD cycle was recorded and compared to the number of days it took the animal to reentrain under control conditions. When subjected to restraint stress, degus and rats took 20–30% longer to reentrain their activity rhythms than did controls that also experienced the shifted light cycle, but not the restraint stress (Mohawk and Lee, 2005; Mohawk, 2006). In a second experiment, suppression of the CORT response in the diurnal degu with either a daily injection of metyrapone (blocks synthesis of CORT from progesterone) or dexamethasone (blocks CORT by negative feedback) for the first five days after the shift in the LD cycle, decreased the number of days required to recover stable entrainment by 33% (Mohawk et al., 2005). We also found that on the third day after the shift in the light cycle, CORT levels were elevated relative to the previous baseline.

We subsequently conducted a series of experiments to examine the immediate HPA axis response to a phase shift, with and without concurrent restraint stress (Mohawk, Pargament and Lee, submitted). We measured CORT and ACTH release in male rats exposed to light, occurring 6 h before the normal lights-on (ZT18). A light pulse resulted in increased CORT and ACTH, and restraint (in the dark) or light (without restraint) induced the same rise in CORT. Interestingly, CORT released in response to both stressors presented simultaneously was approximately 40% greater than the response to either stressor alone. These data support the hypothesis that a photic phase-shift elicits an HPA response, in addition to any sympathetic immediate light-induced effect on the adrenal cortex (Ishida et al., 2005). The magnitude of this rise may play a role in recovery of normal circadian entrainment after destabilization, suggesting that either CORT has an impact on SCN function in this state, or that recovery of stable entrainment of activity, feeding and other endocrine rhythms outside the CNS may be slowed by the need to integrate the signals from the SCN and the adrenal.

Humans experiencing circadian desynchrony suffer from a variety of symptoms (Winget et al. 1984; Katz et al., 2001), which are commonly correlated with glucocorticoid activation (Stephens, 1980; Munck et al., 1984). It could be that the change in the LD cycle itself, as supported by the data presented herein, is partly responsible for the change in glucocorticoid levels and resulting symptoms. Keeping CORT levels constant following an LD shift may result in faster recovery of normal entrainment for humans, as it does in rodents (Sage et al., 2004; Mohawk et al., 2005). The additive effect of restraint and light on CORT levels leads us to speculate that keeping stress at a minimum around the time of a phase-shift could be important for health and reestablishing normal circadian entrainment. It may be possible to employ pharmacological manipulation to reduce CORT to improve entrainment of shift-workers and recovery of transmeridian travelers.

6 Conclusion

The timing of a variety of daily behaviors is controlled directly or indirectly by the SCN. Some behaviors, such as activity or wake cycles and changing EEG patterns during sleep, are controlled primarily be neural outputs from the SCN. However, a variety of behaviors are dependent upon the endocrine state of the animal. In addition to responses to destabilizing events (such as a sudden ingestion of food) that restore homeostasis, most endocrine glands also demonstrate stable circadian release patterns that anticipate daily changes in the behavioral needs of the animal. In this chapter, we discussed four such rhythms that have wide ramifications on behavior and physiology: melatonin, GnRH/LH, prolactin, and glucocorticoids.

Exactly how the circadian mechanism communicates with these neuroendocrine systems has slowly come into focus, with the control of melatonin being best understood. There are still several pieces of information that are unknown. To what extent does each of the secretory cells oscillate independently of the SCN, and are they synchronized by the SCN rather than being passively driven? Clearly, in mammals, the pineal gland has become a completely passive organ; however, the same is not at all clear for GnRH cells, the pituitary, or adrenal cortex. Davidson et al. (2005) argue that most tissues outside the brain that have oscillating clock genes are likely able to oscillate independently, but that the SCN normally maintains the internal synchrony between these tissues and the outside world. Others see this as less likely for the endocrine system than for peripheral tissues that can be synchronized by endocrine tissues (Buijs and Kalsbeek, 2001; Schibler and Brown, 2005). They suggest that the abdication of endogenous oscillation by the pineal might also be true for other endocrine tissues. The research tools are available that allow us to determine which hypothesis is correct.

Lastly, it is important to take note that the maintenance of stable entrainment is no longer unidirectional. That is, while the SCN does act as a conduit for light-controlled circadian rhythms, the oscillations generated throughout the body have an impact on SCN function. Melatonin can be used to entrain circadian rhythms, and the finding of receptors in the SCN of animals sensitive to such entrainment suggests that such feedback may be important. Estrogen released as a function of circadian stimulation of GnRH neurons alters some clock gene-related activity and connections among cells within the SCN at the same time that behavior is altered on the day of estrus. Elevated glucocorticoids during recovery from a light-induced circadian disruption may compete with the SCN for control of peripheral tissues, thereby extending the period of desynchrony. Thus, the stability of circadian neuroendocrine rhythms is the result of a network of bi-directional information between the SCN and the endocrine glands.

Acknowledgements

We wish to acknowledge the assistance of Dr. Megan Mahoney, Dr. Daniel Hummer and soon to be Dr. Jennifer Mohawk, without whose data and assistance this chapter would never have been completed. The authors were supported by NIMH: MH05333 (LS), NSF: IBN-0130977 (LS), NIMH: MH069518 (TML), NSF: IBN-0212322 (TML).

Bibliography

Abe M, Herzog ED, Yamazaki S, Straume M, Tei H, et al. 2002. Circadian rhythms in isolated brain regions. J Neurosci 22: 350-356.

Abrahamson EE, Moore RY. 2001. Suprachiasmatic nucleus in the mouse: retinal innervation, intrinsic organization and efferent projections. Brain Res 916: 172-191.

Abizaid A, Horvath B, Keefe DL, Leranth C, Horvath TL. 2004. Direct visual and circadian pathways target neuroendocrine cells in primates. Eur J Neurosci 20: 2767-2776.

Adler 1981. Neuroendocrinology of Reproduction: Physiology and Behavior. New York: Plenum Press.

Adler Pfaff Goy 1985. Handbook of Behavioral Neurobiology, Vol 7: Reproduction. New York: Plenum Press.

Aguilar-Roblero R, Vega-Gonzalez A. 1993. Splitting of locomotor circadian rhythmicity in hamsters is facilitated by pinealectomy. Brain Res 605: 229-236.

Akema T, Hashimoto R, Kimura F. 1988. Preoptic injection of VIP, but not of secretin or PHI, inhibits LH and stimulates prolactin secretion in the ovariectomized rat. Brain Research 441: 367-370.

Albers HE, Davis FC, Darrow JM, Menaker M. 1981. Gonadal hormones organize and modulate the circadian system of the rat. Am J Physiol 241: R62-R66.

Albers H, Gerall A, Axelson J. 1981. Effects of reproductive state on circadian periodicity in the rat. Physiol Behav 26: 21-25.

Alexander MJ, Clifton DK, Steiner RA. 1985. Vasoactive intestinal polypeptide effects a central inhibition of pulsatile luteinizing hormone secretion in ovariectomized rats. Endocrinology 117: 2134-2139.

Alleva J, Waleski M, Alleva F. 1971. A biological clock controlling the estrous cycle of the hamster. Endocrinology 88: 1368-1379.

Amir S, Lamont EW, Robinson B, Stewart J. 2004. A circadian rhythm in the expression of PERIOD2 protein reveals a novel SCN-controlled oscillator in the oval nucleus of the bed nucleus of the stria terminalis. J Neurosci 24: 781-790.

Andrews RV. 1968. Temporal secretory responses of cultured hamster adrenals. Comp Biochem Physiol 26: 179-193.

Antoni FA. 1993. Vasopressinergic control of pituitary adrenocorticotropin secretion comes of age. Front Neuroendocrinol 14: 76-122.

Aschoff J. 1979. Circadian rhythms: general features and endocrinological aspects. Endocrine Rhythms. Krieger DT, editor. Raven Press; New York: pp. 1-61.

Aschoff 1981. Handbook of Behavioral Neurobiology, Vol 4: Biological Rhythms. New York: Plenum Press.

Atkinson HC, Waddell BJ. 1997. Circadian variation in basal plasma corticosterone and adrenocorticotropin in the rat: sexual dimorphism and changes across the estrous cycle. Endocrinology 138: 3842-3848.

Axelrod J. 1974. The pineal gland: a neurochemical transducer. Science 184: 1341-1348.

Baker JR, Ranson RM. 1932. Facotrs affecting the breeding of the field mouse (*Microtus agrestis*). I Light Proc R Soc Lond (Biol) 110: 313-322.

Balsalobre A, Brown SA, Marcacci L, Tronche F, Kellendonk C, et al. 2000. Resetting of circadian time in peripheral tissues by glucocorticoid signaling. Science 289: 2344-2347.

Barrington J, Jarvis H, Redman JR, Armstrong SM. 1993. Limited effect of three types of daily stress on rat free-running locomotor rhythms. Chronobiol Int 10: 410-419.

Bayliss WM, Starling EH. 1902. The mechanism of pancreatic secretion. J Physiol 28: 325

Beach FA. 1971. Hormonal factors controlling the differentiation, development, and display of copulatory behavior in the ramstergig and related species. The Biopsychology of Development. Tobach E, Aronson LR, Shaw E, editors. Academic Press; New York: pp. 249-296.

Becker JB, Breedlove SM, Crews D, McCarthy MM. (eds.) 2002. Behavioral Endocrinology, 2nd edition. Cambridge, MA: MIT Press.

Berthold AA. 1849. Transplantation der Hoden. Arch Anat Physiol Wiss Med 16: 42.

Bhatnagar S, Dallman M. 1998. Neuroanatomical basis for facilitation of hypothalamic-pituitary-adrenal responses to a novel stressor after chronic stress. Neuroscience 84: 1025-1039.

Bhatnagar S, Dallman MF. 1999. The paraventricular nucleus of the thalamus alters rhythms in core temperature and energy balance in a state-dependent manner. Brain Res 851: 66-75.

Bhatnagar S, Huber R, Nowak N, Trotter P. 2002. Lesions of the posterior paraventricular thalamus block habituation of hypothalamic-pituitary-adrenal responses to repeated restraint. J Neuroendocrinol 14: 403-410.

Bhatnagar S, Meaney MJ. 1995. Hypothalamic-pituitary-adrenal function in chronic intermittently cold-stressed neonatally handled and non handled rats. J Neuroendocrinol 7: 97-108.

Bhatnagar S, Vining C, Denski K. 2004. Regulation of chronic stress-induced changes in hypothalamic-pituitary-adrenal activity by the basolateral amygdala. Ann NY Acad Sci 1032: 315-319.

Bissonette TH. 1932. Modification of mammalian sexual cycles; reactions of ferrets (*Putorius vulgaris*) of both sexes to electric light added after dark in November and December. Proc R Soc Lond (Biol) 110: 322-336.

Bonnefond C, Monnerie R, Richard JP, Martinet L. 1993. Melatonin and the circadian clock in mink: Effects of daily injections of melatonin on circadian rhythm of locomotor activity and autoradiographic localization of melatonin binding sites. J Neruoendocrinol 5: 241-246.

Briski KP, Di Pasquale BM, Gillen E. 1997. Induction of immediate-early gene expression in preoptic and hypothalamic neurons by the glucocorticoid receptor agonist, dexamethasone. Brain Res 768: 185-196.

Bronson F, Vom Saal F. 1979. Control of the preovulatory release of luteinizing hormone by steroids in the mouse. Endocrinology 104: 1247-1255.

Brown-Grant K, JM D, F G. 1973. Induced ovulation in albino rats exposed to constant light. J Endocrin 57: 7-22.

Buijs R, van Eden C, Goncharuk V, Kalsbeek A. 2003. The biological clock tunes the organs of the body: timing by hormones and the autonomic nervous system. Journal of Endocrinology 177: 17-26.

Buijs RM, Kalsbeek A. 2001. Hypothalamic integration of central and peripheral clocks. Nature Rev Neurosci 2: 521-526.

Buijs RM, Wortel J, Van Heerikhuize JJ, Feenstra MG, Ter Horst GJ, et al. 1999. Anatomical and functional demonstration of a multisynaptic suprachiasmatic nucleus adrenal (cortex) pathway. Eur J Neurosci 11: 1535-1544.

Butler J, Donovan B. 1971. The effect of surgical isolation of the hypothalamus upon reproductive function in the femalee guinea pig. J Endocrinol 50: 507-513.

Campmany L, Pol O, Armario A. 1996. The effects of two chronic intermittent stressors on brain monoamines. Pharmacol Biochem Behav 53: 517-523.

Cassone VM. 1992. The pineal gland influences rat circadian activity rhythms in constant light. J Biol Rhythms 7: 27-40.

Cassone VM, Chesworth MJ, Armstrong SM. 1986. Entrainment of rat circadian rhythms by daily injection of melatonin depends upon the hypothalamic suprachiasmatic nuclei. Physiol Behav 36: 1111-1121.

Caufriez A, Moreno-Reyes R, Leproult R, Vertongen F, Van Cauter E, et al. 2002. Immediate effects of an 8-h advance shift of the rest-activity cycle on 24-h profiles of cortisol. Am J Physiol Endocrinol Metab 282: E1147-E1153.

Chappell PE. 2005. Clocks and the black box: Circadian invluences on gonadotropin-releasing hormone secretion. J Neuroendocrin 17: 119-130.

Chappell PE, White RS, Mellon PL. 2003. Circadian gene expression regulates pulsatile gonadotropin-releasing hormone (GnRH) secretory patterns in the hypothalamic GnRH-secreting GT1–7 cell line. J Neurosci 23: 11207-11213.

Charmandari E, Kino T, Chrousos GP. 2004. Glucocorticoids and their actions: an introduction. Ann NY Acad Sci 1024: 1-8.

Cho K. 2001. Chronic 'jet lag' produces temporal lobe atrophy and spatial cognitive deficits. Nat Neurosci 4: 567-568.

Cho K, Ennaceur A, Cole JC, Suh CK. 2000. Chronic jet lag produces cognitive deficits. J Neurosci 20: RC66.

Cullinan WE, Helmreich DL, Watson SJ. 1996. Fos expression in forebrain afferents to the hypothalamic paraventricular nucleus following swim stress. J Comp Neurol 368: 88-99.

Cullinan WE, Herman JP, Battaglia DF, Akil H, Watson SJ. 1995. Pattern and time course of immediate early gene expression in rat brain following acute stress. Neuroscience 64: 477-505.

Cullinan WE, Herman JP, Watson SJ. 1993. Ventral subicular interaction with the hypothalamic paraventricular nucleus: evidence for a relay in the bed nucleus of the stria terminalis. J Comp Neurol 332: 1-20.

D'Agostino J, Vaeth GF, Henning SJ. 1982. Diurnal rhythm of total and free concentrations of serum corticosterone in the rat. Acta Endocrinol (Copenh) 100: 85-90.

Davidson AJ, Tataroglu A, Menaker M. 2005. Circadian effects of timed meals (and other rewards). Methods in Enzymology 393: 509-523.

Davis FC, Darrow JM, Menaker M. 1983. Sex differences in the circadian control of hamster wheel-running activity. Am J Physiol 244: R93-R105.

De Kloet ER, Vreugdenhil E, Oitzl MS, Joels M. 1998. Brain corticosteroid receptor balance in health and disease. Endocr Rev 19: 269-301.

De La Iglesia H, Blaustien J, Bittman E. 1995. The Suprachiasmatic Area in the Female Hamster Projects to Neurons Containing estrogen receptors and GnRH. Neuroreport Abstract.

de la Iglesia HO, Blaustein JD, Bittman EL. 1995. The suprachiasmatic area in the female hamster projects to neurons containing estrogen receptors and GnRH. Neuroreport 6: 1715-1722.

de la Iglesia HO, JMeyer J, Schwartz WJ. 2003. Lateralization of circadian pacemaker output: Activation of left- and right-sided luteinizing hormone-releasing hormone neurons involves a neural rather than a humoral pathway. J Neurosci 23: 7412-7414.

Desir D, Van Cauter E, Fang VS, Martino E, Jadot C, et al. 1981. Effects of "jet lag" on hormonal patterns. I. Procedures, variations in total plasma proteins, and disruption of adrenocorticotropin-cortisol periodicity. J Clin Endocrinol Metab 52: 628-641.

Doan A, Urbanski H. 1994. Diurnal expression of Fos in luteinizing hormone-releasing hormone neurons of Syrian hamsters. Biol Reprod 50: 301-308.

Dobrakovova M, Kvetnansky R, Oprsalova Z, Jezova D. 1993. Specificity of the effect of repeated handling on sympathetic-adrenomedullary and pituitary-adrenocortical activity in rats. Psychoneuroendocrinology 18: 163-174.

Doecke F, Luno D, Rohde W, Dorner G. 1982. Perisuprachiasmatic lesions enhonce the increase in gonadotropin secretion in acutely ovariectomized rats. Endokrinologie 78: 311-314.

Edwards RG. 1981. Test-tube babies, 1981. Nature 293: 253-256.

Elliott JA, Bartness TJ, Goldman BD. 1989. Effect of melatonin infusion duration and frequency on gonad, lipid, and body mass in pinealectomized male Siberian hamsters. J Biol Rhythms 4: 439-455.

Erskine M, Marcus J, Baum M. 1980. Absence of a diurnal rhythm in lordosis behavior induced by oestrogen in gonadectomized rats. J Endocrin 86: 127-134.

Eskes GA. 1984. Neural control of the daily rhythm of sexual behavior in the male golden hamster. Brain Res 293: 127-141.

Everett J, Sawyer C. 1950. A 24 hour periodicity in the LH release apparatus of female rats, disclosed by barbituate sedation. Endocrinology 47: 198-218.

Finkelstein JS, Baum FR, Campbell CS. 1978. Entrainment of the female hamster to reversed photoperiod: Role of the pineal gland. Physiol Behav 21: 105-111.

Fitzgerald K, Zucker I. 1976. Circadian organization of the estrous cycle in the golden hamster. Proc Nat Acad Sci 78: 2923-2927.

Fox M, Smith m. 1984. Postpartum preovualatory surge of gonadotropin secretion in the rat may be initiated by the labor process. Biol Reprod 31: 619-626.

Freeman M, Kanyicska B, Lerant A, Nagy G. 2000. Prolactin: Structure, function, and regulation of secretion. Physiol Rev 80: 1523-1585.

Funabashi T, Shinohara K, Mitsushima D, Kimura F. 2000a. Estrogen increases arginine-vasopression V1a receptor mRNA in the preoptic area of young but not of middle-aged female rats. Neurosci Lett 285: 205-208.

Funabashi T, Shinohara K, Mitsushima D, Kimura F. 2000b. Gonadotropin-releasing hormone exhibits circadian rhythm in phase with arginine-vasopressin in co-cultures of the female rat preoptic area and suprachiasmatic nucleus. J Neuroendocrin 12: 521-528.

Gachon F, Nagoshi E, Brown SA, Ripperger J, Schibler U. 2004. The mammalian circadian timing system: from gene expression to physiology. Chromosoma 113: 103-112.

Gauer F, Masson-Pevet M, Pevet BP. 1994a. Daily variations in melatonin receptor density of rat pars tuberalis and suprachiasmatic nuclei are distinctly regulated. Brain Res 641: 92-98.

Gauer F, Masson-Pevet M, Pevet BP. 1994b. Seasonal regulation of melatonin receptors in rodent pars tuberalis: Correlation with reproductive state. J Neural Transm Gen Sect 96: 187-195.

Gauer F, Masson-Pevet M, Skene DJ, Vivien-Roels B, Pevet BP. 1993. Daily rhythms of melatonin binding-sites in the rat pars tuberalis and suprachiasmatic nuclei: Evidence for a regulation of melatonin receptors by melatonin itself. Neuroendocrinology 57: 120-126.

Gerhold L, Horvath T, Freeman M. 2001. Vasoactive intestinal polypeptide fibers innervate neuroendocrine dopaminergic neurons. Brain Res 919: 48-56.

Gerhold L, Sellix M, Freeman M. 2002. Antagonism of vasoactive intestinal polypeptide mRNA in the suprachiasmatic nucleus disrupts the rhythm of FRAs expression in neuroendocrine dopaminergic neurons. J Comp Neurol 450: 135-143.

Gillespie JMA, Chan BPK, Roy D, Cai F, Belsham DD. 2003. Expression of circadian rhythm genes in GnRH-secreting GT1–7 neurons. Endocrinology 144: 5285-5292.

Goldman BD. 1991. Parameters of the circadian rhythm of pineal melatonin secretion affecting reproductive responses in Siberian hamsters. Steroids 56: 218-225.

Goldman BD, Darrow JM, Yogev L. 1984. Effects of timed melatonin infusions on reproductive development in the Djungarian hamster (*Phodopus sungorus*). Endocrinology 114: 2074-2083.

Goldman BD, Hall V, Hollister C, Reppert S, Roychoudhury P, et al. 1981. Diurnal changes in pineal melatonin content in four rodent species: Relationship to photoperiodism. Biol Reprod 24: 778-783.

Goy RW, Goldfoot DA. 1973. Hormonal influences on sexually dimorphic behavior. Handbook of Physiology, Endocrinology II, Part I. Greep RO, Astwood EB, editors. Williams and Wilkins; Baltimore; pp. 169-186.

Goy RW, McEwen BS. 1980. Sexual Differentiation of the Brain. Cambridge, MA: MIT Press.

Gray GD, Sodersten P, Tallentire D, Davidson JM. 1978a. Effects of lesions in various structures of the suprachiasmatic-preoptic region on LH regulation and sexual behavior in female rats. Neuroendocrinology 25: 174-191.

Gray GD, Sodderstein P, Talentire D, Davidson JM. 1978b. Effects of lesions in various structures of the suprachiasmatic-preoptic region on LH regulation and sexual behavior in female rats. Neuroendocrinology 25: 174-191.

Gu G, Simerly R. 1997. Projections of the sexually dimorphic anteroventral periventricular nucleus in the female rat. J Comp Neurol 384: 142-164.

Han F, Ozawa H, Matsuda K, Nishi M, Kawata M. 2005. Colocalization of mineralocorticoid receptor and glucocorticoid receptor in the hippocampus and hypothalamus. Neurosci Res 51: 371-381.

Hansen S. 1979. A sexually dimorphic rhythmin oestradiol activated lordosis behavior in the rat. J Endocrinol 82: 267-274.

Hansen S, Sodersten P, Srebro B. 1978a. A daily rhythm in the behavioural sensitivity of the female rat to oestradiol. J Endocrinol 77: 381-388.

Hansen S, Sodersten P, Srebro B. 1978b. A daily rhythm in the behavioral sensitivity of the female rat to oestradiol. J Endocrinol 77: 381-388.

Hansen S, Sodersten P, Eneroth P, Srebro B, Hole K. 1979. A sexually dimorphic rhythm in oestradiol-activated lordosis behaviour in the rat. J Endocrinol 83: 267-274.

Harney J, Scarbrough K, Rosewell KL, Wise PM. 1996. In vivo antisense antagonism of vasoactive intestinal peptide in the suprachiasmatic nuclei causes aging-like changes in the estradiol-induced luteinizing hormone and prolactin surges. Endocrinology 137: 3696-3701.

Harper DG, Tornatzky W, Miczek KA. 1996. Stress induced disorganization of circadian and ultradian rhythms: comparisons of effects of surgery and social stress. Physiol Behav 59: 409-419.

Harris GW. 1937. Induction of ovulation in the rabbit by electrical stimulation of the hypothalamo-hypophyseal mechanism. Proc Roy Soc Lond (Suppl. B) 122: 374

Harris GW. 1955. Neural Control of the Pituitary Gland. London: Edward Arnold Ltd.

Hastings MH, Mead SM, Vindlaceravu RR, Ebling FJP, Maywood ES, et al. 1992. Non-photic phase shifting of the circadian activity rhythm of Syrian hamsters: The relative potency of arousal and melatonin. Brain Res 591: 20-26.

Hellman L, Nakada F, Curti J, Weitzman ED, Kream J, et al. 1970. Cortisol is secreted episodically by normal man. Electroencphalogr Clin Neurophysiol 30: 411-422.

Herman JP, Cullinan WE. 1997. Neurocircuitry of stress: central control of the hypothalamo-pituitary-adrenocortical axis. Trends Neurosci 20: 78-84.

Herman JP, Figueiredo H, Mueller NK, Ulrich-Lai Y, Ostrander MM, et al. 2003. Central mechanisms of stress integration: hierarchical circuitry controlling hypothalamo-pituitary-adrenocortical responsiveness. Front Neuroendocrinol 24: 151-180.

Herman JP, Ostrander MM, Mueller NK, Figueiredo H. 2005. Limbic system mechanisms of stress regulation: hypothalamo-pituitary-adrenocortical axis. Prog Neuropsychopharmacol Biol Psych 29: 1201-1013.

Hiebert SM, Thomas EM, Lee TM, Pelz KM, Yellon SM, et al. 2000. Photic entraiment of circannual rhythms in golden-mantled ground squirrels: role of the pineal gland. J Biol Rhythms 15: 126-134.

Hirst JJ, Haluska GJ, Cook MJ, Novy MJ. 1993. Plasma oxytocin and nocturnal uterine activity: Maternal but not fetal concentrations increase progressively during late pregnancy and delivery in rhesus monkeys. Amer J Obstet Gynecol 169: 415-422.

Hoff JD, Quigley ME, Yen SS. 1983. Hormonal dynamics at midcycle: a reevaluation. J Clin Endocrinol Metab 57: 792-796.

Hoffman G, Le W, Abbud R, Lee W, Smith M. 1994. Use of Fos-related antigens (FRAs) as markers of neuronal activity: FRA changes in dopamine neurons during proestrus, pregnancy and lactation. Brain Res 654: 207-215.

Hohlweg W, Junkmann K. 1932. Die hormonal-nervose Regulierung der Funktion des Hypophysenvorderlappens. Klin Wochenschr 11: 231

Horseman ND, Ehret CF. 1982. Glucocorticosteroid injection is a circadian zeitgeber in the laboratory rat. Am J Physiol 243: R373-R378.

Horvath T. 1997. Suprachiasmatic efferents avoid fenestrated capillaries but innervate neuroendocrine cells, including those producing dopamine. Endocrinology 138: 1312-1320.

Horvath TL, Cela V, Beek van der EM. 1998. Gender-specific apposition between vasoactive intestinal peptide-containing axons and gonadotrophin-releasing hormone-producing neurons in the rat. Brain Res 795: 277-281.

Humlova H, Illnerova H. 1992. Resetting of the rat circadian clock after a shift in the light/dark cycle depends on the photoperiod. Neurosci Res 13: 147-153.

Hummer DL, Jechura TJ, Mahoney MM, Lee TM. 2006. Gonadal hormone effects on entrained and free-running circadian activity rhythms in the developing diurnal rodent, *Octodon degus.* Am J Physiol, in press.

Illnerova H, Vanecek J. 1983. Extension of the rat pineal N-acetyltransferase rhythm in continuous darkness and on short photoperiod. Brain Res 261: 176-179.

Illnerova H, Vanecek J, Hoffmann K. 1987. Adjustment of the rat pineal N-acetyltransferase rhythm to eight-hour shifts of the light-dark cycle: Advance of the cycle disturbs the rhythm more than delay. Brain Res 417: 176-171.

Ishida A, Mutoh T, Ueyama T, Bando H, Masubuchi S, et al. (2005). Light activates the adrenal gland: Timing of gene expression and glucocorticoid release. Cell Metab 2: 297-307.

Itoi K, Seasholtz AF, Watson SJ. 1998. Cellular and extracellular regulatory mechanisms of hypothalamic corticotropin-releasing hormone neurons. Endocr J 45: 13-33.

Itoi K, Suda T, Tozawa F, Dobashi I, Ohmori N, et al. (1994). Microinjection of norepinephrine into the paraventricular nucleus of the hypothalamus stimulates corticotropin-releasing factor gene expression in conscious rats. Endocrinology 135: 2177-2182.

Ixart G, Szafarczyk A, Belugou JL, Assenmacher I. 1977. Temporal relationships between the diurnal rhythm of hypothalamic corticotrophin releasing factor, pituitary corticotrophin and plasma corticosterone in the rat. J Endocrinol 72: 113-120.

Jackson G, Thurmon J, Nelson D. 1975. Estrogen-induced release of LH in the ovariectomized ewe: Independence of time of day. Biol Reprod 13: 358-362.

Jechura TJ, Lee TM. 2004 Ovarian hormones influence olfactory cue effects on reentrainment in the diurnal rodent, *Octodon degus.* Horm Behav 46: 349-355.

Johnson JD, O'Connor KA, Deak T, Spencer RL, Watkins LR, et al. (2002). Prior stressor exposure primes the HPA axis. Psychoneuroendocrinology 27: 353-365.

Kallo I, Butler J, Barcovics-Kallo M, Goubillon M, Coen C. 2001. Estrogen receptor beta-immunoreactivity in gonadropin releasing hormone-expressing neurons: Regulation by estrogen. J Neuroendocrinol 13: 741-748.

Kalsbeek A, Buijs R, Engelmann M, Wotjack C, Landgraf R. 1995. In vivo measurement of a diurnal variation in vasopressin release in the rat suprachiasmatic nucleus. Brain Research 682: 75-82.

Kaneko M, Kaneko K, Shinsako J, Dallman MF. 1981. Adrenal sensitivity to adrenocorticotropin varies diurnally. Endocrinology 109: 70-75.

Kant GJ, Bauman RA, Pastel RH, Myatt CA, Closser-Gomez E, et al. (1991). Effects of controllable vs. uncontrollable stress on circadian temperature rhythms. Physiol Behav 49: 625-630.

Kapen S, Boyar R, Hellman L, Weitzman ED. 1973. Episodic release of luteinizing hormone at mid-menstrual cycle in normal adult women. J Clin Endocrinol Metab 36: 724-729.

Katz G, Durst R, Zislin Y, Barel Y, Knobler HY. 2001. Psychiatric aspects of jet lag: review and hypothesis. Med Hypotheses 56: 20-23.

Kawakami M, Arita J. 1981. Effects of lesions in the medial basal part of the suprachiasmatic area on prolactin and gonadotropin surges induced by estrogen and progesterone treatment in ovariectomized rats. Neuroendocrinology 32: 242-247.

Kawakami M, Arita J, Yoshioka E. 1980. Loss of estrogen-induced daily surges of prolactin and gonadotropin by suprachiasmatic nucleus lesions in ovariectomized rats. Endocrinology 106: 1087-1092.

King J, Liu E, Ronsheim P, Slonimski M, Rubin B. 1998. Expression of Fos within luteinizing hormone-releasing hormone neurons, in relation to the steroid-induced luteinizing hormone surge in the guinea pig. Biol Reprod 58: 316-322.

Klein DC, Smoot R, Weller JL, Higa S, Markey SP, 1983. Lesions of the paraventricular nucleus area of the hypothalamus disrupt the suprachiasmatic spinal-cord circuit in the melatonin rhythm generating-system. Brain Res Bull 10: 647-652.

Klein KE, Wegmann HM, Hunt BI. 1972. Desynchronization of body temperature and performance circadian rhythm as a result of outgoing and homegoing transmeridian flights. Aerosp Med 43: 119-132.

Knutsson A. 2003. Health disorders of shift workers. Occup Med (Lond) 53: 103-108.

Krajnak K, Kashon ML, Rosewell KL, Wise PM. 1998. Sex differences in the daily rhythm of vasoactive intestinal polypeptide but not arginine vasopressin messenger ribonucleic acid in the suprachiasmatic nuclei. Endocrinology 139: 4189-4196.

Krey L, Butler W, Knobil E. 1975. Surgical disconnection of the mediobasal hypothalamus and pituitary function in the rhesus monkey. I Gonadotropin secretion Endocrinology 96: 1073-1087.

Krieger 1979. Endocrine Rhythms. New York: Raven Press.

Kriegsfeld LJ, Korets R, Silver R. 2003. Expression of the circadian clock gene Period 1 in neuroendocrine cells: An investigation using mice with a Per1::GFP transgene. European J Neurosci 17: 1-9.

Kriegsfeld LJ, Silver R, Gore AC, Crews D. 2002. Vasoactive intestinal polypeptide contacts on gonadotropin-releasing hormone neurones increase following puberty in female rats. J Neuroendocrinol 14: 685-690.

Labyak SE, Lee TM. 1995. Estrus and estrogen-induced changes in circadian rhythms in a diurnal rodent, *Octodon degus.* Physiol Behav 58: 573-585.

Lee KA, Gay CL. 2004 Sleep in late pregnancy predicts length of labor and type of delivery. Am J Obstet Gynecol 191: 2041-2046.

Lee W, Smith M, Hoffman G. 1990. LHRH neurons express Fos protein during the proestrus surge of luteinizing hormone. Proc Nat Acad Sci USA 87: 5163-5167.

Legan S, Karsch F. 1975. A daily signal for the LH surge in the rat. Endocrinology 96: 57-62.

Le Minh N, Damiola F, Tronche F, Schutz G, Schibler U. 2001. Glucocorticoid hormones inhibit food-induced phase-shifting of peripheral circadian oscillators. EMBO J 20: 7128-7136.

Leproult R, Colecchia EF, L'Hermite-Baleriaux M, Van Cauter E. 2001. Transition from dim to bright light in the morning induces an immediate elevation of cortisol levels. J Clin Endocrinol Metab 86: 151-157.

Lerant a, Freeman M. 1997. Dopaminergic neurons in periventrivular and arcuate nuclei of proestrus and ovariectomized rats: endogenous diurnal rhythm of Fos-related antigens expression. Neuroendocrinology 65: 436-445.

Liu T, Borjigin J. 2005. Free-running rhythms of pineal circadian output. J Biol Rhythms 20: 430-440.

Liu T, Borjigin J. 2005. Reentrainment of the circadian pacemaker through three distinct stages. J Biol Rhythms 20: 441-450.

Lohstroh PN, et al. 2003. Bone resorption is affected by follicular phase length in female rotating shift workers. Environ Health Perspect 111: 618-622.

Mahoney M, Smale L. 2005a. Arginine vasopressin and vasoactive intestinal polypeptide fibers make appositions with gonadotropin-releasing hormone and estrogen receptor cells in the diurnal rodent Arvicanthis niloticus. Brain Res 1049: 156-164.

Mahoney MM, Smale L. 2005b. A daily rhythm in mating behavior in a diurnal murid rodent (Arviucanthis niloticus). Horm Behav 47: 8-13.

Mahoney MM, CSisk C, Ross H, Smale L. 2004. Circadian regulation of gonadotropin-releasing hormone neurons and the preovulatory surge in luteinizing hormone in the diurnal rodent, Arvicanthis niloticus and in a nocturnal rodent Rattus norvegicul. Biol Reprod 70: 1049-1054.

McElhinny TL, Sisk CL, Holekamp KE, Smale, L. 1999. A morning surge in plasma luteinizing hormone coincides with elevated fos expression in gonadotropin-releasing hormone-immunoreactive neurons in the diurnal rodent, Arvicanthis niloticus. Biol Reprod 61: 1115-1122.

Meerlo P, Hoofdakker van den RH, Koolhaas JM, Daan S. 1997. Stress-induced changes in circadian rhythms of body temperature and activity in rats are not caused by pacemaker changes. J Biol Rhythms 12: 80-92.

Meyer-Bernstein EL, Jetton AE, Matsumoto SI, Markuns JF, Lehman MN, et al. 1999a. Effects of suprachiasmatic transplants on circadian rhythms of neuroendocrine function in golden hamsters. Endocrinology 140: 207-218.

Meyer-Bernstein EL, Jetton AE, Matsumoto SI, Markuns JF, Lehman MN, (1999b). Effects of suprachiasmatic transplants on circadian rhythms of neuroendocrine function in golden hamsters. Endocrinology 140: 207-218.

Mohawk JA. 2006. Interactions between glucocorticoids and the desynchronized circadian system. Dissertation completed for Ph.D. at University of Michigan, Ann Arbor.

Mohawk JA, Cashen K, Lee TM. 2005. Inhibiting cortisol response accelerates recovery from a photic phase shift. Am J Physiol 288: R221-R228.

Mohawk JA, Lee TM. 2005. Restraint stress delays reentrainment in male and female diurnal and nocturnal rodents. J Biol Rhythms 20: 245-256.

Mohawk JA, Pargament JM, Lee TM. 2006. Restraint stress alters corticosterone response following a phase-shifting light pulse. Horm Behav submitted.

Moore CR, Price D. 1932. Gonadal hormone functions and the reciprocal influence between gonads and hypophysis with its bearing on the problem of sex hormone antagonism. Amer J Anat Sci 50: 13.

Moore K, Lookingland K. 1995. Dopaminergic neuronal systems in the hypothalamus. New York: Raven Press, Ltd.

Moore RY, Speh JC, Leak RK. 2002. Suprachiasmatic nucleus organization. Cell Tissue Res 309: 89-98.

Moore RY, Eichler VB. 1972. Loss of a circadian adrenal corticosterone rhythm following suprachiasmatic lesions in the rat. Brain Res 42: 201-206.

Moore RY, Klein DC. 1974. Visual pathways and the central neural control of a circadian rhythm in pineal serotonin N-acetyltransferase activity. Brain Res 71: 17-33.

Moore-Ede MC, Sulzman FM, Fuller CA. 1982. The Clocks That Time Us: Physiology of the Circadian Timing System. Cambridge, MA: Harvard University Press.

Morin L. 1980. Effect of ovarian hormones on synchrony of hamster circadian rhythms. Physiol Behav 24: 741-749.

Morin L, Cummings A. 1982. Splitting of wheelrunning rhythms by castrated or steroid treated male and female hamsters. Physiol Behav 29: 665-675.

Morin L, Fitzgerald K, Zucker I. 1977. Estradiol shortens the period of hamster circadian rhythms. Science 196: 305-307.

Morin L, Godless-Sanchez N, Smale L, Moore R. 1994. Projections of the suprachiasmatic nuclei, subparventricular zone and retrochiasmatic area in the golden hamster. Neuroscience 61: 391-410.

Muglia LJ, Jacobson L, Weninger SC, Luedke CE, Bae DS, (1997). Impaired diurnal adrenal rhythmicity restored by constant infusion of corticotropin-releasing hormone in corticotropin-releasing hormone-deficient mice. J Clin Invest 99: 2923-2929.

Munck A, Guyre PM, Holbrook NJ. 1984. Physiological functions of glucocorticoids in stress and their relation to pharmacological actions. Endocr Rev 5: 25-44.

Myer-Bernstein EL, Letton AE, Matsumoto S-I, Markuns JF, Lehman MN, et al. 1999. Effects of suprachiasmatic transplants on circadian rhythms of neuroendocrine function in golden hamsters. Endocrinology 140: 207-218.

Nagano M, Akihito A, Nakahama K-I, Nakamura T, Tamada M, et al. (2003). An abrupt shift in the day/night cycle causes desynchrony in the mammalian circadian center. J Neurosci 23: 6141-6151.

Nakamura T, Moriya T, Inoue S, Shimazoe T, Watanabe S, et al. 2005. Estrogen differentially regulates Expression of Per1 and Per2 genes between central and peripheral clocks and between reproductive and nonreproductive tissues in female rats. J Neurosci Res 82: 622-630.

Nelson RJ. 2000. An Introduction to Behavioral Endocrinology, 2nd edition. Sinauer Assoc, Sunderland, MA.

Nunez AA, Casati MJ. 1979. The role of efferent connections of the suprachiasmatic nucleus in the control of circadian rhythms. Behavioral and Neural Biology 25: 263-267.

Nunez AA, Brown MH, Youngstrom TG. 1985. Hypothalamic circuits involved in the regulation of seasonal and circadian rhythms in male golden hamsters. Brain Res Bull 15: 149-153.

O'Connor LH, Morin LP, Feeder HH. 1985. A diurnal fluctuation in medial basal hypothalamic-preoptic area cytosol estrogen receptors in ovariectomized hamsters. Brain Res 347: 376-380.

Olcese Domagalski R, Bednorz A, Weaver DR, Urbanski HF, Reuss S, et al. 2003. Expression and regulation of mPer1 in immortalized GnRH neurons. NeuroReport 4: 613-618.

Palm I. 2001 Timing of female reproduction: Role of the suprachiasmatic nucleus. Department of Human and Animal Physiology. Amsterdam: Wageningden University.

Palm I, Beek van der E, Wiegant V, Buijs R, Kalsbeek A. 1999a. Vasopressin induces a luteinizing hormone surge in ovariectomized, estradiol-treated rats with lesions of the suprachiasmatic nucleus. Neuroscience 93: 659-666.

Palm I, Beek van der E, Wiegant V, Buijs R, Kalsbeek A. 2001a. The stimulatory effect of vasopressin on the luteinizing hormone surge in ovariectomized, estradiol-treated rats is time-dependent. Brain Res 901: 109-116.

Palm I, Beek van der E, Swarts H, Vliet van der J, Wiegant V, et al. 2001b. Control of estradiol induced prolactin surge by the suprachiasmatic nucleus. Endocrinology 142: 2296-2302.

Palm IF, Van Der Beek EM, Wiegant VM, Buijs RM, Kalsbeek A. 1999b. Vasopressin Induces a Lutenizing Hormone Surge in Ovariectomized, Estradiol-Treated Rats with Lesions of the Suprachiasmatic Nucleus. Neuroscience 93.

Panda S, Hogenesch JB. 2004. It's all in the timing: Many clocks, many outputs. J Biol Rhythms 19: 374-387.

Pfaff DW. 1980. Estrogens and Brain Function: Neural Analysis of a Hormone-Controlled Mammalian Reproductive Behavior. New York: Springer-Verlag.

Phoenix CH, Goy RW, Gerall AA, Young WC. 1959. Organizational action of prenatally administered testosterone propionate on the tissues mediating behavior in the female guinea pig. Endocrinology 65: 369-382.

Pitman DL, Ottenweller JE, Natelson BH. 1988. Plasma corticosterone levels during repeated presentation of two intensities of restraint stress: chronic stress and habituation. Physiol Behav 43: 47-55.

Quay WB. 1970. Precocious entrainment and associated characteristics of activity patterns following pinealectomy and reversal of photoperiod. Physiol Behav 5: 1281-1290.

Redman JR, Armstrong SM, Ng KT. 1983. Free-running activity rhythms in the rat: Entrainment by melatonin. Science 219: 1989-1991.

Reppert S, Weaver D. 2002. Coordination of circadian timing in mammals. Nature 418: 935-941.

Reppert S, Schwartz W, Uhl G. 1987. Arginine vasopressin: a novel peptide rhythm in cerebrospinal fluid. Tr Neurosci 10: 76-80.

Roland BL, Sawchenko PE. 1993. Local origins of some GABAergic projections to the paraventricular and supraoptic nuclei of the hypothalamus in the rat. J Comp Neurol 332: 123-143.

Rosenfeld P, Van Eekelen JAM, Levine S, De Kloet ER. 1988. Ontogeny of the type 2 glucocorticoid receptor in discrete rat brain regions: an immunocytochemical study. Brain Res 470: 119-127.

Rosenfeld P, Van Eekelen JAM, Levine S, De Kloet ER. 1993. Ontogeny of corticosteroid receptors in the brain. Cell Mol Neurobiol 13: 295-319.

Rosenwasser AM, Dwyer SM. 2002. Phase shifting the hamster circadian clock by 15-minute dark pulses. J Biol Rhythms 17: 238-247.

Rosenwasser A, Hollander S, Adler N. 1987. Effects of pregnancy and parturition on free-running circadian activity rhythms in the rat. Chronobiol Int 4: 183-187.

Rusak B, Morin LP. 1976. Testicular responses to photoperiod are blocked by lesions of the suprachiasmatic nuclei in golden hamsters. Biol Reprod 15: 366-374.

Rusak B, Zucker I. 1979a. Neural regulation of circadian rhythms. Physiol Rev 59: 449-524.

Rusak B, Zucker I. 1979b. Neural regulation of circadian rhythms. Physiol Rev 59: 449-522.

Sage D, Ganem J, Guillaumond F, Laforge-Anglade G, Francois-Bellan AM, et al. 2004. Influence of the corticosterone rhythm on photic entrainment of locomotor activity in rats. J Biol Rhythms 19: 144-156.

Salisbury R, Krieg R, Siebel H. 1980. Effects of arginine vasotocin, oxytocin and arginine vasopressin on steroid induced surges of luteinizing hormone and prolactin in ovariectomized rats. Acta Endocrinol 94: 166-173.

Sapolsky RM, Romero LM, Munck AU. 2000. How do glucocorticoids influence stress responses? Integrating permissive, suppressive, stimulatory, and preparative actions. Endocr Rev 21: 55-89.

Schally AV, Arimura A, Bowers CY, Kastin AJ, Sawano S, et al. 1968. Hypothalamic neurohormones regulating anterior pituitary function. Rec Progr Horm Res 24: 497.

Schibler U, Brown SA. 2005. Enlightening the adrenal gland. Cell Metab 2: 278-281.

Schwartz M, Nunez AA, Smale L. 2004. Differences in the suprachiasmatic nucleus and lower subparventricular zone of diurnal and nocturnal rodents. Neuroscience 127: 25-34.

Scott AJ. 2000. Shift work and health. Prim Care 27: 1057-1079.

Scribner KA, Walker CD, Cascio CS, Dallman MF. 1991. Chronic streptozotocin diabetes in rats facilitates the acute stress response without altering pituitary or adrenal responsiveness to secretagogues. Endocrinology 129: 99-108.

Sei H, Fujihara H, Ueta Y, Morita K, Kitahama K, et al. 2003. Single eight-hour shift of light-dark cycle increases brain-derived neurotrophic factor protein levels in the rat hippocampus. Life Sciences 73: 53-59.

Seibel MM, et al. 1982. Biological rhythm of the luteinizing hormone surge in women. Fertil Steril 37: 709-711.

Sellix M, Freeman M. 2003. Circadian rhythms of neuroendocrine dopaminergic neuronal activity in ovariectomized rats. Neuroendocrinology 77: 59-70.

Sellix M, Egli M, Poletini M, McKee D, Bosworth M, et al. 2006. Anatomical and functional characterization of clock gene expression in neuroendocrine dopaminergic neurons. Am J Physiol R-00555–2005: DOI #10.1210/en.2003–0802

Shinohara K, Funabashi T, Nakamura T, Kimura F. 2003. Effects of estrogen and progesterone on the expression of connexin-36 mRNA in the suprachiasmatic nucleus of female rats. Neurosci Lett 309: 37-40.

Shiotsuka R, Jovonovich J, Jovonovich JA. 1974. In vitro data on drug sensitivity: circadian and ultraidan corticosterone rhythms in adrenal organ cultures. Chronobiological Aspects of Endocrinology. Aschoff J, Ceresa F, Halberg F, editors. Stuttgart, Germany: Schattauer-Verlag; pp. 255-267.

Silver R, Le Sauter J. 1995. Localization of pacemaker cells in the hamster SCN:II transplant studies. Soc Neurosci Abstr 21.

Silver R, Le Sauter J, Tresco P, Lehman M. 1996. A diffusible coupling signal from the transplanted suprachiasmatic nucleus controlling circadian locomotor rhythms. Nature 382: 810-813.

Silverman A-J, Livne I, Witkin J. 1994. The gonadotropin-releasing hormone (GnRH), neuronal systems: Immunocytochemistry and in situ hybridization. In The Physiology of Reproduction. Raven Press Ltd; New York: pp. 1683-1709.

Smith M, Jennes L, Weiss P. 2000. Localization of the VIP2 receptor protein on GnRH neurons in the female rat. Neuroendocrinology 141: 4317-4319.

Stephens DB. 1980. Stress and its measurement in domestic animals: a review of behavioral and physiological studies under field and laboratory situations. Adv Vet Sci Comp Med 24: 179-210.

Stetson M, Watson-Whitmyre M. 1976. Nucleus suprachiasmaticus: The biological clock in the hamster. Science 191: 197-200.

Stobie KM, Weick RF. 1989. Vasoactive intestinal peptide inhibits luteinizing hormone secretion: the inhibition is not mediated by dopamine. Neuroendocrinology 49: 597-603.

Sun X, Deng J, Liu T, Borjigin J. 2002. Circadian 5-HT production regulated by adrenergic signaling. Proc Natl Acad Sci USA 99: 4686-4691.

Sun X, Liu T, Deng J, Borjigin J. 2003. Long-term in vivo pineal microdialysis. J Pineal Res 35: 118-124.

Swaab DF, Bao A-M, Lucassen PJ. 2005. The stress system in the human brain in depression and neurodegeneration. Ageing Res Rev 4: 141-194.

Swann J, Turek F. 1985. Multiple circadian oscillators regulate the timing of behavioral and endocrine rhythms in female golden hamsters. Science 22: 898-900.

Takahashi J, Menaker M. 1980. Interaction of estradiol and progesterone: effects on circadian locomotor rhythm in female golden hamsters. Am J Physiol 329: R497-R504.

Tapp WN, Natelson BH. 1989. Circadian rhythms and patterns of performance before and after simulated jet lag. Am J Physiol 257: R796-803.

Terrazzino S, Perego C, De Simoni MG. 1995. Effect of development of habituation to restraint stress on hypothalamic noradrenaline release and adrenocorticotropin secretion. J Neurochem 65: 263-267.

Tersawa E, Yeoman R, Schultz N. 1984. Factors influencing the progesterone-induced luteinizing hormone surge in rhesus monkeys: Diurnal influences and time interval after estrogen. Biol Reprod 31: 732-741.

Tersawa S, Wiegand S, Bridson W. 1980. A role for medial preoptic nucleuson afternoon of proestrus in female rats. Am J Physiol 328: E533-E539.

Testart J, Frydman R, Roger M. 1982. Seasonal influence of diurnal rhythms in the onset of the plasma luteinizing hormone surge in women. J Clin Endocrinol Metab 55: 374-377.

Thind K, Boggan J, Goldsmith P. 1991. Interactions Between Vasopressin- and Gonadotropin-Releasing -Hormone-Containing Neuroendocrine Neurons in the Monkey Supraoptic Nucleus. Neuroendocrinology 39: 287-297.

Turek FW, Campbell CS. 1979. Photoperiodic regulation of neuroendocrine-gonadal activity. Biol Reprod 20: 32-50.

Turner CD, Bagnara JT. 1971. General Endocrinology, 5th edition. Philadelphia, PA: W.B. Saunders Co; pp. 24-73.

van der Beek E. 1996. Circadian control of reproduction in the female rat. Progress in Brain Res 111: 295-320.

van Der Beek E, Swarts H, Wiegant V. 1998. Central Administration of Antiserum to Vasoactive Intestinal Peptide Delays and Reduces Lutenizing Hormone and Prolactin Surges in Ovariectomized, Estrogen-treated Rats. Neuroendocrinology 69: 227-237.

van der Beek /BibAuthorName>van Oudheusden H, Buijs R, Donk van der H, Wiegant V. 1994. Preferential induction of c-fos immunoreactivity in vasoactive intestinal polypeptide-innervated gonadotropin-releasing hormone neourns during a steroid induced luteinizing hormone surge in the female rat. Endocrinology 134: 2636-2644.

van der Beek /BibAuthorName>Horvath T, Wiegant V, Hurk van den R, Buije R. 1997a. Evidence for a direct neuronal pathway from suprachiasmatic nucleus to the gonadotropin-releasing hormone system: Combined tracing and light and electron microscopic immunocytochemical studies. J Comp Neurol 384: 569-579.

van der Beek /BibAuthorName>Swarts HJM, Wiegant VM. 1999. Central administration of antiserum to vasoactive intestinal peptide delays and reduces luteinizing hormone and prolactin surges in ovariectomized, estrogen-treated rats. Neuroendocrinology 69: 227-237.

Van der Beek EM, Wiegant HA, Donk Van der R, Hurk van den HA, Buijs RM. 1993. Lesions of the suprachiasmatic nucleus indicate the presence of a direct vasoactive intestinal polypeptide-containing projection to gonadotropin-releasing hormone neurons in the female rat. J Neuroendocrin 5: 137-144.

van der Beek EM, Horvath TL, Weigant VM, Hurk van den R, Buijs R. 1997b. Evidence for a direct neuronal pathway from the suprachiasmatic nucleus to the gonadotrophin-releasing hormone system: combined tracing and light and electron microscopic immunocytochemical studies. J Comp Neurol 384: 569-579.

Van Reeth O, Hinch D, Tecco JM, Turek FW. 1991. The effects of short periods of immobilization on the hamster circadian clock. Brain Res 545: 208-214.

Viau V, Meaney MJ. 1996. The inhibitory effect of testosterone on hypothalamic-pituitary-adrenal responses to stress is mediated by the medial preoptic area. J Neurosci 16: 1866-1876.

Vijayan E, Samson WK, Said SI, McCann SM. 1979. Vasoactive intestinal peptide: evidence for a hypothalamic site of action to release growth hormone, luteinizing hormone, and prolactin in conscious ovariectomized rats. Endocrinology 104: 53-57.

Waddington Lamont E, Robinson B, Stewart J, Amir S. 2005. The central and basolateral nuclei of the amygdala exhibit opposite diurnal rhythms of expression of the clock protein Period 2. Proc Natl Acad Sci USA 102: 4180-4184.

Watson R, Langub C. 1992. Vasopressinergic synaptic input upon estrogen receptive neurons in the anterior preoptic area of the rat: Suprachiasmatic nucleus origin? Society for Neuroscience Abstracts 18: 113.

Watson R, Langub C, Engle MG, Maley B. 1995. Estrogen-receptive neurons in the anteroventral periventricular nucleus are synaptic targets of the suprachiasmatic nucleus and peri-suprachiasmatic region. Brain Res 689: 254-264.

Watts AG, Swanson LW. 1987. Efferent projections of the suprachiasmatic nucleus: II. Studies using retrograde transport of fluorescent dyes and simultaneous peptide immunohistochemistry in the rat. J Comp Neurol 258: 230-252.

Watts A, Swanson L, Sanchez-Watts G. 1987. Efferent projections of the suprachiasmatic nucleus: I. Studies using anterograde transport of Phaseolus vulgaris leucoagglutinin in the rat. J Comp Neurol 258: 204-229.

Watts AG, Swanson LW. 1987. Efferent projections of the suprachiasmatic nucleus: II. Studies using retrograde transport of fluorescent dyes and simultaneous peptide immunohistochemistry in the rat. J Comp Neurol 258: 230-252.

Weaver DR, Stehle JH, Stopa EG, Reppert SM. 1993. Melatonin receptors in human hypothalamus and pituitary:

Implications for circadian and reproductive responses to melatonin. J Clin Endocrinol Metab 76: 295-301.

Weibel L, Maccari S, Van Reeth O. 2002. Circadian clock functioning is linked to acute stress reactivity in rats. J Biol Rhythms 17: 438-446.

Weick RF, Stobie KM. 1992. Vasoactive intestinal peptide inhibits the steroid-induced LH surge in the ovariectomized rat. J Endocrinol 133: 433-437.

Weitzman ED. 1976. Circadian rhythms and episodic hormone secretion in man. Ann Rev Med 27: 225-243.

Weitzman ED, Fukushima DK, Nogeire C, Roffwarg H, Gallagher TF, et al. 1971. Twenty-four hour pattern of the episodic secretion of cortisol in normal subjects. J Clin Endocrinol Metab 33: 14-22.

Weitzman ED, Hellman L. 1974. Temporal organization of the 24-hour pattern of the hypothalamic-pituitary axis. Biorhythms and Human Reproduction. Ferin M, Halberg F, Richart RM, Vandewiele RL, editors. New York: Wiley Press; pp. 371-395.

Weitzman ED, Schaumburg H, Fishbein W. 1966. Plasma 17-hydroxy-corticosteroid levels during sleep in man. J Clin Endocrinol Metab 26: 121-127.

Weller KL, Smith DA. 1982. Afferent connections to the bed nucleus of the stria terminalis. Brain Res 232: 255-270.

Wiegand S, Tersawa E, Bridson W, Goy R. 1980. Effects of discrete lesions of preoptic and suprachiasmatic structures in the female rat. Neuroendocrinology 31: 137-157.

Wilson M, Rosewell K, Kashon M, Shughrue P, Merchenthaler I, et al. 2002. Age differentially influences estrogen receptor-alpha (ER-alpha) and estrogen receptor-beta (ER-beta) gene expressionin specific regions of the rat brain. Mech Aging Develop 123: 593-601.

Winget CM, De Roshia CW, Markley CL, Holley DC. 1984. A review of human physiological and performance changes associated with desynchronosis of biological rhythms. Aviat Space Environ Med 55: 1085-1096.

Woodfill CJI, Wayne NL, Moenter SM, Karsch FJ. 1994. Photoperiodic synchronization of a circannual reproductive rhythm in sheep: Identification of season-specific time cues. Biol Reprod 50: 965-976.

Yamamoto T, Nakahata Y, Tanaka M, Yoshida M, Soma H, et al. 2005. Acute physical stress elevates mouse period1 mRNA expression in mouse peripheral tissues via a glucocorticoid-responsive element. J Biol Chem 280: 42036-42043.

Yamazaki S, Numano R, Abe M, Hida A, Takahashi R-I, et al. 2000. Resetting central and peripheral circadian oscillators in transgenic rats. Science 288: 682-685.

Yeates NTM. 1949. The breeding season of the sheep with particular reference to its modification by artificial means using light. J Agric Sci 39: 1-43.

Yoo S-H, Yamazaki S, Lowrey PL, Shimomura K, Ko CH, et al. 2004. PERIOD2::LUCIFERASE real-time reporting of circadian dynamics reveals persistent circadian osillations in mouse peripheral tissues. Proc Nat Acad Sci USA 101: 5339-5346.

Yoshida M, Koyanagi S, Matsuo A, Fujioka T, To H, et al. 2005. Glucocorticoid hormone regulates the circadian coordination of micro-opioid receptor expression in mouse brainstem. J Pharmacol Exp Ther 315: 1119-1124.

Ziegler DR, Herman JP. 2002. Neurocircuitry of stress integration: Anatomical pathways regulating the hypothalamo-pituitary adrenocortical axis of the rat. Integr Comp Biol 42: 541-551.

24 Neurochemistry of Sleep

T. Porkka-Heiskanen · L. Alanko · D. Stenberg

Abstract: Neurophysiological and neurochemical mechanisms regulating wakefulness and sleep are reviewed. Wake-active brain areas include the noradrenergic locus coeruleus, the serotonergic dorsal raphe, the dopaminergic ventral tegmental area, cholinergic cells in pontomesencephalic nuclei and basal forebrain (BF), the histaminergic posterior hypothalamic area, and the hypocretin/orexin cells in the lateral hypothalamus. Glutamate is also an important mediator in waking mechanisms. Sleep-active neurons have been found in the ventrolateral preoptic hypothalamic region, and are GABA/galaninergic. These neurons inhibit waking mechanisms. Adenosine is an important homeostatic regulator of sleep, acting on BF neurons to induce sleep. NO production in the BF forebrain through induction of inducible NO synthase has been found to be part of the regulation of recovery sleep after sleep deprivation, involving also local adenosine release. NO influences sleep regulation also through other brain areas. Prostaglandins released in the leptomeninges may also induce sleep through adenosinergic mechanisms. Future research should include studying the role of glial cells in sleep regulation. Also, sleep need is use-dependent, depending especially on cortical neuronal activity, but it is not yet known how this information is conveyed to the homeostatic regulatory mechanisms in the BF.

List of Abbreviations: ACh, acetylcholine; AD, adenosine; AMPA, alpha-amino-3-hydroxy-5-methyl-4-isoxazolepropionic acid, a glutamate agonist; ARAS, ascending reticular activating system; BF, basal forebrain; CRH, corticotropin-releasing hormone; CSF, cerebrospinal fluid; DPR, D type of prostanoid receptor, PGD_2 receptor; DR, dorsal raphe; EEG, electroencephalogram; GABA, gamma-aminobutyric acid; GHRH, growth hormone-releasing hormone; 5-HT, 5-hydroxytryptamine, serotonin; Hcrt, hypocretin = orexin; IL-1, interleukin-1; KO, knockout; LC, locus coeruleus; LDT, laterodorsal tegmental nucleus; L-NAME, NG-nitro-L-arginine methyl ester, a NOS inhibitor; MAO, monoamino-oxidase; NMDA, N-methyl-D-aspartic acid, a glutamate agonist; NOS, nitric oxide synthase (nNOS, eNOS, iNOS); NREM, non-REM (sleep); OX, orexin = hypocretin; PCPA, para-chlorophenylalanine; PGD, prostaglandin, e.g. PGD_2; PGO waves, ponto-geniculo-occipital waves; PPT, pedunculopontine nucleus; PVN, paraventricular nucleus; R, used for receptor (e.g., OX_2R, $5HT_{1A}R$); REM, rapid eye movements (sleep); SCN, suprachiasmatic nucleus; SN, substantia nigra; SWS, slow wave sleep (part of NREM sleep); TM, tuberomamillary; TNFα, tumor necrosis factor alpha; VLPO, ventrolateral preoptic nucleus; VTA, ventral tegmental area

1 Introduction

1.1 What is Sleep?

Sleep is a physiological state during which the subject's ability to react to external stimuli is considerably diminished. Sleep in vertebrates is defined as a specific pattern of electric activity of the brain cortex as measured by electroencephalogram (EEG). During waking, the cortical activity is desynchronized resulting in low-amplitude, high-frequency EEG. On falling asleep the cortical activity gets synchronized, and the amplitude of the EEG waves increases, while frequency decreases (➲ *Figure 24-1*). Sleep consists of two main phases: nonREM (NREM) and REM sleep. NREM is the quiet sleep phase during which EEG waves are high and low-frequent, and depending on species, it is further divided to 2–4 stages (e.g., in humans S1, S2, S3, and S4) describing the intensity of NREM sleep. REM sleep is characterized by high-frequency and low-amplitude EEG (as in waking), but muscle tone, which is high in waking, and decreased in NREM sleep, is further decreased. In the course of night, these states alternate regularly in a highly regulated and composed manner constituting the sleep cycles.

Sleep is homeostatically regulated, which means that a prolonged period of wakefulness is followed by a prolonged period of sleep (Borbély, 1982). The neural substrate(s) and the molecular correlates of sleep, and sleep regulation have been a subject for systematic research for less than 50 years. Considering this, we have managed to reach a considerable understanding of these mechanisms.

1.2 The Emergence of Neurochemistry of Sleep and Wakefulness

The neuroanatomical substrates of sleep–wake regulation were largely known even before contemporary neurochemistry revealed the transmitters involved. The influenza pandemic of 1917–1919 not only killed

Figure 24-1

Human sleep stages. The EEG wave amplitude increases from waking through the sleep stages S1–S4 being largest in stages S3 and S4 (=slow-wave sleep). The frequency of the wave decreases simultaneously. During REM sleep the EEG gets desynchronized as in waking, and is characterized by low-amplitude and high-frequency waves

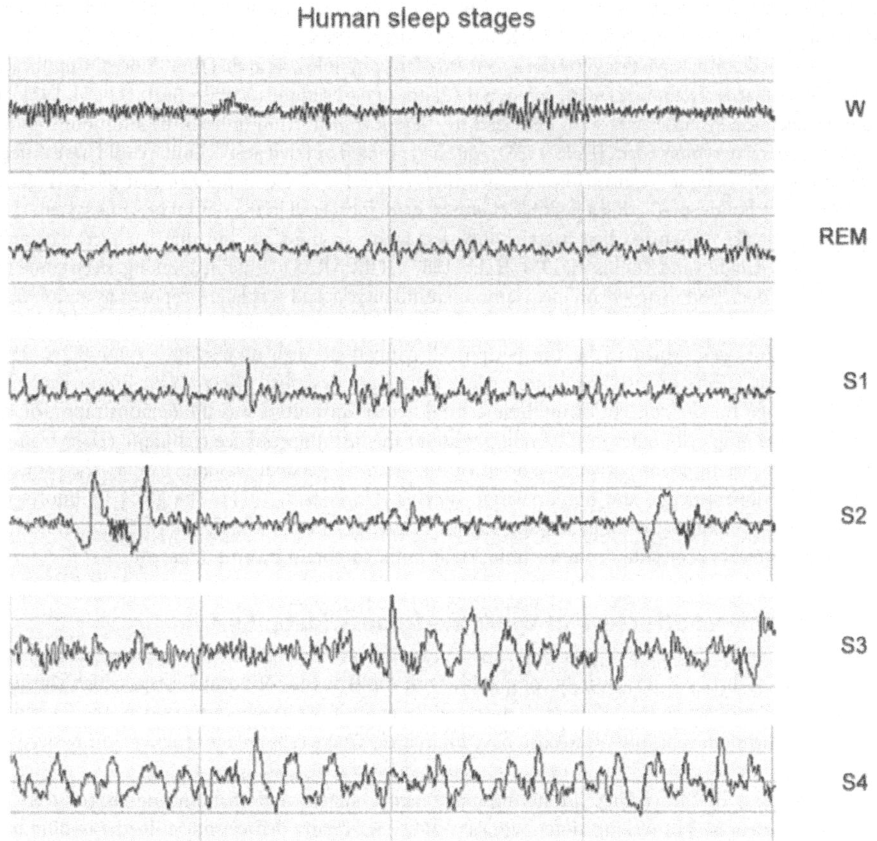

tens of millions of people, but also left some patients in a sleep-like coma, encephalitis lethargica. In a series of studies, von Economo described autopsy findings from these patients (von Economo, 1923, 1930). Loss of viable neurons was found in the posterior hypothalamus and the rostral mesencephalic area. These areas were thus concluded to be necessary for waking, and this view has held. Some patients, on the other hand, displayed insomnia, and their lesion was in the anterior hypothalamus and preoptic area, still considered major sleep-inducing areas.

A major leap was the discovery by Moruzzi and Magoun (1949) that electrical stimulation in cats of the mesencephalic region called "reticular formation" induced waking, and its lesion led to constant sleep. For these studies, a crucial technique was the EEG, first described in animals by Caton (1875) and in further detail in humans by Berger (1929). Although arousal was signaled by low-amplitude fast waves, sleep was characterized by high-amplitude slow waves. The area whose stimulation led to EEG arousal was termed the "ascending reticular activating system" (ARAS), and was assumed to be activated by collaterals of sensory input signals, but lesion studies soon showed that interruption of the afferent sensory input did not prevent the wakefulness, whereas destruction of the ARAS did (Lindsley et al., 1950). Sleep was initially seen simply as deactivation of the ARAS (passive sleep theory), either through fatigue or through the accumulation, during wakefulness, of some unknown chemical substances.

The accumulation in the brain of some sleep-inducing substance(s) was proposed independently by Ishimori (1909) and by Legendre and Piéron (1910). CSF (Legendre and Piéron, 1910) or brain homogenate (Ishimori, 1909) from sleep-deprived dogs was found to induce sleep in recipient animals. As ultrafiltration or heating destroyed the somnogenic action of CSF, the "hypnotoxin" was assumed to be a protein (Legendre and Piéron, 1911). This concept of thinking about sleep regulation and the pioneering findings preceded the modern analysis of sleep-inducing substances, pursued, e.g., in the laboratories of Pappenheimer (Pappenheimer et al., 1975), Krueger (Krueger, 1985), and Inoue (Honda et al., 1984).

Chemical mediation in synapses was not a concern of the physiologists at that time. Since the publication of the evidence that acetylcholine (ACh) acts as a mediator of nerve impulses in the heart (Loewi, 1921), the concept of chemical synapses was soon extended to efferent transmission in both the autonomic and the somatic peripheral nervous system (Dale, 1936). Dale had pointed out that acetylcholine could have different actions at different synapses (muscarinic and nicotinic) (Dale, 1914). He mentioned that the basal ganglia are rich in acetylcholine, predicting a possible chemical mediation even in the central nervous system (Dale, 1936). It was finally recognized that acetylcholine has a role in the CNS and that it affects excitability (Feldberg, 1950; Hughes and Robinson, 1951; Hebb, 1957). If the ARAS maintained waking, sleep could then result from the depletion of acetylcholine. Gamma-aminobutyric acid was later proposed as an inhibitory transmitter (Purpura et al., 1957; Krnjevic and Schwartz, 1966), but not linked to the induction of sleep.

Noradrenaline was shown to be the peripheral sympathetic neurotransmitter (von Euler, 1946), and subsequently effects of catecholamines on the CNS were described (Kety, 1959; Rothballer, 1959). The breakthrough for the concept of multiple central neurotransmitters was the demonstration of catecholamine-containing cell bodies in the brain, by use of the histofluorescence technique (Dahlström and Fuxe, 1964). By placing lesions at various levels of the brain, Ungerstedt was able to map the projection systems of the dopaminergic and noradrenergic systems (Ungerstedt, 1971). The glyoxylic fluorescence technique allowed mapping without making lesions (Lindvall and Björklund, 1974). The complexity of systems and pathways projecting from the brain stem to the forebrain became apparent.

1.3 The General Structure of the Wakefulness/Sleep System

The passive sleep theory is not entirely compatible with the structure of normal sleep, which consists of sleep cycles as sign of active regulation of sleep stages. Recently, it has also become evident that in addition to nuclei that maintain wakefulness, we also have brain areas where cells are more active during sleep than during waking.

The structures of the waking promoting/maintenance system and that putting us to sleep have interesting differences. The waking system consists of several clearly defined nuclei in the midbrain and the pons, with specific neurotransmitters and long projections that reach virtually all parts of the brain (➋ *Figure 24-3*). Neuronal activity in these nuclei is high during waking and decreases during NREM sleep.

The sleep-active neurons lie in the preoptic area and ventrolateral preoptic nucleus (VLPO) of the hypothalamus (Sherin et al., 1996). Analogously with the waking system, also these neurons send widespread projections to distant brain areas, and notably to the wakefulness-regulating nuclei. This structure implies a reciprocal inhibition model for regulation of sleep and wakefulness: during waking the waking-promoting nuclei inhibit the sleep-active neurons and vice versa. An interesting difference from the waking system is that there are several potent sleep-inducing molecules, which originate from diffuse brain structures, and not necessarily only from neurons. These substances have known roles in energy metabolism [e.g., adenosine (AD) and NO], immune function (e.g., cytokines and NO) and hormonal regulation (e.g., GHRH). Whether these molecules act through the sleep-active cells, or by inhibiting the wake-active cells, is presently under investigation.

2 Neurochemistry of Waking

2.1 The Noradrenergic System in Sleep–Wake Regulation

Noradrenaline-containing neurons have their cell bodies in the pons and medulla, in cell groups designed A1–A7, as they did not conform to previously identified anatomical nuclei (Dahlström and Fuxe, 1964).

The major cell group is A6 in the dorsomedial pons, called the locus coeruleus (LC). A4 is close to A6. Also in the pons are the dorsolateral A7 and the ventrolateral A5. In the medulla, A1 is found ventrolaterally and A2 dorsomedially. The noradrenergic cell groups form two projection systems: a dorsal fiber bundle from the locus coeruleus group including A6 and the adjacent A4, projecting to the thalamus, cerebellum, entire neocortex and hippocampus, and a more ventral projection from A1, A2, A5, and A7 to the hypothalamus and preoptic areas (Moore and Bloom, 1979).

Most studies on noradrenergic involvement in sleep have been on neurons in the LC and its immediate vicinity. The firing of LC neurons is state-dependent, decreasing during periods of drowsiness and transitions into sleep (Chu and Bloom, 1974). The bulk of LC neurons fire most intensely in waking, more slowly during slow wave sleep, and cease firing during REMS (Aston-Jones and Bloom, 1981a). Furthermore, changes in their firing rate anticipate the change in vigilance state, pointing to a state-regulatory role. Their activity appears reciprocal to that in neighboring cholinergic cell groups (Hobson et al., 1975).

By comparing the results of pharmacological manipulations and localized brain stem lesions, Jouvet concluded that slow-wave sleep (SWS) and paradoxical (REM) sleep are regulated by different mechanisms, and that tonic and phasic components of REM sleep are also separately regulated, by monoaminergic and cholinergic mechanisms (Jouvet, 1967). Further studies implicated noradrenaline, possibly together with acetylcholine in the maintenance of tonic cortical arousal, and the locus coeruleus area as crucial for the induction of paradoxical (REM) sleep (Jouvet, 1972). It appeared, however, that the effects of lesions are not permanent, but are compensated with time, indicating a more modulatory than imperative role of noradrenaline on sleep–wake states (Jones et al., 1977).

Within waking, LC neuronal firing increases in response to novel sensory stimuli (Foote et al., 1980; Aston-Jones and Bloom, 1981b). Further study has shown that the LC is not only tonically more active during wakefulness, but also phasically activated by stimuli and by behavioral tasks involving selective attention (Rajkowski et al., 1994).

Stress responses are accompanied by increases in noradrenaline turnover and increased LC neuronal activity (Valentino and Foote, 1988). CRH neurons in the paraventricular nucleus (PVN) of the hypothalamus have an important projection to the LC, and intraventricular CRH increases the tonic activity of LC neurons (Valentino and Foote, 1988), as also direct microdialysis of CRH into the LC (Asbach et al., 2001). On the other hand, stress increases the noradrenaline release increases in the PVN, which further activates CRH release (Pacak et al., 1995). Such reciprocal stimulation might serve to increase stress responsivity during chronic or repeated stress, although otherwise repeated stress leads to development of some LC tolerance (Berridge and Waterhouse, 2003).

Noradrenergic projections act on target neurons via a multitude of membrane receptors (Berridge and Waterhouse, 2003). α_1 receptors act on the phosphoinositol system and increase intracellular Ca^{++}. α_2 receptors activate a G_i protein, decreasing cAMP. β_1 receptors act on a G_s protein, increasing cAMP. β_2 receptors have been found presynaptically on noradrenergic neurons and facilitate noradrenaline release from these.

Noradrenaline has a complex modulating effect on target neurons (Berridge and Waterhouse, 2003). In general, the spontaneous firing rate decreases, while the ability to react to afferent stimuli is enhanced, which increases the signal–noise ratio of the response. However, this effect follows an inverted U-type function, as the signal–noise improvement is maximal at a moderate noradrenaline concentration, and decreases with higher or lower noradrenaline. This relationship has been found in the thalamocortical projection cells, in the hippocampus, cerebellum, auditory, and visual cortices, etc. Both excitatory and inhibitory synaptic functions may be enhanced by noradrenaline.

In the context of sleep function, it is interesting that noradrenergic projections appear essential for the memory- and plasticity-related gene expressions (Cirelli and Tononi, 2004). Thus, noradrenaline release during alert wakefulness has a crucial role for the potentiation of synaptic plasticity related to experience and learning.

2.2 The Serotonergic System in Sleep–Wake Regulation

Cell bodies of serotonergic neurons are found in several nuclei in the brain stem midline (raphe), which can be divided into a rostral and a caudal group, depending on whether their projections are mainly ascending to the forebrain or descending to the spinal cord. The rostrally projecting group (B4–B9) comprises the median raphe, pontine raphe, and dorsal raphe nucleus (DR, B7); whereas the caudally projecting group (B1–B3) includes nucleus raphe magnus and nucleus raphe pallidus (Dahlström and Fuxe, 1964). For sleep–wake regulation, DR is the most interesting.

The serotonergic influence on waking was initially thought to be inhibiting, as lesions or pharmacological depletion of serotonin caused insomnia (Jouvet, 1972), and stimulation of raphe nuclei induced behavioral inhibition during waking (Jacobs et al., 1973). This view was difficult to reconcile with studies on unit activity. Like LC neurons, serotonergic DR neurons fire during waking at a regular slow rate of 2–3 spikes/s, more slowly in NREM sleep, and they cease firing in REM sleep (McGinty and Harper, 1976; Trulson and Jacobs, 1979). The neurons are stimulated by sensory stimuli (Trulson and Jacobs, 1979; Heym et al., 1982). Some neurons are activated by specific motor patterns during wakefulness (Jacobs and Fornal, 1991). Serotonin release at the terminals is in parallel with the firing rate, as shown by in vivo microdialysis (Jacobs et al., 1990; Portas et al., 1996). It thus appears that serotonin is released mainly during wakefulness. In addition, local cooling of DR induces NREM sleep and REM sleep, while suppressing waking (Cespuglio et al., 1976).

The early notion that serotonin opposed noradrenaline and favored sleep was favored by experiments showing that pharmacological depletion of serotonin with PCPA caused chronic insomnia (Jouvet, 1972). During this, sustained pontogeniculooccipital (PGO) wave activity supported the absence of serotonin (Petitjean et al., 1985). Repletion of serotonin with 5-HTP immediately caused the PGO waves to disappear, indicating resynthesis of serotonin, but only after a delay of 30 min did NREM sleep, and after 1 h REM sleep reappear, pointing to a delayed action of serotonin. As the reappearance of REM sleep was inhibited by administration of chloramphenicol, it was deduced that at least in the case of REM sleep, the delayed action was mediated by a neurohormone (Jouvet, 1982; Petitjean et al., 1985). A similar indirect effect of serotonin on NREM sleep was also hypothetized.

Serotonin (5-hydroxytryptamine, 5-HT) receptors in the brain include the group of 5-HT_1 receptors (5-HT_{1A}, 5-HT_{1B}, 5-HT_{1D}), which are mainly inhibitory through inhibition of adenylate cyclase, and the group of 5-HT_2 receptors (5-HT_{2A}, 5-HT_{2C}), which are mainly excitatory through the PLC system. Inhibitory autoreceptors are of the 5-HT_1 type. 5-HT_2 receptor antagonists induce SWS (Idzikowski et al., 1987; Dugovic and Wauquier, 1987), some also enhancing EEG slow waves (Landolt et al., 1999) and some not (Dijk et al., 1989). However, 5-HT_{2C} receptor knockout mice have more wakefulness, but also a greater response to sleep deprivation than wild-type mice, which indicates several roles for serotonin in sleep regulation (Frank et al., 2002) or might indicate adaptation in the knockout mice (Adrien et al., 2004).

5-HT_{1A} receptor agonists will inhibit DR neurons (Fornal et al., 1994) and have been shown to favor sleep in man (Seifritz et al., 1996), although large doses will increase waking and inhibit sleep (Monti and Monti, 2000). 5-HT_{1A} receptor antagonists, when infused into DR, will block inhibitory autoreceptors and increase DR activity (Bjorvatn et al., 1998). As suppression of DR activity is a prerequisite for REM sleep, local infusion of 5-HT_{1A} antagonist also suppresses REM sleep (Sorensen et al., 2000). Cataplexy, which resembles the atonia during REM sleep, is also suppressed by 5-HT_{1A} agonists in dogs (Nishino et al., 1995). Mice with a constitutional knockout of either 5-HT_{1A} or 5-HT_{1B} receptors have higher amounts of REM sleep than wild-type mice (Boutrel et al., 1999, 2002). In wild-type mice, either 5-HT_{1A} or 5-HT_{1B} agonist decreased and 5-HT_{1A} or 5-HT_{1B} antagonist increased REM sleep. Thus both receptor types contribute to REM sleep control. The REM sleep-depressing effect of the serotonin reuptake inhibitor citalopram is seen in wild-type and 5-HT_{1B} knockout mice, but not in the 5-HT_{1A} knockouts (Monaca et al., 2003). Serotonin transporter knockouts have increased REM sleep, maybe indicating decreased sensitivity of 5-HT_{1A} and 5-HT_{1B} receptors (Adrien et al., 2004). Monoaminoxidase A knockouts, again, have decreased REM sleep, comparable to the effect of MAO-A inhibitor drugs (Adrien et al., 2004). In constitutional knockouts, adaptive processes often make the interpretation of results complicated.

The serotonergic system is activated in conditions of stress. Considerable release of serotonin takes place in the hypothalamus during immobilization stress of the rat, which is accompanied by REM sleep rebound following an initial period of waking (Cespuglio et al., 1995). Combined with the fact that the immediate effects of serotonin activity is inhibition of sleep, and especially REM sleep, this has led to the concept of serotonin-stimulated events eventually leading to the release of sleep-promoting peptide factors and rebound sleep (Jouvet, 1999). Such sleep-promoting peptides may include pro-opiomelanocortin-derived peptides (Cespuglio et al., 1995).

2.3 The Dopaminergic System in Sleep–Wake Regulation

Cell bodies of dopaminergic brain stem neurons are located in three areas: the substantia nigra (SN, A10), the adjacent ventral tegmental area (VTA, A9), and several small groups in the hypothalamus (A12, A13, A14). The ascending projections are part of the median forebrain bundle; and whereas SN projects mainly to the striatum, VTA projects to the cerebral cortex and basal forebrain (BF) (mesocortical and mesolimbic system). The hypothalamic nuclei have short connections within the hypothalamus and mainly control neurohormone output there and in the hypophysis (a well-known function is the inhibition of prolactin secretion).

The dopaminergic output from SN to striatum mainly facilitates motor activity. Dopaminergic output from VTA to nucleus accumbens is related to reward behavior. However, the neuronal unit activity of dopaminergic neurons in SN and VTA does not appear correlated to sleep–wake state as such (Trulson et al., 1981; Steinfels et al., 1983; Trulson and Preussler, 1984), whereas nondopaminergic neurons fire more during REM sleep and active movement (Miller et al., 1983). Dopaminergic neurons thus differ from other monoaminergic systems. It is also remarkable that a variety of stimuli are ineffective in affecting the firing rate of SN neurons (Strecker and Jacobs, 1985), whereas VTA neurons are significantly excited by stress (Trulson and Preussler, 1984). It would seem that the mesencephalic dopamine system is not primarily a sleep–wake-regulating system. Despite this, pharmacological manipulations cause important changes in vigilance (Monti, 1982; Wauquier, 1983), possibly because of the positive effects on affect and motor activity. Thus, nonspecific dopamine agonists like apomorphine at low doses favor sleep, while large doses prevent sleep. Corresponding opposite biphasic dose effects are seen with dopamine receptor antagonists (Monti, 1982).

2.4 The Cholinergic System in Sleep–Wake Regulation

Two groups of cholinergic neurons are of interest: a posterior one in the ponto-mesencephalic area (Ch5–6), and a rostral one in the basal forebrain (Ch1–4) (Mesulam et al., 1983). These cells are characterized immunohistochemically by their content of cholinacetyltransferase (ChAT) (Vincent and Reiner, 1987) (➲ *Figures 24-2* and ➲ *24-3*).

2.4.1 Pontomesencephalic Nuclei

The laterodorsal tegmental nucleus (LDT, Ch6) is situated immediately medially from the LC, and lateral to the DR. The pedunculopontinenucleus (PPT, Ch5) is situated laterally to the LC, in the peribrachial area (and sometimes called area X). For many purposes, the LDT and PPT can be considered together, although LDT is more involved in muscle atonia during REM sleep, and PPT is more involved in the rapid eye movements.

The dendrites of LDT–PPT neurons extend into neighboring areas (Jones, 1990), and there is considerable reciprocal activity between the cholinergic cells and the noradrenergic and serotonergic ones. It is also apparent that sensory fibers ascending through the brain stem may influence LDT–PPT neurons. The major efferent projections are to the thalamus and into the hypothalamus and basal forebrain (Steriade et al., 1988; Jones and Cuello, 1989; Losier and Semba, 1993; Ford et al., 1995). The posterior cholinergic system thus innervates the rostral one.

Neurons in the PPT area were found to burst in advance to the PGO waves, which are related to REMs, and these neurons were assumed to be PGO wave generators (McCarley et al., 1978). In addition, many nonbursting (tonically firing) neurons in the LDT–PPT area were found to be most active either during

Figure 24-2
The anatomical localization of the basal forebrain (BF) and the ventrolateral preoptic nucleus (VLPO). In a horizontal section of a rat brain, the VLPO and BF are localized in the ventral part of the brain, close to each other. AC = anterior commissure

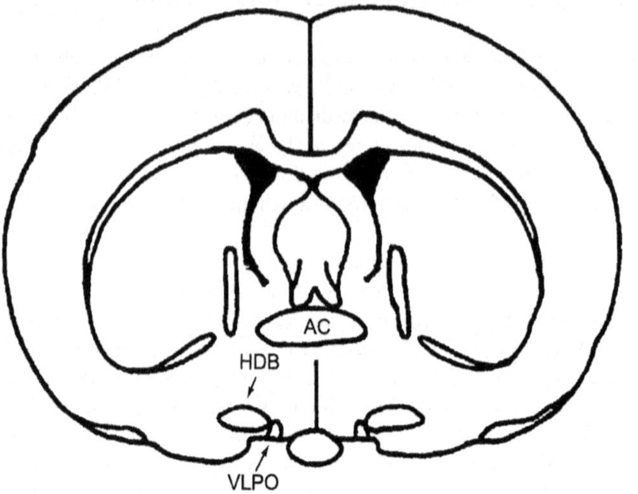

Figure 24-3
The localization of the waking active and sleep-active nuclei in the rat brain. Waking-active: LC = locus coeruleus, DR = dorsal Raphe nucleus, TM = tuberomamillary nucleus, LDT = laterodorsal nucleus; sleep-active: POA = preoptic area; HDB = horizontal limb of the diagonal band of Broca; TH = thalamus; OC = optic chiasm

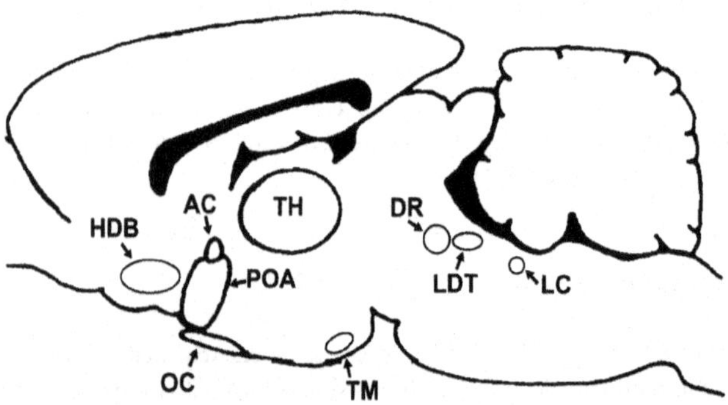

REM sleep or both during waking and REM sleep; their activity increased before the EEG desynchronization (El Mansari et al., 1989; Steriade et al., 1990). This indicated a role for cholinergic brain stem neurons in thalamocortical activation which is different from the noradrenergic, serotonergic, and histaminergic neurons. All of these are active in waking, but silent in REM sleep. The effects on thalamic relay neurons consist in depolarization through both nicotinic and muscarinic receptors (McCormick, 1989), in addition to disinhibition through a hyperpolarization of reticular thalamic neurons (McCormick and Prince, 1986). A curious phenomenon is the production, by systemic administration of the muscarinic receptor blocker atropine, of slow waves and spindles in the EEG, while the animal is still awake (Bradley, 1968).

LDT–PPT neurons may be excited by local injections of glutamate, histamine, adrenaline and noradrenaline (Koyama and Sakai, 2000). Microinjections of glutamate at 0.3 and 1.0 μg into the LDT of cats increased REM sleep without decreasing wakefulness, whereas 3.0 μg increased wakefulness at the expense of both NREM and REM sleep (Datta and Siwek, 1997). A similar dose-relationship was found for glutamate injections into the PPT of the rat (Datta et al., 2001b), and it was found that the cortical activation was mediated though NMDA receptors (Datta et al., 2001a), but REM sleep through kainate receptors, apparently excited by projections from glutamatergic neurons in the pontine reticular formation (Datta et al., 2002). In addition, microinjection of hypocretin/orexin (Hcrt/OX) into the LDT increased the time spent waking and decreased REM sleep in the cat (Xi et al., 2001).

Cholinergic neurons from LDT–PPT project to the medial pontine reticular formation (Shiromani et al., 1988). It has been shown earlier that injections of the cholinergic muscarinic receptor agonist carbachol into the pontine reticular formation induces long-lasting muscle atonia and other REM sleep-like signs in cats, whereas the effect is more variable and involves more brain stem areas in rats (Amatruda et al., 1975; Baghdoyan et al., 1984; Taguchi et al., 1992; Boissard et al., 2002). Carbachol activates G-proteins in the corresponding brain stem areas through several muscarinic receptor types (Capece et al., 1998), and M2 subtype receptors located both postsynaptically and as presynaptical autoreceptors seem to be the most prominent regulators of REM sleep-like signs (Baghdoyan et al., 1998). Similarly, microinjection of the acetylcholine esterase inhibitor neostigmine into the pontine reticular formation enhances REM sleep (Coleman et al., 2004), whereas microinjection of an inhibitor of ACh transport into terminal vesicles decreases REM sleep and prevents the REM sleep-enhancing effect of neostigmine (Capece et al., 1997). During electrical stimulation of the PPT, acetylcholine release was increased in the pontine region where carbachol injection induces REM sleep-like signs (Lydic and Baghdoyan, 1993). Evidently, the cholinergic input from LDT–PPT can induce muscle atonia through pontine reticular glutamatergic neurons, which in turn excite glycinergic, inhibitory neurons in the medullary magnocellular nucleus (Lai and Siegel, 1988). It was shown long ago that this medullary portion of the reticular formation inhibits muscle tone (Magoun, 1950).

PGO waves, which originate in the LDT/PPT area and normally precede REM sleep episodes, are increased by pharmacological agents increasing REM sleep, including cholinomimetics or drugs antagonizing noradrenergic or serotonergic activity (Callaway et al., 1987). Microinjection of carbachol into the peribrachial region of the pontine tegmentum induced sustained PGO activity, followed by a long-lasting REM sleep (Datta et al., 1992; Datta, 1995). The PGO generator area extends more caudally in the pons than the general REM sleep-inducing area (Datta et al., 1998). Several other transmitter sytems modulate the cholinergic PGO wave generator (Datta, 1997).

2.4.2 Basal Forebrain

Basal forebrain is a heterogenous region in the ventral region of forebrain adjacent to preoptic, supraoptic, and rostral infundibular levels of the hypothalamus including the cholinergic cell groups Ch1–4 as defined by Mesulam (Mesulam et al., 1983). The magnocellular cholinergic neurons project widely to the limbic system and the cortex providing most of the cholinergic cortical innervation (Jones, 2004). The cholinergic cells are located interdispersed in several nuclei, from rostral to caudal: the medial septum, the vertical and horizontal diagonal bands of Broca, the magnocellular preoptic are, the substantia innominata and the nucleus basalis of Meynert (Cullinan and Zaborszky, 1991). Even though the BF was originally named according to the cortically projecting magnocellular cholinergic neurons, these neurons are a minority and actually outnumbered by 2:1 the rat BF by GABAergic neurons, which have cortical projections as well. In addition, there is a substantial proportion of unidentified cortically projecting neurons (Gritti et al., 1993; Gritti et al., 1997).

The exact role of the basal forebrain cholinergic neurons in the regulation of sleep is less clear than that of the LDT/PPT neurons, but several experiments suggest that such a role exists. The cholinergic cortically projecting neurons release ACh which excites the cortical neurons and thus stimulates cortical activation (McCormick and Prince, 1986; McCormick, 1992). This ACh release has been shown to be highest during waking and REM sleep (Marrosu et al., 1995). Furthermore, lesions of the BF have been shown to disrupt sleep patterns of both cats and rats (Szymusiak and McGinty, 1986; Buzsaki et al., 1988), and the discharge rates of the presumed cholinergic neurons have been shown to be highest during waking and REM sleep while being lowest during SWS in both cats and rats (Szymusiak and McGinty, 1986; McGinty and Szymusiak, 1989; Alam et al., 1999). The cholinergic cells of the BF are thus waking-active, and are regarded to be important in cortical arousal. The BF is also part of the extrathalamic ventral pathway from the brain stem reticular activating system to the cerebral cortex (Steriade, 1996; Jones, 2004).

The basal forebrain/preoptic area of the hypothalamus is an important target area for endogenous sleep-inducing substances, as discussed later (adenosine, NO, and protaglandins); but presently it is not clear which of these effects, if any, are mediated through cholinergic neurons, as opposed to GABAergic and other neurons in the BF.

2.5 The Histaminergic System in Sleep–Wake Regulation

Histaminergic cell bodies are found exclusively in the tuberomamillary (TM) nuclei in the ventrolateral posterior hypothalamus. von Economo's finding that lesions in the posterior hypothalamus caused somnolence (von Economo, 1923, 1930) is in accordance with present knowledge of the arousing action of histamine, although other types of sleep–wake-related neurons are also found in this area (Steininger et al., 1999).

The histamine neurons receive inputs from many forebrain and brain stem systems, including the ventrolateral posterolateral preoptic area, where GABAergic/galaninergic sleep-active neurons are found (Sherin et al., 1998). Afferent inputs to the tuberomamillary area include cholinergic inputs from the basal forebrain area, dopaminergic hypothalamic input, and various peptidergic hypothalamic inputs (Yoshimoto et al., 1989). The histaminergic neurons project diffusely to all brain regions (Panula et al., 1989). Like other monoaminergic neurons they fire most intensely during wakefulness, less during SWS, and are minimally active during REM sleep (Steininger et al., 1999; Vanni-Mercier et al., 2003).

Postsynaptic receptors mediating arousal are type H_1, and the activity of TM neurons is negatively influenced by H_3 autoreceptors. Thus administration of centrally active H_1 antagonist drugs inhibit waking (Lin et al., 1988; Monti, 1993), whereas H_3 receptor antagonists increase it (Lin et al., 1990; Monti, 1993). These effects have been confirmed in single-unit recordings of TM neurons (Vanni-Mercier et al., 2003).

Efferents of histaminergic TM neurons project to major waking-promoting centers, including noradrenergic neurons of the locus coeruleus and cholinergic neurons in the basal forebrain, as well as to the sleep-promoting preoptic/anterior hypothalamic area (POAH) (Panula et al., 1989). Depletion of histamine in POAH caused a decrease, and histamine injections an increase in waking of cats, and the effect of histamine was blocked by both H_1 and H_2 antagonists (Lin et al., 1994). Using in vivo microdialysis, Strecker et al. (2002) showed that the extracellular levels of histamine in the POAH of cats were highest during waking, intermediate during NREM sleep, and lowest during REM sleep, in accordance with the single neuron activity studies. However, the latter study did not show any increase in histamine levels during sleep deprivation, indicating that histamine does not convey information about sleep pressure to the POAH sleep-promoting area.

In vitro, histamine depolarizes thalamic relay neurons through H_1 receptors, which decrease the potassium current (McCormick and Williamson, 1991). Electrical stimulation of the tuberomamillary nucleus strengthened the processing of visual stimuli in the thalamic lateral geniculate nucleus, indicating promotion of sensory input as one wakefulness-related effect of histamine (Uhlrich et al., 2002). H_2 receptor-mediated inhibition through increased chloride conductance may be related to dampening of sleep-type thalamic activity and facilitate the transition to wakefulness (Lee et al., 2004).

Mice constitutively lacking the histamine-synthesizing enzyme histidine decarboxylase had a slight increase in REM sleep, less theta rhythm in the EEG during waking, and less waking at lights-off (Parmentier et al., 2002). Most interestingly, these mice had less arousal when moved to a new environment, and fell asleep in a few minutes, unlike controls. This indicates that histamine has a role in maintaining the brain in a wake state following behavioral challenge, a role that is not compensated by other systems even when constitutively absent.

2.6 The Hypocretin/Orexinergic System in Sleep–Wake Regulation

The other major waking-promoting system in the posterior hypothalamus consists of perifornical neurons scattered in the lateral posterior hypothamalic regions, containing a peptide called hypocretin (de Lecea et al., 1998) or orexin (Sakurai et al., 1998). The same gene codes for two forms of the peptide: hypocretin 1 and 2 or orexin A and B, the latter of which is degraded very fast and has thus been studied less. Hcrt/OX neurons project diffusely in the brain (Peyron et al., 1998), and innervate other major waking-promoting systems, e.g., the noradrenergic (Horvath et al., 1999) and the basal forebrain cholinergic system (Eggermann et al., 2001). Hcrt/OX also excites neurons of the thalamocortical projection (Bayer et al., 2002) and tuberomamillary histaminergic neurons (Bayer et al., 2001). Single neurons recorded in vivo in the Hcrt/OX neuronal area were wake-active or wake-REM sleep-active (Alam et al., 2002), but they could not be chemically identified.

Two receptor types (OX_1R and OX_2R) mediate the synaptic actions of Hcrt/OX, and both receptors are sensitive to both peptides, although OX_2R is tenfold as sensitive to Hcrt-2/OX-B than to Hcrt-1/OX-A (Sakurai et al., 1998). Both receptors mediate the same type of synaptic action. It was originally found that Hcrt/OX was excitatory on postsynaptic neurons (de Lecea et al., 1998). It raises cytoplasmic calcium via a G-protein mechanism, and can act both pre- and postsynaptically (van den Pol et al., 1998). Interestingly, local glutamate neurons are excited by Hcrt/OX, and in turn serve to excite Hcrt/OX neuons, which provides a means for synchronizing the activity of these latter (Li et al., 2002).

Hcrt/OX was initially found to increase food consumption (van den Pol et al., 1998), but the multiple projections suggested other regulatory actions also (Peyron et al., 1998). It was then found that knockout of the Hcrt/OX gene caused narcolepsy in mice (Chemelli et al., 1999), and that narcolepsy in dogs is caused by a mutation in the OX_2R gene (Lin et al., 1999). Human narcoleptics have an 85–95% reduction in the number of Hcrt/OX neurons (Thannickal et al., 2000). Hypocretin 2-saporin lesions of the posterior hypothalamus increased sleep to an extent correlating with the loss of Hcrt/OX neurons (Gerashchenko et al., 2001). It has become obvious that Hcrt/OX is necessary for the maintenance of consolidated periods of waking, and that its lack fragments waking and increases the propensity for sleep induction (Kilduff and Peyron, 2000). It has been shown that GABAergic neurons in the ventrolateral basal forebrain area control Hcrt/OX neurons, which might be part of the sleep-induction mechanism (Alam et al., 2005).

2.7 Glutamatergic Involvement in Sleep–Wake Regulation

It has been estimated that 75% of the excitatory synapses in the CNS are glutamatergic. It could thus be of no surprise if glutamatergic transmission were crucial in the regulation of wake and sleep. However, its ubiquity would make it difficult to assign specific roles to this transmitter. In addition, glutamatergic neurons may be either local interneurons or projecting neurons.

Short-latency projections to the thalamus from midbrain reticular formation locations, where no noradrenergic or cholinergic neurons are found, indicate possibly amino acids as the transmitter (Steriade and Glenn, 1982; Steriade et al., 1990). Glutamate as a transmitter in brain stem reticular neurons is indicated by immunohistochemical studies (Jones, 1995). Glutamate affects thalamic neurons by exciting thalamocortical relay cells, but inhibiting thalamic reticular cells (McCormick, 1992; Cox and Sherman, 1999). Metabolic glutamate receptors with long-lasting effects are involved in some cases (Cox and Sherman, 1999; Schwarz et al., 2000).

The basal forebrain is also the target of glutamatergic projections from the lower brain stem, including the reticular formation, locus coeruleus, raphe nuclei, and LDT/PPT, as shown by immunohistochemistry (Carnes et al., 1990) Furthermore, glutamatergic projections from the pontine reticular formation mediate muscle atonia during REM sleep (Lai and Siegel, 1988). It is thus apparent that glutamatergic transmission may influence many aspects of wake–sleep regulation.

Some anesthetics apparently act by suppressing glutamatergic transmission, others by enhancing GABAergic transmission, which may have the secondary effect of decreasing glutamate transmission. The anesthetic action of ketamine has been explained by a specific blocking effect on NMDA receptors

(Yamamura et al., 1990). Several other anesthetic agents, including halothane, isoflurane, and thiopental apparently also antagonize ionophoric (AMPA or NMDA) glutamate receptors (Carla and Moroni, 1992). Another cellular target for anesthetics is glutamate uptake into astrocytes, which normally terminate glutamate action by removing it from the synaptic cleft. In vitro, ketamine had no effect on glutamate uptake into astocytes, whereas this was a mechanism of action of halothane, isoflurane, and some other volatile anesthetics (Miyazaki et al., 1997). In conclusion, therefore, decreasing glutamate transmission may impair waking, and this may be a prominent target for the development of anesthetics.

3 Neurochemistry of Sleep

3.1 Adenosine

Adenosine is a ubiquitous molecule, directly associated with energy metabolism, but which also functions in cellular communication and regulation of neural activity. Adenosine mediates its effects via specific receptors, namely A_1, A_{2A}, A_{2B}, and A_3 (Fredholm et al., 2001). The receptors differ in their affinity to adenosine, their distribution and function. The A_1 and A_{2A} are the high-affinity receptors, as they are activated at nanomolar adenosine concentrations, while the A_{2B} and A_3 are activated only at micromolar concentrations. Functionally, the A_1 and A_3 are negatively coupled to adenylate cyclase by $G_{i/o}$ and $G_{q/11}$ proteins and their activation leads to a decrease in the intracellular cAMP levels, while the A_{2A} and A_{2B} are positively coupled to adenylate cyclase by G_s, G_{olf}, $G_{q/11}$, and $G_{15/16}$, and consequently their activation has an opposite effect on the cAMP levels (Fredholm et al., 2000).

The A_1 receptor is the most abundant and widely expressed of the adenosine receptors in the CNS, being most prominent in the hippocampus, cerebellum, and cortex as shown by both mRNA expression studies and immunohistochemical staining (Rivkees et al., 1995). The net effect of A_1 activation is a reduction in transmitter release presynaptically and postsynaptic hyperpolarization (Haas and Selbach, 2000).

The A_{2A} receptor has a more localized expression in the CNS, being concentrated in the striatum, the nucleus accumbens, the olfactory tubercle, and the lateral segment of globus pallidus (Parkinson and Fredholm, 1990) with a low expression throughout the rest of the CNS. The A_{2A} receptor activates the adenylate cyclase pathway and seems to increase neurotransmission, but its role is not only stimulatory (Cunha, 2001). The A_{2B} and A_3 have a low expression in the CNS (Dixon et al., 1996). They are activated only at extreme concentrations of adenosine in pathological conditions, and their physiological role is not clear (Haas and Selbach, 2000).

The first report on the hypnogenic properties of adenosine was that injections of adenosine into the lateral ventricle induced sleep-like behavior in the cat (Feldberg and Sherwood, 1954). Two decades later the same phenomenon was noted in dogs (Haulica et al., 1973). Since then adenosine has been in the focus of sleep research, and its hypnotic properties have been noted repeatedly as both adenosine and its agonist induce sleep in rodents (Virus et al., 1983; Radulovacki et al., 1984; Ticho and Radulovacki, 1991a) and cats (Portas et al., 1997). Furthermore, the most widely used psychoactive stimulant in the world, caffeine, is an adenosine receptor antagonist (Fredholm et al., 1999). It has been shown that unilateral infusion of A_1 receptor antagonist cyclopentyl-1,3-dimethylxanthine (CPT) into the BF of cats and rats increases waking and decreases sleep (Strecker et al., 2000). Most experiments suggest that sleep is induced through A_1 receptors (Ticho and Radulovacki, 1991b; Benington et al., 1995; Schwierin et al., 1996; Portas et al., 1997), while there is also evidence for participation of the A_{2A} receptors (Matsumura et al., 1994; Urade and Hayaishi, 1999; Maquet et al., 1992).

A_{2A} agonists into the subarachnoid space under the rostral basal forebrain induces sleep (Satoh et al., 1996, 1999).

Extracellular adenosine concentrations increase during both increased neuronal activity and increased metabolism (Pull and McIlwain, 1972; Van Wylen et al., 1986; Mitchell et al., 1993). On the other hand, metabolic activity in the brain is higher during wakefulness than sleep, as measured by glucose utilization, cerebral blood flow, and oxygen consumption in humans (Maquet et al.,1990; Madsen et al., 1991; Maquet et al., 1992; Madsen, 1993), and extracellular glucose and lactate concentrations in rats (Netchiporouk et al., 2001).

Thus, it could be expected that adenosine concentrations vary during the sleep–wake cycle. It was first shown that the extracellular adenosine levels were lower during the rest phase and higher during the active phase in the hippocampus and neostriatum of rats (Huston et al., 1996). The first direct evidence for state-dependency was published in 1997, when it was shown using in vivo microdialysis that the extracellular adenosine levels were higher during SWS than during waking in the cholinergic magnocellular BF and the thalamus of cats (Porkka-Heiskanen et al., 1997). Later this phenomenon was shown to be present also in the cortex, the dorsal raphe nucleus, the pedunculopontine tegmental area (PPT) and the preoptic hypothalamic area (Porkka-Heiskanen et al., 2000). Recently, the state-dependency and the circadian variation have been further confirmed in the BF of both young and old rats (Murillo-Rodriguez and Blanco-Centurion, 2004).

As prolonged wakefulness continues to make demands on the neuronal metabolism, it can be expected that adenosine levels would increase. Indeed, increased adenosine levels have been measured in the BF of cats and rats during prolonged wakefulness (Porkka-Heiskanen et al., 1997; Basheer et al., 1999). This effect is site specific since adenosine levels increase only in the BF and transiently in the cortex (Porkka-Heiskanen et al., 2000). Furthermore, it has been shown with combined in vivo microdialysis and unit recording that infusion of adenosine into the BF of rats and cats decreases the unit activity of the cortically projecting wake-related neurons (Alam et al., 1999; Thakkar et al., 2003a). These data strongly suggest that adenosine is an endogenous sleep-regulating substance, acting specifically in the basal forebrain by decreasing the activity of the wake active, presumably cholinergic projection neurons, and thus decreasing cortical activity and promoting sleep.

One scenario to explain the role of adenosine in sleep regulation is through its connection to energy metabolism (Benington and Heller, 1995). The continuous activity of the wakefulness-promoting cells in the basal forebrain imposes a burden on the mitochondrial energy production capacity during wakefulness. When wakefulness is prolonged, these cells start to suffer from an energy deficit, resulting in increased glycolysis and elevations in extracellular concentrations of lactate, pyruvate, and adenosine. The enhanced extracellular concentration of adenosine contributes to the increased sleep propensity and promotes the transition from wakefulness to sleep. This view is supported by the finding that experimentally induced energy depletion in the BF induces sleep that resembles recovery sleep (Kalinchuk et al., 2003). We have proposed that the cholinergic cells of the basal forebrain are particularly vulnerable to the effects of energy depletion, and act as fuses, triggering adenosine release and forcing the brain to sleep (Kalinchuk et al., 2003, 2006a).

The effects of adenosine in the BF appear to be mediated predominantly through A_1 receptors (Basheer and Shiromani, 2001; Basheer et al., 2001a,b; Thakkar et al., 2003b). A_1 agonists decrease and antagonist increased the neuronal discharge rate in the cholinergic mesopontine LDT/PPT by causing hyperpolarization of identified cholinergic cells (Rainnie et al., 1994). Combined microdialysis and unit recording studies have shown that infusion of A_1 agonist into the BF decrease, and infusion of A_1 antagonist increase single unit activity of wake-active neurons in both cats and rats (Alam et al., 1999; Thakkar et al., 2003a). Recently it has been shown that bilateral infusion of antisense to the A_1 receptor mRNA decreases NREM sleep and induces wakefulness (Thakkar et al., 2003a). When considering the evidence presented earlier, the A_1 receptors have an important role in the homeostatic regulation of sleep in the BF.

However, in the immediate vicinity of the BF, effects of adenosine on sleep have been reported to take place through A_{2A} receptors. For details, please, see the ➲ Section 3.5 later in this chapter.

3.2 Cytokines

Cytokines are chemical messengers (low molecular weight proteins), which communicate information between the cells (intercellular messengers). Originally, they were described in connection with immune functions, and the communication within this system is best known. They are secreted through autocrine, paracrine, and endocrine routes, and are involved in the regulation of host defense, cell growth, differentiation, cell death, angiogenesis, and development and repair processes. Injurious stimuli usually initiate the production of cytokines, though they can also be produced constitutively (Vitkovic et al., 2000), and possibly act as neuromodulators.

In addition to their originally defined function as regulatory molecules of the immune system, cytokines regulate a number of other physiological functions, including hormonal secretion and stress. Typically, cytokines rarely act alone, but in concert with other cytokines, forming a network of activity. Of the multitude of cytokines, particularly IL-1 and TNFα have been well assessed in sleep regulation.

Injections of IL-1 beta will increase NREM sleep, while inhibition of IL-1 beta inhibits it [for review see Krueger et al. (2001)]. TNFα induces NREM sleep in several species, including rabbits, mice, sheep, and rats (Shoham et al., 1987; Fang et al., 1997; Dickstein et al., 1999; Kubota et al., 2002), while its inhibition decreases it (Takahashi et al., 1996). In rats, highest levels of both IL-1beta and its mRNA appear in the hypothalamus in the early phase of lights-on (Taishi et al., 1997), when also NREM sleep is maximal. Also, in humans the blood level of IL-1 is maximal at sleep onset (Moldofsky et al., 1986), and its mRNA hypothalamic levels undergo diurnal variation (Bredow et al., 1997). Sleep deprivation decreases IL-1beta mRNA in the hypothalamus (Mackiewicz et al., 1996). Sleep deprivation increases TNFα levels in the hypothalamus (Krueger et al., 2001).

The above cited data suggest a role for cytokines not only in the regulation of the pathologically increased sleep in infection [for review see Opp and Toth (2003)]; but also in the regulation of spontaneous sleep–wake cycle and sleep homeostasis. As the induction of cytokines is part of the early host defense reaction, frequently elicited by immunological challenge, it will be of importance to clarify whether the sleep response is an integral part of such defense. Induction of iNOS during prolonged wakefulness (Kalinchuk et al., 2006b), as well as the induction of heat shock proteins (Terao et al., 2003) and unfolded protein response (Naidoo et al., 2005), imply that this might be the case.

3.3 Nitric Oxide

Nitric oxide (NO) is a gaseous intercellular signaling molecule that regulates both physiological and pathophysiological processes in the CNS (Garthwaite and Boulton, 1995). NO is produced by three nitric oxide synthases (NOS). Neuronal (nNOS) and endothelial (eNOS) types are regulated by intracellular Ca^{2+} levels and they are constitutively expressed, producing small amounts of NO. The third type, iNOS, is regulated at the transcriptional and posttranscriptional levels, and is normally present in the brain only in trace amounts (Nathan and Xie, 1994a). Immunological challenge (Nathan and Xie, 1994b), as well as stressful stimuli including oxygen/glucose deprivation (Moro et al., 1998) and immobilization (Madrigal et al., 2001), activate expression of iNOS, which induces rapid (Harada et al., 1999) and relatively large increases in NO concentration. Excessive synthesis of NO by iNOS has been implicated as a contributing factor for a number of pathological conditions, including acute and chronic neurodegenerative diseases and diabetes (Gross and Wolin, 1995).

nNOS immunopositive neurons are distributed throughout the brain. Dense staining is found in the cortical interneurons, hypothalamic supraoptic and paraventricular nuclei and in the pontine LDT/PPT nuclei and cerebellum. eNOS is predominantly located in the endothelial cells of the blood vessels, and iNOS is induced mainly in the glial cells. The main pathway for NO effects is the cGMP pathway: NO binds to soluble guanylyl cyclase (sGC) activating cGMP production. NO also directly affects the mitochondrial energy metabolism (Brorson and Zhang, 1999).

In the cortex NO release, as measured using voltammetry, is highest during waking, and decreases during NREM and REM sleep (Burlet and Cespuglio, 1997), while in the nucleus raphe dorsalis the changes were opposite (Cespuglio et al., 2004). In the rat thalamus NO levels were highest during waking and REM sleep and lowest during NREM sleep (Williams et al., 1997). In mice with targeted disruptions of nNOS, REM sleep is decreased, while in the iNOS KO mice REM sleep was increased, but NREM in the late dark period was decreased (Chen et al., 2003). nNOS exhibits a circadian rhythm in the rat pineal gland (Spessert and Rapp, 2001) being maximally expressed during the dark phase, and NO production, measured as NO2-, showed a robust circadian rhythm in the SCN (Mitome et al., 2001) and whole brain (Tunctan et al., 2002) with higher NO release during the dark period.

REM sleep deprivation modifies the circadian variation in nNOS expression while there were no changes in iNOS activity (Clement et al., 2004). In recent experiments, total sleep deprivation for 3 h induced iNOS, and when this induction was prevented during sleep deprivation with a specific iNOS inhibitor, recovery sleep was abolished (Kalinchuk et al., 2006b). This suggests that NREM recovery sleep is produced by an increase in NO through iNOS induction.

Most studies using either systemic or icv administration of NOS inhibitors report decreases in sleep (Kapas et al., 1994b; Dzoljic et al., 1996), though increase has also been reported (Dzoljic and de Vries, 1994). NO donors administrated icv have been reported to increase sleep (Kapas and Krueger, 1996). The interpretation of the results obtained by systemic administration is complicated by the voltammetric measurement of NO in the brain after administration of L-NAME, the most frequently used nonspecific NOS inhibitor, showing that the inhibitor has no effect on the brain NO levels (Burlet et al., 1999).

Local microinjections of NOS inhibitors into the LDT/PPT reduce REM sleep (Kapas et al., 1994a; Datta et al., 1997; Leonard and Lydic, 1997; Hars, 1999), while NO donors increase it. Administration of NOS inhibitors into the dorsal raphe nucleus also decreased REM sleep (Burlet et al., 1999; Monti et al., 2001). Interestingly, pretreatment with either a GABA-A or adenosine A1 receptor agonist partially prevented the effect of L-NAME in the dorsal raphe nucleus (Monti et al., 2001).

Taken together these data strongly suggest a role for NO in regulation of sleep and particularly in the regulation of recovery sleep. Of particular interest is the induction of iNOS during prolonged wakefulness in the BF. As iNOS is not expressed in the brain under normal conditions, it suggests that prolonged wakefulness evokes the host defense reaction, and as part of it, sleep is induced.

3.4 GABA

GABA is the main inhibitory neurotransmitter in the brain. GABA neurons are either local interneurons, or project to more distant areas, e.g., GABAergic neurons in the BF. A role for GABA in sleep regulation is suggested by the notion that the most frequently used hypnotics, benzodiazepines, act through GABA binding site, though the exact mechanism of the action has not been clarified (Möhler, 2001)

GABA neurons are widely distributed in the brain, but presently two main sites of action have been proposed to explain the effects of the GABAergic neurons on sleep: the reticular thalamic nucleus, which is part of the reticulo-thalamic circuit, and which has been proposed to be central in the generation of cortical EEG delta (Steriade et al., 1991) and spindle activity (Steriade et al., 1987). A more recent finding, that in the preoptic area of hypothalamus there are sleep active neurons as defined by electrophysiological measurement (Alam et al., 1997) or by using c-fos activation as an indicator of cellular activity (Sherin et al., 1996). These neurons are GABAergic and in the VLPO nucleus they are colocalized with galanin, another inhibitory neuromodulator (Sherin et al., 1998). These neurons send projections to the waking-promoting monoaminergic (locus coeruleus and raphe nuclei) and histaminergic (tuberomamillary nuclei) cell groups (Sherin et al., 1998; Steininger et al., 2001), and they appear to be essential for the production of sleep, since lesion of the VLPO cells reduced the amount of NREM and REM sleep to about half (Lu et al., 2000). VLPO neurons are inhibited by monoamines (Gallopin and Saper, 2000), suggesting a reciprocal inhibition model: during wakefulness the monoamines inhibit the sleep-active VLPO neurons, and during sleep the VLPO neurons inhibit the wake-active monoaminergic neurons (McGinty and Szymusiak, 2000).

GABA concentrations, measured using in vivo microdialysis, were high in the posterior hypothalamus during sleep (Nitz and Siegel, 1997a), while in the dorsal raphe and locus coeruleus the highest GABA levels were measured during REM sleep (Nitz and Siegel, 1997b), suggesting that GABA has a specific role in the shutting off the monoaminergic nuclei during REM sleep.

One neurotransmitter regulating the activity of the VLPO neurons appears to be adenosine, through disinhibition of these neurons (Chamberlin et al., 2003), suggesting a dual role for adenosine in sleep induction in the BF/hypothalamic area: inhibition of the waking-active (presumably cholinergic) neurons through A_1 receptors and disinhibition of the sleep-active GABAergic cells through A_{2A} receptors.

3.5 Prostaglandins

Of the several members of the prostaglandin family, prostaglandin D_2 (PGD_2) has the most sleep-promoting properties. In the brain PGD synthase, a key enzyme in PGD_2 metabolism, is predominantly localized in the leptomeninges (Urade et al., 1993). PGD_2 concentrations are higher during the light period, and particularly during NREM sleep, and sleep deprivation increases it (Ram et al., 1997). PGD_2 injected in the preoptic area (POA) of rats (Ueno et al., 1982), and icv in monkeys (Onoe and Hayaishi, 1988) increased SWS. DPR prostaglandin receptor binding was found to be localized in this area (Yamashita et al., 1983), while immunoreactivity for DPR antibody was found to be most intense in the leptomeninges on the ventral surface of the basal forebrain (Mizoguchi and Hayaishi, 2001).

PGD_2 receptor activation-induced release of adenosine exerts its somnogenic effects via the A_{2A} adenosine receptor in the subarachnoid space below the rostral basal forebrain (Matsumura et al., 1994; Urade and Hayaishi, 1999; Mizoguchi and Hayaishi, 2001; Hayaishi, 2002). Infusion of PGD_2 into the subarachnoid space increased the local extracellular adenosine concentration. Data supporting the somnogenic effects of both PGD_2 and A_{2A} agonists are induction of c-fos immunoreactivity in the ventrolateral preoptic area, this region has been suggested to be involved in promoting sleep by inhibiting the ascending histaminergic arousal system of the tuberomammillary nucleus (Scammell et al., 1998; Sherin et al., 1998; Scammell et al., 2001). Infusion of the A_{2A} agonist CGS 21680 in the subarachnoid space increased SWS, and both A_{2A} agonist- and PGD_2-induced sleep were blocked by the A_{2A} antagonist, KF17837. These data provide the pharmacological evidence for the role of the A_{2A} receptor in mediating the somnogenic effects of PGD_2 (Satoh et al., 1996; Satoh et al., 1999) suggesting that PGD_2-induced release of adenosine in a specific area of subarachnoid space below the rostral forebrain, leads to A_{2A} receptor-mediated effects. The sleep-inducing effects of the PGD_2 and A_{2A} agonists have been localized in the VLPO (Scammell et al., 1998, 2001).

Several experiments thus suggest that PGD_2 has sleep-modulating effects. Interestingly, the effects are localized in the same BF/hypothalamic regions where also other sleep-active substances exert their effects. PGD_2 appears to act through adenosine, particularly through A_{2A} receptors. As also cytokines and NO can increase adenosine release, one of the key questions is whether adenosine is the common key molecule, through which the other sleep-active molecules affect sleep, or whether each of these molecules has a separate mechanism to modulate sleep.

4 Summary

The sleep-promoting substances consist of a variety of endogenous substances, which have regulatory roles in other physiological regulatory systems, as energy metabolism, immune responses and hormonal regulation. These molecules also have a wide interconnecting network, with partly common intracellular signaling pathways. The details of the orchestration of this network to produce the regulation of the spontaneous sleep–wake cycle and the response to prolonged wakefulness are currently one of the most important topics of research within the sleep research community.

5 Questions for Future Research

1. Does the induction of (NREM) sleep take place mainly by activating the sleep-active neurons in the VLPO, or does it take place by inhibiting the wake-active (cholinergic) neurons in the BF, or is it a combination of these?
2. Several substances have (NREM) sleep-inducing effects in the BF/hypothalamic area, and all of them are able to induce adenosine release. Is the effect of these substances on sleep based on their ability to release adenosine, or do they have adenosine-independent mechanism to affect the activity of the sleep/waking active neurons?

3. So far, sleep research has concentrated solely on neuronal activity. However, it is evident that substances like adenosine, nitric oxide and cytokines can be released also by glial cells. The role of glia in regulation of sleep needs to be addressed.
4. Sleep need (the homeostatic component of sleep regulation) appears to be use-dependent, particularly cortical activity-dependent, but the executive mechanism of sleep induction appears to be localized in the basal forebrain/hypothalamic area. How is the information of cortical neuronal activity conveyed to the subcortical areas, and which signaling molecules participate in this?
5. Research on the recently found wake-active cell group, the hypocretin/orexin cells, continues actively.

Acknowledgements

This work was supported by the Academy of Finland, the Finska Läkaresällskapet, EU grant MCRTN-CT-2004-512362, and NIH.

References

Adrien J, Alexandre C, Boutrel B, Popa D. 2004. Contribution of the "knock-out" technology to understanding the role of serotonin in sleep regulations. Arch Ital Biol 142: 369-377.

Alam MN, McGinty D, Szymusiak R. 1997. Thermosensitive neurons of the diagonal band in rats: relation to wakefulness and non-rapid eye movement sleep. Brain Res 752: 81-89.

Alam MN, Szymusiak RS, Gong H, King J, McGinty DJ. 1999. Adenosinergic modulation of rat basal forebrain neurons during sleep and waking: neuronal recording with microdialysis. J Physiol 521: 679-690.

Alam MN, Gong H, Alam T, Jaganath R, McGinty D, et al. 2002. Sleep–waking discharge patterns of neurons recorded in the rat perifornical lateral hypothalamic area. J Physiol 538: 619-631.

Alam MN, Kumar S, Bashir T, Suntsova N, Methippara MM, et al. 2005. GABA-mediated control of hypocretin- but not melanin-concentrating hormone-immunoreactive neurones during sleep in rats. J Physiol 563: 569-582.

Amatruda TT III, Black DA, McKenna TM, McCarley RW, Hobson JA. 1975. Sleep cycle control and cholinergic mechanisms: differential effects of carbachol injections at pontine brain stem sites. Brain Res 98: 501-515.

Asbach S, Schulz C, Lehnert H. 2001. Effects of corticotropin-releasing hormone on locus coeruleus neurons in vivo: a microdialysis study using a novel bilateral approach. Eur J Endocrinol 145: 359-363.

Aston-Jones G, Bloom F. 1981a. Activity of norepinephrine-containing locus coeruleus neurons in behaving rats anticipates fluctuations in the sleep–waking cycle. J Neurosci 1: 876-886.

Aston-Jones G, Bloom FE. 1981b. Nonrepinephrine-containing locus coeruleus neurons in behaving rats exhibit pronounced responses to non-noxious environmental stimuli. J Neurosci 1: 887-900.

Baghdoyan HA, Lydic R, Fleegal MA. 1998. M2 muscarinic autoreceptors modulate acetylcholine release in the medial pontine reticular formation. J Pharmacol Exp Ther 286: 1446-1452.

Baghdoyan HA, Rodrigo-Angulo ML, McCarley RW, Hobson JA. 1984. Site-specific enhancement and suppression of desynchronized sleep signs following cholinergic stimulation of three brainstem regions. Brain Res 306: 39-52.

Basheer R, Shiromani PJ. 2001. Effects of prolonged wakefulness on c-fos and AP1 activity in young and old rats. Mol Brain Res 89: 153-157.

Basheer R, Porkka-Heiskanen T, Stenberg D, McCarley RW. 1999. Adenosine and behavioral state control: adenosine increases c-fos protein and AP1 binding in basal forebrain of rats. Mol Brain Res 73: 1-10.

Basheer R, Halldner L, Alanko L, McCarley RW, Fredholm BB, et al. 2001a. Opposite changes in adenosine A1 and A2a receptor mRNA following sleep deprivation. Neuroreport 12: 1577-1580.

Basheer R, Rainnie DG, Porkka-Heiskanen T, Ramesh V, McCarley RW. 2001b. Adenosine, prolonged wakefulness and A1-activated NF-kB DNA binding in the basal forebrain of the rat. Neuroscience 104: 731-739.

Bayer L, Eggermann E, Serafin M, Saint-Mleux B, Machard D, et al. 2001. Orexins (hypocretins) directly excite tuberomammillary neurons. Eur J Neurosci 14: 1571-1575.

Bayer L, Eggermann E, Saint-Mleux B, Machard D, Jones BE, et al. 2002. Selective action of orexin (hypocretin) on nonspecific thalamocortical projection neurons. J Neurosci 22: 7835-7839.

Benington JH, Heller HC. 1995. Restoration of brain energy metabolism as the function of sleep. Prog Neurobiol 45: 347-360.

Benington JH, Kodali SK, Heller HC. 1995. Stimulation of A1 adenosine receptors mimics the electroencephalographic effects of sleep deprivation. Brain Res 692: 79-85.

Berger H. 1929. über das Elektroenkefalogramm. Arch Psychiat Nervenkrankh 87: 527-570.

Berridge C, Waterhouse B. 2003. The locus coeruleus-noradrenergic system: modulation of behavioral state and state-dependent cognitive processes. Brain Res Rev 42: 33-84.

Bjorvatn B, Fornal CA, Martin FJ, Metzler CW, Jacobs BL. 1998. The 5-HT1A receptor antagonist p-MPPI blocks 5-HT1A autoreceptors and increases dorsal raphe unit activity in awake cats. Eur J Pharmacol 356: 167-178.

Boissard R, Gervasoni D, Schmidt MH, Barbagli B, Fort P, et al. 2002. The rat ponto-medullary network responsible for paradoxical sleep onset and maintenance: a combined microinjection and functional neuroanatomical study. Eur J Neurosci 16: 1959-1973.

Borbély AA. 1982. A two process model of sleep regulation. Hum Neurobiol 1: 195-204.

Boutrel B, Franc B, Hen R, Hamon M, Adrien J. 1999. Key role of 5-HT1B receptors in the regulation of paradoxical sleep as evidenced in 5-HT1B knock-out mice. J Neurosci 19: 3204-3212.

Boutrel B, Monaca C, Hen R, Hamon M, Adrien J. 2002. Involvement of 5-HT1A receptors in homeostatic and stress-induced adaptive regulations of paradoxical sleep: studies in 5-HT1A knock-out mice. J Neurosci 22: 4686-4692.

Bradley PB. 1968. The effect of atropine and related drugs on the EEG and behaviour. Prog Brain Res 28: 3-13.

Bredow S, Guha-Thakurta N, Taishi P, Obal F Jr, Krueger JM. 1997. Diurnal variations of tumor necrosis factor alpha mRNA and alpha-tubulin mRNA in rat brain. Neuroimmunomodulation 4: 84-90.

Brorson JR, Zhang H. 1999. Nitric oxide acutely inhibits neuronal energy production. J Neurosci 19: 147-158.

Burlet S, Cespuglio R. 1997. Voltammetric detection of nitric oxide (NO) in the rat brain: its variations throughout the sleep–wake cycle. Neurosci Lett 226: 131-135.

Burlet S, Leger L, Cespuglio R. 1999. Nitric oxide and sleep in the rat: a puzzling relationship. Neuroscience 92: 627-639.

Buzsáki G, Bickford RG, Ponomareff G, Thal LJ, Mandel RJ, et al. 1988. Nucleus basalis and thalamic control of neocortical activity in the freely moving rat. J Neurosci 8: 4007-4026.

Callaway CW, Lydic R, Baghdoyan HA, Hobson JA. 1987. Pontogeniculooccipital waves: spontaneous visual system activity during rapid eye movement sleep. Cell Mol Neurobiol 7: 105-149.

Capece ML, Baghdoyan HA, Lydic R. 1998. Carbachol stimulates [35S]guanylyl 5'-(gamma-thio)-triphosphate binding in rapid eye movement sleep-related brainstem nuclei of rat. J Neurosci 18: 3779-3785.

Capece ML, Efange SM, Lydic R. 1997. Vesicular acetylcholine transport inhibitor suppresses REM sleep. Neuroreport 8: 481-484.

Carla V, Moroni F. 1992. General anaesthetics inhibit the responses induced by glutamate receptor agonists in the mouse cortex. Neurosci Lett 146: 21-24.

Carnes KM, Fuller TA, Price JL. 1990. Sources of presumptive glutamatergic/aspartatergic afferents to the magnocellular basal forebrain in the rat. J Comp Neurol 302: 824-852.

Caton R. 1875. The electric currents of the brain. Br Med J 2: 278.

Cespuglio R, Debilly G, Burlet S. 2004. Cortical and pontine variations occurring in the voltammetric no signal throughout the sleep–wake cycle in the rat. Arch Ital Biol 142: 551-556.

Cespuglio R, Marinesco S, Baubet V, Bonnet C, el Kafi B. 1995. Evidence for a sleep-promoting influence of stress. Adv Neuroimmunol 5: 145-154.

Cespuglio R, Walker E, Gomez ME, Musolino R. 1976. Cooling of the nucleus raphe dorsalis induces sleep in the cat. Neurosci Lett 3: 221-227.

Chamberlin NL, Arrigoni E, Chou TC, Scammell TE, Greene RW, et al. 2003. Effects of adenosine on GABAergic synaptic inputs to identified ventrolateral preoptic neurons. Neuroscience 119: 913-918.

Chemelli RM, Willie JT, Sinton CM, Elmquist JK, Scammell T, et al. 1999. Narcolepsy in orexin knockout mice: molecular genetics of sleep regulation. Cell 98: 437-451.

Chen L, Majde JA, Krueger JM. 2003. Spontaneous sleep in mice with targeted disruptions of neuronal or inducible nitric oxide synthase genes. Brain Res 973: 214-222.

Chu NS, Bloom FE. 1974. Activity patterns of catecholamine-containing pontine neurons in the dorso-lateral tegmentum of unrestrained cats. J Neurobiol 5: 527-544.

Cirelli C, Tononi G. 2004. Locus ceruleus control of state-dependent gene expression. J Neurosci 24: 5410-5419.

Clement P, Sarda N, Cespuglio R, Gharib A. 2004. Changes occurring in cortical NO release and brain NO-synthases during a paradoxical sleep deprivation and subsequent recovery in the rat. J Neurochem 90: 848-856.

Coleman CG, Lydic R, Baghdoyan HA. 2004. M2 muscarinic receptors in pontine reticular formation of C57BL/6J mouse contribute to rapid eye movement sleep generation. Neuroscience 126: 821-830.

Cox CL, Sherman SM. 1999. Glutamate inhibits thalamic reticular neurons. J Neurosci 19: 6694-6699.

Cullinan WE, Zaborszky L. 1991. Organization of ascending hypothalamic projections to the rostral forebrain with special reference to the innervation of cholinergic projection neurons. J Comp Neurol 306: 631-667.

Cunha RA. 2001. Adenosine as a neuromodulator and as a homeostatic regulator in the nervous system: different roles, different sources and different receptors. J Neurochem 74: 334-338.

Dahlström A, Fuxe K. 1964. Evidence for the existence of monamine-containing neurons in the central nervous system. I. Demonstration of monoamines in the cell bodies of brain stem neurons. Acta Physiol Scand 62: 1-55.

Dale H. 1914. The action of certein esters and ethers of choline and their relation to muscarine. J Pharmacol 6: 147-190.

Dale H. 1936. Some recent extensions of the chemical transmission of the effects of nerve impulses. Nobel lecture.

Datta S. 1995. Neuronal activity in the peribrachial area: relationship to behavioral state control. Neurosci Biobehav Rev 19: 67-84.

Datta S. 1997. Cellular basis of pontine ponto–geniculo–occipital wave generation and modulation. Cell Mol Neurobiol 17: 341-365.

Datta S, Siwek DF. 1997. Excitation of the brain stem pedunculopontine tegmentum cholinergic cells induces wakefulness and REM sleep. J Neurophysiol 77: 2975-2988.

Datta S, Patterson EH, Siwek DF. 1997. Endogenous and exogenous nitric oxide in the pedunculopontine tegmentum induces sleep. Synapse 27: 69-78.

Datta S, Patterson EH, Spoley EE. 2001a. Excitation of the pedunculopontine tegmental NMDA receptors induces wakefulness and cortical activation in the rat. J Neurosci Res 66: 109-116.

Datta S, Spoley EE, Patterson EH. 2001b. Microinjection of glutamate into the pedunculopontine tegmentum induces REM sleep and wakefulness in the rat. Am J Physiol 280: R752-R759.

Datta S, Calvo JM, Quattrochi J, Hobson JA. 1992. Cholinergic microstimulation of the peribrachial nucleus in the cat. I. Immediate and prolonged increases in ponto–geniculo–occipital waves. Arch Ital Biol 130: 263-284.

Datta S, Siwek DF, Patterson EH, Cipolloni PB. 1998. Localization of pontine PGO wave generation sites and their anatomical projections in the rat. Synapse 30: 409-423.

Datta S, Spoley EE, Mavanji VK, Patterson EH. 2002. A novel role of pedunculopontine tegmental kainate receptors: a mechanism of rapid eye movement sleep generation in the rat. Neuroscience 114: 157-164.

de Lecea L, Kilduff TS, Peyron C, Gao X, Foye PE, et al. 1998. The hypocretins: hypothalamus-specific peptides with neuroexcitatory activity. Proc Natl Acad Sci USA 95: 322-327.

Dickstein JB, Moldofsky H, Lue FA, Hay JB. 1999. Intracerebroventricular injection of TNF-alpha promotes sleep and is recovered in cervical lymph. Am J Physiol 276: R1018-R1022.

Dijk DJ, Beersma DG, Daan S, van den Hoofdakker RH. 1989. Effects of seganserin, a 5-HT2 antagonist, and temazepam on human sleep stages and EEG power spectra. Eur J Pharmacol 171: 207-218.

Dixon AK, Gubitz AK, Sirinathsinghji DJ, Richardson PJ, Freeman TJ. 1996. Tissue distribution of adenosine receptor mRNAs in the rat. J Pharmacol 118: 1461-1468.

Dugovic C, Wauquier A. 1987. 5-HT2 receptors could be primarily involved in the regulation of slow-wave sleep in the rat. Eur J Pharmacol 137: 145-146.

Dzoljic MR, de Vries R. 1994. Nitric oxide synthase inhibition reduces wakefulness. Neuropharmacology 33: 1505-1509.

Dzoljic MR, de Vries R, van Leeuwen R. 1996. Sleep and nitric oxide: effects of 7-nitro indazole, inhibitor of brain nitric oxide synthase. Brain Res 718: 145-150.

Eggermann E, Serafin M, Bayer L, Machard D, Saint-Mleux B, et al. 2001. Orexins/hypocretins excite basal forebrain cholinergic neurones. Neuroscience 108: 177-181.

El Mansari M, Sakai K, Jouvet M. 1989. Unitary characteristics of presumptive cholinergic tegmental neurons during the sleep–waking cycle in freely moving cats. Exp Brain Res 76: 519-529.

Fang J, Wang Y, Krueger JM. 1997. Mice lacking the TNF 55 kDa receptor fail to sleep more after TNFalpha treatment. J Neurosci 17: 5949-5955.

Feldberg W. 1950. The role of acetylcholine in the central nervous system. Br Med Bull 6: 312-321.

Feldberg LA, Sherwood PD. 1954. Injections of drugs into the lateral ventricle of the cat. J Physiol 123: 148-167.

Foote SL, Aston-Jones G, Bloom FE. 1980. Impulse activity of locus coeruleus neurons in awake rats and monkeys is a function of sensory stimulation and arousal. Proc Natl Acad Sci USA 77: 3033-3037.

Ford B, Holmes CJ, Mainville L, Jones BE. 1995. GABAergic neurons in the rat pontomesencephalic tegmentum: codistribution with cholinergic and other tegmental neurons projecting to the posterior lateral hypothalamus. J Comp Neurol 363: 177-196.

Fornal CA, Litto WJ, Metzler CW, Marrosu F, Tada K, et al. 1994. Single-unit responses of serotonergic dorsal raphe neurons to 5-HT1A agonist and antagonist drug administration in behaving cats. J Pharmacol Exp Ther 270: 1345-1358.

Frank MG, Stryker MP, Tecott LH. 2002. Sleep and sleep homeostasis in mice lacking the 5-HT2c receptor. Neuropsychopharmacology 27: 869-873.

Fredholm BB, Batting K, Holmen J, Nehling A, Zvartau EE. 1999. Actions of caffeine in the brain with special reference to factors that contribute to its widespread use. Pharmacol Rev 79: 233-243.

Fredholm BB, Ijzerman AP, Jacobson KA, Klotz KN, Linden J. 2001. International union of pharmacology XXV. Nomenclature and classification of adenosine receptors. Pharmacol Rev 53: 527-552.

Fredholm BB, Arslan G, Halldner L, Kull B, Schulte G, et al. 2000. Structure and function of adenosien receptors and their genes. Naunyn Schmiedebergs Arch Pharmacol 362: 364-374.

Gallopin T, Saper CB. 2000. Identification of sleep-promoting neurons in vitro. Nature 404: 992-995.

Garthwaite J, Boulton CL. 1995. Nitric oxide signaling in the central nervous system. Annu Rev Physiol 57: 683-706.

Gerashchenko D, Kohls MD, Greco M, Waleh NS, Salin-Pascual R, et al. 2001. Hypocretin-2-saporin lesions of the lateral hypothalamus produce narcoleptic-like sleep behavior in the rat. J Neurosci 21: 7273-7283.

Gritti I, Mainville L, Jones BE. 1993. Codistribution of GABA-synthesizing with acetylcholine-synthesizing neurons in the basal forebrain of the rat. J Comp Neurol 329: 438-457.

Gritti I, Mainville L, Mancia M, Jones BE. 1997. GABAergic and other noncholinergic basal forebrain neurons, together with cholinergic neurons, project to the mesocortex and isocortex in the rat. J Comp Neurol 383: 163-177.

Gross SS, Wolin MS. 1995. Nitric oxide: pathophysiological mechanisms. Annu Rev Physiol 57: 737-769.

Haas HL, Selbach O. 2000. Functions of neuronal adenosine receptors (Review). Naunyn Schmiedebergs Arch Pharmacol 362: 375-381.

Harada S, Imaki T, Chicada N, Naruse M, Demura H. 1999. Distinct distribution and time-course changes in neuronal nitric oxide synthase and inducible NOS in the paraventricular nucleus following lipopolysaccharide injection. Brain Res 821: 322-332.

Hars B. 1999. Endogenous nitric oxide in the rat pons promotes sleep. Brain Res 816: 209-219.

Haulica I, Ababei L, Branisteanu D, Topoliceanu F. 1973. Letter: preliminary data on the possible hypnogenic role of adenosine. J Neurochem 21: 1019-1020.

Hayaishi O. 2002. Molecular genetic studies on sleep–wake regulation, with special emphasis on the prostaglandin D2 system. J Appl Physiol 92: 863-868.

Hebb C. 1957. Biochemical evidence for the neural function of acetylcholine. Physiol Rev 37: 196-220.

Heym J, Trulson ME, Jacobs BL. 1982. Raphe unit activity in freely moving cats: effects of phasic auditory and visual stimuli. Brain Res 232: 29-39.

Hobson J, McCarley R, Wyzinski P. 1975. Sleep cycle oscillation: reciprocal discharge by two brainstem neuronal groups. Science 189: 55-58.

Honda K, Komoda Y, Nishida S, Nagasaki H, Higashi A, et al. 1984. Uridine as an active component of sleep-promoting substance: its effects on nocturnal sleep in rats. Neurosci Res 1: 243-252.

Horvath TL, Peyron C, Diano S, Ivanov A, Aston-Jones G, et al. 1999. Hypocretin (orexin) activation and synaptic innervation of the locus coeruleus noradrenergic system. J Comp Neurol 415: 145-159.

Hughes R, Robinson F. 1951. Acetylcholine and the central nervous system; a review of neurochemical mechanisms. Yale J Biol Med 24: 35-47.

Huston JP, Haas HL, Pfister M, Decking U, Schrader J, et al. 1996. Extracellular adenosine levels in neostriatum and hippocampus during rest and activity periods of rats. Neuroscience 73: 99-107.

Idzikowski C, Cowen PJ, Nutt D, Mills FJ. 1987. The effects of chronic ritanserin treatment on sleep and the neuroendocrine response to L-tryptophan. Psychopharmacology (Berl) 93: 416-420.

Ishimori K. 1909. True cause of sleep: a hypnogenic substance as evidenced in the brain of sleep-deprived animals. Tokyo Igakkai Zasshi 23: 429-457.

Jacobs BL, Fornal CA. 1991. Activity of brain serotonergic neurons in the behaving animal. Pharmacol Rev 43: 563-578.

Jacobs BL, Asher R, Dement WC. 1973. Electrophysiological and behavioral effects of electrical stimulation of the raphe nuclei in cats. Physiol Behav 11: 489-495.

Jacobs BL, Wilkinson LO, Fornal CA. 1990. The role of brain serotonin. A neurophysiologic perspective. Neuropsychopharmacology 3: 473-479.

Jones BE. 1990. Immunohistochemical study of choline acetyltransferase-immunoreactive processes and cells innervating the pontomedullary reticular formation in the rat. J Comp Neurol 295: 485-514.

Jones BE. 1995. Cytoarchitecture, transmitters and projections. The rat nervous system. Paxinos G, editor. San Diego: Academic Press; pp. 155-171.

Jones BE. 2004. Activity, modulation and role of basal forebrain cholinergic neurons innervating the cerebral cortex. Prog Brain Res 145: 157-169.

Jones BE, Cuello AC. 1989. Afferents to the basal forebrain cholinergic cell area from pontomesencephalic–catecholamine, serotonin, and acetylcholine–neurons. Neuroscience 31: 37-61.

Jones BE, Harper S, Halaris A. 1977. Effects of locus coeruleus lesions upon cerebral monoamine content, sleep–wakefulness states and the response to amphetamine. Brain Res 124: 473-496.

Jouvet M. 1967. Mechanisms of the state of sleep: a neuropharmacological approach. Sleep and altered states of consciousness. Kety SS, Evarts EV, Williams HL, editors. Baltimore: Williams and Wilkins; pp. 86-126.

Jouvet M. 1972. The role of monoamines and acetylcholine-containing neurons in the regulation of the sleep–waking cycle. Ergebn Physiol 64: 166-307.

Jouvet M. 1982. Hypnogenic indolamine-dependent factors and paradoxical sleep rebound. Sleep. Koella WP, editor. Basel: Karger; pp. 2-18.

Jouvet M. 1999. Sleep and serotonin: an unfinished story. Neuropsychopharmacology 21:24S-27S.

Kalinchuk A, Urrila A-S, Alanko L, Heiskanen S, Wigren H-K, et al. 2003. Local energy depletion in the basal forebrain increases sleep. Eur J Neurosci 17: 863-869.

Kalinchuk AV, Lu Y, Stenberg D, Rosenberg PA, Porkka-Heiskanen T. 2006a. Nitric oxide production in the basal forebrain is required ofr recovery sleep. J Neurochem (in press).

Kalinchuk AV, Stenberg D, Rosenberg PA, Porkka-Heiskanen T. 2006b. Inducible and neuronal nitric oxide synthases have complementary roles in recovery sleep induction. Eur J Neurosci (in press).

Kapas L, Krueger JM. 1996. Nitric oxide donors SIN-1 and SNAP promote nonrapid-eye-movement sleep in rats. Brain Res Bull 41: 293-298.

Kapas L, Fang JD, Krueger JM. 1994a. Inhibition of nitric oxide synthesis inhibits rat sleep. Brain Res 664: 189-196.

Kapas L, Shibata M, Kimura M, Krueger JM. 1994b. Inhibition of nitric oxide synthesis suppresses sleep in rabbits. Am J Physiol 266: R151-R157.

Kety SS. 1959. Central actions of catecholamines. Pharmacol Rev 11: 565-566.

Kilduff TS, Peyron C. 2000. The hypocretin/orexin ligand–receptor system: implications for sleep and sleep disorders. Trends Neurosci 23: 359-365.

Koyama Y, Sakai K. 2000. Modulation of presumed cholinergic mesopontine tegmental neurons by acetylcholine and monoamines applied iontophoretically in unanesthetized cats. Neuroscience 96: 723-733.

Krnjevic K, Schwartz S. 1966. Is gamma-aminobutyric acid an inhibitory transmitter? Nature 211: 1372-1374.

Krueger JM. 1985. Somnogenic activity of muramyl peptides. Trends Pharmacol Sci 6: 218-221.

Krueger JM, Obal FJ, Fang J, Kubota T, Taishi P. 2001. The role of cytokines in physiological sleep regulation. Ann NY Acad Sci 933: 211-221.

Kubota T, Li N, Guan Z, Brown RA, Krueger JM. 2002. Intrapreoptic microinjection of TNF-alpha enhances non-REM sleep in rats. Brain Res 932: 37-44.

Lai YY, Siegel JM. 1988. Medullary regions mediating atonia. J Neurosci 8: 4790-4796.

Landolt HP, Meier V, Burgess HJ, Finelli LA, Cattelin F, et al. 1999. Serotonin-2 receptors and human sleep: effect of a selective antagonist on EEG power spectra. Neuropsychopharmacology 21: 455-466.

Lee KH, Broberger C, Kim U, McCormick DA. 2004. Histamine modulates thalamocortical activity by activating a chloride conductance in ferret perigeniculate neurons. Proc Natl Acad Sci USA 101: 6716-6721.

Legendre R, Piéron H. 1910. Le probl'me des facteurs du sommeil. Résultats d'injections vasculaires et intracérebrales de liquides insomniques. CR Soc Biol (Paris) 68: 1077-1108.

Legendre R, Piéron H. 1911. Du développement, au cours de l'insomnie expérimentale, de propriétés hypnotoxiques des humeurs en relation avec le besoin croissant de sommeil. CR Soc Biol (Paris) 70: 190-192.

Leonard TO, Lydic R. 1997. Pontine nitric oxide modulates acetylcholine release, rapid eye movement sleep generation, and respiratory rate. J Neurosci 17: 774-785.

Li Y, Gao XB, Sakurai T, van den Pol AN. 2002. Hypocretin/orexin excites hypocretin neurons via a local glutamate neuron-A potential mechanism for orchestrating the hypothalamic arousal system. Neuron 36: 1169-1181.

Lin JS, Sakai K, Jouvet M. 1988. Evidence for histaminergic arousal mechanisms in the hypothalamus of cat. Neuropharmacology 27: 111-122.

Lin JS, Sakai K, Jouvet M. 1994. Hypothalamo-preoptic histaminergic projections in sleep–wake control in the cat. Eur J Neurosci 6: 618-625.

Lin L, Faraco J, Li R, Kadotani H, Rogers W, et al. 1999. The sleep disorder canine narcolepsy is caused by a mutation in the hypocretin (orexin) receptor 2 gene. Cell 98: 365-376.

Lin JS, Sakai K, Vanni-Mercier G, Arrang JM, Garbarg M, et al. 1990. Involvement of histaminergic neurons in arousal mechanisms demonstrated with H3-receptor ligands in the cat. Brain Res 523: 325-330.

Lindsley DB, Schreiner LH, Knowles WB, Magoun HW. 1950. Behavioral and EEG changes following chronic brain stem lesions in the cat. Electroenceph Clin Neurophysiol 2: 483-498.

Lindvall O, Björklund A. 1974. The organization of the ascending catecholamine neuron systems in the rat brain as revealed by the glyozylic acid fluorescence method. Acta Physiol Scand 412: 1-48.

Loewi O. 1921. Über humorale übertragbarkeit der Herznervenwirkung. Pfluegers Arch ges Physiol 189: 239-242.

Losier BJ, Semba K. 1993. Dual projections of single cholinergic and aminergic brainstem neurons to the thalamus and basal forebrain in the rat. Brain Res 604: 41-52.

Lu J, Greco MA, Shiromani PJ, Saper CB. 2000. Effect of lesions of the ventrolateral preoptic nucleus on NREM and REM sleep. J Neurosci 20: 3830-3842.

Lydic R, Baghdoyan HA. 1993. Pedunculopontine stimulation alters respiration and increases ACh release in the pontine reticular formation. Am J Physiol 264: R544-R554.

Mackiewicz M, Sollars PJ, Ogilvie MD, Pack AI. 1996. Modulation of IL-1b gene expression in the rat CNS during sleep deprivation. Neuroreport 7: 529-533.

Madrigal JLM, Moro MA, Lizasoain I, Lorenzo P, Castrillo A, et al. 2001. Inducible nitric oxide synthase expression in brain cortex after acute restraint stress is regulated by nuclear factor kB-mediated mechanisms. J Neurochem 76: 532-538.

Madsen PL. 1993. Blood flow and oxygen uptake in the human brain during various states of sleep and wakefulness. Acta Neurol Scand 88: U5-U25.

Madsen PL, Holm S, Vorstrup S, Friberg L, Lassen NA, et al. 1991. Human regional cerebral blood flow during rapid-eye-movement sleep. J Cereb Blood Flow Metab 11: 502-507.

Magoun HW. 1950. Caudal and cephalic influences of the brain stem reticular formation. Physiol Rev 30: 459-474.

Maquet P, Dive D, Salmon E, Sadzot B, Franco G, et al. 1990. Cerebral glucose utilization during sleep–wake cycle in man determined by positron emission tomography and [18F]2-fluoro-2- deoxy-D-glucose method. Brain Res 513: 136-143.

Maquet P, Dive D, Salmon E, Sadzot B, Franco G, et al. 1992. Cerebral glucose utilization during stage 2 sleep in man. Brain Res 571: 149-153.

Marrosu F, Portas CM, Mascia MS, Casu MA, Fa M, et al. 1995. Microdialysis measurement of cortical and hippocampal acetylcholine release during sleep–wake cycle in freely moving cats. Brain Res 671: 329-332.

Matsumura H, Matsumura T, Nakajima T, Osaka S, Satoh K, et al. 1994. Prostaglandin D2-sensitive, sleep promoting zone defined in the ventral surface of the rostral basal forebrain. Proc Natl Acad Sci USA 91: 11998-12002.

McCarley RW, Nelson JP, Hobson JA. 1978. Ponto–geniculo–occipital (PGO) burst neurons: correlative evidence for neuronal generators of PGO waves. Science 201: 269-272.

McCormick DA. 1989. Cholinergic and noradrenergic modulation of thalamocortical processing. Trends Neurosci 12: 215-221.

McCormick DA. 1992. Neurotransmitter actions in the thalamus and cerebral cortex and their role in neuromodulation of thalamocortical activity. Prog Neurobiol 39: 337-388.

McCormick DA, Prince DA. 1986. Acetylcholine induces burst firing in thalamic reticular neurones by activating a potassium conductance. Nature 319: 402-405.

McCormick DA, Williamson A. 1991. Modulation of neuronal firing mode in cat and guinea pig LGNd by histamine: possible cellular mechanisms of histaminergic control of arousal. J Neurosci 11: 3188-3199.

McGinty DJ, Harper RM. 1976. Dorsal raphe neurons: depression of firing during sleep in cats. Brain Res 101: 569-575.

McGinty DJ, Szymusiak RS. 1989. The basal forebrain and slow wave sleep: mechanistic and functional aspects. Slow wave sleep. physiological, pathophysiological and functional aspects. Wauquier A, Dugovic C, Radulovacki M, editors. New York: Raven Press; pp. 61-74.

McGinty DJ, Szymusiak RS. 2000. The sleep–wake switch: a neuronal alarm clock. Nat Med 6: 510-511.

Mesulam MM, Mufson EJ, Wainer BH, Levey AI. 1983. Central cholinergic pathways in the rat: an overview based on an alternative nomenclature (Ch1–Ch6). Neuroscience 10: 1185-1201.

Miller JD, Farber J, Gatz P, Roffwarg H, German DC. 1983. Activity of mesencephalic dopamine and non-dopamine neurons across stages of sleep and walking in the rat. Brain Res 273: 133-141.

Mitchell JB, Lupica CR, Dunwiddie TV. 1993. Activity-dependent release of endogenous adenosine modulates synaptic responses in the rat hippocampus. J Neurosci 13: 3439-3447.

Mitome M, Shirakawa T, Oshima S, Nakamura W, Oguchi H. 2001. Circadian rhythm of nitric oxide production in the dorsal region of the suprachiasmatic nucleus in rats. Neurosci Lett 303: 161-164.

Miyazaki H, Nakamura Y, Arai T, Kataoka K. 1997. Increase of glutamate uptake in astrocytes: a possible mechanism of action of volatile anesthetics. Anesthesiology 86: 1359-1366.

Mizoguchi A, Eguchi N, Kimura K, Kiyohara Y, Qu WM, et al. 2001. Dominant localization of prostaglandin D receptors on arachnoid trabecular cells in mouse basal forebrain and their involvement in the regulation of nonrapid eye movement sleep. Proc Natl Acad Sci USA 98: 11674-11679.

Moldofsky H, Lue FA, Eisen J, Keystone E, Gorczynski RM. 1986. The relationship of interleukin-1 and immune functions to sleep in humans. Psychosomatic Med 48: 309-318.

Monaca C, Boutrel B, Hen R, Hamon M, Adrien J. 2003. 5-HT 1A/1B receptor-mediated effects of the selective serotonin reuptake inhibitor, citalopram, on sleep: studies in 5-HT 1A and 5-HT 1B knockout mice. Neuropsychopharmacology 28: 850-856.

Monti JM. 1982. Catecholamines and the sleep–wake cycle. I. EEG and behavioral arousal. Life Sci 30: 1145-1157.

Monti JM. 1993. Involvement of histamine in the control of the waking state. Life Sci 53: 1331-1338.

Monti JM, Monti D. 2000. Role of dorsal raphe nucleus serotonin 5-HT1A receptor in the regulation of REM sleep. Life Sci 66: 1999-2012.

Monti J, Jantos H, Monti D. 2001. Increase of waking and reduction of NREM and REM sleep after nitric oxide synthase inhibition: prevention with GABA-A or adenosine A1 receptor agonists. Behav Brain Res 123: 23-35.

Moore R, Bloom F. 1979. Central catecholamine neuron systems: anatomy and physiology of the norepinephrine and epinephrine systems. Annu Rev Neurosci 2: 113-168.

Moro MA, De Alba J, Leza JC, Lorenzo P, Fernandez AP, et al. 1998. Neuronal expression of inducible nitric oxide synthase after oxygen and glucose deprivation in forebrain slices. Eur J Neurosci 10: 445-456.

Moruzzi G, Magoun H. 1949. Brain stem reticular formation and activation of the EEG. Electroenceph Clin Neurophysiol 1: 455-473.

Murillo-Rodriguez E, Blanco-Centurion C. 2004. The diurnal rhythm of adenosine levels in the basal forebrain of young and old rats. Neuroscience 123: 361-370.

Möhler H. 2001. Functions of GABA-A recpetors: pharmacology and pathophysiology. Pharmacology of GABA and glycine neurotransmission. Möhler H, editor. Heidelberg: Springer; pp. 101-116.

Naidoo N, Giang W, Galante RJ, Pack AI. 2005. Sleep deprivation induces the unfold protein response in mouse cerebral cortex. J Neurochem 92: 1150-1157.

Nathan C, Xie Q. 1994a. Nitric oxide synthases: roles, tolls and controls. Cell 78: 915-918.

Nathan C, Xie Q. 1994b. Regulation of biosynthesis of nitric oxide. J Biol Chem 269: 13725-13728.

Netchiporouk L, Shram N, Salvert D, Cespuglio R. 2001. Brain extracellular glucose assessed by voltammetry throughout the rat sleep–wake cycle. Eur J Neurosci 13: 1429-1434.

Nishino S, Shelton J, Renaud A, Dement WC, Mignot E. 1995. Effect of 5-HT1A receptor agonists and antagonists on canine cataplexy. J Pharmacol Exp Ther 272: 1170-1175.

Nitz D, Siegel JM. 1997a. GABA release in posterior hypothalamus across sleep–wake cycle. Am J Physiol 271: R1707-R1712.

Nitz D, Siegel JM. 1997b. GABA release in the locus coeruleus as a function of sleep/wake state. Neuroscience 78: 795-801.

Onoe H, Ueno R, Fujita I, Nishino H, Oomura U, et al. 1988. Prostaglandin D2, a cerebral sleep-inducing substance in monkeys. Proc Natl Acad Sci USA 85: 4082-4086.

Opp MR, Toth LA. 2003. Neural-immune interactions in the regulation of sleep. Basic science of sleep, Krueger JM, editor. Frontiers in Bioscience, Albertson, NY, Vol. 8, pp. d768-d779.

Pacak K, Palkovits M, Kopin IJ, Goldstein DS. 1995. Stress-induced norepinephrine release in the hypothalamic paraventricular nucleus and pituitary–adrenocortical and sympathoadrenal activity: in vivo microdialysis studies. Front Neuroendocrinol 16: 89-150.

Panula P, Pirvola U, Auvinen S, Airaksinen M. 1989. Histamine-immunoreactive nerve fibers in the rat brain. Neuroscience 28: 585-610.

Pappenheimer JM, Koski G, Fencl V, Karnovsky ML, Krueger J. 1975. Extraction of sleep-promoting factor S from cerebrospinal fluid and from brains of sleep-deprived animals. J Neurophysiol 38: 1299-1311.

Parmentier R, Ohtsu H, Djebbara-Hannas Z, Valatx JL, Watanabe T, et al. 2002. Anatomical, physiological, and pharmacological characteristics of histidine decarboxylase knockout mice: evidence for the role of brain histamine in behavioral and sleep–wake control. J Neurosci 22: 7695-7711.

Parkinson FE, Fredholm BB. 1990. Autoradigraphic evidence for G-protein coupled A2-receptors in the rat neostriatum using [3H]-CGS 21680 as ligand. Naunyn Schmiedebergs Arch Pharmacol 342: 85-89.

Petitjean F, Buda C, Janin M, Sallanon M, Jouvet M. 1985. Insomnie par administration de parachlorophenylalanine: reversibilite par injection peripherique ou centrale de 5-hydroxytryptophane et de serotonine. Sleep 8: 56-67.

Peyron C, Tighe DK, van den Pol AN, de Lecea L, Heller HC, et al. 1998. Neurons containing hypocretin (orexin) project to multiple neuronal systems. J Neurosci 18: 9996-10015.

Porkka-Heiskanen T, Strecker RE, McCarley RW. 2000. Brain site specificity of extracellular concentration changes during sleep deprivation and spontaneous sleep: an in vivo microdialysis study. Neuroscience 99: 507-517.

Porkka-Heiskanen T, Strecker RE, Bjorkum AA, Thakkar M, Greene RW, et al. 1997. Adenosine: a mediator of the sleep-inducing effects of prolonged wakefulness. Science 276: 1265-1268.

Portas CM, Thakkar M, Rainnie D, McCarley RW. 1996. Microdialysis perfusion of 8-hydroxy-2-(di-*n*-propylamino) tetralin (8-OH-DPAT) in the dorsal raphe nucleus decreases serotonin release and increases rapid eye movement sleep in the freely moving cat. J Neurosci 16: 2820-2828.

Portas CM, Thakkar M, Rainnie DG, Greene RW, McCarley RW. 1997. Role of adenosine in behavioral state modulation: a microdialysis study in the freely moving cat. Neuroscience 79: 225-235.

Purpura D, Girado M, Grundfest H. 1957. Selective blockade of excitatory synapses in the cat brain by gamma-aminobutyric acid. Science 125: 1200-1202.

Pull I, McIlwain H. 1972. Metabolism of [14C]adenine and derivatives by cerebral tissues, superfused and electrically stimulated. Biochem J 126: 965-973.

Radulovacki M, Virus RM, Djuricic-Nedelson M, Green RD. 1984. Adenosine analogs and sleep in rats. J Pharm Exp Ther 228: 268-274.

Rainnie DG, Grunze HC, McCarley RW, Greene RW. 1994. Adenosine inhibition of mesopontine cholinergic neurons: implications for EEG arousal. Science 263: 689-692.

Rajkowski J, Kubiak P, Aston-Jones G. 1994. Locus coeruleus activity in monkey: phasic and tonic changes are associated with altered vigilance. Brain Res Bull 35: 607-616.

Ram A, Pandey HP, Matsumura H, Kasahara-Orita K, Nakajima T, et al. 1997. CSF levels of prostaglandins, especially the level of prostaglanding D2, are correlated with increasing propensity towards sleep in rats. Brain Res 751: 81-89.

Rivkees SA, Price SL, Zhou FC. 1995. Immunohistochemical detection of A1 adenosine receptors in rat brain with emphasis on localization in the hippocampal formation, cerebral cortex, cerebellum, and basal ganglia. Brain Res 677: 193-203.

Rothballer A. 1959. The effects of catecholamines on the central nervous system. Pharmacol Rev 11: 494-547.

Sakurai T, Amemiya A, Ishii M, Matsuzaki I, Chemelli RM, et al. 1998. Orexins and orexin receptors: a family of hypothalamic neuropeptides and G protein-coupled receptors that regulate feeding behavior. Cell 92: 573-85.

Satoh S, Matsamura H, Suzuki F, Hayaishi O. 1996. Promotion of sleep mediated by the A_{2a}-adenosine receptor and possible involvement of this receptor in the sleep induced by prostaglandin D_2 in rats. Proc Nat Acad Sci USA 93: 5980-5984.

Satoh S, Matsumura H, Koike N, Tokunaga Y, Maeda T, et al. 1999. Region dependent difference in the sleep-promoting potency of an adenosine A2A receptor agonist. Eur J Neurosci 11: 1587-1597.

Scammell TE, Gerashcenko D, Mochizuki T, McCarthy MT, Estabrooke IV, et al. 2001. An adenosine A_{2a} agonist increases sleep and induces Fos in ventrolateral peroptic neurons. Neuroscience 107: 653-663.

Scammell TE, Gerashcenkko D, Urade Y, Onoe H, Saper CB, et al. 1998. Activation of ventrolateral preoptic neurons by the somnogen prostaglandin D2. Proc Natl Acad Sci USA 95: 7754-7759.

Schwarz DW, Tennigkeit F, Puil E. 2000. Metabotropic transmitter actions in auditory thalamus. Acta Otolaryngol 120: 251-254.

Schwierin B, Borbély AA, Tobler I. 1996. Effects of N^6-cyclopentyladenosine and caffeine on sleep regulation in the rat. Eur J Pharmacol 300: 163-171.

Seifritz E, Moore P, Trachsel L, Bhatti T, Stahl SM, et al. 1996. The 5-HT1A agonist ipsapirone enhances EEG slow wave activity in human sleep and produces a power spectrum similar to 5-HT2 blockade. Neurosci Lett 209: 41-44.

Sherin JE, Elmquist JK, Torrealba F, Saper CB. 1998. Innervation of histaminergic tuberomamillary neurons by GABAergic and galaninergic neurons in the ventrolateral preoptic nucleus of the rat. J Neurosci 18: 4705-4721.

Sherin JE, Shiromani PJ, McCarley RW, Saper CB. 1996. Activation of ventrolateral preoptic neurons during sleep. Science 271: 216-219.

Shiromani PJ, Armstrong DM, Gillin JC. 1988. Cholinergic neurons from the dorsolateral pons project to the medial pons: a WGA–HRP and choline acetyltransferase immunohistochemical study. Neurosci Lett 95: 19-23.

Shoham S, Davenne D, Cady AB, Dinarello CA, Krueger JM. 1987. Recombinant tumor necrosis factor and interleukin 1 enhance slow-wave sleep. Am J Physiol 253: R142-R149.

Sorensen E, Bjorvatn B, Ursin R. 2000. Sleep–wake effects following the selective 5-HT(1A) receptor antagonist p-MPPI in the freely moving rat. Behav Brain Res 114: 31-38.

Spessert R, Rapp M. 2001. Circadian rhythm in NO synthase I transcript expression and its photoperiodic regulation in the rat pineal gland. Neuroreport 12: 781-785.

Steinfels GF, Heym J, Strecker RE, Jacobs BL. 1983. Behavioral correlates of dopaminergic unit activity in freely moving cats. Brain Res 258: 217-228.

Steininger TL, Gong H, McGinty D, Szymusiak R. 2001. Subregional organization of preoptic area/anterior hypothalamic projections to arousal-related monoaminergic cell groups. J Comp Neurol 429: 638-653.

Steininger T, Alam M, Gong H, Szymusiak R, McGinty D. 1999. Sleep–waking discharge of neurons in the posterior lateral hypothalamus of the albino rat. Brain Res 840: 138-147.

Steriade M. 1996. Arousal: revisiting the reticular activation system. Science 272: 225-226.

Steriade M, Glenn LL. 1982. Neocortical and caudate projections of intralaminar thalamic neurons and their synaptic excitation from midbrain reticular core. J Neurophysiol 48: 352-371.

Steriade M, Curró Dossi R, Nuñez A. 1991. Network modulation of a slow intrinsic oscillation of cat thalamocortical neurons implicated in sleep delta waves: cortically induced synchronization and brainstem cholinergic suppression. J Neurosci 11: 3200-3217.

Steriade M, Domich L, Oakson G, Deschenes M. 1987. The deafferented reticular thalamic nucleus generates spindle rhytmicity. J Neurophysiol 57: 260-273.

Steriade M, Pare D, Parent A, Smith Y. 1988. Projections of cholinergic and non-cholinergic neurons of the brainstem core to relay and associational thalamic nuclei in the cat and macaque monkey. Neuroscience 25: 47-67.

Steriade M, Datta S, Pare D, Oakson G, Curro Dossi RC. 1990. Neuronal activities in brain-stem cholinergic nuclei related to tonic activation processes in thalamocortical systems. J Neurosci 10: 2541-2559.

Strecker RE, Jacobs BL. 1985. Substantia nigra dopaminergic unit activity in behaving cats: effect of arousal on spontaneous discharge and sensory evoked activity. Brain Res 361: 339-350.

Strecker RE, Morairty SR, Thakkar MM, Porkka-Heiskanen T, Basheer R, et al. 2000. Adenosinergic modulation of basal forebrain and preoptic/anterior hypothalamic neuronal activity in the control of behavioral state. Behav Brain Res 115: 183-204.

Strecker RE, Nalwalk J, Dauphin LJ, Thakkar MM, Chen Y, et al. 2002. Extracellular histamine levels in the feline preoptic/anterior hypothalamic area during natural sleep–wakefulness and prolonged wakefulness: an in vivo microdialysis study. Neuroscience 113: 663-670.

Szymusiak R, McGinty D. 1986. Sleep suppression following kainic acid-induced lesions of the basal forebrain. Exp Neurol 94: 598-614.

Szymusiak RS, McGinty DJ. 1986. Sleep-related neuronal discharge in the basal forebrain of cats. Brain Res 370: 82-92.

Taguchi O, Kubin L, Pack AI. 1992. Evocation of postural atonia and respiratory depression by pontine carbachol in the decerebrate rat. Brain Res 595: 107-115.

Taishi P, Bredow S, Guha-Thakurta N, Obal F, Krueger JM. 1997. Diurnal variations of interleukin-1 beta mRNA and beta-actin mRNA in rat brain. J Neuroimmunol 75: 69-74.

Takahashi S, Kapas L, Seyer JM, Wang Y, Krueger JM. 1996. Inhibition of tumor necrosis factor attenuates physiological sleep in rabbits. Neuroreport 7: 642-646.

Terao A, Steininger TL, Hyder K, Apte-Deshpande A, Ding J, et al. 2003. Differential increase in the expression of heat shock protein family members during sleep deprivation and during sleep. Neuroscience 116: 187-200.

Thakkar MM, Winston S, McCarley RW. 2003b. A1 receptor and adenosinergic homeostatic regulation of sleep–wakefulness: effects of antisense to the A1 receptor in the cholinergic basal forebrain. J Neurosci 23: 4278-4287.

Thakkar MM, Delgiacco RA, Strecker RE, McCarley RW. 2003a. Adenosinergic inhibition of basal forebrain wakefulness-active neurons: a simultaneous unit recording and microdialysis study in freely behaving cats. Neuroscience 122: 1107-1113.

Thannickal TC, Moore RY, Nienhuis R, Ramanathan L, Gulyani S, et al. 2000. Reduced number of hypocretin neurons in human narcolepsy. Neuron 27: 469-474.

Ticho SR, Radulovacki M. 1991a. Role of adenosine in sleep and temperature regulation in the preoptic area of rats. Pharmacol Biochem Behav 40: 33-40.

Ticho SR, Radulovacki M. 1991b. Role of adenosine in sleep and temperature regulation in the preoptic area of rats. Pharmacol Biochem Behav 40: 33-40.

Trulson ME, Jacobs BL. 1979. Raphe unit activity in freely moving cats: correlation with level of behavioral arousal. Brain Res 163: 135-150.

Trulson ME, Preussler DW. 1984. Dopamine-containing ventral tegmental area neurons in freely moving cats: activity during the sleep–waking cycle and effects of stress. Exp Neurol 83: 367-377.

Trulson ME, Preussler DW, Howell GA. 1981. Activity of substantia nigra units across the sleep–waking cycle in freely moving cats. Neurosci Lett 26: 183-188.

Tunctan B, Weigl Y, Dotan A, Peleg L, Zengil H, et al. 2002. Circadian variation of nitric oxide synthase activity in mouse tissue. Chronobiol Int 19: 393-404.

Ueno R, Ishikawa Y, Nakayama T, Hayaishi O. 1982. Prostaglandin D2 induces sleep when microinjected into the preoptic area of conscious rats. Biochem Biophys Res Commun 109: 576-582.

Uhlrich DJ, Manning KA, Xue JT. 2002. Effects of activation of the histaminergic tuberomammillary nucleus on visual responses of neurons in the dorsal lateral geniculate nucleus. J Neurosci 22: 1098-1107.

Ungerstedt U. 1971. Stereotaxic mapping of the monoamine pathways in the rat brain. Acta Physiol Scand 367: 1-49.

Urade Y, Hayaishi O. 1999. Prostaglandin D2 and sleep regulation. Biochim Biophys Acta 1436: 606-615.

Urade Y, Kitahama K, Ohishi H, Kaneko T, Mizuno N, et al. 1993. Dominant expression of messenger RNA for prostaglandin-D synthase in leptomeninges, choroid plexus, and oligodendrocytes of the adult rat brain. Proc Natl Acad Sci USA 90: 9070-9074.

Valentino RJ, Foote SL. 1988. Corticotropin-releasing hormone increases tonic but not sensory-evoked activity of noradrenergic locus coeruleus neurons in unanesthetized rats. J Neurosci 8: 1016-1025.

van den Pol AN, Gao XB, Obrietan K, Kilduff TS, Belousov AB. 1998. Presynaptic and postsynaptic actions and modulation of neuroendocrine neurons by a new hypothalamic peptide, hypocretin/orexin. J Neurosci 18: 7962-7971.

Vanni-Mercier G, Gigout S, Debilly G, Lin JS. 2003. Waking selective neurons in the posterior hypothalamus and their response to histamine H3-receptor ligands: an electrophysiological study in freely moving cats. Behav Brain Res 144: 227-241.

Van Wylen DG, Park TS, Rubio R, Berne RM. 1986. Increases in cerebral interstitial fluid adenosine concentration during hypoxia, local potassium infusion, and ischemia. J Cereb Blood Flow Metab 6: 522-528.

Vincent SR, Reiner PB. 1987. The immunohistochemical localization of choline acetyltransferase in the cat brain. Brain Res Bull 18: 371-415.

Wauquier A. 1983. Drug effects on sleep–wakefulness patterns in dogs. Neuropsychobiology 10: 60-64.

Virus RM, Djuricic-Nedelson M, Radulovacki M, Green RD. 1983. The effects of adenosine and 2'-deoxycoformycin on sleep and wakefulness in rats. Neuropharmacology 22: 1401-1404.

Vitkovic L, Bockaert J, Jacque C. 2000. "Inflammatory" cytokines: Neuromodulators in normal brain? J Neurochem 74: 457-471.

Williams JA, Vincent SR, Reiner PB. 1997. Nitric oxide production in rat thalamus changes with behavioral state, local depolarization, and brainstem stimulation. J Neurosci 17: 420-427.

von Economo C. 1923. Encephalitis lethargica. Wien Med Wochenschr 73: 777-782.

von Economo C. 1930. Sleep as a problem of localization. J Nervous Mental Dis 71: 248-259.

von Euler U. 1946. A specific sympathomimetic ergone in adrenergic nerve fibres (sympathin) and its relationship to adrenaline and non-adrenaline. Acta Physiol Scand 12: 73-97.

Xi MC, Morales FR, Chase MH. 2001. Effects on sleep and wakefulness of the injection of hypocretin-1 (orexin-A) into the laterodorsal tegmental nucleus of the cat. Brain Res 901: 259-264.

Yamamura T, Harada K, Okamura A, Kemmotsu O. 1990. Is the site of action of ketamine anesthesia the N-methyl-D-aspartate receptor? Anesthesiology 72: 704-710.

Yamashita A, Watanabe Y, Hayaishi O. 1983. Autoradiographic localization of a binding protein(s) specific for prostaglandin D2 in rat brain. Proc Natl Acad Sci USA 80: 6114-6118.

Yoshimoto Y, Sakai K, Luppi PH, Fort P, Salvert D, et al. 1989. Forebrain afferents to the cat posterior hypothalamus: a double labeling study. Brain Res Bull 23: 83-104.

25 Neuroendocrinology of Sleep

A. Steiger

Abstract: Sleep is a time of distinct activity in various endocrine systems. Two major methods of sleep research are the sleep electroencephalogram (EEG) and the assessment of sleep-related endocrine activity (e.g., by collection of hormone profiles). The combination of these methods in human subjects and in animal models helps to elucidate the interaction of sleep EEG and hormones in normal and pathological sleep. Such simultaneous investigations showed a bidirectional interaction between the electrophysiological and the neuroendocrine components of sleep. Certain hormones (neuropeptides and steroids) play a specific role in sleep regulation and some peptides (growth hormone-releasing hormone [GHRH], galanin, ghrelin, neuropeptide Y) promote sleep, at least in males, whereas others (corticotropin-releasing hormone [CRH], somatostatin) impair sleep. A reciprocal interaction of GHRH and CRH plays a keyrole in sleep regulation. GHRH promotes nonrapid-eye-movement sleep (NREMS), at least in males and stimulates GH secretion, whereas CRH maintains wakefulness and enhances the secretion of ACTH and cortisol. Changes in the CRH:GHRH ratio in favor of CRH contribute to shallow sleep, elevated cortisol secretion, and blunted GH levels during depression and normal ageing. However, CRH-like effects of GHRH were found in women, as sleep is impaired, and ACTH and cortisol are elevated after this peptide. Besides peptides, steroids are involved in sleep regulation. Cortisol appears to promote rapid-eye-movement sleep (REMS). This finding points to a REMS-promoting effect of cortisol. Similarly, $GABA_A$ receptors are targets of various neuroactive steroids, which exert specific effects on sleep. The changes of sleep EEG in women after menopause and the beneficial effect of estrogen replacement therapy suggest a role of estrogen in sleep regulation.

List of Abbreviations: ACTH, adrenocorticotropic hormone; CNS, central nervous system; CRH, corticotrophin-releasing hormone; DHEA, dehydroepiandrosterone; DHEAS, dehydroepiandrosterone sulfate; EEG, electroencephalogram; FSH, follicle stimulating hormone; 5-HT-3, 5-hydroxytryptamine type 3; GABA, gamma-aminobutyric acid; CSF, cerebrospinal fluid; GH, growth hormone; GHRH, growth hormone-releasing hormone; GHRP-6, growth hormone-releasing peptide-6; GHS, growth hormone secretagogue; GR, glucorticoid receptor; HPA, hypothalamo-pituitary-adrenocortical; HPS, hypothalamo-pituitary-somatotrophic; IGF-1, insulin-like growth factor-1; iv, intravenous; MR, mineralocorticoid receptor; NREMS, nonrapid-eye-movement sleep; NPS, neuropeptide S; NPSR, neuropeptide S receptor; NPY, neuropeptide Y; OX-A, orexin-A; OX-B, orexin-B; PACAP, pituitary adenylate cyclase activating polypeptide; PRA, plasma renin activity; REM, rapid-eye-movement; REMS, rapid-eye-movement sleep; SWA, slow wave activity; SWS, slow wave sleep; THDOC, deoxycorticosterone 3 alpha, 21-dihydroxy-5-alpha-pregnan-20-one; TRH, thyrotropin-releasing hormone; TSH, thyroid stimulating hormone; VIP, vasoactive intestinal polypeptide

1 Introduction

Sleep is a time of distinct activity in various endocrine systems. Two major methods for the investigation of sleep in various species including humans are the sleep electroencephalogram (EEG) and the assessment of sleep-related endocrine activity (e.g., by collection of hormone profiles) during the night or during 24 h, respectively. The combination of these electrophysiological and neuroendocrinological methods in human subjects and in animal models helps to elucidate the interaction of sleep EEG and hormones in physiological and pathological sleep. Such simultaneous investigations of sleep EEG and hormone secretion were performed in female and male normal control subjects from adulthood to senescence, in patients with psychiatric, particularly affective, with endocrine and sleep disorders, under baseline conditions, after manipulation of the sleep–wake pattern (e.g., sleep deprivation), and after administration of synthetic and endogenous compounds acting at the central nervous system (CNS), particularly neuropeptides and neuroactive steroids. Furthermore, related animal models were studied, including administration of hormones, sleep deprivation, surgery of adrenal glands, and transgenic animals with changes of endocrine activity. These studies showed a bidirectional interaction between the electrophysiological and the neuroendocrine components of sleep. Particularly, it was demonstrated that certain hormones play a specific role in sleep regulation.

Human sleep is characterized by the cyclic occurrence of periods of nonrapid-eye-movement sleep (NREMS) and rapid-eye-movement (REM) sleep (REMS). During the first NREMS period, the major portion of slow wave sleep (SWS) occurs. Correspondingly, in EEG spectral analysis (Borbély et al., 1981; Trachsel et al., 1992; Steiger et al., 1993), the major portion of slow wave activity (SWA) is found in the first sleep cycle. The secretion of various hormones shows distinct patterns as described in detail in the corresponding chapters below. In short, during the first half of the night the growth hormone (GH) surge preponderates, whereas adrenocorticotropic hormone (ACTH) and cortisol levels are low. In contrast, during the second half of the night ACTH and cortisol concentrations are high, whereas GH release is low (➋ *Figure 25-1*) (Weitzman, 1976). This pattern suggests (i) a reciprocal interaction of the hypothalamo-pituitary-somatotrophic (HPS) and the hypothalamo-pituitary-adrenocortical (HPA)

Figure 25-1
Individual hypnograms and patterns of cortisol and growth hormone (GH) secretion in four male subjects (young and old patients with depression and normal controls). From: Steiger, A. 2002. Neuroendocrinology of sleep disorders. In: Textbook of Biological Psychiatry (D'haenen H, den Boer JA, Westenberg H, Willner P, editors.) London: John Wiley & Sons, Ltd; pp. 1229–1246. Copyright John Wiley and Sons Ltd. Reproduced with permission

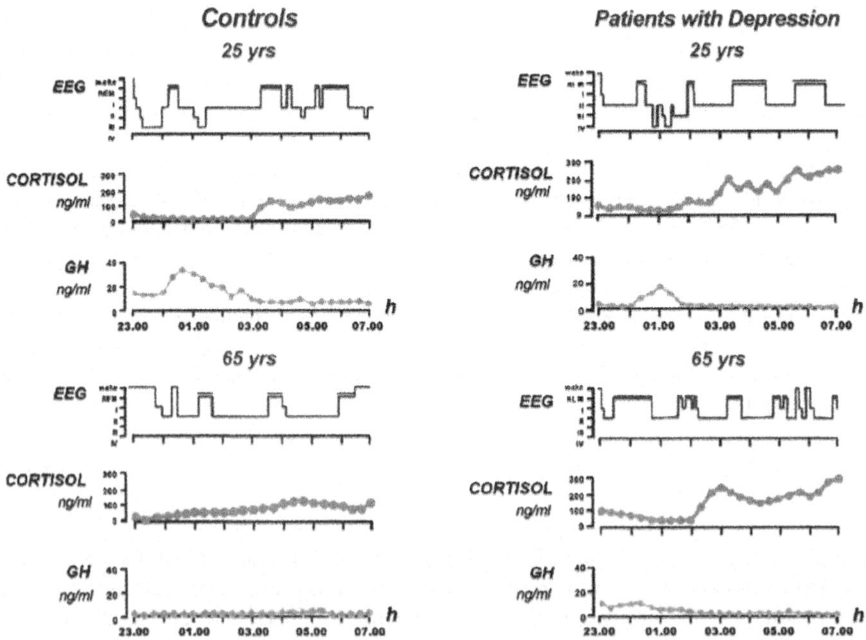

systems (the corresponding peripheral endpoints are GH and cortisol, respectively) and (ii) the existence of common regulating factors of the sleep EEG and the nocturnal hormone secretion. Indeed, there is good evidence, that a reciprocal interaction of the key hormones of the HPS and HPA systems, GH-releasing hormone (GHRH) and corticotropin-releasing hormone (CRH) plays a major role in sleep regulation. This issue is described in detail below.

A sexual dimorphism of sleep-endocrine activities was reported in young normal humans. Cortisol secretion is higher in female than in male subjects. Most men show a single GH peak near to sleep onset, whereas in women characteristically a presleep GH surge and one or more additional GH peaks are found during the second half of the night (Antonijevic et al., 1999a). Sleep EEG (Bliwise, 1993) and nocturnal hormone secretion (Van Coevorden et al., 1991) change throughout the life span. In females, the menopause is a major turning point towards impaired sleep (Ehlers and Kupfer, 1997), whereas in men, the sleep quality declines continuously during ageing.

The one hormone, which is most clearly linked to the NREMS–REMS cycle, is renin. Plasma renin activity (PRA) shows oscillations of about 90 min period strongly linked to the NREMS–REMS cycles. PRA reaches its peak during NREMS and its acrophase during REMS periods (Brandenberger et al., 1988). In particular, a link between SWS and the peaks of renin secretion was observed. Also, in rats, PRA is low in REMS and high in NREMS (Obál et al., 1994).

Leptin, the protein product of the obese (*ob*) gene is released from adipocytes in the periphery. It acts within the hypothalamus and reduces food intake (Tomaszuk et al., 1996). The maximum of serum leptin is found between 0000 and 0400 h. An inverse relationship exists between leptin and cortisol particularly in women (Licinio et al., 1997; Antonijevic et al., 1998). Leptin levels are higher in women than in men (Deuschle et al., 1996; Saad et al., 1997). Ghrelin, which is the first peptide identified as an endogenous ligand of the GH secretagogue (GHS) receptor was delineated as the orexigenic counterpart of leptin in the energy balance (Horvath et al., 2001). For the relationship of ghrelin to sleep, please see ➋ Section 2.3.4.

2 Hypothalamo–Pituitary–Somatotrophic System

2.1 Basic Activity

GH stimulates tissue growth and protein anabolism. These effects are mediated in part by insulin-like growth factor-1 (IGF-1). The synthesis and the secretion of GH is promoted by GHRH and inhibited by somatostatin. Recently, ghrelin was identified as an additional stimulus for GH release (Kojima et al., 1999). Synthetic GHSs were already known before the cloning of the GHS receptor. All these components of the HPS system were shown to participate in sleep regulation.

Whereas about one third of GH is released during daytime in humans, the major peak of GH secretion during 24 h occurs near to sleep onset. This GH surge is associated to the first period of SWS (Quabbe et al., 1966; Takahashi et al., 1968; Steiger et al., 1987). In one study, GH concentrations were determined every 30 sec in normal young male subjects. Maximal GH secretion was reported within minutes after the onset of SWS (Holl et al., 1991). A close temporal relationship was found between GH secretion and SWA (Gronfier et al., 1996). However, GH may be released prior to sleep onset in normal human subjects (Steiger et al., 1987).

The GH surge appears to be widely sleep dependent and is suppressed during sleep deprivation (Sassin et al., 1969; Beck et al., 1975). However, in sleep-deprived, but relaxed normal young male subjects in supine position, an unchanged nocturnal GH peak was observed (Mullington et al., 1996). This finding corroborates the view that lying relaxed is sufficient to trigger the nocturnal GH surge. On the other hand a weak circadian component in the regulation of GH release was delineated. During the second half of the night, GH levels are low. Similarly, it was shown that about one third of SWS periods are not associated with GH secretion (Van Cauter et al., 1992). Already during the third decade of the life span, distinct parallel decreases of SWS, SWA, and GH secretion start. In males near to the onset of the fifth decade, the GH pause occurs. From then on, GH release is nearly absent. In females, the GH pause is related to menopause.

In several animal species (rhesus monkeys, adult rats, goats, cows, and dogs), an episodic release of GH, but no link between sleep and GH secretion was found (reviewed in: Quabbe et al., 1981; Laurentie et al., 1989). However, in calves (Fabry et al., 1982) and in growing pigs (Dubreuil et al., 1988) a relationship between the sleep–wake cycle and GH release was reported. In lambs, the highest GH production rate occurred during SWS and REMS (Laurentie et al., 1989). In contrast, in immature rats a correlation between GH concentration and the duration of sleep during the preceding 10 min was found (Kawakami et al., 1983).

Hypothalamic GHRH mRNA depends on a circadian rhythm. In rats, the highest concentration is found at the onset of the light period when sleep propensity reaches its maximum in these night active animals (Bredow et al., 1996). Furthermore, hypothalamic GHRH contents display sleep-related variations with low levels in the morning, increases in the afternoon (peak at the transition from the light to the dark period) and decreases at night (Gardi et al., 1999). Calcium levels in GABAergic neurons cultured from rat fetal hypothalamus increased when perfused with GHRH (De et al., 2002). It is thought that many hypothalamic GHRH responsive neurons are GABAergic.

2.2 Sleep in Disorders of the HPS System

In patients with isolated GH deficiency, SWS was reduced in comparison to normal control subjects, whereas total sleep time and NREMS stages 1 and 2 increased (Åström and Lindholm, 1990). In addition, a decrease in SWA was found in these patients (Åström and Jochumsen, 1989). In children with psychosocial dwarfism, the amount of SWS was low. After several weeks in a new environment, during recovery of growth, sleep quality improved, and particularly SWS increased (Guilhaume et al., 1982). The multiple sleep latency test, a method to examine daytime sleepiness, was in the normal range in patients with acquired GH deficiency (Schneider et al., 2005).

Excessive GH levels are found in patients with acromegaly. In these patients, obstructive sleep apnea syndrome is frequent due to hyperplasia of their upper airway soft tissue (Hart et al., 1985). However, in patients with acromegaly without sleep apnea, daytime sleepiness and an abnormal sleep structure are also found. One year after adenectomy, REMS and SWS time increased in a sample of these patients (Åström and Trojaborg, 1992). In this study, EEG power spectrum analysis was used to calculate sleep energy. At baseline, REMS and SWS energy were higher before than after adenectomy.

2.3 Effects of HPS Hormone Administration on the Sleep EEG

2.3.1 Growth Hormone-Releasing Hormone

GHRH is an important endogenous sleep-promoting substance. The GHRH receptor gene is found in the mouse in the region of chromosome 13 linked to SWA (Franken et al., 2001). Intracerebroventricular administration of GHRH increases SWS in rats and rabbits (Ehlers et al., 1986; Obál et al., 1988) (➋ *Figure 25-2*). The same effect is found when GHRH is injected into the medial preoptic area in rats (Zhang et al., 1999) or intravenous to rats (Obál et al., 1996).

Similarly, after repetitive hourly iv bolus injections of GHRH between 22:00 and 01:00 h, SWS and GH secretion increased and cortisol levels were blunted in young normal men (Steiger et al., 1992). Mimicking the pulsatile endogenous release appears to be a crucial methodological issue, because sleep remained unchanged after GHRH infusion in normal controls (Marshall et al., 1999) (➋ *Figure 25-3*). Sleep promotion in young male subjects by GHRH was confirmed after iv (Kerkhofs et al., 1993; Marshall et al., 1999) and intranasal (Perras et al., 1999a) administration. The effects of GHRH on human sleep were investigated in three states with a change of the GHRH/CRH ratio in favor of CRH—(i) the second half of the night in young normal men, (ii) in elderly normal men and women, and (iii) in patients with depression. (i) Repetitive iv bolus injections of GHRH during the early morning hours (each hour from 04:00 to 07:00) prompted no major changes of sleep EEG. GH increased whereas HPA hormones remained unchanged (Schier et al., 1997). (ii) At daytime, the response of GH to GHRH is reduced in older men (Iovino et al., 1989). Similarly only a weak sleep-promoting effect of repetitive iv administration of GHRH (hourly from 22:00 to 01:00) was found in healthy elderly women and men. The first NREMS period was prolonged and the number of awakenings decreased. GH levels increased slightly but significantly whereas cortisol levels remained unchanged (Guldner et al., 1997). Furthermore, intranasal administration of GHRH in elderly subjects had only a relatively weak sleep-promoting effect (Perras et al., 1999b). In a pilot study, the hypothesis was tested that after priming (e.g., iv GHRH every 2 days for 12 days) the sleep-promoting effect of GHRH would be restored in the elderly. The study results in two subjects do not support this hypothesis

Figure 25-2

Hourly duration of NREMS (mean ± SEM) in rabbits (n = 6–8) after intracerebroventricular injection of artificial cerebrospinal fluid (open symbols) or GHRH (closed symbols). Arrows: injection. Asterisks: significant differences in hour 1 (paired t-test). Reprinted from *Sleep Medicine Reviews* (2004), 8, Obal F. et al., GHRH and sleep, 367-377, Copyright 2004, with permission from Elsevier

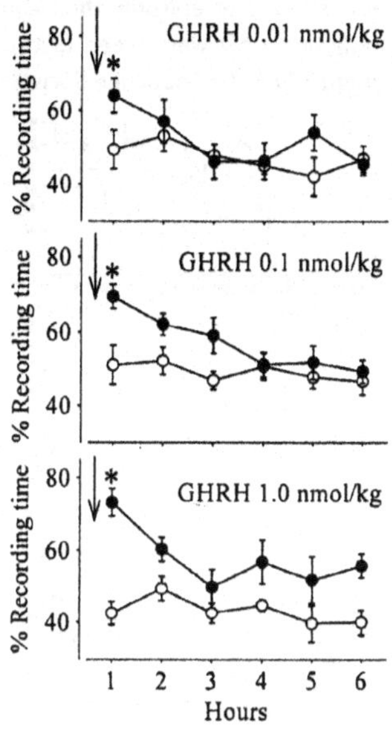

(Murck et al., 1997b). (iii) The influence of pulsatile iv administration of GHRH during the first few hours of the night (22:00 to 01:00) was examined in drugfree patients of both sexes with depression (age range 19–76 years) and in normal controls matched for age and gender. A sexual dimorphism in the response to GHRH was found. In male patients and controls GHRH decreased. ACTH levels during the first half of the night and cortisol levels during the second half of the night decreased. In contrast, these hormones were enhanced in females, regardless of whether they were healthy or depressed. Similarly, NREMS and particularly stage 2 sleep increased and wakefulness decreased in male patients and controls whereas opposite sleep-impairing effects were found in women. These data corroborate a reciprocal antagonism of GHRH and CRH in males (see Section 23.3.4.1), whereas a synergism of GHRH and CRH is suggested in females. The latter issue may contribute to the increased prevalence of mood disorders in women (Antonijevic et al., 2000b, c). In the rat, NREMS decreases and sleep latency increases when GHRH is inhibited by receptor antagonists (Obál et al., 1991). Similarly, NREMS is reduced after intracerebroventricular administration of antibodies to GHRH (Obál et al., 1992b). In humans however, sleep remained unchanged after a GHRH antagonist (Jessup et al., 2004).

Sleep deprivation is a major stimulus for sleep (Borbély et al., 1981; Borbély et al., 1984; Franken et al., 1991). There is good evidence that GHRH mediates this effect. GHRH antibodies antagonized sleep promotion after sleep deprivation in rats (Obál et al., 1992b). Similarly, microinjections of a GHRH antagonist into the area preoptica of rats inhibited the sleep rebound after sleep deprivation (Zhang et al., 1999). Eight hours of sleep deprivation prompted a distinct depletion of hypothalamic GHRH and very low hypothalamic GHRH contents at the end of the experiment (Gardi et al., 1999), whereas hypothalamic GHRH mRNA increased after sleep deprivation in rats (Toppila et al., 1997; Zhang et al., 1999). It is thought that the high rate of release stimulates transcription of GHRH mRNA. The rise in hypothalamic GHRH mRNA concentration is associated with decreases in hypothalamic somatostatin

▪ Figure 25-3

Profiles of Sleep (*top*), cortisol (*middle*), and GH (*bottom*) plasma levels from two representative subjects (upper part, lower part). Left panel, Placebo condition. Middle panel, Episodic GHRH condition (arrows). Right panel, Continuous GHRH condition (hatched bar). Lights were turned off at 2300 h, subjects were awakened at 0700 h. W, Wakefulness; S1,S2,S3,S4, sleep stages 1 to 4; REM, REM sleep. From: Marshall, L., Mölle, M., Böschen, G., Steiger, A., Fehm, H.L., Born, J.: Greater efficacy of episodic than continuous growth hormone releasing hormone (GHRH) administration in promoting slow wave sleep (SWS). *Journal of Clinical Endocrinology & Metabolism*, 81 (1996) 1009-1013. Copyright 1996, The Endocrine Society

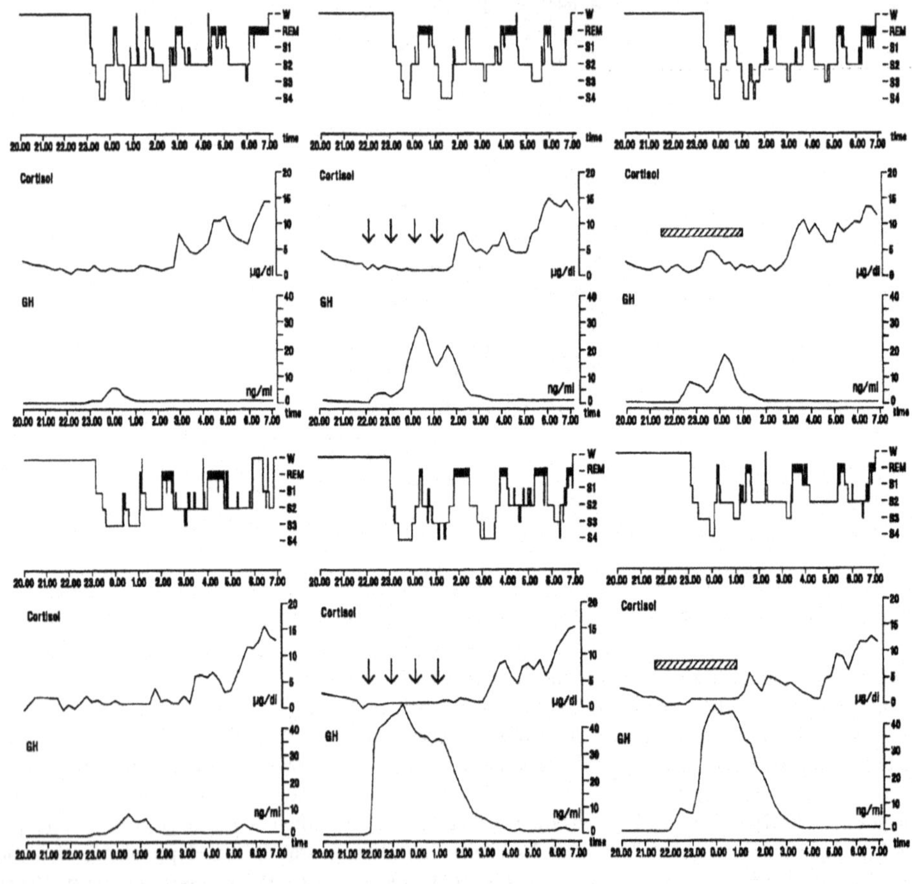

levels (Zhang et al., 1998). GHRH receptor mRNA and GHRH binding declined by 50% in the hypothalamus of rats after 8 h of sleep deprivation. In contrast, pituitary GHRH receptors remained unchanged. Characteristically marked exposure of GHRH receptors to GHRH induces downregulation (Bilezikjian et al., 1986). Therefore, it is thought that a robust intrahypothalamic release explains the downregulation of hypothalamic GHRH receptors. The unchanged pituitary GHRH receptor mRNA and binding suggests that excessive GHRH release into the pituitary vessels does not occur during sleep deprivation (Obál, Jr. and Krueger, 2004).

In mice, viral infections induce excess NREMS. The role of the GHRH receptor in this effect was tested. After influenza A infection, the duration and intensity of NREMS increases in mice with normal phenotype, whereas NREMS and EEG delta power decreases in the GHRH receptor deficient *lit/lit* mice. These findings show that GHRH signaling is involved in the NREMS response to influenza infection (Alt et al., 2003).

2.3.2 Growth Hormone, Insulin-Like Growth Factor-1

Negative feedback inhibition of GHRH after administration of GH in humans (Mendelson et al., 1980), cats (Stern et al., 1975), and rats (Drucker-Colin et al., 1975; Obál, Jr. and Krueger, 2004), or higher dosages of intracerebroventricular IGF-1 (Obál et al., 1999) decrease NREMS. On the other hand, GH antagonism impairs sleep (Obál et al., 1997a). Low dose intracerebroventricular IGF-1 stimulated NREMS in rats (Obál et al., 1998). Sleep remained unchanged after chronic GHRH substitution to patients with acquired GH deficiency (Schneider et al., 2005).

2.3.3 Somatostatin

After, intracerebroventricular somatostatin selective increases of REMS were reported in rats (Danguir, 1986). In contrast, in rats systemic intracerebro-ventricular administration of the somatostatin analogue, octreotide decreased NREMS and GH secretion (Beranek et al., 1999). Similarly, SWS was reduced and intermittent wakefulness was enhanced in young normal men after subcutaneous administration of octreotide (Ziegenbein et al., 2004). Octreotide is known to be long acting and more potent than exogenous somatostatin. This explains why somatostatin impaired sleep in normal elderly female and male controls after repetitive iv administration (Frieboes et al., 1997) (➋ *Figure 25-4*), whereas it had no effect in young normal men after repetitive and single iv injections (Parker et al., 1974a; Steiger et al., 1992; Kupfer et al., 1992). In cats and

Figure 25-4
Sleep pattern in one subject after placebo and SRIF. MT, movement time; REM, rapid eye movement sleep; I-IV, NREMS stages 1-4. From: Frieboes R-M et al.: Somatostatin impairs sleep in elderly human subjects. Neuropsychopharmacology (1997) 16: 339-345. (with permission)

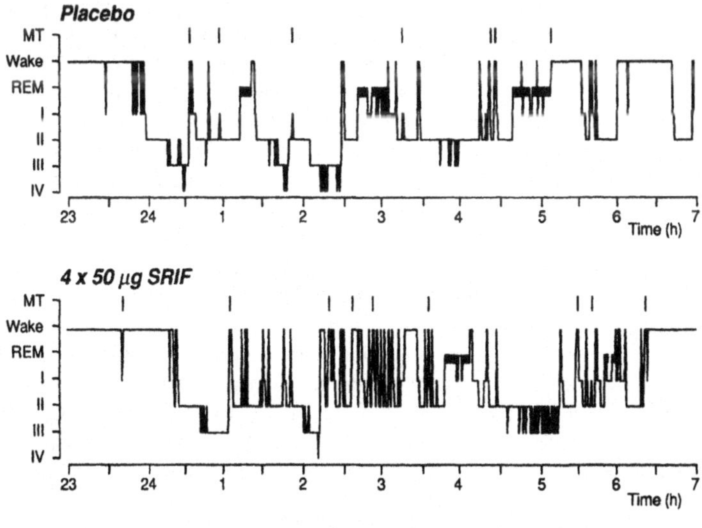

rats, somatostatin inhibits GABAergic transmission in the sensory thalamus via presynaptic receptors (Leresche et al., 2000). It was suggested that this mechanism contributes to the decrease of NREMS after somatostatin. All these data point to a reciprocal interaction of GHRH and somatostatin in sleep regulation similar to their relationship in the regulation of GH release. The same dose of somatostatin, which was not effective in young men impaired sleep in the elderly probably due to a decline of endogenous GHRH during ageing. The finding that infusion of the somatostatin antagonist arginine enhanced SWS in normal elderly male subjects supports this theory. It is thought that in this experiment the action of endogenous GHRH was facilitated (Steiger et al., 2002).

2.3.4 Ghrelin and GH Secretagogues

Similar to the effects of GHRH, repetitive iv administration of ghrelin (22:00 to 0100) enhanced SWS and GH in young normal men (Weikel et al., 2003). In contrast to the effects of GHRH, which blunted cortisol in young men (Steiger et al., 1992), ACTH and cortisol increased, particularly during the first half of the night after ghrelin (Weikel et al., 2003). After the first injection of ghrelin, the response of GH was most distinct and the increase of cortisol was relatively low. In contrast, the fourth injection prompted the highest response of cortisol and the lowest of GH (➋ *Figure 25-5*). The pattern of hormone changes after ghrelin resembles the effects of repetitive iv administration of the synthetic GHSs GH-releasing peptide-6 (GHRP-6) (Frieboes et al., 1995) and hexarelin (Frieboes et al., 2004). These findings suggest that ghrelin acts as an interface between the HPA and the HPS systems. The sleep-EEG effects of GHRP-6 and hexarelin however differed from the changes after ghrelin. After GHRP-6 sleep stage 2 increased (Frieboes et al., 1995), whereas after hexarelin SWS and SWA decreased, probably due to a change of the GHRH/CRH ratio in favor of CRH (Steiger, 2002). In mice, ghrelin promoted NREMS (Obál et al., 2003). An intact GHRH receptor was shown to be the prerequisite for this effect, as in mice with nonfunctional GHRH receptors sleep remained unchanged after ghrelin. Oral administration of the GHS MK-677 for 1 week had a distinct sleep-promoting effect in young men and only a weak effect in elderly controls (Copinschi et al., 1997).

Bodosi et al. (2004) investigated the relationships among plasma ghrelin and leptin concentrations and hypothalamic ghrelin contents and sleep and feeding. In order to do this, rats were examined in three conditions: (i) free-feeding rats with normal diurnal rhythms, (ii) feeding restricted to the 12-h light period, and (iii) 5 h of sleep deprivation at the beginning of the light period. Plasma ghrelin and leptin concentrations displayed diurnal rhythms. (i) The ghrelin peak preceeded and the leptin peak followed, the major daily feeding peak in the first hour after dark onset. (ii) The rats showed vigorous eating in the first hour of the light period after food restriction. The diurnal rhythm of ghrelin and leptin reversed in such a way that they maintained their relationship with respect to one another and to feeding activity. The ghrelin peak continued to preceed the feeding peak. Consequently, the ghrelin maximum was found towards the end of the dark period. The leptin peak followed the major feeding activity. The nocturnal ghrelin peak during restrictive feeding was almost double of that of the diurnal ghrelin peak in rats on normal feeding conditions. REMS decreased throughout the light period and increased during the dark period. (iii) Sleep deprivation did not change leptin concentrations, but it stimulated plasma ghrelin concentrations and induced eating. Hypothalamic ghrelin was highly responsive to sleep deprivation, displaying biphasic variations. Ghrelin contents of the hypothalamus increased during sleep deprivation and returned to baseline after sleep deprivation. These results suggest a strong relationship between feeding and the diurnal rhythm of leptin and a major influence of feeding on the diurnal rhythm of ghrelin. The variations in the hypothalamic ghrelin contents point to an association with the sleep–wake activity in rats.

Furthermore, studies in humans contribute to the understanding of the relationship between sleep–wake-behavior and ghrelin levels. Dzaja et al. (2004) compared interactions of sleep and ghrelin levels in young normal men who were semirecumbent during 24 h. When they were allowed to sleep, they showed a sharp increase of ghrelin levels by sleep onset, which was followed by a decline throughout the night. This nocturnal rise of ghrelin was blunted when the subjects were sleep-deprived. Another study suggests gender differences in the nocturnal secretion of ghrelin. Ghrelin levels were determined between 20:00 and 07:00 in normal female and male control subjects who were active during daytime. Ghrelin concentrations differed during the interval between 20:00 and 23:00, when subjects were allowed to sleep. In males, there was a continuous rise of ghrelin concentrations between 20:00 and 23:00. In females, however, ghrelin levels were at 20:00 already in the same range as during the sleeping period. No relationships between ghrelin and sleep stages or the nocturnal secretion of GH, cortisol, and ACTH was found (Schüssler et al., 2005b). In a large sample of more than 1,000 subjects, the interaction between sleep and ghrelin morning levels was investigated. Short sleep time was associated with higher ghrelin levels

Figure 25-5

Time course of nocturnal plasma hormone concentrations. Nocturnal secretions of growth hormone (GH), adrenocorticotropic hormone (ACTH), cortisol, and prolactin (means ± SE) after iv injection of 4 × 50 µg of ghrelin compared with placebo (n = 7). Arrows, times of injections. From: Weikel, J.C., Wichniak, A., Ising, M., Brunner, H., Friess, E., Held, K., Mathias, S., Schmid, D.A., Uhr, M., Steiger, A.: Ghrelin promotes slow-wave sleep in humans. *American Journal of Physiology, Endocrinology & Metabolism* 284 (2003) E407-E415. Copyright The American Physiological Society

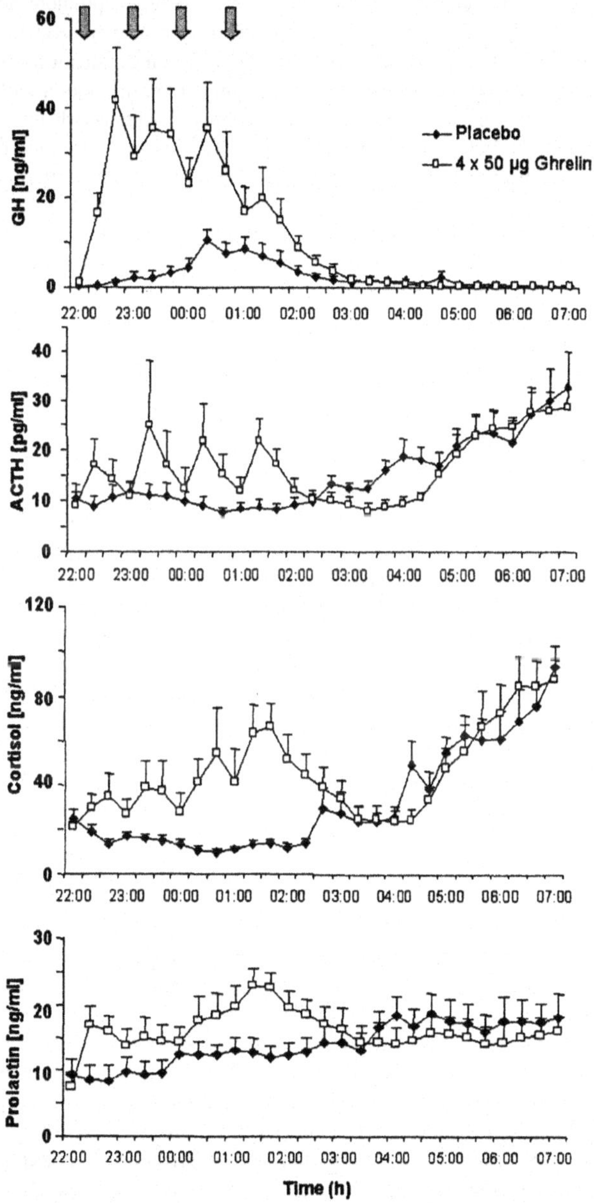

(Taheri et al., 2004). Similarly, the comparison between extended (12 h) and restricted (4 h) sleep time during 2 days each showed at daytime increases in ghrelin levels, self-rated hunger and appetite, and a decrease in leptin after sleep restriction (Spiegel et al., 2004).

2.4 Animal Models of HPS System Changes

In the so-called supermice, the giant transgenic mice, the metalothionine-1 promoter stimulates expression of rat GH (Mt-rGH mice). Therefore, plasma GH is permanently elevated and secretion pulses are absent. Behavioral observation showed that these mice spend more time sleeping than normal mice (Lachmansingh and Rollo, 1994). Sleep recording showed that during the light period, the amount of NREMS is modestly higher and REMS is almost doubled in these mice compared to normal mice, whereas sleep does not differ between groups at night. Also after sleep deprivation, the MT-rGH mice sleep more than normal mice (Hajdu et al., 2002). In dwarf rats with deficits in the central GHRHergic transmission and reduced hypothalamic GHRH, NREMS is reduced compared to control rats (Obál et al., 2001). In dwarf homozygous (*lit/lit*) mice with nonfunctional GHRH receptors, the amounts of NREMS and REMS are lower than in normal control mice. In the dwarf mice, infusion of GH by Alzet minipumps leads to normalization of REMS, but not of NREMS within 9 days. Similar to ghrelin, GHRH, and the somatostatin analogue octreotide exert no effect on sleep EEG in dwarf mice. These results suggest that (i) GHRH deficiency is associated with decreases in NREMS, (ii) decreases in GH lead to decreases in REMS, (iii) the actions of GHRH, ghrelin, and octreotide on sleep EEG require intact GHRH receptor signaling (Obál, Jr. and Krueger, 2004).

2.5 Interactions of HPS System and Orexin

The orexins (OX-A and OX-B) are derived from their common precursor prepro-orexin. Orexins participate in the sleep regulation (see the chapter by Porkka-Heiskanen and Stenberg in this volume). Recent findings suggest an interaction between OX-A, GHRH, and somatostatin in the regulation of sleep, food intake, and GH release. In patients with narcolepsy, who show orexin deficiency, changes of the circadian pattern of GH secretion were reported, pointing to a disruption of GHRH release (Overeem et al., 2003). Lopez et al. (2004) studied the interaction of these peptides in rats. In situ hybridization showed a decrease of GHRH mRNA levels in the paraventricular nucleus of the hypothalamus after OX-A treatment without changes in the arcuate nucleus. The somatostatin mRNA content in the hypothalamus increases GH-dependently after OX-A. In animal models of GH deficiency, hypophysectomized rats and dwarf Lewis rats GHRH mRNA levels in the paraventricular nucleus of the hypothalamus are reduced.

2.6 Summary

All components of the HPA system participate in sleep regulation. GH is widely sleep dependent. GH administration promotes REMS. The influence of GHRH, ghrelin, and somatostatin on sleep EEG corresponds to their role in the regulation of GH release, at least in male subjects. In the rat, hypothalamic GHRH mRNA and hypothalamic GHRH and ghrelin contents display sleep-related changes. GHRH mRNA peaks at the onset of the light period. GHRH promotes NREMS in several species including humans, in male gender. In women, however, GHRH impaired sleep as NREMS decreases and wakefulness increases. Studies on GHRH effects on sleep in female laboratory animals are lacking. Furthermore, GHRH participates in sleep promotion during recovery sleep after sleep deprivation. The sleep-promoting effects of GHRH declines during ageing. In contrast, the impairing effect of somatostatin increases during ageing. These changes appear to contribute to the age-related decline of SWS and GH. Similar to GHRH, ghrelin enhances NREMS in young normal men and in mice. The effects of GHRH and ghrelin on HPA hormones in males are opposite. After GHRH, ACTH and cortisol decrease in males, whereas these hormones increase in females. In women, GHRH enhances ACTH and cortisol levels. Intact GHRH receptors appear to be the prerequisite of the effects of ghrelin and of the somatostatin analogue octreotide on sleep. Decreased GHRH acitivity may explain that SWS is lower in dwarfs than in normal controls. Similarly, feedback inhibition of endogenous GHRH may contribute to the decrease of SWS in patients with acromegaly.

3 Hypothalamo-Pituitary-Adrenocortical System

3.1 Basic Activity

The HPA system mediates the reaction to acute physical and psychological stress. This stress reaction is a prerequisite for survival of the individual. It starts with the release of CRH from the parvocellular neurons of the paraventricular nucleus of the hypothalamus. This results in the secretion of ACTH from the anterior pituitary and finally, in the secretion of cortisol (in humans) or corticosterone (in rats) from the adrenocortex. Various cofactors contribute to this cascade (reviewed in: Holsboer, 1999).

In rats, CRH gene transcription levels increase during the dark period, when the animals are active and decrease in the morning and throughout the light period (Watts et al., 2004). In humans during sleep, both the nadir and the major portion of the secretion of ACTH and cortisol occur. The first few hours of the night contain their quiescent period. Between 02:00 and 03:00 h, the first pulse of cortisol occurs. It is followed by further pulses until awakening (Weitzman, 1976; Born and Fehm, 1998). Whereas ACTH is the prime regulator of nocturnal cortisol secretion in humans, the secretion of ACTH and cortisol may dissociate. A complete lack or a minor ACTH burst before the cortisol morning peak was found in about 40% of normal controls (Fehm et al., 1984). Similarly, the pulse concomitants for nocturnal ACTH with cortisol were 47% and for cortisol with ACTH pulses were 60% (Krishnan et al., 1990). During the quiescent period, the major amounts of SWS and the GH peak both occur.

One approach to examining these interactions in more detail is to analyse the relationships in the course of the NREMS–REMS cycle and of cortisol secretion during normal sleep. An association between a decrease in cortisol levels and REMS periods was observed, particularly during the first four sleep cycles. For the fifth sleep cycle, however, increasing cortisol levels during REMS were reported (Fehm et al., 1993). A novel statistical method, event history analysis, was used to assess the effects of cortisol on the transition between sleep stages in normal human control subjects. The aim of this analysis is to assess the instantaneous probabilities of transitions between sleep stages, provided they are influenced by both various time-dependent factors (e.g., hormone secretion) and the history of the process. High cortisol levels facilitated the transition intensity of (i) waking to sleeping around 2 h after sleep onset, (ii) NREMS to REMS around 6 h later, (iii) sleep stage 1 or 2 to SWS around 2, 4 or 6 h later, and (iv) SWS to sleep stage 1 or 2 about 2 h later. Furthermore, high cortisol concentrations at the beginning of REMS periods favored the change to NREMS, whereas later the influence of cortisol on a change became weaker (Yassouridis et al., 1999).

The relationships between cortisol, EEG spectral activity, and by two measures of heart rate variability, the activity of the autonomic nervous system were examined in normal controls (Gronfier et al., 1999). EEG delta power (SWA) was the greatest during the quiescent period of the HPA system. During the period of pulsatile release, the cortisol secretory variations were concomitant with, or anticipated opposite variations 10–20 min later in SWA. Sleep deepening was associated with decreasing cortisol concentrations and low or decreasing sympathetic tone. In contrast, a lightening of sleep was accompanied by increases in cortisol and sympathetic tone. REMS was associated with a decrease in cortisol secretory rates preceding REMS onset and high autonomic activity. An association between the duration of spontaneous sleep and HPA hormone levels was found in normal male subjects, ACTH and cortisol were higher in subjects with short sleep time when compared to subjects with long sleep time (Späth-Schwalbe et al., 1992).

The genetic factors that contribute to the pattern of cortisol secretion were studied in monozygotic and dizygotic pairs of male, normal twins. Genetic control was found for the timing of the cortisol nadir and for the proportion of overall temporal variability associated with pulsatility. In contrast, for the 24-h mean and the timing of the morning acrophase, environmental effects were identified (Linkowski et al., 1993).

Controversial reports exist on the effect of age on HPA hormones. Elevated and unchanged cortisol levels have both been reported in the elderly. The study including the largest sample of normal human adults over a lifetime reported age-dependent increases of mean cortisol levels and of the nadir and selectively in women, of the acrophase. The rhythmicity of cortisol was preserved in the elderly, but the amplitude was dampened and the morning rise was advanced (Van Cauter et al., 1996).

3.2 Sleep in Disorders with Pathological Changes of HPA Activity

In Addison's disease, the capacity of the adrenal glands to produce corticosteroids is severely reduced. Only a few case reports exist on sleep EEG in these patients. No major disturbances of their sleep were found (Gillin et al., 1974; Krieger and Glick, 1974). Addison's patients were compared intraindividually under two conditions, either continuous hydrocortisone replacement or short-term hydrocortisone withdrawal. After hydrocortisone replacement, REMS latency was shortened, and REMS time and intermittent wake time were increased in comparison to withdrawal. These findings suggest that cortisol may be needed to facilitate the initiation and maintenance of REMS (Garcia-Borreguero et al., 2000). In contrast to Addison's disease, hypercortisolism and disturbed sleep are frequent symptoms in Cushing's disease and in depression. It is very likely that common factors underlie the pathophysiology of these changes. Excessive cortisol levels are produced in Cushing's disease, either of central, or peripheral origin. In these patients, SWS is decreased (Krieger and Glick, 1974; Shipley et al., 1992). In one study, in addition, disturbances of sleep continuity (increased sleep latency, enhanced waketime) and REMS (shortened REMS latency, elevated REM density, a measure for the amount of rapid-eye-movement during REMS) were reported (Shipley et al., 1992).

Similar symptoms are frequently observed in depression, whereas the dysregulation of the HPA system is more subtle in affective disorders. Characteristic sleep-EEG findings in depressed patients are disturbed sleep continuity (prolonged sleep latency, increased number of awakenings, early morning awakening), a decrease of NREMS (decreases of stage 2 sleep and SWS, in younger patients a shift of the major portion of SWS from the first to the second sleep cycle), and REMS disinhibition (shortened REMS latency, prolonged first REMS period, elevated REM density) (reviewed in: Ehlers and Kupfer, 1987); (Steiger et al., 1989). Well-documented endocrine changes include signs of (i) HPA overactivity (reviewed in: Holsboer, 1999) and (ii) HPS dysfunction (reviewed in: (Steiger et al., 1989). Most sleep-endocrine studies in depressed patients report elevated cortisol and ACTH (Linkowski et al., 1987; Steiger et al., 1989; Antonijevic et al., 2000c) throughout the night or throughout day and night (24 h) (Linkowski et al., 1987), respectively, in comparison to normal controls, whereas the circadian pattern of cortisol is preserved. This finding is in contrast to its disappearance due to excessive hypercortisolism in Cushing's disease (Weitzman, 1976). A positive correlation between age and cortisol concentrations was found particularly in female patients with depression (Antonijevic et al., 2000c). Elevated cortisol plasma and norepinephrine cerebrospinal fluid (CSF) levels throughout 30 h, but normal ACTH plasma and CRH CSF levels, were reported in a sample of depressed patients (Wong et al., 2000). GH was blunted in most (Steiger et al., 1989; Jarrett et al., 1990; Voderholzer et al., 1993), but not in all (Linkowski et al., 1987) studies. These findings suggest a causal relationship between shallow sleep, low GH, and HPA hyperactivity in depression. Furthermore, there are similarities in the sleep-endocrine changes during depression and during normal ageing (➋ *Figure 25-1*).

Three studies compared sleep-endocrine activity longitudinally between acute depression and recovery. One study showed a decrease of ACTH and cortisol during 24 h and a normalisation of REMS in adult patients after recovery (Linkowski et al., 1987). As some of the patients received tricyclic antidepressants, which are known to suppress REMS (Steiger, 2002) at the examination during recovery it is difficult to differentiate between the effects of remission and of drugs. Intraindividual comparison of adult patients who were drugfree at least 14 days before each examination (Steiger et al., 1989) confirmed a decrease of cortisol levels after recovery. The pathological sleep EEG and low GH levels, however, remained unchanged. Stage 4 sleep even decreased (Steiger et al., 1989). Both studies corroborate that HPA hypersecretion is a state marker of depression in adult patients. In contrast, sleep EEG and 24 h cortisol profiles of prepubertal depressed children did not differ from children with nonaffective psychiatric disorders and normal controls.

Cortisol hypersecretion was found only in 4 of the 45 young patients with depression. These were restudied after recovery. Only one of them showed persistence of hypersecretion (Puig-Antich et al., 1989). These findings show that abnormalities of sleep EEG and of cortisol secretion are infrequent in prepubertal children with depression. This is in line with the view that ageing and depression exert synergistic effects on sleep-endocrine activity (Antonijevic et al., 2000c). The decrease of cortisol after recovery resembles the normalization of pathological results of challenge tests of the HPA system and of CRH CSF levels after remission (reviewed in: Holsboer, 1999). The persistence of most sleep-EEG (Kupfer et al., 1993) and GH changes (Jarrett et al., 1990) after recovery has been confirmed over a period of 3 years. Obviously, cortisol levels normalize independently from the sleep architecture. Hence, hypercortisolism in depression is not secondary to shallow sleep. Interestingly, patients with primary insomnia had increased nocturnal cortisol and a shorter quiescent period than the controls (Vgontzas et al., 2001; Rodenbeck et al., 2002).

Young individuals with insomnia had higher 24-h ACTH and cortisol secretion than controls. Patients with a high degree of objective sleep disturbance (shortened sleep time) secreted more cortisol than those with a low degree. In another sample, however, nocturnal cortisol levels did not differ between patients with insomnia and controls, whereas their melatonin levels were blunted (Riemann et al., 2002). Salivary cortisol after awakening is decreased in patients with primary insomnia (Backhaus et al., 2004). The authors discussed that this decrease may be due to nocturnal cortisol activation after an increased number of nocturnal awakenings.

It appears likely that the metabolic disturbances during acute depression result in a biological scar as reflected by the persisting changes of sleep EEG and of GH levels in remitted patients. This hypothesis is supported by the results of a study of male patients who survived severe brain injury (Frieboes et al., 1999). Several months later, their cortisol levels did not differ from normal controls, whereas their GH levels and sleep stage 2 time were lower. Obviously, there are similarities between the sleep-endocrine pattern of patients after injury and of remitted depressed patients. Whereas cortisol levels were normal at the time of the examination in this study, it appears likely that either HPA overactivity due to stress under the intensive care situation after brain injury or treatment with glucocorticoids in some patients contributes to the changes of sleep EEG and of GH levels.

Only two studies deal with sleep-endocrine activity in psychiatric disorders other than depression. Hypercortisolism was also found in patients with mania (Linkowski et al., 1994). Enhanced cortisol levels during the first few hours of the night were observed in patients with schizophrenia. In these patients, sleep latency was prolonged and REMS was reduced (Van Cauter et al., 1991b).

In obstructive sleep apnea syndrome, during an episode of apnea upper airway constriction, progressive hypoxemia due to asphyxia, autonomic, and sleep-EEG arousal occur. Buckley and Schatzberg (2005) hypothesized that obstructive sleep apnea causes activation of the HPA system through autonomic activation, awakening, and arousal. This HPA activation may be a risk factor in the development of the metabolic syndrome in untreated obstructive sleep apnea. Furthermore, the authors proposed that HPA hyperactivity may contribute to the pathophysiology of obstructive sleep apnea in hypertension.

3.3 Effects of Changes of Sleep–Wake Behavior on HPA Hormones

Weitzman and colleagues (Weitzman, 1976) did pioneering work as they used nonpharmacological manipulations as sleep deprivation and nocturnal awakenings in studying the interaction between sleep EEG and hormones. Their work led to the still valid conclusion that the pattern of cortisol secretion is widely dependent on a circadian rhythm, whereas manipulation of the sleep–wake pattern prompts subtle changes in HPA secretion. A similar approach is provided by studies on the effects of environmental changes.

In the first investigation by Weitzman and colleagues, normal subjects underwent a repetitive 3 h sleep–wake cycle for 10 consecutive days. Each cycle consisted of 2 h wakefulness and 1 h sleep. Finally, cortisol levels were consistently lower during the interval with sleep than during the period awake in each cycle, independently of circadian time (Weitzman, 1976). In a 4-day protocol, a baseline investigation was followed by sleep deprivation. On day 3, the subject was allowed to sleep 12 h later than usual. On day 4, he slept again at his usual sleeping time. During the first 4 h after sleep onset cortisol levels were reduced, even when sleep took place during the interval when the subject was usually awake (Weitzman et al., 1983). During free running conditions with an average 25 h period, a phase advance of the circadian cortisol rhythm was observed in controls. Cortisol was secreted in large amounts prior to sleep onset. Immediately after sleep onset, cortisol levels decreased sharply for the next 2–3 h (Weitzman et al., 1981).

Various studies investigated HPA activity at several intervals during and after partial and total sleep deprivation. During the night of sleep deprivation, either enhanced or unchanged cortisol concentrations were reported. In the recovery night, after one night of sleep deprivation, cortisol levels were unchanged in young and elderly normal controls compared to baseline conditions (Murck et al., 1999). In one study, patients with depression were investigated during three consecutive nights, before, during, and after sleep deprivation. During the night of sleep deprivation, cortisol levels increased and returned to baseline values during the recovery night (Voderholzer et al., 2004). In the evening of the day after partial or total sleep deprivation, cortisol was elevated in normal controls (Leproult et al., 1997). Similarly, evening cortisol was higher when sleep in normal young men was restricted to 4 h per night for 6 days (Spiegel et al., 1999). In the recovery night, after four nights with restricted sleep, cortisol was blunted during the second half of the night (Follenius et al., 1992). A delayed sleep onset in normal controls prompted a later occurrence of the cortisol rise (Fehm et al., 1993). In rats, 72 h of sleep deprivation resulted in increases in CRH levels in the striatum, limbic areas, and pituitary, whereas hypothalamic CRH was reduced. Significant decreases in CRH binding were reported in the striatum and in the pituitary (Fadda and Fratta, 1997). In vivo microdialysis in rats showed a marked rise in free corticosterone levels in the brain during sleep deprivation (Penalva et al., 2000). Rats were sleep-deprived during intervals up to 92 h. ACTH and corticosterone plasma levels increased from 24 h of sleep deprivation and decreased during the recovery period. After 24 h of recovery, ACTH dropped below baseline levels remaining low until 96 h of recovery (Andersen et al., 2005). All these data point to an enhanced HPA activity during awakenings and sleep deprivation and to a relative decrease of HPA activity during sleep.

The HPA acrophase appears to be linked with the termination of sleep at the morning. The expectation of waking up at a certain time induces a marked increase in ACTH before the end of sleep (Born et al., 1999). This was shown in normal subjects. Lights were switched off at midnight after they had been told that they would be awakened either at 06:00 (once) or at 09:00 (twice). During one of the latter nights, they were allowed to sleep until 09:00. To their surprise the other night, they were awakened at 06:00. When anticipating being awakened at 06:00, distinct increases in ACTH levels before waking were found compared to the condition of unexpected awakening. Arousal from sleep in each of the three conditions prompted increases in ACTH and cortisol concentrations. This change points to an adaptive response to the stress of waking. The increase of ACTH prior to the expected time of waking suggests that anticipation exists also during sleep. This anticipatory increase in ACTH may facilitate spontaneous waking.

The effects of an acute shift of the sleep–wake cycle on cortisol secretion were studied in normal controls. Cortisol secretion was compared during nocturnal and during daytime sleep intraindividually. The 24 h cortisol secretion did not differ between both conditions. The 24 h hormone profiles of the first day were compared to those of the second day of the sleep shift. On this day, the quiescent period was delayed. This finding suggests that an abrupt sleep shift after 2 days can change the circadian pattern of cortisol (Goichot et al., 1998). Cortisol secretion was shown to be relatively stable to environmental changes. After a flight from Europe to the USA, it took 2 weeks for the cortisol pattern of normal control subjects to be totally adapted to the new sleep schedule (Desir et al., 1981). Dissociation of sleep and cortisol rhythm during the first few days after long distance travel are thought to contribute to the symptoms of jet lag.

The interaction between cortisol pulses and SWA was studied in two groups of normal subjects, one group with night sleep (23:00–07:00) and the other with day sleep delayed by 8 h (07:00–15:00). The study showed that SWA may occur in the absence of cortisol pulses, as during the first 3 h of the night. On the other hand, cortisol pulses may occur without any concomitant variations of SWA during wakefulness. When SWA and cortisol are present simultaneously, they are negatively correlated. Cortisol changes precede variations in SWA by about 10 min. This findings show that cortisol and SWA can oscillate independently from each other, but when they occur at the same time, they are oscillating in phase opposition (Gronfier et al., 1998).

In shift workers, even a long-lasting resistance of the cortisol rhythm to adapting totally to an inverted sleep–wake schedule was reported. Young male night workers who had been on night-shift regularly for at least 2 years were compared to matched day-active male controls. In each group, sleep EEG and hormone secretion were investigated during the usual sleep time (07:00–15:00 in the night workers or 23:00–07:00 in the controls respectively). Furthermore, hormone secretion was examined during the usual work time. Whereas sleep EEG did not differ distinctly between groups, cortisol levels were enhanced in the night workers during the usual sleep time. Conversely, during usual work hours, cortisol was blunted in this sample. Obviously, the quiescent period of cortisol persists during the daytime in night workers (Weibel and Brandenberger, 1998).

Light is the major environmental factor which influences the timing of neuroendocrine rhythms via the suprachiasmatic nucleus. This is corroborated by experiments in which 24 h hormone profiles were determined in normal control subjects on two separate occasions, once after they were chronically exposed to simulated short (8 h) "summer nights" and once after they were chronically exposed to simulated long (14 h) "winter nights". During the "winter nights" the duration of the period of rising cortisol levels was longer than during the "summer nights" (Wehr, 1998). It is likely that the physiological responses to seasonal changes in the duration of days and nights are suppressed in people living in modern civilization, in which they are exposed to artificial light after dark or to artificial darkness during the daytime. Bright light was shown to be a stimulus for cortisol in humans. Normal control subjects were kept on bed-rest with enforced wakefulness in dim light. Early morning (05:00–08:00), but not afternoon (13:00–16:00) exposure to bright light resulted in increased cortisol levels (Leproult et al., 2001).

3.4 Effects of HPA Hormone Administration on the Sleep EEG

3.4.1 Corticotropin-Releasing Hormone

Various preclinical and human studies show that the administration of HPA hormones or their antagonists affects sleep. After intracerebroventricular administration of CRH, SWS decreases in rats (Ehlers et al., 1986) and rabbits (Opp et al., 1989). Even following 72 h of sleep deprivation CRH reduces SWS in rats. Furthermore, sleep latency is prolonged and REMS increases (Marrosu et al., 1990). Repetitive hourly iv injections of 4 × 50 μg CRH (22:00–01:00) in young normal men has a similar effect on NREMS as SWS and REMS decrease. Furthermore, the GH surge is blunted, and cortisol levels increase during the first half of the night (Holsboer et al., 1988; Tsuchiyama et al., 1995). Similarly, after an iv dose of 100 μg CRH, SWS decreased and sleep stages 1 and 2 increased in young normal male subjects (Tsuchiyama et al., 1995). As mentioned before, the time interval, the dosage, the age, and the protocol of the administration of peptides appear to be crucial methodological issues. For example, after continuous nocturnal infusion of CRH, the sleep EEG remains unchanged (Fehm et al., 1993). Moreover, hourly iv injections of 10 μg CRH (08:00–18:00) fail to induce sleep-EEG changes during the following night (Kellner et al., 1997), whereas melatonin levels decrease. This finding supports the reciprocal interaction between HPA activity and melatonin secretion. After a single iv bolus of CRH in young healthy men EEG activity in the sigma frequency range increases throughout the first three sleep cycles, both after administration during the first SWS period and during wakefulness (Antonijevic et al., 1999b). This finding is in line with the observations (i) that EEG sigma activity increases throughout the night arbitrarily in parallel to the increase in HPA activity and (ii) that the activities of the HPA system and of the EEG sigma range are greater in young women than in young men (Antonijevic et al., 1999a). A dose of CRH, which was not effective in young normal men, impaired sleep in middle-aged normal men (Vgontzas et al., 2001). This observation suggests that the vulnerability of sleep to CRH increases during ageing, similar to the finding mentioned before, that the same dose of somatostatin impairs sleep in the elderly (Frieboes et al., 1997), but not in young subjects (Steiger et al., 1992).

Two studies investigated sleep-EEG changes after CRH antagonists in rats and reported conflicting results. After administration of two different CRH antagonists, α-helical CRH and astressin, before the dark period, wakefulness is reduced in a dose-related manner, whereas the time courses for these effects differ between drugs (Chang and Opp, 1998). α-helical CRH is effective during the first 2 h after administration. Astressin however is effective during hours 7 to 12 after injection. Administration of the substances before the light period does not prompt any effect. These findings support the view that CRH contributes to the regulation of physiological waking periods. In contrast, in another study (Gonzalez and Valatx, 1997) an effect of α-helical CRH was found only in stressed animals. In these rats REMS was enhanced and decreased to values of the nonstressed condition after the substance was injected. In sleep-deprived rats, α-helical CRH diminished the REMS rebound but not the SWS rebound during recovery sleep. The authors suggested that stress acting via CRH is the major factor inducing the REMS rebound after sleep deprivation (Gonzalez and Valatx, 1998).

In rats, kindling of the amygdala at light onset decreases SWS and REMS and enhances corticosterone plasma levels. Intracerebroventricular administration of the CRH antagonists astressin or alpha-helical CRH antagonized the decrease of SWS after kindling in a dose-dependent manner and blocked the increase of corticosterone (Yi et al., 2004). Central increases of CRH are likely to mediate the changes of SWS and corticosterone after kindling.

Some of this preclinical work (Marrosu et al., 1990; Gonzalez and Valatx, 1997; Gonzalez and Valatx, 1998) supports the hypothesis that CRH promotes REMS. From the study in normal humans, the influence of endogenous CRH on REMS is uncertain, however, as CRH suppresses REMS (Holsboer et al., 1988). Furthermore, in normal male controls alpha-helical CRH prompt CRH-agonistic and CRH-antagonistic effects as well on sleep-endocrine activity (Held et al., 2005). Studies on the sleep-EEG effects of ACTH and cortisol help to differentiate the central and peripherally mediated sleep-EEG changes after CRH in humans (see below). The role of CRH in normal and physiological sleep regulation is further elucidated by a study on the effects of CRH receptor-1-antagonism in patients with depression (Held et al., 2004). After a 4 week trial with the substance R121919, the characteristic sleep-EEG changes in patients with depression were counteracted, as the number of awakenings and REM density decreased and SWS increased (*Figure 25-6*).

Figure 25-6

REM-density analysis of variance (1/min) of the three examinations (baseline, first week, end of treatment) compared for panel A (n = 5) and panel B (n = 5). Panel B showed a significant reduced REM-density at the end of the treatment period ((2)-(0), time × panel*). The significance value was *p < 0.05. Reprinted from Journal of Psychiatric Research, 38, Held K. et al., Treatment with the CRH-receptor-antagonist R121919 improves sleep-EEG in patients with depression, 129-136, Copyright 2004, with permission from Elsevier

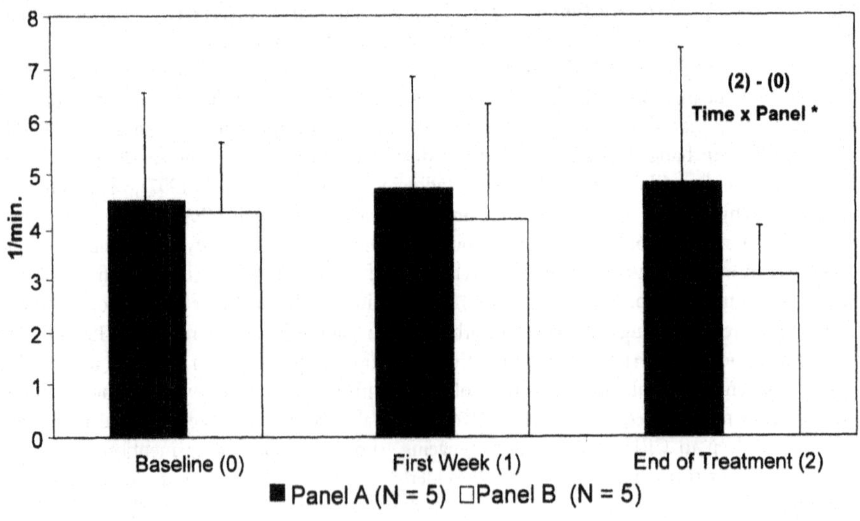

These results support the view that (i) CRH is involved in the pathophysiology of sleep-EEG changes during depression including REMS disinhibition and that (ii) CRH-1-receptor antagonism helps to treat symptoms of depression including impaired sleep.

3.4.2 Vasopressin

The neuropeptide vasopressin is the major cofactor with CRH in the activation of the stress reaction (Holsboer, 1999). Intracerebroventricular administration of vasopressin increases wakefulness in rats (Arnauld et al., 1989). After acute infusion of the peptide to humans stage 2 sleep increased and REMS decreased (reviewed in: Perras et al., 1999b). Studies with pulsatile administration of this peptide are lacking. Surprisingly, chronic intranasal vasopressin for 3 months improves sleep in normal elderly subjects (Perras et al., 1999b). In detail, total sleep time, SWS and, in the second half of the night, REMS increased. The authors suggested that this treatment may compensate for an age-related decrease in vasopressin content of the suprachiasmatic nucleus or that vasopressin could act by stimulating the expression of central corticosteroid receptors.

Finally, the responsiveness of ACTH and cortisol to their secretagogues during sleep and during wakefulness were compared in normal male control subjects. During sleep, particularly during SWS in the first three hours of the night, the HPA stimulating effect of CRH and vasopressin was blunted (Born and Fehm, 1998).

3.4.3 Adrenocorticotropic Hormone

Nocturnal infusions of ACTH suppress REMS in normal controls (Gillin et al., 1974; Fehm et al., 1993), while cortisol and GH increases (Born et al., 1989). Ebiratide is a synthetic ACTH (4–9) analogue. It shares several behavioral effects of ACTH but does not influence peripheral hormone secretion. Accordingly, after repetitive iv administration of ebiratide, GH and cortisol levels remain unchanged in young male controls. A set of sleep-EEG changes is found after ebiratide corresponding to a general CNS activation. Sleep onset increases and during the first third of the night, awake time is elevated and SWS decreases, whereas REMS remains unchanged (Steiger et al., 1991). These findings support the view that the blood–brain interface is no obstacle for CNS effects of iv administered neuropeptides, because ebiratide induces sleep-EEG changes in the absence of effects on peripheral hormone secretion.

3.4.4 Cortisol, Synthetic Glucocorticoid and Mineralocorticoid Receptor Ligands

Since the pioneering work of Gillin, it has been known that certain synthetic and endogenous steroids affect sleep (Friess et al., 1995b). Also continuous nocturnal (23:00–07:00) infusion of cortisol (Born et al., 1991) and pulsatile iv administration (hourly from 17:00–07:00) increases SWS (Friess et al., 1994) and SWA (Friess et al., 2004) and decreases REMS in young normal controls. In a second study, GH levels increased after cortisol. Similarly, SWS, SWA, and GH increase and REMS decreases in analog protocols with pulsatile iv administration of cortisol in normal elderly male subjects (Bohlhalter et al., 1997) and in patients with depression (Schmid et al., 2000). As CRH (Holsboer et al., 1988) and cortisol exert opposite effects on SWS (Born et al., 1991; Friess et al., 1994) and GH levels (Friess et al., 1994; Bohlhalter et al., 1997), it appears unlikely that these effects are mediated by stimulation of cortisol. In contrast, these changes may be explained by negative feedback inhibition of endogenous CRH. Because CRH (Holsboer et al., 1988), ACTH (Born et al., 1989), and cortisol (Born et al., 1991; Friess et al., 1994, 1995b) diminish REMS in contrast to ebiratide, REMS suppression may be mediated by cortisol after each of these hormones. This hypothesis is supported by the observation that the inhibition of cortisol synthesis by metyrapone reduces SWS and cortisol levels in normal controls, whereas REMS is not affected (Jahn et al., 2003). In this experiment, endogenous CRH was probably enhanced, as ACTH was distinctly elevated.

As an alternative to CRH feedback inhibition, action on the central cortical steroid receptors was suggested as a way in which cortisol modulates sleep (Born et al., 1991). Two types of these receptors occur in the CNS, the mineralocorticoid receptor (MR) and the glucocorticoid receptor (GR). Because in normal controls the MR antagonist, canrenoate, reduces SWS, whereas the mixed MR and GR agonist, cortisol, enhances SWS and inhibits REMS, it was hypothesized that the MR regulates SWS, whereas the GR regulates REMS. This theory is challenged however, because (i) in another study MR agonists and antagonists did not modulate sleep EEG (Friess et al., 1995b), (ii) SWS is increased in elderly controls after cortisol, although the number of MRs is known to decrease distinctly during ageing (Bohlhalter et al., 1997), (iii) MRs and GRs coexist frequently in cells as heterodimers (see below).

In contrast to the effects of acute cortisol administration, treatment of female patients with multiple sclerosis with the GR agonist methylprednisolone during 9 days resulted in shortened REMS latency, increased REM density and a shift of the major portion of SWS from the 1st to the 2nd NREMS period. These changes resemble the sleep-EEG disturbances in depression (Antonijevic and Steiger, 2003) (➲ *Figure 25-7*).

◘ Figure 25-7

REM density during corticosteroid treatment. REM density during the first REMS period decreased significantly after acute high dose treatment with glucocorticoids (N2, hatched bars) compared to baseline (N1, open bars), but increased after prolonged corticosteroid treatment (N3, filled bars) compared to N2. For the second REMS period, REM density was not changed after acute treatment (N2, hatched bars), but significantly increased after prolonged treatment (N3, filled bars) compared to both baseline (open bars) and N2. Values are means ± SEM. N1 = before treatment (baseline), N2 = day 2 of high dose corticosteroid treatment, N3 = day 10 of corticosteroid treatment REM period = REMS period. ‡ = significant difference compared to N2, * = significant difference compared to N1. Reprinted from *Psychoneuroendocrinology*, 28, Antonijevic I.A. & Steiger A., Depression-like changes of the sleep-EEG during glucocorticoid treatment in patients with multiple sclerosis, 780-795, Copyright 2003, with permission from Elsevier

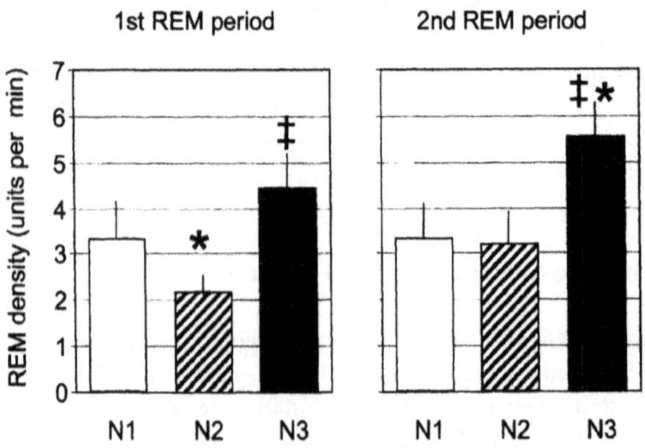

Administration of the mixed GR and progesterone receptor antagonist mifepristone is another approach to modulate the HPA system. In a single case study after oral administration of mifepristone, ACTH and cortisol levels were enhanced in young normal male subjects. Sleep quality was disrupted distinctly with increases of sleep latency and intermittent awake time and decreases of SWS and REMS time. This pilot study was extended in a set of experiments on the effects of GR and MR antagonists in normal controls (Wiedemann et al., 1994). These subjects participated in four protocols: (i) placebo only; in the other protocols before the examination nights dexamethasone was given orally, the following day at 14:00 either (ii) placebo or (iii) the MR antagonist spironolactone, or (iv) mifepristone were administered orally. After pretreatment with dexamethasone sleep EEG remained unchanged. After the combination of dexamethasone with spironolactone, REMS is diminished. Dexamethasone followed by mifepristone results in decreases of SWS and REMS and an increased number of awakenings. Pretreatment with dexamethasone suppresses ACTH and cortisol secretion. Mifepristone, but not spironolactone counteracts this effect.

Finally, the effects of mifepristone, the progesterone receptor agonist megestrol acetate, and placebo were compared (Wiedemann et al., 1998). Mifepristone and megestrol exert opposite effects on hormone levels, but compound their impairing effects on sleep quality. HPA hormones increase in the morning, and GH decreases after mifepristone, whereas megestrol exerts inverse effects. Again, mifepristone disturbs sleep, and megestrol selectively suppresses REMS. The combination of these substances increase wakefulness and shallow sleep and decreases REMS. It is thought that these effects are mediated by an interaction of GR and progesterone receptors. The precise evaluation of the function of corticosteroid receptors becomes puzzling with the discovery that receptor complexes, composed of both MRs and GRs mediate genomic effects on CNS activity (for discussion, see Bohlhalter et al., 1997).

3.5 Animal Models of HPA System Changes

In rats, it was investigated whether changes in the secretion of corticosterone contribute to the circadian variations in sleep–wake behavior. One week before the start of the experiment, rats were adrenalectomized and substituted with subcutaneous 21-day constant release corticosterone pellets, while another group of animals was sham-operated. In rats, after adrenalectomy and corticosterone replacement, the levels of this hormone remain at about 100 ng/ml throughout the day, whereas in the sham-operated control group, corticosterone levels range between peak values of about 240 ng/ml in the evening and minimal concentration of about 20 ng/ml in the morning. Slightly more, but shorter REMS episodes occur in the experimental group compared to the controls. Hence, the tonic levels of corticosterone exert negligible effects on spontaneous sleep–wake behavior (Langebartels and Lancel, 2000).

In another study, the effects of surgical adrenalectomy and subsequent corticosterone replacement were tested in rats under several conditions: before surgery, 14 days after adrenalectomy, which is known to elevate endogenous CRH; after corticosterone replacement in physiological dosage designed to restore CRH levels, or at a supraphysiological dosage suppressing CRH. Adrenalectomy reduces the amplitude of the diurnal rhythms of maximal and average sleep bout lengths. In the EEG power spectra after adrenalectomy, power from 1 to 4 Hz decreases, whereas power from 9 to 12 Hz increases. Physiological corticosterone replacement reverses some of these effects. Supraphysiological corticosterone replacement decreases NREMS. The finding suggests that adrenalectomy and supraphysiological corticosteroid replacement each altered sleep architecture without affecting sleep homeostasis (Bradbury et al., 1998).

In the Lewis rat, the synthesis and the release of CRH is reduced due to a hypothalamic gene defect in comparison to the Fisher 344 and Sprague–Dawley rat strains. Lewis rats spend less time awake and more time in SWS than the intact strains. REMS does not differ between strains. After intracerebroventricular administration of CRH, waking is enhanced similarly in Lewis and Sprague–Dawley rats (Opp, 1997). Obviously, the mechanisms mediating the response to exogenous CRH are not disturbed in the Lewis rats. Vice versa, spontaneous wakefulness of rats is reduced by a CRH antisense oligodeoxynucleotide (Chang and Opp, 2004). A role in the maintenance of wakefulness and sleep-disturbing effects of CRH are confirmed by these studies.

3.6 Summary

In contrast to GH, ACTH, and cortisol are widely dependent on a circadian rhythm. During the night, both their nadir and their acrophase occur in humans. Nocturnal awakenings and sleep deprivation result in increases of HPA hormone levels. CRH and ACTH administration impairs sleep. Various studies in humans and laboratory animals support the view that CRH is involved in the maintenance of wakefulness. Some but not all studies suggest a role of CRH in the promotion of REMS. In addition, cortisol may enhance REMS. HPA overactivity contributes distinctly to the pathophysiology of depression, including the sleep-EEG changes in affective disorders. It is thought that a reciprocal interaction of GHRH and CRH plays a key role in sleep regulation. During depression (CRH overactivity) and during normal ageing (reduced GHRH activity), the GHRH:CRH ratio is changed in favor of CRH. This appears to contribute to the similar changes of sleep-endocrine activity during an episode of depression and during normal ageing.

4 Hypothalamo–Pituitary–Thyroid System

4.1 Basic Activity

The secretion of thyroid stimulating hormone (TSH) and of the thyroid hormone thyroxin is related to circadian rhythm (Chan et al., 1978; Brabant et al., 1987). The minimum levels of TSH occur during the daytime. The level of TSH rises during the night and reaches its maximum by midnight. Then TSH levels decline during the early morning hours. The course of thyroxin release is opposite to that of TSH. Thyroxin concentrations are low during the night and increase during daytime. One study reported declining TSH levels during REMS periods (Follenius et al., 1988).

4.2 Sleep in Hypothyroidism and Hyperthyroidism

Changes of sleep–wake behavior are well-known symptoms of disorders of the thyroid gland. From clinical practice, it is well established that hyperthyroidism is linked with insomnia. In contrast, fatigue occurs frequently in patients with hypothyroidism. Therefore, it is astonishing that only a few data on sleep EEG in these diseases are available. One study reported reduced SWS in patients with hypothyroidism in comparison to normal controls. These changes normalized after therapy (Kales et al., 1967).

4.3 Effects of Thyrotropin-Releasing Hormone Administration

Pulsatile iv administration of thyrotropin-releasing hormone (TRH) decreases sleep efficiency and prompts the advanced occurrence of the cortisol morning rise in young normal male control subjects. All other sleep-EEG, cortisol, and GH variables remained unchanged (Hemmeter et al., 1998).

4.4 Summary

TSH and thyroxin are related to a circadian rhythm. Whereas changes of sleep–wake behavior in hypo- and hyperthyroidism are well known, there is only one study in this field. In patients with hypothyroidism, SWS was reduced. TRH exerts only a weak effect on sleep EEG as sleep efficiency decreases.

5 Insulin

5.1 Basic Activity

Insulin stimulates glucose uptake in adipocytes and skeletal muscles. Both, circadian rhythmicity and sleep, influence the profiles of glucose and the insulin secretion rate throughout 24 h. This results in higher mean levels during nocturnal sleep (Van Cauter et al., 1991a). Furthermore, slow oscillations with a periodicity of 50–150 min occur in animals and humans for both glucose and insulin secretion rate. To determine whether these oscillations are influenced by sleep, a sample of young normal subjects was investigated twice over 24 h during continuous enteral nutrition, once with a normal nocturnal sleep from 23:00 to 07:00 h, and once with a shifted daytime in sleep from 07:00 to 15:00 h. The amplitude of glucose and the insulin secretion rate oscillations increased distinctly during the sleep periods, whatever time they were performed, whereas the influence of the time of the day was not significant. Hence, the increased amplitude of glucose and insulin secretion rate oscillations is related to sleep rather than to the time of the day (Simon et al., 1994).

5.2 Effects of Insulin Administration

Systemic insulin injection (Sangiah et al., 1982), as well as intracerebroventricular insulin infusion during 3 days (Danguir and Nicolaidis, 1984), prompts increases of NREMS in rats. Rats with experimental diabetes mellitus show decreases of NREMS and REMS. Their sleep normalizes after systemic insulin infusion (Danguir, 1984).

6 Prolactin

6.1 Basic Activity

Prolactin is a circulating hormone and a neuroprotein as well. Prolactin is localized particularly in the hypothalamus (review: Roky et al., 1995). In humans, the prolactin level rises after sleep onset and reaches its peak during the second or the last third of the night (Weitzman, 1976). In contrast to various other hormones, prolactin is affected neither by normal ageing (Van Coevorden et al., 1991) nor by an episode of depression (Steiger and Holsboer, 1997b). During the recovery night after sleep deprivation, prolactin levels increase both in young and in elderly normal subjects (Murck et al., 1999).

A relationship between the pattern of prolactin secretion and the NREMS–REMS cycle was found in humans. Prolactin nadirs occur during REMS periods and prolactin levels rise during NREMS periods (Parker et al., 1974b). Furthermore, a temporal relationship between SWA and prolactin secretion was found in young human subjects (Parker et al., 1974b). Decreased dopaminergic inhibition of pituitary prolactin release is thought to lead to the increase of prolactin secretion during NREMS. However, a positive correlation between sleep cycles and plasma prolactin concentrations was not confirmed by another study (Van Cauter et al., 1982). A study on the circadian influences of sleep–wake and light–dark cycles on prolactin release suggests that the nocturnal rise in prolactin is not sleep-associated, but rather is rest dependent (Wehr et al., 1993). In normal human controls, the prolactin secretory rate is elevated throughout the sleep period independently from sleep quality. In this study, experimentally impaired sleep did not change prolactin release (Spiegel et al., 1994).

In Sprague–Dawley and Wistar rats, maximum plasma prolactin concentrations occur at the dark period. In another strain, however, prolactin pulses were observed at the end of the light period (review: Roky et al., 1995). In all, the timing of prolactin release appears to differ between species. In humans, prolactin is released during the night. Most studies suggest that prolactin is not sleep-associated but rather is dependent.

In the twin study mentioned before (➋ Section 3.1), the genetic and environmental influences on prolactin release during waking and sleep were investigated (Linkowski et al., 1998). Baseline daytime prolactin concentrations are partially under genetic influence, whereas the amplitude and overall wave shape of the secretory profile at daytime are genetically determined. The secretory response to a standardized sleep/circadian stimulus is also partly genetically controlled.

6.2 Sleep in Prolactinoma

In patients with hyperprolactinoma, SWS is increased selectively when compared to normal controls (Frieboes et al., 1998). This symptom appears to be unique when compared to other endocrine disorders, which are linked with impaired sleep.

6.3 Effects of Hormone Administration

6.3.1 Prolactin

Prolactin administration promotes REMS in cats, rabbits, and rats. After subcutaneous injection of prolactin, REMS increases in rabbits. The same effect is found after intrahypothalamic injection of prolactin in rats (Roky et al., 1994). In addition, systemic administration of prolactin during the light period stimulates REMS. In contrast, injection of prolactin during the dark period inhibits REMS (Roky et al., 1993). The REMS-promoting effect of prolactin was also found in pontine cats after hypophysectomy (Jouvet et al., 1986). An experimental prolactin-secreting tumor under the kidney capsule leads to long-term hyperprolactinemia in rats, and to an increase in nocturnal REMS, whereas REMS during the day decreases progressively (Valatx et al., 1994). In adult rats bearing juvenile rat anterior grafts under the capsule of the kidney, a distinct increase in REMS and enhanced duration of NREMS with a trend to increased SWA was found (Obál et al., 1992a). Likewise, systemic or intrahypothalamic injection of antiserum to prolactin decreases REMS in rats (Roky et al., 1994; Obál et al., 1997b). Similarly, intrahypothalamic injection of prolactin antiserum to rats decreases REMS. These data point to promotion of REMS after prolactin in rats. The increase of prolactin in these studies differs from the excessive hyperprolactinemia in patients with prolactinoma who showed increased SWS but unchanged REMS (Frieboes et al., 1998). The effects of a moderate increase of prolactin on human sleep have not been investigated so far.

6.3.2 Prolactin-Releasing Peptide

Intracerebroventricular infusion of prolactin-releasing peptide prompts in a dose-dependent manner different effects on sleep and hormone secretion in rats. At 0.1 nmol, REMS and plasma prolactin increase, whereas GH is blunted. After a dose of 1.0 nmol, NREMS and REMS increase, whereas 10.0 nmol enhances only NREMS (Zhang et al., 2001).

6.3.3 Vasoactive Intestinal Polypeptide Interaction with Prolactin and Sleep

In addition, vasoactive intestinal polypeptide (VIP) enhances REMS after intracerebroventricular administration in laboratory animals (Drucker-Colin et al., 1984). When VIP was given to rats during the dark period, NREMS and REMS increased (Riou et al., 1982; Obál et al., 1994). Similarly, VIP microinjections into the pontine reticular tegmentum enhance REMS in rats. This effect lasts up to 8 days (Bourgin et al., 1997). It is thought that interaction with the cholinergic system mediates the enhancement of REMS. This view is corroborated by the observation, that VIP microinjections into the oral pontine tegmentum enhance REMS in rats (Bourgin et al., 1997).

The REMS-promoting effect of systemic VIP was inhibited by immunoneutralisation of circulating prolactin in the rat. Therefore, it is thought that stimulation of prolactin is involved in the promotion of REMS after VIP (Obál et al., 1992a). VIP antibodies neutralized a REMS-promoting substance that accumulated in the CSF of sleep-deprived cats (Drucker-Colin et al., 1988). Subsequently, an increase of VIP was found in the CSF of REMS-deprived cats (Jimenez-Anguiano et al., 1993). After central administration of VIP antibodies (Riou et al., 1982) or a VIP antagonist to rats (Mirmiran et al., 1988), REMS decreases.

In young normal human male subjects, two doses of VIP exert different effects (Murck et al., 1996). After pulsatile administration of 4×10 μg VIP from 22:00 to 01:00, prolactin levels decrease, whereas sleep EEG remains unchanged. In contrast, after the higher dose of 4×50 μg VIP, prolactin levels increase. Furthermore, the NREMS–REMS cycles are decelerated (➋ *Figure 25-8*). Each of the NREMS and REMS periods is prolonged, the cortisol nadir appears advanced, and the GH surge is blunted (Murck et al., 1996). These findings suggest that VIP exerts a specific effect on the temporal organization of the NREMS–REMS cycle and the sleep-related hormone secretion including the timing of the cortisol nadir. VIP appears to

Figure 25-8
Summarized sleep cycles on basis of conventional sleep electroencephalographic analysis under placebo (A) and 4 × 50 μg VIP (B); n = 10. Open bar, NREMS period; solid bar, REMS period. From: Murck, H., Guldner, J., Frieboes, R.M., Schier, T., Wiedemann, K., Holsboer, F., Steiger, A.: VIP decelerates nonREM-REM cycles and modulates hormone secretion during sleep in men. *American Journal of Physiology*, 271 (Regulatory, Integrative and Comparative Physiology 40) (1996) R905-R911. Copyright The American Physiological Society

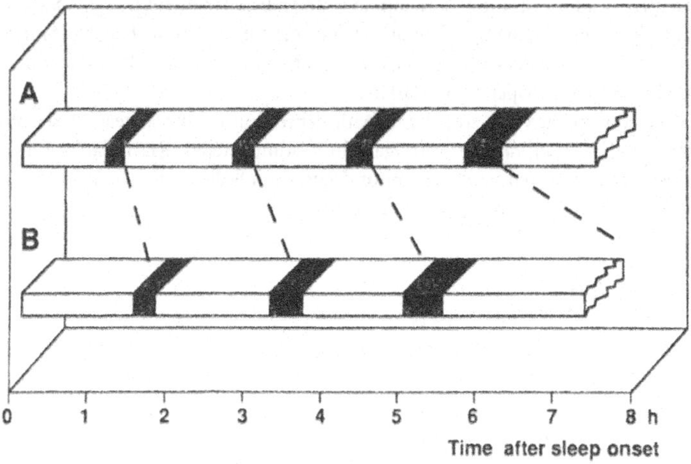

affect the circadian clock, resulting in prolonged sleep cycles, and the early occurrence of the cortisol nadir. The blunted GH surge may be the consequence of advanced elevated HPA activity. It is unclear whether the effects of systemic VIP are the same as the actions of intracerebral administration of VIP. The mechanisms of promotion of REMS by intracerebral VIP may differ from the sleep response to systemic VIP. For example, systemic VIP did not reach the brain stem, whereas it may reach the hypothalamus and modulate activity of the suprachiasmatic nucleus, as suggested by the findings in humans.

6.4 Animal Models of Changed Prolactin Activity

In genetically hypoprolactinemic rats, a decrease of REMS was reported (Valatx and Jouvet, 1988). However, in this study the recordings were done at room temperature. For these hairless rats, this temperature means a cold environment, and cold exposure may cause the decrease of REMS. In a warm environment, sleep durations of these rats did not differ from normal rats, whereas their circadian sleep–wake rhythm was changed. The circadian SWS rhythm remained unchanged, whereas the circadian REMS rhythm was reversed (3/4 during the night, 1/4 during the day) (Roky et al., 1995).

7 Other Peptides

7.1 Cortistatin

The brain-specific peptide cortistatin is highly homologous to somatostatin, whereas it is a product of a gene different from the somatostatin gene (de Lecea et al., 1997; Spier and de Lecea L., 2000). In contrast to somatostatin, which is widely distributed in the brain, cortistatin is localized mainly in the cerebral cortex and in the hippocampus (de Lecea et al., 2002). After intracerebroventricular administration of cortistatin to rats, NREMS increased distinctly (de Lecea et al., 1996). This finding suggests that the effects of cortistatin and somatostatin (see above) on sleep are opposite.

7.2 Galanin

Galanin is a peptide that is widely located in the mammalian brain and coexists in neurons with various peptides and classical neurotransmitters participating in sleep regulation. Furthermore, it stimulates GH via GHRH in man (Davis et al., 1987). Sleep in the rat remains unchanged after intracerebroventricular administration of galanin, whereas REMS deprivation induced galanin gene expression (Toppila et al., 1995). After repetitive iv administration of galanin to young normal male subjects, SWS and the duration of REMS periods increase, whereas the secretion of GH and cortisol remain unchanged (Murck et al., 1997a). A cluster of GABAergic and galaninergic neurons was identified in the ventrolateral preoptic area, which is thought to stimulate NREMS (Saper et al., 2001).

Galanin or placebo was given iv to patients with depression during a steady state of antidepressive therapy with the tricyclic trimipramine. After galanin is administered, REMS latency increases, and the severity of depression as measured by the Hamilton Depression Scale decreases. These findings suggest an acute antidepressive effect of galanin (Murck et al., 2004).

7.3 Neuropeptide Y

In sleep regulation, neuropeptide Y (NPY), besides GHRH, appears to be a physiological antagonist of CRH. NPY exerts a dual action on the HPA system. After intracerebroventricular administration, low doses of NPY suppress corticosterone in the rat, whereas higher doses enhance corticosterone and ACTH (Harfstrand et al., 1987). Similarly, increases of hypothalamic CRH levels after NPY (Haas and George, 1989), as well as a CRH antagonistic action of NPY were described. NPY is found in several neuronal circuits in the CNS, and NPY actions may greatly vary with these action sites.

Similarly, opposite effects of CRH and NPY were found in animal models of anxiety (reviewed in Steiger and Holsboer, 1997a). After intracerebroventricular administration of NPY to rats, EEG spectral activity changes in a similar manner to the effects of benzodiazepines (Ehlers et al., 1997a). The prolongation of sleep latency by CRH is antagonized in a dose-dependent manner by NPY in rats (Ehlers et al., 1997b). In young normal male subjects, repetitive iv administration of NPY decreases sleep latency, the first REMS period, cortisol and ACTH levels, and increases stage 2 sleep and sleep period time (Antonijevic et al., 2000a). In patients with depression of both sexes, with a wide age range and age-matched controls, the sleep latency is shortened, and prolactin levels increase after NPY, whereas cortisol and ACTH levels and the first REMS period remain unchanged (Held et al., 2006). These data suggest that NPY participates in sleep regulation, particularly in the timing of sleep onset as an antagonist of CRH acting via the $GABA_A$ receptor.

7.4 Neuropeptide S

Neuropeptide S (NPS) was recently described as a peptide that potently modulates wakefulness and may also regulate anxiety. NPS acts by activating its cognate receptor (NPSR). The NPSR mRNA is found widely in the brain, including the amygdala and the midline thalamic nuclei. NPS is expressed in a cluster of cells located between the locus coeruleus and Barrington's nucleus. Central administration of NPS in rats increased dose-dependent wakefulness and decreased SWS and REMS in a dose-dependent manner (Xu et al., 2004).

7.5 Pituitary Adenylate Cyclase Activating Polypeptide

Pituitary adenylate cyclase activating polypeptide (PACAP) is a member of the VIP family. Intracerebroventricular injection of PACAP at dark onset enhanced REMS (Fang et al., 1995). In young normal male subjects, however, after pulsatile iv administration of PACAP, the distribution of SWS and REMS changed during the night. SWS decreased during the first and increased during the fourth sleep cycle. The ratio of

REMS to NREMS sleep decreased during the fourth sleep cycle. GH and cortisol remained unchanged, whereas prolactin increased by trend during the first half of the night (Antonijevic et al., 1997).

7.6 Substance P

The neuropeptide substance P is a tachykinin acting as an excitatory neurotransmitter and/or neuromodulator. Microinjection of substance P into the ventrolateral preoptic area of rats enhances SWS. This effect is blocked by a phospholipase C inhibitor and by 3-mercaptopropionic acid, an inhibitor of GABA synthesis and release (Zhang et al., 2004). Whereas this study suggests sleep-promoting effects of substance P, in young normal male subjects, repetitive iv administration of substance P increased wakefulness, stage 1 sleep, REMS latency and cortisol and TSH levels. GH decreased after substance P. It is thought that these results point to a central arousing effect of substance P in humans (Lieb et al., 2002).

8 Melatonin

8.1 Basic Activity

Melatonin secretion is related to the light–dark cycle. Melatonin levels are maximal during the dark period in light active and in dark active species as well. It is thought that melatonin is primarily a neuroendocrine transducer promoting an increased propensity for "dark appropriate" behavior. Hence, it is suggested that melatonin is only a hypnotic in light active species (van den Heuvel et al., 2005).

8.2 Effects of Melatonin Administration

Study results on a beneficial effect of melatonin in young and elderly subjects are ambiguous (Zhdanova et al., 1997; Baskett et al., 2001; Zhdanova, 2005; van den Heuvel et al., 2005). A possible side effect of long-term treatment with melatonin is a blunting of sexual steroids in men and women. So far, there is a lack of sufficient data from clinical studies in order to recommend melatonin as a hypnotic. Some studies suggest that due to phase-shifting properties, melatonin may be helpful in the treatment of rhythm disturbances, like jet lag and disturbed rhythms in blind patients (Zhdanova et al., 1997; Sack et al., 2000).

9 Gonadal Hormones

9.1 Basic Activity

In young females during puberty, highest values of estradiol occur between 14:00 and 16:00 h, and lowest values are found during the night (Boyar et al., 1976). In a small group of adult women, no clear interaction between estradiol levels and sleep was observed (Alford et al., 1973). In males, testosterone rises constantly throughout the night (Weitzman, 1976).

9.2 Sleep in Women

In women, the menstrual cycle, pregnancy, and the menopause reflect distinct changes in endocrine activity and have some impact on sleep regulation. Only few studies have addressed these issues so far. Most studies on sleep regulation and sleep disorders were performed selectively in men or in male animals. One of the reasons why females are not included in such studies is because of the variability of the menstrual cycle (Kimura, 2005). On the other hand, women experience sleep problems more often than men do.

9.2.1 Menstrual Cycle Effects on Sleep

In normal women, sleep EEG was recorded every second night throughout one entire menstrual cycle. The percentage of REMS tended to be higher in the early follicular than in the late luteal phase, and the percentage of NREMS showed higher values in the luteal compared to the follicular phase. In NREMS, EEG power density in the upper frequency range of the sleep spindles (14.25–15.0 Hz) exhibits a large variety across the menstrual cycle, with maximum in the luteal phase (Driver et al., 1996). Normal cycling female rats show significantly less REMS during proestrus nights than during metestrus and diestrus nights, whereas no changes in daytime sleep patterns are found across the estrous cycle (Fang and Fishbein, 1996).

9.2.2 Sleep in Pregnancy

Sleep EEG was recorded in nine healthy women during each trimester of pregnancy. Waking increases from the second to the third trimester, whereas REMS decreases from the first to the second trimester. In NREMS, a progressive reduction of power density occurs (Brunner et al., 1994). Between the third trimester of pregnancy and 1 month postpartum, REMS latency decreases (Lee et al., 2000). In the rat, during pregnancy, nocturnal NREMS increases across the entire period, whereas REMS is enhanced only during the early period. After pregnancy, enhanced sleep returns to baseline (Kimura et al., 1998) (➋ *Figure 25-9*).

9.2.3 Sleep Changes during Menopause

A distinct decline in the sigma frequency range is reported in women during the menopause, whereas in men these changes occurred more gradually (Ehlers and Kupfer, 1997). After menopause, sleep-endocrine changes associated with depression are accentuated. This hypothesis is supported by a comparison of sleep-endocrine activity in pre- and postmenopausal women with depression and in matched controls. Cortisol is enhanced in the postmenopausal patients, whereas it decreases in postmenopausal controls. Sleep-EEG changes namely disturbed sleep continuity is characteristically associated with depression. A decrease in SWS and an increase of REM density are prominent in post- but not in premenopausal patients. An inverse correlation exists between the decline in SWS and in sleep continuity and follicle stimulating hormone (FSH) secretion in patients with depression. These observations suggest a role of menopause for these sleep-EEG changes. In premenopausal patients, however, a shift in SWS and SWA from the first to the second NREMS period was found, which was not related to age or hormone secretion (Antonijevic et al., 2003).

9.3 Effects of Gonadal Hormone Administration

9.3.1 Gonadal Hormone Administration in Adults

Administration of gonadal hormones to adult animals exerts minimal effects on sleep or on sex differences in sleep (Manber and Armitage, 1999). Similarly, only weak effects of high chronic dosages (80–100 mg) of estradiol in transsexual men who underwent cross-gender therapy to women were found. Stage 1 sleep increased in these patients (Künzel et al., 2000).

9.3.2 Hormone Replacement Therapy

In postmenopausal women, estrogen replacement therapy by skin patch (50 μg of estradiol per day) enhances REMS and reduces intermittent wakefulness during the first two sleep cycles. The normal decrease

Figure 25-9
Endocrine changes and sleep patterns during pregnancy in the rat. Upper panel shows circulating concentrations of luteinizing hormone (LH), prolactin (PRL), and progesterone; lower two panels show changes in nocturnal NREMS (nonREM sleep) (*middle*) and REMS (REM sleep) (*bottom*) relative to the baseline in estrous cycle

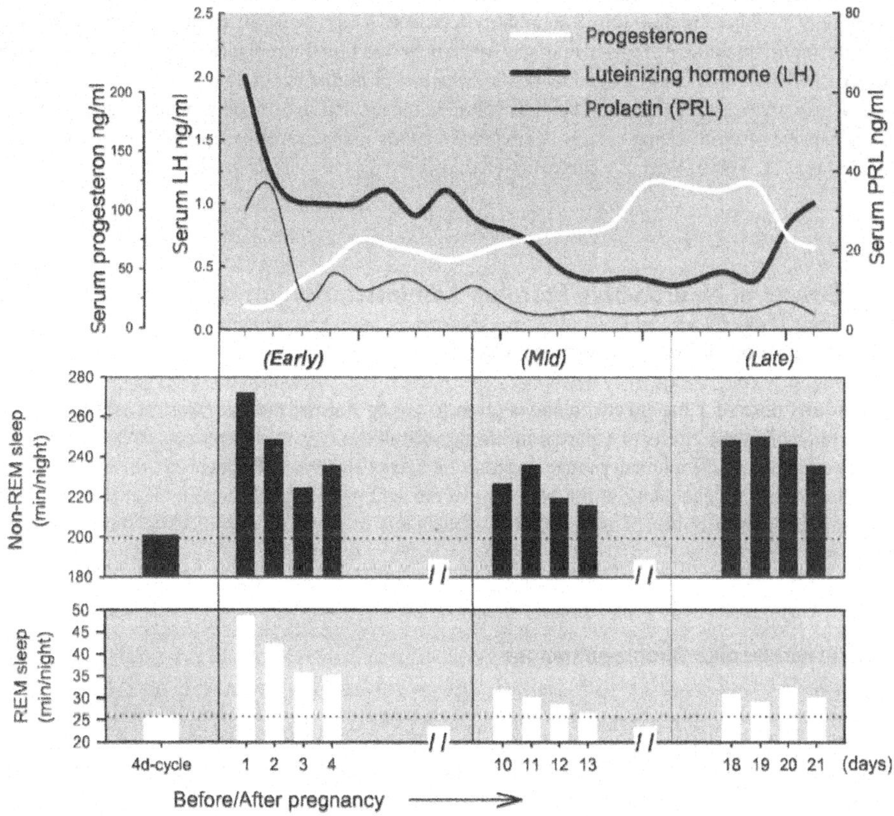

in SWS and SWA from the first to the second cycle is restored (Antonijevic et al., 2000d). These data suggest that estrogen treatment after menopause can help to restore the normal sleep-EEG pattern in women. For the effects of progesterone replacement, please see Section 10.2.2 below.

9.4 Animal Models—Ovariectomy, Castration

REMS is enhanced during the night and is reduced during daytime in adult female rats with ovariectomy, whereas SWS is increased in particular during the night. After subsequent estradiol replacement, the circadian pattern of sleep–wake behavior was restored by reducing the amount of REMS during the night (Colvin et al., 1969). After castration, REMS increases in neonatal mice. This effect is reversed after administration of testosterone (Yang and Fishbein, 1995).

10 Neuroactive Steroids

10.1 Introduction

Certain steroids, so-called neuroactive steroids, exert direct effects only on neuronal membranes and thereby rapidly affect CNS excitability (Paul and Purdy, 1992). It is thought that their effect on neuronal excitability is mediated by the gamma-aminobutyric acid $(GABA)_A$-receptor complex. However, classical steroid hormones including progesterone may also act as functional antagonists at the 5-hydroxytryptamine type 3 (5-HT-3) receptor, a ligand-gated ion channel or at distinct glutamate receptors (Rupprecht, 2003). Neuroactive steroids were found to be involved in the regulation of anxiety, memory, and sleep. Glial cells are capable of synthesizing certain neuroactive steroids independently of peripheral steroid sources (Jung-Testas et al., 1989). Various neuroactive steroids exert specific effects on sleep EEG in humans and rats.

10.2 Effects of Neuroactive Steroids Administration on the Sleep EEG

10.2.1 Pregnenolone, Pregnenolone Sulfate

When an oral dose of 1 mg pregnenolone is given to young normal male controls, it exerts sleep-EEG changes resembling the effects of a partial inverse agonist at the $GABA_A$ receptor, as SWS increases, and EEG power in the spindle frequency range decreases (Steiger et al., 1993) (➋ *Figure 25-10*). Similarly in rats after subcutaneous pregnenolone at the beginning of the light period, SWA increases (Lancel et al., 1994). Intraperitoneal administration of pregnenolone sulfate in rats, however, increases REMS (Darnaudery et al., 1999).

10.2.2 Progesterone, Allopregnanolone

A dose-dependent hypnotic effect of iv progesterone was reported as early as in 1954 (Merryman et al., 1954). After oral progesterone, in normal young men NREMS, especially stage 2 sleep, increases, and SWA decreases (Friess et al., 1997). Furthermore, EEG power in the higher frequency range (>15 Hz) tended to be elevated. In this study, a distinct interindividual variability in the bioavailability of progesterone and consequently in the time course of the concentrations of its metabolite allopregnanolone was found. Therefore, two subgroups were formed, one with an early peak and one with a late peak of allopregnanolone. The time course of this peak is associated with the changes in the EEG power spectra. The initial increase in the EEG activity in the spindle and alpha range in the first few hours of sleep is restricted to the subjects with an early allopregnanolone peak, whereas the decrease of SWA occurred mainly in those showing a later peak of this metabolite. The sleep-EEG changes after administration of progesterone in humans are similar to those induced by agonists at the $GABA_A$ receptor, e.g., benzodiazepines and appear to be mediated in part via the conversion of progesterone into allopregnanolone.

In women, progesterone levels decrease after menopause. Subchronic progesterone replacement increases REMS and decreases intermittent wakefulness in postmenopausal women (Schüssler et al., 2005a). Intraperitoneal administration of three doses of progesterone at the onset of the dark period in rats prompts dose-dependent decreases of NREMS latency, wakefulness and REMS and increases of REMS latency and of preREMS, an intermediate state between NREMS and REMS. Furthermore, EEG activity decreases in the lower frequencies and increases in the higher frequencies. Sleep-EEG effects of two doses of intraperitoneal allopregnanolone itself were studied in rats (Lancel et al., 1997). Both doses reduced NREMS latency and the higher dose increases preREMS. Furthermore, in NREMS EEG activity decreased in the lower frequencies ($\leq$7 Hz) and increased in the higher frequencies ($\geq$13 Hz). These data confirm benzodiazepine-like effects of allopregnanolone on sleep.

Figure 25-10
Sleep-state-specific electroencephalogram (EEG) power spectra after administration of pregnenolone (panel a, left; n = 6) and dehydroepiandrosterone (panel b, right; n = 7) in healthy male control subjects. Bars show deviation (mean ± SEM) from placebo level (=100%) in each frequency bin (0.38 Hz). Panel A; EEG power spectra of NREMS during sleep cycles: solid bars denote significant differences between placebo and pregnenolone (two-sided Bonferroni t-test, nominal $p < 0.05$). Panel B; EEG power spectra of REMS during 2-h periods of night sleep: solid bars denote significant differences between placebo and dehydroepiandrosterone (two-sided Bonferroni MANOVA, nominal $p < 0.05$), shaded bars indicate tendency ($p < 0.01$). Reprinted from Advances in Neuroimmunology, 5, Friess, E., Wiedemann, K., Steiger, A., Holsboer, F., The hypothalamic–pituitary–adrenocortical system and sleep in man, 111-125., Copyright 1995, with permission from Elsevier

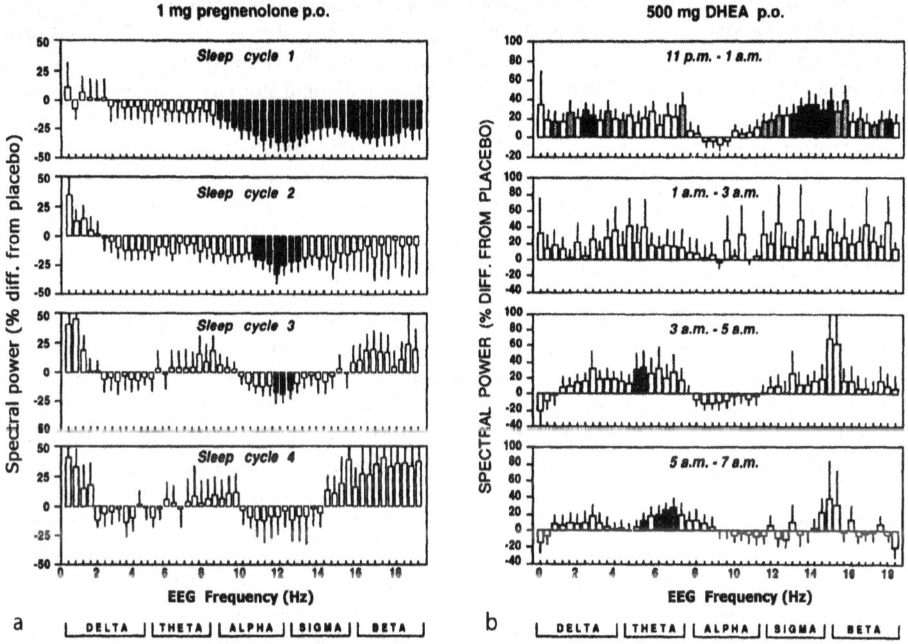

10.2.3 Deoxycorticosterone-3-Alpha, 21-Dihydroxy-5-Alpha-Pregnan-20-One (THDOC)

The ring A reduced metabolite of deoxycorticosterone-3-alpha, 21-dihydroxy-5-alpha-pregnan-20-one (THDOC) is found in the brain. It has been shown to be a barbiturate like ligand of the GABA receptor complex. In the rat, sleep EEG was investigated after THDOC alone and in combination with the benzodiazepine flurazepam. THDOC exerted dose-dependent sleep-inducing effects including shortened sleep latency and increased NREMS. Flurazepam had similar effects and decreased REMS. There were no significant interaction of THDOC and flurazepam except in REMS latency. The findings point to a sedating effect of THDOC (Mendelson et al., 1987).

10.2.4 Dehydroepiandrosterone, DHEA Sulfate

A single oral dose of dehydroepiandrosterone (DHEA) increases selectively REMS in young normal men (Friess et al., 1995a) (*Figure 25-10*). This finding corresponds to a mixed $GABA_A$ agonistic/antagonistic effect. After intraperitoneal DHEA sulfate (DHEAS), a dose-dependent effect on EEG power was observed in rats. A 50 mg/kg DHEAS augmented EEG power in the spindle-frequency range, whereas 100 mg/kg

DHEAS had the opposite effect. Sleep architecture remained unchanged after either dosage of DHEAS (Schiffelholz et al., 2000).

11 Conclusions

The data reviewed in this chapter show that a bidirectional interaction between sleep EEG and endocrine activity is well established. Various hormones (particularly neuropeptides and neuroactive steroids) exert specific effects on the sleep EEG in several species including humans. Vice versa, changes of sleep–wake behavior result in changes of endocrine activity.

In ➋ *Figure 25-11*, a model of peptidergic sleep regulation in humans is proposed. It appears that some peptides (GHRH, galanin, ghrelin) promote sleep, at least in males, whereas others (CRH, somatostatin) impair sleep. The reciprocal interaction of the neuropeptides GHRH and CRH plays a key role in sleep regulation. GHRH promotes NREMS, at least in males, and stimulates GH secretion, whereas CRH maintains wakefulness and enhances the secretion of ACTH and cortisol. Related to its role in the maintenance of wakefulness, it impairs sleep. Changes in the CRH:GHRH ratio in favor of CRH contribute to shallow sleep, elevated cortisol secretion and blunted GH levels during depression and ageing. On the other hand, GHRH participates in sleep promotion after sleep deprivation. Interestingly, recent data suggest CRH-like effects of GHRH in women, as sleep is impaired, and HPA hormones are elevated after GHRH in females. Several findings, particularly the effects of the CRH-1 receptor antagonist, R121919, in patients with depression, and some studies on the effects of CRH antagonists in the rat model suggest that CRH also promotes REMS, whereas other studies failed to support this view. NPY is another endogenous antagonist of CRH. Its major role appears to be in the timing of sleep onset. Similar to their reciprocal role in sleep regulation, GHRH and somatostatin exert opposite effects on sleep EGG, at least in males. Somatostatin is another sleep-impairing peptide. Besides GHRH, galanin and ghrelin were shown to promote SWS. Intact GHRH receptors are the prerequisite for sleep promotion by ghrelin. In contrast to GHRH, ghrelin stimulates HPA hormones in males and may act as an interface between the HPA and HPS systems. Galanin is colocalized with GABA in the ventrolateral preoptic nucleus. Many hypothalamic GHRH responsive neurons are GABAergic. Galanin, ghrelin, and GHRH may either act in a synergistic fashion or these peptides may be part of a cascade resulting in the promotion of NREMS. Probably, GABAergic neurons mediate the effects of these peptides. On the other hand, thalamic GABAergic transmission is thought to be involved in the sleep impairing effect of somatostatin. Whereas the aforementioned peptides either promote or impair sleep, VIP appears to be involved in the temporal organization of sleep. In young normal men after VIP, the NREMS/REMS cycle is decelerated, probably by action at the suprachiasmatic nucleus. In contrast, VIP is thought to promote REMS in animals via prolactin.

Besides peptides, steroids are involved in sleep regulation. Acute administration of cortisol promotes SWS in normal human controls, probably via feedback inhibition of CRH. Furthermore, cortisol suppresses REMS in humans, whereas REMS decreases after short-term withdrawal of hydrocortisone substitution in Addison's patients. This finding points to a REMS-promoting effect of cortisol. Similarly, subchronic administration of the GR agonist methylprednisolone in patients with multiple sclerosis prompted a set of sleep-EEG changes resembling those in patients with depression including REMS disinhibition. These findings suggest that physiological cortisol levels contribute to REMS maintenance and that a synergism of elevated CRH activity and enhanced glucocorticoid levels participates in the development of REMS disinhibition and changes of SWS during depression. $GABA_A$ receptors are also targets of various neuroactive steroids, which exert specific effects on sleep. Pregnenolone promotes SWS in humans and rats. In postmenopausal women, progesterone decreases intermittent wakefulness. Progesterone, most likely via allopregnanolone, acts similarly to benzodiazepine hypnotics in humans and rats. In contrast, DHEA promotes REMS in humans. Finally, the changes of sleep EEG after menopause and the beneficial effect of estrogen replacement therapy point to a role of estrogen in sleep regulation.

The effects of CRH-1 receptor antagonism in depression, of arginine vasopressin in the elderly and of estrogen and progesterone replacement therapy in the menopause are promising hints for a clinical application of research in this exciting area.

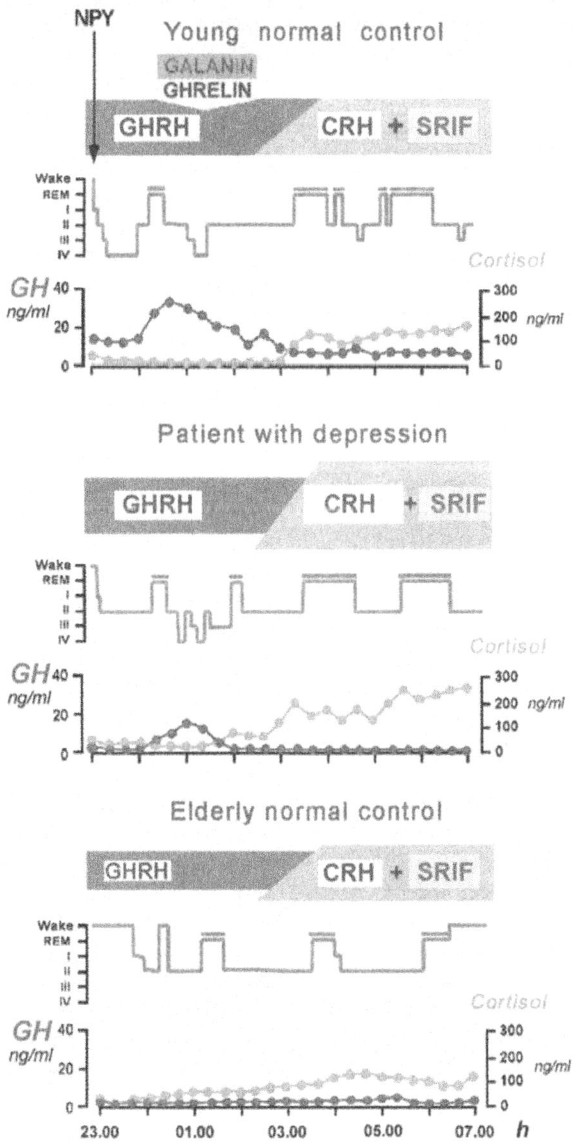

Figure 25-11

Model of peptidergic regulation. CRH, corticotropin-releasing hormone; GHRH, growth hormone-releasing hormone; NPY, neuropeptide Y; SRIF, somatostatin. Characteristic hypnograms and patterns of cortisol and GH secretion are shown in a young and in an elderly control subject and in a depressed patient. It is thought that GHRH is released during the first half of the night, whereas CRH preponderates during the second half of the night. GHRH contributes to the high amounts of GH and SWS after sleep onset, whereas CRH is linked with cortisol release und REM sleep in the morning hours. Neuropeptide Y (NPY) is a signal for sleep onset. In addition to GHRH, galanin and ghrelin are sleep promoting factors, whereas somatostatin is a sleep impairing factor. During depression (CRH overactivity) and during normal ageing, similar changes of sleep-emdocrine activity occur. It is thought that changes in the GHRH/CRH balance in favour of CRH play a key role in these alterations. Nervenarzt (1995), 66: 15-27, Schlafendokrinologie, Axel Steiger, Copyright Springer-Verlag 1995, with kind permission of Springer Science and Business Media

Acknowledgment

Studies from the author's laboratory were supported by grants from the Deutsche Forschungsgemeinschaft (Ste 486/1-2, 5-1, 5-2 and 5-3). I thank Dr. Mayumi Kimura for preparing ➋ *Figure 25-9*.

References

Alford FP, Baker HW, Burger HG, De Kretser DM, Hudson B, et al. 1973. Temporal patterns of integrated plasma hormone levels during sleep and wakefulness. II. Follicle-stimulating hormone, luteinizing hormone, testosterone and estradiol. J Clin Endocrinol Metab 37: 848-854.

Alt JA, Obál F, Jr., Traynor TR, Gardi J, Majde JA, et al. 2003. Alterations in EEG activity and sleep after influenza viral infection in GHRH receptor-deficient mice. J Appl Physiol 95: 460-468.

Andersen ML, Martins PJ, D'Almeida V, Bignotto M, Tufik S. 2005. Endocrinological and catecholaminergic alterations during sleep deprivation and recovery in male rats. J Sleep Res 14:83-90.

Antonijevic IA, Murck H, Bohlhalter S, Frieboes RM, Holsboer F, et al. 2000a. NPY promotes sleep and inhibits ACTH and cortisol release in young men. Neuropharmacology 39: 1474-1481.

Antonijevic IA, Murck H, Frieboes RM, Barthelmes J, Steiger A. 2000b. Sexually dimorphic effects of GHRH on sleep-endocrine activity in patients with depression and normal controls—part I: the sleep EEG. Sleep Res Online 3: 5-13.

Antonijevic IA, Murck H, Frieboes RM, Bohlhalter S, Schier T, et al. 1997. Effects of PACAP on sleep EEG and nocturnal hormone secretion in man. Biol Psychiatry 42: 48S.

Antonijevic IA, Murck H, Frieboes RM, Holsboer F, Steiger A. 1999a. On the gender differences in sleep-endocrine regulation in young normal humans. Neuroendocrinology 70: 280-287.

Antonijevic IA, Murck H, Frieboes RM, Holsboer F, Steiger A. 1999b. Hyporesponsiveness of the pituitary to CRH during slow wave sleep is not mimicked by systemic GHRH. Neuroendocrinology 69: 88-96.

Antonijevic IA, Murck H, Frieboes RM, Horn R, Brabant G, et al. 1998. Elevated nocturnal profiles of serum leptin in patients with depression. J Psychiatr Res 32: 403-410.

Antonijevic IA, Murck H, Frieboes RM, Steiger A. 2000c. Sexually dimorphic effects of GHRH on sleep-endocrine activity in patients with depression and normal controls—part II: hormone secretion. Sleep Res Online 3: 15-21.

Antonijevic IA, Murck H, Frieboes RM, Uhr M, Steiger A. 2003. On the role of menopause for sleep-endocrine alterations associated with major depression. Psychoneuroendocrinology 28: 401-418.

Antonijevic IA, Stalla GK, Steiger A. 2000d. Modulation of the sleep electroencephalogram by estrogen replacement in postmenopausal women. J Obstet Gynecol 182: 277-282.

Antonijevic IA, Steiger A. 2003. Depression-like changes of the sleep-EEG during high dose corticosteroid treatment in patients with multiple sclerosis. Psychoneuroendocrinology 28: 401-418.

Arnauld E, Bibene V, Meynard J, Rodriguez F, Vincent JD. 1989. Effects of chronic icv infusion of vasopressin on sleep-waking cycle of rats. Am J Physiol 256: R674-R684.

Åström C, Jochumsen PL. 1989. Decrease in delta sleep in growth hormone deficiency assessed by a new power spectrum analysis. Sleep 12: 508-515.

Åström C, Lindholm J. 1990. Growth hormone-deficient young adults have decreased deep sleep. Neuroendocrinology 51: 82-84.

Åström C, Trojaborg W. 1992. Effect of growth hormone on human sleep energy. Clin Endocrinol 36: 241-245.

Backhaus J, Junghanns K, Hohagen F. 2004. Sleep disturbances are correlated with decreased morning awakening salivary cortisol. Psychoneuroendocrinology 29: 1184-1191.

Baskett JJ, Wood PC, Broad JB, Duncan JR, English J, et al. 2001. Melatonin in older people with age-related sleep maintenance problems: a comparison with age-matched normal sleepers. Sleep 24: 418-424.

Beck U, Brezinova V, Hunter WM, Oswald I. 1975. Plasma growth hormone and slow wave sleep increase after interruption of sleep. J Clin Endocrinol Metab 40: 812-815.

Beranek L, Hajdu I, Gardi J, Taishi P, Obál F Jr, et al. 1999. Central administration of the somatostatin analog octreotide induces captopril-insensitive sleep responses. Am J Physiol 277: R1297-R1304.

Bilezikjian LM, Seifert H, Vale W. 1986. Desensitization to growth hormone-releasing factor (GRF) is associated with down-regulation of GRF-binding sites. Endocrinology 118: 2045-2052.

Bliwise DL, 1993. Sleep in normal aging and dementia. Sleep 16: 40-81.

Bodosi B, Gardi J, Hajdu I, Szentirmai E, Obál F Jr, et al. 2004. Rhythms of ghrelin, leptin, and sleep in rats: effects of the normal diurnal cycle, restricted feeding, and sleep deprivation. Am J Physiol Regul Integr Comp Physiol 287: R1071-R1079.

Bohlhalter S, Murck H, Holsboer F, Steiger A. 1997. Cortisol enhances non-REM sleep and growth hormone secretion in elderly subjects. Neurobiol Aging 18: 423-429.

Borbély AA, Baumann F, Brandeis D, Strauch I, Lehmann D. 1981. Sleep deprivation: effect on sleep stages and EEG power density in man. Electroencephalogr Clin Neurophysiol 51: 483-495.

Borbély AA, Tobler I, Hanagasioglu M. 1984. Effect of sleep deprivation on sleep and EEG power spectra in the rat. Behav Brain Res 14: 171-182.

Born J, De Kloet ER, Wenz H, Kern W, Fehm HL. 1991. Gluco- and antimineralocorticoid effects on human sleep: a role of central corticosteroid receptors. Am J Physiol 260: E183-E188.

Born J, Fehm HL. 1998. Hypothalamus–pituitary–adrenal activity during human sleep: a coordinating role for the limbic hippocampal system. Exp Clin Endocrinol Diabetes 106: 153-163.

Born J, Hansen K, Marshall L, Mölle M, Fehm HL. 1999. Timing the end of nocturnal sleep. Nature 397: 29-30.

Born J, Späth-Schwalbe E, Schwakenhofer H, Kern W, Fehm HL. 1989. Influences of corticotropin-releasing hormone, adrenocorticotropin, and cortisol on sleep in normal man. J Clin Endocrinol Metab 68: 904-911.

Bourgin P, Lebrand C, Escourrou P, Gaultier C, Franc B, et al. 1997. Vasoactive intestinal polypeptide microinjections into the oral pontine tegmentum enhance rapid eye movement sleep in the rat. Neuroscience 77: 351-360.

Boyar RM, Wu RH, Roffwarg H, Kapen S, Weitzman ED, et al. 1976. Human puberty: 24-hour estradiol in pubertal girls. J Clin Endocrinol Metab 43: 1418-1421.

Brabant G, Brabant A, Ranft U, Ocran K, Köhrle J, et al. 1987. Circadian and pulsatile thyrotropin secretion in euthyroid man under the influence of thyroid hormone and glucocorticoid administration. J Clin Endocrinol Metab 65: 83-88.

Bradbury MJ, Dement WC, Edgar DM. 1998. Effects of adrenalectomy and subsequent corticosterone replacement on rat sleep state and EEG power spectra. Am J Physiol 275: R555-R565.

Brandenberger G, Follenius M, Simon C, Ehrhart J, Libert JP. 1988. Nocturnal oscillations in plasma renin activity and REM–NREM sleep cycles in humans: a common regulatory mechanism? Sleep 11: 242-250.

Bredow S, Taishi P, Obál F Jr Guha-Thakurta N, Krueger JM. 1996. Hypothalamic growth hormone-releasing hormone mRNA varies across the day in rat. NeuroReport 7: 2501-2505.

Brunner DP, Münch M, Biedermann K, Huch R, Huch A, et al. 1994. Changes in sleep and sleep electroencephalogram during pregnancy. Sleep 17: 576-582.

Buckley TM, Schatzberg AF. 2005. On the interactions of the hypothaloamic–pituitary–adrenal (HPA) axis and sleep: Normal HPA axis activity and circadian rhythm, exemplary sleep disorders. J Clin Endocrinol Metab 90: 3106-3114.

Chan V, Jones A, Liendo-Ch P, McNeilly A, Landon J, et al. 1978. The relationship between circadian variations in circulating thyrotrophin, thyroid hormones and prolactin. Clin Endocrinol 9: 337-349.

Chang FC, Opp MR. 1998. Blockade of corticotropin-releasing hormone receptors reduces spontaneous waking in the rat. Am J Physiol 275: R793-R802.

Chang FC, Opp MR. 2004. A corticotropin-releasing hormone antisense oligodeoxynucleotide reduces spontaneous waking in the rat. Regul Pept 117: 43-52.

Colvin GB, Whitmoyer DI, Sawyer CH. 1969. Circadian sleep–wakefulness patterns in rats after ovariectomy and treatment with estrogen. Exp Neurol 25: 616-625.

Copinschi G, Leproult R, Van Onderbergen A, Caufriez A, Cole KY, et al. 1997. Prolonged oral treatment with MK-677, a novel growth hormone secretagogue, improves sleep quality in man. Neuroendocrinology 66: 278-286.

Danguir J, 1986. Intracerebroventricular infusion of somatostatin selectively increases paradoxical sleep in rats. Brain Res 367: 26-30.

Danguir J, 1984. Sleep deficits in diabetic rats: restoration following chronic intravenous or intracerebroventricular infusions of insulin. Brain Res Bull 12: 641-645.

Danguir J, Nicolaidis S. 1984. Chronic intracerebroventricular infusion of insulin causes selective increase of slow wave sleep in rats. Brain Res 306: 97-103.

Darnaudery M, Bouyer JJ, Pallares M, Le Moal M, Mayo W. 1999. The promnesic neurosteroid pregnenolone sulfate increases paradoxical sleep in rats. Brain Res 818: 492-498.

Davis TM, Burrin JM, Bloom SR. 1987. Growth hormone (GH) release in response to GH-releasing hormone in man is 3-fold enhanced by galanin. J Clin Endocrinol Metab 65: 1248-1252.

de Lecea L, Criado JR, Prospero-Garcia O, Gautvik KM, Schweitzer P, et al. 1996. A cortical neuropeptide with neuronal depressant and sleep-modulating properties. Nature 381: 242-245.

de Lecea L, Del Rio JR, Criado JR, Alcantra S, Morales M, et al. 2002. Cortistatin is expressed in a distinct subset of cortical interneurons. J Neurosci 17: 5868-5880.

de Lecea L, Ruiz-Lozano P, Danielson P, Foye P, Peelle-Kirley J, et al. 1997. cDNA cloning, mRNA distribution and chromosomal mapping of mouse and human preprocortistatin. Genomics 42: 499-506.

De A, Churchill L, Obál F, Jr., Simasko SM, Krueger JM. 2002. GHRH and IL1 beta increase cytoplasmic Ca2+ levels in cultured hypothalamic GABAergic neurons. Brain Res 949: 209-212.

Desir D, Van Cauter E, Fang VS, Martino E, Jadot C, et al. 1981. Effects of "jet lag" on hormonal patterns. I. Procedures, variations in total plasma proteins, and disruption of adrenocorticotropin-cortisol periodicity. J Clin Endocrinol Metab 52: 628-641.

Deuschle M, Blum WF, Englaro P, Schweiger U, Weber B, et al. 1996. Plasma leptin in depressed patients and healthy controls. Horm Metab Res 28: 714-717.

Driver HS, Dijk DJ, Werth E, Biedermann K, Borbély AA. 1996. Sleep and the sleep electroencephalogram across the menstrual cycle in young healthy women. J Clin Endocrinol Metab 81: 728-735.

Drucker-Colin R, Bernal-Pedraza J, Fernandez-Cancino F, Oksenberg A. 1984. Is vasoactive intestinal polypeptide (VIP) a sleep factor? Peptides 5: 837-840.

Drucker-Colin R, Prospero-Garcia O, Arankowsky-Sandoval G, Perez-Montfort R. 1988. Gastropancreatic peptides and sensory stimuli as REM sleep factors. Sleep peptides: basic and clinical approaches. Inoué S, Schneider-Helmert D. (editors), Tokyo: Scientific Society and Springer; pp. 73-94.

Drucker-Colin RR, Spanis CW, Hunyadi J, Sassin JF, McGaugh JL. 1975. Growth hormone effects on sleep and wakefulness in the rat. Neuroendocrinology 18: 1-8.

Dubreuil P, Lapierre H, Pelletier G, Petitclerc D, Couture Y, et al. 1988. Serum growth hormone release during a 60-hour period in growing pigs. Domest Anim Endocrinol 5: 157-164.

Dzaja A, Dalal MA, Himmerich H, Uhr M, Pollmächer T, et al. 2004. Sleep enhances nocturnal plasma ghrelin levels in healthy subjects. Am J Physiol Endoc M 286: E963-E967.

Ehlers CL, Kupfer DJ. 1997. Slow-wave sleep: do young adult men and women age differently? J Sleep Res 6: 211-215.

Ehlers CL, Kupfer DJ. 1987. Hypothalamic peptide modulation of EEG sleep in depression: a further application of the S-process hypothesis. Biol Psychiatry 22: 513-517.

Ehlers CL, Reed TK, Henriksen SJ. 1986. Effects of corticotropin-releasing factor and growth hormone-releasing factor on sleep and activity in rats. Neuroendocrinology 42: 467-474.

Ehlers CL, Somes C, Lopez A, Kirby D, Rivier JE. 1997a. Electrophysiological actions of neuropeptide Y and its analogs: new measures for anxiolytic therapy? Neuropsychopharmacology 17: 34-43.

Ehlers CL, Somes C, Seifritz E, Rivier JE. 1997b. CRF/NPY interactions: a potential role in sleep dysregulation in depression and anxiety. Depress Anxiety 6: 1-9.

Fabry J, Oger R, Reynart R. 1982. Rhythme nycthéméral de l'hormone de croissance chez le bovin femelle. Bull Rech Agron Gembloux 17: 331

Fadda P, Fratta W. 1997. Stress-induced sleep deprivation modifies corticotropin releasing factor (CRF) levels and CRF binding in rat brain and pituitary. Pharmacol Res 35: 443-446.

Fang J, Fishbein W. 1996. Sex differences in paradoxical sleep: influences of estrus cycle and ovariectomy. Brain Res 734: 275-285.

Fang J, Payne L, Krueger JM. 1995. Pituitary adenylate cyclase activating polypeptide enhances rapid eye movement sleep in rats. Brain Res 686: 23-28.

Fehm HL, Klein E, Holl R, Voigt KH. 1984. Evidence for extrapituitary mechanisms mediating the morning peak of cortisol secretion in man. J Clin Endocrinol Metab 58: 410-414.

Fehm HL, Späth-Schwalbe E, Pietrowsky R, Kern W, Born J. 1993. Entrainment of nocturnal pituitary–adrenocortical activity to sleep processes in man-a hypothesis. Exp Clin Endocrinol 101: 267-276.

Follenius M, Brandenberger G, Bandesapt JJ, Libert JP, Ehrhart J. 1992. Nocturnal cortisol release in relation to sleep structure. Sleep 15: 21-27.

Follenius M, Brandenberger G, Simon C, Schlienger JL. 1988. REM sleep in humans begins during decreased secretory activity of the anterior pituitary. Sleep 11: 546-555.

Franken P, Chollet D, Tafti M. 2001. The homeostatic regulation of sleep need is under genetic control. J Neurosci 21: 2610-2621.

Franken P, Dijk DJ, Tobler I, Borbely AA. 1991. Sleep deprivation in rats: effects on EEG power spectra, vigilance states, and cortical temperature. Am J Physiol 261: R198-R208.

Frieboes RM, Antonijevic IA, Held K, Murck H, Pollmächer T, et al. 2004. Hexarelin decreases slow-wave sleep and stimulates the sleep-related secretion of GH, ACTH, cortisol and prolactin in healthy volunteers. Psychoneuroendocrinology 29: 851-860.

Frieboes RM, Müller U, Murck H, von Cramon DY, Holsboer F, et al. 1999. Nocturnal hormone secretion and the sleep EEG in patients several months after traumatic brain injury. J Neuropsychiatry Clin Neurosc 11: 354-360.

Frieboes RM, Murck H, Maier P, Schier T, Holsboer F, et al. 1995. Growth hormone-releasing peptide-6 stimulates sleep, growth hormone, ACTH and cortisol release in normal man. Neuroendocrinology 61: 584-589.

Frieboes RM, Murck H, Schier T, Holsboer F, Steiger A. 1997. Somatostatin impairs sleep in elderly human subjects. Neuropsychopharmacology 16: 339-345.

Frieboes RM, Murck H, Stalla GK, Antonijevic IA, Steiger A. 1998. Enhanced slow wave sleep in patients with prolactinoma. J Clin Endocrinol Metab 83: 2706-2710.

Friess E, Tagaya H, Grethe C, Trachsel L, Holsboer F. 2004. Acute cortisol administration promotes sleep intensity in man. Neuropsychopharmacology 29: 598-604.

Friess E, Tagaya H, Trachsel L, Holsboer F, Rupprecht R. 1997. Progesterone-induced changes in sleep in male subjects. Am J Physiol Endoc M 272: E885-E891.

Friess E, Trachsel L, Guldner J, Schier T, Steiger A, et al. 1995a. DHEA administration increases rapid eye movement sleep and EEG power in the sigma frequency range. Am J Physiol 268: E107-E113.

Friess E, von Bardeleben U, Wiedemann K, Lauer C, Holsboer F. 1994. Effects of pulsatile cortisol infusion on sleep-EEG and nocturnal growth hormone release in healthy men. J Sleep Res 3: 73-79.

Friess E, Wiedemann K, Steiger A, Holsboer F. 1995b. The hypothalamic–pituitary–adrenocortical system and sleep in man. Adv Neuroimmunol 5: 111-125.

Garcia-Borreguero D, Wehr TA, Larrosa O, Granizo JL, Hardwick D, et al. 2000. Glucocorticoid replacement is permissive for rapid eye movement sleep and sleep consolidation in patients with adrenal insufficiency. J Clin Endocrinol Metab 85: 4201-4206.

Gardi J, Obál F Jr Fang J, Zhang J, Krueger JM. 1999. Diurnal variations and sleep deprivation-induced changes in rat hypothalamic GHRH and somatostatin contents. Am J Physiol 277: R1339-R1344.

Gillin JC, Jacobs LS, Snyder F, Henkin RI. 1974. Effects of ACTH on the sleep of normal subjects and patients with Addison's disease. Neuroendocrinology 15: 21-31.

Goichot B, Weibel L, Chapotot F, Gronfier C, Piquard F, et al. 1998. Effect of the shift of the sleep–wake cycle on three robust endocrine markers of the circadian clock. Am J Physiol 275: E243-E248.

Gonzalez MMC, Valatx JL. 1998. Involvement of stress in the sleep rebound mechanism induced by sleep deprivation in the rat: use of alpha-helical CRH (9–41). Behav Pharmacol 9: 655-662.

Gonzalez MMC, Valatx JL. 1997. Effect of intracerebroventricular administration of alpha-helical CRH (9–41) on the sleep/waking cycle in rats under normal conditions or after subjection to an acute stressful stimulus. J Sleep Res 6: 164-170.

Gronfier C, Chapotot F, Weibel L, Jouny C, Piquard F, et al. 1998. Pulsatile cortisol secretion and EEG delta waves are controlled by two independent but synchronized generators. Am J Physiol 275: E94-E100.

Gronfier C, Luthringer R, Follenius M, Schaltenbrand N, Macher JP, et al. 1996. Hormones and sleep. A quantitative evaluation of the relationships between growth hormone secretion and delta wave electroencephalographic activity during normal sleep and after enrichment in delta waves. Sleep 19: 817-824.

Gronfier C, Simon C, Piquard F, Ehrhardt J, Brandenberger G. 1999. Neuroendocrine processes underlying ultradian sleep regulation in man. J Clin Endocrinol Metab 84: 2686-2690.

Guilhaume A, Benoit O, Gourmelen M, Richardet JM. 1982. Relationship between sleep stage IV deficit and reversible hGH deficiency in psychosocial dwarfism. Pediatr Res 16: 299-303.

Guldner J, Schier T, Friess E, Colla M, Holsboer F, et al. 1997. Reduced efficacy of growth hormone-releasing hormone in modulating sleep endocrine activity in the elderly. Neurobiol Aging 18: 491-495.

Haas DA, George SR. 1989. Neuropeptide Y-induced effects on hypothalamic corticotropin-releasing factor content and release are dependent on noradrenergic/adrenergic neurotransmission. Brain Res 498: 333-338.

Hajdu I, Obál F Jr Fang J, Krueger JM, Rollo CD. 2002. Sleep of transgenic mice producing excess rat growth hormone. Am J Physiol Regul Integr Comp Physiol 282: R70-R76.

Harfstrand A, Eneroth P, Agnati L, Fuxe K. 1987. Further studies on the effects of central administration of neuropeptide Y on neuroendocrine function in the male rat: relationship to hypothalamic catecholamines. Regul Pept 17: 167-179.

Hart TB, Radow SK, Blackard WG, Tucker HSG, Cooper K.R. 1985. Sleep apnea in active acromegaly. Arch Intern Med 145: 865-866.

Held K, Antonijevic IA, Murck H, Künzel H, Steiger A. 2005. Alpha-helical CRH exerts CRH agonistic effects on sleep-endocrine activity in humans. Neuropsychobiology, 52: 62-67.

Held K, Künzel H, Ising M, Schmid DA, Zobel A, et al. 2004. Treatment with the CRH_1-receptor antagonist R121919 improves sleep EEG in patients with depression. J Psychiatr Res 38: 129-136.

Held K, Murck H, Antonijevic IA, Künzel H, Steiger A, et al. 2006. (NPY) shortens sleep latency and enhances prolactin but does not suppress ACTH and cortisol in depressed patients and controls. Psychoneuroendocrinology, 31: 100-107.

Hemmeter U, Rothe B, Guldner J, Holsboer F, Steiger A. 1998. Effects of thyrotropin-releasing hormone on the sleep EEG and nocturnal hormone secretion in male volunteers. Neuropsychobiology 38: 25-31.

Holl RW, Hartman ML, Veldhuis JD, Taylor WM, Thorner MO. 1991. Thirty-second sampling of plasma growth hormone in man: correlation with sleep stages. J Clin Endocrinol Metab 72: 854-861.

Holsboer F. 1999. The rationale for corticotropin-releasing hormone receptor (CRH-R) antagonists to treat depression and anxiety. J Psychiatr Res 33: 181-214.

Holsboer F, von Bardeleben U, Steiger A. 1988. Effects of intravenous corticotropin-releasing hormone upon sleep-related growth hormone surge and sleep EEG in man. Neuroendocrinology 48: 32-38.

Horvath TL, Diano S, Sotonyi P, Heiman M, Tschöp M. 2001. Minireview: ghrelin and the regulation of energy balance—a hypothalamic perspective. Endocrinology 142: 4163-4169.

Iovino M, Monteleone P, Steardo L. 1989. Repetitive growth hormone-releasing hormone administration restores the attenuated growth hormone (GH) response to GH-releasing hormone testing in normal aging. J Clin Endocrinol Metab 69: 910-913.

Jahn H, Kiefer F, Schick M, Yassouridis A, Steiger A, et al. 2003. Sleep-endocrine effects of the 11-β—Hydroxasteroid-dehydrogenase inhibitor metyrapone. Sleep 26: 823-829.

Jarrett DB, Miewald JM, Kupfer DJ. 1990. Recurrent depression is associated with a persistent reduction in sleep-related growth hormone secretion. Arch Gen Psychiatry 47: 113-118.

Jessup SK, Malow BA, Symons KV, Barkan AL. 2004. Blockade of endogenous growth hormone-releasing hormone receptors dissociates nocturnal growth hormone secretion and slow-wave sleep. Eur J Endocrinol 151: 561-566.

Jimenez-Anguiano A, Baez-Saldana A, Drucker-Colin R. 1993. Cerebrospinal fluid (CSF) extracted immediately after REM sleep deprivation prevents REM rebound and contains vasoactive intestinal peptide (VIP). Brain Res 631: 345-348.

Jouvet M, Buda C, Cespuglio R, Chastrette N, Denoyer M, et al. 1986. Hypnogenic effects of some hypothalamo–pituitary peptides. Clin Neuropharmacol 9 (Suppl. 4): 465-467.

Jung-Testas I, Hu ZY, Baulieu EE, Robel P. 1989. Neurosteroids: biosynthesis of pregnenolone and progesterone in primary cultures of rat glial cells. Endocrinology 125: 2083-2091.

Kales A, Heuser G, Jacobson A, Kales JD, Hanley J, et al. 1967. All night sleep studies in hypothyroid patients, before and after treatment. J Clin Endocrinol Metab 27: 1593-1599.

Kawakami M, Kimura F, Tsai CW. 1983. Correlation of growth hormone secretion to sleep in the immature rat. J Physiol 339: 325-337.

Kellner M, Yassouridis A, Manz B, Steiger A, Holsboer F, et al. 1997. Corticotropin-releasing hormone inhibits melatonin secretion in healthy volunteers—a potential link to low-melatonin syndrome in depression? Neuroendocrinology 16: 339-345.

Kerkhofs M, Van Cauter E, Van Onderbergen A, Caufriez A, Thorner MO, et al. 1993. Sleep-promoting effects of growth hormone-releasing hormone in normal men. Am J Physiol 264: E594-E598.

Kimura M. 2005. Minireview: gender-specific sleep regulation. Sleep Biol Rhythms 3: 75-79.

Kimura M, Zhang SQ, Inoue S. 1998. An animal model for pregnancy-associated sleep disorder. Psychiatry Clin Neurosci 52: 209-211.

Kojima M, Hosoda H, Date Y, Nakazato M, Matsuo H, et al. 1999. Ghrelin is a growth hormone-releasing acylated peptide from stomach. Nature 402: 656-660.

Krieger DT, Glick SM. 1974. Sleep EEG stages and plasma growth hormone concentration in states of endogenous and exogenous hypercortisolemia or ACTH elevation. J Clin Endocrinol Metab 39: 986-1000.

Krishnan KRR, Ritchie JC, Saunders W, Wilson W, Nemeroff CB, et al. 1990. Nocturnal and early morning secretion of ACTH and cortisol in humans. Biol Psychiatry 28: 47-57.

Künzel HE, Murck H, Korali Z, Frieboes RM, Ziegenbein M, et al. 2000. Impaired sleep in male to female transsexual patients treated with estrogens. Exp Clin Endocrinol Diabetes 108 (Suppl. 1): S 93.

Kupfer DJ, Ehlers CL, Frank E, Grochocinski VJ, McEachran AB, et al. 1993. Electroencephalographic sleep studies in depressed patients during long-term recovery. Psychiatry Res 49: 121-138.

Kupfer DJ, Jarrett DB, Ehlers CL. 1992. The effect of SRIF on the EEG sleep of normal men. Psychoneuroendocrinology 17: 37-43.

Lachmansingh E, Rollo CD. 1994. Evidence for a trade-off between growth and behavioural activity in giant "Supermice" genetically engineered with extra growth hormone genes. Can J Zool 72: 2158-2168.

Lancel M, Crönlein TA, Müller-Preuss P, Holsboer F. 1994. Pregnenolone enhances EEG delta activity during non-rapid eye movement sleep in the rat, in contrast to midazolam. Brain Res 646: 85-94.

Lancel M, Faulhaber J, Schiffelholz T, Romeo E, Dimichele F, et al. 1997. Allopregnanolone affects sleep in a benzodiazepine-like fashion. J Pharmacol Exp Ther 282: 1213-1218.

Langebartels A, Lancel M. 2000. Influence of constant corticosterone levels on spontaneous and lipopolysaccharide-induced sleep in the rat. J Sleep Res 9 (Suppl. 1): 109.

Laurentie MP, Barenton B, Charrier J, Garcia-Villar R, Marnet PG, et al. 1989. Instantaneous secretion rate of growth hormone in lambs: relationships with sleep, food intake, and posture. Endocrinology 125: 642-651.

Lee KA, McEnany G, Zaffke ME. 2000. REM sleep and mood state in childbearing women: sleepy or weepy? Sleep 23: 877-885.

Leproult R, Coleccia EF, L'Hermite-Balériaux M, Van Cauter E. 2001. Transition from dim to bright light in the induces an immediate elevation of cortisol levels. J Clin Endocrinol Metab 86: 151-157.

Leproult R, Copinschi G, Buxton O, Van Cauter E. 1997. Sleep loss results in an elevation of cortisol levels the next evening. Sleep 20: 865-870.

Leresche N, Asprodini E, Emri Z, Cope DW, Crunelli V. 2000. Somatostatin inhibits gabaergic transmission in the sensory thalamus via presynaptic receptors. Neuroscience 98: 513-522.

Licinio J, Mantzoros C, Negrao AB, Cizza G, Wong ML, et al. 1997. Human leptin levels are pulsatile and inversely related to pituitary–adrenal function. Nat Med 3: 575-579.

Lieb K, Ahlvers K, Dancker K, Strohbusch S, Reincke M, et al. 2002. Effects of the neuropeptide substance P on sleep, mood, and neuroendocrine measures in healthy young men. Neuropsychopharmacology 27: 1041-1049.

Linkowski P, Kerkhofs M, Van Onderbergen A, Hubain P, Copinschi G, et al. 1994. The 24-hour profiles of cortisol, prolactin, and growth hormone secretion in mania. Arch. Gen. Psychiatry 51: 616-624.

Linkowski P, Mendlewicz J, Kerkhofs M, Leclercq R, Golstein J, et al. 1987. 24-hour profiles of adrenocorticotropin, cortisol, and growth hormone in major depressive illness: effect of antidepressant treatment. J Clin Endocrinol Metab 65: 141-152.

Linkowski P, Spiegel K, Kerkhofs M, L'Hermite-Balériaux M, Van Onderbergen A, et al. 1998. Genetic and environmental influences on prolactin secretion during wake and during sleep. Am J Physiol 274: E909-E919.

Linkowski P, Van Onderbergen A, Kerkhofs M, Bosson D, Mendlewicz J, et al. 1993. Twin study of the 24-h cortisol profile: evidence for genetic control of the human circadian clock. Am J Physiol 264: E173-E181.

Lopez M, Seoane LM, Tovar S, Nogueiras R, Dieguez C, et al. 2004. Orexin-A regulates growth hormone-releasing hormone mRNA content in a nucleus-specific manner and somatostatin mRNA content in a growth hormone-dependent fashion in the rat hypothalamus. Eur J Neurosci 19: 2080-2088.

Manber R, Armitage R. 1999. Sex, steroids and sleep: a review. Sleep 22: 540-555.

Marrosu F, Gessa GL, Giagheddu M, Fratta W. 1990. Corticotropin-releasing factor (CRF) increases paradoxical sleep (PS) rebound in PS-deprived rats. Brain Res 515: 315-318.

Marshall L, Derad L, Strasburger CJ, Fehm HL, Born J. 1999. A determinant factor in the efficacy of GHRH administration in the efficacy of GHRH administration in promoting sleep: high peak concentration versus recurrent increasing slopes. Psychoneuroendocrinology 24: 363-370.

Mendelson WB, Martin JV, Perlis M, Wagner R, et al. 1987. Sleep induction by an adrenal steroid in the rat. Psychopharmacology 93: 226-229.

Mendelson WB, Slater S, Gold P, Gillin JC. 1980. The effect of growth hormone administration on human sleep: a dose-response study. Biol Psychiatry 15: 613-618.

Merryman W, Boiman R, Barnes L, Rothchild I. 1954. Progesterone "anaesthesia" in human subjects. J Clin Endocrinol Metab 14: 1567-1569.

Mirmiran M, Kruisbrink J, Bos NP, Werf Van der D, Boer GJ. 1988. Decrease of rapid-eye-movement sleep in the light by intraventricular application of a VIP-antagonist in the rat. Brain Res 458: 192-194.

Mullington J, Hermann D, Holsboer F, Pollmächer T. 1996. Age-dependent suppression of nocturnal growth hormone levels during sleep deprivation. Neuroendocrinology 64: 233-241.

Murck H, Antonijevic IA, Frieboes RM, Maier P, Schier T, et al. 1997a. Galanin has REM-sleep deprivation-like effects on the sleep EEG in healthy young men. J Psychiatr Res 33: 225-232.

Murck H, Antonijevic IA, Schier T, Frieboes RM, Barthelmes J, et al. 1999. Aging does not affect the sleep endocrine response to total sleep deprivation in humans. Neurobiol Aging 20: 665-668.

Murck H, Frieboes RM, Schier T, Steiger A. 1997b. Long-time administration of growth hormone-releasing hormone (GHRH) does not restore the reduced efficiency of GHRH on sleep endocrine activity in 2 old-aged subjects—a preliminary study. Pharmacopsychiatry 30: 122-124.

Murck H, Guldner J, Colla-Müller M, Frieboes RM, Schier T, et al. 1996. VIP decelerates non-REM-REM cycles and modulates hormone secretion during sleep in men. Am J Physiol 271: R905-R911.

Murck H, Held K, Ziegenbein M, Künzel H, Holsboer F, et al. 2004. Intravenous administration of the neuropeptide galanin has fast antidepressant efficacy and affects the sleep EEG. Psychoneuroendocrinology 29: 1205-1211.

Obál F Jr, Krueger JM. 2004. GHRH and sleep. Sleep Medicine Rev 8: 367-377.

Obál F Jr, Alföldi P, Cady AB, Johannsen L, Sary G, et al. 1988. Growth hormone-releasing factor enhances sleep in rats and rabbits. Am J Physiol 255: R310-R316.

Obál F Jr, Alt J, Taishi P, Gardi J, Krueger JM. 2003. Sleep in mice with non-functional growth hormone-releasing hormone receptors. Am J Physiol Regul Integr Comp Physiol 284: R131-R139.

Obál F Jr, Beranek L, Brandenberger G. 1994. Sleep-associated variations in plasma renin activity and blood pressure in the rat. Neurosci Lett 179: 83-86.

Obál F Jr, Bodosi B, Szilagyi A, Kacsóh B, Krueger JM. 1997a. Antiserum to growth hormone decreases sleep in the rat. Neuroendocrinology 66: 9-16.

Obál F Jr, Fang J, Taishi P, Kacsóh B, Gardi J, et al. 2001. Deficiency of growth hormone-releasing hormone signaling is associated with sleep alterations in the dwarf rat. J Neurosci 21: 2912-2918.

Obál F Jr, Floyd R, Kapás L, Bodosi B, Krueger JM. 1996. Effects of systemic GHRH on sleep in intact and in hypophysectomized rats. Am J Physiol 270: E230-E237.

Obál F Jr, Kacsóh B, Alföldi P, Payne L, Markovic O, et al. 1992a. Antiserum to prolactin decreases rapid eye movement sleep (REM sleep) in the male rat. Physiol Behav 52: 1063-1068.

Obál F Jr, Kacsóh B, Bredow S, Guha-Thakurta N, Krueger JM. 1997b. Sleep in rats rendered chronically hyperprolactinemic with anterior pituitary grafts. Brain Res 755: 130-136.

Obál F Jr, Kapás L, Bodosi B, Krueger JM. 1998. Changes in sleep in response to intracerebral injection of insulin-like growth factor-1 (IGF-1) in the rat. Sleep Research Online 1: 87-91.

Obál F Jr, Kapás L, Gardi J, Taishi P, Bodosi B, et al. 1999. Insulin-like growth factor-1 (IGF-1)-induced inhibition of growth hormone secretion is associated with sleep suppression. Brain Res 818: 267-274.

Obál F Jr, Payne L, Kapás L, Opp M, Krueger JM. 1991. Inhibition of growth hormone-releasing factor suppresses both sleep and growth hormone secretion in the rat. Brain Res 557: 149-153.

Obál F Jr, Payne L, Opp M, Alföldi P, Kapás L, et al. 1992b. Growth hormone-releasing hormone antibodies suppress sleep and prevent enhancement of sleep after sleep deprivation. Am J Physiol 263: R1078-R1085.

Opp M, Obál F Jr, Krueger JM. 1989. Corticotropin-releasing factor attenuates interleukin 1-induced sleep and fever in rabbits. Am J Physiol 257: R528-R535.

Opp MR, 1997. Rat strain differences suggest a role for corticotropin-releasing hormone in modulating sleep. Physiol Behav 63: 67-74.

Overeem S, Kok SW, Lammers GJ, Vein AA, Frolich M, et al. 2003. Somatotropic axis in hypocretin-deficient narcoleptic humans: altered circadian distribution of GH-secretory events. Am J Physiol–Endoc M 284: E641-E647.

Parker DC, Rossman LG, Siler TM, Rivier J, Yen SS, et al. 1974a. Inhibition of the sleep-related peak in physiologic human growth hormone release by somatostatin. J Clin Endocrinol Metab 38: 496-499.

Parker DC, Rossman LG, Vanderlaan EF. 1974b. Relation of sleep-entrained human prolactin release to REM- nonREM cycles. J Clin Endocrinol Metab 38: 646-651.

Paul SM, Purdy RH. 1992. Neuroactive steroids. FASEB J 6: 2311-2322.

Penalva RG, Lancel M, Flachskamm C, Reul JM, Holsboer F, et al. 2000. Effects of sleep dprivation on hippocampal serotonin: a combined in vivo microdialysis/EEG study in rats. J Sleep Res 9 (Suppl. 1): 149.

Perras B, Marshall L, Köhler G, Born J, Fehm HL. 1999a. Sleep and endocrine changes after intranasal administration of growth hormone-releasing hormone in young and aged humans. Psychoneuroendocrinology 24: 743-757.

Perras B, Pannenborg H, Marshall L, Pietrowsky R, Born J, et al. 1999b. Beneficial treatment of age-related sleep disturbances with prolonged intranasal vasopressin. J Clin Psychopharmacol 19: 28-36.

Puig-Antich J, Dahl R, Ryan N, Novacenko H, Goetz D, et al. 1989. Cortisol secretion in prepubertal children with major depressive disorder. Arch Gen Psychiatry 46: 801-809.

Quabbe HJ, Gregor M, Bumke-Vogt C, Eckhof A, Witt I. 1981. Twenty-four-hour pattern of growth hormone secretion in the rhesus monkey: studies including alterations of the sleep/wake and sleep stage cycles. Endocrinology 109: 513-522.

Quabbe HJ, Schilling E, Helge H. 1966. Pattern of growth hormone secretion during a 24-hour fast in normal adults. J Clin Endocrinol Metab 26: 1173-1177.

Riemann D, Klein T, Rodenbeck A, Feige B, Horny A, et al. 2002. Nocturnal cortisol and melatonin secretion in primary insomnia. Psychiatry Res 113: 17-27.

Riou F, Cespuglio R, Jouvet M. 1982. Endogenous peptides and sleep in the rat. III. The hypnogenic properties of vasoactive intestinal polypeptide. Neuropeptides 2: 265-277.

Rodenbeck A, Huether G, Rüther E, Hajak G. 2002. Interactions between evening and nocturnal cortisol secretion and sleep parameters in patients with severe chronic primary insomnia. Neurosci Lett 324: 159-163.

Roky R, Obál F Jr, Valatx JL, Bredow S, Fang J, et al. 1995. Prolactin and rapid eye movement sleep regulation. Sleep 18: 536-542.

Roky R, Valatx JL, Jouvet M. 1993. Effect of prolactin on the sleep–wake cycle in the rat. Neurosci Lett 156: 117-120.

Roky R, Valatx JL, Paut-Pagano L, Jouvet M. 1994. Hypothalamic injection of prolactin or its antibody alters the rat sleep–wake cycle. Physiol Behav 55: 1015-1019.

Rupprecht R. 2003. Neuroactive steroids: mechanisms of action and neuropsychopharmacological properties. Psychoneuroendocrinology 28: 139-168.

Saad MF, Damani S, Gingerich RL, Riad-Gabriel MG, Khan A, et al. 1997. Sexual dimorphism in plasma leptin concentration. J Clin Endocrinol Metab 82: 579-584.

Sack RL, Brandes RW, Kendall AR, Lewy AJ. 2000. Entrainment of free-running circadian rhythms by melatonin in blind people. N Engl J Med 343: 1070-1077.

Sangiah S, Caldwell DF, Villeneuve MJ, Clancy JJ. 1982. Sleep: sequential reduction of paradoxical (REM) and elevation of slow-wave (NREM) sleep by a non-convulsive dose of insulin in rats. Life Sci 31: 763-769.

Saper CB, Chou TC, Scammell TE. 2001. The sleep switch: hypothalamic control of sleep and wakefulness. Trends Neurosci 24: 726-731.

Sassin JF, Parker DC, Mace JW, Gotlin RW, Johnson LC, et al. 1969. Human growth hormone release: relation to slow-wave sleep and sleep-waking cycles. Science 165: 513-515.

Schier T, Guldner J, Colla M, Holsboer F, Steiger A. 1997. Changes in sleep-endocrine activity after growth hormone-releasing hormone depend on time of administration. J Neuroendocrinol 9: 201-205.

Schiffelholz T, Holsboer F, Lancel M. 2000. High doses of systemic DHEA-sulfate do not affect sleep structure and elicit moderate changes in non-REM sleep EEG in rats. Physiol Behav 69: 399-404.

Schmid DA, Brunner H, Holsboer F, Friess E. 2000. Cortisol promotes nonREM sleep in patients with major depression. Int J Neuropsychop 3 (Suppl. 1): S302.

Schneider HJ, Oertel H, Murck H, Pollmächer T, Stalla GK, et al. 2005. Night steep EEG and daytime steep propensity in adult hypopituitary patients with growth hormone deficiency before and after six months of growth hormone replacement. Psychoneuroendocrinology 30: 29-37.

Schüssler P, Kluge M, Dresler M, Yassouridis A, Held K, et al. 2005a. Effects of progesterone on sleep and cognition in healthy postmenopausal women. Pharmacopsychiatry, 38: 219.

Schüssler P, Uhr M, Ising M, Schmid D, Weikel J, et al. 2005b. Nocturnal ghrelin levels-relationship to sleep EEG, the levels of growth hormone, ACTH and cortisol- and gender differences J Sleep Res 14: 329-336.

Shipley JE, Schteingart DE, Tandon R, Starkman MN. 1992. Sleep architecture and sleep apnea in patients with Cushing's disease. Sleep 15: 514-518.

Simon C, Brandenberger G, Saini J, Ehrhart J, Follenius M. 1994. Slow oscillations of plasma glucose and insulin secretion rate are amplified during sleep in humans under continuous enteral nutrition. Sleep 17: 333-338.

Späth-Schwalbe E, Scholler T, Kern W, Fehm HL, Born J. 1992. Nocturnal adrenocorticotropin and cortisol secretion depends on sleep duration and decreases in association with spontaneous awakening in the morning. J Clin Endocrinol Metab 75: 1431-1435.

Spiegel K, Follenius M, Simon C, Saini J, Ehrhart J, et al. 1994. Prolactin secretion and sleep. Sleep 17: 20-27.

Spiegel K, Leproult R, Van Cauter E. 1999. Impact of sleep debt on metabolic and endocrine function. Lancet 354: 1435-1439.

Spiegel K, Tasali E, Penev P, Van Cauter E. 2004. Brief Communication: Sleep curtailment in healthy young men is associated with decreased leptin levels, elevated ghrelin levels, and increased hunger and appetite. Ann Intern Med 141: 846-850.

Spier AD, de Lecea L. 2000. Cortistatin: a member of the somatostatin neuropeptide family with distinct physiological functions. Brain Res Rev 33: 228-241.

Steiger A. 2002. Sleep and the hypothalamo–pituitary–adrenocortical system. Sleep Medicine Rev 6: 125-138.

Steiger A, Guldner J, Hemmeter U, Rothe B, Wiedemann K, et al. 1992. Effects of growth hormone-releasing hormone and somatostatin on sleep EEG and nocturnal hormone secretion in male controls. Neuroendocrinology 56: 566-573.

Steiger A, Guldner J, Knisatschek H, Rothe B, Lauer C, et al. 1991. Effects of an ACTH/MSH(4–9) analog (HOE 427) on the sleep EEG and nocturnal hormonal secretion in humans. Peptides 12: 1007-1010.

Steiger A, Herth T, Holsboer F. 1987. Sleep-electroencephalography and the secretion of cortisol and growth hormone in normal controls. Acta Endocrinol (Copenh) 116: 36-42.

Steiger A, Holsboer F. 1997b. Nocturnal secretion of prolactin and cortisol and the sleep EEG in patients with major endogenous depression during an acute episode and after full remission. Psychiatry Res 72: 81-88.

Steiger A, Holsboer F. 1997a. Neuropeptides and human sleep. Sleep 20: 1038-1052.

Steiger A, Trachsel L, Guldner J, Hemmeter U, Rothe B, et al. 1993. Neurosteroid pregnenolone induces sleep-EEG changes in man compatible with inverse agonistic $GABA_A$-receptor modulation. Brain Res 615: 267-274.

Steiger A, von Bardeleben U, Herth T, Holsboer F. 1989. Sleep EEG and nocturnal secretion of cortisol and growth hormone in male patients with endogenous depression before treatment and after recovery. J Affect Disord 16: 189-195.

Steiger A, Weikel J, Held K, Schmid D, Ziegenbein M. 2002. Arginine improves sleep in elderly men. Sleep 25: A203

Stern WC, Jalowiec JE, Shabshelowitz H, Morgane PJ. 1975. Effects of growth hormone on sleep–waking patterns in cats. Horm Behav 6: 189-196.

Taheri S, Lin L, Austin D, Young T, Mignot E. 2004. Short sleep duration is associated with reduced leptin, elevated ghrelin, and increased body mass index. PLoS Med 1: e62.

Takahashi Y, Kipnis DM, Daughaday WH. 1968. Growth hormone secretion during sleep. J Clin Invest 47: 2079-2090.

Tomaszuk A, Simpson C, Williams G. 1996. Neuropeptide Y, the hypothalamus and the regulation of energy homeostasis. Horm Res 46: 53-58.

Toppila J, Alanko L, Asikainen M, Tobler I, Stenberg D, et al. 1997. Sleep deprivation increases somatostatin and growth hormone-releasing hormone messenger RNA in the rat hypothalamus. J Sleep Res 6: 171-178.

Toppila J, Stenberg D, Alanko L, Asikainen M, Urban JH, et al. 1995. REM sleep deprivation induces galanin gene expression in the rat brain. Neurosci Lett 183: 171-174.

Trachsel L, Edgar DM, Seidel WF, Heller HC, Dement WC. 1992. Sleep homeostasis in suprachiasmatic nuclei-lesioned rat: effects of sleep deprivation and triazolam administration. Brain Res 598: 253-261.

Tsuchiyama Y, Uchimura N, Sakamoto T, Maeda H, Kotorii T. 1995. Effects of hCRH on sleep and body temperature rhythms. Psychiatry Clin Neurosci 49: 299-304.

Valatx JL, Jouvet M. 1988. Circadian rhythms of slow-wave sleep and paradoxical sleep are in opposite phase in genetically hypoprolactinemic rats. CR Acad Sci III 307: 789-794.

Valatx JL, Roky R, Trouillas J, Paut-Pagano L, Jouvet M. 1994. Paradoxical sleep alteration by tumoral hyperprolactinemia. J Sleep Res 3 (Suppl. 1): 260.

Van Cauter E, Blackman JD, Roland D, Spire JP, Refetoff S, et al. 1991a. Modulation of glucose regulation and insulin secretion by circadian rhythmicity and sleep. J Clin Invest 88: 934-942.

Van Cauter E, Desir D, Refetoff S, Spire JP, Neol P, et al. 1982. The relationship between episodic variations of plasma prolactin and REM-non-REM cyclicity is an artifact. J Clin Endocrinol Metab 54: 70-75.

Van Cauter E, Kerkhofs M, Caufriez A, Van Onderbergen A, Thorner MO, et al. 1992. A quantitative estimation of growth hormone secretion in normal man: reproducibility and relation to sleep and time of day. J Clin Endocrinol Metab 74: 1441-1450.

Van Cauter E, Leproult R, Kupfer DJ. 1996. Effects of gender and age on the levels and circadian rhythmicity of plasma cortisol. J Clin Endocrinol Metab 81: 2468-2473.

Van Cauter E, Linkowski P, Kerkhofs M, Hubain P, L'Hermite-Balériaux M, et al. 1991b. Circadian and sleep-related endocrine rhythms in schizophrenia. Arch Gen Psychiatry 48: 348-356.

Van Coevorden A, Mockel J, Laurent E, Kerkhofs M, L'Hermite-Balériaux M, et al. 1991. Neuroendocrine rhythms and sleep in aging men. Am J Physiol 260: E651-E661.

Heuvel van den CJ, Ferguson SA, Macchi MM, Dawson D. 2005. Melatonin as a hypnotic: con. Sleep Medicine Rev 9: 71-80.

Vgontzas AN, Bixler EO, Wittman AM, Zachman K, Lin HM, et al. 2001. Middle-aged men show higher sensitivity of sleep to the arousing effects of corticotropin-releasing hormone than young men: clinical implications. J Clin Endocrinol Metab 86: 1489-1495.

Voderholzer U, Hohagen F, Klein T, Jungnickel J, Kirschbaum C, et al. 2004. Impact of sleep deprivation and subsequent recovery sleep on cortisol in unmedicated depressed patients. Am J Psychiatry 161: 1404-1410.

Voderholzer U, Laakmann G, Wittmann R, Daffner-Bujia C, Hinz A, et al. 1993. Profiles of spontaneous 24-hour and stimulated growth hormone secretion in male patients with endogenous depression. Psychiatry Res 47: 215-227.

Watts AG, Tanimura S, Sanchez-Watts G. 2004. Corticotropin-releasing hormone and arginine vasopressin gene transcription in the hypothalamic paraventricular nucleus of unstressed rats: daily rhythms and their interactions with corticosterone. Endocrinology 145: 529-540.

Wehr TA. 1998. Effect of seasonal changes in daylength on human neuroendocrine function. Horm Res 49: 118-124.

Wehr TA, Moul DE, Barbato G, Giesen HA, Seidel JA, et al. 1993. Conservation of photoperiod-responsive mechanisms in humans. Am J Physiol 265: R846-R857.

Weibel L, Brandenberger G. 1998. Disturbances in hormonal profiles of night workers during their usual sleep and work times. J Biol Rhythms 13: 202-208.

Weikel JC, Wichniak A, Ising M, Brunner H, Friess E, et al. 2003. Ghrelin promotes slow-wave sleep in humans. Am J Physiol–Endoc M 284: E407-E415.

Weitzman ED. 1976. Circadian rhythms and episodic hormone secretion in man. Annu Rev Med 27: 225-243.

Weitzman ED, Czeisler CA, Zimmerman JC, Ronda JM. 1981. Biological rhythms in man: relationship of sleep–wake, cortisol, growth hormone and temperature during temporal isolation. Neurosecretion and brain peptides. Martin JB, Reichlin S, Bick K, editors. New York: Raven Press; pp. 475.

Weitzman ED, Zimmerman JC, Czeisler CA, Ronda J. 1983. Cortisol secretion is inhibited during sleep in normal man. J Clin Endocrinol Metab 56: 352-358.

Wiedemann K, Lauer C, Pollmächer T, Holsboer F. 1994. Sleep-endocrine effects of antigluco- and antimineralocorticoids in healthy males. Am J Physiol 267: E109-E114.

Wiedemann K, Lauer CJ, Hirschmann M, Knaudt K, Holsboer F. 1998. Sleep-endocrine effects of mifepristone and megestrol acetate in healthy men. Am J Physiol–Endoc M 274: E139-E145.

Wong ML, Kling MA, Munson PJ, Listwak S, Licinio J, et al. 2000. Pronounced and sustained central hypernoradrenergic function in major depression with melancholic features: relation to hypercortisolism and corticotropin-releasing hormone. Proc Natl Acad Sci USA 97: 325-330.

Xu YL, Reinscheid RK, Huitron-Resendiz S, Clark SD, Wang Z, et al. 2004. Neuropeptide S: a neuropeptide promoting arousal and anxiolytic-like effects. Neuron 43: 487-497.

Yang SW, Fishbein W. 1995. Castration at birth induces female sleep pattern in mice; neonatal testosterone replacement reverses the effect. Sleep Res 24: 65

Yassouridis A, Steiger A, Klinger A, Fahrmeir L. 1999. Modelling and exploring human sleep with event history analysis. J Sleep Res 8: 25-36.

Yi PL, Tsai CH, Lin JG, Lee CC, Chang FC. 2004. Kindling stimuli delivered at different times in the sleep–wake cycle. Sleep 27: 203-212.

Zhang G, Wang L, Liu H, Zhang J. 2004. Substance P promotes sleep in the ventrolateral preoptic area of rats. Brain Res 1028: 225-232.

Zhang J, Chen Z, Taishi P, Obál J, Fang F Jr, et al. 1998. Sleep deprivation increases rat hypothalamic growth hormone-releasing hormone mRNA. Am J Physiol 275: R1755-R1761.

Zhang J, Obál F Jr, Zheng T, Fang J, Taishi P, et al. 1999. Intrapreoptic microinjection of GHRH or its antagonist alters sleep in rats. J Neurosci 19: 2187-2194.

Zhang SQ, Inoue S, Kimura M. 2001. Sleep-promoting activity of prolactin-releasing peptide (PrRP) in the rat. NeuroReport 12: 3173-3176.

Zhdanova IV. 2005. Reply to the comment on 'Melatonin as a hypnotic: Pro'. Sleep Medicine Rev 9: 69-70.

Zhdanova IV, Lynch HJ, Wurtman RJ. 1997. Melatonin: a sleep-promoting hormone. Sleep 20: 899-907.

Ziegenbein M, Held K, Künzel H, Murck H, Antonijevic IA, et al. 2004. The somatostatin analogue octreotide impairs sleep and decreases EEG sigma power in young male subjects. Neuropsychopharmacology 29: 146-151.

Index

Lightning Source UK Ltd.
Milton Keynes UK
UKOW06n1300171117
312852UK00002B/19/P

9 780387 303628